BRAIN NEUROTRAUMA

MOLECULAR,
NEUROPSYCHOLOGICAL,
and
REHABILITATION ASPECTS

FRONTIERS IN NEUROENGINEERING

Series Editor
Sidney A. Simon, Ph.D.

Published Titles

Brain Neurotrauma: Molecular, Neuropsychological, and Rehabilitation Aspects
Firas H. Kobeissy, University of Florida, Gainesville, Florida

Humanoid Robotics and Neuroscience: Science, Engineering, and Society
Gordon Cheng, ATR Computational Neuroscience Labs, Kyoto, Japan

Methods in Brain Connectivity Inference through Multivariate Time Series Analysis
Koichi Sameshima, University of São Paulo, São Paulo, Brazil
Luiz Antonio Baccala, University São Paulo, São Paulo, Brazil

Neuromorphic Olfaction
Krishna C. Persaud, The University of Manchester, Manchester, UK
Santiago Marco, University of Barcelona, Barcelona, Spain
Agustín Gutiérrez-Gálvez, University of Barcelona, Barcelona, Spain

Indwelling Neural Implants: Strategies for Contending with the *In Vivo* Environment
William M. Reichert, Ph.D., Duke University, Durham, North Carolina

Electrochemical Methods for Neuroscience
Adrian C. Michael, University of Pittsburg, Pennsylvania
Laura Borland, Booz Allen Hamilton, Inc., Joppa, Maryland

BRAIN NEUROTRAUMA

MOLECULAR, NEUROPSYCHOLOGICAL, and REHABILITATION ASPECTS

Edited by
Firas H. Kobeissy, PhD
Psychoproteomics and Nanotechnology Research Center
Department of Psychiatry
University of Florida
Gainsville, Florida, USA

CRC Press
Taylor & Francis Group
Boca Raton London New York

CRC Press is an imprint of the
Taylor & Francis Group, an **informa** business

Cover art designed by Alexdrina Chong, M.S., graphic designer.

First published in paperback 2024

First published 2015 by CRC Press
2385 NW Executive Center Drive, Suite 320, Boca Raton FL 33431

and by CRC Press
4 Park Square, Milton Park, Abingdon, Oxon, OX14 4RN

CRC Press is an imprint of Taylor & Francis Group, LLC

© 2015, 2024 by Taylor & Francis Group, LLC

This book contains information obtained from authentic and highly regarded sources. While all reasonable efforts have been made to publish reliable data and information, neither the author[s] nor the publisher can accept any legal responsibility or liability for any errors or omissions that may be made. The publishers wish to make clear that any views or opinions expressed in this book by individual editors, authors or contributors are personal to them and do not necessarily reflect the views/opinions of the publishers. The information or guidance contained in this book is intended for use by medical, scientific or health-care professionals and is provided strictly as a supplement to the medical or other professional's own judgement, their knowledge of the patient's medical history, relevant manufacturer's instructions and the appropriate best practice guidelines. Because of the rapid advances in medical science, any information or advice on dosages, procedures or diagnoses should be independently verified. The reader is strongly urged to consult the relevant national drug formulary and the drug companies' and device or material manufacturers' printed instructions, and their websites, before administering or utilizing any of the drugs, devices or materials mentioned in this book. This book does not indicate whether a particular treatment is appropriate or suitable for a particular individual. Ultimately it is the sole responsibility of the medical professional to make his or her own professional judgements, so as to advise and treat patients appropriately. The authors and publishers have also attempted to trace the copyright holders of all material reproduced in this publication and apologize to copyright holders if permission to publish in this form has not been obtained. If any copyright material has not been acknowledged please write and let us know so we may rectify in any future reprint.

Except as permitted under U.S. Copyright Law, no part of this book may be reprinted, reproduced, transmitted, or utilized in any form by any electronic, mechanical, or other means, now known or hereafter invented, including photocopying, microfilming, and recording, or in any information storage or retrieval system, without written permission from the publishers.

For permission to photocopy or use material electronically from this work, access www.copyright.com or contact the Copyright Clearance Center, Inc. (CCC), 222 Rosewood Drive, Danvers, MA 01923, 978-750-8400. For works that are not available on CCC please contact mpkbookspermissions@tandf.co.uk

Trademark notice: Product or corporate names may be trademarks or registered trademarks and are used only for identification and explanation without intent to infringe.

Publisher's Note
The publisher has gone to great lengths to ensure the quality of this reprint but points out that some imperfections in the original copies may be apparent.

ISBN: 978-1-4665-6598-2 (hbk)
ISBN: 978-1-03-291771-9 (pbk)
ISBN: 978-0-429-07115-7 (ebk)

DOI: 10.1201/b18126

Visit the Taylor & Francis Web site at
http://www.taylorandfrancis.com

and the CRC Press Web site at
http://www.crcpress.com

Dedication

To the memory of my mom Kawsar Khalil....

To my father Hosni and sister Maha

To my friends and colleagues

To Manal Nahle

-FK-

Contents

SECTION I Neuromechanisms in Brain Injury

SECTION II Management in CNS Trauma

SECTION III Modeling Brain Injury

SECTION IV Imaging and Biomarkers

SECTION VI Neurorehabilitation and Neuroprotection

SECTION VII Mild Brain Injury and Sport Concussion

Series Preface

The Frontiers in Neuroengineering series presents the insights of experts on emerging experimental techniques and theoretical concepts that are or will be at the vanguard of neuroscience. Books in the series cover topics ranging from electrode design methods for neural ensemble recordings in behaving animals to biological sensors. The series also covers new and exciting multidisciplinary areas of brain research, such as computational neuroscience and neuroengineering, and describes breakthroughs in biomedical engineering. The goal is for this series to be the reference that every neuroscientist uses to become acquainted with new advances in brain research.

Each book is edited by an expert and consists of chapters written by leaders in a particular field. Books are richly illustrated and contain comprehensive bibliographies. Chapters provide substantial background material relevant to the particular subject.

We hope that, as the volumes become available, our efforts as well as those of the publisher, the book editors, and the individual authors will contribute to the further development of brain research. The extent to which we achieve this goal will be determined by the utility of these books.

Sidney A. Simon, PhD
Series Editor

Foreword

When asked by Dr. Firas Kobeissy to write the foreword to his proposed textbook on traumatic brain injury, I thought it would be an interesting exercise. Here is a topic that I have worked on in several manners since 1985. Traumatic brain injury is not just a destroyer of worlds, but of many lives beyond those who have sustained the actual brain injury. The effects of traumatic brain injury can be sudden, changing people permanently in an instant, or in the case of repetitive brain injury, bring on prematurely the ravages of neurodegeneration and dementia. However, when I sat down to write this foreword, the typical formula of writing based on published knowledge, which I had used previously in my academic life or in the approved statements of the federal agency in which I am now employed would not suffice. This created a situation where I could actually put into words my thoughts on traumatic brain injury, an idea that can be quite foreboding to a federal employee.

First, we all live and work in our little corner of the research world. In some ways it is almost like describing where you live, for example, Planet Earth, Northern Hemisphere, North America, United States, Virginia, and unincorporated Fairfax County. A lot of us started off in biomedical sciences, neuroscience, brain injury, acute therapeutics, and for me at one time, neurosteroids. Communication among the many disciplines and interests represented in this book is just as critical to the understanding of neurotrauma as is the communication between people located across the world. From my current location in neurotrauma, I believe one of my roles is to foster this type of communication. The program that I administer runs the gambit from animals to humans; acute to chronic; diagnostics to therapies. It is in this position where I met Dr. Kobeissy, who became a regular member of my brain injury scientific review panel. I believe this book opens those lines of communication in a very unique way.

From my former role as a research director at a clinical site and in my current programmatic/review role, I have been lucky to see and learn the different perspectives of brain injury, not only from the scientists and clinicians in the numerous different disciplines of brain injury, but also from those who have had brain injury. The conversation that is initiated in this book is critical to understanding these processes and the impact of these relationships, not only on the scientists investigating neurotrauma, but also for those who have a stake in the outcome of this research. As a student, whether you choose a career as a researcher, a clinician, or some combination of those two, it is your duty not to stove-pipe your career or your interactions in your career.

In my early education on translational neurotrauma research, I was confronted by a thought-provoking statement by my mentor, Dr. Donald Stein. It went something like this: "When you ask a bench researcher what recovery of function is, his response would be more sprouting, greater neuronal survival, better performance on a behavioral task, etc. When you ask a neurosurgeon, her response would be, 'Did the patients regain consciousness, do they respond to commands, are they talking?' etc. When you ask physiatrists or a physical therapists, they judge recovery by how well the person is doing on his/her activities of daily living. Now ask the patients; their concerns lie in areas that we scientists do not generally consider prime areas of research. These people, the stakeholders in our research, their concerns are about incontinence, intimacy with their significant other, going back to work. Finally, you ask the families, and their typical concern is, how long is their love one going to be this way?"

To suggest that every investigator from the bench to the bedside understands the dynamics of brain injury at each level is burdensome; however, transmitting this knowledge between the levels of the researcher, clinician, patient, and caregiver is necessary. Despite living in a period of time with the Internet, where communication and access to information know almost no barriers, this type of cross-disciplinary transmission of knowledge is not occurring at the rate it should. We need better models, more potent therapeutics, improved diagnostics/biomarkers, and outcome measures with greater sensitivity, along with who the stakeholders are in the brain injured community and their needs. After all, when the patients leave the hospital or rehabilitation facility, it will be them and their caregivers who will be dealing with the aftermath of TBI for years to come. Yet, relatively few basic investigators know the ultimate consumers of their research. As a future or present neurotraumatologist, do not forget the bigger picture; clinical researchers need to reach out to their preclinical counterparts and vice versa, to foster translational research, which will advance both the science and healthcare of neurotrauma.

Currently, the research community in the area of TBI has been given a gift of increased visibility. This is due in part to the long-term effects of contact sports in which multiple insults to the brain may result in the early onset of neurodegenerative disease in the form of chronic traumatic encephalopathy. In this situation, the additive effect of mild TBIs and subconcussive "hits" results in the progressive degeneration that begins with behavioral issues and then moves onto early onset dementia. There is now some evidence suggesting that veterans of the recent conflicts may also be at risk for this condition. This area of TBI research has brought about increased realization in the public that brain injury is permanent injury to one's body. Repeated impacts, even without loss of consciousness, can take a toll on the brain. As a research community we need to ascertain the genesis of this condition, determine if particular people are more susceptible, and develop diagnostics and potential therapeutics. The fact that we now have the technology to find minute quantities of molecules that arose from the brain in the blood of those with TBI is impressive. What we also have to remember is that when something can leave, other things can enter. Being able to track not only the presence of these compounds, but how the levels of these compounds change over time, will be essential as the research community develops therapeutics.

The complexity of TBI is truly daunting. Just a quick search of PubMed reveals all the possible pathways that are activated acutely by a brain injury. Combine all these elucidated pathways occurring at the time of injury (whether it is from controlled cortical impact, fluid percussion, blast, etc.) with time post-injury, behavioral changes, individual differences, etc. This continued conversation needs to take place between those who work on models of brain injury from the molecular to neuropsychological to rehabilitation. This textbook begins teaching that interdisciplinary conversation and may show you the way.

Stuart W. Hoffman, PhD
U.S. Department of Veterans Affairs
Washington, DC

Preface

"The important thing in science is not so much to obtain new facts as to discover new ways of thinking about them."

—Sir William Lawrence Bragg

Neurotrauma is a worldwide health problem; it involves the neurological consequences of external forces affecting the brain, spinal cord, or body. The increased frequency and prevalence of central nervous system injuries documented by the Department of Defense (DOD) and the U.S. Department of Veterans Affairs (VA) underscore the increasing concern with these injuries. It has been estimated that at least 1.7 million civilians sustain brain injury in the United States annually, with approximately 50,000 deaths and 200,000 moderate-to-severe injuries resulting in hospitalization, while ~1 million cases are treated in the emergency department. In the United States, approximately $9–10 billion are spent per year on TBI involving injury management, urgent treatment, and rehabilitation coupled with long-term disabilities affecting 50% of moderate brain injury survivors. Medical doctors are currently recognizing that a significant fraction of blast injury patients develop comorbid neuropsychiatric conditions such as posttraumatic stress disorder or depression and are more prone to substance abuse. Repeated mild brain injuries, particularly those stemming from car accidents, falls, and sports-related incidents have become a public health concern due to their long-term consequences and their associated behavioral and cognitive consequences.

Emerging clinical observations and varying pathophysiological presentations of neurotrauma require urgent efforts to develop better assessments and treatments, determine outcomes, and refine disability criteria. Such efforts require robust research initiatives and generous funding. Taken together, these statistics and medical facts have encouraged researchers to develop novel brain injury models and to apply the current understanding of pathology of brain injury to design improved options for treatments and more effective rehabilitation. For these reasons, these topics are discussed in the book.

Over the last decade, the field of neurotrauma has witnessed significant advances, especially at the molecular, cellular, and behavioral levels. This progress is largely due to the introduction of novel techniques, as well as the development of new animal models of central nervous system (CNS) injury. Increasingly sophisticated approaches have proven to be extremely valuable in conducting translational bench-to-bedside studies to decipher the pathological mechanisms that underlie different phases of CNS injury. With the right tools and approaches, many of which are described in this book, scientists are better enabled to identify relevant therapeutic targets to treat the broad spectrum of symptoms and pathologies stemming from traumatic CNS injury.

The general theme of this book evolved during discussions with members of the grant review panel of the VA and members of the National Neurotrauma Society (NNS), many of whom contributed to this work. We are truly fortunate to have valuable contributions from world-renowned experts and leaders in the fields of CNS injury, neurorehabilitation, and biomarker discovery.

This work represents a comprehensive and up-to-date account of current knowledge of key basic and clinical aspects of CNS injury, neurorehabilitative approaches, and novel diagnostic biomarker platforms considered for use in the field of CNS trauma. The book consists of 49 chapters assembled into eight sections that discuss neurotrauma mechanisms, biomarker discovery, neurocognitive and neurobehavioral deficits, and neurorehabilitation and treatment approaches of CNS injury. In addition, a section is devoted to models of milder CNS injury, includes sports injuries.

Topics reviewed in *Brain Neurotrauma: Molecular, Neuropsychological, and Rehabilitation Aspects* include

- Experimental CNS Trauma: A General Overview of Neurotrauma Research
- The Genesis of Brain Injury Research: Combat TBI Overview
- Combat TBI: History, Epidemiology, and Injury Modes
- Neuromechanisms in Brain Injury
- Management in CNS Trauma
- Modeling Brain Injury
- Imaging and Biomarkers
- Neurocognitive and Neurobehavioral Topics in Brain Injury
- Neurorehabilitation and Neuroprotection
- Mild Brain Injury and Sport Concussion
- Drug Abuse and Comorbid Conditions

The first two chapters provide an overview on the genesis of CNS injury and the state of modern translational research. In the first chapter, Dr. Jonathan Lifshitz, PhD, highlights the need for translational research in brain injury, while Professor Ralph DePalma, MD, uses the second chapter to depict the history behind brain injury, providing an elegant historical tour from ancient Greek mythology to our current-day views of CNS injury.

Section I, Neuromechanisms in Brain Injury, consists of eight chapters that recapitulate the major underlying pathophysiological mechanisms involved in the injured nervous system. These include axonal injury, oxidative stress, inflammatory response, and altered signaling pathways. Section II, Management in CNS Trauma, discusses chronic pain, headache, and nonconvulsive seizures, which are still under-investigated topics in the area of CNS injury. Section III, Modeling Brain Injury, describes the currently available experimental brain injury models used by researchers to reproduce the different features of clinical neurotrauma. In this section, comprised of five chapters, several models (including the controlled cortical injury model, fluid percussion injury model, and blast injury model) are thoroughly discussed. In Section IV, Imaging and Biomarkers, biomarker research and the application of imaging techniques are used concurrently to illustrate the bench-to-bedside trajectory of a "perfect" brain injury biomarker identification technique. This section also evaluates the current status of brain injury biomarkers in the area of acute and mild neurotrauma and continues to describe the different "flavors" of available biomarkers (including proteins, gene microRNAs, and autoantibodies) and the utility of systems biology approaches to analyze these biomarkers using high throughput techniques. In Section V, Neurocognitive and Neurobehavioral Topics in Brain Injury, we address the common neurocognitive and neurobehavioral deficits stemming from brain injury. This section aims at improving our general understanding of these behavioral alterations and the available management approaches. Section VI, Neurorehabilitation and Neuroprotection, presents an update of the rehabilitative paradigms in experimental brain injury as well as newly introduced neurotherapeutic molecular approaches and techniques (i.e., stem cell research, noninvasive cranial nerve neuromodulation, and cortical stimulation-induced plasticity). Of particular interest, this section introduces us to the new notion of "rehabilomics" as an emerging area for rehabilitation research. In Section VII, Mild Brain Injury and Sport Concussion, authors including the internationally renowned Professors I. Cernak and V. Koliatsos elaborate on the heterogeneous pathophysiology of blast injury and sport concussion. Finally, the book examines the linkage between brain trauma and substance abuse in Section VIII, Substance Abuse and Comorbid Conditions, with a focus on comorbid alcohol abuse and the controversy surrounding the effect of alcohol on brain injury.

In a general overview of the compiled material, there are a few points that need to be addressed. Due to the broad spectrum of injury manifestation, topics discussed focus on brain injury rather than spinal cord injury (SCI), which we believe requires a separate dedicated resource. To that end, much of what is learned studying the CNS injuries might later be translated to spinal cord injury. Furthermore, in compiling data pertaining to a large field, several pivotal chapters were inevitably missed. These include (1) discussion concerning PTSD and brain injury, (2) alternative animal models of brain injury (penetrating ballistic-like brain injury and Marmarou's impact acceleration model), (3) statistical approaches for experimental design in clinical and basic neurotrauma studies, and (4) clinical trials aiming to evaluate the effectiveness of drugs in head trauma. Finally, while we have sought to minimize this, there is some unavoidable overlap in the contents of chapters. The categorization of each section was selected based on the main theme discussed. Although some chapters may fit a different section from the one selected, we sought to ensure that chapter selection would fulfill the thematic needs of each particular section. We believe that this book, with its diverse coherent content, represents an excellent and reasonably comprehensive reference for individuals interested in the diverse and heterogeneous aspects of CNS pathology and/or rehabilitation needs.

Special thanks go to Barbara Norwitz, Lance Wobus, Kathryn Everett, and Maura Cregan of Taylor & Francis, who were responsible for the development process including book proposal approval, book organization, and material preparation. I would like to thank Lance for allowing the (seemingly endless) deadline extensions asked by several of the authors. To Dr. Stuart Hoffman, PhD, who agreed to write the foreword of this textbook; several contributors were actually insisting on his contribution and anxious to read his foreword. My special thanks and recognition go to the talented graphic designer Alexdrina Chong who agreed—in spite of her endless projects—to design the cover art work that decorates this book. In addition, I want to, sincerely, thank Dr. Noah Walton for his excellence in English and writing skills; Noah edited the final versions of my preface.

To the readers and audience of this work, I hope that you will find this work instrumental to your research and clinical studies. Last, my gratitude goes to all brain injury patients and their families. You are the inspiration and reason for the research reviewed in this book. Our hope is our modest contribution to the literature will benefit brain injury patients and their families and bring them hope, recovery, and peace.

Firas Hosni Kobeissy
University of Florida
Gainesville, Florida

Acknowledgments

First, I would like to thank all the book authors, the ones who contributed as well as those who had urgent circumstances and were not able to submit their chapters; without their encouragement and their contributions this book would never have become a reality. I truly thank each one involved in this project for their patience with the long and repeated emails, reminders, and inquiries sent by me. After all, it is the authors' outstanding expertise and remarkable manuscripts that made this work unique and real.

I wish to thank Professors Ralph DePalma, Jonathan Lifshitz, Ed Dixon, and Stuart Hoffman, whose input facilitated the completion of this work.

I am very grateful to Dr. Noah Walton, PhD, Amaly Nokarri, MS, and Zeinab Dalloul, MS, for their extraordinary editorial help. They generously assisted me in the editing and chapter compilation process.

Thank you.

ABOUT THE COVER DESIGN ARTIST

Alexdrina Chong is a visual designer and a part time design lecturer at a local design program. Before the establishment of SoCo, she spent about a decade in the United States, teaching graphic design at Bowling Green State University for two years, then relocating to Chicago to work as a lead designer on promotional campaigns for major movie studios, among them Warner Brothers and New Line Cinema. Alexdrina came home to Malaysia in July 2008 and started her own private practice, LUCID Collaborative Studio, and in 2011, she decided to put aside most of her commercial work and focus on starting a creative agency working toward a social mission. The result was SoCo, short for Social Collaboration. As an artist, she has exhibited her work in group shows mainly in Chicago, New York and also in Ohio.

Editors

Firas Hosni Kobeissy, PhD, is a research assistant professor in the Department of Psychiatry and part of the McKnight Brain Institute of the University of Florida. He is the associate scientific director of the Center for Neuroproteomics and Biomarkers Research (CNBR). Dr. Kobeissy earned his PhD from the University of Florida in 2007. He has published in the fields of neuroproteomics, brain injury, and systems biology. His current research overlaps the fields of neuroscience and psychiatry with a focus on drug abuse neurotoxicity and traumatic brain injury neuroproteomics. His research interests include brain trauma biomarker discovery utilizing an "omics" approaches, in addition to studying mechanisms of neurotoxicity and drug abuse–induced brain modulation utilizing a systems biology approach and biochemical techniques. He has published more than 60 peer-reviewed papers, reviews, and book chapters and has coauthored five U.S. patents and coedited three books. Dr. Kobeissy serves as an editorial member on several journals related to proteomics, neuroscience, and brain injury. He has served on several review panels for brain trauma and stroke. He is also an actively participating member at the Center of Neuroproteomics and Biomarker Research and at the Center for Traumatic Brain Injury Studies at the McKnight Brain Institute of the University of Florida.

Stuart W. Hoffman, PhD, is the Scientific Program manager of the Brain Injury portfolio in the Rehabilitation Research & Development Service at the U.S. Department of Veterans Affairs (VA). The portfolio is focused on promoting translational research that improves outcomes in veterans living with the chronic effects of stroke, traumatic brain injury, and its comorbidities. Dr. Hoffman also represents the VA Office of Research & Development (ORD) on TBI research. In this role, he is the cochair of the government steering committee for the VA/DOD Chronic Effects of Neurotrauma Consortium, as well as a member of several advisory boards for VA research centers, and serves on the congressionally mandated Traumatic Brain Injury Advisory Committee for the Veterans Health Administration.

Dr. Hoffman earned his doctoral degree in behavioral and molecular neuroscience at Rutgers University in 1995 and completed his postdoctoral training in pharmacology at Virginia Commonwealth University in 1997. His professional career began as an assistant professor in the Department of Emergency Medicine at Emory University from 1998 to 2006, where he was a core member of the research team that translated progesterone from the laboratory to clinical trials for moderate to severe traumatic brain injury. Immediately prior to joining the VA, Dr. Hoffman was the research director for the Defense and Veterans Brain Injury Center in Johnstown, Pennsylvania. He has more than 25 years of translational research experience in brain injury focused on neuroprotection and methods to promote recovery of function after brain injury.

Contributors

DeAnna L. Adkins, PhD
Department of Neuroscience
College of Medicine
Health Sciences and Research
College of Health Professions
and
Center for Biomedical Imaging
Medical University of South Carolina
Charleston, South Carolina

Denes V. Agoston
Department of Anatomy, Physiology, and Genetics
Uniformed Services University
Bethesda, Maryland

Abdullah Shafique Ahmad
Department of Anesthesiology
and
Center for Translational Research in Neurodegenerative Disease
University of Florida College of Medicine
Gainesville, Florida

Elie D. Al-Chaer, PhD, JD
Department of Anatomy, Cell Biology and Physiological Sciences
Faculty of Medicine
American University of Beirut
Beirut, Lebanon

John Anagli
Banyan Biomarkers, Inc.
Alachua, Florida

Carolyn R. Anderson
Minneapolis VA Health Care System
Minneapolis, Minnesota

James W. Bales
Safar Center for Resuscitation Research
Center for Neuroscience
Center for the Neural Basis of Cognition, Neurological Surgery
University of Pittsburgh
Pittsburgh, Pennsylvania

Oneil G. Bhalala, PhD
University of Melbourne Medical School
Melbourne, Victoria, Australia

Erin D. Bigler, PhD
Department of Psychology
Neuroscience Center
Magnetic Resonance Imaging Research Facility
Brigham Young University
Provo, Utah

and

Department of Psychiatry
The Brain Institute of Utah
University of Utah
Salt Lake City, Utah

Amade Bregy, MD, PhD
Department of Neurosurgery
University of Miami Miller School of Medicine
Miami, Florida

Hayde Bolouri
Department of Psychiatry and Neurochemistry
Institute of Neuroscience and Physiology
University of Gothenburg
Gothenburg, Sweden

Corina O. Bondi, PhD
Physical Medicine and Rehabilitation
Safar Center for Resuscitation Research
Center for Neuroscience
University of Pittsburgh
Pittsburgh, Pennsylvania

Prodip Bose, MD, PhD
Malcom Randal VA Medical Center
Brain Rehabilitation Research Center of Excellence
North Florida/South Georgia Veterans Health System
and
Department of Neurology
and
Department of Physiological Sciences
University of Florida
Gainesville, Florida

Helen M. Bramlett, PhD
The Miami Project to Cure Paralysis
Department of Neurosurgery
Bruce W. Carter Department of Veterans Affairs
University of Miami Miller School of Medicine
Miami, Florida

Emilio B. Cagmat, MS
Banyan Biomarkers, Inc.
Alachua, Florida

James P. Caruso, BS
Department of Neurosurgery
Wayne State University School of Medicine
and
John D. Dingell VA Medical Center
Detroit, Michigan

Ibolja Cernak, PhD
Military and Veterans' Clinical Rehabilitation Research
University of Alberta
Edmonton, Alberta, Canada

Namas Chandra, PhD, PE
Center for Injury Biomechanics, Materials, and Medicine
New Jersey Institute of Technology
Newark, New Jersey

Haoxing Chen
University of Maryland School of Medicine
Baltimore, Maryland

Yun Chen
BrightstarTech, Inc.
Clarksburg, Maryland

Jeffrey P. Cheng, BS
Physical Medicine and Rehabilitation
Safar Center for Resuscitation Research
University of Pittsburgh
Pittsburgh, Pennsylvania

Taylor A. Colburn
Department of Child Health
University of Arizona College of Medicine–Phoenix
Phoenix, Arizona

Shlomi Constantini
Department of Pediatric Neurosurgery
Dana Children's Hospital
Tel Aviv Medical Center
Tel Aviv University
Tel Aviv, Israel

Alana C. Conti, PhD
Department of Neurosurgery
Wayne State University School of Medicine
and
John D. Dingell VA Medical Center
Detroit, Michigan

Jennifer A. Creed
MD-PhD Program
Drexel University College of Medicine
Philadelphia, Pennsylvania

Yuri Danilov, PhD
Tactile Communication and Neuromodulation Laboratory
Department of Biomedical Engineering
University of Wisconsin–Madison
Madison, Wisconsin

Nicholas D. Davenport, PhD
Minneapolis VA Health Care System
and
Department of Psychiatry
University of Minnesota–Twin Cities
Minneapolis, Minnestota

Ralph G. DePalma, MD, FACS
VA Office of Research and Development
Washington, DC

and

Uniformed University of the Health Sciences
Bethesda, Maryland

W. Dalton Dietrich, PhD
The Miami Project to Cure Paralysis
Department of Neurosurgery
University of Miami Miller School of Medicine
Miami, Florida

C. Edward Dixon
Safar Center for Resuscitation Research
Department of Neurological Surgery
Brain Trauma Research Center
University of Pittsburgh
and
Veterans Affairs Pittsburgh Healthcare System
Pittsburgh, Pennsylvania

Sylvain Doré, PhD
Center for Translational Research in Neurodegenerative Disease
University of Florida College of Medicine
Gainesville, Florida

Katharine Eakin, PhD
BARROW Neurological Institute
Phoenix Children's Hospital
and
Department of Child Health
University of Arizona College of Medicine
Phoenix, Arizona

Damyan Edwards, BA
Department of Psychology
Concordia University
Montreal, Quebec, Canada

Timothy W. Ellis Jr., MSc
BARROW Neurological Institute
Phoenix Children's Hospital
and
Department of Child Health
University of Arizona College of Medicine–Phoenix
Phoenix, Arizona

and

Biomedical Sciences
College of Health Sciences
Midwestern University
Glendale, Arizona

Megan N. Evilsizor, BS
BARROW Neurological Institute
Phoenix Children's Hospital
Phoenix, Arizona

Michelle D. Failla, BS
Center for Neuroscience
Department of Physical Medicine & Rehabilitation
University of Pittsburgh School of Medicine
Pittsburgh, Pennsylvania

Byron D. Ford, PhD
Department of Neurobiology
Neuroscience Institute
Morehouse School of Medicine
Atlanta, Georgia

Thomas C. Glenn, PhD
Brain Injury Research Center
Department of Neurosurgery
David Geffen School of Medicine at UCLA
University of California–Los Angeles
Los Angeles, California

Justin C. Graves, BS
Veterans Administration Health Centers
Detroit, Michigan

Grace S. Griesbach, PhD
Department of Neurosurgery
Center for Neuro Skills
Bakersfield, California

and

University of California–Los Angeles
Los Angeles, California

Joy D. Guingab-Cagmat, PhD
Banyan Biomarkers, Inc.
Alachua, Florida

Edward D. Hall, PhD
Spinal Cord and Brain Injury Research Center (SCoBIRC)
Department of Anatomy and Neurobiology, Neurosurgery, Neurology, and Physical Medicine and Rehabilitation
University of Kentucky Medical Center
Lexington, Kentucky

Ronald L. Hayes, PhD
Banyan Biomarkers, Inc.
Alachua, Florida

Daniel Hirt, MD
Department of Neurosurgery
Brain Injury Research Center
David Geffen School of Medicine at UCLA
University of California–Los Angeles
Los Angeles, California

Stuart W. Hoffman, PhD
Rehabilitation Research and Development Service
Office of Research and Development
U.S. Department of Veterans Affairs
Washington, DC

Jiamei Hou, MD, PhD
Department of Physiological Sciences
University of Florida
and
Malcom Randal VA Medical Center
Brain Rehabilitation Research Center of Excellence
North Florida/South Georgia Veterans Health System
Gainesville, Florida

Leeanna El Houjeiri, BSc
Department of Biochemistry and Molecular Genetics
American University of Beirut
Beirut, Lebanon

Kurt Kaczmarek
Tactile Communication and Neuromodulation Laboratory
Department of Biomedical Engineering
University of Wisconsin–Madison
Madison, Wisconsin

Alaa Kamnaksh, PhD
Center for Neuroscience and Regenerative Medicine
Uniformed Services University
Bethesda, Maryland

Aida Khodadad, BS
Department of Child Health
University of Arizona College of Medicine–Phoenix
Phoenix, Arizona

and

University of Strasbourg
Strasbourg, France

Mark S. Kindy, PhD
Department of Neurosciences
and
Department of Regenerative Medicine and Cell Biology
Medical University of South Carolina
and
Ralph H. Johnson VA Medical Center
Charleston, South Carolina

and

Department of Bioengineering
Clemson University
Clemson, South Carolina

Anthony E. Kline, PhD
Physical Medicine & Rehabilitation
Safar Center for Resuscitation Research
Center for Neuroscience
Center for the Neural Basis of Cognition, Psychology
Critical Care Medicine
University of Pittsburgh
Pittsburgh, Pennsylvania

Firas Kobeissy, PhD
Department of Psychiatry
Center for Neuroproteomics and Biomarker Research
University of Florida
Gainesville, Florida

and

Department of Biochemistry and Molecular Genetics
Faculty of Medicine
American University of Beirut
Beirut, Lebanon

Vassilis E. Koliatsos, MD
Department of Pathology, Division of Neuropathology
and
Department of Neurology
and
Department of Psychiatry and Behavioral Sciences
Johns Hopkins University School of Medicine
Baltimore, Maryland

Kelsey Korp, BS
Department of Child Health
University of Arizona College of Medicine–Phoenix
Phoenix, Arizona

Jonathan R. Korpon
Safar Center for Resuscitation Research
University of Pittsburgh
Pittsburgh, Pennsylvania

Christian W. Kreipke, PhD
Veterans Administration Health Centers
Detroit, Michigan

Robert A. Laskowski
Program in Neuroscience
Drexel University College of Medicine
Philadelphia, Pennsylvania

Wendy Leung, BS
Neuroprotection Research Laboratory
Massachusetts General Hospital
and
Harvard Medical School
Boston, Massachusetts

Jonathan Lifshitz, PhD
Translational Neurotrauma Research Program
BARROW Neurological Institute
Phoenix Children's Hospital
and
Department of Child Health
University of Arizona College of Medicine–Phoenix
and
Phoenix VA Healthcare System
Phoenix, Arizona

and

Neuroscience Graduate Program
Arizona State University
Tempe, Arizona

Pavel N. Lizhnyak
Department of Anatomy and Neurobiology
Virginia Commonwealth University School of Medicine
Richmond, Virginia

Eng H. Lo, PhD
Neuroprotection Research Laboratory
Departments of Neurology and Radiology
Massachusetts General Hospital
and
Harvard Medical School
Boston, Massachussetts

Josephine Lok
Neuroprotection Research Laboratory
Department of Pediatrics
Division of Pediatric Critical Care Medicine
Massachusetts General Hospital
and
Harvard Medical School
Boston, Massachusetts

Jennifer L. Lowing, MS
Department of Neurosurgery
Wayne State University School of Medicine
and
John D. Dingell VA Medical Center
Detroit, Michigan

Sylvia Lucas, MD, PhD
Department of Neurology
University of Washington Medical Center
Seattle, Washington

Bruce G. Lyeth, PhD
Department of Neurological Surgery
University of California–Davis
Davis, California

Sindhu K. Madathil, PhD
Spinal Cord and Brain Injury Research Center (SCoBIRC)
University of Kentucky
Lexington, Kentucky

Joelle Makoukji, PhD
Neurogenetics Program
Division of Pediatric Neurology
Departments of Pediatrics and Adolescent Medicine, Biochemistry, and Molecular Genetics
American University of Beirut
Beirut, Lebanon

Rebekah Mannix, MD, MPH
Division of Emergency Medicine
Boston Children's Hospital
and
Harvard Medical School
Boston, Massachusetts

William L. Maxwell, PhD, DSc
Veterinary Medicine and Life Sciences
School of Medicine
University of Glasgow
Glasgow, Scotland, United Kingdom

Laura Stone McGuire, BS
The Miami Project to Cure Paralysis
University of Miami Miller School of Medicine
Miami, Florida

William P. Meehan III, MD
Division of Sports Medicine
Boston Children's Hospital
Micheli Center for Sports Injury Prevention
Harvard Medical School
Boston, Massachusetts

Christina M. Monaco, BS
Physical Medicine and Rehabilitation
Safar Center for Resuscitation Research
Philadelphia College of Osteopathic Medicine Graduate Program in Biomedical Sciences
University of Pittsburgh
Pittsburgh, Pennsylvania

Stefania Mondello, MD, PhD
Department of Neurosciences
University of Messina
Messina, Italy

Rabih A. Moshourab, MD, DESA
Department of Anesthesia and Intensive Care
Campus Charité Mitte und Virchow-Klinikum
Charité–Universitätsmedizin
Berlin, Germany

Tarek H. Mouhieddine, MD
Faculty of Medicine
American University of Beirut Medical Center
Beirut, Lebanon

Nathaniel W. Nelson, PhD
Graduate School of Professional Psychology
University of St. Thomas
St. Paul, Minnesota

and

Minneapolis VA Health Care System
Minneapolis, Minnesota

Amaly Nokkari, MS
Department of Biochemistry and Molecular Genetics
American University of Beirut
Beirut, Lebanon

Nicole D. Osier, PhD
School of Nursing
and
Safar Center for Resuscitation Research
University of Pittsburgh
Pittsburgh, Pennsylvania

Andrew K. Ottens, PhD
Department of Anatomy and Neurobiology
Virginia Commonwealth University School of Medicine
Richmond, Virginia

Linda Papa, MD, MSc
Department of Emergency Medicine
Orlando Regional Medical Center
Orlando, Florida

Ramesh Raghupathi, PhD
Department of Neurobiology and Anatomy
Drexel University College of Medicine
Philadelphia, Pennsylvania

Michelle Ramia, MD
Department of Emergency Medicine
Orlando Regional Medical Center
Orlando, Florida

Helen F. Ray-Jones, BS
Department of Child Health
University of Arizona College of Medicine–Phoenix
Phoenix, Arizona

and

Department of Biology and Biochemistry
University of Bath
Bath, United Kingdom

Andrew Rolfe, BS
Department of Neurosurgery
School of Medicine
Virginia Commonwealth University
Richmond, Virginia

Rachel K. Rowe, PhD
BARROW Neurological Institute
Phoenix Children's Hospital
and
Department of Child Health
University of Arizona College of Medicine
Phoenix, Arizona

Kathryn E. Saatman, PhD
Spinal Cord and Brain Injury Research Center (SCoBIRC)
University of Kentucky
Lexington, Kentucky

Mirna Sabra, PhD
Faculty of Sciences
Lebanese University
Beirut, Lebanon

Michael Schäfer
Department of Anesthesia and Intensive Care
Campus Charité Mitte und Virchow-Klinikum
Charité–Universitätsmedizin
Berlin, Germany

Gerry Shaw, PhD
EnCor Biotechnology Inc.
Gainesville, Florida

Justin Sick, MS
The Miami Project to Cure Paralysis
University of Miami Miller School of Medicine
Miami, Florida

Thomas Sick, PhD
Department of Neurology
University of Miami Miller School of Medicine
Miami, Florida

Kimberly Skinner
Tactile Communication and Neuromodulation Laboratory
Department of Biomedical Engineering
University of Wisconsin–Madison
Madison, Wisconsin

Jihane Soueid, PhD
Neurogenetics Program and Division of Pediatric Neurology
Departments of Pediatrics and Adolescent Medicine, Biochemistry, and Molecular Genetics
American University of Beirut
Beirut, Lebanon

and

Faculty of Sciences
Lebanese University
Fanar, Lebanon

Scott R. Sponheim, PhD
Minneapolis VA Health Care System
and
Department of Psychiatry
University of Minnesota–Twin Cities
Minneapolis, Minnesota

Dong Sun, MD, PhD
Department of Neurosurgery
School of Medicine
Virginia Commonwealth University
Richmond, Virginia

Aravind Sundaramurthy, PhD
Center for Injury Biomechanics, Materials, and Medicine
New Jersey Institute of Technology
Newark, New Jersey

Richard L. Sutton, PhD
Department of Neurosurgery
Brain Injury Research Center
David Geffen School of Medicine at UCLA
University of California–Los Angeles
Los Angeles, California

Anna N. Taylor, PhD
Department of Neurobiology
Brain Research Institute and Brain Injury Research Center
David Geffen School of Medicine at UCLA
University of California–Los Angeles
Los Angeles, California

Roya Tehranian-DePasquale, PhD
Physical Medicine and Rehabilitation
Safar Center for Resuscitation Research
University of Pittsburgh
Pittsburgh, Pennsylvania

Theresa Currier Thomas, PhD
Translational Neurotrauma Research Program
BARROW Neurological Institute at Phoenix Children's Hospital
and
Department of Child Health
University of Arizona College of Medicine–Phoenix
and
Phoenix VA Healthcare System
Phoenix, Arizona

Floyd J. Thompson, PhD
Malcom Randal VA Medical Center
Brain Rehabilitation Research Center of Excellence
North Florida/South Georgia Veterans Health System
and
Departments of Physiological Sciences and Neuroscience
University of Florida
and
McKnight Brain Institute
Gainesville, Florida

Hale Zerrin Toklu, PhD
Department of Pharmacology and Therapeutics
University of Florida College of Medicine
and
Geriatric Research Education and Clinical Center
Malcolm Randall Veterans Affairs Medical Center
Gainesville, Florida

Arnold Toth, MD
Department of Neurosurgery
Pécs Medical School
Pécs, Hungary

Nihal Tümer, PhD
Department of Pharmacology and Therapeutics
University of Florida College of Medicine
and
Geriatric Research Education and Clinical Center
Malcolm Randall Veterans Affairs Medical Center
Gainesville, Florida

Mitchell Tyler
Tactile Communication and Neuromodulation Laboratory
Department of Biomedical Engineering
University of Wisconsin–Madison
Madison, Wisconsin

Anatoliy V. Vakulenko, PhD
Banyan Biomarkers, Inc.
Alachua, Florida

Alexey Vertegel, PhD
Department of Bioengineering
Clemson University
Clemson, South Carolina

Amy K. Wagner, MD
Center for Neuroscience
Department of Physical Medicine and Rehabilitation
University of Pittsburgh School of Medicine
and
Safar Center for Resuscitation Research
University of Pittsburgh
Pittsburgh, Pennsylvania

Samantha J. Walas, BS
Neuroprotection Research Laboratory
Massachusetts General Hospital
and
Harvard Medical School
Boston, Massachusetts

Todd E. White, PhD
Department of Neurobiology
Neuroscience Institute
Morehouse School of Medicine
Atlanta, Georgia

Stephanie M. Wolahan, PhD
Brain Injury Research Center
Department of Neurosurgery
David Geffen School of Medicine at UCLA
University of California–Los Angeles
Los Angeles, California

Limin Wu, MD
Neuroprotection Research Laboratory
Massachusetts General Hospital
and
Harvard Medical School
Boston, Massachusetts

and

Neuroscience Center, Department of Neurology
First Hospital of Jilin University
Jilin University
Changchun Jilin, China

Leyan Xu, PhD
Department of Pathology
Division of Neuropathology
Johns Hopkins University School of Medicine
Baltimore, Maryland

Hong Qu Yan
Safar Center for Resuscitation Research
Neurological Surgery
University of Pittsburgh
Pittsburgh, Pennsylvania

Hiyab Yohannes
Department of Anatomy and Neurobiology
Virginia Commonwealth University School of Medicine
Richmond, Virginia

Chenggang Yu, PhD
Henry M. Jackson Foundation for the Advancement of Military Medicine
Bethesda, Maryland

Henrik Zetterberg
Institute of Neuroscience and Physiology
Department of Psychiatry and Neurochemistry
Sahlgrenska Academy at the University of Gothenburg
Mölndal, Sweden

and

UCL Institute of Neurology
London, United Kingdom

Jenna M. Ziebell, PhD
BARROW Neurological Institute
Phoenix Children's Hospital
and
Department of Child Health
University of Arizona College of Medicine–Phoenix
Phoenix, Arizona

1 Experimental CNS Trauma
A General Overview of Neurotrauma Research

Jonathan Lifshitz

CONTENTS

1.1 CLINICAL SIGNIFICANCE OF NEUROTRAUMA RESEARCH: WHAT DOES TRANSLATIONAL MEAN?

Neurotrauma involves the neurological consequences of external forces to the brain, spinal cord, or body. Typically, neurotrauma is associated with car accidents, falls, contact sports, and assault. Diagnosis, treatment, and rehabilitation require a diverse team of knowledgeable experts to deliver excellence in clinical care. To do so, translational research forms the foundation that provides knowledge to advance our understanding and treatment of neurotrauma and thereby empowers clinical providers to deliver the most advanced and effective clinical interventions and treatments. To accomplish translational research, academic scientists and clinicians work collaboratively to generate new knowledge that will ultimately improve patient care. Initially, clinicians help to frame the questions that fill gaps in knowledge that will advance care. Scientists act on these questions and push the envelope of discovery to understand the causes and optimal treatments for neurotrauma. On the other hand, results from basic science investigation shed light on clinical protocols to improve health care delivery. In this way, translational research is the collaborative communication and sharing of knowledge between the laboratory and the clinic.

Traumatic brain injury (TBI) significantly affects the civilian population and is compounded by its prevalence in active duty military personnel. At least 1.4 million civilian TBIs occur in the United States each year, with 50,000–80,000 resulting in death, 200,000–300,000 moderate to severe injuries resulting in hospitalization, and the remaining 1.1 million milder severity injuries treated in the emergency department (Faul, 2010). Today, significant numbers of people are surviving traumatic accidents because of improvements in emergency response, neurological intensive care, and surgical treatments; however, prolonged disability quickly becomes a reality in 20%–50% of mild injuries within weeks of an injury (Faul, 2010; McAllister, 1992). The demographics of TBI are equivalent across race and religion, education, socioeconomic class, health, and genetics. However, young adult males (ages 18–34) and the elderly contribute the greatest to the epidemiology of TBI. It is imperative to implement animal models that replicate the pathophysiology and neurological symptoms to speed the translation of rational therapeutic interventions for chronic neurological dysfunction after injury.

1.2 THE TIME FRAME FOR TBI

TBI is categorized as an acquired brain injury initiated by mechanical forces applied almost instantaneously. This excludes TBIs such as crush injuries, hemorrhages, or intracranial tumors. TBI occurs with the transient application (~20 msec) of a mechanical force (impact or acceleration-deceleration) to the brain that damages cellular membranes, axons, and the vasculature, leaving histopathological hallmarks of TBI across multiple brain regions.

After the primary mechanical injury, secondary molecular, biochemical, and cellular events cause further neuronal, glial, and vascular injuries (Andrews et al., 1990; Miller, 1986). The molecular and cellular processes initiated at the time of injury unfold in an orchestrated series of events, leading to subsequent events further downstream. The rate and distribution of these signaling cascades depend on the parameters of the injury and the injured individual or rodent. In some instances, molecular events start and end rapidly, as would be the case for hemorrhage, membrane permeability, and generation of reactive oxygen species. In other cases, anatomical events progress over prolonged time frames, such as apoptotic cell death and neurogenesis. Evidence points to a nonlinear time course of secondary injury processes between clinical and experimental TBI. The primary injuries occur in the same time frame, followed by immediate secondary injury processes that are likely in sync. However, as the time since the injury progresses, rodent pathology and pathophysiology accelerates ahead of the clinical condition, as indicated by the time course of cerebrometabolic uncoupling (Bergsneider et al., 2000). The principal benefit is that animal models of brain injury permit rapid evaluation of injury-induced pathophysiology, particularly of the long-term consequences.

1.3 BEHAVIORAL DEFICITS AND ENDURING MORBIDITY DEFINE TBI PHASES

TBI leaves individuals with acute injury-induced deficits that can develop into persistent functional morbidity within one month after injury (Langlois et al., 2004; McAllister 1992). After a severe injury, posttraumatic symptoms are injury-induced neurological deficits. However, individuals with mild, diffuse TBI may not receive medical care at the time of the injury and only days, weeks, or even months later begin to articulate transient and mild to ongoing and debilitating posttraumatic morbidity (Faul, 2010; McAllister, 1992).

Injury-induced neurological deficits are a transient consequence of injury-related pathology that impair brain function, including cognitive, motor, and sensory domains, which naturally recover with time after injury. Animals subjected to TBI experience neurological deficits immediately after injury that typically resolves within a few weeks after injury. The cellular mechanisms of injury-induced deficits lie in impaired cellular function because of transient electrical or chemical imbalance along with the ensuing processes to restore homeostasis.

Posttraumatic morbidity is the long-term behavioral and psychological consequence of injury-related pathological processes that impair brain circuit function and activation, including problems with cognition, sensory processing, communication, and behavior/mental health (Chen and D'Esposito, 2010; McAllister, 1992). The multiplicity of symptoms indicates that the organic basis for posttraumatic morbidity that impedes people's return to routines, daily activities, and employment to which they were accustomed preinjury lies scattered throughout the brain. The cellular mechanisms of posttraumatic morbidity lie in impaired circuitry from structural and functional damage in response to injury-related pathological processes. In fact, brain injury survivors exhibit broad and widespread patterns of neural activation compared with control subjects when functionally imaged during cognitive tasks (Christodoulou et al., 2001; Levine et al., 2002), possibly resulting from rewired circuits.

1.4 MILD AND DIFFUSE TBIS ARE DIFFICULT TO DIAGNOSE AND TREAT

Diffuse, and especially mild, brain injury constitutes a majority (more than 80%) of all human TBIs and is a major contributor to posttraumatic morbidity (Graham et al., 2002, 2005; Povlishock and Katz, 2005). Diffuse TBI evolves from acceleration-deceleration forces, typically associated with motor vehicle accidents, automobile-pedestrian contact, and falls. The diffuse forces primarily affect regions where tissue material properties change abruptly, such as gray-white matter interface, the axon hillock, and the blood-brain barrier (Farkas and Povlishock, 2007; Graham et al., 2005; Povlishock and Katz, 2005). It follows that diffuse brain injury mechanically shears axons, vasculature, and membranes. Mechanical disruption promotes cellular signaling, which includes leukocyte infiltration and cerebral exposure to circulating and endogenous cytokines and chemokines (Gentleman et al., 2004; Lenzlinger et al., 2001; Morganti-Kossmann et al., 2007; Passineau et al., 2000; Soares et al., 1995). The ongoing ionic redistribution raises extracellular potassium (Katayama et al., 1990), depolarizes neuronal membranes, and releases large quantities of neurotransmitter. Elevated levels of excitatory amino acids, particularly glutamate (Fei et al., 2005; Goda et al., 2002; Runnerstam et al., 2001), trigger excitotoxic processes that further contribute to cellular damage, including death, axotomy, or synaptic deafferentation (Blinzinger and Kreutzberg, 1968). These processes progress toward neural circuit disconnection coupled with the dysfunction of the entire neurovascular unit (Farkas and Povlishock, 2007; Lo et al., 2004).

Diffuse brain injury establishes a heterogeneous and multifocal neuropathology, interspersed among healthy tissue. The instantaneous nature of TBI produces a synchronized injury in the brain, but the duration and magnitude of injury in different regions may vary. These cellular responses include degenerative and regenerative processes that occur concurrently (Carmichael et al., 2001; Emery et al., 2003). The heterogeneous microscopic pathology avoids detection by routine imaging modalities, making diagnosis more complicated. Thus, part of the strength, but also the challenge, associated with the study of diffuse injury is that, by its very nature, the neuropathology may be indistinguishable or undetectable from spared tissue. In contrast, focal brain injury results in localized tissue destruction often observed in comparison to the contralateral side.

A great challenge lies with mild injury, because far too often no medical advice is sought by those suffering mild TBI. Routinely, mild brain-injured patients present clinically without imaging abnormalities on computed tomography, magnetic resonance imaging, or positron emission tomography scans. In these cases, morbidities emerge after traditional therapeutic windows would have closed, which constrains the available therapeutic approaches. Often, injury-induced morbidities are subtle, such that only under challenging circumstances do the expected responses emerge as attenuated, exaggerated, or inappropriate. Further, limited autopsy specimens, especially of nonfatal diffuse TBI, restrict the ability to explore the cellular basis of uncomplicated posttraumatic morbidity. Plausible mechanisms have to be inferred from clinically relevant experimental models.

Posttraumatic pathology and deafferentation is followed by responsive neuroplastic change in surviving neuronal processes and synaptic terminals (Christman et al., 1997). Plasticity can then promote cellular survival by acquiring trophic support from new local connections and thereby attenuate injury-induced functional deficits (Steward, 1989). However, uncoordinated regenerative responses likely lead to maladaptive circuit reorganization (Povlishock and Katz, 2005), which can have implications for the development of functional morbidity. From weeks to months after injury, a sustained regenerative attempt is evident by increased growth associated protein 43 in regions associated with experimental injury (Christman et al., 1997; Emery et al., 2000; Hulsebosch et al., 1998; Stroemer et al., 1995). Similarly, transcriptional profiling using microarray technology routinely identifies growth factor, cell survival, and plasticity genes (Dash et al., 2004; Rall et al., 2003). Thus, the injured brain is best defined by pathological processes superimposed on reparative processes and, to differing extents, depending on the brain region. Similar neuroplastic responses are proposed to underlie neuropathic pain and spasticity after spinal cord injury (Rabchevsky. 2006; Scholz and Woolf, 2007; Sjolund, 2002; Wang and Thompson, 2008). To this end, for diffuse injury, pharmacological and rehabilitative strategies to combat pathology and regulate repair, rather than surgical intervention, remain as the primary treatment options.

1.5 TBI INCLUDES THE NEUROVASCULAR UNIT, NOT JUST THE NEURONS

The mechanical forces of TBI disrupt the patency of all central nervous system (CNS) compartments, not only damaging neurons. The initial impact and subsequent signaling cascades have effects on neurons, astrocytes, oligodendrocytes, blood vessels, and microglia within the neurovascular units of the brain. Moreover, the compartments and connections within the CNS are stabilized by an intricate interconnected matrix of glycoproteins, proteoglycans, and glycosaminoglycans, known as the extracellular matrix (ECM). All components and compartments of the brain are vulnerable to the mechanical forces and secondary injury processes of TBI.

Diffuse TBI results in the deafferentation of both presynaptic and postsynaptic targets, along with associated components of the neurovascular unit at synaptic sites (e.g., astrocytic processes). During pathological and reparative phases of injury, the extracellular environment must be remodeled to reconstruct circuits in the brain. Soluble molecules, including ECM-degrading enzymes (e.g., matrix metalloproteinases) (Falo et al., 2006; Jia et al., 2010; Wang et al., 2000), redefine the ECM landscape and the associated structural connections between the neural compartments. Therefore, diffuse deafferentation and subsequent repair would disrupt distributed circuits interconnecting central and peripheral systems that control physiological function and higher cortical processing. As an example, thalamocortical projection neurons suffering deafferentation may sprout locally, which would broadly impact information processing toward the cerebral cortices, effectively converting the projection neurons into interneurons (Emery et al., 2000). As such, dramatic behavioral responses and widespread neuronal activation may be the manifestation of maladaptive sprouting in injured brain regions.

1.6 CIRCUITS ARE THE PRINCIPAL CASUALTY OF TBI

Our understanding of neurotrauma can be framed by the way in which the brain repairs itself after injury. The mechanical forces of brain injury disrupt and dismantle brain circuitry. Soon thereafter, endogenous reparative processes proceed to rebuild the circuits, but repairing the injured brain is confounded by ongoing pathology and no coordinated repair strategy. It remains unlikely that injury-induced sprouting responses would reconnect the native circuits in the absence of patterned signals present during development (Emery et al., 2003). Given the complexity of the brain, repaired circuits likely underlie changes in neurological function, including changes in personality, epilepsy, and sensation. Continued investigations are needed into (1) how brain circuits are disrupted by injury, (2) how the brain repairs itself, and (3) ultimately what rehabilitation and pharmacological interventions may promote adaptive reconnection. Regarding treatment, consider that the vast majority of TBIs (>1 million annually in the United States) are sustained by healthy young males who are too proud to seek medical attention, despite uncontrolled secondary processes that can worsen outcome. For these diffuse-injured individuals, self-medication supersedes medical treatment, and they often elect to take over-the-counter anti-inflammatory drugs.

REFERENCES

Andrews, P.J., I.R. Piper, N.M. Dearden, and J.D. Miller. 1990. Secondary insults during intrahospital transport of head-injured patients. *Lancet*. 335:327–330.

Bergsneider, M., D.A. Hovda, S.M. Lee, D.F. Kelly, D.L. McArthur, P.M. Vespa et al. 2000. Dissociation of cerebral glucose metabolism and level of consciousness during the period of metabolic depression following human traumatic brain injury. *Journal of Neurotrauma*. 17:389–401.

Blinzinger, K. and G. Kreutzberg. 1968. Displacement of synaptic terminals from regenerating motoneurons by microglial cells. *Zeitschrift fur Zellforschung und mikroskopische Anatomie (Vienna, Austria: 1948)*. 85:145–157.

Carmichael, S.T., L. Wei, C.M. Rovainen, and T.A. Woolsey. 2001. New patterns of intracortical projections after focal cortical stroke. *Neurobiology of Disease*. 8:910–922.

Chen, A.J. and M. D'Esposito. 2010. Traumatic brain injury: From bench to bedside to society. *Neuron*. 66:11–14.

Christman, C.W., J.B. Salvant, Jr., S.A. Walker, and J.T. Povlishock. 1997. Characterization of a prolonged regenerative attempt by diffusely injured axons following traumatic brain injury in adult cat: A light and electron microscopic immunocytochemical study. *Acta Neuropathologica*. 94:329–337.

Christodoulou, C., J. DeLuca, J.H. Ricker, N.K. Madigan, B.M. Bly, G. Lange et al. 2001. Functional magnetic resonance imaging of working memory impairment after traumatic brain injury. *Journal of Neurology, Neurosurgery, and Psychiatry*. 71:161–168.

Dash, P.K., N. Kobori, and A.N. Moore. 2004. A molecular description of brain trauma pathophysiology using microarray technology: An overview. *Neurochemical Research.* 29:1275–1286.

Emery, D.L., R. Raghupathi, K.E. Saatman, I. Fischer, M.S. Grady, and T.K. McIntosh. 2000. Bilateral growth-related protein expression suggests a transient increase in regenerative potential following brain trauma. *The Journal of Comparative Neurology.* 424:521–531.

Emery, D.L., N.C. Royo, I. Fischer, K.E. Saatman, and T.K. McIntosh. 2003. Plasticity following injury to the adult central nervous system: Is recapitulation of a developmental state worth promoting? *Journal of Neurotrauma.* 20:1271–1292.

Falo, M.C., H.L. Fillmore, T.M. Reeves, and L.L. Phillips. 2006. Matrix metalloproteinase-3 expression profile differentiates adaptive and maladaptive synaptic plasticity induced by traumatic brain injury. *Journal of Neuroscience Research.* 84:768–781.

Farkas, O. and J.T. Povlishock. 2007. Cellular and subcellular change evoked by diffuse traumatic brain injury: A complex web of change extending far beyond focal damage. *Progress in Brain Research.* 161:43–59.

Faul, M., L. Xu, M.M. Wald, and V.G. Coronado. 2010. Traumatic brain injury in the united states: Emergency department visits, hospitalizations and deaths 2002–2006. Atlanta (GA): Centers for Disease Control and Prevention, National Center for Injury Prevention and Control.

Fei, Z., X. Zhang, X.F. Jiang, W.D. Huang, and H.M. Bai. 2005. Altered expression patterns of metabotropic glutamate receptors in diffuse brain injury. *Neuroscience Letters.* 380:280–283.

Gentleman, S.M., P.D. Leclercq, L. Moyes, D.I. Graham, C. Smith, W.S. Griffin et al. 2004. Long-term intracerebral inflammatory response after traumatic brain injury. *Forensic Science International.* 146:97–104.

Goda, M., M. Isono, M. Fujiki, and H. Kobayashi. 2002. Both MK801 and NBQX reduce the neuronal damage after impact-acceleration brain injury. *Journal of Neurotrauma.* 19:1445–1456.

Graham, D.I., T.A. Gennarelli, and T.K. McIntosh. 2002. Trauma. In D.I. Graham and P.L. Lantos (eds.). *Greenfield's Neuropathology.* Vol. 7, pp. 823–898. Arnold Publishers, London.

Graham, D.I., W.L. Maxwell, J.H. Adams, and B. Jennett. 2005. Novel aspects of the neuropathology of the vegetative state after blunt head injury. *Progress in Brain Research.* 150:445–455.

Hulsebosch, C.E., D.S. DeWitt, L.W. Jenkins, and D.S. Prough. 1998. Traumatic brain injury in rats results in increased expression of Gap-43 that correlates with behavioral recovery. *Neuroscience Letters.* 255:83–86.

Jia, F., Y.H. Pan, Q. Mao, Y.M. Liang, and J.Y. Jiang. 2010. Matrix metalloproteinase-9 expression and protein levels after fluid percussion injury in rats: The effect of injury severity and brain temperature. *Journal of Neurotrauma.* 27:1059–1068.

Katayama, Y., D.P. Becker, T. Tamura, and D.A. Hovda. 1990. Massive increases in extracellular potassium and the indiscriminate release of glutamate following concussive brain injury. *Journal of Neurosurgery.* 73:889–900.

Langlois, J.A., W. Rutland-Brown, and K.E. Thomas. 2004. Traumatic brain injury in the United States: Emergency department visits, hopitalizations, and deaths. Centers for Disease Control and Prevention, National Center for Injury Prevention and Control, Atlanta, GA.

Lenzlinger, P.M., M.C. Morganti-Kossmann, H.L. Laurer, and T.K. McIntosh. 2001. The duality of the inflammatory response to traumatic brain injury. *Molecular Neurobiology.* 24:169–181.

Levine, B., R. Cabeza, A.R. McIntosh, S.E. Black, C.L. Grady, and D.T. Stuss. 2002. Functional reorganisation of memory after traumatic brain injury: A study with H(2)(15)0 positron emission tomography. *Journal of Neurology, Neurosurgery, and Psychiatry.* 73:173–181.

Lo, E.H., J.P. Broderick, and M.A. Moskowitz. 2004. tPA and proteolysis in the neurovascular unit. *Stroke.* 35:354–356.

McAllister, T.W. 1992. Neuropsychiatric sequelae of head injuries. *The Psychiatric Clinics of North America.* 15:395–413.

Miller, J.D. 1986. Minor, moderate and severe head injury. *Neurosurgical Review.* 9:135–139.

Morganti-Kossmann, M.C., L. Satgunaseelan, N. Bye, and T. Kossmann. 2007. Modulation of immune response by head injury. *Injury.* 38:1392–1400.

Passineau, M.J., W. Zhao, R. Busto, W.D. Dietrich, O. Alonso, J.Y. Loor et al. 2000. Chronic metabolic sequelae of traumatic brain injury: Prolonged suppression of somatosensory activation. *American Journal of Physiology. Heart and Circulatory Physiology.* 279:H924–931.

Povlishock, J.T. and D.I. Katz. 2005. Update of neuropathology and neurological recovery after traumatic brain injury. *Journal of Head Trauma Rehabilitation.* 20:76–94.

Rabchevsky, A.G. 2006. Segmental organization of spinal reflexes mediating autonomic dysreflexia after spinal cord injury. *Progress in Brain Research.* 152:265–274.

Rall, J.M., D.A. Matzilevich, and P.K. Dash. 2003. Comparative analysis of mRNA levels in the frontal cortex and the hippocampus in the basal state and in response to experimental brain injury. *Neuropathology and Applied Neurobiology.* 29:118–131.

Runnerstam, M., F. Bao, Y. Huang, J. Shi, E. Gutierrez, A. Hamberger et al. 2001. A new model for diffuse brain injury by rotational acceleration: II. Effects on extracellular glutamate, intracranial pressure, and neuronal apoptosis. *Journal of Neurotrauma.* 18:259–273.

Scholz, J. and C.J. Woolf. 2007. The neuropathic pain triad: Neurons, immune cells and glia. *Nature Neuroscience.* 10:1361–1368.

Sjolund, B.H. 2002. Pain and rehabilitation after spinal cord injury: The case of sensory spasticity? *Brain Research. Brain Research Reviews.* 40:250–256.

Soares, H.D., R.R. Hicks, D. Smith, and T.K. McIntosh. 1995. Inflammatory leukocytic recruitment and diffuse neuronal degeneration are separate pathological processes resulting from traumatic brain injury. *The Journal of Neuroscience.* 15:8223–8233.

Steward, O. 1989. Reorganization of neuronal connections following CNS trauma: Principles and experimental paradigms. *Journal of Neurotrauma.* 6:99–152.

Stroemer, R.P., T.A. Kent, and C.E. Hulsebosch. 1995. Neocortical neural sprouting, synaptogenesis, and behavioral recovery after neocortical infarction in rats. *Stroke.* 26:2135–2144.

Wang, G. and S.M. Thompson. 2008. Maladaptive homeostatic plasticity in a rodent model of central pain syndrome: Thalamic hyperexcitability after spinothalamic tract lesions. *The Journal of Neuroscience.* 28:11959–11969.

Wang, X., J. Jung, M. Asahi, W. Chwang, L. Russo, M.A. Moskowitz et al. 2000. Effects of matrix metalloproteinase-9 gene knockout on morphological and motor outcomes after traumatic brain injury. *The Journal of Neuroscience.* 20:7037–7042.

2 Combat TBI
*History, Epidemiology, and Injury Modes**

Ralph G. DePalma

CONTENTS

> All mental processes are biological and so arise from the brain.
>
> —**Franz Joseph Gall, ca 1781**

2.1 INTRODUCTION

Although types and modes of combat injury have changed over the centuries as weapons of war evolved, details about combat traumatic brain injury (TBI) date from the earliest accounts of warfare. This chapter provides a brief historical overview of combat TBI resulting from primitive blunt and penetrating head injuries to current blast-related injuries. Updated numbers of TBI events and injuring mechanisms will be considered. Brain injury causes loss or alteration of consciousness, prograde and retrograde amnesia, and immediate physical and neurological effects ranging from mild to severe. These injuries, in certain cases, cause varying chronic physical, cognitive, and behavioral issues. The most common form of brain injury, acute mild TBI or concussion (mTBI/concussion), has multiple definitions derived from various sources. Vasterling et al. have provided a useful summary of these iterations (Vasterling, 2012).

The operative definition selected for this review includes loss or alteration consciousness for up to 30 minutes at the time of injury, a confused or disoriented state lasting less than 24 hours, memory loss lasting less than 24 hours, and normal structural brain imaging on computed tomographic scanning. Glasgow Coma Scale scores of 13–15 characterize acute mTBI, whereas lower Glasgow Coma Scale scores, 9–12, designate acute moderate TBI. Glasgow Coma Scale scores of 3–8 designate acute severe TBI (Teasdale and Jennett, 1974). Current combat or military TBI/concussions most frequently are classified as mild. Although recovery from mTBI/concussion is said to be the norm, in about 15% (estimates range from 10%–25%) of cases, physical disabilities and symptoms persist beyond three months to become a chronic condition, also known as postconcussion syndrome (Vasterling, 2012). Chronic sequelae of postconcussion syndrome include headache, insomnia, fatigue, sensory, balance, and other neurologic defects as well as cognitive and emotional disorders. Symptoms can be subtle and variable in severity and frequency over time; mTBI and concussion are often used clinically as synonyms. This chapter focuses on mTBI/concussion as a combat injury.

Diagnosis of posttraumatic stress disorder (PTSD), first accepted as a formal diagnosis in 1980 (Horowitz et al., 1980), and other mental illness including depression are reportedly more common in combatants as compared with nondeployed service members during current ongoing military operations (Blakely, 2013). The methods used to obtain estimates affect data concerning numbers of cases of TBI, PTSD, and other mental disorders. Individuals usually are reported only once as a case within a category; data can be presented as the number of diagnoses (prevalence), rate of new diagnoses in a population (incidence), or total number of cases in a population. The total number of diagnoses changes in relation to population size, which for military conditions increases over time with continued combat activities (Blakely, 2013). This chapter uses numbers available from public sources for the Department of Defense (DOD) and updated data through 2013 from the tracking tool used by the Department of Veterans Affairs (VA).

PTSD results from exposure to a traumatic event with risk of serious injury or bodily harm to self or others *and* a response to that event involving intense fear, horror, or helplessness. Symptoms include reexperiencing of the traumatic event, including nightmares and distressing recollections, avoidance

* The views expressed in this chapter are those of the author and do not reflect the position or policy of the Department of Veterans Affairs or the United States government.

of stimuli associated with the trauma with diminished responsiveness and loss of interest in activities, and hyperarousal including irritability, anger, hypervigilance, insomnia, and concentration difficulties. Cognitive and behavioral symptoms of PTSD and depression overlap with those of mTBI; mTBI sustained during the stress of battle is believed to predispose to or accentuate PTSD (Bryant, 2011; Vasterling, 2012).

Historical narratives reveal connections or associations between past and current relationships (Rabins, 2013). Vignettes from past and present conflicts yield insights into the causes and sequelae of combat injuries affecting the brain. Currently, blast injuries predominate among combatants in Operations Enduring Freedom, Iraqi Freedom, and New Dawn (OEF/OIF/OND). Recognition of the importance of mTBI and PTSD relates to enhanced surveillance and clinical guidelines initiated by the DOD and the VA (Management of Concussion/mTBI Working Group 2009). Incidence and prevalence data from both sources provide ongoing estimates of the numbers of service members affected by TBI; the frequency of this particular injury has become a matter of increasing concern. Although the long-term effects of brain damage caused by differing modes of head injury seem to appear identical in the long term (Belanger et al., 2009), recent observations suggest that differing modes of combat injury—for example, blunt as compared with blast injuries—result in differing vestibular-ocular and spinal reflexes (Hoffer et al., 2009) and neural activation responses (Fischer et al., 2013). Behavioral disorders appear to be more common with blast as opposed to blunt injury (Mendez et al., 2013). Repeated injuries, particularly sports-related, have become a public health concern because of their long-term consequences (Jordan, 2013). These emerging observations are important for assessing treatments, outcomes, and disability determinations.

2.2 HISTORY

The Iliad, written by Homer about 800 BCE, described the first casualty of the legendary Trojan War as a combat head injury inflicted upon a warrior, Echepolus, fighting in the front ranks of the defenders:

> *Antilochus, throwing first, struck the horn of the horse head helmet.*
> *And the bronze spear point fixed in his forehead and drove inward*
> *Through the bone, and a mist of darkness clouded his eyes;*
> *Headlong as a tower he fell...*

Blunt and penetrating head injuries from primitive weapons, clubs, spears, and swords caused obvious injuries that were easily understood and described—in contrast, closed-head blast injuries characterize recent combat scenarios. *The Iliad* describes virtually all types of battle wounds and their outcomes in great detail.

Accounts of the epic Trojan War also recorded mental and behavioral disorders in considerable detail. Examples include the rage of Achilles—his savage act of dragging Hector's corpse behind his chariot around the city walls, his refusal to return the body of this noble warrior to his family for burial, and his disputes with the commander of the invading force. The stalwart Achaean, Ajax the Greater, returned to his home, terrorized his wife and child, and slaughtered a herd of sheep he believed to be fellow Greeks. He ended his life by falling upon his sword. Ancient works of art available on the Internet commonly depict these eerily familiar postwar behaviors (Figure 2.1a and b). The duration of the ancient conflict, about 10 years, roughly coincides with the prolonged duration of present hostilities. Publicized similar behaviors by combatants during OEF/OIF/OND have been attributed to PTSD; however, these behaviors could also relate to other mental illnesses or possibly undiagnosed brain trauma. Recent analyses of unethical battlefield conduct suggest that aggression, witnessing war atrocities, and fighting are more strongly associated with unethical conduct than is PTSD (Wilk et al., 2013).

(a)

(b)

FIGURE 2.1 (a) Consequences of the Trojan War: the sorrow of Ajax, with wife and child, upon return home (Asmus Jacob, ca. 1791). (b) Ajax buries the hilt and falls upon his sword. The weapon was a gift from his former foe, Hector of Troy.

Overlapping chronic effects of mTBI, PTSD, and mental illness persist as key issues confronting clinicians and researchers. Symptoms common to each condition complicate estimates of point and lifetime prevalences for each disorder (Carlson et al., 2011) and pose difficulties for diagnosis and management (Bryant, 2011; Hoffman and Harrison, 2009). PTSD and other mental disorders predated use of explosives; these conditions reappear in various forms and guises throughout the history of warfare. The linkage between blast-related mTBI, PTSD, and other mental disorders is a more recent development (Hoge et al., 2008), but conflicting views of the role of each—the dichotomy between mental and physical disorders—persist. For example, in a cohort of deployed combatants referred to a TBI clinic, mTBI was reported to be a predisposing factor for suicidal thoughts and behaviors (Bryan et al., 2013). In sharp contrast, a population-based study reported that suicide risk was associated independently with male gender and mental disorders rather than to military-specific variables (LeardMann et al., 2013). The divergence, a difference between clinical narrative and population-based approaches, invites needed clarification.

Blast TBI, first noted with the use of trinitrotoluene during World War I, related to continual explosive shelling during protracted trench warfare. The introduction of the machine gun had eliminated open battle fields and cavalry charges. Much can be discerned from accounts of a condition then called "shell shock," a term applied to closed-brain blast injury occurring in the absence of overt wounds (DePalma, 2011). An aura of panic attached itself to shell shock. In 1922, the British War Office, based on the report of the Southborough commission (Southborough, 1922), proscribed use of the term, a stricture likely limiting discussion of blast TBI after close of hostilities. Blast TBI reports were published later during the course of World War II; some related to German counterattacks with extensive shelling (Cramer et al., 1949; Draeger et al., 1946; Fulton, 1942). The literature about mTBI/concussion remained relatively silent during and after the Korean Conflict, Vietnam, and the Gulf War.

The contents of the Southborough Commission Report provide insights into blast injury, mTBI/concussion, and probably PTSD along with other mental disorders, some diagnosed as "hysteria." Officially titled *Report of the War Office Committee of Enquiry into "Shell Shock,"* the Right Honorable Lord Southborough chaired the group. Fifteen expert members reviewed fifty-nine witnesses and four patients under the care of the War Ministry suffering from what they called "war neurosis." The charges of the committee were

- Consider the different types of hysteria and traumatic neurosis commonly called shell shock.
- Collate expert knowledge from expert medical authorities derived from war experience.
- Record for future use facts as to its origin, nature, and remedial treatment.
- Advise whether by military training or education, some scientific method of guarding against its occurrence can be devised.

These goals remain relevant today. The committee stipulated that for the purpose of their inquiry, shell shock included

1. a. Commotional disturbances
   b. Emotional disturbances
2. Mental disorders

This group, to its credit, recognized commotional disturbances as a physical entity or syndrome resulting from blast. An appendix to the report lists details of physical examination findings and changes in spinal fluid obtained by sequential lumbar punctures. Spinal fluid analyses compared and contrasted 56 "concussion" cases with 26 "nervous breakdown" cases over time intervals from less than 24 to up to 48 hours or more after blast exposure. In concussion cases, cerebrospinal (CSF) pressure was found to be markedly elevated during the first 24 hours and later fell to normal values. Lymphocytosis occurred along with elevated CSF proteins. In contrast, CSF changes were reported as "slight" in the nervous breakdown group. Clinicians recorded abnormal tendon reflexes in 78 concussion cases compared with minimal changes in 95 nervous breakdown cases. Coordination abnormalities in concussion cases were documented in the medical records; tremor, however, appeared to be more characteristic of nervous breakdown. Clearly, World War I clinicians and committee members were aware of the effects of combat brain injury from blast, including signs and symptoms of the immediate postblast period, if not their long-term effects. Their problem then, as now, remains distinguishing between blast TBI, commotional disturbance, and emotional or mental disorders. In retrospect, CSF findings and focused neurological examinations in the acute postblast period were key to documenting actual brain injury and were used to determine disposition of combat casualties, likely avoiding repeated exposures.

The committee concluded

> a. That concussion or commotion attended by loss of consciousness and evidence of organic lesion of the central nervous system or its adjacent organs (such as rupture of membrana tympani) should be classified as a battle casualty.
> b. That no case of psycho-neurosis or of mental breakdown even when attributable to a shell explosion or the effects thereof, should be classified as a battle casualty....
> c. That in all doubtful cases, it is desirable to have the classification determined by a Board of expert medical officers after observation of the patient in a Neurological Hospital.

The group, or at the least some of its members, grasped the reality of closed blast TBI, notably citing occurrence of ruptured tympanic membranes, recognized as a marker of primary blast exposure (DePalma et al., 2005; Harrison CD et al., 2009; Xydakis et al. 2007). World War I military authorities also endorsed a need for rest after blast exposure, a strategy similar to that occurring during the later stages of the OEF/OIF with the establishment of concussion treatment centers. However, the postwar Southborough Commission likely underestimated and minimized the importance of mental breakdown and its relationship to blast exposure. The committee's final determination can be understood in the context of an overwhelming number of injuries faced by England after World War I. Two and half years after the armistice, approximately 65,000 servicemen were drawing disability pensions for "neurasthenia," among whom 9,000 were undergoing active hospital treatment for mild or moderate blast-related closed TBI and/or PTSD as well as other mental conditions. Delineation and management of these illnesses remain a current problem. The committee made this final recommendation: "Without exception our witnesses condemned the term 'shell shock' and held that it be entirely eliminated from the medical nomenclature."

This opinion was rendered in the context of a remarkably brutal war in which 750,000 British soldiers had died, 400,000 with no known grave. France lost 1.3 million young men and Germany lost 2 million with inscriptions on German War Memorials reading, "Not one too many died for the Fatherland" (Hochschild, 2011). Proscription of the term "shell shock" held until World War II. It is not surprising therefore that neurobehavioral outcomes of mTBI had been overlooked or minimized in the past. On the other hand, Jones et al. (2007), recently reviewing the history of shell shock in World War I and its relationship to current concepts of mTBI, pointed out the dangers of labeling anything as a "signature" injury. Physical disorders crossing a "psychological physical divide," in their opinion, require a nuanced view of diagnosis and treatment. Labels might affect prognosis in that patients who believe that their symptoms have lasting effects will experience more long-lasting disorders. Nonetheless, by 1939 about 120,000 British veterans of World War I had received final awards for primary psychiatric disabilities along with 44,000 getting pensions for "soldier's heart" or "effort-related syndrome." Some disagreed with a compensation policy rewarding functional nervous disabilities (Shephard, 2003).

Better understanding of brain injuries requires timely autopsy studies with detailed neuropathological observations. Autopsy observations of blast effects, apart from those derived from experimental models, are remarkably sparse. In 1916, Mott autopsied cases of "aerial compression" resulting from exposure to high explosives, documenting postmortem punctate hemorrhages and chromatolysis (Mott, 1916). Denny-Brown and Russell (1940) later contested these findings as being due to carbon monoxidemia. Fulton, as mentioned, described brain damage resulting from blast and concussion in World War II (Fulton, 1942). Moore et al. focused attention on the OEF/OIF era, modeling blast concussive effects associated with lethal dose lung injury sustained in the absence of thoracic protection (Moore et al., 2009). They noted that ubiquitous use of body armor made TBI/concussion more common because of survivability and stressed the need for improved head protection.

The pathology and imaging findings of TBI caused by direct skull impact and by vertical or rotational acceleration and deceleration have been documented clinically and experimentally. Typical injuries include brain contusions caused by "coup and contrecoup" brain motion on impact within the skull (Gennerelli, 2005). External forces causing brain rotation or blunt trauma result in more severe brain damage and include subarachnoid hemorrhage, subdural and epidural hematoma, brain contusion and laceration, shearing of nerve fibers, and intracranial hypertension. Understanding these specific brain injury modes in vehicular trauma reduced the effects of TBI by introducing shoulder restraints, seat back extensions, and airbags (Nirula et al., 2003). Overall systemic effects associated with blast exposure include hypovolemia, hypoxia, and sequential inflammatory effects, now blunted to some degree by current use of body armor.

The pathology of repetitive sport injuries has begun to be clarified. Repetitive impact/acceleration injuries cause, in addition to acute concussive effects, chronic traumatic encephalopathy related to deposition of tau and other proteins that promote late neurodegenerative disease (McKee et al., 2009). An autopsy study of 68 males aged 17 to 98 years including 64 athletes and 21 veterans (among whom 86% were athletes) described this unique pathology (McKee et al., 2013). By contrast, the mechanisms and pathology of nonimpact blast-related injury, "the invisible injury," remain ill-defined.

2.3 EPIDEMIOLOGY

Between 2000–2012, service members sustained a total of 255,852 TBIs (DOD Worldwide Numbers for TBI, 2013). These included approximately 212,741 incidences of mTBI, approximately 20,168 classified as moderate TBI, 6,472 incidences of severe TBI/penetrating head injuries, and 16,471 head injuries that remain unclassified. The DOD data include all TBIs seen worldwide, including nonblast TBI caused by vehicle crashes, falls, sports injuries, recreational activity, and military training. Penetrating and other severe head injuries comprised only 3% of this total. Although mechanisms of injury are unavailable for the totality of these occurrences, 80% of them have been estimated to occur in nondeployed settings. However, the diagnosis of TBI in deployed settings nearly doubled between 2010 and 2011 related, in part, to greater focus on identification and treatment of TBIs occurring among deployed service members (MSMR, 2011). The reported incidence of PTSD in service members increased between 2000 to 2011 from approximately 170 diagnoses per 100,000 person years in 2000 to about 1,110 diagnoses per 100,000 person years in 2011, an increase of about 650% (Blakely, 2013), likely related to duration of combat.

The VA screens all presenting OEF/OIF/OND veterans for TBI who have served in a war theater since 2001. This process, mandated by directive, poses four sequential sets of questions: risk of TBI based on event exposure; immediate symptoms following the event; new or worsening symptoms following the event; and current symptoms (VHA Directive 2010-012, 2010). The directive emphasizes that not all those screening positively may have TBI from the presence of PTSD, cervicocranial injury, headaches, and other conditions. A positive response to one or more in each of the four screening questions prompts referral for comprehensive evaluation including detailed inquiries into types, frequency, and mechanisms of combat injuries, and detailed physical and neurologic examinations.

By June 30, 2012, approximately 1,515,700 OEF/OIF/OND veterans left active duty. By December 2012, about 652,445 veterans reported to VA facilities for care and received mandated TBI screening. Approximately 123,000 of these returning veterans screened positive for TBI, predominantly mild TBI/concussion, and were referred for comprehensive specialist evaluations. About 51,900 (~7.95%) of veterans reporting and comprehensively evaluated received a confirmed diagnosis of TBI. More than 95% of these were classified as mTBI. By contrast, by June 30, 2012, approximately 239,200 veterans received a coded diagnosis of PTSD. Among 411,383 OEF/OIF veterans seen in the VA, 30,521 carried a diagnosis of TBI in that year and 72% of those with a TBI also had a PTSD diagnosis in that same year (Taylor et al., 2012). The median additional cost per patient was approximately four times higher for TBI diagnosis as compared with those without TBI ($5,831 vs. $1,547). Within the TBI group, cost increased with diagnostic complexity (Taylor et al., 2012); those with TBI, pain, and PTSD had the highest cost per patient ($7,974).

As of July 31, 2013, data from rehabilitation services indicated that about 760,250 veterans reporting to the VA were screened for TBI. About 3.7% of these presented with a self-reported previous diagnosis of TBI not tracked by the comprehensive screening and comprehensive tracking tools. About 152,248 veterans screened positively for TBI. Among these, 107,635 completed screening and 61,769 (~8.18%) of those reporting and screened as of September 11, 2013, received a confirmed diagnosis of TBI. Most of these, once again, were mTBI.

Increased attention to TBI related to surveillance and clinical follow-up by the VA and the DOD; these joint initiatives aimed to detect and treat TBI/concussion to prevent and possibly to mitigate their persistent effects (Management of Concussion/mTBI Working Group, 2009). PTSD, depression, pain, and substance abuse accompanying mTBI pose challenges for management and disability assessment. A polytrauma study of 613,391 OEF/OIF/OND returning combatants (2009–2011) reported that about 52% of a cohort of returning combatants had one or more of a triad of diagnoses: TBI (9.6%), PTSD (29.4%), and pain (40.7%) (Cifu et al., 2014; Taylor, 2011). Among these estimates of prevalence and coprevalences of combat-associated conditions, about 6% of these veterans had all three diagnoses.

2.4 INJURY MODES: FOCUS ON BLAST EFFECTS

Blast effects fall into four general categories causing bodily injury (DePalma et al., 2005). These are primary (direct effects of shock waves and pressure), secondary (effects from projectiles and debris causing penetrating wounds), tertiary (effects or blunt injuries from a victim being flung through the air and striking an object), and quaternary (burns, hypoxia, and exposure to toxins, toxic inhalants, and other effects). Specific considerations for brain injury focus on the blast shock wave and blast overpressure because these injurious forces encounter the skull. Closed blast TBI has been postulated to relate to vascular surge from the thorax through the neck vessels, air embolism, and piezoelectric currents generated between the skull and the shock wave (Johnson, 2010). Viscoelastic dynamic rippling of the skull secondary to the blast has been postulated based on modeling (Moss et al., 2009). Interactions between the advancing shock wave and blast overpressure, the configuration of the skull, and the brain, including its meninges and cerebrospinal fluid, are complex and cause heterogeneous injury patterns including brain swelling, cerebral vasospasm (Armonda et al., 2006), and diffuse axonal injury (DAI) with disruption across attentional networks (Vakhtin et al., 2013). Brain injury also occurs in conjunction with multiple injuries or polytrauma. For example, a severe combat injury, dismounted complex blast injury, caused by land mines, is characterized by traumatic amputations (Andersen et al., 2012). In these cases, associated TBI relates to several modes of injury: primary blast effects, acceleration, and direct impact from debris. Subsequent combined effects of shock, resuscitation, and air evacuation in these cases can affect brain function and need to be taken into account in severely injured casualties now surviving in unprecedented numbers as compared with prior wars (Kelly et al., 2008).

Kucherov et al. (2012b) postulated a novel mechanism of primary nonimpact blast injury. Calculations show a dramatic shortening the linear scale of the blast shock wave as it passes through brain tissue. The example of a shock wave interacting with water was used, with the assumption that brain tissue's physical properties, on the whole, are quantitatively similar to the properties of water. CSF is even closer to water in its physical characteristics. The proposed mechanism, based upon the dynamic behavior of phonons in water, predicts the length scale of damage to be ~30–200 nm. This phonon-based model recently has been shown to accurately describe failure waves in brittle solids (Kucherov, 2012a). A shock wave traveling through the brain is characterized by a shock front, which is a thermodynamic boundary between shocked and nonshocked states of water. The shock front thickness depends on several parameters and decreases in dimensions relative to the intensity of the shock or blast. For intense shocks, the shock wave front equals the interatomic spacing in the specific medium of propagation. The difference between the two states, the blast wave front and the blast wave pressure, is that some of the energy gets deposited behind the shock front, causing a change in thermodynamic parameters of pressure, volume (density), and temperature. For intense shocks, the change in these parameters becomes pronounced, predicting nanoscale damage occurring within microseconds, in contrast to acceleration injuries having durations measured in milliseconds.

This proposed damage mechanism requires experimental verification using ultrastructural studies. This blast injury mode may account for absence of conventional computed tomographic imaging findings and clinical descriptions of apparent late mTBI progression (Polusny et al., 2011). It predicts that nanoscale cellular damage might cause neuronal apoptosis rather than frank necrosis or DAI from axonal shearing. It also might account for ongoing low-grade inflammation so abundantly demonstrated in experimental models of blast-induced TBI (Cernak and Noble-Haeusslein, 2010). In addition to accounting for blast effects causing the clinical picture of nonimpact mTBI/ concussion, "the invisible injury," the concept of this injury mode may be useful to develop methods to detect, prevent by selective shielding, and possibly to attenuate the effects of nonimpact blast TBI.

Detailed information about specific injury modes and outcomes as previously noted promise novel clinical and research insights. Additionally, DAI after blast exposure has been reported to be different pathologically from that resulting from blunt trauma (Magnuson et al., 2012). Obtaining accurate clinical histories of brain injury mechanisms in austere combat environments pose unique challenges. Nonetheless, better understanding and management of the long-term effects of TBI, given the complexity and heterogeneity of brain injury, particularly mTBI (Rosenbaum and Lipton, 2012), require better characterized longitudinal linkages between original injuries and long-term outcomes.

Federal agencies, including the VA, the DOD, the National Institutes of Health, and the National Institute on Disability and Rehabilitation Research, have focused on diagnosis and understanding of short- and long-term TBI effects, evaluation of existing treatments, and development of novel approaches for rehabilitation and community reintegration. Understanding brain pathology, neurophysiologic responses, and specific treatments for TBI requires autopsy studies, validated imaging techniques revealing changes in brain structure and function, biomarker measurements, functional testing, and prospective randomized treatment trials (Kupersmith et al., 2009). The hope is that these methods will contribute to more objective patient evaluation. Longitudinal clinical assessment of individual responses to immediate treatment strategies remains a crucial need. Because recovery is believed to be the norm, for the present, an episode of mTBI/concussion could be viewed with patients and families as an event rather than as a continuing condition. On the other hand, when disabilities persist as part of a continuing process (Menon et al., 2010), the offer of hope of recovery coexists with provision of financial and social support for returning combatants. We still require better understanding of the injured brain to achieve these goals.

ACKNOWLEDGMENTS

The author thanks David X. Cifu, MD, National Director of Physical Medicine and Rehabilitation, Department of Veterans Affairs, Professor of Rehabilitative Medicine Virginia Commonwealth University, and Douglas Bidelspach, Rehabilitation Specialist, Department of Veterans Affairs, for current Traumatic Brain Injury Data; and Brent Taylor PhD, MPH, Core Investigator for Chronic Disease Outcome Research, for his review of this manuscript and references to the costs of TBI. The author also acknowledges the invaluable assistance of the VA Central Office Library, Ms. Caryl Kazen, Director, and Ms. Vivian Stahl, Reference Librarian, for literature search. I am indebted to my son, Ralph Lawrence DePalma, MD, M.Div., psychiatrist, who long ago pointed out to me the effects of psychotherapy on brain function shown by positron emission and other imaging studies. Franz Joseph Gall's brain-biological insight was cited by Eric R. Kandel: *In Search of Memory: The Emergence of a New Science of the Mind.* 2006; WW Norton & Company, New York and London.

REFERENCES

Andersen, R.C., M. Fleming, J.A. Forsberg, W.T. Gordon, G.P. Nanos, M.T. Charlton et al., 2012. Dismounted complex blast injury. *J Surg Orthop Adv*. 21:2–7.

Armonda, R.A., R.S. Bell, A.H. Vo, G. Ling, T.J. DeGraba, B. Crandall et al., 2006. Wartime traumatic cerebral vasospasm: Recent review of combat casualties. *Neurosurgery*. 59:1215–25; discussion 1225.

Belanger, H.G., T. Kretzmer, R. Yoash-Gantz, T. Pickett, and L.A. Tupler, 2009. Cognitive sequelae of blast-related versus other mechanisms of brain trauma. *J Int Neuropsychol Soc*. 15:1–8.

Blakely, K. and Jansen, D.J., 2013. Post-Traumatic Stress Disorder and Other Mental Health Problems in the Military: Oversight Issues for Congress. Congressional Research Service.

Bryan, C.J., T.A. Clemans, A.M. Hernandez, and M.D. Rudd, 2013. Loss of consciousness, depression, posttraumatic stress disorder, and suicide risk among deployed military personnel with mild traumatic brain injury. *J Head Trauma Rehabil*. 28:13–20.

Bryant, R., 2011. Post-traumatic stress disorder vs traumatic brain injury. *Dialogues Clin Neurosci*. 13:251–62.

Carlson, K.F., S.M. Kehle, L.A. Meis, N. Greer, R. Macdonald, I. Rutks et al., 2011. Prevalence, assessment, and treatment of mild traumatic brain injury and posttraumatic stress disorder: A systematic review of the evidence. *J Head Trauma Rehabil*. 26:103–15.

Cernak, I. and L.J. Noble-Haeusslein. 2010. Traumatic brain injury: An overview of pathobiology with emphasis on military populations. *J Cereb Blood Flow Metab*. 30:255–66.

Cifu, D.X., B.C. Taylor, W.F. Carne, D. Bidelspach, N.A. Sayer, J. Scholten et al., 2014. Traumatic brain injury, posttraumatic stress disorder, and pain diagnoses in OIF/OEF/OND Veterans. *J Rehabil Res Dev*. 50:1169–76.

Cramer, F., S. Paster, and C. Stephenson, 1949. Cerebral injuries due to explosion waves, cerebral blast concussion; a pathologic, clinical and electroencephalographic study. *Arch Neurol Psychiatry*. 61:1–20.

Denny-Brown, D. and W.R. Russell, 1940. Experimental cerebral concussion. *J Physiol*. 99:153.

DePalma, R.G. G.M. Cross, L.B. Beck, and D.W. Chandler, 2011. Epidemiology of mTBI (mild traumatic brain injury) due to blast: History, DOD/VA data bases: Challenges and opportunities. NATO Research Technology Organization. HFM RSY 207 September 2011. MP-HFM-207-01.doc

DePalma, R.G., D.G. Burris, H.R. Champion, and M.J. Hodgson, 2005. Blast injuries. *N Engl J Med*. 352:1335–42.

DOD Worldwide Numbers for TBI, 2013. Defense and Veterans Brain Injury Center, viewed June 15, 2014, from http://dvbic.dcoe.mil/dod-worldwide-numbers-tbi.

Draeger, R.H., J.S. Barr, and W.W. Sager, 1946. Blast injury. *J Am Med Assoc*. 132:762–7.

Fischer, B.L., M. Parsons, S. Durgerian, C. Reece, L. Mourany, M.J. Lowe et al., 2013. Neural activation during response inhibition differentiates blast from mechanical causes of mild to moderate traumatic brain injury. *J Neurotrauma*. 31:169–79.

Fulton, J., 1942. Blast and concussion in the present war. *New Engl J Medi*. 226:1–8.

Gennerelli, T.A. 2005. Neuropathology. In: Silver, J.M., T.W. McAllister, and S.C. Yudofsky, editors. *Textbook of Traumatic Brain Injury.* 2nd ed. American Psychiatric Publishing, Inc., Arlington, VA.

Harrison, C.D., V.S. Barbata, and G.A. Grant, 2009. Tympanic membrane perforation after combat blast exposure in Iraq: A poor biomarker of primary blast injury. *J Trauma*. 67:210–1.

Hochschild, A., 2011. *To End all Wars: A Story of Loyalty and Rebellion, 1914–1918.* Houghton Mifflin Harcourt, Chicago.

Hoffer, M.E., C. Donaldson, K.R. Gottshall, C. Balaban, and B.J. Balough, 2009. Blunt and blast head trauma: Different entities. *Int Tinnitus J*. 15:115–8.

Hoffman, S.W. and C. Harrison, 2009. The interaction between psychological health and traumatic brain injury: A neuroscience perspective. *Clin Neuropsychol*. 23:1400–15.

Hoge, C.W., D. McGurk, J.L. Thomas, A.L. Cox, C.C. Engel, and C.A. Castro, 2008. Mild traumatic brain injury in U.S. soldiers returning from Iraq. *N Engl J Med*. 358:453–63.

Homer. *The Iliad,* transl. S. Butler. Amazon Kindle edition.

Horowitz, M.J., N. Wilner, N. Kaltreider, and W. Alvarez, 1980. Signs and symptoms of posttraumatic stress disorder. *Arch Gen Psychiatry*. 37:85–92.

Johnson, S.C., 2010. Blast induced electromagnetic pulse waves in the brain from bone piezoelectricity. *J Acoustic Soc Am*. 54:S30–6.

Jones, E., N.T. Fear, and S. Wessely, 2007. Shell shock and mild traumatic brain injury: A historical review. *Am J Psychiatry*. 164:1641–5.

Jordan, B.D., 2013. The clinical spectrum of sport-related traumatic brain injury. *Nat Rev Neurol*. 9:222–30.

Kelly, J.F., A.E. Ritenour, D.F. McLaughlin, K.A. Bagg, A.N. Apodaca, C.T. Mallak et al, 2008. Injury severity and causes of death from Operation Iraqi Freedom and Operation Enduring Freedom: 2003–2004 versus 2006. *J Trauma*. 64:S21–6; discussion S26–7.

Kucherov Y., G.K. Hubler, J. Michopoulos, and B. Johnson, 2012a. Acoustic waves excited by phonon decay govern the fracture of brittle materials. *J Appl Phys*. 111:023514.

Kucherov Y., G.K. Hubler, and R.G. DePalma, 2012b. Blast induced mild traumatic brain injury/concussion: A physical Analysis. *J Appl Phys*. 112:104701-1-5.

Kupersmith, J., H.L. Lew, A.K. Ommaya, M. Jaffee, and W.J. Koroshetz. 2009. Traumatic brain injury research opportunities: Results of Department of Veterans Affairs Consensus Conference. *J Rehabil Res Dev*. 46:vii–xvi.

LeardMann, C.A., T.M. Powell, T.C. Smith, M.R. Bell, B. Smith, E.J. Boyko et al., 2013. Risk factors associated with suicide in current and former US military personnel. *JAMA*. 310:496–506.

Magnuson, J., F. Leonessa, and G.S. Ling, 2012. Neuropathology of explosive blast traumatic brain injury. *Curr Neurol Neurosci Rep*. 12:570–9.

Management of Concussion/mTBI Working Group, 2009. VA/DoD Clinical Practice Guideline for Management of Concussion/Mild Traumatic Brain Injury. *J Rehabil Res Dev*. 46:CP1–68.

McKee, A.C., R.C. Cantu, C.J. Nowinski, E.T. Hedley-Whyte, B.E. Gavett, A.E. Budson et al., 2009. Chronic traumatic encephalopathy in athletes: Progressive tauopathy after repetitive head injury. *J Neuropathol Exp Neurol*. 68:709–35.

McKee, A.C., R.A. Stern, C.J. Nowinski, T.D. Stein, V.E. Alvarez, D.H. Daneshvar et al., 2013. The spectrum of disease in chronic traumatic encephalopathy. *Brain*. 136:43–64.

Mendez, M.F., E.M. Owens, E.E. Jimenez, D. Peppers, and E.A. Licht, 2013. Changes in personality after mild traumatic brain injury from primary blast vs. blunt forces. *Brain Inj*. 27:10–8.

Menon, D.K., K. Schwab, D.W. Wright, A.I. Maas; Demographics and Clinical Assessment Working Group of the International and Interagency Initiative toward Common Data Elements for Research on Traumatic Brain and Psychological Health, 2010. Position statement: Definition of traumatic brain injury. *Arch Phys Med Rehabil*. 91:1637–40.

Moore, D.F., A. Jerusalem, M. Nyein, L. Noels, M.S. Jaffee, and R.A. Radovitzky, 2009. Computational biology - modeling of primary blast effects on the central nervous system. *Neuroimage*. 47 Suppl 2:T10–20.

Moss, W.C., M.J. King, and E.G. Blackman, 2009. Skull flexure from blast waves: A mechanism for brain injury with implications for helmet design. *Phys Rev Lett*. 103:108702.

Mott, F., 1916. The effects of high explosives on the central nervous system. *Lancet*. 1:332–315.

MSMR (Monthly Surveillance Medical Report), 2011. External Causes of Traumatic Brain Injury, 2000–2011. Armed Forces Health Surveillance Center. 20(3):13.

Nirula, R., R. Kaufman, and A. Tencer., 2003. Traumatic brain injury and automotive design: Making motor vehicles safer. *J Trauma*. 55:844–8.

Polusny, M.A., S.M. Kehle, N.W. Nelson, C.R. Erbes, P.A. Arbisi, and P. Thuras, 2011. Longitudinal effects of mild traumatic brain injury and posttraumatic stress disorder comorbidity on postdeployment outcomes in National Guard soldiers deployed to Iraq. *Arch Gen Psychiatry*. 68:79–89.

Rabins, P., 2013. Narrative Truth: The Empathic Method. In: *The Why of Things: Causality in Science, Medicine and Life.* Columbia University Press, New York.

Rosenbaum, S.B. and M.L. Lipton, 2012. Embracing chaos: The scope and importance of clinical and pathological heterogeneity in mTBI. *Brain Imaging Behav*. 6:255–82.

Shephard, B., 2003. *A War of Nerves.* Harvard University Press, Harvard..

Southborough, L., 1922. Report of the War Office Committee of Enquiry into "Shell Shock." London, His Majesty's Stationary Office.

Taylor, B.C., 2011. Fiscal Year 2011 VA Utilization Report for Iraq and Afghanistan War Veterans Diagnosed with TBI. Queri, Minneapolis, MN.

Taylor, B.C., E.M. Hagel, K.F. Carlson, D.X. Cifu, A. Cutting, D.E. Bidelspach, and N.A. Sayer, 2012. Prevalence and costs of co-occurring traumatic brain injury with and without psychiatric disturbance and pain among Afghanistan and Iraq War Veteran V.A. users. *Med Care*. 50:342–6.

Teasdale, G. and B. Jennett, 1974. Assessment of coma and impaired consciousness. A practical scale. *Lancet*. 2:81–4.

Vakhtin, A.A., V.D. Calhoun, R.E. Jung, J.L. Prestopnik, P.A. Taylor, and C.C. Ford, 2013. Changes in intrinsic functional brain networks following blast-induced mild traumatic brain injury. *Brain Inj*. 27:1304–10.

Vasterling, J.J., 2012. *PTSD and Mild Traumatic Brain Injury*. The Guilford Press, New York.

VHA Directive 2010-012, 2010. Screening and Evaluation of Possible Traumatic Brain Injury in Operation Enduring Freedom (OEF) and Operation Iraqi Freedom (OIF) Veterans. Department of Veterans Affairs, Washington, DC.

Wilk, J.E., P.D. Bliese, J.L. Thomas, M.D. Wood, D. McGurk, C.A. Castro, and C.W. Hoge. 2013. Unethical battlefield conduct reported by soldiers serving in the Iraq war. *J Nerv Ment Dis*. 201:259–65.

Xydakis, M.S., V.S. Bebarta, C.D. Harrison, J.C. Conner, G.A. Grant, and A.S. Robbins. 2007. Tympanic-membrane perforation as a marker of concussive brain injury in Iraq. *N Engl J Med*. 357:830–1.

Section I

Neuromechanisms in Brain Injury

3 Development of Concepts in the Pathology of Traumatic Axonal and Traumatic Brain Injury

William L. Maxwell

CONTENTS

3.1 INTRODUCTION

The serious and long lasting impact of head injury has been recognized since the paleolithic era, but the potential for treatment of the injured brain has only manifested over the last 50 to 60 years as understanding of cell structure and function within the injured CNS has developed. This chapter reviews the development of ideas concerning the pathobiomechanical response(s) in axons of neurons following transient mechanical loading in central white matter of the CNS. Rather than axons being directly sheared at the time of an insult (primary axotomy), it is now recognized that a complex and incompletely understood series of cellular interactions occurs during hours and days following TBI that may lead to disconnection and disrupted function within and between neuronal networks or circuits, so termed secondary axotomy, following a single, rapid mechanical loading episode. This chapter reviews developments in the experimental literature concerning acute injury to the axolemma and the associated, uncontrolled influx of sodium and calcium ions, intra-axonal release of calcium from mitochondria and the axoplasmic reticulum, and the failure of axonal mitochondria leading to exacerbation of injury to cell membranes. The influx of free calcium into an injured axon mediates proteolysis of the subaxolemma and axonal cytoskeletons, axonal microtubules and neurofilaments. These are reviewed and the progression toward secondary axotomy outlined. Terminal loss of the axonal cytoskeleton by granular degeneration during Wallerian degeneration is reviewed. The possible influences of axonal loss in the post-acute and chronic phases of TBI are reviewed.

Injury to the head and brain has been recognized as a serious clinical scenario since the Paleolithic era (Gross, 2003), and treatment by trepanation of part of the skull was an established and widely practiced therapeutic intervention in Greek and Roman civilizations (Missios, 2007). Hippocrates (460–377 BCE), conventionally described as being the father of medicine, provided a major contribution toward the development of neurosurgery in the classical era by publication of "On Injuries to the Head" written around 400 BCE (Missios, 2007). Hippocrates and his students developed a ranking of five subtypes of injury to the head: (1) a bone contusion with a fissured fracture; (2) a simple bone contusion without fractures; (3) depressed skull fractures; (4) dinted fractures induced by a blow to the head by a weapon leaving an indentation or "hedra" ("εδρα"); and (5) most pertinent to this review, "injuries at distant sites."

Importantly, Hippocrates recognized that injury to the brain may occur in a different part of the brain from a primary wound to the surface of the head. Furthermore, it was widely recognized that wounds to the head were especially dangerous and that death was a high-risk sequel of severe head injury. This chapter will review changes in thinking about the mechanisms and/or causes of injury to the brain developed through the use of experimental models of traumatic brain injury (TBI) and the appreciation at a clinical scenario of an increased understanding of TBI, the ability to monitor and assess progression of pathology within the injured brain, and the outcomes that a patient and their family may anticipate during posttraumatic survival.

3.2 TBI: SIMPLE OR COMPLEX?

Over the past seventy years or so, there have been major changes in our understanding of responses by cells within the brain to a mechanically induced insult. Frequently, analyses had been focused on responses by neurons to specific or particular types of insult to the brain, for example, cerebrovascular and cardiovascular insults or traumatic, mechanical injury. But the brain is composed of not a single population of cells, neurons, but a closely interacting population of different cell types, all of which may respond to a biomechanical or other insult. In 2001, the American Stroke Association recognized such intercellular cross-talk between blood vessels, perivascular glial cells, and neurons and termed the collective the "neurovascular unit" (Figure 3.1). This concept was designed to encourage future research in models of stroke and transient ischemic attacks to investigate the collective properties of the component cells and extracellular matrix of central nervous system (CNS) tissues (Barros et al., 2011). In addition to neurons or parts of neurons, endothelial cells of capillaries in the CNS (Chodobski et al., 2011), perivascular and interstitial astrocytes (Sidoryk-Wegrzynowicz et al., 2011; Sofroniew and Vinters, 2010), oligodendrocytes (Dewar et al., 2003), and microglial cells have been reported to respond to a mechanical insult to the CNS (Venkatesan et al., 2010).

This reductionist approach provided only limited data of value in clinical treatment of a traumatic brain-injured patient because the intimate and complex interactions between neurons, glia, and the immune system were not considered. Recently, however, a consensus within the TBI literature now recognizes that complex, pathophysiological responses occur during and after injury between an integrated range of different cell types ranging from endothelium to all types of CNS glial cells and neurons, the specialized cells of the CNS vascular system, and, recently, the immune system (Bigler, 2013; Johnson et al., 2013a).

A schema to represent the contributing cell types that form the "neurovascular unit" and the recognized interactions between different cell types in a part of the CNS in which axons within central white matter have been exposed to mechanical injury appears in Figure 3.1. Stress induced by mechanical insult results in damage to the plasmalemma of each cell type within the neurovascular unit and results in a loss of homeostatic regulation by that cell. Neurons depolarize and if homeostasis is not reinstated acutely the "calcium cascade" is initiated. Depolarization of astrocytes and oligodendrocytes results in generation of "calcium waves" that extend over possibly hundreds of cubic microns as "spreading depression." When injured axons assume the terminal stage of the axonal, pathological cascade discussed in this chapter,

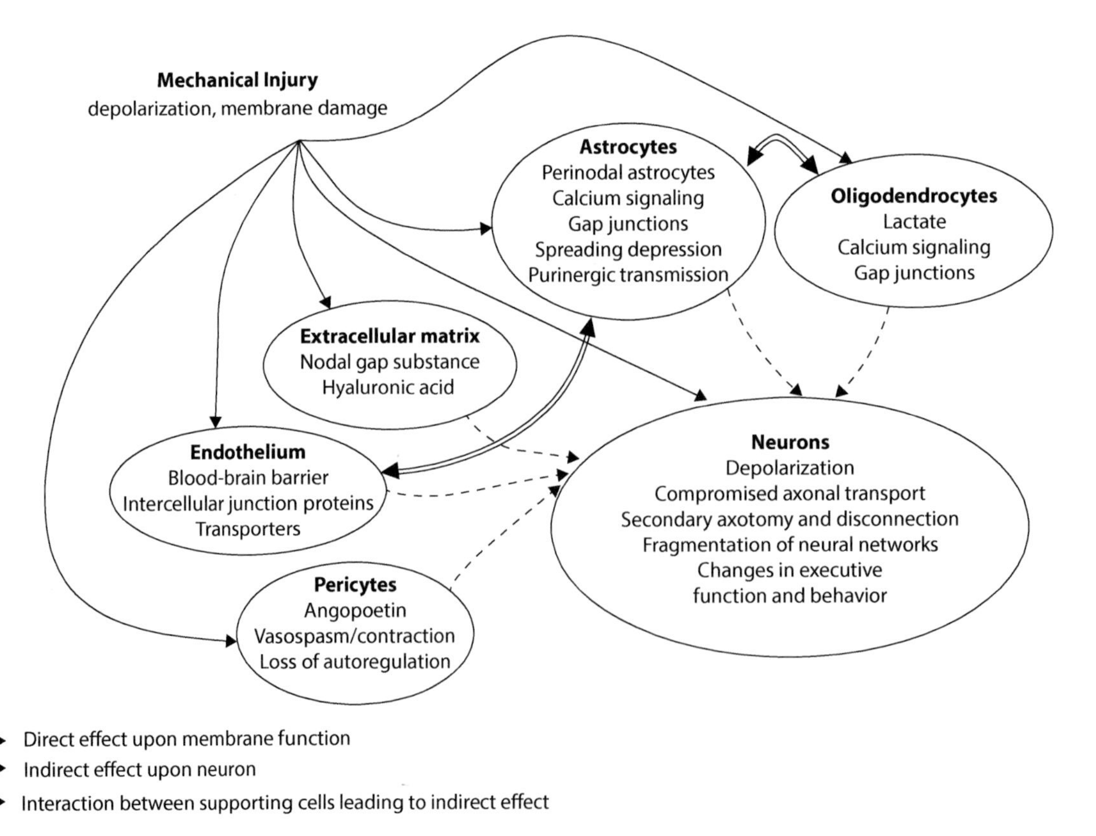

FIGURE 3.1 Components of the "neurovascular unit" and possible interactions between different cell types within the brain neutrophils after direct axonal transient mechanical loading in TBI.

the injured axon fragments disconnect or undergo secondary axotomy. Mechanical loading depolarizes neurons and intercellular signaling between cells (dashed arrows in Figure 3.1) induces Ca^{2+} depolarization of astrocytes and oligodendrocytes through gap junctions and compromises their mitochondrial function, the activity of their transmembrane channels, and induces cellular pathology and axonal degeneration (Figure 3.1). Myelinated axons within the central white matter formerly linking parts of different neural networks disconnect and disrupt the integrity of the said neural networks. Coordination and integration of now separated network nodes is lost and functional integration of cognitive and other neural networks is damaged. The outcome for the patient is a reduction of the efficacy of cognitive, emotional, and memory networks that family members interpret as changes in memory, emotional responses, behavior, and cognitive capacity.

One of the earliest reports in the field of neurotrauma of responses or changes by nonneuronal cells in TBI was made in respect to pericytes (Dore-Duffy et al., 2000) and introduced the concept that coincidental with a traumatic insult to the brain, a transient opening of the blood–brain barrier may occur, followed by activation of the coagulation cascade (Chodobski et al., 2011), in other words, the formation of a blood clot, and that all types of cells at the locus of a mechanical insult to the brain respond in an integrated and coordinated manner. The no-reflow phenomenon that often follows damage to blood vessels may then lead to a significant reduction in blood flow to surrounding parts of the brain and induce either oligemia or, if blood flow through distal vascular branches ceases altogether, focal ischemia. Such reductions in blood flow have been reported in patients after TBI in numerous studies over the past two decades (Schröder et al., 1998; von Oettingen et al., 2002) and now are considered to exercise a considerable influence on development of long-term outcomes in the injured TBI patient.

In conclusion, our understanding of the pathophysiology of injured axons has considerably expanded over the past two decades. Experimental studies have revealed that there is a wide-ranging, complex, and, as yet, poorly understood interaction between an injured axon and its neuronal cell body, between injured axons and all types of glial cell related to that axon, other axons related to these reactive glia, and immune system microglia and blood vessels, all of which potentiate developing pathology in other axons that may possibly extend over decades after a single traumatic episode. This review will attempt to provide an introductory overview of present concepts and how these interact within the injured brain. Contributions in this volume by other investigators will analyze in greater detail specific loci or fields of current investigation.

3.3 HISTOPATHOLOGY AND RESOLUTION OF MODES OF AXONAL DEGENERATION

The major posttraumatic pathology reported in postmortem clinical cases upon whom a neuropathological assessment had been undertaken is the occurrence of so termed "axonal swellings" and "degeneration bulbs" in central white matter (Maxwell et al., 1997; Morrison et al., 2011; Saatman et al., 2011). More recently, the term "axonal varicosities" has been applied to the situation in which a series of swellings interconnected by a contiguous axon has been reported in in vitro studies (Tang-Schomer et al., 2010, 2012).

It was appreciated in the late 1970s or early 1980s that individual axons undergoing pathological degeneration were widely distributed amongst apparently normal or intact nerve fibers in patients who had been exposed to TBI, and, in particular, those diagnosed with diffuse axonal injury (DAI) (Adams et al., 1989; Mendelow and Teasdale, 1983; Saatman et al., 2009). But such axons were difficult to resolve using light microscopy with the Palmgren silver staining technique (Palmgren, 1960). A major advance was reported by Sherriff (Sherriff et al., 1994) in the use of immunocytochemical labeling of foci of accumulation of beta-amyloid precursor protein (β-APP) in degenerating myelinated nerve fibers. Furthermore, use of antibodies against β-APP allowed visualization of damaged nerve fibers in patients that had survived only 0.58 hours after severe TBI (Hortobagyi et al., 2007) rather than the period of 24 hours necessary for identification of injured axons using Palmgren staining.

In addition, β-APP accumulation has been reported in long-term posttraumatic survivors of TBI (Chen et al., 2009; Uryu et al.,, 2007). However, Geddes et al. (2000) cautioned against overinterpretation of such data in central white matter because β-APP may be detected as a result of a derangement of fast axonal transport and this may occur over a range of clinical scenarios. Axonal labeling for β-APP is, therefore, not specific for traumatic injury to axons but may also occur after an ischemic/hypoxic insult, brain swelling, or in a variety of brain diseases. Moreover, axonal bulbs consistently lose positive β-APP labeling after about a week, and no labeling may be obtained if a nerve fiber was injured more than 30 days before death (Geddes et al., 2000). Other immunocytochemical labeling techniques have been reported to identify or label swollen, damaged axons in experimental studies. Examples are labeling for the 68-kD neurofilament subunit core, phosphorylated 200 kDa NF (SMI 31) and nonphosphorylated 150 kDa NF (SMI 32) subunits (Yaghmai and Povlishock, 1992), SMI 32 (Grady et al., 1993; Lee et al., 2011), ubiquitin (Schweitzer et al., 1993), tau (Uryu et al. 2007), compacted neurofilaments (Pettus and Povlishock, 1996; Stone et al., 2001), β–secretase and presenilin-1 (Büki et al., 2000; Chen et al., 2004; Johnson et al., 2013b; Uryu et al., 2007), cytochrome c (Büki et al., 2000), beta amyloid cleaving enzyme (BACE) (Uryu et al., 2007), calpain-mediated spectrin proteolysis (Büki et al., 1999; Saatman et al., 1996, 2003), caspase-3 (Büki et al., 1999; Stone et al., 2002), and caspase breakdown products (Büki et al., 2000).

The use of immunocytochemical techniques to identify injured, damaged, or degenerating nerve fibers has therefore greatly expanded our knowledge about reactive axons over the past twenty years or so and a critical review is beyond the remit of the present chapter. Importantly, however, it should be noted that immunolabeled nerve fibers identified using some or all of these antibodies have been reported also in other types of CNS injury or pathology such as ischemia (Lee et al., 2011; Wang et al., 2012b). There is now a consensus that immunolabeling occurs within nerve fibers wherein loss of the normal spatial relationships of heavy neurofilaments has occurred together with loss of axonal transport but that these immunolabels are not specific markers for nerve fibers injured as a result of traumatic injury.

Nonetheless, development of these techniques has greatly increased the facility to identify and visualize damaged or injured axons within brain deep white matter over a spectrum of both TBI severities and in many degenerative diseases of the human CNS. Because the majority of antibodies now available from commercial sources may be used on either paraffin-embedded or frozen sections obtained from both animal experimental or human pathological material, use of immunocytochemistry has greatly facilitated the identification of injured or reactive nerve fibers in diagnosis of TBI and concussion within the clinicopathological and experimental literature. But, clearly, use of immunocytochemical techniques is limited only to examination of postmortem human material.

A note of caution for consideration in future studies should be raised at this point, however. The reported use of immunocytochemical techniques in experimental and some neuropathological studies has often been limited to use of only one genus of antibody, for example APP and α–beta (Figure 3.2), when these two axonal markers have been used in more than 75% of published experimental studies. However, different antibodies identify or label different proteins, peptides, or epitopes and therefore a differentiation between different pathologies in reactive axons may be resolved. And, perhaps importantly, use of a single marker will not allow a detailed insight as to changes in the number of reactive nerve fibers with posttraumatic survival because individual types of antigens probably identify injured axons at a particular stage of the degenerative posttraumatic pathological cascade. It is suggested that best practice should be use a variety of antibodies against a range of anticipated antigens in an attempt to reduce the risk of obtaining biased or incomplete data in any future quantitative analysis.

Several very important observations or deductions may be drawn from Figure 3.2. First there is a time course for development of axonal pathological change. This was first shown in cats after fluid-percussion injury where reactive axons formed between 1 and 6 hours after injury (Povlishock et al., 1983; Erb and Povlishock, 1988). No evidence in support of the hypothesis that axons sheared or fragmented into smaller axon segments at or close to the time of mechanical insult was obtained. Thus the concept that axons shear in human head injury or DAI (Adams et al., 1989; Strich 1956; Peerless and Rewcastle, 1967) was not supported (Figure 3.2). Neither has support for that concept been obtained in ultrastructural analyses in that at foci of increased axonal calibre the axolemma and myelin sheath are usually morphologically contiguous over at least several hours after injury when discrete examples of axotomy or axonal disconnection have not been obtained until 3 to 4 hours in experimental animals or 12 hours in humans after head injury (Christman et al., 1994).

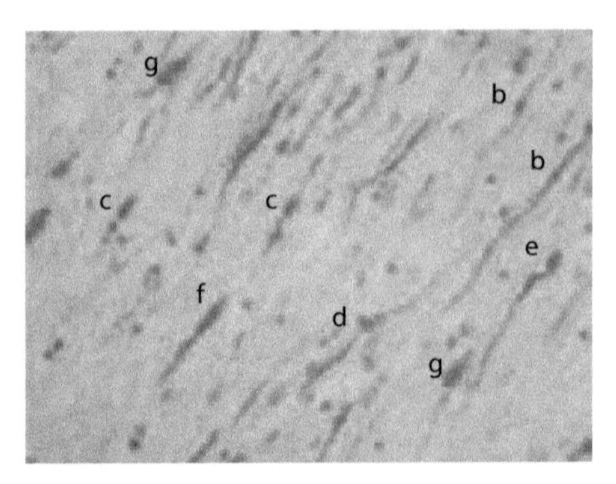

FIGURE 3.2 A colored photomicrograph (immunolabeled for amyloid precursor protein) from the corpus callosum of a patient who died 18 hours after injury. Collectively, these images illustrate the variation in morphology of injured nerve fibers within central white matter over 12 to 18 hours after TBI. Axonal swellings are shown in axons (b) that illustrate foci of increased axonal caliber continuous with adjacent parts of the axon of normal caliber. These were first reported in the late 1980s at one hour after fluid percussion injury in cats (Povlishock et al., 1983; Erb and Povlishock, 1988). Bilobar swellings, frequently separated by short, unlabeled segments of the axon (c) probably represent sites of axonal disconnection or secondary axotomy between two terminal swellings, one on the proximal and the other on the distal axonal fragments of the now disconnected axon. Early studies reported axonal disconnection only at 2 hours' survival in a cat (Povlishock et al., 1983; Erb and Povlishock, 1988). In this section of human corpus callosum obtained at 18 hours' posttrauma comparable axonal profiles are seen (c). That all of the examples illustrated occur within a single section obtained at 18 hours' survival after human TBI provides support for the suggestion that different axons are at different points in the degenerative cascade, with nerve fibers labeled b occurring before axotomy, fibers labeled c representing fibers at the point of axotomy, and fibers d, f, and g representing nerve fiber fragments after secondary axotomy.

3.4 TRANSMISSION ELECTRON MICROSCOPY (TEM)

The great majority of experimental analyses of traumatic axonal injury (TAI) within the TBI literature include use of material examined by TEM. Frequently, immunocytochemical or cytochemical techniques and examination of TEM material has been used together. In this context, two seminal publications were provided by Povlishock et al. (Erb and Povlishock, 1988; Povlishock et al., 1983) in studies using cats. The latter reported axonal responses at a mild level of fluid percussion injury where the pressure change achieved at the surface of intact dura mater during a fluid percussion insult was between 1.8 and 1.9 atmospheres (Povlishock et al., 1983) and in a severe injury (Erb and Povlishock,1988) during which the pressure changes ranged between 2.9 and

3.8 atmospheres. In both studies intraaxonal labeling using horseradish peroxidase identified injured or reactive axons in which the aim was to reproduce in an animal model the morphological changes previously reported in axons within the human corpus callosum and brainstem of patients diagnosed with TBI and DAI (Adams et al., 1989).

After feline axonal injury, discrete intraaxonal pooling of horseradish peroxidase was reported at one-hour survival within longitudinally contiguous axons of the corticospinal, corticorubral, corticoreticular, and cerebellar efferent fibers within the brain stem (Povlishock et al., 1983). Between four and eight labeled nerve fibers were obtained in each 40-µm-thick section of feline brain examined. Clearly, only a small number of injured or reactive nerve fibers were present among a far greater number of intact, unlabeled axons. And with increasing posttraumatic survival, up to 6 hours, intraaxonal peroxidase swellings increased in diameter to 12–15 µm to assume a lobulated profile. And between 12 and 24 hours' survival, labeled swellings achieved 15–40 µm in diameter and formed rounded ball-like and club-like swellings in continuity with their axon only on one side. Povlishock (Povlishock et al., 1983) suggested that at these loci, frank separation or disconnection of an axon had occurred. Comparable examples of axonal pathology are illustrated in Figure 3.2. This specimen was taken from the corpus callosum of a patient who died 18 hours after being exposed to TBI. Damaged axons have been identified by amyloid precursor protein immunolabelling. Povlishock et al. (1983) established a time course for changes is axonal form before, during, and following axotomy, Since all of the above described stages occur in human TBI material obtained from greater than 12 hours survival patients, all of those stages occur in human TBI material obtained from greater than 12 hours' survival patients, it is suggested that the pathological cascade within individual injured nerve fibers may not be initiated only at the time of injury but that there is an ongoing gradual increase in the number of axons entering the pathological cascade with increasing length of posttraumatic survival.

Electron microscopy revealed that mitochondria, axonal microtubules (MTs), and peroxidase-containing smooth endoplasmic or axoplasmic reticulum occurred in axonal swellings (Erb and Povlishock, 1988; Povlishock et al., 1983). Comparable axonal responses have been reported out to 6 hours after stretch-injury to the guinea pig optic nerve (Jafari et al., 1997, 1998). Axonal swellings remained limited by an intact axolemma and contained an enlarged irregular profiled core of axoplasm containing mitochondria, axoplasmic reticulum, and clumped MTs (Figure 3.3b). In some nerve fibers, the axolemma was spatially separated from the internal aspect of the myelin sheath by large, lucent spaces termed *periaxonal spaces* (Figure 3.3a) are suggestive of a change in the caliber of the injured axon, and neurofilaments and MTs assumed a chaotic or nonlinear organization within the axoplasm (Figure 3.3a). Axonal MTs and their interaction with dynein and kinesin are key to fast axonal transport of membranous organelles, and changes in the number and physiology of axonal MTs after traumatic axonal injury results in

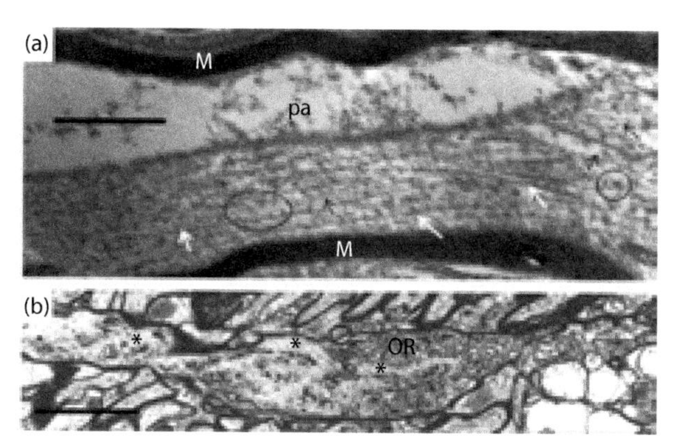

FIGURE 3.3 TEMs of parts of injured axons from the right stretch-injured optic nerve of an adult guinea pig at 2 hours after injury. (a) A longitudinal section of a nerve fiber containing a periaxonal space (pa) that occurs between the contiguous axolemma and the myelin sheath (M) to one side of the axon. Longitudinal sections of axonal microtubules (MTs) (small white arrows) and neurofilaments (NFs) (small black arrows) are seen within the axoplasm. But neither MTs nor NFs follow an organized, parallel linear course. Rather, there is suggestion of a long-pitch spiral orientation and at some points, NFs intersect or cross their neighbors. Marker bar = 0.5 µm. (b) A longitudinal thin section of an axonal swelling within an injured axon at 2 hours after injury. This image contains all the key structural features of an axonal swelling. The axon is limited by an intact myelin sheath and there is no periaxonal space. But the thickness of the myelin sheath appears too thin for the caliber of the nerve fiber because of the accumulation of large numbers of membranous organelles (OR) at foci and the presence of a spiral core of NFs (*) within the axoplasm. Marker bar = 2.0 µm.

accumulation of membranous organelles at foci along the length of injured axons at axonal swellings. Such accumulations of membranous organelles (Figure 3.3b) characterize axonal swellings.

In 1993, Maxwell et al. (1993) tested the hypothesis that examination of injured nerve fibers in animal models of TBI at 1-hour survival may be too late after injury to resolve shearing of central nerve fibers as proposed by the earliest investigators because severed axons reseal their axolemma within minutes of transection (Yawo and Kuno, 1985). Using the lateral head acceleration Penn II device, Maxwell et al. (1993) reported ultrastructural evidence for shearing or tearing of central nerve fibers (Figure 3.4) in nonhuman primates fixed at either 15 or 30 minutes after injury. Examination of nonserial sections through a node and adjacent paranode at 15 minutes after injury (Figure 3.4) illustrates several features.

High-power images of nodal and paranodal regions of injured nerve fibers illustrate disruption of normal nodal and paranodal structure in axons in which the axolemma had been sheared or fragmented (Figure 3.4). At these foci of fragmentation of the axolemma, there is loss of any recognizable axonal cytoskeleton (Figure 3.4). Rather, there is a diffuse, flocculent precipitate, some focal suggestions of linearly organized material, and rounded membrane profiles (small arrows). It has been suggested that the flocculent precipitate represents the breakdown products of the formed elements of the cytoskeleton generated after activation of calcium-activated neutral

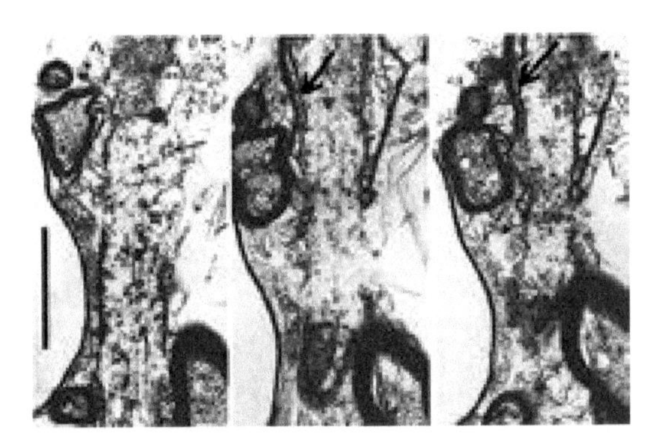

FIGURE 3.4 Nonserial successive, longitudinal TEMs of a node of Ranvier and adjacent paranode from within the corpus callosum of a 20-minute survival nonhuman primate after rotational acceleration of the head. The animal did not regain consciousness before being killed by perfusion fixation for TEM. The nodal axolemma with its characteristic dense undercoating is seen in the left-hand micrograph. But the axolemma does not continue superiorly into the adjacent paranode. Lucent, paranodal glial loops are visible superiorly in the middle and right-hand micrographs (large arrow). There are no longitudinally orientated microtubules (MTs) and neurofilaments (NFs) within the remnants of the axoplasm. Rather there is a diffuse granular precipitate containing several vesicular profiles (small black arrows). These features are suggestive of membrane damage and rapid chemical breakdown of the normal ultrastructural constituents of nodal and paranodal axoplasm as the axon had experienced primary axotomy during rotational acceleration of the brain. Marker bar = 1 μm.

proteases after the immediate influx of free calcium after shearing of the axolemma (Maxwell et al., 1993). The uncontrolled influx of free Ca^{2+} forms the substrate of the "calcium hypothesis" (Tymianski and Tator, 1996), which has been a key part of research studies in TAI for more than 40 years. It is notable that, in the same field, axonal MTs are present within the myelinated fiber to the lower right. This provides evidence that only a proportion of nerve fibers sheared as a result of forces acting during injury (Maxwell et al., 1993).

These findings gave rise to the suggestion that application of transient mechanical loading to central white matter initiates two modes of injury to axons or nerve fibers. One is *primary axotomy,* in which axons are sheared by being stretched rapidly during a mechanical insult to the brain when axons fragment or disconnect at the time of injury and there is rapid proteolysis of the axoplasm (Figure 3.4). *Secondary axotomy* occurs where a probably less severe mechanical loading occurred at the time of a mechanical insult such as a blow to the head and/or uncontrolled acceleration/deceleration of the brain within the cranium. In this paradigm, irregular profiles limited by an intact axolemma extended into the perinodal space at injured or reactive nodes of Ranvier (Figure 3.5a). These were termed nodal blebs (Maxwell, 1996). Membrane-limited vesicles have been reported both at nodal blebs within 15 min after stretch-injury to guinea pig optic nerve fibers (Figure 3.5a) (Maxwell, 1996), and within the internode (Figure 3.6b) at 4 and 6 hours. These membranous vesicles are suggestive of axolemma membrane injury and a cellular response thereto. Tuck and Cavalli (2010) have recently reviewed advances in our understanding of membrane trafficking after axonal injury, and three steps are now recognized. Immediately after injury to the axolemma, influx of calcium ions from the extracellular milieu (Figure 3.5c) allows proteolysis of submembrane spectrin via calcium-dependent proteases or calpains, and accumulation of injury-induced membranous vesicles. The vesicles undergo both exocytosis to provide new-membrane precursors from the cell body, which facilitate membrane repair, and endocytosis, where endosomes are thought to include carriers that convey information back toward the cell soma (Tuck and Cavalli, 2010).

More recent evidence has suggested that the initial calcium peak may originate from intraaxonal sources (Staal et al., 2010), such as the axoplasmic reticulum and mitochondria. Although the axolemma is not directly sheared at the point of injury there is an initial, intraaxonal release of calcium from the axoplasmic reticulum and local mitochondria that potentiates or exacerbates damage to the axolemma. Support for this suggestion has been provided in the guinea pig optic nerve model by evidence for redistribution of transmembrane proteins visualized using freeze-fracture and cytochemistry for loss of or redistribution of membrane pump Ca^{2+} ATPase activity (Figure 3.5b) (Maxwell et al., 1995, 2003) and Na channels (Fehlings and Agrawal, 1995; Wang et al., 2009) in vitro after axonal stretch-injury. An early demonstration of injury-induced calcium influx (Figure 3.5c) was published by Wolf et al. (2001) in an in vitro experiment. However, the literature has significantly expanded over the past decade or more and the interested reader is referred to a recent excellent review by Wang et al. (2012a).

With development of "the calcium hypothesis" (Tymianski and Tator, 1996), a series of detailed analyses of responses by a variety of axonal organelles has been reported and these studies have allowed development of a schema outlining or summarizing a series of intraaxoplasmic responses that culminate in secondary axotomy in an injured central, myelinated nerve fiber. Mechanically induced damage to the nodal axolemma results in disruption of Na^+ channels (Fehlings and Agrawal, 1995; Wolf et al., 2001), depolarization (Ouardouz et al., 2003) and reversed activity of the Na^+/Ca^{2+} exchanger (Agrawal and Fehlings, 1996), loss of transmembrane ionic homeostasis, and uncontrolled influx of free calcium that damages axonal mitochondria and potentiates breakdown of the axonal cytoskeleton.

3.5 INJURY OF AXONAL MITOCHONDRIA

The appearance of swollen, lucent mitochondria within the axoplasm of injured nerve fibers has been reported in almost all ultrastructural studies dating from the 1980s (Erb and Povlishock, 1988; Sherriff et al., 1994; Uryu et al., 2007) to the present day (Figure 3.5e and f). Büki et al. (2000) provided further insight of damage or pathology in axonal mitochondria using the impact acceleration model of TBI. As early as 15 minutes after injury damaged lucent mitochondria lacking cristae were associated with diaminobenzidine labeling of cytochrome-c reaction product either on the

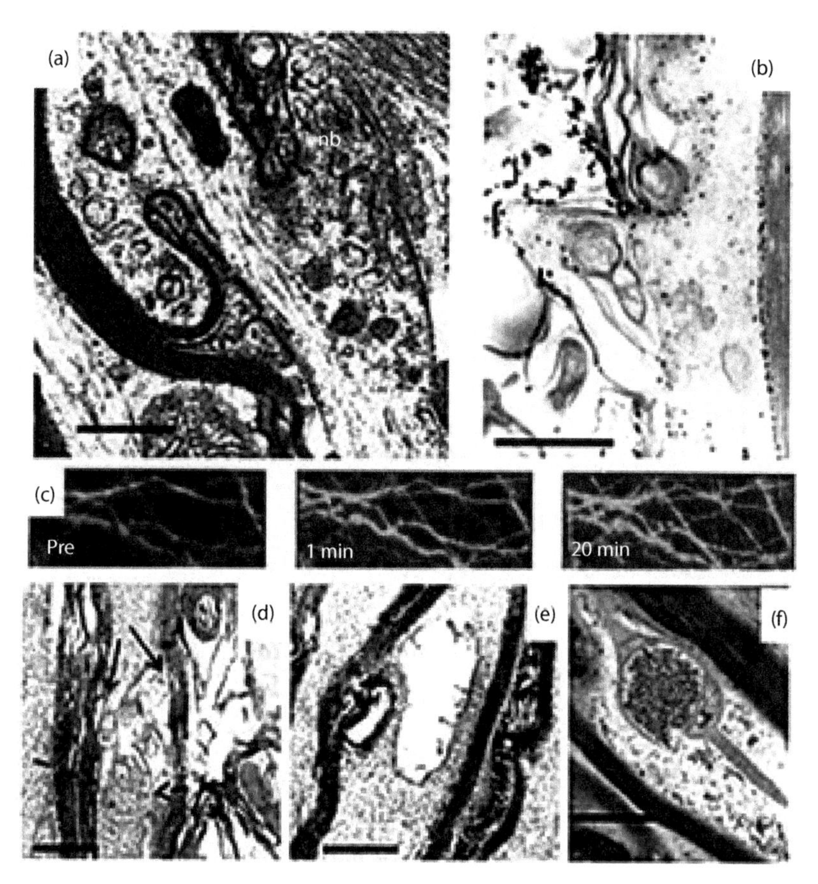

FIGURE 3.5 (a) A longitudinal thin section of a node of Ranvier from the injured right optic nerve of an adult guinea pig at 30 minutes after stretch-injury of the nerve. The ultrastructure differs greatly from that illustrated in Figure 3.4. Paranodal glial loops are adjacent to the external surface of the paranodal axolemma and glial-axonal junctions are discrete. The nodal dense undercoating is visible and several longitudinally orientated microtubules run through the axoplasm. But an axolemma limited nodal bleb (nb) extends from the right side of the node of Ranvier and a similar protrusion of axoplasm extends under and displaces the myelin sheath laterally on the left. Both axoplasmic protrusions contain numerous membranous profiles. Marker bar = 1.0 µm. (b) A longitudinal thin section of the internode of a damaged nerve fiber obtained at 1 hour after stretch-injury. The specimen has been processed for the cytochemical localization of Ca^{2+}-ATPase during fixation. The membrane remnants of the myelin sheath have been disrupted and delaminated on the left-hand side of the figure but remain relatively intact to the right. The axolemma is intact, but has a very irregular profile underlying the locus of disruption of the myelin sheath. On the contrary, the axolemma has a smooth profile immediately subjacent to the relatively intact myelin sheath on the right. The axoplasm contains groups of small membranous vesicles, some of which are decorated with puncta of electron-dense reaction product. It is suggested that these profiles represent an area into which new membrane is being inserted to repair the axolemma that had been damaged at stretch-injury. Marker bar = 0.5 µm. (c) Reprints of figures taken from Wolf et al. (2001) illustrating the time frame over which intraaxoplasmic content of Ca^{2+} increases after stretch-injury to in vitro cultures of axons across a deformable membrane that had been displaced by a puff of air onto one surface. The displacement of the supporting membrane stretched and damaged axons passing over the membrane's surface. The cultured neurons had been incubated with the fluorescent marker for calcium Fuo4 (Wolf et al., 2001) before displacement of their supporting membrane. Before stretch (C, pre) there is no blue fluorescence. At one minute after being stretched, there are numerous fluorescent loci scattered along the length of axons and these fluorescent foci increase in caliber and marker intensity by 20 minutes after injury. There is therefore a gradually increasing influx of free calcium into foci within the axoplasm of stretched axons over real time after a single mechanical insult (Wolf et al., 2001). (d) A TEM obtained from a longitudinal thin section of a 1.0–1.5 µm diameter myelinated axon in stretch-injured guinea pig optic nerve 1 hour after injury. There is a locus of delamination of the myelin sheath (arrows) with membrane profiles extending into the central region of the nerve fiber. Therein the axon is fragmented and contains two lucent mitochondria (dotted arrows). This represents a region of infolded axolemma away from the internal aspect of the myelin sheath. Marker bar = 1 µm. (e, f) Longitudinal thin sections of damaged mitochondria obtained from stretch-injured axons at 1 hour after injury. The section in Figure 3.6e has been routinely processed for thin sectioning, whereas (f) has been processed for pyroantimonate localization of calcium ions. The mitochondrion visible in (e) is swollen and lucent with only a few remnants of profiles of the inner mitochondrial membranes. There is a large locus of pyroantimonate precipitate within the swollen mitochondrion seen in (f). Both micrographs illustrate damage to the mitochondrion within the axoplasm. Marker bars = 0.4 µm.

surface of mitochondria or in close proximity (Figure 3.6a and b) and were closely related to foci of neurofilament compaction and partial digestion (Büki et al., 2000) (Figure 3.6c and d). Moreover, the pathology worsened between 15-minute (Figure 3.6a and c) and 6-hour (Figure 3.6b and d) posttraumatic intervals with an increase in the number of

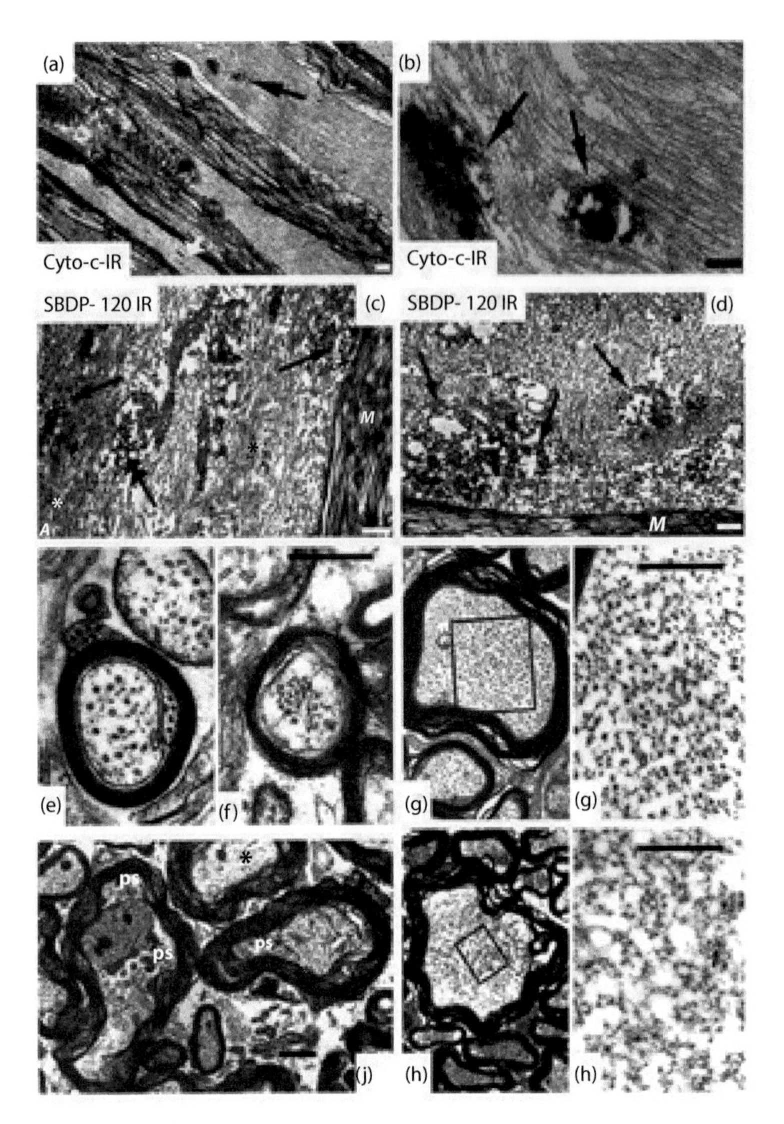

FIGURE 3.6 Copies of transmission electron micrographs published by Büki et al. (2000) with kind permission. (a, b) Low- and high-magnification images from a specimen processed to identify the location of cytochrome c reaction product (Cyto-c-IR) at 15 to 30 minutes after axonal injury. (a) The white arrow indicates an uninjured nerve fiber with compact myelin sheath. The nerve fiber containing the large black arrow is damaged; note the damaged myelin sheath (M). The area indicated by the black arrow is shown at high magnification in (b). Within that micrograph, Cyto-c-IR is localized over or close to two damaged mitochondria indicated by the two black arrows. The right hand labeled mitochondrion also demonstrates the central lucency characteristic of a damaged mitochondrion after axonal stretch-injury and numerous other insults. Notably, the disorganized or nonregular orientation of closely packed neurofilament may also be resolved. The marker bar = 0.5 μm. (e-j) Copies of micrographs published by Jafari et al. (1997) with thanks. (e, g) From control, sham-injured right optic nerves, (f, h) from injured optic nerves at 4 hours after stretch-injury. These micrographs illustrate the ultrastructural reorganization of MTs and neurofilaments within small (0.5–1.0 μm fiber diameter) (e, f) and large (2.0–2.5 μm fiber diameter) nerve fibers (g, h). Uninjured or control nerve fibers illustrate a regular, compact myelin sheath and uniformly spaced MTs and neurofilaments (e). At 4 hours after injury in small nerve fibers (f) although the myelin sheath is still regular in profile, there is an almost complete absence of MT from the axoplasm and neurofilaments form a central core with a reduced spacing between adjacent neurofilaments, termed compaction of neurofilaments. A similar situation pertains in large nerve fibers of the stretch-injured optic nerve (g for controls, h for injured nerve fibers). In the injured nerve fibers, there is an almost complete absence of MTs; one is indicated by the arrow in h, and a nonuniform spacing between adjacent neurofilaments such that lucent lacunae occur between groups of neurofilament, which also have a reduced spacing (h, detail) (Table 3.1). Notably the myelin sheath has an irregular profile with foci of separation of myelin laminae and lucent periaxonal spaces between the axolemma and the internal aspect of the sheath. Marker bar = 1.0 μm. (j) A field of transverse sections of large or intermediate diameter nerve fibers at 6 hours after injury. Some smaller nerve fibers possess normal relations, others possess a lucent axoplasm (*) but retain an intimate relationship to their myelin sheath that, however, has extensive zones of separation of myelin lamellae (j). Two nerve fibers in this field, however, possess extensive periaxonal spaces (ps) and the contained axon has a much reduced caliber (j). The periaxonal spaces (pa) contain loose axoplasmic organelles such as mitochondria and membrane limited profiles. The myelin sheath has an irregular profile and foci of separation of myelin lamellae. The axon appears to have shrunken in diameter in these nerve fibers and the axoplasm has an increased electron density although the occasional mitochondrion appears normal. Examination of the electron dense axoplasm at about 50,000 diameters reveals compaction of neurofilament (Table 3.1) as seen, for example, in Figure 3.7c. Marker bar = 1.0 μm.

immunolabeled damaged axons, and an apparent separation of myelin lamellae within the myelin sheath (labeled M in Figure 3.6). There is now a consensus that a major event in the early stages of axonal responses after TBI is injury to axonal mitochondria after uncontrolled release of or influx of free Ca^{2+} (Figure 3.5f). It is also posited that mitochondrial damage results in a focal loss of production of the essential energy substrate adenosine triphosphate (ATP). ATP content is then rapidly depleted and leads to failure of ATP-dependent membrane pumps close to the locus of injury, loss of ionic homeostasis, and exacerbation of influx of free calcium into the axoplasm that potentiates development of secondary injury paradigms (Hortobagyi et al., 2007).

3.6 RESPONSES BY COMPONENTS OF THE AXONAL CYTOSKELETON

Over the past two decades, a great deal of work has investigated changes to the axonal cytoskeleton in models of TBI. There are, for descriptive purposes, three discrete parts of the axonal cytoskeleton: neurofilaments (10 nm in diameter), neurotubules/MTs (20 nm), and microfilaments (7 nm) of the subaxolemma actin and spectrin complex. The chemistry and spatial relationships of these also differ between nodes of Ranvier, paranodes, internodes, and the initial axonal segment and are associated with a differential distribution of transmembrane proteins that form ions channels and pumps. These ion channels and pumps serve to maintain different concentrations of a variety of ions on opposite sides of the axolemma to maintain the resting membrane potential at −70 mV and allow the passage of the action potential along the length of the axon during depolarization and signal transduction.

3.6.1 NEUROFILAMENTS

Neurofilaments as seen in routine thin sections are linearly organized filamentous proteins forming the unique intermediate filaments of neurons. Neurofilaments are obligate heteropolymers of heavy neurofilaments (NFH), medium (NFM), and light (NFL) neurofilaments. NFM and NFH also possess long C-terminal domains or side arms and at conserved lysine-serine-proline KSXXP regions where phosphorylation may occur, in particular during myelination but also in a number of neurological diseases (Dale and Garcia, 2012). Until very recently, it was thought that phosphorylation of NFM and NFH side arms regulated radial growth of axons. But studies using genetically modified mice have recently questioned this by suggesting that phosphorylation of neurofilaments may regulate, or at least influence, the linkage of neurofilaments to the anterograde motor kinesin and/or the retrograde motor dynein so that neurofilament phosphorylation may, in addition, influence axonal transport (Dale and Garcia, 2012). But the complex interactions between nonphosphorylated and phosphorylated neurofilaments are not well understood and there is accumulating evidence that the functions or roles of NFL, NFM, and NFH differ both between themselves and with stages of development or maturation of a neuron and an organism. For example, deletion of NFM has been reported to result in a reduced radial growth of axons, whereas deletion of NFH resulted in a reduced axonal diameter only in young mice. However, the deletion of either NFH or NFM increased the retrograde rate of transport of neurofilaments toward the cell soma (reviewed in Dale and Garcia, 2012). The complexity of interactions between NFL, NFM, and NFH and their stoichiometry with age at injury, the influence of phosphorylation, and/or dephosphorylation on the interaction between the different neurofilament moieties and changes in possible lowered resistance to trauma or an increased risk of a poorer outcome with aging in both the peripheral nervous system (Verdú et al., 2008) and CNS (Timaru-Kast et al., 2012) are presently poorly reported in the neurotrauma literature in which the majority of experimental studies are carried out on young, mature animals.

A key report in the neurotrauma literature was that neurofilaments within injured axons possess a reduced spacing from their neighboring neurofilaments and were said to be "compacted" (Pettus and Povlishock, 1996). Associated with foci of neurofilament compaction were lucent, mitochondria lacking cristae, low numbers of MTs (Figure 3.6e–h), and separation of the axolemma from the internal aspect of the remnant of the myelin sheath (Figures 3.3a and 3.6j) to form a periaxonal space (Geddes et al., 2000; Jafari et al., 1997, 1998; Schweitzer et al., 1993).

Application of stereological techniques to the guinea-pig optic nerve stretch injury model of TAI has added some insight but also serves to show that understanding of this pathophysiology is still limited. The results of a number of studies (Geddes et al., 2000; Jafari et al., 1997, 1998; Schweitzer et al., 1993; Sulaiman et al., 2011) allow the drawing of the following conclusions. Following TAI, larger diameter nerve fibers underwent secondary axotomy more quickly than smaller fibers (Sulaiman et al., 2011). A dramatic loss of the number of axonal MTs occurred in some axons following TAI/stretch-injury in all sizes/caliber of nerve fibers (Figure 3.6e–h) but there was no change of the spacing between adjacent MT (Table 3.1) when there was partial loss of MT from the axoplasm (Maxwell and Graham, 1997; Sulaiman et al., 2011) (Table 3.1). at the internodes and paranodes of myelinated nerve fibres, neuurofilaments are relatively widely spaced and connected to neighboring neurofilaments by cross-bridges formed by phosphorylated C-terminal side-arms of NFM and NFH.

Neurofilaments at normal nodes of Ranvier are dephosphorylated and are more closely spaced (Chen et al., 2009; Jafari et al., 1997, 1998). After TAI, the side arms of NFH and NFM collapse and the spacing between neighboring neurofilament fall with the correlate that the packing density per unit volume of axoplasm increased (Geddes et al., 2000; Jafari et al., 1997, 1998; Sulaiman et al., 2011) (Figure 3.6e–h) where neurofilament were termed as being "compacted" (Pettus and Povlishock, 1996). Compaction of internodal neurofilaments was associated with a fall in spacing between neighboring neurofilament (Table 3.1) to about a third of

TABLE 3.1
Quantitative Parameters of the Axonal Cytoskeleton in Transverse Sections of Control and Experimental, Seven Days' Survival Stretch-Injured Optic Nerve Axons within 0.5-mm-Wide Bins of the Diameter of Nerve Fibers

Fiber Diameter	0.0–0.5 μm	0.5–1.0 μm		1.0–1.5 μm		1.5–2 μm
		No MTs	With MTs	No MTs	With MTs	
Control G ratio	0.8 ± 0.1	0.8 ± 0.1		0.8 ± 0.09		0.8 ± 0.08
Experimental G ratio	0.59 ± 0.02	0.52 ± 0.11	0.65 ± 0.04	0.66 ± 0.04	0.63 ± 0.05	0.68 ± 0.03
	NS	$p < 0.05$	NS		$p < 0.05$	NS
Control MTs number	19.6 ± 0.8	19.8 ± 0.7		48.6 ± 3.6		68.80 ± 4.95
Experimental MTs number	0	0	8.14± 5.89	0	11.0 ± 3.6	13.4 ± 2.76
	$p = 0.0006$	$p < 0.001$	$p < 0.05$	$p < 0.001$	$p < 0.001$	$p = 0.0004$
Spacing between MTs (nm), control	97.3 ± 5.2	126 ± 27		108 ± 6.9		147.5 ± 7.40
Spacing between MTs (nm), experimental	0		123.03 ± 46.87	0	140.38 ± 41.38	151.31 ± 47.13
	$p < 0.05$	NS		NS		NS
Control NFs number	18.4 ± 2.3	31.4 ± 2.1		30.1 ± 9.8		173.30 ± 34.04
Experimental NFs number	11.5 ± 1.5	29.16 ± 21.59	31.57 ± 14.66	83.75 ± 41.89	91.0 ± 27.48	115.4 ± 39.19
	$p = 0.0037$	$p = 0.87$	$p = 0.98$	NS	$p < 0.05$	$p = 0.124$
Spacing between NFs (nm), control	92.6 ± 0.6	98.7 ± 2.6		143.0 ± 2.7		96.6 ± 12.0
Spacing between NFs (nm), experimental	34.3 ± 3.1	37.4 ± 12.4	52.8 ± 21.8	33.3 ± 6.67	23.9 ±11.5	28.8 ± 3.9
	$p < 0.001$	$p < 0.001$	$p < 0.01$	$p < 0.001$	$p < 0.001$	$p < 0.0001$

NS, not significant.

pre-injury levels. In some nerve fibers compaction of neurofilament was also associated with a fall in the diameter of the reactive axons (Jafari et al., 1997, 1998; Sulaiman et al., 2011) such that periaxonal spaces occurred between the outer aspect of the axolemma and the inner aspect of the myelin sheath (Jafari et al., 1997, 1998; Schweitzer et al., 1993) (Figures 3.6 and 3.7).

TEM examination of longitudinal sections of reactive/injured nerve fibers revealed alternating foci of compacted neurofilaments, about 1 μm in length, (Figure 3.7a, large black arrows; b, double arrows) and zones within which the axoplasm were more electron lucent or pale and contained a disorganized cytoskeleton and lucent/damaged mitochondria (Figure 3.7a, white arrow).

When low-magnification longitudinal sections (Figure 3.7a) were examined, the alternating segments form an appearance like a string of sausages within a myelin sheath of which the caliber was increased at foci of neurofilament compaction (Figure 3.7a and b). When foci of compacted neurofilament are viewed in transverse section (Figure 3.7c), the axolemma is frequently lacking (double black arrows) and the discrete profiles of neurofilaments are replaced by a flocculent ultrastructure (Figure 3.7c, white asterisk). These alternating dark and pale zones differ in size by an order of magnitude from the "string of beads" reported in vitro (Tang-Schomer et al., 2012; Wang et al., 2012a) and were only resolvable at the ultrastructural level in stretch-injured optic nerves.

The development and presence of focal zones of compaction of neurofilament, termed the neurofilamentous core (Figure 3.7d, neurofilamentous core [fcn]), have been suggested to precede secondary axotomy in injured CNS nerve fibers. At the point of secondary axotomy, dissolution of neurofilaments occurred at one end of the fcn (Figure 3.7d). At such a point, there was also loss or fragmentation of the axolemma (Figure 3.7d, black arrows), focal accumulation of injury-induced membranous vesicles (black asterisk), and replacement of the tightly organized linear arrangement of compacted neurofilament (fcn) by a flocculent, dispersed granular precipitate suggestive of lysis/loss or calpain-mediated proteolysis of the neurofilament remnants of the axonal cytoskeleton (Figure 3.7d). It has recently been confirmed that proteolysis of neurofilament is mediated by calpains during Wallerian degeneration after transection injury to both optic and sciatic nerve fibers (Ma et al., 2013). The membranous profiles seen between the lateral margin of the fcn and the electron, dense, relatively intact myelin sheath (ω) are suggested to represent the region of breakdown or damage of the axolemma to provide the membrane fragments indicated by the black asterisk (Figure 3.7d), which are possibly comparable to the injury-induced membrane profiles described by Tuck and Cavalli (2010). It was noteworthy, also, that in the stretch-injury model of TBI, only 24% of axons in the entire injured optic nerve at 1-week survival, 33% at 2 weeks,

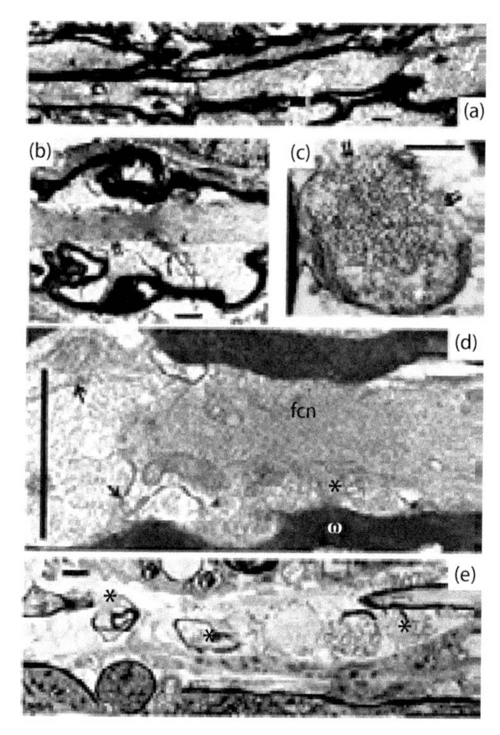

FIGURE 3.7 Reprints of micrographs from Jafari et al. (1997, 1998) (a–d); Figure 3.7e is original. (a) A longitudinal section of a degenerating nerve fiber at 4 hours after stretch-injury to the right optic nerve of adult guinea pig. Repeated zones of neurofilament compaction (large black arrows) are separated by more lucent, sausage-shaped, regions of a degenerating axon that contains swollen, lucent, damaged mitochondria or membrane profiles (white arrows). However, the myelin sheath is relatively intact. At slightly increased magnification of one zone of neurofilament compaction cut in the longitudinal plane (b) and transverse plane (c) the axolemma is lacking or deficient (double black arrows) or reflected into an enlarged periaxonal space (single arrow). There is a central core of compacted neurofilament (c) and a peripheral region containing a flocculent ultrastructure (c, white asterisk) suggestive of calpain-mediated cytoskeletal breakdown. (d) A longitudinal section of a neurofilamentous core (fcn) continuous on the right with a region of compacted neurofilaments. But on the left the fcn terminates abruptly and is replaced by a flocculent precipitate and membrane fragments (black arrows) suggestive of a catastrophic breakdown or dissolution of the neurofilament cytoskeleton and axolemma at the point of fragmentation of the axon within the relatively intact myelin sheath (ω). Small membrane profiles occur close to the black asterisk and are comparable to injury-induced membranous vesicles (Ouardouz et al., 2003). Marker bar = 1.0 μm.

44% at 3 weeks, and 46% at 12 weeks exhibited ultrastructural evidence of axonal pathology or degeneration. This progressive increase in the number of damaged or injured axons provides support for the hypothesis that, after a single mechanical insult to a central myelinated tract, there may be an ongoing recruitment to the number of degenerating axons over a relatively protracted time frame, as has recently been reported in an experimental model (Verdú et al., 2008) in human blunt head injury (Johnson et al., 2013b) and chronic traumatic encephalopathy (McKee et al., 2013).

3.6.2 Microtubules

After stretch-injury to central axons a dramatic, possibly transient, loss of axonal MTs has been reported (Maxwell and Graham, 1997). This loss of MT may lead to loss of fast axonal transport of membranous organelles and result in a lack of delivery by the neuron of mitochondria to replace those destroyed by the uncontrolled influx of free calcium and a failure to transport vesicular profiles to repair foci of damaged membrane or axolemma (Tuck and Cavalli, 2010). Quantitative data for loss of MT (Table 3.1) after stretch-injury in optic nerve axons supported the hypothesis, but also suggested that loss of MT is not an all-or-none response in all nerve fibers of the injured optic nerve; there was a graded response with a reduced number of MT within some larger diameter nerve fibers (Table 3.1) in which, despite loss of some 60%–80% of MT, there was not any change in their spacing within the axoplasm (Jafari et al., 1997, et al., 1998; Maxwell and Graham, 1997). But this observation has still not been explained. Recent studies of beta amyloid labeling of damaged nerve fibers in white matter of surviving patients (Chen et al., 2009; Johnson et al., 2013b) and experimental models (Chung et al., 2005) have extended the hypothesis that ongoing changes in fast axonal transport that is mediated via interaction of membranous organelles with axonal MT may occur over a relatively long time frame after TAI. This is especially pertinent in the context of the diagnostic cautionary publication (Geddes et al., 2000) that stimulates the suggestion that β-amyloid–labeled axons observed months or years after TBI may not have been injured at the injury scenario but at some other point in time during the chronic posttraumatic phase of traumatic head injury and also reflects the current hypothesis that recruitment of increasing numbers of axons to enter the posttraumatic cascade toward secondary axotomy occurs during the chronic phase during survival after TBI.

At relatively mild levels of axonal injury MTs and neurofilaments (see Figure 3.6e–h and Table 3.1), assume a disorganized, nonlinear orientation in respect to the longitudinal axis of the nerve fiber (Figures 3.3a and 3.6e-h). Further insight has been provided after study of stretched axons in vitro (Tang-Schomer et al., 2010, 2012) after culture of primary cortical neurons from embryonic day 18 (E18) Sprague Dawley rats. Axons were exposed to strain levels of 30%, 50%, and 75% by displacement of the supporting medium on which axons were grown (Tang-Schomer et al., 2010). In this in vitro study, there was a rapid but transient assumption of a sine-wave–like appearance by axons termed "undulations" at 50%–75% strain. Undulations then relaxed and by 3 hours after injury were replaced by periodic axonal swellings scattered along the length of axons. The axonal swellings were identified using immunocytochemical labeling against βIII-tubulin, the MT-associated protein tau, the fast-transported protein amyloid precursor protein APP and NF200, and NFH, which is transported via slow axonal transport. Examination using TEM of thin sections revealed breakage of MT, most frequently at the peaks of undulations, at which discrete free ends of MT were reported.

These studies (Tang-Schomer et al., 2010, 2012) confirm that there is a time course to loss of MT within injured axons and provide original evidence that MTs are physically broken immediately after stretch-injury and that axonal swellings form before final destruction of the injured axons. However, and perhaps importantly, the axons formed in culture and exposed to stretch-injury were unmyelinated. It is well established that the components of the axonal cytoskeleton differ between myelinated and unmyelinated axons with high numbers of MT in the latter but neurofilament and a lesser proportion of MT in the former. These differences in the component organelles of the axonal cytoskeleton alter the biophysical properties of axons, with myelinated axons being more resistant to transient tensile strain and responding to that strain within a different time frame compared to unmyelinated axons. Further studies using in vitro methods are therefore necessary.

3.6.3 The Subaxolemma Cytoskeleton or Actin-Spectrin Complex

Underlying the axolemma of CNS myelinated axons is a microfilamentous-supporting cytoskeleton, frequently referred to as the subaxolemma cytoskeleton, formed by a population of microfilaments about 7 nm in diameter. Filamentous proteins, actin, spectrin, ankyrin, and others interact with the axolemma and the core cytoskeleton of neurofilament and MT. The clearest example of the morphology of the subaxolemma cytoskeleton is the so-called "dense undercoating" at nodes of Ranvier (Maxwell, 1996, 2014). The filamentous proteins form a stable and flexible network that binds membrane proteins and the cell cytoskeleton and, in a neuronal axon, provides location and docking and/or anchoring mechanisms to localize and retain a variety of membrane transporters at different, functionally discrete sites such as the axon initial segment, nodes of Ranvier, paranodes, the juxta-paranodes, and internodal parts of the axonal membrane to mitigate its electrical capabilities and function (Xu et al., 2013). Actin filaments bind to tetramers of α and β spectrins to form a flexible, two-dimensional array that binds directly to adhesion proteins of the axolemma such as the neuronal cell adhesion molecule (NCAM) and protein 4.1N as well as ankyrin adaptor proteins (Letourneau, 2009).

Spectrin aggregates are assembled from α and β heterodimers, then into tetramers. Groups of tetramers then link to actin oligomers via protein 4.1N. Within axons, two large ankyrins have been reported, ankyrin-B and ankyrin-G (Bennett and Healey, 2009). Many transmembrane proteins include ion channels, pumps, and exchangers and these ion channels, for example Na^+/K^+-ATPase, voltage-gated Nav 1.6 channels, the Na^+/Ca^{2+} exchanger, $K_v3.1$ channels, KCNQ2/3(K_v7), NMDA receptors, and the neuronal glutamate transporter EAAT4 also chemically bind to spectrin (Letourneau, 2009). The literature about functional groupings of ion channels and components of the spectrin-ankyrin-actin cytoskeleton is vast; the interested reader is referred to Bennett and Healy (2009). In particular, ankyrin-G is essential to stabilize transcellular connections and organize the initial segment of an axon; is essential in the organization of clusters of Nav1.2 channels at developing nodes and clustering of Nav1.6 channels within mature nodes of Ranvier, and for aggregation of KCNQ2/3 potassium channels at nodes. A second large ankyrin, ankyrin-β, associates with Caspr, neurofascin, protein 4.1B, and spectrin βII at paranodes, and with the high density of Kv1.1 and Kv1.2 channels and Caspr2 at juxtaparanodes.

Importantly, ankyrin-spectrin binding is not immutable and may be disrupted under certain physiological conditions. In relation to the TBI literature, ankyrin-spectrin binding may be disrupted when intraaxonal levels of free Ca^{2+} rise after injury to the axolemma and other membranes. Calcium-sensitive neutral proteases or calpains occur within the axoplasm of normal central myelinated nerve fibers but in an inactive form or state (Saatman et al., 1996). Calpain II or m-calpain occurs in glia and axons (Hamakubo et al., 1986), and a major site of its activation is in relation to the spectrin cytoskeleton related to the myelin sheath of myelinated axons. After TBI in rat, using an antibody that recognized spectrin breakdown products (spectrin BDPs) or calpain-mediated proteolysis of spectrin, spectrin BDPs occurred within dendrites of the cerebral cortex at 90 minutes after fluid percussion injury (Saatman et al., 1996). It was also reported that the volume of reactive cortex expanded by 4 hours, increased further by one day, and extended into an even greater volume of cortex and the ipsilateral hippocampus and thalamus by 7 days. Spectrin BDP also occurred in axons within the ipsilateral subcortical white matter at 90 minutes and within focal axonal swellings at 4 hours, and assumed a speckled distribution throughout the subcortical white matter by 7 days (Saatman et al., 1996).

This finding was confirmed and extended in a report describing spectrin BDP–positive axons in patients in the corpus callosum of head injured patients between 4 hours and 12.5 days after TBI (McCracken et al., 1999) and more recently a biphasic activation of calpains was reported after stretch-injury to the optic nerve in C57BL/6 mice (Saatman et al., 2003). An important and novel observation in this latter study was that between 20 to 30 minutes and 1 to 2 hours after injury, there was widespread labeling for calpain-mediated spectrin proteolysis in linearly oriented bands parallel to the longitudinal axis of morphologically intact axons (Figure 3.8, 30 minutes, column A) suggestive of two bands adjacent to the myelin sheath, which confirmed ultrastructural evidence for localization of spectrin-breakdown product just deep to the axolemma at 15 minutes and 6 hours (Figure 3.6c and d) after impact acceleration injury in rat (Büki et al., 1999). In that latter study, calpain-mediated spectrin proteolysis reaction product occurred just internal to the axolemma among microfilaments of the subaxolemma cytoskeleton and related to damaged, lucent mitochondria (Büki et al., 1999). Calpain-mediated spectrin proteolysis reaction product was more widely distributed within the axoplasm at foci of neurofilament compaction at 120 minutes (Buki et al., 1999) suggestive of chemical

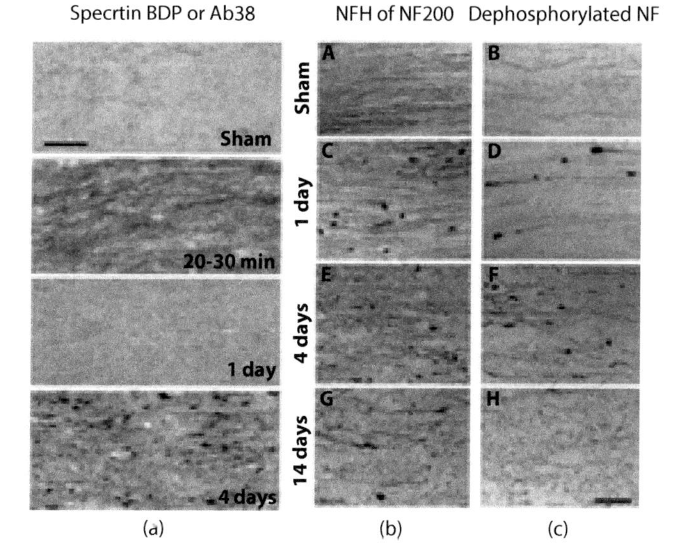

FIGURE 3.8 Copies of immunolabelled micrographs published, with kind permission, by Saatman et al. (2003). These important images illustrate that activation of calpains that break down spectrin microfilaments at two discrete time points after stretch-injury of mouse optic nerve. Shortly after injury, within 20 to 30 minutes, linear arrays of spectrin breakdown products (SBDP) occurred as longitudinal arrays parallel to the long axis of intact nerve fibers (column a, 20 to 30 minutes) possibly related to the internal surface of the myelin sheath at the locus of the subaxolemma compartment as earlier reported (Wang et al., 2011) at 15 minutes after axonal injury in an ultrastructural study (see Figure 3.7). Labeling for SBDP was then lost by 1 day but was regained by 4 days. However, labeling at 4 days differed from that at shorter survivals in that SBDP immunolabeling was localized to axonal swellings or bulbs (4 days, column a) widely distributed throughout an optic nerve. Parallel labeling for phosphorylated heavy neurofilament, its state in normal fibers, (column b) was uniform along the length of axons in sham animals (A, column b), and highlighted small numbers of axonal swellings and bulbs at 1 day (C, column c). More widespread but patchy labeling of axons and axonal bulbs occurred at 4 days (E, column b) and labeled bulbs and axons occurred among cellular debris at 14 days (G, column b). Alternatively, labelling for dephosphorylated HNF (column c, dephosphorylated neurofilament) provided greater discrimination of axonal bulbs and contiguous axons at 1 (D, column c) and 4 days (F, column c), but a lesser intensity of labelling at 14 days (H, column c). Marker bar = 50 μm.

digestion and/or depolymerization of neurofilaments (Dewar et al., 2003). At greater survivals, loss of labeling for spectrin BDP was reported at one day (Figure 3.8, 1 day, column A) with a second, intense expression at 4 days (Figure 3.8, 4 days, column A). Importantly, the spatial distribution of spectrin BDP differed between 4 days and 30 to 120 minutes after injury and was associated with the presence of focal, axonal swellings and fragmented axons (Figure 3.8, 4 days, column A) suggestive of axonal degeneration. Spectrin BDP labeling did not occur within sections at 14 days.

Parallel observations using immunolabeling for phosphorylated and dephosphorylated NFH reported large axonal swellings rich in dephosphorylated neurofilament at 1 day (Figure 3.8d, column C, dephosphorylated NF). By 4 days' survival, there was loss of a parallel orientation of neighboring nerve fibers and greater numbers of dephosphorylated neurofilament-immunolabeled axonal swellings suggestive of more widespread dephosphorylation of neurofilament in injured axons (Figure 3.8f, column C, 4 days), with an increasing proportion of dephosphorylated neurofilament positive axonal swellings and degenerative axonal debris out to 14 days (Figure 3.8f, 14 days, column C). Unfortunately, no experimental study of spectrin BDP or phosphorylated or dephosphorylated NFH has been reported in human brain after TBI although there have been attempts to use the content of spectrin BDP, among other potential biomarkers, within cerebrospinal fluid (Brophy et al., 2009).

In overview, activation of calcium-mediated proteolysis of the subaxolemma and neurofilament components of the axonal cytoskeleton as a result of increased content of free Ca^{2+} leads to loss of normal ionic homeostasis and breakdown of the microfilamentous cytoskeleton and loss of focal concentrations of a variety of transmembrane proteins from nodes of Ranvier, paranode, and juxtaparanode regions of the axolemma. These changes disrupt homeostasis, normal electrical activity, and neurotransmission in the damaged axon and potentiate disconnection of that axon. If recovery of axonal homeostasis is not achieved, the outcome is secondary axotomy that, potentially, may continue for years or tens of years after the initial injury in increasing numbers of axons, which may enter the pathological cascade long after the initial injury. However, experimental characterization of conditions under which axons enter the pathological cascade during the chronic phase of TBI has not yet been investigated. This lack of such investigations has probably occurred because the vast majority of experimental studies of axonal degeneration follow outcomes only out to 14 or 21 days after experimental injury.

3.7 SECONDARY AXOTOMY AND "DIEBACK"

Once the axon had undergone secondary axotomy or disconnection within either a minimum posttraumatic interval of 3 and 4 hours in experimental studies in mice or guinea pig (Maxwell et al., 1997; Staal et al., 2010; Wang et al., 2011), "dieback" of the proximal and distal axonal fragments has been reported. Recently reported morphological criteria have allowed distinction between proximal and distal swellings (Wang et al., 2011) and also provided strong evidence for the withdrawal or dieback of the proximal and distal terminal swellings or bulbs of a fragmented nerve fiber away from the site of disconnection as reported in mouse optic nerve (Wang et al., 2011). By 48 hours after injury, the proximal swelling/bulbs had retracted from the site of secondary axotomy retrogradely toward their respective neuronal cell body by 600–800 μm (Wang et al., 2011). Distal swellings, those on the part of the axon now disconnected from its cell soma, had also withdrawn some 600–800 μm from the site of axotomy/disconnection toward the optic chiasm by 48 hours after TBI (Wang et al., 2011). The proximal and distal swellings had thereby withdrawn or retracted to form a 1200–1600 μm long space between the proximal swelling/bulb and the distal swelling/bulb. Ultrastructural examination of the intermediate zone between the proximal and distal swellings or axonal bulbs revealed a highly vacuolated and organelle deficient region containing remnants of the degenerating myelin sheaths (Wang et al., 2011). But in that study, the occurrence of a band of axonal bulbs across the diameter of the optic nerve possibly suggests induction of a focal lesion because injured axons form two discrete zones in the injured optic nerve after fluid percussion injury rather than being widely distributed within the neuropil.

One zone contained spatially related, numerous proximal swellings in one transverse band and distal swellings in the other transverse zone or band across the diameter of the optic nerve (Wang et al., 2011). An important premise or characteristic of human diffuse brain injury is that small numbers of damaged, degenerating nerve fibers occur among greater numbers of morphologically intact nerve fibers (Adams et al., 1989), although focal lesions or aggregates of degenerating nerve fibers occur either within the splenium of the corpus callosum or the proximal brain stem. In the optic nerve stretch-injury model of traumatic axonal injury, however, myelin figures suggestive of breakdown of myelin fragments after secondary axotomy may be found (Figure 3.7e, black asterisk) intimately related to glial cell processes and morphologically intact nerve fibers (Figure 3.7e). These very small foci of myelin fragments confirm the findings reported (Wang et al., 2011), but probably reflect the pathology in mild TBI and concussion in which the number of axotomized nerve fibers at any one time point after trauma may be small with individual reactive, injured axons scattered among greater numbers of apparently normal or unreactive nerve fibers. The specimen from which Figure 3.7e was obtained was an 8-week survival animal rather than at 3 hours after injury and is suggestive of chronic loss of small numbers of injured nerve fibers over a considerable period after injury.

In mouse optic nerve, proximal swellings or retraction balls had an intact, limiting axolemma, aggregates of axoplasmic reticulum and/or membranous vesicles, intact mitochondria, and a core of disorganized neurofilament with a few MT (Wang et al., 2011). Comparable examples have been obtained in guinea pig injured optic nerve (Figure 3.9a and b). Characteristically, the distal axonal swellings were more electron dense, swollen (Figure 3.9b), of rounded profile, and contained damaged mitochondria lacking recognizable cristae. The specimen illustrated in Figure 3.9c was obtained from an animal 8 weeks after traumatic injury and processed using the pyroantimonate technique to visualize loci with a high concentration of free calcium. Three types of mitochondria occurred with the axoplasm of this distal swelling (Figure 3.9c). Mitochondria were either unremarkable or intact, contained electron lucent spaces within the mitochondrial matrix, or contained aggregates of pyroantimonate crystals (Figure 3.9c). Both of the latter features are indicative of damaged mitochondria incompatible with their long term viability—see discussion earlier.

At 24 hours after injury, proximal, terminal axonal swellings/degeneration bulbs assumed a truncated, rod-like profile limited by a morphologically intact myelin sheath, when the axoplasm contained aggregates of axoplasmic reticulum and intact mitochondria (Wang et al., 2011). Distal swellings, on the other hand, at 24 hours had a spherical or rounded shape and contained examples of damaged mitochondria. These features were confirmed in guinea pig optic nerve after stretch-injury (Figure 3.10b), but also, and perhaps importantly, within specimens obtained at any time point later than 8 hours after injury out to 12 weeks' survival. On the other hand, distal swellings or degeneration bulbs formed large quasi-spherical axonal swellings that contained many damaged mitochondria (Figure 3.9c). Mitochondria contained foci of pyroantimonate precipitate, while autophagic vesicles and lysomal debris in material processed by the pyroantimonate technique (Figure 3.9c). However, the temporal occurrence of distal swellings differed between fluid percussion–injured mouse optic nerve and the stretch-injured nerve in guinea pigs in that examples of proximal and distal bulbs were found at 8, 12, 24, 48 hours, and out to 12 weeks (Figure 3.9c) experimental survival rather than out to just 48 hours (Wang et al., 2011).

3.7.1 Granular Degeneration of the Axon

The final morphologically discrete form of the axonal cytoskeleton encountered in ultrastructural studies of axonal degeneration both in Wallerian degeneration and in models of TBI is loss of discrete neurofilaments and membranous organelles internal to the myelin sheath and their replacement by an amorphous, medium electron dense deposit, the so-called granular degeneration/granular disintegration (Gaudet et al., 2011) of the axonal cytoskeleton.

The granular degenerative remnant of the axon may be viewed both in transverse (Figure 3.10a) and longitudinal (Figure 3.10b) sections of degenerating nerve fibers, and occurred first in the guinea pig optic nerve stretch-injury model between 4 and 6 days after axotomy. But, and perhaps importantly, examples of this final stage of loss of axons continued to occur in small numbers of nerve fibers across the whole diameter of the injured optic nerve out to the longest experimental survival investigated at 12 weeks after injury. It is notable that dark, degenerating nerve fibers occurred immediately adjacent to intact nerve fibers (Table 3.1 and Figure 3.10a), and there was a difference in the G-ratio (axon diameter divided by fiber diameter) between adjacent nerve fibers (Figure 3.10a). The two normal axons, the axoplasm of which contain discrete neurofilament and MT in Figure 3.10a, have a G-ratio ≈ 0.68, whereas the fiber undergoing granular degeneration of the axon (to the top of the figure) has a G-ratio ≈ 0.42 with a significantly increased thickness of the myelin sheath (Figure 3.10a). Notably, a compact, laminated organization of the myelin sheath was still present (Figure 3.10a), closely enveloping the dark axonal remnant.

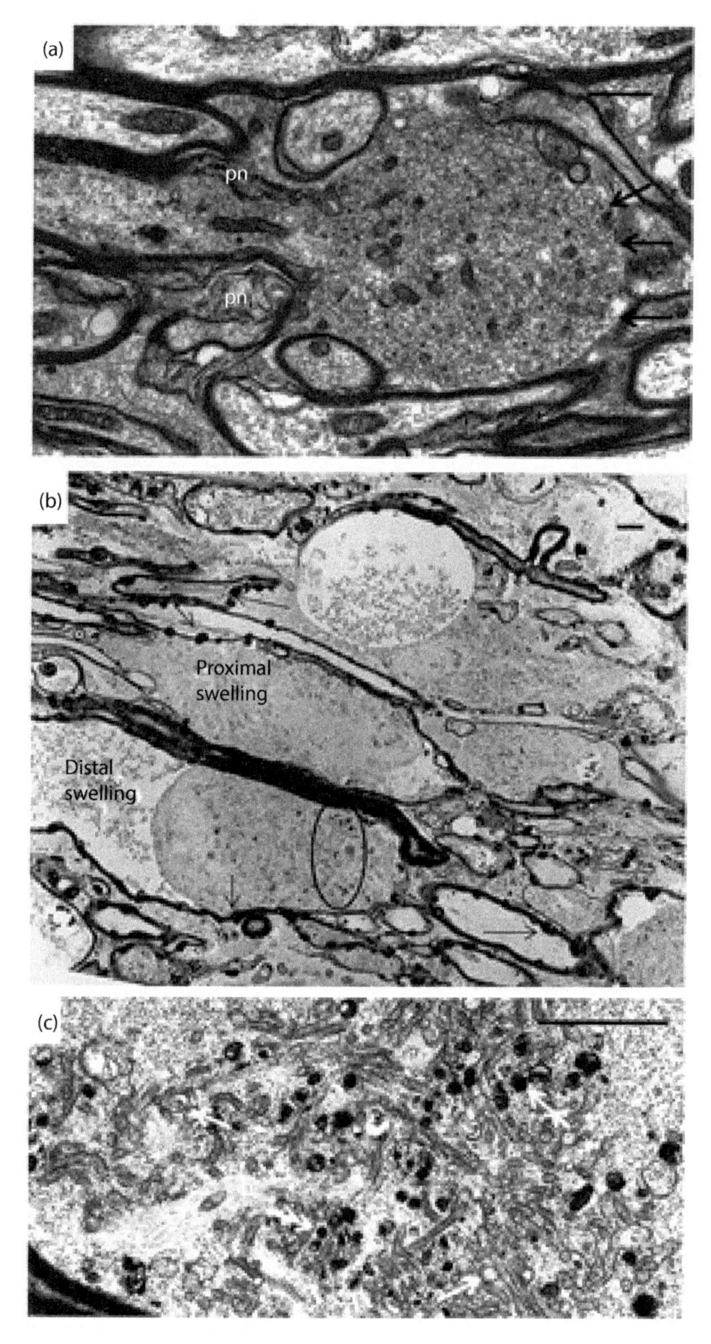

FIGURE 3.9 Longitudinal thin sections of degenerating or disconnected nerve fibers in guinea pig, stretch-injured right optic tract. (a). A longitudinal thin section of an axonal proximal swelling containing a large aggregation of intact mitochondria and membranous vesicular profiles at a remnant of a node of Ranvier defined by the presence of characteristic paranodal myelin glial loops (pn). The axolemma may be seen limiting the distal end of the bulb (arrows). Note that the locus of swelling is proscribed by the paranode glial loops, which even though slightly enlarged are still intimately related to the axolemma. This specimen was obtained from an adult guinea pig 12 hours after injury. Marker bar = 1 μm. (b). A low-magnification field of longitudinal thin sections of examples of truncated proximal swellings and globose distal swelling obtained at 8 weeks after injury. Numerous small-caliber myelinated nerve fibers are also present such that axonal disconnection has occurred in only a relatively small proportion of nerve fibers. This tissue was processed using the pyroantimonate technique for location of free calcium within the specimen. Close examination of the myelin sheath of the labeled proximal swelling and distal swelling, and adjacent small-caliber fibers reveals spherical foci of pyroantimonate deposits across the thickness of the relatively intact myelin sheaths (arrows). The mitochondria within the proximal swelling are mostly unremarkable and intact. But those in the distal swelling (oval outline) are abnormal (shown in Figure 3.10c). Marker bar = 1 μm. (c). A higher power field that is comparable to the area enclosed by the oval outline in Figure 3.10b and is part of a distal axonal swelling. Mitochondria are either intact, contain lucent swollen regions (arrows), or are electron-dense pyroantimonate deposits (double arrows). The latter two features indicate ongoing degeneration of mitochondria and axoplasm in this distal swelling that, clearly, is no longer connected to its appropriate cell soma. The damage to the mitochondria is indicative of the initiation of Wallerian degeneration.

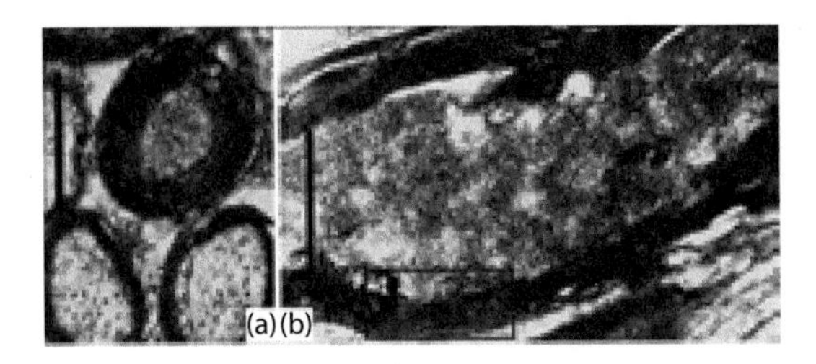

FIGURE 3.10 Transmission electron micrographs of "granular degeneration" of axons within the optic nerve of adult guinea pig following a single mechanical, stretch-injury insult to the optic nerve. (a) Transverse sections obtained at 6 days after stretch-injury. Three nerve fibers are seen. Two of these, toward the base, have the structure of intact nerve fibers. Their myelin sheath is largely intact with closely related myelin lamellae forming an outer "sleeve" enclosing the axon. Within both axons, cross-sections of longitudinally orientated microtubules are discrete within the relatively lucent axoplasm. On the contrary, the nerve fiber within the upper half of the figure has a thickened but regularly organized myelin sheath. The myelin sheath surrounds a circular zone having no discernible structure but containing an amorphous, finely granular appearance with a subtle variation in electron density. There is no evidence for the presence of any defined biological membrane or axolemma limiting the central amorphous material. This is suggestive of a complete dissolution of the axolemma and membranous organelles normally visible within the axoplasm. The central zone represents an area of granular degeneration of an axon. Marker bar = 1 μm. (b) A section through a tubular profile that is formed by an electron dense outer, limiting sleeve enclosing a less electron dense central core that lacks any recognizable ultrastructure. This core has a granular appearance that varies in electron density in a nonuniform manner. There is some suggestion of a laminate substructure within the more electron-dense sleeve within the area outlined by the quadrangle. But that laminate substructure is less discrete than that seen in the upper region of (a). It is suggested to represent early stages of chemical breakdown of the remains of the myelin sheath in a degenerating nerve fiber. The inner, amorphous, less electron-dense material represents amorphous debris resulting from calpain mediated proteolysis of remnants of the axolemma and axoplasmic organelles of a degenerated axon and possibly reflects the known rapid loss of the axon described in a number of studies of Wallerian degeneration. Marker bar = 1 μm.

3.7.2 Overview of Secondary Axotomy and Outcomes

There is a huge and still expanding literature concerning cellular, neural pathways, and behavioral or neurocognitive outcomes in TBI patients, much of which will be discussed in other chapters. But it is of value to place this review material in context in the case of a TBI patient.

1. The concept of mechanical shearing or tearing of large numbers of axons within central white matter of the brain, so-called primary axotomy, has received negligible support within the experimental literature over the past 50 years. There is now a consensus that primary axotomy does not occur within the injured brain of the great majority of TBI patients, with the possible exception of patients who enter coma at the scene of an accident or die there. However, such patients often also suffer such widespread trauma to their bodies that any possibility of surviving the injuries is negligible.
2. At the level of a neuronal axon, if loss of normal neuronal membrane homeostasis occurs, and if recovery of homeostatic mechanisms does not occur within minutes of injury to allow recovery of normal function and return to consciousness, then injured nerve fibers enter the so-called neuropathological/calcium mediated cascade, leading to secondary axotomy and disconnection of neural pathways which mediate normal behavior.
3. There is increasing evidence that the same pathophysiological mechanisms occur in patients exposed to concussion and mild and severe brain injury. The important variable is the relative number of injured axons in terms of the total number within the brain of a patient.
4. Cerebrovascular injury, either extradural, subarachnoid and contusional, or intraparenchymal hemorrhagic lesions, occurs frequently in head-injured patients. The most common scenario is the presence of extradural hematoma when damage to a middle meningeal artery has occurred and this may readily be detected using computed tomography followed by surgical removal of extravasated blood. Changes in function and activity of capillaries within the injured brain are also associated with a number of the microvasculature pathologies reported in TBI (Park et al., 2009) and occur at all levels of brain injury (Bigler and Maxwell, 2012). With increasing severity of injury, there is an associated injury to smaller blood vessels within the neuropil of the brain. Small petechial hemorrhages may occur and the related reduction in the provision of glucose and oxygen to the locus of mechanical injury will compromise potential recovery by injured neurons, which then may be exposed to a combination of mechanical and ischemic insults. The occurrence of many petechial hemorrhages, particularly at the corticomedullary boundary of the cerebral hemispheres, allows magnetic resonance imaging (MRI) and susceptibility-weighted imaging indirect visualization of sites at which axonal injury may be occurring in the living patient. Otherwise, visualization of foci of axonal injury may only be achieved using postmortem neuropathological examination.
5. Mechanical insults to the axolemma of nerve fibers may, especially in the region of nodes of Ranvier of myelinated nerve fibers, damage transmembrane sodium channels and allow an uncontrolled influx of calcium ions into the axoplasm. Calcium may also be released from intraaxonal loci such as the axoplasmic reticulum and mitochondria.

Coincidental reversal of the normal activity of the sodium-calcium exchanger may result in loss of calcium extrusion from an injured neuron and exacerbate the physiological insult to that injured neuron.

6. Injury to and failure of axonal mitochondria close to the locus of axonal injury reduces production of ATP, the essential substrate for many biochemical and physiological processes, and serves to curtail normal function of transmembrane transporters. The presence of lucent, damaged mitochondria lacking cristae correlates strongly with injury to an axon, but is only visible in material examined using TEM of thin sections. An associated release of cytochrome c into the remaining axoplasm hypothetically induces retrograde signaling by "apoptosis-inducing factor" or membranous vesicles to the cell soma of the injured neuron. Programmed cell death and apoptosis of injured neurons may then be initiated.
7. At foci of abnormally high content of free calcium within a damaged axon there is loss of axonal MTs. Traumatic fragmentation of MTs as a result of stretch-injury has been achieved in vitro but only in unmyelinated axons to date. Fast axonal transport is lost and possibly compromises posttraumatic delivery of replacement mitochondria and membranous vesicles for repair of damaged neural membranes.
8. At the same time as damage to MTs, there is also activation of calcium-activated neutral proteases or calpains, phospholipases, and phosphatases. Calpains have been suggested to (1) breakdown microfilaments (ankyrin/spectrin) of the subaxolemma cytoskeleton thereby exacerbating loss of transmembrane proteins and transporters and further damaging cell membranes and (2) mediate dephosphorylation and/or fragmentation of heavy and middle neurofilament side arms causing collapse of the normal spacing between neighboring neurofilaments that form the structurally supporting core within an axon. Repeated foci of compacted neurofilaments occur along the length of a now degenerating axon. This exacerbates disruption of axonal transport.
9. At one or a number of foci of compacted neurofilaments along the length of a degenerating axon, there is also loss of the axolemma adjacent to foci of breakdown of the neurofilament core. The axon has then fragmented into two or more parts and undergone secondary axotomy.
10. The time frame over which secondary axotomy and axonal degeneration occur is shortest in smaller experimental animals such as rodents and longer in humans. In humans, secondary axotomy has been reported not to occur until at least the first 12 hours after injury (Christman et al., 1994).
11. The ends of the two parts of the now-disconnected axon then retract away from the locus of axotomy such that at 48 hours after injury, a gap, some 1200–1600 μm in length, separates the terminal bulb of that part of the axon still connected to its cell body, the so-called proximal bulb, and the terminal bulb at the end of the axonal fragment no longer in continuity with its respective cell body, the so-called distal bulb. The proximal and distal bulbs have recently been characterized as to differences in their size, shape, and content of cellular organelles (Wang et al., 2011)
12. The distal fragment of the disconnected axon undergoes Wallerian degeneration, but the time course of that degeneration is longer than that reported by Waller in his original study using peripheral nerve. A major change in the axon is loss of the remnants of the axolemma, neurofilamentous core, and any remaining membranous organelles. All of these are replaced by an intermediate electron dense or pale gray amorphous precipitate in material processed for TEM. The time frame of this axonal degranulation is only a few hours within individual degenerating nerve fibers (Wang et al., 2012a). The reported occurrence of small numbers of degranulating nerve fibers over weeks and months after TAI in experimental models of TAI (Maxwell et al., 2014) support the hypothesis that the number of degenerating nerve fibers within central white matter tracts following TAI in chronic traumatic encephalopathy may continue to rise with increasing post-traumatic survival.
13. With loss of axons, there is an associated loss of oligodendrocytes and changes in the functional state of astrocytes and microglia. However, understanding of glial responses after traumatic insults is presently relatively poor and warrants future investigation, especially as the complexity of neuronal and glial cell interactions is beginning to be more widely appreciated.

3.7.3 Potential Sequelae within a Patient's Brain

1. Functional and structurally linked areas of cerebral cortex, basal ganglia, and thalamus are disconnected by axotomy.
2. There is thereby disconnection of neural networks within and between the cerebral hemispheres, the cerebellum, and parts of the brain stem. However, the majority of axonal damage and disconnection occurs in commissural nerve fibers (body and splenium of the corpus callosum) and central white matter of the cerebral hemispheres and, in particular, between the frontal and temporal lobes (cingulum, internal capsule, frontothalamic and thalamofrontal radiations, and corticospinal tract among others).
3. In surviving TBI patients, there is an increased degree of loss of central white matter compared with loss associated with increasing age. A key

morphological feature is an apparent increase in the volume of the ventricular system, or ventriculomegaly, and the subarachnoid space. But not only do the ventricles increase in size, there is a parallel loss of the central white matter so that ventriculomegaly is associated with a reduction of the total volume of the brain. The rate of loss of central white matter is relatively high in the acute posttraumatic phase and then falls in the postacute and chronic phases of TBI. It is unknown whether loss of central white matter and associated gray matter during the chronic phase ceases or continues over the rest of the life of a TBI patient.

4. Associated with the observed decrease in central white matter, there is loss in volume and number of neurons within parts of the cerebral cortex (Maxwell et al., 2010), and parts of the thalamus (Maxwell et al., 2004). However, although MRI may provide course evidence of changes in cortical, thalamic, and hippocampal volume, it is not able to be used to obtain data concerning changes in number of neurons or subtypes of neurons. Use of rigorous, stereological techniques should be used, but such data may only be obtained postmortem from long term survivors in post traumatic encephalopathy.
5. Diffusion tensor imaging allows direct visualization of changes in central white matter, but the vast majority of the current literature is largely anecdotal because observations have been made only once in a particular patient. Increased sophistication of analytical techniques and repeated chronological or sequential studies of normal and injured patients are now required.

REFERENCES

Adams, J.H., D. Doyle, I. Ford, T.A. Gennarelli, D.I. Graham, and D.R. McLellan. 1989. Diffuse axonal injury in head injury: Definition, diagnosis and grading. *Histopathology*. 15:49–59.

Agrawal, S.K. and M.G. Fehlings. 1996. Mechanisms of secondary injury to spinal cord axons in vitro: Role of Na+, Na(+)-K(+)-ATPase, the Na(+)-H+ exchanger, and the Na(+)-Ca2+ exchanger. *The Journal of Neuroscience*. 16:545–552.

Barros, C.S., S.J. Franco, and U. Muller. 2011. Extracellular matrix: Functions in the nervous system. *Cold Spring Harbor Perspectives in Biology*. 3:a005108.

Bennett, V. and Healy, J. 2009. Membrane domains based on ankyrin and spectrin associated with cell-cell interactions. *Cold Spring Harbor Perspectives in Biology* 1(6):a003012.

Bigler, E.D. 2013. Neuroinflammation and the dynamic lesion in traumatic brain injury. *Brain*. 136:9–11.

Bigler, E.D. and W.L. Maxwell. 2012. Neuropathology of mild traumatic brain injury: Relationship to neuroimaging findings. *Brain Imaging and Behavior*. 6:108–136.

Brophy, G.M., J.A. Pineda, L. Papa, S.B. Lewis, A.B. Valadka, H.J. Hannay et al. 2009. alphaII-Spectrin breakdown product cerebrospinal fluid exposure metrics suggest differences in cellular injury mechanisms after severe traumatic brain injury. *Journal of Neurotrauma*. 26:471–479.

Büki, A., D. Okonkwo, K.K. Wang, and J.T. Povlishock. 2000. Cytochrome c release and caspases activation in traumatic axonal injury. *Journal of Neuroscience*. 8:2825–2834.

Büki, A., R. Siman, J.Q. Trojanowski, and J.T. Povlishock. 1999. The role of calpain-mediated spectrin proteolysis in traumatically induced axonal injury. *Journal of Neuropathology and Experimental Neurology*. 58:365–375.

Chen, X.H., V.E. Johnson, K. Uryu, J.Q. Trojanowski, and D.H. Smith. 2009. A lack of amyloid beta plaques despite persistent accumulation of amyloid beta in axons of long-term survivors of traumatic brain injury. *Brain Pathology*. 19:214–223.

Chen, X.H., R. Siman, A. Iwata, D.F. Meaney, J.Q. Trojanowski, and D.H. Smith. 2004. Long-term accumulation of amyloid-beta, beta-secretase, presenilin-1, and caspase-3 in damaged axons following brain trauma. *The American Journal of Pathology*. 165:357–371.

Chodobski, A., B.J. Zink, and J. Szmydynger-Chodobska. 2011. Blood-brain barrier pathophysiology in traumatic brain injury. *Translational Stroke Research*. 2:492–516.

Christman, C.W., M.S. Grady, S.A. Walker, K.L. Holloway, and J.T. Povlishock. 1994. Ultrastructural studies of diffuse axonal injury in humans. *Journal of Neurotrauma*. 11:173–186.

Chung, R.S., J.A. Staal, G.H. McCormack, T.C. Dickson, M.A. Cozens, J.A. Chuckowree, et al. 2005. Mild axonal stretch injury in vitro induces a progressive series of neurofilament alterations ultimately leading to delayed axotomy. *Journal of Neurotrauma*. 22:1081–1091.

Dale, J.M. and M.L. Garcia. 2012. Neurofilament phosphorylation during development and disease: Which came first, the phosphorylation or the accumulation? *Journal of Amino Acids*. 2012:382107.

Dewar, D., S.M. Underhill, and M.P. Goldberg. 2003. Oligodendrocytes and ischemic brain injury. *Journal of Cerebral Blood Flow and Metabolism*. 23:263–274.

Dore-Duffy, P., C. Owen, R. Balabanov, S. Murphy, T. Beaumont, and J.A. Rafols. 2000. Pericyte migration from the vascular wall in response to traumatic brain injury. *Microvascular Research*. 60:55–69.

Erb, D.E. and J.T. Povlishock. 1988. Axonal damage in severe traumatic brain injury: An experimental study in cat. *Acta Neuropathologica*. 76:347–358.

Fehlings, M.G. and S. Agrawal. 1995. Role of sodium in the pathophysiology of secondary spinal cord injury. *Spine*. 20:2187–2191.

Gaudet, A.D., P.G. Popovich, and M.S. Ramer. 2011. Wallerian degeneration: Gaining perspective on inflammatory events after peripheral nerve injury. *Journal of Neuroinflammation*. 8:110.

Geddes, J.F., H.L. Whitwell, and D.I. Graham. 2000. Traumatic axonal injury: Practical issues for diagnosis in medicolegal cases. *Neuropathology and Applied Neurobiology*. 26:105–116.

Grady, M.S., M.R. McLaughlin, C.W. Christman, A.B. Valadka, C.L. Fligner, and J.T. Povlishock. 1993. The use of antibodies targeted against the neurofilament subunits for the detection of diffuse axonal injury in humans. *Journal of Neuropathology and Experimental Neurology*. 2:143–152.

Gross, C.G. 2003. Trepanation from the Palaeolithic to the internet. In: Arnott, A., C.U.M. Smith, and S. Finger, editors. *Trepanation: History, Discovery, Theory*. Swets & Zeitlinger, Exton, PA. pp. 307–322.

Hamakubo, T., R. Kannagi, T. Murachi, and A. Matus. 1986. Distribution of calpains I and II in rat brain. *The Journal of Neuroscience*. 6:3103–3111.

Hortobagyi, T., S. Wise, N. Hunt, N. Cary, V. Djurovic, A. Fegan-Earl et al. 2007. Traumatic axonal damage in the brain can be detected using beta-APP immunohistochemistry within 35 min after head injury to human adults. *Neuropathology and Applied Neurobiology*. 33:226–237.

Jafari, S.S., W.L. Maxwell, M. Neilson, and D.I. Graham. 1997. Axonal cytoskeletal changes after non-disruptive axonal injury. *Journal of Neurocytology*. 26:207–221.

Jafari, S.S., M. Nielson, D.I. Graham, and W.L. Maxwell. 1998. Axonal cytoskeletal changes after nondisruptive axonal injury. II. Intermediate sized axons. *Journal of Neurotrauma*. 15:955–966.

Johnson, V.E., J.E. Stewart, F.D. Begbie, J.Q. Trojanowski, D.H. Smith, and W. Stewart. 2013a. Inflammation and white matter degeneration persist for years after a single traumatic brain injury. *Brain*. 136:28–42.

Johnson, V.E., W. Stewart, and D.H. Smith. 2013b. Axonal pathology in traumatic brain injury. *Experimental Neurology*. 246:35–43.

Lee, J.H., J.M. Shin, Y.J. Shin, M.H. Chun, and S.J. Oh. 2011. Immunochemical changes of calbindin, calretinin and SMI32 in ischemic retinas induced by increase of intraocular pressure and by middle cerebral artery occlusion. *Anatomy and Cell Biology*. 44:25–34.

Letourneau, P.C. 2009. Actin in axons: Stable scaffolds and dynamic filaments. *Results and Problems in Cell Differentiation*. 48:65–90.

Ma, M., Ferguson, T.A., Schoch, K.M., Li, J., Qian, Y., Shofer, F.S., Saatman, K.E., and Neumar, R.W. 2013. Calpains mediate axonal cytoskeleton disintegration during Wallerian degeneration. *Neurobiol. Dis.* 56(1), 34–46.

Maxwell, W.L. 1996. Histopathological changes at central nodes of Ranvier after stretch-injury. *Microscopy Research and Technique*. 34:522–535.

Maxwell, W.L. 2014. Nodes of Ranvier. In: *Encyclopedia of the Neurological Sciences*, 2nd edition. Aminoff, M.J. and R.B. Daroff, editors. Elsevier, Oxford.

Maxwell, W.L., A. Domleo, G. McColl, S.S. Jafari, and D.I. Graham. 2003. Post-acute alterations in the axonal cytoskeleton after traumatic axonal injury. *Journal of Neurotrauma*. 20:151–168.

Maxwell, W.L. and D.I. Graham. 1997. Loss of axonal microtubules and neurofilaments after stretch-injury to guinea pig optic nerve fibers. *Journal of Neurotrauma*. 14:603–614.

Maxwell, W.L., M.A. MacKinnon, J.E. Stewart, and D.I. Graham. 2010. Stereology of cerebral cortex after traumatic brain injury matched to the Glasgow outcome score. *Brain*. 133:139–160.

Maxwell, W.L., B.J. McCreath, D.I. Graham, and T.A. Gennarelli. 1995. Cytochemical evidence for redistribution of membrane pump calcium-ATPase and ecto-Ca-ATPase activity, and calcium influx in myelinated nerve fibres of the optic nerve after stretch injury. *Journal of Neurocytology*. 24:925–942.

Maxwell, W.L., K. Pennington, M.A. MacKinnon, D.H. Smith, T.K. McIntosh, J.T. Wilson et al. 2004. Differential responses in three thalamic nuclei in moderately disabled, severely disabled and vegetative patients after blunt head injury. *Brain*. 127:2470–2478.

Maxwell, W.L., J.T. Povlishock, and D.L. Graham. 1997. A mechanistic analysis of nondisruptive axonal injury: A review. *Journal of Neurotrauma*. 14:419–440.

Maxwell, W.L., C. Watt, D.I. Graham, and T.A. Gennarelli. 1993. Ultrastructural evidence of axonal shearing as a result of lateral acceleration of the head in non-human primates. *Acta Neuropathologica*. 86:136–144.

Maxwell,W.L., Bartlet, E., and Morgan, H.L. 2014. Wallerian degeneration in the optic nerve stretch-injury model of TBI: A stereological analysis. *Journal of Neurotrauma* (under peer review).

McCracken, E., A.J. Hunter, S. Patel, D.I. Graham, and D. Dewar. 1999. Calpain activation and cytoskeletal protein breakdown in the corpus callosum of head-injured patients. *Journal of Neurotrauma*. 16:749–761.

McKee, A.C., R.A. Stern, C.J. Nowinski, T.D. Stein, V.E. Alvarez, D.H. Daneshvar et al. 2013. The spectrum of disease in chronic traumatic encephalopathy. *Brain*. 136:43–64.

Mendelow, A.D. and G.M. Teasdale. 1983. Pathophysiology of head injuries. *The British Journal of Surgery*. 70:641–650.

Missios, S. 2007. Hippocrates, Galen, and the uses of trepanation in the ancient classical world. *Neurosurgical Focus*. 23:E11.

Morrison, B., 3rd, B.S. Elkin, J.P. Dolle, and M.L. Yarmush. 2011. In vitro models of traumatic brain injury. *Annual Review of biomedical Engineering*. 13:91–126.

Ouardouz, M., M.A. Nikolaeva, E. Coderre, G.W. Zamponi, J.E. McRory, B.D. Trapp et al. 2003. Depolarization-induced Ca2+ release in ischemic spinal cord white matter involves L-type Ca2+ channel activation of ryanodine receptors. *Neuron*. 40:53–63.

Palmgren, A. 1960. Specific silver staining of nerve fibres. *Acta Zoologica*. 3:239–265.

Park, E., Bell, J.D., Siddiq, I.P., and Baker, A.J. 2009. An analysis of regional microvascular loss and recovery following two grades of fluid percussion trauma: A role for hypoxia-inducible factors in traumatic brain injury. *Journal of Cerebral Blood Flow and Metabolism* 29(3):575–584.

Peerless, S.J. and N.B. Rewcastle. 1967. Shear injuries of the brain. *Canadian Medical Association Journal*. 96:577–582.

Pettus, E.H. and J.T. Povlishock. 1996. Characterization of a distinct subset of intra-axonal ultrastructural changes associated with traumatically induced alteration in axolemma permeability. *Brain Research*. 1–2:1–11.

Povlishock, J.T., D.P. Becker, C.L. Cheng, and G.W. Vaughan. 1983. Axonal change in minor head injury. *Journal of Neuropathology and Experimental Neurology*. 42:225–242.

Saatman, K.E., B. Abai, A. Grosvenor, C.K. Vorwerk, D.H. Smith, and D.F. Meaney. 2003. Traumatic axonal injury results in biphasic calpain activation and retrograde transport impairment in mice. *Journal of Cerebral Blood Flow and Metabolism*. 23:34–42.

Saatman, K.E., D. Bozyczko-Coyne, V. Marcy, R. Siman, and T.K. McIntosh. 1996. Prolonged calpain-mediated spectrin breakdown occurs regionally following experimental brain injury in the rat. *Journal of Neuropathology and Experimental Neurology*. 55:850–860.

Saatman, K.E., G. Serbst, and M.F. Burkhardt. 2009. Axonal damage due to traumatic brain injury. In: Lajtha, A. and D. Johnson, editors. *Handbook of Neurochemistry and Molecular Neurobiology*. Springer, New York. 344–356.

Schröder, M.L., J.P. Muizelaar, P.P. Fatouros, A.J. Kuta, and S.C. Choi. 1998. Regional cerebral blood volume after severe head injury in patients with regional cerebral ischemia. *Neurosurgery*. 42:1276–1280; discussion 1280–1271.

Schweitzer, J.B., M.R. Park, S.L. Einhaus, and J.T. Robertson. 1993. Ubiquitin marks the reactive swellings of diffuse axonal injury. *Acta Neuropathologica*. 85:503–507.

Sherriff, F.E., L.R. Bridges, S.M. Gentleman, S. Sivaloganathan, and S. Wilson. 1994. Markers of axonal injury in post mortem human brain. *Acta Neuropathologica*. 88:433–439.

Sidoryk-Wegrzynowicz, M., M. Wegrzynowicz, E. Lee, A.B. Bowman, and M. Aschner. 2011. Role of astrocytes in brain function and disease. *Toxicologic Pathology*. 39:115–123.

Sofroniew, M.V. and H.V. Vinters. 2010. Astrocytes: Biology and pathology. *Acta Neuropathologica*. 119:7–35.

Staal, J.A., T.C. Dickson, R. Gasperini, Y. Liu, L. Foa, and J.C. Vickers. 2010. Initial calcium release from intracellular stores followed by calcium dysregulation is linked to secondary axotomy following transient axonal stretch injury. *Journal of Neurochemistry*. 112:1147–1155.

Stone, J.R., D.O. Okonkwo, R.H. Singleton, L.K. Mutlu, G.A. Helm, and J.T. Povlishock. 2002. Caspase-3-mediated cleavage of amyloid precursor protein and formation of amyloid Beta peptide in traumatic axonal injury. *Journal of Neurotrauma*. 19:601–614.

Stone, J.R., R.H. Singleton, and J.T. Povlishock. 2001. Intra-axonal neurofilament compaction does not evoke local axonal swelling in all traumatically injured axons. *Experimental Neurology*. 172:320–331.

Sulaimain, A., N. Denman, N. Buchanan, N. Porter, S. Vesi, R. Sharpe et al. 2011. Stereology and ultrastructure of chronic phase axonal and cell soma pathology in stretch-injured central nerve fibers. *Journal of Neurotrauma*. 3:383–400.

Tang-Schomer, M.D., V.E. Johnson, P.W. Baas, W. Stewart, and D.H. Smith. 2012. Partial interruption of axonal transport due to microtubule breakage accounts for the formation of periodic varicosities after traumatic axonal injury. *Experimental Neurology*. 233:364–372.

Tang-Schomer, M.D., A.R. Patel, P.W. Baas, and D.H. Smith. 2010. Mechanical breaking of microtubules in axons during dynamic stretch injury underlies delayed elasticity, microtubule disassembly, and axon degeneration. *FASEB Journal*. 24:1401–1410.

Timaru-Kast, R., C. Luh, P. Gotthardt, C. Huang, M.K. Schafer, K. Engelhard, and S.C. Thal. 2012. Influence of age on brain edema formation, secondary brain damage and inflammatory response after brain trauma in mice. *PloS One*. 7:e43829.

Tuck, E. and V. Cavalli. 2010. Roles of membrane trafficking in nerve repair and regeneration. *Communicative and Integrative Biology*. 3:209–214.

Tymianski, M. and C.H. Tator. 1996. Normal and abnormal calcium homeostasis in neurons: A basis for the pathophysiology of traumatic and ischemic central nervous system injury. *Neurosurgery*. 38:1176–1195.

Uryu, K., X.H. Chen, D. Martinez, K.D. Browne, V.E. Johnson, D.I. Graham et al. 2007. Multiple proteins implicated in neurodegenerative diseases accumulate in axons after brain trauma in humans. *Experimental Neurology*. 208:185–192.

Venkatesan, C., M. Chrzaszcz, N. Choi, and M.S. Wainwright. 2010. Chronic upregulation of activated microglia immunoreactive for galectin-3/Mac-2 and nerve growth factor following diffuse axonal injury. *Journal of Neuroinflammation*. 7:32.

Verdú, E., D. Ceballos, J.J. Viliches, and X. Navarro. 2008. Influence of aging on peripheral nerve function and regeneration. *Journal of the Peripheral Nervous System*. 4:191–208.

von Oettingen, G., B. Bergholt, C. Gyldensted, and J. Astrup. 2002. Blood flow and ischemia within traumatic cerebral contusions. *Neurosurgery*. 50:781–788; discussion 788–790.

Wang, J.A., Lin, W., Morris, T., Banderali, U., Juranka, P.F., and Morris, C.E. 2009. Membrane trauma and Na^+ leak from Nav1.6 channels. *American Journal of Physiology: Cell Physiology*. 297(4) C828-C834.

Wang, J., R.J. Hamm, and J.T. Povlishock. 2011. Traumatic axonal injury in the optic nerve: Evidence for axonal swelling, disconnection, dieback, and reorganization. *Journal of Neurotrauma*. 28:1185–1198.

Wang, J.T., Z.A. Medress, and B.A. Barres. 2012a. Axon degeneration: Molecular mechanisms of a self-destruction pathway. *The Journal of Cell Biology*. 196:7–18.

Wang, Q., R. Vlkolinsky, M. Xie, A. Obenaus, and S.K. Song. 2012b. Diffusion tensor imaging detected optic nerve injury correlates with decreased compound action potentials after murine retinal ischemia. *Investigative Ophthalmology and Visual Science*. 53:136–142.

Wolf, J.A., P.K. Stys, T. Lusardi, D. Meaney, and D.H. Smith. 2001. Traumatic axonal injury induces calcium influx modulated by tetrodotoxin-sensitive sodium channels. *The Journal of Neuroscience*. 21:1923–1930.

Xu, K., G. Zhong, and X. Zhuang. 2013. Actin, spectrin, and associated proteins form a periodic cytoskeletal structure in axons. *Science*. 339:452–456.

Yaghmai, A and J. Povlishock. 1992. Traumatically induced reactive change as visualized through the use of monoclonal antibodies targeted to neurofilament subunits. *Journal of Neuropathology and Experimental Neurology*. 51:158–176.

Yawo, H. and M. Kuno. 1985. Calcium dependence of membrane sealing at the cut end of the cockroach giant axon. *Journal of Neuroscience*. 6:1626–1632.

4 Pathophysiology of Mild TBI
Implications for Altered Signaling Pathways

Robert A. Laskowski, Jennifer A. Creed, and Ramesh Raghupathi

CONTENTS

4.1 INTRODUCTION

Concussions and mild traumatic brain injury (TBI) represent a substantial portion of the annual incidence of TBI aided by the increased reporting of concussions in youth sports, and the increased exposure of soldiers to blast injuries in the war theater. The pathophysiology of concussions and mild TBI consist predominantly of axonal injury at the cellular level and working memory deficits at the behavioral level. Importantly, studies in humans and in animals are making it clear that concussions and mild TBI are not merely a milder form of moderate-severe TBI but represent a separate disease/injury state. Therefore, acute and chronic treatment strategies, both behavioral and pharmacological, need to be implemented based on thorough pre-clinical assessment. The review in this chapter focuses on two under-studied components of the pathophysiology of mild TBI—the role of the c-Jun N-terminal kinase pathway in axonal injury, and the role of the dopaminergic system in working memory deficits.

The growing awareness of the incidence of concussion in contact sports, coupled with the emergence of blast-related injuries in combat fighting, has heightened the urgency to understand the underlying mechanisms of mild brain trauma and devise potential therapeutic interventions. TBI in general, and mild TBI in particular, is considered a "silent epidemic" because many of the acute and enduring alterations in cognitive, motor, and somatosensory functions may not be readily apparent to external observers. Moderate to severe TBI is a major cause of injury-induced death and disability with an annual incidence of approximately 500 in 100,000 people affected in the United States (Sosin et al., 1989; Kraus and McArthur, 1996; Rutland-Brown et al., 2006). However, approximately 80% of all TBI cases are categorized as mild head injuries (Bazarian et al., 2005; Langlois et al., 2006). It is important to note that these approximations are underestimates because they do not account for incidents of TBI in which the person does not seek medical care (Faul et al., 2010). Recent estimates to correct for this underreporting have placed the annual incidence at approximately 3.8 million (Bazarian et al., 2005; Ropper and Gorson, 2007; Halstead and Walter, 2010). The Glasgow Coma Scale (GCS) score, which measures level of consciousness, has been the primary clinical tool for assessing initial brain injury severity in mild (GCS 13–15), moderate (GCS 9–12), or severe (GCS < 8) cases (Teasdale and Jennett, 1974). Although this scoring system serves as a reliable predictor of patient survival (Steyerberg et al., 2008), particularly in the acute phase of trauma and for those patients with more severe head injury (Saatman et al., 2008), it does not necessarily reflect the underlying cerebral pathology because different structural abnormalities can produce a similar clinical picture.

Concussions are a frequent occurrence in contact sports such as football, hockey, lacrosse, and soccer, and increasing evidence suggests that athletes may sustain multiple concussions throughout their career (Bakhos et al., 2010; Bazarian et al., 2005; Grady, 2010; McCrory et al., 2009). Another significant population is soldiers suffering from blast-related injuries, with one in six soldiers returning from combat deployment in Iraq meeting the criteria for concussion (Wilk et al., 2010). Gender factors may also play a role in the epidemiology of concussion. Comparisons of similar sports have yielded the observation that females have nearly twice the rate of concussion compared with males (Dick, 2009; Lincoln et al., 2011). It is important to note that concussed high school males and females self-report different symptoms, with females more often complaining of drowsiness and noise sensitivity, whereas males complain of cognitive deficits and amnesia (Frommer et al., 2011). Furthermore, females also have a higher post-concussion symptom score 3 months postinjury (Bazarian et al., 2010). Two primary complications of concussion are the postconcussion syndrome and second impact syndrome. The postconcussion syndrome is the persistence of concussion-induced symptomatology for greater than 3 months postinjury, presumably because of both neurophysiological and neuropathological processes secondary to the initial concussion (Silverberg and Iverson, 2011).

Second impact syndrome is a condition in which a second head impact is sustained during a "vulnerable period" before the complete symptomatic resolution of the initial impact leading to profound engorgement, massive edema, and increased intracranial pressure within minutes of the impact and resulting in brain herniation, followed by coma and death (Cantu, 1998; Field et al., 2003). It is believed that this vulnerable period is the duration of an injury-induced failure of cerebral blood flow autoregulation (Lam et al., 1997), which can leave the patient highly vulnerable to drastic fluxes and extremes of blood pressure. Second impact syndrome has a morbidity rate of 100% and a mortality rate of 50%, and it is important to note that as of 2001, all reported cases of second impact syndrome had occurred in athletes younger than 20 years of age (McCrory, 2001).

Neurobehavioral symptoms, which often correlate with severity of the TBI, vary in type and duration and are manifested as somatic and/or neuropsychiatric symptoms (reviewed in Riggio and Wong, 2009). Somatic symptoms refer to the physical changes associated with TBI and include headache, dizziness/nausea, fatigue or lethargy, and changes in sleep pattern. Headache is the most commonly reported somatic symptom after mild TBI and is considered acute if resolved within 2 months or chronic if headaches persist for longer than 2 months. Dizziness is another commonly reported symptom of TBI and generally resolves within 2 months but may continue in patients with moderate or severe TBI. Another particularly debilitating symptom is fatigue, likely due to difficulty in initiating or maintaining sleep. Neuropsychiatric sequelae after TBI comprise cognitive deficits and behavioral disorders and are identified in almost all TBI patients for up to 3 months, with a small percentage exhibiting persistent (months–years) symptoms. Cognitive deficits are characterized by impaired attention, memory, and/or executive function and may cause the patient to become irritable, anxious, or depressed. Cognitive deficits in cases of mild TBI generally resolve within days and do not have to be associated with loss of consciousness and posttraumatic amnesia. Behavioral manifestations after TBI include personality changes, depression, and anxiety. Personality changes describe aggression, impulsivity, irritability, emotional lability, and apathy. Major depression is one of the most frequently reported behavioral sequelae of TBI, accounting for approximately 25% to 40% of cases of moderate-to-severe TBI (Riggio and Wong, 2009).

Collectively, these observations underscore the need to develop age-, sex-, and injury severity–appropriate animal models of mild TBI and concussions. The following review describes the current state of knowledge of the pathophysiology of mild TBI/concussions, with particular attention to axonal injury and cognitive deficits.

4.2 EXPERIMENTAL APPROACHES TO STUDYING CONCUSSIONS

The symptomatology associated with concussion appears to be primarily functional in nature because standard neuroimaging studies reveal no structural abnormalities; however, postmortem analyses of brains from patients who had sustained a recent mild TBI, but had died from nontraumatic causes, showed evidence of axonal injury (Blumbergs et al., 1994, 1995). Specialized functional magnetic resonance imaging has revealed decreases in cortical blood flow to the mid-dorsolateral prefrontal cortex during the acute postconcussive period in athletes challenged in a working memory task as well as activation patterns that correlate with symptom severity and recovery (Chen et al., 2004), whereas diffusion tensor imaging has also detected evidence of microstructural white matter and axonal injuries in some cases of prolonged deficits (Arfanakis et al., 2002; Niogi et al., 2008; Smits et al., 2010; Wilde et al., 2008). Furthermore, electroencephalography and transcranial magnetic stimulation studies have determined that acute and long-term electrophysiological changes in brain activity can occur in the absence of overt neuropsychological impairment (De Beaumont et al., 2007a, 2007b; Gosselin et al., 2006).

A concussion may be caused by either a direct blow to the head (contact forces, Figure 4.1a) or by a blow to elsewhere on the body with the forces being subsequently transmitted to the brain (inertial forces, Figure 4.1b)

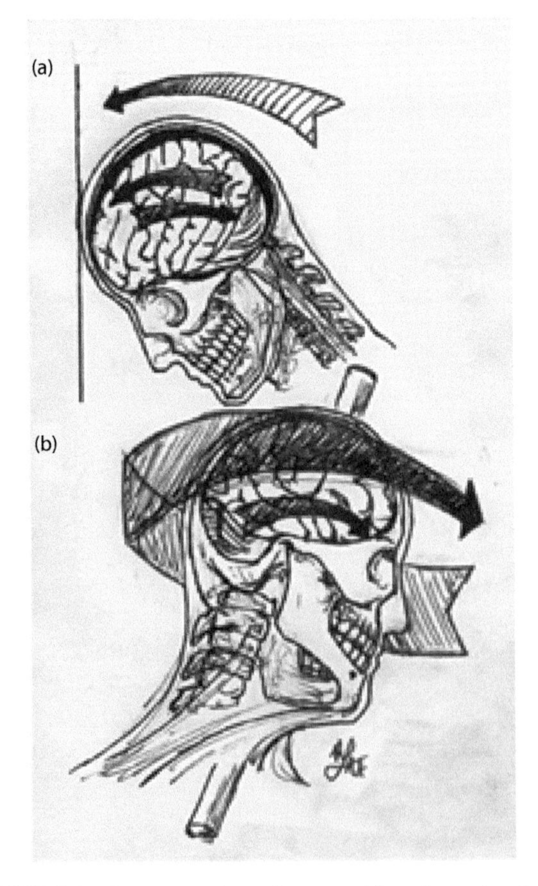

FIGURE 4.1 Representation of contact (a) and rotational forces (b) associated with traumatic brain injury.

(McLean, 1996; Teasdale and Matthew, 1996). Rotational forces around a defined axis are thought to be responsible for damage to deep white matter tracts, resulting in a diffuse axonal injury as well as causing damage to deep gray matter nuclei (McLean, 1996; Thibault and Gennarelli, 1990). A third possible force, the presumable basis of blast trauma, is based on the stereotactile theory, which posits that as a result of the interplay between the spherical shape of the skull and the fact that brain tissue has the same density on concentric planes, the pressure waves created by skull–brain interactions or skull vibrations may propagate through brain tissue as a spherical wave front, resulting in a more focused and direct energy reaching deeper brain structures (Willinger et al., 1996).

Animal models of TBI have been developed in the ferret, cat, pig, and monkey but the most common and developed model is the rodent (Gennarelli, 1994). Two models predominate to elucidate mechanisms of diffuse or concussive brain injury—the midline fluid-percussion model (Dixon et al., 1987) and the impact-acceleration model (Marmarou et al., 1994). Both models were originally characterized in the rat and demonstrate characteristics of human TBI such as cognitive dysfunction (Beaumont et al., 1999; Lyeth et al., 1990) and axonal injury (reviewed in Buki and Povlishock, 2006). More recently, concussive brain injury has been modeled in mice (Laurer et al., 2001; Longhi et al., 2005; Spain et al., 2010; Tang et al., 1997a, 1997b; Zohar et al., 2003). Injury induced by a weight drop, fluid percussion, or a modified cortical impact device resulted in diffuse neurodegeneration in the cortex and hippocampus and βAPP(+) intraaxonal swellings in the thalamus, corpus callosum, and external capsule (Longhi et al., 2005; Spain et al., 2010; Tang et al., 1997b; Tashlykov et al., 2007). Closed-head injury in mice resulted in long-term behavioral dysfunction characterized by learning deficits, depressive behavior, and increased passive avoidance (Milman et al., 2005; Tang et al., 1997a; Spain et al., 2010; Zohar et al., 2003). In contrast, impact to the intact skull using a silicone-tipped indenter only resulted in a transient deficit in motor function with no effect on spatial learning ability (Laurer et al., 2001, 2005). Although these animal models reflect the acute neurochemical, microscopic, and anatomical pathophysiology of concussive brain trauma, they do not appear to model the hallmark of concussion—transient neurologic (cognitive) dysfunction. Impact to the intact skull of mice over the midline suture resulted in spatial learning and working memory deficits only in the first 3 days after trauma (Creed et al., 2011). Traumatic axonal injury was observed up to 3 days postinjury and degenerating axons at 14 days postinjury. These structural alterations in injured axons were accompanied by functional deficits that manifested as reductions in compound action potential and decreased retrograde transport, which were present up to 2 weeks postinjury. Further evidence of diffuse brain injury arose from the observation of cortical edema over the first 24 hours postinjury and neuronal degeneration in the cortex and hippocampus up to 3 days postinjury.

4.3 AXONAL INJURY AFTER MILD TBI

Traumatic axonal injury is triggered by the inertial forces of trauma to the brain, resulting in subsequent structural and subcellular changes within the axon cylinder (Buki and Povlishock, 2006). One of the initial changes is altered axolemmal permeability because of focal microscopic mechanoporation of the axolemma and was first observed as influx of the normally excluded protein, horseradish peroxidase, after head injury (Pettus et al., 1994; Pettus and Povlishock, 1996). These microscopic holes may provide a route for intraaxonal calcium influx, leading to calpain activation (Buki et al., 1999; Saatman et al., 1996). Calpain activation may effect structural alterations to the axonal cytoskeleton leading to disruption of both anterograde and retrograde transport and eventual swellings in contiguous axons and finally secondary axotomy (Buki and Povlishock, 2006; Creed et al., 2011; Shojo and Kibayashi, 2006). Direct evidence of retrograde transport impairment using Fluoro-Gold transport in the brain after a traumatic injury was recently demonstrated (Creed et al., 2011). In part, disruption of axonal transport may be mediated by neurofilament compaction, which occurs as a result of dephosphorylation and has been recognized as another prominent characteristic of axonal injury after TBI (Chen et al., 1999; Christman et al., 1994; Creed et al., 2011; Povlishock et al., 1997).

The c-Jun N-terminal kinases (JNKs) are a subfamily of mitogen-activated protein kinases that play important roles in the central nervous system, in both physiological (neurite outgrowth and extension, brain development and neuronal repair) and pathological conditions (apoptosis, axonal injury) (Herdegen and Waetzig, 2001; Kuan et al., 2003; Waetzig and Herdegen, 2003; Yang et al., 1997). JNK activation has been observed in experimental models of TBI in both neurons (Raghupathi et al., 2003; Ortolano et al., 2009; Otani et al., 2002) and axons (Raghupathi et al., 2003) and in humans (Ortolano et al., 2009).

Their ability to participate in and also be activated by cytoskeletal changes allows JNK to play an important role in dynamic neurite outgrowth and elongation during brain development (Waetzig and Herdegen, 2005). Importantly, JNK activation has been implicated in axonal injury after trauma in vivo (Broude et al., 1997; Raghupathi et al., 2003; Raivich et al., 2004) and in vitro (Cavalli et al., 2005; Verhey et al., 2001). Direct phosphorylation of the kinesin-1 heavy chain subunit by activated JNK in the squid axoplasm led to the inhibition of binding between kinesin-1 and axonal microtubules and subsequent fast axonal transport (Morfini et al., 2006). Interestingly, this disruption in axonal transport appeared to be mediated by the neuron-specific JNK3 isoform (Morfini et al., 2009), which may explain the observed protective effect of genetic deletion of the JNK3 isoform after axotomy of dopaminergic neurons (Brecht et al., 2005).

4.4 WORKING MEMORY DEFICITS AND DOPAMINERGIC SIGNALING IN MILD TBI

Working memory deficits are a major complaint of patients suffering from TBI with transient deficits after mild TBI/concussions and permanent morbidity from severe TBI (Mayers et al., 2011; Gorman et al., 2012; McAllister et al., 2001; Slovarp et al., 2012; Theriault et al., 2011). In rats and mice, working memory deficits have been documented and appear not to be dependent on the location of the impact or the type of model used. Thus, contusive trauma or fluid-percussion injury either over the frontal cortices or the parietal cortex (Hamm et al., 1996; Hoane et al., 2006; Hoskison et al., 2009; Vonder Haar et al., 2011) all resulted in significant long-term working memory deficits in the adult rat. Conversely, closed-head midline cortical contusion injury that impacts the skull midway between Bregma and Lambda is capable of producing a working memory deficit in adult male mice tested on days 1–3 postinjury, but that has resolved by days 7–9 postinjury (Creed et al., 2011).

Working memory is an organism's ability to transiently maintain information in an active and available form over a time delay. It is the mental chalkboard that allows for successful interactions within an ever-changing environment by permitting one to manipulate and actively use the stored information to apply it to a current situation for goal-directed or problem-solving purposes. Working memory relies on the appropriate interactions of a distributed network of brain regions, though the primary region of integration appears to be the prefrontal cortex (PFC). The cellular activity underlying working memory is based on the activity of neurons after the withdrawal of a prior stimulus or event. Neurons within the prefrontal cortex have "memory fields" or the representation of a target stimulus to which a neuron fires maximally (Funahashi et al., 1989). Working memory requires a finely tuned balance of excitatory and inhibitory inputs into and within the PFC. In animals, mild TBI induces a hypoexcitable brain state in which the evoked population excitatory postsynaptic potential is significantly decreased compared with uninjured animals followed by a period of hyperexcitability (Ding et al., 2011). Sanders and colleagues (2001) noted that an fluid-percussion-induced mild TBI over the right parietal cortex of male rats caused reductions of the slope and increases in the latency of vibrissa-evoked potentials 3 days postinjury, whereas alterations in presynaptic neuronal function have also been observed as early as 1 hour postinjury in adult male rats (Reeves et al., 2000).

Pyramidal, excitatory neurons act in concert with inhibitory interneurons; this system is modulated by dopaminergic afferents to the prefrontal cortex from the ventral tegmental area (Durstewitz and Seamans, 2002). These dopamine afferents form symmetric synapses on the dendritic spines of pyramidal neurons, which in turn contain the D1 dopamine receptor subtype (Charuchinda et al., 1987; Lidow et al., 1991; Smiley et al., 1994). Expression of the D1 receptor increased in the PFC as early as 3 hours and remained elevated up to 3 days after contusive brain trauma (Kobori and Dash, 2006). In contrast, in the striatum, the binding properties of the D1 receptor decreased in the acute posttraumatic period but increased in the subacute period, with no concomitant change in the level of expression (Henry et al., 1997; Wagner et al., 2009). Nonspecific dopamine agonists such as methylphenidate (Newsome et al., 2009; Wagner et al., 2007) and amantadine (Dixon et al., 1999; Meythaler et al., 2002) have ameliorated TBI-induced cognitive deficits. In a model of moderate brain trauma, the D1 receptor antagonist SCH23390 attenuated working memory deficits (Kobori and Dash, 2009), whereas after concussive TBI, the efficacy of SCH23390 was augmented by a coadministration of the D2 receptor antagonist sulpiride (Tang et al., 1997a, 1997b). In contrast, in a model of concussion in adolescent rats, we observed that a partial agonist of the D1 receptor (SKF38393) almost completely restored working memory function in brain-injured rats (unpublished observations). These data, while implicating the dopamine system in posttraumatic working memory deficits, underscore the complicated nature of the response of the brain to differing severities of injury.

4.5 CONCLUSION

Concussions and mild TBI represent a significant component of the spectrum of TBI-associated syndromes. Accumulating evidence suggests that the pathophysiology of mild TBI may pose questions not addressed over the years in models of moderate-to-severe TBI. Although the cellular manifestation of axonal injury may be transient in mild TBI, deficits in axonal function may be present over a longer period postinjury. Similarly, alterations in dopaminergic signaling may follow a different trajectory than what has been reported in more severe cases and treatment with dopaminergic agents may have to take into account the severity of the injury. These observations underscore the importance of continued studies in mild TBI.

ACKNOWLEDGMENT

This work is supported, in part, by grants from the National Institutes of Health NS06517 and the Veteran's Administration.

REFERENCES

Arfanakis, K., Haughton, V. M., Carew, J., Rogers, B., Dempsey, R., and Meyerand, M. (2002). Diffusion tensor MR imaging in diffuse axonal injury. *AJNR Am J Neuroradiol*. 23:794–802.

Bakhos, L. L., Lockhart, G. R., Myers, R., and Linakis, J. G. (2010). Emergency department visits for concussion in young child athletes. *Pediatrics*. 126:e550–556.

Bazarian, J. J., McClung, J., Shah, M. N., Cheng, Y. T., Flesher, W., and Kraus, J. (2005). Mild traumatic brain injury in the United States, 1998–2000. *Brain Inj*. 19, 85–91.

Bazarian, J. J., Blyth, B., Mookerjee, S., He, H., and McDermott, M. P. (2010). Sex differences in outcome after mild traumatic brain injury. *J Neurotrauma*. 27:527–539.

Beaumont, A., Marmarou, A., Czigner, A., Yamamoto, M., Demetriadou, K., Shirotani et al. (1999). The impact-acceleration model of head injury: Injury severity predicts motor and cognitive performance after trauma. *Neurol Res.* 21:742–754.

Blumbergs, P. C., Scott, G., Manavis, J., Wainwright, H., Simpson, D. A., and McLean, A. J. (1995). Topography of axonal injury as defined by amyloid precursor protein and the sector scoring method in mild and severe closed head injury. *J Neurotrauma.* 12, 565–572.

Blumbergs, P. C., Scott, G., Manavis, J., Wainwright, H., Simpson, D. A., and McLean, A. J. (1994). Staining of amyloid precursor protein to study axonal damage in mild head injury. *Lancet.* 344, 1055–1056.

Brecht, S., Kirchhof, R., Chromik, A., Willesen, M., Nicolaus, T., Raivich, G. et al. (2005). Specific pathophysiological functions of JNK isoforms in the brain. *Eur J Neurosci.* 21, 363–377.

Broude, E., McAtee, M., Kelley, M. S., and Bregman, B. S. (1997). c-Jun expression in adult rat dorsal root ganglion neurons: Differential response after central or peripheral axotomy. *Exp Neurol.* 148, 367–377.

Buki, A. and Povlishock, J. T. (2006). All roads lead to disconnection? Traumatic axonal injury revisited. *Acta Neurochir (Wien).* 148, 181–93; discussion 193–4.

Buki, A., Siman, R., Trojanowski, J. Q., and Povlishock, J. T. (1999). The role of calpain-mediated spectrin proteolysis in traumatically induced axonal injury. *J Neuropathol Exp Neurol.* 58, 365–375.

Cantu, R. C. (1998). Second-impact syndrome. *Clin Sports Med.* 17:37–44.

Cavalli, V., Kujala, P., Klumperman, J., and Goldstein, L. S. (2005). Sunday Driver links axonal transport to damage signaling. *J Cell Biol.* 168, 775–787.

Charuchinda, C., Supavilai, P., Karobath, M., and Palacios, J. M. (1987). Dopamine D2 receptors in the rat brain: Autoradiographic visualization using high-affinity selective agonist ligand. *J Neurosci.* 7:1352–1360.

Chen, J.-K., Johnston, K., Frey, S., Petrides, M., Worsley, K., and Ptitp, A. (2004). Functional abnormalities in symptomatic concussed athletes: An fMRI study. *NeuroImage.* 22:68–82.

Chen, X. H., Meaney, D. F., Xu, B. N., Nonaka, M., McIntosh, T. K., Wolf, J. A., Saatman, K. E., and Smith, D. H. (1999). Evolution of neurofilament subtype accumulation in axons following diffuse brain injury in the pig. *J Neuropathol Exp Neurol.* 58, 588–596.

Christman, C. W., Grady, M. S., Walker, S. A., Holloway, K. L., and Povlishock, J. T. (1994). Ultrastructural studies of diffuse axonal injury in humans. *J Neurotrauma.* 11, 173–186.

Creed, J. A., DiLeonardi, A. M., Fox, D. P., Tessler, A. R., and Raghupathi, R. (2011). Concussive brain trauma in the mouse results in acute cognitive deficits and sustained impairment of axonal function. *J Neurotrauma.* 28:547–563.

De Beaumont, L., Brisson, B., Lassonde, M., and Jolicoeur, P. (2007a). Long-term electrophysiological changes in athletes with a history of multiple concussions. *Brain Inj.* 21:631–644.

De Beaumont, L., Lassonde, M., Leclerc, S., and Theoret, H. (2007b). Long-term and cumulative effects of sports concussions on motor cortex inhibition. *Neurosurgery.* 61:329–337.

Dick, R. W. (2009). Is there a gender difference in concussion incidence and outcomes? *Br J Sports Med.* 43:i46–50.

Ding, M. C., Wang, Q., Lo, E. H., and Stanley, G. B. (2011). Cortical excitation and inhibition following focal traumatic brain injury. *J Neurosci.* 31:14085–14094.

Dixon, C. E., Lyeth, B. G., Povlishock, J. T., Findling, R. L., Hamm, R. J., Marmarou, A. et al. (1987). A fluid percussion model of experimental brain injury in the rat. *J Neurosurg.* 67:110–119.

Dixon, C.E., Kraus, M.F., Kline, A.E., Ma, X., Yan, H.Q., Griffith, R.G. et al. (1999). Amantadine improves water maze performance without affecting motor behavior following traumatic brain injury in rats. *Restor Neurol Neurosci.* 14:285–294.

Durstewitz, D., Seamans, J. K. (2002). The computational role of dopamine D1 receptors in working memory. *Neural Netw.* 15:561–572.

Faul, M., Xu, L., Wald, M. M., and Coronado, V.G. (2010). Traumatic brain injury in the United States: Emergency department visits, hospitalizations and deaths 2002–2006. Centers for Disease Control and Prevention, National Center for Injury Prevention and Control: Atlanta, GA.

Field, M., Collins, M. W., Lovell, M. R., and Maroon, J. (2003). Does age play a role in recovery from sports-related concussion? A comparison of high school and collegiate athletes. *J Pediatr.* 142:546–553.

Frommer, L. J., Gurka, K. K., Cross, K. M., Ingersoll, C. D., Comstock, R. D., and Saliba, S. A. (2011). Sex differences in concussion symptoms of high school athletes. *J Athl Train.* 46:76–84.

Funahashi, S., Bruce, C. J., and Goldman-Rakic, P. S. (1989). Mnemonic coding of visual space in the monkey's dorsolateral prefrontal cortex. *J Neurophysiol.* 61:331–349.

Gennarelli, T. A. (1994). Animate models of human head injury. *J Neurotrauma.* 11:357–368.

Gorman, S., Barnes, M. A., Swank, P. R., Prasad, M., and Ewing-Cobbs, L. (2012). The effects of pediatric traumatic brain injury on verbal and visual-spatial working memory. *J Int Neuropsychol Soc.* 18:29–38.

Gosselin, N., Theriault, M., Leclerc, S., Montplaisir, J., and Lassonde, M. (2006). Neurophysiological anomalies in symptomatic and asymptomatic concussed athletes. *Neurosurgery.* 58:1151–1161.

Grady, M. W. (2010). Concussion in the adolescent athlete. *Curr Probl Pediatr Adolesc Health Care.* 40:154–169.

Halstead, M. E. and Walter, K. D.(2010). Clinical report—Sport-related concussion in chlidren and adolescents. *Pediatrics.* 126:597–615.

Hamm, R. J., Temple, M. D., Pike, B. R., O, Dell, D. M., Buck, D. L., Lyeth, B. G. (1996). Working memory deficits following traumatic brain injury in the rat. *J Neurotrauma.* 13:317–323.

Henry, J. M., Talukder, N. K., Lee, A. B., and Walker, M. L. (1997). Cerebral trauma-induced changes in corpus striatal dopamine receptor subtypes. *J Invest Surg.* 10:281–286.

Herdegen, T. and Waetzig, V. (2001). The JNK and p38 signal transduction following axotomy. *Restor Neurol Neurosci.* 19, 29–39.

Hoane, M. R., Tan, A. A., Pierce, J. L., Anderson, G. D., and Smith, D. C. (2006). Nicotinamide treatment reduces behavioral impairments and provides cortical protection after fluid percussion injury in the rat. *J Neurotrauma.* 23:1535–1548.

Hoskison, M. M., Moore, A. N., Hu, B., Orsi, S., Kobori, N., and Dash, P. K. (2009). Persistent working memory dysfunction following traumatic brain injury: Evidence for a time-dependent mechanism. *Neuroscience.* 159:483–491.

Kobori, N. and Dash, P. K. (2006). Reversal of brain injury-induced prefrontal glutamic acid decarboxylase expression and working memory deficits by D1 receptor antagonism. *J Neurosci.* 26:4236–4246.

Kraus, J. F. and McArthur, D. L. (1996). Epidemiologic aspects of brain injury. *Neurol Clin.* 14, 435–450.

Kuan, C. Y., Whitmarsh, A. J., Yang, D. D., Liao, G., Schloemer, A. J., Dong, C. et al. (2003). A critical role of neural-specific JNK3 for ischemic apoptosis. *Proc Natl Acad Sci U S A*. 100, 15184–15189.

Lam, J. M., Hsiang, J. N., and Poon, W. S. (1997). Monitoring of autoregulation using Doppler flowmetry in patients with head injury. *J Neurosurg*. 86:438–445.

Laurer, H. L., Bareyre, F. M., Lee, V. M., Trojanowski, J. Q., Longhi, L., Hoover, R. et al. (2001). Mild head injury increasing the brain's vulnerability to a second concussive impact. *J Neurosurg*. 95, 859–870.

Lidow, M. S., Goldman-Rakic, P. S., Gallager, D. W., and Rakic, P. (1991). Distribution of dopaminergic receptors in the primate cerebral cortex: Quantitative autoradiographic analysis using [3H]spiperone and [3H]SCH23390. *Neuroscience*. 40:657–671.

Lincoln, A. E., Caswell, S. V., Almquist, J. L., Dunn, R. E., Norris, J. B., and Hinton, R. Y. (2011). Trends in concussion incidence in high school sports: A prospective 11-year study. *Am J Sports Med*. 39:958–963.

Longhi, L., Saatman, K. E., Fujimoto, S., Raghupathi, R., Meaney, D. F., Davis, J. et al. (2005). Temporal window of vulnerability to repetitive experimental concussive brain injury. *Neurosurgery*. 56, 364–74; discussion 364–74.

Lyeth, B. G., Jenkins, L. W., Hamm, R. J., Dixon, C. E., Phillips, L. L., Clifton, G. L. et al. (1990). Prolonged memory impairment in the absence of hippocampal cell death following traumatic brain injury in the rat. *Brain Res*. 526, 249–258.

Marmarou, A., Foda, M. A., van den Brink, W., Campbell, J., Kita, H., and Demetriadou, K. (1994). A new model of diffuse brain injury in rats. Part I: Pathophysiology and biomechanics. *J Neurosurg*. 80:291–300.

Mayers, L. B., Redick, T. S., Chiffriller, S. H., Simone, A. N., and Terraforte, K. R. (2011). Working memory capacity among collegiate student athletes: Effects of sport-related head contacts, concussions, and working memory demands. *J Clin Exp Neuropsychol*. 33:532–537.

McAllister, T. W., Sparling, M. B., Flashman, L. A., Guerin, S. J., Mamourian, A. C., and Saykin, A. J. (2001). Differential working memory load effects after mild traumatic brain injury. *Neuroimage*. 14:1004–1012.

McCrory, P. (2001). Does second impact syndrome exist? *J Clin Sport Med*. 11:144–149.

McCrory, P., Meeuwisse, W., Johnston, K., Dvorak, J., Aubry, M., Molloy, M. et al. (2009). Consensus statement on Concussion in Sport: The 3rd International Conference on Concussion in Sport held in Zurich, November 2008. *Clin J Sport Med*. 19:185–200.

McLean, A. (1996) Brain injury without head impact? In A. Bandak, R. Eppinger, A. Ommaya (Eds.), *Traumatic Brain Injury: Bioscience and Mechanics*. Mary Ann Liebert, Larchmont, NY.

Meythaler, J. M., Brunner, R. C., Johnson, A., and Novack, T. A. (2002). Amantadine to improve neurorecovery in traumatic brain injury-associated diffuse axonal injury: A pilot double-blind randomized trial. *J Head Trauma Rehabil*. 17:300–313.

Milman, A., Rosenberg, A., Weizman, R., and Pick, C. G. (2005). Mild traumatic brain injury induces persistent cognitive deficits and behavioral disturbances in mice. *J Neurotrauma*. 22, 1003–1010.

Morfini, G. A., You, Y. M., Pollema, S. L., Kaminska, A., Liu, K., Yoshioka, K. et al. (2009). Pathogenic huntingtin inhibits fast axonal transport by activating JNK3 and phosphorylating kinesin. *Nat Neurosci*. 12, 864–871.

Morfini, G., Pigino, G., Szebenyi, G., You, Y., Pollema, S., and Brady, S. T. (2006). JNK mediates pathogenic effects of polyglutamine-expanded androgen receptor on fast axonal transport. *Nat Neurosci*. 9, 907–916.

Newsome, M. R., Scheibel, R. S., Seignourel, P. J., Steinberg, J. L., Troyanskaya, M., Li, X., Levin, H. S. (2009). Effects of methylphenidate on working memory in traumatic brain injury: A preliminary FMRI investigation. *Brain Imaging Behav*. 3:298–305.

Niogi, S. N., Mukherjee, P., Ghajar, J., Johnson, C., Kolster, R. A., Sarkar, R. et al. (2008). Extent of microstructural white matter injury in postconcussive syndrome correlates with impaired cognitive reaction time: A 3T diffusion tensor imaging study of mild traumatic brain injury. *AJNR Am J Neuroradiol*. 29:967–973.

Ortolano, F., Colombo, A., Zanier, E. R., Sclip, A., Longhi, L., Perego, C. et al. (2009). c-Jun N-terminal kinase pathway activation in human and experimental cerebral contusion. *J Neuropathol Exp Neurol*. 68, 964–971.

Otani, N., Nawashiro, H., Fukui, S., Nomura, N., and Shima, K. (2002). Temporal and spatial profile of phosphorylated mitogen-activated protein kinase pathways after lateral fluid percussion injury in the cortex of the rat brain. *J Neurotrauma*. 19, 1587–1596.

Pettus, E. H. and Povlishock, J. T. (1996). Characterization of a distinct set of intra-axonal ultrastructural changes associated with traumatically induced alteration in axolemmal permeability. *Brain Res*. 722, 1–11.

Pettus, E. H., Christman, C. W., Giebel, M. L., and Povlishock, J. T. (1994). Traumatically induced altered membrane permeability: Its relationship to traumatically induced reactive axonal change. *J Neurotrauma*. 11, 507–522.

Povlishock, J. T., Marmarou, A., McIntosh, T., Trojanowski, J. Q., and Moroi, J. (1997). Impact acceleration injury in the rat: Evidence for focal axolemmal change and related neurofilament sidearm alteration. *J Neuropathol Exp Neurol*. 56, 347–359.

Raghupathi, R., Muir, J. K., Fulp, C. T., Pittman, R. N., and McIntosh, T. K. (2003a). Acute activation of mitogen-activated protein kinases following traumatic brain injury in the rat: Implications for posttraumatic cell death. *Exp Neurol*. 183, 438–448.

Raivich, G., Bohatschek, M., Da Costa, C., Iwata, O., Galiano, M., Hristova, M. et al. (2004). The AP-1 transcription factor c-Jun is required for efficient axonal regeneration. *Neuron*. 43, 57–67.

Reeves, T. M., Kao, C. Q., Phillips, L. L., Bullock, M. R., and Povlishock, J. T. (2000). Presynaptic excitability changes following traumatic brain injury in the rat. *J Neurosci Res*. 60:370–379.

Riggio, S. and Wong, M. (2009). Neurobehavioral sequelae of traumatic brain injury. Mt Sinai J Med. 76, 163–172.

Ropper, A. H. and Gorson, K. C. (2007). Clinical practice. Concussion. *N Engl J Med*. 356:166–172.

Rutland-Brown, W., Langlois, J. A., Thomas, K. E., and Xi, Y. L. (2006). Incidence of traumatic brain injury in the United States, 2003. *J Head Trauma Rehabil*. 21:544–548.

Saatman, K. E., Bozyczko-Coyne, D., Marcy, V., Siman, R., and McIntosh, T. K. (1996). Prolonged calpain-mediated spectrin breakdown occurs regionally following experimental brain injury in the rat. *J Neuropathol Exp Neurol*. 55:850–860.

Saatman, K. E., Duhaime, A. C., Bullock, R., Maas, A. I., Valadka, A., Manley, G. T., and Workshop Scientific Team and Advisory Panel Members. (2008). Classification of traumatic brain injury for targeted therapies. *J Neurotrauma*. 25:719–738.

Sanders, M. J., Dietrich, W. D., and Green, E. J. (2001). Behavioral, electrophysiological, and histopathological consequences of mild fluid-percussion injury in the rat. *Brain Res.* 904:141–144.

Shojo, H. and Kibayashi, K. (2006). Changes in localization of synaptophysin following fluid percussion injury in the rat brain. *Brain Res.* 1078:198–211.

Silverberg, N. D. and Iverson, G. L. (2011). Etiology of the post-concussion syndrome: Physiogenesis and psychogenesis revisited. *NeuroRehabilitation*. 29:317–329.

Slovarp, L., Azuma, T., and Lapointe, L. (2012). The effect of traumatic brain injury on sustained attention and working memory. *Brain Inj.* 26:48–57.

Smiley, J. F., Levey, A. I., Ciliax, B. J., and Goldman-Rakic, P. S. (1994). D1 dopamine receptor immunoreactivity in human and monkey cerebral cortex: Predominant and extrasynaptic localization in dendritic spines. *Proc Natl Acad Sci USA*. 91:5720–5724.

Smits, M., Houston, G. C., Dippel, D. W., Wielopolski, P. A., Vernooij, M. W., Koudstaal, P. J. et al. (2011). Microstructural brain injury in post-concussion syndrome after minor head injury. *Neuroradiology*. 53:553–563.

Sosin, D. M., Sacks, J. J., and Smith, S. M. (1989). Head injury-associated deaths in the United States from 1979 to 1986. *JAMA*. 262:2251–2255.

Spain, A., Daumas, S., Lifshitz, J., Rhodes, J., Andrews, P. J., Horsburgh, K., and Fowler, J. H. (2010). Mild fluid percussion injury in mice produces evolving selective axonal pathology and cognitive deficits relevant to human brain injury. *J Neurotrauma.* 27:1429–1438.

Steyerberg, E. W., Mushkudiani, N., Perel, P., Butcher, I., Lu, J., McHugh, G. S. et al. (2008). Predicting outcome after traumatic brain injury: Development and international validation of prognostic scores based on admission characteristics. *PLoS Med.* 5:e165; discussion e165.

Tang, Y. P., Noda, Y., Hasegawa, T., and Nabeshima, T. (1997a). A concussive-like brain injury model in mice (I): Impairment in learning and memory. *J Neurotrauma.* 14:851–862.

Tang, Y. P., Noda, Y., Hasegawa, T., and Nabeshima, T. (1997b). A concussive-like brain injury model in mice (II): Selective neuronal loss in the cortex and hippocampus. *J Neurotrauma.* 14:863–873.

Tashlykov, V., Katz, Y., Gazit, V., Zohar, O., Schreiber, S., and Pick, C. G. (2007). Apoptotic changes in the cortex and hippocampus following minimal brain trauma in mice. *Brain Res.* 1130:197–205.

Teasdale, G. and Jennett, B. (1974). Assessment of coma and impaired consciousness. A practical scale. *Lancet.* 2:81–84.

Teasdale, G. and Matthew, P. (1996). Mechanisms of cerebral concussion, contusion, and other effects of head injury. In J. R. Youmans (Ed.), *Neurological Surgery,* 4th edition. WB Saunders: New York.

Theriault, M., De Beaumont, L., Tremblay, S., Lassonde, M., and Jolicoeur, P. (2011). Cumulative effects of concussions in athletes revealed by electrophysiological abnormalities on visual working memory. *J Clin Exp Neuropsychol.* 33:30–41.

Thibault, L. and Gennarelli, T. (1990). Brain injury: An analysis of neural and neurovascular trauma in the nonhuman primate. Paper presented at: 34th annual proceedings of the Association for the Advancement of Automotive Medicine. Des Plaines, IL.

Verhey, K. J., Meyer, D., Deehan, R., Blenis, J., Schnapp, B. J., Rapoport, T. A. et al. (2001). Cargo of kinesin identified as JIP scaffolding proteins and associated signaling molecules. *J Cell Biol.* 152:959–970.

Vonder Haar, C., Anderson, G. D., and Hoane, M. R. (2011). Continuous nicotinamide administration improves behavioral recovery and reduces lesion size following bilateral frontal controlled cortical impact injury. *Behav Brain Res.* 224:311–317.

Waetzig, V. and Herdegen, T. (2005). Context-specific inhibition of JNKs: Overcoming the dilemma of protection and damage. Trends Pharmacol Sci. 26:455–461.

Wagner, A. K., Kline, A. E., Ren, D., Willard, L. A., Wenger, M. K., Zafonte, R. D. et al. (2007). Gender associations with chronic methylphenidate treatment and behavioral performance following experimental traumatic brain injury. *Behav Brain Res.* 181:200–209.

Wilde, E., McCauley, S., Hunter, J., Bigler, E., Chu, Z., Wang, Z. et al. (2008). Diffusion tensor imaging of acute mild traumatic brain injury in adolescents. *Neurology*. 70:948–955.

Wilk, J. E., Thomas, J. L., McGurk, D. M., Riviere, L. A., Castro, C. A., and Hoge, C. W. (2010). Mild traumatic brain injury (concussion) during combat: Lack of association of blast mechanism with persistent postconcussive symptoms. *J Head Trauma Rehabil.* 25:9–14.

Willinger, R., Taleb, L., and Kopp, C. (1996). Modal and temporal analysis of head mathematical models. In A. Bandak, R. Eppinger, A. Ommaya (Eds.), *Traumatic Brain Injury: Bioscience and Mechanics*. Mary Ann Liebert: Larchmont, NY.

Yang, D. D., Kuan, C. Y., Whitmarsh, A. J., Rincon, M., Zheng, T. S., Davis, R. J. et al. (1997). Absence of excitotoxicity-induced apoptosis in the hippocampus of mice lacking the Jnk3 gene. *Nature.* 389:865–870.

Zohar, O., Schreiber, S., Getslev, V., Schwartz, J. P., Mullins, P. G., and Pick, C. G. (2003). Closed-head minimal traumatic brain injury produces long-term cognitive deficits in mice. *Neuroscience.* 118:949–955.

5 Oxidative Stress, Brain Edema, Blood–Brain Barrier Permeability, and Autonomic Dysfunction from Traumatic Brain Injury

Hale Zerrin Toklu and Nihal Tümer

CONTENTS

5.1 INTRODUCTION

Traumatic brain injury (TBI) contributes to a substantial number of deaths and cases of permanent disability annually. An immune response to head injury gives a timeline to the pathological cascades with multiple cellular, metabolic and immune pathways activated from the moment of injury. In addition to the primary brain injury which refers to an unavoidable brain damage that occurs at the immediate moment of impact, secondary brain injury develops in the latter term, progressively contributing to the worsened neurological outcome. This complex phenomenon is defined by the various neurochemical cascades activated, and the systemic physiological responses which manifest in the afterwards the traumatic event. Microglial activation and macrophage accumulation after diffuse brain injury is observed within 6 to 48 hours post injury. Thus, diffuse brain injury mediated immune responses, blood-brain barrier alterations, oxidative stress and neuroinflammation seem to play an important role in the pathology.

5.2 TRAUMATIC BRAIN INJURY

TBI is the leading cause of death and disability in children and adults from ages 1 to 44. Brain injuries are most often caused by motor vehicle crashes, sports injuries, or simple falls on the playground at work or in the home. Every year, approximately 52,000 deaths result from traumatic brain injury. An estimated 1.5 million head injuries occur every year in the United States, varying from mild to severe. The incidence of sports-related TBIs is estimated between 1.6 and 3.8 million. At least 5.3 million Americans, 2% of the U.S. population, currently live with disabilities resulting from TBI (Langlois, 2006). Long-term effects of TBI can lead to cardiovascular and neurological disorders such as Parkinsonism, Alzheimer disease, dementia, sleep disturbances, anxiety, abnormalities in pain sensation, and muscular dysfunction (Kobeissy, 2014).

Over the past several decades, numerous experimental animal models have been implemented to study the mechanisms of TBI (i.e., overpressure blast injury, controlled cortical impact, and the fluid percussion models). These animal models have been well characterized with predictable neurological, histological, and physiological changes similar to those observed in clinical brain injury and help us to determine the underlying mechanisms of acute injury and establish treatment strategies.

Studies demonstrated that in addition to the primary physical impact of the injury, TBI leads to progressive pathophysiological changes, resulting in a reduction in brain–blood flow and a decrease in tissue oxygen levels leading to ischemia, subsequent secondary injury, blood–brain barrier breakdown, and brain edema (Unterberg et al., 2004). Death of resident cells of the central nervous system (CNS) has traditionally been accepted to take place in two phases: an early necrotic and an ongoing long-term apoptotic phase. Secondary brain injury develops in minutes to months after the original insult, progressively contributing to the worsened neurological impairment. This complex phenomenon is defined by the activation of various neurochemical cascades and the systemic physiological responses that manifest after the traumatic event (Morganti-Kossmann et al., 2007; Werner, 2009). At the cellular level, the biphasic nature of secondary injury is mediated by numerous disturbed pathways that include (1) excitotoxicity caused by an excess of the neurotransmitter glutamate; (2) free-radical generation by mitochondrial dysfunction, causing damage to proteins and phospholipid membranes of neurons and glia; and (3) the neuroinflammatory response that takes place because of both CNS and systemic immunoactivation as shown in Figure 5.1.

5.3 ROLE OF OXIDATIVE STRESS IN TBI

Although the primary injury occurs by the physical impact of the trauma, the tissue injury augmented by the secondary injury. The brain is quite sensitive to free radical damage, although it is secured by the blood–brain barrier (BBB). The high rate of oxidative metabolism in the brain and its elevated levels of polyunsaturated lipids, which are the target of lipid peroxidation, render it particularly vulnerable to oxidative stress.

Previous studies have demonstrated that reactive oxygen species (ROS) such as the superoxide radicals and nitric oxide can form peroxynitrite, a powerful oxidant that impairs cerebral vascular function after TBI (DeWitt and Prough, 2009; Vuceljic et al., 2006). Thus trauma impairs the oxygenation of brain from impaired circulation and ischemia. The reperfusion state after the trauma enables the vitality of the neurons but also increases the amount of ROS generated (Ansari et al., 2008; Cornelius, 2013; Readnower et al., 2010). The generation of free oxygen radicals, superoxide, hydrogen peroxide, nitric oxide, and peroxynitrite cause excitotoxicity and impair the energy metabolism of the cells. The endogenous antioxidant system (i.e., glutathione peroxidase, superoxide dismutase, catalase, and uric acid) aims to convert/neutralize these ROS to less toxic derivatives, thus preventing binding of these to the macromolecules like DNA, RNA, or proteins (Figure 5.2). However, the excessive amount of ROS produced depletes the endogenous antioxidants and the increased peroxidation of membrane lipids or oxidation of proteins lead to DNA fragmentation and inhibits the mitochondrial electron transport system. This process induces apoptosis or necrosis (Tran, 2014).

Because the role of oxidative stress in TBI is shown in different models, a number of therapeutic trials based on the ability of antioxidants to scavenge free radicals have been attempted in both experimental and clinical TBI. Melatonin, alpha tocopherol, ascorbic acid, Tempol, Edaravone, resveratrol, alpha lipoic, and N-acetylcysteine acid have been widely used to protect the brain tissue against TBI-induced oxidative stress (Slemmer, 2008).

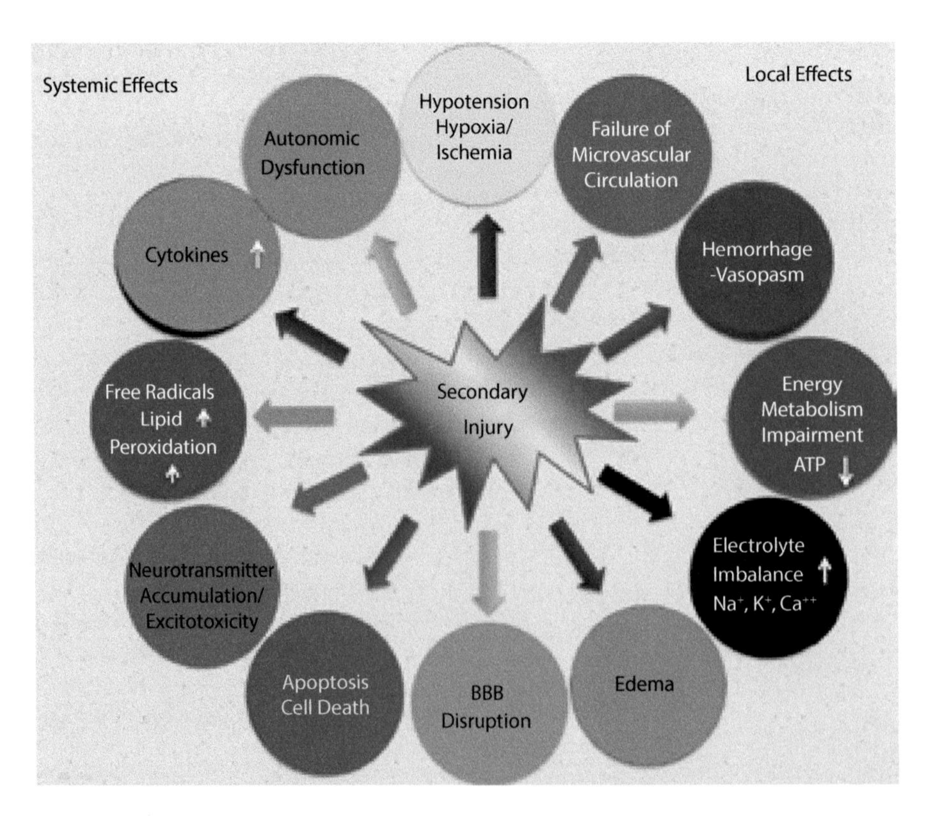

FIGURE 5.1 Mechanisms of brain injury. Brain injury leads to local-systemic secondary changes besides the primary (physical) injury. Failure of the transport system and disturbances in microvascular circulation causes an edema that exacerbates the blood–brain barrier (BBB) disruption and impairs the energy metabolism of the cell. Hypoxia/ischemia and further biochemical changes enhance the generation of free radicals and thus causes immunoactivation and neuroinflammation. This process initiates the pathways through apoptosis and necrosis.

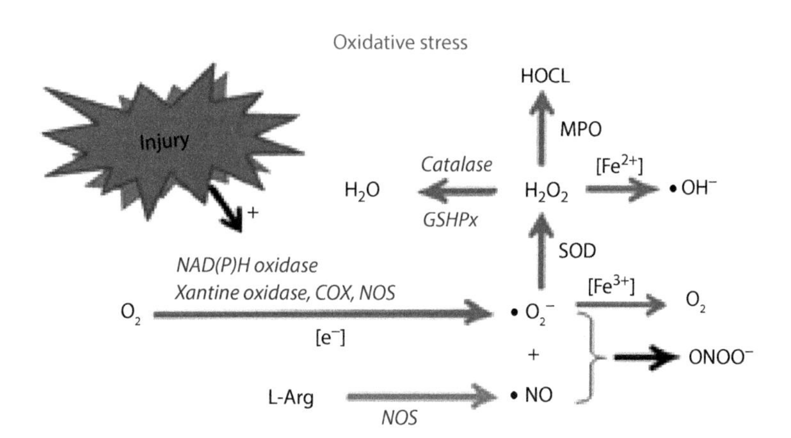

FIGURE 5.2 Oxidative stress and the endogenous antioxidant system. Brain injury causes an increase in oxidative stress. The generated reactive oxygen species have an uncoupled electron and may bind to macromolecules or cause damage in protein or lipid structures within the cell. Superoxide can form a complex with nitric oxide to form peroxynitrite or it can be subject to superoxide dismutase to form hydrogen peroxide, which may further go through Fenton reaction and generate hydroxyl radical. Both peroxynitrite and hydroxyl radicals are highly reactive cytotoxic molecules. The endogenous antioxidant system has a number of enzymes (glutathione peroxidase, superoxide dismutase, catalase) to neutralize/scavenge the reactive metabolites. NOS: nitric oxide synthase; L-arg: L-arginine; SOD: superoxide dismutase; MPO: myeloperoxidase; GSHPx: glutathione peroxidase.

5.4 BRAIN EDEMA AFTER TBI

Brain edema leading to an expansion of brain volume has a crucial impact on morbidity and mortality after TBI because it increases intracranial pressure, impairs cerebral perfusion and oxygenation, and contributes to additional ischemic injuries (Kempski, 2001; Unterberg et al., 2004).

Different types of brain edema and their characteristics are summarized in Table 5.1.

Failure of ion pumps, BBB disruption, neuroinflammation, and oxidative damage are among the major mechanisms that play important roles in the development of cerebral edema after TBI. Generation of free radicals and hypoxia leads to the failure of the Na^+-K^+-ATPase, a membrane-bound enzyme required for cellular transport. Dysfunction of this pump is a common feature in CNS pathologies related to ischemic conditions and TBI. The activity of Na^+-K^+-ATPase pump is very sensitive to free radical reactions and lipid peroxidation. Reductions in this activity can indicate membrane damage indirectly. Thus Na^+-K^+-ATPase is clearly downregulated under low-oxygen conditions that in turn triggers brain edema and enhances the loosening of tight junctions and BBB breakdown. Myeloperoxidase activity, an index for neutrophil infiltration, also increases as evidence of inflammation (Biber et al., 2009). Antiinflammatory agents and antioxidants have both been shown to exert beneficial effects in decreasing tissue injury, but most of them fail to prevent edema formation (Hakan et al, 2010).

5.5 BLOOD BRAIN BARRIER PERMEABILITY AFTER TBI

Diffuse brain injury–mediated immune response; edema, BBB alterations, and neuroinflammation seem to play an important role in the pathology. Microglial activation within the injured area is observed within 6 to 48 hours postinjury. Edema and the increase in BBB permeability were shown to occur immediately after the injury in closed-head injury studies (Ersahin et al., 2010). It was also shown to recover by the third day after the blast exposure (Readnower et al., 2010; Abdul-Muneer et al., 2013). There are several factors (i.e. inflammatory mediators, free radicals, proteases, adhesion molecules, AQP4, VEGF, bradykinin, and arachidonate metabolites) that enhance edema formation and BBB dysfunction (Unterberg et al., 2004). After a blast injury, loosening of the vasculature and perivascular unit was mediated by the activation of matrix metalloproteinases and fluid channel aquaporin-4, promoting edema, enhanced leakiness of the BBB, and progression of neuroinflammation and neuronal degeneration (Abdul-Muneer et al., 2013). Although many studies demonstrate a similar pathophysiologic progression as the conventional TBI, a recent study reported that cerebrovascular injury from a primary blast is distinct from it, suggesting that BBB disruption in blast injury was acute, not resulting from a delayed inflammation as it does in the conventional disruption (Yeoh et al., 2013).

Therapies targeting the restoration of BBB provide the brain homeostasis. Potential agents may be directed to reduce the expression of cell adhesion molecules and/or interfere with signaling of neuroinflammatory mediators of the endothelium. Also neurotrophic factors, such as brain-derived neurotrophic factor and nerve growth factor, can potentially have beneficial effects on functional recovery after injury (Chodobski et al., 2011).

5.6 TRAUMATIC BRAIN INJURY AND AUTONOMIC DYSFUNCTION

One deleterious consequence of brain injury is autonomic nervous system dysregulation and/or dysautonomia. Autonomic nervous system dysfunction has been documented after TBI, but is not well understood. Ninety percent of TBI

TABLE 5.1
Types of Brain Edema and Their Characteristics

	Vasogenic	Cytotoxic	Osmotic	Interstitial
Insult	BBB breakdown **Acute:** trauma, stroke, hemorrhage, arterial hypertension **Chronic:** tumor, encephalitis, abscess	Ischemia, anoxia, intoxication	Metabolic disorders, dialysis, dehydration, diabetic coma	Obstructive hydrocephaly
Mechanism	↑ Permeability in capillary endothelial cells as a result of injury	• Increased cell membrane Na+/K+ permeability • Na+/K+-ATPase pump failure • Uptake of osmotically active solutes	• Osmotic gradient • (plasma→tissue)	• Cerebrospinal efflux ↓
Permeability	↑	-	-	-
Edema fluid	• Rich in protein	• No protein • Electrolytes ↑	• No protein • Electrolytes ↓↑	• Regular cerebrospinal
Morphological findings	• No cell swelling • Interstitial space ↑ • White matter disturbances, secondary swelling in astrocytes	• Cell swelling • Interstitial space ↓	• Cell swelling • Interstitial space ↓	• Interstitial space ↑

Source: Modified from Kempski O., *Semin Nephrol.* 21(3):303–7, 2001; Unterberg, A.W. et al., *Neuroscience.* 129:1021–9, 2004.

patients demonstrate signs of autonomic dysfunction during the first week after injury, with about one-third of the patients developing longer lasting autonomic dysfunction. Autonomic dysregulation is characterized by distinct changes in cardiovascular hyperactivity, sleep function, and specific biomarkers of neural damage. System dysregulation might lead to a range of comorbidities such as hypertension, endothelial dysfunction, and end-organ perfusion abnormalities. Specifically, TBI disruption of autonomic function most often results in sustained sympathoactivation. This sympathetic hyperactivity after TBI remains poorly understood, although sympathetic hyperactivity likely contributes to the high morbidity and mortality associated with TBI. Sympathetic hyperactivity contributes to systemic stress, including neuroinflammation and oxidative stress in the autonomic nervous system. Eventually these disturbances lead to cardiovascular dysfunction (Cernak, 2010) and sleep complications (Viola-Saltzman and Watson, 2012). Systemic stress is associated with activation of the hypothalamic-pituitary-adrenal axis and the hypothalamic sympathoadrenal medullary axis. It is known that TBI activates the hypothalamic-pituitary-adrenal axis (Mcintosh, 1994); however, little is known regarding the TBI-induced activation of the sympathoadrenal medullary axis, and there are limited therapeutic options to treat this sympathoactivation.

We recently demonstrated selective biochemical markers of autonomic function and oxidative stress in male Sprague Dawley rats subjected to head-directed overpressure insult (Tümer et al., 2013). There were increased levels of tyrosine hydroxylase, dopamine-β hydroxylase, and neuropeptide Y in the adrenal medulla along with plasma norepinephrine. In addition, overpressure blast injury (OBI) significantly elevated tyrosine hydroxylase in the nucleus tractus solitarius of the brain stem, whereas AT1 receptor expression and NADPH oxidase activity, a marker of oxidative stress, was elevated in the hypothalamus suggesting that single OBI exposure results in increased sympathoexcitation (Figure 5.3). The mechanism may involve the elevated AT1 receptor expression and NADPH oxidase levels in the hypothalamus. Taken together, such effects may be important factors contributing to pathology of brain injury and autonomic dysfunction associated with the clinical profile of patients after OBI (Tumer et al., 2013). However, insufficient published data are available to describe the long-term effects of TBI on central noradrenergic systems, particularly on neuroplastic adaptations within numerous targets of central noradrenergic projections. In addition, understanding the etiology of these changes may shed new light on the molecular mechanism(s) of injury, potentially offering new strategies for treatment.

5.7 TRAUMATIC BRAIN INJURY AND CEREBROVASCULAR FUNCTION

Recent studies are under way on the basilar artery function. Preliminary results showed that OBI impaired the

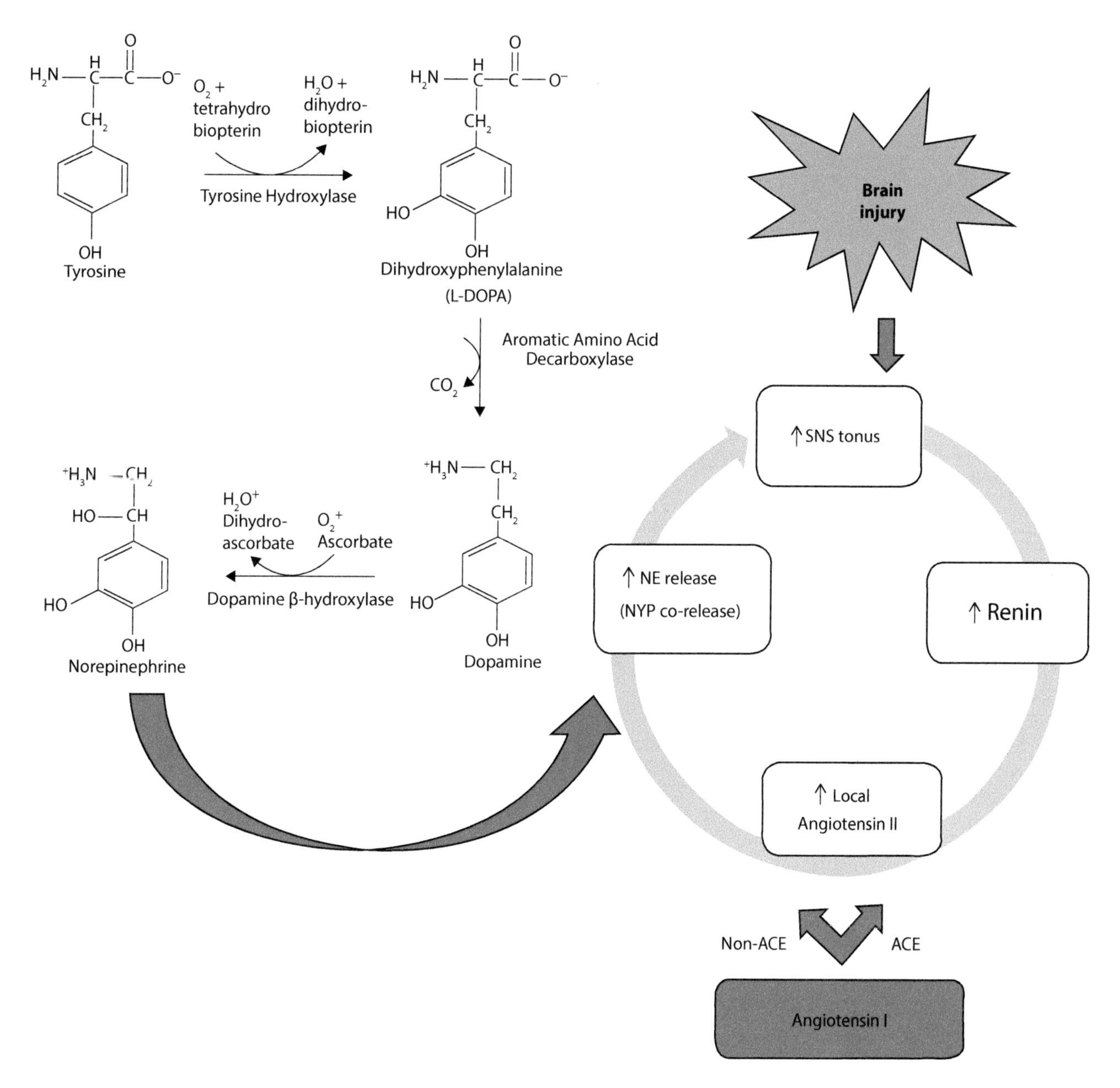

FIGURE 5.3 Brain injury and the sympathetic nervous system. Brain injury increases the sympathetic nervous system (SNS) tonus as a result of the increased norepinephrine (NE) turnover. The levels of the biosynthetic enzymes (i.e., tyrosine hydroxylase [TH], dopamine beta hydroxylase [DBH]) are increased, thus elevating the release of neuropeptide Y that is coreleased with catecholamines. Sympathoactivation also enhances the renin-angiotensin system.

vascular reactivity of the basilar artery. OBI resulted in an increase in the contractile responses to endothelin and a decrease in the relaxant responses to acetylcholine after single or repeated exposure to injury. However, impaired Diethylamine NONO-ate induced dilation and increased arterial wall thickness/lumen ratio were observed only in the repeated injury group (Toklu et al., 2014).

5.8 CONCLUSION

Individuals identified with apparent brain pathology (e.g., edema, hemorrhage, seizures) are well recognized and provided symptomatic treatment. However, the lack of conclusive physical evidence, especially in mild TBI, may not adequately be diagnosed with standard assessment tools. Physical symptoms may resolve rapidly. On the other hand, subtle changes in brain may occur that lead to neurological disorders and cognitive impairment long after the recovery of the primary injury.

Although most of the studies address the acute treatment of TBI, their efficiency in the long term is not clear. Therefore, further studies are needed to elucidate the mechanisms of secondary injury and long-term changes in the brain metabolism and physiological function.

REFERENCES

Abdul-Muneer, P.M, H. Schuetz, F. Wang, M. Skotak, J. Jones, S. Gorantla, MC et al. 2013. Induction of oxidative and nitrosative damage leads to cerebrovascular inflammation in an animal model of mild traumatic brain injury induced by primary blast. *Free Radic Biol Med.* 60:282–291.

Ansari, M.A., K.N. Roberts, and S.W. Scheff. 2008. Oxidative stress and modification of synaptic proteins in hippocampus after traumatic brain injury. *Free Radic Biol Med.* 45:443–52.

Biber, N., H.Z. Toklu, S. Solakoglu, M. Gultomruk, T. Hakan, Z. Berkman, and F.G. Dulger. 2009. Cysteinyl-leukotriene receptor antagonist montelukast decreases blood-brain barrier permeability but does not prevent oedema formation in traumatic brain injury. *Brain Inj.* 23:577–84.

Cernak, I., T. Chang, F.A. Ahmed, M.I. Cruz, R. Vink, B. Stoica, and A.I. Faden. 2010. Pathophysiological response to experimental diffuse brain trauma differs as a function of developmental age. *Develop Neurosci.* 32:442–53.

Chodobski, A., B.J. Zink, and J. Szmydynger-Chodobska. 2011. Blood-brain barrier pathophysiology in traumatic brain injury. *Transl Stroke Res.* 2(4):492–516.

Cornelius, C., R. Crupi, V. Calabrese, A. Graziano, P. Milone, G. Pennisi et al. 2013. Traumatic brain injury: Oxidative stress and neuroprotection. *Antioxid Redox Signal.* 19(8):836–53.

DeWitt, D.S. and D.S. Prough. 2009. Blast-induced brain injury and posttraumatic hypotension and hypoxemia. *J Neurotrauma.* 26:877–87.

Ersahin, M., H.Z. Toklu, C. Erzik, S. Cetinel, D. Akakin, A. Velioglu-Ogunc et al. 2010. The anti-inflammatory and neuroprotective effects of ghrelin in subarachnoid hemorrhage-induced oxidative brain damage in rats. *J Neurotrauma.* 27(6):1143–55.

Hakan, T., H.Z. Toklu, N. Biber, H. Ozevren, S. Solakoglu, P. Demirturk et al. 2010. Effect of COX-2 inhibitor meloxicam against traumatic brain injury-induced biochemical, histopathological changes and blood-brain barrier permeability. *Neurol Res.* 32(6):629–35.

Kempski O. 2001. Cerebral edema. *Semin Nephrol.* 21(3):303–7. Review.

Kobeissy, F., S. Mondello, N. Tümer, H.Z. Toklu, M.A., Whidden, N. Kirichenko et al. 2013. Assessing neuro-systemic and behavioral components in the pathophysiology of blast-related brain injury. *Front Neurol.* 4:186.

Langlois, J., W. Rutland-Brown, and M. Wald. 2006. The epidemiology and impact of traumatic brain injury: A brief overview. *J Head Trauma Rehabil.* 21(5):375–8.

Maduell, P., G. Armengol, M. Llagstera, S. Orduz, and S. Lindow. 2008. *Bacillus thuringiensis* is a poor colonist of leaf surfaces. *Microbiol Ecol.* 55:212–9.

Mcintosh, T.K., T. Yu, and TA Gennarelli. 1994. Alterations in regional brain catecholamine concentrations after experimental brain injury in the rat. *J Neurochem.* 84:1426–33.

Morganti-Kossmann, M.C., L. Satgunaseelan, N. Bye, and T. Kossmann. 2007. Modulation of immune response by head injury. *Injury.* 38:1392–400.

Readnower, R.D., M. Chavko, S. Adeeb, MD Conroy, J.R. Pauly, R.M. McCarron et al. 2010. Increase in blood-brain barrier permeability, oxidative stress, and activated microglia in a rat model of blast-induced traumatic brain injury. *J Neurosci Res.* 88:3530–9.

Slemmer, J.E., J.J. Shacka, M.I. Sweeney, and J.T. Weber. 2008. Antioxidants and free radical scavengers for the treatment of stroke, traumatic brain injury and aging. *Curr Med Chem.* 15(4):404–4.

Tran, L.V. 2014. Understanding the pathophysiology of traumatic brain injury and the mechanisms of action of neuroprotective interventions. *J Trauma Nurs.* 21(1):30–5.

Toklu, H.Z., J. Muller-Delp, Z. Yang, P. Ghosh, K. Strang, P.J. Scarpace et al. 2014. The functional changes in the basilar artery due to overpressure blast injury. *FASEB Journal* 28(1) Suppl 685.1.

Tumer, N., S. Svetlov, M. Whidden, N. Kirichenko, V. Prima, B. Erdos et al. 2013. Overpressure blast-wave induced brain injury elevates oxidative stress in the hypothalamus and catecholamine biosynthesis in the rat adrenal medulla. *Neurosci Lett.* 544:62–7.

Unterberg, A.W., J. Stover, B. Kress, and K.L. Kiening. 2004. Edema and brain trauma. *Neuroscience.* 129:1021–9.

Viola-Saltzman, M. and N.F. Watson. 2012. Traumatic brain injury and sleep disorders. *Neurol Clin.* 30:1299–312.

Vuceljić, M., G. Zunić, P. Romić, and M. Jevtić. 2006. Relation between both oxidative and metabolic-osmotic cell damages and initial injury severity in bombing casualties. *Vojnosanit Pregl.* 63:545–51.

Werner, C. and K. Engelhard. 2007. Pathophysiology of traumatic brain injury. *Br J Anaesth.* 99:4–9.

Yeoh, S., E.D. Bell, and K.L. Monson. 2013. Distribution of blood-brain barrier disruption in primary blast injury. *Ann Biomed Eng.* 41(10):2206–14.

6 The Contributing Role of Lipid Peroxidation and Protein Oxidation in the Course of CNS Injury Neurodegeneration and Neuroprotection

An Overview

Edward D. Hall

CONTENTS

6.1 INTRODUCTION

At present, there are no FDA-approved pharmacological therapies for acute treatment of traumatic brain injury (TBI) patients that are conclusively proven to mitigate the often devastating neurological effects of their injuries. Nevertheless, the possibility of an effective neuroprotective treatment is based upon the fact that even though some of the neural injury is due to the primary mechanical events (i.e., shearing of nerve cells and blood vessels), the majority of post-traumatic neurodegeneration is due to a pathomolecular and pathophysiological secondary cascade that occurs during the first minutes, hours and days following the injury which exacerbates the damaging effects of the primary injury. One of the most validated "secondary injury" mechanisms revealed in experimental TBI studies involves oxygen radical-induced oxidative damage to brain cell lipids and proteins. This chapter outlines the key sources of reactive oxygen species (ROS) including highly reactive (i.e., rapidly oxidizing) free radicals, the mechanisms associated with their neural damage, pharmacological scavenging antioxidants that have been shown to produce neuroprotective effect in TBI models and brief mention of the most widely used methods for studying the extent of lipid and protein oxidative damage in TBI models.

6.1.1 Reactive Oxygen Species and Reactive Nitrogen Species

The term reactive oxygen species (ROS) includes oxygen-derived radicals such as the modestly reactive superoxide radical ($O_2^{\bullet-}$) and the highly reactive hydroxyl (•OH) radicals as well as nonradicals such as hydrogen peroxide (H_2O_2) and peroxynitrite ($ONOO^-$), the latter often referred to as a reactive nitrogen species (RNS). The cascade of post-traumatic oxygen radical reactions begins in response to rapid elevations in intracellular Ca^{2+} immediately after the primary mechanical injury to the brain with the single electron (e^-) reduction of an oxygen molecule (O_2) to produce $O_2^{\bullet-}$, which is considered to be a modestly reactive primordial radical that can potentially react with other molecules to give rise to much more reactive, and thus more potentially damaging radical species. The reason that $O_2^{\bullet-}$ is only modestly reactive is that it can act as either an oxidant stealing an electron from another oxidizable molecule or it can act as a reductant by which it donates its unpaired electron to another radical species, thus acting as a reductant or antioxidant. However, if $O_2^{\bullet-}$ reacts with a proton (H^+) to form a hydroperoxyl radical (HO_2^-), this results in a superoxide form that is much more likely to cause oxidation (i.e., act as an electron stealer).

One of the most important endogenous antioxidants is the enzyme superoxide dismutase (SOD) that rapidly catalyzes the dismutation of $O_2^{\bullet-}$ into H_2O_2 and oxygen. At low pH, $O_2^{\bullet-}$ can dismutate spontaneously. The formation of highly reactive oxygen radicals, which have unpaired electron(s) in their outer molecular orbitals and the propagation of chain reactions, are fueled by nonradical ROS, which do not have unpaired electron(s), but are chemically reactive. For example, •OH radicals are generated in the iron-catalyzed Fenton reaction in which ferrous iron (Fe^{2+}) is oxidized to form •OH in the presence of H_2O_2 ($Fe^{2+} + H_2O_2 \rightarrow Fe^{3+} + \bullet OH + OH-$). Superoxide acting as a reducing agent can actually donate its unpaired electron to ferric iron (Fe^{3+}), cycling it back to the ferrous state in the Haber-Weiss reaction, thus driving subsequent Fenton reactions and increased production of •OH ($O_2^{\bullet-} + Fe^{3+} \rightarrow Fe^{2+} + O_2$). Under physiological conditions, iron is tightly regulated by its transport protein, transferrin, and storage protein, ferritin, both of which bind the ferric (Fe^{3+}) form. This reversible bond of transferrin and ferritin with iron decreases with declining pH (below pH7), as is the case after central nervous system (CNS) injury, resulting in the release of iron and initiation of iron-dependent oxygen radical production. A second source of iron comes from hemoglobin on its release after mechanical-induced hemorrhage.

Although $O_2^{\bullet-}$ itself is less reactive than an •OH radical, its reaction with a nitric oxide (•NO) radical forms the highly reactive oxidizing agent, peroxynitrite (PN: $ONOO^-$). This reaction ($O_2^{\bullet-} + \bullet NO \rightarrow ONOO^-$) occurs at a very high rate constant that outcompetes SOD's ability to convert $O_2^{\bullet-}$ into H_2O_2. Subsequently, at physiological pH, $ONOO^-$ will largely undergo protonation to form peroxynitrous acid (ONOOH) or it can react with carbon dioxide (CO_2) to form nitrosoperoxocarbonate ($ONOOCO_2$). The ONOOH can break down to form highly reactive nitrogen dioxide ($\bullet NO_2$) and •OH ($ONOOH \rightarrow \bullet NO_2 + \bullet OH$). Alternatively, the $ONOOCO_2$ can decompose into $\bullet NO_2$ and carbonate radical ($\bullet CO_3$) ($ONOOCO_2 \rightarrow \bullet NO_2 + \bullet CO_3$).

6.2 LIPID PEROXIDATION

Increased production of reactive free radicals (i.e., "oxidative stress") in the injured brain has been shown to cause "oxidative damage" to cellular lipids and proteins, leading to functional compromise and possibly cell death in both the brain microvascular and parenchymal compartments. Extensive research shows that a major form of radical-induced oxidative damage involves oxidative attack on cell membrane polyunsaturated fatty acids triggering the process of lipid peroxidation (LP). The LP is initiated when highly reactive oxygen radicals (e.g., •OH, $\bullet NO_2$, $\bullet CO_3$) react with polyunsaturated fatty acids such as arachidonic acid, linoleic acid, eicosapentaenoic acid, or docosahexaenoic acid, resulting in disruptions in cellular and membrane integrity. The process sets off a radical chain reaction characterized by three distinct steps: initiation, propagation, and termination (Gutteridge, 1995).

Initiation of LP begins with an ROS-induced hydrogen atom (H^+); its one associated electron is abstracted from an allylic carbon. The basis for the susceptibility of the allylic carbon of the polyunsaturated fatty acid having one of its electrons stolen by a highly electrophilic free radical is that the carbon is surrounded by two relatively electronegative double bonds that tend to pull one of the electrons from the carbon. Consequently, a reactive free radical has an easy time pulling the hydrogen electron off of the carbon because the commitment of the carbon electron to staying paired with it has been weakened by the surrounding double bonds. This results in the original radical being quenched while the polyunsaturated fatty acid (L) becomes a lipid radical (L•) because of its having lost an electron.

In the subsequent propagation step, the unstable L• reacts with O_2 to form a lipid peroxyl radical (LOO•). The LOO• in turn abstracts a hydrogen atom from an adjacent polyunsaturated fatty acid yielding a lipid hydroperoxide (LOOH) and a second L•, which sets off a series of propagation "chain" reactions. These propagation reactions are terminated in the third step when the substrate becomes depleted and a lipid radical reacts with another radical or radical scavenger to yield potentially neurotoxic nonradical end-products. One of those end-products that is often used to measure LP is the 3-carbon-containing malondialdehyde, which is mainly a stable nontoxic compound that, when measured, represents an LP "tombstone."

In contrast, two highly toxic products of LP are 4-hydroxynonenal (4-HNE), the formation of which is illustrated in Figure 6.1, or 2-propenal (acrolein), as shown in Figure 6.2, both of which have been well characterized in CNS injury experimental models (Bains and Hall, 2012; Hall et al., 2010; Hamann and Shi, 2009). These latter two aldehydic LP end-products covalently bind to proteins and amino acids (lysine, histidine, arginine, cysteine), altering their structure and functional properties, the chemistry of which is illustrated in Figure 6.3. Immunohistochemical and immunoblotting (Western, slot, dot) techniques are commonly used to measure 4-HNE or acrolein-modified

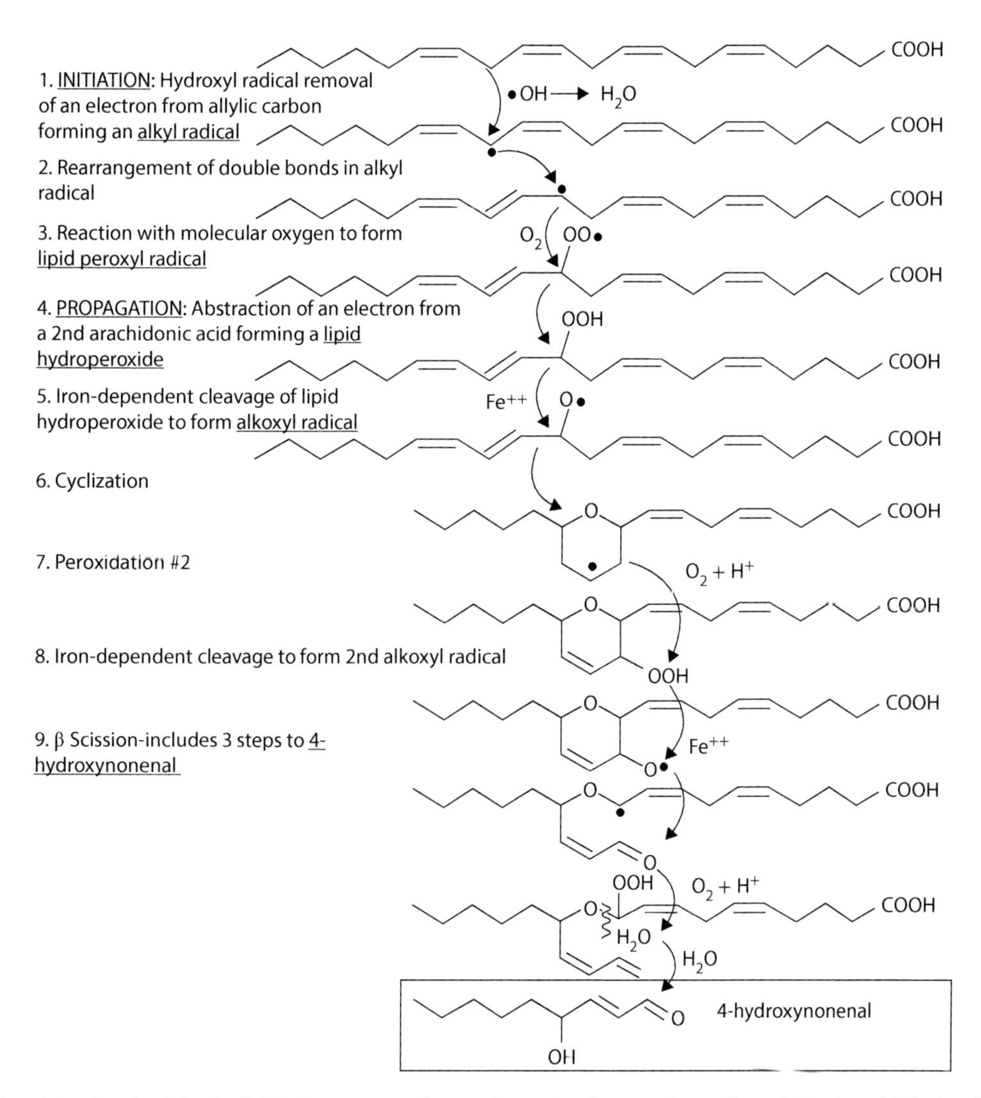

FIGURE 6.1 Chemistry involved in the initiation, propagation, and termination reactions of arachidonic acid during lipid peroxidation with the resulting formation of the aldehydic end-product 4-hydroxynonenal (4-HNE).

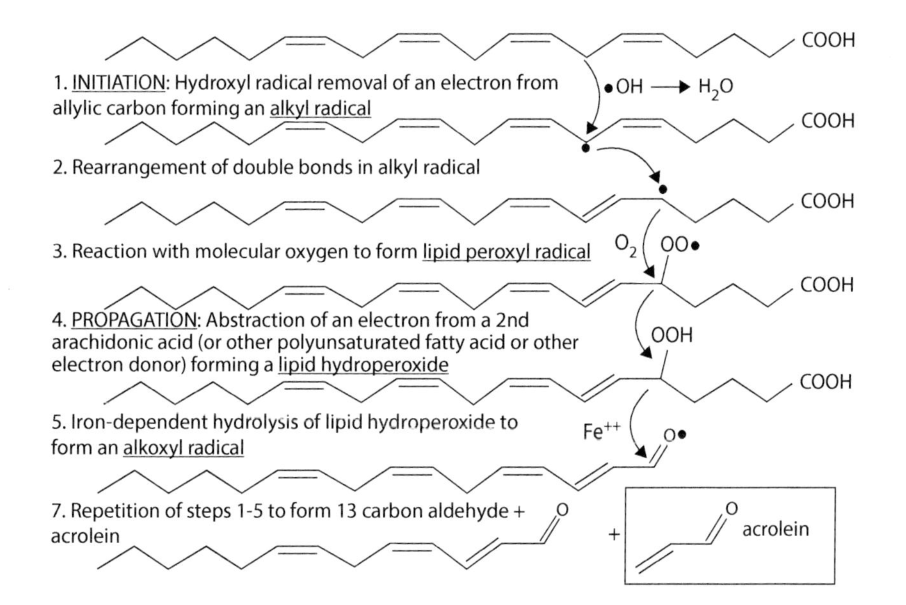

FIGURE 6.2 Chemistry involved in the initiation, propagation, and termination reactions of arachidonic acid during lipid peroxidation with the resulting formation of the aldehydic end-product 2-propenal (acrolein).

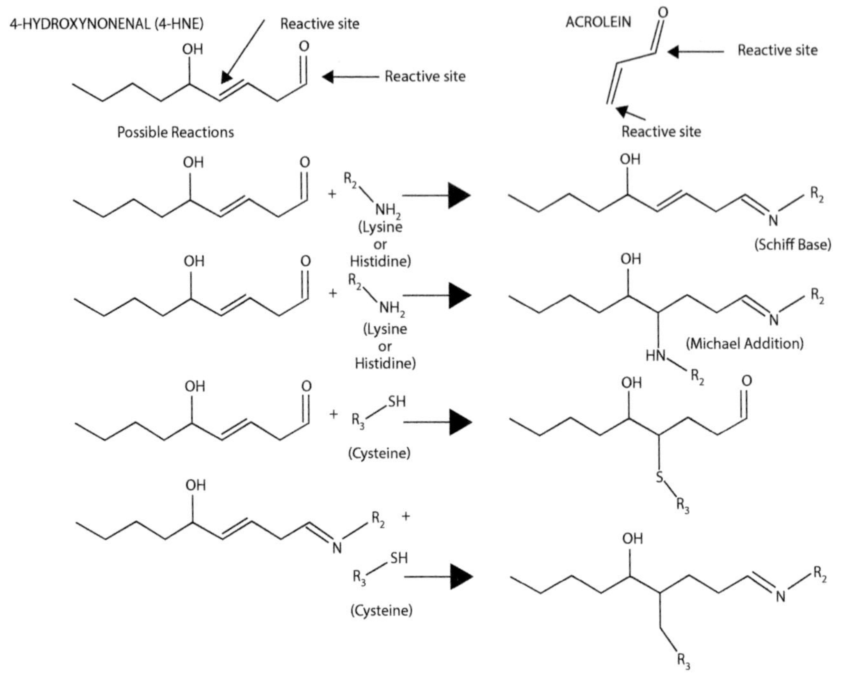

FIGURE 6.3 Chemical reactions of 4-HNE or acrolein with amino acids that lead to impairment of protein structure and function.

proteins in the injured brain. For details on a variety of analytical techniques for measurement of various markers of LP, the reader is referred to the following review (Hall and Bosken, 2009).

6.3 PROTEIN CARBONYLATION AND NITRATION

Free radicals can cause various forms of oxidative protein damage. A major mechanism involves carbonylation by reaction of various free radicals with susceptible amino acids. Figure 6.4 shows the chemistry of •OH-induced histidine oxidation, which results in carbonylation of the 2-position of histidine's imidazole ring. The protein carbonyls thus formed are measurable through immunoblotting after derivatization of the carbonyl groups with diphenylhydrazine. Although the measurement of protein carbonyls by the so-called diphenylhydrazine assay has long been used, a commercially available protein carbonyl Oxyblot assay is widely used by present-day investigators. It should be noted that the carbonyl assay also picks up protein carbonyls that are present because of covalent binding of LP-derived 4-HNE and acrolein to cysteine residues (see Figure 6.3), in addition to those resulting from direct free radical–induced amino acid oxidation. Thus, as a result, the carbonyl assay is as much an indirect index of LP as it is of primary direct protein oxidation.

Second, $\bullet NO_2$ can nitrate the 3-position of tyrosine residues in proteins; 3-NT is a specific footprint of PN-induced cellular damage (see Figure 6.5). Similarly, LOO• can promote tyrosine nitration by producing initial oxidation (loss of an electron), which would enhance the ability of $\bullet NO_2$ to nitrate the phenyl ring as also explained in Figure 6.5. Multiple commercially available polyclonal and monoclonal antibodies are available for immunoblot or immunohistochemical measurement of proteins that have been nitrated by PN.

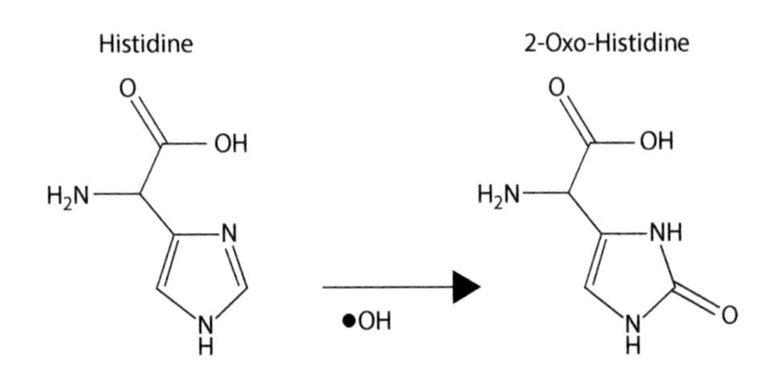

FIGURE 6.4 Chemical reaction of hydroxyl radical (•OH) with a histidine residue resulting in histidine carbonylation.

6.4 INTERACTION OF OXIDATIVE DAMAGE WITH OTHER SECONDARY INJURY MECHANISMS

The impact of ROS/RNS production is heightened when oxygen radicals feed back and amplify other secondary injury pathways, creating a continuous cycle of ion imbalance, Ca^{2+} buffering impairment, mitochondrial dysfunction, glutamate-induced excitotoxicity, and microvascular disruption. One example of ROS-induced ionic disruption arises from LP-induced damage to the plasma membrane adenosine triphosphate (ATP)-driven Ca^{2+} pump (Ca^{++}-ATPase) and Na^+

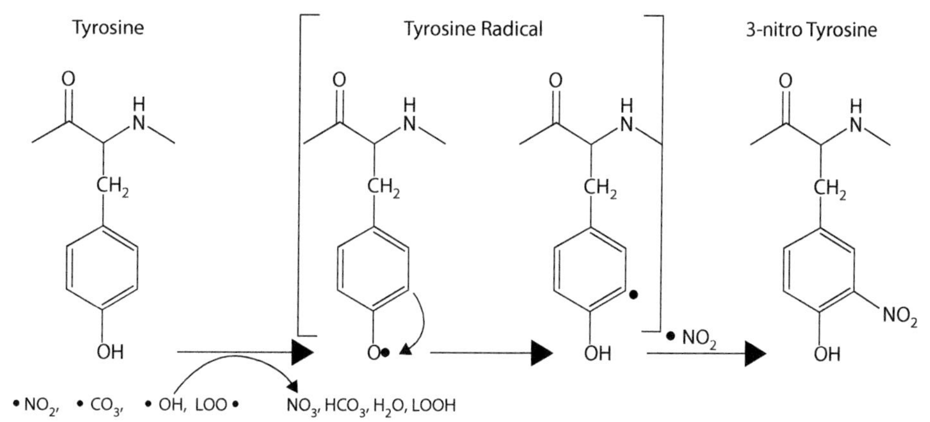

FIGURE 6.5 Mechanism of peroxynitrite-induced nitration of tyrosine residues. The first step involves free radical attack on the phenyl ring pulling away an electron by peroxynitrite-derived radicals or a lipid peroxyl radical (LOO•) that converts tyrosine into a tyrosine radical. The second step involves the reaction of the more reactive tyrosine radical with peroxynitrite-derived $•NO_2$.

pump (Na^+/K^+-ATPase), which contributes to increases in intracellular Ca^{2+} concentrations, mitochondrial dysfunction and additional ROS production. Both Ca^{2+} pump and Na^+/K^+-ATPase disruptions result in further increases in intracellular Ca^{2+} and Na^+ accumulation, respectively (Bains and Hall, 2012), the latter causing reversal of the Na^+/Ca^{++} exchanger (Rohn et al., 1993, 1996).

PN formed from mitochondrial Ca^{2+} overload contribute to mitochondrial dysfunction. Specifically, nitric oxide (•NO), formed from mitochondrial NOS, in turn reacts with $O_2^{•-}$ to produce the highly toxic PN, which impairs respiratory and Ca^{2+} buffering capacity via its derived free radicals (Bringold et al., 2000). Indeed, increased PN-derived 3NT and 4HNE has been detected during the time of mitochondrial dysfunction and correlates with respiratory and Ca^{2+} buffering impairment (Sullivan et al., 2007). Increased synaptosomal 4-HNE content is associated with impaired synaptosomal glutamate and amino acid uptake (Springer et al., 1997; Zhang et al., 1996). Glutamate and N-methyl-D-aspartate induced damage in neuronal cultures is attenuated with LP inhibition, confirming LP and oxidative damage as mediators of glutamate excitotoxicity (Monyer et al., 1990; Pellegrini-Giampietro et al., 1990).

6.5 MECHANISMS FOR PHARMACOLOGICAL INHIBITION OF OXIDATIVE DAMAGE IN TBI

Based on the previous discussion concerning oxidative stress (increased ROS/RNS) and oxidative damage (LP, protein oxidation and nitration), several potential mechanisms for its inhibition are apparent, which fall into three categories. The first category includes compounds that inhibit the initiation of LP and other forms of oxidative damage by preventing the formation of ROS or RNS species. For instance, NOS inhibitors, discussed previously, exert an indirect antioxidant effect by limiting •NO production and thus PN formation. However, they also have the potential to interfere with the physiological roles that •NO is responsible for. These include antioxidant effects that are due to its important role as a scavenger of lipid peroxyl radicals (e.g., LOO• + •NO →LOONO) (Hummel et al., 2006). Another approach to blocking posttraumatic radical formation is the inhibition of the enzymatic (e.g., cyclooxygenase, 5-lipoxygenase) arachidonic acid cascade during which the formation of $O_2^{•-}$ is produced as a by-product of prostanoid and leukotriene synthesis. Kontos and colleagues (Kontos, 1989; Kontos and Wei, 1986) and Hall (Hall, 1986) have shown that cyclooxygenase inhibiting nonsteroidal anti-inflammatory agents (e.g., indomethacin, ibuprofen) are vaso- and neuro-protective in TBI models.

A second indirect LP inhibitory approach involves chemically scavenging the radical species (e.g., $O_2^{•-}$, •OH, $•NO_2$, $•CO_3$) before it has a chance to steal an electron from a polyunsaturated fatty acid and thus initiate LP. The use of pharmacologically administered SOD represents an example of this strategy. Another example concerns the use of the nitroxide antioxidant tempol, which has been shown to catalytically scavenge the PN-derived free radicals $•NO_2$ and $•CO_3$ (Carroll et al., 2000). In either case, a general limitation to these first two approaches is that they would be expected to have a short therapeutic window and would have to be administered rapidly to have a chance to interfere with the initial posttraumatic "burst" of free radical production that has been documented in TBI models (Hall et al., 1993; Kontos and Wei, 1986). Although it is believed that ROS, including PN production, persists several hours after injury, the major portion is an early event that peaks in the first 60 minutes after injury, making it clinically impractical to pharmacologically inhibit, unless the antioxidant compound is already "on board" when the injury occurs or is available for administration immediately after TBI.

In contrast to these indirect-acting antioxidant mechanisms, the third category involves stopping the "chain reaction" propagation of LP once it has begun. The most demonstrated way to accomplish this is by scavenging of

LOO• or lipid alkoxyl (LO•) radicals. The prototype scavenger of these lipid radicals is alpha tocopherol or vitamin E (Vit E) which can donate an electron from its phenolic hydroxyl moiety to quench a LOO•. However, the scavenging process is stoichiometric (1 Vit E can only quench 1 LOO•) and in the process Vit E loses its antioxidant efficacy and becomes Vit E radical (LOO• + Vit E → LOOH + Vit E•). Although Vit E• is relatively unreactive (i.e., harmless), it also cannot scavenge another LOO• until it is reduced back to its active form by receiving an electron from other endogenous antioxidant-reducing agents such as ascorbic acid (vitamin C) or glutathione (GSH). Although this tripartite LOO• antioxidant defense system (Vit E, vitamin C, GSH) works fairly effectively in the absence of a major oxidative stress, numerous studies have shown that each of these antioxidants are rapidly consumed during the early minutes and hours after CNS injury (Hall et al., 1989, 1992). Thus, it has long been recognized that more effective pharmacological LOO• and LO• scavengers are needed. Furthermore, it is expected that compounds that could interrupt the LP process after it has begun would be able to exert a more practical neuroprotective effect (i.e., possess a longer antioxidant therapeutic window).

A second approach to inhibiting the propagation of LP reactions is to chelate free iron, either ferrous (Fe^{++}) or ferric (Fe^{+++}), which potently catalyzes the breakdown of LOOHs, an essential event in the continuation of LP chain reactions in cellular membranes. The prototypical iron-chelating drug that chelates Fe^{+++} is the bacterially (streptomyces pilosus)-derived tri-hydroxamic acid compound deferoxamine.

6.6 NEUROPROTECTIVE EFFECTS OF PHARMACOLOGICAL ANTIOXIDANTS

6.6.1 TBI Clinical Trial Results with PEG-SOD and Tirilazad

During the past 25 years, there has been an intense effort to discover and develop pharmacological agents for acute treatment of TBI. This has included two compounds that possess free radical scavenging/antioxidant properties polyethylene glycol–conjugated SOD (PEG-SOD) (Muizelaar et al., 1995) and the LP inhibitor tirilazad (Langham et al., 2000; Marshall et al., 1998). However, each of these trials was a therapeutic failure in that no overall benefit was documented in severe and/or moderate TBI patient populations, which was the primary goal in each of these trials. These TBI clinical trial failures can be attributed to several factors. Perhaps most importantly, the preclinical assessment of compounds destined for acute TBI trials has often been woefully inadequate in regard to the definition of neuroprotective dose-response relationships, pharmacokinetic–pharmacodynamic correlations, therapeutic window, and optimum dosing regimen and treatment duration. However, a number of other issues related to design of the clinical trials are also believed to be involved (Narayan et al., 2002). The following sections briefly review the TBI histories of PEG-SOD and tirilazad.

6.6.2 PEG-SOD

As mentioned earlier, the earliest studies of free radical scavenging compounds in TBI models were carried out with Cu/Zn SOD based on the work of Kontos and colleagues who showed that posttraumatic microvascular dysfunction was initiated by $O_2^{\bullet-}$ generated as a by-product of the arachidonic acid cascade, which is massively activated during the first minutes and hours after TBI (Kontos, 1989; Kontos and Povlishock, 1986; Kontos and Wei, 1986). Their work showed that administration of SOD prevented the posttraumatic microvascular dysfunction. This led to clinical trials in which the more metabolically stable PEG-conjugated SOD was examined in moderate and severe TBI patients when administered within the first 8 hours after injury. Although an initial small phase II study showed a positive trend, subsequent multicenter phase III studies failed to show a significant benefit in terms of increased survival or improved neurological outcomes (Muizelaar et al., 1995).

Although many explanations for these negative results may be postulated, one reason may be that a large protein like SOD is unlikely to have much brain penetrability and therefore its radical scavenging effects may be limited to the microvasculature. A second reason may be that attempting to scavenge the short-lived inorganic radical $O_2^{\bullet-}$ may be associated with a very short therapeutic window, as suggested previously. As pointed out earlier, the time course of measurable posttraumatic •OH formation in the injured rodent brain has been shown to largely run its course by the end of the first hour after TBI (Hall et al., 1993; Smith et al., 1994).

A more rational strategy would be to inhibit the LP that is triggered by the initial burst of inorganic radicals. A comparison of the time course of LP with that of posttraumatic •OH shows that LP reactions continue to build beyond the first posttraumatic hour (Smith et al., 1994) and may continue for 3 or 4 days (Du et al., 2004; Hall et al., 2012; Miller et al., 2014). Despite the failure of PEG-SOD in human TBI, experimental studies have shown that transgenic mice that overexpress Cu/Zn SOD are significantly protected against post-TBI pathophysiology and neurodegeneration (Chan et al., 1995; Gladstone et al., 2002; Lewen et al., 2000; Mikawa et al., 1996; Xiong et al., 2005). This fully supports the importance of posttraumatic $O_2^{\bullet-}$ in posttraumatic secondary injury, despite the fact that targeting this primordial radical is probably not the best antioxidant strategy for acute TBI compared with trying to stop the downstream LP process that is initiated by the early increases in •OH, $\bullet NO_2$, and $\bullet CO_3$.

6.6.3 Tirilazad

Consistent with that rationale, the 21-aminosteroid LP inhibitor tirilazad (i.e., U74006F) was discovered, which inhibits free radical–induced LP by a combination of LOO• scavenging and a membrane-stabilizing action that limits the propagation of LP reactions between an LOO• and an adjacent polyunsaturated fatty acid. The protective efficacy of tirilazad has been demonstrated in multiple animal models of acute TBI in mice (Hall et al., 1988), rats (McIntosh et al., 1992) and cats (Dimlich et al., 1990). Although the compound is largely localized in the microvascular endothelium, the posttraumatic disruption of the BBB is known to allow the successful penetration of tirilazad into the brain parenchyma, as noted earlier (Hall et al., 1992). Other mechanistic data derived from the rat-controlled cortical impact and the mouse diffuse concussive head injury models have definitively shown that a major effect of tirilazad is to lessen posttraumatic microvascular damage, including BBB opening (Hall et al., 1992; Smith et al., 1994).

Tirilazad was taken into clinical development in the early 1990s and following a small phase II dose-escalation study that demonstrated the drug's safety in TBI patients it was evaluated in two phase III multicenter clinical trials for its ability to improve neurological recovery in moderately and severely injured closed TBI patients. One trial was conducted in North America and the other in Europe. In both trials, TBI patients were treated within 4 hours after injury with either vehicle or tirilazad (2.5 mg/kg intravenously every 6 hours for 5 days). The North American trial was never published because of a major confounding imbalance in the randomization of the patients to placebo or tirilazad in regards to injury severity and pretreatment neurological status. In contrast, the European trial had much better randomization balance and was published (Marshall et al., 1998). The results failed to show a significant beneficial effect of tirilazad in either moderate (Glasgow Coma Scale = 9–12) or severe (Glasgow Coma Scale = 4–8) patient categories. However, a post hoc analysis showed that moderately injured male TBI patients with traumatic subarachnoid hemorrhage (SAH) had significantly less mortality after treatment with tirilazad (6%) compared with placebo (24%, $p < 0.026$). In severely injured males with traumatic SAH tirilazad also lessened mortality from 43% in placebo treated to 34% ($p < 0.071$) Marshall et al., 1998). This result is consistent with the fact that this compound is also highly effective in animal models of SAH (Hall et al., 1994). Nevertheless, additional trials would have been required to establish the neuroprotective utility of tirilazad in certain human TBI subgroups and to gain Food and Drug Administration approval in the United States. However, the sponsoring company, Pharmacia & Upjohn, opted not to continue the compound's development for TBI although tirilazad was successfully approved and marketed for use in aneurysmal SAH in several western European and Australasian countries based on its demonstrated efficacy in phase III SAH trials (Kassell et al., 1996; Lanzino and Kassell, 1999).

6.6.4 Effects of Other Direct and Indirect-Acting Lipid Peroxidation Inhibitors

In addition to tirilazad, several other LP inhibitors have been reported to be effective neuroprotectants in TBI models. These include the LOO• scavenging 2-methylaminochromans U-78517F and U-83836E (Hall et al., 1991), the pyrrolopyrimidine U-101033E (Hall, 1995a, 1995b), OPC-14117 (Mori et al., 1998), the naturally occurring LOO• scavengers curcumin (Sharma et al., 2009; Wu et al., 2006) and resveratrol (Ates et al., 2007; Sonmez et al., 2007), the indoleamine melatonin (Beni et al., 2004; Cirak et al., 1999; Mesenge et al., 1998; Ozdemir et al., 2005a, 2005b), and last, the endogenous antioxidant lipoic acid (Toklu et al., 2009). In the case of curcumin and resveratrol, these are potent LOO• scavengers because they possess multiple phenolic hydroxyl groups that can donate electrons to LOO• radicals. Melatonin also has LOO• scavenging capability (Longoni et al., 1998), but in addition appears to react with PN (Zhang et al., 1998, 1999). Lipoic acid may also have LOO• scavenging effects, but these are more likely to be indirect via the regeneration (i.e., re-reduction) of other endogenous electron-donating antioxidants including Vit E, GSH, and vitamin C.

Among these LP inhibitors, arguably the most potent and effective LOO• scavenging LP inhibitor yet discovered is the 2-methylaminochroman compound U-83836E that combines the LOO• scavenging antioxidant chroman ring structure of Vit E with the bis-pyrrolopyrimidine moiety of tirilazad. The phenolic chroman antioxidant moiety can be rereduced by endogenous ascorbic acid (vitamin C) or GSH after it has donated its phenolic electron to an initial LOO•, making it able to quench a second and then a third LOO•, etc. The bis-pyrrolopyrimidine moiety, on the other hand, can also scavenge multiple moles of LOO• by a true catalytic mechanism (Hall, 1995b; Hall et al., 1991). Thus, U-83836E is a dual functionality LOO• scavenger that is understandably more effective than either Vit E, tirilazad (Hall et al., 1991), and possibly the other naturally occurring LOO• scavengers such as curcumin, resveratrol, melatonin, and lipoic acid. Furthermore, U-83836E possesses a high degree of lipophilicity endowing it with a high affinity for membrane phospholipids where LP takes place. Recent studies from the author's laboratory in the mouse controlled cortical impact TBI model have shown that U-83836E is able to reduce posttraumatic LP and protein nitration and preserve mitochondrial respiratory function in injured cortical tissue and mitochondria (Mustafa et al., 2009).

6.6.5 Effects of Nitroxide Antioxidants and Peroxynitrite Scavengers

In addition to the LOO• radical scavengers, the neuroprotective effects of a family of nitroxide-containing antioxidants have also been examined in experimental TBI models. These are sometimes referred to as "spin-trapping agents" and include α-phenyl-tert-butyl nitrone and its thiol analog NXY-059 and tempol. Both α-phenyl-tert-butyl nitrone and tempol have been shown to be protective in rodent TBI paradigms (Awasthi et al., 1997; Marklund et al., 2001). As mentioned earlier, tempol has been shown by the author and colleagues to catalytically scavenge PN-derived $\bullet NO_2$ and $\bullet CO_3$ (Bonini et al., 2002; Carroll et al., 2000) and to reduce posttraumatic oxidative damage (both LP and protein nitration), preserve mitochondrial function, decrease calcium-activated, calpain-mediated cytoskeletal damage, and reduce neurodegeneration in mice subjected to a severe controlled cortical impact-induced focal TBI (Deng-Bryant et al., 2008). Another laboratory has reported that tempol can reduce posttraumatic brain edema and improve neurological recovery in rat contusion injury model (Beit-Yannai et al., 1996; Zhang et al., 1998). However, the neuroprotective effect of tempol, administered alone, is associated with a therapeutic window of an hour or less in the mouse controlled cortical impact TBI model. Moreover, tempol is not effective at directly inhibiting LP in the latter model (Deng-Bryant et al., 2008).

6.6.6 Effects of the Iron Chelator Deferoxamine

The prototype iron chelator deferoxamine, which binds ferric (Fe^{+++}) iron and thereby would lessen the catalytic effects of iron on LP, has also been reported to have beneficial actions in preclinical TBI or TBI-related models (Gu et al., 2009; Long et al., 1996). However, deferoxamine is hindered by its lack of brain penetration and rapid plasma elimination rate. To partially counter the latter limitation, a dextran-coupled deferoxamine has been synthesized that has been reported to significantly improve early neurological recovery in a mouse diffuse TBI model (Panter et al., 1992). Much of this activity, however, is probably due to microvascular antioxidant protection because of limited brain penetrability. Another caveat to the iron-chelation antioxidant neuroprotective approach that is at least relevant to the ferric iron chelators such as deferoxamine is that they can cause a prooxidant effect in that their binding of Fe^{+++} can actually drive the oxidation of ferrous to ferric iron, which can increase superoxide radical formation in the process ($Fe^{++} + O_2 \rightarrow Fe^{+++} + O_2^{\bullet -}$).

6.6.7 Carbonyl Scavenging as an Approach to Inhibit 4-HNE and Acrolein Binding to Proteins

As pointed out previously (Figure 6.3), the LP-derived aldehydic (carbonyl-containing) breakdown products 4-HNE and acrolein have high affinity for binding to selected protein amino acid residues including histidine, lysine, arginine, and cysteine. These modifications have been shown to inhibit the activities of a variety of enzymatic proteins (Halliwell and Gutteridge, 2008). Several compounds have been recently identified that are able to antagonize this "carbonyl stress" by covalently binding to reactive LP-derived aldehydes. For instance, D-penicillamine has been demonstrated to form an irreversible bond to primary aldehydes. We have previously demonstrated that penicillamine is able to scavenge PN (Althaus et al., 1994) and to protect brain mitochondria from PN-induced respiratory dysfunction in isolated rat brain mitochondria (Singh et al., 2007). This latter action was observed along with an attenuation of 4-HNE–modified mitochondrial proteins (Singh et al., 2007). The PN scavenging action of penicillamine along with its carbonyl scavenging capability may jointly explain our previous findings that acutely administered penicillamine can improve early neurological recovery of mice subjected to moderately severe concussive TBI (Hall et al., 1999).

More recently, it has been demonstrated that a variety of hydrazine-containing compounds such as the antihypertensive agent hydralazine, the antidepressant phenelzine (Figure 6.6), and the antitubercular agent iproniazid can react with the carbonyl moieties of 4-HNE or acrolein that prevents the latter from binding to susceptible amino acids in proteins (Galvani et al., 2008). Consistent with this effect being neuroprotective, others have shown that hydralazine inhibits either compression or acrolein-mediated injuries to ex vivo spinal cord (Hamann et al., 2008). Hydralazine, which is a potent vasodilator, would be difficult to administer in vivo after either spinal cord injury or TBI in which hypotension is already a common pathophysiological problem. However, other hydrazine-containing compounds such as phenelzine and iproniazid do not compromise blood pressure as readily as hydralazine and have a long history of clinical use, although they have never been examined in acute neurotrauma models. Most impressive is that the application of hydrazines can rescue cultured cells from 4-HNE toxicity even when administered after 4-HNE has already covalently bound to cellular proteins (Galvani et al., 2008). Such an effect could be associated with a favorable neuroprotective therapeutic window.

Recently published in vitro studies in our laboratory have documented the ability of the hydrazine-containing phenelzine to protect isolated rat brain mitochondria from the respiratory depressant effects of 4-HNE together with a concentration-related attenuation of the accumulation of 4-HNE–modified mitochondrial proteins. Subsequent in vivo studies in the rat controlled cortical impact TBI model have found that a single 10 mg/kg subcutaneous dose of

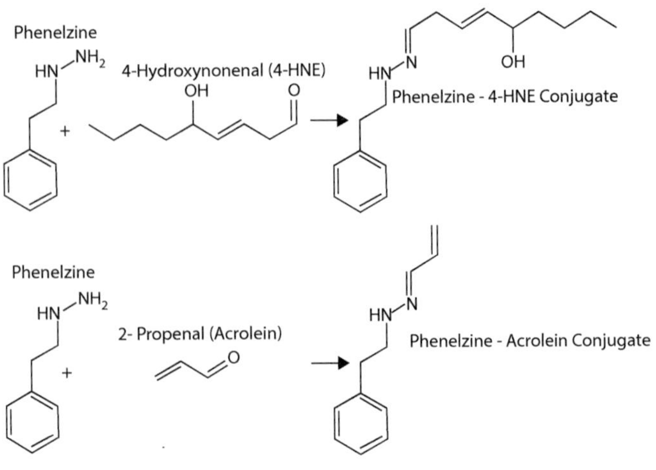

Phenelzine
4-Hydroxynonenal (4-HNE)
Phenelzine - 4-HNE Conjugate
Phenelzine
2- Propenal (Acrolein)
Phenelzine - Acrolein Conjugate

FIGURE 6.6 Chemical scavenging mechanism involved in the reactivity of the hydrazine-containing compound phenelzine with a mole of either 4-HNE or acrolein.

phenelzine can also reduce early (3 hours) posttraumatic mitochondrial respiratory failure as well as reduce cortical lesion volume at 14 days postinjury (Singh et al., 2013).

6.6.8 Small Molecule Nrf2/ARE Signaling Activators

The body's endogenous antioxidant defense system is largely regulated by nuclear factor E2-related factor 2/antioxidant response element (Nrf2/ARE) signaling at the transcriptional level (Kensler et al., 2007; Zhang, 2006). Nrf2 activation and the upregulation of antioxidant and antiinflammatory genes represents a valid neurotherapeutic intervention in CNS injury and has been previously described in various experimental models of stroke and neurodegenerative diseases (Shih et al., 2003). More recently, the role of Nrf2/ARE activation has been explored as a neuroprotective strategy for TBI.

The messenger RNA levels of Nrf2-regulated antioxidant enzymes, heme oxygenase (HO-1) and NAD(P)H:quinonereductase-1 (NQO1) are upregulated 24 hours after TBI (Yan et al., 2008). In addition, Nrf2-knockout mice are susceptible to increased oxidative stress and neurologic deficits after TBI compared with their wild-type counterparts (Hong et al., 2010). Administration of the Nrf2 activator sulforaphane is neuroprotective in an animal model of TBI, reducing cerebral edema and oxidative stress, and improving BBB function and cognitive deficits (Dash et al., 2009). Studies by Chen et al. (2011) demonstrated increased expression of Nrf2 and HO-1 in the cortex of the rat subarachnoid hemorrhage model. Treatment with sulforaphane further increased the expression of Nrf2, HO-1, NQ01, and glutathione S-transferase-α1, resulting in the reduction of brain edema, cortical neuronal death, and motor deficits (Chen et al., 2011). Tert-butyl hydroquinone, another activator of Nrf2, protects against TBI-induced inflammation and damage via reduction in nuclear factor-KB activation and tumor necrosis factor-α and interleukin-1β production after injury in the mouse closed-head injury model (Jin et al., 2011). Collectively, these studies demonstrate a significant neuroprotective role of Nrf2 signaling through the activation of antioxidant enzymes and reduction oxidative secondary injury responses after brain injury. Thus, Nrf2 activation may be a prime candidate for the attenuation of oxidative stress and subsequent neurotoxicity in TBI via the development of small-molecule activators of the Nrf2/ARE pathway.

Recent work in our laboratory has revealed that after controlled cortical impact TBI in mice, there is indeed a progressive activation of the Nrf2-ARE system in the traumatically injured brain as evidenced by an increase in HO-1 messenger RNA and protein that peaks at 72 hours after TBI. However, this effect does not precede, but rather is coincident with, the postinjury increase in LP-related 4-HNE (Miller et al., 2014). Therefore, it is apparent that this endogenous neuroprotective antioxidant response needs to be pharmacologically enhanced and/or sped up if it is to be capable of exerting acute post-TBI neuroprotection. Our laboratory is currently studying another Nrf2-ARE activator natural product, carnosic acid, which has been shown by others to more effectively induce this antioxidant defense system than the prototype sulforaphane (Satoh et al., 2008). We have shown that administration of carnosic acid to non-TBI mice is able to significantly increase the resistance of cortical mitochondria harvested 48 hours later to the respiratory depressant effects of the in vitro applied 4-HNE together with a decrease in 4-HNE modification of mitochondrial proteins (Miller et al., 2013a). Subsequently, we have administered a single dose of carnosic acid to mice at 15 minutes after controlled cortical impact TBI and observed that it is able to significantly reduce the level of 4-HNE in cortical tissue surrounding the injury site (Miller et al., 2013b). Ongoing studies are evaluating the behavioral recovery and tissue protective effects of carnosic acid.

6.7 RATIONALE FOR COMBINATION ANTIOXIDANT TREATMENT OF TBI

Antioxidant neuroprotective therapeutic discovery directed at acute TBI has consistently been focused on attempting to inhibit the secondary injury cascade by pharmacological targeting of a single oxidative damage mechanism. As presented previously, these efforts have included either enzymatic scavenging of superoxide radicals with SOD (Muizelaar et al., 1995) or inhibition of LP with tirilazad (Marshall et al., 1998). Although each of these strategies has shown protective efficacy in animal models of TBI, phase III clinical trials with either compound failed to demonstrate a statistically significant positive effect, although post hoc subgroup analysis suggests that the microvascularly localized tirilazad may have efficacy in moderate and severe TBI patients with traumatic SAH (Marshall et al., 1998). Although many reasons have been identified as possible contributors to the failure, one logical explanation has to with the possible need to interfere at multiple points in the oxidative damage portion of the secondary injury cascade either simultaneously or in a phased manner to achieve a clinically demonstrable level of neuroprotection.

Figure 6.7 summarizes the overall rationale for a multimechanistic antioxidant therapy for TBI. It is anticipated that the combination of two or three antioxidant mechanistic strategies may improve the extent of neuroprotective efficacy, lessen the variability of the effect, and possibly provide a longer therapeutic window of opportunity compared with the window for the individual strategies. If these theoretical combinatorial benefits are confirmed in preclinical TBI models, this should greatly enhance the chance of neuroprotective success in future clinical trials in contrast to previous failures with single antioxidant agents.

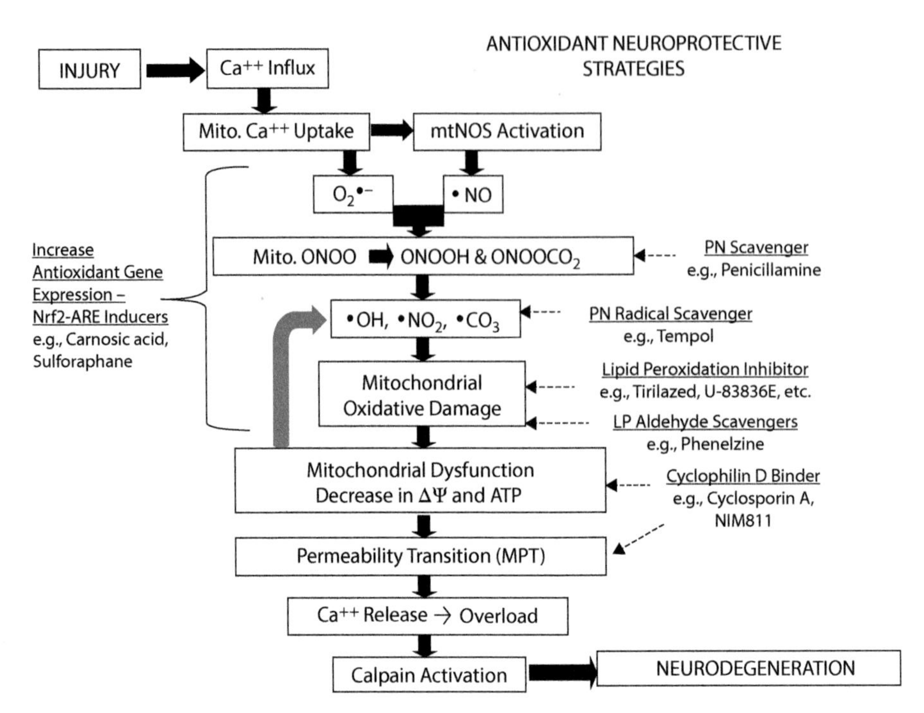

FIGURE 6.7 Rationale for the combination of two or more antioxidant strategies for the more effective protection of the injured brain. Included in the figure, but not discussed in the text, is the immunosuppressant cyclosporin A and its nonimmunosuppressant analog NIM811 that have been demonstrated to inhibit mitochondrial functional collapse by inhibiting formation of the mitochondrial permeability transition pore (MPTP) by binding to cyclophilin D, one of the essential components of MPTP formation. As a result, it has been shown that both compounds indirectly attenuate lipid peroxidative and protein nitrative damage by decreasing mitochondrial ROS/RNS production (Mbye et al., 2009; Mbye et al., 2008; Sullivan et al., 1999).

ACKNOWLEDGMENTS

Portions of the work reviewed in this article were supported by funding from 5R01 NS046566, 5P30 NS051220, and 5P01 NS58484 and from the Kentucky Spinal Cord and Head Injury Research Trust.

REFERENCES

Althaus, J.S., T.T. Oien, G.J. Fici, H.M. Scherch, V.H. Sethy, and P.F. Von Voigtlander. 1994. Structure activity relationships of peroxynitrite scavengers an approach to nitric oxide neurotoxicity. *Res Commun Chem Pathol Pharmacol.* 83:243–254.

Ates, O., S. Cayli, E. Altinoz, I. Gurses, N. Yucel, M. Sener et al. 2007. Neuroprotection by resveratrol against traumatic brain injury in rats. *Mol Cell Biochem.* 294:137–144.

Awasthi, D., D.F. Church, D. Torbati, M.E. Carey, and W.A. Pryor. 1997. Oxidative stress following traumatic brain injury in rats. *Surg Neurol.* 47:575–581; discussion 581–572.

Bains, M. and E.D. Hall. 2012. Antioxidant therapies in traumatic brain and spinal cord injury. *Biochim Biophys Acta.* 1822:675–684.

Beit-Yannai, E., R. Zhang, V. Trembovler, A. Samuni, and E. Shohami. 1996. Cerebroprotective effect of stable nitroxide radicals in closed head injury in the rat. *Brain Res.* 717:22–28.

Beni, S.M., R. Kohen, R.J. Reiter, D.X. Tan, and E. Shohami. 2004. Melatonin-induced neuroprotection after closed head injury is associated with increased brain antioxidants and attenuated late-phase activation of NF-kappaB and AP-1. *Faseb J.* 18:149–151.

Bonini, M.G., R.P. Mason, and O. Augusto. 2002. The mechanism by which 4-hydroxy-2,2,6,6-tetramethylpiperidene-1-oxyl (tempol) diverts peroxynitrite decomposition from nitrating to nitrosating species.. *Chem Res Toxicol.* 15:506–511.

Bringold, U., P. Ghafourifar, and C. Richter. 2000. Peroxynitrite formed by mitochondrial NO synthase promotes mitochondrial Ca2+ release. *Free Radic Biol Med.* 29:343–348.

Carroll, R.T., P. Galatsis, S. Borosky, K.K. Kopec, V. Kumar, J.S. Althaus et al. 2000. 4-Hydroxy-2,2,6,6-tetramethylpiperidine-1-oxyl (Tempol) inhibits peroxynitrite-mediated phenol nitration. *Chem Res Toxicol.* 13:294–300.

Chan, P.H., C.J. Epstein, Y. Li, T.T. Huang, E. Carlson, H. Kinouchi et al. 1995. Transgenic mice and knockout mutants in the study of oxidative stress in brain injury. *J Neurotrauma.* 12:815–824.

Chen, G., Q. Fang, J. Zhang, D. Zhou, and Z. Wang. 2011. Role of the Nrf2-ARE pathway in early brain injury after experimental subarachnoid hemorrhage. *J Neurosci Res.* 89:515–523.

Cirak, B., N. Rousan, A. Kocak, O. Palaoglu, S. Palaoglu, and K. Kilic. 1999. Melatonin as a free radical scavenger in experimental head trauma. *Pediatr Neurosurg.* 31:298–301.

Dash, P.K., J. Zhao, S.A. Orsi, M. Zhang, and A.N. Moore. 2009. Sulforaphane improves cognitive function administered following traumatic brain injury. *Neurosci Lett.* 460:103–107.

Deng-Bryant, Y., I.N. Singh, K.M. Carrico, and E.D. Hall. 2008. Neuroprotective effects of tempol, a catalytic scavenger of peroxynitrite-derived free radicals, in a mouse traumatic brain injury model. *J Cereb Blood Flow Metab.* 28:1114–1126.

Dimlich, R.V., P.A. Tornheim, R.M. Kindel, E.D. Hall, J.M. Braughler, and J.M. McCall. 1990. Effects of a 21-aminosteroid (U-74006F) on cerebral metabolites and edema after severe experimental head trauma. *Adv Neurol.* 52:365–375.

Du, L., H. Bayir, Y. Lai, X. Zhang, P.M. Kochanek, S.C. Watkins et al. 2004. Innate gender-based proclivity in response to cytotoxicity and programmed cell death pathway. *J Biol Chem.* 279:38563–38570.

Galvani, S., C. Coatrieux, M. Elbaz, M.H. Grazide, J.C. Thiers, A. Parini et al. 2008. Carbonyl scavenger and antiatherogenic effects of hydrazine derivatives. *Free Radic Biol Med.* 45:1457–1467.

Gladstone, D.J., S.E. Black, and A.M. Hakim. 2002. Toward wisdom from failure: Lessons from neuroprotective stroke trials and new therapeutic directions. *Stroke.* 33:2123–2136.

Gu, Y., Y. Hua, R.F. Keep, L.B. Morgenstern, and G. Xi. 2009. Deferoxamine reduces intracerebral hematoma-induced iron accumulation and neuronal death in piglets. *Stroke.* 40:2241–2243.

Gutteridge, J.M. 1995. Lipid peroxidation and antioxidants as biomarkers of tissue damage. *Clin Chem.* 41:1819–1828.

Hall, E. 1986. Beneficial effects of acute intravenous ibuprofen on neurological recovery of head injured mice: Comparison of cyclooxygenase inhibition of thromboxane A2 synthetase or 5-lipoxygenase. *CNS Trauma.* 2:75–83.

Hall, E.D. 1995a. The mouse head injury model: Utility in the discovery of acute cerebroprotective agents. In: *Central Nervous System Trauma Research Techniques.* S.T. Ohnishi and T. Ohnishi, editors. CRC Press, Boca Raton, FL. pp. 213–233.

Hall, E.D., P.K. Andrus, and P.A. Yonkers. 1993. Brain hydroxyl radical generation in acute experimental head injury. *J Neurochem.* 60:588–594.

Hall, E.D. and J.M. Bosken. 2009. Measurement of oxygen radicals and lipid peroxidation in neural tissues. *Curr Protoc Neurosci.* Chapter 7:Unit 7 17 11–51.

Hall, E.D., J.M. Braughler, P.A. Yonkers, S.L. Smith, K.L. Linseman, E.D. Means et al. 1991. U-78517F: A potent inhibitor of lipid peroxidation with activity in experimental brain injury and ischemia. *J Pharmacol Exp Ther.* 258:688–694.

Hall, E.D., N.C. Kupina, and J.S. Althaus. 1999. Peroxynitrite scavengers for the acute treatment of traumatic brain injury. *Ann N Y Acad Sci.* 890:462–468.

Hall, E.D., J.M. McCall, and E.D. Means. 1994. Therapeutic potential of the lazaroids (21-aminosteroids) in acute central nervous system trauma, ischemia and subarachnoid hemorrhage. *Adv Pharmacol.* 28:221–268.

Hall, E.D., P.K. Andrus, S.L. Smith, J.A. Oostveen, H.M. Scherch, B.S. Lutzke et al. 1995b. Neuroprotective efficacy of microvascularly-localized versus brain-penetraiting antioxidants. *Acta Neurochir (Suppl).* 66:107–113.

Hall, E.D., R.A. Vaishnav, and A.G. Mustafa. 2010. Antioxidant therapies for traumatic brain injury. *Neurotherapeutics.* 7:51–61.

Hall, E.D., J.A. Wang, and D.M. Miller. 2012. Relationship of nitric oxide synthase induction to peroxynitrite-mediated oxidative damage during the first week after experimental traumatic brain injury. *Exp Neurol.* 238:176–182.

Hall, E.D., P.A. Yonkers, P.K. Andrus, J.W. Cox, and D.K. Anderson. 1992. Biochemistry and pharmacology of lipid antioxidants in acute brain and spinal cord injury. *J Neurotrauma.* 9(Suppl 2):S425–442.

Hall, E.D., P.A. Yonkers, J.M. McCall, and J.M. Braughler. 1988. Effects of the 21-aminosteroid U74006F on experimental head injury in mice. *J Neurosurg.* 68:456–461.

Halliwell, B. and J. Gutteridge. 2008. *Free Radicals in Biology and Medicine.* Oxford University Press.

Hamann, K., G. Nehrt, H. Ouyang, B. Duerstock, and R. Shi. 2008. Hydralazine inhibits compression and acrolein-mediated injuries in ex vivo spinal cord. *J Neurochem.* 104:708–718.

Hamann, K., and R. Shi. 2009. Acrolein scavenging: A potential novel mechanism of attenuating oxidative stress following spinal cord injury. *J Neurochem.* 111:1348–1356.

Hong, Y., W. Yan, S. Chen, C.R. Sun, and J.M. Zhang. 2010. The role of Nrf2 signaling in the regulation of antioxidants and detoxifying enzymes after traumatic brain injury in rats and mice. *Acta Pharmacol Sin.* 31:1421–1430.

Hummel, S.G., A.J. Fischer, S.M. Martin, F.Q. Schafer, and G.R. Buettner. 2006. Nitric oxide as a cellular antioxidant: A little goes a long way. *Free Radic Biol Med.* 40:501–506.

Jin, W., J. Kong, H. Wang, J. Wu, T. Lu, J. Jiang, H. Ni, and W. Liang. 2011. Protective effect of tert-butylhydroquinone on cerebral inflammatory response following traumatic brain injury in mice. *Injury.* 42:714–718.

Kassell, N.F., E.C. Haley, Jr., C. Apperson-Hansen, and W.M. Alves. 1996. Randomized, double-blind, vehicle-controlled trial of tirilazad mesylate in patients with aneurysmal subarachnoid hemorrhage: A cooperative study in Europe, Australia, and New Zealand. *J Neurosurg.* 84:221–228.

Kensler, T.W., N. Wakabayashi, and S. Biswal. 2007. Cell survival responses to environmental stresses via the Keap1-Nrf2-ARE pathway. *Annu Rev Pharmacol Toxicol.* 47:89–116.

Kontos, H.A. 1989. Oxygen radicals in CNS damage. *Chem Biol Interact.* 72:229–255.

Kontos, H.A., and J.T. Povlishock. 1986. Oxygen radicals in brain injury. *Cent Nerv Syst Trauma.* 3:257–263.

Kontos, H.A. and E.P. Wei. 1986. Superoxide production in experimental brain injury. *J Neurosurg.* 64:803–807.

Lanzino, G. and N.F. Kassell. 1999. Double-blind, randomized, vehicle-controlled study of high-dose tirilazad mesylate in women with aneurysmal subarachnoid hemorrhage. Part II. A cooperative study in North America. *J Neurosurg.* 90:1018–1024.

Lewen, A., P. Matz, and P.H. Chan. 2000. Free radical pathways in CNS injury. *J Neurotrauma.* 17:871–890.

Long, D.A., K. Ghosh, A.N. Moore, C.E. Dixon, and P.K. Dash. 1996. Deferoxamine improves spatial memory performance following experimental brain injury in rats. *Brain Res.* 717:109–117.

Longoni, B., M.G. Salgo, W.A. Pryor, and P.L. Marchiafava. 1998. Effects of melatonin on lipid peroxidation induced by oxygen radicals. *Life Sci.* 62:853–859.

Marklund, N., F. Clausen, A. Lewen, D.A. Hovda, Y. Olsson, and L. Hillered. 2001. alpha-Phenyl-tert-N-butyl nitrone (PBN) improves functional and morphological outcome after cortical contusion injury in the rat. *Acta Neurochir (Wien).* 143:73–81.

Marshall, L.F., A.I. Maas, S.B. Marshall, A. Bricolo, M. Fearnside, F. Iannotti et al. 1998. A multicenter trial on the efficacy of using tirilazad mesylate in cases of head injury. *J Neurosurg.* 89:519–525.

Mbye, L.H., I.N. Singh, K.M. Carrico, K.E. Saatman, and E.D. Hall. 2009. Comparative neuroprotective effects of cyclosporin A and NIM811, a nonimmunosuppressive cyclosporin A analog, following traumatic brain injury. *J Cereb Blood Flow Metab.* 29:87–97.

Mbye, L.H., I.N. Singh, P.G. Sullivan, J.E. Springer, and E.D. Hall. 2008. Attenuation of acute mitochondrial dysfunction after traumatic brain injury in mice by NIM811, a non-immunosuppressive cyclosporin A analog. *Exp Neurol.* 209: 243–253.

McIntosh, T.K., M. Thomas, D. Smith, and M. Banbury. 1992. The novel 21-aminosteroid U74006F attenuates cerebral edema and improves survival after brain injury in the rat. *J Neurotrauma.* 9:33–46.

Mesenge, C., I. Margaill, C. Verrecchia, M. Allix, R.G. Boulu, and M. Plotkine. 1998. Protective effect of melatonin in a model of traumatic brain injury in mice. *J Pineal Res.* 25:41–46.

Mikawa, S., H. Kinouchi, H. Kamii, G.T. Gobbel, S.F. Chen, E. Carlson et al. 1996. Attenuation of acute and chronic damage following traumatic brain injury in copper, zinc-superoxide dismutase transgenic mice. *J Neurosurg*. 85:885–891.

Miller, D., J. Wang, A. Buchanan, and E. Hall. 2014. Temporal and spatial dynamics of Nrf2-ARE-mediated gene targets in cortex and hippocampus following controlled cortical impact traumatic brain injury in mice. *J Neurotrauma*. 31:1194–1201.

Miller, D.M., I.N. Singh, J.A. Wang, and E.D. Hall. 2013a. Administration of the Nrf2-ARE activators sulforaphane and carnosic acid attenuates 4-hydroxy-2-nonenal-induced mitochondrial dysfunction ex vivo. *Free Radic Biol Med*. 57:1–9.

Miller, D., I. Singh, J. Wang, and E. Hall. 2013b. The Nrf2-ARE pathway as a therapeutic target for traumatic brain injury: Antioxidant dose-response of the potent activator carnosic acid. *Journal of Neurotrauma*. 30:A166–167.

Monyer, H., D.M. Hartley, and D.W. Choi. 1990. 21-Aminosteroids attenuate excitotoxic neuronal injury in cortical cell cultures. *Neuron*. 5:121–126.

Mori, T., T. Kawamata, Y. Katayama, T. Maeda, N. Aoyama, T. Kikuchi et al. 1998. Antioxidant, OPC-14117, attenuates edema formation, and subsequent tissue damage following cortical contusion in rats. *Acta Neurochir Suppl (Wien)*. 71:120–122.

Muizelaar, J.P., J.W. Kupiec, and L.A. Rapp. 1995. PEG-SOD after head injury. *J Neurosurg*. 83:942.

Mustafa, A.G., I.N. Singh, K.M. Carrico, and E.D. Hall. 2009. Mitochondrial protection after traumatic brain injury by scavenging lipid peroxyl radicals. *J Neurotrauma*. 26:A 93.

Narayan, R.K., M.E. Michel, B. Ansell, A. Baethmann, A. Biegon, M.B. Bracken et al. 2002. Clinical trials in head injury. *J Neurotrauma*. 19:503–557.

Ozdemir, D., K. Tugyan, N. Uysal, U. Sonmez, A. Sonmez, O. Acikgoz et al. 2005a. Protective effect of melatonin against head trauma-induced hippocampal damage and spatial memory deficits in immature rats. *Neurosci Lett*. 385:234–239.

Ozdemir, D., N. Uysal, S. Gonenc, O. Acikgoz, A. Sonmez, A. Topcu et al. 2005b. Effect of melatonin on brain oxidative damage induced by traumatic brain injury in immature rats. *Physiol Res*. 54:631–637.

Panter, S.S., J.M. Braughler, and E.D. Hall. 1992. Dextran-coupled deferoxamine improves outcome in a murine model of head injury. *J Neurotrauma*. 9:47–53.

Pellegrini-Giampietro, D.E., G. Cherici, M. Alesiani, V. Carla, and F. Moroni. 1990. Excitatory amino acid release and free radical formation may cooperate in the genesis of ischemia-induced neuronal damage. *J Neurosci*. 10:1035–1041.

Rohn, T.T., T.R. Hinds, and F.F. Vincenzi. 1993. Ion transport ATPases as targets for free radical damage. Protection by an aminosteroid of the Ca2+ pump ATPase and Na+/K+ pump ATPase of human red blood cell membranes. *Biochem Pharmacol*. 46:525–534.

Rohn, T.T., T.R. Hinds, and F.F. Vincenzi. 1996. Inhibition of Ca2+-pump ATPase and the Na+/K+-pump ATPase by iron-generated free radicals. Protection by 6,7-dimethyl-2,4-DI-1- pyrrolidinyl-7H-pyrrolo[2,3-d] pyrimidine sulfate (U-89843D), a potent, novel, antioxidant/free radical scavenger. *Biochem Pharmacol*. 51:471–476.

Satoh, T., K. Kosaka, K. Itoh, A. Kobayashi, M. Yamamoto, Y. Shimojo et al. 2008. Carnosic acid, a catechol-type electrophilic compound, protects neurons both in vitro and in vivo through activation of the Keap1/Nrf2 pathway via S-alkylation of targeted cysteines on Keap1. *J Neurochem*. 104:1116–1131.

Sharma, S., Y. Zhuang, Z. Ying, A. Wu, and F. Gomez-Pinilla. 2009. Dietary curcumin supplementation counteracts reduction in levels of molecules involved in energy homeostasis after brain trauma. *Neuroscience*. 161:1037–1044.

Shih, A.Y., D.A. Johnson, G. Wong, A.D. Kraft, L. Jiang, H. Erb, J.A. Johnson, and T.H. Murphy. 2003. Coordinate regulation of glutathione biosynthesis and release by Nrf2-expressing glia potently protects neurons from oxidative stress. *J Neurosci*. 23:3394–3406.

Singh, I.N., L.K. Gilmer, D.M. Miller, J.E. Cebak, J.A. Wang, and E.D. Hall. 2013. Phenelzine mitochondrial functional preservation and neuroprotection after traumatic brain injury related to scavenging of the lipid peroxidation-derived aldehyde 4-hydroxy-2-nonenal. *J Cereb Blood Flow Metab*. 33:593–599.

Singh, I.N., P.G. Sullivan, and E.D. Hall. 2007. Peroxynitrite-mediated oxidative damage to brain mitochondria: Protective effects of peroxynitrite scavengers. *J Neurosci Res*. 85:2216–2223.

Smith, S.L., P.K. Andrus, J.R. Zhang, and E.D. Hall. 1994. Direct measurement of hydroxyl radicals, lipid peroxidation, and blood-brain barrier disruption following unilateral cortical impact head injury in the rat. *J Neurotrauma*. 11:393–404.

Sonmez, U., A. Sonmez, G. Erbil, I. Tekmen, and B. Baykara. 2007. Neuroprotective effects of resveratrol against traumatic brain injury in immature rats. *Neurosci Lett*. 420:133–137.

Springer, J.E., R.D. Azbill, R.J. Mark, J.G. Begley, G. Waeg, and M.P. Mattson. 1997. 4-hydroxynonenal, a lipid peroxidation product, rapidly accumulates following traumatic spinal cord injury and inhibits glutamate uptake. *J Neurochem*. 68:2469–2476.

Sullivan, P.G., S. Krishnamurthy, S.P. Patel, J.D. Pandya, and A.G. Rabchevsky. 2007. Temporal characterization of mitochondrial bioenergetics after spinal cord injury. *J Neurotrauma*. 24:991–999.

Sullivan, P.G., M.B. Thompson, and S.W. Scheff. 1999. Cyclosporin A attenuates acute mitochondrial dysfunction following traumatic brain injury. *Exp Neurol*. 160:226–234.

Toklu, H.Z., T. Hakan, N. Biber, S. Solakoglu, A.V. Ogunc, and G. Sener. 2009. The protective effect of alpha lipoic acid against traumatic brain injury in rats. *Free Radic Res*. 43: 658–667.

Wu, A., Z. Ying, and F. Gomez-Pinilla. 2006. Dietary curcumin counteracts the outcome of traumatic brain injury on oxidative stress, synaptic plasticity, and cognition. *Exp Neurology*. 197:309–317.

Xiong, Y., F.S. Shie, J. Zhang, C.P. Lee, and Y.S. Ho. 2005. Prevention of mitochondrial dysfunction in post-traumatic mouse brain by superoxide dismutase. *J Neurochem*. 95:732–744.

Yan, W., H.D. Wang, Z.G. Hu, Q.F. Wang, and H.X. Yin. 2008. Activation of Nrf2-ARE pathway in brain after traumatic brain injury. *Neurosci Lett*. 431:150–154.

Zhang, D.D. 2006. Mechanistic studies of the Nrf2-Keap1 signaling pathway. *Drug Metab Rev*. 38:769–789.

Zhang, H., G.L. Squadrito, R. Uppu, and W.A. Pryor. 1999. Reaction of peroxynitrite with melatonin: A mechanistic study. *Chem Res Toxicol*. 12:526–534.

Zhang, J.R., H.M. Scherch, and E.D. Hall. 1996. Direct measurement of lipid hydroperoxides in iron-dependent spinal neuronal injury. *J Neurochem*. 66:355–361.

Zhang, R., E. Shohami, E. Beit-Yannai, R. Bass, V. Trembovler, and A. Samuni. 1998. Mechanism of brain protection by nitroxide radicals in experimental model of closed-head injury. *Free Radic Biol Med*. 24:332–340.

7 IGF-1/IGF-R Signaling in Traumatic Brain Injury

Impact on Cell Survival, Neurogenesis, and Behavioral Outcome

Sindhu K. Madathil and Kathryn E. Saatman

CONTENTS

7.1 INTRODUCTION

Growing interest in post-traumatic brain plasticity events has fueled investigations of therapeutic approaches that promote endogenous neurorepair. Insulin-like growth factor-1 (IGF-1) is a polypeptide hormone with critical roles in regulating brain plasticity mechanisms. This chapter summarizes literature related to how expression of IGF-1 and its signaling components are altered after traumatic brain injury (TBI). To understand the potential effects of changes in endogenous IGF-1, the major roles of IGF-1 in CNS function are reviewed, with attention to how these IGF-mediated events may impact the response to TBI. In light of the multiplicity of CNS functions mediated by IGF-1, supplementation of endogenous IGF-1 may provide neuroprotection and promote neuronal repair in the injured brain. Coupled with a handful of preclinical studies in TBI, a larger literature in other CNS injuries such as stroke, hypoxic ischemia and spinal cord injury demonstrates potential beneficial effects of IGF-1 following injury.

TBI pathophysiology is multifaceted, including primary and secondary events. Primary injury results from the mechanical forces including acceleration, deceleration, and impact forces at the moment of injury, producing diffuse or focal pathology. This initial phase is characterized by tissue deformation, membrane depolarization, disruption of blood vessels and axons, ischemia, and cell membrane damage (Beauchamp et al., 2008; Dietrich et al., 1994; Gaetz, 2004). Secondary injury evolves from this early damage over a period of hours to days and even weeks to months, characterized by a complex network of biochemical events (Dikmen et al., 2009; Farkas and Povlishock, 2007; McIntosh et al., 1999). Excitatory amino acids and inflammatory cytokines released early in the secondary injury cascade lead to altered calcium homeostasis. Excessive intracellular calcium can signal various biochemical pathways initiating inflammation, free radical generation, and cytoskeletal damage. Increased calcium can activate proteases including calpains and caspases. Once activated, these proteases can cause widespread cell damage via cytoskeletal protein degradation and necrotic or apoptotic cell death pathways initiated within hours and continuing for days after brain injury. Secondary injury responses ultimately culminate in white matter damage and neurodegeneration contributing to behavioral morbidity.

In response to destructive events, the brain also has the capacity to promote cell repair through various compensatory mechanisms commonly referred to as neuroplasticity. Altered growth factor signaling, synaptogenesis, angiogenesis, neurogenesis, and gliogenesis are among these posttrauma brain remodeling events (Kernie and Parent, 2009; Schoch et al., 2012; Stein and Hoffman, 2003; Yu et al., 2008). Expression and release of endogenous neurotrophic factors is altered by various forms of central nervous system (CNS) injuries including TBI. An increase in their expression is considered as one of the mechanisms to promote neuroprotection and neurorepair after damage (Guan et al., 2003). After TBI, expression of growth factors such as neurotrophin 4/5, nerve growth factor, basic fibroblast growth factor, brain-derived neurotrophic factor (BDNF), and IGF-1 are increased (Conte et al., 2003; Madathil et al., 2010; Royo et al., 2006). Many of these growth factors play important roles in brain development and thus their increased expression after brain injury can recapitulate many of the processes involved in brain growth, accelerating neuronal repair.

Despite the improved understanding of TBI pathology, no therapeutic approach for treatment has yet been proved efficacious. Pharmacological approaches under research for TBI can be grouped as either neuroprotective or neuroreparative depending on their mode of action. Neuroprotective strategies that promote neuronal survival are focused mainly on attenuating acute damage from glutamate excitotoxicity, free radicals, or calcium influx. Neurorepair approaches promote neuroregeneration or neuroplasticity events. IGF-1, because of the multiplicity of its actions, provides a combined approach by attenuating cell death and promoting brain repair events (Aberg et al., 2000, 2006; Anderson et al., 2002; Lopez-Lopez et al., 2004).

7.2 IGF-1 SIGNALING AFTER TBI

7.2.1 IGF-1 Signaling

The IGF-1 mature protein is a 70-amino acid peptide with structural similarity to insulin. The IGF-1 peptide is coded by a single IGF-1 gene consisting of six exons (Figure 7.1). Alternate splicing of these exons leads to different IGF-1 protein isoforms, all of which are 70-amino acid peptides and signal through the IGF-1 receptor (IGF-1R) (Sussenbach et al., 1992). In addition to the isoforms, posttranslational processing of IGF-1 protein by acid proteases leads to two biologically active peptides in the brain (Sara et al., 1986) that act through IGF-1R. Another brain-specific cleavage product derived from IGF-1 by acid proteases is the glycyl-prolyl-glutamate tripeptide (GPE) that acts through glutamate receptors rather than the IGF-1R (Cacciatore et al., 2012; Sara et al., 1989).

The liver is the major source of circulating IGF-1, where its synthesis is stimulated by growth hormone (GH). In contrast, synthesis of IGF-1 locally in the brain is not regulated by GH. Brain and systemic IGF-1 tightly bind to IGF binding proteins (IGFBPs) that protect IGF-1 from degradation, prolong its half-life, and deliver them to appropriate receptors (Duan, 2002; Rosenfeld et al., 1999). Among the IGFBP family of six proteins, IGFBPs 2, 4, and 5 are highly expressed in the brain (Duan, 2002; Russo et al., 2005). IGF-1 is expressed predominantly in neurons and its levels are high during brain development and decline with age (Bach et al.,

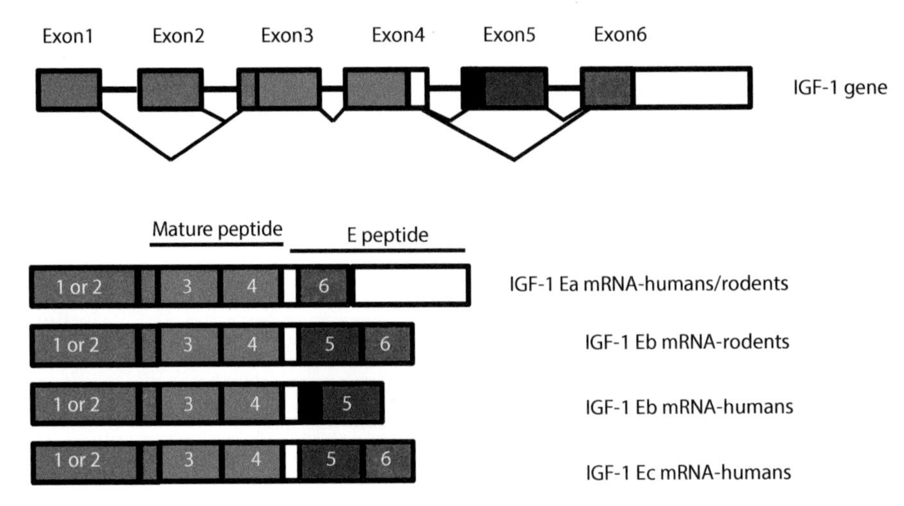

FIGURE 7.1 IGF-1 gene and splicing variants. The IGF-1 gene contains six exons. Alterative splicing can generate multiple mRNAs. Exons 1 and 2 together with part of exon 3 code for the signal peptide. Exons 1 and 2 are leader sequences and are interchangeable. Although exons 3 and 4 code for the mature IGF-1 peptide, exons 5 and 6 are part of the E-peptide. All the variants carry signal peptide and mature IGF-1 (exons 3 and 4). IGF-1 Ea is an mRNA variant generated by the exclusion of exon 5. In humans, IGF-1 Eb mRNA excludes exon 6 from E-peptide. In rodents, IGF-1 Eb (termed as IGF-1 Ec in humans) includes both exon 5 and exon 6 in the E-region. (Modified with permission from Oberbauer, AM, *Front Endocrinol.* 4:39, 2013.)

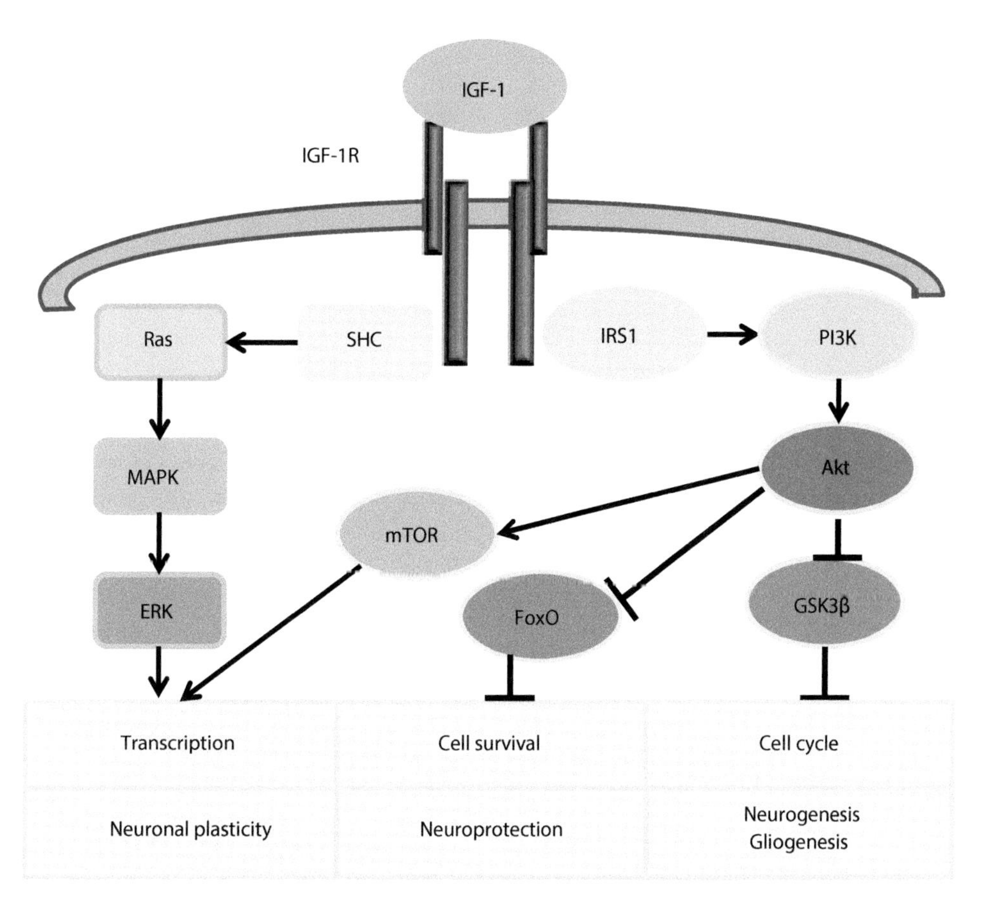

FIGURE 7.2 IGF-1/IGF-1R signaling cascade. IGF-1 is the major ligand for IGF-1R. IGF-1R is a transmembrane tyrosine kinase that consists of two alpha and two beta chains. In the nervous system, IGF-1R signals through two major pathways to mediate its functions. Activation of IGF-1R results in phosphorylation of adaptor proteins belonging to either the insulin-receptor substrate (IRS) or the Src homology 2 domain containing (SHC) family. Activation of SHC leads to signaling through mitogen-activated protein kinase (MAPK), whereas IRS phosphorylation activates PI3K. Both MAPK and PI3K activate or inhibit many downstream signaling proteins and transcription factors that lead to enhanced protein synthesis, cell survival, and proliferation. In the central nervous system, IGF-1 signaling can mediate cell survival and plasticity responses during development and also in response to injury. For simplicity, only major signaling mediators are shown in the pathway illustration.

1991; Beck et al., 1988). Physiological actions of IGF-1 are mediated through the IGF-1R (Czech, 1989). After binding of IGF-1, the IGF-1R can activate multiple pathways including the phosphatidylinositol 3-kinase (PI3K)/Akt and mitogen-activated protein kinase (MAPK) pathways (Cheng et al., 2000; Parrizas et al., 1997; Zheng et al., 2002). Although IGF-1 can act through multiple kinases, the PI3K/Akt pathway predominates in the nervous system (Leinninger et al., 2004a; Sun and D'Ercole, 2006). Activated Akt phosphorylates multiple substrates including GSK3β and mTOR that mediate many of IGF-1's effects in the brain (Figure 7.2).

7.2.2 Changes in IGF-1 Expression after TBI

Changes in serum IGF-1 levels are reported in brain-injured patients. The majority of clinical studies report reduced serum IGF-1 concentrations starting as early as one day that continue even years after injury (Agha et al., 2004a; Olivecrona et al., 2013; Popovic et al., 2004; Sanus et al., 2007; Wagner et al., 2010; Zgaljardic et al., 2011). Although quite a few of the studies included only moderate to severe trauma patients, reduced levels of circulating IGF-1 were also observed after mild TBI (Robles et al., 2009; Wagner et al., 2010). In contrast to the majority of clinical studies showing reduced IGF-1 levels after brain injury, one study reported no change in serum IGF-1 levels when measured months after severe closed-head trauma (Bondanell et al., 2002), whereas another reported elevated serum IGF-1 levels for 14 weeks after injury in patients with polytrauma or TBI alone (Wildburger et al., 2001). Although the reason for the differing observations is unclear at this time, TBI clinical reports commonly include a diverse patient population, with different trauma types, gender, injury severity, patient age, and brain region/regions affected, making comparisons across studies difficult.

Reduced serum IGF-1 levels after TBI may be attributed to pituitary dysfunction (Agha et al., 2004a, 2004b; Aimaretti et al., 2004; Berg et al., 2010; Herrmann et al., 2006). The pituitary gland, residing at the base of the brain, secretes many hormones, including GH. Hypopituitarism after TBI results in reduced levels of GH that in turn affect IGF-1 synthesis by the liver. Approximately 70% of the total circulating

IGF-1 is produced by the liver (Iresjo et al., 2013). Although the brain synthesizes IGF-1 locally, serum IGF-1 also crosses the blood–brain barrier and may affect neural tissue and modulate IGF-1 expression. In liver-specific IGF-1–deficient mice, low circulating IGF-1 induced a decrease in hippocampal IGF-1 expression (Mitschelen et al., 2011). Low levels of serum IGF-1 have been implicated in the development of cognitive dysfunction (Koopmans et al., 2006; Trejo et al., 2004). Furthermore, a meta-analysis study showed a positive correlation between blood IGF-1 levels and cognitive function in aged people (Arwert et al., 2005). Thus it is possible that low levels of circulating IGF-1 is one of the underlying causes of cognitive dysfunction after TBI. Supporting this notion, low serum IGF-1 positively correlated with cognitive impairment in TBI survivors tested a year after their injury (Popovic et al., 2004).

Although limited in number, animal studies are in agreement with the majority of clinical reports that show reduced serum IGF-1 levels in TBI survivors. Serum IGF-1 levels were decreased at 1 week and 1 month after repeated TBI in adolescent rats (Greco et al., 2013). Although single mild TBI did not cause any pituitary dysfunction, repeated TBI induced vascular damage and disrupted the GH/IGF-1 axis (Greco et al., 2013). After weight drop injury in rat pups, serum IGF-1 levels were found to be decreased at 7 days and 3 weeks postinjury (Ozdemir et al., 2012). This study also analyzed cognitive function using a Morris water maze paradigm. Serum IGF-1 was positively correlated with the time spent in the target quadrant (better cognition) and negatively correlated with the time spent on the opposite quadrant (impaired cognition), supporting the role of circulating IGF-1 in mediating cognitive function after trauma. The study also reported a strong negative correlation between serum IGF-1 levels and the number of apoptotic cells in various brain regions, including the hippocampal CA-1 area. Studies in experimental models allows researchers to examine pathophysiological events in the brain in relation to reduced serum IGF-1 levels. Furthermore, serum IGF-1 levels can be altered by exogenous administration or genetic manipulations in animal models to understand therapeutic benefits of IGF-1 in TBI.

In addition to altering peripheral IGF-1 levels, TBI affects IGF-1 expression in the brain. After penetrating or weight drop brain injury in adult rats, IGF-1 messenger RNA (mRNA) levels increase in the cortex within days (Nordqvist et al., 1997; Sandberg Nordqvist et al., 1996; Walter et al., 1997). Consistent with findings in these adult CNS injury models, after a rat pediatric controlled cortical impact (CCI) injury, IGF-1 mRNA was increased from 1 to 14 days, peaking at 3 days postinjury (Schober et al., 2010). When different mRNA variants of IGF-1 were analyzed after CCI brain injury, IGF-1B peaked at 2 days postinjury, whereas IGF-1A peaked at 3 days (Schober et al., 2012). IGF-1B mRNA includes exon 5 and codes for the Eb peptide, whereas IGF-1A mRNA excludes exon 5 and codes for the Ea peptide (Figure 7.1). Muscle gene transfer of either IGF-1 Eb or IGF-1 Ea peptide enhanced motor neuron survival after facial nucleus avulsion injury. However, IGF-1 Eb isoform was more potent in its neuroprotective action compared with IGF-1 Ea peptide (Aperghis et al., 2004). Although TBI differentially upregulated the IGF-1 isoforms, the significance of this finding is still not known. IGF-1 Eb was upregulated after mechanical injury (Cheema et al., 2005), and thus early upregulation in IGF-1 Eb may be in response to mechanical, primary injury. Cortical increases in IGF-1 mRNA after trauma were found to be completely blocked by the NMDA inhibitor MK-801 (Nordqvist et al., 1997), indicating a role of glutamate receptor activation in mediating the posttraumatic response of IGF-1.

IGF-1 protein expression during the acute phase of injury (1 to 7 days postlesion) was increased surrounding the wound area after a penetrating brain injury (Walter et al., 1997). This increase in IGF-1 protein was found to be localized to neurons, astrocytes, and endothelial cells neighboring the injury epicenter. A transient increase in cortical IGF-1 levels was measured by enzyme-linked immunosorbent assay at 1 hour after CCI brain injury in adult mice (Figure 7.3) (Madathil et al., 2010). However, IGF-1 immunostaining revealed a limited increase in IGF-1 staining, confined to contused cortex and periphery, at 1 to 48 hours postinjury, which was mainly localized to neurons (Figure 7.4). A marked increase in IGF-1 expression was noted in the subcortical white matter. Both IGF-1 mRNA and protein expression increase acutely after trauma. Although increased IGF-1 expression may be part of the brain's neuroprotective efforts, it appears to be insufficient in providing protection after TBI.

7.2.3 Changes in IGF-1R and IGFBP Expression after TBI

Upon IGF-1 binding, IGF-1R is autophosphorylated and activates downstream signaling. Thus changes in IGF-1R expression can influence the effectiveness of IGF-1 in regulating processes including cell survival, proliferation, and differentiation. After TBI in pediatric rodents, hippocampal IGF-1R mRNA levels decreased at 1 and 2 days postinjury and later showed a transient increase at 3 days, whereas IGF-1 message levels were higher from 1 to 14 days (Schober et al., 2010). It is possible that low IGF-1R synthesis at early time points can limit the potency of IGF-1. Alternatively, increased IGF-1 production could be a response to low IGF-1R levels to compensate for reduced signaling. Around 1 week after penetrating injury, IGF-1R peptide expression was increased in the neurons, astrocytes, and blood vessels bordering the lesion (Walter et al., 1997). Although no change in IGF-1R expression was observed using Western blot over 1 hour to 3 days after TBI in mice, immunostaining revealed increased vascular expression of IGF-1R in areas neighboring the contusion (Madathil et al., 2010). Proliferating endothelial cells are known to upregulate IGF-1R expression (Chisalita and Arnqvist, 2004) and thus increased vascular expression of IGF-1R after TBI may point to an active angiogenic response after trauma. The functional significance of changes in IGF-1R expression after TBI is not yet known; however, the characterization of IGF-1 and IGF-1R responses will

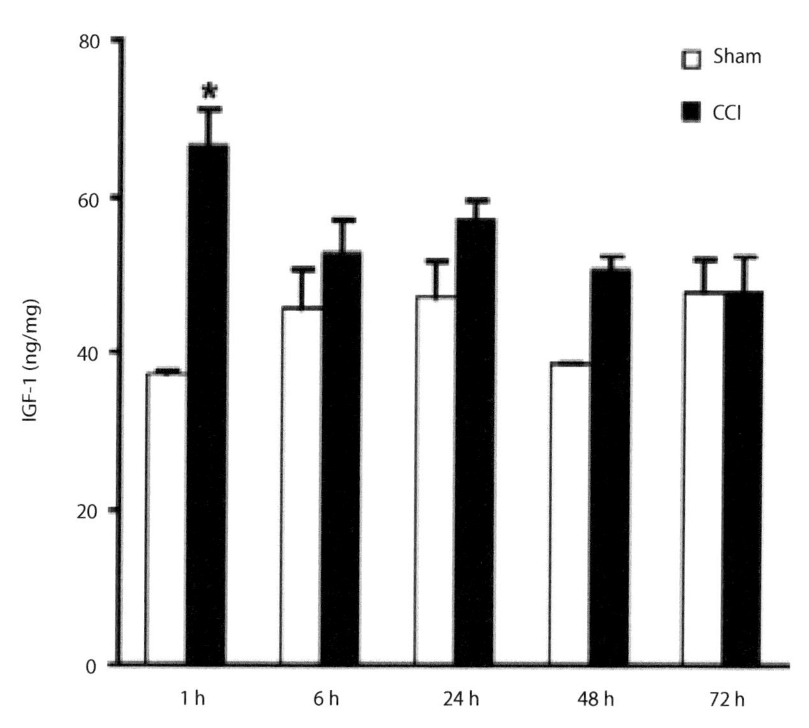

FIGURE 7.3 Quantification of IGF-1 levels after traumatic brain injury. Ipsilateral cortical samples were collected at different time points (1 to 72 hours) from mice that received a 0.5 mm (moderate) CCI brain injury. IGF-1 levels were measured using mouse-specific IGF-1 enzyme-linked immunosorbent assay. A transient increase in IGF-1 level was observed 1 hour after CCI. Data presented as mean ± standard error of the mean. $*p < 0.05$ compared with sham animals. (From Madathil SK et al., *J Neurotrauma*, 27:95–107, 2010. With permission.)

help to design therapeutic interventions involving IGF-1 administration.

Biological effects of IGF-1 are modulated by IGFBPs, expressed in a variety of tissues including brain. mRNA and protein expression of all of the six IGFBPs were increased locally in neurons, astrocytes, and endothelial cells bordering an incisional wound by 5 to 7 days after injury (Walter et al., 1997). After penetrating brain injury, IGFBP-2 and IGFBP-4 mRNA expression was increased in neurons surrounding the cortical lesion (Sandberg Nordqvist et al., 1996). Similar to the IGF-1 response after penetrating brain injury, trauma-induced IGFBP upregulation was also blocked by treatment with the NMDA receptor antagonist MK-801, indicating a role for glutamate receptor activation in IGFBP synthesis after trauma (Nordqvist et al., 1997). The physiological significance of increased IGFBP expression after injury is still unclear. Local increases in IGFBPs after injury may help to maintain IGF-1 levels around the lesion site by protecting it from degradation. Increased IGFBP in endothelial cells after injury could be an adaptive response to bind IGF-1 and facilitate increased transport into the diseased parenchyma from circulation. However, IGFBP-1–overexpressing mice showed decreased gliogenesis after a knife lesion, indicating an inhibitory action on IGF-1's proliferative effects (Ni et al., 1997).

In the CNS, cellular responses initiated by IGF-1 binding to its receptor are mainly mediated by the PI3-kinase/Akt pathway (Brywe et al., 2005; Sun and D'Ercole, 2006; Ye et al., 2010; Zheng et al., 2000). Akt phosphorylation was increased in cortical samples collected acutely (within 24 hours after the injury) from brain-injured patients (Zhang et al., 2006), suggesting the activation of the PI3K/Akt survival pathway. In animal models of CNS injuries, a transient increase in Akt activation has been observed (Janelidze et al., 2001; Madathil et al., 2010; Namura et al., 2000; Noshita et al., 2002; Yano et al., 2001; Zhang et al., 2006), which was associated with increased neuronal survival after TBI in rodents (Noshita et al., 2002). Akt activation was accompanied by phosphorylation of downstream substrates BAD and GSK3b after TBI in mice (Noshita et al., 2002). Although these studies demonstrate activation of the PI3K/Akt pathway after TBI, they do not confirm the involvement of IGF-1. However, increased Akt phosphorylation subsequent to IGF-1 upregulation suggests a role for IGF-1 in initiating PI3K/Akt pathways after TBI (Madathil et al., 2010).

7.3 PHYSIOLOGICAL ROLES OF IGF-1 WITH IMPLICATIONS FOR TBI

7.3.1 Overview

Many reviews are available that summarize the physiological actions of IGF-1 in the CNS (Aber et al., 2006; McMorris et al., 1993; Ye and D'Ercole, 2006). Among the variety of functions regulated by IGF-1, metabolic functions, cell proliferation, survival effects, myelination, and neurite outgrowth are significant in the context of TBI. Although brain injury activates intrinsic neuronal survival and regenerative

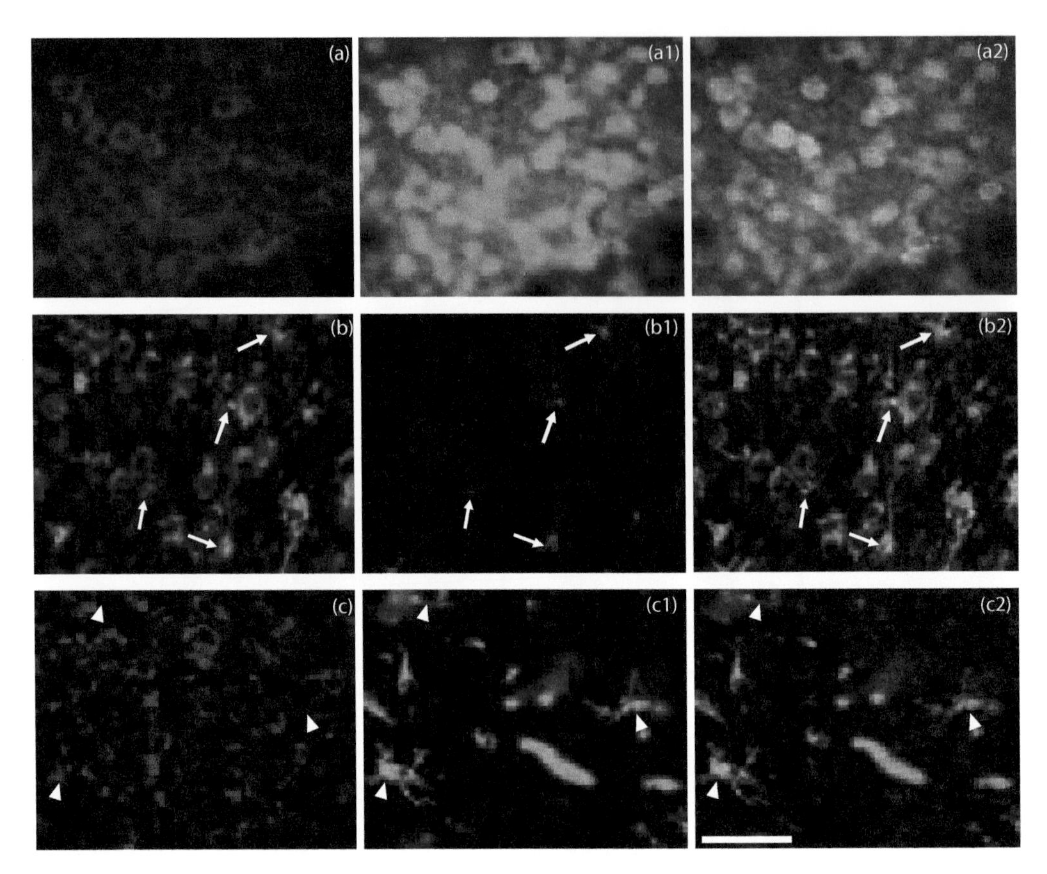

FIGURE 7.4 IGF-1 is expressed in both neurons and astrocytes after brain injury. IGF-1 expression was increased in the cortical penumbra 24 to 48 hours after CCI brain injury in mice. Immunofluorescence staining showed colocalization of IGF-1 (red) and NeuN (green) at the contusion periphery (a, a1, and a2). Some of the IGF-1 (green) positive cells colocalized with glial fibrillary acid protein (GFAP) (red) labeling at the contusion periphery (b, b1, and b2; arrows). IGF-1 (red) did not colocalize with microglia identified by fluorescein isothiocyanate (FITC)-conjugated isolectin B4 (c, c1, and c2; arrowheads) (scale bar in c2 = 50 μm). (From Madathil SK et al., *J Neurotrauma*, 27:95–107, 2010. With permission.)

mechanisms, several vital physiological processes are affected adversely ending in neural tissue loss and behavior complications. Changes in the IGF-1/IGF-1R axis after TBI may affect multiple physiological events as IGF-1 is known to regulate several of these mechanisms. Although increased expression of IGF-1 and its signaling molecules may promote neuronal protection and recovery, low serum IGF-1 or IGF-1R expression after TBI may confound pathology and adversely affect repair mechanisms. Next we discuss major biological functions of IGF-1 in the CNS and how they are important in the setting of brain injury.

7.3.2 Brain Glucose Utilization

Glucose is the brain's primary energy source. Low glucose levels are reported to cause intellectual deficits in diabetic children (Rovet and Ehrlich, 1999), pointing to the importance of glucose in mediating cognitive functions. Glucose utilization is reduced after injury in various animal models of TBI (Hayes et al., 1988; Robertson et al., 2013; Scafidi et al., 2009; Xing et al., 2009; Yoshino et al., 1991) and postinjury administration of glucose provides significant neuroprotection in the cortex and hippocampus (Moro et al., 2013). Neuronal IGF-1, which is more abundant than insulin in the brain, has anabolic functions similar to insulin. Glucose utilization is markedly reduced in the brains of IGF-1 KO mice indicating its insulin-like anabolic functions (Cheng et al., 2000). In a neuronal cell line, IGF-1 enhanced glucose transport and protected cells from low glucose levels (Russo et al., 2004). Ependymal cells with robust expression of glucose transporters and glucokinase serve as cerebral glucose-sensing cells. IGF-1 was more potent than insulin in stimulating glucose uptake in ependymal cells, possibly through regulation of GLUT1 transporters (Verleysdonk et al., 2004), pointing to the predominant role of IGF-1 in regulating brain glucose uptake. IGF-1–mediated Akt activation may play a role in glucose utilization. Translocation of glucose transporters to the plasma membrane is increased after Akt activation. Therefore, one of the mechanisms behind IGF-1's neuroprotective potential may be enhanced glucose uptake in neurons exposed to trauma-induced hypoglycemia. However, to our knowledge, no study has yet confirmed this possibility.

7.3.3 Neuroprotection

Exogenous IGF-1 promotes neuronal survival in both in vitro and in vivo conditions. In a number of cell types including neurons, IGF-1 inhibits apoptosis induced by various stimuli such as hypoxia and excitotoxicity (Chung et al., 2007; Leinninger et al., 2004b; Lu et al., 2008; Lunn et al., 2010; Stohr et al., 2011; Subramaniam et al., 2005; Yang et al., 2013). Many of these studies also report that the cell survival effects of IGF-1 are mediated through PI3K/Akt or MAPK/Erk pathways (Figure 7.2) (Feldman et al., 1997; Parrizas et al., 1997; Russell et al., 1998; Zheng et al., 2000). Although IGF-1 administration reduces neuronal loss in a variety of in vivo CNS injury and neurodegenerative disease models (Bluthe et al., 2005; Jablonka et al., 2011; Kim et al., 2012; Miltiadous et al., 2010, 2011; Quesada et al., 2008; Saenger et al., 2012; Shavali et al., 2003; Traub et al., 2009), relatively few studies have addressed neuroprotective efficacy after TBI. IGF-1, when administered intracerebrally after a penetrating type of brain injury, reduced Hsp70 expression and cell death (Kazanis et al., 2004). IGF-1 overexpression in a transgenic mouse model was effective in reducing acute (3 days postinjury) hippocampal neurodegeneration and promoting neuronal survival 10 days after TBI (Madathil et al., 2013). Moreover, when IGF-1 was overexpressed, phosphorylation of Akt was increased consistent with a role for the PI3K/Akt pathway in mediating IGF-1's neuroprotective effects after TBI. More studies are necessary to validate the neuroprotective efficacy of IGF-1 in various trauma models and also to understand how IGF-1 mediates neuroprotection after brain injury.

7.3.4 Myelination

Traumatic axonal injury in white matter tracts resulting from tensile forces or tissue shear strain is a common occurrence during TBI (Adams et al., 1989; Bramlett et al., 1997; Povlishock et al., 1999; Saatman et al., 2009). White matter damage characterized by features such as microbleeds, axonal transport disruption, axonal swelling, proteolysis, and demyelination persists years after TBI in humans (Johnson et al., 2013a, 2013b). After TBI, demyelination can occur by various mechanisms including mechanical damage, oligodendrocyte death, and myelin degradation by proteases. Myelin increases signal conduction velocity in axons and, when the myelin sheath is damaged or lost as in demyelinating disorders, it impairs neuronal signaling because of reduced speed of electrical signals. Demyelinated axons can be remyelinated by new myelin. However, when oligodendrocytes are lost, remyelination can fail, causing chronic demyelination. Oligodendrocyte loss and subsequent hypomyelination were reported in animal models of TBI (Conti et al., 1998; Flygt et al., 2013; Lotocki et al., 2011). Degradation of structural proteins associated with the myelin sheath may contribute to demyelination or hypomyelination. Myelin basic protein, one of the most abundant myelin-associated proteins, was cleaved into smaller fragments within hours to days after TBI in patients (Su et al., 2012) and experimental animals (Liu et al., 2006), an event that may initiate posttraumatic demyelination. Because there is limited remyelination after TBI, demyelination can be progressive and last months to years (Bramlett and Dietrich, 2002), and may contribute to cognitive dysfunction observed after TBI (Kinnunen et al., 2011; Kraus et al., 2007).

IGF-1 promotes oligodendrocyte proliferation, survival, and differentiation and stimulates myelin synthesis (Cui et al., 2012; D'Ercole et al., 2002; DePaula et al., 2014; McMorris et al., 1993; Palacios et al., 2005; Stangel and Hartung, 2002). Although IGF-1 deficiency in mice induces hypomyelination, constitutive IGF-1 overexpression increases oligodendrocyte numbers, myelin synthesis, and expression of myelin-related genes (Beck et al., 1995; Carson et al., 1993; Mason et al., 2000; Ye et al., 1995, 2002). Furthermore, IGF-1 administration reduces hypomyelination in experimentally induced autoimmune encephalomyelitis (Yao et al., 1996) and in animal models of multiple sclerosis (Chesik et al., 2008). Collectively, these reports suggest the strong influence of IGF-1 in regulating myelin synthesis in the CNS and highlight its therapeutic benefits in demyelinating disorders. Therefore, strategies such as IGF-1 therapy that limit myelin loss either by preventing its degradation or by promoting remyelination by reducing oligodendrocyte loss after TBI carry immense therapeutic potential.

7.3.5 Neurogenesis

The adult CNS is capable of generating new neurons in specific brain regions called neurogenic niches. The hippocampal subgranular zone and the forebrain subventricular zone (SVZ) are the major neurogenic niches in the adult mammalian brain. Brain injury is known to stimulate neurogenesis in both these regions, purportedly to replace lost neurons (Blaiss et al., 2011; Dash et al., 2001, Emery et al., 2005; Kernie et al., 2001; Kernie and Parent, 2009; Rola et al., 2006). Neurons born after injury contribute to posttraumatic functional recovery and tissue recovery (Blaiss et al., 2011; Emery et al., 2005; Kernie and Parent, 2009). Thus, strategies that promote this endogenous regenerative potential of the brain carry significance in TBI therapy. Although IGF-1's mitogenic potential and its ability to promote maturation and differentiation of neural precursors are well described in the context of the developing brain, its influence on adult mammalian neurogenesis is not completely understood (Aberg et al., 2006; Anderson et al., 2002). IGF-1 administration stimulates cell proliferation and neurogenesis in adult and aged rats (Aberg, 2010; Annenkov, 2009; Koltai et al., 2011; Perez-Martin et al., 2010). In mice with conditional IGF-1 overexpression, modest brain overgrowth (approximately 10% more than wild-type littermates) was observed even when IGF-1 overexpressed only after the first 4 weeks of postnatal development (Ye et al., 2004). Together these studies suggest that IGF-1 enhances neurogenesis and cell proliferation in the uninjured adult brain.

Exogenous IGF-1 administration has also been shown to promote neurogenesis in animal models of CNS injury. IGF-1 incorporated microspheres increased numbers of new neurons migrating from the SVZ toward the injured striatum in a mouse model of stroke (Nakaguchi et al., 2012).

Intranasal administration of IGF-1 after hypoxic-ischemic injury increased immature neuron number in neonatal rats (Lin et al., 2009). Postischemic viral delivery of IGF-1 effectively promoted neurogenesis in the SVZ and subcortical white matter (Zhu et al., 2008, 2009). These studies point to neurogenic effects of IGF-1 in the injured adult brain. The contribution of IGF-1 in stimulating posttraumatic neurogenesis is still not known. Recent work in our laboratory using IGF-1 overexpressing mice demonstrates that IGF-1 enhances the density of immature neurons in the injured hippocampus at 10 days after cortical impact TBI without significantly increasing cell proliferation (Carlson et al., 2014). It is possible that IGF-1 promoted immature neuronal survival or increased neuronal differentiation in the hippocampal subgranular zone area or that both these effects contributed to the IGF-1–mediated increase in newborn neuron density. Additional research needs to be done to address these possibilities and to further understand IGF-1's neurogenic potential, including long-term survival and functional integration of newborn neurons into the existing circuitry.

7.3.6 Neurite Outgrowth

Neurite injury and degeneration, which includes both dendritic and axonal damage, are common neuropathological consequences of TBI (Bramlett et al., 1997; Gao et al., 2011; Stone et al., 2004). Reduced dendrite branching complexity and degenerating dendritic spines are observed within days after TBI (Campbell et al., 2012a, 2012b; Gao et al., 2011; Winston et al., 2013). Axonal damage including cytoskeletal damage, transport impairment, and swelling and beading of axons starts acutely and continues for years after TBI (Johnson et al., 2013b; Saatman et al., 2009). Neurites are important in forming functional synapses that mediate neuronal signaling. Damaged dendrites and axons disrupt neuronal connectivity culminating in functional impairment (Gao et al., 2011; Kinnunen et al., 2011; Won et al., 2012). However, during the recovery phase, the brain also stimulates axonal and dendritic sprouting, and enhances functional synaptogenesis (Beck et al., 1993; Campbell et al., 2012b; Wieloch and Nikolich, 2006), a finding that may have implications for functional recovery, but also for complications such as the development of posttraumatic epilepsy.

IGF-1 is known to promote neurite extension in cultured cortical (Shiraishi et al., 2006), retinal (Dupraz et al., 2013), and peripheral (Jones et al., 2003; Kimpinski and Mearow, 2001) neurons. In rat cortical slices, IGF-1 treatment increased both apical and basal dendritic branching (Niblock et al., 2000), and enhanced dendritic arbor complexity to a greater degree than other trophic factors (Figure 7.5). Systemic IGF-1 administration after sciatic nerve crush promoted axon regeneration (Contreras et al., 1995). Brain-specific IGF-1 overexpression promoted a transient increase in hippocampal synaptic densities per neuron in mice (O'Kusky et al., 2000), whereas IGF-1 gene knock out resulted in reduced axonal diameter and peripheral nerve conduction velocities (Gao et al., 1999). Chronic subcutaneous infusion of recombinant IGF-1 to IGF-1–deficient mice restored motor and sensory nerve conduction velocities to normal. Intracerebroventricular infusion of IGF-1 increased dendritic arborization in doublecortin-positive immature neurons after excitotoxic lesion in the hippocampus (Liquitaya-Montiel et al., 2012), an observation whose functional consequence is yet to be explored. In conclusion, IGF-1 appears to be important in maintaining axonal and dendritic morphology, promoting growth of axons and dendrites, and improving synaptic function. Thus, IGF-1 may serve as an effective treatment option to promote neurite outgrowth and synaptogenesis after TBI. However, enhanced neurite outgrowth may not be always desirable after TBI. Aberrant mossy fiber sprouting and mTOR activation contributes to the development of posttraumatic epilepsy (Guo et al., 2013). More research is necessary to understand if the enhanced neurite growth after IGF-1 treatment is targeted appropriately.

7.3.7 Angiogenesis

Angiogenesis in the adult brain occurs by sprouting of new blood vessels from existing ones. Sprouts are formed by endothelial cells proliferating in response to an angiogenic stimulus. Neovascularization in the adult brain usually occurs after certain stimuli such as exercise or brain injury. During the primary injury phase, blood vessels are damaged because of mechanical disruption. In the initial days after TBI, new vessels appear at the contusion periphery and then spread into damaged tissue by one week postinjury (Guo et al., 2009). Brain injury upregulates the expression of vascular endothelial growth factor (VEGF) and its receptor VEGFR1, the major regulators of angiogenesis (Skold et al., 2005), at a time point that parallels the trauma-induced angiogenic response. Moreover, when VEGFR signaling was blocked after TBI, neuronal and glial degeneration was accelerated (Skold et al., 2006), suggesting a neuroprotective role for new blood vessels. Confirming this neuroprotective effect, VEGF administration after closed head injury reduces cortical tissue loss while also stimulating angiogenesis and neurogenesis (Thau-Zuchman et al., 2010). Enhanced vascular perfusion through newly formed blood vessels formation has been shown to improve functional outcome after stroke and TBI (Beck and Plate, 2009; Kreipke et al., 2007; Lu et al., 2004). Experimental studies over years clearly demonstrate the occurrence of angiogenesis after brain injury. New blood vessels formed in and around injured tissue are presumed to be important in restoring adequate perfusion that in turn accelerates recovery mechanisms. More research is needed to explain how postinjury angiogenesis influences outcomes after TBI.

IGF-1 is a well-known angiogenic factor for the developing and adult brain. In patients with genetic defects in IGF-1 signaling, retinal vessel morphology is altered indicating the importance of IGF-1 in the development of proper retinal vasculature (Hellstrom et al., 2002). IGF-1 enhances both human and mice endothelial cell proliferation in culture systems (Li et al., 2009). Age-related declines in

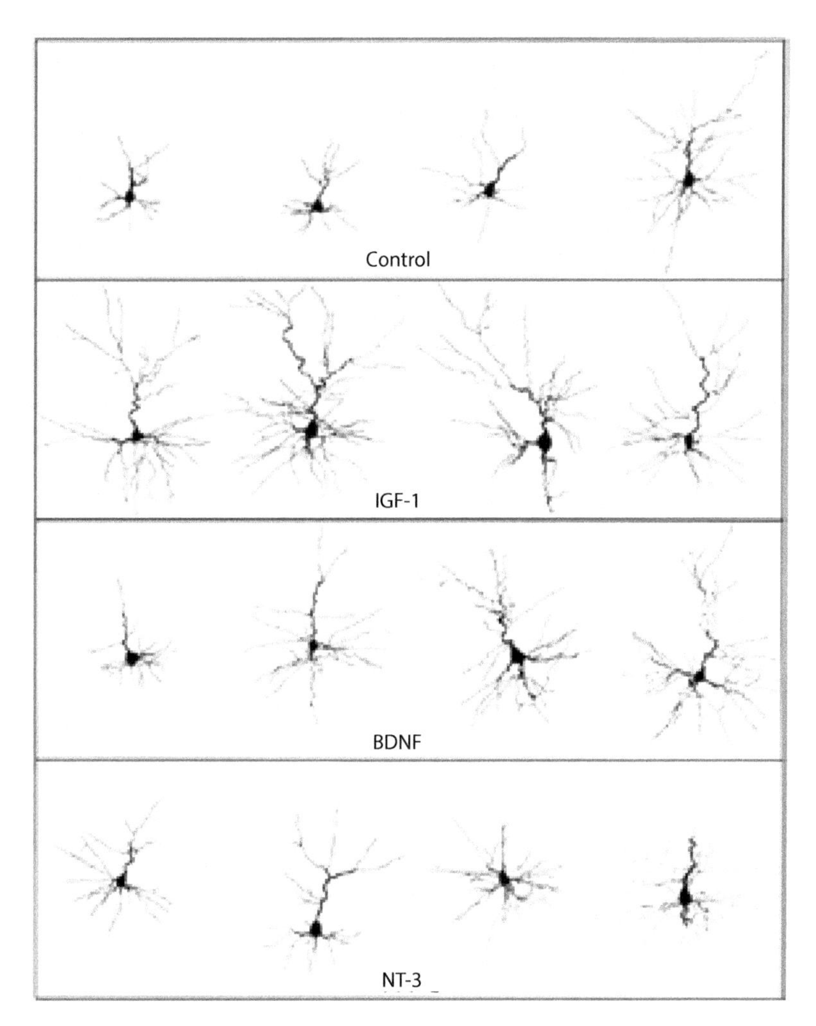

FIGURE 7.5 Dendritic arbor after trophic factor treatment. Brain slices prepared from rat pups (P10) were treated with IGF-1, BDNF, or neurotrophin-3 (NT-3) for 24 hours. Slices were fixed and neurons in slice culture were filled with biotin for visualization using avidin biotin complex/DAB reaction. Labeled cells were reconstructed using a NeuroLucida system. Representative samples of reconstructed neurons are shown. The intricate dendrite branching is most elaborate in IGF-1–treated neurons. (From Niblock MM et al., *J Neurosci.* 20:4165–4176, 2000. With permission.)

serum IGF-1 and GH levels are associated with decreased microvascular density, and administration of GH increases serum IGF-1 levels and enhances cerebral microvascular density in aged rats (Sonntag et al., 1997). Although systemic IGF-1 administration in adult mice increased brain angiogenesis, IGF-1 antibody infusion inhibited new vessel formation (Lopez-Lopez et al., 2004). In an experimental stroke model, postischemic IGF-1 gene transfer enhanced angiogenesis, and improved blood flow (Zhu et al., 2008; Zhu et al., 2009). In summary, IGF-1 represents one of the important regulators of adult brain angiogenesis and may enhance new blood vessel formation after brain injury.

7.4 THERAPEUTIC POTENTIAL OF IGF-1

7.4.1 Overview

Brain injury stimulates the production of several neurotrophins including IGF-1 (Madathil et al., 2010; Schober et al., 2010). However, TBI appears to induce only transient increases in IGF-1 and its signaling molecules, which are probably not sufficient to provide neuroprotection or stimulate subacute repair or regenerative mechanisms. Therefore, exogenous administration of IGF-1 may supplement and extend the actions of endogenous IGF-1. Moreover, IGF-1 has proven to be effective in treating CNS injuries including stroke and spinal cord injury, which share many pathological mechanisms with TBI (Guan et al., 2003; Guan et al., 2001; Hollis et al., 2009; Hung et al., 2007). As described previously, IGF-1 is a pluripotent molecule with both neuroprotective and regenerative potential. Thus IGF-1 is armed with multiple functions and may intervene at various points, halting the progression of TBI pathology and promoting brain repair.

7.4.2 Preclinical Studies Using IGF-1

Few studies have evaluated the therapeutic efficacy of IGF-1 using experimental animal models of TBI. Subcutaneous continuous delivery of recombinant human (rh) IGF-1 to rats for a period of 2 weeks following lateral fluid percussion injury improved both motor and cognitive function (Saatman et al., 1997) (Figure 7.6). Functional improvement in response to IGF-1 has been observed in other TBI animal models as well. Systemic injection of IGF-1 at 24 and 48 hours after a mild closed head injury improved cognitive function in mice (Rubovitch et al., 2010). Brain-specific IGF-1 overexpression in mice attenuated both cognitive and motor impairment after contusion brain injury (Madathil et al., 2013). IGF-1 overexpression also reduced acute hippocampal neurodegeneration and enhanced neuronal survival at 10 days after CCI brain injury (Madathil et al., 2013), (Figure 7.7). Intracranial stereotaxic IGF-1 administration of three equal doses at 15, 45, and 75 minutes after penetrating brain injury increased expression of neurotrophic factors, reduced apoptotic cell death, improved the metabolic status of injured rats, and also improved their motor activity (Kazanis et al., 2003; Kazanis et al., 2004). Systemic IGF-1 administration after weight drop closed head injury resulted in the stimulation of transcription factor C/EBP homologous protein, suggesting the activation of endoplasmic reticulum regulated antiapoptotic mechanisms (Rubovitch et al., 2011). Although limited in number, these studies confirm that IGF-1 has neuroprotective effects and can improve functional outcome, consistent with animal studies in other CNS injury models (Fletcher et al., 2009; Franz et al., 2009; Schabitz et al., 2001; Yao et al., 1995; Zhu et al., 2008). More studies are necessary to understand the therapeutic window of IGF-1 administration, signaling mechanisms, and regenerative potential after TBI.

7.4.3 IGF-1–Derived Peptides and Mimics

Similar to IGF-1, its truncated product GPE is reported to have neuroprotective properties in in vitro and in vivo models of CNS injuries (Guan, 2011; Guan and Gluckman, 2009). However, GPE is reported to have an extremely short plasma half-life (less than 2 minutes) when compared with IGF-1 (10 to 30 minutes), necessitating continuous infusion (Batchelor et al., 2003; Guan, 2011; Guler et al., 1987). To increase the stability of GPE, NNZ-2566, a GPE analogue, was created that is resistant to enzyme degradation but retains the neuroprotective effects of GPE (Guan and Gluckman, 2009; Lu et al., 2009a, 2009b; Wei et al., 2009). Systemic NNZ-2566 administration after a penetrating ballistic-type brain injury attenuated injury-induced upregulation of inflammatory cytokines and improved motor function (Lu et al., 2009a; Wei et al., 2009). NNZ-2566 treatment after penetrating ballistic-type brain injury reduced proapoptotic signaling and increased the expression of transcription factor ATF3, a negative regulator of pro-inflammatory cytokines (Cartagena et al., 2013; Lu et al., 2009a). The findings that NNZ-2566 reduced inflammation and provided functional improvement while providing an increased plasma half-life makes this compound highly desirable for TBI treatment. However, NNZ-2566 does not activate IGF-1R and

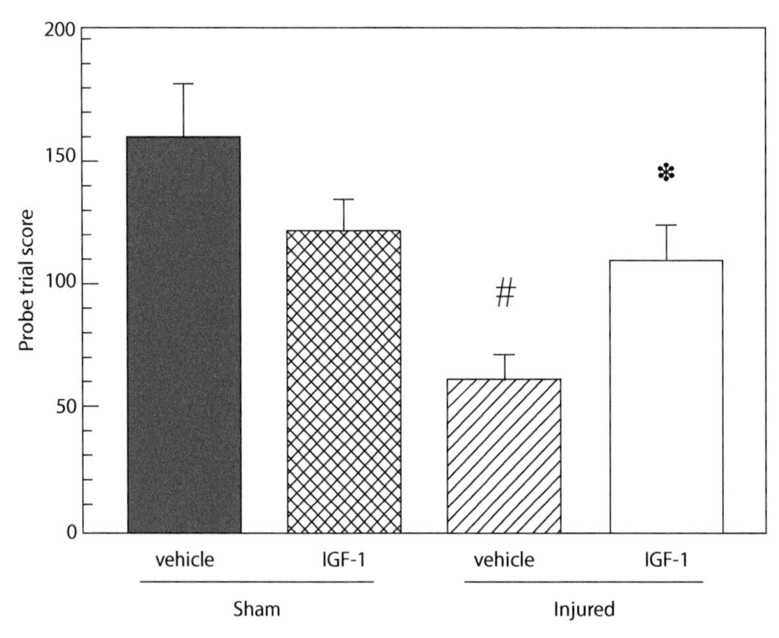

FIGURE 7.6 Systemic IGF-1 administration after brain injury improved cognitive function. IGF-1 was administered subcutaneously at 4 mg/kg/day for 2 weeks following fluid percussion injury in rats. On postinjury days 13 and 14, learning trials were conducted in a Morris water maze (MWM). Probe trials were conducted on day 15 to assess the ability to recall the platform location learned during prior trials. During the probe trial, the platform was removed and animals were scored during a 60-second swimming trial. Swimming patterns were analyzed and scores were assigned based on time spent in specific regions of the pool. High scores mean animals spent more time near the prior platform location. Although vehicle-treated brain-injured rats had impaired memory compared with sham rats (#$p < 0.05$ when compared with either group of sham animals), IGF-1–treated brain-injured rats had significantly improved scores (*$p < 0.01$ when compared with vehicle-treated injured animals), suggesting improved cognitive function in animals that received IGF-1. Data given as mean scores ± standard error of the mean. (From Saatman KE et al., *Exp. Neurol.* 147:418–427, 1997. With permission.)

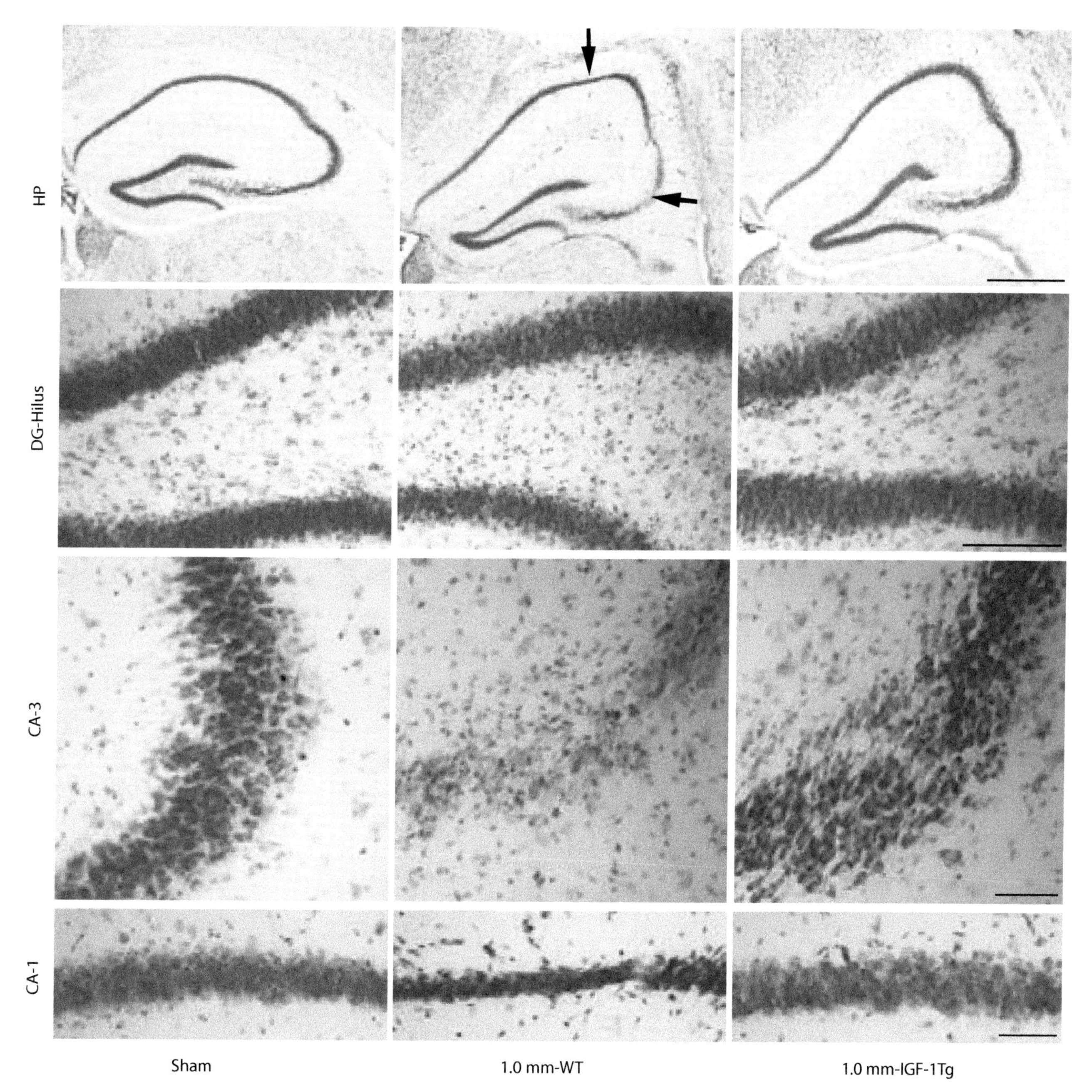

FIGURE 7.7 IGF-1 overexpression promoted hippocampal neuronal survival after traumatic brain injury. Both wild-type (WT) and IGF-1 overexpressing (IGF-1Tg) mice received a severe (1.0 mm) CCI brain injury. Representative hippocampal (HP) images are from Nissl stained brain sections collected 72 hours postinjury. Arrows indicate thinning of the CA-3 and CA-1 areas of pyramidal layer in WT mice, which is less evident in IGF-1Tg mice. Higher magnification images from the DG, CA-3, and CA-1 areas demonstrate hilar and CA-3 neuronal loss and thinning of the CA-1 pyramidal layer. IGF-1Tg mice showed marked neuroprotection in each hippocampal subregion. Scale bars = 500 μm (top HP panel), 100 μm for DG panel, 50 μm for CA-3 and CA-1 panels. (From Madathil SK et al., *PLoS One*. 8:e67204, 2013. With permission.)

its downstream signaling like full-length IGF-1 (Guan and Gluckman, 2009). Thus NNZ-2566 may not possess all the pleiotropic properties of IGF-1. Comparative studies of NNZ-2566 and IGF-1 are necessary to determine the differential effects these compounds may have in TBI. To date, NNZ-2566 has been tested only in a blast injury model of TBI. More studies in multiple trauma models are clearly required to confirm the neuroprotective effects of GPE and its analogue.

7.4.4 IGF-1 Clinical Trials

Despite a shortage of preclinical studies, IGF-1 has been already tested alone and in combination with growth hormone in TBI patients. Starting acutely after injury and lasting for weeks postinjury, patients have low serum IGF-1 and negative caloric balance (Aimaretti et al., 2004; Berg et al., 2010; Clifton et al., 1984; Fruin et al., 1986; Gottardis et al.,

1990; Hatton et al., 1997). Low serum IGF-1 is thought to contribute to inefficient use of protein after TBI. Therefore, supplementation with rhIGF-1 was hypothesized to improve metabolic status for TBI patients. An IGF-1 phase II safety and efficacy trial was conducted in moderate-to-severe TBI patients (Glasgow Coma Scale 4–10). Continuous administration of rhIGF-1 (0.01 mg/kg/hour) for 14 days by intravenous infusion maintained serum IGF-1 concentrations within physiological levels (150–400 ng/mL). However, despite continuous infusion, serum IGF-1 levels greater than 350 ng/mL were maintained for only around 7 days. IGF-1–treated patients gained weight and had higher nitrogen retention than placebo controls despite a low nitrogen intake in treated patients (Hatton et al., 1997). However, systemic IGF-1 administration in brain-injured patients lowered GH and IGFBP-3 concentrations, an effect reported in other IGF-1 clinical trials (Cioffi et al., 1994; Clemmons et al., 1992; Mauras et al., 1992; Turkalj et al., 1992). Because GH administration had been shown to increase levels of IGFBP-3 (Laursen et al., 1995), a subsequent clinical trial investigated the effects of administering IGF-1 in combination with GH in moderate-to-severe brain-injured patients (Rockich et al., 1999), with the goal of achieving higher sustained serum IGF-1 concentrations. IGF-1 (0.01 mg/kg/hour) and GH (0.05 mg/kg/hour) were administered intravenously for 14 days. Patients treated with IGF-1 and GH had significantly elevated IGFBP-3 concentrations compared with control patients (Rockich et al., 1999). Moreover, this study showed that a IGF-1/GH combination was effective in maintaining supraphysiological levels of IGF-1 (>1000 ng/mL) in circulation compared with the infusion of IGF-1 alone, where the plasma IGF-1 concentrations declined over time regardless of continuous IGF-1 administration. In a follow-up study using an identical administration regimen, TBI patients receiving combined IGF-1 and GH showed sustained improvement in nutritional and metabolic status (Hatton et al., 2006). In conclusion, IGF-1 clinical trials in TBI demonstrate that IGF-1 administration either alone or in combination with GH was safe to humans and successful in improving metabolic parameters in moderate-to-severe TBI patients.

7.5 CONCLUSION

TBI initiates a short-lived increase in IGF-1 expression in the brain coupled with a more persistent decrease in serum levels of IGF-1. Altered IGF-1 levels after TBI suggest a role for IGF-1 in modulating TBI pathophysiology. IGF-1 is a pleiotropic molecule that mediates many physiological functions in the developing and adult nervous system. Thus changes in IGF-1 and its signaling components may affect multiple cellular events, contributing to either neuroprotection or neurodegeneration. Administration of exogenous IGF-1 after experimental TBI promotes neuroprotection and improved functional outcomes. Although IGF-1 administration stimulates regenerative events including neurogenesis and angiogenesis in other CNS injury models, its brain repair potential after TBI is not yet known. Importantly, IGF-1 was well tolerated by patients in early-phase clinical trials for TBI and IGF-1–treated patients showed improved metabolic outcome compared to placebo-treated patients. Studies using GPE and GPE analogues open up a new area of investigation as its signaling mechanisms are different from IGF-1.

In conclusion, IGF-1 with its pleiotropic effects appears to be an ideal candidate for TBI therapy, although to move forward with IGF-1 or IGF-1–derived peptide clinical trials, more preclinical studies are absolutely essential. Research is required to delineate the mechanisms of IGF-1–mediated improvements after TBI. Studies are needed in various clinically relevant trauma models to see if IGF-1 is effective over different types of brain trauma. These studies may also help to determine therapeutic window and effective dosage for clinical trials.

ACKNOWLEDGMENT

Supported, in part, by National Institutes of Health grant R01 NS072302.

REFERENCES

Aberg D. (2010). Role of the growth hormone/insulin-like growth factor 1 axis in neurogenesis. *Endocr Develop.* 17:63–76.

Aberg MA, Aberg ND, Hedbacker H, Oscarsson J, and Eriksson PS. (2000). Peripheral infusion of IGF-I selectively induces neurogenesis in the adult rat hippocampus. *J. Neurosci.* 20:2896–2903.

Aberg ND, Brywe KG, and Isgaard J. (2006). Aspects of growth hormone and insulin-like growth factor-I related to neuroprotection, regeneration, and functional plasticity in the adult brain. *Scientific World J.* 6:53–80.

Adams JH, Doyle D, Ford I, Gennarelli TA, Graham DI, and McLellan DR. (1989). Diffuse axonal injury in head injury: Definition, diagnosis and grading. *Histopathology.* 15:49–59.

Agha A, Rogers B, Sherlock M, O'Kelly P, Tormey W, Phillips J, and Thompson CJ. (2004a). Anterior pituitary dysfunction in survivors of traumatic brain injury. *J Clin Endocrinol Metab.* 89:4929–4936.

Agha A, Thornton E, O'Kelly P, Tormey W, Phillips J, and Thompson CJ. (2004b). Posterior pituitary dysfunction after traumatic brain injury. *J Clin Endocrinol Metab.* 89:5987–5992.

Aimaretti G, Ambrosio MR, Benvenga S, Borretta G, De Marinis L, De Menis E et al. (2004). Hypopituitarism and growth hormone deficiency (GHD) after traumatic brain injury (TBI). *Growth Horm IGF Res.* 14 Suppl A:S114–117.

Anderson MF, Aberg MA, Nilsson M, and Eriksson PS. (2002). Insulin-like growth factor-I and neurogenesis in the adult mammalian brain. *Brain Res Dev Brain Res.* 134:115–122.

Annenkov A. (2009). The insulin-like growth factor (IGF) receptor type 1 (IGF1R) as an essential component of the signalling network regulating neurogenesis. *Mol Neurobiol.* 40:195–215.

Aperghis M, Johnson IP, Cannon J, Yang SY, and Goldspink G. (2004). Different levels of neuroprotection by two insulin-like growth factor-I splice variants. *Brain Res.* 1009:213–218.

Arwert LI, Deijen JB, and Drent ML. (2005). The relation between insulin-like growth factor I levels and cognition in healthy elderly: A meta-analysis. *Growth Horm IGF Res.* 15:416–422.

Bach MA, Shen-Orr Z, Lowe WL, Jr., Roberts CT, Jr., and LeRoith D. (1991). Insulin-like growth factor I mRNA levels are developmentally regulated in specific regions of the rat brain. *Brain Res Mol Brain Res.* 10:43–48.

Batchelor DC, Lin H, Wen JY, Keven C, Van Zijl PL, Breier BH et al. (2003). Pharmacokinetics of glycine-proline-glutamate, the N-terminal tripeptide of insulin-like growth factor-1, in rats. *Anal Biochem.* 323:156–163.

Beauchamp K, Mutlak H, Smith WR, Shohami E, and Stahel PF. (2008). Pharmacology of traumatic brain injury: Where is the "golden bullet"? *Mol Med.* 14:731–740.

Beck F, Samani NJ, Byrne S, Morgan K, Gebhard R, and Brammar WJ. (1988). Histochemical localization of IGF-I and IGF-II mRNA in the rat between birth and adulthood. *Development.* 104:29–39.

Beck H, and Plate KH. (2009). Angiogenesis after cerebral ischemia. *Acta Neuropathol.* 117:481–496.

Beck KD, Lamballe F, Klein R, Barbacid M, Schauwecker PE, McNeil TH et al. (1993). Induction of noncatalytic TrkB neurotrophin receptors during axonal sprouting in the adult hippocampus. *J Neurosci.* 13:4001–4014.

Beck KD, Powell-Braxton L, Widmer HR, Valverde J, and Hefti F. (1995). Igf1 gene disruption results in reduced brain size, CNS hypomyelination, and loss of hippocampal granule and striatal parvalbumin-containing neurons. *Neuron.* 14:717–730.

Berg C, Oeffner A, Schumm-Draeger PM, Badorrek F, Brabant G, Gerbert B et al. (2010). Prevalence of anterior pituitary dysfunction in patients following traumatic brain injury in a German multi-centre screening program. *Exp Clin Endocrinol Diabetes.* 118:139–144.

Blaiss CA, Yu TS, Zhang G, Chen J, Dimchev G, Parada LF et al. (2011). Temporally specified genetic ablation of neurogenesis impairs cognitive recovery after traumatic brain injury. *J Neurosci.* 31:4906–4916.

Bluthe RM, Frenois F, Kelley KW, and Dantzer R. (2005). Pentoxifylline and insulin-like growth factor-I (IGF-I) abrogate kainic acid-induced cognitive impairment in mice. *J Neuroimmunol.* 169:50–58.

Bondanelli M, Ambrosio MR, Margutti A, Boldrini P, Basaglia N, Franceschetti P et al. (2002). Evidence for integrity of the growth hormone/insulin-like growth factor-1 axis in patients with severe head trauma during rehabilitation. *Metab Clin Exp.* 51:1363–1369.

Bramlett HM, and Dietrich WD. (2002). Quantitative structural changes in white and gray matter 1 year following traumatic brain injury in rats. *Acta Neuropathol (Berl).* 103:607–614.

Bramlett HM, Kraydieh S, Green EJ, and Dietrich WD. (1997). Temporal and regional patterns of axonal damage following traumatic brain injury: A beta-amyloid precursor protein immunocytochemical study in rats. *J Neuropathol Exp Neurol.* 56:1132–1141.

Brywe KG, Mallard C, Gustavsson M, Hedtjarn M, Leverin AL, Wang X et al. (2005). IGF-I neuroprotection in the immature brain after hypoxia-ischemia, involvement of Akt and GSK3beta? *Eur J Neurosci.* 21:1489–1502.

Cacciatore I, Cornacchia C, Baldassarre L, Fornasari E, Mollica A, Stefanucci A et al. (2012). GPE and GPE analogues as promising neuroprotective agents. *Mini Rev Med Chem.* 12:13–23.

Campbell JN, Low B, Kurz JE, Patel SS, Young MT, and Churn SB. (2012a). Mechanisms of dendritic spine remodeling in a rat model of traumatic brain injury. *J Neurotrauma.* 29:218–234.

Campbell JN, Register D, and Churn SB. (2012b). Traumatic brain injury causes an FK506-sensitive loss and an overgrowth of dendritic spines in rat forebrain. *J Neurotrauma.* 29:201–217.

Carlson SW, Madathil SK, Sama DM, Gao X, Chen J, Saatman KE. (2014). Conditional overexpression of insulin-like growth factor-1 enhances hippocampal neurogenesis and restores immature neuron dendritic processes after traumatic brain injury. *J Neuropathol Exp Neurol.* 73:734–746

Carson MJ, Behringer RR, Brinster RL Carlson S, Madathil SK, Chen J McMorris FA. (1993). Insulin-like growth factor I increases brain growth and central nervous system myelination in transgenic mice. *Neuron.* 10:729–740.

Cartagena CM, Phillips KL, Williams GL, Konopko M, Tortella FC, Dave JR et al. (2013). Mechanism of action for NNZ-2566 anti-inflammatory dffects following PBBI involves upregulation of immunomodulator ATF3. *Neuromol Med.* 15:504–514.

Cheema U, Brown R, Mudera V, Yang SY, McGrouther G Carlson S, Madathil SK, Chen J Goldspink G. (2005). Mechanical signals and IGF-I gene splicing in vitro in relation to development of skeletal muscle. *J Cell Physiol.* 202:67–75.

Cheng CM, Reinhardt RR, Lee WH, Joncas G, Patel SC Carlson S, Madathil SK, Chen J Bondy CA. (2000). Insulin-like growth factor 1 regulates developing brain glucose metabolism. *Proc Natl Acad Sci U S A.* 97:10236–10241.

Cheng HL, Steinway ML, Russell JW Carlson S, Madathil SK, Chen J Feldman EL. (2000). GTPases and phosphatidylinositol 3-kinase are critical for insulin-like growth factor-I-mediated Schwann cell motility. *J Biol Chem.* 275:27197–27204.

Chesik D, De Keyser J Carlson S, Madathil SK, Chen J Wilczak N. (2008). Insulin-like growth factor system regulates oligodendroglial cell behavior: Therapeutic potential in CNS. *J Mol Neurosci.* 35:81–90.

Chisalita SI, and Arnqvist HJ. (2004). Insulin-like growth factor I receptors are more abundant than insulin receptors in human micro- and macrovascular endothelial cells. *Am J Pphysiol Endocrinol Metab.* 286:E896–901.

Chung H, Seo S, Moon M, and Park S. (2007). IGF-I inhibition of apoptosis is associated with decreased expression of prostate apoptosis response-4. *J Endocrinol.* 194:77–85.

Cioffi WG, Gore DC, Rue LW, 3rd, Carrougher G, Guler HP, McManus WF, and Pruitt BA, Jr. (1994). Insulin-like growth factor-1 lowers protein oxidation in patients with thermal injury. *Ann Surg.* 220:310–316; discussion 316–319.

Clemmons DR, Smith-Banks A, and Underwood LE. (1992). Reversal of diet-induced catabolism by infusion of recombinant insulin-like growth factor-I in humans. *J Clin Endocrinol Metab.* 75:234–238.

Clifton GL, Robertson CS, Grossman RG, Hodge S, Foltz R, and Garza C. (1984). The metabolic response to severe head injury. *J Neurosurg.* 60:687–696.

Conte V, Royo C, Shimizu S, Saatman KE, Watson D, Graham DI et al. 2003. Neurotrophic factors pathophysiology and therapeutic applications in traumatic brain injury. *Eur J Trauma.* 335–355.

Conti AC, Raghupathi R, Trojanowski JQ, and McIntosh TK. (1998). Experimental brain injury induces regionally distinct apoptosis during the acute and delayed post-traumatic period. *J Neurosci.* 18:5663–5672.

Contreras PC, Steffler C, Yu E, Callison K, Stong D, and Vaught JL. (1995). Systemic administration of rhIGF-I enhanced regeneration after sciatic nerve crush in mice. *J Pharmacol Exp Ther.* 274:1443–1449.

Cui QL, D'Abate L, Fang J, Leong SY, Ludwin S, Kennedy TE et al. (2012). Human fetal oligodendrocyte progenitor cells from different gestational stages exhibit substantially different potential to myelinate. *Stem Cells Develop.* 21:1831–1837.

Czech MP. (1989). Signal transmission by the insulin-like growth factors. *Cell.* 59:235–238.

D'Ercole AJ, Ye P, and O'Kusky JR. (2002). Mutant mouse models of insulin-like growth factor actions in the central nervous system. *Neuropeptides.* 36:209–220.

Dash PK, Mach SA, and Moore AN. (2001). Enhanced neurogenesis in the rodent hippocampus following traumatic brain injury. *J Neurosci Res.* 63:313–319.

De Paula ML, Cui QL, Hossain S, Antel J, and Almazan G. (2014). The PTEN inhibitor bisperoxovanadium enhances myelination by amplifying IGF-1 signaling in rat and human oligodendrocyte progenitors. *Glia.* 62:64–77.

Dietrich WD, Alonso O, and Halley M. (1994). Early microvascular and neuronal consequences of traumatic brain injury: A light and electron microscopic study in rats. *J Neurotrauma.* 11:289–301.

Dikmen SS, Corrigan JD, Levin HS, Machamer J, Stiers W, and Weisskopf MG. (2009). Cognitive outcome following traumatic brain injury. *J Head Trauma Rehabil.* 24:430–438.

Duan C. (2002). Specifying the cellular responses to IGF signals: Roles of IGF-binding proteins. The *J Endocrinol.* 175:41–54.

Dupraz S, Grassi D, Karnas D, Nieto Guil AF, Hicks D and Quiroga S. (2013). The insulin-like growth factor 1 receptor is essential for axonal regeneration in adult central nervous system neurons. PLoS One. 8:e54462.

Emery DL, Fulp CT, Saatman KE, Schutz C, Neugebauer E, and McIntosh TK. (2005). Newly born granule cells in the dentate gyrus rapidly extend axons into the hippocampal CA3 region following experimental brain injury. *J Neurotrauma.* 22:978–988.

Farkas O, and Povlishock JT. (2007). Cellular and subcellular change evoked by diffuse traumatic brain injury: A complex web of change extending far beyond focal damage. *Prog Brain Res.* 161:43–59.

Feldman EL, Sullivan KA, Kim B, and Russell JW. (1997). Insulin-like growth factors regulate neuronal differentiation and survival. *Neurobiol Dis.* 4:201–214.

Fletcher L, Kohli S, Sprague SM, Scranton RA, Lipton SA, Parra A et al. (2009). Intranasal delivery of erythropoietin plus insulin-like growth factor-I for acute neuroprotection in stroke. Laboratory investigation. *J Neurosurg.* 111:164–170.

Flygt J, Djupsjo A, Lenne F, and Marklund N. (2013). Myelin loss and oligodendrocyte pathology in white matter tracts following traumatic brain injury in the rat. *Eur J Neurosci.* 38:2153–2165.

Franz CK, Federici T, Yang J, Backus C, Oh SS, Teng Q et al. (2009). Intraspinal cord delivery of IGF-I mediated by adeno-associated virus 2 is neuroprotective in a rat model of familial ALS. *Neurobiol Dis.* 33:473–481.

Fruin AH, Taylon C, and Pettis MS. (1986). Caloric requirements in patients with severe head injuries. *Surg Neurol.* 25:25–28.

Gaetz M. (2004). The neurophysiology of brain injury. *Clin Neurophysiol.* 115:4–18.

Gao WQ, Shinsky N, Ingle G, Beck K, Elias KA, and Powell-Braxton L. (1999). IGF-I deficient mice show reduced peripheral nerve conduction velocities and decreased axonal diameters and respond to exogenous IGF-I treatment. *J Neurobiol.* 39:142–152.

Gao X, Deng P, Xu ZC, and Chen J. (2011). Moderate traumatic brain injury causes acute dendritic and synaptic degeneration in the hippocampal dentate gyrus. *PLoS One.* 6:e24566.

Gottardis M, Nigitsch C, Schmutzhard E, Neumann M, Putensen C, Hackl JM, and Koller W. (1990). The secretion of human growth hormone stimulated by human growth hormone releasing factor following severe cranio-cerebral trauma. *Intensive Care Med.* 16:163–166.

Greco T, Hovda D, and Prins M. (2013). The effects of repeat traumatic brain injury on the pituitary in adolescent rats. *J Neurotrauma.* 30:1983–1990.

Guan J. (2011). Insulin-like growth factor-1 (IGF-1) derived neuropeptides, a novel strategy for the development of pharmaceuticals for managing ischemic brain injury. *CNS Neurosci Therapeutics.* 17:250–255.

Guan J, Bennet L, Gluckman PD, and Gunn AJ. (2003). Insulin-like growth factor-1 and post-ischemic brain injury. *Progr Neurobiol.* 70:443–462.

Guan J, and Gluckman PD. (2009). IGF-1 derived small neuropeptides and analogues: A novel strategy for the development of pharmaceuticals for neurological conditions. *Br J Pharmacol.* 157:881–891.

Guan J, Miller OT, Waugh KM, McCarthy DC, and Gluckman PD. (2001). Insulin-like growth factor-1 improves somatosensory function and reduces the extent of cortical infarction and ongoing neuronal loss after hypoxia-ischemia in rats. *Neuroscience.* 105:299–306.

Guler HP, Zapf J, and Froesch ER. (1987). Short-term metabolic effects of recombinant human insulin-like growth factor I in healthy adults. *New Engl J Med.* 317:137–140.

Guo D, Zeng L, Brody DL, and Wong M. (2013). Rapamycin attenuates the development of posttraumatic epilepsy in a mouse model of traumatic brain injury. *PLoS One.* 8:e64078.

Guo X, Liu L, Zhang M, Bergeron A, Cui Z, Dong JF, and Zhang J. (2009). Correlation of CD34+ cells with tissue angiogenesis after traumatic brain injury in a rat model. *J Neurotrauma.* 26:1337–1344.

Hatton J, Kryscio R, Ryan M, Ott L, and Young B. (2006). Systemic metabolic effects of combined insulin-like growth factor-I and growth hormone therapy in patients who have sustained acute traumatic brain injury. *J Neurosurg.* 105:843–852.

Hatton J, Rapp RP, Kudsk KA, Brown RO, Luer MS, Bukar JG et al. (1997). Intravenous insulin-like growth factor-I (IGF-I) in moderate-to-severe head injury: A phase II safety and efficacy trial. *J Neurosurg.* 86:779–786.

Hayes RL, Katayama Y, Jenkins LW, Lyeth BG, Clifton GL, Gunter J et al. (1988). Regional rates of glucose utilization in the cat following concussive head injury. *J Neurotrauma.* 5:121–137.

Hellstrom A, Carlsson B, Niklasson A, Segnestam K, Boguszewski M, de Lacerda L et al. (2002). IGF-I is critical for normal vascularization of the human retina. *J Clin Endocrinol Metab.* 87:3413–3416.

Herrmann BL, Rehder J, Kahlke S, Wiedemayer H, Doerfler A, Ischebeck W et al. (2006). Hypopituitarism following severe traumatic brain injury. *Exp Clin Endocrinol Diabetes.* 114:316–321.

Hollis ER, 2nd, Lu P, Blesch A, and Tuszynski MH. (2009). IGF-I gene delivery promotes corticospinal neuronal survival but not regeneration after adult CNS injury. *Exp Neurol.* 215:53–59.

Hung KS, Tsai SH, Lee TC, Lin JW, Chang CK, and Chiu WT. (2007). Gene transfer of insulin-like growth factor-I providing neuroprotection after spinal cord injury in rats. *J Neurosurg Spine.* 6:35–46.

Iresjo BM, Svensson J, Ohlsson C, and Lundholm K. (2013). Liver-derived endocrine IGF-I is not critical for activation of skeletal muscle protein synthesis following oral feeding. *BMC Physiol.* 13:7.

Jablonka S, Holtmann B, Sendtner M, and Metzger F. (2011). Therapeutic effects of PEGylated insulin-like growth factor I in the pmn mouse model of motoneuron disease. *Exp Neurol.* 232:261–269.

Janelidze S, Hu BR, Siesjo P, and Siesjo BK. (2001). Alterations of Akt1 (PKBalpha) and p70(S6K) in transient focal ischemia. *Neurobiol Dis.* 8:147–154.

Johnson VE, Stewart JE, Begbie FD, Trojanowski JQ, Smith DH, and Stewart W. (2013a). Inflammation and white matter degeneration persist for years after a single traumatic brain injury. *Brain.* 136:28–42.

Johnson VE, Stewart W, and Smith DH. (2013b). Axonal pathology in traumatic brain injury. *Exp Neurol.* 246:35–43.

Jones DM, Tucker BA, Rahimtula M, and Mearow KM. (2003). The synergistic effects of NGF and IGF-1 on neurite growth in adult sensory neurons: Convergence on the PI 3-kinase signaling pathway. *J Neurochem.* 86:1116–1128.

Kazanis I, Bozas E, Philippidis H, and Stylianopoulou F. (2003). Neuroprotective effects of insulin-like growth factor-I (IGF-I) following a penetrating brain injury in rats. *Brain Res.* 991:34–45.

Kazanis I, Giannakopoulou M, Philippidis H, and Stylianopoulou F. (2004). Alterations in IGF-I, BDNF and NT-3 levels following experimental brain trauma and the effect of IGF-I administration. *J Biol Chem.* 186:221–234.

Kernie SG, Erwin TM, and Parada LF. (2001). Brain remodeling due to neuronal and astrocytic proliferation after controlled cortical injury in mice. *J Neurosci Res.* 66:317–326.

Kernie SG and Parent JM. (2009). Forebrain neurogenesis after focal Ischemic and traumatic brain injury. *Neurobiol Dis.* 37:267–274.

Kim Y, Li E, and Park S. (2012). Insulin-like growth factor-1 inhibits 6-hydroxydopamine-mediated endoplasmic reticulum stress-induced apoptosis via regulation of heme oxygenase-1 and Nrf2 expression in PC12 cells. *Int J Neurosci.* 122:641–649.

Kimpinski K and Mearow K. (2001). Neurite growth promotion by nerve growth factor and insulin-like growth factor-1 in cultured adult sensory neurons: Role of phosphoinositide 3-kinase and mitogen activated protein kinase. *J Neurosci Res.* 63:486–499.

Kinnunen KM, Greenwood R, Powell JH, Leech R, Hawkins PC, Bonnelle V et al. (2011). White matter damage and cognitive impairment after traumatic brain injury. *Brain.* 134:449–463.

Koltai E, Zhao Z, Lacza Z, Cselenyak A, Vacz G, Nyakas C et al. (2011). Combined exercise and insulin-like growth factor-1 supplementation induces neurogenesis in old rats, but do not attenuate age-associated DNA damage. *Rejuv Res.* 14:585–596.

Koopmans GC, Brans M, Gomez-Pinilla F, Duis S, Gispen WH, Torres-Aleman I et al. (2006). Circulating insulin-like growth factor I and functional recovery from spinal cord injury under enriched housing conditions. *Eur J Neurosci.* 23:1035–1046.

Kraus MF, Susmaras T, Caughlin BP, Walker CJ, Sweeney JA, and Little DM. (2007). White matter integrity and cognition in chronic traumatic brain injury: A diffusion tensor imaging study. *Brain.* 130:2508–2519.

Kreipke CW, Morgan R, Kallakuri S, and Rafols JA. (2007). Behavioral pre-conditioning enhances angiogenesis and cognitive outcome after brain trauma. *Neurol Res.* 29:388–394.

Laursen T, Jorgensen JO, Jakobsen G, Hansen BL, and Christiansen JS. (1995). Continuous infusion versus daily injections of growth hormone (GH) for 4 weeks in GH-deficient patients. *J Clin Endocrinol Metab.* 80:2410–2418.

Leinninger GM, Backus C, Uhler MD, Lentz SI, and Feldman EL. (2004a). Phosphatidylinositol 3-kinase and Akt effectors mediate insulin-like growth factor-I neuroprotection in dorsal root ganglia neurons. *Faseb J.* 18:1544–1546.

Leinninger GM, Russell JW, van Golen CM, Berent A, and Feldman EL. (2004b). Insulin-like growth factor-I regulates glucose-induced mitochondrial depolarization and apoptosis in human neuroblastoma. *Cell Death Differ.* 11:885–896.

Li W, Yang SY, Hu ZF, Winslet MC, Wang W, and Seifalian AM. (2009). Growth factors enhance endothelial progenitor cell proliferation under high-glucose conditions. *Med Sci Monit.* 15:BR357–363.

Lin S, Fan LW, Rhodes PG, and Cai Z. (2009). Intranasal administration of IGF-1 attenuates hypoxic-ischemic brain injury in neonatal rats. *Exp Neurol.* 217:361–370.

Liquitaya-Montiel A, Aguilar-Arredondo A, Arias C, and Zepeda A. (2012). Insulin growth factor-I promotes functional recovery after a focal lesion in the dentate gyrus. *CNS Neurol Disorders Drug Targets.* 11:818–828.

Liu MC, Akle V, Zheng W, Kitlen J, O'Steen B, Larner SF et al. (2006). Extensive degradation of myelin basic protein isoforms by calpain following traumatic brain injury. *J Neurochem.* 98:700–712.

Lopez-Lopez C, LeRoith D, and Torres-Aleman I. (2004). Insulin-like growth factor I is required for vessel remodeling in the adult brain. *Proc Natl Acad Sci U S A.* 101:9833–9838.

Lotocki G, de Rivero Vaccari JP, Alonso O, Molano JS, Nixon R, Safavi P et al. (2011). Oligodendrocyte vulnerability following traumatic brain injury in rats. *Neurosci Lett.* 499:143–148.

Lu D, Goussev A, Chen J, Pannu P, Li Y, Mahmood A, and Chopp M. (2004). Atorvastatin reduces neurological deficit and increases synaptogenesis, angiogenesis, and neuronal survival in rats subjected to traumatic brain injury. *J Neurotrauma.* 21:21–32.

Lu X, Kambe F, Cao X, Yamauchi M, and Seo H. (2008). Insulin-like growth factor-I activation of Akt survival cascade in neuronal cells requires the presence of its cognate receptor in caveolae. *Exp Cell Res.* 314:342–351.

Lu XC, Chen RW, Yao C, Wei H, Yang X, Liao Z et al. (2009a). NNZ-2566, a glypromate analog, improves functional recovery and attenuates apoptosis and inflammation in a rat model of penetrating ballistic-type brain injury. *J Neurotrauma.* 26:141–154.

Lu XC, Si Y, Williams AJ, Hartings JA, Gryder D, and Tortella FC. (2009b). NNZ-2566, a glypromate analog, attenuates brain ischemia-induced non-convulsive seizures in rats. *J Cereb Blood Flow Metab.* 29:1924–1932.

Lunn JS, Pacut C, Backus C, Hong Y, Johe K, Hefferan M, Marsala M et al. (2010). The pleotrophic effects of insulin-like growth factor-I on human spinal cord neural progenitor cells. *Stem Cells Develop.* 19:1983–1993.

Madathil SK, Carlson SW, Brelsfoard JM, Ye P, D'Ercole AJ, and Saatman KE. (2013). Astrocyte-specific overexpression of insulin-like growth factor-1 protects hippocampal neurons and reduces behavioral deficits following traumatic brain injury in mice. *PLoS One.* 8:e67204.

Madathil SK, Evans HN and Saatman KE. (2010). Temporal and regional changes in IGF-1/IGF-1R signaling in the mouse brain after traumatic brain injury. *J Neurotrauma.* 27:95–107.

Mason JL, Ye P, Suzuki K, D'Ercole AJ, and Matsushima GK. (2000). Insulin-like growth factor-1 inhibits mature oligodendrocyte apoptosis during primary demyelination. *J Neurosci.* 20:5703–5708.

Mauras N, Horber FF, and Haymond MW. (1992). Low dose recombinant human insulin-like growth factor-I fails to affect protein anabolism but inhibits islet cell secretion in humans. *J Clin Endocrinol Metab.* 75:1192–1197.

McIntosh TK, Juhler M, Raghupathi R, Saatman K, and Smith D. (1999). Secondary brain injury: Neurochemical and cellular mediators. In Marion DW, editor. *Traumatic Brain Injury.* Thienne Medical Publishers: New York. pp. 39–54.

McMorris FA, Mozell RL, Carson MJ, Shinar Y, Meyer RD, and Marchetti N. (1993). Regulation of oligodendrocyte development and central nervous system myelination by insulin-like growth factors. *Ann N Y Acad Sci.* 692:321–334.

Miltiadous P, Stamatakis A, Koutsoudaki PN, Tiniakos DG, and Stylianopoulou F. (2011). IGF-I ameliorates hippocampal neurodegeneration and protects against cognitive deficits in an animal model of temporal lobe epilepsy. *Exp Neurol.* 231:223–235.

Miltiadous P, Stamatakis A, and Stylianopoulou F. (2010). Neuroprotective effects of IGF-I following kainic acid-induced hippocampal degeneration in the rat. *Cell Mol Neurobiol.* 30:347–360.

Mitschelen M, Yan H, Farley JA, Warrington JP, Han S, Herenu CB et al. (2011). Long-term deficiency of circulating and hippocampal insulin-like growth factor I induces depressive behavior in adult mice: A potential model of geriatric depression. *Neuroscience.* 185:50–60.

Moro N, Ghavim S, Harris NG, Hovda DA, and Sutton RL. (2013). Glucose administration after traumatic brain injury improves cerebral metabolism and reduces secondary neuronal injury. *Brain Res.* 1535:124–136.

Nakaguchi K, Jinnou H, Kaneko N, Sawada M, Hikita T, Saitoh S et al. (2012). Growth factors released from gelatin hydrogel microspheres increase new neurons in the adult mouse brain. *Stem Cells Int.* 2012:915160.

Namura S, Nagata I, Kikuchi H, Andreucci M, and Alessandrini A. (2000). Serine-threonine protein kinase Akt does not mediate ischemic tolerance after global ischemia in the gerbil. *J Cereb Blood Flow Metab.* 20:1301–1305.

Ni W, Rajkumar K, Nagy JI, and Murphy LJ. (1997). Impaired brain development and reduced astrocyte response to injury in transgenic mice expressing IGF binding protein-1. *Brain Res.* 769:97–107.

Niblock MM, Brunso-Bechtold JK, and Riddle DR. (2000). Insulin-like growth factor I stimulates dendritic growth in primary somatosensory cortex. *J Neurosci.* 20:4165–4176.

Nordqvist AC, Holmin S, Nilsson M, Mathiesen T, and Schalling M. (1997). MK-801 inhibits the cortical increase in IGF-1, IGFBP-2 and IGFBP-4 expression following trauma. *Neuroreport.* 8:455–460.

Noshita N, Lewen A, Sugawara T, and Chan PH. (2002). Akt phosphorylation and neuronal survival after traumatic brain injury in mice. *Neurobiol Dis.* 9:294–304.

O'Kusky JR, Ye P, and D'Ercole AJ. (2000). Insulin-like growth factor-I promotes neurogenesis and synaptogenesis in the hippocampal dentate gyrus during postnatal development. *J Neurosci.* 20:8435–8442.

Oberbauer AM. (2013). The regulation of IGF-1 gene transcription and splicing during development and aging. *Front Endocrinol.* 4:39.

Olivecrona Z, Dahlqvist P, and Koskinen L-OD. (2013). Acute neuro-endocrine profile and prediction of outcome after severe brain injury. *Scand J Trauma Resusc Emerg Med.* 21.

Ozdemir D, Baykara B, Aksu I, Kiray M, Sisman AR, Cetin F et al. (2012). Relationship between circulating IGF-1 levels and traumatic brain injury-induced hippocampal damage and cognitive dysfunction in immature rats. *Neurosci Lett.* 507:84–89.

Palacios N, Sanchez-Franco F, Fernandez M, Sanchez I, and Cacicedo L. (2005). Intracellular events mediating insulin-like growth factor I-induced oligodendrocyte development: Modulation by cyclic AMP. *J Neurochem.* 95:1091–1107.

Parrizas M, Saltiel AR, and LeRoith D. (1997). Insulin-like growth factor 1 inhibits apoptosis using the phosphatidylinositol 3'-kinase and mitogen-activated protein kinase pathways. *J Biol Chem.* 272:154–161.

Perez-Martin M, Cifuentes M, Grondona JM, Lopez-Avalos MD, Gomez-Pinedo U, Garcia-Verdugo JM et al. (2010). IGF-I stimulates neurogenesis in the hypothalamus of adult rats. *Eur J Neurosci.* 31:1533–1548.

Popovic V, Pekic S, Pavlovic D, Maric N, Jasovic-Gasic M, Djurovic B et al. (2004). Hypopituitarism as a consequence of traumatic brain injury (TBI) and its possible relation with cognitive disabilities and mental distress. *J Endocrinol Invest.* 27:1048–1054.

Povlishock JT, Buki A, Koiziumi H, Stone J, and Okonkwo DO. (1999). Initiating mechanisms involved in the pathobiology of traumatically induced axonal injury and interventions targeted at blunting their progression. *Acta Neurochir Suppl.* 73:15–20.

Quesada A, Lee BY, and Micevych PE. (2008). PI3 kinase/Akt activation mediates estrogen and IGF-1 nigral DA neuronal neuroprotection against a unilateral rat model of Parkinson's disease. *Develop Neurobiol.* 68:632–644.

Robertson CL, Saraswati M, Scafidi S, Fiskum G, Casey P, and McKenna MC. (2013). Cerebral glucose metabolism in an immature rat model of pediatric traumatic brain injury. *J Neurotrauma.* 30:2066–2072.

Robles JF, Navarro JEV, Maglinao MLD, Matawaran BJ, Andag-Silva AA et al. (2009). Traumatic injury to the brain and endocrine evaluation of the anterior pituitary a year after the event (The TRIBE Study). *Int J Endocrinol Metab.* 7:72–81.

Rockich KT, Hatton JC, Kryscio RJ, Young BA, and Blouin RA. (1999). Effect of recombinant human growth hormone and insulin-like growth factor-1 administration on IGF-1 and IGF-binding protein-3 levels in brain injury. *Pharmacotherapy.* 19:1432–1436.

Rola R, Mizumatsu S, Otsuka S, Morhardt DR, Noble-Haeusslein LJ, Fishman K et al. (2006). Alterations in hippocampal neurogenesis following traumatic brain injury in mice. *J Biol Chem.* 202:189–199.

Rosenfeld RG, Hwa V, Wilson L, Lopez-Bermejo A, Buckway C, Burren C et al. (1999). The insulin-like growth factor binding protein superfamily: New perspectives. *Pediatrics.* 104:1018–1021.

Rovet JF and Ehrlich RM. (1999). The effect of hypoglycemic seizures on cognitive function in children with diabetes: A 7-year prospective study. *J Pediatr.* 134:503–506.

Royo NC, Conte V, Saatman KE, Shimizu S, Belfield CM, Soltesz KM et al. (2006). Hippocampal vulnerability following traumatic brain injury: A potential role for neurotrophin-4/5 in pyramidal cell neuroprotection. *Eur J Neurosci.* 23:1089–1102.

Rubovitch V, Edut S, Sarfstein R, Werner H, and Pick CG. (2010). The intricate involvement of the Insulin-like growth factor receptor signaling in mild traumatic brain injury in mice. *Neurobiol Dis.* 38:299–303.

Rubovitch V, Shachar A, Werner H, and Pick CG. (2011). Does IGF-1 administration after a mild traumatic brain injury in mice activate the adaptive arm of ER stress? *Neurochem Int.* 58:443–446.

Russell JW, Windebank AJ, Schenone A, and Feldman EL. (1998). Insulin-like growth factor-I prevents apoptosis in neurons after nerve growth factor withdrawal. *J Neurobiol.* 36:455–467.

Russo VC, Gluckman PD, Feldman EL, and Werther GA. (2005). The insulin-like growth factor system and its pleiotropic functions in brain. *Endocr Rev.* 26:916–943.

Russo VC, Kobayashi K, Najdovska S, Baker NL, and Werther GA. (2004). Neuronal protection from glucose deprivation via modulation of glucose transport and inhibition of apoptosis: A role for the insulin-like growth factor system. *Brain Res.* 1009:40–53.

Saatman KE, Contreras PC, Smith DH, Raghupathi R, McDermott KL, Fernandez SC et al. (1997). Insulin-like growth factor-1 (IGF-1) improves both neurological motor and cognitive outcome following experimental brain injury. *Exp Neurol.* 147:418–427.

Saatman KE, Serbest G, and Burkhardt MF. 2009. Axonal damage due to traumatic brain injury. In: Lajtha A, editor. *Handbook of Neurochemistry and Molecular Neurobiology*. Kluwer Academic/Plenum Publishers: Berlin. pp. 343–361.

Saenger S, Holtmann B, Nilges MR, Schroeder S, Hoeflich A, Kletzl H et al. (2012). Functional improvement in mouse models of familial amyotrophic lateral sclerosis by PEGylated insulin-like growth factor I treatment depends on disease severity. *Amyotroph Lat Sclerosis.* 13:418–429.

Sandberg Nordqvist AC, von HH, Holmin S, Sara VR, Bellander BM, and Schalling M. (1996). Increase of insulin-like growth factor (IGF)-1, IGF binding protein-2 and -4 mRNAs following cerebral contusion. *Brain Res Mol Brain Res.* 38:285–293.

Sanus GZ, Tanriverdi T, Coskun A, Hanimoglu H, Is M, and Uzan M. (2007). Cerebrospinal fluid and serum levels of insulin-like growth factor-1 and insulin-like growth factor binding protein-3 in patients with severe head injury. *Turk J Trauma Emerg Surg.* 13:281–287.

Sara VR, Carlsson-Skwirut C, Andersson C, Hall E, Sjogren B, Holmgren A, and Jornvall H. (1986). Characterization of somatomedins from human fetal brain: Identification of a variant form of insulin-like growth factor I. *Proc Natl Acad Sci U S A.* 83:4904–4907.

Sara VR, Carlsson-Skwirut C, Bergman T, Jornvall H, Roberts PJ, Crawford M et al. (1989). Identification of Gly-Pro-Glu (GPE), the aminoterminal tripeptide of insulin-like growth factor 1 which is truncated in brain, as a novel neuroactive peptide. *Biochem Biophys Res Commun.* 165:766–771.

Scafidi S, O'Brien J, Hopkins I, Robertson C, Fiskum G, and McKenna M. (2009). Delayed cerebral oxidative glucose metabolism after traumatic brain injury in young rats. *J Neurochem.* 109 Suppl 1:189–197.

Schabitz WR, Hoffmann TT, Heiland S, Kollmar R, Bardutzky J, Sommer C, and Schwab S. (2001). Delayed neuroprotective effect of insulin-like growth factor-I after experimental transient focal cerebral ischemia monitored with MRI. *Stroke.* 32:1226–1233.

Schober ME, Block B, Beachy JC, Statler KD, Giza CC, and Lane RH. (2010). Early and sustained increase in the expression of hippocampal IGF-1, but not EPO, in a developmental rodent model of traumatic brain injury. *J Neurotrauma.* 27:2011–2020.

Schober ME, Ke XR, Xing BH, Block BP, Requena DF, McKnight R et al. (2012). Traumatic brain injury increased IGF-1B mRNA and altered IGF-1 exon 5 and promoter region epigenetic characteristics in the rat pup hippocampus. *J Neurotrauma.* 29:2075–2085.

Schoch KM, Madathil SK, and Saatman KE. (2012). Genetic manipulation of cell death and neuroplasticity pathways in traumatic brain injury. *Neurotherapeutics.* 9:323–337.

Shavali S, Ren J, and Ebadi M. (2003). Insulin-like growth factor-1 protects human dopaminergic SH-SY5Y cells from salsolinol-induced toxicity. *Neurosci Lett.* 340:79–82.

Shiraishi M, Tanabe A, Saito N, and Sasaki Y. (2006). Unphosphorylated MARCKS is involved in neurite initiation induced by insulin-like growth factor-I in SH-SY5Y cells. *J Cell Physiol.* 209:1029–1038.

Skold MK, Risling M, and Holmin S. (2006). Inhibition of vascular endothelial growth factor receptor 2 activity in experimental brain contusions aggravates injury outcome and leads to early increased neuronal and glial degeneration. *Eur J Neurosci.* 23:21–34.

Skold MK, von Gertten C, Sandberg-Nordqvist AC, Mathiesen T, and Holmin S. (2005). VEGF and VEGF receptor expression after experimental brain contusion in rat. *J Neurotrauma.* 22:353–367.

Sonntag WE, Lynch CD, Cooney PT, and Hutchins PM. (1997). Decreases in cerebral microvasculature with age are associated with the decline in growth hormone and insulin-like growth factor 1. *Endocrinology.* 138:3515–3520.

Stangel M and Hartung HP. (2002). Remyelinating strategies for the treatment of multiple sclerosis. *Progress Neurobiol.* 68:361–376.

Stein DG and Hoffman SW. (2003). Concepts of CNS plasticity in the context of brain damage and repair. *J Head Trauma Rehabil.* 18:317–341.

Stohr O, Hahn J, Moll L, Leeser U, Freude S, Bernard C et al. (2011). Insulin receptor substrate-1 and -2 mediate resistance to glucose-induced caspase-3 activation in human neuroblastoma cells. *Biochim Biophys Acta.* 1812:573–580.

Stone JR, Okonkwo DO, Dialo AO, Rubin DG, Mutlu LK, Povlishock JT et al. (2004). Impaired axonal transport and altered axolemmal permeability occur in distinct populations of damaged axons following traumatic brain injury. *J Biol Chem.* 190:59–69.

Su E, Bell MJ, Kochanek PM, Wisniewski SR, Bayir H, Clark RS et al. (2012). Increased CSF concentrations of myelin basic protein after TBI in infants and children: Absence of significant effect of therapeutic hypothermia. *Neurocrit Care.* 17:401–407.

Subramaniam S, Shahani N, Strelau J, Laliberte C, Brandt R, Kaplan D, and Unsicker K. (2005). Insulin-like growth factor 1 inhibits extracellular signal-regulated kinase to promote neuronal survival via the phosphatidylinositol 3-kinase/protein kinase A/c-Raf pathway. *J Neurosci.* 25: 2838–2852.

Sun LY and D'Ercole AJ. (2006). Insulin-like growth factor-I stimulates histone H3 and H4 acetylation in the brain in vivo. *Endocrinology.* 147:5480–5490.

Sussenbach JS, Steenbergh PH, and Holthuizen P. (1992). Structure and expression of the human insulin-like growth factor genes. *Growth Regul.* 2:1–9.

Thau-Zuchman O, Shohami E, Alexandrovich AG, and Leker RR. (2010). Vascular endothelial growth factor increases neurogenesis after traumatic brain injury. *J Cereb Blood Flow Metab.* 30:1008–1016.

Traub ML, De Butte-Smith M, Zukin RS, and Etgen AM. (2009). Oestradiol and insulin-like growth factor-1 reduce cell loss after global ischaemia in middle-aged female rats. *J Neuroendocrinol.* 21:1038–1044.

Trejo JL, Carro E, Lopez-Lopez C, and Torres-Aleman I. (2004). Role of serum insulin-like growth factor I in mammalian brain aging. *Growth Horm IGF Res.* 14 Suppl A:S39–43.

Turkalj I, Keller U, Ninnis R, Vosmeer S, and Stauffacher W. (1992). Effect of increasing doses of recombinant human insulin-like growth factor-I on glucose, lipid, and leucine metabolism in man. *J Clin Endocrinol Metab.* 75:1186–1191.

Verleysdonk S, Hirschner W, Wellard J, Rapp M, de los Angeles Garcia M, Nualart F et al. (2004). Regulation by insulin and insulin-like growth factor of 2-deoxyglucose uptake in primary ependymal cell cultures. *Neurochem Res.* 29:127–134.

Wagner J, Dusick JR, McArthur DL, Cohan P, Wang C, Swerdloff R, Boscardin WJ, and Kelly DF. (2010). Acute gonadotroph and somatotroph hormonal suppression after traumatic brain injury. *J Neurotrauma.* 27:1007–1019.

Walter HJ, Berry M, Hill DJ, and Logan A. (1997). Spatial and temporal changes in the insulin-like growth factor (IGF) axis indicate autocrine/paracrine actions of IGF-I within wounds of the rat brain. *Endocrinology.* 138:3024–3034.

Wei HH, Lu XC, Shear DA, Waghray A, Yao C, Tortella FC, and Dave JR. (2009). NNZ-2566 treatment inhibits neuroinflammation and pro-inflammatory cytokine expression induced by experimental penetrating ballistic-like brain injury in rats. *J Neuroinflamm.* 6:19.

Wieloch T and Nikolich K. (2006). Mechanisms of neural plasticity following brain injury. *Curr Opin Neurobiol.* 16:258–264.

Wildburger R, Zarkovic N, Leb G, Borovic S, Zarkovic K, and Tatzber F. (2001). Post-traumatic changes in insulin-like growth factor type 1 and growth hormone in patients with bone fractures and traumatic brain injury. *Wien Klin Wochenschr.* 113:119–126.

Winston CN, Chellappa D, Wilkins T, Barton DJ, Washington PM, Loane DJ et al. (2013). Controlled cortical impact results in an extensive loss of dendritic spines that is not mediated by injury-induced amyloid-beta accumulation. *J Neurotrauma.* 30:1966–1972.

Won SJ, Yoo BH, Kauppinen TM, Choi BY, Kim JH, Jang BG et al. (2012). Recurrent/moderate hypoglycemia induces hippocampal dendritic injury, microglial activation, and cognitive impairment in diabetic rats. *J Neuroinflamm.* 9:182.

Xing G, Ren M, Watson WD, O'Neill JT, and Verma A. (2009). Traumatic brain injury-induced expression and phosphorylation of pyruvate dehydrogenase: A mechanism of dysregulated glucose metabolism. *Neurosci Lett.* 454:38–42.

Yang X, Wei A, Liu Y, He G, Zhou Z, Yu Z. (2013). IGF-1 protects retinal ganglion cells from hypoxia-induced apoptosis by activating the Erk-1/2 and Akt pathways. *Mol Vision.* 19:1901–1912.

Yano S, Morioka M, Fukunaga K, Kawano T, Hara T, Kai Y et al. (2001). Activation of Akt/protein kinase B contributes to induction of ischemic tolerance in the CA1 subfield of gerbil hippocampus. *J Cereb Blood Flow Metab.* 21:351–360.

Yao DL, Liu X, Hudson LD, and Webster HD. (1995). Insulin-like growth factor I treatment reduces demyelination and up-regulates gene expression of myelin-related proteins in experimental autoimmune encephalomyelitis. *Proc Natl Acad Sci U S A.* 92:6190–6194.

Yao DL, Liu X, Hudson LD, and Webster HD. (1996). Insulin-like growth factor-I given subcutaneously reduces clinical deficits, decreases lesion severity and upregulates synthesis of myelin proteins in experimental autoimmune encephalomyelitis. *Life Sci.* 58:1301–1306.

Ye P, Carson J, and D'Ercole AJ. (1995). In vivo actions of insulin-like growth factor-1 (IGF-1) on brain myelination: Studies of IGF-1 and IGF binding protein-1 (IGFBP-1) transgenic mice. *J Neurosci.* 15:7344–7356.

Ye P and D'Ercole AJ. (2006). Insulin-like growth factor actions during development of neural stem cells and progenitors in the central nervous system. *J Neurosci Res.* 83:1–6.

Ye P, Hu Q, Liu H, Yan Y, and D'Ercole A J. (2010). beta-catenin mediates insulin-like growth factor-I actions to promote cyclin D1 mRNA expression, cell proliferation and survival in oligodendroglial cultures. *Glia.* 58:1031–1041.

Ye P, Li L, Richards RG, DiAugustine RP, and D'Ercole AJ. (2002). Myelination is altered in insulin-like growth factor-I null mutant mice. *J Neurosci.* 22:6041–6051.

Ye P, Popken GJ, Kemper A, McCarthy K, Popko B, and D'Ercole AJ. (2004). Astrocyte-specific overexpression of insulin-like growth factor-I promotes brain overgrowth and glial fibrillary acidic protein expression. *J Neurosci Res.* 78:472–484.

Yoshino A, Hovda DA, Kawamata T, Katayama Y, and Becker DP. (1991). Dynamic changes in local cerebral glucose utilization following cerebral concussion in rats: Evidence of a hyper- and subsequent hypometabolic state. *Brain Res.* 561:106–119.

Yu TS, Zhang G, Liebl DJ, and Kernie SG. (2008). Traumatic brain injury-induced hippocampal neurogenesis requires activation of early nestin-expressing progenitors. *J Neurosci.* 28:12901–12912.

Zgaljardic DJ, Guttikonda S, Grady JJ, Gilkison CR, Mossberg KA, High WM, Jr. et al. (2011). Serum IGF-1 concentrations in a sample of patients with traumatic brain injury as a diagnostic marker of growth hormone secretory response to glucagon stimulation testing. *Clin Endocrinol.* 74:365–369.

Zhang X, Chen Y, Ikonomovic MD, Nathaniel PD, Kochanek PM, Marion DW et al. (2006). Increased phosphorylation of protein kinase B and related substrates after traumatic brain injury in humans and rats. *J Cereb Blood Flow Metab.* 26:915–926.

Zheng WH, Kar S, Dore S, and Quirion R. (2000). Insulin-like growth factor-1 (IGF-1): A neuroprotective trophic factor acting via the Akt kinase pathway. *J Neural Transm Suppl.* 261–272.

Zheng WH, Kar S, and Quirion R. (2002). Insulin-like growth factor-1-induced phosphorylation of transcription factor FKHRL1 is mediated by phosphatidylinositol 3-kinase/Akt kinase and role of this pathway in insulin-like growth factor-1-induced survival of cultured hippocampal neurons. *Mol Pharmacol.* 62:225–233.

Zhu W, Fan Y, Frenzel T, Gasmi M, Bartus RT, Young WL et al. (2008). Insulin growth factor-1 gene transfer enhances neurovascular remodeling and improves long-term stroke outcome in mice. *Stroke.* 39:1254–1261.

Zhu W, Fan Y, Hao Q, Shen F, Hashimoto T, Yang GY et al. (2009). Postischemic IGF-1 gene transfer promotes neurovascular regeneration after experimental stroke. *J Cereb Blood Flow Metab.* 29:1528–1537.

8 Microglia in Experimental Brain Injury
Implications on Neuronal Injury and Circuit Remodeling

Megan N. Evilsizor, Helen F. Ray-Jones, Timothy W. Ellis Jr., Jonathan Lifshitz, and Jenna M. Ziebell

CONTENTS

8.1 INTRODUCTION

Neurons in the brain are supported by glial cells, including astrocytes and microglia. Together, these cells establish and maintain functional circuits. Often overlooked, microglia function as the immune cells of the central nervous system, monitoring the microenvironment for changes in signaling, pathogens and injury. This chapter discusses the various roles of microglia in the healthy and in diseased brain with a focus on traumatic brain injury. Within the healthy brain, ramified microglia constantly survey the microenvironment playing roles in neurotransmission and maintenance of synaptic integrity. In injury, microglia may interact with neurons to mediate the transition between injury-induced circuit dismantling and subsequent reorganization. Increased understanding of microglial roles could identify therapeutic targets to mitigate the consequences of neurological diseases.

8.1.1 Microglia: Immune Cells of the Central Nervous System

All organs in the body contain resident populations of innate immune cells, such as macrophages and dendritic cells. These cells play a vital role in both the regulation of non-specific inflammation and antigen-specific adaptive immune responses. Within the central nervous system (CNS), microglia are the resident immune cells (Gehrmann et al., 1995; Hanisch, 2002; Lawson et al., 1990). The name is fitting because they are one of the smallest cell types in the brain (Lawson et al., 1990); however, microglia perform essential functions in the healthy and diseased brain. As the resident immune cells of the CNS, microglia contribute to the overall health of nervous tissue. They are the first sentinels of infection and have a role in both the innate and adaptive immune responses of the CNS.

During development microglial precursors undergo three major developmental milestones toward becoming fully integrated microglia. Microglia proliferate and migrate to populate different CNS regions, and then differentiate from the amoeboid microglia into their ramified morphology. Microglia are critical CNS cells with vital roles in both brain development and maintenance. They are also crucial players in synaptic remodeling and clearance of extracellular debris (Domercq et al., 2013; Graeber and Streit, 2010).

Microglia exist in different morphological states and constitute 10–20% of the glial content in the CNS (Lawson et al., 1990). The morphology of microglia is interpreted to reflect their function (Aguzzi et al., 2013). Microglia are typically described in four morphological states: ramified, hyperramified/bushy, active, and amoeboid (Figure 8.1). A fifth morphology is known to exist, rod microglia. The proposed functions for each microglial morphology are discussed in turn, below. However, little research has been done regarding the function of rod microglia (Cao et al., 2012; Graeber, 2010; Graeber and Mehraein,1994; Graeber and Streit,2010; Ziebell et al., 2012).

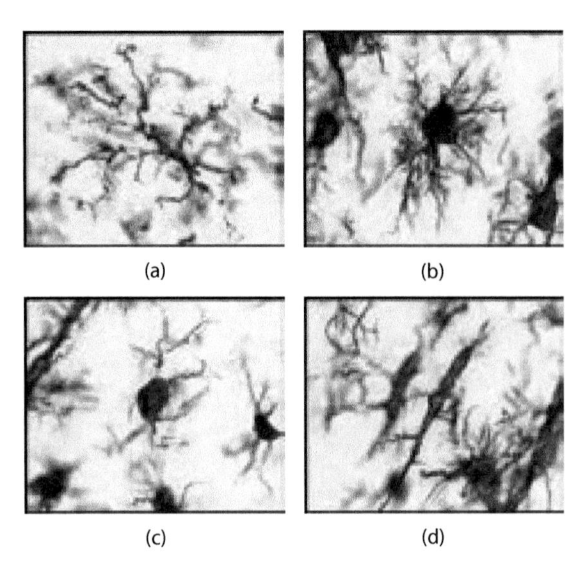

FIGURE 8.1 Morphologies of microglia. Ramified microglia (a) are found throughout the nonpathological CNS; once activated they become hyperramified or bushy (b). At this stage, they begin to retract their processes because activation continues the number of processes decreases (c). When fully activated, they are indistinguishable from bloodborne macrophages. Another morphology of microglia is the rod microglia (d).

Ramified microglia are highly dynamic, continuously monitoring the microenvironment by extension and retraction of their processes (Graeber and Streit, 2010). Ramified microglia may play a role in neurotransmission, particularly synaptic plasticity after injury (Graeber and Streit, 2010). Microglia are being investigated regarding maintenance of synaptic integrity by removing defunct axon terminals while helping neuronal connections stay intact (Tremblay, 2011; Tremblay and Majewska, 2011; Tremblay et al., 2011). The integral role can be attributed, in part, to cell receptors found on the cell surface, several of which take part in adenosine triphosphate and glutamate regulation. Damage to synapses has been shown to elicit changes in the surveying microglia. After these changes, microglia take on a protective, inflammatory role by dividing and phagocytosing cellular debris (Domercq et al., 2013). Ramified microglia are quickly activated after challenges with various stimuli. With activation, microglia proliferate, migrate to the site of insult, and show characteristic morphological alterations, which include the retraction and thickening of processes as well as immunophenotypical and functional change (Kreutzberg, 1996). Fully activated microglia are indistinguishable from bloodborne macrophages and possess an amoeboid morphology. These cells may contribute to brain inflammation by phagocytosing debris, releasing cytotoxins, killing neighboring cells, and secreting growth factors (Giulian, 1987).

Importantly, subsets of microglia are known to exist after injury (Jin et al., 2012). The stimuli that activate microglia subsequently influence receptor expression, which in turn defines their role in injury and repair. For example, M1 microglia are typically designated as proinflammatory in their responses, whereas M2 are activated via anti-inflammatory pathways and act to reduce inflammation.

Lipopolysaccharide (LPS) and interferon gamma promote classically activated microglia that have an M1 phenotype, which produce high levels of proinflammatory cytokines and oxidative metabolites that are essential for host defense and phagocytic activity, but also cause damage to healthy cells and tissue. On the other hand, activation of microglia by interleukin-4 (IL-4) or IL-13 promotes an alternatively activated M2 phenotype. Limited data on resident M2 microglia are available, though it is thought they act similarly to the M2 macrophages. M2 cells can promote angiogenesis, wound healing and tissue repair, and extracellular matrix remodeling as well as suppress immune responses (Colton, 2009). After injury to the brain, it is likely that both M1 and M2 microglia exist in a state of dynamic equilibrium within the injured microenvironment. Whether these cells differentiate into an M1 phenotype that exacerbates tissue injury or into an M2 phenotype that promotes repair depends on local signals within the damaged tissue.

Intracellular signaling responsible for microglia activation is mediated by immune related receptors. Toll-like receptors, scavenger receptors, and cytokine and chemokine receptors all play a role in the neuroimmune response (Kierdorf and Prinz, 2013). The transition from the ramified or surveying phenotype to an activated one is regulated by several factors. Intrinsic factors including runx-1, Irf8, and Pu.1 control the lifecycle and replication of microglia. Runx1 regulates proliferation, transition, and homeostasis of microglia while also contributing to feedback loops regarding other intrinsic factors like Pu.1 (Kierdorf and Prinz, 2013). Pu.1 is regarded as the key intrinsic factor for microglial development and maturation (Scott et al., 1994). It influences Irf8, a transcription factor needed to continue microglial replication. Extrinsic factors such as CD200, CX3CR1, and TREM2 can play a role in both activation and the prevention of activation (Kierdorf and Prinz, 2013). For example, high levels of secreted CX3CL1 are found in healthy individuals. For microglia to remain in a noninflammatory state, they must possess a large number of, or be able to readily upregulate, CX3CR1. It was shown that microglia deficient in this specific receptor remain overactivated after injury (Liu et al., 2008).

8.2 INFLAMMATION IN THE CNS

Inflammation within the CNS, particularly the brain, is an important pathological feature of acute brain injury, immune-mediated disease, and chronic neurodegenerative conditions. In traumatic brain injury (TBI), neuroinflammation is characterized by expression of inflammatory mediators such as cytokines and chemokines as well as glial activation (Perry et al., 2003). Besides blood-derived leukocytes, there are two major contributors to immune responses in the CNS: macroglia and astroglia. Both cell types secrete a cytokines and are important in maintaining homeostasis of the nervous system.

8.2.1 Cytokines

The inflammatory response is mediated by the release of pro- and anti-inflammatory cytokines. Cytokines are polypeptides that are barely detectable in healthy tissue and become rapidly upregulated in response to pathological or stressful challenges (Lucas et al., 2006; Rothwell, 1999). Besides inflammation, cytokines generally function as mediators of cellular communication (Wang and Shuaib, 2002) and play a pivotal role during embryonic development as well as maintenance of tissue homeostasis in the mature organism. Their action is initiated by binding to specific high-affinity cell-surface receptors on target cells. Small secretions of a cytokine can elicit a cascade of intracellular signals, resulting in an amplified and highly specific cellular response. There is significant overlap of function among different cytokines and therefore considerable redundancy (Allan and Rothwell, 2003).

Peripheral immune cells are major producers of cytokines, but microglia, astrocytes, and neurons are also capable of producing cytokines and eliciting an immune response within the CNS. It is thought that neuronally derived cytokines are involved in cellular communication. Cytokines secreted by microglia predominantly mediate neuronal growth and repair and may also direct chronic pathological changes associated with some neurodegenerative diseases (Fan et al., 1995). Under normal physiological conditions, the cytokines IL-6, IL-8, and IL-10 are present in both the plasma and CSF of healthy humans (Maier et al., 2005). Low levels of cytokines including IL-1α, IL-1β, IL-4, IL-6, and tumor necrosis factor (TNF) have also been detected in brain homogenates of healthy rodents (Bye et al., 2007; Fox et al., 2005).

Both pro- and anti-inflammatory cytokines and chemokines are produced in the CNS and may have restorative as well as detrimental effects (Morganti-Kossmann et al., 2007; Schmidt et al., 2004). Specific effects of a cytokine may depend on the target cell and the receptors it expresses. Moreover, cytokine signaling may have a direct effect on the cell or trigger production of additional cytokines or growth factors that could mask the effects of the original signaling (Allan and Rothwell, 2003).

8.2.1.1 Proinflammatory Cytokines

8.2.1.1.1 Interleukin-1 (IL-1)

The IL-1 family of cytokines includes two agonist proteins (IL-1α and IL-1β) and the IL-1 receptor antagonist (IL-1ra) as well as two receptors (IL-1RI and IL-1RII). IL-1 is an important initiator of the inflammatory response and therefore plays a key role in the host's reaction to pathogens. Because of the potency of its inflammatory effects, IL-1 activity is tightly regulated by a complex network of control systems. IL-1–converting enzyme is necessary for cleavage of proIL-1β into the biologically active molecule, IL-1β. Elevated IL-1β has been detected in the CSF and brain parenchyma within hours after brain injury in humans and rodents (Ross et al., 1994; Taupin et al., 1993; Woodroofe et al., 1991).

IL-1ra has been found to attenuate neuronal damage after ischemia or TBI in rats (Loddick and Rothwell, 1996; Relton and Rothwell, 1992). Mice deficient in IL-1–converting enzyme have been shown to have reduced brain damage in cerebral ischemia (Schielke et al., 1998). Increased expression of IL-1β in CSF post-TBI is associated with higher intracranial pressure and unfavorable outcome (reviewed in Woodcock and Morganti-Kossmann, 2013). Together, these data are compelling evidence for the involvement of IL-1 in neurodegeneration.

Although activated microglia are potent producers of IL-1 cytokines after brain injury (Giulian, 1987; Howe et al., 2006; Meme et al., 2006; Raivich et al., 1996; Raivich et al., 1999b), astrocytes (Acarin et al., 2000; Giulian et al., 1986) and neurons (Bandtlow et al., 1990; Tchelingerian et al., 1993) are also known producers. The administration of IL-1 markedly exacerbates neuronal damage following excitotoxic injury, ischemia, and TBI (Lawrence et al., 1998; Loddick and Rothwell, 1996; Relton and Rothwell, 1992; Rothwell, 1999; Taupin et al., 1993), indicating that balanced Il-1 signaling is required to optimize outcome.

8.2.1.1.2 Interleukin-6 (IL-6)

IL-6 is a pleiotropic cytokine that plays an important role in host defense (Hammacher et al., 1994) and has major regulatory effects upon the inflammatory response (Raivich et al., 1999a). In these roles, it displays neurotrophic properties (Benveniste, 1998) and negative effects on the CNS. IL-6 is produced by astrocytes (Aloisi et al., 2000; Morganti-Kossmann and Kossmann, 1995; Veerhuis et al., 2002), microglia (Veerhuis et al., 2002) and neurons (reviewed in Morganti-Kossmann et al., 2007) and has both direct and indirect neurotrophic effects on neurons (Benveniste, 1998).

IL-6 has been shown to increase between 3 and 8 hours post-TBI in rats (Taupin et al., 1993) and has been detected in brain homogenates in several experimental models of TBI (Bye et al., 2007; Shohami et al., 1994; Woodroofe et al., 1991; Ziebell et al., 2011; Ziebell and Morganti-Kossmann, 2010). Increased levels of IL-6 were also reported in human CSF, and to a lesser extent in serum, for up to 3 weeks after TBI (Kossmann et al., 1995).

Mice deficient in IL-6 showed decreased lymphocyte recruitment and reduced activation of astrocytes as well as a slight reduction in microglial activation after brain injury (Raivich et al., 1999b). However, microglia lack receptors for IL-6, indicating that these effects on microglia occur indirectly.

8.2.1.1.3 Tumor Necrosis Factor (TNF)

Brain-derived TNF is mostly synthesized by glial cells in response to pathological stimuli; however, it can also be produced by neurons (Breder et al., 1993). TNF can exert effects directly, but can also function indirectly by further induction of cytokines, namely IL-1 and IL-6. Widespread distribution of TNF receptors has been found in brain homogenates of various brain regions, with maximal levels detected in the brainstem (Kinouchi et al., 1991).

Indeed, TNF is one of several molecules that have been found to activate microglia. Injury-activated endogenous microglia and astrocytes are capable of producing IL-1 and TNF starting at 1 hour and gradually declining within

24 hours postinjury (Csuka et al., 2000). The development of transgenic and knockout animals has allowed more detailed roles for specific cytokines and chemokines to be determined. For example, mice deficient in TNF, IL-1β, or both subjected to a nitric oxide–induced neurotoxicity model showed exacerbated demyelination when TNF was absent (Turrin and Rivest, 2006). This study concluded that the early activation of microglia through TNF had a neuroprotective role, possibly by initiating the removal of debris. On the other hand, data from therapeutic inhibition and knockout studies suggest that TNF may have a toxic effect in the acute phase post-TBI. Experimental focal TBI models have shown that TNF is upregulated in the brain within a few hours of trauma (Knoblach et al., 1999; Shohami et al., 1994; Taupin et al., 1993). TNF was detected in CSF and plasma of human patients within 24 hours of severe head injury (Ross et al., 1994). The detrimental role of TNF has been demonstrated by animal studies whereby recombinant TNF injected into the brain induced cerebral inflammation, breakdown of the blood–brain barrier (BBB), and intracranial leukocyte recruitment (Kim et al., 1992; Ramilo et al., 1990).

The breakdown of the BBB is a common event in traumatic and ischemic brain injuries, and contributes to the transcription of inflammatory cytokines (Streit et al., 1998). Early transient messenger RNA (mRNA) expression of TNF, IL-1, and IL-6 has been shown to precede the appearance of TNF and IL-6 bioactivity in the injured brain regions (Shohami et al., 1997). In addition, the early expression of IL-1 and TNF mRNA occurs before the time of leukocyte infiltration to the injured site, which is usually observed between 24 and 72 hours posttrauma (Riva-Depaty et al., 1994). These data suggest that resident brain cells have the capacity to produce cytokines independent of peripheral immune cell activation.

In the chronic phase, however, when homeostatic and/or repair mechanisms should operate to regain function, the absence of TNF seems to be deleterious. Because of discrepancies across the literature, it is difficult to conclude whether TNF is neuroprotective, neurotoxic, or possesses a combination of both properties after TBI.

8.2.1.2 Anti-inflammatory Cytokines

8.2.1.2.1 Interleukin-4 (IL-4)

IL-4 shows anti-inflammatory effects within the CNS. It is a multifunctional cytokine (Hulshof et al., 2002), with its most well-defined feature being the ability to induce death of activated microglia (Yang et al., 2002). IL-4 levels increase for the first 24 hours after diffuse TBI in rats (Dalgard et al., 2012), which may prevent the sustained production of inflammatory factors including TNF and IL-1 (Howe et al., 2006; Meme et al., 2006). IL-4 also stimulates the synthesis of IL-1ra, the endogenous inhibitor of IL-1 (Dinarello, 1997).

The source of IL-4 within the CNS remains enigmatic because of the inability to consistently detect it without injury or insult. In vitro experiments on human cells have demonstrated a strong immunoreactivity of IL-4 with astrocytes but not microglia. These data were mirrored in brain tissue sections from patients suffering from multiple sclerosis (MS), where IL-4 staining was most intense in reactive astrocytes within areas of chronic lesions (Hulshof et al., 2002). The study concluded that IL-4 had an effect on astrogliosis rather than a specific pathological effect of MS because IL-4 expression was also observed in gliosis induced by cerebral ischemia. Other studies using a mechanical brain injury model demonstrated attenuation of astrocyte proliferation after administration of IL-4 (Estes et al., 1993). The strong immunostaining observed for IL-4 on astrocytic endfeet surrounding blood vessels in the normal white matter as well as in MS lesions suggests a potential role for IL-4 in the maintenance of the BBB (Hulshof et al., 2002). In support of this, IL-4 knockout mice have poorer outcomes than wild-type after experimental stroke (Xiong et al., 2011).

8.2.1.2.2 Interleukin-10 (IL10)

IL-10 is produced by microglia and astrocytes in the CNS and by lymphopoietic cells in the periphery (Aloisi et al., 1999; Mesples et al., 2003; Wu et al., 2005). IL-10 functions by inhibiting activation of microglia and astrocytes, thus decreasing production of proinflammatory cytokines (Knoblach et al., 1999; Kremlev and Palmer, 2005).

It has been proposed that IL-10 therapies should improve neuropathological outcome. After TBI in rats, systemic administration of IL-10 improves neurological recovery (Knoblach and Faden, 1998; Knoblach et al., 1999). Treatment with IL-10 also significantly attenuates the expression of TNF in the injured cortex and IL-1 in the injured hippocampus. Surprisingly, no protection was reported after IL-10 administration in piglets subjected to hypoxic-ischemic insult (Lyng et al., 2005). Indeed, high levels of IL-10 in CSF of children and adults with brain injury have been correlated with adverse outcome (Bell et al., 1997). The data from IL-10 studies suggest that inflammatory processes are required for the restoration of function.

8.2.1.3 Chemokines

Chemokines are chemo-attractant cytokines that serve as cues for cell migration. Chemokines are small proteins that direct leukocyte activation and migration (Lu et al., 2005), playing a fundamental role in development, homeostasis, and immune function. Generally, chemokines are classified into two groups based on the presence and position of conserved cysteine residues. CXC, or α-chemokines, act as neutrophil and lymphocyte chemoattractants whereas C-C, or β-chemokines (e.g., MIP-1α and MCP-1), are chemotactic for monocytes and T cells (Kielian and Hickey, 2000). KC and MIP-2 induce leukocyte recruitment into the CSF in a dose-dependent manner, predominantly influencing neutrophil counts (Zwijnenburg et al., 2003).

8.2.1.3.1 CCL2/monocyte chemo-attractant protein

The monocyte chemoattractant protein CCL2 (formerly MCP-1) is a member of the C-C family of chemokines that attracts monocytes both in vitro and in vivo (Yoshimura et al., 1989). Glabinski and colleagues found upregulation of

CCL2 transcripts within three hours of TBI, before inflammatory cells were evident in the brain tissue, suggesting that resident brain cells are an early source of CCL2 (Glabinski et al., 1996). Expression of CCL2 and its receptor CCR2 are rapidly increased in a wide range of cell types after acute brain injury (Banisadr et al., 2002; Rankine et al., 2006). Activated T cells, astrocytes, and, to a lesser extent, macrophages (or fully reactive microglia) have been found to express CCL2 after ischemia (Gourmala et al., 1997) and mechanical brain injury (Glabinski et al., 1996). On the other hand, its receptor, CCR2, has been shown to colocalize with activated microglia (Galasso et al., 2000), neurons, and astrocytes (Banisadr et al., 2002). In addition to monocyte recruitment, there is direct evidence for the involvement of CCL2 in driving acute inflammatory responses within the CNS. There exists further indication that CCL2 may contribute to the pathogenesis of brain lesion development (Galasso et al., 2000; Hughes et al., 2002; Kurihara et al., 1997; Kuziel et al., 1997; Lu et al., 1998).

8.2.1.3.2 CX3CL1/Fractalkine

CX3CL1, also referred to as fractalkine, is a chemokine responsible for deregulating microglial activity (Prinz and Priller, 2010). It is primarily expressed by neurons, and its receptor, CX3CR1, is present only on microglia (Mattison et al., 2013). The continued release of CX3CL1 from neurons is postulated to keep microglia inactive (Ransohoff and Perry, 2009). Additionally, CX3CL1 has been reported to reduce microglial toxicity and, consequently, neuronal damage after brain injury.

CX3CL1 increases within hours of diffuse TBI, and exhibits increased expression for 3 days in rat models of TBI (Dalgard et al., 2012). The resulting decrease in the release of inflammatory factors (specifically nitric oxide, IL-6, TNF) seems to have a neuroprotective effect in vitro (Mizuno et al., 2003). In models of neurodegeneration CX3CR1 deficiency has been linked to the worst outcomes, possibly because of the lack of fractalkine control of microglia leading to chronic inflammation (Iadecola and Anrather, 2011). On the other hand, experimental models of ischemia that have investigated CX3CR1-deficient mice reported early neuroprotection in the absence of the receptor for CX3CL1 (Fumagalli et al., 2013).

It must be noted that the evidence in vivo using CX3CR1-deficient mice is controversial. CX3CL1-CX3CR1 signaling seems to be neuroprotective in models of amyotrophic lateral sclerosis, Parkinson disease, and MS, but is detrimental in Alzheimer disease, stroke, and spinal cord injury models (Mattison et al., 2013).

8.3 MICROGLIAL ACTIVITY AFTER TBI

Microglial activation has long been documented to be associated with increased inflammation and adverse outcomes. However, microglial activation should not be linked solely to deleterious events. There are instances in which activated microglia may have a protective role in brain injury (Urrea et al., 2007). After injury, resident microglia play a role in tissue repair that is similar to that described for resident macrophages in peripheral organs. Synergistic effects of microglia and astrocytes are needed for tissue remodeling after lesions, as they contribute to the reestablishment of the BBB, thus limiting the invasion of immune cells as well as secreting anti-inflammatory cytokines to suppress the inflammatory cascade (Kreutzberg, 1996).

8.3.1 Activation of Microglia in Response to Diffuse Brain Injury

Diffuse injury is not limited to a specific area of the brain and results in mechanically sheared axons and alterations to the vasculature. Although there is often breakdown of the BBB, it is to a lesser extent than with focal lesions (Adelson et al., 1998). A key component of the diffuse injury cascade is the activation of glial cells, including microglia. In both experimental and clinical cases, the activation of microglia after diffuse brain injury has been reported (Cao et al., 2012; Carthew et al., 2012; Johnson et al., 2013; Ziebell et al., 2012). Indeed, activated microglia and macrophages were observed to span extensive regions of the corpus callosum up to 18 years after a single moderate-severe brain injury in humans (Johnson et al., 2013). Furthermore, in experimental models of diffuse axonal injury activated microglia are reported for weeks postinjury (Cao et al., 2012; Carthew et al., 2012; Hellewell et al., 2010; Ziebell et al., 2012) as well as their proliferation (Carthew et al., 2012).

Limited data on the states of microglial activation (based on gene expression) have been published for diffuse brain injury (Cao et al., 2012). In an experimental model of diffuse brain injury, microglial activation followed by genes for both classically activated and acquired deactivation has been reported (Cao et al., 2012). Analysis of gene expression indicated microglia were predominantly classically activated. However, these genes were upregulated in a time- and region-specific manner, suggesting that microglia in different brain regions have different roles over time. The spatiotemporal distribution of microglial activation adds to the complexity of microglial response to CNS injury.

Furthermore, diffuse brain injury has been reported to induce a fifth rarely described morphology of microglia—rod microglia (Cao et al., 2012; Ziebell et al., 2012). Rod microglia have been known to exist for more than a century; however, little data exist on their role in brain injury and repair. Recently, rod microglia have been reported to align to the same trajectory as neuronal elements (Ziebell et al., 2012). Rod microglia are Iba1 positive and stain intermittently with other known markers of activated microglia. The unique morphology of rod microglia permit conjecture on their role in the injured brain. Reports suggest rod microglia have a role in synaptic reorganization (Graeber, 2010; Ziebell et al., 2012). Whether they strip or remove dysfunctional synapses to allow for new synapses to be built, offer scaffolding for damaged neuronal processes, or simply use neuronal elements to track toward injured areas are not known.

8.3.2 Microglial Response to Focal TBI

Unlike diffuse brain injury, focal TBI results in contusions in discrete regions of the brain (Bye et al., 2007; Hausmann et al., 1998; Semple et al., 2010; Ziebell et al., 2011). At the site of the contusion, there is BBB breakdown and the infiltration of circulating neutrophils, leukocytes, and monocytes, which directly affects neuronal survival (Clark et al., 1994; McIntosh et al., 1998). To restore homeostasis and reduce damage to the brain, there is activation of the resident glial cells. These cells help to reestablish the integrity of the BBB, remove debris and damaged cells, and wall off the contused area.

Macrophage/microglial accumulation has been reported as early as 6 hours after focal lesion (Hausmann et al., 1998). Chemokine mRNA expression preceded the activation and accumulation of microglia and macrophages to the injury site, suggesting that chemokines may promote the activation and migration of these cells.

When CCL2 is knocked out, the accumulation of microglia postinjury was significantly reduced (Semple et al., 2010). Furthermore, the area in which microglia were observed was significantly reduced, suggesting CCL2 is required for the migration of microglia to the injury. Indeed, this decrease in microglia was associated with improved outcomes 1 month postinjury. On the other hand, an early influx of macrophages/amoeboid microglia has been reported in Fas mutant mice after focal TBI, which was associated with improved behavioral outcomes in the first week postinjury (Ziebell et al., 2011). These findings with genetically engineered mice indicate that microglia play an important role following focal brain injury; however, more work is required to determine how microglia can be manipulated to consistently improve outcomes.

8.4 ANTI-INFLAMMATORY TREATMENTS TO REDUCE MICROGLIAL ACTIVATION

Studies on therapeutic interventions following brain injury are primarily aimed at manipulating secondary injury progression. The ability of several agents to attenuate inflammation has been established in various models of brain injury; however, in human clinical trials results have been inconclusive.

One promising treatment was the use of nonsteroidal, anti-inflammatory agents such as ibuprofen. Chronic treatment with ibuprofen improved functional and histopathological outcomes in a mouse model of Alzheimer disease (Breitner et al., 1995; Townsend and Pratico, 2005). The use of ibuprofen in a model of TBI failed to show neuroprotection (Browne et al., 2006). Indeed, the outcome of injured rats chronically treated with ibuprofen was significantly worse than vehicle. Another more recent study reported attenuated inflammatory genes (associated with microglia) with the administration of ibuprofen (Cao et al., 2012). The treatment regime in this study was for 1 week rather than 4 months. These findings suggest that the use of high doses of anti-inflammatory agents for a prolonged period of time after TBI may abolish the neuroprotective effects of inflammatory cascades.

Another common nonsteroidal anti-inflammatory compound is minocycline, a tetracycline derivative. It has consistently been associated with beneficial effects in various models of neurological disease and injury (Maier et al., 2005; Stirling et al., 2004). However, controversy surrounds the effectiveness of minocycline as a therapy for brain injury. Minocycline can transiently improve neurological outcome and lesion volume, with no differences versus vehicle by day 4 (Bye et al., 2007). Although there was no change with the density of apoptotic cells or neutrophils, the amount of activated microglia and macrophages did decrease in the injured cortex, as did the levels of IL-1 and IL-6.

Minocycline and related tetracycline derivatives are also highly potent poly-ADP ribose polymerase (PARP) inhibitors. PARP-1 activation has been identified after brain injury (Besson, 2009; Besson et al., 2005; Fink et al., 2008). Genetic deficiency or enzymatic inhibition of PARP-1 suppresses NF kappaB-dependent gene regulation in microglia and prevents their morphological transformation, proliferation, and migration to injury sites (reviewed in d'Avila et al., 2012). In a model of controlled cortical impact, PARP inhibitor INO-1001 was shown to suppress inflammatory responses. This was associated with increased neuronal survival near the impact site and improved motor performance (d'Avila et al., 2012).

Mounting evidence suggests that mild and transient immunosuppression is more effective in attenuating the detrimental effects of inflammation and microglial activation than full and sustained dosing. Therefore further research is required before blocking microglial activation as a potential treatment for brain injury.

As previously discussed, activation of microglia in neurodegeneration may be regarded as detrimental to recovery because active microglia/macrophages may release reactive oxygen species and cytokines during phagocytosis of myelin (Williams et al., 1994). However, contradictory research shows that activated microglia are essential to repair. For instance, activation of microglia by LPS actually increases brain tissue responsiveness to myelin repair (Glezer et al., 2007).

Myelin debris contains the proteins Nogo (neurite outgrowth inhibitor), MAG (myelin-associated glycoprotein), and OMgp (oligodendrocyte-myelin glycoprotein) that inhibit axon regeneration, and must therefore be removed for successful neuron repair (Gitik et al., 2011). Axon debris as well as myelin debris reduce capacity for axon regrowth following injury. Primary microglia were activated by LPS and interferon-β in vitro, leading to a reduction in axon debris. Additionally, it has been reported that p38 mitogen-activated protein kinase may be essential for activation of phagocytic microglia, but is not required for continued phagocytosis. Once activated, phagocytic microglia facilitate axon regeneration by removal of the growth-inhibitory debris (Tanaka et al., 2009).

Myelin-associated inhibitory proteins mentioned previously (Nogo, MAG, and OMgp) may also play a role in microglial regulation postinjury. Neurite outgrowth inhibitor Nogo-A is located on the innermost membrane and the outer myelin membrane of oligodendrocytes in the postnatal

CNS (Schwab, 2010). Studies have reported upregulation of Nogo-A in oligodendrocytes in the hippocampus of humans with Alzheimer disease and in surviving cells on the edges of chronic demyelinating lesions in MS (Gil et al., 2006; Satoh et al., 2005). After CNS injury, Nogo-A reduces capacity for neuroregeneration and restoration of function.

It is thought that Nogo-A reduces neuronal regeneration and plasticity by interacting with Nogo-66 receptor (NgR) and coreceptors LINGO-1 and TROY found on neurons. This results in the activation of RhoA, leading to a cascade that is associated with the collapse of the growth cone and arrest of neurite outgrowth. In MS, reactive microglia/macrophages reportedly express NgR (Satoh et al., 2005) and subpopulations of microglia/macrophages, astrocytes, and neurons express LINGO-1 and TROY (Satoh et al., 2007). This suggests an interaction between oligodendrocyte-based Nogo-A and microglial-based receptors, the implication being a role for Nogo-A in preventing microglial proliferation and cytokine production in the injured brain. Indeed, NgR-expressing microglia/macrophages have reduced migration and adhesion to myelin in vitro (Fry et al., 2007), suggesting that NgR may have a role in macrophage clearance from injured peripheral nerve. Nogo-66, the 66-amino acid residue extracellular domain of Nogo-A, not only inhibited microglial migration and adhesion, but was also associated with the retraction of microglial processes (Yan et al., 2012). Indeed, Nogo-66 inhibited polarization, further reducing cell motility and converting microglial morphology to an active form (Yan et al., 2012).

8.5 CONCLUSION

Extensive research has been conducted investigating the role microglia play during injury and repair of the CNS. In particular, the signaling cascades involving cytokines and chemokines have been explored to appreciate the activation of the immune system and microglia. Multiple avenues have been explored to interfere with the activation and signaling of microglia; however, disparate results are presented, depending on model, time, and region. These apparently conflicting results suggest that microglial activation cannot be viewed as a solitary event. Rather, the specific roles for microglia in the recovery from injury evolve with the injury. To treat microglial signaling, a series of specific biomarkers are essential, which would indicate the current microglial stage, from which therapeutic decisions can be made.

REFERENCES

Acarin, L., B. Gonzalez, and B. Castellano. 2000. Neuronal, astroglial and microglial cytokine expression after an excitotoxic lesion in the immature rat brain. *The European Journal of Neuroscience*. 12:3505–3520.

Adelson, P.D., M.J. Whalen, P.M. Kochanek, P. Robichaud, and T.M. Carlos. 1998. Blood brain barrier permeability and acute inflammation in two models of traumatic brain injury in the immature rat: A preliminary report. *Acta Neurochirurgica. Supplement*. 71:104–106.

Aguzzi, A., B.A. Barres, and M.L. Bennett. 2013. Microglia: Scapegoat, saboteur, or something else? *Science*. 339:156–161.

Allan, S.M. and N.J. Rothwell. 2003. Inflammation in central nervous system injury. *Philosophical Transactions of the Royal Society of London. Series B, Biological Sciences*. 358:1669–1677.

Aloisi, F., R. De Simone, S. Columba-Cabezas, and G. Levi. 1999. Opposite effects of interferon-gamma and prostaglandin E2 on tumor necrosis factor and interleukin-10 production in microglia: A regulatory loop controlling microglia pro- and anti-inflammatory activities. *Journal of Neuroscience Research*. 56:571–580.

Aloisi, F., F. Ria, and L. Adorini. 2000. Regulation of T-cell responses by CNS antigen-presenting cells: Different roles for microglia and astrocytes. *Immunology Today*. 21:141–147.

Bandtlow, C.E., M. Meyer, D. Lindholm, M. Spranger, R. Heumann, and H. Thoenen. 1990. Regional and cellular codistribution of interleukin 1 beta and nerve growth factor mRNA in the adult rat brain: Possible relationship to the regulation of nerve growth factor synthesis. *The Journal of Cell Biology* 111:1701–1711.

Banisadr, G., F. Queraud-Lesaux, M.C. Boutterin, D. Pelaprat, B. Zalc, W. Rostene et al. 2002. Distribution, cellular localization and functional role of CCR2 chemokine receptors in adult rat brain. *Journal of Neurochemistry*. 81:257–269.

Bell, M.J., P.M. Kochanek, L.A. Doughty, J.A. Carcillo, P.D. Adelson, R.S. Clark et al. 1997. Interleukin-6 and interleukin-10 in cerebrospinal fluid after severe traumatic brain injury in children. *Journal of Neurotrauma*. 14:451–457.

Benveniste, E.N. 1998. Cytokine actions in the central nervous system. *Cytokine & Growth Factor Reviews*. 9:259–275.

Besson, V.C. 2009. Drug targets for traumatic brain injury from poly(ADP-ribose)polymerase pathway modulation. *British Journal of Pharmacology*. 157:695–704.

Besson, V.C., Z. Zsengeller, M. Plotkine, C. Szabo, and C. Marchand-Verrecchia. 2005. Beneficial effects of PJ34 and INO-1001, two novel water-soluble poly(ADP-ribose) polymerase inhibitors, on the consequences of traumatic brain injury in rat. *Brain Research*. 1041:149–156.

Breder, C.D., M. Tsujimoto, Y. Terano, D.W. Scott, and C.B. Saper. 1993. Distribution and characterization of tumor necrosis factor-alpha-like immunoreactivity in the murine central nervous system. *The Journal of Comparative Neurology*. 337:543–567.

Breitner, J.C., K.A. Welsh, M.J. Helms, P.C. Gaskell, B.A. Gau, A.D. Roses et al;. 1995. Delayed onset of Alzheimer's disease with nonsteroidal anti-inflammatory and histamine H2 blocking drugs. *Neurobiology of Aging*. 16:523–530.

Browne, K.D., A. Iwata, M.E. Putt, and D.H. Smith. 2006. Chronic ibuprofen administration worsens cognitive outcome following traumatic brain injury in rats. *Experimental Neurology*. 201:301–307.

Bye, N., M.D. Habgood, J.K. Callaway, N. Malakooti, A. Potter, T. Kossmann, and M.C. Morganti-Kossmann. 2007. Transient neuroprotection by minocycline following traumatic brain injury is associated with attenuated microglial activation but no changes in cell apoptosis or neutrophil infiltration. *Experimental Neurology*. 204:220–233.

Cao, T., T.C. Thomas, J.M. Ziebell, J.R. Pauly, and J. Lifshitz. 2012. Morphological and genetic activation of microglia after diffuse traumatic brain injury in the rat. *Neuroscience*. 225:65–75.

Carthew, H.L., J.M. Ziebell, and R. Vink. 2012. Substance P-induced changes in cell genesis following diffuse traumatic brain injury. *Neuroscience*. 214:78–83.

Clark, R.S., J.K. Schiding, S.L. Kaczorowski, D.W. Marion, and P.M. Kochanek. 1994. Neutrophil accumulation after traumatic brain injury in rats: Comparison of weight drop and controlled cortical impact models. *Journal of Neurotrauma*. 11:499–506.

Colton, C.A. 2009. Heterogeneity of microglial activation in the innate immune response in the brain. *Journal of Neuroimmune Pharmacology: The Official Journal of the Society on NeuroImmune Pharmacology*. 4:399–418.

Csuka, E., V.H. Hans, E. Ammann, O. Trentz, T. Kossmann, and M.C. Morganti-Kossmann. 2000. Cell activation and inflammatory response following traumatic axonal injury in the rat. *Neuroreport*. 11:2587–2590.

d'Avila, J.C., T.I. Lam, D. Bingham, J. Shi, S.J. Won, T.M. Kauppinen et al. 2012. Microglial activation induced by brain trauma is suppressed by post-injury treatment with a PARP inhibitor. *Journal of Neuroinflammation*. 9:31.

Dalgard, C.L., J.T. Cole, W.S. Kean, J.J. Lucky, G. Sukumar, D.C. McMullen et al. 2012. The cytokine temporal profile in rat cortex after controlled cortical impact. *Frontiers in Molecular Neuroscience*. 5:6.

Dinarello, C.A. 1997. Induction of interleukin-1 and interleukin-1 receptor antagonist. *Seminars in Oncology*. 24:S9–81-S89–93.

Domercq, M., N. Vazquez-Villoldo, and C. Matute. 2013. Neurotransmitter signaling in the pathophysiology of microglia. *Frontiers in Cellular Neuroscience*. 7:49.

Estes, M.L., K. Iwasaki, B.S. Jacobs, and B.P. Barna. 1993. Interleukin-4 down-regulates adult human astrocyte DNA synthesis and proliferation. *The American Journal of Pathology*. 143:337–341.

Fan, L., P.R. Young, F.C. Barone, G.Z. Feuerstein, D.H. Smith, and T.K. McIntosh. 1995. Experimental brain injury induces expression of interleukin-1 beta mRNA in the rat brain. *Brain Research. Molecular Brain Research*. 30:125–130.

Fink, E.L., Y. Lai, X. Zhang, K. Janesko-Feldman, P.D. Adelson, C. Szabo et al. 2008. Quantification of poly(ADP-ribose)-modified proteins in cerebrospinal fluid from infants and children after traumatic brain injury. *Journal of Cerebral Blood Flow and Metabolism*. 28:1523–1529.

Fox, C., A. Dingman, N. Derugin, M.F. Wendland, C. Manabat, S. Ji et al. 2005. Minocycline confers early but transient protection in the immature brain following focal cerebral ischemia-reperfusion. *Journal of Cerebral Blood Flow and Metabolism*. 25:1138–1149.

Fry, E.J., C. Ho, and S. David. 2007. A role for Nogo receptor in macrophage clearance from injured peripheral nerve. *Neuron*. 53:649–662.

Fumagalli, S., C. Perego, F. Ortolano, and M.G. De Simoni. 2013. CX3CR1 deficiency induces an early protective inflammatory environment in ischemic mice. *Glia*. 61:827–842.

Galasso, J.M., M.J. Miller, R.M. Cowell, J.K. Harrison, J.S. Warren, and F.S. Silverstein. 2000. Acute excitotoxic injury induces expression of monocyte chemoattractant protein-1 and its receptor, CCR2, in neonatal rat brain. *Experimental Neurology*. 165:295–305.

Gehrmann, J., Y. Matsumoto, and G.W. Kreutzberg. 1995. Microglia: Intrinsic immuneffector cell of the brain. *Brain Research. Brain Research Reviews*. 20:269–287.

Gil, V., O. Nicolas, A. Mingorance, J.M. Urena, B.L. Tang, T. Hirata et al. 2006. Nogo-A expression in the human hippocampus in normal aging and in Alzheimer disease. *Journal of Neuropathology and Experimental Neurology*. 65:433–444.

Gitik, M., S. Liraz-Zaltsman, P.A. Oldenborg, F. Reichert, and S. Rotshenker. 2011. Myelin down-regulates myelin phagocytosis by microglia and macrophages through interactions between CD47 on myelin and SIRPalpha (signal regulatory protein-alpha) on phagocytes. *Journal of Neuroinflammation*. 8:24.

Giulian, D. 1987. Ameboid microglia as effectors of inflammation in the central nervous system. *Journal of Neuroscience Research*. 18:155–171, 132–153.

Giulian, D., T.J. Baker, L.C. Shih, and L.B. Lachman. 1986. Interleukin 1 of the central nervous system is produced by ameboid microglia. *The Journal of Experimental Medicine*. 164:594–604.

Glabinski, A.R., V. Balasingam, M. Tani, S.L. Kunkel, R.M. Strieter, V.W. Yong, and R.M. Ransohoff. 1996. Chemokine monocyte chemoattractant protein-1 is expressed by astrocytes after mechanical injury to the brain. *Journal of Immunology*. 156:4363–4368.

Glezer, I., A.R. Simard, and S. Rivest. 2007. Neuroprotective role of the innate immune system by microglia. *Neuroscience*. 147:867–883.

Gourmala, N.G., M. Buttini, S. Limonta, A. Sauter, and H.W. Boddeke. 1997. Differential and time-dependent expression of monocyte chemoattractant protein-1 mRNA by astrocytes and macrophages in rat brain: Effects of ischemia and peripheral lipopolysaccharide administration. *Journal of Neuroimmunology*. 74:35–44.

Graeber, M.B. 2010. Changing face of microglia. *Science*. 330:783–788.

Graeber, M.B. and P. Mehraein. 1994. Microglial rod cells. *Neuropathology and Applied Neurobiology*. 20:178–180.

Graeber, M.B. and W.J. Streit. 2010. Microglia: Biology and pathology. *Acta Neuropathologica*. 119:89–105.

Hammacher, A., L.D. Ward, J. Weinstock, H. Treutlein, K. Yasukawa, and R.J. Simpson. 1994. Structure-function analysis of human IL-6: Identification of two distinct regions that are important for receptor binding. *Protein Science*. 3:2280–2293.

Hanisch, U.K. 2002. Microglia as a source and target of cytokines. *Glia*. 40:140–155.

Hausmann, E.H., N.E. Berman, Y.Y. Wang, J.B. Meara, G.W. Wood, and R.M. Klein. 1998. Selective chemokine mRNA expression following brain injury. *Brain Research*. 788:49–59.

Hellewell, S.C., E.B. Yan, D.A. Agyapomaa, N. Bye, and M.C. Morganti-Kossmann. 2010. Post-traumatic hypoxia exacerbates brain tissue damage: Analysis of axonal injury and glial responses. *Journal of Neurotrauma*. 27:1997–2010.

Howe, C.L., S. Mayoral, and M. Rodriguez. 2006. Activated microglia stimulate transcriptional changes in primary oligodendrocytes via IL-1beta. *Neurobiology of Disease*. 23:731–739.

Hughes, P.M., P.R. Allegrini, M. Rudin, V.H. Perry, A.K. Mir, and C. Wiessner. 2002. Monocyte chemoattractant protein-1 deficiency is protective in a murine stroke model. *Journal of Cerebral Blood Flow and Metabolism*. 22:308–317.

Hulshof, S., L. Montagne, C.J. De Groot, and P. Van Der Valk. 2002. Cellular localization and expression patterns of interleukin-10, interleukin-4, and their receptors in multiple sclerosis lesions. *Glia*. 38:24–35.

Iadecola, C. and J. Anrather. 2011. The immunology of stroke: From mechanisms to translation. *Nature Medicine*. 17:796–808.

Jin, X., H. Ishii, Z. Bai, T. Itokazu, and T. Yamashita. 2012. Temporal changes in cell marker expression and cellular infiltration in a controlled cortical impact model in adult male C57BL/6 mice. *PloS One*. 7:e41892.

Johnson, V.E., J.E. Stewart, F.D. Begbie, J.Q. Trojanowski, D.H. Smith, and W. Stewart. 2013. Inflammation and white matter degeneration persist for years after a single traumatic brain injury. *Brain*. 136:28–42.

Kielian, T. and W.F. Hickey. 2000. Proinflammatory cytokine, chemokine, and cellular adhesion molecule expression during the acute phase of experimental brain abscess development. *The American Journal of Pathology*. 157:647–658.

Kierdorf, K. and M. Prinz. 2013. Factors regulating microglia activation. *Frontiers in Cellular Neuroscience*. 7:44.

Kim, K.S., C.A. Wass, A.S. Cross, and S.M. Opal. 1992. Modulation of blood-brain barrier permeability by tumor necrosis factor and antibody to tumor necrosis factor in the rat. *Lymphokine and Cytokine Research*. 11:293–298.

Kinouchi, K., G. Brown, G. Pasternak, and D.B. Donner. 1991. Identification and characterization of receptors for tumor necrosis factor-alpha in the brain. *Biochemical and Biophysical Research Communications*. 181:1532–1538.

Knoblach, S.M. and A.I. Faden. 1998. Interleukin-10 improves outcome and alters proinflammatory cytokine expression after experimental traumatic brain injury. *Experimental Neurology*. 153:143–151.

Knoblach, S.M., L. Fan, and A.I. Faden. 1999. Early neuronal expression of tumor necrosis factor-alpha after experimental brain injury contributes to neurological impairment. *Journal of Neuroimmunology*. 95:115–125.

Kossmann, T., V.H. Hans, H.G. Imhof, R. Stocker, P. Grob, O. Trentz, and C. Morganti-Kossmann. 1995. Intrathecal and serum interleukin-6 and the acute-phase response in patients with severe traumatic brain injuries. *Shock*. 4:311–317.

Kremlev, S.G. and C. Palmer. 2005. Interleukin-10 inhibits endotoxin-induced pro-inflammatory cytokines in microglial cell cultures. *Journal of Neuroimmunology*. 162:71–80.

Kreutzberg, G.W. 1996. Microglia: A sensor for pathological events in the CNS. *Trends in Neurosciences*. 19:312–318.

Kurihara, T., G. Warr, J. Loy, and R. Bravo. 1997. Defects in macrophage recruitment and host defense in mice lacking the CCR2 chemokine receptor. *The Journal of Experimental Medicine*. 186:1757–1762.

Kuziel, W.A., S.J. Morgan, T.C. Dawson, S. Griffin, O. Smithies, K. Ley, and N. Maeda. 1997. Severe reduction in leukocyte adhesion and monocyte extravasation in mice deficient in CC chemokine receptor 2. *Proceedings of the National Academy of Sciences of the United States of America*. 94:12053–12058.

Lawrence, C.B., S.M. Allan, and N.J. Rothwell. 1998. Interleukin-1beta and the interleukin-1 receptor antagonist act in the striatum to modify excitotoxic brain damage in the rat. *The European Journal of Neuroscience*. 10:1188–1195.

Lawson, L.J., V.H. Perry, P. Dri, and S. Gordon. 1990. Heterogeneity in the distribution and morphology of microglia in the normal adult mouse brain. *Neuroscience*. 39:151–170.

Liu, C., D. Luo, W.J. Streit, and J.K. Harrison. 2008. CX3CL1 and CX3CR1 in the GL261 murine model of glioma: CX3CR1 deficiency does not impact tumor growth or infiltration of microglia and lymphocytes. *Journal of Neuroimmunology*. 198:98–105.

Loddick, S.A. and N.J. Rothwell. 1996. Neuroprotective effects of human recombinant interleukin-1 receptor antagonist in focal cerebral ischaemia in the rat. *Journal of Cerebral Blood Flow and Metabolism*. 16:932–940.

Lu, B., B.J. Rutledge, L. Gu, J. Fiorillo, N.W. Lukacs, S.L. Kunkel, R. North, C. Gerard, and B.J. Rollins. 1998. Abnormalities in monocyte recruitment and cytokine expression in monocyte chemoattractant protein 1-deficient mice. *The Journal of Experimental Medicine*. 187:601–608.

Lu, W., J.A. Gersting, A. Maheshwari, R.D. Christensen, and D.A. Calhoun. 2005. Developmental expression of chemokine receptor genes in the human fetus. *Early Human Development*. 81:489–496.

Lucas, S.M., N.J. Rothwell, and R.M. Gibson. 2006. The role of inflammation in CNS injury and disease. *British Journal of Pharmacology*. 147 Suppl 1:S232–240.

Lyng, K., B.H. Munkeby, O.D. Saugstad, B. Stray-Pedersen, and J.F. Froen. 2005. Effect of interleukin-10 on newborn piglet brain following hypoxia-ischemia and endotoxin-induced inflammation. *Biology of the Neonate*. 87:207–216.

Maier, B., H.L. Laurer, S. Rose, W.A. Buurman, and I. Marzi. 2005. Physiological levels of pro- and anti-inflammatory mediators in cerebrospinal fluid and plasma: A normative study. *Journal of Neurotrauma*. 22:822–835.

Mattison, H.A., H. Nie, H. Gao, H. Zhou, J.S. Hong, and J. Zhang. 2013. Suppressed pro-inflammatory response of microglia in CX3CR1 knockout mice. *Journal of Neuroimmunology*. 257:110–115.

McIntosh, T.K., K.E. Saatman, R. Raghupathi, D.I. Graham, D.H. Smith, V.M. Lee, and J.Q. Trojanowski. 1998. The Dorothy Russell Memorial Lecture. The molecular and cellular sequelae of experimental traumatic brain injury: Pathogenetic mechanisms. *Neuropathology and Applied Neurobiology*. 24:251–267.

Meme, W., C.F. Calvo, N. Froger, P. Ezan, E. Amigou, A. Koulakoff, and C. Giaume. 2006. Proinflammatory cytokines released from microglia inhibit gap junctions in astrocytes: Potentiation by beta-amyloid. *FASEB Journal*. 20:494–496.

Mesples, B., F. Plaisant, and P. Gressens. 2003. Effects of interleukin-10 on neonatal excitotoxic brain lesions in mice. *Brain Research. Developmental Brain Research*. 141:25–32.

Mizuno, T., J. Kawanokuchi, K. Numata, and A. Suzumura. 2003. Production and neuroprotective functions of fractalkine in the central nervous system. *Brain Research*. 979:65–70.

Morganti-Kossmann, M.C. and T. Kossmann. 1995. The immunology of the brain. In *Molecular and Cellular Neurobiology: Immune Responses in the Nervous System*. N.J. Rothwell, editor. BIOS Scientific Publishers, Oxford.

Morganti-Kossmann, M.C., L. Satgunaseelan, N. Bye, and T. Kossmann. 2007. Modulation of immune response by head injury. *Injury*. 38:1392–1400.

Perry, V.H., T.A. Newman, and C. Cunningham. 2003. The impact of systemic infection on the progression of neurodegenerative disease. *Nature Reviews. Neuroscience*. 4:103–112.

Prinz, M. and J. Priller. 2010. Tickets to the brain: Role of CCR2 and CX3CR1 in myeloid cell entry in the CNS. *Journal of Neuroimmunology*. 224:80–84.

Raivich, G., H. Bluethmann, and G.W. Kreutzberg. 1996. Signaling molecules and neuroglial activation in the injured central nervous system. *The Keio Journal of Medicine*. 45:239–247.

Raivich, G., M. Bohatschek, C.U. Kloss, A. Werner, L.L. Jones, and G.W. Kreutzberg. 1999a. Neuroglial activation repertoire in the injured brain: Graded response, molecular mechanisms and cues to physiological function. *Brain Research. Brain Research Reviews*. 30:77–105.

Raivich, G., L.L. Jones, A. Werner, H. Bluthmann, T. Doetschmann, and G.W. Kreutzberg. 1999b. Molecular signals for glial activation: Pro- and anti-inflammatory cytokines in the injured brain. *Acta Neurochirurgica. Supplement*. 73:21–30.

Ramilo, O., X. Saez-Llorens, J. Mertsola, H. Jafari, K.D. Olsen, E.J. Hansen et al. 1990. Tumor necrosis factor alpha/cachectin and interleukin 1 beta initiate meningeal inflammation. *The Journal of Experimental Medicine*. 172:497–507.

Rankine, E.L., P.M. Hughes, M.S. Botham, V.H. Perry, and L.M. Felton. 2006. Brain cytokine synthesis induced by an intraparenchymal injection of LPS is reduced in MCP-1-deficient mice prior to leucocyte recruitment. *The European Journal of Neuroscience*. 24:77–86.

Ransohoff, R.M. and V.H. Perry. 2009. Microglial physiology: Unique stimuli, specialized responses. *Annual Review of Immunology*. 27:119–145.

Relton, J.K. and N.J. Rothwell. 1992. Interleukin-1 receptor antagonist inhibits ischaemic and excitotoxic neuronal damage in the rat. *Brain Research Bulletin*. 29:243–246.

Riva-Depaty, I., C. Fardeau, J. Mariani, C. Bouchaud, and N. Delhaye-Bouchaud. 1994. Contribution of peripheral macrophages and microglia to the cellular reaction after mechanical or neurotoxin-induced lesions of the rat brain. *Experimental Neurology*. 128:77–87.

Ross, S.A., M.I. Halliday, G.C. Campbell, D.P. Byrnes, and B.J. Rowlands. 1994. The presence of tumour necrosis factor in CSF and plasma after severe head injury. *British Journal of Neurosurgery*. 8:419–425.

Rothwell, N.J. 1999. Annual review prize lecture cytokines - killers in the brain? *The Journal of Physiology*. 514 (Pt 1):3–17.

Satoh, J., H. Onoue, K. Arima, and T. Yamamura. 2005. Nogo-A and nogo receptor expression in demyelinating lesions of multiple sclerosis. *Journal of Neuropathology and Experimental Neurology*. 64:129–138.

Satoh, J., H. Tabunoki, T. Yamamura, K. Arima, and H. Konno. 2007. TROY and LINGO-1 expression in astrocytes and macrophages/microglia in multiple sclerosis lesions. *Neuropathology and Applied Neurobiology*. 33:99–107.

Schielke, G.P., G.Y. Yang, B.D. Shivers, and A.L. Betz. 1998. Reduced ischemic brain injury in interleukin-1 beta converting enzyme-deficient mice. *Journal of Cerebral Blood Flow and Metabolism*. 18:180–185.

Schmidt, O.I., M. Infanger, C.E. Heyde, W. ERtel, and P.F. Stahel. 2004. The role of neuroinflammation in traumatic brain injury. *European Journal of Trauma*. 20:135–149.

Schwab, M.E. 2010. Functions of Nogo proteins and their receptors in the nervous system. *Nature Reviews. Neuroscience*. 11:799–811.

Scott, E.W., M.C. Simon, J. Anastasi, and H. Singh. 1994. Requirement of transcription factor PU.1 in the development of multiple hematopoietic lineages. *Science*. 265:1573–1577.

Semple, B.D., N. Bye, M. Rancan, J.M. Ziebell, and M.C. Morganti-Kossmann. 2010. Role of CCL2 (MCP-1) in traumatic brain injury (TBI): Evidence from severe TBI patients and CCL2-/- mice. *Journal of Cerebral Blood Flow and Metabolism*. 30:769–782.

Shohami, E., R. Gallily, R. Mechoulam, R. Bass, and T. Ben-Hur. 1997. Cytokine production in the brain following closed head injury: Dexanabinol (HU-211) is a novel TNF-alpha inhibitor and an effective neuroprotectant. *Journal of Neuroimmunology*. 72:169–177.

Shohami, E., M. Novikov, R. Bass, A. Yamin, and R. Gallily. 1994. Closed head injury triggers early production of TNF alpha and IL-6 by brain tissue. *Journal of Cerebral Blood Flow and Metabolism*. 14:615–619.

Stirling, D.P., K. Khodarahmi, J. Liu, L.T. McPhail, C.B. McBride, J.D. Steeves et al. 2004. Minocycline treatment reduces delayed oligodendrocyte death, attenuates axonal dieback, and improves functional outcome after spinal cord injury. *The Journal of Neuroscience*. 24:2182–2190.

Streit, W.J., S.L. Semple-Rowland, S.D. Hurley, R.C. Miller, P.G. Popovich, and B.T. Stokes. 1998. Cytokine mRNA profiles in contused spinal cord and axotomized facial nucleus suggest a beneficial role for inflammation and gliosis. *Experimental Neurology*. 152:74–87.

Tanaka, T., M. Ueno, and T. Yamashita. 2009. Engulfment of axon debris by microglia requires p38 MAPK activity. *The Journal of Biological Chemistry*. 284:21626–21636.

Taupin, V., S. Toulmond, A. Serrano, J. Benavides, and F. Zavala. 1993. Increase in IL-6, IL-1 and TNF levels in rat brain following traumatic lesion. Influence of pre- and post-traumatic treatment with Ro5 4864, a peripheral-type (p site) benzodiazepine ligand. *Journal of Neuroimmunology*. 42:177–185.

Tchelingerian, J.L., J. Quinonero, J. Booss, and C. Jacque. 1993. Localization of TNF alpha and IL-1 alpha immunoreactivities in striatal neurons after surgical injury to the hippocampus. *Neuron*. 10:213–224.

Townsend, K.P. and D. Pratico. 2005. Novel therapeutic opportunities for Alzheimer's disease: Focus on nonsteroidal anti-inflammatory drugs. *FASEB Journal*. 19:1592–1601.

Tremblay, M.E. 2011. The role of microglia at synapses in the healthy CNS: Novel insights from recent imaging studies. *Neuron Glia Biology*. 7:67–76.

Tremblay, M.E. and A.K. Majewska. 2011. A role for microglia in synaptic plasticity? *Communicative & Integrative Biology*. 4:220–222.

Tremblay, M.E., B. Stevens, A. Sierra, H. Wake, A. Bessis, and A. Nimmerjahn. 2011. The role of microglia in the healthy brain. *The Journal of Neuroscience*. 31:16064–16069.

Turrin, N.P. and S. Rivest. 2006. Tumor necrosis factor alpha but not interleukin 1 beta mediates neuroprotection in response to acute nitric oxide excitotoxicity. *The Journal of Neuroscience*. 26:143–151.

Urrea, C., D.A. Castellanos, J. Sagen, P. Tsoulfas, H.M. Bramlett, and W.D. Dietrich. 2007. Widespread cellular proliferation and focal neurogenesis after traumatic brain injury in the rat. *Restorative Neurology and Neuroscience*. 25:65–76.

Veerhuis, R., J.J. Hoozemans, I. Janssen, R.S. Boshuizen, J.P. Langeveld, and P. Eikelenboom. 2002. Adult human microglia secrete cytokines when exposed to neurotoxic prion protein peptide: No intermediary role for prostaglandin E2. *Brain Research*. 925:195–203.

Wang, C.X. and A. Shuaib. 2002. Involvement of inflammatory cytokines in central nervous system injury. *Progress in Neurobiology*. 67:161–172.

Williams, K., E. Ulvestad, A. Waage, J.P. Antel, and J. McLaurin. 1994. Activation of adult human derived microglia by myelin phagocytosis in vitro. *Journal of Neuroscience Research*. 38:433–443.

Woodcock, T. and M.C. Morganti-Kossmann. 2013. The role of markers of inflammation in traumatic brain injury. *Frontiers in Neurology*. 4:18.

Woodroofe, M.N., G.S. Sarna, M. Wadhwa, G.M. Hayes, A.J. Loughlin, A. Tinkeret al. 1991. Detection of interleukin-1 and interleukin-6 in adult rat brain, following mechanical injury, by in vivo microdialysis: Evidence of a role for microglia in cytokine production. *Journal of Neuroimmunology*. 33:227–236.

Wu, Z., J. Zhang, and H. Nakanishi. 2005. Leptomeningeal cells activate microglia and astrocytes to induce IL-10 production by releasing pro-inflammatory cytokines during systemic inflammation. *Journal of Neuroimmunology*. 167:90–98.

Xiong, X., G.E. Barreto, L. Xu, Y.B. Ouyang, X. Xie, and R.G. Giffard. 2011. Increased brain injury and worsened neurological outcome in interleukin-4 knockout mice after transient focal cerebral ischemia. *Stroke*. 42:2026–2032.

Yan, J., X. Zhou, J.J. Guo, L. Mao, Y.J. Wang, J. Sun et al. 2012. Nogo-66 inhibits adhesion and migration of microglia via GTPase Rho pathway in vitro. *Journal of Neurochemistry*. 120:721–731.

Yang, M.S., E.J. Park, S. Sohn, H.J. Kwon, W.H. Shin, H.K. Pyo et al. 2002. Interleukin-13 and -4 induce death of activated microglia. *Glia*. 38:273–280.

Yoshimura, T., E.A. Robinson, S. Tanaka, E. Appella, J. Kuratsu, and E.J. Leonard. 1989. Purification and amino acid analysis of two human glioma-derived monocyte chemoattractants. *The Journal of Experimental Medicine*. 169:1449–1459.

Ziebell, J.M., N. Bye, B.D. Semple, T. Kossmann, and M.C. Morganti-Kossmann. 2011. Attenuated neurological deficit, cell death and lesion volume in Fas-mutant mice is associated with altered neuroinflammation following traumatic brain injury. *Brain Research*. 1414:94–105.

Ziebell, J.M. and M.C. Morganti-Kossmann. 2010. Involvement of pro- and anti-inflammatory cytokines and chemokines in the pathophysiology of traumatic brain injury. *Neurotherapeutics*. 7:22–30.

Ziebell, J.M., S.E. Taylor, T. Cao, J.L. Harrison, and J. Lifshitz. 2012. Rod microglia: Elongation, alignment, and coupling to form trains across the somatosensory cortex after experimental diffuse brain injury. *Journal of Neuroinflammation*. 9:247.

Zwijnenburg, P.J., M.M. Polfliet, S. Florquin, T.K. van den Berg, C.D. Dijkstra, S.J. van Deventer et al. 2003. CXC-chemokines KC and macrophage inflammatory protein-2 (MIP-2) synergistically induce leukocyte recruitment to the central nervous system in rats. *Immunology Letters*. 85:1–4.

9 Evaluating the Effects of *APOE4* after Mild Traumatic Brain Injury in Experimental Models

Rebekah Mannix and William P. Meehan III

CONTENTS

9.1 INTRODUCTION

The best known genetic risk factor for poor outcome after traumatic brain injury (TBI) in adults is the E4 allele of the apolipoprotein E (*APOE*) gene. Studying the effect of *APOE* alleles on outcome after mild TBI (mTBI) has proven challenging in clinical studies, due to many factors including heterogeneity of injuries, confounding clinical factors such as age and unmeasured associated genotypes, and general expense of large-scale studies. Studying the effect of *APOE* on outcome after mTBI in experimental models may overcome many of these hurdles, but a thorough understanding of injury and *APOE* models is needed prior to undertaking such studies.

Mild traumatic brain injury is a common injury, with more than 3.8 million new cases annually just in the United States alone. mTBI is the most common form of TBI, with nearly 75% of all TBI being characterized as mild. However, despite the characterization as "mild," mTBI can have long-term and devastating effects. Recently, there has been increased interest in studying short- and long-term effects of mTBI, especially in military and athletic populations.

In studying mTBI, clinicians have observed that outcomes may differ markedly, even among patients with similar mechanisms of injury. These observations have led to the search for genetic risk factors potentially associated with worse outcome after TBI (Nicoll, 1996). The best known genetic risk factor associated with poor outcome after TBI in adults is the E4 allele of the apolipoprotein E (*APOE*) gene (Jordan et al., 1997; Teasdale et al., 1997).

Studying the effect of *APOE* on outcome after mTBI has proven challenging in clinical studies because of many factors including heterogeneity of injuries, confounding clinical factors such as age and unmeasured associated genotypes, and general expense of large-scale studies. Studying the effect of *APOE* on outcome after mTBI in experimental models may overcome many of these hurdles, but a thorough understanding of injury and *APOE* models is needed before undertaking such studies. The purpose of this chapter is to review experimental models of *APOE* and mTBI, starting with the clinical literature that has informed the experimental modeling.

9.1.1 THE EFFECT OF *APOE* ON OUTCOME AFTER TBI IN CLINICAL STUDIES

APOE is the predominant cholesterol and lipid transport protein in brain. Three allelic variants of *APOE* yield three distinct protein isoforms (*APOE2*, *APOE3*, and *APOE4*) *APOE* allele frequencies in Caucasians are 7%, 78%, and 15%, respectively. *APOE* plays an important role in maintaining synaptic intsegrity and function, promoting neural recovery and repair, and modifying inflammatory responses after central nervous system (CNS) injury (Chen et al., 1997; Lynch et al., 2003; Vitek et al., 2009). The efficacy of these functions, however, varies markedly between isoforms. For example, while *APOE3* facilitates amyloid beta (Aβ) transport into microglia for degradation, or across the blood brain–barrier for removal from the brain, *APOE4* promotes Aβ accumulation and is associated with an increased risk of Alzheimer disease (AD) (Kim et al., 2009).

In addition to its association with AD, there is mounting evidence that suggests that the *APOE4* isoform is associated with worse outcome after TBI. The presence of at least one *APOE*4 allele is associated with early mortality (Teasdale et al., 1997), prolonged coma (Sorbi et al., 1995), worse functional outcome (Jordan et al., 1997), and an increased risk of developing AD (Mayeux et al., 1995) after TBI. Importantly, the *APOE*4 allele has been associated with worse outcomes after both moderate-severe TBI (Ariza et al., 2006) and mild TBI (Crawford et al., 2002; Sundstrom et al., 2004).

Although the preponderance of clinical studies indicates that the *APOE* genotype influences outcome after TBI, there have been several notable studies that have not found an association between *APOE* status and outcome. In a long-term follow-up study of 396 patients assessed, on average, 18 years after a TBI, no association was found between *APOE* genotype and neuropsychological function (Millar et al., 2003). Similarly, Chamelian et al. found no association between *APOE* genotype and cognitive and behavioral outcomes after mild to moderate TBI (Chamelian et al., 2004). Another large study found no overall effect of *APOE* status on 6-month outcomes (dichotomized to favorable or unfavorable) but noted that younger *APOE*4 carriers had worse outcomes after TBI than younger non-*APOE*4 carriers (Teasdale et al., 2005), suggesting a possible age-dependent effect of *APOE*4. Other studies, however, have failed to replicate this finding of worse outcome after TBI in young *APOE*4 carriers (Moran et al., 2009). It is also important to note that the effect of *APOE*4 status in non-Caucasian carriers has not been well studied, with one small study finding no statistically significant differences in outcome after traumatic cerebral contusions in an African cohort (Nathoo et al., 2003).

The association of *APOE* with outcome after repetitive head injuries is also not well characterized, though several studies have suggested that *APOE*4 is associated with worse outcome after repetitive injury, including progression to chronic traumatic encephalopathy (CTE) (McKee et al., 2009). Jordan et al. reported that the *APOE*4 allele may be associated with increased severity of chronic neurologic deficits in high-exposure boxers (Jordan et al., 1997). In a small case series of 53 professional football players, Kutner et al. found an association between the *APOE*4 allele and worse cognitive performance (Kutner et al., 2000). Omalu et al., however, failed to find an association of *APOE*4 and CTE in a cohort of professional athletes (Omalu et al., 2011).

Given the sometimes contradictory, underpowered, or confounded results of clinical studies, experimental models of *APOE* and mTBI have promised to offer new insights into the effect of different *APOE* isoforms after injury.

9.2 TRANSGENIC MODELS OF *APOE*

Before developing an experimental model of *APOE*, it is important to understand which cell types in the CNS produce *APOE*, Humans express *APOE* in both glia and, in lower levels, neurons. In contrast, under normal conditions, mouse *APOE* is only expressed in glia, although expression can be induced in neurons under cytotoxic challenge. There is, however, a critical difference in trafficking and cellular localization between receptor-mediated uptake of *APOE* and actual production of *APOE* within the neuron. Thus the potential for significant functional differences in transgenic models is based on glial versus neuronal expression.

Because of its association with AD, *APOE* functioning in the CNS has been well-studied. The human-like neuron expression of *APOE* is found in mouse transgenic models, driven by the human *APOE* promoter (human genomic promoter). Other transgenics driven by cell-type specific promoters have been developed to evaluate which cellular production site is most relevant to form-specific risks of AD. The two most frequently used cell-specific transgenic models have limited expression to neurons using neuron-specific enolase (NSE) (Raber et al., 1998) or to astrocytes using glial fibrillary acidic protein (GFAP) (Sun et al., 1998). Each of these models has been used widely, in both AD and TBI studies, but each is less than ideal, with abnormally high cell-specific expression of *APOE* in the NSE and human genomic promoter transgenic, and nonphysiologic *APOE* regulation in the GFAP transgenics. These caveats are somewhat addressed in newer transgenic models in which the endogenous mouse *APOE* promoter drives expression of human *APOE* isoforms (targeted replacement transgenics, TR), leading to human brain region specific levels of *APOE* without basal neuronal expression.

It is also important to note that all the *APOE* transgenic models (except the targeted replacement transgenics) are maintained on an *APOE*-knockout *(APOE-KO)* background. While *APOE-KO* develop relatively normally through adulthood with only minor cognitive impairments, they begin to show neurodegenerative changes at 5 to 6 months, including defective long-term potentiation (Veinbergs et al., 1998) and destabilization of the synapto-dendritic apparatus (Masliah et al., 1995). Moreover, *APOE-KO* mice have severe peripheral hypercholesterolemia, with potential effects on the integrity of the blood–brain barrier, which is not remediated in the brain cell–specific transgenics (NSE and GFAP). This becomes especially relevant when choosing control mice for studying the effect of a specific *APOE* isoform after TBI because wild-type mice may not be appropriate controls. Indeed, although mouse *APOE* is functionally equivalent to human *APOE*3, the genetic background of the *APOE* transgenic models and resultant peripheral hypercholesterolemia may preclude the use of wild-type controls in many cases. If wild-type controls are used, they should be littermate controls when possible.

9.2.1 Characterizing Transgenic Models of *APOE*

It is important to understand the baseline characteristics of the transgenic *APOE* models before using them to explore the effects of *APOE* isoform on outcome after TBI. Indeed, many of the transgenic models demonstrate isoform-specific differences in baseline functional performance that worsens with aging (Raber et al., 1998; Yao et al., 2004). Baseline differences in toxic intermediaries such as phosphorylated tau and *APOE* fragments are dependent on not only *APOE* isoform, but also *APOE* cellular source (Brecht et al., 2004). In addition, response to toxic insult is specific to isoform and cell source. Using an NSE transgenic model, Buttini et al. demonstrated significant loss of synaptophysin-positive presynaptic terminals and microtubule-associated protein 2-positive neuronal dendrites in the neocortex and hippocampus, and a disruption of neurofilament-positive axons in the hippocampus after kainic acid challenge in *APOE*4 but not *APOE*3 transgenic mice (Buttini et al., 1999). Thus, before undertaking any study using *APOE* transgenics, it is vital to fully characterize the baseline behavioral and histopathological phenotype.

9.3 MILD TBI MODELS USING TRANSGENIC MODELS OF *APOE*

Despite the availability of *APOE* transgenic mice, there is a relative dearth of studies evaluating the effect of *APOE* isoforms on outcome after mTBI. Part of the reason for the limited number of studies is that interest in the field of mTBI has only recently peaked, with prior laboratory-based efforts concentrating on more severe forms of injury. In addition, many experimental models of TBI produce skull fracture, intracranial hemorrhage, or cell death, which does not faithfully replicate the majority of clinical mTBIs, in which no gross or macroscopic structural injury is present. There are several mTBI models in use, including a closed-head projectile concussive impact model (Chen et al., 2012), electromagnetic impaction (Mouzon et al., 2012), modifications of the Marmarou weight drop method (Ucar et al., 2006), lateral fluid percussion methods (Perez-Polo et al., 2013), and blast models (Rubovitch et al., 2011). When choosing between the various models, it is important to note how well the model simulates human mTBI. For example, an experimental model in which mice have prolonged loss of consciousness after injury demonstrates a more severe form of mTBI, since the majority of mTBI patients do not have a prolonged loss of consciousness.

9.3.1 Behavioral Outcomes after mTBI

Before initiating an experimental model of mTBI, behavioral outcomes should be considered. The most widely used test, the Morris Water Maze, assesses spatial memory performance (Morris, 1984). The maze comprises a circular plastic tank partially filled with water, sometimes made nontransparent with milk powder. The perimeter of the pool is then marked with four equally spaced starting positions, dividing the tank into four quadrants. A hidden clear Plexiglas platform is submerged just under the water level in the center of one of the quadrants. Testing is initiated by placing mice at random starting positions, facing the pool wall, and allowing them to swim until they discover the location of the escape platform and climb onto it. The time to find the platform, or latency, is recorded. After multiple trials, mice learn the platform location using the perimeter cues, and latency times decrease. To ensure that the decreased latency times are a result of memory retention, as opposed to alternative, systematic strategies for locating the platform (such as swimming zigzag across the entirety of the tank until the platform is discovered), the platform is then removed from the tank and mice are allowed to free swim for 60 to 90 seconds, with an observer recording the amount of time mice spend swimming in the target quadrant where the platform was previously located. The Morris Water Maze paradigm therefore has a procedural component in which mice learn that they should mount the platform; a spatial memory component, in which mice learn where the platform is located in the tank; and a visual processing component, in which mice interpret the visual cues that allow them to locate the platform, any of which may be compromised by genetic background and/or injury.

Other behavioral paradigms have also been employed in the evaluation of mTBI experimental models, including hind leg flexion reflex, righting reflex, corneal reflex (blinking response), secretory signs (around the mouth and the eye), strength, beam balance task, beam walking coordination task, and exploration and locomotor activity tests as described previously (Pan et al., 2003; Zohar et al., 2003). Tests of basic well-being of the mice include assessments of motor function (staircase test), pain threshold (hot plate test), and anxiety level (elevated plus maze test). Although this list is not exhaustive, it is important to establish prior efforts at behavioral characterization of a given mTBI model before undertaking a new study. Consideration should also be given to clinical outcomes after human mTBI, such as LOC, balance disturbances, sleep problems, and depression. Behavioral analogues of these clinical outcomes should be included in experimental models. Future areas of study may also include fear conditioning, which could be highly clinically relevant as an experimental correlate of posttraumatic stress disorder after mTBI.

9.3.2 Histopathologic Outcomes after mTBI

Prior mTBI experimental models have evaluated a range of histopathological outcomes. Perhaps the most important of these are gross structural outcomes such as skull fracture and intracranial hemorrhage because these outcomes preclude the comparison of the model to the majority of human mTBI, an injury most often described as a disturbance of brain function, as opposed to a gross macroscopic structural injury.

Other frequently employed histopathological outcomes after experimental mTBI include the acute assessment of cerebral edema, blood–brain barrier damage, and cell death, which are notably absent in many models (Kane et al., 2012; Khuman, 2009), although present in blast and lateral fluid percussion models (Perez-Polo et al., 2013). Changes in GFAP and IBA-1 reactivity are sometimes evaluated to better characterize astroglial and microglial response after injury (Kane et al., 2012; Khuman, 2009; Perez-Polo et al., 2013), which is often robust. Several studies have sought to characterize the inflammatory response after mTBI, with particular attention to interleukin-1 alpha/beta and tumor necrosis factor alpha levels and macrophage/microglial and astrocytic activation (Khuman, 2009; Perez-Polo et al., 2013).

One of the most clinically relevant histopathological outputs after mTBI is axonal pathology, which has received increasing attention in the clinical literature (Browne et al., 2011; Inglese et al., 2005; Kasahara et al., 2012; Kirov et al., 2013; Topal et al., 2008). Experimental models of mTBI and repetitive mTBI have also shown axonal pathology with techniques such as amyloid precursor protein immunoreactivity, SMI-31 immunoreactivity, silver staining, and diffusion tensor imaging (Bennett et al., 2012; Mouzon et al., 2012; Shitaka et al., 2011). Of all these techniques, amyloid precursor protein may be the least sensitive, so investigators using this approach are advised to include additional methods for the assessment of axonal injury when possible.

Histopathologic correlates of CTE are also clinically relevant outcomes, including the evaluation of phosphorylated tau, Aβ, and hippocampal cell-less and gross brain atrophy (Rostami et al., 2012). When assessing for tau and Aβ pathology, it is useful to remember that insoluble forms are often not detectable in non-AD transgenic mice, so assessment of soluble forms is recommended.

9.3.3 The Effect of *APOE* on Outcome after mTBI

Of the few mTBI experimental models evaluating the effect of *APOE* isoforms, much of the work to date has focused on better characterizing the role of *APOE* in recovery after mTBI using *APOE-KO* mice. Using a weight drop model of mTBI, Han et al. demonstrated increased hippocampal neuronal degeneration in *APOE*-deficient versus control mice after mTBI (Han and Chung, 2000). In this model, *APOE*-deficient mice also had increased GFAP immunoreactivity. Similarly, in a repetitive mTBI model, Namjoshi et al. also showed worse outcome in *APOE-KO* mice with severe axonal injury and marked Aβ accumulation compared with wild-type controls (Namjoshi et al., 2013).

Although *APOE* has been shown to have an important role in recovery after mTBI, the effect of human *APOE* isoforms has not been well-established. Only one study to date has used *APOE*4 transgenic mice to evaluate the effect of *APOE*4 on outcome after mTBI. Using a weight drop repetitive mTBI injury model, no association was found between immediate or 6 month behavioral outcome in GFAP transgenic *APOE*4 versus WT mice (Mannix and Meehan et al., in press Annals of Neurology). Of note, control mice in this study were wild-type rather than *APOE*3 transgenic mice; further studies are needed to better establish the effect of *APOE*4 in this model.

9.4 CONCLUSION AND PERSPECTIVES

mTBI carries a great public health burden in this country and is the source of increasing attention in both clinical and experimental studies. The best-known genetic risk factor for worse outcome after TBI is the *APOE*4 allele. However, the clinical literature is difficult to interpret and sometimes contradictory, making careful use of experimental models in this setting even more important. Before undertaking a study of the effect of *APOE* status on outcomes after mTBI, investigators need to understand both the genetic background of the *APOE* model and the various choices for mTBI injury platforms. In fact, the effect of *APOE* status has not been well-studied in this setting of experimental models of mTBI, leaving a ripe future area of investigation wide open.

REFERENCES

Ariza, M., R. Pueyo, M. Matarin Mdel, C. Junque, M. Mataro, I. Clemente et al. 2006. Influence of *APOE* polymorphism on cognitive and behavioural outcome in moderate and severe traumatic brain injury. *J Neurol Neurosurg Psychiatry*. 77:1191–3.

Bennett, R.E., C.L. Mac Donald, and D.L. Brody. 2012. Diffusion tensor imaging detects axonal injury in a mouse model of repetitive closed-skull traumatic brain injury. *Neurosci Lett*. 513:160–5.

Brecht, W.J., F.M. Harris, S. Chang, I. Tesseur, G.Q. Yu, Q. Xu et al. 2004. Neuron-specific apolipoprotein e4 proteolysis is associated with increased tau phosphorylation in brains of transgenic mice. *J Neurosci*. 24:2527–34.

Browne, K.D., H. Chen, D.F. Meaney, and D.H. Smith. 2011. Mild traumatic brain injury and diffuse axonal injury in swine. *J Neurotrauma*. 28:1747–55.

Buttini, M., M. Orth, S. Bellosta, H. Akeefe, R.E. Pitas, T. Wyss-Coray et al. 1999. Expression of human apolipoprotein E3 or E4 in the brains of Apoe-/- mice: Isoform-specific effects on neurodegeneration. *J Neurosci*. 19:4867–80.

Chamelian, L., M. Reis, and A. Feinstein. 2004. Six-month recovery from mild to moderate traumatic brain injury: The role of *APOE*-epsilon4 allele. *Brain*. 127:2621–8.

Chen, Y., L. Lomnitski, D.M. Michaelson, and E. Shohami. 1997. Motor and cognitive deficits in apolipoprotein E-deficient mice after closed head injury. *Neuroscience*. 80:1255–62.

Chen, Z., L.Y. Leung, A. Mountney, Z. Liao, W. Yang, C. Lu et al. 2012. A novel animal model of closed-head concussive-induced mild traumatic brain injury: Development, implementation, and characterization. *J Neurotrauma*. 29:268–80.

Crawford, F.C., R.D. Vanderploeg, M.J. Freeman, S. Singh, M. Waisman, L. Michaels et al. 2002. *APOE* genotype influences acquisition and recall following traumatic brain injury. *Neurology*. 58:1115–8.

Han, S.H. and S.Y. Chung. 2000. Marked hippocampal neuronal damage without motor deficits after mild concussive-like brain injury in apolipoprotein E-deficient mice. *Ann N Y Acad Sci*. 903:357–65.

Inglese, M., S. Makani, G. Johnson, B.A. Cohen, J.A. Silver, O. Gonen, and R.I. Grossman. 2005. Diffuse axonal injury in mild traumatic brain injury: A diffusion tensor imaging study. *J Neurosurg*. 103:298–303.

Jordan, B.D., N.R. Relkin, L.D. Ravdin, A.R. Jacobs, A. Bennett, and S. Gandy. 1997. Apolipoprotein E epsilon4 associated with chronic traumatic brain injury in boxing. *JAMA*. 278:136–40.

Kane, M.J., M. Angoa-Perez, D.I. Briggs, D.C. Viano, C.W. Kreipke, and D.M. Kuhn. 2012. A mouse model of human repetitive mild traumatic brain injury. *J Neurosci Meth*. 203:41–9.

Kasahara, K., K. Hashimoto, M. Abo, and A. Senoo. 2012. Voxel- and atlas-based analysis of diffusion tensor imaging may reveal focal axonal injuries in mild traumatic brain injury—Comparison with diffuse axonal injury. *Magn Reson Imaging*. 30:496–505.

Khuman J., W.P. Meehan III, X. Zhu, J. Qiu, U. Hoffmann, J. Zhang, E. Giovannone, E.H. Lo, and M.J. Whalen. 2011. Tumor necrosis factor alpha and Fas receptor contribute to cognitive deficits independent of cell death after concussive traumatic brain injury in mice. *J Cereb Blood Flow Metab.* 31(2):778–89. doi: 10.1038/jcbfm.2010.172. Epub 2010 Oct 13.

Kim, J., J.M. Basak, and D.M. Holtzman. 2009. The role of apolipoprotein E in Alzheimer's disease. *Neuron.* 63:287–303.

Kirov, II, A. Tal, J.S. Babb, Y.W. Lui, R.I. Grossman, and O. Gonen. 2013. Diffuse axonal injury in mild traumatic brain injury: A 3D multivoxel proton MR spectroscopy study. *J Neurol.* 260:242–52.

Kutner, K.C., D.M. Erlanger, J. Tsai, B. Jordan, and N.R. Relkin. 2000. Lower cognitive performance of older football players possessing apolipoprotein E epsilon4. *Neurosurgery.* 47:651–7; discussion 657–8.

Lynch, J.R., W. Tang, H. Wang, M.P. Vitek, E.R. Bennett, P.M. Sullivan et al. 2003. *APOE* genotype and an ApoE-mimetic peptide modify the systemic and central nervous system inflammatory response. *J Biol Chem.* 278:48529–33.

Mannix R., W.P. Meehan, J. Mandeville, P.E. Grant, T. Gray, J. Berglass, J. Zhang, J. Bryant, S. Rezaie, J.Y. Chung, N.V. Peters, C. Lee, L.W. Tien, D.L. Kaplan, M. Feany, and M. Whalen. 2013. Clinical correlates in an experimental model of repetitive mild brain injury. *Ann Neurol.* 74(1):65–75. doi: 10.1002/ana.23858. Epub 2013 Aug 6.

Masliah, E., M. Mallory, N. Ge, M. Alford, I. Veinbergs, and A.D. Roses. 1995. Neurodegeneration in the central nervous system of apoE-deficient mice. *Exp Neurol.* 136:107–22.

Mayeux, R., R. Ottman, G. Maestre, C. Ngai, M. Tang, H. Ginsberg et al. 1995. Synergistic effects of traumatic head injury and apolipoprotein-epsilon 4 in patients with Alzheimer's disease. *Neurology.* 45:555–7.

McKee, A.C., R.C. Cantu, C.J. Nowinski, E.T. Hedley-Whyte, B.E. Gavett, A.E. Budson et al. 2009. Chronic traumatic encephalopathy in athletes: Progressive tauopathy after repetitive head injury. *J Neuropathol Exp Neurol.* 68:709–35.

Millar, K., J.A. Nicoll, S. Thornhill, G.D. Murray, and G.M. Teasdale. 2003. Long term neuropsychological outcome after head injury: Relation to *APOE* genotype. *J Neurol Neurosurg Psychiatry.* 74:1047–52.

Moran, L.M., H.G. Taylor, K. Ganesaligam, J.M. Gastier-Foster, J. Frick, B. Bangert et al. 2009. Apolipoprotein E4 as a predictor of outcomes in pediatric mild traumatic brain injury. *J Neurotrauma.* 26:1489–95.

Morris, R. 1984. Developments of a water-maze procedure for studying spatial learning in the rat. *J Neurosci Methods.* 11:47–60.

Mouzon, B.C., H. Chaytow, G. Crynen, C. Bachmeier, J.E. Stewart, M. Mullan et al. 2012. Repetitive mild traumatic brain injury in a mouse model produces learning and memory deficits accompanied by histological changes. *J Neurotrauma.* 2012:2761–73.

Namjoshi, D.R., G. Martin, J. Donkin, A. Wilkinson, S. Stukas, J. Fan et al. 2013. The liver X receptor agonist GW3965 improves recovery from mild repetitive traumatic brain injury in mice partly through apolipoprotein E. *PLoS One.* 8:e53529.

Nathoo, N., R. Chetry, J.R. van Dellen, C. Connolly, and R. Naidoo. 2003. Apolipoprotein E polymorphism and outcome after closed traumatic brain injury: Influence of ethnic and regional differences. *J Neurosurg.* 98:302–6.

Nicoll, J.A. 1996. Genetics and head injury. *Neuropathol Appl Neurobiol.* 22:515–7.

Omalu, B., J. Bailes, R.L. Hamilton, M.I. Kamboh, J. Hammers, M. Case et al. 2011. Emerging histomorphologic phenotypes of chronic traumatic encephalopathy in American athletes. *Neurosurgery.* 69:173–83; discussion 183.

Pan, W., A.J. Kastin, T. Rigai, R. McLay, and C.G. Pick. 2003. Increased hippocampal uptake of tumor necrosis factor alpha and behavioral changes in mice. *Exp Brain Res.* 149:195–9.

Perez-Polo, J.R., H.C. Rea, K.M. Johnson, M.A. Parsley, G.C. Unabia, G. Xu et al. 2013. Inflammatory consequences in a rodent model of mild traumatic brain injury. *J Neurotrauma.* 30:727–40.

Raber, J., D. Wong, M. Buttini, M. Orth, S. Bellosta, R.E. Pitas et al. 1998. Isoform-specific effects of human apolipoprotein E on brain function revealed in ApoE knockout mice: Increased susceptibility of females. *Proc Natl Acad Sci U S A.* 95:10914–9.

Rostami, E., J. Davidsson, K.C. Ng, J. Lu, A. Gyorgy, J. Walker et al. 2012. A model for mild traumatic brain injury that induces limited transient memory impairment and increased levels of axon related serum biomarkers. *Front Neurol.* 3:115.

Rubovitch, V., M. Ten-Bosch, O. Zohar, C.R. Harrison, C. Tempel-Brami, E. Stein et al. 2011. A mouse model of blast-induced mild traumatic brain injury. *Exp Neurol.* 232:280–9.

Shitaka, Y., H.T. Tran, R.E. Bennett, L. Sanchez, M.A. Levy, K. Dikranian, and D.L. Brody 2011. Repetitive closed skull traumatic brain injury in mice causes persistent multifocal axonal injury and microglial reactivity. *J Neuropathol Exp Neurol.* 70:551–67.

Sorbi, S., B. Nacmias, S. Piacentini, A. Repice, S. Latorraca, P. Forleo, and L. Amaducci. 1995. ApoE as a prognostic factor for post-traumatic coma. *Nat Med.* 1:852.

Sun, Y., S. Wu, G. Bu, M.K. Onifade, S.N. Patel, M.J. LaDu, A.M. Fagan, and D.M. Holtzman. 1998. Glial fibrillary acidic protein-apolipoprotein E (apoE) transgenic mice: Astrocyte-specific expression and differing biological effects of astrocyte-secreted apoE3 and apoE4 lipoproteins. *J Neurosci.* 18:3261–72.

Sundstrom, A., P. Marklund, L.G. Nilsson, M. Cruts, R. Adolfsson, C. Van Broeckhoven, and L. Nyberg. 2004. *APOE* influences on neuropsychological function after mild head injury: Within-person comparisons. *Neurology.* 62:1963–6.

Teasdale, G.M., G.D. Murray, and J.A. Nicoll. 2005. The association between *APOE* epsilon4, age and outcome after head injury: A prospective cohort study. *Brain.* 128:2556–61.

Teasdale, G.M., J.A. Nicoll, G. Murray, and M. Fiddes. 1997. Association of apolipoprotein E polymorphism with outcome after head injury. *Lancet.* 350:1069–71.

Topal, N.B., B. Hakyemez, C. Erdogan, M. Bulut, O. Koksal, S. Akkose et al. 2008. MR imaging in the detection of diffuse axonal injury with mild traumatic brain injury. *Neurol Res.* 30:974–8.

Ucar, T., G. Tanriover, I. Gurer, M.Z. Onal, and S. Kazan. 2006. Modified experimental mild traumatic brain injury model. *J Trauma.* 60:558–65.

Veinbergs, I., M.W. Jung, S.J. Young, E. Van Uden, P.M. Groves, and E. Masliah. 1998. Altered long-term potentiation in the hippocampus of apolipoprotein E-deficient mice. *Neurosci Lett.* 249:71–4.

Vitek, M.P., C.M. Brown, and C.A. Colton. 2009. *APOE* genotype-specific differences in the innate immune response. *Neurobiol Aging.* 30:1350–60.

Yao, J., S.S. Petanceska, T.J. Montine, D.M. Holtzman, S.D. Schmidt, C.A. Parker et al. 2004. Aging, gender and *APOE* isotype modulate metabolism of Alzheimer's Abeta peptides and F-isoprostanes in the absence of detectable amyloid deposits. *J Neurochem.* 90:1011–8.

Zohar, O., S. Schreiber, V. Getslev, J.P. Schwartz, P.G. Mullins, and C.G. Pick. 2003. Closed-head minimal traumatic brain injury produces long-term cognitive deficits in mice. *Neuroscience.* 118:949–55.

10 Cytoprotective Role of Prostaglandin D_2 DP1 Receptor against Neuronal Injury Following Acute Excitotoxicity and Cerebral Ischemia

Sylvain Doré and Abdullah Shafique Ahmad

CONTENTS

10.1 INTRODUCTION

Each year, hundreds of thousands of people experience a new or recurrent stroke. This devastating acute neurologic condition is broadly categorized into ischemic (84%) and hemorrhagic (16%) stroke. Stroke is a leading cause of death and long-term disability around the world. Immediately after an ischemic episode, there is a significant rise in glutamate levels at the synapse, leading to a significant increase in intracellular calcium levels. This leads to a cascade of events ending in acute excitotoxicity and cell death. This process also stimulates the activation of various enzymes, notably calcium-dependent $cPLA_{2\alpha}$, which liberates arachidonic acid from membrane phospholipids, making it a substrate to cyclooxygenase enzymes. At that point, through a coordinated system, all prostaglandin (PG) syntheses generate specific prostanoids, which are members of fatty acids that include PGD_2, PGE_2, $PGF_{2\alpha}$, PGI_2, and TxA_2. These PGs exert their actions through specific high-affinity G protein-coupled receptors named DP1-DP2 (PGD_2 receptor 1-PGD_2 receptor 2), EP1-EP4 (PGE_2 receptor 1-4), FP ($PGF_{2\alpha}$ receptor), IP (PGI_2 receptor or prostacyclin receptor), and TP (thromboxane receptor) receptors, respectively. The focus here is on PGD_2, as it is the most abundant PG in the brain. Although it has been well studied for its role in sleep modulation, recently, the role of PGD_2 and its receptor DP1 has been implicated in acute neurologic conditions. Using genetic and pharmacologic tools, these new results reveal that PGD_2 and its receptor pathway have beneficial effects and protect the brain against acute neurologic conditions such as excitotoxicity and ischemic stroke. Additional studies using compounds that selectively activate the DP1 receptor (PGD_2 receptor 1) might pave the way for therapeutic testing of this target in clinical settings.

This chapter briefly reviews the pathophysiology of ischemic stroke and treatments proposed in preclinical models, and especially focus on PGD_2 and the role of its DP1 receptor in acute ischemic injury. With the understanding that blood flow plays a critical function in ischemic outcomes, a better appreciation of the role of PGD_2/PGD_2 receptor 1 (DP1) in cerebral blood flow will be presented. Finally, there will be a discussion to review whether the PGD_2 DP1 receptor can be used as a target to attenuate brain damage initiated by ischemic stroke or excitotoxicity.

10.2 STROKE PATHOPHYSIOLOGY

Stroke is a medical emergency stemming from acute functional or anatomical neurologic dysfunction caused by a disturbance in the blood supply to discrete areas in the brain. In general, stroke is broken down into two categories: ischemic and hemorrhagic, accounting for approximately 84% and 16% of the total stroke incidents, respectively (Go et al., 2013; Roger et al., 2012). Ischemic stroke is a condition in which the blood supply to a particular region of the brain is permanently or transiently restricted because of the obstruction of a blood vessel by a blood clot or by arterial stenosis, thus depriving the brain cells of the glucose, oxygen, and energy required for their function (Andersen et al., 2009; Chambers et al., 1987; Garcia, 1975; Sutherland et al., 2012). Ischemic stroke is further categorized into thrombotic or embolic stroke. Thrombosis is a condition in which a blood clot is formed at the site of the occlusion, whereas embolism is a condition in which a blood clot is formed in a distant major artery and travels to the occlusion site. Thrombi are formed in the intracranial arteries when the basal lamina is damaged and plaques are formed along the injured site. This process is further facilitated by the activation and aggregation of platelets at the injured site, further facilitating the coagulation cascade. This process by itself can lead to hypoperfusion or occlusion of the blood supply, and certain factors can lead to an increase in shear stress and can dislodge the plaque, which can then lead to occlusion of the blood vessels, causing a significant decrease in the blood flow and leading to an acute ischemic condition. The ischemic episode is also facilitated by atherosclerosis, a condition in which the arterial wall is thickened for various reasons, including the deposition of fatty materials, inflammation, accumulation of macrophages, and formation of multiple thrombi (Aoki et al., 2007; Becker, 1998; Chamorro, 2004; Duvall and Vorchheimer, 2004; Eltzschig and Eckle, 2011; Fenton et al., 1998; Kanematsu et al., 2011; Kargman et al., 1998; Kogure et al., 1996; Mendez et al., 1987). Ultimately, all of this results in reduced blood flow or rupture of the inner arterial wall and the release of the blood clot into the blood stream. This is in contrast to an embolic stroke, in which a blood clot formed at a distant site travels and lodges in a cerebral artery. Some prominent causes of emboli include microemboli breaking away from a plaque in the carotid artery, atrial fibrillation, patent foramen ovale, hypokinetic left ventricle, and vascular or cardiac surgical procedures (Jaillard et al., 1999; Wardlaw et al., 2001; Wolf et al., 1987).

10.3 HYPOXIA, FREE RADICAL DAMAGE, ENERGY DEPLETION, AND NEURONAL DEPOLARIZATION

The brain has a relatively high demand for oxygen and glucose, and similar to the heart, it does not have a high antioxidant system, making it most susceptible to various neurologic conditions, including stroke. As a result, in ischemic/hypoxic conditions, the focal impairment of cerebral blood flow restricts the delivery of vital substrates such as oxygen and glucose. This subsequently imbalances ionic gradients and membrane potentials are lost. Consequently, presynaptic channels are activated and excitatory amino acids such as glutamate are released into the synaptic cleft. Because of energy depletion, the presynaptic reuptake of excitatory amino acids is attenuated, resulting in the accumulation of a significant amount of glutamate in the extracellular space (Ardizzone et al., 2004; Dirnagl et al., 1999; Kemp and McKernan, 2002; Lipton, 1999; Lipton and Rosenberg, 1994; Sutherland et al., 2011; Wang, 2011). As a result, *N*-methyl-*D*-aspartate (NMDA) and metabotropic glutamate receptors are activated and an influx of Ca^{2+} and Na^+ causes excitotoxicity and edema in the postsynaptic neurons (Figure 10.1). Excitotoxicity is one of the primary events propagating immediate cell death, whereas edema is one of the major determinants of the final outcome after stroke (Asano et al., 1985; Bhardwaj et al., 2000; Bounds et al., 1981; Choi, 1992; Dirnagl et al., 1999; Lynch and Guttmann, 2002; Macdonald and Stoodley, 1998; Manley et al., 2000; Urushitani et al., 2001).

10.4 EXCITOTOXICITY, CALCIUM SIGNALING, AND ACTIVATION OF PROSTANOID SYNTHESIS

Neurons in an ischemic brain usually die by means of necrotic or delayed cell death mechanisms. An important insight into the mechanism of penumbral infarction was revealed by the pioneering work of Dr. Brian Meldrum's laboratory showing that the preemptive blockade of the glutamate receptor reduced infarct size (Simon et al., 1984). During ischemic stroke, levels of excitatory amino acids, especially glutamate, increase rapidly within the affected area because of their synaptic release from core neurons undergoing ischemic depolarization. Glutamate regulates the activity of various excitatory metabotropic receptors, including the NMDA receptor, which is a Ca^{2+} membrane channel. Glutamate acts as an NMDA receptor agonist and increases Ca^{2+} influx to the postsynaptic neurons up to a toxic level, causing immediate injury to these neurons (Lipton, 1999; Smith, 2004). The use of a competitive and/or noncompetitive antagonist of the NMDA receptor can reduce experimental infarct volume to a significant level by antagonizing the effects of glutamate (Lipton, 1999). Based on these preclinical reports, a similar approach was further extended in clinical stroke settings; however, no appreciable beneficial effects were observed—most likely from the limitation for timely application of such drugs after the first symptoms of stroke onset.

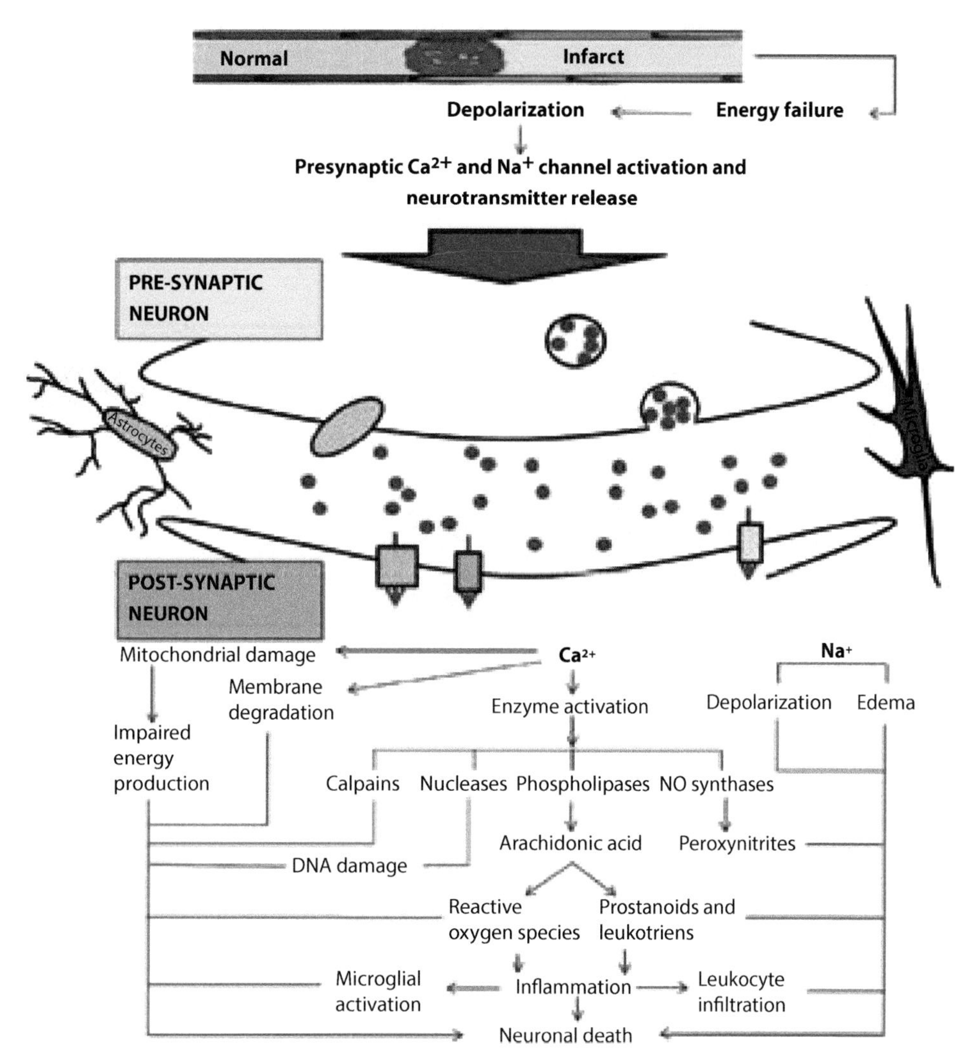

FIGURE 10.1 Some of the major cellular pathophysiologic cascades associated with ischemic stroke. Occlusion of the blood vessel leads to cerebral ischemia, which subsequently leads to energy failure and presynaptic neuronal membrane depolarization. This causes massive release of glutamate into the synaptic cleft, which then activates specific glutamatergic receptors along with other channels and profoundly increases Ca^{2+} and Na^{+} intracellular levels. Increase in Na^{+} results in membrane depolarization and edema (water content in the affected tissue) and finally leads to neuronal cell death. Increased levels of intracellular Ca^{2+} activate various enzymes and initiate membrane degradation and mitochondrial dysfunction. Key enzymes that are activated include calpain, nuclease, phospholipase, and nitric oxide synthase enzymes. Nitric oxide synthase activation can result in peroxynitrite generation, which subsequently leads to cell death. Phospholipase activation results in liberation of AA and the subsequent generation of reactive oxygen species and prostanoids (PGs and thromboxanes). The reactive oxygen species and prostanoids lead to inflammation, which results in leukocyte infiltration and microglial activation, finally leading to cell (neuronal) death.

An increase in the intracellular levels of Ca^{2+} also initiates a cascade of cytoplasmic and nuclear events that dictate the pathophysiology of the ailment. Some of these events include activation of proteolytic enzymes that degrade cytoskeletal and extracellular matrix proteins and activation of phospholipase A_2—notably cytosolic phospholipase A_2 ($cPLA_{2\alpha}$), which initiates a proinflammatory cascade (Adams et al., 1996; Bazan, 2005; Bonventre, 1997; Kishimoto et al., 2010; Phillis et al., 2006; Phillis and O'Regan, 2004; Tanaka et al., 2003). PLA_2 acts on membrane-bound phospholipids and generates arachidonic acid (AA), which is subsequently converted into hydroxyeicosatetraenoic acid (HETE) and leukotriene through lipoxygenase (LOX) pathways, epoxyeicosatrienoic acid (EETs), and HETE through cytochrome P450 (CYP) pathways, or into prostanoids and thromboxanes through cyclooxygenase (COX) and prostaglandin (PG)/thromboxane synthase pathways (Bazan et al., 2002; Funk, 2001; Khanapure et al., 2007; Kudo and Murakami, 2002; Phillis et al., 2006; Wymann and Schneiter, 2008) (Figure 10.2). Thus, the initial insult after ischemic stroke causes a hostile environment with resilient excitotoxicity, the generation of oxygen- and nitrogen-free radicals and proteolytic enzymes, and the activation of an inflammatory cascade that initiates cellular, molecular, and hemodynamic events converting the center of the damage into an ischemic core. These events destroy brain cells in the ischemic core by lipolysis, proteolysis, disaggregation of microtubules, bioenergetic

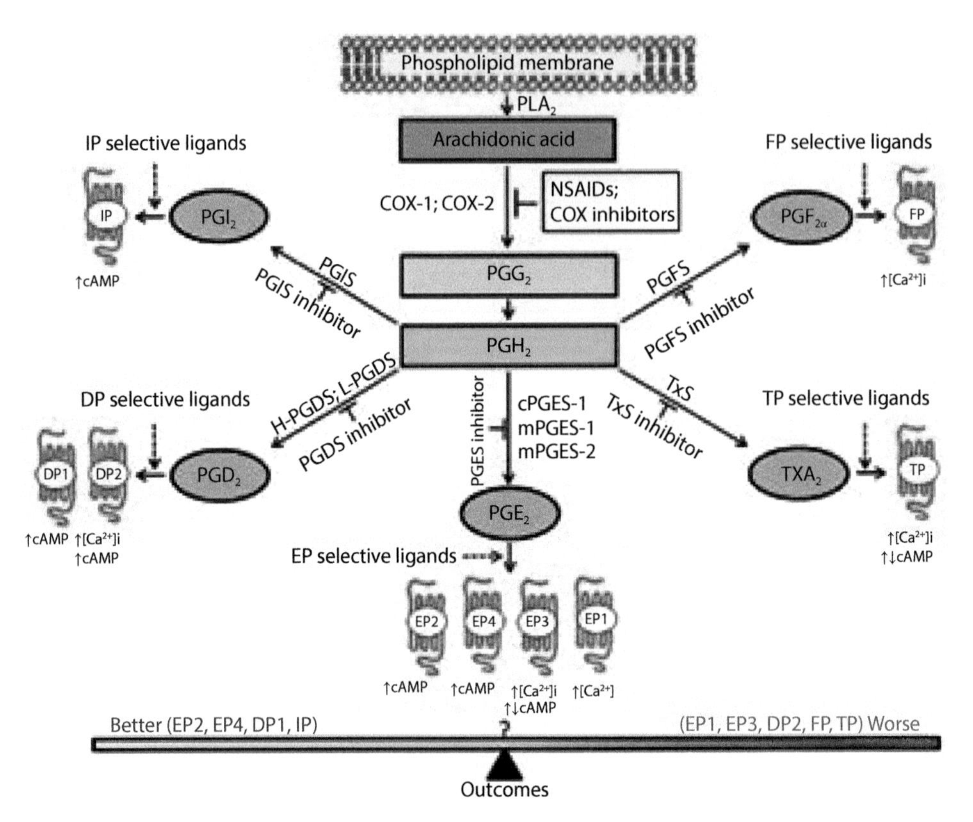

FIGURE 10.2 Simplified schematic illustration of PG syntheses and their action on their high-affinity specific receptors. AAs are generated from membrane phospholipids, and then COX acts on AA to generate intermediate PG endoperoxide H (PGH_2). The tissue-specific PG synthase then converts PGH_2 into respective prostanoids. Use of NSAIDs or COX inhibitors can prevent the generation of prostanoids. Similarly, PG synthase inhibitors can prevent the generation of specific PG, whereas the use of selective receptor ligands can manipulate the outcome related to that specific receptor. According to the generalized hypothesis, those receptors that activate the cAMP pathway tilt the scale toward better outcomes after ischemic stroke, whereas those receptors that activate the Ca^{2+} pathway tilt the scale toward worse outcomes. However, this general rule is not always true, as has been shown recently. *Abbreviations:* cPGES-1, cytosolic PG E synthase-1; mPGES-1 and -2, membrane-associated PG E synthase-1 and -2; H-PGDS, hematopoietic PG D_2 synthase; L-PGDS, lipocalin-type PG D_2 synthase; PGFS, PG $F_{2\alpha}$ synthase; PGIS, PG I_2 (or prostacyclin) synthase; TxS, thromboxane synthase.

failure, and the breakdown of ion homeostasis (del Zoppo et al., 2000; Dirnagl et al., 1999; Iadecola and Alexander, 2001; Lo, 2008; Lo et al., 2003). This ischemic core is surrounded by a partially affected area often referred to as the ischemic penumbra, the peri-infarct area, or the area at risk. Either because of incomplete occlusion, hypoperfusion, or partial compensatory blood flow through collateral circulation, the ischemic penumbra has partially preserved energy metabolism (Bang et al., 2011; Frykholm et al., 2005; Liebeskind, 2005; Lo, 2008). This is the salvageable area; however, if left unattended and without treatment, this area can be overtaken by the advancing ischemic core through ongoing excitotoxicity and/or deleterious secondary cascades such as spreading membrane depolarization, edema, postischemic inflammation, and delayed cell death (del Zoppo et al., 2000; Dirnagl et al., 1999; Iadecola and Alexander, 2001; Lo, 2008; Lo et al., 2003). Although the primary cascade of excitotoxicity has been a target to treat stroke, because of the complex pathophysiology of stroke, many of the poststroke neuroprotective efforts are focused on salvaging the "penumbra" by minimizing secondary events.

10.5 TREATMENTS AVAILABLE FOR STROKE

Although stroke is one of the leading causes of morbidity and mortality in developed nations, effective therapies are still elusive. Essentially, the only approved pharmacologic therapy against stroke is the use of thrombolytic agents such as tissue plasminogen activator (tPA) (del Zoppo and Koziol, 2007; Yamamoto et al., 1991; Zivin et al., 1985). However, because the use of tPA is highly time-dependent, the most beneficial effect of tPA treatment can only be obtained within ~4.5 hours after the onset of ischemia. If given beyond the stroke-onset therapeutic window of ~4.5 hours, tPA can lead to vascular rupture; thereby, transforming an ischemic stroke into a hemorrhagic stroke, a phenomenon known as hemorrhagic transformation (Deb et al., 2010; Rosell et al., 2008; Wang, 2011; Xing et al., 2011). Consequently, a significant amount of research money and effort has been devoted to actively pursuing cutting-edge research and subsequently translating any findings into clinical trials to elicit treatment options or to using other therapies that can potentially minimize tPA side effects and/or extend the therapeutic window of tPA. Some of these approaches include the use of NMDA receptor antagonists, free radical scavengers, vasodilators, matrix metalloproteinase inhibitors, thrombolytics and antithrombotics, platelet inhibitors, COX inhibitors, and many others (Chan, 2001; Cheng et al., 2006; Clark et al., 2001; Cunningham et al., 2005; Fang et al., 2006; Gerriets et al., 2003; Green and Ashwood, 2005; Grossetete and Rosenberg, 2008; Hinkle and Bowman, 2003; Hurley et al., 2002; Maeda et al.; Morita-Fujimura et al., 2001; Rao and Balachandran, 2002; Saleem et al., 2008; Serteser et al., 2002; Vartiainen et al., 2001; Weinberger, 2006; Wilson and Gelb, 2002; Zhao et al., 2006). Although some of these later compounds have had success at certain stages of the preclinical and clinical trials; unfortunately, none of them have been successful in resolving ischemic brain damage (Green, 2002; Ikonomidou and Turski, 2002). It is clear that there is an urgent need to pursue an effective therapeutic option or a combination of drugs that can target the various steps of the stroke cascade over time.

10.6 AA CASCADE

Essential fatty acids (EFAs), a subset of unsaturated fatty acids, play various physiologic and pathophysiologic roles in neural cells by regulating intracellular signaling. Because of the lack of the necessary enzymes that are required for their synthesis, humans obtain EFAs from external sources and then incorporate them into the phospholipids. In the brain, EFAs are released by the action of PLA_2 enzymes and they modulate various physiologic functions, including synaptic function, neuronal toxicity, cerebrovascular tone, neuroinflammation, and oxidative stress (Bazan, 2005; Doré et al., 2003; Kishimoto et al., 2010; Phillis and O'Regan, 2003; Sanchez-Mejia et al., 2008). The PLA_2 enzymes are a family of lipases that are categorized into 12 groups based on various factors such as biological activity, substrate specificity, activating factors, and localization (Burke and Dennis, 2009a, 2009b; Kudo and Murakami, 2002). In the brain, AA and docosahexaenoic acid are present in bilipid layers of brain cells and appear to be among the most important fatty acids because of their action in generating more free fatty acids and their biological activity. AA acts as a substrate for at least three well-studied pathways, which are the LOX, CYP, and COX/PG synthase pathways. These pathways are tissue specific, and often cell specific, and the final outcomes are highly dependent on the nature and site of injury.

10.6.1 LOX Pathway

Briefly, in this pathway, AA is converted into leukotrienes by the action of LOX. Leukotrienes act through specific receptors and play important physiologic and pathophysiologic roles. Recently, some selective and effective compounds targeting the LOX pathway have been developed and are being tested to treat asthma and seasonal allergies (Capra et al., 2007; Ribeiro et al., 2006). These pathways are actively being further explored as novel therapeutic targets, and their roles in various ailments are being defined.

10.6.2 CYP Pathway

The AA in this pathway is acted upon by two different enzymes, namely CYP hydroxylase and CYP epoxygenase, leading to two distinct pathways. The CYP hydroxylase enzymes convert AA into HETE. The main metabolite of this pathway is 20-HETE, which is considered to be proinflammatory, and plays an important vascular function (Ishizuka et al., 2008; Roman, 2002). Various metabolites generated in this pathway are being tested for their efficacy in regulating the outcomes of various disease conditions, including stroke (Miyata et al., 2005; Renic et al., 2009; Roman, 2002). The CYP epoxygenase enzymes generate various EET from AA. It has been proposed that EETs are endothelium-derived hyperpolarizing factors and have an anti-inflammatory action in various animal models of diseases (Campbell et al., 1996; Fleming, 2007; Gross et al., 2008; Imig, 2005).

10.6.3 COX/PG Synthase Pathway

COX isoforms COX-1 and COX-2 are the two most studied and recognized oxygenases (Smith et al., 2000). Once cells are activated by trauma, stress, growth factor, cytokines, etc., a 20-carbon unsaturated fatty acid is released from the lipid bilayer of the plasma membrane and is mobilized by different types of phospholipases such as secretory phospholipase A_2, $cPLA_2$, and calcium-independent phospholipase A_2. The AA generated as a result of PLA_2 action is converted into prostanoids by the sequential action of COX and PG synthase enzymes (Figure 10.2). COXs are considered to be proinflammatory, therefore, drugs such as COX inhibitors, aspirin, nonsteroidal anti-inflammatory drugs (NSAIDs), and other compounds that inhibit COX activity are effective drugs for treating pain, inflammation, and cardiovascular disease (Asanuma et al., 2004; Ballou et al., 2000; Grosser et al., 2006; Hoffmann, 2000; Minghetti, 2004; Park et al., 2007; Simmons et al., 2004; Szekely et al., 2004). The COX inhibitors are usually used acutely due to the side effects. Based on their anti-inflammatory effects, COX inhibitors were promoted as promising candidates for treating stroke and other neurologic conditions; however, this enthusiasm ceased when several clinical trials revealed an increased incidence of acute renal failure, myocardial infarction, and thrombotic stroke in patients treated with COX-2 inhibitors (FitzGerald, 2004; Grosser et al., 2006).

10.7 COX AND NEURODEGENERATION

COX enzymes are rate-limiting enzymes associated with the conversion of AA into PGE_2, PGD_2, $PGF_{2\alpha}$, PGI_2, and TxA_2, collectively known as prostanoids. Two types of COX (COX-1 and COX-2) have been well studied. They are also known as PG endoperoxide H synthases-1 and -2. The two isozymes are quite similar and share ~60% homology in their amino acids. The COX-2 active site has a larger pocket, thus making it more accommodating for using fatty acids (Rouzer and Marnett, 2009; Smith et al., 2000). COX-1 is responsible for normal physiologic or "housekeeping" functions, whereas COX-2 is an inducible isoform involved in inflammation, mitogenesis, and signal transduction (Kam and See, 2000). Although COX-1 is generally considered to be constitutively expressed, depending on which "inducible" PG synthases it is coupled to, it can have effects on various inflammatory and other cellular and organelle pathways.

COX enzymes have catalytic and regulatory subunits and are bound to the endoplasmic reticulum and nuclear membrane. COX-2 immunoreactivity in neurons shows that it is localized to dendritic spines, which are the site of NMDA receptor–mediated neurotransmission (Kaufmann et al., 1996). COX-2 is expressed in the brain at a high basal level selectively in pyramidal neurons of hippocampal and cortical circuits in neurons of the amygdala. COX-2 is an important mediator of inflammation, including neuroinflammation (Feng et al., 1993; Gobbo and O'Mara, 2004). It has been reported that coordinated upregulation of $cPLA_2$ and COX-2 activities contribute to brain damage after cerebral ischemia and neuroinflammation (Bosetti and Weerasinghe, 2003; Sandhya et al., 1998). COX-2 plays an important role in brain injury caused by ischemia (Doré et al., 2003), kainic acid–induced seizures (Baik et al., 1999; Kelley et al., 1999), NMDA toxicity (Hewitt, 2000; Iadecola et al., 2001), Alzheimer disease (Xiang et al., 2002), and Parkinson disease (Teismann et al., 2003), along with other neurologic disorders (Consilvio et al., 2004; Madrigal et al., 2003). The effects of genetic manipulation of COX-2 on focal ischemia have been reported by several groups, including ours (Doré et al., 2003; Iadecola et al., 2001). In COX-2 knockout mice ($COX\text{-}2^{-/-}$), Iadecola and coworkers (Iadecola et al., 2001) found reduced cerebral infarction after middle cerebral artery occlusion (MCAO), whereas in our previous study, we found that overexpression of COX-2 enhanced cerebral infarction (Doré et al., 2003). Consistent with these results, we and others (Doré et al., 2003; Iadecola and Ross, 1997; Nakayama et al., 1998; Nogawa et al., 1997; Resnick et al., 1998) have shown that the use of specific COX-2 inhibitors significantly reduced the infarction in wild-type mice. However, we, for the first time, unveiled that COX-2–specific inhibitors were unable to significantly attenuate the infarction (Doré et al., 2003), notably in conditions in which COX-2 was already overexpressed. These conflicting results suggest that the basal level of COX-1 and other PG synthase enzymes are important factors to consider in addition to COX-2. Moreover, the unique and different roles of various PGs and their receptors in regulation of neuroinflammation and neuronal cell death should not be neglected.

The NSAIDs have been widely used to nonselectively target COX-induced inflammation. Since their discovery (Vane, 1971), NSAIDs have been used to inhibit COX and consequently the synthesis of PGs, and to play a vital role in the physiology and pathophysiology of the cell. Given the anti-inflammatory effects of COX inhibitors, these compounds were proposed and successfully tested to be neuroprotective in diverse neurologic conditions. However, this promising arena essentially ceased when several clinical trials revealed adverse side-effects associated with COX inhibitors. For example, detrimental outcomes were observed in Merck's Adenomatous Polyp PRevention On Vioxx (i.e., APPROVe) study, which resulted in the termination of this clinical trial and withdrawal of the COX-2 inhibitor Vioxx on September 30, 2004, and there was a similar end for Bextra, manufactured and marketed by Pfizer (Davies and Jamali, 2004; Lenzer, 2005; Singh, 2004; Topol, 2004). Essentially, only Celebrex is currently used for various other ailments. This opened a new debate on the pharmacological effects of COX inhibitors and their uses, and the focus of argument shifted toward PGs. Therefore, our laboratory and that of others have been especially interested in the various PG receptors, knowing that the majority of all drugs are targeting G protein–coupled receptors (GPCRs). Therefore, it is imperative to understand the importance of PG syntheses, their relationships, and the role of PGs and their receptors, notably in the cardiovascular and neuronal systems.

10.8 PG SYNTHESES

PGs are bioactive lipids involved in a myriad of diverse and essential homeostatic physiological and pathological functions. PGs are formed by most of the body's cells and can act as autocrine or paracrine agent, or both. Unlike most of the other hormones, PGs are not stored but are synthesized de novo from AA, and are rapidly processed because of their short half-lives. As mentioned previously in the AA cascade section, PGH_2 serves as a common substrate and can be converted into the relatively stable PGs and thromboxanes by the action of specific synthases/isomerases. PGE synthase (cytosolic: cPGES-1; membrane-associated: mPGES-1 and mPGES-2), PG I synthase (PGIS), PG D synthase (hematopoietic-type: H-PGDS, and lipocalin-type: LPGDS), PG F synthase (PGFS), and thromboxane synthase (TxS) into PGE_2, PGI_2, PGD_2, $PGF_2\alpha$, and TxA, respectively (Figure 10.2). Given the important roles that PGs play, PG synthase inhibitors have been tested in various disease conditions to better understand the role of their respective PGs (Kanekiyo et al., 2007; Murakami and Kudo, 2006; Qu et al., 2006; Santovito et al., 2009; Takemiya et al., 2010).

10.9 PGs AND THEIR HIGH-AFFINITY RECEPTORS

Each PG binds with different affinities to various GPCR designated as DP1-DP2 (PGD_2 receptor 1, PGD_2 receptor 2), EP1-EP4 (PGE_2 receptor 1-4), FP ($PGF_{2\alpha}$ receptor), IP (prostacyclin receptor), and TP (thromboxane receptor) for PGD_2, PGE_2, PGF_{2a}, PGI_2, and TxA_2, respectively, and exerts its physiologic and pathophysiologic effects through these receptors (Breyer et al., 2001; Coleman et al., 1995; Jones et al., 2009; Narumiya, 2009; Narumiya et al., 1999; Sharif et al., 1998; Woodward et al., 2011). Although the literature associates these prostanoids as essentially binding only with their receptor family members, in most cases, these receptors cross-react with other receptors. However, this has not been discussed and elaborated on well enough in literature.

In general, EP2, EP4, DP1, and IP are known to cause an increase of intracellular cyclic adenosine 3',5'-monophosphate (cAMP), whereas EP1, FP, and TP cause an increase of intracellular free Ca^{2+} (Wright et al., 2001 and see Mohan et al. for a comprehensive review [Mohan et al., 2012]). These receptors share approximately 20% to 30% of their sequences with each other. We and others have previosuly exposed the "genealogy" of these receptors, which further supports the proposed classification of these receptors and their downstream cell signaling (Hirai et al., 2001; Mohan et al., 2012; Toh et al., 1995). However, caution should be used when impying it as a general rule because there are always exceptions to such rules. Most of the prostanoid receptors interact with a wide range of intracellular signaling pathways and facilitate receptor activation and functions. PGs play significant physiologic and pathophysiologic roles in circulation, vascular regulation, hypertension, sleep regulation, temperature regulation, parturition, ocular pressure, skin inflammation, airway maintenance, bowel syndromes, vasodilation, vasoconstriction, excitotoxicity, and stroke (Chemtob et al., 1990a, 1990b, 1995; Faraci and Heistad, 1998; Wright et al., 2001). In general, it has been shown that the receptors that activate the cAMP pathway lead to better outcomes after ischemic stroke, whereas those that activate the Ca^{2+} pathway lead to worse outcomes. However, this general perception is not true in all conditions, and exceptions have also been reported. Because PGD_2 is the most abundant and often neglected prostanoid, we will briefly expose additional focused information about this prostanoid and its main receptor, DP1. The other PGD_2 receptor was named DP2 once it was found that PGD_2 has some affinity toward this orphan receptor, previously known as $CRTH_2$ receptor (Nagata et al., 1999a). Moreover, this receptor did not evolve from the same phylogenetic tree as the other PG receptors (Mohan et al., 2012).

10.10 PGD_2 AND ITS METABOLISM

PGD_2 is generated from PGH_2 by the action of two distinct PGD synthases: L-PGDS and H-PGDS. PGD_2 was first identified in the late 1960s (Granstrom et al., 1968; Hayashi and Tanouchi, 1973), and a later study provided evidence that this PG is a potent inhibitor of platelet aggregation (Smith et al., 1974). Since then, the role of PGD_2 has been investigated in a wide range of pharmacological activities. The diverse activities of PGD_2 in regulating local blood flow, as a bronchial airway caliber, in leukocyte function, and so on, are mediated by its high-affinity interactions with its GPCRs. However, the half-life of PGD_2 is very short (about 1.5 minutes in blood); therefore, under certain conditions, it is rapidly metabolized. The main products that have been detected in vivo are $\Delta^{12}PGJ_2$ and $9\alpha11\beta PGF_2$ (Heinemann et al., 2003; Sandig et al., 2006b). Other putative metabolites are 13,14-dihydro-15-keto-PGD_2 16, $\Delta^{12}PGD_2$, 15-deoxy- $\Delta^{12,14}PGD_2$, and 15-deoxy-$\Delta^{12,14}PGJ_2$ (Gazi et al., 2005; Monneret et al., 2002) (Figure 10.3). Although the exact mechanism of PGD_2 metabolism is still elusive, it has been postulated that 13,14-dihydro-15-keto-PGD_2, $\Delta^{12}PGD_2$, and 15-deoxy-$\Delta^{12,14}PGJ_2$ are formed locally at the site of inflammation. It is important to note that all of these metabolites confine their activity to DP2 receptor, and their activity on the DP1 receptor has not been reported in the literature. Other metabolites of PGD_2 such as $\Delta^{12}PGJ_2$ and 15-deoxy- $\Delta^{12,13}PGJ_2$ have effects on the resolution of inflammation by inhibiting cytokine production and the induction of apoptosis. These effects are mediated by the peroxisome proliferator-activated receptor-γ (PPARγ)-dependent

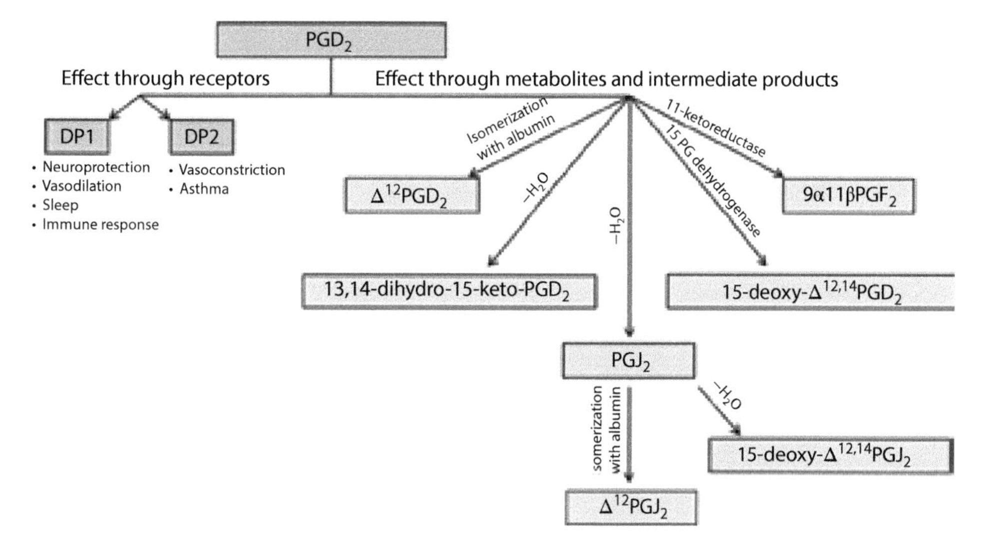

FIGURE 10.3 PGD_2 metabolism and its effects. PGD_2 can act as autocrine or paracrine agent, or both, and exerts its effects mainly through its GPCR DP1 receptor, although some reports on peripheral systems suggest the action of PGD_2 through $CRTH_2$ (or DP2) receptors. The half-life of PGD_2 is very short and it is rapidly metabolized into various stable products. The precise mechanism of PGD_2 metabolism is still obscure. Most of these metabolites have a direct effect or exert their actions through $CRTH_2$ or PPARγ receptors or other intracellular pathways.

or PPARγ-independent mechanisms (Forman et al., 1995; Rossi et al., 2000). Previous work from our laboratory also shows that J series PG ($dPGJ_2$ and 15-deoxy-$\Delta^{12,14}PGJ_2$) treatments upregulate heme oxygenase and attenuate neuroinflammation (Zhuang et al., 2003a, 2003b).

10.11 PGD_2 AND DP RECEPTORS

PGD_2 has peripheral and central physiologic effects. Some of the peripheral effects of PGD_2 include vasodilatation, bronchoconstriction, inhibition of platelet aggregation, glycogenolysis, vasoconstriction, allergic reaction mediation, and intraocular pressure reduction (Angeli et al., 2004; Casteleijn et al., 1988; Darius et al., 1994; Matsugi et al., 1995; Matsuoka et al., 2000; Narumiya and Toda, 1985; Sturzebecher et al., 1989; Whittle et al., 1983). Among all PGs, PGD_2 is the most abundant in the brain. In the central nervous system, under normal physiologic conditions, PGD_2 contributes to sleep induction, modulation of body temperature, olfactory function, hormone release, nociception, and neuromodulation (Eguchi et al., 1999; Gelir et al., 2005; Hayaishi, 2002; Hayaishi and Urade, 2002; Mizoguchi et al., 2001; Urade and Hayaishi, 1999). PGD_2 levels are significantly increased under pathologic conditions (Hata and Breyer, 2004; Hatoum et al., 2005; Luster and Tager, 2004; Naffah-Mazzacoratti et al., 1995; Seregi et al., 1990; Tegtmeier et al., 1990); however, whether this increase is harmful or beneficial is controversial. PGD_2 protects astrocyte cultures by increasing the production of neurotrophins (Angeli et al., 2004). It has also been shown that PGD_2 acts as an inflammatory mediator of allergic reactions, such as asthma and conjunctivitis, which are initiated by the immunoglobulin E–mediated type 1 response. Moreover, there are reports showing some deleterious effects of PGD_2, mainly regulated through the DP2 receptor or through PPARγ pathways (Almishri et al., 2005; Gazi et al., 2005; Kagawa et al., 2011; Nagata and Hirai, 2003; Nagata et al., 1999a).

10.12 DP1 RECEPTOR

PGD_2 exerts its biological action by binding and activating through distinct and specific cell receptors. Recent findings suggest that there are two types of the receptor for PGD_2: the first is known as the DP1 or DP receptor and the second has recently been discovered and is known as chemoattractant receptor homologous molecule expressed on T helper type 2 cells ($CRTH_2$). $CRTH_2$ is also now known as the DP2 receptor; however, it has a different "genealogy" and a different homology compared with other prostanoid receptors (Abe et al., 1999; Hata et al., 2003; Mohan et al., 2012; Toh et al., 1995). $CRTH_2$ signaling is independent of PGD_2 production (Bohm et al., 2004; Sandig et al., 2006a) and has different biological functions and cellular distribution than the DP1 receptor (Hirai et al., 2001; Sawyer et al., 2002). It is located on the 11q12-q13.3, 19A, and 1q43 chromosomes in the human, mouse, and rat, respectively, and is encoded by the genes PTGDR2, Gpr44, and Ptgdr2 in the human, mouse, and rat, respectively. It contains 382 to 403 amino acids (Abe et al., 1999; Hata et al., 2003; Nagata et al., 1999b; Shichijo et al., 2003) and its structure closely resembles that of the formyl peptide receptor (formyl-Met-Leu-Phe) and the B leukotriene receptor. It is expressed on Th2-type lymphocytes, basophils, and eosinophils. DP2 is widely expressed in the brain, heart, stomach, ovary, testes, and skin, and it is biologically involved in eosinophil activation, chemotaxis, degranulation, and cytokine production.

Primarily, the anti-inflammatory and vasoregulatory properties of PGD_2 are associated with DP1 receptors; although, the proinflammatory effects of DP1 have also been reported (Angeli et al., 2004; Hammad et al., 2003; Matsuoka et al., 2000; Spik et al., 2005). DP1 is a GPCR coupled with $G\alpha s$ and is located on the 14q22.1, 14B, and 15p14 chromosomes in the human, mouse, and rat, respectively. The DP1 receptor (encoded by the gene PTGDR) was first cloned in 1994, and it contains 357 to 359 amino acids (Boie et al., 1995; Hirata et al., 1994; Ishikawa et al., 1996; Wright et al., 1999). On the phylogenetic tree, this receptor shows significant sequence homology with IP and EP2 receptors of PGI_2 and PGE_2, respectively, and is distributed throughout the body, including in ciliary epithelial cells, the retinal choroid, the iris, eosinophils, platelets, the colon, and the brain. Physiologically, DP1 receptors are involved in inhibiting platelet aggregation and histamine release, and in the relaxation of myometrium and smooth muscles. Furthermore, any alteration in DP1 receptor results in an asthmatic response, reduced nicotinic-induced vasodilation, and sleep alteration. The human DP1 receptor binds strongly to PGD_2, followed by PGE_2, PGF_2, and PGI_2, and least to thromboxane A_2. These receptors are expressed in various tissues and cells, including bone marrow, small intestine, lung, stomach, mast cells, and brain, and are involved in various pathophysiological functions such as platelet activation, vasodilation (niacin flush), sleep regulation, and immune cell activation. The DP1 receptor stimulates adenylyl cyclase, leading to increased levels of cAMP and decreased platelet aggregation (Hata and Breyer, 2004). Recently, several selective agonists such as BW245C, AS702224, and TS-022, and antagonists such as MK-0524 (or Laropiprant) and BWA868C of the DP1 receptor are being used in various animal models (Cheng et al., 2006; Crider et al., 1999; Hirano et al., 2007; Kabashima and Narumiya, 2003; Van Hecken et al., 2007). The use of these pharmacologic agents and genetic manipulations is extremely helpful in providing insight into the mechanism of action of PGD_2 and the DP1 receptor (Table 10.1). The in vivo role of DP1 in the periphery has been well investigated (Angeli et al., 2004; Hata and Breyer, 2004; Koch et al., 2005), whereas until recently, its role in the brain was limited to studies related to sleep and eye movement (Campbell and Feinberg, 1996; Hayaishi and Urade, 2002; Obal and Krueger, 2003; Urade and Hayaishi, 1999).

TABLE 10.1
List of DP1 Receptor Agonists and Antagonists That Have Been Tested in Various Preclinical and Clinical Studies

Agonist	Route	Dose	Species	Experimental Model
192C86	Intravenous infusion	0.007–0.058 μg/kg/min	Human	Ex vivo platelet aggregation (Gray et al., 1992)
	Topical	2–50 μg	Rabbit, Cat	Intraocular pressure (Matsugi et al., 1995)
AL-6598	Topical	25 μL of 0.01%	Monkey	Intraocular pressure (Toris et al., 2006)
BW 245C	Intrathecal	83 ng/kg	Mouse	Nociception-induced allodynia (Minami et al., 1997)
		100 μM	Mouse	OVA-alum asthma (Hammad et al., 2007)
	Topical	2.5 μg	Rabbit	Intraocular pressure (Goh et al., 1988)
		50 μM	Mouse	Epicutaneous sensitization to OVA (Angeli et al., 2004)
		25 μL 0.01–1%	Rabbit	Intraocular pressure (Woodward et al., 1990)
	Intracerebroventricular	10–50 nmol	Mouse	NMDA-induced acute excitotoxicity (Ahmad et al., 2010)
		10 nmol	Mouse	Ischemia-reperfusion injury (Ahmad et al., 2010)
	Oral	50 and 150 μg	Human	Pharmacodynamics of BW245C (Shah et al., 1984)
	Intraperitoneal	0.02–2.0 mg/kg	Mouse	Cerebral blood flow and hemostasis (Ahmad et al., 2013a, 2013b)
GIF-0173	Intravenous	0.01–0.1 mg/kg	Rat	Photochemically induced thrombosis model of MCAO (Thura et al., 2009)
SQ-27986	Topical	2.5–25 μg	Rabbit	Intraocular pressure (Woodward et al., 1993)
TS-022	Topical	0.25–250 nM	Mouse	NC/Nga mouse model of dermatitis (Arai et al., 2007; Sugimoto et al., 2007)
BW A868C	Intracerebroventricular	1.6 nmol	Mouse	Anxiolytic-like activity (Zhao et al., 2009)
	Intravenous	0.1–1.0 mg/kg	Guinea pig	PGD_2-induced bronchoconstrictor (Hamid-Bloomfield et al., 1990)
Laropiprant (MK-0524)	Oral	20 mg	Human	Niacin-induced flushing (HPS2-THRIVE-Collaborative-Group et al., 2013)
		20 mg	Human	Myocardial infarction (Bregar et al., 2013)
		40–100 mg	Human	Niacin-induced vasodilation (Cheng et al., 2006; Dishy et al., 2009)
	Subcutaneous	0.04–40 mg/kg	Mouse	Niacin-induced vasodilation (Cheng et al., 2006)
S-5751	Oral	3.0–30 mg/kg	Sheep	*Ascaris suum*–induced asthma (Shichijo et al., 2009)

10.13 THE ROLE OF PGD_2 AND DP1 RECEPTOR IN STROKE

PGD_2 has been studied extensively for its role in sleep-awake and thermoregulation studies. However, its importance in regulating brain injuries has recently been recognized. Based on the abundance of PGD_2 in the brain, we wanted to determine the role of PGD_2 and its DP1 receptor in ischemic stroke and excitotoxicity. Our data suggest that genetic deletion of DP1 receptors make mice more susceptible to ischemic brain injury (Saleem et al., 2007). There are several animal models reported to induce cerebral ischemia. In our studies, we used the filament model of transient cerebral ischemia. To induce cerebral ischemia or stroke, mice were anesthetized and a nylon monofilament coated with silicon was inserted into the internal carotid artery and forwarded up to the origin of the middle cerebral artery. With the help of laser Doppler flowmetry, cerebral blood flow was monitored. MCAO by the filament results in a significant (>80%) decrease in cerebral blood flow, confirming the occlusion. After the occlusion, the incision was temporarily sealed and the mice were placed in humidity- and temperature-regulated recovery units. To achieve the reperfusion, the mice were anesthetized, the incision was opened, and the nylon filament was retracted, which resulted in recirculation of the blood. In our study, we found that the neurologic deficit and brain infarction was significantly higher in $DP1^{-/-}$ mice compared with wild-type mice. To confirm that this was through the direct effect of the DP1 receptor, we thoroughly investigated the cerebral vascular anatomy of wild-type and $DP1^{-/-}$ mice. We found that the effect observed in our study was not an artifact from a difference in vascular anatomy.

Thereafter, in a follow-up study, we tested the effect of pharmacologic activation of the DP1 receptor by using a selective DP1 receptor agonist. We found that NMDA-induced excitotoxicity and cerebral ischemic brain damage was significantly attenuated by the DP1 receptor-selective agonist BW245C (Ahmad et al., 2010). In this study, we first showed that intrastriatal microinjection of NMDA resulted in significant brain damage that was significantly aggravated in $DP1^{-/-}$ mice, whereas the DP1 receptor-selective agonist BW245C had a dose-dependent neuroprotective effect. Moreover, we also showed that the minimum effective dose (10 nmol) observed in the NMDA-induced toxicity paradigm was also significantly efficient in minimizing MCAO-induced brain damage. Similarly, the effect of the DP1 agonist BW245C was tested in an in vitro model of ischemia by using the dispersed neuronal culture and hippocampal slice cultures from wild-type mice (Liang et al., 2005). In that study, it was found that the DP1 agonist BW245C attenuated the glutamate-induced cell death and oxygen–glucose deprivation induced hippocampal slice culture degeneration. Recently, in a report investigating a new compound, GIF-0173, it was found that this compound had beneficial effects against cerebral ischemia by regulating the DP1 receptor (Thura et al., 2009). In an animal model of neonatal hypoxic-ischemic encephalopathy (HIE), one of the leading causes of acute mortality and chronic disability in infants and children, a group led by Dr. Masako Taniike found that PGD_2 protects brain against HIE (Taniguchi et al., 2007). Furthermore, they show that the level of PGD_2 is significantly upregulated in the HIE mouse brain. Their data suggest that the beneficial effect of the DP1 receptor could be from its role in endothelial cells and the vasculature.

As we have mentioned in the preceding sections, some of the PG receptors activate calcium signaling, whereas some activate the cAMP pathway. Activation of the cAMP pathway has been reported to have a protective effect and restores normal cellular process (Hanson et al., 1998; Ryde and Greene, 1988; Walton et al., 1999; Walton and Dragunow, 2000). Some of the direct intracellular targets for cAMP are protein kinase A (PKA), the exchange protein activated by cAMP (Epac), and cyclic nucleotide-gated ion channels (Bos, 2006; Craven and Zagotta, 2006; de Rooij et al., 1998; Murray and Shewan, 2008). However, cAMP-mediated activation of PKA is the most widely studied aspect of cAMP signaling (Murray, 2008; Tasken and Aandahl, 2004). There are numerous studies that propose that the activation of cAMP/PKA has a neuroprotective effect through the activation of various transcription factors, including cAMP-responsive element-binding protein (Mantamadiotis et al., 2002; Walton et al., 1999; Walton and Dragunow, 2000). Similarly, it has been reported that DP1 activation significantly augments cAMP levels, which then activate various pathways leading to neuroprotection. Moreover, PGD_2 is substantially produced in blood, and DP1 receptors are significantly present in vascular lining and endothelial cells. Previous reports suggest that DP1 activation by BW245C regulates hemodynamics in rats (Koch et al., 2005). Moreover, DP1 receptor activation has also been implicated in a clinical condition known as niacin flush, where use of niacin increases PGD_2, which then acts on DP1 and induces vasodilation, resulting in hot flushes on a patient's skin. Activation of DP1 by the selective agonist BW245C decreases inflammation in murine models of dermatitis (Angeli et al., 2004). Interestingly, in clinical settings, BW245C was found to increase intraocular pressure (Shah et al., 1984). To further extend our findings on the role of the PGD_2 DP1 receptor in stroke, we tested the effect of the DP1 receptor agonist BW245C on cerebral blood flow before, during, and after ischemic stroke. It was astonishing that this compound was able to improve cerebral blood flow during and after stroke. Moreover, because imbalance hemostasis is also an important trigger of stroke onset or propagation, we tested the effect of BW245C on hemostasis (tail bleeding time and coagulation). To our amazement, we found that BW245C increased tail bleeding in vivo, and inhibited ex vivo blood coagulation. These novel works have been presented at various international scientific meetings and conferences (Ahmad et al., 2013a, 2013b).

Although the potential neuroprotective effect of the DP1 receptor is being reported, the signaling cascade leading to this neuroprotective effect is still unclear. Thus far, by using DP1 selective agents, various researchers are able to show that cAMP/PKA is one of the important pathways thorough which DP1 exerts its effect (Liang et al., 2005). This study also shows that PKA inhibitors were able to inhibit PGD_2 levels as well as PGD_2-mediated neuroprotection. In another study that elegantly used $DP1^{-/-}$ mice, it was shown that wild-type mice, after hypoxic ischemic brain injury, had better angiogenesis compared with $DP1^{-/-}$ mice. From this, the authors suggest that the DP1 receptor–mediated beneficial effect after hypoxic ischemic brain injury could possibly be through its role in the vasculature. Based on these reports and reports from our laboratory, it is evident that activation of the PGD_2 DP1 receptor activates the cAMP/PKA pathway, resulting in the activation of transcription factor(s), cerebral blood flow improvement, and maintenance of hemostasis; however, more experiments and insights are required to fully exploit the role of DP1 in stroke. Therefore, more targeted preclinical studies are required to translate the potential of this treatment in clinical stroke.

10.14 CONCLUSION

The accumulating data thus far show that PGD_2 is substantially produced in the brain, blood, and vascular linings and that the DP1 receptor is expressed in endothelial and neural cells. Under acute or chronic neurodegeneration, this receptor has been shown to have beneficial effects that are achieved by regulating the downstream signaling cascade of cAMP or through regulating physiologic pathways such as blood flow and hemostasis. Although these are preliminary studies, they nevertheless propose the potential of the DP1 receptor as a novel therapeutic target to rescue the brain from neurodegeneration. The availability of selective ligands further facilitates this process and additional rigorous testing in comorbid conditions may pave the way for this target to be tested in clinical settings.

ACKNOWLEDGMENTS

This work was supported by grants from the American Heart Association (0830172N; to A.S.A.) and NIH (R01 NS046400-01; to S.D.). The authors would like to thank Rebecca Astrom for editing the manuscript, and all lab members for constructive feedback on the chapter.

REFERENCES

Abe, H., T. Takeshita, K. Nagata, T. Arita, Y. Endo, T. Fujita et al. 1999. Molecular cloning, chromosome mapping and characterization of the mouse $CRTH_2$ gene, a putative member of the leukocyte chemoattractant receptor family. *Gene*. 227:71–7.

Adams, J., Y. Collaco-Moraes, and J. de Belleroche. 1996. Cyclooxygenase-2 induction in cerebral cortex: An intracellular response to synaptic excitation. *J Neurochem*. 66:6–13.

Ahmad, A.S., M. Ahmad, T. Maruyama, S. Narumiya, and S. Dore. 2010. Prostaglandin D2 DP1 receptor is beneficial in ischemic stroke and in acute exicitotoxicity in young and old mice. *Age (Dordr)*. 32:271–82.

Ahmad, A.S., T. Maruyama, S. Narumiya, and S. Dore. 2013a. Correlation between prostaglandin D2 DP1 receptor effects on cerebral blood flow and stroke outcomes. *Stroke*. 44: International Stroke Conference Poster Abstracts.

Ahmad, A.S., S. Narumiya, and S. Dore. 2013b. Cerebral blood flow is a major contributing factor towards the beneficial effects of PGD2 DP1 activation. *Society for Neuroscience*. 2013 Annual meeting.

Almishri, W., C. Cossette, J. Rokach, J.G. Martin, Q. Hamid, and W.S. Powell. 2005. Effects of prostaglandin D2, 15-deoxy-Delta12,14-prostaglandin J2, and selective DP1 and DP2 receptor agonists on pulmonary infiltration of eosinophils in Brown Norway rats. *J Pharmacol Exp Ther*. 313:64–9.

Andersen, K.K., T.S. Olsen, C. Dehlendorff, and L.P. Kammersgaard. 2009. Hemorrhagic and ischemic strokes compared: Stroke severity, mortality, and risk ractors. *Stroke*. 40:2068–2072.

Angeli, V., D. Staumont, A.S. Charbonnier, H. Hammad, P. Gosset, M. Pichavant et al. 2004. Activation of the D prostanoid receptor 1 regulates immune and skin allergic responses. *J Immunol*. 172:3822–9.

Aoki, T., H. Kataoka, M. Morimoto, K. Nozaki, and N. Hashimoto. 2007. Macrophage-derived matrix metalloproteinase-2 and -9 promote the progression of cerebral aneurysms in rats. *Stroke*. 38:162–169.

Arai, I., A. Takaoka, Y. Hashimoto, Y. Honma, C. Koizumi, N. Futaki et al. 2007. Effects of TS-022, a newly developed prostanoid DP1 receptor agonist, on experimental pruritus, cutaneous barrier disruptions and atopic dermatitis in mice. *Eur J Pharmacol*. 556:207–14.

Ardizzone, T.D., A. Lu, K.R. Wagner, Y. Tang, R. Ran, and F.R. Sharp. 2004. Glutamate receptor blockade attenuates glucose hypermetabolism in perihematomal brain after experimental intracerebral hemorrhage in rat. *Stroke*. 35:2587–91.

Asano, T., O. Gotoh, T. Koide, and K. Takakura. 1985. Ischemic brain edema following occlusion of the middle cerebral artery in the rat. II: Alteration of the eicosanoid synthesis profile of brain microvessels. *Stroke*. 16:110–3.

Asanuma, M., I. Miyazaki, and N. Ogawa. 2004. Neuroprotective effects of nonsteroidal anti-inflammatory drugs on neurodegenerative diseases. *Curr Pharm Des*. 10:695–700.

Baik, E.J., E.J. Kim, S.H. Lee, and C. Moon. 1999. Cyclooxygenase-2 selective inhibitors aggravate kainic acid induced seizure and neuronal cell death in the hippocampus. *Brain Res*. 843:118–29.

Ballou, L.R., R.M. Botting, S. Goorha, J. Zhang, and J.R. Vane. 2000. Nociception in cyclooxygenase isozyme-deficient mice. *Proc Natl Acad Sci U S A*. 97:10272–6.

Bang, O.Y., J.L. Saver, S.J. Kim, G.M. Kim, C.S. Chung, B. Ovbiagele et al. 2011. Collateral flow predicts response to endovascular therapy for acute ischemic stroke. *Stroke*. 42:693–9.

Bazan, N.G. 2005. Lipid signaling in neural plasticity, brain repair, and neuroprotection. *Mol Neurobiol*. 32:89–103.

Bazan, N.G., V. Colangelo, and W.J. Lukiw. 2002. Prostaglandins and other lipid mediators in Alzheimer's disease. *Prostaglandins Other Lipid Mediat*. 68–69:197–210.

Becker, K.J. 1998. Inflammation and acute stroke. *Curr Opin Neurol*. 11:45–9.

Bhardwaj, A., I. Harukuni, S.J. Murphy, N.J. Alkayed, B.J. Crain, R.C. Koehler et al. 2000. Hypertonic saline worsens infarct volume after transient focal ischemia in rats. *Stroke*. 31:1694–701.

Bohm, E., G.J. Sturm, I. Weiglhofer, H. Sandig, M. Shichijo, A. McNamee et al. 2004. 11-Dehydro-thromboxane B2, a stable thromboxane metabolite, is a full agonist of chemoattractant receptor-homologous molecule expressed on TH2 cells (CRTH2) in human eosinophils and basophils. *J Biol Chem*. 279:7663–70.

Boie, Y., N. Sawyer, D.M. Slipetz, K.M. Metters, and M. Abramovitz. 1995. Molecular cloning and characterization of the human prostanoid DP receptor. *J Biol Chem*. 270:18910–6.

Bonventre, J.V. 1997. Roles of phospholipases A2 in brain cell and tissue injury associated with ischemia and excitotoxicity. *J Lipid Mediat Cell Signal*. 17:71–9.

Bos, J.L. 2006. Epac proteins: Multi-purpose cAMP targets. *Trends Biochem Sci*. 31:680–6.

Bosetti, F G.R. Weerasinghe. 2003. The expression of brain cyclooxygenase-2 is down-regulated in the cytosolic phospholipase A2 knockout mouse. *J Neurochem*. 87:1471–7.

Bounds, J.V., D.O. Wiebers, J.P. Whisnant, and H. Okazaki. 1981. Mechanisms and timing of deaths from cerebral infarction. *Stroke*. 12:474–7.

Bregar, U., B. Jug, I. Keber, M. Cevc, and M. Sebestjen. 2013. Extended-release niacin/laropiprant improves endothelial function in patients after myocardial infarction. *Heart and Vessels*:1–7.

Breyer, R.M., C.K. Bagdassarian, S.A. Myers, and M.D. Breyer. 2001. Prostanoid receptors: Subtypes and signaling. *Annu Rev Pharmacol Toxicol*. 41:661–90.

Burke, J.E. and E.A. Dennis. 2009a. Phospholipase A2 biochemistry. *Cardiovasc Drugs Ther*. 23:49–59.

Burke, J.E. and E.A. Dennis. 2009b. Phospholipase A2 structure/function, mechanism, and signaling. *J Lipid Res*. 50 Suppl:S237–42.

Campbell, I.G. and I. Feinberg. 1996. Noncompetitive NMDA channel blockade during waking intensely stimulates NREM delta. *J Pharmacol Exp Ther*. 276:737–42.

Campbell, W.B., D. Gebremedhin, P.F. Pratt, and D.R. Harder. 1996. Identification of epoxyeicosatrienoic acids as endothelium-derived hyperpolarizing factors. *Circ Res*. 78:415–23.

Capra, V., M.D. Thompson, A. Sala, D.E. Cole, G. Folco, and G.E. Rovati. 2007. Cysteinyl-leukotrienes and their receptors in asthma and other inflammatory diseases: Critical update and emerging trends. *Med Res Rev*. 27:469–527.

Casteleijn, E., J. Kuiper, H.C. Van Rooij, J.A. Kamps, J.F. Koster, and T.J. Van Berkel. 1988. Prostaglandin D2 mediates the stimulation of glycogenolysis in the liver by phorbol ester. *Biochem J*. 250:77–80.

Chambers, B.R., J.W. Norris, B.L. Shurvell, and V.C. Hachinski. 1987. Prognosis of acute stroke. *Neurology*. 37:221–5.

Chamorro, A. 2004. Role of inflammation in stroke and atherothrombosis. *Cerebrovasc Dis*. 17 Suppl 3:1–5.

Chan, P.H. 2001. Reactive oxygen radicals in signaling and damage in the ischemic brain. *J Cereb Blood Flow Metab*. 21:2–14.

Chemtob, S., K. Beharry, J. Rex, D.R. Varma, and J.V. Aranda. 1990a. Changes in cerebrovascular prostaglandins and thromboxane as a function of systemic blood pressure. Cerebral blood flow autoregulation of the newborn. *Circ Res*. 67:674–82.

Chemtob, S., K. Beharry, J. Rex, D.R. Varma, and J.V. Aranda. 1990b. Prostanoids determine the range of cerebral blood flow autoregulation of newborn piglets. *Stroke*. 21:777–84.

Chemtob, S., P. Hardy, D. Abran, D.Y. Li, K. Peri, O. Cuzzani, and D.R. Varma. 1995. Peroxide-cyclooxygenase interactions in postasphyxial changes in retinal and choroidal hemodynamics. *J Appl Physiol*. 78:2039–46.

Cheng, K., T.J. Wu, K.K. Wu, C. Sturino, K. Metters, K. Gottesdiener et al. 2006. Antagonism of the prostaglandin D2 receptor 1 suppresses nicotinic acid-induced vasodilation in mice and humans. *Proc Natl Acad Sci U S A*. 103:6682–7.

Choi, D.W. 1992. Excitotoxic cell death. *J Neurobiol*. 23:1261–76.

Clark, W.M., L.G. Rinker, N.S. Lessov, S.L. Lowery, and M.J. Cipolla. 2001. Efficacy of antioxidant therapies in transient focal ischemia in mice. *Stroke*. 32:1000–4.

Coleman, R.A., R.M. Eglen, R.L. Jones, S. Narumiya, T. Shimizu, W.L. Smith, et al. 1995. Prostanoid and leukotriene receptors: A progress report from the IUPHAR working parties on classification and nomenclature. *Adv Prostaglandin Thromboxane Leukot Res*. 23:283–5.

Consilvio, C., A.M. Vincent, and E.L. Feldman. 2004. Neuroinflammation, COX-2, and ALS—A dual role? *Exp Neurol*. 187:1–10.

Craven, K.B and W.N. Zagotta. 2006. CNG and HCN channels: Two peas, one pod. *Annu Rev Physiol*. 68:375–401.

Crider, J.Y., B.W. Griffin, and N.A. Sharif. 1999. Prostaglandin DP receptors positively coupled to adenylyl cyclase in embryonic bovine tracheal (EBTr) cells: Pharmacological characterization using agonists and antagonists. *Br J Pharmacol*. 127:204–10.

Cunningham, L.A., M. Wetzel, and G.A. Rosenberg. 2005. Multiple roles for MMPs and TIMPs in cerebral ischemia. *Glia*. 50:329–39.

Darius, H., J. Michael-Hepp, K.H. Thierauch, and A. Fisch. 1994. Inhibition of human platelets and polymorphonuclear neutrophils by the potent and metabolically stable prostaglandin D2 analog ZK 118.182. *Eur J Pharmacol*. 258:207–13.

Davies, N.M. and F. Jamali. 2004. COX-2 selective inhibitors cardiac toxicity: Getting to the heart of the matter. *J Pharm Pharm Sci*. 7:332–6.

de Rooij, J., F.J. Zwartkruis, M.H. Verheijen, R.H. Cool, S.M. Nijman, A. Wittinghofer et al. 1998. Epac is a Rap1 guanine-nucleotide-exchange factor directly activated by cyclic AMP. *Nature*. 396:474–7.

Deb, P., S. Sharma, and K.M. Hassan. 2010. Pathophysiologic mechanisms of acute ischemic stroke: An overview with emphasis on therapeutic significance beyond thrombolysis. *Pathophysiology*. 17:197–218.

del Zoppo, G., I. Ginis, J.M. Hallenbeck, C. Iadecola, X. Wang, and G.Z. Feuerstein. 2000. Inflammation and stroke: Putative role for cytokines, adhesion molecules and iNOS in brain response to ischemia. *Brain Pathol*. 10:95–112.

del Zoppo, G.J. and J.A. Koziol. 2007. Recanalization and stroke outcome. *Circulation*. 115:2602–5.

Dirnagl, U., C. Iadecola, and M.A. Moskowitz. 1999. Pathobiology of ischaemic stroke: An integrated view. *Trends Neurosci*. 22:391–7.

Dishy, V., F. Liu, D.L. Ebel, G.J. Atiee, J. Royalty, S. Reilley et al. 2009. Effects of aspirin when added to the prostaglandin D2 receptor antagonist laropiprant on niacin-induced flushing symptoms. *J Clin Pharmacol*. 49:416–22.

Doré, S., T. Otsuka, T. Mito, N. Sugo, T. Hand, L. Wu et al. 2003. Neuronal overexpression of cyclooxygenase-2 increases cerebral infarction. *Ann Neurol*. 54:155–62.

Duvall, W.L. and D.A. Vorchheimer. 2004. Multi-bed vascular disease and atherothrombosis: Scope of the problem. *J Thromb Thrombolysis*. 17:51–61.

Eguchi, N., T. Minami, N. Shirafuji, Y. Kanaoka, T. Tanaka, A. Nagata et al. 1999. Lack of tactile pain (allodynia) in lipocalin-type prostaglandin D synthase-deficient mice. *Proc Natl Acad Sci U S A*. 96:726–30.

Eltzschig, H.K and T. Eckle. 2011. Ischemia and reperfusion—From mechanism to translation. *Nat Med*. 17:1391–401.

Fang, Y.C., J.S. Wu, J.J. Chen, W.M. Cheung, P.H. Tseng, K.B. Tam et al. 2006. Induction of prostacyclin/PGI_2 synthase expression after cerebral ischemia-reperfusion. *J Cereb Blood Flow Metab*. 26:491–501.

Faraci, F.M. and D.D. Heistad. 1998. Regulation of the cerebral circulation: Role of endothelium and potassium channels. *Physiol Rev*. 78:53–97.

Feng, L., W. Sun, Y. Xia, W.W. Tang, P. Chanmugam, E. Soyoola et al. 1993. Cloning two isoforms of rat cyclooxygenase: Differential regulation of their expression. *Arch Biochem Biophys*. 307:361–8.

Fenton, J.n., F.A. Ofosu, D.V. Brezniak, and H.I. Hassouna. 1998. Thrombin and antithrombotics [review]. *Semin Thromb Hemost*. 24:87–91.

Fitzgerald, G.A. 2003. COX-2 and beyond: Approaches to prostaglandin inhibition in human disease. *Nat Rev Drug Discov*. 2:879–90.

Fitzgerald, G.A. 2004. Coxibs and cardiovascular disease. *N Engl J Med*. 351:1709–11.

Fleming, I. 2007. DiscrEET regulators of homeostasis: Epoxyeicosatrienoic acids, cytochrome P450 epoxygenases and vascular inflammation. *Trends Pharmacol Sci*. 28:448–52.

Forman, B.M., P. Tontonoz, J. Chen, R.P. Brun, B.M. Spiegelman, and R.M. Evans. 1995. 15-Deoxy-Δ12,14-Prostaglandin J2 is a ligand for the adipocyte determination factor PPARγ. *Cell*. 83:803–812.

Frykholm, P., L. Hillered, B. Langstrom, L. Persson, J. Valtysson, and P. Enblad. 2005. Relationship between cerebral blood flow and oxygen metabolism, and extracellular glucose and lactate concentrations during middle cerebral artery occlusion and reperfusion: A microdialysis and positron emission tomography study in nonhuman primates. *J Neurosurg*. 102:1076–84.

Funk, C.D. 2001. Prostaglandins and leukotrienes: Advances in eicosanoid biology. *Science*. 294:1871–5.

Garcia, J.H. 1975. The neuropathology of stroke. *Hum Pathol*. 6:583–98.

Gazi, L., S. Gyles, J. Rose, S. Lees, C. Allan, L. Xue et al. 2005. Delta12-prostaglandin D2 is a potent and selective CRTH2 receptor agonist and causes activation of human eosinophils and Th2 lymphocytes. *Prostaglandins Other Lipid Mediat*. 75:153–67.

Gelir, E., S.O. Arslan, H. Sayan, and L. Pinar. 2005. Effect of rapid-eye-movement sleep deprivation on rat hypothalamic prostaglandins. *Prostaglandins Leukot Essent Fatty Acids*. 73:391–6.

Gerriets, T., E. Stolz, M. Walberer, M. Kaps, G. Bachmann, and M. Fisher. 2003. Neuroprotective effects of MK-801 in different rat stroke models for permanent middle cerebral artery occlusion: Adverse effects of hypothalamic damage and strategies for its avoidance. *Stroke*. 34:2234–9.

Go, A.S., D. Mozaffarian, V.L. Roger, E.J. Benjamin, J.D. Berry, W.B. Borden et al. 2013. Heart disease and stroke statistics—2013 update: A report from the American Heart Association. *Circulation*. 127:e6-e245.

Gobbo, O.L. and S.M. O'Mara. 2004. Post-treatment, but not pre-treatment, with the selective cyclooxygenase-2 inhibitor celecoxib markedly enhances functional recovery from kainic acid-induced neurodegeneration. *Neuroscience*. 125:317–27.

Goh, Y., M. Nakajima, I. Azuma, and O. Hayaishi. 1988. Effects of prostaglandin D2 and its analogues on intraocular pressure in rabbits. *Jpn J Ophthalmol.* 32:471–80.

Granstrom, E., W.E. Lands, and B. Samuelsson. 1968. Biosynthesis of 9-alpha,15-dihydroxy-11-ketoprost-13-enoic acid. *J Biol Chem.* 243:4104–8.

Gray, S.J., H. Giles, and J. Posner. 1992. The effect of a prostaglandin DP-receptor partial agonist (192C86) on platelet aggregation and the cardiovascular system in healthy volunteers. *Br J Clin Pharmacol.* 34:344–51.

Green, A.R. 2002. Why do neuroprotective drugs that are so promising in animals fail in the clinic? An industry perspective. *Clin Exp Pharmacol Physiol.* 29:1030–4.

Green, A.R. and T. Ashwood. 2005. Free radical trapping as a therapeutic approach to neuroprotection in stroke: Experimental and clinical studies with NXY-059 and free radical scavengers. *Curr Drug Targets CNS Neurol Disord.* 4:109–18.

Gross, G.J., K.M. Gauthier, J. Moore, J.R. Falck, B.D. Hammock, W.B. Campbell, and K. Nithipatikom. 2008. Effects of the selective EET antagonist, 14,15-EEZE, on cardioprotection produced by exogenous or endogenous EETs in the canine heart. *American J physiol.* 294:H2838–44.

Grosser, T., S. Fries, and G.A. Fitzgerald. 2006. Biological basis for the cardiovascular consequences of COX-2 inhibition: Therapeutic challenges and opportunities. *J Clin Invest.* 116:4–15.

Grossetete, M. and G.A. Rosenberg. 2008. Matrix metalloproteinase inhibition facilitates cell death in intracerebral hemorrhage in mouse. *J Cereb Blood Flow Metab.* 28:752–63.

Hamid-Bloomfield, S., A.N. Payne, A.A. Petrovic, and B.J.R. Whittle. 1990. The role of prostanoid TP- and DP-receptors in the bronchoconstrictor effect of inhaled PGD2 in anaesthetized guinea-pigs: Effect of the DP-antagonist BW A868C. *Br J Pharmacol.* 100:761–766.

Hammad, H., H.J. de Heer, T. Soullie, H.C. Hoogsteden, F. Trottein, and B.N. Lambrecht. 2003. Prostaglandin D_2 inhibits airway dendritic cell migration and function in steady state conditions by selective activation of the D prostanoid receptor 1. *J Immunol.* 171:3936–40.

Hammad, H., M. Kool, T. Soullie, S. Narumiya, F. Trottein, H.C. Hoogsteden et al. 2007. Activation of the D prostanoid 1 receptor suppresses asthma by modulation of lung dendritic cell function and induction of regulatory T cells. *J Exp Med.* 204:357–67.

Hanson, M.G., Jr., S. Shen, A.P. Wiemelt, F.A. McMorris, and B.A. Barres. 1998. Cyclic AMP elevation is sufficient to promote the survival of spinal motor neurons in vitro. *J Neurosci.* 18:7361–71.

Hata, A.N. and R.M. Breyer. 2004. Pharmacology and signaling of prostaglandin receptors: Multiple roles in inflammation and immune modulation. *Pharmacol Ther.* 103:147–66.

Hata, A.N., R. Zent, M.D. Breyer, and R.M. Breyer. 2003. Expression and molecular pharmacology of the mouse CRTH2 receptor. *J Pharmacol Exp Therapeutics.* 306:463–470.

Hatoum, O.A., K.M. Gauthier, D.G. Binion, H. Miura, G. Telford, M.F. Otterson et al. 2005. Novel mechanism of vasodilation in inflammatory bowel disease. *Arterioscler Thromb Vasc Biol.* 25:2355–61.

Hayaishi, O. 2002. Molecular genetic studies on sleep-wake regulation, with special emphasis on the prostaglandin D_2 system. *J Appl Physiol.* 92:863–8.

Hayaishi, O. and Y. Urade. 2002. Prostaglandin D2 in sleep-wake regulation: Recent progress and perspectives. *Neuroscientist.* 8:12–5.

Hayashi, M. and T. Tanouchi. 1973. Synthesis of 11-dehydro-13,14-dihydro-PGE1 [prostaglandin E1] and -PGD2 [prostaglandin D2]. *J Organic Chem.* 38:2115–2116.

Heinemann, A., R. Schuligoi, I. Sabroe, A. Hartnell, and B.A. Peskar. 2003. Δ12-Prostaglandin J2, a plasma metabolite of prostaglandin D2, causes eosinophil mobilization from the bone marrow and primes eosinophils for chemotaxis. *J Immunol.* 170:4752–4758.

Hewitt, D.J. 2000. The use of NMDA-receptor antagonists in the treatment of chronic pain. *Clin J Pain.* 16:S73–9.

Hinkle, J.L. and L. Bowman. 2003. Neuroprotection for ischemic stroke. *J Neurosci Nurs.* 35:114–8.

Hirai, H., K. Tanaka, O. Yoshie, K. Ogawa, K. Kenmotsu, Y. Takamori et al. 2001. Prostaglandin D2 selectively induces chemotaxis in T helper type 2 cells, eosinophils, and basophils via seven-transmembrane receptor CRTH2. *J Exp Med.* 193:255–61.

Hirano, Y., M. Shichijo, M. Deguchi, M. Nagira, N. Suzuki, Y. Nishitani et al. 2007. Synergistic effect of PGD2 via prostanoid DP receptor on TNF-[alpha]-induced production of MCP 1 and IL-8 in human monocytic THP-1 cells. *Eur J Pharmacol.* 560:81–88.

Hirata, M., A. Kakizuka, M. Aizawa, F. Ushikubi, and S. Narumiya. 1994. Molecular characterization of a mouse prostaglandin D receptor and functional expression of the cloned gene. *Proc Natl Acad Sci U S A.* 91:11192–6.

Hoffmann, C. 2000. COX-2 in brain and spinal cord implications for therapeutic use. *Curr Med Chem.* 7:1113–20.

HPS2-THRIVE-Collaborative-Group, J. Armitage, and L. Jiang. 2013. HPS2-THRIVE randomized placebo-controlled trial in 25 673 high-risk patients of ER niacin/laropiprant: Trial design, pre-specified muscle and liver outcomes, and reasons for stopping study treatment. *Eur Heart J.* 34:1279–1291.

Hurley, S.D., J.A. Olschowka, and M.K. O'Banion. 2002. Cyclooxygenase inhibition as a strategy to ameliorate brain injury. *J Neurotrauma.* 19:1–15.

Iadecola, C. and M. Alexander. 2001. Cerebral ischemia and inflammation. *Curr Opin Neurol.* 14:89–94.

Iadecola, C., K. Niwa, S. Nogawa, X. Zhao, M. Nagayama, E. Araki et al. 2001. Reduced susceptibility to ischemic brain injury and N-methyl-D-aspartate-mediated neurotoxicity in cyclooxygenase-2-deficient mice. *Proc Natl Acad Sci USA.* 98:1294–9.

Iadecola, C. and M.E. Ross. 1997. Molecular pathology of cerebral ischemia: Delayed gene expression and strategies for neuroprotection. *Ann N Y Acad Sci.* 835:203–17.

Ikonomidou, C. and L. Turski. 2002. Why did NMDA receptor antagonists fail clinical trials for stroke and traumatic brain injury? *Lancet Neurol.* 1:383–6.

Imig, J.D. 2005. Epoxide hydrolase and epoxygenase metabolites as therapeutic targets for renal diseases. *Am J Physiol Renal Physiol.* 289:F496–503.

Ishikawa, T.O., Y. Tamai, J.M. Rochelle, M. Hirata, T. Namba, Y. Sugimoto et al. 1996. Mapping of the genes encoding mouse prostaglandin D, E, and F and prostacyclin receptors. *Genomics.* 32:285–8.

Ishizuka, T., J. Cheng, H. Singh, M.D. Vitto, V.L. Manthati, J.R. Falck et al. 2008. 20-Hydroxyeicosatetraenoic acid stimulates nuclear factor-kappaB activation and the production of inflammatory cytokines in human endothelial cells. *J Pharmacol Exp Ther.* 324:103–10.

Jaillard, A., C. Cornu, A. Durieux, T. Moulin, F. Boutitie, K.R. Lees, and M. Hommel. 1999. Hemorrhagic transformation in acute ischemic stroke. The MAST-E study. MAST-E Group. *Stroke.* 30:1326–32.

Jones, R.L., M.A. Giembycz, and D.F. Woodward. 2009. Prostanoid receptor antagonists: Development strategies and therapeutic applications. *Br J Pharmacol*. 158:104–45.

Kabashima, K. and S. Narumiya. 2003. The DP receptor, allergic inflammation and asthma. *Prostaglandins Leukot Essent Fatty Acids*. 69:187–94.

Kagawa, S., K. Fukunaga, T. Oguma, Y. Suzuki, T. Shiomi, K. Sayama et al. 2011. Role of prostaglandin D2 receptor CRTH2 in sustained eosinophil accumulation in the airways of mice with chronic asthma. *Int Arch Allergy Immunol*. 155 Suppl 1:6–11.

Kam, P.C. and A.U. See. 2000. Cyclo-oxygenase isoenzymes: Physiological and pharmacological role. *Anaesthesia*. 55:442–9.

Kanekiyo, T., T. Ban, K. Aritake, Z.L. Huang, W.M. Qu, I. Okazaki et al. 2007. Lipocalin-type prostaglandin D synthase/beta-trace is a major amyloid beta-chaperone in human cerebrospinal fluid. *Proc Natl Acad Sci U S A*. 104:6412–7.

Kanematsu, Y., M. Kanematsu, C. Kurihara, Y. Tada, T.-L. Tsou, N. van Rooijen et al. 2011. Critical roles of macrophages in the formation of intracranial aneurysm. *Stroke*. 42:173–178.

Kargman, D.E., C. Tuck, L. Berglund, I.F. Lin, R.S. Mukherjee, E.V. Thompson et al. 1998. Lipid and lipoprotein levels remain stable in acute ischemic stroke: The Northern Manhattan Stroke Study. *Atherosclerosis*. 139:391–9.

Kaufmann, W.E., P.F. Worley, J. Pegg, M. Bremer, and P. Isakson. 1996. COX-2, a synaptically induced enzyme, is expressed by excitatory neurons at postsynaptic sites in rat cerebral cortex. *Proc Natl Acad Sci U S A*. 93:2317–21.

Kelley, K.A., L. Ho, D. Winger, J. Freire-Moar, C.B. Borelli, P.S. Aisen, and G.M. Pasinetti. 1999. Potentiation of excitotoxicity in transgenic mice overexpressing neuronal cyclooxygenase-2. *Am J Pathol*. 155:995–1004.

Kemp, J.A. and R.M. McKernan. 2002. NMDA receptor pathways as drug targets. *Nat Neurosci*. 5 Suppl:1039–42.

Khanapure, S.P., D.S. Garvey, D.R. Janero, and L.G. Letts. 2007. Eicosanoids in inflammation: Biosynthesis, pharmacology, and therapeutic frontiers. *Curr Top Med Chem*. 7:311–40.

Kishimoto, K., R.C. Li, J. Zhang, J.A. Klaus, K.K. Kibler, S. Doré et al. 2010. Cytosolic phospholipase A2 alpha amplifies early cyclooxygenase-2 expression, oxidative stress and MAP kinase phosphorylation after cerebral ischemia in mice. *J Neuroinflammation*. 7:42.

Koch, K.A., J.L. Wessale, R. Moreland, G.A. Reinhart, and B.F. Cox. 2005. Effects of BW245C, a prostaglandin DP receptor agonist, on systemic and regional haemodynamics in the anaesthetized rat. *Clin Exp Pharmacol Physiol*. 32:931–5.

Kogure, K., Y. Yamasaki, Y. Matsuo, H. Kato, and H. Onodera. 1996. Inflammation of the brain after ischemia. *Acta Neurochir Suppl*. 66:40–3.

Kudo, I. and M. Murakami. 2002. Phospholipase A2 enzymes. *Prostaglandins Other Lipid Mediat*. 68–69:3–58.

Lenzer, J. 2005. Pfizer is asked to suspend sales of painkiller. *BMJ*. 330:862.

Liang, X., L. Wu, T. Hand, and K. Andreasson. 2005. Prostaglandin D_2 mediates neuronal protection via the DP1 receptor. *J Neurochem*. 92:477–86.

Liebeskind, D.S. 2005. Neuroprotection from the collateral perspective. *IDrugs*. 8:222–8.

Lipton, P. 1999. Ischemic cell death in brain neurons. *Physiol Rev*. 79:1431–568.

Lipton, S.A. and P.A. Rosenberg. 1994. Excitatory amino acids as a final common pathway for neurologic disorders. *N Engl J Med*. 330:613–22.

Lo, E.H. 2008. A new penumbra: Transitioning from injury into repair after stroke. *Nat Med*. 14:497–500.

Lo, E.H., T. Dalkara, and M.A. Moskowitz. 2003. Mechanisms, challenges and opportunities in stroke. *Nat Rev Neurosci*. 4:399–415.

Luster, A.D. and A.M. Tager. 2004. T-cell trafficking in asthma: Lipid mediators grease the way. *Nat Rev Immunol*. 4:711–24.

Lynch, D.R. and R.P. Guttmann. 2002. Excitotoxicity: Perspectives based on N-methyl-D-aspartate receptor subtypes. *J Pharmacol Exp Ther*. 300:717–23.

Macdonald, R.L. and M. Stoodley. 1998. Pathophysiology of cerebral ischemia. *Neurol Med Chir (Tokyo)*. 38:1–11.

Madrigal, J.L., M.A. Moro, I. Lizasoain, P. Lorenzo, A.P. Fernandez, J. Rodrigo et al. 2003. Induction of cyclooxygenase-2 accounts for restraint stress-induced oxidative status in rat brain. *Neuropsychopharmacology*. 28:1579–88.

Maeda, K., R. Hata, and K.A. Hossmann. 1999. Regional metabolic disturbances and cerebrovascular anatomy after permanent middle cerebral artery occlusion in C57black/6 and SV129 mice. *Neurobiol Dis*. 6:101–8.

Manley, G.T., M. Fujimura, T. Ma, N. Noshita, F. Filiz, A.W. Bollen, P. Chan, and A.S. Verkman. 2000. Aquaporin-4 deletion in mice reduces brain edema after acute water intoxication and ischemic stroke. *Nat Med*. 6:159–63.

Mantamadiotis, T., T. Lemberger, S.C. Bleckmann, H. Kern, O. Kretz, A. Martin Villalba et al. 2002. Disruption of CREB function in brain leads to neurodegeneration. *Nat Genet*. 31:47–54.

Matsugi, T., M. Kageyama, K. Nishimura, H. Giles, and E. Shirasawa. 1995. Selective prostaglandin D_2 receptor stimulation elicits ocular hypotensive effects in rabbits and cats. *Eur J Pharmacol*. 275:245–50.

Matsuoka, T., M. Hirata, H. Tanaka, Y. Takahashi, T. Murata, K. Kabashima et al. 2000. Prostaglandin D2 as a mediator of allergic asthma. *Science*. 287:2013–7.

Mendez, I., V. Hachinski, and B. Wolfe. 1987. Serum lipids after stroke. *Neurology*. 37:507–11.

Minami, T., E. Okuda-Ashitaka, M. Nishizawa, H. Mori, and S. Ito. 1997. Inhibition of nociceptin-induced allodynia in conscious mice by prostaglandin D2. *Br J Pharmacol*. 122:605–10.

Minghetti, L. 2004. Cyclooxygenase-2 (COX-2) in inflammatory and degenerative brain diseases. *J Neuropathol Exp Neurol*. 63:901–10.

Miyata, N., T. Seki, Y. Tanaka, T. Omura, K. Taniguchi, M. Doi et al. 2005. Beneficial effects of a new 20-hydroxyeicosatetraenoic acid synthesis inhibitor, TS-011 [N-(3-chloro-4-morpholin-4-yl) phenyl-N'-hydroxyimido formamide], on hemorrhagic and ischemic stroke. *J Pharmacol Exp Ther*. 314:77–85.

Mizoguchi, A., N. Eguchi, K. Kimura, Y. Kiyohara, W.M. Qu, Z.L. Huang et al. 2001. Dominant localization of prostaglandin D receptors on arachnoid trabecular cells in mouse basal forebrain and their involvement in the regulation of non-rapid eye movement sleep. *Proc Natl Acad Sci U S A*. 98:11674–9.

Mohan, S., A.S. Ahmad, A. Glushakov, C. Chambers, and S. Dore. 2012. Putative role of prostaglandin receptor in intracerebral hemorrhage. *Front Neurol*. 3:1–17.

Monneret, G., H. Li, J. Vasilescu, J. Rokach, and W.S. Powell. 2002. 15-Deoxy-Δ12,1412,14-prostaglandins D2 and J2 are potent activators of human eosinophils. *J Immunol*. 168:3563–3569.

Morita-Fujimura, Y., M. Fujimura, T. Yoshimoto, and P.H. Chan. 2001. Superoxide during reperfusion contributes to caspase-8 expression and apoptosis after transient focal stroke. *Stroke*. 32:2356–61.

Murakami, M. and I. Kudo. 2006. Prostaglandin e synthase: A novel drug target for inflammation and cancer. *Curr Pharm Des.* 12:943–54.
Murray, A.J. 2008. Pharmacological PKA inhibition: All may not be what it seems. *Sci Signal.* 1:re4.
Murray, A.J. and D.A. Shewan. 2008. Epac mediates cyclic AMP-dependent axon growth, guidance and regeneration. *Mol Cell Neurosci.* 38:578–88.
Naffah-Mazzacoratti, M.G., M.I. Bellissimo, and E.A. Cavalheiro. 1995. Profile of prostaglandin levels in the rat hippocampus in pilocarpine model of epilepsy. *Neurochem Int.* 27:461–6.
Nagata, K. and H. Hirai. 2003. The second PGD(2) receptor CRTH2: Structure, properties, and functions in leukocytes. *Prostaglandins Leukot Essent Fatty Acids.* 69:169–77.
Nagata, K., H. Hirai, K. Tanaka, K. Ogawa, T. Aso, K. Sugamura et al. 1999a. CRTH2, an orphan receptor of T-helper-2-cells, is expressed on basophils and eosinophils and responds to mast cell-derived factor(s). *FEBS Lett.* 459:195–9.
Nagata, K., K. Tanaka, K. Ogawa, K. Kemmotsu, T. Imai, O. Yoshie et al. 1999b. Selective expression of a novel surface molecule by human Th2 cells in vivo. *J Immunol.* 162:1278–86.
Nakayama, M., K. Uchimura, R.L. Zhu, T. Nagayama, M.E. Rose, R.A. Stetler et al. 1998. Cyclooxygenase-2 inhibition prevents delayed death of CA1 hippocampal neurons following global ischemia. *Proc Natl Acad Sci U S A.* 95:10954–9.
Narumiya, S. 2009. Prostanoids and inflammation: A new concept arising from receptor knockout mice. *J Mol Med.* 87:1015–22.
Narumiya, S., Y. Sugimoto, and F. Ushikubi. 1999. Prostanoid receptors: Structures, properties, and functions. *Physiol Rev.* 79:1193–226.
Narumiya, S. and N. Toda. 1985. Different responsiveness of prostaglandin D2-sensitive systems to prostaglandin D2 and its analogues. *Br J Pharmacol.* 85:367–75.
Nogawa, S., F. Zhang, M.E. Ross, and C. Iadecola. 1997. Cyclooxygenase-2 gene expression in neurons contributes to ischemic brain damage. *J Neurosci.* 17:2746–55.
Obal, F., Jr. and J.M. Krueger. 2003. Biochemical regulation of non-rapid-eye-movement sleep. *Front Biosci.* 8:d520–50.
Park, S.Y., T.H. Kim, H.I. Kim, Y.K. Shin, C.S. Lee, M. Park, and J.H. Song. 2007. Celecoxib inhibits Na+ currents in rat dorsal root ganglion neurons. *Brain Res.* 1148:53–61.
Phillis, J.W., L.A. Horrocks, and A.A. Farooqui. 2006. Cyclooxygenases, lipoxygenases, and epoxygenases in CNS: Their role and involvement in neurological disorders. *Brain Res Rev.* 52:201–43.
Phillis, J.W. and M.H. O'Regan. 2003. The role of phospholipases, cyclooxygenases, and lipoxygenases in cerebral ischemic/traumatic injuries. *Crit Rev Neurobiol.* 15:61–90.
Phillis, J.W. and M.H. O'Regan. 2004. A potentially critical role of phospholipases in central nervous system ischemic, traumatic, and neurodegenerative disorders. *Brain Res Brain Res Rev.* 44:13–47.
Qu, W.M., Z.L. Huang, X.H. Xu, K. Aritake, N. Eguchi, F. Nambu et al. 2006. Lipocalin-type prostaglandin D synthase produces prostaglandin D2 involved in regulation of physiological sleep. *Proc Natl Acad Sci U S A.* 103:17949–54.
Rao, A.V. and B. Balachandran. 2002. Role of oxidative stress and antioxidants in neurodegenerative diseases. *Nutr Neurosci.* 5:291–309.
Renic, M., J.A. Klaus, T. Omura, N. Kawashima, M. Onishi, N. Miyata et al. 2009. Effect of 20-HETE inhibition on infarct volume and cerebral blood flow after transient middle cerebral artery occlusion. *J Cereb Blood Flow Metab.* 29:629–39.
Resnick, D.K., S.H. Graham, C.E. Dixon, and D.W. Marion. 1998. Role of cyclooxygenase 2 in acute spinal cord injury. *J Neurotrauma.* 15:1005–13.
Ribeiro, J.D., A.A. Toro, and E.C. Baracat. 2006. Antileukotrienes in the treatment of asthma and allergic rhinitis. *J Pediatr (Rio J).* 82:S213–21.
Roger, V.L., A.S. Go, D.M. Lloyd-Jones, E.J. Benjamin, J.D. Berry, W.B. Borden et al. 2012. Heart disease and stroke statistics—2012 update: A report from the American Heart Association. *Circulation.* 125:e2-e220.
Roman, R.J. 2002. P-450 metabolites of arachidonic acid in the control of cardiovascular function. *Physiol Rev.* 82:131–85.
Rosell, A., C. Foerch, Y. Murata, and E.H. Lo. 2008. Mechanisms and markers for hemorrhagic transformation after stroke. *Acta Neurochir Suppl.* 105:173–8.
Rossi, A., P. Kapahi, G. Natoli, T. Takahashi, Y. Chen, M. Karin, and M.G. Santoro. 2000. Anti-inflammatory cyclopentenone prostaglandins are direct inhibitors of IkappaB kinase. *Nature.* 403:103–8.
Rouzer, C.A. and L.J. Marnett. 2009. Cyclooxygenases: Structural and functional insights. *J Lipid Res.* 50 Suppl:S29–34.
Rydel, R.E. and L.A. Greene. 1988. cAMP analogs promote survival and neurite outgrowth in cultures of rat sympathetic and sensory neurons independently of nerve growth factor. *Proc Natl Acad Sci U S A.* 85:1257–61.
Saleem, S., H. Zhuang, S. Biswal, Y. Christen, and S. Dore. 2008. Ginkgo biloba extract neuroprotective action is dependent on heme oxygenase 1 in ischemic reperfusion brain injury. *Stroke.* 39:3389–96.
Saleem, S., H. Zhuang, A.J. de Brum-Fernandes, T. Maruyama, S. Narumiya, and S. Doré. 2007. PGD_2 DP1 receptor protects brain from ischemia-reperfusion injury. *Eur J Neurosci.* 26:73–78.
Sanchez-Mejia, R.O., J.W. Newman, S. Toh, G.Q. Yu, Y. Zhou, B. Halabisky et al. 2008. Phospholipase A2 reduction ameliorates cognitive deficits in a mouse model of Alzheimer's disease. *Nat Neurosci.* 11:1311–8.
Sandhya, T.L., W.Y. Ong, L.A. Horrocks, and A.A. Farooqui. 1998. A light and electron microscopic study of cytoplasmic phospholipase A2 and cyclooxygenase-2 in the hippocampus after kainate lesions. *Brain Res.* 788:223–31.
Sandig, H., D. Andrew, A.A. Barnes, I. Sabroe, and J. Pease. 2006a. 9a,11b-PGF_2 and its stereoisomer PGF_{2a} are novel agonists of the chemoattractant receptor, CRTH2. *FEBS Lett.* 580:373–9.
Sandig, H., D. Andrew, A.A. Barnes, I. Sabroe, and J. Pease. 2006b. 9α,11β-PGF2 and its stereoisomer PGF2α are novel agonists of the chemoattractant receptor, CRTH2. *FEBS Letters.* 580:373–379.
Santovito, D., A. Mezzetti, and F. Cipollone. 2009. Cyclooxygenase and prostaglandin synthases: Roles in plaque stability and instability in humans. *Curr Opin Lipidol.* 20:402–8.
Sawyer, N., E. Cauchon, A. Chateauneuf, R.P.G. Cruz, D.W. Nicholson, K.M. Metters et al. 2002. Molecular pharmacology of the human prostaglandin D_2 receptor, CRTH2. *Br J Pharmacol.* 137:1163–72.
Seregi, A., G. Folly, R. Heldt, E.S. Vizi, and G. Hertting. 1990. Differential prostaglandin formation induced by convulsions in the brain of mice susceptible (DBA/2J) and resistant (CFLP) to acoustic stimulation. *Epilepsy Res.* 5:131–6.
Serteser, M., T. Ozben, S. Gumuslu, S. Balkan, and E. Balkan. 2002. The effects of NMDA receptor antagonist MK-801 on lipid peroxidation during focal cerebral ischemia in rats. *Prog Neuropsychopharmacol Biol Psychiatry.* 26:871–7.

Shah, A., H. Pickles, M. Joshi, A. Webster, and J. O'Grady. 1984. Effects of single oral dose administration of a hydantoin prostaglandin analogue BW 245C in man. *Life Sci.* 34:2281–6.

Sharif, N.A., S.X. Xu, G.W. Williams, J.Y. Crider, B.W. Griffin, and T.L. Davis. 1998. Pharmacology of [^{3}H]prostaglandin E_1/[^{3}H]prostaglandin E_2 and [^{3}H]prostaglandin F_{2a} binding to EP_3 and FP prostaglandin receptor binding sites in bovine corpus luteum: Characterization and correlation with functional data. *J Pharmacol Exp Ther.* 286:1094–102.

Shichijo, M., A. Arimura, Y. Hirano, K. Yasui, N. Suzuki, M. Deguchi, and W.M. Abraham. 2009. A prostaglandin D2 receptor antagonist modifies experimental asthma in sheep. *Clin Exp Allergy.* 39:1404–14.

Shichijo, M., H. Sugimoto, K. Nagao, H. Inbe, J.A. Encinas, K. Takeshita et al. 2003. Chemoattractant receptor-homologous molecule expressed on Th2 cells activation in vivo increases blood leukocyte counts and its blockade abrogates 13,14-dihydro-15-keto-prostaglandin D2-induced eosinophilia in rats. *J Pharmacol Exp Ther.* 307:518–25.

Simmons, D.L., R.M. Botting, and T. Hla. 2004. Cyclooxygenase isozymes: The biology of prostaglandin synthesis and inhibition. *Pharmacol Rev.* 56:387–437.

Simon, R.P., J.H. Swan, T. Griffiths, and B.S. Meldrum. 1984. Blockade of N-methyl-D-aspartate receptors may protect against ischemic damage in the brain. *Science.* 226:850–2.

Singh, D. 2004. Merck withdraws arthritis drug worldwide. *BMJ.* 329:816.

Smith, J.B., M.J. Silver, C.M. Ingerman, and J.J. Kocsis. 1974. Prostaglandin D2 inhibits the aggregation of human platelets. *Thromb Res.* 5:291–9.

Smith, W.L., D.L. DeWitt, and R.M. Garavito. 2000. Cyclooxygenases: Structural, cellular, and molecular biology. *Annu Rev Biochem.* 69:145–82.

Smith, W.S. 2004. Pathophysiology of focal cerebral ischemia: A therapeutic perspective. *J Vasc Interv Radiol.* 15:S3–12.

Spik, I., C. Brenuchon, V. Angeli, D. Staumont, S. Fleury, M. Capron et al. 2005. Activation of the prostaglandin D2 receptor DP2/CRTH2 increases allergic inflammation in mouse. *J Immunol.* 174:3703–8.

Sturzebecher, S., B. Nieuweboer, S. Matthes, and E. Schillinger. 1989. Effects of PGD2, PGE1, and PGI2-analogues on PGDF-release and aggregation of human gelfiltered platelets. *Prog Clin Biol Res.* 301:365–9.

Sugimoto, M., I. Arai, N. Futaki, Y. Hashimoto, T. Sakurai, Y. Honma, and S. Nakaike. 2007. The anti-pruritic efficacy of TS-022, a prostanoid DP1 receptor agonist, is dependent on the endogenous prostaglandin D2 level in the skin of NC/Nga mice. *Eur J Pharmacol.* 564:196–203.

Sutherland, B.A., J. Minnerup, J.S. Balami, F. Arba, A.M. Buchan, and C. Kleinschnitz. 2012. Neuroprotection for ischaemic stroke: Translation from the bench to the bedside. *Int J Stroke.* 7:407–18.

Sutherland, B.A., M. Papadakis, R.L. Chen, and A.M. Buchan. 2011. Cerebral blood flow alteration in neuroprotection following cerebral ischaemia. *J Physiol.* 589:4105–14.

Szekely, C.A., J.E. Thorne, P.P. Zandi, M. Ek, E. Messias, J.C. Breitner, and S.N. Goodman. 2004. Nonsteroidal anti-inflammatory drugs for the prevention of Alzheimer's disease: A systematic review. *Neuroepidemiology.* 23:159–69.

Takemiya, T., K. Matsumura, H. Sugiura, M. Maehara, S. Yasuda, S. Uematsu et al. 2010. Endothelial microsomal prostaglandin E synthase-1 exacerbates neuronal loss induced by kainate. *J Neurosci Res.* 88:381–90.

Tanaka, E., S. Niiyama, S. Sato, A. Yamada, and H. Higashi. 2003. Arachidonic acid metabolites contribute to the irreversible depolarization induced by in vitro ischemia. *J Neurophysiol.* 90:3213–23.

Taniguchi, H., I. Mohri, H. Okabe-Arahori, K. Aritake, K. Wada, T. Kanekiyo et al. 2007. Prostaglandin D2 protects neonatal mouse brain from hypoxic ischemic injury. *J Neurosci.* 27:4303–12.

Tasken, K. and E.M. Aandahl. 2004. Localized effects of cAMP mediated by distinct routes of protein kinase A. *Physiol Rev.* 84:137–67.

Tegtmeier, F., C. Weber, U. Heister, I. Haker, D. Scheller, R. Nikolov, and M. Holler. 1990. Eicosanoids in rat brain during ischemia and reperfusion—Correlation to DC depolarization. *J Cereb Blood Flow Metab.* 10:358–64.

Teismann, P., M. Vila, D.K. Choi, K. Tieu, D.C. Wu, V. Jackson-Lewis et al. 2003. COX-2 and neurodegeneration in Parkinson's disease. *Ann N Y Acad Sci.* 991:272–7.

Thura, M., K. Hokamura, S. Yamamoto, M. Maeda, K. Furuta, M. Suzuki et al. 2009. GIF-0173 protects against cerebral infarction through DP1 receptor activation. *Exp Neurol.* 219:481–91.

Toh, H., A. Ichikawa, and S. Narumiya. 1995. Molecular evolution of receptors for eicosanoids. *FEBS Lett.* 361:17–21.

Topol, E.J. 2004. Failing the public health—Rofecoxib, Merck, and the FDA. *N Engl J Med.* 351:1707–9.

Toris, C.B., G.L. Zhan, M.R. Feilmeier, C.B. Camras, and M.A. McLaughlin. 2006. Effects of a prostaglandin DP receptor agonist, AL-6598, on aqueous humor dynamics in a nonhuman primate model of glaucoma. *J Ocul Pharmacol Ther.* 22:86–92.

Urade, Y. and O. Hayaishi. 1999. Prostaglandin D2 and sleep regulation. *Biochim Biophys Acta.* 1436:606–15.

Urushitani, M., T. Nakamizo, R. Inoue, H. Sawada, T. Kihara, K. Honda et al. 2001. N-methyl-D-aspartate receptor-mediated mitochondrial Ca^{2+} overload in acute excitotoxic motor neuron death: A mechanism distinct from chronic neurotoxicity after Ca^{2+} influx. *J Neurosci Res.* 63:377–387.

Van Hecken, A., M. Depré, I. De Lepeleire, C. Thach, M. Oeyen, J. Van Effen et al. 2007. The effect of MK-0524, a prostaglandin D2 receptor antagonist, on prostaglandin D2-induced nasal airway obstruction in healthy volunteers. *Eur J Clin Pharmacol.* 63:135–141.

Vane, J.R. 1971. Inhibition of prostaglandin synthesis as a mechanism of action for aspirin-like drugs. *Nat New Biol.* 231:232–5.

Vartiainen, N., C.Y. Huang, A. Salminen, G. Goldsteins, P.H. Chan, and J. Koistinaho. 2001. Piroxicam and NS-398 rescue neurones from hypoxia/reoxygenation damage by a mechanism independent of cyclo-oxygenase inhibition. *J Neurochem.* 76:480–9.

Walton, M., A.M. Woodgate, A. Muravlev, R. Xu, M.J. During, and M. Dragunow. 1999. CREB phosphorylation promotes nerve cell survival. *J Neurochem.* 73:1836–42.

Walton, M.R. and I. Dragunow. 2000. Is CREB a key to neuronal survival? *Trends Neurosci.* 23:48–53.

Wang, X. 2011. Recent advances in stroke: Molecular mechanisms, approaches, and treatments. *Cent Nerv Syst Agents Med Chem.* 11:80.

Wardlaw, J.M., M.S. Dennis, C.P. Warlow, and P.A. Sandercock. 2001. Imaging appearance of the symptomatic perforating artery in patients with lacunar infarction: Occlusion or other vascular pathology? *Ann Neurol.* 50:208–15.

Weinberger, J.M. 2006. Evolving therapeutic approaches to treating acute ischemic stroke. *J Neurol Sci.* 249:101–9.

Whittle, B.J., S. Moncada, K. Mullane, and J.R. Vane. 1983. Platelet and cardiovascular activity of the hydantoin BW245C, a potent prostaglandin analogue. *Prostaglandins*. 25:205–23.

Wilson, J.X. and A.W. Gelb. 2002. Free radicals, antioxidants, and neurologic injury: Possible relationship to cerebral protection by anesthetics. *J Neurosurg Anesthesiol*. 14:66–79.

Wolf, P.A., R.D. Abbott, and W.B. Kannel. 1987. Atrial fibrillation: A major contributor to stroke in the elderly. The Framingham Study. *Arch Intern Med*. 147:1561–4.

Woodward, D.F., S.B. Hawley, L.S. Williams, T.R. Ralston, C.E. Protzman, C.S. Spada et al. 1990. Studies on the ocular pharmacology of prostaglandin D2. *Invest Ophthalmol Vis Sci*. 31:138–46.

Woodward, D.F., R.L. Jones, and S. Narumiya. 2011. International Union of Basic and Clinical Pharmacology. LXXXIII: Classification of prostanoid receptors, updating 15 years of progress. *Pharmacol Rev*. 63:471–538.

Woodward, D.F., R.A. Lawrence, C.E. Fairbairn, T. Shan, and L.S. Williams. 1993. Intraocular pressure effects of selective prostanoid receptor agonists involve different receptor subtypes according to radioligand binding studies. *J Lipid Mediat*. 6:545–53.

Wright, D.H., D. Abran, M. Bhattacharya, X. Hou, S.G. Bernier, A. Bouayad et al. 2001. Prostanoid receptors: Ontogeny and implications in vascular physiology. *Am J Physiol Regul Integr Comp Physiol*. 281:R1343–60.

Wright, D.H., F. Nantel, K.M. Metters, and A.W. Ford-Hutchinson. 1999. A novel biological role for prostaglandin D2 is suggested by distribution studies of the rat DP prostanoid receptor. *Eur J Pharmacol*. 377:101–15.

Wymann, M.P. and R. Schneiter. 2008. Lipid signalling in disease. *Nat Rev Mol Cell Biol*. 9:162–76.

Xiang, Z., L. Ho, S. Yemul, Z. Zhao, W. Qing, P. Pompl, K. Kelley et al. 2002. Cyclooxygenase-2 promotes amyloid plaque deposition in a mouse model of Alzheimer's disease neuropathology. *Gene Expr*. 10:271–8.

Xing, Y., X. Jiang, Y. Yang, and G. Xi. 2011. Hemorrhagic transformation induced by acute hyperglycemia in a rat model of transient focal ischemia. *Acta Neurochir Suppl*. 111:49–54.

Yamamoto, Y., B.R. Clower, J.L. Haining, and R.R. Smith. 1991. Effect of tissue plasminogen activator on intimal platelet accumulation in cerebral arteries after subarachnoid hemorrhage in cats. *Stroke*. 22:780–4.

Zhao, B.Q., S. Wang, H.Y. Kim, H. Storrie, B.R. Rosen, D.J. Mooney et al. 2006. Role of matrix metalloproteinases in delayed cortical responses after stroke. *Nat Med*. 12:441–445.

Zhao, H., K. Ohinata, and M. Yoshikawa. 2009. Central prostaglandin D(2) exhibits anxiolytic-like activity via the DP(1) receptor in mice. *Prostaglandins Other Lipid Mediat*. 88:68–72.

Zhuang, H., Y.S. Kim, K. Namiranian, and S. Doré. 2003a. Prostaglandins of J series control heme oxygenase expression: Potential significance in modulating neuroinflammation. *Ann N Y Acad Sci*. 993:208–16.

Zhuang, H., S. Pin, X. Li, and S. Doré. 2003b. Regulation of heme oxygenase expression by cyclopentenone prostaglandins. *Exp Biol Med*. 228:499–505.

Zivin, J.A., M. Fisher, U. DeGirolami, C.C. Hemenway, and J.A. Stashak. 1985. Tissue plasminogen activator reduces neurological damage after cerebral embolism. *Science*. 230:1289–92.

Section II

Management in CNS Trauma

11 Chronic Pain in Neurotrauma
Implications on Spinal Cord and Traumatic Brain Injury

Rabih A. Moshourab, Michael Schäfer, and Elie D. Al-Chaer

CONTENTS

11.1 INTRODUCTION

Long-lasting and persistent pain is a frequent consequence of spinal cord and brain injury. Several studies point to a significantly high proportion of patients who experience pain following the trauma. This chapter gives a brief overview of the prevalence, types of persistent pain, animal models that study pain outcomes, and the pertinent mechanisms that underlie the development of neuropathic pain following traumatic spinal cord and brain injury.

Traumatic brain injury (TBI) and spinal cord injury (SCI) impose a high personal, social, and economic burden of disability. Although not as common as low back pain, TBI and SCI combined might have an equivalent economic impact mainly because of the young age of patients, the severity of the associated disability, and the major limitations on daily activity (Ma et al., 2014). Research-based estimates of the prevalence of persistent pain are variable and high in patients with SCI and TBI. Most studies indicate that about two-thirds of patients with either SCI or TBI will experience pain after the injury (Nampiaparampil, 2008; Siddall et al., 2003; Störmer et al., 1997; Uomoto and Esselman, 1993). Pain is consistently rated as one of the most difficult problems associated with these types of injuries (Nepomuceno et al., 1979; Rintala et al., 1998; Stensman, 1994; Westgren and Levi, 1998), hinders the ability to participate in rehabilitation programs (Widerström-Noga et al., 1999), and is difficult to treat. Chronic pain delays the acquisition of an optimal level of activity (Nicholson Perry et al., 2009) and independence and adversely affects the patients' mood (Kennedy et al., 1997; Stroud et al., 2006).

This chapter reviews and summarizes both clinical and experimental studies that focus on chronic pain after TBI and SCI. The emphasis will be in particular on the prevalence of chronic pain in patients with TBI and SCI. We highlight the specific types of pain that occur after injury. A survey of the different experimental animal models of brain and spinal cord injury evaluating pain as an outcome will be discussed. Finally, we will address the mechanisms responsible for the development of chronic pain following SCI and TBI.

11.1.1 Epidemiology of Pain in SCI and TBI

Several studies examined the prevalence of chronic pain after SCI. A systemic review of 42 studies provides a prevalence range between 26% and 96% (Dijkers et al., 2009). The great variability in prevalence might reflect the difficulty and heterogeneity among researchers in classifying pain syndromes (Bryce and Ragnarsson, 2001). Patients with SCI experience multiple types of pain with different characteristics simultaneously in different regions of the body. This highlights the need for international guidelines in the taxonomy of complex pain syndromes to ease the communication and management of pain subtypes. An existing SCI pain taxonomy classifies pain-related syndromes into three different types: musculoskeletal, neuropathic, and visceral pain (Table 11.1) (Widerström-Noga et al., 2008, 2014; Cardenas et al., 2002; Bryce et al., 2012b). The most common type is musculoskeletal pain that arises from muscles (trauma, spasm, inflammation, abnormal use), bones (ectopic bone formation in soft tissue and joints, instability), and joints (swelling due to reduced range of motion) (Biering-Sørensen et al., 2012). In a recent systematic review of eight studies, the prevalence of musculoskeletal pain in spinal cord injury was reported to be around 50% (Michailidou et al., 2013). Perhaps the most debilitating and difficult to treat pain syndromes are neuropathic pain syndromes. In a prospective study over 12 months after onset of injury, neuropathic pain related to SCI afflicted up to 59% of patients, and early hypersensitivity, especially to cold stimuli, might identify patients who will later develope neuropathic pain (Finnerup et al., 2014; Siddall et al., 2003). In addition, research has been inconclusive as to the risk factors that are responsible for the develope of neuropathic pain after SCI. Predictors of neuropathic pain are poorly identified and obscure in that the extent and location of injury do not correlate with the severity of neuropathic pain (McMahon, 2013).

TABLE 11.1
International Association for the Study of Pain Taxonomy of Pain after Spinal Cord Injury

Tier 1	Tier 2	Tier 3
Pain type	Pain subtype	Pain pathology
Nociceptive	Musculoskeletal pain	Muscle: spasms
		Bone: fractures, mechanical instability
		Joints: arthritis, epicondylitis, dislocations
		"Overuse"
	Visceral pain	Bowel impaction,
		Myocardial infarction, cholecystitis,
		Renal calculus
		Urinary tract/bladder infections
	Other	Surgical wounds
Neuropathic	At-level SCI pain	Spinal cord compression, trauma
		Nerve root compression
		Cauda equina compression
		Syringomyelia
	Below-level SCI pain	Ischemia, compression
Other pain		Complex regional pain syndrome, fibromyalgia,
		Headaches

Source: Adapted from Siddall PJ, Yezierski RP, Loeser JD. In: Yezierski RP, Burchiel KJ (eds.), *Spinal Cord Injury Pain: Assessment, Mechanisms, Management. Progress in Pain Research and Management*, volume 23. IASP Press, Seattle, pp. 9–24, 2002.

Similar to SCI, chronic pain is a common complication of TBI, even after minor injuries in both children and adults (Nampiaparampil, 2008; Sherman et al., 2006; Zaloshnja et al., 2008). Eighty percent of TBIs are mild and patients are expected to recover; however, a major subset of patients develop chronic headache and concentration deficits (Weyer Jamora et al., 2013). An analysis of 23 studies yielded a prevalence for chronic posttraumatic headache (PTH) of about 58% in the civilian population and 43% in veterans after mild TBI (Nampiaparampil, 2008). TBI is a hallmark injury in combat veterans, and chronic headache was mainly a sequel of blast-related brain injury. The majority of combat veterans (60%) rated the headache intensity as moderate to severe (Bosco et al., 2013). In severe TBI, about one-third of patients reported chronic pain, and this number should be cautiously interpreted as patients may have cognitive impairments, memory loss, and communication disturbances that make the evaluation of pain difficult (Uomoto and Esselman, 1993; Yamaguchi, 1992). Blume and colleagues found a prevalence of 43–60% of headaches 3 to 12 months after mild TBI in children between 5 and 17 years of age (Blume et al., 2011). In another study involving a pediatric population that suffered mild TBI disability, chronic pain was prevalent in 24% after 36 months (Tham et al., 2013).

11.2 PAIN SYNDROMES AFTER NEUROTRAUMA

11.2.1 Types of Pain in SCI

Clinical studies have revealed specific types of pain syndromes associated with SCI. The SCI Pain Task Force of the International Association for the Study of Pain (IASP) developed a pain taxonomy that achieved widespread consensus and assisted in coding the types of pain (Table 11.1) (Bryce et al., 2012a, 2012b; Siddall et al., 1997, 2002; Widerström-Noga et al., 2008). The classification system consists of three orders or tiers. The first order distinguishes broad types of pain: nociceptive and neuropathic. The second order classifies pain according to broad systems. Nociceptive pain can be classified as visceral or musculoskeletal. Neuropathic pain is classified with reference to the level of SCI: above-level, at-level, or below-level. The third order identifies structures and pathologies pertaining to the etiology of the pain syndrome (e.g., muscle spasm for musculoskeletal pain; syringomyelia for at-level neuropathic pain; see Table 11.1). A redefinition of neuropathic pain has been proposed that requires a demonstration of pathology within the nervous system that can explain the pain (Treede et al., 2008). A complex clinical picture of pain syndromes can develop after spinal trauma. Spinal nerve damage, soft-tissue damage, overuse of skeletal muscle, and visceral dysfunction can accompany chronic neuropathic pain.

An interdisciplinary group comprising experts from IASP, the International Spinal Cord Society, the International Spinal Cord Society, and the American Spinal Injury Association developed the International Spinal Cord Injury Basic Pain Data Set within the framework of the International Spinal Cord Injury Core Data Sets (Widerström-Noga et al., 2008, 2014). The data set consists of core questions that address clinically relevant characteristics of pain in SCI. The questions cover pain severity, pain-intensity rating, temporal pattern of pain for each specific pain problem, and a pain classification. Additionally, the impact of pain on physical, social and emotional function, and sleep is evaluated for each pain. The data set provides standardized collection and reporting of clinically relevant information regarding pain syndromes in SCI among health care professionals. The data set would facilitate a meaningful comparison of study data related to pain in the SCI population, promote collaboration, and increase efficiency of clinical trials.

11.2.2 Nociceptive Pain: Musculoskeletal and Visceral

Chronic musculoskeletal and back pain is very common in patients with SCI (Michailidou et al., 2013). Musculoskeletal pain usually occurs in an area innervated by undamaged nerves (above or below level of injury) that has preserved sensation. Acute nociceptive pain is mainly from injury to the supporting structures of the vertebral column (often not stable). These affected structures include ligaments, muscles, intervertebral discs, and facet joints. Chronic musculoskeletal pain can result from the overuse and abnormal use of extremities and back especially in paraplegics (e.g., transfers using wheelchairs). Pain is suggested by descriptors such as dull or aching, and is typically related to positioning and activity. Examples that contribute to pain include muscle spasms, bone ossification in soft tissue and joints that may lead to joint swelling, and fractures.

The diagnosis of visceral pain can be sometimes challenging in SCI patients because the symptoms of pain are in general located below the level of SCI (thorax, abdomen, and pelvis). The level of injury in SCI can alter pain quality and location. Visceral pain arises from visceral structures, has a delayed onset, and presents in 1–3% of patients with SCI. Pain is dull, tender, and cramping in nature and often accompanied by autonomic dysreflexia. Abdominal pains may be vague in tetraplegics (Finnerup et al., 2014). Bowel impaction, bladder distension, renal and ureteric calculi, and urinary tract infections are common causes of nociceptive visceral pain (Al-Chaer, 2008; Al-Chaer and Traub, 2002). For instance, bladder distension or bowel impaction may present as autonomic dysreflexia and headache in patients with cervical or upper thoracic injury.

11.2.3 Central Neuropathic Pain

Pain syndromes post-SCI are usually permanent and are designated as central neuropathic pain because lesions are central. Neuropathic pain is suggested based on patient history and clinical sensory examination. Pains can be (1) spontaneous and occur independent of peripheral stimulus or (2) evoked and occur in response to a noxious or innocuous stimulus. The constellation of clinical criteria should be well characterized: positive (hyperalgesia and allodynia) and negative (loss of sensation) signs, dermatomal distribution, and a cause related to clinical abnormality (Treede et al., 2008). Spontaneous pain is intermittent and described as shooting, burning, electric-like, and cutting. Typically, the location of pain is in a band-like region at the level of sensory loss. Evoked pain can either be induced by a noxious (hyperalgesia) or innocuous (allodynia) stimulus (Christensen et al., 1996). Allodynia is defined as a painful response to normally innocuous stimuli, and hyperalgesia is an amplified pain response to normally noxious stimuli. There are three major types of chronic neuropathic pain that are specific to SCI: above-level neuropathic pain that occurs in dermatomes cranial to the site of injury; at-level neuropathic pain that occurs in dermatomes at the site of injury; and below-level neuropathic pain that occurs in dermatomes caudal to the neurological lesion (Hulsebosch et al., 2009).

11.2.3.1 At-Level Neuropathic Pain

At-level neuropathic pain denotes pain with a dermatomal or segmental pattern located between the level of neurological injury and three dermatomes below. It can either present at the time of injury or can develop directly afterward. Pain is characterized as shooting, tingling, burning, squeezing, pins and needles, and electric. Allodynia and hyperesthesia are clinical signs associated with at-level neuropathic pain (Widerström-Noga et al., 2008).

A necessary condition is the presence of a lesion in the spinal cord. There are multiple lesions that produce this type of pain: nerve root or spinal cord injury itself. In many cases, pain resulting from nerve root injury is unilateral, spontaneous, and exacerbates upon movement (Yezierski, 2000). Evidence of impingement by facet or disc material is diagnosed by radiographic tests. Sometimes, segmental and radicular pain may occur from damage in the spinal cord, without evidence of nerve root damage. Another potential complication of SCI is syringomyelia (Brodbelt and Stoodley, 2003). This is a disorder in which a cyst (syrinx) forms within the damaged segments of the spinal cord and expands with time, destroying further tissue. Syringomyelia is suspected when new symptoms, such as pain (particularly at-level neuropathic pain), weakness, and sensory loss, appear several months after the initial injury. Cauda equina damage (i.e., compression of the lower region of the spinal canal involving nerve roots L5-S1) is defined as at-level pain referred to the sacral and lumbar dermatomes (Harrop et al., 2004).

11.2.3.2 Below-Level Neuropathic Pain

Below-level neuropathic pain is the most common central pain syndrome after SCI and is difficult to treat. Pain is of gradual onset and has a diffuse distribution located more than three dermatomes below the level of neurological injury. Pain is spontaneous and evoked. It has typically a burning quality and can be triggered by sudden noises and movements (Widerström-Noga et al., 2008). Complete and partial injuries that cause damage to spinothalamic tracts are associated with this type of pain (Finnerup et al., 2007). Tactile allodynia is especially prominent in incomplete injuries, which spare tracts mediating touch sensitivity in affected regions.

11.2.4 Types of Pain in TBI

The occurrence of pain in TBI is not as well systematically evaluated, as is the case for pain after SCI. It presents with a set of diagnostic and treatment challenges that are not adequately addressed. PTH is the most common chronic pain syndrome in brain injury; however, it is poorly characterized (Linder, 2007; Sherman et al., 2006). Neuropathic pain and pain arising from spasticity and fractures are also prevalent in TBI.

11.2.5 Headaches

Headaches are present in 59% of brain injury patients, regardless of trauma severity (Nampiaparampil, 2008). According to the International Headache Society II classification, PTH may present similar to primary headache disorders such as migraine, tension-type, or cluster headaches (Baandrup and Jensen, 2005; Zasler, 2011). PTH usually develops within 7 to 30 days after injury and usually resolves within the expected tissue-healing time of 3 months (Ofek and Defrin, 2007; Walker et al., 2005; Sherman et al., 2006). Persistence of pain beyond 3 months suggests pain chronicity with underlying peripheral and central sensitization in the trigeminal system (Ofek and Defrin, 2007). The average onset time for chronic head and facial pain is about 6 months. Thus, the characteristics of chronic pain after TBI resemble those of chronic central pain from other causes, with painful regions exhibiting allodynia, hyperpathia, and wind-up (Ofek and Defrin, 2007). A 1-year follow-up study on combat veterans with TBI revealed that 54% of those continued to experience headaches; the majority of veterans experienced their headache on a daily basis mostly in the frontal area (Walker et al., 2005). In contrast, in a study that examined the TBI in a civilian population, the occipital area was more involved (Lew et al., 2006). The quality of pain is usually more pricking, throbbing, and pounding, with few complaints of burning pain, which is more frequent in SCI (Defrin et al., 2001; Ofek and Defrin, 2007). Sensory profiles of patients with PTH reveal higher thermal thresholds and lower mechanical pain thresholds in the hands and face suggestive of a central origin of pain (Defrin et al., 2010). Almost all patients with TBI have movement allodynia, which is not common after SCI (Finnerup, 2003). TBI can directly lead to headaches or can worsen preexisting headaches (Jensen and Nielsen, 1990). The literature is scarce on studies that evaluate measures of pain intensity or pain-related behavior among patients with cognitive deficits from moderate to severe TBI. Similarly, no studies compared patients with blast-related headache to patients with other types of headache, or assessed treatments for blast-related headache (Dobscha et al., 2009).

11.2.6 Other Pain Syndromes

Pain can originate from a number of sources after TBI. Muscle spasticity from decerebrate and decorticate posturing may lead to pain. Often overlooked, muscle spasms can contribute to pain from formation of joint contractures, tendinitis, and painful postural abnormalities (Sherman et al., 2006). Other types of nonheadache pains after TBI include low back pain (46%), extremity pains (39%), and complex regional pain syndrome (CRPS, 12%). Interestingly, the incidence of CRPS after severe TBI is comparable to the incidence of CRPS after stroke (12%–25%) (Gellman et al., 1992). Peripheral neuropathy is also observed after TBI.

11.3 ANIMAL MODELS OF PAIN AFTER NEUROTRAUMA

How does chronic neuropathic pain after spinal cord trauma in the clinic compare with the animal models of spinal cord injury in the laboratory setting? For instance, the spinal contusion model, where a device is used to induce a specific spinal contusion injury, can produce above (forelimb), at (trunk), and below (hindlimb) level neuropathic pain highlighted by the presence of mechanical allodynia. This model resembles the alterations in somatosensory profiles found in patients with SCI. Here, the distinct pathological mechanisms—peripheral and central sensitization, intracellular signalling, and microglia activation—responsible for the different types of central neuropathic pain can be investigated.

11.3.1 Overview

Of interest, there exist several animal models that investigate pain in TBI and SCI. These models characterize the anatomical, biochemical, molecular, genetic, physiological, and pharmacological processes associated with chronic pain. Injury to the brain or spinal cord is caused by an external mechanical force (blast waves, crush, projectile penetration, impact, and rapid acceleration/deceleration) or induction of ischemia (Mogil et al., 2010; Vierck et al., 2000; Xiong et al., 2013). The various injury mechanisms can induce a multitude of complex pathophysiological processes with structural damage and functional deficits. The initial or primary injury is the immediate disruption of brain or spinal cord tissue occurring at the time of the event. After the primary injury, secondary injury evolves over several months after the primary injury and is due to a cascade of molecular, cellular, and metabolic derangements leading to inflammation, cell death, and tissue atrophy. The various animal models are each based on a specific component of the primary injury. In contrast, the injury in patients is usually heterogeneous and more likely to involve a mixture of components. Therefore, no single animal model can encompass the whole plethora of pathological changes and this in part explains the sluggish progress in translational studies.

However, animal models are essential to address the molecular and cellular aspects of altered neural plasticity that cannot be studied in the clinical setting (O'Connor et al., 2011; Xiong et al., 2013). Animal models serve the purpose of characterizing the behavioral indices of pain and correlating them with abnormal activity in the central nervous system to elucidate altered nociceptive processing after neurotrauma.

11.3.2 Pain in SCI Animal Models

Several models of SCI are commonly used to investigate central neuropathic pain in rodents. Models that employ excitotoxic, ischemic, transection, and contusive injuries provide specific behavioral and molecular changes involved in the development of chronic pain (Table 11.2) (Nakae et al., 2011; Yezierski, 2005). Not all models involve trauma. For instance, the photochemical model induces an ischemic-like condition. An intravascular injection of a dye that is activated by argon ion laser produces oxygen at the endothelial surface and activates platelets, which results in a thrombus that occludes spinal vessels and leads to parenchymal infarction (Gaviria et al., 2002; Watson et al., 1986). The consequences are behavioral mechanical allodynia and signs of spontaneous pain (Hao et al., 1991). The excitotoxic models use intrathecal or intraspinal injections of excitotoxins (e.g., quisqualic acid or other excitatory amino acids such as glutamate, N-methylaspartate, kainic acid) (Yezierski and Park, 1993; Yezierski et al., 1998). This produces spontaneous pain behavior, mechanical allodynia, and thermal hyperalgesia in both rats and mice. This model simulates the increased levels of excitatory amino acids after spinal cord injury (Panter et al., 1990).

There are various models that employ a defined mechanical or surgical injury to the spinal cord: weight drop or contusion model (Hubscher and Johnson, 2006; Hulsebosch et al., 2000; Lindsey et al., 2000; Nakae et al., 2011; Siddall et al., 1995), spinal lesion models, hemisection, and transection (Christensen et al., 1996; Vierck and Light, 1999). The spinal contusion model is the most widely used in experimental SCI. The spinal cord is injured by a weight drop impactor that drops a 10-gram rod from a certain height to the vertebral column (Gruner, 1992). The injury produces motor and sensory deficits including neuropathic pain (Tanabe et al., 2009). Pain behavior can usually be assessed after recovery of motor dysfunction within 2 to 3 weeks. The procedure has its shortcomings: the impact of the injury is variable and unilateral contusion/paralysis often occurs. There are wide spectra of abnormal sensations to mechanical and thermal stimulations that can be elicited for several weeks after injury. Allodynia can occur in all regions involving the injury site, and thus at-level or below-level neuropathic pain can be poorly controlled with this model (Mills et al., 2001; Nakae et al., 2011).

The transection model is performed after a laminectomy to make a hemitransection or complete transection of the spinal cord. A complete transaction produces at-level and below-level neuropathic pain. Partial transection produces ipsilateral paralysis but allodynia and thermal hyperalgesia are bilaterally observed and, thus, this model provides a window to study neuropathic pain directly after SCI (Densmore et al., 2010). The contralateral side can be assessed directly after partial hemisection injury. The pain phenotype is usually assessed after recovery of motor function (usually as early as 1 to 2 weeks, sometimes 4 weeks). Spinothalamic tract lesions can be assessed after less than a week. Evaluation of pain such as behavior after SCI is based on stimulus-evoked withdrawal responses, and it is often challenging to distinguish pain-like behavior from increased reflex responses. Increased reflexes—which do not correlate with increased behavioral measures of pain—can be mistaken for hypersensitivity and pain behavior. However, there are valid methods that assist in discriminating spinal reflexes from brainstem and cortical responses associated with pain behavior (Baastrup et al., 2010). This is particularly important for the correct diagnosis of at- and below-level neuropathic pain in animal models.

11.3.2.1 Production of At-Level and Below-Level Neuropathic Pain

Not all animal models designed to study pain outcomes after neurotrauma produce similar pain phenotypes (Mills et al., 2001; Nakae et al., 2011; Yezierski, 2005). SCI in animal models causes neuronal loss in a longitudinal (rostrocaudal) direction at the injury site (Gorman et al., 2001). The degree of loss is critically associated with the development of neuropathic pain. Another observation is that often at-level neuropathic pain progresses to below-level. A decisive factor might be the extent of damage to the gray matter and white matter of spinal pathways (Yezierski, 2005). At-level SCI pain is characteristic in animal models with excitotoxic, ischemic, and contusive damages to the spinal gray and white matter (Yezierski et al., 2004; Hulsebosch et al., 2000; Yezierski et al., 1998). Current evidence suggests that strategies that restrict the extent of injury after excitotoxic or ischemic insults to the spinal cord might prevent the development of at-level neuropathic pain (Hulsebosch et al., 2009). Below-level neuropathic pain mainly results from (1) interruption of specific white matter tracts, such as spinothalamic projections to rostral targets; (2) lesions to specific regions of gray matter; and (3) the contusive model in general (Vierck and

TABLE 11.2
TBI and SCI Animal Models Used in Pain Research

Model SCI	Injury Mechanism	Pain Phenotype	Neuropathic Pain	Pathophysiological Features	References*
Contusion	Cervical† to lumbar Weight drop (10-gram rod) on dorsal aspect of vertebral column	Mechanical allodynia Thermal hyperalgesia overgrooming autotomy	At-level and below-level in majority of animals (~80%), can last >6 weeks	Mimics SCI in humans. Lesion grade cannot be well controlled. Motor paralysis 2–4 weeks	1, 5, 6
Transection Complete	Cervical to lumbar Bilateral	Mechanical allodynia Thermal hyperalgesia lasts 4 weeks	At-level Below-level	Motor weakness lasts ~4 weeks	1, 7
Partial/hemi	Unilateral	Mechanical allodynia Thermal hyperalgesia, can last 4 weeks	Below-level (all animals)	Damage to spinal pathways leads to bilateral below-level neuropathic pain	8
Photochemical or ischemic	Thoracic, ischemia through dye injection then argon laser irradiation	Mechanical and cold allodynia (no thermal hyperalgesia), lasts >4 weeks	At-level and below-level (~50% of animals)	Ischemia	1, 9
Excitotoxic	Thoracic, intraspinal/ intrathecal injection of quisqualic acid	Mechanical allodynia Thermal hyperalgesia Overgrooming, lasts 5 weeks	At-level and below-level (50% of animals)	Neuronal loss dorsal horn with sparing of laminae I, increased excitatory amino acids, inflammation, glial scaring	1,2
Spinothalamic tract lesion	Thoracic, anterolateral lesion or cut, electrolytic lesion	Allodynia, mechanical and thermal hyperalgesia for several weeks	At-level Below-level	Damage to spinal pathways leads to bilateral below-level neuropathic pain. Thalamic hyperexcitability. Motor weakness lasts <1 week	1, 11
Avulsion of dorsal root	Unilateral avulsion of T13 and L1 root and lesion to Lissauer's tract, dorsal horn	Mechanical and thermal hyperalgesia	Below-level		12
Clip compression	Thoracic (laminectomy) calibrated clips applied for 1 minute then removed	Mechanical allodynia and thermal hyperalgesia (lasts 4 weeks)	At-level, below-level	Mimics ischemic injuries; 4 weeks of motor paralysis	10
Model TBI	**Injury Mechanism**	**Pain Phenotype**	**Neuropathic Pain**	**Pathophysiological Features**	**References***
Controlled cortical impact	After craniotomy, a device delivers a pneumatic or electromagnetic impact on intact dura	Periorbital allodynia for 4 weeks	Trigeminal system	Mimics concussion, axonal injury, acute subdural hematoma. CGRP and substance P increased in brainstem	3, 4

*1 (Nakae et al., 2011; Vierck Jr et al., 2000); 2 (Yezierski et al., 1998); 3 (Xiong et al., 2013); 4 (Macolino et al., 2014; Elliott et al., 2012); 5 (Gruner, 1992); 6 (Hulsebosch et al., 2000; Lindsey et al., 2000; Siddall et al., 1995; Detloff et al., 2013; Yoon et al., 2004; Scheff et al., 2003); 7 (Scheifer et al., 2002); 8 (Kim et al., 2003; Christensen et al., 1996); 9 (Hao et al., 1991; Watson et al., 1986); 10 (Bruce et al., 2002; Marques et al., 2009); 11 (Finnerup et al., 2007; Koyama et al., 1993; Ovelmen-Levitt et al., 1995; Vierck and Light, 1999; Vierck et al., 1990; Wang and Thompson, 2008); 12 (Wieseler et al., 2010, 2012).

†Cervical hemicontusion by means of a hemilaminectomy is used due to the life-threatening effects of cervical contusion.

Light, 2000; Yezierski, 2005). Therefore, each animal model has characteristic pathophysiological features for at-level and below-level neuropathic pain.

11.3.3 Pain in Animal TBI Models

There is a scarcity of literature that investigates pain in animal models of TBI. Among the many variants of brain injury models, four are widely used in research: fluid percussion injury, impact acceleration injury, controlled cortical impact (CCI), and blast exposure injury. Although functional tests have been conducted for outcomes of sensorimotor function, cognitive dysfunction, and anxiety, the pain parameter is not widely evaluated. The CCI model has been used recently to measure behavioral and biochemical outcomes of pain (Elliott et al., 2012; Macolino et al., 2014). After experimental injury to the brain, periorbital allodynia emerged and closely correlated with elevations in calcitonin gene-related peptide (CGRP) and substance P within the brainstem (Elliott et al., 2012). The findings from these well-controlled studies are intriguing because they seem to directly link injury to the somatosensory cortex with persistent neuropathic pain. It might be very difficult to model the entire complex spectrum of pathological events in an animal. However, animal models can be used to test for hypotheses about the etiologies of pain syndromes.

11.4 MECHANISMS

The experimental animal models and clinical studies on the development of pain in brain and spinal cord injury has provided a lot of insight into the cellular, biochemical, molecular, radiological, electrophysiological, and psychophysical derangements. The pathological sequelae of injury include a wide spectrum and cascade of events that shift the functional integrity of the nervous system to a "new dynamic state." To summarize all these processes requires a three-dimensional perspective: across all levels and scales (social, psychophysical, tissue, cellular, molecular, and genetic levels), temporal aspects (primary and secondary injury) and regional aspects (spinal, supraspinal, cortical). A detailed account is beyond the scope of this chapter. This current section will introduce the basic mechanisms underlying pain conditions, especially neuropathic pain in SCI, which is most disabling and the focus of a large amount of research in recent years. Interestingly, the mechanisms of secondary injury in trauma to the brain and spinal cord share common mechanisms at the cellular level (Xiong et al., 2013; Yezierski, 2005).

11.4.1 Mechanism of Pain in SCI

11.4.1.1 Neurochemical Characteristics of Primary and Secondary Injury

Primary injury denotes the mechanical tissue deformation caused by an impact, which directly results in tissue necrosis, shearing of blood vessels, neurons, and axons initiating secondary injury processes. Tissue destruction and breakdown of cellular membranes activates phospholipases. Within the first minutes, membrane hydrolysis, lipid peroxidation (eicosanoid production), and increased free radical generation cause further damage (Anderson and Hall, 1993; Yezierski, 2005; Xiong et al., 2013). Release of nitrogen and oxygen species further fuels excitotoxic reactions, catecholamine oxidation, and lipid peroxidation (Bains and Hall, 2012; Hall et al., 1992; Hall and Braughler, 1993). Membrane lipid peroxidation also enhances oxidative stress, which perturbs ionic and metabolic homeostasis (malfunction of Ca^{2+} and Na^+/K^+-ATPase ionic pumps, and glucose transporters) and cell signaling pathways (malfunction of receptors and neurotransmitters). The consequences of ionic pump failure are ionic shifts resulting in tissue edema (Kwo et al., 1989). Because of the release of excitatory mediators—glutamate, aspartate, norepinephrine, serotonin—the level of neuronal depolarization increases with massive influx of calcium ions (Andriessen et al., 2010; Liu et al., 1990; Panter et al., 1990). This phenomenon in particular is best studied in the excitotoxic SCI model that simulates the elevated levels of excitatory amino acids (Hao et al., 1991; Yezierski, 2004, 2005). Calcium plays several detrimental roles: increases inflammatory mediators, leads to mitochondrial dysfunction, and augments production of reactive oxygen species and activation of protease systems especially calpains (Kampfl et al., 1997; Saatman et al., 2010). These events pave the way for inflammation, necrosis/apoptosis, and glial cell activation. TBI and SCI upregulate several transcription factors, inflammatory cytokines and chemokines, and gene expression (Raghavendra Rao et al., 2003). In patients with SCI, several inflammatory mediators including interleukins 6 and 8, nerve growth factor, intracellular adhesion molecules, and matrix metalloproteases among others are shown to be upregulated in the cerebrospinal fluid (Tsai et al., 2008).

Rats with at-level neuropathic pain and allodynia after a spinal cord contusion have increased basal and stimulus-evoked levels of c-fos gene expression in the spinal dorsal horn immediately above the level of injury (Siddall et al., 1999). In the same model, activation of transcription factor NF-kappaB could be detected as early as 30 minutes after injury in neuronal, endothelial, and microglial cells (Bethea et al., 1998). NF-kappaB activates around 150 genes involved in inflammation and proliferative and cell death responses. Similarly, p-38a mitogen-activated protein K activation modulated neuronal hyperexcitability and allodynia in a model of SCI (Gwak et al., 2009).

Besides these changes, other signaling systems are compromised. There is an increase in glutamatergic excitatory activity mediated by NMDA, non-NMDA, and metabotropic glutamate receptors (Yu and Yezierski, 2005; Crown et al., 2006). These factors activate signaling cascades and lead to neuronal sensitization and, thus, contribute to pain. The injury cascades not only affect excitatory neurons but also inhibitory neurons. Interference with GABAergic inhibition reduces inhibitory tone and causes secondary amplification of the ascending signals (Gwak et al., 2006; Gwak and Hulsebosch, 2011). The GABAergic dysfunction has direct relation to neuropathic pain in animal models. Another factor contributing to the increased excitability is upregulation of tetrodotoxin-sensitive sodium channels NaV1.3 in second order dorsal horn neurons (Hains et al., 2003). Involvement of the NK-1/SP system has been also linked to the development of at-level neuropathic pain (Yezierski et al., 2004). Several mechanisms have been proposed to explain pain syndromes in SCI: (1) presence of pattern generators (Melzack and Loeser, 1978); (2) loss of inhibitory mechanisms; (3) supraspinal relay nuclei; (4) changes in synaptic plasticity; (5) microglial activation; and (6) changes in cell-signaling pathway in brain and spinal cord (Yezierski, 2009).

11.4.2 Gliopathy and Neuronal Cells

As highlighted previously, secondary injury events culminate in neuronal, endothelial, and glial cell death and white matter degeneration (Bramlett and Dietrich, 2007). Cell death extends over days to months after injury (Raghupathi, 2004). Necrotic and apoptotic cell death have been identified in contused areas, the injury boundary zone, and subcortical regions, and apoptosis coincides with progressive atrophy of gray and white matter after TBI.

In an interesting study using magnetic resonance spectroscopy, Widerström and colleagues compared metabolite concentrations in the anterior cingulated gyrus of SCI patients with severe, mild, and without neuropathic pain. Analysis showed that patients with severe neuropathic pain have altered metabolite profiles than the mild and no neuropathic pain patients (Widerström-Noga et al., 2013). This highlights the existence of a neuronal and glial dysfunction that is characteristic of severe neuropathic pain.

11.4.2.1 Peripheral Sensitization and Hyperexcitability of Spinal Dorsal Horn Neurons after SCI

Peripheral and central sensitization is demonstrated by chronically enhanced responses to stimuli and increased spontaneous activity in recordings from peripheral nociceptors and dorsal horn neurons (Bedi et al., 2010; Carlton et al., 2009). There is evidence that SCI produces alterations in the physiology of primary afferent neurons. Axotomized and injured afferents switch to a hyperfunctional state. This state entails enhanced intrinsic growth and sprouting of low-threshold afferents and peptidergic nociceptive neurons to central targets in dorsal horn and their input may contribute to chronic pain (Bedi et al., 2010, 2012; Walters, 2012). After thoracic injury to the spinal cord, the excitability of all three types of dorsal horn neurons (low-threshold, high-threshold, and wide dynamic range neurons) increases and can be explained by functional changes in glutamate and voltage-gated sodium channels, glial activation, and decreased inhibition of the endogenous inhibitory tone (Gwak and Hulsebosch, 2011). The hyperexcitability that occurs lasts for weeks to months after injury and this may be explained by continued mobilization of microglia (Saab et al., 2006) or a structural reorganization of spinal/supraspinal circuitry via formation of new synaptic connections.

Similarly, studies using animal models have demonstrated that lesions of the spinothalamic tract result in an increase in spontaneous and evoked thalamic neuronal responses linked with changes in thalamic N-methyl-D-aspartate (NMDA) receptor function (Weng et al., 2000), sodium channel expression (Hains and Waxman, 2007), and microglial activation (Zhao et al., 2007). The extent to which these supraspinal changes and pain are dependent on preserved ascending fiber tracts remains unclear. Ascending fiber tracts are surviving tracts that traverse the injury site. Clinical electrophysiological studies suggest that approximately 50% of people with clinically complete injuries still have preserved transmission in spinal pathways (Finnerup et al., 2007, 2004; Wasner et al., 2008). Therefore, preserved spinal cord pathways have been proposed as an important factor underlying the development of neuropathic pain. Based on quantitative sensory testing studies enhanced by chemical sensitization of primary afferents, this would seem to be the case, at least in a proportion of people (Detloff et al., 2012; Wasner et al., 2008). These pathways may transmit information generated by spinal neurons that are sensitized as a result of changes induced by inflammatory mediators, microglial activation, and second-messenger activation below the level of injury (Al-Chaer, 2009; Detloff et al., 2008, 2012). For instance, below-level neuropathic pain has been regarded as dependent upon partial deafferentation of rostral targets of the spinothalamic pathway. Anterolateral quadrant cordotomy, hemisection, spinothalamic tractotomy, and even unilateral lesions produce bilateral deficits and may lead to development of below-level neuropathic pain (Table 11.2) (Finnerup, 2013; Finnerup et al., 2014).

11.4.3 Plastic Changes in Spinal Cord and Brain after Neurotrauma

The degenerative processes of injury and regenerative processes of recovery that ensue after injury have been widely explored in the past decade in animal models, especially with the aid of neurophysiological and neuroimaging techniques (Nardone et al., 2013). The deafferentation that follows SCI can cause alterations, as early as 1 hour, in cortical function, which may play a critical role in the reorganization process (Nardone et al., 2013). This explains the expansion that occurs to the deprived motor and sensory cortical areas. However, reorganization can also have pathological consequences including phantom and neuropathic pain (Henderson et al., 2011; Wrigley et al., 2009). In a study of patients with SCI at the thoracic level, cortical remapping was demonstrated to be responsible for referred sensations and phantom limb pain (Moore et al., 2000). Immediate slowing of spontaneous cortical activity and long-term cortical organization in patients with cervical and thoracic SCI correlate with pathophysiological mechanisms of neuropathic pain after SCI (Hari et al., 2009; Wydenkeller et al., 2009).

A new "state" of activity in the nervous system emerges in patients that suffer pain after SCI. This new state is suggested by a "pain-generating" mechanism that involves several nuclei (dorsal horn and dorsal root ganglion neurons, nuclei in brainstem and thalamus, and cortical neurons) and tracts, which mediate pain sensation. This new emergent pattern of activity in these networks maintains a state of neuropathic pain. A model of low thoracic SCI altered response properties of cervical (above-level), thoracic (at-level), and lumbar (below-level), and thalamic regions (Carlton et al., 2009; Gwak et al., 2010; Hubscher and Johnson, 2006).

11.4.4 Mechanism of Pain in TBI

11.4.4.1 Headaches

Injury to structures such as skull, brain, meninges, and vasculature can cause PTH. The trigeminal system innervates extracranial skin, muscle, meninges, periosteum, and the vasculature and relays nociceptive signals to the trigeminal nuclei located in the brainstem. Few studies have been conducted on PTH with migraine-like features (Faux and Sheedy, 2008; Lieba-Samal et al., 2011). After severe diffuse TBI in adult male rats, neuronal and perivascular substance P immunoreactivity were increased markedly (Donkin et al., 2009). Elliot and colleagues found that nociceptive neuropeptides, CGRP and substance P, are increased in the brainstem in a mouse model of CCI (Elliott et al., 2012). Furthermore, persistent periorbital allodynia was observed in this injury to somatosensory cortex. The glutamate neurotransmitter system in the trigeminal nucleus is known to be involved in pain processing in migraine headaches (Oshinsky and Luo, 2006). In an acute stimulation of the dura by an inflammatory soup, glutamate but not aspartate, GABA, or glutamine levels were elevated in the trigeminal nucleus caudalis.

A combination of neuroinflammation, increased excitability, and disrupted circuitry in the thalamus and cortex resulted in whisker hypersensitivity and increased neuronal activation of the thalamus in a rat model of TBI (Hall and Lifshitz, 2010). Macolino and colleagues found periorbital mechanical allodynia in the CCI model in rats and mice (Macolino et al., 2014). In humans, studies using quantitative sensory testing confirmed the presence of allodynia in 40% of patients with TBI who experienced chronic head and face pain after head injury (Defrin et al., 2010; Ofek and Defrin, 2007). Last, activation of glial cells is an important source of endogenous inflammatory mediators (e.g., cytokines, chemokines, excitatory amino acids) that maintain the sensitized state in central nervous system (Xiong et al., 2013). In conclusion, altered input from sensitized trigeminal nociceptors coupled with central activation is directly involved in maintaining an ongoing state of altered sensation and chronic headache.

11.4.4.2 The Nexus of Pain and Neurotrauma

Severe pain, especially neuropathic type, has a high prevalence, and a significant impact on the quality of life, decreased physical function, and work limitations of individuals with TBI and SCI. Injury to the brain and spinal cord tissue initiates a complex series of events, some of which are local and fast and others that are global and long lasting, which may lead to chronic pain. In this review, we gave an overview of the prevalence, types, animal models, and mechanisms of pain after SCI and TBI. Animal models of pain after neurotrauma have contributed to the understanding of the complex pathophysiological mechanisms and facilitated testing of new therapeutic agents that are much in need because of the difficulty in treating chronic pain in TBI and SCI. It follows that the diverse molecular and neuroplastic changes initiated by neurotrauma, which cannot be mimicked by any single animal model, all serve to promote a persistent state of spontaneous pain, hyperalgesia, and allodynia. Several types of chronic pain conditions with distinctive characteristics such as chronic musculoskeletal, visceral, and neuropathic can afflict patients. It should also be noted that widespread standardized data collection and reporting (e.g., data sets of most relevant clinical information) and treatment guidelines for specific pain types in neurotrauma patients could eventually lead to optimal evaluation of treatment outcomes and enhance the successful management of SCI-related pain in clinical practice and trials.

REFERENCES

Al-Chaer, E.D. 2009. The neuroanatomy of pain and pain pathways. In: R.J. Moore, editor. *Biobehavioral Approaches to Pain*. Springer: New York, pp. 17–44.

Al-Chaer, E.D. 2008. Visceral pain. In: M. Dobrestov and J. M. Zhang, editors. *Mechanisms of Pain in Peripheral Neuropathy*. Springer: New York, pp. 29–45.

Al-Chaer, E.D. and R.J. Traub. 2002. Biological basis of visceral pain: Recent developments. *Pain*. 96:221–225.

Anderson, D.K. and E.D. Hall. 1993. Pathophysiology of spinal cord trauma. *Ann. Emerg. Med.* 22:987–992.

Andriessen, T.M.J.C., B. Jacobs, and P.E. Vos. 2010. Clinical characteristics and pathophysiological mechanisms of focal and diffuse traumatic brain injury. *J. Cell. Mol. Med.* 14:2381–2392. doi:10.1111/j.1582–4934.2010.01164.x.

Baandrup, L. and R. Jensen. 2005. Chronic post-traumatic headache—A clinical analysis in relation to the International Headache Classification 2nd Edition. *Cephalalgia Int. J. Headache*. 25:132–138. doi:10.1111/j.1468–2982.2004.00818.x.

Baastrup, C., C.C. Maersk-Moller, J.R. Nyengaard, T.S. Jensen, and N.B. Finnerup. 2010. Spinal-, brainstem- and cerebrally mediated responses at- and below-level of a spinal cord contusion in rats: Evaluation of pain-like behavior. *Pain*. 151:670–679. doi:10.1016/j.pain.2010.08.024.

Bains, M. and E.D. Hall. 2012. Antioxidant therapies in traumatic brain and spinal cord injury. *Biochim. Biophys. Acta*. 1822:675–684. doi:10.1016/j.bbadis.2011.10.017.

Bedi, S.S., M.T. Lago, L.I. Masha, R.J. Crook, R.J. Grill, and E.T. Walters. 2012. Spinal Cord Injury Triggers an Intrinsic Growth-Promoting State in Nociceptors. *J. Neurotrauma*. 29:925–935. doi:10.1089/neu.2011.2007.

Bedi, S.S., Q. Yang, R.J. Crook, J. Du, Z. Wu, H.M. Fishman et al. 2010. Chronic spontaneous activity generated in the somata of primary nociceptors is associated with pain-related behavior after spinal cord injury. *J. Neurosci.* 30:14870–14882. doi:10.1523/JNEUROSCI.2428–10.2010.

Bethea, J.R., M. Castro, R.W. Keane, T.T. Lee, W.D. Dietrich, and R.P. Yezierski. 1998. Traumatic spinal cord injury induces nuclear factor-kappaB activation. *J. Neurosci. Off. J. Soc. Neurosci.* 18:3251–3260.

Biering-Sørensen, F., A.S. Burns, A. Curt, L.A. Harvey, M. Jane Mulcahey, P.W. Nance et al. 2012. International spinal cord injury musculoskeletal basic data set. *Spinal Cord*. 50:797–802. doi:10.1038/sc.2012.102.

Blume, H.K., M.S. Vavilala, K.M. Jaffe, T.D. Koepsell, J. Wang, N. Temkin et al . 2011. Headache after pediatric traumatic brain injury: A cohort study. *Pediatrics*. 129:e31–e39. doi:10.1542/peds.2011–1742.

Bosco, M.A., J.L. Murphy, and M.E. Clark. 2013. Chronic pain and traumatic brain injury in OEF/OIF service members and veterans. *Headache J. Head Face Pain*. 53:1518–1522.

Bramlett, H.M. and W.D. Dietrich. 2007. Progressive damage after brain and spinal cord injury: Pathomechanisms and treatment strategies. *Prog. Brain Res.* 161:125–141. doi:10.1016/S0079–6123(06)61009–1.

Brodbelt, A.R. and M.A. Stoodley. 2003. Post-traumatic syringomyelia: A review. *J. Clin. Neurosci. Off. J. Neurosurg. Soc. Australas.* 10:401–408.

Bruce, J.C., M.A. Oatway, and L.C. Weaver. 2002. Chronic pain after clip-compression injury of the rat spinal cord. *Exp. Neurol.* 178:33–48. doi:10.1006/exnr.2002.8026.

Bryce, T.N., F. Biering-Sørensen, N.B. Finnerup, D.D. Cardenas, R. Defrin, T. Lundeberg et al. 2012a. International spinal cord injury pain classification: Part I. Background and description. March 6–7, 2009. *Spinal Cord.* 50:413–417. doi:10.1038/sc.2011.156.

Bryce, T.N., E. Ivan, and M. Dijkers. 2012b. Proposed International Spinal Cord Injury Pain (ISCIP) classification. *Top. Spinal Cord Inj. Rehabil.* 18:143–145. doi:10.1310/sci1802–143.

Bryce, T.N. and K.T. Ragnarsson. 2001. Epidemiology and classification of pain after spinal cord injury. *Top. Spinal Cord Inj. Rehabil.* 7:1–17. doi:10.1310/6A5C-354M-681H-M8T2.

Cardenas, D.D., J.A. Turner, C.A. Warms, and H.M. Marshall. 2002. Classification of chronic pain associated with spinal cord injuries. *Arch. Phys. Med. Rehabil.* 83:1708–1714. doi:10.1053/apmr.2002.35651.

Carlton, S.M., J. Du, H.Y. Tan, O. Nesic, G.L. Hargett, A.C. Bopp et al. 2009. Peripheral and central sensitization in remote spinal cord regions contribute to central neuropathic pain after spinal cord injury. *Pain.* 147:265–276. doi:10.1016/j.pain.2009.09.030.

Christensen, M.D., A.W. Everhart, J.T. Pickelman, and C.E. Hulsebosch. 1996. Mechanical and thermal allodynia in chronic central pain following spinal cord injury. *Pain.* 68:97–107.

Crown, E.D., Z. Ye, K.M. Johnson, G.-Y. Xu, D.J. McAdoo, and C.E. Hulsebosch. 2006. Increases in the activated forms of ERK 1/2, p38 MAPK, and CREB are correlated with the expression of at-level mechanical allodynia following spinal cord injury. *Exp. Neurol.* 199:397–407. doi:10.1016/j.expneurol.2006.01.003.

Defrin, R., H. Gruener, S. Schreiber, and C.G. Pick. 2010. Quantitative somatosensory testing of subjects with chronic post-traumatic headache: Implications on its mechanisms. *Eur. J. Pain.* 14:924–931. doi:10.1016/j.ejpain.2010.03.004.

Defrin, R., A. Ohry, N. Blumen, and G. Urca. 2001. Characterization of chronic pain and somatosensory function in spinal cord injury subjects. *Pain.* 89:253–263.

Densmore, V.S., A. Kalous, J.R. Keast, and P.B. Osborne. 2010. Above-level mechanical hyperalgesia in rats develops after incomplete spinal cord injury but not after cord transection, and is reversed by amitriptyline, morphine and gabapentin. *Pain.* 151:184–193. doi:10.1016/j.pain.2010.07.007.

Detloff, M.R., L.C. Fisher, R.J. Deibert, and D.M. Basso. 2012. Acute and chronic tactile sensory testing after spinal cord injury in rats. *J. Vis. Exp. JoVE.* e3247. doi:10.3791/3247.

Detloff, M.R., L.C. Fisher, V. McGaughy, E.E. Longbrake, P.G. Popovich, and D.M. Basso. 2008. Remote activation of microglia and pro-inflammatory cytokines predict the onset and severity of below-level neuropathic pain after spinal cord injury in rats. *Exp. Neurol.* 212:337–347. doi:10.1016/j.expneurol.2008.04.009.

Detloff, M.R., R.E. Wade Jr, and J.D. Houlé. 2013. Chronic at- and below-level pain after moderate unilateral cervical spinal cord contusion in rats. *J. Neurotrauma.* 30:884–890. doi:10.1089/neu.2012.2632.

Dijkers, M., T. Bryce, and J. Zanca. 2009. Prevalence of chronic pain after traumatic spinal cord injury: A systematic review. *J. Rehabil. Res. Dev.* 46:13–29.

Dobscha, S.K., M.E. Clark, B.J. Morasco, M. Freeman, R. Campbell, and M. Helfand. 2009. Systematic review of the literature on pain in patients with polytrauma including traumatic brain injury. *Pain Med. Malden Mass.* 10:1200–1217. doi:10.1111/j.1526–4637.2009.00721.x.

Donkin, J.J., A.J. Nimmo, I. Cernak, P.C. Blumbergs, and R. Vink. 2009. Substance P is associated with the development of brain edema and functional deficits after traumatic brain injury. *J.Cereb. Blood Flow Metab. Off. J. Int. Soc. Cereb. Blood Flow Metab.* 29:1388–1398. doi:10.1038/jcbfm.2009.63.

Elliott, M.B., M.L. Oshinsky, P.S. Amenta, O.O. Awe, and J.I. Jallo. 2012. Nociceptive neuropeptide increases and periorbital allodynia in a model of traumatic brain injury. *Headache.* 52:966–984. doi:10.1111/j.1526–4610.2012.02160.x.

Faux, S. and J. Sheedy. 2008. A prospective controlled study in the prevalence of posttraumatic headache following mild traumatic brain injury. *Pain Med. Malden Mass.* 9:1001–1011. doi:10.1111/j.1526–4637.2007.00404.x.

Finnerup, N.B. 2013. Pain in patients with spinal cord injury. *Pain.* 154(Suppl 1):S71–76. doi:10.1016/j.pain.2012.12.007.

Finnerup, N.B. 2003. Sensory function in spinal cord injury patients with and without central pain. *Brain.* 126:57–70. doi:10.1093/brain/awg007.

Finnerup, N.B., C. Norrbrink, K. Trok, F. Piehl, I.L. Johannesen, J.C. Sørensen et al. 2014. Phenotypes and predictors of pain following traumatic spinal cord injury: A prospective study. *J. Pain Off. J. Am. Pain Soc.* 15:40–48. doi:10.1016/j.jpain.2013.09.008.

Finnerup, N., L. Sorensen, F. Bieringsorensen, I. Johannesen, and T. Jensen. 2007. Segmental hypersensitivity and spinothalamic function in spinal cord injury pain. *Exp. Neurol.* 207:139–149. doi:10.1016/j.expneurol.2007.06.001.

Finnerup, N.B., C. Gyldensted, A. Fuglsang-Frederiksen, F.W. Bach, and T.S. Jensen. 2004. Sensory perception in complete spinal cord injury. *Acta Neurol. Scand.* 109:194–199.

Gaviria, M., H. Haton, F. Sandillon, and A. Privat. 2002. A mouse model of acute ischemic spinal cord injury. *J. Neurotrauma.* 19:205–221. doi:10.1089/08977150252806965.

Gellman, H., M.A. Keenan, L. Stone, S.E. Hardy, R.L. Waters, and C. Stewart. 1992. Reflex sympathetic dystrophy in brain-injured patients. *Pain.* 51:307–311.

Gorman, A.L., C.-G. Yu, G.R. Ruenes, L. Daniels, and R.P. Yezierski. 2001. Conditions affecting the onset, severity, and progression of a spontaneous pain-like behavior after excitotoxic spinal cord injury. *J. Pain.* 2:229–240. doi:10.1054/jpai.2001.22788.

Gruner, J.A. 1992. A monitored contusion model of spinal cord injury in the rat. *J. Neurotrauma.* 9:123–126; discussion 126–128.

Gwak, Y.S. and C.E. Hulsebosch. 2011. GABA and central neuropathic pain following spinal cord injury. *Neuropharmacology.* 60:799–808. doi:10.1016/j.neuropharm.2010.12.030.

Gwak, Y.S., H.K. Kim, H.Y. Kim, and J.W. Leem. 2010. Bilateral hyperexcitability of thalamic VPL neurons following unilateral spinal injury in rats. *J. Physiol. Sci. JPS.* 60:59–66. doi:10.1007/s12576–009–0066–2.

Gwak, Y.S., H.Y. Tan, T.S. Nam, K.S. Paik, C.E. Hulsebosch, and J.W. Leem. 2006. Activation of spinal GABA receptors attenuates chronic central neuropathic pain after spinal cord injury. *J. Neurotrauma.* 23:1111–1124. doi:10.1089/neu.2006.23.1111.

Gwak, Y.S., G.C. Unabia, and C.E. Hulsebosch. 2009. Activation of p-38α MAPK contributes to neuronal hyperexcitability in caudal regions remote from spinal cord injury. *Exp. Neurol.* 220:154–161. doi:10.1016/j.expneurol.2009.08.012.

Hains, B.C., J.P. Klein, C.Y. Saab, M.J. Craner, J.A. Black, and S.G. Waxman. 2003. Upregulation of sodium channel Nav1.3 and functional involvement in neuronal hyperexcitability associated with central neuropathic pain after spinal cord injury. *J. Neurosci.* 23:8881–8892.

Hains, B.C. and S.G. Waxman. 2007. Sodium channel expression and the molecular pathophysiology of pain after SCI. In: *Progress in Brain Research*. Elsevier: New York. pp. 195–203.

Hall, E.D. and J.M. Braughler. 1993. Free radicals in CNS injury. *Res. Publ. Assoc. Res. Nerv. Ment. Dis.* 71:81–105.

Hall, E.D., P.A. Yonkers, P.K. Andrus, J.W. Cox, and D.K. Anderson. 1992. Biochemistry and pharmacology of lipid antioxidants in acute brain and spinal cord injury. *J. Neurotrauma*. 9(Suppl 2):S425–442.

Hall, K.D. and J. Lifshitz. 2010. Diffuse traumatic brain injury initially attenuates and later expands activation of the rat somatosensory whisker circuit concomitant with neuroplastic responses. *Brain Res.* 1323:161–173. doi:10.1016/j.brainres.2010.01.067.

Hao, J.X., X.J. Xu, H. Aldskogius, A. Seiger, and Z. Wiesenfeld-Hallin. 1991. Allodynia-like effects in rat after ischaemic spinal cord injury photochemically induced by laser irradiation. *Pain*. 45:175–185.

Hari, A.R., S. Wydenkeller, P. Dokladal, and P. Halder. 2009. Enhanced recovery of human spinothalamic function is associated with central neuropathic pain after SCI. *Exp. Neurol.* 216:428–430. doi:10.1016/j.expneurol.2008.12.018.

Harrop, J.S., G.E. Hunt Jr, and A.R. Vaccaro. 2004. Conus medullaris and cauda equina syndrome as a result of traumatic injuries: Management principles. *Neurosurg. Focus*. 16:e4.

Henderson, L.A., S.M. Gustin, P.M. Macey, P.J. Wrigley, and P.J. Siddall. 2011. Functional reorganization of the brain in humans following spinal cord injury: Evidence for underlying changes in cortical anatomy. *J. Neurosci. Off. J. Soc. Neurosci.* 31:2630–2637. doi:10.1523/JNEUROSCI.2717-10.2011.

Hubscher, C.H. and R.D. Johnson. 2006. Chronic spinal cord injury induced changes in the responses of thalamic neurons. *Exp. Neurol.* 197:177–188. doi:10.1016/j.expneurol.2005.09.007.

Hulsebosch, C.E., B.C. Hains, E.D. Crown, and S.M. Carlton. 2009. Mechanisms of chronic central neuropathic pain after spinal cord injury. *Brain Res. Rev.* 60:202–213. doi:10.1016/j.brainresrev.2008.12.010.

Hulsebosch, C.E., G.Y. Xu, J.R. Perez-Polo, K.N. Westlund, C.P. Taylor, and D.J. McAdoo. 2000. Rodent model of chronic central pain after spinal cord contusion injury and effects of gabapentin. *J. Neurotrauma*. 17:1205–1217.

Jensen, O.K. and F.F. Nielsen. 1990. The influence of sex and pre-traumatic headache on the incidence and severity of headache after head injury. *Cephalalgia Int. J. Headache*. 10:285–293.

Kampfl, A., R.M. Posmantur, X. Zhao, E. Schmutzhard, G.L. Clifton, and R.L. Hayes. 1997. Mechanisms of calpain proteolysis following traumatic brain injury: Implications for pathology and therapy: Implications for pathology and therapy: A review and update. *J. Neurotrauma*. 14:121–134.

Kennedy, P., H. Frankel, B. Gardner, and I. Nuseibeh. 1997. Factors associated with acute and chronic pain following traumatic spinal cord injuries. *Spinal Cord*. 35:814–817.

Kim, J., Y.W. Yoon, S.K. Hong, and H.S. Na. 2003. Cold and mechanical allodynia in both hindpaws and tail following thoracic spinal cord hemisection in rats: Time courses and their correlates. *Neurosci. Lett.* 343:200–204.

Koyama, S., Y. Katayama, S. Maejima, T. Hirayama, M. Fujii, and T. Tsubokawa. 1993. Thalamic neuronal hyperactivity following transection of the spinothalamic tract in the cat: Involvement of N-methyl-D-aspartate receptor. *Brain Res.* 612:345–350.

Kwo, S., W. Young, and V. Decrescito. 1989. Spinal cord sodium, potassium, calcium, and water concentration changes in rats after graded contusion injury. *J. Neurotrauma*. 6:13–24.

Lew, H.L., P.-H. Lin, J.-L. Fuh, S.-J. Wang, D.J. Clark, and W.C. Walker. 2006. Characteristics and treatment of headache after traumatic brain injury: A focused review. *Am. J. Phys. Med. Rehabil. Assoc. Acad. Physiatr.* 85:619–627. doi:10.1097/01.phm.0000223235.09931.c0.

Lieba-Samal, D., P. Platzer, S. Seidel, P. Klaschterka, A. Knopf, and C. Wöber. 2011. Characteristics of acute posttraumatic headache following mild head injury. *Cephalalgia Int. J. Headache*. 31:1618–1626. doi:10.1177/0333102411428954.

Linder, S.L. 2007. Post-traumatic headache. *Curr. Pain Headache Rep.* 11:396–400.

Lindsey, A.E., R.L. LoVerso, C.A. Tovar, C.E. Hill, M.S. Beattie, and J.C. Bresnahan. 2000. An analysis of changes in sensory thresholds to mild tactile and cold stimuli after experimental spinal cord injury in the rat. *Neurorehabil. Neural Repair*. 14:287–300.

Liu, D.X., V. Valadez, L.S. Sorkin, and D.J. McAdoo. 1990. Norepinephrine and serotonin release upon impact injury to rat spinal cord. *J. Neurotrauma*. 7:219–227.

Ma, V.Y., L. Chan, and K.J. Carruthers. 2014. The incidence, prevalence, costs and impact on disability of common conditions requiring rehabilitation in the US: Stroke, spinal cord injury, traumatic brain injury, multiple sclerosis, osteoarthritis, rheumatoid arthritis, limb loss, and back pain. *Arch. Phys. Med. Rehabil.* 95:986–995. doi:10.1016/j.apmr.2013.10.032.

Macolino, C.M., B.V. Daiutolo, B.K. Alberston, and M.B. Elliott. 2014. Mechanical allodynia induced by traumatic brain injury is independent of restraint stress. *J. Neurosci. Methods*. 226:139–146. doi:10.1016/j.jneumeth.2014.01.008.

Marques, S.A., V.F. Garcez, E.A. Del Bel, and A.M.B. Martinez. 2009. A simple, inexpensive and easily reproducible model of spinal cord injury in mice: Morphological and functional assessment. *J. Neurosci. Methods*. 177:183–193. doi:10.1016/j.jneumeth.2008.10.015.

McMahon, S.B. 2013. *Wall and Melzack's Textbook of Pain*. Elsevier/Saunders, Philadelphia, PA.

Melzack, R., and J.D. Loeser. 1978. Phantom body pain in paraplegics: Evidence for a central "pattern generating mechanism" for pain. *Pain*. 4:195–210.

Michailidou, C., L. Marston, L.H. De Souza, and I. Sutherland. 2013. A systematic review of the prevalence of musculoskeletal pain, back and low back pain in people with spinal cord injury. *Disabil. Rehabil.* 36:705–715. doi:10.3109/09638288.2013.808708.

Mills, C.D., B.C. Hains, K.M. Johnson, and C.E. Hulsebosch. 2001. Strain and model differences in behavioral outcomes after spinal cord injury in rat. *J. Neurotrauma*. 18:743–756. doi:10.1089/089771501316919111.

Mogil, J.S., K.D. Davis, and S.W. Derbyshire. 2010. The necessity of animal models in pain research. *Pain*. 151:12–17. doi:10.1016/j.pain.2010.07.015.

Moore, C.I., C.E. Stern, C. Dunbar, S.K. Kostyk, A. Gehi, and S. Corkin. 2000. Referred phantom sensations and cortical reorganization after spinal cord injury in humans. *Proc. Natl. Acad. Sci. U. S. A.* 97:14703–14708. doi:10.1073/pnas.250348997.

Nakae, A., K. Nakai, K. Yano, K. Hosokawa, M. Shibata, and T. Mashimo. 2011. The animal model of spinal cord injury as an experimental pain model. *J. Biomed. Biotechnol.* 2011:939023. doi:10.1155/2011/939023.

Nampiaparampil, D.E. 2008. Prevalence of chronic pain after traumatic brain injury. *JAMA J. Am. Med. Assoc.* 300:711–719.

Nardone, R., Y. Höller, F. Brigo, M. Seidl, M. Christova, J. Bergmann et al. 2013. Functional brain reorganization after spinal cord injury: Systematic review of animal and human studies. *Brain Res.* 1504:58–73. doi:10.1016/j.brainres.2012.12.034.

Nepomuceno, C., P.R. Fine, J.S. Richards, H. Gowens, S.L. Stover, U. Rantanuabol et al. 1979. Pain in patients with spinal cord injury. *Arch. Phys. Med. Rehabil.* 60:605–609.

Nicholson Perry, K., M.K. Nicholas, and J. Middleton. 2009. Spinal cord injury-related pain in rehabilitation: A cross-sectional study of relationships with cognitions, mood and physical function. *Eur. J. Pain Lond. Engl.* 13:511–517. doi:10.1016/j.ejpain.2008.06.003.

O'Connor, W.T., A. Smyth, and M.D. Gilchrist. 2011. Animal models of traumatic brain injury: A critical evaluation. *Pharmacol. Ther.* 130:106–113. doi:10.1016/j.pharmthera.2011.01.001.

Ofek, H. and R. Defrin. 2007. The characteristics of chronic central pain after traumatic brain injury. *Pain.* 131:330–340. doi:10.1016/j.pain.2007.06.015.

Oshinsky, M.L. and J. Luo. 2006. Neurochemistry of trigeminal activation in an animal model of migraine. *Headache.* 46(Suppl 1):S39–44.

Ovelmen-Levitt, J., J. Gorecki, K.T. Nguyen, B. Iskandar, and B.S. Nashold Jr. 1995. Spontaneous and evoked dysesthesias observed in the rat after spinal cordotomies. *Stereotact. Funct. Neurosurg.* 65:157–160.

Panter, S.S., S.W. Yum, and A.I. Faden. 1990. Alteration in extracellular amino acids after traumatic spinal cord injury. *Ann. Neurol.* 27:96–99. doi:10.1002/ana.410270115.

Raghavendra Rao, V.L., V.K. Dhodda, G. Song, K.K. Bowen, and R.J. Dempsey. 2003. Traumatic brain injury-induced acute gene expression changes in rat cerebral cortex identified by GeneChip analysis. *J. Neurosci. Res.* 71:208–219. doi:10.1002/jnr.10486.

Raghupathi, R. 2004. Cell death mechanisms following traumatic brain injury. *Brain Pathol. Zurich Switz.* 14:215–222.

Rintala, D.H., P.G. Loubser, J. Castro, K.A. Hart, and M.J. Fuhrer. 1998. Chronic pain in a community-based sample of men with spinal cord injury: Prevalence, severity, and relationship with impairment, disability, handicap, and subjective well-being. *Arch. Phys. Med. Rehabil.* 79:604–614.

Saab, C.Y., J. Wang, C. Gu, K.N. Garner, and E.D. Al-Chaer. 2006. Microglia: A newly discovered role in visceral hypersensitivity? *Neuron Glia Biol.* 2:271–277. doi:10.1017/S1740925X07000439.

Saatman, K.E., J. Creed, and R. Raghupathi. 2010. Calpain as a therapeutic target in traumatic brain injury. *Neurother. J. Am. Soc. Exp. Neurother.* 7:31–42. doi:10.1016/j.nurt.2009.11.002.

Scheff, S.W., A.G. Rabchevsky, I. Fugaccia, J.A. Main, and J.E. Lumpp Jr. 2003. Experimental modeling of spinal cord injury: Characterization of a force-defined injury device. *J. Neurotrauma.* 20:179–193. doi:10.1089/08977150360547099.

Scheifer, C., U. Hoheisel, P. Trudrung, T. Unger, and S. Mense. 2002. Rats with chronic spinal cord transection as a possible model for the at-level pain of paraplegic patients. *Neurosci. Lett.* 323:117–120.

Sherman, K.B., M. Goldberg, and K.R. Bell. 2006. Traumatic brain injury and pain. *Phys. Med. Rehabil. Clin. N. Am.* 17:473–490, viii. doi:10.1016/j.pmr.2005.11.007.

Siddall, P., C.L. Xu, and M. Cousins. 1995. Allodynia following traumatic spinal cord injury in the rat: *NeuroReport.* 6:1241–1244. doi:10.1097/00001756-199506090-00003.

Siddall, P., R. Yezierski, and J. Loeser. 2002. Taxonomy and epidemiology of spinal cord injury pain. In: Burcheil KJ, Yezierski RP, editors. *Spinal Cord Injury Pain: Assessment, Mechanisms, Management.* IASP Press: Seattle. pp. 9–24.

Siddall, P.J., J.M. McClelland, S.B. Rutkowski, and M.J. Cousins. 2003. A longitudinal study of the prevalence and characteristics of pain in the first 5 years following spinal cord injury. *Pain.* 103:249–257.

Siddall, P.J., D.A. Taylor, and M.J. Cousins. 1997. Classification of pain following spinal cord injury. *Spinal Cord.* 35:69–75.

Siddall, P.J., C.L. Xu, N. Floyd, and K.A. Keay. 1999. C-fos expression in the spinal cord of rats exhibiting allodynia following contusive spinal cord injury. *Brain Res.* 851:281–286. doi:10.1016/S0006–8993(99)02173–3.

Stensman, R. 1994. Adjustment to traumatic spinal cord injury. A longitudinal study of self-reported quality of life. *Paraplegia.* 32:416–422. doi:10.1038/sc.1994.68.

Störmer, S., H.J. Gerner, W. Grüninger, K. Metzmacher, S. Föllinger, C. Wienke et al. 1997. Chronic pain/dysaesthesiae in spinal cord injury patients: Results of a multicentre study. *Spinal Cord.* 35:446–455.

Stroud, M.W., J.A. Turner, M.P. Jensen, and D.D. Cardenas. 2006. Partner responses to pain behaviors are associated with depression and activity interference among persons with chronic pain and spinal cord injury. *J. Pain Off. J. Am. Pain Soc.* 7:91–99. doi:10.1016/j.jpain.2005.08.006.

Tanabe, M., K. Ono, M. Honda, and H. Ono. 2009. Gabapentin and pregabalin ameliorate mechanical hypersensitivity after spinal cord injury in mice. *Eur. J. Pharmacol.* 609:65–68. doi:10.1016/j.ejphar.2009.03.020.

Tham, S.W., T.M. Palermo, J. Wang, K.M. Jaffe, N. Temkin, D. Durbin, and F.P. Rivara. 2013. Persistent pain in adolescents following traumatic brain injury. *J. Pain.* 14:1242–1249. doi:10.1016/j.jpain.2013.05.007.

Treede, R.-D., T.S. Jensen, J.N. Campbell, G. Cruccu, J.O. Dostrovsky, J.W. Griffin et al. 2008. Neuropathic pain: Redefinition and a grading system for clinical and research purposes. *Neurology.* 70:1630–1635. doi:10.1212/01.wnl.0000282763.29778.59.

Tsai, M.-C., C.-P. Wei, D.-Y. Lee, Y.-T. Tseng, M.-D. Tsai, Y.-L. Shih et al. 2008. Inflammatory mediators of cerebrospinal fluid from patients with spinal cord injury. *Surg. Neurol.* 70:S19–S24. doi:10.1016/j.surneu.2007.09.033.

Uomoto, J.M. and P.C. Esselman. 1993. Traumatic brain injury and chronic pain: Differential types and rates by head injury severity. *Arch. Phys. Med. Rehabil.* 74:61–64.

Vierck, C.J., Jr, J.D. Greenspan, and L.A. Ritz. 1990. Long-term changes in purposive and reflexive responses to nociceptive stimulation following anterolateral chordotomy. *J. Neurosci. Off. J. Soc. Neurosci.* 10:2077–2095.

Vierck, C.J. and A.R. Light. 2000. Allodynia and hyperalgesia within dermatomes caudal to a spinal cord injury in primates and rodents. *Prog. Brain Res.* 129:411–428.

Vierck, C.J., Jr and A.R. Light. 1999. Effects of combined hemotoxic and anterolateral spinal lesions on nociceptive sensitivity. *Pain.* 83:447–457.

Vierck Jr, C.J., P. Siddall, and R.P. Yezierski. 2000. Pain following spinal cord injury: Animal models and mechanistic studies. *Pain.* 89:1–5.

Walker, W.C., R.T. Seel, G. Curtiss, and D.L. Warden. 2005. Headache after moderate and severe traumatic brain injury: A longitudinal analysis. *Arch. Phys. Med. Rehabil.* 86:1793–1800. doi:10.1016/j.apmr.2004.12.042.

Walters, E.T. 2012. Nociceptors as chronic drivers of pain and hyperreflexia after spinal cord injury: An adaptive-maladaptive hyperfunctional state hypothesis. *Front. Physiol.* 3:309. doi:10.3389/fphys.2012.00309.

Wang, G. and S.M. Thompson. 2008. Maladaptive homeostatic plasticity in a rodent model of central pain syndrome: Thalamic hyperexcitability after spinothalamic tract lesions. *J. Neurosci.* 28:11959–11969. doi:10.1523/JNEUROSCI.3296–08.2008.

Wasner, G., B.B. Lee, S. Engel, and E. McLachlan. 2008. Residual spinothalamic tract pathways predict development of central pain after spinal cord injury. *Brain J. Neurol.* 131:2387–2400. doi:10.1093/brain/awn169.

Watson, B.D., R. Prado, W.D. Dietrich, M.D. Ginsberg, and B.A. Green. 1986. Photochemically induced spinal cord injury in the rat. *Brain Res.* 367:296–300.

Weng, H.R., J.I. Lee, F.A. Lenz, A. Schwartz, C. Vierck, L. Rowland, and P.M. Dougherty. 2000. Functional plasticity in primate somatosensory thalamus following chronic lesion of the ventral lateral spinal cord. *Neuroscience.* 101:393–401.

Westgren, N. and R. Levi. 1998. Quality of life and traumatic spinal cord injury. *Arch. Phys. Med. Rehabil.* 79:1433–1439.

Weyer Jamora, C., S.C. Schroeder, and R.M. Ruff. 2013. Pain and mild traumatic brain injury: The implications of pain severity on emotional and cognitive functioning. *Brain Inj.* 27:1134–1140. doi:10.3109/02699052.2013.804196.

Widerström-Noga, E., F. Biering-Sørensen, T. Bryce, D.D. Cardenas, N.B. Finnerup, M.P. Jensen et al. 2008. The international spinal cord injury pain basic data set. *Spinal Cord.* 46:818–823. doi:10.1038/sc.2008.64.

Widerström-Noga, E., F. Biering-Sørensen, T.N. Bryce, D.D. Cardenas, N.B. Finnerup, M.P. Jensen et al. 2014. The International Spinal Cord Injury Pain Basic Data Set (version 2.0). *Spinal Cord.* 52:282–286. doi:10.1038/sc.2014.4.

Widerström-Noga, E., P.M. Pattany, Y. Cruz-Almeida, E.R. Felix, S. Perez, D.D. Cardenas et al. 2013. Metabolite concentrations in the anterior cingulate cortex predict high neuropathic pain impact after spinal cord injury. *Pain.* 154:204–212. doi:10.1016/j.pain.2012.07.022.

Widerström-Noga, E.G., E. Felipe-Cuervo, J.G. Broton, R.C. Duncan, and R.P. Yezierski. 1999. Perceived difficulty in dealing with consequences of spinal cord injury. *Arch. Phys. Med. Rehabil.* 80:580–586.

Wieseler, J., A. Ellis, S.F. Maier, L.R. Watkins, and S. Falci. 2012. Unilateral T13 and L1 dorsal root avulsion: Methods for a novel model of central neuropathic pain. *Methods Mol. Biol. Clifton NJ.* 851:171–183. doi:10.1007/978-1-61779-561-9_12.

Wieseler, J., A.L. Ellis, A. McFadden, K. Brown, C. Starnes, S.F. Maier et al. 2010. Below level central pain induced by discrete dorsal spinal cord injury. *J. Neurotrauma.* 27:1697–1707. doi:10.1089/neu.2010.1311.

Wrigley, P.J., S.R. Press, S.M. Gustin, V.G. Macefield, S.C. Gandevia, M.J. Cousins et al. 2009. Neuropathic pain and primary somatosensory cortex reorganization following spinal cord injury. *Pain.* 141:52–59. doi:10.1016/j.pain.2008.10.007.

Wydenkeller, S., S. Maurizio, V. Dietz, and P. Halder. 2009. Neuropathic pain in spinal cord injury: Significance of clinical and electrophysiological measures. *Eur. J. Neurosci.* 30:91–99. doi:10.1111/j.1460-9568.2009.06801.x.

Xiong, Y., A. Mahmood, and M. Chopp. 2013. Animal models of traumatic brain injury. *Nat. Rev. Neurosci.* 14:128–142. doi:10.1038/nrn3407.

Yamaguchi, M. 1992. Incidence of headache and severity of head injury. *Headache.* 32:427–431.

Yezierski, R.., S. Liu, G.. Ruenes, K.. Kajander, and K. Brewer. 1998. Excitotoxic spinal cord injury: Behavioral and morphological characteristics of a central pain model. *Pain.* 75:141–155. doi:10.1016/S0304-3959(97)00216-9.

Yezierski, R.P. 2000. Pain following spinal cord injury: Pathophysiology and central mechanisms. *Prog. Brain Res.* 129:429–449. doi:10.1016/S0079-6123(00)29033-X.

Yezierski, R.P. 2005. Spinal cord injury: A model of central neuropathic pain. *Neurosignals.* 14:182–193. doi:10.1159/000087657.

Yezierski, R.P. 2009. Spinal cord injury pain: Spinal and supraspinal mechanisms. *J. Rehabil. Res. Dev.* 46:95–107.

Yezierski, R.P. and S.-H. Park. 1993. The mechanosensitivity of spinal sensory neurons following intraspinal injections of quisqualic acid in the rat. *Neurosci. Lett.* 157:115–119. doi:10.1016/0304–3940(93)90656–6.

Yezierski, R.P., C.-G. Yu, P.W. Mantyh, C.J. Vierck, and D.A. Lappi. 2004. Spinal neurons involved in the generation of at-level pain following spinal injury in the rat. *Neurosci. Lett.* 361:232–236. doi:10.1016/j.neulet.2003.12.035.

Yoon, Y.W., H. Dong, J.J.A. Arends, and M.F. Jacquin. 2004. Mechanical and cold allodynia in a rat spinal cord contusion model. *Somatosens. Mot. Res.* 21:25–31. doi:10.1080/0899022042000201272.

Yu, C.-G. and R.P. Yezierski. 2005. Activation of the ERK1/2 signaling cascade by excitotoxic spinal cord injury. *Mol. Brain Res.* 138:244–255. doi:10.1016/j.molbrainres.2005.04.013.

Zaloshnja, E., T. Miller, J.A. Langlois, and A.W. Selassie. 2008. Prevalence of long-term disability from traumatic brain injury in the civilian population of the United States, 2005. *J. Head Trauma Rehabil.* 23:394–400. doi:10.1097/01.HTR.0000341435.52004.ac.

Zasler, N.D. 2011. Pharmacotherapy and posttraumatic cephalalgia. *J. Head Trauma Rehabil.* 26:397–399. doi:10.1097/HTR.0b013e31822721f8.

Zhao, P., S.G. Waxman, and B.C. Hains. 2007. Modulation of thalamic nociceptive processing after spinal cord injury through remote activation of thalamic microglia by cysteine chemokine ligand 21. *J. Neurosci.* 27:8893–8902. doi:10.1523/JNEUROSCI.2209–07.2007.

12 Nonconvulsive Seizures as Secondary Insults in Experimental Traumatic Brain Injury

Laura Stone McGuire, Amade Bregy, Justin Sick,
W. Dalton Dietrich, Helen M. Bramlett, and Thomas Sick

CONTENTS

12.1 INTRODUCTION

Post-traumatic epilepsy is an important consequence of neuronal damage following traumatic brain injury. Recent literature indicates that the majority of seizure events after such traumatic events are non-convulsive seizures. The absence of motor signs makes diagnosis difficult, many cases are under- or mis-diagnosed, and therefore, this disease remains poorly understood. There is very little literature currently available on research in the area of post-traumatic non-convulsive seizures. In this chapter, we will review the state of clinical and experimental research on non-convulsive seizures in post-traumatic epilepsy.

12.1.1 TBI BACKGROUND

Traumatic brain injury (TBI) remains an important cause of morbidity and mortality in the United States, representing a major public health concern and leading to cases of permanent disability. More than 1.7 million TBIs occur every year, contributing to almost one-third (30.5%) of injury-related deaths in the United States. Of these injuries, 1.365 million were managed in the emergency department setting each year, with the number of emergency department visits and hospitalizations increasing annually between the years 2002 and 2006, and the Centers for Disease Control and Prevention further reported that an estimated 52,000 TBIs result in death annually. The vast majority of TBIs, nearly 75%, were classified as mild. Although falls are the most common cause of TBI, motor vehicle accidents are the most lethal. TBIs affect men more commonly than women, and the most frequently affected age groups are children ages 0 to 4, adolescents 15 to 19, and adults 65 years of age and older (CDC, 2003, 2010; Faul et al., 2010).

TBI also poses a significant concern to military personnel. Between 2000 and 2012, the U.S. military reported 30,406 TBIs, including all levels of severity, with mild TBI representing 85.5% or 26,011 cases. Of the remaining reported injuries, 5% were classified as moderate, 0.7% as severe, 0.6% as penetrating, and the remainder as unclassifiable (DVBIC, 2013).

Mild and moderate TBI, although much more common than their severe counterpart, often do not present with obvious clinical sequelae. Mild TBI includes a brief change in mental status lasting less than 24 hours or loss of consciousness for up to 30 minutes with normal results on magnetic resonance imaging (MRI) or computed tomography imaging, whereas severe injury involves an extended period of unconsciousness or amnesia for more than 24 hours after the injury and with either normal or abnormal imaging results.

12.1.2 Epidemiology of Posttraumatic Epilepsy

Clinically, after TBI, the lasting consequences include cognitive deficits, depression, posttraumatic stress disorder, and posttraumatic epilepsy (Hall et al., 1991; Levin et al., 1979). Among these implications, the risk of seizures following injury and the development of posttraumatic epilepsy are well-documented consequences of TBI, supported by both experimental and clinical evidence. The seizure events are classified by their timing following the TBI: impact seizures occur in less than 24 hours after injury, early seizures in less than 1 week, and late seizures in more than 8 days. Posttraumatic epilepsy is characterized by recurrent posttraumatic seizures that can appear up to several years after the inciting trauma and includes spontaneous convulsive and nonconvulsive seizures (NCSs) (Agrawal et al., 2006; Annegers et al., 1980; Vespa et al., 1999). Around 80% of TBI patients suffer from seizures within the first year after the initial trauma, and the rate increases to 90% by the second year (Da Silva et al., 1990). Posttraumatic epilepsy represents 20% of all epilepsy in the general population and up to 50% in military personnel (Annegers and Coan, 2000; Jennett, 1973; Salazar et al., 1985; Semah et al., 1998). In posttraumatic epilepsy, nonconvulsive, also referred to as subclinical, seizures occur more commonly than convulsive seizures. In one study, approximately one-fifth of patients with continuous electroencephalogram (EEG) monitoring were diagnosed with posttraumatic epilepsy after moderate-to-severe TBI; subclinical seizures were more common and occurred in 52% of patients in comparison to overt seizures with an incidence of 22% (Olson, 2004; Vespa et al., 1999).

NCSs encompass nonmotor simple partial, nonmotor complex partial, and absence seizure subtypes. Absence, or petit mal, seizures are characterized by altered awareness, and clinical symptoms typically include confusion and decreased responsiveness, which may be accompanied by personality change, subtle motor activity, myoclonus, persistent eye blinking, staring, aphasia, or other symptoms. Nonmotor simple partial seizures are continuous or repeated focal sensory symptoms or cognitive symptoms without impaired consciousness, whereas complex partial seizures are continuous or repeated episodes of focal sensory or cognitive symptoms with impaired consciousness. Nonconvulsive status epilepticus is defined as NCSs without complete recovery between the seizure events (Drislane, 2000; Kaplan, 1996; Maganti et al., 2007).

The focus of this chapter is to highlight the state of research on posttraumatic epilepsy, specifically NCSs after TBI, and demonstrate future areas of study. The clinical evidence of NCSs in posttraumatic epilepsy will be outlined first, followed by animal models of experimental research in posttraumatic epilepsy. Current EEG recording methods and typical EEG findings in NCSs will be reviewed. Throughout the chapter, evidence available on NCSs will be contrasted to convulsive seizures in posttraumatic epilepsy, and gaps in knowledge and opportunities for future research will be discussed.

12.2 CLINICAL EVIDENCE

12.2.1 Epidemiology of NCSs

Epidemiologically, NCSs present an important clinical entity. Reports of NCSs in neurocritical care patients document incidences of NCSs ranging from 11% to 55% (Classen et al., 2004; Grand'Maison et al., 1991; Jordan, 1993, 1995; Litt et al., 1994; Privitera and Strawsburg, 1994; Vespa et al., 1999, 2007). For instance, the study from Claasen et al. reviewed 570 critically ill patients who underwent continuous EEG monitoring: 19% experienced seizures, of which 92% were nonconvulsive and 88% occurred within the first 24 hours of monitoring (Claasen et al., 2004). NCSs, including posttraumatic, may also present an important cause of unconsciousness in the hospital setting. The study conducted by Alroughani et al. showed that 9.3% of 451 patients with impaired consciousness were found to have NCSs, of which a common cause was intracerebral hemorrhage, including trauma (Alroughani et al., 2009).

The clinical challenge remains the diagnosis of NCSs. Convulsive seizures are easier to diagnose, may be detected through visual recognition of motor events, and can be aggressively treated. In contrast, NCSs are harder to diagnose, require continuous EEG monitoring, and often remain undiagnosed and untreated (Vespa et al., 1999). In the absence of EEG monitoring, common clinical features are sparse and subtle but include remote seizure risk factors and abnormal ocular movements, according to the study by Husain et al. (Husain et al., 2003).

12.2.2 Pathology of NCSs

Clinical evidence demonstrates that NCSs may be a predictor of worse outcome in multiple patient populations, including increased risk of death (Jirsch and Hirsch, 2007). As a complicating second insult after acute brain injury of various etiologies, including intracerebral hemorrhage, cerebral infarction, and TBI, NCSs significantly increase the risk of permanent brain damage and death (Claassen et al., 2007; Hirsch, 2008; Vespa et al., 2003; Vespa, 2005). Furthermore, this risk intensifies with longer duration and a greater delay in the diagnosis of NCSs, and subsequent successful treatment becomes more difficult (Drislane et al., 2008; Young et al., 1996). Following nonconvulsive complex partial seizures, several case reports and series also identify poor outcomes, particularly persistent cognitive and memory deficits (Engel et al., 1978; Krumholz et al., 1995; Treiman, 1995). NCSs may also lead to structural abnormalities: Wasterlain et al. found neuronal loss in the hippocampus after nonconvulsive status epilepticus in three patients without preexisting seizures (Wasterlain et al., 1993). Part of the resulting brain damage is due to an imbalance in neurotransmitters; for example, neuron-specific enolase, an indicator for hypoxic neuronal damage, is elevated following NCSs (DiGeorgio et al., 1996, 1999; Rabinowicz et al., 1994, 1995). Further, NCSs are associated with higher levels of glutamate and glycerol in the brain, indicating excitotoxicity and cell membrane disruption, respectively (Vespa, 2005; Vespa et al., 1998, 2002).

12.2.3 NCSs after TBI

Clinical studies specifically regarding the development of posttraumatic NCSs are less robust, but evidence suggests NCSs pose a significant concern for the patient population suffering from TBI. Specific risk factors for development of NCSs after TBI are depressed skull fracture, penetrating injury, and large cortical contusion/hematomas (Mirski, 2008). In a study of 94 patients with moderate to severe TBI, Vespa et al. found a 22% overall incidence of seizures following injury with a majority (57%) being nonconvulsive, in spite of antiepileptic drug prophylaxis. Another study conducted by Ronne-Engstrom et al. showed that one-third of 70 patients with TBI undergoing continuous EEG monitoring had epileptic seizures, including both convulsive and nonconvulsive (Ronne-Engstrom and Winkler, 2006). In a study of patients with moderate to severe TBI, NCSs exacerbated injury, leading to increased lactate/pyruvate ratio, episodic and sustained increased intracranial pressure, hypoxia, and death (Vespa, 2005; Vespa et al., 2007a, 2007b). The development of posttraumatic epilepsy is associated with a lower Glasgow Outcome Score, behavioral problems, and hippocampal atrophy (Hitiris et al., 2007; Swartz et al., 2006; Thapa et al., 2010; Vespa et al., 2010; Wang et al., 2008).

12.3 EXPERIMENTAL EVIDENCE

Multiple experimental TBI models document the finding of posttraumatic epilepsy, including penetrating ballistic-like brain injury, lateral and parasagittal fluid percussion injury (FPI), controlled cortical impact, weight drop, and undercut models. Of all animal models described, FPI has been most extensively studied for posttraumatic epilepsy. These various animal models show differences in development of spontaneous seizures. Immediate postinjury seizures are documented in weight drop and lateral FPI models. Penetrating ballistic-like brain injury models usually show early seizure events, whereas late seizures are consistent findings in lateral FPI models (Pitkänen et al., 2009). A lowered seizure threshold could be identified in weight drop and controlled cortical impact models, and hyperexcitability was demonstrated in the dentate gyrus in both weight drop and FPI models and CA1 in FPI (Pitkänen et al., 2006).

12.3.1 Animal Models

NCSs in animal TBI models have not been extensively studied. At present, only three studies to date focus on NCSs after TBI (Lu et al., 2011, 2013; Mountney et al., 2013), and the most used TBI model for the study of NCSs is the model of penetrating ballistic-like injury. Research from other posttraumatic epilepsy models generally does not investigate or report the incidence of NCSs. Table 12.1 summarizes clinical symptoms, electrophysiological, metabolic, and histopathological alterations found in NCS. Table 12.2 summarizes published animal studies on posttraumatic seizures and epilepsy, with studies focused on NCS highlighted.

12.3.2 NCS in Penetrating Ballistic-Like Brain Injury

Penetrating ballistic-like brain injury (PBBI) involves unilateral frontal injury that imitates a bullet or shrapnel wound. The procedure involves the stereotactic insertion of a probe into the right frontal cortex, which allows for rapid inflation and deflation of an elastic water balloon and creates a cavity in the brain. The size of the water-filled balloon under control of the hydraulic pressure determines the severity of the injury (Williams et al., 2005).

Lu et al. (2011) investigated three levels of injury severity: 5%, 10%, and 12.5% total rat brain volume, representing mild, moderate, and severe PBBI, respectively. Their study used continuous EEG for a period of 72 hours immediately after TBI and for 2 hours on days 7 and 14 and found that increasing levels of injury severity were associated with higher incidence of spontaneous nonconvulsive seizures and periodic epileptiform seizures. For mild injury, NCSs occurred in 13% of animals and periodic epileptiform discharges in 0%; for moderate injury, NCSs occurred in 39% of animals and periodic epileptiform discharges in 30%; and for severe injury, NCSs occurred in 59% of animals and periodic epileptiform discharges 65%. The study also found that NCSs were not significantly associated with the incidence, frequency, and duration of periodic epileptiform discharges. Further, the authors used EEG power spectral analysis to quantitatively measure the EEG power shift as a function of injury severity and showed that more enhanced, persistent EEG slowing was correlated with PBBI severity.

In the study by Lu et al. (2013), two injury models, middle cerebral artery occlusion (MCAO) injury for cerebral infarction and PBBI at moderate severity in Sprague Dawley rats, were compared for incidence, frequency, and latency of onset of NCSs, as detected by continuous EEG monitoring. Continuous EEG monitoring continued for 72 hours after injury for the PBBI animal group and 24 hours postinjury for MCAO. NCSs occurred spontaneously in both the MCAO and PBBI models. However, in the MCAO model, NCSs occurred more acutely (latency of 0.6 hours vs. 24 hours) and frequently (25 seizures per animal vs. 10) and also at a higher incidence (82% vs. 70%) than in the PBBI model. The study also analyzed other epileptic events among the two injury groups, including periodic epileptiform discharges and intermittent rhythmic delta activities,

TABLE 12.1
Summary of Clinical Symptoms, Electrophysiological, Metabolic, and Histopathological Alterations in Non-Convulsive Seizures

Clinical Symptoms (Kaplan, 1996; Jirsch and Hirsch, 2007)		Electrophysiological Alterations	Neurochemistry/ Metabolic Alterations	Histopathological Alterations	Differential Diagnosis (Maganti et al., 2008; Meierkord and Holtkamp, 2007)
Negative symptoms	***Positive symptoms***	***Experimental posttraumatic nonconvulsive***	***Clinical nonconvulsive***	***Experimental posttraumatic nonconvulsive***	
Anorexia Aphasia/mutism Amnesia Catatonia Coma Confusion Lethargy Staring	Agitation/ aggression Automatisms Blinking Crying Delirium Delusions Echolalia Facial twitching Laughter Nausea/vomiting Nystagmus/eye deviation Perseveration Psychosis Tremulousness	-Increasing severity of injury associated with increased incidence of NCS in PBBI model (Lu et al., 2011) ***Experimental posttraumatic convulsive*** -Hyperexcitability in dentate gyrus in FPI and WD models -Hyperexcitability in CA1 in FPI (Pitkänen et al., 2006) -Lowered seizure threshold in CCI, WD, and FPI models (Bao et al., 2011; Bolkvadze, 2012; Golarai et al., 2001; Kharatishivili, 2010; Mukherjee et al., 2013; Statler et al., 2008; Zanier et al., 2003)	-Increased lactate/ pyruvate ratio (Vespa, 2005; Vespa et al., 2007b) -Episodic and sustained increased intracranial pressure (Vespa, 2005; Vespa et al., 2007b) -Hypoxia (Vespa, 2005; Vespa et al., 2007b) -Elevated neuron-specific enolase (DiGeorgio et al., 1996, 1999; Rabinowicz et al., 1994, 1995) -Increased glutamate and glycerol levels, indicating excitotoxicity and cell membrane damage (Vespa, 2005; Vespa et al., 1998 , 2002) ***Experimental posttraumatic nonconvulsive*** No evidence specifically for NCS ***Experimental post-traumatic convulsive*** -Long-term alterations in GABA(A) and Glutamate receptor subunits in CCI model (Kharlamov et al., 2011) -Increase in GABA(A)-receptor mediated tonic inhibition in dentate granule cells in CCI model (Mtchedlishvili et al., 2010)	No evidence specifically for NCS ***Experimental posttraumatic convulsive*** -Mossy fiber sprouting in ipsilateral hippocampus in FPI model (Kharatishvili et al., 2006) -Loss of hilar neurons in FPI model (Kharatishvili et al., 2006) -Ipsilateral hippocampal surface abnormalities at 1 week post-TBI, FPI (Shultz et al., 2013) -Increased contusional volumes and decreased ipsilateral parietal cortical NeuN reactivity and hippocampal CA3 neurons, FPI (Bao et al., 2011) -Mossy fiber sprouting associated with hyperexcitability, CCI model (Hunt et al., 2009) -More severe and expansive cortical damage, associated with hyperexcitability, CCI model (Kharatishvili et al., 2010) -Early ipsilateral cell loss in hippocampus dentate gyrus and CA3, progressive bilateral sprouting of mossy fibers, persistent bilateral vitro hyperexcitability in the dentate gyrus, WD model (Golarai et al., 2001)	Metabolic encephalopathy Migraine aura Posttraumatic amnesia Prolonged postictal confusion Psychiatric disorders Substance de- or intoxication Transient global amnesia Transient ischemic attack

and found that periodic epileptiform discharges were episodically associated with NCSs, whereas intermittent rhythmic delta activities were independent.

Mountney et al. (2013) investigated two antiepileptic drugs, ethosuximide and phenytoin, in the attenuation and mitigation of NCSs in the moderate PBBI model. Ethosuximide and phenytoin were each tested at four different dosages. Both drugs attenuated the incidence of spontaneously occurring NCSs dose-dependently: at the highest doses, animals experienced a delayed onset of seizures, an incidence of 13%–40%, and a frequency of 1.8–6.2 seizures per rat. In contrast, among the vehicle-tested groups, 69%–73% experienced

TABLE 12.2
Studies Using Traumatic Brain Injury Animal Models That Report Development of Seizures and Posttraumatic Epilepsy

Reference	Model	Animal	Recording Electrodes	Videotape	Seizure Behavioral Assessment	Length of Time Recording/ Observation	Post-TBI Period	Spontaneity	Findings
Lu et al., 2013*	PBBI (10% R frontal) and MAO	SD rat	Continuous EEG, 4 epidural EEG electrodes, and 1 reference electrode	PBBI: 72 hours MAO: first few hours	None	PBBI: 0–72 hours MAO: 0–24 hours	PBBI: 0–72 hours MAO: 0–24 hours	Spontaneous	MAO: 82% incidence NCS, latency = 0.6 hours, frequency = 25/rat PBBI: 70% incidence NCS, latency = 24 hours, frequency = 10/rat
Lu et al., 2011*	PBBI (5, 10, 12.5% R frontal)	SD rat	Continuous EEG, 8 EEG electrodes, 5 ipsilateral, 3 contra	No video recording, only behavioral monitor	None	0–72 hours Day 7: 2 hours Day 14: 2 hours	0–14 days	Spontaneous	Incidence of NCS and PED positively correlated with severity (NCS: 13%, 39%, 59%; PED: 0%, 30%, 65%)
Mountney et al., 2013*	PBBI (10% R frontal) + Phenytoin, Ethosuximide	SD rat	Video EEG, 4 bilateral EEG electrodes, 1 reference electrode	Continuous with EEG monitoring	None	0–72 hours	0–72 hours	Spontaneous	ETX 12.5–187.5 mg/kg and PHT 5–30 mg/kg; VEH 69–73% NCS, 9–10x/rat, 30 hours onset; both attenuated NCS dose-dependently (13–40%, 1.8–6.2x rat, delayed onset); no effect on lesion volume
Williams et al., 2005	PBBI	SD rat	Continuous EEG, 2 epidural screw electrode bilateral, 2 Ag/AgCl electrodes ipsi	No video recording, only behavioral monitor	None	0–72 hours	0–72 hours	Spontaneous	Moderate injury associated with periodic lateralized epileptiform discharges, polymorphic delta activity, recurrent slow waves; delayed persistent cortical spreading discharges
Kharatishvili et al., 2006	LFPI (severe)	SD rat	Video EEG, 1 ipsilateral hippocampal electrode, 1 ipsilateral cortical and 1 contra cortical, 2 ground electrodes cerebellum	Continuous with EEG monitoring	Racine scale 0–5	Begin 7–9 weeks 1 week, 24 hours/day 7-week interval	7–9 weeks to 12 months	Spontaneous	Ex1 43% PTE, Ex2 50% PTE (7 weeks–1 year), 0.3/day, 113 seconds, behavioral seizure severity increased with time
Shulz et al., 2013	LFPI	W rat	Video EEG, 6 extradural electrodes (4 recording, 1 ground)	Continuous with EEG monitoring	Racine scale 0–5, learning, memory, anxiety, depression	At 6 months: 2-wk, 24 hr/d	6 months	Spontaneous	52% PTE or PED, ipsilateral hippocampal structural change at 1 wk predicted PTE

(*Continued*)

TABLE 12.2 (*Continued*)
Studies Using Traumatic Brain Injury Animal Models That Report Development of Seizures and Posttraumatic Epilepsy

Reference	Model	Animal	Recording Electrodes	Videotape	Seizure Behavioral Assessment	Length of Time Recording/ Observation	Post-TBI Period	Spontaneity	Findings
Bolkvadze et al., 2012	LFPI and CCI	C57BL/6S mice	Video EEG, 1 ipsilateral recording electrode, 1 contra recording, 1 ground, 1 reference	Continuous with EEG monitoring	Modified Racine 0–5	3 2-week, 24 hours/day 2 at 6 months (pre/post PTZ), 1 at 9 months	6–9 months	PTZ test, spontaneous	Late spontaneous seizure 9% CCI and 3% LFPI, spontaneous epileptiform spiking 82% CCI and 71% LFPI, spontaneous epileptiform discharge 58% LFPI, severity damage not correlated to seizure
D'Ambrosio et al., 2004	LFPI (severe)	SD rat	Chronic ECoG, 3 epidural montages (1 with 5, 1 with 7, 1 with 8 electrodes)	Synchronous video monitoring	Modified Racine 0–5	Time points at 2–16 weeks, 8 hours	14 days to 16 weeks	Spontaneous, (electrical stimulation for in vitro)	Persistent hyperexcitability at neocortex (in vitro), 92% PTE at 8 weeks
Mukherjee et al., 2013	LFPI (moderate)	C57Bl6 mice	No EEG or ECoG	Videotape only	Modified Racine 1–5	20 minutes	30 days	PTZ test	Increased susceptibility to PTZ test (89% LFPI seizures)
Kharatishvili et al., 2010	LFPI (severe, moderate)	SD rat	Video EEG, no electrode specifications provided	Continuous with EEG monitoring	Unspecified	24 hours/day Ex1 7–14 days every 7–9 weeks, Ex2 7–14 days every 4 weeks	12 months	Spontaneous ex1-2, PTZ ex3-4	Ex1 severe 8/14 PTE, Ex2 severe 7/26 PTE, Ex3-4 moderate 0/14 each PTE but decreased seizure threshold; damage to posterior temporal cortex associated with PTE, severity of cortical injury associated with PTE
Bao et al., 2011	PSFPI (moderate)	SD rat	No EEG or ECoG	None, designated observation only	Racine 1–5	1 hour at 2 week	2 weeks	PTZ test	90% PTZ rats seizure, no VEH/sham rats seizure; contusion volume increased in PTZ rats; decreased ipsi parietal cortical NeuN+ and hippocampal CA3 neurons
Zanier et al., 2003	LFPI (mild-mod)	SD rat	No EEG or ECoG	None, designated observation only	Racine 1–5	Time points at 15 minutes and 4 hours post-KA injection	7 days	Kainic acid 1 hour post-TBI	CA3, CA4, and hilar neurons decreased in LFP+KA > LFP+saline > sham; reduced seizure threshold with KA

Hunt et al., 2009	CCI (mild, severe)	CD-1 Harlan mice	No EEG or ECoG	None, designated observation only	Modified Racine 1–5	11 random 1–2 hours 42–71 days (total 18 hours)	42–71 days	Spontaneous, (electrical stimulation and picrotoxin for in vitro)	Mild 20% PTE, severe 36% PTE
Kharlamov et al., 2010	CCI	SD rat	No EEG or ECoG	Video 06:00–18:00 on recording days	Modified Racine 1–5	12 hours/day, 7 days for every 30 day period	5–11 months	Spontaneous	Long-term changes in GABA-A and glutamate receptor subunits and an HSP
Mtchedlishvili et al., 2010	CCI	SD rat	No EEG or ECoG	Continuous	Racine 1–5	Continuous 7 days post-TBI	90 days	Spontaneous (application ion channel blockers in vitro)	Increase GABA(A)-R mediated tonic inhibition in dentate granule cells
Statler et al., 2008	CCI	SD rat	No EEG or ECoG	None, observation with electrical stimulation only	None	Postnatal days 34–40 and 60–63	Postnatal days 80–85	Electrical stimulation in vivo	TBI no effect on tonic hindlimb extension seizure threshold; lower minimal clonic seizure threshold at maturity; lower partial psychomotor seizure threshold; ipsilateral loss hippocampal neurons
Blaha et al., 2010	WD (mild, moderate, severe)	W rat	No EEG or ECoG	None, immediate observation only	None	Immediate	0–18 hours	Spontaneous	WD 45–100 cm immediate seizure and death, 40 cm 67% seizure, 30 cm 1/3 sz
Golarai et al., 2001	WD	Rat	No EEG or ECoG	None, designated observation only	Racine 1–5	1 hour at 15 weeks	1–5 days, 2 weeks, 8 weeks, 15 weeks	PTZ test, (electrical stimulation in vitro)	Cell loss in DG and CA3, enhanced susceptibility to PTZ at 15 wks, in vitro hyperexcitability in the dentate gyrus following WD TBI at 3 and 15 weeks
Nilsson et al., 1994	WD	Rat	Subcutaneous electrodes bilateral parietal and midline	None, designated observation only	None	2 hours	2 hours	Spontaneous	14/17 generalized seizure for average 59 seconds' duration at average 67 seconds post-TBI; transient increase Asp, taurine, Glu, Gly

Note: *Indicates those studies that focus specifically on the development of nonconvulsive seizures following injury. Knockout studies, solely in vitro studies, and therapeutic studies are not included above.
CCI, controlled cortical impact; ETX, ethosuximide; Ex, experiment; KA, kainic acid; LFPI, lateral fluid percussion injury; MAO, middle cerebral artery occlusion; NCS, nonconvulsive seizures; PED, periodic epileptiform discharges; PHT, phenytoin; PTE, posttraumatic epilepsy; SD, Sprague Dawley; VEH, vehicle; W, Worcester; WD, weight drop.

NCSs at a frequency of 9 to 10 seizures per rat and an onset of 30 hours on average. However, the antiepileptic drugs had no significant effect on lesion volume, and lesion size was not associated with NCS frequency or duration.

Our own group assessed seizures at 1 year after parasagittal FPI in Sprague Dawley rats. Preliminary data presented at the National Neurotrauma Society meeting suggest the presence of subconvulsive seizures in animals. Figure 12.1 represents these findings, showing NCS in animals 1 year post-TBI before pentylenetetrazol testing (Bregy et al., 2013).

12.3.3 NCSs in Other Brain Injury Models

Research involving animal models for other brain injury types showed NCSs, particularly in various ischemic injury models. The incidence of NCSs in the middle cerebral artery occlusion injury model ranges from 78% to 91%, the majority of which occur within the first hour of recording (Hartings et al., 2003; Lu et al., 2009, 2013; Williams et al., 2004). These studies largely investigated the utility of various antiepileptic drugs in experimental ischemic brain injury (Lu et al., 2009; Williams et al., 2004, 2008), which may provide useful data in future research for the use of antiepileptics for NCSs in posttraumatic epilepsy.

12.3.4 Convulsive Seizures

In contrast to NCSs, the physiological and histopathological features of behavioral seizures have been studied more widely. Decreased seizure threshold of the post-TBI has been demonstrated in the FPI model (Bao et al., 2011; Bolkvadze, 2012; Golarai et al., 2001; Kharatishivili, 2010; Mukherjee et al., 2013; Statler et al., 2008; Zanier et al., 2003) and controlled cortical impact model (Bolkvadze, 2012; Statler et al., 2008). Potential MRI biomarkers for increased seizure susceptibility have been identified in the FPI model. Immonen et al. showed that changes in the $T_{1\rho}$ of the perilesional cortex at days 9 to 23 post-TBI at one year after injury and also the average diffusion constant (D_{av}) in the thalamus at 2 months post-TBI predicted seizure susceptibility (Immonen et al., 2013). Shultz et al. showed that ipsilateral hippocampal surface abnormalities by MRI at week 1 predicted development of epilepsy (Shultz et al., 2013). Additionally, structural changes have been identified after TBI in association with the development of posttraumatic epilepsy. Following the FPI model, mossy fiber sprouting in ipsilateral hippocampus and loss of hilar neurons occurred (Kharatishvili et al., 2006), and pentylenetetrazol (PTZ)-susceptible animals had increased contusional volumes and decreased ipsilateral parietal cortical NeuN reactivity and hippocampal CA3 neurons (Bao et al., 2011). In controlled cortical impact, mossy fiber sprouting associated with hyperexcitability (Hunt et al., 2009), and damage to the cortex was more severe and expansive in animals with posttraumatic epilepsy with the extent of damage positively correlated with hyperexcitability (Kharatishvili et al., 2010). In the weight drop model, early ipsilateral cell loss in hippocampus dentate gyrus and CA3, progressive bilateral sprouting of mossy fibers, and persistent bilateral in vitro hyperexcitability in the dentate gyrus has been documented (Golarai et al., 2001). Persistent hyperexcitability at and around the site of injury was shown in FPI (D'Ambrosio et al., 2004). Neurochemical changes were indicated with the controlled cortical impact model, with reports of long-term alterations in GABA(A) and glutamate receptor subunits (Kharlamov et al., 2011) and an increase in GABA(A)-receptor mediated tonic inhibition in dentate granule cells (Mtchedlishvili et al., 2010).

12.4 EEG

Clinically, EEG records from electrodes placed on the patient's scalp reflect correlated synaptic activity caused by postsynaptic potentials of cortical neurons. EEG is the mainstay for confirming a diagnosis of epilepsy and for localizing and monitoring epileptogenic foci as well as characterizing the type of seizure and epileptic disorder (Fisher et al., 2005). Seizures are sudden abnormal electrical discharges in the brain, represented in EEG recordings by frequency changes and increased amplitudes. EEG signals are represented in one of four displays: unipolar, bipolar, Laplacian, and average reference montages. EEG frequently involves single-channel recording; however, algorithms for multiple channels have recently been described (Johnson et al., 2011; Shen et al., 2013). Spectral analysis of EEG data provides information on neural oscillations. Software programs, such as EEGLAB, EEGgui, and the Neurophysiological Biomarker Toolbox, may be used in this process of EEG analysis because these are helpful tools to quantify EEG alterations (Brunner et al., 2013; Sick et al., 2013).

12.4.1 Electrophysiology Recording Methods

In animal models, NCSs can only be detected by continuous EEG or electrocorticographic (ECoG) recordings in vivo. ECoG differs from EEG in the direct placement of electrodes into or on the surface of the cortex. NCSs involve seizure activity on EEG or ECoG monitoring without an accompanying behavioral response. Although simultaneous continuous video recording is not necessary to diagnose NCSs, it aids in the confirmation process. NCSs may be further categorized as generalized or focal in animals with bilateral EEG recording: generalized NCS events are detected from bilateral recording channels, whereas focal NCSs are defined as seizure events detected from single or multiple channels of the ipsilateral hemisphere only.

Published studies on posttraumatic epilepsy in experimental models present a variety of methods for placement of electrodes, the timing after injury, subsequent duration of EEG recording, and behavioral monitoring with or without video. Table 12.2 compares the different recording methodologies. The number of electrodes placed varies from one electrode bilaterally with ground and reference electrodes to up to eight electrodes bilaterally. The depth of electrode placement ranges from subcutaneous to epidural to cortical, and the length of EEG recording time spans from as little as 20 minutes to continuous monitoring for several days.

12.4.2 EEG Findings in TBI

Nonconvulsive or electrographic seizures involve spike-wave discharges at >3 Hz or discharges with any evolving waveform reaching >4 Hz (Hirsch et al., 2005; Jordan, 1999) and occur in the absence of a behavioral or motor component to the seizure. To identify NCSs, several reports indicate the potential benefit in the use of continuous EEG in monitoring neurocritical patients (Claasen et al., 2004, 2013; Friedman et al., 2009; Jordan, 1995; Ronne-Engstrom et al., 2006; Vespa, 2005; Vespa et al., 1999, 2007a). Recently, a technique called inverse localization has been developed to determine the source of epileptiform activity in post-TBI patients by combining MRI imaging with EEG recording (Irimia et al., 2013).

An important clinical consideration for patients suffering TBI is the potential for epileptogenesis. As mentioned previously, epilepsy with accompanying overt seizures occurs in a high proportion of patients after brain injury and the incidence of NCSs is likely to be even higher. Unfortunately, prophylaxis using conventional antiepileptic pharmacotherapy has been ineffective in preventing posttraumatic epilepsy (Temkin et al., 1999). There is a clinical need for a biomarker of posttraumatic epileptogenesis that could be used to test efficacy of antiepileptic treatments. Posttraumatic EEG analysis may serve this purpose.

The underlying basis for epileptiform electrical activity in the brain is the abnormal synchronous firing of neuronal populations. EEG recordings detect these events as the classic "spike and dome" discharges, which are diagnostic of epileptic brain. Examples taken after injection of PTZ and from posttraumatic rat brain are shown in Figure 12.1. In contrast to EEG responses during most other behavioral states in which either low or high frequencies dominate, the epileptiform event contains increased amplitude simultaneously in many frequency domains from synchrony of both slower synaptic potentials and faster action potentials. Power spectral analysis and fast-Fourier transformation of EEG signals are useful methods for identifying epileptiform discharges based on changes in signal power at frequencies characteristic for these events.

We are currently investigating methods to detect epileptiform events and other potentially abnormal events that predict the transition of normal brain to epileptic brain. Our approach has been to compare EEG signals from posttraumatic brain and compare these with "normal" EEG using fast-Fourier transformation and power spectral analysis to detect EEG outliers. An earlier preliminary version of this analysis has been recently published

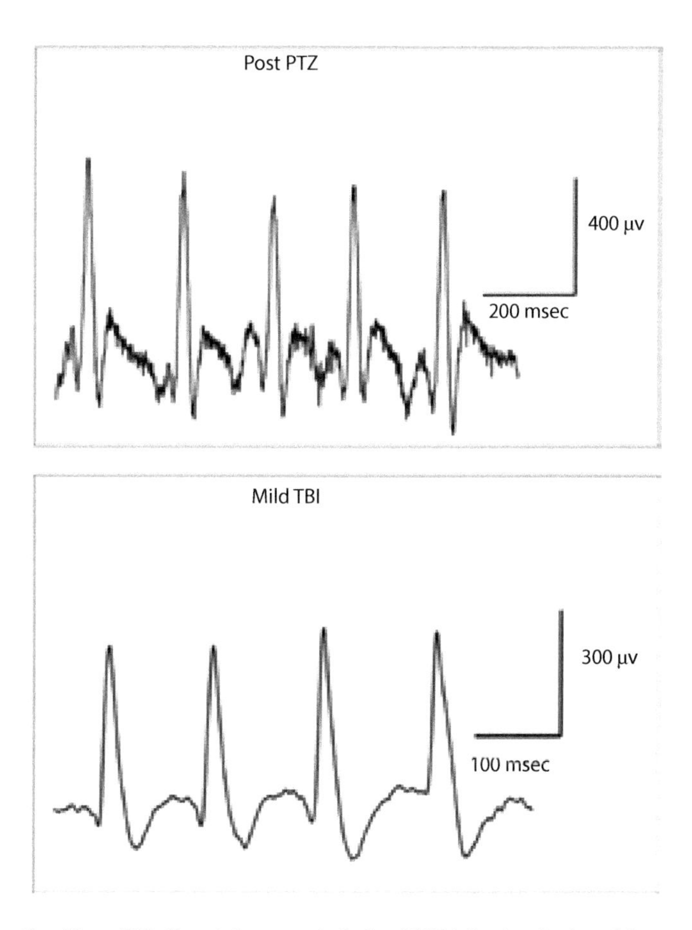

FIGURE 12.1 Examples of rat epileptiform ECoG activity recorded after PTZ injection (top) and 1 year after mild TBI. Waveforms show typical "spike and dome" features consisting of both high and low frequencies. These features allowed detection and quantification of nonconvulsive seizure events.

(Sick et al., 2013). In its first iteration, EEG from awake, normally behaving rats were used as the control. Fast-Fourier transform was used to measure mean and standard deviations of power in the major EEG spectral bands (alpha, beta1, beta2, delta, and gamma) for comparison with power data from rats post-TBI. Abnormal EEG events were defined as changes in power that exceeded four standard deviations from normal EEG in any frequency band. Examples of the analysis are shown in Figure 12.2a from a naive rat, Figure 12.2b from a rat 1 year after mild TBI, and Figure 12.2c from a rat after intraperitoneal injection of PTZ. As expected, epileptiform events were clearly discernible as large increases in power across all frequency bands. However, other "abnormal" events, characterized by heightened power in one or more bands, were also detected. In the future, a similar approach might be used for abnormal EEG event detection in human post-traumatic brain for use as a biomarker of epileptogenesis.

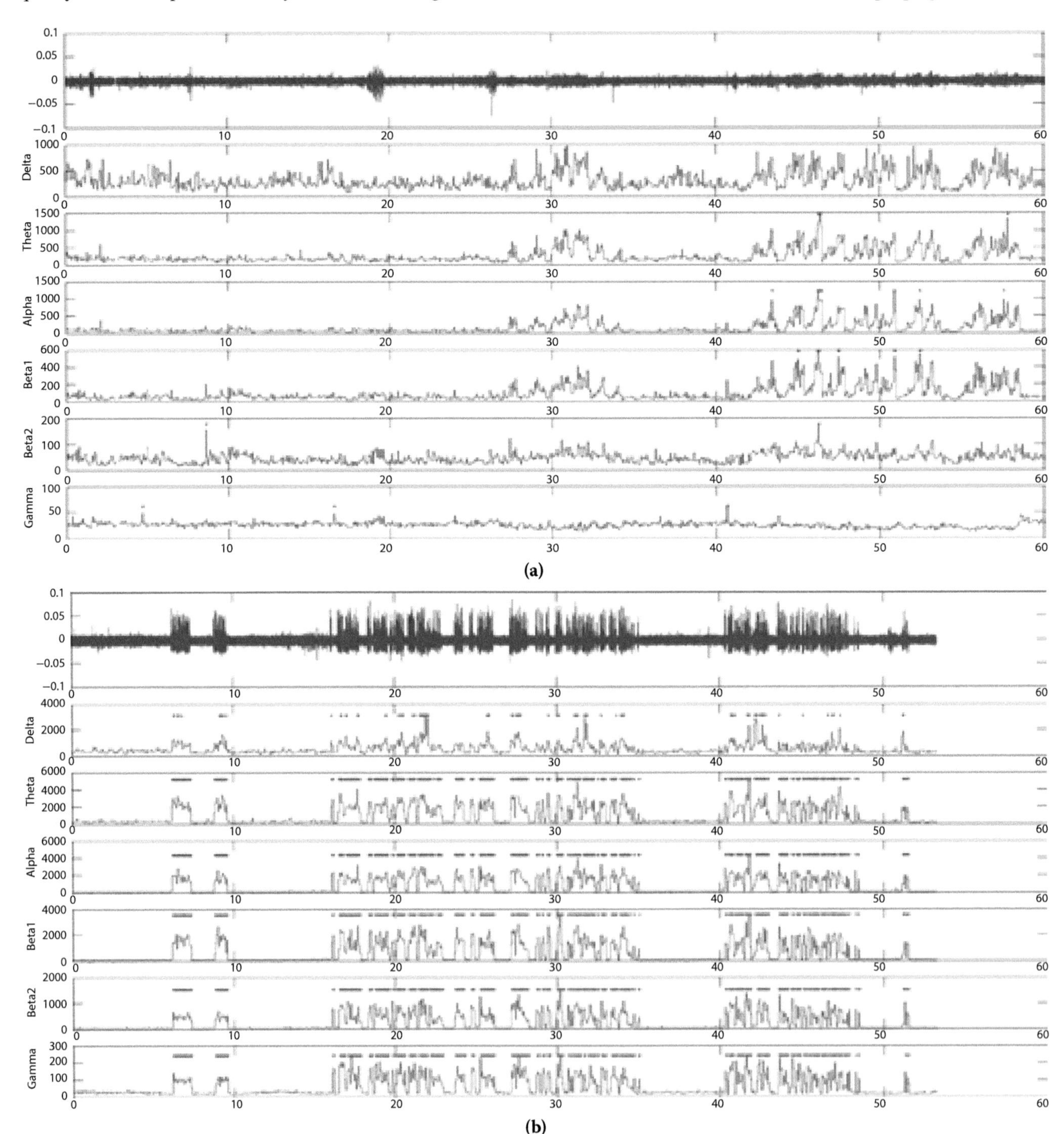

FIGURE 12.2 (a) Naive rat recorded for 1 hour without administration of any drug. Top of the figure shows raw EEG record. Lower traces are changes in power at selected frequencies. Increased power in lower frequencies corresponded to sleeping behavior. (b) An animal 1 year after TBI before injection of PTZ. Spontaneous electrographic seizures are present; however, no behavioral seizures were seen in this animal.

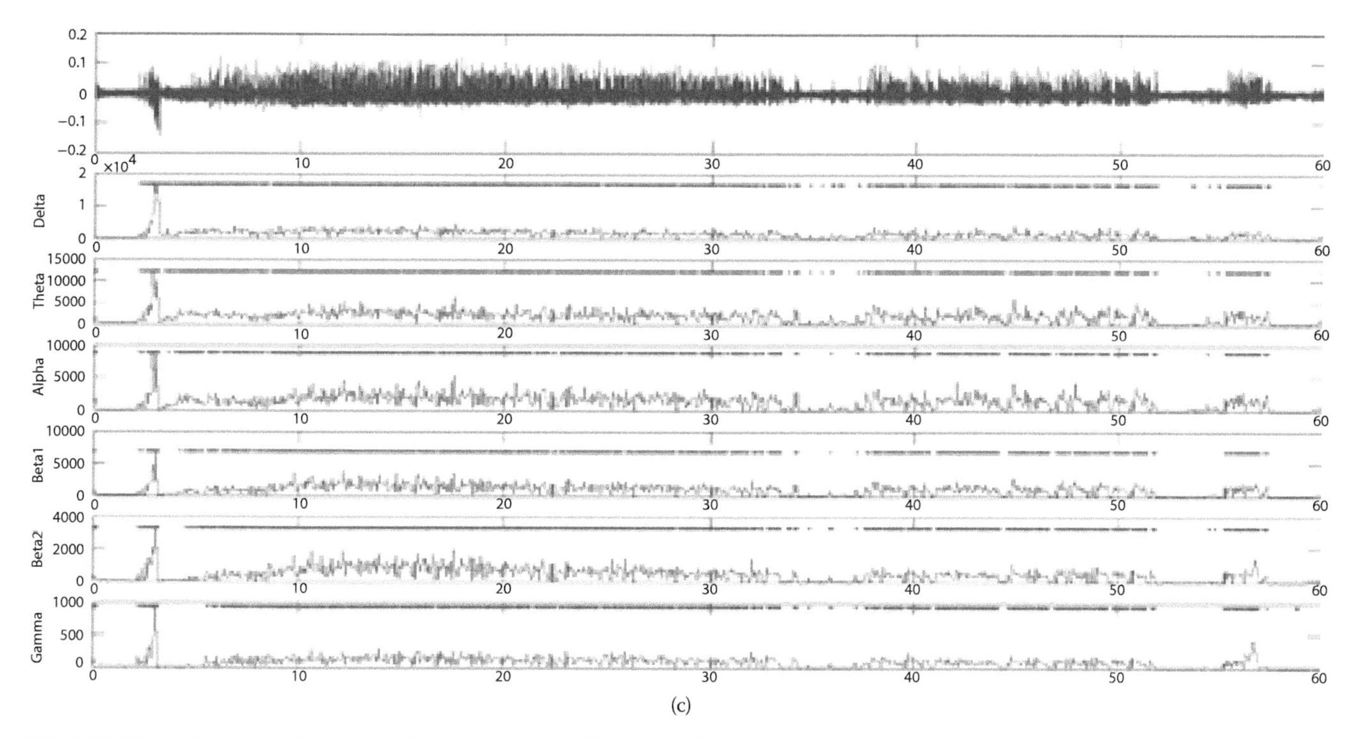

FIUGRE 12.2 *(Continued)* (c) An animal 1 year after TBI after injection of PTZ. Both electrographic and behavioral seizures were detected after PTZ.

12.5 CONCLUSION

Many challenges persist in the diagnosis and treatment of NCSs in posttraumatic epilepsy. In addition to the requirement of continuous EEG monitoring to diagnose NCSs, the duration with which to monitor patients for NCSs remains unclear. Also, the sheer amount of data produced by continuous EEG monitoring may be difficult to adequately interpret in some hospital settings for diagnosing the presence of NCSs. Literature further suggests that NCSs may be misdiagnosed as mental illness, revealing nonconvulsive status epilepticus previously diagnosed as a psychiatric symptom and, therefore, remains undiagnosed and especially untreated (Lopez Arteaga, 2013; Mirsattari et al., 2011). Furthermore, primary antiepileptic drug prophylaxis is frequently unreliable in preventing or suppressing NCSs, as studies have found NCSs to even occur in the presence of antiepileptic drug prophylaxis (Olson, 2004; Temkin, 2009; Vespa, 1999).

NCSs after TBI also represent a significant factor for secondary injury to the brain. NCSs are associated with several dynamic changes in neurochemistry, including increased release of excitatory amino acids, neuron-specific enolase, glutamate, and glycerol levels in the brain (Bullock et al., 1995; DiGeorgio et al., 1996, 1999; Goodman et al., 1996; Rabinowicz et al., 1994, 1995; Vespa, 2005; Vespa et al., 1998, 2002). NCSs post-TBI are also associated with prolonged increased lactate/pyruvate ratio and intracranial pressure, which further increases neuronal damage and worsens outcome (Vespa et al., 2007a, 2007b).

Although clinical evidence for NCSs in posttraumatic epilepsy is abundant and despite substantial experimental evidence for convulsive seizures, presently, experimental research specifically in the field of NCSs posttraumatic brain injury remains lacking. Future areas of research include the characterization of NCSs in other TBI animal models; the pathogenesis of NCS as secondary insult after TBI and subsequent pathophysiology; the attenuation and mitigation of NCS with available antiepileptic agents; the association of NCSs with other disease conditions, such as behavioral seizures; and chronic posttraumatic encephalopathy.

ACKNOWLEDGMENTS

We thank Len Dunikoski for his assistance in acquisition of literature. The experimental work in this manuscript was supported by Veterans Administration BX000521 and NIH NS030291.

REFERENCES

Agrawal A, Timothy J, Pandit L, and Manju M. Post-traumatic epilepsy: An overview. *Clin Neurol Neurosurg.* 2006;108:433–9.

Alroughani R, Javidan M, Qasem A, and Alotaibi N. Non-convulsive status epilepticus: The rate of occurrence in a general hospital. *Seizure.* 2009;18(1):38–42.

Annegers JF, Grabow JD, Groover RV, Laws ER Jr, Elveback LR, and Kurland LT. Seizures after head trauma: A population study. *Neurology.* 1980;30(7):683–9.

Annegers JF and Coan SP. The risks of epilepsy after traumatic brain injury. *Seizure.* 2000;9:453–7.

Bao YH, Bramlett HM, Atkins CM, Truettner JS, Lotocki G, Alonso OF et al. Post-traumatic seizures exacerbate histopathological damage after fluid-percussion brain injury. *J Neurotrauma.* 2011;28(1):35–42.

Blaha M, Schwab J, Vajnerova O, Bednar M, Vajner L, and Michal T. Intracranial pressure and experimental model of diffuse brain injury in rats. *J Korean Neurosurg Soc.* 2010;47(1):7–10.

Bolkvadze T and Pitkänen A. Development of post-traumatic epilepsy after controlled cortical impact and lateral fluid-percussion-induced brain injury in the mouse. *J Neurotrauma.* 2012;29(5):789–812.

Bregy A, Sick J, Bray E, Wick A, Dietrich D, Sick T, and Bramlett H. Subconvulsive epilepsy and changes in PTZ-induced seizure threshold in rats 1 year after moderate fluid percussion injury. Poster session presented at the meeting of the National Neurotrauma Society, Nashville, TN, 2013.

Brunner C, Delorme A, and Makeig S. Eeglab: An open source Matlab toolbox for electrophysiological research. *Biomed Tech (Berl).* 2013 Sep 7. Retrieved from http://www.degruyter.com/view/j/bmte.2013.58.issue-s1-G/bmt-2013-4182/bmt-2013-4182.xml.

Bullock R, Zauner A, Myseros JS, Marmarou A, Woodward JJ, and Young HF. Evidence for prolonged release of excitatory amino acids in severe human head trauma: Relationship to clinical events. *Ann N Y Acad Sci.* 1995;59:290–7.

Centers for Disease Control and Prevention (CDC), National Center for Injury Prevention and Control. Report to Congress on mild traumatic brain injury in the United States: Steps to prevent a serious public health problem. Atlanta (GA): Centers for Disease Control and Prevention; 2003.

Centers for Disease Control and Prevention (CDC), National Center for Injury Prevention and Control. Traumatic brain injury in the United States: Emergency department visits, hospitalizations and deaths 2002–2006. Atlanta (GA): Centers for Disease Control and Prevention; 2010.

Claassen J, Mayer SA, Kowalski RG, Emerson RG, and Hirsch LJ. Detection of electrographic seizures with continuous EEG monitoring in critically ill patients. *Neurology.* 2004;62(10):1743–8.

Claassen J, Jetté N, Chum F, Green R, Schmidt M, Choi H et al. Electrographic seizures and periodic discharges after intracerebral hemorrhage. *Neurology.* 2007;69(13):1356–65

Claassen J, Taccone FS, Horn P, Holtkamp M, Stocchetti N, and Oddo M. Recommendations on the use of EEG monitoring in critically ill patients: Consensus statement from the neurointensive care section of the ESICM. *Intensive Care Med.* 2013;39(8):1337–51.

D'Ambrosio R, Fairbanks JP, Fender JS, Born DE, Doyle DL, and Miller JW. Post-traumatic epilepsy following fluid percussion injury in the rat. *Brain.* 2004;127(2):304–14.

Da Silva AM, Vaz AR, Ribeiro I, Melo AR, Nune B, and Correia M. Controversies in post-traumatic epilepsy. *Acta Neurochir Suppl.* 1990;50:48–51.

Defense and Veterans Brain Injury Center (DVBIC). Department of Defense numbers for traumatic brain injury; 2013. Available from http://www.defense.gov/home/features/2012/0312_tbi/.

DeGiorgio CM, Gott PS, Rabinowicz AL, Heck CN, Smith TD, and Correale JD. Neuron-specific enolase, a marker for acute neuronal injury, is increased in complex partial status epilepticus. *Epilepsia.* 1996;37(7):606–9.

DeGiorgio CM, Heck CN, Rabinowicz AL, Gott PS, Smith T, and Correale J. Serum neuron-specific enolase in the major subtypes of status epilepticus. *Neurology.* 1999;52:746–9.

Drislane FW. Presentation, evaluation, and treatment of nonconvulsive status epilepticus. *Epilepsy Behav.* 2000;1(5):301–14.

Drislane FW, Lopez MR, Blum AS, and Schomer DL. Detection and treatment of refractory status epilepticus in the intensive care unit. *J Clin Neurophysiol.* 2008;25(4):181–86.

Engel J, Ludwig BI, and Fetell M. Prolonged partial complex status epilepticus: EEG and behavioral observations. *Neurology.* 1978;28(9 Pt 1):863–869.

Faul M, Xu L, Wald MM, and Coronado VG. Traumatic brain injury in the United States: Emergency department visits, hospitalizations, and deaths. Atlanta (GA): Centers for Disease Control and Prevention, National Center for Injury Prevention and Control; 2010.

Fisher RS, Boas WVE, Blume W, Elger C, Genton P, Lee P et al. Epileptic seizures and epilepsy: Definitions proposed by the international league against epilepsy (ILAE) and the international bureau for epilepsy (IBE). *Epilepsia.* 2005;46: 470–472.

Friedman D, Claassen J, and Hirsch LJ. Continuous electroencephalogram monitoring in the intensive care unit. *Anesth Analg.* 2009;109:506–523.

Golarai G, Greenwood AC, Feeney DM, and Connor JA. Physiological and structural evidence for hippocampal involvement in persistent seizure susceptibility after traumatic brain injury. *J Neurosci.* 2001;21(21):8523–37.

Goodman JC, Valadka AB, Gopinath SP, Cormio M, and Robertson CS. Lactate and excitatory amino acids measured by microdialysis are decreased by pentobarbital coma in head-injured patients. *J Neurotrauma.* 1996;10:549–56.

Grand'Maison F, Reiher J, and Laduke CP. Retrospective inventory of EEG abnormalities in partial status epilepticus. *Electroencephalogr Clin Neurophysiol.* 1991;79:264–70.

Hall S and Bornstein RA. The relationship between intelligence and memory following minor or mild closed head injury: Greater impairment in memory than intelligence. *J Neurosurg.* 1991;75:378–381.

Hartings JA, Williams AJ, and Tortella FC. Occurrence of nonconvulsive seizures, periodic epileptiform discharges, and intermittent rhythmic delta activity in rat focal ischemia. *Exp Neurol.* 2003;179(2):139–49.

Hirsch LJ, Brenner RP, Drislane FW, So E, Kaplan PW, Jordan KG et al. The ACNS subcommittee on research terminology for continuous EEG monitoring: Proposed standardized terminology for rhythmic and periodic EEG patterns encountered in critically ill patients. *J Clin Neurophysiol.* 2005;22:128–135.

Hirsch LJ. Nonconvulsive seizures in traumatic brain injury: What you don't see can hurt you. *Epilepsy Curr.* 2008;8(4):97–9.

Hunt RF, Scheff SW, and Smith BN. Posttraumatic epilepsy after controlled cortical impact injury in mice. *Exp Neurol.* 2009;215(2):243–52.

Husain A, Horn G, and Jacobson M. Non-convulsive status epilepticus: Usefulness of clinical features in selecting patients for urgent EEG. *J Neuro Neurosurg Psychiatry.* 2003;74(2):189–191.

Immonen R, Kharatishvili I, Gröhn O, and Pitkänen A. MRI biomarkers for post-traumatic epileptogenesis. *J Neurotrauma.* 2013;30(14):1305–9.

Irimia A, Goh SY, Torgerson CM, Stein NR, Chambers MC, Vespa PM et al. Electroencephalographic inverse localization of brain activity in acute traumatic brain injury as a guide to surgery, monitoring and treatment. *Clin Neurol Neurosurg.* 2013;115(10):2159–65.

Jennett B. Trauma as a cause of epilepsy in childhood. *Dev Med Child Neurol.* 1973;15:56–62.

Jirsch J, Hirsch LJ. Nonconvulsive seizures: Developing a rational approach to the diagnosis and management in the critically ill population. *J Clin Neurophys.* 2007;118:1660–70.

Johnson AN, Sow D, and Biem A. A discriminative approach to EEG seizure detection. *AMIA Annu Symp Proc.* 2011;2011:1309–17.

Jordan, KG. Continuous EEG and evoked potential monitoring in the neuroscience intensive care unit. *J Clin Neurophysiol.* 1993;10:445–475.

Jordan KG. Neurophysiologic monitoring in the neuroscience intensive care unit. *Neurol Clin.* 1995;13(3):579–626.

Jordan KG. Nonconvulsive status epilepticus in acute brain injury. *J Clin Neurophysiol.* 1999;16:332–340.

Kaplan PW. Nonconvulsive status epilepticus in the emergency room. *Epilepsia.* 1996;37:643–50.

Kharatishvili I, Nissinen JP, McIntosh TK, and Pitkänen A. A model of posttraumatic epilepsy induced by lateral fluid-percussion brain injury in rats. *Neuroscience.* 2006;140(2):685–97.

Kharatishvili I and Pitkänen A. Association of the severity of cortical damage with the occurrence of spontaneous seizures and hyperexcitability in an animal model of posttraumatic epilepsy. *Epilepsy Res.* 2010;90(1–2):47–59.

Kharlamov EA, Lepsveridze E, Meparishvili M, Solomonia RO, Lu B, Miller ER et al. Alterations of GABA(A) and glutamate receptor subunits and heat shock protein in rat hippocampus following traumatic brain injury and in posttraumatic epilepsy. *Epilepsy Res.* 2011;95(1–2):20–34.

Kilbride RD, Costello DJ, and Chiappa KH. How seizure detection by continuous electroencephalographic monitoring affects the prescribing of antiepileptic medications. *Arch Neurol.* 2009;66(6):723–8.

Krumholz A, Sung GY, Fisher RS, Barry E, Bergey GK, and Grattan LM. Complex partial status epilepticus accompanied by serious morbidity and mortality. *Neurology.* 1995;45(8):1499–1504

Levin HS, Grossman RG, Rose JE, and Teasdale G. Long-term neuropsychological outcome of closed head injury. *J Neurosurg.* 1979;50:412–422.

Litt B, Dizon L, and Ryan D. Fatal nonconvulsive status epilepticus in the elderly. *Epilepsia.* 1994;35:10–7.

López Arteaga T, Amo C, Serrano González C, and Huertas Sánchez D. Nonconvulsive status epilepticus and psychotic symptoms: Case report. *Riv Psichiatr.* 2013;48(3):268–70.

Lu XC, Si Y, Williams AJ, Hartings JA, Gryder D, and Tortella FC. NNZ-2566, a glypromate analog, attenuates brain ischemia-induced non-convulsive seizures in rats. *J Cereb Blood Flow Metab.* 2009;29(12):1924–32.

Lu XC, Hartings JA, Si Y, Balbir A, Cao Y, and Tortella FC. Electrocortical pathology in a rat model of penetrating ballistic-like brain injury. *J Neurotrauma.* 2011;28(1):71–83.

Lu XC, Mountney A, Chen Z, Wei G, Cao Y, Leung LY et al. Similarities and differences of acute nonconvulsive seizures and other epileptic activities following penetrating and ischemic brain injuries in rats. *J Neurotrauma.* 2013;30(7):580–90.

Maganti R, Gerber P, Drees C, and Chung S. Nonconvulsive status epilepticus. *Epilepsy Behav.* 2008;12(4):572–86.

Meierkord H and Holtkamp M. Non-convulsive status epilepticus in adults: Clinical forms and treatment. *Lancet Neurol.* 2007;6(4):329–39.

Mirsattari SM, Gofton TE, and Chong DJ. Misdiagnosis of epileptic seizures as manifestations of psychiatric illnesses. *Can J Neurol Sci.* 2011 May;38(3):487–93.

Mirski MA and Varelas PN. Seizures and status epilepticus in the critically ill. *Crit Care Clin.* 2008;24:115–47.

Mountney A, Shear DA, Potter B, Marcsisin SR, Sousa J, Melendez V et al. Ethosuximide and phenytoin dose-dependently attenuate acute nonconvulsive seizures after traumatic brain injury in rats. *J Neurotrauma.* 2013;30(23):1973–82.

Mtchedlishvili Z, Lepsveridze E, Xu H, Kharlamov EA, Lu B, and Kelly KM. Increase of GABA(A) receptor-mediated tonic inhibition in dentate granule cells after traumatic brain injury. *Neurobiol Dis.* 2010;38(3):464–75.

Mukherjee S, Zeitouni S, Cavarsan CF, and Shapiro LA. Increased seizure susceptibility in mice 30days after fluid percussion injury. *Front Neurol.* 2013;4:28.

Nilsson P, Ronne-Engström E, Flink R, Ungerstedt U, Carlson H, and Hillered L. Epileptic seizure activity in the acute phase following cortical impact trauma in rat. *Brain Res.* 1994;637(1–2):227–32.

Olson S. Review of the role of anticonvulsant prophylaxis following brain injury. *J Clin Neurosci.* 2004;11(1):1–3.

Pandian JD, Cascino GD, So EL, Manno E, and Fulgham JR. Digital video-electroencephalographic monitoring in the neurological–neurosurgical intensive care unit: Clinical features and outcome. *Arch Neurol.* 2004;61:1090–4.

Pitkänen A and McIntosh T. Animal models of post-traumatic epilepsy. *J Neurotrauma.* 2006;23(2):241–61.

Pitkänen A, Immonen RJ, Gröhn OH, and Kharatishvili I. From traumatic brain injury to posttraumatic epilepsy: What animal models tell us about the process and treatment options. *Epilepsia.* 2009;50(2):21–9.

Privitera MD and Strawsburg RH. Electroencephalographic monitoring in the emergency department. *Emerg Med Clin North Am.* 1994;12(4):1089–100.

Rabinowicz AL, Correale JD, Couldwell WT, and DeGeorgio CM. CSF neuron specific enolase after methohexital activation during electrocorticography. *Neurology.* 1994;44(6): 1167–1169.

Rabinowicz AL, Correale JD, Bracht KA, Smith TD, and DeGiorgio CM. Neuron-specific enolase is increased after nonconvulsive status epilepticus. *Epilepsia.* 1995;36(5):475–79.

Ronne-Engstrom E and Winkler T. Continuous EEG monitoring in patients with traumatic brain injury reveals a high incidence of epileptiform activity. *Acta Neurol Scand.* 2006;114(1):47–53.

Salazar AM, Jabbari B, Vance SC, Grafman J, Amin D, and Dillon JD. Epilepsy after penetrating head injury. I. Clinical correlates: A report of the Vietnam Head Injury Study. *Neurology.* 1985;35:1406–14.

Scheuer MJ. Continuous EEG monitoring in the intensive care unit. *Epilepsia.* 2002; 43: 114–27.

Semah F, Picot M-C, Adam C, Broglin D, Arzimanoglou A, Bazin B et al. Is the underlying cause of epilepsy a major prognostic factor for recurrence? *Neurology.* 1998;51:1256–62.

Shen CP, Liu ST, Zhou WZ, Lin FS, Lam AY, Sung HY et al. A physiology-based seizure detection system for multichannel EEG. *PLoS One.* 2013;8(6):e65862.

Shultz SR, Cardamone L, Liu YR, Hogan RE, Maccotta L, Wright DK et al. Can structural or functional changes following traumatic brain injury in the rat predict epileptic outcome? *Epilepsia.* 2013 Jul;54(7):1240–50.

Sick J, Bray E, Bregy A, Dietrich WD, Bramlett HM, and Sick T. EEGgui: A program used to detect electroencephalogram anomalies after traumatic brain injury. *Source Code Biol Med.* 2013;8(1):12.

Statler KD, Swank S, Abildskov T, Bigler ED, and White HS. Traumatic brain injury during development reduces minimal clonic seizure thresholds at maturity. *Epilepsy Res.* 2008;80(2–3):163–70.

Swartz BE, Houser CR, Tomiyasu U, Walsh GO, DeSalles A, Rich JR et al. Hippocampal cell loss in posttraumatic human epilepsy. *Epilepsia.* 2006;47(8):1373–82.

Temkin NR, Dimken SS, Anderson GD, Wilensky AJ, Holmes MD, Cohen W et al. Valproate therapy for prevention of posttraumatic seizures: A randomized trial. *J Neurosurg.* 1999;91(4):593–600.

Temkin NR. Preventing and treating posttraumatic seizures: The human experience. *Epilepsia.* 2009;50(Suppl 2):10–13.

Treiman DM. Electroclinical features of status epilepticus. *J Clin Neurophysiol.* 1995;12(4)343–62.

Vespa PM, Prins M, Ronne-Engstrom E, Caron M, Shalmon E, Hovda DA et al. Increase in extracellular glutamate caused by reduced cerebral perfusion pressure and seizures after human traumatic brain injury: A microdialysis study. *J Neurosurg.* 1998;89:971–82.

Vespa PM, Nuwer MR, Nenov V, Ronne-Engstrom E, Hovda DA, Bergsneider M et al. Increased incidence and impact of nonconvulsive and convulsive seizures after traumatic brain injury as detected by continuous electroencephalographic monitoring. *J Neurosurg.* 1999;91:750–60.

Vespa PM, Boscardin WJ, Hovda DA, McArthur DL, Nuwer MR, Martin Na et al. Early and persistent impaired percent alpha variability on continuous electroencephalography monitoring as predictive of poor outcome after traumatic brain injury. *J Neurosurg.* 2002;97:84–92.

Vespa P, Martin NA, Nenov V, Glenn T, Bergsneider M, Kelly D et al. Delayed increase in extracellular glycerol with post-traumatic electrographic epileptic activity: Support for the theory that seizures induce secondary injury. *Acta Neurochir Suppl.* 2002;81:355–7.

Vespa PM, O'Phelan K, Shah M, Mirabelli J, Starkman S, Kidwell C et al. Acute seizures after intracerebral hemorrhage: A factor in progressive midline shift and outcome. *Neurology.* 2003;60(9):1441–6.

Vespa PM. Continuous EEG monitoring for the detection of seizures in traumatic brain injury, infarction, and intracerebral hemorrhage: "To Detect and Protect." *J Clin Neurophysiol.* 2005;22(2):99–106.

Vespa PM, Miller C, McArthur D, Eliseo M, Etchepare M, Hirt D et al. Nonconvulsive electrographic seizures after traumatic brain injury result in a delayed, prolonged increase in intracranial pressure and metabolic crisis. *Crit Care Med.* 2007a;35(12):2830–6.

Vespa PM, O'Phelan K, McArthur D, Miller C, Eliseo M, Hirt D et al. Pericontusional brain tissue exhibits persistent elevation of lactate/pyruvate ratio independent of cerebral perfusion pressure. *Crit Care Med.* 2007b;35(4):1153–60.

Wasterlain CG, Fujikawa DG, Penix L, and Sankar R. Pathophysiological mechanisms of brain damage from status epilepticus. *Epilepsia.* 1993;34(Suppl 1):37–53.

Williams AJ, Tortella FC, Lu XM, Moreton JE, and Hartings JA. Antiepileptic drug treatment of nonconvulsive seizures induced by experimental focal brain ischemia. *J Pharmacol Exp Ther.* 2004;311(1):220–7.

Williams AJ, Hartings JA, Lu XC, Rolli ML, Dave JR, and Tortella FC. Characterization of a new rat model of penetrating ballistic brain injury. *J Neurotrauma.* 2005;22(2):313–31.

Williams AJ, Tortella FC, Gryder D, and Hartings JA. Topiramate reduces non-convulsive seizures after focal brain ischemia in the rat. *Neurosci Lett.* 2008;430(1):7–12.

Young GB and Doig GS. Continuous EEG monitoring in comatose intensive care patients: Epileptiform activity in etiologically distinct groups. *Neurocrit Care.* 2005;2:5–10.

Young GB and Claassen J. Nonconvulsive status epilepticus and brain damage: Further evidence, more questions. *Neurology.* 2010;75(9):760–1.

Zanier ER, Lee SM, Vespa PM, Giza CC, and Hovda DA. Increased hippocampal CA3 vulnerability to low-level kainic acid following lateral fluid percussion injury. *J Neurotrauma.* 2003;20(5):409–20.

13 Characterization and Management of Headache after Mild Traumatic Brain Injury

Sylvia Lucas

CONTENTS

13.1 INTRODUCTION

In the United States, approximately 1.7 million traumatic brain injuries (TBI) requiring medical attention occur each year. Seventy-five percent of these cases are classified as mild TBI and post traumatic headache (PTH) is the most common physical symptom after injury. PTH is classified as a secondary headache disorder in the International Classification of Headache Disorders (ICHD-3) criteria for headache classification. In recent large, prospective clinical studies after moderate to severe TBI, prevalence can be close to half of the injured population. Cumulative incidence and prevalence have reported to be even higher after mild TBI. A significant risk factor for PTH was found to be a pre-injury history of headache. The most common headache phenotype using ICHD-3 criteria for primary headache disorders was migraine/probable migraine in over 50% of those with headache, followed by tension-type headache. Though the treatment of PTH is largely empiric, one approach to treatment decisions is to use primary headache characterization of the PTH and treat according to recommendations for that headache.

Mild traumatic brain injury (mTBI) has become an extremely important global health issue in the past several years. Considerable attention and interest has focused on several populations susceptible to mTBI: soldiers deployed in war zones, professional athletes and youth involved in school sports activity, and civilians engaged in frequent activities of daily living who may be involved in motor vehicle accidents, falls, or assaults among other injuries.

In the United States, estimates for 1.7 million civilian TBIs occurring each year are for those requiring medical attention. The outcomes of known cases resulted in 53,000 deaths, and 275,000 hospitalizations for nonfatal TBI (Faul et al., 2010). It is estimated that 43% of Americans live with disability 1 year after hospitalization and 3.2 million live with residual disability (Corrigan et al., 2010).

It has been documented that around 75% of TBI is classified as mTBI, and because many of these events do not lead to immediate medical attention, the scope of the problem is likely underestimated. The etiology of mTBI is likely to be different in different age groups. For example, falls are the most common cause of mTBI in children 0 to 4 years old and in adults older than age 75 (Coronado et al., 2011). Civilian deaths have declined in the past decade, likely from industry improvements in protection of occupants with air bags and seat belts in motor vehicles, protective helmet design for use in sports activities and two- or three-wheeled vehicle use, and other public health driven protective measures.

The most common etiology of the additional TBI burden in military personnel in recent Middle East conflicts (Operation Iraqi Freedom and Operation Enduring Freedom) is exposure to combat-related explosions (Eskridge et al., 2012) with approximately 80% of mTBI secondary to blast exposure (Hoge et al., 2008). Injury to military personnel reported by the Congressional Research Service show a total of 253,330 TBI cases between January 1, 2000, and August 20, 2012, with 194,561 mild, 42,083 moderate, and 6,476 severe or penetrating with 10,210 not classifiable (Fischer, 2013).

13.2 POSTTRAUMATIC HEADACHE AND POSTCONCUSSIVE SYMPTOMS

PTH is the most common physical symptom following TBI; however, it may not occur in isolation. Headache may be part of a symptom complex known as postconcussion syndrome

(PCS) comprising physical or somatic, psychological, and cognitive symptoms. One prospective, longitudinal study of symptoms 1 month after TBI reported the most common symptoms as fatigue, headache, dizziness, memory trouble, trouble sleeping, trouble concentrating, irritability, blurred vision, anxiety, increased light, and sound sensitivity (Dikmen et al., 2010). Severity of brain injury was correlated with number of symptoms, and more severe injuries tended to be associated with a greater proportion of cognitive and psychological symptoms in addition to physical symptoms.

PTH has no defining clinical features, and it is classified as a secondary headache disorder in the International Classification of Headache Disorders, third edition-beta (ICHD-3) (Table 13.1) (International Classification of Headache Disorders, 2013). A secondary headache is a headache that occurs in close temporal relation to another disorder that is known to cause headache or fulfills other criteria for causation by that disorder; the new headache is coded as a secondary headache attributed to the causative disorder. This remains true even when the headache has the characteristics of a primary headache (migraine, tension-type headache, cluster headache, or one of the other primary headaches).

The classification of PTH, made on the basis of its close temporal relationship to the head injury thought to imply causation, may be problematic in persons who have preexisting primary headache disorders. If the headaches are worsened in frequency or intensity in close temporal relationship to the injury, then the new or worse headaches are defined as PTH.

Primary headache disorders are described according to a recurring symptom complex and are thought to be genetically acquired syndromes that involve trigeminovascular pathway dysfunction. Although the pathophysiology is not clear, animal models of neurogenic inflammation and the response of migraine to a class of drugs known as the "triptans," which act as agonists at serotonin-1B and 1D receptors, have given us clues into possible physiological pathways of head pain (Olesen and Ramadan, 2006). As a secondary headache, PTH is thought to have a structural or physiological basis that, if corrected, results in the resolution of the secondary headache. The ICHD requirement that PTH occur within 7 days after an injury does not necessarily ensure causation. Many reports have been made of PTH occurring more than 7 days after an injury. In a study of returning war veterans, only 27% of headaches were reported to develop within a week after injury (Theeler and Erickson, 2009). In hospitalized pediatric patients with mTBI, 32% reported headache 2 to 3 weeks after the injury (Blinman et al., 2009). In a prospective study of civilians with complicated mild to moderately severe TBI, approximately 28% of new headaches were reported after 3 months following the injury (Hoffman et al., 2011). If the ICHD classification of PTH is adhered to, then PTH may be underestimated on the basis of the latency constraints.

TABLE 13.1
International Classification of Headache Disorders (ICHD), third edition:
Headache attributed to traumatic injury to the head

Acute post-traumatic headache attributed to mild traumatic injury to the head
Diagnostic criteria:

A. Any headache fulfilling criteria C and D
B. Injury to the head fulfilling both of the following:
   1. Associated with none of the following:
      a) Loss of consciousness for > 30 minutes
      b) Glasgow Coma Scale (GCS) score < 13
      c) Post-traumatic amnesia lasting > 24 hours
      d) Altered level of awareness for > 24 hours
      e) Imaging evidence of traumatic head injury such as intracranial hemorrhage and/or brain contusion
   2. Associated immediately following the head injury with one or more of the following symptoms and/or signs:
      a) Transient confusion, disorientation or impaired consciousness
      b) Loss of memory for events immediately before or after the head injury
      c) Two or more other symptoms suggestive of mild traumatic brain injury: nausea, vomiting, visual disturbances, dizziness and/or vertigo, impaired memory and/or concentration.
C. Headache is reported to have developed within 7 days after one of the following:
   1. The injury to the head
   2. Regaining consciousness following the injury to the head
   3. Discontinuation of medication(s) that impair ability to sense or report headache following the injury to the head
D. Either of the following:
   1. Headache has resolved within 3 months after the injury to the head
   2. Headache has not yet resolved but 3 months have not yet passed since the injury to the head
E. Not better accounted for by another ICHD-3 diagnosis

5.2.2 Persistent headache attributed to mild traumatic injury to the head

Diagnostic criteria: as for 5.1.2 except for D.

D. Headache persists for > 3 months after the injury to the head

Neither the severity of the head injury nor the persistence (whether symptoms last shorter than or longer than 3 months) assists in the determination of PTH treatment. Paradoxically, higher headache rates have been found in those with mTBI, not moderate or severe TBI, in both the clinic and prospective inpatient settings where those with TBI have sustained other, primarily orthopedic injuries (Hoffman et al., 2011; Lucas et al., 2013; Yamaguchi, 1992).

Though headache is the most common symptom after TBI, the prevalence has ranged from 30% to 90% in retrospective studies with 18%–22% lasting more than 1 year (Evans, 2004; Keidel and Deiner, 1997; Lew et al., 2006; Linder, 2007). In a large prospective study of 452 subjects admitted to inpatient rehabilitation units after a moderate to severe TBI, the majority of subjects were men injured in vehicle-related accidents with an average age of 44 (Hoffman et al., 2011). Seventy-one percent reported headache during the first year. Prevalence was 46% at the initial inpatient interview and remained high with 44% reporting headache 1 year after TBI. In another study using a similar headache assessment, a group of 212 individuals admitted to the hospital with mTBI and evaluated within 1 week after TBI had a cumulative incidence of new or worse headache compared with preinjury of 91% over the first year (Lucas et al., 2013). Though prior history of headache was significantly related to PTH regardless of TBI severity, female gender as a risk factor for PTH was much stronger in those sustaining a moderate to severe TBI than after mTBI (Hoffman et al., 2011; Lucas et al., 2013).

After head injury, a variety of headache symptoms may develop without specific location, severity, frequency, duration, or associated features such as nausea, vomiting, photophobia, phonophobia, or presence of aura). In addition, the headache may change features from headache to headache or over time after injury. In an effort to provide a framework for PTH treatment decisions, the classification criteria for primary headache disorders have been used in characterizing PTH (Evans, 2004; Lucas et al., 2013). Similar treatment possibilities despite different causation may be based on overlapping neurochemical changes involving the trigeminovascular system found in models of TBI compared with those found in migraine (Packard, 1999; Packard and Ham, 1997). When using the ICHD-2 diagnostic criteria, migraine/probable migraine was the major headache phenotype in 53% of patients immediately after a mild TBI and in 54% at 1 year. Tension-type headache occurred in up to 40% over the year, with cervicogenic headaches occurring in only 4% at each time point over the year (Lucas et al., 2013). These phenotypes are based on subject report of their worst headache, though many reported more than one type of headache. PTH frequency is higher in those with more severe headache pain. In the mTBI study described previously, up to 27% of subjects reported having headache several times per week to daily (Lucas et al., 2013). For those who had the highest headache frequency, 62% of the headache types were migraine at 1 year. Infrequent headache (fewer than one per month) was most likely to be tension-type headache (59%). In contrast, 4%–5% of those with headache in the general population report chronic daily headache (>15 days of headache per month, greater than 4 hours per day) (Scher et al., 1998).

Very few PTH phenotypes in the civilian population are similar to cluster headache (Walker, 2007), hemicrania continua (Lay and Newman, 1999), chronic paroxysmal hemicrania (Matharu and Goadsby, 2001), and short-lasting, unilateral, neuralgiform headache attacks with conjunctival injection and tearing (Piovesan et al., 1996).

13.3 SPECIAL POPULATIONS

Approximately 20% of all civilian TBI occurs in amateur athletic events at or below the college level in the United States (Coronado et al., 2011; Corrigan et al., 2010). These are primarily mTBI and less than 10% require hospitalization (Blume et al., 2011; Seifert and Evans, 2010). After mTBI, about 90% of athletes are symptom-free within one month, but 10%–20% may continue to have symptoms of PCS including headache (Seifert and Evans, 2010; Blume et al., 2011). Multiple episodes of TBI are more likely to have PTH that persists longer than 1 month (Slobounov et al., 2007).

Military combat–related TBI is complicated by extreme physical and psychological conditions of war (Theeler et al., 2013). Many combat-related PTHs develop after blast exposure, though rarely is this an isolated causative mechanism and it may include blast exposure followed by blunt trauma when hitting a vehicle or the ground (Taber et al., 2006, 2010, 2013). In one report, more than 80% of 978 U.S. Army soldiers reporting headaches after return from deployment were exposed to five or more blasts occurring within 60 feet (Theeler et al., 2013). Similar to the civilian population discussed previously, PTH in this setting also may have a delayed onset beyond the requirements of the ICHD. In recently deployed soldiers, almost 40% of PTH began within the first week after the mTBI, but 20% was reported within the first month and approximately 40% after the first month (Theeler et al., 2010).

Also similar to studies in a civilian population is the high prevalence of the migraine phenotype in PTH as well as a high prevalence of chronic headache syndromes in the military. Recent studies in the military and veteran populations found occurrence of the migraine phenotype in 60%–97% of cases depending on the study population and methodology (Erickson 2011; Lew et al., 2006; Ruff et al., 2008; Theeler et al., 2010). Migraines were 5.4 times more likely in those who sustained a concussion compared with those who did not (Armed Forces Health Surveillance Center, 2009). Although migraine is also the predominant headache phenotype in this population, other headache types did occur (Ruff et al., 2008; Theeler and Erickson, 2009). Among 978 U.S. Army soldiers with deployment-related concussion, 20% reported headaches on 15 or more days per month in the preceding 3 months with a median of 27 headache days per month (Theeler et al., 2013). Similarly, among 100 U.S. Army soldiers with chronic PTHs seen in a headache clinic, the average headache frequency was 17 days per month (Erickson, 2011). Migraine features are present in 70% or more of the chronic PTH disorders in the military studies (Ericson, 2011; Theeler et al., 2013).

In a recent emergency room department prospective cohort of children (0 to 18 years of age) with PTH following mTBI, migraine was the most common headache type seen (Kuczynski et al., 2013).

13.4 MANAGEMENT OF PTH

Treatment of PTH depends on many factors, but those with PTH come to medical attention if the headache causes significant pain, disability from pain, or inability to function or concentrate, all of which may lead to significant work loss and decreased social functioning. Fear of the headache and its unpredictability as well as other reasons may bring people to practitioners. Many people will seek care from primary care providers initially or sports medicine specialists if the injury is sports-related. If prolonged treatment and rehabilitation is necessary for associated orthopedic injuries, a physiatrist may be the specialist providing care. There may be many layers of physicians seen before a headache specialist or practitioner who is knowledgeable regarding PTH is found.

The treatment of PTH is largely empiric. To date, no randomized, blinded, prospective, class I medication treatment trials of PTH have been conducted (Watanabe et al., 2012). One approach to making treatment decisions may be made by using primary headache characterization of the PTH as migraine, probable migraine, or tension-type headache. Because the treatment of migraine and probable migraine is the same, these migraines will be discussed together under the term migraine.

Simplified, useful, diagnostic criteria for migraine and tension-type headaches are presented in Table 13.2. Full diagnostic classification criteria are available in the ICHD-3 (The International Classification of Headache Disorders, 2013). The most important differentiating criteria from a treatment viewpoint is that migraine is a moderate to severe headache, whereas tension-type headache is mild-moderate and non-disabling. Migraine may be unilateral (60% of the time), usually pulsatile with throbbing or pounding described, and mild physical activity or postural change may increase pain. Common accompanying features are nausea and/or vomiting, and photophobia, and/or phonophobia.

TABLE 13.2
Differentiation of Common PTH Phenotypes*

Migraine	Tension-Type
Moderate to severe	Mild to moderate
Unilateral (60%)	Bilateral
Throbbing, pounding, pulsatile	Tight, squeezing, vice-like
Worse with physical movement	No worsening with movement
Nausea and/or vomiting	No nausea or vomiting
Photophobia and phonophobia	Photophobia or phonophobia

*This simplified adaptation is from the classification of the primary headache disorders migraine and tension-type headache found in the International Classification of Headache Disorders, third edition. 2013. *Cephalalgia* 33:629–808.

Acute treatment of a PTH is crucial if the headache is severe or has associated disability. The goal of acute headache treatment is to educate those with headache to treat early with effective therapy. Many patients will wait to see if their headache worsens or choose over-the-counter (OTC) products because of cost or lack of access to medical care. If their usual pattern is that of migraine and if simple analgesic agents or other OTC products do not effectively treat their PTH, then migraine-specific therapy is recommended. Two treatment approaches are used for migraine headache. Acute or abortive therapy treats the headache on an as-needed basis as it occurs. Preventive therapy or prophylaxis is used on a daily basis when attack frequency is high or to treat a suboptimal response to acute therapy (Lucas, 2011; Silberstein, 2000).

13.5 ACUTE PTH TREATMENT

The treatment of acute migraine may be divided into specific and nonspecific therapy. Nonspecific treatment, in general, is less effective in evidence-based reviews of migraine but may be effective for some patients, particularly if used early, when pain is still mild and if the head pain and associated symptoms develop slowly (Silberstein, 2000). Many nonspecific medications exist, are easy to obtain, and are primarily OTC. The mechanism of action of these medications is not well understood. Aspirin, acetaminophen or paracetamol, and nonsteroidal anti-inflammatory drugs such as ibuprofen, naproxen, and diclofenac, may be found OTC, along with combination products such as aspirin-acetaminophen-caffeine, or in some countries, combinations of aspirin or paracetamol, caffeine, and small amounts of codeine. These medications may have antiprostaglandin and antiplatelet activity. Caffeine is an analgesic adjuvant that has vasoconstrictive properties and possibly acts as an adenosine inhibitor. In many cases, the recommended dosing on the label "directions" may be subtherapeutic. For example, the directions for use in brand formulations (such as ALEVE; Bayer Healthcare, Wayne, NJ) and generic formulations of naproxen sodium state to take one tablet (220 mg) two times daily. However, many clinical practitioners empirically recommend approximately 500–550 mg of naproxen sodium per dose for headache treatment. Individual variability in an effective dose is high. Nonsteroidal anti-inflammatory drugs and aspirin can cause gastritis, gastrointestinal bleeding, increased bleeding time, and peptic ulcer disease. These medicines should not be used during the first 24 hours after a head injury to limit the possibility of bleeding. These medications also should not be used if a person with a head injury may require surgery because of the possibility of increasing the bleeding time and causing surgical complications.

Opioid analgesic drugs should be avoided in the treatment of headache if at all possible. These drugs have low effectiveness for migraine phenotypes, the risk of dependency and medication overuse is high, and the chronic use of opioids is not recommended (Ramadan et al., 1999). Opioids may complicate cognitive function, and their sedating effect may be increased in people with TBI.

Triptans, the ergotamines, and dihydroergotamine (DHE) represent a class of treatment medications that are migraine-specific. Triptans (Table 13.3) are serotonin-1B/D agonists that bind to serotonin-1D receptors in meningeal trigeminal afferents to inhibit release of inflammatory peptides, such as calcitonin gene-related peptide, possibly mediators of neurogenic inflammation. They also bind to 1B receptors in middle meningeal artery endothelium and constrict vessels dilated by calcitonin gene-related peptide. Because of their vasoconstrictive potential, they are contraindicated in those with central, coronary, or peripheral vascular disease or in those with significant risk factors for vascular disease. The use of ergotamine has declined since the triptan era, primarily because of poor absorption of the ergots and frequent nausea after taking the oral form. DHE, although highly effective, currently is available only in a nasal spray with inconsistent intranasal absorption and in an injectable form that is inconvenient to use (although self-injection technique may be easily taught in a clinic setting). A convenient, inhaled form of DHE may be available soon and Food and Drug Administration approval is pending. These medications also act at serotonin-1B/D sites but bind at many other sites as well, including other serotonin receptors and adrenergic and dopaminergic sites. They too are vasoconstrictors and have the same contraindication in persons with vascular disease.

Evidence-based guidelines for the treatment of migraine (Silberstein, 2000) have recommended the following for the most optimal treatment response:

1. The goal of treatment is to restore ability to function, and treatment should be stratified based on attack severity and disability; an attempt should be made to match the efficacy of the initial headache therapy to treatment need. For example, if a headache is always severe or escalates rapidly, an OTC product may not be wise, even as an initial medication.
2. Treat a headache attack as soon as possible after onset and if the headache recurs, then retreat with another dose; the triptans, though highly effective with few side effects, are associated with "recurrence." The headache may, in effect, outlast the drug.
3. Minimize the use of backup and rescue medications by making sure that the initial drug is effective. Medication overuse is less likely to happen by using effective medications, avoiding the need for continual retreatment.
4. Optimize self-care by patient education and an effective treatment plan with a patient "toolbox" to avoid urgent care visits or a lapse from physician care because of frustration. The toolbox may be a list of medications to use and how and when to use them. Many patients need written instructions particularly during a headache, for example, an OTC may be effective for slow-onset, mild to moderate

TABLE 13.3
Specific Medications for Acute Migraine Treatment

Generic Name	Brand Name	Formulation and Dose, mg
Triptans		
Almotriptan	Axert (Ortho-McNeil, Janssen, Titusville, NJ)	Tablet 6.25, 12.5
Eletriptan	Relpax (Pfizer, New York, NY)	Tablet 20, 40 (80 outside the United States)
Frovatriptan	Frova (Endo, Newark, DE)	Tablet 2.5
Naratriptan	Amerge (GlaxoSmithKline, Middlesex, UK)	Tablet 1, 2.5
Rizatriptan	Maxalt (Merck, Whitehouse Station, NJ)	Tablet 5, 10 ODT 5, 10
Sumatriptan	Imitrex (in the United States), Imigran (GlaxoSmithKline, Middlesex, UK) Sumavel (Zogenix, San Diego, CA)	Tablet 25, 50, 100 Nasal spray, 5, 20 Subcutaneous 4, 6 Needleless injection 6
Sumatriptan/naproxen	Treximet (GlaxoSmithKline, Middlesex, UK)	Tablet 85/500
Zolmitriptan	Zomig (Astra-Zeneca, London, UK)	Tablet 2.5 5 ZMT 2.5, 5 Nasal spray 5
Dihydroergotamine (DHE)		
Dihydroergotamine mesylate injection	DHE-45 (Bedford Labs, Bedford, OH)	1 mg/mL 1-mL ampules
Dihydroergotamine mesylate spray	Migranal (Valeant, Montreal, Canada)	4 mg/mL

ODT, orally disintegrating tablet; ZMT, zolmitriptan melting tablet.

headaches, but perhaps an injectable formulation is more appropriate for someone who wakes up with a severe headache with nausea.

5. Avoid or minimize adverse effects by choosing medication with good tolerability as initial therapy. However, even medication with good tolerability may have long-term adverse consequences, so for example, acetaminophen may be safe but if used in frequent, high doses, can affect liver function.

13.6 PREVENTIVE PTH TREATMENT

Preventive treatment is headache treatment that is used daily when attack frequency is high. The goal of using preventive treatment is to reduce attack severity, frequency, and duration and improve the response to acute therapy. It can be used when acute therapy is ineffective or contraindicated and if escalating use of acute therapies causes concern for medication overuse headache (Ramadan et al., 1999) Migraine prevention may be pharmacologic or nonpharmacologic. Though daily preventive medication may be considered when headache frequency is more than 4 to 6 days per month, this is arbitrary and may be influenced by how disabling the headaches are or their impact on quality of life. Nonpharmacologic therapies have been shown to be effective in the treatment of primary headache disorders, so cognitive behavioral therapy, biofeedback, and relaxation therapy may be effective in treatment of PTH as well (Olesen and Ramadan, 2006).

When frequent headaches necessitate frequent and usually escalating use of acute medication, this may result in a slow but inexorable transformation of an episodic headache pattern to daily headache. Medication overuse headache (MOH), or rebound headache, can occur when patients use acute medication more than 2 to 3 days per week or approximately 10 days per month, on average. Most MOH results in a daily or near-daily headache frequency (>15 days per month, >4 hours per day). The key to recognition of this phenomenon is that a headache is either there or may occur shortly after awakening, and may have features of tension-type headache with intermittent migraine. The overused medication is less effective over time and the duration of effect is shortened, which necessitates more frequent dosing. Many people will tell providers that the overused medication is the only medication that helps them. Rebound has extensive consequences besides daily headache because it may prevent the effectiveness of migraine-specific therapies and any attempt at preventive therapy. Any acute therapy has a potential to induce MOH, although products that contain opioids, barbiturates, or caffeine appear to be particularly risky. No acute medication is without rebound potential. In many persons, it is difficult to identify a single medication that causes rebound, because the majority of patients may take more than one compound at a time, and each component in a combination analgesic can induce headaches. It is difficult to estimate prevalence of MOH in the general population, because many persons may use OTC products and not come to medical attention. However, in a survey of family physicians, MOH was found to be the third most common cause of headache, and in some specialty headache clinics in North America, medication overuse is present in up to 70% of cases seen in the clinic (Rapoport et al., 1996).

Although there is evidence for, and widespread use of, many preventive medications, the only medications approved by the Food and Drug Administration for use in migraine are propranolol, timolol, valproic acid, and topiramate (methysergide is no longer available in the United States). The most commonly used preventives are the tricyclic antidepressants (TCAs), beta-blockers, calcium-channel blockers, and antiepilepsy medication. Less often used are the selective serotonin reuptake inhibitors and selective norepinephrine reuptake inhibitors (SNRIs). Many patients find herbal, vitamin, and mineral supplements useful. Common medications are listed in Table 13.4; however, it is recommended that consensus guidelines be reviewed (Olesen and Ramadan, 2006).

The mechanism of action of the preventive headache therapies is unknown, but it is likely that all preventives have multiple mechanisms of action. Some likely hypotheses based largely on animal models include inhibition of cortical spreading depression, inhibition of glutamate-dependent mechanisms, and modulation of serotonergic, dopaminergic, and adrenergic pathways and receptors. Choosing a preventive from a long list of medications can appear overwhelming. However, comorbid or cooccurring conditions are frequent in those with PTH or primary headache disorders.

One approach to treatment choice is to choose a medication that can treat both conditions. For example, if a patient with PTH has depression or anxiety, then a selective serotonin reuptake inhibitor or SNRI, perhaps in combination with a low-dose TCA, may be helpful. A sleep disorder may make amitriptyline a good choice. Mood stabilization in those with PTH may be achieved with use of valproic acid or gabapentin. Significant myofascial or somatic pain may warrant treatment with a TCA, tizanidine, or an SNRI. Onabotulinum toxin A (Botox; Allergan Inc, Irvine, CA) is approved for treatment of chronic migraine and may be useful to treat chronic PTH, particularly with associated neck and shoulder pain and spasm. Although a thorough review of potential adverse effects is beyond the scope of this discussion, in general, if cardiac issues are suspected, then a discussion with a patient's primary care physician may be warranted when starting certain medications, such as the TCAs, because of possibility of prolonged QT interval. Concentration or attention difficulty may respond best to cognitive behavioral therapy. Caution must always be used in people who have had an mTBI when medications with significant potential for cognitive adverse effects are used, such as topiramate, because head injury may worsen this adverse effect. After mTBI, children and adolescents may experience different symptom trajectory and recovery time. Optimal management of PTH is discussed in a recent review of concussion symptoms in youths (Blume et al., 2011). Behavioral management is emphasized in this younger population, such as physical and cognitive rest, good hydration, sufficient

TABLE 13.4
Preventive Medications Used in Migraine Treatment

Category	Class	Name	Dose Range, mg	Comments
Antidepressant	TCA	Amitriptyline*	10–100	Sedation, dry mouth, urinary retention
	SSRI	Fluoxetine**	10–60	Weight gain, sexual dysfunction
	SNRI	Duloxetine	30–120	Weight gain, sexual dysfunction, sweating
Antiepilepsy		Valproic acid	500–1500	Weight gain, hair loss, stomach upset
		Topiramate	75–100	Tingling, cognitive dysfunction, weight loss
		Zonisamide	100–400	Weight loss, altered taste
		Gabapentin	900–3600	Sedation, dizziness, poor absorption
		Levetiracetam	500–1000 (divided)	Sedation, dizziness
Cardiovascular	Beta-blockers	Propranolol	40–240	Fatigue, lightheadedness
		Metoprolol	25–100	Fatigue, lightheadedness
		Nadolol	40–120	Fatigue, lightheadedness
	Calcium-channel blockers	Verapamil	120–360	Constipation, ankle edema
		Amlodipine	5–10	Ankle edema
		Flunarizine	10–20	Drowsiness, weight gain
Other		Onabotulinum toxin A	100–155 u	Pain at injection site, muscle weakness
		Cyproheptadine	4–12	Sedation, weight gain
		Magnesium	600 (divided)	Poorly absorbed, laxative effect
		Tizanidine	4–16	Sedation
		Butterbur	100 (divided)	Fatigue, diarrhea

SSRI, selective serotonin reuptake inhibitor.
*Other TCAs such as nortriptyline or protriptyline may be used.
**Other SSRIs such as citalopram or escitalopram may be used.

sleep and regular meals, and stress avoidance; however, acute and/or preventive medication may be required for severe, prolonged, or disabling headache.

For successful preventive treatment and adherence to the management recommendations, the following guidelines are suggested (Ramadan et al., 1999):

1. Begin a preventive medication at a low dose and increase slowly to avoid early dose-dependent adverse events.
2. If the medication is tolerated, then an adequate treatment trial at an effective dose is usually reached with 6 to 8 weeks.
3. Before initiating therapy with a new medication, a thorough review of systems and review of current medications is necessary to avoid drug interactions and contraindications. For example, caution is needed with medications that are metabolized through similar P450 pathways such as combined oral contraceptives and some antiepilepsy medications.
4. Contraception needs to be discussed in an at-risk population because some medications may be associated with birth defects in the fetus if used during pregnancy, such as valproic acid or topiramate.
5. Evaluate potential adverse effects of preventive medication based on lifestyle (for example, avoidance of beta-blockers in athletes).
6. In the setting of comorbid conditions and PTH, choose a medication that may treat both the headache and comorbidity.

13.7 CONCLUSION

PTH is the most frequent physical symptom reported after TBI. PTH is frequent in both military and civilian populations, with recent data from civilian adult, pediatric, and military populations reporting that PTH may be more of a chronic problem than previously thought. Prevalence can be close to half of the injured population. In addition, if ICHD-3 criteria are used to define PTH and strictly adhered to, some PTH may be missed, which underestimates the scope of the problem. New headaches have been reported well after the 7 days required for diagnosis of PTH by ICHD-3 guidelines, but this has been noted in guideline comments, and allows for consideration that latency may be variable, as recent research has shown. A history of headache before head injury is significantly related to occurrence of PTH, which underscores the importance of assessment of preinjury primary headache disorders in this population.

Because of the lack of a consensus approach to PTH treatment, many clinicians are characterizing PTH using primary headache disorder definitions. Reports of successful treatment of PTH with sumatriptan or dihydroergotamine after concussion or mTBI support the use of migraine specific therapies in persons with acute PTH (Erickson, 2011; Gawel et al.,

1993; McBeath and Nanda, 1994). Preventive therapy should be initiated for frequent or refractory headaches, with the choice of preventive based on relevant comorbid conditions.

Future research is needed to evaluate the clinical characteristics of PTH and determine which persons will be at highest risk for development of PTH. Of particular concern is PTH that becomes chronic daily headache. The high incidence, prevalence, and frequency of PTH begs for an evidence-based approach to treatment options for those who sustain brain injury. Double-blind placebo-controlled trials in the PTH population are necessary to see whether similar phenotypes in the primary and secondary headache disorders will respond similarly to treatment. Until blinded treatment trials are completed, it is suggested that, when possible, PTH be treated as one would treat a primary headache disorder of similar phenotype.

REFERENCES

Armed Forces Health Surveillance Center (AFHSC). 2009. Risk factors for migraine after OEF/OIF deployment, active component, U.S. Armed Forces. *Med Surveill Mon Rep.* 12:10–13.

Blinman, T.A., E. Houseknecht, C. Snyder, D.J. Wiebe, and M.L. Nance. 2009. Postconcussive symptoms in hospitalized pediatric patients after mild traumatic brain injury. *J Pediatr Surg.* 44:1223–1228.

Blume, H.K., S. Lucas, and K.R. Bell. 2011. Subacute concussion-related symptoms in youth. *Phys Med Rehabil Clin N Am.* 22:665–681.

Collins, M.S., M.R. Lovell, G.L. Iverson, R.C. Cantu, J.C. Maroon, and M. Field. 2002. Cumulative effects of concussion in high school athletes. *Neurosurgery.* 51:1175–1181.

Coronado, V.G., L. Xu, S.V. Basavaraju, L.C. McGuire, M.M. Wald, M.D. Faul et al. 2011. Surveillance for traumatic brain injury-related deaths-United States, 1997–2007. *MMWR Surveill Summ.* 60:1–32.

Corrigan, J.D., A.W. Selassie, and J.A. Orman. 2010. The epidemiology of traumatic brain injury. *J Head Trauma Rehabil.* 25:72–80.

Dikmen, S., J. Machamer, J.R. Fann, and N.R. Temkin. 2010. Rates of symptom reported following traumatic brain injury. *J Int Neuropsych Soc.* 16:401–411.

Erickson, J.C. 2011. Treatment outcomes of chronic post-traumatic headaches after mild head trauma in US soldiers: An observational study. *Headache.* 51:932–944.

Eskridge, S.L., C.A. Macera, M.R. Galarneau, T.L. Holbrook, A.J. MacGregor, D.J. Morton et al. 2012. Injuries from combat explosions in Iraq: Injury type, location, and severity. *Injury.* 43:1678–1682.

Evans, R.W. 2004. Post-traumatic headaches. *Neurol Clin.* 22:237–249.

Faul, M., L. Xu, M. Wald,and V.G. Coronado. 2010. Traumatic brain injury in the Unites States: Emergency department visits, hospitalizations and deaths 2002–2006. US Department of Health and Human Services Centers for Disease Control and Prevention. Available at http://www.cdc.gov/traumaticbraininjury/pdf/blue_book.pdf.

Fischer, H. 2013. US military casualty statistics: Operation New Dawn, Operation Iraqi Freedom, and Operation Enduring Freedom. Report ID #-RS22452. Congressional Research Service. Available at: http://www.fas.org/sgp/crs/natsec/RS22452.pdf.

Gawel, M.J., R. Rothbart, and H. Jacobs. 1993. Subcutaneous sumatriptan in the treatment of acute episodes of posttraumatic headache. *Headache* 33:96–97.

Hoffman, J.M., S. Lucas, S. Dikmen, C.A. Braden, A.W. Brwon, R. Brunner et al. 2011. Natural history of headache after traumatic brain injury. *J Neurotrauma.* 28:1–8.

Hoge, C.W., D. McGurk, J.L. Thomas, A.L. Cox, C.C. Engel, and C.A. Castro. 2008. Mild traumatic brain injury in U.S. soldiers returning from Iraq. *N Engl J Med.* 358:453–463.

Keidel, M. and H.C. Deiner. 1997. Post-traumatic headache. *Nervenarzt.* 68;769–777

Kuczynski, A., S. Crawford, L. Bodell, D. Dewey, and K.M. Barlow. 2013. Characteristics of post-traumatic headaches in children following mild traumatic brain injury and their response to treatment: A prospective cohort. *Dev Med Child Neurol.* epub http://dx.doi.org/10.1111/dmcn.12145.

Lay, C.L. and L.C. Newman. 1999. Posttraumatic hemicrania continua. *Headache.* 39:275–279.

Lew, H.L., P-H. Lin, J-L Fuh, S.J. Wang, D.J. Clark, and W.C. Walker. 2006. Characteristics and treatment of headache after traumatic brain injury: A focused review. *Am J Phys Med Rehabil.* 85:619–627.

Linder, S.L. 2007. Post-traumatic headache. *Curr Pain Headache Rep.* 11:396–400.

Lucas, S. 2011. Headache management in concussion and mild traumatic brain injury. *PM R.* 3:S406-S412.

Lucas, S., J.M Hoffman, K.R. Bell, W. Walker, and S. Dikmen. 2012. Characterization of headache after traumatic brain injury. *Cephalalgia* 32:600–606.

Lucas, S., J.M. Hoffman, K.R. Bell, and S. Dikmen. 2013. A prospective study of prevalence and characterization of headache following mild traumatic brain injury. *Cephalalgia* 0:1–10.

Matharu, M.J. and P.J. Goadsby. 2001. Post-traumatic chronic paroxysmal hemicrania (CPH) with aura. *Neurology* 56:273–275.

McBeath, J.G. and A. Nanda. 1994. Use of dihydroergotamine in patients with postconcussive syndrome. *Headache* 34: 148–151.

Olesen, J. and N.M. Ramadan. 2006. Migraine mechanisms. In: J. Olesen, P.J. Goadsby, N.M. Ramadan, P. Tfelt-Hansen, and K.M.A. Welch, editors. *The Headaches.* 3rd edition. Lippincott, Williams & Wilkins, Philadelphia, PA. pp. 251–393.

Packard, R.C. and L.P. Ham. 1997. Pathogenesis of posttraumatic headache and migraine: A common headache pathway? *Headache.* 37:142–152.

Packard, R.C. 1999. Epidemiology and pathogenesis of posttraumatic headache. *J Head Trauma Rehabil.* 14:9–21.

Piovesan, E.J., P.A. Kowacs, and L.C. Werneck. 1996. S.U.N.C.T. syndrome: Report of a case preceded by ocular trauma. *Arq Neuropsiquiatr.* 54:494–497.

Ramadan, N., S.D. Silberstein, F.G. Freitag, et al. 1999. Evidence-based guidelines for migraine headache in the primary care setting: Pharmacological management for prevention of migraine. *Neurology*. Available at http://www.neurology.org

Rapoport, A., P. Stand, D.L. Gutterman, R. Cady, H. Markley, R. Weeks et al. 1996. Analgesic rebound headache in clinical practice: Data from a physician survey. *Headache* 36:14–19.

Ruff, R.L., S.S. Ruff, and X.F. Wang. 2008. Headaches among Operation Iraqi Freedom/Operation Enduring Freedom veterans with mild traumatic brain injury associated with exposure to explosions. *J Rehabil Res Dev.* 45:941–952.

Scher, A.I., W.F. Stewart, J. Liberman, and R.B. Lipton. 1998. Prevalence of frequent headache in a population sample. *Headache* 38:497–506.

Seifert, T.D. and R.W. Evans. 2010. Posttraumatic headache: A review. *Curr Pain Headache Rep.* 14:292–298.

Silberstein, S.D. 2000. Practice parameter: Evidence-based guidelines for migraine headache (an evidence based review). Report of the Quality Standards Subcommittee of the American Academy of Neurology. *Neurology.* 55:754–762.

Slobounov, S., E. Slobounov, W. Sebastianelli, C. Cao, and K. Newell. 2007. Differential rate of recovery in athletes after first and second concussion episodes. *Neurosurgery.* 61:338–344.

Taber, K.H., D.L. Warden, and R.A. Hurley. 2006. Blast-related traumatic brain injury: What is known? *J Neuropsychiatry Clin Neurosci.* 18:141–145.

The International Classification of Headache Disorders: Third Edition-beta. 2013. *Cephalalgia.* 33:629–808.

Theeler, B.J. and J. Erickson. 2009. Mild head trauma and chronic headaches in returning US soldiers. *Headache.* 49:529–534.

Theeler, B.J., K. Kenney, O.A. Prokhorenko, U.S. Fideli, W. Campbell, and J.C. Erickson. 2010. Headache triggers in the US military. *Headache.* 50:790–794.

Theeler, B.J., F.G. Flynn, and J.C. Erickson. 2013. Chronic daily headache in U.S. soldiers after concussion. *Headache.* 52:732–738.

Theeler, B.J., F.G. Flynn, and J.C. Erickson. 2010. Headaches after concussion in US soldiers returning from Iraq or Afghanistan. *Headache.* 50:1262–1272.

Theeler, B., S. Lucas, R.G. Reichers, and R.L. Ruff. 2013. Post-traumatic headaches in civilians and military personnel: A comparative, clinical review. *Headache.* 53:881–900.

Walker, R.W. 2007. Cluster headaches and head trauma: Is there an association? *Curr Pain Headache Rep.* 11:137–140

Watanabe, T.K., K.R. Bell, W.C. Walker, and K.S. Schomer. 2012. Systematic review of interventions for post-traumatic headache. *PM R.* 4:129–140.

Yamaguchi, M. 1992. Incidence of headache and severity of head injury. *Headache.* 32:427–431.

14 Traumatic Brain Injury (TBI)-Induced Spasticity

Neurobiology, Treatment, and Rehabilitation

Prodip Bose, Jiamei Hou, and Floyd J. Thompson

CONTENTS

14.1 INTRODUCTION

Traumatic brain injury (TBI) impacts the lives of 1.5 to 2 million new individuals each year; 75,000 to 100,000 of these are classified as severe, and will suffer enduring severe spasticity in addition to cognitive, vestibulomotor (balance), and other motor impairments. Following TBI, the onset of spasticity and associated orthopedic sequellae is rapid, beginning as early as one week following injury. The progressively developing spasticity and other disabilities often represent the most significant barriers for practical re-entry of TBI patients into the community. The lack of sufficient data regarding the neurobiology of TBI-induced spasticity and safety, feasibility and efficacy of early intervention therapy direct the current treatment guidelines to a conservative level. This chapter focuses on several quantitative physiological measures of spasticity, some recent findings regarding a neurobiological basis of spasticity, and finally, a section describing present treatments and the experimental treatments and rehabilitation of TBI-induced spasticity.

Clinically spasticity has been defined as an increased velocity-dependent lengthening resistance of skeletal muscles to passive movement. It is a secondary neurological condition induced by neurological hyperreflexia associated with TBI and spinal cord injury (SCI), stroke, multiple sclerosis (MS), cerebral palsy, amyotrophic lateral sclerosis (ALS), and few other disorders (e.g., anoxic brain damage; some metabolic disorders, such as adrenoleukodystrophy, phenylketonuria). Spasticity is often one of the most troublesome components of upper motor neuron injury (Katz and Rymer, 1989; Ordia et al., 1996) that greatly complicates daily living in individuals with these disorders. Its hallmark feature is altered skeletal muscle tone and spasm, and it is aptly named "spasticity." The word spasm comes from the Greek word "σπασμός" (spasmos), meaning "drawing, pulling." Spasticity symptoms include increased muscle tone (hypertonicity), muscle spasms, increased deep tendon reflexes, clonus, scissoring, and fixed joints. The degree of spasticity varies from mild muscle stiffness to painful, severe uncontrollable muscle spasms. In addition to spasticity symptoms, muscles affected in this way have many other potential features of altered performance, including muscle weakness, decreased movement control, and decreased endurance.

Spasticity is associated with hyperreflexia of the muscle stretch reflexes (Ashby and Verrier, 1980; Bose et al., 2002b; Herman, 1968; Machta and Kuhn, 1948; Thilmann et al., 1991; Toft et al., 1993) that induce velocity-sensitive increased resistance of skeletal muscle lengthening. This dynamic stiffness differentiates spasticity from the changes in passive muscle properties, which are not velocity-sensitive and often seen in patients with spasticity. Although some insights have been made regarding the fundamental neurobiology of spasticity, many aspects of the specific pathophysiology still remain unclear. Therefore, experimental animal models of spasticity have been developed to increase scientific understanding of this clinically troublesome condition. The authors' previous research works have provided evidence of neurophysiological changes in the tibial monosynaptic reflexes that use the neural pathways that subserve hindlimb muscle stretch reflexes (Thompson et al., 1992, 1993, 1998, 1999). These alterations included significant changes in rate-dependent processes that regulate sensory transmission to motoneurons (Thompson et al., 1992, 1998). More recent animal studies have shown that these physiological changes in the muscle stretch reflexes were accompanied by progressive and enduring spastic hypertonia (Bose et al., 2002b, 2012, 2013; Hou et al., 2014). These changes were severe in magnitude and highly relevant to features observed clinically in humans (Schindler-Ivens and Shields, 2000). Human studies have reported that neurophysiological changes in rate-dependent processes that regulate reflex excitability of the stretch reflex pathways also accompany spasticity (Boorman et al., 1992; Brown, 1994; Calancie et al., 1993; Lance, 1981; Nielsen et al., 1993; Schindler-Ivens and Shields, 2000; Thompson et al., 2001a, 2001b).

Rodent spasticity models for SCIs (Bose et al., 2002b, 2012; Hou et al., 2014), TBIs (Bose et al., 2013), experimental allergic encephalitis (EAE) Lewis rat model of MS (Bose et al., 2009), and a rodent white matter stroke model (Thompson et al., 2013) have been developed and studied in the authors' laboratory to increase our scientific understanding about this condition and to find ways to prevent, treat, and diminish disabilities of this condition. There are some distinct differences in the pattern, time course of development, and severity of spasticity that is induced by these different injuries or disease processes. These different patterns of spasticity are presumed to be derived according to the manner in which the central nervous system (CNS) trauma or disease has induced alterations in supraspinal drive, their substrate systems (e.g., serotonergic and noradrenergic innervation of the motoneurons), and associated secondary changes at the cellular level in the spinal cord (e.g., both motor neuron and interneurons) below the lesion. Accordingly, an increased understanding of the mechanisms responsible for these injury/disease-induced neuroplastic changes at supraspinal and spinal levels may be of great importance in relation to the design of the most effective treatment and rehabilitation of spasticity.

14.2 MEASUREMENT OF SPASTICITY

14.2.1 Velocity-Dependent Ankle Torque and Electromyogram: A Measure of Spasticity

Although the clinical symptoms typically categorized as spasticity present as multiple clinical phenomena (Roy and Edgerton, 2012) derived from complex disease processes (Young, 1992), the dynamic features of velocity-dependent exaggeration of lengthening resistance to the skeletal muscle remains a useful defining criteria (Lance, 1981; Nielsen et al., 2007). Because spasticity diminishes the coordination of voluntary movement and gait, it poses significant limitations to motor recovery after CNS injury or disease. Recent studies used a velocity-dependent ankle torque protocol to chart the time course of spasticity development induced by SCI and TBI, and multiple CNS lesions of the EAE MS model (Bose et al., 2002b, 2009, 2012, 2013; Hou et al., 2014). These studies revealed differences in the pattern and severity of spastic hypertonia that was specific to the type of CNS injury or disorder. For example, mid-thoracic spinal cord contusion injury revealed an initial transient pattern of "tonic" spasticity (i.e., a significantly increased resting tone–induced stiffness that was further increased by muscle lengthening velocity). This tonic pattern progressed to a dynamic pattern of velocity-dependent spasticity. This dynamic pattern revealed elevated electromyogram (EMG) amplitudes and associated increased ankle torques only when tested at ankle rotation velocities that were above the threshold velocity of the stretch reflex (in rat, it is 272° per second). This dynamic patterned spasticity was progressive in onset, and once developed, became permanent. The ankle torques recorded at the lower test velocities (49–272° per second) (see Bose et al., 2002b) were not significantly greater than observed in normal controls, nor were these correlated with synchronized EMG activity in the ankle extensor muscles. These low-velocity ankle torques were, therefore, interpreted to be contributed by the passive properties of the muscle and joint tissues.

By contrast, test rotations at the upper test velocities revealed increased stiffness of ankle rotation that was time-locked to the stretch-evoked EMGs recorded from the ankle extensor muscles, indicating resistance contributions from activated ankle extensor muscle stretch reflexes (see also Bose et al., 2002b). In contrast, cervical spinal cord contusion injuries induced a severe pattern of tonic and dynamic spasticity (i.e., a large resting tone with superimposed velocity-dependent spasticity). The elevated EMG amplitudes and ankle torques appeared at the lowest (tonic) range of ankle rotation velocities and were enhanced as a function of velocity throughout the upper (dynamic) range (Bose et al., 2009; Hou et al., 2014). Tonic and superimposed dynamic patterns of spasticity and proportionally elevated EMG amplitudes have been recorded after TBI (Bose et al., 2013), EAE (Bose et al., 2009), and rodent white matter stroke model (Thompson et al., 2013). Because these velocity-dependent changes were quantifiable, robust in magnitude, and of enduring character, they provide a useful and clinically relevant method of measuring spasticity relative to injury, and provide a quantitative outcome measure to assess the influence of experimental treatments on this condition.

14.2.2 Rate-Dependent Inhibition/Depression

During the course of reflex testing, reflex test protocols were developed that revealed quantifiable features of fundamental inhibitory processes that regulate sensory transmission to hindlimb muscle stretch reflexes; this process was reported to be significantly decreased after contusion SCI (Bose et al., 2012; Thompson et al., 1992, 1993, 1998, 2001a). These studies revealed that spasticity developed over a time course that was mirrored by the loss of rate-dependent depression in the reflex pathways that serve the spastic muscles in rats (Bose et al., 2002b; Thompson et al., 1992, 1993, 1998, 2001a) and in humans (Calancie et al., 1993; Ishikawa et al., 1966; Schindler-Ivens and Shields, 2000). It has been proposed that the progressive loss of presynaptic inhibition contributes significantly to the progressive development of hyperreflexia of stretch reflex pathways. Rate-depression is one of three fundamental processes that controls reflex magnitude elicited by repetitive input; facilitation and potentiation are the other two (Mendell, 1984). Rate-sensitive depression has been tested in animal models of SCI to quantify changes in spinal cord reflex excitability and to measure plasticity by experimental interventions.

Reflex repetition at 1.0 Hz elicited by electrical stimulation of the dorsal roots of the L_5 and L_6 lumbar segments (which included Ia's from ankle extensor muscles) in the intact rodent produced a 35% decrease in the reflex amplitude (relative to the 0.3-Hz control). Further increases in frequency were accompanied by additional marked reductions in reflex amplitude (Skinner, 1998; Skinner et al., 1996; Thompson et al., 1992, 1993, 1998). After spinal cord contusion in the adult rat, a progressive reduction in rate-sensitive depression was observed that was permanent once established (Thompson et al., 1992, 1998). Recordings in rodents at 2 months after T_8 contusion injury exhibited significantly less attenuation at each of the test frequencies from 1.0 through 20 Hz (see detail, Thompson et al., 1992, 1998). Rate-sensitive depression has also been tested in humans with CNS lesions (Calancie et al., 1993; Ishikawa et al., 1966; Nielsen and Sinkjaer, 1997; Schindler-Ivens and Shields, 2000; Thompson et al., 2001b; Trimble et al., 1998). This index has distinguished between control subjects, subjects with acute SCI, and subjects with chronic SCI (Calancie et al., 1993; Ishikawa et al., 1966; Schindler-Ivens and Shields, 2000; Trimble et al., 1998). Reduced rate-sensitive depression function after SCI is attributed, in part, to a decrease in presynaptic inhibition on Ia afferents. This reduction in presynaptic inhibition may be due to the injury associated loss of descending paths and alterations in segmental influences converging on inhibitory interneurons that regulate the probability of transmitter release from the Ia afferents (Cardona and Rudomin, 1983; Delwaide, 1973; Nielsen et al., 1995). If depression is reduced, a larger portion of repetitive afferent signals reach spinal reflex paths and contribute to exaggerated levels of reflex activation (Nielsen et al., 1995; Thompson et al., 1993). Accordingly, analysis of rate-depression has provided a sensitive, quantitative probe of segmental reflex excitability of spinal reflex pathways used for locomotion.

Spastic hyperexcitability has been reported to develop over several months after the primary lesion and includes progressive adaptation in the spinal neuronal circuitries caudal to the lesion (Bose et al., 2002; Nielsen et al., 2007; Roy and Edgerton, 2012; Thompson et al., 1998; Thompson et al., 2001a; Thompson et al., 1993). Accordingly, physiological measures of changes in cellular properties have been used to explore mechanisms underlying spastic hyperreflexia. Recently, studies of dendritic potentials called "plateau potentials" have revealed new insights into changes in cellular properties that contribute to hyperreflexia (Bennett et al., 2001), although the relevance of the changes in plateau potentials to human spasticity has not been described clearly because of the difficulty in demonstrating the existence of such intrinsic membrane properties in the intact organism (Nielsen et al., 2007). Plateau potentials are depolarizing potentials recorded in spinal cord motoneurons produced by dendritic persistent inward currents (PICs). Plateau potentials have also been reported in recordings from cortical and hippocampal pyramidal neurons. Spinal cord motoneurons' PICs are normally regulated by descending monoaminergic and reticulospinal pathways, especially through 5-HT and norepinephrine projections. These descending projections modulate the activity of dendritic L-type voltage-dependent calcium channels that elicit the sustained, positive, inward currents that produce the long-lasting depolarization. Once long-lasting depolarization is achieved, the cell fires action potentials independent of synaptic input. There are several reports that argued about a possible contribution of plateau potentials to the development of spasticity in human subjects (Gorassini et al., 2004; Nickolls et al., 2004). The contribution of plateau potentials to the clinical manifestations of the spasticity is based on the fact that plateau potentials were activated during the spasm and appeared to contribute to the occurrence of the spasms in SCI patients (Gorassini et al., 2004)

Moreover, measures of disynaptic reciprocal inhibition, have been suggested as an index of changes in inhibitory processes related to spasticity (Crone et al., 2003). However, the intersubject variability of these mechanisms, and the lack of objective quantitative measures of spasticity has impeded disclosure of a clear causal relationship between the alterations in the inhibitory mechanisms and the stretch reflex hyperexcitability (Nielsen et al., 2007). Therefore, the search continues for physiological measures that can quantitate longitudinal changes in inhibitory mechanism correlated with the development of altered reflex excitability.

14.3 PATHOPHYSIOLOGY OF SPASTICITY

Although some insights have appeared in our understanding of the complex neurobiology of spasticity, the etiology of spasticity remains largely unknown. Because spasticity is a multidimensional and dynamic syndrome (Roy and Edgerton, 2012), there are multiple possible mechanisms for the development of spasticity. Studies in animals and humans suggested that alterations in the biochemical and morphological properties of spinal motor neurons after CNS injury or disease lead to cellular and biochemical dysfunctions as a basis for the development of spasticity. Therefore, multiple processes that contribute to spasticity have been identified and include (1) pre- and postsynaptic changes (loss of presynaptic inhibition of Ia afferents) (Calancie et al., 1993; Faist et al., 1994; Nielsen et al., 1995; Schindler-Ivens and Shields, 2000), alterations in rate-dependent processes that regulate sensory transmission to motoneurons (Bose et al., 2012; Thompson et al., 1992, 1993, 1998), and increase in postsynaptic receptor excitability and upregulation of postsynaptic receptors (Little et al., 1999); (2) changes in neural network (decrease in postactivation depression of the H-reflex) (Nielsen et al., 1995), augmented synaptic inputs and terminal sprouting (Jankowska and Hammar, 2002; Little et al., 1999), and decrease in reciprocal Ia inhibition (Crone et al., 1994, 2003); (3) changes in the motor neurons (alteration in the intrinsic properties of spinal cord motoneurons)

(Bennett et al., 2001; Li et al., 2004), and gap junctions between motoneurons (Yates et al., 2008, 2009), increased excitability of motoneurons (Cope et al., 1986; Heckmann et al., 2005), increased excitability of motoneurons from a decrease in the number of dendritic branches after SCI (Bose et al., 2005; Kitzman, 2005); and (4) alteration in spinal neurotransmission (an increase in excitatory and a decrease in inhibitory spinal neurotransmission) (Shapiro, 1997) mechanisms play an important role in the development of spasticity. Changes in neural networks and their cellular and subcellular substrate systems may take place over a period of months after CNS injury or disease, and thus development of spasticity also follows that time line. The authors view spasticity as the product of unselective, maladaptive changes in the motor and sensory projection systems as part of an unguided attempt for recovery of function. Damage to the CNS (e.g., SCI, TBI, stoke, MS, ALS) alters the regulation of sensory transmission from peripheral nerves.

This change in the regulation of sensory transmission to motor reflex pathways appears to favor excitation and therefore an increased reflex excitability. CNS damage/disease also causes nerve cell membranes (such as changes in dendritic PICs) to have a higher probability of a depolarized state. The combination of decreased regulation of sensory transmission and increased cellular excitability collectively contribute to the increased probability for motoneuron discharge that underlies the resulting spasticity. The majority of the previously mentioned possible mechanisms, however, have only been tested in the setting of SCI. Therefore, it is noteworthy that the contribution of these mechanisms to spasticity resulting from TBI, stroke, and other neurological diseases (e.g., MS, ALS) is yet to be confirmed. In our experience, closed-head TBI (Marmarou et al., 1994) or cortical control impact (Dixon et al., 1991) TBI produces more severe spasticity in the rodents than the magnitude of spasticity after SCI, particularly injury to the thoracic spinal cord (the setting in which most SCI research has been done). Moreover, our preliminary physiological data in our experimental TBI models revealed a different pattern, time course, and neurophysiological mechanisms from that observed following midthoracic SCI. For example, rate-dependent inhibition, an index of a long-acting presynaptic inhibition, was minimally altered in these TBI models. Moreover, a robust permanent tonic and dynamic pattern of spasticity was observed in these TBI models.

Immunocytochemical studies in these animals revealed decreased dopamine beta hydroxylase (DβH), positive norepinephrine (NE) fibers (a surrogate marker for norepinephrine) combined with greatly increased density of serotonin positive fibers in the gray matter of the lumbar spinal cords (Bose et al., 2001; 2002). These observations provoke specific hypotheses regarding monoamine mechanisms that are known to be critically involved in the regulation of stretch reflex excitability. In particular, it has been shown that iontophoretic application of 5-HT enhanced the resting activity of fusimotor (gamma) motoneurons and facilitated the transmission of group II afferent input to these neurons (Jankowska et al., 1998). Sites of noradrenergic and serotonergic modulation may also include intermediate zone interneurons that are pre-motor to gamma motoneurons (Jankowska et al., 2000). In addition, apposition of noradrenergic and serotonergic immunoreactive varicosities is localized to the proximal dendrites and soma of physiologically identified and rhodamine-dextran–labeled gastrocnemius gamma-motoneurons suggesting that they are synapses (Gladden et al., 2000). The number of serotonergic contacts with alpha-motoneurons, themselves, has been estimated to be in excess of the number of synapses from Ia afferents (Alvarez et al., 1998). Serotonin is a metabotrophic neurotransmitter that influences intrinsic membrane conductances and, collectively, these changes increase the excitability of motoneurons by several mechanisms stated previously. In this regard, elegant work by Bennett et al. (2001) showed that the influence of plateau potentials on the firing frequency of the motoneuron was linear with respect to the amplitude of the PIC both during recruitment and derecruitment. Serotonin is known to decrease the threshold and increase the duration of plateau potentials (i.e., eliciting a sustained period of motoneuron firing that greatly outlasts the period of stimulation) (Conway et al., 1988; Hounsgaard et al., 1988). This change has been correlated with a decrease in the slow potassium current and a decrease in the threshold for activation of L-type Ca^{++} channels (Svirskis and Hounsgaard, 1998). Serotonin has also been shown to increase the inward rectifier current (Ih) producing a depolarization that can lead to a strong firing of motoneurons (Takahashi and Berger, 1990). This is a particularly effective mechanism because the Ih operates in a range close to the resting potential. Therefore, the modulation of Ih by serotonin could have a major facilitatory effect on the integration of synaptic potentials as well as on the shaping of spike after potentials (Russo and Hounsgaard, 1999). Therefore, collectively, a robust increase in serotonin expression in the lumbar spinal cord should be correlated with robust increases in motoneuron excitability that can be quantitated by significant changes in the current/frequency curves for tibial motoneurons in TBI-spastic rats.

In this regard, several other recent physiological published reports related to the mechanism of spasticity will be described below. Several studies have been conducted to understand changes in pre- and postsynaptic factors that regulate motoneuron excitability in the normal and the hyperreflexic state following SCI. Accompanying progress in understanding changes in presynaptic mechanisms, substantial progress has occurred in the understanding of postsynaptic mechanisms that regulate the input/output gain of motoneuron discharge (Bennett et al., 1998; Hounsgaard et al., 1988; Kernell, 1979; Lee et al., 2003; Lee and Heckman, 1998a, 1998b, 2000; Schwindt and Crill, 1980a, 1980b). These studies have revealed that the gain of synaptic inputs can be amplified up to a factor of five by brainstem/monoaminergic inputs that regulate dendritic PICs using sodium and calcium channels. The higher the persistent inward current, the higher the synaptic gain and consequent burst rate of the motoneurons. Segmental regulation of PICs occurs through inhibitory mechanisms that regulate afferent inputs. It has been proposed that the acute period of hyporeflexia that follows SCI can be attributed to a reduction in dendritic PICs. Subsequently, after several weeks, recordings have shown that motoneurons reacquired PICs that were indiscriminately initiated by segmental inputs. These unregulated PICs were proposed as a mechanism that significantly contributes to clonus and spasms, and associated amplified bistable properties of motoneurons (Lee and Heckman, 2000).

Furthermore, after a complete transection of the sacral spinal cord (Kitzman, 2005) and after a mid-thoracic spinal cord contusion injury (Bose et al., 2005), a decrease in the number of dendritic branches in the sacrocaudal or soleus motoneurons was reported. These reports suggested that a decrease in the dendritic arborization without a compensatory increase in voltage-gated channels could lead to an increase in motoneuronal excitability that may result spasticity. In this context, segmental inhibitory processes, such as presynaptic inhibition, have an even more important role in the regulation of sensory transmission. Rate-dependent inhibition provides a protocol for testing a type of long-acting presynaptic inhibition that is normally used to regulate the excitability of the muscle stretch reflexes relative to the phase of stepping. Three neurotransmitter systems (serotonergic, GABA, and noradrenergic systems), are known to play critical roles in the modulation of segmental reflex modulation. The authors have observed that the rate-dependent inhibition and velocity-dependent ankle torque are profoundly influenced by $GABA_b$-specific agents (Thompson et al., 2005; Wang et al., 2002) and NE-specific lesions (Bose et al., 2001; Thompson et al., 1999). Specifically, L-baclofen (which acts on $GABA_b$ segmental circuitry) increased rate-dependent inhibition and decreased velocity-dependent ankle torque, whereas selective neurotoxic lesions of NE fibers produced nonspecific increase in reflex excitability. Segmental GABAergic system play an important role in this rate-dependent long-acting presynaptic inhibition. However, there are significant and clear differences between $GABA_a$- and $GABA_b$-mediated presynaptic inhibitions. The findings related to $GABA_a$ in spinal complete transected animals (work from Edgerton's group [Edgerton and Roy, 2010; Khristy et al., 2009; Tillakaratne et al., 2002]) and the recent works from the authors of this chapter (that are related to $GABA_b$ in moderate contusion SCI) provide insight into the differences in mechanism of action. The presynaptic inhibitory (primary afferent depolarization) mechanism related to $GABA_a$ is the action of the $GABA_a$ ionotrophic receptor, which when activated, selectively conducts Cl− through its pore resulting in hypopolarization of the axon terminals.

This unusual presynaptic hypopolarization (instead of hyperpolarization) in axon terminals is due to the increased Cl− concentration in axon terminals that is maintained by the NKCC1 Cl− transporter at a concentration gradient that is greater than that produced by the equilibrium potential (Alvarez-Leefmans, 2007). The issue at odds is related to the reporting of a spinal cord transection–induced downregulation in the potassium-chloride cotransporter-2 (KCC2) in motoneuron membranes; this results in a positive shift in the membrane potential for chloride (Boulenguez et al., 2010). Accordingly, a paradoxical effect could occur at $GABA_a$ receptors on membranes with downregulated KCC2. Presynaptic inhibition associated with $GABA_a$-mediated primary afferent depolarization is a significantly important inhibitory control of afferent transmission to lumbar interneurons and motoneurons. $GABA_a$ receptors, located on presynaptic terminals, are activated by GABAergic interneurons. Although elevated $GABA_a$ and glycine receptors could result in more inhibition, the low KCC2 levels and elevated chloride equilibrium potentials could contribute to the increased excitability leading to spasticity (Roy and Edgerton, 2012). Further, Dr. David Lloyd, Sir John Eccles, Dr. David Curtis, and Dr. P. Rudomin showed that the time course for $GABA_a$-mediated presynaptic inhibition has a rapid onset and a duration (elicited by single shock) of <300 msec. By contrast, $GABA_b$ metabotrophic receptors are also located on presynaptic terminals, adjacent to voltage-gated calcium channels. GABAergic interneurons terminate on, release GABA, and excite these $GABA_b$ receptors. Second messengers released subsequent to $GABA_b$ receptor activation act to decrease the Ca++ conductance of the voltage gated Ca++ channels. This decrease in Ca++ influx, an essential part of the depolarization—Ca influx—staging and adhesion of presynaptic vesicles to the presynaptic membrane, results in a significant decrease in excitatory transmitter release from primary afferents. The time course for these metabotrophic inhibitory processes is slow to activate (i.e., >100 msec) but have the unique signature of long duration (i.e., 1000s of msec). Inhibition associated with these two types of presynaptic inhibition can be selectively tested using afferent volleys initiated at different frequencies. For example, 1 Hz (1000-msec interval between stimulus pulse) induced inhibition is, therefore, primarily $GABA_b$, because $GABA_a$ has a time course ≤300 msec. $GABA_b$ receptors can be selectively activated by L-baclofen or selectively blocked by specific antagonists such as CGP35348. The authors reported that intrathecal administration of $GABA_b$ agonists and antagonists can profoundly alter rate-dependent depression in normal and contusion SCI animals (Thompson et al., 1992, 1993, 1998, 2001a; Wang et al., 2002). These studies reported significant decreases in the magnitude and time course for $GABA_b$-associated rate-dependent depression after contusion SCI (Thompson et al., 1992, 1993, 1998, 2001a; Wang et al., 2002). Moreover, several authors have shown that a similar change occurs in low-frequency (1-Hz) rate-depression–associated changes in spinal reflex excitability after human SCI (Phadke et al., 2009, 2010a, 2010b, 2010c; Thompson et al., 2001a; Trimble et al., 2001) and others (Schindler-Ivens and Shields, 2000). Collectively, these findings contribute to the evidence base for the use of $GABA_b$-specific antispastic medication, intrathecal L-baclofen (ITB) for the treatment of spasticity.

To test the physiological effects of coeruleo-spinal noradrenergic modulation to velocity-dependent ankle torque in hindlimb extensor muscles, the authors injected anti-DβH-saporin, an immunospecific ribosomal toxin to lumbar spinal cerebrospinal fluid. Because DβH is specific to noradrenergic fibers, it produced a selective lesion of noradrenergic neurons via inactivating the ribosomes (Wiley and Kline, 2000). The injected normal animals exhibited a spastic hypertonia that resembled the pattern of TBI-spastic hypertonia observed at one week after TBI. These studies indicated involvement of NE neurobiology in TBI-induced spastic hypertonia and have increased the authors' interest in the possibilities for the therapeutic potential of NE pharmacotherapy in spasticity.

14.4 TREATMENT AND REHABILITATION OF TBI-INDUCED SPASTICITY

Spasticity is one of the most significant challenges associated with treatment and rehabilitation after moderate/severe TBI. Treatment development is hindered by the lack of an understanding of neurobiology of TBI-induced spasticity. As mentioned previously, most of the physiological studies related to the neurobiology of spasticity have focused on thoracic SCI, which has revealed alterations of several physiological parameters of motoneuron firing patterns and associated pre- and postsynaptic inhibition-related mechanisms. However, the lack of fundamental neurobiology that underlies the development and persistence of TBI spasticity prevents focused application of antispasticity medication to specific therapeutic targets. In addition, because data from controlled studies quantifying specific interactions between antispastic medications and cognitive recovery are not available, guidelines regarding initiation of antispastic medications after injury are conservative. Typically, post-injury delay before initiating the most effective antispastic measures (e.g., intrathecal baclofen) is 1 year after injury, although it has been shown that the use of antispastic medications (baclofen), particularly intrathecal baclofen, can decrease the severity of spasticity produced by acquired brain injury (Francisco et al., 2005; Meythaler et al., 1997, 1999, 2001; Saltuari et al., 1989). Therefore, early windows of opportunity to diminish debilitating orthopedic problems induced by hypertonia are usually missed. Current data indicate that of the 75,000–100,000 severe TBIs (total all TBI, 1.5–2.0 million) that occur annually (TBI Model Systems National Database), the incidence of contracture (muscular deformity induced by spasticity) is projected to be as high as 85% (Corrigan et al., 2010; Gottshall, 2011; McGuire, 2011; Thurman et al., 1999). Therefore, current treatment strategies for TBI spasticity include a host of orthopedic and surgical procedures to address musculoskeletal deformities that typically gain a substantial foothold before treatment is initiated. A mild form of spasticity can have beneficial effects, such as improving ambulation and maintaining muscle bulk However, moderate to severe TBI-induced spasticity often interferes with the individual's general functioning, including limitations in rotation of movement, mobility, dysarthria, and dysphagia. In these cases, treatment is often needed.

The currently available treatments for spasticity include (1) oral medications: baclofen, tizanidine, diazepam, and dantrolene sodium, alone or in combination; (2) intrathecal baclofen drug delivery (ITB); (3) orthopedic surgery: such as tendon transfers, tendon lengthening, osteotomy, and bony fusions; (4) neurosurgery: such as percutaneous and open selective dorsal rhizotomy, neurectomy, or myelotomy; (5) injection therapy: such as neurolytic nerve blocks using phenol; anesthetic nerve blocks using procaine and lidocaine, and botulinum toxin (Botox) injections; and (6) rehabilitation therapy: casting, splinting, positioning, electrical stimulation, and rotary movements, etc.

A laboratory TBI-spasticity model recently provided new information relative to the fundamental neurobiology and the treatment of this condition. These new data provided a preclinical platform for safety, feasibility, and efficacy of early ITB intervention after TBI spasticity (Bose et al., 2013). This study was performed in the rodent model to evaluate spasticity, cognitive, vestibulomotor, and locomotor disabilities produced by closed-head TBI and how these disabilities could be modified by Lioresal ITB therapy initiated in an acute setting. The influence of TBI and treatment on the excitability of voluntary motor pathways was assessed using TMS-initiated motor-evoked potentials to activate constituents of the executive motor pathway to lower limb muscles. In addition, spinal cord tissues were studied to assess the expression of immunohistochemical markers for agents known to influence the excitability of neural pathways controlling lower limb motor function, especially spasticity. These data indicated that 1 month of ITB treatment initiated at post-TBI week 1 blocked the early onset of spasticity, significantly attenuated late-onset spasticity, significantly reduced anxiety-like behavior, and produced no significant adverse effects on cognitive and balance performance. More specifically, these data indicated that in the acute treatment group, at 2 weeks post-treatment there was a complete block of spasticity (elevated ankle torques and corresponding EMGs), and at posttreatment week 4, ankle extensor spasticity was reduced by 50% compared with untreated injured controls. By contrast, when ITB was initiated post-TBI 4 weeks, at 2 weeks posttreatment, there was 65% attenuation of spasticity, and at week 4 posttreatment, ankle extensor spasticity was reduced by 42% compared with untreated injured controls. ITB-treated animals in both acute and subchronic treatment groups showed improved scores (trends) for serial learning (probe trail) and improved gait performance compared with untreated injured controls. Although balance performance was not affected in the acute group, a significant deterioration in balance performance was detected in the chronic treatment group. Therefore, acute ITB treatment strategy appeared to be more effective in controlling spasticity (with no impact on balance performance) than observed in the setting of delaying treatment for 1 month (e.g., subchronic ITB treatment). This improved spasticity outcome in acute ITB group was accompanied by marked upregulation of GABA/GABA$_b$, norepinephrine, and BDNF expression in the spinal cord tissues of the treated animals (Bose et al., 2013).

This study was the first to show that early intervention with ITB treatment was safe, feasible, and effective in this closed-head TBI animal model. Collectively, these data provide a strong molecular footprint of the enhanced expression of reflex regulation of factors known to be involved in regulation of sensory transmission (presynaptic inhibition) and postsynaptic excitability. ITB treatment at the acute interval may occur in a setting in which less time has passed postinjury for the progressive development of maladaptive segmental and descending plasticity (Bose et al., 2013). However, in addition to demonstrating that early intervention can be highly effective in reducing TBI spasticity, studies need to be conducted to assess the impact of early ITB versus chronic ITB on cognitive function, balance, sleep, and daily activity, using ITB dosages across a broad range. A comprehensive comparison of acute versus chronic ITB treatments will provide a much needed unequivocal comparison of new versus standard of ITB treatment after TBI.

Although oral baclofen and ITB therapy are partially effective in reducing TBI-induced spasticity in humans, amplification of treatment benefits possibly through the use combination of complementary therapies are needed to produce even more effective treatment benefits for long-term outcomes. The authors' preliminary work in SCI animals revealed that ITB combined with locomotor training (treadmill) produced a profound amplification of long-term spasticity attenuation compared with either of the treatments tested individually. The strategy of combination therapy has also been strongly recommended in recent consensus conferences held to evaluate the disappointing TBI clinical trial outcomes testing single therapies and to encourage combination therapeutic development (Saatman et al., 2008). Because combination therapy potentially represents a paradigm shift in rehabilitation therapy, early intervention combination therapy (ITB and locomotor training) or a combination of ITB and transcranial magnetic stimulation (TMS) for TBI-induced spasticity may be of high importance. To broaden the scope of potential therapeutic application, these therapies need to be evaluated in selected acute and chronic settings. To provide greater confidence in the translation of any of these combination therapies, physiological measures of treatment safety need to be included. A major requirement for the safe application of any new treatment is the quantitative evidence base for safety and efficacy. Although locomotor training has demonstrated positive benefits, the robustness of the single-therapy paradigm (in the setting of human TBI) was not sufficient to yield results significantly greater than conventional therapy. TMS is proposed as a strategy for individual therapy or as an adjuvant to provide an amplification that could move the combined therapy to the next level of effectiveness. The authors' recent report (Hou et al., 2014) suggests the potential for TMS therapy to influence SCI-induced spasticity and gait disabilities that are correlated with neurochemical changes that are consistent with the neurobiology of these processes. Amplification of locomotor training is needed and TMS represents a logical next step in the development of effective therapy for spasticity.

New therapeutic possibilities include the progressive addition of convergent therapies. For example, it is anticipated that the potential for therapeutic exercise and TMS would be enhanced in conjunction with pharmacological therapy that could further induce augmentation of central norepinephrine and segmental $GABA_b$. The authors' recent data suggested that treadmill locomotor training and TMS across the SCI site can be an effective and feasible treatment modality for SCI-induced spasticity and gait impairments, and that the combination of these two therapies was significantly more effective than either treatment tested as individual therapy (Hou et al., 2014). The combination therapy revealed a profound therapeutic reduction of spasticity toward preinjury levels. The therapeutic improvements in spasticity (and gait) with each treatment modality were correlated with significantly amplified expression of the immunohistochemical markers for $GABA_b$ receptors, GAD_{67}, DβH, and BDNF. However, the combination therapy produced a greater expression of these markers. Each treatment modality has its own merit. For example, although TMS may enhance upregulation of trophic factors, without behaviorally relevant guided neural signaling (e.g., treadmill locomotor training), little normalization of gait confirmation was produced by TMS alone. On the other hand, six sessions of TMS therapy provided an increased unguided ascending and descending activity related to pre- and postsynaptic inhibition that resulted in significant improvement of spasticity. This improvement in spasticity was similar to the improvement in spasticity seen after a 7-week program of treadmill locomotor therapy. However, treadmill therapy, as standalone therapy, revealed significant improvement in spasticity and normalization of gait parameters. The treadmill locomotor therapy perhaps increased guided segmental, ascending, and descending plasticity. These findings suggest that task-appropriate therapy was essential for recovery of spasticity with gait confirmation. Therefore, the robustness necessary to make significant functional improvements in spasticity and gait may require amplification of therapeutic impact through the successful combination of complementary individual therapies. Although TMS has been used in a broad range of therapeutic applications since its inception in 1985, many questions remain regarding the mechanisms of its beneficial outcome. In this regard, this SCI study indicated that TMS may enhance the trophic environment through BDNF and GAD_{67} and the upregulation of markers for pre- and postsynaptic inhibition ($GABA_b$ receptors, GAD_{67}, and NE). Similar studies need to be done to test if similar beneficial outcome in spasticity and gait can be achieved in TBI spasticity.

14.5 REHABILITATION OF TBI-INDUCED SPASTICITY AND LONG-TERM BENEFICIAL OUTCOME

Rehabilitation of TBI-induced spasticity may also provide protection against long-term progressive inflammatory cell loss. In addition to TBI-induced acute motor disabilities, long-term functional losses are also threatened by the possibility of enduring injury-induced changes that can produce progressive loss of neural tissue and concomitant exacerbation of disability. One of the most damaging sources for continued loss of tissue is chronic inflammation (Ramlackhansingh et al., 2011). This section will briefly outline that the unusual nature of this problem is that, in addition to TBI-induced increase in signaling factors that induce inflammation, this inflammation is exacerbated by a TBI-induced decrease in the NE-associated regulation of inflammation. Accordingly, rehabilitation procedures that upregulate central NE contribute significantly to the potential for decreasing chronic disability.

In addition to the TBI-induced reduction of NE expression in the spinal cord described previously, it has been reported that TBI results in widespread and long-lasting changes in brain NE turnover (Dunn-Meynell et al., 1994; Fujinaka et al., 2003; Robinson and Justice, 1986). After injury, a short-term phase (minutes) of increased turnover is followed by a prolonged phase (>8 weeks) of substantial decrease in NE metabolism (Fujinaka et al., 2003; Levin et al., 1995). Consistent with these findings, our immunohistochemistry studies (Bose et al., 2002a, 2012) and others (Fujinaka et al., 2003) indicated that TBI produced a marked decrease in DβH labeled locus coeruleus neurons. Accordingly, because this nucleus provides significant NE projections to the CNS, and because these projections influence the excitability of multiple neural networks for cognitive function, anxiety, and balance, we have proposed a TBI multiple-morbidity model based on the TBI-induced injury of the central NE system (Bose et al., 2012). In addition to the important role of monoamines in cellular excitability, NE plays a critical role in the regulation of the CNS immune system. It is known that NE regulates immune signaling (e.g., iCAM, iNOS, NOS2) that is critical for the immune privilege of the CNS (Galea et al., 2003). Locally diffusing NE has been proposed to negatively regulate transcription of inflammatory genes in astrocytes and microglia (Feinstein et al., 2002) and has been proposed to serve as an endogenous anti-inflammatory agent (Mori et al., 2002).

Because TBI induces a cascade of inflammatory signaling factors, a prominent inflammatory response is marshaled via astrocytes and microglia (Ramlackhansingh et al., 2011). The concomitant inflammatory responses (including the secretion of cytokines) may significantly increase neuronal cell loss and enlarge the scope of functional disability. Consequently, injury-related downregulation of the central NE system could disinhibit the inflammatory cascades that are freed to attack intact cells and render them more subject to inflammatory damage. Therefore, a chronic condition of progressing inflammation can potentially inflict a continual erosion of the nervous system and an ever progressing functional degradation.

Accordingly, putative treatments that induce upregulation of NE can potentially result in significant neuronal preservation against the progressive loss resulting from inflammation.

REFERENCES

Alverez-Leefmans, F.J., 2009. Physiology and pathology of chloride transporters and channels in the nervous system. In Alverez-Leefmans F.J., Delpre, E., editors. *Chloride Transporters in Presynaptic Inhibition, Pain, and Neurogenic Inflammation*. Elsevier, Amsterdam. pp. 439–470.

Alvarez, F.J., Pearson, J.C., Harrington, D., Dewey, D., Torbeck, L., and Fyffe, R.E., 1998. Distribution of 5-hydroxytryptamine-immunoreactive boutons on alpha-motoneurons in the lumbar spinal cord of adult cats. *J Compar Neurol*, 393, 69–83.

Aran, S. and Hammond, D.L., 1991. Antagonism of baclofen-induced antinociception by intrathecal administration of phaclofen or 2-hydroxy-saclofen, but not delta-aminovaleric acid in the rat. *J Pharmacol Exp Ther*, 257, 360–368.

Ashby, P. and Verrier, M., 1980. Human motoneuron responses to group 1 volleys blocked presynaptically by vibration. *Brain Res*, 184, 511–516.

Bennett, D.J., Hultborn, H., Fedirchuk, B., and Gorassini, M., 1998. Synaptic activation of plateaus in hindlimb motoneurons of decerebrate cats. *J Neurophysiol*, 80, 2023–2037.

Bennett, D.J., Li, Y., and Siu, M., 2001. Plateau potentials in sacrocaudal motoneurons of chronic spinal rats, recorded in vitro. *J Neurophysiol*, 86, 1955–1971.

Boorman, G., Becker, W.J., Morrice, B.L., and Lee, R.G., 1992. Modulation of the soleus H-reflex during pedalling in normal humans and in patients with spinal spasticity. *J Neurol Neurosurg Psychiatry*, 55, 1150–1156.

Bose, P., Hou, J., Jaiswal, P., Phadke, C., Parmer, R., Hoffman, P.M. et al, 2009. TMS and locomotor therapies improved motor and cognitive disabilities of an experimental autoimmune encephalomyelitis (EAE) rat model for multiple sclerosis (MS), *Soci Neurosci*, 640.4, Chicago, IL.

Bose, P., Hou, J., Nelson, R., Nissim, N., Parmer, R., Keener, J. et al., 2013. Effects of acute intrathecal baclofen in an animal model of TBI-induced spasticity, cognitive, and balance disabilities. *J Neurotrauma*, 30, 1177–1191.

Bose, P., Parmer, R., Reier, P.J., and Thompson, F.J., 2005. Morphological changes of the soleus motoneuron pool in chronic midthoracic contused rats. *Exp Neurol*, 191, 13–23.

Bose, P., Wang, D.C., Parmer, R., Wiley, R.G., and Thompson, F.J., 2001. Monoamine modulation of spinal reflex excitability of the lower limb in the rat: Intrathecal infusion of Anti-DBH saporin toxin - time course for behavior neuroscience. *Soc Neurosci*, 771.3, San Diego, CA.

Bose, P.K., Hou, J., Parmer, R., Reier, P.J., and Thompson, F.J., 2012. Altered patterns of reflex excitability, balance, and locomotion following spinal cord injury and locomotor training. *Front Physiol*, 3, 258.

Bose, P.K., Parmer, R., Parker, J., Hayes, R., and Thompson, F.J., 2002a. Traumatic Brain Injury (TBI)–induced spasticity: Monoamine changes and possible mechanisms. *J. Neurotrauma*, 19, 1305.

Bose, P.K., Parmer, R., and Thompson, F.J., 2002b. Velocity dependent ankle torque in rats after contusion injury of the midthoracic spinal cord: Time course. *J. Neurotrauma,*, 19, 1231–1249.

Boulenguez, P., Liabeuf, S., Bos, R., Bras, H., Jean-Xavier, C., Brocard, C. et al., 2010. Down-regulation of the potassium-chloride cotransporter KCC2 contributes to spasticity after spinal cord injury. *Nat Med*, 16, 302–307.

Brown, P., 1994. Pathophysiology of spasticity. *J Neurol Neurosurg Psychiatry*, 57, 773–777.

Calancie, B., Broton, J.G., Klose, K.J., Traad, M., Difini, J., and Ayyar, D.R., 1993. Evidence that alterations in presynaptic inhibition contribute to segmental hypo- and hyperexcitability after spinal cord injury in man. *Electroenceph Clin Neurophysiol*, 89, 177–186.

Cardona, A. and Rudomin, P., 1983. Activation of brainstem serotoninergic pathways decreases homosynaptic depression of monosynaptic responses of frog spinal motoneurons. *Brain Res*, 280, 373–378.

Conway, B.A., Hultborn, H., Kiehn, O., and Mintz, I., 1988. Plateau potentials in alpha-motoneurones induced by intravenous injection of L-dopa and clonidine in the spinal cat. *J Physiol*, 405, 369–384.

Cope, T.C., Bodine, S.C., Fournier, M., and Edgerton, V.R., 1986. Soleus motor units in chronic spinal transected cats: Physiological and morphological alterations. *J Neurophysiol*, 55, 1202–1220.

Corrigan, J.D., Selassie, A.W., and Orman, J.A., 2010. The epidemiology of traumatic brain injury. *J Head Trauma Rehabil*, 25, 72–80.

Crone, C., Johnsen, L.L., Biering-Sorensen, F., and Nielsen, J.B., 2003. Appearance of reciprocal facilitation of ankle extensors from ankle flexors in patients with stroke or spinal cord injury. *Brain*, 126, 495–507.

Crone, C., Nielsen, J., Petersen, N., Ballegaard, M., and Hultborn, H., 1994. Disynaptic reciprocal inhibition of ankle extensors in spastic patients. *Brain*, 117 (Pt 5), 1161–1168.

Delwaide, P.J., 1973. [Clinical neurophysiology of tendon hyperreflexia]. *Bull Acad R Med Belg*, 128, 271–293.

Dixon, C.E., Clifton, G.L., Lighthall, J.W., Yaghmai, A.A., and Hayes, R.L., 1991. A controlled cortical impact model of traumatic brain injury in the rat. *J Neurosci Methods*, 39, 253–262.

Dunn-Meynell, A., Pan, S., and Levin, B.E., 1994. Focal traumatic brain injury causes widespread reductions in rat brain norepinephrine turnover from 6 to 24 h. *Brain Res*, 660, 88–95.

Edgerton, V.R. and Roy, R.R., 2010. Spasticity: A switch from inhibition to excitation. *Nat Med*, 16, 270–271.

Faist, M., Mazevet, D., Dietz, V., and Pierrot-Deseilligny, E., 1994. A quantitative assessment of presynaptic inhibition of Ia afferents in spastics. Differences in hemiplegics and paraplegics. *Brain*, 117 (Pt 6), 1449–1455.

Feinstein, D.L., Heneka, M.T., Gavrilyuk, V., Dello Russo, C., Weinberg, G., and Galea, E., 2002. Noradrenergic regulation of inflammatory gene expression in brain. *Neurocheml Int*, 41, 357–365.

Francisco, G.E., Hu, M.M., Boake, C., and Ivanhoe, C.B., 2005. Efficacy of early use of intrathecal baclofen therapy for treating spastic hypertonia due to acquired brain injury. *Brain Inj*, 19, 359–364.

Fujinaka, T., Kohmura, E., Yuguchi, T., and Yoshimine, T., 2003. The morphological and neurochemical effects of diffuse brain injury on rat central noradrenergic system. *Neurol Res*, 25, 35–41.

Galea, E., Heneka, M.T., Dello Russo, C., and Feinstein, D.L., 2003. Intrinsic regulation of brain inflammatory responses. *Cell Mole Neurobiol*, 23, 625–635.

Gladden, M.H., Maxwell, D.J., Sahal, A., and Jankowska, E., 2000. Coupling between serotoninergic and noradrenergic neurones and gamma-motoneurones in the cat. *J Physiol*, 527 Pt 2, 213–223.

Gorassini, M.A., Knash, M.E., Harvey, P.J., Bennett, D.J., and Yang, J.F., 2004. Role of motoneurons in the generation of muscle spasms after spinal cord injury. *Brain*, 127, 2247–2258.

Gottshall, K., 2011. Vestibular rehabilitation after mild traumatic brain injury with vestibular pathology. *NeuroRehabilitation*, 29, 167–171.

Heckmann, C.J., Gorassini, M.A., and Bennett, D.J., 2005. Persistent inward currents in motoneuron dendrites: Implications for motor output. *Muscle Nerve*, 31, 135–156.

Herman, R., 1968. Alterations in the dynamic and static properties of the stretch reflex obtained at various rates of stretch during different stages of clinical spasticity. *Electroenceph Clin Neurophysiol*, 25, 408–409.

Hou, J., Nelson, R., Nissim, N., Parmer, R., Thompson, F.J., and Bose, P.K., 2014. Effect of combined treadmill training and magnetic stimulation on spasticity and gait impairments following cervical spinal cord injury (C-SCI). *J Neurotrauma*, 31:1088–1106.

Hounsgaard, J., Hultborn, H., Jespersen, B., and Kiehn, O., 1988. Bistability of alpha-motoneurones in the decerebrate cat and in the acute spinal cat after intravenous 5-hydroxytryptophan. *J Physiol*, 405, 345–367.

Ishikawa, K., Ott, K., Porter, R.W., and Stuart, D., 1966. Low frequency depression of the H wave in normal and spinal man. *Exp Neurol*, 15, 140–156.

Jankowska, E., Gladden, M.H., and Czarkowska-Bauch, J., 1998. Modulation of responses of feline gamma-motoneurones by noradrenaline, tizanidine and clonidine. *J Physiol*, 512 (Pt 2), 521–531.

Jankowska, E. and Hammar, I., 2002. Spinal interneurones; how can studies in animals contribute to the understanding of spinal interneuronal systems in man? *Brain Res*, 40, 19–28.

Jankowska, E., Hammar, I., Chojnicka, B., and Heden, C.H., 2000. Effects of monoamines on interneurons in four spinal reflex pathways from group I and/or group II muscle afferents. *Eur J Neurosci*, 12, 701–714.

Katz, R.T and Rymer, W.Z., 1989. Spastic hypertonia: Mechanisms and measurement. *Arch Phys Med Rehabil*, 70, 144–155.

Kernell, D., 1979. Rhythmic properties of motoneurones innervating muscle fibres of different speed in m. gastrocnemius medialis of the cat. *Brain Res*, 160, 159–162.

Khristy, W., Ali, N.J., Bravo, A.B., de Leon, R., Roy, R.R., Zhong, H. et al., 2009. Changes in GABA(A) receptor subunit gamma 2 in extensor and flexor motoneurons and astrocytes after spinal cord transection and motor training. *Brain Res*, 1273, 9–17.

Kitzman, P., 2005. Alteration in axial motoneuronal morphology in the spinal cord injured spastic rat. *Exp Neurol*, 192, 100–108.

Lance, J.W., 1981. Disordered muscle tone and movement. *Clin Exp Neurol*, 18, 27–35.

Lee, R.H. and Heckman, C.J., 1998a. Bistability in spinal motoneurons in vivo: Systematic variations in persistent inward currents. *J Neurophysiol*, 80, 583–593.

Lee, R.H. and Heckman, C.J., 1998b. Bistability in spinal motoneurons in vivo: Systematic variations in rhythmic firing patterns. *J Neurophysiol*, 80, 572–582.

Lee, R.H. and Heckman, C.J., 2000. Adjustable amplification of synaptic input in the dendrites of spinal motoneurons in vivo. *J Neurosci*, 20, 6734–6740.

Lee, R.H., Kuo, J.J., Jiang, M.C., and Heckman, C.J., 2003. Influence of active dendritic currents on input-output processing in spinal motoneurons in vivo. *J Neurophysiol*, 89, 27–39.

Levin, B.E., Brown, K.L., Pawar, G., and Dunn-Meynell, A., 1995. Widespread and lateralization effects of acute traumatic brain injury on norepinephrine turnover in the rat brain. *Brain Res*, 674, 307–313.

Li, Y., Li, X., Harvey, P.J., and Bennett, D.J., 2004. Effects of baclofen on spinal reflexes and persistent inward currents in motoneurons of chronic spinal rats with spasticity. *J Neurophysiol*, 92, 2694–2703.

Little, J.W., Ditunno, J.F., Jr., Stiens, S.A., and Harris, R.M., 1999. Incomplete spinal cord injury: Neuronal mechanisms of motor recovery and hyperreflexia. *Arch Phys Medic Rehabil*, 80, 587–599.

Macht, M.B. and Kuhn, R.A., 1948. The occurrence of extensor spasm in patients with complete transection of the spinal cord. *N Engl J Med*, 238, 311–314.

Marmarou, A., Foda, M.A., van den Brink, W., Campbell, J., Kita, H., and Demetriadou, K., 1994. A new model of diffuse brain injury in rats. Part I: Pathophysiology and biomechanics. *J Neurosurg*, 80, 291–300.

McGuire, L., 2011. The epidemiology of traumatic brain injury, National Centers for Disease Control and Prevention.

Mendell, L.M., 1984. Modifiability of spinal synapses. *Physiol Rev*, 64, 260–324.

Meythaler, J.M., Guin-Renfroe, S., Brunner, R.C., and Hadley, M.N., 2001. Intrathecal baclofen for spastic hypertonia from stroke. *Stroke*, 32, 2099–2109.

Meythaler, J.M., Guin-Renfroe, S., Grabb, P., and Hadley, M.N., 1999. Long-term continuously infused intrathecal baclofen for spastic-dystonic hypertonia in traumatic brain injury: 1-year experience. *Arch Phys Med Rehabil*, 80, 13–19.

Meythaler, J.M., McCary, A., and Hadley, M.N., 1997. Prospective assessment of continuous intrathecal infusion of baclofen for spasticity caused by acquired brain injury: A preliminary report. *J Neurosurg*, 87, 415–419.

Mori, K., Ozaki, E., Zhang, B., Yang, L., Yokoyama, A., Takeda, I. et al, 2002. Effects of norepinephrine on rat cultured microglial cells that express alpha1, alpha2, beta1 and beta2 adrenergic receptors. *Neuropharmacology*, 43, 1026–1034.

Nickolls, P., Collins, D.F., Gorman, R.B., Burke, D., and Gandevia, S.C., 2004. Forces consistent with plateau-like behaviour of spinal neurons evoked in patients with spinal cord injuries. *Brain*, 127, 660–670.

Nielsen, J., Petersen, N., Ballegaard, M., Biering-Sorensen, F., and Kiehn, O., 1993. H-reflexes are less depressed following muscle stretch in spastic spinal cord injured patients than in healthy subjects. *Exp Brain Res*, 97, 173–176.

Nielsen, J., Petersen, N., and Crone, C., 1995. Changes in transmission across synapses of Ia afferents in spastic patients. *Brain*, 118 (Pt 4), 995–1004.

Nielsen, J.B., Crone, C., and Hultborn, H., 2007. The spinal pathophysiology of spasticity—From a basic science point of view. *Acta Physiol*, 189, 171–180.

Nielsen, J.F. and Sinkjaer, T., 1997. Long-lasting depression of soleus motoneurons excitability following repetitive magnetic stimuli of the spinal cord in multiple sclerosis patients. *Multiple Scler*, 3, 18–30.

Ordia, J.I., Fischer, E., Adamski, E., and Spatz, E.L., 1996. Chronic intrathecal delivery of baclofen by a programmable pump for the treatment of severe spasticity. *J Neurosurg*, 85, 452–457.

Phadke, C.P., Flynn, S.M., Thompson, F.J., Behrman, A.L., Trimble, M.H., and Kukulka, C.G., 2009. Comparison of single bout effects of bicycle training versus locomotor training on paired reflex depression of the soleus H-reflex after motor incomplete spinal cord injury. *Arch Phys Med Rehabil*, 90, 1218–1228.

Phadke, C.P., Klimstra, M., Zehr, E.P., Thompson, F.J., and Behrman, A.L., 2010b. Soleus h-reflex modulation during stance phase of walking with altered arm swing patterns. *Motor Control*, 14, 116–125.

Phadke, C.P., Thompson, F.J., Kukulka, C.G., Nair, P.M., Bowden, M.G., Madhavan, S. et al, 2010c. Soleus H-reflex modulation after motor incomplete spinal cord injury: Effects of body position and walking speed. *J Spinal Cord Med*, 33, 371–378.

Phadke, C.P., Thompson, F.J., Trimble, M.H., Behrman, A.L., and Kukulka, C.G., 2010a. Reliability of soleus H-reflexes in standing and walking post-incomplete spinal cord injury. *Int J Neurosci*, 120, 128–136.

Ramlackhansingh, A.F., Brooks, D.J., Greenwood, R.J., Bose, S.K., Turkheimer, F.E., Kinnunen, K.M. et al., 2011. Inflammation after trauma: Microglial activation and traumatic brain injury. *Ann Neurol*, 70, 374–383.

Robinson, R.G. and Justice, A., 1986. Mechanisms of lateralized hyperactivity following focal brain injury in the rat. *Pharmacol Biochem Behav*, 25, 263–267.

Roy, R.R. and Edgerton, V.R., 2012. Neurobiological perspective of spasticity as occurs after a spinal cord injury. *Exp Neurol*, 235, 116–122.

Russo, R.E. and Hounsgaard, J., 1999. Dynamics of intrinsic electrophysiological properties in spinal cord neurones. *Prog Biophys Mol Biol*, 72, 329–365.

Saatman, K.E., Duhaime, A.C., Bullock, R., Maas, A.I., Valadka, A., and Manley, G.T., 2008. Classification of traumatic brain injury for targeted therapies. *J Neurotrauma*, 25, 719–738.

Saltuari, L., Schmutzhard, E., Kofler, M., Baumgartner, H., Aichner, F., and Gerstenbrand, F., 1989. Intrathecal baclofen for intractable spasticity due to severe brain injury. *Lancet*, 2, 503–504.

Schindler-Ivens, S. and Shields, R.K., 2000. Low frequency depression of H-reflexes in humans with acute and chronic spinal-cord injury. *Exp Brain Res*, 133, 233–241.

Schwindt, P.C. and Crill, W.E., 1980a. Properties of a persistent inward current in normal and TEA-injected motoneurons. *J Neurophysiol*, 43, 1700–1724.

Schwindt, P.C. and Crill, W.E., 1980b. Role of a persistent inward current in motoneuron bursting during spinal seizures. *J Neurophysiol*, 43, 1296–1318.

Shapiro, S., 1997. Neurotransmission by neurons that use serotonin, noradrenaline, glutamate, glycine, and gamma-aminobutyric acid in the normal and injured spinal cord. *Neurosurgery*, 40, 168–176; discussion 177.

Singer, B.J., Jegasothy, G.M., Singer, K.P., Allison, G.T., and Dunne, J.W., 2004. Incidence of ankle contracture after moderate to severe acquired brain injury. *Arch Phys Med Rehabil*, 85, 1465–1469.

Skinner, R.D., Houle, J.D., Reese, N.B., Berry, C.L., and Garcia-Rill, E., 1996. Effects of exercise and fetal spinal cord implants on the H-reflex in chronically spinalized adult rats. *Brain Res*, 729, 127–131.

Skinner, R.D., Houle, J. D., Reese, N.B., Dempster, J., and Garcia-Rill, E, 1998. Amelioration of the H-reflex in chronically spinalized adults rats by exercise and fetal spinal cord implants. *Soc Neurosci* 24, 2105.

Svirskis, G. and Hounsgaard, J., 1998. Transmitter regulation of plateau properties in turtle motoneurons. *J Neurophysiol*, 79, 45–50.

Takahashi, T. and Berger, A.J., 1990. Direct excitation of rat spinal motoneurones by serotonin. *J Physiol*, 423, 63–76.

Thilmann, A.F., Fellows, S.J., and Garms, E., 1991. The mechanism of spastic muscle hypertonus. Variation in reflex gain over the time course of spasticity. *Brain*, 114 (Pt 1A), 233–244.

Thompson, F., Mustafa, G., Hou, J., Nelson, R., Gao, Y., Zhang, Y., Bose, P., 2013. A rodent white matter stroke model: Spasticity and gait disability. *Soc Neurosci*, 338.14, San Diego, CA.

Thompson, F.J., Bose, P., Parmer, R., Vierck, C.J., and Wiley, R.G., 1999. Velocity dependent spasticity of the lower limb of the rat: Monoamine related basic mechanisms. *Soci Neurosci*, Miami Beach, FL.

Thompson, F.J., Jain, R., Parmer, R., Cheng, Y., and Bose, P., 2005. Acute locomotor training and ITB treatment of SCI-spasticity. *Soc Neurosci*, 105.5 Washington, DC.

Thompson, F.J., Parmer, R., and Reier, P.J., 1998. Alteration in rate modulation of reflexes to lumbar motoneurons after mid-thoracic spinal cord injury in the rat. I. Contusion injury. *J Neurotrauma*, 15, 495–508.

Thompson, F.J., Parmer, R., Reier, P.J., Wang, D.C., and Bose, P., 2001a. Scientific basis of spasticity: Insights from a laboratory model. *J Child Neurol*, 16, 2–9.

Thompson, F.J., Reier, P.J., Lucas, C.C., and Parmer, R., 1992. Altered patterns of reflex excitability subsequent to contusion injury of the rat spinal cord. *J Neurophysiol*, 68, 1473–1486.

Thompson, F.J., Reier, P.J., Parmer, R., Lucas, C.C., 1993. Inhibitory control of reflex excitability following contusion injury and neural tissue transplantation. *Adv Neurol*, 59, 175–184.

Thompson, F.J., Reier, P.J., Uthman, B., Mott, S., Fessler, R.G., Behrman, A. et al, 2001b. Neurophysiological assessment of the feasibility and safety of neural tissue transplantation in patients with syringomyelia. *J Neurotrauma*, 18, 931–945.

Thurman, D.J., Alverson, C., Dunn, K.A., Guerrero, J., and Sniezek, J.E., 1999. Traumatic brain injury in the United States: A public health perspective. *J Head Trauma Rehabil*, 14, 602–615.

Tillakaratne, N.J., de Leon, R.D., Hoang, T.X., Roy, R.R., Edgerton, V.R., and Tobin, A.J., 2002. Use-dependent modulation of inhibitory capacity in the feline lumbar spinal cord. *J Neurosci*, 22, 3130–3143.

Toft, E., Sinkjaer, T., Andreassen, S., and Hansen, H.J., 1993. Stretch responses to ankle rotation in multiple sclerosis patients with spasticity. *Electroencephal Clin Neurophysiol*, 89, 311–318.

Trimble, M.H., Behrman, A.L., Flynn, S.M., Thigpen, M.T., and Thompson, F.J., 2001. Acute effects of locomotor training on overground walking speed and H-reflex modulation in individuals with incomplete spinal cord injury. *J Spinal Cord Med*, 24, 74–80.

Trimble, M.H., Kukulka, C.G., and Behrman, A.L., 1998. The effect of treadmill gait training on low-frequency depression of the soleus H-reflex: Comparison of a spinal cord injured man to normal subjects. *Neurosci Lett*, 246, 186–188.

Wang, D.C., Bose, P., Parmer, R., and Thompson, F.J., 2002. Chronic intrathecal baclofen treatment and withdrawal: I. Changes in ankle torque and hind limb posture in normal rats. *J Neurotrauma*, 19, 875–886.

Wiley, R.G. and Kline, I.R., 2000. Neuronal lesioning with axonally transported toxins. *J Neurosci Methods*, 103, 73–82.

Yates, C., Charlesworth, A., Allen, S.R., Reese, N.B., Skinner, R.D., and Garcia-Rill, E., 2008. The onset of hyperreflexia in the rat following complete spinal cord transection. *Spinal Cord*, 46, 798–803.

Yates, C.C., Charlesworth, A., Reese, N.B., Ishida, K., Skinner, R.D., and Garcia-Rill, E., 2009. Modafinil normalized hyperreflexia after spinal transection in adult rats. *Spinal Cord*, 47, 481–485.

Young, W., 1992. Medical treatments of acute spinal cord injury. *J Neurol Neurosurg Psychiatry*, 55, 635–639.

Section III

Modeling Brain Injury

15 Techniques and Methods of Animal Brain Surgery

Perfusion, Brain Removal, and Histological Techniques

Jihane Soueid, Amaly Nokkari, and Joelle Makoukji

CONTENTS

15.1 INTRODUCTION

The brain is a heterogeneous tissue composed of various highly interconnected cell types. Each type has a particular pattern of expression and is differentially located in the brain regions. In some pathological situations such as brain injury, cascades of metabolic, cellular and molecular events ultimately lead to brain cell death, tissue damage and atrophy. To better understand these cellular processes it is important to obtain information regarding the spatial and temporal patterns of gene expression. In this chapter, we will describe the method of intracardial perfusion of a mouse prior to brain removal. We will then describe the detailed methodology for the detection of proteins and mRNA utilizing immunohistochemistry (IHC) and in situ hybridization (ISH) respectively in brain sections; respectively.

The brain is composed of areas of gray and white matter and consists of various regions, including the cerebral cortex, the thalamus, the hypothalamus, the brain stem, and the cerebellum. The sensory areas of the cerebral cortex are involved in perception of sensory information: motor areas control execution of voluntary movements and association areas deal with complex integrative functions such as memory, personality traits, and intelligence. The limbic system promotes a range of emotions including pleasure, pain, affection, fear, and anger. The thalamus relays almost all sensory input to the cerebral cortex: it contributes to motor functions by transmitting information from the cerebellum and basal nuclei to motor areas of the cerebral cortex. It also plays a role in maintaining consciousness. The hypothalamus controls and integrates activities of the autonomic nervous system: it regulates emotional and behavioral patterns and circadian rhythms. The cerebellum smoothes and coordinates contractions of skeletal muscles, regulates posture and balance, and may have a role in cognition and language processing.

The brain is a heterogeneous tissue that contains neurons, neuroglia, and other cell types that vary among anatomical regions. Nonneuronal cell types are broadly categorized into (1) astrocytes, (2) radial glia, (3) oligodendrocytes, (4) ependymal cells, and (5) microglia. The role of each cell type is well defined; moreover, their interaction is essential for the neuronal function of the central nervous system. The cellular communication is substantially involved in the establishment of the majority of neurological disorders (Pham and Gupta, 2009).

Brain function is determined by the communication between electrically excitable neurons and the surrounding glial cells, which perform many tasks in the brain.

Oligodendrocytes are one type of glial cell that form an insulating protective myelin sheath around the axons of neurons that enables saltatory nerve conduction. A loss of myelin in defined areas of brain leads to an impairment of axonal conductance. This is what happens in many forms of myelin disorders, such as multiple sclerosis, and it results in a permanent loss of neuron impulse transmission. It is evident that the demyelinated region contains inflammatory cells such as infiltrating lymphocytes and macrophages and activated microglia. These cells might potentiate or even initiate a damage cascade leading to continuous neurodegeneration.

Microglia are the resident phagocytic cells in the brain, taking part in immune-mediated defense mechanisms and clearing damaged cell debris (Ransohoff and Cardona, 2010; Ransohoff and Perry, 2009). Previously, it was thought that microglia, in their resting state, are relatively quiescent. More recent work suggests that microglia are constantly active and surveying their surroundings (Hughes, 2012; Nimmerjahn et al., 2005). Microglia are now implicated in synapse pruning during both development and throughout adulthood, and therefore play a role in regulating homeostatic synaptic plasticity (Schafer et al., 2012).

Together with astrocytes, another type of glial cell, microglia can release neuromodulatory chemicals that influence neuronal firing and intracellular signaling. When first described, astrocytes were seen merely as structural scaffolding to support and fill the gaps between neurons. However, recent evidence suggests that astrocytes serve as much more than a nutrient supply or supportive scaffolding to protect neural networks (Nedergaard et al., 2003). Astrocytes are highly secretory cells, participating in rapid brain communication by releasing factors that modulate neurotransmission (Haydon and Carmignoto, 2006; Huang et al., 2004; Pascual et al., 2012) and more recently have been suggested to possess their own repertoire of gliotransmitters (Bezzi et al., 2004; Cali et al., 2008; Cali and Bezzi, 2010; Domercq et al., 2006; Jourdain et al., 2007; Prada et al., 2011; Santello et al., 2011). Astrocytes also express a wide variety of functional neurotransmitter receptors essential for sensing neuronal activity (Verkhratsky et al., 1998).

When a local inflammatory reaction is triggered in the brain, the increased levels of proinflammatory mediators such as tumor necrosis factor-alpha and prostaglandin 2 can deeply alter the properties of glial network and thus of neuronal network (Bezzi and Volterra, 2001). The important roles played by glial cells in normal and pathological brain functioning are growing, and a more complete picture of neuron–glia interactions is beginning to emerge.

Cellular behaviors such as proliferation, differentiation, migration, and cell death are studied during brain development and in pathological situations. To better understand the developmental processes involved, it is important to obtain information regarding the spatial and temporal patterns of gene expression.

Over the past three decades, animal models have been developed to replicate the various aspects of human brain injury to better understand the underlying pathophysiology and to explore potential treatments. Among more recent models for traumatic brain injury, four specific models are widely used in research: fluid percussion injury (Dixon et al., 1987), controlled cortical impact injury (Dixon et al., 1991; Lighthall, 1988), weight drop impact acceleration injury (Marmarou et al., 1994), and blast injury (Cernak et al., 1996; Leung et al., 2008). Rodents are mostly used in traumatic brain injury research because of their modest cost, small size, and standardized outcome measurements.

In this chapter, we will describe the method of intracardial perfusion and an appropriate method to dissect and remove the brain of a mouse. We will then describe methods for the detection of protein (immunohistochemistry, IHC) and messenger RNA (mRNA) (in situ hybridization, ISH) in brain sections.

15.2 MATERIALS

15.2.1 Anesthesia, Perfusion, and Brain Removal

1. 1-mL syringe and 27 G needle for anesthesia
2. Anesthetic: ketamine and xylazine
3. Peristaltic pump
4. Scalpel handle and #10 blade
5. Freshly made 4% paraformaldehyde (PFA) 10–150 mL per mouse
6. 0.9% saline (or preferred flush) 8–25 mL per mouse
7. Butterfly catheter (23G) with the needle blunted
8. Straight iris scissors
9. Curved iris scissors
10. Curved narrow pattern forceps
11. Chemical fume hood
12. Container for mouse
13. Corked surface
14. Safety glasses

15.2.2 ISH Buffers

1. 10X salt buffer
   - 2 M NaCL
   - 50 mM EDTA
   - 100 mM Tris-HCl pH 7.5
   - 50 mM $NaH_2PO_4.2H_2O$
   - 50 mM Na_2HPO_4
2. Hybridization buffer
   - 50% desionized formamide
   - 10% (w/v) dextran sulphate
   - 1X Denhart's
   - 0.1 mg/mL yeast transfer RNA
   - 1X salt buffer

3. Posthybridization buffer
   1X SSC
   50% formamide
   0.1% Tween 20
4. MABT buffer
   100 mM Maleic acid pH 7.5
   150 mM NaCl
   0.1% Tween 20
   NaOH
5. NTMT buffer
   100 mM Tris HCl pH 9.5
   100 mM NaCl
   50 mM $MgCl_2$
   0.1% Tween 20
6. Staining buffer
   0.2 mM nitroblueb tetrazolium salt
   0.2 mM 5-bromo-4-chloro-3indolyl-phosphate
   Levamisole 1 M
   NTMT
7. Blocking solution
   MABT
   2% blocking reagent, Roche
   10% heat-inactivated sheep serum

15.2.3 Immunofluorescence Materials and Buffers

1. Coated slides
2. Xylene
3. Ethanol (100%, 95%, 70%, and 50%)
4. Distilled water
5. Formaldehyde 4%
6. TBS 1X
7. Citrate buffer
8. Phosphate-buffered saline (PBS)/bovine serum albumin (BSA) 10%
9. Milk powder
10. Primary antibody
11. Secondary antibody
    A. TBS 1X
       Tris: 121.14 g
       NaCl: 175.32 g
       HCl: to adjust pH 7.5
       H_2O: adjust to 1L
       Add Tris and NaCl in a beaker with 900 mL H_2O
       Shake the solution to dissolve
       Adjust pH 7.5 with HCl
       Add H_2O up to 1L
    B. Citrate buffer

Stock solutions:

- 0.1 M citric acid:
  - 1.92 g citric acid, anhydrous to 100 mL in glass of distilled water
- 0.1 M Na citrate:
  - 14.7 g Na citrate, dihydrate to 500 mL in glass of distilled water

Working solution (store at 4°C):

- 9 mL of 0.1 M citric acid
- 41 mL of 0.1 M Na citrate
- Bring final volume to 500 mL with glass distilled water
- Adjust pH 6 with 5N NaOH

15.3 METHODS

15.3.1 Intracardial Perfusion of Mice

Currently, the recommended techniques for brain fixation require thoracotomy and direct cardiac perfusion with fixative solution under pressure. The goal of perfusion fixation is to use the vascular system of a deeply anesthetized animal to deliver fixatives to the tissues of interest. This is the optimal method of tissue preservation because the tissues are fixed before autolysis begins. The following technique is appropriate for harvesting brain and organs throughout circulation supplied by the left side of the heart. This method combines tissue fixation with euthanasia and can only be performed as a terminal procedure.

15.3.1.1 Anesthesia

1. Inject the mouse intraperitoneally with a ketamine/xylazine mixture. The most widely used dose of ketamine/xylazine for mouse surgery is 100 mg/kg and 10 mg/kg body weight, respectively (Flecknell, 1993).
2. Allow the mouse to rest in a cage by itself in a dark and quiet environment.
3. The withdrawal reflex must be absent in each pelvic limb before the perfusion can begin.

15.3.1.2 Perfusion

1. 4% PFA must be made fresh on the day of the procedure in a chemical fume hood.
2. The perfusion process should be performed in a chemical fume hood for the best personal protection
3. Place the mouse on its back.
4. Check the withdrawal reflex once more to assure adequate depth of anesthesia.
5. Make a midline skin incision from the thoracic inlet to the pelvis (Figure 15.1a).
6. Use scissors to carefully open the abdomen (Figure 15.1b).
7. Grasp the xiphoid, which is the tip of the sternum, with forceps and make an incision through the diaphragm (Figure 15.1c), then down the costal cartilage (Figure 15.1d).
8. Flip the thoracic cage widely enough to visualize the heart.
9. Grasp the heart gently and lift it to the midline and slightly out of the chest.
10. The left ventricle is thicker and lighter pink than the right ventricle.

11. Place butterfly needle into the apex of the left ventricle, and turn on the pump with a flow no higher than 0.5 mL/minute of PBS buffer (Figure 15.1e and f). Then immediately cut the right auricle to allow the perfusate to exit the circulation (Figure 15.1g and h).
12. When the fluid exiting the mouse is clear of blood, perfuse with 4% PFA.
13. Muscle contractions and blanching of the liver and mesenteric blood vessels are signs of good perfusion. Perfusion is complete when all muscle contractions have stopped, the liver and mesenteric vessels are blanched, and the desired amount of preservative has passed through the circulatory system. The mouse should be stiff. A good indication of how well the animal is being fixed is to test tail flexibility.
14. PFA and other fixatives must be collected after the perfusion and stored appropriately as hazardous chemical waste.

For immunofluorescence staining, postfixation can affect the antigen (Ag). Therefore, trials should be done with different times of postfixation. For example, postfixation for any amount of time does not seem to affect SMI-32 (unphosphorylated neurofilaments) immunoreactivity, but prolonged postfixation times will affect most common Ags for visualizing microglia.

15.3.1.3 Troubleshooting

1. A mouse still having a positive withdrawal reflex before the procedure indicates inadequate anesthesia. Administer one-third of the original dose of anesthesia and allow the mouse to rest in a dark, quiet cage until withdrawal reflexes are absent.
2. Fluid dripping from the mouse's nose indicates too much fluid pressure.
3. Fluid not seen dripping in the fourth chamber indicates that the fluid is not flowing into the mouse's circulatory system. The butterfly needle should be repositioned.

15.3.2 Brain Removal

1. Make a midline incision in the skin using the scissors (Figure 15.2a). Flip the skin over the eyes to free the skull. Insert the scissors beneath each eye to cut the optical nerve and extract the eyes (Figure 15.2b).
2. Insert Iris scissors caudally to the interparietal bone and cut along the sagittal suture (Figure 15.2c). Incline the scissors ~45° to avoid cutting through the brain (Figure 15.2d).
3. Tilt one side of the parietal bone with the curved narrow pattern forceps and break it off (Figure 15.2e and f). Do the same with the other side to reveal the brain. Most likely the frontal bone will remain. In that case, make a firm cut through the most anterior part of the skull, between the eyes (frontal bone, Figure 15.2g). This enables removal of the brain more easily.
4. Make a small incision that enables tilting and breaking off this bone plate. Cut the meninges that are beneath the skull.
5. When the brain is freed from meninges, slide the curved narrow pattern forceps (closed) under the anterior part of the brain (olfactory bulb) and tilt the brain gently upward (Figure 15.2h). Gently place the forceps underneath the brain and

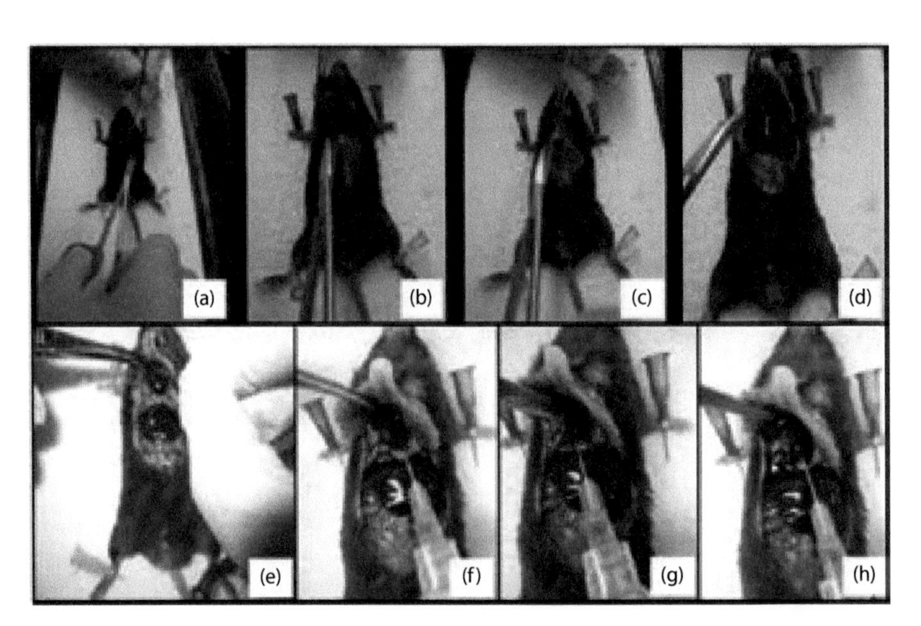

FIGURE 15.1 Intracardial perfusion of mice. (a) A midline skin incision from the thoracic inlet to the pelvis is made. (b) Scissors are used to open the abdomen. (c) The xiphoid process is grasped with forceps and an incision is made through the diaphragm, then down the costal cartilage; (d) the sternum is flipped widely enough to visualize the heart. (e,f) A butterfly needle is placed into the apex of the left ventricle. (g,h) The right auricle is cut to allow the perfusate to exit the circulation.

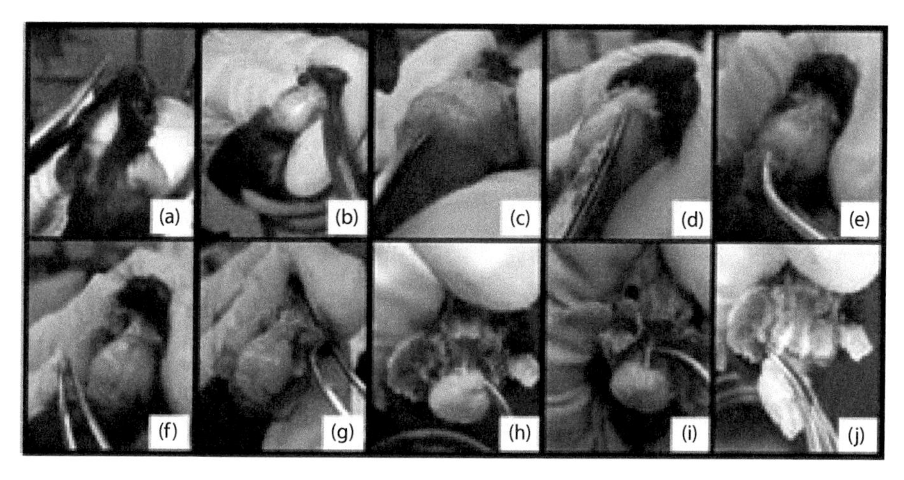

FIGURE 15.2 Brain removal. (a) A midline incision in the skin is made using the scissors. (b) The eyes are removed. (c) Iris scissors are inserted caudally to the interparietal bone (d) and the skull is cut along the sagittal suture. (e,f) One side of the parietal bone is tilted with the curved narrow pattern forceps. (g) The same is done with the other side to reveal the brain. (h) The curved forceps is inserted under the olfactory bulbs and the brain is gently tilted upward. (i) The brain is separated from the underlying tissue, and the optic nerves and other cranial nerves are broken. (j) The brain is lifted out of the skull.

separate it from the underlying tissue. Break the optic nerves and other cranial nerves (Figure 15.2i and j) and gently lift the brain out of the skull (Figure 15.2j).

6. Transfer brain to a petri dish filled with PBS buffer and place on ice to cool down the brain immediately. These steps should be performed within 2 to 3 minutes.

15.3.3 Brain Slicing

To perform histological studies, the brain must be sliced to visualize the brain cytoarchitecture. Different techniques can be used.

The cryosectioning is the slicing of a frozen brain into sections as thin as 10 μm. After fixation, the brain is immersed in 20% sucrose to cryoprotect and prevent freeze artifacts and loss of tissue architecture. The brain is then embeded in fresh optimum cutting temperature compound and immersed in isopentane placed in liquid nitrogen to permit rapid freezing. Cryosections show high preservation of antigenicity and therefore the detection of Ags by microscopy. They can be used in a variety of procedures including immunochemistry, enzymatic detection, and in situ hybridization.

Biological samples often need to be solidified to allow fine sectioning. Thin slices improve the access of dyes, probes, and antibodies. For light microscopy, paraffin wax is most frequently used on hard matrix for cutting. Paraffin sections for light microscopy are typically 5 μm thick. After fixation, the water must be removed from the tissue block, a process called dehydration, usually done by using isopropyl alcohol. Before sectioning, the tissue block must be infiltrated with a material such as paraffin that acts as a support during the sectioning process. Then, the tissue is allowed to solidify in a mold, embedded within a small cube of paraffin. Sectioning is accomplished by using a cutting apparatus called a microtome.

Vibratome sectioning is often used for floating immunostaining. This technique does not require any organic solvents or typical dehydration steps before embedding, which preserves the tissue morphology. The brain is embedded in agarose and a vibrating razor blade is used to cut through the tissue. This technique allows cutting of thick sections (50 μm). The disadvantage of vibratome sections is that the sectioning process is slow and difficult with soft and poorly fixed tissues.

The major planes of sectioning are transverse, coronal, and sagittal sections:

- Transverse sections (also called horizontal sections) (Figure 15.3a) provide a view of the whole brain, from the olfactory lobes, to the cerebellum.
- Coronal sections (Figure 15.3b), if made at specific sites, allow similar areas to be examined, so that comparisons can be made between littermate controls and from mutant animals. Coronal sections of the entire brain can be examined in order to pick up small abnormalities.
- Sagittal sections (Figure 15.3c) are made for easier viewing of each half of the brain, viewed from caudal to rostral aspects, to detect any obvious abnormalities.

15.3.4 In Situ Hybridization

In situ hybridization (Povlishock and Christman, 1995) is a technique that allows direct analysis of gene expression through the localization of specific nucleic acid sequences to individual cells within a morphologically preserved tissue (Guiot and Rahier, 1995). Detection is carried out using nucleic acid probes that are complementary to hybridize with a particular mRNA sequence (Jin and Lloyd, 1997). Because this detection is performed on tissue sections, in situ

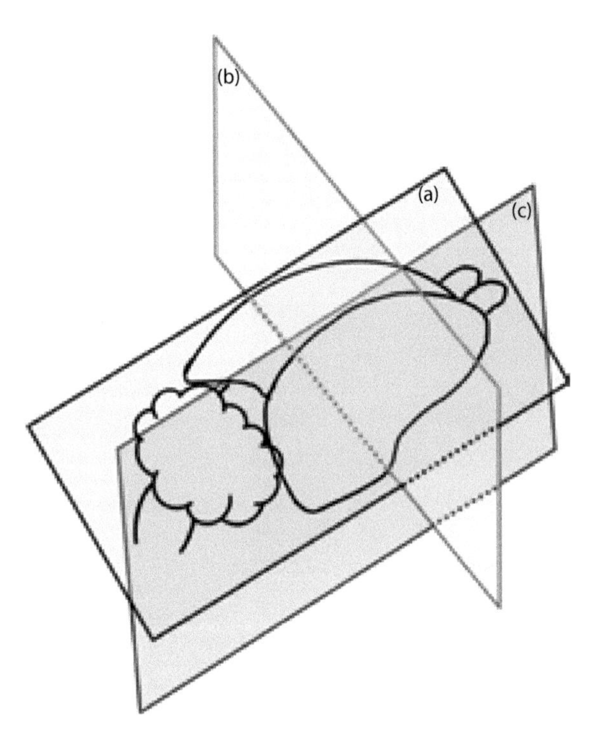

FIGURE 15.3 A schematic diagram showing the three principal section planes of a brain: (a) transverse, (b) coronal, and (c) sagittal.

hybridization provides additional morphological information on the spatial distribution and heterogeneity of gene expression in complex tissues.

Identifying the precise cellular localization of genes of interest is often necessary for understanding the regulation and the function of the genes. Traditionally, ISH depends on the hybridization of the specific RNA sequence in situ to radiolabeled probes (Gall and Pardue, 1969). Currently, digoxigenin-labeled probes are more commonly used in ISH, which can be recognized with antibodies coupled with fluorophore or enzymes such as alkaline phosphatase or peroxidase (Komminoth et al., 1992). ISH can be performed on both frozen sections and paraffin sections, with frozen sections allowing more sensitive detection of weak signals (Wilcox, 1993). The ISH method used in this chapter uses digoxigenin-labeled riboprobes (complementary RNA probes) to detect specific mRNA on frozen brain sections. The procedure includes hybridization of sections with digoxigenin-labeled riboprobes, posthybridization washes, incubation with alkaline phosphatase-conjugated antidigoxigenin antibody (Margulies et al., 2009); colormetric analysis using phosphatase substrate can be used for detection.

15.3.4.1 Protocol

1. Let the slides dry at room temperature 15 minutes to 4 hours.
2. The digoxigenin-labeled probe is diluted (usually 1/200) immediately before use in hybridization buffer, denaturated at 65°C for 5 to 10 minutes.
3. An appropriated volume (usually 150 μL per slide) of diluted probe is placed on each slide.
4. The slides are cover-slipped and placed in a humidified chamber overnight at 65°C.
5. After overnight hybridization, slides are incubated in posthybridization buffer at 65°C in glass Coplin jars until the coverslips slide off.
6. The slides are washed two times for 30 minutes in posthybridization buffer at 65°C.
7. The slides are washed two times for 30 minutes in MABT at room temperature.
8. The slides are transferred to a humidified chamber and incubated in blocking solution for 1 hour at room temperature without a coverslip.
9. Alkaline phosphatase (AP)-conjugated anti-DIG antibodies are diluted 1/2000 in blocking solution. An appropriated volume is placed on each slide.
10. The slides are cover-slipped and placed in a humidified chamber overnight at room temperature.
11. The slides are transferred to Coplin jars and washed five times for 20 minutes at room temperature in MABT buffer.
12. The slides are washed two times for 20 minutes at room temperature in NTMT prestaining buffer.
13. The slides are incubated in the dark at 37°C in the staining buffer until the signal reaches a satisfactory intensity (usually a few hours to overnight).
14. The slides are washed in PBS 1X and mounted.

15.3.5 IHC

The IHC technique is used in the search for cell or tissue Ags ranging from amino acids and proteins to infectious agents and specific cellular populations. IHC is an umbrella term that encompasses many methods used to determine tissue constituents (the Ags) with the use of specific antibodies that can be visualized through staining (Brandtzaeg, 1998; Haines and West, 2005). When used in cell preparations it is called immunocytochemistry.

The fundamental concept behind IHC is the demonstration of Ags within tissue sections by means of specific antibodies (Abs). Once Ag–Ab binding occurs, it is demonstrated with a colored histochemical reaction visible by light microscopy or fluorochromes with ultraviolet light. So, it is based on the binding of Abs to a specific Ag in tissue sections. The most common immunoglobulin (Haydon and Carmignoto, 2006) used in IHC is immunoglobulin G.

Fixation of tissues is necessary to (1) adequately preserve cellular components, including soluble and structural proteins; (2) prevent autolysis and displacement of cell constituents, including Ags and enzymes; (3) stabilize cellular materials against deleterious effects of subsequent procedures; and (4) facilitate conventional staining and immunostaining. Two types of fixatives are used in histopathology: cross-linking (noncoagulating) fixatives and coagulating fixatives.

Formaldehyde is the gold standard of fixatives for routine histology and IHC. Formaldehyde preserves mainly peptides and the general structure of cellular organelles.

It also interacts with nucleic acids, but has little effect on carbohydrates. Many of the formalin substitutes are coagulating fixatives that precipitate proteins by breaking hydrogen bonds in the absence of protein cross-linking. The typical non–cross-linking fixative is ethanol.

Fixation modifies, several times, the tertiary structure of proteins (Ags), making them undetected by specific Abs. This is better understood if one remembers that the reaction between the Ag and the Ab depends on the conformation of the former. One of the challenges of IHC is to develop methods that reverse changes produced during fixation. Antigen retrieval producers reverse at least some of these changes. Antigen retrieval is particularly necessary when tissues are fixed in cross-linking fixatives. Approximately 85% of Ags fixed in formalin require some type of antigen retrieval to optimize the immunoreaction.

The Ag–Ab reaction cannot be seen with the light microscope unless it is labeled. Therefore, labels (reporter molecules) are attached to the primary, secondary, or tertiary Abs of a detection system to allow visualization of the immune reaction. Several labels have been used, including fluorescent compounds, enzymes, and metals. The most commonly used labels are enzymes (e.g., peroxidase, alkaline phosphatase, glucose oxidase). Enzymes in the presence of a specific substrate and a chromogen will produce a colored precipitate at the site of the Ag–Ab reaction. Selection of a detection system is very important, considering that the sensitivity of an immune reaction will depend mostly on the detection system used. Detection systems are classified as direct or indirect methods.

Background is one of the most common problems in IHC, and it can seriously affect the interpretation of the immunologic reaction

15.3.5.1 Protocol

1. Dry paraffin sections (~8–10 μm) on coated slides (eg: HistoBond) overnight at 37°C.
2. Incubate slides at 60°C for 10 minutes.
3. Deparaffinization/rehydration:
   - Put the slides in xylene bath for 5 minutes. Repeat this step twice in a different bath.
   - Put the slides in different ethanol baths (100%, 95%, 70%, and 50%) for 3 minutes each.
   - Put the slides in distilled water for 3 minutes.
4. Fixation:
   - Fix the tissue with 1 mL formaldehyde 4%.
   - Incubate for 20 minutes at room temperature.
5. Aspirate the fixative and wash the slides with TBS 1X for 1 minute.
6. Wash the slides three times with TBS 1X for 3 minutes.
7. Antigen unmasking:
   - Put the slides in a jar containing 10 mM citrate buffer and immerse it in a boiling water bath (99° C) with a glass lid on.
   - Incubate the slides for 10 minutes.
   - Remove the jar from water bath, open the lid, and let it cool on bench for 30 minutes.
8. Wash the slides three times with TBS 1X and 2% milk powder for 5 minutes.
9. Blocking: Incubate the slides with PBS/BSA 10% for 20 minutes at room temperature.
10. Primary antibody:
    - Decant the blocking solution and apply 1 mL of the primary antibody in its working dilution in PBS/BSA 10%.
    - Incubate slides overnight at 4°C in a humid chamber.
11. Remove the primary antibody and wash the slides three times with TBS 1X diluted in 2% milk powder for 5 minutes.
12. Secondary antibody:
    - Apply 1 mL of the secondary antibody in its working dilution in PBS/BSA 10%.
    - Incubate slides 1 hour at room temperature in the dark.
13. Remove the secondary antibody and wash the slides three times with TBS 1X diluted in 2% milk powder for 5 minutes.
14. Wash the slides three times with TBS 1X *without* milk powder for 5 minutes.
15. Mount the slides.

15.4 CONCLUSION

In conclusion, this chapter describes methods and techniques of animal brain surgery and can be applied on experimental brain injury models (mouse model). The first method described is intracardial perfusion. This method consists of dissecting the thoracic cage of the mouse and injecting a fixative solution to fix and preserve the tissues. Importantly, this technique is a terminal procedure because it ends with euthanasia. Then, this chapter presents an appropriate method to dissect and remove the brain with the tools and solutions necessary for an optimal result. This is followed by the description of three different techniques used for brain slicing. The first technique is called cryosectioning and consists of slicing frozen tissues that can be later used for immunochemistry, enzymatic detection, and in situ hybridization. The second technique is microtome sectioning, which consists of solidifying the biological samples with paraffin to allow fine slicing. The third technique is the vibratome sectioning that is used for floating immunostaining and allows thick slicing. In addition, the chapter portrays the three major types of sections: transverse, coronal, and sagittal sections. Afterward, the ISH method used for detection of mRNA in brain sections is assessed. This method allows precise cellular localization of genes of interest by using nucleic acid probes that are complementary to hybridize with a particular mRNA sequence. Of interest, ISH can be used on frozen and paraffin embedded sections. In this chapter the protocol described makes use of digoxigenin-labeled riboprobes to detect specific mRNA on frozen brain sections. Finally, this chapter describes the IHC method used to detect protein in brain sections and provides the most appropriate protocol to be followed. IHC is mainly used to demonstrate the presence of Ags within tissue sections by means of specific Abs. Then, labels, mainly enzymes, attached to the primary, secondary, or tertiary Abs allow visualization of the immune reaction.

REFERENCES

Bezzi, P., V. Gundersen, J.L. Galbete, G. Seifert, C. Steinhauser, E. Pilati et al. 2004. Astrocytes contain a vesicular compartment that is competent for regulated exocytosis of glutamate. *Nature Neuroscience*. 7:613–620.

Bezzi, P. and A. Volterra. 2001. A neuron-glia signalling network in the active brain. *Current Opinion in Neurobiology*. 11:387–394.

Brandtzaeg, P. 1998. The increasing power of immunohistochemistry and immunocytochemistry. *Journal of Immunological Methods*. 216:49–67.

Cali, C. and P. Bezzi. 2010. CXCR4-mediated glutamate exocytosis from astrocytes. *Journal of Neuroimmunology*. 224:13–21.

Cali, C., J. Marchaland, R. Regazzi, and P. Bezzi. 2008. SDF 1-alpha (CXCL12) triggers glutamate exocytosis from astrocytes on a millisecond time scale: Imaging analysis at the single-vesicle level with TIRF microscopy. *Journal of Neuroimmunology*. 198:82–91.

Cernak, I., J. Savic, Z. Malicevic, G. Zunic, P. Radosevic, I. Ivanovic et al. 1996. Involvement of the central nervous system in the general response to pulmonary blast injury. *The Journal of Trauma*. 40:S100–104.

Dixon, C.E., G.L. Clifton, J.W. Lighthall, A.A. Yaghmai, and R.L. Hayes. 1991. A controlled cortical impact model of traumatic brain injury in the rat. *Journal of Neuroscience Methods*. 39:253–262.

Dixon, C.E., B.G. Lyeth, J.T. Povlishock, R.L. Findling, R.J. Hamm, A. Marmarou et al. 1987. A fluid percussion model of experimental brain injury in the rat. *Journal of Neurosurgery*. 67:110–119.

Domercq, M., L. Brambilla, E. Pilati, J. Marchaland, A. Volterra, and P. Bezzi. 2006. P2Y1 receptor-evoked glutamate exocytosis from astrocytes: Control by tumor necrosis factor-alpha and prostaglandins. *The Journal of Biological Chemistry*. 281:30684–30696.

Flecknell, P.A. 1993. Anaesthesia of animals for biomedical research. *British Journal of Anaesthesia*. 71:885–894.

Gall, J.G. and M.L. Pardue. 1969. Formation and detection of RNA-DNA hybrid molecules in cytological preparations. *Proceedings of the National Academy of Sciences of the United States of America*. 63:378–383.

Guiot, Y. and J. Rahier. 1995. The effects of varying key steps in the non-radioactive in situ hybridization protocol: A quantitative study. *The Histochemical Journal*. 27:60–68.

Haines, D.M. and K.H. West. 2005. Immunohistochemistry: Forging the links between immunology and pathology. *Veterinary Immunology and Immunopathology*. 108:151–156.

Haydon, P.G. and G. Carmignoto. 2006. Astrocyte control of synaptic transmission and neurovascular coupling. *Physiological Reviews*. 86:1009–1031.

Huang, Y.H., S.R. Sinha, K. Tanaka, J.D. Rothstein, and D.E. Bergles. 2004. Astrocyte glutamate transporters regulate metabotropic glutamate receptor-mediated excitation of hippocampal interneurons. *The Journal of Neuroscience*. 24:4551–4559.

Hughes, V. 2012. Microglia: The constant gardeners. *Nature*. 485:570–572.

Jin, L. and R.V. Lloyd. 1997. In situ hybridization: Methods and applications. *Journal of Clinical Laboratory Analysis*. 11:2–9.

Jourdain, P., L.H. Bergersen, K. Bhaukaurally, P. Bezzi, M. Santello, M. Domercq et al. 2007. Glutamate exocytosis from astrocytes controls synaptic strength. *Nature Neuroscience*. 10:331–339.

Komminoth, P., F.B. Merk, I. Leav, H.J. Wolfe, and J. Roth. 1992. Comparison of 35S- and digoxigenin-labeled RNA and oligonucleotide probes for in situ hybridization. Expression of mRNA of the seminal vesicle secretion protein II and androgen receptor genes in the rat prostate. *Histochemistry*. 98:217–228.

Leung, L.Y., P.J. VandeVord, A.L. Dal Cengio, C. Bir, K.H. Yang, and A.I. King. 2008. Blast related neurotrauma: A review of cellular injury. *Molecular & Cellular Biomechanics*. 5:155–168.

Lighthall, J.W. 1988. Controlled cortical impact: A new experimental brain injury model. *Journal of Neurotrauma*. 5:1–15.

Margulies, S., R. Hicks, and Combination Therapies for Traumatic Brain Injury Workshop Leaders. 2009. Combination therapies for traumatic brain injury: Prospective considerations. *Journal of Neurotrauma*. 26:925–939.

Marmarou, A., M.A. Foda, W. van den Brink, J. Campbell, H. Kita, and K. Demetriadou. 1994. A new model of diffuse brain injury in rats. Part I: Pathophysiology and biomechanics. *Journal of Neurosurgery*. 80:291–300.

Nedergaard, M., B. Ransom, and S.A. Goldman. 2003. New roles for astrocytes: Redefining the functional architecture of the brain. *Trends in Neurosciences*. 26:523–530.

Nimmerjahn, A., F. Kirchhoff, and F. Helmchen. 2005. Resting microglial cells are highly dynamic surveillants of brain parenchyma in vivo. *Science*. 308:1314–1318.

Pascual, O., S. Ben Achour, P. Rostaing, A. Triller, and A. Bessis. 2012. Microglia activation triggers astrocyte-mediated modulation of excitatory neurotransmission. *Proceedings of the National Academy of Sciences of the United States of America*. 109:E197–205.

Pham, K. and R. Gupta. 2009. Understanding the mechanisms of entrapment neuropathies. Review article. *Neurosurgical Focus*. 26:E7.

Povlishock, J.T. and C.W. Christman. 1995. The pathobiology of traumatically induced axonal injury in animals and humans: A review of current thoughts. *Journal of Neurotrauma*. 12:555–564.

Prada, I., J. Marchaland, P. Podini, L. Magrassi, R. D'Alessandro, P. Bezzi et al. 2011. REST/NRSF governs the expression of dense-core vesicle gliosecretion in astrocytes. *The Journal of Cell Biology*. 193:537–549.

Ransohoff, R.M. and A.E. Cardona. 2010. The myeloid cells of the central nervous system parenchyma. *Nature*. 468:253–262.

Ransohoff, R.M. and V.H. Perry. 2009. Microglial physiology: Unique stimuli, specialized responses. *Annual Review of Immunology*. 27:119–145.

Santello, M., P. Bezzi, and A. Volterra. 2011. TNFalpha controls glutamatergic gliotransmission in the hippocampal dentate gyrus. *Neuron*. 69:988–1001.

Schafer, D.P., E.K. Lehrman, A.G. Kautzman, R. Koyama, A.R. Mardinly, R. Yamasaki et al. 2012. Microglia sculpt postnatal neural circuits in an activity and complement-dependent manner. *Neuron*. 74:691–705.

Verkhratsky, A., R.K. Orkand, and H. Kettenmann. 1998. Glial calcium: Homeostasis and signaling function. *Physiological Reviews*. 78:99–141.

Wilcox, J.N. 1993. Fundamental principles of in situ hybridization. *The Journal of Histochemistry and Cytochemistry*. 41:1725–1733.

16 Controlled Cortical Impact Model

Nicole D. Osier, Jonathan R. Korpon, and C. Edward Dixon

CONTENTS

16.1 INTRODUCTION: DEVELOPMENT AND KEY FEATURES OF CCI

The use of in vivo (i.e., animal) models to study traumatic brain injury (TBI) is far from new; indeed since the late 1800s many models of brain injury have been employed for this purpose (Gennarelli et al., 1982; Govons et al., 1972; Nilsson, Pontén, and Voigt, 1977; Ommaya, Geller, and Parsons, 1971; Ommaya and Gennarelli, 1974; Rinder and Olsson, 1968; Sullivan et al., 1976; Denny-Brown and Russell, 1941; Kramer, 1896; Cannon, 1901; Parkinson, West, and Pathiraja, 1978). In the 1980s, Lighthall and his colleagues developed a pneumatic impactor for the purpose of modeling TBI in ferrets (Lighthall, 1988; Lighthall, Goshgarian, and Pinderski, 1990); this method is now referred to as controlled cortical impact (CCI). The control and reproducibility of CCI led Dixon and colleagues to adapt the model for use in rats (Dixon et al., 1991). Since its development, CCI has become one of the most common models of brain injury in animals and has been applied to numerous species. This chapter will familiarize the reader with the key aspects of CCI (e.g., species applied to; types of devices) the clinical features of brain injury replicated by this model, and considerations for researchers using CCI. It will also provide a standard protocol for pneumatic CCI in rats.

Controlled cortical impact allows for quantitative control over injury force and velocity as well as extent of tissue deformation; thus, it affords control over biomechanical parameters known to be associated with TBI. This independent control over injury parameters across a wide range of contact velocity contributes to the reliability and accuracy of controlled cortical impact (CCI) as a model of TBI. Depending on the goals of the research study, injury can be controlled to produce a range of injury magnitudes, allowing the researcher to produce gradable functional impairment, tissue damage, or both. A thorough review of the literature can provide preliminary guidance regarding how to set the injury parameters for the study. Beyond reviewing the literature, it is encouraged that researchers conduct pilot work to help them fine-tune the CCI parameters to best meet their research goals. Pilot work will also allow the researchers to test the machine and ensure that it is working properly and the velocity and dwell time is consistent with what is set.

Briefly, CCI consists of an anesthetized incision and craniectomy to gain access to the animal's brain for injury induction in the form of deformation of brain tissue using a pneumatic or electromechanical CCI device. Although CCI is an invasive model of head trauma that requires neurosurgery, only one surgical procedure is needed as opposed to the two required when using the standard fluid percussion model of brain injury (Kline and Dixon, 2001). An additional strength of CCI is that it can be scaled to model TBI in different mammalian species ranging from small to large depending on the goals of the research and available facilities. Of the most common models of experimental brain injury, CCI has the broadest applicability across species and has been modified for use in: mice (Fox et al., 1998b; Lee et al., 2012; Smith et al., 1995; Wang et al., 2013), rats (Acosta et al., 2013; Dixon et al., 1991; Vonder Haar et al., 2013; Xing et al., 2012), swine (Costine et al., 2012; Manley et al., 2006; Friess et al., 2007, 2009a), and primates (King et al., 2010). None of the other common models of brain injury (e.g., fluid percussion, weight drop) been used across all the aforementioned animal species.

16.2 CCI DEVICES

The device typically used to produce CCI consists of an impactor tip attached to a shaft that is accelerated by either a pneumatic piston or an electromechanical actuator. Pneumatic CCI devices utilize a small-bore (19.75-mm), reciprocating, double-acting pneumatic piston with a 50-mm travel stroke length. The cylinder is held on a crossbar with variable mounting positions, allowing for both angled and vertical alignment with the surface of the brain (Figure 16.1 illustrates a standard pneumatic CCI device). To ensure uniform injuries, a velocity sensor is mounted on the end of the cylinder, which allows the researcher to control the velocity of the piston by varying the applied gas pressure. To produce a moderately severe TBI, the impactor tip is driven at a velocity of 4.0 m/s, penetrating 2.6–2.8 mm from the dural surface into the dura for a dwell time (duration) of 50–150 ms. The exact parameters may be adjusted by the researcher depending on the research question and the extent of injury desired.

Electromechanical CCI devices are lighter and are typically mounted to stereotaxic manipulator arms (Figure 16.2 is a photograph of a commercially available electromechanical CCI device). Various impactor tips are available including rounded or flat edged with diameters varying from 5 to 6 mm for rats and 3 mm for mice. Some commercially available electromechanical CCI devices come with a variety of tips with additional sizes or replacements available for purchase. The lightweight nature of the electromechanical device makes it a favorable choice when portability is desired, as may be the case when researchers are collaborating or when animal housing and transport rules are stringent.

16.3 CLINICAL RELEVANCE

16.3.1 Overview

TBI is highly variable with respect to underlying cause, pathophysiologic response, and functional outcomes. Thus, to best study TBI and test potential therapeutic strategies to target the myriad of problems associated with it, numerous experimental models of brain injury are needed, including but not limited to CCI. Still, CCI is revered for its ability to produce graded morphological and functional responses to TBI which is useful when evaluating injury severity or the effectiveness of therapeutic agents. Commonly, the goal of CCI research is to conduct preclinical drug trials to evaluate potential novel therapeutic agents; for example, a recent study evaluated the therapeutic window of Edaravone, a free-radical scavenging drug, after TBI (Miyamoto et al., 2013). Such preclinical drug studies have resulted in translating findings from CCI research to clinical trials, as is true for Amantadine, a weak N-methyl-D-aspartate receptor antagonist that releases dopamine (Dixon et al., 1999; Giacino et al., 2012).

Another clinically relevant aspect of CCI is that it can be used to study TBI across the lifespan. For example, CCI has been used to study the effects of TBI on the immature brain (Adelson et al., 2013; Kamper et al., 2013; Pop et al., 2013; Robertson et al., 2013; Russell et al., 2013), providing a preclinical foundation on which later clinical trials can be conducted. Some of this preclinical work includes studies using immature swine (Costine et al., 2012; Friess et al., 2007, 2009b). More commonly, CCI is used in adult rodents and adults of other species. CCI has been used to assess the effects of brain injury in older adult animals. This work includes several CCI studies that used aging or senescence-prone mice (Lee et al. 2012; Sandhir and Berman, 2010; Timaru-Kast et al., 2012; Tran et al., 2011).

A final consideration for the clinical relevance of CCI is that successive CCI procedures can be used to study the consequences of repetitive TBI (Friess et al., 2009a; Vonder Haar et al., 2013), which is relevant to populations at risk for multiple brain injuries (i.e., military personnel and athletes). Overall, CCI represents a clinically relevant model that can be applied to evaluate potential therapies, reproduce key physiological and functional changes that follow TBI, and study the effects of repeated head injuries.

16.3.2 Physiological Outcomes of TBI

Many physiological responses to human TBI can be modeled in animals using CCI. These clinically relevant outcomes of interest to researchers include but are not limited to cortical contusion (Atkins et al., 2013; Adelson et al., 2013; Singleton et al., 2010; Sword et al., 2013), inflammation (Acosta et al., 2013; Gatson et al., 2013; Haber et al., 2013; Khan et al., 2011; Schaible et al. 2013), oxidative stress (Jiang et al., 2013; Khan et al., 2011; Miyamoto et al. 2013; Zhang et al. 2012), axonal injury (Dixon et al. 1991; Goodman et al. 1994; Lighthall 1988; Lighthall et al., 1990; Meaney et al. 1994), apoptosis (Campolo et al., 2013; Chen et al., 2012; Fox et al., 1998b; Kaneko et al., 2012; Schaible et al., 2013; Zhao et al., 2012), and blood–brain barrier disruption (Dhillon et al. 1994; Kochanek et al. 1995; P. Lee et al. 2012). Of additional interest to clinicians and researchers is the lesion size or extent of tissue loss, which is approximated after staining serial coronal sections with hematoxylin and eosin (Acosta et al., 2013; Vonder Haar et al. 2013). Hippocampal neuron loss is often an outcome variable of interest especially when memory and learning are of interest to researchers. Hippocampal cell death can be estimated using semiquantitative methods

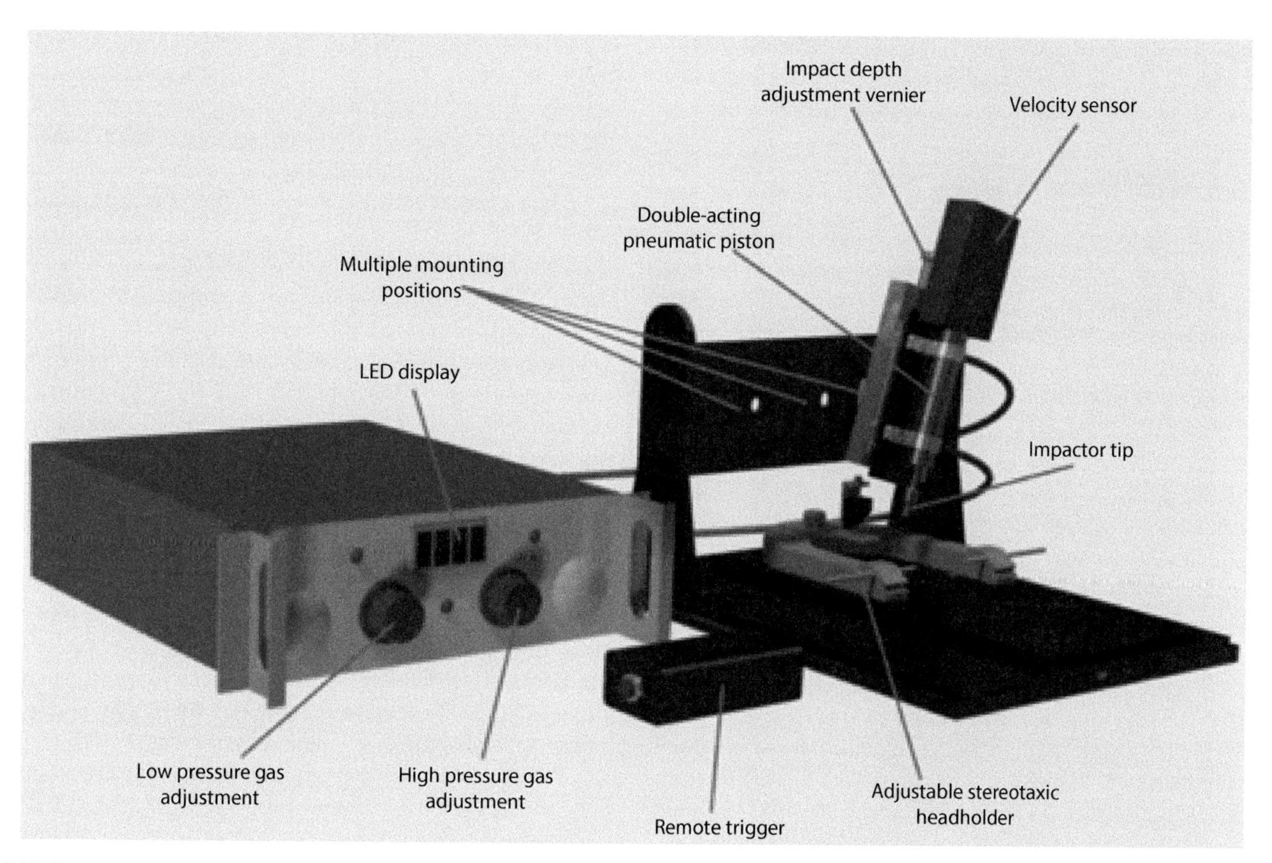

FIGURE 16.1 Diagram of a typical pneumatic device used to produce controlled cortical impact in rats, with all major components labeled.

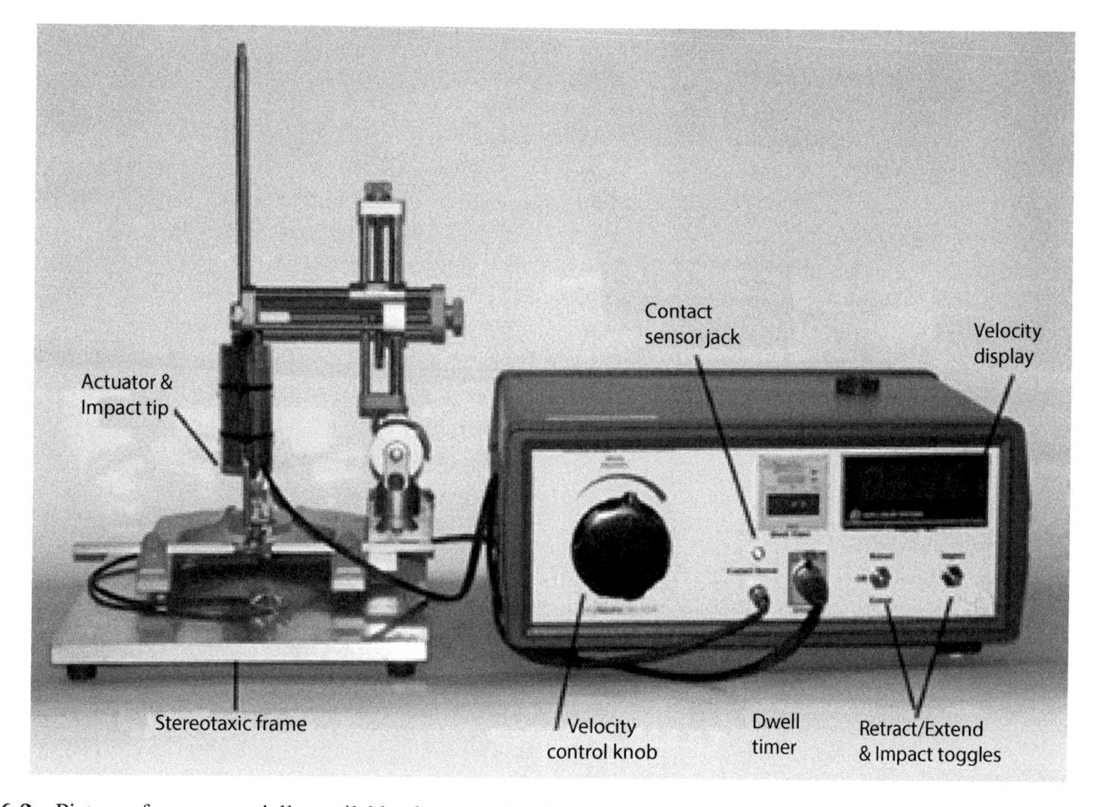

FIGURE 16.2 Picture of a commercially available electromechanical device used to produce controlled cortical impact, with all major components labeled.

based on counting of healthy neurons; conversely, hippocampal neuron survival can be more precisely estimated using unbiased stereological methods (Washington et al., 2012).

16.3.3 Functional Outcomes of Brain Injury

CCI is known to produce many neurobehavioral deficits common in human TBI, including but not limited to acute neurological responses and longer-term problems with cognitive and motor function. When CCI protocol uses anesthesia with a short half-life, acute neurological outcomes can be studied using assessments similar to the Glasgow Coma Scale, which allows for quantification of the magnitude of injury based on extent of functional suppression observed (Teasdale and Jennett, 1974). In preclinical models using CCI, composite neurological scores can be used to evaluate the severity of global neurological deficits post-injury (McIntosh et al., 1989). Similar to the Glasgow Coma Scale, animals are given a total score from 0 (representing complete loss of function) to 4 (representing completely normal function) based on the summed total across four tasks.

One specific area of cognition that is commonly affected by TBI in humans and contributes to disability and decreased independence is memory loss (Corrigan et al., 2010; Ergh et al., 2002; Levin et al., 1979; McKinlay et al., 1981; Schalen et al., 1994). The Morris Water Maze (MWM) is a commonly used test of memory and learning in rodents assessed after CCI (Hamm et al. 1992, 1996; Kline et al. 2002, 2004, 2007; Morris et al., 1982; Smith et al., 1991; Vonder Haar et al, 2013; Washington et al., 2012). The MWM is a well-established test of memory function (Morris, 1981); animals begin each trial in one of four possible start locations (i.e., north, east, south, or west) and are given up to 2 minutes to find a platform randomly placed in one of four quadrants of the maze; during the trial, extra-maze cues within the room are used by the animal in its search for the platform. The MWM has been used to assess rodents after CCI and has resulted in spatial learning deficits (Dixon et al., 1999a, 1999b; Kline et al., 2000). An alternative to the MWM is the Barnes Circular Maze, which is also used to assess cognitive function in rodents after CCI (Fox et al., 1998a; Vonder Haar et al., 2013). The Barnes Circular Maze device consists of a large circular platform with 20–40 holes at the perimeter and an escape tunnel located underneath one of the holes. An emerging assessment of cognitive function after TBI is conditioned fear response (Dash et al., 2002).

Impaired motor function is another debilitating consequence of TBI that can be studied in preclinical models using CCI. To assess gross motor function in rodents (both mice and rats), the Beam Balance Task (BBT) is used to measure the amount of time (i.e., latency) spent on a beam. Typically the task is repeated for three trials, with each trial lasting a maximum of 60 seconds unless the animal falls off prematurely. Scoring on the BBT varies between laboratories, with some protocols using all-or-nothing scoring and other protocols preferring more sophisticated scoring systems (for instance, one that awards more points for animals with both hind limbs on the beam vs. those draped over or dangling from the beam). The width of the beam is adjusted for the species of animal being tested with rats balancing on a wider beam than mice. Also worth noting is the beam's material and shape may differ depending on the laboratory.

When assessment of fine motor function is desired, the Beam Walking Task (BWT) developed by Feeney and colleagues can be used (Feeney et al., 1982). Since the development of the BWT, it has become a common assessment of fine motor function in CCI studies (Kline et al., 2004, 2007; Dixon et al., 2003). The BWT uses negative reinforcement (i.e., aversion to bright light and loud noise) to motivate the animal to traverse a beam to reach a quiet, darkened goal box (i.e., perceived safe zone). Notably, the BWT is primarily used in rat CCI studies, but it has been applied to mice exposed to CCI by counting foot faults (Kline and Dixon, 2001). Also worth noting is that, like the BBT, the scoring of the BWT varies by laboratory protocol, which may contribute to mixed results across studies. Common variations in the BWT include whether the outcome measure is amount of time (latency) to reach the goal box or the number of steps taken. Researchers using the BWT to evaluate fine motor function after CCI should take great care to identify the BWT protocol most appropriate for each research study.

Another option for fine motor assessment is Rotarod performance, which has been used to evaluate both rats (Vonder Haar et al., 2013) and mice (Raghupathi et al., 1998) after CCI. In this test, the animal is placed on a rotating rod or rotating bars and the speed (i.e., number of revolutions per minute) is progressively accelerated over time. Time remaining on the bar (latency) is typically measured as the outcome, with longer latency suggesting less deficit and presumably less neurological damage.

Complex consequences of TBI can also be studied using CCI, including metabolic disturbance. For example, the effects of TBI on the pyruvate-metabolizing enzymes used in glycolysis were evaluated using immunohistochemistry to quantify enzyme levels in the cerebral cortex and real-time polymerase chain reaction to evaluate messenger RNA expression during and after CCI (Xing et al., 2012). The effects of CCI on risk of other conditions such as post-traumatic epilepsy (Bolkvadze and Pitkänen, 2012) and Alzheimer disease (Tran et al., 2011) have also been evaluated. In the case of Alzheimer disease, researchers found that CCI led to increased levels of substances (i.e., amyloid beta peptides and tau proteins) known to be associated with Alzheimer's disease risk (Tran et al. 2011).

16.4 APPLICATIONS OF CCI TO VARIOUS LABORATORY ANIMAL SPECIES

16.4.1 Overview

Generally speaking, scaling across species can be accomplished by keeping the percent tissue volume deformed proportional to total brain volume. When large animals are being used for research, the height of the piston may need to be

TABLE 16.1
Comparison of CCI Parameters by Species of Laboratory Animal

CCI Parameters	Mouse	Rat	Pig	Primate
Impactor tip diameter	3 mm	5–6 mm	15 mm	10 mm
Velocity	4–6 m/ second	4 m/ second	2–4 m/ second	3.5 m/ second
Dwell time	50–250 msec	50–250 msec	50–400 msec	150 msec
Depth setting	0.5–2.0 mm	1–3 mm	12 mm	7 mm
Craniectomy size	4–5 mm	6–8 mm	15 mm	22 mm
Impact site	Parietal cortex	Parietal cortex and midline	Frontal lobe	Frontal lobe

structurally raised to accommodate for the increase in size. A step-by-step protocol for CCI in rats is provided in the next section, but first a brief description of how CCI is applied across the four main species of laboratory animals used in TBI research (rats, mice, swine, and primates) is included to provide a general sense of how CCI can be altered to accommodate a variety of species (Table 16.1 highlights important comparisons in how CCI is used across species).

16.4.2 Rat CCI

As described previously, CCI in the rat is used to produce graded injury with several characteristics similar to humans, including axonal injury, disruption of the blood–brain barrier, as well as alterations in blood flow, edema, inflammation, and other histological changes. In addition to cellular-level outcomes, behavioral effects of brain injury can be studied in rats after CCI. For instance, motor function is assessed using the BBT (gross) or the BWT (fine), whereas memory function can be assessed using the MWM or Novel Object Recognition Task.

16.4.3 Mouse CCI

Traditionally, animal models of TBI have focused on the use of the laboratory rat, but recently, this trend has changed; specifically, the increasing availability of genetically mutant mice (e.g., knockout, transgenic) has allowed researchers to study the effects of various genes on the extent of injury versus recovery after TBI. This led to the modification of the CCI model for use in mice by Smith et al., using a background strain (C57BL6) from which genetic mutants are often made (Smith et al., 1995). Since this initial work, CCI has been further adapted for the study of mice (Fox et al., 1998b, 1999; Hannay et al., 1999). The adaptation of CCI to mice involves scaling down the depth of tissue deformation because mouse cortex is thinner than rat cortex. When CCI is used for mice, impactor tip size is typically 3 mm in diameter. Outcome assessments at the cellular and behavioral levels are similar to those used for rats although some mouse-specific options are available. Notably, in mice the Grip Test adapted by Hall and colleagues can also be used (Hall, 1985). Since it has been developed, CCI has become the most common method of modeling TBI in mice.

16.4.4 Swine CCI

Sometimes studying TBI in larger animals, more specifically on larger, more human-like brains, is necessary to appropriately address a research question. CCI can be applied in such cases by simply increasing the size of the tip used to impact the brain, along with the depth of tissue deformation. Duhaime and colleagues did just that and scaled CCI to piglets based on considerations for the size and dimensions of the growing piglet brain (Duhaime et al., 2000). As is true for mice and rats, CCI produces outcomes that are clinically relevant to human TBI including, but not limited to: neuronal loss, disruption of vasculature, and edema. The pathophysiology and biomechanics of CCI in swine has been evaluated because of the clinical relevance to human TBI (Manley et al., 2006). The effects of CCI on the developing piglet brain have been studied in an attempt to identify relevant biomarkers; researchers found that neuron-specific enolase increases in peripheral blood after CCI in swine (Costine et al., 2012).

Disadvantages of using large animal TBI models include a relatively high expense as well as a lack of normative biochemical and molecular data (Kline and Dixon, 2001). Still, often the rationale for using a large animal (e.g., swine) model is that the brains are similar in mass and biomechanical properties to those of humans. Also worth mentioning is that brain injury tolerances to mechanical stress tend to be easier to calculate in large animal models and often imaging techniques are more readily available.

16.4.5 Primate CCI

CCI has been more recently been developed for nonhuman primate models of TBI (King et al., 2010). Typically, the impact to the brain is over the right frontal cortex, leading to pathophysiology common to human brain injury (e.g., edema, neuron loss, macrophage accumulation). Arguably the most important application of primate CCI is preclinical trials of potential therapeutics for TBI to determine safety before human trials can be conducted. Although controversial, this adaptation of the CCI model is available for research questions that are only possible to answer using nonhuman primate models.

16.5 CONSIDERATIONS FOR RESEARCHERS

16.5.1 Anesthesia Type

As mentioned previously, if the goal of the research is to study the neurological changes in the acute post-TBI period, researchers can choose anesthesia options with short half-lives so that consciousness returns quickly after injury is induced. Shorter-acting options include volatile gases, such as isoflurane and halothane, which are preferred over long-acting agents such as pentobarbital for research aimed at evaluating short-term consequences of TBI. Indeed, Dixon et al. (1991) found that acute neurological assessments could be assessed rapidly after discontinuation of isoflurane; specifically, this study found the righting reflex and hind paw flexion could be assessed within seconds on isoflurane discontinuation in sham animals and within minutes after CCI (Dixon et al., 1991). In addition to the metabolism and half-life of the chosen method of anesthesia, there are other important considerations for researchers. Obviously, it is important that with whatever agent is chosen, a sufficient dose is provided to achieve and maintain a surgical level of anesthesia in the test animal. Commonly, a higher dose is used for anesthesia induction and the dose is reduced to maintain the effects. As part of the effort to reduce animal suffering, in accordance with national and institutional policies, researchers should reassess the animal's level of consciousness (many laboratories test for toe-pinch withdrawal), taking care not to touch the animal after the device has been zeroed but before CCI is induced to avoid personal injury.

Depending on the physiological processes or pathways of interest, the research team can select an appropriate anesthetic by considering the body's response to various agents. It has been established that some types of anesthesia result in hypothermia, others alter the cerebral metabolic rate, whereas some confer neuroprotection in one or more ways. Halothane is known to result in neuroprotection, and this benefit seems to be most pronounced when brain trauma is due to contusion; this has been attributed to the known effects of halothane on vasodilation, which may prevent ischemia (McPherson et al., 1994). Some researchers have set out to evaluate differences in outcome between types of anesthesia and found that compared with fentanyl, isoflurane led to better long-term outcomes at the tissue level as well as reduced deficit on cognitive and motor tasks (Statler et al., 2000). The same research team studied the effects of pre- and postinjury anesthesia with various combinations of isoflurane and/or fentanyl and found that fentanyl was associated with the worst outcomes. Furthermore, they found isoflurane conferred neuroprotection, especially when given pre-TBI (Statler et al., 2006). Although fentanyl is believed to contribute to neural suppression, the beneficial effects of isoflurane might be attributed to its ability to enhance blood flow or reduce excitotoxicity within the brain (Kline and Dixon, 2001). Similarly, the less commonly used ketamine-induced anesthesia may contribute to neuroprotection by antagonizing an important pathway within the brain. Specifically, ketamine acts as an N-methyl-D-aspartate antagonist; this is achieved via inhibition of glutamate and aspartate, two excitatory amino acids.

Whether the research goal consists of studying the effect of brain injury on acute neurological outcomes or assessing the effects of some novel therapeutic agent on longer-term outcomes at the tissue or neurobehavioral level, choice of anesthesia could affect the study results. For instance, if an anesthesia is chosen that suppresses neural functioning, the beneficial effects of treatment may be obscured. Conversely, if an anesthesia is used that contributes to neuroprotection, outcomes could be falsely improved in both injured and sham animals. That is to say that even when evaluation of the effects of anesthesia on brain injury outcome is not the focus of the study, researchers should consider that an anesthesia could influence the study results. This chapter provides some preliminary information that can be used by researchers when selecting an anesthesia to use. A more in-depth review of the literature and pilot work can lead to a more informed decision by researchers.

16.5.2 Craniectomy Location and Method

CCI is an invasive model of brain injury that requires craniectomy before injury induction. To control for the effects of surgery, sham animals are anesthetized and also subjected to craniectomy; consequently, sham animals in CCI studies have greater deficits than sham controls from other less-invasive procedures. The effects of craniectomy have been found to vary depending on the location and method selected. For example, in the past, midline craniectomy was used in CCI protocols but this led to significant sagittal bleeding; to reduce the extent of bleeding, the craniectomy site is now commonly moved laterally (parasagittally). Recently, there has been increased awareness that sham surgery is not benign and can produce a variety of pathological responses (Cole et al., 2011). Consequently, some investigators have chosen to use naive animals as controls. However, without surgical shams, it is not possible to separate the effects of TBI from the surgical effects. Advances in technology, combined with the desire for enhanced control, have contributed to the shift from mechanical trephine use to automatic methods, such as using a pneumatic dental drill (Figure 16.3 depicts the tools used when performing a craniectomy in rats and includes both a trephine and pneumatic dental drill). Vonder Haar and colleagues used bilateral craniectomies to enhance the lateral movement of tissue to produce bilateral cortical contusions.

16.5.3 Tip Size and Geometry

When using the CCI model, researchers can choose impactor tip geometry, with rounded, flat, and beveled options available. Although the original ferret and rat models of CCI used rounded tips (Lighthall, 1988; Dixon et al., 1991) beveled flat tips are currently the norm. Recently, mice have become

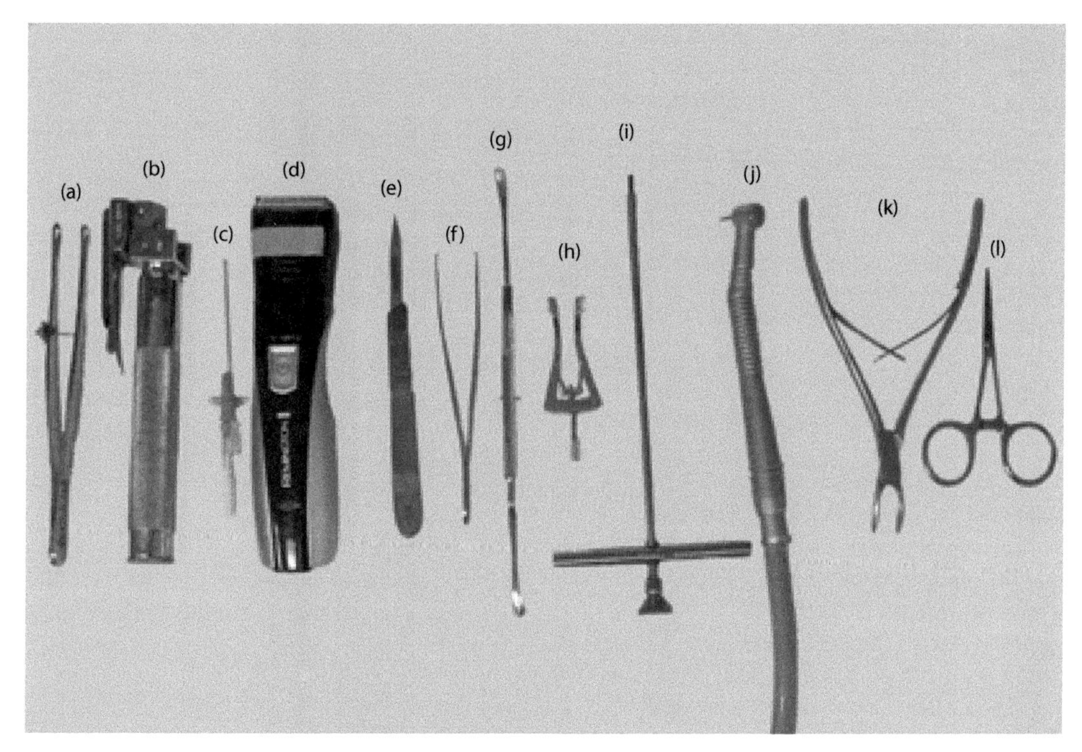

FIGURE 16.3 Picture of the tools used in the surgical procedure preceding induction of CCI using either a pneumatic or electromechanical device. (a) A 6-inch heart-holding forceps, (b) laryngoscope handle and stainless steel blade, (c) 14G catheter (for intubation), (d) cordless hair clipper for small animals, (e) 4-inch scalpel handle and sterile surgical blade #11, (f) 4¾-inch microdissecting forceps, (g) 7-inch periosteal elevator, (h) microdissecting retractor, (i) trephine, (j) pneumatic dental drill (a more modern alternative to trephine), (k) 5½-inch curved bone rongeurs, and (l) 4½-inch needle holder.

increasingly common as the test subjects in CCI studies and in this literature a flat or beveled tip is most common. Despite this convention, little empirical research surrounding the effects of varied shape (i.e., geometry) on CCI has been conducted. One influential study by Pleasant et al. (2011) found that when CCI was modeled in a common wild-type strain of mice (C57BL6J), flat tip impactors caused more extensive cortical hemorrhage and a greater proportion of neuronal loss when compared with the rounded tip.

In this study, the researchers also determined that the rate of neurodegeneration within the neocortex occurred more rapidly with a flat tip, plateauing by 4 hours postinjury, versus 24 hours postinjury when a rounded tip was used. The authors posit that the rounded tip may hold advantages for researchers who desire a longer temporal window to study the mechanisms of TBI. Notably, the overall extent of neurodegeneration within the neocortex was comparable at 9 days postinjury, regardless of tip geometry. Similarly, tip geometry did not result in any significant differences in the extent of axonal injury or blood–brain barrier breakdown or functional assessments of motor or cognitive dysfunction. Notably, at early data collection time points, animals subjected to CCI with a flat impactor tip were found to have a greater amount of hippocampal neurodegeneration; this finding could be relevant to researchers, for instance, if they are interested in specifically studying hippocampal damage (Pleasant et al., 2011).

16.6 STANDARD RAT PROTOCOL USING PNEUMATIC CCI

1. Test the CCI instrument before surgery to ensure that it is set to generate the proper impact velocity and dwell time.
2. Place the animals in a vented anesthesia chamber for 60 seconds and induce anesthesia with inspired isoflurane (IsoFlo, Abbott Laboratories, North Chicago, IL) at a concentration of 4% in 2:1 $N_2O:O_2$.
3. Gently grasp the tongue using the heart-holding forceps and insert the laryngoscope to visualize the trachea. When the trachea can be seen, intubate with a 14G catheter.
4. After endotracheal intubation, the rat is held steady in a stereotaxic frame and mechanically ventilated (Harvard Rodent Ventilator, Model 683, Harvard Apparatus, Inc., Holliston, MA) throughout the surgical procedure.
5. Use a maintenance dose of 2% isoflurane in 2:1 $N_2O:O_2$; assessing the rat to ensure the suppression of pedal response to toe pinch is maintained is helpful in titrating the maintenance dose of anesthesia.
6. Administer sterile ophthalmic ointment to both of the rat's eyes to counteract the decrease in lacrimation while under anesthesia. Note: this step is can

be repeated later if anesthesia is prolonged, such as when the surgery is taking longer than anticipated.

7. Use the animal hair trimmers to shave the scalp; scrub the exposed skin with betadine.
8. Place a sterile drape over the rat in the stereotaxic frame, ensuring there is an opening over the exposed scalp.
9. Ensuring aseptic technique is maintained, create a 20-mm midline scalp incision.
10. Separate the muscle and skull using the periosteal elevator in combination with the microdissecting forceps.
11. Reflect the skin and fascia to expose the underlying skull.
12. Create a craniectomy (approximately 6 mm in diameter) that encircles the bregma and lambda and falls between the sagittal suture and the coronal ridge using either the handheld trephine or a dental drill.
13. Detach and discard the resulting bone flap and, ensuring not to disrupt the dura mater, elongate the craniectomy with rongeurs until there is an adequate opening for the chosen impact tip. *Note: the rationale for discarding the bone flap is that reattachment can augment secondary injury related to pressure on the brain and increase mortality.*
14. Manually extend the impacting shaft and lower the impact tip so that it is centered above the craniectomy; in doing so, the tip will lightly contact the exposed dura mater.
15. Zero the impact tip to the cortical surface. Zeroing the tip while the piston is statically pressurized in the full stroke position increases the accuracy of the zero point.
16. Withdraw the impacting shaft and lower the piston assembly a distance equal to the desired depth of impact (2.8-mm tissue deformation at a velocity of 4 m/second is used for moderately severe injury in our laboratory).
17. Generate the cortical impact, start timers, and discontinue anesthesia administration; note that, depending on user preference, anesthesia can be discontinued before or after the surgical site is sutured.
18. To ensure the discomfort of the animals is minimized, promptly suture the wound shut and spread lidocaine or another topical anesthetic over the incision site.
19. Continue to administer anesthetics postsurgery per institutional and government guidelines surrounding pain management.
20. Remove the rodent from the stereotaxic frame and extubate.
21. Assess acute neurological status (righting reflex latency and limb withdrawal reflex after pinching the footpad).
22. While the rat recovers from anesthesia, keep it in a holding cage; once the effects of anesthesia have subsided (as verified by spontaneous locomotion), the rat can be returned to the colony.
23. Continue to monitor the health status of the animal frequently per institutional guidelines.

16.7 CONCLUSION

CCI is well regarded because it allows researchers to quantify the relationship between measurable engineered parameters (e.g., force, velocity, depth of tissue deformation) and the extent of (either functional and/or tissue) impairment. Thus, CCI affords pristine ability to analyze the biomechanical parameters of injury of interest in TBI research. The ultimate result is that CCI is a reliable and reproducible injury that can be used to study physiological and functional outcomes similar to those seen in human brain injury.

ACKNOWLEDGMENTS

We would like to thank Michael and Marilyn Farmer for their time and assistance with the figures used in this chapter.

REFERENCES

Acosta S.A., Tajiri N., Shinozuka K., Ishikawa H., Grimmig B., Diamond D. et al. 2013. Long-term upregulation of inflammation and suppression of cell proliferation in the Brain of adult rats exposed to traumatic brain injury using the controlled cortical impact model. *PloS One* 8 (1) (January): e53376. doi:10.1371/journal.pone.0053376.

Adelson P.D., Fellows-Mayle W., Kochanek P.M., and Dixon C.E. 2013. Morris Water Maze function and histologic characterization of two age-at-injury experimental models of controlled cortical impact in the immature rat. *Child's Nervous System* 29 (1) (January): 43–53. doi:10.1007/s00381-012-1932-4.

Atkins C.M., Cepero M.L., Kang Y., Liebl D.J., and Dietrich W.D. 2013. Effects of early rolipram treatment on histopathological outcome after controlled cortical impact injury in mice. *Neuroscience Letters* 532 (January 4): 1–6. doi:10.1016/j.neulet.2012.10.019.

Bolkvadze T., and Pitkänen A. 2012. Development of post-traumatic epilepsy after controlled cortical impact and lateral fluid-percussion-induced brain injury in the mouse. *Journal of Neurotrauma* 29 (5) (March 20): 789–812. doi:10.1089/neu.2011.1954.

Campolo M., Ahmad A., Crupi R., Impellizzeri D., Morabito R., Esposito E. et al. 2013. Combination therapy with melatonin and dexamethasone in a mouse model of traumatic brain injury. *The Journal of Endocrinology* 217 (3) (January 26): 291–301. doi:10.1530/JOE-13-0022.

Chen S.F., Tsai H.J., Hung T.H., Chen C.C., Lee C.Y., Wu C.H. et al. 2012. Salidroside improves behavioral and histological outcomes and reduces apoptosis via PI3K/Akt signaling after experimental traumatic brain injury. *PloS One* 7 (9) (January): e45763. doi:10.1371/journal.pone.0045763. http://www.pubmedcentral.nih.gov/articlerender.fcgi?artid=3454376&tool=pmcentrez&rendertype =abstract.

Cole J.T., Yarnell A., Kean W.S., Gold E., Lewis B., Ren M. et al. 2011. Craniotomy: True sham for traumatic brain injury, or a sham of a sham? *Journal of Neurotrauma* 28 (3) (March): 359–69. doi:10.1089/neu.2010.1427.

Corrigan J.D., Selassie A.W., and Orman J.A. 2010. The epidemiology of traumatic brain injury. *J Head Trauma Rehabil* 25 (2): 72–80. doi:10.1097/HTR.0b013e3181ccc8b4.

Costine B.A., Quebeda-Clerkin P.B., Dodge C.P., Harris B.T., Hillier S.C., and Duhaime A.C. 2012. Neuron-specific enolase, but not S100B or myelin basic protein, increases in peripheral blood corresponding to lesion volume after cortical impact in piglets. *Journal of Neurotrauma* 29 (17) (November 20): 2689–95. doi:10.1089/neu.2012.2428.

Dash P.K., Mach S.A., Blum S., and Moore A.N. 2002. Intrahippocampal Wortmannin infusion enhances long-term spatial and contextual memories. *Learning & Memory (Cold Spring Harbor, N.Y.)* 9 (4): 167–77. doi:10.1101/lm.50002.

Dhillon H.S., Donaldson D., Dempsey R.J., and Prasad M.R. 1994. Regional levels of free fatty acids and evans blue extravasation after experimental brain injury. *Journal of Neurotrauma* 11 (4) (August): 405–15.

Dixon C.E., Clifton G.L., Lighthall J.W., Yaghmai A.A., and Hayes R.L. 1991. A controlled cortical impact model of traumatic brain injury in the rat. *Journal of Neuroscience Methods* 39 (3) (October): 253–62.

Dixon C.E., Kochanek P.M., Yan H.Q., Schiding J.K., Griffith R.G., Baum E. et al. 1999a. One-year study of spatial memory performance, brain morphology, and cholinergic markers after moderate controlled cortical impact in rats. *Journal of Neurotrauma* 16 (2) (February): 109–22.

Dixon C.E., Ma X., Kline A.E., Yan H.Q., Ferimer H., Kochanek P.M. et al. 2003. Acute etomidate treatment reduces cognitive deficits and histopathology in rats with traumatic brain injury. *Critical Care Medicine* 31 (8) (August): 2222–7. doi:10.1097/01.CCM.0000080493.04978.73.

Dixon C.E., Kraus M.F., Kline A.E., Ma X., Yan H.Q., Griffith R.G. et al. 1999b. Amantadine improves water maze performance without affecting motor behavior following traumatic brain injury in rats. *Restorative Neurology and Neuroscience* 14 (4) (January): 285–94.

Duhaime A.C., Margulies S.S., Durham S.R., O'Rourke M.M., Golden J.A., Marwaha S. et al. 2000. Maturation-dependent response of the piglet brain to scaled cortical impact. *Journal of Neurosurgery* 93 (3) (September): 455–62. doi:10.3171/jns.2000.93.3.0455.

Ergh T.C., Rapport L.J., Coleman R.D., and Hanks R.A. 2002. Predictors of caregiver and family functioning following traumatic brain injury: Social support moderates caregiver distress. *J Head Trauma Rehabil* 17 (2): 155–74.

Feeney D.M., Gonzalez A., and Law W.A. 1982. Amphetamine, haloperidol, and experience interact to affect rate of recovery after motor cortex injury. *Science (New York, N.Y.)* 217 (4562) (August 27): 855–7.

Fox G.B., Fan L., LeVasseur R.A., and Faden A.I. 1998a. Effect of traumatic brain injury on mouse spatial and nonspatial learning in the barnes circular maze. *Journal of Neurotrauma* 15 (12) (December): 1037–46.

Fox G.B., Fan L., Levasseur R.A., and Faden A.I. 1998b. Sustained sensory/motor and cognitive deficits with neuronal apoptosis following controlled cortical impact brain injury in the mouse. *Journal of Neurotrauma* 15 (8) (August): 599–614. doi:10.1089/neu.1998.15.599.

Fox G.B., LeVasseur R.A., and Faden A.I. 1999. Behavioral responses of C57BL/6, FVB/N, and 129/SvEMS mouse strains to traumatic brain injury: Implications for gene targeting approaches to neurotrauma. *Journal of Neurotrauma* 16 (5) (May): 377–89.

Friess S.H., Ichord R.N., Owens K., Ralston J., Rizol R., Overall K.L. et al. 2007. Neurobehavioral functional deficits following closed head injury in the neonatal pig. *Experimental Neurology* 204 (1) (March): 234–43. doi:10.1016/j.expneurol.2006.10.010.

Friess S.H., Ichord R.N., Ralston J., Ryall K., Helfaer M.A., Smith C. et al. 2009a. Repeated traumatic brain injury affects composite cognitive function in piglets. *Journal of Neurotrauma* 26 (7) (July): 1111–21. doi:10.1089/neu.2008–0845.

Friess S.H., Ichord R.N., Ralston J., Ryall K., Helfaer M.A., Smith C. et al. 2009b. Repeated traumatic brain injury affects composite cognitive function in piglets. *Journal of Neurotrauma* 26 (7) (July): 1111–21. doi:10.1089/neu.2008–0845.

Gatson J.W., Liu M.M., Abdelfattah K., Wigginton J.G., Smith S., Wolf S. et al. 2013. Resveratrol decreases inflammation in the brain of mice with mild traumatic brain injury. *The Journal of Trauma and Acute Care Surgery* 74 (2) (February): 470–4; discussion 474 5. doi:10.1097/TA.0b013e31827e1f51.

Giacino J.T., Whyte J., Bagiella E., Kalmar K., Childs N., Khademi A. et al. 2012. Placebo-controlled trial of amantadine for severe traumatic brain injury. *The New England Journal of Medicine* 366 (9) (March 1): 819–26. doi:10.1056/NEJMoa1102609.

Goodman J.C., Cherian L., Bryan R.M. Jr, and Robertson C.S. 1994. Lateral cortical impact injury in rats: Pathologic effects of varying cortical compression and impact velocity. *Journal of Neurotrauma* 11 (5) (October): 587–97.

Haber M., Abdel Baki S.G., Grin'kina N.M., Irizarry R., Ershova A., Orsi S. et al. 2013. Minocycline plus N-acetylcysteine synergize to modulate inflammation and prevent cognitive and memory deficits in a rat model of mild traumatic brain injury. *Experimental Neurology* 249C (September 10): 169–177. doi:10.1016/j.expneurol.2013.09.002. http://www.ncbi.nlm.nih.gov/pubmed/24036416.

Hall E.D. 1985. High-dose glucocorticoid treatment improves neurological recovery in head-injured mice. *Journal of Neurosurgery* 62 (6) (June): 882–7. doi:10.3171/jns.1985.62.6.0882.

Hamm R.J., Dixon C.E., Gbadebo D.M., Singha A.K., Jenkins L.W., Lyeth B.G. et al. 1992. Cognitive deficits following traumatic brain injury produced by controlled cortical impact. *Journal of Neurotrauma* 9 (1) (January): 11–20.

Hamm R.J., Temple M.D., Pike B.R., O'Dell D.M., Buck D.L., and Lyeth B.G. 1996. Working memory deficits following traumatic brain injury in the rat. *Journal of Neurotrauma* 13 (6) (June): 317–23.

Hannay H.J., Feldman Z., Phan P., Keyani A., Panwar N., Goodman J.C. et al. 1999. Validation of a controlled cortical impact model of head injury in mice. *Journal of Neurotrauma* 16 (11) (November): 1103–14. http://www.ncbi.nlm.nih.gov/pubmed/10595826.

Jiang X., Huang Y., Lin W., Gao D., and Fei Z. 2013. Protective effects of hydrogen sulfide in a rat model of traumatic brain injury via activation of mitochondrial adenosine triphosphate-sensitive potassium channels and reduction of oxidative stress. *The Journal of Surgical Research* 184 (2) (October): e27–35. doi:10.1016/j.jss.2013.03.067.

Kamper J.E., Pop V., Fukuda A.M., Ajao D.O., Hartman R.E., and Badaut J. 2013. Juvenile traumatic brain injury evolves into a chronic brain disorder: Behavioral histological changes over 6 months. *Experimental Neurology* (September 25). doi:10.1016/j.expneurol.2013.09.016.

Kaneko Y., Tajiri N., Yu S., Hayashi T., Stahl C.E., Bae E. et al. 2012. Nestin overexpression precedes caspase-3 upregulation in rats exposed to controlled cortical impact traumatic brain injury. *Cell Medicine* 4 (2) (January): 55–63.

Khan M., Sakakima H., Dhammu T.S., Shunmugavel A., Im Y.B., Gilg A.G. et al. 2011. S-nitrosoglutathione reduces oxidative injury and promotes mechanisms of neurorepair following traumatic brain injury in rats. *Journal of Neuroinflammation* 8 (1) (January): 78. doi:10.1186/1742-2094-8-78. http://www.pubmedcentral.nih.gov/articlerender.fcgi?artid=3158546&tool=pmcentrez&rendertype=abstract.

King C., Robinson T., Dixon C.E., Rao G.R., Larnard D., and Nemoto C.E. 2010. Brain temperature profiles during epidural cooling with the chillerpad in a monkey model of traumatic brain injury. *Journal of Neurotrauma* 27 (10) (October): 1895–903. doi:10.1089/neu.2009.1178.

Kline A.E., Yan H.Q., Bao J., Marion D.W., and Dixon C.E. 2000. Chronic methylphenidate treatment enhances water maze performance following traumatic brain injury in rats. *Neuroscience Letters* 280 (3) (February 25): 163–6.

Kline A.E.,and Dixon C.E. 2001. Contemporary in vivo models of brain trauma and a comparison of injury responses. In LP Miller and RL Hayes (eds.), *Head Trauma: Basic, Preclinical, and Clinical Directions*, , pp. 65–84, John Wiley & Sons, New York.

Kline A.E., Bolinger B.D., Kochanek P.M., Carlos T.M., Yan H.Q., Jenkins L.W. et al. 2002. Acute systemic administration of interleukin-10 suppresses the beneficial effects of moderate hypothermia following traumatic brain injury in rats. *Brain Research* 937 (1–2) (May 24): 22–31.

Kline A.E., Massucci J.L., Dixon C.E., Zafonte R.D., and Bolinger B.D. 2004. The therapeutic efficacy conferred by the 5-HT(1A) receptor agonist 8-hydroxy-2-(di-n-propylamino)tetralin (8-OH-DPAT) after experimental traumatic brain injury is not mediated by concomitant hypothermia. *Journal of Neurotrauma* 21 (2) (February): 175–85. doi:10.1089/089771504322778631.

Kline A.E., Massucci J.L., Ma X., Zafonte R.D., and Dixon C.E. 2004. Bromocriptine reduces lipid peroxidation and enhances spatial learning and hippocampal neuron survival in a rodent model of focal brain trauma. *Journal of Neurotrauma* 21 (12) (December): 1712–22. doi:10.1089/neu.2004.21.1712. http://www.ncbi.nlm.nih.gov/pubmed/15684763.

Kline A.E., Massucci J.L., Zafonte R.D., Dixon C.E., DeFeo J.R., and Rogers E.H. 2007. Differential effects of single versus multiple administrations of haloperidol and risperidone on functional outcome after experimental brain trauma. *Critical Care Medicine* 35 (3) (March): 919–24. doi:10.1097/01.CCM.0000256722.88854.C0.

Kline A.E., Wagner A.K., Westergom B.P., Malena R.R., Zafonte R.D., Olsen A.S. et al. 2007. Acute treatment with the 5-HT(1A) receptor agonist 8-OH-DPAT and chronic environmental enrichment confer neurobehavioral benefit after experimental brain trauma. *Behavioural Brain Research* 177 (2) (February 27): 186–94. doi:10.1016/j.bbr.2006.11.036.

Kochanek P.M., Marion D.W., Zhang W., Schiding J.K., White M., Palmer A.M. et al. 1995. Severe controlled cortical impact in rats: Assessment of cerebral edema, blood flow, and contusion volume. *Journal of Neurotrauma* 12 (6) (December): 1015–25.

Lee P., Kim J., Williams R., Sandhir R., Gregory E., Brooks W.M. et al. 2012. Effects of aging on blood brain barrier and matrix metalloproteases following controlled cortical impact in mice. *Experimental Neurology* 234 (1) (March): 50–61. doi:10.1016/j.expneurol.2011.12.016.

Lee S., Birukov K.G., Romanoski C.E., Springstead JR, Lusis AJ, and Berliner JA. 2012. Role of phospholipid oxidation products in atherosclerosis. *Circulation Research* 111 (6): 778–799. doi:10.1161/circresaha.111.256859.

Levin H.S., Grossman R.G., Rose J.E., and Teasdale G. 1979. Long-term neuropsychological outcome of closed head injury. *Journal of Neurosurgery* 50 (4) (April): 412–22. doi:10.3171/jns.1979.50.4.0412.

Lighthall J.W., Goshgarian H.G., and Pinderski C.R. 1990. Characterization of axonal injury produced by controlled cortical impact. *Journal of Neurotrauma* 7 (2) (January): 65–76.

Lighthall J.W. 1988. Controlled cortical impact: A new experimental brain injury model. *Journal of Neurotrauma* 5 (1) (January): 1–15. doi:10.1089/neu.1988.5.1.

Manley G.T., Rosenthal G., Lam M., Morabito D., Yan D., Derugin N. et al. 2006. Controlled cortical impact in swine: Pathophysiology and biomechanics. *Journal of Neurotrauma* 23 (2) (February): 128–39. doi:10.1089/neu.2006.23.128.

McIntosh T.K., Vink R., Yamakami I., and Faden A.I. 1989. Magnesium protects against neurological deficit after brain injury. *Brain Research* 482 (2): 252–260.

McKinlay W.W., Brooks D.N., Bond M.R., Martinage D.P., and Marshall M.M. 1981. The short-term outcome of severe blunt head injury as reported by relatives of the injured persons. *J Neurol Neurosurg Psychiatry* 44 (6): 527–533.

McPherson R.W., Kirsch J.R., Salzman S.K., and Traystman R.J. 1994. *The Neurobiology of Central Nervous System Trauma.* Oxford University Press, New York.

Meaney D.F., Ross D.T., Winkelstein B.A., Brasko J., Goldstein D., Bilston L.B. et al. 1994. Modification of the cortical impact model to produce axonal injury in the rat cerebral cortex. *Journal of Neurotrauma* 11 (5) (October): 599–612.

Miyamoto K., Ohtaki H., Dohi K., Tsumuraya T., Song D., Kiriyama K. et al. 2013. Therapeutic time window for edaravone treatment of traumatic brain injury in mice. *BioMed Research International* 2013 (January): 379206. doi:10.1155/2013/379206.

Morris G.M. 1981. Spatial localization does not require the presence of local cues. *Learning and Motivation* 12 (2): 239–260.

Morris R.G., Garrud P., Rawlins J.N., and O'Keefe J. 1982. Place navigation impaired in rats with hippocampal lesions. *Nature* 297 (5868) (June 24): 681–3.

Pleasant J.M., Carlson S.W., Mao H., Scheff S.W., Yang K.H., and Saatman K.E. 2011. Rate of neurodegeneration in the mouse controlled cortical impact model is influenced by impactor tip shape: Implications for mechanistic and therapeutic studies. *Journal of Neurotrauma* 28 (11) (November): 2245–62. doi:10.1089/neu.2010.1499.

Pop V., Sorensen D.W., Kamper J.E., Ajao D.O., Murphy M.P., Head E. et al. 2013. Early brain injury alters the blood-brain barrier phenotype in parallel with β-amyloid and cognitive changes in adulthood. *Journal of Cerebral Blood Flow and Metabolism* 33 (2) (February): 205–14. doi:10.1038/jcbfm.2012.154.

Raghupathi R., Fernandez S.C., Murai H., Trusko S.P., Scott R.W., Nishioka W.K. et al. 1998. BCL-2 overexpression attenuates cortical cell loss after traumatic brain injury in transgenic mice. *Journal of Cerebral Blood Flow and Metabolism* 18 (11) (November): 1259–69. doi:10.1097/00004647-199811000-00013.

Robertson C.L., Saraswati M., Scafidi S., Fiskum G., Casey P., and McKenna M.C. 2013. Cerebral glucose metabolism in an immature rat model of pediatric traumatic brain injury. *Journal of Neurotrauma* (September 15). doi:10.1089/neu.2013.3007.

Russell K.L., Berman N.E., and Levant B. 2013. Low brain DHA content worsens sensorimotor outcomes after TBI and decreases TBI-induced Timp1 expression in juvenile rats. *Prostaglandins, Leukotrienes, and Essential Fatty Acids* 89 (2–3) (August): 97–105. doi:10.1016/j.plefa.2013.05.004.

Sandhir R., and Berman N.E. 2010. Age-dependent response of CCAAT/enhancer binding proteins following traumatic brain injury in mice. *Neurochemistry International* 56 (1) (January): 188–93. doi:10.1016/j.neuint.2009.10.002.

Schaible E.V., Steinsträßer A., Jahn-Eimermacher A., Luh C., Sebastiani A., Kornes F. et al. 2013. Single administration of tripeptide α-MSH(11–13) attenuates brain damage by reduced inflammation and apoptosis after experimental traumatic brain injury in mice. *PloS One* 8 (8) (January): e71056. doi:10.1371/journal.pone.0071056.

Schalén W., Hansson L., Nordström G., and Nordström C.H. 1994. Psychosocial outcome 5-8 years after severe traumatic brain lesions and the impact of rehabilitation services. *Brain Inj* 8 (1): 49–64.

Singleton R.H., Yan H.Q., Fellows-Mayle W., and Dixon C.E. 2010. Resveratrol attenuates behavioral impairments and reduces cortical and hippocampal loss in a rat controlled cortical impact model of traumatic brain injury. *Journal of Neurotrauma* 27 (6) (June): 1091–9. doi:10.1089/neu.2010.1291.

Smith D.H., Soares H.D., Pierce J.S., Perlman K.G., Saatman K.E., Meaney D.F. et al. 1995. A model of parasagittal controlled cortical impact in the mouse: Cognitive and histopathologic effects. *Journal of Neurotrauma* 12 (2) (April): 169–78.

Smith D.H., Okiyama K., Thomas M.J., Claussen B., and McIntosh T.K. 1991. Evaluation of memory dysfunction following experimental brain injury using the Morris Water Maze. *Journal of Neurotrauma* 8 (4) (January): 259–269. doi:10.1089/neu.1991.8.259.

Statler K.D., Kochanek P.M., Dixon C.E., Alexander H.L., Warner D.S., Clark R.S. et al. 2000. Isoflurane improves long-term neurologic outcome versus fentanyl after traumatic brain injury in rats. *Journal of Neurotrauma* 17 (12) (December): 1179–89.

Statler K.D., Alexander H., Vagni V., Holubkov R., Dixon C.E., Clark R.S. et al. 2006. Isoflurane exerts neuroprotective actions at or near the time of severe traumatic brain injury. *Brain Research* 1076 (1) (March 3): 216–24. doi:10.1016/j.brainres.2005.12.106.

Sword J., Masuda T., Croom D., and Kirov S.A. 2013. Evolution of neuronal and astroglial disruption in the peri-contusional cortex of mice revealed by in vivo two-photon imaging. *Brain* 136 (Pt 5) (May): 1446–61. doi:10.1093/brain/awt026.

Teasdale G., and Jennett B. 1974. Assessment of coma and impaired consciousness. A practical scale. *Lancet* 2 (7872) (July 13): 81–4.

Timaru-Kast R., Luh C., Gotthardt P., Huang C., Schäfer M.K., Engelhard K. et al. 2012. Influence of age on brain edema formation, secondary brain damage and inflammatory response after brain trauma in mice. *PloS One* 7 (8) (January): e43829. doi:10.1371/journal .pone.0043829.

Tran H.T., LaFerla F.M., Holtzman D.M., and Brody D.L. 2011. Controlled cortical impact traumatic brain injury in 3xTg-AD mice causes acute intra-axonal amyloid-β accumulation and independently accelerates the development of tau abnormalities. *The Journal of Neuroscience* 31 (26) (June 29): 9513–25. doi:10.1523/JNEUROSCI.0858-11.2011.

Vonder Haar C., Friend D.M., Mudd D.B., and Smith J.S. 2013. Successive bilateral frontal controlled cortical impact injuries show behavioral savings. *Behavioural Brain Research* 240 (March 1): 153–9. doi:10.1016/j.bbr.2012.11.029.

Wang G., Zhang J., Hu X., Zhang L., Mao L., Jiang X. et al. 2013. Microglia/macrophage polarization dynamics in white matter after traumatic brain injury. *Journal of Cerebral Blood Flow and Metabolism: Official Journal of the International Society of Cerebral Blood Flow and Metabolism* (August 14). doi:10.1038/jcbfm.2013.146.

Washington P.M., Forcelli P.A., Wilkins T., Zapple D.N., Parsadanian M., and Burns M.P. 2012. The effect of injury severity on behavior: A phenotypic study of cognitive and emotional deficits after mild, moderate, and severe controlled cortical impact injury in mice. *Journal of Neurotrauma* 29 (13) (September): 2283–96. doi:10.1089/neu.2012.2456.

Xing G., Ren M., O'Neill J.T., Verma A., and Watson W.D. 2012. Controlled cortical impact injury and craniotomy result in divergent alterations of pyruvate metabolizing enzymes in rat brain. *Experimental Neurology* 234 (1) (March): 31–8. doi:10.1016/j.expneurol.2011.12.007.

Zhang Q.G., Laird M.D., Han D., Nguyen K., Scott E., Dong Y. et al. 2012. Critical role of NADPH oxidase in neuronal oxidative damage and microglia activation following traumatic brain injury. *PLoS One* 7 (4): e34504. doi:10.1371/journal.pone.0034504.

Zhao S., Fu J., Liu X., Wang T., Zhang J., and Zhao Y. 2012. Activation of Akt/GSK-3beta/beta-catenin signaling pathway is involved in survival of neurons after traumatic brain injury in rats. *Neurological Research* 34 (4) (May): 400–7. doi:10.1179/1743132812Y.0000000025.

17 A Two-Model Approach to Investigate the Mechanisms Underlying Blast-Induced Traumatic Brain Injury

Haoxing Chen, Shlomi Constantini, and Yun Chen

CONTENTS

17.1 INTRODUCTION

Blast-induced traumatic brain injury (TBI) is a signature, invisible wound of wars, sustained with long-lasting neuropsychiatric and neurological symptoms. The mechanism of blast-induced TBI has been controversial for a long time. Direct cranial transmission of blast waves was considered by most investigators as the mechanical mechanism by which the blast wave causes mild TBI. Only few investigators hypothesized that thoraco-abdominal vascular/hydrodynamic transmission of blast waves could be the major cause of blast-induced TBI. To separate direct cranial transmission of blast waves from thoraco-abdominal vascular/hydrodynamic mechanism to blast-induced TBI, two "iron lung"-like protective devices are designed for protection of desired parts of the animal body against blast waves. One "Iron lung"-like protective device allows only the animal head to expose to blast waves, and another makes the animal thorax and abdomen only expose to blast waves. The use of the "iron lung"-like protective devices in blast injury research will lead to new insight into the mechanisms underlying blast-induced TBI. Shock tubes have been employed to investigate blast injuries in animals since 1940s. However, many uncertainties are associated with the results obtained from the animal models using shock tubes because a series of complex shock waves generated by shock tubes affect experimental observations and lead to false-positive results in the studies of blast TBI mechanism. A C4 blast generator that generates blast waves by detonation of C4 charge in a free field can be used as a new experimental tool for blast-induced TBI research. A comparative study between two animal models that use traditional shock tube and novel C4 blast generator respectively to induce TBI, will help develop a reliable and valid experimental approach to identify the mechanism of blast-induced TBI. The physical parameters of blast shock waves and the extent and severity of TBI, which are closely associated with the effects of blast shock waves on the brain, need to be analyzed, assessed and compared comprehensively between the two animal models. The two-model comparative approach will contribute to eliminate knowledge gaps regarding blast-induced TBI and to prioritize future TBI research.

Blast-induced TBI is the signature injury of the wars in Iraq and Afghanistan and has accounted for significant substantial morbidity and disability among U.S. military members (Martin et al., 2008; Warden, 2006). It has become a major public health problem that results in the loss of many years of productive life and incurs large health care costs (Martin et al., 2008). From 2002 through 2010, the U.S. government has spent about $1.5 billion in blast TBI research and about $6 billion on health care expenditures for more than 280,000 veterans with TBI and posttraumatic stress disorder (Congressional Budget Office, 2012). Blast-induced TBI presents a daunting challenge for the military medical community.

The mechanism of blast-induced TBI has been controversial for a long time. However, no satisfying data and results exist to clearly confirm the mechanism of blast TBI. Direct head exposure to blast, skull flexure, and head acceleration were considered by most investigators as the mechanical mechanism by which the blast wave causes mild TBI. Only a few investigators have hypothesized that direct torso impact of blast pressure waves causes noncontact TBI. More recent studies (Assari et al., 2013; Chen et al., 2012; Hue et al., 2013; Sosa et al., 2013; Yeoh et al., 2013) have supported the theory that the major mechanism of blast TBI involves damage to the blood–brain barrier (BBB) and tiny cerebral blood vessels, which is caused by blood surging quickly through large blood vessels from the torso to the brain (Chen and Huang, 2011). Large-scale BBB damage and cerebrovascular insults may be the most important cause of blast-induced TBI after exposure to blast waves.

Because the true mechanism of blast-induced TBI is currently unknown, personal blast protection against TBI is still the most difficult challenge facing medical researchers and body armor engineers. Currently, fielded body armor is unable to properly protect the human body against the impact of blast shock wave. In contrast, it could increase the impact force to the body along with blast shock waves, causing more serious "behind armor blunt trauma" (Chen et al., 2012). Clearly, what is required to prevent and mitigate blast-induced TBI is to employ innovative experimental approaches to investigate the mechanisms of blast-induced TBI. The development of the adequate experimental models with defined blast-wave signatures will help elucidate biomechanisms, pathophysiology, histopathology, and neurological and neuropsychological consequences of blast-induced TBI, thus accelerating the discovery of new protective and therapeutic approaches that can effectively reduce the disabilities and serious complications of TBI.

17.2 THE "IRON LUNG"–LIKE PROTECTIVE DEVICES USED TO INVESTIGATE THE MECHANISMS OF BLAST-INDUCED TBI

Two "iron lung"–like protective devices can be used respectively for investigation of TBIs caused by direct cranial transmission and thoracoabdominal vascular/hydrodynamic transmission of shock waves in rodents (Figure 17.1). The iron lung–like protective device is made of 4–5 mm thick, ultrahigh-strength, crash-resistant steel, and filled with 3-mm-thick hard resilient rubber inside. The ultrahigh-strength, crash-resistant steel will protect the animal body against the rapid impact effects of shock waves. The hard, resilient rubber will prevent the animal body from shaking or moving inside the protective device during exposure to blasts. This protective device will effectively prevent not only the direct damage effects of blast shock waves on the protected object, but also blunt trauma behind the device that is caused by rapid movement or vibrations of the animal body. Iron lung–like protective devices can separate direct cranial transmission of blast waves from thoracoabdominal vascular/hydrodynamic mechanism to blast-induced TBI and can be widely used in any animal models that employ shock tubes or other blast generators to induce blast injuries.

For investigation of the effects of direct cranial transmission of shock waves, the animal head is exposed directly to blast shock waves (Figure 17.2a and Figure 17.5) and other parts of the body will be shielded in the protective device (Figure 17.1a). The animal neck will be protected by rounding rubber. For investigation of the effects of thoracoabdominal vascular/hydrodynamic transmission of shock waves, the animal thorax and abdomen are exposed directly to blast shock waves (Figure 17.2b and Figure 17.6) and other parts of the body, including the head, will be protected by the protective device (Figure 17.1b). The animal's neck and lower abdomen will be rounded by rubber to prevent movement or displacement of the animal body. The use of iron lung–like protective devices will effectively distinguish direct cranial transmission of blast waves from thoracoabdominal vascular/hydrodynamic mechanism to blast-induced TBI. This will help gain new insights into the mechanisms of blast-induced TBI and help evaluate protective effects of body armor, helmets, and other gears on blast-induced TBI.

If no TBI occurs after exposure of the animal head to shock waves, direct cranial transmission, skull flexure, and head acceleration can be ruled out as the mechanism underlying blast-induced TBI. In contrast, if no brain damage occurs after exposure of the animal thorax and abdomen to shock waves, then thoracoabdominal vascular/hydrodynamic mechanisms to blast-induced TBI will not be considered. However, if TBI is observed after exposure of either the head or the thorax and abdomen to shock waves, the major contributor to TBI will be further identified.

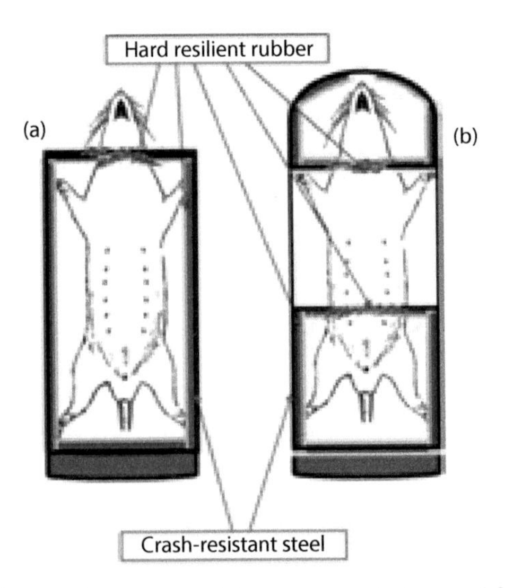

FIGURE 17.1 (a) Iron lung–like protective device used for investigation of blast-induced traumatic brain injury (TBI) caused by direct cranial transmission of shock waves. (b) Iron lung–like protective device used for investigation of thoracoabdominal vascular or hydrodynamic mechanisms to blast-induced TBI.

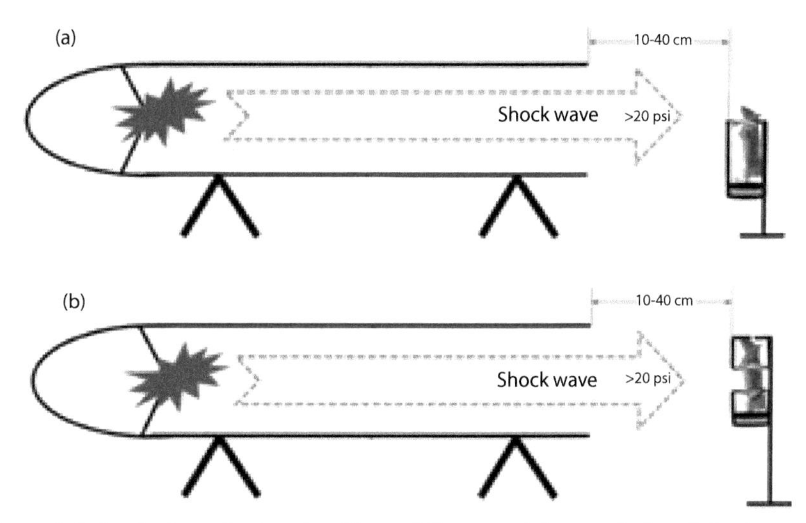

FIGURE 17.2 (a) Experimental design for investigating the effects of direct cranial transmission of shock waves in traumatic brain injury (TBI) induction. (b) Experimental design for investigating thoracoabdominal vascular/hydrodynamic mechanisms to blast-induced TBI.

17.3 THE ANIMAL MODEL FOR STUDYING TBI INDUCED BY SHOCK WAVES GENERATED FROM SHOCK TUBES

17.3.1 Experimental Uncertainties

Over the past several decades, shock tubes (both compressed air-driven and explosive-driven blast shock tubes) have been used as the fundamental research tool for investigating mechanisms of blast injury in rodents and larger animals such as sheep, swine, and dogs (Cooper et al., 1998; Long et al., 2009; Nakagawa et al., 2008; Ritenour and Baskin, 2008). Shock tubes can simulate the small-scale explosive effects on animals by generating and directing blast shock waves. However, because of its design and structure, a shock tube is not able to generate the Friedlander wave (an ideal form of a primary blast wave), that occurs when a powerful explosive detonates in a free field, without nearby surfaces interacting with the wave.

In shock tube experiments, a series of complex shock waves are generated following the lead shock wave (the original shock front) that includes reflected shock waves, a Mach stem, an unsteady turbulent jet, and rarefaction waves (Chen and Constantini, 2013). These waves are 2 to 13 times higher than the lead shock wave and can result in sudden compression or rarefaction effects upon any object encountered in their motion path (Federal Emergency Management Agency, 2003). Upon encounter, the complex shock waves can transfer kinetic energy to the object. This will cause more severe and complex injuries to the animals positioned inside shock tubes that are rarely observed in blast victims, thus leading to false-positive results and incorrect information about the mechanism of blast TBI. Therefore, an adequate experimental design and implementation that can avoid the interference effects of complex shock waves on experimental results and can mimic critical aspects of blast injury sustained in combat or terrorist explosions are of particular importance for investigation of the mechanism of blast-induced TBI.

17.3.2 Reliable Experimental Setup

To avoid severe damaging effects of complex shock waves and to reduce false-positive results, the experimental animals must be placed outside of the shock tube at a distance of approximately 10–40 cm from the mouth of the shock tube (Figure 17.2). Because an unprotected animal body can survive relatively high-incident shock pressure in the free field without experiencing barotraumas, the incident shock pressure that interacts with the animal body should be greater than 20 psi (~138 kPa). Under these conditions, the protective device will be able to shield the shock waves that interact with desired parts (head, neck, torso, or limbs) of the animal body, thus isolating the effects of direct cranial transmission of shock waves from the effects of thoracoabdominal vascular/hydrodynamic transmission of shock waves on the induction of TBI.

17.3.3 Assessment of Extent and Severity of TBI

After exposure to shock waves, the typical parameters that define the extent and severity of TBI and its outcome will be evaluated from 1 hour to 21 days postinjury in the injured animals. Cerebral edema can be evaluated by diffusion magnetic resonance imaging (diffusion tensor imaging) or by measuring the tissue water content in the injured brain. Cerebral blood flow and cerebrovascular function can be measured by functional magnetic resonance imaging or single-photon emission computed tomography. Integrity of the BBB can be evaluated by determining Evans blue dye or immunoglobulin G extravasation. Histopathological changes (tiny cerebrovascular insult, neuronal cell death, neural loss, and axonal degeneration) will be examined using hematoxylin and eosin, van Gieson, or immunohistological staining techniques. Neurological (motor function) deficits will be evaluated using a set of criteria termed the neurological severity score (Chen et al., 1996) that includes reflex test, beam balancing task, and beam walking task. Cognitive and spatial memory functions will be evaluated using Morris Water Maze and Barnes Maze. Real-time blood pressure in the internal carotid artery will be measured during and after explosions using micropressure transducers. Neurotrauma-associated biomarkers in the brain will be detected using Western blot, enzyme-linked immunosorbent assay, or in situ hybridization technique.

17.4 ANIMAL MODEL FOR STUDYING TBI CAUSED BY DETONATION OF SMALL EXPLOSIVE C4 CHARGES IN A FREE FIELD

17.4.1 Salient Features of the C4 Blast Generator

Because shock tubes generate complex shock waves (especially reflected shock waves and an unsteady turbulent jet), some uncertainties may be associated with the results obtained from the animals exposed to shock waves generated in shock tubes. The data on injury mechanisms of blast-induced TBI may be difficult to analyze and compare. To reduce uncertainties in experimental results, a new explosive device, in which blast wave is generated by detonation of a small amount of high-explosive compound (C4) in a free field, can be used as a blast generator to study the mechanisms underlying blast-induced TBI. C4 is a research department explosive plastic explosive. The use of C4 will allow us to measure out any amount of the explosive charge needed from 0.2 to 1100 g. As depicted in Figure 17.3, the explosive device is a semicircle steel shelf to hold the iron lung–like protective device, made of 18–20 mm thick ultrahigh-strength, crash-resistant steel bars. It has a bottom diameter of 1600 mm and a height of 700–720 mm. The salient features of the C4 blast generator are (1) the key parameters (peak pressure, duration, pressure gradient, impulse, etc.) of the blast waves generated by this device approximates that of Friedlander wave; (2) the unique design limits reflected or complex wave formation after explosion; (3) blast injury can be induced simultaneously in six animals during each explosion (Figures 17.5 and 17.6), which will significantly reduce variation in experimental results among animals; (4) the iron lung–like protective devices used in the shock tube experiments can be easily installed in the C4 blast generator to protect desired parts of the animal body against blast waves; and (5) the C4 blast generator can be employed to investigate the effects of both direct cranial transmission and thoracoabdominal vascular/hydrodynamic transmission of shock waves on the induction of TBI.

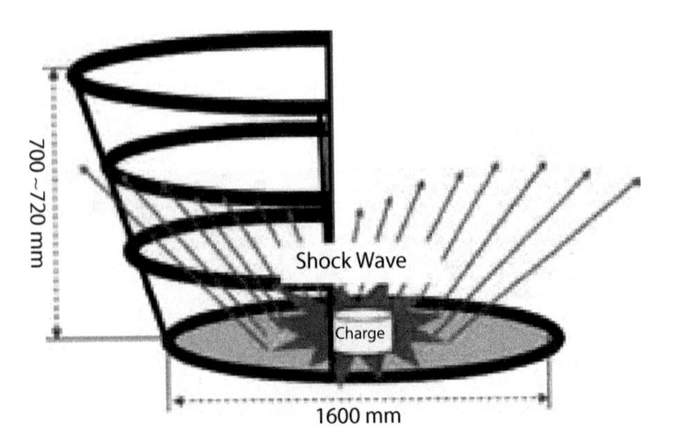

FIGURE 17.3 A semicircle blast generator that generates blast waves by detonating a small amount of C4 charge in a free field and can hold six iron lung–like protective devices to protect desired parts of the animal body against blast waves.

17.4.2 Identification of the Optimal Resultant Peak Pressure and C4 Charge Weight

According to the blast pressure equation created by Olson and Fletcher in 1971 (Olson and Fletcher, 1971), the peak overpressure could be expressed as: $P = 4.9 \times 10^3 (D/W^{1/3})^{-2.15}$, where P is peak overpressure (per square inch [psi]), D is the distance (ft) from the blast, and W is the weight (lb) of explosive charge. Because the peak overpressure decreases exponentially with scaled distance based on the blast pressure equation, the distance from the blast center to the animals tested must be fixed at 2.625 ft (800 mm) through all of the experiments. Three different blast loading levels (light, medium, and heavy) may be applied to induce TBI in the animals. Peak overpressure values for light, medium, and heavy blast loading should be 10.16 psi or 70.07 kPa, 21.77 psi or 150.06 kPa, and 33.37 Psi/230.01 kPa, respectively. The weights of C4 charge for the three different blast loading levels are calculated in the following table.

Blast loading level	Distance from blast	Weight of C4
Light (10.16 Psi/70.07 kPa)	800 mm	145.33 g
Medium (21.77 psi/150.06 kPa)	800 mm	207.06 g
Heavy (33.37 Psi/230.01 kPa)	800 mm	252.56 g

The correlation between blast peak pressure (kPa) and C4 charge weight (g) is shown in Figure 17.4. The peak pressure generated by C4 charge at a distance of 800 mm can be calculated using the formula ($Y = 1.0406X - 143.06$), where Y is peak overpressure (kPa) and X is the weight of C4 charge (g).

To identify the optimal resultant peak pressure and C4 charge weight for this animal model, the animals will be exposed to blast waves generated by detonations of different amounts of C4 charge without protection by the iron lung–like protective devices. The resultant peak pressures applied to the animal bodies will be recorded using pressure sensors during explosions. The extent and severity of blast-induced TBI will be evaluated within 1 hour following blast injury based on mortality rate, lethal dose, 25% (LD_{25}), and anatomical and histopathological abnormalities. A lethal dose is estimated to be LD_{25} in the injured animals after explosions, which may be more appropriately extrapolated to the combat environment (Bell, 2008). Hence, an immediate mortality rate of 25% will be allowed in this animal model resulting from lethal injuries. Thirty percent more animals may be needed for those excluded because of lethal injuries, health problems, and experimental failures. Obvious tissue damage (such as swelling, hyperemia, vasospasm, hemorrhage, and hematoma) should be observed in the brain with lethal dose exposure. An optimal resultant peak pressure applied to the animal body should be in the range between light and medium blast loading levels (10–20 psi). The optimal amount of C4 charge should weigh between 140 and 200 g.

17.4.3 Experimental Setup and Assessment of TBI Extent and Severity

After the optimal C4 amount and peak pressure value are selected, the effects of both direct cranial transmission and thoracoabdominal vascular/hydrodynamic transmission of shock waves on the induction of TBI can be investigated using the calibrated C4 blast generator. To investigate the effects of direct cranial transmission of shock waves in the induction of TBI, the animals will be anesthetized and place in the iron lung–like protective devices. The head is the only part of animal body that is exposed directly to blast waves and will be covered with a wire mesh cylinder tube made of 1.2-mm-diameter steel wire with a 1-mm-square opening (Figure 17.5a). The wire mesh cylinder tube can prevent secondary blast injury caused by fragmentation and other objects propelled by the explosion, thus guaranteeing TBI to be solely induced by blast waves. Six protective devices will be installed on the semicircle steel shelf (Figure 17.5b). To investigate the effects of thoracoabdominal vascular/hydrodynamic transmission of shock waves on the induction of TBI, the animal thorax and abdomen will be left fully exposed to blast waves and other parts of the body will be protected with the iron lung–like protective device. The wire mesh cylinder tube will be also used to protect the thorax and abdomen against secondary blast injury caused by fragmentation and other objects (Figure 17.6a). Six animals are simultaneously exposed to blast waves generated by detonation of a C4 charge (Figure 17.6b).

After exposure to blast waves, the typical parameters that define the extent and severity of TBI and its outcome will be evaluated from 1 hour to 21 days in the injured animals. Cerebral edema, BBB permeability, cerebral blood flow, internal carotid artery pressure, cerebrovascular insults, histopathological abnormalities, motor function, cognitive and spatial memory functions, and biomarkers specific for TBI will be detected. If brain damage cannot be observed in the animal whose head is exposed to blast wave, TBI should be the result of the cerebrovascular insults and the BBB damage caused by the blood surge moving quickly through large blood vessels to the brain from the torso. On the other hand, if no brain damage occurs in the animal whose thorax and abdomen are exposed to blast wave, TBI should be caused directly by direct cranial transmission of blast wave, skull flexure, or head acceleration.

By using both the C4 blast generator and the iron lung–like protective devices, blast-induced polytrauma will be avoided and variation in experimental results among the animals will be significantly reduced. The C4 blast generator will be able to mimic critical aspects of explosive blasts in military operation and terrorist attacks and to reproduce similar posttraumatic sequelae observed in humans. It can help establish a unified experimental model with defined blast-wave signatures in rodents, thus re-creating a real battlefield TBI in the laboratory. The application of the iron lung–like protective devices could help to better understand the mechanisms underlying blast-induced TBI through a comparison study of direct cranial transmission and thoracoabdominal vascular/hydrodynamic transmission of blast waves. The new-generation personal protective equipment (such as body armor and helmets) that provide best protection against blast-induced TBI will be further developed after the transmission characteristics of blast wave in animal body are discovered.

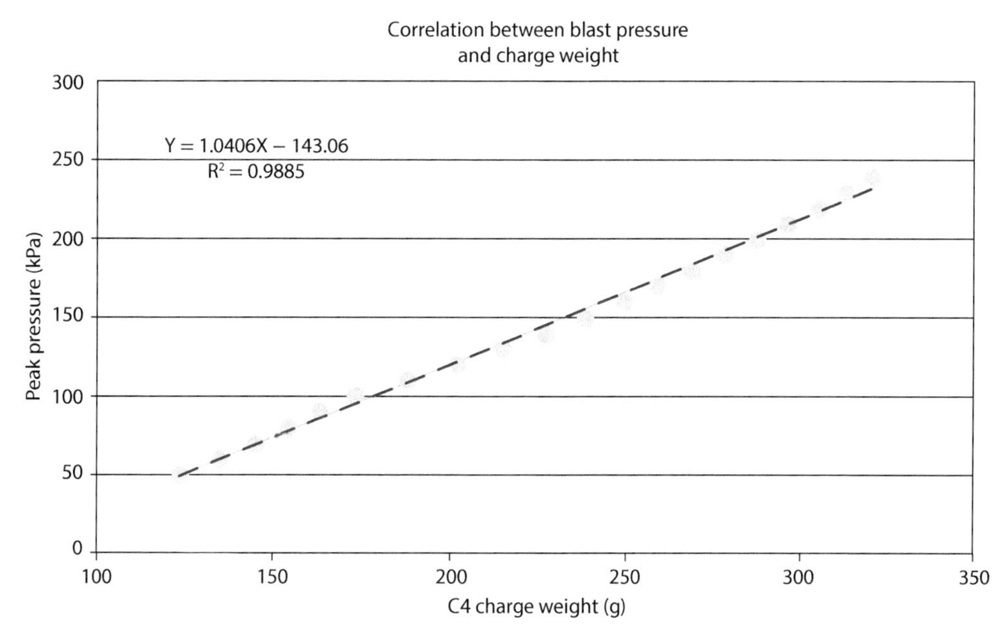

FIGURE 17.4 The correlation between blast peak pressure and C4 charge weight. The peak pressure generated by C4 charge at a distance of 800 mm can be calculated using the formula (Y = 1.0406X – 143.06), where Y is peak overpressure (kPa) and X is the weight of C4 charge (g).

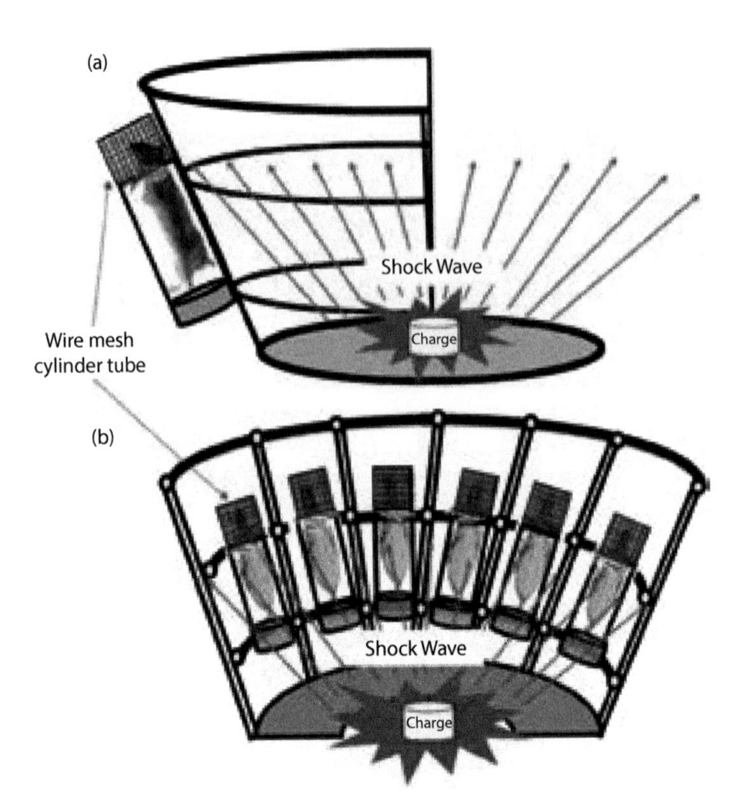

FIGURE 17.5 An animal model that investigates the effects of direct cranial transmission of shock waves in traumatic brain injury (TBI) induction. (a) The animal's head is covered with a wire mesh cylinder tube and then exposed directly to blast waves. (b) Six protective devices are installed simultaneously on the semicircle steel shelf.

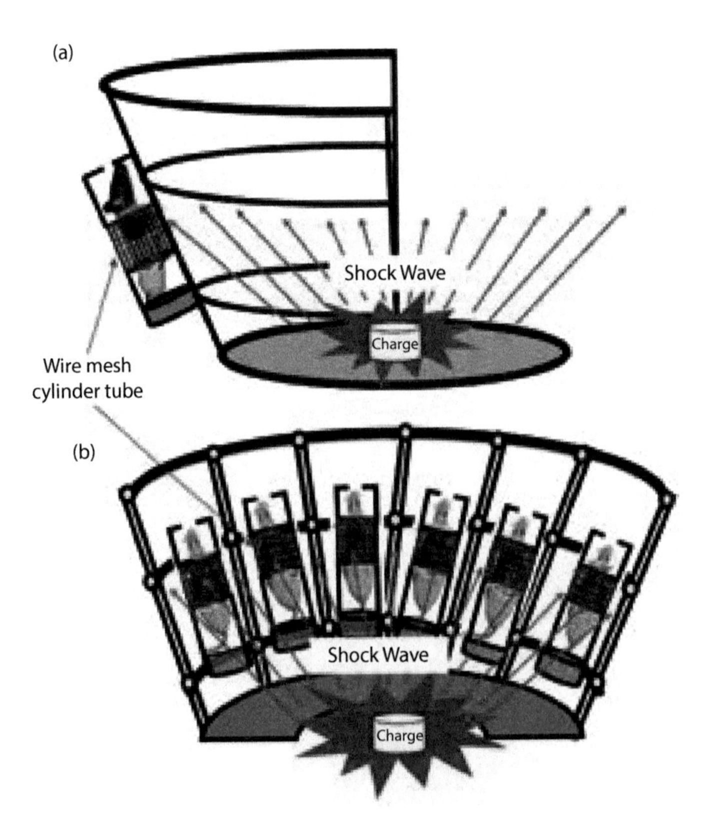

FIGURE 17.6 An animal model that investigates the effects of thoracoabdominal vascular/hydrodynamic transmission of shock waves in traumatic brain injury induction. (a) The animal thorax and abdomen are covered with a wire mesh cylinder tube and then left fully exposed to blast waves. (b) Six animals are simultaneously exposed to blast waves generated by detonation of a C4 charge.

17.5 COMPARATIVE STUDY BETWEEN TWO ANIMAL MODELS

17.5.1 Comprehensive Analysis of Physical Parameters of the Blast Shock Waves

In the two animal models, a shock tube and a C4 blast generator are employed respectively to generate blast shock waves. Blast shock wave is a key contributor to TBI. To mimic TBI that occurs in military operations and terrorist attacks, the characteristics and properties of blast shock waves that are generated by both the shock tube and the C4 blast generator should be the same as or similar to that of the Friedlander wave. The extent and severity of blast injury generally depend on not only overpressure peaks of a shock wave, but also other physical parameters such as duration, pressure gradient, impulse, characteristics of the shear fronts between overpressure peaks, frequency resonance, and rarefaction wave profile. There is general agreement that spalling, implosion, inertia, and pressure differentials are the main mechanisms involved in primary blast injuries caused by shock waves (Treadwell, 1989). Therefore, a comprehensive analysis of physical parameters of the blast shock waves is of particular importance for an in-depth study of blast-induced TBI. The key parameters recorded using pressure sensors or transducers during both positive and negative pressure phases need to be comprehensively analyzed and compared between the two animal models. If there are no optimal physical parameters obtained from an animal model due to uncontrollable physical factors (such as strong reflection of shock wave, Mach stem, or unsteady turbulent jet), the model will not be considered for further TBI research. These complex physical factors will influence the observation of experiments and produce false-positive results and incorrect information about the mechanism of blast TBI.

17.5.2 Evaluation of the Brain Responses to Blast Waves

The brain responses to blast waves can be evaluated by functional neuroimaging, pathophysiological, histopathological, immunochemical, and neurobehavioral approaches. The brain is a very soft, gelatinous tissue that is extremely susceptible to injury and disease (Parent and Carpenter, 1995). The effects of blast wave on the brain may lead to more serious consequences than on other tissues. The initial pathophysiological changes after blast-induced TBI are defined as the pepper-spray pattern: diffuse cerebral edema, hyperemia, vasospasm, and hemorrhage in entire brain (Ling et al., 2009). Neuroimaging technologies (such as three-dimensional computed tomography or magnetic resonance cerebral angiography, single-photon emission computed tomography, and diffusion magnetic resonance imaging) may help detect tiny cerebrovascular insult (hemorrhage), vascular abnormalities (vasospasm), changes in blood flow (hyperemia), and cerebral edema in the animals exposed to blast waves (Chen et al., 2013). Pathophysiological and immunochemical methods can be used for measurements of the BBB permeability, tissue water content in the brain, and neurotrauma-associated biomarkers. Histopathological examinations of the brain tissue can help determine tiny cerebrovascular insults, neuronal cell death, neural loss, and axonal degeneration. Neurobehavioral assessments (such as motor, cognitive, and memory function evaluations) are fundamental to understanding long-term neurobehavioral consequences of TBI. If the pathophysiology, histopathology, and neurobehavioral sequelae of TBI cannot be replicated in any animal model that recapitulates human TBI induced by explosive blasts in military operation and terrorist attacks, the animal model may not be reliable and valid for blast-induced TBI research and will be restricted for further application.

Figure 17.7 shows a flowchart that describes how to respectively induce TBI in the two animal models using shock tubes and the C4 blast generator, to separate direct cranial transmission of blast waves from thoracoabdominal vascular/hydrodynamic mechanisms to blast-induced TBI by using the iron lung–like protective devices, and to compare experimental results between the two animal models. The two-model comparative study will help identify a reliable and valid experimental approach to study the mechanism of TBI caused by exposure to blast waves. It can contribute to eliminate knowledge gaps regarding blast-induced TBI and to prioritize future TBI research, thus accelerating discovery of new therapeutic strategies that can effectively reduce the disability and long-term neuropsychiatric and neurological complications followed by TBI. Importantly, the two-model comparison study can possibly lead to important breakthroughs in the development of new-generation personal protective equipment that can effectively protect against blast threats to civilian and military personnel.

17.6 CONCLUSIONS

Blast-induced TBI is a unique type of brain injury that is caused by blast shock waves in the victims who are exposed to a blast but do not sustain penetrating and blunt impact injuries. The unique injury is characterized by the pepper-spray pattern—diffuse cerebral edema, hyperemia, vasospasm, and hemorrhage—that may be caused by large-scale BBB damage and cerebrovascular insults in entire brain. Patients with blast-induced TBI sustain long-term neuropsychiatric and neurological disorders. There are huge direct and indirect economic costs to society at large for blast-induced TBI through the burden of medical care imposed on the health care system and family members.

Currently no unified experimental model with defined blast-wave signatures is available in laboratory for investigation of blast-induced TBI. Therefore, the development of a reliable and valid research approach to explore the effects of primary blast wave on the brain is critical for understanding the true mechanism of blast-induced TBI. Since the 1940s, shock tubes have been used as the fundamental research tool for investigation of the mechanisms of blast injury in animals.

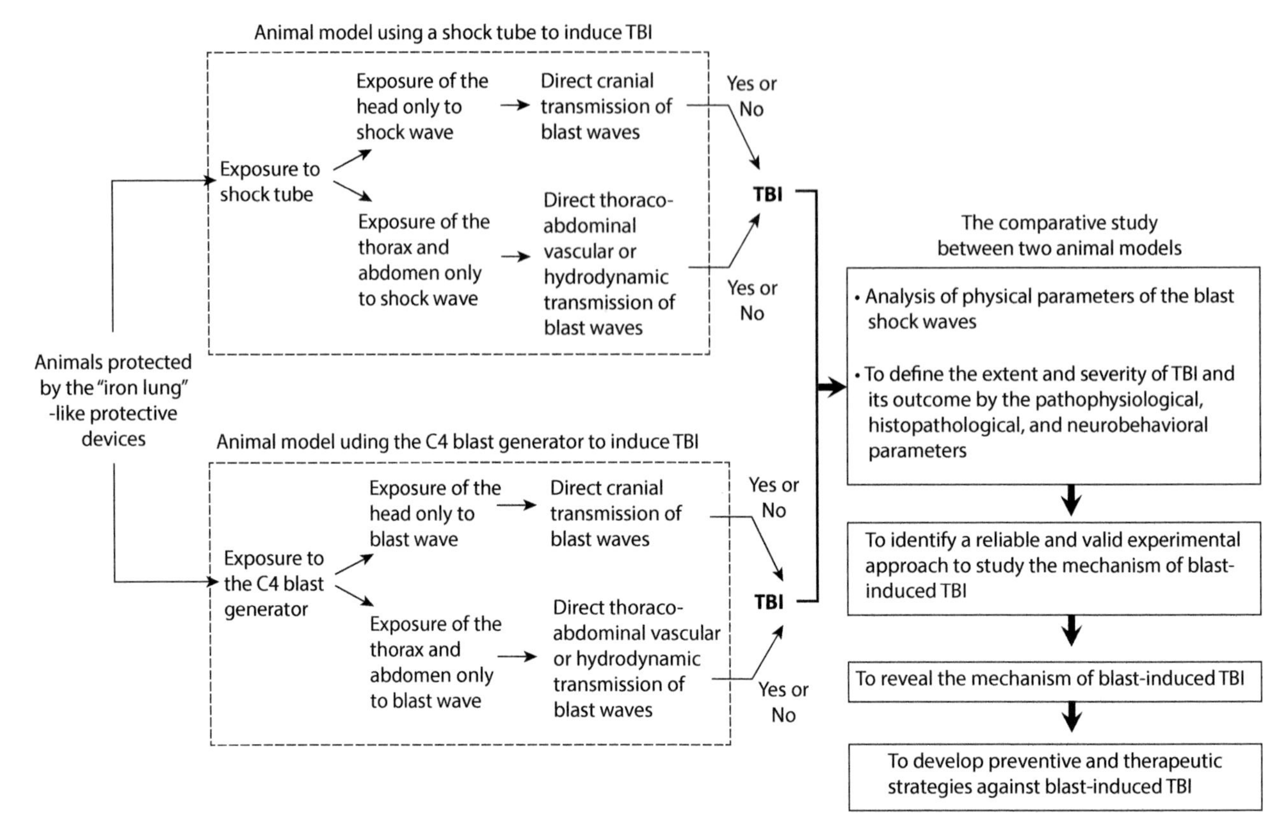

FIGURE 17.7 A flowchart showing the two-model approach to inducing blast-induced traumatic brain injury (TBI) that separates direct cranial transmission of blast waves from a thoracoabdominal vascular/hydrodynamic mechanism and reveals the mechanisms underlying blast-induced TBI.

Because shock tubes generate a series of complex shock waves including reflected shock waves, a Mach stem, an unsteady turbulent jet, and rarefaction waves, many uncertainties are associated with the results obtained from animal models using shock tubes. This means the data on injury mechanisms of shock waves can be difficult to analyze and compare.

A C4 blast generator that generates blast waves by detonation of C4 charge in a free field can be used as a new experimental tool for blast-induced TBI research. The animal model using the C4 blast generator to induce TBI will help yield valuable data from animal experiments and can be compared with the animal model using a shock tube. The use of the iron lung–like protective devices in the two animal models will effectively separate direct cranial transmission of blast waves from thoracoabdominal vascular/hydrodynamic mechanisms to blast-induced TBI, thus gaining a deeper and broader understanding of blast-induced TBI. The comparative study between the two animal models will help identify a reliable and valid experimental approach to study the mechanisms of TBI caused by exposure to blast waves. The two-model approach to investigating blast-induced TBI in animals may lead to important breakthroughs in prevention and treatment of human TBI induced by blasts.

REFERENCES

Assari, S., K. Laksari, M. Barbe, and K. Darvish. 2013. Cerebral blood pressure rise during blast exposure in a rat model of blast-induced traumatic brain injury. 9th Annual Injury Biomechanics Symposium, May 19 –21, 2013, Columbus, Ohio, from http://www.osuibrl.org/symposium/agenda/manuscriptsandposters/AssariS.pdf [viewed October 28, 2013].

Bell, M.K. 2008. Standardized model is needed to study the neurological effects of primary blast wave exposure. *Mil Med.* 173: v–viii.

Chen, Y., S. Constantini, V. Trembovler, M. Weinstock, and E. Shohami. 1996. An experimental model of closed head injury in mice: Pathophysiology, histopathology and cognitive deficits. *J. Neurotrauma* 13:557–568.

Chen, Y., and W. Huang. 2011. Non-impact, blast-induced mild TBI and PTSD: Concepts and caveats. *Brain Inj.* 25:641–650.

Chen, Y., W. Huang, and S. Constantini. 2012. Blast shock wave mitigation using the hydraulic energy redirection and release technology. *PLoS ONE.* 7:e39353. doi: 10.1371/journal.pone.0039353.

Chen, Y., W. Huang, and S. Constantini. 2013. Concepts and strategies for clinical management of blast-induced traumatic brain injury and posttraumatic stress disorder. *J Neuropsychiatry Clin Neurosci* 25:103–110.

Chen, Y. and S. Constantini. 2013. Caveats for using shock tube in blast-induced traumatic brain injury research. *Front. Neurol.* 4:117. doi: 10.3389/fneur.2013.00117.

Congressional Budget Office. 2012. A CBO Study: The Veterans Health Administration's Treatment of PTSD and Traumatic Brain Injury Among Recent Combat Veterans. February 2012. Congressional Budget Office, the Congress of the United States, from http://www.cbo.gov/sites/default/files/cbofiles/attachments/02-09-PTSD.pdf [viewed November 11, 2013].

Cooper, M.F., K.L. Armstrong, N.P. Lawless, M.T. Williams, J.R. Morris, M.J. Topper et al. 1998. Pathological comparison between the effects of single and multiple exposures to blast overpressure in six rat organs. *FASEB J.* 12:4563.

Federal Emergency Management Agency. 2003. Explosive blast. In *Reference Manual to Mitigate Potential Terrorist Attacks Against Buildings*. FEMA Publication Number 426, Washington, DC, from http://www.fema.gov/pdf/plan/prevent/rms/426/fema426_ch4.pdf [viewed November 7, 2013].

Hue, C.D., S. Cao, S.F. Haider, K.V. Vo, G.B. Effgen, E. Vogel III et al. 2013. Blood-brain barrier dysfunction after primary blast injury in vitro. *J Neurotrauma*. 2013 Apr 13. PMID: 23581482.

Ling, G., F. Bandak, R. Armonda, G. Grant, and J. Ecklund. 2009. Explosive blast neurotrauma. *J Neurotrauma.* 26:815–825.

Long, J.B., T.L. Bentley, K.A. Wessner, C. Cerone, S. Sweeney, and R.A. Bauman. 2009. Blast overpressure in rats: Recreating a battlefield injury in the laboratory. *J Neurotrauma.* 26: 827–840.

Martin, E.M., W.C. Lu, K. Helmick, L. French, and D.L. Warden. 2008. Traumatic brain injuries sustained in the Afghanistan and Iraq wars. *J. Trauma Nurs.* 15:94–99.

Nakagawa, A., M. Fujimura, K. Kato, H. Okuyama, T. Hashimoto, K. Takayama et al. 2008. Shock wave-induced brain injury in rat: Novel traumatic brain injury animal model. *Acta Neurochir Suppl.* 102: 421–424.

Olson, J.J., and L.R. Fletcher. 1971. *Airblast-Overpressure Levels From Confined Underground Production Blasts.* Washington, DC: Bureau of Mines Report of Investigation 7574.

Parent, A., and M.B. Carpenter. 1995. *Carpenter's Human Neuroanatomy*. 9th ed. Philadelphia: Lippincott, Williams & Wilkins. pp. 186–192.

Ritenour, A.E., and T.W. Baskin. 2008. Primary blast injury: Update on diagnosis and treatment. *Crit Care Med.* 36:S311–317.

Sosa, M.A., R. De Gasperi, A.J. Paulino, P.E. Pricop, M.C. Shaughness, E. Maudlin-Jeronimo, et al. 2013. Blast overpressure induces shear-related injuries in the brain of rats exposed to a mild traumatic brain injury. *Acta Neuropathol Commun.* 1:51.

Treadwell, I. 1989. Effects of blasts on the human body. *Nurs RSA.* 4:32–36.

Warden, D. 2006. Military TBI during the Iraq and Afghanistan wars. *J. Head Trauma Rehabil.* 21:398–402.

Yeoh, S., E.D. Bell, and K.L. Monson. 2013. Distribution of blood-brain barrier disruption in primary blast injury. *Ann Biomed Eng.* 2013 Apr 9. PMID: 23568152.

18 Acute Pathophysiology of Blast Injury—From Biomechanics to Experiments and Computations

Implications on Head and Polytrauma

Namas Chandra and Aravind Sundaramurthy

CONTENTS

18.1 INTRODUCTION

Traumatic brain injury (TBI) has been a major public health concern in the United States for decades; with an average of 1.4 million sustaining TBI, leading to 50,000 deaths, 235,000 hospitalizations, and 1.1 million emergency department visits. Among the several causes of TBI including fall accidents, motor vehicle crashes, sports collisions, and firearm assaults, explosions-induced TBI, commonly referred to as Blast-Induced Neurotrauma (BINT) has been frequently observed in recent warfare. In this chapter, we will discuss the origin, mechanics, mechanisms of BINT, and focus on the current state-of-the-art in terms of protection, diagnostics, prognostics and therapeutics involved in BINT. In addition, we will outline novel research approaches currently underway in terms of: experiments in laboratory-on animals, surrogate heads, cadaveric heads; as well as computer models of head-brain system. Finally, an analysis on the protective capabilities that have been introduced in BINTs against blast loading will be evaluated. TBI has been a major public health concern in the United States for decades; from 1995 to 2001 there was an average of 1.4 million Americans a year who sustained TBI, leading to 50,000 deaths, 235,000 hospitalizations, and 1.1 million emergency department visits (Langlois et al., 2004). Of these, an estimated 80,000–90,000 experienced the onset of long-term disability (Langlois et al., 2004; Thurman et al., 1999). These TBIs were largely caused by falls, motor vehicle crashes, sports collisions, and firearm assaults (Langlois et al., 2004). However, in this chapter, we are concerned with a new type of brain injury arising from explosions commonly referred to as BINT. We will focus on the origin, mechanics, and mechanisms of BINT, and the current state-of-the-art in terms of protection, diagnostics, prognostics, and therapeutics of the injury. We will outline the research approaches currently under way in terms of experiments in laboratory on animals, surrogate heads, cadaveric heads, computer models of the head–brain system, and finally an analysis on the protective capabilities of the current helmets against blast loading.

BINT has been recognized as a major problem among U.S. service members since Operation Enduring Freedom (OEF) began in Afghanistan in October 2001 and Operation Iraqi

Freedom (OIF) began in Iraq in May 2003. Blasts have been considered a threat to soldiers and civilians since the early 1970s, when researchers mainly considered lung damage and designed better protective body armors. Ironically, these body armors have saved many during explosions, making them more vulnerable to brain injuries. A recent RAND report estimates that 320,000 service members or 20% of the deployed force (total deployed 1.6 million) potentially suffer from TBI (Tanielian, 2008). However, of this population, approximately 60% have never been assessed by a health care provider specifically for TBI. The Department of Defense (DOD) based on data from 2004 to 2006 at selected military installations has estimated that 10%–20% of (the total deployed) OEF/OIF service members potentially sustain mild TBI (mTBI) (Reneer et al., 2011). Other studies also report the occurrence of TBI in OEF/OIF veterans. For example, a recent study has found that 22.8% of soldiers in an Army Brigade Combat Team returning from Iraq had clinically confirmed TBI (Terrio et al., 2009). A survey of OEF/OIF veterans, who had left combat theaters by September 2004, found that about 12% of the 2,235 respondents reported a history consistent with mTBI (Zhu et al., 2010). Among those who have been medically evacuated from theater, the proportions who have suffered TBI are predictably higher. For example, between January 2003 and February 2007, 29% of the patients evacuated from the combat theater to Walter Reed Army Medical Center in Washington, DC, had evidence of TBI (Defense and Veterans Brain Injury Center, 2007). Of 50 OEF/OIF veterans treated at the Tampa Veterans Affairs Polytrauma Rehabilitation Center, 80% had incurred combat-related TBI, with 70% of the injuries caused by improvised explosive devices (IEDs) (Clark et al., 2007).

For active-duty military personnel in war zones, blasts are the primary cause of TBI (Tanielian, 2008). Recent statistics from the conflict in Iraq show that several thousands of active duty U.S. soldiers have sustained TBI, 69% of these as a result of blasts (Warden, 2006). Analysis of data collected (collection period: March 2004–September 2004) from 115 patients from the Navy–Marine Corps that were identified with TBI have found that IEDs were the most common mechanism of injury, responsible for 52% of TBI cases overall. The analysis also showed that intracranial injuries, particularly concussions, were the most common diagnosis, especially among patients with nonbattle injuries (94%). Although multiple TBI-related diagnoses were common, 51% of the patient group had only an intracranial injury with no accompanying head fracture or open wound of the head. It was also found that of 115 patients analyzed, about 63% of patients were wearing a helmet at the time of injury (Clark et al., 2007). In addition to the data reported previously, the DOD, in cooperation with the Armed Forces Health Surveillance Center and Defense and Veterans Brain Injury Center, has consolidated the data of clinically confirmed TBI cases among service members and categorized them based on the severity of injury as shown in Figure 18.1 (DoD, 2012). mTBI contributes to more than 80% of the total reported brain injuries (Figure 18.1) because exposure to repeated low-level blasts is a common feature of the war zone personnel/civilian populations. Indeed, blast-induced mTBI has been identified as the signature injury of OEF and OIF (Jones et al., 2007; Tanielian, 2008; Terrio et al., 2009; Warden, 2006). It should also be noted that the number of TBI cases reported by the DOD (DoD, 2012) is 10 times smaller than that estimated by the RAND (Tanielian, 2008) study. This discrepancy is attributed to methodology used for estimating the numbers. RAND estimates are often based on the subjective response to nonspecific screening questions asked to several hundred individuals; that resulting percentage was then generalized to the entire deployed population. On the contrary, DOD estimates are completed based on clinically confirmed TBI cases.

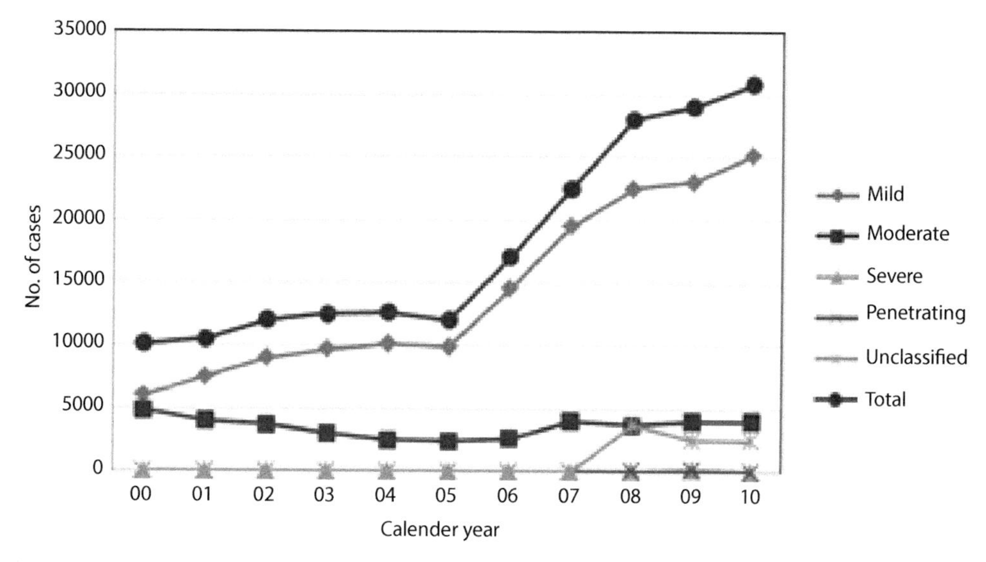

FIGURE 18.1 Blast-induced TBI in U.S. service members between 2000 and 2010. These numbers are based on clinically confirmed TBI cases. mTBI contributes to more than 80% of the total reported brain injuries. (Courtesy of Armed Forces Health Surveillance Center, Department of Defense.)

18.1.1 Definitions of TBI and Blast-Induced TBI

TBI is a silent epidemic with serious consequences that changes the lives of the victims and their families irrevocably (Thurman et al., 1995). The Centers for Disease Control and Prevention defines TBI as a head injury that is associated with decreased levels of consciousness, amnesia, other neurological or neurophysiological abnormalities, skull fracture, diagnosed intracranial lesions, or death. TBIs can be classified as mild, moderate, or severe.

The American Congress of Rehabilitation Medicine has defined blast-induced mTBI as a head injury resulting in at least one of loss of consciousness (LOC) for 30 minutes or less; posttraumatic amnesia (PTA) for less than 24 hours; any alteration in the mental state immediately after the accident; or focal neurological deficit(s) that may or may not be transient. The emergency room physicians upon presentation of the patient routinely use the Glasgow Coma Scale to assess the state of TBI. The 15-point Glasgow Coma Scale defines the severity of injury as mild (13–15), moderate (9–12), severe (3–8), or vegetative state (<3) based on eye, verbal, and motor responses (Teasdale and Jennett, 1974). A new classification specific to blast-related TBI (bTBI) has been also proposed recently, in which a blast-induced mTBI is characterized by loss of consciousness LOC for less than 1 hour and PTA for less than 24 hours after exposure to an explosive blast. Moderate bTBI is characterized by loss of consciousness LOC for 1–24 hours and PTA for 1–7 days, and severe bTBI is characterized by loss of consciousness for more than 24 hours and PTA for more than 7 days (Ling et al., 2009). Mild to moderate cases of bTBI do not involve skull fracture; further, they are not detectable through current neuroimaging techniques. The Centers for Disease Control and Prevention has reported that up to 75% of TBIs that occur each year are mild, and a World Health Organization task force reported that 70%–90% of all treated TBI were mild (Kraus et al., 2007). mTBI is typically not associated with abnormalities in brain imaging (Okie, 2005), and most patients with mTBI recover fully in 4–12 weeks (Alexander, 1995; Kennedy et al., 2007). However, mTBI patients with more severe injuries, such as those who experienced loss of consciousness lasting more than 10 minutes or PTA lasting more than 4–6 hours, may require months to years to recuperate (Alexander, 1995). In addition, some mTBI patients develop postconcussive syndrome, experiencing persistent cognitive, behavioral, and/or somatic symptoms (Alexander, 1995; Heitger et al., 2009; Zhu et al., 2010). Studies have shown that 15%–35% of patients with mTBI experience onset of long-term disability (Alexander, 1995; Heitger et al., 2009; Thurman et al., 1999).

18.1.2 Classification of TBI: Ballistics, Blunt, and Blast

Head and brain injuries are categorized based on the origin of mechanical forces as blast, blunt, or ballistic injuries.

18.1.2.1 Ballistics

In a ballistic impact, the penetrating object pierces the skull and enters the brain parenchyma. Depending on the kinetic energy at the time of contact, a penetrating object may be ejected from the opposite side along with biological tissues or lodged inside the brain. All along the path of propagation of the penetrant, the tissue is dismembered and physical separation occurs, which affects both the neuronal tissues as well as the vasculature. Though ballistic injuries have been significantly prevented with composite armored helmets in the battlefield, this remains a major issue in the civilian population because of increased gun violence.

18.1.2.2 Blunt

In blunt impacts, the head of the victim collides with a stationary or moving object. During the process of collision, the head encounters a directional force in a local region. Furthermore, the head and the brain translate or rotate depending on the magnitude and direction of the impacting force. The energy transfer to the head is governed by the laws of conservation of energy; the peak force, contact area, and the duration of the impact are governed by the principles of conservation of momentum; and the actual state of strain/stress is governed by equilibrium of forces. The severity of injury is a function of the mass and velocity of the impactor, and form, shape, size, and material of the impactor as well as the protective state of the head (helmet type, padding system, impact location, and unprotected). Common types of blunt injuries include epidural/subdural hematoma, subarachnoid hematoma, and contusion and severe hemorrhage, with concomitant increase in intracranial pressure (ICP) (Bolander et al., 2011).

18.1.2.3 Blast

A field blast is extremely complex and the facets of the injuries include components of both blunt and ballistic impact. In general, blast injuries are classified into four main categories: (1) primary (direct effects of overpressure), (2) secondary (effects of projectiles/shrapnel), (3) tertiary (effects from fall from blast winds), and (4) quaternary (burns, asphyxia, and exposure to toxic inhalants). Usually, depending on the strength of explosive, casing, standoff distance, and scenario (urban, theater, indoor/outdoor), victims suffer a combination of these injuries. An illustration of these injuries is shown in Figure 18.2.

18.1.2.3.1 Primary Blast Injury

Once a bomb explodes, it results in a sudden rise in the atmospheric pressure, which represents the shock front followed by an exponential decay resulting from the expansion wave depleting the overpressure. This sudden rise followed by an exponential decay comprises the positive phase of the blast wave and is responsible for primary blast injury. Although the sharp rise constitutes the shock wave, the rest of the pressure pulse is referred to as blast wind. Furthermore, depending on surrounding (urban, indoor/outdoor) structure or an enclosure (such as interior of a bunker or a vehicle), the subjects are also exposed to complex blast waves caused by multiple reflections from the walls, floor, and ceiling and their interactions.

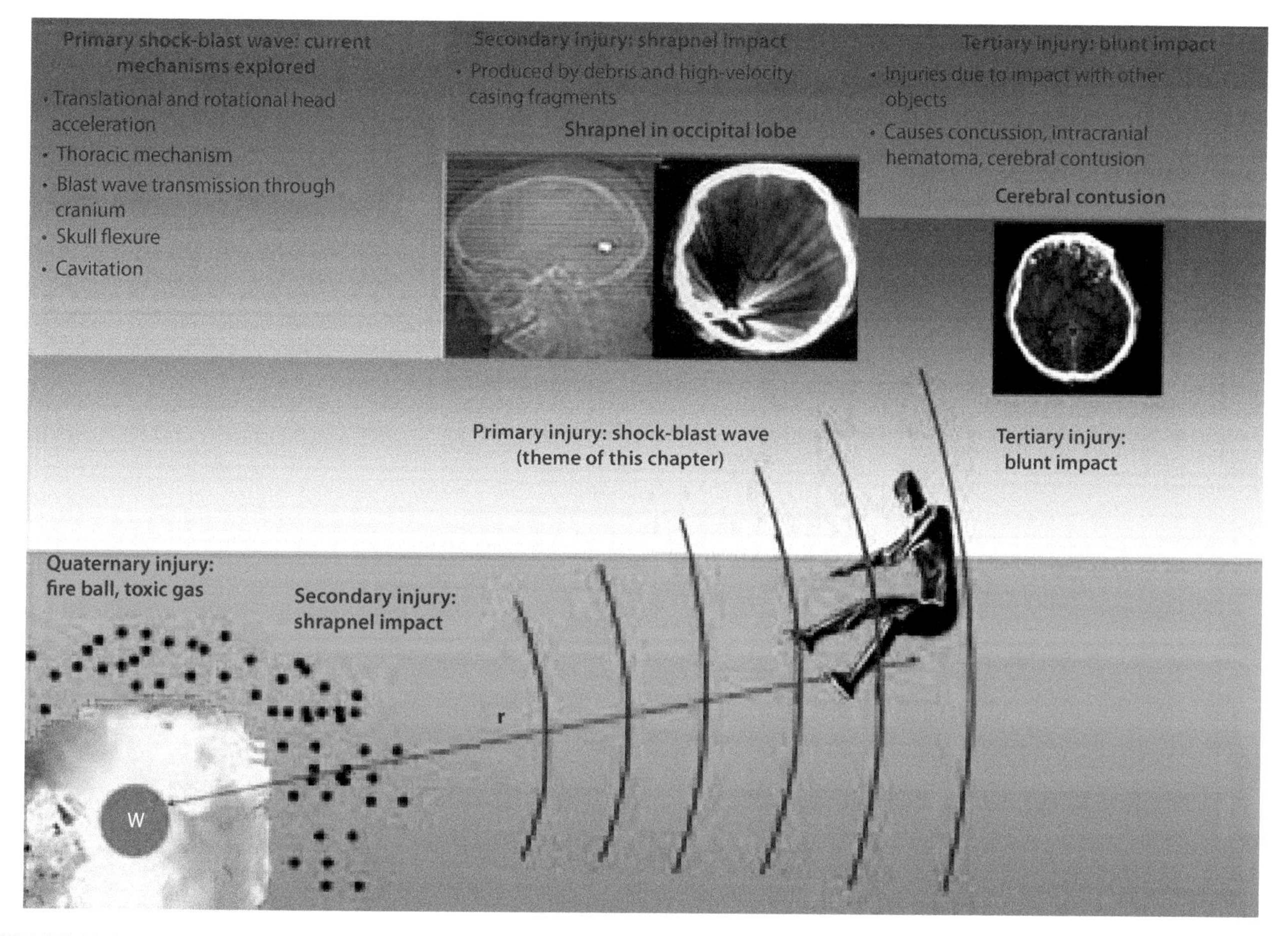

FIGURE 18.2 bTBI classification. In this figure, w is the charge weight and r is the standoff distance.

18.1.2.3.2 *Secondary Blast Injury*

IEDs have metal casings and are usually filled with metal fragments. During the explosion, these metal fragments become highly accelerated projectiles that penetrate the body. The pathology associated with secondary blast injuries is similar to that of the ballistic injury discussed previously.

18.1.2.3.3 *Tertiary Blast Injury*

Tertiary injuries include those with nonpenetrating projectile impacts as well as injuries from a fall. Explosion within a building may result in its collapse, causing impact and crushing of the body, whereas in an outdoor blast, depending on the proximity of the blast epicenter, the whole body might be thrown, resulting in impact with hard surface. When seated inside a vehicle under explosive attacks or in a rollover, the victims are thrown violently inside, leading to an impact-type injury. The pathology of this injury is similar to the blunt impact discussed previously.

18.1.2.3.4 *Quaternary Blast Injury*

Quaternary blast injuries refer to explosion-related injuries that were not included in the primary, secondary, or tertiary injuries (DePalma et al., 2005). Quaternary injuries include burns (chemical or thermal), toxic inhalation, and exposure to radiation.

18.1.3 Mild/Moderate Blast TBI

When subjects are closer to the epicenter of the blast, they will endure injuries with a degree of severity that in most cases will result in fatalities. However, at a sufficiently longer distance from the source (i.e., in the far field range) only primary or pure blast effects are dominant (Reynolds, 1998). It is conjectured that mild to moderate TBI seen in the theater is due to primary blast effects alone because injuries were nonfatal, with neither head fracture nor open wounds (Dewey, 1971; Kleinschmit, 2011; Skotak et al., 2013; Warden, 2006). In 2005, the U.S. military reported 10,953 IED attacks, at an average of 30 per day. As mentioned previously, these explosions were less fatal with a much higher survivability rate than in previous wars, hence mostly resulting in mild or moderate TBI. Hence, in this chapter, we are only concerned with the primary blast injury wherein the victim is exposed to a pure shock-blast wave that eventually results in a mild or moderate TBI.

18.1.4 Current State of the Research

Creating conditions of blast loading and understanding the physics of the blast are fundamental elements in the study of acute and chronic ailments of primary blast injury. Field-testing and shock tubes are the main methods used for creating a primary blast injury. Among them, shock tubes are more commonly and frequently used in primary blast injury research. Table 1 from a review article by Kobeissy and his colleagues shows the recent major studies on primary blast injury (Kobeissy et al., 2013). Among this list, only 8 cases of 49 used field explosive testing as a method of generating primary blast. In 36 other cases, shock tubes were used. Shock tubes are also classified into three types: (1) compressed gas shock tube, (2) detonation shock tube, and (3) combustion shock tube. Among them, compressed gas shock tubes are predominately used as the source for primary blast (in 33 cases).

Understanding the processes of primary blast inflicting loads (i.e., the mechanism by which the injury occurs) is a prerequisite to the design of proper animal model testing and interpretation of results (Benzinger et al., 2009). Over the past few years, several mechanisms of mechanical insult have been suggested. These mechanisms are (1) a thoracic mechanism in which blast waves enter the brain through the thorax and increase brain pressure via vasculature (Cernak et al., 1997, 2001; Courtney and Courtney, 2009), (2) translational and rotational head acceleration (Courtney and Courtney, 2009), (3) blast wave transmission through cranium (Moore et al., 2009; Nyein et al., 2010; Taylor and Ford, 2009), (4) skull flexure (Bolander et al., 2011; Moss et al., 2009), and (5) cavitation (Panzer et al., 2012). Most of these mechanisms are proposed using numerical models alone and experimental evidence is needed to corroborate these proposed mechanisms. Postmortem human specimen (PMHS) heads, which is closest to humans in terms of anthropometry, can be used to study and validate these mechanisms. However, one main drawback with PMHS is that it is not possible to mimic the material and biological aspects of a living brain.

Animal models are essential for studying the biomechanical, cellular, and molecular aspects of human TBI that cannot be addressed using surrogate models. Furthermore, a validated animal model is required for development and characterization of novel therapeutic interventions. Currently, the majority of this work is done using small mammals; very few studies use large mammals (Kobeissy et al., 2013). Animal models are used to investigate the physiological, neuropathological, and neurobehavioral consequences as well as identify biomarkers that are related to brain injury. However, one main issue with the current research in animal models is the inconsistency in specimen location when doing experiments with a shock tube. This leads to two problems: (1) erroneous loading conditions (in some cases) and (2) comparison of the results between different laboratories is virtually impossible. Therefore, developing a standard experimental model that can be validated across laboratories is essential for the interpretation of the mechanisms of blast injury, the identification of biomarkers, and, eventually, the development of strategies for mitigating blast-induced brain injury (Xiong et al., 2013).

Given the difficulties in understanding TBI associated with conducting experiments on PMHS or human volunteers and translating animal models results to human, computational modeling has always been an easier choice. As a result, there have been a large number of models reported in the literature in the past (refer to Table 2.2 in Ganpule, 2013). Computational modeling is currently used for studying the shock-blast interaction with the head (analysis of reflected pressure fields), for flow field analysis to determine flow separations, and for pressure field distribution in the brain and loading mechanisms (skull flexure, direct pressure transmission) as well as to study the effect of helmets. One main drawback in computational modeling is the lack of consistent and accurate material models. Consequently, it is vital to validate the model against an accepted experimental standard to ensure the validity of the predictions in the model.

It is estimated that since 2005, 77% of soldiers who sustained any type of TBI were wearing their helmets at the time of injury (Wojcik et al., 2010). Although the role of helmets in bTBI is critical, there are very few investigations that have studied helmets in blast mitigation (Brown et al., 1993; Grujicic et al., 2011; Nyein et al., 2010; Sogbesan, 2011). These studies concluded that a padded helmet provides only some degree of protection to the head during blast loading. Reductions of pressure were noticed in coup regions, whereas in other areas (e.g., brainstem) the values of ICP, shear stress, and strain remained unchanged compared with the no-helmet case. A comprehensive study using both experiments and numerical models is essential to elucidate the role of helmets in blast mitigation and to gauge the capabilities of the different helmets currently used by military personnel in the theater.

18.1.5 Organization of the Chapter

In this section, we examine the physics of explosion and the characteristics of shock-blast waves as a function of explosive strength and stand-off distance. Because experiments on biological medium require a large number of samples with variations in key parameters, conducting experiments with actual blasts is not an option in terms of time, cost, or safety. Properly designed laboratory-scale shock tubes can replicate field conditions. Unfortunately, many of the current experiments conducted by different experimental groups around the world have not paid attention to this detail; hence, their results should be viewed with caution. If pure primary blast conditions are not achieved during the test conditions, then the specimens are subjected to a combination of blast as well as the loads associated with tertiary injury. Compressed gas-driven shock tubes are very commonly used to replicate field conditions. In this section, we outline the basic components of the shock tube and how each of them interacts to produce the desired shock wave profile. The location of the test specimen within the length of the tube plays a major role.

In Section 18.2 on PMHS blast testing, we examine the basic question of how a blast wave interacts with the human head and causes mechanical loading on the brain. Both the physics of the problem and preliminary experiments clearly show that the brain loading would depend on the geometry (shape and size) of the head as well as the tissue properties of the head, in addition to the intensity of shock wave itself. To represent humans as accurately as possible, cadaveric heads (PMHS) back-filled with ballistic gels were subjected to field-relevant blast conditions. Deformation was measured both on the surface of the head and by measuring ICP variations. These measurements were repeated when the intensity of the external shock was increased. The results showed that there is an increase in pressure excursion in the brain with an increase in the shock wave intensity.

In Section 18.3 on animal model blast testing, we examine the role of animal models in answering the fundamental question of whether the primary blast causes measurable changes at the cellular and tissue level, and also study behavioral patterns. We discuss the observed physiological changes occurring immediately after exposures in terms of heartbeat, pulse, blood oxygen levels, and weight loss over time. We established a dose-response curve between external blast overpressure from 60 kPa to 450 kPa and mortality rates. The response shows a typical sigmoidal curve. Significant pathophysiological changes like plasma membrane integrity loss, blood-brain barrier damage, and elevated levels of reactive oxygen species are observed. The damage in the deeper brain regions distinguishes bTBI from the coup and contrecoup type damage noticed in blunt impacts. Small, but not statistically significant changes are observed in the behavior.

Computer models play a key role in not only understanding individual test results but results across a range of species and loading conditions. This is the subject of Section 18.4. Here we present the basic mathematical formulations involved, geometric modeling techniques used to convert images to numerical models, and material model parameter issues. In Section 18.5, we address the key issue of protection. Though we do not fully understand the blast injury biomechanics, the adequacy or otherwise of current head protection devices still need to be studied. The role of geometry, material, and external loadings on the design of current helmets are discussed in this chapter.

18.1.6 Replicating Primary Blast in the Laboratory

To study mild/moderate BINT, it is vital to replicate field blast conditions associated with the injury. First, we explore the physics of explosion briefly and establish the shock wave parameters (SWPs) that can be measured and controlled. Further, some of the current techniques used for simulating the mild/moderate blast conditions are also outlined. Attention is paid to the compressed gas-driven shock tube because most researchers tend to use this method. The dos and don'ts of the design, construction, and operation of the tubes are also elucidated.

18.1.7 Characteristics of Field Explosion

Explosives are often used by insurgents as improvised weapons. They are usually present in the form of homemade chemical explosives embedded with a variety of shrapnel and typically called IEDs. They are buried under the ground and detonated from a distance or left on the side of the road to injure a vehicle, occupants of the vehicle, pedestrians, or a gathering. The strength of the explosives is measured by kilogram of TNT-equivalent. When a chemical explosive detonates, enormous energy is released in a very short time. This energy expands very quickly and compresses the surrounding air. For a simplified spherical explosive suspended in air, the expansion is spherical in nature; the compression propagates in the form of a fast traveling wave. When the velocity of the wave front exceeds the sonic velocity of air (corresponding to the temperature, altitude, and humidity) then the wave front is termed a shock wave. The shock wave is a very thin layer and the pressure increases from atmospheric to the peak overpressure within a few atomic distances and in a very short time, along the order of microseconds. Therefore, the shock wave is the front part of the blast wave and can be called the shock-blast wave. Though the terms shock and blast are used interchangeably, the shock is the very small portion of the entire blast wave profile.

Explosives can come from different sources (e.g., chemicals, high-pressure gas tanks, nuclear); their strength is expressed as a single unit in terms of TNT-equivalent. One gram of TNT produces a blast energy of 4610 J; 1 ton of TNT produces about 4.61 million kJ. Explosive strength of a given chemical expresses the effectiveness of a given chemical in producing a blast wave comparable to that of the same weight of TNT. For example, pentaerythritol tetranitrate is 1.5 times more powerful than TNT, whereas C4 (composition B of 60% RDX/40% TNT) is 1.34 times more potent than TNT. When an explosive detonates, the matter of the explosive is converted to the heat of explosion (appearing as heat and fireball) and the energy of expanding gas (leading to blast waves). If the explosive mixture contains metallic products (e.g., nails, metallic sheaths), then part of the expansion energy appears as kinetic energy of the fragments acting as projectiles.

The destructive power of the blast wave increases as a function of its velocity of propagation. If a is the velocity of sound in the undisturbed air medium and u is the velocity of the blast wave, then $\frac{u}{a} \equiv M$ is defined as the Mach number. For example, the speed of sound at room temperature and pressure at sea level is approximately 343 m/sec (768 mph); a shock wave with $M = 2$ travels at 686 m/second (1536 mph). The 1-m-long blast wave passes any given point in less than 2 msec, which is such a short time scale the exposed body is unable to react because of its inertial mass.

In the case of an idealized mid-air explosion, the gas products and the fireball expand in a spherical manner and compress the surrounding air. This expanding air continues to increase in velocity at the outer edge and at a certain point, the velocity of the envelope just equals the velocity of sound (i.e., $M = 1$). When this happens, a shock front is formed. The shock front is extremely small, on the order of a few molecular diameters in width. Across this front, there is a sudden change of pressure of the undisturbed medium from atmospheric to a high pressure. There is a sudden change in pressure, velocity, and density across this narrow thickness. Behind the shock front, there is the blast wind in which the particles move at very high velocities. This shock front blast wind continues to expand outward. However, because the expanding gas occupies an increasing volume (spherical radius r), the shock velocity slowly reduces and eventually becomes the same as the sonic velocity of the medium, at which point the shock wave dies leaving behind a low-velocity blast wind.

The strength of the shock wave is measured in terms of overpressure, termed blast overpressure (BOP). BOP can be expressed as

$$\text{BOP} = \frac{7(M^2 - 1)}{6} \text{bars}$$

where M is the Mach number.

Or, equivalently, the shock wave velocity can be found from BOP as

$$M = \sqrt{1 + \frac{6}{7}\text{BOP}}$$

Thus we can see that there is a relationship between shock wave velocity and the blast overpressure. The shock wave velocity decreases as the spherical radius increases. At some distance (standoff distance), this shock degenerates to a sound wave.

18.1.8 Laboratory Reproduction of Shock-Blast Wave

In this section, we consider the characteristics of a spherically expanding blast wave formed by the detonation of an explosive (e.g., an uncased C4 of charge weight W). The effects of the fireball, ground reflections, and other artifacts are not considered because in the far field range of interest these effects are absent (Reynolds, 1998). The impulsive expansion of explosive product initiates first a shock wave propagating spherically outward in the surrounding air and then a family of infinite rarefaction waves propagating in the shock-compressed air. Across the shock front, pressure, density, particle velocity, and temperature rise significantly and rapidly in a few microseconds. As the radial propagation distance r increases, the surface area of the spherical shock front increases as r^2 and consequently its intensity decreases as $1/r^2$.

The propagation of the shock front is supersonic with respect to the ambient air ahead (upstream) but subsonic with respect to the shocked air behind (downstream); the rarefaction waves remain supersonic with respect to the compressed air ahead (upstream) until the air compression is fully released. The sequential arrivals of rarefaction waves at a given r cause decrease of shock compression. Hence, in the intermediate to far range of r, more and more rarefaction waves catch up the shock front, giving rise to erosion of the shock front intensity, evolving nonlinear decay in overpressure p, and even a period of underpressure (negative overpressure) afterward. The shock front diminishes eventually at large r. Therefore, the p–t profiles of the blast waves of interest have a shock front with a peak overpressure P^* followed by a nonlinear decay during positive overpressure with a duration of t^* and then a period of underpressure. This is confirmed through open-field testing conducted by the authors at the Trauma Mechanics Research Facility at the University of Nebraska-Lincoln in conjunction with the U.S. Army. The laboratory and facility has since moved along with the authors to the New Jersey Institute of Technology and is a part of the Center for Injury Bio-mechanics, Materials and Medicine. The test results are further corroborated with conventional weapons effects (ConWep) (Ganpule et al., 2012) (Figure 18.3).

Such p–t curves can be mathematically described by the so-called Friedlander waveform (Kinney and Graham, 1985) given by the following equation:

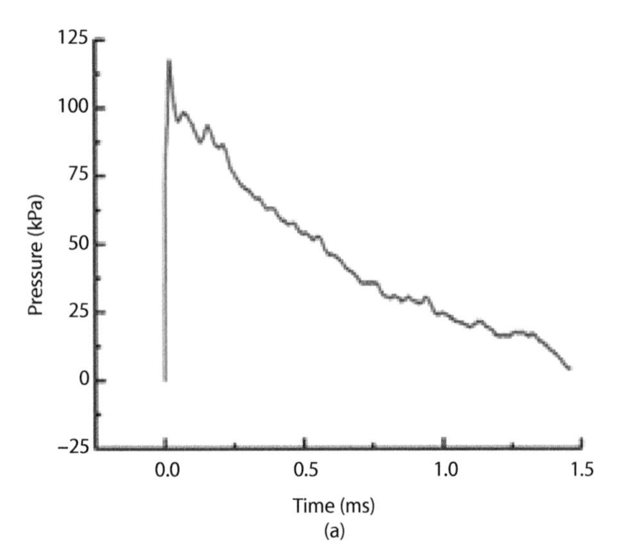

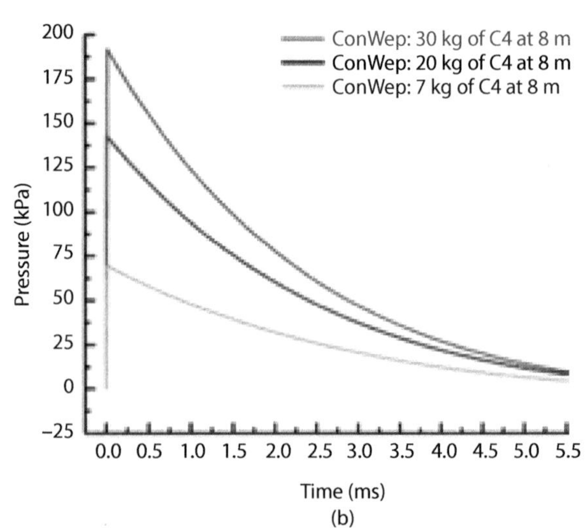

FIGURE 18.3 Pressure time profiles in a free field explosion: (a) IP profiles measured using a pencil gauge by a free field explosion of 1.81 kg of C4 at a distance of 2.8 m. The testing was conducted by the Trauma Mechanics Research Facility at the University of Nebraska-Lincoln in conjunction with the U.S. Army. (b) Pressure profiles obtained using Conventional Weapons Effects calculations using ConWep.

$$p(t) = p^*\left(1 - t/t^*\right)\exp\left(-bt/t^*\right) \quad (18.1)$$

where b is the decay constant.

The Friedlander waveform is characteristic for an open-field blast. A typical Friedlander waveform profile is shown in Figure 18.4. The integration of $p(t)$ over t^* gives the positive impulse per unit area I^*. p^* (overpressure) and t^* (positive time duration) are the two independent parameters along with a decay constant used for describing the essential characteristics of positive portions of the blast waves of interest. The ranges of p^* and t^* that are of interest for bTBI are postulated to be 50–1,000 kPa and 1–8 ms, respectively (Panzer et al., 2012); however, precise ranges are not available from the field data. Measurements conducted in our laboratory indicate that the upper range can be reduced to 400 kPa and the duration restricted to 7 ms. It should be noted that during an actual explosion, the p–t curve could be more complex because of reflections from ground, surrounding building structures (in an urban setting), and explosives that are cased and/or contain shrapnel. Furthermore, there are charge-to-charge variations possible from packing density, shape of the charge (spherical, cylindrical), the purity of explosives (pure or mixed with shrapnel), the type and material of the cover, and the type and location of the trigger. As far as the target is concerned, not only the distance but also the direction of the wave and environmental parameters (temperature, wind, dust, humidity) determine the loading as well as the shape, size, and material property of the target and its conditions (e.g., if human whether subjects are wearing a helmet). To conduct a fundamental scientific study of bTBI, it is critical to isolate these as extraneous effects and generate the blast waves with the Friedlander type wave profile.

Another important feature of the open-field blast waves in the intermediate to far range is that the sizes of wave front are much larger than that of a human body. The interactions of such a blast wave with a human body are influenced strongly by the confinement of an effectively edgeless (no flattop or plateau) wave front. This characteristic represents a planar wave and must be re-created in the laboratory testing to realistically simulate field blast loading.

A spherical wave in the near field while expanding becomes more planar at far field ranges. Therefore, a blast wave in the intermediate to far range can be well approximated with a planar Friedlander wave, as shown in Figure 18.4.

18.1.9 Types of Shock Tubes

Although the free-field blast testing closely replicates real-world blast conditions, there are some significant drawbacks: (1) free-field experiments are expensive and (2) time consuming, and (3) repeatability is difficult to achieve. Furthermore, the by-products of HE testing include potential fireball interactions and penetration from high-speed shrapnel (Bass et al., 2012). This can introduce unnecessary confounds to the experiments in which the objective is to understand the effect of primary blast injury and its subsequent biomechanical and biochemical sequela.

18.1.9.1 Detonation Shock Tube

The idea of a detonation shock tube was first introduced by Clemedson in 1955 to study injuries from blast in animal models (Clemedson and Criborn, 1955). In this case, a tube with one end open and the other end closed is used. A chemical explosive is placed near the closed end of the shock tube and detonated to produce a shock blast wave (Risling et al., 2011; Säljö et al., 2000).

18.1.9.2 Combustion Shock Tube

In a combustion shock tube, a mixture of oxygen and acetylene is filled in the driver section and sealed using a polymer membrane. This mixture is then ignited using an electric match from a safe location. The fuel undergoes combustion and produces carbon dioxide, water vapor, and heat energy. Expanding gases that come from the driver compresses the air in the driven section and initiates a shock-blast wave (Courtney et al., 2012).

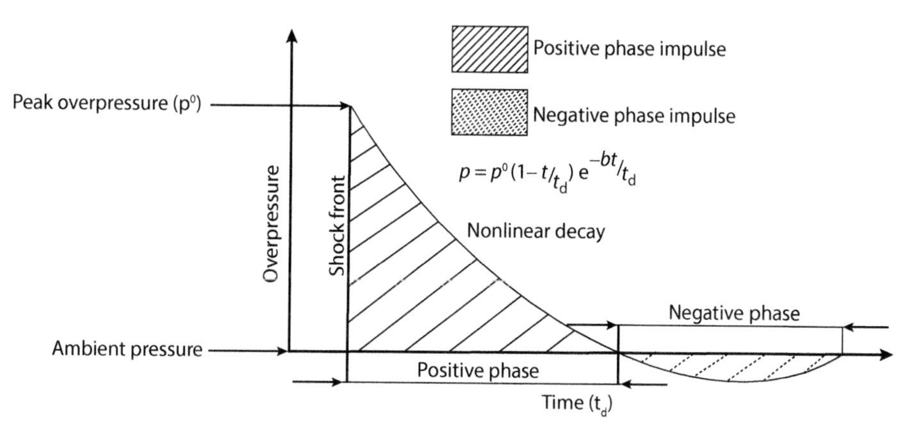

FIGURE 18.4 Mathematical representation of planar Friedlander waveform. The equation in the figure represents instantaneous overpressure $p+$ at given time t, where p^* is the peak overpressure, t^* is positive phase duration, I^* is positive phase impulse, and b is the decay constant. (From Sundaramurthy, A. et al. *Journal of Neurotrauma*. 29:2352, 2012.)

18.1.9.3 Compressed Gas Shock Tube

This is the most popular type of shock-blast generation device, which essentially uses a similar technique as a combustion and detonation type shock tube except here, the sudden release of compressed gas is used for generating the shock-blast wave. A detailed description of the theory and work of a shock tube will be discussed in the following sections. Currently, this technique has been adopted by many researchers who investigate bTBI because of its apparent simplicity, repeatability, and experimental control (Bolander et al., 2011; Leonardi et al., 2011; Long et al., 2009; Sundaramurthy et al., 2012; Zhu et al., 2010).

18.1.9.3.1 Theory of Compressed Gas Shock Tubes

Although individual shock tubes for blast wave simulation may have different features, the essential wave physics can be understood by analyzing the wave propagation in a generic shock tube configuration as shown in Figure 18.5. A typical (compressed gas–driven) shock tube consists of a driver section of pressurized gas and a driven section of air at atmospheric pressure with the two sections separated by a set of membranes. When the membranes burst, the driver gas expands rapidly and compresses the atmospheric air (i.e., driven gas) in front of a shocked state, which propagates forward as an air shock wave. Meanwhile, the driver gas expansion initiates a family of infinite rarefaction waves (expansion fan). These rarefaction waves first travel toward the closed end, are reflected at the closed end, and then travel toward the open end. Their sequential arrivals at a given location of driven section produce a nonlinear decay (see wave profiles a-c of Figure 18.5). The wave profile evolves with propagation distance to that of a Friedlander wave (curve c of Figure 18.5) when the fastest rarefaction wave (which is faster than the shock front) catches the shock front at $x = x^*$, where the shock front intensity is eroded the least by the rarefaction waves. Hence, at $x = x^*$, peak overpressure p^* has the maximum value with a Friedlander wave profile. The time for the nonlinear decay to reach $p = 0$ gives overpressure duration t^*, which has the minimum value at $x = x^*$. Before the initial catchup, $x < x^*$ (curves a and b of Figure 18.5), the blast wave assumes a flattop shape because the rarefaction wave reflected from the closed end has not reached the shock front yet. The flattop duration is given by the difference in the arrival times of the shock front and the fastest rarefaction wave. In the range $x^* < x < 0$ where $x = 0$ represents the shock tube exit, more and more rarefaction waves catch up to the shock front causing decreasing p^* and increasing t^* with increasing x. The pressure-time (p-t) profile near (outside) the exit is shown by curve d of Figure 18.5; notice that the waveform is changed significantly (low p^*, low t^*, followed by jet wind).

18.1.9.4 General Features of Compressed Gas Shock Tube Construction

The design for a shock tube is shown in Figure 18.6. The four main components of the shock tube are the driver section or breech, transition, driven section or straight section, and the catch (or expansion) tank. The straight section includes the test section. The driver contains pressurized gas, which is separated from the transition by several frangible membranes. Upon membrane rupture, a shock wave expands through the transition and develops in the straight section. Subjects are placed in the test section, which is strategically placed to produce a desired shock wave profile. Finally, the shock wave exits the shock tube and enters the catch tank, which reduces the noise intensity. Each of the components is described below.

18.1.9.4.1 Driver Section

The driver section contains the compressed gas (e.g., air, helium, nitrogen) used in the generation of the shock-blast wave. It is a cylindrical tube with one end permanently sealed, whereas the other is sealed with frangible membranes such as Mylar. In some designs, thin sheets of metals (aluminum or steel) have been used. It is critical to strictly control the repeatability of burst characteristics by controlling the material and manufacturing processes of the membranes. Driver gas is filled into the driver section, pressurized, and allowed to burst depending on the membrane material and its thickness. Once the pressure reaches the burst pressure, the membranes rupture releasing the high-pressured gas into the transition.

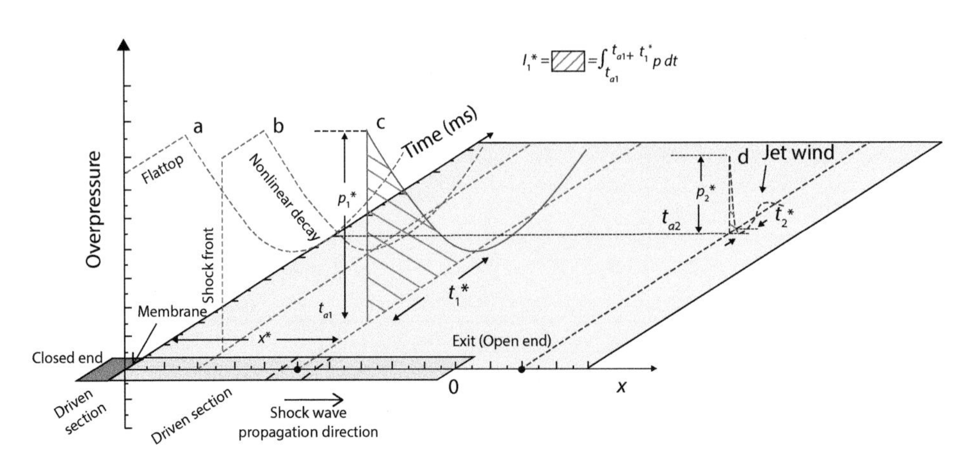

FIGURE 18.5 Evolution of shock wave in a generic shock tube. (From Chandra, N. et al., *Shock Waves*. 22:403–415, 2012.)

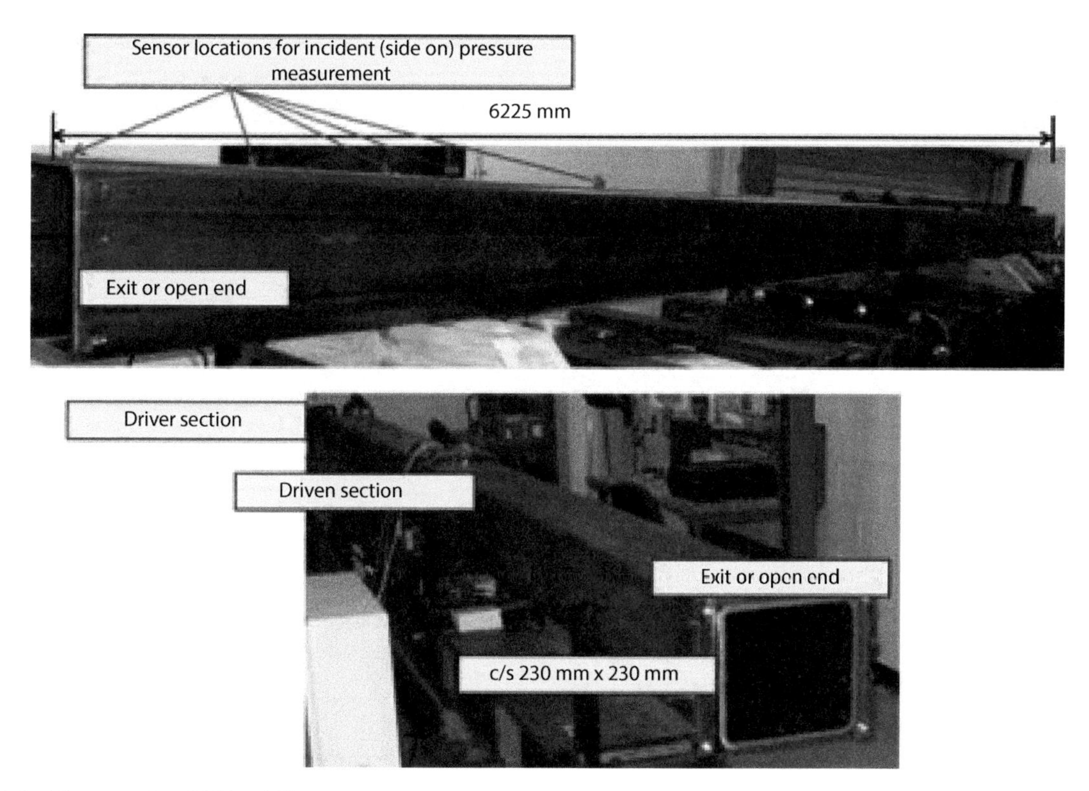

FIGURE 18.6 Photographs of 230 × 230-mm square shock tube used in this work. (From Chandra, N. et al., *Shock Waves*. 22:403–415, 2012; Sundaramurthy, A. et al. *Journal of Neurotrauma*. 29:2352, 2012.)

18.1.9.4.2 Transition

The transition is a design element used to change the cross-section of the tube from a circular cylinder (driver section) to a square (driven sections). The transition was designed with a gradual expansion to minimize flow separation and turbulence associated (reflections from the sidewalls) with abrupt changes in a cross-sectional area. If the driver and driven sections remain the same, there is no need for the transition. However, to produce an effective shock front the driven is usually larger and longer than the driver.

18.1.9.4.3 Driven Section

The driven section or the expansion section is the square section where the fully formed shock-blast wave is generated. In our design, we used a square driver section, which has since been followed by other research groups. The use of a square cross-section as opposed to other geometric sections serves two purposes: (1) uniform expansion from the breech, which is vital to have a planar wave, and (2) to observe and record events in the test section with a high-speed camera (where having a circular section may cause image distortion) (Sundaramurthy et al., 2012). The test section is located within this part; while locating the test section, it is vital to note that the test section should be neither close to the breech nor to the exit. The reasons for this will be discussed in details in the coming sections.

18.1.9.4.4 Catch Tank

Depending on the strength of the shock-blast wave produced by the shock tube, the catch tank may or may not be included in a particular design. Its main purpose is to contain and release the large volume of expanded gas generated from a shot, minimizing blast loading of laboratory structures and reducing noise level. The use of a suddenly changing cross-sectional area was studied and found to successfully mitigate energy (Jiang et al., 1999). The inside of the catch tank is lined with sound absorbing material for reducing the noise.

18.1.10 Test Section Location

Modular design of the shock tube allows the flexibility of altering the distance of the test section from the breech. In the following section, we will discuss the important criteria that have to be considered for test section location and their effect on the shock-blast wave profile. Important criteria include (1) contact surface, (2) planarity, and (3) exit rarefaction.

18.1.10.1 Contact Surface

Upon rapid expansion of the driver gas, the temperature and pressure drop causing a density increase in the driver gas. The location of contact between the driver and driven gas is defined as the contact surface. The shock overpressure across a contact surface does not change, but the density difference causes a discontinuity in kinetic energy. This density difference can cause significant changes in reflected pressure across the contact surface. The contact surface can also be avoided by determining the fully expanded volume of the driver gas, and placing the sample beyond that location. The final volume of the expanded gas can be calculated using the isentropic expansion of an ideal gas, given by the following equation (Cengel et al., 2011):

$$\frac{P_2}{P_1} = \frac{V_1}{V_2}^k$$

where P_2 is the burst pressure, P_1 is the atmospheric pressure (14.5 psi), V_1 is the driver volume, V_2 is the final volume of the gas after complete expansion, and k (1.4 for nitrogen) is the adiabatic index. By knowing the breech volume, type of driver gas, and burst pressure, we can determine the final volume of the completely expanded driver gas. With this, we can theoretically approximate the minimum distance that is required to avoid any interference from the contact surface. However, this methodology does not consider effects from the mixing of driver and driven gases. Usually, the investigator has to be vigilant when analyzing the first 7–10 ms of *p-t* profile of the reflected pressure recorded in the test section to determine sudden changes in the reflected pressure. In such a case, either the test section should be moved downstream or the breech volume has to be reduced.

18.1.10.2 Planarity

In the mild/moderate range depending on the explosive strength and the distance from the epicenter, the shock front is planar. When simulating this type of blast in a laboratory shock tube, this feature is essential to replicate. Experiments can be done in the test section to determine the planarity by placing a bar with a linear array of sensors in the shock tube. The shock front planarity is determined by measuring the arrival time at various locations on the bar, and any deviation of arrival time will show the shock front curvature. With this idea, an experiment was conducted to determine the planarity of the shock front produced in our shock tube.

Along the bar, an array of eight sensors was placed with one sensor in the center and six at 4, 8, and 12 inches from the center in either direction. Assuming the bar is not mounted perfectly perpendicular to the shock tube, averaging of sensors equidistant from the center ensures that nonplanarity measurements caused by a skewed bar are corrected. The sensors at 8 and 12 inches were averaged in the same manner. A final sensor was located 13 inches from the center and was adjusted based on the calculated angle of the sensor bar (Figure 18.7a) (Kleinschmit, 2011).

After the sensor bar was mounted, shock waves were generated using a 11.75-inch nitrogen driver and 10 membranes. This produces shock waves with peak overpressures between 105 and 160 kPa, depending on the burst pressure and distance from the breech. The sensor bar was mounted 48, 98, and 136 inches from the transition exit. The planarity results shown in Figure 18.7b demonstrate nonplanarity with a shock front leading at the edges by approximately 0.15 inches at a location 48 inches from the transition exit.

The curvature corresponds to an approximate diameter of 47 feet, which can be considered planar. The results became increasingly planar to the location 136 inches from the transition exit. The general trends show increasing planarity with increase in the distance from the transition.

18.1.10.3 Exit Rarefaction

Figure 18.8 shows the untailored shock blast wave produced by a shock tube (Ritzel et al., 2011). The secondary decay and the following upstream recompression are the artifacts created by the rarefaction wave from the open end of the shock tube. Therefore, the subject within the test section, if not placed appropriately, will be subjected to a blast wave that does not simulate a field blast. This problem can be resolved by two methods: either having a reflector plate attached to the exit or placing the test section deep inside the driven section. In the case of a reflector plate, the venting gases when impinging on the plate create a compression wave, which essentially nullifies the rarefaction wave generated from venting (Coulter et al., 1983). In the second case, by having the specimen deep inside the driven section the exit rarefaction wave enters the test section only after the passage of the actual shock-blast wave.

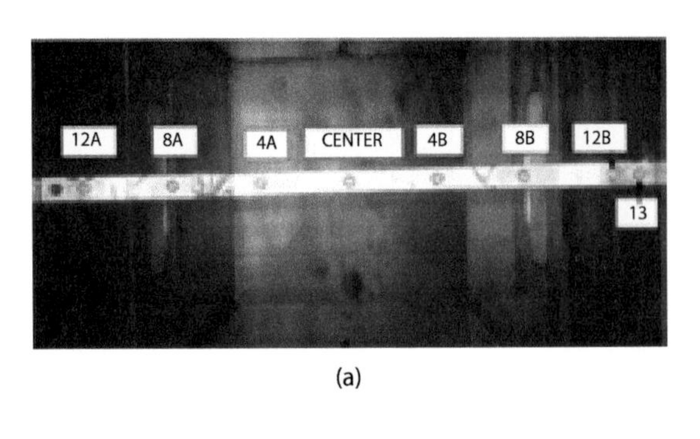

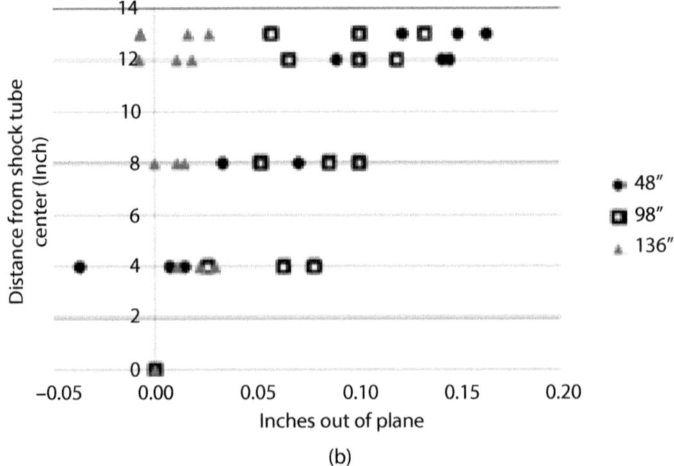

FIGURE 18.7 (a) Sensor bar used for planarity testing in 28-inch shock tube. (b) Shock wave planarity measurements in 28-inch tube taken at 48, 98, and 136 inches from the exit of the transition for 18-psi overpressure shock waves.

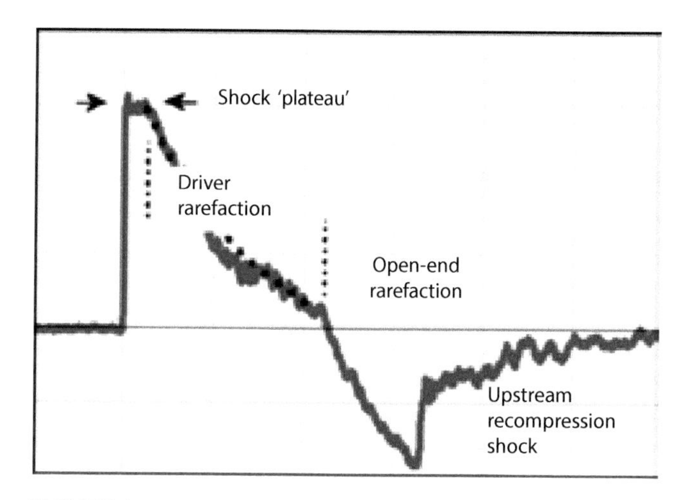

FIGURE 18.8 Untailored shock blast wave profile. (From Ritzel, D. et al., Proceedings HFM-207 NATO Symposium on a Survey of Blast Injury across the Full Landscape of Military Science, 2011.)

18.1.11 Effects of Placing Specimen outside the Shock Tube

Studies on the evolution of the shock wave at the exit or open end have attracted researchers over the years because of numerous flow phenomena occurring at the exit (Arakeri et al., 2004; Honma et al., 2003; Jiang et al., 1999, 2003; Kashimura et al., 2000; Onodera et al., 1998; Setoguchi et al., 1993). It is shown in these studies that at the exit of the shock tube, the shock wave evolves from planar to three-dimensional (3D) spherical with other effects like vortex formation, secondary shock formation, Mach disc, subsonic jet flow, shock-vortex interaction, and impulsive noise. All of these effects may or may not be present depending on the shock wave strength and geometry of the exit. Most of these studies, however, have focused on the flow dynamics aspects with no emphasis on qualitative or quantitative analysis of shock/blast-wave profiles (e.g., *p-t* profiles). This becomes particularly important as many researchers do primary blast-injury testing using animal models and surrogates outside the shock tube. Therefore, to understand the evolution of the shock-blast wave outside the shock tube, and whether it truly represents all aspects of the field blast, a detailed study of the blast wave exiting the shock tube was conducted using numerical simulations. Furthermore, an experimental study was conducted using a cylindrical model as well as an animal model comparing the mechanical responses from outside with the responses obtained inside the shock tube (Chandra et al., 2012; Sundaramurthy et al., 2012)

18.1.11.1 Evolution of Shock-Blast Wave outside the Shock Tube

The flow field at the exit of the shock tube is studied using numerical simulations. Figure 18.9 shows the pressure and velocity (vector) fields at the exit of the shock tube without a cylinder. As the blast wave exits the shock tube, the flow changes from planar to 3D spherical (Figure 18.9a). Rarefaction waves and vorticities at the corners mix with blast and the remaining air ejects as subsonic jet wind, which is evident from the velocity vector field of Figure 18.9b. This jet wind effect is not present deep inside the tube. Further, to clearly demonstrate this, Figure 18.10a shows the nodal velocities at various locations inside and outside the shock tube. Because fixed Eulerian mesh is used for modeling, velocity at a given mesh node corresponds to the instantaneous velocity of the material point coincident at given time '*t*' with the considered node. High-velocity jet wind is recorded in nodal history for locations outside the shock tube. Particle velocity associated with this jet is higher than particle velocity associated with the shock (Figure 18.10b). Locations inside the shock tube that are closer to the exit also show a second peak in velocity because of rarefaction waves moving into the tube, but the magnitude of this second peak is lower than the particle velocity associated with the jet for outside locations. In addition, the magnitude of this second peak gradually reduces as we move inside the shock tube away from the exit (open end). Deep inside the shock tube ($x = -3048$ mm), a second peak is completely absent.

To clearly exhibit transition of the blast wave from planar to 3D spherical, Figure 18.11 shows the pressure distribution at the exit of the shock tube for a sequence of times. The black arrows indicate the (velocity) vector field. In each figure an outer surrogate contour indicates the primary shock wave and an inner green portion indicates the primary vortex loop. The primary shock wave at first appears to be a square shape with rounded corners as shown in Figure 18.11i. These corners become significantly rounded and straight parts at the shock tube walls are shortened (Figure 18.11ii–iv). This indicates that the primary shock wave is planar at the exit (open end) of the shock tube and evolves three dimensionally into a spherical one as time elapses. This process is called a shock wave diffraction and affects the flow expansion behind it (Jiang et al., 2003). Similar arguments can be used to show the 3D nature of the primary vortex loop that is evident from green color of Figure 18.11.

18.1.11.2 Comparison of Inside and Outside Mechanical Response

18.1.11.2.1 Animal Model

In this experiment, five male Sprague Dawley rats of 320–360 g weight were sacrificed and instrumented with a surface mount Kulite sensor (LE-080-250A) on the nose to measure the reflected pressure, and two Kulite probe sensors (XCL-072-500A, diameter: 1.9 mm, length: 9.5 mm) were implanted in the thoracic cavity and in the brain, respectively. For the brain sensor implantation, the tip of the sensor was backfilled with water to ensure good contact with tissue, and the sensor was inserted through the foramen magnum 4–5 mm into the brain tissue. Then the instrumented rat was placed at four different locations twice inside the shock tube and twice outside. Complete details on the experimental setup can be found elsewhere (Sundaramurthy et al., 2012).

Figure 18.12 shows incident pressure as well as pressure in the brain and thoracic cavity corresponding to various

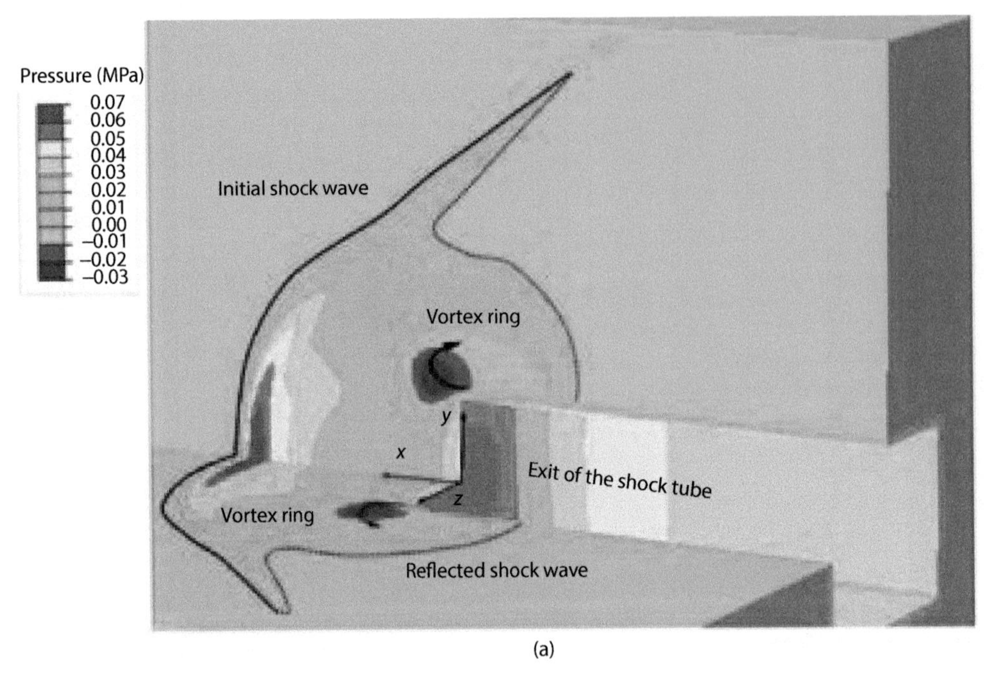

(a)

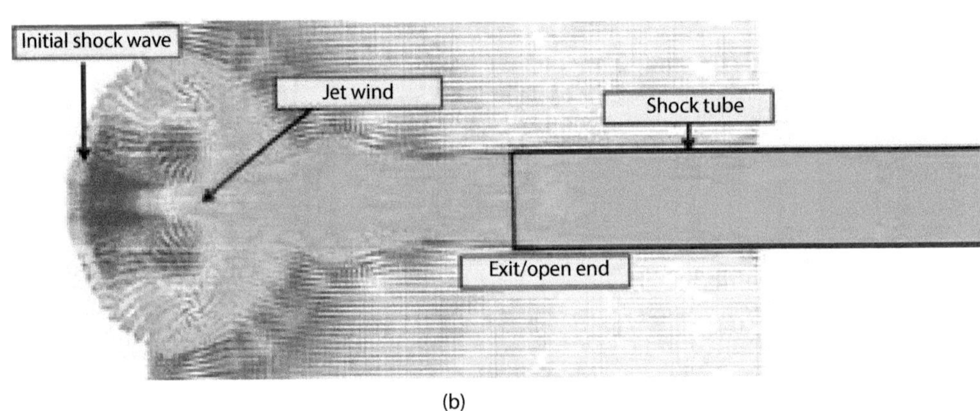

(b)

FIGURE 18.9 (a) Pressure field near the exit of the shock tube. Three-dimensional expansion of shock wave along with vortex formation is seen at the exit. (b) Velocity vector field near the exit of the shock tube. Jet wind is clearly visible in velocity vector field. Representative vector field is shown; jet is also observed at other locations close to the exit at earlier times. (From Chandra, N. et al., *Shock Waves.* 22:403–415, 2012.)

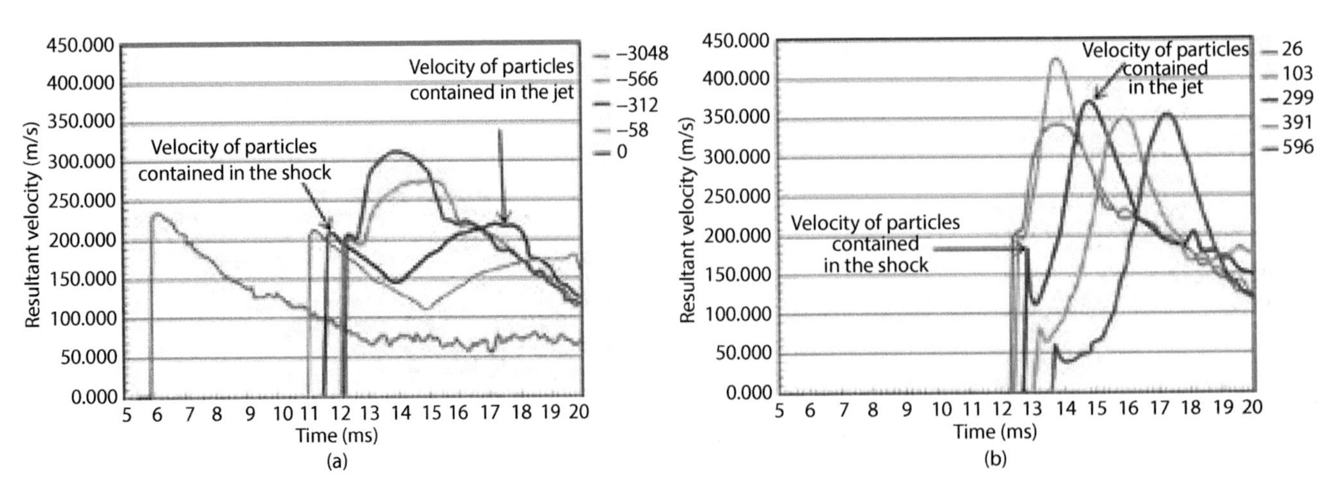

FIGURE 18.10 Nodal velocities at various locations inside and outside the shock tube. Because a fixed Eulerian mesh is used for modeling, velocity at a given mesh node corresponds to the instantaneous velocity of the material point coincident at given time 't' with the considered node. (a) Inside locations, (b) outside locations. (From Chandra, N. et al., *Shock Waves.* 22:403–415, 2012.)

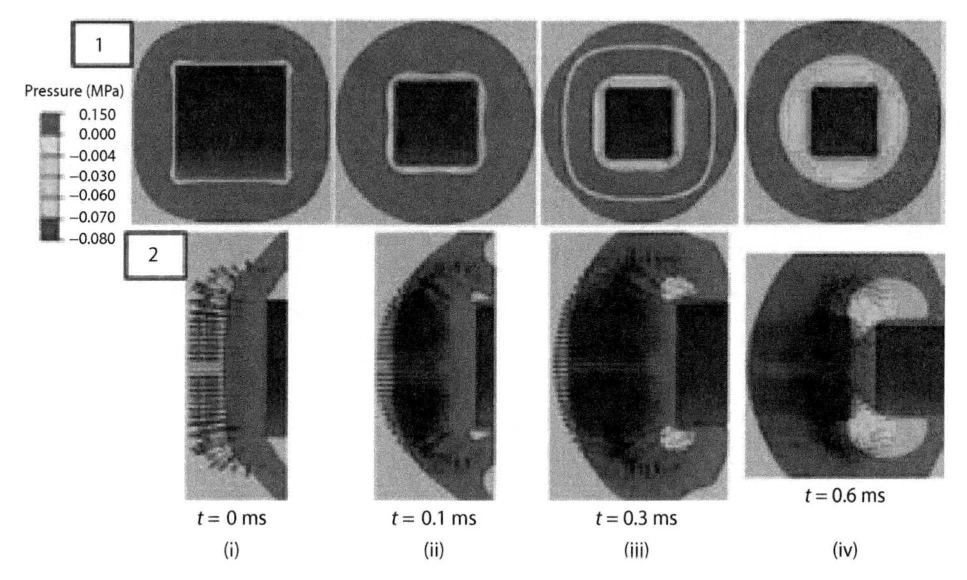

FIGURE 18.11 Flow fields illustrating physics of shock wave diffraction. Row 1 shows the axial view and row 2 shows the top view. Arrival of shock wave at the exit is marked as $t = 0$. (From Chandra, N. et al., *Shock Waves*. 22:403–415, 2012.)

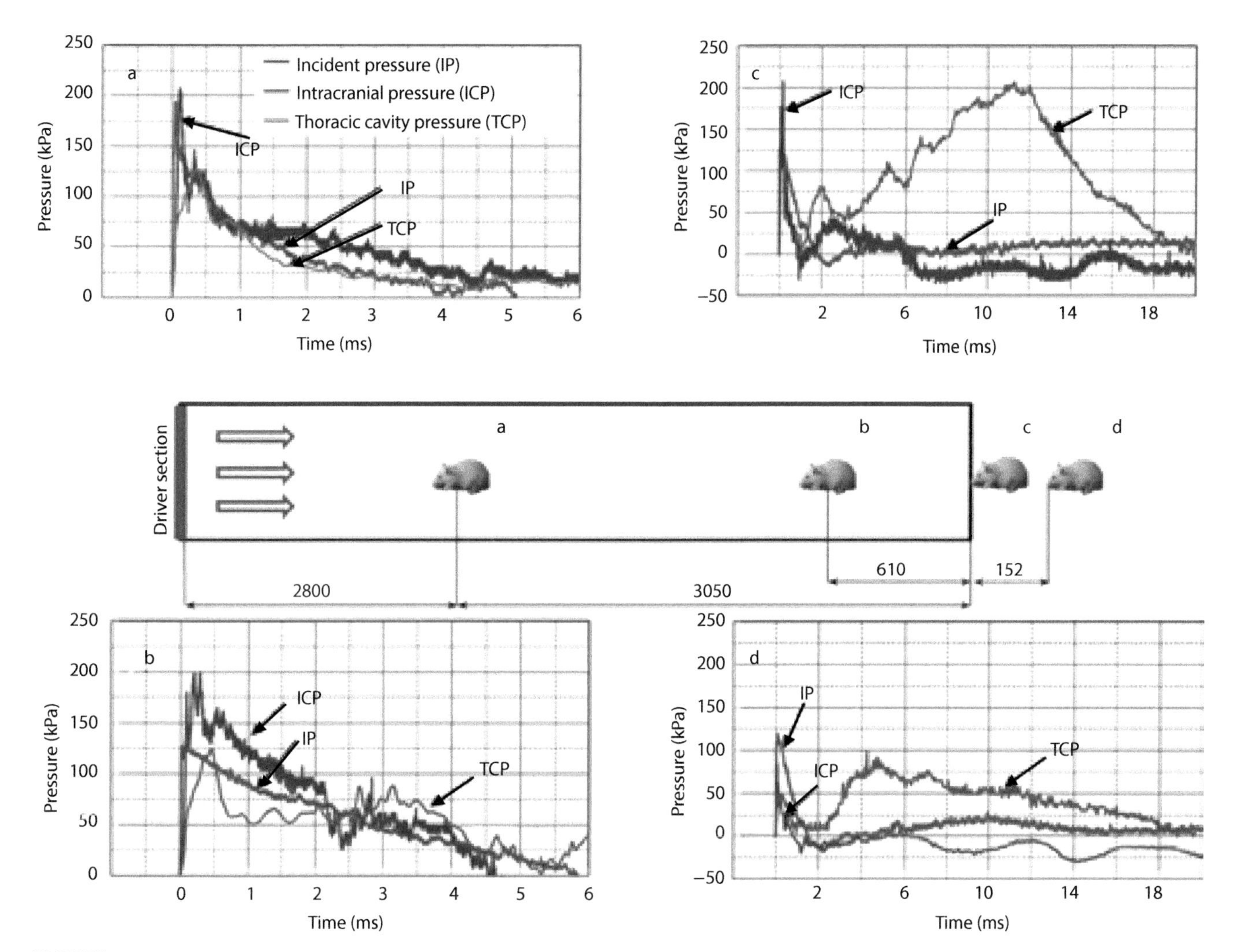

FIGURE 18.12 Measured pressure-time profile in the brain and thoracic cavity with their corresponding incident pressures at all animal placement locations (APLs). At APL (a) and (b), both intracranial and thoracic pressures follow the same behavior as incident pressure; however, in APL (c) and (d) (outside the shock tube), the positive time duration in the brain is reduced drastically and the lung experiences a secondary loading. In this figure, all the dimensions shown are in millimeters. (From Sundaramurthy, A. et al. *Journal of Neurotrauma*. 29:2352, 2012.)

locations along the length of the shock tube. At placement locations a and b, incident pressure profiles follow the Friedlander waveform (Figure 18.4) fairly well. Pressure profiles in the brain and thoracic cavity also have similar profiles (the shape is almost identical) to that of the incident pressure profiles. At these locations, peak pressures recorded in the brain are higher than the incident peak pressure and the peak pressure recorded in the thoracic cavity is equivalent to the incident peak pressure. It is clear, at location c, the incident pressure profile differs significantly from the ideal Friedlander waveform; the overpressure decay is rapid and the positive phase duration is reduced from 5 ms at a to 2 ms at c (Figure 18.12a and c, respectively). The pressure profile in the brain shows a similar trend. The pressure profile in the thoracic cavity shows a secondary loading with higher pressure and longer duration. The pressure profile in d is similar to the pressure profile recorded in c, except the value of the peak pressure reported in the brain is lower than the incident peak pressure.

The biomechanical response of the animal significantly varies with the placement location. Inside the shock tube (a, b, Figure 18.12), the load is due to the pure blast wave, which is evident from the *p-t* profiles (Friedlander type) recorded in the thoracic cavity and brain. For animal placement locations (APLs) at the exit, c and d, *p-t* profiles show sharp decay in pressure after the initial shock front. This decay is due to the interaction between the rarefaction wave from the exit of the shock tube, eliminating the exponentially decaying blast wave, which occurs in a and b. This has two consequences: first, the positive blast impulse (area under the curve) reduces drastically. Second, because the total energy at the exit is conserved, most of the blast energy is converted from supersonic blast wave to subsonic jet wind (Haselbacher et al., 2007). This expansion of blast wave at the exit (subsonic jet) produces an entirely different biomechanical loading effect compared with the blast wave. Consequently, the thoracic cavity experiences secondary loading (i.e., higher pressure and longer positive phase duration). When the animal is constrained, this high-velocity subsonic jet wind exerts severe compression on the tissues in the frontal area (head and neck) that in turn causes pressure increase in the thoracic cavity (lungs, heart). To further illustrate the effect of subsonic jet wind on the rat, experiments a and c without any constraint were performed. Figure 18.13 shows the displacement (motion) of the rat at various time points starting from the moment the blast wave interacts with the animal. At a, the displacement is minimal; however, at c, the rat is tossed away from the bed (motion) because of jet wind. This clearly illustrates the effect of high-velocity subsonic jet wind on the rat when placed outside the shock tube. Consequently, the animal is subjected to extreme compression loading when constrained and subjected to high-velocity (subsonic) wind when free, both of which are not typical of an IED blast. This in turn changes not only the injury type (e.g., brain vs. lung injury) but also the injury severity, outcome (e.g., live vs. dead), and mechanism (e.g., stress wave vs. acceleration).

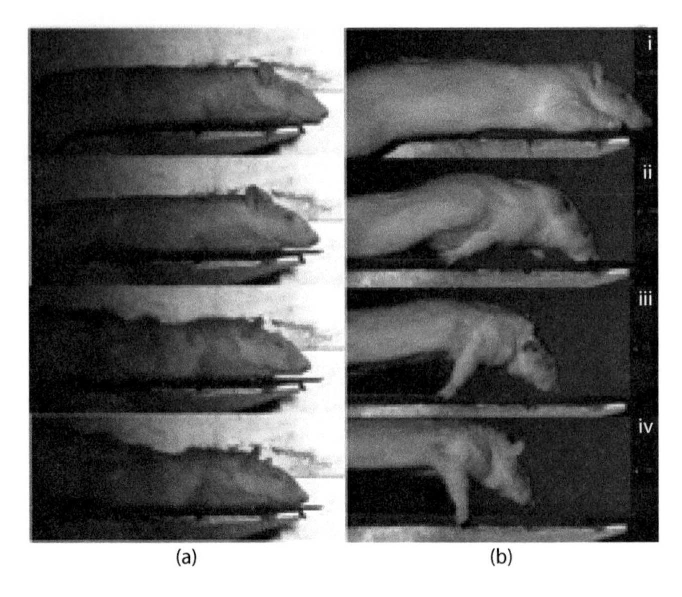

FIGURE 18.13 Motion of unconstrained rat under blast wave loading (a) inside and (b) outside. (i–iv) Time points $t = 0$, 20, 40, and 60 msec, respectively; the rat is thrown out of the bed when placed outside. (From Sundaramurthy, A. et al. *Journal of Neurotrauma*. 29:2352, 2012.)

18.1.11.2.2 Cylindrical Model

The evolution of the shock-blast wave along the length of the shock tube was measured using an aluminum cylinder (length 230 mm, diameter 41.3 mm). In this case, a cylinder was placed along the longitudinal axis of the shock tube at various offset distances from the exit (open end) both outside (+x) and inside (−x) (Figure 18.14). Seven holes were drilled and tapped to locate seven Dytran model 2300V1 piezoelectric pressure sensors used in conjunction with Dytran model 6502 mounting adapters. The location labeled t0 was centered between the two end surfaces of the cylinder, and the rest of the holes were evenly spaced for a total span of 84 mm. The cylinder was mounted (i.e., firmly secured) using brackets made out of flat steel bar. In addition to the gauges mounted on the cylinder, there were a set of gauges (PCB pressure sensor model 134A24) mounted at various locations on the shock tube (along the length) that measure the incident (side-on) pressures. The experiment was repeated three times at each location along the length of the shock tube (N = 3).

Figure 18.15a and b shows the reflected pressure profiles for cylinder placement locations inside and outside of the shock tube. The reflected pressure measures total pressure (both kinetic and potential energy components) at a given point. The reflected pressure profiles for placement locations inside the shock tube show gradual decay in pressure and pressure profiles follow the Friedlander waveform. A small secondary peak in pressure profiles is due to reflection from the walls of the shock tube; however, these wall reflections do not significantly affect pressure profiles. The reflected pressure profiles for placement locations outside the shock tube show rapid pressure decay that does not conform to the Friedlander waveform; the shock front and pressure

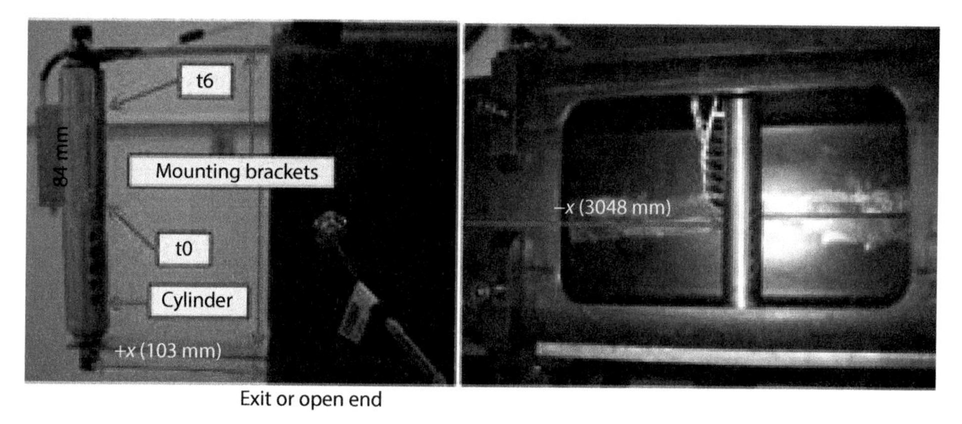

FIGURE 18.14 Experimental setup to measure evolution of the shock wave along the length of the shock tube. Placement of the cylinder at two representative locations along the length of the shock tube is shown. (From Chandra, N. et al., *Shock Waves*. 22:403–415, 2012.)

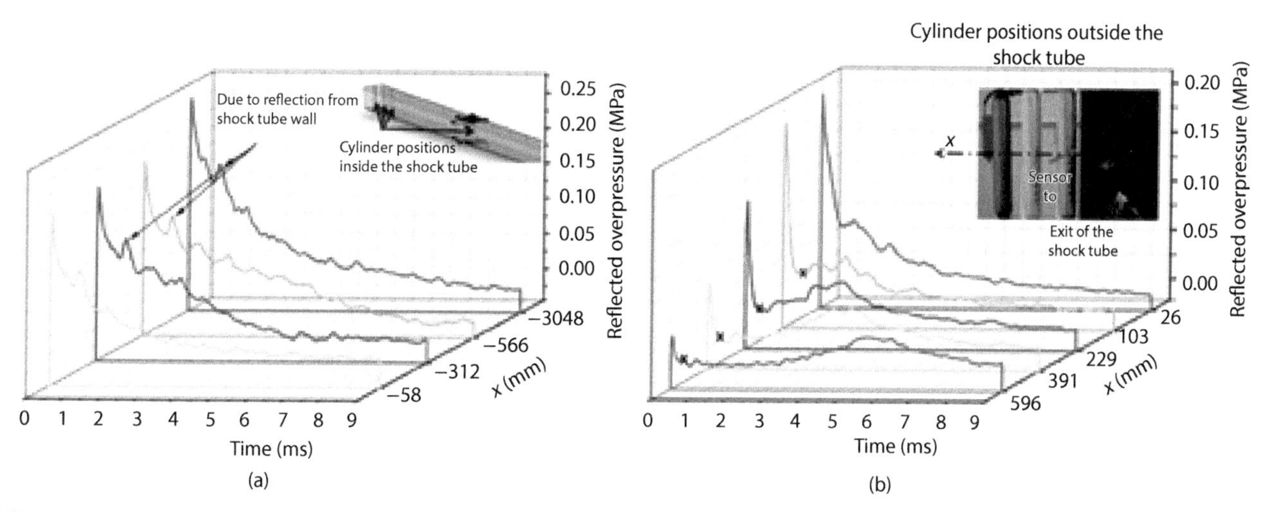

FIGURE 18.15 (a) Experimentally measured *p-t* profiles at various x locations inside the shock tube. *p-t* profiles follow the Friedlander waveform fairly well. (b) Experimentally measured *p-t* profiles at various x locations outside the shock tube. In these profiles, the trends do not follow the Friedlander waveform and peak overpressure drastically reduces as we move away from the exit. The starting points of subsonic jet wind are demarcated by cross symbols. (From Chandra, N. et al., *Shock Waves*. 22:403–415, 2012.)

decay instead look like a delta function. This is followed by a long-duration, relatively constant low-pressure regime (starting points of which are demarcated by cross symbols). This long duration, relatively constant low-pressure regime is referred to as subsonic jet wind in this work. This jet wind is an artifact of the shock tube exit effect and does not occur in free field blast conditions.

Figure 18.16a and b shows the impulse profiles for cylinder placement locations inside and outside of the shock tube. The total impulse is reduced significantly for outside placement locations when compared with inside placement locations. The shape of impulse profiles for placement locations inside the shock tube is relatively constant (i.e., gradual increase) as opposed to nongradual (i.e., with slope changes) for outside placement locations. The contribution of subsonic jet wind to the impulse is high starting points, which are demarcated by cross symbols.

18.1.12 Tailoring the Shock-Blast Wave Profile

In previous sections, we learned about the design, theory, and working of a compressed gas shock tube and then critically analyzed the factors that influence the test section locations for simulating a primary shock-blast wave. In the following sections, we will study the methods to tailor a shock-blast wave profile. To attain this goal, an extensive experimental analysis was carried out to show the methods in which the blast wave profile can be tailored to replicate the field conditions. We study the effects of shock tube–adjustable parameters (SAPs) such as breech length, type of gas, membrane thickness, and measurement location on the SWPs. Furthermore, we characterize the flattop or plateau wave and determine the influence driver gas and breech length has on this phenomenon. Finally, we compare the shock-blast wave profile from the shock tube with the field explosion profiles generated in ConWep.

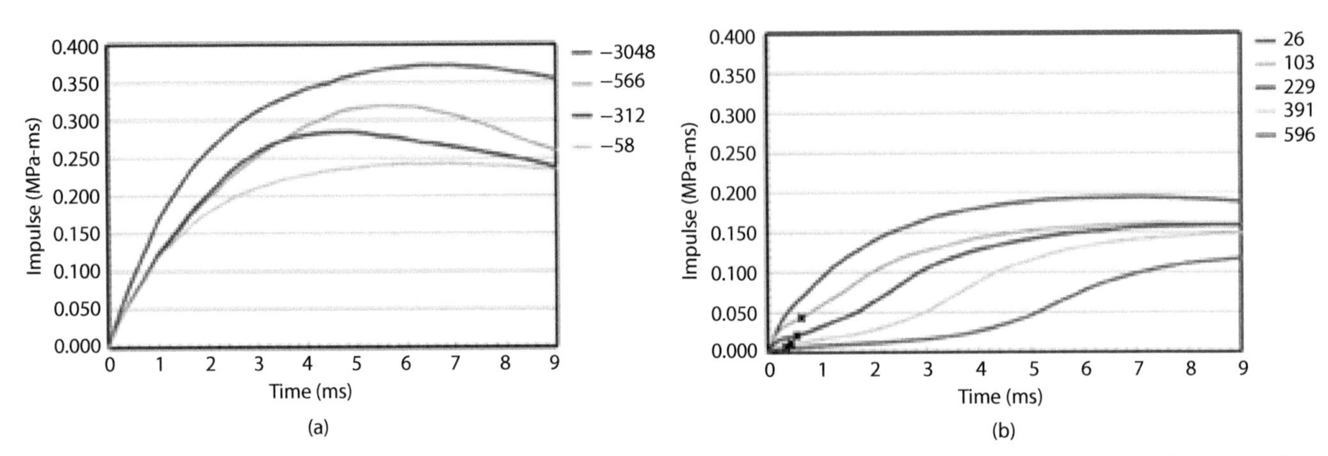

FIGURE 18.16 (a) Impulse profiles at various x locations inside the shock tube obtained by integration of experimentally measured *p-t* profiles. (b) Impulse profiles at various x locations outside the shock tube obtained by integration of experimentally measured *p-t* profiles. Contribution of the subsonic jet wind to the impulse is demarcated by cross symbols. (From Chandra, N. et al., *Shock Waves*. 22:403–415, 2012.)

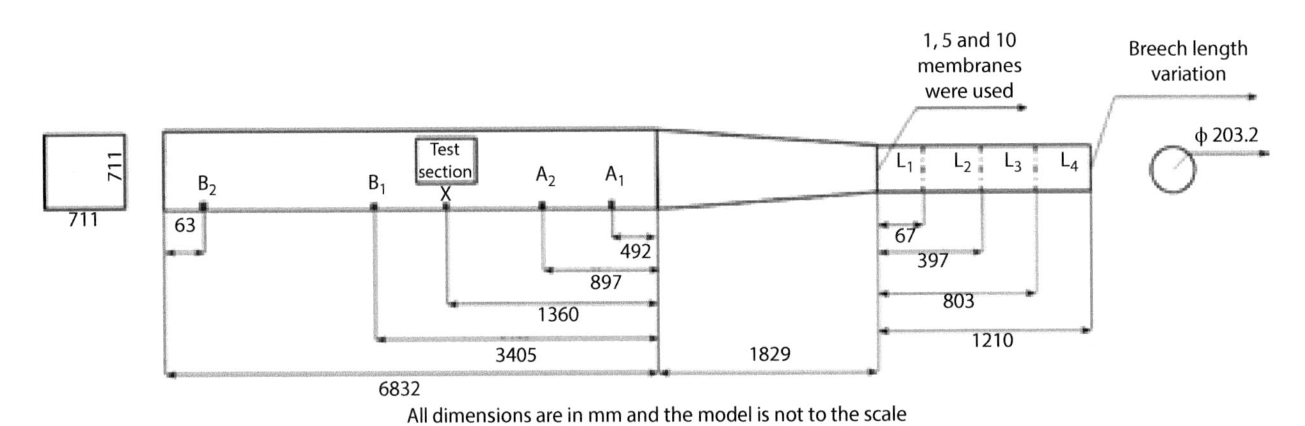

FIGURE 18.17 Experimental variables and sensor location; here A1, A2, X, B1, and B2 are the side-on pressure sensors. (From Sundaramurthy, A. et al. *Journal of Neurotrauma*. 29:2352, 2012.)

Figure 18.17 shows the experimental variables and the sensor locations. The length of the breech is varied between 67 and 1,210 mm. The membrane thickness is varied by varying the number of membranes between 1, 5, and 10 (each membrane is 0.254 mm thick). In this work, both nitrogen and helium were used as the driver gas, and the driven gas was air at ambient laboratory conditions (temperature range of 23 ± 20°C). The evolution of the blast wave along the length of the shock tube was measured using a PCB pressure gauges (model 134A24) mounted on the wall of the shock tube at locations A1, A2, X, B1, and B2 (Figure 18.17). Burst pressure in the driver just before the rupture of the membranes was also recorded.

18.1.12.1 Burst Pressure

Burst pressure is the pressure in the driver section (breech) during the membrane rupture. This highly compressed gas when allowed to expand compresses atmospheric air in the transition section generating a shock front. Burst pressure for different membrane thicknesses and breech lengths were noted and presented in Figure 18.18a. From Figure 18.18a, it can be seen that the burst pressure increases with an increase in the membrane thickness. Furthermore, there is no discernible difference in the burst pressure with respect to increase in breech length for any of the three membrane thicknesses. This is because the membrane rupture is pressure-dependent, and this critical pressure is not influenced by breech volume (i.e., burst pressure that can be achieved at a minimum breech length L_1 can also be achieved at L_2, L_3, or L_4). Therefore, the quantity of membranes used and their thickness is directly proportional to the burst. This result corroborates with the findings from the study conducted by Payman and Shepherd in which they used copper as their membrane. They determined that for the same thickness, the burst pressure does not vary more than ±3%. Similarly, they also determined that membrane thickness has a linear relationship with burst pressure (Payman and Shepherd, 1946).

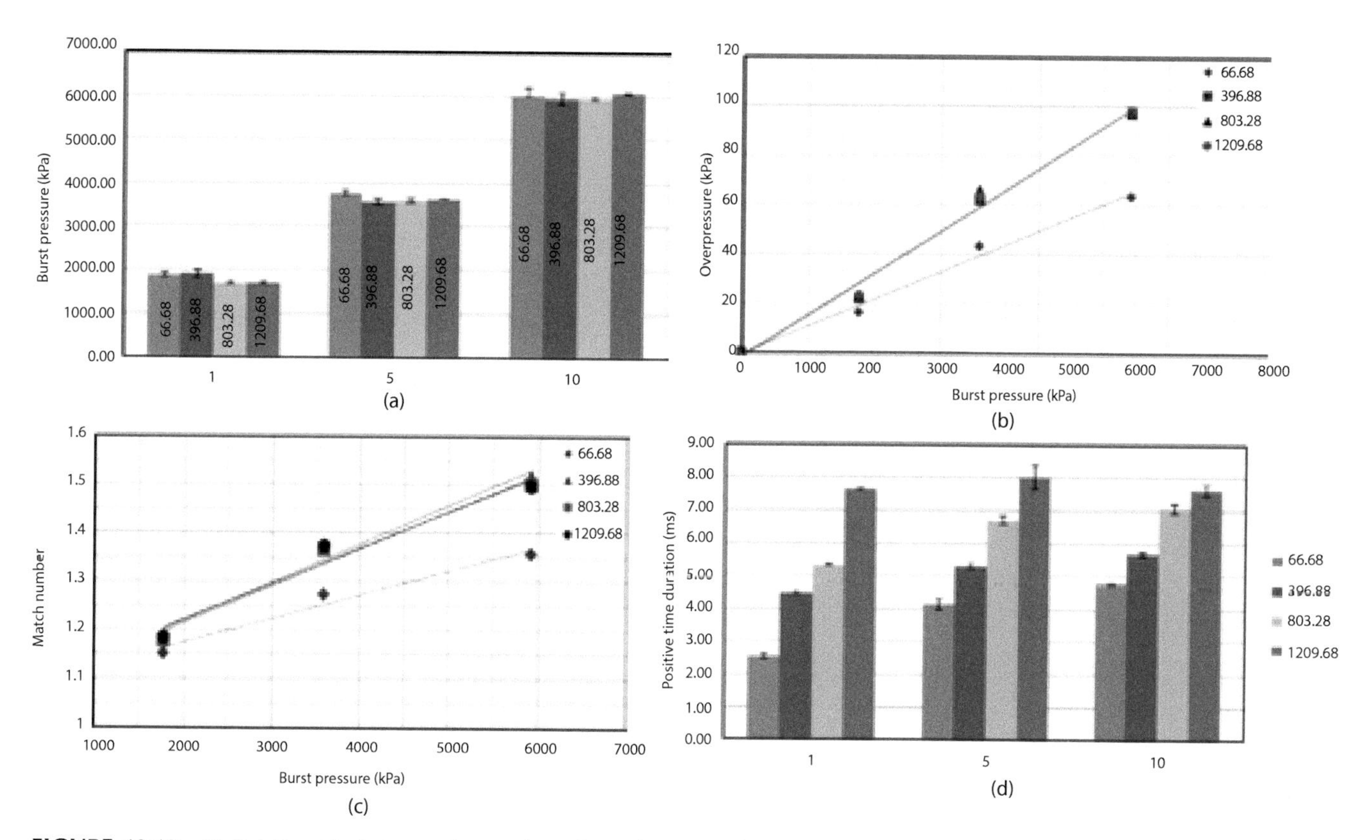

FIGURE 18.18 (a) Relationship between the number of membrane used and burst pressure produced with respect to different breech lengths. (b) Relationship between shock front Mach number and burst pressure; there is a linear relationship between Mach number and burst pressure (with strength of linearity R^2 between 0.96 and 0.98). (c) The relationship between shock tube parameter burst pressure with overpressure measured in the test section for different breech lengths. (d) Relationship between PTD and membrane thickness used for different breech lengths. (From Sundaramurthy, A. et al. ASME 2013 International Mechanical Engineering Congress and Exposition, 2013.)

18.1.12.2 SAPs and Their Influence on the SWPs in the Test Section

By changing the SAPs such as membrane thickness (burst pressure) and breech length, we were able to determine the effect on SWPs such as Mach number, overpressure, and positive time duration (PTD) in the test section. Figure 18.18b and c shows the relationship between overpressure, Mach number, and burst pressure for different breech lengths. Both of these variables have a strong positive relationship with the burst. Furthermore, with increase in the burst pressure both Mach number and overpressure for L_2, L_3, and L_4 increases with a higher rate than L_1.

PTD is the period within which the shock overpressure reaches the atmospheric pressure. Figure 18.18c shows the relationship between PTD and membrane thickness for different breech lengths. For a given membrane thickness, PTD increases with an increase in the breech length. Furthermore, there is an increase in PTD between membrane thicknesses 1 and 5 for breech lengths L_1, L_2, and L_3; however, such an apparent difference is not observed between membrane thicknesses 5 and 10. Finally, for breech length L_4, there is no apparent difference in PTD for different membrane thicknesses.

Controlling overpressure and PTD is essential when replicating a field blast. It is possible in a laboratory shock tube to control the aforementioned variables by manipulating breech length, burst pressure (membrane thickness), type of gas, and test section location (by varying the test section within expansion section). It can be seen within the test section that, with an increase in burst pressure, both the overpressure and the Mach number (strength of shock wave) increases, which implies that both these variables can be increased by increasing the membrane thickness. Similarly, PTD increases with increase in breech length for any given burst pressure. However, at lower breech lengths, both overpressure and PTD are affected by expansion waves (also known as rarefaction waves) that arise from the end of the breech. Furthermore, type of driver gas used also plays a role in the PTD and overpressure. The effect expansion waves and driver gas have on the overpressure and PTD is explained in detail in the following section.

Figure 18.19 shows the x-t wave propagation diagram with shock front, rarefaction head, and tail for two breech lengths. When membranes burst, the driver gas expansion initiates a family of infinite expansion waves toward the closed end. Once the expansion head encounters the closed end, it travels backwards toward the transition. Because of increased density and pressure from the traverse of the shock front, the expansion head accelerates further and catches the shock front. However, because of the expansion, the temperature and pressure in the

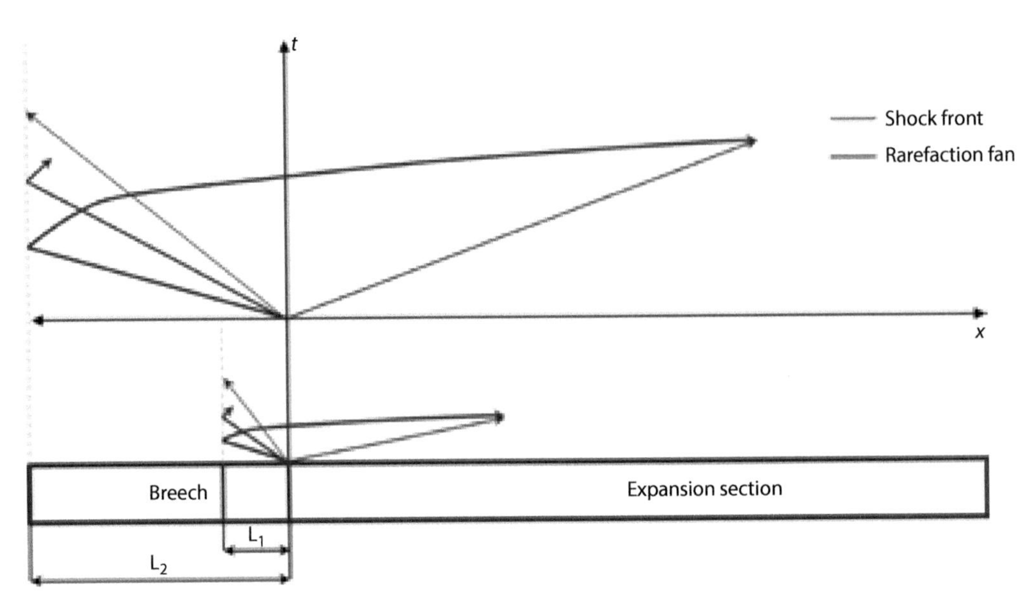

FIGURE 18.19 Ideal breech length x-t diagram for explosive shock wave replication. (From Sundaramurthy, A. et al. ASME 2013 International Mechanical Engineering Congress and Exposition, 2013.)

breech reduces, which reduces the velocity of the successive expansion waves, resulting in a series of waves, which yields the nonlinear decay and ultimately shapes the shock-blast wave (Ritzel et al., 2011). Once the waveform attains the shape shown in the Figure 18.4, expansion waves start to deplete the overpressure and PTD. Similarly, the breech length and driver gas also play major roles in the evolution and interaction of the expansion wave. The expansion wave while traveling toward the closed end of the driver section travels at least the velocity of sound in that medium, which is the driver gas (helium $C_{He} = 972$ m/s and nitrogen $C_N = 353$ m/s). Therefore, for a given breech length and membrane thickness, having helium as a driver gas increases the expansion wave velocity resulting in a Friedlander type wave at an earlier point compared with nitrogen gas. Consequently, by varying the length of the breech in conjunction with using the appropriate driver gas, we can optimize PTD and overpressure.

18.1.12.3 Evolution of the Shock-Blast Wave along the Expansion Section

Figure 18.20(a–c) shows the evolution of the overpressure along the length of the expansion section. From Figure 18.20a, it can be seen that for one membrane there is no discernible change in overpressure for breech lengths L_2, L_3, and L_4. For all cases with breech length L_1, there is a continuous decay in the overpressure downstream of the shock tube. For all the other breech lengths, unique points of overpressure decays are identified along the expansion section, which is illustrated in the following section.

For L_1 (66.68 mm), L_2 (396.88 mm), L_3 (803.28 mm), and L_4 (1,209.68 mm), we observe the following: (1) for any membrane thicknesses, an obvious difference in overpressure is observed between L_1 and other breech lengths (Figure 18.20a–c); (2) beyond 3,000 mm from the breech, for 5 and 10 membranes and breech length L_2, overpressure starts to decay (Figure 18.20b, c); and (3) beyond 5,000 mm from the breech, for 10 membranes and breech lengths L3, the overpressure starts to decay (Figure 18.20c). Finally, for L_4, there is no unique decay point, which implies a flattop wave throughout the expansion section.

In a typical free field blast, the overpressure decreases rapidly with respect to increase in distance from the blast epicenter (Mott et al., 2008). However, overpressure in a shock tube does not show a drastic reduction because of its constant cross section. There is a considerable difference between the overpressures for L_1 and all the other breech lengths. As discussed earlier, this difference arises from the interaction of expansion waves that comes from the breech. This suggests that the expansion waves from the breech for breech length L_1 arrives earlier than all other breech lengths. With an increase in the breech length and burst pressure, distinct points at which the shock blast starts to decay are shown in Figure 18.20b and c, which implies that beyond this point the shock-blast wave has a Friedlander form. Consequently, longer breech lengths that tend to produce a flattop wave in the test section will produce a Friedlander type wave at some point beyond the test section. As discussed in Section 18.1.3, when moving closer to the exit, the rarefaction waves from the exit starts to interact with blast wave creating artifacts, which results in inaccurate blast simulation.

Figure 18.20d–f shows the evolution of the PTD along the length of the shock tube expansion section. For any given breech length and membrane thickness, the PTD remains reasonably constant along the length; however, it decreases drastically toward the exit of the shock tube. Positive impulse (PI) is the area under the shock-blast wave profile. Figure 18.20g–i shows the evolution of the PI along the length of the shock tube expansions section. PI is a function of both overpressure and PTD; hence, it increases with an increase in both membrane thickness and breech length. Because of its relationship with the PTD, the impulse drastically reduces near the exit of the shock tube.

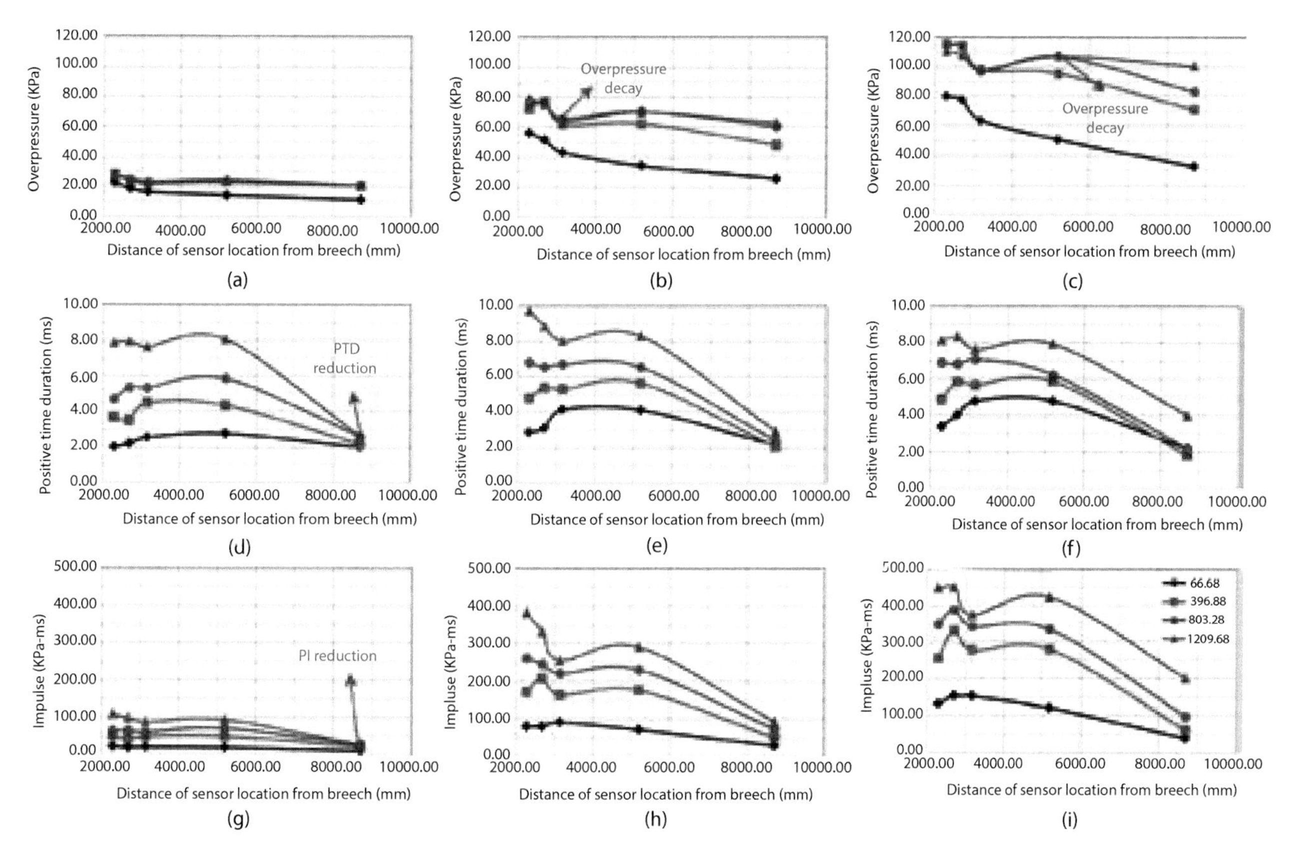

FIGURE 18.20 Variation of shock-blast profile parameters along the length of the shock tube expansion section. All of these experiments were performed for breech lengths 66.68 (black), 396.88 (red), 803.28 (blue), and 1,209.68 (green) mm; (a–c) show the variation of overpressure along the length of the expansion section for burst pressures corresponding to 1, 5, and 10 membranes, respectively; (d–f) show the positive time duration along the expansion section for burst pressures corresponding to 1, 5, and 10 membranes, respectively. (From Sundaramurthy, A. et al. ASME 2013 International Mechanical Engineering Congress and Exposition, 2013.)

As a result, PTD reduces drastically near the exit of the shock tube because of the interaction between shock front and exit expansion waves. This has two consequences: first, the PI (energy of blast wave) reduces drastically (Figure 18.20g–i). Second, because the total energy at the exit is conserved, all the blast energy is converted from supersonic blast wave to subsonic jet wind, which produces erroneous results. The effect of jet wind and specimen placement location along the expansion section for blast simulation using a shock tube is illustrated in Section 18.1.2.2.

18.1.12.4 Flattop or Plateau Wave

A flattop or plateau wave is usually witnessed in a gas-driven shock tube (Reneer et al., 2011). In this case, the shock-blast wave profile, once reaching the peak overpressure, maintains its peak value for a certain period before decay. Longer breech lengths in combination with the use of nitrogen as a driver gas seem to have a strong influence on this phenomenon. Figure 18.21 shows the comparison between the shock-blast wave profile with nitrogen and helium as the driver gas. In both cases, 10 membranes with a breech length of 1,209.68 mm were used. It can be seen that only in the case of nitrogen as the driver gas is a flattop wave observed, whereas in the case of helium, a pure Friedlander wave is witnessed.

There is an inherent relationship between SAPs, such that an optimization of one variable might have a deviating effect on the other variables, resulting in having an nonoptimal shock blast wave (in this case a flattop wave). This particular problem arises depending on (1) breech length and (2) type of driver gas. In the current study, the breech length L_1 is low enough for the expansion waves to reach the shock front and create a Friedlander type wave somewhere before the test section. Therefore, when it arrives at the test section, it has already gone through some overpressure and PTD reduction. Once the breech length is increased, the time taken by the expansion wave to reach the shock front increases. Consequently, for breech lengths L_2, L_3, and L_4, there is no change in the overpressure and PTD at the test section, which implies that it is a flattop wave.

Although nitrogen, because of its low sound speed, has a tendency to produce a longer PTD, using a longer breech length results in a flattop wave. Conversely, helium produces a lower PTD compared with nitrogen but has a sharp decay to the atmospheric pressure. Similar findings are reported by Reneer and Bass in their separate works comparing the wave profiles generated from air (which has a speed of sound close to nitrogen) and helium. They found that using air as a driver gas produces a flattop wave (Bass et al., 2012;

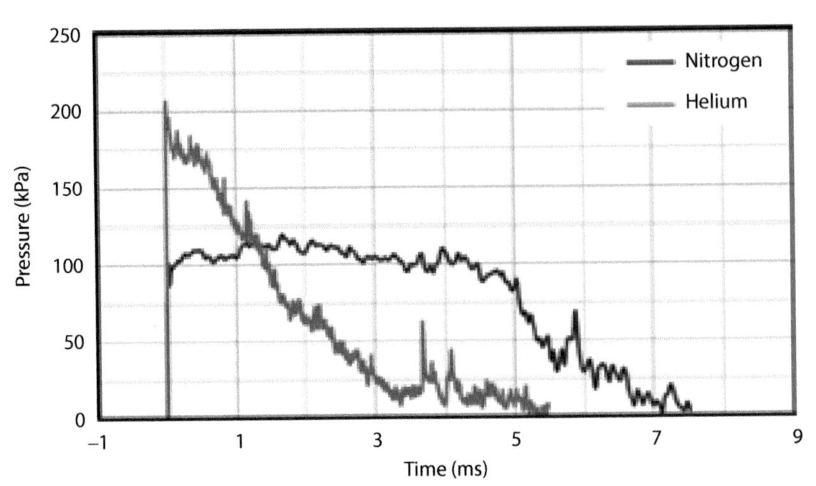

FIGURE 18.21 Comparison of the shock-blast profile for helium and nitrogen with 10 membranes and breech length of 1,209.68 mm; clearly, the wave profile corresponding to helium gas is a Friedlander wave and the wave profile corresponding to nitrogen is a flattop wave.

Reneer et al., 2011). One technique used for avoiding a flattop wave when using long breech length is to place the test section downstream of the expansion section, so that the expansion waves would eventually catch up and produce a Friedlander type wave; nevertheless, this method has its own limitations.

18.1.12.5 Comparison between Field and Laboratory Profiles

Table 18.1 shows the range of IEDs and mines typically used in the field; their explosive capacity is expressed as TNT strength (AEP-55, 2006; Department of Homeland Security, 2009). An important requirement for studying BINT is the ability to produce accurate and repeatable blast loading, which can be related to strengths mentioned in Table 18.1. Using ConWep, we were able to determine the pressure profiles for TNT explosives within the range of strengths described in Table 18.1. A comparison was made between the profiles generated from ConWep with those generated from a shock tube device (Figure 18.22a–c, d). Clearly, there is good match in the results, which indicates that the wave profile generated by the compressed gas shock tube is directly related to relevant field conditions.

18.2 PMHS BLAST TESTING

In this section, the intracranial and surface pressures along with the skull strain response of PMHS heads are studied by subjecting them to blasts of varying peak incident intensities or overpressures (70, 140, and 200 kPa).

18.2.1 Experiments

18.2.1.1 PMHS Testing in the 28-Inch Shock Tube

The PMHS heads were used in conjunction with the Hybrid III neck in these experiments. The head assembly was placed in the test section of the shock tube, as shown in Figure 18.23, and was subjected to frontal blast loading. Experiments were also conducted with helmets mounted on PMHS heads. The shape, overpressure, and duration of the incident blast wave at a given location are known a priori. This is achieved through sample trials in the shock tube, conducted without the surrogate head and the neck.

18.2.1.1.1 PMHS Preparation

Three PMHS heads were used for this purpose. The specimens had no record of osseous disease and preexisting fractures were not present as confirmed by computed tomography (CT) imaging. The age, gender, and basic anthropometry of the specimens are listed in Table 18.2.

PMHS heads that are kept frozen for a substantial period tend to have severe tissue degradation after thawing. Thus, the brain was removed from each PMHS head and intracranial space was backfilled with ballistic gelatin. The brain tissue and dura mater were removed through the foramen magnum using a flat head screwdriver. Twenty percent ballistic gelatin (ballistic gel, from here on) was prepared by dissolving 2 parts of 250 bloom gelatin into 9 parts of warm (40°C) water (by mass), stirring the mixture while pouring in the powdered gelatin. The gelatin was obtained from Gelita USA Inc. (Sioux, IA) in the bloom form. The ballistic gel was poured in the intracranial cavity through the foramen magnum and allowed to settle at room temperature. After the ballistic gel had settled, the entire head was put inside plastic bags and air bubbles were removed using a vacuum cleaner. The foramen magnum was sealed using filler material (Bondo). A Hybrid III neck was attached to the head using a base plate. A base plate was screwed to the bottom of the head.

18.2.1.1.2 Instrumentation

Each PMHS head was instrumented to measure surface pressures, surface strains, and ICPs. Eleven sensor measurements were made on each PMHS head. Surface pressures were measured at two locations, surface strains were measured at four locations, and ICPs were measured at five locations within the head, as shown in Figure 18.23b. CT images of the instrumented head were also taken. CT images were used to verify

TABLE 18.1
Explosive Capacity of the Currently Used IEDs and Mines in the Field

Threat	Explosive Capacity (TNT Equivalent in kg)	Reference
Pipe bomb	2.28	Department of Homeland Security (2009)
Suicide bomber	9	
Briefcase bomb	22.70	
Antipersonnel fragmentation device	0.55	AEP-55 (2006)
Antivehicular blast landmine	6–10	

Explosive capacity of these IEDs are given in terms of TNT equivalents in kilograms.

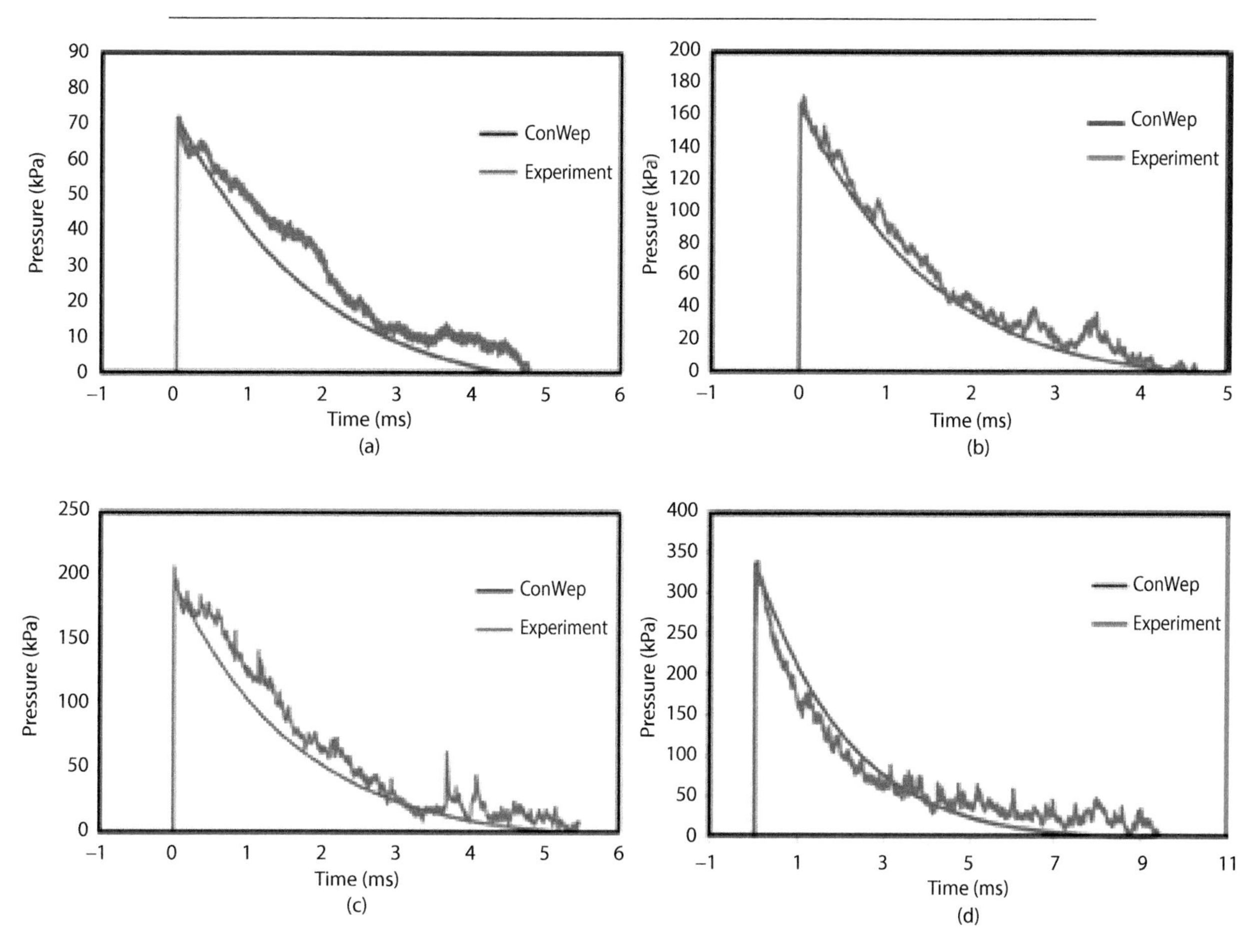

FIGURE 18.22 Comparison of the shock blast profiles from a UNL shock tube device and ConWep simulation software. (a) Comparison between shock blast profile from a 10-membrane, 66.68-mm breech length shot with nitrogen as driver gas, and 2.56 kg of TNT at 5.18 m. (b) Comparison between shock blast profile from an eight-membrane, 752.48-mm breech length shot with helium as driver gas, and 7.68 kg of TNT at 5 m. (c) Comparison between shock blast profile from a 10-membrane, 1,209.68-mm breech length shot with helium as driver gas, and 14.08 kg of TNT at 5.7 m. (d) Comparison between shock blast profile from 15-membrane, 1,209.68-mm breech length shot with helium as driver gas, and 96 kg of TNT at 8.5 m. (From Sundaramurthy, A. et al. ASME 2013 International Mechanical Engineering Congress and Exposition.)

locations of the sensors inside the head (Figure 18.23c). In addition, CT images were also useful in identifying the precise geometry of skull and the face in the vicinity of the sensor; by geometry, we implied anatomical features such as the air sinus, eye socket, and nasal cavity. For example, a huge air sinus was present in front of certain sensors (see Figure 18.23c, nose ICP). Surface pressures were measured using Kulite surface mount sensors (model LE-080-250A) and ICPs are measured using Kulite probe sensors (XCL-072-500A). Surface strains were measured using Vishay strain gauges (model CEA-13-250UN-350). In addition to these sensors, a PCB pressure gauge (model 134A24) was used to measure incident (side-on) pressure of the blast wave. Incident pressure was measured just before (distance = 200 mm) the blast wave encountered the PMHS head. An incident pressure gauge was mounted on the wall of the shock tube.

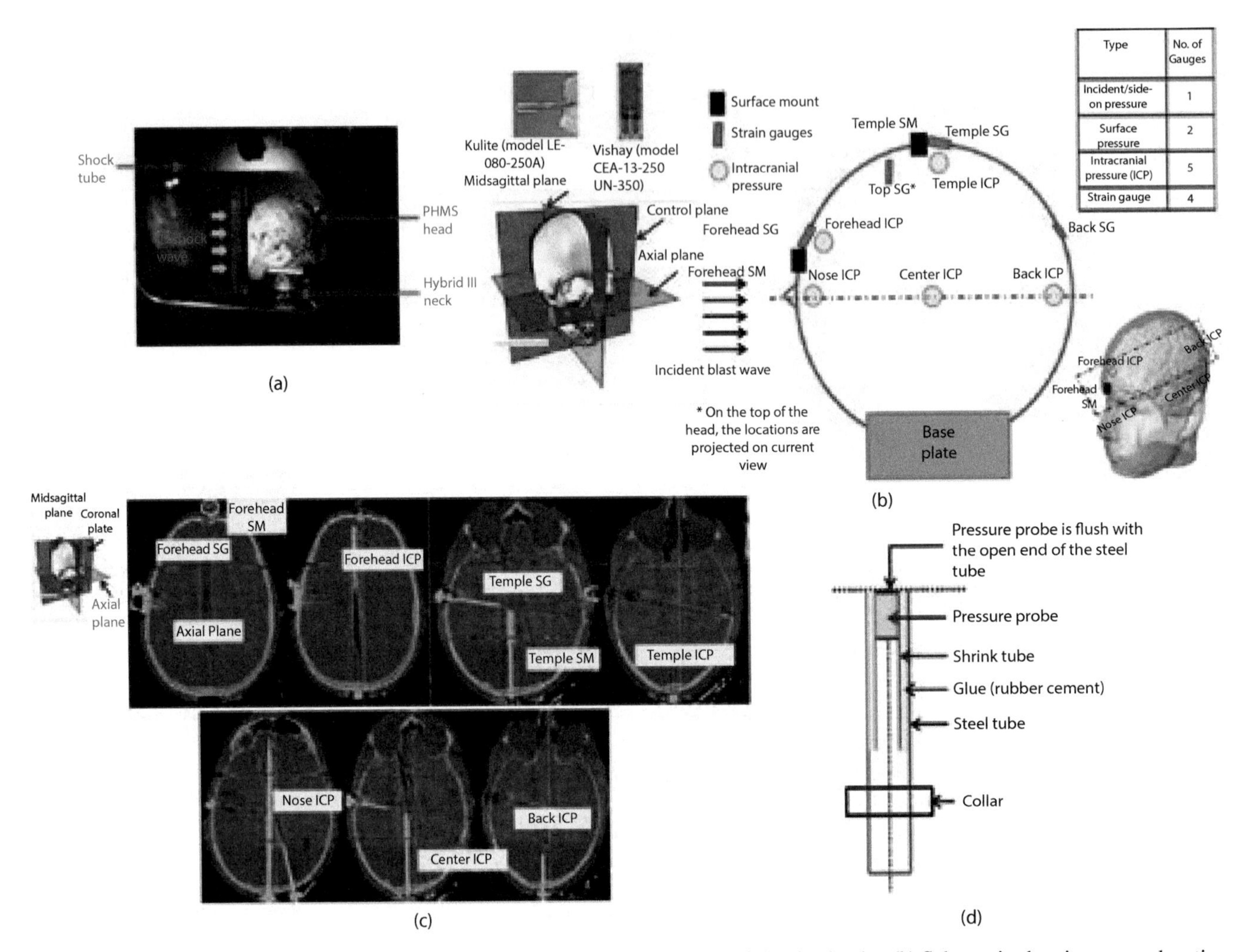

FIGURE 18.23 (a) PMHS heads with hybrid III neck placed in the test section of the shock tube. (b) Schematic showing sensor locations on PMHS head. (c) CT images of instrumented PMHS showing sensor locations. Anthropometric data were also obtained from these CT images. (d) Preparation of pressure probes that are used for ICP measurements.

TABLE 18.2
Characteristics of the Three PMHS Heads Tested in This Study

PMHS No.	PMHS Identification	Sex	Age	Height (cm)	Weight (kg)	Cause of Death
1	421	M	72	175	68	Prostate cancer, diabetes
2	476	M	75	173	79	Cardiovascular disease
3	513	M	65	175	73	Prostate cancer

18.2.1.2 Blast Wave Exposure

All PMHS heads were subjected to blast waves of three different incident intensities or overpressures (70, 140, and 200 kPa). This was achieved by changing the number of membranes, which resulted in different burst pressures. The head was oriented in a frontal direction to the blast. Each intensity was repeated three times for each head, so there were a total of 27 (3 × 3 × 3) shots. The PMHS head was placed in the test section of the shock tube located approximately 2,502 mm from the driver end; the total length of the shock tube was 12,319 mm.

18.2.1.2.1 Sample Pressure-Time Profiles

Figure 18.24a–h shows sample incident pressure (a), surface (reflected) pressure (b, c), and ICP (d–h), respectively. Sample pressure-time profiles presented here are based on the mean of three shots (experiments) for head 1. While presenting sample pressure-time profiles, the time axis was shifted so that arrival of the blast wave at a given sensor location corresponded to t = 0. Incident blast intensity for these profiles was 200 kPa. Raw (pressure) data showed oscillations; thus, the profile was smoothed by performing simple moving average

(Kraus et al., 2007). The number of data points selected for moving average varied from 5 to 20.

Figure 18.24a shows the incident pressure profile. The incident pressure was measured at 200 mm (upstream) from the PMHS head. The incident profile showed a sudden rise in pressure (rise time is 10 μs) followed by nonlinear decay; the peak pressure and positive phase duration were 190 kPa and 5.4 ms, respectively. Secondary peaks were also seen in the incident profile. The incident pressure profile was highly repeatable; shot-to-shot variations in the pressure profile were less than 5%.

Figure 18.24b and c shows pressure profiles on the surface of the head. Surface pressures were measured at the forehead and temple locations. The profile for forehead location showed a sudden rise in pressure (rise time was 30 μs) followed by nonlinear decay. The rate of decay was much faster than that observed in the incident pressure profile. The peak pressure and positive phase duration for this profile were 592 kPa and 5 ms, respectively. The peak pressure was amplified 3.11 times the peak incident pressure but the positive phase duration remained approximately similar at 5 ms. The pressure profile for temple location also showed a sudden rise in pressure (rise time was 30 μs) followed by another spike. This was followed by a sudden (instantaneous) decay in pressure with huge oscillations (from t = 0.25 to t = 0.5). An oscillating profile with much smaller peaks was observed after this time. The peak pressure and positive phase duration for this profile were 295 kPa and 5 ms, respectively. The ratio of peak pressure to peak incident pressure was 1.55.

The shape and positive phase duration of the surface pressure profile at the forehead location was similar to that of the incident pressure profile. However, peak overpressure was significantly higher because of aerodynamic effects and the rate of decay was much faster compared with the incident pressure profile. Temple location had a much smaller peak pressure (50% decrease) compared with the forehead location.

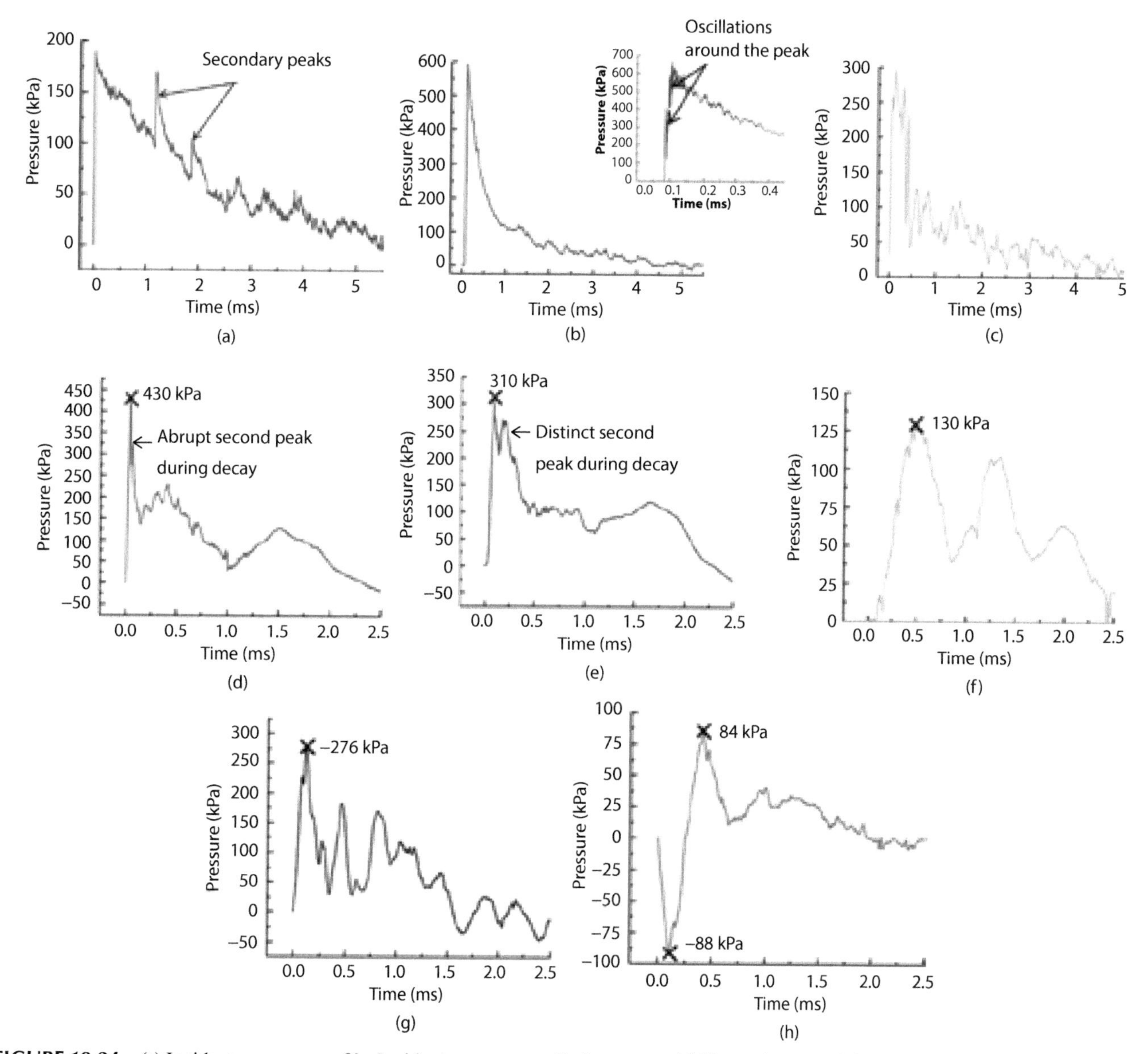

FIGURE 18.24 (a) Incident pressure profile. Incident pressure profile is measured 200 mm (upstream) from the PMHS head. (b, c) Sample pressure profiles on the surface of the head: (b) forehead, (c) temple. (d–h) ICP profiles: (d) forehead ICP, (e) nose ICP (e), center ICP, (f) temple ICP, and (g) back ICP.

These effects can be explained by studying blast wave–head interactions. The flow field around the head is illustrated using numerical simulations. At the beginning of the interaction, as the shock front impinges on the forehead at its most upstream region, a reflected shock propagating in the opposite direction starts to develop. The incident shock starts to propagate around the surface of the head. At the same time, regular reflections occur that propagate radially both in the upstream and downstream directions. These reflections continuously interact with the incoming tail part of the blast wave. The reflections are tensile in nature and hence a compressive pattern of decreasing strength develops as a result of the incident shock reflection over the surface and the forward motion of the shock tail (also known as blast wind) as shown in Figure 18.25a. Thus, the forehead surface gauge recorded a faster decay compared with the incident pressure profile. As the shock wave traversed the head, shock wave diffractions occurred and as a result geometry-induced flow separation took place (Figure 18.25b). This geometry-induced flow separation increased as we moved away from the leading edge (or incident blast site) downstream. Because of this flow separation, the temple location showed a decrease in peak pressure with respect to the forehead location. A similar phenomenon of shock wave diffractions and flow separation over cylindrical objects was seen in studies involving shock wave propagation over cylindrical objects (Bass et al., 2012; Brun, 2009; Gould and Tempo, 1981).

Figure 18.24d–h shows the ICP profiles for sensor locations shown in Figure 18.23b. For ICP profiles, wave action (dynamic events) played out in very short time. Positive phase duration was 2.5 ms with initial (or majority of) intracranial dynamics playing out within 0.5 ms. The forehead (FH) ICP (Figure 18.24d) sensor showed a sharp rise (rise time = 40 μs) in pressure profile. This was followed by a decay in pressure until t = 0.15 ms; an abrupt pressure increase was also seen during this decay at t = 0.07 ms. The decay was not sustained and there was another rise in the pressure. A secondary (loading) pulse, which was similar to a pulse observed in impact loading, was seen after 1.0 ms. The nose ICP sensor (Figure 18.24e) also showed a sharp rise (rise time = 70 μs) in the pressure profile. This was followed by decay in pressure until t = 0.5 ms; during this decay, a distinct pressure increase (like another peak) was seen at t = 0.2 ms. After this decay, the pressure remains approximately constant with small oscillations until t = 1 ms. Secondary pulse, which is similar to the pulse observed in the FH ICP profile, was seen after 1 ms. Figure 18.24f shows the pressure profile for the center ICP. The pressure rise (rise time = 390 μs) is not as sharp compared with the FH and nose ICP profiles. The pressure pulse seemed to repeat itself with damping until the pressure equilibrates. Figure 18.24g shows the pressure profile for temple ICP. The profile has similar features as that of center ICP. It appears that the center and temple ICPs experience several waves that are emanating from different sources during the rise. Rise time for the temple ICP was 115 μs. Figure 18.24h shows the pressure profile for the back ICP. The back ICP shows the negative phase followed by the positive phase. The negative phase had a rise time of 100 μs and duration of 0.25 ms. The positive phase of the back ICP has features similar to the FH and nose ICP profiles. This is followed by decay in pressure after which pressure equilibrates.

Analysis of the peak pressures of ICP profiles has led to some interesting observations. Peak pressures are marked with a black cross on each ICP profile. The highest peak pressure is observed behind the forehead. The peak pressures were decreased as we move away from the coup (impact) site toward the contrecoup (opposite to impact) site. It should also be noted that a significant difference in peak pressure is observed for FH (430 kPa) and nose (310 kPa) ICPs that are in same coronal plane.

The pressure profiles in the brain drastically deviated from surface pressure profiles. ICP dynamics play out in much shorter duration compared with incident or surface pressure profiles. For ICP profiles, the positive phase duration was 2.5 ms with initial (or the majority of) intracranial dynamics playing out within 0.5 ms; this was due to wave propagation in skin–skull–brain parenchyma. The impedance mismatch of the layered system (skin–skull–brain) was such that the magnitude of the pressure wave (or input signal) in the skin–skull–brain parenchyma was either amplified or attenuated as it reflected and transmitted through these layers. The wave traversal times through the skin and the skull were 52.9 μs and 3.62 μs assuming 10-mm thickness for both skin and skull. Within 506 μs (i.e., 0.506 ms) two wave traversals occurred in the skin and 29 wave traversals were possible in the skull. Thus, wave action (reflections and transmissions) happens at a much shorter time scale, and a sharp decay in ICP profiles was seen for FH and nose ICPs. This aspect is further elaborated using a one-dimensional model of skin–skull–brain parenchyma as shown in Figure 18.25c, d. Figure 18.25c shows response of the skull–brain parenchyma when a loading pulse (Heaviside function) of intensity P is applied to the skull. Pressure in the brain was 0.62 (transmission coefficient) times the applied pressure after first transmission from the skull into the brain. Transmission coefficient increased with each transmission and pressure in the brain equilibrated after the fifth transmission from the skull into the brain. Figure 18.25d shows response of the skin–skull–brain parenchyma when a loading pulse of intensity P was applied to the skin. In this setup (or model), thickness of the skull and skin were designed such that ($t_{skull} = 14\ t_{skin}$) reflection/transmission occurs at the same time for skin–skull and skull–brain interfaces. The transmission coefficient was 1.16 after the first transmission from the skull into the brain. However, transmission coefficient drastically decreased to 0.15 after the second transmission. Subsequent transmissions had transmission coefficients of 1.39, 0.42, 1.32, and 0.7. In reality, the thickness of the skin and the skull were almost similar and hence the wave traversal in the skull was ~14 times faster than the skin based on the wave speeds. The transmitted wave from the skull into the brain equilibrated after the fifth transmission (Figure 18.25c). Hence, for wave propagation through skin–skull–brain parenchyma in real scenarios skull (thickness) should not play a major role in the wave amplification or attenuation and brain should ideally see the pressure that is seen by the skull at the skin–skull interface. Even in that case, the transmission coefficient in the brain after the second transmission was 0.25.

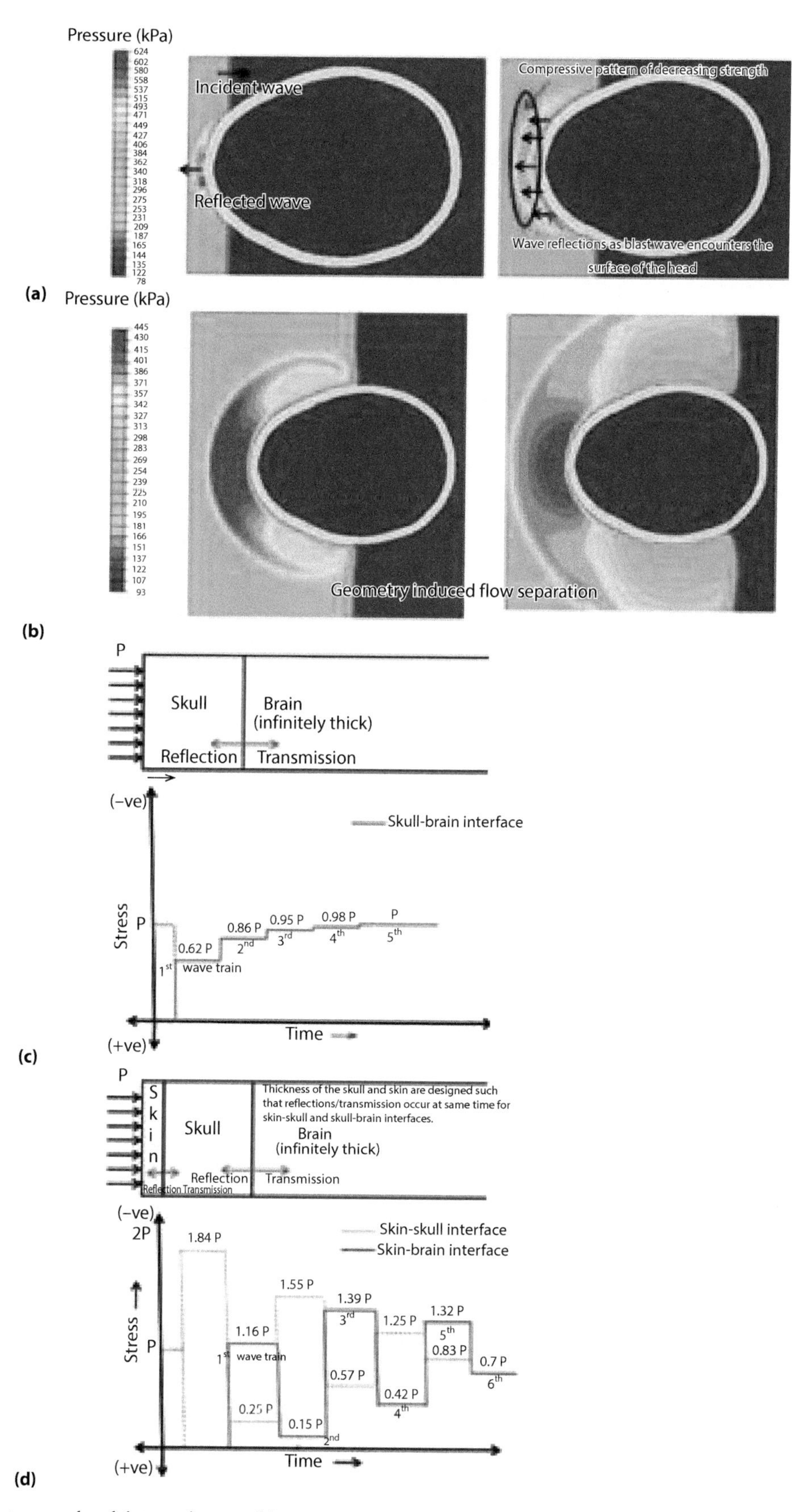

FIGURE 18.25 Blast wave–head interactions as blast wave traverses the head. (a) Blast wave–head interaction at leading edge or incident last site. (b) Illustration of Mach reflection and flow separation as blast wave traverses the head. In all figures, cuts are made along the axial plane. Wave propagation in skin–skull–brain parenchyma using one-dimensional model. (a) Response of the skull–brain parenchyma to the applied loading pulse of intensity P. (b) Response of the skin–skull–brain parenchyma to the applied loading pulse of intensity P.

This explains the sharp decay in ICP profiles for FH and nose ICPs that were closest to the incident blast site wherein initial wave propagation obeys one-dimensional theory fairly well (at least in terms of qualitative trends). For FH ICP, the pressure decreased from 430 kPa (first transmission) to 138 kPa (second transmission) in 117 μs; the ratio of these pressures is 0.32, which is consistent with the one-dimensional theory.

The FH and nose ICPs showed a second peak during the initial decay; the second peak was abrupt in the FH ICP and distinct in the nose ICP. This second peak was due to a delayed wave transmission from the eye socket/eyebrow region as illustrated in Figure 18.26. After these initial phases, wave reflections from the head boundaries dominated the response and it was not possible to delineate these effects because of the complex and highly dynamic nature of the problem. The center and temple ICPs, which are probably located deep inside the brain, experienced many waves emanating from different sources. By the time the wave reached the center ICP, the pressure was attenuated. This attenuation could be due to material damping, wave dispersion over a larger area, and reflections from geometric boundaries and material interfaces. The pressure pattern of the center ICP can be best described as follows. At any given point, the brain experiences a complex set of direct and indirect loadings emanating from different sources (e.g., blast wave transmission, reflections from tissue interfaces, skull deformation) at different time points. These disturbances continuously propagate into the brain as waves. Constructive and destructive interferences of these waves control the pressure history deep inside the brain. The back ICP shows the contrecoup effect (negative pressure) initially because the wave velocity in the skull is higher than the wave velocity in the ballistic gel. Because of this, the skull moves forward; displacement of the brain lags displacement of the skull and hence tension or negative pressure is generated in the brain. If tensile loads/forces are not allowed to transfer, then separation will take place at the skull–brain interface.

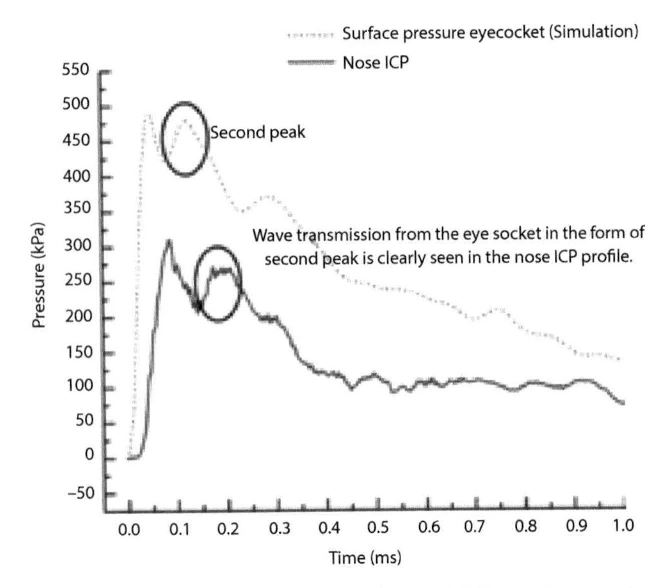

FIGURE 18.26 Comparison of surface and ICP profiles near the eye socket regions.

18.2.1.2.2 Pressure Response as a Function of Incident Blast Intensity

Figure 18.27 shows the positive phase impulse for various sensors as a function of intensity. The positive phase impulse increased as the incident blast intensity increased. The differences were statistically significant ($p < 0.05$) at all sensor locations except the back ICP sensor for all heads and temple ICP for head 1. Head-to-head variations in the positive phase impulse were also significant; impulse variations up to 77% were observed between the heads.

As mentioned earlier, one of the lingering questions facing the medical and scientific communities is whether blast waves cause TBI. We exposed PMHS heads to pure primary blasts of varying intensities and observed statistically significant differences in the peak ICP and total impulse. This finding supported our hypothesis that intracranial response changes with change in incident blast intensity. Thus it is clear that primary blast waves acting alone can cause mechanical insult to the brain. The potential of this mechanical insult in causing the BINT will be assessed in the following section. Over the past few years, several mechanisms (see the previous section) have been proposed based on numerical simulation without experimental backup. Our experimental measurements categorically indicated that the blast wave transmission through the cranium induced mechanical insult to the brain of varying degrees that changed with variations in incident intensity. Thus we propose direct transmission of the blast wave into the intracranial cavity as an essential loading pathway to the brain.

18.2.1.2.3 Sample Strain Profiles

Figure 18.28 shows sample (circumferential) strain profiles for strain gauge locations shown in Figure 18.23b. Incident blast intensity for these profiles was 200 kPa. Negative strain indicated compression and positive strain indicated tension. The front strain gauge showed a compressive phase up to 1 ms; this was followed by small (equilibrium) oscillations for 0.2 ms, followed by another compressive pulse. The right temple strain gauge showed initial tension followed by compression. This compressive phase was sustained for 0.2 ms only. This was followed by a tension-compression phase with higher magnitudes and longer durations. Top and back strain gauges showed several compressive phases; each compressive phase was followed by equilibrium oscillations (i.e., small oscillations around zero). The highest circumferential strain was observed for the front location at 0.06%. The shape of the strain profiles was not consistent across the heads and incident intensities or in some cases, even from experiment to experiment for a given head. The magnitude of the strain however was on the same order of magnitude. Head-to-head variation of ±40% was observed in the peak strain.

18.3 ANIMAL MODEL BLAST TESTING

This section focuses on characterization of biomechanical response and selected pathologies associated with primary blast exposure in an in vivo animal model (rat), in the acute phase (up to 24 hours postexposure). The first part of this study

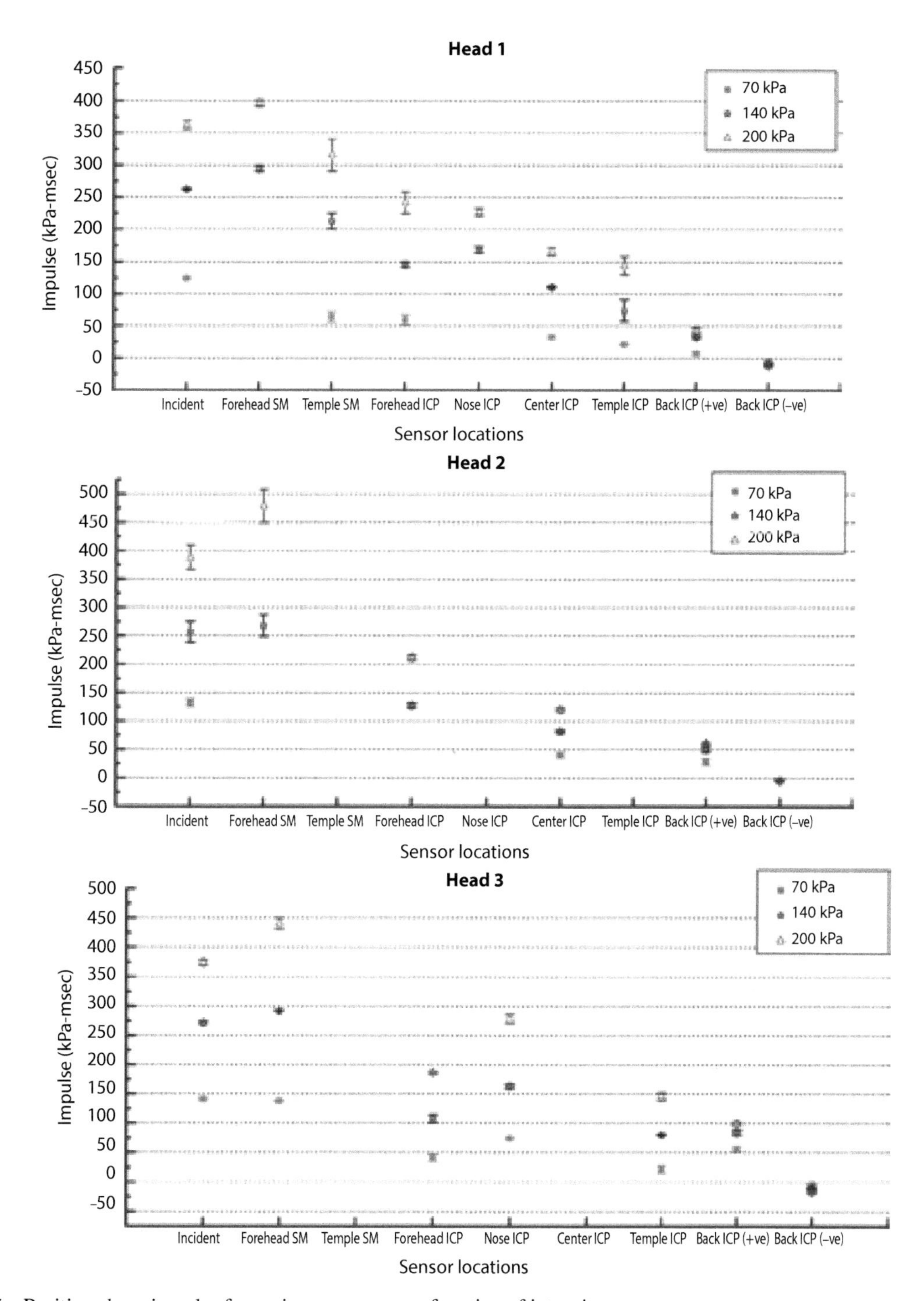

FIGURE 18.27 Positive phase impulse for various sensors as a function of intensity.

is the evaluation of mortality rates in the rodent model at five discrete blast intensities. Concurrently, the primary blast biomechanical loading experienced by the body was studied; that is, a correlation between external load (incident and reflected pressures) and biomechanical response (intracranial and intrathoracic pressures) was established. Furthermore, corresponding response at systemic and organ levels (lung injury, heart rate decrease, and body weight loss) was also studied. Finally, to understand the effect of primary blast exposure on brain, selected pathologies were studied (blood–brain barrier [BBB] damage, immunoglobulin G [IgG] extravasation).

18.3.1 Experiment

Adult 10-week-old male Sprague Dawley rats weighing 320–360 g were used in all studies. Three studies with separate groups of rats were performed. In the first study, five cadaver rats were used to record the pressure on the surface of the nose (reflected pressure), in the lungs, and in the brain. In the second study, rats were anesthetized and exposed to the blast and were sacrificed immediately after blast exposure for gross lung pathology and histology evaluation (27 rats). In the third study, monitoring of physiological vitals such as heart rate, blood oxygen saturation,

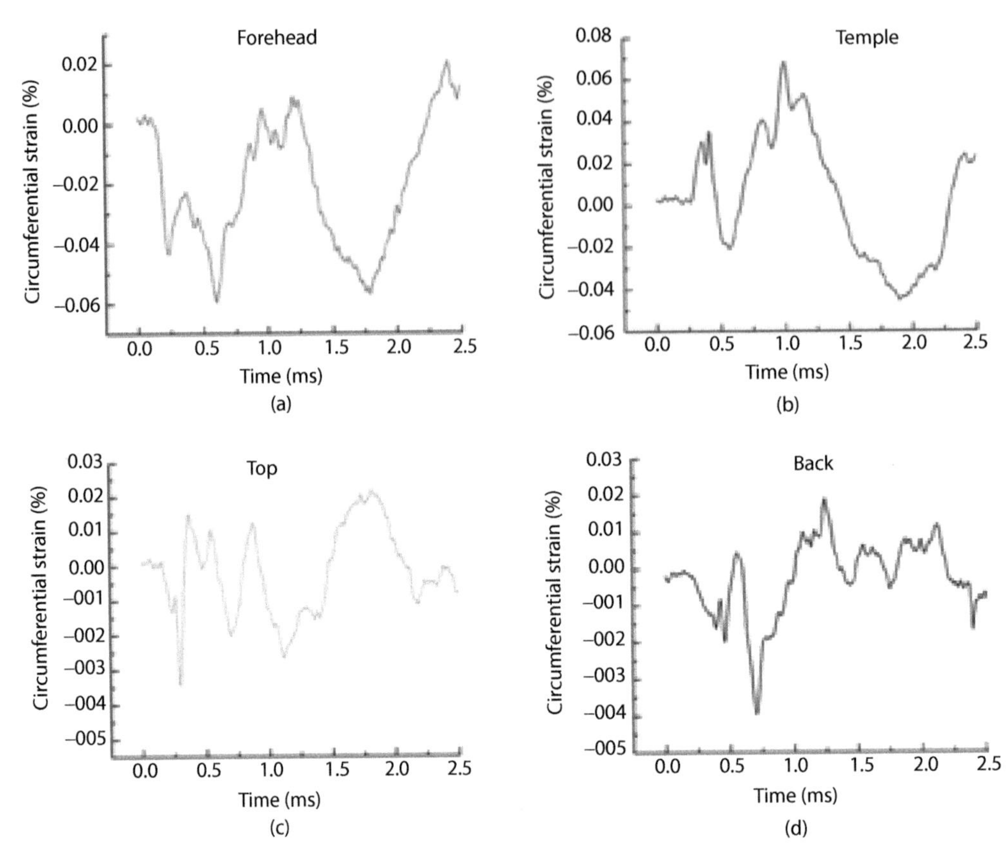

FIGURE 18.28 Strain profiles for various strain gauge locations.

TABLE 18.3
The Peak Overpressure and Impulse Average Values Measured by Side-On Sensor (Incident Pressure)

Rat No.	Group Alias	Peak Overpressure kPa[a]	Impulse (pa.s)[a,b]	Survival (%)	No. of Animals[c]	Lung Injury Score	Number of Animals[d5]
1	130[e]	127 ± 8	184 ± 19	100	18	1.8 ± 0.7	5
2	190	195 ± 19	335 ± 25	70	20	8.5 ± 2.9	6
3	230	223 ± 20	393 ± 44	100	10	10.8 ± 3.0	5
4	250	243 ± 21	437 ± 31	76	17	14.4 ± 2.4	5
5	290[f]	288 ± 17	452 ± 45	0	6	13.5 ± 4.0	6

and perfusion index was performed for two periods of 30 minutes, before and after the blast injury (50 animals). The 14 rats in this group were transcardially perfused 24 hours after the blast exposure, and the brain sections were evaluated by histological and immunohistological methods.

The rats were anesthetized with mixture of ketamine and xylazine (10:1 [100 mg/10 mg/kg], 0.1 mL/100 g) administered via intraperitoneal injection. Rats were exposed to the blast wave in the test section located inside of the shock tube (i.e. 2800 mm from the breech). The tests were performed at five discrete incident peak overpressures detailed in Table 18.3. The incident pressure was controlled by adjusting the number of Mylar membranes, and keeping the breech length constant at 447.7 mm. An aluminum bed was designed and fabricated for holding the rat during the application of blast wave. The aerodynamic riser is attached to the bed to hold the sample in the center of the shock tube. Typically, rats are tested in a prone position and are strapped securely to the bed with a thin cotton cloth wrapped around the body (Figure 18.29a). Sham control rats received anesthesia and noise exposure, but without blast exposure, and naive control rats were given anesthesia only.

18.3.1.1 Biomechanical Loading Evaluation in Cadaver Rats

Five male Sprague Dawley rats were sacrificed using carbon dioxide exposure. After the sacrifice, they were instrumented in the same manner as discussed in previously (Figure 18.30a). Cadaver rats were exposed to five shots per animal at 130 kPa (two rats) and 190, 230, 250, and 290 kPa (two rats). One rat with lungs filled with ballistic gel to ameliorate inconsistent

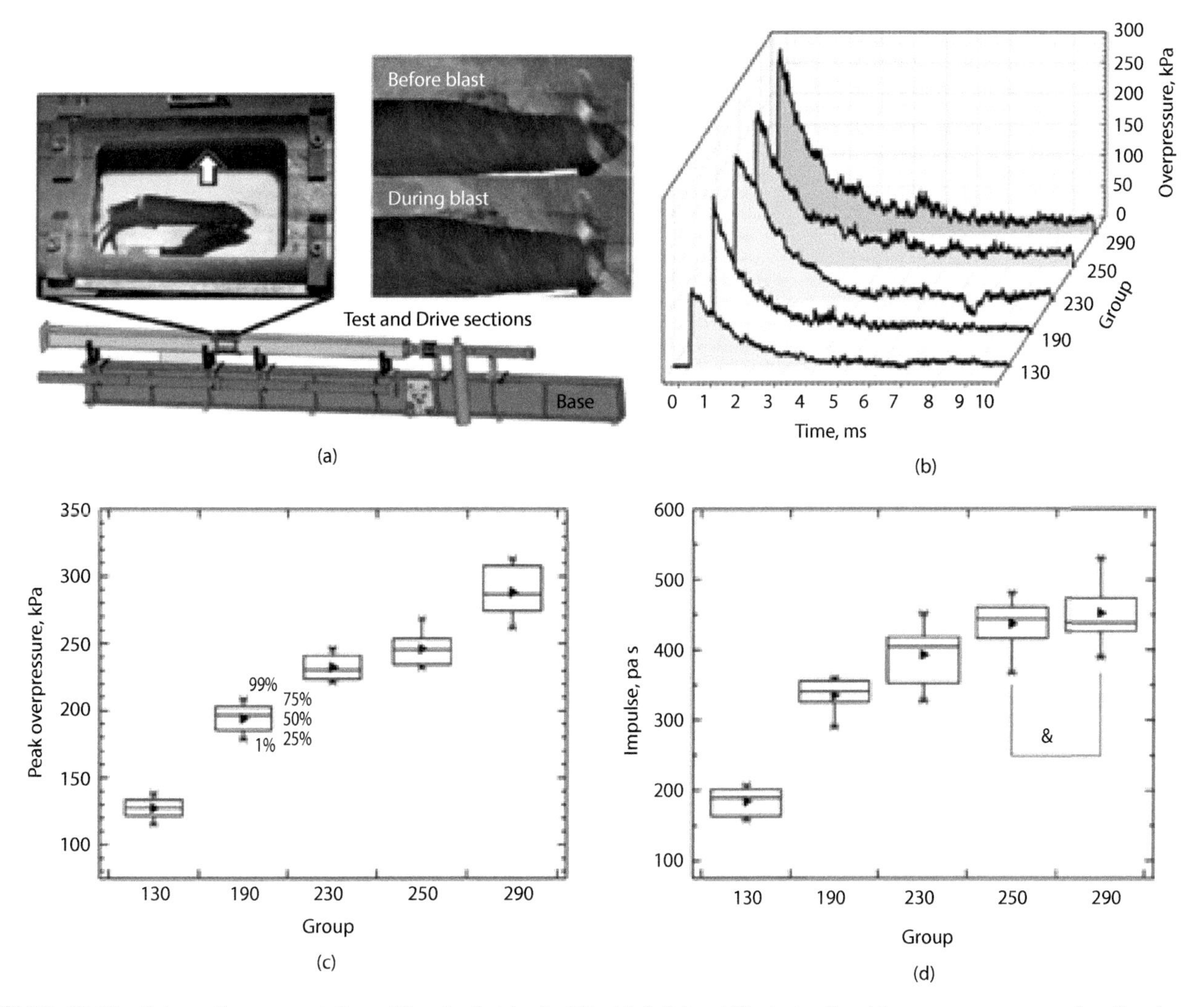

FIGURE 18.29 Schematic representation of the shock tube facility (a): left inset illustrates the side-on pressure sensor location (arrow) and the position and strapping of the animal in the holder; right inset: close-up high-speed video still images demonstrate the typical head displacement during the blast exposure in our model. Representative overpressure profiles (b), side-on overpressure (c), and impulse values (d) of five blast waves with increasing intensities used in our experiments. These data show significant difference in both overpressure values and impulse, except there was no significance between impulse values of groups at 250 and 290 kPa ($p > 0.05$, marked with ampersand). Box range: 25%–75%; whisker range: 1%–99% (median: 50%; mean: ►). (From Skotak, M. et al., *Journal of Neurotrauma*. 30:1147–1160, 2013.)

pressure readings was exposed at five blast overpressures (130–290 kPa). To study the harmonic effects in the ICP signal, fast Fourier transformation (FFT) analysis was used.

18.3.1.2 Live Animal Testing

18.3.1.2.1 Mortality Evaluation

To study the mortality, the anesthetized animals were subjected to various levels of blast, characterized by the peak overpressure ranging from 130 to 290 kPa (Table 18.3). No resuscitation was performed after trauma.

18.3.1.2.2 Physiological Evaluation

Body weight measurements were performed before and 24 hours postexposure. Multiple vital signs were measured noninvasively using clinically validated MouseSTAT, a rodent physiological monitoring system (Kent Scientific Corp., Torrington, CT). Briefly, the anesthetized rat was placed in a supine position on the warming pad attached to the control unit to maintain natural body temperature and prevent hypothermia. The temperature was measured rectally at a rate of 1 Hz over the 30-minute period before and after the blast exposure. Simultaneously, heart rate, blood oxygen saturation, and perfusion index were recorded by a y-clip sensor clamped to the back paws of the rat.

18.3.1.2.3 Injury Evaluation

Immediately after the blast exposure, the lungs were extracted from the thoracic cavity, placed in 30–40 mL of freshly prepared solution of 10% formalin, and stored at 4°C for further evaluation. Lung injury severity associated with the primary blast exposure was evaluated using the Pathology Scoring System for Blast Injuries developed by Yelverton (Yelverton, 1996). The extent of injury is defined

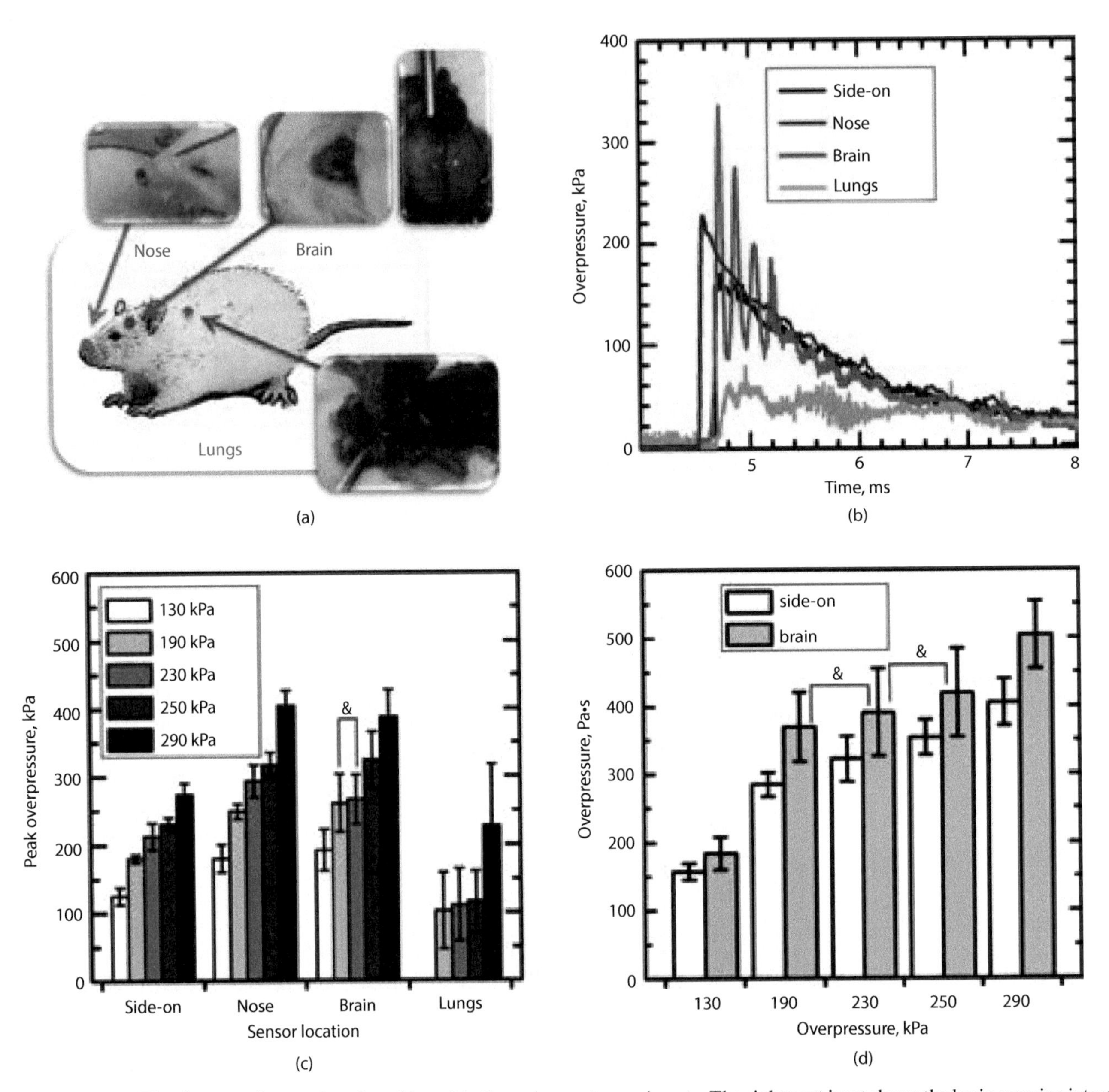

FIGURE 18.30 The diagram of sensor locations (a) used in the cadaver rat experiments. The rightmost inset shows the brain remains intact after the experiments. Representative pressure profiles of the 190-kPa group (b), and average peak overpressures recorded by side-on, nose (reflected pressure), brain, and lungs pressure sensors (c) are presented (no statistically significant differences are noted for 190- and 230-kPa groups, marked with an ampersand). There are no statistical differences between ICP impulse of the 230-kPa group (100% survival, Table 18.1) and ICP impulses of the two lethal groups (190 and 250 kPa) (d). (From Skotak, M. et al., *Journal of Neurotrauma*. 30:1147–1160, 2013.)

by the elements if injury severity (IS) according to the equation

$$IS = (E + G + ST)\,(SD) \tag{18.2}$$

where E is the extent of injury to the lungs (range 0–5); G is the injury grade including the surface area of the lesions (range, 0–4); ST represents the severity type elements, which classify the type of the worst-case lesions (range, 0–4); and SD is the severity depth element, indicating the depth or the degree of disruption of the worst-case lesion (range, 1–4).

18.3.1.2.4 Transcardial Perfusion and Histological Procedures

At 24 hours postexposure, rats were anesthetized with ketamine/xylazine mixture and transcardially perfused with wash solution (0.8% NaCl, 0.4% dextrose, 0.8% sucrose, 0.023% CaCl2, and 0.034% sodium cacodylate), followed by fresh fix solution (4% paraformaldehyde solution supplemented with 4% of sucrose in 1.4% sodium cacodylate buffer, pH = 7.3). Immediately, after perfusion, heads were decapitated and stored for additional 16–18 hours in the excess of fix solution. The brains were subsequently extracted the following day, and stored refrigerated (4°C) in a sodium cacodylate storage buffer. Brains were shipped in phosphate-buffered saline (pH = 7.2) to Neuroscience Associates (Knoxville, TN) for sectioning and staining.

The brains were embedded into one gelatin matrix block using MultiBrain Technology. The block was then allowed to cure and subsequently rapidly frozen in isopentane chilled with crushed dry ice (–70°C). The block was mounted on a freezing stage of a sliding microtome and coronal sections with a thickness of 40 μm were prepared. All sections were collected sequentially into an array of 4 × 6-inch containers that were filled with antigen preserve (buffered 50% ethylene glycol). At the completion of sectioning, each container held a serial set of one of every twelfth section, i.e., single section at every 480-μm interval). These sections were stained with an antibody against rat IgG to assess the BBB damage, and with amino cupric silver staining to detect neurodegeneration.

18.3.2 Results and Discussion

18.3.2.1 Biomechanical Loading Evaluation and ICP Analysis

Typical pressure profiles for all four sensors used (denoted as side-on, nose, brain, and lung) are presented in Figure 18.30b. The intensity of the reflected pressure measured on the rat's nose was higher than the intensity of side-on pressure, and both were characterized by gradually increasing values (Figure 18.30c). The reflected (nose sensor) and ICPs were higher than the incident pressure ($p < 0.001$) as observed in Figure 18.30c, but there are no clear differences between reflected pressure and ICP. Moreover, the ICP not only was higher than the incident pressure, but also showed an oscillatory tendency. Although the reflected pressure showed a monotonic increase with side-on pressure for all of the pressure groups (130, 190, 230, 250, and 290), the same cannot be said for the ICP: there were no statistically significant differences between the 190- and 230-kPa groups ($p = 0.70$) (Figure 18.30c). The impulse values for the brain sensor indicated there were statistically significant correlations between the nonlethal group at 230 kPa and the two lethal groups (i.e., 190 and 250 kPa). This indicates the outcome (mortality) is not a function of the total energy transferred to the brain, but depends on the specific response of the cranium. The mortality rates among animals exposed to different levels of blast overpressures are shown in Table 18.3. Exposure to 130 kPa (n = 18) and 230 kPa (n = 10) resulted in no animal death, whereas exposure to 190 kPa (n = 20), 250 kPa (n = 17), and 290 kPa (n = 6) peak overpressure resulted in 30%, 24%, and 100% mortality, respectively.

To elucidate various modes of blast loading, we performed a set of analyses on the data reported by the brain pressure sensor. Using peak–trough analysis, we correlated biomechanical loading at the early, most violent stages of blast wave–cranium interactions (initial 0.4 ms, Figure 18.31a, b), with the blast-induced mortality (Table 18.3). The monotonically increasing peak overpressure and impulse values measured outside (side-on sensor) or inside of the cranium did not follow the same trend as mortality data in the Table 18.3. However, there is a strong correlation between the sum of the first two peak-to-peak amplitudes and mortality (Figure 18.31c). The ICP data subjected to FFT analysis indicated the oscillations have harmonic characteristics, and the high frequency component (10–20 kHz) is present exclusively in the case of the lethal groups (190, 250, and 290 kPa, Figure 18.31d). Moreover, the maximum ICP frequency increases with increasing blast intensity: peaks at 12.6 kHz (190 kPa), 16.4 kHz (250 kPa), and 19.3 kHz (290 kPa) marked with arrows in Figure 18.31d. In the 290-kPa group, two characteristic frequencies are noted (16.4 and 19.3 kHz). The ICP pressure oscillations are caused by the specific material response of the rat skull, which was also demonstrated in a simplified numerical model of in vivo response to blast proposed by Moss (Moss et al., 2009).

The intrathoracic pressures (lungs) were substantially lower than the incident overpressure and ICPs. However, there were considerable difficulties during pressure measurements in lungs caused by uncontrolled interactions between the pressure sensor and walls of internal organs in the abdomen. It is clearly seen in the experimental data as high variability of peak overpressure values between data sets and relatively high experimental errors (Figure 18.30c). In an attempt to mitigate these issues, we filled up lungs with ballistic gel (20% gelatin), but the recorded pressure profiles in this configuration were of much higher magnitude (i.e., comparable with pressures recorded in the brain).

18.3.2.2 Physiological Parameters

We noted statistically significant weight loss in both 190-kPa and 250-kPa injury groups 24 hours after blast exposure (Figure 18.32). Body weight loss proportional to the blast intensity was reported in mice (Cernak et al., 2011; Wang et al., 2011), and this is typically a manifestation of impairment of the neuroendocrine system (hypothalamic-pituitary-target organ axes) (Brown et al., 1993) associated with the damage to the hypothalamic region of the brain.

Heart rate monitoring was performed over the 30-minute period before and after the blast exposure (Figure 18.33a, b). Observed changes in the heart rate difference between control groups were nonsignificant (Figure 18.33c, $p > 0.05$). The blast groups, as expected, showed an onset of bradycardia occurring immediately after the blast exposure (Figure 18.5c, $p < 0.05$): the heart rates decreased by 40 ± 9 (130 kPa), 62 ± 7 (190 kPa), 62 ± 15 (230 kPa), and 62 ± 10 (250 kPa) bpm. It appears that blast-induced bradycardia (i.e., expressed as the difference of average heart rate measured during 30-minute intervals before and after exposure) follows a simple dose-response model, with an upper value of 62 ± 4 bpm, and inflection point at 126 kPa (Figure 18.33d). However, all the groups of rats exposed to the blast are correlated, with the 130- and 190-kPa groups being borderline correlated (p = 0.065). Animals surviving blast exposure had a significantly decreased heart rate, compared with control (Figure 18.33). At high peak overpressure values (i.e., 190 kPa or higher), the heart rate is independent of the blast intensity (Figure 18.33c, d). The blast-induced bradycardia in the current model is similar to that reported in

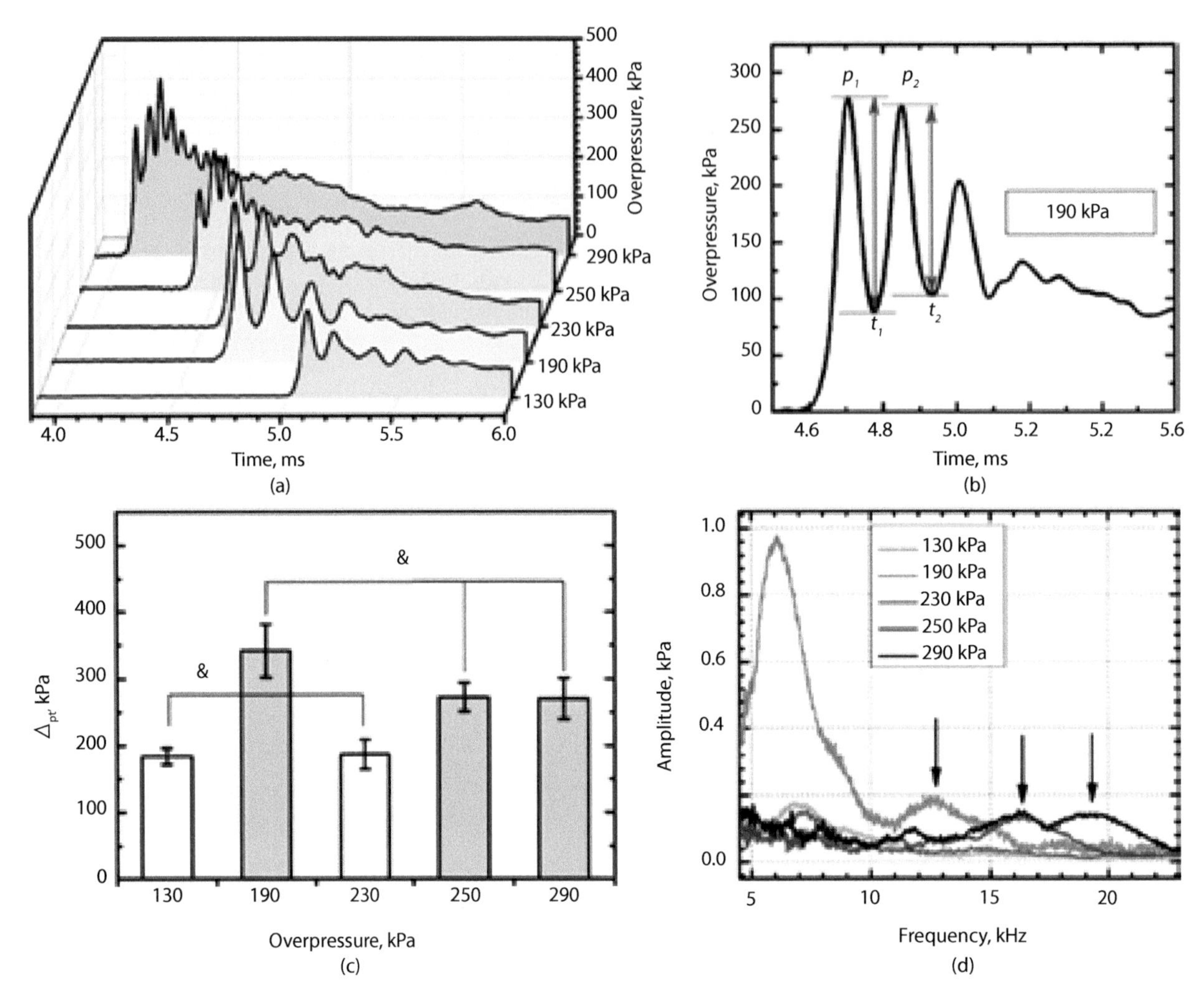

FIGURE 18.31 Representative pressure traces for five exposure groups (a). The signal was smoothed using a FFT band-pass filter for clarity of presentation. (b) Schematic representation of peak–trough analysis on overpressure recorded by brain sensor: Δpt is a sum of the first two peak-to-peak amplitudes. Results of peak–trough analysis: (c) lethal groups (190, 250, and 290 kPa) have significantly higher Δpt ($p < 0.05$) compared with nonlethal groups (130 and 230 kPa). There are no statistically significant differences between Δpt in respective nonlethal and lethal groups ($p > 0.05$, marked with ampersand). (d) FFT analysis of the ICP profiles recorded by sensor implanted in the brain. The high frequency component (peaks marked with arrows, >10 kHz) is present in the signal of lethal groups (190, 250, and 290 kPa). (From Skotak, M. et al., *Journal of Neurotrauma*. 30:1147–1160, 2013.)

other studies using rats (Guy et al., 1998; Irwin et al., 1999; Knöferl et al., 2003). However, in blast trauma mouse models, the opposite relationship was observed: exposure to the mild blast resulted in the heart rate increase immediately after the blast (Cernak et al., 2011). Bilateral cervical vagotomy and intraperitoneal injection of atropine methyl-bromide completely prevented the bradycardia, thus confirming the vasovagal reflex is responsible for bradycardia in the acute phase after the blast exposure (Irwin et al., 1999).

18.3.2.3 Shock Wave–Induced Acute Lung Injury

No animals showed external signs of trauma. At necropsy (performed immediately after the blast exposure), animals subjected to a blast wave showed typical evidence of moderate pulmonary hemorrhage (Figure 18.34), associated with vascular damage, direct alveolar injury, and edema, and generally described as "blast lung" (Brown et al., 1993; Pizov et al., 1999; Sasser et al., 2006; Tsokos et al., 2003). As illustrated in Figure 18.34, animals subjected to a higher blast overpressure (190–290 kPa) were found to have statistically significant pulmonary hemorrhage compared with sham control. The onset of injury in our model takes place at 130 kPa peak overpressure. Modeling of the pulmonary injury as a function of peak overpressure with the dose-response function revealed the inflection point is located at 184.2 ± 15.7 kPa, with a plateau (the highest score) at 14.32 ± 1.85 (Figure 18.34g). This value is rather low: the maximum possible value in Yelverton's blast injury index for lungs is 64 (Yelverton, 1996). Interestingly, pathological evaluation of pulmonary injuries in one rat from the 290-kPa group (100% acute mortality) revealed virtually no signs of lung injury. In the areas most severely affected by hemorrhage, approximately 15%–60% of the alveoli were filled with acute pools of hemorrhage (data not shown). Near these pools, there were respiratory bronchioles, which contained small amounts of edema fluid. The more normal sections of lung have moderate atelectasis and very sparse to absent emphysema.

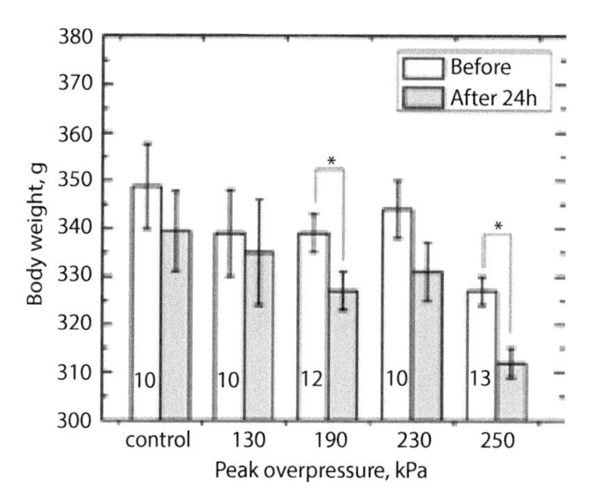

FIGURE 18.32 Exposure to different intensities of primary blast results in body weight decrease in groups 190 and 250 kPa. In these groups, statistically significant differences (marked with asterisk) were established immediately before and 24 hours after the exposure ($p < 0.05$). Numbers of animals used in respective tests are provided in the bars corresponding to each group. (From Skotak, M. et al., *Journal of Neurotrauma*. 30:1147–1160, 2013.)

18.3.2.4 BBB Damage and IgG Uptake by Neurons

Staining to reveal IgG in the brain parenchyma was performed for every twelfth section (each section was 40 μm thick) equally spaced across the entire rat brain (Figure 18.35). The purpose of this staining was twofold: (1) to measure the extent of BBB damage and (2) to visualize the accumulation of IgG in the intracellular space of neurons. Figure 18.35a presents coronal sections where elevated levels of IgG with respect to control samples are obvious. The slides with mounted sections were first digitized, and then we integrated the staining intensity to quantify of the amount of IgG. The olfactory bulb region (initial 12 sections, 5.76 mm from the brain anterior, see inset in Figure 18.35b) was omitted in this analysis. The optical density (OD) of each section was divided by the average OD of three control samples resulting in relative OD (ROD). Figure 18.35b presents RODs of 40 sections as a function of section number for a single rat from the 190-kPa group. At lower blast intensities (190 and 230 kPa), in some rats the ROD had a higher value in the rostral than in the dorsal brain region (Figure 18.35). The ROD

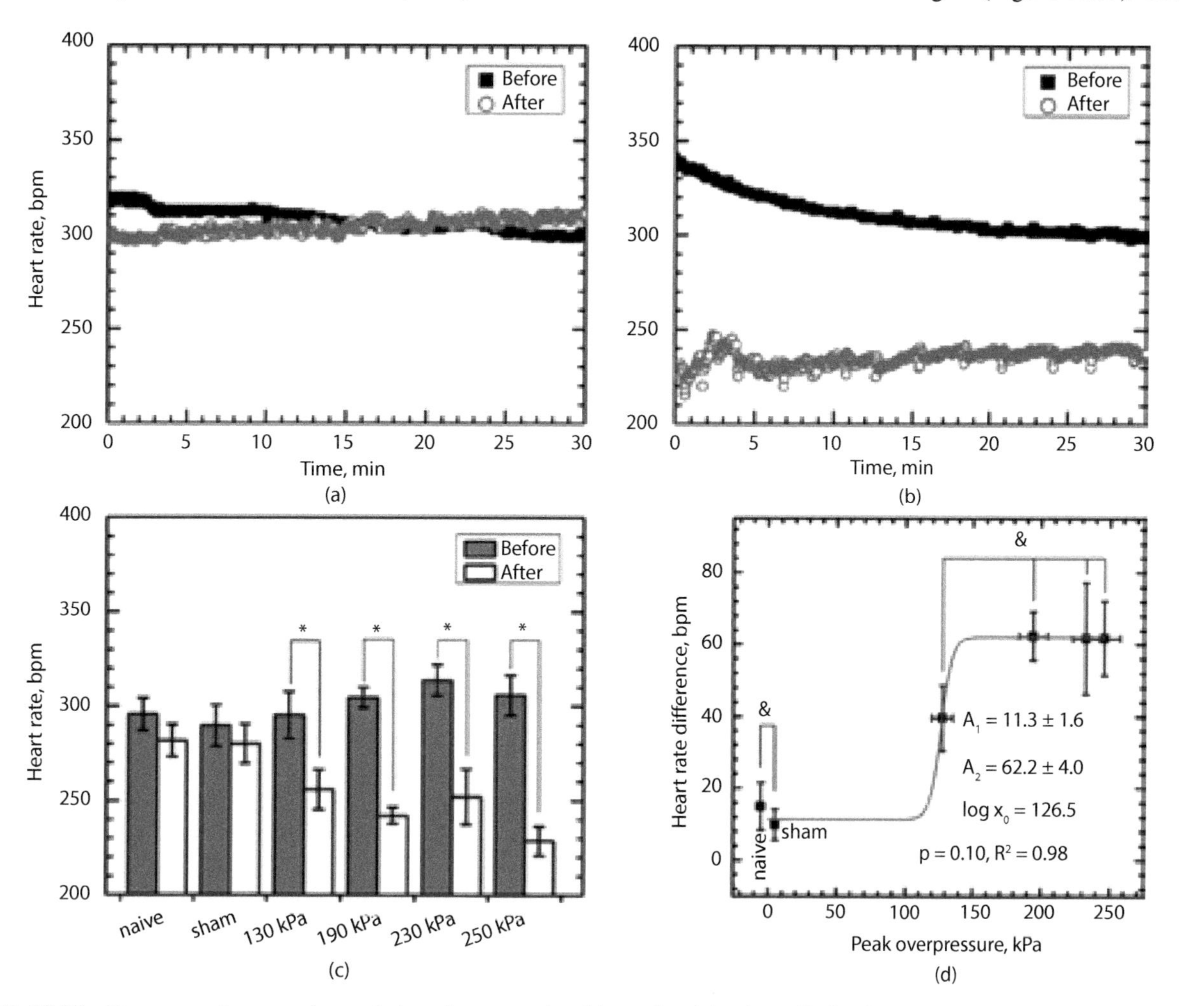

FIGURE 18.33 Representative experimental data demonstrating blast-related bradycardia in the rat model: (a) naive control, (b) rat exposed to 190-kPa peak overpressure (sampling rate 1 Hz in both cases). (c) The heart rate monitoring in the acute phase postexposure revealed the average heartbeat decrease was statistically significant ($p < 0.05$) in animals exposed to blast at peak overpressure of 130 kPa ($\Delta HR = 40 \pm 9$ bpm) and higher (average $\Delta HR = 62 \pm 11$ bpm for 190-, 230-, and 250-kPa groups). (d) The heart rate difference is not statistically significant between the two control and four exposed groups ($p > 0.05$, marked with ampersand) and follows the dose response model in the acute phase postexposure. (From Skotak, M. et al., *Journal of Neurotrauma*. 30:1147–1160, 2013.)

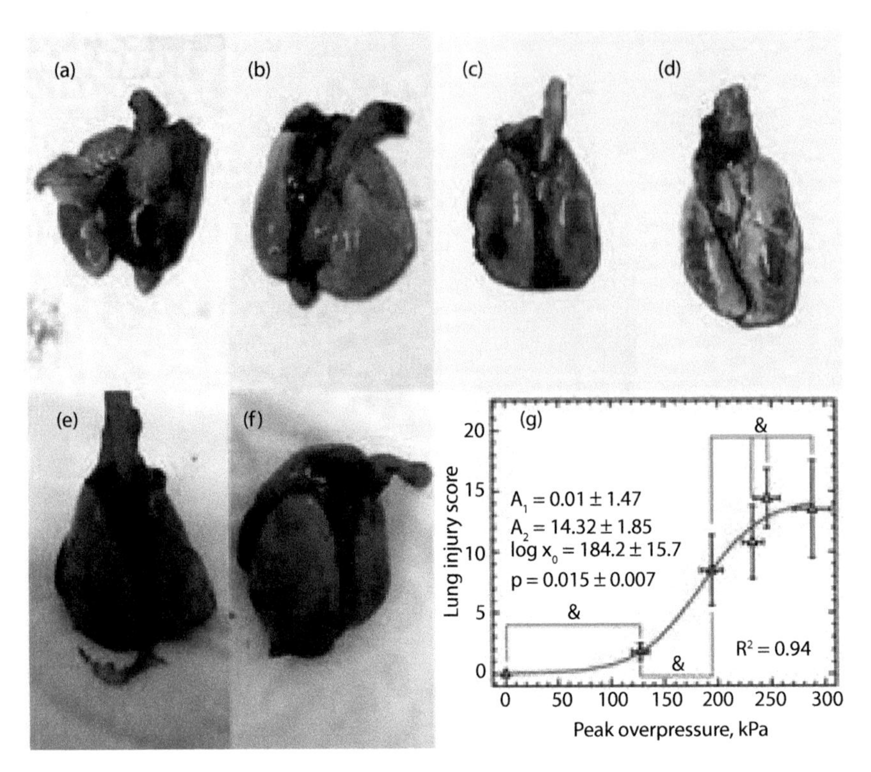

FIGURE 18.34 Blast-induced lung injury immediately postexposure: (a) control, (b) 130 kPa, (c) 190 kPa, (d) 230 kPa, (e) 250 kPa, and (f) 290 kPa. The extent of injury was quantified using the Pathology Scoring System for Blast Injuries (Yelverton, 1996). The extent of injury is defined by the elements of the IS according to Equation 18.1 and was modeled with the dose-response function (g). Groups with no statistically significant differences ($p > 0.05$) are marked with an ampersand. (From Skotak, M. et al., *Journal of Neurotrauma.* 30:1147–1160, 2013.)

gradient was not a general pattern and typically the variations were not substantial. However, in one rat exposed at 250-kPa blast intensity, the highest RODs (2.3–2.7) are noted in slide numbers 28–35 (not shown). The RODs pertaining to the same brain were averaged and expressed as average ROD. The results for rats exposed at three different blast intensities are presented in Figure 18.35c. In the 190 kPa group, two rats have higher average ROD than the control group (average ROD > 1), but two others are comparable. Similar results are noted for the 230-kPa group, but only one rat had a higher average intensity of IgG staining than the rats in the 190-kPa group. The level of BBB damage appears to be correlated with the blast intensity (Figure 18.35c). In animals with compromised BBB, numerous IgG-positive cells of neuronal morphology were noted across brain parenchyma. The IgG accumulated predominantly in the cells localized in various parts of the cerebral cortex (Figure 18.36) and the hippocampus (Figure 18.37). We have also evaluated brain sections of rats exposed to 190-, 230-, and 250-kPa blast intensity using amino cupric silver staining. These tests gave negative results, indicating 24 hours is an insufficient period to induce neuro-degeneration.

18.4 COMPUTER MODELS

In this section, we describe the development of a CT/magnetic resonance imaging (MRI)-based finite element model of human and rat heads. In addition, the computational framework for blast simulations is described.

18.4.1 Development of Human Head Model

The 3D human head model was generated from segmentation of high-resolution MRI data obtained from the Visible Human Project (2009). The MRI data consisted of 192 T1-weighted slices of 256 × 256 pixels taken at 1-mm intervals in a human male head. The image data are segmented into four different tissue types: (1) skin, (2) skull, (3) subarachnoid space (Sasser et al., 2006) and (4) brain (for which material properties are available). Brain included all important sections: frontal, parietal, temporal, and occipital lobes, cerebrum, cerebellum, corpus callosum, thalamus, midbrain, and brain stem. It was not possible to separately segment cerebrospinal fluid and structures such as membranes and bridging veins because of the resolution of the MRI data; as such, they are considered a part of the subarachnoid space (SAS). The segmentation uses 3D image analysis algorithms (voxel recognition algorithms) implemented in Avizo. The segmented 3D head model was imported into the meshing software HyperMesh and was meshed as a triangulated surface mesh. The volume mesh was generated from this surface mesh to generate 10-noded tetrahedrons. Tetrahedron meshing algorithms are more robust than hexahedral meshing algorithms and can model complex head volumes like brain and SAS faster and easier (Baker, 2005; Bourdin et al., 2007; Schneiders, 2000). The modified quadratic tetrahedral element (C3D10M) available in Abaqus is very robust and is as good as hexahedral elements (Abaqus user's manual) as far as accuracy of results is concerned (Cifuentes and Kalbag, 1992; Ramos and Simões, 2006; Wieding et al., 2012).

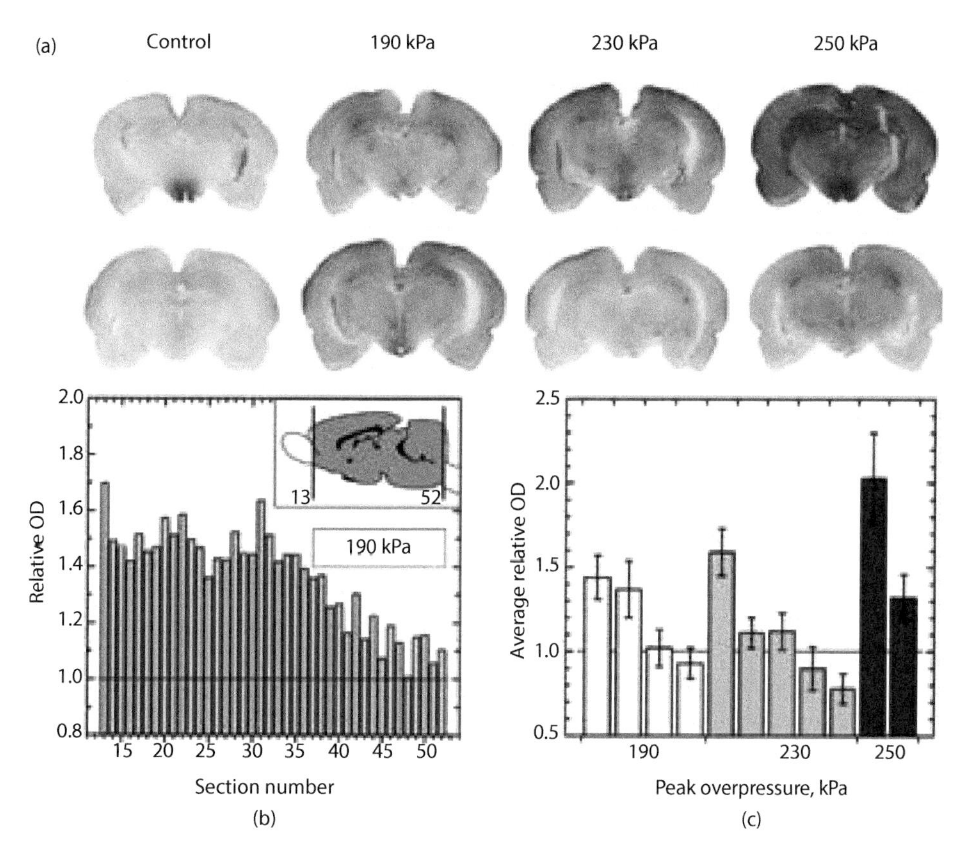

FIGURE 18.35 (a) Immunostaining for rat IgG as an indicator for a compromised BBB. (b) Relative OD of full coronal sections from the brain of a single rat exposed at 190 kPa. The OD of each section was divided by the average OD of three controls (two naive and one sham). Inset illustrates brain region covered in this study. (c) Average relative OD of IgG across the brain parenchyma for rats exposed to primary blast and sacrificed 24 hours postexposure. Error bars are standard deviation. (From Skotak, M. et al., *Journal of Neurotrauma*. 30:1147–1160, 2013.)

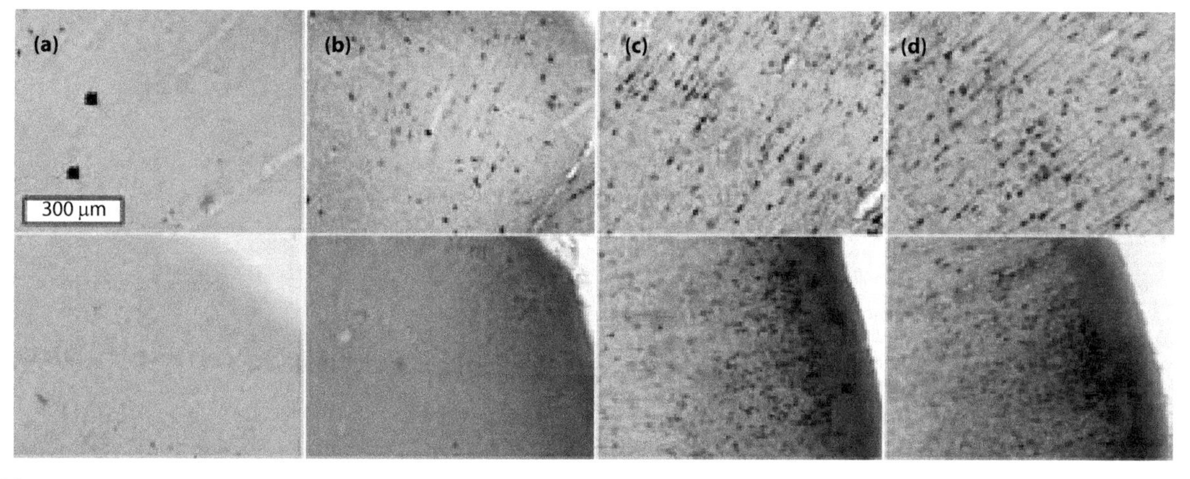

FIGURE 18.36 The IgG uptake in cortical neurons 24 hours after the blast. Coronal sections of the brain were collected from brains of sham control (a) and surviving rats exposed to blast with (b) 190-kPa, (c) 230-kPa, and (d) 250-kPa peak overpressure. The scale bar (300 μm) is the same for all samples. Arrows indicate faintly stained neurons. (From Skotak, M. et al., *Journal of Neurotrauma*. 30:1147–1160, 2013.)

In addition, hexahedral elements can suffer from the problem of volumetric locking for highly incompressible materials like brain. The problem of volumetric locking is not present for a modified quadratic tetrahedral element (C3D10M) (Abaqus user's manual). For these reasons, we chose the modified quadratic tetrahedral element. The use of specialized 3D image processing (Avizo) and meshing software (HyperMesh) allowed for the development of a geometrically accurate finite element (FE) model. FE discretization is schematically shown in Figure 18.38.

18.4.2 Development of Animal Head Model

A very similar approach as described for the human head model was used for developing a 3D rat head model. For development of this model, MRI/CTs of a Sprague Dawley rat

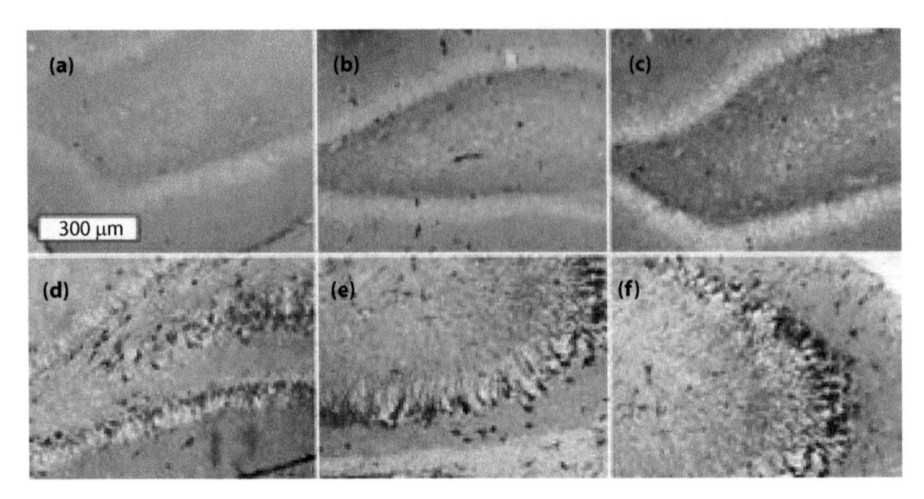

FIGURE 18.37 The IgG-positive cells in hippocampus of sham control (a) and in rats exposed to blast overpressure of 190 (b), 230 (c), and 250 kPa (d–f). The scale bar (300 μm) is the same for all samples. (From Skotak, M. et al., *Journal of Neurotrauma*. 30:1147–1160, 2013.)

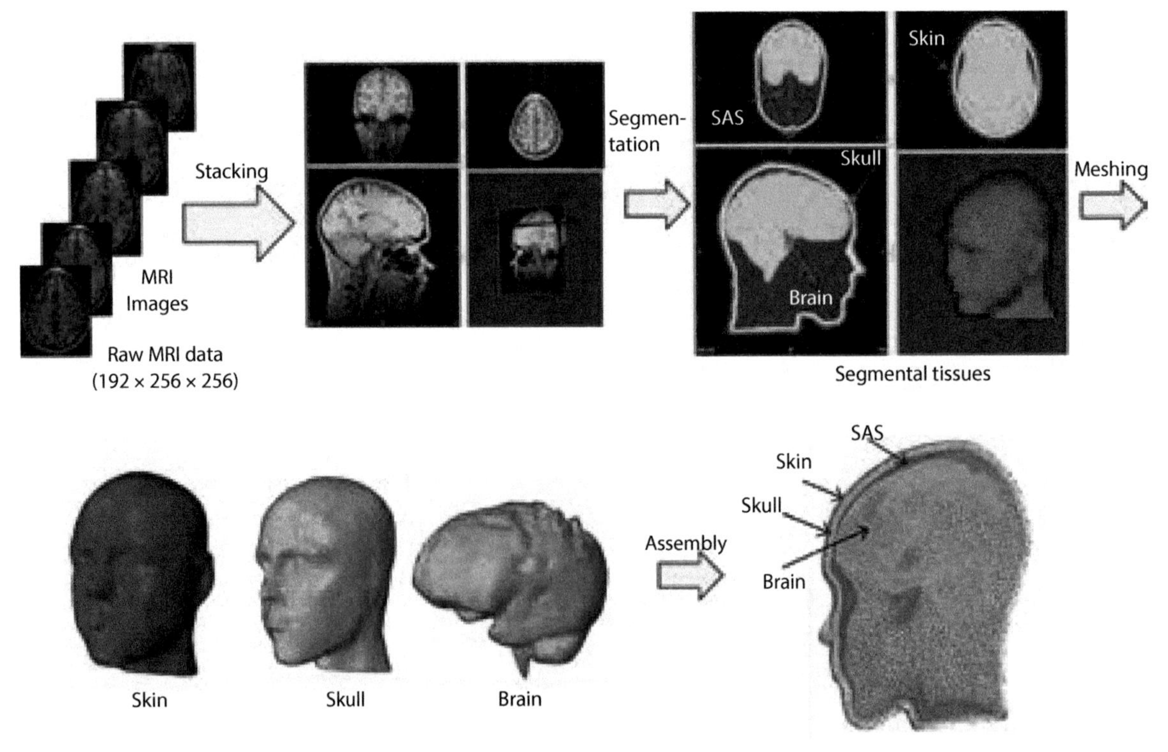

FIGURE 18.38 FE discretization.

was used. Two different T2-weighted MRI scans (one for the muscle skin and other for the brain), and one CT scan (for the skull and the bones) were used. These three different scans were necessary to achieve proper contrast and segmentation of various tissues (i.e., muscle, skin, brain and skull, and bones). The brain MRI has an isotropic resolution of 256 × 256 × 256 pixels, for a field of view of 30 mm in all three directions. The MRI for muscle and skin has an anisotropic resolution, with a pixel size of 512 × 512 × 256, for a field of view of 30, 30, and 50 mm. The three data sets were overlapped, registered, segmented, and triangulated using Avizo 6.2 software. The triangulated mesh (i.e., surface mesh) is imported into HyperMesh and a volume mesh is generated from this surface mesh to generate 10-noded tetrahedrons (C3D10M). The skull, skin, and brain share the node across the interface. These elements are treated as Lagrangian elements.

18.4.3 General Theory of Material Models and Material Parameters Used in the Head Models

In both human and animal models, skin and skull are modeled as linear, elastic, isotropic materials with properties adopted from the literature. Similarly, SAS in the human model is considered elastic. In general, elastic properties are sufficient to capture the wave propagation characteristics for these tissue types and this approach is consistent with other published works (Chafi et al., 2010; Chen and Ostoja-Starzewski, 2010; Grujicic et al., 2011; Moore et al., 2009; Moss et al., 2009; Nyein et al., 2010). For elastic material, stress is related to strain as

$$\sigma_{ij} = \lambda E_{kk}\delta_{ij} + 2\ \ E_{ij} \tag{18.3}$$

where σ is a Cauchy stress, E is Green strain (also known as Green-Lagrange strain), λ and μ are Lame constants, and δ is a Kronecker delta.

The brain is modeled with an elastic volumetric response and viscoelastic shear response. Viscoelastic response is modeled using the Maxwell model. The associated Cauchy stress is computed through

$$\sigma_{ij} = J^{-1}F_{ik} \cdot S_{km} \cdot F^{T}_{mj} \tag{18.4}$$

where F is a deformation gradient, J is a Jacobian, and S is the second Piola-Kirchhoff stress, which is estimated using following integral:

$$S_{ij} = \int_0^t G_{ijkl}(t-\tau)\frac{\partial E_{kl}}{\partial \tau}d\tau \tag{18.5}$$

where E is the Green strain and G_{ijkl} is the tensorial stress relaxation function. The relaxation modulus for an isotropic material can be represented using a Prony series:

$$G(t) = G_{\infty} + \sum_{i=1}^{n} G_i e^{-\beta_i t} \tag{18.6}$$

where G_{∞} is the long-term modulus and β is the decay constant.

18.4.3.1 Material Models for Human Head Model

For material parameters of the brain tissue, the widely accepted bulk modulus value of 2.19 GPa is used in this work. This value is motivated from the works of Stalnaker (Stalnaker, 1969) and McElhaney (McElhaney et al., 1973). The shear properties of the brain tissue are adopted from Zhang et al. (Zhang et al., 2001), who derived the shear modulus from the experimental work of Shuck and Advani (Shuck and Advani, 1972) on human white and gray matter. For material parameters, we relied on widely accepted values in the literature for base simulations. In addition, parametric studies are conducted to account for reported variations in the brain material properties. The material properties of the human head model are summarized in Table 18.4.

18.4.3.2 Material Models for Rat Head Model

Skin and skull are modeled as a homogenous linear elastic isotropic material with properties adopted from the literature (Willinger et al., 1999). Brain tissue is modeled as elastic volumetric response and viscoelastic shear response with properties adopted from previous work (Zhang et al., 2001). Air is modeled as an ideal gas equation of state (EOS). The Mach number of the shock front calculated from our experiments is approximately 1.4 and hence the ideal gas EOS assumption is acceptable; the ratio of specific heats does not change drastically at this Mach number value. The material properties along with longitudinal wave speeds are summarized in Table 18.5.

18.4.4 Computational Framework for Blast Simulations

18.4.4.1 Numerical Approach for Shock Tube

In this section, a detailed description of the computational framework for blast simulations using the Euler-Lagrangian coupling method is discussed. In this method, an Eulerian mesh is used to model shock wave propagation inside the shock tube and a Lagrangian mesh is used for the head. This computational environment allows accurate concurrent simulations of the formation and propagation of the blast wave in air, the fluid–structure interactions between the blast wave and the head models, and the stress wave propagation within the head. The computational framework is shown in Figure 18.39.

The shock tube used in the modeling is based on our experimental shock tube. The head is placed inside the

TABLE 18.4
Material Properties
(a) Elastic Material Properties

Tissue Type	Density (kg/m³)	Young's Modulus (MPa)	Poisson's Ratio
Skin	1,200	16.7	0.42
Skull	1,710	5,370	0.19
SAS	1,000	10	0.49
Neck	2,500	354	0.3

(b) Viscoelastic Material Properties

	Instantaneous Shear Modulus (kPa)	Long-Term Shear Modulus (kPa)	Decay Constant (sec⁻¹)
Brain	41.0	7.8	700

TABLE 18.5
Material Properties for Rat Head
(a) Elastic Material Properties

Material	Young's Modulus (MPa)	Poisson's Ratio
Skin	8	0.42
Skull	100	0.3
Brain	0.123	0.49

(b) Viscoelastic Material Properties

Material	Instantaneous Shear Modulus (kPa)	Long-Term Shear Modulus (kPa)	Decay Constants^{-1}
Brain	41	7.8	700

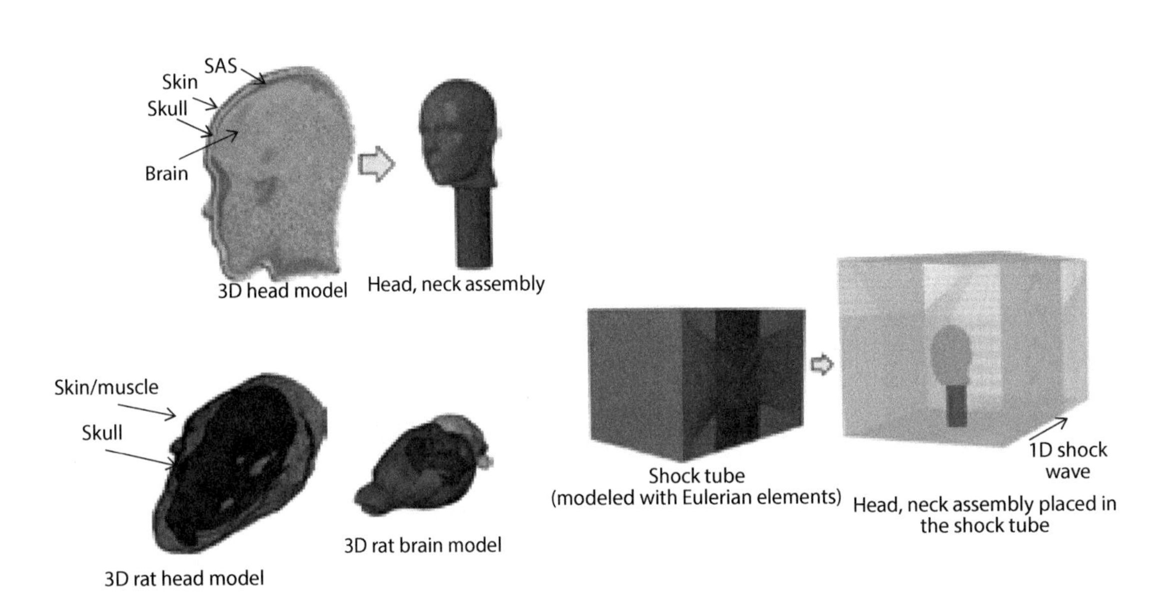

FIGURE 18.39 Computational framework for blast simulations.

shock tube and subjected to the blast wave profile of interest. The Eulerian domain (air inside the shock tube) is meshed with eight-noded brick elements, with appropriate mesh refinement near the regions of solid bodies to capture fluid–structure interaction effects. Parametric studies on mesh size have been performed and it has been found that mesh size of 3 mm is appropriate to capture the flow field around the head (i.e., pressures, velocities) and fluid–structure interaction effects. For Eulerian elements, mesh convergence is achieved at this element size; thus an element size of 3 mm is used near the regions of solid bodies and along the direction of blast wave propagation. Air is modeled as an ideal gas EOS (see Equation 18.7) with the following parameters: density, 1.1607 kg/m^3; gas constant, 287.05 J/(kg-K); and temperature 27°C:

$$P = (\gamma - 1)\frac{\rho}{\rho_0}e \qquad (18.7)$$

where P is the pressure, γ is the constant-pressure to constant-volume specific heat ratio (= 1.4 for air), ρ_0 is the initial air mass density, ρ is the current mass density, and e is the internal volumetric energy density. The Mach number of the shock front from our experiments is approximately 1.4, and hence the ideal gas EOS assumption is acceptable because the ratio of specific heats do not change drastically at this Mach number.

18.4.4.2 Loading, Interface, and Boundary Conditions

To numerically reproduce primary blast loading, there are two possible techniques to impose the boundary conditions. Technique 1 models the entire shock tube, in which driver, transition, and extension sections are included in the model so that events of burst, expansion, and development of a planar of the blast wave are reproduced. Technique 2 is a partial model with the experimentally measured (p-t) history used as the pressure boundary condition, where the numerical model comprises the downstream flow field containing the test specimen. Technique 1 is computationally very expensive. For example, a full scale simulation of a 711 mm × 711 mm cross section, 9,880 mm long shock tube (excluding catch tank) with cylindrical to square transition requires about 5 million eight-noded brick Eulerian elements and takes about 147 CPU hours on 48 dedicated processors. These simulations reach the limits of computing power in terms of memory and simulation time. On the other hand, technique 2 requires about 1.26 million elements with 10 CPU hours. The pressure, velocity, and temperature profiles obtained using technique 2 match well with the profiles that are obtained using the full-scale model (technique 1) at the boundary and downstream locations. Hence technique 2 is capable of capturing the pressure, momentum, and energy of the shock wave. An approach similar to technique 2 has been widely used in shock dynamics studies using shock tubes (Honma et al., 2003; Jiang et al., 2003; Kashimura et al., 2000).

The velocity perpendicular to each face of the Eulerian domain (shock tube) is kept at zero to avoid escaping/leaking of air through these faces. This will maintain a planar shock front traveling in the longitudinal direction with no lateral flow. The bottom of the neck is constrained in all six degrees of freedom to avoid rigid body motion. An enhanced immersed boundary method is used to provide the coupling between the Eulerian and the Lagrangian domains. Here, the Lagrangian region resides fully or partially within the Eulerian region and provides *no-flow* boundary conditions to the fluid in the direction normal to the local surface. Further, the Eulerian region provides the pressure boundary conditions to the Lagrangian region. Thus, a combination of fixed Eulerian mesh and solid–fluid interface modeling through the enhanced immersed boundary method allows for the concurrent simulations of the formation and propagation of a primary blast wave in a fluid medium and accounts for the effects of both fluid–structure interaction and structural deformations once the blast wave encounters a solid. The interactions (contact conditions) between Eulerian (containing air and a propagating blast wave) and Lagrangian regions are defined using the "general contact" feature (card) in Abaqus. In general contact, contact constraints are enforced through the penalty method with a finite sliding contact formulation. Various contact property models are available in general contact. In the present work, frictionless tangential sliding with hard contact is used as the contact property model.

18.4.5 Solution Scheme

The finite element model is solved using the nonlinear transient dynamic procedure with the Euler-Lagrangian coupling method (Abaqus). In this procedure, the governing partial differential equations for the conservation of momentum, mass and energy (Equations 18.8 through 18.10) along with the material constitutive equations (Equations 18.3 through 18.6) and the equations defining the initial and boundary conditions are solved simultaneously.

Conservation of mass (continuity equation):

$$\rho \frac{\partial v_i}{\partial x_i} + \frac{\partial \rho}{\partial t} + v \bullet \quad \rho = 0 \tag{18.8}$$

Conservation of momentum (equation of motion):

$$\frac{\partial \sigma_{ij}}{\partial x_j} + \rho b_i = \rho a_i \tag{18.9}$$

Conservation of energy (energy equation):

$$\rho \frac{\partial e}{\partial t} + v \bullet \quad e = \sigma_{ij} \frac{\partial v_i}{\partial x_j} - \frac{\partial q_i}{\partial x_i} + \rho q_S \tag{18.10}$$

where ρ is a density; x, v, and a are displacement, velocity, and acceleration of a particle, respectively; σ is a Cauchy stress; b is a body force; e is an internal energy per unit mass; q is a heat flow per unit area; and q_S is a rate of heat input per unit mass by external sources.

In the Eulerian–Lagrangian method, we are actually solving the whole model (i.e., both the Eulerian and Lagrangian domains) with the same Lagrangian equations. The notion of a material (solid or fluid) is introduced when specific constitutive assumptions are made. The choice of a constitutive law for a solid or a fluid reduces the equation of motion appropriately (e.g., compressible Navier–Stokes equation, Euler equations). For the Eulerian part/domain in the model, the results are simply mapped back to the original mesh with extensions to allow multiple materials and to support the Eulerian transport phase for Eulerian elements. The Eulerian framework allows for the modeling of highly dynamic events (e.g., shock) which would otherwise induce heavy mesh distortion. In Abaqus, the Eulerian time incrementation algorithm is based on an operator split of the governing equations, resulting in a traditional Lagrangian phase followed by an Eulerian, or transport, phase. This formulation is known as "Lagrange-plus-remap." During the Lagrangian phase of the time increment, nodes are assumed to be temporarily fixed within the material, and elements deform with the material. During the Eulerian phase of the time increment, deformation is suspended, elements with significant deformation are automatically remeshed, and the corresponding material flow between neighboring elements is computed. As material flows through an Eulerian mesh, state variables are transferred between elements by advection. Second-order advection is used in the current analysis. The Lagrangian (solid) body can be a deformable body and can deform based on the forces acting on it; the deformation of the Lagrangian solid influences the Eulerian part/domain.

A typical 3D simulation requires about 7 hours of CPU time on 48 dedicated Opteron parallel processors (processor speed 2.2 GHz, 2 GB memory per processor), for an integration time of 2.5 msec. The simulation time is selected such that the peaks resulting from stress wave action have been established. A time step of the order of 5×10^{-7} seconds is used to resolve and capture wave disturbances of the order of 1 MHz, which increases the overall computational effort for the total simulation time of interest.

18.4.6 Model Validations

In this section, both human and animal models will be validated against the experimental blast results. This is a necessary step in developing a finite element model to show that the model mimics the reality to as close as possible.

18.4.6.1 Validation of the MRI-Based Human Head Model against Blast Experiments

The head model is validated against PMHS experiments, which were previously discussed in this chapter. PMHS specimens (N = 3) were subjected to primary blasts of incident intensities 70, 140, and 200 kPa, respectively. The numerical model is validated against surface (reflected) pressures, surface/skull strains, and ICPs obtained from these experiments. In the PMHS experiments, dura, subarachnoidal spaces, and brain were removed and intracranial contents were backfilled with ballistic gel whose wave speed is calculated at 1,583 ± 118 m/s, which was close to the longitudinal wave speed of water. Thus for head model validation purposes only, SAS and brain were assigned the same bulk modulus value of 2.19 GPa. The validation results are presented for incident intensity of 200 kPa, but similar agreement in the simulation and experimental results are seen for other intensities. The arrival time of the experiment at each sensor location is shifted to match the arrival time of the numerical simulation for ease of comparison of the different features of the *p-t* profile. The experimental profiles are based on the average of three shots (experiments) for head 1.

Figure 18.40a shows a comparison of the surface pressure profile for the front location. For the front location (i.e., FH SM), good agreement is seen between the experiment and the simulation both in terms of peak pressure and nonlinear decay. Figure 18.40b shows the comparison of ICP profiles between the experiment and the simulation. Both experimental and simulation data are filtered using a 10-kHz four-pole Butterworth filter. From the figure, it can be seen that there is a reasonably good agreement between the experiment and numerical ICP profiles both in terms of peak values (maximum% difference in peak ICP value is 17%) and shape of the profiles. In general, simulation ICP profiles show more oscillatory behavior than experimental ICP profiles. This can be attributed to the following: (1) lack of material characterization for ballistic gel. It is possible that the response of ballistic gel to shock loading is much more complex (because of heterogeneity, rate dependent behavior, effect of curing, wave propagation, and dispersion in 3D setting) than assumed here and (2) frequency response of ICP pressure probes (gauges) embedded in the ballistic gel is not known; it is possible that frequency response of the gauge is slower because of which behavior is less oscillatory in the experiments. Because of these artifacts, discrepancies in the oscillatory pattern are to be expected between the simulation and the experiment. It is also common that the repeated experiments will show small variations in the oscillatory pulse patterns from shot to shot. For these reasons, it is improbable if not impossible to match every aspect of experiment with the computational simulation. With these considerations in mind, some aspects of comparison between experimental and simulation ICP profiles are discussed.

For FH ICP, simulation is able to capture major trends including the initial sharp rise (rise time = 40 μs) associated

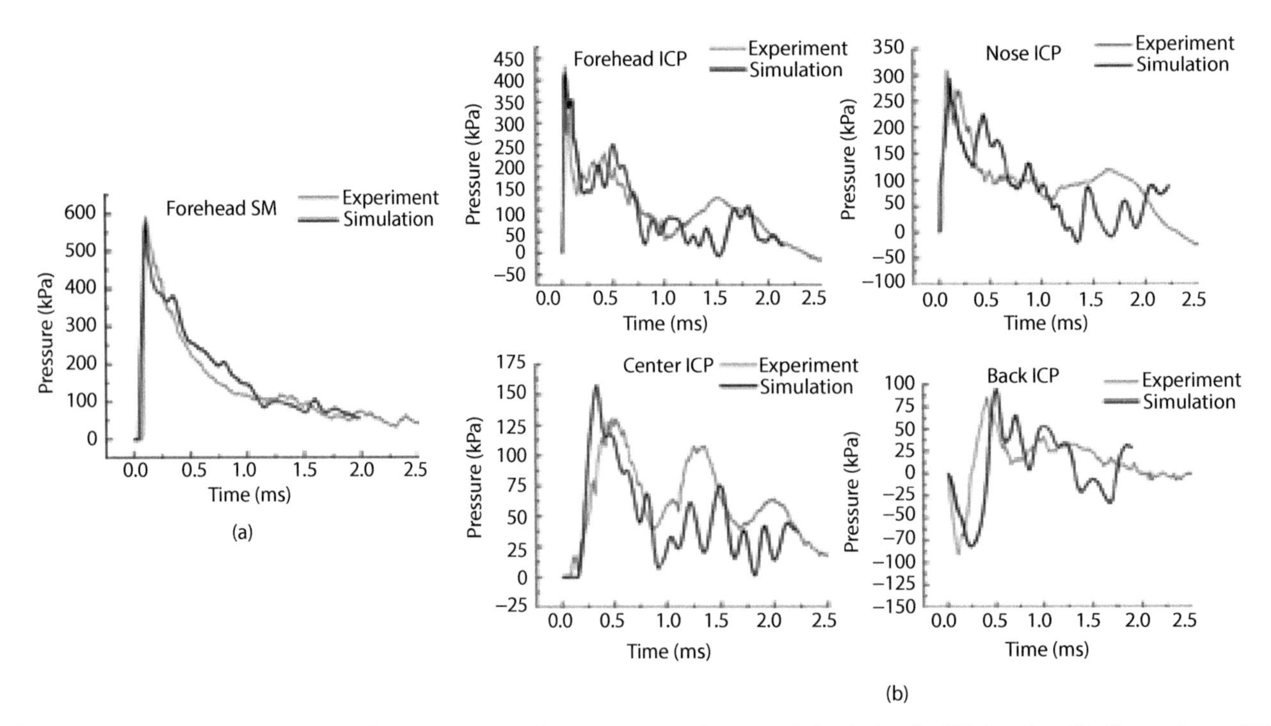

FIGURE 18.40 (a) Comparison of surface pressure profile between experiment and simulation for FH location. (b) Comparison of ICP profiles between experiment and simulation.

with the shock front, initial decay (till t = 0.15 msec), abrupt pressure increase (second peak) during initial decay (at t = 0.07 msec), and subsequent pressure pattern. The secondary (loading) pulse seen in the experiment after 1.0 ms is not seen in the simulation profile for FH ICP. For nose ICP, fair agreement is seen between the experiment and the simulation. The distinct secondary peak seen during the decay (at t = 0.25 msec) is delayed in the simulation. This secondary peak is due to wave transmission from the eye socket. The delay in this transmission between simulation and experiment is attributed to the difference in geometry of the eye socket between the simulation and the experiments. The center ICP also shows fair agreement between the simulation and experiment. The rise time is sharper in the simulation than the experiments. In addition, the peak ICP value is slightly higher (20% difference) in the simulation. These discrepancies are expected considering the complex set of direct and indirect loadings experienced by the center location. The center location experiences a complex set of direct and indirect loadings emanating from different sources (e.g., blast wave transmission, reflections from tissue interfaces, skull deformation) at different points of time. These disturbances continuously propagate into the brain as waves. The constructive and deconstructive interferences of these waves control the pressure history deep inside the brain. We believe the obtained match between experiment and simulation for the center ICP is reasonable considering these complexities. The back ICP also shows reasonably good agreement between experiment and simulation. The contrecoup phase seen in the experiments is replicated in the simulation.

During PMHS experiments (circumferential) skull strains were also measured at four locations, front, temple, top, and back, as shown in figure. Table 18.6 compares the peak skull strains obtained from the simulation with that of experimentally measured skull strains at various locations. The maximum peak strain is seen at the front location and the strain value at this location is less than 0.1%. It should be noted that the standard deviations in experimentally obtained skull strain values are huge. Standard deviations up to 100% of mean values are seen in certain strain measurements. Such standard deviations are normal during strain measurements, especially when strain values are very small. Obtained strain values from the simulations fall within the range of experimentally obtained strain values. The experimental and numerical strain–time profiles are not compared, as the obtained profiles were not highly repeatable during the experiments.

18.4.6.2 Validation of the MRI-Based Rat Head Model against Blast Experiments

The rat head model is validated against rat head experiments, were which previously discussed. In this experiment, five male Sprague Dawley rats of 320–360 g weight were sacrificed and instrumented with a pressure sensor on the nose to measure reflected pressure, and two additional sensors were implanted in the thoracic cavity and brain, respectively. To test the robustness of the model it was validated against experiments both inside and outside of the shock tube. Input loading is simulated using partial model technique in Section 18.4.4.2.

Figure 18.41a, b shows comparisons of *p-t* profiles for the surface mount sensor on the nose inside and outside of the shock tube, respectively. Overpressure and the subsequent decay recorded within 1 ms is slightly lower in the case of simulation in the simulation is slightly lower than that of the overpressure recorded in the experiment. Beyond 1 ms there is an almost perfect match between experiment and simulation. This deviation in the initial part of the pressure profile is due to a coarse Eulerian mesh relative to the size of the Lagrangian body (rat head). A tradeoff was made to get reasonably accurate results for higher model efficiency (time for model simulation) and this did not affect the intracranial measurement drastically, which will be seen in the following section.

Figure 18.41c, d show the ICP comparisons between experiment and simulation for inside and outside, respectively. There is a good agreement between experiment and simulation. Both experimental and simulation data are filtered using a 10-kHz four-pole Butterworth filter. In general, the simulation inside has less oscillations while outside has more oscillations after 0.5 msec. This model also predicts the effect of exit rarefaction depleting a overpressure and resulting in a low-pressure, high-velocity wind (after 0.5 msec).

TABLE 18.6
Comparison of Peak Circumferential Strains at Various Skull Locations

	Peak Strain (%)	
Location	**Experiment (Mean ± S.D.)**	**Simulation**
Front (compressive)	0.09 ± 0.04	0.06
Front (tensile)	0.01 ± 0.01	0.018
Right temple (compressive)	0.04 ± 0.03	0.045
Right temple (tensile)	0.04 ± 0.02	0.028
Top (compressive)	0.04 ± 0.03	0.042
Top (tensile)	0.035 ± 0.02	0.02
Back (compressive)	0.04 ± 0.02	0.03
Back (tensile)	0.02 ± 0.02	0.03

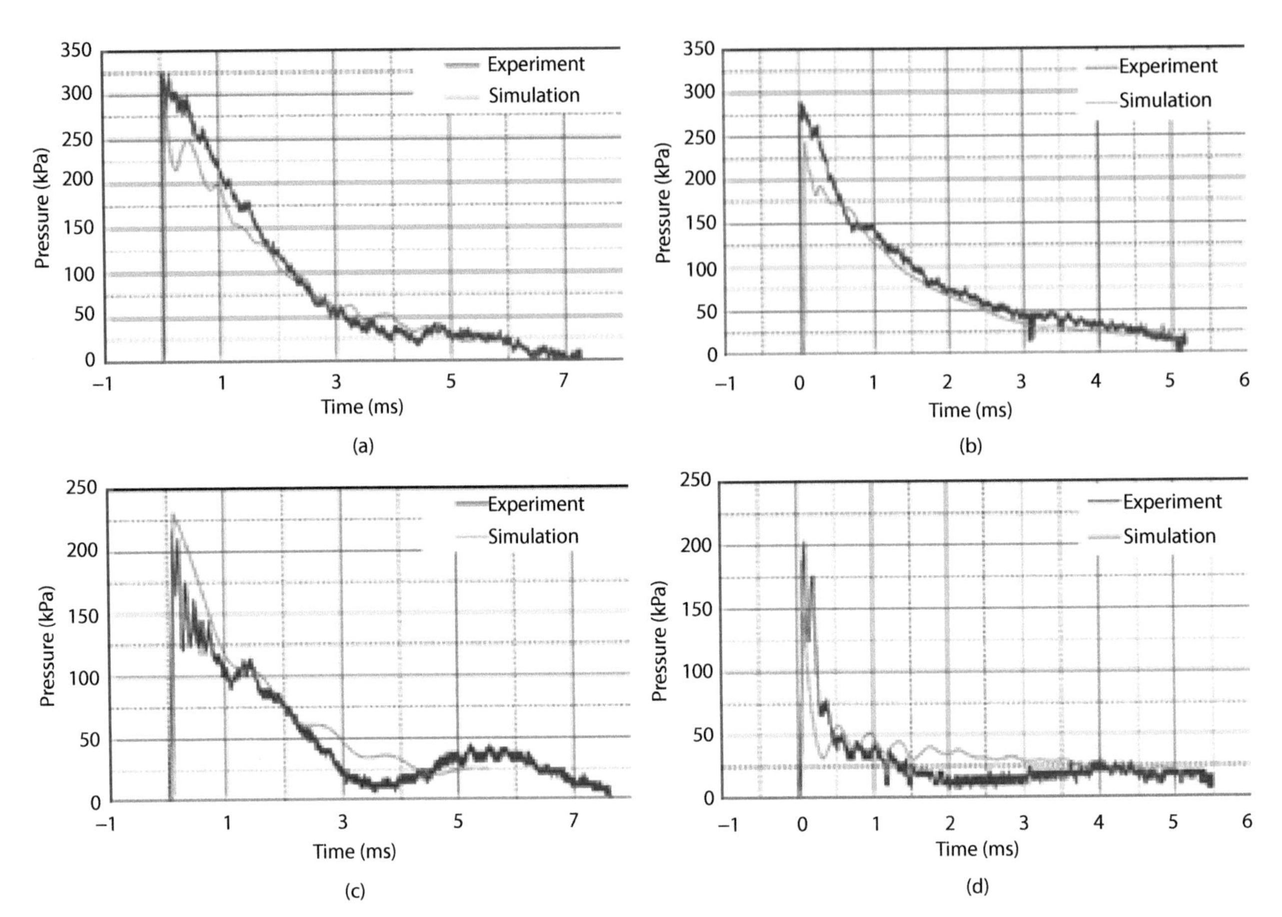

FIGURE 18.41 Comparison between experiments and numerical models both inside and outside the shock tube. (a) Surface pressure measured on the nose. (b) Surface pressure measured on the nose. (c) ICP inside the brain. (d) ICP inside the brain. (From Sundaramurthy, A. et al. *Journal of Neurotrauma*. 29:2352, 2012.)

18.5 MECHANICS OF BLAST WAVE–HUMAN HEAD INTERACTIONS WITH HELMETS

18.5.1 Flow Field on the Surface of the Head with the Helmet

Figure 18.42 shows the pressure-time history on the surface of the head with and without the helmets. In general, rise time and time to peak is increased and rate of pressure decay is decreased with the helmet. Peak pressures are reduced with the padded helmet at all locations compared with the no-helmet case. For the suspension helmet, peak pressure is reduced at the incident blast site (sensor FH); on the contrary, peak pressures are increased on the side away from the incident blast side compared with the no-helmet case. The positive phase impulse either remained equivalent or increased with the suspension and padded helmets as time to peak is increased and rate of pressure decay is decreased with the helmets.

18.5.2 Comparison of Experiments and Numerical Simulations for Helmeted Cases

Numerical simulations are used to understand the mechanics of the flow field around the head with the helmet. Before using the numerical simulations for this purpose, numerical results are compared and validated against helmeteds experiments. Figure 18.43a, b shows a comparison of surface pressures on the surrogate head with suspension and padded helmets, respectively. For the suspension helmet, there is a reasonably good agreement between the experiment and numerical simulation, in terms of peak pressures (maximum difference 26%, minimum difference 0.3%) and nonlinear decay. The simulation is able to capture the majority of the features well, including the arrival of the blast wave at a given location, shock front rise time, and underwash (explained in detail in the next section) beneath the helmet. For the padded helmet, fair agreement is obtained between the experiment and simulation. The huge difference is seen in the values of peak pressure and total impulse. This is because it is very difficult if not impossible to know precise placement of the padded helmet on the surrogate head in the experiments. The blast wave can enter through small gaps, if any, created during the mounting of padded helmet on the surrogate head. In addition, porosity of the foam pads is not modeled in the simulations, which may contribute to the surface pressures.

18.5.3 Underwash Effect of the Helmet

As indicated in previous sections, surface pressures are increased under the suspension helmet on the side away from the incident blast side. This is due to the underwash effect of the helmet. This

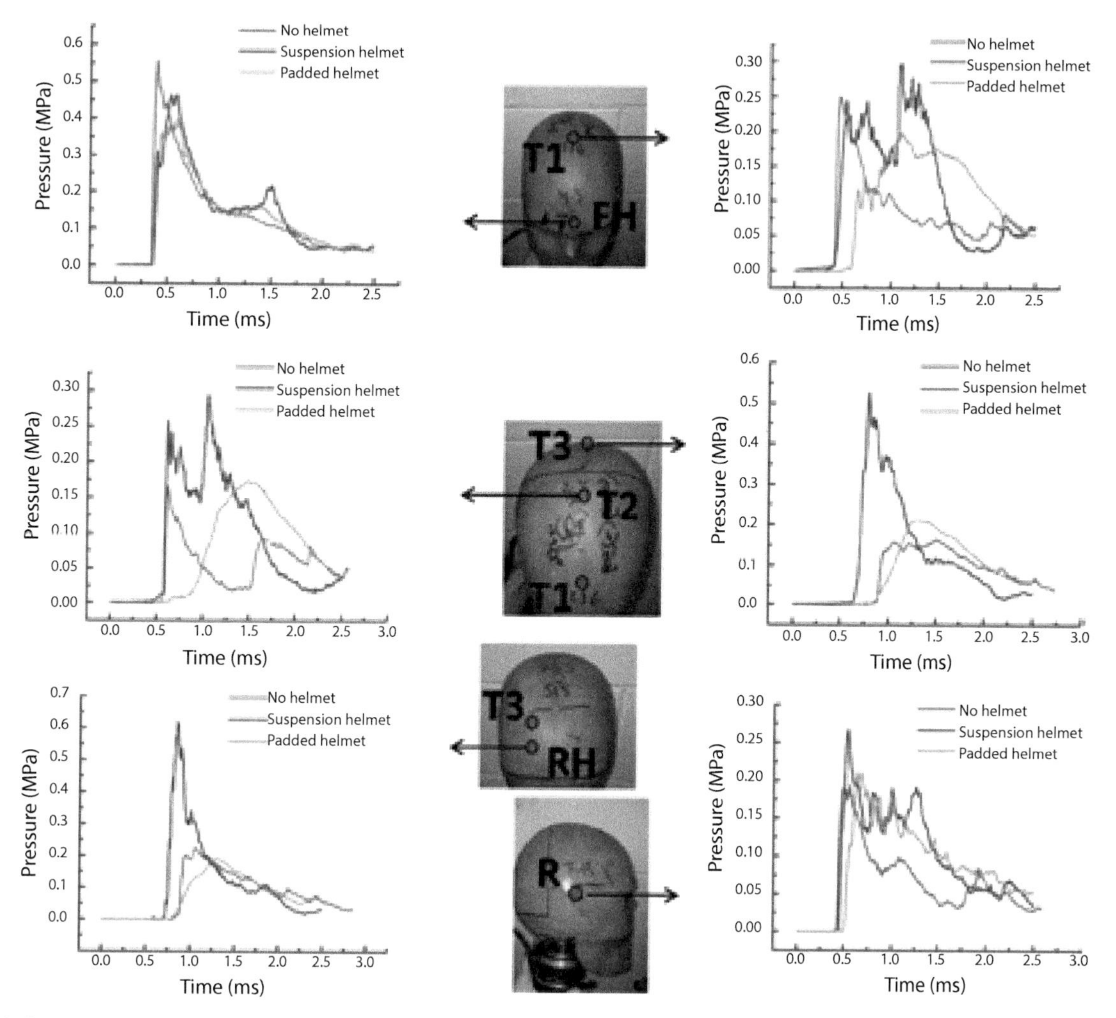

FIGURE 18.42 Pressure-time history on the surface of the head with and without the helmets.

effect is illustrated using numerical simulations (Figure 18.44). The blast front after encountering the head–helmet assembly divides into two fronts: one front travels around the outer perimeter of the helmet whereas the other front penetrates the gap between the head and the helmet (i.e., head–helmet subspace) and travels underneath the helmet toward the back of the head as shown in Figure 18.44a. The blast front traveling outside the helmet reaches the rear of the helmet before the blast front traversing through the gap (Figure 18.44b-i), and eventually when these two blast fronts meet they focus at a region on the back side of the head (Figure 18.44b-ii). This focusing produces higher pressures on the head, away from the incident blast side when the location is shielded by the helmet. After this high pressure is generated, the high pressure air in the head-helmet subspace expands in all directions (Figure 18.44b-iii).

To understand how the underwash influences both the local peak pressure and the impulse, it is postulated that the pressure intensification depends on the shape of the helmet (curvature) and the head–helmet subspace gap size with respect to the oncoming pressure wave and its characteristics (e.g., pressure, velocity, rise/fall time). These aspects are studied in the following section using simplified two-dimensional head models. It should be noted that local peak pressures in the head-helmet subspace and impulse transmitted to the head are analyzed because these quantities determine the effective load on the head.

18.5.3.1 Effect of Curvature, Head–Helmet Gap Size, and Incident Peak Pressure Intensity

To examine the effect of geometry, three different cases are considered. In the first case, the head and the helmet are modeled as cylinders, in the second case the head is cylindrical and the helmet flat, and in the third case both the helmet and the head are flat (Figure 18.45a). In all these cases, there is constant gap of 13 mm between the helmet and the head. Figure 18.45b-i and b-ii shows the pressure and impulse profiles at the back of the head–helmet subspace where the focusing occurs. It is clear from Figure 18.45b that the pressure and impulse are increased when both the shapes are cylindrical in comparison with the other two cases. This trend is the same when the incident overpressure is increased from 0.18 MPa to 0.52 MPa.

Having identified that the cylindrical case offers the most severe loading conditions, this case is used to study the effect of head–helmet gap size and incident peak pressure intensity on the underwash. Figure 18.45c shows the P_{max}/P^* (normalized peak maximum overpressure) in the head–helmet

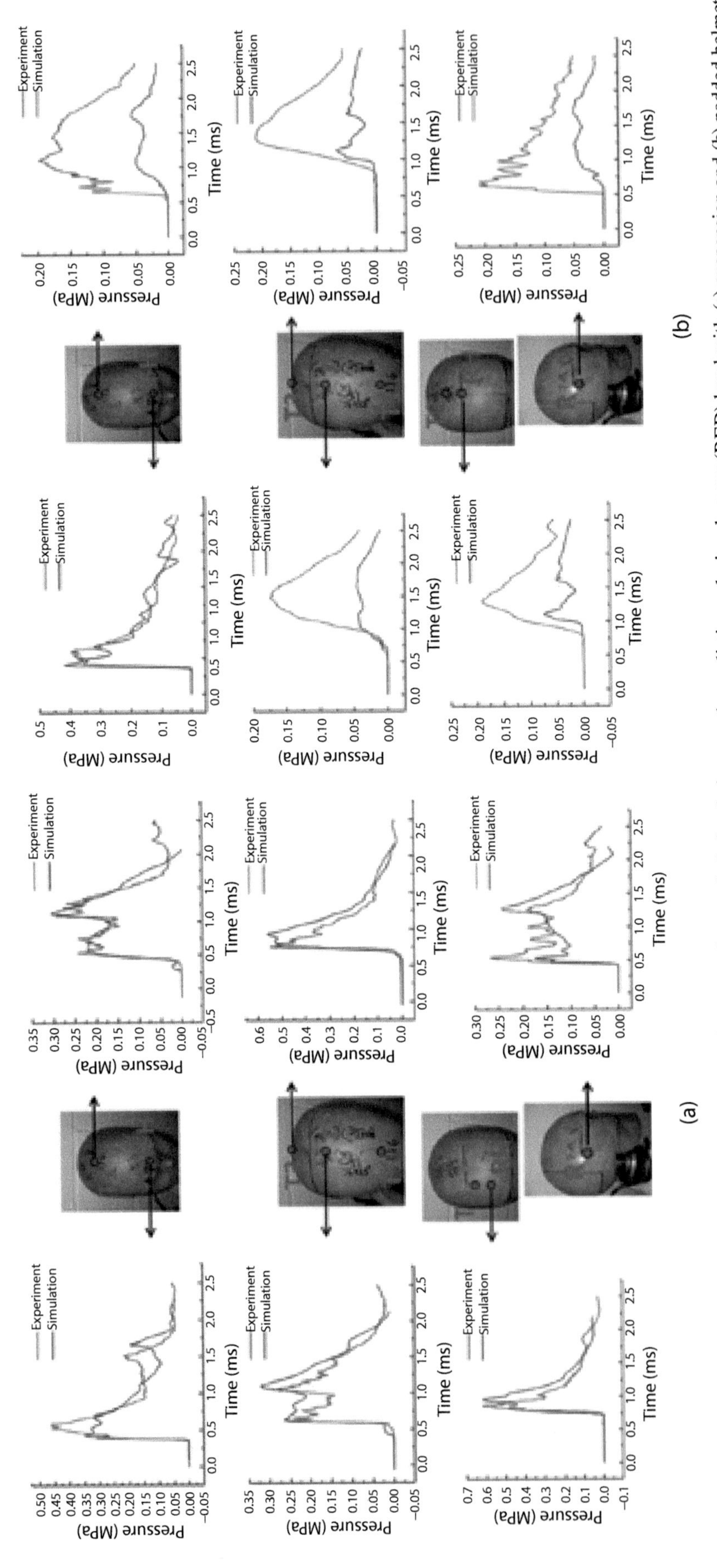

FIGURE 18.43 Comparison of surface pressures from experiments and numerical simulations on the realistic explosive dummy (RED) head with (a) suspension and (b) padded helmet respectively.

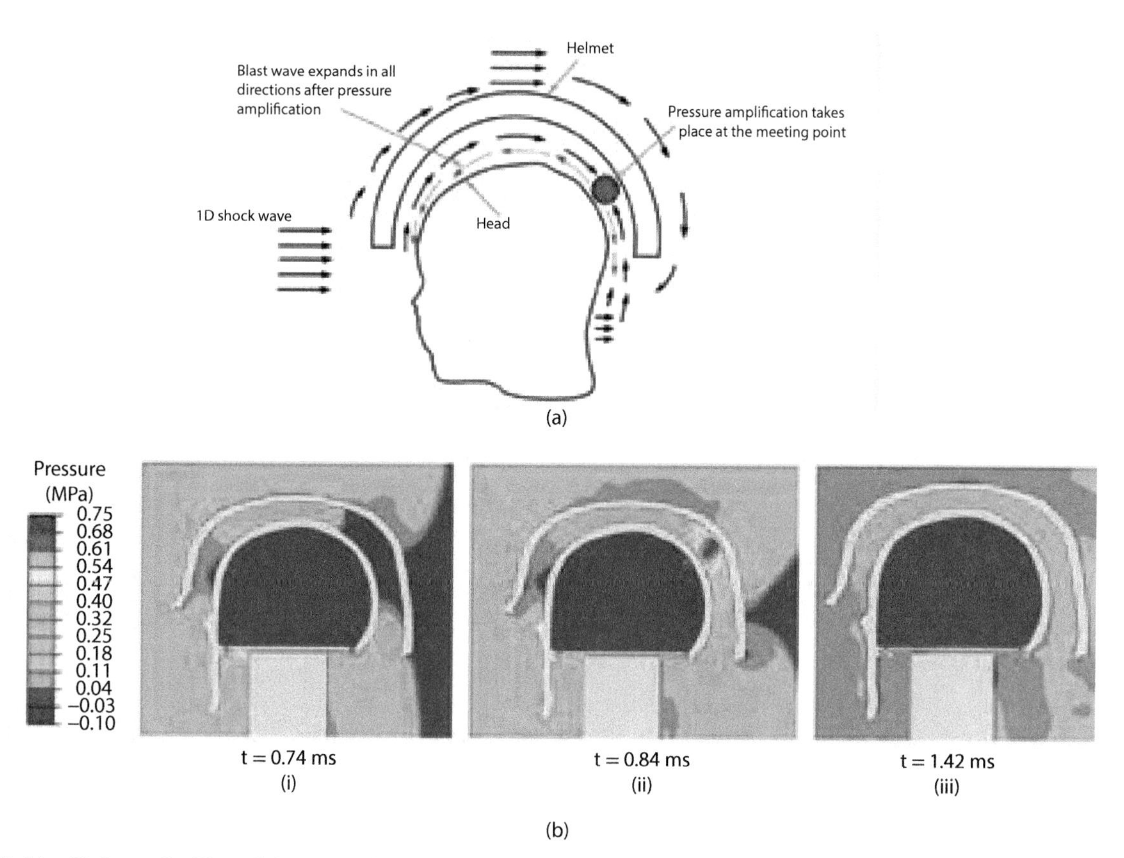

FIGURE 18.44 Underwash effect of the suspension helmet: (a) schematic explaining underwash effect of the helmet; (b) flow field inside and outside of the head–helmet subspace.

subspace as a function of gap size for different incident peak pressure intensities P*. As the gap is reduced, pressure in the gap increases (P α 1/V, V-volume). Thus, the P_{max}/P^* ratio increases as the gap size is reduced until a certain critical gap size. Thereafter, the boundary effects become dominant and P_{max}/P^* ratio decreases because of these boundary effects. The P_{max}/P^* ratio increases as incident peak pressure intensity P* increases. Numerical simulations indicate that for the ranges tested, the angle θ at which P_{max} occurs is between 140 and 155°.

Another quantity of interest is the transmitted impulse, *I*, which depends on the maximum peak pressure, P_{max} and rate of pressure decay (i.e., rate of expansion) once P_{max} is established. The higher P_{max} and lower the rate of pressure decay, the higher is the impulse transmitted. As shown earlier, P_{max} increases as the gap size is reduced until a critical gap size. The rate of pressure decay, however, decreases continuously (no critical gap size) as the gap size is decreased as shown in Figure 18.46a. This is because as the gap size is reduced, there is not enough space for expansion and boundary reflection effects become dominant. Similar observations are reported by Rafaels et al. (2010) from their blast experiments on a helmeted head. From our simulations it was found that, for a given incident peak pressure intensity P*, rate of pressure decay contributes more to impulse transmitted to the head than P_{max}. Hence, for a given incident peak pressure intensity P*, impulse transmitted to the head continuously increases as the gap size is reduced as shown in Figure 18.46b.

18.5.4 Effect of Orientation on Blast Wave–Head Interactions with and without Head Protection

As mentioned earlier, to study the effect of orientation on blast wave–head interactions experiments were conducted on the surrogate head with four different orientations to the blast. These orientations are front, back, side, and 45°. The experiments were conducted with and without helmets and each scenario was repeated three times. Blast wave–head interactions are studied by monitoring surface pressures on the surrogate head and experimental observations are elucidated with the help of validated numerical models. The role of head orientation is studied by understanding the mechanics of the blast wave head interactions for the no helmet, suspension helmet, and padded helmet cases.

18.5.4.1 No-Helmet Case

Figure 18.47a shows the experimentally measured peak pressures for the no-helmet case for each head orientation. Pressure at the incident blast site is amplified ($\Lambda = p_R / p_I$) by 2.40, 2.79, 2.38, and 1.39 times the incident pressure for front, back, side, and 45° orientations, respectively, because of aerodynamic

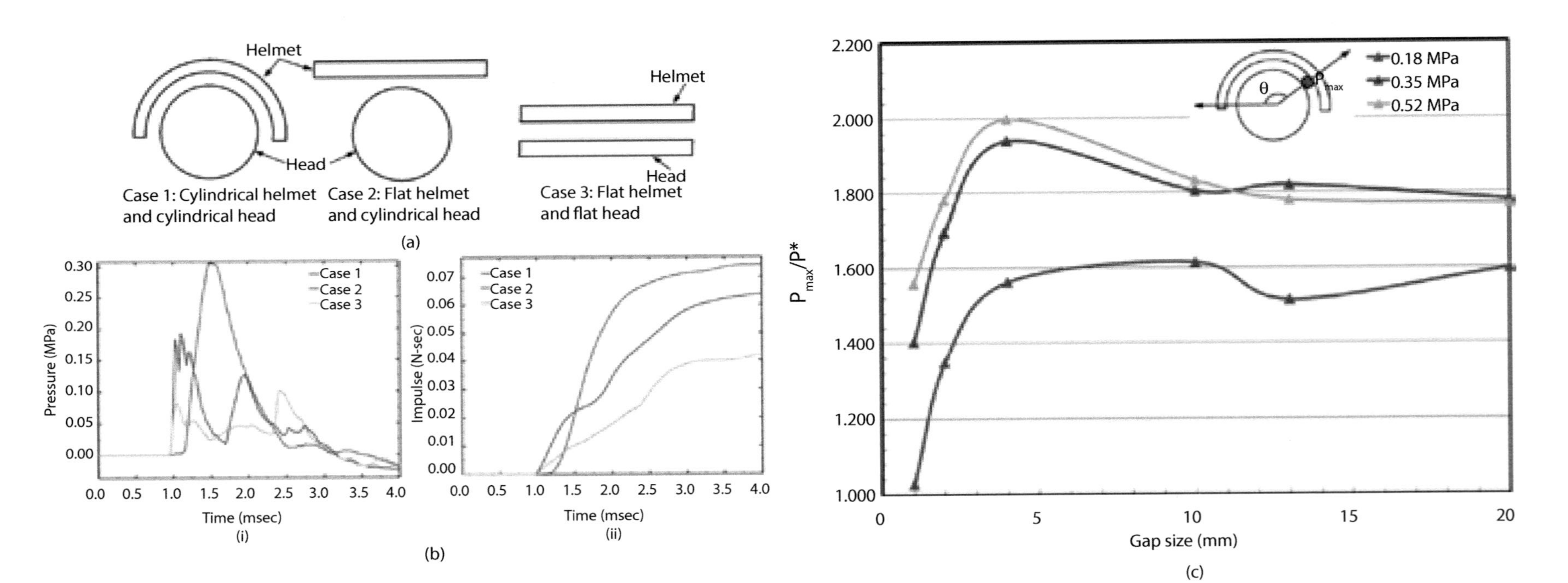

FIGURE 18.45 Effect of curvature of the helmet and the head. (a) Modeling setup for studying curvature effect of the helmet and the head. (b-i) Average pressure in the back region of the head–helmet subspace and (ii) total impulse transmitted to the back region of the head. Incident blast intensity 0.52 Mpa. (c) Normalized maximum peak overpressure in the head–helmet subspace (P_{max}/P^*) as a function of gap size for different incident blast intensities P*. (From Ganpule, S. et al., *Computer Methods in Biomechanics and Biomedical Engineering.* 15:1233–1244, 2011.)

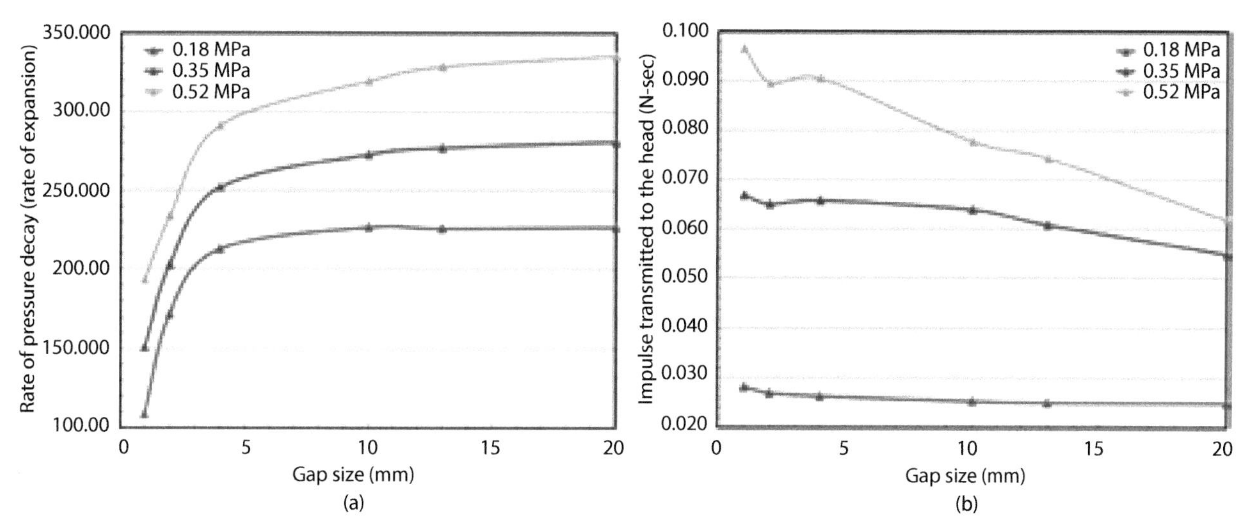

FIGURE 18.46 (a) Rate of pressure decay in head–helmet subspace. (b) Impulse transmitted to the head as a function of gap size for different incident blast intensities P*. (From Ganpule, S. et al., *Computer Methods in Biomechanics and Biomedical Engineering*. 15:1233 -1244, 2011.)

effects. This amplification factor (Λ) is based on the mean values of the incident and reflected pressures for each orientation.

For front orientation, pressure gradually decreases from sensor FH to T1 to T2. Sensors T2 and T3 record equivalent pressures. There is a slight increase in pressure from sensor T3 to sensor RH, which is located on the side opposite to the incident blast side. A similar trend is observed for back orientation but in reverse order (i.e., from sensor RH to sensor FH). For these orientations, sensor R records pressure equivalent (within ±2%) to sensor T2. For the side orientation,a sensor R (i.e., the sensor facing the blast) records the highest pressure; and all sensors in the midsagittal plane record equivalent pressures. Sensor L (i.e. the sensor opposite to the blast side) records marginal pressure. For the 45° orientation, a trend similar to the front orientation is observed, but the flow reunion takes place near sensor T3 (as indicated by increase in pressure) because of tilt.

18.5.4.2 Suspension Helmet Case

Figure 18.47b shows experimentally measured peak pressures for the suspension helmet case for each head orientation. With the suspension helmet, pressure at the incident blast site is amplified by 2.03, 0.94, 1.36, and 2.74 times the incident pressure for front, back, side, and 45° orientations respectively. The pressure field around the head is complex and does not follow any fixed pattern of variation. In general, pressures are increased on the side away from the incident blast side.

18.5.4.3 Padded Helmet Case

Figure 18.47c shows experimentally measured peak pressures for the padded helmet case for each head orientation. With the padded helmet, pressure at the incident blast site is amplified by 1.93, 0.63, 1.26, and 2.68 times the incident pressure for front, back, side, and 45° orientations respectively. Sensors FH and R record higher pressures compared with the other sensors irrespective of the orientation because these sensors are not fully covered by the foam pads. All other sensors record equivalent pressures for a given orientation.

18.5.5 Total Impulse around the Head for Various Head–Helmet Configurations

The total impulse (positive phase, I^+) is obtained by integrating pressure over time ($\equiv P\,dt$). The total impulse shows similar trends as peak surface pressures for all head helmet configurations and for all orientations. The total impulse plots are not shown for brevity.

18.5.6 Results from Numerical Simulations

Numerical simulations are conducted to understand and explain some of the experimental observations. Figure 18.48 shows the flow (pressure) field around the head for various head orientations for the no-helmet case. The flow field around the head is complex. Orientation of the head to the blast wave governs the flow mechanics around the head.

Figure 18.49 shows the pressure field in helmet–head subspace (at the incident blast site) for each orientation for suspension helmet case. The lowest pressures in the helmet–head subspace at the incident blast site are observed for the back orientation.

Figure 18.50 shows pressure field in the helmet–head subspace (away from the incident blast site) for each orientation for the suspension helmet case. From the pressure field it can be seen that pressures are increased under the suspension helmet on the side away from the incident blast side. This also confirms the presence of the underwash effect for back and side orientations. Varying degrees of pressure intensification are observed depending upon the orientation of the head and the helmet to the blast wave. The simulation results are consistent with experimental observations.

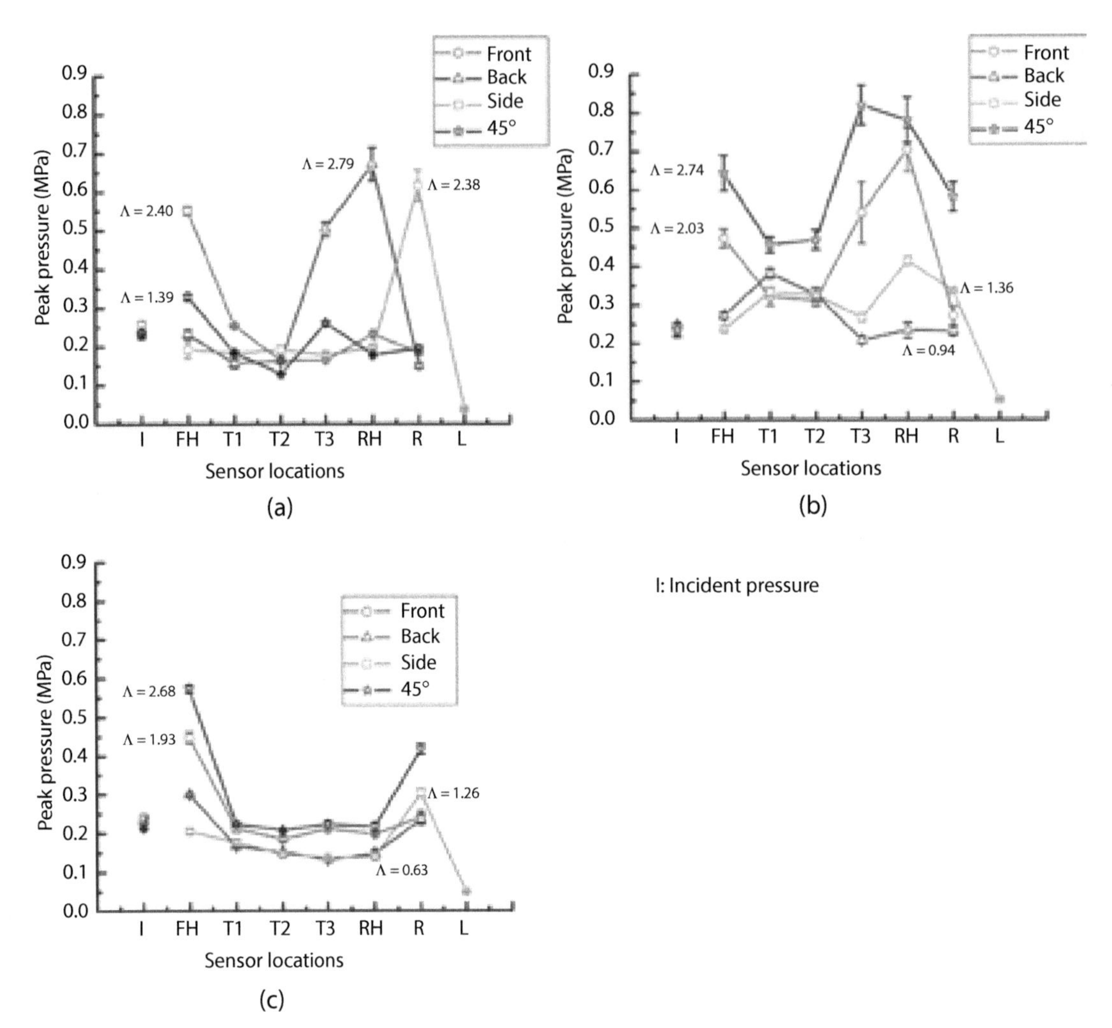

FIGURE 18.47 Peak pressures recorded by the sensors for various orientations: (a) no helmet, (b) suspension helmet, and (c) padded helmet. (Experiment).

18.5.7 Discussion on Orientation-Dependent Blast Wave–Head Interactions

The present results validate both the hypotheses postulated: (1) the external pressure field on the surface of the head depends on whether the wearer has a suspension or padded helmet or no helmet and (2) orientation of the head to the blast wave governs the pressure field experienced by the head, for a given head–helmet configuration. In this section, the results are discussed in the context of these hypotheses.

The blast wave–head interactions are quite complex, as evident from the surface pressure patterns and the values of Λ at the incident blast site for various orientations for various head–helmet configurations (Figure 18.47). For the no-helmet case (Figure 18.47a), statistically similar amplification at the incident blast site is observed for the front and side orientations ($p = 0.81$) and a higher amplification for the back orientation ($p_{max} = 0.019$). The amplification for 45° orientation is lower ($p_{max} = 0.006$) because of flow separation at the face and because a sensor is not present at the exact incident site due to the 45° tilt (Figure 18.48d). The Λ depends on the incident blast intensity, the angle of incidence, the mass and the geometry of the target, and the boundary conditions, and can vary from 2 to 8 (Anderson, 2001; Ganpule et al., 2011). By geometry, we imply geometrical features, such as topology and area of exposure. At the plane of specimen blast wave interaction, the different geometrical features have different effects. For the suspension and padded helmet cases (Figure 18.47b, c), Λ at the incident blast site for each orientation is statistically different. For these cases, Λ is governed by the geometry of the helmet, the head–helmet configuration, and its orientation to the blast.

To better understand surface pressure patterns and hence the flow fields around the head, numerical simulations are used. First, the flow field around the head for the no helmet case is presented. Once the blast wave impinges the head, flow separation occurs, as is evident from the values of the recorded pressure for the sensor next to the incident blast site (Figure 18.47a). For example, pressure reduction of 53.89%, 25%, 67.91%, and 43.94% are observed for front, back, side, and 45° orientations respectively for the sensor next to the incident blast site. Flow separation causes low-pressure zones (e.g., top and sides of the head [Figure 18.48a]); thus,

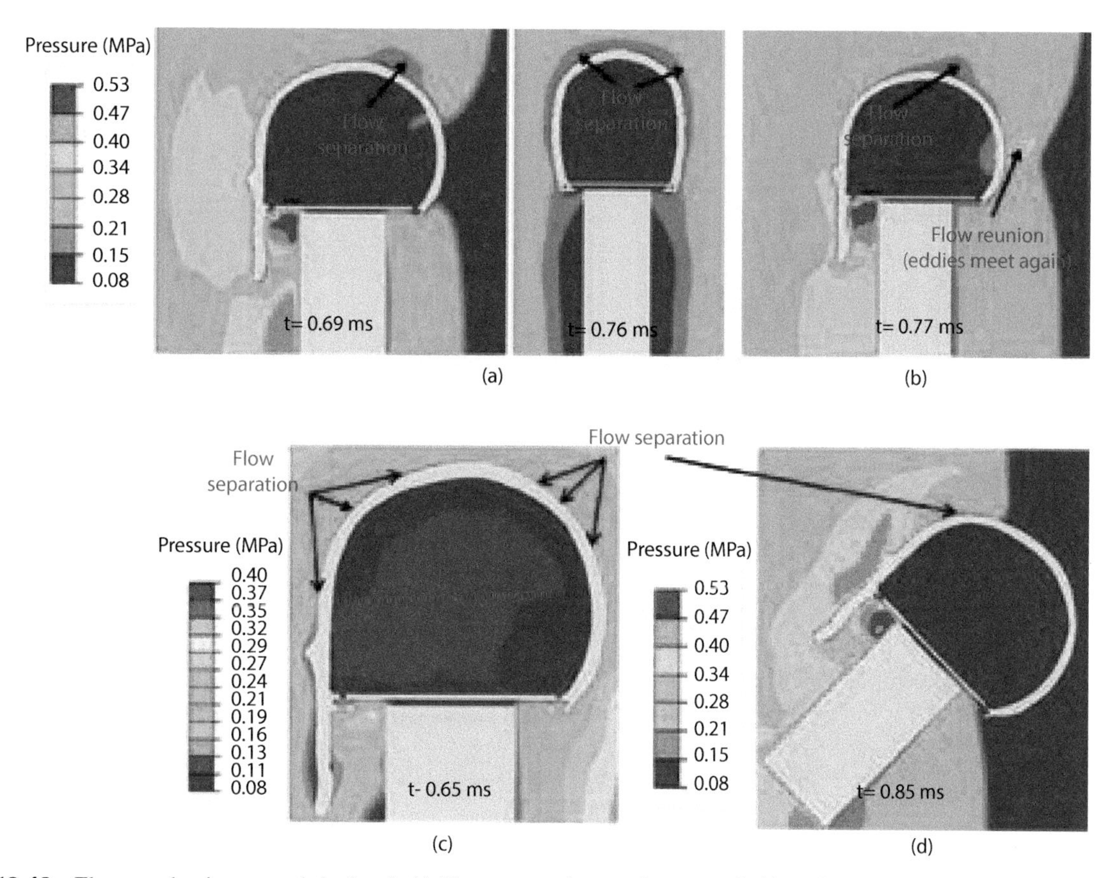

FIGURE 18.48 Flow mechanics around the head. (a) Flow separation on the top and sides of the head for front orientation. (b) Flow reunion on the back of the head for front orientation. (c) Flow separation along the midsagittal plane for side orientation. (d) Flow separation near the face for 45° orientation.

pressures are further reduced as we move away from the incident blast side (Figure 18.48a). The blast wave traversing the head and blast wave traversing the neck reunite on the side opposite to the incident blast side (Figure 18.48b). This flow reunion causes an increase in pressure (e.g., sensor RH for front orientation, sensor FH for back orientation, sensor T3 for 45° orientation) (Figure 18.47a). For the side orientation, flow separation occurs before the blast wave reaches the midsagittal plane (Figure 18.48c), thus all the sensors on the midsagittal plane record similar pressures (Figure 18.48a). This flow separation is further enhanced as the blast wave reaches the side opposite to the incident blast side; hence, the corresponding sensor (sensor L) records very low pressure. Flow separation for the side orientation is attributed to a larger area facing the blast. Numerical simulations clearly show that the surface pressures and the flow field around the head are strongly governed by the geometry of the head. Several other studies (Chavko et al., 2010; Ganpule et al., 2011; Mott DR, 2008; Taylor and Ford, 2009; Zhu et al., 2012) have also shown that the geometry of the head plays an important role in the blast wave head interactions and hence in the biomechanical loading of the brain.

Figure 18.51 shows the peak pressure plots for no-helmet, suspension helmet, and padded helmet cases superimposed on each other for each head orientation. Table 18.7 shows the percentage reduction in peak pressures at the incident blast site for the suspension and padded helmet cases compared with the no-helmet case. By comparing the values of peak pressures (Figure 18.51 and Table 18.7) at the incident blast site, it can be seen that varying degrees of pressure reduction at the incident blast site are observed for suspension and padded cases compared with the no-helmet case. Back and side orientations show statistically significant reduction ($p < 0.05$) in pressure under the helmet, but only marginal reduction is seen under the helmet for front orientation ($p > 0.05$). For front orientation, part of the oncoming blast wave contributes to the pressure as the helmet does not cover the forehead completely (Figure 18.49). A 45° orientation does not show reduction in peak pressure at the incident blast side (i.e., sensor FH). In contrast, for 45° orientation, peak pressures are increased by 95.13% and 73.94% for the suspension and padded helmet cases respectively compared with the no-helmet case. This is because (1) there is flow separation at the face for the no helmet case (Figure 18.48) and (2) in contrast to flow separation for the no-helmet case, the blast wave is directed to the head–helmet subspace for the suspension and padded helmentt cases (Figure 18.49). Of all orientations, the maximum reduction in pressure (65.18% and 77.98%, respectively, for suspension and padded helmet cases) at the incident blast site compared with no-helmet case is observed for the back orientation (Figure 18.51 and Table 18.7). The helmet has a larger area and height on the back than the front

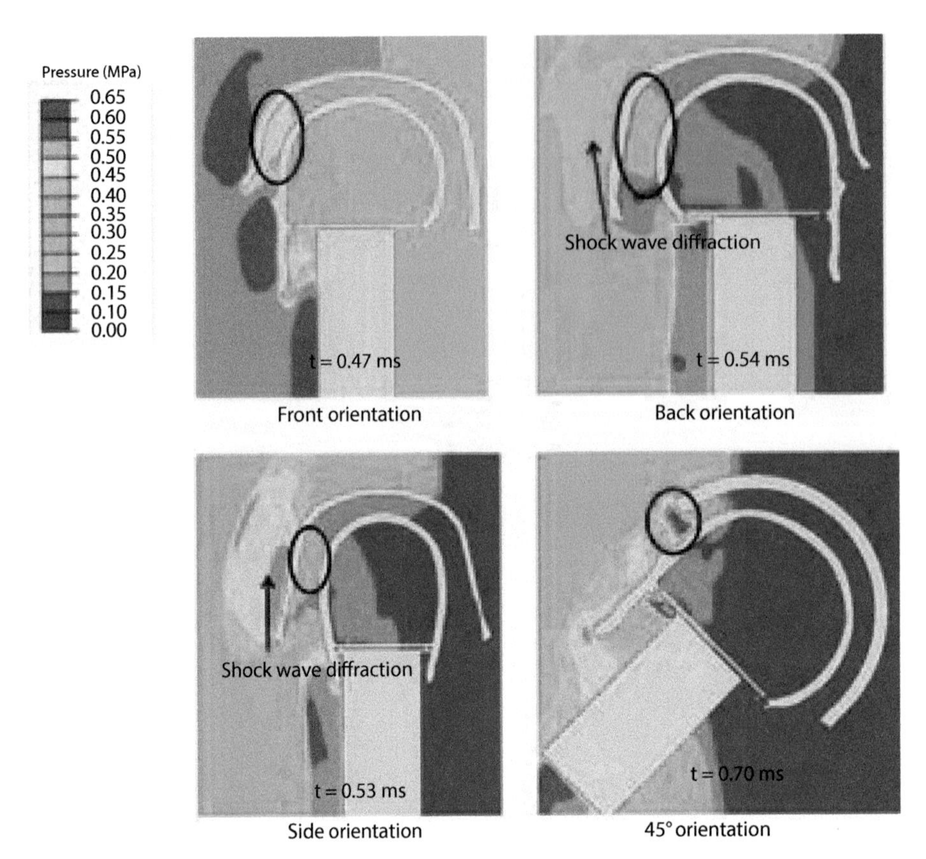

FIGURE 18.49 Pressure contours in the helmet–head subspace at the incident blast site for each orientation. The lowest pressures in the helmet–head subspace at the incident blast site are observed for the back orientation because of the shock wave diffraction around the outer surface of the helmet. The suspension helmet case is used for illustration.

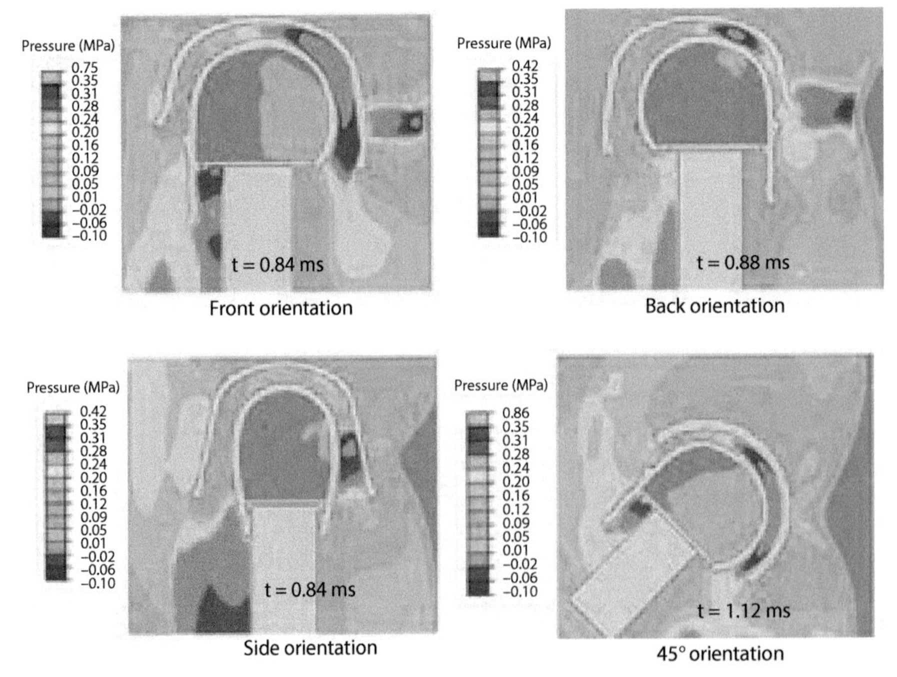

FIGURE 18.50 Pressure intensification on the side away from the incident blast side for the suspension helmet. Varying degrees of intensification are observed for various orientations from the geometric effects that govern the flow field within the head–helmet subspace. Maximum intensification is observed for 45° orientation.

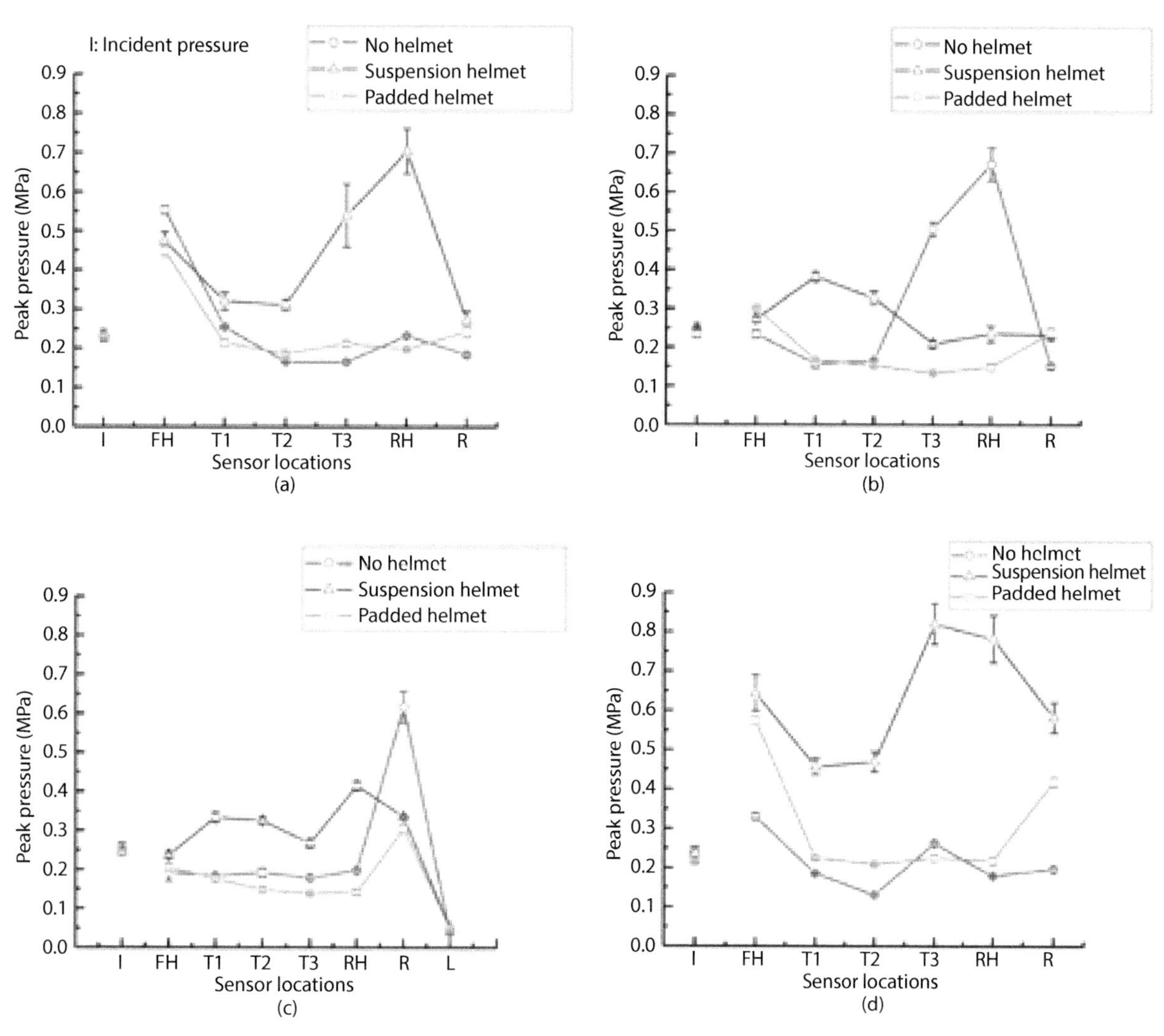

FIGURE 18.51 Peak pressures for no helmet, suspension helmet, and padded helmet cases superimposed on each other for various orientations: (a) front, (b) back, (c) side, and (d) 45°.

TABLE 18.7
Percent Reduction in Peak Pressures at Incident Blast Site for Suspension and Padded Helmet Cases as Compared with a No-Helmet Case

	Peak Pressure Incident Blast Site (MPa)			Percent Reduction in Peak Pressure	
Orientation	**No Helmet**	**Suspension Helmet**	**Padded Helmet**	**Suspension Helmet**	**Padded Helmet**
Front	0.55	0.47	0.45	14.47	18.99
Back	0.67	0.23	0.15	65.18	77.98
Side	0.62	0.34	0.30	45.54	50.73
45°	0.33	0.64	0.57	−95.13[a]	−73.94[a]

Peak pressures are based on mean values.
[a] Negative number suggests increase in pressure compared with a no-helmet case.

or the side. Thus the helmet diffracts and blocks the oncoming blast wave, offering maximum protection as shown in Figure 18.49. Zhang and Makwana (2011) have also found maximum reduction in peak ICP for the back orientation from their numerical simulations.

Pressures are increased under the suspension helmet on the side away from the incident blast side (Figure 18.51). This is due to the underwash effect of the helmet. For the orientations studied, the maximum underwash (i.e., pressure intensification) under the suspension helmet is observed for the

45° orientation followed by the front orientation (see sensor RH, Figure 18.51). This is mainly for two reasons: (1) orientation of the head–helmet configuration to the blast and (2) the geometry of the helmet. For the 45° orientation, the blast wave penetrates the head–helmet subspace (i.e., gap) more effectively from both sides (Figure 18.50). The blast wave is continuously directed in the head–helmet subspace from the face because of tilt. As mentioned earlier, the back of the helmet has a larger area and height than the front. Thus, for the front orientation, the blast wave traversing the neck and blast wave traversing outside the helmet, after reaching the head–helmet subspace, engulfs the head–helmet subspace (due to geometric effects) and causes higher intensification. For other orientations, this engulfment is less intense (Figure 18.50) because of the shorter height of the helmet in the corresponding regions. It should be noted that, for front and 45° scenarios, the maximum surface pressure recorded by the surrogate head with the suspension helmet (P_{max} = 0.76 and 0.87 for front and 45° orientations, respectively) exceeds the maximum surface pressure recorded in the no-helmet case (P_{max} = 0.56 and 0.34 for front and 45° orientations, respectively).

The underwash effect is not seen for the padded helmet case, as evident from Figure 18.51. However, equivalent pressures are seen on the top region of the head (sensors T1–T3) compared with the no-helmet case (Figure 18.51). This indicates that additional pathways/modes of energy transfer exist under the padded helmet. Thus, performance of the foam pads under the blast-loading conditions needs further investigation to identify these pathways/modes.

18.5.8 Role of Helmet in Mechanics of the Blast Wave–Head Interactions: Effect on ICP Response

In published work (Ganpule et al., 2011; Ropper, 2011), the role of the helmet in the mechanics of the blast wave–head interactions has been studied in terms of pressure field experienced on the surface of the surrogate dummy head. From the previous study done on surrogates, it was shown that peak surface pressures were reduced at the incident blast site (i.e., site closest to impact) for both padded and suspension helmet configurations and these reductions were statistically significant. At other locations, surface pressures were marginally reduced for the padded helmet configuration and increased for the suspension helmet configuration. Increases in surface pressures under the suspension helmet were due to focusing of the blast wave under the suspension helmet. Impulse values were either equivalent or were marginally reduced for both padded and suspension helmet configurations. In addition, for the suspension helmet, a statistically significant increase in

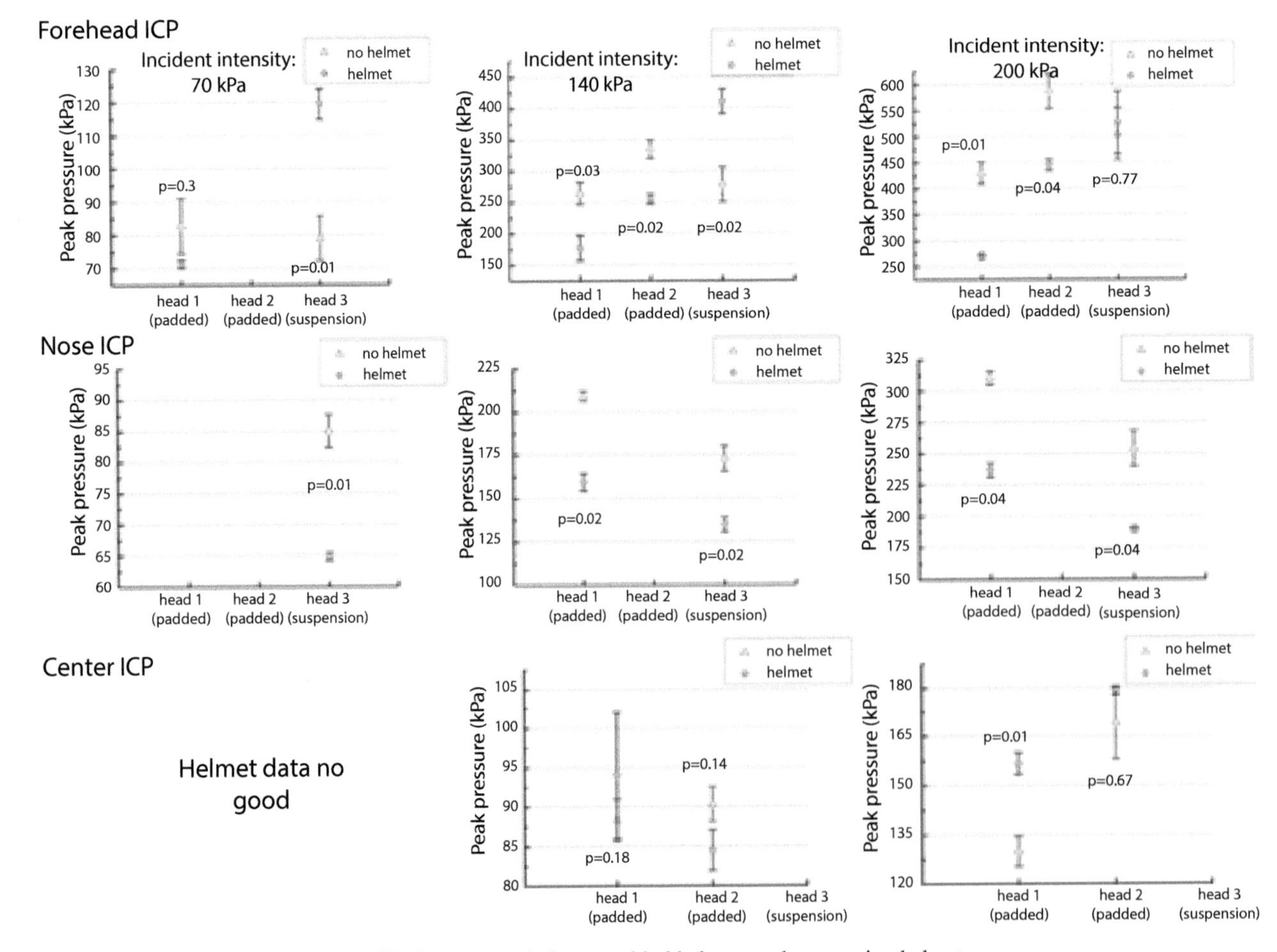

FIGURE 18.52 Comparison of peak ICPs between no helmet, padded helmet, and suspension helmet cases.

impulse was seen for locations that were directly below the focused region. How the external pressure field measured on the surface of the head translates to the intracranial contents is currently not known. Some of the interesting questions include the following: What dose increase in surface pressures under a suspension helmet imply in terms of brain injury? Does blast wave focusing affect injury outcome and injury severity? In this section, an attempt is made to address these questions by studying ICP response with and without helmets.

For these experiments, the cadaver heads used in Section 18.2 were used with the same array of instrumentation used for the bare head tests. Padded helmets were used for heads 1 and 2 and a suspension helmet was used for head 3. Figure 18.52 shows the comparison of peak ICPs between no-helmet and helmeted cases for FH, nose, and center ICP. Peak ICP values are reduced with the padded helmet for FH and nose locations compared with the no-helmet case; these reductions are statistically significant for most of the cases ($p < 0.05$). Center and back (not shown) locations show equivalent peak ICP values with the padded helmet, compared with the no-helmet case. With the suspension helmet, peak ICP value were increased (statistically significant increase) for the FH location at incident blast intensities of 70 kPa and 140 kPa and remained equivalent for an incident blast intensity of 200 kPa. With the suspension helmet, a statistically significant reduction in peak nose ICP is seen at all intensities. Center and back locations show equivalent peak ICP values with the suspension helmet.

Figure 18.53 shows the ICP impulse comparison between the no-helmet and helmeted cases for FH, nose, and center ICP. Equivalent ICP impulse values are seen with the padded helmet at all locations compared with no helmet, but there is no fixed pattern of variation across the heads and/or intensities, making it difficult to draw any concrete conclusions. Increased ICP impulse values are seen with the suspension helmet at all locations compared with no-helmet scenarios. This increase in impulse under suspension helmet is statistically significant for most of the scenarios.

For the padded helmet, peak ICP values at the FH, nose, and temple locations are reduced and these reductions are statistically significant; center and back locations do not show significant reduction in peak ICP values. ICP impulse values are either equivalent or marginally reduced with the padded helmet (Figure 18.54a). This suggests that the face is an important pathway of load transfer to the brain. The area of the face is ~25%–30% that of the area of the forehead. Thus,

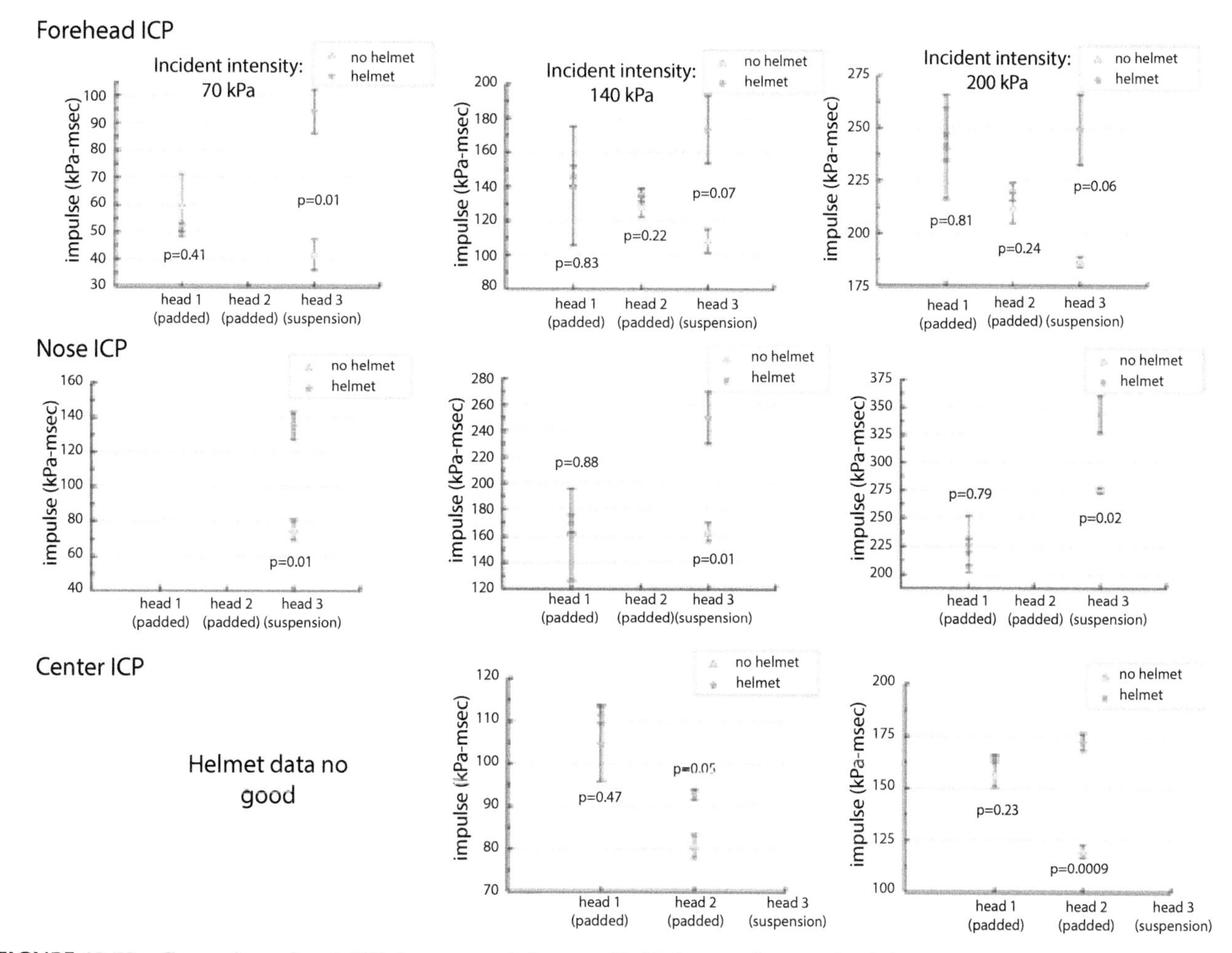

FIGURE 18.53 Comparison of peak ICPs between no helmet, padded helmet, and suspension helmet cases.

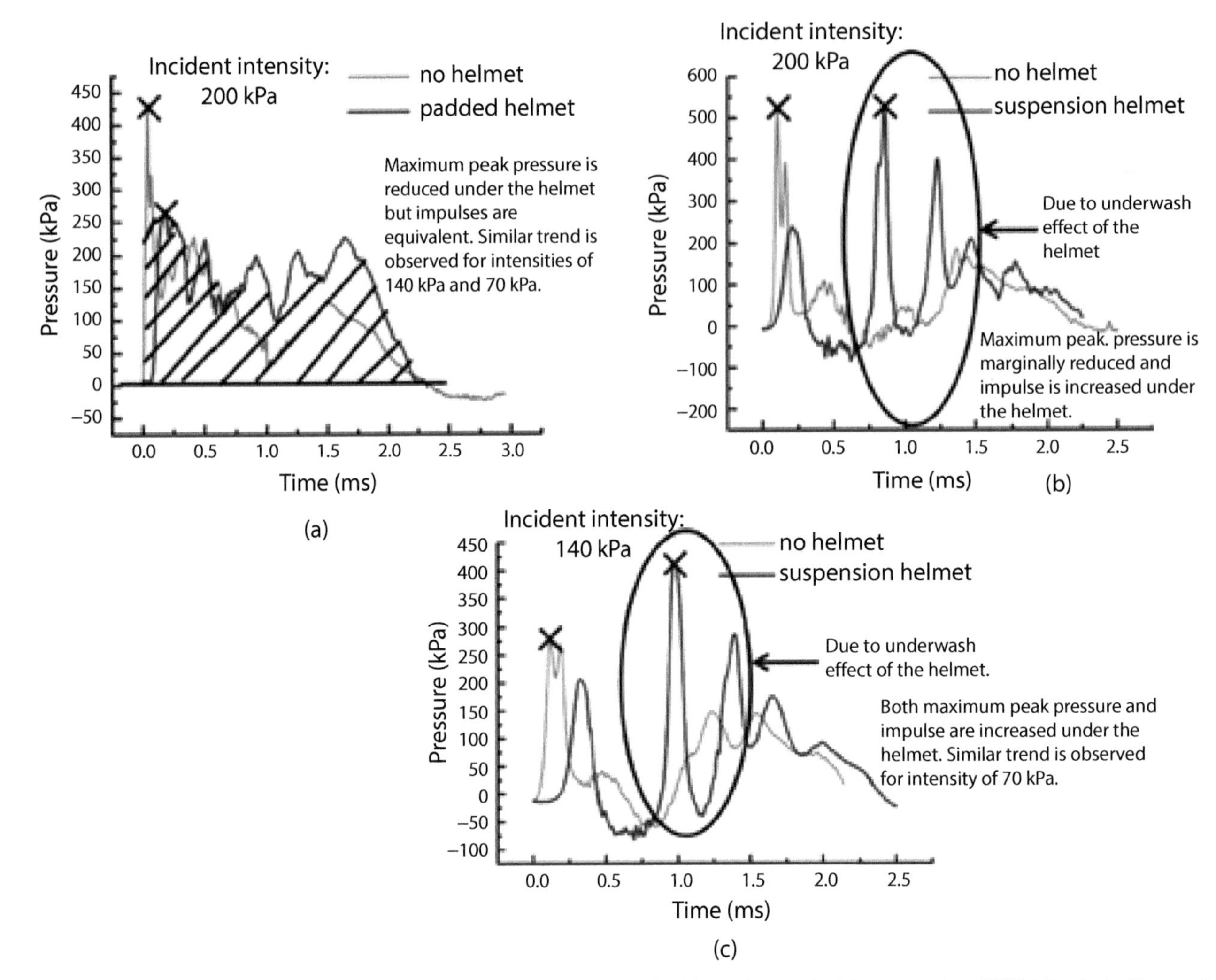

FIGURE 18.54 Pressure time histories with and without the helmet. (a) Padded helmet at incident intensity of 200 kPa. A similar trend is seen for intensities of 70 kPa and 140 kPa. (b) Suspension helmet at incident intensity of 200 kPa. (c) Suspension helmet at incident intensity of 140 kPa. A similar trend is seen at incident intensity of 70 kPa.

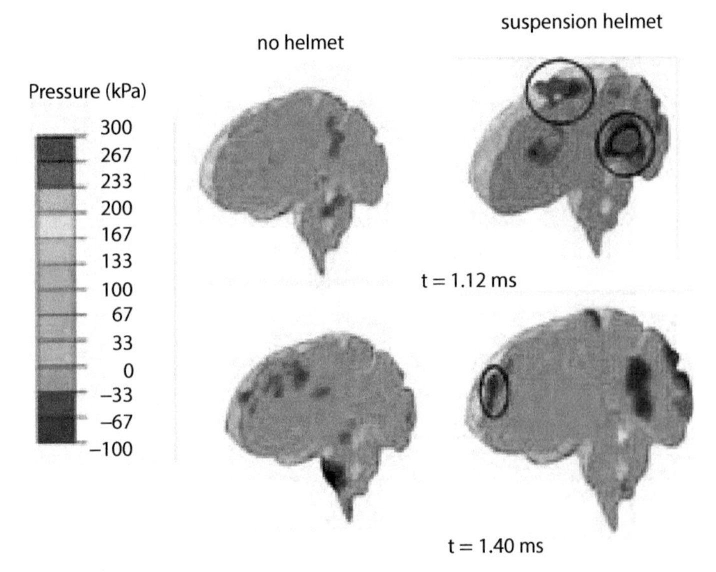

FIGURE 18.55 Focusing effect seen under the suspension helmet translates to the intracranial contents. Intracranial regions with increased pressure with respect to the no-helmet counterpart are highlighted with an ellipse.

a significant amount of impulse/energy is transferred through the face; this is further confirmed using numerical simulations. The face has been identified as an important pathway of load transfer by Nyein et al. (2010) and a face shield has been proposed for blast mitigation; peak pressure reductions up to 80% were proposed with the face shield. However, some of the major limitations of Nyein et al.'s study were total simulation time and military/field relevance of their input data. The authors simulated the entire blast event for 0.76 msec only; this time period is not sufficient to truly evaluate the role of a face shield in blast mitigation as a face shield delays the blast wave transmission into the intracranial cavity. In addition, incident blast pressures and durations used as an input for numerical simulations were not representative of a realistic blast. In addition to the role the face plays in blast wave transmission, marginal impulse reduction seen with the padded helmet from our experiments also highlight the role that structure/geometry of the head plays in governing the ICP response.

With the suspension helmet, peak ICP values are reduced for nose ICP, and these reductions are statistically significant. On the contrary, peak values are increased for FH ICP at incident blast intensities of 70 and 140 kPa and remained equivalent for blast intensity of 200 kPa because of the focusing/underwash effect as shown in Figure 18.54b, c. In addition, temple, center, and back locations show equivalent peak ICP values with the suspension helmet. Statistically significant increases in impulse values are seen with the suspension helmet at all locations compared with no-helmet scenarios. Thus it is clear that the focusing effect seen under the suspension helmet translates to the intracranial contents and adversely affects the ICP response (Figure 18.55). This does imply that for a suspension helmet configuration, wearing a helmet can be worse than not wearing the helmet for primary blasts in which blast waves can potentially focus under the helmet. However, this does not preclude the use of helmets that provide critical protection against blunt and penetrating conditions.

18.6 CONCLUSION

In this chapter, a comprehensive review of the current state of the art in BINT research is presented. From the field data, it is very clear that BINT, especially mild bTBIs, is a serious concern to defense personnel during and beyond their service. Although the mechanics of BINT is different from that of blunt and ballistic injuries, pathophysiological and neurobiological aspects of mild TBI are possibly related. BINT in general and mild bTBI in particular have been the subject of investigations by many groups around the world using a variety of models and we are still far away in having a clear understanding of the problem and hence their solution.

In this chapter, we have addressed a few of those challenges including (1) issues in replication of primary blasts in laboratory and methods used for accurate simulation of the primary blasts using compressed gas driven shock tubes, (2) loading mechanisms on the head that have to be considered during a blast event, (3) injury mechanisms associated with primary blasts, (4) procedure for developing anatomically accurate human and animal models, and finally (5) the role of protective equipment in mitigation of the primary blast. Some of the key points of this chapter are

- Before performing experiments with a shock tube, it is essential to tune the location of the test section. If located too close to the breech, it will not produce a Friedlander wave and instead will have a flattop wave; if located too close to the exit, it will interact with vorticities or jet winds that are also not representative of a field blast. In any case, it is critical to measure the blast overpressure and impulse impacting the specimen using surrogates. Further, it is necessary to relate these parameters to field relevant values because otherwise we will be exploring injuries in the wrong loading domains.
- Through PMHS testing, we were able to determine that two important loading mechanisms of BINT are (1) direct transmission of pressure, which was evident from the increase in the ICP and (2) indirect transmission through skull flexure, which was evident from the strain data recorded in the skull.
- Through an animal model, we were able to determine the mortality as well as gross pathology and BBB damage from blast exposure. Furthermore, we also determined that within 24 hours of blast exposure, there was no significant neurodegeneration.
- Methods for developing anatomically accurate human and rat models along with validations were shown.
- Finally, the role of helmets in mitigation of blasts is discussed. It was shown through numerical models and experimental testing that use of a suspension helmet, though it works well against ballistic impact, actually aggravates the blast injury.

ACKNOWLEDGEMENTS

The majority of the work reported by the authors was carried out when the senior author was a faculty member at University of Nebraska-Lincoln and was the director of the Army Center on Trauma Mechanics. A number of faculty and students were involved during that period. Among them, we wish to acknowledge the work of and discussions with Ruqiang Feng, Shailesh Ganpule, Maciej Skotak, Fang Wang, Aaron Holmberg, Nick Kleinschmit, Veera Selvan, Aaron Alai, and James Haorah. Financial support was provided by the U.S. Army (project monitors: Bruce Lamattina and Larry Russell).

REFERENCES

AEP-55. 2006. Procedures for evaluating the protection level of logistic and light armoured vehicle: For mine threat, volume 2. NATO/Partnership for Peace.

Alexander, M.P. 1995. Mild traumatic brain injury: Pathophysiology, natural history, and clinical management. *Neurology*. 45:1253–60.

Anderson, J. 2001. *Fundamentals of Aerodynamics*. McGraw-Hill, New York.

Arakeri, J.H., D. Das, A. Krothapalli, and L. Lourenco. 2004. Vortex ring formation at the open end of a shock tube: A particle image velocimetry study. *Physics of Fluids*. 16:1008–1019.

Baker, T.J. 2005. Mesh generation: Art or science? *Progress in Aerospace Sciences*. 41:29–63.

Bass, C.R., M.B. Panzer, K.A. Rafaels, G. Wood, J. Shridharani, and B. Capehart. 2012. Brain injuries from blast. *Annals of Biomedical Engineering*. 40:1–18.

Benzinger, T.L., D. Brody, S. Cardin, K.C. Curley, M.A. Mintun, S.K. Mun, K.H. Wong, and J.R. Wrathall. 2009. Blast-related brain injury: Imaging for clinical and research applications: Report of the 2008 St. Louis workshop. *Journal of Neurotrauma*. 26:2127–2144.

Bolander, R., B. Mathie, C. Bir, D. Ritzel, and P. VandeVord. 2011. Skull flexure as a contributing factor in the mechanism of injury in the rat when exposed to a shock wave. *Annals of Biomedical Engineering*. 39:1–10.

Bourdin, X., P. Beillas, P. Petit, and X. Troseille. 2007. Comparison of tetrahedral and hexahedral meshes for human finite element modelling: An application to kidney impact. Paper No.

07-0424-W. In: 20th Enhanced Safety of Vehicles Conference: Innovations for safety: Opportunities and challenges. Lyon, France.

Brown, R., G. Cooper, and R.L. Maynard. 1993. The ultrastructure of rat lung following acute primary blast injury. *International Journal of Experimental Pathology*. 74:151.

Brun, R. 2009. Shock Tubes and Shock Tunnels: Design and Experiments. *In:* NATO-OTAN. Vol. RTO-EN-AVT-162.

Cengel, Y.A., M.A. Boles, and M. Kanoğlu. 2011. *Thermodynamics: An Engineering Approach*. McGraw-Hill, New York.

Cernak, I., A.C. Merkle, V.E. Koliatsos, J.M. Bilik, Q.T. Luong, T.M. Mahota et al. 2011. The pathobiology of blast injuries and blast-induced neurotrauma as identified using a new experimental model of injury in mice. *Neurobiology of Disease*. 41:538–551.

Cernak, I., J. Savic, and J. Lazarov. 1997. Relations among plasma prolactin, testosterone, and injury severity in war casualties. *World Journal of Surgery*. 21:240–246.

Cernak, I., Z.G. Wang, J.X. Jiang, X.W. Bian, and J. Savic. 2001. Ultrastructural and functional characteristics of blast injury-induced neurotrauma. *Journal of Trauma-Injury Infection and Critical Care*. 50:695–706.

Chafi, M., G. Karami, and M. Ziejewski. 2010. Biomechanical assessment of brain dynamic responses due to blast pressure waves. *Annals of Biomedical Engineering*. 38:490–504.

Chandra, N., S. Ganpule, N. Kleinschmit, R. Feng, A. Holmberg, A. Sundaramurthy, V. Selvan, and A. Alai. 2012. Evolution of blast wave profiles in simulated air blasts: Experiment and computational modeling. *Shock Waves*. 22:403–415.

Chavko, M., T. Watanabe, S. Adeeb, J. Lankasky, S.T. Ahlers, and R.M. McCarron. 2010. Relationship between orientation to a blast and pressure wave propagation inside the rat brain. *Journal of Neuroscience Methods*. 195:61–66.

Chen, Y., and M. Ostoja-Starzewski. 2010. MRI-based finite element modeling of head trauma: Spherically focusing shear waves. *Acta Mechanica*. 213:155–167.

Cifuentes, A.O., and A. Kalbag. 1992. A performance study of tetrahedral and hexahedral elements in 3-D finite element structural analysis. *Finite Elements in Analysis and Design*. 12:313–318.

Clark, M.E., M.J. Bair, C. Buckenmaier, R.J. Gironda, and R.L. Walker. 2007. Pain and combat injuries in soldiers returning from Operations Enduring Freedom and Iraqi Freedom: Implications for research and practice. *Journal of Rehabilitation Research and Development*. 44:179.

Clemedson, C., and C. Criborn. 1955. A detonation chamber for physiological blast research. *The Journal of Aviation Medicine*. 26:373.

Coulter, G.A., G. Bulmash, and C.N. Kingery. 1983. Experimental and Computational Modeling of Rarefaction Wave Eliminators Suitable for the BRL 2.44 m Shock Tube. DTIC Document.

Courtney, A.C., L.P. Andrusiv, and M.W. Courtney. 2012. Oxy-acetylene driven laboratory scale shock tubes for studying blast wave effects. *Review of Scientific Instruments*. 83:045111–045111-7.

Courtney, A.C., and M.W. Courtney. 2009. A thoracic mechanism of mild traumatic brain injury due to blast pressure waves. *Medical Hypotheses*. 72:76–83.

Defense and Veterans Brain Injury Center. 2007. Operation Iraqi Freedom (OIF)/Operation Enduring Freedom (OEF) Fact Sheet.

DePalma, R.G., D.G. Burris, H.R. Champion, and M.J. Hodgson. 2005. Blast Injuries. *New England Journal of Medicine*. 352:1335–1342.

Department of defense (DoD). 2012. DoD worldwide numbers for traumatic brain injury. http://www.health.mil/Research/TBI_Numbers.aspx. Accessed January 2012.

Department of Homeland Security. 2009. Bomb threat stand-off chart, from https://www.llis.dhs.gov/sites/default/files/DHS-BombThreatChart-6-5-09.pdf.

Dewey, J.M. 1971. The properties of a blast wave obtained from an analysis of the particle trajectories. *Proceedings of the Royal Society of London. Series A, Mathematical and Physical Sciences*. 324:275–299.

Ganpule, S., A. Alai, E. Plougonven, and N. Chandra. 2012. Mechanics of blast loading on the head models in the study of traumatic brain injury using experimental and computational approaches. *Biomechanics and Modeling in Mechanobiology*. 12:511–531.

Ganpule, S., L. Gu, A. Alai, and N. Chandra. 2011. Role of helmet in the mechanics of shock wave propagation under blast loading conditions. *Computer Methods in Biomechanics and Biomedical Engineering*. 15:1233–1244.

Ganpule, S.G. 2013. Mechanics of blast loading on post-mortem human and surrogate heads in the study of Traumatic Brain Injury (TBI) using experimental and computational approaches. *In:* Mechanical engineering. University of Nebraska-Lincoln, Lincoln.

Gould, K.E., and K. Tempo. 1981. High-Explosive Field Tests, Explosion Phenomena and Environmental Impacts. DTIC Document.

Grujicic, M., W. Bell, B. Pandurangan, and P. Glomski. 2011. Fluid/structure interaction computational investigation of blast-wave mitigation efficacy of the advanced combat helmet. *Journal of Materials Engineering and Performance*. 20:877–893.

Guy, R.J., E. Kirkman, P.E. Watkins, and G.J. Cooper. 1998. Physiologic responses to primary blast. *The Journal of Trauma and Acute Care Surgery*. 45:983–987.

Haselbacher, A., S. Balachandar, and S.W. Kieffer. 2007. Open-ended shock tube flows: Influence of pressure ratio and diaphragm position. *AIAA Journal*. 45:1917–1929.

Heitger, M.H., R.D. Jones, A.D. Macleod, D.L. Snell, C.M. Frampton, and T.J. Anderson. 2009. Impaired eye movements in post-concussion syndrome indicate suboptimal brain function beyond the influence of depression, malingering or intellectual ability. *Brain*. 132:2850–2870.

Honma, H., M. Ishihara, T. Yoshimura, K. Maeno, and T. Morioka. 2003. Interferometric CT measurement of three-dimensional flow phenomena on shock waves and vortices discharged from open ends. *Shock Waves*. 13:179–190.

Irwin, R.J., M.R. Lerner, J.F. Bealer, P.C. Mantor, D.J. Brackett, and D.W. Tuggle. 1999. Shock after blast wave injury is caused by a vagally mediated reflex. *The Journal of Trauma and Acute Care Surgery*. 47:105–110.

Jiang, Z., O. Onodera, and K. Takayama. 1999. Evolution of shock waves and the primary vortex loop discharged from a square cross-sectional tube. *Shock Waves*. 9:1–10.

Jiang, Z., C. Wang, Y. Miura, and K. Takayama. 2003. Three-dimensional propagation of the transmitted shock wave in a square cross-sectional chamber. *Shock Waves*. 13:103–111.

Jones, E., N.T. Fear, and S. Wessely. 2007. Shell shock and mild traumatic brain injury: A historical review. *American Journal of Psychiatry*. 164:1641–1645.

Kashimura, H., T. Yasunobu, H. Nakayama, T. Setoguchi, and K. Matsuo. 2000. Discharge of a shock wave from an open end of a tube. *Journal of Thermal Science*. 9:30–36.

Kennedy, J.E., M.S. Jaffee, G.A. Leskin, J.W. Stokes, F.O. Leal, and P.J. Fitzpatrick. 2007. Posttraumatic stress disorder and posttraumatic stress disorder-like symptoms and mild traumatic brain injury. *J Rehabil Res Dev*. 44:895–920.

Kinney, G.F., and K.J. Graham. 1985. *Explosive Shocks in Air.* Springer-Verlag, New York.

Kleinschmit, N.N. 2011. A shock tube technique for blast wave simulation and studies of flow structure interactions in shock tube blast experiments, from http://digitalcommons.unl.edu/cgi/viewcontent.cgi?article=1022&context=engmechdiss.

Knöferl, M.W., U.C. Liener, D.H. Seitz, M. Perl, U.B. Brückner, L. Kinzl, and F. Gebhard. 2003. Cardiopulmonary, histological, and inflammatory alterations after lung contusion in novel mouse model of blunt chest trauma. *Shock*. 19:519–525.

Kobeissy, F.H., S. Mondello, N. Tumer, H.Z. Toklu, M.A. Whidden, N. Kirichenko et al. 2013. Assessing neuro-systemic & behavioral components in the pathophysiology of blast-related brain injury. *Frontiers in Neurology*. 4:186.

Kraus, M.F., T. Susmaras, B.P. Caughlin, C.J. Walker, J.A. Sweeney, and D.M. Little. 2007. White matter integrity and cognition in chronic traumatic brain injury: A diffusion tensor imaging study. *Brain*. 130:2508–2519.

Langlois, J.A., W. Rutland-Brown, and K.E. Thomas. 2004. Traumatic brain injury in the United States: Emergency department visits, hospitalizations, and deaths. Department of Health and Human Services, Centers for Disease Control and Prevention, Division of Acute Care, Rehabilitation Research and Disability Prevention, National Center for Injury Prevention and Control.

Leonardi, A.D.C., C.A. Bir, D.V. Ritzel, and P.J. VandeVord. 2011. Intracranial pressure increases during exposure to a shock wave. *Journal of Neurotrauma*. 28:85–94.

Ling, G., F. Bandak, R. Armonda, G. Grant, and J. Ecklund. 2009. Explosive blast neurotrauma. *Journal of Neurotrauma*. 26:815–825.

Lockhart, P.A. 2010. Primary blast injury of the head: Numerical prediction and evaluation of protection. *In:* Mechanical Engineering. University of Waterloo, Waterloo, Ontario, Canada, p. 136.

Long, J.B., T.L. Bentley, K.A. Wessner, C. Cerone, S. Sweeney, and R.A. Bauman. 2009. Blast overpressure in rats: Recreating a battlefield injury in the laboratory. *Journal of Neurotrauma*. 26:827–840.

McElhaney, J., J.W. Melvin, V.L. Roberts, and H.D. Portnoy. 1973. Dynamic characteristics of the tissues of the head. In: Kenedi, R.M. (ed.). *Perspectives in Biomedical Engineering*. Baltimore: University Park Press; pp. 215–222.

Moore, D.F., A. Jerusalem, M. Nyein, L. Noels, M.S. Jaffee, and R.A. Radovitzky. 2009. Computational biology - modeling of primary blast effects on the central nervous system. *Neuroimage*. 47:T10–T20.

Moss, W.C., M.J. King, and E.G. Blackman. 2009. Skull flexure from blast waves: A mechanism for brain injury with implications for helmet design. *Physical Review Letters*. 103:108702.

Mott, D., D. Schwer, T. Young, J. Levine, J.-P. Dionne, A. Makris, and G. Hubler. 2008. Blast-induced pressure fields beneath a military helmet for non-lethal threats. *Bulletin of the American Physical Society*. 53.

Mott DR, S.D., Young TR, Levine J, Dionne JP, Makris A, Hubler G. 2008. Blast-induced pressure fields beneath a military helmet. 20th International Symposium on Military Aspects of Blast and Shock. September 1–5, 2008. Oslo, Norway.

N.Kleinschmit. 2011. A shock tube technique for blast wave simulation and studies of flow structure interactions in shock tube blast experiments. Master's thesis. Department of Engineering Mechanics, University of Nebraska-Lincoln, Lincoln.

Nyein, M.K., A.M. Jason, L. Yu, C.M. Pita, J.D. Joannopoulos, D.F. Moore, and R.A. Radovitzky. 2010. In silico investigation of intracranial blast mitigation with relevance to military traumatic brain injury. *Proceedings of the National Academy of Sciences*. 107:20703–20708.

Okie, S. 2005. Traumatic brain injury in the war zone. *New England Journal of Medicine*. 352:2043–2047.

Onodera, O., Z.L. Jiang, and K. Takayama. 1998. Holographic interferometric observation of shock waves discharged from an open-end of a square cross-sectional shock tube. *JSME International Journal Series B-Fluids and Thermal Engineering*. 41:408–415.

Panzer, M.B., B.S. Myers, B.P. Capehart, and C.R. Bass. 2012. Development of a finite element model for blast brain injury and the effects of CSF cavitation. *Ann Biomed Eng*. 40:1530–1544.

Payman, W., and W.C.F. Shepherd. 1946. Explosion waves and shock waves. VI. The disturbance produced by bursting diaphragms with compressed air. *Proceedings of the Royal Society of London. Series A. Mathematical and Physical Sciences*. 186:293–321.

Pizov, R., A. Oppenheim-Eden, I. Matot, Y.G. Weiss, L.A. Eidelman, A.I. Rivkind, and C.L. Sprung. 1999. Blast lung injury from an explosion on a civilian bus. *CHEST Journal*. 115:165–172.

Ramos, A., and J.A. Simões. 2006. Tetrahedral versus hexahedral finite elements in numerical modelling of the proximal femur. *Medical Engineering & Physics*. 28:916–924.

Reneer, D.V., R.D. Hisel, J.M. Hoffman, R.J. Kryscio, B.T. Lusk, and J.W. Geddes. 2011. A multi-mode shock tube for investigation of blast-induced traumatic brain injury. *Journal of Neurotrauma*. 28:95–104.

Reynolds, G.T. 1998. Walker Bleakney February 8, 1901–January 15, 1992. *Biographical Memoirs*. 73:87.

Risling, M., S. Plantman, M. Angeria, E. Rostami, B.-M. Bellander, M. Kirkegaard, U. et al. 2011. Mechanisms of blast induced brain injuries, experimental studies in rats. *Neuroimage*. 54:S89–S97.

Ritzel, D., S. Parks, J. Roseveare, G. Rude, and T. Sawyer. 2011. Experimental blast simulation for injury studies. Proceedings HFM-207 NATO Symposium on a Survey of Blast Injury Across the Full Landscape of Military Science.

Ropper, A. 2011. Brain injuries from blasts. *New England Journal of Medicine*. 364:2156–2157.

Säljö, A., F. Bao, K.G. Haglid, and H.-A. Hansson. 2000. Blast exposure causes redistribution of phosphorylated neurofilament subunits in neurons of the adult rat brain. *Journal of Neurotrauma*. 17:719–726.

Sasser, S.M., R.W. Sattin, R.C. Hunt, and J. Krohmer. 2006. Blast lung injury. *Prehospital Emergency Care*. 10:165–172.

Schneiders, R. 2000. Algorithms for quadrilateral and hexahedral mesh generation. Proceedings of the VKI Lecture Series on Computational Fluid Dynamics.

Setoguchi, T., K. Matsuo, F. Hidaka, and K. Kaneko. 1993. Impulsive noise induced by a weak shock wave discharged from an open end of a tube: Acoustic characteristics and its passive control. Proceedings of the 1993 ASME Winter Meeting, November 28, 1993 December 3, 1993. Vol. 170. New Orleans, ASME; pp. 57–64.

Shuck, L.Z., and S.H. Advani. 1972. Rheological response of human brain-tissue in shear. *Journal of Basic Engineering*. 94:905–911.

Skotak, M., F. Wang, A. Alai, A. Holmberg, S. Harris, I. Switzer, Robert C, and N. Chandra. 2013. Rat injury model under controlled field-relevant primary blast conditions: Acute response to a wide range of peak overpressures. *Journal of Neurotrauma*. 30:1147–1160.

Sogbesan, E.A. 2011. Design and Analysis of Blast Induced Traumatic Brain Injury Mechanism Using a Surrogate Headform: Instrumentation and Outcomes. DTIC Document.

Stalnaker, R.L. 1969. Mechanical properties of the head. Ph.D. dissertation. West Virginia University.

Sundaramurthy, A., A. Alai, S. Ganpule, A. Holmberg, lE. Plougonven, and N. Chandra. 2012. Blast-induced biomechanical loading of the rat: An experimental and anatomically accurate computational blast injury model. *Journal of Neurotrauma*. 29:2352.

Sundaramurthy, A., R. K. Gupta, and N. Chandra. 2013. Design considerations for compression gas driven shock tube to replicate field relevant primary blast condition. ASME 2013 International Mechanical Engineering Congress and Exposition. American Society of Mechanical Engineers, 2013.

Tanielian, T., and Jaycox, L.H. 2008. Invisible wounds of war. RAND Corporation, Santa Monica, CA.

Taylor, P.A., and C.C. Ford. 2009. Simulation of blast-induced early-time intracranial wave physics leading to traumatic brain injury. *Journal of Biomechanical Engineering-Transactions of the ASME*. 131:061007.

Teasdale, G., and B. Jennett. 1974. Assessment of coma and impaired consciousness: A practical scale. *The Lancet*. 304:81–84.

Terrio, H., L.A. Brenner, B.J. Ivins, J.M. Cho, K. Helmick, K. Schwab et al. 2009. Traumatic brain injury screening: Preliminary findings in a US Army Brigade Combat Team. *Journal of Head Trauma Rehabilitation*. 24:14–23.

The Visible Human Project. 2009. U.S. National Library of Medicine, Washington, DC.

Thurman, D.J., C. Alverson, K.A. Dunn, J. Guerrero, and J.E. Sniezek. 1999. Traumatic brain injury in the United States: A public health perspective. *Journal of Head Trauma Rehabilitation*. 14:602–615.

Thurman, D.J., J.E. Sniezek, and D. Johnson. 1995. Guidelines for Surveillance of Central Nervous System Injury. Centers for Disease Control and Prevention, Atlanta.

Tsokos, M., F. Paulsen, S. Petri, B. Madea, K. Puschel, and E.E. Turk. 2003. Histologic, immunohistochemical, and ultrastructural findings in human blast lung injury. *American Journal of Respiratory and Critical Care Medicine*. 168:549–555.

Wang, Y., Y. Wei, S. Oguntayo, W. Wilkins, P. Arun, M. Valiyaveettil, J. Song, J.B. Long, and M.P. Nambiar. 2011. Tightly coupled repetitive blast-induced traumatic brain injury: Development and characterization in mice. *Journal of Neurotrauma*. 28:2171–2183.

Warden, D. 2006. Military TBI during the Iraq and Afghanistan wars. *The Journal of Head Trauma Rehabilitation*. 21:398–402.

Wieding, J., R. Souffrant, A. Fritsche, W. Mittelmeier, and R. Bader. 2012. Finite element analysis of osteosynthesis screw fixation in the bone stock: An appropriate method for automatic screw modelling. *PLoS ONE*. 7:e33776.

Willinger, R., H.-S. Kang, and B. Diaw. 1999. Three-dimensional human head finite-element model validation against two experimental impacts. *Annals of Biomedical Engineering*. 27:403–410.

Wojcik, B.E., C.R. Stein, K. Bagg, R.J. Humphrey, and J. Orosco. 2010. Traumatic brain injury hospitalizations of US army soldiers deployed to Afghanistan and Iraq. *American Journal of Preventive Medicine*. 38:S108–S116.

Xiong, Y., A. Mahmood, and M. Chopp. 2013. Animal models of traumatic brain injury. *Nature Reviews Neuroscience*. 14:128–142.

Yelverton, J.T. 1996. Pathology scoring system for blast injuries. *The Journal of Trauma and Acute Care Surgery*. 40:111S–115S.

Zhang, L., and R. Makwana. 2011. Comparison of the head response in blast insult with and without combat helmet. RTO Human Factors and Medicine Panel (HFM) Symposium Halifax, Canada.

Zhang, L.Y., K.H. Yang, and A.I. King. 2001. Comparison of brain responses between frontal and lateral impacts by finite element modeling. *Journal of Neurotrauma*. 18:21–30.

Zhu, F., H. Mao, A. Dal Cengio Leonardi, C. Wagner, C. Chou, X. Jin et al. 2010. Development of an FE model of the rat head subjected to air shock loading. *Stapp Car Crash Journal*. 54:211–25.

Zhu, F., C. Wagner, A. Dal Cengio Leonardi, X. Jin, P. VandeVord, C. Chou et al. 2012. Using a gel/plastic surrogate to study the biomechanical response of the head under air shock loading: A combined experimental and numerical investigation. *Biomechanics and Modeling in Mechanobiology*. 11:341–353.

19 Modeling Fluid Percussion Injury
Relevance to Human Traumatic Brain Injury

Katharine Eakin, Rachel K. Rowe, and Jonathan Lifshitz

CONTENTS

19.1 INTRODUCTION

Fluid percussion injury (FPI) is capable of producing diffuse (midline FPI) and mixed (focal and diffuse; lateral FPI) injury pathologies that are frequently seen following human traumatic brain injury (TBI). The injuries produced with this model are scalable and can produce mild, moderate, and severe injury severities. The surgical procedures and assessment techniques described in this chapter are specific to adult rat and mouse models. However, this injury model can be used with subjects of any age and across different species. By following the procedures detailed in this chapter, it is possible to reproduce behavioral and pathophysiological characteristics of TBI in laboratory animals.

19.1.1 Fluid Percussion Injury Model

Experimental brain injury models must generate similar injuries to those observed after human TBI. Regardless of the physiological, behavioral, or anatomical outcome measures used to evaluate the injury response, it is important that the results be reproducible, quantifiable, and produce a continuum of injury severities (Lighthall, 1989). No single model can replicate the complex mechanisms that occur after human TBI, but several preclinical models of TBI have been developed and implemented to properly characterize its underlying pathology. This chapter focuses on the methods, techniques, and clinical relevance of the fluid percussion model of TBI. The procedures described are based on the authors' years of training and experience using this injury model.

FPI is one of the most commonly used and well-characterized models of TBI. It has been successfully applied in several animal models including rabbit, cat, rat, mouse, and pig. The injury can be applied centrally (midline FPI), over the sagittal suture midway between bregma and lambda, or laterally (lateral FPI), over the parietal cortex. This model is capable of producing graded neurological, histological, and cognitive outcomes that are similar to those seen in human TBI (Dixon, 1987). The injury is produced by applying a fluid pulse (~20 msec) directly onto the surface of the dura via a craniotomy, producing a brief deformation of the brain tissue. By altering the severity of injury and location of the injury site, the FPI model can reproduce neurological impairments and neuropathology associated with both diffuse and focal injury.

Other aspects of human TBI are reproduced by FPI, such as hemorrhaging at the gray–white matter interface, acute hypertension, bradycardia, increased plasma glucose levels, and suppression of electroencephalogram amplitude, which is related to the magnitude of the head injury (Cortez, 1989; Dixon, 1988). In rodent models, FPI has been shown to produce cognitive deficits that can last for weeks to months postinjury (Hamm, 1993; Pierce, 1998), providing avenues to evaluate therapeutic efficacy.

19.2 MATERIALS

19.2.1 Animals

The procedures described in this chapter focus on adult male Sprague Dawley rats (approximately 300–400 g) and 8-week-old adult male C57BL/6 mice (approximately 25–35 g).

19.2.2 Equipment

19.2.2.1 Injury Device

- Fluid percussion injury device (Figure 19.1)
  Custom Design and Fabrication
  Virginia Commonwealth University
  http://www.radiology.vcu.edu/research/customdesign/fpi.html

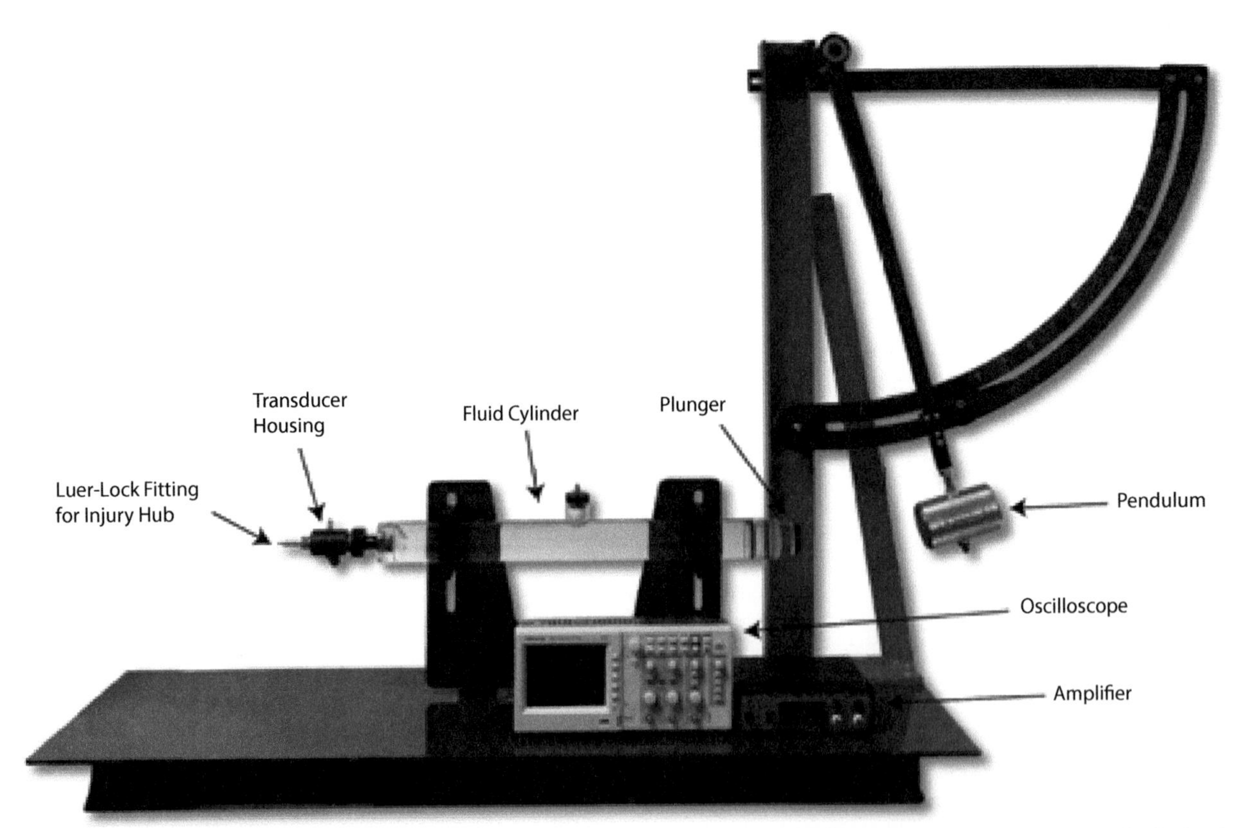

FIGURE 19.1 Fluid percussion injury device. Injury is induced by a fluid pulse injected onto the exposed dura via a surgically implanted injury hub. The fluid pulse is generated by the pressure wave produced when the weighted end of the pendulum arm strikes the piston at the end of the fluid cylinder. The force of the pulse is detected by a transducer and the signal amplified before being sent to the oscilloscope. The oscilloscope output is in mV and needs to be converted to atmospheres of pressure (atm).

Product information including assembly manual, operation manual, and product brochure are provided on the website.

Note: The plunger impact pad on the fluid cylinder should be replaced every 8–12 months.

Information and instructions for the setup, cleaning, and maintenance of the FPI device can be found in the FPI Operation Manual: http://www.radiology.vcu.edu/docs/FPIOperationManual.pdf.

- Recording oscilloscope (recommended: Tektronix, Model 1001B)
- Rubberized shelf liner or industrial Velcro to prevent the FPI device from moving on the bench
- High-vacuum grease (e.g., Fisher Scientific, #14-635-5D)
- Dishwashing solution to clean fluid cylinder
- Jet Dry finishing rinse to minimize air bubbles in the fluid cylinder

19.2.2.2 Anesthesia

- Vaporizer for delivery of inhaled anesthesia
- Tubing/petcocks
- Induction chamber
- Isoflurane
- Oxygen
- Rodent nose cone for inhaled anesthetic that is compatible with the stereotaxic frame

  When using an inhaled anesthetic, it is recommended that all procedures are performed in a well-ventilated area, on a downdraft or similar table, or in a type II biosafety cabinet to minimize anesthesia exposure to the surgeon (current Occupational Safety and Health Administration recommendation for halogenated gasses is <2 ppm).

19.2.2.3 Surgical Supplies

- Sterilization pouches, sterilization wrap (e.g., CSR Wrap 18 in. × 18 in.; Butler Schein, #1004683), sterilization indicator strips, autoclave confirmation tape
- Gauze sponges, 2 in. × 2 in. and 4 in. × 4 in.
- Cotton tip applicators (recommended: 6-in. long, rounded tip, and 3-in. long, pointed tip)
- Heating pad (recommended: Deltaphase isothermal heating pad, Braintree Scientific, #39DP)
- Absorbent blue pads (chux)
- Three autoclavable cups for intraoperative cleaning of surgical instruments: one for chlorhexidine solution and two for saline rinses (e.g., Griffin Nalgene beakers, 150–250 mL or SENTRY Silicone Instrument Cup, Roboz, #RS-9908)

  Cylinder-shaped desk organizers for small items can be used to stabilize the Nalgene cups and prevent them from tipping over (e.g., black wire mesh Large doodad cup, Staples, #385747).
- Chlorhexidine solution (e.g., Evsco Pharmaceutical Corp., #9880085)
- Sterilized water to dilute chlorhexidine to clean/disinfect instruments during surgery
- 0.9% sterile sodium chloride, irrigation and injectable, USP
- 20-gauge needles (recommended: 1-in. length)
- 1-mL syringes, slip tip
- ≥10-mL syringes, Luer-Lock tip
- Laboratory timer or stopwatch
- Small animal trimmer for fur removal (e.g., Wahl, Mini Arco Animal Trimmer)
- Ophthalmic ointment to prevent drying of eyes during surgery
- 4% chlorhexidine solution (or Betadine scrub) for prepping the surgery site
- 70% ethanol
- Surgical drape for animal (recommended: Glad Press'n Seal is a sterile and inexpensive option for a surgical drape and allows the surgeon to observe, unobstructed, the status of the animal throughout the procedure) http://research.utsa.edu/files/larc/RodentSurgeryApplicationhandouts.pdf. Sterilized Press'n Seal in pre- packaged ETO-exposed rolls can be purchased from SAI-Infusion Technologies, http://www.sai-infusion.com/products/press-n-seal
- Gelfoam
- Cyanoacrylate (e.g., Krazy Glue or Super Glue)
- Methyl methacrylate (dental acrylic; e.g., Perm Reline and Repair Resin, liquid and powder)
- Metal spatula for mixing and shaping dental acrylic
- Cups for mixing dental acrylic (e.g., small Dixie cups)
- Optional: Applicator for dental acrylic 3-mL syringe (Luer-Lock tip) and a 200-µL pipette tip
- Antibiotic ointment
- Glass bead sterilizer to sterilize instruments between surgeries

19.2.2.4 Surgical Instruments

- Small animal stereotaxic frame
- Surgery light (recommended: bifurcated gooseneck illuminator light)
- Scalpel handle and blade (recommended: #3 handle, #15 blade)
- Delicate bone scraper
- Molt elevator
- #5 forceps, 45° (0.10 × 0.06-mm or 0.05 × 0.01-mm tip size)
- Graefe forceps, long serrated slight curve (recommended: 0.8-mm tip)
- Needle holder and scissors for suturing (recommended: Olsen-Hegar combination scissor and needle holder)

19.2.2.4.1 Rat Surgical Instruments

- Rat nosecone for anesthesia delivery
- Retractor (recommended: Mixter hemostatic forceps, curved extra delicate (quantity, 4) or bulldog clips (quantity, 4)

- Pin vise and drill bit (e.g., pin vise, 0.64–1.9 mm, tapered collet, Grainger Industrial Supply, #2ZVH6; Jobber Drill, #50, 135 Deg., Grainger Industrial Supply, #1K016)
- Michelle Trephine (Roboz, #RS-9200)
- Small screwdriver
- Stainless steel machine screws (thread size must be compatible with drill bit)
- 4-0 suture

19.2.2.4.2 Mouse Surgical Instruments

- Mouse nosecone for anesthesia delivery
- Stereo surgical microscope with a swing/flex arm (e.g., Meiji EMZ-5 stereo surgical microscope)
- Micro bulldog clamp, serrated jaw, curved, 3-cm long (quantity, 4)
- Custom trephine (3-mm diameter)
  Research Instrument Shop
  University of Pennsylvania, Philadelphia, PA
  http://www.uphs.upenn.edu/biocbiop/ris/ris.html
- Weed whacker line for cranial discs (1.7-mm diameter) to use as a guide for the trephine
- Graefe forceps: serrated slight curve extra delicate 0.5-mm tip (e.g., Roboz, #RS-5136)
- 5-0 suture

19.2.2.5 Injury Hub

Supplies for making the injury hub are shown in Figure 19.2. Hub preparation for FPI is illustrated in Figures 19.3 through 19.5.

- 20-gauge needle
- 1-mL syringes, slip tip
- Gauze sponge
- Razor blades
- Calipers or ruler
- Optional:
- Sand paper (fine or very fine grit) or emery board

19.2.2.5.1 Rat Injury Hub

- Handheld eyebrow pencil sharpener to bevel the tip of the hub

19.2.2.5.2 Mouse Injury Hub

Mouse injury hubs do not need a beveled tip.

19.2.2.6 Fluid Percussion Injury

- Laboratory timer/stopwatch
- Heating pad
- Absorbent blue pads (chux)
- Scalpel handle and blade
- Gauze sponges, 2 in. × 2 in.
- Cotton tipped applicators (with rounded tip)
- Graefe forceps, serrated slight curve 0.8-mm tip (e.g., Roboz, #RS-5135)

19.2.2.6.1 Rat Model

- 4-0 suture
- Optional suture method:
- 20-gauge needles (three to four per rat) threaded with suture (e.g., Braunamid white suture USP 4-0)
  Optional Luer-Lock adapter:
  To keep blood from entering the injury device: Male Luer X Female Luer Coupler, PVDF (Cole-Parmer, #YO-45512-82)

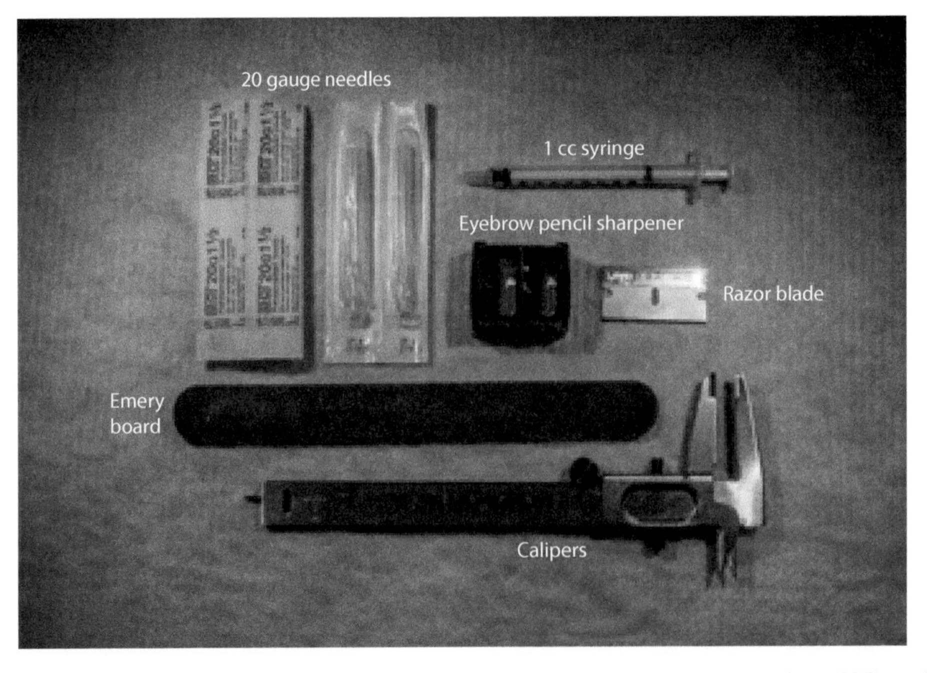

FIGURE 19.2 Materials for making the injury hub. Materials needed to make the injury hub: 1-mL syringe, 20G needles (1–1.5 in.), razor blade, 4 × 4-in. gauze sponge (not shown), handheld eyebrow pencil sharpener, calipers, and emery board or sandpaper (optional).

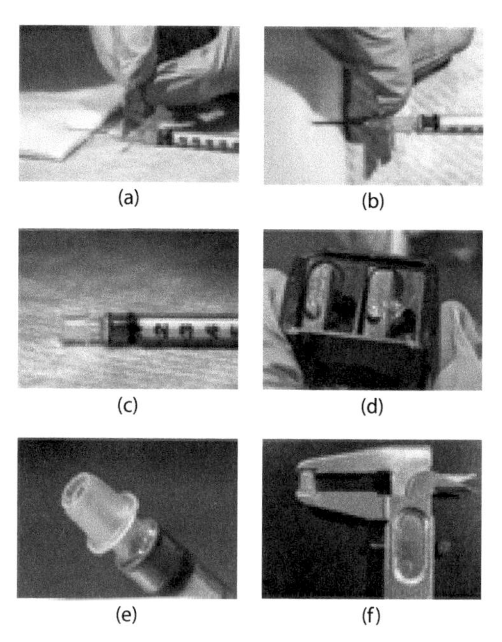

FIGURE 19.3 Crafting the injury hub. Firmly attach a 20G needle onto a 1-mL syringe. Insert the needle tip into gauze (or something similar) to prevent the needle tip from becoming a projectile after it is cut (a). Use a razor blade to cut off the tip of the needle just before the (a, b). The cut end of the injury hub should be flat and even, parallel to the Luer-Lock plane. Use a handheld eyebrow pencil sharpener to bevel the tip of the injury hub (d, e). The final length of the hub should be approximately 7 mm (f). If using a mouse model of fluid percussion injury, do not bevel the tip of the hub.

FIGURE 19.4 Optional shaping of the injury hub. If the end is cut at an angle, an emery board (or fine grit sandpaper) can be used to sand it down (a, b). However, any plastic "pulp" (a, white arrow) should be removed from the end of the hub before implantation. Pulp on the outside of the hub can be removed with the pencil sharpener. The corner of a razor blade can be used to remove any residual plastic from inside the hub.

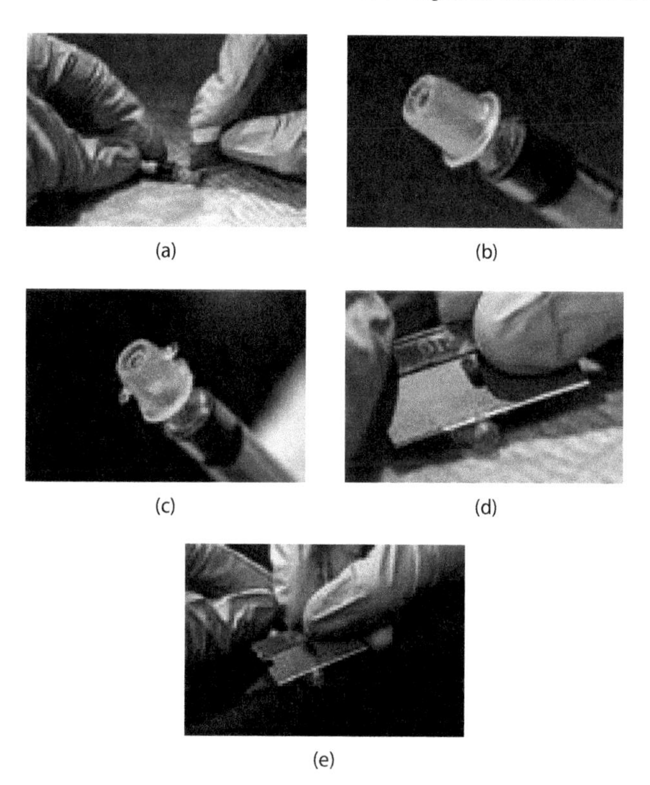

FIGURE 19.5 Scoring the hub. Score the exterior of the hub (a) at a 45° down to the base at even intervals around the hub, approximately four to six times (a, b). The grooves make a path for the glue to follow down to the skull. An alternative method to facilitate a secure bond between the methyl methacrylate and hub is to make four flanges on the hub, resembling a fleur-de-lis (c). Use a razor blade to make superficial cuts into the side of the hub (d). To make the fourth tab, use the edge of a table to provide stability while cutting without flattening the opposing tab (e).

19.2.2.6.2 Mouse Model

- Baxter extension tubing (Baxter, #2C5643)
- 5-0 suture

19.3 PROCEDURE

All experimental procedures in this chapter follow the guidelines established in the *Guide for the Care and Use of Laboratory Animals* (U.S. Department of Health and Human Services). A useful guide for rodent surgical technique can be found on the University of Texas at San Antonio, Laboratory Animal Resources Center website: Rodent Surgery: Application of Aseptic Technique and Perioperative Care (Perret-Gentil, 2013).

19.3.1 Record Keeping

A standard surgery sheet should be used to record information pertaining to the preoperative condition of the animal, surgical procedure, injury, and both immediate and long-term postoperative care. Separate sheets for postoperative observations and treatment for each animal should be maintained in the animal room in the vivarium. This should include notes about the general condition of the animal and any supportive care the animal received (e.g., saline, mash [chow + water], nutritional supplements).

19.3.2 General Preparation for Surgery and Injury

- Autoclave the Nalgene beakers, distilled water, and surgical packs. A typical surgery pack would include the following:
  - Gauze sponges
  - Cotton tip applicators
  - Surgical instruments, screws, hubs (200-μL pipette tips, optional)

Note: The pin vise and screwdriver will oxidize after autoclaving; to prevent transfer onto other instruments, autoclave them in a separate pouch.

19.3.3 Preparatory Surgery for Fluid Percussion Injury

19.3.3.1 Preoperative Preparation

1. Appropriate protective personal equipment should be worn: clean lab coat or scrubs, gloves, face mask, hair covering, and protective eyewear.
2. Assess the animal for any signs of pain, distress, or disease and record this information on the surgery/injury data sheet (e.g., abnormal posture, movement, poor grooming, evidence of porphyrin accumulation on eyes, nose, or fur).
3. Clean the operating area and stereotaxic frame with 70% ethanol.
4. Place a heating pad on the stereotaxic frame and cover it using an absorbent blue pad (chux) or other absorbent pad.
5. Pour irrigation saline into two of the Nalgene beakers, ~50 mL each.
6. Pour ~1 mL of chlorhexidine solution into the remaining Nalgene beaker and add ~30 mL of sterilized water.
7. Confirm that surgical supplies (packs and instruments) have been properly sterilized.

19.3.3.2 Rat Preparatory Surgery

All surgical procedures should be performed using aseptic or sterile technique. Instructions on making the injury hub are described in Figures 19.3 through 19.5.

1. Anesthetize the animal in the induction chamber under 5% isoflurane mixed with 100% oxygen at a rate of 1 L/minute for 5 minutes.
2. Shave the surgical area.
3. Position the animal in the stereotaxic frame and adjust the isoflurane level to between 2% and 2.5%. The back of the upper teeth should be flush with the bite bar, without any tension applied to the teeth. Check the condition of the teeth after the surgery; note any injury or pathology.
4. Depth of anesthesia is determined by suppression of the toe pinch reflex.
   If you observe mouth breathing, check the positioning of the teeth over the bite bar and/or reposition the nose cone. Make sure there is adequate space for the animal's nose to allow normal respiration.
5. Apply ophthalmic ointment liberally to the eyes to prevent drying, reapply as needed throughout the procedure.
6. Prepare the surgical area (top of head to base of neck) with 4% chlorhexidine solution followed by 70% ethanol wipes. Wipe in a circular pattern, starting in the middle and moving to the edge of the surgical site. Do this three times.
   Betadine scrub can be used in place of chlorhexidine; however, according to recent reports it is less effective in preventing surgical-site infection (Darouiche, 2010; UTSA website).
7. Place surgical drape over the animal.
8. Don clean procedural gloves.
9. Using sterile technique, open the surgical pack and instruments for preparatory surgery, including scalpel blade, syringe, and 20-gauge needle for saline.
10. Prepare a ≥10 mL syringe filled with sterile 0.9% saline to use for irrigation and filling of the injury hub.
11. Make a superficial midline sagittal incision to the scalp, extending from between the eyes to the base of the skull.
12. Make a midline sagittal incision to the periosteum.
13. Use a periosteal elevator or bone scraper to laterally retract the periosteum to the lateral ridge, exposing the skull.

14. If greater exposure is needed, stretch the skin by applying pressure in a rostral direction and rostrally extend the periosteal incision.
    To prevent excessive bleeding, avoid cutting the underlying muscle at the base of the skull.
15. Grasp the skin with forceps and dissect the subcutaneous tissue laterally.
16. Use hemostatic forceps or bulldog clips to grasp the periosteal tissue bilaterally at the level of the coronal and lambdoid sutures to expose the surgical site. It is important to have a clear view of the sutures and the craniometrical points, bregma and lambda.
17. Use the pin vise to manually drill screw holes at the specified locations for either lateral or midline FPI; see Figure 19.6. The skull screws are used to secure the injury hub to the skull.
    a. For midline FPI, position the first screw hole 1 mm lateral to bregma and 1 mm rostral to the coronal suture on the right side; position the second screw hole 2 mm lateral to lambda and 1 mm caudal to the lambdoid suture.
    b. For lateral FPI, position the first screw hole 1 mm lateral to bregma and 1 mm rostral to the coronal suture on the ipsilateral side to the craniotomy; position the second hole midway between bregma and lambda and 1 mm lateral to the central suture contralateral to the craniotomy.
18. Use the pin vise to manually drill a shallow divot at the center of the intended craniotomy location to provide stability for the trephine's guide pin (see white arrow in Figure 19.7a, lateral preparation). Use the trephine to make the craniotomy without disrupting the underlying dura; see Figure 19.7b.
19. Remove the bone piece and gently clear any blood from the craniotomy site.
20. Loosely secure the skull screws into the screw holes (approximately 1½ turns); the screw should be able to withstand gentle side-to-side pressure using the screwdriver.
    Do not overtighten the screws or the hub will not easily come off after the injury.
21. Clean and dry the injury site by gently rolling a round-tip cotton swab over the opening.

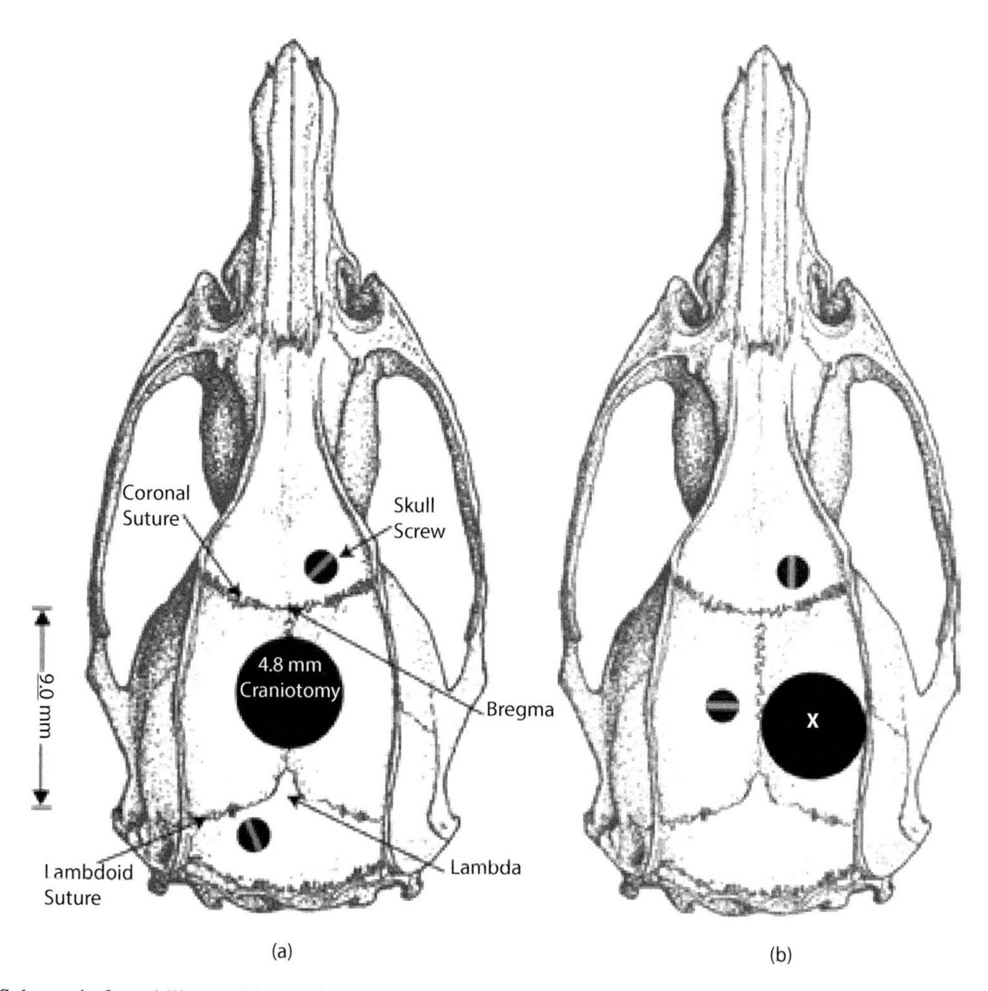

FIGURE 19.6 Schematic for midline and lateral injury preparatory surgery. This illustration shows the screw and hub placement for midline (a) and lateral (b) fluid percussion injury (FPI) in the rat. Similar positioning for the craniotomy can be used in the mouse FPI model. (Modified from Paxinos, G., Watson, C., *The Rat Brain in Stereotaxic Coordinates*. Academic Press/Elsevier, Amsterdam, 2007.)

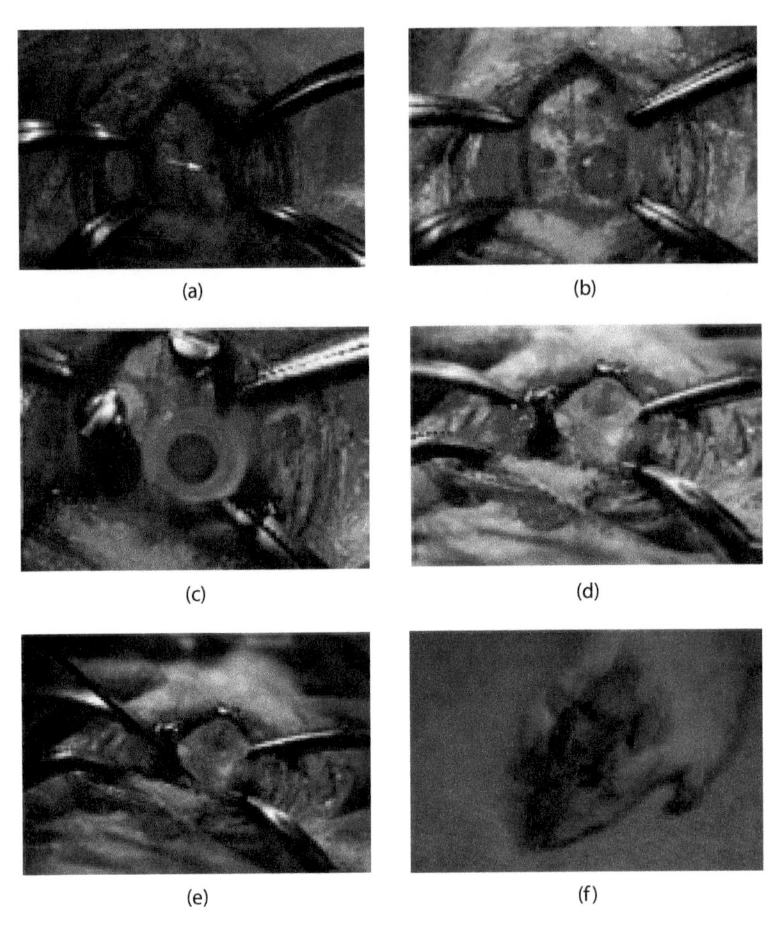

FIGURE 19.7 Surgical preparation for lateral fluid percussion injury (FPI) in the rat. These images show the surgical preparation for lateral FPI in the rat. The scalp is incised and the underlying fascia removed from the bone. The coronal and lambdoid sutures should be visible. Use a pin vise to make a superficial divot in the skull (white arrow) to stabilize the trephine guide pin (a). The position of the two screw holes and 4.8-mm craniotomy are shown before (b) and after (c) placement of the screws and injury hub. The brain should be clearly visible and the dura intact. Fill the hub with sterile saline to ensure there is a watertight seal (d) and apply cyanoacrylate with a needle or applicator stick around the perimeter of the hub to provide additional stability (e). Prepare the methyl methacrylate to the appropriate consistency and pour it around the injury hub and so that it covers the skull screws. Use a metal spatula to shape the methyl methacrylate and keep it away from the animal's skin. Allow the methyl methacrylate to dry completely before suturing the scalp closed (f).

22. Apply a small amount of cyanoacrylate to the cut end of the hub and evenly distribute it around the outside at the cut end. The wooden stick end of a cotton swab is a useful tool for evenly distributing the glue around the tip of the hub.
23. Blot any excess glue from the hub using the back of your gloved hand until the resulting ring of glue is similar to the width of the hub and there is an even amount of transferred glue.
24. Gently place the hub inside the 4.8-mm craniotomy and apply light pressure for approximately 3–5 seconds. Figure 19.7c shows placement of the hub and skull screws.

    Graefe forceps are a useful tool for gripping and placing the hub. While maintaining your grip on the hub with the forceps, resting the wooden stick end of a cotton swab across the top of the hub can provide stability during placement within the perimeter of the craniotomy. Once the hub is in place, release the forceps while maintaining light, even pressure with the wooden stick. To ensure evenly distributed downward pressure, rest the Graefe forceps on top of and perpendicular to the wooden stick, such that both curved ends are pointing upward.
25. Fill the hub with sterile saline to ensure a watertight seal; see Figure 19.7d. Make sure the skull remains dry. Once the hub is filled with saline you should still be able to clearly visualize the underlying cortex. It is important that there is no glue covering the dura.

    If any at any point saline leaks from the base or if the interior of the hub appears opaque and whitish in color, remove the syringe hub and replace it with a new one.
26. Apply additional cyanoacrylate to the skull at the base of the hub; see Figure 19.7e.

    The tip of a 20–gauge needle or 200 µL pipette tip is useful for application of the cyanoacrylate.
27. In a hood, prepare a small amount of enough to surround the hub and cover both skull screws.

    Do not let glue or get into the hub or cover the dura. This will obstruct the fluid pulse and may either dampen the force of the injury or if partially obstructed, may produce a more intense and focal injury.

28. Use the metal spatula to shape the and push it away from the skin.
29. Once the methyl methacrylate has completely hardened, gently stretch the skin over the hub and suture the scalp; see Figure 19.7f.

 If the methyl methacrylate around the hub is attached to the skin it may dislodge the hub during suturing.

 Tip: Use microdissecting forceps (e.g., 45° #5 forceps with a fine tip) to gently separate the skin from where it is attached to the methyl methacrylate.
30. Apply antibiotic ointment to the surgery site and place the animal in a warmed recovery area.
31. Postoperative observations should be performed every 15 minutes until the animal regains normal ambulatory behavior and the information recorded on the surgery-injury data sheet.

19.3.3.3 Mouse Preparatory Surgery

All surgical procedures should be performed using aseptic or sterile technique. Images showing the surgery procedures are in Figure 19.8.

1. Anesthetize the animal in the induction chamber, under 5% isoflurane mixed with 100% oxygen at a rate of 1 L/min for 5 minutes.
2. Position the animal in stereotaxic frame and adjust the isoflurane level to 2–2.5%. Depth of anesthesia is determined by suppression of the toe pinch reflex. If you observe mouth breathing, check the positioning of the teeth/tongue and/or reposition the nose

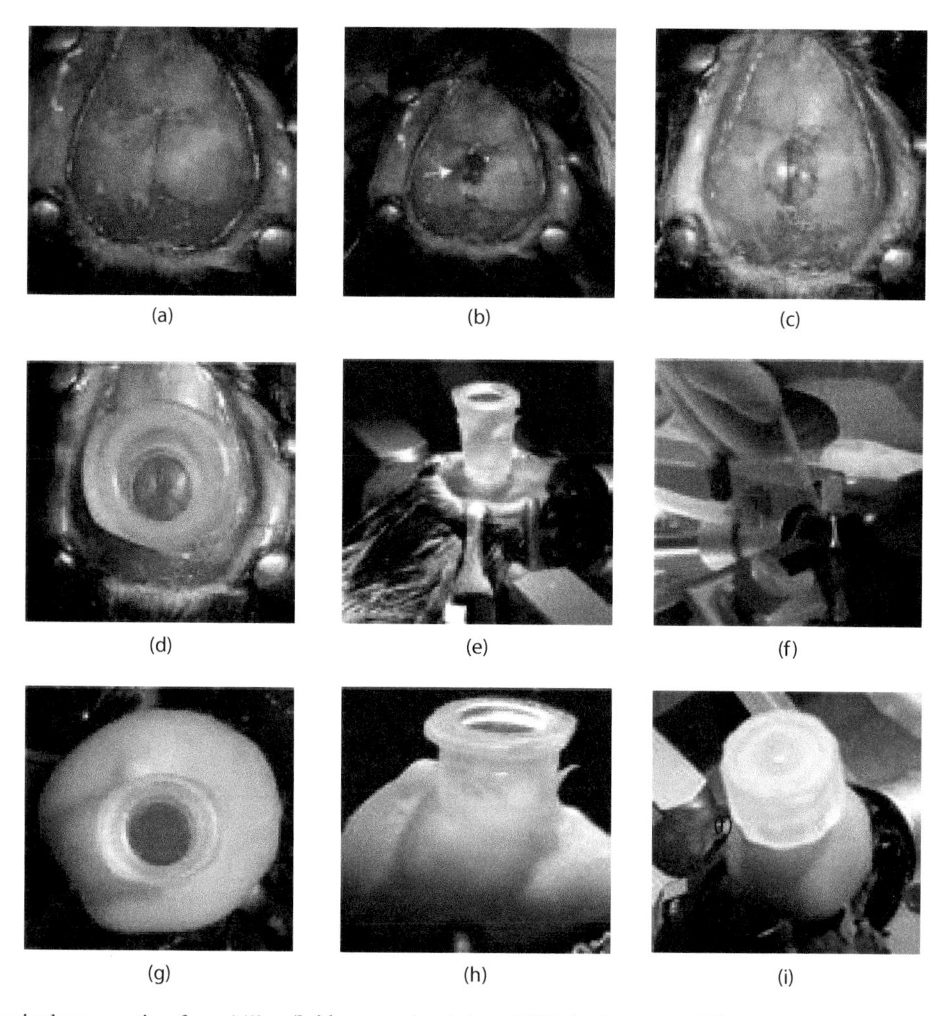

FIGURE 19.8 Surgical preparation for midline fluid percussion injury (FPI) in the mouse. These images illustrate the surgical preparation for midline FPI in the mouse. The scalp is incised and underlying fascia removed (a). Four bulldog clips are used to expose the surgical site. The central and lambdoid suture should be visible. Place a small amount of cyanoacrylate on the disc created from weed whacker line (white arrow in b) and affix it to the skull midway between bregma and lambda over the central suture (b). This will be used as a guide for the trephine. Carefully remove the bone flap without disrupting the underlying dura (c). Apply cyanoacrylate to the bottom of the hub and place it on the skull, over the burr hole (overhead [d] and lateral [e] images are shown). Use an applicator (e.g., sharpened wooden stick, 20G needle, 200-µL pipette tip) to apply additional cyanoacrylate around the base of the hub (f). Prepare the methyl methacrylate to the desired consistency and apply it around the hub using a 1-mL syringe (g), leaving the top third of the hub exposed (h). Secure a Luer-Lock cap to the hub to protect the opening (i).

cone. Make sure there is adequate space for the animal's nose to allow normal respiration.

3. Apply ophthalmic ointment liberally to the eyes to prevent drying; reapply as needed throughout the procedure.
4. Prepare the surgical area (top of head to base of neck) with 4% chlorhexidine solution followed by 70% ethanol wipes. Wipe in a circular pattern, starting in the middle and moving to the edge of the surgical site. Do this three times.
   Betadine scrub can be used in place of chlorhexidine; however, according to recent reports it is less effective in preventing surgical-site infection (Darouiche, 2010; Perrett-Gentil, 2013).
5. Place surgical drape over the animal.
6. Don clean procedural gloves.
7. Using sterile technique, open the surgical pack and instruments for preparatory surgery, including scalpel blade, syringe, and 20-gauge needle for saline.
8. Prepare a ≥10-mL syringe filled with sterile 0.9% saline to use for irrigation and filling of the injury hub.
9. Make a superficial midline sagittal incision to the scalp, extending from between the eyes to the base of the skull.
10. Make a midline sagittal incision to the periosteum.
11. Use a periosteal elevator or bone scraper to laterally retract the periosteum to the lateral ridge, exposing the skull.
    If greater exposure is needed, stretch the skin by applying pressure in a rostral direction and rostrally extend the periosteal incision. To prevent excessive bleeding, avoid cutting the underlying muscle at the base of the skull.
12. Use hemostatic forceps or bulldog clips to grasp the periosteal tissue bilaterally at the level of the coronal and lambdoid sutures to expose the surgical site.
13. It is important to have a clear view of the sutures and the craniometrical points, bregma and lambda.
14. Trim (shave) weed whacker line with a razor blade or scalpel to make a thin circular cranial disk. This will be used as a guide for the trephine.
15. Use Vetbond (Henry Schein, #700-3449) to secure the weed whacker line in place.
16. For midline FPI, position the weed whacker line centered between bregma and lambda.
17. For lateral FPI, position the weed whacker line in the center of the desired craniotomy location.
18. Use the trephine to drill through the skull without disrupting the underlying dura.
19. Remove the bone piece and gently clear any blood from the craniotomy site
    A Wedelstaedt chisel, scalpel, or forceps is best for removing the piece of bone. This step should be done under magnification. If the bone piece from the craniotomy is attached at points to the skull, use the chisel or forceps to work around the circumference of the craniotomy, without disturbing the dura, until the bone piece can be easily removed.
20. Clean and dry the injury site by gently rolling a round-tip cotton swab over the opening.
21. Apply a small amount of cyanoacrylate to the cut end of the hub and evenly distribute it around the outside at the cut end. The wooden stick end of a cotton swab is a useful tool for evenly distributing the glue around the tip of the hub.
22. Blot any excess glue from the hub using the back of your gloved hand until the resulting ring of glue is similar to the width of the hub and there is an even amount of transferred glue.
23. Gently place the hub on the skull over the craniotomy.
    Graefe forceps are a useful tool for gripping and placing the hub. While maintaining your grip on the hub with the forceps, resting the wooden stick end of a cotton swab across the top of the hub can provide stability during placement over the perimeter of the craniotomy. Once the hub is in place, release the forceps while maintaining light, even pressure with the wooden stick. To ensure evenly distributed downward pressure, rest the Graefe forceps on top of and perpendicular to the wooden stick, such that both curved ends are pointing upward.
24. Fill the hub with sterile saline to ensure a watertight seal. Make sure the skull remains dry. Once the hub is filled with saline, you should still be able to clearly visualize the underlying cortex. It is important that there is no glue covering the dura.
25. If any at any point saline leaks from the base or if the interior of the hub appears opaque and whitish in color, remove the syringe hub and replace it with a new one.
26. Apply additional cyanoacrylate to the skull at the base of the hub.
27. The tip of a 20-gauge needle or 200-µL pipette tip are useful for application of the cyanoacrylate.
28. In a hood, prepare a small amount of methyl methacrylate, enough to surround the hub.
    Do not let glue or methyl methacrylate get into the hub or cover the dura. This will obstruct the fluid pulse and may either dampen the force of the injury or, if partially obstructed, may produce a more intense and focal injury.
29. Use the metal spatula to shape the methyl methacrylate and push it away from the skin.
30. Apply antibiotic ointment to the surgery site and place the animal in a warmed recovery area.
31. Postoperative observations should be performed every 15 minutes until the animal regains normal ambulatory behavior and the information recorded on the surgery/injury data sheet.

19.3.4 Fluid Percussion Injury

Before using the injury device, check that when the weighted pendulum arm is hanging in a neutral position, at a 0° angle, that it is flush and centered on the foam pad at the end of the plunger. It is useful to mark the correct position of the plunger's O-rings on the outside of the fluid cylinder to use as a guide.

19.3.4.1 Setup

1. Microwave the Deltaphase heating pad (or if using a heating pad, set it to the lowest heat setting) and place the heating pad on the injury frame.
2. Position an absorbent blue pad (chux) over the heating pad and tape the corners to the injury frame.
3. Using aseptic technique, open the autoclaved injury pack and surgical instruments.
4. Before anesthetizing the animal, ensure that the injury level as indicated by the oscilloscope is within the parameters of your study.
5. Convert mV from the oscilloscope to atmospheres of pressure (atm).
6. Place the animal in the induction chamber under 5% isoflurane for 5–6 minutes.
7. While the animal is in the induction chamber, prime the injury device by releasing the pendulum once every 30 seconds; this will produce more consistent pressure at the time of injury.

19.3.4.2 Injury Procedure

Images of the FPI procedure for the rat and mouse are shown in Figures 19.9 and 19.10, respectively.

1. Place the anesthetized animal on the covered heating pad.
2. Using aseptic technique, reopen the scalp incision to expose the injury hub.
3. Fill the hub with sterile saline until a bead of water is formed by surface tension, take care to remove any air bubbles from inside the hub (see Figure 19.9a).
4. Remove the syringe from the end of the fluid cylinder.
5. Press the plunger so that a drop of fluid from is produced at the end of the transfer tube attached to the transducer housing.
6. Join the saline bubble from the injury hub with the fluid drop from the transfer tube and connect the animal via the injury hub to the injury device. Images from rat FPI (Figure 19.9b–d) and mouse FPI (Figure 19.10a, b) are shown.
7. Release the pendulum to injure the animal. This will produce a period of reduced responsiveness to stimuli (mouse, Figure 19.10c, d) and, depending on the injury severity, may produce a brief to extended period of postinjury apnea.
8. Immediately after the injury, start a timer to measure the duration of the suppression of the righting reflex (see step 11). Additionally, record the occurrence and duration of any physiological indicators of injury severity, such as apnea and seizure.
9. Remove the injury hub. Note the condition/appearance the surgical site and the brain tissue beneath the injury site (e.g., active bleeding, brain swelling, currant jelly clot).

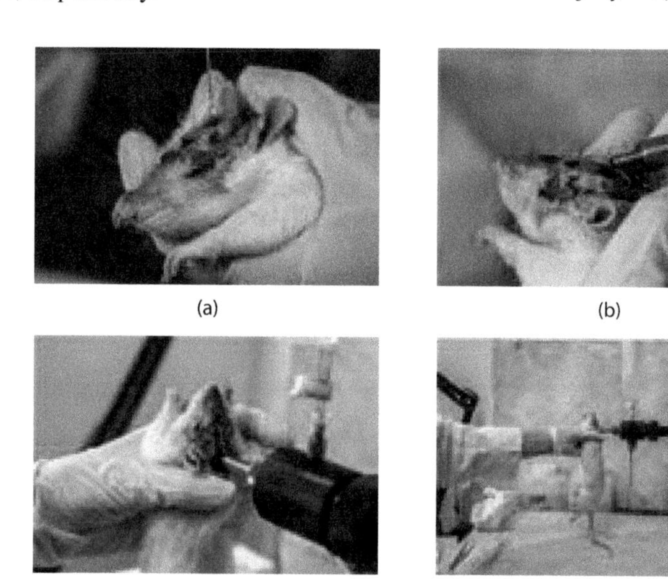

FIGURE 19.9 Fluid percussion injury (FPI) in the rat model. Fill the injury hub with sterile saline until a bead of saline is visible at the top (a). Push on the plunger of the injury device until a small bead of water is visible at the opposite end. Hold the animal at a 45° angle to the Luer-Lock end of the injury device so that the fluids from the device and injury hub are touching. Connect the injury hub to the FPI device and hold the rat in place. Release the pendulum arm to produce the injury (c, d). After the injury, hold the rat in place with the injury hub still attached at the Luer-Lock end of the FPI device. Start a timer. To remove the injury hub, angle the rat's body upward, similar to removing a bottle cap. Suture the scalp closed and lay the rat supine. Record the injury pressure and latency for the animal to regain the righting reflex.

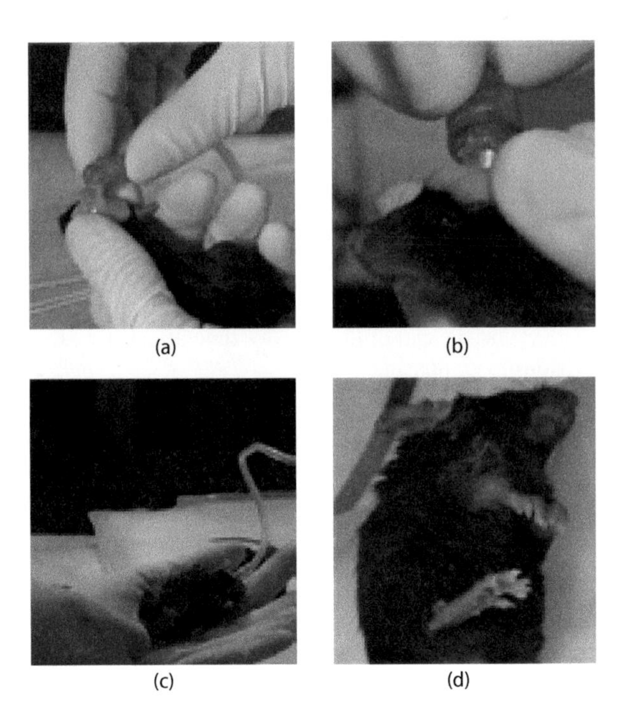

FIGURE 19.10 Fluid percussion injury (FPI) in the mouse model. Attach one end of the tubing to the end of the FPI device, push on the plunger until a small bead of water is visible at the other end and all air bubbles are removed. Fill the injury hub with sterile saline until a bead of saline is visible at the top. Create a watertight seal between the tubing and the injury hub (a). Screw the Luer-Lock female end of the tubing to the male injury hub (b). While holding the mouse, release the pendulum arm on the fluid percussion device to induce the injury (c). Remove the injury hub and suture the scalp closed. Position the mouse supine and record the injury pressure and latency for the animal to regain the righting reflex. An indicator of a concussive brain injury is the fencing response (d), with characteristic extension of one or both forelimbs.

10. Quickly close the wound (i.e., suture or staple) and place the animal supine to assess the righting reflex.

    Tip: apply a thin strip of antibiotic ointment on a 2-in. × 2-in. gauze sponge before injury; after suturing the scalp, place the gauze with ointment over the incision, and lay the animal supine on the heating pad. This will cushion the head and allow better penetration of the antibiotic ointment at the incision site.
11. The criterion for the return of the righting reflex is that the animal must roll completely over three consecutive times. After each attempt at righting, the animal is returned to the supine position by the experimenter until the animal is able to right itself three consecutive times. The righting time is then recorded on the injury sheet along with the atm produced by the injury.
12. Once the animal has righted, it is placed in a designated recovery area, such as an animal Intensive Care Unit (e.g., Braintree Scientific Inc., Lyon Technologies Inc.). Once the animal regains normal ambulatory behavior, it can be returned to its home cage. During the recovery period, postinjury observations should be made every 10–15 minutes and recorded on the surgery/injury data sheet.
13. Additional sterile water (or fill solution) may need to be added via syringe on the injury device to replace any lost during the procedure.

19.4 INJURY OUTCOME ASSESSMENT TECHNIQUES

To assess the injury severity level, there are multiple outcome measures that can be used (e.g., neurological, physiological, neurobehavioral, histological techniques). Based on resource availability (i.e., availability of reagents, having the necessary equipment to perform analyses, established behavioral models) and expertise of the research staff, it is possible to select the most efficient study design that has the greatest probability of success.

19.4.1 Neurological Assessment

One way to assess injury severity is to perform a neurological evaluation of the animal immediately after FPI by timing the duration of the reflex suppression (e.g., pinna, cornea, righting). The pinna reflex is produced by touching the inside of the pinna with the wooden end of a cotton-tip applicator. A positive response will include a head shake and twitching of the ear. The corneal reflex is present if there is a blink response after gentle touching of the cornea with a saline-moistened wisp of cotton. The righting reflex is assessed by placing the animal on its back and recording the latency for the animal to roll over onto their feet three consecutive times (Schmidt, 1993). Of these reflexes, the righting reflex is non-invasive and has been shown to correlate with injury severity (Hamm, 2001; Morehead et al., 1994). The atm produced by the injury can vary between devices and laboratories; therefore, reporting the righting time with the atm provides an additional reference point for comparing data in the literature. However, the duration of reflex suppression can be effected by the type of anesthesia used and if the animal is intubated. Postinjury apnea may coincide with reflex suppression and correlates with injury severity. Some groups provide postinjury ventilation support using room air until a normal, regular rate of respiration is restored, which may mitigate apnea-induced effects.

19.4.2 Postinjury Weight Loss

The amount of weight loss after TBI is correlated with the degree of injury severity. Mild injury is associated with 0%–10% weight loss from the preinjury weight. A moderate injury level-can produce weight loss between 15% and 20% that can go up to 25%–30% with increasing injury severity. If by the second day postinjury, there is no change or continued weight loss, the animal will likely require fluid injections (0.9% sterile saline) to prevent dehydration. The animal should also be examined for non–TBI-related causes for weight loss, such as malocclusion. Injury to the teeth can also occur during the preparatory surgery if the teeth are not positioned properly in the bite bar on the stereotaxic frame. It is beneficial to prophylactically provide mash (chow + water) and/or place normal rat chow on the floor of the home cage to facilitate weight gain. Hydration packs can also be used. Providing high-calorie supplements to the diet, such as Nutri-Cal or Suplical, will help animals gain weight; however, alterations in postinjury glucose levels have been shown to influence injury outcome (Kokiko-Cochran, 2008).

19.4.3 Neurobehavioral Assessment

There is an abundance of normative data showing graded motor and cognitive deficits after different severity levels of injury. Injured animals will have some degree of motor impairment that typically resolves within 7 days (Hamm, 2001). There are several means of evaluating motor impairment after TBI. Motor deficits can be tested using the beam balance, beam walk, and Rotarod tests. Both midline and lateral FPI models are capable of producing cognitive deficits via damage to the hippocampus, a region of the brain that is known to be selectively vulnerable in human TBI. The two models differ in the type of damage inflicted on the hippocampus. Midline FPI does not produce the same magnitude of cell loss that is typically seen after lateral FPI; however, it does produce hippocampal damage (Hamm, 1993; Lyeth, 1990; Thompson, 2005). The memory impairment observed after midline FPI is not believed to be due to cell death but instead from neuronal dysfunction in the hippocampus (Hayes, 1992; Lyeth, 1990). The position of the craniotomy, particularly in lateral FPI, may alter the degree of cognitive impairment (Floyd, 2002; Vink, 2001).

19.5 TROUBLESHOOTING

19.5.1 Air Bubbles

Air bubbles in the FPI device can prevent an accurate measurement of the injury magnitude. When air is present in the device, the oscilloscope reading will have many jagged peaks instead of a smooth curve with one peak. The syringe ports can be used to remove any air that enters the device. One way for air to become trapped in the fluid cylinder is after cleaning of the cylinder. This can be minimized by rinsing with a spot remover solution for the dishwasher (e.g., Jet Dry). Degassing the water before filling the cylinder will also help to minimize air bubbles.

If an air bubble is found inside the fluid cylinder, adjust the angle of the tube by loosening one of the braces used to mount the cylinder to the metal frame of the FPI device and extract the trapped air using one of the fill ports. To remove air bubbles introduced at the male Luer-Lock end of the device during normal use, reattach the three-way petcock with the attached syringe and open the valve between the FPI device and syringe. With one hand, stabilize the three-way petcock, syringe, and Luer-Lock end of the device. Using the other hand, forcefully pull the syringe plunger to suck up any trapped air. When pulling the fluid into the syringe, make sure that the plunger at the end of the fluid cylinder is not pulled completely inside the cylinder. Slowly, depress the syringe plunger to reintroduce the degassed fluid into the cylinder. Close the petcock valve to the injury device and retest the device. Repeat this degassing step until the waveform produced on the oscilloscope is smooth (not jagged).

19.5.2 Inconsistent Pressure Pulse Waveforms

The atm corresponding to the height of the pendulum arm should be tested before starting and in between injuries. Overnight, the vacuum grease around the plunger will cause it to stick to the wall of the fluid cylinder. To loosen the plunger, manually rotate it inside the cylinder, then use the syringe attached to the fill port to remove and inject fluid into the cylinder. When loosening the plunger or adjusting it for the injury, always remove fluid first to avoid injecting an unseen air bubble further into the device.

The force associated with the fluid pulse can decrease within minutes without changing the angle of the hammer. To control for this, prime the device by performing a "test hit" every 30 seconds before performing the injury. With one hand, brace the syringe and petcock on the end of the device and release the weighted pendulum arm with the other. Make sure the plunger at the end of the fluid cylinder is adequately lubricated with vacuum grease and able to move freely within the cylinder.

19.5.3 Inconsistent Righting Reflex Times

Inconsistent righting times despite similar atmospheres of pressure generated by the injury device can result from several factors such as obstruction (complete or partial) at the injury site, lack of a watertight seal between the injury hub and skull, unintended disruption of the dura, and the amount of time required to close the scalp incision after FPI. The first thing to consider is an obstruction over the injury site, such as glue or blood clot. If there is a complete obstruction of the craniotomy opening, it will impede the fluid pulse from making contact with the dura and potentially result in reduced righting times. If the craniotomy site is partially obstructed, it may focus the fluid pulse and intensify the injury effect and possibly increase righting time. Lack of a watertight seal between the skull and the injury hub will likely result in a less severe injury to the animal than what was intended because the force of the injury is not restricted to the craniotomy opening. Disruption of the dura underlying the screw holes or injury site can cause a second insult to the brain and potentially increase the righting time. Righting time can also be influenced by the amount of time it takes to close the incision after the injury. Use the minimum number of staples or stitches in combination with precise and efficient technique to minimize this potential confounding variable.

19.5.4 Lack of Injury-Induced Neurobehavioral Deficits

A lack of injury-induced effects on behavioral task performance between sham and injured animals is more likely to be an issue when using a mild or moderate injury severity level. The first thing to consider is the placement of the craniotomy. It has been shown that even small changes in the placement of the craniotomy can dramatically affect physiological and cognitive outcome measures (Floyd, 2002; Vink, 2001). These studies illustrate the importance of precise surgical technique in obtaining reliable data.

The sensitivity of the selected outcome measure may not be sufficient to detect differences between sham and injured animals. Some outcome measures have greater sensitivity; for example, the Rotarod is a more sensitive measure of motor deficits as compared to beam balance and beam walk tasks (Hamm, 2001). The Morris Water Maze, a cognitive task that measures spatial navigation and memory, may not be as sensitive as the swim T-maze in detecting differences between sham animals and those injured using a mild severity level FPI. In these instances, if a particular test must be used, a larger sample size may also be necessary to detect subtle between-group differences.

19.6 VARIATIONS

Because of the extensive characterization of the fluid percussion injury model, it is easier to compare findings between different laboratories. However, variations do exist between laboratories. The FPI model has been used in different species and the injury severity can be adjusted by either raising or lowering the angle of the pendulum. Given the variability across species, it is necessary to establish the injury parameters that will produce the desired mortality rate and suppression of the righting reflex. Minor adjustments to the surgery and injury procedures may be necessary to account for interspecies differences in relative size as well as variations in the characteristics of the scalp, skull, and dura. Other variations include the location of the craniotomy, anesthetic agent used, duration of time between completion of the preparatory surgery and injury, animal gender, and animal age at injury.

All anesthetic agents will, to some degree, alter the progression of the secondary injury effects (Rowe, 2013). The rapid clearance rate of inhaled anesthetics, such as isoflurane, can minimize these side effects. The injury can be delivered immediately after the preparatory surgery, following a delay between 60 and 90 minutes, or 24 hours later.

REFERENCES

Cortez, S. C., T. K. McIntosh, and L. J. Noble. 1989. Experimental fluid percussion brain injury: Vascular disruption and neuronal and glial alterations. *Brain Res* 482(2):271–82.

Darouiche, R. O., M. J. Wall Jr., K. M. Itain, M. F. Otterson, A. L. Webb, M. M. Carrick, et al. 2010. Chlorhexidine-alcohol versus povidone-iodine for surgical-site antisepsis. *N Engl J Med* 362(1):18–26.

Dixon, C. E., J. W. Lighthall, and T. E. Anderson. 1988. Physiologic, histopathologic, and cineradiographic characterization of a new fluid-percussion model of experimental brain injury in the rat. *J Neurotrauma* 5 (2):91–104.

Dixon, C. E., B. G. Lyeth, J. T. Povlishock, R. L. Findling, R. J. Hamm, A. Marmarou et al. 1987. A fluid percussion model of experimental brain injury in the rat. *J Neurosurg* 67(1):110–9.

Floyd, C. L., K. M. Golden, R. T. Black, R. J. Hamm, and B. G. Lyeth. 2002. Craniectomy position affects Morris water maze performance and hippocampal cell loss after parasagittal fluid percussion. *J Neurotrauma* 19(3):303–16.

Hamm, R. J. 2001. Neurobehavioral assessment of outcome following traumatic brain injury in rats: An evaluation of selected measures. *J Neurotrauma* 18(11):1207–16.

Hamm, R. J., B. G. Lyeth, L. W. Jenkins, D. M. O'Dell, and B. R. Pike. 1993. Selective cognitive impairment following traumatic brain injury in rats. *Behav Brain Res* 59 (1–2):169–73.

Hayes, R. L., L. W. Jenkins, and B. G. Lyeth. 1992. Neurotransmitter-mediated mechanisms of traumatic brain injury: Acetylcholine and excitatory amino acids. *J Neurotrauma* 9(Suppl 1): S173–87.

Kokiko-Cochran, O. N., M. P. Michaels, and R. J. Hamm. 2008. Delayed glucose treatment improves cognitive function following fluid-percussion injury. *Neurosci Lett* 436(1):27–30.

Lighthall, J. W., C. E. Dixon, and T. E. Anderson. 1989. Experimental models of brain injury. *J Neurotrauma* 6(2):83–97.

Lyeth, B. G., L. W. Jenkins, R. J. Hamm, C. E. Dixon, L. L. Phillips, G. L. Clifton et al. 1990. Prolonged memory impairment in the absence of hippocampal cell death following traumatic brain injury in the rat. *Brain Res* 526(2):249–58.

Morehead, M., R. T. Bartus, R. L. Dean, J. A. Miotke, S. Murphy, J. Sall et al. 1994. Histopathologic consequences of moderate concussion in an animal model: Correlations with duration of unconsciousness. *J Neurotrauma* 11(6):657–67.

Paxinos, G., Watson, C. (2007) *The Rat Brain in Stereotaxic Coordinates*. Academic Press/Elsevier, Amsterdam.

Perrett-Gentil, M.I. 2013. Rodent urgery: Application of aseptic technique and perioperative care, from http://research.utsa.edu/files/larc/RodentSurgeryApplicationhandouts.pdf.

Pierce, J. E. S., D. H. Smith, J. Q. Trojanowski, and T. K. McIntosh. 1998. Enduring cognitive, neurobehavioral and histopathological changes persist for up to one year following severe experimental brain injury in rats. *Neuroscience* 87(2):359–369.

Rowe, R.K., Harrison, J.L., Thomas, T.C., Pauly, J.R., Adelson, P.D., and Lifshitz, J. 2013. Using anesthetics and analgesics in experimental traumatic brain injury. *Lab Anim (NY)* 42(8):286–91.

Schmidt, R. H., and M. S. Grady. 1993. Regional patterns of blood-brain barrier breakdown following central and lateral fluid percussion injury in rodents. *J Neurotrauma* 10 (4):415–30.

Section IV

Imaging and Biomarkers

20 CNS Trauma Biomarkers and Surrogate Endpoints Pipeline from Bench to Bedside

A Translational Perspective

Tarek H. Mouhieddine, Leeanna El Houjeiri, Mirna Sabra, Ronald Hayes, and Stefania Mondello

CONTENTS

20.1 INTRODUCTION

A biomarker is a qualitative and quantitative biological substance or characteristic that defines a certain pathological condition and may give indications on disease severity and the type of therapy that should be administered to the patient. Traumatic brain injury (TBI) is an ever-growing public health concern, where the annual number of TBI patients reaches a staggering number of around 1.7 million cases in the United States alone. It is important that we continue the pursuit for the ideal TBI biomarker in order to decrease the number of fatalities secondary to TBI, and its direct and indirect costs. However, the path towards reaching a surrogate endpoint biomarker is not easy but rather long and needs time and effort. In this chapter, we discuss the process of assessment and validation through which a biomarker passes through in order to be finally established as a TBI biomarker and approved by the FDA. We also provide a brief history on TBI biomarkers, discuss the different types and sources of biomarkers and give a summary of TBI biomarkers currently in use and their potential clinical application.

The term "biomarker" refers to objective and quantifiable characteristics of biological processes that can define a normal or a pathogenic state. Biomarkers can be identified as a broad subcategory of medical signs, which, by accurate measures and reproducibility, can provide an objective suggestion of the medical state examined from outside the patient. With evolving technology, we are able to better detect, identify, and quantify biomarkers and validate them in the clinical setting.

The use of the term biomarker dates back to as early as 1980 (Atkinson et al., 2001). Many precise definitions of biomarkers are found in the literature, and their contexts overlap. In 1998, the National Institutes of Health Biomarkers Definitions Working Group defined a biomarker as "a characteristic that is objectively measured and evaluated as an indicator of normal biological processes, pathogenic processes, or pharmacologic responses to a therapeutic intervention" (Strimbu and Tavel, 2010). The National Academy of Sciences defines a biomarker as an indicator that signals events in biological samples or systems (Committee on Biological Markers of the National Research Council, 1987). In addition, the International Program on Chemical Safety, led by the World Health Organization (WHO) and in coordination with the United Nations and the International Labor Organization, has defined a biomarker as "any substance, structure, or process that can be measured in the body, or its products and influences or predicts the incidence of outcome or disease" (WHO, 2001). In its report on the validity of biomarkers in the environmental risk assessment, the WHO went even more broadly with its definition, stating that "a biomarker includes almost any measurement reflecting an interaction between a biological system and a potential hazard, which may be chemical, physical, or biological. The measured response may be functional, physiological and biochemical at the cellular level, or a molecular interaction" (WHO, 1993).

All these terms, definitions, and characteristics were proposed to describe a biomarker, indicating that it may have the greatest value in early efficacy and safety evaluations, such as in vitro studies in tissue samples, in vivo studies in animal models, and early-phase clinical trials, to establish "proof of concept" (Atkinson et al., 2001). Yet, biomarkers may have many other important applications in detecting diseases and monitoring the health status of a patient. As Atkinson et al. stated, these applications include the use of such biomarkers as a diagnostic tool to identify patients with a possible disease or abnormal condition, such as hyperglycemia for the diagnosis of diabetes mellitus (Atkinson et al., 2001). Moreover, biomarkers can be used as a method to predict the stage of the disease and its severity such as measuring the concentration of prostate-specific antigen (PSA) in the blood to detect the level of tumor growth and metastasis, even though different clinical trials are currently having conflicting results about the efficacy of this biomarker (Acimovic et al., 2013; Haines and Gabor Miklos, 2013; Izumi et al., 2014; Pataky et al., 2014). Also, they can be used as an indicator of disease prognosis or a predictor of clinical response to an intervention whereby, for example, biomarkers can be measured to detect tumor shrinkage or regression in certain cancers (Everson et al., 2014; Okegawa et al., 2014) or determine the risk of heart disease (Eapen et al., 2014; Ligthart et al., 2014).

20.2 BIOMARKERS VERSUS CLINICAL ENDPOINTS

Despite many studies having explored the ability of different biomarkers to aid in decision-making and the management of patients with different diseases, it is important to note that a biomarker may or may not correlate perfectly with clinical efficacy/toxicity, and as such, does not necessarily associate with a patient's experience and sense of well-being. On the contrary, a clinical endpoint is a variable that reflects how a patient feels, functions, or survives (Temple, 1995). It represents one of the most credible characteristic assessments of how a patient benefits, which is manifested by the effect of a therapeutic intervention. For example, survival is considered as the gold-standard clinical endpoint for anticancer agents. However, many clear, well-defined clinical variables have also been used as clinical endpoints in appropriate circumstances such as stroke, myocardial infarction, and occurrence of predefined opportunistic infections. These variables have the potential to indicate the effectiveness of the intervention, as well as its safety, and could be used for internal decision-making within a pharmaceutical company. Such clinical endpoints, however, can be replaced with a faster, more sensitive evaluation of the effect of an experimental treatment: a surrogate endpoint (Boone and Kelloff, 1993).

20.3 SURROGATE ENDPOINT

Originating from the Latin word "subrogare," the term "surrogate marker" was introduced in the late 1980s (Brotman and Prince, 1988), preceded by a few years by the term biomarker (Paone et al., 1980), and was succeeded and replaced by yet another term: "surrogate endpoint" (Boone and Kelloff, 1993). When used as outcomes in clinical trials, biomarkers are considered to be surrogate endpoints; that is, they act as surrogates or substitutes for clinically meaningful endpoints. Surrogate endpoints are a subset of biomarkers. Although all surrogate endpoints can be considered biomarkers, it is likely that only a few biomarkers will achieve surrogate endpoint status. To be considered a surrogate endpoint, a biomarker should show solid epidemiological, therapeutic, and/or pathophysiological evidence that consistently and accurately predicts a clinical outcome, whether it be benefit or harm (Strimbu and Tavel, 2010) (Table 20.1).

TABLE 20.1
Summary of the Definitions of Biomarkers, Clinical Endpoints, and Surrogate Endpoints

Term	Definition
Biomarker	A characteristic that is objectively measured and evaluated as an indicator of normal biological processes, pathogenic processes, or pharmacologic responses to a therapeutic intervention (Biomarkers Definitions Working Group, 2001).
Clinical endpoint	Measurement providing information on how a patient feels, functions, or survives (Temple, 1995).
Surrogate endpoint	Measurement providing early and accurate prediction of both a clinical endpoint and the effects of treatment on this endpoint (Boone and Kelloff, 1993).

20.4 VALIDITY OF A BIOMARKER AS A SURROGATE ENDPOINT

Clinical relevance of a biomarker is essential for its identification as a surrogate endpoint. As such, a biomarker should provide clinically relevant information and answers to the questions asked by health providers and officials. Atkinson et al. have compiled a table of biomarkers that have found varying degrees of clinical utility (Table 20.2) (Atkinson et al., 2001).

In addition, a biomarker needs to be identified as a valid marker through its degree of effectiveness as a surrogate endpoint. Unfortunately, there are no strict criteria for the validation of a surrogate endpoint, but there exist general guidelines that provide a framework for understanding the nature of the evidence that is provided by surrogate endpoints (Bucher et al., 1999). As Bucher et al. describe, for its validation, a surrogate endpoint should show a strong, independent, consistent association with its respective clinical endpoint (Bucher et al., 1999). Also, there should be evidence from randomized trials in other drug classes showing that improvement in the surrogate endpoint has consistently led to improvement in the target outcome. Moreover, large, precise, and lasting treatment effects should exist, accompanied by benefits that are worth the potential harms and costs. Stated in another way, "if there is no clearly established causal link between (biomarkers of) a given pharmacologic action and clinical outcome, then no amount of evidence confirming that the pharmacologic action is linked to treatment will increase confidence that the clinical outcome is also linked to treatment" (Peck et al., 2003).

Nonetheless, it is important to note that some researchers have replaced the term "validation" with "evaluation," considering the former unsuitable for studying biomarkers because it suggests that there could be a complete biological understanding of the relationship between a given biomarker and a clinical endpoint (Biomarkers Definitions Working Group, 2001). Evaluation refers to the ongoing process of studying the biomarkers' success in acting as surrogates for individual clinical endpoint. In this case, treatment development via biomarkers could help reduce the uncertainty about the relationship between an intervention, a biomarker, and a clinical endpoint rather than conclusively proving so.

TABLE 20.2
Examples of Biomarkers Used as Surrogate Endpoints with Their Clinical Relevance/Endpoints

Biomarker/Surrogate Endpoint	Clinical Relevance/Endpoint
Cholesterol	Coronary artery disease, heart attacks
Blood pressure	Stroke, atherosclerosis, heart failure
Viral RNA	Survival, decrease in infections
CD4+ T cells	Sustained reduction in viral RNA
HbA1C, glucose	Diabetic neuropathy
Intraocular pressure	Preservation of peripheral vision
Bone mineral density	Fracture rate
Magnetic resonance imaging scans	Decrease in rate of progression disease
Computed tomography scans/X-rays	Survival, cancer
Intraocular pressure	Glaucoma
Fever	Infection
Increased viral load	HIV infection
Reduced absolute neutrophil count	Risk for infection
Reduced platelet count	Risk of bleeding
Increased serum creatinine phosphokinase	Rhabdomyolysis
Increased serum creatinine	Damage to the kidney

20.5 ADVANTAGES AND CAPABILITIES OF BIOMARKERS AS SURROGATE ENDPOINTS

Novel biomarkers have the ability to modify the current health care system, shifting from a "one-size-fits-all" approach to one that is increasingly more "personalized," through which biomarker-based tests can be used by practitioners to optimize treatment strategies for patients.

There exist many reasons for the increasing interest in using biomarkers as surrogate endpoints for targeting clinical endpoints in pharmaceutical research. Surrogate endpoints may offer the potential for improved compliance (Chuang-Stein and DeMasi, 1998) because the trial duration for each patient is reduced (Weir and Walley, 2006). In addition, it becomes possible to recruit less severely ill patients when the clinical endpoint is disease-related mortality (Albert et al., 1998), leading to an improvement in cost-effectiveness (DeGruttola et al., 2001) and trials becoming more ethical (Biomarkers Definitions Working Group, 2001) by avoiding waiting for a potentially fatal final clinical endpoint. Moreover, surrogate endpoints may assist in dose selection (Rolan, 1997) and understanding the route of drug action (Lonn, 2001). Currently, there are few surrogates that are accepted by regulatory authorities. Biomarkers have been approved by U.S Food and Drug Administration (FDA) regulation for use as surrogate endpoints in the treatment development process (Lesko and Atkinson, 2001). For example, blood pressure is an accepted surrogate endpoint for antihypertensive agents because it predicts cardiovascular disease, heart failure, stroke, and kidney failure (Chobanian, 1983). As such, a surrogate endpoint is a biomarker that can be trusted to serve as a stand-in for, but not as a replacement of, a clinical endpoint (Strimbu and Tavel, 2010).

The role of biomarkers in the health care system is clear and significant; however, there exist many drawbacks and challenges that remain to be overcome to achieve a widespread adoption of biomarkers in personalized medicine and effective disease prediction.

20.6 DRAWBACKS AND RISKS OF BIOMARKER USE

As with any diagnostic tool, biomarkers have been associated with many risks and disadvantages. An ideal biomarker used as a surrogate endpoint should reflect all relevant clinical information, not just information relating to efficacy (Prentice, 1989). An example of this problem is illustrated in a trial that compared the effect of different cholesterol-lowering agents on the reduction of atherosclerosis (Nissen et al., 2004). In this trial, Lipitor was compared with Pravachol in a blinded manner. When matched for degree of cholesterol lowering, Lipitor produced a significantly larger benefit in reversing atherosclerotic disease, suggesting that measuring cholesterol does not confine all of the relevant clinical information. As such, a surrogate endpoint might be useful for predicting benefit, but might not be satisfactory for determining degree of benefit. Thus, one should be cautious in using a biomarker as the main predictor of an outcome instead of clinical parameters; this entails an obvious need for developing appropriate standards and requirements to facilitate the sensitive evaluation of biomarkers before integrating in clinical systems.

Lesko and Atkinson have described many risks of biomarker usage that could arise for some of the following reasons (Lesko and Atkinson, 2001). A new molecular entity, such as a drug or intervention produced as a response to a biomarker, could affect the biomarker itself but can have no clinical outcome results. In this case, the biomarker is nonspecific. As such, in choosing this specific biomarker in early-phase clinical development, the pharmaceutical company could end up wasting time and huge amounts of money by relying on an inappropriate biomarker. In addition, the new molecular entity could affect both the biomarker and the clinical outcome; however, such an effect might be to a different extent in each. In such a case, even if a correlation exists between a biomarker and the clinical outcome, the former could account for a small portion of the clinical benefit and thus create a false-negative impression that fools the pharmaceutical company into making the wrong decision of discontinuing the development of a potentially good drug (Blyth et al., 2011).

20.7 BIOMARKERS OF TRAUMATIC BRAIN INJURY

Detecting biomarkers of TBI is a difficult process because of the complex pathophysiological changes that take place in the brain after TBI. TBI may include the activation of apoptotic, necrotic, or even necroptotic pathways that include using multiple molecular pathways that induce oxidative stress, inflammation, vascular dysfunction, and secondary tissue injury (Wang et al., 2012). In this chapter we will discuss the long process that a biological substance or characteristic passes through before finally being established as a biomarker and a surrogate endpoint for a certain disease and being used in the clinical setting.

20.7.1 Selection of TBI Biomarkers

After TBI, a vast array of cellular pathways are activated, yielding hundreds of molecular products that are differentially expressed, depending on the type and severity of the injury that in turn dictate which pathways are activated (Guingab-Cagmat et al., 2012; Loov et al., 2013). This is the first challenge of the TBI biomarker selection process. The release of hundreds of products into the cerebrospinal fluid (CSF) and blood requires a lot of time and money to invest in identifying and studying each substance alone. Thus, it is imperative to narrow down the list of markers to be studied before further assay development. Other factors that come into play in deciding the appropriateness of the TBI marker are the level of expression of the biomarker and the specificity toward TBI with respect to other disease states and neurological disorders. Fortunately, methods of identification and quantification are evolving (e.g., proteomics, genomics, lipidemics, metabolomics, degradomics) and we are able to better isolate and quantify molecules that are expressed in negligible amounts. This will allow us to update the list of TBI biomarkers to include substances that we are currently unable to detect in biological fluids because of technological limitations.

20.7.2 Assay Optimization

One of the longest and most cumbersome processes of biomarker discovery is optimization of the selection, identification, and quantification methodologies. Even though not much credit is given to the optimization process, it entails a major part of translational research without which we could end up with nonreplicable results or false positives and negatives. Assay optimization may range from selecting the type of assay—such as enzyme-linked immunosorbent assay (ELISA), Luminex, immunofluorescence, chemiluminescence, polymerase chain reaction, microarray, next-generation sequencing, and mass spectroscopy—to choosing the appropriate temperature, concentration, and incubation time of reagents. Different parameters and conditions are applied and repeated several times, and the conditions with the most consistent results are usually adopted. Noteworthy is that the same experiment may need different parameters when done in two different laboratories. This could be due to multiple factors such as the company from which the laboratory instruments are obtained, the way the laboratory users handle the instruments and biological fluids, the laboratory temperature and location, etc. Thus, even though there is a general protocol to be followed for a certain experiment, an assay optimization will always bring at least some minor changes into the process. The ultimate goal of assay optimization is to create a very specific and sensitive protocol that practically abolishes or at least diminishes the chance of getting false results.

20.7.3 Model and Biomarker Validation

It is of the utmost importance to validate the model we are using for seeking TBI biomarkers. The factors to be considered for model validity assessment are what the FDA seeks in every newly introduced model, including sensitivity, specificity, precision, accuracy, stability, and reproducibility. More important is the ability of the newly identified biomarker to fulfill the same factors that are needed for the validity of the model to reach appropriate cutoff points. The sensitivity (detecting true positives) and specificity (detecting true negatives) of a biomarker are usually measured and their relationship calculated through the receiver operating characteristic curve (Bossuyt et al., 2003; Obuchowski et al., 2004). The receiver operating characteristic curve enables us to assess the discriminative predictive ability of the biomarker through the area under the curve where 50% represents random guessing, whereas 100% represents a perfect prediction (Obuchowski et al., 2004). Moreover, the ultimate goal of the biomarker is to be used in the clinical setting, and thus it is only logical that we validate the biomarker through clinical research (prospective or retrospective) and clinical trials. We should achieve a statistically significant correlation between a biomarker expression and a specific malady (TBI in this case) through strictly controlled clinical trials that compare different TBI states and severities with controls and healthy people from the general population. By that, we would validate that the TBI biomarker is only absent or present at a high level of expression in TBI patients only. Furthermore, we should monitor any changes in expression over long periods through constant follow-ups. Such studies could also provide us with an assessment of various factors (such as additional morbidities in a patient) and their contribution to the production of the TBI biomarker. Thus, tightly controlled clinical trials are one of the pillars of reaching the ideal TBI biomarker and applying it in the clinic, which will be discussed in the following section.

20.7.4 Assay Development

The desire for more accurate and precise instrumentation for biomarker detection has thrived in the industry of assay platform development. Traditional methodologies have been enhanced such as Luminex (an enhanced version of ELISA), which is able to detect 100 more analytes than the traditional ELISA and at a higher sensitivity. Another example is chemiluminescence, which has the same function as ELISA but can detect analytes at a much higher sensitivity and specificity. Furthermore, we currently have the cytometric bead array, which represents a new, sensitive method to detect analytes using flow cytometry. What we need in the field of TBI is the ability to measure acute biomarkers in a point-of-care (POC) setting and not just in the laboratory. A POC represents the setting in which the injury took place, such as a car accident site, a battlefield, or a sports field. We need POC handheld devices that can directly and accurately measure TBI biomarkers, similar to the handheld dextrometer of diabetic patients that measure glucose level within seconds (Bressan et al., 2014; Haselwood and La Belle, 2014). One of the prominent POC handheld devices is the i-STAT System (produced by Abbott Point of Care Laboratories) that provides a wide selection of cartridges that can be used for different disease states and that can accurately test for a wide array of substances in a few drops of blood (Bock et al., 2008; Donaldson et al., 2010; Nanduri et al., 2012). What is special about POC handheld devices is the ease of use and portability, and that there is no need for any kind of sample preparation. POC handheld devices can include different formats, including small strips that drag substrates by capillary or magnetic action, polymerase chain reaction–based amplifications, or fluid reagents in cartridges that are mixed once the assay is initiated. The realm of a POC handheld device seems to be a very promising one that is on its way to becoming one of the gold standards of medical practice because of its numerous benefits.

20.7.5 History of TBI Biomarkers

Studies to determine biomarkers for TBI have been exponentially increasing, especially during the past decade. The tens of thousands of studies on TBI date back to as far as the early and mid-1800s (Brodie, 1828; May, 1850; Mynter, 1894), but the concurrent studies of TBI and its biomarkers emerged in 1971 and started on guinea pigs and not humans (Shirinian, 1971a, 1971b). The first studies on human TBI looked for subsequent biomarkers in the CSF, such as the increasing levels of cyclic adenosine monophosphate (Fleischer et al., 1977; Rudman et al., 1976), cytokines (Bell et al., 1997; McClain et al., 1991; McClain et al., 1987; Whalen et al., 2000) and some cellular effector molecules (Clark et al., 2000; Robertson et al., 2001; Satchell et al., 2005). The studies either investigated markers of structural tissue damage or biochemical repair cascades after TBI. The difference in the type of investigated biomarkers dictated different methodologies of exploration and opened up the field to various options and ways to treat TBI. Today, the search for TBI biomarkers involves various body fluids and tissue lysates and not just the CSF, and new technological instruments and neuroimaging have allowed us to accurately identify and assess a new and larger panel of biomarkers and even better predict the progress and outcome of the injury (Agoston et al., 2009; Guingab-Cagmat et al., 2014; Strathmann et al., 2014).

20.7.6 The Ideal TBI Biomarker

To classify a molecule as an ideal biomarker, there are certain criteria that should be taken into consideration. The ideal candidate TBI biomarker should be an easily measured, sensitive, brain-specific marker that could be detected in the blood as soon as the TBI symptoms ensue (Bettermann and Slocomb, 2012). The level of the TBI biomarker should correlate with the size and location of the lesion, the severity of the brain injury (i.e., Glasgow Coma Scale), and the clinical outcome (i.e., Extended Glasgow Outcome Scale), and respond to the administered treatment (Bettermann and Slocomb, 2012; Ottens et al., 2007). Furthermore, the biomarker signal should not be confused with other neurological states that are of non-TBI etiology (Ottens et al., 2007). Thus, it is important to continuously improve our instruments to have better sensitivity and specificity to accommodate the required criteria for discovering new biomarkers that would enable us to reliably diagnose a TBI case even if it had no apparent symptoms.

20.7.7 Types of TBI Biomarkers

In the field of TBI, there are different types of biomarkers, including molecular biomarkers and biomarkers of neuroimaging. Discovering molecular biomarkers in TBI can be further divided into finding a single marker of injury via traditional methods (Western blot) or screening for a panel of markers via genomic (Birnie et al., 2013; White et al., 2013), proteomic (Guingab-Cagmat et al., 2012; Loov et al., 2013; Wu et al., 2013), metabolomic (Glenn et al., 2013; Viant et al., 2005; Walsh et al., 2012; Wang et al., 2013b), and lipidomic (Fonteh et al., 2006; Kiebish et al., 2008) approaches. Furthermore, the field of neuroimaging greatly enriched the search for TBI biomarkers because TBIs are heterogeneous in nature and need different methodologies to detect the variable TBI manifestations. This could be translated into the fact that computed tomography scans can easily and rapidly detect TBI manifestations in patients with severe head trauma to directly provide the appropriate diagnosis and treatment (Toyama et al., 2005), whereas magnetic resonance imaging scans are better at detecting changes in the white matter of the brain (Garnett et al., 2001) and can be provided after stabilizing the patient. The different neuroimaging modalities include ultrasound (comprising transcranial Doppler ultrasonography and contrast-enhanced ultrasound) (Bouzat et al., 2011; Gura et al., 2012; He et al., 2013), magnetic resonance imaging (Caeyenberghs et al., 2013; Fozouni et al., 2013; Takayanagi et al., 2013), computed tomography (Sarkar et al., 2014; Sarubbo et al., 2014), positron emission tomography (Mendez et al., 2013; Selwyn et al., 2013; Shultz et al., 2013), and single photon emission computed tomography (Koizumi et al., 2010; Koizumi et al., 2013; Newberg et al., 2014). The different neuroimaging techniques were further enhanced by the introduction of fractional anisotropy, diffusion tensor imaging, and fiber tractography (Fox et al., 2013).

20.7.8 Sources of TBI Biomarkers

Biological fluids (blood, CSF, and urine) are the most analyzed sources for biomarkers because of the ease of acquiring them with minimal invasiveness and the ability to undergo high throughput screening. The CSF is highly preferred for diagnostic analysis due to its direct contact with the central nervous system, thus giving a direct picture on what's happening there (Yuan and Desiderio, 2005). On the downside, CSF can only be acquired from rats and mice in small volumes (50–150 μL and 25–30 μL, respectively) and at low protein concentrations (less than 1 μg/L), whereas in humans the volume is larger (2–5 mL) but still negligible compared with other body fluid volumes such as blood and urine (Ottens et al., 2006). Unfortunately, the different methods of obtaining CSF (such as lumbar puncture and ventriculotomy) from different locations leads to differential protein expression in the CSF that has nothing to do with the injury (Ottens et al., 2006).

Once the brain is inflicted by trauma, the blood–brain barrier (BBB) could be breached and molecules, secreted by the brain because of the injury, would leak into the blood (Romner, 2000). However, the concentrations of those substrates could be very low compared to the other highly expressed proteins in the blood and thus would be difficult to detect even via extremely sensitive instruments such as mass spectroscopy. Moreover, the BBB may not be breached after TBI after all, and any potential biomarkers would not be secreted and detected in the blood. Thus, even though blood seems to be a very appealing source of biomarkers, it still is not an accurate diagnostic body fluid. Urine and saliva are also body fluids that can be obtained noninvasively; however, the drawback lies in the fact that any biomarkers that would be found in these body fluids, including blood, would be highly diluted and complex and would necessitate the use of analytical tools that are much more powerful than the ones we currently have (Fang et al., 2009; Wang et al., 2013a).

Therefore, it seems that the best source of TBI biomarkers is brain tissue, which can only be obtained from postmortem patients. Even then, the biomarkers found in the tissues would not be very representative because of the rapid deterioration of the brain samples from the open air, reagents and collecting techniques themselves (Kobeissy et al., 2008; Taurines et al., 2011). Even though there is a general protocol to be followed for detecting biomarkers, one still has to adjust the protocol according to the type of biological samples (Ottens et al., 2007). Furthermore, with more heterogeneous samples, there will be more biological variability that can reach a coefficient of variation of as much as 26% for mammalian cell line cultures and around 50% in primary animal cell cultures (Molloy et al., 2003).

20.7.9 Clinical Application of TBI Biomarkers

The ultimate aim behind finding TBI specific biomarkers is to use them for timely and accurate diagnosis of TBI in the clinical setting. Instead of undergoing plain observation that could be fooled by hidden symptoms of TBI, or expensive neuroimaging, a rapid and inexpensive test for an ideal biomarker can be done to determine whether the patient has TBI or not and to determine the severity of the injury. Another advantage of biomarkers is the ability to predict the progress of TBI symptoms and any potential complications that may arise (such as edema and ischemia) that could prevent patients from safely resuming their day-to-day activities (Bermpohl et al., 2007; Harris et al., 2012). If appropriate biomarkers were used in the clinical setting, the degree of severity of TBI and its forthcoming complications could be reduced or even prevented with a minimal chance of error and at a faster pace than what we are capable of today. Furthermore, TBI biomarkers could provide us with an assessment of the patients' response to TBI treatment or can even specify and guide the patients' therapy by informing us of any secondary insults that might have taken place in the patients. There is an estimated annual number of 1.7 million TBI cases in the United States, of which 52,000 people die (Faul et al., 2010), and the total direct medical and indirect TBI cost was $60 billion in 2000 (Finkelstein et al., 2006). With the use of TBI biomarkers, it would be possible to decrease that huge number of deaths and the ever-increasing costs of TBI.

TABLE 20.3
A Summary of the Most Prominent Candidate TBI Biomarkers

Biomarker	Abbreviation	Source	Location	Physiologic Role	Mechanism/Target	References
Neuron-specific enolase	NSE	CSF and serum	Cytoplasm of neurons, platelets, red blood cells	Increases chloride levels in neurons upon activation	Glycolytic enzyme, marker of axonal injury	Bohmer et al., 2011; Ondruschka et al., 2013; Wolf et al., 2013
S100B	N/A	CSF and serum	Astroglia, bone marrow, rat, skeletal muscle	Calcium-binding protein with low affinity. Possibly involved in the inhibition of synaptic plasticity	Glial and BBB dysfunction marker	Egea-Guerrero et al., 2013a, 2013b; Thelin et al., 2013a, 2013b
Myelin-basic protein	MBP	CSF and serum	Myelin	A protein found in abundance in myelin	Demyelination marker	Ajao et al., 2012; Rostami et al., 2012; Su et al., 2012
Cleaved tau	C-tau	CSF and serum	Axons of the central nervous system	Usually cleaved proteolytically, associated with axonal structural proteins	Axonal injury marker	Bazarian et al., 2006; Gabbita et al., 2005; Huber et al., 2013
Glial fibrillary acidic protein	GFAP	Serum	Astroglial skeleton	Skeletal intermediate filament protein in glial cells	Gliosis marker	Diaz-Arrastia et al., 2014; Okonkwo et al., 2013
Ubiquitin C-terminal hydrolase	UCH-L1	CSF and serum	Neuron cell body	Deubiquitinating enzyme	Neural cell breakdown	Berger et al., 2012; Brophy et al., 2011; Mondello et al., 2012; Papa et al., 2010, 2012
Neurofilament-L and neurofilament-H	NFL and NFH	CSF and serum	Axons	Axonal structural support	Axonal injury marker	Ahmed et al., 2012; Anderson et al., 2008; Vajtr et al., 2012; Zurek et al., 2011; Zurek and Fedora, 2012
Brain fatty acid binding protein	B-FABP	Serum	Glial cells	Carrier proteins for fatty acids and other lipophilic molecules	Glial cell break-down marker	Hulscher et al., 2014; Sharifi et al., 2011
αII-spectrin break-down products	SBDP150, SBDP145, SBDP120	CSF and serum	Axons	Cytoskeletal protein associated with the intracellular side of plasma membranes	Neural necrosis/apoptosis, axonal injury	Brophy et al., 2009; Cardali and Maugeri, 2006; Mondello et al., 2010; Pineda et al., 2007; Siman et al., 2013; Valiyaveettil et al., 2014
Amyloid-beta (1-42)	Aβ1-42	Serum	Generated in the neuronal axonal membranes	Not well understood	Neural necrosis/apoptosis, axonal injury	Chen et al., 2009; Cheng et al., 2013; Gatson et al., 2013; Marklund et al., 2014; Tajiri et al., 2013

20.7.10 Prominent TBI Biomarkers

Currently, a wide panel of TBI biomarkers has been established of which only a specific few are being studied and used extensively (Table 20.3). One of the most heavily studied TBI biomarkers is the S100B, which is a glial, low-affinity, calcium-binding protein that is mainly expressed and secreted by astrocytes, in addition to being present in neurons, microglia, chondrocytes, adipocytes, and bone marrow cells (Donato, 2001). The exact function of S100B is still ambiguous but it may be involved in regulating cellular activities, morphology, and viability depending on its concentration (Goncalves et al., 2008; Kleindienst and Ross Bullock, 2006). S100B is present in negligible amounts in healthy individuals, but after TBI, it rises rapidly with a serum half-life of about 30–90 minutes (Jonsson et al., 2000). Another glial TBI biomarker is the glial fibrillary acidic protein, a monomeric intermediate filament protein present in the astroglial skeleton (Missler et al., 1999; Vos et al., 2004) that has been correlated with TBI (Nylen et al., 2006; Vos et al., 2004), stroke (Herrmann et al., 2000), and neurodegenerative disorders (Baydas et al., 2003).

TBI biomarkers also include a large set of neuronal markers such as the neuron-specific enolase that is a cytoplasmic glycolytic pathway enzyme found in neurons, oligodendrocytes, erythrocytes, platelets, and neuroendocrine cells (Kovesdi et al., 2010). Neuron-specific enolase may also indicate neuronal cell death and its increasing concentration in the CSF and serum correlates with the severity of injury and clinical outcome (Herrmann et al., 2000; Ross et al., 1996; Selakovic et al., 2005). Another neuronal cytoplasmic enzyme TBI marker is the ubiquitin C-terminal hydrolase (UCH-L1) that is involved in ubiquitin addition or removal (Laser et al., 2003). UCH-L1 is responsible for removing excessive, misfolded, or oxidized proteins during both normal and neuropathological situations, including neurodegenerative disorders (Kobeissy et al., 2008). CSF levels of UCH-L1 are associated with the severity of TBI, clinical outcome, and abnormal BBB permeability (Blyth et al., 2011; Papa et al., 2010; Papa et al., 2012). A fifth biomarker is the αII-spectrin that is found in neurons (especially axons and presynaptic terminals) (Riederer et al., 1986) and is broken down into spectrin breakdown products (SBDPs) of molecular weights 150 kDa (SBDP150) and 145 kDa (SBDP145) via calpain and 120-kDa product (SBDP120) via caspase-3 (Mondello et al., 2010). Because calpain and caspase-3 execute apoptotic and necrotic cell death, SBDPs not only reveal the severity of TBI in patients but also give an idea on the underlying cell death mechanisms (Mondello et al., 2010). The list of TBI biomarkers includes myelin basic protein, phosphorylated neurofilaments (neurofilament L and H), brain fatty acid binding protein, amyloid β_{1-42}, and cleaved tau protein.

20.8 LIMITATIONS AND CHALLENGES

The main challenge in developing a panel of TBI biomarkers lies in the methodology of validating the biomarkers that were found in cell culture and animal model studies in the clinical field. That is mainly done by comparing the identified biomarkers in preclinical studies to those found in clinical trials on human subjects. The major drawback in that comparison is our inability to simulate and integrate all the risk factors, which are normally found in human subjects, in preclinical studies. Such risk factors may include genetics, age, gender, race, secondary insults, and environmental factors such as occupation and smoking. Furthermore, those risk factors may also contribute to the differential results when comparing different studies on human subjects. In addition to that, current assays for assessing the severity of TBI, such as the Glasgow Coma Scale, are not specific for TBI but rather for other conditions such as coma. Moreover, different preclinical and clinical studies have used different methods and techniques with different specificities and sensitivities for isolating and identifying TBI biomarkers, which may be behind the discrepancies between the results of different studies. These in turn push us to conduct more studies with larger sample sizes and develop better, standardized methods with proper normalization techniques and higher sensitivities and specificities to accurately identify biomarkers and reduce variability in the results.

20.9 STRATEGY FOR REGULATORY APPROVAL BY FDA

In the United States, compliance with FDA regulations is required for bringing a biomarker test (classified as an in vitro diagnostic device) into the market for clinical use. According to the Medical Device Amendments Act (Greenberg, 1976), an in vitro diagnostic device must prove safety and effectiveness for its intended use. In addition, a new assay is required to demonstrate an adequate analytical performance (appropriate accuracy and precision, among others) and clinical performance (sensitivity, specificity, and some indication of clinical utility).

There are two primary regulatory pathways used by the FDA depending on the test classification (e.g., the risk to the patient that arises when the test provides incorrect results): 510(k) premarketing clearance or premarket approval (PMA).

The 510(k) applications are used to approve moderate risk (class II) tests through the premarket notification. Specifically, the sponsor informs the FDA that it will be introducing the test into the market in 90 days, unless the FDA raises any concerns before the 90 days are up. Under the 510(k) process, the sponsor demonstrates that there is a "substantially equivalence" to a predicate device (e.g., a class II device that has been previously cleared by the FDA or that was in commercial distribution before the introduction of the 510(k) program in 1976) and tests are cleared, not approved, by the agency. The new test does not need to prove safety and effectiveness for its intended use but has to demonstrate that its performance is substantially equivalent to the performance of the predicate. In addition, the sponsor must show characterization of analytical capability of the test (e.g., specificity and accuracy, precision and linearity by correlating patient studies against the predicate device) (FDA, 2014).

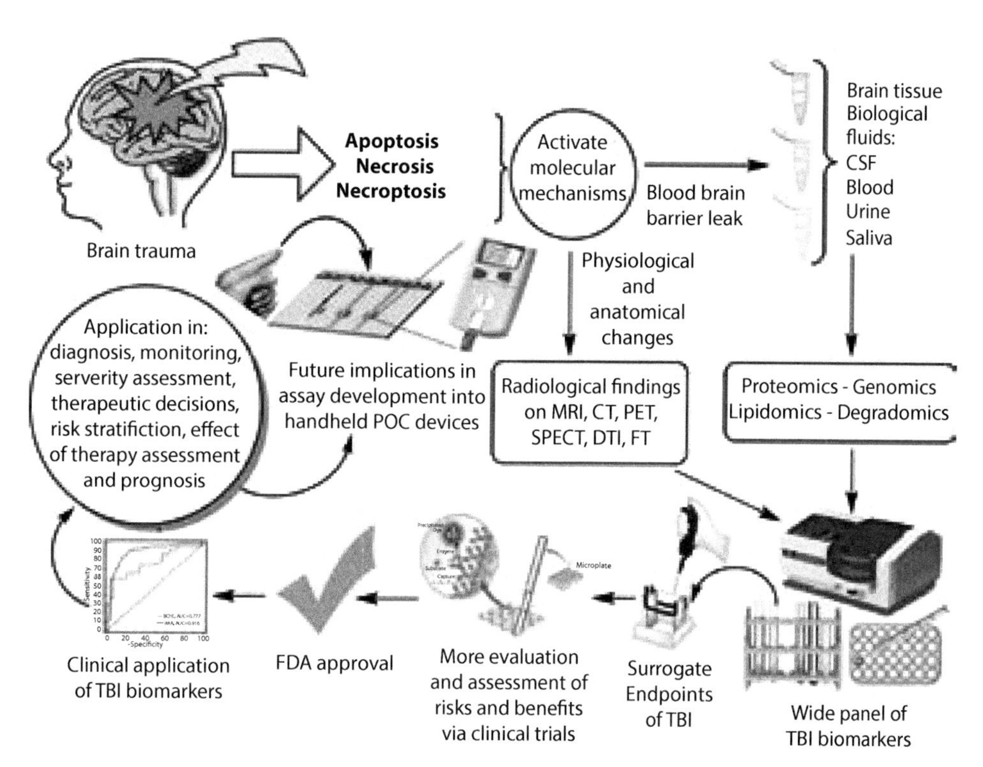

FIGURE 20.1 A schematic of the trajectory of identifying and validating a potential TBI biomarker from bench to bedside. The figure represents a pipeline starting from the point of TBI infliction, activating cellular mechanisms involved in apoptosis, necrosis, and necroptosis. This is followed by a leakage of cellular components through the BBB to be found in biological fluids and brain tissue via different detection techniques, and physiological and anatomical changes detected on radiological tests. After assay optimization, a group of surrogate endpoints are selected from the wide panel of detected biomarkers. These undergo further evaluation through clinical trials to be finally approved by the FDA. Finally, we are able to use these surrogate endpoints to diagnose and assess the severity of TBI, guide our therapeutic decisions, and even use them in POC handheld devices.

On the other hand, the PMA process is used when the test is classified as class III; that is, either it is associated with high risk (e.g., when the outcome determines cancer diagnosis or treatment) or the clinical utility of the marker or the technology of the measurement are novel and no predicate device can be identified. The standard for premarket approval is that the sponsor (the firm submitting the PMA application) must demonstrate, through analytical and clinical performance studies, that the test is safe and effective for use in patient care (FDA, 2014).

For some newly discovered analytes that do not have safety concerns (class II) but where no appropriate predicate device exists, the FDA created a special process known as the de novo pathway. This process allows a new biomarker to be regulated as in a 510(k), but requires the demonstration of clinical effectiveness. The advantage of the de novo pathway for manufacturers is that it permits novel biomarker tests to reach the market more quickly and with fewer expenses.

Finally, a class I test (the lowest risk test) does not require a premarket submission to the FDA; however, the sponsor is still required to register the test with the agency. Importantly, the FDA and other international regulatory bodies have identified the development of biomarkers as a high priority. In this regard, biomarkers are considered a substantial component of the Critical Path Initiative introduced by the FDA with the intent of modernizing drug development (Figure 20.1).

20.10 CONCLUSION

Blood biomarkers have a potential high diagnostic value in the context of TBI. Demonstration of clinical utility and compliance with regulatory requirements is critical for the commercialization of novel biochemical markers but also formidable, uncertain, and costly. Nevertheless, a biomarker panel approach elucidating mechanisms of injury and allowing the identification of potentially complex pathophysiological alterations is likely to provide a significant added value when combined with clinical and radiological information to guiding therapies and improving patient outcome, and we appear to be on the verge of a breakthrough with the use of these markers in TBI.

REFERENCES

Acimovic, M., K. Dabic-Stankovic, T. Pejcic, Z. Dzamic, D. Rafailovic, and J. Hadzi-Djokic. 2013. Preoperative Gleason score, percent of positive prostate biopsies and PSA in predicting biochemical recurrence after radical prostatectomy. *Journal of B.U.ON.: Official journal of the Balkan Union of Oncology.* 18:954–960.

Agoston, D.V., A. Gyorgy, O. Eidelman, and H.B. Pollard. 2009. Proteomic biomarkers for blast neurotrauma: Targeting cerebral edema, inflammation, and neuronal death cascades. *J Neurotrauma*. 26:901–911.

Ahmed, F., A. Gyorgy, A. Kamnaksh, G. Ling, L. Tong, S. Parks et al. 2012. Time-dependent changes of protein biomarker levels in the cerebrospinal fluid after blast traumatic brain injury. *Electrophoresis*. 33:3705–3711.

Ajao, D.O., V. Pop, J.E. Kamper, A. Adami, E. Rudobeck, L. Huang et al. 2012. Traumatic brain injury in young rats leads to progressive behavioral deficits coincident with altered tissue properties in adulthood. *J Neurotrauma*. 29:2060–2074.

Albert, J.M., J.P. Ioannidis, P. Reichelderfer, B. Conway, R.W. Coombs, L. Crane et al. 1998. Statistical issues for HIV surrogate endpoints: Point/counterpoint. An NIAID workshop. *Statis Med*. 17:2435–2462.

Anderson, K.J., S.W. Scheff, K.M. Miller, K.N. Roberts, L.K. Gilmer, C. Yang et al. 2008. The phosphorylated axonal form of the neurofilament subunit NF-H (pNF-H) as a blood biomarker of traumatic brain injury. *J Neurotrauma*. 25:1079–1085.

Atkinson, A.J., W.A. Colburn, V.G. DeGruttola, D.L. DeMets, G.J. Downing, D.F. Hoth et al. 2001. Biomarkers and surrogate endpoints: Preferred definitions and conceptual framework. *Clin Pharmacol Therapeutics*. 69:89–95.

Baydas, G., V.S. Nedzvetskii, M. Tuzcu, A. Yasar, and S.V. Kirichenko. 2003. Increase of glial fibrillary acidic protein and S-100B in hippocampus and cortex of diabetic rats: Effects of vitamin E. *Eur J Pharmacol*. 462:67–71.

Bazarian, J.J., F.P. Zemlan, S. Mookerjee, and T. Stigbrand. 2006. Serum S-100B and cleaved-tau are poor predictors of long-term outcome after mild traumatic brain injury. *Brain Inj*. 20:759–765.

Bell, M.J., P.M. Kochanek, L.A. Doughty, J.A. Carcillo, P.D. Adelson, R.S. Clark et al. 1997. Interleukin-6 and interleukin-10 in cerebrospinal fluid after severe traumatic brain injury in children. *J Neurotrauma*. 14:451–457.

Berger, R.P., R.L. Hayes, R. Richichi, S.R. Beers, and K.K. Wang. 2012. Serum concentrations of ubiquitin C-terminal hydrolase-L1 and alphaII-spectrin breakdown product 145 kDa correlate with outcome after pediatric TBI. *J Neurotrauma*. 29:162–167.

Bermpohl, D., Z. You, E.H. Lo, H.H. Kim, and M.J. Whalen. 2007. TNF alpha and Fas mediate tissue damage and functional outcome after traumatic brain injury in mice. *Journal of cerebral blood flow and metabolism: Official journal of the International Society of Cerebral Blood Flow and Metabolism*. 27:1806–1818.

Bettermann, K., and J.E. Slocomb. 2012. Chapter 1: Clinical relevance of biomarkers for traumatic brain injury. In: D. Thurston (ed.), *Biomarkers for Traumatic Brain Injury*. The Royal Society of Chemistry, Cambridge, UK. 1–18.

Biomarkers Definitions Working Group. 2001. Biomarkers and surrogate endpoints: Preferred definitions and conceptual framework. *Clin Pharmacol Therapeutics*. 69:89–95.

Birnie, M., R. Morrison, R. Camara, and K.I. Strauss. 2013. Temporal changes of cytochrome P450 (Cyp) and eicosanoid-related gene expression in the rat brain after traumatic brain injury. *BMC Genom*. 14:303.

Blyth, B.J., A. Farahvar, H. He, A. Nayak, C. Yang, G. Shaw, and J.J. Bazarian. 2011. Elevated serum ubiquitin carboxy-terminal hydrolase L1 is associated with abnormal blood-brain barrier function after traumatic brain injury. *J Neurotrauma*. 28:2453–2462.

Bock, J.L., A.J. Singer, and H.C. Thode, Jr. 2008. Comparison of emergency department patient classification by point-of-care and central laboratory methods for cardiac troponin I. *American journal of clinical pathology*. 130:132–135.

Bohmer, A.E., J.P. Oses, A.P. Schmidt, C.S. Peron, C.L. Krebs, P.P. Oppitz et al. 2011. Neuron-specific enolase, S100B, and glial fibrillary acidic protein levels as outcome predictors in patients with severe traumatic brain injury. *Neurosurgery*. 68:1624–1630; discussion 1630–1621.

Boone, C.W., and G.J. Kelloff. 1993. Intraepithelial neoplasia, surrogate endpoint biomarkers, and cancer chemoprevention. *J Cell Biochem Suppl*. 17F:37–48.

Bossuyt, P.M., J.B. Reitsma, D.E. Bruns, C.A. Gatsonis, P.P. Glasziou, L.M. Irwig et al. 2003. The STARD statement for reporting studies of diagnostic accuracy: Explanation and elaboration. *Clin Chem*. 49:7.

Bouzat, P., G. Francony, P. Declety, C. Genty, A. Kaddour, P. Bessou et al. 2011. Transcranial Doppler to screen on admission patients with mild to moderate traumatic brain injury. *Neurosurgery*. 68:1603–1609; discussion 1609–1610.

Bressan, S., M. Daverio, F. Martinolli, D. Dona, F. Mario, I.P. Steiner, and L.D. Dalt. 2014. The use of handheld near-infrared device (Infrascanner) for detecting intracranial haemorrhages in children with minor head injury. *Child's nervous system: ChNS: Official journal of the International Society for Pediatric Neurosurgery*. 30:477–484.

Brodie, B. 1828. Pathological and surgical observations relating to injuries of the brain: Part I. *Med Chir Trans*. 14:325–423.

Brophy, G.M., S. Mondello, L. Papa, S.A. Robicsek, A. Gabrielli, J. Tepas, 3rd et al. 2011. Biokinetic analysis of ubiquitin C-terminal hydrolase-L1 (UCH-L1) in severe traumatic brain injury patient biofluids. *J Neurotrauma*. 28:861–870.

Brophy, G.M., J.A. Pineda, L. Papa, S.B. Lewis, A.B. Valadka, H.J. Hannay et al. 2009. alphaII-Spectrin breakdown product cerebrospinal fluid exposure metrics suggest differences in cellular injury mechanisms after severe traumatic brain injury. *J Neurotrauma*. 26:471–479.

Brotman, B., and A.M. Prince. 1988. Gamma-glutamyltransferase as a potential surrogate marker for detection of the non-A, non-B carrier state. *Vox Sanguinis*. 54:144–147.

Bucher, H.C., G.H. Guyatt, D.J. Cook, A. Holbrook, and F.A. McAlister. 1999. Users' guides to the medical literature: XIX. Applying clinical trial results. A. How to use an article Measuring the effect of an intervention on surrogate endpoints. Evidence-Based Medicine Working Group. *JAMA*. 282:771–778.

Caeyenberghs, K., A. Leemans, I. Leunissen, K. Michiels, and S.P. Swinnen. 2013. Topological correlations of structural and functional networks in patients with traumatic brain injury. *Front Hum Neurosci*. 7:726.

Cardali, S., and R. Maugeri. 2006. Detection of alphaII-spectrin and breakdown products in humans after severe traumatic brain injury. *J Neurosurg Sci*. 50:25–31.

Chen, X.H., V.E. Johnson, K. Uryu, J.Q. Trojanowski, and D.H. Smith. 2009. A lack of amyloid beta plaques despite persistent accumulation of amyloid beta in axons of long-term survivors of traumatic brain injury. *Brain Pathol*. 19:214–223.

Cheng, S.X., S. Zhang, H.T. Sun, and Y. Tu. 2013. Effects of mild hypothermia treatment on rat hippocampal beta-amyloid expression following traumatic brain injury. *Therapeutic Hypotherm Temp Manage*. 3:132–139.

Chobanian, A.V. 1983. The influence of hypertension and other hemodynamic factors in atherogenesis. *Progr Cardiovasc Dis*. 26:177–196.

Chuang-Stein, C., and R. DeMasi. 1998. Surrogate endpoints in AIDS drug development: Current status. *Drug Information J*. 32:439–448.

Clark, R.S., P.M. Kochanek, P.D. Adelson, M.J. Bell, J.A. Carcillo, M. Chen, S.R et al. 2000. Increases in bcl-2 protein in cerebrospinal fluid and evidence for programmed cell death in infants and children after severe traumatic brain injury. *J Pediatr*. 137:197–204.

Committee on Biological Markers of the National Research Council. 1987. Biological markers in environmental health research. *Environ Health Perspect*. 74:3–9.

De Gruttola, V.G., P. Clax, D.L. DeMets, G.J. Downing, S.S. Ellenberg, L. Friedman et al. 2001. Considerations in the evaluation of surrogate endpoints in clinical trials. Summary of a National Institutes of Health workshop. *Control Clin trials*. 22:485–502.

Diaz-Arrastia, R., K.K. Wang, L. Papa, M.D. Sorani, J.K. Yue, A.M. Puccio et al. 2014. Acute biomarkers of traumatic brain injury: Relationship between plasma levels of ubiquitin C-terminal hydrolase-L1 and glial fibrillary acidic protein. *J Neurotrauma*. 31:19–25.

Donaldson, M., J. Sullivan, and A. Norbeck. 2010. Comparison of International Normalized Ratios provided by two point-of-care devices and laboratory-based venipuncture in a pharmacist-managed anticoagulation clinic. *American journal of health-system pharmacy: AJHP: Official journal of the American Society of Health-System Pharmacists*. 67:1616–1622.

Donato, R. 2001. S100: A multigenic family of calcium-modulated proteins of the EF-hand type with intracellular and extracellular functional roles. *Int J Biochem Cell Biol*. 33:637–668.

Eapen, D.J., P. Manocha, N. Ghasemzedah, R.S. Patel, H. Al Kassem, M. Hammadah, E. Veledar, N.A. Le, T. Pielak, C.W. Thorball, A. Velegraki, D.T. Kremastinos, S. Lerakis, L. Sperling, and A.A. Quyyumi. 2014. Soluble urokinase plasminogen activator receptor level is an independent predictor of the presence and severity of coronary artery disease and of future adverse events. *Journal of the American Heart Association*. 3:e001118.

Egea-Guerrero, J., F. Murillo-Cabezas, E. Gordillo-Escobar, A. Rodríguez-Rodríguez, J. Enamorado-Enamorado, J. Revuelto-Rey et al. 2013a. S100B protein may detect brain death development after severe traumatic brain injury. *J Neurotrauma*. 30:1762–1769.

Egea-Guerrero, J.J., J. Revuelto-Rey, E. Gordillo-Escobar, A. Rodriguez-Rodriguez, J. Enamorado-Enamorado, Z. Ruiz de Azua Lopez et al. 2013b. Serologic behavior of S100B protein in patients who are brain dead: Preliminary results. *Transplant Proc*. 45:3569–3572.

Everson, R.G., R.M. Jin, X. Wang, M. Safaee, R. Scharnweber, D.N. Lisiero, H. Soto, L.M. Liau, and R.M. Prins. 2014. Cytokine responsiveness of CD8(+) T cells is a reproducible biomarker for the clinical efficacy of dendritic cell vaccination in glioblastoma patients. *Journal for immunotherapy of cancer*. 2:10.

Fang, X., B.M. Balgley, and C.S. Lee. 2009. Recent advances in capillary electrophoresis-based proteomic techniques for biomarker discovery. *Electrophoresis*. 30:3998–4007.

Faul, M., L. Xu, M. Wald, and V. Coronado. 2010. Traumatic Brain Injury in the United States: Emergency Department Visits, Hospitalizations and Deaths 2002–2006. Atlanta (GA): Centers for Disease Control and Prevention, National Center for Injury Prevention and Control.

Finkelstein, E., P. Corso, and T. Miller. 2006. *The Incidence and Economic Burden of Injuries in the United States*. Oxford University Press, New York.

Fleischer, A.S., D.R. Rudman, C.B. Fresh, and G.T. Tindall. 1977. Concentration of 3',5' cyclic adenosine monophosphate in ventricular CSF of patients following severe head trauma. *J Neurosurg*. 47:517–524.

Fonteh, A.N., R.J. Harrington, A.F. Huhmer, R.G. Biringer, J.N. Riggins, and M.G. Harrington. 2006. Identification of disease markers in human cerebrospinal fluid using lipidomic and proteomic methods. *Dis Markers*. 22:39–64.

Food and Drug Administration (FDA). 2014. Medical Device User Fee and Modernization Act (MDUFMA).

Fox, W.C., M.S. Park, S. Belverud, A. Klugh, D. Rivet, and J.M. Tomlin. 2013. Contemporary imaging of mild TBI: The journey toward diffusion tensor imaging to assess neuronal damage. *Neurol Res*. 35:223–232.

Fozouni, N., M. Chopp, S.P. Nejad-Davarani, Z.G. Zhang, N.L. Lehman, S. Gu et al. 2013. Characterizing brain structures and remodeling after TBI based on information content, diffusion entropy. *PLoS One*. 8:e76343.

Gabbita, S.P., S.W. Scheff, R.M. Menard, K. Roberts, I. Fugaccia, and F.P. Zemlan. 2005. Cleaved-tau: A biomarker of neuronal damage after traumatic brain injury. *J Neurotrauma*. 22:83–94.

Garnett, M., T. Cadoux-Hudson, and P. Styles. 2001. How useful is magnetic resonance imaging in predicting severity and outcome in traumatic brain injury? *Curr Opin Neurol*. 14:753–757.

Gatson, J.W., V. Warren, K. Abdelfattah, S. Wolf, L.S. Hynan, C. Moore et al. 2013. Detection of beta-amyloid oligomers as a predictor of neurological outcome after brain injury. *J Neurosurg*. 118:1336–1342.

Glenn, T.C., D. Hirt, G. Mendez, D.L. McArthur, R. Sturtevant, S. Wolahan et al. 2013. Metabolomic analysis of cerebral spinal fluid from patients with severe brain injury. *Acta Neurochir Suppl*. 118:115–119.

Goncalves, C.A., M.C. Leite, and P. Nardin. 2008. Biological and methodological features of the measurement of S100B, a putative marker of brain injury. *Clin Biochem*. 41:755–763.

Greenberg, R.B. 1976. Medical Device Amendments of 1976. *Am J Hosp Pharmacy*. 33:1308–1311.

Guingab-Cagmat, J.D., E.B. Cagmat, F.H. Kobeissy, and J. Anagli. 2014. Uses and challenges of bioinformatic tools in mass spectrometric-based proteomic brain perturbation studies. *International journal of bioinformatics research and applications*. 10:27–42.

Guingab-Cagmat, J.D., K. Newsom, A. Vakulenko, E.B. Cagmat, F.H. Kobeissy, S. Zoltewicz et al. 2012. In vitro MS-based proteomic analysis and absolute quantification of neuronal-glial injury biomarkers in cell culture system. *Electrophoresis*. 33:3786–3797.

Gura, M., G. Silav, N. Isik, and I. Elmaci. 2012. Noninvasive estimation of cerebral perfusion pressure with transcranial Doppler ultrasonography in traumatic brain injury. *Turkish Neurosurg*. 22:411–415.

Haines, I.E., and G.L. Gabor Miklos. 2013. Prostate-specific antigen screening trials and prostate cancer deaths: The androgen deprivation connection. *Journal of the National Cancer Institute*. 105:1534–1539.

Harris, J.L., H.W. Yeh, I.Y. Choi, P. Lee, N.E. Berman, R.H. Swerdlow, S.C. Craciunas, and W.M. Brooks. 2012. Altered neurochemical profile after traumatic brain injury: (1) H-MRS biomarkers of pathological mechanisms. *Journal of cerebral blood flow and metabolism: Official journal of the International Society of Cerebral Blood Flow and Metabolism*. 32:2122–2134.

Haselwood, B.A., and J.T. La Belle. 2014. Development of electrochemical methods to enzymatically detect traumatic brain injury biomarkers. *Biosensors & bioelectronics*. Sep 17, p. ii: S0956-5663(14)00719-2. doi: 10.1016/j.bios.2014.09.032.

He, W., L.S. Wang, H.Z. Li, L.G. Cheng, M. Zhang, and C.G. Wladyka. 2013. Intraoperative contrast-enhanced ultrasound in traumatic brain surgery. *Clinl Imaging*. 37:983–988.

Herrmann, M., P. Vos, M.T. Wunderlich, C.H. de Bruijn, and K.J. Lamers. 2000. Release of glial tissue-specific proteins after acute stroke: A comparative analysis of serum concentrations of protein S-100B and glial fibrillary acidic protein. *Stroke*. 31:2670–2677.

Huber, B.R., J.S. Meabon, T.J. Martin, P.D. Mourad, R. Bennett, B.C. Kraemer et al. 2013. Blast exposure causes early and persistent aberrant phospho- and cleaved-tau expression in a murine model of mild blast-induced traumatic brain injury. *JAD*. 37:309–323.

Hulscher, J.B., B.H. Vervliet, N. Wilczak, and J. van der Naalt. 2014. The diagnostic value of brain-fatty acid binding protein in traumatic brain injury. *J Neurotrauma*. 31:411.

Izumi, K., W.J. Lin, H. Miyamoto, C.K. Huang, A. Maolake, Y. Kitagawa, Y. Kadono, H. Konaka, A. Mizokami, and M. Namiki. 2014. Outcomes and predictive factors of prostate cancer patients with extremely high prostate-specific antigen level. *Journal of cancer research and clinical oncology*. 140:1413–1419.

Jonsson, H., P. Johnsson, P. Hoglund, C. Alling, and S. Blomquist. 2000. Elimination of S100B and renal function after cardiac surgery. *J Cardiothor Vasc Anesth*. 14:698–701.

Kiebish, M.A., X. Han, H. Cheng, J.H. Chuang, and T.N. Seyfried. 2008. Cardiolipin and electron transport chain abnormalities in mouse brain tumor mitochondria: Lipidomic evidence supporting the Warburg theory of cancer. *J Lipid Res*. 49:2545–2556.

Kleindienst, A., and M. Ross Bullock. 2006. A critical analysis of the role of the neurotrophic protein S100B in acute brain injury. *J Neurotrauma*. 23:1185–1200.

Kobeissy, F.H., S. Sadasivan, M.W. Oli, G. Robinson, S.F. Larner, Z. Zhang et al. 2008. Neuroproteomics and systems biology-based discovery of protein biomarkers for traumatic brain injury and clinical validation. *Proteomics Clin Appl*. 2:1467–1483.

Koizumi, H., H. Fujisawa, T. Kurokawa, E. Suehiro, H. Iwanaga, J. Nakagawara et al. 2010. Recovered neuronal viability revealed by Iodine-123-iomazenil SPECT following traumatic brain injury. *J Cereb Blood Flow Metab*. 30:1673–1681.

Koizumi, H., H. Fujisawa, E. Suehiro, H. Iwanaga, J. Nakagawara, and M. Suzuki. 2013. Cortical damage following traumatic brain injury evaluated by iomazenil SPECT and in vivo microdialysis. *Acta Neurochir Suppl*. 118:121–123.

Kovesdi, E., J. Luckl, P. Bukovics, O. Farkas, J. Pal, E. Czeiter et al. 2010. Update on protein biomarkers in traumatic brain injury with emphasis on clinical use in adults and pediatrics. *Acta Neurochir (Wien)*. 152:1–17.

Laser, H., T.G. Mack, D. Wagner, and M.P. Coleman. 2003. Proteasome inhibition arrests neurite outgrowth and causes "dying-back" degeneration in primary culture. *J Neurosci Res*. 74:906–916.

Lesko, L., and A.J. Atkinson. 2001. Use of biomarkers and surrogate endpoints in drug development and regulatory decision making: Criteria, validation, strategies. *Annu Rev Pharmacol Toxicol*. 41:347–366.

Ligthart, S., S. Sedaghat, M.A. Ikram, A. Hofman, O.H. Franco, and A. Dehghan. 2014. EN-RAGE: A Novel Inflammatory Marker for Incident Coronary Heart Disease. Arteriosclerosis, thrombosis, and vascular biology. Oct. 23, p. ii: ATVBAHA.114.304306.

Lonn, E. 2001. The use of surrogate endpoints in clinical trials: Focus on clinical trials in cardiovascular diseases. *Pharmacoepidem Drug Safe*. 10:497–508.

Loov, C., G. Shevchenko, A. Geeyarpuram Nadadhur, F. Clausen, L. Hillered, M. Wetterhall et al. 2013. Identification of injury specific proteins in a cell culture model of traumatic brain injury. *PLoS One*. 8:e55983.

Marklund, N., N. Farrokhnia, A. Hanell, E. Vanmechelen, P. Enblad, H. Zetterberg et al. 2014. Monitoring of beta-amyloid dynamics after human traumatic brain injury. *J Neurotrauma*. 31:42–55.

May, G. 1850. Illustrations of the tolerance of injuries occasionally exhibited by the brain. *Prov Med Surg J*. 14:8–10.

McClain, C., D. Cohen, R. Phillips, L. Ott, and B. Young. 1991. Increased plasma and ventricular fluid interleukin-6 levels in patients with head injury. *J Lab Clin Med*. 118:225–231.

McClain, C.J., D. Cohen, L. Ott, C.A. Dinarello, and B. Young. 1987. Ventricular fluid interleukin-1 activity in patients with head injury. *J Lab Clin Med*. 110:48–54.

Mendez, M.F., E.M. Owens, G. Reza Berenji, D.C. Peppers, L.J. Liang, and E.A. Licht. 2013. Mild traumatic brain injury from primary blast vs. blunt forces: Post-concussion consequences and functional neuroimaging. *NeuroRehabilitation*. 32:397–407.

Missler, U., M. Wiesmann, G. Wittmann, O. Magerkurth, and H. Hagenstrom. 1999. Measurement of glial fibrillary acidic protein in human blood: Analytical method and preliminary clinical results. *Clin Chem*. 45:138–141.

Molloy, M.P., E.E. Brzezinski, J. Hang, M.T. McDowell, and R.A. VanBogelen. 2003. Overcoming technical variation and biological variation in quantitative proteomics. *Proteomics*. 3:1912–1919.

Mondello, S., A. Linnet, A. Buki, S. Robicsek, A. Gabrielli, J. Tepas et al. 2012. Clinical utility of serum levels of ubiquitin C-terminal hydrolase as a biomarker for severe traumatic brain injury. *Neurosurgery*. 70:666–675.

Mondello, S., S.A. Robicsek, A. Gabrielli, G.M. Brophy, L. Papa, J. Tepas et al. 2010. alphaII-spectrin breakdown products (SBDPs): Diagnosis and outcome in severe traumatic brain injury patients. *J Neurotrauma*. 27:1203–1213.

Mynter, H. 1894 III. Contribution to the Study of head injuries, and of the results of trephining for subdural haemorrhage, abscess of brain, and epilepsy. *Ann Surg*. 19:539–545.

Nanduri, S., A.H. Tayal, G.G. Hegde, J. Shang, and A. Venkat. 2012. An analysis of discrepancy between point-of-care and central laboratory international normalized ratio testing in ED patients with cerebrovascular disease. *The American journal of emergency medicine*. 30:2025–2029.

Newberg, A.B., M. Serruya, A. Gepty, C. Intenzo, T. Lewis, D. Amen et al. 2014. Clinical comparison of (99m)Tc exametazime and (123)I ioflupane SPECT in patients with chronic mild traumatic brain injury. *PLoS One*. 9:e87009.

Nissen, S.E., E.M. Tuzcu, P. Schoenhagen, B.G. Brown, P. Ganz, R.A. Vogel, T. Crowe, G. Howard, C.J. Cooper, B. Brodie, C.L. Grines, and A.N. DeMaria. 2004. Effect of intensive compared with moderate lipid-lowering therapy on progression of coronary atherosclerosis: A randomized controlled trial. *Jama*. 291:1071–1080.

Nylen, K., M. Ost, L.Z. Csajbok, I. Nilsson, K. Blennow, B. Nellgard et al. 2006. Increased serum-GFAP in patients with severe traumatic brain injury is related to outcome. *J Neurol Sci*. 240:85–91.

Obuchowski, N.A., M.L. Lieber, and F.H.W. Jr. 2004. ROC curves in clinical chemistry: Uses, misuses, and possible solutions. *Clin Chem*. 50:1118.

Okegawa, T., N. Itaya, H. Hara, M. Tambo, and K. Nutahara. 2014. Circulating Tumor Cells as a Biomarker Predictive of Sensitivity to Docetaxel Chemotherapy in Patients with Castration-resistant Prostate Cancer. *Anticancer research*. 34:6705–6710.

Okonkwo, D.O., J.K. Yue, A.M. Puccio, D.M. Panczykowski, T. Inoue, P.J. McMahon et al. 2013. GFAP-BDP as an acute diagnostic marker in traumatic brain injury: Results from the prospective transforming research and clinical knowledge in traumatic brain injury study. *J Neurotrauma*. 30:1490–1497.

Ondruschka, B., D. Pohlers, G. Sommer, K. Schober, D. Teupser, H. Franke et al. 2013. S100B and NSE as useful postmortem biochemical markers of traumatic brain injury in autopsy cases. *J Neurotrauma*. 30:1862–1871.

Ottens, A.K., F.H. Kobeissy, B.F. Fuller, M. Chen Liu, M.W. Oli, R.L. Hayes et al. 2007. Novel neuroproteomic approaches to studying traumatic brain injury. *Prog Brain Res*. 161:401–418.

Ottens, A.K., F.H. Kobeissy, E.C. Golden, Z. Zhang, W.E. Haskins, S.S. Chen et al. 2006. Neuroproteomics in neurotrauma. *Mass Spectrom Rev*. 25:380–408.

Paone, J.F., T.P. Waalkes, R.R. Baker, and J.H. Shaper. 1980. Serum UDP-galactosyl transferase as a potential biomarker for breast carcinoma. *J Surg Oncol*. 15:59–66.

Papa, L., L. Akinyi, M.C. Liu, J.A. Pineda, J.J. Tepas, 3rd, M.W. Oli et al. 2010. Ubiquitin C-terminal hydrolase is a novel biomarker in humans for severe traumatic brain injury. *Crit Care Med*. 38:138–144.

Papa, L., L.M. Lewis, S. Silvestri, J.L. Falk, P. Giordano, G.M. Brophy et al. 2012. Serum levels of ubiquitin C-terminal hydrolase distinguish mild traumatic brain injury from trauma controls and are elevated in mild and moderate traumatic brain injury patients with intracranial lesions and neurosurgical intervention. *J Trauma Acute Care Surg*. 72:1335–1344.

Pataky, R., R. Gulati, R. Etzioni, P. Black, K.N. Chi, A.J. Coldman, T. Pickles, S. Tyldesley, and S. Peacock. 2014. Is prostate cancer screening cost-effective? A microsimulation model of prostate-specific antigen-based screening for British Columbia, Canada. International journal of cancer. *Journal international du cancer*. 135:939–947.

Peck, C.C., D.B. Rubin, and L.B. Sheiner. 2003. Hypothesis: A single clinical trial plus causal evidence of effectiveness is sufficient for drug approval. *Clin Pharmacol Therapeutics*. 73:481–490.

Pineda, J.A., S.B. Lewis, A.B. Valadka, L. Papa, H.J. Hannay, S.C. Heaton et al. 2007. Clinical significance of alphaII-spectrin breakdown products in cerebrospinal fluid after severe traumatic brain injury. *J Neurotrauma*. 24:354–366.

Prentice, R.L. 1989. Surrogate endpoints in clinical trials: Definition and operational criteria. *Stat Med*. 8:431–440.

Riederer, B.M., I.S. Zagon, and S.R. Goodman. 1986. Brain spectrin(240/235) and brain spectrin(240/235E): Two distinct spectrin subtypes with different locations within mammalian neural cells. *J Cell Biol*. 102:2088–2097.

Robertson, C.L., N. Minamino, R.A. Ruppel, K. Kangawa, S.R. Wisniewski, T. Tsuji et al. 2001. Increased adrenomedullin in cerebrospinal fluid after traumatic brain injury in infants and children. *J Neurotrauma*. 18:861–868.

Rolan, P. 1997. The contribution of clinical pharmacology surrogates and models to drug development—A critical appraisal. *Br J Clin Pharmacol*. 44:219–225.

Romner, B. 2000. Traumatic brain damage: Serum S-100 protein measurements related to neuroradiological findings. *J Neurotrauma*. 17:641– 647.

Ross, S.A., R.T. Cunningham, C.F. Johnston, and B.J. Rowlands. 1996. Neuron-specific enolase as an aid to outcome prediction in head injury. *Br J Neurosurg*. 10:471–476.

Rostami, E., J. Davidsson, K.C. Ng, J. Lu, A. Gyorgy, J. Walker et al. 2012. A model for mild traumatic brain injury that induces limited transient memory impairment and increased levels of axon related serum biomarkers. *Front Neurol*. 3:115.

Rudman, D., A. Fleischer, and M.H. Kutner. 1976. Concentration of 3', 5' cyclic adenosine monophosphate in ventricular cerebrospinal fluid of patients with prolonged coma after head trauma or intracranial hemorrhage. *N Engl J Med*. 295:635–638.

Sarkar, K., K. Keachie, U. Nguyen, J.P. Muizelaar, M. Zwienenberg-Lee, and K. Shahlaie. 2014. Computed tomography characteristics in pediatric versus adult traumatic brain injury. *J Neurosurg Pediatr*. 13;307–314.

Sarubbo, S., F. Latini, S. Ceruti, A. Chieregato, C. d'Esterre, T.Y. Lee et al. 2014. Temporal changes in CT perfusion values before and after cranioplasty in patients without symptoms related to external decompression: A pilot study. *Neuroradiology*. 56: 237–243.

Satchell, M.A., Y. Lai, P.M. Kochanek, S.R. Wisniewski, E.L. Fink, N.A. Siedberg et al. 2005. Cytochrome c, a biomarker of apoptosis, is increased in cerebrospinal fluid from infants with inflicted brain injury from child abuse. *J Cereb Blood Flow Metab*. 25:919–927.

Selakovic, V., R. Raicevic, and L. Radenovic. 2005. The increase of neuron-specific enolase in cerebrospinal fluid and plasma as a marker of neuronal damage in patients with acute brain infarction. *J Clin Neurosci*. 12:542–547.

Selwyn, R., N. Hockenbury, S. Jaiswal, S. Mathur, R.C. Armstrong, and K.R. Byrnes. 2013. Mild traumatic brain injury results in depressed cerebral glucose uptake: An (18)FDG PET study. *J Neurotrauma*. 30:1943–1953.

Sharifi, K., Y. Morihiro, M. Maekawa, Y. Yasumoto, H. Hoshi, Y. Adachi et al. 2011. FABP7 expression in normal and stab-injured brain cortex and its role in astrocyte proliferation. *Histochem Cell Biol*. 136:501–513.

Shirinian, E. 1971a. Catecholamine metabolism in guinea pig tissues in craniocerebral injuries. *Biull Eksp Biol Med*. 71:46–49.

Shirinian, E. 1971b. The effect of craniocerebral trauma on catecholamine metabolism in the heart, liver and adrenals of guinea pigs. *Vopr Med Khim*. 17:640–644.

Shultz, S., L. Cardamone, Y. Liu, R. Hogan, L. Maccotta, D. Wright et al. 2013. Can structural or functional changes following traumatic brain injury in the rat predict epileptic outcome? *Epilepsia*. 54:1240–1250.

Siman, R., N. Giovannone, G. Hanten, E.A. Wilde, S.R. McCauley, J.V. Hunter et al. 2013. Evidence that the blood biomarker SNTF predicts brain imaging changes and persistent cognitive dysfunction in mild TBI patients. *Front Neurol*. 4:190.

Strathmann, F.G., S. Schulte, K. Goerl, and D.J. Petron. 2014. Blood-based biomarkers for traumatic brain injury: Evaluation of research approaches, available methods and potential utility from the clinician and clinical laboratory perspectives. *Clin Biochem*. 47:876–888.

Strimbu, K., and J. Tavel. 2010. What are biomarkers? *Curr Opin HIV AIDS*. 5:463–466.

Su, E., M.J. Bell, P.M. Kochanek, S.R. Wisniewski, H. Bayir, R.S. Clark et al. 2012. Increased CSF concentrations of myelin basic protein after TBI in infants and children: Absence of significant effect of therapeutic hypothermia. *Neurocrit Care*. 17:401–407.

Tajiri, N., S.L. Kellogg, T. Shimizu, G.W. Arendash, and C.V. Borlongan. 2013. Traumatic brain injury precipitates cognitive impairment and extracellular Abeta aggregation in Alzheimer's disease transgenic mice. *PLoS One*. 8:e78851.

Takayanagi, Y., G. Gerner, M. Takayanagi, V. Rao, T.D. Vannorsdall, A. Sawa et al. 2013. Hippocampal volume reduction correlates with apathy in traumatic brain injury, but not schizophrenia. *J Neuropsychiatry Clin Neurosci*. 25:292–301.

Taurines, R., E. Dudley, J. Grassl, A. Warnke, M. Gerlach, A.N. Coogan, and J. Thome. 2011. Proteomic research in psychiatry. *J Psychopharmacol*. 25:151–196.

Temple, R. 1995. A regulatory authority's opinion about surrogate endpoints. In: Nimmo, W.S. and G.T. Tucker (eds.), *Clinical Measurement in Drug Evaluation*, Wiley, New York.

Thelin, E.P., L. Johannesson, D. Nelson, and B.M. Bellander. 2013a. S100B is an important outcome predictor in traumatic brain injury. *J Neurotrauma*. 30:519–528.

Thelin, E.P., D.W. Nelson, and B.M. Bellander. 2013b. Secondary peaks of S100B in serum relate to subsequent radiological pathology in traumatic brain injury. *Neurocrit Care*. 20:217–219.

Toyama, Y., T. Kobayashi, Y. Nishiyama, K. Satoh, M. Ohkawa, and K. Seki. 2005. CT for acute stage of closed head injury. *Radiat Medi*. 23:309–316.

Vajtr, D., O. Benada, P. Linzer, F. Samal, D. Springer, P. Strejc et al. 2012. Immunohistochemistry and serum values of S-100B, glial fibrillary acidic protein, and hyperphosphorylated neurofilaments in brain injuries. *Soud Lek*. 57:7–12.

Valiyaveettil, M., Y.A. Alamneh, Y. Wang, P. Arun, S. Oguntayo, Y. Wei et al. 2014. Cytoskeletal protein alpha-II spectrin degradation in the brain of repeated blast exposed mice. *Brain Res*. 1549:32–41.

Viant, M.R., B.G. Lyeth, M.G. Miller, and R.F. Berman. 2005. An NMR metabolomic investigation of early metabolic disturbances following traumatic brain injury in a mammalian model. *NMR Biomed*. 18:507–516.

Vos, P.E., K.J. Lamers, J.C. Hendriks, M. van Haaren, T. Beems, C. Zimmerman et al. 2004. Glial and neuronal proteins in serum predict outcome after severe traumatic brain injury. *Neurology*. 62:1303–1310.

Walsh, B.H., D.I. Broadhurst, R. Mandal, D.S. Wishart, G.B. Boylan, L.C. Kenny et al. 2012. The metabolomic profile of umbilical cord blood in neonatal hypoxic ischaemic encephalopathy. *PLoS One*. 7:e50520.

Wang, C., X. Fang, and C.S. Lee. 2013a. Recent advances in capillary electrophoresis-based proteomic techniques for biomarker discovery. *Methods Mol Biol*. 984:1–12.

Wang, Y.Q., L. Wang, M.Y. Zhang, T. Wang, H.J. Bao, W.L. Liu, D.K. Dai, L. Zhang, P. Chang, W.W. Dong, X.P. Chen, and L.Y. Tao. 2012. Necrostatin-1 suppresses autophagy and apoptosis in mice traumatic brain injury model. *Neurochemical research*. 37:1849–1858.

Wang, Y., Y. Wang, M. Li, P. Xu, T. Gu, T. Ma et al. 2013b. (1)H NMR-based metabolomics exploring biomarkers in rat cerebrospinal fluid after cerebral ischemia/reperfusion. *Mole bio-Sys*. 9:431–439.

Weir, C.J., and R.J. Walley. 2006. Statistical evaluation of biomarkers as surrogate endpoints: A literature review. *Stat Medi*. 25:183–203.

Whalen, M.J., T.M. Carlos, P.M. Kochanek, S.R. Wisniewski, M.J. Bell, R.S. Clark et al. 2000. Interleukin-8 is increased in cerebrospinal fluid of children with severe head injury. *Crit Care Med*. 28:929–934.

White, T.E., G.D. Ford, M.C. Surles-Zeigler, A.S. Gates, M.C. Laplaca, and B.D. Ford. 2013. Gene expression patterns following unilateral traumatic brain injury reveals a local pro-inflammatory and remote anti-inflammatory response. *BMC Genom*. 14:282.

WHO. 1993. *Biomarkers and Risk Assessment: Concepts and Principles*, from http://apps.who.int/iris/handle/10665/39037.

WHO. 2001. *Biomarkers in Risk Assessment: Validity and Validation*, from http://apps.who.int/iris/handle/10665/42363.

Wolf, H., S. Frantal, G.S. Pajenda, O. Salameh, H. Widhalm, S. Hajdu et al. 2013. Predictive value of neuromarkers supported by a set of clinical criteria in patients with mild traumatic brain injury: S100B protein and neuron-specific enolase on trial: Clinical article. *J Neurosurg*. 118:1298–1303.

Wu, P., Y. Zhao, S.J. Haidacher, E. Wang, M.O. Parsley, J. Gao et al. 2013. Detection of structural and metabolic changes in traumatically injured hippocampus by quantitative differential proteomics. *J Neurotrauma*. 30:775–788.

Yuan, X., and D.M. Desiderio. 2005. Proteomics analysis of prefractionated human lumbar cerebrospinal fluid. *Proteomics*. 5:541–550.

Zurek, J., L. Bartlova, and M. Fedora. 2011. Hyperphosphorylated neurofilament NF-H as a predictor of mortality after brain injury in children. *Brain Inj*. 25:221–226.

Zurek, J., and M. Fedora. 2012. The usefulness of S100B, NSE, GFAP, NF-H, secretagogin and Hsp70 as a predictive biomarker of outcome in children with traumatic brain injury. *Acta Neurochir (Wien)*. 154:93–103; discussion 103.

21 The Use and Potential of pNF-H as a General Blood Biomarker of Axonal Loss

An Immediate Application for CNS Injury

Gerry Shaw

CONTENTS

21.1 INTRODUCTION

This chapter describes the unusual protein chemical properties of the phosphorylated axonal form of the heavy neurofilament subunit NF-H (pNF-H). I show how these properties conspire to make this protein an excellent biomarker of ongoing chronic and acute axonal loss useful in clinical and research contexts. pNF-H can be detected by ELISA not only in cerebrospinal fluid but also in plasma and serum. I summarize neurological damage and disease states in which this protein has been measured in elevated and informative levels in blood. Finally, I use this biomarker to discuss some of the problems, questions and opportunities associated with the biomarker field in general, and highlight some areas which require further research on this and other potential biomarkers.

There has been much interest in the development of assays that can reflect the seriousness and type of central nervous system (CNS) injury. Ideally, such assays could rapidly inform a researcher or a clinician about ongoing challenges to the CNS of an individual. Such challenges may result from chronic or acute events, and different kinds of assays would inform about different types of events. Much recent thought has focused on sports-related injuries, the issue being whether a sportsman should carry on playing following a mild head injury. As other examples, diseases such as multiple sclerosis (MS) tend to be variable in different patients and several different therapies are available. Understanding which therapy is effective would aid in patient treatment and could be assessed from an appropriate test. Potential rapid assays include advanced magnetic, X-ray and radiochemical-based imaging methods that will not be discussed here. Rather, I will focus on the detection of substances in biological fluids with the potential to provide information about CNS status, response to therapy, and prognosis. So how does this field of research stand? There are many excellent and comprehensive reviews in this area (Dash et al., 2011; Kovesdi et al., 2010; Kuhle et al., 2011; Mondello et al., 2011), but I will write on just one particular biomarker, the phosphorylated axonal form of the major neurofilament subunit NF-H (pNF-H). This biomarker has some particularly favorable properties that allow detection not only in cerebrospinal fluid (CSF), but also in blood. I will also attempt to raise important issues that have had little exposure to date and that may point the way to future advances with this particular biomarker and by analogy to other biomarkers.

21.2 WHAT IS A GOOD BIOMARKER?

First, which biofluids are convenient sources for biomarker measurements? Ideally a biomarker should be detectable in blood samples because these are taken routinely in experimental and clinical practice. Other potential sources of biomarkers are CSF, saliva, tears, hair, skin biopsy, feces, and urine. However, each of these has significant disadvantages, so that the only widely used alternative to blood has been the use of CSF. CSF is easy to work with, being normally very low in protein content, typically in the range of 150–600 μg/mL. Biomarkers released from the damaged or diseased brain or spinal cord readily enter the CSF and can be collected soon after their release. As a result, biomarkers with short half-lives or that are present in only low amounts may be detected relatively easily in CSF samples, and substances with no ability to pass from CSF to blood may also be detected. The problems with CSF sampling are that it is impractical to do on a sports field or in an ambulance, it is rather invasive and painful, it cannot normally be repeated over shorter periods, and it carries some patient risk of injury, headaches, and infection. As a result, taking CSF is either impractical or tends to be avoided in routine clinical practice, with liability issues being a particular problem, especially in the United States. However, we should not overlook the advantages of CSF biomarkers. The typical volume of CSF in an adult human is 150 mL with a typical protein content of about 300 μg/mL. In contrast, a typical adult has about 5 L of blood, with a protein concentration of about 60–80 mg/mL. So blood protein content is about 200 times that of CSF. Blood is almost a totally saturated protein solution, and is also an extremely complex mix of substances that is variable between individuals and even in the same individual on a daily or even hourly basis (Adkins et al., 2002; Farrah et al., 2011; Jacobs et al., 2005; Shen et al., 2005). So if, for example, 10 μg of a potential biomarker enters the CSF, assuming no degradation or sequestration and assuming perfect mixing, the concentration will equilibrate to about 10/150 μg/mL, or about 60 ng/mL, against a total protein background only about 5,000-fold higher.

The same amount of this biomarker in blood will be diluted to 10/5,000 μg/mL final concentration, or about 2 ng/mL. In this case, the biomarker is found against a background of about 80 mg/mL protein, effectively 1 part in about 4,000,000, or 800 times more diluted than in the CSF example. Clearly it is likely to be much harder to detect a CNS-derived protein in blood than in CSF. Blood is not only concentrated and complex in composition but also contains large amounts of human antibodies and potential antibody binding substances such as complement factors, which can interfere with antibody-based assays. In addition, 99% of the content of blood is accounted for by only 22 highly abundant proteins, and even a low-affinity interaction of an assay reagent or the analyte with one or more of these can prevent an assay from working correctly. Blood also contains many active proteases, most of them poorly characterized, which may degrade potential blood biomarkers (Wildes and Wells, 2010). All of these factors combine to mean that a biomarker assay, if it is to work on blood samples, must have the ability to bind the analyte with enough affinity and avidity to pull a small amount of it specifically and avidly out of the concentrated and complex background of blood.

As a result, detection of biomarkers in blood is far more technically challenging than looking for the same substance in CSF; it should also be mentioned that the 10 μg of any particular biomarker discussed previously is a lot of material, and most biomarkers studied to date are not released in such large amounts except in cases of acute and extensive CNS damage. As a result, the publication of a reliable method for detecting a CNS-derived biomarker in blood should be regarded as an important procedural breakthrough.

21.3 CHARACTERISTICS OF GOOD POTENTIAL BIOMARKERS

Bearing these issues in mind, which proteins are likely to be useful biomarkers of CNS damage and disease? There are a few simple observations that lead directly to a panel of usable biomarkers and a particularly strong candidate.

The potential for detecting a biomarker is greatly increased if the biomarker in question is normally an abundant component of the CNS, so that it is released in quantity following damage of disease. Obviously, the more abundant a protein is in CNS tissue, the more likely it is to be reliably detected after release. The most abundant components of cells and tissues are typically cytoskeletal proteins and cytosolic enzymes. Interestingly, all of the currently most promising CNS biomarker proteins are abundant and already well known many years ago, so that exhaustive and expensive searches for novel biomarkers may have been unnecessary.

Another feature is that the biomarker in question should be informative. For example, glial fibrillary acidic protein (GFAP) and S100β are major proteins localized primarily in astrocytes, so release of these proteins reflects damage primarily to these cells, important components of the blood–brain barrier (BBB). Neurofilament proteins, microtubule-associated protein 2, and ubiquitin C-terminal hydrolase 1 (UCHL1) are all found specifically and in large amounts in neuronal perikarya and dendrites, and so their detection reflects neuronal loss.

A third feature is that it is helpful if the biomarker is relatively stable because this will increase the chances of it being detectable. Proteins such as neurofilament subunits, GFAP, and others are components of the cytoskeleton and their normal functioning requires long half-lives, slow turnover rates, and some resistance to proteases because providing a robust framework in the cell is their normal cellular role. Similarly many cytoplasmic enzymes have relatively long half lives. Searching for proteins that are resistant to proteases might be beneficial, and one of the approaches that identified UCHL1 as a potential CNS injury and disease biomarker was open-ended searches for protease-resistant brain-derived molecules (Lewis et al., 2010).

A fourth feature is that the potential biomarker should be freely diffusible from the site of injury. This should favor soluble proteins, so heavily charged proteins, which do not associate with membranes would a priori be viable biomarker candidates. S100β is a low-molecular-weight, strongly negatively charged protein lacking hydrophobic residues and thus has this feature. Neurofilament subunits are cytoplasmic proteins that are not membrane localized and are also unusually highly charged, and so might readily migrate away from a site of injury or degeneration.

A final feature is that a protein may have unusual properties that render it particularly usable as a biomarker. The best example of this is the neurofilament subunit NF-H, the subject of this review. The NF-H protein has an unusual sequence of lysine-serine-proline (KSP) repeats that are the major axonal phosphorylation sites (Lee et al., 1988). The number of these repeats is variable across species and across alleles, and the most common human allele has 44 versions of these repeats (Figure 21.1). The axonal form of NF-H, pNF-H, is phosphorylated on most if not all serine residues in the KSP sequences, which is distinct from perikaryal and dendritic forms that are not normally phosphorylated on those sites (Goldstein et al., 1987; Strong et al., 2001). This region of the molecule is particularly immunogenic in the phosphorylated form, so that many commercial and in-house–developed antibodies, both monoclonal and polyclonal, bind preferentially to this region (Carden et al., 1985). The repeated nature of this region would be expected to allow for unusually efficient multiepitope capture and detection. All of these features led us to hypothesize that pNF-H was likely to be detectable in blood after CNS injury and disease associated with axonal loss (Shaw et al., 2005). As is well known, axons are peculiarly sensitive to both mechanical and metabolic stress (Buki and Povlishock, 2006; Stys, 2005), so measuring levels of pNF-H in blood would be expected to give valuable surrogate information about ongoing axonal damage and degeneration. As outlined in the following section, this simple hypothesis was correct and there are now many examples that show that this protein has great potential utility as a surrogate blood biomarker of axonal loss.

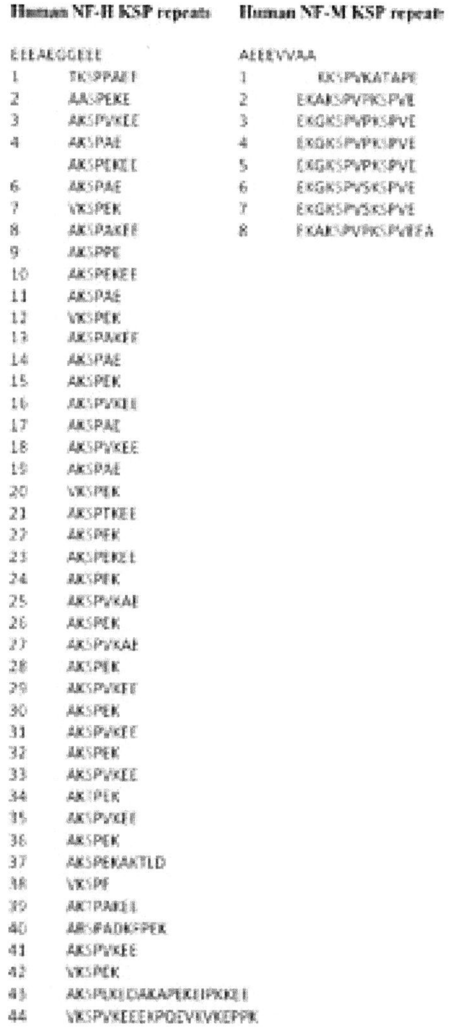

Human NF-H KSP repeats		Human NF-M KSP repeats	
EEEAEGGEEE		AEEEVVAA	
1	TKSPPAEE	1	EKSPVKATAPE
2	AASPEKE	2	EKAKSPVPKSPVE
3	AKSPVKEE	3	EKGKSPVPKSPVE
4	AKSPAE	4	EKGKSPVPKSPVE
	AKSPEKEE	5	EKGKSPVPKSPVE
6	AKSPAE	6	EKGKSPVSKSPVE
7	VKSPEK	7	EKGKSPVSKSPVE
8	AKSPAKEE	8	EKAKSPVPKSPVEEA
9	AKSPPE		
10	AKSPEKEE		
11	AKSPAE		
12	VKSPEK		
13	AKSPAKEE		
14	AKSPAE		
15	AKSPEK		
16	AKSPVKEE		
17	AKSPAE		
18	AKSPVKEE		
19	AKSPAE		
20	VKSPEK		
21	AKSPTKEE		
22	AKSPEK		
23	AKSPEKEE		
24	AKSPEK		
25	AKSPVKAE		
26	AKSPEK		
27	AKSPVKAE		
28	AKSPEK		
29	AKSPVKEE		
30	AKSPEK		
31	AKSPVKEE		
32	AKSPEK		
33	AKSPVKEE		
34	AKSPEK		
35	AKSPVKEE		
36	AKSPEK		
37	AKSPEKAKTLD		
38	VKSPE		
39	AKTPAKEE		
40	ARSPADKFPEK		
41	AKSPVKEE		
42	VKSPEK		
43	AKSPLKEDAKAPEKEIPKKEE		
44	VKSPVKEEEKPQEVKVKEPPK		

FIGURE 21.1 Amino acid sequences of the KSP regions of the most common allele of human NF-H (left) and NF-M (right). The NF-H repeats are organized in tandem in one continuous segment of the molecule as shown. Variants of KSP such as RSP, KTP, and ASP have been counted as part of this repeat. In the case of NF-M peptide number 1 is in isolation whereas peptides 2–8 are present in a continuous tandem arrangement in another part of the molecule as shown. The serine residues are the sites of phosphorylation and are highlighted in light gray in both sequences.

21.4 PROBLEMS WITH BIOMARKER ASSAYS

Biomarker protein levels are usually determined by the enzyme linked immunosorbent assay (ELISA) technique, in which typically one antibody acts as a capture reagent and another acts as a detection reagent. The antibodies may be purified polyclonals or specific monoclonals, and in some cases defined mixtures of monoclonals may be used. There are many uncertainties associated with this kind of assay, particularly applied to blood samples. Although commercial antibodies may work as advertised on the pure protein or peptide immunogen, may stain the appropriate cell type immunohistologically and may pick up one clean band of the expected size on Western blots, all of these assays require different and distinct properties of the antibody–antigen interaction. As a result, many antibodies that are of otherwise good quality do not work well in ELISA-type assays, particularly on blood samples. Of course it is well known that many commercial antibodies are not of high quality. As a result, there is a need for well-validated ELISAs known to detect the appropriate protein in blood samples. In the case of pNF-H, the unusual phosphorylated KSP repeat sequences have long been known to be unusually immunogenic, so that even polyclonal antibodies to purified pNF-H fail to show appreciable binding to nonphosphorylated versions of this protein (Lee et al., 1988; Shaw et al., 1986).

The sequence of the mammalian NF-H KSP repeats fall into three groups based on the peptides AKSPAE, AKSPEK, and AKSPVKEE and their minor variants (Lees et al., 1988) (Figure 21.1). These peptides are obviously highly charged in the native form and in addition are heavily and perhaps completely serine phosphorylated in axonal neurofilaments (Strong et al., 2001). These three kinds of related sequences are phosphorylated by different protein kinases, and in that form human pNF-H presents 44 similar but not identical epitopes. As a result, antibodies may be specific for one or another kind of sequence, for pairs or more of these sequences, or may cross-react with all three types of sequences (Lee et al., 1988). In the case of pNF-H, we would therefore expect to obtain antibodies to all three kinds of phosphorylated repeats alone and in combination, which would allow unusually efficient multiepitope capture and detection. We took this approach with our first publication in this area, using an in-house–generated affinity-purified chicken antibody in capture and an in-house–generated affinity-purified rabbit antibody in detection, the final signal being detected with a commercial goat anti-mouse enzyme conjugate (Shaw et al., 2005). In this article, we were able to show that pNF-H could be reliably detected in blood of rats given experimental brain and spinal cord injuries, and that the levels detected reflect the seriousness of injury. We have used this assay in several other studies (Anderson et al., 2008; Guy et al., 2008; Lewis et al., 2010; Lewis et al., 2008; Pasol et al., 2010; Petzold and Shaw, 2007; Prasad et al., 2012; Ringger et al., 2011; Talla et al., 2013). The assay has also been used by collaborators on independent projects, in which our contribution was only in providing reagents, running assays, interpreting the results, or aiding in the preparation of the final article (Allahtavakoli et al., 2011; Blyth et al., 2011; Douglas-Escobar et al., 2011; Ganesalingam et al., 2011; Gresle et al., 2012; Jokubaitis et al., 2013; Menkhorst et al., 2011). Slightly different versions of this assay, all using the same pair of antibodies, have been commercialized by EnCor Biotechnology, BioVendor, and Millipore-EMD and have resulted in a series of further publications (Ganesalingam et al., 2011; Gnanapavan et al., 2013; Hayakawa et al.; Hayakawa et al., 2012; Lehnert et al., 2014; Levine et al., 2012; Matsushige et al., 2008; Matsushige et al., 2009; Matsushige et al., 2012; Obeid et al., 2011; Singh et al., 2011; Stratford et al., 2013; Ueno et al., 2010; Zurek et al., 2011; Zurek and Fedora, 2012).

A collaborator of ours, Dr. Axel Petzold, has independently recognized the utility of assays to pNF-H (Petzold, 2005; Petzold et al., 2003) and has made use of various mouse monoclonal antibodies of the SMI series, many of which also bind phospho-KSP type sequences. These antibodies were originally developed in the laboratory of Ludwig and Nancy Sternberger (Sternberger and Sternberger, 1983) and are now marketed through Covance. Although most of Petzold's collaborative studies have been performed on CSF, an increasing number of articles use these antibodies to look at pNF-H in blood samples (Lu et al., 2012; Rundgren et al., 2012; Sellner et al., 2011; Sellner et al., 2009; Wieske et al., 2014; Wild et al., 2007). We have also developed a second-generation pNF-H assay using two in-house–generated monoclonal antibodies (Boylan et al., 2009), which is now commercially available from EnCor and is being used in newer studies (Boylan et al., 2013). Finally, we have developed a third-generation assay on the Mesoscale Discovery electrochemiluminescent detection platform, which allows for somewhat increased speed and efficiency. This is now being used for a series of ongoing collaborations and will shortly, in collaboration with Iron Horse DX of Phoenix, AZ, be used as part of a Clinical Laboratory Improvement Amendments–certified amyotrophic lateral sclerosis (ALS) diagnostic assay (Ganesalingam et al., 2011, 2013). The use of this assay for clinical purposes has required the detailed characterization of important parameters such as the stability of reagents, the limit of assay detection, linearity, parallelism, and inter- and intraassay variability. Such parameters are generally available for assays developed for research use only, but clinical use requires a layer of regulation that forces much more rigorous study and documentation of these parameters. This emerging body of data, from many laboratories and with several different assays, shows that measurement of pNF-H in blood has considerable promise in the monitoring of axonal loss in a variety of research and clinical contexts.

In summary, pNF-H has a unique combination of characteristics that make it unusually detectable not only in CSF but also in blood and the levels found are expected to reflect the level of axonal loss. Of course, axonal loss is generally a bad thing, so in any particular patient group higher levels of blood pNF-H predict a poor outcome, whereas lower levels predict recovery.

21.5 OTHER NEUROFILAMENT SUBUNITS

The neurofilament is a structure defined by ultrastructural morphology as an electron dense ~10 nm in diameter fiber with fine filaments protruding from its side, which is found only in neurons. These fine filaments keep the filament about 30 nm apart in axons and differentiate neurofilaments from other types of 10-nm-diameter filaments such as GFAP and the keratins. The subunits of neurofilaments of the adult consist of four major proteins—NF-L, NF-M, NF-H, and α-internexin—with smaller amounts of peripherin being also present in some regions. Human NF-M has a total of 14 KSP sequences in one tandem set of repeats (Figure 21.1), and a single KSP repeat in another part of the molecule (Myers et al., 1987); these are also phosphorylated in axonal neurofilaments (Hill et al., 1991). The 14 tandem human NF-M KSP repeats are somewhat different in sequence from those of NF-H, having the consensus EKGKSPVPKSPVE, which is repeated seven times in tandem. The pair of KSP sequences in each of these peptides is separated by proline residues, so this region would be predicted to adopt a conformation quite different from the KSP region of NF-H. This 14 KSP repeat arrangement is actually somewhat unique to humans, so that antibodies specific to this region of human NF-M may be unreactive with NF-M from other species (Hill et al., 1991). To date, such antibodies have not been used to attempt to capture axonal pNF-M from biological fluids. Because the KSP phosphorylated forms of NF-M, like those of NF-H, are only normally found in axons these phosphoforms are also potentially useful blood biomarkers of axonal loss. It remains to be seen if assays to pNF-M would complement in some fashion information obtained from assays to pNF-H.

NF-L is argued to be the core component and most abundant component of the neurofilament and so might be a useful potential biomarker also. However, the absolute amounts of NF-L, NF-M, and NF-H in any particular region of the nervous system are rather comparable, and NF-L does not contain any of the unusual KSP type repeats, which makes NF-M and particularly NF-H unusually favorable targets for ELISA-type assays. An NF-L ELISA using two in-house–generated mouse monoclonal antibodies, 47:3 and 2:1, which bind to distinct sites on the NF-L molecule, has been described (Norgren et al., 2002). This assay has been studied in multiple centers (Petzold et al., 2010) and is now marketed by Uman Diagnostics of Umea, Sweden. This assay has worked very well when applied to CSF samples, but the commercial version apparently does not work with blood samples. However, a new article shows that the same antibody pair will produce a usable signal if the signal is quantified using the Mesoscale Discovery electrochemiluminescence method instead of the colorimetric method as in a standard ELISA (Gaiottino et al., 2013). Another interesting recent article shows that NF-L can be detected at elevated levels in blood using a different NF-L ELISA after cardiac arrest in humans (Rana et al., 2013). These authors made use of a commercially available monoclonal antibody to NF-L, clone MCA-DA2 (EnCor), for capture and a rabbit polyclonal for detection, and found that serum NF-L levels were significantly higher in patients who either died or recovered poorly compared with those who recovered well. These two studies show that informative levels of NF-L, as with pNF-H, can enter the blood after CNS damage and disease. Whether the NF-L is associated with pNF-H or some other proteins, whether it is proteolyzed, phosphorylated, or modified in other ways has not been addressed. Presumably, α-internexin and peripherin, the other major components of neurofilaments, may also enter the blood under similar circumstances, and thus could also be detected with appropriate assays.

21.6 QUESTIONS ABOUT BIOMARKER ASSAYS

Antibodies do not primarily react against intact proteins, but against peptides digested from proteins, which have been internalized in antigen-presenting cells (APC). The cleaved proteins must then be fitted into the groove in major histocompatibility complex (MHC) class II proteins on the APC, and it is in this form that an antigen is presented on the cell surface to antibody-producing lymphocytes (Harlow and Lane, 1988). The MHC class II groove is about the same width and length as a single 18–20 amino acid polypeptide chain, so it should not be surprising to find that antibody epitopes are mostly short linear peptide sequences. It is therefore quite possible to raise an antibody against a protein or peptide that is unable to bind the parent molecule because the peptide in the parent may be masked, in a different conformation, or posttranslationally modified. Presenting the peptide in a particular conformation depends on the exact structure of the MHC class II groove in the relevant animal. MHC class II molecules are present in many allelic variants and it may be impossible to present a particular peptide in a particular conformation in many individual animals, meaning that no antibody to that particular conformation can be obtained. Getting the antibody to work immunocytochemically can provide good confirmation that it behaves as expected but requires that it binds both to the APC conformation and whatever form of the protein is present in the processed tissue.

Processing may be done by rapid freezing, formalin fixation, glutaraldehyde fixation, methanol or acetone denaturation, or other methods, all of which are expected to change the conformation of a molecule in different and in some cases unsubtle ways. Similarly, Western blotting will only show a signal if the immunogen in question can still be recognized by the antibody after denaturation typically in sodium dodecyl sulfate and β-mercaptoethanol, followed by binding to a nitrocellulose or polyvinylidene difluoride membrane and then limited and usually uncontrolled renaturation. It is frequently the case that a monoclonal antibody will work on Western blots and not immunocytochemically or vice versa, but researchers generally only study antibodies that work in both kinds of assay. The most important reagent in an ELISA-type assay is the capture antibody, which must have unusually high affinity and avidity for the relevant target molecule. Identifying an antibody that works well as a capture reagent seems to be particularly challenging, requiring not only clean and specific binding but also a high on rate and a low off rate, two characteristics of antibody–antigen interactions that are rarely measured. These parameters can be obtained using various approaches, the commonest being using biolayer interferometry, surface plasmon resonance, or other biosensors (Rich and Myszka, 2007). Unfortunately, most commercially available antibodies were made primarily for use in Western blotting or immunocytochemical studies and their ability to work in ELISA, in particular as capture reagents, has usually not been examined. In our experience, many antibodies of apparently good quality do not work well as ELISA capture reagents, and the best capture antibodies are found by a directed search for them.

This can be done either by purchasing and testing all available commercial reagents or by generating novel reagents screened specifically for this purpose. Because these two approaches cost about the same in the short term and producing antibodies in house has considerable long-term advantages, we generally make our own antibodies. We now screen directly for the ability of novel antibodies to pull the potential biomarker out of blood or blood mimetic solutions, and developed novel pNF-H monoclonals selected specifically to work well as antigen capture reagents (Boylan et al., 2009). Such antibodies can then be characterized by Western blotting and immunocytochemistry, and, if they behave as expected, used in an ELISA-type assay.

In the case of pNF-H antibodies, the phospho-KSP type repeats are known to be structurally quite different from the nonphosphorylated forms of the protein (Chang et al., 2009; Julien and Mushynski, 1982). Because the antibodies are known to be phosphate-specific, they must bind either directly to epitopes containing one or more phosphate groups, or to regions of the peptides whose conformation depend on the presence of the phosphate groups. The structure of the KSP sequences has not been rigorously studied but, given the high content of charged amino acids is expected to be extended and flexible. In fact, these types of highly charged sequence, often targets for phosphorylation, are found in a large family of CNS proteins that include growth-associated protein 43 and the microtubule associated proteins 2 and tau. These in turn all belong to a superfamily of polypeptides referred to as "intrinsically unstructured proteins." Such proteins include regions lacking defined α-helical or β-sheet structure, many of which only adopt a defined conformation on binding to a structurally defined substrate (Wright and Dyson, 1999). Possibly this lack of defined structure in the phospho-KSP repeat regions is one reason why they are so immunogenic, allowing them to readily adopt conformations to fit into a variety of MHC class II molecules on APC cells.

21.6.1 What Is the Form of the Biomarker Signal?

The previous paragraph raises the obvious question of what the ELISA is actually detecting in a blood or CSF sample. Although the protein standard for an ELISA is a pure protein, the genuine biomarker in blood may not be a single intact molecule. Many potential biomarkers exist as multimeric complexes, so that, for example, neuron-specific enolase (NSE) and S100β are present as dimers in vivo. The pNF-H molecule is known to form dimers, tetramers, and higher order complexes with itself and also NF-L, NF-M, α-internexin, and peripherin, interactions mediated by the homologous α-helical coiled coil region common to all five proteins (Perrot and Eyer, 2009). So capturing pNF-H may at least result in multiple pNF-H signals and also possibly signals from these other molecules, because it appears that any heterotetrameric combination of the various subunits can be formed. Intermediate filaments also associate with intermediate filament cross-linking proteins such as plectin, dystonin, and ACF7, which might also be found as part of the pNF-H signal (Errante et al., 1994). These issues have been little addressed to date, even though whether a protein is a single molecule or part of a multimeric complex would have a profound impact on the signal obtained by an ELISA-type assay. A preliminary simple experiment would be to see what size the biomarker signal is using during gel filtration. Our data suggest that much of the pNF-H signal in the blood of a G93A SOD1 mouse model of ALS (Boylan et al., 2009) is quite large, with a peak of material eluting at a native size of at least 600 kDa. There was also some material eluting at a lower apparent molecular weight, suggesting some heterogeneity in the signal. These so-far unpublished findings suggest that much of the blood pNF-H signal in this disease model is not monomeric, but part of a larger complex, possibly tetrameric, and potentially containing other neurofilament subunits. It is also possible that pNF-H or fragments of it complex with other endogenous blood components.

Another complexity is that the protein may be partially degraded or posttranslationally modified differently from the form used as a protein standard. This may alter the signal in unpredictable ways. Some evidence suggests that the blood pNF-H signal may be significantly increased by treatment of samples with urea, which may help unmask hidden epitopes (Lu et al., 2011). In some pathological situations, proteins may be cleaved in specific ways by proteases involved in either apoptosis or necrosis. Such specific forms, if they can be specifically detected by appropriate immunoreagents, might provide supplemental information about the type of degeneration in progress. However, antibodies that recognize specific cleaved forms of molecules are rather difficult to develop. Some have been described for calpain and caspase cleaved forms of αII-spectrin (Siman et al., 1984; Wang et al., 1998), but this work has not so far translated into workable commercial assays. Recently, Petzold and coworkers showed that pNF-H can be detected in microdialysates of TBI patients and in these patients can be cleaved by enterokinase, a protease endogenous to the brain and other tissues (Petzold et al., 2011). These authors also characterized two resulting pNF-H breakdown products, both of which include the phospho-KSP sites and that would be predicted to be highly soluble. Independently, we have characterized one calpain site in NF-M, which was observed both in vitro and in appropriately treated cells of human origin in tissue culture and which would also be predicted to be highly soluble (Shaw et al., 2004). Both sets of findings suggest that it may be possible to develop antibodies to these specific breakdown products, though this has not been done yet. NF-L and NF-H are also digested by calpains (Greenwood et al., 1993), but the sites have not been characterized. None of the neurofilament sequences contain obvious variants of the DE(T/S/A)D-type consensus sequences predicted to be cleaved by caspase enzymes (Fischer et al., 2003), although this tentative conclusion has not been experimentally confirmed as far as is known. The calpain consensus cleavage site is very loosely defined (Tompa et al., 2004), so the identity of such sites in the various neurofilament subunits can be hypothesized but must be confirmed experimentally.

The issue of proteolysis raises some other pertinent questions. As noted previously, in most cases an antibody being used in ELISA capture and/or detection binds to a short peptide derived from the immunogen. Such short peptides may be accessible to antibody in the intact molecule, in the partially proteolyzed protein, and in the fully proteolyzed protein, but are likely to bind most strongly to the digested peptide. That being the case, what is the ELISA actually measuring? Some ELISAs based on two antibodies with neighboring epitopes would likely work well on even small fragments of the intact protein, whereas an ELISA based on two reagents binding distant epitopes might be exquisitely sensitive to proteolysis. As a result, the same amount of an analyte could give completely different ELISA results depending on the state of proteolysis. In the case of pNF-H, we could envision segments of the phosphorylated KSP segment alone being detected as well or possibly even better than the intact molecule. Again the issue of whether the protein forms multimers becomes important, as does the issue of whether the proteolyzed fragments of a protein dissociate or not. These considerations are important issues that are rarely addressed because the epitopes for the majority of the antibodies used in ELISA-type assays are frequently not accurately known.

A difficulty is that ELISA is significantly more sensitive than Western blotting, so that it is problematic to show what, in Western blotting terms, an ELISA is actually detecting. With sufficient biomarker-rich sample and antibody it should be possible to obtain this information but this is rarely done and has not been done for pNF-H. The ELISA could be effectively scaled up by using, instead of an ELISA plate, either immunoprecipitation or an affinity column to which the relevant antibody is bound. The specifically bound material could then be eluted and probed by biochemical and/or proteomic methods. This approach, if done carefully and with appropriate controls, would give at least a clue as to the composition of the biomarker signal. In addition, it would be relatively easy to characterize the native size of the biomarker signal, as outlined previously for the pNF-H signal. Another approach would be to see if the captured biomarker signal is detectable by antibodies to other plausible candidates. For example, it would be straightforward to probe the captured pNF-H signal from particular blood samples with antibodies to spanning all regions of the NF-L, NF-M and other molecules that might plausibly be part of the same complex. Similarly, αII-spectrin has been reported to associate with not only various β-spectrins in vivo, but also with calmodulin, actin, and several other proteins, and it would be relatively simple to see if the αII-spectrin biomarker signal has this property also, assuming good reagents capable of capturing this molecule from blood were to become widely available. The better understanding of the exact composition of each biomarker signal is of more than academic interest because different types of signal are likely be associated with different types of damage and disease state, and different associated proteins may give further cell type and cell region–specific information. Taking all of these issues in mind, it should not be surprising that different assays to the same analyte can give different results.

In summary, it is expected that different proteolytic fragments, different protein complexes, and differently phosphorylated, oxidized, reduced, methylated, transglutaminated, and other forms of biomarkers are associated specifically with particular damage and disease states. Understanding this will allow the development of yet more specific and selective next-generation assays, an exciting possibility.

21.6.2 How Does the Biomarker Get into Blood and CSF?

This is a topic that has so far mostly been the subject of speculation. It is assumed that proteins such as neuron specific enolase (NSE) and UCHL1 are released after TBI and diffuse away from the site of injury because they are of relatively low molecular weight and are normally soluble. As noted previously, it is not clear in what form most other biomarker proteins are present in blood. In some cases, such as S100β, a small amount of the protein is normally present in CSF and blood. Such proteins may leave healthy brain tissue either by mechanically induced leakage, direct secretion, or in exosomes and enter the interstitial fluid. From there, they would enter the CSF, but to get into blood would have to cross the BBB. The BBB consists of several types of cell including neuronal processes, astrocyte endfeet, pericytes, and endothelia joined by tight junctions (Hawkins and Davis, 2005). This barrier is not as intact as often thought, and hydrophilic dyes and enzymes such as horseradish peroxidase can enter and leave the brain in the various circumventricular organs (Horsburgh and Massoud, 2013).

These circumventricular organ regions, such as the area postrema, neurohypophysis, the subfornical organ, and the pineal, allow the brain to come into more direct contact with the blood. This allows CNS tissue to detect blood components and also allows the secretion of specific CNS-derived components into blood. These regions lack the typical tight junctions responsible for the BBB in most parts of the CNS and could therefore allow the direct transitioning of CNS-derived biomarkers into the blood. However, these regions of contact are relatively small and localized in deeper brain regions, so it seems unlikely that significant amounts of, for example, cortical-derived biomarkers would leak into the blood by this route. There are other possible means of biomarker egress especially in the various subarachnoid spaces. Reiber has argued that in the healthy CNS, the barrier in these regions works as a sizing filter in the blood to CSF direction (Reiber, 2003). It is permeable even to blood molecules as large as immunoglobulin M, with a molecular weight of about 1,000 kDa, but with low efficiency, there being a linear relationship between molecular size and entry efficiency. A major factor preventing the entry of blood proteins into the CNS is the outward flow of CSF that antagonizes the inward diffusion of blood proteins. Clearly, one would expect the outward flowing CSF to augment the tendency of even quite large CNS proteins to cross the BBB.

Of course, TBI or spinal cord injury would be expected to allow egress of large amounts of CNS-derived material directly into the blood through a damaged BBB. For example, after TBI and spinal cord injury significant amounts of pNF-H were detected in peripheral circulation (Anderson et al., 2008; Blyth et al., 2011; Shaw et al., 2005). Importantly, many, if not most, chronic CNS disorders are associated with compromise of the BBB, presumably allowing enhanced egress of potential biomarkers (Hawkins and Davis, 2005; Weiss et al., 2009). For example, BBB compromise is seen in MS (Weiss et al., 2009), ALS (Rodrigues et al., 2012), Alzheimer disease (Erickson and Banks, 2013), and ischemia (Schoknecht and Shalev, 2012). Animal models of MS and ALS in fact show elevated blood levels of pNF-H, which are in both cases indicative of disease progression (Boylan et al., 2009; Gresle et al., 2008; Lu et al., 2012; Talla et al., 2013). Several studies suggest that inflammation of the CNS may result in compromise of the BBB (Simka, 2009). Finally the BBB may leak idiopathically providing another potential avenue for biomarker release, although this is itself a potentially serious clinical problem.

There seems to be no reason why pNF-H or other biomarker signals should be limited to protein fragments, single proteins, or relatively small protein complexes. In traumatic injury and ischemia, small pieces of cytoplasm and even small pieces of tissue might leave a damaged or degenerating brain region. Finally it has recently been realized that exosomes are continually being shed by all types of cells, and that these tiny vesicles can contain a microcosm of the proteins and microRNAs (MIR) of their cells of origin (Makino et al., 2013). It remains to be demonstrated whether meaningful amounts of neurofilament subunits leave damaged or diseased neurons by this method, but, if so, perhaps they could provide information in combination with studies of exosomal MIRs (Rao et al., 2013). MIRs are newly recognized as potentially very informative biomarkers, because they can be amplified using polymerase chain reaction, providing for great sensitivity from small amounts of starting material. Exosomes are known to contain abundant membrane associated and cytosolic proteins such as GAPDH, actin, and so on, and exosomes originating from, for example, neurons would be expected to contain neuron specific cytoplasmic proteins such as pNF-H, UCHL1, NSE, and others. Therefore, with appropriate techniques it might be possible to not only measure levels of particular MIRs in blood, but also to stratify from which cell type or cell region the MIR originated. To date, this has not been performed.

Another issue is that we do not know the half-life of most of the brain injury and degeneration biomarkers in blood. This is true of the pNF-H signal in blood and brings another unexplored area for future research. Because biomarkers are released in a continuous fashion from typical damage or disease states, it is difficult to deduce the half-life from the release profile. In principle, the half-life can be determined as it would be for a drug by injecting the purified protein and then measuring the level in blood samples over a suitable time course. However, performing this experiment presupposes that the protein leaves the tissue in the intact form similar to the purified form injected and not complexed with other proteins or subjected to posttranslational modifications which, as discussed previously, we do not currently understand.

21.6.3 Why Are Some Biomarkers Detectable in CSF but Not Blood?

Much interest has been engendered by the search for "point of care" (POC) devices that would allow a paramedic to, for example, assess if a brain injury had occurred from a blood sample. Such a device would allow an informed decision about whether a player should or should not return to a football game after a head injury. Such POC devices may use filters, fluidics in combination with rapid colorimetric, fluorescent, or chemiluminescent detection, but are, like ELISA, critically dependent on high avidity and affinity antibodies for both capture and detection. To date pNF-H, NSE, GFAP, UCHL1, and S100β have apparently been reliably detected in not only CSF but also blood, and so could in principle be used in such POC assays. However, many other potential analytes, including αII-spectrin, have been difficult or impossible to detect in blood. Why is this? There are several possible reasons. One is that proteins detectable in CSF are unstable in blood, being degraded by proteases, being sequestered rapidly in the kidney, or bound up by one or other of the numerous blood proteins. A second is that the protein in question may not readily enter the blood because of it being part of too large a molecular complex. Third, it is possible that a protein has disadvantageous adhesive, hydrophobic, or charge properties that hinder it diffusivity. Fourth, it is quite possible that assays for the protein in question are simply not of high enough affinity to capture the small amount of a particular protein from the complex milieu of blood.

21.7 CONFOUNDING FACTORS

It is important to know if antibodies to any of the potential biomarker proteins are already present in the blood of some individuals because such antibodies may bind the biomarker protein and reduce or prevent it from being detected efficiently by ELISA. There is quite old literature showing autoantibodies to pNF-H both in the vast majority of humans, irrespective of disease state, and in patients with Kuru, MS, and other serious neurodegenerative diseases (Braxton et al., 1989; Ehling et al., 2004; Sotelo et al., 1980; Stefansson et al., 1985). Although the presence of these antibodies may present another variable in this area of research, they have not been studied in this context. Presumably pNF-H and other neurofilament proteins may leak into the blood as a part of normal aging or after trauma and may raise antibody responses there. Key issues to understand are which individuals have these antibodies and what is their titer. Most studies suggest primarily an immunoglobulin M response and relatively low titer, so it is quite possible that these antibodies will have only a marginal effect on pNF-H biomarker detection, though this remains to be demonstrated. It is possible that the so-called "hook effect," meaning that blood samples rich in pNF-H biomarker gave a less strong ELISA signal than expected, is due partly or wholly to such antibodies (Lu et al., 2011). These authors proposed that this effect was due to aggregation reducing the access to antibodies, and dealt with this by treating samples with urea. However, they did not see the hook effect in pNF-H–rich CSF samples that would not normally contain circulating antibodies (Lu et al., 2011).

Finally, and as noted previously, blood is by far not an ideal fluid in which to detect very small amounts of a protein efficiently. This may be partly due to endogenous human anti-mouse antibodies (Bruggink et al., 2013; Kricka, 1999), which may cross-link or block binding and hence give a signal independent of the presence of the analyte, a problem that can be addressed. Less easily dealt with are the fortuitous binding of one or another antibody to some serum or plasma component, another possible cause of the hook effect. As a result, much work goes into obtaining antibody pairs that are able to capture and detect the analyte in the presence of human plasma or serum. This work typically involves painstaking studies with control and patient blood samples and the use of different diluents and detergents, and may involve addition of mild denaturants. Our experience is that antibodies that seem ideal by every other criterion may show a dramatic loss of binding ability, usually for unknown reasons, in the presence of human serum or plasma.

21.8 ENDOGENOUS LEVELS OF BIOMARKER PROTEINS

Certain potential biomarker proteins are known to be present at low levels in blood and CSF in healthy individuals, but in most cases this has not been studied. S100β is known to leak into both CSF and blood in small amounts (Reiber, 2003) and has been shown to bind the receptor for advanced glycation end products and certain Toll-like receptors (TLRs). Increased amounts of S100β may therefore actually function as a signal that tissue has been damaged and so S100β appears to be a damage-associated molecular profile (DAMP) (Kaczmarek et al., 2013). DAMPs bind to endogenous receptors in a similar fashion to the much better studied pathogen-associated molecular profiles. Binding of pathogen-associated molecular profiles or DAMPs to TLRs and other receptors induces activation of nuclear factor-κB leading to an inflammatory and presumably generally beneficial response. A similar mechanism may be at work with other biomarker proteins, and it is possible that pNF-H and/or other neurofilament subunits may bind to receptors for advanced glycation end products, TLR, and similar receptors and help induce an inflammatory response after injury, we can speculate perhaps eliciting neuroprotective mechanisms. By analogy to S100β, small amounts of these proteins would normally be secreted and increased amounts would activate the DAMP receptors and so induce inflammation. Such a mechanism would also explain the generation of neurofilament autoantibodies as described previously. ELISA-type assays always have a small background signal particularly when tested on blood samples. We noted that this small signal increased slightly as a function of age of control individuals (Boylan et al., 2009), perhaps suggestive of a small amount of endogenous pNF-H in blood. This could be addressed by biochemical means, such as affinity purification or immunoprecipitation with pNF-H antibodies followed by mass spectroscopy or other biochemical analysis, but this has not been performed to date. There is some direct proteomic evidence of the presence of low levels of peptides derived from NF-H and other neurofilament subunits in normal and TBI patient blood samples (Haqqani et al., 2007) and in normal and MS patient blood samples (Gresle et al., in press). The origin of these peptides in the blood of healthy individuals is currently unknown. In the case of the MS samples, levels of the neurofilament-derived peptides were elevated in patients compared with controls, as expected from ELISA data and presumably indicating release of pNF-H from the sites of MS pathology. This is therefore another area that requires more study to more fully understand the significance of the pNF-H blood biomarker.

21.9 CONCLUSION

Several studies from a variety of laboratories have provided compelling evidence that pNF-H can be reliably and sensitively detected in the blood of a wide variety of damage and disease states associated with axonal injury and degeneration. As a result this biomarker is now being developed for potential clinical use. Further studies of this unique and promising biomarker therefore seem warranted.

REFERENCES

Adkins, J.N., S.M. Varnum, K.J. Auberry, R.J. Moore, N.H. Angell, R.D. Smith et al. 2002. Toward a human blood serum proteome: Analysis by multidimensional separation coupled with mass spectrometry. *Molecular & Cellular Proteomics*. 1:947–955.

Allahtavakoli, M., and B. Jarrott. 2011. Sigma-1 receptor ligand PRE-084 reduced infarct volume, neurological deficits, pro-inflammatory cytokines and enhanced anti-inflammatory cytokines after embolic stroke in rats. *Brain Research Bulletin*. 85:219–224.

Anderson, K.J., S.W. Scheff, K.M. Miller, K.N. Roberts, L.K. Gilmer, C. Yang et al. 2008. The phosphorylated axonal form of the neurofilament subunit NF-H (pNF-H) as a blood biomarker of traumatic brain injury. *Journal of Neurotrauma*. 25:1079–1085.

Blyth, B.J., A. Farahvar, H. He, A. Nayak, C. Yang, G. Shaw et al. 2011. Elevated serum ubiquitin carboxy-terminal hydrolase l1 is associated with abnormal blood-brain barrier function after traumatic brain injury. *Journal of Neurotrauma*. 28:2453–2462

Boylan, K., C. Yang, J. Crook, K. Overstreet, M. Heckman, Y. Wang et al. 2009. Immunoreactivity of the phosphorylated axonal neurofilament H subunit (pNF-H) in blood of ALS model rodents and ALS patients: Evaluation of blood pNF-H as a potential ALS biomarker. *Journal of Neurochemistry*. 111:1182–1191.

Boylan, K.B., J.D. Glass, J.E. Crook, C. Yang, C.S. Thomas, P. Desaro et al. 2013. Phosphorylated neurofilament heavy subunit (pNF-H) in peripheral blood and CSF as a potential prognostic biomarker in amyotrophic lateral sclerosis. *Journal of Neurology, Neurosurgery, and Psychiatry*. 84:467–472.

Braxton, D.B., M. Williams, D. Kamali, S. Chin, R. Liem, and N. Latov. 1989. Specificity of human anti-neurofilament autoantibodies. *Journal of Neuroimmunology*. 21:193–203.

Bruggink, K.A., W. Jongbloed, E.A. Biemans, R. Veerhuis, J.A. Claassen, H.B. Kuiperij et al. 2013. Amyloid-beta oligomer detection by ELISA in cerebrospinal fluid and brain tissue. *Analytical Biochemistry*. 433:112–120.

Buki, A., and J. Povlishock. 2006. All roads lead to disconnection?—Traumatic axonal injury revisited. *Acta Neurochirurgica*. 148:181–193.

Carden, M.J., W.W. Schlaepfer, and V.M. Lee. 1985. The structure, biochemical properties, and immunogenicity of neurofilament peripheral regions are determined by phosphorylation state. *The Journal of Biological Chemistry*. 260:9805–9817.

Chang, R., Y. Kwak, and Y. Gebremichael. 2009. Structural properties of neurofilament sidearms: Sequence-based modeling of neurofilament architecture. *Journal of Molecular Biology*. 391:648–660.

Dash, P.K., S. Gorantla, H.E. Gendelman, J. Knibbe, G.P. Casale, E. Makarov et al. 2011. Loss of neuronal integrity during progressive HIV-1 infection of humanized mice. *The Journal of Neuroscience*. 31:3148–3157.

Douglas-Escobar, M., C. Yang, J. Bennett, J. Shuster, D. Theriaque, A. Leibovici et al. 2011. A pilot study of novel biomarkers in neonates with hypoxic-ischemic encephalopathy. *Pediatric Research*. 68:531–536

Ehling, R., A. Lutterotti, J. Wanschitz, M. Khalil, C. Gneiss, F. Deisenhammer et al. 2004. Increased frequencies of serum antibodies to neurofilament light in patients with primary chronic progressive multiple sclerosis. *Multiple Sclerosis*. 10:601–606.

Erickson, M.A., and W.A. Banks. 2013. Blood-brain barrier dysfunction as a cause and consequence of Alzheimer's disease. *Journal of Cerebral Blood Flow and Metabolism*. 33:1500–1513.

Errante, L.D., G. Wiche, and G. Shaw. 1994. Distribution of plectin, an intermediate filament-associated protein, in the adult rat central nervous system. *Journal of Neuroscience Research*. 37:515–528.

Farrah, T., E.W. Deutsch, G.S. Omenn, D.S. Campbell, Z. Sun, J.A. Bletz et al. 2011. A high-confidence human plasma proteome reference set with estimated concentrations in PeptideAtlas. *Molecular & Cellular Proteomics* 10:M110 006353.

Fischer, U., R.U. Janicke, and K. Schulze-Osthoff. 2003. Many cuts to ruin: A comprehensive update of caspase substrates. *Cell Death and Differentiation*. 10:76–100.

Gaiottino, J., N. Norgren, R. Dobson, J. Topping, A. Nissim, A. Malaspina et al. 2013. Increased neurofilament light chain blood levels in neurodegenerative neurological diseases. *PloS One*. 8:e75091.

Ganesalingam, J., J. An, R. Bowser, P.M. Andersen, and C.E. Shaw. 2013. pNfH is a promising biomarker for ALS. *Amyotrophic Lateral Sclerosis & Frontotemporal Degeneration*. 14:146–149.

Ganesalingam, J., J. An, C.E. Shaw, G. Shaw, D. Lacomis, and R. Bowser. 2011. Combination of neurofilament heavy chain and complement C3 as CSF biomarkers for ALS. *Journal of Neurochemistry*. 117:528–537.

Gnanapavan, S., D. Grant, S. Morant, J. Furby, T. Hayton, C.E. Teunissen et al. 2013. Biomarker report from the phase II lamotrigine trial in secondary progressive MS - neurofilament as a surrogate of disease progression. *PloS One*. 8:e70019.

Goldstein, M.E., L.A. Sternberger, and N.H. Sternberger. 1987. Varying degrees of phosphorylation determine microheterogeneity of the heavy neurofilament polypeptide (Nf-H). *Journal of Neuroimmunology*. 14:135–148.

Greenwood, J.A., J.C. Troncoso, A.C. Costello, and G.V. Johnson. 1993. Phosphorylation modulates calpain-mediated proteolysis and calmodulin binding of the 200-kDa and 160-kDa neurofilament proteins. *Journal of Neurochemistry*. 61:191–199.

Gresle, M.M., E. Alexandrou, Q. Wu, G. Egan, V. Jokubaitis, M. Ayers et al. 2012. Leukemia inhibitory factor protects axons in experimental autoimmune encephalomyelitis via an oligodendrocyte-independent mechanism. *PloS One*. 7:e47379.

Gresle, M.M., G. Shaw, B. Jarrott, E.N. Alexandrou, A. Friedhuber, T.J. Kilpatrick et al. 2008. Validation of a novel biomarker for acute axonal injury in experimental autoimmune encephalomyelitis. *Journal of Neuroscience Research*. 86:3548–3555.

Guy, J., G. Shaw, F.N. Ross-Cisneros, P. Quiros, S.R. Salomao, A. Berezovsky, V. Carelli et al. 2008. Phosphorylated neurofilament heavy chain is a marker of neurodegeneration in Leber hereditary optic neuropathy (LHON). *Molecular Vision*. 14:2443–2450.

Haqqani, A.S., J.S. Hutchison, R. Ward, and D.B. Stanimirovic. 2007. Biomarkers and diagnosis; protein biomarkers in serum of pediatric patients with severe traumatic brain injury identified by ICAT-LC-MS/MS. *Journal of Neurotrauma*. 24:54–74.

Harlow, E., and D. Lane. 1988. *Antibodies: A Laboratory Manual*. Cold Spring Harbor Laboratory Press, New York.

Hawkins, B.T., and T.P. Davis. 2005. The blood-brain barrier/neurovascular unit in health and disease. *Pharmacological Reviews*. 57:173–185.

Hayakawa, K., T. Itoh, H. Niwa, T. Mutoh, and G. Sobue. 1998. NGF prevention of neurotoxicity induced by cisplatin, vincristine and taxol depends on toxicity of each drug and NGF treatment schedule: In vitro study of adult sympathetic ganglion explants. *Brain Research*. 794:313–319.

Hayakawa, K., R. Okazaki, K. Ishii, T. Ueno, N. Izawa, Y. Tanaka et al. 2012. Phosphorylated neurofilament subunit NF-H as a biomarker for evaluating the severity of spinal cord injury patients, a pilot study. *Spinal Cord*. 50:493–496.

Hill, W.D., V.M. Lee, H.I. Hurtig, J.M. Murray, and J.Q. Trojanowski. 1991. Epitopes located in spatially separate domains of each neurofilament subunit are present in Parkinson's disease Lewy bodies. *The Journal of Comparative Neurology*. 309:150–160.

Horsburgh, A., and T.F. Massoud. 2013. The circumventricular organs of the brain: Conspicuity on clinical 3T MRI and a review of functional anatomy. *Surgical and Radiologic Anatomy*. 35:343–349.

Jacobs, J.M., J.N. Adkins, W.J. Qian, T. Liu, Y. Shen, D.G. Camp, 2nd, and R.D. Smith. 2005. Utilizing human blood plasma for proteomic biomarker discovery. *Journal of Proteome Research*. 4:1073–1085.

Jokubaitis, V.G., M.M. Gresle, D.A. Kemper, W. Doherty, V.M. Perreau, T.L. Cipriani et al. 2013. Endogenously regulated Dab2 worsens inflammatory injury in experimental autoimmune encephalomyelitis. *Acta Neuropathologica Communications*. 1:32.

Julien, J.P., and W.E. Mushynski. 1982. Multiple phosphorylation sites in mammalian neurofilament polypeptides. *The Journal of Biological Chemistry*. 257:10467–10470.

Kaczmarek, A., P. Vandenabeele, and D.V. Krysko. 2013. Necroptosis: The release of damage-associated molecular patterns and its physiological relevance. *Immunity*. 38:209–223.

Kovesdi, E., P. Bukovics, V. Besson, J. Nyiradi, J. Luckl, J. Pal et al. 2010. A novel PARP inhibitor L-2286 in a rat model of impact acceleration head injury: An immunohistochemical and behavioral study. *International Journal of Molecular Sciences*. 11:1253–1268.

Kricka, L.J. 1999. Human anti-animal antibody interferences in immunological assays. *Clinical Chemistry*. 45:942–956.

Kuhle, J., D. Leppert, A. Petzold, A. Regeniter, C. Schindler, M. Mehling et al. 2011. Neurofilament heavy chain in CSF correlates with relapses and disability in multiple sclerosis. *Neurology*. 76:1206–1213.

Lee, V.M., L. Otvos, Jr., M.J. Carden, M. Hollosi, B. Dietzschold, and R.A. Lazzarini. 1988. Identification of the major multiphosphorylation site in mammalian neurofilaments. *Proceedings of the National Academy of Sciences of the United States of America*. 85:1998–2002.

Lees, J.F., P.S. Shneidman, S.F. Skuntz, M.J. Carden, and R.A. Lazzarini. 1988. The structure and organization of the human heavy neurofilament subunit (NF-H) and the gene encoding it. *The EMBO Journal*. 7:1947–1955.

Lehnert, S., J. Costa, M. de Carvalho, J. Kirby, M. Kuzma-Kozakiewicz, C. Morelli et al. 2014. Multicentre quality control evaluation of different biomarker candidates for amyotrophic lateral sclerosis. *Amyotrophic Lateral Sclerosis & Frontotemporal Degeneration*. Feb 28.

Levine, T.D., R. Bowser, N.C. Hank, S. Gately, D. Stephan, D.S. Saperstein et al. 2012. A pilot trial of pioglitazone HCl and tretinoin in ALS: Cerebrospinal fluid biomarkers to monitor drug efficacy and predict rate of disease progression. *Neurology Research International*. 2012:582075.

Lewis, S.B., R. Wolper, Y.Y. Chi, L. Miralia, Y. Wang, C. Yang et al. 2010. Identification and preliminary characterization of ubiquitin C terminal hydrolase 1 (UCHL1) as a biomarker of neuronal loss in aneurysmal subarachnoid hemorrhage. *Journal of Neuroscience Research*. 88:1475–1484.

Lewis, S.B., R.A. Wolper, L. Miralia, C. Yang, and G. Shaw. 2008. Detection of phosphorylated NF-H in the cerebrospinal fluid and blood of aneurysmal subarachnoid hemorrhage patients. *Journal of Cerebral Blood Flow and Metabolism*. 28:1261–1271.

Lu, C.H., B. Kalmar, A. Malaspina, L. Greensmith, and A. Petzold. 2011. A method to solubilise protein aggregates for immunoassay quantification which overcomes the neurofilament "hook" effect. *Journal of Neuroscience Methods*. 195:143–150.

Lu, C.H., A. Petzold, B. Kalmar, J. Dick, A. Malaspina, and L. Greensmith. 2012. Plasma neurofilament heavy chain levels correlate to markers of late stage disease progression and treatment response in SOD1(G93A) mice that model ALS. *PloS One*. 7:e40998.

Makino, D.L., F. Halbach, and E. Conti. 2013. The RNA exosome and proteasome: Common principles of degradation control. *Nature Reviews. Molecular Cell Biology*. 14:654–660.

Matsushige, T., T. Ichiyama, B. Anlar, J. Tohyama, K. Nomura, Y. Yamashita et al. 2008. CSF neurofilament and soluble TNF receptor 1 levels in subacute sclerosing panencephalitis. *Journal of Neuroimmunology*. 205:155–159.

Matsushige, T., T. Ichiyama, M. Kajimoto, M. Okuda, S. Fukunaga, and S. Furukawa. 2009. Serial cerebrospinal fluid neurofilament concentrations in bacterial meningitis. *Journal of the Neurological Sciences*. 280:59–61.

Matsushige, T., H. Inoue, S. Fukunaga, S. Hasegawa, M. Okuda, and T. Ichiyama. 2012. Serum neurofilament concentrations in children with prolonged febrile seizures. *Journal of the Neurological Sciences*. 321:39–42.

Menkhorst, E., J.G. Zhang, N.A. Sims, P.O. Morgan, P. Soo, I.J. Poulton et al. 2011. Vaginally administered PEGylated LIF antagonist blocked embryo implantation and eliminated non-target effects on bone in mice. *PloS One*. 6:e19665.

Mondello, S., U. Muller, A. Jeromin, J. Streeter, R.L. Hayes, and K.K. Wang. 2011. Blood-based diagnostics of traumatic brain injuries. *Expert Review of Molecular Diagnostics*. 11:65–78.

Myers, M.W., R.A. Lazzarini, V.M. Lee, W.W. Schlaepfer, and D.L. Nelson. 1987. The human mid-size neurofilament subunit: A repeated protein sequence and the relationship of its gene to the intermediate filament gene family. *The EMBO Journal*. 6:1617–1626.

Norgren, N., J.E. Karlsson, L. Rosengren, and T. Stigbrand. 2002. Monoclonal antibodies selective for low molecular weight neurofilaments. *Hybrid Hybridomics*. 21:53–59.

Obeid, R., J. Schlundt, N. Umanskaya, W. Herrmann, and M. Herrmann. 2011. Folate is related to phosphorylated neurofilament-H and P-tau (Ser396) in rat brain. *Journal of Neurochemistry*. 117:1047–1054.

Pasol, J., W. Feuer, C. Yang, G. Shaw, R. Kardon, and J. Guy. 2010. Phosphorylated neurofilament heavy chain correlations to visual function, optical coherence tomography, and treatment. *Multiple Sclerosis International*. 2010:542691.

Perrot, R., and J. Eyer. 2009. Neuronal intermediate filaments and neurodegenerative disorders. *Brain Research Bulletin*. 80:282–295.

Petzold, A. 2005. Neurofilament phosphoforms: Surrogate markers for axonal injury, degeneration and loss. *Journal of the Neurological Sciences*. 233:183–198.

Petzold, A., A. Altintas, L. Andreoni, A. Bartos, A. Berthele, M.A. Blankenstein et al. 2010. Neurofilament ELISA validation. *Journal of Immunological Methods*. 352:23–31.

Petzold, A., G. Keir, A.J. Green, G. Giovannoni, and E.J. Thompson. 2003. A specific ELISA for measuring neurofilament heavy chain phosphoforms. *Journal of Immunological Methods*. 278:179–190.

Petzold, A., and G. Shaw. 2007. Comparison of two ELISA methods for measuring levels of the phosphorylated neurofilament heavy chain. *Journal of Immunological Methods*. 319:34–40.

Petzold, A., M.M. Tisdall, A.R. Girbes, L. Martinian, M. Thom, N. Kitchen et al. 2011. in vivo monitoring of neuronal loss in traumatic brain injury: A microdialysis study. *Brain*. 134:464–483.

Prasad, A., Q.S. Xue, V. Sankar, T. Nishida, G. Shaw, W. Streit et al. 2012. Comprehensive characterization of tungsten microwires in chronic neurocortical implants. Conference proceedings: Annual International Conference of the IEEE Engineering in Medicine and Biology Society. IEEE Engineering in Medicine and Biology Society Conference. 2012:755–758.

Rana, O.R., J.W. Schroder, J.K. Baukloh, E. Saygili, K. Mischke, J. Schiefer et al. 2013. Neurofilament light chain as an early and sensitive predictor of long-term neurological outcome in patients after cardiac arrest. *International Journal of Cardiology*. 168:1322–1327.

Rao, P., E. Benito, and A. Fischer. 2013. MicroRNAs as biomarkers for CNS disease. *Frontiers in Molecular Neuroscience*. 6:39.

Reiber, H. 2003. Proteins in cerebrospinal fluid and blood: Barriers, CSF flow rate and source-related dynamics. *Restorative Neurology and Neuroscience*. 21:79–96.

Rich, R.L., and D.G. Myszka. 2007. Survey of the year 2006 commercial optical biosensor literature. *Journal of Molecular Recognition*. 20:300–366.

Ringger, N.C., S. Giguere, P.R. Morresey, C. Yang, and G. Shaw. 2011. Biomarkers of brain injury in foals with hypoxic-ischemic encephalopathy. *Journal of Veterinary Internal Medicine/American College of Veterinary Internal Medicine*. 25:132–137.

Rodrigues, M.C., D.G. Hernandez-Ontiveros, M.K. Louis, A.E. Willing, C.V. Borlongan, P.R. Sanberg et al. 2012. Neurovascular aspects of amyotrophic lateral sclerosis. *International Review of Neurobiology*. 102:91–106.

Rundgren, M., H. Friberg, T. Cronberg, B. Romner, and A. Petzold. 2012. Serial soluble neurofilament heavy chain in plasma as a marker of brain injury after cardiac arrest. *Critical Care*. 16:R45.

Schoknecht, K., and H. Shalev. 2012. Blood-brain barrier dysfunction in brain diseases: Clinical experience. *Epilepsia*. 53(Suppl 6):7–13.

Sellner, J., A. Patel, P. Dassan, M.M. Brown, and A. Petzold. 2011. Hyperacute detection of neurofilament heavy chain in serum following stroke: A transient sign. *Neurochemical Research*. 36:2287–2291.

Sellner, J., A. Petzold, S. Sadikovic, L. Esposito, M.S. Weber, P. Heider ealt. 2009. The value of the serum neurofilament protein heavy chain as a biomarker for peri-operative brain injury after carotid endarterectomy. *Neurochemical Research*. 34:1969–1974.

Shaw, G., M. Osborn, and K. Weber. 1986. Reactivity of a panel of neurofilament antibodies on phosphorylated and dephosphorylated neurofilaments. *European Journal of Cell Biology*. 42:1–9.

Shaw, G., C. Yang, R. Ellis, K. Anderson, J. Parker Mickle, S. Scheff et al. 2005. Hyperphosphorylated neurofilament NF-H is a serum biomarker of axonal injury. *Biochemical and Biophysical Research Communications*. 336:1268–1277.

Shaw, G., C. Yang, L. Zhang, P. Cook, B. Pike, and W.D. Hill. 2004. Characterization of the bovine neurofilament NF-M protein and cDNA sequence, and identification of in vitro and in vivo calpain cleavage sites. *Biochemical and Biophysical Research Communications*. 325:619–625.

Shen, Y., J. Kim, E.F. Strittmatter, J.M. Jacobs, D.G. Camp, 2nd, R. Fang et al. 2005. Characterization of the human blood plasma proteome. *Proteomics*. 5:4034–4045.

Siman, R., M. Baudry, and G. Lynch. 1984. Brain fodrin: Substrate for calpain I, an endogenous calcium-activated protease. *Proceedings of the National Academy of Sciences of the United States of America*. 81:3572–3576.

Simka, M. 2009. Blood brain barrier compromise with endothelial inflammation may lead to autoimmune loss of myelin during multiple sclerosis. *Current Neurovascular Research*. 6:132–139.

Singh, P., J. Yan, R. Hull, S. Read, J. O'Sullivan, R.D. Henderson et al. 2011. Levels of phosphorylated axonal neurofilament subunit H (pNfH) are increased in acute ischemic stroke. *Journal of the Neurological Sciences*. 304:117–121.

Sotelo, J., C.J. Gibbs, Jr., and D.C. Gajdusek. 1980. Autoantibodies against axonal neurofilaments in patients with Kuru and Creutzfeldt-Jakob disease. *Science*. 210:190–193.

Stefansson, K., L.S. Marton, M.E. Dieperink, G.K. Molnar, W.W. Schlaepfer, and C.M. Helgason. 1985. Circulating autoantibodies to the 200,000-dalton protein of neurofilaments in the serum of healthy individuals. *Science*. 228:1117–1119.

Sternberger, L.A., and N.H. Sternberger. 1983. Monoclonal antibodies distinguish phosphorylated and nonphosphorylated forms of neurofilaments in situ. *Proceedings of the National Academy of Sciences of the United States of America*. 80:6126–6130.

Stratford, C.H., A. Pemberton, L. Cameron, and B.C. McGorum. 2013. Plasma neurofilament pNF-H concentration is not increased in acute equine grass sickness. *Equine Veterinary Journal*. 45:254–255.

Strong, M.J., W.L. Strong, H. Jaffe, B. Traggert, M.M. Sopper, and H.C. Pant. 2001. Phosphorylation state of the native high-molecular-weight neurofilament subunit protein from cervical spinal cord in sporadic amyotrophic lateral sclerosis. *Journal of Neurochemistry*. 76:1315–1325.

Stys, P.K. 2005. General mechanisms of axonal damage and its prevention. *Journal of Neurological Sciences*. 233:3–13.

Talla, V., C. Yang, G. Shaw, V. Porciatti, R.D. Koilkonda, and J. Guy. 2013. Noninvasive assessments of optic nerve neurodegeneration in transgenic mice with isolated optic neuritis. *Investigative Ophthalmology & Visual Science*. 54:4440–4450.

Tompa, P., P. Buzder-Lantos, A. Tantos, A. Farkas, A. Szilagyi, Z. Banoczi et al. 2004. On the sequential determinants of calpain cleavage. *The Journal of Biological Chemistry*. 279:20775–20785.

Ueno, T., Y. Ohori, J. Ito, S. Hoshikawa, S. Yamamoto, K. Nakamura et al. 2010. Hyperphosphorylated neurofilament NF-H as a biomarker of the efficacy of minocycline therapy for spinal cord injury. *Spinal Cord*. 49:333–336.

Wang, K.K., R. Posmantur, R. Nath, K. McGinnis, M. Whitton, R.V. Talanian et al. 1998. Simultaneous degradation of alphaII- and betaII-spectrin by caspase 3 (CPP32) in apoptotic cells. *The Journal of Biological Chemistry*. 273:22490–22497.

Weiss, N., F. Miller, S. Cazaubon, and P.O. Couraud. 2009. The blood-brain barrier in brain homeostasis and neurological diseases. *Biochimica et Biophysica Acta*. 1788:842–857.

Wieske, L., E. Witteveen, A. Petzold, C. Verhamme, M.J. Schultz, I.N. van Schaik et al. 2014. Neurofilaments as a plasma biomarker for ICU-acquired weakness: An observational pilot study. *Critical Care*. 18:R18.

Wild, E.J., A. Petzold, G. Keir, and S.J. Tabrizi. 2007. Plasma neurofilament heavy chain levels in Huntington's disease. *Neuroscience Letters*. 417:231–233.

Wildes, D., and J.A. Wells. 2010. Sampling the N-terminal proteome of human blood. *Proceedings of the National Academy of Sciences of the United States of America*. 107:4561–4566.

Wright, P.E., and H.J. Dyson. 1999. Intrinsically unstructured proteins: Re-assessing the protein structure-function paradigm. *Journal of Molecular Biology*. 293:321–331.

Zurek, J., L. Bartlova, and M. Fedora. 2011. Hyperphosphorylated neurofilament NF-H as a predictor of mortality after brain injury in children. *Brain Injury*. 25:221–226.

Zurek, J., and M. Fedora. 2012. The usefulness of S100B, NSE, GFAP, NF-H, secretagogin and Hsp70 as a predictive biomarker of outcome in children with traumatic brain injury. *Acta Neurochirurgica*. 154:93–103; discussion 103.

22 Exploring Serum Biomarkers for Mild Traumatic Brain Injury

Linda Papa, Damyan Edwards, and Michelle Ramia

CONTENTS

22.1 INTRODUCTION

The diagnosis of traumatic brain injury (TBI) in the acute setting is based on neurological examination and neuroimaging tools such as CT scanning and MRI. However, CT scanning has low sensitivity to diffuse brain damage and confers exposure to radiation. On the other hand, MRI can provide information on the extent of diffuse injuries but its widespread application is restricted by cost, the limited availability of MRI in many centers, and the difficulty of performing it in physiologically unstable patients. Although some patients with Mild traumatic brain injury (mTBI) may be admitted to the hospital overnight, the vast majority are treated and released from emergency departments with basic discharge instructions. This group of TBI patients represents the greatest challenges to accurate diagnosis and outcome prediction. The lack of clinical tools to detect the deficits that affect daily function, have left these individuals with little or no treatment options. The injury is often seen as "not severe" and subsequently therapies have not been aggressively sought for MTBI. The diagnostic and prognostic tools for risk stratification of TBI patients are therefore limited in the early stages after injury. Unlike other organ-based diseases where rapid diagnosis employing biomarkers from blood tests are clinically essential to guide diagnosis and treatment, there are no rapid, definitive diagnostic blood tests for TBI. Over the last decade there has been a myriad of studies exploring many promising biomarkers. Despite the large number of published studies there is still a lack of any FDA-approved biomarkers for clinical use in adults and children. There is now an important need to validate and introduce them into the clinical setting. This chapter will review some of the most widely studied biomarkers for TBI in the clinical setting, with an emphasis on those that have been evaluated in MTBI.

Mild traumatic brain injury is also known as a concussion and is a traumatic force to the brain leading to a disruption of brain function. This disruption may seem transient, but could have long-lasting effects. It can manifest as an alteration in mental status such as confusion, amnesia, or loss of consciousness. There is a misconception among many that loss of consciousness must occur to have mTBI or concussion. As a result, many people with mTBI do not seek help, and many health care professionals do not recognize that an mTBI has occurred. There are an estimated 10 million people affected annually by TBI across the globe (Hyder et al., 2007). However, this is likely an underestimate given that many patients sustain mTBI but do not seek medical care. According to the World Health Organization, TBI will surpass many diseases as the major cause of death and disability by the year 2020 (Hyder et al., 2007).

Although TBI is often categorized into mild, moderate, and severe based on the Glasgow Coma Scale (GCS) score, it really represents a spectrum of injury. The GCS is a 15-point neurological scale used to characterize severity of TBI and was originally intended to provide an easy-to-use assessment tool and to facilitate communication between care providers on rotating shifts (Teasdale and Jennett, 1974). A GCS equal to or less than 8 is considered a "severe" TBI, a GCS of 9–12 is a "moderate" TBI, and a GCS of 13–15 is considered mTBI. The term "mild TBI" is actually a misnomer.

Individuals who incur a TBI and have an initial GCS score of 13–15 are acutely at risk for intracranial bleeding and diffuse axonal injury (Stein et al., 2009). Additionally, a significant proportion is at risk for impairment of physical, cognitive, and psychosocial functioning (Alexander, 1995; Alves et al., 1993; Barth et al., 1983; Millis et al., 2001; Rimel et al., 1981).

An important neuropathological consequence of TBI is axonal injury, termed diffuse axonal injury (DAI) and more recently called traumatic axonal injury (TAI) (Povlishock, 1992). DAI/TAI can be found after severe, moderate, and mild TBI, and can occur after rapid acceleration and deceleration forces that can occur following motor vehicle accidents. DAI/TAI involves a number of abnormalities from direct damage to the axonal cytoskeleton to secondary damage from disruption of transport, proteolysis, and swelling (Johnson et al., 2012). For instance, ionic imbalances, through an efflux of potassium and influx of sodium, lead to calcium influx into cells, creating mitochondrial damage and impaired oxidative metabolism with lactate production (Buki et al., 2003; Maxwell et al., 2003).

The diagnosis of TBI in the acute setting is based on neurological examination and neuroimaging tools such as computed tomography (CT) scanning and magnetic resonance imaging (MRI). However, CT scanning has low sensitivity to diffuse brain damage and confers exposure to radiation. MRI can provide information on the extent of diffuse injuries, but its widespread application is restricted by cost, the limited availability of MRI in many centers, and the difficulty of performing it in physiologically unstable patients. In particular, the recognition of DAI/TAI is even more difficult and standard neuroimaging techniques may not detect TBI (Metting et al., 2012). Diffusion tensor imaging (DTI) is a promising neuroimaging technique that may help to identify axonal injury after mTBI (Bazarian et al., 2007; Huang et al., 2009). However, the role of MRI and DTI in the acute clinical management of TBI patients has not been established (Jagoda et al., 2008; Kesler, 2000). Although some patients with mTBI may be admitted to the hospital overnight, the vast majority are treated and released from emergency departments with basic discharge instructions. This group of TBI patients represents the greatest challenge to accurate diagnosis and outcome prediction. The lack of clinical tools to detect the deficits that affect daily function have left these individuals with little or no treatment options. The injury is often seen as "not severe" and subsequently therapies have not been aggressively sought for mTBI.

The diagnostic and prognostic tools for risk stratification of TBI patients are therefore limited in the early stages after injury. Unlike other organ-based diseases in which rapid diagnosis employing biomarkers from blood tests are clinically essential to guide diagnosis and treatment, such as for myocardial ischemia or kidney and liver dysfunction, there are no rapid, definitive diagnostic blood tests for TBI. Over the past decade, there have been myriad studies exploring many promising biomarkers. Despite the large number of published studies (Kochanek et al., 2008; Papa, 2012), there is still a lack of any Food and Drug Administration–approved biomarkers for clinical use in adults and children (Papa, 2012; Papa et al., 2013). There is now a strong need to validate and introduce them into the clinical setting.

This chapter will review some of the most widely studied biomarkers for TBI in the clinical setting, with an emphasis on those that have been evaluated with mTBI. Figure 22.1 shows the neuroanatomical locations of the biomarkers that will be reviewed.

22.2 BIOFLUID BIOMARKERS OF ASTROGLIAL INJURY

22.2.1 S100β

S100β is the major low-affinity calcium-binding protein in astrocytes (Xiong et al., 2000) that helps to regulate intracellular levels of calcium; it is considered a marker of astrocyte injury or death. It can also be found in nonneural cells such as adipocytes, chondrocytes, and melanoma cells (Olsson et al., 2011; Zimmer et al., 1995). S100β is one of the most extensively studied biomarkers (Berger et al., 2002; Haimoto,1987; Jonsson, 2000; Korfias et al., 2007; Missler, 1997; Raabe et al., 1999; Romner et al., 2000; Usui, 1989; Vos et al., 2010; Woertgen et al., 1997; Ytrebo, 2001). Elevated S100β levels in serum have been associated with increased incidence of postconcussive syndrome and problems with cognition (Ingebrigtsen and Romner, 1997; Waterloo et al., 1997). Moreover, studies have reported that serum levels of S-100β are associated with MRI abnormalities and with neuropsychological examination disturbances after mTBI (Ingebrigtsen and Romner, 1996; Ingebrigtsen et al., 1999). Several studies have found significant correlations between elevated serum levels of S100β and CT abnormalities (Biberthaler et al., 2006; Ingebrigtsen et al., 2000; Muller et al., 2007). It has been suggested that adding the measurement of S100β concentration to clinical decision tools for mTBI patients could potentially reduce the number of CT scans by 30% (Biberthaler et al., 2006). However, these results have not been consistently reproduced and many other investigators have failed to detect associations between S100β with CT abnormalities (Bechtel et al., 2009; Phillips et al., 1980; Piazza et al., 2007; Rothoerl et al., 1998).

There is also a concern about the brain specificity of S100β. Although S100β remains promising as an adjunctive marker, its utility in the setting of multiple trauma remains controversial because it is also elevated in trauma patients without head injuries (Anderson et al., 2001; Pelinka et al., 2004b; Romner and Ingebrigtsen, 2001; Rothoerl and Woertgen, 2001).

22.2.2 Glial Fibrillary Acid Protein

Glial fibrillary acidic protein (GFAP) is a monomeric intermediate protein found in astroglial skeleton that was first isolated by Eng et al. in 1971 (Eng et al., 1971). GFAP is found in white and gray brain matter and is strongly upregulated during astrogliosis (Duchen, 1984). Current evidence indicates that serum GFAP might be a useful marker for various types of brain damage, including neurodegenerative

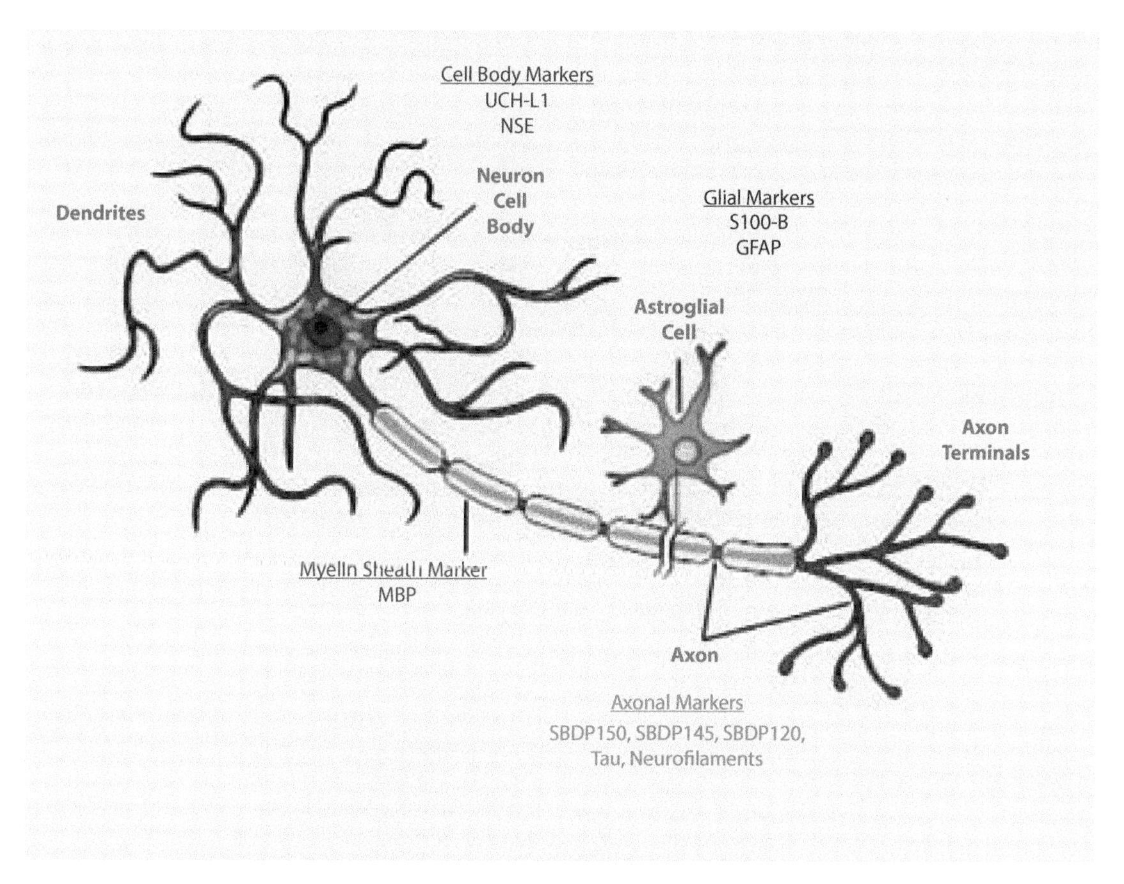

FIGURE 22.1 Neuroanatomical locations of the TBI biomarkers. S100β is the major low-affinity calcium binding protein in astrocytes that helps to regulate intracellular levels of calcium. GFAP is a monomeric intermediate protein found in astroglial skeleton that is found in white and gray brain matter and is strongly upregulated during astrogliosis. NSE is one of the five isozymes of the glycolytic enzyme enolase found in central and peripheral neuronal cell bodies. UCH-L1 is highly abundant in neurons and was previously used as a histological marker for neurons. Alpha-II spectrin (280 kDa) is the major structural component of the cortical membrane cytoskeleton and is particularly abundant in axons and presynaptic terminals. Tau is an intracellular, microtubule-associated protein that is highly enriched in axons. Neurofilaments are heteropolymeric components of the neuron cytoskeleton.

disorders (Baydas et al., 2003; Mouser et al., 2006), stroke (Herrmann et al., 2000), and severe TBI (Missler et al., 1999; Mondello et al., 2011; Nylen et al., 2006; Pelinka et al., 2004a; Pelinka et al., 2004b; vanGeel et al., 2002). In 2010, Vos et al. described serum GFAP profile in severe and moderate TBI with GCS <12 and found an association with unfavorable outcome at 6 months (Vos et al., 2010). More recently, Metting et al. found GFAP to be elevated in patients with axonal injury on MRI in patients with mTBI at 3 months postinjury, but it was not predictive of global outcome at 6 months (Metting et al., 2012). Furthermore, they demonstrated that serum GFAP was increased in patients with an abnormal CT after mTBI. In this study, GFAP was also compared to S100β. S100β was poorly correlated with outcome and with neuroimaging, including CT scan.

In a study by Papa et al. in 2012, GFAP was detectable in serum less than 1 hour after a concussion and was able to distinguish concussion patients from other trauma patients without head injury. Two groups of controls—single-limb orthopedic injuries and motor vehicle crash occupants without blunt head injury—had very low levels of GFAP when compared with mTBI patients (Papa et al., 2012a). In this same study, serum GFAP was significantly higher in mTBI patients with intracranial lesions on CT compared with those without lesions. GFAP was also able to predict patients who required neurosurgical intervention (Papa et al., 2012a). These studies suggest that GFAP has a good specificity for brain injury acutely after injury.

22.3 BIOFLUID BIOMARKERS OF NEURONAL INJURY

22.3.1 Neuron-Specific Enolase

Neuron-specific enolase (NSE) is one of the five isozymes of the glycolytic enzyme enolase found in central and peripheral neuronal cell bodies and has been shown be elevated after cell injury (Skogseid et al., 1992). It is also present in erythrocytes and endocrine cells and has a biological half-life of 48 hours (Schmechel et al., 1978). This protein is passively released into the extracellular space only under pathological conditions during cell destruction. Several reports on serum NSE measurements of mTBI have been published (Ergun et al., 1998; Fridriksson et al., 2000; Ross et al., 1996; Skogseid et al., 1992; Yamazaki et al., 1995). Many of these studies either contained inadequate control groups or concluded that serum NSE had limited utility as a marker of neuronal damage. Early levels of NSE and MBP concentrations have been correlated with outcome in children, particularly those younger than 4 years of age (Bandyopadhyay et al., 2005; Berger et al., 2005; Berger et al., 2007; Varma et al., 2003). In the setting of DAI/TAI in severe TBI, levels of NSE at 72 hours after injury have shown an association with unfavorable outcome (Chabok et al., 2012). One of the limitations of NSE is the occurrence of false-positive results in the setting of hemolysis (Johnsson et al., 2000; Ramont et al., 2005).

22.3.2 Ubiquitin C-Terminal Hydrolase

A promising candidate biomarker for TBI currently under investigation is ubiquitin C-terminal hydrolase-L1 (UCH-L1). This protein is involved in the addition and removal of ubiquitin from proteins that are destined for metabolism (Tongaonkar et al., 2000). It has an important role in the removal of excessive, oxidized, or misfolded proteins during both normal and pathological conditions in neurons (Gong and Leznik, 2007). Because of its high abundance and specific expression in neurons, UCH-L1 was previously used as a histological marker for neurons (Jackson and Thompson, 1981). Clinical studies in humans with severe TBI have confirmed, using enzyme-linked immunosorbent analysis, that the UCH-L1 protein is significantly elevated in human cerebrospinal fluid (CSF; Papa et al., 2010; Siman et al., 2009) and is detectable very early after injury. It remains significantly elevated for at least 1 week postinjury (Papa et al., 2010). Studies in severe TBI patients have demonstrated a very good correlation between CSF and serum levels (Brophy et al., 2011). Increases in serum UCH-L1 have also been found in children with moderate and severe TBI (Berger et al., 2012). Most recently, UCH-L1 was detected in the serum of mild and moderate TBI patients within an hour of injury (Papa et al., 2012b). Serum levels of UCH-L1 discriminated concussion patients from uninjured and non–head-injured trauma control patients that had orthopedic injuries or motor vehicle trauma without head injury. Notably, initial levels taken within 4 hours of injury were significantly higher in those with intracranial lesions on CT than those without lesions. Accordingly, levels were much higher in patients who eventually required neurosurgical intervention (Papa et al., 2012b). In a study by Berger et al., UCH-L1 was correlated with Glasgow Outcome Score (GOS), and this correlation was stronger than the correlations with NSE, S100β, or myelin basic protein (Berger et al., 2012).

22.4 BIOFLUID BIOMARKERS OF AXONAL INJURY

22.4.1 Alpha-II Spectrin Breakdown Products

Alpha-II spectrin (280 kDa) is the major structural component of the cortical membrane cytoskeleton and is particularly abundant in axons and presynaptic terminals (Goodman et al., 1995; Riederer et al., 1986). It is also a major substrate for both calpain and caspase-3 cysteine proteases (McGinn et al., 2009; Wang et al., 1998). A hallmark feature of apoptosis and necrosis is an early cleavage of several cellular proteins by activated caspases and calpains. A signature of caspase-3 and calpain-2 activation is cleavage of several common proteins such as cytoskeletal αII spectrin (Pike et al., 2004; Ringger et al., 2004). Levels of spectrin breakdown products (SBDPs) have been reported in CSF from adults with severe TBI and they have shown a significant relationship with severity of injury and clinical outcome (Cardali and Maugeri, 2006; Farkas et al., 2005; Mondello et al., 2010; Papa et al., 2004; Papa et al., 2006; Papa et al., 2005; Pineda et al., 2007). The time course of calpain-mediated SBDP150 and SBDP145 (markers of necrosis) differs from that of caspase-3–mediated SBDP120 (a marker of apoptosis). Average SBDP values measured in CSF early after injury have been shown to correlate with severity of injury, CT scan findings, and outcome at 6 months postinjury (Brophy et al., 2009).

Serum SBDP145 has also been measured in serum in children with TBI (Berger et al., 2012). Levels were significantly greater in subjects with moderate and severe TBI than in control patients, but were not significantly different in mTBI patients. Levels were correlated with dichotomized GOS at 6 months. This correlation did not hold true for mTBI. More recently, however, serum levels of SBDP150 has been examined in patients with mTBI and have shown a significant association with acute measures of injury severity, such as GCS score, intracranial injuries on CT, and neurosurgical intervention (Papa et al., 2012c). In this study, serum SBDP150 levels were much higher in patients with mTBI/concussion than other trauma patients who did not have a head injury (Papa et al., 2012c).

22.4.2 Tau Protein

Tau is an intracellular, microtubule-associated protein that is highly enriched in axons and is involved with assembling axonal microtubule bundles and participating in anterograde axoplasmic transport (Teunissen et al., 2005). Because tau is preferentially localized in the axon, tau lesions are apparently related to axonal disruption (Higuchi et al., 2002; Kosik and Finch, 1987). A supposedly cleaved form of tau, c-tau, has been investigated as a potential biomarker of CNS injury. After TBI, tau is proteolytically cleaved (c-tau) and gains access to CSF and serum. In a study by Shaw et al., CSF levels of c-tau were significantly elevated in TBI patients compared with control patients, and these levels correlated with clinical outcome (Shaw et al., 2002). In a similar study, initial elevated CSF c-tau levels in severe TBI patients were significant predictors of intracranial pressure and clinical outcome (Zemlan et al., 2002). However, these findings did not hold true when c-tau was measured in peripheral blood or in mTBI. Although levels of c-tau were also elevated in plasma from patients with severe TBI, there was no correlation between plasma levels and clinical outcome (Chatfield et al., 2002). Two studies assessed whether serum c-tau could predict postconcussion syndrome in adults after mTBI at 3 months postinjury. C-tau was a poor predictor of CT lesions and also a poor predictor of postconcussion syndrome (Bazarian et al., 2006; Ma et al., 2008).

Total tau protein is highly expressed in thin, nonmyelinated axons of cortical interneurons (Trojanowski et al., 1989) and thus may be indicative of axonal damage in gray matter neurons. It has been found to be correlated with severity of injury in severe TBI (Franz et al., 2003; Marklund et al., 2009; Ost et al., 2006; Sjogren et al., 2001). Ost et al. found that total tau measured in CSF on days 2–3 discriminated between TBI and controls (normal pressure hydrocephalus) and also between good and bad outcome at 1 year per dichotomized GOS score (Ost et al., 2006). However, total tau was not detected in serum throughout the study.

22.4.3 Neurofilaments

Neurofilaments are heteropolymeric components of the neuron cytoskeleton that consist of a 68-kDa light neurofilament subunit (NF-L) backbone with either 160-kDa medium (NF-M) or 200-kDa heavy subunit (NF-H) side arms (Julien and Mushynski, 1998). After TBI, calcium influx into the cell contributes to a cascade of events that activates calcineurin, a calcium-dependent phosphatase that dephosphorylates neurofilament side arms, presumably contributing to axonal injury (Buki and Povlishock, 2006). Phosphorylated NF-H has been found to be elevated in the CSF of adult patients with severe TBI compared with controls (Siman et al., 2009). Similarly, hyperphosphorylated NF-H has also been correlated with severity of brain injury in children (Zurek et al., 2012). In a study by Zurek et al., NF-H levels taken on the second to fourth day remained significantly higher in patients with poor outcome in comparison to patients with good outcome. Additionally, NF-H was significantly higher in those children with diffuse axonal injury on initial CT scan (Zurek et al., 2012). Accordingly, Vajtr et al. compared 10 patients with DAI/TAI with 28 patients with focal injuries and found that serum NF-H was much higher in patients with DAI/TAI over 10 days after admission. Serum NF-H levels were highest from the fourth to the tenth day in both groups (Vajtr et al., 2013).

22.5 CONCLUSION

Although research in the field of TBI biomarkers has increased exponentially over the past 20 years (Kochanek et al., 2008; Papa, 2012; Papa et al., 2013), most of the studies have focused on severe TBI. Because more than 80% of patients have a mild brain injury (Consensus conference, 1999; Vollmer and Dacey, 1991; Yealy and Hogan, 1991), there is a great need to explore biomarkers in this population. Biomarkers could potentially facilitate diagnosis and risk stratification of mTBI and impact management of these patients. Therapies for mTBI have not been well explored, so there are major opportunities for improving the conduct of clinical research for mTBI patients by providing early information about injury mechanism(s) and possible drug targets. Because of significant advances in proteomic techniques that are more specific and selective than traditional methods, the detection of proteins amid complex high-protein-content biofluids such as serum or plasma has improved. These developments are bringing neurobiomarkers a step closer to the bedside.

REFERENCES

Alexander, M.P. 1995. Mild traumatic brain injury: Pathophysiology, natural history, and clinical management. *Neurology*. 45:1253–1260.

Alves, W., S. Macciocchi, and J.T. Barth. 1993. Postconcussive symptoms after uncomplicated mild head injury. *J Head Trauma Rehabil*. 8:48–59.

Anderson, R.E., L.O. Hansson, O. Nilsson, R. Dijlai-Merzoug, and G. Settergen. 2001. High serum S100B levels for trauma patients without head injuries. *Neurosurgery*. 49:1272–1273.

Bandyopadhyay, S., H. Hennes, M.H. Gorelick, R.G. Wells, and C.M. Walsh-Kelly. 2005. Serum neuron-specific enolase as a predictor of short-term outcome in children with closed traumatic brain injury. *Acad Emerg Med*. 12:732–738.

Barth, J.T., S.N. Macciocchi, B. Giordani, R. Rimel, J.A. Jane, and T.J. Boll. 1983. Neuropsychological sequelae of minor head injury. *Neurosurgery*. 13:529–533.

Baydas, G., V.S. Nedzvetskii, M. Tuzcu, A. Yasar, and S.V. Kirichenko. 2003. Increase of glial fibrillary acidic protein and S-100B in hippocampus and cortex of diabetic rats: Effects of vitamin E. *Eur J Pharmacol*. 462:67–71.

Bazarian, J.J., F.P. Zemlan, S. Mookerjee, and T. Stigbrand. 2006. Serum S-100B and cleaved-tau are poor predictors of long-term outcome after mild traumatic brain injury. *Brain Inj*. 20:759–765.

Bazarian, J.J., J. Zhong, B. Blyth, T. Zhu, V. Kavcic, and D. Peterson. 2007. Diffusion tensor imaging detects clinically important axonal damage after mild traumatic brain injury: A pilot study. *J Neurotrauma*. 24:1447–1459.

Bechtel, K., S. Frasure, C. Marshall, J. Dziura, and C. Simpson. 2009. Relationship of serum S100B levels and intracranial injury in children with closed head trauma. *Pediatrics*. 124:e697–704.

Berger, R.P., P.D. Adelson, M.C. Pierce, T. Dulani, L.D. Cassidy, and P.M. Kochanek. 2005. Serum neuron-specific enolase, S100B, and myelin basic protein concentrations after inflicted and noninflicted traumatic brain injury in children. *J Neurosurg*. 103:61–68.

Berger, R.P., S.R. Beers, R. Richichi, D. Wiesman, and P.D. Adelson. 2007. Serum biomarker concentrations and outcome after pediatric traumatic brain injury. *J Neurotrauma*. 24:1793–1801.

Berger, R.P., R.L. Hayes, R. Richichi, S.R. Beers, and K.K. Wang. 2012. Serum concentrations of ubiquitin C-terminal hydrolase-L1 and alphaII-spectrin breakdown product 145 kDa correlate with outcome after pediatric TBI. *J Neurotrauma*. 29:162–167.

Berger, R.P., M.C. Pierce, S.R. Wisniewski, P.D. Adelson, and P.M. Kochanek. 2002. Serum S100B concentrations are increased after closed head injury in children: A preliminary study. *J Neurotrauma*. 19:1405–1409.

Biberthaler, P., U. Linsenmeier, K.J. Pfeifer, M. Kroetz, T. Mussack, K.G. Kanz et al. 2006. Serum S-100B concentration provides additional information for the indication of computed tomography in patients after minor head injury: A prospective multicenter study. *Shock*. 25:446–453.

Brophy, G., S. Mondello, L. Papa, S. Robicsek, A. Gabrielli, J. Tepas J 3rd et al. 2011. Biokinetic Analysis of ubiquitin C-terminal hydrolase-L1 (Uch-L1) in severe traumatic brain injury patient biofluids. *J Neurotrauma*. 28:861–870.

Brophy, G.M., J.A. Pineda, L. Papa, S.B. Lewis, A.B. Valadka, H.J. Hannay et al. 2009. alphaII-Spectrin breakdown product cerebrospinal fluid exposure metrics suggest differences in cellular injury mechanisms after severe traumatic brain injury. *J Neurotrauma*. 26:471–479.

Buki, A., O. Farkas, T. Doczi, and J.T. Povlishock. 2003. Preinjury administration of the calpain inhibitor MDL-28170 attenuates traumatically induced axonal injury. *J Neurotrauma*. 20:261–268.

Buki, A., and J.T. Povlishock. 2006. All roads lead to disconnection?—Traumatic axonal injury revisited. *Acta Neurochir (Wien)*. 148:181–193; discussion 193–184.

Cardali, S., and R. Maugeri. 2006. Detection of alphaII-spectrin and breakdown products in humans after severe traumatic brain injury. *J Neurosurg Sci*. 50:25–31.

Chabok, S.Y., A.D. Moghadam, Z. Saneei, F.G. Amlashi, E.K. Leili, and Z.M. Amiri. 2012. Neuron-specific enolase and S100BB as outcome predictors in severe diffuse axonal injury. *J Trauma Acute Care Surg*. 72:1654–1657.

Chatfield, D.A., F.P. Zemlan, D.J. Day, and D.K. Menon. 2002. Discordant temporal patterns of S100beta and cleaved tau protein elevation after head injury: A pilot study. *Br J Neurosurg*. 16:471–476.

Consensus conference. 1999. Rehabilitation of persons with traumatic brain injury. NIH Consensus Development Panel on Rehabilitation of Persons With Traumatic Brain Injury. *JAMA*. 282:974–983.

Duchen, L.W. 1984. General pathology of neurons and neuroglia. In J.A. Adams, J.A.N. Corsellis, and L.W. Duchen (edis.). *Greenfield's Neuropathology*. pp. 1–52. Edward Arnold, London.

Eng, L.F., J.J. Vanderhaeghen, A. Bignami, and B. Gerstl. 1971. An acidic protein isolated from fibrous astrocytes. *Brain Res*. 28:351–354.

Ergun, R., U. Bostanci, G. Akdemir, E. Beskonakli, E. Kaptanoglu, F. Gursoy et al. 1998. Prognostic value of serum neuron-specific enolase levels after head injury. *Neurol Res*. 20:418–420.

Farkas, O., B. Polgar, J. Szekeres-Bartho, T. Doczi, J.T. Povlishock, and A. Buki. 2005. Spectrin breakdown products in the cerebrospinal fluid in severe head injury—Preliminary observations. *Acta Neurochir (Wien)*. 147:855–861.

Franz, G., R. Beer, A. Kampfl, K. Engelhardt, E. Schmutzhard, H. Ulmer, and F. Deisenhammer. 2003. Amyloid beta 1–42 and tau in cerebrospinal fluid after severe traumatic brain injury. *Neurology*. 60:1457–1461.

Fridriksson, T., N. Kini, C. Walsh-Kelly, and H. Hennes. 2000. Serum neuron-specific enolase as a predictor of intracranial lesions in children with head trauma: A pilot study. *Acad Emerg Med*. 7:816–820.

Gong, B., and E. Leznik. 2007. The role of ubiquitin C-terminal hydrolase L1 in neurodegenerative disorders. *Drug News Perspect*. 20:365–370.

Goodman, S.R., W.E. Zimmer, M.B. Clark, I.S. Zagon, J.E. Barker, and M.L. Bloom. 1995. Brain spectrin: Of mice and men. *Brain Res Bull*. 36:593–606.

Haimoto, H.H., S; Kato, K. 1987. Differential distribution of immunoreactive S100-a and S100-b proteins in normal nonnervous human tissues. *Lab Invest*. 57:489–498.

Herrmann, M., P. Vos, M.T. Wunderlich, C.H. de Bruijn, and K.J. Lamers. 2000. Release of glial tissue-specific proteins after acute stroke: A comparative analysis of serum concentrations of protein S-100B and glial fibrillary acidic protein. *Stroke*. 31:2670–2677.

Higuchi, M., V.M. Lee, and J.Q. Trojanowski. 2002. Tau and axonopathy in neurodegenerative disorders. *Neuromolecular Med*. 2:131–150.

Huang, M.X., R.J. Theilmann, A. Robb, A. Angeles, S. Nichols, A. Drake et al. 2009. Integrated imaging approach with MEG and DTI to detect mild traumatic brain injury in military and civilian patients. *J Neurotrauma*. 26:1213–1226.

Hyder, A.A., C.A. Wunderlich, P. Puvanachandra, G. Gururaj, and O.C. Kobusingye. 2007. The impact of traumatic brain injuries: A global perspective. *NeuroRehabilitation*. 22:341–353.

Ingebrigtsen, T., and B. Romner. 1996. Serial S-100 protein serum measurements related to early magnetic resonance imaging after minor head injury. Case report. *J Neurosurg*. 85:945–948.

Ingebrigtsen, T., and B. Romner. 1997. Management of minor head injuries in hospitals in Norway. *Acta Neurol Scand*. 95:51–55.

Ingebrigtsen, T., B. Romner, S. Marup-Jensen, M. Dons, C. Lundqvist, J. Bellner et al. 2000. The clinical value of serum S-100 protein measurements in minor head injury: A Scandinavian multicentre study. *Brain Inj*. 14:1047–1055.

Ingebrigtsen, T., K. Waterloo, E.A. Jacobsen, B. Langbakk, and B. Romner. 1999. Traumatic brain damage in minor head injury: Relation of serum S-100 protein measurements to magnetic resonance imaging and neurobehavioral outcome. *Neurosurgery*. 45:468–475; discussion 475–466.

Jackson, P., and R.J. Thompson. 1981. The demonstration of new human brain-specific proteins by high-resolution two-dimensional polyacrylamide gel electrophoresis. *J Neurol Sci*. 49:429–438.

Jagoda, A.S., J.J. Bazarian, J.J. Bruns, Jr., S.V. Cantrill, A.D. Gean, P.K. Howard et al. 2008. Clinical policy: Neuroimaging and decisionmaking in adult mild traumatic brain injury in the acute setting. *Ann Emerg Med*. 52:714–748.

Johnson, V.E., W. Stewart, and D.H. Smith. 2012. Axonal pathology in traumatic brain injury. *Exp Neurol*. 246:35–43.

Johnsson, P., S. Blomquist, C. Luhrs, G. Malmkvist, C. Alling, J.O. Solem, and E. Stahl. 2000. Neuron-specific enolase increases in plasma during and immediately after extracorporeal circulation. *Ann Thorac Surg*. 69:750–754.

Jonsson H, J.P., Hoglund P, Alling C, Blomquist S. 2000. The elimination of S-100b and renal function after cardiac surgery. *J Cardiothorac Vasc Aneth*. 14:698–701.

Julien, J.P., and W.E. Mushynski. 1998. Neurofilaments in health and disease. *Prog Nucleic Acid Res Mol Biol*. 61:1–23.

Kesler, e.a. 2000. APECT, MR and quantitative MR imaging: Correlates with neuropsycholgical. *Brain Injury*. 14:851–857.

Kochanek, P.M., R.P. Berger, H. Bayr, A.K. Wagner, L.W. Jenkins, and R.S. Clark. 2008. Biomarkers of primary and evolving damage in traumatic and ischemic brain injury: Diagnosis, prognosis, probing mechanisms, and therapeutic decision making. *Curr Opin Crit Care*. 14:135–141.

Korfias, S., G. Stranjalis, E. Boviatsis, C. Psachoulia, G. Jullien, B. Gregson, A.D. Mendelow et al. 2007. Serum S-100B protein monitoring in patients with severe traumatic brain injury. *Intensive Care Med*. 33:255–260.

Kosik, K.S., and E.A. Finch. 1987. MAP2 and tau segregate into dendritic and axonal domains after the elaboration of morphologically distinct neurites: An immunocytochemical study of cultured rat cerebrum. *J Neurosci*. 7:3142–3153.

Ma, M., C.J. Lindsell, C.M. Rosenberry, G.J. Shaw, and F.P. Zemlan. 2008. Serum cleaved tau does not predict post-concussion syndrome after mild traumatic brain injury. *Am J Emerg Med*. 26:763–768.

Marklund, N., K. Blennow, H. Zetterberg, E. Ronne-Engstrom, P. Enblad, and L. Hillered. 2009. Monitoring of brain interstitial total tau and beta amyloid proteins by microdialysis in patients with traumatic brain injury. *J Neurosurg*. 110:1227–1237.

Maxwell, W.L., A. Domleo, G. McColl, S.S. Jafari, and D.I. Graham. 2003. Post-acute alterations in the axonal cytoskeleton after traumatic axonal injury. *J Neurotrauma*. 20:151–168.

McGinn, M.J., B.J. Kelley, L. Akinyi, M.W. Oli, M.C. Liu, R.L. Hayes et al. 2009. Biochemical, structural, and biomarker evidence for calpain-mediated cytoskeletal change after diffuse brain injury uncomplicated by contusion. *J Neuropathol Exp Neurol*. 68:241–249.

Metting, Z., N. Wilczak, L.A. Rodiger, J.M. Schaaf, and J. van der Naalt. 2012. GFAP and S100B in the acute phase of mild traumatic brain injury. *Neurology*. 78:1428–1433.

Millis, S.R., M. Rosenthal, T.A. Novack, M. Sherer, T.G. Nick, J.S. Kreutzer et al. 2001. Long-term neuropsychological outcome after traumatic brain injury. *J Head Trauma Rehabil*. 16:343–355.

Missler, U., M. Wiesmann, G. Wittmann, O. Magerkurth, and H. Hagenstrom. 1999. Measurement of glial fibrillary acidic protein in human blood: Analytical method and preliminary clinical results. *Clin Chem*. 45:138–141.

Missler, U. 1997. S-100 protein and neuron-specific enolase concentrations in blood as indicators of infarction volume and prognosis in acute ischemic stroke. *Stroke*. 28:1956–1960.

Mondello, S., L. Papa, A. Buki, R. Bullock, E. Czeiter, F. Tortella et al. 2011. Neuronal and glial markers are differently associated with computed tomography findings and outcome in patients with severe traumatic brain injury: A case control study. *Crit Care*. 15:R156.

Mondello, S., S.A. Robicsek, A. Gabrielli, G.M. Brophy, L. Papa, J. Tepas et al. 2010. alphaII-spectrin breakdown products (SBDPs): Diagnosis and outcome in severe traumatic brain injury patients. *J Neurotrauma*. 27:1203–1213.

Mouser, P.E., E. Head, K.H. Ha, and T.T. Rohn. 2006. Caspase-mediated cleavage of glial fibrillary acidic protein within degenerating astrocytes of the Alzheimer's disease brain. *Am J Pathol*. 168:936–946.

Muller, K., W. Townend, N. Biasca, J. Unden, K. Waterloo, B. Romner et al. 2007. S100B serum level predicts computed tomography findings after minor head injury. *J Trauma*. 62:1452–1456.

Nylen, K., M. Ost, L.Z. Csajbok, I. Nilsson, K. Blennow, B. Nellgard et al. 2006. Increased serum-GFAP in patients with severe traumatic brain injury is related to outcome. *J Neurol Sci*. 240:85–91.

Olsson, B., H. Zetterberg, H. Hampel, and K. Blennow. 2011. Biomarker-based dissection of neurodegenerative diseases. *Prog Neurobiol*. 95:520–534.

Ost, M., K. Nylen, L. Csajbok, A.O. Ohrfelt, M. Tullberg, C. Wikkelso et al. 2006. Initial CSF total tau correlates with 1-year outcome in patients with traumatic brain injury. *Neurology*. 67:1600–1604.

Papa, L. 2012. Exploring the role of biomarkers for the diagnosis and management of traumatic brain injury patients. In T.K. Man and R.J. Flores (eds.). *Poteomics - Human Diseases and Protein Functions*. In Tech Open Access Publisher, Rijeka, Croatia.

Papa, L., L. Akinyi, M.C. Liu, J.A. Pineda, J.J. Tepas, 3rd, M.W. Oli et al. 2010. Ubiquitin C-terminal hydrolase is a novel biomarker in humans for severe traumatic brain injury. *Crit Care Med*. 38:138–144.

Papa, L., D. D'Avella, M. Aguennouz, F.F. Angileri, O. de Divitiis, A. Germano et al. 2004. Detection of alpha-II spectrin and breakdown products in humans after severe traumatic brain injury (abstract). *Acad Emerg Med*. 11:515–516.

Papa, L., L.M. Lewis, J.L. Falk, Z. Zhang, S. Silvestri, P. Giordano et al. 2012a. Elevated levels of serum glial fibrillary acidic protein breakdown products in mild and moderate traumatic brain injury are associated with intracranial lesions and neurosurgical intervention. *Ann Emerg Med*. 59:471–483.

Papa, L., L.M. Lewis, S. Silvestri, J.L. Falk, P. Giordano, G.M. Brophy et al. 2012b. Serum levels of ubiquitin C-terminal hydrolase distinguish mild traumatic brain injury from trauma controls and are elevated in mild and moderate traumatic brain injury patients with intracranial lesions and neurosurgical intervention. *J Trauma Acute Care Surg*. 72:1335–1344.

Papa, L., S.B. Lewis, S. Heaton, J.A. Demery, J.J. Tepas III, K.K.W. Wang et al. 2006. Predicting early outcome using alpha-II spectrin breakdown products in human CSF after severe traumatic brain injury (abstract). *Acad Emerg Med*. 13.

Papa, L., J. Pineda, K.K.W. Wang, S.B. Lewis, J.A. Demery, S. Heaton et al. 2005. Levels of alpha-II spectrin breakdown products in human CSF and outcome after severe traumatic brain injury (abstract). *Acad Emerg Med*. 12.

Papa, L., M.M. Ramia, J.M. Kelly, S.S. Burks, A. Pawlowicz, and R.P. Berger. 2013. Systematic review of clinical research on biomarkers for pediatric traumatic brain injury. *J Neurotrauma*. 30:324–338.

Papa, L., K.W. Wang, G.B. Brophy, J.A. Demery, S. Silvestri, P. Giordano et al. 2012c. Serum levels of spectrin breakdown product 150 (SBDP150) distinguish mild traumatic brain injury from trauma and uninjured controls and predict intracranial injuries on CT and neurosurgical intervention. *J Neurotrauma*. 29:A28.

Pelinka, L.E., A. Kroepfl, M. Leixnering, W. Buchinger, A. Raabe, and H. Redl. 2004a. GFAP versus S100B in serum after traumatic brain injury: Relationship to brain damage and outcome. *J Neurotrauma*. 21:1553–1561.

Pelinka, L.E., A. Kroepfl, R. Schmidhammer, M. Krenn, W. Buchinger, H. Redl et al. 2004b. Glial fibrillary acidic protein in serum after traumatic brain injury and multiple trauma. *J Trauma*. 57:1006–1012.

Phillips, J.P., H.M. Jones, R. Hitchcock, N. Adama, and R.J. Thompson. 1980. Radioimmunoassay of serum creatine kinase BB as index of brain damage after head injury. *Br Med J*. 281:777–779.

Piazza, O., M.P. Storti, S. Cotena, F. Stoppa, D. Perrotta, G. Esposito et al. 2007. S100B is not a reliable prognostic index in paediatric TBI. *Pediatr Neurosurg*. 43:258–264.

Pike, B.R., J. Flint, J.R. Dave, X.C. Lu, K.K. Wang, F.C. Tortella et al. 2004. Accumulation of calpain and caspase-3 proteolytic fragments of brain-derived alphaII-spectrin in cerebral spinal fluid after middle cerebral artery occlusion in rats. *J Cereb Blood Flow Metab*. 24:98–106.

Pineda, J.A., S.B. Lewis, A.B. Valadka, L. Papa, H.J. Hannay, S.C. Heaton et al. 2007. Clinical significance of alphaII-spectrin breakdown products in cerebrospinal fluid after severe traumatic brain injury. *J Neurotrauma*. 24:354–366.

Povlishock, J.T. 1992. Traumatically induced axonal injury: Pathogenesis and pathobiological implications. *Brain Pathol*. 2:1–12.

Raabe, A., C. Grolms, and V. Seifert. 1999. Serum markers of brain damage and outcome prediction in patients after severe head injury. *Br J Neurosurg*. 13:56–59.

Ramont, L., H. Thoannes, A. Volondat, F. Chastang, M.C. Millet, and F.X. Maquart. 2005. Effects of hemolysis and storage condition on neuron-specific enolase (NSE) in cerebrospinal fluid and serum: Implications in clinical practice. *Clin Chem Lab Med*. 43:1215–1217.

Riederer, B.M., I.S. Zagon, and S.R. Goodman. 1986. Brain spectrin(240/235) and brain spectrin(240/235E): Two distinct spectrin subtypes with different locations within mammalian neural cells. *J Cell Biol*. 102:2088–2097.

Rimel, R.W., B. Giordani, J.T. Barth, T.J. Boll, and J.A. Jane. 1981. Disability caused by minor head injury. *Neurosurgery*. 9:221–228.

Ringger, N.C., B.E. O'Steen, J.G. Brabham, X. Silver, J. Pineda, K.K. Wang et al. 2004. A novel marker for traumatic brain injury: CSF alphaII-spectrin breakdown product levels. *J Neurotrauma*. 21:1443–1456.

Romner, B., and T. Ingebrigtsen. 2001. High serum S100B levels for trauma patients without head injuries. *Neurosurgery*. 49:1490; author reply 1492–1493.

Romner, B., T. Ingebrigtsen, P. Kongstad, and S.E. Borgesen. 2000. Traumatic brain damage: Serum S-100 protein measurements related to neuroradiological findings. *J Neurotrauma*. 17:641–647.

Ross, S.A., R.T. Cunningham, C.F. Johnston, and B.J. Rowlands. 1996. Neuron-specific enolase as an aid to outcome prediction in head injury. *Br J Neurosurg*. 10:471–476.

Rothoerl, R.D., and C. Woertgen. 2001. High serum S100B levels for trauma patients without head injuries. *Neurosurgery*. 49:1490–1491; author reply 1492–1493.

Rothoerl, R.D., C. Woertgen, M. Holzschuh, C. Metz, and A. Brawanski. 1998. S-100 serum levels after minor and major head injury. *J Trauma*. 45:765–767.

Schmechel, D., P.J. Marangos, and M. Brightman. 1978. Neurone-specific enolase is a molecular marker for peripheral and central neuroendocrine cells. *Nature*. 276:834–836.

Shaw, G.J., E.C. Jauch, and F.P. Zemlan. 2002. Serum cleaved tau protein levels and clinical outcome in adult patients with closed head injury. *Ann Emerg Med*. 39:254–257.

Siman, R., N. Toraskar, A. Dang, E. McNeil, M. McGarvey, J. Plaum et al. 2009. A panel of neuron-enriched proteins as markers for traumatic brain injury in humans. *J Neurotrauma*. 26:1867–1877.

Sjogren, M., M. Blomberg, M. Jonsson, L.O. Wahlund, A. Edman, K. Lind et al. 2001. Neurofilament protein in cerebrospinal fluid: A marker of white matter changes. *J Neurosci Res*. 66:510–516.

Skogseid, I.M., H.K. Nordby, P. Urdal, E. Paus, and F. Lilleaas. 1992. Increased serum creatine kinase BB and neuron specific enolase following head injury indicates brain damage. *Acta Neurochir (Wien)*. 115:106–111.

Stein, S.C., A. Fabbri, F. Servadei, and H.A. Glick. 2009. A critical comparison of clinical decision instruments for computed tomographic scanning in mild closed traumatic brain injury in adolescents and adults. *Ann Emerg Med*. 53:180–188.

Teasdale, G., and B. Jennett. 1974. Assessment of coma and impaired consciousness. A practical scale. *Lancet*. 2:81–84.

Teunissen, C.E., C. Dijkstra, and C. Polman. 2005. Biological markers in CSF and blood for axonal degeneration in multiple sclerosis. *Lancet Neurol*. 4:32–41.

Tongaonkar, P., L. Chen, D. Lambertson, B. Ko, and K. Madura. 2000. Evidence for an interaction between ubiquitin-conjugating enzymes and the 26S proteasome. *Mol Cell Biol*. 20:4691–4698.

Trojanowski, J.Q., T. Schuck, M.L. Schmidt, and V.M. Lee. 1989. Distribution of tau proteins in the normal human central and peripheral nervous system. *J Histochem Cytochem*. 37:209–215.

Usui A, K.K., Abe T, Murase M, Tanaka M, Takeuchi E. 1989. S-100ao protein in blood and urine during open-heart surgery. *Clin Chem*. 35:1942–1944.

Vajtr, D., O. Benada, P. Linzer, F. Samal, D. Springer, P. Strejc et al. 2013. Immunohistochemistry and serum values of S-100B, glial fibrillary acidic protein, and hyperphosphorylated neurofilaments in brain injuries. *Soud Lek*. 57:7–12.

van Geel, W.J., H.P. de Reus, H. Nijzing, M.M. Verbeek, P.E. Vos, and K.J. Lamers. 2002. Measurement of glial fibrillary acidic protein in blood: An analytical method. *Clin Chim Acta*. 326:151–154.

Varma, S., K.L. Janesko, S.R. Wisniewski, H. Bayir, P.D. Adelson, N.J. Thomas et al. 2003. F2-isoprostane and neuron-specific enolase in cerebrospinal fluid after severe traumatic brain injury in infants and children. *J Neurotrauma*. 20:781–786.

Vollmer, D.G., and R.G. Dacey, Jr. 1991. The management of mild and moderate head injuries. *Neurosurg Clin N Am*. 2:437–455.

Vos, P.E., B. Jacobs, T.M. Andriessen, K.J. Lamers, G.F. Borm, T. Beems et al. 2010. GFAP and S100B are biomarkers of traumatic brain injury: An observational cohort study. *Neurology*. 75:1786–1793.

Wang, K.K., R. Posmantur, R. Nath, K. McGinnis, M. Whitton, R.V. Talanian et al. 1998. Simultaneous degradation of alphaII- and betaII-spectrin by caspase 3 (CPP32) in apoptotic cells. *J Biol Chem*. 273:22490–22497.

Waterloo, K., T. Ingebrigtsen, and B. Romner. 1997. Neuropsychological function in patients with increased serum levels of protein S-100 after minor head injury. *Acta Neurochir (Wien)*. 139:26–31; discussion 31–22.

Woertgen, C., R.D. Rothoerl, M. Holzschuh, C. Metz, and A. Brawanski. 1997. Comparison of serial S-100 and NSE serum measurements after severe head injury. *Acta Neurochir (Wien)*. 139:1161–1164; discussion 1165.

Xiong, H., W.L. Liang, and X.R. Wu. 2000. [Pathophysiological alterations in cultured astrocytes exposed to hypoxia/reoxygenation]. *Sheng Li Ke Xue Jin Zhan*. 31:217–221.

Yamazaki, Y., K. Yada, S. Morii, T. Kitahara, and T. Ohwada. 1995. Diagnostic significance of serum neuron-specific enolase and myelin basic protein assay in patients with acute head injury. *Surg Neurol*. 43:267–270; discussion 270–261.

Yealy, D.M., and D.E. Hogan. 1991. Imaging after head trauma. Who needs what? *Emerg Med Clin North Am*. 9:707–717.

Ytrebo L.M., G.I. Nedredal, C. Korvald, O.J. Holm Nielsen, T. Ingebrigtsen, B. Romner et al. 2001. Renal elimination of protein S-100beta in pigs with acute encephalopathy. *Scand J Clin Lab Invest*. 61:217–225.

Zemlan, F.P., E.C. Jauch, J.J. Mulchahey, S.P. Gabbita, W.S. Rosenberg, S.G. Speciale et al. 2002. C-tau biomarker of neuronal damage in severe brain injured patients: Association with elevated intracranial pressure and clinical outcome. *Brain Res*. 947:131–139.

Zimmer, D.B., E.H. Cornwall, A. Landar, and W. Song. 1995. The S100 protein family: History, function, and expression. *Brain Res Bull*. 37:417–429.

Zurek, J., L. Bartlova, and M. Fedora. 2012. Hyperphosphorylated neurofilament NF-H as a predictor of mortality after brain injury in children. *Brain Inj*. 25:221–226.

23 Modeling the Neurobehavioral Consequences of Blast-Induced Traumatic Brain Injury Spectrum Disorder and Identifying Related Biomarkers

Denes V. Agoston and Alaa Kamnaksh

CONTENTS

23.1 INTRODUCTION

Blast induced traumatic brain injury (bTBI) has become the most common type of military head injuries affecting close to 500,000 service members. bTBI is a spectrum disorder ranging from the severe form that is frequently comorbid with polytrauma to the mild form that shares symptoms and/or comorbid with post-traumatic stress disorder (PTSD). The epidemiologic scale and complexity of bTBI and closely related neuropsychiatric conditions present especially significant short- and long-term challenges for the military health care system. Due to the complexity of physical forces generated by explosive blast combined with our limited understanding how these forces interact with the biological entity (physical-to-biological coupling), modeling bTBI poses special challenges. The goal of this chapter is to: 1) familiarize readers with experimental modeling of bTBI, including physical and biological considerations toward high-fidelity modeling; 2) provide a brief overview of bTBI animal models; 3) present some of the neurobehavioral consequences of bTBI; and 4) identify related biomarkers for the diagnosis and monitoring of the disease.

Since the advent of the military conflicts in 2001, approximately 1.5 million service members were deployed to Afghanistan and Iraq where 15%–30% suffered traumatic brain injuries (TBIs) (http://www.dcoe.mil/). Approximately 80% of the injuries sustained during those wars were caused by explosive blast, resulting in a specific form of TBI called bTBI (Ling et al., 2009). Using criteria similar to other forms of TBI, bTBI is also classified as severe, moderate, or mild (Ling and Ecklund, 2011). Based on traditional assessments of injury severity, approximately 5% of documented bTBI cases (10,000–20,000 soldiers) were severe and approximately 10% (20,000–40,000 soldiers) were moderate. The overwhelming majority of bTBI cases, approximately 85% affecting 170,000–340,000 service members, were mild (Masel et al., 2012). Thus, there may be as many as 300,000 individuals who have sustained bTBIs and will be in need of some level of medical and/or rehabilitative care over the next several decades.

Depending on the severity of the injury, functional changes in bTBI span an entire spectrum of neurobehavioral symptoms and deficits: some have an acute onset and are transient in nature, whereas others develop over time and may become chronic (Marion et al., 2011). At the different ends of the spectrum, there are two important factors that confound underlying injury mechanisms, patient care, and outcome. In severe bTBI, the majority (60%–80%) of individuals sustain serious penetrating injuries to the head as well as to the extremities and body (Bass et al., 2012; Hicks et al., 2010; Ling et al., 2009). These polytrauma patients present with highly complex and critical physical and functional deficits depending on the brain region(s) (and associated circuitries) damaged by the blast (Ling and Ecklund, 2007, 2011; Ling and Marshall, 2008; Ling et al., 2010; Ling and Neal, 2005). Emotional and cognitive impairments are almost always detectable in severe bTBI.

In mild bTBI, the physical damage is minimal, but there is a significant psychological stress component that plays an important role as cofactor in the development of the condition (Chen and Huang, 2011; Elder et al., 2010; Kennedy et al., 2010b; Rosenfeld and Ford, 2010; Thompson et al., 2008). In this form of bTBI, mood, learning, and memory are primarily impaired but executive function, sleep, and social interactions are also frequently affected. Accordingly, the neurobehavioral symptoms of mild bTBI (especially when it is repeated) greatly overlap with those of PTSD, thus complicating the pathobiology, differential diagnosis, and, more importantly, the treatment of mild bTBI (Chen and Huang, 2011; Jaffee and Meyer, 2009; Peterson et al., 2011).

The behavioral symptoms observed in bTBI reflect similarly complex structural, cellular, and molecular changes in the brain. Some of these changes can be detected noninvasively by advanced imaging techniques (Benzinger et al., 2009). Various magnetic resonance imaging (MRI) modalities, predominantly diffusion tensor imaging (DTI), have identified structural anomalies in a number of brain regions, including white matter changes in the cerebrum and cerebellum. The limited number of experimental bTBI imaging studies have recapitulated some of these clinical findings. Furthermore, the nature and extent of the cellular and molecular damage can be determined by monitoring time-dependent changes in the serum and/or cerebrospinal fluid (CSF) concentrations of select protein biomarkers (Agoston and Elsayed, 2012). Several putative biomarkers have been identified in experimental bTBI, indicating complex and dynamically changing pathologies that include oxidative stress, axonal and vascular damage, and inflammation.

The epidemiologic scale and complexity of bTBI and closely related neuropsychiatric conditions have and will continue to present very significant challenges for the military health care system on a short-term as well as a long-term basis. The aim of this chapter is to (1) familiarize readers with experimental modeling of bTBI, including physical and biological considerations toward high-fidelity modeling; (2) provide a brief overview of bTBI animal models; (3) present some of the neurobehavioral consequences of bTBI; and (4) identify related biomarkers for the diagnosis and monitoring of the disease.

23.2 EXPERIMENTAL MODELING OF bTBI

23.2.1 Physical Considerations

Explosive blast creates a highly complex physical environment that extends beyond the scope of our discussion. Nonetheless, it is important to note that the inherent complexity of blast environments is due to variations in a number of factors, such as the size and shape of the charge, the presence of casing and accelerants (or the lack thereof), and the initiation of the explosive charge (Masel et al., 2012). These factors are further affected by meteorological conditions (e.g., temperature, humidity) and the nature of the explosion site (open vs. closed space) among others. As a result, blast flow conditions differ from near-field to mid- and far-field.

In the near-field (i.e., within the expansion area of the fireball where incident blast overpressure is higher than 10 atm), blast conditions predominantly include high energy supersonic shock wave(s), toxic combustion gases, extreme pressure and temperature, and ejecta depending on the detonation site. The propagation of the primary and secondary shock waves continues into mid- and far-fields (i.e., beyond the area of maximum fireball expansion where blast overpressure ranges from 10-1 atm).

Shock waves generated by an idealized explosive blast (uncased spherical charge; open field detonation) are spherically symmetric and decay with distance. Therefore, when a blast wave encounters an obstacle (e.g., a human body) it reflects and diffracts resulting in a wide range and types of stress waves. The direct interaction between a rapidly propagating shock wave and viscoelastic biological structures, such as the head and the body, triggers an array of extremely complex structural, cellular, and molecular changes in what is referred to as primary blast injury (Cernak and Noble-Haeusslein, 2010; Cernak et al., 1999). The nature and the extent of the resulting biological damage is impacted by many of the abovementioned blast conditions as well as the orientation of the subject relative to the explosion, distance from the charge, and the presence of body protection.

23.2.2 Biological Considerations

In the case of bTBI, when a high-velocity, high-energy blast wave encounters a human head, the blast wave reflects and diffracts and its energy dissipates upon interacting with various hard and soft structures (e.g., the scalp, the skull, the dura and pia mater, gray and white matter structures) (Wolf et al., 2009). The spherical structure of the head results in spatially nonuniform loading on the skull where reflected and diffracted pressures cause minute deflections of varying magnitudes (Ganpule et al., 2013; Selvan et al., 2013). Higher intensity shockwaves result in greater loading and larger deflections in the flexible skull bone. Such flexures create transient, albeit highly complex stress conditions in the brain (e.g., intracranial pressure, cellular disruption). Therefore, the primary cause of damage to biological structures is caused by the sudden (millisecond range) increase in pressure followed by a similarly rapid decrease (Cernak, 2010; Chavko et al., 2011; Ganpule et al., 2013; Nakagawa et al., 2011).

Biological materials are especially sensitive to extreme rates of change (i.e., pressurization and de-pressurization) due to highly regulated physical and chemical properties. The head and the brain contain substantial amounts of fluid. High-energy pressure waves interact differently with incompressible liquids such as blood and CSF than with soft, viscoelastic materials like the brain parenchyma (Bolander et al., 2011; Leonardi et al., 2011). White matter has significantly different physicochemical properties (e.g., tissue density, distribution, energy consumption) than gray matter, and is affected differently by the physical components of blast (Chafi et al., 2010). In fact, clinical and experimental data have shown that a significant proportion of damage occurs at the interface of differing biological materials (Cernak, 2010; Chavko et al., 2011; Ganpule et al., 2013; Nakagawa et al., 2011).

Primary blast injury also affects the gyrencephalic brains of humans, nonhuman primates, and pigs very differently than the lissencephalic brains of rodents (Saatman et al., 2008). Based on cadaver studies, it has been suggested that the human head has a much higher biomechanical response to blast compared with the head of a rat (Masel et al., 2012). This is due to a number of factors that include head size and shape as well as skull thickness and elasticity. It is critical to bear in mind that blast affects the entire body, not only the head (Cernak and Noble-Haeusslein, 2010). The energy from blast waves can travel to the brain from other body parts via (large) blood vessels. This "vascular load" can also contribute to the pathogenesis of bTBI (Cernak, 2010). Accordingly, systemic changes transmitted by the vasculature and the autonomic nervous system play a critical role in the overall pathomechanism of blast as described in Chapter 45.

23.3 METHODOLOGY

Re-creating the physical components of explosive blast environments under controlled conditions is logistically challenging for reasons of safety, fidelity, and reproducibility (Mediavilla Varas et al., 2011). Moreover, the complexity and variability of explosive blast environments introduce a plethora of confounding factors (e.g., noxious gases, hypoxic conditions, secondary traumas from blast wind, penetrating injuries) to the study of primary blast injury (Bauman et al., 2009; Masel et al., 2012). For reproducing the direct interactions between blast waves and viscoelastic materials such as the skull and the brain, the modeling of explosive blast flow conditions is restricted to the generation of the blast overpressure wave.

23.3.1 Experimental Modeling of Blast Physics

The most frequently used and accepted models, the shock and blast tubes, re-create a cut-out segment of spherical blast (Bauman et al., 2009; Masel et al., 2012). The shock tube consists of a compression chamber and an expansion chamber separated by a Mylar diaphragm of varying thickness (Andersen and Louie, 1978; Celander et al., 1955; Chen and Constantini, 2013; Mediavilla Varas et al., 2011). The compression chamber is either connected to a high-capacity air compressor that pressurizes room air or to high-pressure gas cylinders of helium or another driver gas. When the compression chamber is pressurized above the threshold of the selected membrane thickness the Mylar membrane ruptures, resulting in a high-velocity pressure wave inside the expansion chamber.

The advantage of this relatively simple device is that the thickness of the Mylar membrane dictates the peak overpressure that is generated, thus guaranteeing a high level of reproducibility and scalability (Long et al., 2009). In some shock

TABLE 23.1
Summary of Experimental bTBI Models and Their Reported Outcomes

Method Used to Induce bTBI	Species and Brain			
	Mouse (Lissencephalic)	Rat (Lissencephalic)	Porcine (Gyrencephalic)	NHP (Gyrencephalic)
Open field; explosives	*Acute*: NM *Subchronic*: B *Chronic*: B; I; H Rubovitch et al., 2011	NM	*Acute*: PBB; PCSF *Subchronic*: B; P; H; PBT; PBB; PCSF Baumann et al., 2009	*Acute*: B; I *Subchronic*: B *Chronic*: I; H Lu et al., 2012
Three-wall structure; explosives	NM	NM	*Acute*: PBB; PC *Subchronic*: B; P; PBT; PBB; PCSF Ahmed et al., 2012 Ahmed et al., 2014	NM
Blast tube; explosives	NM	*Acute*: H; D Risling et al., 2011	*Acute*: PBB *Subchronic*: PBB *Chronic*: H de Lanerolle et al., 2011; Gyorgy et al., 2011	NM
Shock tube; air or helium	*Acute*: B; H; PBB; I; P; PBT. *Subchronic*: B; H; PBB; I; P; PBT. *Chronic*: B; H; PBB; I; P; PBT. Ahmed et al., 2014 (submitted); Cernak et al., 2011; Koliatsos et al., 2011	*Acute*: B; H; PBB; I; P; PBT *Chronic*: B; H; PBB; I; P; PBT Ahmed et al., 2012, 2013; Cernak et al., 2001; Kamnaksh et al., 2011; Kwon et al., 2011; Long et al., 2009	NM	NM

Time points: Acute, <72 hours; subchronic, ±2 weeks; chronic, >4 weeks.
Outcome measures: B, behavior; D, DNA microarray; H, histopathology; I, imaging; P, physiology; PBB, proteomics blood-based biomarkers; PBT, proteomics brain tissue; PCSF, proteomics CSF; NM, not measured.

tube models, compressed air has been replaced by helium, which results in an improved profile of the Friedlander curve because helium gas behaves more ideally (Koliatsos et al., 2011). The blast tube uses an explosive charge as the driver to simulate real blast environments (Risling and Davidsson, 2012; Risling et al., 2011). However, logistical, safety, and regulatory matters rendered the compressed air–driven shock tube the most widely used model. Table 23.1 summarizes the methods most commonly used to induce bTBI as well as tested outcome measures in the different bTBI animal models.

23.3.2 Animal Models of bTBI

The vast majority of blast-induced TBI modeling utilizes rodents, mostly rats. As previously mentioned, there is evidence that blast waves interact differently with gyrencephalic brains compared with lissencephalic brains (Nakagawa et al., 2011; Saatman et al., 2008). This finding alone makes scalability (i.e., the relationship between the measurable physical components of blast and the ensuing biological response) very difficult across species. Nonetheless, several factors have tilted the balance toward rodents as a biological model for bTBI. These include the cost-effectiveness of the rodent model, the availability of well-established behavioral assays, and the ability to compare blast-induced changes with outcomes observed in other forms of TBI.

A limited number of bTBI studies have been performed with other species. The large bTBI animal models with gyrencephalic brains, predominantly swine and nonhuman primates (NHPs), offer significantly improved scalability and high-fidelity physical-to-biological coupling—two critical factors in bTBI (Nakagawa et al., 2011; Saatman et al., 2008). In addition, these models are amenable to testing various blast scenarios, such as real open field exposure, closed structures (e.g., building, Humvee surrogate), and blast/shock tube-simulated open field conditions (Masel et al., 2012). As a narrative to Table 23.1, we provide a brief overview of the different animal models in the following sections.

23.3.2.1 Rat Model

Exposing rats to various forms of blast overpressure has recapitulated some of the salient features of human bTBI, including the neurobehavioral consequences of the injury (Kamnaksh et al., 2012; Kovesdi et al., 2012b; Long et al., 2009). As described elsewhere in this chapter, general locomotion as well as anxiety- and depression-related behaviors can be measured using the open field test; anxiety can be alternatively assessed by the elevated plus maze. Learning and memory functions can be measured using the Barnes maze. Blast overpressure exposure in rats results in complex structural, cellular, and molecular changes that implicate metabolic and vascular abnormalities, neuronal and glial damage, gliosis, inflammation (Kamnaksh et al., 2012; Kanemura et al., 1999), and axonal pathologies (Garman et al., 2011) in bTBI. As described later in detail, some of these changes can be detected by using various imaging modalities, histopathology, and proteomic analyses of blood, CSF, and brain tissue.

23.3.2.2 Mouse Model

Using mice in experimental bTBI confers many of the same advantages as the rat model: cost-effectiveness, logistical convenience (e.g., housing, handling, transportation), comparability of results across different forms of TBI, and the availability of an impressive array of neurobehavioral tests tailored to rodents (Koliatsos et al., 2011). Another advantage is the availability of genetically modified mouse strains, which enable researchers to study the role specific genes play in the pathobiology and outcome of bTBI (Rubovitch et al., 2011). However, using the mouse model of bTBI has the same disadvantages as the rat model on account of these models' limited scalability and obvious anatomical differences between rodents and humans (Nakagawa et al., 2011; Saatman et al., 2008).

23.3.2.3 Porcine Model

The limited studies using the porcine model of bTBI have found pathological changes similar to those identified in the rat model (Ahmed et al., 2012, 2013; Bauman et al., 2009; de Lanerolle et al., 2011; Gyorgy et al., 2011; Kamnaksh et al., 2012). The swine model offers several important advantages over rodent models in bTBI research. These include a human-like gyrencephalic brain, similar biochemical and cellular processes, and comparable physiologies. The swine model is also amenable for using techniques similar to those used in clinical neurointensive care units, including physiological monitoring, imaging, and serial blood/CSF sampling for metabolic and proteomic analyses (Cairns et al., 2010; Manley et al., 2006; Nakagawa et al., 2011; Saatman et al., 2008). These are very important considerations for bTBI modeling given the complexity of blast physics and biomechanics. Despite these key advantages, the costs and logistics associated with the bTBI swine model are rather prohibitive.

23.3.2.4 NHP Model

NHPs are the most (if not the only) high-fidelity models for human TBI (Margulies et al., 1990) including bTBI (Lu et al., 2012). In addition to their critical anatomical and physiological similarities to humans, primates have advanced cognitive abilities that render them far superior to any other animal model for the study of the neurobehavioral consequences of bTBI. Moreover, the availability of sophisticated behavioral testing, imaging techniques, and other clinically relevant outcome measures enable translational research of the highest fidelity. Although the use of NHPs in research is ethically controversial and extensively regulated, bTBI is one of the very few conditions in which their use is probably justifiable considering the spectrum of neurobehavioral symptoms and chronic nature of the disease. The most critical application of the NHP model would be late-stage preclinical testing of candidate drugs for bTBI.

23.4 SPECIAL CONSIDERATIONS FOR bTBI MODELING

Various ethical, economical, and logistical issues made the rat the animal of choice for bTBI studies. Similarly, practical and safety considerations made the shock tube the most frequently used physical model of blast. When using any animal model, there are multiple factors that must be taken into account to ensure that experimental conditions mimic the intended real-life scenario as closely as possible. Below, we summarize some of these considerations for experimental bTBI using the rat model in a shock tube.

23.4.1 Severity of bTBI

The main determinant of bTBI severity when using the shock tube is the peak pressure (Long et al., 2009; Skotak et al., 2013). This can be adjusted by selecting the appropriate Mylar membrane thickness. There is an empirical correlation between the thickness of the membrane and the peak pressure generated after the membrane ruptures as measured by pressure sensors placed inside the tube. However, it should be noted that the geometry of the tube (diameter to length) can significantly modify the actual physical parameters that interact with the experimental animal.

23.4.2 Position of the Animal

To mimic the effect of explosive blast exposure as closely as possible, the animal must be placed inside the shock tube. The shock tube re-creates a cut-out segment of the (open field) spherical blast wave and enables highly reproducible blast overpressure exposure (Masel et al., 2012). However, as the shock wave exits the end of the tube, it rapidly dissipates in different directions creating a highly non-uniform flow field. If the animal is placed outside the shock tube, it is exposed to the blast wind (not the blast overpressure waves). The blast wind not only has a different velocity and wave characteristics, but it also induces an entirely different type of injury (i.e., coup-contrecoup tertiary blast injury as opposed to primary blast injury). (Cernak and Noble-Haeusslein, 2010; Chodobski et al., 2003). Experimental data have demonstrated that injury severity, including lethality, is affected by the position of the animal relative to the shockwave. Severity and lethality is significantly higher if the animal is in the supine position versus the prone position (Cernak et al., 2011). The differences in lethality are likely due to a greater extent of lung injury sustained in the supine position. To account for these and other anatomical differences between rodent and human heads/skulls, we typically expose rats in the prone position with the lateral (right) side of the head facing the blast.

23.4.3 Anesthesia and Environmental Cofactors

Anesthesia, required not only for ethical reasons but also to minimize stress, pain, and mobility-induced complications that can confound the topic of study, is in itself a factor that significantly affects outcome after bTBI. All anesthetics used in experimental TBI (halothane, ketamine, and isoflurane) are neuroprotective (Matchett et al., 2009; Yan and Jiang, 2013). Conversely, environmental factors such as blast sounds, vibrations, and infrasound can cause significant behavioral and molecular changes without direct exposure to blast, especially when modeling mild and repeated mild bTBI (Kamnaksh et al., 2011). Therefore, one would need to design mild bTBI and repeated mild bTBI experiments with these issues in mind. This can be achieved by controlling for and reducing the effect of various experimental manipulations, which can be both logistically challenging (e.g., eliminating blast sounds) and biologically undesirable (e.g., anesthesia). Thus, the inclusion of the appropriate sham groups is especially important due to the partly overlapping symptomatology of PTSD and mild bTBI.

23.4.4 Body Protection

Exposing animals without body protection significantly increases lethality and poorly mimics real field scenarios in which soldiers wear body armor (Long et al., 2009). However, head-only exposures (typically performed outside the shock tube) result in an unrealistic pathophysiology not only because of the substandard simulation of physical blast conditions discussed previously, but also because the pathobiology of bTBI—unique to other TBIs—has a substantial systemic component (Cernak, 2010).

23.4.5 Fidelity and Scalability

Scalability is a greater concern in bTBI than in any other form of TBI. This is because, although we can measure and mostly know the physical parameters of explosive blast, we still have very little knowledge of how and to what extent the various physical components of blast interact with a biological structure of a highly complex mechanical and biochemical nature such as the brain. From a biological viewpoint, there are obviously huge differences in size and anatomy between human and rodent bodies, heads, and brains. Given the complexity of explosive blast physics, this should be a major concern when using rodents in bTBI. Although rodent models have provided important insights into the pathobiology of bTBI, they provide little (if any) useful information about the relationship between the physical forces and biological responses that are relevant to human scenarios. Peak pressure and duration, the two physical parameters frequently measured in experimental bTBI, do not directly correlate with the extent of the damage sustained after blast. Additional factors, including the geometry of the tube and the position of the animal, can substantially modify the original physical forces that interact with the organism. Moreover, studies have varyingly used head-only or whole-body exposure(s) with and without body protection, making the comparison of the physical forces and the respective biological responses nearly impossible to compare. If one intends to study the biological effects of explosive blast as close as possible to real-life scenarios, then whole-body exposure with torso protection must be used. As experimental bTBI studies have shown, blast triggers a systemic response that cannot be recapitulated with a head-only exposure.

23.4.6 Repeated Mild bTBI

The overwhelming majority of military bTBI cases are mild. Soldiers who test negative for neurocognitive impairments return to duty and are frequently exposed to a second, third, or more mild blasts (http://www.dcoe.mil/). Modeling this very important condition, repeated mild bTBI, has additional considerations. When designing repeated mild bTBI experiments, the temporal differences in rodent and human pathologies must be taken into account (Agoston et al., 2012). As available experimental evidence suggests, the period of increased cerebral vulnerability (ICV), a critical issue in repeated mild bTBI, is measured in hours in rodents compared with days to weeks in humans (Povlishock, 2013). Beyond the apparent anatomical differences, the physiology (and pathophysiology) of rodents versus large (gyrencephalic) animals vastly differs in the timescale of biological processes, both pathological and restorative. Metabolically-speaking, a rodent day/week/month is not the equivalent of a human, NHP, or porcine day/week/month. Accordingly, the interval between repeat exposures ranges from a single minute to a full day in experimental bTBI (Agoston, 2013). Even though the exact conversion rate between rat and human time is still unknown, the cumulative effect of mild bTBI was seen when the interval between exposures was less than 24 hours (Ahlers et al., 2012; Ahmed et al., 2013; Huang et al., 2013; Kamnaksh et al., 2012).

23.5 NEUROBEHAVIORAL CONSEQUENCES OF bTBI

The exposure to blast results in a wide spectrum of neurobehavioral consequences that range from transient confusion to coma depending on injury severity (Chew and Zafonte, 2009; French et al., 2012; Iverson et al., 2011; Kelly et al., 2012; Kennedy et al., 2010a; Meterko et al., 2012). bTBI is classified as severe if the Glasgow Coma Scale (GCS) score is <9, loss of consciousness is >24 hours, and amnesia is >7 days (http://www.dcoe.mil/). Imaging, standard MRI and computed tomography (CT), is positive, and most if not all severe bTBI victims suffer penetrating head injuries and polytrauma (Ling et al., 2009; Ling and Ecklund, 2011). If the patient survives, there are always significant and lasting neurological and neuropsychiatric impairments in severe bTBI. In moderate bTBI, the GCS score is 9–12, loss (or altered state) of consciousness is relatively short (>30 minutes to 24 hours), and amnesia lasts between 1 and 7 days (Dalle Lucca et al., 2012;

Levin et al., 2010). Imaging shows minimal and transient changes in the brain. These acute symptoms are followed by moderate levels of chronic neurological and neurobehavioral impairments that adversely affect cognition, memory, and mood. The overwhelming majority of bTBI cases (75–85%) are mild with GCS scores of 13–15. In mild bTBI, there is no or <30-minute loss (or altered state) of consciousness, no or <24 hours of amnesia, and standard imaging is always negative (Brenner et al., 2010; Elder et al., 2010; Lange et al., 2012; Luethcke et al., 2011; Mendez et al., 2013; Petrie et al., 2013; Thompson et al., 2008; Vanderploeg et al., 2012).

Many individuals who suffer a mild bTBI recover within days to weeks after the injury. However, a significant number develop minor, albeit lasting neurobehavioral impairments that can adversely affect their performance and overall quality of life. The most prevalent symptoms associated with mild bTBI are impaired learning and memory, including forgetfulness, poor concentration/attention, and slowed thinking, in addition to emotional and mood imbalances, such as increased anxiety and depression (Belanger et al., 2011; Elder et al., 2012; Levin et al., 2010; Schultz et al., 2011; Vanderploeg et al., 2012). Because these symptoms are subjective and relatively minor, most soldiers return to duty where many of them are exposed to additional mild blasts. Repeated exposure to mild blast overpressure, especially within the period of ICV that follows an insult, increases the probability of developing more severe acute symptoms as well as long-term adverse outcomes. This is similar to what is observed in civilian mild TBI (i.e., repeated concussions in athletes such as boxers and football players) (Fujita et al., 2012; Povlishock, 2013; Prins et al., 2012). Unfortunately, in the absence of objective biomarkers, the exact period of ICV after a mild TBI is currently unknown.

A major challenge associated with the diagnosis of mild bTBI is that many of the neurobehavioral symptoms are also observed in PTSD (Chen and Huang, 2011; Jaffee and Meyer, 2009; Liberzon and Sripada, 2008; McAllister, 2009; Peterson et al., 2011; Simmons and Matthews, 2012; Trudeau et al., 1998). PTSD is by definition a condition triggered by the exposure to extreme psychological stress/trauma in the absence of physical damage. Because of the lack of objective outcome measures (e.g., imaging, molecular biomarkers), the partly overlapping symptomatology and subjective nature of testing makes the differential diagnosis between mild bTBI and PTSD very challenging. Furthermore, the overlapping symptomatology fuels the debate about the validity of mild bTBI as a disease (Hoge et al., 2008; Wilk et al., 2010, 2012).

23.6 NEUROBEHAVIORAL OUTCOMES IN EXPERIMENTAL AND CLINICAL bTBI

Some of the neurobehavioral impairments observed in clinical bTBI, particularly those related to learning, memory, anxiety, and depression, can be modeled in experimental bTBI (Elder et al., 2012; Kamnaksh et al., 2011; Kamnaksh et al., 2012; Kovesdi et al., 2012a; Kovesdi et al., 2012b; Kwon et al., 2011). Others (such as headaches) are among the most common symptoms in veterans with a history of bTBI, but they are not possible to test in animal models (Table 23.2).

TABLE 23.2
Summary of the Neurobehavioral Symptoms Observed in Clinical and Experimental Mild bTBI (Including Repeated Mild bTBI)

	Clinical		Experimental	
Neurobehavioral Symptom	**Acute <72 hours[a]**	**Chronic (1.2–7.1 years)[b]**	**Acute <72 hours[c,d]**	**Chronic (72 days)[e,f]**
Dazed, confused, disoriented	+++	+	NA	NA
Headaches	+++	+++	NA	NA
Sleep disturbances	+	+++	NM	NM
Anxiety and depression	+	+++	+ (EPM and OF)	+ (EPM and OF)
Memory loss	+	+++	+ (BM)	+++ (BM)
Imbalance	++	+	NM	NM
Impaired executive function	+	+++	NM	NM
Altered sensory sensitivities and functions (auditory and visual)	+	+++	NA	NA

[a] Luethcke et al., 2011
[b] Petrie et al., 2013
[c] Kamnaksh et al., 2011
[d] Kamnaksh et al., 2012
[e] Kwon et al., 2011
[f] Kovesdi et al., 2012a, 2012b
Relative occurrence of symptoms: +, rarely observed; ++, typically observed; +++, frequently observed.
EPM, elevated plus maze; OF, open field; BM, Barnes maze; NA, not applicable; NM, not measured.

Increased anxiety and impaired learning and memory have been detected in the rodent model of moderate and mild bTBI (Kovesdi et al., 2012a, 2012b). In animal studies, the severity of these neurobehavioral abnormalities showed significant changes over time. At 2 weeks post-injury, anxiety levels were only slightly higher in injured animals than in their sham controls. At approximately 6 weeks post-injury, anxiety levels were significantly elevated in injured animals after which they normalized at 10 weeks post-injury. These findings suggest that in the rodent model of bTBI anxiety is transient, and it can return to normal levels. Once again, the caveat is that the rodent and human biological and pathobiological clocks run on substantially different scales (Seok et al., 2013). Therefore, the increase in anxiety levels observed in veterans at the study time point of 1.2–7.1 years post-injury may correspond to the rodent midterm (~6 weeks) time point. The dynamic changes in neurobehavioral outcomes observed in the animal model of bTBI underlines the critical need to perform longitudinal neurobehavioral assessments in bTBI veterans. Studies of this nature should be similar to the Vietnam Veterans Head Injury Study (Raymont et al., 2011), a gold mine of knowledge about the long-term neurobehavioral effects of (penetrating) TBI, the role of gene polymorphism in recovery, and more.

Unlike the temporal pattern of anxiety in the rodent model of bTBI, the temporal pattern of memory impairment showed similarities to human observations. At 2 weeks post-injury, only minimal impairments were detected; injured animals required a significantly longer time to learn the task and recall the location of the escape box in the Barnes maze, but after the first day of testing they performed just as well as their sham counterparts (Ahmed et al., 2013; Kamnaksh et al., 2011, 2012; Kovesdi et al., 2012a, 2012b; Kwon et al., 2011). At 6 weeks post-injury, injured animals had higher latency times each day but the last day of testing, suggesting an increased difficulty to learn and recall (spatial) information. At 10 weeks post-injury, injured animals spent two to three times longer to find the escape box every day of the 5-day testing period, indicating a progressively worsening memory that is similar to what is observed in blast-injured veterans (Belanger et al., 2011; Kennedy et al., 2010a).

The main lesson for both clinical and experimental bTBI studies is that the neurobehavioral consequences of blast exposure need to be tested at several post-injury time points in long-term studies. Short-term and single detection point studies—both clinical and experimental—run the risk of producing temporary findings or false negative outcomes. They also overlook the temporal nature of behavioral changes in bTBI.

23.6.1 Special Considerations for Behavioral Assessments in Experimental bTBI

Testing the neurobehavioral consequences of human bTBI in an experimental setting poses several challenges. Although rodent behavioral testing has played an important role in determining injury severity in experimental TBI studies (Gold et al., 2013), some of the leading neurocognitive changes observed in human bTBI (e.g., headaches, disorientation, impaired executive function) are impossible to assess in animal models. Anxiety- and depressive-like behaviors, memory deficits, and select sensory functions are more amenable for animal testing. However, the comparison and interpretation of experimental data in the context of human data can still be difficult.

Many of the existing clinical tests, such as the Military Acute Concussion Evaluation (MACE) and the Automated Neuropsychological Assessment Metrics (ANAM), are predominantly based on self-reporting. Such subjective tools are valid for assessing the acute effects of mild blast exposure on neurobehavioral functions under real-life, field conditions with limited resources. The few corresponding tests used in experimental bTBI (e.g., Barnes maze for spatial memory), while highly simplified, are objective.

Another important issue that is often overlooked is the frequency of neurobehavioral testing in clinical vs. experimental bTBI. Most clinical assessments are administered at multiple post-injury time points, thus providing information about functional improvement or decline. Animal behavioral testing is typically performed at a single post-injury time point. If we take into account the poorly understood temporal relationship between human and rodent physiology, the comparison and interpretation of clinical and experimental data becomes even more convoluted. Despite these innate differences, conducting longitudinal studies with several neurobehavioral assessments would provide invaluable information about the dynamics of functional change in bTBI..

23.7 BIOMARKERS IN bTBI

Biomarkers are objectively measurable indicators of pathological (or normal physiological) processes. The two classes of clinically-relevant biomarkers in bTBI are 1) structural, obtained by various imaging modalities, and 2) molecular, obtained by analyzing biosamples (e.g., blood, CSF) using a proteomic, genomic, or metabolomic approach. Due to the specific and non-/minimally invasive nature of biomarkers, they can be tested repeatedly. Therefore, they can help identify underlying pathological mechanisms important for developing evidence-based therapeutic interventions, and they can monitor disease progression and/or regression, respectively. Unfortunately, there is a significant gap between clinical and experimental bTBI in the use of imaging and molecular biomarkers (Agoston et al., 2012). Next, we summarize the current knowledge about imaging and molecular biomarkers in clinical and experimental bTBI.

23.8 IMAGING BIOMARKERS IN CLINICAL bTBI

CT imaging is the current standard of care, and virtually all severe bTBI victims have been examined by CT (Benzinger et al., 2009). However, CT-based "biomarkers" are only relevant in severe and moderate bTBI cases—a minority of all bTBI cases. CT scans can identify changes such as skull fractures, hemorrhage, edema, shrapnel, and foreign bodies that require immediate neurointensive or neurosurgical interventions. No CT imaging has been used in experimental bTBI studies.

The overwhelming majority (80%–90%) of bTBI patients have normal CT findings, which when coupled with normal GCS scores places them in the mild bTBI category. Various MRI modalities have identified structural lesions in mild bTBI patients (Benzinger et al., 2009); of these modalities, DTI is the most frequently used. Altered fractional anisotropy (FA) reflects altered directional water diffusion indicative of the status of axonal membranes and myelin sheaths. Injury-induced changes in FA values, thus reflect damage to myelinated axonal tracts and (indirectly) to axonal function. Accordingly, DTI can identify subtle changes, and altered FA values have the potential to serve as an imaging biomarker for mild bTBI for cases with negative CT scans (Aoki et al., 2012; Gardner et al., 2012; Lipton et al., 2008; Sotak, 2002). Keeping MRI systems fully operational under field conditions has not proven possible, thus we have no clinical information about acute changes in FA in mild bTBI. Routine MRI scans performed at later post-injury time points failed to detect structural abnormalities in mild bTBI cases. However, some DTI studies examining veterans with a history of mild bTBI found white matter changes (Jorge et al., 2012; Levin et al., 2010; Mac Donald et al., 2011, 2013; Mendez et al., 2013; Petrie et al., 2013). The detected changes in FA values were multifocal, frequently involving cerebellar tracts, and (depending on the study) either correlated with injury-induced neuropsychiatric changes or lacked such a correlation. These conflicting findings can be explained by: 1) the different post-injury testing time points, 2) differences in DTI data analyses, 3) inter-patient variability, and 4) the control populations that were scanned.

A limited number of studies have used more specific and sensitive imaging modalities, such as functional MRI (fMRI) (Fischer et al., 2013; Han et al., 2013, 2014; Matthews et al., 2011), positron emission tomography (PET) (Peskind et al., 2011; Petrie et al., 2013), and magnetic resonance spectroscopy imaging (MRSI) (Hetherington et al., 2013), that identified various functional, structural, and molecular abnormalities in mild bTBI. Findings included increased amygdalar activation to fear in veterans with major depressive disorder as a consequence of bTBI, white matter abnormalities, and reduced cerebral glucose metabolism in several cortical areas in veterans with a history of mild bTBI (especially after repeated mild bTBI). Importantly, the detected abnormalities were found to be specific to mild bTBI and not related to PTSD comorbidity (Peskind et al., 2013; Petrie et al., 2013). A 7-T, high-resolution MRSI study showed reduced right hippocampal volume and decreased N-acetylaspartate levels in veterans with a history of mild bTBI, as well as cognitive impairment measured at various time points (1 and 10 years) after single or repeated (up to 10) mild blast injury. In this small study (25 veterans and 20 controls), however, there were no significant differences between subjects with or without PTSD (Hetherington et al., 2013).

Veterans exhibiting behavioral changes also displayed (chronically) altered cerebral glucose metabolism as indicated by fludeoxyglucose positron emission tomography (FDG-PET), and white matter abnormalities as indicated by MRI using FA and macromolecular proton fraction (MPF) mapping (Petrie et al., 2013). The white matter abnormalities were mainly localized at the corpus callosum, various cortical/subcortical white matter tracts, and gray matter/white matter border regions signifying axonal injury as one of the main underlying pathologies in bTBI. Importantly, the study found that PTSD comorbidity does not affect structural abnormalities, thus it supports the idea that mild bTBI is a distinct disease entity. In addition to decreased FA values in veterans with a history of bTBI, MPF mapping showed decreased values in multiple brain regions including the superior frontal gyrus, subcallosal gyrus, and medial orbital gyrus. The extent of these abnormalities correlated with the number of mild blast exposures. Veterans with 20–100 blast exposures had the lowest MPF values. FDG-PET imaging showed significant reductions in cerebral glucose metabolism mostly in the right and the left parietal cortices, the left somatosensory, and right visual cortices. Given the average 4-year interval between the last blast exposure and imaging study, these findings indicate chronic metabolic abnormalities as a consequence of bTBI. Veterans with more than 20 exposures displayed a single spot within the parahippocampal gyrus with especially low cerebral glucose metabolism. Again, PTSD as a comorbidity did not affect any of these imaging findings.

Other studies found no significant differences in FA values between veterans with and without a history of bTBI (Levin et al., 2010; Matthews et al., 2011; Mendez et al., 2013). Potential explanations for these discrepancies include differences in imaging protocols and time elapsed before the imaging studies were conducted, which varied between 14 days and 10 years from the last injury. An additional contributing factor is the complexity of real-life blast exposure, from a physical as well as a biological point of view.

Although imaging has provided some of the most important insights into the pathobiology of bTBI, the summarized findings underline key issues that remain unsolved. These include 1) our limited understanding of how altered FA values relate to axonal function, including injury-induced axonal damage. This can be at least partially addressed by using magnetoencephalography to measure axonal functionality in addition to DTI (Matthews et al., 2011). Another issue is (2) the lack of information about early structural changes, and the absence of longitudinal studies as a function of neurobehavioral performance, which can be addressed by retesting previously tested populations at several time points.

23.9 IMAGING BIOMARKERS IN EXPERIMENTAL bTBI

Using imaging modalities identical to those used on humans, animal studies can—in principle—address some of the main issues that confound human imaging studies. Experimental imaging studies allow protocol standardization and impart much needed homogeneity to the subject population. Therefore, animal imaging, in combination with behavioral testing and other clinically-relevant outcome measures (e.g., blood-based biomarkers), can provide critical information about how imaging findings relate to functional outcomes and molecular changes. Despite these clear needs, very few imaging studies have been performed in experimental bTBI.

Consistent with the abovementioned clinical data indicating axonal damage, animal imaging studies found altered FA values in the cerebellum and several subcortical white matter structures after single and repeated mild bTBI (Budde et al., 2013; Calabrese et al., 2014; Kamnaksh et al., 2014). Importantly, the existing studies were performed ex vivo (i.e., by scanning perfusion-fixed brains to reduce logistical and technical challenges). Although the ex vivo approach has its own technical issues, primarily related to scanning a fixed organ, findings can be verified by histology.

The available studies using similar rat models of bTBI have shown significant changes in FA values after single and repeated blast (delivered at different intervals) at various survival time points (Calabrese et al., 2014; Kamnaksh et al., 2014). Both studies found minor changes in FA values and axial diffusivity values at 2 hours, 72 hours, and 42 days after a single injury. Multiple mild injuries, however, resulted in substantially increased FA values in the cerebellum and several subcortical structures. The changes in the rodent model of repeated mild bTBI were transient and detectable within the acute period of injury (2 and 72 hours) but not at 42 days after the last exposure. In the absence of acute data, these findings cannot be directly compared to clinical studies where which other factors such as the intensity of blast and intervals between exposures are unknown. Nevertheless, these are important first steps toward identifying imaging biomarkers in bTBI and determining how structural changes relate to injury intensities. The imaging findings in bTBI are summarized in Table 23.3.

23.9.1 Special Considerations for Imaging Biomarkers in Experimental bTBI

Human imaging studies were able to identify some important structural and functional changes even though they were hampered by the heterogeneity of the patient population, challenges with the appropriate controls, and the lack of acute post-injury information and longitudinal data. Animal studies can help address these issues. However, imaging, small rodents, has its own—mostly technical—issues. One is the resolution, which is becoming less and less of a limitation with the advent of larger magnets and advances in data analysis software. Another issue is scanning time; to obtain sufficient spatial resolution, rodent brains need to be scanned for several hours. This can only be performed under anesthesia, which introduces another confounding factor. Ex vivo DTI imaging enables extended scanning times but the fixation process itself can introduce other artifacts, including reduced DTI sensitivity. The main issue, which also hampers human imaging, is the lack of standardized imaging and analytical protocols, making the comparison between imaging studies challenging.

23.10 THE IMPORTANCE OF CORRELATING BETWEEN NEUROBEHAVIOR AND IMAGING

The importance of correlating structural and functional/behavioral changes in bTBI cannot be overstated. However, the logistical, ethical, and technical challenges are formidable, as reflected in the small number of clinical and experimental bTBI studies. The few existing integrated studies failed to identify associations between structural and functional changes (Petrie et al., 2013). Most imaging studies have used DTI, but how altered FA values relate to altered neurobehavioral functions is still unknown. Another problem is that there can be substantial differences in the spatial associations of altered FA values. So far, no experimental bTBI studies have successfully combined high-fidelity blast modeling with high-resolution imaging and neurobehavioral monitoring.

Similarly, the associations between neurobehavioral impairment and altered glucose metabolism are still elusive. Civilian TBI, sustained during sports or motor vehicle accidents, results in transient cerebral glucose hypometabolism; levels typically normalize within a few weeks after injury. In contrast, cerebral glucose metabolism remained decreased in veterans with a history of bTBI even 4 years after the (last) injury (Petrie et al., 2013). This observation is very important because it implicates cerebral glucose hypometabolism in the pathobiology of bTBI.

23.11 BLOOD- AND CSF-BASED PROTEIN BIOMARKERS

The interactions between the body and the various physical components of blast are extremely complex and still poorly understood (Cernak et al., 2011; Cernak and Noble-Haeusslein, 2010; Masel et al., 2012). The observed neurobehavioral consequences of bTBI are the result of dynamic changes at the structural, cellular, and molecular levels triggered by blast exposure. At the molecular level, the transferred energy can profoundly alter cellular metabolism, and disrupt or alter cell–cell adhesions between the various cells populating the central nervous system. At the intracellular level, the cytoskeleton, including the axoskeleton as well as transport networks, can be directly or indirectly affected by the exposure to blast. Each of these molecular changes can trigger a myriad of downstream processes, some reactive and others restorative. Many of these changes can be detected by analyzing serum samples prepared from the peripheral blood (Agoston and Elsayed, 2012; Agoston et al., 2009).

Blood-based biomarkers are essential diagnostic tools in medicine. Obtaining blood is easy, minimally invasive, and

TABLE 23.3
Summary of Imaging Findings in Clinical and Experimental bTBI

Imaging Modality	Clinical Injury Severity, Number of Exposures; Postinjury Time Point, and Anatomical Locations			Experimental Injury Severity, Number of Exposures; Postinjury Time Point, and Anatomical Locations		
	Acute (<2 hours)	Subacute and Subchronic (1–90 days)	Chronic (>12 months)	Acute (<2–72 hours)	Subchronic (42 days)	Chronic (>72 days)
DTI		*1x mbTBI*; Cib; right orbitofrontal WM; Cb peduncles[a,b]	*1x and xx mbTBI*; Cib; right orbitofrontal WM; Cb peduncles[a,b]	*1x and xx mbTBI*; Th, Cb[c]	*1x and xx mbTBI* No changes[c]	
			1x month and mbTBI; no change[e]	*1x and xx mbTBI*; Cb, Hb[d]		
			xx mbTBI; "potholes" in WH, CC[f]			
			mbTBI; CC, Cr, slf[g]			
fMRI			*1x mbTBI*; AMG[g]			
fMRI			*xx mbTBI*; HIP[h]			
MRSI			*xx mbTBI*; Pcx, Vcx, Scx[i]			
FDG-PET DTI, MPF			*xx mbTBI*; CC, subgyral and subcortical WM[i]			

[a] MacDonald et al., 2011
[b] MacDonald et al., 2013
[c] Kamnaksh et al., 2014
[d] Calabrese et al., 2014
[e] Levin et al., 2010
[f] Jorge et al., 2012
[g] Matthews et al., 2011
[h] Hetherington et al., 2013
[i] Petrie et al., 2013

1x, single exposure; AMG, amygdala; CC, corpus callosum; Cb, cerebellum; Cib, cingulum bundles; Cr, corona radiata; DTI, diffusion tensor imaging > fractional anisotropy; fMRI, functional magnetic resonance imaging > neuronal activity; FDG-PET, fludeoxyglucose positron emission tomography > cerebral glucose metabolism; Hb, hindbrain; HIP, hippocampus; mbTBI, mild blast-induced TBI; mobTBI, moderate blast-induced TBI; MPF, macromolecular proton fraction mapping > white matter abnormalities; MRSI, magnetic resonance spectroscopy imaging > N-acetyl aspartate; NA, not applicable/measurable; NM, not measured; Pcx, parietal cortex; Scx, somatosensory cortex; slf, superior longitudinal fasciculus; Th, thalamus; Vcx, visual cortex; WM, white matter tracts; xx, multiple exposures.

repeatable. Moreover, time-dependent changes in serum biomarker levels can reflect the progression or regression of a disease. For these reasons, serum-based biomarkers are of special value in bTBI. In the absence of sophisticated diagnostic tools such as MRI, their value is even higher. Despite these facts, there is currently no publicly available information about the identity of serum-based biomarkers in human bTBI. Experimental bTBI work, however, has identified several candidate biomarkers that are capable of determining the severity of the injury, monitoring the progression of the disease, predicting outcome, and identifying ongoing pathological changes that can aid in designing evidence-based therapeutic interventions. The current state of knowledge is summarized and discussed in the following section.

23.11.1 bTBI Specificity of Biomarkers

In the absence of comparative studies employing blast and nonblast models of TBI, we can only speculate about the specificity of blood-based biomarkers in bTBI (Agoston and Elsayed, 2012; Agoston et al., 2009). The different physical components of blast suggest that some biological structures and molecules are more sensitive to blast than others. The three main physical mechanisms—spallation, implosion, and the inertial effect of blast—can exert unique damage (Cernak et al., 2011; Cernak and Noble-Haeusslein, 2010; Masel et al., 2012) (see Chapter 45). The inertial effect that occurs at the interface of different tissue densities accelerates the lighter density component more than the denser one (e.g., gray matter vs. white matter). Then there are frequency-dependent primary blast effects that can disrupt different structures at the molecular level. The high-frequency (~1 kHz), low-amplitude wave component of the blast disrupts distinct sets of biological structures, such as cell–cell adhesions, by "loosening" the junctional structures between cells, and shedding cell surface and junctional proteins. This is supported by substantially elevated serum levels of cell surface molecules, gap junction proteins, and vascular proteins in experimental bTBI. The cyto- and axoskeletal networks can be altered by the other type of stress waves (low frequency and high amplitude), as evidenced by elevated serum levels of specific markers such as GFAP, Tau, and NF (Agoston and Elsayed, 2012; Agoston et al., 2009).

As of today, there are no serum-based biomarkers that are specific to bTBI and it is unlikely that a single molecule will emerge as "the" bTBI-specific biomarker. The identification of such markers can be accomplished by: 1) determining the qualitative and quantitative changes in select biomarkers at different post-injury time points; and 2) comparing the temporal pattern of these changes in bTBI with changes detected after other forms of TBI. In doing so, we will be able to establish the "bTBI specificity" of biomarkers associated with various pathological processes, such as axonal injury, inflammation, and so forth (Agoston and Elsayed, 2012; Agoston et al., 2009). Furthermore, comparing the onset, extent, and duration of changes in the blood/serum levels of markers specific to the different pathologies (e.g., changes in Tau and NF proteins as markers of axonal injury relative to changes in chemokines and cytokines as markers of inflammation) can reveal the molecular fingerprint of bTBI. Conducting such analyses, especially if combined with neurobehavioral and imaging studies of the same subjects, would provide a vastly improved understanding of the pathobiology of bTBI and simultaneously identify potential molecular targets for intervention. The experimental evidence showing that the exposure to blast triggers a complex systemic response, and that systemic inflammation appears to be a significant component of bTBI pathobiology makes such an approach even more important as only a handful of the protein biomarkers are central nervous system–specific. Current findings are summarized in Table 23.4.

Metabolic changes (including oxidative stress) are a hallmark of different forms of mild TBI (i.e., concussions) in civilians (Giza and Hovda, 2001; Hovda, 2007; Prins et al., 2012; Slobounov et al., 2013). Importantly, both clinical and experimental TBI studies have demonstrated that metabolic depression is a key component of ICV after mild TBI. 4-HNE, HIF1α, and ceruloplasmin are protein biomarkers of oxidative stress. 4-HNE is a byproduct of lipid peroxidation and a marker of (systemic) oxidative stress. HIF1α, a transcription factor, is part of organisms' adaptive response to noxious insults involving hypoxia (e.g., TBI, stroke) (Agoston and Elsayed, 2012; Ahmed et al., 2014b). Cerebral edema and vasospasm are hallmarks of severe bTBI, indicating major changes in water transport (Lu et al., 2011; Manley et al., 2006; McMahon et al., 2014; Nakagawa et al., 2011; Papadopoulos et al., 2004) and altered vascular endothelial/vascular reactivity after TBI (Fujita et al., 2012). Of the various water channel proteins, AQP1 is broadly expressed by various cell types in the body. In the brain, AQP1 is expressed by the ependymal cells that line the choroid plexus and are involved in CSF production (Badaut et al., 2011; Benga and Huber, 2012; Papadopoulos et al., 2002; Venero et al., 2001). AQP4, the main water channel of the brain, transports water between the blood and the brain parenchyma and is mostly expressed on astrocytic foot processes and endothelial cells. vWF is an endothelium-specific glycoprotein; elevated vWF serum levels indicate endothelial activation in response to bTBI, which can lead to vascular damage, increased permeability, and even microbleeding (Lenting et al., 2012; Lip and Blann, 1997; Rauch et al., 2013). Elevated VEGF levels also indicate endothelial damage in bTBI (Agoston and Elsayed, 2012; Ahmed et al., 2014b).

Significant increases in the serum and CSF levels of various inflammatory markers have been observed in bTBI (Agoston and Elsayed, 2012; Agoston et al., 2009; Ahmed et al., 2014b). MCP-1, a member of the C-C class of the beta-chemokine family, is involved in attracting monocytes and regulating the permeability of the blood–brain barrier (Yadav et al., 2010). OPN is an extracellular matrix protein with multiple immunological functions. Elevated OPN levels have been associated with various diseases, including multiple sclerosis (Butler, 1989; Shin, 2012). Increased levels of CINC-1α, CD53 (OX-44), and TLR-9 further implicate inflammation in the pathobiology of bTBI. CINC-1α, a member of the CXC chemokine family, is produced by astrocytes in response to oxidative stress (Callcut et al., 2012). CD53 is a cell-surface protein expressed in various cells (including microglia) that mediates cell growth and immune response. Therefore, elevated OX44 levels in the CSF suggest a robust neuroinflammatory process (Woodcock and Morganti-Kossmann, 2013). Elevated serum levels of fibrinogen, a plasma glycoprotein, are also indicative of inflammation. An increase in fibrinogen affects blood viscosity as well as vascular permeability through the ERK-1/2 signaling pathways. Elevated fibrinogen levels can link inflammation and vascular damage to longer-term neuronal dysfunctions, such as neurodegeneration, as suggested by studies showing a correlation between elevated fibrinogen precursor proteins in the CSF and the severity and progression of Alzheimer's disease (Lee et al., 2010). MIP1α, a proinflammatory chemokine produced by a variety of immune cells including macrophages and microglia (McManus et al., 1998), is indicative of a systemic response to injury when its levels are elevated in serum (Hsieh et al., 2008). TLR9, a member of the Toll-like receptor family, is involved in mediating innate immunity, inflammation, and adaptive immunity; increased TLR-9 levels also suggest an immune response after blast (Barton and Kagan, 2009; Vollmer, 2006). As previously discussed, intercellular function can be especially sensitive to blast as evidenced by changes in several markers related to cell adhesion.

Elevated serum levels of neuronal markers, such as NSE and CK-BB, are indicative of neuronal damage and/or loss

TABLE 23.4
Summary of Protein Biomarker Findings in Various Experimental Models of bTBI

Model		**Mouse**			**Rat**				**Swine**		
Biomaterial		Blood			Blood			Blood			CSF	
Time point(s)	A	S	C	A	S	C	A	S	C	A	S	C
Metabolic function												
4-HNE	+	+	+	+	NM	+	+	+	+	+	+	+
HIF-1α	+	+	+	−	NM	−	+	+	+	+	+	+
Ceruloplasmin	+	+	+	+	NM	+	+	+	+	+	+	+
Vascular function												
VEGF	+	+	+	+	NM	−	+	+	+	+	+	+
FLK-1/VEGFR2	+	+	+	−	NM	NM	−	−	−	+	+	+
vWF	+	+	+	NM	NM	+	+	+	+	+	+	+
AQP1	+	+	+	NM	NM	NM	+	+	+	+	+	+
AQP4	+	+	+	NM	NM	NM	+	+	+	+	+	+
Claudin-5	−	−	−	NM	NM	NM	+	+	+	+	+	+
Inflammation												
MMP8	−	−	−	−	NM	−	+	+	+	+	+	+
MCP-1	−	−	−	NM	NM	NM	−	−	−	+	+	+
MIP-1α	+	+	+	NM	NM	NM	+	+	+	+	+	+
CINC-1α	+	+	+	NM	NM	NM	+	+	+	+	+	+
Fibrinogen	+	+	+	NM	NM	NM	+	+	+	+	+	+
CCR5	−	−	−	−	NM	+	+	+	+	+	+	+
CRP	−	−	−	NM	NM	NM	+	+	+	+	+	+
OPN	+	+	+	NM	NM	NM	−	−	−	+	+	+
TLR-9	−	−	−	+	NM	−	+	+	+	+	+	+
FPR-1	−	−	−	+	NM	−	+	+	+	−	−	−
P38	+	+	+	NM	NM	−	+	+	+	−	−	−
CD53 (OX-44)	+	+	+	NM	NM	NM	+	+	+	+	+	+
Galectin-1	−	−	−	NM	NM	NM	+	+	+	−	−	−
Cell adhesion												
Integrin-6α	+	+	+	NM	NM	NM	+	+	+	+	+	+
TIMP-1	−	−	−	NM	NM	NM	−	−	−	−	−	−
TIMP-4	+	+	+	NM	NM	NM	+	+	+	+	+	+
Connexin-43	+	+	+	NM	NM	NM	−	−	−	+	+	+
N-cadherin	+	+	+	NM	NM	NM	+	+	+	+	+	+
Neuronal damage												
NSE	+	+	+	+	NM	NM	+	+	+	+	+	+
CK-BB	+	+	+	NM	NM	NM	+	+	+	+	+	+
Axonal damage												
NF-H	+	+	+	+	NM	+	+	+	+	+	+	+
Tau	+	+	+	NM	NM	NM	−	−	−	+	+	+
Astroglia and myelin damage												
GFAP	+	+	+	+	NM	+	+	+	+	+	+	+
S100β	+	+	+	NM	NM	NM	+	+	+	+	+	+
MBP	+	+	+	−	NM	+	+	+	+	+	+	+
Other												
BDNF	−	−	−	NM	NM	NM	+	+	+	+	+	+

References: Agoston et al., 2009; Gyorgy et al., 2011; Kamnaksh et al., 2011; Agoston and Elsayed, 2012; Kamnaksh et al., 2012; Kovesdi et al., 2012; Ahmed et al., 2013, 2014a, 2014b.

Post-injury time point(s): A, acute; S, subchronic; C, chronic.

List of blood- and CSF-based biomarkers associated with bTBI: *Metabolic function:* 4-HNE, 4-hydroxy-2-noneal; HIF1α, hypoxia-inducible factor 1α; ceruloplasmin. *Vascular function*: VEGF, vascular endothelial growth factor; Flk-1/VEGFR2, fetal liver kinase 1/VEGF receptor 2; vWF, von Willebrand factor; AQP1, aquaporin 1; AQP4, aquaporin 4; claudin-5. *Inflammation:* MMP8, matrix metalloproteinase 8; MCP-1, monocyte chemotactic protein 1; MIP-1α, macrophage inflammatory protein 1α; CINC-1α, cytokine-induced neutrophil chemoattractant 1α; fibrinogen, CCR5, chemokine (C-C motif) receptor 5; CRP, C-reactive protein; OPN, osteopontin; TLR-9,Toll-like receptor-9; FPR-1, formyl peptide receptor 1; p38, p38 mitogen-activated protein kinase; CD53 (OX-44); galectin-1. *Cell adhesion:* integrin-6α, TIMP1, TIMP metallopeptidase inhibitor 1; TIMP4, TIMP metallopeptidase inhibitor 4; connexin-43, N-cadherin. *Axonal damage*: NF-H, neurofilament-heavy chain; Tau. *Neuronal damage*: NSE, neuron-specific enolase; CK-BB, creatine kinase-brain type. *Astroglia and myelin damage*: S100β, S100 calcium binding protein β. *Other:* BDNF, brain-derived neurotrophic factor.

(Agoston and Elsayed, 2012; Agoston et al., 2009; Ahmed et al., 2014a, 2014b). Both markers have been extensively studied in experimental and clinical TBI. In some studies, elevated serum levels have been correlated with the severity of various central nervous system injuries. The axonal markers NF-H and tau have been extensively used in clinical and experimental TBI, however, there is no consensus regarding their specificity as indicators of TBI severity and disease outcome (Begaz et al., 2006; Blennow et al., 2012; Kovesdi et al., 2010). It has been shown that an early peak in serum NF-H levels correlates with severe bTBI (and poor outcome), whereas a gradual increase reflects a mild to moderate injury (Gyorgy et al., 2011). Exposure to a single moderate blast results in a robust increase in the serum levels of GFAP, S100β, and MBP (Agoston et al., 2009). GFAP and S100β are the two most studied and established serum biomarkers in both experimental and clinical TBI. Injury-induced increases in serum S100β levels have been associated with injury severity and outcome (Thelin et al., 2013a; Thelin et al., 2013b). However, early increases in serum S100β levels can be due to nonneuronal sources (Rainey et al., 2009; Vos et al., 2010). MBP has been routinely used to monitor white matter damage after TBI. A previous study also showed significant increases in serum MBP levels after bTBI, which correlates with the white matter damage found by using DTI in clinical and experimental TBI studies (Calabrese et al., 2014; Johnson et al., 2013; Jorge et al., 2012; Lipton et al., 2008; Mac Donald et al., 2013).

23.11.2 Special Considerations for Blood- and CSF-Based Biomarkers in Experimental bTBI

Experimental bTBI studies have identified several candidate blood- and CSF-based protein biomarkers that have yet to be validated in clinical studies (Agoston and Elsayed, 2012; Agoston et al., 2009). The validation of these markers will be particularly challenging considering the post-injury time points of blood and/or CSF sampling. Similar to imaging studies, biosamples are mostly obtained from veterans with a multi-year history of bTBI. Experimental bTBI studies should seek to mimic those time points as closely as possible. Accordingly, experimental bTBI studies should be longitudinal, and blood (and/or CSF) samples should be collected at numerous acute, subacute, and chronic post-injury time points. Such a design can identify markers specific to the different post-injury phases in experimental bTBI, and establish their role in disease propagation or mitigation. In order to validate these markers as a function of time and disease progression, first we have to establish the temporal patterns of the different pathological processes in human bTBI victims and animal bTBI models, then compare the clinical and experimental data (Agoston, 2013). Therefore, the importance of obtaining blood (and/or CSF) samples from soldiers and veterans for biomarker analysis cannot be overemphasized.

23.12 BIOMARKER IDENTIFICATION AND CHOICE OF ASSAY PLATFORM

The use of discovery style, mass spectroscopy–based proteomics platforms is not practical in experimental as well as in clinical bTBI biomarker studies for a number of reasons discussed elsewhere (Agoston and Elsayed, 2012; Agoston et al., 2009). There are numerous antibody-based platforms that are both sensitive and specific. Of these tests, the enzyme-linked immunosorbent assay (ELISA) is the most frequently used despite its disadvantages, which include a limited selection of markers, the capacity to measure a single marker at a time, and variations in sensitivity and dynamic range between markers. Platforms that can measure multiple markers in a sensitive, reproducible, and parallel fashion are preferable. These systems include Luminex, Mesoscale, and reverse-phase protein microarray. Each has distinct advantages and disadvantages (Agoston and Elsayed, 2012; Agoston et al., 2009; Dudal et al., 2013). Reverse-phase protein microarray is the most sensitive (Gyorgy et al., 2010) while the Luminex is the most standardized, albeit limited in the number and types of markers it can measure. The Mesoscale performs in a multiplex manner but its sensitivity is suboptimal for detecting changes in minute quantities of biomaterials such as those available in TBI research.

23.13 CONCLUDING REMARKS

Blast-induced TBI, especially its mild form, is and will remain a significant military health issue for decades to come. There may be as many as 300,000 soldiers and veterans with some form of bTBI-related condition. The long-term effects of bTBI on afflicted individuals' quality of life, and the costs associated with treatment, rehabilitation, and care are currently unknown. However, the number of affected individuals and their ~40-year life expectancy suggests a very substantial long-term financial pressure on the military health care system.

The past decade of bTBI research yielded significant and novel findings on the physics and the biology of bTBI. However, the bench-to-bed and bed-to-bench translations in bTBI research have not been optimal. There have been several promising findings in the bTBI field, but practically none of the experimentally successful diagnostics and therapeutics have made it to clinical practice. In addition, real-life clinical needs continue to be poorly represented in experimental bTBI research.

There are two critical aspects of bTBI research that have not been adequately addressed. One is an in-depth knowledge and understanding of the acute consequences of bTBI for the purpose of translating them into noninvasive "fieldable" diagnostics. This entails identifying the immediate consequences of blast on serum biomarker levels and monitoring the temporal pattern of these changes, as well as determining blast-induced changes in brain activity using quantitative electroencephalogram. The other aspect is to understand the long-term consequences of mild bTBI, especially repeated mild bTBI. There is a reasonable likelihood that in a decade or so there will be thousands, possibly tens of thousands, of veterans with chronic traumatic encephalopathy or other neurodegenerative conditions. Such a scenario would impart an enormous emotional and financial toll on all parties involved if diagnosis and treatment are not addressed quickly and adequately.

High-fidelity models are essential for generating high-value experimental data. This is especially significant in experimental bTBI research because of the complexity of blast physics. Consequently, there should be clear guidelines for what qualifies as a blast model in relation to high-fidelity simulation of the physical forces of blast. High-fidelity biological modeling is equally important in bTBI. One of the main unanswered questions is: What is the rodent equivalent of the physical force/energy that causes the same type and extent of damage in the human brain? Although the physics are scalable, the bigger challenge is how similarly or dissimilarly rodent and human heads/brains react to the same exact physical force. Moreover, how rodent and human times compare when it comes to complex biological and pathological processes such as inflammation is still unknown (Seok et al., 2013). Addressing such issues will likely require the use of gyrencephalic models such as NHPs. The use of NHPs would better address the issue of scalability not only in terms of physical-to-biological coupling, but also in terms of biological/pathological fidelity.

Experimental bTBI research should place great emphasis on using the same or similar outcome measures to those used in clinical bTBI (Agoston et al., 2012). These non-invasive or minimally invasive technologies—imaging, quantitative electroencephalogram, physiological monitoring, and blood-based biomarkers—are readily available for experimental research. Finally, employing a systems biology approach to connect the vast quantities of existing experimental and clinical data will generate the much-needed level of understanding toward a differential diagnosis and treatment for bTBI.

REFERENCES

Agoston, D.V. 2013. Of timescales, animal models, and human disease: The 50th anniversary of *C. elegans* as a biological model. *Front Neurol*. 4:129.

Agoston, D.V., and M. Elsayed. 2012. Serum-based protein biomarkers in blast-induced traumatic brain injury spectrum disorder. *Front Neurol*. 3:107.

Agoston, D.V., A. Gyorgy, O. Eidelman, and H.B. Pollard. 2009. Proteomic biomarkers for blast neurotrauma: Targeting cerebral edema, inflammation, and neuronal death cascades. *J Neurotrauma*. 26:901–11.

Agoston, D.V., M. Risling, and B.M. Bellander. 2012. Bench-to-bedside and bedside back to the bench; coordinating clinical and experimental traumatic brain injury studies. *Front Neurol*. 3:3.

Ahlers, S.T., E. Vasserman-Stokes, M.C. Shaughness, A.A. Hall, D.A. Shear, M. Chavko et al. 2012. Assessment of the effects of acute and repeated exposure to blast overpressure in rodents: Toward a greater understanding of blast and the potential ramifications for injury in humans exposed to blast. *Front Neurol*. 3:32.

Ahmed, F., I. Cernak, and D.V. Agoston. 2014a. Time dependent changes of protein biomarker levels in the serum in a mouse model of mild blast traumatic brain injury. *Journal of Neurotrauma*. Submitted.

Ahmed, F., M. Elsayed, A. Kamnaksh, G. Ling, L. Tong, W. Taylor et al. 2014b. Time-dependent changes of protein biomarker levels in the serum and cerebrospinal fluid in a large, gyrencephalic animal model of blast-induced mild traumatic brain injury *Journal of Neurotrauma*. Submitted.

Ahmed, F., A. Gyorgy, A. Kamnaksh, G. Ling, L. Tong, S. Parks, and D. Agoston. 2012. Time-dependent changes of protein biomarker levels in the cerebrospinal fluid after blast traumatic brain injury. *Electrophoresis*. 33:3705–11.

Ahmed, F.A., A. Kamnaksh, E. Kovesdi, J.B. Long, and D.V. Agoston. 2013. Long-term consequences of single and multiple mild blast exposure on select physiological parameters and blood-based biomarkers. *Electrophoresis*. 34:2229–33.

Andersen, W.H., and N.A. Louie. 1978. Shock tube for simulating nuclear blast durations. *Rev Sci Instrum*. 49:1729.

Aoki, Y., R. Inokuchi, M. Gunshin, N. Yahagi, and H. Suwa. 2012. Diffusion tensor imaging studies of mild traumatic brain injury: A meta-analysis. *J Neurol Neurosurg Psychiatry*. 83:870–6.

Badaut, J., S. Ashwal, and A. Obenaus. 2011. Aquaporins in cerebrovascular disease: A target for treatment of brain edema? *Cerebrovasc Dis*. 31:521–31.

Barton, G.M., and J.C. Kagan. 2009. A cell biological view of Toll-like receptor function: Regulation through compartmentalization. *Nat Rev Immunol*. 9:535–42.

Bass, C.R., M.B. Panzer, K.A. Rafaels, G. Wood, J. Shridharani, and B. Capehart. 2012. Brain injuries from blast. *Ann Biomed Eng*. 40:185–202.

Bauman, R.A., G. Ling, L. Tong, A. Januszkiewicz, D. Agoston, N. Delanerolle et al. 2009. An introductory characterization of a combat-casualty-care relevant swine model of closed head injury resulting from exposure to explosive blast. *J Neurotrauma*. 26:841–60.

Begaz, T., D.N. Kyriacou, J. Segal, and J.J. Bazarian. 2006. Serum biochemical markers for post-concussion syndrome in patients with mild traumatic brain injury. *J Neurotrauma*. 23:1201–10.

Belanger, H.G., Z. Proctor-Weber, T. Kretzmer, M. Kim, L.M. French, and R.D. Vanderploeg. 2011. Symptom complaints following reports of blast versus non-blast mild TBI: Does mechanism of injury matter? *Clin Neuropsychol*. 25:702–15.

Benga, O., and V.J. Huber. 2012. Brain water channel proteins in health and disease. *Mol Aspects Med*. 33:562–78.

Benzinger, T.L., D. Brody, S. Cardin, K.C. Curley, M.A. Mintun, S.K. Mun et al. 2009. Blast-related brain injury: Imaging for clinical and research applications: Report of the 2008 st. Louis workshop. *J Neurotrauma*. 26:2127–44.

Blennow, K., J. Hardy, and H. Zetterberg. 2012. The neuropathology and neurobiology of traumatic brain injury. *Neuron*. 76:886–99.

Bolander, R., B. Mathie, C. Bir, D. Ritzel, and P. VandeVord. 2011. Skull flexure as a contributing factor in the mechanism of injury in the rat when exposed to a shock wave. *Ann Biomed Eng*. 39:2550–9.

Brenner, L.A., H. Terrio, B.Y. Homaifar, P.M. Gutierrez, P.J. Staves, J.E. Harwood et al. 2010. Neuropsychological test performance in soldiers with blast-related mild TBI. *Neuropsychology*. 24:160–7.

Budde, M.D., A. Shah, M. McCrea, W.E. Cullinan, F.A. Pintar, and B.D. Stemper. 2013. Primary blast traumatic brain injury in the rat: Relating diffusion tensor imaging and behavior. *Front Neurol.* 4:154.

Butler, W.T. 1989. The nature and significance of osteopontin. *Connect Tissue Res.* 23:123–36.

Cairns, C.B., R.V. Maier, O. Adeoye, D. Baptiste, W.G. Barsan, L. Blackbourne et al. 2010. NIH Roundtable on Emergency Trauma Research. *Ann Emerg Med.* 56:538–50.

Calabrese, E., F. Du, R.H. Garman, G.A. Johnson, C. Riccio, L.C. Tong et al. 2014. Diffusion tensor imaging reveals white matter injury in a rat model of repetitive blast-induced traumatic brain injury. *J Neurotrauma.* 15:938–950.

Callcut, R.A., D.J. Hanseman, P.D. Solan, K.S. Kadon, N.K. Ingalls, G.R. Fortuna et al. 2012. Early treatment of blunt cerebrovascular injury with concomitant hemorrhagic neurologic injury is safe and effective. *J Trauma.* 72:338–46.

Celander, H., C.J. Clemedson, U.A. Ericsson, and H.I. Hultman. 1955. The use of a compressed air operated shock tube for physiological blast research. *Acta Physiol Scand.* 33:6–13.

Cernak, I. 2010. The importance of systemic response in the pathobiology of blast-induced neurotrauma. *Front Neurol.* 1:151.

Cernak, I., A.C. Merkle, V.E. Koliatsos, J.M. Bilik, Q.T. Luong, T.M. Mahota et al. 2011. The pathobiology of blast injuries and blast-induced neurotrauma as identified using a new experimental model of injury in mice. *Neurobiol Dis.* 41:538–51.

Cernak, I., and L.J. Noble-Haeusslein. 2010. Traumatic brain injury: An overview of pathobiology with emphasis on military populations. *J Cereb Blood Flow Metab.* 30:255–66.

Cernak, I., J. Savic, D. Ignjatovic, and M. Jevtic. 1999. Blast injury from explosive munitions. *J Trauma.* 47:96–103; discussion 103–4.

Chafi, M.S., G. Karami, and M. Ziejewski. 2010. Biomechanical assessment of brain dynamic responses due to blast pressure waves. *Ann Biomed Eng.* 38:490–504.

Chavko, M., T. Watanabe, S. Adeeb, J. Lankasky, S.T. Ahlers, and R.M. McCarron. 2011. Relationship between orientation to a blast and pressure wave propagation inside the rat brain. *J Neurosci Methods.* 195:61–6.

Chen, Y., and S. Constantini. 2013. Caveats for using shock tube in blast-induced traumatic brain injury research. *Front Neurol.* 4:117.

Chen, Y., and W. Huang. 2011. Non-impact, blast-induced mild TBI and PTSD: Concepts and caveats. *Brain Inj.* 25:641–50.

Chew, E., and R.D. Zafonte. 2009. Pharmacological management of neurobehavioral disorders following traumatic brain injury—A state-of-the-art review. *J Rehabil Res Dev.* 46:851–79.

Chodobski, A., I. Chung, E. Kozniewska, T. Ivanenko, W. Chang, J.F. Harrington et al. 2003. Early neutrophilic expression of vascular endothelial growth factor after traumatic brain injury. *Neuroscience.* 122:853–67.

Dalle Lucca, J.J., M. Chavko, M.A. Dubick, S. Adeeb, M.J. Falabella, J.L. Slack et al. 2012. Blast-induced moderate neurotrauma (BINT) elicits early complement activation and tumor necrosis factor alpha (TNFalpha) release in a rat brain. *J Neurol Sci.*

de Lanerolle, N.C., F. Bandak, D. Kang, A.Y. Li, F. Du, P. Swauger etal. 2011. Characteristics of an explosive blast-induced brain injury in an experimental model. *J Neuropathol Exp Neurol.* 70:1046–57.

Dudal, S., D. Baltrukonis, R. Crisino, M.J. Goyal, A. Joyce, K. Osterlund et al. 2013. Assay formats: Recommendation for best practices and harmonization from the global bioanalysis consortium harmonization team. *AAPS J.* 16:194–205.

Elder, G.A., N.P. Dorr, R. De Gasperi, M.A. Gama Sosa, M.C. Shaughness, E. Maudlin-Jeronimo et al. 2012. Blast exposure induces post-traumatic stress disorder-related traits in a rat model of mild traumatic brain injury. *J Neurotrauma.* 29:2564–75.

Elder, G.A., E.M. Mitsis, S.T. Ahlers, and A. Cristian. 2010. Blast-induced mild traumatic brain injury. *Psychiatr Clin North Am.* 33:757–81.

Fischer, B.L., M. Parsons, S. Durgerian, C. Reece, L. Mourany, M.J. Lowe et al. 2013. Neural activation during response inhibition differentiates blast from mechanical causes of mild to moderate traumatic brain injury. *J Neurotrauma.* 31:169–179.

French, L.M., R.T. Lange, G.L. Iverson, B. Ivins, K. Marshall, and K. Schwab. 2012. Influence of bodily injuries on symptom reporting following uncomplicated mild traumatic brain injury in US military service members. *J Head Trauma Rehabil.* 27:63–74.

Fujita, M., E.P. Wei, and J.T. Povlishock. 2012. Intensity- and interval-specific repetitive traumatic brain injury can evoke both axonal and microvascular damage. *J Neurotrauma.* 29:2172–80.

Ganpule, S., A. Alai, E. Plougonven, and N. Chandra. 2013. Mechanics of blast loading on the head models in the study of traumatic brain injury using experimental and computational approaches. *Biomech Model Mechanobiol.* 12:511–31.

Gardner, A., F. Kay-Lambkin, P. Stanwell, J. Donnelly, W.H. Williams, A. Hiles et al. 2012. A systematic review of diffusion tensor imaging findings in sports-related concussion. *J Neurotrauma.* 29:2521–38.

Garman, R.H., L.W. Jenkins, R.C. Switzer, 3rd, R.A. Bauman, L.C. Tong, P.V. Swauger et al. 2011. Blast exposure in rats with body shielding is characterized primarily by diffuse axonal injury. *J Neurotrauma.* 28:947–59.

Giza, C.C., and D.A. Hovda. 2001. The neurometabolic cascade of concussion. *J Athl Train.* 36:228–235.

Gold, E.M., D. Su, L. Lopez-Velazquez, D.L. Haus, H. Perez, G.A. Lacuesta et al. 2013. Functional assessment of long-term deficits in rodent models of traumatic brain injury. *Regen Med.* 8:483–516.

Gyorgy, A., G. Ling, D. Wingo, J. Walker, L. Tong, S. Parks et al. 2011. Time-dependent changes in serum biomarker levels after blast traumatic brain injury. *J Neurotrauma.* 28:1121–6.

Gyorgy, A.B., J. Walker, D. Wingo, O. Eidelman, H.B. Pollard, A. Molnar et al. 2010. Reverse phase protein microarray technology in traumatic brain injury. *J Neurosci Methods.* 192:96–101.

Han, K., C.L. Mac Donald, A.M. Johnson, Y. Barnes, L. Wierzechowski, D. Zonies et al. 2013. Disrupted modular organization of resting-state cortical functional connectivity in U.S. military personnel following concussive 'mild' blast-related traumatic brain injury. *Neuroimage.* 84C:76–96.

Han, K., C.L. Mac Donald, A.M. Johnson, Y. Barnes, L. Wierzechowski, D. Zonies et al. 2014. Disrupted modular organization of resting-state cortical functional connectivity in U.S. military personnel following concussive 'mild' blast-related traumatic brain injury. *Neuroimage.* 84:76–96.

Hetherington, H.P., H. Hamid, J. Kulas, G. Ling, F. Bandak, N.C. de Lanerolle et al. 2013. MRSI of the medial temporal lobe at 7 T in explosive blast mild traumatic brain injury. *Magn Reson Med.* 71:1358–1367.

Hicks, R.R., S.J. Fertig, R.E. Desrocher, W.J. Koroshetz, and J.J. Pancrazio. 2010. Neurological effects of blast injury. *J Trauma.* 68:1257–63.

Hoge, C.W., D. McGurk, J.L. Thomas, A.L. Cox, C.C. Engel, and C.A. Castro. 2008. Mild traumatic brain injury in U.S. soldiers returning from Iraq. *N Engl J Med.* 358:453–63.

Hovda, D.A. 2007. Oxidative need and oxidative capacity following traumatic brain injury. *Crit Care Med.* 35:663–4.

Hsieh, C.H., M. Frink, Y.C. Hsieh, W.H. Kan, J.T. Hsu, M.G. Schwacha et al. 2008. The role of MIP-1 alpha in the development of systemic inflammatory response and organ injury following trauma hemorrhage. *J Immunol.* 181:2806–12.

Huang, L., J.S. Coats, A. Mohd-Yusof, Y. Yin, S. Assaad, M.J. Muellner et al. 2013. Tissue vulnerability is increased following repetitive mild traumatic brain injury in the rat. *Brain Res.* 1499:109–20.

Iverson, K.M., A.M. Hendricks, R. Kimerling, M. Krengel, M. Meterko, K.L. Stolzmann et al. 2011. Psychiatric diagnoses and neurobehavioral symptom severity among OEF/OIF VA patients with deployment-related traumatic brain injury: A gender comparison. *Womens Health Issues.* 21:S210–7.

Jaffee, M.S., and K.S. Meyer. 2009. A brief overview of traumatic brain injury (TBI) and post-traumatic stress disorder (PTSD) within the Department of Defense. *Clin Neuropsychol.* 23:1291–8.

Johnson, V.E., J.E. Stewart, F.D. Begbie, J.Q. Trojanowski, D.H. Smith, and W. Stewart. 2013. Inflammation and white matter degeneration persist for years after a single traumatic brain injury. *Brain.* 136:28–42.

Jorge, R.E., L. Acion, T. White, D. Tordesillas-Gutierrez, R. Pierson, B. Crespo-Facorro et al. 2012. White matter abnormalities in veterans with mild traumatic brain injury. *Am J Psychiatry.* 169:1284–91.

Kamnaksh, A., M.D. Budde, E. Kovesdi, J.B. Long, J.A. Frank, and D. Agoston. 2014. Diffusion tensor imaging reveals acute subcortical changes after mild blast-induced traumatic brain injury. *Scientific Reports.* Submitted.

Kamnaksh, A., E. Kovesdi, S.K. Kwon, D. Wingo, F. Ahmed, N.E. Grunberg et al. 2011. Factors affecting blast traumatic brain injury. *J Neurotrauma.* 28:2145–53.

Kamnaksh, A., S.K. Kwon, E. Kovesdi, F. Ahmed, E.S. Barry, N.E. Grunberg et al. 2012. Neurobehavioral, cellular, and molecular consequences of single and multiple mild blast exposure. *Electrophoresis.* 33:3680–92.

Kanemura, Y., S. Hiraga, N. Arita, T. Ohnishi, S. Izumoto, K. Mori et al. 1999. Isolation and expression analysis of a novel human homologue of the Drosophila glial cells missing (gcm) gene. *FEBS Lett.* 442:151–6.

Kelly, J.C., E.H. Amerson, and J.T. Barth. 2012. Mild traumatic brain injury: Lessons learned from clinical, sports, and combat concussions. *Rehabil Res Pract.* 2012:371970.

Kennedy, J.E., M.A. Cullen, R.R. Amador, J.C. Huey, and F.O. Leal. 2010a. Symptoms in military service members after blast mTBI with and without associated injuries. *NeuroRehabilitation.* 26:191–7.

Kennedy, J.E., F.O. Leal, J.D. Lewis, M.A. Cullen, and R.R. Amador. 2010b. Posttraumatic stress symptoms in OIF/OEF service members with blast-related and non-blast-related mild TBI. *NeuroRehabilitation.* 26:223–31.

Koliatsos, V.E., I. Cernak, L. Xu, Y. Song, A. Savonenko, B.J. Crain et al. 2011. A mouse model of blast injury to brain: Initial pathological, neuropathological, and behavioral characterization. *J Neuropathol Exp Neurol.* 70:399–416.

Kovesdi, E., A.B. Gyorgy, S.K. Kwon, D.L. Wingo, A. Kamnaksh, J.B. Long et al. 2012a. The effect of enriched environment on the outcome of traumatic brain injury; a behavioral, proteomics, and histological study. *Front Neurosci.* 5:42.

Kovesdi, E., A. Kamnaksh, D. Wingo, F. Ahmed, N.E. Grunberg, J.B. Long et al. 2012b. Acute minocycline treatment mitigates the symptoms of mild blast-induced traumatic brain injury. *Front Neurol.* 3:111.

Kovesdi, E., J. Luckl, P. Bukovics, O. Farkas, J. Pal, E. Czeiter et al. 2010. Update on protein biomarkers in traumatic brain injury with emphasis on clinical use in adults and pediatrics. *Acta Neurochir (Wien).* 152:1–17.

Kwon, S.K., E. Kovesdi, A.B. Gyorgy, D. Wingo, A. Kamnaksh, J. Walker et al. 2011. Stress and traumatic brain injury: A behavioral, proteomics, and histological study. *Front Neurol.* 2:12.

Lange, R.T., S. Pancholi, T.A. Brickell, S. Sakura, A. Bhagwat, V. Merritt et al. 2012. Neuropsychological outcome from blast versus non-blast: Mild traumatic brain injury in U.S. military service members. *J Int Neuropsychol Soc.* 18:595–605.

Lee, Y.J., S.B. Han, S.Y. Nam, K.W. Oh, and J.T. Hong. 2010. Inflammation and Alzheimer's disease. *Arch Pharm Res.* 33:1539–56.

Lenting, P.J., C. Casari, O.D. Christophe, and C.V. Denis. 2012. von Willebrand factor: The old, the new and the unknown. *J Thromb Haemost.* 10:2428–37.

Leonardi, A.D., C.A. Bir, D.V. Ritzel, and P.J. VandeVord. 2011. Intracranial pressure increases during exposure to a shock wave. *J Neurotrauma.* 28:85–94.

Levin, H.S., E. Wilde, M. Troyanskaya, N.J. Petersen, R. Scheibel, M. Newsome et al. 2010. Diffusion tensor imaging of mild to moderate blast-related traumatic brain injury and its sequelae. *J Neurotrauma.* 27:683–94.

Liberzon, I., and C.S. Sripada. 2008. The functional neuroanatomy of PTSD: A critical review. *Prog Brain Res.* 167:151–69.

Ling, G., F. Bandak, R. Armonda, G. Grant, and J. Ecklund. 2009. Explosive blast neurotrauma. *J Neurotrauma.* 26:815–25.

Ling, G., and J. Ecklund. 2007. Neuro-critical care in modern war. *J Trauma.* 62:S102.

Ling, G.S., and J.M. Ecklund. 2011. Traumatic brain injury in modern war. *Curr Opin Anaesthesiol.* 24:124–30.

Ling, G.S., and S.A. Marshall. 2008. Management of traumatic brain injury in the intensive care unit. *Neurol Clin.* 26:409–26, viii.

Ling, G.S., S.A. Marshall, and D.F. Moore. 2010. Diagnosis and management of traumatic brain injury. *Continuum (Minneapolis Minn).* 16:27–40.

Ling, G.S., and C.J. Neal. 2005. Maintaining cerebral perfusion pressure is a worthy clinical goal. *Neurocrit Care.* 2:75–81.

Lip, G.Y., and A. Blann. 1997. von Willebrand factor: A marker of endothelial dysfunction in vascular disorders? *Cardiovasc Res.* 34:255–65.

Lipton, M.L., E. Gellella, C. Lo, T. Gold, B.A. Ardekani, K. Shifteh et al. 2008. Multifocal white matter ultrastructural abnormalities in mild traumatic brain injury with cognitive disability: A voxel-wise analysis of diffusion tensor imaging. *J Neurotrauma.* 25:1335–42.

Long, J.B., T.L. Bentley, K.A. Wessner, C. Cerone, S. Sweeney, and R.A. Bauman. 2009. Blast overpressure in rats: Recreating a battlefield injury in the laboratory. *J Neurotrauma.* 26:827–40.

Lu, D.C., Z. Zador, J. Yao, F. Fazlollahi, and G.T. Manley. 2011. Aquaporin-4 reduces post-traumatic seizure susceptibility by promoting astrocytic glial scar formation in mice. *J Neurotrauma.*

Lu, J., K.C. Ng, G. Ling, J. Wu, D.J. Poon, E.M. Kan et al. 2012. Effect of blast exposure on the brain structure and cognition in Macaca fascicularis. *J Neurotrauma.* 29:1434–54.

Luethcke, C.A., C.J. Bryan, C.E. Morrow, and W.C. Isler. 2011. Comparison of concussive symptoms, cognitive performance, and psychological symptoms between acute blast-versus non-blast-induced mild traumatic brain injury. *J Int Neuropsychol Soc*. 17:36–45.

Mac Donald, C., A. Johnson, D. Cooper, T. Malone, J. Sorrell, J. Shimony et al. 2013. Cerebellar white matter abnormalities following primary blast injury in US military personnel. *PLoS One*. 8:e55823.

Mac Donald, C.L., A.M. Johnson, D. Cooper, E.C. Nelson, N.J. Werner, J.S. Shimony et al. 2011. Detection of blast-related traumatic brain injury in U.S. military personnel. *N Engl J Med*. 364:2091–100.

Manley, G.T., G. Rosenthal, M. Lam, D. Morabito, D. Yan, N. Derugin et al. 2006. Controlled cortical impact in swine: Pathophysiology and biomechanics. *J Neurotrauma*. 23:128–39.

Margulies, S.S., L.E. Thibault, and T.A. Gennarelli. 1990. Physical model simulations of brain injury in the primate. *J Biomech*. 23:823–36.

Marion, D.W., K.C. Curley, K. Schwab, and R.R. Hicks. 2011. Proceedings of the military mTBI Diagnostics Workshop, St. Pete Beach, August 2010. *J Neurotrauma*. 28:517–26.

Masel, B.E., R.S. Bell, S. Brossart, R.J. Grill, R.L. Hayes, H.S. Levin et al. 2012. Galveston Brain Injury Conference 2010: Clinical and experimental aspects of blast injury. *J Neurotrauma*. 29:2143–71.

Matchett, G.A., M.W. Allard, R.D. Martin, and J.H. Zhang. 2009. Neuroprotective effect of volatile anesthetic agents: Molecular mechanisms. *Neurol Res*. 31:128–34.

Matthews, S.C., I.A. Strigo, A.N. Simmons, R.M. O'Connell, L.E. Reinhardt, and S.A. Moseley. 2011. A multimodal imaging study in U.S. veterans of Operations Iraqi and Enduring Freedom with and without major depression after blast-related concussion. *Neuroimage*. 54(Suppl 1):S69–75.

McAllister, T.W. 2009. Psychopharmacological issues in the treatment of TBI and PTSD. *Clin Neuropsychol*. 23:1338–67.

McMahon, P.J., A.J. Hricik, J.K. Yue, A.M. Puccio, T. Inoue, H. Lingsma et al. 2014. Symptomatology and functional outcome in mild traumatic brain injury: Results from the prospective TRACK-TBI Study. *J Neurotrauma*. 31:26–33.

McManus, C.M., C.F. Brosnan, and J.W. Berman. 1998. Cytokine induction of MIP-1 alpha and MIP-1 beta in human fetal microglia. *J Immunol*. 160:1449–55.

Mediavilla Varas, J., M. Philippens, S.R. Meijer, A.C. van den Berg, P.C. Sibma, J.L. van Bree et al. 2011. Physics of IED blast shock tube simulations for mTBI Research. *Front Neurol*. 2:58.

Mendez, M.F., E.M. Owens, G. Reza Berenji, D.C. Peppers, L.J. Liang, and E.A. Licht. 2013. Mild traumatic brain injury from primary blast vs. blunt forces: Post-concussion consequences and functional neuroimaging. *NeuroRehabilitation*. 32:397–407.

Meterko, M., E. Baker, K.L. Stolzmann, A.M. Hendricks, K.D. Cicerone, and H.L. Lew. 2012. Psychometric assessment of the Neurobehavioral Symptom Inventory-22: The structure of persistent postconcussive symptoms following deployment-related mild traumatic brain injury among veterans. *J Head Trauma Rehabil*. 27:55–62.

Nakagawa, A., G.T. Manley, A.D. Gean, K. Ohtani, R. Armonda, A. Tsukamoto et al. 2011. Mechanisms of primary blast-induced traumatic brain injury: Insights from shock-wave research. *J Neurotrauma*. 28:1101–19.

Papadopoulos, M.C., S. Krishna, and A.S. Verkman. 2002. Aquaporin water channels and brain edema. *Mt Sinai J Med*. 69:242–8.

Papadopoulos, M.C., S. Saadoun, D.K. Binder, G.T. Manley, S. Krishna, and A.S. Verkman. 2004. Molecular mechanisms of brain tumor edema. *Neuroscience*. 129:1011–20.

Peskind, E.R., D. Brody, I. Cernak, A. McKee, and R.L. Ruff. 2013. Military- and sports-related mild traumatic brain injury: Clinical presentation, management, and long-term consequences. *J Clin Psychiatry*. 74:180–8; quiz 188.

Peskind, E.R., E.C. Petrie, D.J. Cross, K. Pagulayan, K. McCraw, D. Hoff et al. 2011. Cerebrocerebellar hypometabolism associated with repetitive blast exposure mild traumatic brain injury in 12 Iraq war Veterans with persistent post-concussive symptoms. *Neuroimage*. 54(Suppl 1):S76–82.

Peterson, A.L., C.A. Luethcke, E.V. Borah, A.M. Borah, and S. Young-McCaughan. 2011. Assessment and treatment of combat-related PTSD in returning war veterans. *J Clin Psychol Med Settings*. 18:164–75.

Petrie, E.C., D.J. Cross, V.L. Yarnykh, T. Richards, N.M. Martin, K. Pagulayan et al. 2013. Neuroimaging, behavioral, and psychological sequelae of repetitive combined blast/impact mild traumatic brain injury in Iraq and Afghanistan War veterans. *J Neurotrauma*.

Povlishock, J.T. 2013. The window of risk in repeated head injury. *J Neurotrauma*. 30:1.

Prins, M.P.D., D. Alexander, C.C. Giza, and D. Hovda. 2012. Repeat mild traumatic brain injury: Mechanisms of cerebral vulnerability. *J Neurotrauma*. 30:30–38.

Rainey, T., M. Lesko, R. Sacho, F. Lecky, and C. Childs. 2009. Predicting outcome after severe traumatic brain injury using the serum S100B biomarker: Results using a single (24h) time-point. *Resuscitation*. 80:341–5.

Rauch, A., N. Wohner, O.D. Christophe, C.V. Denis, S. Susen, and P.J. Lenting. 2013. On the versatility of von Willebrand factor. *Mediterr J Hematol Infect Dis*. 5:e2013046.

Raymont, V., A.M. Salazar, F. Krueger, and J. Grafman. 2011. "Studying injured minds" - the Vietnam head injury study and 40 years of brain injury research. *Front Neurol*. 2:15.

Risling, M., and J. Davidsson. 2012. Experimental animal models for studies on the mechanisms of blast-induced neurotrauma. *Front Neurol*. 3:30.

Risling, M., S. Plantman, M. Angeria, E. Rostami, B.M. Bellander et al. 2011. Mechanisms of blast induced brain injuries, experimental studies in rats. *Neuroimage*. 54(Suppl 1):S89–97.

Rosenfeld, J.V., and N.L. Ford. 2010. Bomb blast, mild traumatic brain injury and psychiatric morbidity: A review. *Injury*. 41:437–43.

Rubovitch, V., M. Ten-Bosch, O. Zohar, C.R. Harrison, C. Tempel-Brami, E. Stein et al. 2011. A mouse model of blast-induced mild traumatic brain injury. *Exp Neurol*. 232:280–9.

Saatman, K.E., A.C. Duhaime, R. Bullock, A.I. Maas, A. Valadka, and G.T. Manley. 2008. Classification of traumatic brain injury for targeted therapies. *J Neurotrauma*. 25:719–38.

Schultz, B.A., D.X. Cifu, S. McNamee, M. Nichols, and W. Carne. 2011. Assessment and treatment of common persistent sequelae following blast induced mild traumatic brain injury. *NeuroRehabilitation*. 28:309–20.

Selvan, V., S. Ganpule, N. Kleinschmit, and N. Chandra. 2013. Blast wave loading pathways in heterogeneous material systems-experimental and numerical approaches. *J Biomech Eng*. 135:61002–14.

Seok, J., H.S. Warren, A.G. Cuenca, M.N. Mindrinos, H.V. Baker, W. Xu et al. 2013. Genomic responses in mouse models poorly mimic human inflammatory diseases. *Proc Natl Acad Sci U S A*. 110:3507–12.

Shin, T. 2012. Osteopontin as a two-sided mediator in acute neuroinflammation in rat models. *Acta Histochem*. 114:749–54.

Simmons, A.N., and S.C. Matthews. 2012. Neural circuitry of PTSD with or without mild traumatic brain injury: A meta-analysis. *Neuropharmacology*. 62:598–606.

Skotak, M., F. Wang, A. Alai, A. Holmberg, S. Harris, R.C. Switzer III et al. 2013. Rat injury model under controlled field-relevant primary blast conditions: Acute response to a wide range of peak overpressures. *J Neurotrauma*. 30:1147–60.

Slobounov, S., J. Bazarian, E. Bigler, R. Cantu, M. Hallett, R. Harbaugh et al. 2013. Sports-related concussion: Ongoing debate. *Br J Sports Med*. 48:75–6.

Sotak, C.H. 2002. The role of diffusion tensor imaging in the evaluation of ischemic brain injury—A review. *NMR Biomed*. 15:561–9.

Thelin, E.P., L.K. Johannesson, D.W. Nelson, and B.M. Bellander. 2013a. S100B is an important outcome predictor in traumatic brain injury. *J Neurotrauma*. 30:519–528.

Thelin, E.P., D.W. Nelson, and B.M. Bellander. 2013b. Secondary peaks of S100B in serum relate to subsequent radiological pathology in traumatic brain injury. *Neurocrit Care*. 20:217–29.

Thompson, J.M., K.C. Scott, and L. Dubinsky. 2008. Battlefield brain: Unexplained symptoms and blast-related mild traumatic brain injury. *Can Fam Physician*. 54:1549–51.

Trudeau, D.L., J. Anderson, L.M. Hansen, D.N. Shagalov, J. Schmoller, S. Nugent et al. 1998. Findings of mild traumatic brain injury in combat veterans with PTSD and a history of blast concussion. *J Neuropsychiatry Clin Neurosci*. 10:308–13.

Vanderploeg, R.D., H.G. Belanger, R.D. Horner, A.M. Spehar, G. Powell-Cope, S.L. Luther et al. 2012. Health outcomes associated with military deployment: Mild traumatic brain injury, blast, trauma, and combat associations in the Florida National Guard. *Arch Phys Med Rehabil*. 93:1887–95.

Venero, J.L., M.L. Vizuete, A. Machado, and J. Cano. 2001. Aquaporins in the central nervous system. *Prog Neurobiol*. 63:321–36.

Vollmer, J. 2006. TLR9 in health and disease. *Int Rev Immunol*. 25:155–81.

Vos, P.E., B. Jacobs, T.M. Andriessen, K.J. Lamers, G.F. Borm, T. Beems et al. 2010. GFAP and S100B are biomarkers of traumatic brain injury: An observational cohort study. *Neurology*. 75:1786–93.

Wilk, J.E., R.K. Herrell, G.H. Wynn, L.A. Riviere, and C.W. Hoge. 2012. Mild traumatic brain injury (concussion), posttraumatic stress disorder, and depression in U.S. soldiers involved in combat deployments: Association with postdeployment symptoms. *Psychosom Med*. 74:249–57.

Wilk, J.E., J.L. Thomas, D.M. McGurk, L.A. Riviere, C.A. Castro, and C.W. Hoge. 2010. Mild traumatic brain injury (concussion) during combat: Lack of association of blast mechanism with persistent postconcussive symptoms. *J Head Trauma Rehabil*. 25:9–14.

Wolf, S.J., V.S. Bebarta, C.J. Bonnett, P.T. Pons, and S.V. Cantrill. 2009. Blast injuries. *Lancet*. 374:405–15.

Woodcock, T., and M.C. Morganti-Kossmann. 2013. The role of markers of inflammation in traumatic brain injury. *Front Neurol*. 4:18.

Yadav, A., V. Saini, and S. Arora. 2010. MCP-1: Chemoattractant with a role beyond immunity: A review. *Clin Chim Acta*. 411:1570–9.

Yan, J., and H. Jiang. 2013. Dual effects of ketamine: Neurotoxicity versus neuroprotection in anesthesia for the developing brain. *J Neurosurg Anesthesiol*. 26:155–60.

24 Magnetic Resonance Imaging Application in the Area of Mild and Acute Traumatic Brain Injury

Implications for Diagnostic Markers?

Arnold Toth

CONTENTS

24.1 INTRODUCTION

Mild traumatic brain injury (mTBI) causes brain damage generally invisible for conventional imaging methods. Its diagnosis mostly relies on the patient's history, subjective complaints and neuropsychological status. Long-term complication development is just scarcely linked to these clinical factors. Imaging markers would contribute not only to the diagnosis and prognosis of mTBI, but to the understanding of its pathomechanisms as well. Advanced Magnetic Resonance Imaging (MRI) methods offer new insights to the background of mTBI. The microscopic scale white matter disease following mTBI can be evaluated by Diffusion Tensor Imaging (DTI) and Susceptibility Weighted Imaging (SWI). It's possible to detect subtle atrophy using advanced volumetric analyses of submillimeter resolution images, while biochemical aspects can be assessed by MR spectroscopy. Functional MRI (fMRI) provides information of altered and compensational brain activity due to injury. Further advanced MRI techniques and perspectives are discussed as well in the chapter.

Mild traumatic brain injury is a special field calling for advanced imaging methods, first of all magnetic resonance imaging (MRI) methods. The numerous definitions for mTBI and inconsistent nomenclature (e.g., concussion, minor head injury, minor brain injury, minor head trauma) show that the confinement of this clinical category is challenging. Diagnoses are mostly based on symptoms and self-reported history, yet no generally deployable objective marker exists; however, recent attempts for both imaging and biomarkers are promising. The most widely accepted criteria for mTBI are blunt trauma, Glasgow Coma Scale of 13–15, brief period (<30 minutes) of loss of consciousness, and brief period (<24 hours) of posttraumatic amnesia (Carroll et al., 2004; Mild Traumatic Brain Injury Committee, 1993). Inclusion of cases in which computed tomography (CT) scans show trauma-related pathology is debated; mostly these cases are excluded (i.e., mTBI is considered to be CT scan–negative). Indeed, CT scans are normal in about 90% of the cases fulfilling the aforementioned criteria; this fact is somewhat paradoxical considering the sometimes alarming neuropsychological signs and symptoms of these patients. The categorization of CT-positive mTBI cases as "complicated mTBI" (e.g., finding of focal contusion) seems to be useful because these cases generally deserve extra attention acutely; however, focal lesions may not predict the outcome 3 months after injury (Lannsjo et al., 2013). Clinical variables together with age may be stronger predictors for outcome (Jacobs et al., 2010).

Beyond the issues with definitions and diagnostic criteria, the greater problem from a clinical point of view is that the severity of the complaints or neuropsychological deficits at admission are only very scarcely linked to the prognosis and true severity of the injury. This means that, for example, loss of consciousness or the length of posttraumatic amnesia is not necessarily associated with the actual mechanical force suffered or the chance of developing persistent posttraumatic complaints. It is important to keep in mind that mTBI can be interpreted as "mild" only when compared with moderate or severe TBI, which is known to be life-threatening. In itself, mTBI is also potentially dangerous because in 10%–30% of cases it may lead to serious long-term complications significantly worsening life quality and disabling work or social interactions (see the following section). Additionally, considering its extremely high incidence (up to 500/100,000), mTBI deserves to be called a public health problem. Long-term complications may include such persisting acute symptoms as headache, dizziness, nausea, or concentration/memory problems, although new complaints may also develop in time such as depression, sleeping disorders, or anxiety. Patients suffering from repetitive mTBIs are especially exposed to long-term complications. This makes the decision of letting one return to work or return to play (in case of sports concussion) really serious. Still, without enough objective information of mTBI-related mechanisms, the background of long-term complications is not fully understood. It is debated whether these complications are a result of psychological or organic factors.

Because mTBI-related pathomechanisms remain elusive, therapeutic possibilities are also going to be limited. Presently, the only widely accepted treatment is rest, both physical and cognitive. Medications used merely serve as symptomatic treatment and their use is generally based on local anecdotal evidence (Meehan, 2011).

One reason why identification of the details of related mechanisms has been held up is that, generally, histopathological examination is not possible. Human histopathological observations are very scarce and are from the rare cases of mTBI accompanying fatal conditions. The vast majority of histological information and data about pathomechanism have been obtained from animal (mostly rodent) mTBI models (for an overview on neuropathology in mTBI, see Bigler and Maxwell, 2012). These models allow an infinite range of controlled observations on different elements of brain injury and have provided irreplaceable findings. However, all mTBI animal models suffer from the problem that mTBI can only be interpreted truthfully at the level of the human brain's complexity. Most of the neuropsychological deficits characterizing this condition are hardly transposable to animals. For example, a mainstay element of the mTBI definition is posttraumatic amnesia. To be simplistic, mTBI, grossly, is the damage of a theoretical fraction of the human brain that an animal does not even have.

These concerns regarding mTBI diagnosis, prognosis evaluation, and pathomechanism have together called for noninvasive, highly sensitive contemporary imaging tools. Among these are single photon emission computed tomography, positron emission tomography, and MRI, the latter of which has become the most widely applied in mTBI studies because it is the most accessible, multimodal, and the least harmful because no ionizing radiation is used and, generally, no contrast agent has to be administered. Multimodality in MRI means that this method, depending on actual acquisition parameters, can provide different insights to the complex pathology of the damaged brain, from detailed microstructural to functional components. Unlike classic neuroradiological scan evaluation, assessment of advanced MRI data is based often on quantitative and statistical methods. This means that although visible images are created, the true information is held in the underlying numbers that allow objective, often group-wise analytical processes.

One of the most promising methods of the field is diffusion tensor imaging (DTI) that is able to detect change in water microcompartments due to microstructural pathology as axonal deformation and swelling. Focal microscopic bleeds developing as part of diffusion axonal imaging are most successfully detectable by susceptibility-weighted imaging (SWI), a method exploiting the magnetic property of iron. Recent efforts seem to validate the clinical importance of these methods (see the following section). High-resolution, three-dimensional, T1-weighted images allow precise volumetric analyses to be performed shedding light on subtle changes in the brain macrostructure because of, for example, edematic and atrophic mechanisms after injury. Beyond the advanced investigation of brain structure, magnetic resonance spectroscopy (MRS) offers information of the metabolic state of the brain by measuring specific magnetic signals from mainly the 1H nuclei in different metabolites. Getting to the functional level, the effect of injury on brain functions such as perception or cognitive tasks (memory and concentration functions are typically affected) can be investigated by functional MRI (fMRI). The following sections provide a brief overview of the benefits and also challenges of using these methods in the mTBI field.

24.2 ROUTINE MRI METHODS IN mTBI

Here, routine MRI refers to magnetic resonance techniques such as T1-, T2-, and spin density–weighted imaging and also fluid-attenuated inversion recovery assessed conventionally by a neuroradiologist. These modalities hold similar information on the injured brain as CT scans, so traumatic pathology such as epidural or subdural bleeding, contusion, or skull fracture can be identified. Nevertheless, CT is more appropriate for this purpose because it is more widely accessible, faster, and less expensive; furthermore, acute bleeding and skull fractures are better outlined. Hence it can be stated that CT is still the imaging method to be chosen to disclose traumatic conditions requiring neurosurgical intervention. The only exceptions may be cases of children or young women where ionizing irradiation has to be avoided. Then MRI may be considered as the first-line imaging tool. However, if the question is the presence of more subtle injuries such as small contusions or microscopic bleeding (hemorrhagic axonal injury), routine MRI is the preferable tool because it is far more sensitive to such lesions (Yuh et al.,

2013). Unfortunately, to date, the clinical value of these focal lesions is debated; no general conclusions can be drawn on how these lesions can be attributed to injury severity within the spectrum of mTBI or the outcome (Hughes et al., 2004). However, a recent study drew attention to the significance of routine MRI features by showing that lesion number accompanied by proper controlling for demographic, socioeconomic, and clinical features improved outcome predictions (Yuh et al., 2013). Although MRI is not yet considered cost beneficial for mTBI, it may become so in the future by developing cheaper magnetic resonance instruments (e.g., head-only MRI).

24.3 ADVANCED MRI METHODS IN mTBI

24.3.1 Overview

Thanks to the nonstop technical development of MRI, newer and newer methods are becoming available every year. Some of these are decades old but are often called advanced because their capabilities are still not fully discovered and generally need special operation and evaluation. Each can reveal certain special features of brain structure, function, and pathology, for example MRS shows metabolic profile, or DTI reveals microstructural condition. The object for research groups is to explore the clinical effectiveness of these modalities or to find new components of pathomechanism and their correlations in the field of mTBI. This section of the chapter discuss the most important advanced methods for mTBI such as DTI, SWI, fMRI, volumetric analysis, and MRS.

24.3.2 DTI: The "Fingerprint" of White Matter

DTI measures Brownian movement of water molecules and applies at least six diffusion gradient directions and thus is able to provide information of both extent and directionality of diffusion (Pierpaoli et al., 1996). Fractional anisotropy (FA) refers to the degree of directionality, calculated from the ratio of eigenvalues of the diffusion tensor, while mean diffusivity (MD) or the synonym apparent diffusion coefficient refers to the overall, directionally indifferent mobility of water molecules. The character of diffusivity in brain is widely accepted to be associated with fiber tracts (i.e., axons and myelin sheath) (Beaulieu, 2002). In the classic theory, in a direction parallel to axons, diffusion is greater than in the direction perpendicular to them, because cell membranes and other structures restrict diffusion. This is why white matter tracts can be visualized by DTI tractography. Though this concept has never been exactly confirmed, tremendous empirical data show that it works quite well. FA and MD are very sensitive parameters indicating subtle alterations of white matter because they are likely to be influenced by axon density, diameter, and continuity; myelin content; myelin sheath thickness; and interstitial water content. This way, DTI can reveal differences among healthy subjects as well (i.e., gender, aging, or education are known to have an effect on DTI parameters).

It is not surprising therefore that DTI is able to detect axonal pathology in severe TBI; however, it might be surprising that to date it has been accepted as fact that DTI finds white matter abnormalities in mTBI as well. This pathology includes axonal disintegration related to shear-strain deformation of the fiber structure as a mild version of diffuse axonal injury. Changes in water microcompartments because of vasogen or cytotoxic edema may also be present and are visible for DTI (Peled, 2007).

A large cohort of studies investigated diffusion in mTBI focusing on several different relationships (i.e., age, acute or chronic phase, clinical symptoms or neuropsychological tests, recovery, or sport- and combat-related injuries).

Many investigations on mTBI found reduced FA or elevated MD (apparent diffusion coefficient) in mildly injured patients (Figure 24.1) and often interpreted the findings as reduced integrity and misalignment of axonal and myelin structures because of shear-strain forces, including local expansion of axonal cylinder or axonal disconnection (Arfanakis et al., 2002; Inglese et al., 2005; Lipton et al., 2008; Miles et al., 2008; Nakayama et al., 2006). Later studies observed oppositely elevated FA or reduced MD acutely after mild injury over several white matter regions (Bazarian et al., 2007; Chu et al., 2010; Mayer et al., 2010). The suggested underlying mechanism was cytotoxic edema; in this condition, the injury altered function of gated ion channels, resulting in intracellular swelling and decreased extracellular water that may cause reduced radial diffusivity (Peled, 2007; Rosenblum, 2007; Wilde et al., 2008). The output yielded by

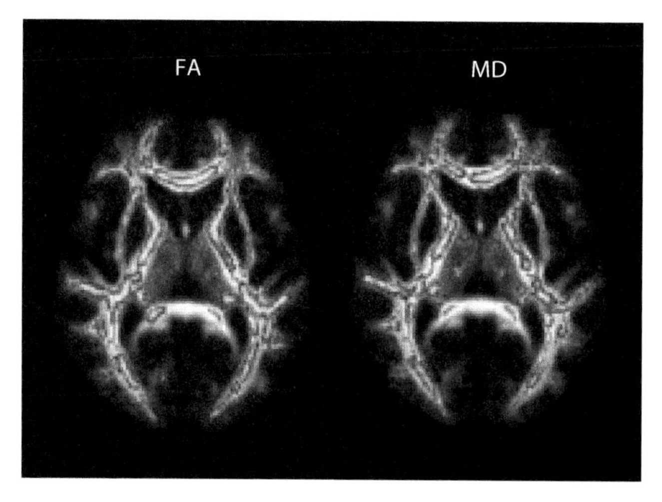

FIGURE 24.1 Voxel-wise statistical comparison of DTI parameters between an acute-phase mTBI group (n = 15) and an age-, sex-, and education-matched healthy control group. Red-yellow indicates voxels of significant (corrected $p < 0.05$) difference between the groups: FA refers to fractional anisotropy, which was decreased in the mTBI group compared with controls; MD refers to mean diffusivity, which was increased in the mTBI group compared with controls. Green voxels indicate white matter tract midlines where no significant difference was found. Background image is an average FA map of the two groups. To achieve these data and images, tract-based spatial statistics, which is part of the FSL software library, was used. (Courtesy of FMRIB Oxford).

DTI may show a summarized effect of the two basic mechanisms: microstructural disintegration and cytotoxic edema. A recent study also found bidirectional irregularities in DTI parameters after injury (Lipton et al., 2012). The actual dominance of these substantial mechanisms in the white matter may theoretically depend on temporal and spatial factors, attributes of the patient, and the circumstances of injury (Obenaus et al., 2007). Future research should shed light on the proper interpretation of the different diffusion indices possibly by focusing also on less robust parameters such as eigenvectors and eigenvalues and studying patient groups by well-homogenized study parameters.

Findings of follow-up studies are also various; some longitudinal studies revealed partial normalization of DTI indices after different periods (Arfanakis et al., 2002; Mayer et al., 2010; Rutgers et al., 2008), whereas other investigations indicated traumatic microstructural alteration to be more permanent (Bendlin et al., 2008) or even to evolve over time (MacDonald et al., 2011). It is likely that DTI parameters change quite dynamically after injury; this has been shown in a study comparing the acute and subacute phases that found dramatic differences in both FA and MD values (Toth et al., 2013). There are promising observations on the relationship of DTI findings with cognitive or psychological dysfunction (Miles et al., 2008; Niogi et al., 2008) and clinical outcome (Messe et al., 2011), especially in moderate to severe cases (Rutgers et al., 2008; Sidaros et al., 2008).

Besides temporal features of white matter changes, spatial characteristics also imply several questions. Though axonal pathology is regarded as mostly diffuse in mTBI, it is clear that some regions must be more vulnerable because of general mechanical and anatomical rules even if considering subject variability. For instance, posterior corpus callosum seems to be the most susceptible to mild injuries (Aoki et al., 2012). In some cases, injury of a certain white matter tract is obviously associated with accompanying complaints. In other situations, the complaint can be linked to the extent of overall injury. Based on clinical history, it is impossible to exactly draw the model of biomechanical forces and thus predict the predominant site (fascicle) of injury (this can be performed on sports concussion cases where video recordings are available and are analyzed by specialized computer algorithms). However, DTI offers retrospective assessment of the manifestation peak sites of axonal injury that can be correlated with occurring signs and symptoms. For example, in case of damaged tracts beginning from the hippocampal areas, impaired memory functions may be more easily understood. If imaging is performed in the chronic phase, it is challenging to decide if DTI abnormalities are a cause or a result of the clinical disorders because mTBI-independent disorders (e.g., depression) themselves may also be associated with DTI abnormalities (Maller et al., 2012). The specificity of posttraumatic neuropsychological testing for DTI resulting in mTBI is a topic of debate because non-TBI factors may affect both (Larrabee et al., 2013).

Although DTI brings up numerous issues to be solved—most importantly to be able to provide clinically useful information at subject level—it possibly could be the first advanced MRI method involved in the clinical arsenal for mTBI.

24.3.3 SWI Deployed to Find Microscopic Bleeding

SWI is particularly sensitive in detecting both intravascular venous deoxygenated blood and extravascular blood products (Haacke et al., 2009; Reichenbach et al., 1997). This method exploits the magnetic property of heme iron: iron causes local magnetic field distortion altering both T2 star relaxation times and phase data that are measurable and can be visualized by proper MRI sequencing. Anatomical structures do not appear well on these images, for example contrast is low between the cortex and white matter or cerebrospinal fluid. In turn, iron content, most importantly bleeding, is shown pronouncedly as hypodense (black) lesions (Figure 24.2).

SWI was shown to be the most sensitive modality for detecting microhemorrhage, primarily in pediatric TBI of mixed severity (Babikian et al., 2005; Tong et al., 2008). SWI does not only reveal more focal lesions in a certain patient than other MRI modalities such as T2-weighted imaging, fluid-attenuated inversion recovery, gradient-recalled echo, or CT does, but SWI hemorrhagic lesions are more unmistakable than lesions on other imaging modalities. This is supported by interrater-reliability data (Geurts et al., 2012).

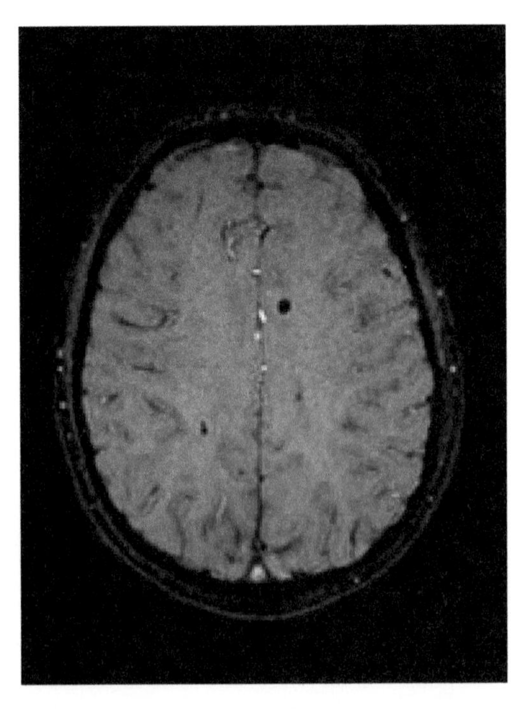

FIGURE 24.2 SWI of an mTBI patient. Hypodense lesions (black dots) indicate microscopic bleeding (i.e., hemorrhagic axonal injury). This patient had a Glasgow Coma Scale score of 14 at admission, reported loss of consciousness, and also posttraumatic amnesia.

It was possible to explore correlation of SWI lesion number, volume, and location with neuropsychological functioning (Babikian et al., 2005) or with outcome (Tong et al., 2008) in children. A pediatric patient can be reliably placed in the spectrum of mild-to-severe TBI based on SWI, with a better prediction of cognitive outcome (Beauchamp et al., 2013).

In contrast, adult data and especially studies focusing strictly on mTBI are limited. A study proved the superiority of SWI over CT and conventional MRI in sensitivity to microhemorrhage in a group of adults with dominantly severe TBI patients (Akiyama et al., 2009).

Microhemorrhages do not seem that frequent in mTBI. Based on a study of amateur and professional boxers (Hasiloglu et al., 2011) and on experiences of SWI using diagnostic centers, SWI lesions in mTBI occur in about 1 or 2 of 10 patients. Large amounts of observations were needed to draw correlations with clinical parameters, such as outcomes besides this lesion occurrence. An attempt to do was successfully done by Yuh et al., who found that four hemorrhagic lesions detected early after injury can be regarded as the threshold for predicting poorer 3-month outcomes (Yuh et al., 2013).

24.3.4 Atrophy, Edema Revealed by Volumetric Analysis

Brain volume changes such as edema and chronically developing atrophy are known mechanisms in severe TBI. These dramatic volume disorders can be evaluated well by classical neuroradiological methods such as manual morphometric measurements. Different manifestations of brain atrophy after injury were identified in a large number of morphometric studies conducted on mixed (mainly moderate to severe) TBI populations (Bigler et al., 1997; Fearing et al., 2008; Kim et al., 2008; Wilde et al., 2005). Injury severity or cognitive function was correlated with atrophy rate (Bigler et al., 2002; Levine et al., 2008); in one group, outcome was found to be independent from atrophy (Bendlin et al., 2008). The association of posttraumatic stress disorder with atrophy of whole brain (Woodward et al., 2007), corpus callosum (Villarreal et al., 2004), anterior cingulum (Kitayama et al., 2006), and hippocampus (Villarreal et al., 2002) was presented in some studies.

However, in mTBI, the volume changes are not that apparent (i.e., if present at all, they are too subtle for routine neuroradiology methods to detect them), hence far fewer data are available. Furthermore, an important point is that when comparing two healthy subjects' brain volumes, even if they are normalized for total intracranial volumes, we can find great differences between structure volumes. For instance, if we compare the normalized ventricular volume of two healthy subjects of the same age, gender, or education, a two-fold or even larger difference can be found.

Hence, it is clear that detecting a volume change of a few percentages and regarding it as a trustworthy consequence of mTBI is quite problematic. For such investigations, structural

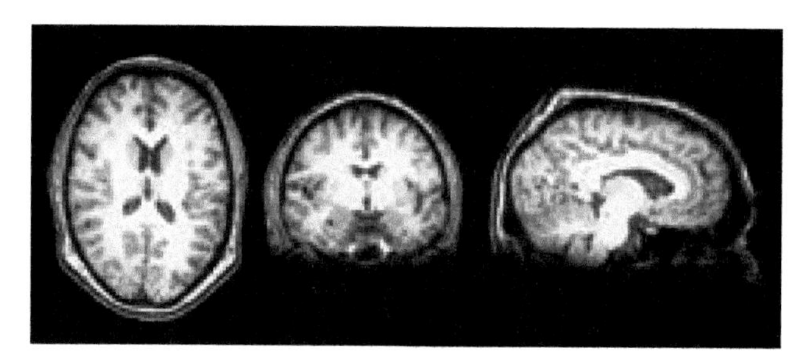

FIGURE 24.3 This figure demonstrates brain tissue edge displacement of an mTBI patient occurring between the acute (48 hours) and subacute phase (32 days) after injury. Blue indicates brain volume decrease, whereas red indicates brain volume increase along tissue borders. It can be seen that, around the ventricles, virtually only blue can be observed. This means that more than 1 month after injury, this patient's ventricles were expanding. This can be a result of recovering initial edema or developing atrophy as well. These images were generated using the "SIENA" two time point estimation tool, part of the FSL software library. These volume changes can be quantified when necessary (Courtesy of FMRIB Oxford).

images of the highest possible resolution are needed, in addition to proper quantitative-automatic volumetric analysis algorithms and a high-enough subject number (Figure 24.3). Follow-up arrangements are also advantageous (Ross, 2011). Among the few volumetric studies focusing on homogenous mTBI groups, MacKenzie et al. found global atrophy developed in 3 months in a group of mild and moderate injured patients that was correlated with length of consciousness (MacKenzie et al., 2002); the presence of atrophy in mTBI was supported by others as well (Zhou et al., 2013). Gray matter atrophy was detected years after injury by Cohen et al. in an mTBI group (Cohen et al., 2007). Messe et al. showed gray matter volume to be decreased, but not to be predictive for outcome (Messe et al., 2011). A study concentrating on the acute-subacute phase of mTBI detected cortical gray matter and ventricle volume changes over the first month after injury proposed to be due to recovery from an initial subtle edema (Toth et al., 2013).

Although group-wise studies providing valuable data of mTBI volumetric changes are gathering, the single time point volumetric assessment at the subject level is not informative enough in mTBI. However, follow-up volumetric analysis was shown to possibly be beneficial in single cases as well (Ross et al., 2013).

24.3.5 Functional MRI: Closer to Understanding Cognitive Disorders

fMRI detects local hemodynamic changes following increased metabolic rate in neural activity, by measuring the blood oxygen-level dependent (BOLD) contrast (Nair, 2005; Ogawa et al., 1992). Specific cognitive, motor, memory tasks, or sensory stimulation are repeated and the associated BOLD signals are compared (Moonen, 2000). Functional connectivity investigation reveals brain areas with correlated fluctuations (i.e., coupled functionality) during an experimental task or resting state (in the absence of any active task or external stimulus) (Rogers et al., 2007).

Cognitive disorders such as impaired processing speed, concentration, and memory problems are typical in mTBI. fMRI is hence a plausible tool in mTBI to better understand underlying neural function abnormalities and plasticity or to detect specific mTBI-related functional patterns (McDonald et al., 2012). An outstanding advance of fMRI is the opportunity to measure actual task performance of a patient simultaneously with functional imaging. For example, during a memory task, the number of correct answers or reaction speed can be quantified. This helps the observer interpret the functional imaging findings (e.g., additional activations in an injured patient performing as well as a control subject is likely to mean a compensational neural recruitment).

Most studies concentrated on memory functions, especially working memory (Chen et al., 2004; McAllister et al., 1999; McAllister et al., 2001; Pardini et al., 2010; Perlstein et al., 2004; Smits et al., 2009). The altered activation of primarily the dorsolateral prefrontal cortex was suggested to underlie working memory dysfunctions. A few studies focused on spatial working memory or declarative/episodic memory (Figure 24.4) (Russell et al., 2011; Slobounov et al., 2010; Stulemeijer et al., 2010).

In these studies, various injury-related changes of BOLD signal levels and distribution were detected. Some investigators found attenuated activation in mTBI patients compared with healthy subjects that may be a result of an injured neural network (Chen et al., 2004; Gosselin et al., 2011; Lovell et al., 2007; Perlstein et al., 2004). Others reported increased or additional activation (i.e., involvement of new, normally silent areas) (McAllister et al., 1999; Slobounov et al., 2010; Smits et al., 2009). The latter findings are generally considered to be an effect of neural reorganization or functional accommodation. The discrepancy across these findings of hypo- and hyperactivation patterns in mTBI memory tasks was somewhat resolved by a recent study pointing out the significance of a working memory task being considered a continuous or a discrete task (Bryer et al., 2013).

Correlation between BOLD signal changes and neuropsychological findings or task performance was proposed by some studies (Lovell et al., 2007; Smits et al., 2009; Stulemeijer et al., 2010); however, the alteration of BOLD signal distribution was observed independently of clinical complaints or performance as well (McAllister et al., 1999; Slobounov et al., 2010). Because fMRI can reveal abnormal memory functional activity beside normal behavioral performance of the patients, it was suggested to be a more sensitive tool for neuropsychological evaluation than classical tests (Chen et al., 2012). A relatively low number of longitudinal studies showed the cessation of symptoms over time to be associated with the normalization of cortical patterns (Chen et al., 2008; Lovell et al., 2007).

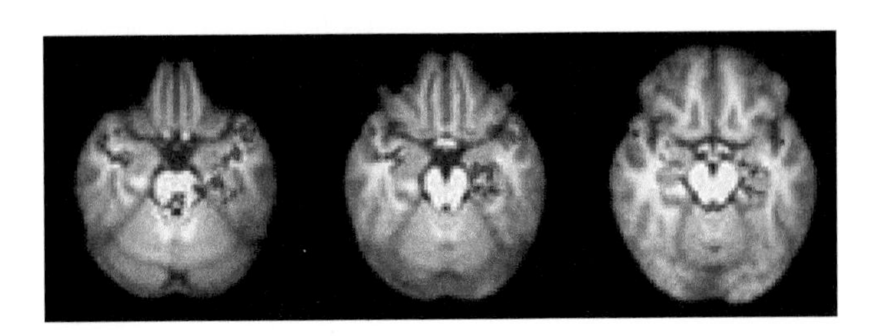

FIGURE 24.4 fMRI group analysis reveals attenuated activation during a spatial retrieval memory task (Roland's Hometown Walking task) in mTBI patients (n = 12) compared with age- and sex-matched healthy subjects. Red-yellow indicates significantly lower activations, which can be seen at the left parahippocampus and the temporal poles.

The recent wave of resting state fMRI studies on mTBI patients provided further important insights into the functionality of the injured brain. In this method, the brain's intrinsic connections (functional connectivity) are mapped by the analysis of low-frequency fluctuations. An important network of a resting (or deactivation) state of the brain is called the default mode network. The default mode network involves brain areas such as the medial prefrontal cortex and parietal and retrosplenial areas (Deco et al., 2011). The extent of these areas and their connectivity strength may be altered because of injury. Depending on explored areas, both decreased and increased connectivity were registered (Johnson et al., 2012b; Mayer et al., 2011; Slobounov et al., 2011; Tang et al., 2011). Alterations in the default mode network connectivity were suggested to be predictive for acute neuropsychological complaints (Johnson et al., 2012b; Mayer et al., 2011) and for later developing postconcussion syndrome as well (Messe et al., 2013).

Abnormal functional patterns in mTBI can be interpreted both as a cause or a consequence of neuropsychological malfunction that is a challenging theoretical, near-philosophical question. If certain cortical areas or linked axonal pathways are injured and cause complaints, those may appear as altered function. However, it is also possible that injury causes a more general and complex abnormality and a certain local BOLD signal alteration is its only mark. Integration of structural and functional connectivity data may be a subsequent step to elucidate these dilemmas (Sharp and Ham, 2011).

Because of the inherent heterogeneity of mTBI, future fMRI studies should strain after larger and more characterized cohorts by means of injury type, age, psychosocial factors, and image acquisition timing after injury to fully exploit the possibility of understanding cognitive sequelae of mTBI held by fMRI.

24.3.6 "Virtual Biopsy" of mTBI: MRS

By measuring chemical compound-specific magnetic signals from 1H nuclei, MRS offers metabolic information of the brain tissue in vivo. This method can be used to detect and characterize altered metabolism in mTBI. Metabolic disorder is believed to start with the neural shear-strain deformation leading to abnormal ion flow through cell membranes triggering excitatory factors (e.g., glutamate) followed by compensatory mechanisms (increased glycolysis and ATP generation). Manifestation of cell death and inflammation may also interlard the metabolic picture.

The main peaks of a proton MRS spectrum refer to metabolites revealing important data about the brain's injured state: N-acetylaspartate (NAA) is a marker of neuronal integrity and viability; choline is a membrane marker altered in membrane damage (e.g., from diffusion axonal imaging or inflammation and also during proliferation); myo-inositol is regarded a glial marker; lactate is attributed to ischemic /hypoxic conditions; creatine (Cr) and phosphocreatine are related to energy metabolism but are often assumed to be relatively constant so are widely used as an internal reference for other peaks; and glutamate and glutamine (glx when combined) are important neurotransmitters or metabolites (glutamine).

Decreased levels of NAA seem to be convincing in characterizing the acute phase of mTBI in adults, which is reasonable considering the lower NAA turnover capacity of the injured neurons (Cecil et al., 1998; Henry et al., 2011; Maugans et al., 2012; Vagnozzi et al., 2010; Yeo et al., 2011). NAA decrease may be postulated a marker for impact severity, when the spectrum is obtained soon after the trauma. However the later (subacute to chronic) fate of NAA levels is more contradictory. Some longitudinal studies have shown NAA levels return to normal over a few weeks (Henry et al., 2011; Vagnozzi et al., 2010). This implies NAA is also able to reflect recovery, at least of the neural tissue—this is not necessary linked to the patients' neuropsychological status; nevertheless, a study elucidated significant associations between NAA levels and neuropsychological test results in the subacute phase (Govind et al., 2010). On the other hand, a considerable amount of studies have found abnormal NAA levels while still in the chronic phase (Cecil et al., 1998; Cohen et al., 2007; Garnett et al., 2000; Govindaraju et al., 2004; Kirov et al., 2007), indicating mTBI can potentially cause persistent alteration. Yet it is hard to evaluate the clinical significance of these definitive changes because many of these studies had no information concerning the patients' neuropsychological state or the available clinical correlations were inconsistent. Some studies support that NAA levels are sensitive to postconcussive symptoms (Kirov et al., 2013; Sarmento et al., 2009), whereas others state that MRS may detect metabolic abnormalities even after the patient's clinical recovery (Johnson et al., 2012a). Interestingly, premorbid intelligence may also affect the neurometabolite normalization rate (Yeo et al., 2011).

A factor definitely worsening both clinical outcome and neurometabolite abnormalities is if head injury is repetitive. The classical theory holding that recurrent brain injury has a cumulative effect has been supported by MRS studies, by means of extraneurometabolite alterations when compared with a single mTBI episode (Vagnozzi et al., 2008). These observations are promising in predicting or evaluating chronic traumatic encephalopathy.

Some studies indicated that the instability of Cr levels in mTBI has important connotations (Gasparovic et al., 2009; Yeo et al., 2011). First, this means that Cr may not work well as an internal reference for metabolite ratios in brain injuries, which, at least partially, explains inconsistencies among MRS studies on mTBI. Second, altered Cr levels as energy markers may be attributed to hypo- or hypermetabolic state (Castillo et al., 1996), as known from different diseases as well (Hattingen et al., 2008).

Although MRS is a relatively time-consuming magnetic resonance method because of its noninvasiveness, it is a unique tool for the longitudinal metabolic description of mTBI. Therefore it holds great promise in better characterizing the injured brain and also as a clinical tool. The provided data may be particularly important adjuncts when interpreting other MRI modality findings.

24.3.7 Further Evolving Advanced MRI Methods

An "upgraded" version of DTI, diffusion spectrum imaging, is believed to resolve crossing fibers (unlike conventional DTI). This method recently provided novel insights into the human white matter microstructure (Hagmann et al., 2008; Wedeen et al., 2008). This, together with novel connectomic techniques (Irimia et al., 2012) and functional connectivity data, may revolutionize understanding alterations in mTBI at the brain network level. Some preliminary data already indicate distinct structural connectivity alterations.

Iron deposition detection and quantification was proposed to be advantageous in mTBI based on previous animal study observations of nonfocal hemoglobin degradation products resulting from oxidative stress and blood–brain barrier dysfunction. For this, magnetic field correlation was used by a group that found abnormal iron accumulation in deep gray matter (Raz et al., 2011).

Beyond function-related neurovascular abnormalities presented by fMRI studies, general hemodynamic (i.e., perfusion disorders) are also likely to occur in mTBI. MRI offers different methods to assess perfusion, such as arterial spin labeling or dynamic contrast-enhanced perfusion imaging. A few studies on mTBI revealed quite specific regional cerebral blood flow deficits attributable for neuropsychological malfunctions (Ge et al., 2009; Liu et al., 2013).

24.4 CONCLUSION AND PERSPECTIVES

Over just more than a decade, advanced MRI studies provided results that brought reappraisal of mTBI. It turned out that the so-called mild injury is characterized by a rather complex pathophysiology that was previously not recognized, or was only hypothesized. Both structural and functional components have become objectively and noninvasively examinable. The multimodality of MRI offers different insights into mTBI sequelae within one patient at a certain time point, and divergent data may be integrated to better interpret results. Beyond the theoretical mapping of this pathological state, contemporary research is very close to yielding clinically useful MRI markers. This means that in cases where routine imaging such as CT does not indicate pathology calling for urgent care, more advanced methods may gain ground and become a useful adjunct in diagnosis, prognosis, and follow-up of mTBI.

Presently, the main limiting factors of such deployment are the heterogeneity of mTBI and the problematic standardization of these methods. Advanced MRI methods are generally quite sensitive but in turn not specific enough for different etiologies. Normal intersubject variability may be in some cases bigger than mTBI-related alteration of a MRI parameter. Similar alterations may be caused by numerous other pathologies, but for instance many parameters are altered due to normal aging or education as well. Yet, the main findings are drawn from group analyses. Future research should enable the clinically feasible application of advanced MRI methods at the subject level.

REFERENCES

Akiyama, Y., K. Miyata, K. Harada, Y. Minamida, T. Nonaka, I. Koyanagi et al. 2009. Susceptibility-weighted magnetic resonance imaging for the detection of cerebral microhemorrhage in patients with traumatic brain injury. *Neurol Med Chir (Tokyo)*. 49:97–99; discussion 99.

Aoki, Y., R. Inokuchi, M. Gunshin, N. Yahagi, and H. Suwa. 2012. Diffusion tensor imaging studies of mild traumatic brain injury: A meta-analysis. *J Neurol Neurosurg Psychiatry*. 83:870–876.

Arfanakis, K., V.M. Haughton, J.D. Carew, B.P. Rogers, R.J. Dempsey, and M.E. Meyerand. 2002. Diffusion tensor MR imaging in diffuse axonal injury. *AJNR Am J Neuroradiol*. 23:794–802.

Babikian, T., M.C. Freier, K.A. Tong, J.P. Nickerson, C.J. Wall, B.A. Holshouser et al. 2005. Susceptibility weighted imaging: Neuropsychologic outcome and pediatric head injury. *Pediatr Neurol*. 33:184–194.

Bazarian, J.J., J. Zhong, B. Blyth, T. Zhu, V. Kavcic, and D. Peterson. 2007. Diffusion tensor imaging detects clinically important axonal damage after mild traumatic brain injury: A pilot study. *J Neurotrauma*. 24:1447–1459.

Beauchamp, M.H., R. Beare, M. Ditchfield, L. Coleman, F.E. Babl, M. Kean et al. 2013. Susceptibility weighted imaging and its relationship to outcome after pediatric traumatic brain injury. *Cortex*. 49:591–598.

Beaulieu, C. 2002. The basis of anisotropic water diffusion in the nervous system—A technical review. *NMR Biomed*. 15:435–455.

Bendlin, B.B., M.L. Ries, M. Lazar, A.L. Alexander, R.J. Dempsey, H.A. Rowley et al. 2008. Longitudinal changes in patients with traumatic brain injury assessed with diffusion-tensor and volumetric imaging. *Neuroimage*. 42:503–514.

Bigler, E.D., C.V. Anderson, and D.D. Blatter. 2002. Temporal lobe morphology in normal aging and traumatic brain injury. *AJNR Am J Neuroradiol*. 23:255–266.

Bigler, E.D., D.D. Blatter, C.V. Anderson, S.C. Johnson, S.D. Gale, R.O. Hopkins et al. 1997. Hippocampal volume in normal aging and traumatic brain injury. *AJNR Am J Neuroradiol*. 18:11–23.

Bigler, E.D., and W.L. Maxwell. 2012. Neuropathology of mild traumatic brain injury: Relationship to neuroimaging findings. *Brain Imaging Behav*. 6:108–136.

Bryer, E.J., J.D. Medaglia, S. Rostami, and F.G. Hillary. 2013. Neural recruitment after mild traumatic brain injury is task dependent: A meta-analysis. *J Int Neuropsychol Soc*:1–12.

Carroll, L.J., J.D. Cassidy, L. Holm, J. Kraus, and V.G. Coronado. 2004. Methodological issues and research recommendations for mild traumatic brain injury: The WHO Collaborating Centre Task Force on Mild Traumatic Brain Injury. *J Rehabil Med*:113–125.

Castillo, M., L. Kwock, and S.K. Mukherji. 1996. Clinical applications of proton MR spectroscopy. *AJNR Am J Neuroradiol*. 17:1–15.

Cecil, K.M., E.C. Hills, M.E. Sandel, D.H. Smith, T.K. McIntosh, L.J. Mannon et al. 1998. Proton magnetic resonance spectroscopy for detection of axonal injury in the splenium of the corpus callosum of brain-injured patients. *J Neurosurg*. 88:795–801.

Chen, C.J., C.H. Wu, Y.P. Liao, H.L. Hsu, Y.C. Tseng, H.L. Liu et al. 2012. Working memory in patients with mild traumatic brain injury: Functional MR imaging analysis. *Radiology*. 264:844–851.

Chen, J.K., K.M. Johnston, S. Frey, M. Petrides, K. Worsley, and A. Ptito. 2004. Functional abnormalities in symptomatic concussed athletes: An fMRI study. *Neuroimage*. 22:68–82.

Chen, J.K., K.M. Johnston, M. Petrides, and A. Ptito. 2008. Recovery from mild head injury in sports: Evidence from serial functional magnetic resonance imaging studies in male athletes. *Clin J Sport Med*. 18:241–247.

Chu, Z., E.A. Wilde, J.V. Hunter, S.R. McCauley, E.D. Bigler, M. Troyanskaya et al. 2010. Voxel-based analysis of diffusion tensor imaging in mild traumatic brain injury in adolescents. *AJNR Am J Neuroradiol*. 31:340–346.

Cohen, B.A., M. Inglese, H. Rusinek, J.S. Babb, R.I. Grossman, and O. Gonen. 2007. Proton MR spectroscopy and MRI-volumetry in mild traumatic brain injury. *AJNR Am J Neuroradiol*. 28:907–913.

Deco, G., V.K. Jirsa, and A.R. McIntosh. 2011. Emerging concepts for the dynamical organization of resting-state activity in the brain. *Nat Rev Neurosci*. 12:43–56.

Fearing, M.A., E.D. Bigler, E.A. Wilde, J.L. Johnson, J.V. Hunter, L. Xiaoqi et al. 2008. Morphometric MRI findings in the thalamus and brainstem in children after moderate to severe traumatic brain injury. *J Child Neurol*. 23:729–737.

Garnett, M.R., A.M. Blamire, B. Rajagopalan, P. Styles, and T.A. Cadoux-Hudson. 2000. Evidence for cellular damage in normal-appearing white matter correlates with injury severity in patients following traumatic brain injury: A magnetic resonance spectroscopy study. *Brain*. 123 (Pt 7):1403–1409.

Gasparovic, C., R. Yeo, M. Mannell, J. Ling, R. Elgie, J. Phillips et al. 2009. Neurometabolite concentrations in gray and white matter in mild traumatic brain injury: An 1H-magnetic resonance spectroscopy study. *J Neurotrauma*. 26:1635–1643.

Ge, Y., M.B. Patel, Q. Chen, E.J. Grossman, K. Zhang, L. Miles et al. 2009. Assessment of thalamic perfusion in patients with mild traumatic brain injury by true FISP arterial spin labelling MR imaging at 3T. *Brain Inj*. 23:666–674.

Geurts, B.H., T.M. Andriessen, B.M. Goraj, and P.E. Vos. 2012. The reliability of magnetic resonance imaging in traumatic brain injury lesion detection. *Brain Inj*. 26:1439–1450.

Gosselin, N., C. Bottari, J.K. Chen, M. Petrides, S. Tinawi, E. de Guise et al. 2011. Electrophysiology and functional MRI in post-acute mild traumatic brain injury. *J Neurotrauma*. 28:329–341.

Govind, V., S. Gold, K. Kaliannan, G. Saigal, S. Falcone, K.L. Arheart et al. 2010. Whole-brain proton MR spectroscopic imaging of mild-to-moderate traumatic brain injury and correlation with neuropsychological deficits. *J Neurotrauma*. 27:483–496.

Govindaraju, V., G.E. Gauger, G.T. Manley, A. Ebel, M. Meeker, and A.A. Maudsley. 2004. Volumetric proton spectroscopic imaging of mild traumatic brain injury. *AJNR Am J Neuroradiol*. 25:730–737.

Haacke, E.M., S. Mittal, Z. Wu, J. Neelavalli, and Y.C. Cheng. 2009. Susceptibility-weighted imaging: Technical aspects and clinical applications, part 1. *AJNR Am J Neuroradiol*. 30:19–30.

Hagmann, P., L. Cammoun, X. Gigandet, R. Meuli, C.J. Honey, V.J. Wedeen et al. 2008. Mapping the structural core of human cerebral cortex. *PLoS Biol*. 6:e159.

Hasiloglu, Z.I., S. Albayram, H. Selcuk, E. Ceyhan, S. Delil, B. Arkan elt. 2011. Cerebral microhemorrhages detected by susceptibility-weighted imaging in amateur boxers. *AJNR Am J Neuroradiol*. 32:99–102.

Hattingen, E., P. Raab, K. Franz, H. Lanfermann, M. Setzer, R. Gerlach et al. 2008. Prognostic value of choline and creatine in WHO grade II gliomas. *Neuroradiology*. 50:759–767.

Henry, L.C., S. Tremblay, S. Leclerc, A. Khiat, Y. Boulanger, D. Ellemberg et al. 2011. Metabolic changes in concussed American football players during the acute and chronic post-injury phases. *BMC Neurol*. 11:105.

Hughes, D.G., A. Jackson, D.L. Mason, E. Berry, S. Hollis, and D.W. Yates. 2004. Abnormalities on magnetic resonance imaging seen acutely following mild traumatic brain injury: Correlation with neuropsychological tests and delayed recovery. *Neuroradiology*. 46:550–558.

Inglese, M., S. Makani, G. Johnson, B.A. Cohen, J.A. Silver, O. Gonen et al. 2005. Diffuse axonal injury in mild traumatic brain injury: A diffusion tensor imaging study. *J Neurosurg*. 103:298–303.

Irimia, A., M.C. Chambers, C.M. Torgerson, and J.D. Horn. 2012. Circular representation of human cortical networks for subject and population-level connectomic visualization. *Neuroimage*. 60:1340–1351.

Jacobs, B., T. Beems, M. Stulemeijer, A.B. van Vugt, T.M. van der Vliet, G.F. Borm et al. 2010. Outcome prediction in mild traumatic brain injury: Age and clinical variables are stronger predictors than CT abnormalities. *J Neurotrauma*. 27:655–668.

Johnson, B., M. Gay, K. Zhang, T. Neuberger, S.G. Horovitz, M. Hallett et al. 2012a. The use of magnetic resonance spectroscopy in the subacute evaluation of athletes recovering from single and multiple mild traumatic brain injury. *J Neurotrauma*. 29:2297–2304.

Johnson, B., K. Zhang, M. Gay, S. Horovitz, M. Hallett, W. Sebastianelli et al. 2012b. Alteration of brain default network in subacute phase of injury in concussed individuals: Resting-state fMRI study. *Neuroimage*. 59:511–518.

Kim, J., B. Avants, S. Patel, J. Whyte, B.H. Coslett, J. Pluta et al. 2008. Structural consequences of diffuse traumatic brain injury: A large deformation tensor-based morphometry study. *Neuroimage*. 39:1014–1026.

Kirov, II, A. Tal, J.S. Babb, J. Reaume, T. Bushnik, T. Ashman et al. 2013. Proton MR spectroscopy correlates diffuse axonal abnormalities with post-concussive symptoms in mild traumatic brain injury. *J Neurotrauma*. 30:1200–1204.

Kirov, I., L. Fleysher, J.S. Babb, J.M. Silver, R.I. Grossman, and O. Gonen. 2007. Characterizing 'mild' in traumatic brain injury with proton MR spectroscopy in the thalamus: Initial findings. *Brain Inj*. 21:1147–1154.

Kitayama, N., S. Quinn, and J.D. Bremner. 2006. Smaller volume of anterior cingulate cortex in abuse-related posttraumatic stress disorder. *J Affect Disord*. 90:171–174.

Lannsjo, M., M. Backheden, U. Johansson, J.L. Af Geijerstam, and J. Borg. 2013. Does head CT scan pathology predict outcome after mild traumatic brain injury? *Eur J Neurol*. 20:124–129.

Larrabee, G.J., L.M. Binder, M.L. Rohling, and D.M. Ploetz. 2013. Meta-analytic methods and the importance of non-TBI factors related to outcome in mild traumatic brain injury: Response to Bigler et al. (2013). *Clin Neuropsychol*. 27:215–237.

Levine, B., N. Kovacevic, E.I. Nica, G. Cheung, F. Gao, M.L. Schwartz et al. 2008. The Toronto traumatic brain injury study: Injury severity and quantified MRI. *Neurology*. 70:771–778.

Lipton, M.L., E. Gellella, C. Lo, T. Gold, B.A. Ardekani, K. Shifteh et al. 2008. Multifocal white matter ultrastructural abnormalities in mild traumatic brain injury with cognitive disability: A voxel-wise analysis of diffusion tensor imaging. *J Neurotrauma*. 25:1335–1342.

Lipton, M.L., N. Kim, Y.K. Park, M.B. Hulkower, T.M. Gardin, K. Shifteh et al. 2012. Robust detection of traumatic axonal injury in individual mild traumatic brain injury patients: Intersubject variation, change over time and bidirectional changes in anisotropy. *Brain Imaging Behav*. 6:329–342.

Liu, W., B. Wang, R. Wolfowitz, P.H. Yeh, D.E. Nathan, J. Graner et al. 2013. Perfusion deficits in patients with mild traumatic brain injury characterized by dynamic susceptibility contrast MRI. *NMR Biomed*. 26:651–663.

Lovell, M.R., J.E. Pardini, J. Welling, M.W. Collins, J. Bakal, N. Lazar et al. 2007. Functional brain abnormalities are related to clinical recovery and time to return-to-play in athletes. *Neurosurgery*. 61:352–359; discussion 359–360.

Mac Donald, C.L., A.M. Johnson, D. Cooper, E.C. Nelson, N.J. Werner, J.S. Shimony et al. 2011. Detection of blast-related traumatic brain injury in U.S. military personnel. *N Engl J Med*. 364:2091–2100.

MacKenzie, J.D., F. Siddiqi, J.S. Babb, L.J. Bagley, L.J. Mannon, G.P. Sinson et al. 2002. Brain atrophy in mild or moderate traumatic brain injury: A longitudinal quantitative analysis. *AJNR Am J Neuroradiol*. 23:1509–1515.

Maller, J.J., R.H. Thomson, K. Pannek, S.E. Rose, N. Bailey, P.M. Lewis et al. 2012. The (Eigen)value of diffusion tensor imaging to investigate depression after traumatic brain injury. *Hum Brain Mapp*. 35:227–237.

Maugans, T.A., C. Farley, M. Altaye, J. Leach, and K.M. Cecil. 2012. Pediatric sports-related concussion produces cerebral blood flow alterations. *Pediatrics*. 129:28–37.

Mayer, A.R., J. Ling, M.V. Mannell, C. Gasparovic, J.P. Phillips, D. Doezema et al. 2010. A prospective diffusion tensor imaging study in mild traumatic brain injury. *Neurology*. 74:643–650.

Mayer, A.R., M.V. Mannell, J. Ling, C. Gasparovic, and R.A. Yeo. 2011. Functional connectivity in mild traumatic brain injury. *Hum Brain Mapp*. 32:1825–1835.

McAllister, T.W., A.J. Saykin, L.A. Flashman, M.B. Sparling, S.C. Johnson, S.J. Guerin et al. 1999. Brain activation during working memory 1 month after mild traumatic brain injury: A functional MRI study. *Neurology*. 53:1300–1308.

McAllister, T.W., M.B. Sparling, L.A. Flashman, S.J. Guerin, A.C. Mamourian, and A.J. Saykin. 2001. Differential working memory load effects after mild traumatic brain injury. *Neuroimage*. 14:1004–1012.

McDonald, B.C., A.J. Saykin, and T.W. McAllister. 2012. Functional MRI of mild traumatic brain injury (mTBI): Progress and perspectives from the first decade of studies. *Brain Imaging Behav*. 6:193–207.

Meehan, W.P., 3rd. 2011. Medical therapies for concussion. *Clin Sports Med*. 30:115–124, ix.

Messe, A., S. Caplain, G. Paradot, D. Garrigue, J.F. Mineo, G. Soto Ares ealt. 2011. Diffusion tensor imaging and white matter lesions at the subacute stage in mild traumatic brain injury with persistent neurobehavioral impairment. *Hum Brain Mapp*. 32:999–1011.

Messe, A., S. Caplain, M. Pelegrini-Issac, S. Blancho, R. Levy, N. Aghakhani et al. 2013. Specific and evolving resting-state network alterations in post-concussion syndrome following mild traumatic brain injury. *PLoS One*. 8:e65470.

Mild Traumatic Brain Injury Committee. 1993. Head Injury Interdisciplinary Special Interest Group of the American Congress of Rehabilitation Medicine. Definition of mild traumatic brain injury. *J Head Trauma Rehabil*:86–87.

Miles, L., R.I. Grossman, G. Johnson, J.S. Babb, L. Diller, and M. Inglese. 2008. Short-term DTI predictors of cognitive dysfunction in mild traumatic brain injury. *Brain Inj*. 22:115–122.

Moonen, C.T., Bandettini, P., eds. 2000. *Functional MRI*. Springer Verlag, Berlin, Germany.

Nair, D.G. 2005. About being BOLD. *Brain Res Brain Res Rev*. 50:229–243.

Nakayama, N., A. Okumura, J. Shinoda, Y.T. Yasokawa, K. Miwa, S.I. Yoshimura et al. 2006. Evidence for white matter disruption in traumatic brain injury without macroscopic lesions. *J Neurol Neurosurg Psychiatry*. 77:850–855.

Niogi, S.N., P. Mukherjee, J. Ghajar, C. Johnson, R.A. Kolster, R. Sarkar et al. 2008. Extent of microstructural white matter injury in postconcussive syndrome correlates with impaired cognitive reaction time: A 3T diffusion tensor imaging study of mild traumatic brain injury. *AJNR Am J Neuroradiol*. 29:967–973.

Obenaus, A., M. Robbins, G. Blanco, N.R. Galloway, E. Snissarenko, E. Gillard et al. 2007. Multi-modal magnetic resonance imaging alterations in two rat models of mild neurotrauma. *J Neurotrauma*. 24:1147–1160.

Ogawa, S., D.W. Tank, R. Menon, J.M. Ellermann, S.G. Kim, H. Merkle et al. 1992. Intrinsic signal changes accompanying sensory stimulation: Functional brain mapping with magnetic resonance imaging. *Proc Natl Acad Sci U S A*. 89:5951–5955.

Pardini, J.E., D.A. Pardini, J.T. Becker, K.L. Dunfee, W.F. Eddy, M.R. Lovell et al. 2010. Postconcussive symptoms are associated with compensatory cortical recruitment during a working memory task. *Neurosurgery*. 67:1020–1027; discussion 1027–1028.

Peled, S. 2007. New perspectives on the sources of white matter DTI signal. *IEEE Trans Med Imaging*. 26:1448–1455.

Perlstein, W.M., M.A. Cole, J.A. Demery, P.J. Seignourel, N.K. Dixit, M.J. Larson et al. 2004. Parametric manipulation of working memory load in traumatic brain injury: Behavioral and neural correlates. *J Int Neuropsychol Soc*. 10:724–741.

Pierpaoli, C., P. Jezzard, P.J. Basser, A. Barnett, and G. Di Chiro. 1996. Diffusion tensor MR imaging of the human brain. *Radiology*. 201:637–648.

Raz, E., J.H. Jensen, Y. Ge, J.S. Babb, L. Miles, J. Reaume etal. 2011. Brain iron quantification in mild traumatic brain injury: A magnetic field correlation study. *AJNR Am J Neuroradiol*. 32:1851–1856.

Reichenbach, J.R., R. Venkatesan, D.J. Schillinger, D.K. Kido, and E.M. Haacke. 1997. Small vessels in the human brain: MR venography with deoxyhemoglobin as an intrinsic contrast agent. *Radiology*. 204:272–277.

Rogers, B.P., V.L. Morgan, A.T. Newton, and J.C. Gore. 2007. Assessing functional connectivity in the human brain by fMRI. *Magn Reson Imaging*. 25:1347–1357.

Rosenblum, W.I. 2007. Cytotoxic edema: Monitoring its magnitude and contribution to brain swelling. *J Neuropathol Exp Neurol*. 66:771–778.

Ross, D.E. 2011. Review of longitudinal studies of MRI brain volumetry in patients with traumatic brain injury. *Brain Inj*. 25:1271–1278.

Ross, D.E., C. Castelvecchi, and A.L. Ochs. 2013. Brain MRI volumetry in a single patient with mild traumatic brain injury. *Brain Inj*. 27:634–636.

Russell, K.C., P.M. Arenth, J.M. Scanlon, L.J. Kessler, and J.H. Ricker. 2011. A functional magnetic resonance imaging investigation of episodic memory after traumatic brain injury. *J Clin Exp Neuropsychol*. 33:538–547.

Rutgers, D.R., P. Fillard, G. Paradot, M. Tadie, P. Lasjaunias, and D. Ducreux. 2008. Diffusion tensor imaging characteristics of the corpus callosum in mild, moderate, and severe traumatic brain injury. *AJNR Am J Neuroradiol*. 29:1730–1735.

Sarmento, E., P. Moreira, C. Brito, J. Souza, C. Jevoux, and M. Bigal. 2009. Proton spectroscopy in patients with post-traumatic headache attributed to mild head injury. *Headache*. 49:1345–1352.

Sharp, D.J., and T.E. Ham. 2011. Investigating white matter injury after mild traumatic brain injury. *Curr Opin Neurol*. 24:558–563.

Sidaros, A., A.W. Engberg, K. Sidaros, M.G. Liptrot, M. Herning, P. Petersen et al. 2008. Diffusion tensor imaging during recovery from severe traumatic brain injury and relation to clinical outcome: A longitudinal study. *Brain*. 131:559–572.

Slobounov, S.M., M. Gay, K. Zhang, B. Johnson, D. Pennell, W. Sebastianelli et al. 2011. Alteration of brain functional network at rest and in response to YMCA physical stress test in concussed athletes: RsFMRI study. *Neuroimage*. 55:1716–1727.

Slobounov, S.M., K. Zhang, D. Pennell, W. Ray, B. Johnson, and W. Sebastianelli. 2010. Functional abnormalities in normally appearing athletes following mild traumatic brain injury: A functional MRI study. *Exp Brain Res*. 202:341–354.

Smits, M., D.W. Dippel, G.C. Houston, P.A. Wielopolski, P.J. Koudstaal, M.G. Hunink, and A. van der Lugt. 2009. Postconcussion syndrome after minor head injury: Brain activation of working memory and attention. *Hum Brain Mapp*. 30:2789–2803.

Stulemeijer, M., P.E. Vos, S. van der Werf, G. van Dijk, M. Rijpkema, and G. Fernandez. 2010. How mild traumatic brain injury may affect declarative memory performance in the post-acute stage. *J Neurotrauma*. 27:1585–1595.

Tang, L., Y. Ge, D.K. Sodickson, L. Miles, Y. Zhou, J. Reaume et al. 2011. Thalamic resting-state functional networks: Disruption in patients with mild traumatic brain injury. *Radiology*. 260:831–840.

Tong, K.A., S. Ashwal, A. Obenaus, J.P. Nickerson, D. Kido, and E.M. Haacke. 2008. Susceptibility-weighted MR imaging: A review of clinical applications in children. *AJNR Am J Neuroradiol*. 29:9–17.

Toth, A., N. Kovacs, G. Perlaki, G. Orsi, M. Aradi, H. Komaromy et al. 2013. Multi-modal magnetic resonance imaging in the acute and sub-acute phase of mild traumatic brain injury: Can we see the difference? *J Neurotrauma*. 30:2–10.

Vagnozzi, R., S. Signoretti, L. Cristofori, F. Alessandrini, R. Floris, E. Isgro et al. 2010. Assessment of metabolic brain damage and recovery following mild traumatic brain injury: A multicentre, proton magnetic resonance spectroscopic study in concussed patients. *Brain*. 133:3232–3242.

Vagnozzi, R., S. Signoretti, B. Tavazzi, R. Floris, A. Ludovici, S. Marziali et al. 2008. Temporal window of metabolic brain vulnerability to concussion: A pilot 1H-magnetic resonance spectroscopic study in concussed athletes—Part III. *Neurosurgery*. 62:1286–1295; discussion 1295–1286.

Villarreal, G., D.A. Hamilton, D.P. Graham, I. Driscoll, C. Qualls, H. Petropoulos et al. 2004. Reduced area of the corpus callosum in posttraumatic stress disorder. *Psychiatry Res*. 131:227–235.

Villarreal, G., D.A. Hamilton, H. Petropoulos, I. Driscoll, L.M. Rowland, J.A. Griego et al. 2002. Reduced hippocampal volume and total white matter volume in posttraumatic stress disorder. *Biol Psychiatry*. 52:119–125.

Wedeen, V.J., R.P. Wang, J.D. Schmahmann, T. Benner, W.Y. Tseng, G. Dai et al. 2008. Diffusion spectrum magnetic resonance imaging (DSI) tractography of crossing fibers. *Neuroimage*. 41:1267–1277.

Wilde, E.A., J.V. Hunter, M.R. Newsome, R.S. Scheibel, E.D. Bigler, J.L. Johnson et al. 2005. Frontal and temporal morphometric findings on MRI in children after moderate to severe traumatic brain injury. *J Neurotrauma*. 22:333–344.

Wilde, E.A., S.R. McCauley, J.V. Hunter, E.D. Bigler, Z. Chu, Z.J. Wang et al. 2008. Diffusion tensor imaging of acute mild traumatic brain injury in adolescents. *Neurology*. 70:948–955.

Woodward, S.H., D.G. Kaloupek, C.C. Streeter, M.O. Kimble, A.L. Reiss, S. Eliez et al. 2007. Brain, skull, and cerebrospinal fluid volumes in adult posttraumatic stress disorder. *J Trauma Stress*. 20:763–774.

Yeo, R.A., C. Gasparovic, F. Merideth, D. Ruhl, D. Doezema, and A.R. Mayer. 2011. A longitudinal proton magnetic resonance spectroscopy study of mild traumatic brain injury. *J Neurotrauma*. 28:1–11.

Yuh, E.L., P. Mukherjee, H.F. Lingsma, J.K. Yue, A.R. Ferguson, W.A. Gordon, A.B. Valadka et al. 2013. Magnetic resonance imaging improves 3-month outcome prediction in mild traumatic brain injury. *Ann Neurol*. 73:224–235.

Zhou, Y., A. Kierans, D. Kenul, Y. Ge, J. Rath, J. Reaume et al. 2013. Mild traumatic brain injury: Longitudinal regional brain volume changes. *Radiology*. 267:880–890.

25 Translational Metabolomics of Head Injury

Exploring Dysfunctional Cerebral Metabolism with Ex Vivo NMR Spectroscopy-Based Metabolite Quantification

Stephanie M. Wolahan, Daniel Hirt, and Thomas C. Glenn

CONTENTS

25.1 INTRODUCTION TO METABOLOMICS

There are four biochemical components that control biological systems by serving as building blocks and as information databases: genes, transcripts, proteins, and metabolites. The study of these four components have become entire fields of biological study and have often been referred to collectively as the omics, including genomics, transcriptomics, proteomics, and metabolomics. The ability to study each of these biological components in great detail and to study the relationship between them has led to significant advances in medical discovery and understanding. The goal of medical systems biology is to integrate all biological information to understand mechanistic information about cellular events and functions that may contribute to disease propensity, development, progression, diagnosis, and/or treatment.

Having a systems perspective on human biology is desirable, where details of various system components can be integrated with increasing complexity to better understand properties of the entire system. The systems-oriented approach requires extensive and complex datasets; reliable analytical techniques; thoughtful data integration across platforms; and advanced biostatistical methods. Medical systems biology necessitates an unbiased and comprehensive approach when interpreting experimental results and biological interpretations need to be carefully explained, justified by the data, and tested on larger data sets.

Traumatic brain injury (TBI) patients would benefit from a medical systems biology understanding of the systemic dysregulation and cellular changes that follow an insult to the head. A subspecialty in the critical care environment, neurocritical care, evolved from the acceptance that recovery from the primary injury to the brain tissue is affected by systemic alterations that can result in secondary injuries to the brain. The neurological intensive care unit (ICU) has realized significant improvements in patient outcomes due to protocols to address and prevent secondary injuries and due to neurointensivist-led teamwork, both aided by modern technological advances in multimodality neuromonitoring (Elf et al., 2002, Le Roux et al., 2012, Varelas et al., 2006).

Considering the notable advances achieved through incorporating a systems-level approach to treating head injury and improving outcomes, in this review we discuss metabolomics applied to TBI. First, we will introduce metabolomics for readers not familiar with the field. Second, we summarize research on the metabolic changes following TBI to highlight what information has been translated to the clinic and what treatments exist. Finally, we discuss metabolomics techniques applied to TBI metabolism, reviewing the examples in the literature, and offering the authors' suggestions for using NMR spectroscopy to study biofluids from head injured patients. As researchers and clinicians report and validate metabolomics findings, building a medical systems biology perspective on post-TBI metabolic dysfunction is likely to aid in informing physicians' decisions and in integrating treatments into daily practice.

Metabolomics refers to the study of the metabolome, which has been defined as "the quantitative complement of metabolites in a biological system" (Dunn et al., 2011). A metabolome, estimated to contain thousands of compounds, is organism-specific and sample type–specific. The human serum metabolome has been reported to contain 4,229 unique compounds, detection of which involved the use of several analytic techniques, and is still not considered exhaustive (Psychogios et al., 2011). Metabolomics studies aim to discriminate pathological metabolic profiles from that of a normal physiological state and to predict class assignment based on this set of metabolite biomarkers (Baker, 2011; Holmes et al., 2008; Nicholson et al., 2012).

The field of metabolomics research consists of several investigative methods. First, there is a distinction to be made between targeted and exploratory metabolomics studies (Lenz and Wilson, 2007). In the latter, the goal is to generate a metabolomic fingerprint for each case and to use multivariate analysis to probe class-specific patterns. Generally, the focus of such studies is not to identify and quantify metabolites nor to propose mechanistic explanations of the results, but rather to predict class assignment based on the metabolomic fingerprint.

Targeted metabolomics studies aim to identify and quantify specific metabolites. These metabolites may be hypothesized to be biomarkers of disease progression or may be considered an indicator of the severity of a physiological state. Targeted metabolomics studies may use the same multivariate statistical techniques as the metabolome fingerprint-type studies, but also typically include more traditional univariate and multivariate analyses on the metabolite concentrations. Targeted studies can be targeted to a set of endogenous metabolites or can be targeted to study an exogenous substance, including labeled tracer metabolites or a pharmaceutical.

Blood plasma, blood serum, urine, and cerebrospinal fluid (CSF) have been extensively investigated in the metabolomics literature. These biofluids are readily available and are interpreted as an average representation of the surrounding tissue. Researchers working with animal models have access to tissue after sacrifice, which is considerably rarer in human studies. As the field has grown, online metabolite databases containing biological, structural, and experimental information have been developed and are a key tool for metabolomics researchers (Ulrich et al., 2008; Wishart et al., 2007).

The term *metabolomics* resulted from research in the 1980s and 1990s (Nicholson et al., 1999), yet the concept behind metabolomics was a focus of research for several decades prior. What distinguishes contemporary metabolomics studies from past studies on metabolic changes is the technology available for analyzing such biofluid samples and, therefore, the extent and accuracy of the metabolome quantified. In addition to the larger data set, there have also been computational and statistical advances that make the prospect of drawing meaningful conclusions from thousands of metabolites and the changes that occur between classes possible. With improvements in technology, metabolomics research has reached a level of complexity requiring a multidisciplinary team and has made providing biological rationale for the findings challenging because of data set complexity. The Institute of Medicine of the National Academies published a report on translational omics that issued recommendations for improving the overall quality of the metabolomics research and for translating these findings to the clinical setting (Committee on the Review of Omics-Based Tests for Predicting Patient Outcomes in Clinical Trials, 2012).

The use of mass spectrometry (MS)-based and nuclear magnetic resonance (NMR)-based quantification are the most common in the metabolomics literature. Both of these analytical instruments are reliable, accurate, and widely available. There are advantages and disadvantages associated with each, some of which will be briefly mentioned, and the reader is referred to a number of excellent metabolomics review articles (Dunn et al., 2011; Lenz and Wilson, 2007; Nicholson et al., 1999). Because an individual's metabolome is highly influenced by environment and diet, population studies require a large number of subjects, and the reliability and reproducibility of these analytical techniques is key. The focus of this review is NMR-based metabolomics applied to TBI, but both analytical methods will be described. The reader is referred to extensive review articles focused on the application of MS and/or NMR to metabolomics (Dettmer et al., 2007; Zhang et al., 2010).

MS detects compounds in the picomolar concentration range that become ionized after injection into the mass spectrometer; the readout is the mass-to-charge ratio of the detectable compounds in solution. MS-based metabolomics have used gas chromatography MS and liquid chromatography MS. Preparing samples for MS analysis requires extraction of metabolites and may require derivitization, which can be a labor-intensive process. Metabolite extraction involves a series of experimental steps in which metabolite loss can occur and where additional sample-to-sample variability may be introduced. The high sensitivity of MS-based quantification makes it a powerful tool in targeted metabolomics studies. In metabolome fingerprinting studies, it is challenging to measure all compounds with the same efficiency and accuracy for technical reasons.

NMR spectroscopy is used to identify and quantify compounds in solution containing elements that are magnetic resonance–detectable (i.e., elemental isotopes that will absorb photons when placed in a magnetic field). NMR is considerably less sensitive than MS and is able to detect concentrations in the micromolar concentration range, but does not destroy the sample in the process of measurement. Application of a radiofrequency field at a known frequency and power excites the spin of the magnetic resonance–detectable isotopes. Spin is a fundamental property of elements akin to mass and charge and both the absorption and emission of radiofrequency photons is nondestructive and noninvasive. Each unique chemical structure in a molecule will resonate in the magnetic field at a specific frequency as the spins relax to equilibrium alignment with the magnetic field. The signal collected by the NMR spectrometer is then Fourier transformed into a NMR spectrum with spectral peaks at specific frequencies corresponding to the molecular structure of the compound being measured. The integrated area of the spectral peaks is proportional to the concentration of the compound. All compounds in solution above a certain concentration will be detected, unlike the variable efficiency of MS-based quantification. There is minimal sample preparation required when compared with MS. There are a number of biologically relevant isotopes that can be measured, including ^{1}H, ^{13}C, ^{31}P, and ^{15}N. ^{1}H is the most abundant isotope of hydrogen (99.99%) and, because biologically relevant molecules contain hydrogen, ^{1}H NMR is widely used. NMR spectrometers are standard equipment in research environments and increased spectral resolution is possible due to the prevalence of high-field spectrometers with field strengths ≥400 MHz (9.4 T). High-resolution magic angle spinning spectroscopy is able to quantify metabolites in intact tissue using solid-state NMR spectrometers (Beckonert et al., 2010).

Another aspect of modern metabolomics research is application of multivariate statistical approaches. Unsupervised multivariate techniques such as principal component analysis (PCA) reduce the number of variables to a few principal components. Principal components are orthogonal to one another, are linear combinations of the original data, and can reduce hundreds of input variables to three or four. There are many NMR-based metabolomics fingerprint-type studies that use the complete NMR spectrum as the set of variables. Some metabolomics studies are designed to build a prediction model with supervised multivariate techniques, for example partial least squares (PLS) or PLS-discriminant analysis (PLS-DA) among others (Bylesjo et al., 2006). Most metabolomics studies generate a PCA model of the data to test whether the groups can be reasonably separated based on metabolic information. To build a predictive model, validation is vital and the data set is randomly separated into a larger training set and a smaller test set; the model generated from the training set is then tested on the test set.

In reality, metabolomics studies generally quantify fewer than 100 metabolites per sample. Several advances are required to achieve high-throughput quantification of the entire metabolome and to translate metabolomics to the clinical setting. The steps following data collection, including processing and statistical analyses, will be discussed later in this chapter within the context of metabolomics of TBI.

25.2 METABOLIC CHANGES AFTER TBI AND CURRENT TREATMENT RECOMMENDATIONS

TBI results from a strong force applied to the head and, although the primary injury may penetrate and/or cause physical strain on the cerebral tissue requiring surgery, the primary injury also initiates cellular metabolic changes and hemodynamic dysregulation. Experimental models of head trauma have characterized what is known about pathophysiological changes after TBI. The primary injury initiates indiscriminate excitatory neurotransmitter release (primarily glutamate and aspartate) and an increase in extracellular potassium (Katayama et al., 1990). Flux of calcium ions from intracellular stores into the cytoplasm leads to mitochondrial damage and membrane microporation (Verweij et al., 2000). The cellular response to these changes, such as operation of membrane ion pumps to restore the ionic gradient, requires large amounts of adenosine triphosphate (ATP). There is a notable increase in glucose uptake during the acute period without a concomitant increase in oxygen uptake. Increased anaerobic metabolism of glucose, termed hyperglycolysis, is followed by a prolonged period of depressed glucose metabolism with respect to oxygen metabolism. Hyperglycolysis has been interpreted as the cerebral response to the high energy demands required to respond to the metabolic disruption and can lead to increased lactate production (De Salles et al., 1987).

Several of the metabolic changes after TBI, including hyperglycemia, high cerebral glutamate, and high cerebral lactate/pyruvate ratio (LPR), have been extensively characterized in the literature (Goodman et al., 1999; Soustiel and Sviri, 2007; Vespa et al., 2003, 2005) because restoring metabolic homeostasis becomes a major focus of clinical treatment after surgical intervention. A metabolic crisis has been defined as a high LPR, high glutamate, and low glucose concentration in the extracellular space of cerebral tissue, as measured with cerebral microdialysis. Stein and colleagues demonstrated that almost three-quarters of patients with brain injury suffer from a metabolic crisis during the acute postinjury period (Stein et al., 2012). The deleterious effects of these metabolic derangements have been correlated with poor outcomes (De Salles et al., 1986; Marcoux et al., 2008; Obrist et al., 1979; Stein et al., 2012; Xu et al., 2010). In this section, we will discuss these changes, focusing on glucose because it is routinely monitored in the clinic. Additionally, we will discuss current treatment recommendations for TBI.

25.2.1 GLUCOSE

After TBI, patients present in a state of hyperglycemia and of depressed cerebral metabolic rate of oxygen ($CMRO_2$) compared with normal levels. It has been shown that high admission glucose levels correlate with poor outcomes (Lam et al., 1991; Rovlias and Kotsou, 2000; Young et al., 1989) and the clinical team works to lower glucose to the normal range. The frequency of hyperglycemia simultaneous with depressed $CMRO_2$ led researchers to extensively study glucose and oxygen in the acute period after injury. Positron emission tomography (PET) imaging studies confirmed acute hyperglycolysis in TBI patients, where the cerebral metabolic rate of glucose is significantly elevated above normal levels. Additional PET studies on TBI patients in the months after injury revealed hyperglycolysis is followed by a period of depressed cerebral metabolic rate of glucose compared with normal levels, whereas depressed $CMRO_2$ persists throughout (Ackermann and Lear, 1989; Andersen and Marmarou, 1989; Bergsneider et al., 1997, 2001). The purpose and effects of the upregulation of glycolysis, whether deleterious or beneficial, are not fully understood. However, numerous studies have demonstrated a clear relationship between hyperglycemia and poor outcomes after TBI (Cochran et al., 2003; De Salles et al., 1987; Lam et al., 1991; Rovlias and Kotsou, 2000).

After identifying hyperglycemia as a contributor to poor outcomes, a surge of studies emerged addressing glucose management in the intensive care unit setting. Initial studies suggested a benefit of fewer neurologic complications associated with tight glycemic control, when blood glucose is targeted to a relatively low and narrow range and controlled by insulin infusion (Van den Berghe et al., 2001, 2005). However, subsequent investigations cast controversy on this subject, concluding intensive glucose management either had no effect on neurological outcome or contributed to worse outcomes (Meier et al., 2008; Van den Berghe et al., 2006).

The appropriate glucose target remains controversial and there are many aspects of glucose metabolism after brain injury that remain poorly understood. Increased glucose levels are clearly correlated with poor outcome, yet the conclusions on strict glucose control are mixed. An explanation was offered by Meier et al. who demonstrated in a retrospective study that aggressive glucose management in the intensive care unit setting increases the likelihood of and subsequently the frequency of hypoglycemic episodes (Green et al., 2010; Meier et al., 2008). Patients with even one episode of hypoglycemia had worse outcomes than patients without. Later studies demonstrated that tight glucose control results in lower extracellular glucose as measured by microdialysis as well as an increased incidence of markers of cellular distress, such as high cerebral glutamate and LPR (Meierhans et al., 2010; Vespa et al., 2006). Recently, a prospective, randomized within-subject crossover trial of tight versus loose glycemic control, where blood glucose in maintained within the 80–110 or 120–150 mg/dL ranges, respectively, revealed increased glucose uptake under tight control with fluorodeoxyglucose-PET imaging (Vespa et al., 2012). Additionally, microdialysis measurements showed an increased incidence of metabolic crisis in patients under tight control than under loose control. This study suggests, somewhat counterintuitively, that lowering systemic glucose increases glucose uptake. Although delivering more glucose to the energy-hungry brain might be beneficial, the microdialysis results suggest the increased uptake leads to increased damaging metabolic processes. This study supports previous evidence that strict glycemic control may be inappropriate for TBI patients.

The ideal glucose range in the acute postinjury period remains undefined. The general recommendations are based on the above as well as trials conducted in a general critical care patient population, such as the NICE-SUGAR trial, the VISEP trial, and the Glucontrol trial (Brunkhorst et al., 2008; Finfer et al., 2009; Preiser et al., 2009). Based on the findings of these studies and trials, it is recommended to avoid the extremes of hypo- and hyperglycemia and to maintain a broad glucose range of up to 140 mg/dL or even as high as 180 mg/dL.

The mechanism of and physiological reason for these pathological changes are unknown and these conflicting results on glucose add confusion. Yet, an alternative to the question of whether glucose is maintained at a low, medium, or high level is to ask how glucose is metabolized and whether systemic glucose levels could control the mechanism of glucose metabolism. Alternative mechanisms of glucose metabolism, called alternative glucose utilization, and/or alternative fuel sources have also been studied to test whether outcome is improved or diminished by generating ATP through other biochemical pathways (Bartnik et al., 2005, 2007; Glenn et al., 2003).

Dusick et al. performed a ^{13}C-labeled glucose infusion study in humans and demonstrated increased activation of the pentose phosphate shunt after brain injury compared with non–brain-injured subjects (Dusick et al., 2007), as was previously demonstrated in animal models (Bartnik et al., 2005). Glucose may be shunted into alternate pathways to assist in cellular repair; the pentose phosphate pathway plays an important role in the neutralization of free radicals and DNA repair. Therefore, moderate hyperglycolysis and/or higher systemic glucose concentrations may be necessary to promote the increased activation of this pathway. Alternatively, a recent analysis of CSF using ^{1}H NMR demonstrated the presence of propylene glycol at higher concentrations in TBI patients compared with noninjured controls (Glenn et al., 2013). Propylene glycol is a by-product of the methylglyoxal pathway; this pathway is considered deleterious because advanced glycation end-products are produced that are catastrophic to the nervous system. In another research study, the microdialysis catheter in TBI patients was used to infuse ^{13}C-labeled metabolites directly into cerebral tissue and, by pooling fluid from 24 hours of monitoring, samples were analyzed with ^{13}C NMR spectroscopy (Gallagher et al., 2009). Although not using quantitative analysis, this metabolomics-like study showed labeled acetate and labeled lactate converted to glutamine, demonstrating tricarboxylic acid cycle activity in the human brain after injury. The mystery of metabolic pathway activation and/or suppression remains to be solved, but there is growing evidence supporting the importance of the biochemical activity.

This introduction to post-TBI glucose metabolism is in no way meant to be complete, but, when applying metabolomics to study head injury, there are several avenues the interdisciplinary field of metabolomics seems poised to investigate, including the various biochemical pathways that are activated, the physiological states that upregulate beneficial and downregulate deleterious pathways, and the impact of alternative metabolism on outcome.

25.2.2 LPR and Glutamate

Both glutamate and the LPR are elevated in a state of metabolic crisis and can therefore be valuable markers of a patient's condition in the clinical setting (Hillered and Enblad, 2008; Hillered et al., 2006; Vespa et al., 2005). Circumstances that lead to a metabolic crisis and therefore the elevation in these markers are numerous, but are ultimately linked to mitochondrial dysfunction as well as increased metabolic demand (Lindquist and LeRoy, 1942; Vespa et al., 2005, 2007). This vicious cycle, in which energy sources are depleted in the context of increased demand, results in the failure of electrolyte transporters. Electrolytes accumulate within cells and mitochondria, leading to electrolyte imbalance and water influx. The water influx furthers mitochondrial dysfunction because swelling lowers energy production and perpetuates this deleterious cycle (Carre et al., 2013; Soustiel and Larisch, 2010; Unterberg et al., 2004).

Metabolic crisis is associated with poor outcome (Stein et al., 2012; Vespa et al., 2003), but it is inherently difficult to treat. Protocols for the treatment of TBI patients, including the management of glucose, intracranial pressure, mean arterial pressure, and/or cerebral perfusion pressure, rarely and inconsistently improve markers of metabolic distress (Bor-Seng-Shu et al., 2010; Stein et al., 2012; Vespa et al., 2007). At this point, no successful treatments exist to address or improve these markers in TBI patients.

Several decades of research and trials have yielded important insights into the complex paradigm of cerebral metabolism after TBI. However, some basic questions, such as an ideal glucose management range, remain unanswered to this day. Further research is necessary to assist the treating physicians of these critically ill patients. Understanding pathologically related production and consumption of metabolites, describing the network of biochemical pathways activated in TBI, and discovering treatments that control these biochemical processes and improve outcomes are promising avenues for metabolomics studies to investigate.

25.3 METABOLOMICS OF HEAD INJURY

^{1}H NMR-based metabolomics has been used to study diabetes, Alzheimer's disease, cancer, stroke, and subarachnoid hemorrhage (Fan et al., 2009; Floegel et al., 2013; Jung et al., 2011; Lanza et al., 2010; Srivastava et al., 2009; Tukiainen et al., 2008). In this section, we review examples in the literature of metabolomics applied to head injury not previously mentioned. There is a large body of the head injury animal model literature that uses NMR (^{1}H primarily, ^{31}P to study phosphate containing metabolites such as ATP, ^{13}C for tracer studies) to quantify metabolites in cerebral tissue extract. Many of these studies fall into the targeted metabolomics study category previously described, using either endogenous or exogenous metabolites. These metabolomics studies are not reviewed here because our focus is to highlight modern human metabolomics studies with a systems biology-level perspective of TBI.

To the best of the authors' knowledge, there is one example of ^{1}H NMR tissue analysis combined with multivariate PCA in the TBI literature. Using a lateral fluid percussion rat model of head injury, Viant et al. compared metabolite levels 1 hour postinjury in the hippocampus and cortex cerebral tissue and in the plasma (Viant et al., 2005). The authors report discernible changes in cerebral tissue metabolites linked to oxidative stress (ascorbate), excitotoxic damage (glutamate), membrane disruption (choline-containing metabolites), and neuronal injury (N-acetylaspartate), but are not able to discern these changes with plasma alone, perhaps because of systemic dilution of cerebral metabolites. There are a growing number of research studies that use in vivo magnetic resonance spectroscopy to measure cerebral metabolites after injury (Harris et al., 2012; Lin et al., 2012).

NMR-based analysis of CSF collected from TBI patients revealed several changes in the injured population when compared with controls and studied these changes over the first 2 weeks of injury (Toczylowska et al., 2006). The authors reported increased lactate and pyruvate compared to controls, with no discernible difference in the CSF LPR between injured and noninjured patients, which persisted for the 2 weeks studied. The changes observed over the 2 weeks are described, but little is offered as a biological interpretation, and these changes are not studied in the context of secondary injuries. As previously mentioned, ^{1}H NMR analysis of CSF from TBI patients compared with CSF from noninjured controls showed increased propylene glycol and decreased creatinine in the injured cohort (Glenn et al., 2013). Correlation analysis revealed different correlative patterns between metabolites in the two groups. Generalized linear models and multimodel inference revealed a subset of the metabolites measured to be the strongest predictors of outcome and clinical measures of oxygen metabolism and intracranial pressure.

There are two examples in the literature of metabolomics applied to traumatic injuries that exclude head injury. An outcome study on trauma patients, excluding isolated head-injured patients, using blood collected within the first 24 hours of admission was able to discriminate survivors from nonsurvivors with blood lipid biomarkers (Cohen et al., 2010). Study of the metabolomic network altered by a porcine model of trauma, hemorrhagic shock, and resuscitation supports a dual-phase metabolic response to trauma and hemorrhagic shock (Lusczek et al., 2013). It would be interesting to test these findings in a head injured population.

25.4 EXPERIMENTAL PERSPECTIVE ON NMR-BASED METABOLOMICS APPLIED TO HEAD INJURY

In this section, we highlight specific steps in metabolomics protocols where samples from head-injured patients or animals differ from other populations. Additionally, we have a few recommendations from personal experience about metabolomics studies that we hope prove to be useful to researchers.

Samples are stored frozen from the time of collection until the time of NMR sample preparation. There are a variety of techniques to prepare biological samples for NMR analysis that aim to remove proteins and other compounds that reduce the reliability of quantification, while leaving metabolites in solution. There have been concerns about metabolite loss during solvent removal, methods including drying under a flow of nitrogen gas, freeze drying, using a centrifugal concentrator. In our experience, the NMR spectral quality is significantly improved by solvent removal (and replacement with deuterium oxide: ^2H_2O or D_2O) and there is no significant metabolite loss. It should be kept in mind that samples in D_2O will, over time, exchange NMR-invisible deuterium with metabolite hydrogen, so it is best to collect NMR spectra soon after sample preparation. The primary advantage of replacing the solvent with D_2O is reducing the large solvent peak before NMR data collection. To detect the low concentration metabolite peaks in solution, the solvent peak must be suppressed, but spectral techniques for water suppression are known to influence nearby peak quantification. For this reason, reducing the solvent peak before NMR data collection improves quantification.

Spectral peak location is significantly influenced by solvent, temperature, ionic strength, and pH. We have found the use of a 0.1 M phosphate-buffered solution in D_2O, in lieu of checking sample pH before NMR measurements, is advantageous to save time and to keep solvent conditions consistent. Quantification with NMR spectroscopy requires an internal standard in solution at a known concentration. We prepare the phosphate-buffered solution containing the NMR internal standard used for quantification, minimizing variability between samples.

A common internal NMR standard in metabolomics is sodium 3-trimethylsilyltetradeuteriopropionate (TSP) (Pohl and Eckle, 1969). TSP and other NMR internal standards can bind to proteins and/or protein-like macromolecules (Bell et al., 1989; Nowick et al., 2003) that lead to line broadening because of the relatively slow rotational dynamics of macromolecules in solution when compared with the small hydrocarbons of interest. A broad internal standard peak leads to inaccuracies in setting the chemical shift and in metabolite quantification. 4,4-dimethyl-4-silapentane-1-amomonium trifluoroacetate (DSA) was recently introduced, with a claim that this molecule does not interact with peptides in solution as much as other common NMR internal standards do (Nowick et al., 2003), but results in our laboratory do not support this claim.

We recommend ^{13}C-labeled formate as an internal standard for aqueous solutions. Although formate is an endogenous metabolite with a sharp, single spectral peak at 8.44 ppm, ^{13}C-labeled formate does not overlap with endogenous formate because the universal ^{13}C-labeling splits the signal in ^1H NMR spectra (^{13}C satellite peaks located at 8.25 and 8.64 ppm). Unpublished research from our laboratory shows formate does not suffer from line broadening, even in very protein-rich biofluids such as plasma.

Blood serum and blood plasma contain larger quantities of proteins than urine and CSF. NMR spectra of protein-rich biofluids are characterized by broad spectral peaks from protein and lipid macromolecules. Overlap between the sharp metabolite resonances of interest and the broad macromolecule resonances makes quantification difficult and inaccurate, may hide some low concentration peaks under the macromolecule peak envelope, and makes true baseline correction impossible.

Common practice in metabolomics is to change the method of NMR acquisition for protein-rich biofluids; instead of using a simple excitation-acquisition pulse sequence (e.g., *zgpr* on Bruker spectrometers), researchers use the Carr-Purcell-Meiboom-Gill (CPMG) pulse sequence, which includes a relaxation delay between excitation and acquisition (*cpmgprld* on Bruker spectrometers). The idea behind the relaxation filter is to remove the broad protein peaks without significantly affecting metabolite spectral peaks; the slow rotational dynamics of large macromolecules make their NMR signal decay to zero at a faster rate than small, low-molecular-weight molecules.

Modifying the pulse sequence is an extremely straightforward solution to this problem, but we would caution our readers to two potential problems. First, the combination of TSP or DSA internal standards with CPMG spectral acquisition of protein-rich biofluids leads to extremely inaccurate quantification. The broad TSP or DSA peak in these solutions is a sign that the relaxation rate of the internal standard is increased above normal. CPMG spectra acquired under these conditions are characterized by a TSP or DSA peak that is nearly zero and will lead to vast overestimation of metabolite concentrations in solution. Figure 25.1 plots three spectra acquired on a plasma NMR sample and, although there is a noticeable decrease in the DSA peak at 0 ppm in the CPMG spectra, there is very little

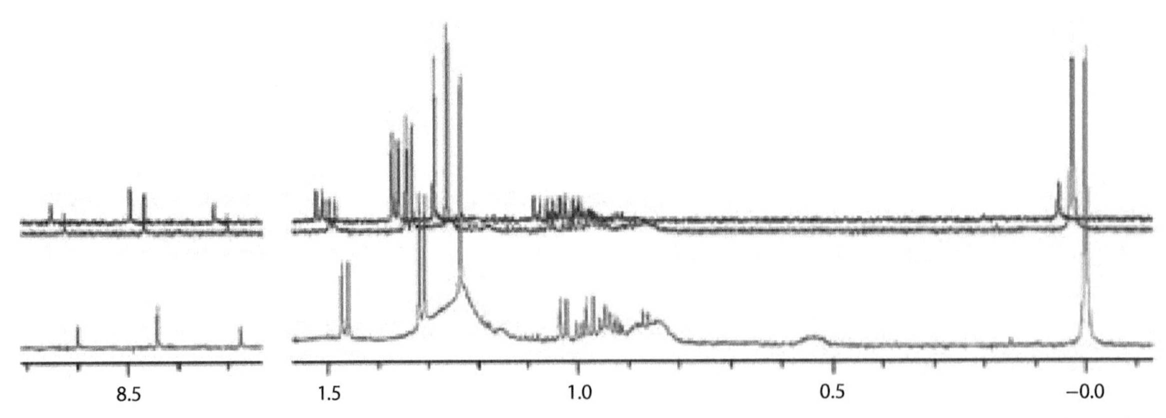

FIGURE 25.1 NMR spectra acquired on a sample prepared from 230 μL blood plasma in 0.1 M phosphate-buffered D_2O solution containing 0.2 mM DSA and 0.1 mM ^{13}C-labeled formate. The three spectra were acquired with a simple excitation-acquisition pulse sequence (green), a CPMG pulse sequence with a 50-ms relaxation delay (red), and a CPMG pulse sequence with a 200-ms relaxation delay (blue). The red and blue CPMG spectra are not rescaled but are shifted upfield from the green spectra and the x-axis so that the peaks do not overlap for visual clarity.

change in the ^{13}C-labeled formate peaks at 8.25 and 8.64 ppm. Additionally, the broad macromolecule peaks overlapping metabolite peaks between 1.5 and 0.8 ppm are reduced as the CPMG relaxation delay increases. The ^{13}C-labeled formate spectral peaks are not changed significantly by acquisition with the CPMG pulse sequence because the relaxation rate of formate is closer to the relaxation rates of metabolites in solution. Second, CSF from TBI patients can contain a relatively high level of blood from cerebral hemorrhage, from disruption of the blood–CSF barrier, from clearance of blood from the brain, and/or from contamination during sample collection through the ventricular drain. Metabolomics of CSF from TBI patients potentially requires some of the techniques typically applied to blood plasma or serum.

An important aspect of the multivariate methods applied in metabolomics, including PCA and/or PLS-DA, is that the input variables have a normal distribution and are homoscedastic. Data transformation and scaling are aspects of multivariate analysis that should be justified before application. In the authors' experience, metabolite variable distributions are highly non-normal as determined by the Shapiro P test of normality, but discussion in the literature of how researchers address data's non-normality is rare.

Because of the complexity of human biology and of experimental designs, there is a strong danger of bias in metabolomics studies. It is human nature to seek a pattern when there is none and to pay close attention to aspects of the experiments that directly relate to hypotheses. Even though researchers are trained take steps to avoid bias and false discoveries, it may still occur unknowingly. There was a metabolomics study published in 2002 that reported ^{1}H NMR metabolomics analysis of human serum accurately diagnosed coronary heart disease (Brindle et al., 2002). A PLS-DA model generated from 80% of the data (the training set) predicted, with a sensitivity of 92% and a specificity of 93%, the presence of coronary heart disease in the test set; the spectral regions that led to class separation correspond to lipids. Subsequent analysis of blood serum identified confounding factors, such as gender and drug treatments, that were not considered in the original publication and that significantly reduced the predictive value of ^{1}H NMR-metabolomics analysis of blood serum in diagnosing coronary heart disease (Kirschenlohr et al., 2006). Avoiding bias in scientific experiments is important, as all scientists know, but because of the interdisciplinary nature and complexity of the omics fields, is worth repeating. There is an excellent opinion piece on the threat of bias to biomarker discovery in cancer research that is applicable to all metabolomics and head injury research (Ransohoff, 2005). We also highly recommend a review article on metabolomics statistical strategies to avoid false discovery (Broadhurst and Kell, 2006).

25.5 CONCLUSION

Prevention of secondary brain injury after trauma is the single most important task of the treating physician. Clinical management of acute, aberrant metabolic changes most certainly represents the cornerstone of what metabolomics has to offer TBI research. Therefore, this endeavor is likely to become one of the most important and relevant forms of post-TBI metabolism investigation.

Metabolic crisis has been associated with poor outcome, yet few treatment options exist that address this significant problem. As a result, exhaustive metabolic analyses, followed by exhaustive validation, will need to be conducted to further delineate the complex metabolic composition of the injured brain and to deepen our understanding of the biochemical pathways involved. It is our belief that only under these conditions will treatment options and interventions emerge. Ultimately, measuring and monitoring postinjury metabolic profiles could assist not only in the treatment of metabolic crisis, but also in predicting an individual's clinical parameters that may deteriorate in the days after injury and in determining long-term prognosis. Metabolomics represents one of the most important tools in the arsenal of the modern-day TBI scientist.

REFERENCES

Ackermann, R. and J. Lear. 1989. Glycolysis-induced discordance between glucose metabolic rates measured with radiolabeled fluorodeoxyglucose and glucose. *J. Cerebr. Blood Flow Metab.* 9: 774–785.

Andersen, B. and A. Marmarou. 1989. Isolated stimulation of glycolysis following traumatic brain injury. In J.T. Hoff and A.L. Betz (eds.). *Intracranial Pressure VII*, pp. 575–580. Springer-Verlag/Berlin, Germany.

Baker, M. 2011. Metabolomics: From small molecules to big ideas. *Nat. Methods.* 8: 117–121.

Bartnik, B., D. Hovda, and P. Lee. 2007. Glucose metabolism after traumatic brain injury: Estimation of pyruvate carboxylase and pyruvate dehydrogenase flux by mass isotopomer analysis. *J. Neurotrauma.* 24: 181–194.

Bartnik, B., R. Sutton, M. Fukushima, N. Harris, D. Hovda, and S.Lee. 2005. Upregulation of pentose phosphate pathway and preservation of tricarboxylic acid cycle flux after experimental brain injury. *J. Neurotrauma.* 22: 1052–1065.

Beckonert, O., M. Coen, H. Keun, Y. Wang, T. Ebbels, E. Holmes et al. 2010. High-resolution magic-angle-spinning NMR spectroscopy for metabolic profiling of intact tissues. *Nat. Protoc.* 5: 1019–1032.

Bell, J., J. Brown, and P. Sadler. 1989. NMR studies of body fluids. *NMR Biomed.* 2: 245–256.

Bergsneider, M., D. Hovda, D. McArthur, M. Etchepare, S.-C. Huang, N. Sehati et al. 2001. Metabolic recovery following human traumatic brain injury based on FDG-PET: Time course and relationship to neurological disability. *J. Head Trauma Rehabil.* 16: 135–148.

Bergsneider, M., D. Hovda, E. Shalmon, D. Kelly, P. Vespa, N. Martin et al. 1997. Cerebral hyperglycolysis following severe traumatic brain injury in humans: A positron emission tomography study. *J. Neurosurg.* 86: 241–251.

Bor-Seng-Shu, E., M. de Lima Oliveira, and M. Teixeira. 2010. Traumatic brain injury and metabolism. *J. Neurosurg.* 112: 1351–1353.

Brindle, J., H. Antti, E. Holmes, G. Tranter, J. Nicholson, H. Bethell et al. 2002. Rapid and noninvasive diagnosis of the presence and severity of coronary heart disease using ^{1}H-NMR-based metabonomics. *Nat. Med.* 8: 1439–1444.

Broadhurst, D. and D. Kell. 2006. Statistical strategies for avoiding false discoveries in metabolomics and related experiments. *Metabolomics.* 2: 171–196.

Brunkhorst, F., C. Engel, F. Bloos, A. Meier-Hellmann, M. Ragaller, N. Weiler et al. 2008. Intensive insulin therapy and pentastarch resuscitation in severe sepsis. *N. Engl. J. Med.* 358: 125–139.

Bylesjo, M., M. Rantalaineen, O. Cloarec, J. Nicholson, E. Holmes, and J. Trygg. 2006. OPLS discriminant analysis: Combining the strength of PLS-DA and SIMCA classification. *J. Chemometrics.* 20: 341–351.

Carre, E., M. Ogier, H. Boret, A. Montcriol, L. Bourdon, and J.-J. Risso. 2013. Metabolic crisis in severely head-injured patients: Is ischemia just the tip of the iceberg? *Front. Neurol.* 11: 146.

Cochran, A., E. Scaife, K. Hansen, and E. Downey. 2003. Hyperglycemia and outcomes from pediatric traumatic brain injury. *J Trauma.* 55: 1035–1038.

Cohen, M., N. Serkova, J. Wiener-Kronish, J.-F. Pittet, and C. Niemann. 2010. ^{1}H-NMR-based metabolic signatures of clinical outcomes in trauma patients—beyond lactate and base deficit. *J. Trauma* 69: 31–40.

Committee on the Review of Omics-Based Tests for Predicting Patient Outcomes in Clinical Trials. 2012. *Evolution of Translational Omics: Lessons Learned and the Path Forward.* The National Academies Press, Washington, DC.

De Salles, A., H. Kontos, D. Becker, M. Yang, J. Ward, R. Moulton et al. 1986. Prognostic significance of ventricular CSF lactic acidosis in severe head injury. *J. Neurosurg.* 65: 615–624.

De Salles, A., J. Muizelaar, and H. Young. 1987. Hyperglycemia, cerebrospinal fluid lactic acidosis, and cerebral blood flow in severely head-injured patients. *Neurosurgery.* 21: 45–50.

Dettmer, K., P. Aronov, and B. Hammock. 2007. Mass spectrometry-based metabolomics. *Mass Spectrom. Rev.* 26: 51–78.

Dunn, W., D. Broadhurst, H. Atherton, R. Goodacre, and J. Griffin. 2011. Systems level studies of mammalian metabolomes: The roles of mass spectrometry and nuclear magnetic resonance spectroscopy. *Chem. Soc. Rev.* 40: 387–426.

Dusick, J., T. Glenn, W. Lee, P. Vespa, D. Kelly, S. Lee et al. 2007. Increased pentose phosphate pathway flux after clinical traumatic brain injury: A [1,2-$^{13}C_2$]glucose labeling study in humans. *J. Cerebr. Blood Flow Metab.* 27: 1593–1602.

Fan, T., A. Lane, R. Higashi, M. Farag, H. Gao, M. Bousamra et al. 2009. Altered regulation of metabolic pathways in human lung cancer discerned by ^{13}C stable isotope-resolved metabolomics (SIRM). *Mol. Cancer.* 8: 41.

Finfer, S., D. Chittock, S. Su, D. Blair, D. Foster, V. Dhingra et al. 2009. Intensive versus conventional glucose control in critically ill patients. *N. Engl. J. Med.* 360: 1283–1297.

Floegel, A., N. Stefan, Z. Yu, K. Muhlenbruch, D. Drogan, H.-G. Joost et al. 2013. Identification of serum metabolites associated with risk of type 2 diabetes using a targeted metabolomic approach. *Diabetes.* 62: 639–648.

Gallagher, C., K. Carpenter, P. Grice, D. Howe, A. Mason, I. Timofeev et al. 2009. The human brain utilizes lactate via the tricarboxylic acid cycle: A ^{13}C-labelled microdialysis and high-resolution nuclear magnetic resonance study. *Brain.* 132: 2839–2849.

Glenn, T., D. Hirt, G. Mendez, D. McArthur, R. Sturtevant, S. Wolahan et al. 2013. Metabolomic analysis of cerebral spinal fluid from patients with severe brain injury. *Acta Neurochir. Suppl.* 118: 115–119.

Glenn, T., D. Kelly, W. Boscardin, D. McArthur, P. Vespa, M. Oertel et al. 2003. Energy dysfunction as a predictor of outcome after moderate or severe head injury: Indices of oxygen, glucose, and lactate metabolism. *J. Cerebr. Blood Flow Metab.* 23: 1239–1250.

Goodman, J., A. Valadka, S. Gopinath, M. Uzura, and C. Robertson. 1999. Extracellular lactate and glucose alterations in the brain after head injury measured by microdialysis. *Crit. Care Med.* 27: 1965–1973.

Green, D., K. O'Phelan, S. Bassin, C. Chang, T. Stern, and S. Asai. 2010. Intensive versus conventional insulin therapy in critically ill neurologic patients. *Neurocrit. Care.* 13: 299–306.

Harris, J., H.-W. Yeh, I.-Y. Choi, P. Lee, P. Lee, N. Berman et al. 2012. Altered neurochemical profile after traumatic brain injury: ^{1}H-MRS biomarkers of pathological mechanisms. *J. Cerebr. Blood Flow Metab.* 32: 2122–2134.

Hillered, L. and P. Enblad. 2008. Nonischemic energy metabolic crisis in acute brain injury. *Crit. Care Med.* 36: 2952–2953.

Hillered, L., L. Persson, P. Nilsson, E. Ronne-Engstrom, and P. Enblad. 2006. Continuous monitoring of cerebral metabolism in traumatic brain injury: A focus on cerebral microdialysis. *Curr. Opin. Crit. Care.* 12: 112–118.

Holmes, E., I. Wilson, and J. Nicholson. 2008. Metabolic phenotyping in health and disease. *Cell.* 134: 714–717.

Jung, J., H.-S. Lee, D.-G. Kang, N. Kim, M. Cha, O.-S. Bang et al. 2011. ^{1}H-NMR-based metabolomics study of cerebral infarction. *Stroke.* 42: 1282–1288.

Katayama, Y., D. Becker, T. Tamura, and D. Hovda. 1990. Massive increases in extracellular potassium and the indiscriminate release of glutamate following concussive brain injury. *J. Neurosurg.* 73: 889–900.

Kirschenlohr, H., J. Griffin, S. Clarke, R. Rhydwen, A. Grace, P. Schofield et al. 2006. Proton NMR analysis of plasma is a weak predictor of coronary artery disease. *Nat. Med.* 12: 705–710.

Lam, A., H. Winn, B. Cullen, and N. Sundling. 1991. Hyperglycemia and neurological outcome in patients with head injury. *J. Neurosurg.* 75: 545–551.

Lanza, I., S. Zhang, L. Ward, H. Karakelides, D. Raftery, and K. Nair. 2010. Quantitative metabolomics by 1H-NMR and LC-MS/MS confirms altered metabolic pathways in diabetes. *PLoS ONE.* 5: e10538.

Lenz, E. and I. Wilson. 2007. Analytical strategies in metabonomics. *J. Proteome Res.* 6: 443–458.

Lin, A., H. Liao, S. Merugumala, S. Prabhu, W. Meehan, and B. Ross. 2012. Metabolic imaging of mild traumatic brain injury. *Brain Imaging Behav.* 6: 208–223.

Lindquist, J. and G. LeRoy. 1942. Studies of cerebral oxygen consumption following experimental head injury. *Surg. Gynecol. Obstet.* 75: 28–33.

Lusczek, E., D. Lexcen, N. Witowski, K. Mulier, and G. Beilman. 2013. Urinary metabolic network analysis in trauma, hemorrhagic shock, and resuscitation. *Metabolomics.* 9: 223–235.

Marcoux, J., D. McArthur, C. Miller, T. Glenn, P. Villablanca, N. Martin et al. 2008. Persistent metabolic crisis as measured by elevated cerebral microdialysis lactate-pyruvate ratio predicts chronic frontal lobe brain atrophy after traumatic brain injury. *Crit. Care Med.* 36: 2871–2877.

Meier, R., S. Bechir, M ad Ludwig, J. Sommerfeld, M. Keel, P. Steiger, R. Stocker et al. 2008. Differential temporal profile of lowered blood glucose levels (3.5 to 6.5 mmol/l versus 5 to 8 mmol/l) in patients with severe traumatic brain injury. *Crit. Care.* doi:10.1186/cc6974.

Meierhans, R., M. Bechir, S. Ludwig, J. Sommerfeld, G. Brandi, C. Haberthur et al. 2010. Brain metabolism is significantly impaired at blood glucose below 6 mM and brain glucose below 1 mM in patients with severe traumatic brain injury. *Crit. Care.* doi:10.1186/cc8869.

Nicholson, J., E. Holmes, J. Kinross, A. Darzi, Z. Takats, and J. Lindon. 2012. Metabolic phenotyping in clinical and surgical environments. *Nature.* 491: 384–392.

Nicholson, J., J. Lindon, and E. Holmes. 1999. 'Metabonomics': Understanding the metabolic responses of living systems to pathophysiological stimuli via multivariate statistical analysis of biological NMR spectroscopic data. *Xenobiotica.* 29: 1181–1189.

Nowick, J., O. Khakshoor, M. Hashemzadeh, and J. Brower. 2003. DSA: A new internal standard for NMR studies in aqueous solution. *Org. Lett.* 5: 3511–3513.

Obrist, W., T. Gennarelli, H. Segawa, C. Dolinskas, and T. Langfitt. 1979. Relation of cerebral blood flow to neurological status and outcome in head-injured patients. *J. Neurosurg.* 51: 292–300.

Pohl, L. and M. Eckle. 1969. Sodium 3-trimethylsilyltetradeuteriopropionate, a new water-soluble standard for ^{1}H-NMR. *Angew. Chem. Int. Ed.* 5: 381.

Preiser, J., P. Devos, S. Ruiz-Santana, C. Melot, D. Annane, J. Groeneveld et al. 2009. A prospective randomised multicentre controlled trial on tight glucose control by intensive insulin therapy in adult intensive care units: The Glucontrol study. *Intensive Care Med.* 35: 1738–1748.

Psychogios, N., D. Hau, A. Guo, R. Mandal, S. Bouatra, I. Sinelnikov et al. 2011. The human serum metabolome. *PLoS ONE.* 6: e16957.

Ransohoff, D. 2005. Bias as a threat to the validity of cancer molecular-marker research. *Nat. Rev.* 5: 142–149.

Rovlias, A. and S. Kotsou. 2000. The influence of hyperglycemia on neurological outcome in patients with severe head injury. *Neurosurgery.* 46: 335–342.

Soustiel, J. and S. Larisch. 2010. Mitochondrial damage: A target for new therapeutic horizons. *Neurotherapeutics.* 7: 13–21.

Soustiel, J. and G. Sviri. 2007. Monitoring of cerebral metabolism: Non-ischemic impairment of oxidative metabolism following severe traumatic brain injury. *Neuro. Res.* 29: 654–660.

Srivastava, N., S. Pradhan, G. Gowda, and R. Kumar. 2009. In vitro, high-resolution ^{1}H and ^{31}P NMR based analysis of the lipid components in the tissue, serum, and CSF of the patients with primary brain tumors: One possible diagnostic view. *NMR Biomed.* 23: 113–122.

Stein, N., D. McArthur, M. Etchepare, and P. Vespa. 2012. Early cerebral metabolic crisis after TBI influences outcome despite adequate hemodynamic resuscitation. *Neurocrit. Care* 17: 49–57.

Toczylowska, B., M. Chalimoniuk, M. Wodowska, and E. Mayzner-Zawadzka. 2006. Changes in concentration of cerebrospinal fluid components in patients with traumatic brain injury. *Brain Res.* 1104: 183–189.

Tukiainen, T., T. Tynkkynen, V.-P. Makinen, P. Jylanki, A. Kangas, J. Hokkanen et al. 2008. A multi-metabolite analysis of serum by ^{1}H NMR spectroscopy: Early systemic signs of Alzheimer's disease. *Biochem. Biophys. Res. Commun.* 375: 356–361.

Ulrich, E., H. Akutsu, J. Doreleijers, Y. Harano, Y. Ioannidis, J. Lin et al. 2008. BioMagResBank. *Nucleic Acids Res.* 36: D402–D408.

Unterberg, A., J. Stover, B. Kress, and K. Kiening. 2004. Edema and brain trauma. *Neuroscience.* 129: 1021–1029.

Van den Berghe, G., K. Schoonheydt, P. Becx, F. Bruyninckx, and P. Wouters. 2005. Insulin therapy protects the central and peripheral nervous system of intensive care patients. *Neurology.* 64: 1348–1353.

Van den Berghe, G., A. Wilmer, G. Hermans, W. Meersseman, P. Wouters, I. Milants et al. 2006. Intensive insulin therapy in the medical ICU. *N. Engl. J. Med.* 354: 449–461.

Van den Berghe, G., P. Wouters, F. Weekers, C. Verwaest, F. Bruyninckx et al. 2001. Intensive insulin therapy in critically ill patients. *N. Engl. J. Med.* 345: 1359–1367.

Verweij, B., J. Muizelaar, F. Vinas, P. Peterson, Y. Xiong, and C. Lee. 2000. Impaired cerebral mitochondrial function after traumatic brain injury in humans. *J. Neurosurg.* 93: 815–820.

Vespa, P., M. Bergsneider, N. Hattori, H.-M. Wu, S.-C. Huang, N. Martin et al. 2005. Metabolic crisis without brain ischemia is common after traumatic brain injury: A combined microdialysis and positron emission tomography study. *J. Cerebr. Blood Flow Metab.* 25: 763–774.

Vespa, P., R. Boonyaputthikul, D. McArthur, C. Miller, M. Etchepare, M. Bergsneider et al. 2006. Intensive insulin therapy reduces microdialysis glucose values without altering glucose utilization or improving the lactate/pyruvate ratio after traumatic brain injury. *Crit. Care Med.* 34: 850–856.

Vespa, P., D. McArthur, K. O'Phelan, T. Glenn, M. Etchepare, D. Kelly et al. 2003. Persistently low extracellular glucose correlates with poor outcome 6 months after human traumatic brain injury despite a lack of increased lactate: A microdialysis study. *J. Cerebr. Blood Flow Metab.* 23: 865–877.

Vespa, P., D. McArthur, N. Stein, S.-C. Huang, W. Shao, M. Filippou et al. 2012. Tight glycemic control increases metabolic distress in traumatic brain injury: A randomized controlled within-subjects trial. *Crit. Care Med.* 40: 1923–1929.

Vespa, P., C. Miller, D. McArthur, M. Eliseo, M. Etchepare, D. Hirt et al. 2007. Nonconvulsive electrographic seizures after traumatic brain injury result in a delayed, prolonged increase in intracranial pressure and metabolic crisis. *Crit. Care Med.* 35: 2830–2836.

Vespa, P., K. O'Phelan, D. McArthur, C. Miller, M. Eliseo, D. Hirt et al. 2007. Pericontusional brain tissue exhibits persistent elevation of lactate/pyruvate ratio independent of cerebral perfusion pressure. *Crit. Care Med.* 35: 1153–1160.

Viant, M., B. Lyeth, M. Miller, and R. Berman. 2005. An NMR metabolomic investigation of early metabolic disturbances following TBI in a mammalian model. *NMR Biomed.* 18: 507–516.

Wishart, D., D. Tzur, C. Knox, R. Eisner, A. Guo, N. Young et al. 2007. HMDB: The Human Metabolome Database. *Nucleic Acids Res.* 35: D521–D526.

Xu, Y., D. McArthur, J. Alger, M. Etchepare, D. Hovda, T. Glenn, S. Huang, I. Dinov, and P. Vespa. 2010. Early nonischemic oxidative metabolic dysfunction leads to chronic brain atrophy in traumatic brain injury. *J. Cerebr. Blood Flow Metab.* 30: 883–894.

Young, B., L. Ott, R. Dempsey, D. Haack, and P. Tibbs. 1989. Relationship between admission hyperglycemia and neurologic outcome of severely brain-injured patients. *Ann. Surg.* 210: 466–472.

Zhang, S., G. Gowda, T. Ye, and D. Raftery. 2010. Advances in NMR-based biofluid analysis and metabolite profiling. *Analyst.* 135: 1490–1498.

26 The Emerging Impact of microRNAs in Neurotrauma Pathophysiology and Therapy

Oneil G. Bhalala

CONTENTS

26.1 INTRODUCTION

Neurotrauma results in significant morbidity and mortality, due in large part to the vast array of cellular changes occurring in the central nervous system (CNS). Elucidating the complex nature of such transformations is critical in understanding the resulting pathogenesis. MicroRNAs (miRNAs) are 20–24 nucleotide long RNA molecules that regulate cellular function epigenetically. As they are implicated in various CNS injuries, identifying how they are affected by neurotrauma can provide insight into the molecular networks regulating cellular responses. In this chapter, the diverse roles of miRNAs in stroke and traumatic brain injury (TBI) obtained from human and animal studies are explored. Their potential as biomarkers for the type and severity of injury is also considered. Finally, the promise of miRNA-based therapeutics in improving outcomes for stroke and TBI is discussed.

Neurotrauma, such as stroke and TBI, are common pathologies that affect more than 10 million people in the United States alone (Coronado et al., 2011; National Stroke Association, 2014). These diseases are a significant cause of morbidity and mortality, primarily because of a lack of effective diagnostic tools and therapies. A major impediment for successful outcomes is the incomplete understanding of mechanisms underpinning these pathologies because cellular and molecular changes are dynamic and complex. Therefore, it is imperative to develop sound insights into stroke and TBI pathogenesis.

Capturing a snapshot of the multitude of changes after injury can be accomplished by using microarrays to identify altered expression levels of the transcriptome (Munro and Perreau, 2009). These studies have shown that microRNAs (miRNAs) are as important as messenger RNAs (mRNAs) in regulating both central nervous system (CNS) homeostasis and pathology (Qureshi and Mehler, 2012; Salta and DeStrooper, 2012). miRNAs are a family of short noncoding RNAs (ncRNAs)—in contrast to long protein-coding mRNAs—that epigenetically regulate gene function by inhibiting mRNA translation (Carthew and Sontheimer, 2009). A single miRNA can antagonize numerous distinct mRNAs and, conversely, a single mRNA can be targeted by a number of different miRNAs. Because of these types of interactions, miRNAs are thought of as potent regulators of gene networks (Gurtan and Sharp, 2013), and have been implicated in neural development and disease (Bian and Sun, 2011; Meza-Sosa et al., 2012). Hence they have the potential to serve as important markers of as well as therapeutic for CNS injury.

In this chapter, the biogenesis of miRNAs and their roles in CNS development are summarized. Studies describing the changes and effects of miRNAs in patients with—and in animal models of—stroke and TBI as well as their implications for use as biomarkers and diagnostics tools are extensively discussed. Furthermore, common themes between these injuries are identified. Finally, the development and promise of miRNA-based therapies is highlighted.

26.2 miRNA BIOGENESIS AND FUNCTION

miRNAs are single-stranded ncRNA molecules composed of 20–24 nucleotides (Bartel, 2004). The biogenesis of miRNAs begins with transcription, mostly by RNA polymerase II, from genomic DNA, yielding primary miRNA transcripts (pri-miRNA) that can be thousands of nucleotides in length (Figure 26.1) (Lee et al., 2002). Similar to mRNAs, transcription of pri-miRNAs is governed by adjacent DNA sequences, allowing for upstream transcription control (Lee et al., 2004). However, unlike mRNAs, pri-miRNAs form functional secondary structures, termed stem-loops. A stem-loop structure contains a RNA sequence of approximately 24 nucleotides that are separated from a nearly complementary sequence of 24 nucleotides by a few nucleotides. These complementary sequences base-pair with one another, forming the stem-loop structure (Zeng and Cullen, 2003). These loops are significant because they are recognized by the RNAse III endonuclease Drosha, which liberates stem-loops from the pri-miRNA to yield precursor-miRNA (pre-miRNA) (Basyuk et al., 2003; Han et al., 2004).

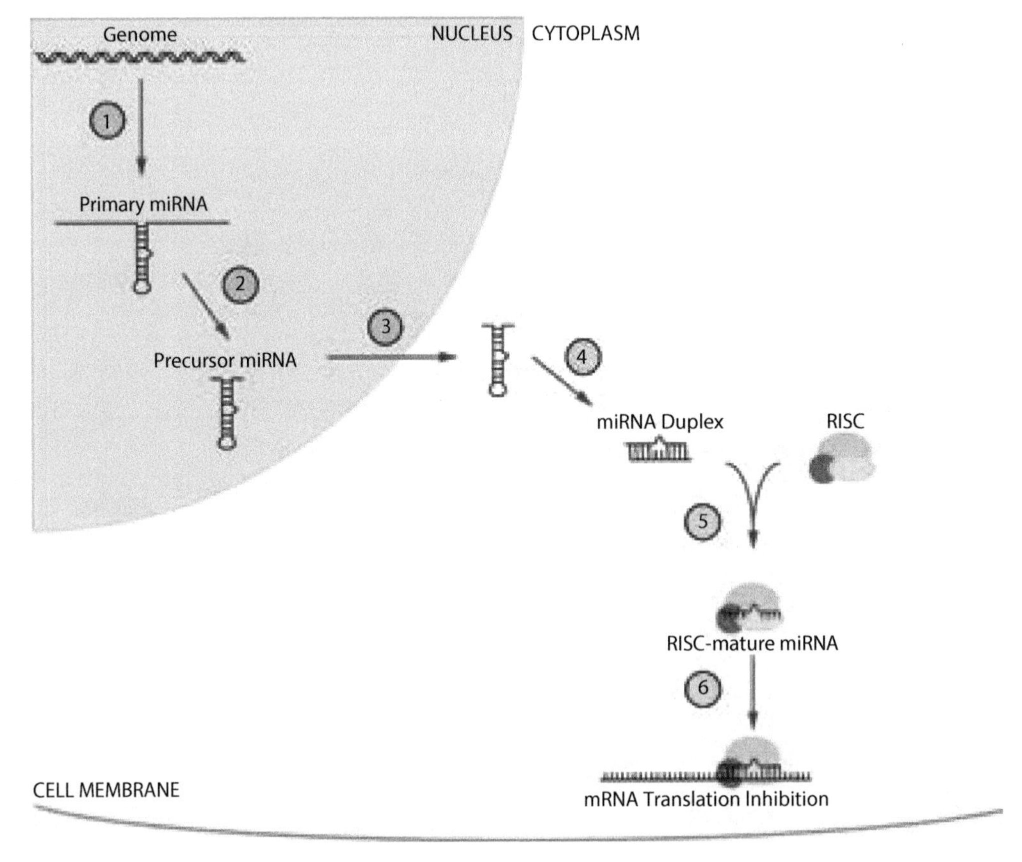

FIGURE 26.1 miRNA biogenesis. (1) Primary miRNA transcript is transcribed from genomic DNA by RNA polymerase II. (2) The precursor miRNA is liberated from primary miRNA by DROSHA. (3) Precursor miRNA is exported from the nucleus by exportin-5. (4) Dicer-TRBP enzyme complex cleaves the 20- to 24-nucleotide miRNA duplex. (5) RISC associates with the mature miRNA strand, which then targets mRNAs with complementary miRNA binding sites (6) to inhibit translation.

Pre-miRNAs are transported from the nucleus into the cytoplasm by exportin-5 (Lund et al., 2004; Yi et al., 2003), where they are processed by Dicer and TAR RNA-binding protein (TRBP) (Chendrimada et al., 2005; Hutvagner et al., 2001; Lau et al., 2001). This processing yields a double-stranded RNA molecule composed of the 20–24 nucleotide-long mature miRNA strand and the complementary strand of similar length (designated miRNA*). Characteristically, this duplex molecule contains a two-nucleotide overhang on the 3' terminus of each strand and a 5' phosphate group (Schwarz et al., 2003). The mature miRNA strand typically exerts biological activity, whereas the miRNA* strand is inert and degraded. However, exceptions to this rule are becoming increasingly common, to the point where both strands of the duplex may actually be functional (Long and Lahiri, 2012; Nass et al., 2009).

Mature miRNAs exert their effects by binding to mRNA molecules. These interactions are mediated by the RNA-induced silencing complex (RISC), which minimally consists of Dicer, TRBP, and the Argonaute family of proteins (Hammond et al., 2000; Martinez et al., 2002; Peters and Meister, 2007). Human and mouse genomes encode for four Argonaute proteins (Ago1–4) that are involved in the miRNA pathway (Hock and Meister, 2008), where they facilitate miRNA loading into RISC. Interestingly, Ago2 is the only member of the family that exhibits endonuclease activity, thereby allowing it to act as a surrogate for Dicer function (Liu et al., 2004).

The net result of the RISC-miRNA association is targeting of mRNAs to reduce protein expression. Targeting is not indiscriminate, but is, instead, governed by nucleotides 2–7 at the 5' end of the miRNA, termed the "seed sequence" (Lewis et al., 2003). The seed sequence binds to complementary regions on the mRNA, which are typically conserved and found in the 3' untranslated region (UTR) (Bartel, 2009). However, other locations for miRNA binding sites have been described. The association of RISC-miRNA with the mRNA diminishes protein production because of the combination of translation inhibition and mRNA destabilization, with destabilization having a greater role (Baek et al., 2008; Fabian and Sonenberg, 2012; Guo et al., 2010).

Given the nature of the miRNA–mRNA interaction, there are many ways this system can control protein levels (Bartel, 2009). For example, mRNAs can contain multiple copies of a given seed sequence as well as sequences for multiple different miRNAs. Also, many different miRNAs can target similar mRNAs if they contain the same seed sequence. The impact of miRNA–mRNA bindings has been revealed through quantitative proteomic analyses, where it was found that a single miRNA can downregulate expression of hundreds of proteins under physiologic conditions (Selbach et al., 2008). Interestingly, on average, the magnitude of the change is less than four-fold. However, not all mRNAs are targeted by miRNAs because it is estimated that only one-third of mRNAs contain miRNA binding sites (Esquela-Kerscher and Slack, 2006). Transcripts that escape miRNA-mediated repression because of absent or short 3'UTRs have been found to regulate key cellular processes such as DNA repair, protein synthesis, and proliferation, demonstrating a difference between housekeeping and cell-specific genes (Sandberg et al., 2008; Stark et al., 2005).

Understanding the extent of miRNA effects depends on which genes are targeted. Identifying these genes in vivo or in vitro is an arduous task because the number of possible miRNA–mRNA combinations may be prohibitively large, requiring significant amounts of time and reagents (Chi et al., 2009; Hafner et al., 2010). However, the use of in silico algorithms makes this more feasible because the putative target list can be substantially shortened. Because miRNA–mRNA interactions are governed by Watson-Crick base-pairing, programs have been developed to predict the strength of miRNA–mRNA binding (Lhakhang and Chaudhry, 2012). These predictions are based on various characteristics such as the degree of complementarity between the miRNA (particularly in the seed sequence) and mRNA, thermodynamic stability of the duplex, evolutionary conservation of binding sites, the number of binding sites, and 3' UTR secondary structures. Certain programs, such as PicTar, miRanda, and TargetScan emphasize site conservation, whereas others (PITA and rna22) place more weight on thermodynamics (Witkos et al., 2011). Not surprisingly, the output of mRNA targets can be vastly different for a given miRNA. Online resources have been developed to integrate results from various prediction algorithms to provide a consensus list (Lhakhang and Chaudhry, 2012). However, this is not always better because consensus results have been shown to perform worse than individual programs at predicting miRNA targets (Alexiou et al., 2009). An important consideration for all algorithms is that mRNA targets need to be validated in vitro and/or in vivo because many of these predicted targets are cell- and context-dependent. However, a benefit of such tools is that gene-ontological analyses can be performed with mRNA targets, providing a rough indication of which molecular pathways are implicated. Therefore, although not perfect, miRNA in silico prediction tools are useful in narrowing the scope of analysis so that meaningful results can be attained expediently.

26.3 THE DYNAMIC ROLES OF miRNAs IN CNS DEVELOPMENT

lin-4 was the first identified miRNA in *Caenorhabditis elegans* where it targeted the 3'UTR of *lin-14* (Lee et al., 1993; Wightman et al., 1993). Posttranscriptional regulation of the mRNA was suggested when overexpression of *lin-4* led to a decrease in *lin-14* protein without a change in *lin-14* mRNA levels. The second miRNA to be discovered, *let-7*, was equally as significant because homologs in other species exist, indicating a conserved phenomenon (Pasquinelli et al., 2000; Reinhart et al., 2000; Slack et al., 2000). Subsequently, large-scale sequencing studies have identified numerous miRNAs expressed in various tissues conserved amongst many species (Lagos-Quintana et al., 2001, 2002, 2003; Lau et al., 2001; Lee and Ambros, 2001).

The ubiquitous expression of miRNAs suggests that they are critical for organism development and physiology. Strong evidence was provided when complete knock-out of Dicer, leading to a global depletion of miRNAs, resulted in mouse embryos arrested at embryonic day 7.5 (Bernstein et al., 2003). A more CNS-specific role for miRNAs was observed when Dicer was ablated from Emx1- and CamKII-expressing and derived cells, yielding smaller cortices (Davis et al., 2008; De Pietri Tonelli et al., 2008). Conditional deletion of Dicer in Wnt1-derived cells resulted in loss of neural crest–derived cells as well as malformations of the midbrain and cerebellum (Huang et al., 2010; Zehir et al., 2010). Furthermore, whole-embryo loss of Ago2 prevented neural tube closure (Liu et al., 2004). These studies implicate miRNAs as being essential throughout development.

Because the CNS contains a heterogeneous population of cells, it comes as no surprise that many miRNAs are expressed in lineage and cell-type specific patterns (Bak et al., 2008; Lagos-Quintana et al., 2003; Landgraf et al., 2007; Miska et al., 2004; Smith et al., 2010; Zheng et al., 2010). For example, *let*-7 is expressed in neural stem cells (NSCs), where it targets the nuclear receptor TLX and the cell cycle regulator cyclin D1 to modulate the balance between NSC proliferation and differentiation (Zhao et al., 2010a). Similarly, miRNA-9 (miR-9) is found to be expressed in neurogenic regions and regulates NSC proliferation by binding the 3'UTR of TLX. Neuronal miRNAs include miR-124, whose expression increases upon differentiation and reaches its maximal levels in mature neurons (Cheng et al., 2009; Lagos-Quintana et al., 2002). Consequently, it is one of the most abundant neural miRNAs. Targets of miR-124 include Dlx2 (a neuronal-subtype regulating transcription factor), Jag1 (a Notch ligand receptor), and Sox9 (a transcription factor involved in adult neurogenesis) (Cheng et al., 2009; Liu et al., 2011). miR-128 is another abundant neuronal miRNA that has been shown to regulate membrane excitability and motor behavior (Smirnova et al., 2005; Tan et al., 2013a).

Oligodendrocytic miRNAs include miR-219 and -338 (He et al., 2012), which promote differentiation by targeting PDGFRα, Sox6, and Hes5 proteins involved in precursor cell proliferation (Dugas et al., 2010; Zhao et al., 2010b). Contrastingly, proliferation of oligodendrocyte precursor cells is promoted by miR-17-92 cluster-mediated repression of AKT pathway inhibitors (Budde et al., 2010). Overexpression of miR-23a in vivo increases CNS myelination in mice, providing a possible mechanism for combating myelin diseases (Lin et al., 2013). Differentiation of glial precursor cells into oligodendrocytes or astrocytes is, in part, regulated by miR-184, which downregulates astrocytic differentiating proteins (Letzen et al., 2010). miR-21, -23, -26, and -29a have been found to be expressed in astrocytes (Bhalala et al., 2012; Sahni et al., 2010; Smirnova et al., 2005). Dicer ablation from astrocytic lineage confirms the importance of miRNAs because this results in aberrant astrogliosis and astrocytic function (Tao et al., 2011; Zheng et al., 2010, 2012).

26.4 miRNAs AS A MEANS TO UNDERSTAND NEUROTRAUMA PATHOPHYSIOLOGY

As discussed previously, miRNA functions shed light on pathways important for CNS development and physiology. Similarly, miRNAs also have a significant role in CNS injuries, where they modulate molecular networks. In the subsequence sections, miRNAs' roles in stroke and TBI as well as in the healing process are examined.

26.4.1 Contribution of miRNAs to Stroke Pathology

26.4.1.1 Epidemiology of Stroke

Stroke is a global burden, affecting one in six individuals in their lifetime (Bushnell, 2008a). In the United States, deaths from stroke have steadily decreased from 60.9 per 100,000 in 2000 to 38.9 in 2009 (Go et al., 2013). Despite this decreasing trend in the United States and in other developed countries, lower- and middle-income countries are experiencing a large increase in prevalence because 85% of stroke cases are occurring in the latter countries. Moreover, of the 72 million disability-adjusted life years lost to stroke each year, these countries account for nearly 90% of them (Lopez et al., 2006). The costs are also staggering at $38.6 billion in the United States alone in 2009; there, the mean lifetime cost to a stroke patient is approximately $140,000 (Go et al., 2013).

26.4.1.2 Stroke miRNA Microarray Studies

At an organ level, the effects of a stroke can be significant and long-lasting, with extensive tissue damage. Analysis of how and which miRNAs are affected in a stroke can provide insight into the cellular and molecular changes that transpire. miRNA microarrays are very powerful tools as they allow for a high-level analysis of global changes that occur after injury. In a rat middle cerebral artery occlusion (MCAO) model for stroke, 32 miRNAs were found to be significantly altered in the acute phase (within 6 hours) and the late phase (7–14 days after injury) (Table 26.1) (Gubern et al., 2013). Interestingly, some of the levels for specific miRNAs stayed either elevated or depressed, whereas others were altered one direction acutely and then the other direction thereafter. In another rat MCAO microarray, 334 miRNAs were found to be differentially regulated, with most of them decreased in the first 24 hours followed by an increase over the next 6 days (Liu et al., 2013a). Gene-ontology analyses of the miRNA targets revealed that genes involved in energy regulation were initially targeted in the first 6 hours, followed by those involved in oxidative stress and excitotoxicity within the first 24 hours. In the later phase of recovery, miRNAs repressing the transforming growth factor-β pathway were downregulated, allowing for fibrosis and healing to proceed. Changes in the rat hippocampus after MCAO were also observed, where, of the 266 miRNAs detected on the microarray chip, 23 were upregulated and 32 were

TABLE 26.1
Summary of Stroke miRNA Microarray Studies

Study	Tissue Analyzed	Number of miRNAs Increased at Time Points Studied	Number of miRNAs Decreased at Time Points Studied	Bioinformatics-Predicted Pathways
Liu et al., 2011	Rat brain	0 HPI: 3 3 HPI: 5 6 HPI: 5 12 HPI: 6 1 DPI: 17 2 DPI: 26 3 DPI: 26 5 DPI: 31 7 DPI: 30	0 HPI: 11 3 HPI: 7 6 HPI: 10 12 HPI: 23 1 DPI: 27 2 DPI: 28 3 DPI: 28 5 DPI: 28 7 DPI: 29	Cell organization Cellular metabolism DNA replication Intracellular signaling GABAergic synaptic function
Gubern et al., 2013	Rat cortex	30 MPI: 9 6 HPI: 7 7 DPI: 4 14 DPI: 4	30 MPI: 8 6 HPI: 6 7 DPI: 4 14 DPI: 3	Nervous system development and function Organ morphology Cellular development Cellular movement Cell cycle Cell organization
Dharap et al., 2009	Rat cortex	3 HPI: 3 6 HPI: 8 12 HPI: 8 1 DPI: 17 3 DPI: 24	3 HPI: 9 6 HPI: 12 12 HPI: 9 1 DPI: 21 3 DPI: 22	Antioxidation Inflammation Ion channels Neurotransmitter receptors Transcription factors Neuroprotection
Yuan et al., 2010	Rat hippocampus	30 MPI: 23 1 DPI: 40	30 MPI: 32 1 DPI: 31	Not reported
Jeyaseelan et al., 2008	Rat brain and blood	*Brain* 1 DPI: 42 2 DPI: 33 *Blood* 1 DPI: 7 2 DPI: 10	*Brain* 1 DPI: 64 2 DPI: 49 *Blood* 1 DPI:13 2 DPI: 15	Aquaporin proteins Cellular regeneration Extracellular matrix remodeling
Liu et al., 2010	Rat hippocampus and blood	*Hippocampus* 1 DPI: 32 *Blood* 1 DPI: 22	*Hippocampus* 1 DPI: 19 *Blood* 1 DPI: 72	Cell cycle Apoptosis Cell morphology Gene expression
Tan et al., 2009	Human peripheral blood	6–18 months postinjury: 138	6–18 months postinjury: 19	Angiogenesis Hematopoietic regulation Immune/lymphocyte regulation Hypoxia Metabolism

DPI, days postinjury; HPI, hours postinjury; MPI, minutes postinjury.

downregulated (Yuan et al., 2010). These studies indicate that modulation of miRNAs following a stroke is dynamic and nonrandom because particular pathways seem to be involved at various times.

An important consideration in these microarray analyses is the timing of the sample collection postinjury because miRNA levels are quickly changing. This was illustrated in another rat MCAO model in which changes to 32 miRNAs were detected 8 hours postinjury compared with only 8 miRNAs at 48 hours postinjury (Jeyaseelan et al., 2008). This may be partially due to modifications of miRNA processing proteins. Although extensive studies in CNS tissue have not yet been carried out, results from other organ systems shed light on how ischemia regulates miRNA biogenesis and stability. In pulmonary artery smooth muscle cells, hypoxic conditions increase levels of type 1 collagen prolyl-4-hydroxylase, which leads to hydroxylation of Ago2 and increased association of hydroxylated-Ago2 with HSP90

(Wu et al., 2011). This is significant because HSP90 facilitates Ago2 translocation into RISC, resulting in increased miRNA formation. Epidermal growth factor receptor has been shown to negatively regulate Ago2 by phosphorylating it at tyrosine 293 (Shen et al., 2013). The consequence of this posttranslational modification of Ago2 is reduced binding with Dicer, and hence, decreased miRNA function. Dicer mRNA and protein is also affected directly by chronic hypoxia, where levels were decreased in human umbilical vein endothelial cells (HUVEC) and in mouse tissue, including in the CNS (Ho et al., 2012). HUVEC also respond to hypoxia by upregulating the staphylococcal nuclease domain-containing protein Tudor-SN, which incorporates into RISC to reduce miRNA processing of the miR-17-92a cluster (Heinrich et al., 2013). The observation that most tissues contain unprocessed miRNA precursors (Obernosterer et al., 2006), in conjunction with the above studies, indicate that there must be significant regulation (both in homeostatic and pathological states) of miRNA processing machinery.

26.4.1.3 Role of Specific miRNAs in Stroke

Although microarrays provide for a general understanding of miRNA changes after stroke, studies of individual miRNAs are needed to decipher their specific effects. In one such study, levels of the brain-enriched miRNA, miR-124, was increased in the penumbra, but not in the ischemic core, of a mouse MCAO model (Liu et al., 2013c). The inhibitor of apoptosis-stimulating protein of p53 (iASPP) was identified as a target of miR-124 because it contains two binding sites in the 3'UTR. This was confirmed by attaching the binding sites 3' to a luciferase-encoding transcript and observing reduced luciferase activity in the presence of miR-124. Moreover, luciferase activity returned to normal when the miR-124 binding sites were mutated as to no longer be complementary to miR-124. iASPP is an inhibitor of p53-mediated apoptosis, so its repression by miR-124 makes cells vulnerable to death, as seen in neurons within the penumbra. Interestingly, injecting a miR-124 inhibitor into the ventricles increased iASPP levels and reduced infarct size. In another MCAO study, miR-124 levels were found to be reduced in the infarcted tissue (Zhu et al., 2014). Here, miR-124 was found to target Ku70, a protein that inhibits Bax-mediated apoptosis. Although Ku70 mRNA levels were found to be elevated after MCAO, inhibition of miR-124 further increased Ku70 mRNA and protein expression as well as reduced infarct volume. This collection of studies suggests that, in addition to its functions in neuronal development, miR-124 plays an important role in neuronal survival.

Another miRNA that has differential expression in the ischemic core compared with the penumbra is miR-181a. In contrast to miR-124, miR-181a was found to be increased in the ischemic core and reduced in the penumbra (Ouyang et al., 2012). Grp78, a chaperone protein and a validated target of miR-181a, is found to be decreased in the core compared with the penumbra. In vivo antagonism of miR-181a increased Grp78 and reduced infarct size. This may further help shed light on mechanisms that explain differences in propensity of neurons to die in the two locations.

The subventricular zone (SVZ) of the lateral ventricle is greatly affected after a stroke, where neural progenitor cells (NPCs) proliferate and migrate into the injured tissue (Zhang et al., 2009; Zhang et al., 2006). As miRNAs are involved in physiologic SVZ neurogenesis, it is not surprising they are also involved in this process after a stroke (Lopez-Ramirez and Nicoli, 2013). In fact, the miR-17-92 cluster was found to be increased in SVZ NPCs collected 7 days after a mouse MCAO (Liu et al., 2013d). This cluster consists of six miRNAs (miR-17, -18a, -19a, -20a, -19b-1, and -92a-1) that are transcribed as a single polycistronic unit (Xiao et al., 2008). Expression of this cluster is induced by the Shh pathway (Liu et al., 2013d; Sims et al., 2009). The miR-17-92 cluster, in turn, targets PTEN, a protein that inhibits proliferation and enhances apoptosis, suggesting a mechanism for poststroke neurogenesis. miR-124 was also found to expand the SVZ NPC pool after stroke because its poststroke decrease in the SVZ increases Notch signaling by the derepression of Jag1 (Liu et al., 2011).

miR-210, which is increased in ischemic tissue after a rat MCAO, is another miRNA found to exert its effects through the Notch pathway (Lou et al., 2012). Though its targets have not been elucidated, miR-210 increases Notch1 protein levels in HUVECs, resulting in the formation of capillary-like structures. These findings suggest that Notch is a central pathway modulated after stroke because multiple miRNAs mediate their effects through this signaling to regulate processes such as neurogenesis and angiogenesis.

Neurons are particular vulnerable to death after stroke, in part from changes in miRNAs that regulate survival and apoptosis. In an MCAO model in mouse and oxygen–glucose deprivation model in neurons, miR-29b was increased and found to target the anti-apoptotic protein Bcl2L2 (Shi et al., 2012). miR-181a, which is increased in the CA1 region of the hippocampus within 24 hours after rat forebrain ischemia, is another miRNA that targets the antiapoptotic Bcl-2 family member Bcl-2 (Moon et al., 2013). miR-497 was increased in both a mouse MCAO model and an oxygen–glucose deprivation model (Yin et al., 2010b), where it was found to target Bcl-2 and Bcl-w. Antagonism of this miRNA in vivo reduced infarct size and improved neurological outcomes. Bcl-2 is also targeted by miR-15a in vascular endothelial cells, through the ischemia-mediated reduction of PPARδ (Yin et al., 2010a). Regulation of miR-15a by PPAR may be mediated by Kruppel-like factor 11, which associates with PPARγ to repress miR-15a expression (Yin et al., 2013). Neuronal apoptosis is also inhibited by miR-21, whose upregulation in the penumbra reduces protein expression of the proapoptotic Fas ligand (Buller et al., 2010). Another mechanism of neuronal death is excitotoxicity resulting from excessive extracellular glutamate levels after stroke (Lai et al., 2013). Neuronal regulation of levels for two of the receptors mediating this toxicity, GluR2 and

NR2B, has been observed via miR-223 (Harraz et al., 2012). This exciting finding may identify a novel pathway for modulating neuronal death after stroke.

Neuronal apoptosis can also be triggered by the increase in reactive oxygen species that is observed after ischemia. Proteins, such as superoxide dismutase 2 (SOD2), help protect neurons by quenching reactive oxygen species reactivity (Raha and Robinson, 2001). Interestingly, SOD2 is repressed by miR-145, which is found to be elevated after ischemia (Dharap et al., 2009). In a mouse MCAO model, infusion of miR-145 antagonists into the cerebrospinal fluid (CSF) led to an increase in neuronal SOD2 expression. Moreover, this reduced infarct volume identifies this miRNA as a strong candidate for therapeutic modulation as a means to minimize stroke damage.

The repressor element-1 silencing transcription factor (REST) plays a critical role in neuronal differentiation (Chen et al., 1998). Not surprisingly, REST is reactivated after global ischemia (Calderone et al., 2003). As levels of REST protein increase, they interact with three REST-binding sites upstream of pri-miR-29c, resulting in reduced miR-29c expression (Pandi et al., 2013). Physiological levels of miR-29c bind to and repress expression of DNA methyltransferase 3a (DNMT3a), which creates de novo methylations (Chedin, 2011). Therefore, after stroke, as REST levels increase, miR-29c decreases, leading to increased DNMT3a expression. This may explain some of the neuronal damage that is seen after stroke because cells cannot cope with new and potentially aberrant methylation patterns. In fact, increasing miR-29c decreases both DNMT3a levels and the infarct volume.

Cerebral edema can be a severe consequence of stroke, leading to further damage after the ischemic event. Aquaporin proteins, which function as channels for water movement, increase in expression after a stroke (Bernstein et al., 2003). However, it unclear if this family of proteins augments or attenuates the injury (Manley et al., 2000; Papadopoulos et al., 2004). In a rodent MCAO model, levels of miR-30a-3p and -383, which are predicted to target aquaporin 4 (Aqp4), are decreased as Aqp4 levels are increased after ischemia (Jeyaseelan et al., 2008). Another miRNA that targets Aqp1 and Aqp4 is miR-320a, which is also decreased after an ischemic event (Sepramaniam et al., 2010). Taken together, these studies highlight the role miRNAs can have in stroke pathology by modulating cerebral edema via aquaporin protein levels.

One of the interesting aspects of stroke recovery is the sex differences in outcomes, with women having worse prognoses (Bushnell, 2008b; Lang and McCullough, 2008). A potential mediator of these differences may be X-linked inhibitor of apoptosis (XIAP), an inhibitor of caspases. XIAP is a target of miR-23a, which is increased in female mice after a stroke (Siegel et al., 2011). Surprisingly, miR-23a levels are unchanged in male mice. Consequently, XIAP levels are decreased in females compared with males, resulting in increased caspase activity and neuronal death. Another mediator of sex differences may be the neuroprotective protein insulin-like growth factor-1 (IGF-1) (Selvamani and Sohrabji, 2010). IGF-1 is targeted by two miRNAs, *let-7f* and miR-1, which increase in age in female rats (Selvamani et al., 2012). Furthermore, antagonism of *let-7f* and miR-1 increases IGF-1 levels in the CNS and reduces infarct volumes only in female.

26.4.1.4 Stroke-Associated miRNA Single Nucleotide Polymorphisms

Being able to predict an individual's risk for developing a stroke is a powerful tool to facilitate early interventions and prevent such an occurrence. For example, examining an individual's smoking status, exercise levels, and diet are some of the ways the likelihood of a stroke can be assessed. Another important risk factor for stroke is family history because embedded in family history is a person's genomic architecture (Markus, 2012). With the growth in high-throughput DNA sequencing technology, genome-wide association studies have been used to identify genetic motifs that increase stroke predisposition. Single nucleotide polymorphisms (SNPs) are one way of identifying base-pair changes in genomic sequences that may mark disease risk. Studies in Asian populations have been conducted to identify miRNA-related SNPs that contribute to stroke. In one Chinese study, analysis of 340 stroke patients and 309 healthy controls revealed that SNP rs1056628, which is located in the 3'UTR of matrix metalloproteinase 9 (MMP-9), increased stroke risk by 1.653 (95% confidence interval [CI]: 1.252–2.183, $p < 0.01$) with a C allele compared with an A allele (Yuan et al., 2013). Significantly, this A > C change was located in the binding site for miR-491, where the C-nucleotide reduced miR-491 repression of MMP-9 expression. This may be of clinical significance because increased MMP-9 is associated with increased atherosclerotic plaque instability and ischemic stroke (Newby, 2008). In another study of 296 ischemic stroke patients and 391 healthy Chinese individuals, SNP rs3746444 was identified as having a 1.509 increased risk of stroke with the G allele compared with A (95% CI: 1.151–1.978, $p = 0.003$) (Liu et al., 2013e). This SNP is located in the mature strand of miR-499, which may be implicated in thrombosis and/or circulatory system inflammation.

The A > G SNP in miR-499 was also found in a study of South Koreans (678 ischemic stroke patients and 553 age- and sex-matched controls) (Jeon et al., 2013). This SNP, when present with three other SNPs in miR-146a, -149, and -196a2, increased the stroke risk by 2.480 (95% CI: 1.207–5.096). Moreover, homozygous expression of the G allele compared with the C allele in miR-146a alone yielded a 1.583 risk increase (95% CI: 1.104–2.270, $p = 0.013$). Bioinformatics analysis indicated that these miRNAs may be associated with the vascular damage response. Interestingly, three of the miRNAs identified in the South Korean study (miR-146a, -149, and -196a2) were found not to be significant in the Chinese study mentioned previously (Liu et al., 2013e), suggesting population-specific miRNA SNPs may exist.

A different miRNA was identified in a Taiwanese study of 657 stroke patients and 687 matched controls (Liu et al., 2013b). Here, the CT heterozygous alleles at SNP rs3735590 were found to be protective (compared with homozygous CC alleles) by 0.72 ($p = 0.036$). This SNP is located in the 3'UTR of paraoxonase 1 (PON1) gene, which inhibits low- and high-density lipoprotein oxidation. miR-616 binds to the PON1 mRNA at this SNP; however, the T allele reduces binding affinity compared with the C allele, thereby increasing PON1 expression and potentially reducing plaque formation and embolus. Studies such as these demonstrate that SNPs in either miRNAs or miRNA binding sites affect stroke risk. However, analyses with larger sample sizes are crucial in order to increase the detection power for other such SNPs. In addition, other ethnic groups need to be examined to identity both common and population-specific SNPs.

26.4.2 Using miRNAs to Diagnose Stroke

The studies described previously clearly demonstrate that numerous miRNA expression levels are altered after a stroke. The next step would be to use this information about miRNA changes to determine a stroke diagnosis and prognosis. However, obtaining brain tissue in a suspected patient is not practical because it may lead to further damage. Contrastingly, obtaining peripheral blood samples is more feasible, and may provide an insight into CNS changes. Analysis of blood 1 day post-MCAO in rats revealed that changes in miRNAs can be detected and, importantly, that these changes mirror those observed in the brain (Table 26.2) (Liu et al., 2010). In fact, whole blood analysis of young stroke patients between 6 and 18 months after the event revealed an increase in 138 miRNAs and a decrease in 19 miRNAs (Tan et al., 2009). These preliminary studies suggest that serum miRNAs may have the potential to serve as stroke biomarkers.

Not surprisingly, the brain-enriched miR-124 is detected in rat blood after a stroke, with plasma levels increasing as early as 6 hours after the injury (Weng et al., 2011). Although these levels stayed elevated until 48 hours after, there was no correlation between plasma concentration and infarct size, suggesting that miR-124 may be used as a maker of a stroke event, but not of the magnitude. In a small study of 11 ischemic stroke patients and 14 healthy controls, miR-145 was found to be to two-fold higher ($p = 0.022$) in patient blood (Gan et al., 2012). Plasma levels of miR-424 were lower in another study of 11 stroke patients and 11 matched controls (Zhao et al., 2013). Interestingly, miR-424 concentrations were positively correlated with the patients' performance in activities of daily living (Barthel Index). This raises the possibility of using therapies to increase miR-424 levels in stroke patients and help improve their Barthel Index.

A larger study of 167 stroke patients and 157 healthy controls found miR-21 and miR-221 as strong plasma biomarkers (Tsai et al., 2013). Log-fold elevation in miR-21 increased the likelihood of having had a stroke 6.16-fold (95% CI: 2.82 – 14.64, $p < 0.0001$); conversely, log-fold reduction in miR-221 increased stroke likelihood by 10.38 (95% CI: 4.52–26.45, $p < 0.0001$). miR-30a and -126 were identified as other serum miRNA markers that could identify stroke because they were decreased in patient blood for up to 24 weeks after the injury (Long et al., 2013).

In addition to using circulating miRNA levels to determine if a stroke has occurred, their expression profile can also be used to distinguish the stroke subtype (Tan et al., 2009). Examining peripheral blood in stroke patients aged 18 to 49 years revealed similar expression patterns in 21 miRNAs irrespective of the stroke type (Tan et al., 2013b). However, five miRNAs were found to be upregulated only in large-artery strokes, whereas cardioembolic strokes demonstrated a reduction in three different miRNAs (Table 26.2). *let-7b* is another miRNA that can discriminate between subtypes because it is reduced in large-vessel stroke but higher in all other types (Long et al., 2013). In a rat model, miRNAs could be used to distinguish between ischemic and hemorrhagic strokes (Liu et al., 2010). Similar findings were made in a study of plasma from 15 patients with hemorrhagic stroke, 16 patients with ischemic stroke, and eight healthy volunteers, where 13 miRNAs were found to be specifically altered in hemorrhagic stroke compared to ischemic stroke and no stroke (Guo et al., 2013). These studies highlight the future direction of analyzing peripheral miRNAs as biomarkers to detect the occurrence of a stroke as well as to differentiate stroke etiology.

26.4.3 Contribution of miRNAs to TBI Pathology

26.4.3.1 Epidemiology of TBI

TBI is a significant cause of morbidity and mortality in the general population. In the United States, TBI accounts for more than 50,000 deaths annually (Coronado et al., 2011). However, TBI significantly affects a larger proportion of those serving in the military. A survey of soldiers returning from combat in Iraq indicated that up to 20% of them suffered from at least one TBI (Shively and Perl, 2012). Studies have found that even mild TBI can increase the likelihood of developing posttraumatic stress disorder (Kontos et al., 2013). TBI has been labeled the "invisible" wounds of military service as diagnosis and management are not fully defined yet (Tanielian et al., 2008). Therefore, it is imperative to develop a more comprehensive understanding of the pathophysiology and possible treatment options for TBI (Rosenfeld et al., 2013). Analysis of miRNAs involved in TBI will aid in this understanding because it will provide insights into injury-related gene networks.

26.4.3.2 TBI miRNA Microarray Studies

miRNA microarrays of brain tissue provide a broad view of the changes occurring after TBI. Examination of rat injured cortical tissue revealed the dynamic temporal regulation of miRNAs, with the number of downregulated and upregulated miRNAs peaking at 24 and 72 hours after injury,

TABLE 26.2
Potential miRNA Biomarkers for Neurotrauma

Injury	miRNA Biomarker	Study
Stroke	miR-124 levels elevated >100-fold in plasma from 6 to 48 hours after injury in rat MCAO model	Weng et al., 2011
	miR-10a, -182, -200b, and -298 were elevated and miR-155, -210, -223, and -362-3p were reduced in both brain and blood by >1.5-fold in rat ischemic stroke model compared with control	Liu et al., 2010
	miR-107, -200b, -331–5p, and -672 were elevated and miR-155 and -188-5p were reduced in both brain and blood by >1.5-fold in rat hemorrhagic stroke model compared with control	
	miR-145 levels elevated approximately two-fold in plasma from ischemic stroke patients (n = 32) compared with healthy controls (n = 14)	Gan et al., 2012
	miR-424 levels reduced approximately four-fold in plasma within 72 hours from ischemic stroke patients (n = 11) compared with healthy controls (n = 11)	Zhao et al., 2013
	miR-30a and -126 levels were reduced in blood for up to 24 weeks from ischemic stroke patients (n = 197) compared with healthy controls (n = 50)	Long et al., 2013
	let-7b levels were reduced in patients with large-vessel atherosclerosis compared with other causes of stroke	
	miR-21 levels elevated and miR-221 reduced in serum from ischemic stroke patients (n = 147) compared with healthy controls (n = 150)	Tsai et al., 2013
	miR-130b, -29b, -301a, -339–5p, -532–5p, -634, and -886-5p levels elevated by approximately two-fold in small artery strokes (n = 3) compared with large artery strokes (n = 8) in blood from stroke patients	Tan et al., 2009
	miR-208a, -519d, -605, -634, and -99b* levels elevated in blood from patients with ischemic stroke compared with that from cardioembolic or lacunar stroke patients	Tan et al., 2013b
	miR-377, -767-5p, and -875-3p were reduced in blood from patients with cardioembolic stroke compared with other stroke etiologies	
	miR-27a, -29b, -134, -150*, -197,- 365, -874, -526b, -574-5p, -575, -654-5p, -671-5p, and -1225-5p were elevated in plasma from hemorrhagic stroke patients (n = 15) compared with healthy controls (n = 8)	Guo et al., 2010
TBI	*let-7i* was elevated by >1.5 in serum with 3 hours from a rat TBI model	Balakathiresan et al., 2012
	miR-16 and -92a were reduced and miR-765 were elevated in plasma within 72 hours from patients with severe TBI (n = 10) compared with healthy controls (n = 10)	Redell et al., 2010
	miR-16 and -92a were elevated in plasma within 10 hours from patients with mild TBI (n = 11) compared with healthy control (n = 10)	
	miR-9 and -451 were elevated in CSF from patients with TBI (n = 11) compared with healthy controls (n = 17)	Patz et al., 2013

respectively (Table 26.3) (Lei et al., 2009). The total number of miRNAs expressed at different time points was 136 at 6 hours, 118 at 24 hours, 149 at 48 hours, and 203 at 72 hours, indicating that miRNA modulation continues well after the initial injury.

The hippocampus is exquisitely sensitive to TBI; structural changes were found in 59% of patients (Orrison et al., 2009). Significantly, changes such as atrophy can be predictive of long-term cognitive and memory deficits (Cohen et al., 2007). The effects of TBI on rat hippocampi have been assessed where of the 444 assayed miRNAs, 35 were upregulated and 50 were downregulated within the first 24 hours (Redell et al., 2009). Moreover, 12 of the upregulated miRNAs were not detected in sham-control animals, suggesting that these miRNAs may serve injury-specific functions. Gene-ontological analyses of all altered miRNAs identified signal transduction, transcription, proliferation, and differentiation as regulated pathways. Analysis of hippocampi at later time points found 21 miRNAs with altered expression levels at 24 hours (Hu et al., 2012); at 7 days postinjury, only 10 miRNAs were affected, of which five were also previously affected at 24 hours. Bioinformatics analyses of targets suggest that the temporal changes to the miRNA profile have various effects at different stages after injury. Targeted genes at 24 hours were involved with inhibition of cell death and protection from energy exhaustion; those at 7 days were involved with cell repair and structural remodeling. These studies illuminate some of the consequences of dynamic changes to miRNAs profiles.

26.4.3.3 Role of Specific miRNAs in TBI

So far, the roles of two miRNAs have been described in hippocampi after TBI. The antiapoptotic miRNA miR-21 was found to be upregulated in CA3 and dentate gyrus (Redell et al., 2011). Its levels peaked 3 days after injury and

TABLE 26.3
Summary of TBI miRNA Microarray Studies

Study	Tissue Analyzed	Number of miRNAs Increased at Time Points Studied	Number of miRNAs Decreased at Time Points Studied	Bioinformatics-Predicted Pathways
Lei et al., 2009	Rat cortex	6 HPI: 13 1 DPI: 4 2 DPI: 16 3 DPI: 19	6 HPI: 14 1 DPI: 23 2 DPI: 11 3 DPI: 5	Not reported
Redell et al., 2009	Rat hippocampus	3 HPI: 27 1 DPI: 24	3 HPI: 46 1 DPI: 28	Cell differentiation Cell morphology Cell proliferation Transcription Cell cycle
Hu et al., 2012	Rat hippocampus	1 DPI: 8 7 DPI: 3	1 DPI: 13 7 DPI: 7	Apoptosis Inflammation Energy conservation Cell organization Cell remodeling and repair
Balakathiresan, 2012	Rat serum and CSF	3 HPI: 41 1 DPI: 35	3 HPI: 0 1 DPI: 4	Axon guidance Cell adhesion Intracellular signaling Cell organization
Redell et al., 2012	Human plasma	3 DPI: 27	3 DPI: 33	Cell proliferation Cell cycle Apoptosis Angiogenesis

DPI, days postinjury; HPI, hours postinjury.

returned to baseline by 15 days. Similar to its role in cancer, miR-21 targeted PDCD4, a pro-apoptotic protein, suggesting a mechanism by which hippocampal cells try to remain viable after injury. miR-107 is another miRNA that is altered in CA3 as well as in CA1 (Wang et al., 2010). In contrast to miR-21, miR-107 levels are repressed after injury, allowing its target, granulin, to increase in expression. Granulin functions in wound healing and cellular repair, suggesting another mechanism by which the tissue tries to repair itself after TBI.

The cholinergic system is known to be an important regulator of neuroinflammation after TBI (Pavlov and Tracey, 2005), as anti-acetylcholinesterase inhibitors are beneficial in treating patients with blast-induced TBI (Tenovuo, 2005). Identifying miRNAs that can modulate cholinergic signaling can aid in understanding the mechanisms and how they can be manipulated to improve TBI outcomes. Analysis of miRNAs in mice cerebellum, which are particularly vulnerable to blast injuries (Wang et al., 2011), revealed miR-132 was significantly decreased after injury (Valiyaveettil et al., 2013). One of the targets of miR-132 is anti-acetylcholinesterase, thereby explaining the increase in cholinergic signaling. Moreover, this may help shed light on why inhibition of this enzyme is beneficial in TBI treatment (Shaked et al., 2009).

26.4.4 Using miRNAs to Diagnose TBI

A significant impediment to effective TBI therapy is the difficulty in diagnosing an injury as physical and cognitive symptoms may not manifest until weeks after, when the therapeutic window has substantially narrowed. However, as discussed previously with stroke diagnosis, peripheral blood may serve as a window into what is occurring in the brain. Analysis of plasma from 10 healthy controls and 10 patients with severe TBI identified three miRNAs that were significantly altered within the first 72 hours after injury: miR-16 and miR-92a were decreased, whereas miR-765 was increased (Table 26.2) (Redell et al., 2010). A miRNA profile in the first 24 hours after a mild TBI was also established, with an increase in miR-16 and -92a observed.

CSF communicates with both CNS tissue and the circulatory system. Therefore, it is a strong candidate to carry miRNA biomarkers. In a rat model of TBI, *let-7i* was found to be increased in CSF as early as 3 hours after injury (Balakathiresan et al., 2012). Prediction algorithms suggested that this miRNA may modulate S100B and UCH-L1, which have been linked to TBI pathology (Geyer et al., 2009; Papa et al., 2010; Svetlov et al., 2010). Analysis of human CSF found miR-9 and miR-451 increased in microparticles (membrane-sheathed structures containing proteins and nucleic acids) from patients with brain injury compared with noninjured controls (Patz et al., 2013). Although larger studies are needed to confirm these miRNA biomarkers as well as discover new ones, these findings support the goal of identifying diagnostic miRNAs that are both easy to obtain and robust at indicating injury.

26.5 OVERLAP OF STROKE AND TBI miRNA PROFILES

Microarray studies performed on stroke and TBI tissues reveal the kinetics of miRNA modulation. As illustrated in Tables 26.1 and 26.3, many miRNA levels are significantly altered well within the first 24 hours after injury, indicating that mechanisms regulating miRNAs levels are at least partially established within the cell and do not wholly rely on de novo transcription and translation of various signaling molecules and effector proteins. This early change in miRNAs is consistent with in vitro models of CNS injuries (Shih et al., 2011; Ziu et al., 2011). Regulation of miRNA expression also appears to be spatially controlled as changes are detected at the injury site and adjacent structures, indicating strong paracrine and autocrine signaling, although there may be some systemic signaling given the involvement of the immune system. Evidence of miRNAs in CSF and blood suggests that those affected by stroke and TBI may have long-range effects at distant tissues (Liang et al., 2013).

From a network perspective, the patterns of miRNA changes suggest some conserved pathways after injury: apoptosis, inflammation, cell proliferation, and cell differentiation (Tables 26.1 and 26.3). Although the pathways in general may be common between stroke and TBI, some of the exact miRNAs and the associated gene targets mediating these changes appear to be distinct. For example, peripheral serum biomarkers for determining stroke occurrence are different than those used for detecting TBI (Table 26.2). Moreover, miR-499 appears to regulate inflammation in stroke patients, whereas this may be mediated by miR-132 in TBI patients (Liu et al., 2013e; Valiyaveettil et al., 2013). However, some miRNAs are affected in both types of injury, such as miR-21, which may serve as a common cell-death regulator. Interestingly, many of the pathways and miRNAs altered in stroke and TBI are also dysregulated in other CNS injuries, such as spinal cord injury (Bhalala et al., 2013; Liu and Xu, 2011). This suggests that there may be a common set of pathways activated after CNS injury, regardless of etiology.

26.6 THE THERAPEUTIC POTENTIAL OF miRNAs

Improving therapies for CNS injuries is paramount as current strategies yield only modest outcomes. The involvement of miRNAs in current drug mechanisms is detailed in the following section. Moreover, the use of miRNAs as a therapeutic modality is discussed.

26.6.1 Augmenting Stroke Therapy with miRNAs

Poststroke therapy is critical in minimizing damage while maximizing recovery. Histone deacetylase inhibitors, such as valproic acid (VPA), are emerging as a class of drugs that improve stroke outcomes, in part, by reducing infarct volumes and enhancing neurogenesis (Langley et al., 2009). Although the exact mechanism of histone deacetylase inhibitors has not been completely defined, potential mediators have been identified. Administering VPA postinjury in a rat MCAO model increased miR-331 and decreased miR-885-3p (Hunsberger et al., 2012). Bioinformatics analyses suggest that these miRNAs modulate apoptosis, necrosis, and synaptic transmission, providing insight into putative protective mechanisms that are activated by VPA.

Stroke pathology is also mediated by Toll-like receptors (TLRs), which are found to be increased in neurons after injury (Tang et al., 2007). TLRs appear to be detrimental as mice deficient in these proteins have improved outcome after stroke (Cao et al., 2007). Bortezomib, a proteasome inhibitor, is neuroprotective by inhibiting TLR signaling through increasing miR-146a expression (Zhang et al., 2012). This miRNA targets IRAK1, an intracellular mediator of TLR signaling. Implicating miRNAs in the mechanism of action for stroke drugs identifies more putative loci for therapeutic modulation (see the following section).

Another mechanism by which injury can be attenuated, though theoretical in humans, is ischemic preconditioning (IPC) (Eltzschig and Eckle, 2011). IPC involves transient exposure of tissue to nonlethal ischemia before a prolonged ischemic insult to activate protective pathways. However, the mechanism of IPC is not fully understood. Microarray analysis has found that, during IPC, miRNAs that target the mitogen-activated protein kinase and mammalian target of rapamycin signaling pathways are increased, whereas those targeting WNT signaling are decreased, resulting in increased cytoprotection and cellular regeneration (Dharap and Vemuganti, 2010). Another global regulator of cell function and mediator of IPC is methyl CpG binding protein 2, which is a target of many repressed miRNAs (Lusardi et al., 2010). The protein kinase C (PKC) family is also an important mediator of IPC (Bright and Mochly-Rosen, 2005), and are targets of 19 miRNAs found to be differentially expressed by IPC (Liu et al., 2012). PKC interacting proteins, such as Grp78 (a heat shock protein) and UCH-L1 (involved in the ubiquitin system), are also targets of IPC-affected miRNAs, such as miR-181b (Peng et al., 2013). miR-181b is repressed by IPC, thereby allowing further activation of PKC signaling. Other involved miRNAs include the miR-200 and -182 families, which were found to be increased within 3 hours of IPC (Lee et al., 2010). These miRNAs provide protection by targeting prolyl hydroxylase 2, which negatively regulates the pro-survival protein HIF1a.

Adding to its already increasing roles in CNS function, miR-124 has been shown to facilitate stroke recovery. Overexpression of miR-124 transcript in mouse striatum leads to significantly reduced infarct volumes while increasing neuronal density at 4 and 8 weeks postinjury (Doeppner et al., 2013). The 3'UTR of Usp14, a deubiquitinating enzyme, is targeted by miR-124, resulting in lower Usp14 protein levels. Importantly, this leads to REST degradation, as Usp14 facilitates REST protein stability. As REST is a pro-death factor in neurons, overexpression of miR-124 ultimately results in lower levels of REST and increased number of neurons.

26.6.2 Augmenting TBI Therapy with miRNAs

Hypothermia is a promising and powerful treatment for TBI because reducing body temperature attenuates neurological damage and improves functional outcomes (Choi et al., 2012; Dietrich et al., 2009; Marion and Bullock, 2009). Interestingly, some miRNAs that are affected by TBI are also temperature-sensitive. miR-9 is one such miRNA whose increased levels after TBI are reduced by hypothermic conditions (Truettner et al., 2011). This miRNA is predicted to target genes involved in the maintenance of cytoskeletal and membrane architecture. Therefore, derepressing these targets may allow cells to maintain their cellular structure and survive. miR-874 is another miRNA reduced by hypothermia and is predicted to inhibit genes involved in intracellular transport, membrane organization, and ion channels. Transfecting this miRNA, along with two other temperature-sensitive ones (miR-34a and -451), in primary neuronal cultures demonstrated augmented vulnerability to injury (Truettner et al., 2013). Stretching the transfected neurons increased levels of tumor necrosis factor-α, caspase 11, and interleukin-1β, suggesting that reduced expression of these miRNAs can attenuate susceptibility to death from cell stress. However, miR-34a is involved in NPC proliferation through its regulation of Notch signaling. As hypothermia reduces miR-34a levels, expansion of NPCs is reduced, thereby limiting one of the natural repair mechanisms that occur after injury (Wang et al., 2012b). Therefore, it is imperative to better understand the specific mediators of hypothermia. Examining this process at the level of miRNAs allows for identifying which ones can be exploited to maximize the beneficial aspects of hypothermia while minimizing the adverse effects.

26.6.3 Therapeutic Avenues to Modulate miRNA Function

These studies indicate that miRNA expression is greatly altered after neurotrauma and that these changes have profound effects on recovery. Therefore, restoring miRNAs back to homeostatic levels represents a therapeutic opportunity to improve outcomes in injured patients. This can be achieved through either enhancing or reducing miRNA activity, or by creating artificial miRNAs, as discussed in the following sections.

26.6.3.1 miRNA Agonism

Because many miRNAs levels are reduced after injury, it seems appropriate to increase levels of protective miRNAs. One option is to use miRNA mimics, which are modified double-stranded RNA molecules that contain one strand sequence that is identical to the mature miRNA of interest. Because uridine bases and uridine-guanine base pairs activate TLRs and induce an immune response when introduced exogenously (Judge et al., 2005), chemical modifications have to be done on miRNA mimic molecules. The alternative, base-pair changes to adenosine or cytosine, are not effective as miRNA potency is sequence-dependent. Mimic modifications include replacing guanine with cyclopentyl-guanine at the 2'-ribose position, which demonstrates reduced immune activation with strong silencing activity (Peacock et al., 2011). Other modifications, such as to the 3' overhang of the miRNA duplex, are also used to prevent endonuclease-mediated degradation and to increase miRNA stability (Akao et al., 2010).

Pri-miRNAs and pre-miRNAs can be used as another means of increasing miRNA levels. DNA-based plasmid vectors containing pri- or pre-sequences can be targeted to particular cells and tissues, where the cells' machinery produces mature miRNAs from the vector. This method has been demonstrated in vivo with an intraventricular infusion of pri-miR-181 to improve stroke outcomes in a rodent model (Ouyang et al., 2012). An advantage of this method over directly using miRNA mimics is that a large number of mature miRNA molecules can be amplified because of constant transcription of the plasmid vector compared with the fixed number of transfected mimics. However, as described previously, neurotrauma can alter expression of miRNA processing machinery, thereby limiting the effectiveness of the pri- and pre-miRNAs vectors (Heinrich et al., 2013; Ho et al., 2012; Lee et al., 2011; Wu et al., 2011).

26.6.3.2 miRNA Antagonism

A few methods for miRNA inhibition have also been developed. First-generation inhibitors include antagomiRs that bind with the mature miRNAs of interest to prevent miRNA repression of mRNA translation (Krutzfeldt et al., 2005). Although antagomiRs contain many chemical modifications, such as 3'-end conjugated cholesterol groups to increase in vivo stability and bioavailability, including to the CNS, they may have limited use as high doses are needed for sufficient potency (Broderick and Zamore, 2011). Locked nucleic acids (LNAs) are another miRNA inhibitor, but differ from antagomiRs in that LNAs are significantly more potent. LNA molecules contain locked ribose rings that both minimize endonuclease degradation and increase miRNA inhibition (Grunweller and Hartmann,2007; Orom et al., 2006; Stenvang et al., 2008). Moreover, LNAs have been demonstrated to have less toxicity and more function in vivo than antagomiRs (Elmen et al., 2008a, 2008b). In fact, LNAs targeting miR-122 are being used in phase II clinical trials to treat chronic hepatitis C; as inhibition of this miRNA by the LNA reduced circulating levels of hepatitis C virus (Janssen et al., 2013; Lanford et al., 2010; Multiple Ascending Dose Study of Miravirsen in Treatment-Naive Chronic Hepatitis C Subjects, 2014).

Although antagomiRs and LNAs are very potent inhibitors, the half-life of these molecules can be quite short, being functional in serum for only a few days (Elmen et al., 2005; Wang et al., 2012a). One technique to sidestep this issue is to use miRNA-sponge constructs, which are artificial RNA transcripts containing multiple repeats of the miRNA-binding site (Ebert et al., 2007). Therefore, miRNA sponges serve as competitive inhibitors of miRNAs, sequestering the mature miRNAs from endogenous mRNA targets. In fact, this mimics a natural phenomenon where circular RNA molecules found in mammalian cells act as miRNA sponges (Hansen et al., 2013). miRNA sponge constructs can be integrated into the genome of CNS cells (Bhalala et al., 2012; Gentner et al., 2009; Luikart et al., 2011), allowing for constant expression of the sponge as well as persistent inhibition of the miRNA of interest.

Target protectors, also known as miRNA masks, have also been used to inhibit miRNA function. These are synthetic RNA molecules that contain equivalent sequences as mature miRNAs, thereby binding to the same mRNA targets (Choi et al., 2007). However, a major difference is that target protectors are chemically modified so as not to interact with RISC, thereby preventing mRNA translation inhibition. As with miRNA sponges, endogenous target protectors have also been found, where a long ncRNA binds to BACE1 transcripts (which are dysregulated in Alzheimer disease) to prevent function of miR-485-5p, a miRNA that represses BACE1 protein levels (Faghihi et al., 2010).

26.6.3.3 Artificial miRNAs

miRNAs exert their effect by simultaneously targeting a network of genes based on binding site similarities that have been conserved throughout evolution (Bartel, 2009; Gurtan and Sharp, 2013). This idea can be exploited to regulate genes of interest that have no common naturally occurring miRNA binding sites through the use of artificial miRNAs. Artificial miRNAs are designed miRNAs that bind to novel seed sequences, identified through the use of computer algorithms, in a set of genes of interest (DeGuire et al., 2010; Zeng et al., 2002). This approach may be advantageous as microarray studies have identified a multitude of transcripts that are upregulated after injury, many of which are involved in deleterious processes such as apoptosis and inflammation (Kotipatruni et al., 2011; Nagel et al., 2012). Although many of these transcripts may be miRNA targets, using endogenous miRNAs to target these mRNAs may not be desirable because it may require too many miRNAs for effective modulations. Moreover, there is an increased chance of miRNA off-target binding of nonnatural target mRNAs with use of large amounts of miRNA mimics. Artificial miRNAs may be more useful because they can be adjusted to target only intended transcripts and designed to maximize the number of mRNAs of interest that are targeted. Therefore, fewer miRNAs need to be introduced into the CNS to efficiently modulate postinjury recovery.

26.6.3.4 miRNA Delivery

The utility of miRNA modulators faces a major barrier for success as an effective means of therapy: delivery. Direct delivery of these miRNA modulators into ventricles or parenchyma is possible, but is limited by the potential of further tissue damage and immune-system activation. Because the CNS is a privileged location, successful access of the tissue needs to overcome some technical limitations, including the blood–brain barrier. Exosomes are endogenous lipid-membrane vesicles that are secreted by the CNS and can enter the systemic circulation (Liang et al., 2013; Schorey and Bhatnagar, 2008; Simpson et al., 2008). Because these exosomes have also been found to carrier miRNAs, they can also be used to shuttle RNA molecules in the opposite direction, where systemic injection of nucleotide-carrying exosomes has been shown to enter and deliver material into the CNS (Alvarez-Erviti et al., 2011). Another method of getting targeted CNS delivery is the use of recombinant adeno-associated viruses (rAAVs), which can carry miRNA encoding genes. Although these viruses can cross the blood–brain barrier and enter the CNS, they also have been found to infect liver and muscle cells (Foust et al., 2009). However, to restrict the expression to the CNS, rAAVs have been engineered to contain binding sites to non–CNS-specific miRNAs, thereby effectively silencing the virus in non-CNS tissues (Xie et al., 2011). These studies highlight the forefront of miRNA-based therapeutics and their amazing potential. However, substantially more work is needed to develop safe and efficient delivery vehicles for miRNA modulators before they can be used to treat brain injuries.

26.7 CONCLUDING REMARKS

Neurotrauma can be life-changing, creating a substantial burden on both the individual patient and society. Understanding the cellular and molecular changes that occur after injury provides a foundation for developing effective treatments. Identifying miRNAs that are altered in stroke and TBI provides a high-level network perspective on global patterns of such changes. Microarray and specific analyses of miRNAs demonstrate that many are involved in apoptosis, inflammation, and cell proliferation. Studies have also found that miRNAs may be used as biomarkers for injury, to either diagnose a patient before any symptoms have become evident or to differentiate between various etiologies of disease.

Advances in miRNA modulators show promise as a potential therapeutic modality. However, significantly more work is needed to translate findings from animal models to human patients. Moreover, targets of miRNAs need to be confirmed because prediction algorithms may not capture the complex regulation of miRNA-mRNA interactions within cells, especially after injury. Nonetheless, the success of miRNA-based therapies has been demonstrated in phase II clinical trials for anti–miR-122 in chronic hepatitis C. Coupling the increasingly evident diverse utility of miRNAs with powerful bioinformatics and prediction algorithms and rapid pharmaceutical development, neurotrauma may eventually be recognized and treated quickly, thereby restoring functionality and normalcy.

REFERENCES

Akao, Y., Y. Nakagawa, I. Hirata, A. Lio, T. Itoh, K. Kojima et al. 2010. Role of anti-oncomirs miR-143 and -145 in human colorectal tumors. *Cancer Gene Therapy*. 17:398–408.

Alexiou, P., M. Maragkakis, G.L. Papadopoulos, M. Reczko, and A.G. Hatzigeorgiou. 2009. Lost in translation: An assessment and perspective for computational microRNA target identification. *Bioinformatics*. 25:3049–3055.

Alvarez-Erviti, L., Y. Seow, H. Yin, C. Betts, S. Lakhal, and M.J. Wood. 2011. Delivery of siRNA to the mouse brain by systemic injection of targeted exosomes. *Nature Biotechnology*. 29:341–345.

Baek, D., J. Villen, C. Shin, F.D. Camargo, S.P. Gygi, and D.P. Bartel. 2008. The impact of microRNAs on protein output. *Nature*. 455:64–71.

Bak, M., A. Silahtaroglu, M. Moller, M. Christensen, M.F. Rath, B. Skryabin et al. 2008. MicroRNA expression in the adult mouse central nervous system. *RNA*. 14:432–444.

Balakathiresan, N., M. Bhomia, R. Chandran, M. Chavko, R.M. McCarron, and R.K. Maheshwari. 2012. MicroRNA let-7i is a promising serum biomarker for blast-induced traumatic brain injury. *Journal of Neurotrauma*. 29:1379–1387.

Bartel, D.P. 2009. MicroRNAs: Target recognition and regulatory functions. *Cell*. 136:215–233.

Bartel, D.P. 2004. MicroRNAs: Genomics, biogenesis, mechanism, and function. *Cell*. 116:281–297.

Basyuk, E., F. Suavet, A. Doglio, R. Bordonne, and E. Bertrand. 2003. Human let-7 stem-loop precursors harbor features of RNase III cleavage products. *Nucleic Acids Research*. 31:6593–6597.

Bernstein, E., S.Y. Kim, M.A. Carmell, E.P. Murchison, H. Alcorn, M.Z. Li et al. 2003. Dicer is essential for mouse development. *Nature Genetics*. 35:215–217.

Bhalala, O.G., L. Pan, V. Sahni, T.L. McGuire, K. Gruner, W.G. Tourtellotte et al. 2012. microRNA-21 regulates astrocytic response following spinal cord injury. *The Journal of Neuroscience*. 32:17935–17947.

Bhalala, O.G., M. Srikanth, and J.A. Kessler. 2013. The emerging roles of microRNAs in CNS injuries. *Nature Reviews. Neurology*. 9:328–339.

Bian, S., and T. Sun. 2011. Functions of noncoding RNAs in neural development and neurological diseases. *Molecular Neurobiology*. 44:359–373.

Bright, R., and D. Mochly-Rosen. 2005. The role of protein kinase C in cerebral ischemic and reperfusion injury. *Stroke*. 36:2781–2790.

Broderick, J.A., and P.D. Zamore. 2011. MicroRNA therapeutics. *Gene Therapy*. 18:1104–1110.

Budde, H., S. Schmitt, D. Fitzner, L. Opitz, G. Salinas-Riester, and M. Simons. 2010. Control of oligodendroglial cell number by the miR-17-92 cluster. *Development*. 137:2127–2132.

Buller, B., X. Liu, X. Wang, R.L. Zhang, L. Zhang, A. Hozeska-Solgot et al. 2010. MicroRNA-21 protects neurons from ischemic death. *The FEBS Journal*. 277:4299–4307.

Bushnell, C.D. 2008a. Stroke and the female brain. *Nature Clinical Practice. Neurology*. 4:22–33.

Bushnell, C.D. 2008b. Stroke in women: Risk and prevention throughout the lifespan. *Neurologic Clinics*. 26:1161–1176, xi.

Calderone, A., T. Jover, K.M. Noh, H. Tanaka, H. Yokota, Y. Lin et al. 2003. Ischemic insults derepress the gene silencer REST in neurons destined to die. *The Journal of Neuroscience*. 23:2112–2121.

Cao, C.X., Q.W. Yang, F.L. Lv, J. Cui, H.B. Fu, and J.Z. Wang. 2007. Reduced cerebral ischemia-reperfusion injury in Toll-like receptor 4 deficient mice. *Biochemical and Biophysical Research Communications*. 353:509–514.

Carthew, R.W., and E.J. Sontheimer. 2009. Origins and mechanisms of miRNAs and siRNAs. *Cell*. 136:642–655.

Chedin, F. 2011. The DNMT3 family of mammalian de novo DNA methyltransferases. *Progress in Molecular Biology and Translational Science*. 101:255–285.

Chen, Z.F., A.J. Paquette, and D.J. Anderson. 1998. NRSF/REST is required in vivo for repression of multiple neuronal target genes during embryogenesis. *Nature Genetics*. 20:136–142.

Chendrimada, T.P., R.I. Gregory, E. Kumaraswamy, J. Norman, N. Cooch, K. Nishikura et al. 2005. TRBP recruits the Dicer complex to Ago2 for microRNA processing and gene silencing. *Nature*. 436:740–744.

Cheng, L.C., E. Pastrana, M. Tavazoie, and F. Doetsch. 2009. miR-124 regulates adult neurogenesis in the subventricular zone stem cell niche. *Nature Neuroscience*. 12:399–408.

Chi, S.W., J.B. Zang, A. Mele, and R.B. Darnell. 2009. Argonaute HITS-CLIP decodes microRNA-mRNA interaction maps. *Nature*. 460:479–486.

Choi, H.A., N. Badjatia, and S.A. Mayer. 2012. Hypothermia for acute brain injury—mechanisms and practical aspects. *Nature Reviews. Neurology*. 8:214–222.

Choi, W.Y., A.J. Giraldez, and A.F. Schier. 2007. Target protectors reveal dampening and balancing of Nodal agonist and antagonist by miR-430. *Science*. 318:271–274.

Cohen, A.S., B.J. Pfister, E. Schwarzbach, M.S. Grady, P.B. Goforth, and L.S. Satin. 2007. Injury-induced alterations in CNS electrophysiology. *Progress in Brain Research*. 161:143–169.

Coronado, V.G., L. Xu, S.V. Basavaraju, L.C. McGuire, M.M. Wald, M.D. Faul et al., Centers for Disease Control and Prevention. 2011. Surveillance for traumatic brain injury-related deaths—United States, 1997–2007. *Morbidity and Mortality Weekly Report. Surveillance Summaries*. 60:1–32.

Davis, T.H., T.L. Cuellar, S.M. Koch, A.J. Barker, B.D. Harfe, M.T. McManus et al. 2008. Conditional loss of Dicer disrupts cellular and tissue morphogenesis in the cortex and hippocampus. *The Journal of Neuroscience*. 28:4322–4330.

De Guire, V., M. Caron, N. Scott, C. Menard, M.F. Gaumont-Leclerc, P. Chartrand et al. 2010. Designing small multiple-target artificial RNAs. *Nucleic Acids Research*. 38:e140.

De Pietri Tonelli, D., J.N. Pulvers, C. Haffner, E.P. Murchison, G.J. Hannon, and W.B. Huttner. 2008. miRNAs are essential for survival and differentiation of newborn neurons but not for expansion of neural progenitors during early neurogenesis in the mouse embryonic neocortex. *Development*. 135:3911–3921.

Dharap, A., K. Bowen, R. Place, L.C. Li, and R. Vemuganti. 2009. Transient focal ischemia induces extensive temporal changes in rat cerebral microRNAome. *Journal of Cerebral Blood Flow and Metabolism*. 29:675–687.

Dharap, A., and R. Vemuganti. 2010. Ischemic pre-conditioning alters cerebral microRNAs that are upstream to neuroprotective signaling pathways. *Journal of Neurochemistry*. 113:1685–1691.

Dietrich, W.D., C.M. Atkins, and H.M. Bramlett. 2009. Protection in animal models of brain and spinal cord injury with mild to moderate hypothermia. *Journal of Neurotrauma*. 26:301–312.

Doeppner, T.R., M. Doehring, E. Bretschneider, A. Zechariah, B. Kaltwasser, B. Muller et al. 2013. MicroRNA-124 protects against focal cerebral ischemia via mechanisms involving Usp14-dependent REST degradation. *Acta Neuropathologica*. 126:251–265.

Dugas, J.C., T.L. Cuellar, A. Scholze, B. Ason, A. Ibrahim, B. Emery et al. 2010. Dicer1 and miR-219 are required for normal oligodendrocyte differentiation and myelination. *Neuron*. 65:597–611.

Ebert, M.S., J.R. Neilson, and P.A. Sharp. 2007. MicroRNA sponges: Competitive inhibitors of small RNAs in mammalian cells. *Nature Methods*. 4:721–726.

Elmen, J., M. Lindow, S. Schutz, M. Lawrence, A. Petri, S. Obad et al. 2008a. LNA-mediated microRNA silencing in non-human primates. *Nature*. 452:896–899.

Elmen, J., M. Lindow, A. Silahtaroglu, M. Bak, M. Christensen, A. Lind-Thomsen et al. 2008b. Antagonism of microRNA-122 in mice by systemically administered LNA-antimiR leads to up-regulation of a large set of predicted target mRNAs in the liver. *Nucleic Acids Research*. 36:1153–1162.

Elmen, J., H. Thonberg, K. Ljungberg, M. Frieden, M. Westergaard, Y. Xu et al. 2005. Locked nucleic acid (LNA) mediated improvements in siRNA stability and functionality. *Nucleic Acids Research*. 33:439–447.

Eltzschig, H.K., and T. Eckle. 2011. Ischemia and reperfusion—from mechanism to translation. *Nature Medicine*. 17:1391–1401.

Esquela-Kerscher, A., and F.J. Slack. 2006. Oncomirs - microRNAs with a role in cancer. *Nature Reviews. Cancer*. 6:259–269.

Fabian, M.R., and N. Sonenberg. 2012. The mechanics of miRNA-mediated gene silencing: A look under the hood of miRISC. *Nature Structural & Molecular Biology*. 19:586–593.

Faghihi, M.A., M. Zhang, J. Huang, F. Modarresi, M.P. Van der Brug, M.A. Nalls et al. 2010. Evidence for natural antisense transcript-mediated inhibition of microRNA function. *Genome Biology*. 11:R56.

Foust, K.D., E. Nurre, C.L. Montgomery, A. Hernandez, C.M. Chan, and B.K. Kaspar. 2009. Intravascular AAV9 preferentially targets neonatal neurons and adult astrocytes. *Nature Biotechnology*. 27:59–65.

Gan, C.S., C.W. Wang, and K.S. Tan. 2012. Circulatory microRNA-145 expression is increased in cerebral ischemia. *Genetics and Molecular Research*. 11:147–152.

Gentner, B., G. Schira, A. Giustacchini, M. Amendola, B.D. Brown, M. Ponzoni etal. 2009. Stable knockdown of microRNA in vivo by lentiviral vectors. *Nature Methods*. 6:63–66.

Geyer, C., A. Ulrich, G. Grafe, B. Stach, and H. Till. 2009. Diagnostic value of S100B and neuron-specific enolase in mild pediatric traumatic brain injury. *Journal of Neurosurgery. Pediatrics*. 4:339–344.

Go, A.S., D. Mozaffarian, V.L. Roger, E.J. Benjamin, J.D. Berry, W.B. Borden et al.; American Heart Association Statistics Committee and Stroke Statistics Subcommittee. 2013. Heart disease and stroke statistics—2013 update: A report from the American Heart Association. *Circulation*. 127:e6-e245.

Grunweller, A., and R.K. Hartmann. 2007. Locked nucleic acid oligonucleotides: The next generation of antisense agents? *BioDrugs*. 21:235–243.

Gubern, C., S. Camos, I. Ballesteros, R. Rodriguez, V.G. Romera, R. Canada et al. 2013. miRNA expression is modulated over time after focal ischaemia: Up-regulation of miR-347 promotes neuronal apoptosis. *The FEBS Journal*. 280:6233–6246.

Guo, D., J. Liu, W. Wang, F. Hao, X. Sun, X. Wu et al. 2013. Alteration in abundance and compartmentalization of inflammation-related miRNAs in plasma after intracerebral hemorrhage. *Stroke*. 44:1739–1742.

Guo, H., N.T. Ingolia, J.S. Weissman, and D.P. Bartel. 2010. Mammalian microRNAs predominantly act to decrease target mRNA levels. *Nature*. 466:835–840.

Gurtan, A.M., and P.A. Sharp. 2013. The role of miRNAs in regulating gene expression networks. *Journal of Molecular Biology*. 425:3582–3600.

Hafner, M., M. Landthaler, L. Burger, M. Khorshid, J. Hausser, P. Berninger et al. 2010. Transcriptome-wide identification of RNA-binding protein and microRNA target sites by PAR-CLIP. *Cell*. 141:129–141.

Hammond, S.M., E. Bernstein, D. Beach, and G.J. Hannon. 2000. An RNA-directed nuclease mediates post-transcriptional gene silencing in Drosophila cells. *Nature*. 404:293–296.

Han, J., Y. Lee, K.H. Yeom, Y.K. Kim, H. Jin, and V.N. Kim. 2004. The Drosha-DGCR8 complex in primary microRNA processing. *Genes & Development*. 18:3016–3027.

Hansen, T.B., T.I. Jensen, B.H. Clausen, J.B. Bramsen, B. Finsen, C.K. Damgaard et al. 2013. Natural RNA circles function as efficient microRNA sponges. *Nature*. 495:384–388.

Harraz, M.M., S.M. Eacker, X. Wang, T.M. Dawson, and V.L. Dawson. 2012. MicroRNA-223 is neuroprotective by targeting glutamate receptors. *Proceedings of the National Academy of Sciences of the United States of America*. 109:18962–18967.

He, X., Y. Yu, R. Awatramani, and Q.R. Lu. 2012. Unwrapping myelination by microRNAs. *The Neuroscientist*. 18:45–55.

Heinrich, E.M., J. Wagner, M. Kruger, D. John, S. Uchida, J.E. Weigand, B. Suess, and S. Dimmeler. 2013. Regulation of miR-17-92a cluster processing by the microRNA binding protein SND1. *FEBS Letters*. 587:2405–2411.

Ho, J.J., J.L. Metcalf, M.S. Yan, P.J. Turgeon, J.J. Wang, M. Chalsev et al. 2012. Functional importance of Dicer protein in the adaptive cellular response to hypoxia. *The Journal of Biological Chemistry*. 287:29003–29020.

Hock, J., and G. Meister. 2008. The Argonaute protein family. *Genome Biology*. 9:210.

Hu, Z., D. Yu, C. Almeida-Suhett, K. Tu, A.M. Marini, L. Eiden et al. 2012. Expression of miRNAs and their cooperative regulation of the pathophysiology in traumatic brain injury. *PloS One*. 7:e39357.

Huang, T., Y. Liu, M. Huang, X. Zhao, and L. Cheng. 2010. Wnt1-cre-mediated conditional loss of Dicer results in malformation of the midbrain and cerebellum and failure of neural crest and dopaminergic differentiation in mice. *Journal of Molecular Cell Biology*. 2:152–163.

Hunsberger, J.G., E.B. Fessler, Z. Wang, A.G. Elkahloun, and D.M. Chuang. 2012. Post-insult valproic acid-regulated microRNAs: Potential targets for cerebral ischemia. *American Journal of Translational Research*. 4:316–332.

Hutvagner, G., J. McLachlan, A.E. Pasquinelli, E. Balint, T. Tuschl, and P.D. Zamore. 2001. A cellular function for the RNA-interference enzyme Dicer in the maturation of the let-7 small temporal RNA. *Science*. 293:834–838.

Janssen, H.L., H.W. Reesink, E.J. Lawitz, S. Zeuzem, M. Rodriguez-Torres, K. Patel et al. 2013. Treatment of HCV infection by targeting microRNA. *The New England Journal of Medicine*. 368:1685–1694.

Jeon, Y.J., O.J. Kim, S.Y. Kim, S.H. Oh, D. Oh, O.J. Kim et al. 2013. Association of the miR-146a, miR-149, miR-196a2, and miR-499 polymorphisms with ischemic stroke and silent brain infarction risk. *Arteriosclerosis, Thrombosis, and Vascular Biology*. 33:420–430.

Jeyaseelan, K., K.Y. Lim, and A. Armugam. 2008. MicroRNA expression in the blood and brain of rats subjected to transient focal ischemia by middle cerebral artery occlusion. *Stroke*. 39:959–966.

Judge, A.D., V. Sood, J.R. Shaw, D. Fang, K. McClintock, and I. MacLachlan. 2005. Sequence-dependent stimulation of the mammalian innate immune response by synthetic siRNA. *Nature Biotechnology*. 23:457–462.

Kontos, A.P., R.S. Kotwal, R.J. Elbin, R.H. Lutz, R.D. Forsten, P.J. Benson, and K.M. Guskiewicz. 2013. Residual effects of combat-related mild traumatic brain injury. *Journal of Neurotrauma*. 30:680–686.

Kotipatruni, R.R., V.R. Dasari, K.K. Veeravalli, D.H. Dinh, D. Fassett, and J.S. Rao. 2011. p53- and Bax-mediated apoptosis in injured rat spinal cord. *Neurochemical Research*. 36:2063–2074.

Krutzfeldt, J., N. Rajewsky, R. Braich, K.G. Rajeev, T. Tuschl, M. Manoharan et al. 2005. Silencing of microRNAs in vivo with 'antagomirs'. *Nature*. 438:685–689.

Lagos-Quintana, M., R. Rauhut, W. Lendeckel, and T. Tuschl. 2001. Identification of novel genes coding for small expressed RNAs. *Science*. 294:853–858.

Lagos-Quintana, M., R. Rauhut, J. Meyer, A. Borkhardt, and T. Tuschl. 2003. New microRNAs from mouse and human. *RNA*. 9:175–179.

Lagos-Quintana, M., R. Rauhut, A. Yalcin, J. Meyer, W. Lendeckel, and T. Tuschl. 2002. Identification of tissue-specific microRNAs from mouse. *Current Biology*. 12:735–739.

Lai, T.W., S. Zhang, and Y.T. Wang. 2013. Excitotoxicity and stroke: Identifying novel targets for neuroprotection. *Progress in Neurobiology*. 115:157–188.

Landgraf, P., M. Rusu, R. Sheridan, A. Sewer, N. Iovino, A. Aravin et al. 2007. A mammalian microRNA expression atlas based on small RNA library sequencing. *Cell*. 129:1401–1414.

Lanford, R.E., E.S. Hildebrandt-Eriksen, A. Petri, R. Persson, M. Lindow, M.E. Munk et al. 2010. Therapeutic silencing of microRNA-122 in primates with chronic hepatitis C virus infection. *Science*. 327:198–201.

Lang, J.T., and L.D. McCullough. 2008. Pathways to ischemic neuronal cell death: Are sex differences relevant? *Journal of Translational Medicine*. 6:33.

Langley, B., C. Brochier, and M.A. Rivieccio. 2009. Targeting histone deacetylases as a multifaceted approach to treat the diverse outcomes of stroke. *Stroke*. 40:2899–2905.

Lau, N.C., L.P. Lim, E.G. Weinstein, and D.P. Bartel. 2001. An abundant class of tiny RNAs with probable regulatory roles in Caenorhabditis elegans. *Science*. 294:858–862.

Lee, R.C., and V. Ambros. 2001. An extensive class of small RNAs in Caenorhabditis elegans. *Science*. 294:862–864.

Lee, R.C., R.L. Feinbaum, and V. Ambros. 1993. The C. elegans heterochronic gene lin-4 encodes small RNAs with antisense complementarity to lin-14. *Cell*. 75:843–854.

Lee, S.T., K. Chu, W.S. Im, H.J. Yoon, J.Y. Im, J.E. Park et al. 2011. Altered microRNA regulation in Huntington's disease models. *Experimental Neurology*. 227:172–179.

Lee, S.T., K. Chu, K.H. Jung, H.J. Yoon, D. Jeon, K.M. Kang et al. 2010. MicroRNAs induced during ischemic preconditioning. *Stroke*. 41:1646–1651.

Lee, Y., K. Jeon, J.T. Lee, S. Kim, and V.N. Kim. 2002. MicroRNA maturation: Stepwise processing and subcellular localization. *The EMBO Journal*. 21:4663–4670.

Lee, Y., M. Kim, J. Han, K.H. Yeom, S. Lee, S.H. Baek et al. 2004. MicroRNA genes are transcribed by RNA polymerase II. *The EMBO Journal*. 23:4051–4060.

Lei, P., Y. Li, X. Chen, S. Yang, and J. Zhang. 2009. Microarray based analysis of microRNA expression in rat cerebral cortex after traumatic brain injury. *Brain Research*. 1284:191–201.

Letzen, B.S., C. Liu, N.V. Thakor, J.D. Gearhart, A.H. All, and C.L. Kerr. 2010. MicroRNA expression profiling of oligodendrocyte differentiation from human embryonic stem cells. *PloS One*. 5:e10480.

Lewis, B.P., I.H. Shih, M.W. Jones-Rhoades, D.P. Bartel, and C.B. Burge. 2003. Prediction of mammalian microRNA targets. *Cell*. 115:787–798.

Lhakhang, T.W., and M.A. Chaudhry. 2012. Current approaches to micro-RNA analysis and target gene prediction. *Journal of Applied Genetics*. 53:149–158.

Liang, H., F. Gong, S. Zhang, C.Y. Zhang, K. Zen, and X. Chen. 2013. The origin, function, and diagnostic potential of extracellular microRNAs in human body fluids. *Wiley Interdisciplinary Reviews. RNA*. 5:285–300.

Lin, S.T., Y. Huang, L. Zhang, M.Y. Heng, L.J. Ptacek, and Y.H. Fu. 2013. MicroRNA-23a promotes myelination in the central nervous system. *Proceedings of the National Academy of Sciences of the United States of America*. 110:17468–17473.

Liu, C., Z. Peng, N. Zhang, L. Yu, S. Han, D. Li, and J. Li. 2012. Identification of differentially expressed microRNAs and their PKC-isoform specific gene network prediction during hypoxic pre-conditioning and focal cerebral ischemia of mice. *Journal of Neurochemistry*. 120:830–841.

Liu, D.Z., Y. Tian, B.P. Ander, H. Xu, B.S. Stamova, X. Zhan et al. 2010. Brain and blood microRNA expression profiling of ischemic stroke, intracerebral hemorrhage, and kainate seizures. *Journal of Cerebral Blood Flow and Metabolism*. 30:92–101.

Liu, F.J., K.Y. Lim, P. Kaur, S. Sepramaniam, A. Armugam, P.T. Wong et al. 2013a. microRNAs involved in regulating spontaneous recovery in embolic stroke model. *PloS One*. 8:e66393.

Liu, J., M.A. Carmell, F.V. Rivas, C.G. Marsden, J.M. Thomson, J.J. Song et al. 2004. Argonaute2 is the catalytic engine of mammalian RNAi. *Science*. 305:1437–1441.

Liu, M.E., Y.C. Liao, R.T. Lin, Y.S. Wang, E. Hsi, H.F. Lin et al. 2013b. A functional polymorphism of PON1 interferes with microRNA binding to increase the risk of ischemic stroke and carotid atherosclerosis. *Atherosclerosis*. 228:161–167.

Liu, N.K., and X.M. Xu. 2011. MicroRNA in central nervous system trauma and degenerative disorders. *Physiological Genomics*. 43:571–580.

Liu, X., F. Li, S. Zhao, Y. Luo, J. Kang, H. Zhao et al. 2013c. MicroRNA-124-mediated regulation of inhibitory member of apoptosis-stimulating protein of p53 family in experimental stroke. *Stroke*. 44:1973–1980.

Liu, X.S., M. Chopp, X.L. Wang, L. Zhang, A. Hozeska-Solgot, T. Tang et al. 2013d. MicroRNA-17-92 cluster mediates the proliferation and survival of neural progenitor cells after stroke. *The Journal of Biological Chemistry*. 288:12478–12488.

Liu, X.S., M. Chopp, R.L. Zhang, T. Tao, X.L. Wang, H. Kassis et al. 2011. MicroRNA profiling in subventricular zone after stroke: MiR-124a regulates proliferation of neural progenitor cells through Notch signaling pathway. *PloS One*. 6:e23461.

Liu, Y., Y. Ma, B. Zhang, S.X. Wang, X.M. Wang, and J.M. Yu. 2013e. Genetic polymorphisms in pre-microRNAs and risk of ischemic stroke in a Chinese population. *Journal of Molecular Neuroscience*. 52:493–480.

Long, G., F. Wang, H. Li, Z. Yin, C. Sandip, Y. Lou et al. 2013. Circulating miR-30a, miR-126 and let-7b as biomarker for ischemic stroke in humans. *BMC Neurology*. 13:178.

Long, J.M., and D.K. Lahiri. 2012. Advances in microRNA experimental approaches to study physiological regulation of gene products implicated in CNS disorders. *Experimental Neurology*. 235:402–418.

Lopez-Ramirez, M.A., and S. Nicoli. 2013. Role of miRNAs and epigenetics in neural stem cell fate determination. *Epigenetics*. 9:90–100.

Lopez, A.D., C.D. Mathers, M. Ezzati, D.T. Jamison, and C.J. Murray. 2006. Global and regional burden of disease and risk factors, 2001: Systematic analysis of population health data. *Lancet*. 367:1747–1757.

Lou, Y.L., F. Guo, F. Liu, F.L. Gao, P.Q. Zhang, X. Niu et al. 2012. miR-210 activates notch signaling pathway in angiogenesis induced by cerebral ischemia. *Molecular and Cellular Biochemistry*. 370:45–51.

Luikart, B.W., A.L. Bensen, E.K. Washburn, J.V. Perederiy, K.G. Su, Y. Li et al. 2011. miR-132 mediates the integration of newborn neurons into the adult dentate gyrus. *PloS One*. 6:e19077.

Lund, E., S. Guttinger, A. Calado, J.E. Dahlberg, and U. Kutay. 2004. Nuclear export of microRNA precursors. *Science*. 303:95–98.

Lusardi, T.A., C.D. Farr, C.L. Faulkner, G. Pignataro, T. Yang, J. Lan et al. 2010. Ischemic preconditioning regulates expression of microRNAs and a predicted target, MeCP2, in mouse cortex. *Journal of Cerebral Blood Flow and Metabolism*. 30:744–756.

Manley, G.T., M. Fujimura, T. Ma, N. Noshita, F. Filiz, A.W. Bollen et al. 2000. Aquaporin-4 deletion in mice reduces brain edema after acute water intoxication and ischemic stroke. *Nature Medicine*. 6:159–163.

Marion, D., and M.R. Bullock. 2009. Current and future role of therapeutic hypothermia. *Journal of Neurotrauma*. 26:455–467.

Markus, H.S. 2012. Stroke genetics: Prospects for personalized medicine. *BMC Medicine*. 10:113.

Martinez, J., A. Patkaniowska, H. Urlaub, R. Luhrmann, and T. Tuschl. 2002. Single-stranded antisense siRNAs guide target RNA cleavage in RNAi. *Cell*. 110:563–574.

Meza-Sosa, K.F., D. Valle-Garcia, G. Pedraza-Alva, and L. Perez-Martinez. 2012. Role of microRNAs in central nervous system development and pathology. *Journal of Neuroscience Research*. 90:1–12.

Miska, E.A., E. Alvarez-Saavedra, M. Townsend, A. Yoshii, N. Sestan, P. Rakic let. 2004. Microarray analysis of microRNA expression in the developing mammalian brain. *Genome Biology*. 5:R68.

Moon, J.M., L. Xu, and R.G. Giffard. 2013. Inhibition of microRNA-181 reduces forebrain ischemia-induced neuronal loss. *Journal of Cerebral Blood Flow and Metabolism*. 33:1976–1982.

Multiple Ascending Dose Study of Miravirsen in Treatment-Naive Chronic Hepatitis C Subjects. 2014. Study NCT01200420. US National Institutes of Health, from http://clinicaltrials.gov/ct2/show/NCT01200420?term=NCT01200420&rank=1.

Munro, K.M., and V.M. Perreau. 2009. Current and future applications of transcriptomics for discovery in CNS disease and injury. *Neuro-Signals*. 17:311–327.

Nagel, S., M. Papadakis, K. Pfleger, C. Grond-Ginsbach, A.M. Buchan, and S. Wagner. 2012. Microarray analysis of the global gene expression profile following hypothermia and transient focal cerebral ischemia. *Neuroscience*. 208:109–122.

Nass, D., S. Rosenwald, E. Meiri, S. Gilad, H. Tabibian-Keissar, A. Schlosberg et al. 2009. MiR-92b and miR-9/9* are specifically expressed in brain primary tumors and can be used to differentiate primary from metastatic brain tumors. *Brain Pathology*. 19:375–383.

National Stroke Association. 2014. Stroke 101: Fast facts on stroke. http://www.stroke.org/site/DocServer/STROKE101_2009.pdf?docID=4541.

Newby, A.C. 2008. Metalloproteinase expression in monocytes and macrophages and its relationship to atherosclerotic plaque instability. *Arteriosclerosis, Thrombosis, and Vascular Biology*. 28:2108–2114.

Obernosterer, G., P.J. Leuschner, M. Alenius, and J. Martinez. 2006. Post-transcriptional regulation of microRNA expression. *RNA*. 12:1161–1167.

Orom, U.A., S. Kauppinen, and A.H. Lund. 2006. LNA-modified oligonucleotides mediate specific inhibition of microRNA function. *Gene*. 372:137–141.

Orrison, W.W., E.H. Hanson, T. Alamo, D. Watson, M. Sharma, T.G. Perkins et al. 2009. Traumatic brain injury: A review and high-field MRI findings in 100 unarmed combatants using a literature-based checklist approach. *Journal of Neurotrauma*. 26:689–701.

Ouyang, Y.B., Y. Lu, S. Yue, L.J. Xu, X.X. Xiong, R.E. White et al. 2012. miR-181 regulates GRP78 and influences outcome from cerebral ischemia in vitro and in vivo. *Neurobiology of Disease*. 45:555–563.

Pandi, G., V.P. Nakka, A. Dharap, A. Roopra, and R. Vemuganti. 2013. MicroRNA miR-29c down-regulation leading to derepression of its target DNA methyltransferase 3a promotes ischemic brain damage. *PloS One*. 8:e58039.

Papa, L., L. Akinyi, M.C. Liu, J.A. Pineda, J.J. Tepas, 3rd, M.W. Oli et al. 2010. Ubiquitin C-terminal hydrolase is a novel biomarker in humans for severe traumatic brain injury. *Critical Care Medicine*. 38:138–144.

Papadopoulos, M.C., G.T. Manley, S. Krishna, and A.S. Verkman. 2004. Aquaporin-4 facilitates reabsorption of excess fluid in vasogenic brain edema. *FASEB Journal*. 18:1291–1293.

Pasquinelli, A.E., B.J. Reinhart, F. Slack, M.Q. Martindale, M.I. Kuroda, B. Maller et al. 2000. Conservation of the sequence and temporal expression of let-7 heterochronic regulatory RNA. *Nature*. 408:86–89.

Patz, S., C. Trattnig, G. Grunbacher, B. Ebner, C. Gully, A. Novak et al. 2013. More than cell dust: Microparticles isolated from cerebrospinal fluid of brain injured patients are messengers carrying mRNAs, miRNAs, and proteins. *Journal of Neurotrauma*. 30:1232–1242.

Pavlov, V.A., and K.J. Tracey. 2005. The cholinergic anti-inflammatory pathway. *Brain, Behavior, and Immunity*. 19:493–499.

Peacock, H., R.V. Fucini, P. Jayalath, J.M. Ibarra-Soza, H.J. Haringsma, W.M. Flanagan et al. 2011. Nucleobase and ribose modifications control immunostimulation by a microRNA-122-mimetic RNA. *Journal of the American Chemical Society*. 133:9200–9203.

Peng, Z., J. Li, Y. Li, X. Yang, S. Feng, S. Han et al. 2013. Downregulation of miR-181b in mouse brain following ischemic stroke induces neuroprotection against ischemic injury through targeting heat shock protein A5 and ubiquitin carboxyl-terminal hydrolase isozyme L1. *Journal of Neuroscience Research*. 91:1349–1362.

Peters, L., and G. Meister. 2007. Argonaute proteins: Mediators of RNA silencing. *Molecular Cell*. 26:611–623.

Qureshi, I.A., and M.F. Mehler. 2012. Emerging roles of non-coding RNAs in brain evolution, development, plasticity and disease. *Nature Reviews. Neuroscience*. 13:528–541.

Raha, S., and B.H. Robinson. 2001. Mitochondria, oxygen free radicals, and apoptosis. *American Journal of Medical Genetics*. 106:62–70.

Redell, J.B., Y. Liu, and P.K. Dash. 2009. Traumatic brain injury alters expression of hippocampal microRNAs: Potential regulators of multiple pathophysiological processes. *Journal of Neuroscience Research*. 87:1435–1448.

Redell, J.B., A.N. Moore, N.H. Ward, 3rd, G.W. Hergenroeder, and P.K. Dash. 2010. Human traumatic brain injury alters plasma microRNA levels. *Journal of Neurotrauma*. 27:2147–2156.

Redell, J.B., J. Zhao, and P.K. Dash. 2011. Altered expression of miRNA-21 and its targets in the hippocampus after traumatic brain injury. *Journal of Neuroscience Research*. 89:212–221.

Reinhart, B.J., F.J. Slack, M. Basson, A.E. Pasquinelli, J.C. Bettinger, A.E. Rougvie et al. 2000. The 21-nucleotide let-7 RNA regulates developmental timing in Caenorhabditis elegans. *Nature*. 403:901–906.

Rosenfeld, J.V., A.C. McFarlane, P. Bragge, R.A. Armonda, J.B. Grimes, and G.S. Ling. 2013. Blast-related traumatic brain injury. *Lancet Neurology*. 12:882–893.

Sahni, V., A. Mukhopadhyay, V. Tysseling, A. Hebert, D. Birch, T.L. McGuire et al. 2010. BMPR1a and BMPR1b signaling exert opposing effects on gliosis after spinal cord injury. *The Journal of Neuroscience*. 30:1839–1855.

Salta, E., and B. De Strooper. 2012. Non-coding RNAs with essential roles in neurodegenerative disorders. *Lancet Neurology*. 11:189–200.

Sandberg, R., J.R. Neilson, A. Sarma, P.A. Sharp, and C.B. Burge. 2008. Proliferating cells express mRNAs with shortened 3' untranslated regions and fewer microRNA target sites. *Science*. 320:1643–1647.

Schorey, J.S., and S. Bhatnagar. 2008. Exosome function: From tumor immunology to pathogen biology. *Traffic*. 9:871–881.

Schwarz, D.S., G. Hutvagner, T. Du, Z. Xu, N. Aronin, and P.D. Zamore. 2003. Asymmetry in the assembly of the RNAi enzyme complex. *Cell*. 115:199–208.

Selbach, M., B. Schwanhausser, N. Thierfelder, Z. Fang, R. Khanin, and N. Rajewsky. 2008. Widespread changes in protein synthesis induced by microRNAs. *Nature*. 455:58–63.

Selvamani, A., P. Sathyan, R.C. Miranda, and F. Sohrabji. 2012. An antagomir to microRNA Let7f promotes neuroprotection in an ischemic stroke model. *PloS One*. 7:e32662.

Selvamani, A., and F. Sohrabji. 2010. The neurotoxic effects of estrogen on ischemic stroke in older female rats is associated with age-dependent loss of insulin-like growth factor-1. *The Journal of Neuroscience*. 30:6852–6861.

Sepramaniam, S., A. Armugam, K.Y. Lim, D.S. Karolina, P. Swaminathan, J.R. Tan et al. 2010. MicroRNA 320a functions as a novel endogenous modulator of aquaporins 1 and 4 as well as a potential therapeutic target in cerebral ischemia. *The Journal of Biological Chemistry*. 285:29223–29230.

Shaked, I., A. Meerson, Y. Wolf, R. Avni, D. Greenberg, A. Gilboa-Geffen, and H. Soreq. 2009. MicroRNA-132 potentiates cholinergic anti-inflammatory signaling by targeting acetylcholinesterase. *Immunity*. 31:965–973.

Shen, J., W. Xia, Y.B. Khotskaya, L. Huo, K. Nakanishi, S.O. Lim et al. 2013. EGFR modulates microRNA maturation in response to hypoxia through phosphorylation of AGO2. *Nature*. 497:383–387.

Shi, G., Y. Liu, T. Liu, W. Yan, X. Liu, Y. Wang, J. Shi, and L. Jia. 2012. Upregulated miR-29b promotes neuronal cell death by inhibiting Bcl2L2 after ischemic brain injury. *Experimental Brain Research*. 216:225–230.

Shih, J.D., Z. Waks, N. Kedersha, and P.A. Silver. 2011. Visualization of single mRNAs reveals temporal association of proteins with microRNA-regulated mRNA. *Nucleic Acids Research*. 39:7740–7749.

Shively, S.B., and D.P. Perl. 2012. Traumatic brain injury, shell shock, and posttraumatic stress disorder in the military—past, present, and future. *The Journal of Head Trauma Rehabilitation*. 27:234–239.

Siegel, C., J. Li, F. Liu, S.E. Benashski, and L.D. McCullough. 2011. miR-23a regulation of X-linked inhibitor of apoptosis (XIAP) contributes to sex differences in the response to cerebral ischemia. *Proceedings of the National Academy of Sciences of the United States of America*. 108:11662–11667.

Simpson, R.J., S.S. Jensen, and J.W. Lim. 2008. Proteomic profiling of exosomes: Current perspectives. *Proteomics*. 8:4083–4099.

Sims, J.R., S.W. Lee, K. Topalkara, J. Qiu, J. Xu, Z. Zhou et al. 2009. Sonic hedgehog regulates ischemia/hypoxia-induced neural progenitor proliferation. *Strok*. 40:3618–3626.

Slack, F.J., M. Basson, Z. Liu, V. Ambros, H.R. Horvitz, and G. Ruvkun. 2000. The lin-41 RBCC gene acts in the C. elegans heterochronic pathway between the let-7 regulatory RNA and the LIN-29 transcription factor. *Molecular Cell*. 5:659–669.

Smirnova, L., A. Grafe, A. Seiler, S. Schumacher, R. Nitsch, and F.G. Wulczyn. 2005. Regulation of miRNA expression during neural cell specification. *The European Journal of Neuroscience*. 21:1469–1477.

Smith, B., J. Treadwell, D. Zhang, D. Ly, I. McKinnell, P.R. Walker et al. 2010. Large-scale expression analysis reveals distinct microRNA profiles at different stages of human neurodevelopment. *PloS One*. 5:e11109.

Stark, A., J. Brennecke, N. Bushati, R.B. Russell, and S.M. Cohen. 2005. Animal microRNAs confer robustness to gene expression and have a significant impact on 3'UTR evolution. *Cell*. 123:1133–1146.

Stenvang, J., A.N. Silahtaroglu, M. Lindow, J. Elmen, and S. Kauppinen. 2008. The utility of LNA in microRNA-based cancer diagnostics and therapeutics. *Seminars in Cancer Biology*. 18:89–102.

Svetlov, S.I., V. Prima, D.R. Kirk, H. Gutierrez, K.C. Curley, R.L. Hayes et al. 2010. Morphologic and biochemical characterization of brain injury in a model of controlled blast overpressure exposure. *The Journal of Trauma*. 69:795–804.

Tan, C.L., J.L. Plotkin, M.T. Veno, M. von Schimmelmann, P. Feinberg, S. Mann et al. 2013a. MicroRNA-128 governs neuronal excitability and motor behavior in mice. *Science*. 342:1254–1258.

Tan, J.R., K.S. Tan, Y.X. Koo, F.L. Yong, C.W. Wang, A. Armugam et al. 2013b. Blood microRNAs in Low or No Risk Ischemic Stroke Patients. *International Journal of Molecular Sciences*. 14:2072–2084.

Tan, K.S., A. Armugam, S. Sepramaniam, K.Y. Lim, K.D. Setyowati, C.W. Wang et al. 2009. Expression profile of MicroRNAs in young stroke patients. *PloS One*. 4:e7689.

Tang, S.C., T.V. Arumugam, X. Xu, A. Cheng, M.R. Mughal, D.G. Jo et al. 2007. Pivotal role for neuronal Toll-like receptors in ischemic brain injury and functional deficits. *Proceedings of the National Academy of Sciences of the United States of America*. 104:13798–13803.

Tanielian, T.L., L. Jaycox, and Rand Corporation. 2008. *Invisible Wounds of War: Psychological and Cognitive Injuries, Their Consequences, and Services to Assist Recovery*. RAND, Santa Monica, CA.

Tao, J., H. Wu, Q. Lin, W. Wei, X.H. Lu, J.P. Cantle et al. 2011. Deletion of astroglial Dicer causes non-cell-autonomous neuronal dysfunction and degeneration. *The Journal of Neuroscience*. 31:8306–8319.

Tenovuo, O. 2005. Central acetylcholinesterase inhibitors in the treatment of chronic traumatic brain injury-clinical experience in 111 patients. *Progress in Neuro-psychopharmacology & Biological Psychiatry*. 29:61–67.

Truettner, J.S., O.F. Alonso, H.M. Bramlett, and W.D. Dietrich. 2011. Therapeutic hypothermia alters microRNA responses to traumatic brain injury in rats. *Journal of Cerebral Blood Flow and Metabolism*. 31:1897–1907.

Truettner, J.S., D. Motti, and W.D. Dietrich. 2013. MicroRNA overexpression increases cortical neuronal vulnerability to injury. *Brain Research*. 1533:122–130.

Tsai, P.C., Y.C. Liao, Y.S. Wang, H.F. Lin, R.T. Lin, and S.H. Juo. 2013. Serum microRNA-21 and microRNA-221 as potential biomarkers for cerebrovascular disease. *Journal of Vascular Research*. 50:346–354.

Valiyaveettil, M., Y.A. Alamneh, S.A. Miller, R. Hammamieh, P. Arun, Y. Wang et al. 2013. Modulation of cholinergic pathways and inflammatory mediators in blast-induced traumatic brain injury. *Chemico-biological Interactions*. 203:371–375.

Wang, H., M. Chiu, Z. Xie, Z. Liu, P. Chen, S. Liu et al. 2012a. Synthetic microRNA cassette dosing: Pharmacokinetics, tissue distribution and bioactivity. *Molecular Pharmaceutics*. 9:1638–1644.

Wang, W.X., B.R. Wilfred, S.K. Madathil, G. Tang, Y. Hu, J. Dimayuga et al. 2010. miR-107 regulates granulin/progranulin with implications for traumatic brain injury and neurodegenerative disease. *The American Journal of Pathology*. 177:334–345.

Wang, Y., F. Guo, C. Pan, Y. Lou, P. Zhang, S. Guo, J. Yin, and Z. Deng. 2012b. Effects of low temperatures on proliferation-related signaling pathways in the hippocampus after traumatic brain injury. *Experimental Biology and Medicine*. 237:1424–1432.

Wang, Y., Y. Wei, S. Oguntayo, W. Wilkins, P. Arun, M. Valiyaveettil et al. 2011. Tightly coupled repetitive blast-induced traumatic brain injury: Development and characterization in mice. *Journal of Neurotrauma*. 28:2171–2183.

Weng, H., C. Shen, G. Hirokawa, X. Ji, R. Takahashi, K. Shimada et al. 2011. Plasma miR-124 as a biomarker for cerebral infarction. *Biomedical Research*. 32:135–141.

Wightman, B., I. Ha, and G. Ruvkun. 1993. Posttranscriptional regulation of the heterochronic gene lin-14 by lin-4 mediates temporal pattern formation in C. elegans. *Cell*. 75:855–862.

Witkos, T.M., E. Koscianska, and W.J. Krzyzosiak. 2011. Practical aspects of microRNA target prediction. *Current Molecular Medicine*. 11:93–109.

Wu, C., J. So, B.N. Davis-Dusenbery, H.H. Qi, D.B. Bloch, Y. Shi et al. 2011. Hypoxia potentiates microRNA-mediated gene silencing through posttranslational modification of Argonaute2. *Molecular and Cellular Biology*. 31:4760–4774.

Xiao, C., L. Srinivasan, D.P. Calado, H.C. Patterson, B. Zhang, J. Wang et al. 2008. Lymphoproliferative disease and autoimmunity in mice with increased miR-17–92 expression in lymphocytes. *Nature Immunology*. 9:405–414.

Xie, J., Q. Xie, H. Zhang, S.L. Ameres, J.H. Hung, Q. Su et al. 2011. MicroRNA-regulated, systemically delivered rAAV9: A step closer to CNS-restricted transgene expression. *Molecular Therapy*. 19:526–535.

Yi, R., Y. Qin, I.G. Macara, and B.R. Cullen. 2003. Exportin-5 mediates the nuclear export of pre-microRNAs and short hairpin RNAs. *Genes & Development*. 17:3011–3016.

Yin, K.J., Z. Deng, M. Hamblin, Y. Xiang, H. Huang, J. Zhang et al. 2010a. Peroxisome proliferator-activated receptor delta regulation of miR-15a in ischemia-induced cerebral vascular endothelial injury. *The Journal of Neuroscience*. 30:6398–6408.

Yin, K.J., Z. Deng, H. Huang, M. Hamblin, C. Xie, J. Zhang et al. 2010b. miR-497 regulates neuronal death in mouse brain after transient focal cerebral ischemia. *Neurobiology of Disease*. 38:17–26.

Yin, K.J., Y. Fan, M. Hamblin, J. Zhang, T. Zhu, S. Li et al. 2013. KLF11 mediates PPARgamma cerebrovascular protection in ischaemic stroke. *Brain*. 136:1274–1287.

Yuan, M., Q. Zhan, X. Duan, B. Song, S. Zeng, X. Chen et al. 2013. A functional polymorphism at miR-491-5p binding site in the 3'-UTR of MMP-9 gene confers increased risk for atherosclerotic cerebral infarction in a Chinese population. *Atherosclerosis*. 226:447–452.

Yuan, Y., J.Y. Wang, L.Y. Xu, R. Cai, Z. Chen, and B.Y. Luo. 2010. MicroRNA expression changes in the hippocampi of rats subjected to global ischemia. *Journal of Clinical Neuroscience*. 17:774–778.

Zehir, A., L.L. Hua, E.L. Maska, Y. Morikawa, and P. Cserjesi. 2010. Dicer is required for survival of differentiating neural crest cells. *Developmental Biology*. 340:459–467.

Zeng, Y., and B.R. Cullen. 2003. Sequence requirements for micro RNA processing and function in human cells. *RNA*. 9:112–123.

Zeng, Y., E.J. Wagner, and B.R. Cullen. 2002. Both natural and designed micro RNAs can inhibit the expression of cognate mRNAs when expressed in human cells. *Molecular Cell*. 9:1327–1333.

Zhang, L., M. Chopp, X. Liu, H. Teng, T. Tang, H. Kassis et al. 2012. Combination therapy with VELCADE and tissue plasminogen activator is neuroprotective in aged rats after stroke and targets microRNA-146a and the toll-like receptor signaling pathway. *Arteriosclerosis, Thrombosis, and Vascular Biology*. 32:1856–1864.

Zhang, R.L., M. Chopp, S.R. Gregg, Y. Toh, C. Roberts, Y. Letourneau et al. 2009. Patterns and dynamics of subventricular zone neuroblast migration in the ischemic striatum of the adult mouse. *Journal of Cerebral Blood Flow and Metabolism*. 29:1240–1250.

Zhang, R.L., Z.G. Zhang, M. Lu, Y. Wang, J.J. Yang, and M. Chopp. 2006. Reduction of the cell cycle length by decreasing G1 phase and cell cycle reentry expand neuronal progenitor cells in the subventricular zone of adult rat after stroke. *Journal of Cerebral Blood Flow and Metabolism*. 26:857–863.

Zhao, C., G. Sun, S. Li, M.F. Lang, S. Yang, W. Li et al. 2010a. MicroRNA let-7b regulates neural stem cell proliferation and differentiation by targeting nuclear receptor TLX signaling. *Proceedings of the National Academy of Sciences of the United States of America*. 107:1876–1881.

Zhao, H., J. Wang, L. Gao, R. Wang, X. Liu, Z. Gao et al. 2013. MiRNA-424 protects against permanent focal cerebral ischemia injury in mice involving suppressing microglia activation. *Stroke*. 44:1706–1713.

Zhao, X., X. He, X. Han, Y. Yu, F. Ye, Y. Chen et al. 2010b. MicroRNA-mediated control of oligodendrocyte differentiation. *Neuron*. 65:612–626.

Zheng, K., H. Li, H. Huang, and M. Qiu. 2012. MicroRNAs and glial cell development. *The Neuroscientist*. 18:114–118.

Zheng, K., H. Li, Y. Zhu, Q. Zhu, and M. Qiu. 2010. MicroRNAs are essential for the developmental switch from neurogenesis to gliogenesis in the developing spinal cord. *The Journal of Neuroscience*. 30:8245–8250.

Zhu, F., J.L. Liu, J.P. Li, F. Xiao, Z.X. Zhang, and L. Zhang. 2014. MicroRNA-124 (miR-124) regulates Ku70 expression and is correlated with neuronal death induced by ischemia/reperfusion. *Journal of Molecular Neuroscience: MN*. 52:148–155.

Ziu, M., L. Fletcher, S. Rana, D.F. Jimenez, and M. Digicaylioglu. 2011. Temporal differences in microRNA expression patterns in astrocytes and neurons after ischemic injury. *PloS One*. 6:e14724.

27 Neuroproteome Dynamics in Modeled Brain Injury

A Systems Neurobiology Perspective

Pavel N. Lizhnyak, Hiyab Yohannes, and Andrew K. Ottens

CONTENTS

27.1 INTRODUCTION TO NEUROPROTEOMICS IN MODELED BRAIN INJURY

Neuroproteomics enables the holistic interrogation of functional changes down at the molecular level following brain injury. Results are a window into the protein-driven mechanisms underlying cellular and anatomical degradation as well as regeneration. In this chapter, we discuss the merits and limitations of neuroproteomics in the context of studying brain injury, and discuss important consideration for study design. We present results from our recent studies on how this new level of detail informs our understanding of the fundamental processes governing regeneration. Placed in context with observed anatomical restructuring, neuroproteomic results can be instrumental in guiding targeted interventions of regeneration. Delineating the temporal series of molecular events would facilitate time-dependent therapeutic administration as well as precision endpoints with which to decipher efficacy. Ultimately, deep-neuroproteomic analysis can enhance our understanding of the complex interactive-nature of the molecular underpinnings of brain injury pathobiology and help delivery on the promise of improved care and treatment.

Traumatic brain injury (TBI) broadly encompasses any damage from a mechanical insult to the brain, accounting for more than 1.7 million incidents annually in the United States (Faul et al., 2010). As a neurodegenerative disease, TBI requires unique consideration when employing model systems. TBI initiates with an acute mechanical event, providing a clear start to the pathobiology that can be reasonably approximated in the laboratory. However, a multitude of external and individual factors critically influence the ensuing damage and progression of disease, instilling heterogeneity that is difficult to account for in model systems (Adamczak et al., 2012; Roozenbeek et al., 2012). Secondary biochemical insults soon follow that challenge targeted molecular studies. Following the course of acute degeneration are the onset of repair mechanisms as well as chronic degradative processes. Thus, there are many facets and caveats to understanding the evolving molecular and anatomical picture of TBI.

Systems neurobiology aspires to the holistic comprehension of neural networks underlying brain function, and, in application to TBI, its dysfunction. The discipline takes into account the biochemical dynamics that facilitate integration, maturation, and plastic organization of neural networks, extending the bounds of molecular neuroscience beyond the process of neurotransmission. Neuroproteomics is a logical component in systems neurobiological studies, by which complex and dynamic posttranslational mechanisms of action can be interrogated under developing, mature, and pathological conditions. Although other omic disciplines importantly study nucleotide, lipid, and small molecule constituents, proteomics is perhaps the most functionally informative on acute injury, given that proteins and their products are the chief actors within and between cells. TBI models provide researchers with the means to interrogate the functional relationships between the neuroproteomic, anatomic, and neurologic domains. Thereby, an integrated understanding of TBI pathobiology is poised to provide a deeper reflection of dysfunction and how to minimize or therapeutically reverse the consequences.

27.2 STUDY DESIGN IN UNBIASED NEUROPROTEOMICS OF TBI

In designing a neuroproteomics experiment, one must carefully consider what molecular change are to be assessed. The neuroproteome exhibits a multidimensional response well beyond that exhibited with messenger RNA in its reaction to brain injury. For example, the posttranslational actions of enzymes result in protein modifications that alter normal function in response to the traumatic disruption of homeostasis. Some actions by design induce protective mechanisms, whereas others are uncontrolled consequences of environmental change that result in rapid degradation of proximal structures. A host of posttranslational changes, from signaling motifs, such as phosphorylation and acetylation, to cleavage events by a host of proteases may be assessed within the neuroproteome. Posttranslational events occur in parallel and on an evolving serial timescale. Modifications can also initiate protein translocation within and between cells, with functional relevance often dictated by how much of a protein is in any given location. Thus, it is essential to consider the combination of states by which any given protein may exist within the neural circuitry in its dynamic response to TBI.

Yet, it remains a significant analytical challenge to account for the complexity within the neuroproteome, limiting our interpretation of its response to TBI. Much of the diversity within the neuroproteome remains unknown. For example, protein databases comprise a core of well-annotated proteins along with a larger number of predicted protein variants. The latter are suggested translational products from genomic screens that remain unconfirmed. In our neurodevelopmental and acute insult research alike, we routinely observe sequences specific to unconfirmed variants. Immunological-based approaches may be used to resolve isoform-specific epitopes within a target protein family, but remain impractical for consideration of the full complement of predicted protein variants across the neuroproteome. Alternative nontargeted methods are hence needed to fill a critical gap in our understanding of the translational and posttranslational response to modeled TBI.

Mass spectrometry offers a flexible platform to perform both targeted (Ottens et al., 2007) and broad-scale studies of the TBI neuroproteome (Cortes et al., 2012). Nontargeted, also known as *unbiased*, approaches evaluate peptide sequences from thousands of proteins without a priori selection. However, the TBI research must appreciate limits to mass spectrometric approaches in order to properly design and interpret a study as many variables influence result generation. For example, methods are limited by the sensitivity and dynamic range of the mass spectrometer and the overall separation space within which analytes are measured. Biospecimens contain proteins over a wide array of concentrations, yet mass spectrometers are limited in their dynamic range between 2.5 and 4.5 orders in magnitude. Unlike with familiar nucleotide analyses, in proteomics research detection of low abundant species cannot be achieved through sequence amplification. Enrichment preparations do exist to allow for more effective characterization of a subset of the neuroproteome (e.g., subcellular, immunoaffinity, or ligand affinity enrichment). Another consideration is to broaden the analytical capacity of the mass spectrometric method by incorporating multiple dimensions of molecular separation ahead of detection. Liquid- and gas-phase chromatographic methods can often be performed in tandem to more effectively resolve sample components and provide more comprehensive detection. Fundamentally this occurs not by extending the detection range of the mass spectrometer but by separating lesser abundant analytes from higher concentration species that would otherwise interfere with detection. There is, however, a time penalty paid with each successive chromatographic method added to the study design. In practice, non-targeted methods are generally limited to detection of approximately 2000 more abundant proteins, which is estimated to be about one-half of the full neuroproteome. Yet, molecular separations and mass spectrometry for proteomics is expected to continue at a rapid pace, warranting optimism for improved coverage in the next five years.

Unbiased methods rely on the generation of selective mass information to confidently match a peptide sequence. Recent improvements in mass measurement accuracy and precision have lessened ambiguity in assigning an appropriate sequence to a measured peptide. However, tandem mass spectra remain an essential source of selective information for differentiating peptides that overlap in mass to within a few thousandths of a Dalton. Yet, the natural variety of posttranslational modifications is perhaps the most challenging aspect to unbiased proteomics. Consider the important signaling motif of phosphorylation. Lacking a full account as to which of a protein's residues can be phosphorylated, proteomic software must consider the addition of phosphate (80 Daltons) to all serine, threonine, and tyrosine residues, along with the possibility that there are multiple phosphates on a given peptide. Thus, the list of possible masses to consider swells several-fold with just this one type of modification, necessitating further advancements in informatic approaches for deciphering the posttranslational modified neuroproteome. One must also recognize that experimental conditions bias which peptides may be detected. Digestion with restrictive proteases produces a multitude of small peptides that are two short in sequence to selectively associate with any given protein, whereas other peptides may be too large to accurately assign a mass and fragmentation pattern.

Neuroproteomic study design must carefully balance sample complexity with appropriate analytical capacity. Simpler subproteomes, such as derived through immunoprecipitation, can be effectively analyzed with most liquid chromatography–tandem mass spectrometry configurations. Analysis of whole-tissue lysates requires multidimensional separations, perhaps including subcellular fractionation and multiple steps of liquid and gas phase separations. Protein extraction procedures must also be carefully considered; for example, the addition of surfactants and other denaturants will provide accessibility to integral membrane and membrane bound proteins. Yet, these factors should be considered carefully because they may interfere with subsequent digestion and separation stages. It is also critical to control solution pH while processing proteomic samples; for example, iodoacetamide, a common alkylation agent used to protect reactive thiols, requires basic conditions to avoid unintended reactions with free amines. Restrictive proteases such as trypsin also require controlled pH conditions to maximize reactivity.

Lastly, neuroproteome sampling time is a particularly critical variable for modeled TBI research. Acute responses occur in minutes to hours, whereas postacute cell survival and death processes extend out days from injury. Those processes are gradually superseded by regenerative mechanisms lasting weeks to months. Thus, a faster sampling rate is essential in the acute phase after injury, with a lengthened period during the postacute and chronic phases of TBI.

27.3 QUANTITATIVE ANALYSIS

The complexities of the neuroproteome discussed previously must also be factored into procedures for quantitative analysis. A key consideration is that different peptide measures of the same protein family can often reflect independent responses to TBI, a fact that has gone underappreciated by the casual proteomics user. Consider the hypothetical scenario of kinase activation after TBI rapidly phosphorylating 20% more of protein X within the cytosol, causing a proportional translocation of protein X to the mitochondrial membrane. First, only 20% of the targeted motif changes its phosphorylation state, such the phosphorylated and non-phosphylated forms of the associated peptide are altered in abundance in complementary amounts that are distinct from any other peptides to protein X. Second, the total abundance of protein X remains unchanged; however, were the experiment to quantify cytosolic and mitochondrial-associated amounts of the protein separately, then the critical redistribution of protein X could be discerned. As another example, consider protein Y with two potential isoforms and only two peptides cover each of two alternatively spliced regions. In this case, the selective peptides to isoform 1 may respond oppositely to TBI than those of isoform 2, all the while peptides common to both may appear to exhibit no significant change. All told, it is instrumental to consider independent dynamics of peptides in order to reveal more specific information regarding the response to TBI.

With tens of thousands of peptides quantified in any given study, one must give consideration to the statistical testing and appropriate power required to generate meaningful results. A group size of three may be tempting given its common use in the literature; however, as with all quantitative studies, the number of replicates should be set based on the expected variance and minimum effect size of interest. Per this point, an effect size of two-fold is biochemically arbitrary, and its use can underpower your observations. For example, in our research, we often find that a minimum group size of five replicates provides sufficient power to differentiate a median effect size of 30% change in analysis of roughly 10,000 peptides between two experimental groups, which often is necessary to inform on post-translational change after TBI. Adding additional study factors or levels (e.g., multiple time points post-TBI) would require an increased sample size. As with other omic methods, one must also consider the problem of multiple testing (e.g., 10,000 comparisons between two or more experimental groups). Put another way, 750 peptides would be falsely identified as significantly responsive to TBI (type I errors) at a customary alpha of 0.05. Several statistical approaches exist to account for multiple comparisons—our preference is to calculate a false discovery rate (which provides a greater power than a false-positive rate) by calculating q values from the distribution of p values. One must also carefully consider the appropriate use of statistical testing. For example, it is inappropriate to use analysis of variance on raw mass spectrometric intensity measures; however, the data may meet the assumption of a normal distribution of values following a log-transformation.

27.4 NEUROPROTEOME DYNAMICS AND DISCLOSING NEW KNOWLEDGE

Unbiased approaches offer us the potential to elucidate as-yet unexplored dynamic aspects of TBI pathobiology relevant to functional outcomes and effective interventions. Neuroproteomics offers new capability to study the temporal evolution of the biochemical mechanisms underlying not only degeneration but also repair and regeneration. For example, few studies have looked at critical perturbation to the inhibitory synaptic network following modeled TBI (Bonislawski et al., 2007; Pavlov et al., 2011). Inhibitory network plasticity is critical to counteracting aberrant signaling for effective restoration of function. Deficits in inhibitory signaling are further linked with post-TBI epileptic activity in the post-acute and chronic phases. Yet, little is known regarding the mechanisms governing restoring functional plasticity to the inhibitory synaptic network following TBI.

In recent neuroproteomic studies of the temporal response to modeled focal contusion injury, we observed a conspicuous and prolonged reduction in the activated, membrane-bound level of Rab3b (Figure 27.1a) within the perilesional neocortex. This *Rab3* isoform selectively reflects the loss of inhibitory synaptic transmission (Huusko et al., 2013). These findings were consistent with a loss in inhibitory synaptic vesicle staining within the perilesional region using the marker synaptotagmin 2 (Syt2) as early as 2 days after TBI (Figure 27.1b). *Rab3's* are guanosine-5'-triphosphate–binding proteins that regulate synaptic vesicle membrane trafficking (Armstrong, 2000; Pavlos and Jahn, 2011; Sudhof, 2004; Tuvim et al., 2001; Tsetsenis et al., 2011). A *Rab3* soluble pool remains available for potential recruitment after a proper molecular signal; however, the bound/integral proportion of Rab3b found diminished here reflects a loss in vesicle association as used to recruit vesicles for docking to the presynaptic membrane and facilitating priming and release of neurotransmitter (Figure 27.2). The functional differences between isoforms are not fully understood (Schluter et al., 2004). Yet, we critically know although Rab3a is involved in maintaining glutamatergic synaptic transmission, Rab3b is selectively associated with inhibitory synapses in the neocortex (Feliciano et al., 2013). Thus, these findings emphasize two earlier points: that it is essential to consider independent dynamics among isoforms and their distribution within the cell.

Further understanding of the molecular mechanisms involved in restoration of the inhibitory network is needed given that its dysfunction is known to result in cognitive deficits as well as other behavioral and neuropsychiatric impairments (Sun and Feng, 2014). Adhesion proteins are critical during synaptogenesis in guiding the axonal

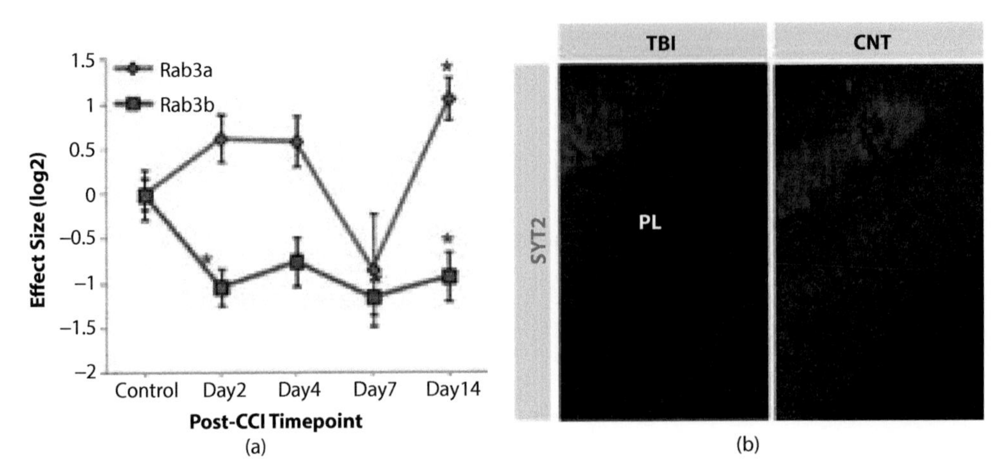

FIGURE 27.1 Loss of inhibitory synaptic transmission in perilesional (PL) neocortex. (a) Temporal pattern of activated/bound rab3 isoforms reflecting a differential response within the 2 weeks following modeled TBI. Rab3b results reflect a shutdown in synaptic vesicle trafficking and inhibitory neurotransmission throughout the postacute period. Mean ± standard error; * $p \leq 0.01$, relative to control; n = 6 animals/time point. (b) PL staining of depicting the loss of inhibitory synaptic vesicles with synaptotagmin-2 (Syt2) immunofluorescence at 2 days post-TBI relative to naive control (CNT).

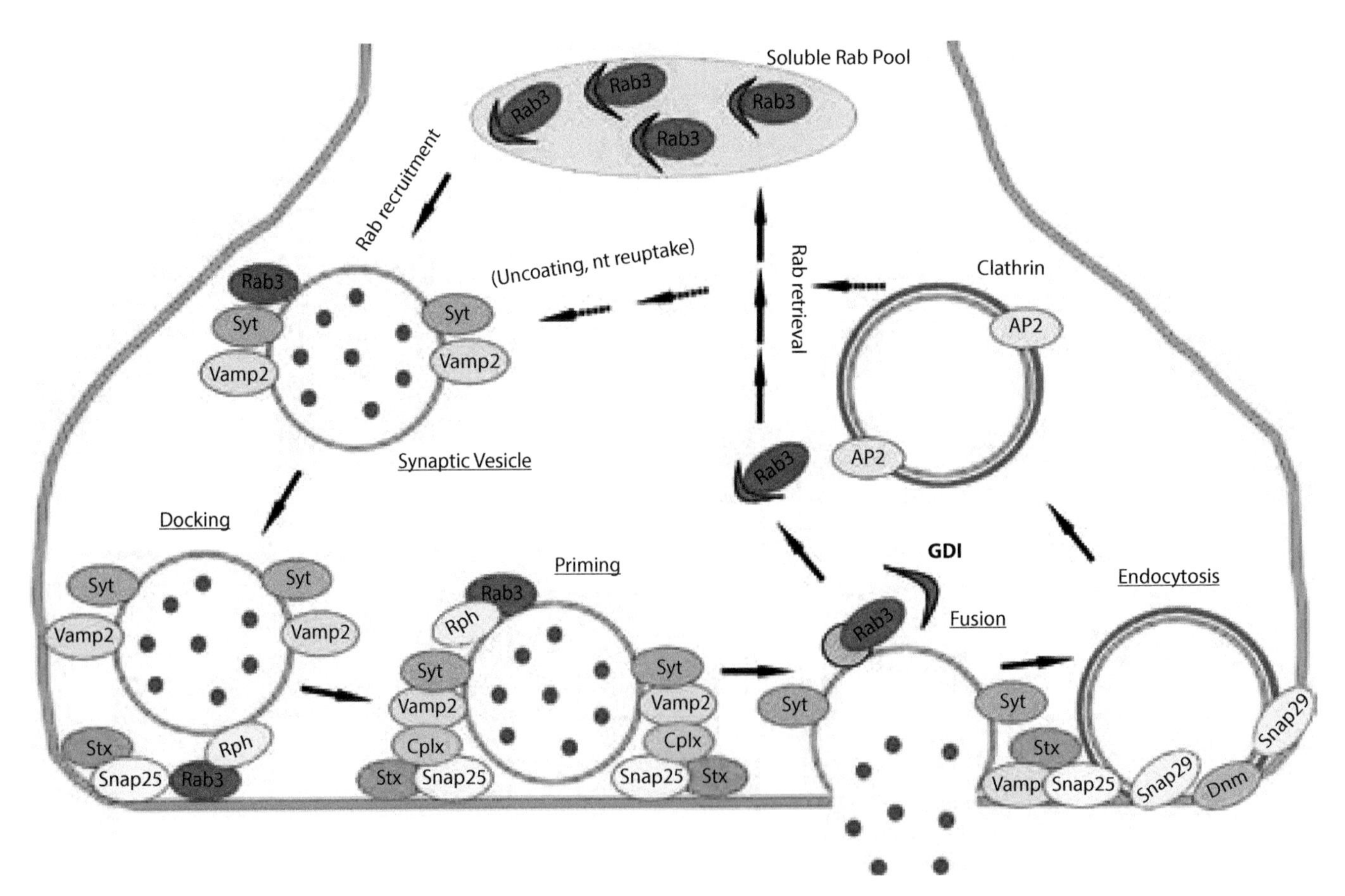

FIGURE 27.2 Rab3s are essential GTPases involved in trafficking synaptic vesicles and regulated neurotransmitter release. A reserve Rab3 soluble pool is tapped for activity-dependent vesicle binding. Different isoforms are responsible for excitatory (Rab3a) and inhibitory (Rab3b) vesicle trafficking at the postsynaptic terminal.

segment, anchoring the inhibitory synapse, and maintaining the synaptic machinery involved in the initiation and the progression of neurotransmission (Figure 27.3). Neuroligin-2 in particular is specific to the inhibitory postsynaptic differentiation, anchoring both pre- and postsynaptic aspects (Poulopoulos et al., 2009). Our data show that membrane-bound neuroligin-2 is significantly diminished by 2 days post-TBI (Figure 27.4a), denoting early loss of the inhibitory synapse–anchoring mechanism as well as the decrease of VGAT (Blundell et al., 2009). Our data then show recovery of neuroligin-2 levels with an increased membrane presentation as is consistent with initiation of synatogenesis. We further identified a distinct temporal profile (Figure 27.4b) for a known neuroligin-2 phosphopeptide containing S714 (Tweedie-Cullen et al., 2009; Wisniewski et al., 2010). These results show that neuroligin-2 was significantly

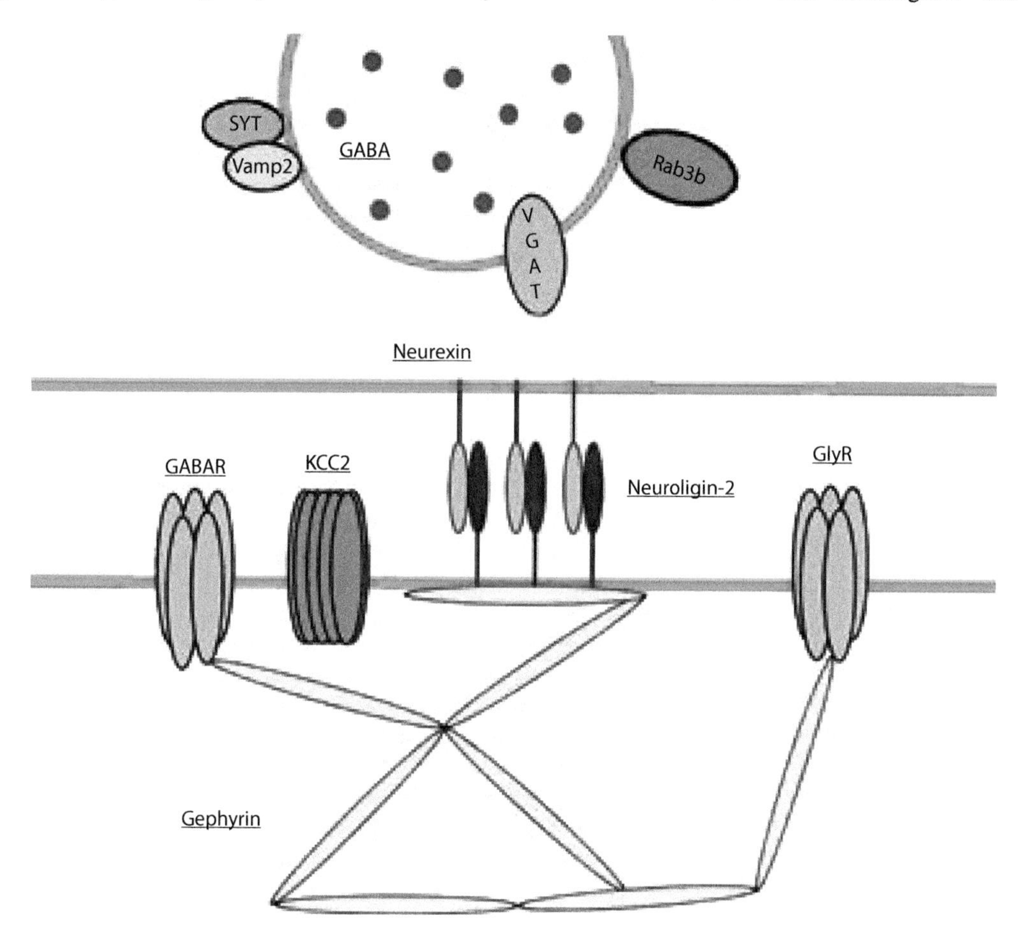

FIGURE 27.3 Molecular components underlying the formation and stabilization of inhibitory synapses. Rab3b regulates trafficking of GABA-containing synaptic vesicles. Neuroligin-2 guides synaptogenesis and anchoring of inhibitory synapses. KCC2 modulates postsynaptic chloride homeostasis, controlling the developmental excitatory/inhibitory switch. Gephyrin provides a polymeric scaffold with which to secure components of the inhibitory postsynaptic differentiation.

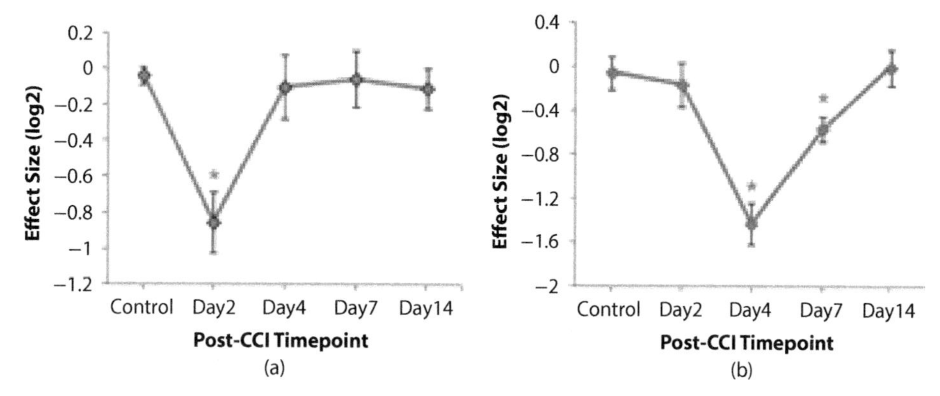

FIGURE 27.4 Post-TBI dynamics of neuroligin-2. (a) Temporal profile of membrane-bound neuroligin-2 levels after TBI. Reduced binding of neuroligin-2 denotes deafferentation of inhibitory synapses. (b) Temporal profile of phosphorylated state of neuroligin-2 at S714, reflecting a shift in the functional state at day 4 in time with the recovery of membrane-bound neuroligin-2. Mean ± standard error; $*p \leq 0.01$; n = 6 animals/time point.

dephosphorylated at day 4, paralleling the observed recovery of membrane-bound neuroligin levels, with both measures returning to control levels by day 14. These data point to a time-specific neuroligin-2 dephosphorylation event that may be a critical signal in initiating inhibitory synaptogenesis.

Importantly, gamma aminobutyric acid (GABA) transmission initially serves an excitatory function with a high chloride concentration inside the cell during synaptogenesis, and GABA receptor activation leading to chloride efflux and depolarization of the cell (Lee et al., 2005). The switch from depolarizing to hyperpolarizing function occurs with an increase in the type 2 K^+-Cl^- symporter (KCC2), which maintains a reduced chloride gradient in mature inhibitory neurons (Ben-Ari, 2002; Kahle et al., 2008; Papp et al., 2008; Payne et al., 2003). The excitatory functionality of GABA is, further, developmentally critical to morphological maturation of cortical neurons (Cancedda et al., 2007). Our TBI data reveal that KCC2 levels are significantly reduced starting around day 4 with recovery 2 weeks after TBI (Figure 27.5a). These results agree with an earlier report of reduced KCC2 levels in hippocampus after TBI, but provide additional insight into the transient nature of this signal as is relevant to restoration of the inhibitory network. (Bonislawski et al., 2007). A loss of KCC2 connotes a transient intracellular buildup of chloride resetting nascent GABA transmission to a depolarizing state. The purported chloride switch is poised to be a critical event in the regenerative mechanism of the inhibitory network after TBI.

From neurodevelopment, we also know that there is a precise order of events for establishing inhibitory synapses: an early presentation of neuroligin-2 along the postsynaptic membrane followed by increased KCC2 expression to guide recruitment of additional players to the synaptic cleft (Sun et al., 2013). Yet, our data appear to be the first to outline this progression of events during the regenerative phase after brain injury. We conclude that the depolarizing function of GABA is needed for proper regeneration and renewal of the damaged neuronal network during the postacute period of TBI. Yet another important temporal event in inhibitory synaptogenesis involves the scaffolding protein gephyrin, which interacts with neuroligin-2 and provides the structural support for the postsynaptic differentiation (Michels and Moss, 2007; Fritschy et al., 2008; Jacob et al., 2005). Gephyrin polymerization is critical to the clustering of GABA and glycine receptors (Gonzalez, 2013; Levi et al., 2004; Yu et al., 2007). We observe a loss of the multimeric gephyrin scaffold at day 7, which is consistent with the remodeling of the postsynaptic neuronal membrane (Figure 27.5b). Polymerized gephyrin levels begin to recover by day 14 post-TBI and reflect formation of postsynaptic complex simultaneously with the reexpression of the KCC2 cotransporter. All together, these findings reveal a parallel course of molecular events as seen in neurodevelopment, initiating more than 2 days after injury and extending on through the first 2 weeks. These findings are illustrative of the powerful discovery platform of unbiased neuroproteomics, which is poised to extend our understanding of the temporal course of critical steps in underlying post-TBI regeneration as well as other aspects of pathobiology.

27.5 CONCLUSION

Omic methods are a critical complement to targeted approaches in that they facilitate out-of-the-box examination of a complex biochemical response. Careful consideration must be paid when designing a study and interpreting results, as discussed in this chapter. However, the potential for novel findings are immense as exemplified by the presented results elucidating the dynamic progression of molecular events initiating restoration of the inhibitory network after TBI. Such detail is instrumental in engineering a combinatorial approach to TBI therapeutics. It is critical to assess when a drug should be administered, for example, in the use of diazepam to restore GABAergic function at the right point after injury (Gibson et al., 2010; O'Dell et al., 2000; Mtchedlishvili et al., 2010). With neuroproteomic data as a template, other treatments may be developed to promote more effective initiation and/or progression of the assessed synaptogenesis mechanism, as needed to minimize functional deficits and potential chronic epileptogenesis (Avramescu et al., 2009; Bissonette et al., 2014; Hashimoto et al., 2008).

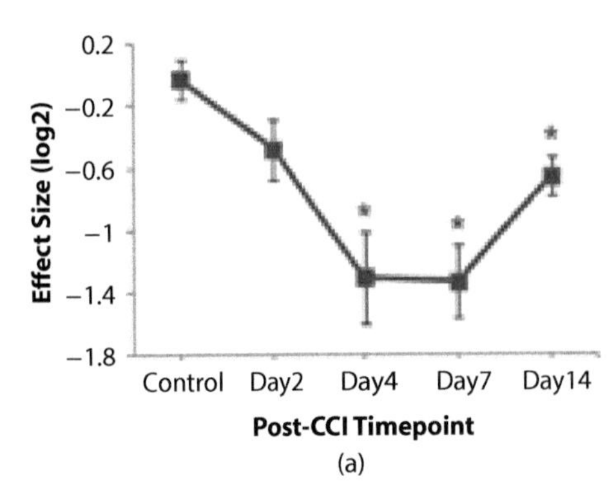

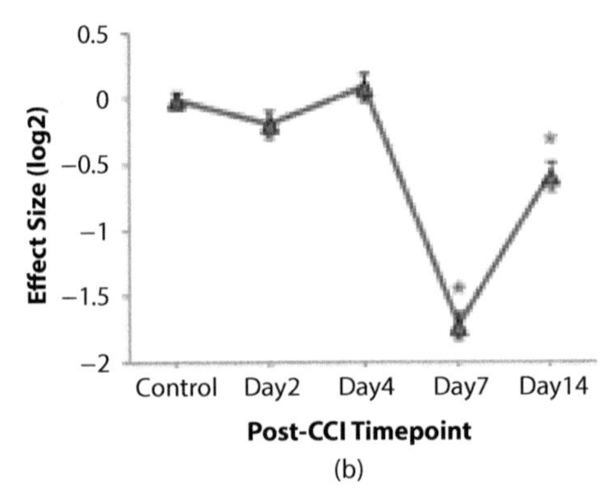

FIGURE 27.5 Dynamics involved in reestablishment of the inhibitory network in perilesional neocortex after TBI. (a) Temporal profile of KCC2 levels after TBI. KCC2 is essential to maintenance of chloride levels within inhibitory neurons, which is diminished in early stages of synaptogenesis, resulting in excitatory activation of GABA transmission. (b) Temporal profile of membrane-bound multimeric gephyrin after TBI. Reduction in levels denotes depolymerization of the gephyrin scaffold at the postsynaptic differentiation. Mean ± standard error; $*p \leq 0.01$; n = 6 animals/time point.

REFERENCES

Adamczak, S., G. Dale, J. P. de Rivero Vaccari, M. R. Bullock, W. D. Dietrich, and R. W. Keane. 2012. Inflammasome proteins in cerebrospinal fluid of brain-injured patients as biomarkers of functional outcome: Clinical article. *Journal of Neurosurgery* 117 (6) (Dec): 1119–25.

Armstrong, J. 2000. How do rab proteins function in membrane traffic? *The International Journal of Biochemistry & Cell Biology* 32 (3) (Mar): 303–7.

Avramescu, S., D. A. Nita, and I. Timofeev. 2009. Neocortical post-traumatic epileptogenesis is associated with loss of GABAergic neurons. *Journal of Neurotrauma* 26 (5) (May): 799–812.

Ben-Ari, Y. 2002. Excitatory actions of GABA during development: The nature of the nurture. *Nature Reviews Neuroscience* 3 (9) (Sep): 728–39.

Bissonette, G. B., M. H. Bae, T. Suresh, D. E. Jaffe, and E. M. Powell. 2014. Prefrontal cognitive deficits in mice with altered cerebral cortical GABAergic interneurons. *Behavioural Brain Research* 259 (Feb 1): 143–51.

Blundell, J., K. Tabuchi, M. F. Bolliger, C. A. Blaiss, N. Brose, X. Liu et al. 2009. Increased anxiety-like behavior in mice lacking the inhibitory synapse cell adhesion molecule neuroligin 2. *Genes, Brain, and Behavior* 8 (1) (Feb): 114–26.

Bonislawski, D. P., E. P. Schwarzbach, and A. S. Cohen. 2007. Brain injury impairs dentate gyrus inhibitory efficacy. *Neurobiology of Disease* 25 (1) (Jan): 163–9.

Cancedda, L., H. Fiumelli, K. Chen, and M. M. Poo. 2007. Excitatory GABA action is essential for morphological maturation of cortical neurons in vivo. *The Journal of Neuroscience* 27 (19) (May 9): 5224–35.

Cortes, D. F., M. K. Landis, and A. K. Ottens. 2012. High-capacity peptide-centric platform to decode the proteomic response to brain injury. *Electrophoresis* 33 (24) (Dec): 3712–9.

Faul, M., L. Xu, M. M. Wald, and V. G. Coronado. 2010. Traumatic brain injury in the United States: Emergency department visits, hospitalizations and deaths 2002–2006. Centers for Disease Control and Prevention, National Center for Injury Prevention and Control, Atlanta, GA.

Feliciano, P., R. Andrade, and M. Bykhovskaia. 2013. Synapsin II and Rab3a cooperate in the regulation of epileptic and synaptic activity in the CA1 region of the hippocampus. *The Journal of Neuroscience* 33 (46) (Nov 13): 18319–30.

Fritschy, J. M., R. J. Harvey, and G. Schwarz. 2008. Gephyrin: Where do we stand, where do we go? *Trends in Neurosciences* 31 (5) (May): 257–64.

Gibson, C. J., R. C. Meyer, and R. J. Hamm. 2010. Traumatic brain injury and the effects of diazepam, diltiazem, and MK-801 on GABA-A receptor subunit expression in rat hippocampus. *Journal of Biomedical Science* 17 (May 18): 38,0127–17–38.

Gonzalez, M. I. 2013. The possible role of GABAA receptors and gephyrin in epileptogenesis. *Frontiers in Cellular Neuroscience* 7: 113.

Hashimoto, T., H. H. Bazmi, K. Mirnics, Q. Wu, A. R. Sampson, and D. A. Lewis. 2008. Conserved regional patterns of GABA-related transcript expression in the neocortex of subjects with schizophrenia. *The American Journal of Psychiatry* 165 (4) (Apr): 479–89.

Huusko, N., C. Romer, X. E. Ndode-Ekane, K. Lukasiuk, and A. Pitkanen. 2013. Loss of hippocampal interneurons and epileptogenesis: A comparison of two animal models of acquired epilepsy. *Brain Structure & Function* (Oct 6).

Jacob, T. C., Y. D. Bogdanov, C. Magnus, R. S. Saliba, J. T. Kittler, P. G. Haydon et al. 2005. Gephyrin regulates the cell surface dynamics of synaptic GABAA receptors. *The Journal of Neuroscience* 25 (45) (Nov 9): 10469–78.

Kahle, K. T., K. J. Staley, B. V. Nahed, G. Gamba, S. C. Hebert, R. P. Lifton et al. 2008. Roles of the cation-chloride cotransporters in neurological disease. *Nature Clinical Practice Neurology* 4 (9) (Sep): 490–503.

Lee, H., C. X. Chen, Y. J. Liu, E. Aizenman, and K. Kandler. 2005. KCC2 expression in immature rat cortical neurons is sufficient to switch the polarity of GABA responses. *The European Journal of Neuroscience* 21 (9) (May): 2593–9.

Levi, S., S. M. Logan, K. R. Tovar, and A. M. Craig. 2004. Gephyrin is critical for glycine receptor clustering but not for the formation of functional GABAergic synapses in hippocampal neurons. *The Journal of Neuroscience* 24 (1) (Jan 7): 207–17.

Michels, G., and S. J. Moss. 2007. GABAA receptors: Properties and trafficking. *Critical Reviews in Biochemistry and Molecular Biology* 42 (1) (Jan-Feb): 3–14.

Mtchedlishvili, Z., E. Lepsveridze, H. Xu, E. A. Kharlamov, B. Lu, and K. M. Kelly. 2010. Increase of GABAA receptor-mediated tonic inhibition in dentate granule cells after traumatic brain injury. *Neurobiology of Disease* 38 (3) (Jun): 464–75.

O'Dell, D. M., C. J. Gibson, M. S. Wilson, S. M. DeFord, and R. J. Hamm. 2000. Positive and negative modulation of the GABA(A) receptor and outcome after traumatic brain injury in rats. *Brain Research* 861 (2) (Apr 10): 325–32.

Ottens, A. K., F. H. Kobeissy, B. F. Fuller, M. C. Liu, M. W. Oli, R. L. Hayes et al. 2007. Novel neuroproteomic approaches to studying traumatic brain injury. *Progress in Brain Research* 161: 401–18.

Papp, E., C. Rivera, K. Kaila, and T. F. Freund. 2008. Relationship between neuronal vulnerability and potassium-chloride cotransporter 2 immunoreactivity in hippocampus following transient forebrain ischemia. *Neuroscience* 154 (2) (Jun 23): 677–89.

Pavlos, N. J., and R. Jahn. 2011. Distinct yet overlapping roles of rab GTPases on synaptic vesicles. *Small GTPases* 2 (2) (Mar): 77–81.

Pavlov, I., N. Huusko, M. Drexel, E. Kirchmair, G. Sperk, A. Pitkanen et al. 2011. Progressive loss of phasic, but not tonic, GABAA receptor-mediated inhibition in dentate granule cells in a model of post-traumatic epilepsy in rats. *Neuroscience* 194 (Oct 27): 208–19.

Payne, J. A., C. Rivera, J. Voipio, and K. Kaila. 2003. Cation-chloride co-transporters in neuronal communication, development and trauma. *Trends in Neurosciences* 26 (4) (Apr): 199–206.

Poulopoulos, A., G. Aramuni, G. Meyer, T. Soykan, M. Hoon, T. Papadopoulos et al. 2009. Neuroligin 2 drives postsynaptic assembly at perisomatic inhibitory synapses through gephyrin and collybistin. *Neuron* 63 (5) (Sep 10): 628–42.

Roozenbeek, B., Y. L. Chiu, H. F. Lingsma, L. M. Gerber, E. W. Steyerberg, J. Ghajar et al. 2012. Predicting 14-day mortality after severe traumatic brain injury: Application of the IMPACT models in the brain trauma foundation TBI-trac(R) New York state database. *Journal of Neurotrauma* 29 (7) (May 1): 1306–12.

Schluter, O. M., F. Schmitz, R. Jahn, C. Rosenmund, and T. C. Sudhof. 2004. A complete genetic analysis of neuronal Rab3 function. *The Journal of Neuroscience* 24 (29) (Jul 21): 6629–37.

Sudhof, T. C. 2004. The synaptic vesicle cycle. *Annual Review of Neuroscience* 27: 509–47.

Sun, C., L. Zhang, and G. Chen. 2013. An unexpected role of neuroligin-2 in regulating KCC2 and GABA functional switch. *Molecular Brain* 6 (May 12): 23.

Sun, Z. L., and D. F. Feng. 2014. Biomarkers of cognitive dysfunction in traumatic brain injury. *Journal of Neural Transmission* 121 (1) (Jan): 79–90.

Tsetsenis, T., T. J. Younts, C. Q. Chiu, P. S. Kaeser, P. E. Castillo, and T. C. Sudhof. 2011. Rab3B protein is required for long-term depression of hippocampal inhibitory synapses and for normal reversal learning. *Proceedings of the National Academy of Sciences of the United States of America* 108 (34) (Aug 23): 14300–5.

Tuvim, M. J., R. Adachi, S. Hoffenberg, and B. F. Dickey. 2001. Traffic control: Rab GTPases and the regulation of interorganellar transport. *News in Physiological Sciences* 16 (Apr): 56–61.

Tweedie-Cullen, R. Y., J. M. Reck, and I. M. Mansuy. 2009. Comprehensive mapping of post-translational modifications on synaptic, nuclear, and histone proteins in the adult mouse brain. *Journal of Proteome Research* 8 (11) (Nov): 4966–82.

Wisniewski, J. R., N. Nagaraj, A. Zougman, F. Gnad, and M. Mann. 2010. Brain phosphoproteome obtained by a FASP-based method reveals plasma membrane protein topology. *Journal of Proteome Research* 9 (6) (Jun 4): 3280–9.

Yu, W., M. Jiang, C. P. Miralles, R. W. Li, G. Chen, and A. L. de Blas. 2007. Gephyrin clustering is required for the stability of GABAergic synapses. *Molecular and Cellular Neurosciences* 36 (4) (Dec): 484–500.

28 Gene Interaction Hierarchy Analysis Can Be an Effective Tool for Managing Big Data Related to Unilateral Traumatic Brain Injury

Todd E. White and Byron D. Ford

CONTENTS

28.1 INTRODUCTION

This chapter discusses the utility of gene interaction hierarchy (GIH) analysis for the management of large datasets. This step-wise analysis combines Ingenuity Pathways Analysis (IPA®) with other offline techniques to identify genes that are central to gene expression patterns in the data. Specifically, this example uses IPA®'s functional analysis capabilities to identify genes that are associated with cell growth and proliferation (CGP) following unilateral traumatic brain injury (TBI). Gene expression patterns reveal distinct ipsilateral and contralateral responses. The majority of CGP genes increase ipsilateral to the injury while the inverse is true on the contralateral side of the brain. Network analysis in IPA® identified genes of interest. Subsequently, these genes were interconnected based on direct (first order) interactions in the Ingenuity knowledge base. The number of interactions in that network determined the gene's placement in the resultant GIH. After microarray analysis, the original datasets numbered 31099 genes. Out of that large number, the process presented here identified 22 primary and 30 secondary genes on the ipsilateral side and 9 primary and 17 secondary genes on the contralateral side of the brain that were central to the CGP gene response following unilateral TBI. This GIH methodology can also be applied to any dataset where the potential for molecular interaction exists.

Traumatic brain injury (TBI) initiates multiple molecular cascades within the brain that, in parallel, lead to exacerbation of the initial mechanical injury and attempt to protect surviving brain tissue. A widely used experimental model of TBI is controlled cortical impact (CCI). For this model, a craniotomy is performed to expose the brain, which is impacted directly by a mechanically actuated piston (Dixon and Kline, 2009). Unilateral CCI results in distinct gene expression patterns ipsilateral and contralateral to the injury (White et al., 2013), allowing for isolated molecular analysis of the injured brain both proximal and distal to the site of impact. Determining how the proximal and distal molecular responses differ in the respective presence and absence of frank neural damage may provide clues for the development of therapies to counteract deleterious processes and augment neuroprotection.

The development of gene microarray technology has allowed for examination of thousands of genes in one assay. The task of determining which genes within the big data sets generated are most important can be a daunting task. This contribution discusses one method that we have developed to manage these data sets and identify genes for further study.

Invariably, arbitrary thresholds must be used to manage large data sets. These thresholds include expression value, fold change, and statistical significance (p value) among others. Making these thresholds stringent enough to get the number of genes into a manageable range can possibly result in the elimination of important genes from the analysis. The process described here uses an advanced bioinformatics analysis tool to reduce the number of genes based on biological significance.

The Ingenuity Pathway Analysis (IPA) software program uses a database built from published scientific literature to draw direct and indirect interactions between genes and to assign genes to specific biological functions, canonical pathways, and networks (IPA, 2011). IPA also features a strong network building component that allows for the creation and analysis of networks composed of any genes of interest (GOIs). We devised a method for using the initial information that IPA provides and subsequent network analysis to determine which genes are most significant to the inflammatory response following neuronal injury unilateral CCI in the rat (White et al., 2013). Although this previous analysis pooled the genes from both sides of the brain, it has become apparent that it is more informative to analyze each side of the brain separately. The analysis described in this contribution results in gene interaction hierarchies (GIHs) where genes of interest are ranked based on the number of interactions they have with each other. The theory behind the analysis is that a gene that interacts with more genes in a particular set of genes has greater potential to influence that set of genes.

Although other analyses may be referenced for discussion purposes, this contribution is not intended to be a comprehensive review of similar analyses. The purpose here is to present one method of analysis that we have found useful and that may be useful to others doing similar work.

28.2 DETERMINING GOIs

The goal for this analysis was to have approximately 75–150 genes put into ranked order in our final analysis. In order to identify GOI, a step-wise process was followed to reduce the number of genes into this range.

28.2.1 Present Sort, Fold Change, and IPA Gene Reduction

The Affymetrix Rat Genome 230 2.0 GeneChip reports the expression values for 31,099 gene transcripts. At the time of this analysis, the Affymetrix expression console output predicted detection values for every gene in each sample, assigning values of present, marginal, or absent. For a gene to be kept in the study, it had to be "present" in all samples of either ipsilateral, contralateral, or naive brain tissue. Sorting the genes with this restriction reduced the number of genes to 20,195.

Fold changes were next calculated to create ipsilateral versus naive (TBI-I) and contralateral versus naive (TBI-C) data sets. Genes with fold changes ≥2 were kept in the study. This reduced our numbers to 3145 TBI-I genes and 2636 TBI-C genes; these data sets were uploaded to IPA.

Some Affymetrix transcript identifications remain unmapped by IPA; these genes were eliminated from the study. This left 2255 TBI-I genes and 1837 TBI-C genes that were mapped by IPA; however, this number still included some duplicate gene transcripts because of the redundancy of the GeneChip. Core analyses were performed on these data sets within IPA, which eliminated any duplicate gene transcripts remaining and brought our numbers to 1999 TBI-I and 1623 TBI-C "analysis-ready" genes. A total of 1149 genes were expressed on both sides of the brain (common genes), leaving 850 unique TBI-I and 474 unique TBI-C genes.

28.2.2 Initial Functional Analysis

Investigating only the genes related to a specific biological function is another way to further narrow our focus. Analysis of the top 10 molecular and cellular functions associated with the TBI-I and TBI-C data sets in IPA showed that cellular growth and proliferation (CGP) was the highest ranked TBI-I function and ranked fourth for TBI-C (Figure 28.1a, b).

28.2.3 CGP Gene Expression Patterns

By cross-referencing all the CGP genes in IPA with our analysis-ready data sets, we identified 717 TBI-I and 519 TBI-C CGP genes. Of these genes, 340 CGP genes changed uniquely on the ipsilateral side of the brain. A total of 311 of those genes (91%) increased, whereas 29 genes (9%) decreased in expression (Figure 28.2a). A total of 142 CGP genes changed uniquely on the contralateral side of the brain and, in contrast to what we observed on the ipsilateral side, only 42 genes (30%) increased, whereas 100 genes (70%) decreased in expression (Figure 28.2b).

There were 377 common CGP genes. To determine whether these common genes changed differently on one side of the brain compared with the other, we calculated the ratio of the TBI-I fold change to the TBI-C fold change. Those genes that had a TBI-I/TBI-C ratio >2 were determined to have changed differently. We observed that 320 of the common CGP genes (85%) changed similarly (TBI-I/TBI-C ratio <2; Figure 28.2c). Of the genes that changed similarly, 234 genes (62%) increased and 86 genes (23%) decreased in expression. The remaining 57 common CGP genes (15%) changed differently (TBI-I/TBI-C ratio >2) (Figure 28.2c). Table 28.1 shows the 57 common CGP genes that changed differently. These genes span all cellular compartments (extracellular space, plasma

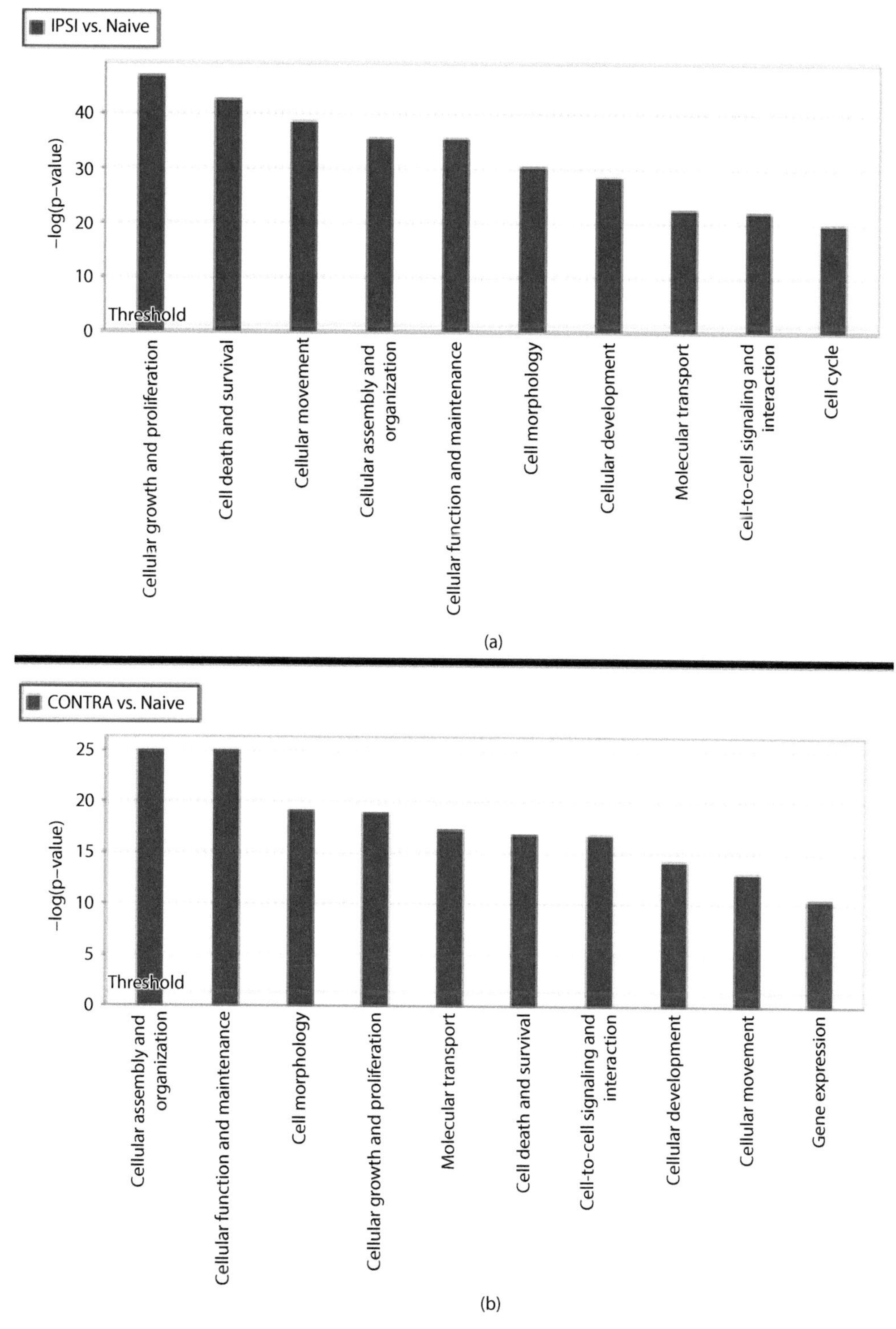

FIGURE 28.1 Analysis of the top 10 molecular and cellular functions determined by IPA for the TBI-I data set (a) and the TBI-C data set (b) showed that cellular growth and proliferation was a top-ranked function on both sides of the brain.

membrane, cytoplasm, and nucleus) with diverse molecule types. The expression of all these genes was lower on the contralateral side of the brain. Because of their different expression patterns, these 57 genes became our first group of GOI (Table 28.1). When these 57 genes were added to the uniquely expressed CGP genes, the potential GOI total was 397 for TBI-I and 199 for TBI-C.

28.2.4 Gene Network Analysis

Gene networks are generated de novo in IPA based on the list of genes that are imported. IPA takes "seed" molecules from the gene list, searches the Ingenuity Knowledge Base, and uses a network algorithm to draw connections between molecules based on biological function (IPA, 2013). Gene network

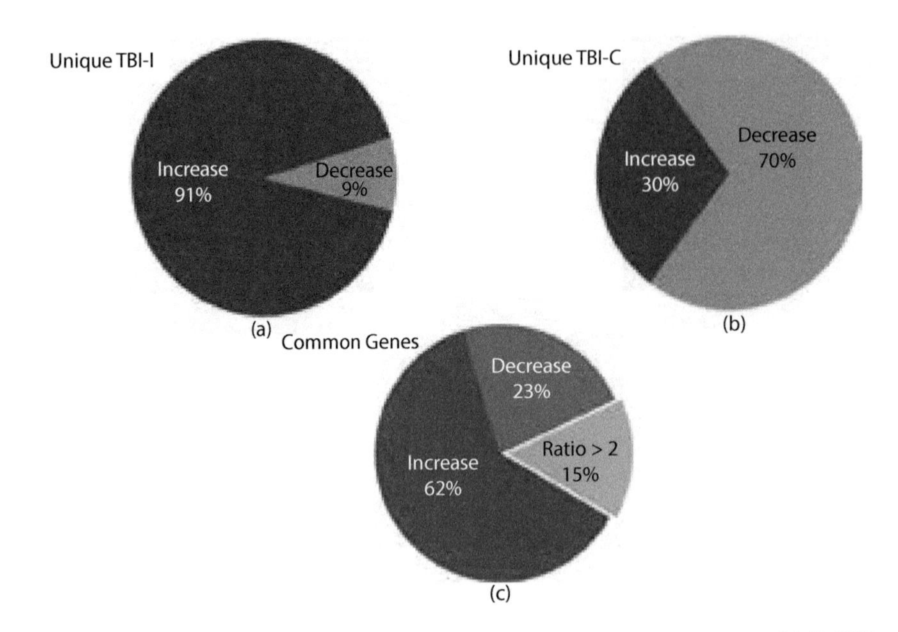

FIGURE 28.2 Breakdown of CGP genes based on up- and downregulation in expression. (b) A total of 340 CGP genes changed uniquely on the ipsilateral side of the brain and 91% (311 genes) of those increased in expression. (b) A total of 142 IR genes changed uniquely on the contralateral side of the brain and 70% (100 genes) of those decreased in expression. (c) There were 377 genes that changed more than two-fold on both sides of the brain. Eighty-five percent of them (320 genes) changed similarly, whereas the remaining 15% (57 genes) changed differently (TBI-I/TBI-C ratio >2; see text).

TABLE 28.1
Genes that Change Differently on Each Side of the Brain

Gene Symbol	Entrez Gene Name	TBI-I Fold Change	TBI-C Fold Change	TBI-I/TBI-C Ratio	Molecular Type
Extracellular space					
SPP1	Secreted phosphoprotein 1	37.905	2.37	15.99367089	Cytokine
ESM1	Endothelial cell-specific molecule 1	20.805	5.748	3.619519833	Growth factor
FGL2	Fibrinogen-like 2	16.793	4.017	4.180482947	Peptidase
LCN2	Lipocalin 2	71.824	3.895	18.44005135	Transporter
TIMP1	TIMP metallopeptidase inhibitor 1	38.486	2.101	18.31794384	Other
Plasma membrane					
CD44	CD44 molecule (Indian blood group)	15.558	2.399	6.485202168	Enzyme
EHD4	EH-domain containing 4	2.361	−2.056	4.854216	Enzyme
MGLL	Monoglyceride lipase	−7.853	−18.145	2.310581943	Enzyme
SDC1	Syndecan 1	13.681	2.566	5.331644583	Enzyme
CAMK2N1	Calcium/calmodulin-dependent protein kinase II inhibitor 1	−11.813	−23.824	2.016761195	Kinase
EGFR	Epidermal growth factor receptor	6.773	2.374	2.852990733	Kinase
EPHA5	EPH receptor A5	−2.945	−8.441	2.866213922	Kinase
KCNN4	Potassium intermediate/small conductance calcium-activated channel, subfamily N, member 4	3.088	−9.429	29.116752	Ion channel
PTPRF	Protein tyrosine phosphatase, receptor type, F	−6.365	−20.492	3.21948154	Phosphatase
IL6ST	Interleukin-6 signal transducer (gp130, oncostatin M receptor)	2.307	−3.283	7.573881	Transmembrane receptor
AQP4	Aquaporin 4	6.58	2.986	2.203616879	Transporter
PMEPA1	Prostate transmembrane protein, androgen induced 1	2.682	−2.937	7.877034	Other
Cytoplasm					
CYP1B1	Cytochrome P450, family 1, subfamily B, polypeptide 1	10.998	4.808	2.287437604	Enzyme
KIF3A	Kinesin family member 3A	−5.083	−11.754	2.312413929	Enzyme
MX1	Myxovirus (influenza virus) resistance 1, interferon-inducible protein p78 (mouse)	28.177	7.326	3.846164346	Enzyme
RND3	Rho family GTPase 3	2.864	−2.971	8.508944	Enzyme

TABLE 28.1 (*Continued*)
Genes that Change Differently on Each Side of the Brain

Gene Symbol	Entrez Gene Name	TBI-I Fold Change	TBI-C Fold Change	TBI-I/TBI-C Ratio	Molecular Type
CARD11	Caspase recruitment domain family, member 11	7.343	2.892	2.539073306	Kinase
CSNK2A1	Casein kinase 2, alpha 1 polypeptide	2.992	−2.75	8.228	Kinase
PPP1R12A	Protein phosphatase 1, regulatory subunit 12A	−3.911	−10.97	2.80490923	Phosphatase
EIF3C/EIF3CL	Eukaryotic translation initiation factor 3, subunit C	−4.369	−9.072	2.0764477	Translation regulator
RASA1	RAS p21 protein activator (GTPase activating protein) 1	2.392	−2.105	5.03516	Transporter
ACTN1	Actinin, alpha 1	2.483	−2.088	5.184504	Other
CMIP	c-Maf inducing protein	−3.778	−13.763	3.642932769	Other
KIFAP3	Kinesin-associated protein 3	−2.281	−7.831	3.433143358	Other
LCP1	Lymphocyte cytosolic protein 1 (L-plastin)	6.082	2.799	2.1729189	Other
PHLDA1	Pleckstrin homology-like domain, family A, member 1	5.129	2.16	2.374537037	Other
Nucleus					
ARID4A	AT rich interactive domain 4A (RBP1-like)	−3.263	−7.033	2.155378486	Transcription regulator
ARID5B	AT rich interactive domain 5B (MRF1-like)	−5.312	−16.105	3.031814759	Transcription regulator
BTG2	BTG family, member 2	−2.22	−5.803	2.613963964	Transcription regulator
KDM5A	Lysine (K)-specific demethylase 5A	−2.984	−19.489	6.53116622	Transcription regulator
KLF13	Kruppel-like factor 13	−2.006	−4.582	2.284147557	Transcription regulator
NFIX	Nuclear factor I/X (CCAAT-binding transcription factor)	−2.548	−8.112	3.183673469	Transcription regulator
PA2G4	Proliferation-associated 2G4, 38kDa	−2.702	−5.783	2.140266469	Transcription regulator
SMARCA4	SWI/SNF-related, matrix-associated, actin-dependent regulator of chromatin, subfamily a, member 4	2.521	−7.712	19.441952	Transcription regulator
STAT3	Signal transducer and activator of transcription 3 (acute-phase response factor)	4.219	−3.771	15.909849	Transcription regulator
TCF4	Transcription factor 4	−2.216	−4.625	2.087093863	Transcription regulator
BRD4	Bromodomain containing 4	−3.528	−15.202	4.308956916	Kinase
CDK11A/ CDK11B	Cyclin-dependent kinase 11B	−4.29	−14.872	3.466666667	Kinase
GSK3B	Glycogen synthase kinase 3 beta	−2.733	−6.635	2.42773509	Kinase
SRPK2	SRSF protein kinase 2	−5.614	−23.589	4.201816886	Kinase
MCM3	Minichromosome maintenance complex component 3	10.812	3.363	3.214986619	Enzyme
TOP2A	Topoisomerase (DNA) II alpha 170kDa	2.26	−2.406	5.43756	Enzyme
THRA	Thyroid hormone receptor, alpha	−2.799	−11.518	4.115041086	Ligand-dependent nuclear receptor
DUSP8	Dual specificity phosphatase 8	−6.392	−17.059	2.668804756	Phosphatase
TPR	Translocated promoter region, nuclear basket protein	2.212	−2.728	6.034336	Transporter
CCND1	Cyclin D1	2.152	−2.027	4.362104	Other
CDT1	Chromatin licensing and DNA replication factor 1	3.098	−2.295	7.10991	Other
CEBPD	CCAAT/enhancer binding protein (C/EBP), delta	11.271	2.037	5.533136966	Other
GADD45G	Growth arrest and DNA-damage-inducible, gamma	3.191	−2.384	7.607344	Other
ZBTB20	Zinc finger and BTB domain containing 20	4.12	−2.757	11.35884	Other
Other					
PHF14	PHD finger protein 14	3.602	−6.323	22.775446	Other
RASSF4	Ras association (RalGDS/AF-6) domain familymember 4	4.289	2.106	2.036562203	Other

TBI-I/TBI-C ratio: gene increased on both sides (TBI-I > TBI-C): ratio = (TBI-I)/(TBI-C); gene decreased on both sides (TBI-I > TBI-C): ratio = 1/[(TBI-I)/(TBI-C)]; gene decreased on both sides (TBI-I < TBI-C): ratio = −1/[(TBI-C)/(TBI-I)]; gene increased ipsilaterally and decreased contralaterally: ratio = (TBI-I)/−[1/(TBI-C)]; gene decreased ipsilaterally and increased contralaterally: ratio = (TBI-C)/[1/(TBI-I)].

analysis in IPA was used in this analysis to identify genes in our data sets that were potentially most relevant to the observed CGP gene response. To generate the networks, we performed an IPA core analysis on the TBI-I and TBI-C CGP data sets. IPA scores the networks to rank them according to their degree of relevance to the network eligible molecules in the data set (IPA, 2013). The top six scoring networks for each data set were used to identify GOI. Previous analyses suggested that using six networks would yield the desired number of genes to bring the total number of GOI into the desired range (White et al., 2013).

Figure 28.3 shows networks 2 and 6 (Table 28.2) as examples of the TBI-I CGP analysis. Figure 28.4 shows networks 3 and 5 (Table 28.3) as examples of the TBI-C CGP analysis. All relevant gene families, groups, and complexes

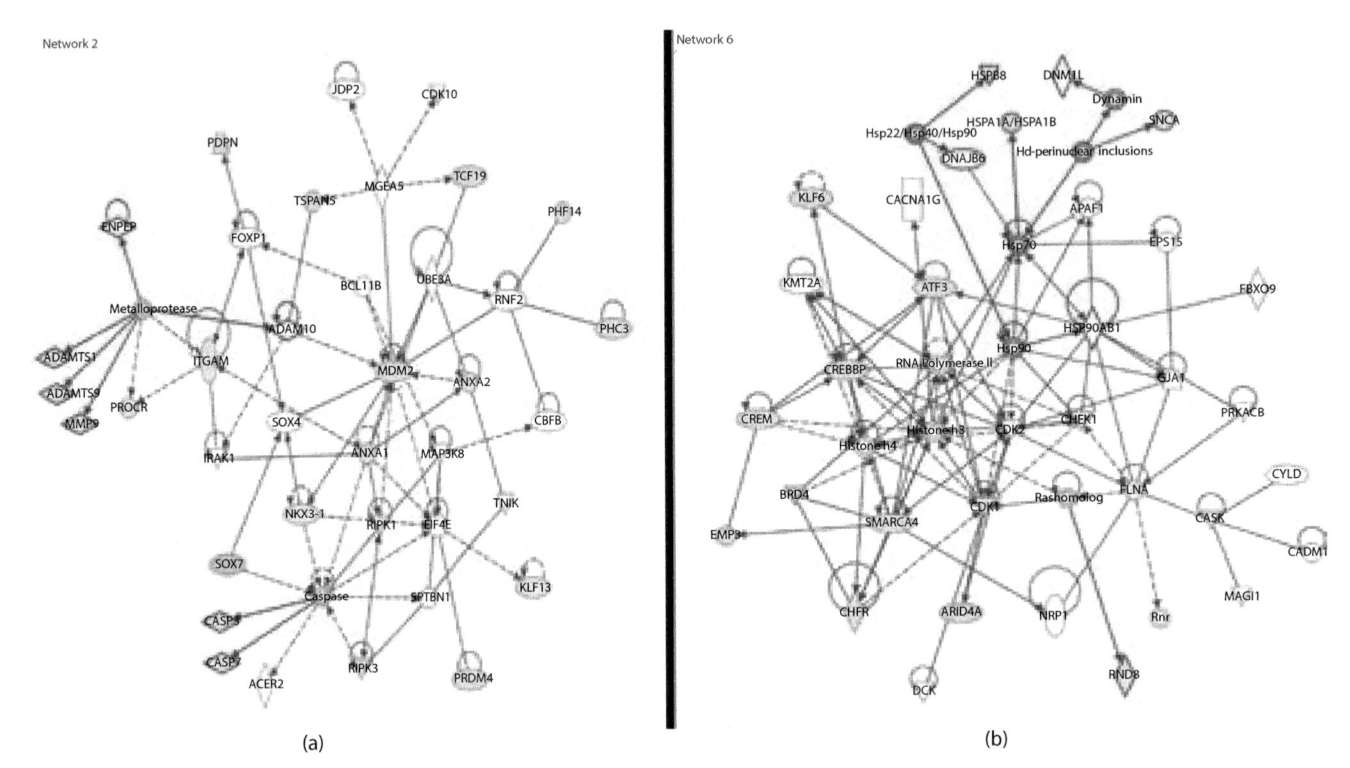

FIGURE 28.3 Examples of TBI-I network analysis. TBI-I CGP networks 2 (a) and 2 (b) (see Table 28.2) with all gene families, groups, and complexes expanded to show the member genes and showing the relative expression values of potential GOI for TBI-I. Red: relative increase in expression; green: relative decrease in expression; white: no change in expression; purple connections and outlines: expansion of gene families, groups, and complexes in the original network.

TABLE 28.2
Top Six Gene Networks Associated with the TBI-I CGP Data Set

Network ID	Molecules in Network	Score	Focus Molecules	Top Functions
1	Alp, AMOT, BTG2, CDKN1A, CDKN2AIP, CNOT6L, COPS2, CUL5, EIF1AY, EIF2AK2, FUS, GTPBP4, histone, HNRNPA2B1, HSPA1A/HSPA1B, MAB21L2, PA2G4, PMEPA1, PRKAA1, PRMT1, PTBP1, PURA, RNF20, SIN3A, SMAD7, SRPK2, TCEB3, THRA, TNRC6A, TRA2A, TWIST1, WHSC1, WWTR1, XRN2, ZMYM2	49	33	Cellular growth and proliferation, cellular development, tissue development
2	ACER2, ADAM10, ANXA1, ANXA2, BCL11B, caspase, CBFB, CDK10, EIF4E, ENPEP, FOXP1, IRAK1, ITGAM, JDP2, KLF13, MAP3K8, MDM2, metalloprotease, MGEA5, NKX3-1, PDPN, PHC3, PHF14, PRDM4, PROCR, RIPK1, RIPK3, RNF2, SOX4, SOX7, SPTBN1, TCF19, TNIK, TSPAN5, UBE3A	48	33	Cellular growth and proliferation, gene expression, cancer
3	ADM, ARID5B, BSG, CA3, CIRBP, CXCL16, focal adhesion kinase, GADD45B, GIT1, GTF2I, H2AF, HEYL, ING2, ITGAV, LGALS1, LITAF, MTPN, NCSTN, Notch, NRARP, NREP, POR, pro-inflammatory cytokine, PRPS2, RORA, S100A8, S100A9, SDC1, SDC4, SLC39A6, SPP1, TAF4B, TAGLN, TGFB3, TGM2	47	31	Cellular growth and proliferation, cellular movement, cardiovascular system development and function

TABLE 28.2 (*Continued*)
Top Six Gene Networks Associated with the TBI-I CGP Data Set

Network ID	Molecules in Network	Score	Focus Molecules	Top Functions
4	**ANGPT2,** Ap1, **BCL2L1, CASP7, COL1A1, CSF1, ENAH,** estrogen receptor, **ETV5, EZR, FGF2, FGFR1, FN1, HMOX1, ITGA6, ITGB1, MAP3K1, MLLT4,** p85 (pik3r), **PGR, PLAU, PPAP2B, PRDM2, PTGS2, SGPL1, SMAD4,** SRC (family), **STX6, TFRC, TJP2, UNC5B, USP9X, VANGL1, VASP, YTHDF2**	45	31	Cellular movement, cellular growth and proliferation, cell death and survival
5	**A2M, ALB, ARL11, AVP, BCL6B, CASP3, CCND1,** CD3, **CD38, CDKN1B, CEBPB, CREB1, EGR2, ELK4, EMP1, FBXW7, GRIA3, HBEGF,** IKK (complex), Mmp, **NEO1, NTRK2,** P110, **PAK1, PLAT, PSEN1, RAC2, RASGRP1, RUNX1, RUNX1T1, SCAMP2, TAC1, TGIF1, ZBTB18, ZFP36**	43	31	Cellular growth and proliferation, cellular development, cell death and survival
6	**APAF1, ARID4A, ATF3, BRD4, CACNA1G, CADM1, CASK, CDK1, CDK2, CHEK1, CHFR, CREBBP, CREM, CYLD, DCK, DNAJB6, EMP3, EPS15, FBXO9, FLNA, GJA1,** histone h3, histone h4, Hsp70, Hsp90, **HSP90AB1, KLF6, KMT2A, MAGI1, NRP1, PRKACB,** Ras homolog, RNA polymerase II, Rnr, **SMARCA4**	40	28	Cellular growth and proliferation, cell death and survival, cellular development

Bold = gene included in the data set.
Note: Some of the nodes in the original networks represent gene groups, complexes, or families that, when expanded, contain more potential GOI.

TABLE 28.3
Top Six Gene Networks Associated with the TBI-C CGP Data Set

Network ID	Molecules in Network	Score	Focus Molecules	Top Functions
1	**ACIN1, APAF1, BTG2, CADM1, CAPRIN1, CBFB, CNOT6L, CSNK2A1, CUL2, CUL5, DNAJB6, ENC1, FBXO9, FBXW7, FUS,** Gsk3, hemoglobin, **HNRNPA2B1, HSP90AA1, HSP90AB1, MAB21L2, MAFG, MBNL1, PHF14, PRMT1, RBM5, RNF2, SMAD7, SRPK2, TNKS2, TNRC6B, TRA2A, UBE3A, ZBTB20, ZMYM2**	57	33	Cellular growth and proliferation, RNA posttranscriptional modification, posttranslational modification
2	14-3-3, **ACTN1,** alpha tubulin, **APC, ATXN2, CACNA1A, CMIP, FYN, GABARAPL1, GFRA1, GIT1, KIF3A, KIFAP3, KRT23, LASP1, MAP3K1, MCAM, PAFAH1B1, PDPK1, PHF6,** PP2A, **PRKCI, RANBP2, RBFOX1, RBFOX2,** Rnr, **SPTBN1, TNIK, TOP2A, TPR, TUBB4B,** tubulin, **USP9X, VCL, ZYX**	49	30	Cellular growth and proliferation, cell-to-cell signaling and interaction, nervous system development and function
3	26s proteasome, **ABCC1, ADC, AMFR, APBB2, App, ATF2,** ATPase, **BRD4,** calmodulin, **CDK10, CDK11A/CDK11B, CLDN1,** E3 RING, **ELF1, GADD45A,** histone, **HSF2, IDE,** IL-1R, **JDP2, KLF15, KMT2E, MAGI1, MGEA5, MMP14, NFYA, OGT, PDS5B,** proinsulin, **SMARCA4, STAG2, TOPBP1, TTR,** ubiquitin	39	27	Cellular growth and proliferation, carbohydrate metabolism, cell cycle
4	ACAC, **ACACA, AMOT,** APC-AXIN-GSK3β, **APPL1, ARID4A, C3, CDH13,** Ck2, collagen type ix, **COPS2, CXCL14,** estrogen receptor, **GSK3B, ING2, MED1,** mediator, **MTF2,** N-cor, **NCOR1,** Nr1h, **NR2F1, NR2F2, PGR, PIAS1, PMEPA1, PPARGC1B, RNF4, RNF20, SMURF2,** T3-TR-RXR, **THRA, TRH, UIMC1, ZMYND11**	38	26	Cellular growth and proliferation, gene expression, endocrine system development and function
5	**AKAP12, ALDH1A2, ARHGAP1, ARL6IP5, CD2AP,** Cg, **CSF1, DUSP8, DUSP10,** FSH, **GBX2,** Gαq, **KIDINS220,** Lh, **MAFB,** NMDA receptor, Pdgf (complex), PLA2, **PPP1R12A, PTK2B, PTP4A1,** PTPase, **PTPRF, PTPRJ, PTPRS, RAB27A, RET, RPRM,** Sfk, **SIRPA,** SRC (family), **SSTR2, SYMPK, TIMP1, YTHDF2**	37	25	Cellular growth and proliferation, renal and urological system development and function, reproductive system development and function
6	**BAX, BCL11B, BRINP1, CAV1, CCND1,** CD3, **CDK2, CYLD, DNM1L, FOLR1, FOXO1, FOXP1, GJA1,** GNRH, histone h3, Hsp70, Hsp90, IKK (complex), **IRAK1, MAP2K4, MAP3K8, MAPK8, MAPK9, MAPT,** p85 (pik3r), **PAK1, PAK2, PPP2R1A, PRLR,** RNA polymerase II, **RND3, STUB1, TAC1, TSN, ZBTB18**	36	27	Cellular growth and proliferation, cell death and survival, cellular movement

Bold = gene included in the data set.
Note: Some of the nodes in the original networks represent gene groups, complexes, or families that, when expanded, contain more potential GOI.

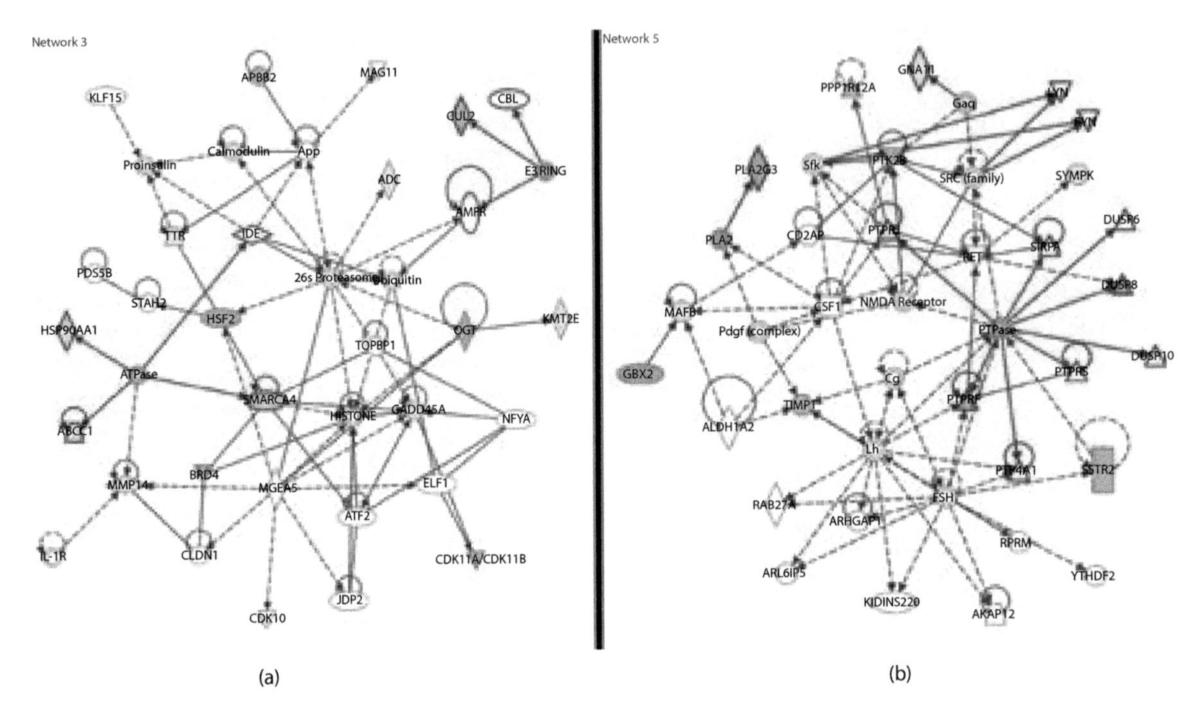

FIGURE 28.4 Examples of TBI-C network analysis. TBI-C CGP networks 3 (a) and 5 (b) (see Table 28.3) with all gene families, groups, and complexes expanded to show the member genes and showing the relative expression values of potential GOI for TBI-C. Red: relative increase in expression; green: relative decrease in expression; white: no change in expression; purple connections and outlines: expansion of gene families, groups, and complexes in the original network.

were expanded to show the member genes included in the CGP data sets. The relative TBI-I (Figure 28.3) and TBI-C (Figure 28.4) gene expression values of potential GOIs were overlaid on these networks and additional GOIs were identified. Tables 28.4 and 28.5 shows the resulting GOI that were identified through this analysis. For TBI-I, a total of 115 GOI were found in these networks, 19 of which were previously identified (common genes or previous network) (Table 28.4). Thus, 96 additional GOIs were identified for TBI-I. For TBI-C, 29 of the 72 GOIs found in the networks had been previously identified, leaving 43 additional GOIs (Table 28.5). Combining these network-identified GOIs with the common GOIs in Table 28.1 gives a total 153 TBI-I CGP and 100 TBI-C CGP GOIs. These numbers fall in the desired range for moving on to the next part of the analysis.

28.3 COMPILING THE GIH

Now that the GOIs have been identified, the next step is to rank them based on the number of interactions they have with each other. To do this, we make a network in IPA consisting of the GOIs using only direct (first-order) connections. The hierarchy is divided into four tiers: primary, secondary, peripheral, and orphan. Genes having direct connections with more than 10% of the other genes within the main GOI network were ranked in the primary tier. Genes having direct connections with 5%–10% of the other genes were ranked in the secondary tier. Genes with direct connections with less than 5% of the other genes were ranked in the peripheral tier. Orphan genes have no connections in the main GOI network.

28.3.1 TBI-I

Of the 153 TBI-I CGP GOIs, 127 of the GOIs formed an interconnected network, leaving 26 orphan genes. Therefore, based on the criteria provided previously, the tier breakdown is as follows: primary tier: >12 connections (22 genes; see Figure 28.5 for an example); secondary tier: seven to 12 connections (30 genes); and peripheral tier: less than connections (75 genes). The resultant GIH is displayed in Table 28.6. Twenty-two genes were ranked in the primary tier, 30 genes in the secondary tier, and 75 in the peripheral tier.

28.3.2 TBI-C

Of the 100 TBI-I CGP GOIs, 73 of the GOIs formed an interconnected network, leaving 27 orphan genes. The tier breakdown is as follows: primary tier: more than seven connections (nine genes; see Figure 28.6 for an example); secondary tier: four to seven connections (17 genes); and peripheral tier: less than four connections (47 genes). The resultant GIH is displayed in Table 28.7. Nine genes were ranked in the primary tier, 17 genes in the secondary tier, and 47 in the peripheral tier.

28.3.3 Effects of GIH on Functional Analysis

We know that, in this example, all the genes at this point are categorized as CGP genes. However, these genes also have additional biological functions. To determine the effect of the GIH rankings on the significance of other biological functions, we ran an IPA core analysis on the unranked GOIs and the top two tiers (primary and

TABLE 28.4
Identification of GOIs from TBI-I CGP Network Analysis

Network ID	GOIs Found	Total No. of GOIs	Overlap with Previous Analyses	Net No. of GOIs	Top Molecular Types
1	*BTG2,* **CDKN1A, CDKN2AIP, EIF2AK2, HSPA1A/HSPA1B,** *PA2G4, PMEPA1,* **PTBP1, SIN3A,** *SRPK2,* **TCEB3,** *THRA,* **TWIST1, WHSC1, WWTR1**	15	5	10	Transcription regulators and kinases
2	**ADAMTS1, ADAMTS9, ANXA1, ANXA2, CASP3, CASP7, ITGAM,** *KLF13,* **MDM2, MMP9, NKX3–1, PDPN, PHC3,** *PHF14,* **PRDM4, PROCR, RIPK1, RIPK3, SOX7, TCF19, TSPAN5**	21	2	19	Transcription regulators and peptidases
3	**ADM,** *ARID5B,* **CXCL16, GADD45B, IL1B, ITGAV, LGALS1, LITAF, MTPN, NRARP, NREP, PRPS2, RORA, S100A8, S100A9,** *SDC1,* **SDC4,** *SPP1,* **TAGLN, TGFB3, TGM2**	21	3	18	Transcription regulators and cytokines
4	**ANGPT2,** *CASP7,* **ETV5, EZR, FGF2, FN1, FOS, HCK, HMOX1, PIK3R1, PLAU, PRDM2, PTGS2, SGPL1, STX6, TFRC, TJP2, UNC5B, VANGL1, VASP**	20	1	19	Regulators and kinases
5	**ALB, ARL11, BCL6B,** *CASP3, CCND1,* **CDKN1B, CEBPB, CREB1, EGR2, ELK4, EMP1, HBEGF, IKBKB, MMP3,** *MMP9,* **PLAT, RAC2, RUNX1, SCAMP2, TGIF1, ZFP36**	21	3	18	Transcription regulators and peptidases
6	*ARID4A,* **ATF3,** *BRD4,* **CASK, CDK1, CHEK1, CHFR, CREBBP, CREM, EMP3, FLNA,** *HSPA1A/HSPA1B,* **HSPB8, KLF6,** *RND3, SMARCA4,* **SNCA**	17	5	12	Transcription regulators and kinases

Italics = GOIs also found in a previous analysis.

TABLE 28.5
Identification of Genes of Interest from TBI-C CGP Network Analysis

Network ID	GOIs Found	Total No. of GOIs	Overlap with Previous Analyses	Net No. of GOIs	Top Molecular Types
1	**ACIN1,** *BTG2, CSNK2A1,* **CUL2,** *GSK3B,* **HSP90AA1, MAFG,** *PHF14, SRPK2, ZBTB20*	10	6	4	Enzymes, kinases, and transcription regulators
2	*ACTN1,* **ATXN2, CACNA1A,** *CMIP, KIF3A, KIFAP3,* **KRT23, PAFAH1B1, RBFOX1,** *TOP2A, TPR*	11	6	5	Enzymes
3	**ABCC1, ADC, APBB2,** *BRD4, CDK11A/CDK11B, CUL2,* **GADD45A, HSF2,** *HSP90AA1,* **KMT2E, MMP14, OGT,** *SMARCA4,* **TTR**	14	5	9	Enzymes, kinases, transcription regulators, and transporters
4	*ARID4A,* **CDH13, CDK19,** *CSNK2A1,* **CXCL14,** *GSK3B,* **MED1, NCOR1, NR2F1, NR2F2,** *PMEPA1,* **SMURF2, SP1,** *THRA,* **TRH, ZMYND11**	16	5	11	Transcription regulators, kinases, and ligand-dependent nuclear receptors
5	*DUSP8,* **GBX2, GNA11, MAFB, PLA2G3,** *PPP1R12A,* **PTK2B,** *PTPRF,* **PTPRJ, SSTR2, SYMPK,** *TIMP1*	12	4	8	Phosphatases and enzymes
6	**BAX,** *CCND1,* **FOXO1,** *HSP90AA1,* **MAPK8, MAPT, PIK3R2, PRLR,** *RND3*	9	3	6	Kinases and enzymes

Italics = GOIs also found in a previous analysis.

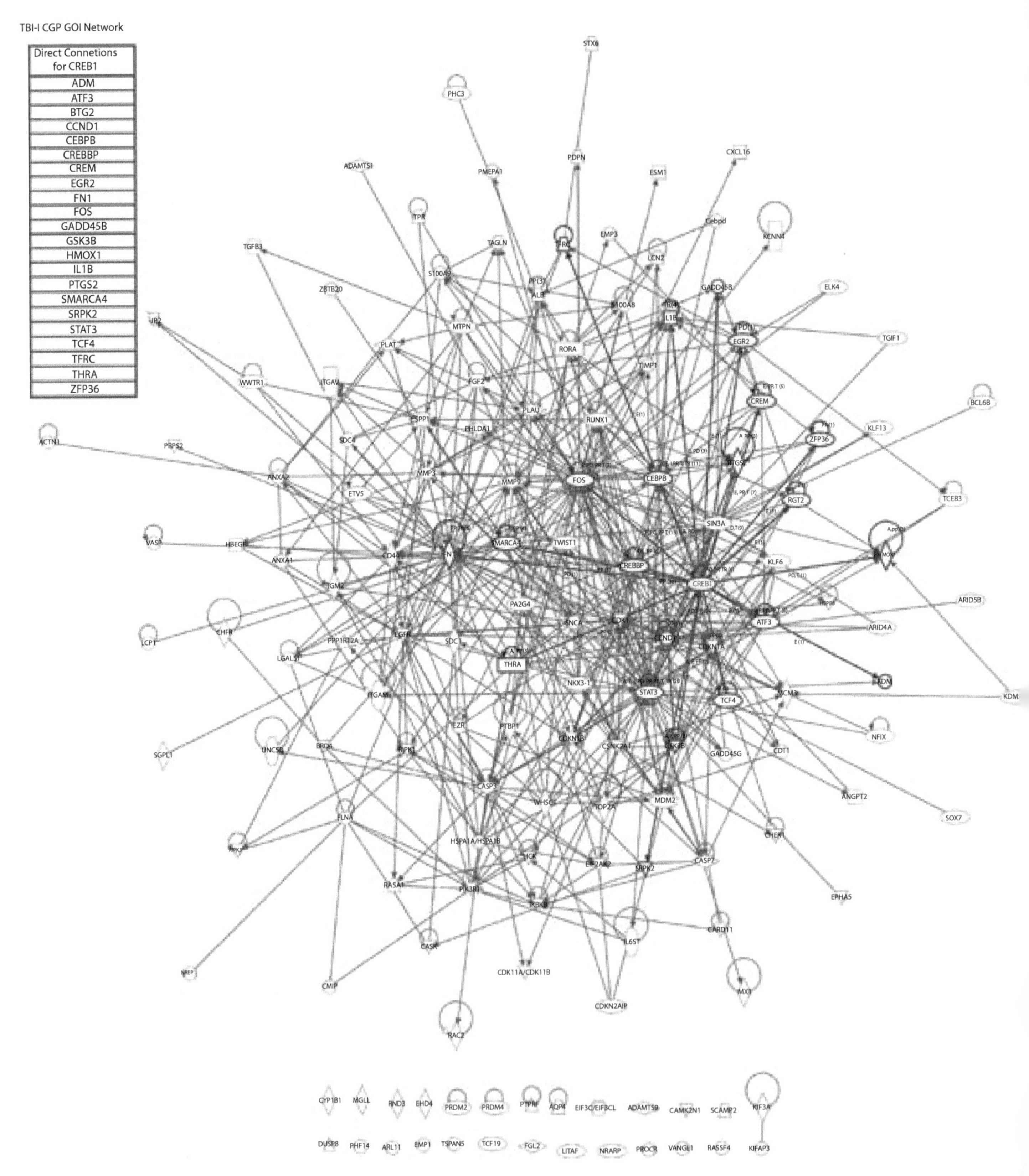

FIGURE 28.5 An example of how we calculated the number of direct connections for a gene in our TBI-I CGP GOI network. In IPA, the gene in question was selected (CREB1 in this example). Then, its direct connections were selected by right-clicking on CREB1 and using the "select nearest neighbors" option (highlighted in blue). A list of the selected genes was exported and CREB1 was removed from the list (upper left corner). The remaining genes were counted (22 in this example) and CREB1 was ranked in the TBI-I CGP gene interaction hierarchy (Table 28.6; primary tier) by this number.

TABLE 28.6
TBI-I CGP GIH

Gene Symbol	Entrez Gene Name	Fold Change	Cellular Compartment	Molecular Type
Primary				
ATF3	Activating transcription factor 3	12.027	Nucleus	Transcription regulator
CEBPB	CCAAT/enhancer binding protein (C/EBP), beta	3.366	Nucleus	Transcription regulator
CREB1	cAMP-responsive element binding protein 1	2.666	Nucleus	Transcription regulator
CREBBP	CREB binding protein	2.421	Nucleus	Transcription regulator
FOS	FBJ murine osteosarcoma viral oncogene homolog	2.832	Nucleus	Transcription regulator
MDM2	MDM2 oncogene, E3 ubiquitin protein ligase	2.01	Nucleus	Transcription regulator
RUNX1	Runt-related transcription factor 1	2.534	Nucleus	Transcription regulator
SIN3A	SIN3 transcription regulator family member A	2.42	Nucleus	Transcription regulator
SMARCA4	SWI/SNF-related, matrix-associated, actin-dependent regulator of chromatin, subfamily a, member 4	2.521	Nucleus	Transcription regulator
STAT3	Signal transducer and activator of transcription 3 (acute-phase response factor)	4.219	Nucleus	Transcription regulator
CDK1	Cyclin-dependent kinase 1	2.105	Nucleus	Kinase
CDKN1A	Cyclin-dependent kinase inhibitor 1A (p21, Cip1)	2.675	Nucleus	Kinase
CSNK2A1	Casein kinase 2, alpha 1 polypeptide	2.992	Cytoplasm	Kinase
EGFR	Epidermal growth factor receptor	6.773	Plasma membrane	Kinase
GSK3B	Glycogen synthase kinase 3 beta	−2.733	Nucleus	Kinase
PIK3R1	Phosphoinositide-3-kinase, regulatory subunit 1 (alpha)	2.136	Cytoplasm	Kinase
CASP3	Caspase 3, apoptosis-related cysteine peptidase	2.535	Cytoplasm	Peptidase
MMP3	Matrix metallopeptidase 3 (stromelysin 1, progelatinase)	4.045	Extracellular space	Peptidase
MMP9	Matrix metallopeptidase 9 (gelatinase B, 92-kDa gelatinase, 92kDa type IV collagenase)	7.178	Extracellular space	Peptidase
CD44	CD44 molecule (Indian blood group)	15.558	Plasma membrane	Enzyme
FN1	Fibronectin 1	3.97	Extracellular space	Enzyme
CCND1	Cyclin D1	2.152	Nucleus	Other
Secondary				
BTG2	BTG family, member 2	−2.22	Nucleus	Transcription regulator
CREM	cAMP responsive element modulator	2.165	Nucleus	Transcription regulator
EGR2	Early growth response 2	2.271	Nucleus	Transcription regulator
KLF6	Kruppel-like factor 6	2.51	Nucleus	Transcription regulator
MTPN	Myotrophin	2.763	Nucleus	Transcription regulator
NKX3–1	NK3 homeobox 1	−2.046	Nucleus	Transcription regulator
PA2G4	Proliferation-associated 2G4, 38kDa	−2.702	Nucleus	Transcription regulator
TCF4	Transcription factor 4	−2.216	Nucleus	Transcription regulator
TWIST1	Twist basic helix-loop-helix transcription factor 1	2.1	Nucleus	Transcription regulator
CDKN1B	Cyclin-dependent kinase inhibitor 1B (p27, Kip1)	3.732	Nucleus	Kinase
EIF2AK2	Eukaryotic translation initiation factor 2-alpha kinase 2	2.18	Cytoplasm	Kinase
IKBKB	Inhibitor of kappa light polypeptide gene enhancer in B-cells, kinase beta	2.127	Cytoplasm	Kinase
RIPK1	Receptor (TNFRSF)-interacting serine-threonine kinase 1	2.723	Plasma membrane	Kinase
MCM3	Minichromosome maintenance complex component 3	10.812	Nucleus	Enzyme
PTGS2	Prostaglandin-endoperoxide synthase 2 (prostaglandin G/H synthase and cyclooxygenase)	3.106	Cytoplasm	Enzyme
WHSC1	Wolf-Hirschhorn syndrome candidate 1	2.407	Nucleus	Enzyme
IL1B	Interleukin 1, beta	5.166	Extracellular space	Cytokine
SPP1	Secreted phosphoprotein 1	37.905	Extracellular space	Cytokine
RORA	RAR-related orphan receptor A	2.504	Nucleus	Ligand-dependent nuclear receptor
THRA	Thyroid hormone receptor, alpha	−2.799	Nucleus	Ligand-dependent nuclear receptor
CASP7	Caspase 7, apoptosis-related cysteine peptidase	2.579	Cytoplasm	Peptidase
PLAU	Plasminogen activator, urokinase	5.946	Extracellular space	Peptidase
FGF2	Fibroblast growth factor 2 (basic)	2.387	Extracellular space	Growth factor
ALB	Albumin	−3.125	Extracellular space	Transporter
ANXA2	Annexin A2	3.441	Plasma membrane	Other
EZR	Ezrin	2.31	Plasma membrane	Other
FLNA	Filamin A, alpha	3.45	Cytoplasm	Other

(*Continued*)

TABLE 28.6 (*Continued*)
TBI-I CGP GIH

Gene Symbol	Entrez Gene Name	Fold Change	Cellular Compartment	Molecular Type
HSPA1A/ HSPA1B	Heat shock 70-kDa protein 1A	3.137	Cytoplasm	Other
S100A9	S100 calcium binding protein A9	19.725	Cytoplasm	Other
SNCA	Synuclein, alpha (non-A4 component of amyloid precursor)	−2.169	Cytoplasm	Other
Peripheral				
ARID4A	AT rich interactive domain 4A (RBP1-like)	−3.263	Nucleus	Transcription regulator
ARID5B	AT rich interactive domain 5B (MRF1-like)	−5.312	Nucleus	Transcription regulator
BCL6B	B-cell CLL/lymphoma 6, member B	−2.23	Nucleus	Transcription regulator
CDKN2AIP	CDKN2A interacting protein	2.033	Nucleus	Transcription regulator
ELK4	ELK4, ETS-domain protein (SRF accessory protein 1)	2.532	Nucleus	Transcription regulator
ETV5	ets variant 5	−2.163	Nucleus	Transcription regulator
KDM5A	Lysine (K)-specific demethylase 5A	−2.984	Nucleus	Transcription regulator
KLF13	Kruppel-like factor 13	−2.006	Nucleus	Transcription regulator
NFIX	Nuclear factor I/X (CCAAT-binding transcription factor)	−2.548	Nucleus	Transcription regulator
PHC3	Polyhomeotic homolog 3 (Drosophila)	2.017	Nucleus	Transcription regulator
SOX7	SRY (sex determining region Y)-box 7	19.788	Nucleus	Transcription regulator
TCEB3	Transcription elongation factor B (SIII), polypeptide 3 (110 kDa, elongin A)	3.053	Nucleus	Transcription regulator
TGIF1	TGFB-induced factor homeobox 1	2.331	Nucleus	Transcription regulator
WWTR1	WW domain containing transcription regulator 1	3.243	Nucleus	Transcription regulator
ZFP36	ZFP36 ring finger protein	7.278	Nucleus	Transcription regulator
BRD4	Bromodomain containing 4	−3.528	Nucleus	Kinase
CARD11	Caspase recruitment domain family, member 11	7.343	cytoplasm	Kinase
CASK	Calcium/calmodulin-dependent serine protein kinase (MAGUK family)	2.439	Plasma Membrane	Kinase
CDK11A/ CDK11B	Cyclin-dependent kinase 11B	−4.29	Nucleus	Kinase
CHEK1	Checkpoint kinase 1	2.191	Nucleus	Kinase
EPHA5	EPH receptor A5	−2.945	Plasma membrane	Kinase
HCK	Hemopoietic cell kinase	3.887	Cytoplasm	Kinase
HSPB8	Heat shock 22-kDa protein 8	4.112	Cytoplasm	Kinase
PRPS2	Phosphoribosyl pyrophosphate synthetase 2	2.087	Other	Kinase
RIPK3	Receptor-interacting serine-threonine kinase 3	7.55	Plasma membrane	Kinase
SRPK2	SRSF protein kinase 2	−5.614	Nucleus	Kinase
TJP2	Tight junction protein 2	2.552	Plasma membrane	Kinase
CHFR	Checkpoint with forkhead and ring finger domains, E3 ubiquitin protein ligase	2.055	Nucleus	Enzyme
HMOX1	Heme oxygenase (decycling) 1	9.778	Cytoplasm	Enzyme
MX1	Myxovirus (influenza virus) resistance 1, interferon-inducible protein p78 (mouse)	28.177	Cytoplasm	Enzyme
PTBP1	Polypyrimidine tract binding protein 1	2.212	Nucleus	Enzyme
RAC2	ras-related C3 botulinum toxin substrate 2 (rho family, small GTP binding protein Rac2)	2.989	Cytoplasm	Enzyme
SDC1	Syndecan 1	13.681	Plasma membrane	Enzyme
SGPL1	Sphingosine-1-phosphate lyase 1	3.108	Cytoplasm	Enzyme
TGM2	Transglutaminase 2	3.574	Cytoplasm	Enzyme
TOP2A	Topoisomerase (DNA) II alpha 170 kDa	2.26	Nucleus	Enzyme
LCN2	Lipocalin 2	71.824	Extracellular space	Transporter
PDPN	Podoplanin	7.911	Plasma membrane	Transporter
RASA1	RAS p21 protein activator (GTPase activating protein) 1	2.392	Cytoplasm	Transporter
STX6	Syntaxin 6	2.294	Cytoplasm	Transporter
TFRC	Transferrin receptor	2.094	Plasma membrane	Transporter
TPR	Translocated promoter region, nuclear basket protein	2.212	Nucleus	Transporter
ANGPT2	Angiopoietin 2	2.979	Extracellular space	Growth factor
ESM1	Endothelial cell-specific molecule 1	20.805	Extracellular space	Growth factor
HBEGF	Heparin-binding EGF-like growth factor	7.073	Extracellular space	Growth factor
TGFB3	Transforming growth factor, beta 3	2.74	Extracellular space	Growth factor

TABLE 28.6 (*Continued*)
TBI-I CGP GIH

Gene Symbol	Entrez Gene Name	Fold Change	Cellular Compartment	Molecular Type
IL6ST	Interleukin 6 signal transducer (gp130, oncostatin M receptor)	2.307	Plasma membrane	Transmembrane receptor
ITGAM	Integrin, alpha M (complement component 3 receptor 3 subunit)	4.739	Plasma membrane	Transmembrane receptor
UNC5B	unc-5 homolog B (*C. elegans*)	2.067	Plasma membrane	Transmembrane receptor
ITGAV	Integrin, alpha V	2.09	Plasma membrane	Ion channel
KCNN4	Potassium intermediate/small conductance calcium-activated channel, subfamily N, member 4	3.088	Plasma membrane	Ion channel
ADAMTS1	ADAM metallopeptidase with thrombospondin type 1 motif, 1	5.238	Extracellular space	Peptidase
PLAT	Plasminogen activator, tissue	2.25	Extracellular space	Peptidase
CXCL16	Chemokine (C-X-C motif) ligand 16	3.69	Extracellular space	Cytokine
PPP1R12A	Protein phosphatase 1, regulatory subunit 12A	−3.911	Cytoplasm	Phosphatase
ACTN1	Actinin, alpha 1	2.483	Cytoplasm	Other
ADM	Adrenomedullin	2.142	Extracellular space	Other
ANXA1	Annexin A1	3.535	Plasma membrane	Other
CDT1	Chromatin licensing and DNA replication factor 1	3.098	Nucleus	Other
CEBPD	CCAAT/enhancer binding protein (C/EBP), delta	11.271	Nucleus	Other
CMIP	c-Maf inducing protein	−3.778	Cytoplasm	Other
EMP3	Epithelial membrane protein 3	4.346	Plasma membrane	Other
GADD45B	Growth arrest and DNA-damage-inducible, beta	2.349	Cytoplasm	Other
GADD45G	Growth arrest and DNA-damage-inducible, gamma	3.191	Nucleus	Other
LCP1	Lymphocyte cytosolic protein 1 (L-plastin)	6.082	Cytoplasm	Other
LGALS1	Lectin, galactoside-binding, soluble, 1	3.2	Extracellular space	Other
NREP	Neuronal regeneration-related protein	−2.259	Cytoplasm	Other
PHLDA1	Pleckstrin homology-like domain, family A, member 1	5.129	Cytoplasm	Other
PMEPA1	Prostate transmembrane protein, androgen-induced 1	2.682	Plasma membrane	Other
S100A8	S100 calcium-binding protein A8	35.214	Cytoplasm	Other
SDC4	Syndecan 4	2.427	Plasma membrane	Other
TAGLN	Transgelin	7.353	Cytoplasm	Other
TIMP1	TIMP metallopeptidase inhibitor 1	38.486	Extracellular space	Other
VASP	Vasodilator-stimulated phosphoprotein	2.927	Plasma membrane	Other
ZBTB20	Zinc finger and BTB domain containing 20	4.12	Nucleus	Other
Orphan				
CYP1B1	Cytochrome P450, family 1, subfamily B, polypeptide 1	10.998	Cytoplasm	Enzyme
EHD4	EH-domain containing 4	2.361	Plasma membrane	Enzyme
KIF3A	Kinesin family member 3A	−5.083	Cytoplasm	Enzyme
MGLL	Monoglyceride lipase	−7.853	Plasma membrane	Enzyme
RND3	Rho family GTPase 3	2.864	Cytoplasm	Enzyme
LITAF	Lipopolysaccharide-induced TNF factor	2.224	Nucleus	Transcription regulator
NRARP	NOTCH-regulated ankyrin repeat protein	2.215	Nucleus	Transcription regulator
PRDM2	PR domain containing 2, with ZNF domain	3.677	Nucleus	Transcription regulator
PRDM4	PR domain containing 4	2.168	Nucleus	Transcription regulator
TCF19	Transcription factor 19	2.289	Nucleus	Transcription regulator
ADAMTS9	ADAM metallopeptidase with thrombospondin type 1 motif, 9	3.11	Extracellular space	Peptidase
FGL2	Fibrinogen-like 2	16.793	Extracellular space	Peptidase
DUSP8	Dual specificity phosphatase 8	−6.392	Nucleus	Phosphatase
PTPRF	Protein tyrosine phosphatase, receptor type, F	−6.365	Plasma membrane	Phosphatase
AQP4	Aquaporin 4	6.58	Plasma membrane	Transporter
SCAMP2	Secretory carrier membrane protein 2	2.333	Cytoplasm	Transporter
CAMK2N1	Calcium/calmodulin-dependent protein kinase II inhibitor 1	−11.813	Plasma membrane	kinase
EIF3C/EIF3CL	Eukaryotic translation initiation factor 3, subunit C	−4.369	Cytoplasm	Translation regulator
ARL11	ADP-ribosylation factor-like 11	3.143	Other	Other
EMP1	Epithelial membrane protein 1	9.353	Plasma membrane	Other
KIFAP3	Kinesin-associated protein 3	−2.281	Cytoplasm	Other

(*Continued*)

TABLE 28.6 (*Continued*)
TBI-I CGP GIH

Gene Symbol	Entrez Gene Name	Fold Change	Cellular Compartment	Molecular Type
PHF14	PHD finger protein 14	3.602	Other	Other
PROCR	Protein C receptor, endothelial	6.923	Plasma membrane	Other
RASSF4	Ras association (RalGDS/AF-6) domain family member 4	4.289	Other	Other
TSPAN5	Tetraspanin 5	2.67	Plasma membrane	Other
VANGL1	VANGL planar cell polarity protein 1	2.009	Cytoplasm	Other

Primary: >12 connections in GOI network (see text); secondary: 7–12 connections in GOI network
Peripheral: <7 connections in GOI network; orphan: no connections in GOI network
Italics = gene changes on both sides of the brain

TABLE 28.7
TBI-C CGP GIH

Gene Symbol	Entrez Gene Name	Fold Change	Cellular Compartment	Molecular Type
Primary				
NCOR1	Nuclear receptor corepressor 1	−9.305	Nucleus	Transcription regulator
SMARCA4	SWI/SNF-related, matrix-associated, actin-dependent regulator of chromatin, subfamily a, member 4	−7.712	Nucleus	Transcription regulator
SP1	Sp1 transcription factor	−2.076	Nucleus	Transcription regulator
STAT3	Signal transducer and activator of transcription 3 (acute-phase response factor)	−3.771	Nucleus	Transcription regulator
CSNK2A1	Casein kinase 2, alpha 1 polypeptide	−2.75	Cytoplasm	Kinase
EGFR	Epidermal growth factor receptor	2.374	Plasma membrane	Kinase
GSK3B	Glycogen synthase kinase 3 beta	−6.635	Nucleus	Kinase
HSP90AA1	Heat shock protein 90-kDa alpha (cytosolic), class A member 1	−4.843	Cytoplasm	Enzyme
CCND1	Cyclin D1	−2.027	Nucleus	Other
Secondary				
CD44	CD44 molecule (Indian blood group)	2.399	Plasma membrane	Enzyme
MCM3	Minichromosome maintenance complex component 3	3.363	Nucleus	Enzyme
OGT	O-linked N-acetylglucosamine (GlcNAc) transferase	2.296	Cytoplasm	Enzyme
TOP2A	Topoisomerase (DNA) II alpha 170kDa	−2.406	Nucleus	Enzyme
MAPK8	Mitogen-activated protein kinase 8	2.102	Cytoplasm	Kinase
MAPT	Microtubule-associated protein tau	−2.366	Plasma membrane	Kinase
PTK2B	Protein tyrosine kinase 2 beta	2.15	Cytoplasm	Kinase
FOXO1	Forkhead box O1	−3.329	Nucleus	Transcription regulator
MED1	Mediator complex subunit 1	−4.011	Nucleus	Transcription regulator
TCF4	Transcription factor 4	−4.625	Nucleus	Transcription regulator
THRA	Thyroid hormone receptor, alpha	−11.518	Nucleus	Ligand-dependent nuclear receptor
MMP14	Matrix metallopeptidase 14 (membrane-inserted)	−3.034	Extracellular space	Peptidase
BAX	BCL2-associated X protein	−3.306	Cytoplasm	Transporter
ATXN2	Ataxin 2	−2.479	Nucleus	Other
GADD45A	Growth arrest and DNA-damage-inducible, alpha	−2.656	Nucleus	Other
TIMP1	TIMP metallopeptidase inhibitor 1	2.101	Extracellular space	Other
TRH	Thyrotropin-releasing hormone	−2.211	Extracellular space	Other
Peripheral				
ACIN1	Apoptotic chromatin condensation inducer 1	−2.515	Nucleus	Enzyme
CUL2	Cullin 2	2.033	Nucleus	Enzyme

TABLE 28.7 (*Continued*)
TBI-C CGP GIH

Gene Symbol	Entrez Gene Name	Fold Change	Cellular Compartment	Molecular Type
CYP1B1	Cytochrome P450, family 1, subfamily B, polypeptide 1	4.808	Cytoplasm	Enzyme
KIF3A	Kinesin family member 3A	−11.754	Cytoplasm	Enzyme
KMT2E	Lysine (K)-specific methyltransferase 2E	−3.465	Nucleus	Enzyme
MX1	Myxovirus (influenza virus) resistance 1, interferon-inducible protein p78 (mouse)	7.326	Cytoplasm	Enzyme
PAFAH1B1	Platelet-activating factor acetylhydrolase 1b, regulatory subunit 1 (45 kDa)	−2.856	Cytoplasm	Enzyme
SDC1	Syndecan 1	2.566	Plasma membrane	Enzyme
SMURF2	SMAD-specific E3 ubiquitin protein ligase 2	−3.458	Cytoplasm	Enzyme
BRD4	Bromodomain containing 4	−15.202	Nucleus	Kinase
CARD11	Caspase recruitment domain family, member 11	2.892	Cytoplasm	Kinase
CDK11A/ CDK11B	Cyclin-dependent kinase 11B	−14.872	Nucleus	Kinase
CDK19	Cyclin-dependent kinase 19	−2.191	Nucleus	Kinase
EPHA5	EPH receptor A5	−8.441	Plasma membrane	Kinase
PIK3R2	Phosphoinositide-3-kinase, regulatory subunit 2 (beta)	2.332	Cytoplasm	Kinase
SRPK2	SRSF protein kinase 2	−23.589	Nucleus	Kinase
ARID4A	AT rich interactive domain 4A (RBP1-like)	−7.033	Nucleus	Transcription regulator
ARID5B	AT rich interactive domain 5B (MRF1-like)	−16.105	Nucleus	Transcription regulator
BTG2	BTG family, member 2	−5.803	Nucleus	Transcription regulator
KDM5A	Lysine (K)-specific demethylase 5A	−19.489	Nucleus	Transcription regulator
MAFG	v-maf avian musculoaponeurotic fibrosarcoma oncogene homolog G	−2.632	Nucleus	Transcription regulator
NFIX	Nuclear factor I/X (CCAAT-binding transcription factor)	−8.112	Nucleus	Transcription regulator
DUSP8	Dual specificity phosphatase 8	−17.059	Nucleus	Phosphatase
PPP1R12A	Protein phosphatase 1, regulatory subunit 12A	−10.97	Cytoplasm	Phosphatase
PTPRJ	Protein tyrosine phosphatase, receptor type, J	−3.838	Plasma membrane	Phosphatase
LCN2	Lipocalin 2	3.895	Extracellular space	Transporter
RASA1	RAS p21 protein activator (GTPase activating protein) 1	−2.105	Cytoplasm	Transporter
TTR	Transthyretin	−2.058	Extracellular space	Transporter
NR2F1	Nuclear receptor subfamily 2, group F, member 1	2.659	Nucleus	Ligand-dependent nuclear receptor
NR2F2	Nuclear receptor subfamily 2, group F, member 2	−2.674	Nucleus	Ligand-dependent nuclear receptor
IL6ST	Interleukin 6 signal transducer (gp130, oncostatin M receptor)	−3.283	Plasma membrane	Transmembrane receptor
PRLR	Prolactin receptor	−3.192	Plasma membrane	Transmembrane receptor
SPP1	Secreted phosphoprotein 1	2.37	Extracellular space	Cytokine
SSTR2	Somatostatin receptor 2	2.068	Plasma membrane	G-protein–coupled receptor
CACNA1A	Calcium channel, voltage-dependent, P/Q type, alpha 1A subunit	−2.235	Plasma membrane	Ion channel
ACTN1	Actinin, alpha 1	−2.088	Cytoplasm	Other
APBB2	Amyloid beta (A4) precursor protein-binding, family B, member 2	2.046	Cytoplasm	Other
CDT1	Chromatin licensing and DNA replication factor 1	−2.295	Nucleus	Other
GADD45G	Growth arrest and DNA-damage-inducible, gamma	−2.384	Nucleus	Other
KIFAP3	Kinesin-associated protein 3	−7.831	Cytoplasm	Other
LCP1	Lymphocyte cytosolic protein 1 (L-plastin)	2.799	Cytoplasm	Other
PHLDA1	Pleckstrin homology-like domain, family A, member 1	2.16	Cytoplasm	Other
PMEPA1	Prostate transmembrane protein, androgen induced 1	−2.937	Plasma membrane	Other
RBFOX1	RNA binding protein, fox-1 homolog (C. elegans) 1	−2.109	Cytoplasm	Other
SYMPK	Symplekin	−2.13	Cytoplasm	Other

(*Continued*)

TABLE 28.7 (*Continued*)
TBI-C CGP GIH

Gene Symbol	Entrez Gene Name	Fold Change	Cellular Compartment	Molecular Type
ZBTB20	Zinc finger and BTB domain containing 20	−2.757	Nucleus	Other
ZMYND11	Zinc finger, MYND-type containing 11	−2.211	Nucleus	Other
Orphan				
ADC	Arginine decarboxylase	−2.112	Cytoplasm	Enzyme
EHD4	EH-domain containing 4	−2.056	Plasma membrane	Enzyme
GNA11	Guanine nucleotide binding protein (G protein), alpha 11 (Gq class)	−2.104	Plasma membrane	Enzyme
MGLL	Monoglyceride lipase	−18.145	Plasma membrane	Enzyme
PLA2G3	Phospholipase A2, group III	2.484	Extracellular space	Enzyme
RND3	Rho family GTPase 3	−2.971	Cytoplasm	Enzyme
GBX2	Gastrulation brain homeobox 2	2.59	Nucleus	Transcription regulator
HSF2	Heat shock transcription factor 2	2.159	Nucleus	Transcription regulator
KLF13	Kruppel-like factor 13	−4.582	Nucleus	Transcription regulator
PA2G4	Proliferation-associated 2G4, 38kDa	−5.783	Nucleus	Transcription regulator
ABCC1	ATP-binding cassette, sub-family C (CFTR/MRP), member 1	−3.109	Plasma membrane	Transporter
AQP4	Aquaporin 4	2.986	Plasma membrane	Transporter
TPR	Translocated promoter region, nuclear basket protein	−2.728	Nucleus	Transporter
CXCL14	Chemokine (C-X-C motif) ligand 14	2.126	Extracellular space	Cytokine
ESM1	Endothelial cell–specific molecule 1	5.748	Extracellular space	Growth factor
KCNN4	Potassium intermediate/small conductance calcium–activated channel, subfamily N, member 4	−9.429	Plasma membrane	Ion channel
CAMK2N1	Calcium/calmodulin-dependent protein kinase II inhibitor 1	−23.824	Plasma membrane	Kinase
FGL2	Fibrinogen-like 2	4.017	Extracellular space	Peptidase
PTPRF	Protein tyrosine phosphatase, receptor type, F	−20.492	Plasma membrane	Phosphatase
EIF3C/EIF3CL	Eukaryotic translation initiation factor 3, subunit C	−9.072	Cytoplasm	Translation regulator
CDH13	Cadherin 13	−2.692	Plasma membrane	Other
CEBPD	CCAAT/enhancer-binding protein (C/EBP), delta	2.037	Nucleus	Other
CMIP	c-Maf inducing protein	−13.763	Cytoplasm	Other
KRT23	Keratin 23 (histone deacetylase inducible)	−3.062	Other	Other
MAFB	v-maf avian musculoaponeurotic fibrosarcoma oncogene homolog B	−2.018	Nucleus	Other
PHF14	PHD finger protein 14	−6.323	Other	Other
RASSF4	Ras association (RalGDS/AF-6) domain family member 4	2.106	Other	Other

Primary: >7 connections in GOI network (see text); secondary: 4–7 connections in GOI network
Peripheral: <4 connections in GOI network; orphan: no connections in GOI network
Italics = gene changes on both sides of the brain

secondary) of the respective GIHs. Figure 28.7 shows that the top 10 molecular and cellular for TBI-I are reordered in terms of significance. Cell cycle, gene expression, and cell-to-cell signaling and interaction move up in significance, whereas cellular movement and cell morphology move down (Figure 28.7a, b). The effect for TBI-C is more dramatic. In addition to a significant reordering of the functions, molecular transport and small molecule biochemistry are replaced in the top 10 by DNA replication, recombination, and repair and cellular function and maintenance (Figure 28.7c and d). Although in a slightly different order of significance, the top functions are similar for both TBI-I and TBI-C after GIH analysis (Figure 28.7b and d). However, the near-inverse expression patterns (Figure 28.2) demonstrate significantly different priorities for these functions on each side of the brain after TBI.

28.4 OPTIMIZING THE GIH ANALYSIS

Having now completed several of these analyses, it has become clear that some modifications to the process could improve the outcome. This section presents potential ways to augment the significance the GIH analysis.

28.4.1 Origin of Genetic Material

The GIH is built by ranking genes in a virtual network of GOI based on their interactions with each other. To increase

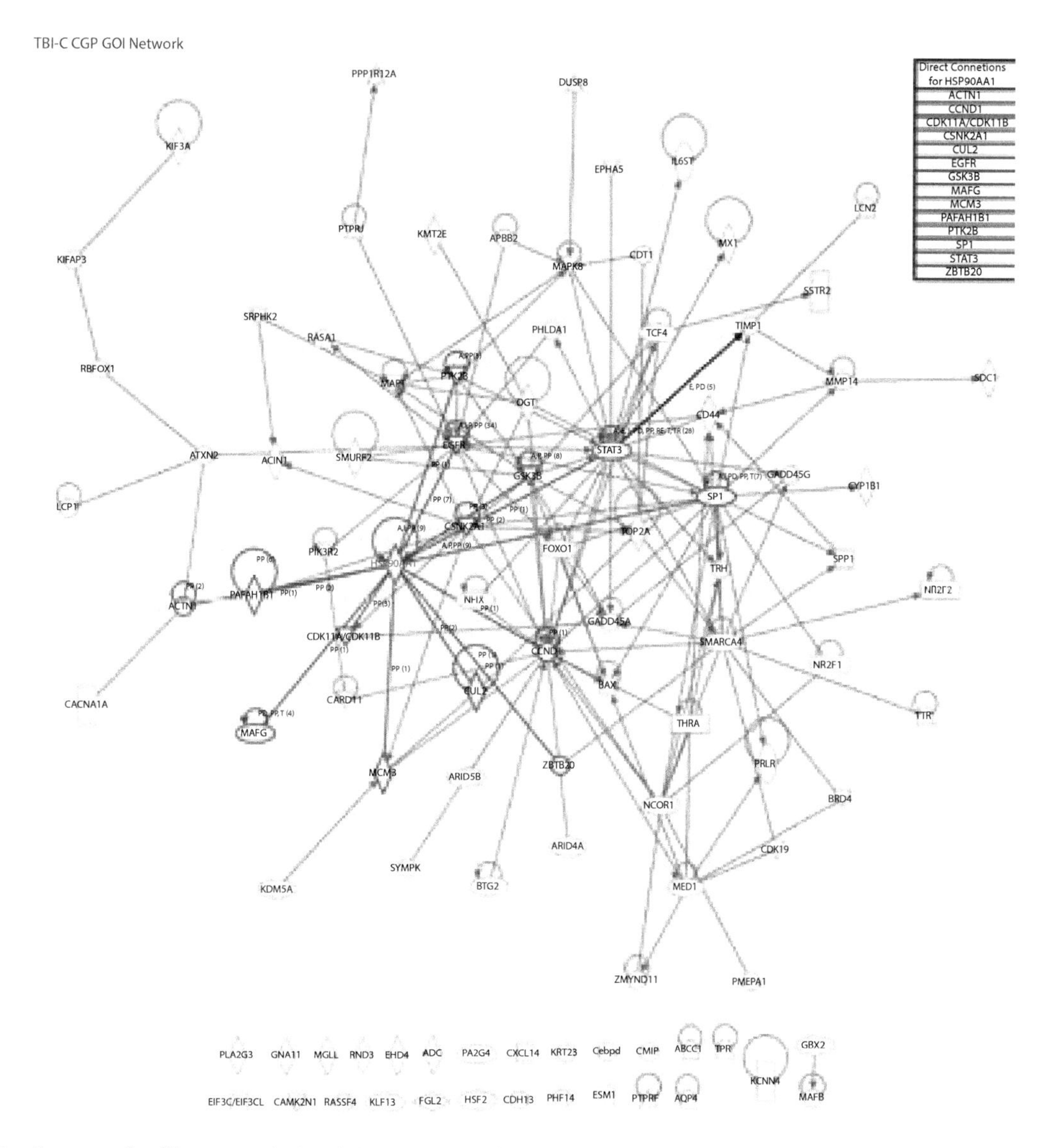

FIGURE 28.6 An example of how we calculated the number of direct connections for a gene in our TBI-C CGP GOI network. In IPA, the gene in question was selected (HSP90AA1 in this example). Then, its direct connections were selected by right-clicking on HSP90AA1 and using the "select nearest neighbors" option (highlighted in blue). A list of the selected genes was exported and HSP90AA1 was removed from the list (upper right corner). The remaining genes were counted (14 in this example) and HSP90AA1 was ranked in the TBI-C CGP gene interaction hierarchy (Table 28.7; primary tier) by this number.

the likelihood that these genes actually would interact, it would be best to isolate your RNA from a very discrete area. This would improve the reproducibility of the analysis and aid in the targeting of confirmation studies.

28.4.2 Weighting Connections

In the current analysis, all interactions are weighted equally. It may be sensible to give more weight to an interaction that "acts on" another molecule as opposed to one where the molecule is "acted upon." IPA has more than 20 main types of interactions for molecules. Several of these main types have several subtypes of interactions. Depending on a particular biological function or process, certain molecule interactions may be more significant than others and thus weighted more in the analysis. This would require case-by-case weighting that could hinder reproducibility of complex analyses.

28.4.3 Identifying and Eliminating Bias

Another issue that is related to evenly weighted interactions is an inherent bias toward or away from specific molecular types. Molecules such as cytokines and growth factors that bind specifically to their respective receptors can have a near one-to-one relationship with those receptors and, therefore, are limited in the number of interactions they may exhibit. Conversely, a single transcription factor can modulate transcription for many molecules and interact with other molecules both upstream and downstream to support transcription (Cheng et al., 2011). Eliminating these biases will require

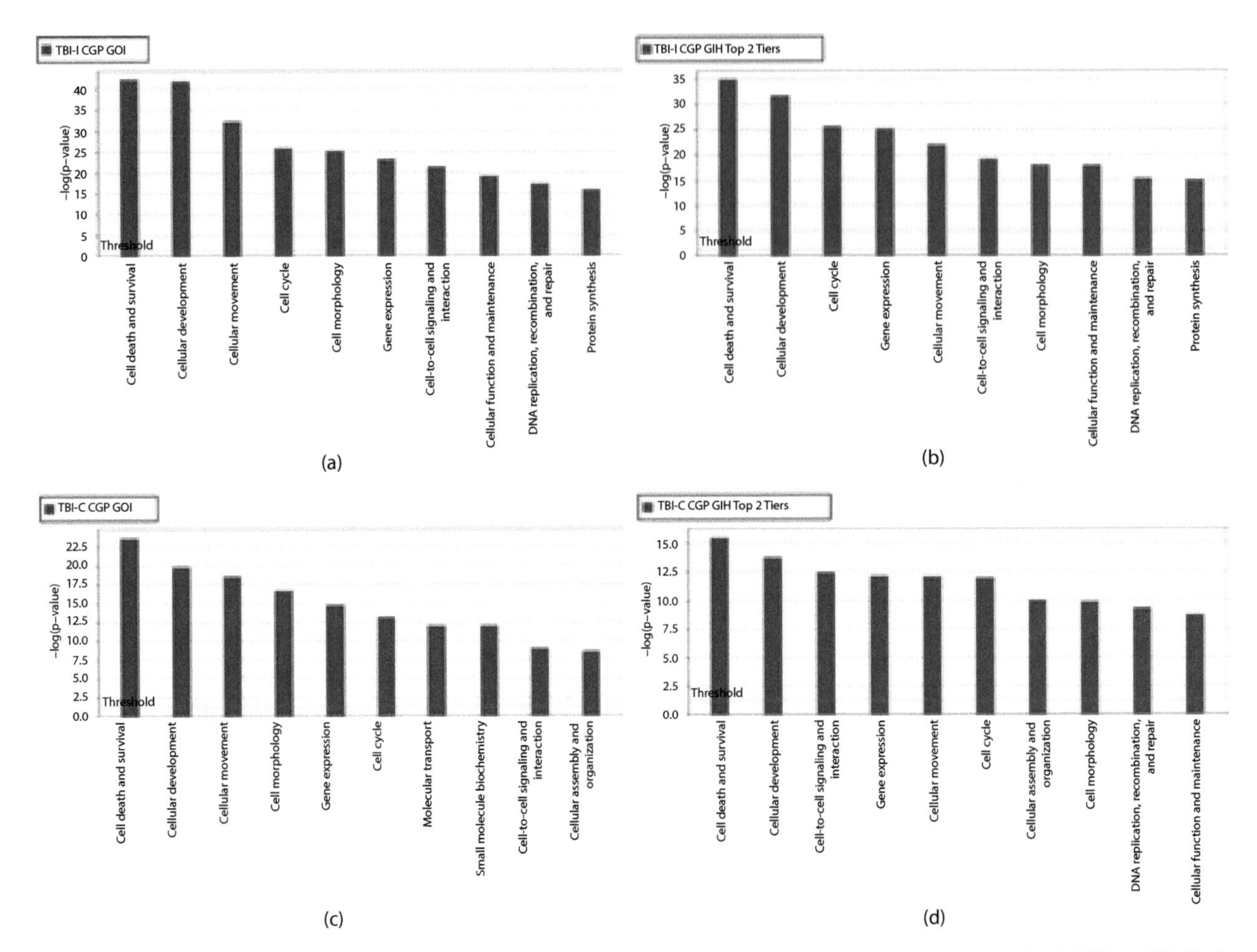

FIGURE 28.7 The top 10 molecular and cellular functions determined by IPA to be associated with the unranked GOI for TBI-I (a) and TBI-C (c) and the primary and secondary tiers of the TBI-I (b) and TBI-C (d) GIHs. Side-by-side comparison allowed for visualization of how functions changed in significance order once the genes were put into a ranked order. Notably, two functions were replaced in the top 10 for TBI-C. The cellular growth and proliferation category was removed from this analysis because all genes were initially selected from that functional category.

advanced modeling to determine the correct way to normalize interactions across molecular types.

28.5 DISCUSSION

This chapter presents a method to manage the large data sets that can be produced by high throughput analyses. In particular, we developed this method to manage gene microarray data generated from unilateral TBI. As stated previously, optimization of this analysis is an ongoing process. As it stands, it has been a valuable tool making our large data sets manageable and giving us direction for further study of the genes that were ranked highest in the GIH.

A key aspect of the unilateral TBI injury model was that there were significant gene changes both locally and remote to the injury site. This necessitated the development of an analysis that was both specific to the region being analyzed and objective in nature so that the individual results could be compared. IPA's robust network component filled these requirements because it allowed for identification of GOI that were specific to the regional genes imported and free from investigator bias as the networks were generated based on IPA network generation rules (IPA, 2013).

Although IPA was used in our analysis as the source for biological function analysis, molecule interactions, and network analysis, it is possible to use other tools to approximate the combined analysis we present. These and other alternatives have been reviewed previously (Selvaraj and Natarajan, 2011). Briefly, gene ontologies can be used to assign genes to biological functions. Molecule interactions can be found in several international databases including the Biomolecular Interaction Database, the Database of Interacting Proteins, the Molecular Interactions Database (Rome, Italy), and the Search Tool for the Retrieval of Interacting Genes/Proteins (EMBL, Heidelberg, Germany). Pathway Studio and Cytoscape, among others, can be used for network generation and analysis. We chose IPA because we had access to the program and it had the ability to do all the necessary steps of the analysis. As part of our ongoing optimization, other analysis tools will be explored for their utility.

Hierarchical ranking based on network connections has been demonstrated previously (Camargo and Azuaje, 2008), but their ranking was based on arbitrary numbers of connections in the network, building on previous studies of hubs in networks (Han et al., 2004; Patil and Nakamura, 2006; Lu et al., 2007). Here we determined that using percentages of the total number of molecules in the network was best for determining hierarchical ranking ranges and allowed for application to networks of varying sizes.

We believe that our gene interaction hierarchy methodology can also be applied to any molecular data set in which the potential for molecular interaction exists. Although this analysis is effective for ranking genes/molecules in one region or a control situation, there seems to be added value when investigating the molecular response in two or more regions to a singular initiating event. How various regions of the brain respond molecularly to a single focal injury will be a continuing focus of our research and the GIH analysis will be useful in deciphering the data.

28.6 METHODS

The methods for the animal surgery and microarray processing have been published previously (White et al., 2013). Only the methods required for the understanding of this chapter are included here.

28.6.1 Controlled Cortical Impact

Under isoflurane anesthesia, rats received a unilateral controlled cortical impact (CCI/TBI) using the Pittsburgh Precision Instruments, Inc., device. A craniotomy was made with the center 4 mm posterior and 3–4 mm lateral to bregma using a 6-mm-diameter trephan drill bit. The impact was done at an angle of 15° from vertical with a velocity of 3 m/s to a depth of 2 mm using a 5-mm diameter impact tip. The rats were sacrificed 24 hours postinjury and the brains were removed for RNA isolation.

28.6.2 Microarray Data Analysis

The data were analyzed in Microsoft Excel for calculation of fold change and whether the genes were confirmed as present in the tissue sample (as determined by the Affymetrix software). Genes in the injured brain that increased or decreased in expression by two-fold or more compared with controls and were present in either all three ipsilateral samples or all three contralateral samples were identified. The gene data sets that were generated were TBI-I and TBI-C fold changes.

28.6.3 IPA

The gene data sets were analyzed between December 23, 2013 and December 31, 2013 using IPA (Ingenuity Systems, www.ingenuity.com) and overlaid onto a global molecular network developed from information contained in the Ingenuity Knowledge Base.

ACKNOWLEDGMENTS

The authors thank Michelle C. LaPlaca, PhD, Gregory D. Ford, PhD, Monique C. Surles-Zeigler, Alicia S. Gates, Benem Davids, and Timothy Distel for their assistance with this work. This work was supported by The National Institutes of Health (NIH), The Department of Defense, The W.M. Keck Foundation, The Howard Hughes Medical Institute, and The Clinical and Translational Science Award program from The National Center for Research Resources, a component of NIH. Its contents are solely the responsibility of the authors and do not necessarily represent the official views of the above stated agencies.

REFERENCES

Camargo, A. and F. Azuaje, 2008. Identification of dilated cardiomyopathy signature genes through gene expression and network data integration. *Genomics,* 92 (6), 404–13.

Cheng, C., K.K. Yan, W. Hwang, J. Qian, N. Bhardwaj, J. Rozowsky et al., 2011. Construction and analysis of an integrated regulatory network derived from high-throughput sequencing data. *PLoS Comput Biol,* 7 (11), e1002190.

Dixon, C.E. and A. Kline, 2009. Controlled cortical impact injury model. In Chen, J., Xu, Z., Xu, X.-M. and Zhang, J. (eds.). *Animal Models of Acute Neurological Injuries.* pp. 385–391. Humana Press, New York.

Han, J.D., N. Bertin, T. Hao, D.S. Goldberg, G.F. Berriz, L.V. Zhang et al., 2004. Evidence for dynamically organized modularity in the yeast protein-protein interaction network. *Nature,* 430 (6995), 88–93.

IPA, 2011. *Ingenuity Systems website,* viewed December 2013, from www.ingenuity.com

IPA, 2013. *IPA network generation*, viewed December 2013, from http://ingenuity.force.com/ipa/IPATutorials?id=kA250000000TNBZCA4.

Lu, X., V.V. Jain, P.W. Finn, and D.L. Perkins, 2007. Hubs in biological interaction networks exhibit low changes in expression in experimental asthma. *Mol Syst Biol,* 3, 98.

Patil, A. and H. Nakamura, 2006. Disordered domains and high surface charge confer hubs with the ability to interact with multiple proteins in interaction networks. *FEBS Lett,* 580 (8), 2041–5.

Selvaraj, S. and J. Natarajan, 2011. Microarray data analysis and mining tools. *Bioinformation,* 6 (3), 95–9.

White, T.E., G.D. Ford, M.C. Surles-Zeigler, A.S. Gates, M.C. Laplaca, and B.D. Ford, 2013. Gene expression patterns following unilateral traumatic brain injury reveals a local pro-inflammatory and remote anti-inflammatory response. *BMC Genomics,* 14 (1), 282.

29 Autoantibodies in CNS Trauma and Neuropsychiatric Disorders

A New Generation of Biomarkers

Firas Kobeissy and Rabih A. Moshourab

CONTENTS

29.1 INTRODUCTION

Central nervous system (CNS) trauma is a growing public health concern resulting from various types of cerebral insults, leading to acute neurological and non-neurological manifestations that can leave life-long consequences. To date, there are no standardized therapeutic and management protocols dealing with brain trauma. Current research is uncovering novel biomarkers that can aid in diagnosis, management and therapy. Current status of brain injury biomarkers includes the presence, absence or altered expression levels of certain neural (neuronal astrocytic or glial) related genes/proteins, protein degradation products and microRNAs which are discussed in different chapters of this book. Recently, there has been an increased interest in the new emerging role of autoantibodies—which have been long identified—as new generation biomarkers in the areas of neurotrauma, neuropsychiatric disorders and neurotoxicity. In this chapter, we will discuss the genesis and implications of autoantibodies in neurotrauma; focusing on the area of spinal cord injury (SCI) and shedding light on recent application in traumatic brain injury (TBI). In addition, the potential pathogenic mechanistic role of autoantibodies in the areas of Autism spectrum disorder (ASD) and neurotoxicity will be evaluated as this may reflect on the neural injury observed in brain trauma. The key value of these new generation biomarkers is that—unlike their corresponding autoantigens that may serve as acute markers of injury—these identified autoantibodies represent long-lasting, chronic signature biomarkers that can be associated with advanced chronic stages of injury sequelae. Such work has the potential to be applied in the fields of neurotrauma and neuropsychiatric fields that may reflect underlying mechanisms and can be utilized for diagnosis, staging and treatment guidance as well as be the target for therapy.

Autoimmune diseases, characterized by the presence of autoantibodies, affect about 5%–7% of the world's population; 3% of these are brain reactive autoantibodies with no overt effects (Diamond et al., 2013; Fairweather and Rose, 2004; National Institutes of Health Autoimmune Diseases Coordinating Committee Report, 2002). These brain-specific autoantibodies have a restricted access to our brain tissues unless under pathologic conditions (Diamond et al., 2013). Autoimmune response mechanisms have been observed in a number of CNS disorders involving multiple sclerosis (MS), paraneoplastic syndromes, brain trauma, and dementia-related disorders (Cross et al., 2001; Popovic et al., 1998; Sjogren and Wallin, 2001). A number of neurological disorders are associated with blood–brain barrier (BBB) disruption or increased permeability observed in Alzheimer disease (AD), stroke, TBI, and schizophrenia (Fazio et al., 2004; Marchi et al., 2003, 2004; Neuwelt et al., 2011). Injury to the BBB such as in brain injury may lead to the release of intracellular proteins either intact or proteolytic fragments from protease activation into the cerebrospinal fluid (CSF) or blood stream. The leakage of such entities into the circulation may lead to the formation of autoantibodies that have been defined as brain-reactive antibodies that recognize self- (auto-) antigens i.e., an antigen that is normally found in a subject tissue or cell or organelles. For a schematic representation of the above mechanism, please refer to Figure 29.1.

Several hypotheses have been proposed for the development of these brain-specific autoantibodies and it has been argued whether their presence contributes to the pathogenic outcome of the disease in question or maybe they are epiphenomenal in nature. Recent studies by Davies and Skoda have indicated that patients with SCI or TBI would develop autoantibodies that target a number of CNS self-antigens including GM1 gangliosides, myelin-associated glycoprotein, α-amino-3-hydroxy-5-methyl-4-isoxazolepropionic acid (AMPA) and N-methyl-D-aspartate (NMDA) glutamate receptors, and β-III-tubulin and nuclear antigens (Davies et al., 2007; Skoda et al., 2006). Based on the data presented by Ankeny in the area of SCI and by Zhang and Marchi in the areas of TBI (Ankeny and Popovich, 2010; Marchi et al., 2013a; Zhang et al., 2014), it is reasonable to regard the presence of an actual anti-brain reactivity as a potential threat to brain tissue integrity (Rudehill et al., 2006).

As such, there is an increased interest in this newly discovered mine of biomarkers for several reasons. Autoantibodies can be correlated to disease activity/severity and are shown to be related to particular clinical manifestation or tissue injury presenting years before disease onset and may constitute potential biomarkers of the disease. Autoantibodies act as a predictive marker of disease occurrence and valuable indicators for therapeutic response to biologics as well as to side effects. These autoantibodies can be a useful tool for diagnosis and management relevant to organ-specific or non–organ-specific disorders (Tron, 2014). Several brain-derived autoantibodies are presented in Table 29.1.

29.2 BBB, IMMUNE SYSTEM, MOLECULAR MIMICRY, AND CNS INTERACTION

The CNS has a delicate interaction with the immune system. The CNS immune homeostasis is in the quiescence and self-tolerance state. Specific anatomical structures, including the BBB, limit interactions between the CNS and the systemic immune system (Diamond et al., 2013). The BBB is a dynamic system that can be regulated in vivo by stimuli that can modulate—in specific circumstances—the entry of typically nonpermeating molecules (antibodies) into brain tissue (Banks, 2005). The relative quiescence state involves that low neuronal major histocompatibility complex antigens expression exist to limit inflammatory damage in an organ with limited regenerative capacity and extraordinary ongoing metabolic demand (Diamond et al., 2013). The CNS immune environment suggests that the brain immune system components are involved in different processes in the CNS different than in other tissues (Diamond et al., 2013). For example, complement components in the CNS exert important effects

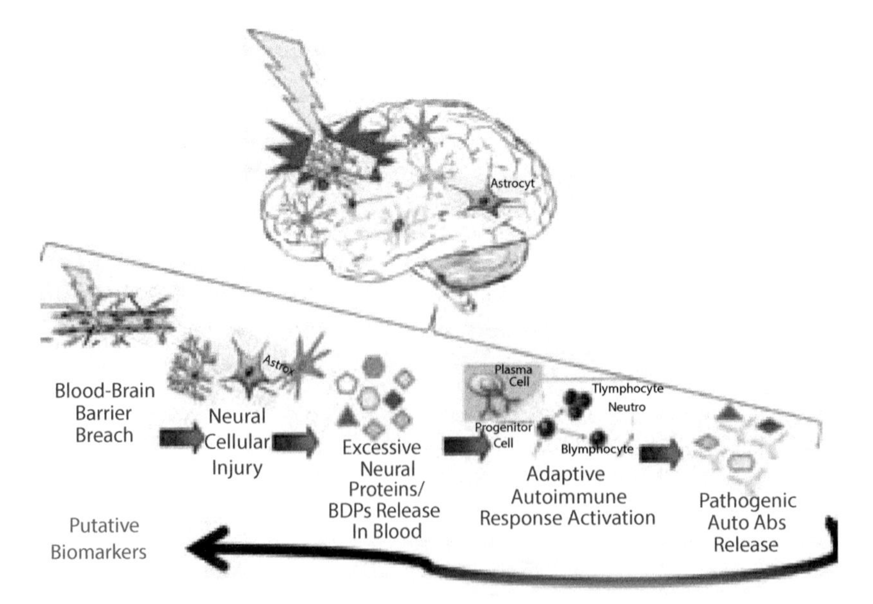

FIGURE 29.1 The trajectory and genesis of autoantibodies as potential biomarkers. Upon neurotrauma to the CNS, several pathologic events are activated, including a breach in the BBB coupled with the activation of injurious pathways (e.g., excitotoxicity, necrosis, apoptosis, protease activation) leading to neural cell injury. Molecular components (DNA, lipids, proteins, receptors) of these neural structures are released into the circulation where they are encountered by our adaptive immune response (B and T lymphocytes) that may recognize these as non–self-antigens mounting an autoimmune response that can have several implications as pathogenic as in SCI or as a neuroprotective. The long-lasting existence of these autoantibodies may represent an excellent target as novel biomarkers indicative of injury severity and may serve as potential guide in drug therapy applied in CNS trauma.

TABLE 29.1
Summary of Autoantibodies and their Targets in Relation to Neurological and Nonneurological Diseases

Autoantibody ID	Neuropathology	Location of Corresponding Protein	Class of Corresponding Protein	Place Identified (CSF, Serum)	References
Anti-microtubule–associated protein tau protein	Alzheimer disease	Nerve axon, glial cells of CNS	Structural	Serum, CSF	Bartos et al., 2012
Anti-microtubule–associated protein 2 (MAP2)	PANDAS, Tourette syndrome, CNS injury	Neuronal dendrite	Structural	Serum	Abou-Donia et al., 2013; Morris et al., 2009
Anti-synapsin 1	MS	Nerve terminal in CNS	Vesicular	Serum	Bitsch et al., 2004; Gitlits et al., 2001
Anti-tubulin beta 3	Neurological and nonneurological disorders	Nerve microtubule	Structural	Serum, CSF	Bansal et al., 2009; Op De Beeck et al., 2011
Anti-neurofilament	Neurological disorders and injuries	Nerve axon	Structural	Serum, CSF	Abou-Donia et al., 2013; Fialova et al., 2013a; Fialova et al., 2013b; Jones et al., 2014
Anti-neuronal antibody	Cerebellar damage, epilepsy	NA	NA	Serum	Darnell et al., 1991; Ekizoglu et al., 2014
Anti-myelin basic protein	Neuropathy, MS, and CNS injury	Myelin of nerve axon	Tissue antigen	Serum	Hedegaard et al., 2009
Anti-GFAP	AD, MS, and CNS injury	Glial cells	Structural	Serum	El-Fawal and McCain, 2008; Storoni et al., 2011; Zhang et al., 2014
Anti-nerve growth factor	Lupus, arthritis, sensorineural loss	Neuronal milieu	Neuronal maintenance	Serum	Dicou et al., 1993; Zolotova and Grebeniuk, 2010
Anti-alpha internexin	Neuropsychiatric SLE	Nerve axon	Structural	Serum, CSF	Lu et al., 2010
Anti-NMDA	Myelitis, encephalitis	Transmembrane receptor	Signal conduction	Serum, CSF	Gresa-Arribas et al., 2014; Outteryck et al., 2013; Thomas et al., 2014
Anti-acetyl choline receptor	Schizophrenia, myasthenia gravis, CNS injury	Transmembrane receptor	Signal conduction	Serum	Jones et al., 2014; Sorokina et al., 2011; Wang et al., 2014
Anti-desmin	ALS	Nerve axon	Structural	Serum, CSF	Niebroj-Dobosz et al., 2006
Anti-actin	Systemic autoimmune diseases	Cytoskeleton	Structural	Serum	Porcelli et al., 2013; Zachou et al., 2012
Anti-vimentin	Neurologic ischemia and systemic autoimmune diseases	Cytoskeleton	Structural	Serum	Joachim et al., 2011; Zahran et al., 2013
Anti-MOG	MS, dementia	Oligodendrocyte and myelin sheath	Tissue antigen	Serum, CSF	Maetzler et al., 2011; Tsuburaya et al., 2014; Wang et al., 2008
Anti-MAG	MS, autoimmune polyneuropathy	Myelinating cells, myelin internode	Tissue antigen	Serum, CSF	Baig et al., 1991; Sotgiu et al., 2009
Anti-spectrin	Neurodegeneration, autoimmune disease	Cytoskeleton	Structural	Serum, CSF, saliva	Fernandez-Shaw et al., 1997; Moody et al., 2001
Anti-GABA receptor	Autism, encephalitis, neuropsychiatric SLE	Transmembrane receptor	Signal conduction	Serum, CSF	Mundiyanapurath et al., 2013; Tsuchiya et al., 2014; Wills et al., 2011a
Anti-glutamate receptor	Autoimmune diseases	Transmembrane receptor	Signal conduction	Serum, CSF	Murakami et al., 2013; Trippe et al., 2014

on neuronal synapse development (Stephan et al., 2012; Veerhuis et al., 2011), whereas cytokines, which are proinflammatory in peripheral organs, regulate synaptic activity in the CNS (Pavlowsky et al., 2010; Piton et al., 2008; Stellwagen and Malenka, 2006). Similarly, microglia, the CNS-resident myeloid cells, are characterized by their resting state with a limited capacity for antigen presentation and phagocytosis (Prinz et al., 2011).

The concept of BBB and its selective permeability plays a major role in biomarker research under normal circumstances and pathological conditions (CNS trauma). The BBB is composed of endothelial cells and astroglial cellular barriers with tight junctions. These barriers are composed of the vascular endothelial cells and the tight junctions that effectively orchestrate the traffic of brain chemicals. The selective permeability exerts regulated control in excluding most macromolecules from entering the brain and prevents neurotransmitters from leaking to the circulation. Most of the CNS (spinal cord and brain)-derived proteins are relatively sequestered in their localization due to the BBB control (Reiber, 2001, 2003). However, when the BBB is compromised, these macromolecules of proteins and their breakdown products are released in the CSF and may leak into the peripheral circulation comprising attractive candidates as signature biomarkers indicative of injury modalities (see Chapter 20).

The concept of the immune privilege of the CNS refers strictly to the parenchyma, which has limited interaction with blood vessels, meninges, or ventricles and is not localized within a circumventricular organ (Galea et al., 2007). The CNS parenchyma either expresses many antigens that are not found to an appreciable degree in other tissues or expresses a given antigen in a form not present in other tissues. The presence of CNS-reactive antibodies in circulation is, in general, harmless to the host; however, under pathologic conditions, when the BBB integrity is compromised, one would expect antibody-mediated pathology to occur (Diamond et al., 2013). Once they encounter their cognate antigens, B cells differentiate via antigen-specific T-cell interaction into a germinal center response in which immunoglobulin genes undergo class-switch recombination and somatic hypermutation. Under regular conditions, the BBB selectivity is permeable to some systemic circulating antibodies (Bard et al., 2000). However, in some cases of BBB disruption, as in brain trauma, a high influx of circulating antibodies and other serum proteins, leukocytes, macrophages, and red blood cells gain access to the brain vicinity and to injured lesions (Baloyannis and Gonatas, 1979; Soares et al., 1995). Thus, it is expected that neural cells (neurons and glial cells) will be exposed to circulating antibodies and serum proteins after brain trauma (Aihara et al., 1994).

Among the major mechanisms proposed that might explain the self-antigen–autoantibody response is the molecular mimicry mechanism. Molecular mimicry usually involves a foreign antigen mimicking a self-antigen and mounting an aberrant autoimmune response against self-antigens because of cross-reactivity. Under normal conditions, a low level of autoimmunity is necessary for normal function (Davidson and Diamond, 2001). Substantial evidence highlights the role of free radical generation and nonenzymatic lipid peroxidation as a posttranslational protein modification (Keller and Mattson, 1998). This covalent modification can alter the immunogenicity of these proteins (Uchida and Stadtman, 1992; Yahya et al., 1996). It has been shown that autoantibodies recognize proteins cross-linked to lipid peroxidation that have been identified in normal mice and humans (Kergonou et al., 1988; Yahya et al., 1996). As such, lipid peroxidation in injured neural cells may elicit circulating immunoglobulins that can recognize these proteins. In mounting immune response, the primary antibody secretion exhibits a lag phase of 2 to 6 days for the immunoglobulin G (IgG) secretion after antigen occurrence (Dudley, 1992; Poletaev and Boura, 2011), that is the time needed for antigen presentation and clonal expansion and generation of B cells.

29.3 AUTOANTIBODIES GENESIS AND MECHANISMS IN THE CNS: AGING PROCESS

The concept of autoantibodies can be traced back to the early 1970s (Ingram et al., 1974). In a very interesting report, Ingram et al. discussed the role of autoantibodies in aging. They proposed that, with advancing age, there is a natural increase in the prevalence of autoantibodies against neurons of the brain. The earliest site of binding for these autoantibodies is localized in the occipital lobe (visual cortex). However, the binding sites can vary between individuals (Ingram et al., 1974). For the origin of these autoantibodies and their genesis, they debated that our CNS constituents are naturally not antigenic but may be converted to be immunogens by a number of mechanisms including (1) release of brain debris into circulation from normally sequestered immunoprivileged sites, (2) formation of complexes with other substances such as viruses or environmental components, (3) alteration of configuration (Behan et al., 1973), or via (4) exposure of the neural proteins to antibodies as a result of cellular injury or death. Concerning the latter, the "free solution form" theory has been proposed to occur after necrotic cell death where autosensitization is coupled with the release of autoantigens from intact organs into peripheral circulation.

This is observed in several conditions as in stroke or TBI where a massive cell death with necrotic mechanisms followed by apoptotic cell death sequelae is observed. For the first option, it has been proposed that aging induces BBB disruption (Nandy, 1972; Threatt et al., 1971), coupled with the exposure of immunological privileged sites (Levy et al., 1972). Regarding formation of complexes with a foreign microorganism, it involves the presence of common antigens with the neural components leading to the development of autoantibodies capable of interacting with both entities. Of interest, there is another possibility of cross-reactivity arising from the fact that the CNS appears to possess normal self-antigens coexisting and shared with other organs (e.g., the thymus tissue) (Murphy et al., 1970). Such a condition is known as "a progressive qualitative or quantitative change" in the neural tissue. As to the alteration of configuration, cytoplasmic structural changes such as the development of neurofibrillary tangles are suggestive of misfolded proteins in addition to other modifications (Terry and Wisnewski, 1972). Interestingly, these studies go back to the early 1970s and still hold today and offer some explanation to what is observed in SCI and TBI records, as will be discussed later.

In cases of aging and senile dementia conditions, the brain is characterized by cellular loss and neuronal damage sustained over long periods and is associated with astrogliosis and high levels of gliofibrillar acid protein (GFAP) in the CNS (Bjorklund et al., 1985). GFAP is a major intermediate filament protein in the CNS, and its levels are elevated in cases of inflammatory response of reactive astrocytes (Eng et al., 1971). It has been shown to be a biomarker of cerebral infarctions, head injuries, and other conditions associated with reactive gliosis (Hayakawa et al., 1979; Mori et al., 1978).

In their work, Mecocci et al. evaluated serum GFAP autoantibodies in patients (n = 108) divided into vascular dementia, presenile AD, and senile AD and were compared with controls (Mecocci et al., 1992). Autoantibodies against GFAP were found elevated in vascular dementia, which is correlated with an increase in BBB permeability with no prognostic value to AD (Mecocci et al., 1992). Vascular dementia, compared with other types of dementia, allowed immune competent cells to gain access to the CNS antigens. These autoantibodies can be considered a secondary phenomenon to BBB disruption in which the exact role of anti-GFAP antibodies in the pathogenesis of AD is not clear. In addition, the presence of altered immune response in the case of degenerative dementia such as in AD; autoantibodies against cholinergic neurons, astrocytes, neurofilaments, or GFAP proteins have been identified in degenerative dementia with potential diagnostic importance (Bahmanyar et al., 1983; Kingsley et al., 1988; Mcraedegueurce et al., 1987; Tanaka et al., 1988, 1989).

29.4 NEUROPROTECTIVE ROLE OF AUTOANTIBODIES

The neuroprotective role of natural autoreactive monoclonal antibodies has been discussed by Wright et al. (2009). They proposed a new role of natural IgM autoantibodies that can bind to the surface of neurons stimulating neurite outgrowth and inhibiting neuronal apoptosis; this has a therapeutic potential in CNS neurodegenerating diseases such as amyotrophic lateral sclerosis, stroke, and SCI (Wright et al., 2009). The binding of these autoantibodies would activate intracellular signaling promoting glial and neuronal survival, which may involve crossing the BBB and accumulating at injured sights (Howe et al., 2004; Paz Soldan et al., 2003; Pirko et al., 2004). This signaling pathway involves antibody–protein glycolipid interactions leading to a slight calcium influx activating mitogen-activated protein (MAP) kinases inducing a cascade that downregulates caspase-3 activation.

29.5 AUTOANTIBODIES IN TBI

Among the dominant pathologic features of TBI is the occurrence of primary mechanical shear stress and secondary mediated cell death mechanisms involving the hyperactivation of the caspase and the calpain proteases (Arbour, 2013; Wang, 2000; Xing et al., 2009a, 2009b; Zheng et al., 2008). Furthermore, in the pathophysiology of TBI, there is major compromise in the BBB integrity that contributes to the release of neural proteins into the CSF along with the breakdown products (BDPs) representing protease substrates that may leak into the peripheral circulation. The proteins include UCH-L1, MAP-2, NF-H, αII-spectrin BDPs, and synaptotagmin, which have been proposed as putative biomarkers (discussed fully in Chapters 20, 21, and 22). Recently, the concept of autoimmunity has been revisited in the area of neurotrauma, a discipline that is not an autoimmune disease per se, but involves the production of autoantibodies. The role of these antibodies has been debated as neuroprotective or correlated to pathogenic conditions as discussed in the following section. These autoantibodies have been "newly" exploited in the field of neurotrauma biomarker research. Recently, two main TBI studies have been published by Zhang et al. and Marchi et al. that reflect the true utility of autoantibodies as TBI biomarkers (Marchi et al., 2013b; Zhang et al., 2014).

In one recent study on football players, Marchi et al. tested the hypothesis whether BBB disruption is coupled with increased influx of S100β astrocytic protein in blood, which may lead to the production of autoantibodies (Marchi et al., 2013b). Diffusion tensor imaging was performed to see if these events result in white matter disruption. A total of 67 football players were assessed before and after games and compared with the number of repeated head hits. Moreover, levels of S100β were evaluated and correlated to the diffusion tensor imaging scan data (pre- and postseason) that were performed for a 6-month interval coupled with cognitive and functional assessment. Serum levels of S100β correlated with the magnetic resonance imaging/diffusion tensor imaging findings showed abnormal cognitive changes. Elevated levels of autoantibodies against S100β were correlated to repeated subconcussive events characterized by BBB disruption.

Complementary to this work, a recent study by Zhang et al. evaluated the role of systemic autoantibodies in TBI (Zhang et al., 2014). A systematic analysis of human TBI serum was performed to identify serum autoantibody responses to brain-specific proteins. It was found that human autoantibodies showed prominent immune reactivity to a cluster of proteins in the region of 38–50 kDa identified as GFAP and GFAP BDPs, which increased at 7–10 days postinjury of the IgG subtype. Interestingly, these results were translated into an experimental model of rat TBI showing that human TBI autoantibodies colocalized in injured rat brain and in primary rat astrocytes, which is suggestive that these autoantibodies enter living astroglial cells compromising their survival. Among the major findings from this work is that in vitro digestion indicated that calpain was responsible for fragmenting GFAP protein yielding a 38-kDa BDP fragment. On the clinical level, a global neuroproteomics approach was used on 53 patients with severe TBI and age-matched healthy controls. Sera of subjects were collected on days 0–10 postinjury and screened against control brain tissue. TBI patients showed an average of 3.77-fold increase in anti-GFAP autoantibody levels from early (0–1 days) to late (7–10 days) postinjury. The presence of these autoantibodies was inversely correlated to Glasgow Coma Scale after a 6-month follow-up, suggesting that TBI patients with greater anti-GFAP immune responses had worse outcomes. Taken together, these data suggest that anti-GFAP autoantibodies represent excellent putative markers that can be used to monitor and assess brain injury in chronic conditions of human TBI.

In another study by Ngankam et al., CSF of 100 TBI patients were assessed for myelin-basic protein (MBP) and phospholipids (PL) autoantibodies on the first, 10th, and 21st days postinjury. Interestingly, autoantibodies against MBP and PL were elevated in TBI groups, whereas MBP autoantibodies were shown to correlate with Glasgow Coma Scale in the first days and the level of recovery on the 21st day. PL autoantibodies were correlated to the severity of vascular complications of trauma (Ngankam et al., 2011).

In another study by the Sorokina et al., autoantibodies against different fragments of α7-subunit of acetylcholine receptor were evaluated in children with craniocerebral trauma with different severities. Brain injury severity correlated with the titers of α7-subunit and the autoantibody levels mounted against them during the first week of injury (Sorokina et al., 2011). They hypothesized that inflammatory response coupled with disruption of the BBB increase the release of structural neuronal and glial proteins components in the circulation, which led to the mounting of immune response (autoantibodies) against them. These autoantibodies can hold compensatory role by binding to the damaged receptors providing compensatory-adaptive response (Sorokina et al., 2009). Other autoantibodies can contribute to the pathologic profile by increasing edema, inflammation, and calcium influx, increasing the secretion of proinflammatory cytokines (Sorokina et al., 2006; Whitney and McNamara, 2000). On the compensatory role of α7-subunit, signal transduction of the α7-subunit of AChR is known to support neuronal plasticity (Berg and Conroy, 2002; Drisdel and Green, 2000). In addition, it is suggested that these receptors carry protective roles in suppressing the secretion of the proinflammatory tumor necrosis factor α (TNF-α) mediated via cholinergic pathways, the vagus nerve, and the spleen leading to macrophage activation (source of TNF-α) (Borovikova et al., 2000). Thus, the α7-types of ACHR have been proposed to be evaluated for their diagnostic value in assessing severity of brain injury.

In another study, Sorokina et al. assessed levels of S100β and its cognate autoantibodies in the serum of children with different severity and outcomes of TBI (1–75 days) in chronic phases (15–75 days) after TBI was stratified into five groups (complete recovery, moderate disability, high disability, vegetative state, and fatal outcome). Glasgow Coma Scale was used to assess severity. The maximal level of S100β protein and its autoantibody were observed in patients with fatal outcome. In groups 1–3, the changes of S100β in the blood serum did not depend on the severity of brain damage. The S100β protein levels went up the first day and then declined, whereas S100β autoantibodies were elevated at days 3 and 5 (Sorokina et al., 2010).

In another study by Kamchatnov et al., neuron-specific enolase, GFAP, S100β proteins, and their associated autoantibodies were evaluated in 42 patients with acute ischemic stroke. A reverse correlation between concentrations of neuron-specific enolase and its autoantibodies was observed. Similarly, there was a positive correlation between the restoration of lost functions (from the first to 21st days) and GFAP autoantibody levels (Kamchatnov et al., 2009). In another study by Pinelis et al., acute craniocerebral trauma period is characterized by a marked change in the levels of nitric oxide metabolites and antibodies to two subtypes of glutamate receptor, the AMPA and NMDA receptors. Unfavorable outcome of craniocerebral trauma was associated with the lowest level of autoantibodies and high nitric oxide levels. It was suggested that antibodies to glutamate receptors and receptor hyperstimulation play an important role in the pathogenesis of hypoxia (Pinelis and Sorokina, 2008).

In an elegant work by Stein et al., it was clearly demonstrated that circulating IgG autoantibodies bind to dying neurons in the time frame of 4 hours to 7 days after visual cortex lesion (Stein et al., 2002). They proposed that this autoantibody binding may participate in the phagocytosis and clearance of injured neurons; IgG-positive neurons were in advanced stages of degeneration. The major finding their work is that injured neurons are associated only with the IgG class immunoglobulins that have entered the cortical lesion binding specifically to the injured cells. These autoantibodies are naturally occurring autoantibodies because of the time frame of detection (4 hours). They may be involved in microglial-dependent opsonization initiating phagocytosis because of the Fc receptors that enable binding of IgG opsonized targets (Ulvestad et al., 1994). Finally, in one study by Rudehill et al., serum autoantibodies against neurons, basal lamina, and astrocytic cells were assessed after experimental brain contusion (Rudehill et al., 2006). These autoantibodies were elevated in injured rats at different time points (antineuronal IgG and antivascular basal lamina IgG were detected at 2 weeks and 3 months postinjury, respectively). These autoantibodies were detected in the rat serum but not as tissue bound, 2 weeks postinjury. This is suggestive that the BBB reformation prevented the chronic passage of these autoantibodies to the brain, which is supported by Stein et al.'s findings that IgG bound to injured neurons 4–7 days postinjury but not a 14-day period. The presence of these IgG antibodies suggests a potential pathological role to neural integrity (Rudehill et al., 2006).

Finally, among the devastating outcomes observed in TBI is that it may result in pituitary dysfunction; the exact mechanisms for such an outcome have not been elucidated (Kelly et al., 2000; Tanriverdi et al., 2006). Several mechanisms have been proposed, including direct mechanical injury to the hypothalamus, compression from hemorrhage, edema, or increased intracranial pressure along with an autoimmune response (Tanriverdi et al., 2008b; Yuan and Wade, 1991). In one study by Tanriverdi, antipituitary autoantibodies (APA) post-TBI and pituitary functions were assessed in TBI patients 3 years after TBI (29 TBI patients vs. 60 controls) (Tanriverdi et al., 2008a). Of interest, there was a significant association between APA positivity and hypopituitarism from TBI. APAs were detected in 13 of the 29 TBI patients but none in control. Similarly, pituitary dysfunction development ratio was significantly higher in APA-positive patients (46.2%) compared with APA-negative patients (12.5%). A follow-up study by the same group was performed on amateur boxers assessing anti-hypothalamus (AHA) and APA antibodies in chronic repetitive head trauma (Tanriverdi et al., 2010). Sixty-one actively competing (n = 44) and retired (n = 17) male boxers were assessed for AHA and APA and compared with 60 normal controls. AHAs and APAs were detected in 13 and in 14 of 61 boxers, respectively. Pituitary dysfunction was significantly higher in AHA-positive boxers than in AHA-negative boxers. These results demonstrate the presence of AHAs and APAs in boxers with repeated head injuries; however, the functional value of these autoantibodies requires further studying to establish a causal relation with the injury.

29.6 SCI, B LYMPHOCYTES, AND THE PERPLEXING ROLE OF AUTOANTIBODIES

Among the hallmarks of CNS trauma in the context of SCI and TBI is the activation of several interrelated pathologic responses involving the necrotic, excitotoxic, and inflammatory responses in the primary phase followed by apoptotic cell death in the secondary phase; hence contributing to the worsening of the neurological outcomes. The neuroinflammatory response observed, though it is part of the natural wound-healing process, contributes to the increased damage to the neural tissue specifically in the area of SCI (Dekaban and Thawer, 2009).

Neuroinflammation cascade is a key mechanism occurring post-SCI involving the activation of antibody-producing B cells, in addition to the already well-studied activated neutrophils, monocytes/macrophages, and T lymphocytes (Blight, 1993, 1994; Fleming et al., 2006; Kigerl et al., 2006; Popovich et al., 1999; Sroga et al., 2003). Ankeny et al. discuss elegantly the impact of B-cell activation and its associated autoantibodies in the realm of SCI pathology genesis and exacerbation offering new venues for neurotherapeutic targets for patients with SCI (Ankeny et al., 2006, 2009; Ankeny and Popovich, 2009, 2010). B lymphocytes arise from bone marrow hematopoietic stem cells as immature B-cell stage cells (Dalakas, 2008a, 2008b).

Upon the entry of a "non-self" foreign antigen, our system mounts a host immune response in which mature B cells coupled with T-cell are activated. However, when the encountered antigen is host-derived (DNA, peptide, or protein), a state of autoimmune response is elicited (Ankeny et al., 2009). Usually, during development, negative selection would eliminate highly reactive lymphocytes while positive selection would keep "subthreshold" stimulation of lymphocytes that recognize self-host antigens and increase sensitivity to foreign antigens (Stefanova et al., 2002). This mechanism of positive selection plays a major role in regulating the immune response. Nevertheless, when the threshold level is bypassed, we have a pathologic state of autoimmunity (Stefanova et al., 2002). In the case of CNS trauma, T-dependent and T-independent self-antigens induce an adaptive immune response that has several implications (Ankeny et al., 2006, 2009; Schwartz and Kipnis, 2001).

Upon activation in the presence of its cognate antigens, B cells differentiate into antibody-secreting plasma cells and later into long-lived, antibody-secreting plasma cells (Dalakas, 2008a, 2008b). In addition to secreting antibody, B cells can act as antigen-presenting cells (Dalakas, 2008b; Waubant, 2008). These activated B cells migrate to the secondary lymphoid tissue, bone marrow, and to the CNS (Dalakas, 2008b). The CNS recruitment and B-cell traffic is upregulated in CNS autoimmune disorders as in multiple sclerosis (Corcione et al., 2004b; Dalakas, 2008b).

Activated B lymphocytes and autoantibodies have been shown to be the main players in a number of neurological disorders including systemic lupus erythematosus (SLE), MS, and experimental autoimmune encephalomyelitis (EAE) (Genain et al., 1999; Kowal et al., 2004; Raine et al., 1999). In SLE, antinuclear and anti-DNA antibodies cross-react with neural antigens in the brain, causing excitotoxicity coupled with cognitive decline (DeGiorgio et al., 2001). Similarly, in MS, levels of myelin-reactive antibodies are elevated in CSF and are associated with progressive demyelination and neurological dysfunction (Lyons et al., 1999; Willenborg and Prowse, 1983), whereas in EAE, B-cell activation and autoantibody synthesis are the main triggers to cause demyelination and neuropathology. The role of autoantibodies involves cytotoxic effects on neurons and glia via complement activation and phagocytosis stimulation with cytokine release and protease activation from microglia and macrophages (Abdul-Majid et al., 2002; Beuche and Friede, 1986; Griot-Wenk et al., 1991; Mosley and Cuzner, 1996). In their work, Ankeny et al. show that experimental spinal contusion injury elicits chronic systemic and intraspinal B-cell activation. Immunoblots of sera from injured mice showed reactivity against multiple CNS proteins, including autoantibodies that can bind nuclear antigens (DNA and RNA), as observed in EAE. These autoantibodies show similar neurotoxic potentials via cross-reacting with glutamine receptors, causing neuronal excitotoxicity (Ankeny et al., 2006). They propose that the mechanisms leading to B-cell activation involve SCI-dependent stimulation of cognate B and T cells in response to autoantigens liberated by SCI.

A major finding in human SCI is the elevation of myelin-reactive antibodies in serum and CSF that recognize CNS proteins as autoantigens (Hayes et al., 2002; Kil et al., 1999; Mizrachi et al., 1983). By inducing experimental SCI, long-lasting B-cell activation was observed in both spleen and bone marrow with increased levels of IgG and IgM (Ankeny et al., 2006). Activated B lymphocytes are shown to dwell in the injured spinal cord associated with de novo expression of mRNA that encodes a range of autoantibodies (Ankeny and Popovich, 2009).

29.7 PATHOGENIC ROLE OF SCI AUTOANTIBODIES AND SCI LEVEL DEPENDENCE

Ankeny showed that in a mouse model of moderate severity spinal cord contusion at the mid-thoracic level (T9), there is an induction of a pathologic B-cell response coupled with pathogenic antibody secretion (Ankeny et al., 2009). Functional recovery testing via coordinated stepping of the four limbs applied on SCI mice was demonstrated in 88% of B-cell knockout compared with 35% in wild-type (WT) mice after a 9-week period. Similarly, the neuropathology was less pronounced in the B-cell knockout SCI mice. The pathogenic antibodies from the SCI mice injected in WT mice induced a similar type of neurotoxicity as in WT SCI mice. Interestingly, the pathogenic neurotoxic phenotype observed in SCI was achieved if the injury occurred in the lower half of the spinal cord (T9–T10), whereas at higher levels (T4–T5), a profound immune suppression and diminished B-cell activation was observed (Lucin et al., 2007). This is attributed to the cholinergic anti-inflammatory pathway disruption at high SCI due to the removal of the sympathetic contribution as it injures the intermediolateral column sympathetic fibers that regulate systemic inflammation (Rosas-Ballina and Tracey, 2009). In a further study, Ankeny et al. highlighted the integral role of B cell and autoantibody secretion in mediating axonal and myelin pathology and motor function impairment. Antibody-mediated pathology involved complement activation Fc-receptors bearing cells (e.g., microglia/macrophages) in the spinal cord (Ankeny et al., 2009). It was shown that in mice with normal B-cell function, large deposits of antibody and complement component 1q (C1q) accumulated at sites of axon pathology and demyelination. Of high interest, mice that were B cell–deficient were incapable of antibody production, and lesion pathology was reduced coupled with spontaneous recovery of locomotor function. Furthermore, injection of purified antibodies from SCI mice into naive/uninjured spinal cord led to paralysis and spinal cord pathology (Ankeny et al., 2009). The profile observed by injecting these pathogenic autoantibodies mimicked what is observed in models of muscle and gut ischemia/reperfusion injury, which are complement C1q- and FcRs-dependent (Zhang et al., 2006, 2008). Taken together, these results suggest that controlling B-cell activation and/or plasmapheresis would be among the therapeutic options considered in treating SCI.

SCI initiates a cascade of altered immunological responses from the primary injury phase through the secondary injury phase and it even passes through the chronic rehabilitative phase (Cruse et al., 2000; Kliesch et al., 1996; Yang et al., 2004). On the other hand, the immunological response of patients in the postacute (2–52 weeks) or chronic (>52 weeks) stages is not well-characterized (Davies et al., 2007). It is proposed that due to the disrupted sympathetic innervations of the lymphoid tissue coupled with altered neuroendocrine responses and the dysregulation of the afferent input to immunoregulatory neurons would create a state of immune suppression observed in SCI (Iversen et al., 2000). It has been reported that within 2 weeks, there was a decrease in natural and adaptive immune responses, in which the function of both natural killer cells and T cells dropped significantly. In addition to that, the hematopoietic cell lineages including dendritic cells of the decentralized bone marrow were also shown to be reduced (Iversen et al., 2000).

Taken together, the level and severity of SCI influence the kinetics and magnitude of B-cell activation and autoantibody synthesis. Finally, it is hypothesized that the accumulated intraspinal autoantibodies contribute to pathology via activating complement binding to antigen–antibody immune complex; this in turn activates other complement proteins. This will result in activation and recruitment of myeloid lineage cells (e.g., microglia/macrophages) bearing complement receptors.

In one study by Davies et al., inflammatory serum cytokines were assessed in SCI subjects (n = 56) with varying clinical presentations compared with control subjects. Several end points were assessed including levels of the proinflammatory cytokines (interleukin [IL]-1, IL-6, TNF-α), the anti-inflammatory cytokines (IL-4 and IL-10), the regulatory cytokines (IL-2), the IL-1 receptor antagonist (IL-1RA), and autoantibodies against myelin-associated glycoprotein, and GM1 ganglioside (anti-GM1) immunoglobulin (IgG and IgM) (Davies et al., 2007). Furthermore, findings from this work showed that there is an elevation in circulating proinflammatory cytokines and autoantibodies. However, these were present in SCI subjects with or without complications (e.g., pain, ulcers, pressure). These findings may be indicative of a protective autoimmunity or may be due to evident infection.

29.8 B-LYMPHOCYTES AS THERAPEUTIC TARGETS POST-SCI

During B-cell activation, maturation, and antibody secretion process, a number of factors and receptors that can be therapeutic targets in the activated immune response in SCI (Dalakas, 2008a, 2008b) have been identified. Among these, B cell–activating factor (BAFF), lymphotoxin-β, and "a proliferation-inducing ligand" (APRIL) contribute to B-cell survival, maturation, and activation (Dalakas, 2008a, 2008b). Moreover, it is observed that B-cell "follicle-like" structures (as detected in MS) (Corcione et al., 2004a) are present in the lesion area. This reflects that these pathogenic antibodies are in part derived from the lesion and in part systemically introduced due to disruption of the blood–spinal cord barrier (Ankeny et al., 2009). The "follicle-like" structures are partly formed by the BAFF and APRIL factors derived from the infiltrating monocyte/macrophages and proliferating astrocytes. Based on these results, the pathogenic parameters discussed, including the activated B cells and the secreted autoantibodies, render them potential targets for therapy in SCI (Dekaban and Thawer, 2009).

Intravenous immunoglobulin administration therapy has shown to block both autoantibody binding and complement activation, which would be an optimal neuroprotective target in SCI (Gold et al., 2007). Another alternative would be using B-cell depletion as a neuroprotective option, which has been tried as a therapy in a number of CNS autoimmune conditions with the ability to remove activated B cells (Matsushita and Tedder, 2009; Waubant, 2008). Alternatively, the use of anti-CD20 antibody–mediated depletion of B cells can be applied; however, preexisting plasma cells with pathogenic antibodies will not be eliminated by anti-CD20 therapy, which raises some concerns (Dalakas, 2008c; Waubant, 2008). Other options involve the use of drugs that can target factors involved in B cell growth/differentiation (e.g., BAFF, APRIL), limiting the accumulation of pathogenic autoantibodies (Dorner et al., 2009; Levesque and St Clair, 2008). This can be delivered within the intraspinal or intracranial space that can target B cell follicles in the chronically injured CNS (Ankeny et al., 2006, 2009; Ankeny and Popovich, 2009; Katz, 1980).

29.9 AUTOANTIBODIES IN AUTISM SPECTRUM DISORDER

ASDs are a group of heterogeneous neurodevelopmental disorders affecting approximately one in 88 children in the United States (Braunschweig and Van de Water, 2012; Lord et al., 1997). Current diagnosis is based entirely on behavioral testing and the analysis of medical and developmental history (LeCouteur et al., 2008; Lord et al., 2000a; Lord et al., 2000b; Lord et al., 1994). ASD is characterized by core deficits in social interaction and stereotypical movements (Lord et al., 1997). The pathology and etiology of these disorders remain unclear with some implications of genetic, neurological, and environmental factors (Pardo et al., 2005). In addition, ASD has been implicated with altered immune response, which is an important factor contributing to the development of some cases of ASD. These immune response alterations include immune dysfunction (van Gent et al., 1997; Warren et al., 1987, 1996), peripheral immune abnormalities (Ashwood and Van de Water, 2004b; Ashwood et al., 2006), and an ongoing inflammatory response (Croonenberghs et al., 2002; Vargas et al., 2005). Furthermore, there is evidence showing that autism may be due to an ongoing autoimmune process (Ashwood and Van de Water, 2004a; Braunschweig et al., 2013). Several studies have already identified a number of autoantibodies in individuals with autism mounted against various CNS proteins, including glial and neuron-axon filament proteins (Kirkman et al., 2008), MBP (Singh et al., 1993), serotonin receptor (Todd et al., 1988), cerebellar peptides (Vojdani et al., 2004), brain-derived neurotrophic factor (Connolly et al., 2006), and brain endothelial cells (Connolly et al., 1999). Several ongoing studies are under way to describe the identification of several of brain-specific autoantibodies and their relation to ASD development and severity. However, the debate of whether or not these autoantibodies represent putative markers of ASD is still under investigation and most of the described studies that follow are correlative in nature, with some hints of biomarker implication as discussed.

Dysregulation in immune response as well as neuroinflammation coupled with the presence of maternal autoantibodies mounted against brain tissue support the role of the immune system in some of the ASD cases observed (Braunschweig et al., 2008, 2012; Goines and Van de Water, 2010; Vargas et al., 2005). It has been shown that in mothers of ASD children, there is immunoreactivity to proteins of 37- and 73-kDa antigens that were correlated to increased severity of language deficits in the offspring (Braunschweig et al., 2008, 2012). A potential role of these maternal autoantibodies is proposed because of the accessibility of maternal IgG during pregnancy. Maternal IgG is detected in circulation as early as 13 weeks' gestation. In addition, by 30 weeks' gestation, the levels of circulating autoantibodies reach 50% of that of the mother and at birth, the levels of these autoantibodies exceed the maternal IgG (Simister, 2003). In addition, the BBB is still in the developmental stages and permits IgG during this period (Bake et al., 2009; Malek et al., 1996). It has been shown that rodents and nonhuman primates exhibit ASD-like behaviors in offspring born to dams exposed to passive transfusion of human IgG from mothers with brain-reactive autoantibodies compared with control serum (Dalton et al., 2003; Martin et al., 2008; Singer et al., 2009). These maternal autism-related autoantibodies were investigated by Braunschweig et al. through proteomics approaches (Braunschweig et al., 2013). Reactivity to specific antigen combinations was noted in 23% of mothers with ASD children compared with 1% in controls. These seven autoantibodies included lactate dehydrogenase A and B, cypin, stress-induced phosphoprotein 1, collapsin response mediator proteins 1 and 2, and Y-box-binding protein, which compose what is known as maternal autoantibody-related (MAR) autism. ASD children from mothers with specific reactivity to these MAR autoantibodies had elevated stereotypical behaviors compared with siblings with no MAR occurrence. The identification of these specific significant biomarkers assists in understanding the etiologic mechanisms and therapeutic potentials for MAR autism. Thus, identifying these MAR autoantibodies provides one potential understanding for the development of the traits observed in the ASD cases. Moreover, the high specificity of the MAR autoantibody profiles may constitute the first biomarker panel predicting ASD risk.

A study by Goines et al. explored the relationship between the presence of brain-specific autoantibodies and several behavioral characteristics of autism in 227 ASD children. It was found that autoantibodies that target pairs of fetal brain proteins at 37/73 kDa and 39/73 kDa are directed against cerebellar proteins. The autoantibodies specific for a 45-kDa cerebellar protein in children were associated with a diagnosis of autism, whereas autoantibodies directed toward a 62-kDa protein were associated with diagnosis of broader diagnosis of ASD. Children with elevated levels of autoantibodies showed lower adaptive and cognitive functions relative to children without the autoantibodies in sera (Goines et al., 2011). Another implication of the contribution of autoantibodies to ASD development comes from maternal IgG experiments. During pregnancy, maternal IgG is passed across the placenta to fetal circulation (Simister, 2003). It was shown that animals exposed to these fetal brain-directed autoantibodies exhibited altered ASD behavior, thus indicative that maternal autoantibodies may be of pathologic significance contributing to the occurrence of ASD (Martin et al., 2008; Singer et al., 2006).

Wills et al. showed that plasma from children with ASD demonstrated immunoreactivity directed against a 52-kDa human cerebellar protein in 21% of subjects with autism, whereas it was present in only 2% of the typically developing controls; these autoantibodies stained positive for Golgi cells of the cerebellum (Wills et al., 2009). This was validated using immunohistochemical staining of sections from Macaca fascicularis monkey cerebellum. Autism has been implicated in a number of brain regions in addition to the cerebellum (Amaral et al., 2008) and the question arises whether the autoantibodies present in autistic individuals identify a broader class of neurons that are distributed throughout the brain with distinct morphological features. Thus, in a different study, plasma from the same individuals who demonstrated positive cerebellar Golgi cell immunoreactivity were applied to stain tissue sections from the full rostrocaudal extent of the macaque monkey brain (Wills et al., 2011b). Significantly, it was found that children with ASD demonstrated autoantibodies are reactive to a subset of GABAergic interneurons throughout the brain. Colocalization studies with GABA antibodies revealed that ASD children recognize GABAergic interneurons in V1 layers. This study raises some major questions about the identity of the autoantigen that these autoantibodies recognize and the reason for a potential pathologic role of these autoantibodies or it is merely an epiphenomenon.

In another study by Mostafa et al., neurokinin-A, a proinflammatory neuropeptide and a main component in the neurogenic inflammatory pathway, was evaluated in 70 autistic children (Mostafa and Al-Ayadhi, 2011). It was shown that neurokinin-A titer was elevated and was correlated to the severity of the autistic phenotype. Similarly, antiribosomal P protein autoantibodies were also associated with the occurrence of autism in these children. Several possible pathogenic mechanisms have been proposed for these autoantibodies, including inducing cytotoxic effects, cellular invasion into living cells, and initiating apoptosis (Ben-Ami et al., 2010). Previously, antiribosomal P protein antibody occurrence had been associated with the neuropsychiatric manifestations for SLE involving psychosis, mood disorders, anxiety, and cognitive dysfunction (Ben-Ami et al., 2010; Mostafa et al., 2010). However, a clear mechanistic analysis of the formation of these specific autoantibodies was not described and the question of whether these autoantibodies can be considered a marker of ASD was not evaluated (Mostafa and Al-Ayadhi, 2011).

Finally, other studies have associated the presence of ASD with the presence of folate receptor autoantibodies and their relation to maternal folate receptor-α (FRα) autoantibodies (the blocking or the binding type) (Rothenberg et al., 2004). It has been shown that in ASD cases, either the blocking or the binding type of the FRα autoantibody was detected in 75% of the ASD children and high-dose folinic acid supplementation improved the core symptoms of autism in these children (Frye et al., 2013). Of interest, in a study by Ramaekers et al., ASD children were shown to develop these autoantibodies postnatally (Ramaekers et al., 2013). However, in ASD children with negative autoantibodies, either one or both parents had positivity for the autoantibodies, which may suggest that parental FRα autoantibodies may contribute to the development of autism in their offspring (Ramaekers et al., 2013).

29.10 AUTOANTIBODIES IN THE AREA OF NEUROTOXICITY

Another area of autoantibody investigation is the area of neurotoxicity, involving the exposure of neurotoxins that have been shown to exert an autoimmune response detected in circulation and found to be proportional to injury severity (Abou-Donia et al., 2013; El-Fawal and McCain, 2008; El-Fawal et al., 1999; Evans, 1995; Mason et al., 2013; Moneim et al., 1999). In his review, Evans discusses the role of neurotoxicity and autoantibody response and proposes that debris in damaged CNS cells from neurotoxins may present as novel neoantigens, giving rise to autoantibodies detected at chronic phases of injury (Evans, 1995). These autoantibodies can be considered noninvasive organ-localized markers of valuable importance to evaluate the underlying mechanisms involved (Evans, 1995). It has been shown that toxicant-induced neuronal injury might cause autoantibody formation (Lotti, 1995). These environmental agents can act as autoantigens with autoantibodies that can be measured as a biomarker of neurotoxicity and may provide evidence of neurotoxicity that has disappeared from the brain (Lotti, 1995). This is of very high importance. Several obstacles exist in surveying real-time biomarkers of neurotoxicity including the physical protection of the brain by the skull and the CSF cushion. These features prevent direct monitoring of marker changes and indirect measurement of neurotransmission (metabolites) as in cases of neurotoxicity lacking sensitivity, reproducibility, and reliability (Kanada et al., 1994; Lotti, 1995; Mailman and Lewis, 1987; Moretto et al., 1994)

Several schemes are available explaining the appearance of autoantibodies: (1) changes of cell-specific protein levels as in cases of astrogliosis associated with elevated GFAP (Brock and O'Callaghan, 1987); (2) protein fragmentation from injured cells that can leak into the CSF and then to the periphery, which involves disruption of the BBB and impairment of the scavenging properties of microglia (Bressler and Goldstein, 1991; Kida et al., 1993; Skouen et al., 1993; Sundstrom and Kalimo, 1987; Sundstrom et al., 1985); and (3) the early appearance and long survival of autoantibodies to these proteins might permit practical surveillance of exposure and toxicity levels (Vlajkovic and Jankovic, 1991). However, one must be cautious to make sure that there are low levels of autoantibodies in controls, which may reflect a long-past injury that needs to be normalized when assessing these autoantibodies as biomarkers (Evans, 1995).

In one study by Abou-Donia, a panel of CNS autoantibodies was assessed in the sera of 12 healthy controls and a group of 34 flight crew members who experienced adverse effects after exposure to air emissions containing gaseous, vapor, and particulate contaminants (Abou-Donia et al., 2013). These have adverse effects on the CNS dysfunction (Abou-Donia et al., 2013; Winder and Balouet, 2002). The autoantibodies selected represent several types of neural protein affected by neuronal degeneration. These included triplet proteins, tubulin, microtubule-associated tau proteins (tau), MAP-2, myelin basic protein (MBP), GFAP, and glial S100β protein. After exposure, damaged BBB would allow neuronal and glial proteins to leak from the CNS or damaged peripheral nervous system into the circulation and act as autoantigens that will react with B lymphocytes. These B lymphocytes would convert the short-lived nervous system–specific proteins in the serum into long-term biomarkers for neurological damage (Abou-Donia et al., 2013). Data from this study demonstrated that there is a temporal relationship between exposure to air emissions, clinical condition, and level of serum neural autoantibodies (Abou-Donia et al., 2013).

In another study by al El-Fawal et al., neuronal protein autoantibodies were assessed in organophosphorus-induced delayed neuropathy in a hen model treated with phenylsaligenin phosphate (PSP) (El-Fawal and McCain, 2008). Serum autoantibodies against neuronal cytoskeletal proteins (e.g., neurofilament triplet [NF]) and glial proteins MBP and GFAP proteins as biomarkers of neurotoxicity were assessed as markers of neurotoxicity. A subgroup of hens was treated with calcium channel blocker verapamil 4 days before PSP treatment to monitor its amelioration. IgG against all neural proteins were detected on days 7 and 21, with titer levels being significantly higher in sera of hens receiving PSP only. NF (Neurofilament light [NF-L], Neurofilament medium [NF-M], and Neurofilament heavy [NF-H]), MBP and GFAP, and anti-GFAP and anti-MBP were highest and correlated with clinical scores showing neuro-axonal degradation accompanied by myelin loss at days 7 and 21 as markers of neuropathy (El-Fawal and McCain, 2008). Thus, CNS autoantibodies monitoring may be used to assess neuropathogenesis from occupational exposures and assess treatment intervention options.

29.11 AUTOANTIBODIES IN THE AREA OF PARANEOPLASTIC SYNDROMES

Another area of autoantibody investigation is the paraneoplastic syndromes of the CNS, which involve neurological disorders associated with cancers and are independent of metastasis or tumor compression of the brain tissue (Posner and Dalmau, 1997a, 1997b; Younger et al., 1991). The role of autoantibodies can be traced to the 1970s in the areas of neuromuscular disorders, which later advanced to recognize specific onconeural antibodies in patients with paraneoplastic neurological disorders (Dalmau and Rosenfeld, 2008). These disorders involve autoimmune pathogenesis where they exhibit autoantibodies directed against antigens shared by the neural cells/components and the cancer tissue (Floyd et al., 1998). The hypothesis proposed is that these autoantibodies share protective roles aimed at stopping the spread of tumor to the brain because these cancerous cells share antigens of neural homology. However, these autoantibodies may lead to the impairment of the neurological functions and may injure neuronal cells, leading to their death. Interestingly, the onset of neurological dysfunction in the paraneoplastic syndromes often precedes the clinical manifestation of the tumor. Thus, detecting these autoantibodies may be of clinical use, leading to the early detection of an occult tumor (Folli et al., 1993). In an article by Vincent et al., they discussed the association of several CNS disorders with the occurrence of autoantibodies; these disorders are often associated with tumors and the use of autoantibody assays can help in their diagnosis (Vincent et al., 2011). These autoantibodies span a number of protein families, including voltage-gated potassium channel complexes, NMDA receptor, AMPA receptors, GABA type B receptors, and glycine receptors (Vincent et al., 2011).

In their work, Berghs et al. have identified autoantibodies against two isoforms of β-IV spectrin, which are enriched at axonal areas and nodes of Ranvier in a patient with paraneoplastic lower motor neuron symptoms (LMNS) and breast cancer (Berghs et al., 2000, 2001). An immune response toward these structural components would induce neuronal cell death causing functional impairment, and eventually the death of motoneurons. Other autoantibodies directed against a surface antigen(s) that is enriched at axon initial segments was identified and hypothesized to contribute to the pathogenesis of LMNS. Along the same line, removal of the breast tumor actually stopped the rapid progression of LMNS and was coupled with significant neurological improvement, concomitant with a decrease in the concentration of the autoantibody titers of β-IV spectrin and surface autoantigens. These autoantibodies may provide new diagnostic capabilities in the area of paraneoplastic neurological disorders.

29.12 CONCLUSION

As discussed, the area of autoantibodies has been recently investigated as new-generation biomarkers in neuropsychiatric disorders, CNS trauma, and neurotoxicity. The importance of detecting autoantibodies and their corresponding cognate proteins as potential biomarkers cannot be overstressed. The latter may serve as acute injury markers, whereas their corresponding autoantibodies represent long-lasting, chronic molecular signature biomarkers associated with advanced chronic stages of injury sequelae. Thus, the timely detection of these composite biomarkers may provide a theragnostic window that may help in diagnosing pathological progression of CNS trauma and help in follow-up prognosis (Sorokina et al., 2009).

Finally, at this stage, one should be cautious in drawing conclusions about the exact utility of autoantibodies, since there are a lot of justifiable questions that haven't been answered yet. Even upon identifying these autoantibodies, the exact roles of the elevated titers are being argued as naturally occurring epiphenomenon markers or actually representing culprit pathogenic contributors. In addition, it is also questionable whether autoantibodies are able or not to translate the pathologic state into the control subjects if transferred or this is more dependent on the neuropathological condition in question. Other valid queries include whether these autoantibodies trigger irreversible damage to the brain and whether their clearance is accompanied with a cessation of symptomatology. Finally, as demonstrated in SCI studies, the mechanism of how these autoantibodies initiate injury and at what titers or cut offs they induce pathology should be critically evaluated. These considerations must be thoroughly investigated in light of the heterogeneous profile of autoantibodies-related neurological disorders (Diamond et al., 2013).

ACKNOWLEDGMENT

We extend our thanks to Abeer Naser Eddine (PhD) for proofreading this chapter.

REFERENCES

Abdul-Majid, K.B., A. Stefferl, C. Bourquin, H. Lassmann, C. Linington, T. Olsson et al. 2002. Fc receptors are critical for autoimmune inflammatory damage to the central nervous system in experimental autoimmune encephalomyelitis. *Scandinavian Journal of Immunology*. 55:70–81.

Abou-Donia, M.B., M.M. Abou-Donia, E.M. ElMasry, J.A. Monro, and M.F. Mulder. 2013. Autoantibodies to nervous system-specific proteins are elevated in sera of flight crew members: Biomarkers for nervous system injury. *Journal of Toxicology and Environmental Health. Part A*. 76:363–380.

Aihara, N., H. Tanno, J.J. Hall, L.H. Pitts, and L.J. Noble. 1994. Immunocytochemical localization of immunoglobulins in the rat brain: Relationship to the blood-brain barrier. *The Journal of Comparative Neurology*. 342:481–496.

Amaral, D.G., C.M. Schumann, and C.W. Nordahl. 2008. Neuroanatomy of autism. *Trends in Neurosciences*. 31:137–145.

Ankeny, D.P., Z. Guan, and P.G. Popovich. 2009. B cells produce pathogenic antibodies and impair recovery after spinal cord injury in mice. *The Journal of Clinical Investigation*. 119:2990–2999.

Ankeny, D.P., K.M. Lucin, V.M. Sanders, V.M. McGaughy, and P.G. Popovich. 2006. Spinal cord injury triggers systemic autoimmunity: Evidence for chronic B lymphocyte activation and lupus-like autoantibody synthesis. *Journal of Neurochemistry*. 99:1073–1087.

Ankeny, D.P., and P.G. Popovich. 2009. Mechanisms and implications of adaptive immune responses after traumatic spinal cord injury. *Neuroscience*. 158:1112–1121.

Ankeny, D.P., and P.G. Popovich. 2010. B cells and autoantibodies: Complex roles in CNS injury. *Trends in Immunology*. 31:332–338.

Arbour, R.B. 2013. Traumatic brain injury pathophysiology, monitoring, and mechanism-based care. *Crit Care Nurs Clin*. 25:297–319.

Ashwood, P., and J. Van de Water. 2004a. Is autism an autoimmune disease? *Autoimmunity Reviews*. 3:557–562.

Ashwood, P., and J. Van de Water. 2004b. A review of autism and the immune response. *Clinical and Developmental Immunology*. 11:165–174.

Ashwood, P., S. Wills, and J. Van de Water. 2006. The immune response in autism: A new frontier for autism research. *Journal of Leukocyte Biology*. 80:1–15.

Bahmanyar, S., M.C. Moreau-Dubois, P. Brown, F. Cathala, and D.C. Gajdusek. 1983. Serum antibodies to neurofilament antigens in patients with neurological and other diseases and in healthy controls. *Journal of Neuroimmunology*. 5:191–196.

Baig, S., T. Olsson, J. Yu-Ping, B. Hojeberg, M. Cruz, and H. Link. 1991. Multiple sclerosis: Cells secreting antibodies against myelin-associated glycoprotein are present in cerebrospinal fluid. *Scandinavian Journal of Immunology*. 33:73–79.

Bake, S., J.A. Friedman, and F. Sohrabji. 2009. Reproductive age-related changes in the blood brain barrier: Expression of IgG and tight junction proteins. *Microvascular Research*. 78:413–424.

Baloyannis, S.J., and N.K. Gonatas. 1979. Distribution of anti-HRP antibodies in the central nervous system of immunized rats after disruption of the blood brain barrier. *J Neuropathology and Experimental Neurology*. 38:519–531.

Banks, W.A. 2005. Blood-brain barrier transport of cytokines: A mechanism for neuropathology. *Current Pharmaceutical Design*. 11:973–984.

Bansal, D., F. Herbert, P. Lim, P. Deshpande, C. Becavin, V. Guiyedi et al. 2009. IgG autoantibody to brain beta tubulin III associated with cytokine cluster-II discriminate cerebral malaria in central India. *PLoS One*. 4:e8245.

Bard, F., C. Cannon, R. Barbour, R.L. Burke, D. Games, H. Grajeda et al. 2000. Peripherally administered antibodies against amyloid beta-peptide enter the central nervous system and reduce pathology in a mouse model of Alzheimer disease. *Nature Medicine*. 6:916–919.

Bartos, A., L. Fialova, J. Svarcova, and D. Ripova. 2012. Patients with Alzheimer disease have elevated intrathecal synthesis of antibodies against tau protein and heavy neurofilament. *Journal of Neuroimmunology*. 252:100–105.

Behan, P.O., L.M. Lowenstein, M. Stilmant, and D.S. Sax. 1973. Landry-Guillain-Barre-Strohl syndrome and immune-complex nephritis. *Lancet*. 1:850–854.

Ben-Ami, S.D., M. Blank, and A. Altman. 2010. [The clinical importance of anti-ribosomal-P antibodies]. *Harefuah*. 149:794–810.

Berg, D.K., and W.G. Conroy. 2002. Nicotinic alpha 7 receptors: Synaptic options and downstream signaling in neurons. *Journal of Neurobiology*. 53:512–523.

Berghs, S., F. Ferracci, E. Maksimova, S. Gleason, N. Leszczynski, M. Butler et al. 2001. Autoimmunity to beta IV spectrin in paraneoplastic lower motor neuron syndrome. *Proceedings of the National Academy of Sciences of the United States of America*. 98:6945–6950.

Berghs, S.A., D. Aggujaro, R. Dirkx, E. Maksimova, P. Stabach, J.M. Hermel et al. 2000. beta IV spectrin, a new spectrin localized at axon initial segments and nodes of Ranvier. *The Journal of Cell Biology*. 151:985–1002.

Beuche, W., and R.L. Friede. 1986. Myelin phagocytosis in Wallerian degeneration of peripheral nerves depends on silica-sensitive, bg/bg-negative and Fc-positive monocytes. *Brain Research*. 378:97–106.

Bitsch, A., A. Dressel, K. Meier, T. Bogumil, F. Deisenhammer, H. Tumanial et. 2004. Autoantibody synthesis in primary progressive multiple sclerosis patients treated with interferon beta-1b. *Journal of Neurology*. 251:1498–1501.

Bjorklund, H., M. Eriksdotter-Nilsson, D. Dahl, G. Rose, B. Hoffer, and L. Olson. 1985. Image analysis of GFA-positive astrocytes from adolescence to senescence. *Experimental Brain Research*. 58:163–170.

Blight, A.R. 1993. Remyelination, revascularization, and recovery of function in experimental spinal cord injury. *Advances in Neurology*. 59:91–104.

Blight, A.R. 1994. Effects of silica on the outcome from experimental spinal cord injury: Implication of macrophages in secondary tissue damage. *Neuroscience*. 60:263–273.

Borovikova, L.V., S. Ivanova, M. Zhang, H. Yang, G.I. Botchkina, L.R. Watkins et al. 2000. Vagus nerve stimulation attenuates the systemic inflammatory response to endotoxin. *Nature*. 405:458–462.

Braunschweig, D., P. Krakowiak, P. Duncanson, R. Boyce, R.L. Hansen, P. Ashwood et al. 2013. Autism-specific maternal autoantibodies recognize critical proteins in developing brain. *Translational Psychiatry*. 3:e277.

Braunschweig, D., P. Duncanson, R. Boyce, R. Hansen, P. Ashwood, I.N. Pessah, I. Hertz-Picciotto, and J. Van de Water. 2012. Behavioral correlates of maternal antibody status among children with autism. *Journal of Autism and Developmental Disorders*. 42:1435–1445.

Braunschweig, D., and J. Van de Water. 2012. Maternal autoantibodies in autism. *Archives of Neurology*. 69:693–699.

Braunschweig, D., P. Ashwood, P. Krakowiak, I. Hertz-Picciotto, R. Hansen, L.A. Croen, I.N. Pessah et al. 2008. Autism: Maternally derived antibodies specific for fetal brain proteins. *Neurotoxicology*. 29:226–231.

Bressler, J.P., and G.W. Goldstein. 1991. Mechanisms of lead neurotoxicity. *Biochemical Pharmacology*. 41:479–484.

Brock, T.O., and J.P. O'Callaghan. 1987. Quantitative changes in the synaptic vesicle proteins synapsin I and p38 and the astrocyte-specific protein glial fibrillary acidic protein are associated with chemical-induced injury to the rat central nervous system. *Journal of Neuroscience*. 7:931–942.

Connolly, A.M., M. Chez, E.M. Streif, R.M. Keeling, P.T. Golumbek, J.M. Kwon et al. 2006. Brain-derived neurotrophic factor and autoantibodies to neural antigens in sera of children with autistic spectrum disorders, Landau-Kleffner syndrome, and epilepsy. *Biological Psychiatry*. 59:354–363.

Connolly, A.M., M.G. Chez, A. Pestronk, S.T. Arnold, S. Mehta, and R.K. Deuel. 1999. Serum autoantibodies to brain in Landau-Kleffner variant, autism, and other neurologic disorders. *The Journal of Pediatrics*. 134:607–613.

Corcione, A., S. Casazza, E. Ferretti, D. Giunti, E. Zappia, A. Pistorio et al. 2004a. Recapitulation of B cell differentiation in the central nervous system of patients with multiple sclerosis. *Proceedings of the National Academy of Sciences of the United States of America*. 101:11064–11069.

Corcione, A., S. Casazza, E. Ferretti, D. Giunti, E. Zappia, A. Pistorio et al. 2004b. Recapitulation of B cell differentiation in the central nervous system of patients with multiple sclerosis. *Proceedings of the National Academy of Sciences of the United States of America*. 101:11064–11069.

Croonenberghs, J., E. Bosmans, D. Deboutte, G. Kenis, and M. Maes. 2002. Activation of the inflammatory response system in autism. *Neuropsychobiology*. 45:1–6.

Cross, A.H., J.L. Trotter, and J.A. Lyons. 2001. B cells and antibodies in CNS demyelinating disease. *Journal of Neuroimmunology*. 112:1–14.

Cruse, J.M., R.E. Lewis, D.L. Roe, S. Dilioglou, M.C. Blaine, W.F. Wallace et al. 2000. Facilitation of immune function, healing of pressure ulcers, and nutritional status in spinal cord injury patients. *Experimental and Molecular Pathology*. 68:38–54.

Dalakas, M.C. 2008a. B cells as therapeutic targets in autoimmune neurological disorders. *Nature Clinical Practice. Neurology*. 4:557–567.

Dalakas, M.C. 2008b. Invited article: Inhibition of B cell functions - Implications for neurology. *Neurology*. 70:2252–2260.

Dalakas, M.C. 2008c. Invited article: Inhibition of B cell functions: Implications for neurology. *Neurology*. 70:2252–2260.

Dalmau, J., and M.R. Rosenfeld. 2008. Paraneoplastic syndromes of the CNS. *Lancet Neurology*. 7:327–340.

Dalton, P., R. Deacon, A. Blamire, M. Pike, I. McKinlay, J. Stein et al. 2003. Maternal neuronal antibodies associated with autism and a language disorder. *Annals of Neurology*. 53:533–537.

Darnell, R.B., H.M. Furneaux, and J.B. Posner. 1991. Antiserum from a patient with cerebellar degeneration identifies a novel protein in Purkinje cells, cortical neurons, and neuroectodermal tumors. *Journal of Neuroscience*. 11:1224–1230.

Davidson, A., and B. Diamond. 2001. Advances in immunology - Autoimmune diseases. *New England Journal of Medicine*. 345:340–350.

Davies, A.L., K.C. Hayes, and G.A. Dekaban. 2007. Clinical correlates of elevated serum concentrations of cytokines and autoantibodies in patients with spinal cord injury. *Archives of Physical Medicine and Rehabilitation*. 88:1384–1393.

DeGiorgio, L.A., K.N. Konstantinov, S.C. Lee, J.A. Hardin, B.T. Volpe, and B. Diamond. 2001. A subset of lupus anti-DNA antibodies cross-reacts with the NR2 glutamate receptor in systemic lupus erythematosus. *Nature Medicine*. 7:1189–1193.

Dekaban, G.A., and S. Thawer. 2009. Pathogenic antibodies are active participants in spinal cord injury. *The Journal of Clinical Investigation*. 119:2881–2884.

Diamond, B., G. Honig, S. Mader, L. Brimberg, and B.T. Volpe. 2013. Brain-reactive antibodies and disease. *Annual Review of Immunology*. 31:345–385.

Dicou, E., D. Hurez, and V. Nerriere. 1993. Natural autoantibodies against the nerve growth factor in autoimmune diseases. *Journal of Neuroimmunology*. 47:159–167.

Dorner, T., A. Radbruch, and G.R. Burmester. 2009. B-cell-directed therapies for autoimmune disease. *Nature Reviews. Rheumatology*. 5:433–441.

Drisdel, R.C., and W.N. Green. 2000. Neuronal alpha-bungarotoxin receptors are alpha7 subunit homomers. *Journal of Neuroscience*. 20:133–139.

Dudley, D.J. 1992. The immune-system in health and disease. *Bailliere's Clinical Obstetrics and Gynecology*. 6:393–416.

Ekizoglu, E., E. Tuzun, M. Woodhall, B. Lang, L. Jacobson, S. Icoz et al. 2014. Investigation of neuronal autoantibodies in two different focal epilepsy syndromes. *Epilepsia*. 55:414–422.

El-Fawal, H.A., and W.C. McCain. 2008. Antibodies to neural proteins in organophosphorus-induced delayed neuropathy (OPIDN) and its amelioration. *Neurotoxicology and teratology*. 30:161–166.

El-Fawal, H.A., S.J. Waterman, A. De Feo, and M.Y. Shamy. 1999. Neuroimmunotoxicology: Humoral assessment of neurotoxicity and autoimmune mechanisms. *Environmental Health Perspectives*. 107 Suppl 5:767–775.

Eng, L.F., J.J. Vanderhaeghen, A. Bignami, and B. Gerstl. 1971. An acidic protein isolated from fibrous astrocytes. *Brain Research*. 28:351–354.

Evans, H.L. 1995. Markers of neurotoxicity - from behavior to autoantibodies against brain proteins. *Clinical Chemistry*. 41:1874–1881.

Fairweather, D., and N.R. Rose. 2004. Women and autoimmune diseases. *Emerging Infectious Diseases*. 10:2005–2011.

Fazio, V., S.K. Bhudia, N. Marchi, B. Aumayr, and D. Janigro. 2004. Peripheral detection of S100beta during cardiothoracic surgery: What are we really measuring? *The Annals of Thoracic Surgery*. 78:46–52; discussion 52–43.

Fernandez-Shaw, C., A. Marina, P. Cazorla, F. Valdivieso, and J. Vazquez. 1997. Anti-brain spectrin immunoreactivity in Alzheimer's disease: Degradation of spectrin in an animal model of cholinergic degeneration. *Journal of Neuroimmunology*. 77:91–98.

Fialova, L., A. Bartos, J. Svarcova, D. Zimova, and J. Kotoucova. 2013a. Serum and cerebrospinal fluid heavy neurofilaments and antibodies against them in early multiple sclerosis. *Journal of Neuroimmunology*. 259:81–87.

Fialova, L., A. Bartos, J. Svarcova, D. Zimova, J. Kotoucova, and I. Malbohan. 2013b. Serum and cerebrospinal fluid light neurofilaments and antibodies against them in clinically isolated syndrome and multiple sclerosis. *Journal of Neuroimmunology*. 262:113–120.

Fleming, J.C., M.D. Norenberg, D.A. Ramsay, G.A. Dekaban, A.E. Marcillo, A.D. Saenz et al. 2006. The cellular inflammatory response in human spinal cords after injury. *Brain*. 129:3249–3269.

Floyd, S., M.H. Butler, O. Cremona, C. David, Z. Freyberg, X. Zhang et al. 1998. Expression of amphiphysin I, an autoantigen of paraneoplastic neurological syndromes, in breast cancer. *Molecular Medicine*. 4:29–39.

Folli, F., M. Solimena, R. Cofiell, M. Austoni, G. Tallini, G. Fassetta et al. 1993. Autoantibodies to a 128-Kd synaptic protein in 3 women with the Stiff-Man Syndrome and breast-cancer. *New England Journal of Medicine*. 328:546–551.

Frye, R.E., J.M. Sequeira, E.V. Quadros, S.J. James, and D.A. Rossignol. 2013. Cerebral folate receptor autoantibodies in autism spectrum disorder. *Molecular Psychiatry*. 18:369–381.

Galea, I., I. Bechmann, and V.H. Perry. 2007. What is immune privilege (not)? *Trends in Immunology*. 28:12–18.

Genain, C.P., B. Cannella, S.L. Hauser, and C.S. Raine. 1999. Identification of autoantibodies associated with myelin damage in multiple sclerosis. *Nature Medicine*. 5:170–175.

Gitlits, V.M., J.W. Sentry, L.S. Matthew, A.I. Smith, and B.H. Toh. 2001. Synapsin I identified as a novel brain-specific autoantigen. *Journal of Investigative Medicine*. 49:276–283.

Goines, P., L. Haapanen, R. Boyce, P. Duncanson, D. Braunschweig, L. Delwiche et al. 2011. Autoantibodies to cerebellum in children with autism associate with behavior. *Brain, Behavior, and Immunity*. 25:514–523.

Goines, P., and J. Van de Water. 2010. The immune system's role in the biology of autism. *Current Opinion in Neurology*. 23:111–117.

Gold, R., M. Stangel, and M.C. Dalakas. 2007. Drug Insight: The use of intravenous immunoglobulin in neurology - therapeutic considerations and practical issues. *Nature Clinical Practice. Neurology*. 3:36–44.

Gresa-Arribas, N., M.J. Titulaer, A. Torrents, E. Aguilar, L. McCracken, F. Leypoldt et al. 2014. Antibody titres at diagnosis and during follow-up of anti-NMDA receptor encephalitis: A retrospective study. *Lancet Neurology*. 13:167–177.

Griot-Wenk, M., C. Griot, H. Pfister, and M. Vandevelde. 1991. Antibody-dependent cellular cytotoxicity in antimyelin antibody-induced oligodendrocyte damage in vitro. *Journal of Neuroimmunology*. 33:145–155.

Hayakawa, T., Y. Ushio, T. Mori, N. Arita, T. Yoshimine, Y. Maeda et al. 1979. Levels in stroke patients of CSF astroprotein, an astrocyte-specific cerebroprotein. *Stroke*. 10:685–689.

Hayes, K.C., T.C. Hull, G.A. Delaney, P.J. Potter, K.A. Sequeira, K. Campbell et al. 2002. Elevated serum titers of proinflammatory cytokines and CNS autoantibodies in patients with chronic spinal cord injury. *Journal of Neurotrauma*. 19:753–761.

Hedegaard, C.J., N. Chen, F. Sellebjerg, P.S. Sorensen, R.G. Leslie, K. Bendtzen et al. 2009. Autoantibodies to myelin basic protein (MBP) in healthy individuals and in patients with multiple sclerosis: A role in regulating cytokine responses to MBP. *Immunology*. 128:e451–461.

Howe, C.L., A.J. Bieber, A.E. Warrington, L.R. Pease, and M. Rodriguez. 2004. Antiapoptotic signaling by a remyelination-promoting human antimyelin antibody. *Neurobiology of Disease*. 15:120–131.

Ingram, C.R., K.J. Phegan, and H.T. Blumenthal. 1974. Significance of an aging-linked neuron binding gamma globulin fraction of human sera. *Journal of Gerontology*. 29:20–27.

Iversen, P.O., N. Hjeltnes, B. Holm, T. Flatebo, I. Strom-Gundersen, W. Ronning et al. 2000. Depressed immunity and impaired proliferation of hematopoietic progenitor cells in patients with complete spinal cord injury. *Blood*. 96:2081–2083.

Joachim, S.C., M.B. Wax, N. Boehm, D.R. Dirk, N. Pfeiffer, and F.H. Grus. 2011. Upregulation of antibody response to heat shock proteins and tissue antigens in an ocular ischemia model. *Investigative Ophthalmology andand Visual Science*. 52:3468–3474.

Jones, A.L., B.J. Mowry, D.E. McLean, B.X. Mantzioris, M.P. Pender, and J.M. Greer. 2014. Elevated levels of autoantibodies targeting the M1 muscarinic acetylcholine receptor and neurofilament medium in sera from subgroups of patients with schizophrenia. *Journal of Neuroimmunology*. 269:68–75.

Kamchatnov, P.R., N.Y. Ruleva, S.F. Dugin, L.I. Buriachkovskaya, A.V. Chugunov, N.A. Mikhailova et al. 2009. Neurospecific proteins and autoantibodies in serum of patients with acute ischemic stroke. *Zhurnal Neurologii i Psikhiatrii*. 109:69–72.

Kanada, M., M. Miyagawa, M. Sato, H. Hasegawa, and T. Honma. 1994. Neurochemical profile of effects of 28 neurotoxic chemicals on the central nervous system in rats (1). Effects of oral administration on brain contents of biogenic amines and metabolites. *Industrial Health*. 32:145–164.

Katz, D.H. 1980. Adaptive differentiation of lymphocytes: Theoretical implications for mechanisms of cell— cell recognition and regulation of immune responses. *Advances in Immunology*. 29:137–207.

Keller, J.N., and M.P. Mattson. 1998. Roles of lipid peroxidation in modulation of cellular signaling pathways, cell dysfunction, and death in the nervous system. *Reviews in the Neurosciences*. 9:105–116.

Kelly, D.F., I.T. Gonzalo, P. Cohan, N. Berman, R. Swerdloff, and C. Wang. 2000. Hypopituitarism following traumatic brain injury and aneurysmal subarachnoid hemorrhage: A preliminary report. *Journal of Neurosurgery*. 93:743–752.

Kergonou, J.F., I. Pennacino, C. Lafite, and R. Ducousso. 1988. Immunological relevance of malonic dialdehyde (Mda). 2. Precipitation reactions of human-sera with Mda-crosslinked proteins. *Biochemistry International*. 16:835–843.

Kida, S., A. Pantazis, and R.O. Weller. 1993. CSF drains directly from the subarachnoid space into nasal lymphatics in the rat. Anatomy, histology and immunological significance. *Neuropathology and Applied Neurobiology*. 19:480–488.

Kigerl, K.A., V.M. McGaughy, and P.G. Popovich. 2006. Comparative analysis of lesion development and intraspinal inflammation in four strains of mice following spinal contusion injury. *The Journal of Comparative Neurology*. 494:578–594.

Kil, K., Y.C. Zang, D. Yang, J. Markowski, G.S. Fuoco, G.C. Vendetti et al. 1999. T cell responses to myelin basic protein in patients with spinal cord injury and multiple sclerosis. *Journal of Neuroimmunology*. 98:201–207.

Kingsley, B.S., F. Gaskin, and S.M. Fu. 1988. Human antibodies to neurofibrillary tangles and astrocytes in Alzheimer's disease. *Journal of Neuroimmunology*. 19:89–99.

Kirkman, N.J., J.E. Libbey, T.L. Sweeten, H.H. Coon, J.N. Miller, E.K. Stevenson et al. 2008. How relevant are GFAP autoantibodies in autism and Tourette Syndrome? *Journal of Autism and Developmental Disorders*. 38:333–341.

Kliesch, W.F., J.M. Cruse, R.E. Lewis, G.R. Bishop, B. Brackin, and J.A. Lampton. 1996. Restoration of depressed immune function in spinal cord injury patients receiving rehabilitation therapy. *Paraplegia*. 34:82–90.

Kowal, C., L.A. DeGiorgio, T. Nakaoka, H. Hetherington, P.T. Huerta, B. Diamond et al. 2004. Cognition and immunity; antibody impairs memory. *Immunity*. 21:179–188.

Le Couteur, A., G. Haden, D. Hammal, and H. McConachie. 2008. Diagnosing autism spectrum disorders in pre-school children using two standardised assessment instruments: The ADI-R and the ADOS. *Journal of Autism and Developmental Disorders*. 38:362–372.

Levesque, M.C., and E.W. St Clair. 2008. B cell-directed therapies for autoimmune disease and correlates of disease response and relapse. *The Journal of Allergy and Clinical Immunology*. 121:13–21; quiz 22–33.

Levy, N.L., M.S. Mahaley, Jr., and E.D. Day. 1972. In vitro demonstration of cell-mediated immunity to human brain tumors. *Cancer Research*. 32:477–482.

Lord, C., E.H. Cook, B.L. Leventhal, and D.G. Amaral. 2000a. Autism spectrum disorders. *Neuron*. 28:355–363.

Lord, C., A. Pickles, J. McLennan, M. Rutter, J. Bregman, S. Folstein et al. 1997. Diagnosing autism: Analyses of data from the Autism Diagnostic Interview. *Journal of Autism and Developmental Disorders*. 27:501–517.

Lord, C., S. Risi, L. Lambrecht, E.H. Cook, Jr., B.L. Leventhal, P.C. DiLavore et al. 2000b. The autism diagnostic observation schedule-generic: A standard measure of social and communication deficits associated with the spectrum of autism. *Journal of Autism and Developmental Disorders*. 30:205–223.

Lord, C., M. Rutter, and A. Le Couteur. 1994. Autism Diagnostic Interview-Revised: A revised version of a diagnostic interview for caregivers of individuals with possible pervasive developmental disorders. *Journal of Autism and Developmental Disorders*. 24:659–685.

Lotti, M. 1995. Cholinesterase inhibition: Complexities in interpretation. *Clinical Chemistry*. 41:1814–1818.

Lu, X.Y., X.X. Chen, L.D. Huang, C.Q. Zhu, Y.Y. Gu, and S. Ye. 2010. Anti-alpha-internexin autoantibody from neuropsychiatric lupus induce cognitive damage via inhibiting axonal elongation and promote neuron apoptosis. *PLoS One*. 5:e11124.

Lucin, K.M., V.M. Sanders, T.B. Jones, W.B. Malarkey, and P.G. Popovich. 2007. Impaired antibody synthesis after spinal cord injury is level dependent and is due to sympathetic nervous system dysregulation. *Experimental Neurology*. 207:75–84.

Lyons, J.A., M. San, M.P. Happ, and A.H. Cross. 1999. B cells are critical to induction of experimental allergic encephalomyelitis by protein but not by a short encephalitogenic peptide. *European Journal of Immunology*. 29:3432–3439.

Maetzler, W., D. Berg, M. Synofzik, K. Brockmann, J. Godau, A. Melms et al. 2011. Autoantibodies against amyloid and glial-derived antigens are increased in serum and cerebrospinal fluid of Lewy body-associated dementias. *Journal of Alzheimer's disease*. 26:171–179.

Mailman, R.B., and M.H. Lewis. 1987. Neurotoxicants and central catecholamine systems. *Neurotoxicology*. 8:123–139.

Malek, A., R. Sager, P. Kuhn, K.H. Nicolaides, and H. Schneider. 1996. Evolution of maternofetal transport of immunoglobulins during human pregnancy. *American Journal of Reproductive Immunology*. 36:248–255.

Marchi, N., J.J. Bazarian, V. Puvenna, M. Janigro, C. Ghosh, J. Zhong et al. 2013a. Consequences of repeated blood-brain barrier disruption in football players. *PloS One*. 8:e56805.

Marchi, N., J.J. Bazarian, V. Puvenna, M. Janigro, C. Ghosh, J.H. Zhong et al. 2013b. Consequences of repeated blood-brain barrier disruption in football players. *PloS One*. 8:e56805.

Marchi, N., M. Cavaglia, V. Fazio, S. Bhudia, K. Hallene, and D. Janigro. 2004. Peripheral markers of blood-brain barrier damage. *Clinica Chimica Acta*. 342:1–12.

Marchi, N., P. Rasmussen, M. Kapural, V. Fazio, K. Kight, M.R. Mayberg et al. 2003. Peripheral markers of brain damage and blood-brain barrier dysfunction. *Restorative Neurology and Neuroscience*. 21:109–121.

Martin, L.A., P. Ashwood, D. Braunschweig, M. Cabanlit, J. Van de Water, and D.G. Amaral. 2008. Stereotypies and hyperactivity in rhesus monkeys exposed to IgG from mothers of children with autism. *Brain, Behavior, and Immunity*. 22:806–816.

Mason, D.L., M.M. Assimon, J.R. Bishop, and H.A. El-Fawal. 2013. Nervous system autoantibodies and vitamin D receptor gene polymorphisms in hemodialysis patients. *Hemodialysis International*. 17:3–11.

Matsushita, T., and T.F. Tedder. 2009. B-lymphocyte depletion for the treatment of multiple sclerosis: Now things really get interesting. *Expert Review of Neurotherapeutics*. 9:309–312.

Mcraedegueurce, A., S. Booj, K. Haglid, L. Rosengren, J.E. Karlsson, I. Karlsson et al. 1987. Antibodies in cerebrospinal-fluid of some Alzheimer-disease patients recognize cholinergic neurons in the rat central nervous-system. *Proceedings of the National Academy of Sciences of the United States of America*. 84:9214–9218.

Mecocci, P., L. Parnetti, R. Donato, C. Santucci, A. Santucci, D. Cadini et al. 1992. Serum autoantibodies against glial fibrillary acidic protein in brain aging and senile dementias. *Brain, Behavior, and Immunity*. 6:286–292.

Mizrachi, Y., A. Ohry, A. Aviel, R. Rozin, M.E. Brooks, and M. Schwartz. 1983. Systemic humoral factors participating in the course of spinal cord injury. *Paraplegia*. 21:287–293.

Moneim, I.A., M.Y. Shamy, R.M. el-Gazzar, and H.A. El-Fawal. 1999. Autoantibodies to neurofilaments (NF), glial fibrillary acidic protein (GFAP) and myelin basic protein (MBP) in workers exposed to lead. *The Journal of the Egyptian Public Health Association*. 74:121–138.

Moody, M., M. Zipp, and I. Al-Hashimi. 2001. Salivary anti-spectrin autoantibodies in Sjogren's syndrome. *Oral Surgery, Oral Medicine, Oral Pathology, Oral Radiology, and Endodontics*. 91:322–327.

Moretto, A., M. Bertolazzi, and M. Lotti. 1994. The phosphorothioic acid O-(2-chloro-2,3,3-trifluorocyclobutyl) O-ethyl S-propyl ester exacerbates organophosphate polyneuropathy without inhibition of neuropathy target esterase. *Toxicology and Applied Pharmacology*. 129:133–137.

Mori, T., K. Morimoto, T. Hayakawa, Y. Ushio, H. Mogami, and K. Sekiguchi. 1978. Radioimmunoassay of astroprotein (an astrocyte-specific cerebroprotein) in cerebrospinal fluid and its clinical significance. *Neurologia Medico-chirurgica*. 18:25–31.

Morris, C.M., C. Pardo-Villamizar, C.D. Gause, and H.S. Singer. 2009. Serum autoantibodies measured by immunofluorescence confirm a failure to differentiate PANDAS and Tourette syndrome from controls. *Journal of the Neurological Sciences*. 276:45–48.

Mosley, K., and M.L. Cuzner. 1996. Receptor-mediated phagocytosis of myelin by macrophages and microglia: Effect of opsonization and receptor blocking agents. *Neurochemical Research*. 21:481–487.

Mostafa, G.A., and L.Y. Al-Ayadhi. 2011. The possible link between the elevated serum levels of neurokinin A and anti-ribosomal P protein antibodies in children with autism. *Journal of Neuroinflammation*. 8:180.

Mostafa, G.A., D.H. Ibrahim, A.A. Shehab, and A.K. Mohammed. 2010. The role of measurement of serum autoantibodies in prediction of pediatric neuropsychiatric systemic lupus erythematosus. *Journal of Neuroimmunology*. 227:195–201.

Mundiyanapurath, S., S. Jarius, C. Probst, W. Stocker, B. Wildemann, and J. Bosel. 2013. GABA-B-receptor antibodies in paraneoplastic brainstem encephalitis. *Journal of Neuroimmunology*. 259:88–91.

Murakami, H., S. Iijima, M. Kawamura, Y. Takahashi, and H. Ichikawa. 2013. [A case of acute cerebellar ataxia following infectious mononucleosis accompanied by intrathecal anti-glutamate receptor delta2 antibody]. *Rinsho shinkeigaku*. 53:555–558.

Murphy, W.H., M.R. Tam, R.L. Lanzi, M.R. Abell, and C. Kauffman. 1970. Age dependence of immunologically induced central nervous system disease in C58 mice. *Cancer Research*. 30:1612–1622.

Nandy, K. 1972. Brain-reactive antibodies in mouse serum as a function of age. *Journal of Gerontology*. 27:173–177.

Neuwelt, E.A., B. Bauer, C. Fahlke, G. Fricker, C. Iadecola, D. Janigro et al. 2011. Engaging neuroscience to advance translational research in brain barrier biology. *Nature Reviews. Neuroscience*. 12:169–182.

Ngankam, L., N.V. Kazantseva, and M.M. Gerasimova. 2011. Immunological markers of severity and outcome of traumatic brain injury. *Zhurnal Neurologii i Psikhiatrii*. 111:61–65.

Niebroj-Dobosz, I., D. Dziewulska, and P. Janik. 2006. Autoantibodies against proteins of spinal cord cells in cerebrospinal fluid of patients with amyotrophic lateral sclerosis (ALS). *Folia Neuropathologica*. 44:191–196.

National Institutes of Health Autoimmune Diseases Coordinating Committee Report. 2002. *Autoimmune Diseases Research Plan*. National Institutes of Health, Bethesda, MD.

Op De Beeck, K., K. Van den Bergh, S. Vermeire, S. Decock, R. Derua, E. Waelkens et al. 2011. Immune reactivity to beta-tubulin isotype 5 and vesicular integral-membrane protein 36 in patients with autoimmune gastrointestinal disorders. *Gut*. 60:1601–1602.

Outteryck, O., G. Baille, J. Hodel, M. Giroux, A. Lacour, J. Honnorat et al. 2013. Extensive myelitis associated with anti-NMDA receptor antibodies. *BMC Neurology*. 13:211.

Pardo, C.A., D.L. Vargas, and A.W. Zimmerman. 2005. Immunity, neuroglia and neuroinflammation in autism. *International Review of Psychiatry*. 17:485–495.

Pavlowsky, A., A. Gianfelice, M. Pallotto, A. Zanchi, H. Vara, M. Khelfaoui et al. 2010. A postsynaptic signaling pathway that may account for the cognitive defect due to IL1RAPL1 mutation. *Current Biology*. 20:103–115.

Paz Soldan, M.M., A.E. Warrington, A.J. Bieber, B. Ciric, V. Van Keulen, L.R. Pease et al. 2003. Remyelination-promoting antibodies activate distinct Ca2+ influx pathways in astrocytes and oligodendrocytes: Relationship to the mechanism of myelin repair. *Molecular and Cellular Neurosciences*. 22:14–24.

Pinelis, V.G., and E.G. Sorokina. 2008. [Autoimmune mechanisms of modulation of the activity of glutamate receptors in children with epilepsy and craniocerebral injury]. *Vestnik Rossiiskoi akademii meditsinskikh nauk/Rossiiskaia akademiia meditsinskikh nauk*. 44–51.

Pirko, I., B. Ciric, J. Gamez, A.J. Bieber, A.E. Warrington, A.J. Johnson et al. 2004. A human antibody that promotes remyelination enters the CNS and decreases lesion load as detected by T2-weighted spinal cord MRI in a virus-induced murine model of MS. *FASEB Journal*. 18:1577–1579.

Piton, A., J.L. Michaud, H. Peng, S. Aradhya, J. Gauthier, L. Mottron et al. 2008. Mutations in the calcium-related gene IL1RAPL1 are associated with autism. *Human Molecular Genetics*. 17:3965–3974.

Poletaev, A., and P. Boura. 2011. The immune system, natural autoantibodies and general homeostasis in health and disease. *Hippokratia*. 15:295–298.

Popovic, M., M. Caballero-Bleda, L. Puelles, N. Popovic, and N. Popovic. 1998. Importance of immunological and inflammatory processes in the pathogenesis and therapy of Alzheimer's disease. *International Journal of Neuroscience*. 95:203–236.

Popovich, P.G., Z. Guan, P. Wei, I. Huitinga, N. van Rooijen, and B.T. Stokes. 1999. Depletion of hematogenous macrophages promotes partial hindlimb recovery and neuroanatomical repair after experimental spinal cord injury. *Experimental Neurology*. 158:351–365.

Porcelli, B., F. Ferretti, C. Vindigni, C. Scapellato, and L. Terzuoli. 2013. Detection of autoantibodies against actin filaments in celiac disease. *Journal of Clinical Laboratory Analysis*. 27:21–26.

Posner, J.B., and J. Dalmau. 1997a. Paraneoplastic syndromes. *Current Opinion in Immunology*. 9:723–729.

Posner, J.B., and J.O. Dalmau. 1997b. Paraneoplastic syndromes affecting the central nervous system. *Annual Review of Medicine*. 48:157–166.

Prinz, M., J. Priller, S.S. Sisodia, and R.M. Ransohoff. 2011. Heterogeneity of CNS myeloid cells and their roles in neurodegeneration. *Nature Neuroscience*. 14:1227–1235.

Raine, C.S., B. Cannella, S.L. Hauser, and C.P. Genain. 1999. Demyelination in primate autoimmune encephalomyelitis and acute multiple sclerosis lesions: A case for antigen-specific antibody mediation. *Annals of Neurology*. 46:144–160.

Ramaekers, V.T., E.V. Quadros, and J.M. Sequeira. 2013. Role of folate receptor autoantibodies in infantile autism. *Molecular Psychiatry*. 18:270–271.

Reiber, H. 2001. Dynamics of brain-derived proteins in cerebrospinal fluid. *Clinica Chimica Acta*. 310:173–186.

Reiber, H. 2003. Proteins in cerebrospinal fluid and blood: Barriers, CSF flow rate and source-related dynamics. *Restorative Neurology and Neuroscience*. 21:79–96.

Rosas-Ballina, M., and K.J. Tracey. 2009. Cholinergic control of inflammation. *Journal of Internal Medicine*. 265:663–679.

Rothenberg, S.P., M.P. da Costa, J.M. Sequeira, J. Cracco, J.L. Roberts, J. Weedon et al. 2004. Autoantibodies against folate receptors in women with a pregnancy complicated by a neural-tube defect. *The New England Journal of Medicine*. 350:134–142.

Rudehill, S., S. Muhallab, A. Wennersten, C. von Gertten, F. Al Nimer, A.C. Sandberg-Nordqvist et al. 2006. Autoreactive antibodies against neurons and basal lamina found in serum following experimental brain contusion in rats. *Acta Neurochirurgia*. 148:199–205.

Schwartz, M., and J. Kipnis. 2001. Protective autoimmunity: Regulation and prospects for vaccination after brain and spinal cord injuries. *Trends in Molecular Medicine*. 7:252–258.

Simister, N.E. 2003. Placental transport of immunoglobulin G. *Vaccine*. 21:3365–3369.

Singer, H.S., C. Morris, C. Gause, M. Pollard, A.W. Zimmerman, and M. Pletnikov. 2009. Prenatal exposure to antibodies from mothers of children with autism produces neurobehavioral alterations: A pregnant dam mouse model. *Journal of Neuroimmunology*. 211:39–48.

Singer, H.S., C.M. Morris, P.N. Williams, D.Y. Yoon, J.J. Hong, and A.W. Zimmerman. 2006. Antibrain antibodies in children with autism and their unaffected siblings. *Journal of Neuroimmunology*. 178:149–155.

Singh, V.K., R.P. Warren, J.D. Odell, W.L. Warren, and P. Cole. 1993. Antibodies to myelin basic protein in children with autistic behavior. *Brain, Behavior, and Immunity*. 7:97–103.

Sjogren, M., and A. Wallin. 2001. Pathophysiological aspects of frontotemporal dementia - emphasis on cytoskeleton proteins and autoimmunity. *Mechanisms of Ageing and Development*. 122:1923–1935.

Skoda, D., K. Kranda, M. Bojar, L. Glosova, J. Baurle, J. Kenney et al. 2006. Antibody formation against beta-tubulin class III in response to brain trauma. *Brain Research Bulletin*. 68:213–216.

Skouen, J.S., J.L. Larsen, and S.E. Vollset. 1993. Cerebrospinal fluid proteins as indicators of nerve root compression in patients with sciatica caused by disc herniation. *Spine*. 18:72–79.

Soares, H.D., R.R. Hicks, D. Smith, and T.K. McIntosh. 1995. Inflammatory leukocytic recruitment and diffuse neuronal degeneration are separate pathological processes resulting from traumatic brain injury. *Journal of Neuroscience*. 15:8223–8233.

Sorokina, E.G., J.B. Semenova, O.K. Granstrem, O.V. Karaseva, S.V. Meshcheryakov, V.P. Reutov et al. 2010. S100B protein and autoantibodies to S100B protein in diagnostics of brain damage in craniocerebral trauma in children. *Zhurnal Neurologii i Psikhiatrii*. 110:30–35.

Sorokina, E.G., Zh.B. Semenova, N.A. Bazarnaya, S.V. Meshcheryakov, V.P. Reutov, A.V. Goryunova et al. 2009. Autoantibodies to glutamate receptors and products of nitric oxide metabolism in serum in children in the acute phase of craniocerebral trauma. *Neuroscience and Behavioral Physiology*. 39:329–334.

Sorokina, E.G., T.P. Storozhevykh, Y.E. Senilova, O.K. Granstrem, V.P. Reutov, and V.G. Pinelis. 2006. Effect of antibodies against AMPA glutamate receptors on brain neurons in primary cultures of the cerebellum and hippocampus. *Bulletin of Experimental Biology and Medicine*. 142:51–54.

Sorokina, E.G., O.M. Vol'pina, B. Semenova Zh, O.V. Karaseva, D.O. Koroev, A.V. Kamynina et al. 2011. [Autoantibodies to alpha7-subunit of neuronal acetylcholine receptor in children with traumatic brain injury]. *Zhurnal Nevrologii i Psikhiatrii*. 111:56–60.

Sotgiu, S., A. Giua, M.R. Murrighile, and R. Ortu. 2009. A case of anti-myelin-associated glycoprotein polyneuropathy and multiple sclerosis: One disease instead of two? *BMJ Case Reports*. 2009.

Sroga, J.M., T.B. Jones, K.A. Kigerl, V.M. McGaughy, and P.G. Popovich. 2003. Rats and mice exhibit distinct inflammatory reactions after spinal cord injury. *The Journal of Comparative Neurology*. 462:223–240.

Stefanova, I., J.R. Dorfman, and R.N. Germain. 2002. Self-recognition promotes the foreign antigen sensitivity of naive T lymphocytes. *Nature*. 420:429–434.

Stein, T.D., J.P. Fedynyshyn, and R.E. Kalil. 2002. Circulating autoantibodies recognize and bind dying neurons following injury to the brain. *Journal of Neuropathology and Experimental Neurology*. 61:1100–1108.

Stellwagen, D., and R.C. Malenka. 2006. Synaptic scaling mediated by glial TNF-alpha. *Nature*. 440:1054–1059.

Stephan, A.H., B.A. Barres, and B. Stevens. 2012. The complement system: An unexpected role in synaptic pruning during development and disease. *Annual Review of Neuroscience*. 35:369–389.

Storoni, M., A. Petzold, and G.T. Plant. 2011. The use of serum glial fibrillary acidic protein measurements in the diagnosis of neuromyelitis optica spectrum optic neuritis. *PloS One*. 6:e23489.

Sundstrom, R., and H. Kalimo. 1987. Extracellular edema and glial response to it in the cerebellum of suckling rats with low-dose lead encephalopathy. An electron microscopic and immunohistochemical study. *Acta Neuropathologica*. 75:116–122.

Sundstrom, R., K. Muntzing, H. Kalimo, and P. Sourander. 1985. Changes in the integrity of the blood-brain barrier in suckling rats with low dose lead encephalopathy. *Acta Neuropathologica*. 68:1–9.

Tanaka, J., K. Murakoshi, M. Takeda, Y. Kato, K. Tada, S. Hariguchi et al. 1988. A high-level of anti-GFAP autoantibody in the serum of patients with Alzheimer's disease. *Biomedical Research-Tokyo*. 9:209–216.

Tanaka, J., K. Nakamura, M. Takeda, K. Tada, H. Suzuki, H. Morita et al. 1989. Enzyme-linked immunosorbent-assay for human autoantibody to glial fibrillary acidic protein - higher titer of the antibody is detected in serum of patients with Alzheimer's disease. *Acta Neurologica Scandinavia*. 80:554–560.

Tanriverdi, F., A. De Bellis, M. Battaglia, G. Bellastella, A. Bizzarro, A.A. Sinisi et al. 2010. Investigation of antihypothalamus and antipituitary antibodies in amateur boxers: Is chronic repetitive head trauma-induced pituitary dysfunction associated with autoimmunity? *European Journal of Endocrinology*. 162:861–867.

Tanriverdi, F., A. De Bellis, A. Bizzarro, A.A. Sinisi, G. Bellastella, E. Pane et al. 2008a. Antipituitary antibodies after traumatic brain injury: Is head trauma-induced pituitary dysfunction associated with autoimmunity? *European Journal of Endocrinology/European Federation of Endocrine Societies*. 159:7–13.

Tanriverdi, F., H. Senyurek, K. Unluhizarci, A. Selcuklu, F.F. Casanueva, and F. Kelestimur. 2006. High risk of hypopituitarism after traumatic brain injury: A prospective investigation of anterior pituitary function in the acute phase and 12 months after trauma. *The Journal of Clinical Endocrinology and Metabolism*. 91:2105–2111.

Tanriverdi, F., H. Ulutabanca, K. Unluhizarci, A. Selcuklu, F.F. Casanueva, and F. Kelestimur. 2008b. Three years prospective investigation of anterior pituitary function after traumatic brain injury: A pilot study. *Clinical Endocrinology*. 68:573–579.

Terry, R.D., and H. Wisnewski. 1972. Ultrastructure of senile dementia and of experimental analogs. In *Aging and the Brain*. Plenum Press,, New York,.

Thomas, L., A. Mailles, V. Desestret, F. Ducray, E. Mathias, V. Rogemond et al. 2014. Autoimmune N-methyl-D-aspartate receptor encephalitis is a differential diagnosis of infectious encephalitis. *The Journal of Infection*. 68:419–425.

Threatt, J., K. Nandy, and R. Fritz. 1971. Brain-reactive antibodies in serum of old mice demonstrated by immunofluorescence. *Journal of Gerontology*. 26:316–323.

Todd, R.D., J.M. Hickok, G.M. Anderson, and D.J. Cohen. 1988. Antibrain antibodies in infantile autism. *Biological Psychiatry*. 23:644–647.

Trippe, J., K. Steinke, A. Orth, P.M. Faustmann, M. Hollmann, and C.G. Haase. 2014. Autoantibodies to glutamate receptor antigens in multiple sclerosis and Rasmussen's encephalitis. *Neuroimmunomodulation*. 21:189–194.

Tron, F. 2014. [Autoantibodies as biomarkers]. *Presse Medicále*. 43:57–65.

Tsuburaya, R.S., N. Miki, K. Tanaka, T. Kageyama, K. Irahara, S. Mukaida et al. 2014. Anti-myelin oligodendrocyte glycoprotein (MOG) antibodies in a Japanese boy with recurrent optic neuritis. *Brain and Development*. Feb 28.

Tsuchiya, H., S. Haga, Y. Takahashi, T. Kano, Y. Ishizaka, and A. Mimori. 2014. Identification of novel autoantibodies to GABAB receptors in patients with neuropsychiatric systemic lupus erythematosus. *Rheumatology (Oxford)*. 53:1219–1228.

Uchida, K., and E.R. Stadtman. 1992. Modification of histidine-residues in proteins by reaction with 4-hydroxynonenal. *Proceedings of the National Academy of Sciences of the United States of America*. 89:4544–4548.

Ulvestad, E., K. Williams, R. Matre, H. Nyland, A. Olivier, and J. Antel. 1994. Fc-receptors for IgG on cultured human microglia mediate cytotoxicity and phagocytosis of antibody-coated targets. *J Neuropathol Exp Neurol*. 53:27–36.

van Gent, T., C.J. Heijnen, and P.D. Treffers. 1997. Autism and the immune system. *Journal of child Psychology and Psychiatry, and Allied Disciplines*. 38:337–349.

Vargas, D.L., C. Nascimbene, C. Krishnan, A.W. Zimmerman, and C.A. Pardo. 2005. Neuroglial activation and neuroinflammation in the brain of patients with autism. *Annals of Neurology*. 57:67–81.

Veerhuis, R., H.M. Nielsen, and A.J. Tenner. 2011. Complement in the brain. *Molecular Immunology*. 48:1592–1603.

Vincent, A., C.G. Bien, S.R. Irani, and P. Waters. 2011. Autoantibodies associated with diseases of the CNS: New developments and future challenges. *Lancet Neurology*. 10:759–772.

Vlajkovic, S., and B.D. Jankovic. 1991. Experimental epilepsy in vitro: Neuromodulating activity of anti-brain autoantibodies from rats exposed to electroconvulsive shock. *International Journal of Neuroscience*. 59:205–211.

Vojdani, A., M. Bazargan, E. Vojdani, J. Samadi, A.A. Nourian, N. Eghbalieh et al. 2004. Heat shock protein and gliadin peptide promote development of peptidase antibodies in children with autism and patients with autoimmune disease. *Clinical and Diagnostic Laboratory Immunology*. 11:515–524.

Wang, H., K.L. Munger, M. Reindl, E.J. O'Reilly, L.I. Levin, T. Berger et al. 2008. Myelin oligodendrocyte glycoprotein antibodies and multiple sclerosis in healthy young adults. *Neurology*. 71:1142–1146.

Wang, K.K.W. 2000. Calpain and caspase: Can you tell the difference? *Trends in Neurosciences*. 23:59–59.

Wang, Y.Z., F.F. Tian, M. Yan, J.M. Zhang, Q. Liu, J.Y. Lu et al. 2014. Delivery of an miR155 inhibitor by anti-CD20 single-chain antibody into B cells reduces the acetylcholine receptor-specific autoantibodies and ameliorates experimental autoimmune myasthenia gravis. *Clinical and Experimental Immunology*. 176:207–221.

Warren, R.P., A. Foster, and N.C. Margaretten. 1987. Reduced natural killer cell activity in autism. *Journal of the American Academy of Child and Adolescent Psychiatry*. 26:333–335.

Warren, R.P., V.K. Singh, R.E. Averett, J.D. Odell, A. Maciulis, R.A. Burger et al. 1996. Immunogenetic studies in autism and related disorders. *Molecular and Chemical Neuropathology*. 28:77–81.

Waubant, E. 2008. Spotlight on anti-CD20. *International MS Journal/MS Forum*. 15:19–25.

Whitney, K.D., and J.O. McNamara. 2000. GluR3 autoantibodies destroy neural cells in a complement-dependent manner modulated by complement regulatory proteins. *Journal of Neuroscience*. 20:7307–7316.

Willenborg, D.O., and S.J. Prowse. 1983. Immunoglobulin-deficient rats fail to develop experimental allergic encephalomyelitis. *Journal of Neuroimmunology*. 5:99–109.

Wills, S., M. Cabanlit, J. Bennett, P. Ashwood, D.G. Amaral, and J. Van de Water. 2009. Detection of autoantibodies to neural cells of the cerebellum in the plasma of subjects with autism spectrum disorders. *Brain, Behavior, and Immunity*. 23:64–74.

Wills, S., C.C. Rossi, J. Bennett, V.M. Cerdeño, P. Ashwood, D.G. Amaral et al. 2011a. Further characterization of autoantibodies to GABAergic neurons in the central nervous system produced by a subset of children with autism. *Molecular Autism*. 2:5.

Wills, S., C.C. Rossi, J. Bennett, V. Martinez Cerdeno, P. Ashwood, D.G. Amaral et al. 2011b. Further characterization of autoantibodies to GABAergic neurons in the central nervous system produced by a subset of children with autism. *Molecular Autism*. 2:5.

Winder, C., and J.C. Balouet. 2002. The toxicity of commercial jet oils. *Environmental Research*. 89:146–164.

Wright, B.R., A.E. Warrington, D.E. Edberg, and M. Rodriguez. 2009. Cellular mechanisms of central nervous system repair by natural autoreactive monoclonal antibodies. *Archives of Neurology*. 66:1456–1459.

Xing, G.Q., M. Ren, W.A. Watson, J.T. O'Neil, and A. Verma. 2009a. Traumatic brain injury-induced expression and phosphorylation of pyruvate dehydrogenase: A mechanism of dysregulated glucose metabolism. *Neuroscience Letters*. 454:38–42.

Xing, G.Q., M. Ren, W.D. Watson, J.T. O'Neill, and A. Verma. 2009b. Traumatic brain injury-induced expression and phosphorylation of pyruvate dehydrogenase: A mechanism of dysregulated glucose metabolism. *Neuroscience Letters*. 463:258–258.

Yahya, M.D., J.L. Pinnas, G.C. Meinke, and C.C. Lung. 1996. Antibodies against malondialdehyde (MDA) in MRL/lpr/pr mice: Evidence for an autoimmune mechanism involving lipid peroxidation. *Journal of Autoimmunity*. 9:3–9.

Yang, L., P.C. Blumbergs, N.R. Jones, J. Manavis, G.T. Sarvestani, and M.N. Ghabriel. 2004. Early expression and cellular localization of proinflammatory cytokines interleukin-1beta, interleukin-6, and tumor necrosis factor-alpha in human traumatic spinal cord injury. *Spine*. 29:966–971.

Younger, D.S., L.P. Rowland, N. Latov, A.P. Hays, D.J. Lange, W. Sherman et al. 1991. Lymphoma, motor neuron diseases, and amyotrophic lateral sclerosis. *Annals of Neurology*. 29:78–86.

Yuan, X.Q., and C.E. Wade. 1991. Neuroendocrine abnormalities in patients with traumatic brain injury. *Frontiers in Neuroendocrinology*. 12:209–230.

Zachou, K., K. Oikonomou, Y. Renaudineau, A. Chauveau, N. Gatselis, P. Youinou et al. 2012. Anti-alpha actinin antibodies as new predictors of response to treatment in autoimmune hepatitis type 1. *Alimentary Pharmacology and Therapeutics*. 35:116–125.

Zahran, W.E., M.I. Mahmoud, K.A. Shalaby, and M.H. Abbas. 2013. Unique correlation between mutated citrullinated vimentine IgG autoantibodies and markers of systemic inflammation in rheumatoid arthritis patients. *Indian Journal of Clinical Biochemistry*. 28:272–276.

Zhang, M., E.M. Alicot, and M.C. Carroll. 2008. Human natural IgM can induce ischemia/reperfusion injury in a murine intestinal model. *Molecular Immunology*. 45:4036–4039.

Zhang, M., K. Takahashi, E.M. Alicot, T. Vorup-Jensen, B. Kessler, S. Thiel et al. 2006. Activation of the lectin pathway by natural IgM in a model of ischemia/reperfusion injury. *Journal of Immunology*. 177:4727–4734.

Zhang, Z., J.S. Zoltewicz, S. Mondello, K.J. Newsom, Z. Yang, B. Yang et al. 2014. Human traumatic brain injury induces autoantibody response against glial fibrillary acidic protein and its breakdown products. *PloS One*. 9:e92698.

Zheng, W.R., M.C. Liu, R.L. Hayes, and K.K.W. Wang. 2008. Calpain and caspase mediated proteolysis of alpha II-spectrin and beta II-spectrin in rat cerebrocortical culture under oncotic, apoptotic and excitotoxic challenges. *Journal of Neurotrauma*. 25:900–900.

Zolotova, T.V., and I.E. Grebeniuk. 2010. [Serum level of anti-neurotrophine antibodies and activity of proteolytic enzymes in patients with sensorineural loss of hearing]. *Vestnik Otorinolaringologii* 25–28.

30 Systems Biology Applications to Decipher Mechanisms and Novel Biomarkers in CNS Trauma

Chenggang Yu and Firas Kobeissy

CONTENTS

30.1 INTRODUCTION

Systems biology approaches have become indispensable in processing huge and heterogeneous biomedical data, extracting essential information and forming novel hypotheses from various experimental results. In the field of neurotrauma studies, decades of efforts in animal experiments and clinical studies accumulated wealthy knowledge on the pathophysiology of traumatic brain injury (TBI). Application of systems biology strategies to holistically analyze the complex molecular pathways and networks of brain trauma can be of high importance due to its impact on biomarker research. In this chapter, we discuss current available systems biology strategies, databases, and tools and their applications to existing TBI data sets that can be used to identify new biomarker candidates investigating the different underlying molecular mechanisms and pathways of TBI responses.

The emerging large-scale and high-throughput technologies in molecular biology allowed rapid accumulation of vast biomedical data and its documentation in open access databases. This entailed unraveling sequences of whole genomes for thousands of species, identification of new genes and proteins, and studying biological function of these interacting biomolecules. This has provided scientists with the tools and material to investigate complex molecular mechanisms of biological systems (Kitano, 2002; Wang et al., 2010). However, researchers could be easily overwhelmed by this enormous amount of data and by the intricacy of the large number of biomolecules that are intertwined in a system. In fact, the complexity of biological processes is challenging the traditional reductionist thinking of biomedical studies investigating a few biomolecules for their effects under well-controlled experimental conditions (Strange, 2005). Although small-scale experiments are essential in accumulating knowledge and validating hypotheses, they are not effective in incubating novel ideas and hypotheses of the molecular mechanisms of complicated biological phenomenon. This instigated the birth of the new field of systems biology.

Systems biology emerged from molecular biology as an interdisciplinary subject that employs mathematics and computer science to study complex molecular mechanisms of biological phenomenon in a systematic and holistic way. In contrast to reductionist thinking that decomposes a biological system into small components and studies them separately, systems biology explores the molecular basis of the whole biological system by studying its construction, operation, different components' coordination, and system responses to external perturbations (Strange, 2005). By focusing on the interactions of multiple different biomolecules, systems biology approaches have been developed to incorporate heterogeneous biomedical data and knowledge to explain experimental results, provide insights into the molecular basis of complex diseases, prioritize genes and proteins from large-scale screenings, and discover molecular biomarkers and potential drug targets (Kaimal et al., 2011; Noorbakhsh et al., 2009; Ptitsyn et al., 2008).

Today, systems biology approaches have become indispensable in processing huge and heterogeneous biomedical data, extracting essential information and forming novel hypotheses from various experimental results. In the field of neurotrauma studies, decades of efforts in animal experiments and clinical studies accumulated wealthy knowledge on the pathophysiology of TBI. Meanwhile, we recognize that the complexity of the molecular mechanisms underpinning the primary and the secondary injury induced by traumatic insult are still poorly understood (Albert-Weissenberger and Sirén, 2010). This significantly hindered the discovery of molecular biomarkers for diagnosis and prognosis of TBI and the development of neurotrauma therapeutics. As more neurotrauma data are garnered and become available to the whole community, systems biology is in immense need for researchers in the field to incorporate these experimental data and knowledge so as to decipher molecular mechanisms and discover novel molecular biomarkers. Actually, systems biology has been applied in some neurotrauma studies and a couple of review articles have been published recently (Feala et al., 2013; Zhang et al., 2010). However, the full potential of systems biology is not yet recognized by the whole neurotrauma community.

In this chapter, we will introduce systems biology basics and methodologies. We will focus on three groups of approaches: (1) gene set analysis; (2) pathway analysis; and (3) protein-protein interaction (PPI) analysis. All these methods can be applied in neurotrauma studies, where they help gain insights into molecular mechanisms or infer novel biomarkers. The actual application of systems biology in neurotrauma studies will be briefly reviewed in the second part of this chapter. This chapter provides general guidance and inspires more neurotrauma studies that exploit the predictive power of systems biology.

30.2 THE APPROACHES OF SYSTEMS BIOLOGY

30.2.1 Overview

Systems biology studies biosystems by incorporating the knowledge of the functions of individual biomolecules, their biological roles in well-studied biological processes, and their binding partners and regulators. The information helps link individual biomolecules together to construct a systematic understanding of complex biological processes. Several databases have been developed to provide such information. Table 30.1 lists a collection of resources that are available over the Internet. These resources could be roughly categorized into three types: function category (also called gene set), pathway, and PPI network. Figure 30.1 illustrates the three types of data.

Classifying genes into different functional groups is an important approach in the analysis of large numbers of genes (as well as proteins), which will yield major functional

TABLE 30.1
Knowledge Bases for Systems Biology Studies

Name	URL	Reference
	Function category	
Pfam	http://www.sanger.ac.uk/resources/databases/pfam.html	Punta et al., 2012
Panther	http://www.pantherdb.org/	Thomas et al., 2003
UniProt	http://www.uniprot.org/	Magrane and Consortium, 2011
Gene Ontology	http://www.geneontology.org/	Ashburner et al., 2000
DAVID	http://david.abcc.ncifcrf.gov/	Huang et al., 2009a
	Pathway	
KEGG	http://www.genome.jp/kegg/pathway.html	Kanehisa et al., 2010
Reactome	http://www.reactome.org/	Croft, 2013
Biocarta	http://www.biocarta.com	Nishimura, 2001
WikiPathway	http://www.wikipathways.org/	Kelder et al., 2012
PID	*http://pid.nci.nih.gov/*	Schaefer et al., 2009
CPDB	http://cpdb.molgen.mpg.de/	Kamburov et al., 2009
MsigDB	http://www.broadinstitute.org/gsea/msigdb/	Subramanian et al., 2005
	Protein–protein interaction	
Interactome	http://interactome.dfci.harvard.edu/	Rual et al., 2005
BIND	http://bond.unleashedinformatics.com/	Bader et al., 2003
BioGRID	http://thebiogrid.org/	Breitkreutz et al., 2008
HPRD	http://www.hprd.org/	Mathivanan et al., 2008
IntAct	http://www.ebi.ac.uk/intact/	Kerrien et al., 2012
MIPS	http://mips.helmholtz-muenchen.de/proj/ppi/	Pagel et al., 2005
MINT	http://mint.bio.uniroma2.it/mint/Welcome.do	Licata et al., 2012

BIND, Biomolecular Object Network Database; CPDB, Consensus Pathway Database; DAVID, Database for Annotation, Visualization and Integrated Discovery; HPRD, Human Protein Reference Database; MsigDB, Molecular Signatures Database; PID, Pathway Interaction Database.

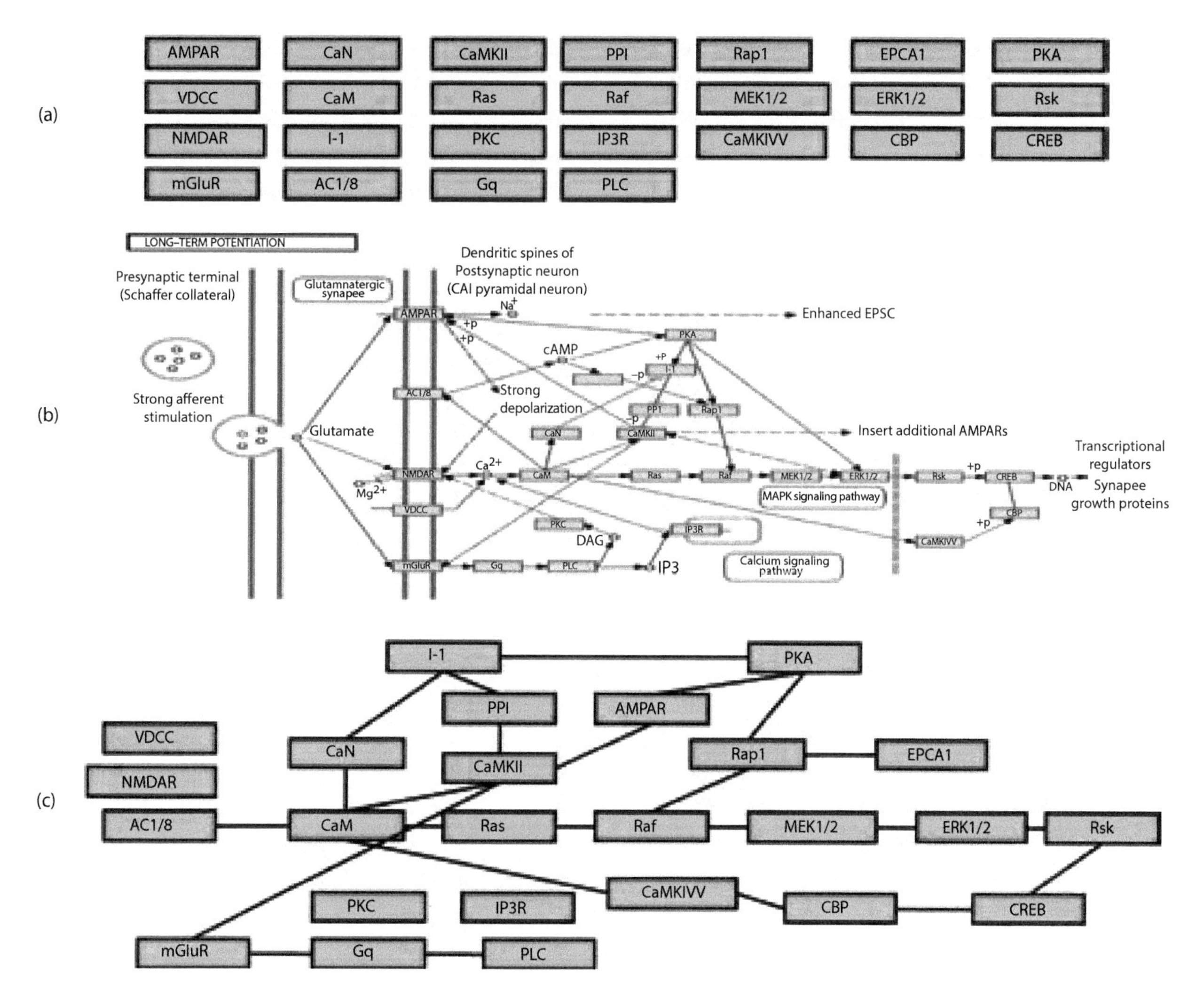

FIGURE 30.1 An illustration of a gene set (a), a KEGG pathway (b), and a protein–protein interaction (PPI) network (c). The gene set and the PPI network are derived from the KEGG pathway by either ignoring interactions among proteins or by ignoring detail information that describes the interactions. Because a PPI network only describes direct interactions among proteins, those interactions via small molecules in the KEGG pathway are ignored in a PPI network (c), which results in four proteins that do not connect to others.

groups that might be essential in a study. Although various function annotations, such as Pfam (Punta et al., 2012), PANTHER (Thomas et al., 2003), and UniProt (Magrane and Consortium, 2011) could be used to categorize genes, gene ontology (GO) is probably the most widely used function categorization scheme (Ashburner et al., 2000). The GO project aims at standardizing the annotation of genes and gene products by providing a controlled vocabulary of terms. These GO terms are grouped in three main categories: molecular function, biological process, and subcellular localization. In each category, several GO terms are organized in a hierarchical fashion to describe gene function in different granularity. For example, a protein kinase could be annotated by a set of GO terms, including generic terms such as "macromolecule modification," "transferase activity," and more specific terms such as "kinase activity" and "protein kinase activity." This makes for easy and flexible categorization of genes into groups with controlled functional granularity. In addition to GO, the Database for Annotation, Visualization and Integrated Discovery Bioinformatics Resources integrates a number of databases, including GO, in one warehouse to provide comprehensive function annotation, categorization, and systems biology analysis for user-uploaded data (Huang et al., 2009a).

A molecular pathway is a wiring diagram that illustrates the molecular mechanism of a particular biological process. The diagram uses blocks of different shapes to represent proteins and other molecules, and uses lines and arrows that link blocks to represent interactions among them. The diagram also illustrates cellular locations of pathway members and labels the types of interactions. A pathway diagram is usually constructed by field experts according to experimental data, literature, and established theories or hypotheses. The size of pathways varies from a few proteins to hundreds of proteins, depending on the complexity of a defined biological process, and the details included in the pathway (Croft, 2013). A big number of pathways has been created by different authors and organizations. A selection of pathway resources is shown in Table 30.1. Kyoto Encyclopedia of Genes and Genomes (KEGG) pathways are constructed and regularly updated by experts from the Bioinformatics Center and the Human

Genome Center at the University of Tokyo (Kanehisa et al., 2010). These pathways are organized into categories including metabolism, genetic and environmental information processing, cellular process, organismal systems, and human diseases. KEGG provides pathways for different species, which include more than 400 human pathways consisting of around 5,000 proteins. KEGG pathways for different species could be derived from the same generalized pathway, called the canonical pathway, which represents common properties of a biological process. They may not be suitable in the study of species-specific mechanisms. Reactome is an open source, open access, manually created, and peer-reviewed pathway database that contains a quality pathway constructed by expert biologists in collaboration with Reactome editorial staff (Croft, 2013; D'Eustachio, 2013). Reactome pathways are developed for 21 species. These include more than 1,400 human pathways consisting of more than 7,000 proteins, which almost doubles the total number of proteins covered by the KEGG human pathways. In addition, Reactome pathways distinguish protein isoforms, whereas KEGG pathways use a single protein-coding gene to represent all isoforms. Thus Reactome may be more suitable and optimal than KEGG for proteomic studies that may assist in distinguishing protein isoforms and their functional differences.

In contrast to KEGG and Reactome, which have their own staff to develop pathways, BioCarta (Nishimura, 2001) and WikiPathways (Kelder et al., 2012) make an open platform that allows everyone to construct his or her own pathways and publish on their websites. This strategy can effectively aggregate the intelligence of individuals to quickly update and expand the collection of molecular pathways. However, the quality of these pathways can be a concern. To ensure the quality, BioCarta recruits a group of "gurus" who are recognized as experts of their research fields to curate newly submitted pathways, whereas WikiPathways employs a peer-review strategy to correct false information in its pathways. Nevertheless, how many human experts participate is the key for the quality of the two databases. Recently, pathway databases that incorporate different resources also have emerged. These include Pathway Interaction Database (PID) (Schaefer et al., 2009), which imported BioCarta and Reactome human pathways, MsigDB (Subramanian et al., 2005), and the Consensus Pathway Database (Kamburov et al., 2009), which includes pathways from 3 and 32 resources, respectively. Although the coverage of these database warehouses is considerably increased, users should be aware of the quality of those pathways. Knowing the pathways' original resources and whether these pathways have been professionally curated is important for users for an appropriate data analysis. For example, in addition to pathways from BioCarta and Reactome, PID also contains a set of National Cancer Institute–Nature Curated pathways that were created by Nature Publishing Group editors and were reviewed by experts in the field (Schaefer et al., 2009).

Although biological pathways provide essential and abundant information on how biomolecules and their interactions contribute to particular biological processes, only a fraction of proteins in a whole proteome could be covered by these pathways. For example, combining KEGG and Reactome human pathways together only covers about 8000 unique proteins, which is less than 40% of the total proteins in a human proteome. The protein–protein interaction (PPI) network is a complementary resource that can cover many more proteins that are not presented in any pathways. A PPI network uses nodes and edges to represent proteins and their interactions. The interaction here means physical binding of two proteins, which is usually determined through high-throughput assays, such as yeast two-hybrid (Y2H) screening (Young, 1998) and the tandem affinity purification (TAP) (Puig et al., 2001). Several online resources provide PPI experimental data, some of which are listed in Table 30.1. Integration of PPI data from multiple resources usually yields a PPI network with maximal coverage of a proteome. However, the quality of PPI data from different resources and different experiments from manually curated interactions and interactions that were reported independently by multiple experiments can be considered high-quality data.

One may also pay attention to the types of experiments because of their inherent bias. For example, because the abundance of two proteins in a Y2H experiment could be higher than their actual level in living cells, their interaction observed in the Y2H experiment may never happen in a real cellular environment. The TAP experiment reports the identification of multiprotein complexes instead of protein pairs. The extraction of pair-wise protein interaction data from TAP experiments could be erroneous. Recently, Yu et al. developed a statistical method, termed interaction detection based on shuffling, to determine high-quality protein interactions from raw experimental data (Yu et al., 2012). This method can be used to construct a high-confidence PPI network in case the number of false interactions needs to be controlled. However, the tradeoff here is the size of the PPI network, which is reduced as a result.

30.2.2 Gene Set Analysis

The application of high-throughput technology such as gene microarray in experimental studies can easily screen tens of thousands of genes and identify hundreds of genes that may be relevant to a studied biological condition. These identified genes usually convey a variety of molecular functions and are involved in many different biological processes. Thus the identification of those molecular functions and biological processes that are prevalent in the identified genes is essential in understanding the molecular mechanisms of a biological condition. Gene set analysis, also known as gene set enrichment analysis or gene function enrichment analysis, aims to distill critical functions for genes observed from experimental studies, as illustrated in Figure 30.2. A gene set in this context refers to a group of genes that have the same function or that are involved in the same biological process.

Many gene set analysis approaches are based on an overrepresentation principle, which assumes that the function of a gene set is relevant and essential if it is overrepresented

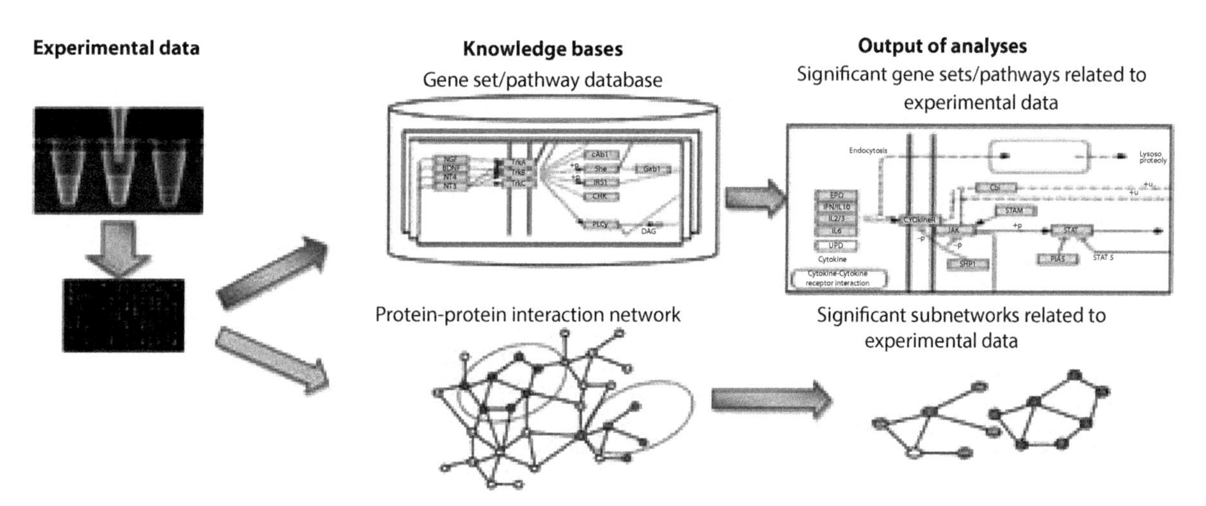

FIGURE 30.2 Systems biology analyses of experimental data. Experimental data, such as differential expression of genes, can be superimposed onto gene sets in a database such that the sets that are enriched with differential expressed genes are identified as significant gene sets that are related to the experimental condition. Significant pathways are identified in a similar way. However, interactions among gene products in pathways are taken into account in the analysis. Different from gene sets/pathway analysis, analysis based on protein–protein interaction (PPI) network is used to identify subnetworks enriched with proteins whose coding genes are either upregulated (red) or downregulated (green).

in genes identified from a study. The hypergeometric test, also known as Fisher's exact test, is the standard procedure in determining overrepresented functions (i.e., gene sets). Let us assume that an experiment tests N genes, which include N_f genes that are associated with a function F. Now the experiment identifies m differentially expressed genes, which include m_f genes of function F. To determine if function F is overrepresented in the m genes, we calculate the probability to obtain m_f or more genes of function F by randomly picking up m genes from the N genes, which is

$$p(X > m_f) = \sum_{x=m_f}^{\min(m,N_f)} \binom{N_f}{x} \binom{N-N_f}{m-x} \Big/ \binom{N}{m}$$

A smaller probability means that it is less likely that m_f or more genes of function F could be observed by chance. Therefore, it is more likely that function F is actually relevant to the experimental study. This analysis is also termed overrepresentation analysis (ORA). A caveat of ORA is that one has to use some cutoff threshold of some measurement to choose a group of genes (called significant genes) from all screened genes. The measurement could be a fold change or p value that reflects a gene's differential expression and the cutoff threshold is, to some extent, arbitrary. The clear arbitrary cutoff inevitably separates genes whose measurements are around the threshold level and just slightly different into different categories. On the other hand, all genes in the same category receive the same treatment regardless of the difference of their differentially expression measurements, which should have revealed different significance of these genes. This issue was addressed in the gene set enrichment analysis (GSEA) method (Subramanian et al., 2005). The method ranks all genes by the measurements of their differential expression and a weighted Kolmogorov-Smirnov test is applied to calculate a score for each gene set. Generally, if genes of a gene set locate closer to each other on the ranked gene list, the gene set will acquire a higher score. If genes of a gene set are randomly scattered in the ranked list, the gene set will acquire a very low score. A score is eventually converted into statistical significance through a randomization test, which is then used to determine if a gene set (or its corresponding function) is overrepresented.

Clearly, the GSEA method determines a significant gene set based on whether its member genes have similar changes in expression level, rather than the number of arbitrarily identified significant genes. Another solution to avoid setting an arbitrary cutoff threshold is through computing gene set aggregate score, which could be the mean, median, or other statistics of differential expression measurements (e.g., fold change, p value) for all genes in a gene set. A randomization test then converts the aggregate scores to gene set significance. This idea has been employed in gene set analysis tools, such as parametric analysis of gene set enrichment (Kim and Volsky, 2005) and generalized random set (Freudenberg et al., 2011).

Over the past 10 years, many gene set analysis tools have been developed. A review paper in 2008 listed approximately 68 tools (Huang et al., 2009b). However, many of these tools are based on similar approaches and all of them can be described by a general modular framework that was proposed by Ackermann and Strimmer (2009). The framework consists of five components: gene-level statistics, an optional transformation of gene-level statistics, gene set–level statistics, a choice of null hypothesis, and significance assessment. The first two components analyze individual genes and the last two components estimate gene set significance according to gene set–level statistics. The selection of gene set–level statistics is believed to be more important for the performance of a gene set analysis approach. This framework facilitates the systematic comparison of different gene set analysis tools

TABLE 30.2
Currently Available Systems Biology Approaches

Name	Source	Reference
	Gene set analysis	
ORA		
GSEA	http://www.broadinstitute.org/gsea/index.jsp	Subramanian et al., 2005
GRS	http://ClusterAnalysis.org/	Freudenberg et al., 2011
PAGE	From author upon request	Kim and Volsky, 2005
RS	http://www.stat.wisc.edu/~newton/	Newton et al., 2007
GLAPA	From author upon request	Maglietta et al., 2007
PADOG	http://bioinformaticsprb.med.wayne.edu/PADOG/	Tarca et al., 2012
DAVID	http://david.abcc.ncifcrf.gov/	Huang et al., 2009a
	Pathway analysis	
SPIA	http://vortex.cs.wayne.edu/ontoexpress/	Tarca et al., 2009
PathNet	http://www.bhsai.org/downloads/pathnet/	Dutta et al., 2012
GANPA	http://cran.r-project.org/web/packages/GANPA/	Fang et al., 2012
ACST	Article's online supplement	Mieczkowski et al., 2012
DEAP	Article's online supplement	Haynes et al., 2013
Clipper	http://romualdi.bio.unipd.it/	Martini et al., 2013
TEAK	http://code.google.com/p/teak/	Judeh et al., 2013
	Network analysis	
Aggregate z-score	http://www.cytoscape.org/	Ideker et al., 2002
MATISSE	http://www.cs.tau.ac.il/~rshamir/matisse/	Ulitsky and Shamir, 2007
COSINE	http://cran.r-project.org/web/packages/COSINE/	Ma et al., 2011
DEGAS	http://www.cs.tau.ac.il/~rshamir/matisse/	Ulitsky et al., 2010
NIPD	http://www.broadinstitute.org/~sroy/condspec/	Roy et al., 2011
	Commercial software	
IPA	http://www.ingenuity.com/	
MetaCore	http://thomsonreuters.com/metacore/	
Pathway Studio	http://www.elsevier.com/online-tools/pathway-studio/	

ASCT, analysis of consistent signaling transduction; COSINE, condition-specific subnetwork; DEAP, differential expression analysis for pathways; DEGAS, DysrEgulated Gene set Analysis via Subnetworks; GANPA, gene association network-based pathway analysis; GLAPA, Gene List Analysis with Prediction Accuracy; GRS, generalized random set; MATISSE, module analysis via topology of interactions and similarity sets; NIPD, network inference with pooling data; PADOG, pathway analysis with down-weighting of overlapping genes; PAGE, parametric analysis of gene set enrichment; TEAK, Topology Enrichment Analysis framework. See Table 30.1 for other definitions.

and helps the construction of customized tools for particular analysis. However, it may not benefit end users who rely on ready-to-use software and need practical guidance on the selection of right software.

Recent comparative studies provide some assistance on the selection of the right gene set analysis methods. Abatangelo et al. compared the performance of four methods: Fisher's exact test (i.e., ORA), GSEA, random set (RS) (Newton et al., 2007), and Gene List Analysis with Prediction Accuracy (Maglietta et al., 2007) using simulation data and nine real data sets (Abatangelo et al., 2009). They found that the performance of ORA was inferior to the other three methods. However, they could not determine the most suitable method from the three since no one could outperform the others in all data sets. Nevertheless, the authors still recommended GSEA in cases with no computational constraints. This method was also believed to have higher sensitivity in a recent comparative study (Hung et al., 2012). A selection of gene set analysis tools is listed in Table 30.2, which also includes pathway analysis with down-weighting of overlapping genes (Tarca et al., 2012). Even though it is claimed as a pathway analysis tool. Actually, pathways can be considered gene sets and analyzed by gene set analysis tools. Here we define pathway analysis tools as those that analyze pathways using pathway topological information; they are introduced in the next section.

30.2.3 Pathway Analysis

Although the identification of molecular pathways that are significantly activated or suppressed from experimental data is generally called pathway analysis, many such studies treat pathways as a set of genes and employ gene set analysis methods to find significant pathways. Some recent review articles on pathway analysis approaches also included these gene set analysis methods (Khatri et al., 2012). Unfortunately, these methods completely ignored the pathway's topological information (i.e., those paths and branches defined by the interactions of a series of proteins and other molecules). This information not only distinguishes a pathway from a simple gene set, but also conveys important information on the roles of different proteins and the ways that they interact to accomplish a cellular function. While gene set analysis treats all set members equally, a practical pathway analysis treats pathway members differently according to their locations in the pathway. In this section, we will introduce pathway analysis approaches that consider pathway topological information.

The early efforts to incorporate pathway topological information still used the framework of gene set overrepresentation analysis. Pathway topological information was used to adjust the significance of individual genes in a pathway. For example, Tarca et al. developed a signaling pathway impact analysis (SPIA) approach that measures the significance of a gene by two factors: (1) the differential expression of the gene and (2) the differential expression of upstream genes connected to the gene through paths in a pathway, which is termed perturbation factor (Tarca et al., 2009). The introduction of perturbation factor makes SPIA different from the standard gene set overrepresentation analysis. The SPIA adjustment of gene significance favors a pathway with differentially expressed genes located in the upstream of a pathway or along linear paths in the pathway, comparing it with a pathway with the same number of differentially expressed genes located randomly in it. One should expect that the SPIA-favored pathway is more likely to be activated, which yields better specificity and more sensitivity for SPIA. This was claimed by its authors after comparing the method with two gene set analysis methods (i.e., the standard overrepresentation analysis and GSEA, using three gene expression data sets and simulation data).

One potential issue for SPIA is that it perhaps gives too much weight to upstream genes because they will affect the significance of all downstream genes. This cannot be well justified by experimental studies. Therefore, the PathNet method (Dutta et al., 2012) took a similar framework as SPIA but developed a different way to compute the perturbation factor, which is called indirect evidence in PathNet. The statistical significance of a gene according to its differential expression is termed direct evidence in PathNet. A gene's significance is a combination of its direct evidence and indirect evidence that is computed by the direct evidence of all its interaction partners (i.e., genes that directly interact with that gene). A gene becomes more significant if more interaction partners are significant according to their direct evidence. This strategy favors a pathway with differentially expressed genes that directly interact with each other, whereas no additional weight is given to upstream genes. The authors of PathNet compared their method with SPIA, GSEA, and the standard overrepresentation analysis in the analysis of two gene microarray data sets of Alzheimer disease. They found that PathNet was able to identify the Alzheimer disease pathway and other biologically relevant pathways in multiple scenarios, whereas the other three methods oftentimes failed to do so (Dutta et al., 2012).

The overrepresentation principle adopted by both SPIA and PathNet requires an arbitrary threshold to determine significant genes. This issue has been discussed in the section of gene set analysis. In pathway analysis, the approach implies that those genes that are determined insignificant have no contribution to the activation of a pathway. Actually, multiple genes with a minor change of expression may have a collective impact to a pathway. This idea began to be adopted in more recently developed pathway analysis methods. For example, the gene association network-based pathway analysis (Fang et al., 2012) used a weighted mean of the statistical significance of all genes in a pathway to quantify a pathway's activation. Pathway topology is used through the computation of the weight assigned to each gene in a pathway, which is proportional to the fraction of interactions between a gene and other genes present in the pathway over the total number of interactions that involve the gene.

In more developed pathway analysis methods, the direct use of pathway topology information and the search of activated/suppressed subpathways inside a predefined pathway are adopted. For example, the analysis of the consistent signaling transduction method (Mieczkowski et al., 2012) searches a pathway topology for a maximal consistent subgraph, which is a part of the pathway consisting of connected genes that have the same direction of impact to downstream genes. For example, in a part of pathway, genes that activate downstream genes are upregulated, whereas genes that inhibit downstream genes are downregulated such that all genes' effect to downstream genes is activation. Thus this part of the pathway forms a consistent subgraph. A pathway is scored by the expression of all genes in its maximal consistent subgraph and the subgraph's distance to the downstream target gene of the pathway. Significant pathways are then determined via a statistical test on whether a pathway's score is significantly higher than the score of a pathway with randomly expressed genes. A similar method was developed by Haynes et al. and termed differential expression analysis for pathways (Haynes et al., 2013). Differential expression analysis for pathways searches all linear paths for a pathway and uses a different formula to score linear paths and estimate their significance. Unlike analysis of consistent signaling transduction, a linear path does not require its member genes to be consistently expressed. However, inconsistent expression would reduce the total score of a linear path, making it less favorable as compared to linear paths with consistent expression. Recently developed pathway analysis approaches that search for subpathways also include Clipper and the Topology Enrichment Analysis framework (Judeh et al., 2013; Martini et al., 2013). Clipper converts pathway topology into a Gaussian graphical model and tests differences in mean and in covariance matrices between two experimental conditions, whereas the Topology Enrichment

Analysis framework method uses a Gaussian Bayesian network to rank and estimate the significance of subpathways.

In summary, although pathways were often analyzed by gene set analysis approaches with pathway topological information completely ignored, recent developments have offered a number of tools designed especially for pathway analysis. These tools, which are complicated and might require more computational resources, usually outperform those gene set analysis approaches in sensitivity and specificity. More biological meaningful results would be expected from these specialized pathway analysis tools. More importantly, the identification of subpathways from a big predefined pathway would help biologists focus on particular molecular processes that are relevant to their study and thus help in proposing new hypotheses for subsequent studies.

30.2.4 Protein–Protein Interaction Network Analysis

Similar to pathway analysis that explores experimental data and a group of predefined pathways to determine the pathways that are relevant to an experiment, PPI network analysis explores experimental data and a single protein interaction network to identify network regions that are relevant to a study. As one of the major subjects of systems biology studies, PPI networks have been intensively studied and used in a variety of research. This section will focus on the PPI network analysis used in discovering novel molecular mechanisms from high-throughput experimental data. Table 30.2 lists some of the currently existing tools.

The procedure of PPI network analysis begins with superimposing high-throughput experimental data, such as differential expression of genes from a gene microarray experiment or the change of protein abundance from a proteomic study, onto a PPI network. The abundant changed proteins are expected to concentrate in some regions of the network instead of randomly and uniformly scattering through the whole network. These regions, termed network modules or subnetworks, are assumed to represent the underlying molecular mechanisms of a study. A key part of PPI network analysis is employing a subnetwork search algorithm to find all subnetworks. Such an algorithm has at least two components: (1) a scoring function used to evaluate the subnetwork and (2) a searching algorithm to obtain an optimal subnetwork that has the maximal score. The scoring function is formulated based on a biological hypothesis about the characteristics of a subnetwork that would represent a particular biological pathway. This is very important in choosing PPI network analysis tools for a particular study. Regarding the searching algorithm, theoretically, given any scoring function, a greedy search will find the optimal subnetwork. However, it has been proved that finding the maximal-score subnetwork is nondeterministic polynomial-time (NP) hard, which means that the process is extremely time-consuming. Therefore in practice, heuristic methods aiming to obtain suboptimal solutions have been developed and are broadly used.

In their seminal work, Ideker et al. formulated their PPI network analysis method by scoring a subnetwork with a normalized aggregate z-score (i.e., a sum of z-scores of all genes in the network) and employing a heuristic simulated annealing algorithm to find suboptimal subnetworks (Ideker et al., 2002). Their search of a subnetwork begins with a working subnetwork that is a randomly selected protein and ends after a number of expansions and updating of the working subnetwork. The scoring function and the search algorithm that was designed to increase the score of a subnetwork actually retrieve subnetworks that were enriched with differentially expressed genes because the z-score of a gene measures its differential expression. The authors validated their method by showing that they could identify subnetworks that were consistent with known regulatory circuits. This method has been frequently cited over the past decade and inspired many similar methods aiming to derive biological pathways from a PPI network.

For example, Guo et al. developed a method that used a scoring function that measures the coexpression of genes in a subnetwork (Guo et al., 2007). A raw score of a subnetwork takes a sum of edge scores, which is the covariance of the expression of every two genes that form an edge in the subnetwork. The final score is calculated as the raw score normalized by the mean and the standard deviation of raw scores of randomly formed subnetworks. A similar scoring function was also formulated in program module analysis via topology of interactions and similarity sets, which measures coexpression of genes in a subnetwork while allowing the presence of uncorrelated genes (Ulitsky and Shamir, 2007). Module analysis via topology of interactions and similarity sets also allows assignment of different priors to different genes to represent their coexpression probability in a subnetwork. This is justified by the findings that only a fraction of genes in a pathway are usually coherently regulated under a particular condition. Corresponding to the edge score, a score such as Ideker's z-score for an individual gene is called the node score. Recently, an idea that combines node score and edge score was proposed in a method termed condition-specific subnetwork (Ma et al., 2011). The method used the F-statistic to measure the differential expression of a gene and used the expected conditional F (ECF)-statistic to measure the differential gene–gene coexpression.

The score of a subnetwork is then defined as the sum of normalized node scores (F-statistic) and edge scores (ECF-statistic). A scoring function of a subnetwork can also be defined implicitly in particular studies. For example, in the search of disease dysregulated pathways, the DysrEgulated Gene set Analysis via Subnetworks method determines dysregulated genes in each of multiple cases from a case-control study of a disease (Ulitsky et al., 2010). A gene is dysregulated in one case if the expression value is significantly different from the gene's expression values in the control cohort. The score of a subnetwork is then defined by the number of dysregulated genes in the subnetwork. When probabilistic graphic models are applied in the identification of subnetworks, a scoring function is replaced by a joint distribution of a set of random variables. Nevertheless, the underlying assumption is usually that connected genes tend to be coregulated, which is the most important thing that an end-user should know. The methods that used probabilistic graphic models include the method developed by Segal et al. and the most recent method, network inference with pooling data (Roy et al., 2011; Segal et al., 2003).

Whether the characteristics of a subnetwork are enriched with differentially expressed genes or with highly coexpressed genes, it is determined by a scoring function. Actually, the efficiency in getting the optimal or suboptimal subnetwork is dependent on the subnetwork searching algorithms. In addition to the simulated annealing algorithm, a genetic algorithm (Ma et al., 2011), variants of greedy search strategies (Nacu et al., 2007; Sohler et al., 2004), and a more efficient algorithm based on integer-linear programming (Dittrich et al., 2008) have been proposed. Nevertheless, for end-users in the neurotrauma community, selection of PPI network analysis tools with a proper scoring function is more important than tools implemented with effective searching algorithms.

30.3 SYSTEMS BIOLOGY IN NEUROTRAUMA STUDIES

30.3.1 Deciphering Molecular Mechanisms of Neurotrauma

Neurotrauma studies have accumulated a number of high-throughput genomic and proteomics data, which enable the application of systems biology approaches to gain insights into molecular mechanisms of neurotrauma. Feala et al. listed 10 of these data sets in their recent review paper (Feala et al., 2013). Early studies usually clustered identified genes into predefined functional categories as the first simple procedure to understand genomic responses that involve multiple genes. For example, Matzilevich et al. categorized more than 500 genes that were differentially expressed in a TBI rat model into 12 broad functional groups and found genes associated to categories of cell cycle, growth factors, inflammation, and neuropeptides were exclusively upregulated, indicating the activation of these processes after TBI (Matzilevich et al., 2002). Natale et al. classified differentially expressed genes identified in two TBI murine models into functional categories defined in Gene Ontology (Natale et al., 2003). They found that more than 50% of genes were associated with three categories: inflammation, transcription regulation, and cell adhesion/extracellular matrix. Gene set enrichment analysis methods have been applied in recent neurotrauma studies. By determining functional groups that are statistically overrepresented by differentially expressed genes, these methods make statistically sound hypotheses on the underlying biological processes induced in neurotrauma studies. For example, Babikian et al. employed a hypergeometric test in gene set enrichment analysis for hundreds of differentially expression genes from a juvenile rat TBI model (Babikian et al., 2010). They used the Database for Annotation, Visualization and Integrated Discovery online resource to perform the analysis and found genes associated with inflammation/immune processes and cytokine activity were consistently upregulated, and genes associated with neurotransmission/plasticity, development, and metabolism were downregulated.

Gene set enrichment analysis was also conducted by Shojo et al. in a study of apoptosis and the causal relationship between inflammation and apoptosis in TBI brain (Shojo et al., 2010). They employed the GSEA program and molecular signature databases MSigDB and GenMAPP to identify apoptosis-related genes from a time-series gene expression data set. Through the systems biology analysis and subsequently an immunohistochemical experiment, they concluded that TBI-induced apoptosis was mediated by an inflammation signaling pathway that was activated shortly after TBI. More recently, advanced pathway analysis tools and network-based analysis have been recognized by the neurotrauma research community. For example, the SPIA pathway analysis tool was used by Kochanek et al. in determining KEGG signaling pathways that could be involved in a blast-induced mild TBI rat model (Kochanek et al., 2013). From gene expression data at 2 hours postinjury, they determined the ribosome pathway and pathways for three neurodegenerative diseases: Parkinson's disease, Huntington's disease, and Alzheimer's disease. More pathways related to nervous systems were determined from gene expression data at 24 hours postinjury, including long-term potentiation, calcium-signaling, and neurodegenerative disease pathways. Boutte et al. conducted PPI network–based analyses on differential proteomic data from a TBI experiment using a rat penetrating ballistic-like brain injury model (Boutté et al., 2012). They used Pathway Studio tools to superimpose more than 300 proteins with differential abundance in penetrating ballistic-like brain injury and sham brains onto a PPI network and found that the two subnetworks having the greatest degree of connectivity were associated to neurite outgrowth and cellular differentiation. Other biological processes obtained from the global network analysis include brain plasticity, neuronal death inflammation, and proteolysis.

In spite of these applications, systems biology is still new to the whole community of neurotrauma researchers. Its potential in generating testable hypotheses on molecular mechanisms is not yet fully explored. Most of the findings from systems biology have been reported earlier in multiple studies. Novel hypotheses have not yet been proposed and tested in subsequent experiments. However, these efforts demonstrate its capability in systematically explaining large genomic data from neurotrauma studies by incorporating the knowledge of field experts that has been effectively collected and represented in various systems biology databases. Systems biology study of neurotrauma is moving toward revealing the complex molecular processes induced by brain trauma. It provides indispensable tools in comprehending the complicated roles of a large number of biomolecules in a system that involves immunological and inflammatory responses, neuroprotective and neurodegenerative processes, cell signaling, metabolic alterations, and so on. Although at present field experts are still needed to filter out meaningful results from gene set enrichment analysis, pathway analysis, and PPI network–based analysis, we can expect in the near future that novel testable hypothesis will be effectively generated from systems biology as new and advanced approaches are actively being pursued.

30.3.2 Inferring Molecular Biomarkers

Molecular biomarkers for traumatic brain injury have been vigorously searched over the past decade. The traditional strategies in discovering molecular biomarkers can be categorized as knowledge-driven or discovery-driven methods, which are also called "top-down" and "bottom-up" methods (Noorbakhsh et al., 2009). The knowledge-driven strategy infers biomarkers through understanding disease pathology, molecular mechanism, and the involved biomolecules. The strategy is restricted by our knowledge of diseases and is less effective in the search of TBI biomarkers because of the lack of understanding of the molecular mechanisms of action of TBI. Because, theoretically, all biomolecules could become biomarker candidates, the discovery-driven strategy employs high-throughput technologies to screen a large number of genes and proteins to determine those whose change in abundance could indicate TBI. The caveat of this strategy is that usually a large number of candidates could pass the screening, which may include a significant number of false positives from inherent noise and the semiquantification nature of the high-throughput technologies.

Systems biology could mitigate the limitation of discovery-driven strategy by filtering high-throughput data and identifying genes and proteins that might be relevant to neurotrauma. This idea was proposed earlier (Kobeissy et al., 2008) and discussed again in recent reviews of the discovery of TBI biomarkers (Feala et al., 2013; Guingab-Cagmat et al., 2013), TBI drug targets (Zhang et al., 2010), and biomarkers in neuropsychiatry (Alawieh et al., 2012), Kobeissy et al. used Pathway Studio to construct a functional interaction map that linked 59 proteins that showed significant abundant change post-TBI in a previous study (Kobeissy et al., 2006, 2008) to altered pathways that were associated with inflammation, cell survival/proliferation, and synaptic plasticity. A study of this interaction map would help the downselection of candidate biomarkers for subsequent confirmation and validation. Another advantage that systems biology might offer is to identify genes or proteins that might have been missed by experimental analysis, which was proposed by Kobeissy et al. and was supported in a recent study of systems biology approaches for discovering TBI biomarkers (Kobeissy et al., 2008).

In one study, Feala et al. analyzed 32 TBI biomarker candidates from the literature and found their associations with four KEGG pathways were statistically significant (Feala et al., 2013). Three of the pathways were relevant to TBI or nervous systems: the apoptosis pathway, amyotrophic lateral sclerosis pathway, and Alzheimer disease pathway. A PPI network analysis showed that these 32 TBI biomarker candidates were tightly connected to each other on a PPI network of more than 10,000 proteins. There were 15 interactions among these biomarker candidates, which is a sharp contrast to the average 0.39 interactions among 32 randomly selected proteins. This suggests that proteins that interact with the 32 candidates might also be potential TBI biomarker candidates. The authors determined seven such proteins, among which they found that the protein kinase, ABL1, was the most interesting one because of its known association with Alzheimer's disease.

To some extent, the intention of searching for a TBI biomarker candidate from TBI-relevant pathways or interaction networks is against the principle of systems biology, which believes that complex biological processes, such as TBI secondary injury, are governed by the interactions of multiple genes and proteins. Thus a panel of biomolecules serving as a TBI biomarker should be suggested by systems biology. Actually, this idea has been proposed and a study of using two proteins GFAP and UCH-L1 together as a TBI biomarker has been studied (Mondello et al., 2012). Considering the huge number of possible combinations of multiple proteins, systems biology will become pivotal in efficiently identifying the most effective combinations of proteins for TBI biomarker panels.

30.4 CONCLUSION

Today's biotechnologies are accumulating immense experimental data and knowledge from laboratories all over the world. Our knowledge update via literature browsing is no longer able to keep up with the accumulation of biomedical data. In the meantime, high-throughput technologies enable an exhaustive survey of biomolecules that could be involved in complex biomedical conditions. Manual comprehension of vast quantities of the output becomes formidable, if not completely impossible. Systems biology provides a unique computational solution to explore knowledge buried in vast heterogeneous data to explain specific high-throughput experimental results. The advancement in this field keeps renovating and inventing powerful computational tools for life scientists in broad research areas. Because of the complexity of molecular mechanisms underpinning traumatic brain injury, systems biology approaches are indispensable in gaining insights into the disease and in discovering molecular biomarkers for diagnostics. This was demonstrated by the pioneering works introduced here. It is expected that systems biology will manifest its full potential when more advanced approaches are explored.

REFERENCES

Abatangelo, L., R. Maglietta, A. Distaso, A. D'Addabbo, T.M. Creanza, S. Mukherjee, et al. 2009. Comparative study of gene set enrichment methods. *BMC Bioinform.* 10:275.

Ackermann, M., and K. Strimmer. 2009. A general modular framework for gene set enrichment analysis. *BMC Bioinform.* 10:47.

Alawieh, A., F.A. Zaraket, J.-L. Li, S. Mondello, A. Nokkari, M. Razafsha et al. 2012. Systems biology, bioinformatics, and biomarkers in neuropsychiatry. *Front. Neurosci.* 6:187.

Albert-Weissenberger, C., and A.-L. Sirén. 2010. Experimental traumatic brain injury. *Exp. Transl. Stroke Med.* 2:16.

Ashburner, M., C.A. Ball, J.A. Blake, D. Botstein, H. Butler, J.M. Cherry et al. 2000. Gene ontology: Tool for the unification of biology. The Gene Ontology Consortium. *Nat. Genet.* 25:25–9.

Babikian, T., M.L. Prins, Y. Cai, G. Barkhoudarian, I. Hartonian, D.A. Hovda et al. 2010. Molecular and physiological responses to juvenile traumatic brain injury: Focus on growth and metabolism. *Dev. Neurosci.* 32:431–41.

Bader, G.D., D. Betel, and C.W. V Hogue. 2003. BIND: The Biomolecular Interaction Network Database. *Nucleic Acids Res.* 31:248–50.

Boutté, A.M., C. Yao, F. Kobeissy, X.-C. May Lu, Z. Zhang, K.K. Wang et al. 2012. Proteomic analysis and brain-specific systems biology in a rodent model of penetrating ballistic-like brain injury. *Electrophoresis.* 33:3693–704.

Breitkreutz, B.-J., C. Stark, T. Reguly, L. Boucher, A. Breitkreutz, M. Livstone et al. 2008. The BioGRID Interaction Database: 2008 update. *Nucleic Acids Res.* 36:D637–40.

Croft, D. 2013. Building models using reactome pathways as templates. *Methods Mol. Biol.* 1021:273–83.

D'Eustachio, P. 2013. Pathway databases: Making chemical and biological sense of the genomic data flood. *Chem. Biol.* 20:629–35.

Dittrich, M.T., G.W. Klau, A. Rosenwald, T. Dandekar, and T. Müller. 2008. Identifying functional modules in protein-protein interaction networks: An integrated exact approach. *Bioinformatics.* 24:i223–31.

Dutta, B., A. Wallqvist, and J. Reifman. 2012. PathNet: A tool for pathway analysis using topological information. *Source Code Biol. Med.* 7:10.

Fang, Z., W. Tian, and H. Ji. 2012. A network-based gene-weighting approach for pathway analysis. *Cell Res.* 22:565–80.

Feala, J.D., M.D.M. Abdulhameed, C. Yu, B. Dutta, X. Yu, K. Schmid et al. 2013. Systems biology approaches for discovering biomarkers for traumatic brain injury. *J. Neurotrauma.* 30:1101–16.

Freudenberg, J.M., S. Sivaganesan, M. Phatak, K. Shinde, and M. Medvedovic. 2011. Generalized random set framework for functional enrichment analysis using primary genomics datasets. *Bioinformatics.* 27:70–7.

Guingab-Cagmat, J.D., E.B. Cagmat, R.L. Hayes, and J. Anagli. 2013. Integration of proteomics, bioinformatics, and systems biology in traumatic brain injury biomarker discovery. *Front. Neurol.* 4:61.

Guo, Z., L. Wang, Y. Li, X. Gong, C. Yao, W. Ma et al. 2007. Edge-based scoring and searching method for identifying condition-responsive protein-protein interaction sub-network. *Bioinformatics.* 23:2121–8.

Haynes, W.A., R. Higdon, L. Stanberry, D. Collins, and E. Kolker. 2013. Differential expression analysis for pathways. *PLoS Comput. Biol.* 9:e1002967.

Huang, D.W., B.T. Sherman, and R.A. Lempicki. 2009a. Systematic and integrative analysis of large gene lists using DAVID bioinformatics resources. *Nat. Protoc.* 4:44–57.

Huang, D.W., B.T. Sherman, and R.A. Lempicki. 2009b. Bioinformatics enrichment tools: Paths toward the comprehensive functional analysis of large gene lists. *Nucleic Acids Res.* 37:1–13.

Hung, J.-H., T.-H. Yang, Z. Hu, Z. Weng, and C. DeLisi. 2012. Gene set enrichment analysis: Performance evaluation and usage guidelines. *Brief. Bioinform.* 13:281–91.

Ideker, T., O. Ozier, B. Schwikowski, and A.F. Siegel. 2002. Discovering regulatory and signalling circuits in molecular interaction networks. *Bioinformatics.* 18 Suppl 1:S233–40.

Judeh, T., C. Johnson, A. Kumar, and D. Zhu. 2013. TEAK: Topology enrichment analysis framework for detecting activated biological subpathways. *Nucleic Acids Res.* 41:1425–37.

Kaimal, V., D. Sardana, E.E. Bardes, R.C. Gudivada, J. Chen, and A.G. Jegga. 2011. Integrative systems biology approaches to identify and prioritize disease and drug candidate genes. *Methods Mol. Biol.* 700:241–59.

Kamburov, A., C. Wierling, H. Lehrach, and R. Herwig. 2009. ConsensusPathDB—a database for integrating human functional interaction networks. *Nucleic Acids Res.* 37:D623–8.

Kanehisa, M., S. Goto, M. Furumichi, M. Tanabe, and M. Hirakawa. 2010. KEGG for representation and analysis of molecular networks involving diseases and drugs. *Nucleic Acids Res.* 38:D355–60.

Kelder, T., M.P. van Iersel, K. Hanspers, M. Kutmon, B.R. Conklin, C.T. Evelo et al. 2012. WikiPathways: Building research communities on biological pathways. *Nucleic Acids Res.* 40:D1301–7.

Kerrien, S., B. Aranda, L. Breuza, A. Bridge, F. Broackes-Carter, C. Chen et al. 2012. The IntAct molecular interaction database in 2012. *Nucleic Acids Res.* 40:D841–6.

Khatri, P., M. Sirota, and A.J. Butte. 2012. Ten years of pathway analysis: Current approaches and outstanding challenges. *PLoS Comput. Biol.* 8:e1002375.

Kim, S.-Y., and D.J. Volsky. 2005. PAGE: Parametric analysis of gene set enrichment. *BMC Bioinformatics.* 6:144.

Kitano, H. 2002. Systems biology: A brief overview. *Science.* 295:1662–4.

Kobeissy, F.H., A.K. Ottens, Z. Zhang, M.C. Liu, N.D. Denslow, J.R. Dave et al. 2006. Novel differential neuroproteomics analysis of traumatic brain injury in rats. *Mol. Cell. Proteomics.* 5:1887–98.

Kobeissy, F.H., S. Sadasivan, M.W. Oli, G. Robinson, S.F. Larner, Z. Zhang et al. 2008. Neuroproteomics and systems biology-based discovery of protein biomarkers for traumatic brain injury and clinical validation. *Proteomics. Clin. Appl.* 2:1467–83.

Kochanek, P.M., C.E. Dixon, D.K. Shellington, S.S. Shin, H. Bayır, E.K. Jackson et al. 2013. Screening of biochemical and molecular mechanisms of secondary injury and repair in the brain after experimental blast-induced traumatic brain injury in rats. *J. Neurotrauma.* 30:920–37.

Licata, L., L. Briganti, D. Peluso, L. Perfetto, M. Iannuccelli, E. Galeota et al. 2012. MINT, the molecular interaction database: 2012 update. *Nucleic Acids Res.* 40:D857–61.

Ma, H., E.E. Schadt, L.M. Kaplan, and H. Zhao. 2011. COSINE: COndition-SpecIfic sub-NEtwork identification using a global optimization method. *Bioinformatics.* 27:1290–8.

Maglietta, R., A. Piepoli, D. Catalano, F. Licciulli, M. Carella, S. Liuni et al. 2007. Statistical assessment of functional categories of genes deregulated in pathological conditions by using microarray data. *Bioinformatics.* 23:2063–72.

Magrane, M., and U. Consortium. 2011. UniProt Knowledgebase: A hub of integrated protein data. *Database (Oxford).* 2011:bar009.

Martini, P., G. Sales, M.S. Massa, M. Chiogna, and C. Romualdi. 2013. Along signal paths: An empirical gene set approach exploiting pathway topology. *Nucleic Acids Res.* 41:e19.

Mathivanan, S., M. Ahmed, N.G. Ahn, H. Alexandre, R. Amanchy, P.C. Andrews et al. 2008. Human Proteinpedia enables sharing of human protein data. *Nat. Biotechnol.* 26:164–7.

Matzilevich, D.A., J.M. Rall, A.N. Moore, R.J. Grill, and P.K. Dash. 2002. High-density microarray analysis of hippocampal gene expression following experimental brain injury. *J. Neurosci. Res.* 67:646–63.

Mieczkowski, J., K. Swiatek-Machado, and B. Kaminska. 2012. Identification of pathway deregulation—gene expression based analysis of consistent signal transduction. *PLoS One.* 7:e41541.

Mondello, S., A. Jeromin, A. Buki, R. Bullock, E. Czeiter, N. Kovacs et al. 2012. Glial neuronal ratio: A novel index for differentiating injury type in patients with severe traumatic brain injury. *J. Neurotrauma.* 29:1096–104.

Nacu, S., R. Critchley-Thorne, P. Lee, and S. Holmes. 2007. Gene expression network analysis and applications to immunology. *Bioinformatics.* 23:850–8.

Natale, J.E., F. Ahmed, I. Cernak, B. Stoica, and A.I. Faden. 2003. Gene expression profile changes are commonly modulated across models and species after traumatic brain injury. *J. Neurotrauma.* 20:907–27.

Newton, M., F. Quintana, J. Den Boon, S. Sengupta, and P. Ahlquist. 2007. Random-set methods identify distinct aspects of the enrichment signal in gene-set analysis. *Ann. Appl. Stat.* 1:85–106.

Nishimura, D. 2001. BioCarta. *Biotech Softw. Internet Rep.* 2:117–120.

Noorbakhsh, F., C.M. Overall, and C. Power. 2009. Deciphering complex mechanisms in neurodegenerative diseases: The advent of systems biology. *Trends Neurosci.* 32:88–100.

Pagel, P., S. Kovac, M. Oesterheld, B. Brauner, I. Dunger-Kaltenbach, G. Frishman et al. 2005. The MIPS mammalian protein-protein interaction database. *Bioinformatics.* 21:832–4.

Ptitsyn, A.A., M.M. Weil, and D.H. Thamm. 2008. Systems biology approach to identification of biomarkers for metastatic progression in cancer. *BMC Bioinformatics.* 9 Suppl 9:S8.

Puig, O., F. Caspary, G. Rigaut, B. Rutz, E. Bouveret, E. Bragado-Nilsson et al. 2001. The tandem affinity purification (TAP) method: A general procedure of protein complex purification. *Methods.* 24:218–29.

Punta, M., P.C. Coggill, R.Y. Eberhardt, J. Mistry, J. Tate, C. Boursnell et al. 2012. The Pfam protein families database. *Nucleic Acids Res.* 40:D290–301.

Roy, S., M. Werner-Washburne, and T. Lane. 2011. A multiple network learning approach to capture system-wide condition-specific responses. *Bioinformatics.* 27:1832–8.

Rual, J.-F., K. Venkatesan, T. Hao, T. Hirozane-Kishikawa, A. Dricot, N. Li et al. 2005. Towards a proteome-scale map of the human protein-protein interaction network. *Nature.* 437:1173–8.

Schaefer, C.F., K. Anthony, S. Krupa, J. Buchoff, M. Day, T. Hannay et al. 2009. PID: The Pathway Interaction Database. *Nucleic Acids Res.* 37:D674–9.

Segal, E., H. Wang, and D. Koller. 2003. Discovering molecular pathways from protein interaction and gene expression data. *Bioinformatics.* 19 Suppl 1:i264–71.

Shojo, H., Y. Kaneko, T. Mabuchi, K. Kibayashi, N. Adachi, and C. V Borlongan. 2010. Genetic and histologic evidence implicates role of inflammation in traumatic brain injury-induced apoptosis in the rat cerebral cortex following moderate fluid percussion injury. *Neuroscience.* 171:1273–82.

Sohler, F., D. Hanisch, and R. Zimmer. 2004. New methods for joint analysis of biological networks and expression data. *Bioinformatics.* 20:1517–21.

Strange, K. 2005. The end of "naive reductionism": Rise of systems biology or renaissance of physiology? *Am. J. Physiol. Cell Physiol.* 288:C968–74.

Subramanian, A., P. Tamayo, V.K. Mootha, S. Mukherjee, B.L. Ebert, M.A. Gillette et al. 2005. Gene set enrichment analysis: A knowledge-based approach for interpreting genome-wide expression profiles. *Proc. Natl. Acad. Sci. U.S.A.* 102:15545–50.

Tarca, A.L., S. Draghici, G. Bhatti, and R. Romero. 2012. Down-weighting overlapping genes improves gene set analysis. *BMC Bioinformatics.* 13:136.

Tarca, A.L., S. Draghici, P. Khatri, S.S. Hassan, P. Mittal, J.-S. Kim et al. 2009. A novel signaling pathway impact analysis. *Bioinformatics.* 25:75–82.

Thomas, P.D., M.J. Campbell, A. Kejariwal, H. Mi, B. Karlak, R. Davermanet al. 2003. PANTHER: A library of protein families and subfamilies indexed by function. *Genome Res.* 13:2129–41.

Ulitsky, I., A. Krishnamurthy, R.M. Karp, and R. Shamir. 2010. DEGAS: De novo discovery of dysregulated pathways in human diseases. *PLoS One.* 5:e13367.

Ulitsky, I., and R. Shamir. 2007. Identification of functional modules using network topology and high-throughput data. *BMC Syst. Biol.* 1:8.

Wang, C., C.M. Sanders, Q. Yang, H.W. Schroeder, E. Wang, F. Babrzadeh et al. 2010. High throughput sequencing reveals a complex pattern of dynamic interrelationships among human T cell subsets. *Proc. Natl. Acad. Sci. U. S. A.* 107:1518–23.

Young, K.H. 1998. Yeast two-hybrid: So many interactions, (in) so little time... *Biol. Reprod.* 58:302–11.

Yu, X., A. Wallqvist, and J. Reifman. 2012. Inferring high-confidence human protein-protein interactions. *BMC Bioinformatics.* 13:79.

Zhang, Z., S.F. Larner, F. Kobeissy, R.L. Hayes, and K.K.W. Wang. 2010. Systems biology and theranostic approach to drug discovery and development to treat traumatic brain injury. *Methods Mol. Biol.* 662:317–29.

Section V

Neurocognitive and Neurobehavioral Topics in Brain Injury

31 Neuropathology of Mild Traumatic Brain Injury

Correlation to Neurocognitive and Neurobehavioral Findings

Erin D. Bigler

CONTENTS

31.1 INTRODUCTION

The use of computed tomography (CT) and magnetic resonance imaging (MRI) in mild traumatic brain injury (mTBI) is overviewed in this chapter. Although in the majority of mTBI cases no abnormality will be shown, the common neuropathological changes that may be identified on CT and/or MRI are highlighted with an emphasis that such abnormalities provide only a macroscopic perspective of the pathology that may be viewed. Emphasis is placed on understanding the subtle nature of neuropathology that may accompany mTBI, the potential for dynamic changes that vary with time postinjury and that detection depends on which neuroimaging method is used. The role of advanced neuroimaging techniques that provide quantitative information about potential network-level damage using diffusion tensor imaging (DTI) and resting state functional MRI is overviewed with numerous examples provided that illustrate neuroimaging techniques that detect mTBI abnormalities.

The common structural neuroimaging methods and findings in mTBI will be overviewed. The field of neuroimaging is expansive and the basics of neuroimaging will not be covered in this review. For the reader who would like additional background information in neuroimaging of TBI, Wilde et al. (2012) provide such a synopsis. The chapter will conclude with a section that relates the macroscopically identified pathologies using conventional and advanced neuroimaging techniques with the ultrastructure and underlying pathophysiology of mTBI.

The neuroimaging investigation by Yuh et al. (2013) represents a comprehensive investigation of the common, visibly detected abnormalities observed in mTBI using day-of-injury (DOI) computed tomography (CT) followed by magnetic resonance imaging (MRI) during the subacute timeframe. In the Yuh et al. study, 135 mTBI patients were evaluated for acute head injury in three separate level 1 trauma centers in the United States and all were enrolled through an emergency department (ED) for prospective 3-month neurobehavioral outcome, assessed by the Extended Glasgow Outcome Scale (GOS-E). Although DOI CT imaging was done acutely, MRI on average was performed within 2 weeks postinjury. The National Institutes of Health has established the TBI Common Data Elements (TBI-CDEs; Haacke et al., 2010; Yue et al., 2013) for classifying both acute as well as chronic abnormalities, with all scan abnormalities identified by CDE criteria. CDE guidelines for pathoanatomical TBI findings on DOI CT or early MRI include skull fracture, hematoma (either epidural and/or subdural), traumatic axonal injury (defined as one to three foci), and diffuse axonal injury (DAI; defined as at least four foci). DOI CT foci are typically characterized as visibly identified contusions or intraparenchymally identified petechia. On MRI, such foci may take the form of white matter (WM) signal abnormality (hyperintense) and/or characteristic signal changes (hypointense) that reflect prior hemorrhage, often at the gray matter (GM)-WM interface. All of these types of macroscopic pathologies will be depicted in this chapter. Importantly in the Yuh et al. investigation, TBI-CDE features of more severe TBI such as midline shift ≥5 mm and partial/complete basal cistern effacement were not observed in any of the mTBI patients as part of that study. This is understandable and highlights that the visible abnormalities in mTBI do not reach the threshold associated with more severe TBI; nonetheless, very significant parenchymal injury may accompany mTBI.

The 2013 Yuh et al. investigation was a subset of a much larger investigation (McMahon et al., 2013) that prospectively followed 375 mTBI patients at 3, 6, and 12 months. The McMahon et al. (2013) study found that by 1 year, 22.4% of mTBI patients were still below functional status as measured by the GOS-E. Although there was an association of positive CT findings with poorer 3-month outcome, by 1 year whether DOI CT was abnormal or not did not predict outcome. Clearly, mTBI results in lasting sequelae for some, but this is not necessarily predicted by DOI CT findings. As will be shown in this chapter, advanced neuroimaging studies provide additional information and insight into the neuropathological effects of mTBI potentially useful in better understanding mTBI sequelae as well as providing additional information in the assessment and treatment of mTBI.

Figure 31.1 summarizes the findings of Yuh et al., which show that 44% of all mTBIs in this cohort of ED assessed individuals with mTBI had at least an identifiable neuroimaging abnormality. Clearly MRI was superior to CT in identifying abnormalities, especially those neuroimaging markers that infer axonal pathology. In fact, 27% of mTBI patients with normal head CTs had abnormal MRIs that were otherwise "missed" by DOI CT imaging. Of the 135 mTBI patients assessed in the Yu et al. investigation, only one had a Glasgow Coma Scale (GCS) of 13, with 26/135 (19%) assessed with a GCS of 14 and 108/135 (80%) with a GCS of 15. As such, the majority had a classically defined *maximum* GCS score, yet almost half had some positive neuroimaging finding. This observation underscores the frequency with which MRI may identify structural pathology in mTBI, even with a GCS of 15. In terms of the frequency of CDE findings and their relation to outcome, presence of any type of CDE-identified TBI abnormality increased the likelihood of lower GOS-E at 3 months, supporting the importance of identifying neuroimaging-based MRI abnormalities because of increased sensitivity in detecting gross pathology (Bigler, 2013a,b). However, the majority scanned had negative conventional imaging. For those with DOI CT, presence of subarachnoid hemorrhage was associated with poorer 3-month GOS-E. For those with positive MRI findings in the subacute time frame, presence of contusion or DAI was found to predict lower GOS-E.

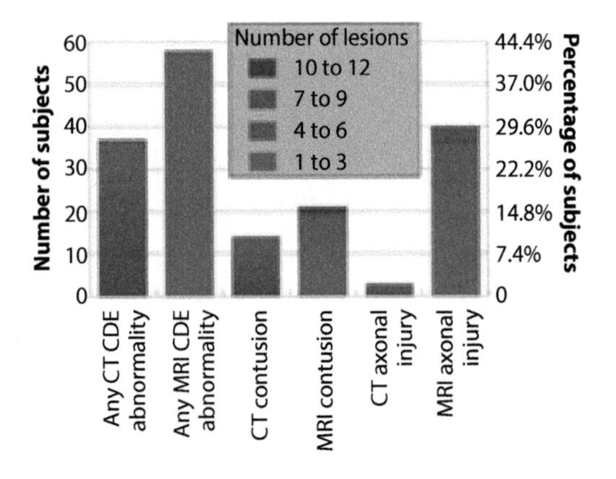

FIGURE 31.1 Incidence of CT versus MRI traumatic brain injury CDE abnormalities in 135 study participants. For MRI evidence of contusion and MRI evidence of hemorrhagic axonal injury, progressively darker shades of red indicate larger numbers of lesions. Study participants with CT evidence of brain contusion had, in most cases, evidence of one or two hemorrhagic contusions, with no CT demonstrating more than three convincing brain contusions. CT showed evidence of hemorrhagic axonal injury in 3 of 135 study participants, all with one to three foci of injury. (Color figure can be viewed at www.annalsofneurology.org.) (From Yuh, E.L. et al. Magnetic resonance imaging improves 3-month outcome prediction in mild traumatic brain injury. *Annals of Neurology*. 2013. 73(2):224–235. Copyright Wiley-VCH Verlag GmbH & Co. KGaA. Reproduced with permission.)

Figure 31.2 from the Yuh et al. investigation depicts some common CT and MRI findings in mTBI during the acute to early subacute timeframe. What is depicted in this illustration shows many of the classic observable, macroscopic lesion types in mTBI, which will be discussed more fully throughout this chapter.

Impressively, the Yuh et al. study shows that greater than 40% of mTBI patients initially evaluated within the ED will have positive CDE-identified abnormalities. However, the CDE technique requires visual identification of the abnormality from conventional clinical imaging (CT and/or MRI) studies and does not incorporate advanced magnetic resonance (MR) techniques to be discussed in subsequent sections. Nonetheless, it is important to understand what constitutes early neuroimaging identified abnormalities and how such findings relate to underlying neuropathology.

Conventional CT and MRI clinical studies configure anatomical images with millimeter resolution, meaning they detect gross pathology at a similar level, although submillimeter MR resolution is now possible (Yassa et al., 2010; Heidemann et al., 2012). In contrast, the fundamental pathological changes that occur from TBI happen at the micron and nanometer cellular level (Bigler and Maxwell, 2011, 2012), with only the largest of lesions being visible with contemporary neuroimaging (Bigler, 2013b). This means for brain injuries in the mild range, with the subtlest of neural injury that the macroscopic lesions will not be observed. However, as will be discussed in this chapter, this does not mean that those with negative imaging have no underlying pathology.

31.2 CT AND MRI IN mTBI

Once the mTBI patient is prepared and in the scanner, CT imaging can be completed within seconds to minutes.

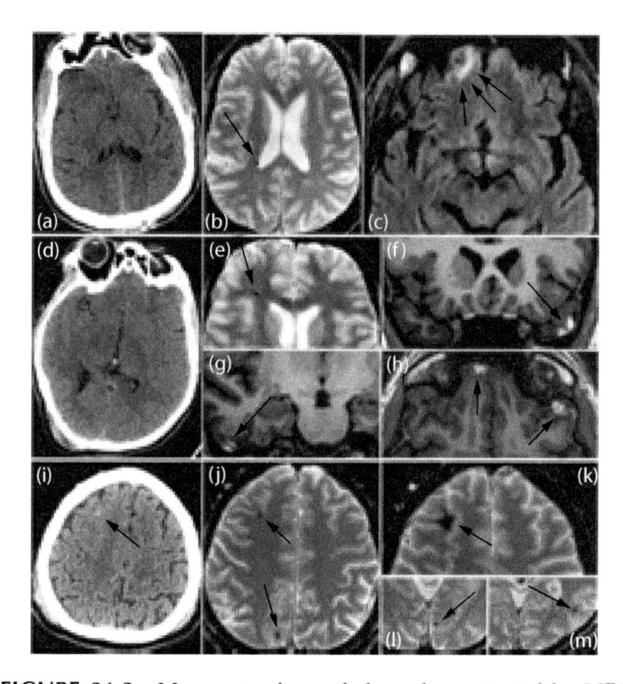

FIGURE 31.2 More extensive pathology demonstrated by MRI compared with CT in the study participants. (a–c) Fifty-year-old assaulted man. (a) Initial head CT was normal. MRI at 7 days postinjury demonstrated (b) hemorrhagic axonal injury along the right lateral ventricle (yellow arrow, axial T2*-weighted gradient echo) and (c) a right frontal contusion (red arrows, axial T2-weighted FLAIR). (d–h) Nineteen-year-old woman in motor vehicle collision. (e) Head CT showed no intracranial hemorrhage. (e) MRI at 12 days postinjury demonstrated hemorrhagic axonal injury in the deep right frontal white matter (yellow arrow, axial T2*-weighted gradient echo) and (f, h) four unsuspected hemorrhagic contusions (red arrows, three-dimensional T1-weighted inversion recovery spoiled gradient echo). (i–m) Fifty-four-year-old man after falling off a bicycle. (i) CT was interpreted as demonstrating trace right frontal subarachnoid hemorrhage (red arrow). (j–m) T2*-weighted gradient echo MRI demonstrated numerous discrete foci of subcortical white matter signal loss (yellow arrows), consistent with hemorrhagic axonal injury and with the TBI-CDE definition of diffuse axonal injury (≥4 visible discrete areas of axonal injury). (Color figure can be viewed in the online issue, which is available at www.annalsofneurology.org.) (From Yuh, E.L. et al. Magnetic resonance imaging improves 3-month outcome prediction in mild traumatic brain injury. *Annals of Neurology*. 2013. 73(2):224–235. Copyright Wiley-VCH Verlag GmbH & Co. KGaA. Reproduced with permission.)

Because CT imaging uses x-ray beam technology, objects with paramagnetic properties including life support and other medical assist devices are not precluded as in MRI. Likewise, metallic fragments from injury that may be paramagnetic can be imaged without concern about displacement by the strong magnetic fields generated by MRI, although the images will still be influenced by artifacts. Excellent contrast between bone and brain parenchyma can be achieved with CT, where CT clearly has the advantage over MRI in demonstrating presence and location of skull fractures, common sequelae with head injury, including mTBI. CT also provides methods for examining cerebrovasculature and inflammation in TBI, where potential perfusion technology may have application in identifying mTBI patients with more permanent sequelae (Metting et al., 2009, 2010, 2013).

It should be emphasized again that the clinical abnormalities as identified by CDE standards must be visible. To be visible on CT, there must be sufficient contrast between normal appearing parenchyma and damaged or abnormal tissue, typically observed in the form of hemorrhage or edema. On CT imaging, hemorrhage because of the quick clotting action of blood makes the lesion appear hyperdense in comparison to normal appearing parenchyma, whereas with edema the abnormality might appear less dense (i.e., more water content) or the typically distinct GM-WM boundaries are lost and/or there is loss of sulcal definition from swelling (Gean and Fischbein, 2010; Kim and Gean, 2011). However, these findings may not define where other shear/strain effects have occurred in the brain and likely reflect only a minimalist view of the pathology. To best understand TBI, the more advanced neuroimaging methods provide the tools to uncover additional neuropathology missed by the traditional "lesion analysis/identification" approach of conventional CT and MRI. Furthermore it is probably a fruitless endeavor to seek the prototypical "lesion" in mTBI because it likely does not exist. The macroscopic abnormalities identified in the CDE, especially with the DOI CT, may only indicate "tip-of-the-iceberg" phenomena and are not proportional to the total pathological effects of mTBI at the histological level. This is probably why the presence and location of a DOI CT abnormality does not necessarily predict outcome, because there is more than just the visibly identifiable lesion.

The common CT identified TBI-induced surface contusions typically occur at the brain–skull interface, whereas petechial hemorrhages often occur at the GM-WM interface (Gean and Fischbein, 2010; Kim and Gean, 2011). Both of these lesion patterns may be associated with focal or more diffuse patterns of edema. Presence of petechial hemorrhage in TBI is considered a marker of DAI, including in mTBI (Scheid et al., 2003, 2006); two examples are shown in Figure 31.3. Although skull fracture occurs externally to the brain, the biomechanical forces that resulted in the fracture do affect the brain. Skull fractures are associated with subdural, epidural, and subarachnoid hemorrhage. These hemorrhagic abnormalities may in turn be associated with brain displacement, mid-line shift, and associated edema. Interestingly, although hemorrhage and edema may evolve into a life-threatening medical emergency in mTBI, especially an epidural hematoma, once treated or removed, follow-up imaging may appear normal as shown in Figure 31.4. Figure 31.4 also demonstrates that dramatic acute abnormalities may occur in mTBI with subsequent "normal" appearing gross anatomy, which should not necessarily be interpreted as representing an entirely normal underlying microstructure, because it cannot actually be seen. Histological studies in TBI may show underlying pathology in what macroscopically appears to be normal tissue (Budde et al., 2011). This is also suggested by the fact that presence of extraaxial traumatic abnormalities such as skull fracture, epidural, subdural, and subarachnoid

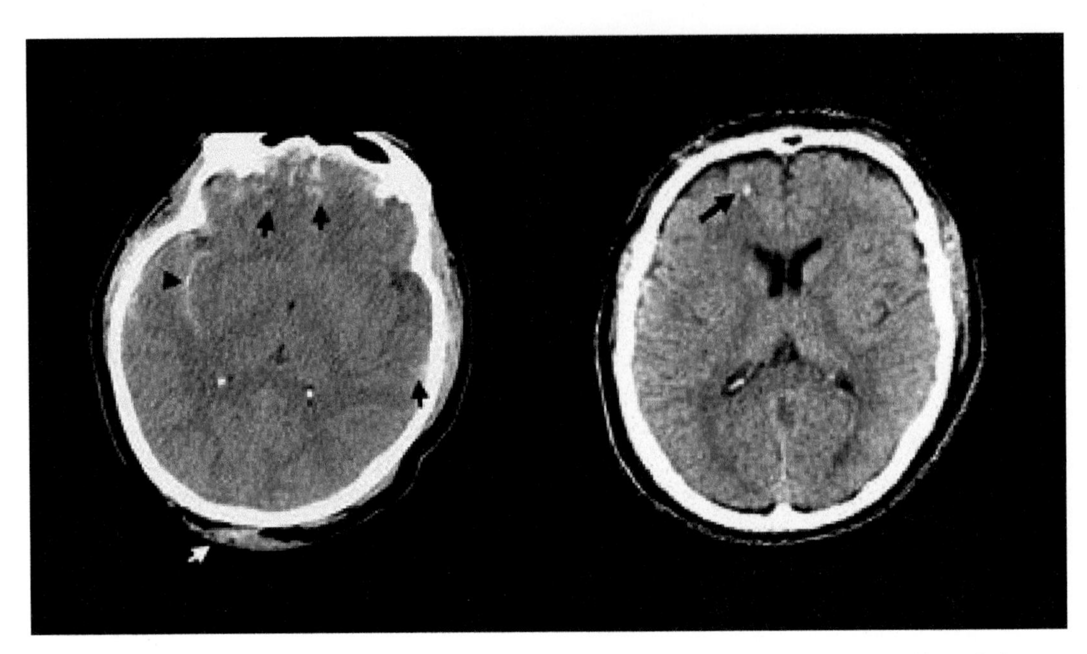

FIGURE 31.3 (Left) Hemorrhagic contusions as noted on DOI CT after a blow to the occiput, as indicated by soft-tissue swelling (white arrow) with characteristic contrecoup hemorrhagic contusions are seen in the inferior frontal and temporal lobes (black arrows). Also note the subarachnoid hemorrhage (arrowhead) in the right Sylvian fissure. (Right) Single punctate hemorrhagic lesion (bright white dot) in the frontal region right at the GM-WM interface. (From Kim, J.J. and A.D. Gean. *Neurotherapeutics* 8(1):39–53, 2011. With permission.)

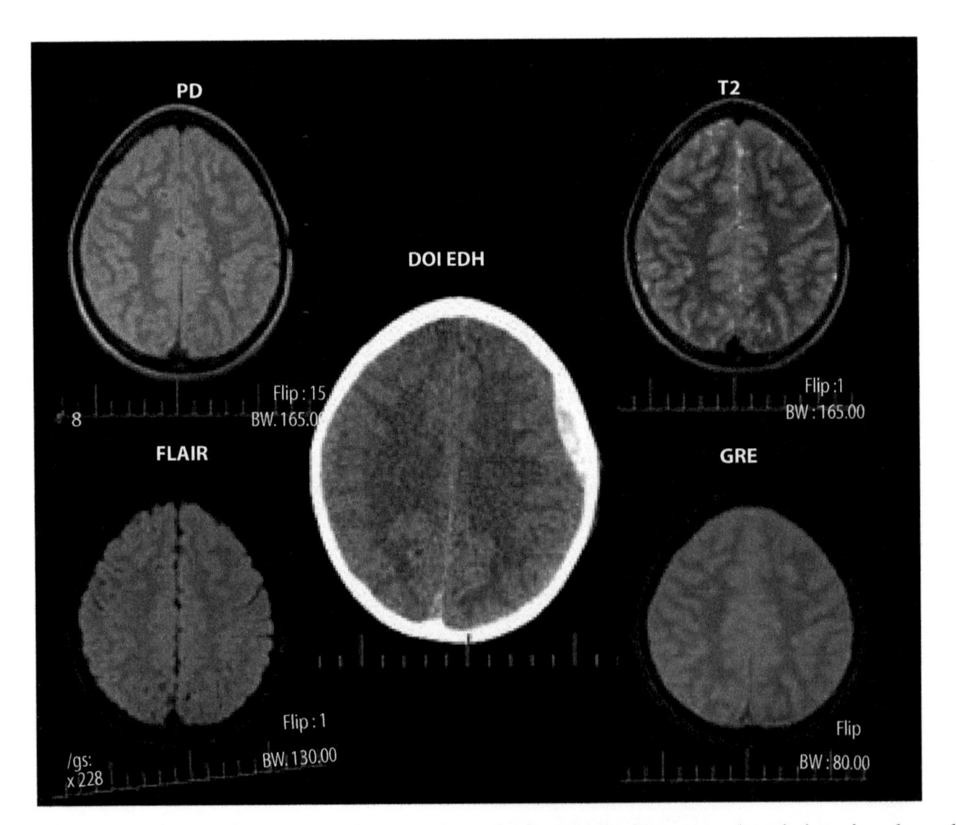

FIGURE 31.4 The DOI CT revealed an epidural hematoma as shown in the middle. However, chronic imaging showed no distinct parenchymal abnormality, regardless of the imaging sequence used. This illustrates that the pathology observed in the DOI scan may not result in identifiable pathology on follow-up imaging in mTBI.

hemorrhages all increased the odds of poorer outcome in the Yuh et al. mTBI study, with simple skull fracture having the least influence and subarachnoid hemorrhage the most. As such, an identifiable intraparenchymal lesion does not have to be present in mTBI to be associated with residual impairments implicating pathology below visible detection.

Presence of an abnormality on the DOI CT is the basis for the classification of "complicated mild TBI." However, given contemporary advances that identify mTBI abnormalities that simply are not detected by CT imaging, this classification is mostly meaningless. As stated in the Yuh et al. investigation, close to one-third of mTBI patients who had no DOI CT abnormality in fact had underlying pathology identified with MRI. Figure 31.1 demonstrated this point as well as the case shown in Figure 31.15. Currently, the superior MRI method for detecting hemorrhagic shear lesions in mTBI is susceptibility-weighted imaging (SWI) (Spitz et al., 2013), although as shown in Figure 31.1, the T2*-weighted gradient echo readily detects hemosiderin as well. The fluid attenuated inversion recovery (FLAIR) sequence is best suited for visibly detecting WM abnormalities in mTBI (Benson et al., 2012), as will be shown in the cases highlighted in Figures 31.11 and 31.15. In both of these cases, the WM pathology detected during MRI follow-up was completely undetected in the DOI CT.

31.3 EMPIRICALLY DERIVED QUANTITATIVE MRI ABNORMALITIES

The common anatomic images generated from MR technology, as with those shown in the various figures up to this point, are all derived from the underlying MR physics that constitute the image, which also form the basis for MR metrics. Diffusion tensor imaging (DTI) is one such metric related to assessing water anisotropy. As an example of how water diffusion metrics relate to axon integrity, and hence WM integrity, Figure 31.5 shows a DTI sequence with its associated color map in comparison to T1- and T2-weighted MR anatomical sequences. Two common metrics derived from DTI are referred to as fractional anisotropy (FA) and the apparent diffusion coefficient or ADC. Figure 31.5 provides a DTI schematic depicting the relationship of FA and ADC to axon integrity and putatively what happens with axon damage. These DTI metrics assess the microstructure of WM and are based on how water molecules are influenced by cell membranes and myelin characteristics including myelin sheath thickness. Healthy axonal membranes that are tightly compacted constrain the free movement and direction of water. Consequently, water molecules tend to move faster in parallel to nerve fibers rather than perpendicular to them. This characteristic, which is referred to as anisotropic diffusion, is measured by FA, which is influenced by the thickness of the axons, their myelin sheaths, aggregate compactness, and orientation with other axons. FA ranges from 0 to 1, where 0 represents maximal isotropic diffusion (e.g., free diffusion in perfect sphere) and 1 represents maximal anisotropic diffusion, i.e., diffusion in one direction (e.g., a long cylinder of minimal diameter). Diffusion anisotropy varies across WM regions, presumed to reflect differences in fiber myelination, fiber diameter, and directionality.

Using a technique referred to as tractography, the aggregate fiber tracts of an entire brain can be derived from DTI, as shown in Figure 31.6. In TBI, DTI may demonstrate a loss of fiber tract integrity, reflected as a thinning out of the number

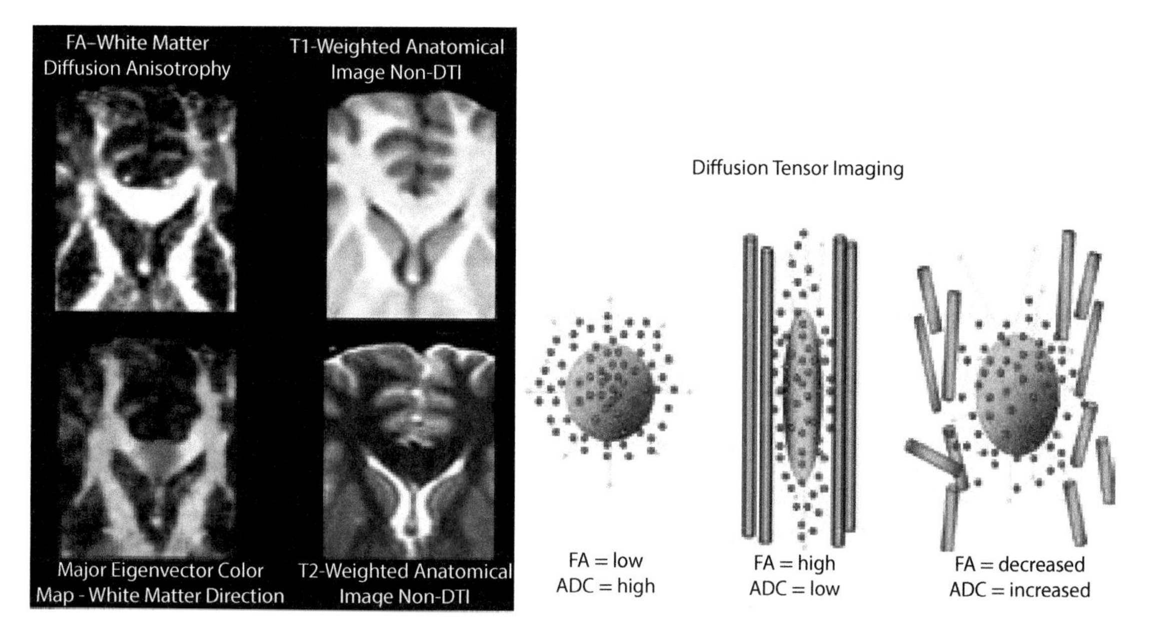

FIGURE 31.5 The images on the left show the comparison T1- and T2-weighted imaging sequence cropped at the level of the anterior corpus callosum (forceps minor) compared with the DTI sequence (upper left) and diffusion color map that reflects directionality of tracts (blue: vertically oriented tracts, green: anteroposterior oriented tracts, red-orange: laterally, side-to-side oriented tracts). The left image in the illustration on the right shows the random unrestricted water diffusion with spherical dispersion in contrast to directional movement imposed by the vertically oriented barrier that symbolizes an axon membrane. However, in injury or disease, if the axon membrane breaks down, water dispersion approaches the unrestricted spherical movement.

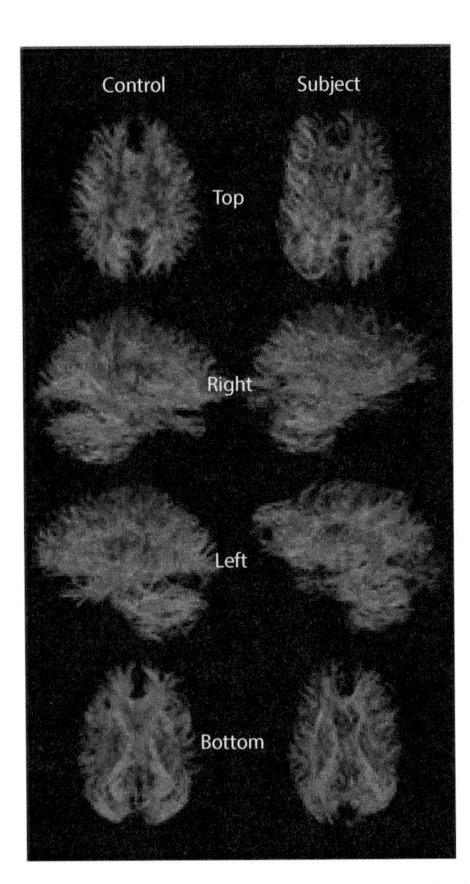

FIGURE 31.6 DTI tractography is shown in a control subject age-matched to a TBI case with severe injury showing the loss of tracts throughout the brain regardless of perspective. This illustration nicely demonstrates the complexity of the connections that create the networking matric of the brain.

of aggregate tracts. In mTBI, this is typically not dramatic and as readily visible as that observed in severe TBI, shown in Figure 31.6. Nonetheless, if even a few of the aggregate pathways depicted in Figure 31.6 are affected in mTBI, DTI along with other advanced neuroimaging techniques provide methods for examining WM pathology associated with brain connectivity (Caeyenberghs et al., 2013; Vakhtin et al., 2013; Yeh et al., 2013). Although DTI tractography may indicate where tracts are damaged and even their absence, DTI metrics may also reflect inflammation and other findings about WM integrity (Filley, 2011; Voss and Schiff, 2009; Zappala et al., 2012). As such, some of the DTI metric is not so much a marker of a lesion in the traditional sense but an indication of WM health. Viewing network damage/disruption in the broad sense of WM integrity (and not just where a "lesion" may reside) provides an improved framework to better understand the effects of brain injury and its influence on cognition and behavior. Furthermore, these types of neuroimaging methods provide techniques for viewing the complexity of neural pathways in three-dimensional images, which assists in inspecting where lesion/abnormalities may be located within the network. Accordingly, DTI and other methods that provide network analyses will likely play a much more substantial role in the future in identifying abnormalities associated with mTBI.

There are numerous methods for analyzing DTI metrics (Shenton et al., 2012). One method, referred to as tract-based spatial statistics, is used to show where significant group differences may be observed by comparing the TBI group with a matched control sample. In Figure 31.7 from Wada et al. (2012), significant FA reductions associated with mTBI were found most prominently in the superior longitudinal fasciculus, region of the forceps minor of the corpus callosum, superior frontal gyrus, insula, and fornix during the chronic stage. These changes along with others also related to reduced cognitive performance. Although the areas of abnormal FA as shown in Wada et al. are unique to that sample, these areas of reported FA changes associated with mTBI are common regions consistent with what others have found in mTBI (Benson et al., 2012; Bigler, 2013a,b; Hellyer et al., 2012; Hulkower et al., 2013; Lipton et al., 2012; Messe et al., 2011). Furthermore, these are the brain regions assumed to classically experience the greatest stress/strain/shear and rotational forces during head trauma, including mTBI (Ropper and Gorson, 2007).

Trauma induced edematous reactions in the brain compress parenchyma, which in turn may influence water diffusion potentially detected by DTI. Using the FA metric, increases in FA beyond some normal baseline may signify edema, whereas low FA may occur when axon degradation, membrane abnormalities increase water diffusion, or actual degeneration has occurred, which increases extracellular water (Wilde et al., 2008). Accordingly, some aspects of DTI findings reflect different pathological features that are time dependent, such as early swelling with subsequent parenchymal loss (Rosenbaum and Lipton, 2012). Because TBI may induce dynamic changes over time, differences in FA over acute, subacute, and chronic time frames postinjury may differ as well. When axons degenerate, the increased space frees extracellular water, resulting in lower FA. Thus, in mTBI, low FA may reflect WM degeneration, whereas increased FA may reflect neuroinflammation. Rosenbaum and Lipton (2012) and Shenton et al. (2012) provide much more elaborate discussions concerning FA markers of pathology in mTBI as well as other DTI metrics. Sometimes in mTBI dilated perivascular, so-called Virchow-Robin spaces will be detected, especially on T2-weighted MR sequences (Inglese et al., 2005, 2006). How these findings relate to DTI in mTBI has not been examined.

31.4 MICROSTRUCTURE EFFECTS OF mTBI

Intuitively, mTBI must involve more subtle pathology than moderate-to-severe TBI and therefore fewer abnormalities including those visibly identified on conventional imaging (Bigler, 2013b). Diagrammatically, some of the potential differences on axon morphology between mild and severe TBI are depicted in Figure 31.8 from Smith et al. (2013). As shown in this illustration, there are not only morphological effects of shear/strain injury at the axonal level but ionic. The axon membrane that regulates ionic movement is but a few nanometers in thickness and similarly measured in nanometers

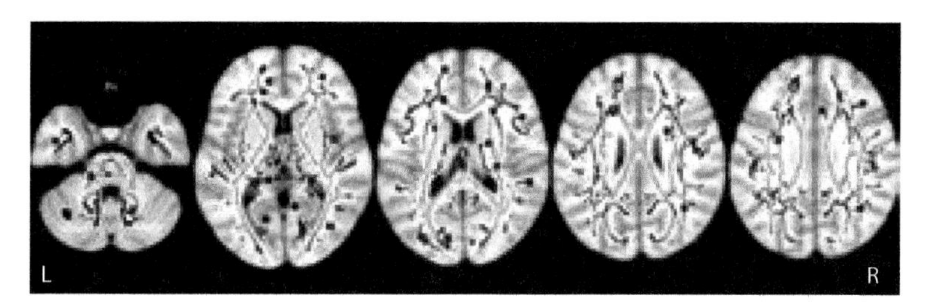

FIGURE 31.7 Tract-based spatial statistics analysis of the white matter skeleton. Voxels demonstrating significantly ($p < 0.01$) decreased FA values for the subjects with mTBI compared with the control group are shown in red-yellow. Voxels are thickened into local tracts and overlaid on the white matter skeleton (green). Further cluster details are given in Wada et al., as outlined in Table 2 of the original article. From Wada, T., Y. Asano, et al. (2012). Decreased fractional anisotropy evaluated using tract-based spatial statistics and correlated with cognitive dysfunction in patients with mild traumatic brain injury in the chronic stage. (From Wada, T., Y. Asano, et al. *American Journal of Neuroradiology* 33(11):2117–2122, 2012.)

is the cytoskeleton that provides the scaffolding for normal axon morphology (Maxwell, 2013). If the injury, even at the mild level, is sufficient to induce physiological change, it may be insufficient to produce morphological change, but may still produce changes in neural transmission and function. If the injury is time-limited and but transient disruption in neuronal cellular function recovery may be complete with no lasting effect. However, as shown in Figure 31.8, if a breakdown in axon integrity occurs, this has the potential to affect water diffusion properties because of membrane degradation or dissolution (Figure 31.5). If a sufficient number of axons becomes regionally affected, this may result in detectable differences with MRI techniques. In the classic chronic injury where loss of axonal fibers has occurred, this would typically result in reduced anisotropy, hence decreased FA and increased ADC.

As already mentioned, finite element biomechanical modelling of mTBI over the past decade has refined our understanding of the most commonly vulnerable WM pathways involved in the stretch/shear mechanism associated with mTBI, including pathways coursing through the corona radiata, internal capsule, cerebral peduncle, and corpus callosum (Bayly et al., 2012; Ji et al., 2013; McAllister et al., 2012). In more severe injury, these regions also exhibit classic atrophic changes concomitant with DTI-identified abnormalities (Bigler, 2013b; Dinkel et al., 2013), so clearly these are particularly vulnerable areas in head injury. Accordingly, given the likelihood of significant stretch/strain influences occurring within these regions of interest in mTBI (Ropper and Gorson, 2007), it is also most likely that within these regions of interest is where the most frequent pathology will be detected with advanced MRI methods, including DTI. This is especially true in individuals younger than age 50 without prior neuropsychiatric disorder because WM abnormalities in these regions occur infrequently as just incidental findings (Bigler, 2013b).

What is also so important about where these maximal strain fields occur is that axon projection through densely compacted WM occurs with slightly different trajectory for each neuron depending on where in the trajectory the strain field influences the axon. No two neurons as adjacent partners in their origin will have an identical terminus, meaning that their respective axonal projections differ. Each axon contributes unique connectivity to the network. As such, a maximal strain field will not necessarily affect all axons equally, as shown in Figure 31.9 from Kraft et al. (2012). This in turn sets up an important dynamic in mTBI—a sufficient number of neural cells must be affected in aggregate for there to be detection with advanced MRI metrics and depending on the number of cells affected, neural networks will differentially be affected. But not all neurons may be affected, adding to the uniqueness of each injury.

As already pointed out, the CDE-identified abnormalities associated with mTBI are visibly recognized. If 500 small diameter (<5 microns) axons were damaged in a particular strain field that would not necessarily be detected by conventional MRI because the "lesion" physically would be less than 0.5 mm. As is well known from epilepsy research, a small number of aberrant pathways may wreak havoc with neural communication (Devinsky et al., 2013; Izhikevich and Edelman, 2008). Accordingly, when either conventional or advanced neuroimaging techniques are found, they likely represent significant aberrations in underlying neuronal integrity.

Advanced neuroimaging techniques provide a much more fruitful approach over just visible "lesion" detection in understanding mTBI. Microstructure aberration that affects membrane integrity, metabolic functioning including hemodynamic responsivity along with synaptic, and/or neurotransmitter dysfunction may all be the source of impaired neural function in mTBI, which would not be associated with an identifiable visible lesion. Whether an mTBI abnormality is structural, functional, or a combination ranging from a classic shear lesion to some physiological aberration associated with membrane or synaptic dysfunction, how such abnormalities disrupt neural networks is more important than where a "lesion" may reside. This leads to a discussion of network damage and how to use neuroimaging to identify where in the network damage/disruption may have occurred (Bigler, 2013b).

31.5 NETWORK DAMAGE AND DISRUPTION IN mTBI

Over the past two decades, neuroimaging techniques have provided a more refined view of the anatomical loci of certain neural networks that underlie behavior and cognition. Advanced neuroimaging methods, especially application of

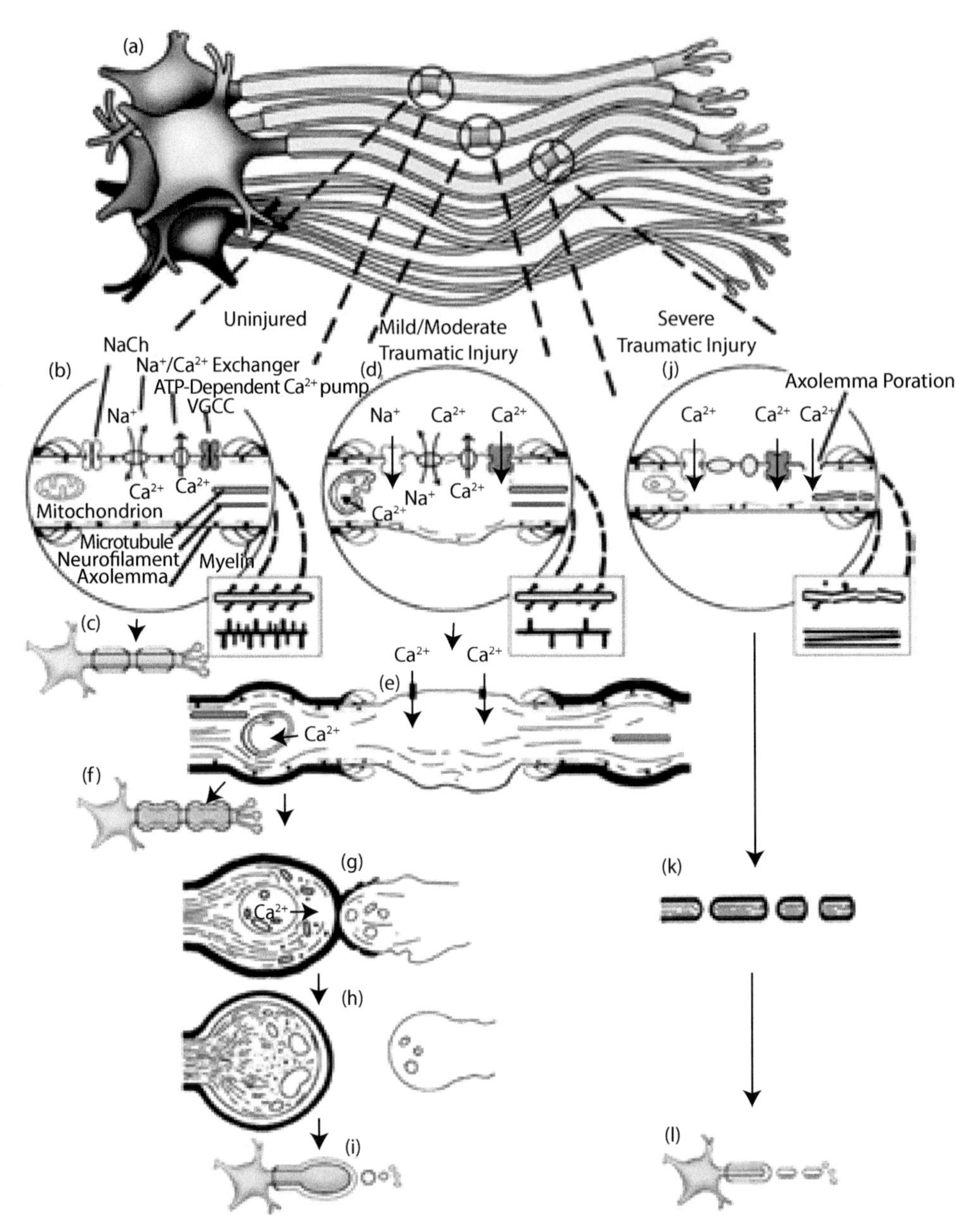

FIGURE 31.8 Evolving pathophysiology of traumatic injury in myelinated axons as reproduced from Smith (2013). In this figure, the authors attempt, in an abbreviated fashion, to illustrate some of the key events believed to be involved in the pathobiology of traumatic axonal injury and, thereby, identify potential therapeutic targets. Although framed in the view of primary nodal involvement (a), this focus does not preclude comparable change ongoing in other regions of the axon. (b, c) Normal axonal detail including the paranodal loops and the presence of intraaxonal mitochondria, microtubules, and neurofilaments, together with the presence of multiple axolemmal channels localized primarily to the nodal domain. Mild to moderate traumatic brain injury (d) is observed to involve a mechanical dysregulation of the voltage-sensitive sodium channels, which contribute to increased calcium influx via reversal of the sodium calcium exchanger and the opening of voltage-gated calcium channels. This also impacts on the proteolysis of sodium channel inactivation that contributes further to local calcium dysregulation. Microtubular loss, neurofilament impaction, and local mitochondrial damage can follow, that, if unabated, collectively alters/impairs axonal transport illustrated (e). Alternatively, if these abnormalities do not progress, recovery is possible (r). When progressive, these events not only impair axonal transport but also lead to rapid intra-axonal change in the paranodal and perhaps internodal domains that elicit the collapse of the axolemma and its overlying myelin sheath resulting in lobulated and disconnected axonal segments (g) that, over the next 15 minutes to 2 hours, fully detach (h). The proximal axonal segment in continuity with the cell body of origin now continues to swell from the delivery of vesicles and organelles via anterograde transport, whereas the downstream fiber undergoes Wallerian change (i). Last, with the most severe forms of injury, the previously identified calcium-mediated destructive cascades are further augmented by the poration of the axolemma, again primarily at the nodal region (j). The resulting calcium surge, together with potential local microtubular damage and disassembly, pose catastrophic intra-axonal change that converts anterograde to retrograde axonal transport, precluding continued axonal swelling, whereas the distal axonal segment fragments and disconnects (k), with Wallerian degeneration ensuing downstream (l). (From Smith, D. H., R. Hicks, et al., *Journal of Neurotrauma* 30(5):307–323, 2013. With permission.)

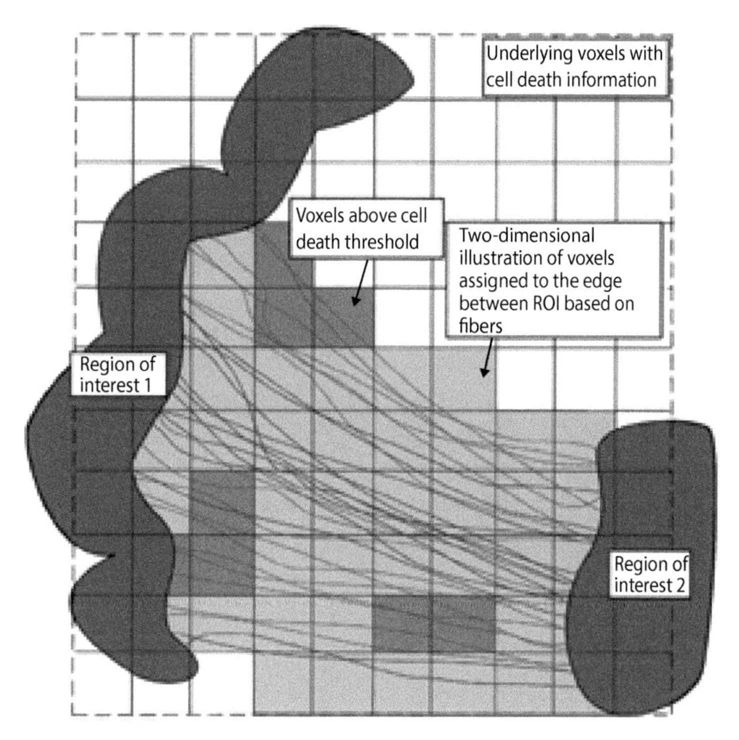

FIGURE 31.9 This schematic shows how different axon trajectories may or may not be vulnerable to injury. As can be seen, there are only certain sectors where the biomechanical deformation sufficiently alters brain parenchyma to damage axons. Lines represent hypothetical axon projections from one gray matter structure to another. Note that even though all connect region of interest (ROI) 1 with ROI 2, and that out of all sectors where these hypothetical axons project, only eight of the sectors experienced sufficient deformation to damage axons, because of the differences in crossing routes and trajectories numerous axons were affected. (From Kraft, R. H et al., *PLoS Computational Biology* 8(8):e1002619, 2012. With permission.)

DTI and resting state (rs) functional connectivity (fc) mapping of the brain using functional MRI (fMRI) technology. Figure 31.10 is a simplified depiction of a network from the work of van den Heuvel and Sporns (2011). As depicted in this figure, there are both short and long coursing pathways within a hemisphere (intrahemispheric tracts) as well as across the two hemispheres (interhemispheric tracts) that integrate the brain via either gateway hubs that appear to be along main neural highways as signals pass from one region to another, or lesser hubs and nodes, with some of the smallest nodes out toward the periphery of the network. From this analogy, even a small lesion that affected a major hub could be very disruptive to the network, with lesser disruption if the injury only affected a peripheral node. Note that disruption to a hub could occur by damage to pathways remote to the hub but highly connected to it. Importantly, van de Heuvel and Sporns show that even minor hubs typically are but a single link away from a critical hub, which they refer to as a "rich club" (van den Heuvel and Sporns, 2013).

Figures 31.11 and 31.12 present two mTBI cases with different lesion patterns that show how dissimilar lesions very likely differentially influence distinctly different networks.

In Figure 31.11, this individual with mTBI (at-the-scene GCS of 14, with a 15 at ED evaluation) had a solitary DOI CT finding of a small bleed within the basal ganglia on the left. This mTBI patient was observed overnight and the scan repeated, revealing no change, and discharged the next day. In the mTBI case (GCS never below 15) presented in Figure 31.12, a small left frontal surface contusion occurred, which resulted in a region of focal encephalomalacia. Based on the conventional imaging performed, both of these cases involved left hemisphere injury, but the network effects of what the mild injury induced in the case in Figure 31.11 would likely be greater that what occurred in Figure 31.12. Returning to the network diagram of van den Heuvel and Sporns, the patient in Figure 31.11 would most likely have central network disruption affecting a major hub, whereas the patient in Figure 31.12 would most likely have only a minor node affected.

The problem with the overly simplistic view just presented in these cases is that it does not take into account what is occurring below conventional lesion detection because it uses the old "lesion-localization" approach to image analysis. For example, the case in Figure 31.11 shows the distinct hemorrhage in the DOI CT scan, but SWI and FLAIR imaging during the chronic phase show much more widespread and frequent lesions than the DOI CT. The distribution and multifocal nature of the shear injuries demonstrated by the SWI and FLAIR imaging would reinforce the likelihood of greater key hub as well as nonspecific network disruption.

So beyond the lesion, how can network dysfunction be demonstrated in mTBI? As mentioned previously, one of

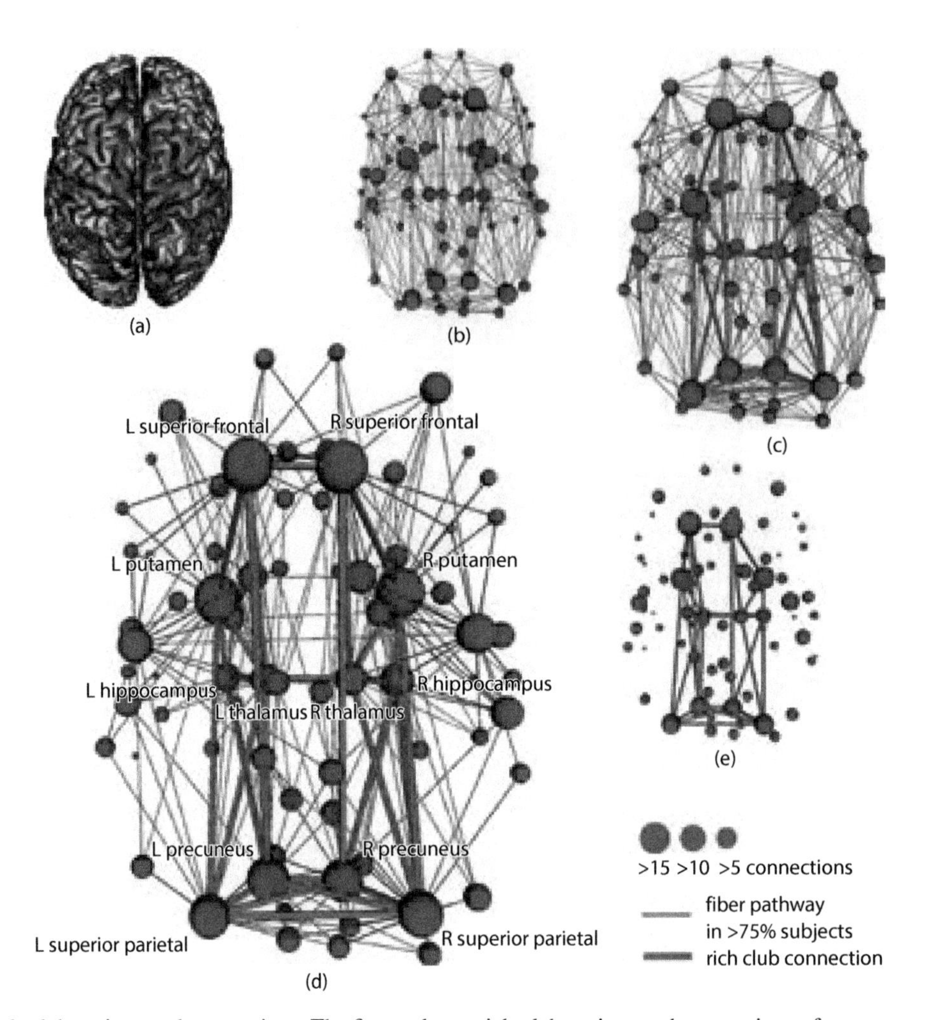

FIGURE 31.10 Rich-club regions and connections. The figure shows rich-club regions and connections of a group-averaged connectome (unweighted, $k = 17$; Figure 3a from the original article). The size of nodes reflects their number of connections, with bigger nodes representing more densely connected regions. (a) Anatomical perspective. (b) Group-averaged connectome. (c) Group connectome with rich-club connections marked in dark blue. (d) Connections between rich-club regions (dark blue) and connections from rich-club nodes to the other regions of the brain network (light blue). The figure shows that almost all regions of the brain have at least one link directly to the rich club. (e) Rich-club connections. Used with permission from the Society for Neuroscience.

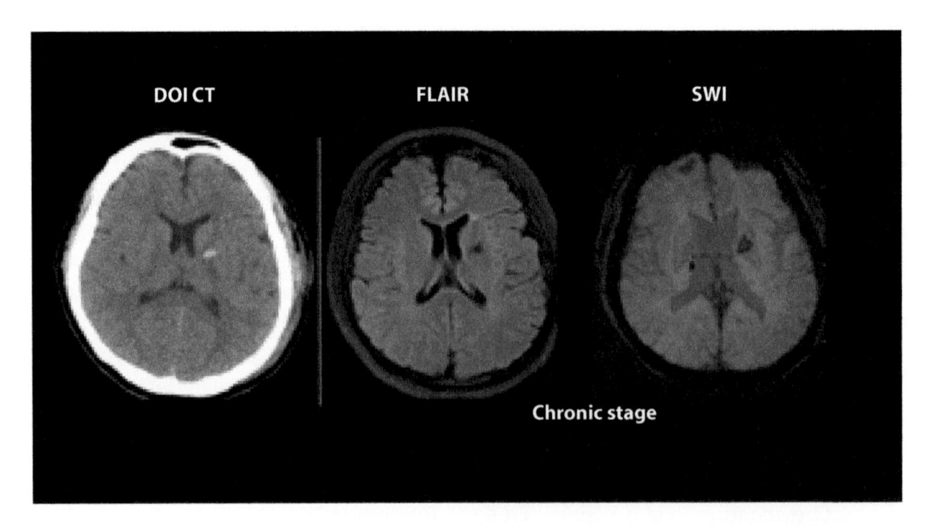

FIGURE 31.11 The DOI CT scan on the left revealed a small hemorrhage in the region of the globus pallidus-putamen as a solitary lesion. However, on follow-up MRI multiple other areas of prior hemorrhage are noted along with foci of abnormal hyperintense WM signal on the FLAIR sequence. This case illustrates how the DOI scan may underestimate the number and extent of identifiable macroscopic lesions associated with mTBI.

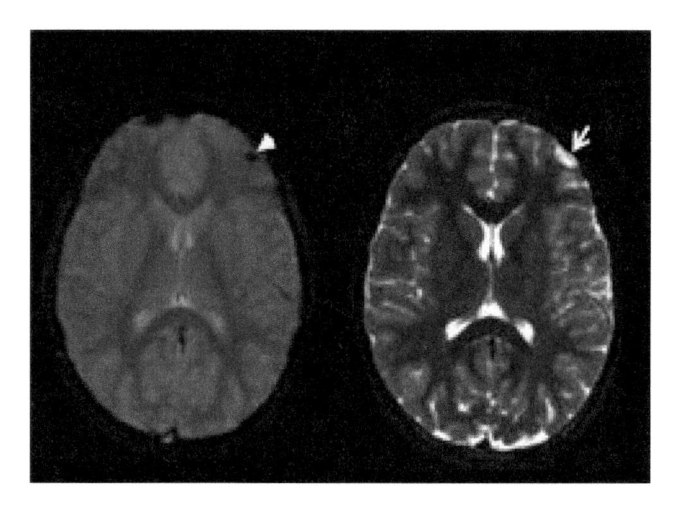

FIGURE 31.12 This child sustained an mTBI resulting in a left frontal contusion with residual hemosiderin deposition and an area of focal encephalomalacia that is back-filled with a pocket of CSF. In contrast to the mTBI case presented in Figure 31.11, note the entirely different location of pathology.

the most common areas of WM injury arising from mTBI occurs within long-coursing pathways within the corona radiata like the superior longitudinal fasciculus. An extensively studied pathway associated with the superior longitudinal fasciculus is the default mode network (DMN). This network involves interconnectiveness of the parietal and frontal lobes as well as between each hemisphere that participate in regulation of attention and working memory, two common deficits associated with mTBI (Mayer et al., 2011). The likely pathological consequence of superior longitudinal fasciculus injury in mTBI would be to weaken network coherence. Interestingly, the inefficiency of the network connections may actually lead to increased recruitment within parts of the network or other networks to maintain functionality (McAllister et al., 2001; Perlstein et al., 2004). The recent rs-fcMRI study by Zhou et al. (2012) demonstrates some of these points where they found increased frontal connectivity and decreased posterior connectivity of the DMN in mTBI patients compared with controls. This is depicted in Figure 31.13. Importantly, these findings occurred in mTBI with no conventional abnormalities on clinical MRI. So, this aberration within the DMN occurred as a result of microstructural pathology below the threshold of conventional detection. Bigler and Maxwell (2011, 2012) have shown that a likely correlate of disrupted axonal integrity is axon membrane damage that would alter neural transmission. Because there were no conventional lesions or abnormalities visible in this cohort of mTBI patients, a network analysis using conventional imaging would have yielded no findings.

The study by Johnson et al. (2012) provides a different approach in examining the DMN in concussed athletes during the subacute recovery phase. Also using fMRI-based functional connectivity mapping, as shown in Figure 31.14, they created maps akin to what was displayed in Figure 31.10. These investigators examined athletes who had sustained concussive head injuries from sports contact. mTBI resulted in less network complexity with the magnitude of connections significantly reduced.

31.6 HETEROGENEITY OF THE LESION

As pointed out in the CDE and previously discussed, a variety of conventional abnormalities may occur in mTBI, some of which have been shown Figures 31.2 through 31.4, 31.11, and 31.12. What should be clearly evident in viewing these figures is the heterogeneity of the "lesion" patterns. In a study with 41 mTBI children with confirmed DOI abnormalities, there were no identical lesions or overlap in lesion distribution (Bigler, Abildskov, Petrie et al., 2013). At the macroscopic level, if lesions do not overlap, this would certainly be true for underlying microstructure changes. All of this would implicate heterogeneous outcome because as lesions/abnormalities vary with each patient, neurobehavioral and neurocognitive outcome would likely vary as well.

Heterogeneity in lesion type is also influenced by the metabolic cascade associated with concussion and a variety of pathophysiological events that may occur over time (Signorett et al., 2011). The interplay between structural and pathophysiological events associated with mTBI is dynamic in the first days to weeks postinjury, with individual time frames for recovery or persistence in defect (Bigler, 2013). For example, Figure 31.15 shows an mTBI case with "normal" DOI CT but because of persistence in symptoms, this patient underwent MRI scanning several weeks postinjury, which revealed significant findings of hemosiderin deposition within the frontal white matter along with associated FLAIR abnormalities.

Contrasting the case in Figure 31.15, in which DOI CT is negative but with follow-up MRI depicting abnormalities, with the case in Figure 31.4, in which DOI CT was distinctly abnormal but follow-up conventional MRI negative, shows the range of potential dynamic changes in neuroimaging of mTBI. Although not possible in human mTBI studies, animal investigations that model mTBI where acute neuroimaging abnormalities become nondetectable over time nonetheless may show histological pathology (Dewitt et al., 2013; Hylin et al., 2013). This has also been shown in a postmortem human mTBI case (Bigler, 2004). In this case study, the patient sustained a well-documented mTBI with positive loss of consciousness (LOC) and an initial GCS of 14, but DOI CT was negative. Several months postinjury, the patient died from a cardiac arrest, at which time an autopsy was performed and showed no gross pathology but microscopic hemosiderin deposition within the WM. This injury occurred in 2000, a time before improvements in MR gradient sequences that detect hemosiderin in vivo. As such, it is likely that cases where DOI CT demonstrates petechia but follow-up does not detect the prior hemorrhage may, in fact, have underlying microscopic residual hemosiderin, as shown in this case. Furthermore, this mTBI case showed macrophages within the WM, suggesting that even in the chronic stage of injury inflammatory reactions were still occurring (Bigler, 2013).

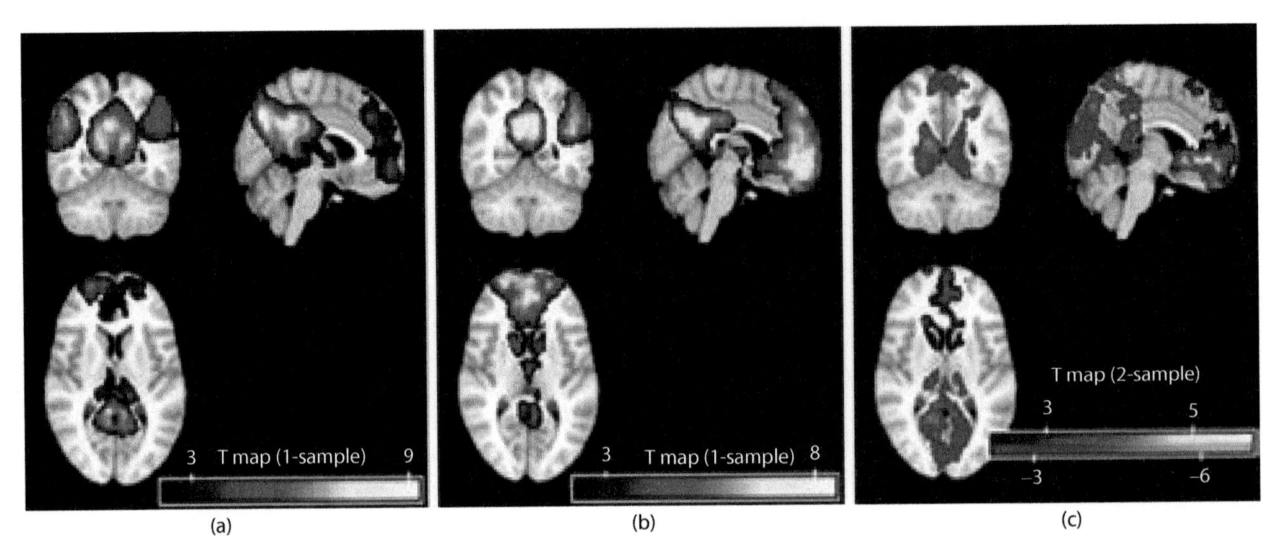

FIGURE 31.13 These fMRI activation plots are taken from Zhou et al. (2012). DMN templates obtained with the hybrid ICA seed method in patients and control subjects (corrected, $p < 0.05$; $K \geq 20$). (a) Typical but enhanced connectivity pattern of the DMN was identified in the healthy control group. (b) Disrupted DMN pattern is shown in the patient group. (c) Group-level voxel-wise image of hybrid DMN template differences comparing MTBI with control groups. Red, increased functional connectivity; blue, decreased functional connectivity in patients compared with control subjects. These changes in DMN pattern were consistent with the single-subject independent component analysis (ICA) results by means of visual inspection. (From Zhou, Y., M. P. Milham et al., *Radiology* 265(3):882–892, 2012. With permission.)

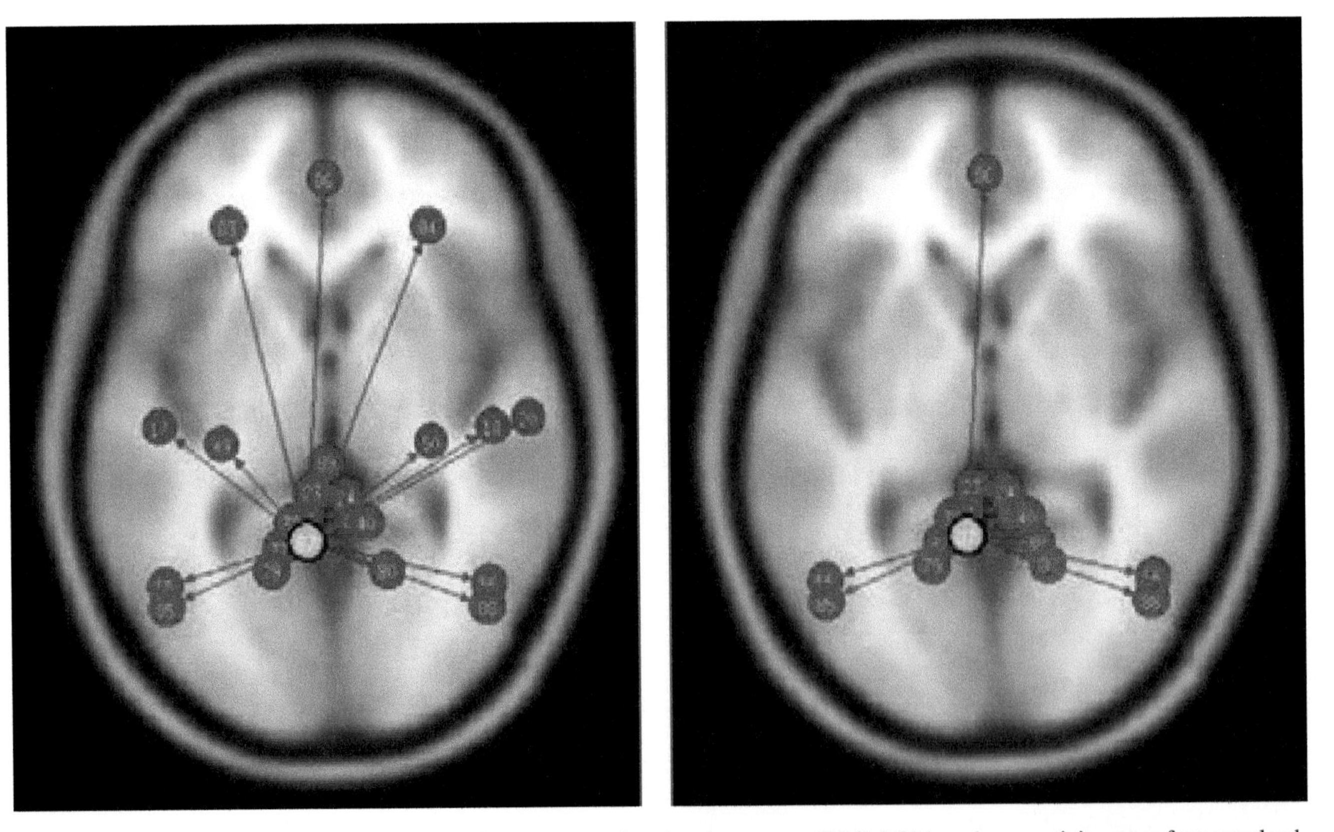

FIGURE 31.14 The study by Johnson et al. (2012) shows posterior cingulate cortex (PCC) ROI-based connectivity maps for normal volunteers (left) and mTBI (right) subjects. The green dot designates the PCC ROI seed and red arrows and dots indicate significant positively correlated ($p < 0.05$ false discovery rate [FDR]) brain regions (note numbers do not represent Brodmann's area numbers). Left and right inferior temporal gyrus, left and right parahippocampal gyrus, left and right angular gyrus, left and right dorsolateral prefrontal cortex, and medial prefrontal cortex all show differences.

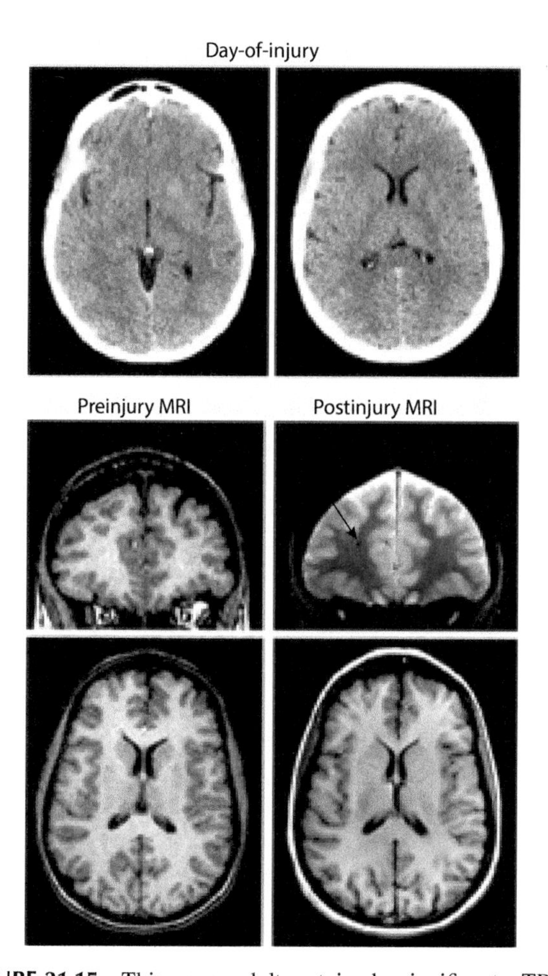

FIGURE 31.15 This young adult sustained a significant mTBI in an auto-pedestrian injury where she had positive LOC, but the DOI CT revealed no abnormality. However, as symptoms persisted, this patient was assessed with MRI that revealed hemosiderin and focal white matter hyperintensities. Interestingly, this patient had participated as a research subject before the injury, confirming no prior brain abnormalities, as shown in the preinjury MRI, although only a T1-weighted MRI had been performed.

31.7 UNIQUENESS OF EACH MTBI "LESION" TO DISRUPT THE NETWORK

There is probably a simple explanation for the heterogeneity of abnormalities that may occur with mTBI: each injury is a unique circumstance of biomechanical forces affecting an individual with unique genetic and developmental environments that apply *only* to that individual. Although two brains may appear similar, neuronal development is, in part, experience-dependent; therefore, aspects of connectivity naturally will be unique for each. Furthermore, as shown in this chapter, lesion characteristics are diverse and change over time. As such, individual differences combined with heterogeneity in mechanism of injury and the unique nature of each brain to the effects of injury means no singular uniformity to outcome would occur across all aspects of mTBI.

As depicted in Figure 31.2, only certain WM tracts are actually damaged within a particular biomechanical strain field (Kraft et al., 2012). A neuron's cell body is tightly held within the neuropil but axons intertwine and course is various directions within WM parenchyma, the direction of which may influence the strain field after impact. Watanabe et al. (2012) have shown how, with each individual impact injury, unique influences occur from the biomechanical movement of the brain within the cranium. These unique individual differences, when coupled with the fact that neural tissue has different elastic properties that are region- and structure-dependent (Feng et al., 2013; Mao et al., 2013), demonstrates that no two injuries from mTBI will ever produce identical pathology detectable by neuroimaging. Although a particular injury at the mild level may never be identical, some general aspects of brain morphology do result in certain WM regions (corpus callosum and corona radiata in particular and others as previously mentioned) that become especially vulnerable to stretch, strain, and tensile effects after the mechanical deformation that occurs with impact and/or acceleration/deceleration injury (Bayly et al., 2012; Feng et al., 2013).

31.8 TIME SEQUENCE OF NEUROPATHOLOGY ASSOCIATED WITH mTBI

There are primary and secondary effects associated with TBI. Some of the primary effects are immediate, the consequences of which play out within hours of the mTBI, whereas secondary effects have a slower time course (Bigler and Maxwell, 2011, 2012). In mTBI, peak symptoms may not occur for hours to days postinjury (Duhaime et al., 2012; Prichep et al., 2012). After the immediate effects of injury, there are complex pathophysiological effects that occur at the vascular and neuroinflammatory response level (Stahel et al., 1998) that may be partially detected with neuroimaging methods in mTBI (Metting et al., 2009, 2010, 2013).

Wilde et al. (2012) examined a group of mTBI patients repeatedly assessed with DTI metrics and neurocognitive examination over the first 8 days postinjury. Cognitive and neuroimaging findings fluctuated over time in mTBI. All of these mTBI patients had experienced an "uncomplicated" mTBI meaning that no abnormalities were identified in the DOI CT scan, almost all had an ED identified GCS of 15, and all were the result of some type of motor vehicle accident. The first assessment was completed within 2 days of injury, and serially at days 3–4, 5–6, and 7–8 post injury. Alternate forms of the Hopkins Verbal Learning Test were administered at each time point where memory performance typically dipped between days 3 and 6, suggesting the confluence of primary and secondary effects from mTBI reaching their apex on cognitive functioning at this time point in this cohort of mTBI patients. DTI metrics also fluctuated but not necessarily in concert with cognitive performance. Increased FA, which may reflect neuroinflammation (Wilde et al., 2008), was found in several mTBI patients with FA peaks observed between days 3–4 and 7–8. Decrease in FA may reflect axon damage, but without preinjury neuroimaging to know precisely each individual's FA baseline, it is difficult to fully interpret these findings. However, from a memory performance perspective, almost all showed a decrease after days 1–2, with postconcussive symptoms reaching their peak around day 3. This does suggest that the variability in FA during this acute/subacute timeframe may reflect instability of WM microstructure associated with the injury.

Although a positive LOC in mTBI is abrupt and an obvious indicator of TBI, by definition for mTBI LOC has to be brief and transient or otherwise, the injury would no longer be considered "mild." The evolution of symptoms/problems associated with the initial injury likely has much to do with complex cellular responses to the mechanical deformation of brain parenchyma following injury as overviewed by Stahel et al. (1998). Although mTBI is initiated by an event involving traumatic deformation of neural tissue, the event does not induce a singular pathological event that can be identified by neuroimaging methods at this time, but initiates a most complex array of structural and physiological changes in brain parenchyma that play out over time, potentially with permanent changes. If the biomechanical deformation is minimal, only transient disruption in neuron integrity occurs (Biasca and Maxwell, 2007; Magdesian et al., 2012), but as already shown in Figure 31.8 from Smith et al. (2013), various structural pathologies involving axon damage may occur and take time to develop.

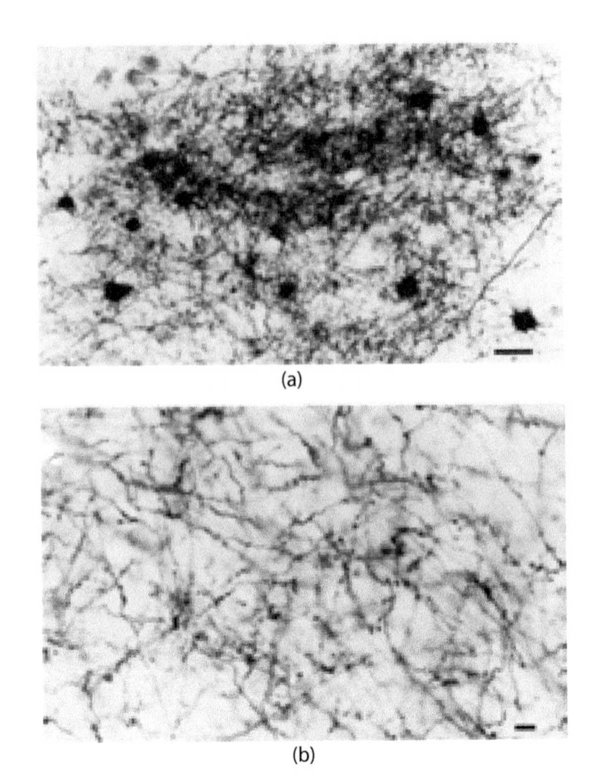

FIGURE 31.16 The photomicrograph from Kultas-Ilinsky et al. (2003) is taken from a study involving the histology of thalamocortical connection of a rhesus monkey (Kultas-Ilinsky et al., 2003). (a) Photomicrographs of dense plexuses (cell bodies dark) within the ventrolateral nucleus of the thalamus. (b) Type 2 axon plexus in the centromedian nucleus with a few type 1 axons in the background. Two basic types of corticothalamic axons were identified: small to medium width, type 1 axons that formed large terminal fields with small boutons, and thick, type 2 axons that formed small terminal fields with large boutons Scale bars: 50 μm in (a), 10 μm in (b). When brain deformation is viewed in mTBI, it is important to remember that this delicate and rather intestinally small ultrastructure is what is being deformed. (From Kultas-Ilinsky, K. E. et al., *The Journal of Comparative Neurology* 457(2):133–158, 2003.)

31.9 CELLULAR BASIS OF mTBI NEUROPATHOLOGY

As repeatedly mentioned throughout this chapter, what is observed in mTBI with contemporary neuroimaging even with advanced MR techniques still represents a macro perspective of brain injury. The CDE definition of traumatic brain injury (TBI) is "... an alteration in brain function, or other evidence of brain pathology, caused by an external force" (Menon et al., 2010). External force induces brain injury via deformation of neural tissue that surpasses tolerance limits for normal displacement or strain that accompanies movement such as jumping, rapid turning of the head, simple bumps to the head, etc. So, at the most fundamental level of injury, cellular deformation disrupts anatomy and physiology sufficient to at least transiently impair function when the threshold for mTBI has been reached.

Too often, neural cells are viewed schematically as an artist's rendition of what a neural cell looks like, typically with a large-appearing singular axon that looks sturdy. However, artistic schematics detract from the true complexity and delicate nature of what really constitutes neural tissue. For example, primary motor cortex-thalamic connections are depicted in two photomicrographs as shown in Figure 31.16 taken from the ventrolateral nucleus of the thalamus in a rhesus monkey (Kultas-Ilinsky et al., 2003). Disruption of corticothalamic and thalamocortical connections has been implicated in mTBI by DTI and rs-fcMRI studies (Little et al., 2010; Tang et al., 2011), including mild TBI (Zhou et al., 2013), so the axon elements depicted in this illustration are representative of neurons that could be affected by mechanical deformation in mTBI. Histological labeling was performed to track axons and their projections between cortex and within the thalamus. As viewed in Figure 31.16, axon projections are not straight-lined but complexly interdigitated. Note also the micron scaling indicating that most axons in this photomicrograph are <3 μ in diameter and that these views are merely two-dimensional of a three-dimensional structure. Given their size, axons are extremely delicate. Differences in angulation of the projecting axons would further influence which axons would be most affected depending on the orientation of the head at time of impact or angular acceleration/deceleration.

Amid the complex intertwining of axons, all with multiple branches forming terminal boutons and synaptic connections, as shown in Figure 31.16, any misalignment from traumatically induced deformation would likely affect axonal function and synaptic integrity. Likewise, if the axon membrane is disrupted, membrane permeability will directly impact neuronal function and axon potential propagation. Only one strategic axon segment need be affected to disrupt neural transmission for the entire axon. A variety of finite element and various methods for re-creating the motion that displaces brain parenchyma in mTBI have been performed, mostly using sports concussion models. For example, Viano et al. (2005) showed on average in the typical sports-related concussion that the brain displaces between 4 and 8 mm in regions such as the corpus callosum, midbrain, medial temporal lobe, and fornix. Viewing Figure 31.16 from the perspective of this amount of deformation, noting that the photomicrograph depicts axons <0.5 mm in length, even deformation of a few millimeters would reflect a massive distortion of neurons this size and their axonal projections.

From what has been outlined in the previous paragraph and in Figure 31.16, looking back on the macroscopic pathology within the images presented in this chapter, essentially assessing tissue with millimeter-level resolution, even with all of the neuroimaging advancements, only the coarsest associations may be demonstrated via neuroimaging techniques with what may be occurring at the microstructure level with mTBI.

Blood vessels are just as delicate as neural tissue, especially at the capillary level. Each neuron is dependent on receiving a continuous source of glucose and oxygen with the smallest capillaries large enough for just a single red blood cell to traverse the capillary to deliver its oxygen and glucose (Bigler and Maxwell, 2011, 2012). As such, blood vessels are just as susceptible to the shear–strain biomechanics of head injury as are neurons (Madri, 2009). Subtle vascular changes in mTBI could also be the source for neural dysregulation (Pomschar et al., 2013), altering the hemodynamic response necessary for normal cellular function.

Understanding these cellular effects of mTBI and their potential relation to biomarkers of brain injury, combined with neuroimaging, will hopefully lead to furthering the diagnostic utility of advance neuroimaging methods, including DTI in assessing mTBI. It may be that blood biomarkers combined with tracking DTI metrics over time will reveal diagnostic features of mTBI (Fo et al., 2013; Kou et al., 2013).

31.10 VOLUMETRY FINDINGS IN mTBI

Based on the biomechanics of injury, as discussed earlier in this chapter, if atrophic changes associated with mTBI were to occur, they would most likely be found within those regions associated with the greatest likelihood for shear/strain and deformation injury. Indeed, several studies that have prospectively examined mTBI subjects have demonstrated this regional and whole brain atrophy (Benson et al., 2012; Lannsjo et al., 2013; MacKenzie et al., 2002; Ross et al., 2013; Tot et al., 2013; Zhou et al., 2013). For example, Zhou et al. demonstrated that by establishing a baseline in mTBI patients within the acute to early subacute time frame, when they were assessed with various volumetric techniques 1 year later, significant volume loss was observed in the anterior cingulum, cingulate gyrus, and scattered regions within the frontal lobes. Interestingly, they observed volume loss in the cuneus and precuneus regions as well. The volume loss within the cuneus and precuneus, both posterior brain regions, may actually be the result of Wallerian degeneration from the more focal frontal loss disrupting long coursing frontoparietal connections particularly vulnerable to stretch and shearing effects (Biasca and Maxwell, 2007, 2011, 2012).

31.11 CONCLUSION

Various neuroimaging methods provide a variety of techniques to detect underlying neuropathology that results from mTBI. The most common visible abnormalities are in the form of focal encephalomalacia, hemosiderin deposition, and/or white matter signal abnormalities. Several quantitative MRI methods have demonstrated techniques for the detection of underlying pathology associated with mTBI, which differ depending on the time postinjury that the scan is performed. Despite these advances neuroimaging remains but a coarse view of the microstructure pathology associated with mTBI.

REFERENCES

Bayly, P. V., E. H. Clayton, et al. (2012). Quantitative imaging methods for the development and validation of brain biomechanics models. *Annual Review of Biomedical Engineering* 14:369–396.

Benson, R. R., R. Gattu, et al. (2012). Detection of hemorrhagic and axonal pathology in mild traumatic brain injury using advanced MRI: Implications for neurorehabilitation. *NeuroRehabilitation* 31(3):261–279.

Biasca, N. and W. L. Maxwell (2007). Minor traumatic brain injury in sports: A review in order to prevent neurological sequelae. *Progress in Brain Research* 161:263–291.

Bigler, E. D. (2013a). Neuroinflammation and the dynamic lesion in traumatic brain injury. *Brain* 136(Pt 1): 9–11.

Bigler, E. D. (2013b). Neuroimaging biomarkers in mild traumatic brain injury (mTBI). *Neuropsychology Review* 23(3):169–209.

Bigler, E. D., T.J., Abildskov, J., Petrie, M., Dennis, N., Simic, H.G. Taylor, et al. (2013). Heterogeneity of brain lesions in pediatric traumatic brain injury. *Neuropsychology*. 27(4):438–451.

Bigler, E. D. and W. L. Maxwell (2011). Neuroimaging and neuropathology of TBI. *NeuroRehabilitation* 28(2):63–74.

Bigler, E. D. and W. L. Maxwell (2012). Neuropathology of mild traumatic brain injury: Relationship to neuroimaging findings. *Brain Imaging and Behavior* 6(2):108–136.

Bigler, E. D. (2004). Neuropsychological results and neuropathological findings at autopsy in a case of mild traumatic brain injury. *Journal of the International Neuropsychological Society* 10(5):794–806.

Budde, M. D., L. Janes, et al. (2011). The contribution of gliosis to diffusion tensor anisotropy and tractography following traumatic brain injury: Validation in the rat using Fourier analysis of stained tissue sections. *Brain* 134(Pt 8):2248–2260.

Caeyenberghs, K., A. Leemans, et al. (2013). Topological correlations of structural and functional networks in patients with traumatic brain injury. *Frontiers in Human Neuroscience* 7:726.

Devinsky, O., A. Vezzani, et al. (2013). Glia and epilepsy: Excitability and inflammation. *Trends in Neurosciences* 36(3):174–184.

Dewitt, D. S., R. Perez-Polo, et al. (2013). Challenges in the development of rodent models of mild traumatic brain injury. *Journal of Neurotrauma* 30(9):688–701.

Dinkel, J., A. Drier, et al. (2013). Long-term white matter changes after severe traumatic brain injury: A 5-year prospective cohort. *AJNRAmerican Journal of Neuroradiology* 35:23–29.

Duhaime, A. C., J. G. Beckwith, et al. (2012). Spectrum of acute clinical characteristics of diagnosed concussions in college athletes wearing instrumented helmets: Clinical article. *Journal of Neurosurgery* 117(6):1092–1099.

Feng, Y., E. H. Clayton, et al. (2013). Viscoelastic properties of the ferret brain measured in vivo at multiple frequencies by magnetic resonance elastography. *Journal of Biomechanics* 46(5):863–870.

Feng, Y., R. J. Okamoto, et al. (2013). Measurements of mechanical anisotropy in brain tissue and implications for transversely isotropic material models of white matter. *Journal of the Mechanical Behavior of Biomedical Materials* 23C:117–132.

Filley, C. M. (2011). White matter: Beyond focal disconnection. *Neurologic Clinics* 29(1):81–97, viii.

Fox, W. C., M. S. Park, et al. (2013). Contemporary imaging of mild TBI: The journey toward diffusion tensor imaging to assess neuronal damage. *Neurological Research* 35(3):223–232.

Gean, A. D. and N. J. Fischbein (2010). Head trauma. *Neuroimaging Clinics of North America* 20(4):527–556.

Haacke, E. M., A. C. Duhaime, et al. (2010). Common data elements in radiologic imaging of traumatic brain injury. *Journal of Magnetic Resonance Imaging* 32(3): 516–543.

Heidemann, R. M., D. Ivanov, et al. (2012). Isotropic submillimeter fMRI in the human brain at 7 T: Combining reduced field-of-view imaging and partially parallel acquisitions. *Magnetic Resonance in Medicine* 68(5):1506–1516.

Hellyer, P. J., R. Leech, T. E. Ham, and D. J. Sharp. (2012). Individual prediction of white matter injury following traumatic brain injury. *Annals of Neurology* 73:489–499.

Hulkower, M. B., D. B. Poliak, et al. (2013). A decade of DTI in traumatic brain injury: 10 years and 100 articles later. *AJNR. American Journal of Neuroradiology* 34(11):2064–2074.

Hylin, M. J., S. A. Orsi, et al. (2013). Behavioral and histopathological alterations resulting from mild fluid percussion injury. *Journal of Neurotrauma* 30(9):702–715.

Inglese, M., E. Bomsztyk, et al. (2005). Dilated perivascular spaces: Hallmarks of mild traumatic brain injury. *AJNR. American Journal of Neuroradiology* 26(4):719–724.

Inglese, M., R. I. Grossman, et al. (2006). Clinical significance of dilated Virchow-Robin spaces in mild traumatic brain injury. *Brain Injury* 20(1):15–21.

Izhikevich, E. M. and G. M. Edelman (2008). Large-scale model of mammalian thalamocortical systems. *Proceedings of the National Academy of Sciences of the United States of America* 105(9):3593–3598.

Ji, S., H. Ghadyani, R. P. Bolander, J. G. Beckwith, J.C. Ford, T. W. McAlllister et al. (2013). Parametric comparisons of intracranial mechanical responses from three validated finite element models of the human head. *Annals of Biomedical Engineering* 42:11–24.

Johnson, B., K. Zhang, et al. (2012). Alteration of brain default network in subacute phase of injury in concussed individuals: Resting-state fMRI study. *NeuroImage* 59(1):511–518.

Kim, J. J. and A. D. Gean (2011). Imaging for the diagnosis and management of traumatic brain injury. *Neurotherapeutics* 8(1):39–53.

Kou, Z., R. Gattu, et al. (2013). Combining biochemical and imaging markers to improve diagnosis and characterization of mild traumatic brain injury in the acute setting: Results from a pilot study. *PloS One* 8(11):e80296.

Kraft, R. H., P. J. McKee, et al. (2012). Combining the finite element method with structural connectome-based analysis for modeling neurotrauma: Connectome neurotrauma mechanics. *PLoS Computational Biology* 8(8):e1002619.

Kultas-Ilinsky, K., E. Sivan-Loukianova, et al. (2003). Reevaluation of the primary motor cortex connections with the thalamus in primates. *The Journal of Comparative Neurology* 457(2):133–158.

Lannsjo, M., R. Raininko, et al. (2013). Brain pathology after mild traumatic brain injury: An exploratory study by repeated magnetic resonance examination. *Journal of Rehabilitation Medicine* 45(8):721–728.

Lipton, M. L., N. Kim, et al. (2012).Robust detection of traumatic axonal injury in individual mild traumatic brain injury patients: Intersubject variation, change over time and bidirectional changes in anisotropy. *Brain Imaging and Behavior* 6(2):329–342.

Little, D. M., M. F. Kraus, et al. (2010). Thalamic integrity underlies executive dysfunction in traumatic brain injury. *Neurology* 74(7):558–564.

MacKenzie, J. D., F. Siddiqi, et al. (2002). Brain atrophy in mild or moderate traumatic brain injury: A longitudinal quantitative analysis. *AJNR. American Journal of Neuroradiology* 23(9):1509–1515.

Madri, J. A. (2009). Modeling the neurovascular niche: Implications for recovery from CNS injury. *Journal of Physiology and Pharmacology* 60 Suppl 4:95–104.

Magdesian, M. H., F. S. Sanchez, et al. (2012). Atomic force microscopy reveals important differences in axonal resistance to injury. *Biophysical Journal* 103(3):405–414.

Mao, H., B. S. Elkin, V. V. Genthikatti, B. Morrison 3rd, and K. H. Yang. (2013). Why is CA3 more vulnerable than CA1 in experimental models of controlled cortical impact-induced brain injury? *Journal of Neurotrauma* 30:1521–1230.

Maxwell, W. L. (2013). Damage to myelin and oligodendrocytes: A role in chronic outcomes following traumatic brain injury? *Brain Sciences* 3(3):1374–1394.

Mayer, A. R., M. V. Mannell, et al. (2011). Functional connectivity in mild traumatic brain injury. *Human Brain Mapping* 32(11):1825–1835.

McAllister, T. W., J. C. Ford, et al. (2012). Maximum principal strain and strain rate associated with concussion diagnosis correlates with changes in corpus callosum white matter indices. *Annals of Biomedical Engineering* 40(1):127–140.

McAllister, T. W., M. B. Sparling, et al. (2001). Differential working memory load effects after mild traumatic brain injury. *NeuroImage* 14(5):1004–1012.

McMahon, P., A. Hricik, J. K. Yue, A. M. Puccio, T. Inoue, H. F. Lingsma et al. (2013). Symptomatology and functional outcome in mild traumatic brain injury: Results from the prospective TRACK-TBI Study. *Journal of Neurotrauma* 31:26–33.

Menon, D. K., K. Schwab, et al. (2010). Position statement: Definition of traumatic brain injury. *Archives of Physical Medicine and Rehabilitation* 91(11):1637–1640.

Messe, A., S. Caplain, et al. (2011). Diffusion tensor imaging and white matter lesions at the subacute stage in mild traumatic brain injury with persistent neurobehavioral impairment. *Human Brain Mapping* 32(6):999–1011.

Metting, Z., L. Cerliani, et al. (2013). Pathophysiological concepts in mild traumatic brain injury: Diffusion tensor imaging related to acute perfusion CT imaging. *PloS One* 8(5):e64461.

Metting, Z., L. A. Rodiger, et al. (2010). Acute cerebral perfusion CT abnormalities associated with posttraumatic amnesia in mild head injury. *Journal of Neurotrauma* 27(12):2183–2189.

Metting, Z., L. A. Rodiger, et al. (2009). Perfusion computed tomography in the acute phase of mild head injury: Regional dysfunction and prognostic value. *Annals of Neurology* 66(6):809–816.

Perlstein, W. M., M. A. Cole, et al. (2004). Parametric manipulation of working memory load in traumatic brain injury: Behavioral and neural correlates. *Journal of the International Neuropsychological Society* 10(5):724–741.

Pomschar, A., I. Koerte, S. Lee, R. P. Laubender, A. Straube, F. Heinen, B. Ertl-Wagner, N. Alperin (2013). MRI evidence for altered venous drainage and intracranial compliance in mild traumatic brain injury. *PloS One* 8(2):e55447.

Prichep, L. S., M. McCrea, W. Barr, M. Powell, and R. J. Chabot. (2012). Time course of clinical and electrophysiological recovery after sport-related concussion. *The Journal of Head Trauma Rehabilitation* 28:266–273.

Ropper, A. H. and K. C. Gorson (2007). Clinical practice. Concussion. *The New England Journal of Medicine* 356(2):166–172.

Rosenbaum, S. B. and M. L. Lipton (2012). Embracing chaos: The scope and importance of clinical and pathological heterogeneity in mTBI. *Brain Imaging and Behavior* 6(2):255–282.

Ross, D. E., C. Castelvecchi, et al. (2013). Brain MRI volumetry in a single patient with mild traumatic brain injury. *Brain Injury* 27(5):634–636.

Scheid, R., C. Preul, et al. (2003). Diffuse axonal injury associated with chronic traumatic brain injury: Evidence from T2*-weighted gradient-echo imaging at 3 T. *AJNR. American Journal of Neuroradiology* 24(6):1049–1056.

Scheid, R., K. Walther, et al. (2006). Cognitive sequelae of diffuse axonal injury. *Archives of Neurology* 63(3):418–424.

Shenton, M. E., H. M. Hamoda, et al. (2012). A review of magnetic resonance imaging and diffusion tensor imaging findings in mild traumatic brain injury. *Brain Imaging and Behavior* 6(2):137–192.

Signoretti, S., G. Lazzarino, B. Tavazzi, and R. Vagnozzi. (2011). The pathophysiology of concussion. *PM & R: The Journal of Injury, Function, and Rehabilitation* 3(10 Suppl 2):S359–368.

Smith, D. H., R. Hicks, et al. (2013). Therapy development for diffuse axonal injury. *Journal of Neurotrauma* 30(5):307–323.

Spitz, G., J. J. Maller, et al. (2013). Detecting lesions after traumatic brain injury using susceptibility weighted imaging: A comparison with fluid-attenuated inversion recovery and correlation with clinical outcome. *Journal of Neurotrauma* 30(24):2038–2050.

Stahel, P. F., M. C. Morganti-Kossmann, et al. (1998). The role of the complement system in traumatic brain injury. *Brain Research. Brain Research Reviews* 27(3):243–256.

Tang, L., Y. Ge, et al. (2011). Thalamic resting-state functional networks: Disruption in patients with mild traumatic brain injury. *Radiology* 260(3):831–840.

Toth, A., N. Kovacs, et al. (2013). Multi-modal magnetic resonance imaging in the acute and sub-acute phase of mild traumatic brain injury: Can we see the difference? *Journal of Neurotrauma* 30(1):2–10.

Vakhtin, A. A., V. D. Calhoun, et al. (2013). Changes in intrinsic functional brain networks following blast-induced mild traumatic brain injury. *Brain Injury* 27(11):1304–1310.

van den Heuvel, M. P. and O. Sporns (2011). Rich-club organization of the human connectome. *The Journal of Neuroscience* 31(44):15775–15786.

van den Heuvel, M. P. and O. Sporns (2013). Network hubs in the human brain. *Trends in Cognitive Sciences* 17(12): 683–696.

Viano, D. C., I. R. Casson, et al. (2005). Concussion in professional football: Brain responses by finite element analysis: Part 9. *Neurosurgery* 57(5):891–916; discussion 891–916.

Voss, H. U. and N. D. Schiff (2009). MRI of neuronal network structure, function, and plasticity. *Progress in Brain Research* 175:483–496.

Wada, T., Y. Asano, et al. (2012). Decreased fractional anisotropy evaluated using tract-based spatial statistics and correlated with cognitive dysfunction in patients with mild traumatic brain injury in the chronic stage. *AJNR. American Journal of Neuroradiology* 33(11):2117–2122.

Watanabe, R., T. Katsuhara, et al. (2012). Research of the relationship of pedestrian injury to collision speed, car-type, impact location and pedestrian sizes using human FE model (THUMS Version 4). *Stapp Car Crash Journal* 56:269–321.

Wilde, E. A., J. V. Hunter, et al. (2012). A primer of neuroimaging analysis in neurorehabilitation outcome research. *Neuro Rehabilitation* 31(3):227–242.

Wilde, E. A., S. R. McCauley, et al. (2012). Serial measurement of memory and diffusion tensor imaging changes within the first week following uncomplicated mild traumatic brain injury. *Brain Imaging and Behavior* 6(2):319–328.

Wilde, E. A., S. R. McCauley, et al. (2008). Diffusion tensor imaging of acute mild traumatic brain injury in adolescents. *Neurology* 70(12):948–955.

Yassa, M. A., L. T. Muftuler, et al. (2010). Ultrahigh-resolution microstructural diffusion tensor imaging reveals perforant path degradation in aged humans in vivo. *Proceedings of the National Academy of Sciences of the United States of America* 107(28):12687–12691.

Yeh, P. H., B. Wang, T. R. Oakes, L. M. French, H. Paner, J. Graner et al. (2013). Postconcussional disorder and PTSD symptoms of military-related traumatic brain injury associated with compromised neurocircuitry. *Human Brain Mapping* 35:2652–2672.

Yue, J. K., M. J. Vassar, et al. (2013). Transforming research and clinical knowledge in traumatic brain injury pilot: Multicenter implementation of the common data elements for traumatic brain injury. *Journal of Neurotrauma* 30(22):1831–1844.

Yuh, E. L., P. Mukherjee, et al. (2013). Magnetic resonance imaging improves 3-month outcome prediction in mild traumatic brain injury. *Annals of Neurology* 73(2):224–235.

Zappala, G., M. Thiebaut de Schotten, et al. (2012). Traumatic brain injury and the frontal lobes: What can we gain with diffusion tensor imaging? *Cortex* 48(2):156–165.

Zhou, Y., A. Kierans, D. Kenul, Y. Ge, J. Rath, J. Reaume et al. (2013). Mild traumatic brain injury: Longitudinal regional brain volume changes. *Radiology* 267:880–890.

Zhou, Y., Y. W. Lui, X. N. Zuo, M. P. Milham, J. Reaume, R. I. Grossman et al. (2013). Characterization of thalamo-cortical association using amplitude and connectivity of functional MRI in mild traumatic brain injury. *Journal of Magnetic Resonance Imaging* 39:1558–1568.

Zhou, Y., M. P. Milham, et al. (2012). Default-mode network disruption in mild traumatic brain injury. *Radiology* 265(3):882–892.

32 Blast-Related Mild Traumatic Brain Injury
Neuropsychological Evaluation and Findings

Nathaniel W. Nelson, Nicholas D. Davenport, Scott R. Sponheim, and Carolyn R. Anderson

CONTENTS

32.1 INTRODUCTION

Blast exposure is often reported by military personnel who have been deployed to recent wars in Iraq and Afghanistan. Difficulties of returning personnel with reintegrating into civilian society have in part been attributed to brain injury that was caused by blast concussions. In this chapter, we describe the challenges of evaluating the long-term cognitive impact of blast concussion through neuropsychological evaluation of self-reported events in military and veteran samples. We offer strategies of ascertaining whether a history of blast concussion or other combat-related condition, such as emotional distress, might represent the source of cognitive limitations. The chapter also reviews some of the latest neuropsychological findings in blast concussion samples and identifies areas of crucial need with respect to future research, ultimately in the interest of ensuring that veterans receive care that is appropriate to their injuries.

Blast-related traumatic brain injury (TBI) is among the most frequent injuries sustained by soldiers and other personnel who have served in recent wars in Iraq and Afghanistan (Eskridge et al., 2012; McCrea et al., 2009; MacGregor et al., 2011). Estimates of the prevalence of blast-related TBI in military personnel deployed to Iraq (Operation Iraqi Freedom, OIF; now Operation New Dawn) and Afghanistan (Operation Enduring Freedom; OEF) have been as high as 19%–23% (Tanielian and Jaycox, 2008; Terrio et al., 2009; Polusny et al., 2011). Reports of blast-related TBI among OEF/OIF military personnel may be unprecedented relative to military personnel in any previous war or conflict. As a consequence, research devoted to blast as a unique mechanism of concussive injury is greater now than in any other time in history.

The aim of the present chapter is to describe unique challenges and complexities that surround research conducted in OEF/OIF veteran samples with self-reported histories of blast-related TBI and provide a summary of some of the latest neuropsychological findings. Experimental neuroimaging studies in these samples will also be reviewed to the extent that they may potentially provide an objective and biological reference point for neuropsychological assessment. We begin by providing an overview of how TBI severity is conventionally assessed and diagnosed in civilian concussion samples. We also identify those features of concussion research that are essential to inform cause-and-effect relationships as well as expectations about short- and long-term recovery. It is our hope that a review of this literature will elucidate how researchers have been able to effectively document convergent findings regarding long-term recovery after nonblast forms of concussion and identify how military blast concussion is in many ways different and confounded by more factors relative to civilian concussion. We then summarize the methodology and rationale underlying a recent strategy that was developed at the Minneapolis Veterans Administration (VA) Health Care System that aims to maintain uniformity in assessing and diagnosing blast-related TBI on the basis of self-report in OEF/OIF veterans. Next, we present a summary of recent neuropsychology outcomes in blast-related TBI and preliminary results of experimental neuroimaging techniques that have been applied to blast concussion samples (most notably diffusion tensor imaging [DTI]). We conclude by highlighting how future research can most effectively enhance understanding of the true natural history of recovery after blast concussion, as has been established in conventional mechanisms of concussive brain injury (i.e., impact). It is reasonable to presume that the blast component of a TBI in most instances results in a mild injury to the brain because recent modeling indicates pressure waves at about 50% lethality because of lung rupture (an unprotected air-filled organ) would result in a mild TBI (Moore et al., 2009). Greater pressure waves are more often than not fatal. Thus, our discussion will focus on blast-related mild TBI (mTBI).

32.2 CONVENTIONAL (NONBLAST) mTBI (CONCUSSION)

Before turning to issues related to the assessment and diagnosis of blast-related mTBI as well as recent findings from studies using neuropsychology and neuroimaging techniques, it is essential to review what is known about conventional forms (i.e., nonblast) of concussive injury, how concussion is assessed and diagnosed, regardless of mechanism, and what the literature suggests with respect to recovery expectations after uncomplicated concussion, issues that may well pertain to blast-related mTBI. Familiarity with what is known about conventional concussion will also underline the major challenges and complexities that exist in assessing blast-related concussion in soldiers and veterans and highlight important areas of weakness to be addressed in future blast concussion research.

32.2.1 Assessment and Diagnosis

A variety of concussion assessment schemes has been developed in recent decades, such as those developed by the American Congress of Rehabilitation Medicine (ACRM; Kay et al., 1993), World Health Organization (Holm et al., 2005), and American Academy of Neurology (1997). Although subtle variations exist across these schemes, a fundamental tenet is that a diagnosis of concussion is made on the basis of *acute stage* injury parameters, not on the basis of retrospective self-report of symptoms alone. The ACRM criteria have proven to be especially influential and continue to be relied upon with regularity in both civilian and military/veteran samples. For example, the ACRM criteria were instrumental in the development of the later World Health Organization (2005) criteria. More recently, the VA and Department of Defense (DOD) relied upon the ACRM criteria to define combat-related concussion. According to the VA/DOD Clinical Practice Guideline for Management of Concussion/mTBI (U.S. Department of Veterans Affairs/Department of Defense), the ACRM criteria are "the most widely accepted criteria for mild TBI" in the United States.

Table 32.1 presents the criteria that are relied upon to define TBI severity within the VA and DOD (2009). The criteria included in the mild severity category are essentially the same as those included within the ACRM convention. According to the ACRM, an mTBI consists of "a traumatically induced physiological disruption of brain function" that is accompanied by (1) any period of loss of consciousness (LOC) persisting for approximately 30 minutes or less; (2) any loss of memory for events immediately before or after the accident (posttraumatic amnesia [PTA]), but not greater than 24 hours; (3) any alteration of mental state (e.g., dazed, disoriented, confused); and (4) focal neurological deficit(s) that may or may not be transient. To be regarded as mild traumatic brain injuries, they must result in Glasgow Coma Scale (GCS) status that is not less than 13 (i.e., between 13 and 15). Importantly, these injuries are the result of the head

TABLE 32.1
Definition of Traumatic Brain Injury by Severity

Criteria	Mild	Moderate	Severe
Structural imaging	Normal	Normal or abnormal	Normal or abnormal
Loss of consciousness (LOC)	0–30 minutes	>30 minutes and <24 hours	>24 hours
Alteration of consciousness/mental state*	A moment up to 24 hours	>24 hours; severity based on other criteria.	>24 hours; severity based on other criteria.
Posttraumatic amnesia	0–1 day	>1 and <7 days	>7 days
Glasgow Coma Scale (best available score in first 24 hours)	13–15	9–12	<9

Source: Adapted from U.S. Department of Veterans Affairs. (2009, April). *Clinical Practice Guideline: Management of Concussion/Mild Traumatic Brain Injury,* http://www.healthquality.va.gov/guidelines/Rehab/mtbi/concussion_mtbi_full_1_0.pdf., April 2009; and ACRM (Kay, T. et al. *J Head Trauma Rehabil,* 8, 86–87, 1993) criteria.

* (U.S. Department of Veterans Affairs/Department of Defense, 2009, p. 17): "Alteration of mental status must be immediately related to trauma to the head. Typical symptoms would be looking and feeling dazed and uncertain of what is happening, confusion, difficulty thinking clearly or responding appropriately to mental status questions, and being unable to describe events immediately before or after the trauma event."

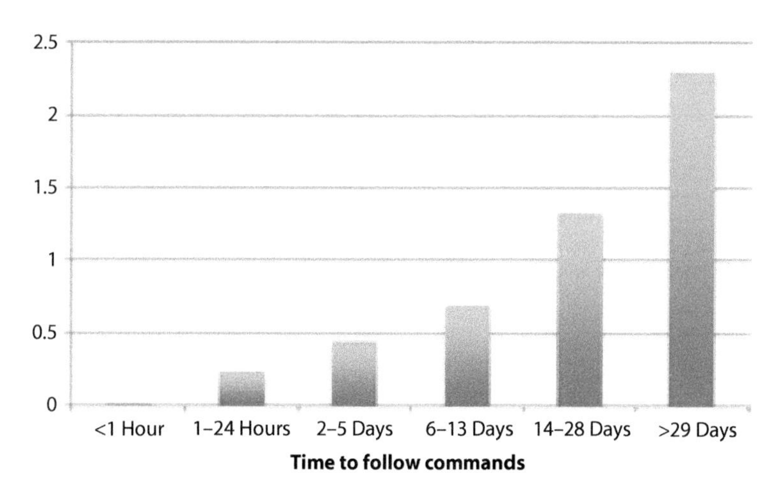

FIGURE 32.1 Overall impairment in neuropsychological performance at 1 year postinjury by TBI severity (the amount of time after injury until victim could follow verbal commands). Results from Dikmen et al. (1995) as summarized by Larrabee (2012). Effect sizes represented as Cohen's d with positive values representing greater impairment.

being struck or striking an object or acceleration/deceleration, which excludes other acquired conditions, such as stroke, anoxia, tumor, or encephalitis.

Why is it essential to rely upon acute-stage injury parameters, rather than late-stage self-reported symptoms, to define TBI severity? First, an extensive neuropsychology literature has shown that LOC, PTA, and GCS (e.g., time to follow commands) are strongly predictive of long-term recovery outcomes; injuries of greater severity typically result in worse outcomes relative to injuries of lesser severity as defined by these parameters. Indeed, there is strong empirical support for a dose-response relationship between such variables as GCS, and LOC/PTA even more so (Dikmen, Machamer, and Temkin, 2009), and long-term cognitive and adaptive functioning.

Consider, for example, the findings of one of the highest quality prospective studies that has been carried out in civilian TBI samples to date (Dikmen et al., 1995). The authors conducted a prospective examination of neuropsychological outcomes in a group of 436 adults who underwent hospital admission related to varying severities of TBI and compared performances against a group of 121 general-trauma control participants hospitalized for injuries other than TBI. The latter group did not differ significantly to the TBI group with respect to age, educational attainment, or gender, which allowed the authors to control for the potential effects of socioeconomic status as well as the various effects of physical injury itself independent of TBI (e.g., pain, psychosocial stress) on neuropsychological performance. The TBI group was classified according to six different injury severities on the basis of time to follow commands after injury as follows: (1) <1 hour; (2) 1–24 hours; (3) 25 hours to 6 days; (4) 7–13 days; (5) 14–28 days; and (6) >29 days. Performances were then compared across groups at 1 month postinjury and 1 year later.

Results suggested a general dose-response relationship between injury severity and cognition over time across the six TBI groups on a comprehensive neuropsychological test battery, with progressively higher levels of persisting impairment noted by TBI injury severity at 1 year. By contrast, participants who sustained minor brain injury (i.e., <1 hour time to follow command) demonstrated far less evidence of cognitive impairment at 1 year postinjury. In fact, cognitive performances in the minor brain injury group were not significantly different from the non-TBI injury controls. Summarizing the Dikmen et al. results through development of composite cognitive effect sizes, Larrabee (2012) concluded that "the effect size for the Dikmen et al. MTBI group, tested 1 year post-trauma, is essentially zero, consistent with complete overlap of the MTBI and trauma control group score distributions" (p. 235). (See Figure 32.1 for a review of overall effect sizes in cognitive performance in the Dikmen et al., 1995, study between the TBI and injury control groups, as summarized by Larrabee, 2012.)

A second reason why acute-stage injury parameters are relied upon to define TBI diagnosis, rather than late stage self-reported symptoms, relates to the highly nonspecific quality of the physical, cognitive, and emotional symptoms that may be reported in the postacute phase of recovery. Factors other than concussion itself often better account for the symptoms/impairments that individuals report/exhibit in the months and years after concussions are sustained. Indeed, this issue was emphasized by the developers of the ACRM criteria; although the authors suggested that various physical (e.g., nausea, vomiting, dizziness, headache), cognitive (e.g., inattention, memory difficulty), and behavioral changes (e.g., irritability, emotional lability) could potentially be used as "additional evidence" of mTBI, the developers emphasized that symptoms could be used to support a mTBI diagnosis only to the extent that these symptoms *could not be accounted for by causes other than mTBI itself* (e.g., peripheral injury, psychological reaction to physical or emotional stress).

The various symptoms included in the so-called "postconcussion syndrome" (PCS) (*Diagnostic and Statistical Manual of Mental Disorders* IV), for example, are not at all specific to remote history of concussion and cannot be relied upon, in isolation, to infer a history of traumatic brain injury of any severity. Meares et al. (2011) compared symptom reports of 62 mTBI participants and 58 non–brain-injured trauma controls (i.e., physical injuries other than mTBI). Whereas mTBI status was not predictive of PCS symptoms, the authors found that preinjury psychiatric symptoms (depression, anxiety), and acute posttraumatic stress were significantly predictive of PCS symptoms after physical trauma. Pain, regardless of mTBI status, was also predictive of PCS. These and similar findings reported by other independent researchers (Dikmen et al., 1995; Ponsford et al., 2012) highlight the importance of comparing outcomes of concussion samples with those of non-TBI trauma controls to account for various non-TBI confounds (e.g., pain, psychosocial stress) that may impact performances and/or symptom reports following concussion.

The nonspecific nature of PCS symptoms is also illustrated by the fact that a significant proportion of individuals with alternate conditions, but not TBI, also endorse these symptoms with great regularity. PCS symptoms are frequently endorsed in psychiatric groups (Iverson, 2006; Lange et al., 2011), chronic pain patients (Iverson and McCracken, 1997), and healthy community samples (Iverson and Lange, 2003). Moreover, researchers have demonstrated that non–TBI-related factors, such as cognitive bias (Ozen and Fernandes, 2011; Suhr and Gunstad, 2002, 2005) and false expectations regarding recovery after mTBI (Mittenberg et al., 1992), may of themselves reinforce persisting symptoms among those who continue to endorse symptoms in the late stage of recovery. The very belief that concussion results in lasting impairments, rather than concussion itself, may for some individuals maintain symptoms/impairments.

Secondary gain and response validity issues are also known to moderate self-reported symptoms as well as objective "impairments" on performance-based tasks in the months/years that follow concussive injury. In their survey of clinical neuropsychologists practicing within the United States, for example, Mittenberg et al. (2003) found that rates of probable malingering among those who underwent neuropsychological evaluation in the context of litigated mTBI were as high as 39%. In their widely cited meta-analytic review of neuropsychological outcomes in mTBI samples, Belanger et al. (2005b) found litigation status to be the most significant moderator of persisting cognitive limitations. The World Health Organization concluded that litigation and compensation issues were very influential factors to consider in the context of persisting symptoms/impairments following mTBI (Carroll et al., 2004). Recent studies of OEF/OIF concussion samples have also shown that rates of insufficient effort (Nelson et al., 2010) and symptom exaggeration (Nelson et al., 2011a,b) increase when veterans are assessed in a disability context or show indication of active disability claims. Thus, whether a civilian or veteran, response validity assessment is essential to the neuropsychological evaluation of any individual with a claim of disabling concussion, especially in the presence of known incentives for appearing more impaired than may in fact be the case.

32.2.2 "Natural History" of Recovery and Long-Term Neuropsychological Outcomes

If the primary task of concussion assessment is to establish that a given injury event likely resulted in concussion as defined by previously mentioned acute-stage injury parameters, or in other words surpassed a "minimum biomechanical threshold" (McCrea, 2008) of concussive injury, a secondary task is to establish the specific effects caused by the concussion, particularly with respect to *onset* and *course* of post-injury symptoms/impairments over time.

David Hume (1739–40, 1748; as cited by Field, 2013) suggested that to infer cause and effect (1) cause and effect must transpire contiguously (close together in time), (2) the cause must transpire before the effect, and (3) the effect should never occur in the absence of the cause. Considering concussion, researchers must demonstrate that (1) a temporal relationship exists between concussion and postinjury symptoms/impairments (contiguity), (2) the concussion transpires before the reported symptoms/impairments (i.e., were not present before the injury event), and (3) symptoms/impairments do not occur in the absence of concussion. Arguably, these conditions of causation have been most closely satisfied in studies conducted among athletes who undergo longitudinal assessments before (i.e., preseason baseline) and at multiple time points after (acute, subacute, and postacute) the times of their sports-related injuries.

As nicely described by McCrea (2008), there are various "unique advantages" inherent to conducting concussion research among athletes, and several researchers (Belanger and Vanderploeg, 2005a; McCrea et al., 2003, 2013) have capitalized on these advantages to elucidate the "natural history" of recovery that follows concussive injury. First, athletes represent a group that is at elevated risk of sustaining concussive injury within a well-defined time period (sports season). Close surveillance of potential concussive injuries can be conducted throughout the period of risk, and researchers are able to conduct their assessments very soon after the time of injury, thus allowing researchers to observe the necessary condition of contiguity. Related, sports-related concussions are most often witnessed (e.g., by teammates, spectators, game film), and this independent observation of concussion allows researchers to summarize acute injury characteristics (e.g., LOC, PTA) with far greater precision relative to retrospective self-report alone. Sports concussion researchers are also able to maintain a continuity of care that allows for direct comparison of acute-stage symptoms/impairments with baseline functioning, thus identifying the magnitude of effect that is attributable to the concussion itself. Systematic follow-up assessments conducted at various postinjury time points elucidate the usual patterns of recovery in subsequent hours, days, weeks, and months.

Furthermore, athletes represent a relatively "clean sample" that is typically healthy and free of premorbid physical (e.g., migraine headache), cognitive (e.g., attention-deficit), and psychiatric (e.g., depression, anxiety) symptoms that might confound an understanding of symptoms/impairments that may follow during the postinjury assessment phase. Sports concussion researchers also have ready access to a matched noninjury control group to further control for any symptoms/impairments that might naturally occur in the absence of concussion. Issues of secondary gain and response validity, which have been identified as important moderators of persisting symptoms/impairments in certain concussion samples (Belanger et al., 2005a; Carroll et al., 2004), are less likely relevant to athletes, a sample that tends to be very motivated to participate in sport (i.e., they typically have less incentive to exaggerate symptoms and/or subvert cognitive performances relative to other civilian concussion samples). By eliminating, or at least minimizing the degree to which these premorbid and postinjury variables have on recovery after concussion, sports concussion researchers are also able to conduct studies that largely uphold Hume's second (i.e., ensure that the symptoms/impairments were present only after the causal concussion) and third (i.e., control for any symptoms/impairments that might occur in the absence of concussion) conditions of causation.

To illustrate, consider the approach of McCrea et al. (2003), who conducted a prospective cohort study of 1,631 college football players to understand recovery expectations following sports-related concussion. From this sample, the symptoms/performances of 94 athletes with verified concussion (as defined by American Academy of Neurology, 1997, criteria) were compared against those of 56 noninjured controls at eight independent postinjury assessment points: immediately, 3 hours, and 1, 2, 3, 5, 7, and 90 days after injury. The authors conducted various assessments across these time points, including measures of self-reported postconcussive symptoms, postural stability/balance, and cognition (i.e., Standardized Assessment of Concussion) (McCrea et al., 2000). Relative to controls, individuals who sustained sports concussion endorsed significantly more symptoms, imbalance, and cognitive impairment immediately after concussion, which implicates an abrupt onset of considerable brain disturbance. However, these significant symptoms/impairments were followed by a rapid and very favorable course of improvement in subsequent days. Overall, postconcussion symptoms resolved by 7 days, balance difficulties resolved between 3–5 days, and cognitive functioning improved to baseline within 5–7 days. No significant differences were observed across symptom- or performance-based outcomes at 90 days. This pattern of acute-stage disturbance and exponential rate of recovery is remarkably consistent with the pattern of physiological disturbance (e.g., ionic shifts, reduced glucose metabolism) and recovery that has been reported in neuroscience research (Giza and Hovda, 2001).

In a more recent study, which represents the largest prospective study conducted to date on the incidence and course of "prolonged" recovery in sports concussion samples, McCrea et al. (2013) examined outcomes of 570 concussed athletes and 166 controls obtained from an overall sample of 18,531 athletes who had been monitored over a 10-year time frame. Using the same measures of postconcussion symptoms, balance, and cognitive functioning as McCrea et al. (2003), athletes were assessed immediately, 3 hours, and 1, 2, 3, 5, 7, and 45 or 90 days after concussion, and results were compared against preseason baseline. The concussion group was further subdivided according to those who sustained "typical" (i.e., recovery within 7 days; n = 513) and "prolonged" (>7 days; n = 57) recovery (i.e., the authors found that approximately 10% of the sample showed prolonged recovery patterns beyond the usual expectation of recovery within 7 days). The typical recovery group endorsed significantly more symptoms than the control group at day 3, and symptoms were no longer significant at day 5 postinjury. The typical recovery group also showed significantly greater performance-based impairments than controls immediately after injury, but no differences were found at day 2. By contrast, the prolonged recovery group endorsed significantly more symptoms than the typical recovery and control groups at every postinjury recovery point, including the 45/90-day period. The prolonged recovery group also showed significantly greater cognitive impairment relative to controls through the first 7 days postinjury, but no significant differences were observed in cognitive performance (or balance) at the 45/90-day mark relative to controls. Predictors of prolonged recovery in this study included LOC, PTA, and more severe acute-stage symptoms. Only 2.3% of the full injury sample reported symptoms at the 6- to 12-week assessment point. The authors concluded that only a small percentage of concussed athletes experience symptoms and cognitive impairment beyond 1 week, and prolonged recovery was most associated with acute-stage indices that were suggestive of more severe injury.

Results of well-controlled sports concussion studies have detailed the "natural history" of concussive injury (McCrea, 2008), and recovery outcomes following sports concussion. In their meta-analytic review of neuropsychological functioning following sports-related concussion, Belanger and Vanderploeg (2005) found acute-stage (i.e., first 24 hours) cognitive impairments to be greatest, particularly in the areas of learning/memory and global cognition measures. However, the authors found that the acute effect of concussion was "essentially zero beyond 7 days post injury (10 days for delayed memory)" (p. 352). Other meta-analytic reviews (Belanger et al., 2005b; Binder et al., 1997; Frencham et al., 2005; Iverson et al., 2005; Rohling et al., 2011; Schretlen and Shapiro, 2003), including civilian samples that have not been exclusively devoted to sports concussion, have reported very similar findings with respect to long-term outcomes after concussion. Rather than describe these in detail, we refer the reader to Figure 32.2, which summarizes the results of five meta-analyses reporting neuropsychological outcomes at approximately 90 days after concussion. As can be seen, there is no evidence of meaningful cognitive impairment at this 3-month assessment point. This contrasts dramatically with cognitive effects in the case of moderate-severe traumatic brain injury, which may result in significant impairments at 2 years postinjury and beyond (Schretlen and Shapiro, 2003).

Two primary lessons can be derived from these studies of neuropsychological outcomes following concussion in primarily civilian samples. First, impairments following a single, uncomplicated (i.e., not accompanied by any form of demonstrable intracranial disturbance), non–blast-related mTBI are most pronounced immediately after the time of injury and may include fairly significant impairments within the first week postinjury (e.g., d = −0.41) (Schretlen et al., 2004). Second, the vast majority of individuals who sustain concussion show a rapid trajectory of recovery; many if not most attain baseline function within 7–10 days, and there is no evidence of persisting impairment at approximately 90

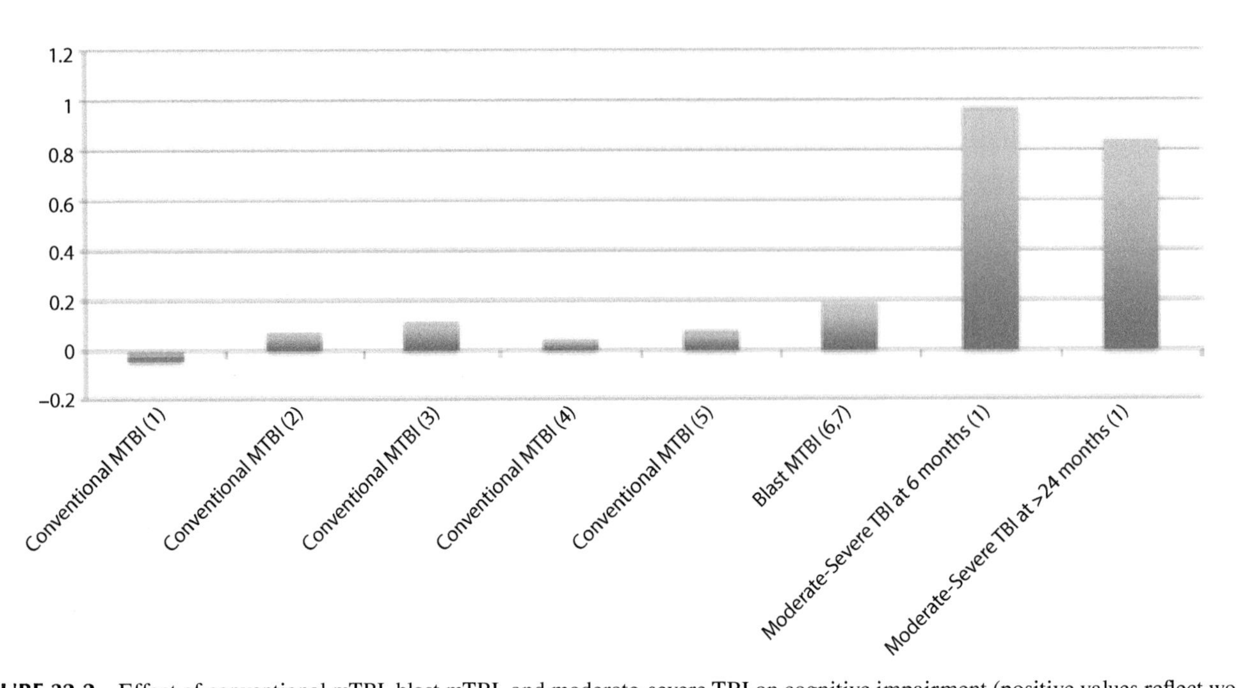

FIGURE 32.2 Effect of conventional mTBI, blast mTBI, and moderate-severe TBI on cognitive impairment (positive values reflect worse functioning). Effect sizes, represented as Cohen's d, for conventional mTBI and moderate-severe TBI are derived from the following meta-analyses that compared cognitive performances of TBI and comparison samples: (1) Schretlen and Shapiro (2003); (2) Binder et al. (1997); (3) Frencham et al. (2005); (4) Belanger et al. (2005); (5) Rohling et al (2011). Effect size for "Blast mTBI" is adapted from (6) Nelson et al. (2012) and (7) Shandera-Ochser (2013).

days postinjury at most. In a review of findings like these, Iverson (2005, p. 311) properly concluded "under most circumstances, we should anticipate good recovery following mTBI. Patients and athletes should be reassured."

32.3 BLAST-RELATED mTBI (CONCUSSION)

32.3.1 Blast as a Unique Injury Mechanism

Blast represents an injury mechanism that is qualitatively distinct from conventional (i.e., non–blast-related) forms of injury. As summarized by Taber et al. (2006), explosive blast is accompanied by an extraordinary change in atmospheric pressure that involves an initial period of peak overpressure, and then a subsequent negative blast phase (underpressure) that may cause various forms of physical injury, including TBI. Theoretically, blast may result in four types of independent mechanisms of injury: primary, secondary, tertiary, and quaternary effects (DePalma et al., 2005; Mayorga et al., 1997; Taber et al., 2006).

Primary blast injury refers to the direct effect of the blast itself. Air-filled organs, such as the lungs, have been identified as especially vulnerable to primary blast effect, and it is not uncommon for primary blast exposure to cause pulmonary hemorrhages or other forms of internal bleeding. The tympanic membrane (TM) is also regarded as especially vulnerable to primary blast, and has in fact been identified as a potential proxy for concussive injury (Xydakis et al., 2007). There is debate as to whether concussion likely results from the primary blast effect itself (Saljo et al., 2011) in the absence of TM perforation or injury to other bodily regions known to be more vulnerable to pressure wave effects than the brain, which is relatively well-protected within the skull. There are several theorized means by which concussion might result from primary blast exposure, such as thoracic, translational/rotational, and direct cranial entry mechanisms (Courtney and Courtney, 2011). Secondary blast effects, which refer to injuries that are sustained as a result of debris being projected toward the body, include such injuries as penetrating injuries from projectiles or fragments (DePalma et al., 2005). Tertiary blast injuries occur as a result of the body being displaced as a result of blast, and in the case of TBI, may occur when an individual is thrown against a wall or other stationary object. Quaternary blast injury refers to burns, toxic exposure, and other potential effects of the explosion. In short, the nature of the blast mechanism itself is novel. As such, some researchers (Cernak et al., 2001) have suggested that blast-related concussion may be accompanied by a pattern of neuropsychological recovery that is distinct relative to more conventional (i.e., non–blast-related) concussion.

Blast-related concussion often occurs in life-threatening circumstances beyond the blast (e.g., military combat). Thus, psychological distress may interact with blast concussion to complicate long-term neuropsychological recovery. Uncomplicated, non–blast-related concussion typically results in rapid and favorable patterns of cognitive recovery (Belanger et al., 2005b; Binder et al., 1997; Frencham et al., 2005; Iverson et al., 2005; Rohling et al., 2011; Schretlen and Shapiro, 2003), but also psychological and emotional recovery. A meta-analytic review of psychological adjustment after conventional concussion (Panayiotou et al., 2010) revealed that mTBI was associated with small overall effect sizes across the domains of depression, anxiety, coping, and psychosocial disability; the authors concluded, "mTBI had a small to negligible effect on emotional symptom reporting" (p. 463).

An unpredicted blast event that results in the death or severe injury of a fellow soldier in a war zone, for example, in many instances represents a more traumatic event relative to a concussion sustained as a result of a sports injury. Indeed, exposure to explosive blast among OEF/OIF personnel has been identified as increasing risk of developing trauma symptoms relative to those who sustain non–blast-related injuries (Belanger et al., 2009). Blasts frequently result in burns and other physical injuries in addition to concussion (Edwards et al., 2012) and also contribute to trauma symptoms. Lopes Cardozo et al. (2012) conducted a mental health survey of Cambodian civilians who survived landmine explosions as part of war. The authors found very high rates of mental health difficulties, including depression, anxiety, and posttraumatic stress disorder (PTSD), despite the fact that only a minority of the sample sustained comorbid head injuries. There is the possibility that some of these psychological consequences were at least partially associated with the general situation of the Cambodian war (i.e., not necessarily a direct effect of blast exposure itself). Nevertheless, findings like these illustrate that further research is needed to identify not only the degree to which blast concussion itself may impact long-term recovery, but also potential independent and interaction effects due to the blast event being a significant psychosocial stressor.

32.3.2 Challenges and Complexities Associated with Assessment and Diagnosis

Although a great deal of research has been devoted to the study of long-term outcomes after blast-related concussion in recent years, it is far less research than conducted in civilian samples (Lamberty et al., 2013). Moreover, blast concussion researchers do not enjoy the previously described "unique advantages" (McCrea, 2008) that exist in conducting research in sports concussion samples (Nelson and Keenan, in press). Like sports concussion samples, OEF/OIF soldiers and other military personnel represent a group that is at increased risk of sustaining mTBI, and researchers can also follow them during a specified period of mTBI risk exposure (e.g., defined period of deployment). However, unlike sports concussion samples, blast concussions sustained by OEF/OIF personnel are only occasionally witnessed, assessments are not typically conducted in a standardized manner and a controlled environment immediately after the blast, and only rarely are personnel systematically followed during acute, subacute, and postacute phases as has been conducted in sports concussion samples (e.g., McCrea et al., 2003, 2013). Most notably, blast concussion researchers do not typically have access to acute-stage injury findings that corroborate the previously mentioned indices of LOC, PTA, GCS, and other parameters that are so essential to establishing injury severity and predicting recovery.

The Military Acute Concussion Evaluation (MACE) (DVBIC, 2006) is a useful tool that may improve assessment of blast concussion. The MACE, which was meant to be completed by medics or other emergency medical personnel as soon as feasible after injury, consists of a history (e.g., incident description, cause of injury), neurological screening (eye, verbal, motor responsivity), and cognitive performance sections. The latter section was modeled after the Standardized Assessment of Concussion (McCrea et al., 2000) and allows the examiner to obtain rapid evaluation of orientation, immediate memory, concentration, and delayed recall performances that result in a composite score ranging from 0 to 30. From this information examiners are to offer an impression of whether a concussion was sustained and whether it resulted in LOC. Integration of MACE results with the accounts provided by OEF/OIF veterans months or years later could be helpful in determining the reliability of postdeployment accounts years after blast events. For example, it would be difficult to support a diagnosis of blast concussion in a veteran if MACE results (i.e., information documenting acute-stage concussion indices) were inconsistent with concussive injury. Unfortunately, researchers rarely have access to MACE information or other records documenting acute-stage injury characteristics following blast exposure. More often than not, researchers who work with soldiers and veterans are forced to rely upon retrospective self-report or results of "TBI screening" instruments months or years postinjury to inform diagnosis of blast concussion.

Concerns have been noted about the reliability of retrospective self-report of blast events and TBI screening in OEF/OIF samples (Belanger et al., 2012; Donnelly et al., 2011; Nelson et al., in press; Van Dyke et al., 2010). Test-retest reliability of the VA TBI Screening Instrument (VATBISI or TBI Clinical Reminder; GAO, 2008), for example, was described as "sobering" in a study conducted by Van Dyke et al. The authors repeatedly administered the VATBISI to 44 OEF/OIF veterans at average 6-month assessment points, and found inconsistency on such important parameters as LOC (36% inconsistency) and amnesia (32% inconsistency). Other more recent studies have raised serious concerns related to false-positive identifications of TBI using screening instruments like the VATBISI (Belanger et al., 2012; Donnelly et al., 2011), and it appears that PTSD and physical symptoms reported during the postdeployment are predictive of inconsistent reports of combat-related TBI over time.

Amidst these challenges and complexities, blast concussion researchers and clinicians are much more restricted in their ability to uphold the conditions necessary to establish cause and effect (Hume, 1739–40, 1748; as cited by Field, 2013) relative to what has been reported in sports concussion samples. With few exceptions (Kennedy et al., 2013; Luethcke et al., 2011), evaluators of blast concussion are not able to corroborate acute-stage injury characteristics and cognitive performances immediately after blast concussions, which impede their ability to uphold the condition of contiguity. Moreover, because of the frequent lack of control for premorbid and postmorbid physical and psychiatric conditions in many blast concussion studies, evaluators are typically unable to uphold the second and third of Hume's conditions of causality—absent before event and not present when the event does not occur.

Thus, evaluators of blast concussion confront a dilemma: blast concussion has been identified as a common hazard of the recent wars in Iraq and Afghanistan, but the elucidation of the severity, frequency, and anticipated outcomes associated with historical blast concussions is obscured by the unknown reliability and validity of self-report information obtained through contemporary TBI screening methodologies. With these limitations in mind, evaluators are forced to "work with what we have" when determining whether an historical blast exposure likely resulted in blast concussion.

32.3.3 Semistructured Interview Strategies

In light of the important limitations that have been documented of the VA TBI clinical reminder and other contemporary TBI screening instruments, several research groups have developed semistructured interviews for TBI to arrive at a more informed or "clinician-confirmed" assessment and diagnosis of historical blast-related concussion. Examples of such interview approaches include the Structured Interview for TBI Diagnosis (Donnelly et al., 2011), VA TBI Identification Clinical Interview (Vanderploeg et al., 2012), Boston Assessment of Traumatic Brain Injury – Lifetime (Brawn Fortier et al., 2012), and the Minnesota Blast Exposure Screening Tool (MN-BEST) (Nelson et al., 2011). To be clear, none of these methods has been developed with complementary acute-stage injury information for cross-validation, leaving significant questions regarding their own psychometric properties and utility in classifying concussion. But there is little question that these more extended interview-based approaches represent an improvement relative to TBI screening instruments alone, which as mentioned result in unacceptably high false-positive identification rates (Belanger et al., 2012). Here, we briefly describe the rationale and development of just one of these interview-based approaches, the MN-BEST, to illustrate how researchers of the Minneapolis Veterans Administration Health Care System "work with what they have" when confronting the challenge of assessing and classifying reports of blast concussion in the absence of corroborating information from the combat zone.

The MN-BEST, which is reproduced in Appendix 32.A, was developed in 2009 to be used in conjunction with results of the VATBISI (GAO, 2008) and TBI Secondary Evaluation that is the current procedure of screening and further evaluating OEF/OIF veterans who endorse a history of combat-related TBI. Early in our research, it was recognized that many of our OEF/OIF participants reported a history of many previous blast exposures that may or may not have resulted in blast-related concussion. Therefore, a primary rationale in developing the MN-BEST was to develop a composite index of blast-related concussion that could be used as a single indicator of previous blast concussion history. It was also our anecdotal observation that physicians and other clinical providers varied widely in their approach to the assessment and diagnosis of self-reported blast concussion. Review of clinical records revealed that providers did not necessarily assess such parameters as LOC and PTA to the same level of specificity. Variability was also noted in the number of blast-related concussions that were explored with veterans during the TBI interview process. Thus, another rationale for the development of the MN-BEST was to develop a method of maintaining a uniform, standardized approach to the assessment and diagnosis of every case of blast-related concussion that was reported among our participants. More specifically, the MN-BEST was developed to maintain a consistent strategy of assessing the *frequency, severity,* and *plausibility* of the three most significant concussions sustained as a result of blast.

Regular consensus meetings are held by a team of experienced clinical neuropsychologists who review self-reported information (and corroborating information if available) among study participants. To administer the MN-BEST, the evaluator first invites participants to estimate the sheer number of blast exposures that they confronted during combat, regardless of whether they perceived that these events resulted in blast concussion. This gives the assessor a general sense of relative risk that each individual participant may have had of sustaining blast concussion. An explosive ordnance disposal worker whose task it is to locate and intentionally detonate improvised explosive devices (IEDs), for example, would be regarded to be at greater risk of sustaining blast-related concussion than a veteran who never received combat training and was never located near the combat zone.

Next, the evaluator invites the veteran to describe the three most significant blast events that were confronted during combat. For each of the three events, it is the assessor's task to essentially re-create the injury event with as much detail as possible to establish whether it is more likely than not that a given blast *exposure* resulted in a blast *concussion* according to a standardized set of diagnostic criteria. We were interested in identifying whether there might be meaningful differences within the mild spectrum of injury (e.g., whether an individual who did not sustain any LOC as a result of blast may show outcomes that are meaningfully different relative to an individual with 20 minutes of LOC). Specifically, the MN-BEST includes four concussion severity classifications, labeled as types 0, I, II, and III (refer to Appendix 32.A) that were adapted from both the ACRM (Kay et al., 1993) criteria and standards proposed by Ruff and Richardson (1999). As shown in Appendix 32.A, type 0 injuries involve no indication of LOC/PTA and are restricted to one or more neurologic symptoms or signs. Type I injuries involve altered mental state or unclear LOC, very brief PTA (1–60 seconds), and one or more neurologic symptoms/signs. Type II concussions are defined by definite LOC persisting less than 5 minutes, PTA of 60 seconds to 12 hours, and one or more neurologic symptoms/signs. Finally, type III injuries are accompanied by LOC from 5 to no more than 30 minutes in duration, 12–24 hours of PTA, and one or more neurologic symptoms or signs. Each concussive injury is then converted into a metric that, if determined to be plausible, may contribute to an overall composite of blast concussion. Type 0 injuries are assigned a severity score of 1, type I injuries a 2, type II injuries a 3, and type III injuries a 4, resulting in a range of 0–12 for the overall composite of blast concussion. The same process is also conducted for all non–blast-related events during the respondent's lifetime (including those that transpired before, during, or after deployment), allowing for the control of non–blast-related concussions on outcomes of interest.

The classification of blast-related concussions during the frequency and severity rating stages is determined by taking the self-reported symptoms/signs at face value. After classifying the frequency and severity of blast- and non–blast-related concussions on the basis of face validity alone, the final step in the MN-BEST assessment process is to offer an opinion regarding the *plausibility* that the blast event as described resulted in concussive injury; is it "more likely than not" on the basis of available information that the blast event surpassed the minimal biomechanical threshold of concussive injury? Only those injuries that are determined to be plausible are included in the overall composites of concussion (blast or nonblast).

It is during this final plausibility rating process that the researcher must exercise clinical judgment and knowledge of blast effects to determine whether the blast accounts are coherent and plausible. For example, the level of *specificity* surrounding reported blast events may determine whether a given event was consistent with concussion. Thus, descriptions of blast events that cannot be located in place or time may be regarded as less plausible relative to events that are described in a specified city on a specified month, date, and year. Other detailed information, such as personnel who accompanied the respondent at the time of the blast, the precise activities being conducted at the time of the blast, type of weaponry (e.g., IED, rocket-propelled grenade), and conveyances (e.g., riding in a Humvee), may also assist the assessor to contextualize the event.

Another common factor that evaluators may rely upon to inform the likelihood that a reported blast exposure resulted in blast concussion relates to the *level of anticipation* that was associated with the blast event. For instance, a rather common scenario among veterans who report historical blast exposures in our work in Minneapolis is the report of blast exposure as a result of intentional detonation of an IED. More often than not, the consensus team concludes that these intentional detonations are less likely than not to have resulted in concussive injury. Unlike blasts encountered during combat, blast exposures that result from these intentional detonations are known to transpire at predetermined times, which allows soldiers to take steps to insure their safety (e.g., standing behind protective barrier an extended distance away, making use of protective gear).

A veteran's reported *proximity from blast* represents another example of an indicator that might inform the plausibility of blast concussion. Howe (2009) indicates that the blast wave dissipates by a cubed root of one's distance from the blast source; thus, an individual who is 10 feet from a given blast source is exposed to nine times more overpressure than an individual who is 20 feet from the blast source. Thus, a participant who reports having been 1,500 m from a controlled detonation might be regarded as less likely to have sustained blast concussion relative to an individual who reports having encountered a rocket-propelled grenade that was 5 m away. Conversely, the reliability of an event involving the detonation of an IED "1 foot" away from one's person might be brought into question if he or she denies that this exposure resulted in any immediate concussive effects or physical injury.

The *other physical injuries* that a given blast event contributed to may further inform the likelihood of blast concussion. As noted previously, the tympanic membrane is especially vulnerable to blast overpressure and may even represent a reliable proxy for concussive injury (Xydakis et al., 2007). Thus, a blast event that contributes to TM perforation may be plausibly regarded as an event that also resulted in concussion. Related, consideration of the *injuries of peers* who were in the same vicinity of the respondent at the time of the blast may also assist the researcher to determine the credibility of a given blast event and inform the likelihood of concussive injury. Case studies have shown that individuals who are exposed to blast at comparable distances show "strikingly similar" physical injuries (Commandeur et al., 2012). Thus, accounts suggesting that the respondents' peers sustained similar injuries as the respondent may be more plausible than accounts in which a blast resulted in injury to the respondent only (in spite of the report that his or her peers were located the same distance from the blast).

A benefit of the MN-BEST is that mTBI can be either dimensionally or categorically characterized, as well as collapsing across blast and conventional mTBIs to achieve a total cumulative mTBI for an individual's lifetime. Given the sources of information and variability in assessor ratings MN-BEST scores are unlikely to be numerically precise but nonetheless provide quantitative estimate of the cumulative degree of brain injury from traumatic events. Therefore, the MN-BEST overall blast concussion composite can be used as a continuous measure to explore questions of interest that are better suited for dimensional data (e.g., dose-response relationship between blast concussion and emotional distress as defined by select Minnesota Multiphasic Personality Inventory-2 clinical scales). Although preliminary work suggests that the inter-rater reliability of the MN-BEST is strong (>0.90 per Nelson et al., 2011), to date there is no validation of the instrument with known injury characteristics obtained during the acute phase of injury. Ideally, blast-related concussion would be assessed on the basis of acute-stage injury characteristics, but in the vast majority of cases this is not possible with military samples. We recognize the MN-BEST as only one potential strategy of enhancing the reliability of self-report information in OEF/OIF veterans with self-reported histories of concussion, but analysis of white matter integrity (number of voxels with low FA) in 116 OEF/OIF military personnel revealed that individuals classified on the MN-BEST as having blast-related mTBI tended to fall in the

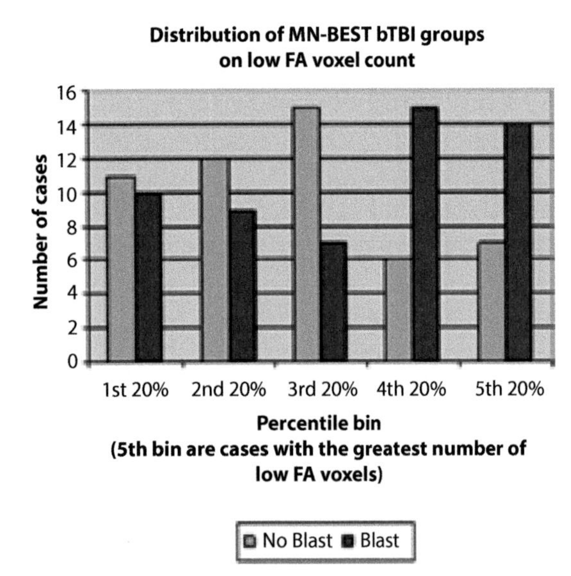

FIGURE 32.3 Number of individuals with and without blast-related mTBI (i.e., bTBI) as determined by the MN-BEST within each 20th percentile bin for a measure of white matter integrity (count of number of white matter voxels with low FA) in 116 individuals who had been deployed to war zones in Iraq and Afghanistan. See Davenport et al. (2012) for further explanation of the FA voxel count index.

upper two-fifths of subjects on the low FA voxel count index (Figure 32.3). Thus, there is some preliminary neuroimaging evidence to support use of the MN-BEST.

32.3.4 Neuropsychological Outcomes

As noted, no research group to date has effectively implemented the same longitudinal approach to the assessment of blast concussion as has been conducted in sports concussion samples (e.g., McCrea et al., 2003, 2013). This dramatically limits the ability to reliably determine the "natural history" of recovery after blast-related concussion. Some concussion researchers have examined cognitive performance within the acute stage of combat-related concussion (Coldren et al., 2010, 2012; Kennedy et al., 2012), but investigators have not systematically followed performance trajectories beyond the acute stage to understand long-term recovery patterns. Further, some of these acute stages of injury blast concussion studies (e.g., Kennedy et al., 2012) have mixed blast concussion samples with other forms of combat-related concussion, which eliminates the possibility of understanding differences between blast and nonblast mechanisms of injury.

Researchers have recently studied predeployment cognitive performance as a control for postdeployment cognitive function (Roebeck-Spencer et al., 2012; Vasterling et al., 2012), yet these have not included assessments during the acute stage of concussive injury. Some of these pre-post studies have also mingled blast with other mechanisms of injury (e.g., only 63% of the concussion sample in Roebeck-Spencer et al. were blast-related), which again limits the ability to identify whether blast or other combat-related concussions account for reported findings. Interestingly, in both studies, the majority of individuals who reported having sustained mTBI during deployment failed to show lasting cognitive impairments. However, psychological and emotional difficulties were found to contribute to longer term neuropsychological impairments. Vasterling et al. concluded that "milder TBI reported by deployed service members typically has limited lasting neuropsychological consequences; PTSD and depression are associated with more enduring cognitive compromise" (p. 186). These findings are very much consistent with previous studies showing favorable course of recovery in the case of non–blast-related concussion (Belanger et al., 2005) as well as the impact that psychological conditions, such as PTSD (Oien et al., 2011) and depression (Zakzansis et al., 1999), have on cognitive performance.

Of particular interest to the current review are two threads of neuropsychology literature: (1) studies that have compared blast and nonblast concussion samples on cognitive performances and self-reported psychological symptoms (Belanger et al., 2009, 2011; Cooper et al., 2012; Kontos et al., 2013; Lange et al., 2012; Lippa et al., 2010; Luethcke et al., 2011) and (2) studies that have compared blast concussion samples with deployment controls and psychiatric comparison groups (Nelson et al., 2012; Shandera-Ochsner et al., 2013). With respect to contrasts of blast and nonblast samples, we generated neuropsychological effect size differences across five studies that examined performances in blast and nonblast concussion groups at various postinjury recovery stages (Belanger et al., 2009; Cooper et al., 2012; Kontos et al., 2013; Lange et al., 2012; Luethcke et al., 2011). As shown in Figure 32.4, the composite effect size for neuropsychological performance between blast (n = 1,006) and nonblast concussion (n = 1,806) samples in these studies was minimal ($d = -0.06$). The similarity of blast and nonblast concussion samples was demonstrated at 72 hours (Luethcke et al., 2011; $d = 0.19$), 4 months (Lange et al., 2012; $d = 0.14$), 27 weeks (Cooper et al., 2012; $d = -0.22$), and more than a year (Belanger et al., 2009; $d = 0.13$) after blast concussions were sustained. Taking this literature together, there is minimal evidence that blast results in cognitive changes that are distinct from nonblast mechanisms of concussive injury.

To contrast psychological consequences of blast and nonblast concussive injury we also generated effect size differences across six studies that included self-report measures of psychological and emotional functioning (Belanger et al., 2011; Cooper et al., 2012; Kontos et al., 2013; Lippa et al., 2010; Luethcke et al., 2011, Lange et al., 2012). As shown in Figure 32.4, the composite effect size of self-reported psychological symptoms between blast (n = 1,404) and non-blast (n = 1,939) samples was small in magnitude ($d = 0.28$) by conventional standards (Cohen, 1988), though certainly more sizeable than what was observed of neurocognitive outcomes in blast versus nonblast samples. Effect sizes for psychological symptoms were generally small across assessments conducted 72 hours (Luethcke et al., 2011; $d = 0.10$), and approximately 1 month (Lippa et al., 2010; $d = 0.31$), 4 months (Lange et al., 2012; $d = 0.15$), 6 months (Cooper et al., 2012; d = 0.10), and 12 months (Belanger et al., 2011; $d = 0.24$) after blast

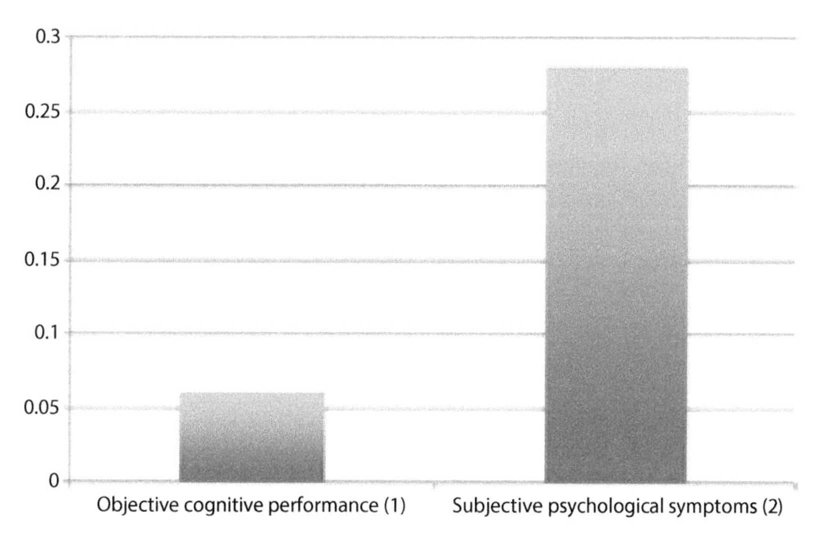

FIGURE 32.4 Composite effect sizes of blast versus non–blast-related concussion. Effect sizes, represented as Cohen's d, with positive values reflecting greater prevalence in blast samples. Effect size (1) derived from cognitive performances obtained in blast versus nonblast concussion samples: Belanger et al., 2009; Cooper et al., 2012; Kontos et al., 2013; Lange et al., 2012; Luethcke et al., 2011. Effect size (2) derived from self-reported symptoms obtained in blast versus nonblast concussion samples: Belanger et al., 2011; Lippa et al., 2010; Kontos et al., 2013; Lange et al., 2012; Luethcke et al., 2011.

concussions were sustained. Thus, individuals who sustain concussion as a result of blast might be at somewhat greater risk of developing trauma symptoms and other forms of emotional difficulty than those who sustain concussion by other means (Belanger et al., 2009).

Studies have also compared OEF/OIF veterans with reported histories of blast concussion to deployment controls on neuropsychological performance (Nelson et al., 2012; Shandera-Ochsner et al., 2013). Nelson et al. (2012) examined neuropsychological outcomes in a sample of 118 OEF/OIF veterans with varied combat histories. Using the Clinician-Administered PTSD Scale (CAPS) and Structured Clinical Interview for DSM Disorders (SCID) to ascertain axis I diagnosis, and the MN-BEST to define self-reported history of blast concussion, the authors compared neuropsychological performances across four groups: (1) mTBI only (n = 18), (2) axis I diagnosis only (n = 24), (3) comorbid mTBI/axis I (n = 34), and (4) deployment control (n = 28). Although a main effect was found for axis I diagnosis on neuropsychological performance, no main effect was found for mTBI status.

Shandera-Ochsner et al. (2013) aimed to extend and replicate the Nelson et al. (2012) study by examining neuropsychological performances in a sample of 81 OEF/OIF veterans in four groups: (1) mTBI only (n = 20), (2) PTSD only (n = 19), (3) comorbid PTSD/ mTBI (*n* = 21), and (4) deployment control (n = 21). Similar to Nelson et al., the mTBI-only group demonstrated neuropsychological performances that were comparable to the deployment control group, whereas both PTSD groups demonstrated significantly worse cognitive performances relative to deployment controls. The authors concluded that "PTSD seems to be an important variable affecting neuropsychological profiles in the post-deployment time period" (p. 881). They also concluded that the lack of effect between the mTBI and control group was consistent with findings from the civilian mTBI literature, which suggests that mTBI does not of itself contribute to enduring cognitive impairments.

Taken together, the results of Nelson et al. (2012) and Shandera-Ochsner et al. (2013) support two primary conclusions. First, when comparing blast concussion samples with no ongoing indication of psychological and emotional symptoms to deployment controls, there is no evidence that a remote history of blast concussion alone contributes to lasting cognitive impairments months or years later. In spite of the noteworthy limitations of these two studies (e.g., small sample sizes; some proportion of the Nelson et al. mTBI group showed ongoing signs of alcohol dependency), the overall effect size between the 38 OEF/OIF veterans with mTBI (and without axis I conditions) and OEF/OIF controls in these two studies was small ($d = 0.19$), and comparable to effect sizes obtained in meta-analytic reviews of concussion (Figure 32.2). Second, the results of Nelson et al. (2012) and Shandera-Ochsner et al. (2013) clearly illustrate that PTSD and other forms of axis I psychopathology clearly impact cognitive performances during the postdeployment phase, and their findings comport with those of other researchers who have found psychological distress as having a significant effect on cognitive function in OEF/OIF personnel (e.g., Vasterling et al., 2012). As shown in Figure 32.5, regardless of blast concussion history, the effect sizes of the two samples with ongoing psychopathology from the Nelson et al. and Shandera-Ochsner et al. studies are quite consistent with the moderate effect sizes reported in two meta-analyses that examined neuropsychological functioning among individuals with major depression (Zakzanis et al., 1999) and PTSD (Oien et al., 2011).

In sum, despite various weaknesses in neuropsychological studies of blast-related mTBI, several preliminary conclusions can be made. First, although blast has been shown to represent a unique injury mechanism (DePalma et al., 2005; Taber et al., 2006), there is essentially no evidence that neuropsychological outcomes vary by injury mechanism. Studies have generally failed to show meaningful differences between blast and non–blast-related mTBI at various points after injury. Second, there is very little evidence that blast concussion per se results in enduring neuropsychological impairment. Available research appears to suggest that long-term recovery after a single blast concussion is quite favorable, consistent with what has been reported in the civilian literature among those who sustain non–blast-related concussion. Third, blast concussion does appear to increase risk of developing PTSD and other forms of psychological distress compared with other causes of concussion, though effect sizes for differences in psychological symptoms are fairly modest based upon studies reviewed here. Finally, PTSD, depression, and other forms of psychological distress, independent of blast concussion history, contribute to cognitive impairments of at least moderate magnitude in OEF/OIF samples, consistent with what has been reported in previous meta-analytic reviews of neuropsychological functioning in PTSD and major depression.

32.4 STRUCTURAL NEUROIMAGING OF BLAST-RELATED mTBI

Although neuroimaging measures offer objective characterization of how the brain may be altered by mTBI, the data are limited in the same way as neuropsychological measures in terms of determining cause and effect (at least per Hume's criteria). The hope is that these direct measures of brain structure are more sensitive to possible effects of concussion on the brain and thus can detail the neural degeneration that might occur because of one or more concussions suffered by an individual. Structural measures are also clinically practical because they do not require difficult to standardize experimental manipulations as is typically required for functional magnetic resonance imaging.

32.4.1 Clinical Neuroimaging Strategies

In civilian settings, neuroimaging is frequently used to determine the presence and extent of any cerebrovascular ruptures (e.g., hematoma) that may require immediate attention and/or acute signs of brain damage. Both X-ray computed tomography (CT) and magnetic resonance imaging (MRI) are able to provide this information, though CT is somewhat preferred in emergency room settings because of its lower cost, shorter duration, and overall higher safety for persons with altered or reduced consciousness who cannot reliably report contraindications to MRI (e.g., implanted medical devices). Although CT can reliably be used to detect gross structural damage, mild TBI is not typically associated with CT abnormalities (Borg et al., 2004). In contrast, MRI is much more sensitive to traumatic lesions such as diffuse axonal injury and cerebral contusions, with up to 77% of mTBI individuals demonstrating abnormalities on MRI; however, the presence of abnormal findings on MRI relates poorly to cognitive impairments or long-term outcomes (Hofman et al., 2001; Hughes et al., 2004; Lee et al., 2008). As David Hughes and colleagues (2004) stated, "Although non-specific abnormalities are frequently seen, standard MRI techniques are not helpful in identifying patients with MTBI who are likely to have delayed recovery" (p. 550). Therefore, the search for neuroimaging measures that can provide a clinically useful, objective marker of mTBI has focused on more specific measures than overall structure.

32.4.2 DTI

Although conventional MRI scans merely reflect the *type* of tissue (e.g., gray matter, white matter, fat) present at various locations in the brain, DTI has received attention over the past decade for being able to provide information about the underlying microstructural properties of that tissue based on

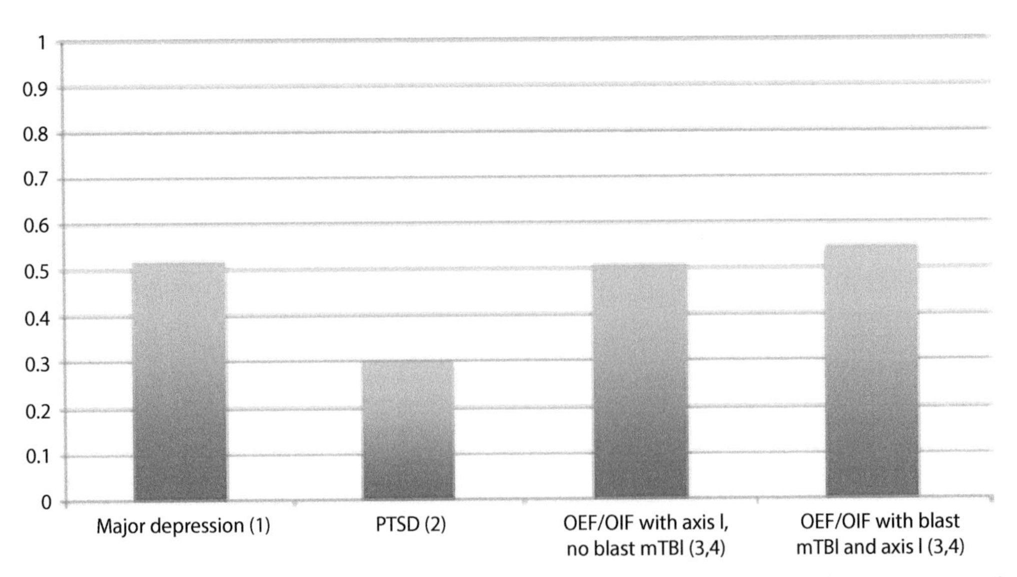

FIGURE 32.5 Effect of major depression, PTSD, and other axis I conditions on cognitive impairment. Effect sizes for major depression and PTSD are derived from the following meta-analyses that compared cognitive performances of psychiatric and comparison samples: (1) Zakzanis et al. (1998) and (2) Oien et al. (2011). Effect sizes for OEF/OIF groups are adapted from (3) Nelson et al. (2012) and (4) Shandera-Ochser (2013).

patterns of water diffusion (Basser, 1995; Beaulieu, 2002). By measuring the magnitude of Brownian motion of water (i.e., diffusion) in each of several directions, DTI summarizes the overall diffusion pattern within each volume element (i.e., "voxel") as a set of three perpendicular vectors of varying magnitudes (Basser et al., 1994). In white matter, diffusion along the orientation of the axon (axial diffusivity) is much more substantial than diffusion perpendicular to the axon (radial diffusivity) because the latter is constrained by the axon membrane, myelin, and cytoskeletal elements (Beaulieu, 2002). This anisotropic (i.e., nonspherical) diffusion pattern is dependent on multiple characteristics of healthy white matter (e.g., high axon density, myelination), and the degree of anisotropy is generally considered a valid measure of overall white matter integrity (Beaulieu, 2002). Fractional anisotropy (FA), the most commonly used DTI measure, varies between 0 (perfect sphere) and 1 (infinite cylinder) to represent the shape of the tensor independent of its size, whereas the complementary measure of mean diffusivity (MD) represents the overall size of the tensor independent of its shape (Basser and Pierpaoli, 1996). Reduced FA and increased MD (i.e., reduced constraints to diffusion) are generally considered to be associated with white matter damage or poor integrity.

In civilian populations, some researchers have reported mTBI to be associated with lower FA and higher MD, especially in frontal association pathways and posterior corpus callosum (Aoki et al., 2012; Niogi and Mukherjee, 2010). However, the nature of military mTBI in general, and those involving exposure to blast in particular, limits the generalizability of the civilian literature. Indeed, among the seven studies that directly compared DTI measures of soldiers who experienced mTBI during their most recent deployment to those of deployed veterans without mTBI (Bazarian et al., 2012; Davenport et al., 2012; Jorge et al., 2012; Levin et al., 2010; MacDonald et al., 2011; Morey et al., 2013; Sorg et al., 2013), only two reported group differences in individual regions (MacDonald et al., 2011) or voxels (Morey et al., 2013), and the affected regions in these studies did not overlap. However, even the studies with negative results have provided valuable insight into the complexity of studying this population. The first DTI study of blast-related mTBI, conducted by Levin and colleagues (2010), compared a group of veterans previously diagnosed with mild or moderate TBI to a group of veterans with neither exposure to blast nor mTBI during deployment, thus maximizing group differences in both mTBI and exposure to blast. As expected, individuals with mTBI had higher rates of PTSD, depression, and postconcussive symptoms as well as higher levels of global distress, but there were no group differences in quality of life or overall cognitive function. No group differences in FA or MD were found for any region of interest (ROI) or in a voxel-wise analysis, indicating that even moderate TBI may not be associated with systematic reductions of white matter integrity of particular regions. This is consistent with a lack of regional FA differences between veterans with mTBI and healthy civilians despite differences in coherence of electrophysiological activity among brain regions (Sponheim et al., 2011). One potential interpretation for the failure to detect group differences in FA using traditional ROI and voxel-wise analyses is that the structural effects of mTBI are spatially heterogeneous across individuals (Ptak et al., 2003). Consistent with this hypothesis, veterans with blast mTBI have been reported to have more ROIs (MacDonald et al., 2011) or voxels (Davenport et al., 2012; Jorge et al., 2012) with abnormally low anisotropy (i.e., several standard deviations below the mean), raising the suggestion of white matter findings associated with mTBI even if no single region is consistently affected.

Given the general lack of differences in DTI measures between mTBI and veteran control groups, several studies have instead focused on subsets of veterans with mTBI. Interestingly, although Jorge and colleagues (2012) reported that the number of voxels with abnormally low FA was similarly elevated in veterans with evidence of LOC or PTA (i.e., probable mTBI) as in veterans with "vague symptoms of confusion and dazedness" (i.e., possible mTBI), Sorg and colleagues (2013) reported that only mTBI involving LOC was associated with lower FA in ventral prefrontal white matter. Moreover, Matthews and colleagues (2012) reported lower FA associated with LOC, compared to alteration of consciousness, in 14 widespread regions of white matter. These results underscore the potential importance of acute symptoms (e.g., LOC, PTA) on long-term neurological changes, making the collection of this information even more critical to the continued study of blast mTBI.

Overall, select experimental investigations that have included DTI have raised the possibility that white matter disruptions associated with blast mTBI are present in the chronic stage after injury and most likely vary in location across individuals. There is also preliminary evidence that certain features of the mTBI event (e.g., loss vs. alteration of consciousness) affect the neurological outcomes, though this does not necessarily comport with clinical observations (e.g., Dikman et al., 1995). Additional studies that can better characterize these acute injury symptoms may lead to the ability for DTI to inform assessments of injury severity and, eventually, reduce the reliance on retrospective self-report.

32.5 CONCLUSION AND FUTURE DIRECTIONS

Neuropsychological recovery after mTBI (concussion) has been a topic of great interest to researchers and clinicians who work with civilian and military/veteran samples alike. Most of what is known of neuropsychological recovery following concussion follows from studies conducted in civilian samples, and we have pointed to the sports concussion literature to illustrate not only how concussion researchers have gone about identifying the natural history of recovery following a specific type of non–blast-related concussion, but to also emphasize the many complexities and challenges that exist in understanding blast concussion. The heavy reliance on retrospective self-report alone to define blast concussion status represents perhaps the most significant limitation of the current blast concussion literature, and in spite of the efforts of certain research groups to improve upon the lack of reliability that accompanies contemporary TBI screening instruments, the reality is that the degree to which these interview strategies improve the reliability of mTBI assessment and diagnosis remains unclear because of the absence of acute-stage injury information for blast events.

Nevertheless, working with the information that is available to them, blast concussion researchers have attempted to implement experimental neuroimaging strategies (e.g., DTI) to identify signs of blast-related trauma. It is essential to recognize that the use of DTI and other novel imaging methodologies remain *experimental* in nature, and although select preliminary findings from the neuroimaging literature might suggest that DTI and other methods show promise in future detection of blast-related neurotrauma, blast-related (and non–blast-related) concussion remains a *clinical* diagnosis.

From a neuropsychological perspective, it is important to recognize that no single blast concussion research group has implemented the same longitudinal approach that has been conducted in sports concussion samples to establish the "natural history" of recovery after blast-related concussion. However, neuropsychological outcomes have been reported by independent blast concussion researchers at various postinjury stages (e.g., acute-stage only, postdeployment only), and using alternate methodologies (e.g., pre-post deployment outcomes, cross-sectional comparison of blast vs. no blast conditions). Overall, despite the many weaknesses of these studies, the available blast concussion literature in neuropsychology suggests a remarkable degree of consistency with literature on conventional concussion samples. The blast concussion literature published to date suggests that when controlling for important confounding factors, such as premorbid psychiatric difficulties, postdeployment adjustment issues (e.g., chronic PTSD, depression), and secondary gain issues, there is very little evidence that a single, blast-related concussion results in lasting cognitive effects beyond the acute stage of recovery. Injury severity (mild, moderate, severe) appears to be a far more important determining factor than injury mechanism (blast vs. nonblast) with respect to long-term recovery. Comorbid psychopathology is also recognized as having an untoward effect on cognitive function, and to date it appears that PTSD, depression, and other independent psychiatric conditions play a more significant role in postdeployment cognitive outcomes than blast concussion per se. There are unclear interaction effects between blast concussion and psychopathology, though it does appear that blast concussion increases risk of developing PTSD and other psychiatric difficulties, which may then in turn contribute to cognitive limitations beyond the acute stage of concussive injury.

At present, we perceive that there are two primary needs that should be the focus of future blast concussion research. First, it is essential that researchers and clinicians integrate self-report information pertaining to historical concussion with acute-stage injury information whenever available to enhance self-reported accounts offered by military and veteran samples in the months and years that follow deployment. A VA/DOD collaboration, for example, that allowed researchers to examine in-theater MACE findings and compare them with postdeployment accounts of blast injury would inform whether the recently developed semi-structured interviews (e.g., MN-BEST) are of any utility in diagnosing blast concussion long after the injuries are sustained. Defining groups by acute-stage injury information (e.g., LOC, PTA) included on the MACE or other external records will also serve as a more reliable gold standard of blast concussion than retrospective self-report alone, and will enhance researchers' ability to uphold necessary standards to establish cause-and-effect relationships between blast concussion and various late-stage outcomes (e.g., neuropsychological functioning, white matter integrity). Similar to what has been modeled in sports concussion literature, future studies would ideally involve larger longitudinal investigations that incorporate premorbid, acute, subacute, and long-term outcome data. Longitudinal studies would enable researchers to better characterize the full course of recovery and identify premorbid risk factors for complicated recovery (i.e., persistent postconcussive symptoms) as well as factors that contribute to resilience following deployment-related mTBI.

Second, researchers are encouraged to continue to investigate the possible dose-response relationship that may exist between recurrent blast concussion and neuropsychological functioning. Most blast concussion studies published to date have included samples with reported histories of relatively few self-reported blast concussions. Although it is unclear whether recurrent concussion results in recovery patterns that are distinct relative to a single concussive injury (Belanger et al., 2011), certain researchers have raised the possibility of a dose-response relationship in the case of multiple blast concussions (Kontos et al., 2013). Discrepant findings across blast concussion studies point to the need for better study designs using both objective and subjects measures of blast effects. Such work would enhance our understanding of blast concussion and allow better management and improved education of individuals in the acute stage of injury. Ultimately, such research would lead to the development of prevention models and more effective interventions to minimize the impact of blast concussion on those serving in military conflicts.

APPENDIX 32.A MINNESOTA BLAST EXPOSURE SCREENING TOOL (MN-BEST)

Minnesota Blast Exposure Screening Tool (MN-BEST)

Patient Name ______ Date of Clinical Interview ______

A. Blast Exposures

A1. Estimated number of blast exposures (i.e., times the participant felt pressure wave from an explosion) ______________________

A2. Worst three blast exposures (i.e., greatest likelihood of injury to brain): **complete attached Table 32.2**

A3. Estimated total number of probable or definite blast-related mTBIs (complete after Table 32.2 is finished) __________________

A4. Estimated total number of possible, probable, or definite blast-related mTBIs (complete after Table 32.2 is finished) ___________

A5. Estimated total number of unlikely, possible, probable, or definite blast-related mTBIs (complete after Table 32.2 is finished) ___________

A6. Estimated total number of probable or definite blast related TBIs (**moderate or severe**) (complete after Table 32.2 is finished) ___________

TABLE 32.2
Worst Three Blast Exposures (i.e., Greatest Likelihood of Injury to Brain)

Event Description	Date	LOC (duration)	PTA (duration)	Neurological Sign(s)	TBI Severity	mTBI Type	Rating of Certainty	Severity Score (*if* Rating of Certainty ≥2)
1. Deployed: Y/N				Headache ___ Dizzy/Disoriented __ Trouble tracking ___ Tinnitus ___ Nauseous ___ Sensitive to light/ noise ___ Other ____________ Comments:	Type: ___ Rationale: Severity score: __ Mild = 0 Moderate* = 15 Severe = 30	Type: ___ Rationale: Severity score: __ Type 0 = 1 Type I = 2 Type II = 3 Type III = 4	Rating: ___ Rationale: 0 = unlikely 1 = less likely than not 2 = more likely than not 3 = likely	Rating: ___ Rationale:
2. Deployed: Y/N				Headache ___ Dizzy/Disoriented __ Trouble tracking ___ Tinnitus ___ Nauseous ___ Sensitive to light/ noise ___ Other ____________ Comments:	Type: ___ Rationale: Severity score: __ Mild = 0 Moderate* = 15 Severe = 30	Type: ___ Rationale: Severity score: __ Type 0 = 1 Type I = 2 Type II = 3 Type III = 4	Rating: ___ Rationale: 0 = unlikely 1 = less likely than not 2 = more likely than not 3 = likely	Rating: ___ Rationale:
3. Deployed: Y/N				Headache ___ Dizzy/Disoriented __ Trouble tracking ___ Tinnitus ___ Nauseous ___ Sensitive to light/ noise ___ Other ____________ Comments:	Type: ___ Rationale: Severity score: __ Mild = 0 Moderate* = 15 Severe = 30	Type: ___ Rationale: Severity score: __ Type 0 = 1 Type I = 2 Type II = 3 Type III = 4	Rating: ___ Rationale: 0 = unlikely 1 = less likely than not 2 = more likely than not 3 = likely	Rating: ___ Rationale:

Classification of mTBI (Code TBI According to Symptom of Greatest Severity)

	Type 0	Type I	Type II	Type III
LOC	Definite no LOC	Altered state (including dazed, confused, disoriented) or transient loss and unsure LOC	Definite loss with time unknown or <5 minutes	Loss 5–30 minutes
PTA	Definite no PTA	1–60 seconds	60 seconds–12 hours	>12 hours
Neurological symptoms	One or more	One or more	One or more	One or more

* Includes complicated mTBI.

Total blast-related TBI score: (0–90):_______

ACKNOWLEDGMENT

This work was supported by grants from the Congressionally Directed Medical Research Program (W81XWH-08-2-0038: Sponheim), the Department of Veterans Affairs, Rehabilitation R&D Program (I01RX000622: Sponheim; IK2RX000709: Davenport), and the Minnesota Veterans Medical Research and Education Foundation (mnvets.org) (Sponheim; Nelson)

REFERENCES

Aoki, Y., Inokuchi, R., Gunshin, M., Yahagi, N., and Suwa, H. (2012). Diffusion tensor imaging studies of mild traumatic brain injury: A meta-analysis. *J Neurol Neurosurg Psychiatry*, 83(9), 870–6.

Basser, P. J. (1995). Inferring microstructural features and the physiological state of tissues from diffusion-weighted images. *NMR in Biomedicine*, 8(7–8), 333–44.

Basser, P. J., Mattiello, J., and LeBihan, D. (1994). Estimation of the effective self-diffusion tensor from the NMR spin echo. *J Magn Reson. Series B*, 103(3), 247–254.

Basser, P. J., and Pierpaoli, C. (1996). Microstructural and physiological features of tissues elucidated by quantitative-diffusion-tensor MRI. *J Magn Reson Series B*, 111(3), 209–19.

Bazarian, J. J., Donnelly, K., Peterson, D. R., Warner, G. C., Zhu, T., and Zhong, J. (2012). The relation between posttraumatic stress disorder and mild traumatic brain injury acquired during Operations Enduring Freedom and Iraqi Freedom. *J Head Trauma Rehabil*, 28(1), 1–12.

Beaulieu, C. (2002). The basis of anisotropic water diffusion in the nervous system - a technical review. *NMR Biomed*, 15, 435–455.

Belanger, H. G., Kretzmer, T., Vanderploeg, R. D., and French, L. M. (2010). Symptom complaints following combat-related traumatic brain injury: relationship to traumatic brain injury symptoms and posttraumatic stress disorder. *J Int Neuropsychol Soc*, 16, 194–199.

Belanger, H. G., Kretzmer, T., Yoash-Gantz, R., Pickett, T., and Tupler, L. A. (2009a). Cognitive sequelae of blast-related versus other mechanisms of brain trauma. *J Int Neuropsychol Soc*, 15, 1–8.

Belanger, H. G., Proctor-Weber, Z., Kretzmer, T., Kim, M., French, L. M., and Vanderploeg, R. D. (2011). Symptom complaints following reports of blast versus non-blast mild TBI: Does mechanism of injury matter? *Clin Neuropsychol*, 25, 702–715.

Belanger, H. G., Spiegel, E., and Vanderploeg, R. D. (2010). Neuropsychological performance following a history of multiple self-reported concussions: a meta-analysis. *J Int Neuropsycholo So*, 162, 62–267.

Belanger, H. G., Uomoto, J. M., and Vanderploeg, R. D. (2009b). The Veterans Health Administration system of care for mild traumatic brain injury: Costs, benefits, and controversies. *J Head Trauma Rehabil*, 24, 4–13.

Belanger, H. G., and Vanderploeg, R. D. (2005). The neuropsychological impact of sports-related concussion: A meta-analysis. *J Int Neuropsychol Soc*, 11, 345–357.

Belanger, H. G., Vanderploeg, R. D., Soble, J. R., Richardson, M., and Groer, S. (2012). Validity of the Veterans Health Administration's traumatic brain injury screen. *Arch Phys Med Rehabil*, 931234–39.

Binder, L. M., Rohling, M. L., and Larrabee, G. J. (1997). A review of mild head trauma. Part I: Meta-analytic review of neuropsychological studies. *J Clin Exp Neuropsychol*, 19,421–431.

Borg, J., Holm, L., Cassidy, J. D., Peloso, P. M., Carroll, L. J., von Holst, H., and Erickson, K (2004). WHO Collaborating Centre Task Force on Mild Traumatic Brain Injury. Diagnostic procedures in mild traumatic brain injury: Results of the WHO Collaborating Centre Task Force on Mild Traumatic Brain Injury. *J Rehabil Med*, 43 Suppl, 61–75.

Brawn Fortier, C., Amick, M. M., Grande, L., McGlynn, S., Kenna, A., Morra, L. et al. (2013). The Boston Assessment of Traumatic Brain Injury – Lifetime (BAT-L) Semistructured Interview: Evidence of research utility and validity. *J Head Trauma Rehabil*, 29:89–98.

Cernak, J., Wang, Z., Jiang, J., Bian, X., and Savic, J. (2001a). Cognitive deficits following blast injury-induced neurotrauma: Possible involvement of nitric oxide. *Brain Injury*, 15, 593–612.

Coldren, R. L., Kelly, M. P., Parish, R. V., Dretsch, M., and Russell, M. L. (2010). Evaluation of the military acute concussion evaluation for use in combat operations more than 12 hours after injury. *Mil Med*, 175, 477–481.

Coldren, R. L., Russell, M. L., Parish, R. V., Dretsch, M., and Kelly, M. P. (2012). The ANAM lacks utility as a diagnostic or screening tool for concussion more than 10 days following injury. *Mil Med*, 177, 179–183.

Commandeur, J., Derksen, R. J., MacDonald, D., and Breederveld, R. (2012). Identifical fracture patterns in combat vehicle blast injuries due to improvised explosive devices: A case series. *BMC Emerg Med*, 12, 12.

Cooper, D. B., Chau, P. M., Armistead-Jehle, P., Vanderploeg, R. D., and Bowles, A. O. (2012). Relationship between mechanism of injury and neurocognitive functioning in OEF/OIF service members with mild traumatic brain injuries. *Mil Med*, 177, 1157–1160.

Cooper, D. B., Kennedy, J. E., Cullen, M. A., Critchfield, E., Amador, R. R., and Bowles, A. O. (2011). Association between combat stress and post-concussive symptom reporting in OEF/OIF service members with mild traumatic brain injuries. *Brain Inj*, 25, 1–7.

Cooper, D. B., Nelson, L., Armistead-Jehle, P., and Bowles, A. O. (2011). Utility of the mild brain injury atypical symptoms scale as a screening measure for symptom over-reporting in Operation Enduring Freedom/Operation Iraqi Freedom service members with post-concussive complaints. *Arch Clin Neuropsychol*, 26, 718–727.

Courtney, M. W., and Courtney, A. C. (2011). Working toward exposure thresholds for blast-induced thoracic and acceleration mechanisms. *NeuroImage*, 54, S55–S61.

Davenport, N. D., Lim, K. O., Armstrong, M. T., and Sponheim, S. R. (2012). Diffuse and spatially variable white matter disruptions are associated with blast-related mild traumatic brain injury. *NeuroImage*, 59(3), 2017–24.

DePalma, R. G., Burris, D. G., Champion, H. R., and Hodgson, M. J. (2005). Blast injuries. *The New England Journal of Medicine*, 352, 1335–1342.

Defense and Veterans Brain Injury Center [DVBIC] (2007). Military Acute Concussion Evaluation, from www.dvbic.org.

Dikmen, S. S., Machamer, J. E., Winn, H. R., and Temkin, N. R. (1995). Neuropsychological outcome at 1-year post head injury. *Neuropsychology*, 9, 80–90.

Donnelly, K. T., Donnelly, J. P., Dunnam, M., Warner, G. C., Kittelson, C. J., Constance, J. E., et al. (2011). Reliability, sensitivity, and specificity of the VA traumatic brain injury screening tool. *J Head Trauma Rehabil*, 26, 439–453.

Edwards, M. J., Lustik, M., Eichelberger, M. R., Elster, E., Azarow, K., and Coppola, C. (2012). Blast injury in children: An analysis from Afghanistan and Iraq, 2002–2010. *J Trauma Acute Care Surg,* 73, 1278–1283.

Eskridge, S. L., Macera, C. A., Galarneau, M. R., Holbrook, T. L., Woodruff, S. I., MacGregor, A. J. et al. (2012). Injuries from combat explosions in Iraq: Injury type, location, and severity. *Injury,* 43, 1678–168.

Field, A. (2013). *Discovering Statistics Using IBM SPSS Statistics*, 4th ed. Sage, Los Angeles.

Frencham, K. A., Fox, A. M., and Maybery, M. T. (2005). Neuropsychological studies of mild traumatic brain injury: A meta-analytic review of research since 1995. *J Clin Exp Neuropsychol, 27*, 334–351.

Giza, C. C., and Hovda, D. A. (2001). The neurometabolic cascade of concussion. *J Athletic Train,* 36, 228–235.

General Accounting Office. (2008, February). VA Health Care. Mild traumatic brain injury screening and evaluation implemented for OEF/OIF veterans, but challenges remain. GAO-08-276. General Accounting Office, Washington, DC, from www.gao.gov/new.items/d08276.pdf.

Hofman, P. A., Stapert, S. Z., van Kroonenburgh, M. J., Jolles, J., de Kruijk, J., and Wilmink, J. T. (2001). MR imaging, single-photon emission CT, and neurocognitive performance after mild traumatic brain injury. *AJNR. Am J Neuroradiol, 22* (3), 441–9.

Howe, L. L. S. (2009). Giving context to post-deployment post-concussive-like symptoms: Blast-related potential mild traumatic brain injury and comorbidities. *Clin Neuropsychol,* 23, 1315–1337.

Hughes, D. G., Jackson, A., Mason, D. L., Berry, E., Hollis, S., and Yates, D. W. (2004). Abnormalities on magnetic resonance imaging seen acutely following mild traumatic brain injury: Correlation with neuropsychological tests and delayed recovery. *Neuroradiology*, 46(7), 550–8.

Iverson, G. L. (2005). Outcome from mild traumatic brain injury. *Curr Opin Psychiatry,* 18, 301–317.

Iverson, G. L. (2006). Misdiagnosis of the persistent postconcussion syndrome in patients with depression. *Arch Clin Neuropsychol,* 21, 303–310.

Iverson, G. L., and Lange, R. T. (2003). Examination of "postconcussion-like" symptoms in a healthy sample. *Appl Neuropsychol,* 10, 137–144.

Iverson, G. L., Langlois, J. A., McCrea, M. A., and Kelly, J. P. (2009). Challenges associated with post-deployment screening for mild traumatic brain injury in military personnel. *Clin Neuropsychol,* 23, 1299–1314.

Iverson, G. L., and McCracken, L. M. (1997). 'Postconcussive' symptoms in persons with chronic pain. *Brain Inj,* 11, 783–790.

Jorge, R. E., Acion, L., White, T., Tordesillas-Gutierrez, D., Pierson, R., Crespo-Facorro, B., and Magnotta, V. A. (2012). White matter abnormalities in veterans with mild traumatic brain injury. *Am J Psychiatry*, 169(12), 1284–91.

Kay, T., Harrington, D.E., Adams, R., Anderson, T., Berrol, S., et al. (1993). Definition of mild traumatic brain injury. *J Head Trauma Rehabil,* 8, 86–87.

Kennedy, C. H., Evans, J. P., Chee, S., Moore, J. L., Barth, J. T., and Stuessi, K. A. (2012). Return to combat duty after concussive blast injury. *Arch Clin Neuropsychol,* 27, 817–827.

Kontos, A. P., Kotwal, R. S., Elbin, R. J., Lutz, R. H., Forsten, R. D., Benson, P. J et al. (2013). Residual effects of combat-related mild traumatic brain injury. *J Neurotrauma,* 30, 680–686.

Lamberty, G. J., Nelson, N. W., and Yamada, T. (2013). Effects and outcomes in civilian and military traumatic brain injury: Similarities, differences, and forensic implications. *Behav Sci Law.* 31:814–821.

Lange, R. T., Iverson, G. L., and Rose, A. (2011). Depression strongly influences postconcussion symptom reporting following mild traumatic brain injury. *J Head Trauma Rehabil,* 26, 127–137.

Lange, R. T., Pancholi, S., Brickell, T. A., Sakura, S., Bhagwat, A., Merritt, V. et al. (2012). Neuropsychological outcome from blast versus non-blast: Mild traumatic brain injury in U.S. military service members. *J Neuropsychol Soc,* 18, 595–605.

Larrabee, G. J. (2012). Mild traumatic brain injury. In G. J. Larrabee (Ed.) *Forensic Neuropsychology A Scientific Approach*, 2nd ed. Oxford, New York.

Lee, H., Wintermark, M., Gean, A. D., Ghajar, J., Manley, G. T., and Mukherjee, P. (2008). Focal lesions in acute mild traumatic brain injury and neurocognitive outcome: CT versus 3T MRI. *J Neurotrauma*, 25(9), 1049–56.

Levin, H. S., Wilde, E., Troyanskaya, M., Petersen, N. J., Scheibel, R., Newsome, M. et al. (2010). Diffusion tensor imaging of mild to moderate blast-related traumatic brain injury and its sequelae. *Journal of Neurotrauma*, 27(4), 683–94.

Lippa, S. M., Pastorek, N. J., Benge, J. F., and Thornton, G. M. (2010). Postconcussive symptoms after blast and nonblast-related mild traumatic brain injuries in Afghanistan and Iraq War veterans. *J Int Neuropsychol Soc,* 16, 856–866.

Lopes Cardozo, B., Blanton, C., Zalewski, T., Tor, S., McDonald, L., Lavelle, J. et al. (2012). Mental health survey among landmine survivors in Siem Reap province, Cambodia. *Med Conflict Survival,* 28, 161–181.

Luethcke, C. A., Bryan, C. J., Morrow, C. E., and Isler, W. C. (2011). Comparison of concussive symptoms, cognitive performance, and psychological symptoms between acute blast-versus non-blast-induced mild traumatic brain injury. *J Int Neuropsychol So,* 17, 36–45.

MacDonald, C. L., Johnson, A. M., Cooper, D., Nelson, E. C., Werner, N. J., Shimony, J. S., ... Brody, D. L. (2011). Detection of blast-related traumatic brain injury in U.S. military personnel. *New England Journal of Medicine* 364 (22), 2091–2100. Retrieved from http://www.nejm.org/doi/full/10.1056/nejmoa1008069.

MacGregor, A. J., Dougherty, A. L., and Galarneau, M. R. (2011). Injury-specific correlates of combat-related traumatic brain injury in Operation Iraqi Freedom. *J Head Trauma Rehabil,* 26, 312–318.

Matthews, S. C., Spadoni, A. D., Lohr, J. B., Strigo, I. A., and Simmons, A. N. (2012). Diffusion tensor imaging evidence of white matter disruption associated with loss versus alteration of consciousness in warfighters exposed to combat in Operations Enduring and Iraqi Freedom. *Psychiatry Res Neuroimaging*, 204(2–3), 149–54.

Mayorga, M. A. (1997). The pathology of primary blast overpressure injury. *Toxicology,* 121, 17–28.

McCrea, M. A. (2008). *Mild Traumatic Brain Injury and Postconcussion Syndrome.* Oxford University Press, New York.

McCrea, M., Guskiewicz, K. M., Marshall, S. W., Barr, W., Randolph, C. et al. (2003). Acute effects and recovery time following concussion in collegiate football players. *JAMA,* 290, 2556–2563.

McCrea, M., Guskiewicz, K., Randolph, C., Barr, W. B., Hammeke, T. A., Marshall S. W. et al. (2013). Incidence, clinical course, and predictors of prolonged recovery time following sport-related concussion in high school and college athletes. *J the I Neuropsychol So,* 19, 22–33.

McCrea, M., Iverson, G. L., McAllister, T. W., et al. (2009). An integrated review of recovery after mild traumatic brain injury (MTBI): Implications for clinical management. *Clin Neuropsychol,* 23, 1368–1390.

Meares, S., Shores, E. A., Bryant, R. A., Taylor, A. J., Batchelor, J. et al. (2011). The prospective course of postconcussion syndrome: The role of mild traumatic brain injury. *Neurosychology,* 25, 454–465.

Mittenberg, W., Canyock, E. M., Condit, D., and Patton, C. (2001). Treatment of post-concussion syndrome following mild head injury. *J Clin Exp Neuropsychol,* 23, 829–836.

Mittenberg, W., DiGiulio, D. V., Perrin, S., and Bass, A. E. (1992). Symptoms following mild head injury: Expectation as aetiology. *J Neurol Neurosurg Psychiatry,* 55, 200–204.

Moore, D. F., Jerusalem, A., Nyein, M., Noels, L., Jaffee, M. S., and Radovitzky, R. A. (2009). Computational biology – modeling of primary blast effects on the central nervous system. *Neuroimage,* 47, S2, 10–20.

Morey, L. C. (1991). *The Personality Assessment Inventory.* Psychological Assessment Inventory, Odessa, FL.

Morey, R. A., Haswell, C. C., Selgrade, E. S., Massoglia, D., Liu, C., Weiner, J. et al. (2013). Effects of chronic mild traumatic brain injury on white matter integrity in Iraq and Afghanistan war veterans. *Hum Brain Map,* 34(11), 2986–99.

Nelson, N. W., Anderson, C. R., Thuras, P., Kehle-Forbes, S. M., Arbisi, P. A., Erbes, C. R. et al. (in press). Factors associated with inconsistency in self-reported mild traumatic brain injuury over time among U.S. National Guard soldiers deployed to Iraq. *British Journal of Psychiatry.*

Nelson, N. W., Hoelzle, J. B., Doane, B. M., McGuire, K. A., Ferrier-Auerbach, A. G., Charlesworth, M. J. et al. (2012a). Neuropsychological outcomes of U.S. veterans with report of remote blast concussion and current psychopathology. *J Int Neuropsychol Soc,* 18, 845–55.

Nelson, N. W., Hoelzle, J. B., McGuire, K. A., Ferrier-Auerbach, A. G., Charlesworth, M. J., and Sponheim, S. R. (2010). Evaluation context impacts neuropsychological performance of OEF/OIF veterans with reported combat-related concussion. *Arch Clin Neuropsychol,* 25, 713–723.

Nelson, N. W., Hoelzle, J. B., McGuire, K. A., Ferrier-Auerbach, A. G., Charlesworth, M. J., and Sponheim, S. R. (2011a). Neuropsychological evaluation of blast-related concussion: Illustrating the challenges and complexities through OEF/OIF case studies. *Brain Inj,* 25, 511–525.

Nelson, N. W., Hoelzle, J. B., McGuire, K. A., Sim, A. H., Goldman, D. J., Ferrier-Auerbach, A. G. et al. (2011b). Self-report of psychological function among OEF/OIF personnel who also report combat-related concussion. *Clin Neuropsychol,* 25, 716–740.

Nelson, N. W., and Keenan, P. (in press). Blast-related traumatic brain injury: review and update. In: S. Koffler, J. Morgan, M. Greiffenstein, and B. Marcopoulos, eds. *Neuropsychology: Science and Practice V.* Oxford University Press, Oxford.

Niogi, S. N., and Mukherjee, P. (2010). Diffusion tensor imaging of mild traumatic brain injury. *J Head Trauma Rehabil,* 25(4), 241–55.

Oien, M. L., Nelson, N. W., Lamberty, G. J., and Arbisi, P. A. (June, 2011). Neuropsychological function in posttraumatic stress disorder: A meta-analytic review. Poster session presented at the annual conference of the American Academy of Clinical Neuropsychology, Washington, D.C..

O'Neil, M. E., Carlson, K. F., Storzbach, D., Brenner, I. A., Freeman, M., Quiñones, A. et al. Complications of Mild Traumatic Brain Injury in Veterans and Military Personnel: A Systematic Review. VA-ESP Project #05-225; 2012

Ozen, L. J., and Fernandes, M. A. (2011). Effects of 'diagnosis threat' on cognitive and affective functioning long after mild head injury. *J Int Neuropsychol Soc,* 17, 219–229.

Panayiotou, A., Jackson, M., and Crowe, S. F. (2010). A meta-analytic review of the emotional symptoms associated with mild traumatic brain injury. *JJ Clin Exp Neuropsychol,* 32, 463–473.

Ptak, T., Sheridan, R. L., Rhea, J. T., Gervasini, A. A., Yun, J. H., Curran, M. A. et al. (2003). Cerebral fractional anisotropy score in trauma patients: A new indicator of white matter injury after trauma. *Am J Roentgenol,* 181(5), 1401–1407.

Polusny, M. A., Kehle, S. M., Nelson, N. W., Erbes, C. R., Arbisi, P. A., and Thuras, P. (2011). Longitudinal effects of mild TBI and PTSD comorbidity on post-deployment outcomes in National Guard soldiers deployed to Iraq. *Arch Gen Psychiatry,* 68, 79–89.

Randolph, C. (1998). *Repeatable Battery for the Assessment of Neuropsychological Status (RBANS) Manual.* The Psychological Corporation, San Antonio, TX.

Roebuck-Spencer, T. M., Vincent, A. S., Twillie, D. A., Logan, B. W., Lopez, M., Friedl, K. E. et al. (2012). Cognitive change associated with self-reported mild traumatic brain injury sustained during the OEF/OIF conflicts. *Clin Neuropsychol,* 26, 473–489.

Rohling, M. L., Binder, L. M., Demakis, G. J., Larrabee, G. J., Ploetz, D. M., and Langhinrichsen-Rohling, J. (2011). A meta-analysis of neuropsychological outcome after mild traumatic brain injury: Re-analyses and reconsiderations of Binder et al. (1997), Frencham et al. (2005), and Pertab et al. (2009). *The Clin Neuropsychol,* 25, 608–623.

Ruff R., Richardson A. M. (1999). Mild traumatic brain injury. In: Sweet, J. J., ed. *Forensic Neuropsychology; Fundamentals and Practice.* Swets and Zeitlinger, Lisse, Hollan., pp 315–338.

Saljo, A., Mayorga, M., Bolouri, H., Svensson, B., and Hamberger, A. (2011). Mechanisms and pathophysiology of the low-level blast brain injury in animal models. *NeuroImage,* 54, S83–S88.

Schretlen, D. J., and Shapiro, A. M. (2003). A quantitative review of the effects of traumatic brain injury on cognitive functioning. *Int Rev Psychiatry,* 15, 341–349.

Shandera-Ochsner, A. L., Berry, D. T. R., Harp, J. P., Edmundson, M., Graue, L. O., Roach, A. et al. (2013). Neuropsychological effects of self-reported deployment-related mild TBI and current PTSD in OIF/OEF veterans. *Clin Neuropsychol,* 27, 881–907.

Sorg, S. F., Delano-Wood, L., Luc, N., Schiehser, D. M., Hanson, K. L., Nation et al. (2013). White matter integrity in veterans with mild traumatic brain injury: associations with executive function and loss of consciousness. *J Head Trauma Rehabil,* 29:21–32.

Sponheim, S. R., Mcguire, K. A., Suk, S., Davenport, N. D., Aviyente, S., Bernat, E. M. et al. (2011). Evidence of disrupted functional connectivity in the brain after combat-related blast injury. *Neuroimage,* 54, S21–S29.

Suhr, J. A., and Gunstad, J. (2002). "Diagnosis Threat": The effect of negative expectations on cognitive performance in head injury. *J Clin Exp Neuropsychol,* 24, 448–457.

Suhr, J. A., and Gunstad, J. (2005). Further exploration of the effect of "diagnosis threat" on cognitive performance in individuals with mild head injury. *J the Int Neuropsychol Society,* 11, 23–29.

Taber, K. H., Warden, D. L., and Hurley, R. A. (2006). Blast-related traumatic brain injury: What is known? *J Neuropsychiatry Clin Neurosci,* 18, 141–145.

Tanielian, T., and Jaycox, L. H. (2008). *Invisible Wounds of War: Psychological and Cognitive Injuries, Their Consequences, and Services to Assist Recovery*. Rand Corporation, Santa Monica, CA.

Terrio, H., Brenner, L. A., Ivins, B. J., Cho, J. M., Schwab, K., Scally, K. et al. (2009). Traumatic brain injury screening: Preliminary findings in a U.S. Army Brigade combat team. *J Head Trauma Rehabil,* 24, 14–23.

U.S. Department of Veterans Affairs. (2009, April). *Clinical Practice Guideline: Management of Concussion/Mild Traumatic Brain Injury,* from http://www.healthquality.va.gov/guidelines/Rehab/mtbi/concussion_mtbi_full_1_0.pdf.

Van Dyke, S. A., Axelrod, B. N., and Schutte, C. (2010). Test-retest reliability of the traumatic brain injury screening instrument. *Mi Med,* 175, 947–49.

Vanderploeg, R. D., Groer, S., and Belanger, H. G. (2012). The initial developmental process of a VA Semi-Structured Clinical Interview for TBI identification. *J Rehabil Res Develop,* 49, 545–556.

Vanderploeg, R. D., and Belanger, H. G. (2013). Screening for a remote history of mild traumatic brain injury: When a good idea is bad. *JJ Head Trauma Rehabil,* 28, 211–218.

Vasterling, J. J., Brailey, K., Proctor, S. P., Kane, R., Heeren, T., and Franz, M. (2012). Neuropsychological outcomes of mild traumatic brain injury, post-traumatic stress disorder, and depression in Iraq-deployed U.S. Army soldiers. *Br J Psychiatry,* 201, 186–192.

Xydakis, M. S., Bebarta, V. S., Harrison, C. D., Conner, J. C., and Grant, G. A. (2007). Tympanic-membrane perforation as a marker of concussive brain injury in Iraq. *New Engl J Med,* 35, 7830–831.

Zakzanis, K.K., Leach, L., and Kaplan, E. (1998). On the nature and pattern of neurocognitive function in major depressive disorder. *Neuropsychiatry Neuropsychol Behav Neurol,* 11, 111–119.

33 Persistent Cognitive Deficits
Implications of Altered Dopamine in Traumatic Brain Injury

Hong Qu Yan, Nicole D. Osier, Jonathan Korpon, James W. Bales, Anthony E. Kline, Amy K. Wagner, and C. Edward Dixon

CONTENTS

33.1 INTRODUCTION

Traumatic brain injury (TBI) represents a significant cause of death and disability in industrialized countries. Of particular importance to patients is the chronic effect that TBI has on cognitive function. Therapeutic strategies have been difficult to evaluate because of the complexity of injuries and variety of patient presentations within a TBI population. However, pharmacotherapies targeting dopamine (DA) have consistently shown benefits in attention, behavioral outcome, executive function, and memory. Ongoing research in animal models has begun to elucidate the pathophysiology of DA alterations after TBI. The purpose of this review is to discuss the potential role of dopamine in persistent brain injury disabilities and the role that dopaminergic therapies have in improving recovery of function for TBI.

33.1.1 Traumatic Brain Injury: A Significant Public Health Problem

TBI is a significant worldwide health problem, affecting individuals across the lifespan (NIH, 1998). Approximately 1.7 million people sustain a TBI in the United States each year (Faul et al., 2010), and approximately 2% live with long-term deficits (Thurman et al., 1999). TBI is characterized by acute, subacute, and chronic pathology that contributes to the heterogeneous nature of brain injury and the complexity in clinically managing TBI patients (DeKosky et al., 1998; Kochanek 1993; Park et al., 2008). Indeed, a myriad of diverse problems are known to follow TBI, including direct mechanical damage, and secondary problems such as oxidative stress, metabolic disturbance, excitotoxicity, and neurotransmitter imbalance. These processes, in turn, contribute in complex ways to affect the severity and type of functional deficit as well as the extent of recovery.

The United States spends approximately $9–10 billion annually on TBI care ranging from acute emergent treatment to rehabilitation services; however, long-term disability remains common, affecting 50% of moderate TBI survivors (Kraus et al., 2005). Researchers have sought to apply the current understanding of brain injury consequences to develop viable treatment options that will significantly improve outcomes of TBI survivors. Such efforts include studies using experimental models of brain injury (Lighthall 1988; Dixon et al., 1991; 1999; Kline et al., 2007; Wagner et al., 2007; Xiong et al., 2013) and to a lesser-extent clinical trials of TBI patients (Cantor et al., 2013; Nelson et al., 2013; S. Wang et al., 2013; Zafonte et al., 2012; Giacino et al., 2012). Barriers to translation of findings to clinical practice have been discussed elsewhere (Tolias and Bullock, 2004) and include confounding effects of multi-system injury (Capone-Neto and Rizoli, 2009; Ladanyi and Elliott, 2008), adverse drug reactions and toxicity (Muir 2006), and disruption of the blood–brain barrier (Folkersma et al., 2009; Whalen et al., 1998). Overall, there is a need for additional research to identify safe, effective, and feasible therapies for TBI patients. There is some promising evidence suggesting a role for DA within the central nervous system (CNS) after TBI. Thus, DA systems are a potential therapeutic target.

33.1.2 Cognitive Dysfunction after Brain Injury

Of the various functional deficits experienced post-TBI, changes in cognitive status are among the most common and distressing to TBI survivors and their families. A plethora of cognitive problems have been observed after TBI including deficits in frontal executive functioning (e.g., impulse control, problem solving), memory (short- and long-term), attention, learning/information processing, and language abilities (Sun and Feng, 2013). In addition to being distressful, cognitive dysfunction can be life-altering, affecting survivors' ability to live independently, readapt to social life, resume work, and return to preinjury family role(s). The extent of cognitive dysfunction tends to increase with injury severity, but other important factors have been identified. One of the factors that may impact post-TBI cognitive dysfunction is complex changes within the DA system (Bales et al., 2009, 2010). Preclinical animal models of brain injury can produce clinically relevant cognitive dysfunction in rodents (Dixon et al., 1987, 1991; Fox et al., 1998; Hamm et al., 1994, 1996; Kline et al., 2007; Wagner et al., 2013). Chronic dysfunction within areas of the brain known to be important for cognitive functioning, such as the hippocampus and thalamus (Vertes, 2006), have been observed in animal (Bramlett and Dietrich, 2002; Lifshitz et al., 2007) and human studies (Langfitt et al., 1986; Fontaine et al., 1999).

Currently, little is known about the underlying cause of this long-term pathology or the best way to prevent or treat cognitive decline post-TBI. However, one popular hypothesis is that DA plays a role in cognition and may affect TBI outcomes. The aims of this chapter are to (1) summarize the role of DA in cognitive functioning; (2) briefly describe important factors that affect DA signaling; (3) summarize important research related to the role of DA in TBI or its application as a therapeutic for TBI patients; and (4) highlight areas for future research regarding DA in the context of TBI. The reader should be aware that this chapter is meant to be a preliminary introduction to the role of DA in post-TBI cognition. More in-depth reviews of the roles of DA in cellular function (Verheij and Cools, 2008) and cognition (Bales et al., 2009; El-Ghundi et al., 2007) have been published.

33.2 EVIDENCE FOR A ROLE OF DA IN BRAIN INJURY

DA can act as both a neurotransmitter and neuromodulator within the CNS. Two important consequences of TBI are diffuse axonal injury in the tracts of white matter as well as damage to gray matter (Bramlett and Dietrich, 2002; Meythaler et al., 2001; Reeves et al., 2007; Smith, et al, 2003); these consequences subsequently impact the functioning of various neurotransmitter systems including the dopaminergic (DAergic) system. In addition to generalized brain damage, it is well-established that multiple brain regions are affected by acute TBI including, but not limited to the frontal cortex (FC) (Lighthall et al., 1989; Dixon et al., 1991), striatum (Dietrich et al., 1994), thalamus (Vertes 2006), and hippocampus (Sanders et al., 2001; Hicks et al., 1993). Notably, neuroprotection within the thalamus and hippocampus depends on interactions between DA and glutamate (Centonze et al., 1999; Chen et al., 2007; Granado et al., 2008; O'Carroll et al., 2006). Moreover, the previously mentioned brain regions are known to play important roles in cognition including executive function, attention, memory, and learning (Baron et al., 1985; Chudasama and Robbins, 2006; McDonald et al., 2002; Seeman et al., 1978). These domains of cognition (executive function, memory, and learning) affect quality of life and are often impaired after TBI (Binder 1987; Levin 1990; Levin et al., 1990; McDonald et al., 2002; McDowell et al., 1997; Millis et al., 2001).

Research has demonstrated vulnerability of the substantia nigra to TBI (van Bregt et al., 2012; Hutson et al., 2011). Moreover, loss of DAergic fibers in the substantia nigra and ventral tegmental area in experimental Parkinson's disease models alters dendrites and synapses and leads to adverse consequences for brain functioning (Blanchard et al., 1995; Mura and Feldon, 2003; Onn et al., 1986). This evidence supported the evaluation of two classes of drugs known to affect DA activity (i.e., antipsychotics and stimulants), which can improve dendritic and synaptic structures in a variety of CNS disorders (Bütefisch 2003; Tanaka 2006; Viggiano et al., 2003). Though evidence specific to TBI is more limited, similarities in underlying pathology make it plausible that drugs targeting the DAergic system would result in similar benefits for this population.

Notably, the structure of the DAergic system is not changed in response to TBI, but post-TBI functional changes within the DAergic system have been reported. Specifically, (1) the rate-limiting enzyme in DA synthesis (tyrosine hydroxylase [TH]) is increased after TBI within the substantia nigra as well as the infralimbic and prelimbic cortices (Kobori et al., 2006); however, TH activity (and subsequent DA release) is decreased in rat brain after experimental TBI (Shin et al., 2011). (2) Cortical DA synthesis is enhanced post-TBI, as evidenced by elevated

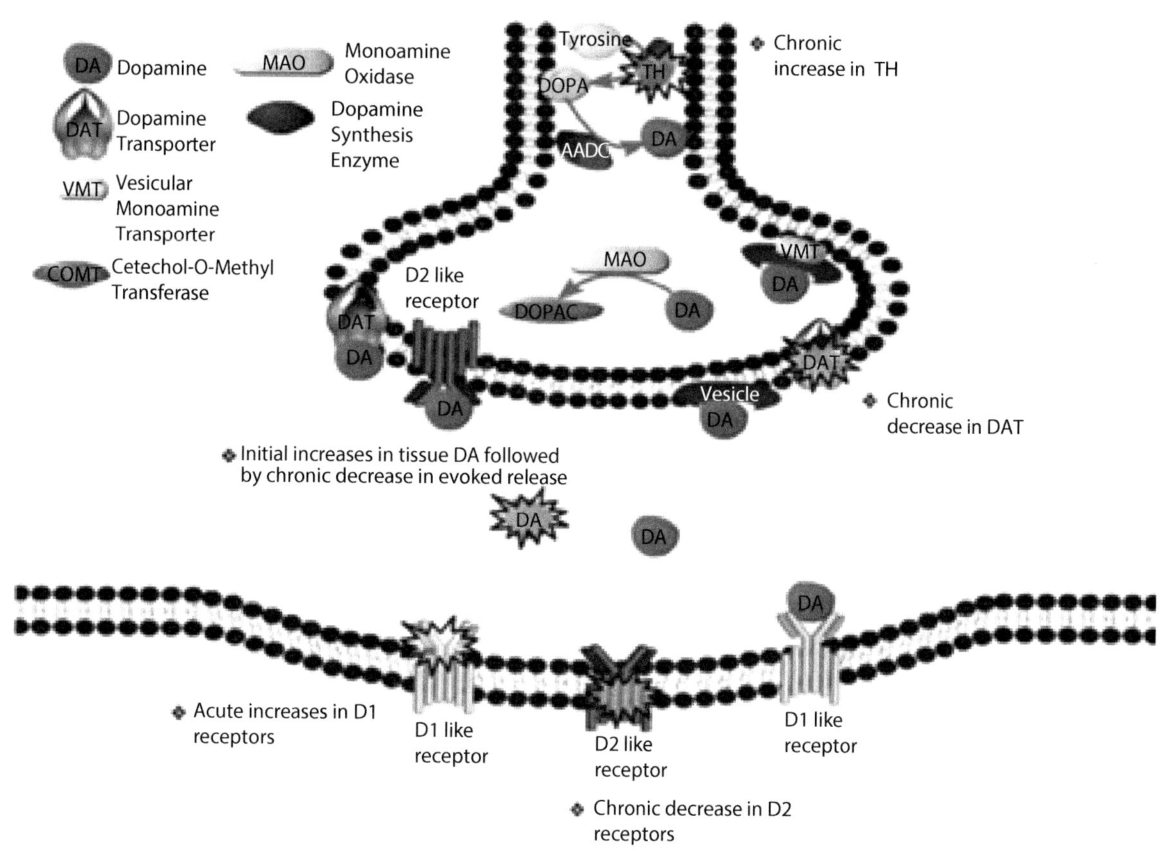

FIGURE 33.1 Summary of changes observed within the dopaminergic system after TBI including alterations in DA receptors, DAT, and TH. (From Bales, J.W. et al., *Neurosci Biobehav Rev* 33(7), 2009. With permission.)

levels of phospho-TH (Kobori et al., 2006). (3) There is an early transient increase in DA levels in several brain regions (Huger and Patrick, 1979; Massucci et al., 2004; McIntosh et al., 1994), followed by decrease in DA levels at 1 hour that persisted out to 2 weeks post-injury (McIntosh et al., 1994). Finally, (4) both electrically evoked DA release and reuptake components of presynaptic neurotransmission in dorsal striatum are functionally impaired as measured using fast scan cyclic voltammetry (FSCV) (Wagner et al., 2005).

DA receptors will be discussed in more detail later, but it is worth noting that one team of researchers (Henry et al., 1997) found that DA binding of DA to DA_1 receptors transiently decreases immediately post-TBI. This decrease was not evident in the chronic period (Henry et al., 1997). There are no DA_2 receptor expression decreases reported at 2 weeks post-TBI in rodents (Wagner et al., 2005; Wagner et al., 2009), however acute time point studies to evaluate expression changes early after TBI have not been conducted.

Acute alterations of DA metabolism have also been observed within the striatum as evidenced by ratios of dihydrophenylacetic acid to DA (Massucci et al., 2004). Altered DA metabolism has also been reported in the hippocampus starting within 24 hours of injury and persisting for upward of 2 weeks in hippocampal microglia (Redell and Dash, 2007). Recent advances have led to new ways to monitor DA and its metabolites (Suominen et al., 2013). Also worth noting is that DA transporter (DAT) binding is altered post-TBI (Wagner et al., 2014; Donnemiller et al., 2000); in this study, DAT binding in the striatum was reduced at 4–5 months after severe TBI, even in the absence of direct injury to the striatum. Also, regional decreases in total DAT expression have been observed following experimental models of TBI that are proportional to reductions in DA clearance observed with FSCV, though sex differences in DAT expression may exist, (Wagner et al., 2005, 2009) and FSCV studies examining sex differences in striatal neurotransmission are underway. Notably, in these studies DA clearance was also impaired after experimental TBI.

Overall it is known that DAergic structures throughout the brain impact cognition and are vulnerable to TBI. Although no structural changes within the DAergic system after brain injury have been observed, changes in DA synthesis, activity, metabolism, and neurotransmission have been reported. A summary of the changes observed within the dopaminergic system following TBI (specifically with respect to DAT, DA receptors, and TH) is depicted in Figure 33.1.

33.3 IMPLICATIONS OF ALTERED DA

33.3.1 Cell Death

In addition to DA being necessary for normal hippocampal, FC, and striatal functioning (Alexander and Crutcher, 1990; Meredith et al., 1990; Graybiel, 1990), dysregulation of DA can lead to cell dysfunction and death within the CNS (Williams and Castner, 2006) that can be associated with the symptoms of schizophrenia and Parkinson's disease (Goldberg et al., 1991; Schneider and Roeltgen, 1993) as well as significant disruption of learning and working memory (Arnsten and Goldman-Rakic, 1998; Morrow et al., 2000). In contrast to these roles, DA can be a potent excitotoxic agent promoting calcium influx (Chapman et al., 1989; Filloux and Wamsley 1991; Olney et al., 1990). DA also can lead to oxidative stress if an excessive amount oxidized to DA semiquinone/quinine and then further transformed to hydrogen peroxide or superoxide (Brunmark and Cadenas, 1988; Hastings 1995; Sinet et al., 1980).

There are also other mechanisms by which DA signaling can contribute to activation of cellular death. The first is a calcium-dependent process: DA signaling involving the D_2 receptor can increase intracellular Ca^{2+}, which subsequently releases kinases and phosphatases that play key roles in cell death cascades (Azdad et al., 2009; Hernandez-Lopez et al., 2000; So et al., 2009). The second involves the modulation of excitotoxicity within the striatum by DAergic fibers (Palmer et al., 1993; Rao et al., 1999). Some researchers posit that the excitotoxic and oxidative damage to DAergic cells and alterations in DA kinetics and reduction in evoked DA release may be precipitated by the initial spike in DA observed post-TBI (Wagner et al., 2007). A 500-fold increase in striatal DA concentration has been observed after ischemia (Globus et al., 1988); notably, striatal ischemia occurs after experimental TBI (Dietrich et al., 1994). Additional evidence that DA can be neurotoxic comes from studies showing that depletion of striatal DAergic projections confers neuroprotection if done before ischemic insult (Globus et al., 1987).

33.3.2 Acute Cellular Dysfunction

DA has significant effects within the hippocampus, FC, and striatum, in ways that are relevant to TBI pathology. For instance, DAergic signaling involving DA_1 and DA_2 receptors influences N-methyl-D-aspartate (NMDA) receptor regulation; this is accomplished via glutamate-mediated activation, which activates striatal-enriched protein and subsequently reduces extracellular regulated kinase signaling (Paul et al., 2003; Snyder et al., 1998). This process may represent a possible strategy for targeting brain injury. The rationale for this potential target is that hypoxia/ischemia can increase both expression and phosphorylation of one of the subunits (NR1) of the NMDA receptor; this change happens at the DA-dependent serine-897 site (Yang et al., 2007). The expression of two NMDA receptor subunits (NR1 and NR2) can decrease at 6–12 hours post-TBI (Kumar et al., 2002). Additional research in this area is needed and may lead to therapeutic attempts to target NMDA receptors post-TBI.

Another potential target for DA-based therapy within the context of brain injury is DARPP-32. It has been established that calcineurin activity is involved in DARPP-32 regulation, specifically the phosphorylation of this protein, which contributes to the overall control of protein phosphatase 1 activity (Nishi et al., 2002). Protein phosphatase 1 is known to control nuclear transcription activity and phosphorylation of important proteins including the NMDA receptor NR1 subunit as well as the Na/K ATPase, which is important to cellular energy production (Flores-Hernández et al., 2002; Snyder et al., 1998). There are potential therapeutic applications acting upon calcineurin, which is known to be induced and altered by TBI, and DA activation of PKA may be a key contributor (Bales et al., 2009; Kurz et al., 2005a, 2005b;). Overall, DA contributes to necessary intracellular signaling within neurons of the hippocampus, striatum, and FC, leading many to believe therapies that regulate DAergic activity may hold promise for care of TBI.

33.3.3 CNS Inflammation

Although some evidence suggests that reduction of post-TBI neuronal inflammation may contribute to neuronal sparing and beneficial outcomes (Clausen et al., 2009; Ekdahl et al., 2009; Kelso et al., 2009), there is concern that inflammatory cells may play an important neuroprotective role that may be disrupted if inflammation is treated (Ekdahl et al., 2009; Kriz, 2006). DA is known to be a potent promoter of inflammation by modifying cytokines within the CNS, which contributes to the pathophysiology of Parkinson's disease (Färber and Kettenmann, 2005). In the context of TBI, blocking inflammatory factors like interleukin-1β can have beneficial results within the CNS (Clausen et al., 2009). Also worth noting is that drugs known to act on the DAergic system (e.g., bupropion) may reduce the extent of CNS inflammation after TBI (Brustolim et al., 2006).

33.4 COGNITIVE IMPLICATIONS OF ALTERED DA

33.4.1 Overview of DA and Cognition

As discussed earlier in this chapter and more extensively elsewhere (Bales et al., 2009; Binder, 1986, 1987; Levin et al. 1988a, 1988b; McMillan 1997; Millis et al., 2001), TBI leads to serious and often persistent deficits in cognition. Cognitive symptoms are among the most troubling and life-altering for TBI survivors and their families (Binder, 1986). Efforts have been made to design and test therapies that might improve cognitive function after TBI; DA is one promising target because it is known to play important roles in the physiological events within the hippocampus, striatum, and FC believed to underlie cognition.

33.4.2 DAergic Systems

The ascending DAergic pathways can be divided into two systems: (1) the mesocorticolimbic pathway and (2) the nigrostriatal pathway (Alexander and Crutcher, 1990; Graybiel, 1990). Both of these DAergic pathways impact cognition with the mesocorticolimbic pathway believed to modulate motivation (Baldo and Kelley, 2007; Mitchell and Gratton, 1994; Salamone 1994), addiction and drug reinforcement (Berridge, 2006; Di Chiara and Bassareo, 2007; Ikemoto, 2007; Salamone et al., 2005; Schultz, 2004; Sutton and Beninger, 1999), and the consolidation of memory (Cools et al., 1993; Ploeger et al., 1991, 1994; Setlow and McGaugh, 1998). The mesocorticolimbic system also plays a role in stress and the development of various neuropsychiatric disorders (Sonuga-Barke, 2005; Tidey and Miczek, 1996; Viggiano et al., 2003).

The other of the two main DAergic systems, the nigrostriatal system, plays a role in spatial learning/memory (Mura and Feldon, 2003) and reward processing (Wickens et al., 2007). Specifically, Mura and Feldon (2003) suggest that lesions in the nigrostriatal DAergic system impact the selection and maintenance of behavioral strategies involved in spatial navigation, as assessed using the Morris water maze. Regional variation in DA signaling is cited as a potential key player in reinforcement of habits (Wickens et al., 2007). Moreover, cognitive dysfunction in Parkinson's disease has been linked to dysfunction within the nigrostriatal system (Ridley et al., 2006; Tamaru, 1997).

33.4.3 Effects of DA in the Striatum

In the context of various neurological conditions, DAergic activity within the striatum affects cognition. In Parkinson's disease, positron emission tomography (PET) studies suggest that DA depletion affects performance on neuropsychiatric assessments (Broussolle et al., 1999; Brück et al., 2001; Duchesne et al., 2002; Marié et al., 1999). Similarly, individuals affected by Huntington's disease have shown decreased binding to striatal DA receptors (specifically D_2 receptors) affects cognitive performance in the areas of attention, working memory, and executive function (Bäckman et al., 1997; Lawrence et al., 1998). Notably, both the striatum and other brain structures related to cognition are vulnerable to damage from TBI (Dietrich et al., 1994; Ding et al., 2001; Dunn-Meynell and Levin, 1997; Fontaine et al., 1999), which affects memory performance (Ricker et al., 2001; Sánchez-Carrión et al., 2008). Medium spiny neurons within the striatum also receive input from DAergic cells and are vulnerable to the effects of TBI and ischemia (Pennartz and Kitai, 1991; Meredith et al., 1990).

33.4.4 Effects of DA on Cognition

Outside of the striatum, hippocampal DA receptors have been identified and are hypothesized to play a role in the physiologic basis of cognition (Lemon and Manahan-Vaughan, 2006; Li et al., 2003). Specifically, DA is reported to regulate synaptic plasticity and ultimately affect storage of unpredicted information within the CA_1 subregion of the hippocampus. It is well-established that the hippocampus is vulnerable to TBI and that TBI patients experience problems with attentional processing manifesting in confusion, distraction, and memory problems (Draper and Ponsford, 2008; Gentilini et al., 1989). DAergic systems focus attention (Brennan and Arnsten, 2008; Cohen and Servan-Schreiber, 1992; Wise et al., 1996) as evidenced in the context of ADHD (Solanto, 1998). Although this evidence is not sufficient to definitively prove the role for DAergic hippocampally mediated memory dysfunction in TBI, nor rule out other possible mechanisms, it does suggest plausibility sufficient to warrant future research.

33.4.5 Effects of DA on Emotion, Affect, and Behavior

TBI patients have long been known to suffer emotional disturbances and behavioral changes (Arciniegas et al., 2000; Dyer et al., 2006; Oddy et al., 1985). The rate of depression is elevated after TBI (Jorge et al., 2004; Moldover et al., 2004; Seel et al., 2003). One study found that TBI survivors had worse verbal aggressiveness and impulsivity compared with individuals with spinal cord injury or noninjured controls (Dyer et al., 2006). DA is implicated in these types of symptoms because the mesocorticolimbic DAergic pathway plays a role in neuropsychiatric disturbances (e.g., depression, other mood disorders) and thus may represent a target for therapy (Mega and Cummings, 1994). Notably, neuropsychiatric disorders like schizophrenia are also related to problems in mesocorticolimbic DA signaling (Abi-Dargham and Moore, 2003). However, TBI is more complex and includes pathology such as inflammation, white matter changes, and excitotoxicity that could theoretically impact the DAergic system. Thus, devoting time and effort into developing an enhanced understanding of how DAergic signaling affects brain function, especially post-TBI, may lead to better ability to treat brain injury.

33.5 TARGETING DA DIRECTLY

33.5.1 Overview

Studies of DAergic agonists within the context of TBI have largely been limited to studies of chronic TBI patients, where DAergic agonists have shown some benefit to clinical outcomes (Gualtieri and Evans, 1988; Kaelin et al., 1996; Bales et al., 2009; Zweckberger et al., 2010). For instance, methylphenidate (MPD), a pharmaceutical that inhibits the DAT has been found to have beneficial effects on cognitive function (e.g., attention and memory) when administered chronically to TBI patients (Gualtieri and Evans, 1988; Kim et al., 2006; Lee et al., 2005; Pavlovskaya et al., 2007). MPD administration within the first month of injury (i.e., the subacute period) is associated with beneficial effects on attention and memory (Kaelin et al., 1996).

Other drugs have been tried during the recovery period post-TBI and have been shown to positively affect cognition, including bromocriptine (McDowell et al., 1998; McAllister et al., 2011) and amantadine hydrochloride (Kraus et al., 2005). Amantadine is a drug known to influence DA (Gianutsos et al., 1985; Von Voigtlander and Moore, 1971). Amantadine can improve neurological outcome as assessed by measures of patient agitation (Chandler et al., 1988), and other cognitive outcomes (Giacino et al., 2012). Although relatively few studies have explored the role of therapies targeting DA on cognitive outcomes after TBI, many of those that have been conducted to date have yielded promising results.

The Neurotrauma Foundation (NTF) recognized the benefits of DA enhancing drugs in their 2006 publication where they reviewed clinical recommendations for acute- and rehabilitative-TBI management (Warden et al., 2006). This review mentioned DAergic drugs and their potential roles in cognitive recovery post-TBI. Both MPD and amantadine were acknowledged by the NTF for their beneficial effect on generalized cognitive functioning post-TBI, with MPD also acknowledged as playing a role in attention and processing speed. Bromocriptine, on the other hand, is recognized in the NTF review for its role in improving executive function. Taken together, the evidence from the NTF and other publications suggest that targeting DA directly postinjury may be beneficial.

33.5.2 DA Agonists

The majority of available evidence of the efficacy of DA agonists comes from clinical and laboratory studies of DAergic therapies that were begun within the first day of TBI and continued chronically through the time of behavioral outcome assessment. One undesirable consequence of this research design is that the effects of acute administration are unclear and warrant additional research. Still, there is evidence that DA therapy during the acute phase after experimental TBI may be beneficial when administered chronically. One agent that has been found clinically to have positive therapeutic properties is MPD (Volz, 2008); this drug has been found to exert its action via DAT blockade in clinical and pre-clinical studies (Volkow et al., 1998, 2002a, 2002b). The effective increase in DA observed after MPD or amphetamine (AMPH) treatment has been found to correlate with better cognitive function in experimental models (Schiffer et al., 2006; Volkow et al., 2004).

Experimentally, MPD preserves cognitive function after both cortical ablation and cortical impact injury (Kline et al., 1994, 2000). The specific types of function that were preserved were motor skills, assessed via beam walking (Kline et al., 1994). In addition to reduced post-TBI motor deficits, MPD therapy started as late as 24 hours after injury in rodent models enhanced spatial memory (Kline et al., 2000).

AMPH can accelerate recovery after experimental TBI in animal models (Barbay and Nudo, 2009; Chudasama et al., 2005; Dhillon et al., 1998; Feeney et al., 1981; Hovda et al., 1989; M'Harzi et al., 1988). AMPH affects DA release and reuptake (Kahlig and Galli, 2003). Moreover, clinical trials have found beneficial effects of MPD treatment on neuropsychological outcomes following TBI (Kaelin et al., 1996; McDowell et al., 1997; Plenger et al., 1996; Whyte et al., 2004). Conversely, other studies have failed to find a beneficial effect of MPD therapy on cognitive outcomes (Mooney and Haas, 1993; Speech et al., 1993; Tiberti et al., 1998). Methamphetamine has also been evaluated in experimental TBI with mixed results (Rau et al., 2012; Shen et al., 2011).

At the cellular level, AMPH therapy reduces accumulation of unwanted substances (e.g., lactate, free fatty acids) within the hippocampus and cortex (Dhillon et al., 1998) and also attenuates decreases in cerebral glucose utilization after experimental TBI (Queen et al., 1997). AMPH may produce benefits through its ability to induce hippocampal brain-derived neurotrophic factor, as was observed in a rat model of controlled cortical impact (Griesbach et al., 2008). Although exciting, this finding was unsurprising, as AMPH was already known to induce synaptogenesis and plasticity affter TBI (Goldstein 2003; Ramic et al., 2006). Notably, AMPH treatment is complicated by a number of factors; for instance-AMPH has effects on other monoamines (Fleckenstein et al., 2007). Moreover, DA_2 antagonists (e.g., haloperidol) that may also be part of the treatment regimen may counteract the desirable effects of AMPH treatment (Feeney et al., 1982; Hovda and Feeney, 1985).

Another relevant drug is the DA receptor agonist bromocriptine. This drug is a D_2 receptor–specific agonist that protects cortical neurons from glutamate-induced toxicity (Kihara et al., 2002). When evaluated after experimental TBI, bromocriptine enhanced working and spatial memory, which was associated with a decrease in oxidative damage (Kline et al., 2002, 2004).

33.5.3 DA Antagonists

In addition to neuroprotection conferred by DAergic agonists, DAergic antagonists can be neuroprotective in ischemia, excitotoxicity, and DA lesion models (Armentero et al., 2002; Okada et al., 2005; Yamamoto et al., 1994; Zou et al., 2005). In the TBI literature, early administration of a DA2 receptor antagonist can improve cognitive outcome; though there was no effect of a DA1 antagonism alone, there was an additive effect when DA1 and DA2 antagonists were given in combination (Tang et al., 1997). However, there is some concern regarding the potential use of DA antagonists as a therapy for TBI. Support for this concern comes from rodent studies using drugs with D_2 antagonist properties such as risperidone and haloperidol after TBI, which have been associated with poorer outcomes (Hoffman et al., 2008; Hovda and Feeney, 1985) and the induction of Parkinson-like symptoms after post-TBI risperidone therapy (Kang and Kim, 2013). Among the undesirable outcomes of haloperidol treatment for TBI is delayed motor recovery, which has been observed in pre-clinical studies of adult rodents (Goldstein and Bullman, 2002; Feeney et al., 1982). Moreover, haloperidol, but not olanzapine was found to impair cognitive recovery following experimental TBI (Wilson et al. 2003). Other studies showed impaired cognitive and motor function with antipsychotic drug administration (Hoffman et al., 2008; Wilson et al., 2003). Notably, these drugs have been prescribed to TBI patients; commonly, antipsychotics are used for sedation, and less commonly to treat the post-TBI agitation or psychosis that can occur after injury (Stanislav, 1997).

33.6 INDIRECTLY TARGETING DA

Beyond the drugs discussed above that directly target DAergic signaling, other drugs that indirectly target this pathway can have neuroprotective effects in the context of Parkinson's (Schapira, 2009). For example, other DA agonists (e.g., ropinirole, pramipexole) can protect against TBI-induced oxidative stress in a way that is partly mediated by binding with DA receptors (Nair and Sealfon, 2003). Of additional interest is that these drugs have other desirable qualities including upregulating neurotrophic factors (Du et al., 2005). Another study found pramipexole reduced neuronal death within the nigrostriatal region after ischemic injury (Hall et al., 1996). In low-response pediatric TBI survivors, pramipexole started 1 month after injury was associated with improved outcomes (Patrick et al., 2006).

Another type of drug found to indirectly impact DAergic signaling and, subsequently, brain injury outcomes are monoamine-oxidase B (MAO-B) inhibitors. One example of a MAO-B inhibitor is selegiline, a drug that inhibits breakdown of DA within the synapse and subsequently raise DA levels. Selegiline protects cells in culture against toxicity induced by 1-methyl-4-phenylpyridinium (Tatton et al., 1994). In animal models, this MAO-B inhibitor attenuated loss of DAergic cells (Tatton and Greenwood, 1991). Another desirable property of selegiline is its ability to reduce DA-induced free-radical generation (Wu et al., 1993). This drug is also clinically relevant where it can improve cognitive performance (Zhu et al., 2000) and reduce TBI-related apathy in adults (Newburn and Newburn, 2005). Other MAO-B inhibitor drugs reportedly reduce excitotoxic damage and improve motor and memory function (Huang et al., 1999; Mandel et al., 2005).

33.7 OPPORTUNITIES FOR ADDITIONAL RESEARCH

The available evidence surrounding the role of DA and appropriateness of DA therapies remains limited. There are many opportunities for future research in this area. First and foremost, additional research regarding the temporal sequence of events surrounding disruptions of DA signaling in needed. This investigation should include both a precise examination of the early events after TBI as well as long-term tracking through the acute, recovery, and chronic periods. A sophisticated understanding of the molecular events that follow brain injury will help researchers design therapies as well as critically evaluate the successes and failures of past efforts. A second area worth devoting research effort to is optimizing the drug regimen of DA therapy; this approach would include considerations for the appropriate dosage, therapeutic window, and possible benefits of combination therapy. The distinct mechanisms associated with the different dopamine receptors (e.g., D_1 and D_2) represent one possible avenue for combination therapy. It is also possible that differential approaches to DA therapy in the acute and chronic periods after TBI may result in improved outcomes. One study found that DA administration along with intrinsic characteristics (e.g., sex, DAT genotype) affected levels of DA and its metabolites in the cerebrospinal fluid of severe TBI patients; the effects of these and other intrinsic characteristics on DA should be further researched (Wagner et al., 2007). These and other considerations regarding therapeutic regimen of DA-targeting therapies are needed. Finally, efforts will be necessary to identify which patients are most likely to benefit from DA therapy. This will include considerations regarding how sex, hormones, and injury characteristics as a way to stratify patients. Sex differences in post-TBI outcomes have been reported (Ratcliff et al., 2007; Slewa-Younan et al., 2004). Such intrinsic characteristics may affect response to treatment since sex-specific differences in response to treatment with the DA agonist methylphenidate for TBI have also been reported (Wagner et al., 2007); the effect of sex and other intrinsic factors on other therapies are worth exploring.

In light of available genetic technologies, considerations regarding genetic markers and DAT expression profiles (Failla et al., 2013; Zhuang et al., 2001) may also be relevant to personalized medicine approaches. Such considerations will help to increase the rate at which findings can be translated to clinical practice. Part of this effort should include additional work exploring the potential effects of environmental enrichment on gene expression, DA signaling, and TBI outcomes (Shin et al., 2013; Wagner et al., 2005). Other specific areas for future research include further evaluation into the role of DARPP-32 in the DAergic system and the effects of DA receptor expression. Preliminary research suggests that DARPP-32 phosphorylation is decreased by TBI with the potential consequence of altered DA signaling. Research in this area will be complicated by the complex interplay of DA receptors with other receptor systems.

33.8 CONCLUSION

The current state-of-the-science suggests a complex role for the DAergic system in TBI pathology. Moreover, the current literature suggests additional research is warranted surrounding the potential therapeutic effects of targeting the DAergic system post-TBI. Evidence for the cognitive benefits of therapies which directly or indirectly effect DA signaling comes from both animal models of brain injury (Armstead et al., 2013) and, to a lesser extent, data from patients after TBI. Taken together, the evidence suggests that therapies that impact the DA system may have beneficial effects on cognitive factors including memory, learning, and executive function. These benefits are clinically relevant because these problems are known to be distressful and life-altering for TBI survivors and their families. For a summary of the clinical effects of altered dopamine and the potential therapeutic implications see Table 33.1.

TABLE 33.1
Effects of Alterations in Dopamine Tone and Potential Therapeutic Targets

Dopamine Tone	Decreased	Normal	Increased
Commonly observed in	Chronic TBI	Healthy individuals	Acute TBI
Effects on cognition	Problems with memory, attention, and executive function	Normal	Possible effects on mood/affect
Other effects	Dysfunction in LTP and LTD	Normal long-term potentiation and long-term depression; oxidative stress = antioxidants	Increased oxidative stress Induction of inflammatory signaling pathways
Potential therapies	Increase DA tone -increase synapse time -increase synthesis -target receptor	N/A	Decrease DA tone -inhibit receptors Decrease metabolism -oxidative metabolites

REFERENCES

Abi-Dargham, A. and H. Moore. 2003. Prefrontal DA transmission at D1 receptors and the pathology of schizophrenia. *Neuroscientist* 9(5):404–16.

Alexander, G.E. and M.D. Crutcher. 1990. Functional architecture of basal ganglia circuits: Neural substrates of parallel processing. *Trends Neurosci* 13(7):266–71.

Arciniegas, D.B., Topkoff, J., and J.M. Silver. 2000. Neuropsychiatric aspects of traumatic brain injury. *Curr Treat Opin Neurol* 2(2):169–186.

Armentero, M-T., Fancellu, R., Nappi, G., and F. Blandini. 2002. Dopamine receptor agonists mediate neuroprotection in malonate-induced striatal lesion in the rat. *Exp Neurol* 178(2):301–5.

Armstead, W.M., Riley J., and M.S. Vavilala. 2013. Dopamine prevents impairment of autoregulation after traumatic brain injury in the newborn pig through inhibition of up-regulation of endothelin-1 and extracellular signal-regulated kinase mitogen-activated protein kinase. *Pediatr Crit Care Med* 14(2):e103–11.

Arnsten, A.F., and P.S. Goldman-Rakic. 1998. Noise stress impairs prefrontal cortical cognitive function in monkeys: Evidence for a hyperdopaminergic mechanism. *Arch Gen Psychiatry* 55(4):362–8.

Azdad, K., Gall, D., Woods, A.S., Ledent, C., Ferré, S., and S.N. Schiffmann. 2009. Dopamine D2 and adenosine A2A receptors regulate NMDA-mediated excitation in accumbens neurons through A2A-D2 receptor heteromerization. *Neuropsychopharmacology* 34(4):972–86.

Bäckman, L., Robins-Wahlin, T.B., Lundin, A., Ginovart, N., and L. Farde. 1997. Cognitive deficits in Huntington's Disease are predicted by dopaminergic PET markers and brain volumes. *Brain* 120(Pt 12):2207–17.

Baldo, B.A., and A.E. Kelley. 2007. Discrete neurochemical coding of distinguishable motivational processes: Insights from nucleus accumbens control of feeding. *Psychopharmacology (Berl.)* 191(3):439–59.

Bales, J.W., Yan, H.Q., Ma, X., Li, Y., Samarasinghe, R., and C.E. Dixon. 2011. The dopamine and cAMP regulated phosphoprotein, 32 kDa (DARPP-32) signaling pathway: A novel therapeutic target in traumatic brain injury. *Exp Neurol* 229(2):300–7.

Bales, J.W., Kline A.K., Wagner, A.K., and C.E. Dixon. 2010. Targeting dopamine in acute traumatic brain injury. *Open Drug Discov J* 2:119–128.

Bales, J.W., Wagner, A.K., Kline, A.E., and C.E. Dixon. 2009. Persistent cognitive dysfunction after traumatic brain injury: A dopamine hypothesis. *Neurosci Biobehav Rev* 33(7):981–1003.

Barbay, S., and R.J. Nudo. 2009. The effects of amphetamine on recovery of function in animal models of cerebral injury: A critical appraisal. *NeuroRehabilitation* 25(1):5–17.

Baron, J.C., Comar, D., Zarifian, E., Agid, Y., Crouzel, C., Loo, H. et al. 1985. Dopaminergic receptor sites in human brain: Positron emission tomography. *Neurology* 35(1):16–24.

Bayir, H., Marion, D.W., Puccio, A.M., Wisniewski, S.R., Janesko, K.L., Clark, R.S. et al. 2004. Marked gender effect on lipid peroxidation after severe traumatic brain injury in adult patients. *J Neurotrauma* 21(1):1–8.

Berridge, C.W. 2006. Neural substrates of psychostimulant-induced arousal. *Neuropsychopharmacology* 31(11):2332–40.

Binder, L.M. 1986. Persisting symptoms after mild head injury: A review of the postconcussive syndrome. *J Clin Exp Neuropsychol* 8(4):323–46.

Binder, L.M. 1987. Neurobehavioral recovery after mild head injury. *J Neurosurg* 67(5):785–7.

Blanchard, V., Chritin, M., Vyas, S., Savasta, M., Feuerstein, C., Agid, Y. et al. 1995. Long-term induction of tyrosine hydroxylase expression: Compensatory response to partial degeneration of the dopaminergic nigrostriatal system in the rat brain. *J Neurochem* 64(4):1669–79.

Bramlett, H.M., and W.D. Dietrich. 2002. Quantitative structural changes in white and gray matter 1 year following traumatic brain injury in rats. *Acta Neuropathol* 103(6):607–14.

Brennan, A.R., and A.F. Arnsten. 2008. Neuronal mechanisms underlying attention deficit hyperactivity disorder: The influence of arousal on prefrontal cortical function. *Ann NY Acad Sci* 1129:236–45.

Broussolle, E., Dentresangle, C., Landais, P., Garcia-Larrea, L., Pollak, P., Croisile, B. et al. 1999. The relation of putamen and caudate nucleus 18F-Dopa uptake to motor and cognitive performances in Parkinson's disease. *J Neurol Sci* 166(2):141–51.

Brück, A., Portin, R., Lindell, A., Laihinen, A., Bergman, J., Haaparanta, M. et al. 2001. Positron emission tomography shows that impaired frontal lobe functioning in Parkinson's disease is related to dopaminergic hypofunction in the caudate nucleus. *Neurosci Lett* 311(2):81–4.

Brunmark, A., and E. Cadenas. 1988. Oxidation of quinones by H2O2: Formation of epoxy- and hydroxyquinone adducts and electronically excited states. *Basic Life Sci* 49:81–6.

Brustolim, D., Ribeiro-dos-Santos, R., Kast, R.E., Altschuler, E.L., and M.B. Soares. 2006. A new chapter opens in anti-inflammatory treatments: The antidepressant bupropion lowers production of tumor necrosis factor-alpha and interferon-gamma in mice. *Int Immunopharmacol* 6(6):903–7.

Bütefisch, C.M. 2003. Modulation of use-dependent plasticity by D-amphetamine. *Suppl Clin Neurophysiol* 56:242–5.

Cantor, J., Ashman, T., Dams-O'Connor, K., Dijkers, M.P., Gordon, W., Spielman, L. et.al. 2014. Evaluation of the short-term executive plus intervention for executive dysfunction after traumatic brain injury: A randomized controlled trial with minimization. *Arch Phys Med Rehabil* 95(1):1–9.e3.

Capone-Neto, A., and S.B. Rizoli. 2009. Linking the chain of survival: Trauma as a traditional role model for multisystem trauma and brain injury. *Curr Opin Crit Care* 15(4):290–4.

Centonze, D., Gubellini, P., Picconi, B., Calabresi, P., Giacomini, P., and G. Bernardi. 1999. Unilateral dopamine denervation blocks corticostriatal LTP. *J Neurophysiol* (Dec), 82(6):3575–9.

Chandler, M.C., Barnhill, J.L., and C.T. Gualtieri. 1988. Amantadine for the agitated head-injury patient. *Brain Inj* 2(4):309–11.

Chapman, A.G., Dürmuller, N., Lees, G.J., and B.S. Meldrum. 1989. Excitotoxicity of NMDA and kainic acid is modulated by nigrostriatal dopaminergic fibres. *Neurosci Lett* 107(1–3):256–60.

Chen, L., Bohanick, J.D., Nishihara, M., Seamans, J.K., and C.R. Yang. 2007. Dopamine D1/5 receptor-mediated long-term potentiation of intrinsic excitability in rat prefrontal cortical neurons: Ca2+-dependent intracellular signaling. *J Neurophysiol* 97(3):2448–64.

Chudasama, Y., Nathwani, F., and T.W. Robbins. 2005. D-Amphetamine remediates attentional performance in rats with dorsal prefrontal lesions. *Behav Brain Res* 158(1):97–107.

Chudasama, Y., and T.W. Robbins. 2006. Functions of frontostriatal systems in cognition: Comparative neuropsychopharmacological studies in rats, monkeys and humans. *Biol Psychol* 73(1):19–38.

Clausen, F., Hanell, A., Bjork, M., Hillered, L., Mir, A.K., Gram H. et al. 2009. Neutralization of interleukin-1beta modifies the inflammatory response and improves histological and cognitive outcome following traumatic brain injury in mice. *Eur J Neursci* 30(3):385–96.

Cohen, J.D., and D. Servan-Schreiber. 1992. Context, cortex, and dopamine: A connectionist approach to behavior and biology in schizophrenia. *Psychol Rev* 99(1):45–77.

Cools, A.R., Ellenbroek, B., Heeren, D., and L. Lubbers. 1993. Use of high and low responders to novelty in rat studies on the role of the ventral striatum in radial maze performance: Effects of intra-accumbens injections of sulpiride. *Can J Physiol Pharmacol* 71(5–6):335–42.

DeKosky, S.T., Kochanek, P.M., Clark, R.S., Ciallella, J.R., and C.E. Dixon. 1998. Secondary injury after head trauma: Subacute and long-term mechanisms. *Semin Clin Neuropsychiatry* 3(3):176–85.

Dhillon, H.S., Dose, J.M., and R.M. Prasad. 1998. Amphetamine administration improves neurochemical outcome of lateral fluid percussion brain injury in the rat. *Brain Res* 804(2):231–7.

Di Chiara, G., and V. Bassareo. 2007. Reward system and addiction: What dopamine does and doesn't do. *Curr Opin Phamacol* 7(1):69–76.

Dietrich, W.D., Alonso, O., and M. Halley. 1994. Early microvascular and neuronal consequences of traumatic brain injury: A light and electron microscopic study in rats. *J Neurotrauma* 11(3):289–301.

Ding, Y., Yao, B., Lai, Q., and J.P. McAllister. 2001. Impaired motor learning and diffuse axonal damage in motor and visual systems of the rat following traumatic brain injury. *Neurol Res* 23(2–3):193–202.

Dixon, C.E., Clifton, G.L., Lighthall, J.W., Yaghmai, A.A., and R.L. Hayes. 1991. A controlled cortical impact model of traumatic brain injury in the rat. *J Neurosci Methods* 39(3):253–62.

Dixon, C.E., Lyeth, B.G., Povlishock, J.T., Findling, R.L., Hamm, R.J., Marmarou, A. et al. 1987. A fluid percussion model of experimental brain injury in the rat. *J Neurosurg* 67(1):110–9.

Dixon, C.E., Kraus, M.F., Kline, A.E., Ma, X., Yan, H.Q., Griffith, R.G. et al. 1999. Amantadine improves water maze performance without affecting motor behavior following traumatic brain injury in rats. *Restor Neurol Neurosci* 14(4):285–94.

Donnemiller, E., Brenneis, C., Wissel, J., Scherfler, C., Poewe, W., Riccabona, G. et al. 2000. Impaired dopaminergic neurotransmission in patients with traumatic brain injury: A SPECT study using 123I-beta-CIT and 123I-IBZM. *Eur J Nucl Med* 27(9):1410–4.

Draper, K., and J. Ponsford. 2008. Cognitive functioning ten years following traumatic brain injury and rehabilitation. *Neuropsychology* 22(5):618–25.

Du, F., Li, R., Huang, Y., Li, X., and W. Le. 2005. Dopamine D3 Receptor-preferring agonists induce neurotrophic effects on mesencephalic dopamine neurons. *Eur J Neurosci* 22(10):2422–30.

Duchesne, N., Soucy, J.P., Masson, H., Chouinard, S., and M.A. Bédard. 2002. Cognitive deficits and striatal dopaminergic denervation in Parkinson's disease: A single photon emission computed tomography study using 123iodine-beta-CIT in patients on and off levodopa. *Clin Neuropharmacol* 25(4):216–24.

Dunn-Meynell, A.A., and B.E. Levin. 1997. Histological markers of neuronal, axonal and astrocytic changes after lateral rigid impact traumatic brain injury. *Brain Res* 761(1):25–41.

Dunn-Meynell, A., Pan, S., and B.E. Levin. 1994. Focal traumatic brain injury causes widespread reductions in rat brain norepinephrine turnover from 6 to 24 H. *Brain Res* 660(1):88–95.

Dyer, K.F., Bell, R., McCann, J., and R. Rauch. 2006. Aggression after traumatic brain injury: Analysing socially desirable responses and the nature of aggressive traits. *Brain Inj* 20(11):1163–73.

Ekdahl, C.T., Kokaia, Z., and O. Lindvall. 2009. Brain inflammation and adult neurogenesis: The dual role of microglia. *Neuroscience* 158(3):1021–9.

El-Ghundi, M., O'Dowd, B.F., and S.R. George. 2007. Insights into the role of dopamine receptor systems in learning and memory. *Rev Neurosci* 18(1):37–66.

Failla, M.D., Burkhardt, J.N., Miller, M.A., Scanlon, J.M., Conley, Y.P., Ferrell, R.E. et al. 2013. Variants of SLC6A4 in depression risk following severe TBI. *Brain Inj* 27(6):696–706.

Färber, K., and H. Kettenmann. 2005. Physiology of microglial cells. *Brain Res Brain Res Rev* 48(2):133–43.

Faul, M., Xu, L., Wald, M.M., and V.G. Coronado. 2010. Traumatic brain injury in the United States: Emergency department visits, hospitalizations, and deaths. Centers for Disease Control and Prevention, National Center for Injury Prevention and Control, Atlanta, GA.

Feeney, D.M., Gonzales, A., and W.A. Law. 1981. Amphetamine restores locomotor function after motor cortex injury in the rat. *Proc West Pharmacol Soc* 24:15–7.

Feeney, D.M., Gonzalez, A., and W.A. Law. 1982. Amphetamine, haloperidol, and experience interact to affect rate of recovery after motor cortex injury. *Science (New York, N.Y.)* 217(4562): 855–7.

Filloux, F., and J.K. Wamsley. 1991. Dopaminergic modulation of excitotoxicity in rat striatum: Evidence from nigrostriatal lesions. *Synapse* 8(4):281–8.

Fleckenstein, A.E., Volz, T.J., Riddle, E.L., Gibb, J.W., and G.R. Hanson. 2007. New insights into the mechanism of action of amphetamines. *Annu Rev Pharmacol Toxicol* 47:681–98.

Flores-Hernández, J., Cepeda, C., Hernández-Echeagaray, E., Calvert, C.R., Jokel, E.S., Fienberg, A.A. et al. 2002. Dopamine enhancement of NMDA currents in dissociated medium-sized striatal neurons: Role of D1 receptors and DARPP-32. *J Neurophysiol* 88(6):3010–20.

Folkersma, H., Boellaard, R., Vandertop, W.P., Kloet, R.W., Lubberink M., Lammertsma, A.A. et al.. 2009. Reference tissue models and blood-brain barrier disruption: Lessons from (R)-[11C]PK11195 in traumatic brain injury. *J Nucl Med* 50(12):1975–9.

Fontaine, A., Azouvi, P., Remy, P., Bussel, B., and Y. Samson. 1999. Functional anatomy of neuropsychological deficits after severe traumatic brain injury. *Neurology* 53(9):1963–8.

Fox, G.B., Fan, L., LeVasseur, R.A., and A.I. Faden. 1998. Sustained sensory/motor and cognitive deficits with neuronal apoptosis following controlled cortical impact brain injury in the mouse. *J Neurotrauma* 15(8):599–614.

Frenette, A.J., Kanji, S., Rees L., Williamson, D.R., Perreault, M.M., Turgeon, A.F. et al. 2012. Efficacy and safety of dopamine agonists in traumatic brain injury: A systematic review of randomized controlled trials. *J Neurotrauma* 29(1):1–18.

Fridman, E.A., Krimchansky, B.Z., Bonetto, M., Galperin, T., Gamzu, E.R., Leiguarda R.C. et al. 2010. Continuous subcutaneous apomorphine for severe disorders of consciousness after traumatic brain injury. *Brain Inj* 24(4):636–41.

Gentilini, M., Barbieri, C., De Renzi, E., and P. Faglioni. 1989. Space exploration with and without the aid of vision in hemisphere-damaged patients. *Cortex* 25(4):643–51.

Giacino, J.T., Whyte, J., Bagiella, E., Kalmar, K., Childs, N., Khademi, A. et al. 2012. Placebo-controlled trial of amantadine for severe traumatic brain injury. *N Engl J Med* 366(9):819–26.

Gianutsos, G., Chute, S., and J.P. Dunn. 1985. Pharmacological changes in dopaminergic systems induced by long-term administration of amantadine. *Eur J Pharmacol* 110(3):357–61.

Globus, M.Y., Busto, R., Dietrich, W.D., Martinez, E., Valdes, I., and M.D. Ginsberg. 1988. Effect of ischemia on the in vivo release of striatal dopamine, glutamate, and gamma-aminobutyric acid studied by intracerebral microdialysis. *J Neurochem* 51(5):1455–64.

Globus, M.Y., Ginsberg, M.D., Dietrich, W.D., Busto, R., and P. Scheinberg. 1987. Substantia nigra lesion protects against ischemic damage in the striatum. *Neurosci Lett* 80(3):251–6.

Goldberg, T.E., Bigelow, L.B., Weinberger, D.R., Daniel, D.G., and J.E. Kleinman. 1991. Cognitive and behavioral effects of the coadministration of dextroamphetamine and haloperidol in schizophrenia. *Am J Psychiatry* 148(1):78–84.

Goldstein, L.B. 2003. Neuropharmacology of TBI-induced plasticity. *Brain Inj* 17(8):685–94.

Goldstein, L.B., and S. Bullman. 2002. Differential effects of haloperidol and clozapine on motor recovery after sensorimotor cortex injury in rats. *Neurorehabil Neural Repair* 16(4):321–5.

Granado, N., Ortiz, O., Suárez, L.M., Martín, E.D., Ceña, V., Solís, J.M. et al. 2008. D1 but not D5 dopamine receptors are critical for LTP, spatial learning, and LTP-induced arc and Zif268 expression in the hippocampus. *Cereb Cortex* 18(1):1–12.

Graybiel, A.M. 1990. Neurotransmitters and neuromodulators in the basal ganglia. *Trends Neurosci* 13(7):244–54.

Griesbach, G.S., Hovda, D.A., Gomez-Pinilla, F., and R.L. Sutton. 2008. Voluntary exercise or amphetamine treatment, but not the combination, increases hippocampal brain-derived neurotrophic factor and synapsin I following cortical contusion injury in rats. *Neuroscience* 154(2):530–40.

Gualtieri, C.T., and R.W. Evans. 1988. Stimulant treatment for the neurobehavioural sequelae of traumatic brain injury. *Brain Inj* 2(4):273–90.

Hall, E.D., Andrus, P.K., Oostveen, J.A., Althaus, J.S., and P.F. VonVoigtlander. 1996. Neuroprotective effects of the dopamine D2/D3 agonist pramipexole against postischemic or methamphetamine-induced degeneration of nigrostriatal neurons. *Brain Res* 742(1–2):80–8.

Hamm, R.J., Pike, B.R., O'Dell, D.M., and B.G. Lyeth. 1994. Traumatic brain injury enhances the amnesic effect of an NMDA antagonist in rats. *J Neurosurg* 81(2):267–71.

Hamm, R.J., Temple, M.D., O'Dell, D.M., Pike, B.R., and B.G. Lyeth. 1996. Exposure to environmental complexity promotes recovery of cognitive function after traumatic brain injury. *J Neurotrauma* 13(1):41–7.

Hastings, T.G. 1995. Enzymatic oxidation of dopamine: The role of prostaglandin H synthase. *J Neurochem* 64(2):919–24.

Henry, J.M., Talukder, N.K., Lee, A.B., and M.L. Walker. 1997. Cerebral trauma-induced changes in corpus striatal dopamine receptor subtypes. *J Invest Surg* 10(5):281–6.

Hernandez-Lopez, S., Tkatch, T., Perez-Garci, E., Galarraga, E., Bargas, J., Hamm, H. et al. 2000. D2 dopamine receptors in striatal medium spiny neurons reduce L-type Ca2+ currents and excitability via a novel PLC[beta]1-IP3-calcineurin-signaling cascade. *J Neurosci* 20(24):8987–95.

Hicks, R.R., Smith, D.H., Lowenstein, D.H., Saint Marie, R., and T.K. McIntosh. 1993. Mild experimental brain injury in the rat induces cognitive deficits associated with regional neuronal loss in the hippocampus. *J Neurotrauma* 10(4):405–14.

Hoffman, A.N., Cheng, J.P., Zafonte, R.D., and A.E. Kline. 2008. Administration of haloperidol and risperidone after neurobehavioral testing hinders the recovery of traumatic brain injury-induced deficits. *Life Sci* 83(17–18):602–7.

Hovda, D.A., and D.M. Feeney. 1985. Haloperidol blocks amphetamine induced recovery of binocular depth perception after bilateral visual cortex ablation in cat. *Proc West Pharmacol Soc* 28:209–11.

Hovda, D.A., Sutton, R.L., and D.M. Feeney. 1989. Amphetamine-induced recovery of visual cliff performance after bilateral visual cortex ablation in cats: Measurements of depth perception thresholds. *Behav Neurosci* 103(3):574–84.

Huang, W., Chen, Y., Shohami, E., and M. Weinstock. 1999. Neuroprotective effect of rasagiline, a selective monoamine oxidase-B inhibitor, against closed head injury in the mouse. *Eur J Pharmacol* 366(2–3):127–35.

Huger, F., and G. Patrick. 1979. Effect of concussive head injury on central catecholamine levels and synthesis rates in rat brain regions. *J Neurochem* 33(1):89–95.

Hutson, C.B., Lazo, C.R., Mortazavi, F., Giza, C.C, Hovda, D., and M.F.Chesselet 2011. Traumatic brain injury in adult rats causes progressive nigrostriatal dopaminergic cell loss and enhanced vulnerability to the pesticide paraquat. *J Neurotrauma* 28(9):1783–801.

Ikemoto, S. 2007. Dopamine reward circuitry: Two projection systems from the ventral midbrain to the nucleus accumbens-olfactory tubercle complex. *Brain Res Rev* 56(1):27–78.

Jorge, R.E., Robinson, R.G., Moser, D., Tateno, A., Crespo-Facorro, B., and S. Arndt. 2004. Major depression following traumatic brain injury. *Arch Gen Psychiatry* 61(1):42–50.

Kaelin, D.L., Cifu, D.X., and B. Matthies. 1996. Methylphenidate effect on attention deficit in the acutely brain-injured adult. *Arch Phys Med Rehabil* 77(1):6–9.

Kahlig, K.M., and A. Galli. 2003. Regulation of dopamine transporter function and plasma membrane expression by dopamine, amphetamine, and cocaine. *Eur J Pharmacol* 479(1–3):153–8.

Kang, S.H., and D.K. Kim. 2013. Drug induced Parkinsonism caused by the concurrent use of donepezil and risperidone in a patient with traumatic brain injuries. *Ann Rehabil Med* 37(1):147–50.

Kelso, M.L., Scheff, S.W., Pauly, J.R., and C.D. Loftin. 2009. Effects of genetic deficiency of cyclooxygenase-1 or cyclooxygenase-2 on functional and histological outcomes following traumatic brain injury in mice. *BMC Neurosci* 10:108.

Kihara, T., Shimohama, S., Sawada, H., Honda, K., Nakamizo, T., Kanki, R. et al. 2002. Protective effect of dopamine D2 agonists in cortical neurons via the phosphatidylinositol 3 kinase cascade. *J Neurosci Res* 70(3):274–82.

Kim, Y.H., Ko, M.H., Na, S.Y., Park, S.H., and K.W. Kim. 2006. Effects of single-dose methylphenidate on cognitive performance in patients with traumatic brain injury: A double-blind placebo-controlled study. *Clin Rehabil* 20(1):24–30.

Kline, A.E, Chen, M.J Tso-Olivas, D.Y and D.M Feeney. 1994. Methylphenidate treatment following ablation-induced hemiplegia in rat: Experience during drug action alters effects on recovery of function. *Pharmacol Biochem Behav* 48(3):773–9.

Kline, A.E., Yan, H.Q., Bao, J., Marion, D.W., and C.E. Dixon. 2000. Chronic methylphenidate treatment enhances water maze performance following traumatic brain injury in rats. *Neurosci Lett* 280(3):163–6.

Kline, A.E., Massucci, J.L., Ma, X., Zafonte, R.D., and C.E. Dixon. 2004. Bromocriptine reduces lipid peroxidation and enhances spatial learning and hippocampal neuron survival in a rodent model of focal brain trauma. *J Neurotrauma* 21(12):1712–22.

Kline, A.E., Massucci, J.L., Marion, D.W., and C.E. Dixon. 2002. Attenuation of working memory and spatial acquisition deficits after a delayed and chronic bromocriptine treatment regimen in rats subjected to traumatic brain injury by controlled cortical impact. *J Neurotrauma* 19(4):415–25.

Kline, A.E., Massucci, J.L., Zafonte, R.D., Dixon, C.E., DeFeo, J.R., and E.H. Rogers. 2007. Differential effects of single versus multiple administrations of haloperidol and risperidone on functional outcome after experimental brain trauma. *Crit Care Med* 35(3):919–24.

Kline, A.E., Wagner, A.K., Westergom, B.P., Malena, R.R, Zafonte, R.D., Olsen, A.S. et al. 2007. Acute treatment with the 5-HT(1A) receptor agonist 8-OH-DPAT and chronic environmental enrichment confer neurobehavioral benefit after experimental brain trauma. *Behav Brain Res* 177(2):186–94.

Kobori, N., Clifton, G.L., and P.K. Dash. 2006. Enhanced catecholamine synthesis in the prefrontal cortex after traumatic brain injury: Implications for prefrontal dysfunction. *J Neurotrauma* 23(7):1094–102.

Kochanek, P.M. 1993. Ischemic and traumatic brain injury: Pathobiology and cellular mechanisms. *Crit Care Med* 21(9 Suppl):S333–5.

Kraus, M.F., Smith, G.S., Butters, M., Donnell, A.J., Dixon, E., Yilong, C. et al. 2005. Effects of the dopaminergic agent and NMDA receptor antagonist amantadine on cognitive function, cerebral glucose metabolism and D2 receptor availability in chronic traumatic brain injury: A study using positron emission tomography (PET). *Brain Inj* 19(7):471–9.

Kriz, J. 2006. Inflammation in ischemic brain injury: Timing is important. *Crit Rev Neurobiol* 18(1–2):145–57.

Kumar, A., Zou, L., Yuan, X., Long, Y., and K. Yang. 2002. N-methyl-D-aspartate receptors: Transient loss of NR1/NR2A/NR2B subunits after traumatic brain injury in a rodent model. *J Neurosci Res* 67(6):781–6.

Kurz, J.E., Hamm, R.J., Singleton, R.H., Povlishock, J.T., and S.B. Churn. 2005a. A persistent change in subcellular distribution of calcineurin following fluid percussion injury in the rat. *Brain Res* 1048(1–2):153–60.

Kurz, J.E., Parsons, J.T., Rana, A., Gibson C.J., Hamm, R.J., and S.B. Churn. 2005b. A significant increase in both basal and maximal calcineurin activity following fluid percussion injury in the rat. *J Neurotrauma* 22(4):476–90.

Ladanyi, S., and D. Elliott. 2008. Traumatic brain injury: An integrated clinical case presentation and literature review. Part I: Assessment and initial management. *Aust Crit Care* 21(2):86–95.

Langfitt, T.W., Obrist, W.D., Alavi, A., Grossman, R.I., Zimmerman, R., J Jaggi et al. 1986. Computerized tomography, magnetic resonance imaging, and positron emission tomography in the study of brain trauma. Preliminary observations. *J Neurosurg* 64(5):760–7.

Lawrence, A.D., Weeks, R.A., Brooks, D.J., Andrews, T.C., Watkins, L.H., Harding, A.E. et al. 1998. The relationship between striatal dopamine receptor binding and cognitive performance in Huntington's disease. *Brain* 121(Pt 7):1343–55.

Lee, H., Kim, S.W., Kim, J.M., Shin, I.S., Yang, S.J., and J.S. Yoon. 2005. Comparing effects of methylphenidate, sertraline and placebo on neuropsychiatric sequelae in patients with traumatic brain injury. *Hum Phychopharmacol* 20(2):97–104.

Lemon, N., and D. Manahan-Vaughan. 2006. Dopamine D1/D5 receptors gate the acquisition of novel information through hippocampal long-term potentiation and long-term depression. *J Neurosci* 26(29):7723–9.

Levin, H.S. 1990. Memory deficit after closed head injury. *J Clin Exp Neuropsychol* 12(1):129–53.

Levin, H.S., Gary, H.E., Eisenberg, H.M., Ruff, R.M., Barth, J.T., Kreutzer, J. et al. 1990. Neurobehavioral outcome 1 year after severe head injury. experience of the traumatic coma data bank. *J Neurosurg* 73(5):699–709.

Levin, H.S., Goldstein, F.C., High, W.M., and H.M. Eisenberg. 1988a. Disproportionately severe memory deficit in relation to normal intellectual functioning after closed head injury. *J Neurol Neurosurg Psychiatry* 51(10):1294–301.

Levin, H.S., Goldstein, F.C., High, W.M., and D. Williams. 1988b. Automatic and effortful processing after severe closed head injury. *Brain Cogn* 7(3):283–97.

Li, S., Cullen, W.K., Anwyl, R., and M.J. Rowan. 2003. Dopamine-dependent facilitation of LTP induction in hippocampal CA1 by exposure to spatial novelty. *Nat Neurosci* 6(5):526–31.

Lifshitz, J., Kelley, B.J., and J.T. Povlishock. 2007. Perisomatic thalamic axotomy after diffuse traumatic brain injury is associated with atrophy rather than cell death. *J Neuropathol Exp Neurol* 66(3):218–29.

Lighthall, J.W. 1988. Controlled cortical impact: A new experimental brain injury model. *J Neurotrauma* 5(1):1–15.

Lighthall, J.W., Dixon, C.E., and T.E. Anderson. 1989. Experimental models of brain injury. *J Neurotrauma* 6(2):83–97.

M'Harzi, M., Willig, F., Costa, J.C., and J. Delacour. 1988. d-Amphetamine enhances memory performance in rats with damage to the fimbria. *Physiol Behav* 42(6):575–9.

Mandel, S., Weinreb, O., Amit, T., and M.B.Youdim. 2005. Mechanism of neuroprotective action of the anti-Parkinson drug rasagiline and its derivatives. *Brain Res Brain Res Rev* 48(2):379–87.

Marié, R.M., Barré, L., Dupuy, B., Viader, F., Defer, G., and J.C. Baron. 1999. Relationships between striatal dopamine denervation and frontal executive tests in Parkinson's disease. *Neurosci Lett* 260(2):77–80.

Massucci, J.L., Kline, A.E., Ma, X., Zafonte, R.D., and C.E. Dixon. 2004. Time dependent alterations in dopamine tissue levels and metabolism after experimental traumatic brain injury in rats. *Neurosci Lett* 372(1–2):127–31.

McAllister, T.W., Flashman, L.A., McDonald, B.C., Ferrell, R.B., Tosteson, T.D., Yanofsky, N.N. et al. 2011. Dopaminergic challenge with bromocriptine one month after mild traumatic brain injury: Altered working memory and BOLD response. *J Neuropsychiatry Clin Neurosci* 23(3):277–86.

McDonald, B.C., Flashman, L.A., and A.J. Saykin. 2002. Executive dysfunction following traumatic brain injury: Neural substrates and treatment strategies. *NeuroRehabilitation* 17(4):333–44.

McDowell, S., Whyte, J., and M. D'Esposito. 1997. Working memory impairments in traumatic brain injury: Evidence from a dual-task paradigm. *Neuropsychologia* 35(10):1341–53.

McDowell, S., Whyte, J., and M. D'Esposito. 1998. Differential effect of a dopaminergic agonist on prefrontal function in traumatic brain injury patients. *Brain* 121(Pt 6):1155–64.

McIntosh, T.K., Yu, T., and T.A. Gennarelli. 1994. Alterations in regional brain catecholamine concentrations after experimental brain injury in the rat. *J Neurochem* 63(4):1426–33.

McMillan, T.M. 1997. Minor head injury. *Curr Opin Neurol* 10(6):479–83.

Mega, M.S., and J.L. Cummings. 1994. Frontal-subcortical circuits and neuropsychiatric disorders. *J Neuropsychiatry Clin Neurosci* 6(4):358–70.

Meredith, G.E., Wouterlood, F.G., and A. Pattiselanno. 1990. Hippocampal fibers make synaptic contacts with glutamate decarboxylase-immunoreactive neurons in the rat nucleus accumbens. *Brain Res* 513(2):329–34.

Meythaler, J.M., Peduzzi, J.D., Eleftheriou, E., and T.A. Novack. 2001. Current concepts: Diffuse axonal injury-associated traumatic brain injury. *Arch Phys Med Rehabil* 82(10):1461–71.

Millis, S.R., Rosenthal, M., Novack, T.A., Sherer, M., Nick, T.G., Kreutzer, J.S. et al. 2001. Long-term neuropsychological outcome after traumatic brain injury. *J Head Trauma Rehabil* 16(4):343–55.

Mitchell, J.B., and A. Gratton. 1994. Involvement of mesolimbic dopamine neurons in sexual behaviors: Implications for the neurobiology of motivation. *Rev Neurosci* 5(4):317–29.

Moldover, J.E., Goldberg, K.B., and M.F. Prout. 2004. Depression after traumatic brain injury: A review of evidence for clinical heterogeneity. *Neuropsychol Rev* 14(3):143–54.

Mooney, G.F., and L.J. Haas. 1993. Effect of methylphenidate on brain injury-related anger. *Arch Phys Med Rehabil* 74(2):153–60.

Morrow, B.A., Roth, R.H., and J.D. Elsworth. 2000. TMT, a predator odor, elevates mesoprefrontal dopamine metabolic activity and disrupts short-term working memory in the rat. *Brain Res Bull* 52(6):519–23.

Muir, K.W. 2006. Glutamate-based therapeutic approaches: Clinical trials with NMDA antagonists. *Curr Opin Phamacol* 6(1):53–60.

Mura, A., and J. Feldon. 2003. Spatial learning in rats is impaired after degeneration of the nigrostriatal dopaminergic system. *Mov Disord* 18(8): 860–71.

Nair, V.D., and S.C. Sealfon. 2003. Agonist-specific transactivation of phosphoinositide 3-kinase signaling pathway mediated by the dopamine D2 receptor. *J Biol Chem* 278(47):47053–61.

Nelson, L.A., Macdonald, M., Stall, C., and R. Pazdan. 2013. Effects of interactive metronome therapy on cognitive functioning after blast-related brain injury: A randomized controlled pilot trial. *Neuropsychology* 27(6):666–79.

Newburn, G., and D. Newburn. 2005. Selegiline in the management of apathy following traumatic brain injury. *Brain Inj* 19(2):149–54.

NIH. 1998. Rehabilitation of persons with traumatic brain injury. *NIH Consens Statement* 16(1):1–41.

Nishi, A., Bibb, J.A., Matsuyama, S., Hamada, M., Higashi, H., Nairn, A.C. et al. 2002. Regulation of DARPP-32 dephosphorylation at PKA- and Cdk5-sites by NMDA and AMPA receptors: Distinct roles of calcineurin and protein phosphatase-2A. *J Neurochem* 81(4):832–41.

O'Carroll, C.M., Martin, S.J., Sandin, J., Frenguelli, B., and R.G. Morris. 2006. Dopaminergic modulation of the persistence of one-trial hippocampus-dependent memory. *Learn Mem* 13(6):760–9..

Oddy, M., Coughlan, T., Tyerman, A., and D. Jenkins. 1985. Social adjustment after closed head injury: A further follow-up seven years after injury. *J Neurol Neurosurg Psychiatry* 48(6): 564–8.

Okada, Y., Sakai, H., Kohiki, E., Suga, E., Yanagisawa, Y., Tanaka, K. et al. 2005. A dopamine D4 receptor antagonist attenuates ischemia-induced neuronal cell damage via upregulation of neuronal apoptosis inhibitory protein. *J Cereb Blood Flow Metab* 25(7):794–806.

Olney, J.W., Zorumski, C.F., Stewart, G.R., Price, M.T., Wang, G.J., and J. Labruyere. 1990. Excitotoxicity of L-dopa and 6-OH-dopa: Implications for Parkinson's and Huntington's diseases. *Exp Neurol* 108(3):269–72.

Onali, P., Olianas, M.C., and G.L. Gessa. 1985. Characterization of dopamine receptors mediating inhibition of adenylate cyclase activity in rat striatum. *Mol Pharmacol* 28(2):138–45.

Onn, S.P., Berger, T.W. Stricker, E.M. and M.J. Zigmond. 1986. Effects of intraventricular 6-hydroxydopamine on the dopaminergic innervation of striatum: Histochemical and neurochemical analysis. *Brain Res* 376(1)):8–19.

Palmer, A.M., Marion, D.W., Botscheller, M.L., Swedlow, P.E., Styren, S.D., and S.T. DeKosky. 1993. Traumatic brain injury-induced excitotoxicity assessed in a controlled cortical impact model. *J Neurochem* 61(6):2015–24.

Park, C., Cho, I.H., Kim, D., Jo, E.K., Choi, S.Y., Oh, S.B. et al. 2008. Toll-like receptor 2 contributes to glial cell activation and heme oxygenase-1 expression in traumatic brain injury. *Neurosci Lett* 431(2):123–8.

Patrick, P.D., Blackman, J.A., Mabry, J.L., Buck, M.L., Gurka, M.J., and M.R. Conaway. 2006. Dopamine agonist therapy in low-response children following traumatic brain injury. *J Child Neurol* 21(10):879–85.

Paul, S., Nairn, A.C., Wang, P., and P.J. Lombroso. 2003. NMDA-mediated activation of the tyrosine phosphatase STEP regulates the duration of ERK signaling. *Nat Neurosci* 6(1):34–42.

Pavlovskaya, M., Hochstein, S., Keren, O., Mordvinov, E., and Z. Groswasser. 2007. Methylphenidate effect on hemispheric attentional imbalance in patients with traumatic brain injury: A psychophysical study. *Brain Inj* 21(5):489–97.

Pennartz, C.M., and S.T. Kitai. 1991. Hippocampal inputs to identified neurons in an in vitro slice preparation of the rat nucleus accumbens: Evidence for feed-forward inhibition. *J Neurosci* 11(9):2838–47.

Plenger, P.M., Dixon, C.E., Castillo, R.M., Frankowski, R.F., Yablon, S.A., and H.S. Levin. 1996. Subacute methylphenidate treatment for moderate to moderately severe traumatic brain injury: A preliminary double-blind placebo-controlled study. *Arch Phys Med Rehabil* 77(6):536–40.

Ploeger, G.E., Spruijt, B.M., and A.R. Cools. 1994. Spatial localization in the Morris water maze in rats: Acquisition is affected by intra-accumbens injections of the dopaminergic antagonist haloperidol. *Behav Neurosci* 108(5):927–34.

Ploeger, G.E., Willemen, A.P., and A.R. Cools. 1991. Role of the nucleus accumbens in social memory in rats. *Brain Res Bull* 26(1):23–7.

Queen, S.A., Chen, M.J., and D.M. Feeney. 1997. d-Amphetamine attenuates decreased cerebral glucose utilization after unilateral sensorimotor cortex contusion in rats. *Brain Res* 777(1–2):42–50.

Ramic, M., Emerick, A.J., Bollnow, M.R., O'Brien, T.E., Tsai, S.Y., and G.L. Kartje. 2006. Axonal plasticity is associated with motor recovery following amphetamine treatment combined with rehabilitation after brain injury in the adult rat. *Brain Res* 1111(1):176–86.

Rao, V.L., Dogan, A., Bowen, K.K., and R.J. Dempsey. 1999. Traumatic injury to rat brain upregulates neuronal nitric oxide synthase expression and L-[3H]nitroarginine binding." *J Neurotrauma* 16(10):865–77.

Ratcliff, J.J., Greenspan, A.I., Goldstein, F.C., Stringer, A.Y., Bushnik, T., Hammond, F.M. et al. 2007. Gender and traumatic brain injury: Do the sexes fare differently? *Brain Inj* 21(10):1023–30.

Rau, T.F., Kothiwal, A.S., Rova, A.R., Brooks, D.M., and D.J. Poulsen. 2012. Treatment with low-dose methamphetamine improves behavioral and cognitive function after severe traumatic brain injury. *J Trauma Acute Care Surg* 73(2 Suppl 1):S165–72.

Redell, J.B., and P.K. Dash. 2007. Traumatic brain injury stimulates hippocampal catechol-O-methyl transferase expression in microglia. *Neurosci Lett* 13(1):36–41.

Reeves, T.M., Phillips, L.L., Lee, N.N., and J.T. Povlishock. 2007. Preferential neuroprotective effect of tacrolimus (FK506) on unmyelinated axons following traumatic brain injury. *Brain Res* 1154:225–36.

Ricker, J.H., Hillary, F.G., and J. DeLuca. 2001. Functionally activated brain imaging (O-15 PET and fMRI) in the study of learning and memory after traumatic brain injury. *J Head Trauma Rehabil* 16(2):191–205.

Ridley, R.M., Cummings, R.M., Leow-Dyke, A., and H.F. Baker. 2006. Neglect of memory after dopaminergic lesions in monkeys. *Behav Brain Res* 166(2):253–62.

Salamone, J.D. 1994. The involvement of nucleus accumbens dopamine in appetitive and aversive motivation. *Behav Brain Res* 61(2):117–33.

Salamone, J.D., Correa, M., Mingote, S.M., and S.M. Weber. 2005. Beyond the reward hypothesis: Alternative functions of nucleus accumbens dopamine. *Curr Opin Phamacol* 5(1):34–41.

Sánchez-Carrión, R., Gómez, P.V., Junqué, C., Fernández-Espejo, D., Falcon, C., Bargalló, N. et al. 2008. Frontal hypoactivation on functional magnetic resonance imaging in working memory after severe diffuse traumatic brain injury. *J Neurotrauma* 25(5):479–94.

Sanders, M.J., Dietrich, W.D., and E.J. Green. 2001. Behavioral, electrophysiological, and histopathological consequences of mild fluid-percussion injury in the rat. *Brain Res* 904(1):141–4.

Saniova, B., Drobny, M., Kneslova, L., and M. Minarik. 2004. The outcome of patients with severe head injuries treated with amantadine sulphate. *J Neural Transm* 111(4):511–4.

Schapira, A.H. 2009. Neuroprotection in Parkinson's disease. *Parkinsonism Relat Disord* 15 Suppl 4:S41–3.

Schiffer, W.K., Volkow, N.D., Fowler, J.S., Alexoff, D.L., Logan, J., and S.L. Dewey. 2006. Therapeutic doses of amphetamine or methylphenidate differentially increase synaptic and extracellular dopamine. *Synapse* 59(4):243–51.

Schneider, J.S., and D.P. Roeltgen. 1993. Delayed matching-to-sample, object retrieval, and discrimination reversal deficits in chronic low dose MPTP-treated monkeys. *Brain Res* 615(2):351–4.

Schultz, W. 2004. Neural coding of basic reward terms of animal learning theory, game theory, microeconomics and behavioural ecology. *Curr Opin Neurobiol* 14(2):139–47.

Seel, R.T., Kreutzer, J.S., Rosenthal, M., Hammond, F.M., Corrigan, J.D., and K. Black. 2003. Depression after traumatic brain injury: A National Institute on Disability and Rehabilitation Research Model Systems multicenter investigation. *Arch Phys Med Rehabil* 84(2):177–84.

Seeman, P., Tedesco, J.L., Lee, T., Chau-Wong, M., Muller, P., Bowles, J. et al. 1978. Dopamine receptors in the central nervous system. *Fed Proc* 37(2):131–6.

Setlow, B., and J.L. McGaugh. 1998. Sulpiride infused into the nucleus accumbens posttraining impairs memory of spatial water maze training. *Behav Neurosci* 112(3):603–10.

Shen, H., Harvey, B.K., Chiang, Y.H., Pick, C.G., and Y. Wang. 2011. Methamphetamine potentiates behavioral and electrochemical responses after mild traumatic brain injury in mice. *Brain Res* 1368:248–53.

Shin, S.S., Bales, J.W., Yan, H.Q., Kline, A.K., Wagner, A.K., Lyons-Weiler, J., and C.E. Dixon. 2013. The effect of environmental enrichment on substantia nigra gene expression after traumatic brain injury in rats. *J Neurotrauma* 30(4):259–70.

Shin, S.S., Bray, E.R., Zhang, C.Q., and C.E. Dixon. 2011. Traumatic brain injury reduces striatal tyrosine hydroxylase activity and potassium-evoked dopamine release in rats. *Brain Res* 1369:208–15.

Shin, S.S., and C.E. Dixon. 2011. Oral fish oil restores striatal dopamine release after traumatic brain injury. *Neurosci Lett* 496 (3): 168–71.

Sinet, P.M., Heikkila, R.E., and G. Cohen. 1980. Hydrogen peroxide production by rat brain in vivo. *J Neurochem* 34(6):1421–8.

Slewa-Younan, S., Green, A.M., Baguley, I.J., Gurka, J.A., and J.E. Marosszeky. 2004. Sex differences in injury severity and outcome measures after traumatic brain injury. *Arch Phys Med Rehabil* 85(3):376–9.

Smith, D.H., Meaney, D.F., and W.H. Shull, 2003. Diffuse axonal injury in head trauma. *J Head Trauma Rehabil* 18(4):307–16.

Snyder, G.L., Fienberg, A.A., Huganir, R.L., and P. Greengard. 1998. A dopamine/D1 receptor/protein kinase A/dopamine- and cAMP-regulated phosphoprotein (Mr 32 kDa)/protein phosphatase-1 pathway regulates dephosphorylation of the NMDA receptor. *J Neurosci* 18(24):10297–303.

So, C.H., Verma, V., Alijaniaram, M., Cheng, R., Rashid, A.J., O'Dowd, B.F., and S.R. George. 2009. Calcium signaling by dopamine D5 receptor and D5-D2 receptor hetero-oligomers occurs by a mechanism distinct from that for dopamine D1-D2 receptor hetero-oligomers. *Mol Pharmacol* 75(4):843–54.

Solanto, M.V. 1998. Neuropsychopharmacological mechanisms of stimulant drug action in attention-deficit hyperactivity disorder: A review and integration. *Behav Brain Res* 94(1):127–52.

Sonuga-Barke, E.J. 2005. Causal models of attention-deficit/hyperactivity disorder: From common simple deficits to multiple developmental pathways. *Biol Psyciatry* 57(11):1231–8.

Speech, T.J., Rao, S.M., Osmon, D.C., and L.T. Sperry. 1993. A double-blind controlled study of methylphenidate treatment in closed head injury. *Brain Inj* 7(4):333–8.

Stanislav, S.W. 1997. Cognitive effects of antipsychotic agents in persons with traumatic brain injury. *Brain Inj.* 11, 335–42.

Sun, Z.L., and D.F. Feng. 2014. Biomarkers of cognitive dysfunction in traumatic brain injury. *J Neural Transm* 121(1):79–90.

Suominen, T., Uutela, P., Ketola, R.A., Bergquist, J., Hillered, L., Finel, M. et al. 2013. Determination of serotonin and dopamine metabolites in human brain microdialysis and cerebrospinal fluid samples by UPLC-MS/MS: Discovery of intact glucuronide and sulfate conjugates. *PloS One* 8(6):e68007.

Sutton, M.A., and R.J. Beninger. 1999. Psychopharmacology of conditioned reward: Evidence for a rewarding signal at D1-like dopamine receptors. *Psychopharmacology* 144(2):95–110.

Tamaru, F. 1997. Disturbances in higher function in Parkinson's disease. *Eur Neurol* 38 Suppl 2:33–6.

Tanaka, S. 2006. Dopaminergic control of working memory and its relevance to schizophrenia: A circuit dynamics perspective. *Neuroscience* 139(1):153–71.

Tang, Y.P., Noda, Y., and T. Nabeshima. 1997. Involvement of activation of dopaminergic neuronal system in learning and memory deficits associated with experimental mild traumatic brain injury. *Eur J Neurosci* 9(8):1720–7.

Tatton, W.G., and C.E. Greenwood. 1991. Rescue of dying neurons: A new action for deprenyl in MPTP Parkinsonism. *J Neurosci Res* 30(4):666–72.

Tatton, W.G., Ju, W.Y., Holland, D.P., Tai, C., and M. Kwan. 1994. (-)-Deprenyl reduces PC12 cell apoptosis by inducing new protein synthesis. *J Neurochem* 63(4):1572–5.

Tauscher, J., Hussain, T., Agid, O., Verhoeff, N.P., Wilson, A.A., Houle, S. et al. 2004. Equivalent occupancy of dopamine D1 and D2 receptors with clozapine: Differentiation from other atypical antipsychotics. *Am J Psychiatry* 161(9):1620–5.

Thurman, D.J., Alverson, C., Dunn, K.A., Guerrero, J., and J.E. Sniezek. 1999. Traumatic brain injury in the United States: A public health perspective. *J Head Trauma Rehabil* 14(6):602–15.

Tiberti, C., Sabe, L., Jason, L., Leiguarda, R., and S. Starkstein. 1998. A randomized, double-blind, placebo-controlled study of methylphenidate in patients with organic amnesia. *Eur J Neurol* 5(3):297–99.

Tidey, J.W., and K.A. Miczek. 1996. Social defeat stress selectively alters mesocorticolimbic dopamine release: An in vivo microdialysis study. *Brain Res* 721(1–2):140–9.

Tolias, C.M., and M.R. Bullock. 2004. Critical appraisal of neuroprotection trials in head injury: What have we learned? *NeuroRx* 1(1):71–9.

van Bregt, D.R., Thomas, T.C., Hinzman, J.M., Cao, T., Liu, M., Bing, G. et al. 2012. Substantia nigra vulnerability after a single moderate diffuse brain injury in the rat. *Exp Neurol* 234(1):8–19.

Verheij, M.M., and A.R. Cools. 2008. Twenty years of dopamine research: Individual differences in the response of accumbal dopamine to environmental and pharmacological challenges. *Eur J Pharmacol* 585(2–3):228–44.

Vertes, R.P. 2006. Interactions among the medial prefrontal cortex, hippocampus and midline thalamus in emotional and cognitive processing in the rat. *Neuroscience* 142(1):1–20.

Viggiano, D., Vallone, D., Ruocco, L.A., and A.G. Sadile. 2003. Behavioural, pharmacological, morpho-functional molecular studies reveal a hyperfunctioning mesocortical dopamine system in an animal model of attention deficit and hyperactivity disorder. *Neurosci Biobehav Rev* 27(7):683–9.

Volkow, N.D., Fowler, J.S., Wang, G., Ding, Y., and S.J. Gatley. 2002a. Mechanism of action of methylphenidate: Insights from PET imaging studies. *J Atten Disord* 6 Suppl 1:S31–43.

Volkow, N.D., Wang, G.J., Fowler, J.S., Gatley, S.J., Logan, J., Ding, Y.S. et al. 1998. Dopamine transporter occupancies in the human brain induced by therapeutic doses of oral methylphenidate. *Am J Psychiatry* 155(10):1325–31.

Volkow, N.D., Wang, G.J., Fowler, J.S., Logan, J., Franceschi, D., Maynard, L. et al. 2002b. Relationship between blockade of dopamine transporters by oral methylphenidate and the increases in extracellular dopamine: Therapeutic implications. *Synapse* 43(3):181–7.

Volkow, N.D., Wang, G.J., Fowler, J.S., Telang, F., Maynard, L., Logan, J. et al. 2004. Evidence that methylphenidate enhances the saliency of a mathematical task by increasing dopamine in the human brain. *Am J Psychiatry* 161(7):1173–80.

Volz, T.J. 2008. Neuropharmacological mechanisms underlying the neuroprotective effects of methylphenidate. *Curr Neuropharmacol* 6(4):379–85.

Von Voigtlander, P.F., and K.E. Moore. 1971. Dopamine: Release from the brain in vivo by amantadine. *Science* 174(4007):408–10.

Wagner, A.K., Ren, D., Conley, Y.P., Ma, X., Kerr, M.E., Zafonte, R.D. et al. 2007. Sex and genetic associations with cerebrospinal fluid dopamine and metabolite production after severe traumatic brain injury. *J Neurosurg* 106(4):538–47.

Wagner, A.K., Chen, X., Kline, A.E., Li, Y., Zafonte, R.D., and C.E. Dixon. 2005. Gender and environmental enrichment impact dopamine transporter expression after experimental traumatic brain injury. *Exp Neurol* 195(2):475–83.

Wagner, A.K., Sokoloski, J.E., Ren, D., Chen, X., Khan, A.S., Zafonte, R.D. et al. 2005. Controlled cortical impact injury affects dopaminergic transmission in the rat striatum. *J Neurochem* 95(2):457–65.

Wagner, A.K., Brayer, S.W., Hurwitz, M., Niyonkuru, C., Zou, H., Failla, M. et al. 2013. Non-spatial pre-training in the water maze as a clinically relevant model for evaluating learning and memory in experimental TBI. *Neurobiol Learn Mem* 106:71–86.

Wagner, A.K., Drewencki, L.L., Chen, X., Santos, F.R., Khan, A.S., Harun, R. et al. 2009. Chronic methylphenidate treatment enhances striatal dopamine neurotransmission after experimental traumatic brain injury. *J Neurochem* 108(4):986–97.

Wagner, A.K., Kline, A.E., Ren, D., Willard, L.A., Wenger, M.K., Zafonte, R.D. et al. 2007. Gender associations with chronic methylphenidate treatment and behavioral performance following experimental traumatic brain injury. *Behav Brain Res* 181(2):200–9.

Wagner, A.K., Willard, L.A., Kline, A.E., Wenger, M.K., Bolinger, B.D., Ren, D. et al. 2004. Evaluation of estrous cycle stage and gender on behavioral outcome after experimental traumatic brain injury. *Brain Res* 998(1):113–21.

Wagner, A.K., Scanlon, J.M., Niyonkuru, C., Dixon, C.E., Conley, Y.P., Becker, C., and Price, J. 2014 (Epub). Genetic variation in the dopamine transporter gene and the D2 receptor gene influences striatal DAT binding in adults with severe TBI. *JCBFM* 34(8):1328–39.

Wang, S., Cheng, H., Dai, G., Wang, X., Hua, R., Liu, X. et al. 2013. Umbilical cord mesenchymal stem cell transplantation significantly improves neurological function in patients with sequelae of traumatic brain injury. *Brain Res* 1532:76–84.

Wang, T., Huang, X.J., Van, K.C., Went, G.T., Nguyen, J.T., and B.G. Lyeth. 2013. Amantadine improves cognitive outcome and increases neuronal survival after fluid percussion traumatic brain injury in rats. *J Neurotrauma* 31:370–377.

Warden, D.L., Gordon, B., McAllister, T.W., Silver, J.M., Barth, J.T., Bruns, J. et al. [Neurobehavioral Guidelines Working Group]. 2006. Guidelines for the pharmacologic treatment of neurobehavioral sequelae of traumatic brain injury. *J Neurotrauma* 23(10):1468–501.

Weiss, S., Sebben, M., Garcia-Sainz, J.A., and J. Bockaert. 1985. D2-dopamine receptor-mediated inhibition of cyclic AMP formation in striatal neurons in primary culture. *Mol Pharmacol* 27(6):595–9.

Whalen, M.J., Carlos, T.M., Kochanek, P.M., and S. Heineman. 1998. Blood-brain barrier permeability, neutrophil accumulation and vascular adhesion molecule expression after controlled cortical impact in rats: A preliminary study. *Acta Neurochir Suppl* 71:212–4.

Whyte, J., Hart, T., Vaccaro, M., Grieb-Neff, P., Risser, A., Polansky, M. et al. 2004. Effects of methylphenidate on attention deficits after traumatic brain injury: A multidimensional, randomized, controlled trial. *Am J Phys Med Rehabil* 83(6):401–20.

Wickens, J.R., Budd, C.S., Hyland, B.I., and G.W. Arbuthnott. 2007. Striatal contributions to reward and decision making: Making sense of regional variations in a reiterated processing matrix. *Ann NY Acad Sci* 1104:192–212.

Williams, G.V., and S.A. Castner. 2006. Under the curve: Critical issues for elucidating D1 receptor function in working memory. *Neuroscience* 139(1):263–76.

Wilson, M.S., Gibson, C.J., and R.J. Hamm. 2003. Haloperidol, but not olanzapine, impairs cognitive performance after traumatic brain injury in rats. *Am J Phys Med Rehabil* 82(11):871–9.

Wise, S.P., Murray, E.A., and C.R. Gerfen. 1996. The frontal cortex-basal ganglia system in primates. *Crit Rev Neurobiol* 10(3–4): 317–56.

Writer, B.W., and J.E. Schillerstrom. 2009. Psychopharmacological treatment for cognitive impairment in survivors of traumatic brain injury: A critical review. *J Neuropsychiatry Clin Neurosci* 21(4):362–70.

Wu, R.M., Chiueh, C.C., Pert, A., and D.L. Murphy. 1993. Apparent antioxidant effect of l-deprenyl on hydroxyl radical formation and nigral injury elicited by MPP+ in vivo. *Eur J Pharmacol* 243(3):241–7.

Xiong, Y., Mahmood, A., and M. Chopp. 2013. Animal models of traumatic brain injury. *Nat Rev Neurosci* 14(2):128–42.

Yamamoto, Y., Tanaka, T., Shibata, S., and S. Watanabe. 1994. Involvement of D1 dopamine receptor mechanism in ischemia-induced impairment of CA1 presynaptic fiber spikes in rat hippocampal slices. *Brain Res* 665(1):151–4.

Yang, Z.J., Torbey, M., Li, X., Bernardy, J., Golden, W.C., Martin, L.J. et al. 2007. Dopamine receptor modulation of hypoxic-ischemic neuronal injury in striatum of newborn piglets. *J Cereb Blood Flow Metab* 27(7):1339–51.

Zafonte, R.D., Bagiella, E., Ansel, B.M., Novack, T.A., Friedewald, W.T., Hesdorffer, D.C. et al. 2012. Effect of citicoline on functional and cognitive status among patients with traumatic brain injury: Citicoline Brain Injury Treatment Trial (COBRIT). *JAMA* 308(19):1993–2000.

Zhu, J., Hamm, R.J., Reeves, T.M., Povlishock, J.T., and L.L. Phillips. 2000. Postinjury administration of L-deprenyl improves cognitive function and enhances neuroplasticity after traumatic brain injury. *Exp Neurol* 166(1):136–52.

Zhuang, X., Oosting, R.S., Jones, S.R., Gainetdinov, R.R., Miller, G.W., Caron, M.G. et al. 2001. Hyperactivity and impaired response habituation in hyperdopaminergic mice. *Proc Natl Acad Sci USA* 98(4):1982–7.

Zou, S., Li, L., Pei, L., Vukusic, B., Van Tol, H.H., Lee, F.J. et al. 2005. Protein-protein coupling/uncoupling enables dopamine D2 receptor regulation of AMPA receptor-mediated excitotoxicity. *J Neurosci* 25(17):4385–95.

Zweckberger, K., Simunovic, F., Kiening, K.L., Unterberg, A.W., and O.W. Sakowitz. 2010. Anticonvulsive effects of the dopamine agonist lisuride maleate after experimental traumatic brain injury. *Neurosci Lett* 470(2):150–4.

Section VI

Neurorehabilitation and Neuroprotection

34 Rehabilitative Paradigms after Experimental Brain Injury

Relevance to Human Neurotrauma

Corina O. Bondi, Roya Tehranian-DePasquale, Jeffrey P. Cheng, Christina M. Monaco, Grace S. Griesbach, and Anthony E. Kline

CONTENTS

34.1 INTRODUCTION

The goal of this chapter is to describe four relatively noninvasive rehabilitative paradigms that may have clinical relevance following brain injury. Specifically, the benefits and limitations of environmental enrichment, exercise, low-level laser therapy, and constraint-induced movement therapy will be discussed. Timing issues (e.g., best time to initiate treatment as well as duration of treatment) and the advantage of adjunct therapies (i.e., can they further improve functional outcome) will also be discussed. Overall, the literature suggests that each of the aforementioned therapies confer significant behavioral improvement after experimental brain trauma. Hence, we propose that they should be considered for implementation in clinical rehabilitation.

34.1.1 Statement of the Problem

More than 10 million people worldwide sustain a traumatic brain injury (TBI) each year. Of those, approximately 2 million reside in the United States and suffer long-term and often permanent neurological dysfunction (Bales et al., 2009; Centers for Disease Control and Prevention [CDC] National Center for Injury Prevention and Control, 2003; Faul, 2010; Garcia et al., 2011; Laker, 2011; Moore et al., 2006; Selassie et al., 2008; Summers et al., 2009). In an effort to combat the deleterious effects of TBI, numerous preclinical therapeutic approaches, some of which are rather invasive, have been evaluated (Cheng et al., 2008; Clark et al., 1996; Clifton et al., 1991; Kline et al., 2002a, 2002b; Koizumi and Povlishock, 1998; McDermott et al., 1997; McDowell et al., 1998; Yan et al., 2000). Although several potential candidate therapies have emerged from the laboratory, their translation to the clinic has been relatively ineffective (Doppenberg et al., 2004; Menon, 2009), thus prompting the evaluation of noninvasive approaches that mimic clinical rehabilitation post-TBI, such as environmental enrichment (EE), exercise, low-level laser therapy (LLLT), and constraint-induced movement therapy (CIMT). These therapeutic manipulations can be effectively, effortlessly, and consistently applied to patients. Hence, the goal of this review is to describe the benefits and limitations of these rehabilitative paradigms while validating their relevance to human neurotrauma.

34.2 ENVIRONMENTAL ENRICHMENT

Environmental enrichment is a specialized environment that consists of a living situation that is significantly larger than standard (STD) laboratory cages and incorporates group housing as well as various toys (e.g., balls, blocks, tubes) and nesting materials (e.g., paper towels). The features of the EE promote socialization and ample opportunity for exploration and interaction of novel stimuli, which collectively provide physical and cognitive stimulation (Sozda et al., 2010). This paradigm induces neuroplasticity and confers significant behavioral performance in normal rodents (Bruel-Jungerman et al., 2005; Doulames et al., 2014; Veena et al., 2009). Furthermore, EE enhances learning and memory and improves motor performance after controlled cortical impact (CCI) injury and fluid percussion (FP) brain injury (Cheng et al., 2012; deWitt et al., 2011; Hamm et al., 1996; Hoffman et al., 2008; Kline et al., 2007, 2010, 2012; Matter et al., 2011; Passineau et al., 2001; Sozda et al., 2010), and therefore it is considered a reasonable rodent correlate of human neurorehabilitation (Cheng et al., 2012; deWitt et al., 2011; Kline et al., 2007, 2010, 2012; Matter et al., 2011; Sozda et al., 2010; Will et al., 2004). Despite also demonstrating improvements in models of Alzheimer's disease (Cracchiolo et al., 2007; Jankowsky et al., 2005), Parkinson's disease (Bezard et al., 2003; Faherty et al., 2005; Jadavji et al., 2006), spinal cord injury (Berrocal et al., 2007; Fischer and Peduzzi, 2007), and stroke (Buchhold et al., 2007; Nygren and Wieloch, 2005; Yu et al., 2013; Zai et al., 2011; Zhao et al., 2000), the focus of this brief EE review is on its efficacy after TBI, as are the subsequent reviews on exercise, LLLT, and CIMT.

34.2.1 EE as a Preclinical Model of Neurorehabilitation

Will and colleagues provided the first report that EE is protective and promotes cognitive benefits after brain injury in both neonatal and young adult rats (Will et al., 1976, 1977). Subsequent confirmations came from cortical ablation studies by Held and colleagues (1985), Gentile and coworkers (1987), and Rose et al. (1987). By demonstrating significant benefits of EE, the cortical ablation studies provided the impetus for evaluating the potential efficacy of EE in models of TBI, such as CCI and FP injury that have greater clinical relevance because they mimic many of the pathophysiological features of human TBI (Dixon et al., 1987, 1991; Kline and Dixon, 2001; Statler et al., 2001). Moreover, both CCI and FP injuries produce significant and long-lasting cognitive dysfunction (Bramlett and Dietrich, 2002; Dixon et al., 1999; Lindner et al., 1998; Pierce et al., 1998; Smith et al., 1997), which parallels clinical TBI (Binder, 1986; Levin et al., 1987).

Numerous studies have shown that enriched rats exhibit significant improvement in motor ability and spatial learning when compared to their nonenriched counterparts after CCI or FP injury (Hamm et al., 1996; Held et al., 1985; Hicks et al., 2002; Hoffman et al., 2008; Kline et al., 2007; Passineau et al., 2001; Sozda et al., 2010; Will et al., 1977). Specifically, Hamm and colleagues (1996) showed that TBI rats exposed to early EE located the escape platform in a Morris water maze (MWM) task (Morris, 1984) significantly quicker than STD-housed controls. Moreover, the enriched rats performed as well as the uninjured sham controls, which demonstrates robust EE-induced neurocognitive protection (Hamm et al., 1996). In addition to cognitive benefits, Passineau and colleagues (2001) reported a reduction in cortical lesion size. Several other studies from independent laboratories have also reported EE-induced benefits after CCI or FP injury in rats using traditional EE (Briones et al., 2013; Hicks et al., 2002; Smith et al., 2007) or a paradigm augmented with multimodal early onset stimulation (MEOS), which consists of incorporating auditory and olfactory stimuli to the EE setting (Gruner and Terhaag, 2000; Lippert-Gruner et al., 2007; Lippert-Gruner et al., 2002; Maegele et al., 2005a,b). EE has also been reported to be beneficial after blast TBI (Kovesdi et al., 2011; Lippert-Gruner et al., 2011). Although the overall consensus is that EE exerts beneficial effects, one study reported a paradoxical effect as rats exposed to enriched housing before injury exhibited larger contusion cavities and a greater initial deficit in forelimb use (Kozlowski et al., 2004).

Further support for EE as a preclinical model of neurorehabilitation comes from a recent study by Monaco and colleagues (2013), who showed that females also benefit significantly from EE, which is in contrast to a previous report (Wagner et al., 2002). Briefly, female rats received a CCI injury at various phases of the estrous stage, which was intentionally designed as such to mimic real-world experiences, and then were placed in either STD or enriched housing for 3 weeks. Motor and cognitive performance was assessed while rehabilitation (i.e., EE) was ongoing. EE improved both gross and fine motor performance as well as spatial learning. Moreover, EE conferred significant histological protection as evidenced by increased hippocampal $CA_{1/3}$ cell survival and decreased cortical lesion size (Monaco et al., 2013). This finding replicates the studies in male rats from several independent laboratories.

The behavioral benefits of EE extend beyond adult male and females as demonstrated in a study by Giza and coworkers, where post-natal day-19 rats underwent a lateral FP injury followed by rearing in EE or STD housing (Giza et al., 2005). Delayed EE led to enhanced spatial learning and memory in both sham and injured animals. However, when EE was provided immediately after TBI, there was no benefit in the injured rats. Taken together these findings suggest that EE-induced benefits after TBI are time-sensitive (Giza et al., 2005).

34.2.2 Transitioning to a Viable Preclinical Model of Neurorehabilitation

Although the studies described thus far have consistently shown that utilizing EE as a rehabilitative tool enhances neurobehavior and attenuates histopathology after various models of TBI, they were not designed to address long-term efficacy. However, knowledge regarding whether the benefits acquired by a treatment can be maintained after its discontinuation is clinically important. Moreover, demonstrating longevity would strengthen the validity of EE as a preclinical model of neurorehabilitation. Thus, to address this issue, Cheng et al. (2012) evaluated motor and cognitive ability in CCI or sham-injured rats that initially received 3 weeks of EE, and then were subsequently returned to STD housing. The experiment was designed to mimic real-world rehabilitation where patients receive a finite amount of rehabilitation and then are discharged home. Both the EE groups, regardless of whether having continuous or limited rehabilitation performed at the same level, which was significantly better than the nonenriched controls. The findings from this study indicate that the benefits of EE are maintained for up to 6 months after cessation (Cheng et al., 2012).

Another important question regarding the clinical relevance of EE is whether early initiation after TBI is necessary to confer benefits or if delayed exposure can also be effective. This is an important topic because patients will not be prescribed rehabilitation until after critical care has ended, which may take days to weeks. Two studies using the CCI injury model were conducted to address this issue. In the first study, which evaluated only a subset of possible rehabilitation time periods, it was found that early (immediately after surgery) and continuous EE postinjury facilitated motor recovery and spatial learning relative to STD-housed injured controls. Importantly, cognitive enhancement was also observed when delaying the initiation of EE by a week after TBI (Hoffman et al., 2008). Expanding on this first study by assessing additional temporal permutations consisting of various delays and durations of EE during different states of behavioral assessments, Matter et al. (2011) showed again that early EE enhanced motor recovery, whereas late EE facilitated cognitive recovery. Moreover, if the rats received enrichment for a total of 14 continuous days (1 week before and during maze training), they performed significantly better than rats that also received 14 days of cumulative EE, but that were not in succession. This finding suggests there is a threshold of approximately 2 weeks of contiguous EE required to elicit cognitive benefits after TBI (Matter et al., 2011).

In the clinical setting, rehabilitation therapy is typically provided for short periods (Blackerby, 1990; Shiel et al., 2001; Vanderploeg et al., 2008; Zhu et al., 2007). Hence, another important question to address when considering EE as a preclinical model of neurorehabilitation is whether limited amounts of daily exposure can provide benefits. In a study designed to address this question, de Witt et al. (2011) showed that a 6-hour dose of EE conferred significant improvements in motor and cognitive performance. Moreover, the benefits did not differ from the groups receiving EE around the clock, which suggests that continuous EE (i.e., rehabilitation) may not be essential for optimal functional benefits after moderate TBI in rats. No benefits were observed with the TBI groups receiving only 2 or 4 hours of EE (de Witt et al., 2011), which suggests that a minimum threshold of daily exposure may be required to benefit functional recovery after CCI injury, as previously described in the EE timing study by Matter et al. (2011). However, evidence that brief exposure (i.e., 1 hour daily) is sufficient to enhance behavior is provided from a study employing FP injury (Gaulke et al., 2005). The differences in injury severity between CCI and FP injury models may account for the amount of rehabilitation needed postinjury.

34.2.3 Combinational Therapies

The following section describes findings from TBI studies where EE was combined with pharmacotherapies in an attempt to confer additive benefits. This combinational paradigm is clinically relevant as TBI patients in rehabilitation will, more often than not, also be prescribed some form of pharmacotherapy. Moreover, the limitation with translation from the bed to bedside may be attributable to the use of single therapies, as opposed to a more translatable approach of combined treatments (Kline et al., 2010; Margulies et al., 2009). In one of the first combinational studies of EE and a pharmacotherapy, Kline and colleagues (2007) showed that acute treatment with EE or the serotonin$_{1A}$ (5-HT$_{1A}$) receptor agonist 8-hydroxy-2-(di-*n*-propyl amino) tetralin (8-OH-DPAT) promoted cognitive recovery and reduced histopathology after CCI injury. However, there was no additional benefit when combining the two therapeutic strategies (Kline et al., 2007). Similar behavioral outcomes were observed when EE was combined with chronic 8-OH-DPAT given daily for 3 weeks postinjury (Kline et al., 2010). A difference between the acute and chronic studies is that the latter reduced TBI-induced choline acetyltransferase medial septal cell loss (Kline et al., 2010). A plausible explanation for the lack of additive behavioral effects could be that EE is quite robust and thus produces a ceiling effect that precludes 8-OH-DPAT from conferring additional improvement.

Subsequent studies from Kline and colleagues evaluated the potential efficacy of the 5-HT$_{1A}$ receptor agonist buspirone combined with EE. Buspirone was selected based on experiments demonstrating efficacy with 5-HT$_{1A}$ receptor agonists and well as a dose response study by Olsen et al. (2012) showing that it facilitates spatial learning after TBI. Moreover, buspirone is used clinically for anxiety and depression (Chew and Zafonte, 2009). The finding from this combination study showed that buspirone and EE alone conferred significant behavioral recovery in CCI-injured rats, which is comparable to the 8-OH-DPAT studies (Kline et al., 2010; Kline et al., 2007). Also reminiscent of the 8-OH-DPAT studies, the combination of EE and buspirone did not provide additive or synergistic effects (Kline et al., 2012).

34.2.4 Potential Mechanisms for the EE-Induced Benefits after TBI

Seminal studies evaluating the effects of EE demonstrated that noninjured adult rats develop heavier and thicker cerebral cortices (Bennett et al., 1964) as well as increases in the number and length of glia (Diamond et al., 1966) relative to STD-housed controls. Subsequent studies have shown EE-induced increases in synapse size (Rosenzweig and Bennett, 1996) as well as increased densities of synaptic vesicles (Nakamura et al., 1999). EE also stimulates neurogenesis in rats (Nilsson et al., 1999) and mice (Kempermann et al., 1997, 1998; Kempermann and Gage, 1999). Furthermore, EE increases pre- and postsynaptic plasticity markers, such as synaptophysin and postsynaptic density protein-95, respectively (Nithianantharajah et al., 2004), which may be potential mechanisms mediating the benefits observed after TBI.

Several studies using TBI models have reported significant changes in the expression of neurotrophins when EE was presented after TBI (Chen et al., 2005; Hicks et al., 1997, 1998, 1999, 2002; Yang et al., 1996). After FP injury, brain-derived neurotrophic factor (BDNF) messenger RNA (mRNA) was increased in the CA_3 and dentate gyrus at various time points (Hicks et al., 1997). Chen and colleagues reported that BDNF protein levels were not affected by EE exposure after CCI injury (Chen et al., 2005). In the uninjured hippocampus, there were qualitative decreases in BDNF expression for both males and females not receiving enrichment, whereas significant increases in BDNF protein expression was observed in EE females (Chen et al., 2005). In addition to neurotrophins serving as potential mediating factors for the benefits conferred by EE, Briones and colleagues identified other potential mechanisms by demonstrating that EE reduces the proinflammatory cytokines interleukin-1β and tumor necrosis factor-alpha (TNF-α) in the prefrontal cortex and hippocampus (Briones et al., 2013).

34.3 EXERCISE

It should be noted that the literature describing randomized exercise studies in the TBI patient population is greatly limited (Rimmer et al., 2010). Fortunately, animal studies have allowed us to obtain reliable biomarkers for neuroplasticity that have been used to study exercise in experimental TBI models.

34.3.1 Beneficial Effects of Exercise

Exercise is increasingly being reported as producing beneficial effects on the brain (Hillman et al., 2008; vanPraag, 2009). For example, human and animal studies have linked exercise with improved learning and memory (Farmer et al., 2004; Fordyce and Wehner, 1993; Pereira et al., 2007; Rhyu et al., 2010) as well as prevention of cognitive decay and lowered risks of Alzheimer's disease (Friedland et al., 2001; Schuit et al., 2001). In addition, exercise has been shown to increase synaptic strength, improve motor function, and confer neuroprotection in animal models of Parkinson's disease (Tajiri et al., 2010; VanLeeuwen et al., 2009) and stroke (Ding et al., 2006a; Stummer et al., 1994). Several mechanisms play a role on the effects of exercise on synaptic plasticity and neural protection. One such mechanism is the decrease of oxidative stress markers (Navarro et al., 2004; Pan et al., 2007; Wu et al., 2004). Oxidative stress occurs after TBI and results in the production of reactive oxygen species that, when excessive, compromise cell functioning and ultimately lead to cell death (Hall et al., 2004; Shohami et al., 1999). An additional neuroprotective mechanism of exercise is the reduction of cytokines. Cytokines are intercellular messengers with pro- or anti-inflammatory properties. Both early and delayed exercise after CCI injury has been reported to reduce neuroinflammation (Chen et al., 2013; Piao et al., 2013). Cytokine overexpression is involved in several neurodegenerative disorders (Szelenyi, 2001). Animal studies have indicated that exercise decreases TNF-α, the cytokine found in ischemic injuries (Ang et al., 2004; Ding et al., 2006b). Elevations in TNF-α have been linked with neuronal damage in brain injury (Barone et al., 1997; Feuerstein et al., 1994).

Moreover, voluntary exercise in rodents increases growth factors (Fabel et al., 2003; Gomez-Pinilla et al., 1997; Neeper et al., 1996; Trejo et al., 2001), angiogenic factors (Ding et al., 2004), and hippocampal neurogenesis (vanPraag et al., 2005), thus promoting dendritic complexity and axonal connectivity. Among the growth factors increased with exercise, BDNF is prominently implicated in the beneficial effects of exercise. BDNF is well-recognized for its potent neuroplasticity and trophic effects. It plays a key role in hippocampal neuronal plasticity by facilitating long-term potentiation (Farmer et al., 2004), a cellular mechanism underlying learning and memory (Berninger, 1999; Levine et al., 1998; Tyler and Pozzo-Miller, 2001). Unlike most growth factors, BDNF expression is in large part activity dependent (Thoenen, 1995; Zafra et al., 1992). Accordingly, studies in rodents using a running wheel have shown that increases in BDNF are positively correlated to the amount of exercise (Griesbach et al., 2004b; Neeper et al., 1995). Other molecular substrates of learning that are increased with exercise and are functionally linked to BDNF are synapsin I, cyclic-AMP response-element-binding protein (CREB) and calcium-calmodulin-dependent protein kinase II (CAMKII) (Vaynman et al., 2003). Synapsin I is a phosphoprotein that is involved in synaptic vesicle release and the establishment of a synaptic vesicle reserve pool (Jovanovic, 2000; Li et al., 1995; Llinas et al., 1991; Melloni et al., 1994). CREB is a transcriptional regulator thought to play an important role in long-term potentiation (Bailey et al., 1996; Finkbeiner et al., 1997; Silva et al., 1998). CAMKII plays key roles in neurotransmission, gene expression and the regulation of glutamate receptors and calcium channels (Bronstein et al., 1993) and has a vital role in the formation of long-term memories (Cammarota et al., 2002; Fukunaga and Miyamoto, 1999).

34.3.2 Effects of Exercise after TBI Are Time-Dependent

Experimental data support the use of exercise as a treatment for TBI. Rats that are exercised in a voluntary running wheel after FP injury show increases in BDNF, CREB, and synapsin I within the hippocampus (Griesbach et al., 2004b; Hicks, 1998). Exercised animals also show cognitive improvement when compared with sedentary FP or CCI injured counterparts (Griesbach et al., 2004b; Itoh et al., 2011; Kim et al., 2010; Piao et al., 2013). Animals receiving a CCI injury and subsequently exercised demonstrate increased neuron survival (Seo et al., 2010), protection against oxidative damage (Griesbach et al., 2008), and reductions in caspase-3 and TUNEL-positive cells (Kim et al., 2010).

As indicated previously, exercise can improve outcome after TBI. However, this may not always be the case if certain TBI-related factors are not considered when using exercise as a therapeutic tool for recovery. Despite the neuroprotective effects of exercise, premature postconcussive exercise has been linked with an exacerbation of postconcussive symptomatology such as depression and cognitive deterioration (Guskiewicz et al., 2003, 2007; Jordan, 2000). Likewise, a retrospective cohort study in a population of student athletes revealed that high levels of daily cognitive and physical postinjury activity had a negative effect on neurocognitive function and concussive symptoms (Majerske et al., 2008). Accordingly, animal studies indicate that, in contrast to delayed posttraumatic exercise, subacute exercise after a FP injury, does not increase BDNF. FP-injured rats that underwent subacute voluntary exercise also showed more pronounced learning and memory deficits compared to unexercised injured rats (Griesbach et al., 2004b). Other animal studies have shown that forced exercise, when administered acutely following a unilateral sensorimotor cortex injury, increases the brain lesion size and results in a behavioral impairment (Humm et al., 1998; Kozlowski, 1996). In addition, premature exercise appears to suppress a neuroplastic response to brain injury that has been observed in humans (Chu et al., 2000; Weiller et al., 1992) and animals (Dunn-Meynell and Levin,1995; Ip et al., 2002; Kozlowski, 1998). Compensatory and/or restorative responses may be observed in areas of the brain that have endured a lesser amount of harm, and exertion after brain injury may disrupt these restorative processes (Kleim et al., 2003; Kozlowski, 1996). Along these lines, injury-induced cortical increases of synapsin, CREB, and CAMKII are disrupted when injured rats are allowed to exercise within 24 hours after injury (Griesbach et al., 2004a).

34.3.3 Potential Mechanisms behind the Negative Effects of Subacute Exercise

It is probable that some of the key molecular players implicated in exercise will be influenced by the dynamic nature of TBI. Acutely, there is a metabolic cascade from stretching of neuronal cell membranes and axons induced by mechanical trauma that results in ionic disequilibrium leading to the accumulation of intracellular calcium and compromised mitochondrial respiration and protease activation (Raghupathi, 2004; Sullivan et al., 1998). Furthermore, restoration attempts of ionic equilibrium augment local cerebral glucose demand (Yoshino et al., 1991). The previously mentioned ionic shifts and the downstream effects of indiscriminate neurotransmitter release, in conjunction with an increase of cerebral glucose demand, lead to an acute energy crisis. Although the mechanisms underlying the inability of acute exercise to upregulate BDNF after FP injury are not fully understood, it is likely that the disruption of cellular function from compromised cerebral metabolism after a TBI contributes to the delayed time window for exercise. It is possible that premature stimulation to the brain, through exercise, imposes an energetic demand at a time when the brain is metabolically compromised. Accordingly, cortical stimulation to the brain after FP injury elicits a metabolic response that may act as a secondary injury by increasing cortical degeneration (Ip et al., 2003). Metabolic and ionic disturbances will also affect function of proteins such as BDNF (Falkenberg, 1992). In turn, alterations in BDNF will also influence brain activation given its role on synaptic facilitation (Tyler et al., 2006) and neurotransmitter release (Lessmann et al., 1994; Levine et al., 1995; Takei et al., 1997).

Of particular relevance is the interaction between exercise and stress. Stress is a perturbation of either psychological or physiological homeostasis. There is an abundance of studies demonstrating the ill effects of prolonged or repeated stress. Among these effects is inhibition of BDNF and CREB phosphorylation (Gronli et al., 2006; Schaaf et al., 1998). Indeed, corticosteroids regulate BDNF expression via activation of glucocorticoid receptors, which can act directly as transcription factors (Hansson et al., 2000). Glucocorticoid receptors are widespread throughout the brain and are predominantly expressed within the hippocampus thus regulating the magnitude and duration of the stress response (McEwen, 1999). Given the high concentration of GRs within the hippocampus, pronounced stress has been associated with decreased levels of hippocampal BDNF mRNA (Nibuya et al., 1995; Smith, 1995; Ueyama, 1997). The stress response involves the activation of the hypothalamic-pituitary-adrenal axis (HPA), resulting in the release of corticotrophin-releasing hormone by paraventricular neurons and the subsequent release of adrenocorticotrophic hormone (ACTH) from pituitary cells. ACTH stimulates the adrenal gland to release glucocorticoids, which leads to a negative feedback to restrain release of corticotrophin-releasing hormone, ACTH, and corticosterone. Corticosterone is the main circulating glucocorticoid in rats and is equivalent to cortisol in humans.

Recent findings from rodent studies indicate that there is a hyperresponsiveness to stress following a mild FP injury. This hyperresponsiveness was observed during the first postinjury weeks (Griesbach et al., 2011) and is likely to be associated with disruptions in glucocorticoid regulation following TBI. Excitatory changes within the hippocampus will thus have an effect on glucocorticoid negative feedback. Accordingly, it appears that hippocampal glucocorticoid receptors' response to the stressful stimulus is altered during the subacute period after a FP injury (Griesbach et al., 2012a, 2012b). Glucocorticoid release is regulated by the hypothalamus, a region receiving multiple suprahypothalamic inputs with notable projections from the limbic system. TBI-related changes in brain activation and connectivity have an influence on the regulation of the HPA axis. Thus, alterations in function and connectivity are likely to exert a substantial influence on HPA regulation. Efficient HPA regulation is necessary for neuronal vitality and function. The duration of TBI-induced neuroendocrine abnormalities appears to be severity-dependent (Woolf et al., 1990). These clinical findings have been supported by animal studies indicating TBI-induced changes in HPA regulation (Griesbach et al., 2011; Taylor et al., 2010).

34.3.4 Alterations in Autonomic Regulation Become Evident with Subacute Exercise

Changes in heart rate (HR) and core body temperature (CBT) may be detrimental to TBI recovery. Elevations in CBT after TBI are concerning. It is well-known that fever worsens TBI outcomes in humans and animals (Dietrich et al., 1996; Kilpatrick et al., 2000). Moreover, even mild temperature variations can negatively affect outcome. A recent preclinical study of mild TBI demonstrated that peritraumatic mild hyperthermia increased cortical contusion size (Sakurai et al., 2012). Hyperthermia has also been associated with increases of extracellular glutamate (Sharma, 2006; Suehiro et al., 1999) and inflammatory processes (Chatzipanteli et al., 2000). Clinically, irregularities in HR are reported during the subacute period (Biswas et al., 2000; Keren et al., 2005). A recent study reported that both CBT and HR disruptions become evident when exercise is introduced (Griesbach et al., 2013). FP-injured animals have more pronounced increases in HR as a result of voluntary exercise compared to uninjured controls, despite the controls having exercised more. Likewise, higher HRs has been reported in concussed athletes as a response to exercise compared with matched controls (Gall et al., 2004a,b). This study also showed that impaired thermoregulation was evident when FP-injured rats were exposed to exercise during the subacute period. In other words, exercise-induced elevations in CBT took longer to recover after exercise. Prolonged exercise normally increases CBT in healthy individuals (Hayashi et al., 2006). However, these increases are of significant concern if thermoregulation is impaired.

Autonomic effects appear to be dependent on the characteristics of exercise. A recent study compared the effects of forced and voluntary exercise, via a running wheel, in FP-injured rats. Given the differences between forced and voluntary exercise paradigms, it is likely that different motivational mechanisms are involved. Accordingly, the stress response also differs when comparing these forms of exercise. A forced running wheel was found to be a potent stimulus for the HPA axis, unlike voluntary exercise (Griesbach et al., 2012a). In addition, forced exercise did not increase BDNF levels irrespective of injury. It is unknown if the pronounced increase in CORT inhibited BDNF, and whether increases in BDNF would have been observed if forced exercise continued for more days, thus allowing animals to fully habituate. What became evident was that, whereas the stress response to forced exercise decreased over time in control rats, FP-injured rats exposed to the forced exercise subacutely showed a delay in habituation. Interestingly, injured rats exposed to forced exercise were also more responsive to experimental manipulations as indicated by increases in CBT and HR (Griesbach et al., 2013). These studies suggest that certain exercise regimens with stronger stress responses may be particularly counterproductive during the early postinjury period.

34.3.5 When Is It Best to Exercise after TBI?

Both animal and humans studies support the concept of post-concussive vulnerability. All the previously mentioned post-concussive physiological changes have been shown to increase the brain's vulnerability to further injury during the subacute post injury period (Laurer et al., 2001; Longhi et al., 2005; Schulz et al., 2004). Given all the concerns regarding early return to exercise in conjunction with the high incidence of concussions in the athletic and military settings (Terrio et al., 2009; Thurman et al., 1998), guidelines have been formed in which it is recommended that symptoms be allowed to resolve before reinstating sports-related activities (Aubry et al., 2002; McCrory et al., 2009). Recently, because of the increasing concern of mild TBI in the military, guidelines were also suggested for military members (Weightman et al., 2010). For these guidelines, physicians have generally relied on cognitive neuropsychological testing and the evaluation of concussive symptoms, including vestibular dysfunction and impaired balance (Ingebrigtsen et al., 1998; Maroon et al., 2000; McCrory et al., 2005). However, symptom resolution may not always imply that there is full cerebral recovery. There is evidence that cerebral metabolism is still engaged in restoring energetic functions after symptom resolution (Vagnozzi et al., 2008, 2010). Unfortunately, assessing injury severity to determine time for recovery or rest, particularly with more mild brain injuries, is not straightforward. Because injury recovery varies greatly among patients, having fixed waiting guidelines may be too conservative or not sufficient. This concern has given rise to recommendations of tailoring restriction periods to individualized evaluation of cognitive function and clinical profile (Makdissi et al., 2010).

Recent animal studies have addressed this issue, indicating that the recovery of the stress response coincides with responsiveness to voluntary exercise after TBI (Griesbach et al., 2014). Earlier studies indicated that the appropriate time-window for voluntary exercise following FP injury was dependent on injury severity, where an increase in injury severity prolonged the period of nonresponsiveness to voluntary exercise (Griesbach et al., 2007). In accordance with the temporal component of rehabilitative exercise, the disruption in cellular functioning, observed after TBI, is severity-dependent both in humans and animal models of TBI (Dietrich et al., 1994; Glenn et al., 2003; Rogatsky et al., 2003; Thompson et al., 2005). Another factor that should be considered is that a history of previous concussions is likely to influence the waiting period for return to activities and rehabilitative exercise.

In summary, the studies in experimental brain injury models indicate that awareness of temporal brain vulnerability should be considered during rehabilitative intervention, particularly those that are likely to elicit a pronounced stress response and/or are likely to place further energetic demands on the brain. Injury-induced alterations on metabolism may limit energy in such a manner that neuronal functioning and signaling are jeopardized, ultimately affecting neuroendocrine regulation and behavioral outcome. Moreover, despite the therapeutic capabilities of BDNF, its role as a therapeutic agent for TBI has been diminished because of its low blood–brain barrier permeability and short half-life in systemic circulation (Kroll and Neuwelt, 1998; Morse et al., 1993; Poduslo and Curran, 1996; Sakane and Pardridge, 1997). However, exercise circumvents these difficulties by endogenously upregulating BDNF and associated molecular systems and making it particularly advantageous. Lastly, exercise can be easily implemented in the rehabilitative setting.

34.4 LOW-LEVEL LASER THERAPY

Another area of rehabilitative research that is gaining momentum as a treatment for TBI and other neurological disorders is LLLT or photo-biostimulation. Photo-biostimulation is a novel method of noninvasive neural stimulation that can, at specific wavelengths (600–1,000 nm) and at power densities in the range of 1–5 W/cm^2, safely penetrate the brain and produce beneficial cellular and physiological effects (Desmet et al., 2006). LLLT uses coherent lasers, or noncoherent light-emitting diodes. During LLLT stimulation, infrared/near-infrared photons are absorbed by cytochrome C oxidase in the mitochondrial respiratory chain causing an increase in cellular respiration at times of reduced energy. This interaction ultimately leads to improved mitochondrial function and increased ATP production (Byrnes et al., 2005; Karu et al., 2005).

LLLT has been used in many clinical applications, specifically for soft tissue damage, as well as in preclinical models. In experimental stroke models, LLLT has been shown to attenuate neurological deficits (Oron et al., 2007) and reduce levels of inducible nitric oxide synthase expression while increasing transforming growth factor beta-1 (Leung et al., 2002). LLLT has also been reported to improve functional outcome in a rat model of spinal cord injury (Byrnes et al., 2005) and has also been applied to models of TBI.

In the first study assessing the potential efficacy of LLLT in TBI, Oron and colleagues (2007) subjected mice to a cortical weight-drop injury and 4 hours later presented a single application of two doses of LLLT (10 or 20 mW/cm^2; 1.2–2.4 J/cm^2) or no laser therapy. A significant improvement in neurological severity scores was observed on postoperative days 5–28 as well as sparing of brain tissue after treatment, regardless of dose (Oron et al., 2007). In a subsequent study, Oron et al. (2012) reported that near infrared transcranial laser therapy (10 mW/cm^2) applied to separate groups of closed head–injured mice at 4 and 6 hours postinjury produced a beneficial effect on long-term neurological outcome and smaller infarct lesion volumes as determined by magnetic resonance imaging (Oron et al., 2012). In a mouse CCI model, Khuman and colleagues (2012) applied LLLT (810 nm, 500 mW/cm^2, 60J/cm^2, for 2 minutes) either directly to the contused brain parenchyma or transcranially at 60–80 minutes postinjury. A modest improvement in MWM performance was observed. Moreover, the beneficial effects of LLLT were associated with reduced microglial activation at 48 hours. Furthermore, marginal or no effect of LLLT was observed on cognitive function with a 4-hour postinjury or a repeated 7-day postinjury administration. This study suggests that LLLT might be a therapeutic option in improving spatial learning and memory and reducing neuroinflammation when applied acutely post-TBI. Using the rat CCI model, Quirk and associates found that two light-emitting diode treatments (670 nm, 50 mW/cm^2, 15 J/cm^2, 5 minutes) per day for 3 or 10 days postinjury improved motor activity. Furthermore, LLLT also decreased levels of the pro-apoptotic protein, Bax, in the contused cortical hemisphere (Quirk et al., 2012). Using a mouse CCI model, Xuan and colleagues (2013) explored the effect of repeated laser treatments (810 nm, 25 mW/cm^2, 18 J/cm^2) on neurological performance. LLLT was delivered either as a single treatment (4 hours post-TBI) or as 3 and 14 daily treatments. Mice receiving 1 or 3 daily treatments exhibited significant improvements in neurological severity score, wire grip, and motion test. In contrast, 14 daily treatments provided no benefit compared with the TBI control group that received no laser treatment. Mice treated with 1 or 3 daily laser treatments also showed smaller lesion volumes and fewer degenerating neurons when stained with fluoro-jade at 28 days postinjury. These findings suggest that transcranial laser therapy may be beneficial given the optimal treatment regimen after TBI (Xuan et al., 2013). Lastly, Ando et al. (2011) assessed the therapeutic effectiveness between continuous wave (810 nm) and pulsed laser irradiation administered 4 hours post-CCI injury in mice and found that the 810 nm (50mW/cm^2, 36 J/cm^2, 12 minutes) pulse was effective as it improved neurological severity score and spared cortical tissue loss at 15 days as well as 4 weeks post-TBI. The same study showed that LLLT had an antidepressant effect in the forced swim and tail suspension tests (Ando et al., 2011).

Clinical applications of LLLT have been sparse. In a recent study, Naeser and associates showed that transcranial LLLT in two chronic TBI patients improved cognitive function (Naeser et al., 2011). In one patient continuous treatment over an extended period (9 months) resulted in significant improvement in executive function and memory as well as a reduction in posttraumatic stress disorder. In a randomized clinical stroke trial, LLLT was shown to improve functional outcome when initiated 24 hours after stroke onset (Lampl et al., 2007).

Overall, the research, albeit still limited, indicates that LLLT is effective in improving functional outcome and attenuating histological damage after experimental TBI. Further preclinical studies are warranted to optimize parameters such as irradiation doses, duration, and therapeutic windows. Such assessments could potentially lead to more translatable studies and augment the clinical report of improved cognitive function.

34.5 CONSTRAINT-INDUCED MOVEMENT THERAPY

Whereas EE, exercise, and LLLT may be considered relatively noninvasive therapeutic strategies for TBI, CIMT, which is increasingly gaining attention in both experimental and clinical rehabilitation, is more invasive by design. Briefly, CIMT involves a series of techniques developed by Taub and colleagues (Taub, 1980; Taub et al., 1999), in which "learned nonuse" (Knapp et al., 1958) is counterconditioned by restraining the unaffected limb, and in essence forcing use of the affected/injured limb with repetitive movements (Taub, 1980). Simply stated, the rationale of CIMT is to prevent avoidance-induced compensatory behaviors, particularly using the unaffected limb simply because it is easier to do so, and encourage use of the impaired limb. The CIMT paradigm induces neuroplasticity that can be correlated with a therapeutic effect (Taub, 2012a). Taub and colleagues (Knapp et al., 1958; Shaw et al., 2005; Taub, 1980; Taub, 1999, 2012a, 2012b) are the pioneers of CIMT as a rehabilitative approach, and over the past 50 years have applied it to numerous neurological disorders. There are more than 300 studies that report positive findings utilizing CIMT for stroke (Taub, 2012a). Moreover, the findings from a multisite randomized controlled trial by Wolf et al. (2006) showed that CIMT was effective when compared with usual and customary care, which ranged from no treatment to application of orthotics, home and clinic-based occupational therapy as well as physiotherapy. CIMT, however, has not been as thoroughly investigated in other CNS injuries, specifically TBI. Nonetheless, given its success after stroke (Stevenson et al., 2012), hemiplegia (Cimolin et al., 2012), cerebral palsy (Klingels et al., 2013), and neurological motor disorders and amblyopia (Taub, 2012b) as well as in various age groups, such as adults (Shaw et al., 2005; Sterr et al., 2002; Stevenson et al., 2012) and pediatric patients (Brady and Garcia, 2009; Cimolin et al., 2012; Klingels et al., 2013; Mancini et al., 2013; Pedlow et al., 2013; Taub, 2012a), we briefly introduce it in this chapter as an additional potentially efficacious therapeutic approach for TBI.

Animal studies have been conducted extensively during the early development of the underlying principles of CIMT and are critical for the progress and advancement of clinically relevant rehabilitative paradigms. Kozlowski and colleagues illustrated the complexity of TBI modality of injury and recovery as putatively different from recovery mechanisms seen after stroke, therefore highlighting the need for detailed assessment of functional organization of movement control following TBI in particular (Kozlowski et al., 2013).

Successful employment of CIMT in animal models originated with pioneering work by Knapp and Taub and colleagues using behavioral counterconditioning after deafferentation in monkeys (Knapp et al., 1958; Taub, 1980, 2012a), which later produced replicable and effective clinical results (Wolf, 2006). Although the seminal work applying CIMT in preclinical models involved primarily stroke as the model of CNS injury, recent forays researching the experimental models of TBI using modified forms of CIMT have been mixed. Use-dependent plasticity of cortical structures have been investigated in studies using lesion (Humm et al., 1998, 1999; Jones and Schallert, 1994) and CCI (Jones et al., 2012) models of injury. In these models, early immobilization of the intact limb may have exacerbated functional deficits (Humm et al., 1998; Jones and Schallert, 1994). Moreover, time-dependent alterations after injury have differential effects on dendritic arborization and pruning, in relation to timing of immobilization of the intact limb, which may result in long-term effects on functional outcomes (Jones and Schallert, 1994). This early immobilization rendering a worsening of injury is also seen in a devascularization injury model (DeBow et al., 2004). Additionally, it has been proposed that an interaction between lesion/injury and behavior is required for morphological dendritic changes to occur (Jones and Schallert, 1994). These findings, although not always positive, are encouraging and suggest that manipulating behavior at appropriate times after injury could promote recovery via alterations of plasticity at the site of injury. The complexity of the injury–behavior interaction needs to be further elucidated when describing future use-dependent therapies such as CIMT. Further delineation of the complexity and disconnect between TBI and stroke has been reported (Jones et al., 2012). It may be beneficial to use pharmacotherapies concurrently with CIMT to augment functional recovery. Studies with concomitant pharmacotherapies such as MK-801 (Humm et al., 1999) and more recently, minocycline (Lam et al., 2013), have shown some promising effects using lesion and CCI models, respectively. The latter suggests a role for reduced inflammation when CIMT is applied as a rehabilitation paradigm (Lam et al., 2013).

Although this rehabilitative paradigm presents some intrinsic difficulties in the utilization of the affected extremity on a daily basis (Mancini et al., 2013), CIMT may also show clinical benefits after TBI (Cimolin et al., 2012; Lillie and Mateer, 2006). Cimolin and colleagues showed that pediatric TBI patients presenting with a Glasgow Coma Scale score of ≤8 and a diagnosis of hemiplegia had significant improvements in movement efficiency and upper limb kinematics following a CIMT regimen (Cimolin et al., 2012).

34.6 CONCLUSION

The current chapter highlighted a series of therapeutic manipulations relevant to clinical rehabilitation post-TBI, such as environmental enrichment, exercise, low-level laser therapy, and constraint-induced movement therapy. These noninvasive rehabilitative approaches were shown to promote significant behavioral, histological, and neuroplasticity-related benefits across a variety of brain injury models. Moreover, as discussed, beneficial effects of these rehabilitative paradigms have also been reported throughout the rather limited clinical literature, rendering it essential to further implement and optimize these noninvasive and easily applicable therapeutic manipulations, alone or in combination with clinically relevant pharmacotherapies, into the acute and long-term clinical care settings. Successful translation of these therapeutic strategies would significantly benefit the millions of TBI survivors, and their caregivers, in both the United States and abroad as they endure the path to recovery to regain physical and mental capabilities similar to those pre-injury, which is typically a long, difficult, and costly process.

REFERENCES

Ando, T., W. Xuan, T. Xu, T. Dai, S.K. Sharma, G.B. Kharkwal et al. 2011. Comparison of therapeutic effects between pulsed and continuous wave 810-nm wavelength laser irradiation for traumatic brain injury in mice. *PloS One*. 6:e26212.

Ang, E.T., P.T. Wong, S. Moochhala, and Y.K. Ng. 2004. Cytokine changes in the horizontal diagonal band of Broca in the septum after running and stroke: A correlation to glial activation. *Neuroscience*. 129:337–347.

Aubry, M., R. Cantu, J. Dvorak, T. Graf-Baumann, K.M. Johnston, J. Kelly et al. Concussion in sport. 2002. Summary and agreement statement of the 1st International Symposium on Concussion in Sport, Vienna 2001. *Clin J Sport Medi*. 12:6–11.

Bailey, C.H., D. Bartsch, and E.R. Kandel. 1996. Toward a molecular definition of long-term memory storage. *Proc Natl Acad Sci U S A*. 93:13445–13452.

Bales, J.W., A.K. Wagner, A.E. Kline, and C.E. Dixon. 2009. Persistent cognitive dysfunction after traumatic brain injury: A dopamine hypothesis. *Neurosci Biobehav Rev*. 33:981–1003.

Barone, F.C., B. Arvin, R.F. White, A. Miller, C.L. Webb, R.N. Willette et al. 1997. Tumor necrosis factor-alpha. A mediator of focal ischemic brain injury. *Stroke*. 28:1233–1244.

Bennett, E.L., M.C. Diamond, D. Krech, and M.R. Rosenzweig. 1964. Chemical and anatomical plasticity brain. *Science*. 146:610–619.

Berninger, B., A.F. Schinder, and M.M Poo. 1999. Synaptic reliability correlates with reduced suceptibility to synaptic potentiation by brain-derived neurotrfophic factor. *Learn Memory*. 6:232–242.

Berrocal, Y., D.D. Pearse, A. Singh, C.M. Andrade, J.S. McBroom, R. Puentes et al. 2007. Social and environmental enrichment improves sensory and motor recovery after severe contusive spinal cord injury in the rat. *J Neurotrauma*. 24:1761–1772.

Bezard, E., S. Dovero, D. Belin, S. Duconger, V. Jackson-Lewis, S. Przedborski et al. 2003. Enriched environment confers resistance to 1-methyl-4-phenyl-1,2,3,6-tetrahydropyridine and cocaine: Involvement of dopamine transporter and trophic factors. *J Neurosci*. 23:10999–11007.

Binder, L.M. 1986. Persisting symptoms after mild head injury: A review of the postconcussive syndrome. *J Clin Exp Neuropsychol*. 8:323–346.

Biswas, A.K., W.A. Scott, J.F. Sommerauer, and P.M. Luckett. 2000. Heart rate variability after acute traumatic brain injury in children. *Crit Care Med*. 28:3907–3912.

Blackerby, W.F. 1990. Intensity of rehabilitation and length of stay. *Brain Inj*. 4:167–173.

Brady, K., and T. Garcia. 2009. Constraint-induced movement therapy (CIMT): Pediatric applications. *Develop Disabil Res Rev*. 15:102–111.

Bramlett, H.M., and W.D. Dietrich. 2002. Quantitative structural changes in white and gray matter 1 year following traumatic brain injury in rats. *Acta Neuropathol*. 103:607–614.

Briones, T.L., J. Woods, and M. Rogozinska. 2013. Decreased neuroinflammation and increased brain energy homeostasis following environmental enrichment after mild traumatic brain injury is associated with improvement in cognitive function. *Acta Neuropathol Commun*. 1:57.

Bronstein, J.M., D.B. Farber, and C.G. Wasterlain. 1993. Regulation of type-II calmodulin kinase: Functional implications. *Brain Res*. 18:135–147.

Bruel-Jungerman, E., S. Laroche, and C. Rampon. 2005. New neurons in the dentate gyrus are involved in the expression of enhanced long-term memory following environmental enrichment. *Eur J Neurosci*. 21:513–521.

Buchhold, B., L. Mogoanta, Y. Suofu, A. Hamm, L. Walker, C. Kessler et al. 2007. Environmental enrichment improves functional and neuropathological indices following stroke in young and aged rats. *Restorative Neurol Neurosci*. 25:467–484.

Byrnes, K.R., R.W. Waynant, I.K. Ilev, X. Wu, L. Barna, K. Smith et al. 2005. Light promotes regeneration and functional recovery and alters the immune response after spinal cord injury. *Lasers Surg Med*. 36:171–185.

Cammarota, M., L.R. Bevilaqua, H. Viola, D.S. Kerr, B. Reichmann, V. Teixeira et al. 2002. Participation of CaMKII in neuronal plasticity and memory formation. *Cell Molec Neurobiol*. 22:259–267.

Centers for Disease Control and Prevention (CDC) National Center for Injury Prevention and Control. 2003. Report to Congress on mild traumatic brain injury in the United States: Steps to prevent a serious public health problem. Centers for Disease Control and Prevention, Atlanta, GA.

Chatzipanteli, K., O.F. Alonso, S. Kraydieh, and W.D. Dietrich. 2000. Importance of posttraumatic hypothermia and hyperthermia on the inflammatory response after fluid percussion brain injury: Biochemical and immunocytochemical studies. *J Cerebr Blood Flow Metab*. 20:531–542.

Chen, M.F., T.Y. Huang, Y.M. Kuo, L. Yu, H.I. Chen, and C.J. Jen. 2013. Early postinjury exercise reverses memory deficits and retards the progression of closed-head injury in mice. *J Physiol*. 591:985–1000.

Chen, X., Y. Li, A.E. Kline, C.E. Dixon, R.D. Zafonte, and A.K. Wagner. 2005. Gender and environmental effects on regional brain-derived neurotrophic factor expression after experimental traumatic brain injury. *Neuroscience*. 135:11–17.

Cheng, J.P., A.N. Hoffman, R.D. Zafonte, and A.E. Kline. 2008. A delayed and chronic treatment regimen with the 5-HT_{1A} receptor agonist 8-OH-DPAT after cortical impact injury facilitates motor recovery and acquisition of spatial learning. *Behav Brain Res*. 194:79–85.

Cheng, J.P., K.E. Shaw, C.M. Monaco, A.N. Hoffman, C.N. Sozda, A.S. Olsen et al. 2012. A relatively brief exposure to environmental enrichment after experimental traumatic brain injury confers long-term cognitive benefits. *J Neurotrauma*. 29:2684–2688.

Chew, E., and R.D. Zafonte. 2009. Pharmacological management of neurobehavioral disorders following traumatic brain injury—a state-of-the-art review. *J Rehabil Res Develop*. 46:851–879.

Chu, D., P.R. Huttenlocher, D.N. Levin, and V.L. Towle. 2000. Reorganization of the hand somatosensory cortex following perinatal unilateral brain injury. *Neuropediatrics*. 31:63–69.

Cimolin, V., E. Beretta, L. Piccinini, A.C. Turconi, F. Locatelli, M. Galli, and S. Strazzer. 2012. Constraint-induced movement therapy for children with hemiplegia after traumatic brain injury: A quantitative study. *J Head Trauma Rehabil*. 27:177–187.

Clark, R.S., P.M. Kochanek, D.W. Marion, J.K. Schiding, M. White, A.M. Palmer, and S.T. DeKosky. 1996. Mild posttraumatic hypothermia reduces mortality after severe controlled cortical impact in rats. *J Cerebr Blood Flow Metab*. 16:253–261.

Clifton, G.L., J.Y. Jiang, B.G. Lyeth, L.W. Jenkins, R.J. Hamm, and R.L. Hayes. 1991. Marked protection by moderate hypothermia after experimental traumatic brain injury. *J Cerebr Blood Flow Metab*. 11:114–121.

Cracchiolo, J.R., T. Mori, S.J. Nazian, J. Tan, H. Potter, and G.W. Arendash. 2007. Enhanced cognitive activity—over and above social or physical activity—is required to protect Alzheimer's mice against cognitive impairment, reduce Abeta deposition, and increase synaptic immunoreactivity. *Neurobiol Learn Memory*. 88:277–294.

de Witt, B.W., K.M. Ehrenberg, R.L. McAloon, A.H. Panos, K.E. Shaw, P.V. Raghavan et al. 2011. Abbreviated environmental enrichment enhances neurobehavioral recovery comparably to continuous exposure after traumatic brain injury. *Neurorehabil Neural Repair*. 25:343–350.

DeBow, S.B., J.E. McKenna, B. Kolb, and F. Colbourne. 2004. Immediate constraint-induced movement therapy causes local hyperthermia that exacerbates cerebral cortical injury in rats. *Can J Physiol Pharmacol*. 82:231–237.

Desmet, K.D., D.A. Paz, J.J. Corry, J.T. Eells, M.T. Wong-Riley, M.M. Henry et al. 2006. Clinical and experimental applications of NIR-LED photobiomodulation. *Photomed Laser Surg*. 24:121–128.

Diamond, M.C., F. Law, H. Rhodes, B. Lindner, M.R. Rosenzweig, D. Krech et al. 1966. Increases in cortical depth and glia numbers in rats subjected to enriched environment. *J Comparative Neurol*. 128:117–126.

Dietrich, W.D., O. Alonso, and M. Halley. 1994. Early microvascular and neuronal consequences of traumatic brain injury: A light and electron microscopic study in rats. *J Neurotrauma*. 11:289–301.

Dietrich, W.D., O. Alonso, M. Halley, and R. Busto. 1996. Delayed posttraumatic brain hyperthermia worsens outcome after fluid percussion brain injury: A light and electron microscopic study in rats. *Neurosurgery*. 38:533–541; discussion 541.

Ding, Y.H., J. Li, W.X. Yao, J.A. Rafols, J.C. Clark, and Y. Ding. 2006a. Exercise preconditioning upregulates cerebral integrins and enhances cerebrovascular integrity in ischemic rats. *Acta Neuropathol*. 112:74–84.

Ding, Y.H., X.D. Luan, J. Li, J.A. Rafols, M. Guthinkonda, F.G. Diaz, and Y. Ding. 2004. Exercise-induced overexpression of angiogenic factors and reduction of ischemia/reperfusion injury in stroke. *Curr Neurovasc Res*. 1:411–420.

Ding, Y.H., M. Mrizek, Q. Lai, Y. Wu, R. Reyes, Jr., J. Li, W.W. Davis, and Y. Ding. 2006b. Exercise preconditioning reduces brain damage and inhibits TNF-alpha receptor expression after hypoxia/reoxygenation: An in vivo and in vitro study. *Curr Neurovasc Res*. 3:263–271.

Dixon, C.E., G.L. Clifton, J.W. Lighthall, A.A. Yaghmai, and R.L. Hayes. 1991. A controlled cortical impact model of traumatic brain injury in the rat. *J Neurosci Methods*. 39:253–262.

Dixon, C.E., P.M. Kochanek, H.Q. Yan, J.K. Schiding, R.G. Griffith, E. Baum et al. 1999. One-year study of spatial memory performance, brain morphology, and cholinergic markers after moderate controlled cortical impact in rats. *J Neurotrauma*. 16:109–122.

Dixon, C.E., B.G. Lyeth, J.T. Povlishock, R.L. Findling, R.J. Hamm, A. Marmarou et al. 1987. A fluid percussion model of experimental brain injury in the rat. *J Neurosurg*. 67:110–119.

Doppenberg, E.M., S.C. Choi, and R. Bullock. 2004. Clinical trials in traumatic brain injury: Lessons for the future. *J Neurosurg Anesthesiol*. 16:87–94.

Doulames, V., S. Lee, and T.B. Shea. 2014. Environmental enrichment and social interaction improve cognitive function and decrease reactive oxidative species in normal adult mice. *Int J Neurosci*. 124:369–376

Dunn-Meynell, A.A., and B.E. Levin. 1995. Lateralized effect of unilateral somatosensory cortex contusion on behavior and cortical reorganization. *Brain Res*. 675:143–156.

Fabel, K., K. Fabel, B. Tam, D. Kaufer, A. Baiker, N. Simmons et al. 2003. VEGF is necessary for exercise-induced adult hippocampal neurogenesis. *Eur J Neurosci*. 18:2803–2812.

Faherty, C.J., K. Raviie Shepherd, A. Herasimtschuk, and R.J. Smeyne. 2005. Environmental enrichment in adulthood eliminates neuronal death in experimental Parkinsonism. *Brain Res*. 134:170–179.

Falkenberg, T., Mohammed, A.K., Henriksson, B., Persson, H., Winblad, B., and Lindefors, N. 1992. Increased expression of brain-derived neurotrophic factor mRNA in rat hippocampus is associated with improved spatial memory and enriched environment. *Neuroscience Lett*. 138:153–156.

Farmer, J., X. Zhao, H. van Praag, K. Wodtke, F.H. Gage, and B.R. Christie. 2004. Effects of voluntary exercise on synaptic plasticity and gene expression in the dentate gyrus of adult male Sprague Dawley rats in vivo. *Neuroscience*. 124:71–79.

Faul, M.X., L.; Wald, MM.; Coronado, VG. 2010. Traumatic brain injury in the United States: Emergency department visits, hospitalizations and deaths 2002–2006. Centers for Disease Control and Prevention, National Center for Injury Prevention and Control, Atlanta, Ga.

Feuerstein, G.Z., T. Liu, and F.C. Barone. 1994. Cytokines, inflammation, and brain injury: Role of tumor necrosis factor-alpha. *Cerebrovasc Brain Metab Rev*. 6:341–360.

Finkbeiner, S., S.F. Tavazoie, A. Maloratsky, K.M. Jacobs, K.M. Harris, and M.E. Greenberg. 1997. CREB: A major mediator of neuronal neurotrophin responses. *Neuron*. 19:1031–1047.

Fischer, F.R., and J.D. Peduzzi. 2007. Functional recovery in rats with chronic spinal cord injuries after exposure to an enriched environment. *J Spinal Cord Med*. 30:147–155.

Fordyce, D.E., and J.M. Wehner. 1993. Physical activity enhances spatial learning performance with an associated alteration in hippocampal protein kinase C activity in C57BL/6 and DBA/2 mice. *Brain Res*. 619:111–119.

Friedland, R.P., T. Fritsch, K.A. Smyth, E. Koss, A.J. Lerner, C.H. Chen et al. 2001. Patients with Alzheimer's disease have reduced activities in midlife compared with healthy control-group members. *Proc Natl Acad Sci U S A*. 98:3440–3445.

Fukunaga, K., and E. Miyamoto. 1999. Current studies on a working model of CaM kinase II in hippocampal long-term potentiation and memory. *Jap J Pharmacol*. 79:7–15.

Gall, B., W. Parkhouse, and D. Goodman. 2004a. Heart rate variability of recently concussed athletes at rest and exercise. *Med Sci Sports Exerc*. 36:1269–1274.

Gall, B., W.S. Parkhouse, and D. Goodman. 2004b. Exercise following a sport induced concussion. *Br J Sports Med*. 38:773–777.

Garcia, A.N., M.A. Shah, C.E. Dixon, A.K. Wagner, and A.E. Kline. 2011. Biologic and plastic effects of experimental traumatic brain injury treatment paradigms and their relevance to clinical rehabilitation. *PM & R*. 3:S18–27.

Gaulke, L.J., P.J. Horner, A.J. Fink, C.L. McNamara, and R.R. Hicks. 2005. Environmental enrichment increases progenitor cell survival in the dentate gyrus following lateral fluid percussion injury. *Brain Res*. 141:138–150.

Gentile, A.M., Z. Beheshti, and J.M. Held. 1987. Enrichment versus exercise effects on motor impairments following cortical removals in rats. *Behav Neural Biol*. 47:321–332.

Giza, C.C., G.S. Griesbach, and D.A. Hovda. 2005. Experience-dependent behavioral plasticity is disturbed following traumatic injury to the immature brain. *Behav Brain Res*. 157:11–22.

Glenn, T.C., D.F. Kelly, W.J. Boscardin, D.L. McArthur, P. Vespa, M. Oertel et al. 2003. Energy dysfunction as a predictor of outcome after moderate or severe head injury: Indices of oxygen, glucose, and lactate metabolism. *J Cerebr Blood Flow Metab*. 23:1239–1250.

Gomez-Pinilla, F., L. Dao, and V. So. 1997. Physical exercise induces FGF-2 and its mRNA in the hippocampus. *Brain Res.* 764:1–8.

Griesbach, G.S., F. Gomez-Pinilla, and D.A. Hovda. 2004a. The upregulation of plasticity-related proteins following TBI is disrupted with acute voluntary exercise. *Brain Res.* 1016:154–162.

Griesbach, G.S., F. Gomez-Pinilla, and D.A. Hovda. 2007. Time window for voluntary exercise-induced increases in hippocampal neuroplasticity molecules after traumatic brain injury is severity dependent. *J Neurotrauma.* 24:1161–1171.

Griesbach, G.S., D.A. Hovda, F. Gomez-Pinilla, and R.L. Sutton. 2008. Voluntary exercise or amphetamine treatment, but not the combination, increases hippocampal BDNF and synapsin I following cortical contusion injury in rats. *Neuroscience.* 154:530–540.

Griesbach, G.S., D.A. Hovda, R. Molteni, A. Wu, and F. Gomez-Pinilla. 2004b. Voluntary exercise following traumatic brain injury: Brain-derived neurotrophic factor upregulation and recovery of function. *Neuroscience.* 125:129–139.

Griesbach, G.S., D.A. Hovda, D.L. Tio, and A.N. Taylor. 2011. Heightening of the stress response during the first weeks after a mild traumatic brain injury. *Neuroscience.* 178:147–158.

Griesbach, G.S., D.L. Tio, S. Nair, and D. Hovda. 2013. Temperature and heart rate responses to exercise following mild traumatic brain injury. *J Neurotrauma.* 30:281–291.

Griesbach, G.S., D.L. Tio, J. Vincelli, D.L. McArthur, and A.N. Taylor. 2012a. Differential effects of voluntary and forced exercise on stress responses after traumatic brain injury. *J Neurotrauma.* 29:1426–1433.

Griesbach, G.S., J. Vincelli, D.L. Tio, and D.A. Hovda. 2012b. Effects of acute restraint-induced stress on glucocorticoid receptors and brain-derived neurotrophic factor after mild traumatic brain injury. *Neuroscience.* 210:393–402.

Griesbach G.S., D. Tio, S. Nair, and A. Hovda. 2014. Recovery of stress response coincides with responsiveness to voluntary exercise after traumatic brain injury. *J Neurotrauma.* 31:674-682.

Gronli, J., C. Bramham, R. Murison, T. Kanhema, E. Fiske, B. Bjorvatn et al. 2006. Chronic mild stress inhibits BDNF protein expression and CREB activation in the dentate gyrus but not in the hippocampus proper. *Pharmacol Biochem Behav.* 85:842–849.

Gruner, M.L., and D. Terhaag. 2000. Multimodal early onset stimulation (MEOS) in rehabilitation after brain injury. *Brain Inj.* 14:585–594.

Guskiewicz, K.M., S.W. Marshall, J. Bailes, M. McCrea, H.P. Harding, Jr., A. Matthews et al. 2007. Recurrent concussion and risk of depression in retired professional football players. *Med Sci Sports Exerc.* 39:903–909.

Guskiewicz, K.M., M. McCrea, S.W. Marshall, R.C. Cantu, C. Randolph, W. Barr et al. 2003. Cumulative effects associated with recurrent concussion in collegiate football players: The NCAA Concussion Study. *JAMA.* 290:2549–2555.

Hall, E.D., M.R. Detloff, K. Johnson, and N.C. Kupina. 2004. Peroxynitrite-mediated protein nitration and lipid peroxidation in a mouse model of traumatic brain injury. *J Neurotrauma.* 21:9–20.

Hamm, R.J., M.D. Temple, D.M. O'Dell, B.R. Pike, and B.G. Lyeth. 1996. Exposure to environmental complexity promotes recovery of cognitive function after traumatic brain injury. *J Neurotrauma.* 13:41–47.

Hansson, A.C., A. Cintra, N. Belluardo, W. Sommer, M. Bhatnagar, M. Bader et al. 2000. Gluco- and mineralocorticoid receptor-mediated regulation of neurotrophic factor gene expression in the dorsal hippocampus and the neocortex of the rat. *Eur J Neurosci.* 12:2918–2934.

Hayashi, K., Y. Honda, T. Ogawa, N. Kondo, and T. Nishiyasu. 2006. Relationship between ventilatory response and body temperature during prolonged submaximal exercise. *J Appl Physiol.* 100:414–420.

Held, J.M., J. Gordon, and A.M. Gentile. 1985. Environmental influences on locomotor recovery following cortical lesions in rats. *Behav Neurosci.* 99:678–690.

Hicks, R.R., Boggs, A., Leider, D., Kraemer, P., Brown, R., Scheff, S.W. et al. 1998. Effects of exercise following lateral fluid percussion brain injury in rats. *Restorative Neurol Neurosci.* 12:41–47.

Hicks, R.R., V.B. Martin, L. Zhang, and K.B. Seroogy. 1999. Mild experimental brain injury differentially alters the expression of neurotrophin and neurotrophin receptor mRNAs in the hippocampus. *Exp Neurol.* 160:469–478.

Hicks, R.R., S. Numan, H.S. Dhillon, M.R. Prasad, and K.B. Seroogy. 1997. Alterations in BDNF and NT-3 mRNAs in rat hippocampus after experimental brain trauma. *Brain Res.* 48:401–406.

Hicks, R.R., L. Zhang, A. Atkinson, M. Stevenon, M. Veneracion, and K.B. Seroogy. 2002. Environmental enrichment attenuates cognitive deficits, but does not alter neurotrophin gene expression in the hippocampus following lateral fluid percussion brain injury. *Neuroscience.* 112:631–637.

Hicks, R.R., L. Zhang, H.S. Dhillon, M.R. Prasad, and K.B. Seroogy. 1998. Expression of trkB mRNA is altered in rat hippocampus after experimental brain trauma. *Brain Res.* 59:264–268.

Hillman, C.H., K.I. Erickson, and A.F. Kramer. 2008. Be smart, exercise your heart: Exercise effects on brain and cognition. *Nat Rev. Neurosci.* 9:58–65.

Hoffman, A.N., R.R. Malena, B.P. Westergom, P. Luthra, J.P. Cheng, H.A. Aslam et al. 2008. Environmental enrichment-mediated functional improvement after experimental traumatic brain injury is contingent on task-specific neurobehavioral experience. *Neurosci Lett.* 431:226–230.

Humm, J.L., D.A. Kozlowski, S.T. Bland, D.C. James, and T. Schallert. 1999. Use-dependent exaggeration of brain injury: Is glutamate involved? *Exp Neurol.* 157:349–358.

Humm, J.L., D.A. Kozlowski, D.C. James, J.E. Gotts, and T. Schallert. 1998. Use-dependent exacerbation of brain damage occurs during an early post-lesion vulnerable period. *Brain Res.* 783:286–292.

Ingebrigtsen, T., K. Waterloo, S. Marup-Jensen, E. Attner, and B. Romner. 1998. Quantification of post-concussion symptoms 3 months after minor head injury in 100 consecutive patients. *J Neurol.* 245:609–612.

Ip, E.Y., C.C. Giza, G.S. Griesbach, and D.A. Hovda. 2002. Effects of enriched environment and fluid percussion injury on dendritic arborization within the cerebral cortex of the developing rat. *J Neurotrauma.* 19:573–585.

Ip, E.Y., E.R. Zanier, A.H. Moore, S.M. Lee, and D.A. Hovda. 2003. Metabolic, neurochemical, and histologic responses to vibrissa motor cortex stimulation after traumatic brain injury. *J Cerebr Blood Flow Metab.* 23:900–910.

Itoh, T., M. Imano, S. Nishida, M. Tsubaki, S. Hashimoto, A. Ito et al. 2011. Exercise inhibits neuronal apoptosis and improves cerebral function following rat traumatic brain injury. *J Neural Transmission.* 118:1263–1272.

Jadavji, N.M., B. Kolb, and G.A. Metz. 2006. Enriched environment improves motor function in intact and unilateral dopamine-depleted rats. *Neuroscience.* 140:1127–1138.

Jankowsky, J.L., T. Melnikova, D.J. Fadale, G.M. Xu, H.H. Slunt, V. Gonzales et al. 2005. Environmental enrichment mitigates cognitive deficits in a mouse model of Alzheimer's disease. *J Neurosci.* 25:5217–5224.

Jones, T.A., D.J. Liput, E.L. Maresh, N. Donlan, T.J. Parikh, D. Marlowe, and D.A. Kozlowski. 2012. Use-dependent dendritic regrowth is limited after unilateral controlled cortical impact to the forelimb sensorimotor cortex. *J Neurotrauma.* 29:1455–1468.

Jones, T.A., and T. Schallert. 1994. Use-dependent growth of pyramidal neurons after neocortical damage. *J Neurosci.* 14:2140–2152.

Jordan, B.D. 2000. Chronic traumatic brain injury associated with boxing. *Semin Neurol.* 20:179–185.

Jovanovic, J.N., A.J. Czernik, A.A. Fienberg, P. Greengard, and T.S. Sihra. 2000. Synapsins as mediators of BDNF-enhanced neurotransmitter release. *Nat Neurosci.* 3:323–329.

Karu, T.I., L.V. Pyatibrat, and N.I. Afanasyeva. 2005. Cellular effects of low power laser therapy can be mediated by nitric oxide. *Lasers Surg Med.* 36:307–314.

Kempermann, G., E.P. Brandon, and F.H. Gage. 1998. Environmental stimulation of 129/SvJ mice causes increased cell proliferation and neurogenesis in the adult dentate gyrus. *Curr Biol.* 8:939–942.

Kempermann, G., and F.H. Gage. 1999. Experience-dependent regulation of adult hippocampal neurogenesis: Effects of long-term stimulation and stimulus withdrawal. *Hippocampus.* 9:321–332.

Kempermann, G., H.G. Kuhn, and F.H. Gage. 1997. More hippocampal neurons in adult mice living in an enriched environment. *Nature.* 386:493–495.

Keren, O., S. Yupatov, M.M. Radai, R. Elad-Yarum, D. Faraggi, S. Abboud et al. 2005. Heart rate variability (HRV) of patients with traumatic brain injury (TBI) during the post-insult sub-acute period. *Brain Inj.* 19:605–611.

Khuman, J., J. Zhang, J. Park, J.D. Carroll, C. Donahue, and M.J. Whalen. 2012. Low-level laser light therapy improves cognitive deficits and inhibits microglial activation after controlled cortical impact in mice. *J Neurotrauma.* 29:408–417.

Kilpatrick, M.M., D.W. Lowry, A.D. Firlik, H. Yonas, and D.W. Marion. 2000. Hyperthermia in the neurosurgical intensive care unit. *Neurosurgery.* 47:850–855; discussion 855–856.

Kim, D.H., I.G. Ko, B.K. Kim, T.W. Kim, S.E. Kim, M.S. Shinal et . 2010. Treadmill exercise inhibits traumatic brain injury-induced hippocampal apoptosis. *Physiol Behav.* 101:660–665.

Kleim, J.A., T.A. Jones, and T. Schallert. 2003. Motor enrichment and the induction of plasticity before or after brain injury. *Neurochem Res.* 28:1757–1769.

Kline, A.E., and C.E. Dixon. 2001. Contemporary in vivo models of brain trauma and a comparison of injury responses. In *Head Trauma: Basic, Preclinical, and Clinical Directions.* Miller, R.H., Hayes, R.L., and Newcomb, J.K., eds. John Wiley & Sons, Inc., New York, NY. pp. 65–84.

Kline, A.E., J.L. Massucci, D.W. Marion, and C.E. Dixon. 2002a. Attenuation of working memory and spatial acquisition deficits after a delayed and chronic bromocriptine treatment regimen in rats subjected to traumatic brain injury by controlled cortical impact. *J Neurotrauma.* 19:415–425.

Kline, A.E., R.L. McAloon, K.A. Henderson, U.K. Bansal, B.M. Ganti, R.H. Ahmed et al. 2010. Evaluation of a combined therapeutic regimen of 8-OH-DPAT and environmental enrichment after experimental traumatic brain injury. *J Neurotrauma.* 27:2021–2032.

Kline, A.E., A.S. Olsen, C.N. Sozda, A.N. Hoffman, and J.P. Cheng. 2012. Evaluation of a combined treatment paradigm consisting of environmental enrichment and the 5-HT_{1A} receptor agonist buspirone after experimental traumatic brain injury. *J Neurotrauma.* 29:1960–1969.

Kline, A.E., A.K. Wagner, B.P. Westergom, R.R. Malena, R.D. Zafonte, A.S. Olsen et al. 2007. Acute treatment with the 5-HT_{1A} receptor agonist 8-OH-DPAT and chronic environmental enrichment confer neurobehavioral benefit after experimental brain trauma. *Behav Brain Res.* 177:186–194.

Kline, A.E., J. Yu, J.L. Massucci, R.D. Zafonte, and C.E. Dixon. 2002b. Protective effects of the 5-HT_{1A} receptor agonist 8-hydroxy-2-(di-n-propylamino)tetralin against traumatic brain injury-induced cognitive deficits and neuropathology in adult male rats. *Neurosci Lett.* 333:179–182.

Klingels, K., H. Feys, G. Molenaers, G. Verbeke, S. Van Daele, J. Hoskens, K. Desloovere, and P. De Cock. 2013. Randomized trial of modified constraint-induced movement therapy with and without an intensive therapy program in children with unilateral cerebral palsy. *Neurorehabil Neural Repair.* 27:799–807.

Knapp, H.D., E. Taub, and A.J. Berman. 1958. Effect of deafferentation on a conditioned avoidance response. *Science.* 128:842–843.

Koizumi, H., and J.T. Povlishock. 1998. Posttraumatic hypothermia in the treatment of axonal damage in an animal model of traumatic axonal injury. *J Neurosurg.* 89:303–309.

Kovesdi, E., A.B. Gyorgy, S.K. Kwon, D.L. Wingo, A. Kamnaksh, J.B. Long et al. 2011. The effect of enriched environment on the outcome of traumatic brain injury; a behavioral, proteomics, and histological study. *Front Neurosci.* 5:42.

Kozlowski, D. and T. Schallert. 1998. Relationship between dendritic pruning and behavioral recovery following sensorimotor cortex lesions. *Behav Brain Res.* 97:89–98.

Kozlowski, D.A., D.C. James and T. Schallert. 1996. Use-dependent exageration of neuronal injury after unilateral sensorimotor cortex lesions. *J Neurosci.* 16:4776–4786.

Kozlowski, D.A., J.L. Leasure, and T. Schallert. 2013. The control of movement following traumatic brain injury. *Comprehens Physiol.* 3:121–139.

Kozlowski, D.A., B.V. Nahed, D.A. Hovda, and S.M. Lee. 2004. Paradoxical effects of cortical impact injury on environmentally enriched rats. *J Neurotrauma.* 21:513–519.

Kroll, R.A., and E.A. Neuwelt. 1998. Outwitting the blood-brain barrier for therapeutic purposes: Osmotic opening and other means. *Neurosurgery.* 42:1083–1099; discussion 1099–1100.

Laker, S.R. 2011. Epidemiology of concussion and mild traumatic brain injury. *PM & R.* 3:S354–358.

Lam, T.I., D. Bingham, T.J. Chang, C.C. Lee, J. Shi, D. Wang, S. Massa, R.A. Swanson, and J. Liu. 2013. Beneficial effects of minocycline and botulinum toxin-induced constraint physical therapy following experimental traumatic brain injury. *Neurorehabil Neural Repair.* 27:889–899

Lampl, Y., J.A. Zivin, M. Fisher, R. Lew, L. Welin, B. Dahlof et al. 2007. Infrared laser therapy for ischemic stroke: A new treatment strategy: Results of the NeuroThera Effectiveness and Safety Trial-1 (NEST-1). *Stroke.* 38:1843–1849.

Laurer, H.L., F.M. Bareyre, V.M. Lee, J.Q. Trojanowski, L. Longhi, R. Hoover et al. 2001. Mild head injury increasing the brain's vulnerability to a second concussive impact. *J Neurosurg.* 95:859–870.

Lessmann, V., K. Gottmann, and R. Heumann. 1994. BDNF and NT-4/5 enhance glutamatergic synaptic transmission in cultured hippocampal neurones. *Neuroreport.* 6:21–25.

Leung, M.C., S.C. Lo, F.K. Siu, and K.F. So. 2002. Treatment of experimentally induced transient cerebral ischemia with low energy laser inhibits nitric oxide synthase activity and up-regulates the expression of transforming growth factor-beta 1. *Lasers Surg Med.* 31:283–288.

Levin, H.S., S. Mattis, R.M. Ruff, H.M. Eisenberg, L.F. Marshall, K. Tabaddor, W.M. High, Jr., and R.F. Frankowski. 1987. Neurobehavioral outcome following minor head injury: A three-center study. *J Neurosurg*. 66:234–243.

Levine, E.S., R.A. Crozier, I.B. Black, and M.R. Plummer. 1998. Brain-derived neurotrophic factor modulates hippocampal synaptic transmission by increasing N-methyl-D-aspartic acid receptor activity. *Proc Natl Acad Sci U S A*. 95:10235–10239.

Levine, E.S., C.F. Dreyfus, I.B. Black, and M.R. Plummer. 1995. Brain-derived neurotrophic factor rapidly enhances synaptic transmission in hippocampal neurons via postsynaptic tyrosine kinase receptors. *Proc Natl Acad Sci U S A*. 92:8074–8077.

Li, L., L.S. Chin, O. Shupliakov, L. Brodin, T.S. Sihra, O. Hvalby et al. 1995. Impairment of synaptic vesicle clustering and of synaptic transmission, and increased seizure propensity, in synapsin I-deficient mice. *Proc Natl Acad Sci U S A*. 92:9235–9239.

Lillie, R., and C.A. Mateer. 2006. Constraint-based therapies as a proposed model for cognitive rehabilitation. *J Head Trauma Rehabil*. 21:119–130.

Lindner, M.D., M.A. Plone, C.K. Cain, B. Frydel, J.M. Francis, D.F. Emerich, and R.L. Sutton. 1998. Dissociable long-term cognitive deficits after frontal versus sensorimotor cortical contusions. *J Neurotrauma*. 15:199–216.

Lippert-Gruner, M., M. Maegele, J. Pokorny, D.N. Angelov, O. Svestkova, M. Wittner et al. 2007. Early rehabilitation model shows positive effects on neural degeneration and recovery from neuromotor deficits following traumatic brain injury. *Physiol Res*. 56:359–368.

Lippert-Gruner, M., M. Magele, O. Svestkova, Y. Angerova, T. Ester-Bode, and D.N. Angelov. 2011. Rehabilitation intervention in animal model can improve neuromotor and cognitive functions after traumatic brain injury: Pilot study. *Physiol Res*. 60:367–375.

Lippert-Gruner, M., C. Wedekind, R.I. Ernestus, and N. Klug. 2002. Early rehabilitative concepts in therapy of the comatose brain injured patients. *Acta Neurochirurg Suppl*. 79:21–23.

Llinas, R., J.A. Gruner, M. Sugimori, T.L. McGuinness, and P. Greengard. 1991. Regulation by synapsin I and Ca(2+)-calmodulin-dependent protein kinase II of the transmitter release in squid giant synapse. *J Physiol*. 436:257–282.

Longhi, L., K.E. Saatman, S. Fujimoto, R. Raghupathi, D.F. Meaney, J. Davis et al. 2005. Temporal window of vulnerability to repetitive experimental concussive brain injury. *Neurosurgery*. 56:364–374; discussion 364–374.

Maegele, M., M. Lippert-Gruener, T. Ester-Bode, J. Garbe, B. Bouillon, E. Neugebauer et al. 2005a. Multimodal early onset stimulation combined with enriched environment is associated with reduced CNS lesion volume and enhanced reversal of neuromotor dysfunction after traumatic brain injury in rats. *Eur J Neurosci*. 21:2406–2418.

Maegele, M., M. Lippert-Gruener, T. Ester-Bode, S. Sauerland, U. Schafer, M. Molcanyi et al. 2005b. Reversal of neuromotor and cognitive dysfunction in an enriched environment combined with multimodal early onset stimulation after traumatic brain injury in rats. *J Neurotrauma*. 22:772–782.

Majerske, C.W., J.P. Mihalik, D. Ren, M.W. Collins, C.C. Reddy, M.R. Lovell et al. 2008. Concussion in sports: Postconcussive activity levels, symptoms, and neurocognitive performance. *J Athletic Training*. 43:265–274.

Makdissi, M., D. Darby, P. Maruff, A. Ugoni, P. Brukner, and P.R. McCrory. 2010. Natural history of concussion in sport: Markers of severity and implications for management. *Am J Sports Med*. 38:464–471.

Mancini, M.C., M.B. Brandao, A. Dupin, A.F. Drummond, P.S. Chagas, and M.G. Assis. 2013. How do children and caregivers perceive their experience of undergoing the CIMT protocol? *Scand J Occup Therapy*. 20:343–348.

Margulies, S., R. Hicks, and L. Combination Therapies for Traumatic Brain Injury Workshop. 2009. Combination therapies for traumatic brain injury: Prospective considerations. *J Neurotrauma*. 26:925–939.

Maroon, J.C., M.R. Lovell, J. Norwig, K. Podell, J.W. Powell, and R. Hartl. 2000. Cerebral concussion in athletes: Evaluation and neuropsychological testing. *Neurosurgery*. 47:659–669; discussion 669–672.

Matter, A.M., K.A. Folweiler, L.M. Curatolo, and A.E. Kline. 2011. Temporal effects of environmental enrichment-mediated functional improvement after experimental traumatic brain injury in rats. *Neurorehabil Neural Repair*. 25:558–564.

McCrory, P., K. Johnston, W. Meeuwisse, M. Aubry, R. Cantu, J. Dvorak et al. 2005. Summary and agreement statement of the 2nd International Conference on Concussion in Sport, Prague 2004. *Br J Sports Med*. 39:196–204.

McCrory, P., W. Meeuwisse, K. Johnston, J. Dvorak, M. Aubry, M. Molloy et al. 2009. Consensus Statement on Concussion in Sport: The 3rd International Conference on Concussion in Sport held in Zurich, November 2008. *Br J Sports Med*. 43 Suppl 1:i76–90.

McDermott, K.L., R. Raghupathi, S.C. Fernandez, K.E. Saatman, A.A. Protter, S.P. Finklestein et al. 1997. Delayed administration of basic fibroblast growth factor (bFGF) attenuates cognitive dysfunction following parasagittal fluid percussion brain injury in the rat. *J Neurotrauma*. 14:191–200.

McDowell, S., J. Whyte, and M. D'Esposito. 1998. Differential effect of a dopaminergic agonist on prefrontal function in traumatic brain injury patients. *Brain*. 121 (Pt 6):1155–1164.

McEwen, B.S. 1999. Stress and hippocampal plasticity. *Annu Rev Neurosci*. 22:105–122.

Melloni, R.H., Jr., P.J. Apostolides, J.E. Hamos, and L.J. DeGennaro. 1994. Dynamics of synapsin I gene expression during the establishment and restoration of functional synapses in the rat hippocampus. *Neuroscience*. 58:683–703.

Menon, D.K. 2009. Unique challenges in clinical trials in traumatic brain injury. *Crit Care Med*. 37:S129–135.

Monaco, C.M., V.V. Mattiola, K.A. Folweiler, J.K. Tay, N.K. Yelleswarapu, L.M. Curatolo et al. 2013. Environmental enrichment promotes robust functional and histological benefits in female rats after controlled cortical impact injury. *Exp Neurol*. 247:410–418.

Moore, E.L., L. Terryberry-Spohr, and D.A. Hope. 2006. Mild traumatic brain injury and anxiety sequelae: A review of the literature. *Brain Inj*. 20:117–132.

Morris, R. 1984. Developments of a water-maze procedure for studying spatial learning in the rat. *J Neurosci Methods*. 11:47–60.

Morse, J.K., S.J. Wiegand, K. Anderson, Y. You, N. Cai, J. Carnahan et al. 1993. Brain-derived neurotrophic factor (BDNF) prevents the degeneration of medial septal cholinergic neurons following fimbria transection. *J Neurosci*. 13:4146–4156.

Naeser, M.A., A. Saltmarche, M.H. Krengel, M.R. Hamblin, and J.A. Knight. 2011. Improved cognitive function after transcranial, light-emitting diode treatments in chronic, traumatic brain injury: Two case reports. *Photomed Laser Surg*. 29:351–358.

Nakamura, H., S. Kobayashi, Y. Ohashi, and S. Ando. 1999. Age-changes of brain synapses and synaptic plasticity in response to an enriched environment. *J Neurosci Res*. 56:307–315.

Navarro, A., C. Gomez, J.M. Lopez-Cepero, and A. Boveris. 2004. Beneficial effects of moderate exercise on mice aging: Survival, behavior, oxidative stress, and mitochondrial electron transfer. *Am J Physiol.* 286:R505–511.

Neeper, S.A., F. Gomez-Pinilla, J. Choi, and C. Cotman. 1995. Exercise and brain neurotrophins. *Nature.* 373:109.

Neeper, S.A., F. Gomez-Pinilla, J. Choi, and C.W. Cotman. 1996. Physical activity increases mRNA for brain-derived neurotrophic factor and nerve growth factor in rat brain. *Brain Res.* 726:49–56.

Nibuya, M., S. Morinobu, and R.S. Duman. 1995. Regulation of BDNF and trkB mRNA in rat brain by chronic electroconvulsive seizure and antidepressant drug treatments. *J Neurosci.* 15:7539–7547.

Nilsson, M., E. Perfilieva, U. Johansson, O. Orwar, and P.S. Eriksson. 1999. Enriched environment increases neurogenesis in the adult rat dentate gyrus and improves spatial memory. *J Neurobiol.* 39:569–578.

Nithianantharajah, J., H. Levis, and M. Murphy. 2004. Environmental enrichment results in cortical and subcortical changes in levels of synaptophysin and PSD-95 proteins. *Neurobiol Learning Memory.* 81:200–210.

Nygren, J., and T. Wieloch. 2005. Enriched environment enhances recovery of motor function after focal ischemia in mice, and downregulates the transcription factor NGFI-A. *J Cerebr Blood Flow Metab.* 25:1625–1633.

Olsen, A.S., C.N. Sozda, J.P. Cheng, A.N. Hoffman, and A.E. Kline. 2012. Traumatic brain injury-induced cognitive and histological deficits are attenuated by delayed and chronic treatment with the 5-HT_{1A} receptor agonist buspirone. *J Neurotrauma.* 29:1898–1907.

Oron, A., U. Oron, J. Streeter, L. de Taboada, A. Alexandrovich, V. Trembovler et al. 2007. low-level laser therapy applied transcranially to mice following traumatic brain injury significantly reduces long-term neurological deficits. *J Neurotrauma.* 24:651–656.

Oron, A., U. Oron, J. Streeter, L. De Taboada, A. Alexandrovich, V. Trembovler et al. 2012. Near infrared transcranial laser therapy applied at various modes to mice following traumatic brain injury significantly reduces long-term neurological deficits. *J Neurotrauma.* 29:401–407.

Pan, Y.X., L. Gao, W.Z. Wang, H. Zheng, D. Liu, K.P. Patel, I.H. Zucker, and W. Wang. 2007. Exercise training prevents arterial baroreflex dysfunction in rats treated with central angiotensin II. *Hypertension.* 49:519–527.

Passineau, M.J., E.J. Green, and W.D. Dietrich. 2001. Therapeutic effects of environmental enrichment on cognitive function and tissue integrity following severe traumatic brain injury in rats. *Exp Neurol.* 168:373–384.

Pedlow, K., S. Lennon, and C. Wilson. 2013. Application of constraint-induced movement therapy in clinical practice: An online survey. *Arch Phys Med Rehabil.* 95:276–282.

Pereira, A.C., D.E. Huddleston, A.M. Brickman, A.A. Sosunov, R. Hen, G.M. McKhann et al. 2007. An in vivo correlate of execise-induced neurogenesis in the adult dentate gyrus. *Proc Natl Acad Sci U S A.* 104:5638–5643.

Piao, C.S., B.A. Stoica, J. Wu, B. Sabirzhanov, Z. Zhao, R. Cabatbat et al. 2013. Late exercise reduces neuroinflammation and cognitive dysfunction after traumatic brain injury. *Neurobiol Dis.* 54:252–263.

Pierce, J.E., D.H. Smith, J.Q. Trojanowski, and T.K. McIntosh. 1998. Enduring cognitive, neurobehavioral and histopathological changes persist for up to one year following severe experimental brain injury in rats. *Neuroscience.* 87:359–369.

Poduslo, J.F., and G.L. Curran. 1996. Permeability at the blood-brain and blood-nerve barriers of the neurotrophic factors: NGF, CNTF, NT-3, BDNF. *Brain Res.* 36:280–286.

Quirk, B.J., M. Torbey, E. Buchmann, S. Verma, and H.T. Whelan. 2012. Near-infrared photobiomodulation in an animal model of traumatic brain injury: Improvements at the behavioral and biochemical levels. *Photomed Laser Surg.* 30:523–529.

Raghupathi, R. 2004. Cell death mechanisms following traumatic brain injury. *Brain Pathol.* 14:215–222.

Rhyu, I.J., J.A. Bytheway, S.J. Kohler, H. Lange, K.J. Lee, J. Boklewski et al. 2010. Effects of aerobic exercise training on cognitive function and cortical vascularity in monkeys. *Neuroscience.* 167:1239–1248.

Rimmer, J.H., M.D. Chen, J.A. McCubbin, C. Drum, and J. Peterson. 2010. Exercise intervention research on persons with disabilities: What we know and where we need to go. *Am J Phys Med Rehabil.* 89:249–263.

Rogatsky, G.G., J. Sonn, Y. Kamenir, N. Zarchin, and A. Mayevsky. 2003. Relationship between intracranial pressure and cortical spreading depression following fluid percussion brain injury in rats. *J Neurotrauma.* 20:1315–1325.

Rose, F.D., M.J. Davey, S. Love, and P.A. Dell. 1987. Environmental enrichment and recovery from contralateral sensory neglect in rats with large unilateral neocortical lesions. *Behav Brain Res.* 24:195–202.

Rosenzweig, M.R., and E.L. Bennett. 1996. Psychobiology of plasticity: Effects of training and experience on brain and behavior. *Behav Brain Res.* 78:57–65.

Sakane, T., and W.M. Pardridge. 1997. Carboxyl-directed pegylation of brain-derived neurotrophic factor markedly reduces systemic clearance with minimal loss of biologic activity. *Pharmaceut Res.* 14:1085–1091.

Sakurai, A., C.M. Atkins, O.F. Alonso, H.M. Bramlett, and W.D. Dietrich. 2012. Mild hyperthermia worsens the neuropathological damage associated with mild traumatic brain injury in rats. *J Neurotrauma.* 29:313–321.

Schaaf, M.J., J. de Jong, E.R. de Kloet, and E. Vreugdenhil. 1998. Downregulation of BDNF mRNA and protein in the rat hippocampus by corticosterone. *Brain Res.* 813:112–120.

Schuit, A.J., E.J. Feskens, L.J. Launer, and D. Kromhout. 2001. Physical activity and cognitive decline, the role of the apolipoprotein e4 allele. *Med Sci Sports Exerc.* 33:772–777.

Schulz, M.R., S.W. Marshall, F.O. Mueller, J. Yang, N.L. Weaver, W.D. Kalsbeek et al. 2004. Incidence and risk factors for concussion in high school athletes, North Carolina, 1996–1999. *Am J Epidemiol.* 160:937–944.

Selassie, A.W., E. Zaloshnja, J.A. Langlois, T. Miller, P. Jones, and C. Steiner. 2008. Incidence of long-term disability following traumatic brain injury hospitalization, United States, 2003. *J Head Trauma Rehabil.* 23:123–131.

Seo, T.B., B.K. Kim, I.G. Ko, D.H. Kim, M.S. Shin, C.J. Kim et al. 2010. Effect of treadmill exercise on Purkinje cell loss and astrocytic reaction in the cerebellum after traumatic brain injury. *Neurosci Lett.* 481:178–182.

Sharma, H.S. 2006. Hyperthermia influences excitatory and inhibitory amino acid neurotransmitters in the central nervous system. An experimental study in the rat using behavioural, biochemical, pharmacological, and morphological approaches. *J Neural Trans.* 113:497–519.

Shaw, S.E., D.M. Morris, G. Uswatte, S. McKay, J.M. Meythaler, and E. Taub. 2005. Constraint-induced movement therapy for recovery of upper-limb function following traumatic brain injury. *J Rehabil Res Develop.* 42:769–778.

Shiel, A., J.P. Burn, D. Henry, J. Clark, B.A. Wilson, M.E. Burnett, and D.L. McLellan. 2001. The effects of increased rehabilitation therapy after brain injury: Results of a prospective controlled trial. *Clin Rehabil*. 15:501–514.
Shohami, E., I. Gati, E. Beit-Yannai, V. Trembovler, and R. Kohen. 1999. Closed head injury in the rat induces whole body oxidative stress: Overall reducing antioxidant profile. *J Neurotrauma*. 16:365–376.
Silva, A.J., J.H. Kogan, P.W. Frankland, and S. Kida. 1998. CREB and memory. *Annu Rev Neurosci*. 21:127–148.
Smith, D.H., X.H. Chen, J.E. Pierce, J.A. Wolf, J.Q. Trojanowski, D.I. Graham et al. 1997. Progressive atrophy and neuron death for one year following brain trauma in the rat. *J Neurotrauma*. 14:715–727.
Smith, J.M., P. Lunga, D. Story, N. Harris, J. Le Belle, M.F. James et al. 2007. Inosine promotes recovery of skilled motor function in a model of focal brain injury. *Brain*. 130:915–925.
Smith, M.A., S. Makino, R. Kvetnansky, and R.M. Post. 1995. Effects of stress on neurotrophic factor expression in the rat brain. *Ann NY Acad, Sci*. 771:234–239.
Sozda, C.N., A.N. Hoffman, A.S. Olsen, J.P. Cheng, R.D. Zafonte, and A.E. Kline. 2010. Empirical comparison of typical and atypical environmental enrichment paradigms on functional and histological outcome after experimental traumatic brain injury. *J Neurotrauma*. 27:1047–1057.
Statler, K.D., L.W. Jenkins, C.E. Dixon, R.S. Clark, D.W. Marion, and P.M. Kochanek. 2001. The simple model versus the super model: Translating experimental traumatic brain injury research to the bedside. *J Neurotrauma*. 18:1195–1206.
Sterr, A., T. Elbert, I. Berthold, S. Kolbel, B. Rockstroh, and E. Taub. 2002. Longer versus shorter daily constraint-induced movement therapy of chronic hemiparesis: An exploratory study. *Arch Phys Med Rehabil*. 83:1374–1377.
Stevenson, T., L. Thalman, H. Christie, and W. Poluha. 2012. Constraint-induced movement therapy compared to dose-matched interventions for upper-limb dysfunction in adult survivors of stroke: A systematic review with meta-analysis. *Physiother Canada*.64:397–413.
Stummer, W., K. Weber, B. Tranmer, A. Baethmann, and O. Kempski. 1994. Reduced mortality and brain damage after locomotor activity in gerbil forebrain ischemia. *Stroke*. 25:1862–1869.
Suehiro, E., H. Fujisawa, H. Ito, T. Ishikawa, and T. Maekawa. 1999. Brain temperature modifies glutamate neurotoxicity in vivo. *J Neurotrauma*. 16:285–297.
Sullivan, P.G., J.N. Keller, M.P. Mattson, and S.W. Scheff. 1998. Traumatic brain injury alters synaptic homeostasis: Implications for impaired mitochondrial and transport function. *J Neurotrauma*. 15:789–798.
Summers, C.R., B. Ivins, and K.A. Schwab. 2009. Traumatic brain injury in the United States: An epidemiologic overview. *Mt Sinai J Med*. 76:105–110.
Szelenyi, J. 2001. Cytokines and the central nervous system. *Brain Res Bull*. 54:329–338.
Tajiri, N., T. Yasuhara, T. Shingo, A. Kondo, W. Yuan, T. Kadota et al. 2010. Exercise exerts neuroprotective effects on Parkinson's disease model of rats. *Brain Res*. 1310:200–207.
Takei, N., K. Sasaoka, K. Inoue, M. Takahashi, Y. Endo, and H. Hatanaka. 1997. Brain-derived neurotrophic factor increases the stimulation-evoked release of glutamate and the levels of exocytosis-associated proteins in cultured cortical neurons from embryonic rats. *J Neurochem*. 68:370–375.
Taub, E. 1980. Somatosensory deafferentation research with monkeys: Implications for rehabilitation medicine. In *Behavioral Psychology in Rehabilitation Medicine: Clinical Applications*. L.P. Ince, ed. Williams & Wilkins, New York. pp. 371–401.
Taub, E. 2012a. The behavior-analytic origins of constraint-induced movement therapy: An example of behavioral neurorehabilitation. *Behav Analyst*. 35:155–178.
Taub, E. 2012b. Parallels between use of constraint-induced movement therapy to treat neurological motor disorders and amblyopia training. *Develop Psychobiol*. 54:274–292.
Taub, E., G. Uswatte, and R. Pidikiti. 1999. Constraint-Induced Movement Therapy: A new family of techniques with broad application to physical rehabilitation—a clinical review. *J Rehabil Res Develop*. 36:237–251.
Taylor, A.N., S.U. Rahman, D.L. Tio, S.M. Gardner, C.J. Kim, and R.L. Sutton. 2010. Injury severity differentially alters sensitivity to dexamethasone after traumatic brain injury. *J Neurotrauma*. 27:1081–1089.
Terrio, H., L.A. Brenner, B.J. Ivins, J.M. Cho, K. Helmick, K. Schwab, K. Scally, R. Bretthauer, and D. Warden. 2009. Traumatic brain injury screening: Preliminary findings in a US Army Brigade Combat Team. *J Head Trauma Rehabil*. 24:14–23.
Thoenen, H. 1995. Neurotrophins and neuronal plasticity. *Science*. 270:593–598.
Thompson, H.J., J. Lifshitz, N. Marklund, M.S. Grady, D.I. Graham, D.A. Hovda et al. 2005. Lateral fluid percussion brain injury: A 15-year review and evaluation. *J Neurotrauma*. 22:42–75.
Thurman, D.J., C.M. Branche, and J.E. Sniezek. 1998. The epidemiology of sports-related traumatic brain injuries in the United States: Recent developments. *J Head Trauma Rehabil*. 13:1–8.
Trejo, J.L., E. Carro, and I. Torres-Aleman. 2001. Circulating insulin-like growth factor I mediates exercise-induced increases in the number of new neurons in the adult hippocampus. *J Neurosci*. 21:1628–1634.
Tyler, W.J., and L.D. Pozzo-Miller. 2001. BDNF enhances quantal neurotransmitter release and increases the number of docked vesicles at the active zones of hippocampal excitatory synapses. *J Neurosci*. 21:4249–4258.
Tyler, W.J., X.L. Zhang, K. Hartman, J. Winterer, W. Muller, P.K. Stanton et al. 2006. BDNF increases release probability and the size of a rapidly recycling vesicle pool within rat hippocampal excitatory synapses. *J Physiol*. 574:787–803.
Ueyama, T.K., K. Nemoto, M. Sekimoto, S. Tone, and E. Senba. 1997. Immobilization stress reduced the expression of neurotrohins and their receptors in the rat brain. *Neurosci Res*. 28:103–110.
Vagnozzi, R., S. Signoretti, L. Cristofori, F. Alessandrini, R. Floris, E. Isgro et al. 2010. Assessment of metabolic brain damage and recovery following mild traumatic brain injury: A multicentre, proton magnetic resonance spectroscopic study in concussed patients. *Brain*. 133;3232–3242.
Vagnozzi, R., S. Signoretti, B. Tavazzi, R. Floris, A. Ludovici, S. Marziali et al. 2008. Temporal window of metabolic brain vulnerability to concussion: A pilot 1H-magnetic resonance spectroscopic study in concussed athletes—part III. *Neurosurgery*. 62:1286–1295; discussion 1295–1286.
van Praag, H. 2009. Exercise and the brain: Something to chew on. *Trends Neurosci*. 32:283–290.
van Praag, H., T. Shubert, C. Zhao, and F.H. Gage. 2005. Exercise enhances learning and hippocampal neurogenesis in aged mice. *J Neurosci*. 25:8680–8685.

Vanderploeg, R.D., K. Schwab, W.C. Walker, J.A. Fraser, B.J. Sigford, E.S. Date et al, and Defense and Veterans Brain Injury Center Study Group. 2008. Rehabilitation of traumatic brain injury in active duty military personnel and veterans: Defense and Veterans Brain Injury Center randomized controlled trial of two rehabilitation approaches. *Arch Phys Med Rehabil.* 89:2227–2238.

VanLeeuwen, J.E., G.M. Petzinger, J.P. Walsh, G.K. Akopian, M. Vuckovic, and M.W. Jakowec. 2009. Altered AMPA receptor expression with treadmill exercise in the 1-methyl-4-phenyl-1,2,3,6-tetrahydropyridine-lesioned mouse model of basal ganglia injury. *J Neurosci Res.* 88:650–668.

Vaynman, S., Z. Ying, and F. Gomez-Pinilla. 2003. Interplay between brain-derived neurotrophic factor and signal transduction modulators in the regulation of the effects of exercise on synaptic-plasticity. *Neuroscience.* 122:647–657.

Veena, J., B.N. Srikumar, K. Mahati, V. Bhagya, T.R. Raju, and B.S. Shankaranarayana Rao. 2009. Enriched environment restores hippocampal cell proliferation and ameliorates cognitive deficits in chronically stressed rats. *J Neurosci Res.* 87:831–843.

Wagner, A.K., A.E. Kline, J. Sokoloski, R.D. Zafonte, E. Capulong, and C.E. Dixon. 2002. Intervention with environmental enrichment after experimental brain trauma enhances cognitive recovery in male but not female rats. *Neurosci Lett.* 334:165–168.

Weightman, M.M., R. Bolgla, K.L. McCulloch, and M.D. Peterson. 2010. Physical therapy recommendations for service members with mild traumatic brain injury. *J Head Trauma Rehabil.* 25:206–218.

Weiller, C., F. Chollet, K.J. Friston, R.J. Wise, and R.S. Frackowiak. 1992. Functional reorganization of the brain in recovery from striatocapsular infarction in man. *Ann Neurol.* 31:463–472.

Will, B., R. Galani, C. Kelche, and M.R. Rosenzweig. 2004. Recovery from brain injury in animals: Relative efficacy of environmental enrichment, physical exercise or formal training (1990–2002). *Prog Neurobiol.* 72:167–182.

Will, B.E., M.R. Rosenzweig, and E.L. Bennett. 1976. Effects of differential environments on recovery from neonatal brain lesions, measured by problem-solving scores and brain dimensions. *Physiol Behav.* 16:603–611.

Will, B.E., M.R. Rosenzweig, E.L. Bennett, M. Hebert, and H. Morimoto. 1977. Relatively brief environmental enrichment aids recovery of learning capacity and alters brain measures after postweaning brain lesions in rats. *J Comp Physiol Psychol.* 91:33–50.

Wolf, S., Winstein, C., Miller, J., Taub, E., Uswatte, G., Morris, D., et al. 2006. Effect of constraint-induced movement therapy on upper extremity function 3–9 months after stroke: The EXCITE randomized clinical trial. *JAMA.* 296:2095–2104.

Woolf, P.D., C. Cox, M. Kelly, D. Nichols, J.V. McDonald, and R.W. Hamill. 1990. The adrenocortical response to brain injury: Correlation with the severity of neurologic dysfunction, effects of intoxication, and patient outcome. *Alcohol Clin Exp Res.* 14:917–921.

Wu, A., Z. Ying, and F. Gomez-Pinilla. 2004. The interplay between oxidative stress and brain-derived neurotrophic factor modulates the outcome of a saturated fat diet on synaptic plasticity and cognition. *Eur J Neurosci.* 19:1699–1707.

Xuan, W., F. Vatansever, L. Huang, Q. Wu, Y. Xuan, T. Dai et al. 2013. Transcranial low-level laser therapy improves neurological performance in traumatic brain injury in mice: Effect of treatment repetition regimen. *PloS One.* 8:e53454.

Yan, H.Q., J. Yu, A.E. Kline, P. Letart, L.W. Jenkins, D.W. Marion et al. 2000. Evaluation of combined fibroblast growth factor-2 and moderate hypothermia therapy in traumatically brain injured rats. *Brain Res.* 887:134–143.

Yang, K., J.R. Perez-Polo, X.S. Mu, H.Q. Yan, J.J. Xue, Y. Iwamoto et al. 1996. Increased expression of brain-derived neurotrophic factor but not neurotrophin-3 mRNA in rat brain after cortical impact injury. *J Neurosci Res.* 44:157–164.

Yoshino, A., D.A. Hovda, T. Kawamata, Y. Katayama, and D.P. Becker. 1991. Dynamic changes in local cerebral glucose utilization following cerebral conclusion in rats: Evidence of a hyper- and subsequent hypometabolic state. *Brain Res.* 561:106–119.

Yu, K., Y. Wu, Y. Hu, Q. Zhang, H. Xie, G. Liu et al. 2013. Neuroprotective effects of prior exposure to enriched environment on cerebral ischemia/reperfusion injury in rats: The possible molecular mechanism. *Brain Res.* 1538:93–103.

Zafra, F., D. Lindholm, E. Castren, J. Hartikka, and H. Thoenen. 1992. Regulation of brain-derived neurotrophic factor and nerve growth factor mRNA in primary cultures of hippocampal neurons and astrocytes. *J Neurosci.* 12:4793–4799.

Zai, L., C. Ferrari, C. Dice, S. Subbaiah, L.A. Havton, G. Coppola et al. 2011. Inosine augments the effects of a Nogo receptor blocker and of environmental enrichment to restore skilled forelimb use after stroke. *J Neurosci.* 31:5977–5988.

Zhao, L.R., B. Mattsson, and B.B. Johansson. 2000. Environmental influence on brain-derived neurotrophic factor messenger RNA expression after middle cerebral artery occlusion in spontaneously hypertensive rats. *Neuroscience.* 97:177–184.

Zhu, X.L., W.S. Poon, C.C. Chan, and S.S. Chan. 2007. Does intensive rehabilitation improve the functional outcome of patients with traumatic brain injury (TBI)? A randomized controlled trial. *Brain Inj.* 21:681–690.

35 Models of Posttraumatic Brain Injury Neurorehabilitation

Michelle D. Failla and Amy K. Wagner

CONTENTS

35.1 INTRODUCTION TO EXPERIMENTAL MODELS OF NEUROREHABILITATION

The effects of traumatic brain injury (TBI) are heterogeneous and have limited predictability, particularly in relation to neurorehabilitation. To address and treat the diverse sequelae that occur post-TBI, several experimental models have been designed to model human TBI. Yet, there is a growing gap between experimental therapeutic treatment studies and their translation to clinical TBI. As experimental models are vital to advance our understanding, management and treatment of TBI, this review aims to describe the role of experimental models in post-TBI neurorehabilitation research. We discuss important themes to consider in experimental rehabilitation modeling. Variations in models, treatment timing and chronicity, and factors like genetics, age, and gender can influence rehabilitation response and benefits. We propose more rigorous experimental rehabilitation models, integrated with a Rehabilomics-framework to examine how individual factors interact with injury to influence response to rehabilitation and outcomes following TBI, will yield important neurorehabilitation research breakthroughs and improved clinical care.

TBI is a major cause of disability and death worldwide (Maas et al., 2008). The effects of TBI are heterogeneous, and in addition to damage from the direct injury, TBI facilitates the development of secondary pathologic processes in the brain, including inflammation, excitotoxicity, ischemia, edema, and many chronic secondary signaling changes (Park et al., 2008). To examine the diverse sequelae post-TBI, several experimental models, primarily in rodents, have been designed to produce injuries reflecting many components of those observed in human TBI.

Experimental models are vital to the advancement of our understanding, management, and treatment of TBI. A recent example of the importance of animal models lies within work examining genetic risk factors for mortality after clinical TBI. Failla and colleagues (Failla et al., 2014) showed that genetic risk factors within the gene coding for brain-derived neurotrophic factor (BDNF) are associated with mortality risk after TBI. However, variations within the *BDNF* gene that reportedly increase secretion of the normally pro-survival neurotrophin were associated with reduced survival probabilities, especially in older individuals. Although this result seems to contradict reports of BDNF's dominantly pro-survival signaling in uninjured animals, a recent study by Rostami and colleagues (2013) showed that there may actually be an injury-induced increase in BDNF's pro-apoptotic signaling capabilities after experimental TBI. To evaluate and model this balance in BDNF signaling, and its impact on recovery post-TBI, it will be imperative to use both clinical and experimental paradigms to further contextualize and interpret the clinical genetic biomarker associations reported.

While furthering our understanding of TBI mechanisms relevant to clinical care is important, there are limitations with current experimental TBI approaches. Several therapies can be neuroprotective in experimental TBI, yet most of these agents have consistently failed to translate into successful treatments in the clinic (Ikonomidou and Turski, 2002; Narayan et al., 2002). In fact, after severe TBI, only about 40% of patients have a favorable outcome (MRC CRASH Trial Collaborators et al., 2008), suggesting there is a wide gap between experimental therapeutic treatment studies and their translation to clinical TBI. This apparent disconnect may be due to heterogeneity in injury as well as a lack of consideration for covariates. Similarly, it is unclear how concurrent extracerebral injuries may influence clinical trials results (Aarabi and Simard, 2009; Kochanek and Yonas, 1999; Narayan et al., 2002). One additional limitation of successfully translating experimental clinical trials to consider is the lack of understanding of TBI as a chronic, yet rehabilitation sensitive, disease.

To date, there has been a far more limited emphasis in the literature on mechanisms of chronic dysfunction after TBI. Contemporary research now suggests that TBI, in some instances including a subset of those with mild TBI (mTBI), is not simply a transient and static syndrome from which people recover, but rather a chronic and evolving disease state. Our own work suggests that TBI pathology has a dynamic time course in which secondary complications and symptoms can arise and require ongoing management (Failla et al., 2014). Within this chronic disease framework, therapeutic plasticity and recovery mechanisms interplay with ongoing neurodegenerative and other chronic state pathology to affect symptoms, complications, and function (Dixon et al., 1999; Niogi et al., 2008; Sidaros et al., 2008). Common functional impairment and disability after TBI can vary and lead to faster recovery patterns (months for motor function) (Katz et al., 1998) compared with longer recovery patterns (years for cognitive/mood recovery) (Hammond et al., 2004). This evolving understanding of the dynamic nature of TBI requires a paradigm shift in experimental TBI studies and a clear understanding of chronic states of TBI pathology and associated sequelae.

However, there are numerous advantages to using established experimental models to further elucidate the mechanisms and treatment of chronic TBI through neurorehabilitation approaches. One of the most important issues is a relative lack of established neurorehabilitation methodology within the experimental literature. There are several important issues to consider when modeling neurorehabilitation strategies after experimental TBI. Many therapeutic interventions are administered briefly in experimental models, yet in clinical practice, treatments reflect a more chronic intervention period. Similarly, rehabilitation models may allow for more controlled administration compared with variable practice parameters for common rehabilitation interventions in the clinical population. Importantly, much of the experimental TBI model literature has focused on acute neuroprotection and/or management, with little consideration or understanding of how acute secondary injury and acute care practices may influence chronic TBI and neurorehabilitation effects. This lack of understanding could have important clinical implications. For example, studies suggest now that treatment with haloperidol in a limited fashion immediately following TBI may not impede recovery, but delayed administration can greatly reduce motor and cognitive recovery (Kline et al., 2007a).

Given the utility of experimental models, and the need for a greater understanding of neurorehabilitation mechanisms as well as considerations and caveats for experimental rehabilitation model use, this review focuses on the current state of neurorehabilitation research in experimental TBI. We (1) provide insight into important rehabilitation-centric considerations in common TBI models, (2) summarize the current literature pairing current animal models with rehabilitation-specific interventions, and (3) identify areas of importance, impact, and improvement as a developmental guide for future integration of neurorehabilitation concepts in animal models of TBI.

35.2 TBI ANIMAL MODELS: CONSIDERATIONS FOR REHABILITATION

Several models have been developed that further our understanding of pharmacological and physiological changes after TBI and aid in development of new therapeutic avenues for study. Development of experimental TBI models began in larger animals like nonhuman primates (Gennarelli et al., 1981) and ferrets (Lighthall, 1988), but experimental TBI and therapeutics research progress has blossomed in rodent models. Many models were later developed/adapted for mice with the advent of and increased access to transgenic animals. To model critical components of neurorehabilitation in experimental animals, several important factors need to be considered. The models must be relevant to clinical TBI, reproducible across laboratories and investigators, and allow for severity manipulation. Parameters within each of the described models below can be manipulated to produce a continuum of injury severity (O'Connor et al., 2011), yet no model can capture the full breadth of clinical TBI, including those mechanisms uniquely associated with some injury types such as those seen with military TBI. Newer models relevant to military TBI are emerging (both blast and penetrating ballistic-like injury models) (Cunningham et al., 2013; Kochanek et al., 2011; Zoltewicz et al., 2013), as noted in the subsequent section on new avenues for rehabilitation models. In this section, we describe common rodent models and compare how they contribute to the understanding of specific chronic TBI sequelae and neurorehabilitation concepts (Table 35.1).

35.3 OVERVIEW OF COMMON ANIMAL MODELS IN TBI

One of the major models of experimental TBI is the controlled cortical impact (CCI) model. The CCI model procedures include a craniotomy and uses a pneumatic or electromagnetic device to deliver a blow to the exposed dura. CCI was developed in 1988 in the ferret (Lighthall, 1988), was later developed and characterized in the rat (Dixon et al., 1991a) and mouse (Smith et al., 1995), and now has become one of the most prominent models in neurorehabilitation. Across species, CCI mimics several aspects of human TBI: cognitive and motor deficits, blood–brain barrier disruption, and subdural hematoma (Dixon et al., 1991b; Lighthall, 1988; Smith et al., 1995). Damage is reported in both gray and white matter though considerably less in the latter) (Fox et al., 1998; Goodman et al., 1994; Scheff et al., 1997), evolving up to 1 year postinjury (Dixon et al., 1999). As such, there is evidence of motor deficits following CCI for up to 1 year postinjury in mice (Shear et al., 2004). Cognitive deficits also persist up to 1 year after CCI (Dixon et al., 1999; Fox et al., 1999; Lindner et al., 1998; Shear et al., 2004). Importantly, most applications of CCI result in damage to hippocampus, thalamus, striatum, and amygdala, important for both implicit and explicit components of effective spatial navigation in the Morris Water Maze (MWM) (Packard, 2009; Packard and Knowlton, 2002; Wolff et al., 2008). The CCI model is a favorable model for evaluation of neurorehabilitation but lacks some aspects of human TBI (i.e., coma, mortality, and diffuse injury) that may complicate its translation to neurorehabilitation studies.

The fluid percussion injury (FPI) model, initially designed for use in rabbits, has been refined in rats (Dixon et al., 1987). After anesthesia, the animal's skull is removed, and a pendulum swings from a known height to impact a saline-filled tube, inducing a fluid-filled bolus onto the intact dura. Movement of the injury location from midline to a relatively lateral position can produce varying types of histological injuries. As such, injury involvement within the contralateral hemisphere is dependent on how close to midline the craniotomy is performed (Vink et al., 2001), although this injury evolves into a cortical contusion near the impact site having both focal and diffuse characteristics (Thompson et al., 2005). There is evidence of cell death in the hippocampus, thalamus, and cortex (Hicks et al., 1996) with cavity expansion and neuronal death observed up to 1 year postinjury (Pierce et al., 1998; Smith et al., 1997). Eight weeks following FPI in rats, there is evidence of tissue damage, gliosis, and

TABLE 35.1
Summary of TBI Models Relevant in Rehabilitation

Model Name	Summary	Rehabilitation Considerations	References
		Classic Models	
CCI	After a craniotomy, a pneumatic/electromagnetic device delivers a blow to exposed dura	• TBI-induced damage and cognitive deficits evolving to 1 year postinjury • Damage to hippocampus, thalamus, striatum, and amygdala • Contrary to clinical presentation, CCI produces no significant coma or diffuse injury • Ongoing histological/behavioral changes 1 year post-injury	Dixon et al., 1999; Shear et al., 2004; Wolffet al., 2008
Fluid percussion injury	After a craniotomy, a pendulum impacts a saline-filled tube, inducing a fluid-filled bolus onto intact dura	• Both focal and diffuse characteristics • Cavity expansion and neuronal death observed up to 1 year postinjury • Motor and cognitive deficits reported up to 1 year postinjury	Pierce et al., 1998; Smith et al., 1997; Thompson et al., 2005
Weight-drop	A small weight is dropped through a guiding apparatus to deliver a blow directly to the skull	• No craniotomy; most reminiscent to human closed-head TBI • Altered inflammatory state 2 months after injury • Memory/learning deficits at 3 months postinjury	Schwulst et al., 2013; Zohar et al., 2003
Impact acceleration	Similar to weight-drop, with addition of a steel protection plate against the skull; head movement is allowed after impact	• Reduces the possibility of skull fracture (compared with weight-drop) • Significantly more diffuse model • Produces coma and subarachnoid hemorrhage • Limited understanding of chronic sequelae	Gennarelli et al., 1981; Marmarou et al., 1994
		Emerging Models	
Blast	Exposure to explosions, compression air-driven shock tubes, and detonation-shock tubes	• Little is known about how current therapies translate in blast-induced TBI • Models in swine show promise mirroring the polytrauma of blast TBI	Bauman et al., 2009
Repetitive	Repeated injury, applied across certain time intervals, in many different models	• Increased exposures in military and interest in multiple sports concussions • Experimental TBI models show exacerbated tissue damage and outcomes with repetitive injuries • Subconcussive blows difficult to capture	Longhi et al., 2005; Shitaka et al., 2011
Mild	Most classic models have been adapted to produce mild injuries	• Current prognostication of complications post-mTBI are lacking • Difficulty in development of experimental models from no measureable tissue damage • Evidence for anxiety-like behaviors and altered metabolism post-mTBI	DeWitt et al., 2013; Prins et al., 2013; Shultz et al., 2011
Combination	Primary injury with the addition of secondary injury (e.g., hypoxia) or additional TBI-related pathology	• No single model can encompass all neurorehabilitation • Combined injury models may better reflect TBI + polytrauma • In experimental TBI models, secondary insults in combination models can increase deficits chronically	Hemerka et al., 2012; Sauerbeck et al., 2012

reactive astrocytosis in the thalamus, hippocampus, and cortex with enlarged lateral ventricles (Bramlett et al., 1997a). Importantly, this model greatly impacts cerebral blood flow and produces uncoupling of cerebral glucose metabolism (Ginsberg et al., 1997) similar to that observed with human TBI (Bergsneider et al., 2000). Motor and cognitive deficits have been reported following severe lateral FPI up to 1 year postinjury (Pierce et al., 1998).

One of the only models that does not involve a craniotomy is the weight-drop method. This closed injury model produces a contusion by dropping a small weight through a guiding apparatus to deliver a blow directly to the skull

(Feeney et al., 1981). Thus, the weight-drop model has the advantage of being the most reminiscent to human closed-head TBI. Without the need for a craniotomy, this model also allows shams to more closely resemble control populations in human TBI. Although the majority of the studies using the weight-drop method have focused on acute methodology (Morales et al., 2005), recent work has begun to characterize more chronic deficits from mild injuries. The weight-drop method is easily adaptable to produce concussive-like injures (Tang et al., 1997). As such, Schwulst and colleagues observed an altered inflammatory state in mice 2 months after weight-drop injury (Schwulst et al., 2013), and Zohar and colleagues show memory and learning deficits at 3 months following a mild weight-drop injury in mice (Zohar et al., 2003).

The impact acceleration model of experimental TBI is a much more diffuse model, reproducing diffuse axonal injury observed in a large subset of clinical TBI patients. This model employs a weight-drop, but rodents are outfitted with a steel protection plate designed to distribute the force of the impact across the skull, minimizing the possibility of skull fracture. Similarly, the anesthetized animal's head is placed on a specialized foam that allows for some movement after the weight-drop (Marmarou et al., 1994). This model is significantly more diffuse, producing a coma, axonal swelling, and subarachnoid hemorrhage acutely (Gennarelli et al., 1981; Marmarou et al., 1994). Although literature is sparse regarding how this model impacts chronic behavioral recovery, one study suggests this model produces both anxiety and depression-like behaviors (Pandey et al., 2009).

35.4 CHALLENGES IN TRANSLATION BETWEEN ANIMAL MODELS AND CLINICAL TBI

Similar to the caveats with modeling acute TBI pathology, there is no one model that captures the wide range of pathology associated with clinical TBI especially when modeling the chronic disease components associated with TBI. There are several challenges present that are difficult to quantify, yet are integral to consider when evaluating translation between animal models and human disease. First, there is always the issue of translation in signaling pathways that likely differ in subtle ways. Similarly, it is unclear how specific neurotransmission and signaling pathways may impact chronic recovery after TBI differently in animal models compared with clinical populations (see Section 35.7).

Another consideration with experimental TBI in general is the use of anesthesia during the injury. As such, it is important to consider the impact anesthesia may have on animal models. Importantly, anesthetic drugs like isoflurane may even be neuroprotective (Statler et al., 2006), thus influencing injury cascades being modeled, and subsequently, their effects on recovery.

The rodent brain has important differences in structure and function that can limit translation and interpretability of these models. With a different organizational structure compared with humans, experimental rodent TBI models can differ from humans in terms of injury biomechanics. In fact, the presence of gyri in the human brain, compared with the smooth rodent brain, likely has implications with brain movement and biomechanics during an injury. Similarly, there are important functional neuroanatomical differences. For example, rodents lack specific histological patterns and organizational structure within the prefrontal cortex that makes injury to analogous structures like the dorsolateral prefrontal cortex difficult to interpret (Öngür and Price, 2000; Uylings et al., 2003). Since executive function is a central component to cognitive and clinical recovery post-TBI, this species-specific difference within frontal cortex is an important caveat to consider when assessing the value and translational potential for studies involving experimental TBI. Rodents also have a much shorter lifespan compared with humans, and in the context of long-term rehabilitation treatment research design, species-specific differences in aging pose an important timing issue. The time frames associated with acute versus chronic stages of injury are dynamic in humans, but in experimental TBI, timing issues are even more complex. In this review, we generally consider chronic time frames post-TBI in rodent models as beginning approximately 3–4 weeks postinjury. However, it is unclear if time frames designated for chronicity should be different based on the specific experimental TBI model.

There are other important issues inherent in neurorehabilitation, for which experimental modeling options are limited. Adherence/participation issues, voluntary versus forced interventions, and constant versus intermittent rehabilitation dosing make the idea of using rodents to model complex rehabilitation paradigms seem daunting. We consider, later in this review, some ideas on minimizing the impact of some of these issues.

35.5 ANATOMICAL-BEHAVIORAL CORRELATES

Because TBI is a highly heterogeneous disease, predicting behavioral and functional outcomes following human TBI is difficult. Injury parameters (i.e., location, severity, specific types) do not always predict specific behavioral or functional changes. In the clinic, outcome prediction can be complicated by other factors like heterogeneity in medical care, genetic variation, preexisting conditions, and social support after the injury. In animal studies, many of these variables are controlled and/or are irrelevant. Given the relative simplicity that experimental models afford, it seems logical to hypothesize that there would be strong correlations between injury parameters and behavioral/functional complications after experimental TBI. However, the identification of structure-function relationships have been mixed, suggesting there are either methodological issues (i.e., inconsistencies in injury parameters or behavioral tasks) or that there are important injury-related factors (like alterations in signaling pathways) that are not being fully considered.

To determine anatomical-behavioral correlates, many studies examine injury severity parameters in relation to behavioral performance measures to determine how injury severity might relate to recovery. In this regard, many studies examine contusion volume, or some method of tissue loss, to compare with behavior. Many studies measure contusion volume, quantifying the volume of tissue loss produced by the impact, but interestingly, these values are not always consistent across laboratories even when using similar injury parameters (Dixon et al., 1999; Goodman et al., 1994; Kochanek et al., 1995). Some studies quantify residual tissue area/volume present by comparing the ipsilateral and contralateral hemispheres (Dixon et al., 1999; Longhi et al., 2009; Marklund et al., 2001), or calculating the ratio or percentage of ipsilateral tissue volume to contralateral tissue volume remaining (Marklund et al., 2001; Zhang et al., 1998). Both methods use the uninjured contralateral hemisphere for a control because of significant interanimal variation in brain size (Marklund et al., 2009). Yet, these measures of injury severity also do not always correlate with behavioral performance after TBI (Clausen and Hillered, 2005; Marklund et al., 2009).

The experimental TBI literature shows fairly consistent associations between MWM performance and severity of injury (Markgraf et al., 2001; Smith et al., 1991; Yu et al., 2009), though the majority of these studies test animals within 3 weeks of the injury. Thus, a loss of memory function in severely injured animals result in slower recovery time that may not necessarily represent permanent dysfunction akin to the clinical chronic presentation. However, some longitudinal observations actually suggest stability of MWM performance in the presence of increasing contusion volume and ventricular expansion across the first year after moderate CCI (Dixon et al., 1999). This continuing evolution of the injury is consistent with dynamic tissue changes in both gray (Bendlin et al., 2008) and white (Sidaros et al., 2008) matter across the first year of recovery in clinical TBI. In contrast to injury severity studies, the previous work discussed demonstrates that histological measures do not always yield correlations with behavioral performance. One possibility for this poor correlation may be due to the redundancy of structures on the uninjured and injured hemispheres that allow for reorganization of function in the brain.

Although some of the lack of observable correlations may be due to methodological issues in contusion volume assessment, as suggested by Goodman et al. (1994) and Kochanek et al. (1995), the disparity between structure and function may be related to the neurobiological complexities contributing to behavioral or functional outcome. Although there is evidence of anatomical correlations to cognitive or motor functional recovery in experimental models (Markgraf et al., 2001; Smith et al., 1991; Yu et al., 2009), the literature is lacking with regard to anatomical correlates for behaviors associated with depression or anxiety after the injury (Washington et al., 2012). Similarly, functional and cognitive recovery metrics show strong correlations clinically to severity of injury; however, the relationship between mood and severity is still controversial. The lack of anatomical relationships to mood complications may be due to the paucity of research specifically assessing mood complications post-TBI, but evidence is mounting to suggest there is no relationship between injury severity and development or severity of depression or anxiety (Washington et al., 2012).

Some experimental behavioral modeling methods may not be sensitive enough to interpret TBI severity-related differences in behavior or other symptoms. As in clinical populations, injury severity and injury parameters usually only explain a fraction of behavioral complications (Novack et al., 2001). The inability to use functional imaging paradigms in rodent models is another translational limitation when considering applicability of these models to clinical behavioral outcomes. Although histological methods have become more sophisticated, and methods such as awake in vivo electrochemical monitoring are accessible for use with neurotrauma models, measures that truly assess what is specifically responsible for alterations in performance are limited.

35.6 DIFFUSE VERSUS FOCAL INJURY MODELS

Most of the current animal models produce primarily a diffuse or focal injury, whereas clinical presentation of TBI usually contains elements of both. The major animal models used (CCI, FPI, and weight-drop) all show predominantly focal rather than diffuse injuries, though there are certainly components of a diffuse injury in FPI and weight-drop. For example, in FPI there is more involvement of the brainstem, suggesting a possible explanation for an observed lack of pressure increases contralaterally in CCI and weight-drop but not FPI (Clausen and Hillered, 2005). This reduced capacity for diffuse injury may be one reason these models do not consistently induce a coma after the injury, but more diffuse models like the acceleration impact model can induce prolonged unconsciousness after anesthesia (Marmarou et al., 1994). Similarly, the location of the injury can affect the diffuse nature of the injury, as many models suggest that movement of the injury location from respectively lateral to more medial areas can increase the diffuse injury sustained (Thompson et al., 2005).

The diffuse elements of an experimental injury model can have a great impact on how behavioral studies are interpreted. For example, examination of MWM performance, a hippocampal-dependent task, may not correlate to histological measures of damage induced by a parasagittal CCI model (Lyeth et al., 1990). Since the injury is unilateral, and the contralateral hemisphere is still mostly intact, function may be reorganized after TBI. Thus, the contralateral side may be exhibiting learning and memory functions that the injured hippocampus was originally responsible for executing. Similarly, there is increasing evidence for involvement of thalamic nuclei in focal injury models like CCI. For example, when CCI is applied to the medial prefrontal cortex, there is evidence of increased cell death in thalamic nuclei that is related to performance in the MWM (Goss et al., 2003; He et al., 2004). Understanding the diffuse nature of the injury produced is important to fully interpreting deficits.

35.6.1 Involvement of the Hippocampus

Hippocampal atrophy is common after clinical TBI (Bigler et al., 1997, Jorge et al., 2007). Importantly, most experimental models show volume reductions in the hippocampus after TBI (Bramlett et al., 1997a; Smith et al., 1997). The hippocampus and its associated pathways are common regions of interest after experimental TBI because of their involvement in learning and memory processes. This point is not surprising because direct hippocampal lesions result in deficits in learning and memory (Scoville and Milner, 1957; Squire, 1992). Common applications of the CCI model results in damage to structures important for both implicit and explicit components of effective spatial navigation in the MWM, including the hippocampus, thalamus, striatum and amygdala (Packard, 2009; Packard and Knowlton, 2002; Wolff et al., 2008). Thus, from a behavioral perspective, commonly performed applications of the CCI model yields injury patterns where there is difficulty mastering the spatial components of the task and a reduced capability to learn the general strategies needed to locate the platform (Wolff et al., 2008).

35.7 IMPORTANT SIGNALING CASCADES IN NEUROREHABILITATION

35.7.1 Inflammation

Several studies have documented the immediate and large inflammatory response after experimental TBI (Morganti-Kossmann et al., 2007). However, a growing number of studies suggest that chronic inflammation is an important consideration after clinical TBI (Kumar et al., 2014; Ramlackhansingh et al., 2011). For example, we have shown acute inflammation can predict chronic depressive-like symptoms in clinical populations (Juengst et al., 2014). Experimental studies have begun to explore chronic inflammation and its role in neurogenesis and cognitive recovery. In our group, we have shown increased cytokine expression at 3 weeks post-CCI in the hippocampus (Zou et al., 2013). There is increased microglial activation 60 days out from CCI (Acosta et al., 2013). Interestingly, there are also increased inflammatory markers in subcortical regions like the thalamus and striatum, distal from the CCI contusion (Acosta et al., 2013). In the weight-drop injury method, there is evidence of an altered inflammatory state in mice 2 months postinjury (Schwulst et al., 2013). Similarly, clinical studies show increased inflammation decades after TBI in the thalamus that correlates with cognitive recovery (Ramlackhansingh et al., 2011). There are reductions in neurogenesis and gliogenesis 4 weeks after CCI in mice, with accompanying microglial activation (Rola et al., 2006). However, using the diffuse model of acceleration-impact injury, Bye and colleagues showed increased cell proliferation, though there was not a significant increase in surviving neurons at 8 weeks postinjury (Bye et al., 2011).

Chronic inflammation is an important consideration in rehabilitative therapies as interventions such as exercise (Wu et al., 2007) may interact with inflammatory states to influence recovery. Importantly, it may be the case that chronic inflammation could delay or reduce efficacy of certain therapies, necessitating management of the inflammatory state before application of certain interventions and/or monitoring of inflammatory biomarkers during other neurorehabilitation therapies.

35.7.2 Neurotrophic Support

The neurotrophin BDNF is important for synaptic plasticity, neurogenesis, and neuronal survival (Chen et al., 2004) and is relevant to TBI recovery. In clinical TBI, serum BDNF is acutely decreased, correlating with severity of injury (Kalish and Phillips, 2010). Hippocampal BDNF is chronically decreased in experimental TBI (Chen et al., 2005). BDNF is ubiquitous in the hippocampus, where it plays a critical role in synaptogenesis and maintenance, particularly in long-term potentiation with activity-dependent secretion of BDNF (Kovalchuk et al., 2002). Similarly, BDNF may be an underlying substrate for persistent long term memory storage (Bekinschtein et al., 2008a, 2008b). Increased BDNF levels also inhibit neuronal degeneration and apoptotic cell death (Itoh et al., 2011a). BDNF has also shown to be an important component to exercise therapies post-TBI (Wu et al., 2007).

BDNF signals through the tyrosine-related kinase-B (TrkB) receptor, full-length (TrkB.FL), and truncated (TrkB.T), as well as the p75^{NTR} receptor, activating antagonistic signaling cascades that are dependent on receptor milieu. Studies suggest that TrkB.FL/TrkB.T/p75^{NTR} expression ratios may vary after ischemia (Gomes et al., 2012) and that dynamic expression changes in these receptors after experimental TBI (Rostami et al., 2013). BDNF's role in TBI recovery may be dependent on the relative balance of these target receptors. Increased BDNF levels can inhibit neuronal degeneration and apoptotic cell death (Itoh et al., 2011a). Recent work in our group suggests genetic variation in the *BDNF* gene may be related to mortality after TBI (Failla et al., 2014).

Importantly, therapies thought to increase BDNF expression in the brain, like environmental enrichment (Chen et al., 2005) and exercise (Griesbach et al., 2009) are promising therapies for cognitive recovery post-TBI. Based on these studies and the aforementioned work, BDNF may be a viable therapeutic target and biomarker for long-term complications such as mood and cognitive impairments throughout TBI recovery.

35.7.3 Hypothalamus-Pituitary-Adrenal Axis

Stress is likely an important factor in recovery after TBI. The psychosocial adjustments and enduring cognitive/physical impairments after injury can create a chronically stressful environment for those with TBI. The hypothalamus-pituitary-adrenal (HPA) axis modulates the body's reaction to both physical and emotional stress. Cortisol (in humans) or corticosterone (rodents), or CORT, is a known stress hormone and an end-product of the HPA axis that binds to brain glucocorticoid receptors (Joëls, 2008). The stress of major trauma results in increase adrenocorticotropin and CORT (Woolf, 1992). Our published work suggests that serum CORT is elevated for many subjects after clinical TBI and linked to other adrenal/peripheral hormone production (estradiol) that is associated with outcome (Wagner et al., 2011). Cerebrospinal fluid increases in CORT early after TBI are even more striking after TBI and associated with poor outcomes (Santarsieri et al., 2013). Elevated cortisol is a known biological characteristic of depression. Although there are limited data on cortisol levels and cognitive function in depressed subjects, several lines of evidence suggest cortisol has effects on cognition, including performance on memory testing (Hinkelmann et al., 2009; Newcomer et al., 1994).

Experimental TBI studies show important relationships to stress response after TBI. One study showed that stressed and injured rats showed increased levels of CORT up to 2 months after blast TBI that was related to altered anxiety-like behaviors (Kwon et al., 2011). Forced exercise therapies increase CORT in rodents and result in reduced benefits of exercise after experimental TBI (Griesbach et al., 2012a). Importantly, the benefits of exercise in experimental TBI also seem to be dependent on recovery of the stress response before induction of exercise (Griesbach et al., 2013). These studies suggest some understanding that the stress response will greatly impact experimental neurorehabilitation research design and study interpretation.

35.8 MODELING MOTOR AND BEHAVIORAL DEFICITS

35.8.1 Motor

Nearly all experimental TBI paradigms have characterized motor deficits, even if this is not their primary outcome of interest. Motor assessments can be used to validate severity of the injury, monitor spontaneous recovery in the acute phase post-TBI, or establish recovery of motor function before other behavioral tests that require baseline motor function to interpret the findings (i.e., behavioral or cognitive tests such as the MWM).

The beam walking or beam balance tests assess vestibulomotor function after experimental TBI. This beam balance test examines the ability of an animal to remain upright on a narrow, elevated beam, usually for up to 60 seconds. Lateral brain injuries with motor cortex involvement induce transient hemiparesis, causing the animal to slip to one side. The beam-walking test examines the amount of time it takes the animal to traverse the beam and enter a goal box on the other side of the beam. Longer latencies to enter the goal box can indicate motor slowing or reduced balance.

In addition to the beam tests, the Rotarod test has also been used to evaluate vestibulomotor function post-TBI. For this test, animals are placed on a rotating rod and asked to maintain balance as the speed of rotation increases. This task has been used to measure severity of motor and balance deficits after application of a number of TBI models (Hamm et al., 1994).

Another way to assess functional recovery is the grip strength test. In the wire grip test, the animal is placed onto a wire grid and the animal is allowed to attach its forelimbs. The animal is then judged on its ability to remain on the wire grid, usually using a scale of 0–5. Animals with vestibulomotor dysfunction after TBI will show difficulty remaining on the wire grid (Sinz et al., 1999). The wire grip test is especially sensitive to effects on neuromuscular function, either from injury, recovery, or pharmacological interventions.

Often vestibulomotor tests (i.e., Rotarod) are assessed early in recovery and are shown to be transient in most experimental models. After a short recovery period, even some severely injured animals show minimal motor differences from sham animals (Washington et al., 2012). It is not clear if there are fluctuations in vestibulomotor function chronically after experimental TBI with these motor function tests. Importantly, skilled motor tasks like Rotarod may show more persistent deficits after experimental TBI compared to less skilled tests such as the grip strength test. Other factors such as sex can also influence performance on beam tasks (Wagner et al., 2004). Also, female rodents tend to perform better on the Rotarod task compared with male rodents (O'Connor et al., 2003). Although not typically considered, in neurorehabilitation research and modeling, it is important to understand motor function across sex and other demographics factors (i.e., age) (Cekic et al., 2011; Wali et al., 2011) that may also impact recovery.

35.8.2 Mood and Motivation in Exploratory and Forced Tasks

There are several common neurobehavioral complications after TBI. In fact, depression after TBI, or posttraumatic depression (PTD), and anxiety are both quite common after TBI (Bombardier et al., 2010; Jorge et al., 1993) and can greatly affect recovery post-TBI. Our data show individuals with PTD have more cognitive problems and report lower satisfaction with life scores. Further, those with PTD are at increased risk for suicide (León-Carrión et al., 2001). Also, studies suggest that treatments used for major depressive disorders are not necessarily effective in the context of PTD (Warden et al., 2006).

After a TBI, depression and anxiety can overlap with cognitive and somatic symptoms that can be attributed to the brain injury. Apathy, fatigue, and attentional difficulties can mask or influence identification of mood complications in the clinic. To address this type of behavior in experimental models, it is important to also examine motivation and apathy as well as depression or anxiety-like behaviors. As such, in experimental models, the distinction of forced versus voluntary tasks is important. Certain tasks put the animal in a forced situation and examine behaviors such as learned helplessness (forced swim test [FST]), yet others examine voluntary exploratory behavior (open field) or anhedonia (sucrose preference test [SPT]). Similarly, habituation to the experimental apparatus is also an important component of examining anxiety behaviors in tests such as open field (Bolivar et al., 2000). TBI may specifically affect habituation and thus examination of behavior during habituation may also provide insight into these behaviors post-TBI. As such, depression and anxiety-like behaviors will need to be conceptualized in the context of post-TBI to increase the validity of these studies in experimental TBI.

35.8.3 Depression

Although not commonly used as a behavioral component of experimental TBI, depressive behaviors in rodents are usually evaluated using the tail suspension test (TST) (Castagné et al., 2011), the FST (Slattery and Cryan, 2012), or the SPT (Willner et al., 1987). The TST involves suspension of the animal by its tail, upside down. Depressed animals spend less active time struggling, suggesting learned helplessness. The FST is used by placing an animal in a large beaker of water with no escape platform. Animals will swim or struggle to escape via the sides of the beakers. Floating behaviors or lack of escape attempts suggest depressive symptomology. The sucrose preference test is used in an animal's home cage. Animals are given two water bottles, one with water and another with a sucrose–water solution. Healthy animals greatly prefer the sucrose solution, yet animals with depressive-like behavior show no preference for sucrose. Both the TST and FST measure learned helplessness as a marker for depression, but the SPT is more reflective of the anhedonic aspect of depression. There is evidence in experimental models that depressive-like symptoms are present early after injury (Cope et al., 2011; Fenn et al., 2013) and may persist (Washington et al., 2012) as well, though the investigation into depressive-like symptoms in the context of rehabilitation is preliminary.

35.8.4 Anxiety

Anxiety can be measured using a number of tasks, either directly or indirectly within tests designed to evaluate other aspects of behavior. The open field task involves placing an animal in an open arena with no obstacles or stimuli and observing the animal for behavior after a period of acclimation. Anxious animals will not venture into the middle of the field, where less anxious animals will cross the center of the field in exploratory behavior (Prut and Belzung, 2003). The elevated plus and zero mazes are also used to evaluate anxiety-like behaviors in rodents. The elevated plus maze uses a raised platform with two closed arms perpendicular to two open arms. The elevated zero maze uses a raised platform in a circle formation with two nonadjacent closed portions, equal to the open portions. Because rodents prefer dark areas in a novel environment (Montgomery, 1955), animals will spend more time in the closed areas of these mazes compared with open areas, but more anxious animals will show an even greater preference for closed areas (Pellow et al., 1985; Shepherd et al., 1994). Similarly, thigmotaxic behavior in the MWM can be indicative of anxiety-like behaviors (Treit and Fundytus, 1988). Thigmotaxic behavior in the MWM is described as an animal that follows the outside of the pool, never venturing away from the wall to attempt to find the platform. There is mounting evidence in experimental models that anxiety-like symptoms are present after injury (Wagner et al., 2007a) up to 3 months postinjury (Jones et al., 2008; Liu et al., 2010), though some studies report reduced anxiety (Washington et al., 2012) after experimental TBI. Similarly, some studies have shown treatment with progesterone reduces anxiety-like behaviors after TBI (Cutler et al., 2006).

35.8.5 Learning and Memory

Cognitive dysfunction after TBI is the most common complaint cited by caregivers, even years after TBI (Hoofien et al., 2001). Multiple neuropsychological domains of cognition are impaired after TBI, including attention, arousal, memory, and executive control. Cognitive deficits mirror the severity of injury as indicated by the extent of diffuse axonal injury, generalized atrophy, and the location and extent of focal injuries (Katz and Alexander, 1994; Wilson et al., 1995). The neurobiological basis of cognitive function and TBI-induced cognitive dysfunction are not wholly understood, but are informed to some degree by experimental TBI studies.

Memory problems, both explicit and implicit, are common after TBI. Explicit memory involves the conscious acquisition and retention of new information, whereas implicit learning involves skill or knowledge acquisition without explicit awareness of how the material is learned. Explicit learning-memory is hippocampal dependent (Brandeis et al., 1989), and hippocampal damage is associated with reproducible deficits in spatial and temporal memory processing. Explicit learning-memory is vulnerable to the effects of TBI, whereas implicit learning-memory is hippocampal independent and involves both cortical and subcortical regions (Buckley, 2005; Vakil, 2005).

A frequently used method for assessing learning and memory after TBI is the MWM (D'Hooge and De Deyn, 2001; Vorhees and Williams, 2006). The MWM has been used extensively in experimental TBI for evaluation for cognitive and memory deficits after TBI (Scheff et al., 1997; Smith et al., 1991). In this task, the rodent is placed in a pool and given a finite amount of time to find a platform hidden below the surface of the water. On average, an animal can use three different strategies in finding the hidden platform. Taxis is defined as the strategy that uses proximal cues located within the maze. Praxis involves remembering a previously used sequence of movements. Finally, the spatial strategy involves explicit awareness and incorporation of the extra-maze cues to find a hidden platform (Brandeis et al., 1989). Although implicitly learned strategies like taxis and praxis are considered to be hippocampal independent (Bramlett et al., 1997b; Cain et al., 2006), the spatial mapping required to solve the task and the spatial strategy used is hippocampal dependent (Wolff et al., 2008). Wagner and colleagues has shown that implicit and explicit learning and memory are dissociable in the MWM, and that preinjury training can be leveraged to study each of these systems in the context of TBI (Wagner et al., 2013).

35.8.6 Attention and Executive Function

One of the most common cognitive complaints after TBI is attentional deficits (Mathias and Wheaton, 2007). Attentional deficits are important in rehabilitation because they can influence performance in other cognitive domains. Dopaminergic signaling is critically important for attention and cognitive control (Braver and Cohen, 2000). Attentional networks are likely vulnerable after TBI from the hypodopaminergic state that occurs post-TBI (Bales et al., 2009). Similarly, areas involved in attentional control (such as the dorsolateral prefrontal cortex) often show relative damage in diffuse TBI that may correlate to attentional difficulties in clinical TBI (Kraus et al., 2007).

Attention in rodents has been evaluated using the five-choice serial reaction time task (5-CSRTT) (Robbins, 2002). In the 5-CSRTT, the animal is in an apparatus with five nose-poke holes and a food magazine. The animal pushes the food magazine to initiate a trial then a light stimulus appears in one of the five nose-poke holes. The animal then pokes its nose in the appropriate hole to receive a food reward. Response latency is the most common measure on this task, and responses can be characterized by premature responses (before the light stimulus), correct response, incorrect response, or omission. The animal is usually asked to perform this task for up to 100 trials or 30 minutes to measure sustained attention. Performance on the 5-CSRTT is modulated by injury to attentional networks, specifically in the prefrontal cortex (Maddux et al., 2007) as well as subcortical regions like the habenula (Lecourtier and Kelly, 2005). One model of penetrating brain injury has shown a significant attention deficit during acute recovery using the 5-CSRTT (Plantman et al., 2012). Yet, overall, the examination of attention in experimental models is lacking in TBI research.

Difficulties in executive function are common after clinical TBI and remain one of the more complex issues in recovery. Specifically, many individuals experience a loss in behavioral flexibility and strategy planning (Bonatti et al., 2008; Brooks et al., 1999; Lipsky et al., 2005). To evaluate executive function deficits in experimental TBI, Bondi and colleagues (2014) have recently adapted the attentional set-shifting test (AST) for use in experimental TBI. In the AST, an animal is given access to two digging pots with each pot containing two cues in two dimensions (digging medium and odor). In each trial, one of the pots contained a food reward. Animals are expected to learn the cue associated with food reward. However, as trials increase, experimenters can vary the "rule" as to what stimulus is associated with the food reward. An inability to switch rules is reflective of a lack of behavioral flexibility. Bondi and colleagues showed that rats subjected to CCI injury show severity-dependent deficits on the AST. This suggests that the AST may be a valuable new tool in examining executive function in rehabilitation.

35.9 PRECLINICAL MODELS AND NONPHARMACOLOGICAL REHABILITATION INTERVENTIONS AND CONSTRUCTS

35.9.1 Introduction to Nonpharmacological Rehabilitation in Animal Models

Pharmacological intervention is not always ideal or practical after TBI. In the clinical setting, patients with TBI are treated for multiple symptomologies with numerous drugs that can have serious interactions and side effects. Similarly, the effects of traditional drug treatments are not always as efficacious in TBI, thus pointing to the need for alternative avenues of treatment. In this section, we will detail nonpharmacological, promising therapeutic avenues being investigated in experimental TBI and their potential impact on rehabilitation relevant components of recovery.

35.9.2 Forced versus Natural or Voluntary Behavioral Interventions

One important caveat in modeling neurorehabilitation is examining the difference between forced and voluntary therapies in animals to better mimic the clinical population. For example, in animal studies evaluating the impact of exercise, forced training immediately after TBI can result in detrimental cognitive effects as well as reduced neuroplasticity (Griesbach et al., 2004). In contrast, delayed, voluntary exercise can mitigate some of the cognitive deficits induced by experimental TBI (Griesbach et al., 2009). The detrimental effects of interventions such as forced exercise may be due to induction of stress, either physical or psychological, which could impede TBI rehabilitation. In this regard, animal models using a forced paradigm of exercise therapy could have negative effects on TBI recovery compared to voluntary paradigms. This example illustrates the importance of understanding the voluntary nature of therapies in relevance to clinical neurorehabilitation.

35.9.3 Exercise

Physical activity (exercise) enhances mood and cognitive function in both healthy and neurologically impaired populations (Hillman et al., 2008). Exercise has historically been a core rehabilitative tool that may be used alone or in conjunction with cognitive, occupational, and pharmaceutical interventions to facilitate recovery (Archer et al., 2012). Animal studies report both neuroprotective and neuroenhancing effects of exercise documenting improvement in cognitive ability, inhibition of neuronal degeneration and apoptotic cell death, stimulation of corticotrophic axes, restoration of homeostatic regulation, and improved intactness of the blood–brain barrier (Hillman et al., 2008).

Exercise increases expression and secretion of BDNF, which in turn can elevate learning, memory, and mood (Erickson et al., 2011). Studies show an immediate increase in BDNF protein during the first occurrence of exercise activity; the duration and amount of exercise was significantly and positively correlated to hippocampal increases in BDNF levels (Oliff et al., 1998). Modifications in serotonin and inflammatory cytokines (e.g., interleukin-6) are also instrumental in explaining the positive effects of exercise on brain health. Yet, animal studies have shown that exercise may rescue adverse effects of inflammation through increased BDNF levels (Wu et al., 2007). Preclinical TBI work shows that exercise improves cognitive recovery in animal models (Griesbach et al., 2004), and that this improvement is mediated by BDNF signaling (Griesbach et al., 2009). Griesbach and colleagues found that rats exposed to voluntary exercise after a delayed period recovered almost full functional ability on the MWM (compared with preinjury performance) (Griesbach et al., 2007). Exercise-induced elevations in BDNF may facilitate synaptic strength within existing neural networks as well as increase hippocampal neurogenesis, possibly leading to increased cognitive recovery post-TBI (Archer, 2012). Researchers have shown BDNF increases (Crane et al., 2012; Griesbach et al., 2007), reductions in inflammation, and increased neurogenesis (Piao et al., 2013) correlate to improved cognitive recovery after experimental TBI (Crane et al., 2012; Griesbach et al., 2007). Exercise has also shown to increase stem-cell proliferation after a focal brain injury (Itoh et al., 2011b). Although exercise therapies certainly show great promise in TBI, there are important caveats in this research.

35.9.4 Timing

Two factors can alter the effectiveness of exercise in treating patients—the therapeutic window in which exercise is introduced and the willingness of the patient to perform activities (Griesbach, 2011). Animal model studies suggest that when introduced too early in the recovery period, exercise acts as a stressor, exacerbating symptoms, decreasing levels of BDNF, and slowing the overall recovery process (Griesbach et al., 2013). Forced exercise early in recovery also increases corticosterone in animals after FPI, indicating an increased stress response to the forced exercise (Griesbach et al., 2012b). Similarly, forced exercise immediately after TBI can result in detrimental cognitive effects (Piao et al., 2013) as well as reduced neuroplasticity (Griesbach et al., 2004). In contrast, delayed exercise can mitigate some of the cognitive deficits induced by experimental TBI (Griesbach et al., 2009). Most studies define the therapeutic window as 12 days postinjury, although more recent studies suggest that the time window for exercise-induced benefits is severity dependent; more severe injuries warrant further delays before the introduction of exercise (Crane et al., 2012; Griesbach 2011; , 2012b). It is also not clear how early interventions impact chronic recovery and how long the intervention must be to create lasting effects on recovery.

35.9.5 Comparisons of Clinical and Animal Literature

Two clinical studies have examined exercise as a possible therapeutic in a TBI population and have reported promising and stable remittance of depressive symptomology (Hoffman et al., 2010; Wise et al., 2012). The mechanisms and timing for this response have not been explored in experimental animal models. Experimental TBI work shows that exercise improves cognitive recovery in animal models (Griesbach et al., 2004), and that this improvement is mediated by BDNF signaling (Griesbach et al., 2009). However, the dose-response of cognitive recovery to exercise in clinical TBI populations is not known.

Importantly, these studies have also shown that certain combinatorial interventions may not be advantages. One recent study showed that D-amphetamine administration for the first 7 days after CCI in rats interfered with benefits from voluntary exercise (Griesbach et al., 2008). In another study, docosahexaenoic acid administration in conjunction with exercise show synergistic effects (Wu et al., 2013). These combinatory studies are especially important given the fact that most of the TBI population is on medications for other issues; thus, combinatorial studies with experimental rehabilitation models are highly warranted.

35.10 ENVIRONMENTAL ENRICHMENT

Another treatment that has shown promise in animal models of TBI is environmental enrichment (EE). EE involves increasing the typical living space, allowing for interaction with continuously novel stimuli, and facilitating the opportunity for social interactions, all to increase cognitive and physical stimulation. EE has been difficult to study because of the wide range of potentially therapeutic, but poorly measured, components to the intervention that cause disparities in meta-analysis and limit applicability of this animal research to a clinical or rehabilitation setting. These factors include housing density, ages of rats, genders of rats, duration of environmental enrichment, frequencies of cleaning and rearranging housing environments, types of cages, and types of enrichment objects. In fact, a number of experimental studies have examined particular parts of EE to determine the lowest "dose" of EE; in experimental TBI; studies suggest having all of the components is necessary to reap the benefits of EE (Sozda et al., 2010).

A large body of work in experimental TBI is beginning to form with regard to EE as a possible therapeutic for TBI recovery. After experimental TBI, EE can improve motor recovery (Gentile et al., 1987; Held et al., 1985) and performance on the MWM (Passineau et al., 2001). After CCI, EE has shown motor and cognitive benefits with attenuated hippocampal CA3 cell loss (Kline et al., 2007b). After severe FPI, animals showed increased cognitive recovery as well as reduced lesion volume (Passineau et al., 2001).

Interestingly, sex has been shown to interact with EE after experimental TBI. In one study, males, but not females, benefited from EE after CCI (Wagner et al., 2002). Interestingly, there are sex-specific responses in dopamine transporter expression, although it is not clear if these mediate the differences in EE benefits (Wagner et al., 2005). In a follow-up study, this group showed sex-specific, region-specific differences in BDNF expression after CCI and environmental enrichment, though, in this study, BDNF did not correlate with EE-mediated cognitive recovery (Chen et al., 2005).

35.10.1 Translational Considerations with Environmental Enrichment

TBI is a chronic disease, and treatments need to be evaluated across recovery spectrum post-TBI. EE has mostly been administered early after recovery, and it is not clear how initiation in a more chronic period might impact recovery. Interestingly, however, acute and relatively brief exposure to EE in early recovery conferred long-term benefits (up to 6 months after CCI) (Cheng et al., 2012). It is still unclear if or how delayed EE would benefit after TBI. Similarly, it is still unclear if intermittent exposure to EE, as opposed to animals living in EE housing, would show similar benefits. Because many clinical environments try to minimize stimulation early in recovery, it would be important to examine exactly how EE might translate into a clinical population with regard to timing as well as selection and implementation of potentially enriching components.

35.11 COGNITIVE REHABILITATION

Many individuals experience significant cognitive complications after TBI. In the clinical population, cognitive rehabilitation (CR) paradigms have shown significant promise but have not been subjected to the rigor of experimental studies to characterize their mechanisms of action, effectiveness, or capacity to enhance efficacy of other interventions (Cicerone et al., 2011). Although implicit tasks are more resistant to external factors such as anxiety or fatigue, interference with explicit feedback during implicit learning can adversely affect task performance after TBI (Vakil, 2005). Implicitly learned information is often considered relatively "inflexible," where carryover of implicitly learned knowledge can be limited (Schmitter-Edgecombe, 2006). Given the importance of dissociating memory networks to study implicit-explicit learning-memory impairments and training approaches, recent work by Wagner et al. evaluated the effects of CCI on implicitly learned, nonspatial components of the MWM place learning task as well as spatial mapping components of the task by subjecting some animals to a preinjury course of nonspatial training in the MWM (Wagner et al., 2013). Injured rats that received this nonspatial, implicit training performed similarly to trained and untrained shams in the MWM void of extra-maze cues, suggesting that implicit learning and memory is intact 2 weeks after CCI. Moreover, in the cued condition, preinjury training reduced thigmotaxis, improved place learning, and largely eliminated the apparent injury-induced deficits typically observed between untrained CCI and sham rats (Smith et al., 1991).

Although CR interventions appear to improve attention, memory, social communication skills, and executive function in humans (Cicerone et al., 2011), it is unknown to what degree a paradigm of this nature can affect learning and memory in experimental models. Based on the experimental findings above with preinjury nonspatial training, and because of the benefits observed in clinical studies with CR, Wagner and colleagues (2013) have developed a rodent model of implicit cognitive training. The Wagner model of CR uses the knowledge that retention for implicitly learned material after clinical TBI, particularly automatized tasks, is similar to uninjured controls (Schmitter-Edgecombe, 2006).

The Wagner model used an implicit learning paradigm postinjury that was similar to the previously published pretraining paradigm. Animals were able to use this nonspatial strategy to improve MWM performance after the injury (Brayer et al., 2014). The current data with this CR model suggest that implicit training provided postinjury can ameliorate deficits in the MWM in a manner similar to preinjury implicit training (Wagner et al., 2013). Future work will address other avenues to enhance CR effects, such as visual priming (Vorhees and Williams, 2006) and errorless learning (Clare and Jones, 2008). Additional work will focus on potential performance variance with regard to sex, age, timing, and alternative injury models. This work will pave the way for studying CR in combination with other rehabilitation relevant therapies.

35.12 PRECLINICAL MODELS AND PHARMACOLOGICAL MANIPULATION OF REHABILITATION CONSTRUCTS

35.12.1 Overview

One of the significant advantages of animal models for rehabilitation research is their use in preclinical pharmacological interventions. In experimental TBI, drug treatments show improvement in behavioral (Clausen and Hillered, 2005; Marklund et al., 2001; Longhi et al., 2009), physiological (Xiong et al., 2012), and morphological (Clausen and Hillered, 2005; Longhi et al., 2009; Marklund et al., 2001; Zhang et al., 1998) markers of recovery. In this section, we will highlight several pharmacological studies that are important for rehabilitation research.

One of the most important approaches in experimental TBI research has been the development of neuroprotectants that could be administered immediately after the injury to attenuate secondary injury cascades. Several classes of drugs have been used with experimental TBI to address this line of research. Loane and Faden (2010) discuss a number of these possible drugs (statins, progesterone, cyclosporine A, cell-cycle inhibitors, and PARP inhibitors) in their review of neuroprotectants that target a wide range of secondary injury cascades including excitotoxicity, apoptosis, inflammation, edema, and oxidative stress.

From a rehabilitation perspective, treatments provided early after injury and are considered neuroprotectants and are important when evaluating all of the contributors to recovery. Further, early interventions may influence rehabilitation treatment effectiveness and/or interact with response to rehabilitation focused treatment. However, preclinical testing of rehabilitation focused drugs is far less common compared with early intervention studies; for the field of translational TBI research to continue to evolve and progress, rehabilitation relevant treatment designs are an important contribution to the field. Among the important rehabilitation research design concepts are considerations of timing, dosing, and the posttreatment withdrawal effects of drugs commonly used during rehabilitation.

35.12.2 Rehabilitation-Focused Pharmacological Studies

35.12.2.1 Stimulants

The use of psychostimulants in treating cognitive deficits after TBI initially led to the hypothesis that DA systems were altered post-TBI (McAllister et al., 2004). Although we do not provide a comprehensive review of the hypo-dopaminergic hypothesis (see Bales et al., 2009 for review), it is important to understand the mounting evidence for reduced dopaminergic signaling post-TBI. Importantly, dopaminergic signaling reductions can have a wide range of effects on brain function. In this section, we highlight some of the major uses of stimulants targeting dopaminergic signaling in neurorehabilitation post-TBI (for a comprehensive review, see Harun and Wagner, 2013).

Three common pharmacological interventions clinically, methylphenidate (MPH), D-amphetamine (AMPH), and methamphetamine, target some combination of dopamine (DA), norepinephrine, and/or serotonin neurotransmission by acting primarily on their respective transporters. MPH blocks the DA transporter, while amphetamine is internalized after binding to DA transporter and promotes efflux of dopamine through a reverse transport mechanism. AMPH also inhibits degradation of monoamines through inhibition of monoamine oxidase, depleting DA/NE vesicles (Fleckenstein et al., 2007). Methamphetamine reverses the directionality of all three neurotransmitter transporters, in addition to blocking reuptake of neurotransmitters (Fleckenstein et al., 2007). MPH and AMPH are used almost interchangeably for disorders like attention deficit hyperactivity disorder (ADHD), and the considerable similarities between TBI and ADHD in terms of fatigue, mental slowing, and distractibility have led to off-label use of psychostimulants in TBI. However, dosing strategies differ clinically, with lower dosing ranges commonly observed relative to dosing ranges used for ADHD (Warden et al., 2006).

Animal models of TBI have been critical in understanding the utility of psychostimulants after TBI and the mechanisms by which they confer beneficial effects. Beyond the short-term stimulant effects, some experimental work demonstrates the plasticity and neurorestorative capabilities associated with MPH administration. Using an electrically evoked DA release paradigm and fast-scan cyclic voltammetry, Wagner and colleagues have shown that daily MPH administration for 14 days post-CCI restores many of the observed alterations in striatal DA neurotransmission (Wagner et al., 2009a). Further work in this area has shown chronic administration of MPH can restore the responsiveness to an acute MPH challenge after CCI (Wagner et al., 2009b), and normalization of D2 autoreceptor function in mediating these restorative effects is being evaluated.

Behaviorally, this same treatment paradigm of chronic MPH treatment improved spatial learning after CCI in male but not female rats (Kline et al., 2000). Adding to the literature on sex-specific responses to dopaminergic drugs, female rats were more sensitive to the stimulant effects of daily MPH therapy and had increased locomotor activity in an open field chamber on an established behaviorally restorative dose of MPH treatment for male rats (Wagner et al., 2007b). These studies suggest chronic MPH treatment improves behavioral recovery; however, they also suggest further work evaluating sex-specific sensitivity and efficacy in MPH treatment will be important to better personalize clinical dosing (Bales et al., 2009). Because most of these studies have focused on initiating MPH acutely, it is important to note there are a number of clinical studies to suggest efficacy of MPH in the chronic phase to treat both cognitive (Whyte et al., 2004) and mood (Warden et al., 2006) issues; thus, subsequent animal studies, beginning in chronic phases of TBI, are warranted. Given the plastic effects noted in experimental models with MPH, future work should also establish what lasting benefits remain after withdrawal of the drug.

The efficacy of AMPH and methamphetamine is less clear in the TBI population. In a clinical population, AMPH has been shown in one study to improve participation in rehabilitation (Hornstein et al., 1996) and to increase attentional capacity and working memory (Bleiberg et al., 1993) in chronic TBI. A recent study showed that AMPH administration for the first 7 days after CCI in rats increases hippocampal BDNF (Griesbach et al., 2008), though this finding has not been shown to affect chronic recovery. Interestingly, a single, acute, low dose of methamphetamine may positively influence recovery after FPI (Rau et al., 2012).

35.12.2.2 Anticonvulsants

Anticonvulsants are commonly used in prevention of acute post-TBI seizures (<1 week after trauma). However, therapeutic evidence supporting prolonged prophylactic use of anticonvulsants clinically is absent, and the effects of prolonged antiepileptic drugs on neurobehavioral recovery have not been well characterized. However, some recent, experimental models suggest chronic phenytoin administration decreases hippocampal cell survival and impedes cognitive and motor recovery (Darrah et al., 2011). These findings are consistent with clinical studies where prolonged use of phenytoin has been associated poor cognitive and motor performance (Dikmen et al., 2004; Pulliainen and Jokelainen, 1995). Conversely, daily treatment with Levetiracetam after experimental TBI resulted in multiple improvements in behavior, cell sparing, chronic inflammation, and excitatory modulation (Zou et al., 2013), but a clinically relevant brief acute treatment paradigm did not confer any beneficial effects on behavior or histological endpoints (Zou et al., 2014). Together, this work suggests that more systematic work is needed to define if/how this drug affects epileptogenesis when administered over the recovery continuum and establish support for potential long-term treatment trials to prevent epileptogenesis and promote neurorecovery.

35.12.2.3 Antipsychotics

Because signaling through postsynaptic dopamine D2 receptors modulates arousal and stress responses, D2 antagonists like haloperidol and risperidone are commonly used to treat agitation or aggression in acute TBI. As such, haloperidol is the most commonly used agent to treat posttraumatic agitation (Vincent et al., 1986). These drugs are also often used for their sedative rather than antipsychotic properties. Antipsychotics should also be used with caution within the TBI population because they can interfere with cognitive recovery. Similar to clinical studies (Rao et al., 1985), rodent models of TBI have shown that haloperidol and risperidone, both of which have high affinity for the dopamine D2 receptor, impair spatial learning in rodent models of TBI (Kline et al., 2007a, 2008; Wilson et al., 2003). Thus, experimental models suggest acute management of aggression or agitation with antipsychotic agents can greatly impact chronic cognitive recovery.

35.12.3 Challenges with Experimental Rehabilitation Models

35.12.3.1 Chronicity and Dosing of Rehabilitation

Much of the focus on rehabilitation studies in animal models has focused on interventions that are administered chronically and systematically. In the future, it will be important to determine therapeutic windows. For example, it may be that there are treatments that can be administered for a limited exposure that will have lasting effects or are these chronic treatments that will be long-term adjustments. Similarly, the understanding of how multiple treatments interact with each other to affect recovery and susceptibility to complications will be important to evaluate. It may be that treatment timing and order are necessary to maximize patient benefits. Further, it will be important to look at pharmacological interventions with non-pharmacological interventions in both combined and time-sensitive studies in neurorehabilitation.

There is a need for specific dosing experiments to determine minimal regimens for translation into clinical use. For example, EE studies often involve housing the animal consistently in the EE cage. In thinking of translatability, it is important to understand if EE done for a few hours a day, a few times a week, would produce significant improvements in recovery as this is a more feasible structure in the clinic. Interestingly, relatively shorter exposures to EE confer some benefit in experimental TBI (Cheng et al., 2012). Similarly, determination of when to begin these interventions will be crucial to translate into clinical practice. Of course, these studies will need to be performed in comparative effectiveness or clinical trials in order to determine dosing and timing in the human.

35.12.4 Genetic Effects on Rehabilitation Models

Several human studies have begun to evaluate the impact of genetic variation in TBI recovery and biosusceptibility to complications, both in the acute care (Failla et al., 2014; Wagner et al., 2007c, 2010) and chronic rehabilitation phases (Darrah et al., 2012; Failla et al., 2013; Hoh et al., 2010; Krueger et al., 2011; Wagner et al., 2010, 2012). Many of these studies have focused on genetic variation within pathways discussed in this review (i.e., neurotrophin signaling, inflammation). As mentioned in the introduction, however, human studies are correlative and need to be supplemented by important work in experimental animal models. Tracking gene expression patterns and regulation after TBI in rodent models will likely identify potential molecular mechanisms involved in recovery. Molecular biology techniques, including transgenic animals, allow researchers to examine how particular genetic changes can alter injury recovery as well as response to treatments. Although transgenic animals allow for important genetic variation studies in rehabilitation, there are important considerations. For example, transgenic animals often have gene-specific behavioral baselines and thus, proper genetic and sham controls will be crucial to interpretation of these models in experimental TBI. Importantly, work in transgenic animals will likely lead to important pharmacogenetic studies elucidating individual responses to targeted pharmacological interventions post-TBI. Eventually, an understanding of each individual patient's genetic variation would allow for treatment to be tailored to the individual rather than generalizing treatment to type and severity of injury.

35.12.5 Age and Sex Considerations in Rehabilitation Models

As noted throughout this review, age and sex are important considerations in rehabilitation as both factors have been shown to impact recovery and intervention after TBI. Age is a consistent predictor of outcome/survival after TBI (Brooks et al., 2013). A large proportion of TBI incidence each year, older adults often have worse outcomes despite similar injury types and severity (Susman et al., 2002). Also, younger age groups can carry some unique risk for poor outcome (Majerske et al., 2008; Shein et al., 2012). Because age interacts with many of the secondary injury cascades (Berger et al., 2010; Wagner et al., 2004a, 2011) it is an important factor to consider in neurorehabilitation. For example, in experimental TBI models, older animals show reduced neuronal survival (Onyszchuk et al., 2008) and impaired immune states (Kumar et al., 2013) postinjury. In the clinical population, Failla and colleagues (2014) show age can interact with genetic variants to improve mortality predictions, and age is an important factor to consider in neurorehabilitation.

Although the clinical literature is mixed, sex can affect response to TBI and its treatments. For example, studies have shown important differences in TBI recovery for men and women, including sex differences in postconcussive syndrome (Bazarian et al., 1999; Broshek et al., 2009), depression (Schopp et al., 2001), and return to gender-based roles (Gutman, 2000). Similarly, in experimental models, there are differences in response to environmental enrichment (Wagner et al., 2002) and pharmacological therapies

(Wagner et al., 2007b), behavior, and learning strategies in female compared with male rodents after TBI. Importantly, this work suggests sex differences may inform rehabilitation treatment decisions. Because women make up a significant portion of the clinical population, sex should routinely be considered in experimental rehabilitation research design.

35.12.6 New Avenues for Rehabilitation Models

35.12.6.1 Blast TBI

With the increased number of blast-related TBI, particularly among military populations, there has been significant recent focus of new research in this field. Clinically, much of how blast TBI (bTBI)-related sequelae are treated is drawn from knowledge about TBI treatments from conventional mechanisms such as fall or motor vehicle collision. Since little is known about how these therapies translate to blast-induced TBI, there are major gaps in the literature about how to treat individuals after bTBI.

Three broad techniques have been implemented to model and study blast injuries: swine exposure to explosions, compression air-driven shock tubes, and detonation-shock tubes. Compression air-driven shock tubes use a long, cylindrical tube with two compartments of pressurized air. When a high-pressure compartment ruptures, a shock wave is created, passes through the low-pressure compartment, and travels down the length of the metal tube where it affects either a silico model or an animal model. A detonation-shock tube also uses a long, cylindrical tube but uses an explosive to generate a shock wave. It is unclear how relevant these models are to the clinical population because neither of these two models is capable of producing a Friedlander wave, the wave most commonly associated with primary TBI (Chen and Constantini, 2013). The detonation-shock tube is potentially a more realistic model because it encompasses all four levels of polytraumatic blast injury. Swine exposure to an explosion is a third model currently being used (Bauman et al., 2009). Swine models are more representative of polytraumatic injuries because the projectile motion of the swine body is similar to that in a human subject.

Several studies in bTBI have begun to show diffuse injuries (Garman et al., 2011) that lead to cognitive (Ning et al., 2013) and behavioral (Elder et al., 2012) phenotypes similar to the clinical presentation of bTBI (Magnuson et al., 2012). Because of the complex nature of bTBI, utilization of these models to better understand the neurorestorative properties of nonpharmacological treatments, such as exercise and cognitive rehabilitation paradigms, are warranted as are pharmacological neurotransmission enhancement studies.

35.12.6.2 mTBI

Since nearly three-fourths of reported TBIs each year are classified as mild (Bruns and Jagoda, 2009), mTBI is an important subset of clinical TBI that is in need of improved diagnosis and treatment, particularly because a significant proportion of this population can have persistent symptoms and conditions associated with their injury (Suffoletto et al., 2013). Current paradigms for prognostication of complications after mTBI are lacking, and treatment is often reliant upon a patient's self-report of injury and symptoms. Importantly, mTBI often results in a loss of function for patients, including depression and cognitive deficits, years out from injury (Vanderploeg et al., 2005).

For this reason, the use of animal models in studying mild TBI is important, but the neurotrauma field has encountered many obstacles with developing and implementing relevant models. To validate mTBI, one looks to the clinical diagnosis first. Mild TBI is considered to have the following criteria (Cassidy et al., 2004): normal structural brain imaging, loss of consciousness <30 minutes, alteration of consciousness <24 hours, posttraumatic amnesia <24 hours, and/or an initial Glasgow Coma Scale of 13–15. Many of these criteria are hard to translate into a rodent model. With no structural differences, researchers must look to the other criteria to determine validity. Alterations in consciousness and posttraumatic amnesia are difficult to assess in rodents. Pre- and postinjury tests (such as the MWM (Vorhees and Williams, 2006)) can examine retrograde amnesia, if conducted in a suitable time frame consistent with mTBI, but to date, there have been no studies in experimental mTBI that meet these criteria. Similarly, loss of consciousness criteria need to be adapted to animal studies. As such, it has been suggested that the righting reflex suppression after unconsciousness in rodents is an analog to human loss of consciousness. Dewitt and colleagues (2013) suggest durations of righting reflex suppression of 5–15 minutes correlate to mTBI. They conclude mTBI models that meet this criteria show similar histological patterns to human mTBI.

Most animal models (including CCI, FPI, weight-drop, and impact acceleration) have been adapted in some aspect to produce mild injuries. Mild FPI and CCI show overlapping deficits in a number of signaling pathways immediately after injury (Redell et al., 2013). Mild FPI produces reduced histopathological tissue loss (Gurkoff et al., 2006), but still results in selected gene/protein alterations and behavioral consequences (Hylin et al., 2013) in impaired spatial memory, neuroendocrine dysfunction (Greco et al., 2013) and increased anxiety-like behaviors (Abrous et al., 1999; Gurkoff et al., 2006; Shultz et al., 2011). Similarly, other mTBI models show altered glucose metabolism after the injury as well (Prins et al., 2013). Although these models do hold promise and mirror some of the above criteria for mTBI, they lack validation in chronic mTBI sequelae like executive function (Brooks et al., 1999) and attention deficits (Vanderploeg et al., 2005). Additional work needed for postconcussion model development includes high repetition subconcussive blows to better model sports-related concussion (Bailes et al., 2013) and some forms of mild military TBI (Kelly et al., 2012; Mac Donald et al., 2013).

35.12.7 Repetitive TBI

Repetitive injuries have become increasingly relevant for clinical TBI. With increased exposures to multiple blast injuries in the military (Galarneau et al., 2008), and with the gaining insight into repetitive concussions in sports, there is a great demand for experimental repetitive TBI (rTBI) models. Experimental rTBI has shown consistently that a repeated injury (applied across certain intervals) can exacerbate both tissue damage and functional/behavioral outcomes (Longhi et al., 2005; Shitaka et al., 2011), consistent with clinical rTBI (Belanger et al., 2010). Even with a mild injury, when there was little to no histological damage, a second injury 24 hours later can produce profound motor impairment and blood–brain barrier disruption compared with a single injury (Laurer et al., 2001).

Because many individuals with mild TBI often report multiple concussions in their clinical history (Belanger et al., 2010), experimental rTBI data are very salient. Further, the compounding of multiple injuries may explain why 10%–20% of individuals with mTBI do not fully recover. Importantly, this work also suggests that rehabilitation providers need to be cognizant of the possibility of multiple head injuries when developing their treatment plans, and treatments that may reduce the vulnerable period after a single TBI could reduce the effects of a second injury. Characterizing long-term effects of repetitive mTBI, and identifying relevant therapeutic targets, will be an important goal for neurorehabilitation research. The addition of multiple injuries will likely need evaluation in terms of dosing and timing of rehabilitation, as well. Many of these factors are addressable using the current rTBI models with some variation on protocols.

35.12.8 Nonrodent Animal Models for Rehabilitation

Although rodent models have been instrumental in the current state of neurorehabilitation, there are other animal models that may provide certain advantages from an experimental rehabilitation perspective. There are specific species differences in anatomy, physiology, or behavior that can greatly impact selection of animal experimental models in rehabilitation. For example, certain biochemical factors involved in secondary-injury cascades (i.e., neurogenesis) show important species-specific differences (Johansson et al., 2002). Next we describe two nonrodent models that may be able bridge specific gaps in preclinical experimental rehabilitation research.

35.12.8.1 Swine

While rodent models have greatly advanced TBI research, there is still a gap between preclinical studies and clinical trials. Some researchers suggest swine models may be able to bridge this gap. Important for TBI, the pig brain shows greater complexity in both chemistry and anatomy compared with the rodent (Schook et al., 2005). For example, gyri are present in swine compared with the smooth rodent brain, which makes it more representative of the human brain. For certain models, such as blast TBI, the larger body size and increased skull thickness increase the relevance of this model to human blast TBI (Bauman et al., 2009). Similarly, the inertial acceleration brain injury model has been further developed in the mini-pig rather than the rodent as the physics of this model make reliability in rodents difficult because of their decreased body mass (Morales et al., 2005).

However, as with any model, there are disadvantages: swine are more expensive to care for, require larger housing, and are increasingly difficult to obtain. Importantly, there are fewer behavioral data available in swine (compared with rodents) that diminishes their use studying certain rehabilitation-relevant problems such as cognition, depression, or anxiety. However, because the pigs' cardiovascular system is similar to humans (Swindle, 1992) (more than rats), they could be important tools in rehabilitation paradigms such as exercise. Similarly, their similar cardiovascular system to humans makes pigs more appropriate for certain combined TBI models such as hemorrhagic shock combined with TBI (Imam et al., 2013). In this vein, the use of swine in TBI and neurorehabilitation research will likely increase our understanding of the limitations in rodent experimental models and improve translation as its own preclinical model.

35.12.8.2 Drosophila

Recently, Katzenberger et al. (2013) have developed a Drosophila model that provides a less expensive and more accessible way to study TBI because of the abundance of Drosophila, reduced cost of daily care, and their short life span. Katzenberger's recent publications revealed the similarities between Drosophila and human models both in macro-level responses and cellular-level responses. In this model, Drosophila were placed in a small tube and subjected to a rapid acceleration and deceleration that resulted in a TBI when the fly impacted the sides of the home tube. Similar to humans, Drosophila experienced "temporary incapacitation, ataxia, activation of the innate immune response, neurodegeneration, and death" (Katzenberger et al., 2013). Katzenberger proposed that the cellular and molecular pathways involved in the recovery of Drosophila models parallels the pathways of humans. One major disadvantage of the Drosophila model is the exposure to polytrauma in this injury paradigm. Because of their size and anatomy, it is difficult to isolate a TBI without injuring the rest of the body. Another potential disadvantage is the short life cycle of the fly, thus chronic studies may be difficult to conduct and interpret. However, there are several behavioral and functional studies that can be conducted to examine such important TBI sequelae like seizures (Parker et al., 2011) and cognitive (Margulies et al., 2005) recovery. Similarly, there are a multitude of genetic manipulations available in the fly, with numerous genetic homologs to humans, affording many possibilities in terms of therapeutic development. Though more research is needed into the ability of Drosophila as a translational model in TBI and rehabilitation, it is certainly an attractive avenue for high-throughput and genetic studies.

35.12.9 Combination Injury Models for Rehabilitation

Because no single model can encompass all of TBI or neurorehabilitation, one movement to address this problem is combined injury models. Combined injury models have developed across species (in the mouse, Ditelberg et al., 1996; rat, Bramlett et al., 1999; cat, Zauner et al., 2002; and pig, Imam et al., 2013) and include a number of secondary injuries such as hypoxia, ischemia, and hypotension frequently observed in experimental and clinical TBI (Chesnut, 1995, Manley et al., 2001). The combination of these secondary insults with primary injury usually leads to increased mortality or exacerbation of deficits in the clinical population, but it has also been shown in experimental TBI models that these combinations can increase deficits chronically (Hemerka et al., 2012).

However, acute secondary injuries related to the trauma are not the only "combined injury" to be cognizant of in the TBI population. Experimental TBI models are now incorporating other neurological injury models such as Parkinson models or Alzheimer models to examine how pathology relevant to those conditions may impact TBI pathology. For example, animal models of Alzheimer disease show alterations in pathology after experimental TBI (Abrahamson et al., 2013; Bennett et al., 2013). Alternatively, some studies examine how TBI in combination with other factors might predispose animals to Parkinson-like pathology (Sauerbeck et al., 2012). These and the aforementioned types of combined studies can be valuable experimental TBI rehabilitation tools that increase our understanding of the complicated, long-term clinical picture that can arise with TBI and develop new areas of understanding with how TBI-induced consequences and mechanisms of secondary injury affect long-term sequelae and recovery.

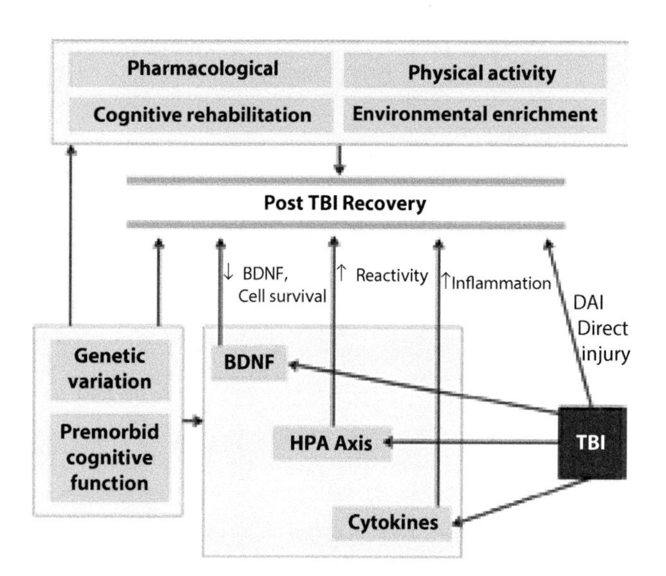

FIGURE 35.1 Schematic of overall model of neurorehabilitation after TBI. Data from both animal and clinical studies in neurorehabilitation suggest TBI affect multiple signaling pathways that can alter recovery (BDNF, HPA axis, and inflammation) that interact with individual genetic variation and premorbid conditions to influence post-TBI recovery. Similarly, response to rehabilitation (either pharmacological or nonpharmacological) is modulated by these signaling pathways and their interaction with genetic and premorbid conditions. It is likely that a "Rehabilomics" approach, which includes all of these factors, will be necessary in rehabilitation research. It is likely to be especially important to incorporate this framework in animal models of rehabilitation.

35.13 CONCLUSIONS: REHABILOMICS IN NEUROREHABILITATION RESEARCH

The effects of TBI are widely varied and heterogeneous compared with other clinical diseases. In this review, we have discussed common themes to consider in experimental rehabilitation modeling: variations in models, the importance of timing and chronicity of treatments, and important factors like genetics, age, and sex that can influence rehabilitation. To improve neurorehabilitation, we must examine the diverse sequelae post-TBI within a framework that considers these wide-ranging factors. A new framework called Rehabilomics has been proposed (Wagner 2010; Wagner and Zitelli, 2012). The Rehabilomics framework aims to understand how a wide-range of individual factors (like genetics, age, and sex) interact with injury parameters to influence response to rehabilitation and alter outcomes after TBI (Figure 35.1).

Within this Rehabilomics framework, it will be important to use a wide range of experimental models to examine important issues such as how acute care can influence response to chronic rehabilitation, focusing on any unintended consequences of acute management of injuries. Experimentally, the Rehabilomics model can be applied to study sex, age, and genetic variation with regard to treatment response and to understand how neurotrophin status, plasticity, inflammation, and stress hormone pathology can be modeled and mapped to the clinical condition. However, one of the major challenges of translational rehabilitation research is the ability to model complex, higher order behaviors, and deficits after TBI. For example, affective disorders are highly detrimental to TBI recovery, and thus, modeling them in experimental TBI research is important for understanding these complications. However, language or cultural barriers, social support, and participation in rehabilitation are important factors in the clinic that cannot be addressed in experimental models.

Mapping portions of this holistic Rehabilomics approach to an experimental rehabilitation paradigm will likely accelerate the pace in which rehabilitation research can translate into individualized treatment plans while yielding important insights into rehabilitation-relevant mechanisms. However, there is still is a relative lack of established neurorehabilitation methodology within the experimental literature that needs to be addressed. It is imperative for experimental models to evaluate rigorously important factors in rehabilitation such as timing, chronicity of treatment, maintenance of treatment effects, and influences of acute management on chronic care. With an integrative approach, the Rehabilomics framework combined with rigorous experimental models of neurorehabilitation will likely yield rapid improvements in the current state of neurorehabilitation research and clinical care.

REFERENCES

Aarabi, B. and Simard, J.M., 2009. Traumatic brain injury. *Current Opinion in Critical Care*, 15 (6), 548–553.

Abrahamson, E.E., Foley, L.M., Dekosky, S.T., Kevin Hitchens, T., Ho, C., Kochanek, P.M., and Ikonomovic, M.D., 2013. Cerebral blood flow changes after brain injury in human amyloid-beta knock-in mice. *Journal of Cerebral Blood Flow and Metabolism*, 33 (6), 826–833.

Abrous, D.N., Rodriguez, J., le Moal, M., Moser, P.C., and Barnéoud, P., 1999. Effects of mild traumatic brain injury on immunoreactivity for the inducible transcription factors c-Fos, c-Jun, JunB, and Krox-24 in cerebral regions associated with conditioned fear responding. *Brain Research*, 826 (2), 181–192.

Acosta, S.A., Tajiri, N., Shinozuka, K., Ishikawa, H., Grimmig, B., Diamond, D. et al., 2013. Long-term upregulation of inflammation and suppression of cell proliferation in the brain of adult rats exposed to traumatic brain injury using the controlled cortical impact model. *PLoS ONE*, 8 (1), e53376.

Archer, T., 2012. Influence of physical exercise on traumatic brain injury deficits: Scaffolding effect. *Neurotoxicity Research*, 21 (4), 418–434.

Archer, T., Svensson, K., and Alricsson, M., 2012. Physical exercise ameliorates deficits induced by traumatic brain injury. *Acta Neurologica Scandinavica*, 125 (5), 293–302.

Bailes, J.E., Petraglia, A.L., Omalu, B.I., Nauman, E., and Talavage, T., 2013. Role of subconcussion in repetitive mild traumatic brain injury. *Journal of Neurosurgery*, 119 (5), 1235–1245.

Bales, J.W., Wagner, A.K., Kline, A.E., and Dixon, C.E., 2009. Persistent cognitive dysfunction after traumatic brain injury: A dopamine hypothesis. *Neuroscience and Biobehavioral Reviews*, 33 (7), 981–1003.

Bauman, R.A., Ling, G., Tong, L., Januszkiewicz, A., Agoston, D., Delanerolle, N. et al., 2009. An introductory characterization of a combat-casualty-care relevant swine model of closed head injury resulting from exposure to explosive blast. *Journal of Neurotrauma*, 26 (6), 841–860.

Bazarian, J.J., Wong, T., Harris, M., Leahey, N., Mookerjee, S., and Dombovy, M., 1999. Epidemiology and predictors of post-concussive syndrome after minor head injury in an emergency population. *Brain Injury*, 13 (3), 173–189.

Bekinschtein, P., Cammarota, M., Izquierdo, I., and Medina, J.H., 2008a. Reviews: BDNF and memory formation and storage. *The Neuroscientist*, 14 (2), 147 –156.

Bekinschtein, P., Cammarota, M., Katche, C., Slipczuk, L., Rossato, J.I., Goldin, A. et al, 2008b. BDNF is essential to promote persistence of long-term memory storage. *Proceedings of the National Academy of Sciences*, 105 (7), 2711–2716.

Belanger, H.G., Spiegel, E., and Vanderploeg, R.D., 2010. Neuropsychological performance following a history of multiple self-reported concussions: A meta-analysis. *Journal of the International Neuropsychological Society*, 16 (2), 262–267.

Bendlin, B.B., Ries, M.L., Lazar, M., Alexander, A.L., Dempsey, R.J., Rowley, H.A. et al., 2008. Longitudinal changes in patients with traumatic brain injury assessed with diffusion-tensor and volumetric imaging. *NeuroImage*, 42 (2), 503–514.

Bennett, R.E., Esparza, T.J., Lewis, H.A., Kim, E., Mac Donald, C.L., Sullivan, P.M. et al., 2013. Human apolipoprotein E4 worsens acute axonal pathology but not amyloid-β immunoreactivity after traumatic brain injury in 3xTG-AD mice. *Journal of Neuropathology and Experimental Neurology*, 72 (5), 396–403.

Berger, R.P., Bazaco, M.C., Wagner, A.K., Kochanek, P.M., and Fabio, A., 2010. Trajectory analysis of serum biomarker concentrations facilitates outcome prediction after pediatric traumatic and hypoxemic brain injury. *Developmental Neuroscience*. 32, 396–405.

Bergsneider, M., Hovda, D.A., Lee, S.M., Kelly, D.F., McArthur, D.L., Vespa, P.M. et al., 2000. Dissociation of cerebral glucose metabolism and level of consciousness during the period of metabolic depression following human traumatic brain injury. *Journal of Neurotrauma*, 17 (5), 389–401.

Bigler, E., Blatter, D., Anderson, C., Johnson, S., Gale, S., Hopkins, R. et al., 1997. Hippocampal volume in normal aging and traumatic brain injury. *AJNR American Journal of Neuroradiology*, 18 (1), 11–23.

Bleiberg, J., Garmoe, W., Cederquist, J., and Reeves, D., 1993. Effect of dexedrine on performance consistency following brain injury: A double-blind placebo crossover case study. *Neuropsychiatry, Neuropsychology, and Behavioral Neurology*

Bolivar, V.J., Caldarone, B.J., Reilly, A.A., and Flaherty, L., 2000. Habituation of activity in an open field: A survey of inbred strains and F1 hybrids. *Behavior Genetics*, 30 (4), 285–293.

Bombardier, C.H., Fann, J.R., Temkin, N.R., Esselman, P.C., Barber, J., and Dikmen, S.S., 2010. Rates of major depressive disorder and clinical outcomes following traumatic brain injury. *JAMA*, 303 (19), 1938–1945.

Bonatti, E., Zamarian, L., Wagner, M., Benke, T., Hollosi, P., Strubreither, W. et al., 2008. Making decisions and advising decisions in traumatic brain injury. *Cognitive and Behavioral Neurology*, 21 (3), 164–175.

Bondi, C.O., Cheng, J.P., Tennant, H.M., Monaco, C.M., and Kline, A.E., 2014. Old dog, new tricks: The attentional set-shifting test as a novel cognitive behavioral task after controlled cortical impact injury. *Journal of Neurotrauma*, 31 (10), 926–937.

Bramlett, H.M., Green, E.J., and Dietrich, W.D., 1999. Exacerbation of cortical and hippocampal CA1 damage due to posttraumatic hypoxia following moderate fluid-percussion brain injury in rats. *Journal of Neurosurgery*, 91 (4), 653–659.

Bramlett, H.M., Dietrich, W.D., Green, E.J., and Busto, R., 1997a. Chronic histopathological consequences of fluid-percussion brain injury in rats: Effects of post-traumatic hypothermia. *Acta Neuropathologica*, 93 (2), 190–199.

Bramlett, H.M., Green, E.J., and Dietrich, W.D., 1997b. Hippocampally dependent and independent chronic spatial navigational deficits following parasagittal fluid percussion brain injury in the rat. *Brain Research*, 762 (1), 195–202.

Brandeis, R., Brandys, Y., and Yehuda, S., 1989. The use of the Morris water maze in the study of memory and learning. *International Journal of Neuroscience*, 48 (1–2), 29–69.

Braver, T.S. and Cohen, J.D., 2000. On the control of control: The role of dopamine in regulating prefrontal function and working memory. *Control of Cognitive Processes: Attention and Performance XVIII*, 713–737.

Brayer, S., Zou, H., Hurwitz, M., Ketcham S., Henderson C., Fuletra, J., Kumar K., Skidmore, E., Thiels, E., Wagner, A.K. 2014. Developing a clinically relevant model of cognitive training after experimental traumatic brain injury. In Press, *Neurorehabilitation and Neural Repair*.

Brooks, J., Fos, L.A., Greve, K.W., and Hammond, J.S., 1999. Assessment of executive function in patients with mild traumatic brain injury. *The Journal of Trauma*, 46 (1), 159–163.

Brooks, J.C., Strauss, D.J., Shavelle, R.M., Paculdo, D.R., Hammond, F.M., and Harrison-Felix, C.L., 2013. Long-term disability and survival in traumatic brain injury: Results from the National Institute on Disability and Rehabilitation Research Model Systems. *Archives of Physical Medicine and Rehabilitation*, 94 (11), 2203–2209.

Bruns, J.J., Jr. and Jagoda, A.S., 2009. Mild traumatic brain injury. *The Mount Sinai Journal of Medicine, New York*, 76 (2), 129–137.

Buckley, M.J., 2005. The role of the perirhinal cortex and hippocampus in learning, memory, and perception. *The Quarterly Journal of Experimental Psychology. B, Comparative and Physiological Psychology*, 58 (3–4), 246–268.

Bye, N., Carron, S., Han, X., Agyapomaa, D., Ng, S.Y., Yan, E. et al., 2011. Neurogenesis and glial proliferation are stimulated following diffuse traumatic brain injury in adult rats. *Journal of Neuroscience Research*, 89 (7), 986–1000.

Cain, D.P., Boon, F., and Corcoran, M.E., 2006. Thalamic and hippocampal mechanisms in spatial navigation: A dissociation between brain mechanisms for learning how versus learning where to navigate. *Behavioural Brain Research*, 170 (2), 241–256.

Cassidy, J.D., Carroll, L.J., Peloso, P.M., Borg, J., von Holst, H., Holm, L. et al., and WHO Collaborating Centre Task Force on Mild Traumatic Brain Injury, 2004. Incidence, risk factors and prevention of mild traumatic brain injury: Results of the WHO Collaborating Centre Task Force on Mild Traumatic Brain Injury. *Journal of Rehabilitation Medicine*, (43 Suppl), 28–60.

Castagné, V., Moser, P., Roux, S., and Porsolt, R.D., 2011. Rodent models of depression: Forced swim and tail suspension behavioral despair tests in rats and mice. *Current Protocols in Neuroscience*, 55, 11–18.

Cekic, M., Cutler, S.M., VanLandingham, J.W., and Stein, D.G., 2011. Vitamin D deficiency reduces the benefits of progesterone treatment after brain injury in aged rats. *Neurobiology of Aging*, 32 (5), 864–874.

Chen, X., Li, Y., Kline, A.E., Dixon, C.E., Zafonte, R.D., and Wagner, A.K., 2005. Gender and environmental effects on regional brain-derived neurotrophic factor expression after experimental traumatic brain injury. *Neuroscience*, 135 (1), 11–17.

Chen, Y. and Constantini, S., 2013. Caveats for using shock tube in blast-induced traumatic brain injury research. *Frontiers in Neurology*, 4.

Chen, Z.-Y., Patel, P.D., Sant, G., Meng, C.-X., Teng, K.K., Hempstead, B.L. et al., 2004. Variant brain-derived neurotrophic factor (BDNF) (Met66) alters the intracellular trafficking and activity-dependent secretion of wild-type BDNF in neurosecretory cells and cortical neurons. *Journal of Neuroscience*, 24 (18), 4401–4411.

Cheng, J.P., Shaw, K.E., Monaco, C.M., Hoffman, A.N., Sozda, C.N., Olsen, A.S. et al., 2012. A relatively brief exposure to environmental enrichment after experimental traumatic brain injury confers long-term cognitive benefits. *Journal of Neurotrauma*, 29 (17), 2684–2688.

Chesnut, R.M., 1995. Secondary brain insults after head injury: Clinical perspectives. *New Horizons (Baltimore, Md.)*, 3 (3), 366–375.

Cicerone, K.D., Langenbahn, D.M., Braden, C., Malec, J.F., Kalmar, K., Fraas, M. et al., 2011. Evidence-based cognitive rehabilitation: Updated review of the literature from 2003 through 2008. *Archives of Physical Medicine and Rehabilitation*, 92 (4), 519–530.

Clare, L. and Jones, R., 2008. Errorless learning in the rehabilitation of memory impairment: A critical review. *Neuropsychology Review*, 18 (1), 1–23.

Clausen, F. and Hillered, L., 2005. Intracranial pressure changes during fluid percussion, controlled cortical impact and weight drop injury in rats. *Acta Neurochirurgica*, 147 (7), 775–780; discussion 780.

Cope, E.C., Morris, D.R., Scrimgeour, A.G., VanLandingham, J.W., and Levenson, C.W., 2011. Zinc supplementation provides behavioral resiliency in a rat model of traumatic brain injury. *Physiology & Behavior*, 104 (5), 942–947.

Crane, A.T., Fink, K.D., and Smith, J.S., 2012. The effects of acute voluntary wheel running on recovery of function following medial frontal cortical contusions in rats. *Restorative Neurology and Neuroscience*, 30 (4), 325–333.

Cunningham, T.L., Cartagena, C.M., Lu, X.-C.M., Konopko, M., Dave, J.R., Tortella, F.C. et al., 2013. Correlations between blood-brain barrier disruption and neuroinflammation in an experimental model of penetrating ballistic-like brain injury. *Journal of Neurotrauma*, 31 (5), 505–514.

Cutler, S.M., Vanlandingham, J.W., and Stein, D.G., 2006. Tapered progesterone withdrawal promotes long-term recovery following brain trauma. *Experimental Neurology*, 200 (2), 378–385.

D'Hooge, R. and De Deyn, P.P., 2001. Applications of the Morris water maze in the study of learning and memory. *Brain Research Reviews*, 36 (1), 60–90.

Darrah, S.D., Chuang, J., Mohler, L.M., Chen, X., Cummings, E.E. et al., 2011. Dilantin therapy in an experimental model of traumatic brain injury: Effects of limited versus daily treatment on neurological and behavioral recovery. *Journal of Neurotrauma*, 28 (1), 43–55.

Darrah, S.D., Miller, M.A., Ren, D., Hoh, N.Z., Scanlon, J.M., Conley, Y.P. et al., 2012. Genetic variability in glutamic acid decarboxylase genes: Associations with post-traumatic seizures after severe TBI. *Epilepsy Research* 103 (2–3), 180–194.

DeWitt, D.S., Perez-Polo, R., Hulsebosch, C.E., Dash, P.K., and Robertson, C.S., 2013. Challenges in the development of rodent models of mild traumatic brain injury. *Journal of Neurotrauma*, 30 (9), 688–701.

Dikmen, S.S., Bombardier, C.H., Machamer, J.E., Fann, J.R., and Temkin, N.R., 2004. Natural history of depression in traumatic brain injury. *Archives of Physical Medicine and Rehabilitation*, 85 (9), 1457–1464.

Ditelberg, J.S., Sheldon, R.A., Epstein, C.J., and Ferriero, D.M., 1996. Brain injury after perinatal hypoxia-ischemia is exacerbated in copper/zinc superoxide dismutase transgenic mice. *Pediatric Research*, 39 (2), 204–208.

Dixon, C., Kochanek, P.M., Yan, H.Q., Schiding, J.K., Griffith, R.G., Baum, E. et al., 1999. One-year study of spatial memory performance, brain morphology, and cholinergic markers after moderate controlled cortical impact in rats. *Journal of Neurotrauma*, 16 (2), 109–122.

Dixon, C.E., Clifton, G.L., Lighthall, J.W., Yaghmai, A.A., and Hayes, R.L., 1991a. A controlled cortical impact model of traumatic brain injury in the rat. *Journal of Neuroscience Methods*, 39 (3), 253–262.

Dixon, C.E., Clifton, G.L., Lighthall, J.W., Yaghmai, A.A., and Hayes, R.L., 1991b. A controlled cortical impact model of traumatic brain injury in the rat. *Journal of Neuroscience Methods*, 39 (3), 253–262.

Dixon, C.E., Lyeth, B.G., Povlishock, J.T., Findling, R.L., Hamm, R.J., Marmarou, A. et al., 1987. A fluid percussion model of experimental brain injury in the rat. *Journal of Neurosurgery*, 67 (1), 110–119.

Broshek, D.K., Kaushik, T., Freeman, J.R., Erlanger, D., Webbe, F., and Barth, J.T., 2009. Sex differences in outcome following sports-related concussion. *Journal of Neuroscience*, 102 (5), 856–863.

Elder, G.A., Dorr, N.P., De Gasperi, R., Gama Sosa, M.A., Shaughness, M.C., Maudlin-Jeronimo, E. et al., 2012. Blast exposure induces post-traumatic stress disorder-related traits in a rat model of mild traumatic brain injury. *Journal of Neurotrauma*, 29 (16), 2564–2575.

Erickson, K.I., Voss, M.W., Prakash, R.S., Basak, C., Szabo, A., Chaddock, L. et al., 2011. Exercise training increases size of hippocampus and improves memory. *Proceedings of the National Academy of Sciences of the United States of America*, 108 (7), 3017–3022.

Failla, M.D., Burkhardt, J.N., Miller, M.A., Scanlon, J.M., Conley, Y.P., Ferrell, R.E. et al., 2013. Variants of SLC6A4 in depression risk following severe TBI. *Brain Injury*, 27 (6), 696–706.

Failla, M.D., Kumar, R., Conley, YP., Ferrell, R.P., Wagner, A.K. 2014. Interactions with BDNF gene variation and age in mortality following severe TBI. *Neurorehabilitation and Neural Repair*, pii: 1545968314542617. [Epub ahead of print] PMID: 25063686.

Feeney, D.M., Boyeson, M.G., Linn, R.T., Murray, H.M., and Dail, W.G., 1981. Responses to cortical injury: I. Methodology and local effects of contusions in the rat. *Brain Research*, 211 (1), 67–77.

Fenn, A.M., Gensel, J.C., Huang, Y., Popovich, P.G., Lifshitz, J., and Godbout, J.P., 2013. Immune activation promotes depression 1 month after diffuse brain injury: A role for primed microglia. *Biological Psychiatry*. October 25.

Fleckenstein, A.E., Volz, T.J., Riddle, E.L., Gibb, J.W., and Hanson, G.R., 2007. New insights into the mechanism of action of amphetamines. *Annual Review of Pharmacology and Toxicology*, 47 (1), 681–698.

Fox, G.B., Fan, L., Levasseur, R.A., and Faden, A.I., 1998. Sustained sensory/motor and cognitive deficits with neuronal apoptosis following controlled cortical impact brain injury in the mouse. *Journal of Neurotrauma*, 15 (8), 599–614.

Fox, G.B., LeVasseur, R.A., and Faden, A.I., 1999. Behavioral responses of C57BL/6, FVB/N, and 129/SvEMS mouse strains to traumatic brain injury: Implications for gene targeting approaches to neurotrauma. *Journal of Neurotrauma*, 16 (5), 377–389.

Galarneau, M.R., Woodruff, S.I., Dye, J.L., Mohrle, C.R., and Wade, A.L., 2008. Traumatic brain injury during Operation Iraqi Freedom: Findings from the United States Navy-Marine Corps Combat Trauma Registry. *Journal of Neurosurgery*, 108 (5), 950–957.

Garman, R.H., Jenkins, L.W., Switzer, R.C., 3rd, Bauman, R.A., Tong, L.C., Swauger, P.V. et al., 2011. Blast exposure in rats with body shielding is characterized primarily by diffuse axonal injury. *Journal of Neurotrauma*, 28 (6), 947–959.

Gennarelli, T.A., Adams, J.H., and Graham, D.I., 1981. Acceleration induced head injury in the monkey. I. The model, its mechanical and physiological correlates. In Jellinger, K., Gullotta, F., and Mossakowski, M., eds. *Experimental and Clinical Neuropathology*. Springer, New York, pp. 23–25.

Gentile, A.M., Beheshti, Z., and Held, J.M., 1987. Enrichment versus exercise effects on motor impairments following cortical removals in rats. *Behavioral and Neural Biology*, 47 (3), 321–332.

Ginsberg, M.D., Zhao, W., Alonso, O.F., Loor-Estades, J.Y., Dietrich, W.D., and Busto, R., 1997. Uncoupling of local cerebral glucose metabolism and blood flow after acute fluid-percussion injury in rats. *American Journal of Physiology-Heart and Circulatory Physiology*, 272 (6), H2859–H2868.

Gomes, J.R., Costa, J.T., Melo, C.V., Felizzi, F., Monteiro, P., Pinto, M.J. et al., 2012. Excitotoxicity downregulates TrkB.FL signaling and upregulates the neuroprotective truncated TrkB receptors in cultured hippocampal and striatal neurons. *The Journal of Neuroscience*, 32 (13), 4610–4622.

Goodman, J.C., Cherian, L., Bryan, R.M., Jr., and Robertson, C.S., 1994. Lateral cortical impact injury in rats: Pathologic effects of varying cortical compression and impact velocity. *Journal of Neurotrauma*, 11 (5), 587–597.

Goss, C.W., Hoffman, S.W., and Stein, D.G., 2003. Behavioral effects and anatomic correlates after brain injury: A progesterone dose–response study. *Pharmacology Biochemistry and Behavior*, 76 (2), 231–242.

Greco, T., Hovda, D., and Prins, M., 2013. The effects of repeat traumatic brain injury on the pituitary in adolescent rats. *Journal of Neurotrauma*, 30 (23), 1983–1990.

Griesbach, G.S., 2011. Exercise after traumatic brain injury: Is it a double-edged sword? *PM & R*, 3 (6 Suppl 1), S64–72.

Griesbach, G.S., Gómez-Pinilla, F., and Hovda, D.A., 2007. Time window for voluntary exercise-induced increases in hippocampal neuroplasticity molecules after traumatic brain injury is severity dependent. *Journal of Neurotrauma*, 24 (7), 1161–1171.

Griesbach, G.S., Hovda, D.A., and Gomez-Pinilla, F., 2009. Exercise-induced improvement in cognitive performance after traumatic brain injury in rats is dependent on BDNF activation. *Brain Research*, 1288, 105–115.

Griesbach, G.S., Hovda, D.A., Gomez-Pinilla, F., and Sutton, R.L., 2008. Voluntary exercise or amphetamine treatment, but not the combination, increases hippocampal brain-derived neurotrophic factor and synapsin I following cortical contusion injury in rats. *Neuroscience*, 154 (2), 530–540.

Griesbach, G.S., Hovda, D.A., Molteni, R., Wu, A., and Gomez-Pinilla, F., 2004. Voluntary exercise following traumatic brain injury: Brain-derived neurotrophic factor upregulation and recovery of function. *Neuroscience*, 125 (1), 129–139.

Griesbach, G.S., Tio, D.L., Nair, S., and Hovda, D., 2013. Recovery of stress response coincides with responsiveness to voluntary exercise after traumatic brain injury. *Journal of Neurotrauma*, 31 (17), 674–682.

Griesbach, G.S., Tio, D.L., Vincelli, J., McArthur, D.L., and Taylor, A.N., 2012a. Differential effects of voluntary and forced exercise on stress responses after traumatic brain injury. *Journal of Neurotrauma*, 29 (7), 1426–1433.

Griesbach, G.S., Tio, D.L., Vincelli, J., McArthur, D.L., and Taylor, A.N., 2012b. Differential effects of voluntary and forced exercise on stress responses after traumatic brain injury. *Journal of Neurotrauma*, 29 (7), 1426–1433.

Gurkoff, G.G., Giza, C.C., and Hovda, D.A., 2006. Lateral fluid percussion injury in the developing rat causes an acute, mild behavioral dysfunction in the absence of significant cell death. *Brain Research*, 1077 (1), 24–36.

Gutman, S.A., 2000. *Brain Injury and Gender Role Strain: Rebuilding Adult Lifestyles after Injury*. Routledge, Florence, KY.

Hamm, R.J., Pike, B.R., O'Dell, D.M., Lyeth, B.G., and Jenkins, L.W., 1994. The Rotarod test: An evaluation of its effectiveness in assessing motor deficits following traumatic brain injury. *Journal of Neurotrauma*, 11 (2), 187–196.

Hammond, F.M., Grattan, K.D., Sasser, H., Corrigan, J.D., Rosenthal, M., Bushnik, T., and Shull, W., 2004. Five years after traumatic brain injury: A study of individual outcomes and predictors of change in function. *NeuroRehabilitation*, 19 (1), 25–35.

Harun, R. and Wagner, A.K., 2013. The neurobiological basis of pharmacological approaches for patients with traumatic brain injury. In *Understanding Traumatic Brain Injury: Current Research and Future Directions*. Levin, H., Shum, D., and Chan, R. (eds.). Oxford University Press, Oxford, UK.

He, J., Hoffman, S.W., and Stein, D.G., 2004. Allopregnanolone, a progesterone metabolite, enhances behavioral recovery and decreases neuronal loss after traumatic brain injury. *Restorative Neurology and Neuroscience*, 22 (1), 19–31.

Held, J.M., Gordon, J., and Gentile, A.M., 1985. Environmental influences on locomotor recovery following cortical lesions in rats. *Behavioral Neuroscience*, 99 (4), 678–690.

Hemerka, J.N., Wu, X., Dixon, C.E., Garman, R.H., Exo, J.L., Shellington, D.K. et al., 2012. Severe brief pressure-controlled hemorrhagic shock after traumatic brain injury exacerbates functional deficits and long-term neuropathological damage in mice. *Journal of Neurotrauma*, 29 (12), 2192–2208.

Hicks, R., Soares, H., Smith, D., and McIntosh, T., 1996. Temporal and spatial characterization of neuronal injury following lateral fluid-percussion brain injury in the rat. *Acta Neuropathologica*, 91 (3), 236–246.

Hillman, C.H., Erickson, K.I., and Kramer, A.F., 2008. Be smart, exercise your heart: Exercise effects on brain and cognition. *Nature Reviews. Neuroscience*, 9 (1), 58–65.

Hinkelmann, K., Moritz, S., Botzenhardt, J., Riedesel, K., Wiedemann, K., Kellner, M. et al., 2009. Cognitive impairment in major depression: Association with salivary cortisol. *Biological Psychiatry*, 66 (9), 879–885.

Hoffman, J.M., Bell, K.R., Powell, J.M., Behr, J., Dunn, E.C., Dikmen, S. et al., 2010. A randomized controlled trial of exercise to improve mood after traumatic brain injury. *PM & R*, 2 (10), 911–919.

Hoh, N.Z., Wagner, A.K., Alexander, S.A., Clark, R.B., Beers, S.R., Okonkwo, D.O. et al., 2010. BCL2 genotypes: Functional and neurobehavioral outcomes after severe traumatic brain injury. *Journal of Neurotrauma*, 27 (8), 1413–1427.

Hoofien, D., Gilboa, A., Vakil, E., and Donovick, P.J., 2001. Traumatic brain injury (TBI) 10–20 years later: A comprehensive outcome study of psychiatric symptomatology, cognitive abilities and psychosocial functioning. *Brain Injury*, 15 (3), 189–209.

Hornstein, A., Lennihan, L., Seliger, G., Lichtman, S., and Schroeder, K., 1996. Amphetamine in recovery from brain injury. *Brain Injury*, 10 (2), 145–148.

Hylin, M.J., Orsi, S.A., Zhao, J., Bockhorst, K., Perez, A., Moore, A.N., and Dash, P.K., 2013. Behavioral and histopathological alterations resulting from mild fluid percussion injury. *Journal of Neurotrauma*, 30 (9), 702–715.

Ikonomidou, C. and Turski, L., 2002. Why did NMDA receptor antagonists fail clinical trials for stroke and traumatic brain injury? *Lancet Neurology*, 1 (6), 383–386.

Imam, A.M., Jin, G., Sillesen, M., Duggan, M., Jepsen, C.H., Hwabejire, J.O. et al., 2013. Early treatment with lyophilized plasma protects the brain in a large animal model of combined traumatic brain injury and hemorrhagic shock. *The Journal of Trauma and Acute Care Surgery*, 75 (6), 976–983.

Itoh, T., Imano, M., Nishida, S., Tsubaki, M., Hashimoto, S., Ito, A., and Satou, T., 2011a. Exercise inhibits neuronal apoptosis and improves cerebral function following rat traumatic brain injury. *Journal of Neural Transmission*, 118 (9), 1263–1272.

Itoh, T., Imano, M., Nishida, S., Tsubaki, M., Hashimoto, S., Ito, A. et al., 2011b. Exercise increases neural stem cell proliferation surrounding the area of damage following rat traumatic brain injury. *Journal of Neural Transmission*, 118 (2), 193–202.

Joëls, M., 2008. Functional actions of corticosteroids in the hippocampus. *European Journal of Pharmacology*, 583 (2–3), 312–321.

Johansson, C.B., Lothian, C., Molin, M., Okano, H., and Lendahl, U., 2002. Nestin enhancer requirements for expression in normal and injured adult CNS. *Journal of Neuroscience Research*, 69 (6), 784–794.

Jones, N.C., Cardamone, L., Williams, J.P., Salzberg, M.R., Myers, D., and O'Brien, T.J., 2008. Experimental traumatic brain injury induces a pervasive hyperanxious phenotype in rats. *Journal of Neurotrauma*, 25 (11), 1367–1374.

Jorge, R.E., Acion, L., Starkstein, S.E., and Magnotta, V., 2007. Hippocampal volume and mood disorders after traumatic brain injury. *Biological Psychiatry*, 62 (4), 332–338.

Jorge, R.E., Robinson, R.G., Starkstein, S.E., and Arndt, S.V., 1993. Depression and anxiety following traumatic brain injury. *The Journal of Neuropsychiatry and Clinical Neurosciences*, 5 (4), 369–374.

Juengst, S.B., Kumar, R., Failla, M.D., Goyal, A., Wagner, A.K. 2014. Acute inflammatory associations with depression after severe TBI. *J. Head Trauma Rehabil.* [Epub ahead of print] PMID: 24590155.

Kalish, H. and Phillips, T.M., 2010. Analysis of neurotrophins in human serum by immunoaffinity capillary electrophoresis (ICE) following traumatic head injury. *Journal of Chromatography. B, Analytical Technologies in the Biomedical and Life Sciences*, 878 (2), 194–200.

Katz, D.I. and Alexander, M.P., 1994. Traumatic brain injury. Predicting course of recovery and outcome for patients admitted to rehabilitation. *Archives of Neurology*, 51 (7), 661–670.

Katz, D.I., Alexander, M.P., and Klein, R.B., 1998. Recovery of arm function in patients with paresis after traumatic brain injury. *Archives of Physical Medicine and Rehabilitation*, 79 (5), 488–493.

Katzenberger, R.J., Loewen, C.A., Wassarman, D.R., Petersen, A.J., Ganetzky, B., and Wassarman, D.A., 2013. A Drosophila model of closed head traumatic brain injury. *Proceedings of the National Academy of Sciences*, 201316895.

Kelly, J.C., Amerson, E.H., and Barth, J.T., 2012. Mild traumatic brain injury: Lessons learned from clinical, sports, and combat concussions. *Rehabilitation Research and Practice*, 2012, 371970.

Kline, A.E., Hoffman, A.N., Cheng, J.P., Zafonte, R.D., and Massucci, J.L., 2008. Chronic administration of antipsychotics impede behavioral recovery after experimental traumatic brain injury. *Neuroscience Letters*, 448 (3), 263–267.

Kline, A.E., Massucci, J.L., Dixon, C.E., DeFeo, J.R., and Rogers, E.H., 2007a. Differential effects of single versus multiple administrations of haloperidol and risperidone on functional outcome after experimental brain trauma. *Critical Care Medicine*, 35 (3), 919–924.

Kline, A., Wagner, A.K., Westergom, B.P., Malena, R.R., Zafonte, R.D., Olsen, A.S. et al., 2007b. Acute treatment with the 5-HT(1A) receptor agonist 8-OH-DPAT and chronic environmental enrichment confer neurobehavioral benefit after experimental brain trauma. *Behavioural Brain Research*, 177 (2), 186–194.

Kline, A.E., Yan, H.Q., Bao, J., Marion, D.W., and Dixon, C.E., 2000. Chronic methylphenidate treatment enhances water maze performance following traumatic brain injury in rats. *Neuroscience Letters*, 280 (3), 163–166.

Kochanek, P., Marion, D.W., Zhang, W., Schiding, J.K., White, M., Palmer, A.M. et al., 1995. Severe controlled cortical impact in rats: Assessment of cerebral edema, blood flow, and contusion volume. *Journal of Neurotrauma*, 12 (6), 1015–1025.

Kochanek, P. and Yonas, H., 1999. Subarachnoid hemorrhage, systemic immune response syndrome, and MODS: Is there cross-talk between the injured brain and the extra-cerebral organ systems? Multiple organ dysfunction syndrome. *Critical Care Medicine*, 27 (3), 454–455.

Kochanek, P.M., Bramlett, H., Dietrich, W.D., Dixon, C.E., Hayes, R.L., Povlishock, J. et al., 2011. A novel multicenter preclinical drug screening and biomarker consortium for experimental traumatic brain injury: Operation brain trauma therapy. *The Journal of Trauma*, 71 (1 Suppl), S15–24.

Kovalchuk, Y., Hanse, E., Kafitz, K.W., and Konnerth, A., 2002. Postsynaptic induction of BDNF-mediated long-term potentiation. *Science*, 295 (5560), 1729–1734.

Kraus, M.F., Susmaras, T., Caughlin, B.P., Walker, C.J., Sweeney, J.A., and Little, D.M., 2007. White matter integrity and cognition in chronic traumatic brain injury: A diffusion tensor imaging study. *Brain*, 130 (10), 2508 –2519.

Krueger, F., Pardini, M., Huey, E.D., Raymont, V., Solomon, J., Lipsky, R.H. et al., 2011. The role of the Met66 brain-derived neurotrophic factor allele in the recovery of executive functioning after combat-related traumatic brain injury. *The Journal of Neuroscience*, 31 (2), 598–606.

Kumar, A., Stoica, B.A., Sabirzhanov, B., Burns, M.P., Faden, A.I., and Loane, D.J., 2013. Traumatic brain injury in aged animals increases lesion size and chronically alters microglial/macrophage classical and alternative activation states. *Neurobiology of Aging*, 34 (5), 1397–1411.

Kumar, R., Boles, J.A., Failla, M.D., Wagner, A.K. 2014. Chronic inflammation characterization after severe TBI and associations with outcome. *J. Head Trauma Rehabil.* [Epub ahead of print] PMID: 24901329.

Kwon, S.-K.C., Kovesdi, E., Gyorgy, A.B., Wingo, D., Kamnaksh, A., Walker, J., Long, J.B., and Agoston, D.V., 2011. Stress and traumatic brain injury: A behavioral, proteomics, and histological study. *Frontiers in Neurology*, 2.

Laurer, H.L., Bareyre, F.M., Lee, V.M., Trojanowski, J.Q., Longhi, L., Hoover, R. et al., 2001. Mild head injury increasing the brain's vulnerability to a second concussive impact. *Journal of Neurosurgery*, 95 (5), 859–870.

Lecourtier, L. and Kelly, P.H., 2005. Bilateral lesions of the habenula induce attentional disturbances in rats. *Neuropsychopharmacology*, 30 (3), 484–496.

León-Carrión, J., De Serdio-Arias, M.L., Cabezas, F.M., Roldán, J.M., Domínguez-Morales, R., Martín, J.M. et al., 2001. Neurobehavioural and cognitive profile of traumatic brain injury patients at risk for depression and suicide. *Brain Injury*, 15 (2), 175–181.

Lighthall, J.W., 1988. Controlled cortical impact: A new experimental brain injury model. *Journal of Neurotrauma*, 5 (1), 1–15.

Lindner, M.D., Plone, M.A., Cain, C.K., Frydel, B., Francis, J.M., Emerich, D.F., and Sutton, R.L., 1998. Dissociable long-term cognitive deficits after frontal versus sensorimotor cortical contusions. *Journal of Neurotrauma*, 15 (3), 199–216.

Lipsky, R.H., Sparling, M.B., Ryan, L.M., Xu, K., Salazar, A.M., Goldman, D. et al., 2005. Association of COMT Val158Met genotype with executive functioning following traumatic brain injury. *The Journal of Neuropsychiatry and Clinical Neurosciences*, 17 (4), 465–471.

Liu, Y.R., Cardamone, L., Hogan, R.E., Gregoire, M.-C., Williams, J.P., Hicks, R.J. et al., 2010. Progressive metabolic and structural cerebral perturbations after traumatic brain injury: An in vivo imaging study in the rat. *Journal of Nuclear Medicine*, 51 (11), 1788–1795.

Loane, D.J. and Faden, A.I., 2010. Neuroprotection for traumatic brain injury: Translational challenges and emerging therapeutic strategies. *Trends in Pharmacological Sciences*, 31 (12), 596–604.

Longhi, L., Perego, C., Ortolano, F., Zanier, E.R., Bianchi, P., Stocchetti, N. et al., 2009. C1-inhibitor attenuates neurobehavioral deficits and reduces contusion volume after controlled cortical impact brain injury in mice. *Critical Care Medicine*, 37 (2), 659–665.

Longhi, L., Saatman, K.E., Fujimoto, S., Raghupathi, R., Meaney, D.F., Davis, J. et al, T.K., 2005. Temporal window of vulnerability to repetitive experimental concussive brain injury. *Neurosurgery*, 56 (2), 364–374; discussion 364–374.

Lyeth, B.G., Jenkins, L.W., Hamm, R.J., Dixon, C.E., Phillips, L.L., Clifton, G.L. et al., 1990. Prolonged memory impairment in the absence of hippocampal cell death following traumatic brain injury in the rat. *Brain Research*, 526 (2), 249–258.

Maas, A.I., Stocchetti, N., and Bullock, R., 2008. Moderate and severe traumatic brain injury in adults. *The Lancet Neurology*, 7 (8), 728–741.

Mac Donald, C., Johnson, A., Cooper, D., Malone, T., Sorrell, J., Shimony, J. et al., 2013. Cerebellar white matter abnormalities following primary blast injury in US military personnel. *PloS One*, 8 (2), e55823.

Maddux, J.-M., Kerfoot, E.C., Chatterjee, S., and Holland, P.C., 2007. Dissociation of attention in learning and action: Effects of lesions of the amygdala central nucleus, medial prefrontal cortex, and posterior parietal cortex. *Behavioral Neuroscience*, 121 (1), 63.

Magnuson, J., Leonessa, F., and Ling, G.S.F., 2012. Neuropathology of explosive blast traumatic brain injury. *Current Neurology and Neuroscience Reports*, 12 (5), 570–579.

Majerske, C.W., Mihalik, J.P., Ren, D., Collins, M.W., Reddy, C.C., Lovell, M.R. et al., 2008. Concussion in sports: Postconcussive activity levels, symptoms, and neurocognitive performance. *Journal of Athletic Training*, 43 (3), 265–274.

Manley, G., Knudson, M.M., Morabito, D., Damron, S., Erickson, V., and Pitts, L., 2001. Hypotension, hypoxia, and head injury: Frequency, duration, and consequences. *Archives of Surgery (Chicago, Ill.: 1960)*, 136 (10), 1118–1123.

Margulies, C., Tully, T., and Dubnau, J., 2005. Deconstructing memory in Drosophila. *Current Biology*, 15 (17), R700–713.

Markgraf, C.G., Clifton, G.L., Aguirre, M., Chaney, S.F., Knox-Du Bois, C., Kennon, K. et al., 2001. Injury severity and sensitivity to treatment after controlled cortical impact in rats. *Journal of Neurotrauma*, 18 (2), 175–186.

Marklund, N., Clausen, F., McIntosh, T.K., and Hillered, L., 2001. Free radical scavenger posttreatment improves functional and morphological outcome after fluid percussion injury in the rat. *Journal of Neurotrauma*, 18 (8), 821–832.

Marklund, N., Morales, D., Clausen, F., Hånell, A., Kiwanuka, O., Pitkänen, A. et al., 2009. Functional outcome is impaired following traumatic brain injury in aging Nogo-A/B-deficient mice. *Neuroscience*, 163 (2), 540–551.

Marmarou, A., Foda, M.A., van den Brink, W., Campbell, J., Kita, H., and Demetriadou, K., 1994. A new model of diffuse brain injury in rats. Part I: Pathophysiology and biomechanics. *Journal of Neurosurgery*, 80 (2), 291–300.

Mathias, J.L. and Wheaton, P., 2007. Changes in attention and information-processing speed following severe traumatic brain injury: A meta-analytic review. *Neuropsychology*, 21 (2), 212.

McAllister, T.W., Flashman, L.A., Sparling, M.B., and Saykin, A.J., 2004. Working memory deficits after traumatic brain injury: Catecholaminergic mechanisms and prospects for treatment—a review. *Brain Injury*, 18 (4), 331–350.

Montgomery, K.C., 1955. The relation between fear induced by novel stimulation and exploratory drive. *Journal of Comparative and Physiological Psychology*, 48 (4), 254.

Morales, D.M., Marklund, N., Lebold, D., Thompson, H.J., Pitkanen, A., Maxwell, W.L. et al., 2005. Experimental models of traumatic brain injury: Do we really need to build a better mousetrap? *Neuroscience*, 136 (4), 971–989.

Morganti-Kossmann, M.C., Satgunaseelan, L., Bye, N., and Kossmann, T., 2007. Modulation of immune response by head injury. *Injury*, 38 (12), 1392–1400.

MRC CRASH Trial Collaborators, Perel, P., Arango, M., Clayton, T., Edwards, P., Komolafe, E., Poccock, S. et al., 2008. Predicting outcome after traumatic brain injury: Practical prognostic models based on large cohort of international patients. *BMJ (Clinical research ed.)*, 336 (7641), 425–429.

Narayan, R.K., Michel, M.E., Ansell, B., Baethmann, A., Biegon, A., Bracken, M.B., Bullock, M.R. et al., 2002. Clinical trials in head injury. *Journal of Neurotrauma*, 19 (5), 503–557.

Newcomer, J., Craft, S., Hershey, T., Askins, K., and Bardgett, M., 1994. Glucocorticoid-induced impairment in declarative memory performance in adult humans. *The Journal of Neuroscience*, 14 (4), 2047–2053.

Ning, Y.-L., Yang, N., Chen, X., Xiong, R.-P., Zhang, X.-Z., Li, P. et al., 2013. Adenosine A2A receptor deficiency alleviates blast-induced cognitive dysfunction. *Journal of Cerebral Blood Flow and Metabolism*, 33 (11), 1789–1798.

Niogi, S.N., Mukherjee, P., Ghajar, J., Johnson, C., Kolster, R.A., Sarkar, R. et al., 2008. Extent of microstructural white matter injury in postconcussive syndrome correlates with impaired cognitive reaction time: A 3T diffusion tensor imaging study of mild traumatic brain injury. *AJNR American Journal of Neuroradiology*, 29 (5), 967–973.

Novack, T.A., Bush, B.A., Meythaler, J.M., and Canupp, K., 2001. Outcome after traumatic brain injury: Pathway analysis of contributions from premorbid, injury severity, and recovery variables. *Archives of Physical Medicine and Rehabilitation*, 82 (3), 300–305.

O'Connor, C.A., Cernak, I., and Vink, R., 2003. Interaction between anesthesia, gender, and functional outcome task following diffuse traumatic brain injury in rats. *Journal of Neurotrauma*, 20 (6), 533–541.

O'Connor, W.T., Smyth, A., and Gilchrist, M.D., 2011. Animal models of traumatic brain injury: A critical evaluation. *Pharmacology & Therapeutics*, 130 (2), 106–113.

Oliff, H.S., Berchtold, N.C., Isackson, P., and Cotman, C.W., 1998. Exercise-induced regulation of brain-derived neurotrophic factor (BDNF) transcripts in the rat hippocampus. *Molecular Brain Research*, 61 (1), 147–153.

Öngür, D. and Price, J.L., 2000. The organization of networks within the orbital and medial prefrontal cortex of rats, monkeys and humans. *Cerebral Cortex*, 10 (3), 206–219.

Onyszchuk, G., He, Y.-Y., Berman, N.E.J., and Brooks, W.M., 2008. Detrimental effects of aging on outcome from traumatic brain injury: A behavioral, magnetic resonance imaging, and histological study in mice. *Journal of Neurotrauma*, 25 (2), 153–171.

Packard, M.G., 2009. Exhumed from thought: Basal ganglia and response learning in the plus-maze. *Behavioural Brain Research*, 199 (1), 24–31.

Packard, M.G. and Knowlton, B.J., 2002. Learning and memory functions of the Basal Ganglia. *Annual Review of Neuroscience*, 25, 563–593.

Pandey, D.K., Yadav, S.K., Mahesh, R., and Rajkumar, R., 2009. Depression-like and anxiety-like behavioural aftermaths of impact accelerated traumatic brain injury in rats: A model of comorbid depression and anxiety? *Behavioural Brain Research*, 205 (2), 436–442.

Park, E., Bell, J.D., and Baker, A.J., 2008. Traumatic brain injury: Can the consequences be stopped? *Canadian Medical Association Journal*, 178 (9), 1163–1170.

Parker, L., Padilla, M., Du, Y., Dong, K., and Tanouye, M.A., 2011. Drosophila as a model for epilepsy: Bss is a gain-of-function mutation in the para sodium channel gene that leads to seizures. *Genetics*, 187 (2), 523–534.

Passineau, M.J., Green, E.J., and Dietrich, W.D., 2001. Therapeutic effects of environmental enrichment on cognitive function and tissue integrity following severe traumatic brain injury in rats. *Experimental Neurology*, 168 (2), 373–384.

Pellow, S., Chopin, P., File, S.E., and Briley, M., 1985. Validation of open: Closed arm entries in an elevated plus-maze as a measure of anxiety in the rat. *Journal of Neuroscience Methods*, 14 (3), 149–167.

Piao, C.-S., Stoica, B.A., Wu, J., Sabirzhanov, B., Zhao, Z., Cabatbat, R. et al., 2013. Late exercise reduces neuroinflammation and cognitive dysfunction after traumatic brain injury. *Neurobiology of Disease*, 54, 252–263.

Pierce, J.E.S., Smith, D.H., Trojanowski, J.Q., and McIntosh, T.K., 1998. Enduring cognitive, neurobehavioral and histopathological changes persist for up to one year following severe experimental brain injury in rats. *Neuroscience*, 87 (2), 359–369.

Plantman, S., Ng, K.C., Lu, J., Davidsson, J., and Risling, M., 2012. Characterization of a novel rat model of penetrating traumatic brain injury. *Journal of Neurotrauma*, 29 (6), 1219–1232.

Prins, M.L., Alexander, D., Giza, C.C., and Hovda, D.A., 2013. Repeated mild traumatic brain injury: Mechanisms of cerebral vulnerability. *Journal of Neurotrauma*, 30 (1), 30–38.

Prut, L. and Belzung, C., 2003. The open field as a paradigm to measure the effects of drugs on anxiety-like behaviors: A review. *European Journal of Pharmacology*, 463 (1–3), 3–33.

Pulliainen, V. and Jokelainen, M., 1995. Comparing the cognitive effects of phenytoin and carbamazepine in long-term monotherapy: A two-year follow-up. *Epilepsia*, 36 (12), 1195–1202.

Ramlackhansingh, A.F., Brooks, D.J., Greenwood, R.J., Bose, S.K., Turkheimer, F.E., Kinnunen, K.M. et al., 2011. Inflammation after trauma: Microglial activation and traumatic brain injury. *Annals of Neurology*, 70 (3), 374–383.

Rao, N., Jellinek, H.M., and Woolston, D.C., 1985. Agitation in closed head injury: Haloperidol effects on rehabilitation outcome. *Archives of Physical Medicine and Rehabilitation*, 66 (1), 30–34.

Rau, T.F., Kothiwal, A.S., Rova, A.R., Brooks, D.M., and Poulsen, D.J., 2012. Treatment with low-dose methamphetamine improves behavioral and cognitive function after severe traumatic brain injury. *The Journal of Trauma and Acute Care Surgery*, 73 (2 Suppl 1), S165–172.

Redell, J.B., Moore, A.N., Grill, R.J., Johnson, D., Zhao, J., Liu, Y. et al., 2013. Analysis of functional pathways altered after mild traumatic brain injury. *Journal of Neurotrauma*, 30 (9), 752–764.

Robbins, T.W., 2002. The 5-choice serial reaction time task: Behavioural pharmacology and functional neurochemistry. *Psychopharmacology*, 163 (3–4), 362–380.

Rola, R., Mizumatsu, S., Otsuka, S., Morhardt, D.R., Noble-Haeusslein, L.J., Fishman, K. et al., 2006. Alterations in hippocampal neurogenesis following traumatic brain injury in mice. *Experimental Neurology*, 202 (1), 189–199.

Rostami, E., Krueger, F., Plantman, S., Davidsson, J., Agoston, D., Grafman, J., and Risling, M., 2013. Alteration in BDNF and its receptors, full-length and truncated TrkB and p75 NTR following penetrating traumatic brain injury. *Brain Research*, 1542, 195–205.

Santarsieri, M., Niyonkuru, C., McCullough, E., Dobos, J., Dixon, C.E., Berga, S. et al., 2013. CSF cortisol and progesterone profiles and outcomes prognostication after severe TBI. *Journal of Neurotrauma*, 31 (8), 699–712.

Sauerbeck, A., Hunter, R., Bing, G., and Sullivan, P.G., 2012. Traumatic brain injury and trichloroethylene exposure interact and produce functional, histological, and mitochondrial deficits. *Experimental Neurology*, 234 (1), 85–94.

Scheff, S.W., Baldwin, S.A., Brown, R.W., and Kraemer, P.J., 1997. Morris water maze deficits in rats following traumatic brain injury: Lateral controlled cortical impact. *Journal of Neurotrauma*, 14 (9), 615–627.

Schmitter-Edgecombe, M., 2006. Implications of basic science research for brain injury rehabilitation: A focus on intact learning mechanisms. *The Journal of Head Trauma Rehabilitation*, 21 (2), 131–141.

Schook, L., Beattie, C., Beever, J., Donovan, S., Jamison, R., Zuckermann, F. et al., 2005. Swine in biomedical research: Creating the building blocks of animal models. *Animal Biotechnology*, 16 (2), 183–190.

Schopp, L.H., Shigaki, C.L., Johnstone, B., and Kirkpatrick, H.A., 2001. Gender differences in cognitive and emotional adjustment to traumatic brain injury. *Journal of Clinical Psychology in Medical Settings*, 8 (3), 181–188.

Schwulst, S.J., Trahanas, D.M., Saber, R., and Perlman, H., 2013. Traumatic brain injury-induced alterations in peripheral immunity. *Journal of Trauma and Acute Care Surgery November 2013*, 75 (5), 780–788.

Scoville, W.B. and Milner, B., 1957. Loss of recent memory after bilateral hippocampal lesions. *Journal of Neurology, Neurosurgery, and Psychiatry*, 20 (1), 11.

Shear, D.A., Tate, M.C., Archer, D.R., Hoffman, S.W., Hulce, V.D., Laplaca, M.C. et al., 2004. Neural progenitor cell transplants promote long-term functional recovery after traumatic brain injury. *Brain Research*, 1026 (1), 11–22.

Shein, S.L., Bell, M.J., Kochanek, P.M., Tyler-Kabara, E.C., Wisniewski, S.R., Feldman, K. et al., 2012. Risk factors for mortality in children with abusive head trauma. *The Journal of Pediatrics*, 161 (4), 716–722.e1.

Shepherd, J.K., Grewal, S.S., Fletcher, A., Bill, D.J., and Dourish, C.T., 1994. Behavioural and pharmacological characterisation of the elevated 'zero-maze' as an animal model of anxiety. *Psychopharmacology*, 116 (1), 56–64.

Shitaka, Y., Tran, H.T., Bennett, R.E., Sanchez, L., Levy, M.A., Dikranian, K. et al., 2011. Repetitive closed-skull traumatic brain injury in mice causes persistent multifocal axonal injury and microglial reactivity. *Journal of Neuropathology and Experimental Neurology*, 70 (7), 551–567.

Shultz, S.R., MacFabe, D.F., Foley, K.A., Taylor, R., and Cain, D.P., 2011. A single mild fluid percussion injury induces short-term behavioral and neuropathological changes in the Long-Evans rat: Support for an animal model of concussion. *Behavioural Brain Research*, 224 (2), 326–335.

Sidaros, A., Engberg, A.W., Sidaros, K., Liptrot, M.G., Herning, M., Petersen, P. et al., 2008. Diffusion tensor imaging during recovery from severe traumatic brain injury and relation to clinical outcome: A longitudinal study. *Brain*, 131 (2), 559 –572.

Sinz, E.H., Kochanek, P.M., Dixon, C.E., Clark, R.S.B., Carcillo, J.A., Schiding, J.K. et al., 1999. Inducible nitric oxide synthase is an endogenous neuroprotectant after traumatic brain injury in rats and mice. *Journal of Clinical Investigation*, 104 (5), 647–656.

Slattery, D.A. and Cryan, J.F., 2012. Using the rat forced swim test to assess antidepressant-like activity in rodents. *Nature Protocols*, 7 (6), 1009–1014.

Smith, D.H., Chen, X.-H., Pierce, J.E., Wolf, J.A., Trojanowski, J.Q., Graham, D.I. et al., 1997. Progressive atrophy and neuron death for one year following brain trauma in the rat. *Journal of Neurotrauma*, 14 (10), 715–727.

Smith, D.H., Okiyama, K., Thomas, M.J., Claussen, B., and McIntosh, T.K., 1991. Evaluation of memory dysfunction following experimental brain injury using the Morris Water Maze. *Journal of Neurotrauma*, 8 (4), 259–269.

Smith, D.H., Soares, H.D., Pierce, J.S., Perlman, K.G., Saatman, K.E., Meaney, D.F. et al., 1995. A model of parasagittal controlled cortical impact in the mouse: Cognitive and histopathologic effects. *Journal of Neurotrauma*, 12 (2), 169–178.

Sozda, C.N., Hoffman, A.N., Olsen, A.S., Cheng, J.P., Zafonte, R.D., and Kline, A.E., 2010. Empirical comparison of typical and atypical environmental enrichment paradigms on functional and histological outcome after experimental traumatic brain injury. *Journal of Neurotrauma*, 27 (6), 1047–1057.

Squire, L.R., 1992. Memory and the hippocampus: A synthesis from findings with rats, monkeys, and humans. *Psychological Review*, 99 (2), 195.

Statler, K.D., Alexander, H., Vagni, V., Holubkov, R., Dixon, C.E., Clark, R.S.B. et al., 2006. Isoflurane exerts neuroprotective actions at or near the time of severe traumatic brain injury. *Brain Research*, 1076 (1), 216–224.

Suffoletto, B., Wagner, A.K., Arenth, P.M., Calabria, J., Kingsley, E., Kristan, J. et al., 2013. Mobile phone text messaging to assess symptoms after mild traumatic brain injury and provide self-care support: A pilot study. *The Journal of Head Trauma Rehabilitation*, 28 (4), 302–312.

Susman, M., DiRusso, S.M., Sullivan, T., Risucci, D., Nealon, P., Cuff, S. et al., 2002. Traumatic brain injury in the elderly: Increased mortality and worse functional outcome at discharge despite lower injury severity. *The Journal of Trauma*, 53 (2), 219–223; discussion 223–224.

Swindle, M.M., 1992. *Swine as Models in Biomedical Research.* Iowa State University Press, Iowa City, IA.

Tang, Y.P., Noda, Y., Hasegawa, T., and Nabeshima, T., 1997. A concussive-like brain injury model in mice (II): Selective neuronal loss in the cortex and hippocampus. *Journal of Neurotrauma*, 14 (11), 863–873.

Thompson, H.J., Lifshitz, J., Marklund, N., Grady, M.S., Graham, D.I., Hovda, D.A. et al., 2005. Lateral fluid percussion brain injury: A 15-year review and evaluation. *Journal of Neurotrauma*, 22 (1), 42–75.

Treit, D. and Fundytus, M., 1988. Thigmotaxis as a test for anxiolytic activity in rats. *Pharmacology Biochemistry and Behavior*, 31 (4), 959–962.

Uylings, H., Groenewegen, H.J., and Kolb, B., 2003. Do rats have a prefrontal cortex? *Behavioural Brain Research*, 146 (1), 3–17.

Vakil, E., 2005. The effect of moderate to severe traumatic brain injury (TBI) on different aspects of memory: A selective review. *Journal of Clinical and Experimental Neuropsychology*, 27 (8), 977–1021.

Vanderploeg, R.D., Curtiss, G., and Belanger, H.G., 2005. Long-term neuropsychological outcomes following mild traumatic brain injury. *Journal of the International Neuropsychological Society*, 11 (3), 228–236.
Vincent, F.M., Zimmerman, J.E., and Van Haren, J., 1986. Neuroleptic malignant syndrome complicating closed head injury. *Neurosurgery*, 18 (2), 190–193.
Vink, R., Mullins, P.G., Temple, M.D., Bao, W., and Faden, A.I., 2001. Small shifts in craniotomy position in the lateral fluid percussion injury model are associated with differential lesion development. *Journal of Neurotrauma*, 18 (8), 839–847.
Vorhees, C.V. and Williams, M.T., 2006. Morris water maze: Procedures for assessing spatial and related forms of learning and memory. *Nature Protocols*, 1 (2), 848–858.
Wagner, A.K., Brayer, S.W., Hurwitz, M., Niyonkuru, C., Zou, H., Failla, M. et al., 2013. Non-spatial pre-training in the water maze as a clinically relevant model for evaluating learning and memory in experimental TBI. *Neurobiology of Learning and Memory*, 106, 71–86.
Wagner, A.K., Hatz, L.E., Scanlon, J.M., Niyonkuru, C., Miller, M.A., Ricker, J.H. et al., 2012. Association of KIBRA rs17070145 polymorphism and episodic memory in individuals with severe TBI. *Brain Injury*, 26 (13–14), 1658–1669.
Wagner, A.K. and Zitelli, K.T., 2012. A Rehabilomics focused perspective on molecular mechanisms underlying neurological injury, complications, and recovery after severe TBI. *Pathophysiology,* 20 (1), 39–48.
Wagner, A.K., McCullough, E.H., Niyonkuru, C., Ozawa, H., Loucks, T.L., Dobos, J.A. et al., 2011. Acute serum hormone levels: Characterization and prognosis after severe traumatic brain injury. *Journal of Neurotrauma*, 28 (6), 871–888.
Wagner, A.K., 2010. TBI translational rehabilitation research in the 21st century: Exploring a Rehabilomics research model. *European Journal of Physical and Rehabilitation Medicine*, 46 (4), 549–556.
Wagner, A.K., Miller, M.A., Scanlon, J., Ren, D., Kochanek, P.M., and Conley, Y.P., 2010. Adenosine A1 receptor gene variants associated with post-traumatic seizures after severe TBI. *Epilepsy Research*, 90 (3), 259–272.
Wagner, A.K., Drewencki, L.L., Chen, X., Santos, F.R., Khan, A.S., Harun, R. et al., 2009a. Chronic methylphenidate treatment enhances striatal dopamine neurotransmission after experimental traumatic brain injury. *Journal of Neurochemistry*, 108 (4), 986–997.
Wagner, A.K., Sokoloski, J.E., Chen, X., Harun, R., Clossin, D.P., Khan, A.S. et al., 2009b. Controlled cortical impact injury influences methylphenidate-induced changes in striatal dopamine neurotransmission. *Journal of Neurochemistry*, 110 (3), 801–810.
Wagner, A.K., Postal, B.A., Darrah, S.D., Chen, X., and Khan, A.S., 2007a. Deficits in novelty exploration after controlled cortical impact. *Journal of Neurotrauma*, 24 (8), 1308–1320.
Wagner, A.K., Kline, A.E., Ren, D., Willard, L.A., Wenger, M.K., Zafonte, R.D. et al., 2007b. Gender associations with chronic methylphenidate treatment and behavioral performance following experimental traumatic brain injury. *Behavioural Brain Research*, 181 (2), 200–209.
Wagner, A.K., Ren, D., Conley, Y.P., Ma, X., Kerr, M.E., Zafonte, R.D. et al., 2007c. Sex and genetic associations with cerebrospinal fluid dopamine and metabolite production after severe traumatic brain injury. *Journal of Neurosurgery*, 106 (4), 538–547.
Wagner, A.K., Chen, X., Kline, A.E., Li, Y., Zafonte, R.D., and Dixon, C.E., 2005. Gender and environmental enrichment impact dopamine transporter expression after experimental traumatic brain injury. *Experimental Neurology*, 195 (2), 475–483.
Wagner, A.K., Bayir, H., Ren, D., Puccio, A., Zafonte, R.D., and Kochanek, P.M., 2004a. Relationships between cerebrospinal fluid markers of excitotoxicity, ischemia, and oxidative damage after severe TBI: The impact of gender, age, and hypothermia. *Journal of Neurotrauma*, 21 (2), 125–136.
Wagner, A.K., Willard, L.A., Kline, A.E., Wenger, M.K., Bolinger, B.D., Ren, D. et al., 2004. Evaluation of estrous cycle stage and gender on behavioral outcome after experimental traumatic brain injury. *Brain Research*, 998 (1), 113–121.
Wagner, A.K., Kline, A.E., Sokoloski, J., Zafonte, R.D., Capulong, E., and Dixon, C.E., 2002. Intervention with environmental enrichment after experimental brain trauma enhances cognitive recovery in male but not female rats. *Neuroscience letters*, 334 (3), 165–168.
Wali, B., Sayeed, I., and Stein, D.G., 2011. Improved behavioral outcomes after progesterone administration in aged male rats with traumatic brain injury. *Restorative Neurology and Neuroscience*, 29 (1), 61–71.
Warden, D.L., Gordon, B., McAllister, T.W., Silver, J.M., Barth, J.T., Bruns, J. et al., 2006. Guidelines for the pharmacologic treatment of neurobehavioral sequelae of traumatic brain injury. *Journal of Neurotrauma*, 23 (10), 1468–1501.
Washington, P.M., Forcelli, P.A., Wilkins, T., Zapple, D.N., Parsadanian, M., and Burns, M.P., 2012. The effect of injury severity on behavior: A phenotypic study of cognitive and emotional deficits after mild, moderate, and severe controlled cortical impact injury in mice. *Journal of Neurotrauma*, 29 (13), 2283–2296.
Whyte, J., Hart, T., Vaccaro, M., Grieb-Neff, P., Risser, A., Polansky, M., and Coslett, H.B., 2004. Effects of methylphenidate on attention deficits after traumatic brain injury: a multidimensional, randomized, controlled trial. *American Journal of Physical Medicine & Rehabilitation*, 83 (6), 401–420.
Willner, P., Towell, A., Sampson, D., Sophokleous, S., and Muscat, R., 1987. Reduction of sucrose preference by chronic unpredictable mild stress, and its restoration by a tricyclic antidepressant. *Psychopharmacology*, 93 (3), 358–364.
Wilson, J.T., Hadley, D.M., Wiedmann, K.D., and Teasdale, G.M., 1995. Neuropsychological consequences of two patterns of brain damage shown by MRI in survivors of severe head injury. *Journal of Neurology, Neurosurgery, and Psychiatry*, 59 (3), 328–331.
Wilson, M.S., Gibson, C.J., and Hamm, R.J., 2003. Haloperidol, but not olanzapine, impairs cognitive performance after traumatic brain injury in rats. *American Journal of Physical Medicine & Rehabilitation*, 82 (11), 871–879.
Wise, E.K., Hoffman, J.M., Powell, J.M., Bombardier, C.H., and Bell, K.R., 2012. Benefits of exercise maintenance after traumatic brain injury. *Archives of Physical Medicine and Rehabilitation*, 93 (8), 1319–1323.
Wolff, M., Gibb, S.J., Cassel, J.-C., and Dalrymple-Alford, J.C., 2008. Anterior but not intralaminar thalamic nuclei support allocentric spatial memory. *Neurobiology of Learning and Memory*, 90 (1), 71–80.
Woolf, P.D., 1992. Hormonal responses to trauma. *Critical Care Medicine*, 20 (2), 216–226.
Wu, A., Ying, Z., and Gomez-Pinilla, F., 2013. Exercise facilitates the action of dietary DHA on functional recovery after brain trauma. *Neuroscience*, 248, 655–663.
Wu, C.-W., Chen, Y.-C., Yu, L., Chen, H., Jen, C.J., Huang, A.-M. et al., 2007. Treadmill exercise counteracts the suppressive effects of peripheral lipopolysaccharide on hippocampal neurogenesis and learning and memory. *Journal of Neurochemistry*, 103 (6), 2471–2481.

Xiong, Y., Mahmood, A., Meng, Y., Zhang, Y., Zhang, Z.G., Morris, D.C., and Chopp, M., 2012. Neuroprotective and neurorestorative effects of thymosin β4 treatment following experimental traumatic brain injury. *Annals of the New York Academy of Sciences*, 1270, 51–58.

Yu, S., Kaneko, Y., Bae, E., Stahl, C.E., Wang, Y., van Loveren, H., Sanberg, P.R. et al., 2009. Severity of controlled cortical impact traumatic brain injury in rats and mice dictates degree of behavioral deficits. *Brain Research*, 1287, 157–163.

Zauner, A., Clausen, T., Alves, O.L., Rice, A., Levasseur, J., Young, H.F., and Bullock, R., 2002. Cerebral metabolism after fluid-percussion injury and hypoxia in a feline model. *Journal of Neurosurgery*, 97 (3), 643–649.

Zhang, C., Raghupathi, R., Saatman, K.E., Smith, D.H., Stutzmann, J.M., Wahl, F. et al., 1998. Riluzole attenuates cortical lesion size, but not hippocampal neuronal loss, following traumatic brain injury in the rat. *Journal of Neuroscience Research*, 52 (3), 342–349.

Zohar, O., Schreiber, S., Getslev, V., Schwartz, J.., Mullins, P.., and Pick, C.., 2003. Closed-head minimal traumatic brain injury produces long-term cognitive deficits in mice. *Neuroscience*, 118 (4), 949–955.

Zoltewicz, J.S., Mondello, S., Yang, B., Newsom, K.J., Kobeissy, F., Yao, C. et al., 2013. Biomarkers track damage after graded injury severity in a rat model of penetrating brain injury. *Journal of Neurotrauma*, 30 (13), 1161–1169.

Zou, H., Brayer, S.W., Hurwitz, M., Niyonkuru, C., Fowler, L.E., and Wagner, A.K., 2013. Neuroprotective, neuroplastic, and neurobehavioral effects of daily treatment with levetiracetam in experimental traumatic brain injury. *Neurorehabilitation and Neural Repair*, 1545968313491007.

Zou, H., Hurwitz, M., Fowler, L., Wagner, A.K. 2014. Abbreviated levetiracetam treatment effects on behavioral and histological outcomes after experimental TBI. In press, *Brain Injury*.

36 Translational Considerations for Behavioral Impairment and Rehabilitation Strategies after Diffuse Traumatic Brain Injury

Theresa Currier Thomas, Taylor A. Colburn, Kelsey Korp, Aida Khodadad, and Jonathan Lifshitz

CONTENTS

36.1 INTRODUCTION

Millions of diffuse traumatic brain injuries (TBI) occur annually through sports-related incidents, motor vehicle accidents, falls and violence; many of which are never reported. Post-concussive symptoms can last from days to years post-injury and include cognitive, behavioral, motor, emotional and social deficits. Some symptoms manifest long after acute symptoms have subsided; hindering a distinct connection between the symptoms and brain injury event. These chronic and debilitating post-concussive symptoms can negatively impact the patient's ability to maintain pre-injury skills required for employment and relationships; significantly reducing their quality of life. At this time, rehabilitation has not been extensively evaluated in an experimental model of diffuse traumatic brain injury (dTBI). This chapter discusses clinical and experimental approaches of rehabilitation after brain injury that should be considered in the evaluation of experimental models of diffuse TBI. The consolidation of this knowledge can streamline and optimize rehabilitation approaches to minimize or compensate for chronic post-concussive symptoms.

Diffuse traumatic brain injury is caused by external mechanical forces causing rapid acceleration and deceleration of the brain within the skull, including both rotational and linear forces (Harmon et al., 2013). This movement causes axonal shearing and tissue distortion resulting in diffuse axonal injury (DAI), the hallmark of dTBI, and damage to the microvasculature. This transient disturbance in brain function is commonly referred to as a concussion (Harmon et al., 2013). The primary injury causes immediate structural and functional (ionic, metabolic, and pathophysiological) damage that initiates a cascade of complex secondary processes that continue to alter neurons and glia over minutes and months postinjury (Xiong et al., 2013). This extended time frame for secondary injury progression may leave the brain vulnerable to secondary insults, but also provides opportunity for reorganization and strengthening of remaining networked neurons using rehabilitative interventions that may prevent or mitigate long-term deficits (Xiong et al., 2013).

Survivors of TBI frequently experience long-term disabling changes in cognition, sensorimotor function, and personality (Mott et al., 2012). Preventative and long-term rehabilitative strategies for acute and chronic clinical care depend on a number of factors, many of which are still being tested. This chapter will focus on the types of dTBI that result in behavioral impairment, animal models of dTBI, the immediate circuit damage and recovery that underlie aspects of these impairments, and the available rehabilitation approaches. The complexity of rehabilitation after dTBI will be emphasized in a discussion on what to consider when choosing the specific type of rehabilitation and variables that may influence or confound recovery. The approach to rehabilitation therapy after diffuse TBI in experimental models is novel and many of the approaches are adapted from focal injury models. Throughout this chapter, ongoing experimental focal and diffuse TBI research will be highlighted in support or opposition of constantly evolving rehabilitative strategies in diffuse TBI.

36.2 TYPES OF CLINICAL dTBI AND RESULTING BEHAVIORAL IMPAIRMENTS

Media have recently highlighted the severe behavioral impairments resulting from repetitive head trauma in sports injury and blast-related trauma in the military; however, long-term behavioral impairment can result from a single mild, moderate, or severe incident and be influenced by the rotational or linear forces causing the DAI. The multiple variables that influence the acute injury can often effect the resulting behavioral impairments or the approach to rehabilitation. Most patients with a mild-moderate TBI or concussion (the most common form of TBI) improve over several days to weeks; however, 5%–38% of these patients continue to experience postconcussive symptoms for extended periods (Iverson, 2005; Meares et al., 2011; Mott et al., 2012).

Repetitive dTBIs are associated with contact sports (football, rugby, hockey) where repeated head injury occurs more frequently and over an extended period. The Centers for Disease Control and Prevention estimate 1.6–3.8 million occurrences of sports-related concussions are reported annually in the United States, whereas substantially more go unreported (Langlois et al., 2006). At 1 week postinjury, immediate acute deficits have normally subsided and players are allowed to return to play (Harmon et al., 2013); however, both neuropsychological testing in patients and electrochemical and histological testing in experimental animal models of diffuse brain injury indicate ongoing postconcussive symptoms and neuropathological events for weeks to months after a single dTBI (Hall and Lifshitz, 2010; Harmon et al., 2013; Lifshitz and Lisembee, 2011; McCrea et al., 2003; McNamara et al., 2010b; Thomas et al., 2012, 2013). Clinical and animal studies both support that multiple concussions during the postconcussion recovery can exacerbate metabolic changes in neurons and contribute to more substantial cognitive morbidities. This includes symptoms of chronic traumatic encephalopathy, which can be verified postmortem (Baugh et al., 2012; Harmon et al., 2013; McKee et al., 2009; Prins et al., 2010; Tavazzi et al., 2007; Vagnozzi et al., 2007, 2008).

Military blast injuries take on another level of complexity, given the highly stressful environment in which the injuries occur, the circumstances of the acute recovery phase and the opportunity for repetitive injury (Bogdanova and Verfaellie, 2012). In addition to concussion, the blast can cause a fall, resulting in the head striking a solid object (Bogdanova and Verfaellie, 2012). In addition, the occurrence of other physical injury (polytrauma) as a result of the blast could also influence the outcome (Bogdanova and Verfaellie, 2012). As a result of multiple deployments, soldiers, as with athletes, are also at a risk for repeated injury, increasing the likelihood of persistent symptoms (Corrigan and Deutschle, 2008). In the Veterans Administration database, Carlson et al. showed that 85% of veterans with a reported TBI had at least one psychiatric diagnosis and 64% had at least two (Carlson et al., 2010).

The behavioral impairments resulting from TBI are similar between types of injury, but can vary in magnitude as a function of injury severity, number of injuries, circumstances of injury, premorbid conditions, age, and gender. These symptoms include cognitive (attention difficulties, memory problems, executive dysfunction), somatic (depression, sensory, pain, anxiety, diarrhea with no biological basis), behavioral (irritability, mood, sleep disturbances, fatigue) (Mottet al., 2012), motor (fine motor skills, speed, coordination, balance) (Guskiewicz et al., 1996; Haaland et al., 1994), and social (disinhibition, reception, processing and communication) (Struchen et al., 2011) deficits, which reduce patients' ability to reintegrate into society and overall compromises their quality of life and life expectancy. The occurrence of dTBI-induced endocrine dysfunction is also gaining recognition in the literature for the high correlation with postconcussion symptoms (Ciancia, 2012; Edwards and Clark,1986; Gasco et al., 2012; Glynn and Agha,2013; Kelly et al., 1998; Klose et al., 2007).

36.3 ANIMAL MODELS OF dTBI AND ASSOCIATED BEHAVIORAL DEFICITS

Experimental models of dTBI are designed to reproduce DAI while controlling for age, sex, medication, drugs of abuse, genetics, and premorbid conditions. Additionally, the type of injury and injury parameters can be controlled, providing the ability to evaluate structural, functional, cellular, and molecular sequelae and rehabilitative approaches over time that cannot otherwise be addressed in a clinical setting. Experimental models simplify the complex issues involved with human TBI in exchange for a reproducible model that allows for the assessment of acute and chronic structural and functional consequences. Chronic structural and functional consequences may be conducive to rehabilitative strategies to mitigate long-term deficits. Several species have been selected for use in animal models, including mice, rats, cats, dogs, sheep, swine, and nonhuman primates. However, rodents are the predominantly used species based on their small size, accelerated life span, modest cost, and extensive normative data (Morganti-Kossmann et al., 2010; Xiong et al., 2013).

The models and evidence of their translational contributions are highlighted in the following section. There are three widely used models of experimental dTBI to replicate aspects of the human condition: fluid percussion injury, weight-drop impact acceleration, and blast injury. Each has been modified over time to increase validity toward specific aspects of the human condition, with concussion, contusion, DAI, hemorrhage, and skull fracture being the primary pathological features of interest (Xiong et al., 2013). Because experimental models of dTBI have become increasingly better characterized, the evaluation of behavioral impairment and therapeutic approaches will continually evolve from previously qualified approaches in focal injury. So although the presence of a contusion or hemorrhage (focal injury) along with the DAI is not strictly a diffuse experimental model, rehabilitation approaches from these models will be evaluated with consideration for their potential translation to dTBI rehabilitation.

36.3.1 Fluid Percussion Injury

The fluid percussion injury (FPI) model delivers a fluid pressure pulse against the intact dura through a craniotomy fitted with a head cap. The fluid pulse causes deformation of the underlying tissue. The extent of pathology can be modified by adjusting the fluid pressure force and moving the craniotomy from midline between lambda and bregma to a more lateral position; midline FPI causes reproducible DAI in both hemispheres without a focal contusion (McNamara et al., 2010b; Thomas et al., 2012) and, depending on the location, lateral FPI produces a mixed injury with both DAI and a focal contusion at the injury site (Thompson et al., 2005). However, when the lateral contusion is moved 3.5 mm or greater from midline (sagittal suture) the focal contusion does not develop (Vink et al., 2001). Petechial hemorrhage, edema, and progressive white and gray matter damage are also typical of FPI. Injury severity can also be influenced by the size of the craniotomy, level of anesthesia, weight, skull size, and skull thickness (Thompson et al., 2005). FPI is highly reproducible and allows for modification of injury severity; however, the preparation requires a craniotomy and can have a mortality rate of 5%–10% in rats (Xiong et al., 2013). FPI has been demonstrated to cause dTBI in several animals when compared with other experimental dTBI models, including rabbits, cats, dog, sheep, swine, rats, and mice (Xiong et al., 2013). Pathology and behavioral deficits need to be verified in each species because of differences between brain structure and geometry (Wang and Ma, 2010).

36.3.2 Weight-Drop Injury

Weight-drop models mimic an impact acceleration injury by dropping a free-falling, guided weight onto either the exposed skull (Shohami model), exposed skull with a metal disc glued or cemented in place with dental acrylic (Marmarou model), or through a craniotomy with the dura intact (Feeney model). The animal, normally a mouse or rat, is anesthetized and placed on a hard surface or foam during the impact. Injury severity in these models can be changed by varying the mass of the weight and the height from which it falls. Other variations include where the injury is delivered (midline, lateral, anterior as in the Maryland model) and the foundation on which the animal is placed during the hit (foam or hard surface). Aside from the different types of methods used for this technique, variation in some of the outcome measures can result from the type of foam used, interaction of the weight with the tube during the fall, rebounding of the weight against the skull to induce a second injury, and skull fracture (which is reduced by the metal helmet). The continued popularity of this model is due to a relatively inexpensive setup, ease of operation, and ability to induce graded levels of diffuse axonal injury (Morganti-Kossmann et al., 2010; O'Connor et al., 2011; Wang and Ma, 2010; Xiong et al., 2013).

In rodents, aside from widespread DAI, common pathology that can result from the weight-drop model is petechial hemorrhages (especially in white matter tracts); damage of neurons, axons, dendrites, and microvasculature; compromised blood–brain barrier; and loss of righting reflex. In the Feeney and Shohami models, a necrotic cavity often forms; however, the Marmarou and Maryland models show only DAI without cavitation or significant cell loss (Morganti-Kossmann et al., 2010; O'Connor et al., 2011; Wang and Ma, 2010; Xiong et al., 2013). If considering a model of purely DAI, Feeney and Shohami models should be excluded based on the tissue loss caused by the injury. Behavioral impairment in the variations of DAI models has been extensively researched. Neurological severity score assessments, performed to evaluate neurological impairment in motor function, alertness, and seeking behavior, indicate that the score highly correlates with the severity of brain injury. Neurobehavioral deficits, activation of microglia and astrocytes, neurodegeneration, and morphological changes, each demonstrate that weight-drop models can resemble the clinical condition in humans (Xiong et al., 2013).

36.3.3 Blast Injury

Another model for DAI in animals is the blast injury model. The models are developed to mimic real blast scenarios commonly reported in military conflicts. Effects of the primary blast wave result from the propagation of a supersonic pressure transient and are characterized by factors such as peak pressure, duration, and shape of the wave. The secondary effects of a blast, caused by the impact of flying objects or the head hitting the ground, are relevant as second, repetitive injuries (see the following section) but not necessarily comparable to single injury models with DAI only. The tertiary effects of a blast result from acceleration movements, which may result in tissue shearing and diffuse injuries, such as DAI. The quaternary effects of a blast result from heat, smoke, or emission of electromagnetic pulses (Kirkman et al., 2011; Risling and Davidsson, 2012). With the four clinical effects of blast injury considered, the parameters for first peak pressure, duration, and wave shape must be standardized for their ability to cause DAI when the position of the rat, protection from polytrauma, and the type of explosion are controlled.

To establish the effects of primary blast waves on the central nervous system, rodent and swine models of blast TBI have been established (Cernak et al., 2011; Risling and Davidsson, 2012; Xiong et al., 2013). Parameters of the primary blast can be modified and impact the severity and type of injury induced. Blasts can be simulated using explosives, compressed air, and gas in either blast or shock tubes. The length and diameter of the tube as well as the distance the animal is restrained from the initial charge can also impact severity and add variability that must be controlled for. The position of the animal (parallel vs. perpendicular to the force, inside, outside, or near the exit of the tube), the presence of objects in the tube that would reflect the wave, and any type of protective gear for the animals would also influence the primary location and severity of the injury (Risling and Davidsson, 2012; Xiong et al., 2013).

DAI is the most prominent neuropathological feature resulting from blast injury, with loss of responsiveness, and hemorrhage often replicated in the animal models (Xiong et al., 2013). Provisions should be made to control for systemic events that influence the outcome of the dTBI, including hypotension and hypoxemia, which are possibly evoked by blast-induced lung injury and/or hemorrhage (Xiong et al., 2013). Body shielding with Kevlar vests over the thorax and abdominal region of the animals exposed to the blast have been able to reduce mortality and axonal injury and behavioral deficits (Koliatsos et al., 2011; Long et al., 2009). The variety of approaches and the lack of continuity between peak pressure, duration, rise time, positive and negative phase, and calibration of sensors together make it very difficult to compare models, much less re-create a comparable blast injury tube at an alternate research institution. The blast injury models have only been around for approximately 10 years, compared with FPI and weight-drop models, which have been actively used for 25–30 years. Although much more standardization is required from this model to produce meaningful translational research (Xiong et al., 2013), this newer more military relevant model is making exceptional progress.

36.3.4 Behavioral Impairment after Experimental dTBI

Behavioral impairment is evident in injured rats, regardless of the injury source. Behavioral tests can be used to evaluate the efficacy of rehabilitation, depending on the model chosen and the parameters identified for the injury. Neuromotor deficits have been identified in the days and weeks after injury using composite neuroscores (McIntosh et al., 1989) and behavioral tests such as the rotating pole, Rotarod, balance beam, and rope grip (Floyd et al., 2002; Hamm, 2001; Mattiasson et al., 2000; Thompson et al., 2005). Sensory processing deficits have been identified using sticky paper, limb placement, the whisker nuisance task, and a stationary shock zone on a rotating arena (Baki et al., 2009; McNamara et al., 2010b; O'Dell et al., 2000; Riess et al., 2001). Cognitive assessment is most often accomplished using the Morris Water Maze, modified for either memory or learning; however deficits have also been reported for the Lashley radial arm and Barnes mazes (Griesbach et al., 2004; Hicks et al., 1993; Hoover et al., 2004; Lyeth et al., 2001; Pierce et al., 1998; Piot-Grosjean et al., 2001; Sanders et al., 1999; Sanderson et al., 1999; Smith et al., 1991). Cognition in rats has also been documented in the freezing response test and object recognition tests (Fujimoto et al., 2004). Anxiety-like tests that have been indicative of deficits after DAI are the elevated-plus maze, exploratory activity, and open field tests (Xiong et al., 2013).

36.4 EVIDENCE FOR CIRCUIT DAMAGE AND RECOVERY FROM EXPERIMENTAL dTBI

In experimental models of diffuse brain injury, damage to neurons results in short- and long-term changes in circuit function and structure that likely evolves over time. Damage to neuronal processes are hypothesized to be the substrate for late-onset and chronic injury-induced deficits because of their ability to structurally reorganize, impacting circuit function and thus behavior (Hall and Lifshitz, 2010; Thomas et al., 2012). Diffuse-injured axons upregulate growth-associated proteins (e.g., GAP-43) and reorganize their cytoskeleton, which indicates regenerative responses (Christman et al., 1997; Emery et al., 2000; Hulsebosch et al., 1998; Stroemer et al., 1998). Similarly, transcriptional profiling using microarray technology routinely identifies growth factor, cell survival, and plasticity gene upregulation after experimental TBI (Dash et al., 2004; Rall et al., 2003). Injury-induced sprouting may promote cellular survival by acquiring trophic support from new local connections, whereas creating maladaptive neural circuits in the absence of guidance cues (Emery et al., 2003; Povlishock et al., 1992). Adaptive regrowth returns acute and subacute symptoms to preinjury status. Maladaptive regrowth results in late-onset, gain-of-function chronic morbidities. The timing and effectiveness required to guide reorganization events and strengthen remaining networks will likely be impacted by the onset, duration, intensity, and type of the rehabilitation approach in both animals and their clinical counterparts.

36.4.1 The Need for a Translational Approach to Rehabilitation after Experimental dTBI

In the clinical setting, rehabilitation approaches that are initiated to restore normal function include gait and balance training (Bland et al., 2011), community integration (Kim and Colantonio, 2010), strength training (Killington et al., 2010), aerobic therapy, splinting and casting, and sensory stimulation (Hellweg and Johannes, 2008; Sanders et al., 2001). Though these approaches are supported by research, no ideal time frame or intervention mechanism for TBI rehabilitation has been indicated to achieve the optimal impact in regard to ongoing plasticity and structural reorganization (Bland et al., 2011; Hellweg and Johannes, 2008; Killington et al., 2010). Animal models of DAI providing evidence of plasticity and reorganization over a time course postinjury may prove useful for testing time and modality of rehabilitation strategies. One example is the whisker system in the rat model of DAI that has been used as an in vivo model of plasticity and structural reorganization. After DAI using midline FPI, there is structural and functional neuropathology in the whisker circuit that progresses over time and parallels with injury-induced, late-onset sensory sensitivity to manual whisker stimulation (Hall and Lifshitz, 2010; Lifshitz and Lisembee, 2011; McNamara et al., 2010a; Thomas et al., 2012). Models such as this would provide the opportunity to test the translational relevance of clinical rehabilitation over time. Consideration for the current types of rehabilitation that are clinically relevant can aid in the choice of translational rehabilitation approaches that could rapidly impact clinical choices.

36.5 TYPES OF REHABILITATION

There are two phases of injury after a traumatic event to the brain. The first phase is the mechanical injury such as shearing, lacerations, or skull impact. The second phase is an inflammatory response, which can cause intracranial pressure, cell apoptosis, and degradation of neurons (Bingham et al., 2013). TBI treatment is largely focused on advanced life support to treat the initial insult followed by measures to limit damage to the brain caused by inflammation. Although experimental models have indicated some level of efficacy for very specific rehabilitative approaches, few experimental models of chronic cognitive, behavioral, motor, emotional, and/or social deficits have been identified to fully test approaches. This section addresses clinical treatment and measures taken after the patient has been stabilized and is ready to resume life post-TBI. These aspects should be taken into consideration in experimental models of rehabilitation after dTBI.

36.5.1 Cognitive

In general, cognitive deficits and impairments refer to any deficiency in cognitive processes that impact intellectual performance and abilities in comparison to preinjury status. After diffuse TBI, memory, information processing, attention, and executive function are most often affected (Carr and Shepherd, 1998).

The most common rehabilitative approaches are memory aids and strategies (e.g., notes, electronic reminders) to cope with cognitive impairment (Hart et al., 2004). Some studies indicate that spaced repetition can benefit patients with mild to moderate TBI. "Impairments in recall, recognition and rates of forgetting following TBI are more likely a result of insufficient acquisition rather than deficient retrieval from long-term storage" (Hillary et al., 2003). Based on this assumption, introducing information more than once has been shown to improve memory and recall of TBI patients; for example, consider stating a name before and after a conversation in comparison to repeating a name twice in a row. Patients are also instructed to perform one task at a time to compensate for attention deficits. It has been shown through functional magnetic resonance imaging that after TBI, cerebral activation from memory-related tasks was similar in injured and uninjured patients, but that injured patients had more regionally dispersed activation (Christodoulou et al., 2001). These methods of rehabilitation can be beneficial assuming the patient is willing to participate in cognitive rehabilitative measures. A challenge concerning cognitive rehabilitation is the prevalence of depression and lack of social relationships (Gomez-Hernandez et al., 1997) after dTBI. Social relationships can foster participation and improvement, whereas depression lessens the patient's desire to participate in social interactions, including rehabilitation.

36.5.2 Behavioral

Two types of behavioral changes have been observed in patients suffering from TBI: impaired drive (e.g., apathy, lethargy) or dyscontrol (aggressiveness, anger control, impulsiveness). These changes adversely impact every aspect of the patient's life and should be recognized as an impairment that needs to be addressed, just like any physical injury. Rehabilitation involves "a comprehensive lifestyle change, including construction of a meaningful role of personal value" (Ylvisaker et al., 2005). Because of the lack of drive and control, a structured and focused lifestyle will avoid stagnation of rehabilitation and the onset of depression. Simplifying the environment and having a consistent schedule can also have this effect by reducing patient frustration. Some approaches avoid triggers for bad behavior and offer rewards (vocal or other) for good behavior.

36.5.3 Motor

Depending on the severity of brain injury, varied levels of motor rehabilitation are necessary for patient rehabilitation. Severity of motor deficits can range from being in a coma state to loss of fine articulation of peripheral limbs. Coma patients need passive stretching for at-risk muscles to avoid muscle atrophy and shortening that can occur in stationary limbs. After awakening from a coma, reestablishment of swallowing, unassisted breathing, and communicating is necessary (Carr and Shepherd, 1998). Besides overcoming these extremes of surviving TBI, less severe but still crippling impairments may also hinder motor performance. Rehabilitation for functional performance includes treating primary sensorimotor impairments (weakness, loss of coordination) as well as physical injuries that may have also occurred (fractures or nerve injury). Task-oriented motor training is used to reinstate simple everyday activities of daily life. More physically demanding means of rehabilitating motor skills are used in patients with less severe symptoms such as gait and stair-climbing or pivoting deficits. This more aggressive therapy takes the form of resistance exercise with increasing repetitions or stretches conducted alone or assisted. Augmented exercise therapy is the addition of repetitions or extended time in exercise therapy. The increase in physical therapy has shown to improve long-term motor recovery by continually strengthening the desired muscles as well as strengthening the neuronal pathways associated with the affected limb movement (Griesbach et al., 2011). Another method is constraint-induced movement therapy used in conjunction with physical therapy. The method is to restrain the nonimpaired limb, forcing the impaired limb to be used exclusively, with the desire to accelerate and improve range, movement, and articulation of that limb (Wolf et al., 2006).

36.5.4 Emotional

Emotional therapy refers to mitigation of apathy, anger, aggression, and depression that can be brought about by adapting to injury-induced deficits. Frequent frustration has a huge emotional toll on rehabilitation and psychosocial interactions during rehabilitation are noted and addressed by therapists. There is also an increased risk for deficits in emotional and social perception (McDonald and Flanagan, 2004). Studies have shown that "individuals with TBI frequently talk better than they communicate" (Milton and Wertz, 1986). This implies that for some patients the recall of words, grammar, and ability to form sentences or hold a conversation are not impaired, but the act of forming abstract ideas ("What if?" Or "in the future…") or of expressing and interpreting emotion may be impaired (Kwa et al., 1996; McDonald and Flanagan, 2004). For patients' families, the act of simplifying language, avoiding patronization, and redirecting agitated behavior may be helpful for improvement.

36.5.5 Social

Interpersonal relationships between family members and friends that the patient previously fostered may be affected by the cognitive and emotional changes caused by injury. The patient's family should be properly informed and trained to deal with possible personality changes. For instance, a task that the patient performed on a regular basis (doing dishes or balancing the checkbook) may no longer be within their cognitive or motor abilities postinjury. Often, the patient realizes that this task is something that took little effort before the injury, which can increase the frustration level of the patient. In combination with other deficits, the inability to communicate emotional frustration effectively can augment patient stress. As a consequence, the patient's interactions with people will be deeply impacted. The caretaker and family need to understand that these changes are injury-induced deficits that may require assistance and encouragement to overcome. A novel approach of social and emotional rehabilitation is emphasized by resilience training, which teaches skills for adaptive responses to social and emotional cues (Kent et al., 2013).

Another step in social reintegration is returning to work. Introduction of a more enriching and challenging environment has been reported to decrease recovery times (Hamm et al., 1996). Although returning to work encompasses more rehabilitative opportunities, it can also be frustrating to the patient because of physical restraints, cognitive deficits, and emotional lability, depending on the demand of the work (Carr and Shepherd, 1998). Emotional lability, otherwise known as emotionalism, has been documented by clinicians and associated with disorders of the central nervous system (House et al., 1989) including TBI, and can be a deterrent toward returning to work. In most cases, physical and emotional deficits because of TBI usually have to be resolved or compensated for before resuming employment.

36.6 CONSIDERATIONS WHEN CHOOSING CLINICAL REHABILITATION

Rehabilitation, as discussed in this chapter, is defined as any therapy (including pharmacological) aimed at restoring neurological function lost or diminished as a result of injury. When choosing a clinically relevant rehabilitation strategy, many factors must be considered, including intensity, type, timing, impact of polytrauma, and willing participation. Also, there needs to be an understanding of the possible outcomes of the rehabilitation therapy, including recovery of function, compensation, no recovery, or a combination. This is often assessed through the ability of the patient to complete a goal-directed task. The intensity of the rehabilitation (strenuousness, duration of time in a single bout, and repetitions in a week) will likely need to be adjusted to incorporate factors that would influence recovery of function. Research is inconclusive in regard to a single optimal approach. Just as no two people are alike, no two brain injuries are alike. Recovery is typically lengthy, from months to years, because the brain requires time to heal and repair. In addition, it can be affected by different factors such as age at injury, gender, brain injury severity, premorbid conditions, concurrent dementia, preexisting psychiatric conditions, and social, demographic, and cultural factors that interact with injury circumstances. The following section will highlight these aspects of rehabilitation for consideration in experimental models of diffuse TBI.

To properly mediate the chronic and acute complications that arise after TBI, a variety of rehabilitative actions must be taken, depending on the location and severity of an injury. Rehabilitation methods can be broken down into two categories: intensity and timing. Intensity can be further categorized into mild, moderate, and intense, with a combination of intensity that represents the level of recovery desired for each deficit. Mild rehabilitation is defined as behavioral/environmental management, nutritional and/or medicinal regiments, and cognitive therapies. Moderate treatment consists of major nutritional and/or medicinal regiments in addition to physical and speech therapies, whereas intense rehabilitation consists of surgical interventions and physical therapy to treat specific gross motor dysfunction. The optimal timing includes the onset of rehabilitation and the duration of therapies. The goal of rehabilitation is to help the body heal and assists the brain in relearning processes so that an individual recovers as quickly and efficiently as possible. Rehabilitation will also help the person with TBI learn new ways to do things if any previous abilities have been lost.

The severity of the injury impacts the type, intensity, onset, and duration of rehabilitation. In the case of focal injuries, the size, symmetry, and quantity of lesions play a pivotal role in severity (Gentile et al., 1980; Glees and Cole, 1950; Kennard, 1936; Travis and Woolsey, 1956). Kennard showed that small lesions in specific regions of the motor cortex resulted in a shorter recovery period than did the removal of the whole cortical region. She also showed that lesions of both the motor and premotor cortices caused slower recovery in nonhuman primates (NHP) than lesions in either of the areas individually (Kennard, 1936). Additionally, Glees and Cole explored the effects of varying lesion sizes on the thumb, fingers, and facial regions of the motor cortex of macaques, demonstrating that small lesions resulted in more rapid recoveries (Glees and Cole, 1950). This principle, that the smaller a lesion, the shorter the recovery period, was later supported by Travis and Woolsey, who conducted a similar study using NHPs (Travis and Woolsey, 1956), as well as Gentile et al., who were able to confirm it in rats (Gentile et al., 1980). Travis and Woolsey also showed the significance of lesion symmetry on recovery time, finding that bilaterally symmetrical lesions in the motor cortex result in a shorter locomotive recovery period when compared to asymmetrical motor cortex lesions (Travis and Woolsey, 1956). If the size and location of the focal lesion predicts recovery time, then the intensity, onset, and duration of rehabilitation could be adapted to accommodate for these predictions. Also, if recovery and rehabilitation of focal injuries is similar to dTBI recovery, then we may speculate that mild dTBI would result in more rapid recovery in comparison to severe. The diffuse nature of dTBI increases the complexity of symptoms and modalities of necessary rehabilitation, and adaption of our knowledge from focal injury models may be useful in facilitating more rapid and effective rehabilitation paradigms.

36.6.1 Factors Affecting Recovery Can Impact Rehabilitation Plan

36.6.1.1 Location of Injury

Certain injury locations will also call for the implementation of rehabilitative therapies sooner rather than later. After having the forelimb area of the motor cortex unilaterally removed, NHPs were exposed to three different types of rehabilitation: training of contralateral arm, ipsilateral arm, or both. These rehabilitations were implemented at two time points: immediately after surgery and 4 months after surgery (Black et al., 1975). Recovery was shown to be less complete in the late training groups when compared with the recovery of the early training groups (Black et al., 1975). Although Schallert et al. found that immediate rehabilitation training of a limb impaired by lesioning led to permanent changes in behavior and a greater number of dead neurons (Schallert et al., 1997), Nudo et al. demonstrated that a 5-day latency in the implementation of rehabilitation leads to a more thorough representation of the limb in tissue proximal to the damage, and facilitates some behavioral recovery (Nudo et al., 1996). These studies indicate that the location of the injury and the modality of rehabilitative approaches must be strategically chosen based on individual assessment of each subject (whether animal or patient) when focal injuries are present.

36.6.1.2 Gender

There is mounting evidence of gender difference in regard to brain organization and neurochemistry, with growing support for the role of hormonal influences on recovery after TBI (Bazarian et al., 2010). Females (patients and rats) were observed to recover better than males (patients and rats) after focal injury (Stein, 2001). They found that progesterone, a hormone involved in menstruation, pregnancy, and fetal development, has neuroprotective effects. In rodents, progesterone reduced amounts of edema present in the brain postinjury, and also improved mortality and functional recovery after contusions in the frontal cortex (Roof et al., 1993). While progesterone is native to both female and male physiology, circulating levels of progesterone tend to be higher in females, predisposing females to improved neuroprotection. However, progesterone treatment in both male and female patients has shown effectiveness in clinical trials (Stein, 2013).

36.6.1.3 Age

Much research has focused on the importance of age on recovery after TBI. According to the Centers for Disease Control and Prevention, there are four TBI patient classifications: youngest (<16), young (16–26), intermediate (27–39), and oldest (>40). Children younger than age 16 are at the greatest risk for TBI. The results of recovery after injury show that disability rating scores for the younger groups have been improved significantly from year 1 to year 5, with the highest level of improvement in disability seen among the youngest group. In contrast, the oldest group had the least level of improvement as indicated by disability rating scores (delaPlata et al., 2008). Although all these results suggest that the younger the patient, the better the chances of recovering from TBI, recent studies have demonstrated that the youngest TBI patients, especially the youngest group (younger than age 5), recover with increased cognitive disabilities many years after injury. This is potentially because of the immaturity of the nervous system at the time of injury. Therefore, after injury, their nervous systems must go through both the challenges of recovery after injury, and the challenges of development and maturation, which can make the rehabilitation of TBI more complicated for children than for adults (Yen and Wong, 2007).

In experimental models, Smith et al. showed that the adult and very young rats (10 days old; clinically relevant to 0- to 1-year-old infants) had the worst and longest lasting deficits after unilateral cerebellar damage, and that the neonates (15 and 21 day olds; ~1- to 2-year-old children) experienced the most rapid and complete recoveries (Andreollo et al., 2012; Smith et al., 1974). In the same rats, lesions were made to the motor cortex after the animals recovered from the cerebellar damage. The adult and 21-day-old neonates had the least number of deficits. This indicates a potentially more significant degree of plasticity in the neonatal rats (Smith et al., 1974). This developmental plasticity, according to Milner, allows for the recovery of speech in human patients younger than 6 years old who have suffered from lesions to the left hemisphere, but at the cost of diminished visuospatial functionality (Milner, 1974). Additionally, Stein and Firl showed that aged rats were capable of learning the delayed spatial alternation task in roughly the same amount of time, regardless of frontal cortex lesions, whereas the task results for the lesioned young rats suggested more severe impairments (Stein and Firl, 1976). Together, this research indicates a pronounced age-dependent recovery response to brain injury that likely translates to experimental dTBI models.

36.6.1.4 Pre- and Postinjury Conditions

Premorbid conditions have been addressed as another explanation for poorer outcome of recovery after TBI. For instance, history of any head injury before the time of TBI, drug or alcohol abuse, and psychological diseases have been shown to have an impact on the process of recovery of TBI by increasing the time of rehabilitation (Ponsford et al., 2000). In addition, preexisting personality characteristics such as overachievement, perfectionism, and dependency can lengthen the rehabilitation (Kay et al., 1992).

In addition to preinjury conditions, postinjury conditions such as depression, anxiety, headache, dizziness, and insomnia have been reported for failure in recovery after TBI. Many studies found that TBI patients are at great risk of depression after TBI. Many studies showed that all together neuroanatomic, neurochemical, and psychosocial factors are responsible for the depression after TBI (Rosenthal et al., 1998). It has also been suggested that depression in TBI patients can be related to unemployment and poor financial situation (Seel et al., 2003). But the treatment of depression that develops following TBI is still unclear because it is likely to be the result of a complex variety of interacting factors.

There are conflicting reports of the impact of alcohol consumption in regard to brain injury recovery. In a clinical report, blood alcohol levels during blunt trauma causing DAI did not correlate with patient outcome and acquired disabilities (Matsukawa et al., 2013). However, pre-TBI "risky" use of alcohol was an indicator of posttraumatic psychiatric disturbances (Hart et al., 2014). One animal study showed that preinjury intoxication with ethanol can reduce the neuroinflammatory response to injury and consequently improve the rehabilitation (Goodman et al., 2013). In regard to alcohol consumption during rehabilitation, alcohol abuse has been indicated to lengthen the recovery period after TBI, impairing cognition and balance, which increases a TBI survivor's chance of reinjury (Corrigan, 1995). In light of the conflicting reports in both experimental and clinical settings, alcohol abuse during recovery and rehabilitation is more likely to cause setbacks rather than improve time and recovery of function.

36.6.1.5 Nutrition

According to a report by the Institute of Medicine, nutrition appears to play an important role in improving the outcome of TBI, especially if it is administered soon after the injury occurs. Other studies suggest nutritional approaches as part of TBI management, including very strict avoidance of extreme fluid or glucose shifts, in addition to consumption of adequate calories and protein in a TBI patient's diet (Cook et al., 2008). The impact of exercise and a diet rich in the omega-3 fatty acid, docosahexaenoic acid (DHA), after TBI was indicated by Wu et al. as being beneficial to the recovery process (Wu et al., 2013). Voluntary exercise increased and stabilized DHA levels (which drop after injury), reduced oxidative stress, and enhanced membrane homeostasis (as measured by levels of the iPLA2 enzyme and STX-3 protein) for up to 12 days (Wu et al., 2013). These reports highlight that nutrition and exercise improve outcomes after TBI and may be a relevant consideration as part of the rehabilitation strategy.

36.6.1.6 Cultural

Cultural differences can be another factor to have impact on recovery after TBI. Human studies demonstrate that minority populations such as Hispanic and African-American populations in the United Stated are at higher risk for sustaining a TBI in comparison to the white population. These populations are known to have a higher level of disability and are less likely to return to work. Functional outcome at 1 year postinjury shows that disability rating scale and functional independence measures are worse in minority populations than in white populations, which may be due to educational and economic differences (Arango-Lasprilla et al., 2007). Although we have only addressed educational and economic differences, many ethnic and local cultures use complementary or alternative medicine in the treatment of brain injury (Gau et al., 2012; Purohit et al., 2013). These include both physical and biological approaches, including massage, mediation, and herbal remedies. Recently, a traditional Chinese medicine, "Shengyu" decoction, was found to have anti-inflammatory effects after dTBI using the Feeney weight-drop model in rodents (Zhao et al., 2014). As approaches to rehabilitation progress, evaluation in established experimental models of dTBI can validate efficacy.

36.7 CONCLUSION

We have briefly discussed the types of dTBI that result in behavioral impairment, animal models of dTBI, the immediate circuit damage, and recovery that is predicted to underlie aspects of these impairments, the rehabilitation strategies available for TBI patients, and important considerations when choosing a rehabilitation plan. Rehabilitation strategies in the clinic and in experimental models have provided evidence that rehabilitation should take into account the type of injury, gender, age, pre- and postmorbid conditions, nutrition, and the level of determination of each patient. Animal models of focal TBI support an individualized plan based on the location of injury. However, changing variables of the injury or rehabilitation could influence long-term recovery. Many of these approaches have *not* been tested in an experimental model of dTBI.

The heterogeneous nature of the dTBI may support a marginally more generalized treatment approach, where well-controlled experimental models of dTBI can provide definitive support for the timing, duration, and modality of rehabilitation. Also, because typical types of clinical rehabilitation do not always translate to rodents, it is important to continue to consider formidable methods of rehabilitation for our experimental counterparts. Of particular interest is whether repeated behavioral testing could be considered a form of rehabilitation. As our understanding of the dTBI-induced chronic structural, functional, and behavioral deficits continues to expand, the evaluation of the effectiveness of rehabilitation from cellular to behavioral outcome measures continues to improve in experimental models with the potential for a truly profound impact on clinical rehabilitation.

REFERENCES

Andreollo, N.A., E.F. Santos, M.R. Araujo, and L.R. Lopes. 2012. Rat's age versus human's age: What is the relationship? *Arquivos Brasileiros de Cirurgia Digestiva*. 25:49–51.

Arango-Lasprilla, J.C., M. Rosenthal, J. Deluca, E. Komaroff, M. Sherer, D. Cifu, and R. Hanks. 2007. Traumatic brain injury and functional outcomes: Does minority status matter? *Brain Injury*. 21:701–708.

Baki, S.G.A., H.Y. Kao, E. Kelemen, A.A. Fenton, and P.J. Bergold. 2009. A hierarchy of neurobehavioral tasks discriminates between mild and moderate brain injury in rats. *Brain Research*. 1280:98–106.

Baugh, C.M., J.M. Stamm, D.O. Riley, B.E. Gavett, M.E. Shenton, A. Lin et al. 2012. Chronic traumatic encephalopathy: Neurodegeneration following repetitive concussive and subconcussive brain trauma. *Brain Imaging and Behavior*. 6:244–254.

Bazarian, J.J., B. Blyth, S. Mookerjee, H. He, and M.P. McDermott. 2010. Sex differences in outcome after mild traumatic brain injury. *Journal of Neurotrauma*. 27:527–539.

Bingham, D., C.M. John, J. Levin, S.S. Panter, and G.A. Jarvis. 2013. Post-injury conditioning with lipopolysaccharide or lipooligosaccharide reduces inflammation in the brain. *Journal of Neuroimmunology*. 256:28–37.

Black, P., R.S. Markowitz, and S.N. Cianci. 1975. Recovery of motor function after lesions in motor cortex of monkey. *Ciba Foundation Symposium* 34:65–83.

Bland, D.C., C. Zampieri, and D.L. Damiano. 2011. Effectiveness of physical therapy for improving gait and balance in individuals with traumatic brain injury: A systematic review. *Brain Injury*. 25:664–679.

Bogdanova, Y., and M. Verfaellie. 2012. Cognitive sequelae of blast-induced traumatic brain injury: Recovery and rehabilitation. *Neuropsychology Review*. 22:4–20.

Carlson, K.F., D. Nelson, R.J. Orazem, S. Nugent, D.X. Cifu, and N.A. Sayer. 2010. Psychiatric diagnoses among Iraq and Afghanistan war veterans screened for deployment-related traumatic brain injury. *Journal of Traumatic Stress*. 23:17–24.

Carr, J.H. and R.B. Shepherd. 1998. *Neurological Rehabilitation: Optimizing Motor Performance*. Butterworth-Heinemann, Oxford.

Cernak, I., A.C. Merkle, V.E. Koliatsos, J.M. Bilik, Q.T. Luong, T.M. Mahota et al. 2011. The pathobiology of blast injuries and blast-induced neurotrauma as identified using a new experimental model of injury in mice. *Neurobiological Disease*. 41:538–551.

Christman, C.W., J.B. Salvant, Jr., S.A. Walker, and J.T. Povlishock. 1997. Characterization of a prolonged regenerative attempt by diffusely injured axons following traumatic brain injury in adult cat: A light and electron microscopic immunocytochemical study. *Acta Neuropathology*. 94:329–337.

Christodoulou, C., J. DeLuca, J.H. Ricker, N.K. Madigan, B.M. Bly, G. Lange et al. 2001. Functional magnetic resonance imaging of working memory impairment after traumatic brain injury. *Journal of Neurology, Neurosurgery, and Psychiatry*. 71:161–168.

Ciancia, S. 2012. [Pituitary insufficiency after traumatic brain injury: Consequences? Screening?]. *Annales Francaises D'anesthesie et de Reanimation*. 31:e117–124.

Cook, A.M., A. Peppard, and B. Magnuson. 2008. Nutrition considerations in traumatic brain injury. *Nutrition in Clinical Practice*. 23:608–620.

Corrigan, J.D. 1995. Substance abuse as a mediating factor in outcome from traumatic brain injury. *Archives of Physical Medicine and Rehabilitation*. 76:302–309.

Corrigan, J.D., and J.J. Deutschle, Jr. 2008. The presence and impact of traumatic brain injury among clients in treatment for co-occurring mental illness and substance abuse. *Brain Injury*. 22:223–231.

Dash, P.K., A.E. Hebert, and J.D. Runyan. 2004. A unified theory for systems and cellular memory consolidation. *Brain Research. Brain Research Reviews*. 45:30–37.

de la Plata, C.A.M., T. Hart, F.H. Hammond, A.B. Frol, A. Hudak, C.R. Harper et al. 2008. Impact of age on long-term recovery from traumatic brain injury. *Archives of Physical Medicine and Rehabilitation*. 89:896–903.

Edwards, O.M., and J.D. Clark. 1986. Post-traumatic hypopituitarism. Six cases and a review of the literature. *Medicine*. 65:281–290.

Emery, D.L., R. Raghupathi, K.E. Saatman, I. Fischer, M.S. Grady, and T.K. McIntosh. 2000. Bilateral growth-related protein expression suggests a transient increase in regenerative potential following brain trauma. *Journal of Comprehensive Neurology*. 424:521–531.

Emery, D.L., N.C. Royo, I. Fischer, K.E. Saatman, and T.K. McIntosh. 2003. Plasticity following injury to the adult central nervous system: Is recapitulation of a developmental state worth promoting? *Journal of Neurotrauma*. 20:1271–1292.

Floyd, C.L., K.M. Golden, R.T. Black, R.J. Hamm, and B.G. Lyeth. 2002. Craniectomy position affects Morris water maze performance and hippocampal cell loss after parasagittal fluid percussion. *Journal of Neurotrauma*. 19:303–316.

Fujimoto, S.T., L. Longhi, K.E. Saatman, V. Conte, N. Stocchetti, and T.K. McIntosh. 2004. Motor and cognitive function evaluation following experimental traumatic brain injury. *Neuroscience and Biobehavioral Reviews*. 28:365–378.

Gasco, V., F. Prodam, L. Pagano, S. Grottoli, S. Belcastro, P. Marzullo et al. 2012. Hypopituitarism following brain injury: When does it occur and how best to test? *Pituitary*. 15:20–24.

Gau, B.S., H.L. Yang, S.J. Huang, and M.F. Lou. 2012. The use of complementary and alternative medicine for patients with traumatic brain injury in Taiwan. *BMC Complementary and Alternative Medicine*. 12:211.

Gentile, A.M., J.M. Held, and R. Musii. 1980. Recovery of locomotion following symmetrical and asymmetrical serial lesions in rats. *Society for Neuroscience Abstracts*.650.

Glees, P. and J. Cole. 1950. Recovery of skilled motor functions after small repeated lesions of motor cortex in macaque. *Journal of Neurophysiology*. 13:137–148.

Glynn, N., and A. Agha. 2013. Which patient requires neuroendocrine assessment following traumatic brain injury, when and how? *Clinical Endocrinology*. 78:17–20.

Gomez-Hernandez, R., J.E. Max, T. Kosier, S. Paradiso, and R.G. Robinson. 1997. Social impairment and depression after traumatic brain injury. *Archives of Physical Medicine and Rehabilitation*. 78:1321–1326.

Goodman, M.D., A.T. Makley, E.M. Campion, L.A.W. Friend, A.B. Lentsch, and T.A. Pritts. 2013. Preinjury alcohol exposure attenuates the neuroinflammatory response to traumatic brain injury. *The Journal of Surgical Research*. 184:1053–1058.

Griesbach, G.S., D.A. Hovda, R. Molteni, A. Wu, and F. Gomez-Pinilla. 2004. Voluntary exercise following traumatic brain injury: Brain-derived neurotrophic factor upregulation and recovery of function. *Neuroscience*. 125:129–139.

Griesbach, G.S., D.A. Hovda, D.L. Tio, and A.N. Taylor. 2011. Heightening of the stress response during the first weeks after a mild traumatic brain injury. *Neuroscience*. 178:147–158.

Guskiewicz, K.M., D.H. Perrin, and B.M. Gansneder. 1996. Effect of mild head injury on postural stability in athletes. *Journal of Athletic Training*. 31:300–306.

Haaland, K.Y., N. Temkin, G. Randahl, and S. Dikmen. 1994. Recovery of simple motor skills after head injury. *Journal of Clinical and Experimental Neuropsychology*. 16:448–456.

Hall, K.D., and J. Lifshitz. 2010. Diffuse traumatic brain injury initially attenuates and later expands activation of the rat somatosensory whisker circuit concomitant with neuroplastic responses. *Brain Res*. 1323:161–173.

Hamm, R.J. 2001. Neurobehavioral assessment of outcome following traumatic brain injury in rats: An evaluation of selected measures. *Journal of Neurotrauma*. 18:1207–1216.

Hamm, R.J., M.D. Temple, D.M. O'Dell, B.R. Pike, and B.G. Lyeth. 1996. Exposure to environmental complexity promotes recovery of cognitive function after traumatic brain injury. *Journal of Neurotrauma*. 13:41–47.

Harmon, K.G., J. Drezner, M. Gammons, K. Guskiewicz, M. Halstead, S. Herring et al. 2013. American Medical Society for Sports Medicine position statement: Concussion in sport. *Clinical Journal of Sport Medicine*. 23:1–18.

Hart, T., E.K. Benn, E. Bagiella, P. Arenth, S. Dikmen, D.C. Hesdorffer et al. 2014. Early trajectory of psychiatric symptoms after traumatic brain injury: Relationship to patient and injury characteristics. *Journal of Neurotrauma*. 31:610–617.

Hart, T., R. Buchhofer, and M. Vaccaro. 2004. Portable electronic devices as memory and organizational aids after traumatic brain injury: A consumer survey study. *Journal of Head Trauma Rehabilitation*. 19:351–365.

Hellweg, S., and S. Johannes. 2008. Physiotherapy after traumatic brain injury: A systematic review of the literature. *Brain Injyury*. 22:365–373.

Hicks, R.R., D.H. Smith, D.H. Lowenstein, R.S. Marie, and T.K. Mcintosh. 1993. Mild experimental brain injury in the rat induces cognitive deficits associated with regional neuronal loss in the hippocampus. *Journal of Neurotrauma*. 10:405–414.

Hillary, F.G., M.T. Schultheis, B.H. Challis, S.R. Millis, G.J. Carnevale, T. Galshi et al. 2003. Spacing of repetitions improves learning and memory after moderate and severe TBI. *Journal of Clinical and Experimental Neuropsychology*. 25:49–58.

Hoover, R.C., M. Motta, J. Davis, K.E. Saatman, S.T. Fujimoto, H.J. Thompson et al. 2004. Differential effects of the anticonvulsant topiramate on neurobehavioral and histological outcomes following traumatic brain injury in rats. *Journal of Neurotrauma*. 21:501–512.

House, A., M. Dennis, A. Molyneux, C. Warlow, and K. Hawton. 1989. Emotionalism after stroke. *British Medical Journal*. 298:991–994.

Hulsebosch, C.E., D.S. DeWitt, L.W. Jenkins, and D.S. Prough. 1998. Traumatic brain injury in rats results in increased expression of Gap-43 that correlates with behavioral recovery. *Neuroscience Letters*. 255:83–86.

Iverson, G.L. 2005. Outcome from mild traumatic brain injury. *Current Opinion in Psychiatry*. 18:301–317.

Kay, T., B. Newman, M. Cavallo, O. Ezrachi, and M. Resnick. 1992. Toward a neuropsychological model of functional disability after mild traumatic brain injury. *Neuropsychology*. 6:371–384.

Kelly, M.A., M. Rubinstein, T.J. Phillips, C.N. Lessov, S. Burkhart-Kasch, G. Zhang et al. 1998. Locomotor activity in D2 dopamine receptor-deficient mice is determined by gene dosage, genetic background, and developmental adaptations. *Journal of Neuroscience*. 18:3470–3479.

Kennard, M.A. 1936. Age and other factors in motor recovery from precentral lesions in monkeys. *American Journal of Physiology*. 115:138–146.

Kent, M., M.C. Davis, and J.W. Reich, editors. 2013. *The Resilience Handbook: Approaches to Stress and Trauma*. Routledge, Florence, KY.

Killington, M.J., S.F. Mackintosh, and M.B. Ayres. 2010. Isokinetic strength training of lower limb muscles following acquired brain injury. *Brain Injury*. 24:1399–1407.

Kim, H., and A. Colantonio. 2010. Effectiveness of rehabilitation in enhancing community integration after acute traumatic brain injury: A systematic review. *The American Journal of Occupational Therapy*.64:709–719.

Kirkman, E., S. Watts, and G. Cooper. 2011. Blast injury research models. *Philosophical Transactions of the Royal Society of London. Series B, Biological Sciences*. 366:144–159.

Klose, M., A. Juul, J. Struck, N.G. Morgenthaler, M. Kosteljanetz, and U. Feldt-Rasmussen. 2007. Acute and long-term pituitary insufficiency in traumatic brain injury: A prospective single-centre study. *Clinical Endocrinology*. 67:598–606.

Koliatsos, V.E., I. Cernak, L. Xu, Y. Song, A. Savonenko, B.J. Crain et al. 2011. A mouse model of blast injury to brain: Initial pathological, neuropathological, and behavioral characterization. *Journal of Neuropathology and Experimental Neurology*. 70:399–416.

Kwa, V.I., M. Limburg, and R.J. de Haan. 1996. The role of cognitive impairment in the quality of life after ischaemic stroke. *Journal of Neurology*. 243:599–604.

Langlois, J.A., W. Rutland-Brown, and M.M. Wald. 2006. The epidemiology and impact of traumatic brain injury: A brief overview. *The Journal of Head Trauma Rehabilitation*. 21:375–378.

Lifshitz, J., and A.M. Lisembee. 2011. Neurodegeneration in the somatosensory cortex after experimental diffuse brain injury. *Brain Structure and Function*. 217:49–61.

Long, J.B., T.L. Bentley, K.A. Wessner, C. Cerone, S. Sweeney, and R.A. Bauman. 2009. Blast overpressure in rats: Recreating a battlefield injury in the laboratory. *Journal of Neurotrauma*. 26:827–840.

Lyeth, B.G., Q.Z. Gong, S. Shields, J.P. Muizelaar, and R.F. Berman. 2001. Group I metabotropic glutamate antagonist reduces acute neuronal degeneration and behavioral deficits after traumatic brain injury in rats. *Experimental Neurology*. 169:191–199.

Matsukawa, H., M. Shinoda, M. Fujii, O. Takahashi, A. Murakata, and D. Yamamoto. 2013. Acute alcohol intoxication, diffuse axonal injury and intraventricular bleeding in patients with isolated blunt traumatic brain injury. *Brain Injury*. 27:1409–1414.

Mattiasson, G.J., M.F. Philips, G. Tomasevic, B.B. Johansson, T. Wieloch, and T.K. McIntosh. 2000. The rotating pole test: Evaluation of its effectiveness in assessing functional motor deficits following experimental head injury in the rat. *Journal of Neuroscience Methods*. 95:75–82.

McCrea, M., K.M. Guskiewicz, S.W. Marshall, W. Barr, C. Randolph, R.C. Cantu et al. 2003. Acute effects and recovery time following concussion in collegiate football players: The NCAA Concussion Study. *JAMA*. 290:2556–2563.

McDonald, S., and S. Flanagan. 2004. Social perception deficits after traumatic brain injury: Interaction between emotion recognition, mentalizing ability, and social communication. *Neuropsychology*. 18:572–579.

McIntosh, T.K., R. Vink, L. Noble, I. Yamakami, S. Fernyak, H. Soares et al. 1989. Traumatic brain injury in the rat: Characterization of a lateral fluid-percussion model. *Neuroscience*. 28:233–244.

McKee, A.C., R.C. Cantu, C.J. Nowinski, E.T. Hedley-Whyte, B.E. Gavett, A.E. Budson et al. 2009. Chronic traumatic encephalopathy in athletes: Progressive tauopathy after repetitive head injury. *Journal of Neuropathology and Experimental Neurology*. 68:709–735.

McNamara, K.C., A.M. Lisembee, and J. Lifshitz. 2010a. The whisker nuisance task identifies a late-onset, persistent sensory sensitivity in diffuse brain-injured rats. *Journal of Neurotrauma*. 27:695–706.

McNamara, K.C., A.M. Lisembee, and J. Lifshitz. 2010b. The whisker nuisance task identifies a late-onset, persistent sensory sensitivity in diffuse brain-injured rats. *Journal of Neurotrauma*. 27:695–706.

Meares, S., E.A. Shores, A.J. Taylor, J. Batchelor, R.A. Bryant, I.J. Baguley et al. 2011. The prospective course of postconcussion syndrome: The role of mild traumatic brain injury. *Neuropsychology*. 25:454–465.

Milner, B. 1974. Sparing of language function after early unilateral brain damage. *Neurosciences Research Program Bulletin*.12:213–217.

Milton, S., and R. Wertz. 1986. Management of persisting communication deficits in patients with traumatic brain injury. In B. Uzzell and Y. Gross, editors. *Clinical Neuropsychology of Intervention*. Springer US, New York. pp. 223–256.

Morganti-Kossmann, M.C., E. Yan, and N. Bye. 2010. Animal models of traumatic brain injury: Is there an optimal model to reproduce human brain injury in the laboratory? *Injury*. 41 Suppl 1:S10–13.

Mott, T.F., M.L. McConnon, and B.P. Rieger. 2012. Subacute to chronic mild traumatic brain injury. *American Family Physician*. 86:1045–1051.

Nudo, R.J., B.M. Wise, F. SiFuentes, and G.W. Milliken. 1996. Neural substrates for the effects of rehabilitative training on motor recovery after ischemic infarct. *Science*. 272:1791–1794.

O'Connor, W.T., A. Smyth, and M.D. Gilchrist. 2011. Animal models of traumatic brain injury: A critical evaluation. *Pharmacology and Therapy*. 130:106–113.

O'Dell, D.M., J.K. Muir, C. Zhang, F.M. Bareyre, K.E. Saatman, R. Raghupathi et al. 2000. Lubeluzole treatment does not attenuate neurobehavioral dysfunction or CA3 hippocampal neuronal loss following traumatic brain injury in rats. *Restorative Neurology and Neuroscience*. 16:127–134.

Pierce, J.E.S., D.H. Smith, J.Q. Trojanowski, and T.K. McIntosh. 1998. Enduring cognitive, neurobehavioral and histopathological changes persist for up to one year following severe experimental brain injury in rats. *Neuroscience*. 87:359–369.

Piot-Grosjean, O., F. Wahl, O. Gobbo, and J.M. Stutzmann. 2001. Assessment of sensorimotor and cognitive deficits induced by a moderate traumatic injury in the right parietal cortex of the rat. *Neurobiology of Disease*. 8:1082–1093.

Ponsford, J., C. Willmott, A. Rothwell, P. Cameron, A.M. Kelly, R. Nelms et al. 2000. Factors influencing outcome following mild traumatic brain injury in adults. *Journal of the International Neuropsychological Society*. 6:568–579.

Povlishock, J.T., D.E. Erb, and J. Astruc. 1992. Axonal response to traumatic brain injury: Reactive axonal change, deafferentation, and neuroplasticity. *Journal of Neurotrauma*. 9 Suppl 1:S189-S200.

Prins, M.L., A. Hales, M. Reger, C.C. Giza, and D.A. Hovda. 2010. Repeat traumatic brain injury in the juvenile rat is associated with increased axonal injury and cognitive impairments. *Developmental Neuroscience*. 32:510–518.

Purohit, M.P., R.E. Wells, R.D. Zafonte, R.B. Davis, and R.S. Phillips. 2013. Neuropsychiatric symptoms and the use of complementary and alternative medicine. *PM&R*. 5:24–31.

Rall, J.M., D.A. Matzilevich, and P.K. Dash. 2003. Comparative analysis of mRNA levels in the frontal cortex and the hippocampus in the basal state and in response to experimental brain injury. *Neuropathology and Applied Neurobiology*. 29:118–131.

Riess, P., F.M. Bareyre, K.E. Saatman, J.A. Cheney, J. Lifshitz, R. Raghupathi et al. 2001. Effects of chronic, post-injury Cyclosporin A administration on motor and sensorimotor function following severe, experimental traumatic brain injury. *Restorative Neurology and Neuroscience*. 18:1–8.

Risling, M., and J. Davidsson. 2012. Experimental animal models for studies on the mechanisms of blast-induced neurotrauma. *Frontiers in Neurology*. 3:30.

Roof, R.L., R. Duvdevani, and D.G. Stein. 1993. Gender influences outcome of brain injury - progesterone plays a protective role. *Brain Research*. 607:333–336.

Rosenthal, M., B.K. Christensen, and T.P. Ross. 1998. Depression following traumatic brain injury. *Archives of Physical Medicine and Rehabilitation*. 79:90–103.

Sanders, M.J., W.D. Dietrich, and E.J. Green. 1999. Cognitive function following traumatic brain injury: Effects of injury severity and recovery period in a parasagittal fluid-percussive injury model. *Journal of Neurotrauma*. 16:915–925.

Sanders, M.J., W.D. Dietrich, and E.J. Green. 2001. Behavioral, electrophysiological, and histopathological consequences of mild fluid-percussion injury in the rat. *Brain Research*. 904:141–144.

Sanderson, K.L., R. Raghupathi, K.E. Saatman, D. Martin, G. Miller, and T.K. McIntosh. 1999. Interleukin-1 receptor antagonist attenuates regional neuronal cell death and cognitive dysfunction after experimental brain injury. *Journal of Cerebral Blood Flow and Metabolism*. 19:1118–1125.

Schallert, T., D.A. Kozlowski, J.L. Humm, and R.R. Cocke. 1997. Use-dependent structural events in recovery of function. *Advances in Neurology*. 73:229–238.

Seel, R.T., J.S. Kreutzer, M. Rosenthal, F.M. Hammond, J.D. Corrigan, and K. Black. 2003. Depression after traumatic brain injury: A National Institute on Disability and Rehabilitation Research Model Systems multicenter investigation. *Archives of Physical Medicine and Rehabilitation*. 84:177–184.

Smith, D.H., K. Okiyama, M.J. Thomas, B. Claussen, and T.K. McIntosh. 1991. Evaluation of memory dysfunction following experimental brain injury using the Morris Water Maze. *Journal of Neurotrauma*. 8:259–269.

Smith, R.L., T. Parks, and G. Lynch. 1974. A comparison of the role of the motor cortex in recovery from cerebellar damage in young and adult rats. *Behavioral Biology*. 12:177–198.

Stein, D.G. 2001. Brain damage, sex hormones and recovery: A new role for progesterone and estrogen? *Trends in Neurosciences*. 24:386–391.

Stein, D.G. 2013. A clinical/translational perspective: Can a developmental hormone play a role in the treatment of traumatic brain injury? *Hormones and Behavior*. 63:291–300.

Stein, D.G., and A.C. Firl. 1976. Brain-damage and reorganization of function in old-age. *Experimental Neurology*. 52:157–167.

Stroemer, R.P., T.A. Kent, and C.E. Hulsebosch. 1998. Enhanced neocortical neural sprouting, synaptogenesis, and behavioral recovery with D-amphetamine therapy after neocortical infarction in rats. *Stroke*. 29:2381–2393; discussion 2393–2385.

Struchen, M.A., M.R. Pappadis, A.M. Sander, C.S. Burrows, and K.A. Myszka. 2011. Examining the contribution of social communication abilities and affective/behavioral functioning to social integration outcomes for adults with traumatic brain injury. *The Journal of Head Trauma Rehabilitation*. 26:30–42.

Tavazzi, B., R. Vagnozzi, S. Signoretti, A.M. Amorini, A. Belli, M. Cimatti et al. 2007. Temporal window of metabolic brain vulnerability to concussions: Oxidative and nitrosative stresses—Part II. *Neurosurgery*. 61:390–395; discussion 395–396.

Thomas, T.C., J.M. Hinzman, G.A. Gerhardt, and J. Lifshitz. 2012. Hypersensitive glutamate signaling correlates with the development of late-onset behavioral morbidity in diffuse brain-injured circuitry. *Journal of Neurotrauma*. 29:187–200.

Thomas, T.C., H.F. Ray-Jones, R.P. Hammer, P.D. Adelson, and J. Lifshitz. 2013. Morphological changes in neurons along a diffuse-injured circuit associated with the development of late-onset morbidity in rats. *Journal of Neurotrauma*. 30:A51-A51.

Thompson, H.J., J. Lifshitz, N. Marklund, M.S. Grady, D.I. Graham, D.A. Hovda et al. 2005. Lateral fluid percussion brain injury: A 15-year review and evaluation. *Journal of Neurotrauma*. 22:42–75.

Travis, A.M., and C.N. Woolsey. 1956. Motor performance of monkeys after bilateral partial and total cerebral decortications. *American Journal of Physical Medicine*. 35:273–310.

Vagnozzi, R., S. Signoretti, B. Tavazzi, R. Floris, A. Ludovici, S. Marziali et al. 2008. Temporal window of metabolic brain vulnerability to concussion: A pilot 1H-magnetic resonance spectroscopic study in concussed athletes—Part III. *Neurosurgery*. 62:1286–1295; discussion 1295–1286.

Vagnozzi, R., B. Tavazzi, S. Signoretti, A.M. Amorini, A. Belli, M. Cimatti et al. 2007. Temporal window of metabolic brain vulnerability to concussions: Mitochondrial-related impairment—Part I. *Neurosurgery*. 61:379–388; discussion 388–379.

Vink, R., P.G. Mullins, M.D. Temple, W. Bao, and A.I. Faden. 2001. Small shifts in craniotomy position in the lateral fluid percussion injury model are associated with differential lesion development. *Journal of Neurotrauma*. 18:839–847.

Wang, H.C., and Y.B. Ma. 2010. Experimental models of traumatic axonal injury. *Journal of Clinical Neuroscience*. 17:157–162.

Wolf, S.L., C.J. Winstein, J.P. Miller, E. Taub, G. Uswatte, D. Morris et al. 2006. Effect of constraint-induced movement therapy on upper extremity function 3 to 9 months after stroke: The EXCITE randomized clinical trial. *JAMA*. 296:2095–2104.

Wu, A., Z. Ying, and F. Gomez-Pinilla. 2013. Exercise facilitates the action of dietary DHA on functional recovery after brain trauma. *Neuroscience*. 248:655–663.

Xiong, Y., A. Mahmood, and M. Chopp. 2013. Animal models of traumatic brain injury. *Nature Reviews. Neuroscience*. 14:128–142.

Yen, H.L., and J.T.Y. Wong. 2007. Rehabilitation for traumatic brain injury in children and adolescents. *Annals of the Academy of Medicine, Singapore*. 36:62–66.

Ylvisaker, M., L.S. Turkstra, and C. Coelho. 2005. Behavioral and social interventions for individuals with traumatic brain injury: A summary of the research with clinical implications. *Seminars in Speech and Language*. 26:256–267.

Zhao, G.W., Y. Wang, Y.C. Li, Z.L. Jiang, L. Sun, X. Xi et al. 2014. The neuroprotective effect of modified "Shengyu" decoction is mediated through an anti-inflammatory mechanism in the rat after traumatic brain injury. *Journal of Ethnopharmacology*. 151:694–703.

37 Endothelin, Cerebral Blood Flow, and Traumatic Brain Injury

Implications for a Future Therapeutic Target

Justin C. Graves and Christian W. Kreipke

CONTENTS

37.1 INTRODUCTION

An early event in the pathotrajectory of traumatic brain injury (TBI) is reduction in cerebral blood flow (CBF). This decrease in CBF likely inhibits the brain's ability to recover following injury. Until recently, mechanisms underlying dysfunctional CBF have been unknown. However, recent research implicates the endothelin system. Endothelin, working through its receptor A (ETrA), exerts a prolonged period of vasoconstriction throughout brain tissue and contributes to vasospasm. This chapter will not only elucidate the mechanisms by which endothelin affects CBF and ultimate outcome following TBI, but will also offer insight into developments in treatments centered around antagonizing ETrA. It is the hope that this work will inspire future research into pharmacological therapeutics aimed at mitigating the devastating effects of TBI.

37.1.1 Historical Perspectives Situating Cerebral Blood Flow in the Pathotrajectory of Traumatic Brain Injury

TBI, whether an open- (such as a penetration wound) or closed-head injury (such as acceleration/deceleration of the brain), is characterized by both primary and secondary injuries. Primary injuries may include diffuse axonal injury and direct vascular disruption. These phenomena lead to secondary injuries such as metabolic disturbances, edema, and dysfunctional autoregulation of CBF (reviewed in Betrus, 2013). Arguably, these secondary injuries contribute most to the ensuing pathotrajectory of TBI and present the most complex challenges for attempting to mitigate suffering associated with brain injury. For instance, lack of proper oxygen and metabolite delivery to the brain can lead to long-term alterations in membrane permeability, improper release of excitatory neurotransmitters (Bullock et al., 1998), and ion imbalances that ultimately cause axon degeneration (Bullock et al., 1998; Tekkok et al., 2007).

Although diffuse axonal injury and edema after TBI have been studied extensively for decades, disruption of CBF has only recently gained the attention of the research community. This is, in part, due to an earlier misconception that changes in CBF underlined brain swelling. Earlier notions that CBF actually would increase after a brain injury necessarily led to the belief that this increase would lead to increased intracranial pressure (ICP) and brain swelling. However, later data showed that, in fact, CBF decreases after TBI and that brain swelling is the result of edema and not changes in CBF (Betrus, 2013).

Changes in CBF after TBI became a hotbed for discovery during the latter half of the 1970s and throughout the 1980s. Miller et al. was one of the first laboratories to show that, contrary to popular belief, after a brain injury, CBF did not increase (Miller et al., 1975). Using liquid nitrogen to induce injury to the brain in baboons, Miller and colleagues determined that in some cases autoregulation of microvascular tone is disrupted following initial injury. In 1977, Langfitt and colleagues published the results of a clinical study that provided strong evidence for a role of decreased CBF in the pathologic outcome of TBI (Langfitt et al., 1977). Twenty-four post-TBI patients were analyzed for CBF as a function of outcome. It was determined that there was a strong correlation between CBF and outcome. Not only did those patients that showed low CBF exhibit the worst outcome (including death), but as patients recovered, CBF increased.

During the 1980s and early 1990s, great emphasis was being placed on producing animal models of TBI that accurately mimicked typical human head injuries. The open-skull, lesion-producing models that may have recapitulated open-head wounds were no longer sufficient for extrapolating data to closed-head injuries. Through the development of better models such as the weight-drop and fluid percussion models of inducing TBI, the impact of CBF to the pathotrajectory of TBI has become more clear. A flurry of studies showed that, after increased or decreased cerebral perfusion pressure (CPP), CBF autoregulation is either impaired or abolished (Czosnyka et al., 2001; Lang et al., 2003; Mudaliar et al., 2003; Obrist et al., 1984; Panerai et al., 2004). Yuan et al., using a rat fluid percussion model of TBI, showed that TBI caused significant decreases in cortical CBF (Yuan et al., 1988). More importantly, given that this model of TBI is considered a moderate injury, this study suggested that dysfunctional CBF may play an important role in more than just severe injuries.

Clinical data during the 1980s and 1990s were consistent with that which was elucidated through basic science research. Muizelaar et al. (1984) reported that in many cases severe head trauma was characterized by a decrease in CBF and loss of normal vascular autoregulation. It was also shown that patients who exhibited dysfunctional autoregulation were less responsive to mannitol-induced decreases in ICP. Overall, this study concluded that vasoconstriction is effective in reducing ICP. However, later studies demonstrate that sustained vasoconstriction exacerbates injury (reviewed in Graves, 2013).

A seminal clinical study was published in 1997 that studied CBF in 54 patients that suffered a TBI. Hemispheric CBF was measured for 5 days after injury. The results showed that only 3 of the 54 patients exhibited CBF considered at ischemic levels. This suggested that CBF *contributes* to the pathologic signs and symptoms of TBI (Kelly et al., 1997). Therefore, it may not be necessary for CBF to drop to ischemic levels for damage to ensue. Likely, lowered CBF exacerbates secondary injury by causing dampened oxygen and metabolic delivery to brain tissue, which is vital for its proper function and survival.

37.2 UNDERSTANDING CBF, ITS REGULATION, AND ITS IMPORTANCE TO OUTCOME

After a brief history of situating CBF in the pathotrajectory of TBI, it is important to understand the relationship of CBF to ultimate outcome after head trauma. CBF must be tightly regulated to ensure constant delivery of oxygen and metabolites vital for proper brain function. During periods of brain activity, this delivery must increase to supply the increased demand on that particular brain tissue.

CBF is tightly regulated through a homeostatic process in which arterioles dilate and constrict to maintain CBF nearly constant over a range of blood pressures (reviewed in Armstead, 2013). Even in instances in which CPP rises, autoregulation maintains a constant CBF (Lassen, 1959; Paulson et al., 1990). Thus, in theory, even in the midst of increased CPP as a consequence of TBI, normal CBF should be maintained. However, as previously outlined, both basic science and clinical research show that this is not always the case, thus suggesting that TBI causes disruption in normal autoregulatory processes. Clinical work supports this notion. Vavilala et al. (2008) showed that up to 40% of pediatric TBI patients exhibit impaired cerebral autoregulation. Furthermore, this impaired autoregulation is associated with poor outcome (Vavilala et al., 2006, 2007).

37.3 THE ROLE OF ENDOTHELIN IN AUTOREGULATION

Autoregulation is achieved via a myriad of vasoconstrictors and vasodilators (Table 37.1), which act locally on vascular smooth muscle cells that regulate the media layer of blood vessel walls and, ultimately, control vascular luminal area (Martinez-Lemus and Galinanes, 2011). Vasoconstriction leads to decreased luminal area and decreased blood flow, whereas vasodilation facilitates an increase in luminal area and consequent increased blood flow.

Endothelin, first discovered in the venom of the African snake species, *Atractaspis engaddensis,* is a highly potent, long-lasting vasoconstrictor that is ubiquitously expressed throughout the brain. Working through its receptors, ETA and ETB, endothelin contributes to the maintenance of CBF and is considered a chief autoregulatory peptide. Endothelin signaling includes G-protein activation, cytosolic inositol triphosphate generation, and subsequent release of calcium stores that ultimately leads to contraction of vascular smooth muscle cells. More recently, it was shown that endothelin can also mediate vasoconstriction via protein tyrosine kinases (reviewed in Kropinski, 2013). Cerebral circulation decreases in a dose-dependent manner with application of endothelin, these decreases being long-lasting (Faraci, 1989; Kobayashi et al., 1991; Shigeno et al., 1995; Willette and Sauermelch, 1990; Zimmermann and Seifert, 1998).

Although endothelin is widely accepted as a potent vasoconstrictor, its true role in autoregulation appears more complex. Through binding to ETB, endothelin has been shown to exhibit vasodilatory effects as well. Numerous studies show that ETB stimulation leads to vasodilation (DeNucci et al., 1988; Kropinski, 2013). However, it is not fully understood whether this is via a direct mechanism or whether ETB stimulation causes release of endothelium-derived vasodilators.

TABLE 37.1
Examples of Typical Endogenous Vasoactive Substances

Vasoconstrictors	Vasodilators
Endothelin	Epinephrine
Thromboxane	Adenosine
Angiotensin II	L-arginine
Muscarinic agonists	EDHF
ATP	Nitric oxide
Antidiuretic hormone	CO_2
Asymmetric dimethylarginine	Substance P

Nonetheless, endothelin, via activation of its receptors, plays a critical role in the tightly regulated autoregulatory processes of brain microcirculation.

37.4 THE ROLE OF ENDOTHELIN IN DYSFUNCTIONAL AUTOREGULATION AFTER TBI

A role for endothelin in the pathotrajectory of TBI is widely supported (Armstead, 2004; Grasso, 2004; Sato and Noble, 1998). Increased levels of ET-1 have been found in both plasma and cerebrospinal fluid of trauma patients and have been correlated with occurrence of cerebral vasospasm (Beuth et al., 2001; Lampl et al., 1997; Salonia et al., 2010). Numerous findings from animal models of TBI have linked endothelin to decreased CBF and poor outcome following injury. As early as 1996, Armstead's laboratory posited that endothelin is a primary contributor to pial arterial vasoconstriction after focal brain injury. These results were supported using a weight-impact model of TBI. Armstead and Kreipke (2011) reported similar increases in endothelin in pig and rat models of TBI that were sustained for a significant period after injury. These increases correlated with decreased CBF (Rafols et al., 2007) (Figure 37.1). Additionally, the increased endothelin and resulting decreased CBF correlated with poor behavioral outcome (Kreipke et al., 2007) (Figure 37.2). Taken together, these results suggest that TBI causes dysregulation of autoregulatory processes through increased endothelin-mediated vasoconstriction that decreases arterial luminal areas, causing decreased CBF and, ultimately, poor cognitive outcome.

37.5 THE ROLE OF ENDOTHELIN RECEPTORS IN DYSFUNCTIONAL CBF AFTER TBI

Recent evidence has suggested a more complex role of endothelin signaling in the pathotrajectory of TBI. In addition to research that supports an increase in the ligand, several laboratories have shown that ETA is also increased after TBI. Zhang et al. reported an increase in ETA messenger RNA within brain endothelial cells of traumatized rabbits (Zhang et al., 2000). Dore-Duffy et al. (2011) measured ETA protein levels in brain vascularization after TBI and showed that ETA was significantly elevated as early as 4 hours post-TBI (coincident with decreased CBF). Curiously this same report showed that ETB levels peak at 24 and 48 hours post-TBI, the period in which CBF begins to normalize. This suggests that ETA likely contributes to the ensuing decreased CBF and that, perhaps, ETB increases help to restore autoregulation by allowing more vasodilation to equal the balance of constriction/dilation after injury. Thus, it appears that endothelin exerts its effects of CBF after injury via ETA. ETA is, therefore, a possible target for therapeutic intervention.

To suggest that ETA is associated with pathology of brain disorders is not novel. Endothelin and its receptors have been implicated in stroke, subarachnoid hemorrhage, migraines, and brain tumors (reviewed by Kropinski, 2013). ETA and mixed ETA/B antagonists have been used in multiple animal studies and some clinical studies with mixed results in trying to mitigate damage associated with subarachnoid hemorrhage and stroke (Davenport and Morton, 1991; Lopez-Farre et al., 1991; Suzuki et al., 1992). Unfortunately, endothelin-based therapies have shown to have discouraging results (Macdonald, 2008; Roux et al., 1999).

37.6 ETA: A THERAPEUTIC TARGET FOR MITIGATING THE DELETERIOUS EFFECTS OF TBI

In summary thus far, TBI appears to cause dysregulation of autoregulatory processes through upregulation of endothelin and its receptor, ETA. This causes decreased CBF that, in turn, leads to poorer outcome after injury. Is there evidence to suggest that manipulating ETA may be beneficial after TBI?

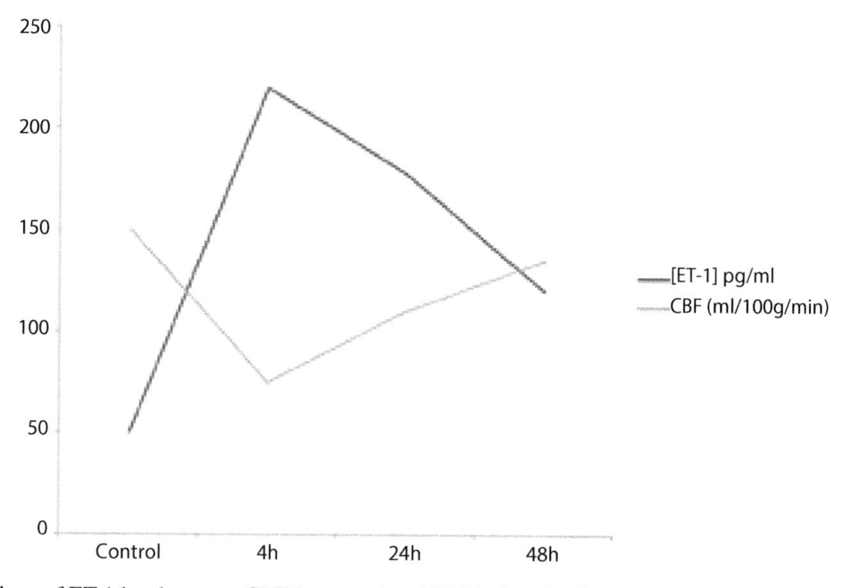

FIGURE 37.1 Comparison of ET-1 levels versus CBF in control and TBI-induced animals (up to 48 hours postinjury).

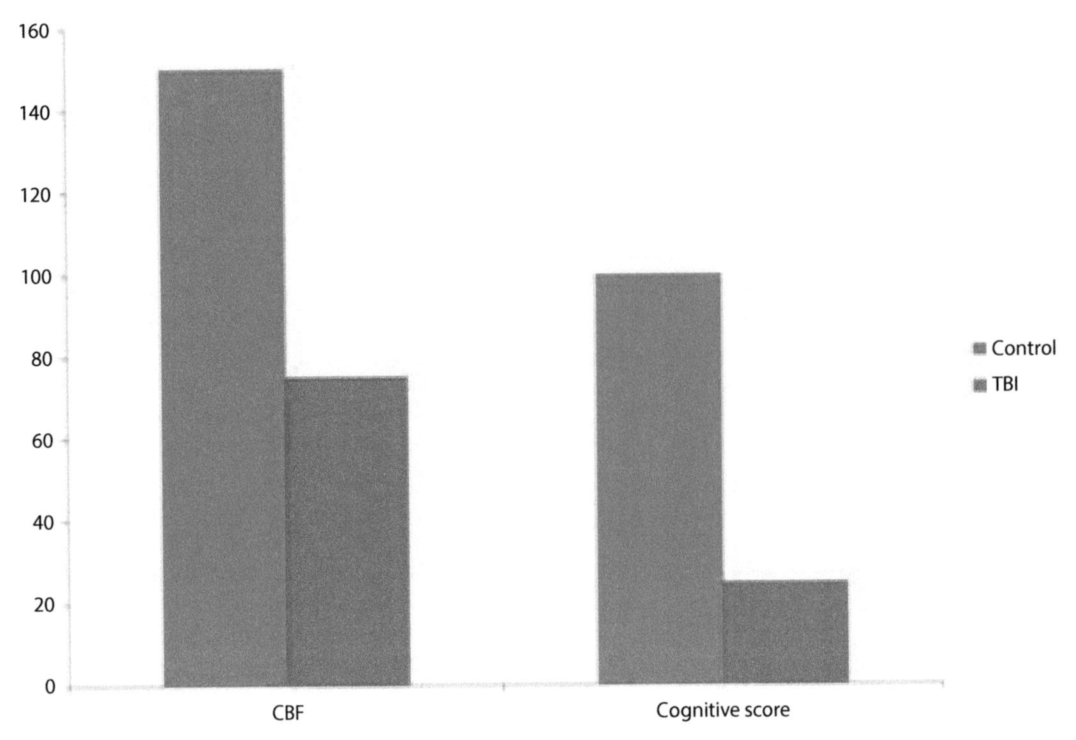

FIGURE 37.2 Comparison of CBF to cognitive score. As shown, CBF was decreased in TBI animals, as was cognition.

Zhang et al. first published a report in 2000 showing that BQ-123, a selective ETA antagonist, reduces the extent of hypoperfusion in the brain after trauma to rabbit brains (Zhang et al., 2000). This report was followed by a study conducted by Barone and colleagues with similar results—chiefly, that selective ETA antagonism reduces TBI-induced hypoperfusion (Barone et al., 2000). In 2010, Kreipke et al. (2010) published that BQ-123 mitigated hypoperfusion in a rodent weight-acceleration impact model of TBI. This was followed by two reports from the same laboratory that showed that BQ-123 also decreased the extent of histopathologic and behavioral deficits after TBI (Reynolds et al., 2011). All three of these works also concluded that ETB antagonism is either ineffective or exacerbates injury. Taken together, these data provided sound rationale that ETA is not only a likely target for developing TBI therapies, but also that selectively blocking this receptor while leaving ETB intact could prove to be an effective therapy to improve cognitive outcome following TBI.

As reviewed in Kreipke et al. (2013), a likely candidate for clinical trial aimed at antagonizing ETA to mitigate secondary injury and ensuing behavioral deficits after TBI already exists. The use of clazosentan, a highly selective ETA antagonist, has been attempted to improve outcome after subarachnoid hemorrhage. Although clazosentan reduced the extent of vasospasm after subarachnoid hemorrhage, it did not improve overall outcome as predicted (Macdonald, 2012). There appears to be some controversy as to whether this is due to possible side effects or to an overly ambitious study design that did not discriminate for the heterogeneity of injury.

In preclinical studies, clazosentan not only improved blood flow as effectively as BQ-123, it also improved behavioral outcome when administered at a variety of time points postinjury. Most effective was administration at 2 and 24 hours postinjury in which cognition was measured to be almost identical to sham-operated animals and was dramatically improved versus animals that received no drug (Kreipke et al., 2013). Given that clazosentan, in the low doses used in these preclinical studies, was shown to have little to no side effects in human subjects (Vajkoczy et al., 2005; van Giersbergen and Dingemanse, 2007), it is not surprising that there is great hope for clazosentan undergoing clinical trial for treatment of TBI. As noted, a careful study design will have to be implemented so as not to negate any potential efficacy.

37.7 CONCLUSION

CBF is tightly regulated in the brain via autoregulatory mechanisms that use a complex set of vasoreactive substances, including endothelin. TBI appears to disrupt this delicate system, which leads to decreased CBF and worsened histopathologic and behavioral outcomes. The precise mechanism underlying this dysfunctional situation may include upregulation of endothelin and its receptor, ETA. Attempts to mitigate the changes in CBF may be achieved through selective antagonism of ETA, which has been shown to improve not only CBF, but also cellular and cognitive outcomes. Therefore, future clinical trials may be developed using ETA as the target; these clinical trials may prove useful in mitigating the ensuing deficits caused by secondary injury in the brain after TBI.

REFERENCES

Armstead, W.M. 2004. Endothelins and the role of endothelin antagonists in the management of posttraumatic vasospasm. *Current Pharmaceutical Design*. 10:2185–2192.

Armstead, W.M., and M.S. Vavilala. 2013. Age and sex differences in cerebral blood flow and autoregulation after pediatric traumatic brain injury. In Kreipke, C.W. and Rafols, J.A., eds. *Cerebral Blood Flow, Metabolism, and Head Trauma*. Springer, New York, pp. 135–153.

Armstead, W.M., and C.W. Kreipke. 2011. Endothelin-1 is upregulated after traumatic brain injury: A cross-species, cross-model analysis. *Neurological Research*. 33:133–136.

Barone, F.C., E.H. Ohlstein, A.J. Hunter, C.A. Campbell, S.H. Hadingham, A.A. Parsons et al. 2000. Selective antagonism of endothelin-A-receptors improves outcome in both head trauma and focal stroke in rat. *Journal of Cardiovascular Pharmacology*. 36:S357–361.

Betrus, C., and C.W. Kreipke. 2013. Historical perspectives in understanding traumatic brain injury and in situating disruption in CBF in the pathotrajectory of head trauma. In Kreipke, C.W. and Rafols, J.A., eds. *Cerebral Blood Flow, Metabolism, and Head Trauma*. Springer, New York, pp. 135–153.

Beuth, W., H. Kasprzak, M. Kotschy, B. Wozniak, A. Kulwas, and M. Sniegocki. 2001. [Endothelin in the plasma and cerebrospinal fluid of patients after head injury]. *Neurologia i Neurochirurgia Polska*. 35 Suppl 5:125–129.

Bullock, R., A. Zauner, J.J. Woodward, J. Myseros, S.C. Choi, J.D. Ward et al. 1998. Factors affecting excitatory amino acid release following severe human head injury. *Journal of Neurosurgery*. 89:507–518.

Czosnyka, M., P. Smielewski, S. Piechnik, L.A. Steiner, and J.D. Pickard. 2001. Cerebral autoregulation following head injury. *Journal of Neurosurgery*. 95:756–763.

Davenport, A.P., and A.J. Morton. 1991. Binding sites for 125I ET-1, ET-2, ET-3 and vasoactive intestinal contractor are present in adult rat brain and neurone-enriched primary cultures of embryonic brain cells. *Brain Research*. 554:278–285.

De Nucci, G., R. Thomas, P. D'Orleans-Juste, E. Antunes, C. Walder, T.D. Warner, and J.R. Vane. 1988. Pressor effects of circulating endothelin are limited by its removal in the pulmonary circulation and by the release of prostacyclin and endothelium-derived relaxing factor. *Proceedings of the National Academy of Sciences of the United States of America*. 85:9797–9800.

Dore-Duffy, P., S. Wang, A. Mehedi, V. Katyshev, K. Cleary, A. Tapper et al. 2011. Pericyte-mediated vasoconstriction underlies TBI-induced hypoperfusion. *Neurological Research*. 33:176–186.

Faraci, F.M. 1989. Effects of endothelin and vasopressin on cerebral blood vessels. *The American Journal of Physiology*. 257:H799–803.

Grasso, G. 2004. An overview of new pharmacological treatments for cerebrovascular dysfunction after experimental subarachnoid hemorrhage. *Brain Research. Brain Research Reviews*. 44:49–63.

Graves, J.C., C. Betrus, and J.A. Rafols. 2013. Situating cerebral blood flow in the pathotrajectory of head trauma. In Kreipke, C.W. and Rafols, J.A., eds. *Cerebral Blood Flow, Metabolism, and Head Trauma*. Springer, New York.

Kelly, D.F., N.A. Martin, R. Kordestani, G. Counelis, D.A. Hovda, M. Bergsneider et al. 1997. Cerebral blood flow as a predictor of outcome following traumatic brain injury. *Journal of Neurosurgery*. 86:633–641.

Kobayashi, H., M. Hayashi, S. Kobayashi, M. Kabuto, Y. Handa, H. Kawano et al. 1991. Cerebral vasospasm and vasoconstriction caused by endothelin. *Neurosurgery*. 28:673–678; discussion 678–679.

Kreipke, C.W., A. Kropinski, J. Graves, D. Tiesma, M. Kaufman, S. Schafer et al. 2013. New frontiers in clinical trials aimed at improving outcome following traumatic brain injury. In Kreipke, C.W. and Rafols, J.A., eds. *Cerebral Blood Flow, Metabolism, and Head Trauma*. Springer, New York.

Kreipke, C.W., R. Morgan, G. Roberts, M. Bagchi, and J.A. Rafols. 2007. Calponin phosphorylation in cerebral cortex microvessels mediates sustained vasoconstriction after brain trauma. *Neurological Research*. 29:369–374.

Kreipke, C.W., P.C. Schafer, N.F. Rossi, and J.A. Rafols. 2010. Differential effects of endothelin receptor A and B antagonism on cerebral hypoperfusion following traumatic brain injury. *Neurological Research*. 32:209–214.

Kropinski, A., P. Dore-Duffy, and C.W. Kreipke. 2013. Situation the endothelin system in th pathotrajectory of TBI-induced changes in hemodynamics. In Kreipke, C.W. and Rafols, J.A., eds. *Cerebral Blood Flow, Metabolism, and Head Trauma*. Springer, New York.

Lampl, Y., G. Fleminger, R. Gilad, R. Galron, I. Sarova-Pinhas, and M. Sokolovsky. 1997. Endothelin in cerebrospinal fluid and plasma of patients in the early stage of ischemic stroke. *Stroke*. 28:1951–1955.

Lang, E.W., J. Lagopoulos, J. Griffith, K. Yip, A. Yam, Y. Mudaliar et al. 2003. Cerebral vasomotor reactivity testing in head injury: The link between pressure and flow. *Journal of Neurology, Neurosurgery, and Psychiatry*. 74:1053–1059.

Langfitt, T.W., W.D. Obrist, T.A. Gennarelli, M.J. O'Connor, and C.A. Weeme. 1977. Correlation of cerebral blood flow with outcome in head injured patients. *Annals of Surgery*. 186:411–414.

Lopez-Farre, A., D. Gomez-Garre, F. Bernabeu, and J.M. Lopez-Novoa. 1991. A role for endothelin in the maintenance of post-ischaemic renal failure in the rat. *The Journal of Physiology*. 444:513–522.

Macdonald, R.L. 2008. Clazosentan: An endothelin receptor antagonist for treatment of vasospasm after subarachnoid hemorrhage. *Expert Opinion on investigational Drugs*. 17:1761–1767.

Macdonald, R.L. 2012. Endothelin antagonists in subarachnoid hemorrhage: What next? *Critical Care*. 16:171.

Martinez-Lemus, L.A., and E.L. Galinanes. 2011. Matrix metalloproteinases and small artery remodeling. *Drug Discovery Today. Disease Models*. 8:21–28.

Miller, J.D., J. Garibi, J.B. North, and G.M. Teasdale. 1975. Effects of increased arterial pressure on blood flow in the damaged brain. *Journal of Neurology, Neurosurgery, and Psychiatry*. 38:657–665.

Muizelaar, J.P., H.A. Lutz, 3rd, and D.P. Becker. 1984. Effect of mannitol on ICP and CBF and correlation with pressure autoregulation in severely head-injured patients. *Journal of Neurosurgery*. 61:700–706.

Obrist, W.D., T.W. Langfitt, J.L. Jaggi, J. Cruz, and T.A. Gennarelli. 1984. Cerebral blood flow and metabolism in comatose patients with acute head injury. Relationship to intracranial hypertension. *Journal of Neurosurgery*. 61:241–253.

Panerai, R.B., V. Kerins, L. Fan, P.M. Yeoman, T. Hope, and D.H. Evans. 2004. Association between dynamic cerebral autoregulation and mortality in severe head injury. *British Journal of Neurosurgery*. 18:471–479.

Rafols, J.A., R. Morgan, S. Kallakuri, and C.W. Kreipke. 2007. Extent of nerve cell injury in Marmarou's model compared to other brain trauma models. *Neurological Research.* 29:348–355.

Reynolds, C.A., S. Schafer, R. Pirooz, A. Marinica, A. Chbib, C. Bedford et al. 2011. Differential effects of endothelin receptor A and B antagonism on behavioral outcome following traumatic brain injury. *Neurological Research.* 33:197–200.

Roux, S., V. Breu, S.I. Ertel, and M. Clozel. 1999. Endothelin antagonism with bosentan: A review of potential applications. *Journal of Molecular Medicine (Berlin).* 77:364–376.

Salonia, R., P.E. Empey, S.M. Poloyac, S.R. Wisniewski, M. Klamerus, H. Ozawa et al. 2010. Endothelin-1 is increased in cerebrospinal fluid and associated with unfavorable outcomes in children after severe traumatic brain injury. *Journal of Neurotrauma.* 27:1819–1825.

Sato, M., and L.J. Noble. 1998. Involvement of the endothelin receptor subtype A in neuronal pathogenesis after traumatic brain injury. *Brain Research.* 809:39–49.

Shigeno, T., M. Clozel, S. Sakai, A. Saito, and K. Goto. 1995. The effect of bosentan, a new potent endothelin receptor antagonist, on the pathogenesis of cerebral vasospasm. *Neurosurgery.* 37:87–90; discussion 90–81.

Suzuki, R., H. Masaoka, Y. Hirata, F. Marumo, E. Isotani, and K. Hirakawa. 1992. The role of endothelin-1 in the origin of cerebral vasospasm in patients with aneurysmal subarachnoid hemorrhage. *Journal of Neurosurgery.* 77:96–100.

Tekkok, S.B., Z. Ye, and B.R. Ransom. 2007. Excitotoxic mechanisms of ischemic injury in myelinated white matter. *Journal of Cerebral Blood Flow and Metabolism.* 27:1540–1552.

Vajkoczy, P., B. Meyer, S. Weidauer, A. Raabe, C. Thome, F. Ringel et al. 2005. Clazosentan (AXV-034343), a selective endothelin A receptor antagonist, in the prevention of cerebral vasospasm following severe aneurysmal subarachnoid hemorrhage: Results of a randomized, double-blind, placebo-controlled, multicenter phase IIa study. *Journal of Neurosurgery.* 103:9–17.

van Giersbergen, P.L., and J. Dingemanse. 2007. Effect of gender on the tolerability, safety and pharmacokinetics of clazosentan following long-term infusion. *Clinical Drug Investigation.* 27:797–802.

Vavilala, M.S., S. Muangman, N. Tontisirin, D. Fisk, C. Roscigno, P. Mitchell et al. 2006. Impaired cerebral autoregulation and 6-month outcome in children with severe traumatic brain injury: Preliminary findings. *Developmental Neuroscience.* 28:348–353.

Vavilala, M.S., S. Muangman, P. Waitayawinyu, C. Roscigno, K. Jaffe, P. Mitchell et al. 2007. Neurointensive care; impaired cerebral autoregulation in infants and young children early after inflicted traumatic brain injury: A preliminary report. *Journal of Neurotrauma.* 24:87–96.

Vavilala, M.S., N. Tontisirin, Y. Udomphorn, W. Armstead, J.J. Zimmerman, R. Chesnut et al. 2008. Hemispheric differences in cerebral autoregulation in children with moderate and severe traumatic brain injury. *Neurocritical Care.* 9:45–54.

Willette, R.N., and C.F. Sauermelch. 1990. Abluminal effects of endothelin in cerebral microvasculature assessed by laser-Doppler flowmetry. *The American Journal of Physiology.* 259:H1688–1693.

Yuan, X.Q., D.S. Prough, T.L. Smith, and D.S. Dewitt. 1988. The effects of traumatic brain injury on regional cerebral blood flow in rats. *Journal of Neurotrauma.* 5:289–301.

Zhang, Y., Y. Zou, M. Xu, P. Zhu, and Z. Wang. 2000. Effect of endothelin and endothelin A receptors on regional cerebral blood flow after traumatic brain injury in rabbits. *Chinese Journal of Traumatology.* 3:185–188.

Zimmermann, M., and V. Seifert. 1998. Endothelin and subarachnoid hemorrhage: An overview. *Neurosurgery.* 43:863–875; discussion 875–866.

38 Application of Novel Therapeutic Agents for CNS Injury

NAAG Peptidase Inhibitors

Bruce G. Lyeth

CONTENTS

38.1 INTRODUCTION

Glutamate excitotoxicity is a significant determinant of traumatic brain injury (TBI) pathophysiology. Elevated glutamate can damage neurons through activation of various glutamate receptors. Excessive extracellular glutamate also initiates astrocyte pathology through excessive uptake of Na^+ through sodium-glutamate co-transporters. This chapter discusses a novel strategy for reducing glutamate toxicity following TBI by inhibiting the cleavage of N-acetylaspartylglutamate (NAAG) into N-acetylaspartate (NAA) and glutamate. NAAG is an abundant peptide found in the brain and is released from neurons after intense depolarization and functions as a selective agonist at metabotropic glutamate receptor subtype 3 (mGluR3), which is located on both neurons and astrocytes. NAAG is catabolized into NAA and glutamate by a specific carboxypeptidases. NAAG could play a significant role in modulating glutamate excitotoxicity if its rapid catabolism can be inhibited thereby conferring protection to the traumatized brain in several ways. First, NAAG reduces excessive glutamate release by activation of presynaptic mGluR3 autoreceptors. Second, by inhibiting the cleavage of NAAG into NAA and glutamate a secondary source of synaptic glutamate could be diminished. Third, activation of mGluR3 on astrocytes increases the expression of glutamate transporters thereby facilitating removal of excess glutamate from the synapse. Fourth, inhibiting the accumulation of the NAAG cleavage product, NAA, reduces NAA-Na^+ co-transport and subsequent astrocyte Na^+ overload. Overload of $[Na^+]i$ can initiate astrocyte pathology that subsequently impacts negatively on surrounding neurons. Thus, inhibition of carboxypeptidases represents a novel strategy for reducing glutamate excitotoxicity following TBI through multiple mechanisms. NAAG peptidase inhibition could provide new and important insights into glutamate excitotoxicity and lead to important insights into the dynamics of neuron-astrocyte interactions in TBI pathophysiology.

An estimated 1.7 million persons per year sustain a TBI in the United States resulting in more than 230,000 hospitalizations and 50,000 deaths (Faul et al., 2010). Survivors of TBI are often left with long-term disability (approximately 85,000 new cases per year) encompassing a broad spectrum of physical, neurological, and psychological dysfunctions such as memory impairment and difficulties with attention and concentration (Thurman and Guerrero, 1999; Thurman et al., 1999). The annual economic cost to society for the care of head-injured patients has been estimated to exceed $25 billion (Goldstein, 1990). TBI research to date, including glutamate excitotoxicity pathways, has focused almost exclusively on the pathophysiology of injured neurons with very little attention paid to glial cells or the interaction between glia and neurons. Astrocytes, which play an important role in normal brain function, increasingly appear to play a critical role in perturbed brain functions associated with seizures and ischemia brain injury (Vernadakis 1996; Vesce et al., 1999). This chapter examines a novel method of reducing the damaging effects of elevated extracellular glutamate by manipulating a putative endogenous peptide neurotransmitter, N-acetylaspartylglutamate (NAAG). The role of NAAG in exogenous and endogenous brain cellular protection involves the interdependent nature of neuronal glial interactions and the critical involvement of glutamate excitotoxicity initiated by TBI.

Previous therapeutic strategies targeting excitotoxicity in central nervous system (CNS) injury have focused on pharmacological blockade of various ionotropic and metabotropic glutamate receptors. Although laboratory results have often been impressive, translation into clinical applications have been overwhelmingly disappointing (Bullock et al., 1999; Narayan et al., 2002). However, a general perception from laboratory experiments is that excessive glutamate is indeed damaging to neurons (Bullock et al., 1999). This chapter discusses a novel strategy to reduce glutamate toxicity by simultaneously enhancing levels of an endogenous neurotransmitter peptide, NAAG, while reducing its cleavage products, N-acetylaspartate (NAA) and glutamate, through application of NAAG peptidase inhibitors. This approach is designed to reduce excitotoxicity through multiple mechanisms acting in concert.

Three pathways of the NAAG neuropeptide story are depicted in Figure 38.1. The first pathway (highlighted in green) depicts how NAAG reduces glutamate release by activation of presynaptic metabotropic glutamate receptor subtype 3 (mGluR3) autoreceptors. This is based on data demonstrating that inhibition of NAAG cleavage elevates levels of extracellular NAAG and reduces extracellular levels of glutamate after TBI (Zhong et al., 2005). Because NAAG is hydrolyzed into NAA and glutamate, the first pathway also depicts how NAAG cleavage provides a secondary source of extracellular glutamate.

A second pathway (highlighted in red in the figure) depicts elevated concentrations of NAAG peptide increasing glutamate uptake in astrocytes after TBI as a result of mGluR3 receptor activation on astrocytes. There is an established relationship between NAAG activation of mGluR3 on astrocytes and increased glutamate transport (Aronica et al., 2003; Yao et al., 2005). The expression and kinetics of glutamate transporters could be potentially increased by NAAG activating mGluR3 on astrocytes (Aronica et al., 2003). Thus, the extent of cellular protection afforded by NAAG peptidase inhibitors may not simply be because of reduced glutamate release, but also because of NAAG-induced enhancement of glutamate transport into astrocytes.

A third pathway (highlighted in blue in the figure) depicts the role of NAA (a product of NAAG cleavage by NAAG peptidase) in astrocyte pathology after TBI. NAAG is exclusively stored and released in neurons, whereas NAAG peptidase is exclusively found on the surface of astrocytes (Neale et al., 2000). NAA is preferentially and overwhelmingly taken up by sodium-dependent transporters located on astrocytes (Sager et al., 1999b). The excessive NAA/Na^+ cotransport into astrocytes after TBI initiates a pathological cascade in astrocytes related to elevated $[Na^+]_i$. Glutamate/

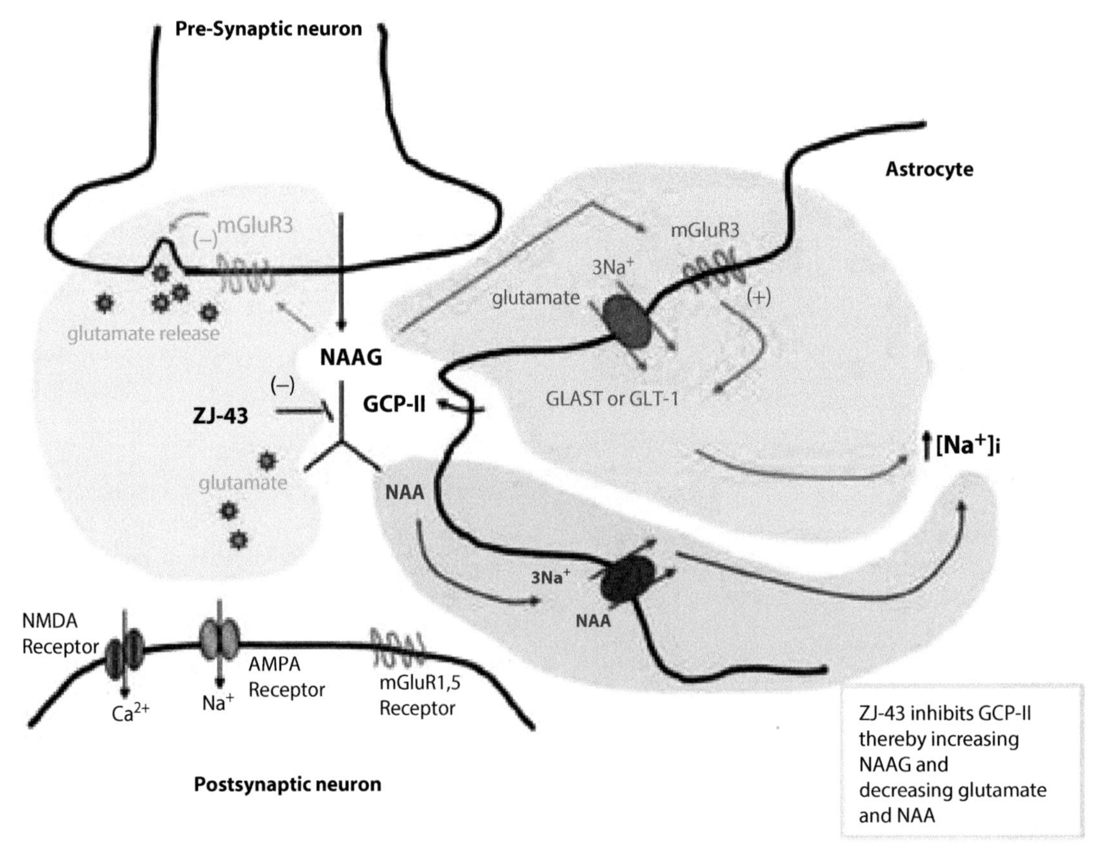

FIGURE 38.1 Actions of NAAG and NAAG peptidase inhibition at glutamatergic synapse. This diagram depicts a glutamatergic synapse with pre- and postsynaptic neurons and surrounding astrocyte. NAAG neuropeptide is coreleased along with glutamate during intense stimulation associated with TBI. NAAG preferentially stimulates mGluR3 inhibitory autoreceptors that reduce glutamate release. ZJ-43 inhibits the NAAG peptidase, GCP-II thereby increasing and sustaining the level of NAAG within the synapse, which subsequently dampens glutamate release. See the main text for more details.

Na^+ cotransporters (GLT-1 and GLAST) located on astrocytes also initiates a Na^+-dependent pathological cascade in astrocytes that eventually impacts upon survival of nearby neurons (Floyd et al., 2005).

NAAG likely functions as an endogenous protective peptide through multiple pathways involving both neurons and astrocytes. However, NAAG is short-lived in the synapse because of its rapid catabolism by NAAG peptidases (glutamate carboxypeptidase II [GCP-II] and GCP-III), which diminishes the beneficial effects of NAAG. Application of NAAG peptidase inhibitors represents a novel method of enhancing the endogenous protective effects of NAAG by enhancing and sustaining concentrations of NAAG in the synapse.

38.2 EXCESSIVE GLUTAMATE CONTRIBUTES TO TBI PATHOPHYSIOLOGY

Glutamate excitotoxicity is a significant determinant of TBI pathophysiology. Numerous laboratories (Faden et al., 1989; Katayama et al., 1990; Nilsson et al., 1990; Zhong et al., 2005) and several clinical studies (Bullock et al., 1995a, 1995b, 1998a; Koura et al., 1998; Vespa et al., 1998; Zauner et al., 1996) have documented an excessive glutamate release after TBI. The elevated extracellular concentration of glutamate results in excitotoxicity from excessive activation of ion channel–linked (Hayes et al., 1988; Hicks et al., 1994; McIntosh et al., 1998) and G-protein–linked glutamate receptors (Faden et al., 1997; Lyeth et al., 2001; Mukhin et al., 1996, 1997; Zwienenberg et al., 2001). Experimental cerebral ischemia also is associated with a large, rapid increase in extracellular glutamate (Choi and Rothman, 1990). During normal neuronal activity, extracellular glutamate in the synaptic cleft must be rapidly cleared to optimize the signal-to-noise ratio as well as to prevent neuronal damage from excessive glutamate receptor activation. Glutamate is removed from the synapse by glutamate transporters located in the plasma membrane of neurons and adjacent astrocytes (Kanner and Schuldiner, 1987). The major glutamate removal mechanism is through astrocyte specific sodium-dependent glutamate transporters (GLT-1 and GLAST) (Danbolt, 1994; Takahashi et al., 1997). Data indicate that elevated intracellular Na^+ from excessive glutamate transport can be lethal to astrocytes (Floyd et al., 2005). Thus, astrocytes as well as neurons are likely adversely affected by excitotoxic glutamate release.

38.3 PEPTIDE NEUROTRANSMITTER: NAAG

The neurotransmitter NAAG is an acidic dipeptide that was first discovered in the mid-1960s and found in high concentrations (millimolar) in the brain and spinal cord (Curatolo et al., 1965). NAAG was not considered to be a neurotransmitter because the concentrations in nervous tissue far exceeded other putative neurotransmitters (except glutamate) and at the time, peptides were not considered a significant category of neurotransmitters. Furthermore, since NAAG did not directly alter membrane potential, it was thought to have only a metabolic function. In the 1980s, Coyle's laboratory provided further evidence of NAAG's presence in the nervous system along with evidence that it was acting as a neurotransmitter at the glutamate receptor (Koller and Coyle, 1984; Robinson et al., 1987; Serval et al., 1990; Zaczek et al., 1983). Substantial data now substantiate that NAAG not only serves as a neurotransmitter but that it is the most prevalent and widely distributed peptide neurotransmitter in the mammalian CNS (Neale et al., 2000). NAAG has a wide distribution throughout the brain including the forebrain and areas that use a variety of amine transmitters such as glutamate, GABA, acetylcholine, serotonin, and dopamine (Moffett and Namboodiri, 1995; Moffett et al., 1994; Renno et al., 1997). Ultrastructural and immunohistochemical analyses indicate that NAAG is concentrated in synaptic vesicles and has a strictly neuronal localization (Renno et al., 1997; Williamson and Neale, 1988). NAAG is released from neurons after depolarization in a calcium-dependent manner (Pittaluga et al., 1988; Zollinger et al., 1988, 1994).

NAAG acts as a highly selective and potent agonist at the group II mGluR3, but not at other cyclic AMP (cAMP)-coupled group II or group III mGluRs (Wroblewska et al., 1997, 1993). Several subtypes of mGluRs have been described and grouped based on amino acid sequence homology, signal transduction pathways, and agonist selectivity (Nakanishi, 1994; Schoepp and Conn, 1993). Group I includes mGluR1 and mGluR5 subtypes, which are coupled through phosphoinositide-specific phospholipase C to phosphoinositide hydrolysis and intracellular Ca^{++} mobilization and are sensitive to quisqualate and the cyclic glutamate analogue, (1S,3R)-ACPD. Group II includes mGluR2 and mGluR3 subtypes that are coupled to the inhibition of cAMP formation and are also activated by (1S,3R)-ACPD but are less sensitive to quisqualate. Group III includes mGluR4, mGluR6, mGluR7, and mGluR8 subtypes, which are also negatively coupled to cAMP and are activated by L-AP4 but are insensitive to (1S,3R)-ACPD and quisqualate (Roberts, 1995). To a lesser extent, activation of mGluRs can also lead to the activation of phospholipases A_2 and D as well as changes in ion channel function (Schoepp and Conn, 1993). Similar to other group II mGluR agonists (DCG-IV or glutamate), NAAG decreases voltage-activated calcium currents and decreases cAMP levels in neurons containing mGluR3s (Bischofberger and Schild, 1996; Sayer et al., 1992). NAAG also acts as a low-potency agonist at N-methyl-D-aspartate (NMDA) receptors but not at kainate or AMPA receptors. NAAG displaces the competitive NMDA antagonist, CGS-19755, with a half maximal inhibitory concentration (IC_{50}) value of 8.8 μM, which is 20-fold less potent than glutamate displacement of CGS-19755 (IC_{50} of 0.4 μM) (Valivullah et al., 1994). Thus, NAAG can act as a mixed agonist/antagonist at NMDA receptors depending upon the concentration of glutamate present at the synapse.

NAAG is hydrolyzed by an extracellular peptidase, GCP-II, into NAA and glutamate (Figure 38.2). Previous literature (Robinson et al., 1987) used the nomenclature, N-acylated-α-linked l-aminodipeptidase (NAALADase) for

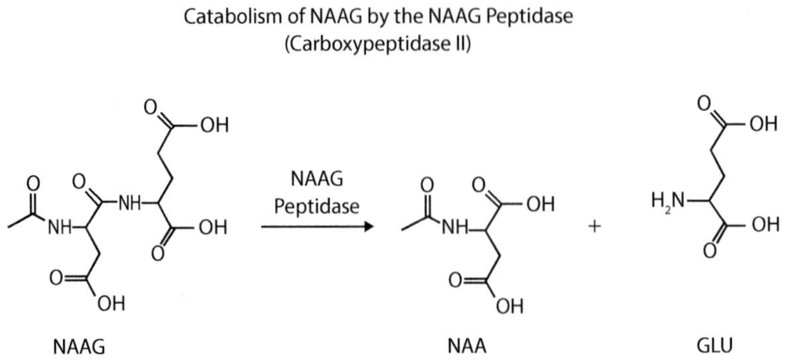

Catabolism of NAAG by the NAAG Peptidase
(Carboxypeptidase II)

NAAG Peptidase

NAAG + NAA GLU

FIGURE 38.2 Catabolism of the brain neuropeptide NAAG. Carboxypeptidase II is a Cl-dependent, membrane-bound, metallopeptidase that cleaves the endogenous brain neuropeptide NAAG into glutamate and NAA.

what is now commonly termed GCP-II. NAAG peptidase activity is widely distributed throughout the nervous system, consistent with the distribution of NAAG (Fuhrman et al., 1994). NAAG peptidase is concentrated in, if not exclusively located on, the extracellular face of astrocytes (Slusher et al., 1992) and is not found in neurons (Berger et al., 1999). NAAG peptidase is most likely synthesized exclusively in astrocytes because in situ hybridization studies reveal that the NAAG peptidase messenger RNA is localized in glial cells (Luthi-Carter et al., 1998).

A second brain enzyme with NAAG peptidase activity, GCP III, has been cloned in mouse and characterized with similar pharmacological properties and affinity for NAAG (Bacich et al., 2002; Bzdega et al., 2004). NAAG peptidase activity is potently inhibited by 2-(phosphonomethyl)pentanedioic acid (2-PMPA) with a K_i in the low nanomolar range (Jackson et al., 1996). However, 2-PMPA has limited ability to penetrate the blood–brain barrier because of strong hydrophilicity (Rong et al., 2002). A urea-based NAAG analogue, ZJ-43, was synthesized and characterized that had improved lipophilicity but also retained nanomolar inhibitory activity (Kozikowski et al., 2004; Olszewski et al., 2004). ZJ-43 was found to be a potent inhibitor of NAGG peptidase with an IC_{50} of 2.4 nm and a K_i of 0.8 nm. In the same assay system, 2-PMPA had an IC_{50} of 4.1 nm and a K_i of 1.4 nm. When tested against GCP III, ZJ-43 had a K_i of 23 nm (Kozikowski et al., 2004; Olszewski et al., 2004). In summary, accumulating data indicate that NAAG peptidase functions both to terminate any synaptic NAAG neuromodulatory activity and to liberate glutamate from NAAG, which subsequently acts at the various glutamate receptor subtypes (Slusher et al., 1999).

38.4 NAAG PEPTIDASE INHIBITORS, NAAG, AND EXCITOTOXICITY

Several studies have shown that NAAG peptidase inhibitors provide benefit in a number of pathological models, including amyotrophic lateral sclerosis (Ghadge et al., 2003), diabetic neuropathy (Zhang et al., 2002), allodynia (Yamamoto et al., 2001), neuropathic and inflammatory pain (Kozikowski et al., 2004), and phencyclidine-induced schizophrenia-like behaviors (Olszewski et al., 2004). High levels of extracellular glutamate significantly also contribute to delayed neuronal cell death after TBI. Glutamate-mediated delayed neuronal death is reduced by administration of group II mGluR agonists in ischemia (Bond et al., 1999) and TBI (Allen et al., 1999; Bruno et al., 1997; Movsesyan and Faden, 2006; Zwienenberg et al., 2001). Application of NAAG to cultured mouse cortical cells provided neuroprotection from a toxic NMDA insult (Bruno et al., 1998). Prestroke administration of the NAAG peptidase inhibitor, 2-PMPA, decreased extracellular glutamate, decreased lesion volume, and improved behavioral performance after transient middle cerebral artery occlusion in rats (Slusher et al., 1999). The ability of NAAG to activate presynaptic mGluRs and inhibit glutamate release (Sanabria et al., 2004; Xi et al., 2002) supports the hypothesis that elevated levels of NAAG after peptidase inhibition decrease glutamate levels after trauma. These observations provide strong evidence that NAAG peptidase inhibition (and subsequent activation of mGluR3s by NAAG) is likely to be neuroprotective in animal models of TBI. Thus, administration of NAAG peptidase inhibitors after TBI may represent a viable therapeutic strategy to augmenting the brain's natural endogenous system of controlling glutamate release via presynaptic feedback on mGluR3 by the NAAG peptide. This strategy represents a novel method of reducing the damaging glutamate release associated with TBI.

Recent data demonstrate that application of the group II mGluR agonist, DCG-IV, produces a significant increase in expression of both GLT-1 and GLAST receptor proteins in human astrocyte cultures (Aronica et al., 2003). Because astrocytes express mGluR3 group II receptors (Aronica et al., 2003), stimulation of mGluR3 by NAAG is also likely to increase the expression of astrocytic glutamate transporters. This becomes an important factor when NAAG peptidase inhibitors are administered during TBI, driving up extracellular levels of NAAG. From a neuronal perspective, this would be protective because the increased expression of glutamate transporters should increase the removal of excessive synaptic glutamate after TBI. Furthermore, the elevated NAAG levels would also likely reduce presynaptic glutamate release. From an astrocyte perspective the increased glutamate transporter expression and subsequent increase in glutamate uptake may actually be harmful because transport of elevated glutamate may have lethal consequences for astrocytes because of resultant elevations in $[Na^+]_i$ via excessive cotransport of Na^+ with glutamate (Floyd et al., 2005). Thus, the strategy of inhibiting NAAG peptidase after TBI may act as a "double-edged sword"—protective to neurons and damaging to astrocytes. However, recent data indicate that both neurons and astrocytes benefit from NAAG peptidase inhibition after TBI (Feng et al., 2012a). At least two possibilities may begin to explain the benefit to astrocytes in spite of possible increased glutamate (and obligatory Na^+) uptake. First, NAAG peptidase inhibition may reduce presynaptic glutamate release to an extent that the glutamate uptake "burden" is greatly diminished. Second, the other by-product of NAAG cleavage, NAA, may be substantially reduced via inhibition of NAAG cleavage. As discussed in the following section, NAA is cotransported with Na^+ into astrocytes with the same stoichiometry as the glutamate transporters, GLT-1 and GLAST. It is to consider this issue regarding the relative benefits and liabilities of NAAG peptidase inhibition before translation to clinical trials.

38.5 NAAG PEPTIDASE INHIBITORS APPLIED TO TBI

A series of experimental TBI studies examined the effects of NAAG peptidase inhibition in the rat fluid percussion model. A series of experiments examined the urea-based NAAG analogue, ZJ-43, which is a potent inhibitor of NAAG peptidases, GCP II and GCP III (Bzdega et al., 2004; Olszewski et al., 2004). Using in vivo microdialysis, concentrations of extracellular NAAG and glutamate from the dorsal hippocampus were elevated after TBI (Zhong et al., 2006). In the same study, TBI animals treated with ZJ-43 had higher and persistently elevated extracellular concentrations of NAAG, whereas extracellular concentrations of glutamate were significantly reduced. Furthermore, coadministration of the group II mGluR antagonist, LY341495 partially blocked the effects of ZJ-43 on dialysate glutamate. The results of this study provided the first evidence that inhibition of NAAG peptidase increases NAAG and reduces glutamate elevations associated with TBI and that NAAG effects are mediated through group II mGluR activation.

Subsequent studies demonstrated that systemic administration of ZJ-43 significantly reduced neuronal degeneration and astrocyte damage in the hippocampus 24 hours after lateral fluid percussion TBI (Feng et al., 2012a; Zhong et al., 2005). Moreover, the group II mGluR selective antagonist, LY341495, successfully abolished the protective effects of ZJ-43. These initial finding supported the hypothesis that the effects of blockade of NAAG peptidase activity are mediated by activation of group II mGluRs.

Further TBI studies examined post-injury administration of PGI-02776, a novel di-ester prodrug modified from the urea-based NAAG peptidase inhibitor ZJ-43. This new compound attached a di-ester to the parent compound, ZJ-43, to enhance lipophilicity to increase blood–brain barrier penetration. The strategy was to increase penetration of the prodrug into the brain where esterases would cleave the attached di-esters, uncovering the active parent compound, ZJ-43. PGI-02776 administered 30 minutes after lateral fluid percussion TBI in rats significantly reduced acute neuronal degeneration in the hippocampus assessed at 24 hours postinjury (Feng et al., 2011). In chronic behavioral experiments, a single administration of PGI-02776 administered 30 minutes postinjury improved cognitive performance (Feng et al., 2011)

NAAG peptidase inhibitors have also been examined in experimental TBI with imposed secondary hypoxia/hypotension. Severe TBI is often associated with a number of deleterious secondary insults including hypoxia (Davis et al., 2004, 2009; Manley et al., 2001; Miller et al., 1978; Schmoker et al., 1992). Indeed, as many as one-third of the severe TBI patients arriving in the emergency department have significant hypoxia and hypotension (Manley et al., 2001). Secondary insults such as hypoxemia can compound the accumulation of extracellular glutamate, increasing glutamate concentrations for hours after the primary insult (Bullock et al., 1998b). Experimental studies have demonstrated that secondary insults of hypoxia can exacerbate TBI by increasing neuronal damage (Bauman et al., 2000; Clark et al., 1997; Feng et al., 2012b; Nawashiro et al., 1995), exacerbating axonal pathology and neuro-inflammatory response (Goodman et al., 2011; Hellewell et al., 2010), and exacerbating sensorimotor and cognitive deficits (Bauman et al., 2000; Clark et al., 1997). In a clinically relevant model of TBI with a secondary insult of 30 minutes of hypoxia/hypotension, a single dose of PGI-02776 was administered immediately after the cessation of hypoxia. PGI-02776 treatment significantly reduced neuronal and astrocytic cell death across all regions of the hippocampus (Feng et al., 2012). In follow-up studies, a single administration of PGI-02776 at 30 minutes post-TBI significantly reduced motor and cognitive deficits over a 2-week period post-TBI (Gurkoff et al., 2013). The effectiveness of PGI-02776 in ameliorating the effects of secondary hypoxia indicate the potential utility of this therapeutic strategy for treating secondary insults that may occur hours after the primary TBI.

38.6 ROLE OF NAA IN CNS INJURY

N-acetylaspartate (NAA) is the second-most abundant amino acid in the adult brain and is one of the two by-products (glutamate being the other) of NAAG cleavage by NAAG peptidases. Although NAA is one of the major organic constituents of the brain, little is known about its function and metabolism (Tsai and Coyle, 1995). Stores of NAA are found only in neurons in the fully developed CNS (Urenjak et al., 1992, 1993) and are released from neurons by an unknown mechanism (Sager et al., 1999a). Because NAA is found almost exclusively in neurons (Miyake et al., 1981), reduction of NAA is generally considered to reflect neuronal injury and death in various neuronal diseases and injuries (Dunlop et al., 1992; Plaitakis and Constantakakis, 1993).

Intraneuronal concentrations of NAA (10–15 mM) are much higher than in the brain interstitial space (80–100 μM) (Sager et al., 1997; Taylor et al., 1994). The enzyme responsible

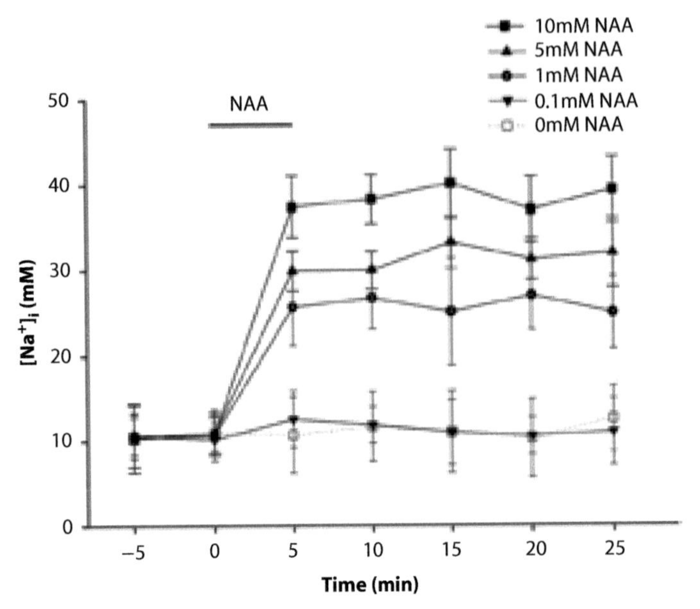

FIGURE 38.3 NAA increases intracellular sodium in cultured astrocytes. A range of concentrations of NAA (0, 0.1, 1, 5, or 10 mM) was applied to uninjured cortical astrocytes and $[Na^+]_i$ was measured using live-cell imaging techniques with sodium-binding benzofuran isophthalate (SBFI-AM). NAA was bath applied for 5 minutes in HEPES-buffered saline (pH 7.2). Three to six cells were measured in three independent experiments. NAA concentration-dependently increased intracellular sodium with 1-mM NAA producing a modest elevation and 10-mM NAA producing a more robust elevation in intracellular sodium. The transport process is electrogenic with a Na^+:NAA stoichiometry of 3:1 (Huang et al., 2000). NAA concentrations that produced robust elevations in $[Na^+]_i$ were also associated with significant astrocyte cell death measured with uptake of PrI. These data indicate that NAA/Na^+ cotransport elevates $[Na^+]_i$ and may contribute to astrocyte pathology. Data are means ± standard error of the mean.

for synthesis of NAA is located exclusively in neurons while the enzyme responsible for degradation of NAA is predominantly in glia cells (Benuck and D'Adamo, 1968; Goldstein, 1976). Studies of NAA transport indicate that NAA is taken up exclusively by astrocytes in a sodium-dependent manner (Huang et al., 2000; Sager et al., 1999b). NAA has two negative charges at pH 7.3 (Patel and Clark, 1979) and thus has an outward directed charge gradient, making it likely that Na^+ cotransport is critical for transport of NAA into glia. Subsequent studies identified the NAA transporter as a dicarboxylate structurally related to succinate termed NaDC3 that cotransports Na^+ with NAA (Huang et al., 2000). The transport process is electrogenic with a Na^+:NAA stoichiometry of 3:1 (Huang et al., 2000). NAA undergoes cleavage after transport into astrocytes by the enzyme aspartoacyclase II into acetate and aspartate (Tsai and Coyle, 1995). It is generally accepted that once transported into astrocytes, the cleavage products of NAA participate in myelin synthesis (Tsai and Coyle, 1995). Evidence points to dysfunction of aspartoacylase II in Canavan disease in which there is impaired synthesis of myelin (Kaul et al., 1991). NAA is also greatly reduced in children with demyelinating disorders (Peden et al., 1990). NAA is detected with a very prominent peak on ^{1}H-NMR spectroscopy and has been examined in a number of CNS disorders including stroke and TBI (Alessandri et al., 2000; Schuhmann et al., 2003). Stroke studies generally report a reduction in NAA when measured well after the event (Birken and Oldendorf, 1989; Tsai and Coyle, 1995). However, one group has reported acute elevations of NAA after ischemia (Sager, et al., 1997, 1999a). Acute measurements of NAA after TBI in rats indicate a rapid decrease in NAA measured by ^{1}H-NMR spectroscopy (Schuhmann et al., 2003) or microdialysis and high-performance liquid chromatography (Signoretti et al., 2001). ^{1}H-NMR spectroscopy measurements of NAA do not differentiate between cellular and extracellular NAA, whereas microdialysis sampling of NAA should give an indication of extracellular NAA. Because extracellular NAA is rapidly transported into astrocytes, extracellular measures may not reflect a true assessment of NAA dynamics.

Although NAA measurements using MRI spectroscopy are typically used as a marker for neuronal damage or loss (Narayanan et al., 1997), NAA/Na^+ cotransport into astrocytes likely results in detrimental Na^+ accumulation in astrocytes. Indeed, recent data from our laboratory demonstrate that NAA application to cultured astrocytes increases intracellular Na^+ in a dose-dependent manner (Figure 38.3). Thus, inhibiting NAAG cleavage will eliminate a significant pool of extracellular NAA and thus reduce NAA/Na^+ transport into astrocytes. Because both glutamate and NAA transporters can potentially contribute to elevated $[Na^+]_i$ in astrocytes, the dynamic interactions between astrocytes and neurons represent a novel strategy for understanding and reducing pathology to both cell types after TBI.

38.7 ELEVATED INTRACELLULAR NA+ ADVERSELY AFFECTS ASTROCYTES

In the mid-1980s, Plum (1983) and later Largo and colleagues (1996) suggested that the degree of infarction from focal ischemia might depend upon the viability of astrocytes.

Recent in vivo studies provide compelling evidence of early astrocyte loss after subdural hematoma-induced ischemia (Jiang et al., 2000) or permanent middle cerebral artery occlusion (Liu et al., 1999). Immunohistochemical evidence indicates that astrocytes are damaged and lost in selectively vulnerable brain regions within hours after experimental TBI (Zhao et al., 2003). These results are consistent with recent reports of early astrocyte loss after cerebral ischemia (Liu et al., 1999). Furthermore, this astrocyte loss appears to precede neuronal cell loss in vulnerable brain regions.

Intracellular Na^+ in mechanically injured cultured astrocytes rises in an injury magnitude-dependent manner (Floyd et al., 2005). Application of glutamate to astrocyte cultures also increases $[Na^+]_i$ in a glutamate dose-dependent manner (Floyd et al., 2005). The increased $[Na^+]_i$ observed after either mechanical injury or glutamate application is similar to that previously reported in astrocytes following exogenous glutamate application (Rose and Ransom, 1996) or ischemia (Bondarenko and Chesler, 2001a, 2001b; Rose et al., 1998). Additionally, injury-induced Na^+ influx has been implicated as a mechanism of neuronal injury including mechanical stretch injury to axons (Wolf et al., 2001), glutamate toxicity to dendrites (Hasbani et al., 1998), and neuronal ischemic injury (Fern and Ransom, 1997; LoPachin and Lehning, 1997; Nishizawa, 2001; Vornov et al., 1999). Moreover, astrocyte swelling after severe CNS injuries is likely related to Na^+-dependent glutamate uptake (Hansson et al., 1997; Kimelberg et al., 1995). Astrocyte cell death can be significantly reduced by inhibiting Na^+ entry into mechanically injured or glutamate exposed astrocytes by blocking either Na^+-glutamate cotransport or Na^+-proton exchange (Floyd et al., 2005). Application of NAAG peptidase inhibitors after TBI results in a reduction in concentrations of extracellular glutamate and NAA and should therefore decrease subsequent co-transport of Na^+ into astrocytes. Thus, the actions of NAAG and the consequences of NAAG peptidase inhibition could have significant therapeutic consequences for both neuronal and astrocytic pathology after TBI.

38.8 CONCLUSIONS AND PERSPECTIVE

There is a well-established role of elevated extracellular glutamate and excitotoxicity in TBI. Reducing extracellular glutamate and restoring homeostasis remains a prime target for pharmacological intervention. However, a major limitation of traditional antiglutamatergic therapeutic targets including receptor blockade is that both physiological and pathophysiological activity is blocked. NAAG peptidase inhibitors were designed to enhance an endogenous neuropeptide transmitter system. NAAG peptidase inhibiotors increase synaptic levels of NAAG and increase activation of mGluR3 that ultimately reduces glutamate release. One of the advantages of this strategy is that NAAG activation of mGluR3 only plays a substantial role when excessive glutamate is released and thus does not disrupt physiological neurotransmission of information via glutamate release. Because NAAG is only released during intense neuronal signaling, side effects of NAAG peptidase inhibition should be minimal.

The application of NAAG peptidase inhibitors represents a novel departure from traditional approaches of studying excitotoxicity in the context of TBI pathophysiology. The abundant neuropeptide, NAAG, could potentially confer a multifaceted approach to reduce glutamatergic excitotoxic processes associated with TBI. Compelling evidence is accumulating that elevating levels of extracellular NAAG (via inhibition of its cleavage) significantly reduces acute cellular TBI pathology in neurons and astrocytes, improves functional outcome, and reduces the deleterious effects of secondary insults. The accumulating experimental data lend support for the continued development of NAAG peptidase inhibitors as therapeutics to reduce glutamate-induced excitotoxicity after TBI.

REFERENCES

Alessandri B, al-Samsam R, Corwin F, Fatouros P, Young HF, Bullock RM. (2000) Acute and late changes in N-acetyl-aspartate following diffuse axonal injury in rats: An MRI spectroscopy and microdialysis study. *Neurol Res* 22:705–712.

Allen JW, Ivanova SA, Fan L, Espey MG, Basile AS, Faden AI. (1999) Group II metabotropic glutamate receptor activation attenuates traumatic neuronal injury and improves neurological recovery after traumatic brain injury. *J Pharmacol Exp Therapeutics* 290:112–120.

Aronica E, Gorter JA, Ijlst-Keizers H, Rozemuller AJ, Yankaya B, Leenstra S et al. (2003) Expression and functional role of mGluR3 and mGluR5 in human astrocytes and glioma cells: Opposite regulation of glutamate transporter proteins. *Eur J Neurosci* 17:2106–2118.

Bacich DJ, Ramadan E, O'Keefe DS, Bukhari N, Wegorzewska I, Ojeifo O et al. (2002) Deletion of the glutamate carboxypeptidase II gene in mice reveals a second enzyme activity that hydrolyzes N-acetylaspartylglutamate. *J Neurochem* 83:20–29.

Bauman RA, Widholm JJ, Petras JM, McBride K, Long JB. (2000) Secondary hypoxemia exacerbates the reduction of visual discrimination accuracy and neuronal cell density in the dorsal lateral geniculate nucleus resulting from fluid percussion injury. *J Neurotrauma* 17:679–693.

Benuck M, D'Adamo AF, Jr. (1968) Acetyl transport mechanisms. Metabolism of N-acetyl-L-aspartic acid in the non-nervous tissues of the rat. *Biochim Biophys Acta* 152:611–618.

Berger UV, Luthi-Carter R, Passani LA, Elkabes S, Black I, Konradi C et al. (1999) Glutamate carboxypeptidase II is expressed by astrocytes in the adult rat nervous system. *J Comp Neurol* 415:52–64.

Birken DL, Oldendorf WH. (1989) N-acetyl-L-aspartic acid: A literature review of a compound prominent in 1H-NMR spectroscopic studies of brain. *Neurosci Biobehav Rev* 13:23–31.

Bischofberger J, Schild D. (1996) Glutamate and N-acetylaspartylglutamate block HVA calcium currents in frog olfactory bulb interneurons via an mGluR2/3-like receptor. *J Neurophysiol* 76:2089–2092.

Bond A, Ragumoorthy N, Monn JA, Hicks CA, Ward MA, Lodge D, O'Neill MJ. (1999) LY379268, a potent and selective Group II metabotropic glutamate receptor agonist, is neuroprotective in gerbil global, but not focal, cerebral ischaemia. *Neurosci Lett* 273:191–194.

Bondarenko A, Chesler M. (2001a) Calcium dependence of rapid astrocyte death induced by transient hypoxia, acidosis, and extracellular ion shifts. *Glia* 34:143–149.

Bondarenko A, Chesler M. (2001b) Rapid astrocyte death induced by transient hypoxia, acidosis, and extracellular ion shifts. *Glia* 34:134–142.

Bruno V, Sureda FX, Storto M, Casabona G, Caruso A, Knopfel T, Kuhn R, Nicoletti F. (1997) The neuroprotective activity of group-II metabotropic glutamate receptors requires new protein synthesis and involves a glial-neuronal signaling. *J.Neurosci* 17:1891–1897.

Bruno V, Wroblewska B, Wroblewski JT, Fiore L, Nicoletti F. (1998) Neuroprotective activity of N-acetylaspartylglutamate in cultured cortical cells. *Neuroscience* 85:751–757.

Bullock MR, Lyeth BG, Muizelaar JP. (1999) Current status of neuroprotection trials for traumatic brain injury: Lessons from animal models and clinical studies. *Neurosurgery* 45:207–217.

Bullock MR, Zauner A, Myseros JS, Marmarou A, Woodward JJ, Young HF. (1995a) Evidence for prolonged release of excitatory amino acids in severe human head trauma. *Neuroprotective Agents*. New York Academy of Science, New York, pp. 290–297.

Bullock MR, Zauner A, Woodward J, Young HF. (1995b) Massive persistent release of excitatory amino acids following human occlusive stroke. *Stroke* 26:2187–2189.

Bullock R, Zauner A, Woodward JJ, Myseros J, Choi SC, Ward JD et al. (1998a) Factors affecting excitatory amino acid release following severe human head injury. *J Neurosurg* 89:507–518.

Bullock R, Zauner A, Woodward JJ, Myseros J, Choi SC, Ward JD et al. (1998b) Factors affecting excitatory amino acid release following severe human head injury. *J Neurosurg* 89:507–518.

Bzdega T, Crowe SL, Ramadan ER, Sciarretta KH, Olszewski RT, Ojeifo OA et al. (2004) The cloning and characterization of a second brain enzyme with NAAG peptidase activity. *J Neurochem* 89:627–635.

Choi DW, Rothman SM. (1990) The role of glutamate neurotoxicity in hypoxic-ischemic neuronal death. *Annu Rev Neurosci* 13:171–182.

Clark RS, Kochanek PM, Dixon CE, Chen M, Marion DW, Heineman S et al. (1997) Early neuropathologic effects of mild or moderate hypoxemia after controlled cortical impact injury in rats. *J Neurosurg* 14:179–189.

Curatolo A, D'Arcangelo P, Lino A, Brancati A. (1965) Distribution of N-acetyl-aspartic and N-acetyl-aspartyl-glutamic acids in nervous tissue. *J Neurochem* 12:339–342.

Danbolt NC. (1994) The high affinity uptake system for excitatory amino acids in the brain. *Prog Neurobiol* 44:377–396.

Davis DP, Dunford JV, Poste JC, Ochs M, Holbrook T, Fortlage D et al. (2004) The impact of hypoxia and hyperventilation on outcome after paramedic rapid sequence intubation of severely head-injured patients. *J Trauma* 57:1–8; discussion 8–10.

Davis DP, Meade W, Sise MJ, Kennedy F, Simon F, Tominaga G et al. (2009) Both hypoxemia and extreme hyperoxemia may be detrimental in patients with severe traumatic brain injury. *J Neurotrauma* 26:2217–2223.

Dunlop DS, Mc Hale DM, Lajtha A. (1992) Decreased brain N-acetylaspartate in Huntington's disease. *Brain Res* 580:44–48.

Faden AI, Demediuk P, Panter SS, Vink R. (1989) The role of excitatory amino acids and NMDA receptors in traumatic brain injury. *Science* 244:798–800.

Faden AI, Ivanova SA, Yakovlev AG, Mukhin AG. (1997) Neuroprotective effects of group III mGluR in traumatic neuronal injury. *J Neurotrauma* 14:885–895.

Faul M, Xu L, Wald MM, Coronado VG. (2010) Traumatic brain injury in the United States: Emergency department visits, hospitalizations and deaths 2002–2006. Centers for Disease Control and Prevention, National Center for Injury Prevention and Control, Atlanta, Ga., U.S. Department of Health and Human Services.

Feng JF, Gurkoff GG, Van KC, Song M, Lowe DA, Zhou J, Lyeth BG. (2012a) NAAG peptidase inhibitor reduces cellular damage in a model of TBI with secondary hypoxia. *Brain Res* 1469:144–152.

Feng JF, Van KC, Gurkoff GG, Kopriva C, Olszewski RT, Song M et al. (2011) Post-injury administration of NAAG peptidase inhibitor prodrug, PGI-02776, in experimental TBI. *Brain Res* 1395:62–73.

Feng JF, Zhao X, Gurkoff GG, Van KC, Shahlaie K, Lyeth BG. (2012b) Post-traumatic hypoxia exacerbates neuronal cell death in the hippocampus. *J Neurotrauma* 29:1167–1179.

Fern R, Ransom BR. (1997) Ischemic injury of optic nerve axons: The nuts and bolts. *Clin Neurosci* 4:246–250.

Floyd CL, Gorin FA, Lyeth BG. (2005) Mechanical strain injury increases intracellular sodium and reverses Na+/Ca2+ exchange in cortical astrocytes. *Glia* 51:35–46.

Fuhrman S, Palkovits M, Cassidy M, Neale JH. (1994) The regional distribution of N-acetylaspartylglutamate (NAAG) and peptidase activity against NAAG in the rat nervous system. *J Neurochem* 62:275–281.

Ghadge GD, Slusher BS, Bodner A, Canto MD, Wozniak K, Thomas AG et al. (2003) Glutamate carboxypeptidase II inhibition protects motor neurons from death in familial amyotrophic lateral sclerosis models. *Proc Natl Acad Sci U S A* 100:9554–9559.

Goldstein FB. (1976) Amidohydrolases of brain; enzymatic hydrolysis of N-acetyl-L-aspartate and other N-acyl-L-amino acids. *J Neurochem* 26:45–49.

Goldstein M. (1990) Traumatic brain injury: A silent epidemic. *Ann Neurol* 27:237–237.

Goodman MD, Makley AT, Huber NL, Clarke CN, Friend LA, Schuster RM et al. (2011) Hypobaric hypoxia exacerbates the neuroinflammatory response to traumatic brain injury. *J Surg Res* 165:30–37.

Gurkoff GG, Feng JF, Van KC, Izadi A, Ghiasvand R, Shahlaie K et al. (2013) NAAG peptidase inhibitor improves motor function and reduces cognitive dysfunction in a model of TBI with secondary hypoxia. *Brain Res* 1515:98–107.

Hansson E, Blomstrand F, Khatibi S, Olsson T, Ronnback L. (1997) Glutamate induced astroglial swelling—methods and mechanisms. *Acta Neurochir Suppl (Wien)* 70:148–151.

Hasbani MJ, Hyrc KL, Faddis BT, Romano C, Goldberg MP. (1998) Distinct roles for sodium, chloride, and calcium in excitotoxic dendritic injury and recovery. *Exp Neurol* 154:241–258.

Hayes RL, Jenkins LW, Lyeth BG, Balster RL, Robinson SE, Clifton GL et al. (1988) Pretreatment with phencyclidine, an N-methyl-D-aspartate antagonist, attenuates long-term behavioral deficits in the rat produced by traumatic brain injury. *J Neurotrauma* 5:259–274.

Hellewell SC, Yan EB, Agyapomaa DA, Bye N, Morganti-Kossmann MC. (2010) Post-traumatic hypoxia exacerbates brain tissue damage: Analysis of axonal injury and glial responses. *J Neurotrauma* 27:1997–2010.

Hicks RR, Smith DH, Gennarelli TA, Mcintosh TK. (1994) Kynurenate is neuroprotective following experimental brain injury in the rat. *Brain Res* 655:91–96.

Huang W, Wang H, Kekuda R, Fei YJ, Friedrich A, Wang J et al. (2000) Transport of N-acetylaspartate by the Na(+)-dependent high-affinity dicarboxylate transporter NaDC3 and its relevance to the expression of the transporter in the brain. *J Pharmacol Exp Ther* 295:392–403.

Jackson PF, Cole DC, Slusher BS, Stetz SL, Ross LE, Donzanti BA et al. (1996) Design, synthesis, and biological activity of a potent inhibitor of the neuropeptidase N-acetylated alpha-linked acidic dipeptidase. *J Med Chem* 39:619–622.

Jiang ZW, Gong QZ, Di X, Zhu J, Lyeth BG. (2000) Dicyclomine, an M1 muscarinic antagonist, reduces infarct volume in a rat subdural hematoma model. *Brain Res* 852:37–44.

Kanner BI, Schuldiner S. (1987) Mechanism of transport and storage of neurotransmitters. *CRC Crit Rev Biochem* 22:1–38.

Katayama Y, Becker DP, Tamura T, Hovda DA. (1990) Massive increases in extracellular potassium and the indiscriminate release of glutamate following concussive brain injury. *J Neurosurg* 73:889–900.

Kaul R, Casanova J, Johnson AB, Tang P, Matalon R. (1991) Purification, characterization, and localization of aspartoacylase from bovine brain. *J Neurochem* 56:129–135.

Kimelberg HK, Rutledge E, Goderie S, Charniga C. (1995) Astrocytic swelling due to hypotonic or high K+ medium causes inhibition of glutamate and aspartate uptake and increases their release. *J Cereb Blood Flow Metab* 15:409–416.

Koller KJ, Coyle JT. (1984) Characterization of the interactions of N-acetyl-aspartyl-glutamate with [3H]L-glutamate receptors. *Eur J Pharmacol* 98:193–199.

Koura SS, Doppenberg EM, Marmarou A, Choi S, Young HF, Bullock R. (1998) Relationship between excitatory amino acid release and outcome after severe human head injury. *Acta Neurochir Suppl* 71:244–246.

Kozikowski AP, Zhang J, Nan F, Petukhov PA, Grajkowska E, Wroblewski JT et al. (2004) Synthesis of urea-based inhibitors as active site probes of glutamate carboxypeptidase II: Efficacy as analgesic agents. *J Med Chem* 47:1729–1738.

Largo C, Cuevas P, Herreras O. (1996) Is glia disfunction the initial cause of neuronal death in ischemic penumbra? *Neurol Res* 18:445–448.

Liu D, Smith CL, Barone FC, Ellison JA, Lysko PG, Li K, Simpson IA. (1999) Astrocytic demise precedes delayed neuronal death in focal ischemic rat brain. *Brain Res* 68:29–41.

LoPachin RM, Lehning EJ. (1997) Mechanism of calcium entry during axon injury and degeneration. *Toxicol Appl Pharmacol* 143:233–244.

Luthi-Carter R, Berger UV, Barczak AK, Enna M, Coyle JT. (1998) Isolation and expression of a rat brain cDNA encoding glutamate carboxypeptidase II. *Proc Natl Acad Sci U S A* 95:3215–3220.

Lyeth BG, Gong QZ, Shields S, Muizelaar JP, Berman RF. (2001) Group I metabotropic glutamate antagonist reduces acute neuronal degeneration and behavioral deficits after traumatic brain injury in rats. *Exp Neurol* 169:191–199.

Manley G, Knudson MM, Morabito D, Damron S, Erickson V, Pitts L. (2001) Hypotension, hypoxia, and head injury: Frequency, duration, and consequences. *Arch Surg* 136:1118–1123.

McIntosh TK, Juhler M, Wieloch T. (1998) Novel pharmacologic strategies in the treatment of experimental traumatic brain injury: 1998. *J Neurotrauma* 15:731–769.

Miller JD, Sweet RC, Narayan R, Becker DP. (1978) Early insults to the injured brain. *JAMA* 240:439–442.

Miyake M, Kakimoto Y, Sorimachi M. (1981) A gas chromatographic method for the determination of N-acetyl-L-aspartic acid, N-acetyl-alpha-aspartylglutamic acid and beta-citryl-L-glu tamic acid and their distributions in the brain and other organs of various species of animals. *J Neurochem* 36:804–810.

Moffett JR, Namboodiri MA. (1995) Differential distribution of N-acetylaspartylglutamate and N-acetylaspartate immunoreactivities in rat forebrain. *J Neurocytol* 24:409–433.

Moffett JR, Palkovits M, Namboodiri A, Neale JH. (1994) Comparative distribution of N-acetylaspartylglutamate and GAD67 in the cerebellum and precerebellar nuclei of the rat utilizing enhanced carbodiimide fixation and immunohistochemistry. *J Comp Neurol* 347:598–618.

Movsesyan VA, Faden AI. (2006) Neuroprotective effects of selective group II mGluR activation in brain trauma and traumatic neuronal injury. *J Neurotrauma* 23:117–127.

Mukhin A, Fan L, Faden AI. (1996) Activation of metabotropic glutamate receptor subtype mGluR1 contributes to post-traumatic neuronal injury. *J Neurosci* 16:6012–6020.

Mukhin AG, Ivanova SA, Faden AI. (1997) mGluR modulation of post-traumatic neuronal death: Role of NMDA receptors. *Neuroreport* 8:2561–2566.

Nakanishi S. (1994) Metabotropic glutamate receptors: Synaptic transmission, modulation, and plasticity. *Neuron* 13:1031–1037.

Narayan RK, Michel ME, Ansell B, Baethmann A, Biegon A, Bracken MB et al. (2002) Clinical trials in head injury. *J Neurotrauma* 19:503–557.

Narayanan S, Fu L, Pioro E, De Stefano N, Collins DL, Francis GS et al. (1997) Imaging of axonal damage in multiple sclerosis: Spatial distribution of magnetic resonance imaging lesions. *Ann Neurol* 41:385–391.

Nawashiro H, Shima K, Chigasaki H. (1995) Selective vulnerability of hippocampal CA3 neurons to hypoxia after mild concussion in the rat. *Neurol Res* 17:455–460.

Neale JH, Bzdega T, Wroblewska B. (2000) N-Acetylaspartylglutamate: The most abundant peptide neurotransmitter in the mammalian central nervous system. *J Neurochem* 75:443–452.

Nilsson P, Hillered L, Ponten U, Ungerstedt U. (1990) Changes in cortical extracellular levels of energy-related metabolites and amino acids following concussive brain injury in rats. *J Cereb Blood Flow Metab* 10:631–637.

Nishizawa Y. (2001) Glutamate release and neuronal damage in ischemia. *Life Sci* 69:369–381.

Olszewski RT, Bukhari N, Zhou J, Kozikowski AP, Wroblewski JT, Shamimi-Noori S et al. (2004) NAAG peptidase inhibition reduces locomotor activity and some stereotypes in the PCP model of schizophrenia via group II mGluR. *J Neurochem* 89:876–885.

Patel TB, Clark JB. (1979) Synthesis of N-acetyl-L-aspartate by rat brain mitochondria and its involvement in mitochondrial/cytosolic carbon transport. *Biochem J* 184:539–546.

Peden CJ, Cowan FM, Bryant DJ, Sargentoni J, Cox IJ, Menon DK et al. (1990) Proton MR spectroscopy of the brain in infants. *J Comput Assist Tomogr* 14:886–894.

Pittaluga A, Barbeito L, Serval V, Godeheu G, Artaud F, Glowinski J et al. (1988) Depolarization-evoked release of N-acetyl-L-aspartyl-L-glutamate from rat brain synaptosomes. *Eur J Pharmacol* 158:263–266.

Plaitakis A, Constantakakis E. (1993) Altered metabolism of excitatory amino acids, N-acetyl-aspartate and N-acetyl-aspartyl-glutamate in amyotrophic lateral sclerosis. *Brain Res Bull* 30:381–386.

Plum F. (1983) What causes infarction in ischemic brain? The Robert Wartenberg Lecture. *Neurology* 33:222–233.

Renno WM, Lee JH, Beitz AJ. (1997) Light and electron microscopic immunohistochemical localization of N-acetylaspartylglutamate (NAAG) in the olivocerebellar pathway of the rat. *Synapse* 26:140–154.

Roberts PJ. (1995) Pharmacological tools for the investigation of metabotropic glutamate receptors (mGluRs): Phenylglycine derivatives and other selective antagonists - an update. *Neuropharmacology* 34:813–819.

Robinson MB, Blakely RD, Couto R, Coyle JT. (1987) Hydrolysis of the brain dipeptide N-acetyl-L-aspartyl-L-glutamate. Identification and characterization of a novel N-acetylated alpha-linked acidic dipeptidase activity from rat brain. *J Biol Chem* 262:14498–14506.

Rong SB, Zhang J, Neale JH, Wroblewski JT, Wang S, Kozikowski AP. (2002) Molecular modeling of the interactions of glutamate carboxypeptidase II with its potent NAAG-based inhibitors. *J Med Chem* 45:4140–4152.

Rose CR, Ransom BR. (1996) Mechanisms of H+ and Na+ changes induced by glutamate, kainate, and D-aspartate in rat hippocampal astrocytes. *J Neurosci* 16:5393–5404.

Rose CR, Waxman SG, Ransom BR. (1998) Effects of glucose deprivation, chemical hypoxia, and simulated ischemia on Na+ homeostasis in rat spinal cord astrocytes. *J Neurosci* 18:3554–3562.

Sager TN, Fink-Jensen A, Hansen AJ. (1997) Transient elevation of interstitial N-acetylaspartate in reversible global brain ischemia. *J Neurochem* 68:675–682.

Sager TN, Laursen H, Fink-Jensen A, Topp S, Stensgaard A, Hedehus M et al. (1999a) N-Acetylaspartate distribution in rat brain striatum during acute brain ischemia. *J Cereb Blood Flow Metab* 19:164–172.

Sager TN, Thomsen C, Valsborg JS, Laursen H, Hansen AJ. (1999b) Astroglia contain a specific transport mechanism for N-acetyl-L-aspartate. *J Neurochem* 73:807–811.

Sanabria ER, Wozniak KM, Slusher BS, Keller A. (2004) GCP II (NAALADase) inhibition suppresses mossy fiber-CA3 synaptic neurotransmission by a presynaptic mechanism. *J Neurophysiol* 91:182–193.

Sayer RJ, Schwindt PC, Crill WE. (1992) Metabotropic glutamate receptor-mediated suppression of L-type calcium current in acutely isolated neocortical neurons. *J Neurophysiol* 68:833–842.

Schmoker JD, Zhuang J, Shackford SR. (1992) Hemorrhagic hypotension after brain injury causes an early and sustained reduction in cerebral oxygen delivery despite normalization of systemic oxygen delivery. *J Trauma* 32:714–720; discussion 721–712.

Schoepp DD, Conn PJ. (1993) Metabotropic glutamate receptors in brain function and pathology. *Trends Pharmacol. Sci.* 14:13–20.

Schuhmann MU, Stiller D, Skardelly M, Bernarding J, Klinge PM, Samii A, Samii M, Brinker T. (2003) Metabolic changes in the vicinity of brain contusions: A proton magnetic resonance spectroscopy and histology study. *J Neurotrauma* 20:725–743.

Serval V, Barbeito L, Pittaluga A, Cheramy A, Lavielle S, Glowinski J. (1990) Competitive inhibition of N-acetylated-alpha-linked acidic dipeptidase activity by N-acetyl-L-aspartyl-beta-linked L-glutamate. *J Neurochem* 55:39–46.

Signoretti S, Marmarou A, Tavazzi B, Lazzarino G, Beaumont A, Vagnozzi R. (2001) N-Acetylaspartate reduction as a measure of injury severity and mitochondrial dysfunction following diffuse traumatic brain injury. *J Neurotrauma* 18:977–991.

Slusher BS, Tsai G, Yoo G, Coyle JT. (1992) Immunocytochemical localization of the N-acetyl-aspartyl-glutamate (NAAG) hydrolyzing enzyme N-acetylated alpha-linked acidic dipeptidase (NAALADase). *J Comp Neurol* 315:217–229.

Slusher BS, Vornov JJ, Thomas AG, Hurn PD, Harukuni I, Bhardwaj A et al. (1999) Selective inhibition of NAALADase, which converts NAAG to glutamate, reduces ischemic brain injury. *Nat Med* 5:1396–1402.

Takahashi M, Billups B, Rossi D, Sarantis M, Hamann M, Attwell D. (1997) The role of glutamate transporters in glutamate homeostasis in the brain. *J Exp Biol.* 200:401–409.

Taylor DL, Davies SE, Obrenovitch TP, Urenjak J, Richards DA, Clark JB et al. (1994) Extracellular N-acetylaspartate in the rat brain: In vivo determination of basal levels and changes evoked by high K+. *J Neurochem* 62:2349–2355.

Thurman D, Guerrero J. (1999) Trends in hospitalization associated with traumatic brain injury. *JAMA* 282:954–957.

Thurman DJ, Alverson C, Dunn KA, Guerrero J, Sniezek JE. (1999) Traumatic brain injury in the United States: A public health perspective. *J Head Trauma Rehabil* 14:602–615.

Tsai G, Coyle JT. (1995) N-acetylaspartate in neuropsychiatric disorders. *Prog Neurobiol* 46:531–540.

Urenjak J, Williams SR, Gadian DG, Noble M. (1992) Specific expression of N-acetylaspartate in neurons, oligodendrocyte-type-2 astrocyte progenitors, and immature oligodendrocytes in vitro. *J Neurochem* 59:55–61.

Urenjak J, Williams SR, Gadian DG, Noble M. (1993) Proton nuclear magnetic resonance spectroscopy unambiguously identifies different neural cell types. *J Neurosci* 13:981–989.

Valivullah HM, Lancaster J, Sweetnam PM, Neale JH. (1994) Interactions between N-acetylaspartylglutamate and AMPA, kainate, and NMDA binding sites. *J Neurochem* 63: 1714–1719.

Vernadakis A. (1996) Glia-neuron intercommunications and synaptic plasticity. *Prog Neurobiol* 49:185–214.

Vesce S, Bezzi P, Volterra A. (1999) The active role of astrocytes in synaptic transmission. *Cell Mol Life Sci* 56:991–1000.

Vespa P, Prins M, Ronne-Engstrom E, Caron M, Shalmon E, Hovda DA, Martin NA, Becker DP. (1998) Increase in extracellular glutamate caused by reduced cerebral perfusion pressure and seizures after human traumatic brain injury: A microdialysis study. *J Neurosurg* 89:971–982.

Vornov JJ, Wozniak K, Lu M, Jackson P, Tsukamoto T, Wang E, Slusher B. (1999) Blockade of NAALADase: A novel neuroprotective strategy based on limiting glutamate and elevating NAAG. *Ann N Y Acad Sci* 890:400–405.

Williamson LC, Neale JH. (1988) Ultrastructural localization of N-acetylaspartylglutamate in synaptic vesicles of retinal neurons. *Brain Res* 456:375–381.

Wolf JA, Stys PK, Lusardi T, Meaney D, Smith DH. (2001) Traumatic axonal injury induces calcium influx modulated by tetrodotoxin-sensitive sodium channels. *J Neurosci* 21:1923–1930.

Wroblewska B, Wroblewski JT, Pshenichkin S, Surin A, Sullivan SE, Neale JH. (1997) *N*-acetylaspartylglutamate selectively activates mGluR3 receptors in transfected cells. *J Neurochem.* 69:174–181.

Wroblewska B, Wroblewski JT, Saab OH, Neale JH. (1993) N-acetylaspartylglutamate inhibits forskolin-stimulated cyclic AMP levels via a metabotropic glutamate receptor in cultured cerebellar granule cells. *J Neurochem* 61:943–948.

Xi ZX, Baker DA, Shen H, Carson DS, Kalivas PW. (2002) Group II metabotropic glutamate receptors modulate extracellular glutamate in the nucleus accumbens. *J Pharmacol Exp Ther* 300:162–171.

Yamamoto T, Nozaki-Taguchi N, Sakashita Y. (2001) Spinal N-acetyl-alpha-linked acidic dipeptidase (NAALADase) inhibition attenuates mechanical allodynia induced by paw carrageenan injection in the rat. *Brain Res* 909:138–144.

Yao HH, Ding JH, Zhou F, Wang F, Hu LF, Sun T, Hu G. (2005) Enhancement of glutamate uptake mediates the neuroprotection exerted by activating group II or III metabotropic glutamate receptors on astrocytes. *J Neurochem* 92:948–961.

Zaczek R, Koller K, Cotter R, Heller D, Coyle JT. (1983) N-acetylaspartylglutamate: An endogenous peptide with high affinity for a brain "glutamate" receptor. *Proc Natl Acad Sci U S A* 80:1116–1119.

Zauner A, Bullock R, Kuta AJ, Woodward J, Young HF. (1996) Glutamate release and cerebral blood flow after severe human head injury. *Acta Neurochir Suppl*, 67:40–44.

Zhang W, Slusher B, Murakawa Y, Wozniak KM, Tsukamoto T, Jackson PF et al. (2002) GCPII (NAALADase) inhibition prevents long-term diabetic neuropathy in type 1 diabetic BB/Wor rats. *J Neurol Sci* 194:21–28.

Zhao X, Ahram A, Berman RF, Muizelaar JP, Lyeth BG. (2003) Early loss of astrocytes after experimental traumatic brain injury. *Glia* 44:140–152.

Zhong C, Zhao X, Sarva J, Kozikowski A, Neale JH, Lyeth BG. (2005) NAAG peptidase inhibitor reduces acute neuronal degeneration and astrocyte damage following lateral fluid percussion TBI in rats. *J Neurotrauma* 22:266–276.

Zhong C, Zhao X, Van KC, Bzdega T, Smyth A, Zhou J et al. (2006) NAAG peptidase inhibitor increases dialysate NAAG and reduces glutamate, aspartate and GABA levels in the dorsal hippocampus following fluid percussion injury in the rat. *J Neurochem* 97:1015–1025.

Zollinger M, Amsler U, Do KQ, Streit P, Cuenod M. (1988) Release of N-acetylaspartylglutamate on depolarization of rat brain slices. *J Neurochem* 51:1919–1923.

Zollinger M, Brauchli-Theotokis J, Gutteck-Amsler U, Do KQ, Streit P, Cuenod M. (1994) Release of N-acetylaspartylglutamate from slices of rat cerebellum, striatum, and spinal cord, and the effect of climbing fiber deprivation. *J Neurochem* 63:1133–1142.

Zwienenberg M, Gong QZ, Berman RF, Muizelaar JP, Lyeth BG. (2001) The effect of groups II and III metabotropic glutamate receptor activation on neuronal injury in a rodent model of traumatic brain injury. *Neurosurgery* 48:1119–1126; discussion 1126–1117.

39 Neuregulin-1 and Neurovascular Protection

Limin Wu, Samantha J. Walas, Wendy Leung, Eng H. Lo, and Josephine Lok

CONTENTS

39.1 INTRODUCTION

Neuregulin-1 (NRG1) is a growth factor that belongs to a family of polypeptide growth factors encoded by four neuregulin genes. NRG1 signals are transduced through the ErbB family of receptor protein tyrosine kinases, in which NRG1 binding results in phosphorylation and dimerization of these receptors, leading to activation of downstream pathways. NRG1/erbB signaling is involved in cellular processes including cell survival, proliferation, migration, and differentiation. In the CNS, NRG1/ErbB4 signaling also affects dendritic spine maturation and the formation of inhibitory synapses. Neuregulin-1/ErbB signaling has been shown to have beneficial effects on many of the different types of cells in the brain, including neurons, astrocytes, oligodendrocytes, oligodendrocyte precursor cells, endothelial cells, and microglia. Many of these actions may have neuroprotective potential in the setting of traumatic brain injury, and will be further discussed in this chapter.

Traumatic brain injury (TBI) is a significant public health problem, affecting individuals of all ages daily. Survivors are often left with motor and cognitive deficits, ranging from mild to devastating. Because there is no definitive treatment beyond supportive care in the acute postinjury period, there is much focus on searching for neuroprotective agents that will attenuate the pathological events that arise after the injury. The process is challenging because of the heterogeneity of TBI. An attempt to classify TBI highlights the diverse array of injuries—for example, focal or diffuse, mild or severe, presence or absence of hemorrhage, a concussion or a contusion, or a single hit or repeated hits. It is likely that no single agent will be efficacious for the different types of TBI, and it is necessary to understand the pathophysiology of the various injuries and investigate appropriate therapies for each type of injury. Here the term "traumatic brain injury" is used as a general term, with the understanding that it is a simplification of the complex nature of TBI.

In recent years, there has been an increasing amount of research and interest in NRG1 an endogenous growth factor with multiple functions in the central nervous system (CNS), and now in clinical trials for the treatment of patients with chronic heart failure. The actions of NRG1 intersect with many cellular pathways, thus it may have applicability to many of the pathological processes in TBI. In this chapter, we will review pathophysiological processes after TBI that intersect with NRG1/ErbB signaling as well as experimental data in models of brain trauma suggesting that NRG1 may be neuroprotective. Because some of the brain injury pathways in TBI are similar to those in ischemic brain injury (Bramlett and Dietrich, 2004), literature on the effects of NRG1 in models of ischemic stroke will also be discussed.

39.2 NRG1/ErbB SIGNALING

NRG1 is a growth factor that belongs to a family of polypeptide growth factors encoded by four genes—NRG1, NRG2, NRG3, and NRG4. NRG1 was identified in parallel by several groups working in different areas of biology—cancer biology, Schwann cell proliferation, and acetylcholine receptor synthesis (Falls et al., 1993; Holmes et al., 1992; Marchionni et al., 1993; Peles et al., 1992; Wen et al., 1992). The neuregulin proteins isolated by each of these groups are encoded by the gene now known as NRG1. The NRG1 gene is a large and complex gene transcribing more than 20 different transmembrane proteins with a single membrane-spanning domain, and generating a large number of isoforms in tissue- and cell type–specific patterns (Buonanno and Fischbach, 2001; Falls, 2003a; Tan et al., 2007).

NRG1 can be classified as type I, II, or III based on the type of amino-terminal sequence, and further characterized as type α or β according to the characteristics of the epidermal growth factor (EGF)-like domain. Type I consists of secreted isoforms that contain an immunoglobulin (Ig)-like domain (NRG1-Ig) with a carbohydrate-spacer region in their extracellular regions. Type II isoforms have an Ig-like domain without a carbohydrate-spacer region. Type III isoforms contain a cysteine-rich domain (NRG1-CRD) and are membrane bound. All three types of NRG1 proteins share high-sequence homology in their EGF-like domain, which is sufficient to bind and activate ErbB receptors to dimerize, phosphorylate tyrosine kinase, and activate downstream signaling pathways. The β-type EGF-like domain has more potent activity compared with the α-type. Therefore, many studies have been performed using a NRG1-β1 fragment containing only the EGF-like domain. Soluble NRG is produced through limited proteolysis of the membrane-bound protein, or by RNA splice variation that produces a secreted form (Buonanno and Fischbach, 2001; Falls, 2003b; Tan et al., 2007).

NRG1 signals are transduced through the ErbB family of receptor protein tyrosine kinases: ErbB1 (EGFR and HER1), ErbB2 ($p185^{Her2/neu}$ and HER2), ErbB3 (HER3), and ErbB4 (HER4). The basic structure of the receptors consists of an extracellular amino terminus and two cysteine-rich domains that function in ligand binding, a transmembrane domain, a short intracellular tyrosine kinase domain, and a carboxyl terminal tail. Interestingly, although the study of ErbB2 contributed to the discovery of NRG1, ErbB2 does not actually bind NRG1—it has strong kinase activity but lacks ligand-binding activity. ErbB3, on the other hand, binds ligand but lacks kinase activity, whereas ErbB4 possesses both. NRG1 binding to ErbB3 or ErbB4 results in phosphorylation and dimerization of the receptors. Heterodimers containing ErbB2 or ErbB4 or homodimers of ErbB4 may effect signal transduction intracellularly. The downstream signals involve SHC-adaptor protein, growth factor receptor–bound protein 2, the regulatory subunit of phosphatidylinositol-3-kinase, Ca^{2+}-dependent protein kinase C, and NFAT activity. Activation of these pathways is followed by cellular responses including proliferation, migration, differentiation, and cell survival (Citri and Yarden, 2006; Flames et al., 2004; Jones et al., 2006; Pawson, 2004; Stein and Staros, 2006; Yarden and Sliwkowski, 2001).

NRG1/ErbB pathways are important in embryonic CNS development and postnatal brain function. Of the ErbB receptors, ErbB4 is the predominant one in the brain. In the developing mouse nervous system, ErbB4 mRNA is first detected at around E9, mostly in the hindbrain and the forebrain. As development proceeds, ErbB4 expression becomes more widespread within the brain and spinal cord. NRG1/ErbB4 signaling is involved in many processes in the CNS, including neuronal migration, dendritic spine maturation, and the formation of inhibitory synapses onto excitatory pyramidal neurons (Buonanno, 2010; Buonanno and Fischbach, 2001; Citri and Yarden, 2006; Falls, 2003b; Flames et al., 2004; Jones et al., 2006; Pawson, 2004; Rio et al., 1997; Stein and Staros, 2006; Tan et al., 2012; Yarden and Sliwkowski, 2001).

39.3 BLOOD–BRAIN BARRIER PERMEABILITY

In the setting of TBI, one of the most important early events is the increase in blood–brain barrier (BBB) permeability (Adelson et al., 1998; Baskaya et al., 1997; Rosenberg, 2012; Shapira et al., 1993). Compromise of BBB integrity exposes the brain parenchyma to substances normally in the vascular space; additionally, it disrupts the ionic gradients necessary for proper neuronal function and may lead to acute and long-term CNS dysfunction (Pop and Badaut, 2011; Pop et al., 2013; Shlosberg et al., 2010; Tomkins et al., 2011; Zlokovic, 2008). Additionally, BBB disruption may contribute to brain edema, leading to increased intracranial pressure, compromised perfusion, ischemia, brain herniation, and death. The increase in BBB permeability is often associated with increased inflammation and with a local immune response, including the activation of astrocytes and microglia (Aihara et al., 1995; Stamatovic et al., 2006). In some clinical series, patients who have disrupted BBB appear to have an increased risk of developing acute seizures, delayed epilepsy, cognitive impairment, and Alzheimer disease (Rosenberg, 2012; Shlosberg et al., 2010; Tomkins et al., 2011; Zlokovic, 2008).

NRG1's actions in the brain microvasculature include the ability to protect endothelial barrier function (Lok et al., 2009, 2012). In the in vitro experiments in which incubation with the inflammatory cytokine interleukin (IL)-1β increases permeability through a monolayer of brain microvascular endothelial cells, addition of NRG1 ameliorates the permeability increase (Lok et al., 2012). In addition to its effects on endothelial permeability, NRG1 promotes the survival of endothelial cells in an in vitro model of oxidative stress with H_2O_2-induced injury (Lok et al., 2009). Data from in vivo experiments using a controlled cortical impact model of brain trauma in mice also showed a beneficial effect of NRG1 on BBB barrier function. Administration of NRG1 by intravenous injection immediately after trauma was associated with a reduction in acute BBB permeability (Lok et al., 2012). In the in vivo setting, it is possible that NRG1 effects are extended to other types of cells that comprise the BBB, including the astrocytes and pericytes.

39.4 AXONAL INJURY

Along with injuring neurons, brain trauma also damages the axons, often resulting in diffuse axonal injury. Axonal injury is one of the most important pathological features after TBI, with a spectrum of severity ranging from mild disruption of the axonal cytoskeleton to major disruptions of axonal function (Johnson et al., 2013). Clinically, these changes may be manifested acutely as loss of consciousness or confusion and persist as coma and/or cognitive dysfunction (Johnson et al., 2013).

It is well-known that NRG1 plays an integral role in axon physiology in the peripheral nervous system, influencing the survival and migration of Schwann cell precursors as well as remyelination and functional recovery after injury (Carroll et al., 1997; Viehover et al., 2001). In regard to myelination in the CNS, NRG1 signaling affects oligodendrocyte specification, differentiation, function, and survival. NRG1 plays a role in oligodendrocyte development, and oligodendrocyte precursors fail to develop in mice with a deletion of the NRG1 gene (Calaora et al., 2001; Kim et al., 2003; Park et al., 2001; Roy et al., 2007; Sussman et al., 2005; Vartanian et al., 1994, 1999). In in vitro experiments using oxygen-glucose deprivation to induce cell death, exogenous NRG1 reduced apoptosis in oligodendrocytes via ErbB4-dependent activation of PI3-kinase/Akt (Xu et al., 2012). In an animal model of spinal cord injury, endogenous NRG1 levels were decreased after injury. Exogenous NRG1 enhanced the proliferation and differentiation of spinal neural precursor cells into oligodendrocytes, enhanced axonal preservation, and attenuated astrogliosis and tissue degeneration (Gauthier et al., 2013).

39.5 POSTTRAUMA ISCHEMIC BRAIN INJURY

In the acute period after TBI, a transient period of decreased blood flow may occur (Bouma et al., 1992; Kunz et al., 2010). Additionally, cerebral edema, if present, may limit cerebral perfusion and contribute to brain ischemia. Therefore, TBI may involve cell death pathways seen in ischemic brain injury. NRG1 effect on ischemic brain injury has been investigated in MCAO models of stroke, in which several groups reported that the administration of NRG1-β significantly reduced cortical infarct volume. Additionally, NRG1-treated rats had significantly improved neurological scores compared with vehicle-treated rats, with fewer degenerating nerve fibers, and a lower amount of TUNEL staining in the cortex and the striatum (Guo et al., 2006; Shyu et al., 2004; Xu et al., 2004).

39.6 POSTTRAUMA INFLAMMATION

After brain trauma, an acute inflammatory state occurs. Peripherally derived immune cells, including neutrophils, macrophages and T cells, along with activated astrocytes and microglia, contribute to secondary pathological injury and increase tissue destruction (Bethea and Dietrich, 2002). Release of inflammatory cytokines, such as tumor necrosis factor, IL-1, IL-6, and IL-10, exacerbate the inflammatory condition. Interaction of activated platelets with the injured endothelium further contributes to a proinflammatory, prothrombotic microenvironment (del Zoppo et al., 2000). Many of these changes are mediated through the altered expression of surface adhesion molecules (Feuerstein et al., 1998). Acute inflammation may increase tissue damage, although in the delayed phase, inflammatory responses may contribute to the recovery process (Bethea and Dietrich, 2002; Kerschensteiner et al., 1999).

NRG1 has been reported to have anti-inflammatory actions. In cell culture, recombinant human NRG1 decreases the production of superoxide and nitrite by stimulated N9 microglial cells (Dimayuga et al., 2003). In a rat model of stroke, NRG1 administered before middle cerebral artery occlusion is associated with a decrease in microglial activation and interleukin-1 mRNA expression, indicating a possible anti-inflammatory effect in the brain. NRG1 also attenuated the expression of many inflammatory genes, including genes associated with leukocyte migration and activation. In parallel in vitro studies, NRG1 suppressed inflammatory gene expression in activated macrophages (Xu et al., 2005).

39.7 POSTTRAUMA EPILEPSY

Posttrauma epilepsy (PTE) is a complication that occurs in a significant number of patients who have suffered brain trauma. Not only does PTE adversely affect the recovery process, it is often unresponsive to conventional antiepilepsy drugs, making its treatment problematic (Jensen, 2009; Lowenstein, 2009). Preventing the development of PTE would contribute substantially to the recovery of patients with TBI.

Of the many factors that contribute to posttrauma epileptogenesis, an important one is an alteration in the normal balance between excitatory and inhibitory circuits in the hippocampus (Santhakumar et al., 2001). Inhibitory signals from GABAergic interneurons are thought to be important in controlling excitatory output of pyramidal cells, and dysfunction of GABAergic signaling is a factor in the pathogenesis of epilepsy (Thind et al., 2010). NRG1 interacts with parvalbumin (PV+) GABAergic interneurons, which highly express ErbB4. Interestingly, ErbB4 expression is decreased in human epileptogenic tissue (Li et al., 2012). In experimental models of epilepsy, mice with specific deletion of ErbB4 in PV+ interneurons are more susceptible to seizures (Li et al., 2012). In another experimental model, limbic epileptogenesis was inhibited by infusing NRG1 intracerebrally, but exacerbated by neutralizing endogenous NRG1, inhibiting ErbB4 activation or deleting the ErbB4 gene. Depletion of ErbB4 in parvalbumin-expressing interneurons abolished NRG1-mediated inhibition of epileptogenesis and promoted kindling progression, resulting in increased spontaneous seizures and exuberant mossy fiber sprouting (Tan et al., 2012). These findings suggest that NRG1 activation of ErbB4 in PV+ interneurons may provide an endogenous negative-feedback mechanism to suppress limbic epileptogenesis.

39.8 POSTTRAUMATIC COGNITIVE DEFICITS AND NEUROGENESIS

Long-term cognitive deficits, including difficulty with learning and memory, are common morbidities associated with TBI. This may be significant morbidity even after mild TBI. NRG1 is connected with a multitude of processes that may impact cognition (Andersson et al., 2012; Furth et al., 2013; Makinodan et al., 2012; Roy et al., 2007; Schillo et al., 2005; Taylor et al., 2012). Additionally, there is evidence implicating NRG1 as a schizophrenia susceptibility gene (Buonanno, 2010; Deng et al., 2013; Stefansson et al., 2003a, 2003b). In regard to its impact on cognitive deficits after TBI, NRG1 administration was associated with improved spatial memory in mice after controlled cortical impact (Lok et al., 2007).

Increased hippocampal neurogenesis has been reported in rodent models of TBI and has been postulated to play a role in improving motor and cognitive function (Chauhan and Gatto, 2010; Han et al., 2011; Petraglia et al., 2010; Ramaswamy et al., 2005; Wu et al., 2008). Because NRG1 is known to interact with processes affecting neuronal progenitor cells in embryonic as well as postnatal life, it may have the potential to enhance neurogenesis after TBI. The NRG1 receptor erbB4 is expressed prominently by neuroblasts in the subventricular zone and the rostral migratory stream and modulates neuroblast migration and placement (Anton et al., 2004). NRG1 affects the configuration of neuronal progenitor cells in the subventricular zone (Ghashghaei et al., 2006). Peripherally administered exogenous NRG1 was reported to have pro-proliferative and neurogenic properties (Mahar et al., 2011). In addition to its effect on neuronal progenitor cells, NRG1 may affect the development of radial glia. In a scratch-injury model, mature astrocytes revert to a radial glial progenitor cell phenotype in a process associated with activation of ErbB2 receptors (Yang et al., 2011). Thus, further studies of NRG1 actions in post-TBI neurogenesis and recovery are warranted.

39.9 NRG1 IN CLINICAL TRIALS

There have been several clinical trials investigating the use of NRG1 in the treatment of cardiac failure. It is known that the NRG1/ErbB signaling plays a major role in cardiac development and in maintaining the function of cardiomyocytes (Odiete et al., 2012). In one study, administration of a 10-day course of recombinant human NRG1 to patients with chronic heart failure was associated with improved cardiac function when the patients were evaluated at 30 days (Gao et al., 2010). In another phase II study in which recombinant human NRG-1 was infused over 11 days, the improvement in cardiac function was sustained for 12 weeks (Jabbour et al., 2011). As progress is made in the understanding of NRG1 actions in the setting of chronic cardiac failure, it would be important to determine the CNS effects of NRG1 at the relevant dosages.

39.10 BIVALENT NRG1 DERIVATIVES

There is a concern that NRG1 may have pro-neoplastic effects, because of its ability to activate ErbB2 receptors that are expressed in some types of cancer. To address this issue, Jay et al. engineered bivalent ligands, formed by the synthetic linkage of two NRG1-β moieties (NN). When ErbB receptors are activated by NN rather than NRG1, the interaction favors a lower rate of ErbB2-containing heterodimers compared with ErbB3-containing heterodimers. When tested in cancer cells, NN signaling led to a decrease in migration and proliferation and an increase in apoptosis (Jay et al., 2011, 2013). When tested in cultured cardiomyocytes, the engineered covalent ligand

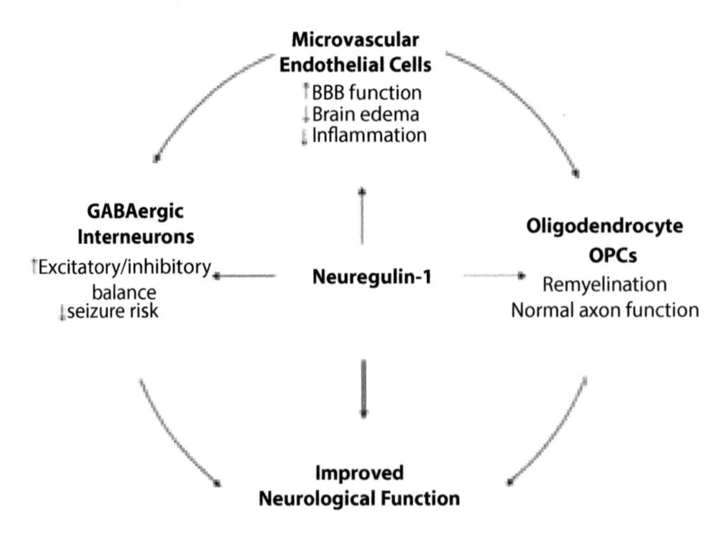

FIGURE 39.1 Neuroprotective effects of NRG1 after brain trauma. NRG1's actions on cells in the neurovascular unit contribute to its neuroprotective potential after brain trauma. NRG1 protects BBB function by preventing pathological increases in endothelial permeability, providing a normal environment for neuronal function. Additionally, NRG1 provides a pro-survival effect to GABAergic interneurons, which may play a role in preventing PTE, and to oligodendrocytes and their precursor cells, which enable remyelination to take place.

retained the ability to prevent doxorubicin-induced toxicity (Jay et al., 2013). This is an important development in the investigation of NRG1 as a therapy for heart failure and for brain injury.

39.11 CONCLUSION

NRG1/ErbB signaling as discussed has beneficial actions on endothelial, oligodendrocyte, and interneuron physiology, with effects that may have neuroprotective potential as illustrated in Figure 39.1. Further investigations on NRG1/ErbB pathways in the setting of TBI are merited.

ACKNOWLEDGMENTS

The authors are supported by NS037074 (E.H.L.), NS076694 (E.H.L.), and NS K08N5057339 (J.L.). The authors do not have any conflicts of interest.

REFERENCES

Adelson, P.D., M.J. Whalen, P.M. Kochanek, P. Robichaud, and T.M. Carlos. 1998. Blood brain barrier permeability and acute inflammation in two models of traumatic brain injury in the immature rat: A preliminary report. *Acta Neurochir Suppl.* 71:104–106.

Aihara, N., J.J. Hall, L.H. Pitts, K. Fukuda, and L.J. Noble. 1995. Altered immunoexpression of microglia and macrophages after mild head injury. *J Neurotrauma.* 12:53–63.

Andersson, R.H., A. Johnston, P.A. Herman, U.H. Winzer-Serhan, I. Karavanova, D. Vullhorst et al. 2012. Neuregulin and dopamine modulation of hippocampal gamma oscillations is dependent on dopamine D4 receptors. *Proc Natl Acad Sci U S A.* 109:13118–13123.

Anton, E.S., H.T. Ghashghaei, J.L. Weber, C. McCann, T.M. Fischer, I.D. Cheung et al. 2004. Receptor tyrosine kinase ErbB4 modulates neuroblast migration and placement in the adult forebrain. *Nat Neurosci.* 7:1319–1328.

Baskaya, M.K., A.M. Rao, A. Dogan, D. Donaldson, and R.J. Dempsey. 1997. The biphasic opening of the blood-brain barrier in the cortex and hippocampus after traumatic brain injury in rats. *Neurosci Lett.* 226:33–36.

Bouma, G.J., J.P. Muizelaar, W.A. Stringer, S.C. Choi, P. Fatouros, and H.F. Young. 1992. Ultra-early evaluation of regional cerebral blood flow in severely head-injured patients using xenon-enhanced computerized tomography. *J Neurosurg.* 77:360–368.

Bramlett, H.M., and W.D. Dietrich. 2004. Pathophysiology of cerebral ischemia and brain trauma: Similarities and differences. *J Cereb Blood Flow Metab.* 24:133–150.

Buonanno, A. 2010. The neuregulin signaling pathway and schizophrenia: From genes to synapses and neural circuits. *Brain Res Bull.* 83:122–131.

Buonanno, A., and G.D. Fischbach. 2001. Neuregulin and ErbB receptor signaling pathways in the nervous system. *Curr Opin Neurobiol.* 11:287–296.

Calaora, V., B. Rogister, K. Bismuth, K. Murray, H. Brandt, P. Leprince et al. 2001. Neuregulin signaling regulates neural precursor growth and the generation of oligodendrocytes in vitro. *J Neurosci.* 21:4740–4751.

Carroll, S.L., M.L. Miller, P.W. Frohnert, S.S. Kim, and J.A. Corbett. 1997. Expression of neuregulins and their putative receptors, ErbB2 and ErbB3, is induced during Wallerian degeneration. *J Neurosci.* 17:1642–1659.

Chauhan, N.B., and R. Gatto. 2010. Synergistic benefits of erythropoietin and simvastatin after traumatic brain injury. *Brain Res.* 1360:177–192.

Citri, A., and Y. Yarden. 2006. EGF-ERBB signalling: Towards the systems level. *Nat Rev Mol Cell Biol.* 7:505–516.

Deng, C., B. Pan, M. Engel, and X.F. Huang. 2013. Neuregulin-1 signalling and antipsychotic treatment: Potential therapeutic targets in a schizophrenia candidate signalling pathway. *Psychopharmacology (Berl).* 226:201–215.

Dimayuga, F.O., Q. Ding, J.N. Keller, M.A. Marchionni, K.B. Seroogy, and A.J. Bruce-Keller. 2003. The neuregulin GGF2 attenuates free radical release from activated microglial cells. *J Neuroimmunol.* 136:67–74.

Falls, D.L. 2003a. Neuregulins and the neuromuscular system: 10 years of answers and questions. *J Neurocytol.* 32:619–647.

Falls, D.L. 2003b. Neuregulins: Functions, forms, and signaling strategies. *Exp Cell Res.* 284:14–30.

Falls, D.L., K.M. Rosen, G. Corfas, W.S. Lane, and G.D. Fischbach. 1993. ARIA, a protein that stimulates acetylcholine receptor synthesis, is a member of the neu ligand family. *Cell.* 72:801–815.

Flames, N., J.E. Long, A.N. Garratt, T.M. Fischer, M. Gassmann, C. Birchmeier et al. 2004. Short- and long-range attraction of cortical GABAergic interneurons by neuregulin-1. *Neuron.* 44:251–261.

Furth, K.E., S. Mastwal, K.H. Wang, A. Buonanno, and D. Vullhorst. 2013. Dopamine, cognitive function, and gamma oscillations: Role of D4 receptors. *Front Cell Neurosci.* 7:102.

Gao, R., J. Zhang, L. Cheng, X. Wu, W. Dong, X. Yang et al. 2010. A Phase II, randomized, double-blind, multicenter, based on standard therapy, placebo-controlled study of the efficacy and safety of recombinant human neuregulin-1 in patients with chronic heart failure. *J Am Coll Cardiol.* 55:1907–1914.

Gauthier, M.K., K. Kosciuczyk, L. Tapley, and S. Karimi-Abdolrezaee. 2013. Dysregulation of the neuregulin-1-ErbB network modulates endogenous oligodendrocyte differentiation and preservation after spinal cord injury. *Eur J Neurosci.* 38:2693–2715.

Ghashghaei, H.T., J. Weber, L. Pevny, R. Schmid, M.H. Schwab, K.C. Lloyd et al. 2006. The role of neuregulin-ErbB4 interactions on the proliferation and organization of cells in the subventricular zone. *Proc Natl Acad Sci U S A.* 103:1930–1935.

Guo, W.P., J. Wang, R.X. Li, and Y.W. Peng. 2006. Neuroprotective effects of neuregulin-1 in rat models of focal cerebral ischemia. *Brain Res.* 1087:180–185.

Han, X., J. Tong, J. Zhang, A. Farahvar, E. Wang, J. Yang et al. 2011. Imipramine treatment improves cognitive outcome associated with enhanced hippocampal neurogenesis after traumatic brain injury in mice. *J Neurotrauma.* 28:995–1007.

Holmes, W.E., M.X. Sliwkowski, R.W. Akita, W.J. Henzel, J. Lee, J.W. Park et al. 1992. Identification of heregulin, a specific activator of p185erbB2. *Science.* 256:1205–1210.

Jabbour, A., C.S. Hayward, A.M. Keogh, E. Kotlyar, J.A. McCrohon, J.F. England et al. 2011. Parenteral administration of recombinant human neuregulin-1 to patients with stable chronic heart failure produces favourable acute and chronic haemodynamic responses. *Eur J Heart Fail.* 13:83–92.

Jay, S.M., E. Kurtagic, L.M. Alvarez, S. de Picciotto, E. Sanchez, J.F. Hawkins et al. 2011. Engineered bivalent ligands to bias ErbB receptor-mediated signaling and phenotypes. *J Biol Chem.* 286:27729–27740.

Jay, S.M., A.C. Murthy, J.F. Hawkins, J.R. Wortzel, M.L. Steinhauser, L.M. Alvarez et al. 2013. An engineered bivalent neuregulin protects against doxorubicin-induced cardiotoxicity with reduced proneoplastic potential. *Circulation.* 128:152–161.

Jensen, F.E. 2009. Introduction. Posttraumatic epilepsy: Treatable epileptogenesis. *Epilepsia*. 50 Suppl 2:1–3.

Johnson, V.E., W. Stewart, and D.H. Smith. 2013. Axonal pathology in traumatic brain injury. *Exp Neurol*. 246:35–43.

Jones, R.B., A. Gordus, J.A. Krall, and G. MacBeath. 2006. A quantitative protein interaction network for the ErbB receptors using protein microarrays. *Nature*. 439:168–174.

Kim, J.Y., Q. Sun, M. Oglesbee, and S.O. Yoon. 2003. The role of ErbB2 signaling in the onset of terminal differentiation of oligodendrocytes in vivo. *J Neurosci*. 23:5561–5571.

Kunz, A., U. Dirnagl, and P. Mergenthaler. 2010. Acute pathophysiological processes after ischaemic and traumatic brain injury. *Best Pract Res Clin Anaesthesiol*. 24:495–509.

Li, K.X., Y.M. Lu, Z.H. Xu, J. Zhang, J.M. Zhu, J.M. Zhang et al. 2012. Neuregulin 1 regulates excitability of fast-spiking neurons through Kv1.1 and acts in epilepsy. *Nat Neurosci*. 15:267–273.

Lok, J., S.P. Sardi, S. Guo, E. Besancon, D.M. Ha, A. Rosell et al. 2009. Neuregulin-1 signaling in brain endothelial cells. *J Cereb Blood Flow Metab*. 29:39–43.

Lok, J., H. Wang, Y. Murata, H.H. Zhu, T. Qin, M.J. Whalen, and E.H. Lo. 2007. Effect of neuregulin-1 on histopathological and functional outcome after controlled cortical impact in mice. *J Neurotrauma*. 24:1817–1822.

Lok, J., S. Zhao, W. Leung, J.H. Seo, D. Navaratna, X. Wang et al. 2012. Neuregulin-1 effects on endothelial and blood-brain-barrier permeability after experimental injury. *Transl Stroke Res*. 3:S119-S124.

Lowenstein, D.H. 2009. Epilepsy after head injury: An overview. *Epilepsia*. 50 Suppl 2:4–9.

Mahar, I., S. Tan, M.A. Davoli, S. Dominguez-Lopez, C. Qiang, A. Rachalski et al. 2011. Subchronic peripheral neuregulin-1 increases ventral hippocampal neurogenesis and induces antidepressant-like effects. *PLoS One*. 6:e26610.

Makinodan, M., K.M. Rosen, S. Ito, and G. Corfas. 2012. A critical period for social experience-dependent oligodendrocyte maturation and myelination. *Science*. 337:1357–1360.

Marchionni, M.A., A.D. Goodearl, M.S. Chen, O. Bermingham-McDonogh, C. Kirk, M. Hendricks et al. 1993. Glial growth factors are alternatively spliced erbB2 ligands expressed in the nervous system. *Nature*. 362:312–318.

Odiete, O., M.F. Hill, and D.B. Sawyer. 2012. Neuregulin in cardiovascular development and disease. *Circ Res*. 111:1376–1385.

Park, S.K., R. Miller, I. Krane, and T. Vartanian. 2001. The erbB2 gene is required for the development of terminally differentiated spinal cord oligodendrocytes. *J Cell Biol*. 154:1245–1258.

Pawson, T. 2004. Specificity in signal transduction: From phosphotyrosine-SH2 domain interactions to complex cellular systems. *Cell*. 116:191–203.

Peles, E., S.S. Bacus, R.A. Koski, H.S. Lu, D. Wen, S.G. Ogden et al. 1992. Isolation of the neu/HER-2 stimulatory ligand: A 44 kd glycoprotein that induces differentiation of mammary tumor cells. *Cell*. 69:205–216.

Petraglia, A.L., A.H. Marky, C. Walker, M. Thiyagarajan, and B.V. Zlokovic. 2010. Activated protein C is neuroprotective and mediates new blood vessel formation and neurogenesis after controlled cortical impact. *Neurosurgery*. 66:165–171; discussion 171–162.

Pop, V., and J. Badaut. 2011. A neurovascular perspective for long-term changes after brain trauma. *Transl Stroke Res*. 2:533–545.

Pop, V., D.W. Sorensen, J.E. Kamper, D.O. Ajao, M.P. Murphy, E. Head et al. 2013. Early brain injury alters the blood-brain barrier phenotype in parallel with beta-amyloid and cognitive changes in adulthood. *J Cereb Blood Flow Metab*. 33:205–214.

Ramaswamy, S., G.E. Goings, K.E. Soderstrom, F.G. Szele, and D.A. Kozlowski. 2005. Cellular proliferation and migration following a controlled cortical impact in the mouse. *Brain Res*. 1053:38–53.

Rio, C., H.I. Rieff, P. Qi, T.S. Khurana, and G. Corfas. 1997. Neuregulin and erbB receptors play a critical role in neuronal migration. *Neuron*. 19:39–50.

Rosenberg, G.A. 2012. Neurological diseases in relation to the blood-brain barrier. *J Cereb Blood Flow Metab*. 32:1139–1151.

Roy, K., J.C. Murtie, B.F. El-Khodor, N. Edgar, S.P. Sardi, B.M. Hooks et al. 2007. Loss of erbB signaling in oligodendrocytes alters myelin and dopaminergic function, a potential mechanism for neuropsychiatric disorders. *Proc Natl Acad Sci U S A*. 104:8131–8136.

Santhakumar, V., A.D. Ratzliff, J. Jeng, Z. Toth, and I. Soltesz. 2001. Long-term hyperexcitability in the hippocampus after experimental head trauma. *Ann Neurol*. 50:708–717.

Schillo, S., V. Pejovic, C. Hunzinger, T. Hansen, S. Poznanovic, J. Kriegsmann et al. 2005. Integrative proteomics: Functional and molecular characterization of a particular glutamate-related neuregulin isoform. *J Proteome Res*. 4:900–908.

Shapira, Y., D. Setton, A.A. Artru, and E. Shohami. 1993. Blood-brain barrier permeability, cerebral edema, and neurologic function after closed head injury in rats. *Anesth Analg*. 77:141–148.

Shlosberg, D., M. Benifla, D. Kaufer, and A. Friedman. 2010. Blood-brain barrier breakdown as a therapeutic target in traumatic brain injury. *Nat Rev Neurol*. 6:393–403.

Shyu, W.C., S.Z. Lin, M.F. Chiang, H.I. Yang, P. Thajeb, and H. Li. 2004. Neuregulin-1 reduces ischemia-induced brain damage in rats. *Neurobiol Aging*. 25:935–944.

Stamatovic, S.M., O.B. Dimitrijevic, R.F. Keep, and A.V. Andjelkovic. 2006. Inflammation and brain edema: New insights into the role of chemokines and their receptors. *Acta Neurochir Suppl*. 96:444–450.

Stefansson, H., J. Sarginson, A. Kong, P. Yates, V. Steinthorsdottir, E. Gudfinnsson et al. 2003a. Association of neuregulin 1 with schizophrenia confirmed in a Scottish population. *Am J Hum Genet*. 72:83–87.

Stefansson, H., T.E. Thorgeirsson, J.R. Gulcher, and K. Stefansson. 2003b. Neuregulin 1 in schizophrenia: Out of Iceland. *Mol Psychiatry*. 8:639–640.

Stein, R.A., and J.V. Staros. 2006. Insights into the evolution of the ErbB receptor family and their ligands from sequence analysis. *BMC Evol Biol*. 6:79.

Sussman, C.R., T. Vartanian, and R.H. Miller. 2005. The ErbB4 neuregulin receptor mediates suppression of oligodendrocyte maturation. *J Neurosci*. 25:5757–5762.

Tan, G.H., Y.Y. Liu, X.L. Hu, D.M. Yin, L. Mei, and Z.Q. Xiong. 2012. Neuregulin 1 represses limbic epileptogenesis through ErbB4 in parvalbumin-expressing interneurons. *Nat Neurosci*. 15:258–266.

Tan, W., Y. Wang, B. Gold, J. Chen, M. Dean, P.J. Harrison et al. 2007. Molecular cloning of a brain-specific, developmentally regulated neuregulin 1 (NRG1) isoform and identification of a functional promoter variant associated with schizophrenia. *J Biol Chem*. 282:24343–24351.

Taylor, A.R., S.B. Taylor, and J.I. Koenig. 2012. The involvement of type II neuregulin-1 in rat visuospatial learning and memory. *Neurosci Lett*. 531:131–135.

Thind, K.K., R. Yamawaki, I. Phanwar, G. Zhang, X. Wen, and P.S. Buckmaster. 2010. Initial loss but later excess of GABAergic synapses with dentate granule cells in a rat model of temporal lobe epilepsy. *J Comp Neurol*. 518:647–667.

Tomkins, O., A. Feintuch, M. Benifla, A. Cohen, A. Friedman, and I. Shelef. 2011. Blood-brain barrier breakdown following traumatic brain injury: A possible role in posttraumatic epilepsy. *Cardiovasc Psychiatry Neurol*. 2011:765923.

Vartanian, T., G. Corfas, Y. Li, G.D. Fischbach, and K. Stefansson. 1994. A role for the acetylcholine receptor-inducing protein ARIA in oligodendrocyte development. *Proc Natl Acad Sci U S A*. 91:11626–11630.

Vartanian, T., G. Fischbach, and R. Miller. 1999. Failure of spinal cord oligodendrocyte development in mice lacking neuregulin. *Proc Natl Acad Sci U S A*. 96:731–735.

Viehover, A., R.H. Miller, S.K. Park, G. Fischbach, and T. Vartanian. 2001. Neuregulin: An oligodendrocyte growth factor absent in active multiple sclerosis lesions. *Dev Neurosci*. 23:377–386.

Wen, D., E. Peles, R. Cupples, S.V. Suggs, S.S. Bacus, Y. Luo et al. 1992. Neu differentiation factor: A transmembrane glycoprotein containing an EGF domain and an immunoglobulin homology unit. *Cell*. 69:559–572.

Wu, H., D. Lu, H. Jiang, Y. Xiong, C. Qu, B. Li et al. 2008. Simvastatin-mediated upregulation of VEGF and BDNF, activation of the PI3K/Akt pathway, and increase of neurogenesis are associated with therapeutic improvement after traumatic brain injury. *J Neurotrauma*. 25:130–139.

Xu, C., L. Lv, G. Zheng, B. Li, L. Gao, and Y. Sun. 2012. Neuregulin1beta1 protects oligodendrocyte progenitor cells from oxygen glucose deprivation injury induced apoptosis via ErbB4-dependent activation of PI3-kinase/Akt. *Brain research*. 1467:104–112.

Xu, Z., G.D. Ford, D.R. Croslan, J. Jiang, A. Gates, R. Allen et al. 2005. Neuroprotection by neuregulin-1 following focal stroke is associated with the attenuation of ischemia-induced proinflammatory and stress gene expression. *Neurobiol Dis*. 19:461–470.

Xu, Z., J. Jiang, G. Ford, and B.D. Ford. 2004. Neuregulin-1 is neuroprotective and attenuates inflammatory responses induced by ischemic stroke. *Biochem Biophys Res Commun*. 322:440–446.

Yang, H., W. Ling, A. Vitale, C. Olivera, Y. Min, and S. You. 2011. ErbB2 activation contributes to de-differentiation of astrocytes into radial glial cells following induction of scratch-insulted astrocyte conditioned medium. *Neurochem Int*. 59:1010–1018.

Yarden, Y., and M.X. Sliwkowski. 2001. Untangling the ErbB signalling network. *Nat Rev Mol Cell Biol*. 2:127–137.

Zlokovic, B.V. 2008. The blood-brain barrier in health and chronic neurodegenerative disorders. *Neuron*. 57:178–201.

40 Potential Use of Calpain Inhibitors as Brain Injury Therapy

Emilio B. Cagmat, Joy D. Guingab-Cagmat, Anatoliy V. Vakulenko, Ronald L. Hayes, and John Anagli

CONTENTS

40.1 INTRODUCTION

Existing therapies for traumatic brain injuries (TBI) only relieve the symptoms of patients but do not treat the underlying causes that lead to debilitating effects of the trauma. One of the well understood molecular mechanism of TBI is the hyperactivation of the calpain enzyme. The need for drugs to inhibit the effect of the enzyme is viewed as one of the neurotherapies for traumatic brain injury. In this chapter, we summarize one of the strategies in finding small molecules to slow down or even eliminate the devastating effect of calpain hyperactivation.

The brain is a well-protected organ. Because of the very fragile nature of the brain, the organ is encased in a series of bone, membrane, and fluid-based protection. But despite all the privileges the brain is bestowed with, the organ is still very vulnerable to damage. A car accident, a football tackle, and an explosive's shock wave from a blast are all blows to the head that can set off a chain reaction, rendering the victim at least incapacitated or at worst, dead. At the molecular level, it is now well-documented that part of the damage is from the hyperactivation of calpain.

Calpain is a tightly regulated Ca^{2+}-dependent proteolytic enzyme that plays a critical role in important signaling pathways, including synaptic function and memory formation. Calpain exerts its biological function through limited proteolysis of its substrates. However, in some pathological conditions, such as after a TBI, calpain is dysregulated and participates in cell injury and death.

In TBI, neuronal calpains are hyperactivated within minutes after the initial trauma event, causing early damage to specific areas of the brain (Huh et al., 2007; Kampfl et al., 1996; Liu et al., 2006; Pike et al., 1998, 2001; Posmantur et al., 1997; Thompson et al., 2006; Upadhya et al., 2003; Warren et al., 2005; Zhang et al., 2007; Zhao et al., 1998a, 1998b, 2000). In multiple models of TBI, sustained calpain activation is associated with neuronal and axonal cell death. Calpain is not just linked to TBI but also to a handful of neurogenerative diseases. These neurogenerative pathologies include cerebral ischemia (Blomgren et al., 1999; Garcia-Bonilla et al., 2006; Gutierrez et al., 2007; Hong et al., 1994; Johnson et al., 1997; Kambe et al., 2005; Liebetrau et al., 1999; Minger et al., 1998; Neumar et al., 1996; Tsubokawa et al., 2006), Alzheimer's disease (DiRosa et al., 2002; Garg et al., 2011; Getz, 2012; Grynspan et al., 1997; Iwamoto et al., 1991; Medeiros et al., 2012; Tsuji et al., 1998), multiple sclerosis (Das et al., 2008; Shields et al., 1999), Huntington's disease (Bizat et al., 2003; Gafni and Ellerby, 2002; Gladding et al., 2012), prion-related encephalopathy (Gray et al., 2006; Hachiya et al., 2011; Wang et al., 2005; Yadavalli et al., 2004), and Parkinson's disease (Alvira et al., 2008; Mouatt-Prigent et al., 1996; Ray et al., 2000; Samantaray et al., 2008).

Enzyme inhibition is a well-known strategy for therapy development, and based on our understanding of calpain in the pathology of brain injury, this enzyme is an attractive therapeutic target. Calpain inhibitors probably came out at the same time as the discovery of the said enzyme 40 years ago. However, these inhibitors were general protease inhibitors, lacking selectivity and therefore inhibiting other proteases such as the proteasome and cathepsins. Interest in making selective calpain inhibitors sprung from the need to pinpoint the specific role(s) of calpain in the body. Subsequent years of research in calpain inhibitors led to a variety of compounds, most of which are small molecules (average MW ~500 g/mole).

As drug candidates, small-molecule compounds have some advantages over peptides or proteins, such as the ability to penetrate cells and resistance to proteolytic degradation. However, peptide and protein drug candidates exhibit higher selectivity toward their targets. A desirable treatment strategy in a clinical setting is through oral administration with a convenient dosing schedule, probably twice a day frequency of ingestion. These desirable pharmacological properties of a drug that can be taken orally and dosed once or twice daily are all met by small molecule compounds. In addition, small molecules are simpler: only a small portion of the enzyme, the active site is targeted to render the enzyme unable to perform its function. Compared with the whole enzyme, the active site is relatively small in comparison to the total volume of the macromolecule. This is akin to poking a small hole in the gas tank of a car so that the whole automobile cannot function.

The active site is precisely arranged in a three-dimensional structure to facilitate the catalysis of substrates. This is generally where the action occurs, and rendering the active site unable to perform its duty (or partially incapacitated) is one of the goals of small molecule inhibitors.

Designing small molecules as inhibitors is a strategy that is common in the pharmaceutical industry. In the market today, small-molecule enzyme inhibitors comprise half of all drugs. Pharmaceutical companies will still continue to follow in that direction, developing small molecules to inhibit dysregulated enzymes. And there are relatively a lot of success stories in drug development that targets enzymes. An example of this is the angiotensin-converting enzyme (ACE) inhibitor for treating hypertension. ACE converts angiotensin I into angiotensin II, resulting in blood vessel constriction and eventually increasing blood pressure. Blocking ACE lowers blood pressure. Pfizer's Viagra, a drug that blocks the enzyme cyclic guanosine monophosphate–specific phosphodiesterase type 5 to combat erectile dysfunction, is another blockbuster success. There has been a focus on designing new inhibitors of proteases that have been identified as disease-causing proteolytic enzymes. In addition to the successfully inhibited enzymes mentioned earlier, several inhibitors of a number of proteases are believed to progress from purely research to clinical trials. Bayer has developed a cathepsin K inhibitor (AAE581) to treat osteoporosis (Falgueyret et al., 2005). We saw the development of a serine aminopeptidase inhibitor, vildagliptin, as a therapy for type 2 diabetes (Deacon and Holst, 2006), and another ACE inhibitor (Aliskiren) that is an alternative to the current ACE inhibitor drugs (Stanton, 2003).

In this chapter, we will address the issues and challenges in drug discovery, specifically for the viability of small molecule inhibitors to regulate the destructive effects of calpain hyperactivation during or after TBI. Broadly, the enzyme (2) calpain will be discussed, followed by (3) calpain inhibitors, (4) challenges in realizing a therapy for brain injury, and then THE (5) conclusion. We hope that this chapter will provide students and scientists in brain injury research an overview of the potential of small-molecule calpain inhibitors in TBI drug development.

40.2 THE TARGET: CALPAIN

Calpains (Clan CA, family C02, EC 3.4.22.17) are a class of intracellular cysteine proteases that participate in several intracellular signaling pathways (NC-IUB, 1981). These pathways are regulated by Ca^{2+}. Although the precise function of the enzyme is not completely understood, dysregulation of the enzyme, as mentioned previously, is implicated in several pathological conditions.

In 1964, the existence of a Ca^{2+}-dependent neutral proteinase in rat brain was first reported (Guroff, 1964). Four years later, it was identified again in skeletal muscle as an enzyme that was similar to what was reported in 1964 (Huston and Krebs, 1968). In the 1970s, calpain was reidentified both as a Z-line hydrolyzing enzyme and a protein kinase C–activating factor (Busch et al., 1972).

Although calpain was first purified to homogeneity in 1978 (Ishiura et al., 1978), the breakthrough in studying calpain in terms of production and purification came in 1984, when cloning of the cDNA for the large subunit of a chicken calpain occurred (Ohno et al., 1984). The name calpain was derived from the enzymes calcium-binding protein calmodulin and the cysteine protease papain (Croall and DeMartino, 1991). This was after an extensive characterization of the structure and function of calpain that showed a calmodulin-like Ca^{2+}-binding domain (*-cal*) in addition to some similarity of its enzymatic activity to that of papain (*-pain*).

Calpain is an intracellular Ca^{2+}-activated protease found in all mammalian tissues. To date, 15 calpain genes have been found in humans and numbered *CAPN1–3* and *CAPN5–16,* but the most extensively studied gene products are calpain-1 (also called calpain-I or μ-CANP) and calpain-2 (also called calpain-II or m-CANP), the two main isoforms with varying sensitivity to calcium. These two calpains were originally labeled μ-calpain and m-calpain, the prefix indicating the Ca^{2+} concentration needed to activate the enzyme in vitro. These two calpains are known also as the conventional, ubiquitous, or the classical calpains, and we will limit our discussion to these calpains only.

Under normal physiological conditions, the enzyme is hypothesized to be part of Ca^{2+}-regulated signal transduction and cell physiology modulating processes. These biological situations are tightly regulated by Ca^{2+} and calpastatin (Betts and Anagli, 2004; Betts et al., 2003; Crawford et al., 1993; Kawasaki et al., 1993; Kenessey et al., 1990; Murachi, 1980, 1984), a protein that is the natural inhibitor of calpain. Because of its potency and specificity, a polypeptide derived from calpain's endogenous inhibitor is currently being studied as a calpain inhibitor drug candidate by some drug designers.

40.3 STRUCTURE-FUNCTIONAL ASPECTS OF CONVENTIONAL CALPAINS

Calpain-1 and calpain-2 are heterodimers and composed of two subunits. The small, 28-kDa regulatory subunit is identical in calpain-1 and calpain-2, whereas the large catalytic subunits (~80 kDa) are distinct in the two conventional calpains and share around 55%–65% homology. In humans, the two subunits of calpain-1 and calpain-2 are products of different genes (genes on chromosome 1 and 11, respectively) (Ohno et al., 1990). The catalytic and regulatory subunits of the conventional calpains are divided into four and two regions/domains, respectively (Figure 40.1).

Advances in X-ray crystallography have allowed structural biologists to peek into the detailed and high resolution image of the protease core (PC) and the active site of calpain (Cuerrier et al., 2006, 2007; Li et al., 2006; Moldoveanu et al., 2004). The calpain protease (CysPc) domain is the catalytic site containing the cysteine, histidine, asparagine triad. In the absence of Ca^{2+}, the CysPc domain is divided into PC domains PC1 and PC2, which are folded into one domain upon Ca^{2+} binding. Calpain free of Ca^{2+} is inactive, and from the standpoint of structure and function, this is because without Ca^{2+}, Cys-105 of calpain-2 is 10.5 Å away from His-282 and Asn-386, too far away from each other to be catalytically functioning (Goll et al., 2003). Ca^{2+} ion induces conformational

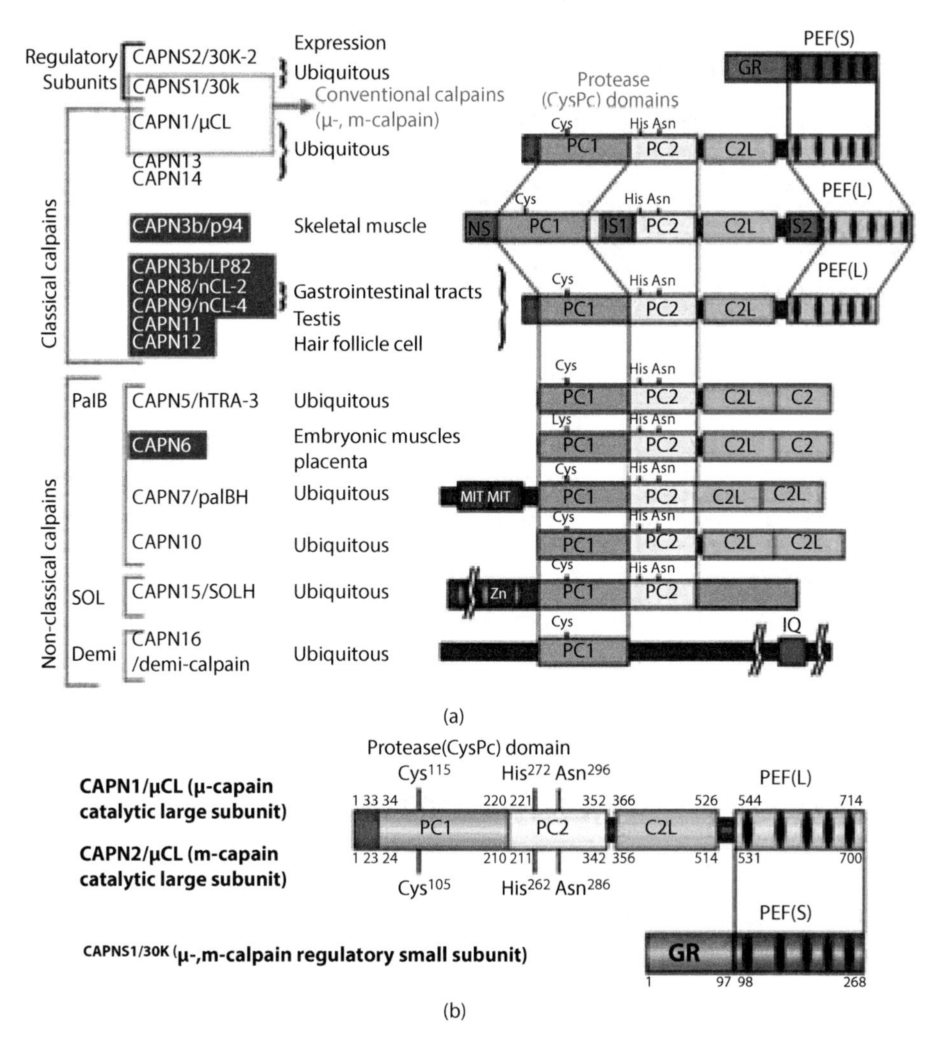

FIGURE 40.1 Schematic structures of human calpains and their related molecules. (a) Schematic structures of 15 human calpains. Names for calpains are by the general nomenclature of gene product (e.g., *CAPN1* > CAPN1), and their previous names, if any, are shown after the general names. Black and reversed letters indicate ubiquitous and tissue/organ-specific calpains, respectively. PC1 and PC2, protease core domains 1 and 2 in the calpain protease (CysPc) domain; C2L, C2-domain-like domain; PEF(L/S), penta-EF-hand domains in the large(L)/small(S) subunit; GR, glycine-rich hydrophobic domain; MIT, microtubule interacting and transport motif; C2, C2 domain; Zn, zinc-finger motif; SOH, SOL-homology domain; IQ, a motif interactive with calmodulin; NS/IS1/IS2, CAPN3/p94-characteristic sequences. (b) Domain architecture of human calpain 1 (μ-calpain) and calpain 2 (m-calpain). Calpains 1 and 2 (μ- and m-calpains) are heterodimers composed of a common regulatory small subunit (CAPNS1/30K) and each catalytic large subunit (CAPN1/μCL and CAPN2/mCL, respectively). See (a) for abbreviations of domain names. (With kind permission from Springer Science+Business Media: *Structure and Function*, Calpains in health and disease, 2013, 395–431, Anagli, J. et al.)

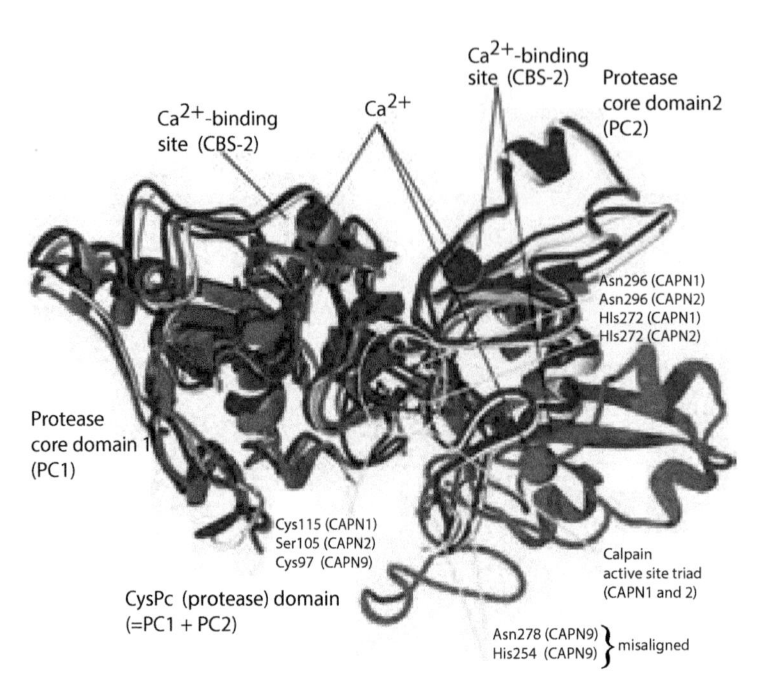

FIGURE 40.2 Superimposed schematic three-dimensional (3D) structures of CysPc domains of the active CAPN1/μCL, CAPN2/mCL, and CAPN9/nCL-4. Schematic 3D ribbon structures of the active (Ca^{2+}-bound) forms of CysPc domains of CAPN1/μCL (blue), CAPN2/mCL (white), and CAPN9/nCL-4 (red); their PC1 domains are superimposed by Deli server (PDB data: 2ARY, 1MDW, and 1ZIV) (Davis et al., 2007; Moldoveanu et al., 2003). The active sites are circled in yellow. PC1 subdomains of the three CysPc domains fit very well (root-mean-square deviation [RMSD] = 1~4 Å). CAPN1/μCL and CAPN2/mCL show overall fitting with very low RMSD (1.0 Å), including Ca^{2+}, but with one significant difference at Trp116 (CAPN1/μCL) and Trp106 (CAPN2/mCL). On the other hand, the PC2 domains of CAPN1/μCL and CAPN9/nCL-4 are markedly displaced. Note that, in the CAPN9/nCL-4 structure, His254 and Asn278 are too far away from Cys97 to form an active triad.

changes in calpain, moving the distance between the amino acids responsible for catalysis close to 3.7 Å. From the best evidence acquired, one result of the conformational change is that Trp-288 from PC2 rotates for catalysis to proceed (Arthur et al., 1995).

Figure 40.2 shows Ca^{2+} ions bound to calpain. Domains PC1, PC2, calpain-type beta-sandwich, PEF(L), and PEF(S) (penta-EF hand domains in the large(L)/small(S) subunit) bind at least one Ca^{2+} with varying affinities. Although we are armed with this information, the detailed step by step mechanism of the catalysis of calpain is still unknown.

As with any program in drug development, medicinal chemists have been exploiting the three-dimensional X-ray crystallography structure of calpain's active site to design inhibitors. By analyzing the binding pocket, including the electronic properties of the active site, with the help of interactive graphics, ligands, or compounds can be designed. This method is called structure-based design, and later in the chapter we will learn that in the drug development process, structure-based drug design is used in parallel with mechanism-based design.

40.4 CALPAIN INHIBITOR DRUG DEVELOPMENT

The process of drug development is a lengthy one. On average, a drug is developed in 15–17 years, from the discovery process to Food and Drug Administration (FDA) approval (Figure 40.3). Before a drug is available for doctors to prescribe, several million dollars are spent, starting from the discovery process that involves screening chemically diverse compounds (synthetic or natural sources), including the preclinical testing of animal models, to clinical trials, and then FDA approval.

The discovery process, which will be the process this chapter is concerned with, usually takes 3–5 years. In our case, peptidomimetic calpain inhibitor drug candidates are developed and assayed for their potency and selectivity in blocking the activity of the target calpain compared with a panel of mechanistically diverse group of proteolytic enzymes.

Preclinical testing is the phase in which pharmaceutical companies perform studies in the laboratory and then in animals, generally testing for biological activity for compounds against a disease. It is in this phase also that the safety of the compound is evaluated.

Before the start of the clinical trials, an application for an Investigational New Drug (IND) is usually filed with the FDA. The IND application provides previous experimental results relevant to proceed to the next studies, which will involve human testing. Outlined also in the IND application are the structures of the investigated compounds, the inner workings of the drug, the toxic effects, and the detailed manufacturing process.

In phase 1 clinical trials, around 80 healthy volunteers are tested for safety, establishing the dosage and the duration of

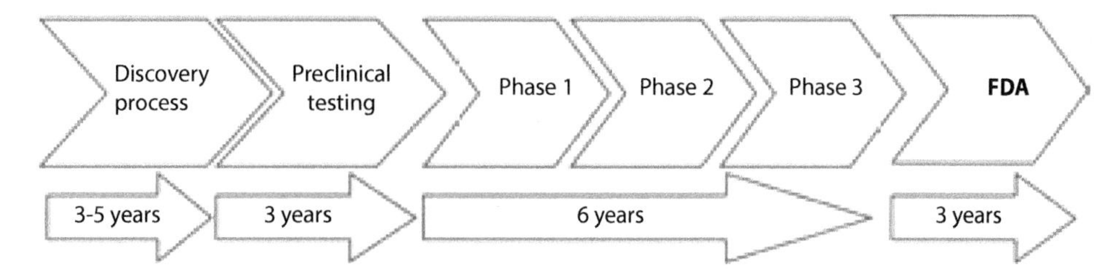

FIGURE 40.3 Typical duration of drug discovery, discovery process, preclinical testing, phases 1–3, and FDA approval.

the compound's action. In this particular study, the main concerns are the compound's absorption, distribution, metabolism, and excretion.

Phase 2 clinical trials move the drugs to testing more volunteers (~300 patients). These volunteers are known to have the targeted disease. It is in this phase in which the effectiveness of the drug is evaluated.

In the final phase of the clinical trials, the pharmaceutical company will rigorously analyze the data from the previous phases, especially concerning effectiveness and safety. The company will then apply for New Drug Application (NDA) with the FDA. For a minimum of 6 months, the FDA will evaluate the data of the clinical trials contained in the NDA.

If the process is smooth, the drug will be available for prescription; however, the FDA still requires periodic reports regarding the drug's safety. The process as presented, as we all know, is an ideal situation. Sometimes, in the clinical trials stage, the drug is found to have adverse side effects and the study has to be ceased. If the drug is found to have a serious adverse event the drug has to be recalled for safety issues.

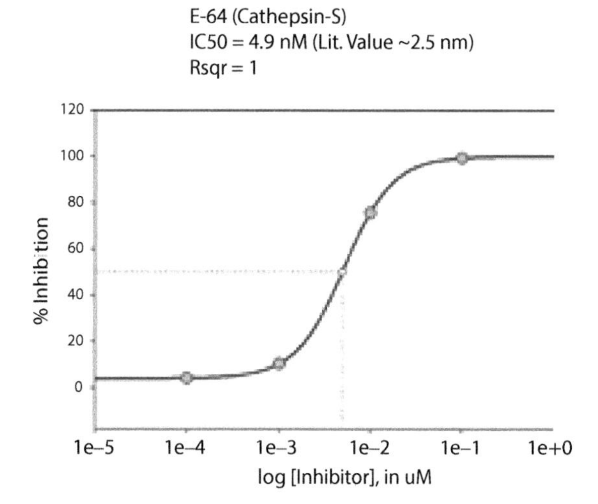

FIGURE 40.4 Inhibition of cathepsin S by E-64. The figure shows a dose-response curve for the inhibition of cathepsin S by the protease inhibitor E-64. The half-maximal inhibitory concentration (IC_{50} value), the concentration of the drug required for 50% of the maximal inhibition of the enzyme, is derived from the dose-response curve.

40.4.1 Early-Phase Drug Discovery

During the early stage of the drug discovery campaign, a computational group screens compounds virtually using simulation programs. Computational chemists begin by docking structures of potential inhibitors to the active site of the free enzyme using computer-modeling software (Ali et al., 2011; Cosconati et al., 2010; Mortier et al., 2012). In the absence of virtual screening, pharmaceutical companies screen for inhibitors from their library of compounds, or buy compounds from third-party suppliers to test against their target. These assays are referred to as in vitro assays, the term referring to biochemical and/or cell-based assays. In vitro assays require the test inhibitor compound to bind to the target protein, in solution, and inhibit the enzyme's activity. A typical in vitro assay is shown in Figure 40.4, quantifying the small-molecule's capability of inhibiting the enzyme (IC_{50}).

High-throughput assays can also be applied to the compound library, to generate "hit" compounds. As mentioned earlier, hit compounds are identified in the laboratory as structures that exhibit the desired biological activity. These are drug-like entities that possess inhibitory activities to the biological system that is being studied. Figure 40.5 shows a typical flow of a drug discovery project, although not all follow these flowcharts.

In this initial stage, rounds of testing of the compounds are performed to quantify binding affinity. Once binding activity is validated in vitro assays, lead optimization follows. Lead optimization is an iterative process. In this step, a series of compounds that are closely related (or structural analogues) are designed, synthesized, and assayed. The systematic variation of the structures leads to a series of structurally related compounds with their corresponding activity quantified. This is the basis of structure-activity relationship (SAR) analysis. The goal is to clearly understand the structural basis of the ligand's binding affinity to the target enzyme. Selectivity is also evaluated in addition to establishing the mechanism of inhibition. This is followed by biochemical and biological evaluation. Subsequent goals, such as optimization of membrane permeability, oral bioavailability, and pharmacokinetics, are important to continue to the next stage of the drug discovery effort but beyond the scope of this chapter.

Once the SARs are established, patterns may emerge that can be used to develop hypotheses with regards to the interaction of the inhibitors with the binding pocket of the target. In the next section, a discussion of calpain inhibitors will be presented, including the rationale of the drug designs and a summary of SAR findings from the literature.

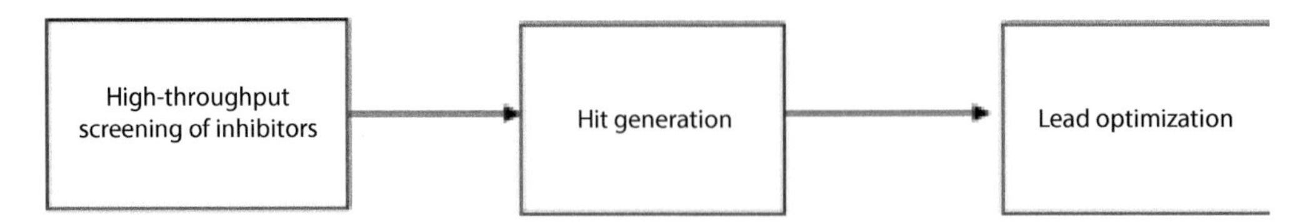

FIGURE 40.5 Typical flowchart of the early-stage drug discovery effort.

40.5 CALPAIN INHIBITORS

Since the discovery of calpain in the mid-1960s (Guroff, 1964; Huston and Krebs, 1968), scientists have sought methods to inhibit its activity. Early attempts used calcium chelators, but as knowledge of calpain substrate specificity grew, and inhibitors from natural products screening programs became available, more selective inhibitors were designed and synthesized. Most small-molecule calpain inhibitors reported are modified peptides containing reactive functional groups that interact with the cysteine thiol in the calpain active site. Most were obtained by incorporating calpain preferring P1 and P2 residues. Unfortunately, they lack selectivity for calpain; they also inhibit other cysteine proteases such as cathepsins B and L, and the proteasome, which leads to cellular apoptosis. Throughout the literature, low-molecular-weight calpain inhibitors were classified in various ways. Initially, these compounds were classified as *active-site directed* and *allosteric effectors*. However, detailed studies of the kinetics of inhibition revealed much more detailed categories such as *mechanism-based, tight binding, slow binding, affinity labels, suicide substrates, transition state analogs,* and *dead-end inhibitors* (Otto and Schirmeister, 1997; Vicik et al., 2006). Allosteric calpain inhibitors are known not to target the active site but most likely interact with other sites (allosteric sites) that are involved in catalysis and activation. An example of this type of compound is PD150606, which came out from the laboratory of Parke-Davis (Wang et al., 1996). The main focus however of this chapter is not on allosteric inhibitors but rather on active site–directed inhibitors that are drug-like compounds with a peptide or peptidomimetic backbone as its skeletal structure. In the next section, we will discuss the address regions of the active site and the theory of active site–directed inhibitors.

40.5.1 Active Site–Directed Inhibitors

As the name implies, these calpain inhibitors interact directly with the active site, unlike the allosteric effectors that target a site that is far away from the active center of the enzyme. These types of compounds are usually peptidomimetics of calpain substrates, composed of modified amino acids that are recognized by the enzyme. An electrophilic functional group called the *warhead* is added to the design to interact directly with the active site, specifically binding to the thiol of the cysteine residue. The warhead's bonding to the cysteine residue can be covalent/noncovalent and reversible/irreversible. To differentiate between reversible and irreversible inhibitors, one can perform an experiment of lowering the inhibitor concentration. A reversible inhibitor will exhibit an increase in the enzyme activity because the inhibitors are not attached to the active site anymore. The consensus is that noncovalent inhibition is also a reversible interaction, except for cases of covalently bonded inhibitors that are labile and can be easily detached by water.

Examples of reversible calpain inhibitors include warheads with the functional group of aldehydes, α-keto heterocycles, and α-ketocarbonyls. The α-ketocarbonyls can be subdivided into α-ketoacids, α-diketones, α-ketophosphorus (Tao et al., 1998), α-ketoamides (Angelastro et al., 1990), and α-ketoesters. The problem with aldehydes as the electrophilic warhead is their reactivity. Because of the highly reactive aldehyde warhead, researchers have been exploring the less electrophilic peptidyl α-ketoamides (Carragher, 2006). In addition to the synthetic reversible inhibitors listed previously, there are also examples that come from natural sources; these are calpastatin and leupeptin.

As mentioned earlier, irreversible inhibitors are referred to as inactivators; they bind to the site covalently and permanently. An example of this type of inhibitor is E-64 (epoxy succinyl) (Chatterjee et al., 2005), a compound that was isolated from *Aspergillus japonicas*. E-64 is a potent calpain inhibitor; however, one of the drawbacks that plagued E-64 is that it is not cell-permeable, thus forcing medicinal chemists to redesign the compound, resulting in numerous derivatives. By modifying E-64, E-64d was formed, a derivative that is cell-permeable. Once E-64d is inside the cells, it can convert to its active form (E-64c) in vivo.

In addition to epoxides, ketomethylenes, disulfide linkages, vinyl sulfones, and methylsulfonium salts as warheads are known functional groups that are irreversible calpain inhibitor warheads (Donkor, 2000). One may ask why it is important to understand the mode of inhibition. In the early stage of drug discovery, the value lies in quantitative comparison of binding affinity (K_i) of different compounds against the target. But it is not rational to compare the affinity of two compounds that differ in their inhibition modalities; it should be apples to apples or oranges to oranges. And in the long run, knowing the modalities could have clinical implications.

In general, however, the practice of differentiating between reversible or irreversible is difficult. In some cases, a reversible inhibitor binds very tightly to the target enzyme and only dissociates very slowly, thus showing qualities of being irreversible. These compounds are known as *tight-binding* inhibitors. Readers who wish to delve more into this topic are invited to read books on enzyme inhibition kinetics.

40.5.2 Active Site's Address

Although drug discovery campaigns now use high-throughput screens using the database of a pharmaceutical company, most of the drugs that are used in the clinic today function as competitive enzyme inhibitors discovered and optimized by the complementary approach of *mechanism-based* and *structure-based* drug design.

Mechanism-based drug design depends on the knowledge of the enzyme's catalytic reaction mechanism. Compounds designed in this way use structures that mimic the substrate, transition state, or product state of the enzyme as the initial structure. On the other hand, structure-based drug designs use the structural information of the enzyme's active site. If the X-ray structure of a particular enzyme is not available, homology models based on the structures of related proteins can be used. If available, X-ray crystal structures and nuclear magnetic resonance data can be used. Based on the detailed structure (atomic resolution) of the active site, small molecules with steric and electronic features are designed to fit into the active site pocket. Fortunately, the X-ray structures of calpain and its active site have already been solved and are available in public domains. Drug designers have used this information to optimize lead compounds.

To facilitate protease inhibition research in general, a convention was proposed by Schecter and Berger (Abramowitz et al., 1967; Schechter and Berger, 1967). The convention designates the regions of the inhibitor as P_1, P_2, etc., and the corresponding regions in the active site of the enzyme as S_1, S_2, etc. These S designations are regions that refer to the pockets in the active site, the binding positions in the enzyme corresponding to that of the substrate, which are P. In other words, the S_1 is where the P_1 amino acid from the inhibitor is supposed to bind. Illustrated in Figure 40.6, the address regions of a substrate or the inhibitor are subdivided into P_1 to P_3, and the position to the right after the active site is designated as P_1', P_2', etc. The same thing can be said for the corresponding regions of the active site (e.g., S_1', S_2', S_3').

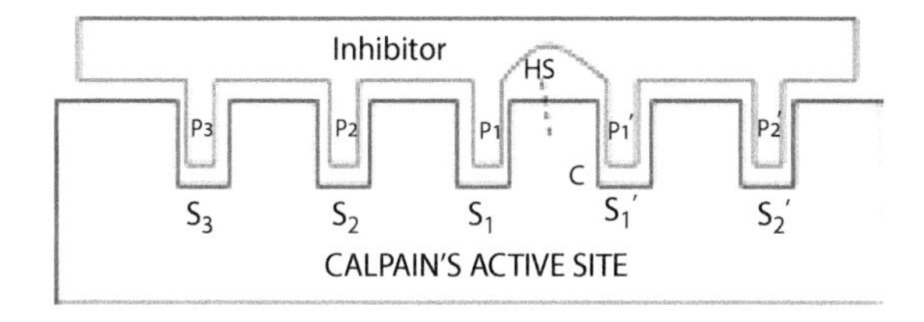

FIGURE 40.6 A depiction of the address regions of the enzyme's active site.

Figure 40.7 shows a depiction of the address regions of the enzyme's active site. The pocket where the inhibitor or substrate can bind can be above, below, or on the side (not depicted in the illustration) of the ligand. The scissile bond of the substrate is right after P_1, a bond that is acted upon by the enzyme.

What may help is a real example of a calpain inhibitor with its designated addresses. This may be well illustrated in an example of a real calpain inhibitor. Shown in Figure 40.8 is calpain inhibitor I with its corresponding addresses in colored bonds.

The aldehyde warhead of calpain inhibitor I was shown previously. Warheads are designed to interact directly with the active site, and changing the warhead into other different functional group changes the modality of interaction between the thiol group of the active site and the warhead (reversible or irreversible). Some of the known warheads were summarized earlier in the text. The last region that needs a little bit of attention is the P_3 position, known also as the N-terminal capping. Although it is an additional element for the enzyme to recognize, its primary job is to protect the inhibitor from degradation in vivo. As we explore SAR studies of calpain inhibitors in the next section, we will see later on that there is more to S_3 than just being a cap.

40.6 COUNTERSCREENS

Before we go on exploring the SAR studies as it relates to potency of drug candidates in inhibiting calpain, it is worthwhile to emphasize that potency is not the only criterion for drugs to advance to the next phase of drug discovery. Potency against calpain is important, but equally important is its selectivity to other structurally or mechanistically related enzymes. The aim of setting up counterscreen assays is to make sure that the inhibitor is only inhibiting calpain and not related enzymes.

Because the active site of proteases in general evolved to interact with the peptide backbone structure, the task of finding a selective inhibitor remains a challenge for medicinal

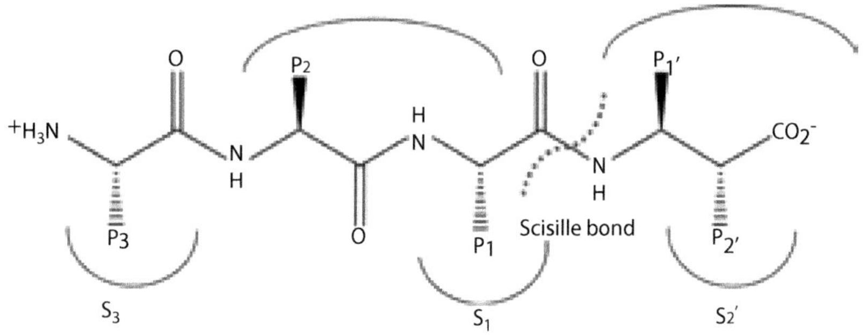

FIGURE 40.7 Regions of the calpain's active site pocket and the corresponding substrate/inhibitor regions. Also shown is the amide scissile bond of the substrate.

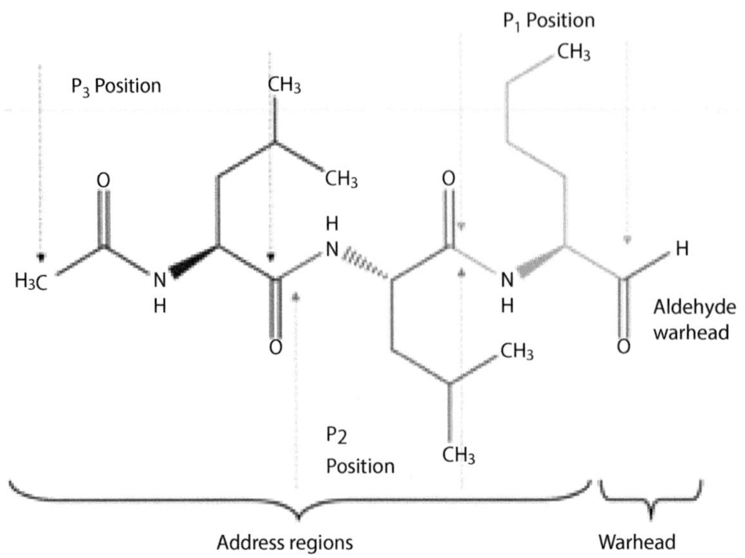

FIGURE 40.8 Calpain inhibitor I with its corresponding address regions and depicting an aldehyde warhead.

chemists. As mentioned previously, the active site of calpain is composed of the catalytic triad Cys, His, and Asn, and this catalytic triad is not unique to calpain. This is also a characteristic of cysteine proteases such as papain, and cathepsins B, L, S, and K. The sequence homology of these proteases is not the same however, suggesting that the ancestral genes that they evolved from were different (Goll et al., 2003). Nevertheless, in vitro assays should be set up also to test these similar enzymes, to avoid adverse side effects in the future. In subsequent cell assays of drug candidates, compounds that inhibit calpain and at the same time proteasome are known to cause the cell to die by apoptosis. In our laboratory, we have set up assays for cathepsin B, L, K, S, MMP-2, MMP-9, trypsin, chymotrypsin, and proteasome. The ratio of the K_i of a counterscreen enzyme over the K_i of calpain is usually the measure of the fold-selectivity. It is important and a requirement in our laboratory to have more than 30 times fold-selectivity.

40.6.1 Structure–Activity Relationship of Current Calpain Inhibitors

Based on previous research on small-molecule protease inhibitors, a considerable amount of information has been obtained and used as requirements for designing inhibitors of typical calpains.

40.6.1.1 P_1 Position

Although it is known that the S_1 pocket of the calpain active site can accept a wide variety of amino acids (Donkor, 2000), current SAR studies have revealed that amino acids with side chains that are capable of hydrogen bonding are less well-tolerated in this position. Examples of these are Ser, Thr, Gln, and Tyr.

On the other hand, notably good potency was achieved when drug designers incorporated aliphatic and aromatic amino acids at the P1 position of calpain inhibitors (Iqbal, 1997). Aliphatic substituents that are well-tolerated include Abu, Cha, Leu, Nle, Met, and Val, whereas the aromatic residues include Tyr(O-Benzyl) and Phe. Furthermore, a comparison between aromatic and aliphatic amino acids at the P1 position revealed that aliphatics are superior in terms of potency. Improvements in potency are achieved however when the P_1 residue is replaced with relatively hydrophobic groups.

40.6.1.2 P_2 Position

What we know about the residues preferred at the P_2 position of inhibitors is that the S_2 subsite of calpain has a preference for aliphatic side chains and, of course, amino acid side chains that are capable of hydrogen bonding reduce the binding and/or inhibitory potency. It follows also that acidic and basic amino acids do not increase potency and are even detrimental, based on previous SAR studies found in the literature (Donkor, 2000; Iqbal, 1997).

40.6.1.3 P_3 Position

This position is also known as the N-terminal capping, a region in the inhibitor in which the drug designers usually tweak to improve and optimize the selectivity and cell permeability of their lead compounds. SAR studies show that bulky groups can be positioned in this region too. Known N-terminal capping substituents are numerous and are not limited to the list here: pyridine ethanol, naphthol, thianonaptalene, fluorine, xanthines, benzyloxycarbonyl, substituted benzoyl, acetyl, aryl, alkyl, and sulfonyl (Chatterjee et al., 1998; Iqbal, 1997). Figure 40.9 is a pictorial representation of all the available SAR data gathered from the literature.

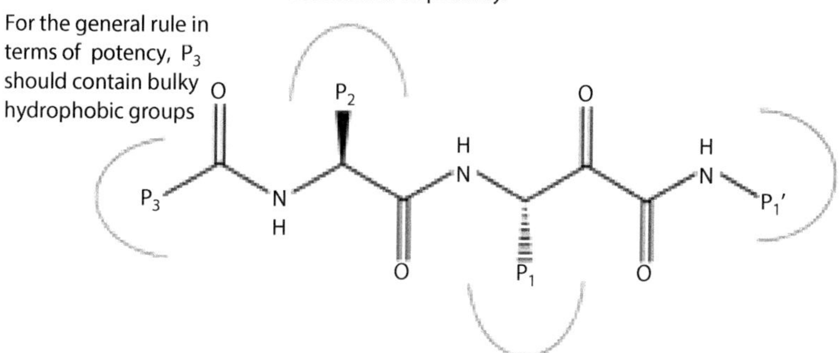

FIGURE 40.9 Visual summary of SAR studies of calpain inhibitors.

40.7 CHALLENGES IN MAKING CALPAIN INHIBITOR TO THE BRAIN

The so-called first generation of calpain inhibitors was inundated with problems such as permeability into the cells and selectivity. Early constructs of calpain inhibitors were non–cell-permeable and were compromised when these inhibitors react with closely related enzymes.

Technological advances made most of the protease inhibitors cell-permeable and in the area of calpain inhibitors, new inhibitors coming out from the patent office show several folds of selectivity to calpain. However, the challenge of making more selective compounds is still one of the daunting tasks drug designers focused on. After the early discovery phase, as with any other drug, toxicity is the next criterion that needs to be addressed by the drug designer.

However, granted that a safe drug is available, one formidable challenge specific to the brain as a target is how to deliver these drugs to the brain. Though this problem is not fully addressed, novel delivery models such as nasal sprays and encapsulating drugs in nanoparticles are some of the new routes researchers are using.

40.8 CONCLUSION

There is a clinical need to treat patients of TBI. Several inhibitors of calpain coming from the industry and academia are proof that calpains can be inhibited. Small-molecule inhibition of calpain has also been demonstrated in animal models. Medicinal chemists are still in the process of making more potent and selective calpain inhibitors, making the structures explored more diverse and numerous. There are plenty of structures that are disclosed in the literature and patent offices, from active site–directed peptidomimetics and nonpeptidomimetics, to novel calpain inhibitors that do not target the active site (allosteric site). This increase can be attributed also to the availability of more refined X-ray crystal structures of the enzyme that are available for free in public databases, such as the Protein Data Bank.

REFERENCES

Abramowitz, N., I. Schechter, and A. Berger. 1967. On the size of the active site in proteases. II. Carboxypeptidase-A. *Biochemical and Biophysical Research Communications*. 29:862–867.

Ali, H.I., T. Nagamatsu, and E. Akaho. 2011. Structure-based drug design and AutoDock study of potential protein tyrosine kinase inhibitors. *Bioinformation*. 5:368–374.

Alvira, D., I. Ferrer, J. Gutierrez-Cuesta, B. Garcia-Castro, M. Pallas, and A. Camins. 2008. Activation of the calpain/cdk5/p25 pathway in the girus cinguli in Parkinson's disease. *Parkinsonism & Related Disorders*. 14:309–313.

Anagli, J., K.W. Wang, Y. Ono, H. Sorimachi. 2013. Calpains in health and disease. In *Proteases: Structure and Function*, K. Brix and W. Stöcker eds., 395–431. Springer-Verlag, Vienna DOI 10.1007/978-3-7091-0885-7.

Angelastro, M.R., S. Mehdi, J.P. Burkhart, N.P. Peet, and P. Bey. 1990. Alpha-diketone and alpha-keto ester derivatives of N-protected amino acids and peptides as novel inhibitors of cysteine and serine proteinases. *Journal of Medicinal Chemistry*. 33:11–13.

Arthur, J.S., S. Gauthier, and J.S. Elce. 1995. Active site residues in m-calpain: Identification by site-directed mutagenesis. *FEBS Letters*. 368:397–400.

Betts, R., and J. Anagli. 2004. The beta- and gamma-CH2 of B27-WT's Leu11 and Ile18 side chains play a direct role in calpain inhibition. *Biochemistry*. 43:2596–2604.

Betts, R., S. Weinsheimer, G.E. Blouse, and J. Anagli. 2003. Structural determinants of the calpain inhibitory activity of calpastatin peptide B27-WT. *The Journal of Biological Chemistry*. 278:7800–7809.

Bizat, N., J.M. Hermel, F. Boyer, C. Jacquard, C. Creminon, S. Ouary et al. 2003. Calpain is a major cell death effector in selective striatal degeneration induced in vivo by 3-nitropropionate: Implications for Huntington's disease. *The Journal of Neuroscience*. 23:5020–5030.

Blomgren, K., U. Hallin, A.L. Andersson, M. Puka-Sundvall, B.A. Bahr, A. McRae et al. 1999. Calpastatin is up-regulated in response to hypoxia and is a suicide substrate to calpain after neonatal cerebral hypoxia-ischemia. *Journal of Biological Chemistry*. 274:14046–14052.

Busch, W.A., M.H. Stromer, D.E. Goll, and A. Suzuki. 1972. Ca 2+-specific removal of Z lines from rabbit skeletal muscle. *The Journal of Cell Biology*. 52:367–381.

Carragher, N.O. 2006. Calpain inhibition: A therapeutic strategy targeting multiple disease states. *Current Pharmaceutical Design*. 12:615–638.

Chatterjee, P.K., Z. Todorovic, A. Sivarajah, H. Mota-Filipe, P.A. Brown, K.N. Stewart et al. 2005. Inhibitors of calpain activation (PD150606 and E-64) and renal ischemia-reperfusion injury. *Biochemical Pharmacology*. 69:1121–1131.

Chatterjee, S., M. Iqbal, S. Mallya, S.E. Senadhi, T.M. O'Kane, B.A. McKenna et al. 1998. Exploration of the importance of the P2-P3-NHCO-moiety in a potent di- or tripeptide inhibitor of calpain I: Insights into the development of nonpeptidic inhibitors of calpain I. *Bioorganic and Medicinal Chemistry*. 6:509–522.

Cosconati, S., S. Forli, A.L. Perryman, R. Harris, D.S. Goodsell, and A.J. Olson. 2010. Virtual screening with AutoDock: Theory and practice. *Expert Opinion on Drug Discovery*. 5:597–607.

Crawford, C., N.R. Brown, and A.C. Willis. 1993. Studies of the active site of m-calpain and the interaction with calpastatin. *The Biochemical Journal*. 296 (Pt 1):135–142.

Croall, D.E., and G.N. DeMartino. 1991. Calcium-activated neutral protease (calpain) system: Structure, function, and regulation. *Physiological Reviews*. 71:813–847.

Cuerrier, D., T. Moldoveanu, R.L. Campbell, J. Kelly, B. Yoruk, S.H. Verhelst et al. 2007. Development of calpain-specific inactivators by screening of positional scanning epoxide libraries. *The Journal of Biological Chemistry*. 282:9600–9611.

Cuerrier, D., T. Moldoveanu, J. Inoue, P.L. Davies, and R.L. Campbell. 2006. Calpain inhibition by alpha-ketoamide and cyclic hemiacetal inhibitors revealed by X-ray crystallography. *Biochemistry*. 45:7446–7452.

Das, A., M.K. Guyton, J.T. Butler, S.K. Ray, and N.L. Banik. 2008. Activation of calpain and caspase pathways in demyelination and neurodegeneration in animal model of multiple sclerosis. *CNS and Neurological Disorders Drug Targets*. 7:313–320.

Davis, T.L., J.R. Walker, P.J. Finerty, Jr., F. Mackenzie, E.M. Newman, and S. Dhe-Paganon. 2007. The crystal structures of human calpains 1 and 9 imply diverse mechanisms of action and auto-inhibition. *Journal of Molecular Biology*. 366:216–229.

Deacon, C.F., and J.J. Holst. 2006. Dipeptidyl peptidase IV inhibitors: A promising new therapeutic approach for the management of type 2 diabetes. *The International Journal of Biochemistry and Cell Biology*. 38:831–844.

Di Rosa, G., T. Odrijin, R.A. Nixon, and O. Arancio. 2002. Calpain inhibitors: A treatment for Alzheimer's disease. *Journal of Molecular Neuroscience*. 19:135–141.

Donkor, I.O. 2000. A survey of calpain inhibitors. *Current Medicinal Chemistry*. 7:1171–1188.

Falgueyret, J.P., S. Desmarais, R. Oballa, W.C. Black, W. Cromlish, K. Khougaz et al. 2005. Lysosomotropism of basic cathepsin K inhibitors contributes to increased cellular potencies against off-target cathepsins and reduced functional selectivity. *Journal of Medicinal Chemistry*. 48:7535–7543.

Gafni, J., and L.M. Ellerby. 2002. Calpain activation in Huntington's disease. *The Journal of Neuroscience*. 22:4842–4849.

Garcia-Bonilla, L., J. Burda, D. Pineiro, I. Ayuso, M. Gomez-Calcerrada, and M. Salinas. 2006. Calpain-induced proteolysis after transient global cerebral ischemia and ischemic tolerance in a rat model. *Neurochemical Research*. 31:1433–1441.

Garg, S., T. Timm, E.M. Mandelkow, E. Mandelkow, and Y. Wang. 2011. Cleavage of Tau by calpain in Alzheimer's disease: The quest for the toxic 17 kD fragment. *Neurobiology of Aging*. 32:1–14.

Getz, G.S. 2012. Calpain inhibition as a potential treatment of Alzheimer's disease. *The American Journal of Pathology*. 181:388–391.

Gladding, C.M., M.D. Sepers, J. Xu, L.Y. Zhang, A.J. Milnerwood, P.J. Lombroso et al. 2012. Calpain and STriatal-Enriched protein tyrosine phosphatase (STEP) activation contribute to extrasynaptic NMDA receptor localization in a Huntington's disease mouse model. *Human Molecular Genetics*. 21:3739–3752.

Goll, D.E., V.F. Thompson, H. Li, W. Wei, and J. Cong. 2003. The calpain system. *Physiological Reviews*. 83:731–801.

Gray, B.C., P. Skipp, V.M. O'Connor, and V.H. Perry. 2006. Increased expression of glial fibrillary acidic protein fragments and mu-calpain activation within the hippocampus of prion-infected mice. *Biochemical Society Transactions*. 34:51–54.

Grynspan, F., W.R. Griffin, A. Cataldo, S. Katayama, and R.A. Nixon. 1997. Active site-directed antibodies identify calpain II as an early-appearing and pervasive component of neurofibrillary pathology in Alzheimer's disease. *Brain Research*. 763:145–158.

Guroff, G. 1964. A neutral, calcium-activated proteinase from the soluble fraction of rat brain. *The Journal of Biological Chemistry*. 239:149–155.

Gutierrez, M., M. Alonso de Lecinana, M. Salinas, J. Masjuan, I. Ayuso, and E. Diez-Tejedor. 2007. Calpain activation plays a role in the pathophysiology of focal cerebral ischaemia and determines loss of tissue viability at the ischaemic penumbra. *Journal of Neurolgy*. 254:169–169.

Hachiya, N., Y. Komata, S. Harguem, K. Nishijima, and K. Kaneko. 2011. Possible involvement of calpain-like activity in normal processing of cellular prion protein. *Neuroscience Letters*. 490:150–155.

Hong, S.C., Y. Goto, G. Lanzino, S. Soleau, N.F. Kassell, and K.S. Lee. 1994. Neuroprotection with a calpain inhibitor in a model of focal cerebral-ischemia. *Stroke*. 25:663–669.

Huh, J.W., A.G. Widing, and R. Raghupathi. 2007. Basic science; repetitive mild non-contusive brain trauma in immature rats exacerbates traumatic axonal injury and axonal calpain activation: A preliminary report. *Journal of Neurotrauma*. 24:15–27.

Huston, R.B., and E.G. Krebs. 1968. Activation of skeletal muscle phosphorylase kinase by Ca2+. II. Identification of the kinase activating factor as a proteolytic enzyme. *Biochemistry*. 7:2116–2122.

Iqbal, M., Messina, P., Freed, B., and Das, M.,. 1997. Subsite requirements for peptide aldehyde inhibitors of human calpain 1. *Bioorganic and Medicinal Chemistry Letters*. 7:6.

Ishiura, S., H. Murofushi, K. Suzuki, and K. Imahori. 1978. Studies of a calcium-activated neutral protease from chicken skeletal muscle. I. Purification and characterization. *Journal of Biochemistry*. 84:225–230.

Iwamoto, N., W. Thangnipon, C. Crawford, and P.C. Emson. 1991. Localization of calpain immunoreactivity in senile plaques and in neurones undergoing neurofibrillary degeneration in Alzheimer's disease. *Brain Research*. 561:177–180.

Johnson, M.P., C.G. Markgraf, N.L. Velayo, D.R. McCarty, P.A. Chmielewski, J.R. Koehl et al. 1997. Calpain inhibition protects neurons when therapy is delayed for 6 hours after focal cerebral ischemia. *Stroke*. 28:115–115.

Kambe, A., M. Yokota, T.C. Saido, I. Satokata, H. Fujikawa, S. Tabuchi et al. 2005. Spatial resolution of calpain-catalyzed proteolysis in focal cerebral ischemia. *Brain Research*. 1040:36–43.

Kampfl, A., R. Posmantur, R. Nixon, F. Grynspan, X. Zhao, S.J. Liu et al. 1996. mu-calpain activation and calpain-mediated cytoskeletal proteolysis following traumatic brain injury. *Journal of Neurochemistry*. 67:1575–1583.

Kawasaki, H., Y. Emori, and K. Suzuki. 1993. Calpastatin has two distinct sites for interaction with calpain—effect of calpastatin fragments on the binding of calpain to membranes. *Archives of Biochemistry and Biophysics*. 305:467–472.

Kenessey, A., M. Banay-Schwartz, T. DeGuzman, and A. Lajtha. 1990. Calpain II activity and calpastatin content in brain regions of 3- and 24-month-old rats. *Neurochemical Research*. 15:243–249.

Li, Q., R.P. Hanzlik, R.F. Weaver, and E. Schonbrunn. 2006. Molecular mode of action of a covalently inhibiting peptidomimetic on the human calpain protease core. *Biochemistry*. 45:701–708.

Liebetrau, M., B. Staufer, E.A. Auerswald, D. Gabrijelcic-Geiger, H. Fritz, C. Zimmermann et al. 1999. Increased intracellular calpain detection in experimental focal cerebral ischemia. *Neuroreport*. 10:529–534.

Liu, M.C., V. Akle, W. Zheng, J. Kitlen, B. O'Steen, S.F. Larner et al. 2006. Extensive degradation of myelin basic protein isoforms by calpain following traumatic brain injury. *Journal of Neurochemistry*. 98:700–712.

Medeiros, R., M. Kitazawa, M.A. Chabrier, D. Cheng, D. Baglietto-Vargas, A. Kling et al. 2012. Calpain inhibitor A-705253 mitigates Alzheimer's disease-like pathology and cognitive decline in aged 3xTgAD mice. *The American Journal of Pathology*. 181:616–625.

Minger, S.L., J.W. Geddes, M.L. Holtz, S.D. Craddock, S.W. Whiteheart, R.G. Siman et al. 1998. Glutamate receptor antagonists inhibit calpain-mediated cytoskeletal proteolysis in focal cerebral ischemia. *Brain Research*. 810:181–199.

Moldoveanu, T., R.L. Campbell, D. Cuerrier, and P.L. Davies. 2004. Crystal structures of calpain-E64 and -leupeptin inhibitor complexes reveal mobile loops gating the active site. *Journal of Molecular Biology*. 343:1313–1326.

Moldoveanu, T., C.M. Hosfield, D. Lim, Z. Jia, and P.L. Davies. 2003. Calpain silencing by a reversible intrinsic mechanism. *Nature Structural Biology*. 10:371–378.

Mortier, J., C. Rakers, R. Frederick, and G. Wolber. 2012. Computational tools for in silico fragment-based drug design. *Current Topics in Medicinal Chemistry*. 12:1935–1943.

Mouatt-Prigent, A., J.O. Karlsson, Y. Agid, and E.C. Hirsch. 1996. Increased M-calpain expression in the mesencephalon of patients with Parkinson's disease but not in other neurodegenerative disorders involving the mesencephalon: A role in nerve cell death? *Neuroscience*. 73:979–987.

Murachi, T. 1984. Calcium-dependent proteinases and specific inhibitors: Calpain and calpastatin. *Biochemical Society Symposium*. 49:149–167.

Murachi, T., K. Tanaka, M. Hatanaka, and T. Murakami. 1980. Intracellular Ca2+-dependent protease (calpain) and its high-molecular-weight endogenous inhibitor (calpastatin). *Advances in Enzyme Regulation*. 19:407–424.

NC-IUB. 1981. Nomenclature Committee of the International Union of Biochemistry. Enzyme nomenclature. Recommendations 1978. Supplement 2: Corrections and additions. *European Journal of Biochemistry/FEBS*. 116:423–435.

Neumar, R.W., S.M. Hagle, D.J. DeGracia, G.S. Krause, and B.C. White. 1996. Brain mu-calpain autolysis during global cerebral ischemia. *Journal of Neurochemistry*. 66:421–424.

Ohno, S., Y. Emori, S. Imajoh, H. Kawasaki, M. Kisaragi, and K. Suzuki. 1984. Evolutionary origin of a calcium-dependent protease by fusion of genes for a thiol protease and a calcium-binding protein? *Nature*. 312:566–570.

Ohno, S., S. Minoshima, J. Kudoh, R. Fukuyama, Y. Shimizu, S. Ohmi-Imajoh et al. 1990. Four genes for the calpain family locate on four distinct human chromosomes. *Cytogenetics and Cell Genetics*. 53:225–229.

Otto, H.H., and T. Schirmeister. 1997. Cysteine proteases and their inhibitors. *Chemical Reviews*. 97:133–172.

Pike, B.R., J. Flint, S. Dutta, E. Johnson, K.K. Wang, and R.L. Hayes. 2001. Accumulation of non-erythroid alpha II-spectrin and calpain-cleaved alpha II-spectrin breakdown products in cerebrospinal fluid after traumatic brain injury in rats. *Journal of Neurochemistry*. 78:1297–1306.

Pike, B.R., X. Zhao, J.K. Newcomb, R.M. Posmantur, K.K. Wang, and R.L. Hayes. 1998. Regional calpain and caspase-3 proteolysis of alpha-spectrin after traumatic brain injury. *Neuroreport*. 9:2437–2442.

Posmantur, R., A. Kampfl, R. Siman, J. Liu, X. Zhao, G.L. Clifton, and R.L. Hayes. 1997. A calpain inhibitor attenuates cortical cytoskeletal protein loss after experimental traumatic brain injury in the rat. *Neuroscience*. 77:875–888.

Ray, S.K., G.G. Wilford, S.F. Ali, and N.L. Banik. 2000. Calpain upregulation in spinal cords of mice with 1-methyl-4-phenyl-1,2,3,6-tetrahydropyridine (MPTP)-induced Parkinson's disease. *Annals of the New York Academy of Sciences*. 914:275–283.

Samantaray, S., S.K. Ray, and N.L. Banik. 2008. Calpain as a potential therapeutic target in Parkinson's disease. *CNS and Neurological Disorders Drug Targets*. 7:305–312.

Schechter, I., and A. Berger. 1967. On the size of the active site in proteases. I. Papain. *Biochemical and Biophysical Research Communications*. 27:157–162.

Shields, D.C., K.E. Schaecher, T.C. Saido, and N.L. Banik. 1999. A putative mechanism of demyelination in multiple sclerosis by a proteolytic enzyme, calpain. *Proceedings of the National Academy of Sciences of the United States of America*. 96:11486–11491.

Stanton, A. 2003. Therapeutic potential of renin inhibitors in the management of cardiovascular disorders. *American Journal of Cardiovascular Drugs*. 3:389–394.

Tao, M., R. Bihovsky, G.J. Wells, and J.P. Mallamo. 1998. Novel peptidyl phosphorus derivatives as inhibitors of human calpain I. *Journal of Medicinal Chemistry*. 41:3912–3916.

Thompson, S.N., T.R. Gibson, B.M. Thompson, Y. Deng, and E.D. Hall. 2006. Relationship of calpain-mediated proteolysis to the expression of axonal and synaptic plasticity markers following traumatic brain injury in mice. *Experimental Neurology*. 201:253–265.

Tsubokawa, T., I. Solaroglu, H. Yatsushige, J. Cahill, K. Yata, and J.H. Zhang. 2006. Cathepsin and calpain inhibitor E64d attenuates matrix metalloproteinase-9 activity after focal cerebral ischemia in rats. *Stroke*. 37:1888–1894.

Tsuji, T., S. Shimohama, J. Kimura, and K. Shimizu. 1998. m-Calpain (calcium-activated neutral proteinase) in Alzheimer's disease brains. *Neuroscience Letters*. 248:109–112.

Upadhya, G.A., S.A. Topp, R.S. Hotchkiss, J. Anagli, and S.M. Strasberg. 2003. Effect of cold preservation on intracellular calcium concentration and calpain activity in rat sinusoidal endothelial cells. *Hepatology*. 37:313–323.

Vicik, R., M. Busemann, K. Baumann, and T. Schirmeister. 2006. Inhibitors of cysteine proteases. *Current Topics in Medicinal Chemistry*. 6:331–353.

Wang, K.K., R. Nath, A. Posner, K.J. Raser, M. Buroker-Kilgore, I. Hajimohammadreza et al. 1996. An alpha-mercaptoacrylic acid derivative is a selective nonpeptide cell-permeable calpain inhibitor and is neuroprotective. *Proceedings of the National Academy of Sciences of the United States of America*. 93:6687–6692.

Wang, X., F. Wang, M.S. Sy, and J. Ma. 2005. Calpain and other cytosolic proteases can contribute to the degradation of retrotranslocated prion protein in the cytosol. *The Journal of Biological Chemistry*. 280:317–325.

Warren, M.W., F.H. Kobeissy, M.C. Liu, R.L. Hayes, M.S. Gold, and K.K. Wang. 2005. Concurrent calpain and caspase-3 mediated proteolysis of alpha II-spectrin and tau in rat brain after methamphetamine exposure: A similar profile to traumatic brain injury. *Life Sciences*. 78:301–309.

Yadavalli, R., R.P. Guttmann, T. Seward, A.P. Centers, R.A. Williamson, and G.C. Telling. 2004. Calpain-dependent endoproteolytic cleavage of PrPSc modulates scrapie prion propagation. *The Journal of Biological Chemistry*. 279:21948–21956.

Zhang, Z., A.K. Ottens, S. Sadasivan, F.H. Kobeissy, T. Fang, R.L. Hayes et al. 2007. Calpain-mediated collapsin response mediator protein-1, -2, and -4 proteolysis after neurotoxic and traumatic brain injury. *Journal of Neurotrauma*. 24:460–472.

Zhao, X., J.K. Newcomb, B.R. Pike, and R.L. Hayes. 2000. Casein zymogram assessment of mu-calpain and m-calpain activity after traumatic brain injury in the rat in vivo. *Methods in Molecular Biology*. 144:117–120.

Zhao, X., J.K. Newcomb, R.M. Posmantur, K.K. Wang, B.R. Pike, and R.L. Hayes. 1998a. pH dependency of mu-calpain and m-calpain activity assayed by casein zymography following traumatic brain injury in the rat. *Neuroscience Letters*. 247:53–57.

Zhao, X., R. Posmantur, A. Kampfl, S.J. Liu, K.K. Wang, J.K. Newcomb et al. 1998b. Subcellular localization and duration of mu-calpain and m-calpain activity after traumatic brain injury in the rat: A casein zymography study. *Journal of Cerebral Blood Flow and Metabolism*. 18:161–167.

41 Nanoparticles for Neurotherapeutic Drug Delivery in Neurodegenerative Disorders

Application in Neurotrauma

Mark S. Kindy and Alexey Vertegel

CONTENTS

41.1 INTRODUCTION

Stroke is among the top three leading causes of death and disability in the United States and in the world. In the United States, approximately 700,000 strokes lead to 165,000 deaths every year. Mortality for survivors is high, with as low as 50% 5-year survival in some studies. All surviving stroke patients suffer from at least some degree of disability. Prognosis for the survivors largely depends on the severity of the brain injury. In the long term, this injury is caused by both the primary stroke event and secondary neuronal death; the latter is produced by a series of cellular and molecular events initiated by the primary trauma. While it is practically impossible to reduce the damage from the primary stroke event after it has happened, factors responsible for secondary damage can be alleviated. However, current therapeutic approaches focus primarily on prevention, with limited or no treatment options currently available for patients after the stroke onset. Thus, developing a new therapy to address secondary neuronal death in stroke is highly significant as it is expected to improve the long term mortality and reduce the degree of disability for large number of stroke patients. This chapter will address the impact of oxidative stress in stroke and the potential therapeutic approach of using nanoparticles containing antioxidants to protect that brain from the secondary effects.

41.1.1 STROKE AND OXIDANT STRESS

Stroke is one of the leading causes of disability and death in the United States (American Heart Association and Heart Disease and Stroke Statistics, 2012). The outcome and infarction size after focal cerebral ischemia are determined by both "necrotic" cell death and by delayed neuronal cell loss in the border zone of ischemia (programmed cell death or apoptosis) (Ekshyyan and Aw, 2004). Recent therapies have emerged to treat ischemic stroke; however, these treatments do not address neuroprotection, reduction of behavioral deficit, or brain infarct volume once the neuronal cell death cycle has been triggered, and mostly deal with dissolving the blood clot (Grupper et al., 2007). Past and current neuroprotective strategies have been successful in animal models but have failed significantly in clinical trials as such there is a need for better therapeutics (Arumugam et al., 2008; Young et al., 2007). Understanding the basic mechanisms that influence cell loss helps to design drugs that are targeted to reduce cell death associated with ischemic injury and improve functional outcome (Leker and Shohami, 2002; Watt, 1962). Cells in the ischemic penumbra are subject to various pathological processes that can lead to their own death and to risk the survival of their neighbors. These death-promoting mechanisms are shared by both ischemic and traumatic brain injury and include oxidative stress, excitotoxicity, and inflammation (Poeckel et al., 2006).

Stroke is a complex disorder precipitated by genetic and environmental factors. The clinical variability of stroke, mainly in terms of duration, localization, and severity of ischemia as well as the patients' age and coexisting systemic diseases, poses huge challenges for clinical research, because very large patient groups are needed to avoid the confounding effects of disease diversity. Although experimental longitudinal clinical studies are of major importance for understanding stroke, having provided strong evidence for the existence of genetic and environmental influence factors, they are seldom helpful in understanding disease mechanisms and their results can be misleading (Hachinski, 2005). On the other hand, in vitro systems alone cannot thoroughly evaluate stroke and its consequences, given the importance of the brain vasculature to study the effects of abnormal brain perfusion and given the involvement of other systems as well (Graham et al., 2004). For these reasons, animal models are required to reproduce specific aspects of human disease, to understand the complex pathophysiology of stroke, and to find efficacious preventive and therapeutic approaches (Carmichael, 2005; Hainsworth and Markus, 2008).

Studies over the years have indicated that various risk factors contribute to the pathogenic events that culminate in stroke (American Heart Association and Heart Disease and Stroke Statistics, 2012; Roger et al., 2012). Recent studies have shown that stroke is the result of multiple risk factors (Osmond et al., 2009). Inflammation has been hypothesized as a critical factor in the contribution to and the generalized outcomes associated with stroke (Auriel et al., 2014; Chiba and Umegaki, 2013; Koellhoffer and McCullough, 2014; Miao and Liao, 2014; van Rooy and Pretorius, 2014). Oxidative stress is a key factor in the pathogenesis of ischemic/stroke injury and the subsequent secondary injury that persists.

Oxidative stress arises as a result of an imbalance between the prooxidant and antioxidant entities and consequent overproduction of reactive oxygen species. Reactive oxygen species (ROS) are biphasic, because they participate in normal physiological cellular processes and are also involved in a number of disease processes, whereby they facilitate impairment to cellular structures, comprising lipids, membranes, proteins, RNA, and DNA (Allen and Bayraktutan, 2009). The cerebral vasculature and parenchyma are major targets of oxidative stress participating in a critical role culminating in the pathogenesis of ischemic brain injury after a cerebrovascular attack. Superoxide, the primary ROS, and its derivatives are implicated in vasodilatation via altered vascular reactivity, breakdown of the blood–brain barrier (BBB) and lesions in animal models of ischemic stroke. Therefore, understanding the role of ROS and targeting for therapeutic intervention is an important aspect of recovery for ischemic stroke.

41.1.2 Nanoparticles for Targeting Neurological Disorders

Recently, nanoparticles (NPs) and liposomes attracted much attention as potentially efficient systems for targeted drug delivery to a desired location. Applications of nanotechnology in basic and clinical neuroscience, however, are only in the early stages of development. This is in part because of some of the unique challenges imposed by central nervous system (CNS) such as restricted and difficult anatomical access, an extremely heterogeneous cellular and molecular environment, and complexity of interaction between these components. Despite these challenges, the potential benefits of nanotechnology for the treatment of both peripheral and CNS disorders are tremendous and may eventually offer the patient and clinician novel therapeutic choices that simply do not exist now (Silva, 2006).

One of significant achievements was demonstration that functionalized nanoparticles can penetrate through the BBB and be used for drug and gene delivery to CNS (Begley, 2004). To achieve this, various materials and approaches are being investigated. Oligonucleotides have been delivered to CNS in gels of cross-linked poly(ethylene glycol) and polyethylenimine (Vinogradov et al., 2004). Several cell-penetrating peptides, including transferrin and transactivating-transduction peptide, were found to be capable of penetrating BBB and their attachment to the surface of liposomes and nanoparticles was used to facilitate internalization of these nanostructures by the CNS (Lewin et al., 2000; Torchilin et al., 2000). Poly(butyl)cyanoacrylate (PBCA) nanoparticles have also been used to deliver drugs to the CNS with a good degree of success (Kreuter et al., 2002). PBCA nanoparticles were loaded with drug by incorporating the drug during the polymerization process. The particles were then coated with Tween 80. Upon intravenous injection, the surface of PBCA particles becomes further coated with adsorbed plasma proteins, especially apolipoprotein E. It is believed that this final product is mistaken for low-density lipoprotein particles by the cerebral endothelium and is internalized by the low-density lipoprotein uptake system (Kurakhmaeva et al., 2009). Other surfactants are active in producing protein binding by the nanoparticles, but only Tween 80 preferentially absorbs apolipoprotein E. Enzymes and proteins, including β-galactosidase, rhoG, growth factors, and mouse anti–α-synuclein monoclonal antibody H3C adsorbed onto PBCA NPs were also shown to be delivered to neurons in cell cultures without loss of function (Hasadsri et al., 2009; Kurakhmaeva et al., 2009).

In conclusion, current research on targeted delivery to CNS is primarily focused on penetration through the BBB. Although BBB penetration is essential, in stroke it can be assisted to some extent by the biphasic opening during cerebral ischemia and reperfusion injury (Kuroiwa et al., 1985; Spatz, 2010). Further improvement of efficacy can be achieved by better retention of the antioxidant agent at the injury site. However, much less attention has been paid to specific targeting to the injured sites of the brain. Thus, there is an apparent need for development of approaches, which would allow retention of the desired therapeutic agent at the injury site rather than a uniform distribution of the drug within the brain.

41.1.3 Impact of Antioxidant Nanoparticles in Stroke

41.1.3.1 Nanomaterials in Neuroprotection

Applications of nanotechnology for neuroprotection have focused on limiting the damaging effects of free radicals generated after injury. Carboxyfullerenes have been shown to have antioxidant properties. They function as free radical scavengers, which can lead to a reduction in the cell damage in secondary injury (Dugan et al., 2001). Other fullerene derivatives, polyhydroxyfullerenes, have been shown to reduce glutamate-induced neuronal apoptosis (Jin et al., 2000). However, poor biodegradability of fullerenol derivatives and possibility of their accumulation in CNS is a concern. Nanocrystalline ceria has also been shown to be neuroprotective in both cell culture and in vivo models (Estevez et al., 2011; Kim et al., 2012). However, literature data on their antioxidant efficacy are controversial: although some researchers suggest that they are neuroprotective because of their ability to degrade free radicals, others observed prooxidant long-term effect of ceria NPs in brain in a rat model (Das et al., 2007; Hardas et al., 2012; Kim et al., 2012). Although ceria is biocompatible and is currently used in a clinical trial on treatment of glaucoma, its lack of biodegradability would in the long term be a safety concern for treatment of stroke (Hardas et al., 2010). Also, ceria NPs were found to induce adverse effects in liver because of oxidative stress (Tseng et al., 2012). This could be another concern for clinical applications of nanocrystalline ceria because majority of systemically administered ceria NPs, similarly to most other NPs, will end up in the liver and will reside there for a long time because of a lack of biodegradability or elimination pathways.

41.1.3.2 Superoxide Dismutase and its Mimetics in Stroke Treatment

Biological free radical scavengers, such as superoxide dismutase (SOD) or catalase, as well as their mimetics, could be superior to other agents because of their extremely high ROS scavenging ability. Both free SOD and its mimetics were tested as neuroprotectants in rodent stroke models (Margaill et al., 2005). In the case of free enzymes, lack of BBB penetration, and quick degradation in the bloodstream (half-life of ~7 minutes) required continuous intravenous infusion at high doses (1,200 U per animal over 3 hours) to achieve efficacy in a rat ischemia/reperfusion model (Francis et al., 1997). SOD mimetics, being low-molecular-weight compounds, degrade much slower than SOD and show better penetration through BBB. Several SOD mimetics including AEOL 10150, AEOL 10113, EUK-134, and M40401 were found to be efficient in reducing infarct volume in rodent models of ischemic stroke (Baker et al., 1998; Mackensen et al., 2001; Sheng et al., 2002; Shimizu et al., 2003). However, high doses and, in many cases, continuous infusion were still required to achieve this efficacy, and neurological side effects were noticed in both mouse and rat models. For example, achievement of a long-term therapeutic effect at 8 weeks, as reflected by improved neurological scores, required intracerebroventricular continuous infusion of AEOL 10150 (Sheng et al., 2009). Limitations of therapeutic efficacy of free SOD mimetics were linked to their limited ability to penetrate BBB and poor availability at the site of injury (Sheng et al., 2012). More lipophilic analog was therefore developed and tested in a rat ischemia/reperfusion model, and was found to be efficient a low doses (Rajic et al., 2012). However, high toxicity of this compound required development of novel, less toxic, lipophilic SOD mimetics, which are yet to be tested in a stroke model (Rajic et al., 2012). Although search for new more potent and less toxic compounds in an important research direction, an alternative approach could be to improve brain delivery of the existing antioxidants.

The required dose could be reduced considerably if prolonged release of SOD or SOD mimetic in brain were achieved. Towards this end, SOD-loaded nanoparticles have been studied by several groups. Reddy et al. used poly(lactide-co-glycolide) (PLGA) nanoparticles loaded with SOD to protect human neurons against action of peroxide in cell culture (Reddy et al., 2008). Such nanoparticles showed neuroprotective efficacy in neuronal cell cultures. The response to treatment by SOD-loaded nanoparticles was found to be highly dose- and time-dependent. It should be noted that although demonstration of efficacy of SOD-loaded nanoparticles against the oxidation stress caused by peroxide is important and encouraging, studies of efficacy against much more toxic superoxide, a natural SOD substrate associated with secondary neuronal injury, would be of value for further development of this approach. More recent in vivo study by the same group evaluated neuroprotective efficacy of SOD-loaded PLGA nanoparticles in a rat focal cerebral ischemia-reperfusion injury model (Reddy et al., 2009). The authors studied three routes of administration: intravenous (tail vein), intrajugular, and internal carotid arterial routes. They found that only ~0.1% of nanoparticles were taken up by the brain in the case of intravenous and intrajugular administration and ~1.5% of nanoparticles were taken up by the brain in the case of administration via the carotid artery. The nanoparticles administered via the carotid arterial route improved the survival rate (75% vs. 0% in controls) and showed good neuroprotective efficacy in this in vivo model. Free enzyme was not found to be efficient at these doses, in agreement with the previous reports. It appears that improved delivery to the brain was critical for in vivo efficacy.

We recently published a paper describing the impact of various nanoparticles (liposomes, PBCA, or PLGA) that encased active SOD enzyme in the mouse model of cerebral ischemia and reperfusion injury (Yun et al., 2013). We also developed nanoparticles with unconjugated or conjugated nonspecific antibodies or antibodies directed against the N-methyl-D-aspartate (NMDA) receptor 1 (Figure 41.1). We showed that the nanoparticles containing SOD not only protected primary neurons in vitro from oxygen-glucose deprivation, but protected against ischemia and reperfusion injury in vivo. The targeted nanoparticles elicited a 50%–60% reduction in infarct volume, attenuated inflammatory markers, and improved behavior in the mouse model (Figure 41.2). In addition to the enhanced protection, they also showed localization to the CA regions of the hippocampus. Whereas the nanoparticles alone were not effective in altering the various parameters associated with ischemic injury, these studies showed that targeted nanoparticles containing SOD or other antioxidant enzymes/protective factors may be viable candidates for the treatment of stroke.

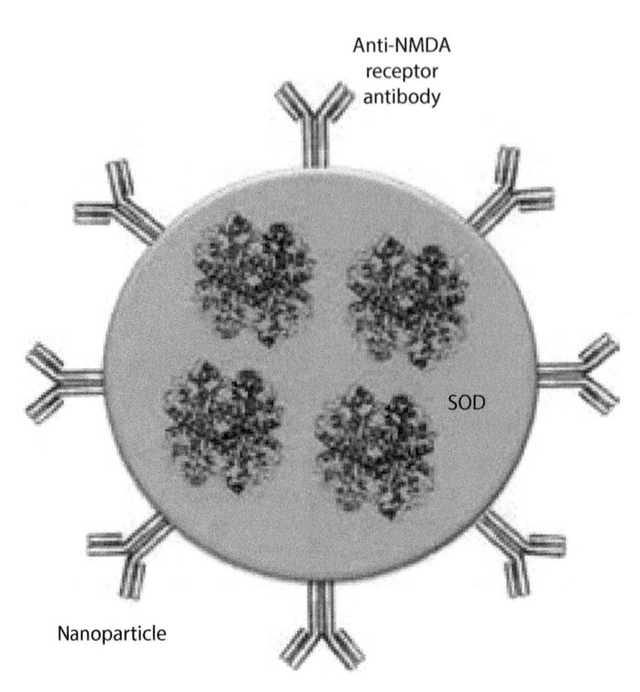

FIGURE 41.1 Depiction of nanoparticle with SOD buried in the nanoparticle with anti-NMDA antibodies coating the outside of the particle. The nanoparticle can be made of various compounds (liposomes, PBCA, or PLGA).

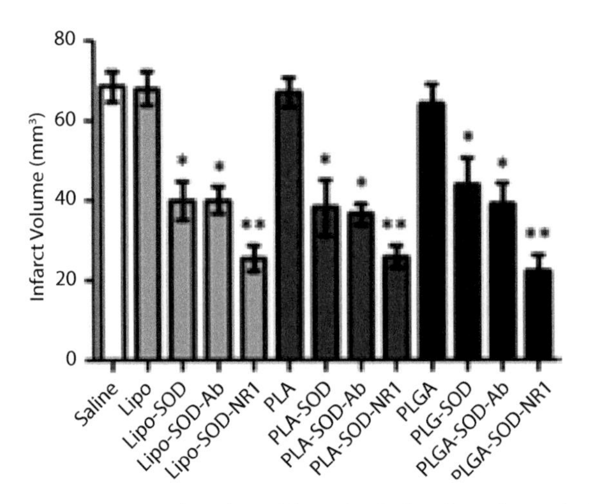

FIGURE 41.2 Effect of nanoparticles on infarct volume in mice subjected to ischemia and reperfusion injury (IRI). Mice were subjected to 1 hour of ischemia and 24 hours of reperfusion with treatment of nanoparticles immediately after ischemia. Animals were administered nanoparticles and infarct volumes determined at 24 hours. n = 10 per group. $*p < 0.01$; $**p < 0.01$ compared with SOD and SOD-Ab for each group. Lipo, liposomes; PLA, polylactic acid; Ab, nonspecific antibody; NR-1, NMDA receptor antibody. (From Yun, X. et al., *J. Cereb. Blood Flow Metab.*, 33, 583–592, 2013. With permission.)

In conclusion, results of the prior studies indicate that free radical scavengers reduce harmful effects of secondary neuronal injury in stroke models, both in vitro and in vivo. However, high doses are required because of poor delivery to the brain, leading to neurological side effects and/or making the therapy impractical. Outcomes of the in vivo studies, in particular, indicate that delivery of antioxidant enzymes or their mimetics to the brain plays a critical role in establishing therapeutic efficacy. On the other hand, several approaches have been developed to achieve targeted delivery of nanoparticulate systems to the brain; however, none of these approaches has been tested in conjunction with delivery of neuroprotective agents for stroke treatment. Future studies are needed to fill this gap. If successful, it will lead to the development of a novel therapeutic approach for reducing harmful effects of secondary neuronal injury in stroke and will accumulate the in vivo data necessary for its translation toward clinical testing. Although the primary focus was on models of stroke, it is clear that general methodology developed could have broader implications because it can be expanded, with relatively minor amendments, to address a number of conditions involving any secondary neuronal injury in CNS, such as brain trauma or secondary spinal cord injury.

41.2 CONCLUSIONS

The research presented here builds upon the results of prior studies, which showed that enzymatic free radical scavengers can reduce harmful effects of secondary neuronal injury in stroke models. Analysis of the results of these background studies indicates that therapeutic efficacy in vivo is largely dependent upon successful delivery of antioxidants to the brain. Although several approaches have been developed recently to achieve targeted delivery of therapeutic agents to CNS, none have been used for delivery of neuroprotective agents. The current research studies attempt to fill in this gap by focusing on targeted delivery of neuroprotective agents to the site of the injury. The use of nanoparticles to target therapeutics is the primary scientific novelty of the work.

Secondly, use of anti-NMDA receptor antibody (anti-NR1) or other neuronal-specific makers as the targeting ligand will be innovative. Prior research on targeted delivery to the brain has focused primarily on penetration of BBB. This approach generally results in a rather uniform distribution of the drug within the brain. In stroke, where BBB is already compromised by the primary insult, it is critical to achieve maximized drug retention at the injury site. It appears that use of anti-NR1 antibody as the targeting ligand leads to the preferential uptake by the injured parenchyma. This new targeting agent can potentially be used broadly, for delivery of many other drugs to the CNS.

REFERENCES

AHA and Heart Disease and Stroke Statistics, 2012. http://circ.ahajournals.org/content/125/1/e2.

Allen CL, Bayraktutan U (2009) Oxidative stress and its role in the pathogenesis of ischaemic stroke. *Int J Stroke* 4:461–70.

Arumugam TV, Selvaraj PK, Woodruff TM, Mattson MP (2008) Targeting ischemic brain injury with intravenous immunoglobulin. *Expert Opin Ther Targets* 12:19–29.

Auriel E, Regev K, Korczyn AD (2014) Nonsteroidal anti-inflammatory drugs exposure and the central nervous system. *Handb Clin Neurol* 119:577–84.

Baker K, Marcus CB, Huffman K, Kruk H, Malfroy B, Doctrow SR (1998) Synthetic combined superoxide dismutase/catalase mimetics are protective as a delayed treatment in a rat stroke model: A key role for reactive oxygen species in ischemic brain injury. *J Pharmacol Exp Ther* 284:215–21.

Begley DJ (2004) Delivery of therapeutic agents to the central nervous system: The problems and the possibilities. *Pharmacol Therap* 104:29–45.

Carmichael ST (2005) Rodent models of focal stroke: Size, mechanism, and purpose. *NeuroRx* 2:396–409.

Chiba T, Umegaki K (2013) Pivotal roles of monocytes/macrophages in stroke. *Mediators Inflamm* 2013:759103.

Das M, Patil S, Bhargava N, Kang JF, Riedel LM, Seal S et al (2007) Auto-catalytic ceria nanoparticles offer neuroprotection to adult rat spinal cord neurons. *Biomaterials* 28:1918–1925.

Dugan LL, Lovett EG, Quick KL, Lotharius J, Lin TT, O'Malley KL (2001) Fullerene-based antioxidants and neurodegenerative disorders. *Parkinsonism Relat Disord* 7:243–46.

Ekshyyan O, Aw TY (2004) Apoptosis: A key in neurodegenerative disorders. *Curr Neurovasc Res* 1: 355–71.

Estevez AY, Pritchard S, Harper K, Aston JW, Lynch A, Lucky JJ et al (2011) Neuroprotective mechanisms of cerium oxide nanoparticles in a mouse hippocampal brain slice model of ischemia. *Free Radic Biol Med* 51:1155–63.

Francis JW, Ren JM, Warren L, Brown RH, Finklestein SP (1997) Postischemic infusion of Cu/Zn superoxide dismutase or SOD: Tet451 reduces cerebral infarction following focal ischemia/reperfusion in rats. *Exp Neurol* 146:435–443.

Graham SM, McCullough LD, Murphy SJ (2004) Animal models of ischemic stroke: Balancing experimental aims and animal care. *Comp Med* 54:486–96.

Grupper M, Eran A, Shifrin A (2007) Ischemic stroke, aortic dissection, and thrombolytic therapy—the importance of basic clinical skills. *J Gen Intern Med* 22:1370–2.

Hachinski V (2007) The 2005 Thomas Willis Lecture: Stroke and vascular cognitive impairment: A transdisciplinary, translational and transactional approach. *Stroke* 38:1396.

Hainsworth AH, Markus HS (2008) Do in vivo experimental models reflect human cerebral small vessel disease? A systematic review. *J Cereb Blood Flow Metab* 28:1877–91.

Hardas SS, Sultana R, Warrier G, Dan M, Florence RL, Wu P et al (2012) Rat brain pro-oxidant effects of peripherally administered 5 nm ceria 30 days after exposure. *Neurotoxicology* 33:1147–55.

Hardas SS, Butterfield DA, Sultana R, Tseng MT, Dan M, Florence RL et al (2010) Brain distribution and toxicological evaluation of a systemically delivered engineered nanoscale ceria. *Toxicol Sci* 116:562–76.

Hasadsri L, Kreuter J, Hattori H, Iwasaki T, George JM (2009) Functional protein delivery into neurons using polymeric nanoparticles. *J Biol Chem* 284:6972–6981.

Jin H, Chen WQ, Tang XW, Chiang LY, Yang CY, Schloss JV et al (2000) Polyhydroxylated C-60, fullerenols as glutamate receptor antagonists and neuroprotective agents. *J Neurosci Res* 62:600–7.

Kim CK, Kim T, Choi IY, Soh M, Kim D, Kim YJ et al (2012) Ceria nanoparticles that can protect against ischemic stroke. *Angew Chem Int Ed Engl* 51:11039–43.

Koellhoffer EC, McCullough LD (2013) The effects of estrogen in ischemic stroke. *Transl Stroke Res* 4:390–401.

Kreuter J, Shamenkov D, Petrov V, Ramge P, Cychutek K, Koch-Brandt C et al (2002) Apolipoprotein-mediated transport of nanoparticle-bound drugs across the blood-brain barrier. *J Drug Targ* 10:317–325.

Kurakhmaeva KB, Djindjikhashvili IA, Petrov VE, Balabanyan VU, Voronina TA, Trofimov SS et al (2009) Brain targeting of nerve growth factor using poly(butyl cyanoacrylate) nanoparticles. *J Drug Target* 17:564–74.

Kuroiwa T, Ting P, Martinez H, Klatzo I (1985) The biphasic opening of the blood-brain barrier to proteins following temporary middle cerebral artery occlusion. *Acta Neuropathol* 68:122–9.

Leker RR, Shohami E (2002) Cerebral ischemia and trauma-different etiologies yet similar mechanisms: Neuroprotective opportunities. *Brain Res Brain Res Rev* 39:55–73.

Lewin M, Carleso N, Tung C-H, Tang X-W, Cory D, Scadden DT et al (2000) Tat peptide-derivatized magnetic nanoparticles allow in vivo tracking and recovery of progenitor cells. *Nat Biotech* 18:410.

Mackensen GB, Patel M, Sheng H, Calvi CL, Batinic-Haberle I, Day BJ et al (2001) Neuroprotection from delayed postoschemic administration of a metalloporphyrin catalytic antioxidant. *J Neurosci* 21:4582–92.

Margaill I, Plotkine M, Lerouet D (2005) Antioxidant strategies in the treatment of stroke. *Free Rad Biol Med* 39:429–43.

Miao Y, Liao JK (2014) Potential serum biomarkers in the pathophysiological processes of stroke. *Expert Rev Neurother* 14:173–85.

Osmond JM, Mintz JD, Dalton B, Stepp DW (2009) Obesity increases blood pressure, cerebral vascular remodeling, and severity of stroke in the Zuckerrat. *Hypertension* 53:381–6.

Poeckel, D., Tausch, L., Kather, N., Jauch, J., Werz, O (2006) Boswellic acids stimulate arachidonic acid release and 12-lipoxygenase activity in human platelets independent of Ca2+ and differentially interact with platelet-type 12-lipoxygenase. *Mol Pharmacol* 70:1071–8.

Rajic Z, Tovmasyan A, Spasojevic I, Sheng H, Lu M, Li AM et al (2012) A new SOD mimic, Mn(III) ortho N-butoxyethylpyridylporphyrin, combines superb potency and lipophilicity with low toxicity. *Free Radic Biol Med* 52:1828–34.

Reddy MK, Wu L, Kou W, Ghorpade A, Labhasetwar V (2008) Superoxide dismutase-loaded PLGA nanoparticles protect cultured human neurons under oxidative stress. *Appl Biochem Biotech* 151:565–77.

Reddy MK, Labhasetwar V (2009) Nanoparticle-mediated delivery of superoxide dismutase to the brain: An effective strategy to reduce ischemia-reperfusion injury. *FASEB J* 23:1384–95.

Roger VL, Go AS, Lloyd-Jones DM, Benjamin EJ, Berry JD, Borden WB et al. (2012) Heart disease and stroke statistics—2012 update: A report from the American Heart Association. *Circulation* 125:e2–220.

Sheng H, Yang W, Fukuda S, Tse HM, Paschen W, Johnson K et al (2009) Long-term neuroprotection from a potent redox-modulating metalloporphyrin in the rat. *Free Radic Biol Med* 47:917–23.

Sheng H, Spasojevich I, Tse HM, Jung JY, Hong J, Zhang Z et al (2011) Neuroprotective efficacy from a lipophilic redox-modulating Mn(III) N-hexylpyridylporphyrin, MnTnHex-2-PyP: Rodent models of ischemic stroke and subarachnoid hemorrhage. *J Pharmacol Exp Ther* 338:906–16.

Sheng H, Enghild JJ, Bowler R, Patel M, Batinic-Haberle I, Calvi CL et al (2002) Effects of metalloporphyrin catalytic antioxidants in experimental brain ischemia. *Free Radic Biol Med* 33:947–61.

Shimizu K, Rajapakse N, Horiguchi T, Payne RM, Busija DW (2003) Protective effect of a new nonpeptidyl mimetic of SOD, M40401, against focal cerebral ischemia in the rat. *Brain Res* 963:8–14.

Silva GA (2006) Neuroscience nanotechnology: Problems, opportunities, and challenges. Nat *Rev Neurosci* 7:65.

Spatz M (2010) Past and recent BBB studies with particular emphasis on changes in ischemic brain edema: Dedicated to the memory of Dr Igor Klatzo. *Acta Neurochir Suppl* 106:21–7.

Torchilin VP, Rammohan R, Weissig V, Leuchenko TS (2000) TAT peptide on the surface of liposomes affords their efficient intracellular delivery even at low temperature and in the presence of metabolic inhibitors. *Proc Natl Acad Sci* 98:8786.

Tseng MT, Lu X, Duan X, Hardas SS, Sultana R, Wu P et al (2012) Alteration of hepatic structure and oxidative stress induced by intravenous nanoceria. *Toxicol Appl Pharmacol* 260:173–82.

van Rooy M, Pretorius E (2014) Obesity, hypertension and hypercholesterolemia as risk factors for atherosclerosis leading to ischemic events. *Curr Med Chem* 21:2121–9.

Vinogradov SV, Batrakova EV, Kabanov AV (2004) Nanogels for oligonucleotide delivery to the brain. *Bioconj Chem* 15:50–60.

Watt JM (1962) T*he Medicinal/Poisonous Plants of Southern/ Eastern Africa*, 2nd ed. Livingstone, Edinburgh.

Young AR, Ali C, Duretête A, Vivien D (2007) Neuroprotection and stroke: Time for a compromise. *J Neurochem* 103:1302–9.

Yun X, Maximov V, Yu J, Zhu H, Vertegel A, Kindy MS (2013) Nanoparticles for targeted delivery of antioxidant enzymes to the brain following cerebral ischemia and reperfusion injury. *J Cereb Blood Flow Metab* 33:583–92.

42 Stem Cell Therapy in Brain Trauma

Implications for Repair and Regeneration of Injured Brain in Experimental TBI Models

Andrew Rolfe and Dong Sun

CONTENTS

42.1 INTRODUCTION

Traumatic brain injury (TBI) is a major health problem worldwide. Currently, there is no effective treatment to improve neural structural repair and functional recovery of patients in clinic. Recent studies suggest that adult neural stem/progenitor cells residing in the neurogenic regions in the adult mammalian brain may play regenerative and reparative roles in response to CNS injuries or diseases. Alternatively, cell transplantation is a potential strategy to repair and regenerate the injured brain. This chapter will discuss the potential of neural stem cells to repair the injured brain with emphasize on modulating endogenous adult neurogenesis to promote regeneration following TBI. The potential of neural stem cells for neural transplantation to repair the injured brain will also been discussed.

Approximately 350,000 individuals in the United States are affected annually by severe and moderate TBI that may result in long-term disability. This rate of injury has produced more than 3 million disabled citizens in the United States alone. Despite generally improving rates of survival after TBI, approximately 80,000 individuals in the United States annually sustain TBIs that result in significant long-term disability. These impairments involve both memory and behavior and can result in a total vegetative state. Most of these 3 million survivors depend upon others for daily care. Many clinical and animal model studies have now shown that severe and even moderate TBI is characterized by both neuronal and white matter loss with resultant brain atrophy and functional neurological impairment. Injury may be in the form of focal damage because it typically occurs after acute subdural hematoma, or it may be diffuse with widespread delayed neuronal loss as it typically occurs after diffuse axonal injury. To date, there is no effective treatment for TBI. Current therapies are primarily focused on reducing the extent of secondary insult rather than repairing the damage from the primary injury. After TBI, the hippocampus is particularly vulnerable to the secondary insults. Hippocampal injury associated to learning and memory deficits are the hallmarks of brain trauma and are the most enduring and debilitating of TBI deficits because they prevent reintegration of patients into a normal lifestyle by impairing employment and social interactions. Spontaneous cognitive improvement is not uncommon but is greatly limited and not normally seen past the second year postinjury (Schmidt et al., 1999). This natural recovery, however, does suggest that innate mechanisms for repair and regeneration are present within the brain.

Recent findings reveal that multipotent neural stem cells/progenitor cells (NSCs/NPCs) persist in selected regions of the brain throughout the life span of an animal, rendering the brain capable of generating new neurons and glia (Gage et al., 1998; Lois and Alvarez-Buylla, 1993). Furthermore, increasing evidence indicates that these endogenous NSCs/NPCs may play regenerative and reparative roles in response to central nervous system (CNS) injuries or diseases. In support of this notion, heightened levels of cell proliferation and neurogenesis have been observed in response to brain trauma or insults suggesting that the brain has the inherent potential to restore populations of damaged or destroyed neurons. This raises the possibility of developing therapeutic strategies aiming at harnessing this neurogenic capacity to repopulate and repair the damaged brain. Recent experimental successes in cell replacement in models of Parkinson disease and other neurodegenerative diseases have inspired TBI researchers to investigate this approach for treating the injured brain. The therapeutic prospects of cell transplantation are based on the potential for transplanted cells to differentiate into region-specific cells and integrate into the host tissue to replace lost cells in the injured brain; alternatively, transplanted cells could provide neurotransmitters or trophic support to the host tissue to facilitate survival or regeneration.

These two approaches, through modulating endogenous NSCs or using exogenous stem cells, are gaining increasing attention in the field of neural regeneration. This chapter will review recent understanding and progress in experimental TBI therapeutic development with endogenous neurogenesis and neural transplantation.

42.2 POTENTIAL OF ENDOGENOUS NSCs FOR BRAIN REPAIR

42.2.1 Extent of Endogenous Neurogenesis in the Normal Brain

The mature mammalian brain is traditionally considered an organ without regenerative capacity. Recently, this statement was revised after the discovery of multipotent NSCs that are capable of generating neurons and glial cells residing in the mature mammalian brain. The region of neurogenesis in the mature brain is primarily confined to the subventricular zone (SVZ) surrounding the lateral ventricle and the dentate gyrus (DG) of the hippocampus (Altman and Das, 1965; Lois and Alvarez-Buylla, 1993). The majority of the SVZ progeny are neuroblasts that undergo chain migration along the rostral migratory stream to the olfactory bulb, where they differentiate into olfactory interneurons (Doetsch and Alvarez-Buylla, 1996). Another subpopulation of these cells migrate into cortical regions for reasons yet to be identified, but evidence suggests they may be involved in repair or cell renewal mechanisms (Parent, 2002). Likewise, the newly generated cells of the DG migrate laterally into the granule cell layer and exhibit properties of fully integrated mature dentate granule neurons (Kempermann and Gage, 2000; van Praag et al., 2002). Most importantly, the newly generated DG granule neurons form synapses and extend axons into their correct target area, the CA3 region (Hastings and Gould, 1999).

Multiple studies have quantified the degree of cytogenesis occurring in these regions and have clearly shown that large numbers of new cells are regularly produced (Lois and Alvarez-Buylla, 1993; Cameron and McKay, 2001). Specifically, the rat dentate gyrus produces ~9,000 new cells per day, which equates to ~270,000 cells per month (Cameron and McKay, 2001). Considering that the total granule cell population in the rat is 1–2 million cells, this degree of new cell addition is certainly large enough to affect network function. A more recent study has found that in the olfactory bulb almost the entire granule cell population in the deep layer and half of the super layer was replaced by new neurons over a 12-month period (Imayoshi et al., 2008). The same study also reported that in the hippocampus, the adult-generated neurons comprised about 10% of the total number of dentate granule cells and they were equally present along the anteroposterior axis of the DG (Imayoshi et al., 2008). However, studies have also found that in normal adult rodent brains, many newly generated neurons in the DG and nonolfactory-bound SVZ cells have a transient existence of 2 weeks or less (Gould et al., 2001). Although this interval is long enough for supportive glial roles, neuron formation and integration into an existing network takes approximately 10–14 days (Alvarez-Buylla and Nottebohm, 1988; Kirn et al., 1999). It must be noted, however, that a small population of these cells are sustained for months to years (Altman and Das, 1965; Eriksson et al., 1998; Gould et al., 2001), strongly supporting the theory of network integration. Furthermore, this dramatic loss of newly generated cells might be a recapitulation of network pruning seen in early mammalian development. Whether the limited life span represents network pruning or merely distinct cell-specific roles is yet to be understood.

42.2.2 Functions of Adult-Generated Neurons

In the normal hippocampus, the newly generated granular cells in the adult DG can become functional neurons by displaying passive membrane properties, generating action potentials and functional synaptic inputs as seen in mature DG neurons (van Praag et al., 2002). Increasing evidence has also shown that adult hippocampal neurogenesis is involved in learning and memory function (Clelland et al., 2009; Deng et al., 2009). For example, mouse strains with genetically low levels of neurogenesis perform poorly on learning tasks when compared with those with higher level of baseline neurogenesis (Kempermann et al., 1997, 1998). Conversely, physical activity stimulates a robust increase in the generation of new neurons and subsequently enhances spatial learning and long-term potentiation (van Praag et al., 1999a, 1999b). Additionally, diminished hippocampal neurogenesis, as observed after the administration of antimitotic drugs such as methylazoxymethanol acetate, cytosine-β-D-arabinofuranoside, by irradiation or by genetic manipulation, was associated with worse performance on hippocampus-dependent trace eye blink conditioning (Shors et al., 2001), contextual fear conditioning (Saxe et al., 2006; Shors et al., 2002), and long-term spatial memory function tests (Rola et al., 2004; Snyder et al., 2005). Collectively, these studies provide compelling evidence that adult born neurons in the hippocampus play a critical role in many important hippocampal-dependent functions in normal adult brain. Compared with the evident role of hippocampal neurogenesis in hippocampal-dependent functions, the function of SVZ-olfactory neurogenesis is less certain. Thus far, limited studies have found that adult-generated neurons in the olfactory bulb have a critical role in olfactory tissue maintenance and are involved in olfactory discrimination and olfactory perceptual learning functions (Gheusi et al., 2000; Kageyama et al., 2012; Moreno et al., 2009).

The proliferation and maturational fate of cells within the SVZ and DG is modulated by a number of physical and chemical cues. For example, biochemical factors such as serotonin, glucocorticoids, ovarian steroids, and growth factors tightly regulate the proliferative response, suggesting that cell proliferation within these regions have physiologic importance (Banasr et al., 2001; Cameron and Gould, 1994; Kuhn et al., 1997; Tanapat et al., 1999). In addition, certain physical stimuli produce alterations in cell production suggesting a role in network adaptation (Gould et al., 1997; Kempermann et al., 1997b; van Praag et al., 1999b). For example, environments that are cognitively and physically enriched increase cell proliferation and neurogenesis in both the SVZ and DG, whereas stress reduces this type of cellular response (Gould and Tanapat, 1999). Nevertheless, a functional role for these new cells is dependent upon a significant number of cells being generated, and their survival, differentiation, and ultimate integration into existing neuronal circuitry.

42.2.3 Neurogenesis in the Human Brain

Compared with rodent brains, the degree of adult neurogenesis in human brain is less clear. The most well-characterized neurogenic region in the adult human brain is the SVZ lining the lateral ventricle, where a ribbon of SVZ astrocytes have been identified that proliferate in vivo and behave as multipotent progenitor cells in vitro (Sanai et al., 2004). In rodents and primates, neurons born in the SVZ migrate in chains through the rostral migratory stream to replace interneurons of the olfactory bulb currently. In contrast, there is no evidence for chains of migrating neuroblasts in the human SVZ (Sanai et al., 2004). It has been estimated that in normal humans less than 1% of astrocytes within the SVZ ribbon are undergoing cell division and although these endogenous NSCs can be expanded in culture, their response to injury in patients has not been studied. In another neurogenic region in humans, the hippocampal DG neurogenesis in vivo was demonstrated on histological sections obtained in patients who had died of cancer but for which BrdU staining was used for diagnostic purposes (Eriksson et al., 1998). A recent study has found that the generation and migration of new neurons is very much limited to the early childhood (Curtis et al., 2012). Less well-characterized in the human brain are proliferating NPCs in the hippocampus, white matter, and other regions, where cells isolated from the adult human brain are capable of generating both neurons and glia under culture conditions (Kukekov et al., 1999; Murrell et al., 2013; Nunes et al., 2003).

42.2.4 Response of Endogenous NSCs to Brain Injury and the Role of These Cells for Brain Repair

The regenerative capacity of the SVZ and DG is of particular interest with regard to TBI. Because adult-generated neurons from both regions have functional roles, harnessing this endogenous population of stem cells to repopulate the damaged brain is an attractive strategy to repair and regenerate the injured brain. In the injured brain, studies from our laboratory and others have shown that TBI significantly increases cell proliferation in both the SVZ and DG in adult mice and rats in various TBI models including diffuse and focal injury models (Chirumamilla et al., 2002; Rice et al., 2003). We have also found that the juvenile brain has more robust neurogenic response after injury than the adult and aged brain (Sun et al., 2005). Such increased levels of cell proliferation with increased generation of new neurons likely contribute to the better functional recovery in juvenile animals after TBI. Furthermore, we and others have found that injury-induced newly generated granular cells integrate into the existing hippocampal circuitry (Emery et al., 2005; Sun et al., 2007), and this endogenous neurogenesis is associated to the innate cognitive recovery after injury (Sun et al., 2007). In human brain specimens, a recent study has found an increased number of cells expressing NSCs/NPCs markers in the perilesion cortex in the injured brain (Zheng et al., 2013). These studies strongly indicated the inherent attempts of the brain to repair and regenerate after injury through endogenous NSCs.

The degree of endogenous neurogenesis can be enhanced via exogenous means and augmentation of endogenous neural stem cells could be a potential therapy for treating the injured brain. So far, many factors have been shown to enhance neurogenesis particularly in the hippocampus. Studies have found that various types of growth factors and drugs can enhance neurogenesis and improve functional recovery of the injured brain after trauma. For example, studies from our laboratory have shown that intraventricular administration of growth factors basic fibroblast growth factor or epidermal growth factor can significantly enhance TBI-induced cell proliferation in the hippocampus and the SVZ, and drastically improve cognitive functional recovery of the injured adult animals (Sun et al., 2009, 2010). Other studies have found that infusion of S100β or vascular endothelial growth factor can also enhance neurogenesis in the hippocampus and improve the functional recovery of animals after TBI (Kleindienst et al., 2005; Lee and Agoston, 2010; Thau-Zuchman et al., 2010). Several drugs that are currently used in clinical trials for treating TBI or other conditions have shown effects in enhancing neurogenesis and cognitive function in TBI animals including statins (Lu et al., 2007b), erythropoietin (Lu et al., 2005; Xiong et al., 2010), progesterone (Barha et al., 2011), and the antidepressant imipramine (Han et al., 2011). Other strategies that have beneficial effects for TBI such as hypothermia and environment enrichment are also shown enhanced hippocampal neurogenesis in injured animals (Bregy et al., 2012; Kovesdi et al., 2011). Collectively, these studies suggest the therapeutic potential of augmenting the endogenous repair response for treating TBI.

42.3 STEM CELLS AS CELL SOURCE FOR NEURAL TRANSPLANTATION FOR BRAIN REPAIR AND REGENERATION

Because of the limited capacity of the injured brain to repair and replace the damaged neurons, neural transplantation is a prospective therapy for TBI as transplanted cells may differentiate into region-specific cells and integrate into the host tissue to replace the lost cells in the injured brain. Additionally, transplanted cells could provide trophic support to the host tissue to facilitate regeneration. Over the past few decades, researchers have explored a wide array of cell sources for neural transplantation. These cells include embryonic stem cells isolated to fetal or embryonic tissue, mesenchymal stromal cells such as bone marrow stromal cells and umbilical cord cells, adult NSCs, and more recently, induced pluripotent stem cells (iPSCs). The following section will discuss the application of these cell types in the setting of TBI.

42.3.1 Embryonic Stem Cells

Embryonic stem (ES) cells are pluripotent stem cells that have unlimited capacity of self-renewal and can give rise to cells of all three primary germ layers. Because of their high plasticity, ES cells are the ideal cell source for neural transplantation. When transplanted into normal or damaged CNS, human ES cells can differentiate, migrate, and make innervations (Hentze et al., 2007). Thus, ES cells derived from human or mice fetal brains have been tested as a transplantation cell source for TBI treatment in animal studies in different TBI models with different results reported.

NSCs from human ES cells isolated from fetal brain were capable of surviving for an extended period of 6 weeks, migrating to the contralateral cortex, and differentiating into neurons and astrocytes when transplanted into the injured brain after a cortical contusion injury (Wennersten et al., 2004). Gao et al. (2006) have reported that NSCs from human ES cells survived and differentiated to neurons after transplantation into the injured brain when examined at 2 weeks after cell injection, and the injured animals with cell transplantation had improved cognitive functional recovery. In a more recent study, Skardelly et al. (2011) transplanted predifferentiated human fetal ES cells into injured rat brain after a severe controlled cortical injury. They observed a transient increase in angiogenesis and reduced astrogliosis together with improved long-term motor functional improvement, brain injury lesion volume reduction, and increased neuronal survival in the border zone of the lesion. Shear et al. (2004) assessed the long-term survival, migration, differentiation, and functional significance of NSCs derived from mice fetal brain after transplantation into the injured brain up to 1 year posttransplantation. They found that the injured animals receiving transplants showed significant improvement in motor and spatial learning functions, and the transplanted cells migrated widely in the injured brain, with the majority of transplanted cells expressing NG2, an oligodendrocyte progenitor cell marker, but not neuronal markers. Post-TBI neural transplantation of immortalized fetal ES-derived NSCs (C17.2 cells) has also shown improved motor function with the transplanted cells surviving for up to 13 weeks and differentiating into mature neurons and glial cells (Riess et al., 2002; Boockvar et al., 2005). In vitro–modified ES cells either predifferentiated into mature neurons expressing neurotransmitters or with overexpression of growth factors such as glial cell line–derived and brain-derived neurotrophic factor showed beneficial effects when transplanted into the injured animals by promoting motor and cognitive improvement of the injured animals concomitant with better graft survival and neuronal differentiation (Bakshi et al., 2006; Becerra et al., 2007; Ma et al., 2012).

Taken together, these data suggest that post-TBI transplantation using ES-derived cells can restore motor and cognitive functions of the injured animals. However, the beneficial effect of the transplanted cells may be associated with the neural trophic effect of the transplanted cells rather than direct neural replacement as long-term survival and neuronal differentiation is rather limited. Further studies are needed to improve survival and functional neural replacement by modulating the injured host environment. Caution must be exercised when working with multipotent ES cells as undifferentiated ES cells have a potential risk of tumor formation (Riess et al., 2007).

42.3.2 Adult NSCs

Recent findings show that the mature mammalian CNS harbors multipotent stem cells capable of differentiation into a variety of specialized cells throughout life (Gage et al., 1998; Lois and Alvarez-Buylla, 1993). In the adult mammalian CNS, the NSCs/NPCs are primarily confined to the SVZ surrounding the lateral ventricle and the DG of the hippocampus (Altman and Das, 1965; Lois and Alvarez-Buylla, 1993). Aside from these major neurogenic regions, adult neurogenesis in rodents has also been reported in other regions in the CNS including the striatum, the substantia nigra, the cortex, and the spinal cord (Lie et al., 2002; Palmer et al., 1999; Weiss et al., 1996). These adult-derived NSCs express low levels of the major histocompatibility complex antigens (Klassen et al., 2003), display high survival rates, and become region-specific cells when transplanted into normal adult rat brains (Gage et al., 1995; Richardson et al., 2005; Zhang et al., 2003). When transplanted into the injured brain in a rat experimental TBI model, we found that the adult derived NSCs can survive for an extended period in the injured brain. Many cells migrated out of the injection site into the surrounding areas expressing markers for mature astrocytes or oligodendrocytes. Electrophysiological studies showed that the transplanted cells possessed typical mature glial cell properties demonstrating that adult-derived NSCs became region-specific functional cells (Sun et al., 2011) (Figure 42.1).

In humans, multipotent stem/progenitors cells have been identified and successfully isolated from various regions of adult human brain including the hippocampus, SVZ, neocortex, and subcortical white matters from neurosurgical resection tissues (Arsenijevic et al., 2001; Brunet et al., 2002, 2003; Kukekov et al., 1999; Nunes et al., 2003; Richardson et al., 2006; Roy et al., 2000; Windrem et al., 2002). These raise the possibility of using such cells as an autologous cell source for transplantation therapy. Indeed, Brunet and colleagues have demonstrated that adult monkey NSCs/NPCs derived from cortical biopsy survived for at least 3 months and displayed a neuronal phenotype after reimplantation into the normal or ibotenic acid excitotoxic-lesioned motor cortex of the donor brains (Brunet et al., 2005). These cells may also restore the anatomy and function of the injured CNS as shown in a study after grafting adult human NSCs/NPCs into the demyelinated rat spinal cord (Akiyama et al., 2001).

To date, very few studies have attempted to examine the behavior of adult-derived human NSCs/NPCs in the injured mature CNS. Olstorn and colleagues recently reported that a small portion (4 ± 1%) of adult human NSCs/NPCs can survive for 16 weeks after transplantation into the posterior periventricular region in normal adult rats or rats with hippocampal CA1 ischemic injury (Olstorn et al., 2007). Although the results of this study are promising, questions remain whether these cells become anatomically and functionally integrated into the injured brain and whether the proportions of surviving cells can be increased by transplanting NSCs/NPC's at a different developmental stage.

42.3.3 Bone Marrow Stromal Cells

Because of ethical and immunological concerns as well as the risk of tumorigenesis, the translational value of using ES cells for clinic application is limited. Autologous transplantation of NSCs isolated from neurosurgical removal of brain tissue from TBI patients is an attractive strategy; however, so far, the success of long-term cell survival and functional outcomes of these cells in the treatment of experimental TBI is rather limited. Because of these limitations, adult-derived

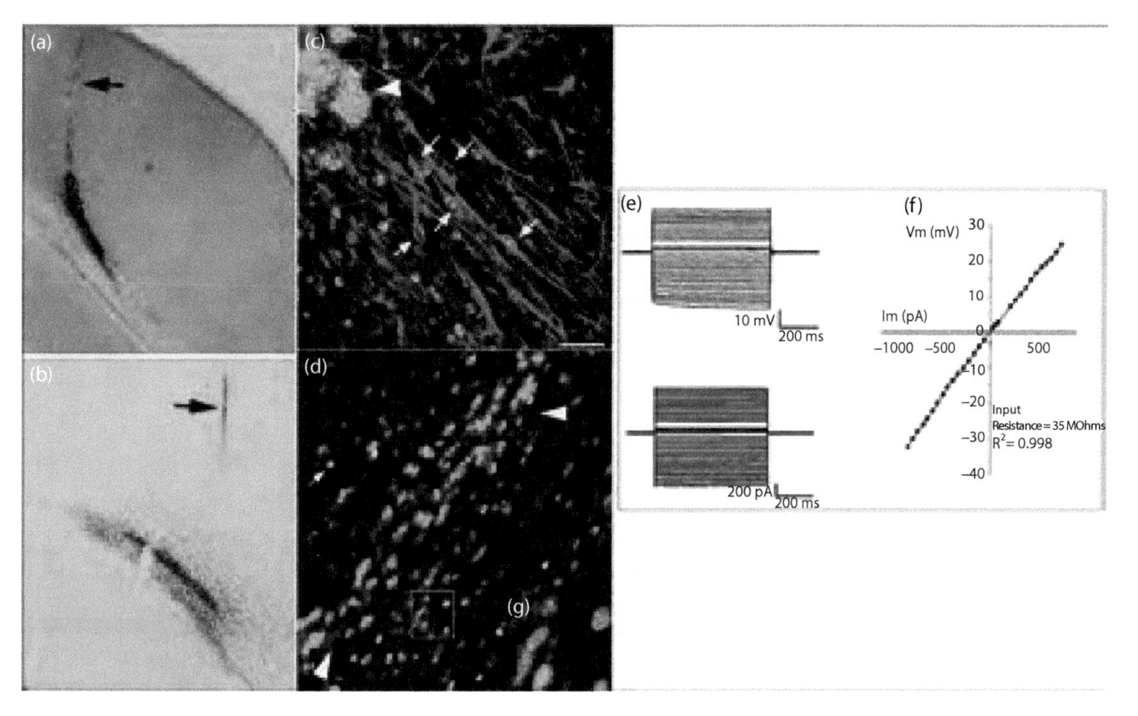

FIGURE 42.1 Survival and functional differentiation of transplanted adult neural stem and progenitor cells after TBI. Cultured adult rat NSCs/NPCs were labeled with BrdU in vitro 3 days before being used for transplantation and subsequently identified with BrdU immunostaining. (a) Coronal section from an animal sacrificed 2 weeks after transplantation. Grafted cells were mostly located at the cortex–white matter interface. (b) Coronal sections taken from an animal sacrificed at 4 weeks after transplantation. Compared to 2 weeks, at 4 weeks many BrdU-labeled cells migrated out of the injection site to the surrounding areas. Arrow indicates the injection needle tract. (c) Confocal micrograph showing that many BrdU-labeled transplanted cells (green) migrated away from the injection center (arrowhead) along white matter tracts and differentiated into spindle-shaped glial fibrillary acidic protein–labeled astrocytes (red, arrows). Bar scale = 50 μm. (d) Arrows indicate many BrdU+ transplanted cells (green) away from the injection center (arrowhead) were colabeled with mature oligodendrocyte marker Olig2 (red). Merged image shows colocalization of BrdU and Olig2 (f). (g) Enlarged image of the boxed area in (f) showing colabeling of BrdU and Olig2. (e) Electrophysiological properties of transplanted cells showing membrane potential response (top) to current injection (bottom). (f) The amplitude of the membrane potential responses is plotted against the current injected. The input resistance was low and the current-voltage relationship was linear, consistent with mature glial cell but not neuronal electrophysiological properties.

mesenchymal cells, particularly bone marrow stromal cells, (BMSCs) have received much attention.

BMSCs are undifferentiated cells with mixed cell population including stem and progenitor cells. These cells can be easily isolated from the mononuclear fraction of patients' bone marrow and be expanded in culture without ethical and technical concerns. Another advantage of considering BMSCs for cell transplantation is the low antigenicity because of their low expression of the major histocompatibility complex antigens (class II) (Le and Ringden, 2005). In addition, these cells produce high levels of growth factors, cytokines, and extracellular matrix molecules that could have potential neurotrophic or neuroprotective effect in the injured brain. As a matter of fact, all studies using BMSCs for neural transplantation have demonstrated that the beneficial effects of BMSCs are attributed to their neurotrophic or neuroprotective effect rather than direct cell replacement (Li and Chopp, 2009).

The potential of BMSCs for treating TBI have been extensively assessed in experimental TBI models. Cells were delivered either focally to the injured brain, or systemically through intravenous or intraarterial injections at the acute or subacute phase after TBI and significant reduction of neurological deficits including motor and cognitive deficits were reported. For example, intracranial injection of rat BMSCs into the brain region adjacent to the brain lesion site or intravenous injection of cells at 24 hours after a controlled cortical contusion injury in rats and it was found that the injured animals had improved sensory motor functional (Lu et al., 2001; Mahmood et al., 2001, 2003). When human BMSCs were combined with collagen scaffolds and transplanted into the injury cavity at 4 or 7 days after TBI, animals had significantly improved sensorimotor and spatial learning functions together with reduced brain lesion volume and enhanced focal brain angiogenesis (Lu et al., 2007a; Xiong et al., 2009). The effect of BMSCs in improving sensorimotor function of injured animals was reported even when delivered at 2 months after TBI (Bonilla et al., 2009). Further studies have demonstrated that the beneficial effort of BMSCs in the injured brain is due largely by their production of bioactive factors, which facilitates the endogenous plasticity and remodeling of the host brain thus promoting functional recovery (Li and Chopp, 2009). Although a low number of BMSCs can be found in the injured brain expressing neuronal or glial markers (Mahmood et al., 2001, 2003), no study has demonstrated that BMSCs can become fully differentiated functional neurons in vivo. Taken together, extensive experimental studies have demonstrated the beneficial effects of BMSCs in the injured brain and highlighted the potential use of BMSCs

42.3.4 Other Potential Types of Stem Cells for Cell Replacement Therapy

Apart from the previously mentioned stem cells, researchers have recently explored several other types of stem or stem-like cells for TBI application. Published data have reported that the use of human amnion–derived multipotent progenitor cells can significantly attenuate axonal degeneration and improve neurological function and brain tissue morphology of the injured rats (Chen et al., 2009; Yan et al., 2013). Intravenous administration of human adipose–derived stem cells or the derived culture medium into a controlled cortical impact rat model significantly improved motor and cognitive functions and reduced focal tissue damage and hippocampal cell loss (Tajiri et al., 2014).

Human umbilical cord blood is an abundant source of multiple stem cells, including hematopoietic stem cells, mesenchymal stem cells, unrestricted somatic stem cells, and embryonic-like stem cells. These cells can be easily harvested without ethical controversy and could be an attractive source of stem cells for brain repair. These studies have shown that these cells can survive in injury sites and promote survival of local host neurons in ischemic and spinal cord injury animal models (Sun and Ma, 2013). In a recent study, Wang et al. have conducted a small-scale clinical trial using these cells for treating TBI patients. The authors reported that the patients treated with umbilical cord stem cells had improved neurological function and self-care compared with control group (Wang et al., 2013).

Recent development of somatic cell reprogramming that generates iPSCs provides prospects for novel neural replacement strategies. Human iPSCs possess the dual properties of unlimited self-renewal and the pluripotent potential to differentiate into multilineage cells without ethical concerns. More importantly, patient-specific iPSCs can serve as an autologous cell source for transplantation without encountering graft rejection. These unique properties of iPSCs have raised the widespread hope that many neurological diseases including TBI might be cured or treated. Thus far rapid progress has been made in the field of reprogramming; however, the optimal source of somatic cells used for applications in neurological disorders has not been identified yet.

42.4 CONCLUSION AND PERSPECTIVES

The existence of multipotent stem cells in the mammalian brain and other organs has raised high enthusiasm for using these cells to treat the injured brain. Extensive studies have shown the potential brain repair through endogenous NSCs or through cell replacement strategies using varying types of stem cells. However, to successfully repair and regenerate the injured brain with stem cells, many challenges must be overcome. One major challenge is generation of sufficient functional neurons capable of integrating into existing neural circuitry in the injured brain. Another major challenge, which is particularly important for stem cell transplantation, is the focal microenvironment of the site of injury. After TBI, primary brain damage together with secondary tissue loss induced by ischemia, excitotoxicity, oxidative stress, and inflammation create a hostile environment preventing the survival and integration of the transplanted cells. So far, ample studies have supported the notion that the in vivo fate of transplanted cells is regulated by the intrinsic properties of grafted cells and the local environmental cues in the host. These challenges must be overcome in experimental TBI studies before moving forward with stem cell therapies for treating the injured brain clinically.

REFERENCES

Akiyama Y, Honmou O, Kato T, Uede T, Hashi K, Kocsis JD (2001) Transplantation of clonal neural precursor cells derived from adult human brain establishes functional peripheral myelin in the rat spinal cord. *Exp Neurol* 167:27–39.

Altman J, Das GD (1965) Autoradiographic and histological evidence of postnatal hippocampal neurogenesis in rats. *J Comp Neurol* 124:319–335.

Alvarez-Buylla A, Nottebohm F (1988) Migration of young neurons in adult avian brain. *Nature* 335:353–354.

Arsenijevic Y, Villemure JG, Brunet JF, Bloch JJ, Deglon N, Kostic C et al. (2001) Isolation of multipotent neural precursors residing in the cortex of the adult human brain. *Exp Neurol* 170:48–62.

Bakshi A, Shimizu S, Keck CA, Cho S, LeBold DG, Morales D et al. (2006) Neural progenitor cells engineered to secrete GDNF show enhanced survival, neuronal differentiation and improve cognitive function following traumatic brain injury. *Eur J Neurosci* 23:2119–2134.

Banasr M, Hery M, Brezun JM, Daszuta A (2001) Serotonin mediates oestrogen stimulation of cell proliferation in the adult dentate gyrus. *Eur J Neurosci* 14:1417–1424.

Barha CK, Ishrat T, Epp JR, Galea LA, Stein DG (2011) Progesterone treatment normalizes the levels of cell proliferation and cell death in the dentate gyrus of the hippocampus after traumatic brain injury. *Exp Neurol* 231:72–81.

Becerra GD, Tatko LM, Pak ES, Murashov AK, Hoane MR (2007) Transplantation of GABAergic neurons but not astrocytes induces recovery of sensorimotor function in the traumatically injured brain. *Behav Brain Res* 179:118–125.

Bonilla C, Zurita M, Otero L, Aguayo C, Vaquero J (2009) Delayed intralesional transplantation of bone marrow stromal cells increases endogenous neurogenesis and promotes functional recovery after severe traumatic brain injury. *Brain Inj* 23:760–769.

Boockvar JA, Schouten J, Royo N, Millard M, Spangler Z, Castelbuono D et al. (2005) Experimental traumatic brain injury modulates the survival, migration, and terminal phenotype of transplanted epidermal growth factor receptor-activated neural stem cells. *Neurosurgery* 56:163–171.

Bregy A, Nixon R, Lotocki G, Alonso OF, Atkins CM, Tsoulfas P et al. (2012) Posttraumatic hypothermia increases doublecortin expressing neurons in the dentate gyrus after traumatic brain injury in the rat. *Exp Neurol* 233:821–828.

Brunet JF, Pellerin L, Arsenijevic Y, Magistretti P, Villemure JG (2002) A novel method for in vitro production of human glial-like cells from neurosurgical resection tissue. *Lab Invest* 82:809–812.

Brunet JF, Pellerin L, Magistretti P, Villemure JG (2003) Cryopreservation of human brain tissue allowing timely production of viable adult human brain cells for autologous transplantation. *Cryobiology* 47:179–183.

Brunet JF, Rouiller E, Wannier T, Villemure JG, Bloch J (2005) Primate adult brain cell autotransplantation, a new tool for brain repair? *Exp Neurol* 196:195–198.

Cameron HA, Gould E (1994) Adult neurogenesis is regulated by adrenal steroids in the dentate gyrus. *Neuroscience* 61:203–209.

Cameron HA, McKay RD (2001) Adult neurogenesis produces a large pool of new granule cells in the dentate gyrus. *J Comp Neurol* 435:406–417.

Chen Z, Tortella FC, Dave JR, Marshall VS, Clarke DL, Sing G et al. (2009) Human amnion-derived multipotent progenitor cell treatment alleviates traumatic brain injury-induced axonal degeneration. *J Neurotrauma* 26:1987–1997.

Chirumamilla S, Sun D, Bullock MR, Colello RJ (2002) Traumatic brain injury induced cell proliferation in the adult mammalian central nervous system. *J Neurotrauma* 19:693–703.

Clelland CD, Choi M, Romberg C, Clemenson GD, Jr., Fragniere A, Tyers P et al. (2009) A functional role for adult hippocampal neurogenesis in spatial pattern separation. *Science* 325:210–213.

Curtis MA, Low VF, Faull RL (2012) Neurogenesis and progenitor cells in the adult human brain: A comparison between hippocampal and subventricular progenitor proliferation. *Dev Neurobiol* 72:990–1005.

Deng W, Saxe MD, Gallina IS, Gage FH (2009) Adult-born hippocampal dentate granule cells undergoing maturation modulate learning and memory in the brain. *J Neurosci* 29:13532–13542.

Doetsch F, Alvarez-Buylla A (1996) Network of tangential pathways for neuronal migration in adult mammalian brain. *Proc Natl Acad Sci USA* 93:14895–14900.

Emery DL, Fulp CT, Saatman KE, Schutz C, Neugebauer E, McIntosh TK (2005) Newly born granule cells in the dentate gyrus rapidly extend axons into the hippocampal CA3 region following experimental brain injury. *J Neurotrauma* 22:978–988.

Eriksson PS, Perfilieva E, Bjork-Eriksson T, Alborn AM, Nordborg C, Peterson DA et al. (1998) Neurogenesis in the adult human hippocampus. *Nat Med* 4:1313–1317.

Gage FH, Coates PW, Palmer TD, Kuhn HG, Fisher LJ, Suhonen JO et al. (1995) Survival and differentiation of adult neuronal progenitor cells transplanted to the adult brain. *Proc Natl Acad Sci USA* 92:11879–11883.

Gage FH, Kempermann G, Palmer TD, Peterson DA, Ray J (1998) Multipotent progenitor cells in the adult dentate gyrus. *J Neurobiol* 36:249–266.

Gao J, Prough DS, McAdoo DJ, Grady JJ, Parsley MO, Ma L et al. (2006) Transplantation of primed human fetal neural stem cells improves cognitive function in rats after traumatic brain injury. *Exp Neurol* 201:281–292.

Gheusi G, Cremer H, McLean H, Chazal G, Vincent JD, Lledo PM (2000) Importance of newly generated neurons in the adult olfactory bulb for odor discrimination. *Proc Natl Acad Sci USA* 97:1823–1828.

Gould E, McEwen BS, Tanapat P, Galea LA, Fuchs E (1997) Neurogenesis in the dentate gyrus of the adult tree shrew is regulated by psychosocial stress and NMDA receptor activation. *J Neurosci* 17:2492–2498.

Gould E, Tanapat P (1999) Stress and hippocampal neurogenesis. *Biol Psychiatry* 46:1472–1479.

Gould E, Vail N, Wagers M, Gross CG (2001) Adult-generated hippocampal and neocortical neurons in macaques have a transient existence. *Proc Natl Acad Sci USA* 98:10910–10917.

Han X, Tong J, Zhang J, Farahvar A, Wang E, Yang J et al. (2011) Imipramine treatment improves cognitive outcome associated with enhanced hippocampal neurogenesis after traumatic brain injury in mice. *J Neurotrauma* 28:995–1007.

Hastings NB, Gould E (1999) Rapid extension of axons into the CA3 region by adult-generated granule cells. *J Comp Neurol* 413:146–154.

Hentze H, Graichen R, Colman A (2007) Cell therapy and the safety of embryonic stem cell-derived grafts. *Trends Biotechnol* 25:24–32.

Imayoshi I, Sakamoto M, Ohtsuka T, Takao K, Miyakawa T, Yamaguchi M et al. (2008) Roles of continuous neurogenesis in the structural and functional integrity of the adult forebrain. *Nat Neurosci* 11:1153–1161.

Kageyama R, Imayoshi I, Sakamoto M (2012) The role of neurogenesis in olfaction-dependent behaviors. *Behav Brain Res* 227:459–463.

Kempermann G, Brandon EP, Gage FH (1998) Environmental stimulation of 129/SvJ mice causes increased cell proliferation and neurogenesis in the adult dentate gyrus. *Curr Biol* 8:939–942.

Kempermann G, Gage FH (2000) Neurogenesis in the adult hippocampus. *Novartis Found Symp* 231:220–235.

Kempermann G, Kuhn HG, Gage FH (1997a) Genetic influence on neurogenesis in the dentate gyrus of adult mice. *Proc Natl Acad Sci USA* 94:10409–10414.

Kempermann G, Kuhn HG, Gage FH (1997b) More hippocampal neurons in adult mice living in an enriched environment. *Nature* 386:493–495.

Kirn JR, Fishman Y, Sasportas K, Alvarez-Buylla A, Nottebohm F (1999) Fate of new neurons in adult canary high vocal center during the first 30 days after their formation. *J Comp Neurol* 411:487–494.

Klassen H, Imfeld KL, Ray J, Young MJ, Gage FH, Berman MA (2003) The immunological properties of adult hippocampal progenitor cells. *Vision Res* 43:947–956.

Kleindienst A, McGinn MJ, Harvey HB, Colello RJ, Hamm RJ, Bullock MR (2005) Enhanced hippocampal neurogenesis by intraventricular S100B infusion is associated with improved cognitive recovery after traumatic brain injury. *J Neurotrauma* 22:645–655.

Kovesdi E, Gyorgy AB, Kwon SK, Wingo DL, Kamnaksh A, Long JB et al. (2011) The effect of enriched environment on the outcome of traumatic brain injury; a behavioral, proteomics, and histological study. *Front Neurosci* 5:42.

Kuhn HG, Winkler J, Kempermann G, Thal LJ, Gage FH (1997) Epidermal growth factor and fibroblast growth factor-2 have different effects on neural progenitors in the adult rat brain. *J Neurosci* 17:5820–5829.

Kukekov VG, Laywell ED, Suslov O, Davies K, Scheffler B, Thomas LB et al. (1999) Multipotent stem/progenitor cells with similar properties arise from two neurogenic regions of adult human brain. *Exp Neurol* 156:333–344.

Le BK, Ringden O (2005) Immunobiology of human mesenchymal stem cells and future use in hematopoietic stem cell transplantation. *Biol Blood Marrow Transplant* 11:321–334.

Lee C, Agoston DV (2010) Vascular endothelial growth factor is involved in mediating increased de novo hippocampal neurogenesis in response to traumatic brain injury. *J Neurotrauma* 27:541–553.

Li Y, Chopp M (2009) Marrow stromal cell transplantation in stroke and traumatic brain injury. Neurosci Lett 456:120–123.

Lie DC, Dziewczapolski G, Willhoite AR, Kaspar BK, Shults CW, Gage FH (2002) The adult substantia nigra contains progenitor cells with neurogenic potential. *J Neurosci* 22:6639–6649.

Lois C, varez-Buylla A (1993) Proliferating subventricular zone cells in the adult mammalian forebrain can differentiate into neurons and glia. *Proc Natl Acad Sci USA* 90:2074–2077.

Lu D, Mahmood A, Qu C, Goussev A, Schallert T, Chopp M (2005) Erythropoietin enhances neurogenesis and restores spatial memory in rats after traumatic brain injury. *J Neurotrauma* 22:1011–1017.

Lu D, Mahmood A, Qu C, Hong X, Kaplan D, Chopp M (2007a) Collagen scaffolds populated with human marrow stromal cells reduce lesion volume and improve functional outcome after traumatic brain injury. *Neurosurgery* 61:596–602.

Lu D, Mahmood A, Wang L, Li Y, Lu M, Chopp M (2001) Adult bone marrow stromal cells administered intravenously to rats after traumatic brain injury migrate into brain and improve neurological outcome. *Neuroreport* 12:559–563.

Lu D, Qu C, Goussev A, Jiang H, Lu C, Schallert T et al. (2007b) Statins increase neurogenesis in the dentate gyrus, reduce delayed neuronal death in the hippocampal CA3 region, and improve spatial learning in rat after traumatic brain injury. *J Neurotrauma* 24:1132–1146.

Ma H, Yu B, Kong L, Zhang Y, Shi Y (2012) Neural stem cells over-expressing brain-derived neurotrophic factor (BDNF) stimulate synaptic protein expression and promote functional recovery following transplantation in rat model of traumatic brain injury. *Neurochem Res* 37:69–83.

Mahmood A, Lu D, Lu M, Chopp M (2003) Treatment of traumatic brain injury in adult rats with intravenous administration of human bone marrow stromal cells. *Neurosurgery* 53:697–702.

Mahmood A, Lu D, Yi L, Chen JL, Chopp M (2001) Intracranial bone marrow transplantation after traumatic brain injury improving functional outcome in adult rats. *J Neurosurg* 94:589–595.

Moreno MM, Linster C, Escanilla O, Sacquet J, Didier A, Mandairon N (2009) Olfactory perceptual learning requires adult neurogenesis. *Proc Natl Acad Sci USA* 106:17980–17985.

Murrell W, Palmero E, Bianco J, Stangeland B, Joel M, Paulson L et al. (2013) Expansion of multipotent stem cells from the adult human brain. *PLoS One* 8:e71334.

Nunes MC, Roy NS, Keyoung HM, Goodman RR, McKhann G, Jiang L et al. (2003) Identification and isolation of multipotential neural progenitor cells from the subcortical white matter of the adult human brain. *Nat Med* 9:439–447.

Olstorn H, Moe MC, Roste GK, Bueters T, Langmoen IA (2007) Transplantation of stem cells from the adult human brain to the adult rat brain. *Neurosurgery* 60:1089–1098.

Palmer TD, Markakis EA, Willhoite AR, Safar F, Gage FH (1999) Fibroblast growth factor-2 activates a latent neurogenic program in neural stem cells from diverse regions of the adult CNS. *J Neurosci* 19:8487–8497.

Parent JM (2002) The role of seizure-induced neurogenesis in epileptogenesis and brain repair. *Epilepsy Res* 50:179–189.

Rice AC, Khaldi A, Harvey HB, Salman NJ, White F, Fillmore H et al. (2003) Proliferation and neuronal differentiation of mitotically active cells following traumatic brain injury. *Exp Neurol* 183:406–417.

Richardson RM, Broaddus WC, Holloway KL, Sun D, Bullock MR, Fillmore HL (2005) Heterotypic neuronal differentiation of adult subependymal zone neuronal progenitor cells transplanted to the adult hippocampus. *Mol Cell Neurosci* 28:674–682.

Richardson RM, Holloway KL, Bullock MR, Broaddus WC, Fillmore HL (2006) Isolation of neuronal progenitor cells from the adult human neocortex. *Acta Neurochir (Wien)*148:773–777.

Riess P, Molcanyi M, Bentz K, Maegele M, Simanski C, Carlitscheck C et al. (2007) Embryonic stem cell transplantation after experimental traumatic brain injury dramatically improves neurological outcome, but may cause tumors. *J Neurotrauma* 24:216–225.

Riess P, Zhang C, Saatman KE, Laurer HL, Longhi LG, Raghupathi R et al. (2002) Transplanted neural stem cells survive, differentiate, and improve neurological motor function after experimental traumatic brain injury. *Neurosurgery* 51:1043–1052.

Rola R, Otsuka S, Obenaus A, Nelson GA, Limoli CL, VandenBerg SR et al. (2004) Indicators of hippocampal neurogenesis are altered by 56Fe-particle irradiation in a dose-dependent manner. *Radiat Res* 162:442–446.

Roy NS, Benraiss A, Wang S, Fraser RA, Goodman R, Couldwell WT et al. (2000) Promoter-targeted selection and isolation of neural progenitor cells from the adult human ventricular zone. *J Neurosci Res* 59:321–331.

Sanai N, Tramontin AD, Quinones-Hinojosa A, Barbaro NM, Gupta N, Kunwar S et al. (2004) Unique astrocyte ribbon in adult human brain contains neural stem cells but lacks chain migration. *Nature* 427:740–744.

Saxe MD, Battaglia F, Wang JW, Malleret G, David DJ, Monckton JE et al. (2006) Ablation of hippocampal neurogenesis impairs contextual fear conditioning and synaptic plasticity in the dentate gyrus. *Proc Natl Acad Sci USA* 103:17501–17506.

Schmidt RH, Scholten KJ, Maughan PH (1999) Time course for recovery of water maze performance and central cholinergic innervation after fluid percussion injury. *J Neurotrauma* 16:1139–1147.

Shear DA, Tate MC, Archer DR, Hoffman SW, Hulce VD, LaPlaca MC et al. (2004) Neural progenitor cell transplants promote long-term functional recovery after traumatic brain injury. *Brain Res* 1026:11–22.

Shors TJ, Miesegaes G, Beylin A, Zhao M, Rydel T, Gould E (2001) Neurogenesis in the adult is involved in the formation of trace memories. *Nature* 410:372–376.

Shors TJ, Townsend DA, Zhao M, Kozorovitskiy Y, Gould E (2002) Neurogenesis may relate to some but not all types of hippocampal-dependent learning. *Hippocampus* 12:578–584.

Skardelly M, Gaber K, Burdack S, Scheidt F, Hilbig H, Boltze J et al. (2011) Long-term benefit of human fetal neuronal progenitor cell transplantation in a clinically adapted model after traumatic brain injury. *J Neurotrauma* 28:401–414.

Snyder JS, Hong NS, McDonald RJ, Wojtowicz JM (2005) A role for adult neurogenesis in spatial long-term memory. *Neuroscience* 130:843–852.

Sun D, Bullock MR, Altememi N, Zhou Z, Hagood S, Rolfe A et al. (2010) The effect of epidermal growth factor in the injured brain after trauma in rats. *J Neurotrauma* 27:923–938.

Sun D, Bullock MR, McGinn MJ, Zhou Z, Altememi N, Hagood S et al. (2009) Basic fibroblast growth factor-enhanced neurogenesis contributes to cognitive recovery in rats following traumatic brain injury. *Exp Neurol* 216:56–65.

Sun D, Colello RJ, Daugherty WP, Kwon TH, McGinn MJ, Harvey HB et al. (2005) Cell proliferation and neuronal differentiation in the dentate gyrus in juvenile and adult rats following traumatic brain injury. *J Neurotrauma* 22:95–105.

Sun D, Gugliotta M, Rolfe A, Reid W, McQuiston AR, Hu W et al. (2011) Sustained survival and maturation of adult neural stem/progenitor cells after transplantation into the injured brain. *J Neurotrauma* 28:961–972.

Sun D, McGinn MJ, Zhou Z, Harvey HB, Bullock MR, Colello RJ (2007) Anatomical integration of newly generated dentate granule neurons following traumatic brain injury in adult rats and its association to cognitive recovery. *Exp Neurol* 204:264–272.

Sun T, Ma QH (2013) Repairing neural injuries using human umbilical cord blood. *Mol Neurobiol* 47:938–945.

Tajiri N, Acosta SA, Shahaduzzaman M, Ishikawa H, Shinozuka K, Pabon M et al. (2014) Intravenous transplants of human adipose-derived stem cell protect the brain from traumatic brain injury-induced neurodegeneration and motor and cognitive impairments: Cell graft biodistribution and soluble factors in young and aged rats. *J Neurosci* 34:313–326.

Tanapat P, Hastings NB, Reeves AJ, Gould E (1999) Estrogen stimulates a transient increase in the number of new neurons in the dentate gyrus of the adult female rat. *J Neurosci* 19:5792–5801.

Thau-Zuchman O, Shohami E, Alexandrovich AG, Leker RR (2010) Vascular endothelial growth factor increases neurogenesis after traumatic brain injury. *J Cereb Blood Flow Metab* 30:1008–1016.

van Praag H, Christie BR, Sejnowski TJ, Gage FH (1999a) Running enhances neurogenesis, learning, and long-term potentiation in mice. *Proc Natl Acad Sci USA* 96:13427–13431.

van Praag H, Kempermann G, Gage FH (1999b) Running increases cell proliferation and neurogenesis in the adult mouse dentate gyrus. *Nat Neurosci* 2:266–270.

van Praag H, Schinder AF, Christie BR, Toni N, Palmer TD, Gage FH (2002) Functional neurogenesis in the adult hippocampus. *Nature* 415:1030–1034.

Wang S, Cheng H, Dai G, Wang X, Hua R, Liu X et al. (2013) Umbilical cord mesenchymal stem cell transplantation significantly improves neurological function in patients with sequelae of traumatic brain injury. *Brain Res* 26:76–84.

Weiss S, Dunne C, Hewson J, Wohl C, Wheatley M, Peterson AC et al. (1996) Multipotent CNS stem cells are present in the adult mammalian spinal cord and ventricular neuroaxis. *J Neurosci*16:7599–7609.

Wennersten A, Meier X, Holmin S, Wahlberg L, Mathiesen T (2004) Proliferation, migration, and differentiation of human neural stem/progenitor cells after transplantation into a rat model of traumatic brain injury. *J Neurosurg* 100:88–96.

Windrem MS, Roy NS, Wang J, Nunes M, Benraiss A, Goodman R, McKhann GM, Goldman SA (2002) Progenitor cells derived from the adult human subcortical white matter disperse and differentiate as oligodendrocytes within demyelinated lesions of the rat brain. *J Neurosci Res* 69:966–975.

Xiong Y, Mahmood A, Meng Y, Zhang Y, Qu C, Schallert T et al. (2010) Delayed administration of erythropoietin reducing hippocampal cell loss, enhancing angiogenesis and neurogenesis, and improving functional outcome following traumatic brain injury in rats: Comparison of treatment with single and triple dose. *J Neurosurg* 113:598–608.

Xiong Y, Qu C, Mahmood A, Liu Z, Ning R, Li Y et al. (2009) Delayed transplantation of human marrow stromal cell-seeded scaffolds increases transcallosal neural fiber length, angiogenesis, and hippocampal neuronal survival and improves functional outcome after traumatic brain injury in rats. *Brain Res* 1263:183–191.

Yan ZJ, Zhang P, Hu YQ, Zhang HT, Hong SQ, Zhou HL et al. (2013) Neural stem-like cells derived from human amnion tissue are effective in treating traumatic brain injury in rat. *Neurochem Res* 38:1022–1033.

Zhang RL, Zhang L, Zhang ZG, Morris D, Jiang Q, Wang L et al. (2003) Migration and differentiation of adult rat subventricular zone progenitor cells transplanted into the adult rat striatum. *Neuroscience* 116:373–382.

Zheng W, Zhuge Q, Zhong M, Chen G, Shao B, Wang H et al. (2013) Neurogenesis in adult human brain after traumatic brain injury. *J Neurotrauma* 30:1872–1880.

43 Cortical Stimulation-Induced Structural Plasticity and Functional Recovery after Brain Damage

DeAnna L. Adkins

CONTENTS

43.1 INTRODUCTION

Experimental treatments for traumatic brain injury (TBI) often focus on reducing cognitive disorders; however, motor impairments are also common following TBI, but have received less attention. Over the last several decades, there has been growing support for the use of electric or magnetic stimulation of the cortex (CS), or cortical pathways, to enhance recovery of function after other brain injuries, such as stroke. CS studies in stroke patients and in animal models provide compelling evidence that CS can alter brain plasticity and that this likely supports the improved functional motor recovery. While there are only a few studies directly examining the use of CS following TBI, these studies suggest that CS is safe and may also be effective in experimental TBI models. This chapter will cover the evidence supporting the use of CS following brain damage as a means to drive functional recovery and that it can result in structural and functional plasticity of remaining brain areas following injury.

TBI commonly results in physical, cognitive, emotional, and behavioral impairments (Hellawell et al., 1999; Vogenthaler, 1987). The Centers for Disease Control and Prevention has estimated that as a result of TBI, at least 5.3 million people in the United States need long-term or life-long aid in performing activities of daily living (Thurman et al., 1999).

Many experimental treatments for TBI focus on neuroprotection during the acute period after injury. Although this approach holds promise, many people may either miss the small treatment window or may not fully benefit from it. Rehabilitative therapies and treatments that target the post-acute period provide a larger window of opportunity. Although many experimental and clinical studies focus mainly on improving cognitive disorders in this period (Constantinidou et al., 2008; Goldstein, 1990; Griesbach et al., 2004; Kline et al., 2002; Mateer and Sira, 2006; Prins et al., 2003), TBI also frequently results in motor impairments (Langlois et al., 2006; Marshall et al., 2007; Pickett et al., 2007; Teasell et al., 2007; Walker and Pickett, 2007) and these have received less research attention. Rodent models of motor rehabilitative training are relatively well-developed, focusing on animal models of TBI with motor impairments provides a great opportunity to investigate the effects of rehabilitative training on behavioral recovery and the ability for adjuvant treatments, such as cortical stimulation, to enhance training efficacy and drive supportive brain reorganization/plasticity.

Over the past several decades, cortical electrical and magnetic stimulation of the cortex, or cortical pathways, has gained support as a promising therapeutic tool to enhance recovery of function after brain damage and enhance structural and functional brain plasticity. Motor impairments are common but understudied in TBI patients. However, there is a wealth of both human and animal data supporting the efficacy of CS over the motor cortex as a safe and effective tool to improve motor skill learning in intact individuals and as a treatment to improve motor function after brain damage. This chapter will cover the evidence supporting that motor rehabilitation can drive functional recovery and that CS can further enhance behavioral recovery and result in structural and function plasticity of remaining brain areas after brain damage. Although there are only a few studies that have explored the use of brain stimulation techniques after TBI, there is a rapidly expanding body of evidence from stroke recovery research that lays the foundation for the safety and efficacy of using CS as an adjunctive treatment after TBI.

43.2 EFFECTIVENESS OF MOTOR REHABILITATION

A large body of motor rehabilitation research in the past couple of decades has focused on treatment approaches that promote recovery of function, primarily after stroke, by enhancing experience-dependent plasticity as a means to drive reorganization of the remaining brain tissue, often via task-specific motor practice (reviewed in Johansson, 2000; Jones et al., 2013; Kleim and Jones, 2008). Research suggests that motor rehabilitative training can drive the injured brain to alter neural function and structure in tissue that subserves the reinstatement of, or compensation for, lost motor function (reviewed in Adkins et al., 2006a; Nudo, 2003, 2006; Nudo, 2007). In fact, the brain sometimes actually becomes more sensitive to activity-dependent neural plasticity after stroke-like damage (Jones, 2009; Jones et al., 2008; Kleim and Jones, 2008; Nudo, 2003).

Long-term motor dysfunction is common after TBI, yet there are few treatments available. It is assumed that motor rehabilitative therapy after TBI will also improve motor function through similar neural plastic mechanisms seen after stroke damage. Yet, recent findings suggest that injury-induced adaptive plasticity after TBI is restricted, at least in some aspects, compared with injury-induced plasticity found in stroke models (Jones et al., 2012; Kozlowski et al., 2013). Because adaptive plasticity is reduced, it likely will require a great deal of effort to promote recovery after TBI. Additionally, even in these stroke models, behavioral manipulations often have varying degrees of success (Jones, 2009; Jones et al., 2008; Wolf et al., 2008), and significant improvements can take weeks or months of effortful training. Therefore, motor recovery after TBI may be greatly improved by combining appropriate experience-dependent practice with treatments that have been shown to alter brain activity, such as CS of the motor cortex. Combining CS and motor practice after TBI injury may facilitate greater neural plasticity because CS has been found to do in stroke survivors and thus may enhance greater motor recovery.

43.3 IMPLANTABLE CS DEVICES ENHANCE MOTOR FUNCTION RECOVERY AND DRIVE NEURAL PLASTICITY

Extradural electrical stimulation over the motor cortex was initially used as an experimental treatment for intractable central pain in humans after stroke. In several studies, a subset of stroke subjects reported improvements in motor and sensory impairments, including reductions in dyskinesia, tremor, and/or weakness after CS when the electrodes were placed over the motor cortex (Brown and Pilitsis, 2005; Franzini et al., 2000; Garcia-Larrea et al., 1999; Katayama et al., 1997, 1998, 2001; Tsubokawa et al., 1991, 1993). In clinical phase 1–2 trials, implantable epidural bipolar electrodes were placed over the motor cortex of stroke survivors. In these trials, chronic stroke survivors received epidural bipolar CS or no stimulation during physical therapy. In phase 1 and 2 studies, CS was delivered to the hand area of the motor cortex at 50 Hz or 101 Hz. These two studies demonstrated the safety and efficacy of epidural CS combined with motor practice to improve motor function as assessed on the Upper Extremity Fugl Meyer (Brown et al., 2006; Levy et al., 2008a), Stroke Impact Scale (Brown et al., 2006), and Arm Motor Ability Test (Levy et al., 2008a) compared with physical therapy alone. Although these earlier studies were promising, the larger phase 3 trial failed to improve motor function to meet the primary motor end points, when all subjects were included in the analysis (Levy et al., 2008b). However, follow-up analysis of the phase 3 data revealed that a subset of subjects actually showed major motor improvements (Plow et al., 2009). The subjects with the motor improvements were the ones in which CS was delivered in a manner consistent with the parameters found to be effective in the smaller phase 1 and 2 clinical trials and in the preceding animal studies. In the previous animal studies, short bursts of CS were used to determine the lowest current necessary to evoke involuntary motor movements or the movement threshold. CS was then delivered during physical therapy between 50% and 70% of these thresholds.

In all of the clinical trials, CS current was also set at 50% of movement thresholds. In those in which movement could not be evoked, subjects were stimulated at a predefined maximum current of 6.5 mA. The earlier clinical trials differed from phase 3 in the proportion of subjects in which movements could be evoked. In the phase 3 trial, movement thresholds were undefined or unobtainable in the majority of the participants. This difference is likely the reason for the lack of overall effect in the phase 3 trial because in only 16% of subjects completing the phase 3 study were movements evoked and this was the subset of subjects that were found to show major improvements in Fugl-Meyer (upper extremity Fugl-Meyer; Northstar Inc.) (Plow et al., 2009). In contrast, movement thresholds were evoked in 100% and 42% of subjects in the phase 1 and 2 trials, respectively. These data suggest that a minimum level of motor system integrity or activation threshold of the descending motor pathways must exist for CS to be effective. Effective poststroke CS likely also requires that the stimulation be focused over the appropriate cortical region(s). Thus, it is likely that the phase 3 trial failed because it did not use the parameters tested to be most effective in preclinical trials, or at least not in all the subjects. Despite the phase 3 clinical trial findings, CS remains a promising adjunctive therapy based upon the combination early clinical studies with implantable electrodes, the strong preclinical data, and promising results from noninvasive stimulation approaches.

In rats and monkeys, bipolar, cathodal, or anodal CS delivered via epidural or subdural electrodes over peri-ischemic motor cortex (MC) combined with rehabilitative motor training significantly enhances motor performance on skilled reaching tasks (Adkins-Muir and Jones, 2003; Adkins et al., 2006b, 2008; Boychuk et al., 2011; Kleim et al., 2003b; Moon et al., 2009; Plautz et al., 2003; Teskey et al., 2003a; Zheng et al., 2013; Zhou et al., 2010). CS combined with training increases dendritic and synaptic densities (Figure 43.1) (Adkins-Muir and Jones, 2003; Adkins et al., 2006, 2008; Zheng et al., 2013), forelimb movement representation area (Kleim et al., 2003a;

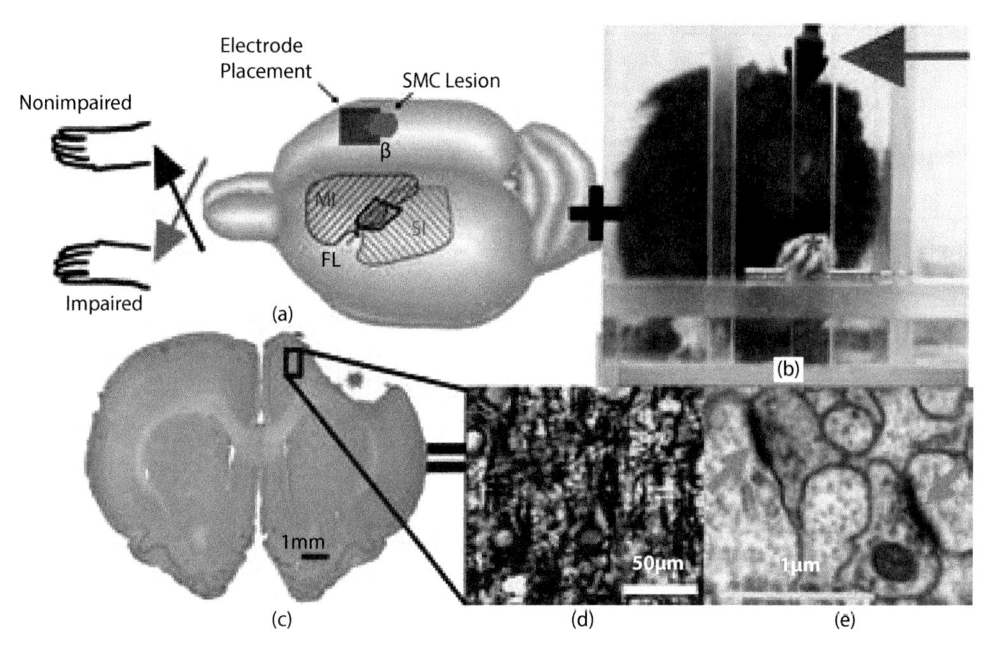

FIGURE 43.1 Illustration of extradural CS following an experimental model of ischemic damage and resulting brain plasticity. (a) Illustration of a rat brain demonstrating the relative location of the extradural electrode and ischemic lesion/damage of the SMC. (b) Image of a rat reaching through a small window to retrieve a pellet in the single-pellet reaching task. When CS is delivered via implanted electrode (purple arrow) concurrent with the impaired forelimb (red asterisk) practice on a skilled reaching task, animals significantly improve their reaching performance compared with animals undergoing training without CS. (c) Example of ischemic damage (red starburst) to the SMC shown in a coronal Nissl-stained section. The black box represents the sampling area within the remaining motor cortex, where CS plus training effects on neuronal plasticity are found. (d) Photomicrograph representative image of microtubule-2 immunoreactive (MAP2-IR) dendritic processes in layer V of the lesion hemisphere's remaining motor cortex. Three weeks of subdural 50-Hz bipolar CS concurrent with impaired limb reach training significantly increased the density of MAP2-IR processes compared to training alone (Adkins-Muir and Jones, 2003). (e) Electron microscopy photomicrograph of layer V of remaining motor cortex following unilateral ischemic lesion. Cathodal 100HZ CS combined with reach training significantly increased the density of synapses (orange arrows) compared with reach training alone (Adkins et al., 2008).

Plautz et al., 2003), and motor cortical evoked potentials (Teskey et al., 2003b) in peri-infarct MC. CS combined with training also reduces peri-injury gliosis (Zheng, 2013) and, although not always demonstrated, may in some cases promote anti-apoptotic pathways (Zhou et al., 2010). Extradural CS has been found to be most beneficial when paired concurrently with task-specific practice and does not appear to generalize to other motor behaviors performed nonconcurrent with CS (Adkins-Muir and Jones, 2003; Adkins et al., 2006b, 2008). Electrical stimulation effects likely occur by delivery current across neuronal membrane and driving activity (Histed et al., 2009) and are likely polarity, frequency, and timing dependent. For example, while epidural cathodal and anodal CS both improve the efficacy of motor rehabilitative treatment in animal models of stroke, cathodal current resulted in earlier improvements during the treatment period than did anodal current and may preserve a greater density of layer V penumbra neurons compared to anodal CS and rehabilitative training (RT) alone (Adkins et al., 2006b).

43.4 NONINVASIVE CS IMPROVES MOTOR FUNCTION AFTER STROKE AND ALTERS BRAIN FUNCTION

Further support that CS will likely improve motor performance in survivors of TBI comes from studies using noninvasive transcranial stimulation to modulate cortical activity in stroke subjects (Pape et al., 2006). Transcranial magnetic stimulation (TMS) has been used both to investigate normal and injured brain function, including after TBI, and as a tool to alter brain excitability to treat motor impairments, aphasia, and depression (Bernabeu et al., 2009; Chistyakov et al., 1999, 2001; Lapitskaya et al., 2013; reviewed in Dayan et al., 2013; George et al., 2013; Schulz et al., 2010; Shah et al., 2013). TMS coils, which use magnetic force to generate electric current, when held over the motor cortex can drive involuntary muscle movements and repeated TMS (rTMS) can either facilitate neural activity using high-frequency pulse trains (5–20 Hz) or inhibit neural activation via low-frequency trains (~0.2–1 Hz; Figure 43.2a) (Dayan et al., 2013; Shah et al., 2013). Although the exact mechanism of TMS effects are not fully known, the current hypothesis is that TMS likely modulates ion channels on axons and can directly activate neurons (Huerta and Volpe, 2009). TMS has been found to alter cortical excitation for a period after stimulation has ceased (Chen et al., 1997; Fregni and Pascual-Leone, 2007; Hummel and Cohen, 2005) and thus may facilitate learning off-line when followed by rehabilitative training treatments. Currently, the size and bulkiness of TMS makes it less ideal to provide CS during most rehabilitative treatment regiments.

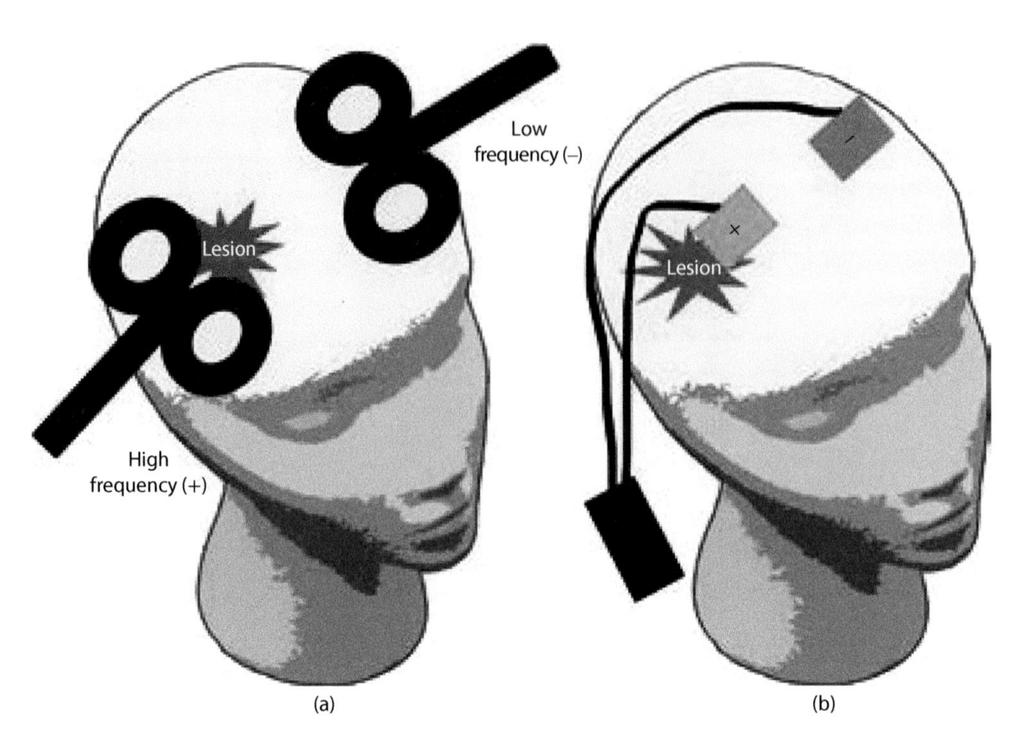

FIGURE 43.2 Illustration of noninvasive stimulation techniques. (a) Diagram of TMS coils over the motor cortex relative to side of brain injury. High-frequency rTMS over the lesion hemisphere has been found to be excitatory and can increase motor performance after stroke, whereas low-frequency rTMS is inhibitory and may reduce overactivation of the nonlesion hemisphere (Dayan et al., 2013; Shah et al., 2013). (b) Diagram of tDCS electrode placement relative to side of injury. Generally, the anodal electrode is placed ipsilateral to injury and is excitatory, whereas the cathode electrode can induce inhibition of the nonlesion hemisphere (Monte-Silva et al., 2013). Rarely are the electrodes placed in this configuration, but are done so in the image for illustrative purposes.

Similarly, transcranial direct current (tDCS) applied over the cortex can alter neural activity. Compared with TMS, tDCS modulates brain activity through relatively weak electric currents (~1–2 mA), likely by modulating resting membrane activity via ion channel activation (Fregni and Pascual-Leone, 2007; Nitsche et al., 2008). The strength of current, length of stimulation, and the electrode positioning of the anodal and cathode electrode contacts on the skull determines the direction of modulation, such that anodal leads are excitatory and cathodal stimulation is inhibitory (Figure 43.2b) (Monte-Silva et al., 2013; Nitsche and Paulus, 2000). CS with tDCS is thought to shift neural excitation longer than that found after repeated TMS (Paulus, 2003).

Excitatory CS has been shown to enhance motor learning and "relearning" after brain damage, likely by strengthening synaptic connections through long-term potentiation (LTP)-like molecular changes (Nitsche et al., 2003). Recent studies support this hypothesis. Direct stimulation of mouse slices is N-methyl-D-aspartate receptor dependent and when combined with repetitive, low-frequency synaptic activation enhances secretion of brain-derived neurotrophic factor and TrkB activation (Fritsch et al., 2010). Further, anodal tDCS over the primary motor cortex induces long-lasting LTP-like effects that are N-methyl-D-aspartate and calcium dependent in healthy human cortex (Monte-Silva et al., 2013).

CS may also be beneficial following brain damage by restoring interhemispheric balance between the damaged and nondamaged hemispheres or other connected brain regions. After stroke and brain injury, several studies have revealed that cortical damage can lead to disruption of the balance between excitatory and inhibitory neural transmission, such that, for example, the contralesional hemisphere is overly excited, whereas in the damaged hemisphere excitation is suppressed (Butefisch et al., 2008; Domann et al., 1993; Mittmann et al., 1994; Nair et al., 2007; Neumann-Haefelin and Witte, 2000; Shimizu et al., 2002). Noninvasive CS has successfully been used to alter the balance of postinjury excitation by either exciting the injured hemisphere or inhibiting the contrainjury cortex. Using high-frequency TMS or anodal tDCS over the stroke-affected motor cortex (Fregni et al., 2005, 2006; Hummel et al., 2005; Khedr et al., 2005; Kim et al., 2006; Mansur et al., 2005) can acutely improve paretic hand and limb performance. These findings also are likely related to facilitatory activation of the neurons to drive motor movements. Similar improvements are also found by stimulating with low-frequency TMS or cathodal tDCS stimulation of the contralesional hemisphere (Di Lazzaro et al., 2008; Fregni et al., 2005; Hummel and Cohen, 2006).

Poststroke grasping function was improved in subcortical stroke subjects after ipsilesional repeated excitatory TMS (Nowak et al., 2008). Repeated high-frequency TMS has also been found to increase motor evoked potential amplitudes in acute stroke (Di Lazzaro et al., 2008) and reduce spasticity and hemiparesis in chronic stroke subjects (Mally and Dinya, 2008). However, many of these improvements are only evident for a short period after stimulation. Motor improvements are found to be more enduring, however, if stimulation is coupled to motor training sessions (Boggio et al., 2007; Kim et al., 2006; Nair, 2007). For example, Nair et al. (2007) found that tDCS delivered during occupational therapy after stroke resulted in more persistent improvements than therapy alone.

43.5 CORTICAL STIMULATION AFTER TBI

Thus, stimulation of motor cortex after stroke has been repeatedly shown to be a promising adjunctive therapy and, in a handful of human and many animal studies, CS further improved motor functional recovery when combined with motor rehabilitation. In animal models, we have consistently found that motor cortical electrical stimulation coupled with motor rehabilitative therapy improves motor function after ischemic stroke (Adkins et al., 2006b, 2008; Adkins-Muir and Jones, 2003). Although these studies suggest that cortical stimulation is linked to plasticity of neural structures and a reorganization of motor cortex after ischemic damage, it is relatively untested whether after TBI. Further, previous research suggests that pathological cellular sequelae in stroke models are quite different from those after TBI, even when the region of damage is similar (Jones et al., 2012). CS efficacy will depend, in part, on its interaction with the TBI-specific neurobiological pathology.

43.6 MODELING MOTOR IMPAIRMENTS AND RECOVERY IN ANIMAL MODELS OF TBI

Although not all animal models of TBI produce direct injury to the sensorimotor cortex (SMC), those that do affect the SMC can result in forelimb and hindlimb deficits. Contusion impact injury to the frontal lobe produces a cavity in the sensorimotor cortex (SMC) and damage to the pyramidal tract and dorsolateral striatum (Whishaw et al., 2004). After contusions, rats display a hyperreliance upon the limb opposite of the injury during postural support and reduced reaching performance on the single-pellet reaching task and tray-reaching task with both limbs, but more markedly with the contralesional limb (Whishaw et al., 2004). Using a detailed, qualitative assessment of reaching movement analysis, the ipsilesional limb was found to be significantly impaired in aiming, supination, and food pellet release. Additionally rats had postural impairments that contributed to these abnormal reaching movements.

Unilateral TBI models can also be used to model hemiparesis, as seen following many moderate to severe TBIs in humans. Controlled cortical impact (CCI) models provide a means to precisely control impact duration and depth of penetration into the brain, producing relatively consistent injuries that reproduce many aspects of human TBI, including cortical contusions and distal axonal injury (Dixon et al., 1991; Kozlowski et al., 2013). Unilateral CCI over the SMC induces long-term impairments with both contrainjury forelimbs and hindlimbs as assessed in tasks such as the pegged beam, narrow beam, and grid platform tasks that measure coordinated limb use and alterations in postural support with the forelimbs. Soblosky et al. (1996) demonstrated that over the first 7–10 days postinjury, rats showed rapid improvement in performance on these tasks, although they do not return to postinjury levels. Over the proceeding weeks, these rats show a more gradual reduction in motor impairments (Soblosky et al., 1996) and do not return to preinjury levels. The authors argued that it is likely that the first stage of rapid improvement was due to resolution of acute pathological consequences of the brain injury (i.e. reduction in edema) or may reflect a functional recovery of less severely affected neurons or improvements due to rapid reorganization of other motor-related brain regions that take over some of the lost function.

Although the actual mechanism of the early recovery processes are unknown, Soblosky et al. (1996) provide behavioral data supporting that much of the early recovery is likely not due to a true restoration of function, but more because of behavioral compensation. Further, the timing of improvements on the different tasks were not the same, such that better motor performance on one task (nonpegged beam) did not transfer to improvements on a similar task (pegged beam) at the same time frame postinjury. Taken together, these data suggest, as has been found in many motor rehabilitation tasks, that many behavioral improvements may be task-specific, linked to experience or learning that occurs during the performance or practice of that task, as is often reported in human and animal models of poststroke rehabilitation research (Jones et al., 2013).

Similarly, following more focal unilateral CCI over the forelimb area of the sensorimotor cortex (Fl-SMC), adult male rats show impairments in coordinate locomotion on the grid platform task and a hyperreliance upon the unimpaired forelimb for postural support compared to sham animals, out to 30 days postinjury (Jones et al., 2012). Although these motor impairments are similar to those also seen after electrolytic and ischemic lesion to the Fl-SMC, the resulting experience-dependent brain changes are quite different. After electrolytic or ischemic lesion of the Fl-SMC, the hyperreliance upon the forelimb opposite the injury results in a time- and experience-dependent increase in dendritic and synaptic plasticity in the contralesional SMC compared with sham operates (Adkins et al., 2004; Bury et al., 2000; Jones and Schallert, 1992, 1994). However, after CCI to the Fl-SMC, which results in similar behavioral impairments, there is a reduction in experience-dependent plasticity (Jones et al., 2012). After TBI to Fl-SMC, the cortex undergoes neuronal structural degradation in both hemispheres out to 28 days postinjury, which likely indicates that the pathological sequence of TBI-linked injury limits use-dependent plasticity and may limit the effectiveness of therapeutic interventions, such as rehabilitative training and CS.

43.7 POSSIBLE LIMITATIONS OF CS AFTER TBI

There are already identified limitations of CS when used to enhance rehabilitative training after stroke and these may also interfere with the efficacy when used with TBI patients. Animal studies demonstrate that injury severity can reduce the efficacy of both epidural CS (Adkins et al., 2008) and tDCS in animal models of stroke (Moon et al., 2013). For example, CS coupled with rehabilitative training reduces lesion-induced impairments in forelimb reaching in moderately impaired subjects, but not in more severely impaired animals (Adkins et al., 2008); however, both the moderate and severely impaired subjects did improve in the overall quality of reaching compared with training alone, and thus more severely impaired animals may require longer treatment to fully benefit from CS (Adkins et al., 2008). Additionally, the timing of CS treatment after initial injury may alter efficacy. In animal studies, CS initiated 3 months after ischemic damage to the SMC does not enhance motor function compared with training alone, even though training alone did have some benefit during this period (O'Bryant and Jones, 2008). Thus, this last study suggests that, at least in a stroke model, CS may only significantly enhance training effects if administered within an appropriate time window after injury.

43.8 CONCLUSION

In rat models of TBI, CS has been shown to drive an expansion of forelimb representation of the forelimb representational maps (Nishibe et al., 2010), improve behavior on a reaching task, and induce greater neuronal activity as visualized by greater c-Fos immunoreactivity (Yoon et al., 2012). Thus, CS has been demonstrated to be a safe and effective modulator of post-TBI behavioral and cortical function. Studies using CS to further enhance effectiveness of post-TBI motor rehabilitative training requires more study to determine the proper parameters and limitations.

REFERENCES

Adkins-Muir, D.L., and T.A. Jones. 2003. Cortical electrical stimulation combined with rehabilitative training: Enhanced functional recovery and dendritic plasticity following focal cortical ischemia in rats. *Neurol Res*. 25:780–8.

Adkins, D.L., J. Boychuk, M.S. Remple, and J.A. Kleim. 2006a. Motor training induces experience-specific patterns of plasticity across motor cortex and spinal cord. *J Appl Physiol*. 101:1776–82.

Adkins, D.L., P. Campos, D. Quach, M. Borromeo, K. Schallert, and T.A. Jones. 2006b. Epidural cortical stimulation enhances motor function after sensorimotor cortical infarcts in rats. *Exp Neurol*. 200:356–70.

Adkins, D.L., J.E. Hsu, and T.A. Jones. 2008. Motor cortical stimulation promotes synaptic plasticity and behavioral improvements following sensorimotor cortex lesions. *Exp Neurol*. 212:14–28.

Adkins, D.L., A.C. Voorhies, and T.A. Jones. 2004. Behavioral and neuroplastic effects of focal endothelin-1 induced sensorimotor cortex lesions. *Neuroscience*. 128:473–86.

Bernabeu, M., A. Demirtas-Tatlidede, E. Opisso, R. Lopez, J.M. Tormos, and A. Pascual-Leone. 2009. Abnormal corticospinal excitability in traumatic diffuse axonal brain injury. *J Neurotrauma*. 26:2185–93.

Boggio, P.S., A. Nunes, S.P. Rigonatti, M.A. Nitsche, A. Pascual-Leone, and F. Fregni. 2007. Repeated sessions of noninvasive brain DC stimulation is associated with motor function improvement in stroke patients. *Restor Neurol Neurosci*. 25:123–9.

Boychuk, J.A., D.L. Adkins, and J.A. Kleim. 2011. Distributed versus focal cortical stimulation to enhance motor function and motor map plasticity in a rodent model of ischemia. *Neurorehabil Neural Repair*. 25:88–97.

Brown, J.A., H.L. Lutsep, M. Weinand, and S.C. Cramer. 2006. Motor cortex stimulation for the enhancement of recovery from stroke: A prospective, multicenter safety study. *Neurosurgery*. 58:464–73.

Brown, J.A., and J.G. Pilitsis. 2005. Motor cortex stimulation for central and neuropathic facial pain: A prospective study of 10 patients and observations of enhanced sensory and motor function during stimulation. *Neurosurgery*. 56:290–7; discussion 290–7.

Bury, S.D., D.L. Adkins, J.T. Ishida, C.M. Kotzer, A.C. Eichhorn, and T.A. Jones. 2000. Denervation facilitates neuronal growth in the motor cortex of rats in the presence of behavioral demand. *Neurosci Lett*. 287:85–8.

Butefisch, C.M., M. Wessling, J. Netz, R.J. Seitz, and V. Homberg. 2008. Relationship between interhemispheric inhibition and motor cortex excitability in subacute stroke patients. *Neurorehabil Neural Repair*. 22:4–21.

Chen, R., J. Classen, C. Gerloff, P. Celnik, E.M. Wassermann, M. Hallett, and L.G. Cohen. 1997. Depression of motor cortex excitability by low-frequency transcranial magnetic stimulation. *Neurology*. 48:1398–403.

Chistyakov, A.V., H. Hafner, J.F. Soustiel, M. Trubnik, G. Levy, and M. Feinsod. 1999. Dissociation of somatosensory and motor evoked potentials in non-comatose patients after head injury. *Clin Neurophysiol*. 110:1080–9.

Chistyakov, A.V., J.F. Soustiel, H. Hafner, M. Trubnik, G. Levy, and M. Feinsod. 2001. Excitatory and inhibitory corticospinal responses to transcranial magnetic stimulation in patients with minor to moderate head injury. *J Neurol Neurosurg Psychiatry*. 70:580–7.

Constantinidou, F., R.D. Thomas, and L. Robinson. 2008. Benefits of categorization training in patients with traumatic brain injury during post-acute rehabilitation: Additional evidence from a randomized controlled trial. *J Head Trauma Rehabil*. 23:312–28.

Dayan, E., N. Censor, E.R. Buch, M. Sandrini, and L.G. Cohen. 2013. Noninvasive brain stimulation: From physiology to network dynamics and back. *Nat Neurosci*. 16:838–44.

Di Lazzaro, V., F. Pilato, M. Dileone, P. Profice, F. Capone, F. Ranieri et al. 2008. Modulating cortical excitability in acute stroke: A repetitive TMS study. *Clin Neurophysiol*. 119:715–23.

Di Lazzaro, V., F. Pilato, M. Dileone, P. Profice, A. Oliviero, P. Mazzone et al. 2008. Low frequency rTMS suppresses specific excitatory circuits in the human motor cortex. *J Physiol*. 586:4481–7.

Dixon, C.E., G.L. Clifton, J.W. Lighthall, A.A. Yaghmai, and R.L. Hayes. 1991. A controlled cortical impact model of traumatic brain injury in the rat. *J Neurosci Methods*. 39:253–62.

Domann, R., G. Hagemann, M. Kraemer, H.J. Freund, and O.W. Witte. 1993. Electrophysiological changes in the surrounding brain tissue of photochemically induced cortical infarcts in the rat. *Neurosci Lett*. 155:69–72.

Franzini, A., P. Ferroli, D. Servello, and G. Broggi. 2000. Reversal of thalamic hand syndrome by long-term motor cortex stimulation. *J Neurosurg*. 93:873–5.

Fregni, F., P.S. Boggio, C.G. Mansur, T. Wagner, M.J. Ferreira, M.C. Lima et al. 2005. Transcranial direct current stimulation of the unaffected hemisphere in stroke patients. *Neuroreport*. 16:1551–5.

Fregni, F., P.S. Boggio, A.C. Valle, R.R. Rocha, J. Duarte, M.J. Ferreira et al. 2006. A sham-controlled trial of a 5-day course of repetitive transcranial magnetic stimulation of the unaffected hemisphere in stroke patients. *Stroke*. 37:2115–22.

Fregni, F., and A. Pascual-Leone. 2007. Technology insight: Noninvasive brain stimulation in neurology-perspectives on the therapeutic potential of rTMS and tDCS. *Nat Clin Pract Neurol*. 3:383–93.

Fritsch, B., J. Reis, K. Martinowich, H.M. Schambra, Y. Ji, L.G. Cohen et al. 2010. Direct current stimulation promotes BDNF-dependent synaptic plasticity: Potential implications for motor learning. *Neuron*. 66:198–204.

Garcia-Larrea, L., R. Peyron, P. Mertens, M.C. Gregoire, F. Lavenne, D. Le Bars et al. 1999. Electrical stimulation of motor cortex for pain control: A combined PET-scan and electrophysiological study. *Pain*. 83:259–73.

George, M.S., J.J. Taylor, and E.B. Short. 2013. The expanding evidence base for rTMS treatment of depression. *Curr Opin Psychiatry*. 26:13–8.

Goldstein, M. 1990. Traumatic brain injury: A silent epidemic. *Ann Neurol*. 27:327.

Griesbach, G.S., D.A. Hovda, R. Molteni, A. Wu, and F. Gomez-Pinilla. 2004. Voluntary exercise following traumatic brain injury: Brain-derived neurotrophic factor upregulation and recovery of function. *Neuroscience*. 125:129–39.

Hellawell, D.J., R.T. Taylor, and B. Pentland. 1999. Cognitive and psychosocial outcome following moderate or severe traumatic brain injury. *Brain Inj*. 13:489–504.

Histed, M.H., V. Bonin, and R.C. Reid. 2009. Direct activation of sparse, distributed populations of cortical neurons by electrical microstimulation. *Neuron*. 63:508–22.

Huerta, P.T., and B.T. Volpe. 2009. Transcranial magnetic stimulation, synaptic plasticity and network oscillations. *J Neuroeng Rehabil*. 6:7.

Hummel, F., P. Celnik, P. Giraux, A. Floel, W.H. Wu, C. Gerloff et al. 2005. Effects of non-invasive cortical stimulation on skilled motor function in chronic stroke. *Brain*. 128:490–9.

Hummel, F., and L.G. Cohen. 2005. Improvement of motor function with noninvasive cortical stimulation in a patient with chronic stroke. *Neurorehabil Neural Repair*. 19:14–9.

Hummel, F.C., and L.G. Cohen. 2006. Non-invasive brain stimulation: A new strategy to improve neurorehabilitation after stroke? *Lancet Neurol*. 5:708–12.

Johansson, B.B. 2000. Brain plasticity and stroke rehabilitation. The Willis lecture. *Stroke*. 31:223–30.

Jones, T.A., R.P. Allred, D.L. Adkins, J.E. Hsu, A. O'Bryant, and M.A. Maldonado. 2008. Remodeling the brain with behavioral experience after stroke. *Stroke*. In press.

Jones, T.A., R.P. Allred, S.C. Jefferson, A.L. Kerr, D.A. Woodie, S.Y. Cheng, and D.L. Adkins. 2013. Motor system plasticity in stroke models: Intrinsically use-dependent, unreliably useful. *Stroke*. 44:S104–6.

Jones, T.A., Allred, R. P., Adkins, D. L., Hsu, J. E., O'Bryant, A., Maldonado, M. A. 2009. Remodeling the brain with behavioral experience after stroke. *Stroke*. 40:S136–8.

Jones, T.A., D.J. Liput, E.L. Maresh, N. Donlan, T.J. Parikh, D. Marlowe et al. 2012. Use-dependent dendritic regrowth is limited after unilateral controlled cortical impact to the forelimb sensorimotor cortex. *J Neurotrauma*. 29:1455–68.

Jones, T.A., and T. Schallert. 1992. Overgrowth and pruning of dendrites in adult rats recovering from neocortical damage. *Brain Res*. 581:156–60.

Jones, T.A., and T. Schallert. 1994. Use-dependent growth of pyramidal neurons after neocortical damage. *J Neurosci*. 14:2140–52.

Katayama, Y., C. Fukaya, and T. Yamamoto. 1997. Control of post-stroke involuntary and voluntary movement disorders with deep brain or epidural cortical stimulation. *Stereotact Funct Neurosurg*. 69:73–9.

Katayama, Y., C. Fukaya, and T. Yamamoto. 1998. Poststroke pain control by chronic motor cortex stimulation: Neurological characteristics predicting a favorable response. *J Neurosurg*. 89:585–91.

Katayama, Y., T. Yamamoto, K. Kobayashi, M. Kasai, H. Oshima, and C. Fukaya. 2001. Motor cortex stimulation for post-stroke pain: Comparison of spinal cord and thalamic stimulation. *Stereotact Funct Neurosurg*. 77:183–6.

Khedr, E.M., M.A. Ahmed, N. Fathy, and J.C. Rothwell. 2005. Therapeutic trial of repetitive transcranial magnetic stimulation after acute ischemic stroke. *Neurology*. 65:466–8.

Kim, Y.H., S.H. You, M.H. Ko, J.W. Park, K.H. Lee, S.H. Jang et al. 2006. Repetitive transcranial magnetic stimulation-induced corticomotor excitability and associated motor skill acquisition in chronic stroke. *Stroke*. 37:1471–6.

Kleim, J.A., R. Bruneau, P. VandenBerg, E. MacDonald, R. Mulrooney, and D. Pocock. 2003a. Motor cortex stimulation enhances motor recovery and reduces peri-infarct dysfunction following ischemic insult. *Neurol Res*. 25:789–93.

Kleim, J.A., R. Bruneau, P. VandenBerg, E. MacDonald, R. Mulrooney, and D. Pocock. 2003b. Motor cortex stimulation enhances motor recovery and reduces peri-infarct dysfunction following ischemic insult. *Neurol Res*. 25:789–93.

Kleim, J.A., and T.A. Jones. 2008. Principles of experience-dependent neural plasticity: Implications for rehabilitation after brain damage. *J Speech Lang Hear Res*. 51:S225–39.

Kline, A.E., J.L. Massucci, D.W. Marion, and C.E. Dixon. 2002. Attenuation of working memory and spatial acquisition deficits after a delayed and chronic bromocriptine treatment regimen in rats subjected to traumatic brain injury by controlled cortical impact. *J Neurotrauma*. 19:415–25.

Kozlowski, D.A., J.L. Leasure, and T. Schallert. 2013. The control of movement following traumatic brain injury. *Comprehens Physiol*. 3:121–39.

Langlois, J.A., W. Rutland-Brown, and M.M. Wald. 2006. The epidemiology and impact of traumatic brain injury: A brief overview. *J Head Trauma Rehabil*. 21:375–8.

Lapitskaya, N., S.K. Moerk, O. Gosseries, J.F. Nielsen, and A.M. de Noordhout. 2013. Corticospinal excitability in patients with anoxic, traumatic, and non-traumatic diffuse brain injury. *Brain Stim*. 6:130–7.

Levy, R., S. Ruland, M. Weinand, D. Lowry, R. Dafer, and R. Bakay. 2008a. Cortical stimulation for the rehabilitation of patients with hemiparetic stroke: A multicenter feasibility study of safety and efficacy. *J Neurosurg*. 108:707–14.

Levy, R.M., R.R. Benson, and C.J. Winstein. 2008b. Cortical stimulation for upper-extremity hemiparesis from ischemic stroke: Everest Study Primary Endpoint Results. *International Stroke Conference*

Mally, J., and E. Dinya. 2008. Recovery of motor disability and spasticity in post-stroke after repetitive transcranial magnetic stimulation (rTMS). *Brain Res Bull*. 76:388–95.

Mansur, C.G., F. Fregni, P.S. Boggio, M. Riberto, J. Gallucci-Neto, C.M. Santos et al. 2005. A sham stimulation-controlled trial of rTMS of the unaffected hemisphere in stroke patients. *Neurology*. 64:1802–4.

Marshall, S., R. Teasell, N. Bayona, C. Lippert, J. Chundamala, J. Villamere et al. 2007. Motor impairment rehabilitation post acquired brain injury. *Brain Inj*. 21:133–60.

Mateer, C.A., and C.S. Sira. 2006. Cognitive and emotional consequences of TBI: Intervention strategies for vocational rehabilitation. *NeuroRehabilitation*. 21:315–26.

Mittmann, T., H.J. Luhmann, R. Schmidt-Kastner, U.T. Eysel, H. Weigel, and U. Heinemann. 1994. Lesion-induced transient suppression of inhibitory function in rat neocortex in vitro. *Neuroscience*. 60:891–906.

Monte-Silva, K., M.F. Kuo, S. Hessenthaler, S. Fresnoza, D. Liebetanz, W. Paulus et al. 2013. Induction of late LTP-like plasticity in the human motor cortex by repeated non-invasive brain stimulation. *Brain Stim.* 6:424–32.

Moon, S.K., Y.I. Shin, H.I. Kim, H. Kim, J.O. Lee, and M.C. Lee. 2009. Effect of prolonged cortical stimulation differs with size of infarct after sensorimotor cortical lesions in rats. *Neurosci Lett.* 460:152–5.

Nair, D.G., Hamelin, S., Pascual-Leone, A., and Schlaug, G.,. 2007. Direct current stimulation in combination with occupational therapy for 5 consecutive days improves motor function in chronic stroke patients. *Stroke.* 38:517.

Nair, D.G., S. Hutchinson, F. Fregni, M. Alexander, A. Pascual-Leone, and G. Schlaug. 2007. Imaging correlates of motor recovery from cerebral infarction and their physiological significance in well-recovered patients. *NeuroImage.* 34:253–63.

Neumann-Haefelin, T., and O.W. Witte. 2000. Periinfarct and remote excitability changes after transient middle cerebral artery occlusion. *J Cerebral Blood Flow Metabol.* 20:45–52.

Nishibe, M., S. Barbay, D. Guggenmos, and R.J. Nudo. 2010. Reorganization of motor cortex after controlled cortical impact in rats and implications for functional recovery. *J Neurotrauma.* 27:2221–32.

Nitsche, M.A., L.G. Cohen, E.M. Wassermann, A. Priori, N. Lang, A. Antal, W. Paulus, F. Hummel, P.S. Boggio, F. Fregni, and A. Pascual-Leone. 2008. Transcranial direct current stimulation: State of the art 2008. *Brain Stim.* 1:206–23.

Nitsche, M.A., K. Fricke, U. Henschke, A. Schlitterlau, D. Liebetanz, N. Lang et al. 2003. Pharmacological modulation of cortical excitability shifts induced by transcranial direct current stimulation in humans. *J Physiol.* 553:293–301.

Nitsche, M.A., and W. Paulus. 2000. Excitability changes induced in the human motor cortex by weak transcranial direct current stimulation. *J Physiol.* 527 Pt 3:633–9.

Nowak, D.A., C. Grefkes, M. Dafotakis, S. Eickhoff, J. Kust, H. Karbe et al. 2008. Effects of low-frequency repetitive transcranial magnetic stimulation of the contralesional primary motor cortex on movement kinematics and neural activity in subcortical stroke. *Arch Neurol.* 65:741–7.

Nudo, R.J. 2003. Adaptive plasticity in motor cortex: Implications for rehabilitation after brain injury. *J Rehabil Med*:7–10.

Nudo, R.J. 2006. Plasticity. *NeuroRx.* 3:420–7.

Nudo, R.J. 2007. Postinfarct cortical plasticity and behavioral recovery. *Stroke.* 38:840–5.

O'Bryant, A., and T.A. Jones. 2008. Effects of delayed onset of motor rehabilitative training and cortical stimulation on functional deficits after unilateral cortical infarcts in rats. *Society for Neuroscience Abstracts,* 148.20.

Pape, T.L., J. Rosenow, and G. Lewis. 2006. Transcranial magnetic stimulation: A possible treatment for TBI. *J Head Trauma Rehabil.* 21:437–51.

Paulus, W. 2003. Transcranial direct current stimulation (tDCS). *Suppl Clin Neurophysiol.* 56:249–54.

Pickett, T.C., L.S. Radfar-Baublitz, S.D. McDonald, W.C. Walker, and D.X. Cifu. 2007. Objectively assessing balance deficits after TBI: Role of computerized posturography. *J Rehabil Res Dev.* 44:983–90.

Plautz, E.J., S. Barbay, S.B. Frost, K.M. Friel, N. Dancause, E.V. Zoubina et al. 2003. Post-infarct cortical plasticity and behavioral recovery using concurrent cortical stimulation and rehabilitative training: A feasibility study in primates. *Neurol Res.* 25:801–10.

Plow, E.B., J.R. Carey, R.J. Nudo, and A. Pascual-Leone. 2009. Invasive cortical stimulation to promote recovery of function after stroke: A critical appraisal. *Stroke.* 40:1926–31.

Prins, M.L., J.T. Povlishock, and L.L. Phillips. 2003. The effects of combined fluid percussion traumatic brain injury and unilateral entorhinal deafferentation on the juvenile rat brain. *Brain Res Dev Brain Res.* 140:93–104.

Schulz, J.M., P. Redgrave, and J.N. Reynolds. 2010. Cortico-striatal spike-timing dependent plasticity after activation of subcortical pathways. *Front Synaptic Neurosci.* 2:23.

Shah, P.P., J.P. Szaflarski, J. Allendorfer, and R.H. Hamilton. 2013. Induction of neuroplasticity and recovery in post-stroke aphasia by non-invasive brain stimulation. *Fronti Hum Neurosci.* 7:888.

Shimizu, T., A. Hosaki, T. Hino, M. Sato, T. Komori, S. Hirai et al. 2002. Motor cortical disinhibition in the unaffected hemisphere after unilateral cortical stroke. *Brain.* 125:1896–907.

Soblosky, J.S., M.A. Matthews, J.F. Davidson, S.L. Tabor, and M.E. Carey. 1996. Traumatic brain injury of the forelimb and hindlimb sensorimotor areas in the rat: Physiological, histological and behavioral correlates. *Behav Brain Res.* 79:79–92.

Teasell, R., N. Bayona, S. Marshall, N. Cullen, M. Bayley, J. Chundamala et al. 2007. A systematic review of the rehabilitation of moderate to severe acquired brain injuries. *Brain Inj.* 21:107–12.

Teskey, G.C., C. Flynn, C.D. Goertzen, M.H. Monfils, and N.A. Young. 2003a. Cortical stimulation improves skilled forelimb use following a focal ischemic infarct in the rat. *Neurol Res.* 25:794–800.

Teskey, G.C., C. Flynn, C.D. Goertzen, M.H. Monfils, and N.A. Young. 2003b. Cortical stimulation improves skilled forelimb use following a focal ischemic infarct in the rat. *Neurol Res.* 25:794–800.

Thurman, D.J., C. Alverson, K.A. Dunn, J. Guerrero, and J.E. Sniezek. 1999. Traumatic brain injury in the United States: A public health perspective. *J Head Trauma Rehabil.* 14:602–15.

Tsubokawa, T., Y. Katayama, T. Yamamoto, T. Hirayama, and S. Koyama. 1991. Chronic motor cortex stimulation for the treatment of central pain. *Acta Neurochir Suppl (Wien).* 52:137–9.

Tsubokawa, T., Y. Katayama, T. Yamamoto, T. Hirayama, and S. Koyama. 1993. Chronic motor cortex stimulation in patients with thalamic pain. *J Neurosurg.* 78:393–401.

Vogenthaler, D.R. 1987. An overview of head injury: Its consequences and rehabilitation. *Brain Inj.* 1:113–27.

Walker, W.C., and T.C. Pickett. 2007. Motor impairment after severe traumatic brain injury: A longitudinal multicenter study. *J Rehabil Res Dev.* 44:975–82.

Whishaw, I.Q., D.M. Piecharka, F. Zeeb, and D.G. Stein. 2004. Unilateral frontal lobe contusion and forelimb function: Chronic quantitative and qualitative impairments in reflexive and skilled forelimb movements in rats. *J Neurotrauma.* 21:1584–600.

Wolf, S.L., C.J. Winstein, J.P. Miller, P.A. Thompson, E. Taub, G. Uswatte et al. 2008. Retention of upper limb function in stroke survivors who have received constraint-induced movement therapy: The EXCITE randomised trial. *Lancet Neurol.* 7:33–40.

Yoon, Y.S., K.P. Yu, H. Kim, H.I. Kim, S.H. Kwak, and B.O. Kim. 2012. The effect of electric cortical stimulation after focal traumatic brain injury in rats. *Ann Rehabil Med.* 36:596–608.

Zheng, J., L. Liu, X. Xue, H. Li, S. Wang, Y. Cao et al. 2013. Cortical electrical stimulation promotes neuronal plasticity in the peri-ischemic cortex and contralesional anterior horn of cervical spinal cord in a rat model of focal cerebral ischemia. *Brain Res.* 1504:25–34.

Zhou, Q., Q. Zhang, X. Zhao, Y.Y. Duan, Y. Lu, C. Li et al. 2010. Cortical electrical stimulation alone enhances functional recovery and dendritic structures after focal cerebral ischemia in rats. *Brain Research.* 1311:148–57.

44 Cranial Nerve Noninvasive Neuromodulation

New Approach to Neurorehabilitation

Yuri Danilov, Kurt Kaczmarek, Kimberly Skinner, and Mitchell Tyler

CONTENTS

44.1 INTRODUCTION

Cranial-nerve non-invasive neuromodulation (CN-NINM) is a multi-targeted rehabilitation therapy that initiates the recovery of multiple damaged or suppressed brain functions affected by neurological disorders. It is deployable as a simple, home-based device (portable tongue neurostimulator, PoNSTM) and targeted training regimen following initial patient training in an outpatient clinic. It may be easily combined with all existing rehabilitation therapies, and may reduce or eliminate need for more aggressive invasive procedures or decrease the total medication intake. CN-NINM uses sequenced patterns of electrical stimulation on the tongue. CN-NINM induces neuroplasticity by noninvasive stimulation of two major cranial nerves: trigeminal, CN-V, and facial, CN-VII. This stimulation excites a natural flow of neural impulses to the brainstem (pons varolli and medulla), and cerebellum, to effect changes in the function of these targeted brain structures. CN-NINM represents a synthesis of a new non-invasive brain stimulation technique with applications in physical medicine, cognitive, and affective neurosciences.

It is difficult to find a more challenging problem in rehabilitation medicine than recovery of neurological function, either fully or partially, that is lost through injury, disease, or aging. Despite intensive scientific effort, the need for greater understanding of both physiological and psychiatric brain dysfunction and effective clinical applications has thus far overwhelmed existing rehabilitative tools and methods. This situation has grown particularly acute with increasing life spans and dramatically increasing survival rates for major diseases and trauma, leading to ever increasing numbers of people with substantial neurological dysfunction.

At present, the primary approach for treating traumatic brain injury (TBI) is through physical medicine, yet a systematic and unified approach has not been universally established. Recent studies have demonstrated that neuroplasticity and functional benefits are stimulated through specific forms of motor-behavioral interventions (Nudo et al., 2001; Weiller et al., 1993), whereas others have determined that early, intensive, targeted cognitive skill training after injury, even in older adults, is clinically effective (Cicerone et al., 2000; Rohling et al., 2009). Concomitant scientific efforts have identified the potential for noninvasive electrical stimulation of the brain to improve patients' ability to learn and retain a motor task (Reis et al., 2009).

Additionally, there has been an emerging call for an integrative approach to developing clinical applications that draw on the latest understanding of the interactive factors affecting the potential for functional physical, cognitive and psychological recovery from TBI (Bach-y-Rita, 1990, 2001, 2003a, 2003b; Bédard, 2003; Taub et al., 2002).

In this chapter, we will introduce the emerging neurorehabilitation technology developed at the University of Wisconsin-Madison, Tactile Communication and Neurorehabilitation Laboratory (TCNL).

44.2 ACQUIRED BRAIN INJURY

A primary goal of TCNL research and development is to improve and optimize the treatment of acquired brain injury, including stroke, and mild and moderate traumatic brain injury (mTBI) from whiplash, sport, and blunt or blast concussive injuries, by using CN-NINM. This new method enhances the brain's natural ability to repair damage, form new functional pathways, assist the recovery process, and increase and accelerate functional neurorehabilitation.

44.2.1 Epidemiology of Brain Injury

TBI is a common occurrence in the United States in both the civilian and military populations. The Centers for Disease Control and Prevention (CDC) estimates that at least 1.7 million people receive urgent medical care for TBI each year (Coronado et al., 2011). Of these, 80%–90% likely experienced a mild TBI, or concussion, based on symptoms at the time of injury (Binder et al., 2005; Coronado et al., 2005; Guerrero et al., 2000). Unfortunately 12%–35% of the military personnel who have deployed to recent combat operations (2.4 million as of June 30, 2012) have sustained at least one mTBI (Rigg and Mooney, 2011).

Approximately 290,000 individuals are hospitalized and discharged annually, most of whom have moderate or severe injuries. More than 53,000 individuals die before hospitalization or before hospital discharge (McMillan et al., 2011). About 43% of TBI subjects experience long-lasting disabilities (Selassie et al., 2008; Zaloshnja et al., 2008).

Precise measures of prevalence within the population are unknown; however, the CDC estimates that approximately 5.3 million Americans are living with a TBI-related disability, with about 125,000 new individuals annually requiring assistance with activities of daily living at least a year after their injuries (Rutland-Brown et al., 2008; Rutland-Brown et al., 2006; Thurman and Guerrero, 1999; Thurman et al., 1998).

Studies in civilian populations suggest that 10%–30% of individuals experiencing a single mild TBI may develop some long-term implications (Ponsford et al., 2013; Silver et al., 2001; Spitz et al., 2013; Tate et al., 2012; Wilde et al., 2008), and recent studies are now proposing possible long-term adverse health effects emerging years after mild TBI, including the development of chronic traumatic encephalopathy

(CTE), thus increasing the importance of prompt and accurate identification of injuries and appropriate intervention (Chen et al., 2011; Goldstein et al., 2012; McKee et al., 2013; McMillan et al., 2011; Omalu et al., 2011).

The economic and social impact of TBI is considered enormous but has not been extensively researched to date. The total lifetime cost for all people who sustained TBI in the United States was $37.8 billion in 1985 (Max, 1991), $56.3 billion in 1995 (Thurman, 2001), and in the year 2000, it was $60.4 billion (Corso et al., 2006). Considering loss of income, and economic impact on the family for caregiving, other researchers estimated that the lifetime costs of TBI in year 2009 dollars totaled more than $221 billion (Langlois Orman, 2011). Charges for acute care and rehabilitation alone in the Model System database averaged about $120,000 per patient, excluding physician charges. In addition to the medical costs, TBI creates a strain in intimate relationships, affects role functioning, fosters economic hardship, and creates a great burden on the family (Cifu et al., 1999; McKinlay et al., 1981; Ponsford et al., 2013; Taylor et al., 2012; Wehman et al., 1994, 2003).

Concussions or mTBIs are a common occurrence in athletes, with an estimated 1.6–3.8 million sport-related concussions annually in the United States (McKee et al., 2009). DeKosky et al. reported that each year more than 1.5 million Americans have mTBI with no loss of consciousness and no need for hospitalization as well as an equal number with consciousness-impairing trauma that is insufficiently severe to require long-term hospitalization (DeKosky et al., 2010). In a 2009 review of CTE, McKee et al. found that of 51 neuropathologically diagnosed cases of CTE, 46 (90%) occurred in athletes. Specifically, athletes participating in American football, boxing, soccer, and hockey comprise the majority of cases. (Baugh et al., 2012; Gavett et al., 2011).

The broader scope of brain injury also includes cerebrovascular accidents, or strokes. A stroke is another mechanism for a brain injury. Those who have had a stroke often experience similar neural disabilities as those with a TBI. Stroke is the leading cause of chronic disability, affecting 500,000 individuals per year. Only 50% of those affected return to the workforce without residual disability. The annual combined cost of health care and lost productivity from stroke is estimated to be between $30 and $40 billion (Baker and Epstein, 1991; Coronado et al., 2005; McGregor and Pentland, 1997; Pai et al., 2012; Schneier et al., 2006; Taylor et al., 2012; Wehman et al., 2003).

44.2.2 The Essence of Brain Injury

The brain is made up of billions of cells that interconnect and communicate. The neuron is the main functional cell of the brain and nervous system, consisting of a cell body, single efferent or output part (axon), and many afferent or input parts—projections of the cell body called dendrites. The axons travel in tracts or clusters throughout the brain, providing extensive interconnections between brain areas via numerous contact synapses.

Each function of the human body and mind is based on highly organized complex neuronal networks that include numerous interrelated structures (cortices, nuclei, neuronal clusters, and pathways) located in multiple levels of the brain and spinal cord. Cooperative and synchronized work of all components of such complex networks provide optimal human performance in behavioral, cognitive, and autonomic functions. This tight integration is especially important in such complex sensory and motor functions as vision, hearing, balance, gait, and speech. Sensory-motor integration is an example of a high level neurophysiological process that needs optimal function in order to reach peak performance in critical situations, sports, and even everyday life.

There are many types of brain damage. The most typical are

- Concussion: The most minor and the most common type of TBI, involving possible brief loss of consciousness in response to a head injury, but in common language the term has come to mean any minor injury to the head or brain.
- Contusion: An area of swollen brain tissue mixed with blood released from a broken vascular system.
- Contrecoup: A contusion that occurs in response to the shaking of the brain back and forth within the confines of the skull. This injury often occurs in car accidents after high-speed stops, and in shaken baby syndrome.
- Diffuse axonal injury (shearing): Involves damage to individual nerve cells (neurons) and loss of connections among neurons. This damage causes a series of reactions that eventually leads to swelling of the axon and disconnection from the cell body of the neuron. This form of neurotrauma is frequently associated with exposure to explosive blasts.
- Hematoma: Heavy bleeding into or around the brain tissue.
- Anoxia: A condition in which there is an absence of oxygen supply to an organ's tissues, even if there is adequate blood flow to the tissue. Without oxygen, the cells of the brain die within several minutes. This type of injury is often seen in near-drowning victims, in heart attack victims, or in people who suffer significant blood loss from other injuries that decrease blood flow to the brain.
- Neurotoxicity: Occurs when the neurons that communicate with other neurons degenerate and release toxic levels of neurotransmitters into the synapse, damaging neighboring neurons through a secondary neuroexcitatory cascade.

Neurologic changes may occur as a result of head trauma, cerebrovascular accident, or postsurgical complications, ultimately leading to a physical breakdown of overall communication among neurons in the brain and/or to neural cell death (apoptosis).

In spite of the wide spectrum of different classes of TBI and types of stroke, the residual damage to the brain and debilitation of the patient looks very similar. For example, the pattern of residual dysfunction that corresponds to a focal lesion caused by a bullet passing through the brain is similar to that caused by a similar focal lesion resulting from a stroke.

In addition to the primary brain injury, there is secondary brain damage that is frequently undetectable by all contemporary diagnostic methods, and therefore "invisible." In secondary damage, neurons appear alive, but physical damage and structural changes in dendrites, cell body, and axons are so small that they are below the resolution of computed tomography, functional magnetic resonance imaging (fMRI), or defusion tensor imaging. The remaining neurons, even if they appear physically healthy, are nonetheless functionally damaged due to the loss of major connections, the loss of input from neighboring dead cells, or the lack of appropriate signals through broken connections. Disruption of sensory and/or motor action potentials in the complex network immediately leads to abnormalities in the supported function, disturbances in spatio-temporal interactions, and loss of sensory-motor integration (Barroso-Chinea and Bezard, 2010; D'Angelo, 2011; Price and Drevets, 2012; Watson et al., 2013; Zikopoulos and Barbas, 2013). As a result, physical manifestation of hypo- or hyper-function (dystonia or spasm and rigidity), and desynchronization and imbalance of inhibition and excitation in the neural network (tremor, seizures, oscillopsia) will occur, for which there is no corroborating evidence.

44.3 DEMANDS OF NEUROREHABILITATION

A deficiency in multiple functional networks in brain injury presents as a wide spectrum of symptoms. In acquired brain injury, the brain may be injured not only in one specific location, but many small injuries may be diffused throughout the brain. It is this indefinite nature of brain injury that makes the pattern of injury and the consequent presentation unique for each individual.

Symptoms of a TBI can be mild, moderate, or severe, depending on the extent of the damage to the brain. Some symptoms are evident immediately, while others do not surface until several days or even weeks after the initial injury. Typical symptoms of mild TBI include headache, confusion, lightheadedness, dizziness, blurred vision or tired eyes, ringing in the ears, bad taste in the mouth, fatigue or lethargy, a change in sleep patterns, behavioral or mood changes, and trouble with memory, concentration, attention, or thinking.

A person with a moderate or severe TBI may show these same symptoms, but may also have a headache that gets worse or does not go away, repeated vomiting or nausea, convulsions or seizures, inability to awaken from sleep, dilation of one or both pupils of the eyes, slurred speech, weakness, numbness in the extremities, loss of coordination, and/or increased confusion, restlessness, or agitation.

If a brain injury results in blindness, deafness, loss of balance or limb control, in the best of circumstances, the neurology is defined, the site of damage specified by imaging, and rehabilitative strategies pursued to restore, substitute, or boost function, and the impaired individual taught to adapt, compensate, and survive.

Less well known is that the remnant sensory cortex can be induced to learn new things, providing that not too much volume—circuits and connectivity—have been lost. For example, the somatosensory cortex can learn to process information from a prosthetic "eye" (Bach-y-Rita, 1969b, 1972, 1983, 2003) long after vision has been lost. People with a damaged vestibular system can improve their moment-to-moment posture, balance, and bipedal movement (Danilov, 2004; Wildenberg et al., 2010) using the sensory circuitry of the tongue.

Even with the knowledge that TBI can result in a plethora of symptoms, neurorehabilitation of brain injury is arguably the field in medicine with one of the greatest number of unmet needs. In response to this demand, we have developed CN-NINM. The objective of our research is to develop and demonstrate an innovative and integrated therapeutic regimen for treating the multiple symptoms of TBI.

44.4 BRIEF HISTORY OF CN-NINM TECHNOLOGY

To understand how we arrived at our current technology, it is helpful to know a brief history of our research.

44.4.1 Sensory Substitution

The TCNL at the University of Wisconsin-Madison was established in 1992 by the late Paul Bach-y-Rita. The laboratory's original mission was to advance the development of tactile displays for sensory substitution initiated by Bach-y-Rita (1969a, 1983; Danilov et al., 2007). The research program has evolved from pioneering studies beginning in 1963 that resulted in the development of the Tactile Vision Substitution System (TVSS). The TVSS delivered information from a TV camera to arrays of mechanical stimulators in contact with the skin of one of several parts of the body, including the abdomen, back, thigh, forehead, and fingertip. After sufficient training with the TVSS, blind subjects reported experiencing the images in space, instead of on the skin. They learned to make perceptual judgments using visual means of analysis, such as perspective, parallax, looming and zooming, and depth judgments. The TVSS was sufficient to allow users to perform complex perception and eye-hand coordination tasks. These included facial recognition, accurate judgment of speed and direction of a rolling ball with more than 95% accuracy in batting the ball as it rolls over a table edge, and complex inspection-assembly tasks (Bach-y-Rita, 1969b, 1983; Lambert et al., 2004; Sampaio et al., 2001; Segond et al., 2005, 2013).

44.4.2 Why the Tongue? Part I

Electrotactile stimulation supplanted the vibrotactile stimulation because it is simpler, lighter, and consumes less energy, and it is easier to control the stimulus. Various improvements have led to the current system. It is an example of a new generation of sensory substitution devices based on computer-controlled electrical stimulation of the human skin in the most densely innervated tactile areas: the tongue and the fingers. The tongue was preferable because it affords a better environment (constant acidity level [pH], constant temperature and humidity, and low excitability thresholds) in comparison with a fingertip (variable hydration, thickness of the skin, surface contaminants, relatively limited and highly curved surface area available for stimulation, and high excitability thresholds).

44.4.3 Evolution of the PoNS Device

44.4.3.1 Vision Substitution via the Tongue

Initially a stationary, shoebox-sized device, the tongue display unit (TDU, Figure 44.1a), was the basis for our research and development. It was the first electrotactile visual substitution system (ETVSS), which we successfully tested in our laboratory and independently applied in different laboratories around the globe. Using ETVSS as a research tool, Canadian scientist Maurice Ptito verified that stimulation from the tongue was capable of unmasking hidden pathways to visual areas of the human brain in people who were blind. He demonstrated that information delivered through the tongue was capable of reaching the main visual cortical and subcortical areas. The system encoded an electrotactile image on the tongue surface using information provided through a video camera in real time. Using the system, a blind person was capable of orienting themselves in three-dimensional space, navigating in building corridors, avoiding obstacles, recognizing simple objects on tables and walls, or reading simple words (Chebat et al., 2007, 2011; Kupers and Ptito, 2013; Matteau et al., 2010; Ptito et al., 2005). The research demonstrated effective use of brain neuroplasticity to substitute for the missing visual sense.

The portable version of the ETVSS system, the BrainPort Vision Device (Figure 44.1b, commercialized by Wicab, Inc., under license from the Wisconsin Alumni Research Foundation at the University of Wisconsin-Madison), became our first electrotactile system for the blind. It was intensively tested on blind veterans in the United States and Europe and in 2013 received the CE mark. (As a historical note, the concept for vision substitution via electrotactile stimulation of the tongue was conceived as early as the first part of the 20th century) (Machts, 1920).

44.4.3.2 Vestibular Sensory Substitution Systems

In 2001, we developed the electrotactile vestibular substitution system (EVSS), wherein tactile sensations produced by an electrode matrix placed on the tongue provide head orientation information (relative to gravity) normally obtained from the vestibular complex of the inner ear. We demonstrated that the brain is able to process the head-orientation information presented on the tongue to make appropriate postural corrections and dramatically improve balance in affected individuals.

The first population we tested with EVSS was people who had suffered peripheral bilateral vestibular damage (BVD), typically from ototoxicity. As a result of training with the EVSS, subjects demonstrated significant improvement in balance control. Moreover, we found that training with the EVSS system induced several different levels of balance recovery. The effects of EVSS training were both immediate and retained.

The immediate effect was observed in BVD subjects after 5–10 minutes of familiarization with EVSS and included the ability to control stable vertical posture and body alignment during use. Even for well-compensated BVD subjects,

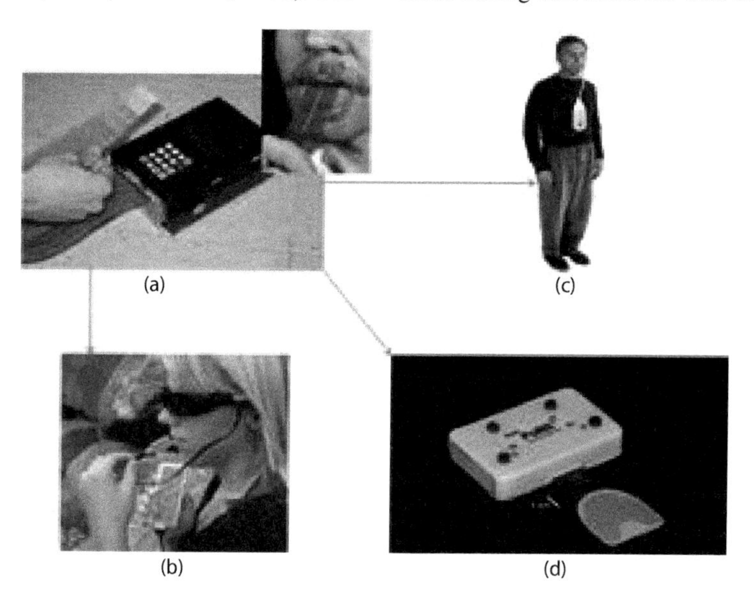

FIGURE 44.1 Electrotactile tongue stimulation technology. The TDU (a), developed at the University of Wisconsin-Madison (UW), became the platform technology leading to the BrainPort Vision (b) and Balance (c) devices marketed by Wicab, Inc., as well as the Portable Neuromodulation Stimulator (d), also developed at UW.

standing on soft or uneven surfaces or stances with a limited base such as tandem Romberg position, was challenging, and unthinkable with closed eyes. To our surprise, using the EVSS, these subjects readily acquired the ability to control balance and body alignment in challenging positions with their eyes closed (Tyler et al., 2003).

44.4.4 Retention Effects

Residual balance effects also were observed in all tested BVD subjects; however, the strength and extent of effects significantly varied from subject to subject depending on the severity of vestibular damage, the time of subject recovery, and the length and intensity of EVSS training.

At least three groups of residual balance effects were noted: short-term residual effects (sustained for a few minutes), long-term residual effects (sustained for 1–12 hours), and a rehabilitation effect that we observed during several months of training in a single subject. All residual effects were observed after complete removal of EVSS from the subject's mouth.

44.4.4.1 Short-Term Aftereffect

This effect was usually observed during the initial stages of EVSS training. Subjects were able to maintain their balance for some time, without immediately developing an abnormal sway. Moreover, the length of short-term aftereffects was almost linearly dependent on the time of EVSS exposure. After 100 seconds of EVSS exposure, stabilization continued for 30–35 seconds; after 200 seconds of EVSS exposure, for 65–70 seconds; and after 300 seconds, the subject was able to maintain balance for more than 100 seconds. The short-term aftereffect continued to be approximately 30%–50% of the EVSS exposure time (Tyler et al., 2003).

44.4.4.2 Long-Term Aftereffect

After longer (up to 40 minutes) sessions of EVSS training long-term aftereffects developed and were sustained for a several hours. The relative duration of the balance improvement was much longer than the observed short-term aftereffect. Instead of the expected 7 minutes of stability (if one were to linearly extrapolate the 30% rule based on 20-minute trials), we observed from 1 to 6 hours of improved stability (Danilov, 2004).

After 20 minutes of training, the BVD subjects were able to not only stand steadily with good alignment on a hard or soft surface, but were also able to accomplish completely different kinds of balance-challenging activities, such as walking on a beam, standing on one leg, and even riding a bicycle or dancing. After a few hours, however, all the symptoms returned, although frequently not as severe as before. We noted that the strength of the long-term aftereffects was also dependent on the time of EVSS exposure: 10-minute trials were much less effective than 20-minute trials, and 40-minute trials had about the same efficiency as 20 minutes. Consequently, we concluded that 20 minutes was sufficient for standing trials with closed eyes.

44.4.4.3 Rehabilitation Effect

We were capable of repeating two and sometimes three 20-minute EVSS trials with a single subject during 1 day. After the second trial, the effects continued about 6 hours on average (Figure 44.2). In total, after two 20-minute EVSS

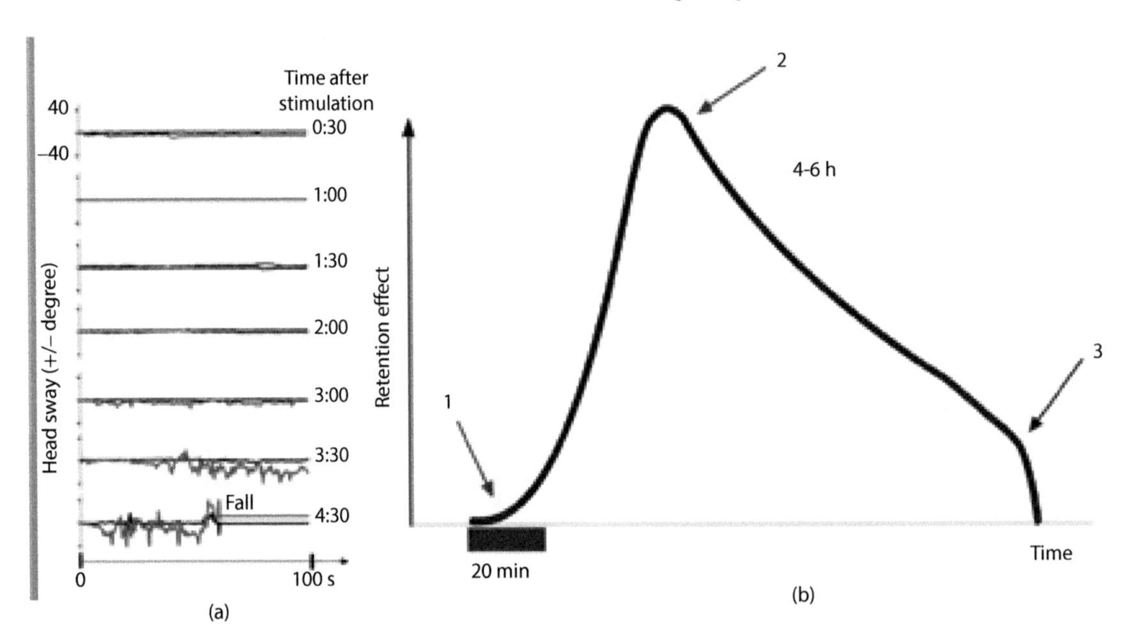

FIGURE 44.2 Long-term retention effect. (a) Head stabilogram of a subject with bilateral vestibular loss, standing with closed eyes test during 100 seconds. Subject could not stand in test position before stimulation. The head stabilograms were recorded every 0.5 and 1 hour until the stability effect disappeared (after approximately 4 hours). (b) Conceptual diagram of the retention effect dynamic. Note: Positive effects start to develop after approximately 10 minutes of stimulation (1), reaching a maximum at 30–60 minutes after the end of stimulation (2), and beneficial effects continue for up to 4–6 hours after initial training sessions, after which they rapidly diminish (3).

stabilization trials, BVD subjects reported feeling and behaving "normal" for up to 10–14 hours a day. Our conclusion was that consistent, consecutive training produced both systematic improvement and gradual increase in duration of the long-term aftereffect.

Moreover, we found that repetitive EVSS training produced both accumulated improvement in balance control, and global recovery of the central mechanisms of the vestibular system (Danilov, 2004). The improvement in most individuals' balance continued even after the sensory substitution ceased. These subjects exhibited balance retention and functional transfer of postural stability to gait and other activities of daily living (Danilov et al., 2006, 2007).

We have tested more than 200 subjects with peripheral, central, or idiopathic vestibular loss, Parkinson disease, and other sensory-motor control impairments such as stroke and TBI, all with the primary symptom of moderate to severe balance dysfunction. The great majority of subjects showed significant short- and long-term functional improvement when scored by multiple standardized objective measures, regardless of etiology and age (Danilov et al., 2007).

Similar to our vision substitution technology, the portable version of the EVSS was commercialized as the BrainPort Balance device (Figure 44.1c). It received the CE mark and was tested in the United States, Europe, Canada, Brazil, and Russia, mainly reproducing our core results. (Badke et al., 2011; Barros et al., 2010; Bittar and Barros, 2011; Borisoff et al., 2010; Danilov et al., 2007; Robinson et al., 2009; Vuillerme and Cuisinier, 2008; Vuillerme et al., 2008). Unfortunately, Wicab, Inc., decided to postpone BrainPort Balance device development, instead focusing on the BrainPort Vision device, which received CE mark approval in 2013.

It was already becoming clear by 2005 that this technology was also effective with TBI subjects, where balance and gait problems are "secondary" to primary symptoms of brain damage. Our first experience in helping moderate and severe TBI subjects was very encouraging, as was the recovery of balance in people who had suffered a stroke, and functional improvement in people with spinal cord injuries (Badke et al., 2011; Borisoff et al., 2010). In addition to balance recovery effects, we found multiple effects directly or indirectly related to other areas, for example improvement of vision (oscillopsia, visual neglect, and eye movement control), cognitive function, tinnitus, headache, sleep, and mood.

44.4.5 CN-NINM Technology

After the initial successful application of our new portable BrainPort Balance Device system in central and peripheral vestibular affected subjects, it became obvious that we needed to reconsider our main scientific concept. Multiple independent and consistent results led us to believe that these long-lasting retention effects cannot be attributed to sensory substitution alone (translation of meaningful environmental information) because the signal is no longer present, but rather are induced by the mechanism of delivery—the patterned, high-frequency electrical stimulation of the lingual nerve and chorda tympani (Tyler et al., 2003). The source and driving force for neuroplasticity induced by our intervention is the electrotactile stimulation itself, delivered to the brain by two cranial nerves directly to the brainstem.

The new technology, CN-NINM, was born as direct implementation of that idea—noninvasive electrotactile stimulation of the tongue producing strong brain activation and exciting the self-rehabilitation processes. Our stimulation appears to facilitate the state of neuroplasticity simultaneously in many areas of the brain, improving many affected functions in parallel. We also recognized that combining stimulation with intensive rehabilitation therapy methods would enhance the stimulation effects.

44.5 CONCEPTUAL FRAMEWORK

There are a several widely accepted axioms about rehabilitation after brain injury. Our approach differs from these in several ways.

First of all, although conventional physical rehabilitation therapy does employ retraining with the intention to return the patient to normal function, this occurs primarily during the acute and postacute period after trauma (typically up to 1 year). CN-NINM technology is oriented primarily on rehabilitation during chronic stages (years after traumatic incident), when conventional thinking assumes that there is no further capacity for change.

In situations when patient internal resources are not sufficient to activate the neurorehabilitation process (naturally or with help of physical therapy), our technology can provide an additional "boost," reinforcing the process of neurorehabilitation. Another distinctive feature of our technology is its integrative nature and multidirectional rehabilitation effect. The physical medicine (the primary therapeutic pathway) has not, until recently, recognized or given significant attention to the profound and interrelated cognitive and emotional impact of TBI and even less to rehabilitation of them. Most practitioners also consider the oculomotor consequences of brain injury untreatable.

For many years, Paul Bach-y-Rita was a lone voice pointing out the obvious inconsistencies in the contemporary view of neurorehabilitation. His pioneering journey, and those of others like him (e.g., Edward Taub) is captured in a book titled *The Brain That Changes Itself* (Doidge, 2007). They called for a more holistic, integrated, and targeted approach to treating brain injury, and gave special attention to the need for cognitive and psychosocial training in conjunction with the physical rehabilitation.

Integrated CN-NINM therapy intends to restore physiological and cognitive functions affected by brain injury beyond traditionally expected limits, by employing both newly developed and novel therapeutic mechanisms for progressive physical and cognitive training, while simultaneously applying brain stimulation through a device we call the Portable NeuroModulation Stimulator (PoNS, Figure 44.1d). Based on our previous research and recent pilot data, we believe a rigorous in-clinic CN-NINM training program, followed by regular at-home exercises also performed with

PoNS, simultaneously enhances, accelerates, and extends recovery from multiple impairments from brain injury (e.g., movement, vision, speech, memory, attention, mood), based on divergent, but deeply interconnected neurophysiological mechanisms (Wildenberg et al., 2010, 2011a, 2011b, 2013).

44.6 TECHNICAL DESCRIPTION OF THE PoNS DEVICE

The PoNS device evolved from a related set of technologies aimed at stimulation of the human nervous system to achieve beneficial results. These include information delivery as well as broad activation and modulation of large-scale neural networks. This section describes how this technological evolution parallels the conceptual and theoretical evolution of applications summarized in Section 44.3.

44.6.1 Purposeful Neurostimulation

The PoNS device was designed to provide optimized neurostimulation via the tongue specifically to induce neuromodulation as part of CN-NINM therapy. In this sense it belongs to two broad categories of technologies. The first category includes devices that electrically activate the nervous system. For example, cochlear implants induce action potentials that propagate to the end of the nerve fiber, and that once there, are indistinguishable from APs of natural origin, resulting in auditory sensations. Similarly, transcutaneous electrical nerve stimulation introduces stimulation for pain relief. The PoNS electrical stimulation functions in the same manner in that it too is applied to a body surface, the tongue.

The PoNS device also belongs to a category of technologies intended to induce broad modulation of brain activity. For example, transcranial magnetic stimulation (TMS) can activate, upregulate, or downregulate limited regions of brain tissue with volumes ranging from mm^3 to cm^3. Based on the principal of electromagnetic induction, TMS actively evokes action potentials in the underlying neural tissue, mainly in axons oriented orthogonal to the magnetic field. Transcranial direct current stimulation (TDCS) applies an external electric current to the head surface, in this case spreading stimulation across the entire skull. TDCS does not cause neurons to produce spikes, but rather modulates the spontaneous firing rate by changing the membrane potentiation level. In both cases, the direct neural activation (or suppression) occurs under the influence of external electromagnetic factors, changing the excitability of neuronal membranes.

Unlike both TMS and TDCS, however, the PoNS device achieves changes through natural neural pathways. The electrode array of the PoNS device induces an electric field in the tongue epithelia that we believe, based on the relevant anatomy and sensory percept, activates sensory fibers (predominantly mechano, thermo, and free nerve endings) to a depth of approximately 300–400 μm. This creates a massive flow of action potentials that are perceived by the subject as a "buzzing" or "champagne bubble" sensation. Here, the stimulation is a flow of neural impulses filling the brainstem nuclei through the trigeminal and facial nerve fibers. Activation of primary targets—these brainstem nuclei neurons—happens through existing synaptic connections, initiating a cascade of activation through multiple neural circuitries.

44.6.2 Electrotactile Stimulation

Controlled electrotactile stimulation, a particular form of purposeful neurostimulation (others include, for example, mechanical, thermal, chemical, and electromagnetic), was first described as early as 1879 (Tschiriew and Watteville, 1879) and has since been applied to a wide variety of applications, most of which involve information display to replace lost senses or to create entirely new ones (Kaczmarek et al., 1985, 1991). Electrotactile stimulation became the technological basis for Bach-y-Rita's collaborative efforts to make the ETVSS more portable as well as a number of related applications, including feedback of touch information from artificial pressure sensors mounted on an insensate or prosthetic hand, auditory sensory substitution, and aircraft cockpit augmentative information displays (Collins and Madey, 1974).

In the 1990s, Bach-y-Rita continued this line of research at the University of Wisconsin-Madison, this time in collaboration with the present authors, discovering that the fingertips (Haase and Kaczmarek, 2005; Kaczmarek and Haase, 2003a, 2003b; Kaczmarek et al., 1994) and tongue (Bach-y-Rita and Kaczmarek, 2002; Bach-y-Rita et al., 1998) were also suitable sites for high-resolution electrotactile information display. These studies resulted in the TDU, an experimental tool for programmable electrical stimulation of the tongue (Figure 44.1a). The principles of operation of this device have been described in detail (Kaczmarek, 2011), and were subsequently instantiated commercially as the BrainPort Vision and Balance Devices described in Section 44.3. With the subsequent discovery that information-free electrotactile stimulation of the tongue enhanced neurorehabilitation, the concept for the PoNS device was born, likewise derived from the TDU.

44.6.3 PoNS Device

The current-generation PoNS device (Figure 44.1d) achieves localized electrical stimulation of afferent nerve fibers on the dorsal surface of the tongue via small surface electrodes. Because of the resulting tactile sensation, which, depending on stimulation waveform, typically feels like vibration, tingle, or pressure, it is certain that tactile nerve fibers are activated. Taste sensations are infrequently reported, although it is not known whether gustatory afferents are in fact stimulated, given the nonphysiological patterns of activation likely to result from PoNS-induced stimulation of these fibers.

All electrotactile systems, including the PoNS device, must adhere to a set of core principles to ensure comfortable and controllable tactile percepts, as well as safe operation. As these have already been extensively reviewed (Gibson, 1968; Kaczmarek and Bach-y-Rita, 1995; Kaczmarek et al., 1991; Rollman, 1973; Schaning and Kaczmarek, 2008; Szeto and Riso, 1990; Szeto and Saunders, 1982), we will focus on application of these principles specifically to the PoNS device. An expanded discussion of the waveform, electrode, and safety features appropriate for tongue stimulation has been previously published (Kaczmarek, 2011).

44.6.3.1 Physical Construction

The PoNS device is held lightly in place by the lips and teeth around the neck of a tab that goes into the mouth and rests on the anterior, superior part of the tongue. The paddle-shaped tab of the system has a hexagonally patterned array of 143 gold-plated circular electrodes (1.50-mm diameter on 2.34-mm centers) that is created by a photolithographic process used to make printed circuit boards. The board is an industry-standard polyimide composite that is USP Class VI compliant and meets ISO 10993 biocompatibility standards. The edges and nonelectrode surface of the array tab are coated with a rugged USP Class VI biocompatible epoxy. Therefore, the only materials that contact oral tissues are the gold electrodes and the biocompatible polymers. The remainder of the PCB and all electronic components, including battery, are in a sealed Delrin (USP Class VI compliant) enclosure that remains outside the mouth. Although the PoNS device is built using biocompatible materials, it is investigational and not approved by any regulatory agency. Device function is user-controlled by four buttons: On, Off, Intensity "Up," and Intensity "Down." The PoNS device is powered by an internal battery that may be recharged via an external power supply that plugs into a 120-V or 240-V AC electric mains outlet, similarly to a mobile phone.

44.6.3.2 Electrical Stimulation

The tongue electrodes deliver 19-V positive voltage-controlled pulses that are capacitively coupled both to limit maximal charge delivered under any rare circuit failure and also to ensure zero DC to the electrodes, minimizing potential tissue irritation from electrochemical reactions. (Such irritation has never been observed or reported.) Tongue sensitivity to positive pulses is greater than that for negative pulses. The pulse width is adjustable in 64 unequal steps from 0.3 to 60 μs by the intensity buttons. This intensity control scheme takes advantage of the steep section of the strength–duration relationship for electrical stimulation of neural tissue (Reilly, 1998). These pulses repeat at a rate of 200 per second, within the typical physiological firing rate for tactile afferents. Because of the neural refractory period, and extrapolated from earlier single-fiber median nerve response to similar electrotactile stimuli on a rhesus monkey fingerpad (Kaczmarek et al., 2000), it is presumed that at most one action potential results in any given afferent fiber for each stimulation pulse. To minimize sensory adaptation (Kaczmarek and Tyler, 2000) and to ensure a good quality of sensation (Kaczmarek et al., 1992), every fourth pulse is removed from the pulse train, so that each electrode delivers a burst of three pulses every 20 ms (Figure 44.3). This combination of pulse amplitude and width results in an electrotactile stimulus that may be varied by the user from well below sensory threshold to a perceived sensation at the upper limit of comfortability.

44.6.3.3 Electrode Array and Pulse Sequencing

The PoNS electrode array, irregularly shaped to take advantage of the most sensitive regions of the tongue, comprises 143 electrodes nominally organized into nine 16-electrode sectors. Within each sector, one electrode is active at any moment (pulse beginnings staggered by 312.5 μs), with unstimulated electrodes serving as the return current path. The nine sectors present simultaneous stimulation, with the intensity of each sector adjusted to compensate for the variability of tongue sensitivity to electrotactile stimuli (Tyler et al., 2009). The sensation produced by the array has been described as similar to the feeling of drinking a carbonated beverage.

The impedance of the electrode–skin interface presents as a resistive component of approximately 1 kΩ, in series with

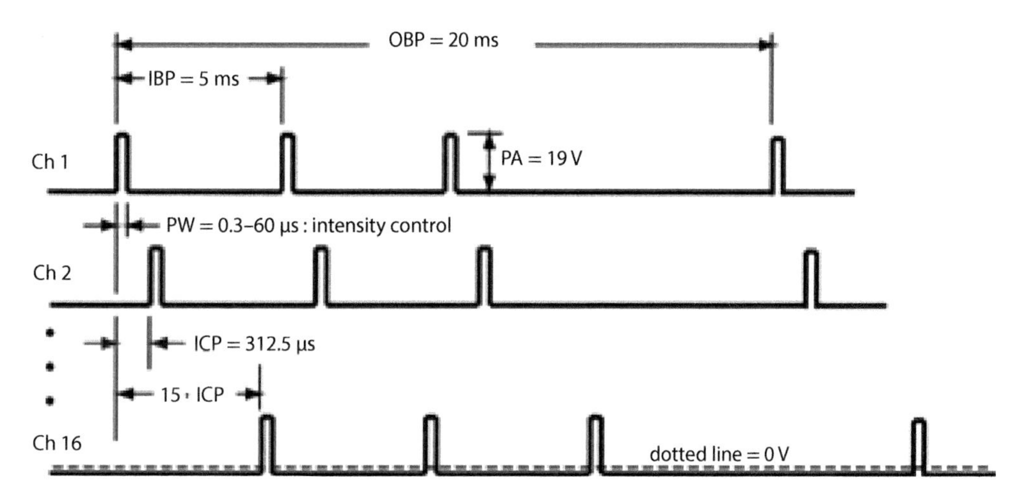

FIGURE 44.3 The PoNS stimulation waveform. This incorporates pulse timing to achieve a comfortable neuromodulation stimulus to the tongue. Sixteen individual stimulation channels each activate nine electrodes on the electrode array.

a resistive–capacitive network of 4–6 kΩ in parallel with 0.5 nF (Kaczmarek, 2011). The pulse current therefore contains a brief leading spike followed by an exponential decay to a plateau current of approximately 3 mA. Voltage control is used rather than the current control of typical electrotactile systems because the tongue electrode impedance is relatively stable compared with cutaneous loci otherwise used (abdomen, fingertip). Use of voltage control affords circuit simplicity and therefore component, space, and battery economy.

The electrode size and geometry were chosen to achieve a reasonable balance between number of electrodes that may be packed into the array area and the comfortability and controllability of the electrotactile percept (Kaczmarek and Tyler, 2000). The overall result of this stimulation is the comfortable and convenient presentation of almost 26 million stimulation pulses to the tongue during a typical 20-minute therapy session. How many action potentials are propagated to the brain as a result of this surface stimulation is at this point unknown.

44.7 HOW IT WORKS

In brief, CN-NINM uses sequenced patterns of electrical stimulation on the anterior dorsal surface of the tongue to stimulate the trigeminal and facial nerves.

44.7.1 Why the Tongue? Part II

The anterior dorsal surface of the tongue is a unique patch of the human skin with a unique innervation pattern. The relatively thin (in comparison to other areas) epithelium (300–400 μm) is saturated by different kinds of mechano, thermo, and taste receptors in addition to free nerve endings, stratified in its depth. It is the area with maximal density of mechanoreceptors, and (e.g., the fovea in the retina) have the minimal two-point discrimination threshold: 0.5–1 mm for mechanical stimulation (Vallbo and Johansson, 1984; Vallbo et al., 1984) and 0.25–0.5 mm for electrotactile stimulation (unpublished data). The physical density, spatial distribution, size of the receptive fields and their overlapping coefficient, and spatial and temporal summation properties are largely unknown, especially for electrotactile stimulation (Johansson and Vallbo, 1979a, 1979b).

The two major nerves from the tip of the tongue deliver information streams directly to the brainstem—the lingual nerve (texture of food) and chorda tympani (taste of food). According to our approximation, approximately 20–25 thousand neural fibers are delivering neural impulses from this area (about 7.5 cm^2) covered by our electrode array (Heasman and Beynon, 1986).

44.7.2 Hypothesis

Our hypothesis is that CN-NINM induces neuroplasticity by noninvasive stimulation of two major cranial nerves: trigeminal (CN-V) and facial (CN-VII). This stimulation excites a natural flow of neural impulses to the brainstem (pons varolii and medulla) and cerebellum via the lingual branch of the cranial nerve (CN-Vc), and chorda tympani branch of CN-VII, to effect changes in the function of these targeted brain structures (Ptito et al., 2005; Wildenberg et al., 2010).

The spatiotemporal trains of neural activation induced in these nerves eventually produce changes of neural activity in corresponding nuclei of the brainstem—at least in the sensory and spinal nuclei of trigeminal nuclei complex (the largest nuclei in the brainstem, extending from the midbrain to the nuclei of the descending spinal tracts), and the caudal part of the nucleus tractus solitarius where both stimulated nerves have direct projections.

Changes in neural activity were evidenced in the results of our pilot study, wherein we also developed a new fMRI signal processing method to yield high-resolution images of the pons, brainstem, and cerebellum beyond that previously reported, allowing observation of changes in functional activity in all of the regions of interest (Wildenberg et al., 2010, 2011a, 2011b, 2013). We are particularly interested in these specific changes (in pons, brainstem, and cerebellum) because these neural structures are the major sensory integration and movement control centers of the brain and therefore primary targets for neuromodulation.

We postulate that the intensive activation of these structures initiates a sequential cascade of changes in neighboring and/or connected nuclei by direct projections and collateral connections, by activation of brainstem interneuron circuitries (reticular formation of the brainstem), and/or by passive transmission of biochemical compounds in the intercellular space (release of neurotransmitters in the synaptic gaps). The stream of neural impulses lead to activation of corresponding neural networks and massive release of neurotransmitters that eventually activate the glial networks of the brainstem (responsible for maintenance of neuronal environment and synaptic gaps). This activation includes the trigeminal nuclei complex and nucleus tractus solitarius, vestibular nuclei complex, cochlear, facial, and hypoglossal nerve nuclei, and vermis of the cerebellum. This in turn, causes radiating therapeutic neurochemical and neurophysiological changes affecting both synaptic and extrasynaptic circuitries affecting information processing of afferent and efferent neural signals involved in movement control, including the cerebellum and nuclei of spinal motor pathways.

The temporal pattern of our observed retention effects is strikingly similar to the process well known in neuroscience literature for several decades as long term potentiation (LTP) and depression. Both processes were tested and verified in multiple animal models by analyzing changes in brain tissue samples, and both are in intensive use in different models of human processes of learning and memory as a basic mechanism of the synaptic plasticity of the brain (Abraham and Bear, 1996; Bear and Malenka, 1994; Dudek and Bear, 1993; Kirkwood et al., 1995; Larkman and Jack,1995; Levenes et al., 1998).

In brief, synaptic plasticity is a natural manifestation of using activity-dependent processes affecting structure and function of neuronal connections. As a result of such processes, multiple consequential adaptive changes are happening on different levels of brain organization (molecular, cellular, regional, and systemic), with different temporal patterns and dynamics (short and long) that reflects on multiple sensory and motor functions, cognitive performance, and behavior (Bi et al., 2006; Gisabella et al., 2005; Li et al., 2013; Shin et al., 2006, 2010; Tully and Bolshakov, 2010; Tully et al., 2007a, 2007b).

Intensive repetitive stimulation of neurons leads to corresponding activation of synaptic contacts on the axonal tree, including the whole complex of pre- and postsynaptic neurochemical mechanisms. Multiphasic fluctuations of postsynaptic potentials, frequently described as short-term activity-dependent synaptic plasticity (in the range of milliseconds, seconds, and minutes) has been shown are capable to enhance synaptic transmissions (Morice et al., 2008; Zucker, 1999; Zucker and Regehr, 2002).

In contrast, LTP is the phenomena of synaptic structural remodeling and formation of new synaptic contacts that is activated by high frequency stimulation (Buchs and Muller, 1996; Calverley and Jones, 1990; Engert and Bonhoeffer, 1999; Geinisman et al., 1991, 1993, 2000, 2001; Jones and Calverley, 1991; Jones et al., 1991; Maletic-Savatic et al., 1999). After 10–40 minutes of high-frequency stimulation (50–400 Hz, range of frequencies used in animal research) the number of synapses and proportion of multiple spine boutons can increase the efficiency of neural connections (Toni et al., 1999). Effects of LTP can continue during several hours and even days (Bliss and Gardner-Medwin, 1973; Bliss and Lomo, 1973).

In our experiments, using the PoNS device, prolonged and repetitive activation (20 minutes or more) of functional neuronal circuits (balance, gait) can initiate long-lasting processes of neuronal reorganization (similar to LTP), that we can see and measure in subjects' behavior. The functional improvement after initial training sessions continues for several hours (see Figure 44.2). Multiple regular sequential training sessions lead to consistent increase of improved symptom duration and cumulative enhancement of affected functions.

This regular excitation may also increase the receptivity of numerous other neural circuitries and/or affect internal mechanisms of homeostatic self-regulation, according to contemporary concepts of synaptic plasticity. We cannot exclude also that this induces simultaneous activation of serotoninergic and noradrenalinergic regulation systems of the brain as well.

The result of this intervention is essentially brain plasticity on demand—a priming or upregulating of targeted neural structures to develop new functional pathways, which is the goal of neurorehabilitation and a primary means of functional recovery from permanent physical damage caused by stroke or trauma.

44.8 CN-NINM TRAINING WITH THE PoNS DEVICE

CN-NINM training is an advanced level of neurorehabilitation therapy. It combines brain stimulation using the PoNS device with neuromuscular reeducation. Stimulation from the PoNS device enhances the brain's innate capability toward improved function. The intervention is intended to produce beneficial neuroplastic changes by retraining mind-body connections.

The goal of CN-NINM training is to recover normal movement control. By combining brain activation with targeted physical training, we believe we are affecting neural pathways directly related to the task. Through experimentation in multiple studies with various populations (TBI, stroke, multiple sclerosis, Parkinson disease), we have found that the most effective way to train using this technology involves five main components:

I. Movement training
II. Balance training
III. Gait training
IV. Cognitive training
V. Breathing and awareness training (BAT)

Each component requires focused attention on tasks, with a goal of challenging the individual without pushing them beyond the point where they cannot continue training. Finding the "just right" challenge point is a key factor in achieving success with this intervention. In the presence of excessive fatigue, or absent adequate challenge, the PoNS stimulation may be less effective because the underlying, natural drivers of neuroplasticity are suboptimal.

We have achieved the best results using uninterrupted 20- to 25-minute training sessions with the PoNS device. We experimented with longer periods of continuous stimulation (e.g., 30, 40, 60 minutes), but found that these did not produce better results when compared with the 20-minute sessions.

An example of a day of training is found in Table 44.1. Figure 44.4 shows how the different training components can be combined temporally.

In our studies, individuals are trained in the clinic initially for 1 to 2 weeks (Monday through Friday). As they improve, they are challenged with harder tasks in order to progress. After the clinical training period, they continue training at home, performing the same components of CN-NINM training that they learned in the clinic. Individuals return to the

TABLE 44.1
Daily Training Session Sample

Morning	**Movement training—warm-up exercises**	
	Balance training with PoNS	20 minutes
	Gait training with PoNS	20 minutes
	Cognitive training with PoNS	20 minutes
Break		3–4 hours
Afternoon	Movement control exercises with PoNS	20 minutes
	Balance training with PoNS	20 minutes
	Gait training with PoNS	20 minutes
	Cognitive training with PoNS	20 minutes
Evening	BAT with PoNS	20 minutes

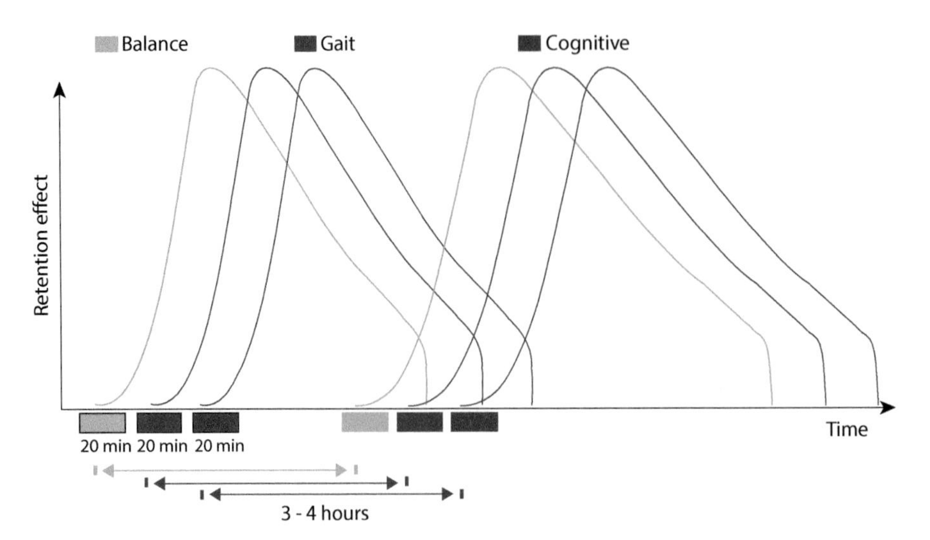

FIGURE 44.4 Timeline of combined training. Different functions can be exercised consecutively. There should be a minimum of 2 hours before repeating the same type of training.

clinic approximately at weekly and monthly intervals to review training. They must receive frequent and progressive upgrades to the physical challenge in their training in order to improve and retain therapeutic benefit. The following sections illustrate the types of training and how each is progressed for training.

44.8.1 Movement Training

Movement training has two components: warm-up exercises and movement control exercises. Warm-up exercises are performed at the beginning of each training session. The goal of the warm up is two-fold. First, the warm-up prepares the system for work. The body responds best when the systems are primed. This is similar to an athlete warming up before tasking the body with physical demands. Second, the warm-up exercises are targeted towards areas of abnormal movement and physical limitations. The warm-up exercises help to "wake up" the body's pathways and relieve stress on tight areas. This helps the body prepare to work toward maximum performance. The PoNS device is not used during the warm-up exercises. Focused attention is given to the quality of the movements. All exercises emphasize neutral posture and good alignment.

The second component of movement training is movement control exercises. Human movement occurs as a result of the synergy of three interrelated systems: the muscular system (soft-tissue structures; i.e., muscles, tendons, ligaments, and fascia), nervous system (recruitment of muscle synergies during movement), and articular system (functional joint motion) (Neuman, 2009). In an unaffected individual, structural alignment, neuromuscular control, and movement are optimal, as long as no impairments are present. When any of these is impaired, the result is abnormal movement. In people with neurological disorders, the movement synergies are affected. As a result, they typically present with abnormal movement patterns and often develop compensatory strategies. The movement control exercises are targeted towards the individual's abnormal movements. A critical piece of the CN-NINM intervention is to properly challenge the individual's movement control patterns in a manner specific to their needs.

The movement control exercises are designed to retrain movements for improved neuromuscular control, alignment, and mobility. This is achieved by:

- Isolating body segments
- Focusing on quality of movement
- Moving joints within normal range by balancing muscle contraction and relaxation

Sample exercises include chin tucks, shoulder circles, scapular circles, pelvic tilts, and pelvic figure 8s. Each exercise starts out simply and builds in complexity as the individual demonstrates mastery. To illustrate how to progress an exercise, we will use leg circles. The starting point may be standing on one leg while holding a support and making a circle in space with the other leg. The leg circle can be performed in front of the body, then in back of the body; progress by connecting the two circles and making a figure 8; the figure 8 progresses to a four-leaf clover. Further progress this exercise by turning the foot inward and outward (add greater internal and external hip rotation). Increase the challenge more by performing the exercise without holding a support. Each exercise can be developed in a similar fashion, progressing from single to multiple planes, and simple to more complex movements. The goal is to demonstrate coordinated and fluid movement through the maximum range of motion of the joints. Movement control exercises are performed with the PoNS device. The focus is on correcting abnormal movement patterns. Using a mirror when learning and practicing the movements facilitates visual feedback, which allows the individual to self-correct. The treating clinician helps determine which exercises are important to address the individual's needs.

44.7.2 Balance Training

The goal of balance training is to create body awareness, correct postural alignment, and improve balance. The individual uses the PoNS device during balance training while getting feedback from the trainer, which allows proprioceptive and vestibular inputs to recalibrate.

44.8.2.1 Training Positions

Various sitting and standing positions are used for balance training. Finding the proper starting point is key. To find the starting point, the trainer tests the individual until they find the point at which the person must work in order to maintain balance control with proper alignment, but not work so hard that they continuously struggle or lose their balance.

Sitting and standing positions are affected by external and internal variables. External variables that affect seated stability include sitting surface, back support, arm position, base of support, use of touch point, and eyes open or closed (Table 44.2). External variables that affect stability in standing include standing surface, base of support, foot position, shoes, use of touch point, and eyes open or closed (Table 44.3). In a person with a neurological dysfunction, internal variables are related to the lack of communication in the neurological network, which includes the lack of firing or misfiring of neurons, manifesting outwardly as abnormal posture and instability.

To find a starting point, the trainer tests the individual in the various positions, adding challenges. For example, the individual starts by standing in a comfortable position with eyes open, then in the same position with the eyes closed. If the individual sways significantly or appears unstable, this is where the training begins. If this position can be done easily, changing the variables (i.e., base of support, foot position, or use of soft foam) increases the difficulty. Each position starts with eyes open and progresses to eyes closed. When on soft foam, the individual starts with feet apart, increasing the difficulty by bringing them closer together. The trainer continues to assess the individual in this manner to find a position where he or she is challenged, but does not continuously lose his or her balance. This is where the individual begins their balance training.

44.8.2.2 Performing Balance Training

To perform balance training, the individual stands close to a table for support. With the stimulus intensity set at a comfortable level, the individual holds the device in his or her mouth, closes his or her eyes, and focuses on maintaining good posture in this position. The trainer provides standby assistance and monitors posture. Verbal cues, such as "unlock your knees" or "imagine a string pulling at the top of your head," help create the awareness necessary for the individual to make and maintain postural corrections. The goal is for the individual to be able to adjust his or her posture internally without cues. Individuals need to work at the highest level possible where they do not need assistance. The subject's balance is challenged with each training session, as they demonstrate improvement.

TABLE 44.2
External Variables That Affect Sitting Balance Training

	Easier	More Challenging
Sitting surface	Chair	Stability ball
Back	Supported	Unsupported
Arm position	Supported	Unsupported
Base of support	Wide (feet apart)	Narrow (feet together)
Touch point	Using touch point for support	No touch point
Eyes	Open	Closed

TABLE 44.3
External Variables That Affect Standing Balance Training

	Easier	More Challenging
Standing surface	Floor	Soft foam
Base of support	Wide (feet apart)	Narrow (feet together)
Foot position	Side by side	Staggered stance
Shoes	On	Off
Touch point	Using touch point for support	No touch point
Eyes	Open	Closed

44.8.3 Gait Training

Individuals with a neurological lesion may lack selective motor control, which prevents them from controlling the timing and intensity of muscle action (Perry, 1994). As a result, they exhibit an abnormal gait. The goal of gait training is to retrain movement patterns to achieve normal gait. Similar to balance training, a trainer uses verbal and tactile cues to correct abnormal patterns as the patient walks while using the PoNS device.

Gait training is more easily performed on the treadmill, eventually transferring to overground walking. As in balance, it is important to determine an appropriate starting point. This is relative to the individual's baseline performance. Variables for gait training on the treadmill include speed, grade, and use of handrails (Table 44.4).

To find the starting point, the trainer has the individual walk on the treadmill (no incline) at a comfortable pace and observes their gait pattern to identify areas that need work.

TABLE 44.4
Variables That Affect Treadmill Training

	Easier	More Challenging
Speed	Slower (may be harder for some)	Faster
Grade	Level surface	Inclined
Handrails	Use of handrails	No use of handrails

Training typically begins at a pace slower than normal. This allows the subject to work on the areas identified by the trainer.

Gait training sessions are 20 minutes and performed with the PoNS device. Working on the treadmill, the first 5 minutes are done at a slow pace. For the second 5 minutes, the speed should be increased 0.1–0.2 mph. In the third 5 minutes, speed should increase again and should be the most challenging. The last 5 minutes, the individual should work at a comfortable speed that he/she can maintain while exhibiting good walking posture. At each level, the individual is working to adjust their movement pattern as instructed. Change in grade and use of handrails can be used to vary the difficulty. If the individual's performance starts to break down, the challenge is too hard.

- Speed: Generally the individual walks at the highest speed that they are able until the observed quality of performance breaks down. The optimal speed is just below this point.
- Grade: An incline increases effort and affects the work of ankle dorsiflexors, knee, and hip flexors.
- Handrails: Using the handrails allows greater stability. Without using handrails allows work on arm swing.

It is important for the trainer to cue proper walking posture, focusing on areas identified as abnormal. Cues may be needed to equalize step length, decrease foot drag, emphasize heel strike, straighten a knee, relax the shoulders, or to modify arm swing. Additional activities can be incorporated into gait training, such as backwards walking or use of a metronome. Backwards walking can improve upright posture and foot clearance. A metronome can help with pacing. The trainer's observations and expertise should guide the treatment.

After treadmill training, individuals practice walking on the ground, incorporating the cues practiced on the treadmill. In addition, the individual can practice walking on a straight line, or staying in between two lines that are 12 inches apart. The goal is to eventually exhibit normal gait when walking without the PoNS device.

44.8.4 Cognitive Training

Cognitive functions such as memory and attention are often affected with a brain injury. Just as movement control can be retrained with CN-NINM training, so can cognitive abilities. Subjects report improved mental clarity, and demonstrate improved scores on memory, attention, and visuospatial tests after regular cognitive training with the PoNS device.

A cognitive training session is also 20 minutes. Individuals work at tasks that challenge their memory and attention while using the PoNS device. This can be in the form of games such as Simon, playing card games, solving puzzles, or playing challenging computer games designed specifically for cognitive training such as those found in Lumosity or Posit Science. Individuals should be provided with guidance as to the type of games that would be best for their needs. As they improve, they need to be challenged by progressing to higher levels.

44.8.5 BAT

The goal of breathing and awareness training is for the individual to develop relaxed and mindful breathing and awareness. A BAT session, also referred to as relaxation training, can have a significant positive impact on training progress. It requires 20 minutes of continuous uninterrupted PoNS use with eyes closed with attention toward relaxed muscles, breathing, and concentration.

The concept of attention to conscious breathing and relaxation helps the individual derive maximum benefit from PoNS training. Visualization and breathing are the focus points. Verbal cues are used to train the individual in how to perform a BAT session. What follows are some of the cues that we have used:

- Sit unsupported (back not touching back of chair) with hands resting on thighs.
- Distribute weight evenly.
- Good sitting posture with head and shoulders in good alignment.
- Place one hand on abdomen to feel the movement of the diaphragm.
- Bring attention to the breath.
- Before each breath, check in—"How am I doing?"
- Slowly inhale through the nose, starting from the belly.
- Feel the spine straighten and the chest expand with each inhale.
- Exhale slowly through the mouth and allow spine to curl gently (flexion).
- With each breath, check in with a body region.
- Scan from head to feet ("body scan").
- Release tension, sensation, thoughts, judgment as you breathe.
- Visualization: Pair a positive experience or outcome with inhalation and a negative experience or outcome with exhalation. Use positive memory and imagery (e.g., visualize an ice cube melting or waves at the seaside).

Breathing and awareness training is introduced in the first or second day of clinical training. During the rest of the training, individuals perform a BAT session on their own at the end of the day, approximately 1–2 hours before bedtime. BAT sometimes is used during the clinical training as an alternative to a balance, gait, or cognitive session if the individual is fatigued.

44.8.6 Continued Research

We continue to experiment to find the optimal training protocol for each type of patient population. Similar to other therapies, the final determinant of the training is the individual.

The greatest success with CN-NINM training is achieved when the training is targeted to each individual according to his/her presentation, progressed appropriately so that they continue to be challenged during their training, and conducted at a level of intensity that allows for adjustment based on their level of fatigue.

44.9 SELECTED RESULTS

44.9.1 Application to Symptoms of Acquired Brain Injury

44.9.1.1 Gait

44.9.1.1.1 *Four Subject TBI Cohort Dynamic Gait Index Results*

The results presented below represent the changes over a 5-day period of CN-NINM intervention in subjects with a TBI. Four female subjects (mean age: 48.3) presented with sustained and significant balance and gait deficits from moderate closed-head, nonpenetrating, concussive TBI (9–13 on Glasgow Coma Scale) at initial diagnosis. All were approximately 5 years postinjury and had previously completed rehabilitative therapy programs at their respective primary care facilities.

The results of the gait testing for four subjects that participated in the pilot study are summarized in Figure 44.5. The Dynamic Gait Index (DGI) is a clinician-scored index of eight facets of gait. Scores range from 0 (worst) to 24 (normal). A score change of 3.0 is generally considered clinically significant. The DGI scores indicate significant improvements in stability and gait that are retained for as much as 6 hours after completion of the second intervention session of the day.

44.9.1.1.2 *Single TBI Subject Electromyelogram Results*

Additional quantitative gait analysis using electromyography was performed on one of these subjects (D). At baseline, it revealed desynchronization of muscular activity—early activation of the left soleus during stance, and delayed activation of the left vastus lateralis, creating an abnormal gait pattern. This is shown in Figure 44.6. After 1 week of CN-NINM rehabilitation, much more normal phasing of both these muscles is present when the subject walked at the same speed. The medial hamstrings and medial gastrocnemius were not substantially affected, exhibiting similar phasing both before and after treatment.

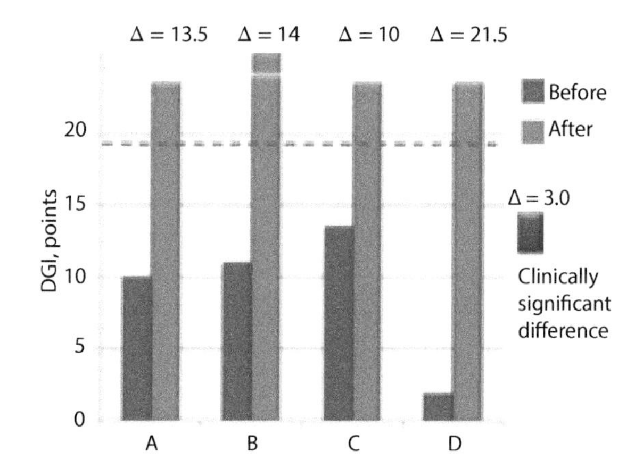

FIGURE 44.5 Summary of results of Dynamic Gait Index for four subjects with nonpenetrating TBI.

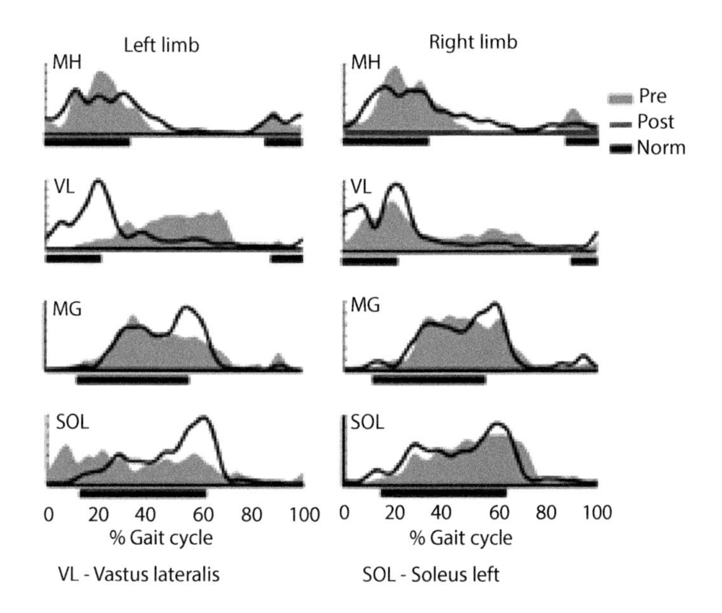

FIGURE 44.6 Muscle activation pattern was recorded before (pre) and after (post) training. (a) Head stabilogram of a subject with bilateral vestibular loss, standing with closed eyes test during 100 seconds. Subject could not stand in test position before stimulation. The head stabilograms were recorded every 0.5 and 1 hour until the stability effect disappeared (after approximately 4 hours). (b) Conceptual diagram of the retention effect dynamic. Note: Positive effects start to develop after approximately 10 minutes of stimulation (1); reaching a maximum at 30–60 minutes after the end of stimulation (2); and beneficial effects continue for up to 4–6 hours after initial training sessions, after which they rapidly diminish (3). (Data recorded at the UW Neuromuscular Biomechanics Lab by Darryl G. Thelen, PhD.)

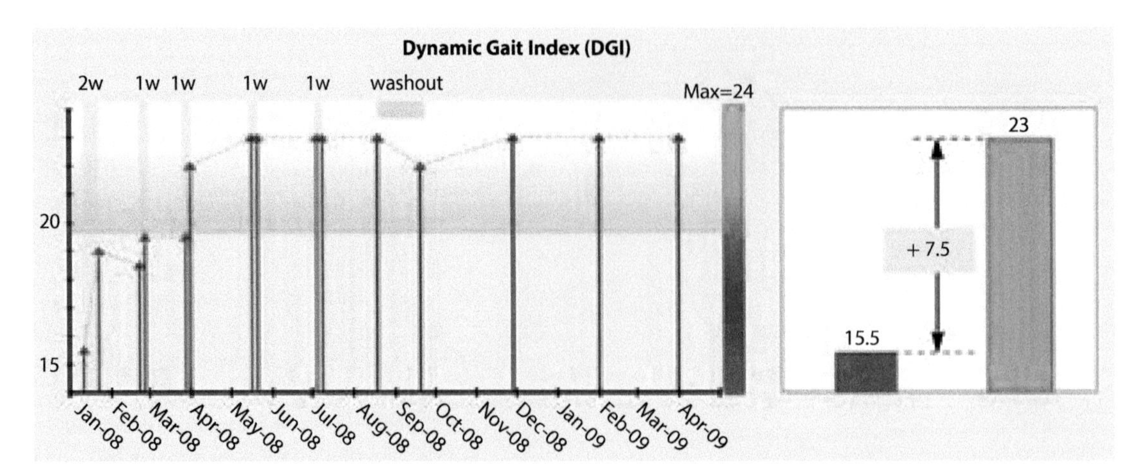

FIGURE 44.7 DGI recovery timeline. Stroke subject, 81 years of age, 4 years after cerebrovascular accident. Note: The major improvements occurred during in-laboratory training (blue stripes) and slight decrease of performance during "washout" period (purple). Subject trained at home between in-laboratory training sessions.

44.9.1.1.3 *Stroke Subject DGI Results*

Careful analysis of gait improvement in a stroke subject revealed a very important feature of CN-NINM training. The training protocol included balance, gait, and movement training (see previous section) during the initial 2 weeks in the laboratory, and an additional 5-day retraining and adjustment every month. In between the laboratory training sessions, the subject was instructed to continue the training at home. In this particular case, measurements of gait performance were conducted before and after every in-laboratory training period. Results are presented in Figure 44.7, which shows that the subject's gait performance improved 48% over 6 months. However, development of such performance was not smooth and continuous, but looks stepwise. Maximum improvement shift was achieved during in-laboratory training (initial 2 weeks and two sequential follow-up periods). At home, however, the subject only maintained achieved performance without noticeable improvement. Most likely, this reflects the difference in level of "challenge" and effort that was applied during balance and gait exercises at home versus in the laboratory (in this particular case, subject used regular walking at home instead of specialized treadmill exercises in the laboratory). After the third training period, when gait performance was near maximal (23/24 of DGI score), the additional in-laboratory training did not change gait performance further and the subject successfully maintained this level of function by training at home. After 6 months, subject was asked to stop any training for 1 month (washout period). As a result, we noticed minimal performance drop, which was easily reversed when subject returned back to regular in-home exercises.

44.9.2 Balance

The four TBI subjects were tested on the NeuroCom CDP Sensory Organization Test (SOT) before and after the week of twice-daily interventions. The SOT is an objective, automated measure of sensory-motor integration that evaluates the functional contribution of the somatosensory, visual, and vestibular components of balance. A composite score is calculated and compared with a database normalized for age and height. Sample pre- and postintervention scores are shown in Figure 44.8 for the latter two subjects in this cohort. The composite scores for all four subjects are summarized in Figure 44.9.

44.9.3 Cognitive Functions

Additionally, TBI subjects C and D were tested for changes in cognitive function, memory, attention, and mood both before the 5-day intervention began, and within 24 hours of completing the training. Their primary indications and scores on the Brief Repeatable Battery of Neuropsychological Tests (Boringa et al., 2001; Portaccio et al., 2010; Strauss, 2006) are summarized in Tables 44.5 and 44.6, and the attendant lists below each indicate the symptoms and changes.

1. Subject given 12 words in 24 seconds, then must recall them in any order within 60 seconds.
2. Recall of original 12 words 15 minutes later.
3. Recall position of 10 checkers on 6 x 6 board after viewing for 10 seconds.
4. Number of correctly matched digits to unfamiliar systematic symbolic code in 90 seconds.
5. Paced arithmetic serial addition test; successive entry every 3 seconds.
6. Paced arithmetic serial addition test; successive entry every 2 seconds.
7. Spontaneous production of names of a given category within 90 seconds.
8. Beck Depression Inventory; self-report that measures symptoms of depression.

Symptoms: pre-CN-NINM intervention

- Difficulties with concentration, emotions, and physical fatigue
- Slow and stuttered speech (significant dysarthria and expressive aphasia)

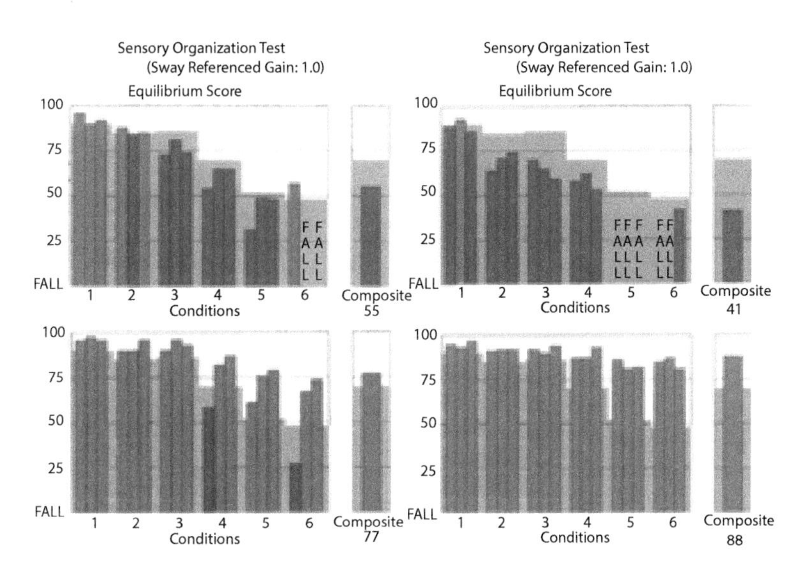

FIGURE 44.8 SOT scores before (upper) and after 1 week of training (lower) for two TBI subjects. Note that the greatest functional improvement occurred in the most dynamic and challenging tasks: condition 5: eyes *closed* with platform-induced anterior/posterior sway; condition 6, eyes open with platform and visually induced anterior/posterior sway.

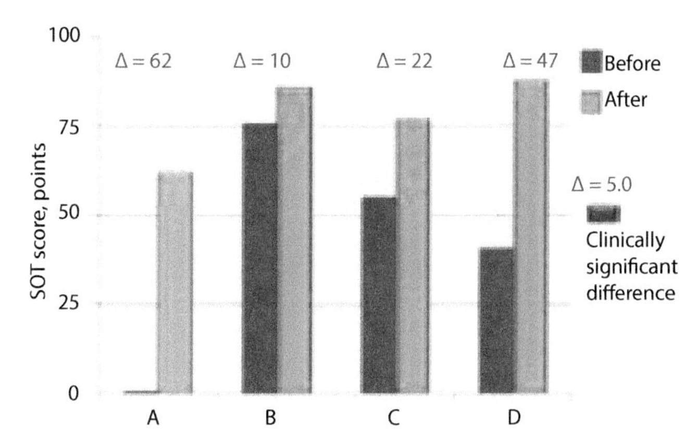

FIGURE 44.9 Summary of results from Computerized Dynamic Posturography SOT. Changes greater than 5 points are considered clinically significant improvement. SOT score below the age corrected norm marked in red color, SOT above the norm is marked in green color. Blue color demonstrates that patient A (initially incapable of performing this test at all, with continued falling in all the tests) improved dramatically (the largest improvement in score), but was still below the age corrected norm.

TABLE 44.5
Cognitive Tests, Subject C

Test Cycle	Word Span[1]	Word[2] Storage	Spatial Recall[3]	Symbol-Digit[4]	PASAT[5]	PASAT[6]	Word Generation[7]	Depression Scale[7]
	(12)	(11 minutes)	(10/36)	(90 seconds)	(3 seconds)	(2 seconds)	(60 seconds)	
Baseline	6	7	6.5	18	34	28	27	4 (mild)
2 weeks	9	9	8	42	34	35	32	2
1 month	10	11	9	53	52	36	33	2

- Diplopia and blurred vision
- Frequent vertigo and loss of balance in crowded places and between store aisles
- Problems with attention span, sustained and divided attention (low scores on word span on Symbol-Digit and Paced Audio Serial Addition Tests)
- Depression and emotional instability from chronic disability

Results: posttraining

- Showed statistically and clinically relevant improvement on most of the tests after 1 month.
- Verbal and visual were retrieval intact.
- After almost 5 years staying at home unable to do anything but simple tasks, she was able to go shopping, go to appointments, cook meals, and participate in family conversations.

TABLE 44.6
Cognitive Tests, Subject D

Test Cycle	Word Span	Word Storage	Spatial Recall	Symbol-Digit	PASAT	PASAT	Word Generation	Depression Scale
	(12)	(11 minutes)	(10/36)	(90 seconds)	(3 seconds)	(2 seconds)	(60 seconds)	
Baseline	5	7	9	30	29	23	27	12 (mod)
2 weeks	7	9	9	46	50	31	38	5 (mild)
1 month	11	11	10	59	58	42	38	2

Symptoms: preintervention

- Hypersensitivity to sounds, light, motion, or to objects or people moving in environment.
- Felt very tired, constant muscle aches because of cocontraction and aberrant neuromotor pattern, frequently had to lie down in complete isolation for at least 30 minutes.
- Showed significant attention impairment (low numbers on all tests).
- Moderate depression on depression scale, frequent crying.
- Both receptive and expressive aphasia, difficulty participating in conversations.
- Dysarthria; difficulty pronouncing even single-syllable words, stutters.

Results: posttraining

- After very first balance training with PoNS regained balance; after 5 days of intensive training was able to jog, walk, and stand with eyes closed.
- Showed significant improvement on all tests, no signs of depression.
- Now working again nearly full-time as a nurse-midwife, and is biking, jogging, singing, cooking, and traveling with her family.

All subjects anecdotally reported progressive improvements in eye movement, short-term memory, executive function, and mood elevation. Additionally, subjects C and D noted significant and progressive reductions in both expressive aphasia and social anxiety.

44.9.4 Eye Movement

Beginning with our first studies with rehabilitation of peripheral and central balance disorders, we noticed striking effects of CN-NINM training on the recovery of visual dysfunctions (oscillopsia, abnormal nystagmus, color perception, visual acuity, light and dark adaptation, limits of visual field). Similar and even stronger effects were observed during studies with stroke, traumatic brain injury, multiple sclerosis, and Parkinson subjects. Two examples of eye movement abnormalities and rehabilitation effects of CN-NINM therapy are presented in Figures 44.10 and 44.11.

The quantitative eye movement recording tests were performed with binocular videonystagmograph VisualEyes (Micromedical Technologies).

The first case demonstrates problematic eye stabilization during a fixation task in a subject with a sport concussive injury 7 months after the incident. The subject also reported continuous hypersensitivity to light and sound, difficulty with reading books and computer screens, and attention, concentration, and short-term memory problems.

The diagram in Figure 44.10a demonstrates binocular fixation precision (distance from the target cross) and stability (size of the color spots) in a normal, healthy subject. Each spot represents the contour map of bivariate distribution of 3,000 eye position measurements recorded during 15 seconds for each target position. In Figure 44.10b (mTBI subject) there is a striking difference in a key parameter: the deviation from target position (enlarged size of spots, elongated form, and appearance of two separate clusters). After 5 days of balance and gait training with the mTBI subject (without special eye movement exercises), we found significant improvement in muscular stabilization, evidenced by more condensed spots that are more appropriately aligned both spatially and axially (Figure 44.10c). This "shrinking" of the spots reflects increased stability of the extrinsic eye muscles.

The second example, shown in Figure 44.11, demonstrates changes in dynamic eye movement control during a smooth pursuit test performed by an 82-year-old stroke subject, 4 years after incident.

44.10 CONCLUSION

CN-NINM technology is a new noninvasive brain stimulation technique with potential applications in physical medicine, cognitive, and affective neurosciences (Bach-y-Rita, 2003a,2003b; Bach, 2004, 2005). This stimulation method appears to be a promising treatment of a full spectrum of movement disorders, and probably for multiple other dysfunctions associated with TBI.

Based on CN-NINM technology, our training program is designed to noninvasively induce multitargeted neural plasticity that, with consistent progressive application, affects

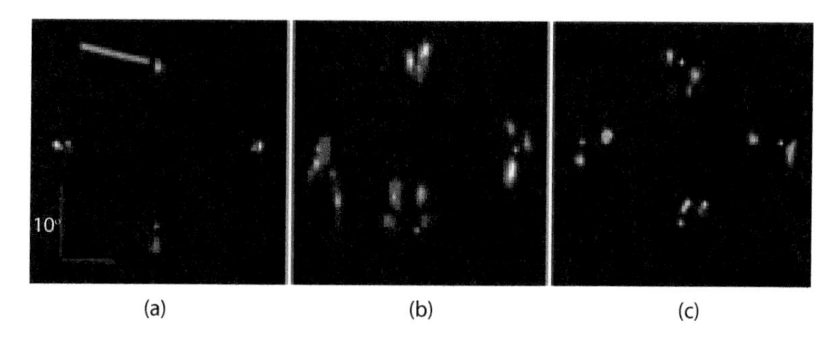

FIGURE 44.10 Density plot of targeted eye fixation. Eye position were recorded during a 15 second (3,000 data points) interval. Crosses, position of the targets. Position, size, and shape of each blob reflect precision and stability of fixation point. Bright colors correspond to higher density and white spots to the point of maximal stability of the eye. (a) Normal subject: left (red) and right (blue) eye position density plots. (b) mTBI subject, 17 years of age, 7 months after injury. Left (orange) and right (purple) eye, before CN-NINM training. (c) The same subject after 5 days of CN-NINM training. Left (blue) and right (green) eye. Note changes in eyes stabilization.

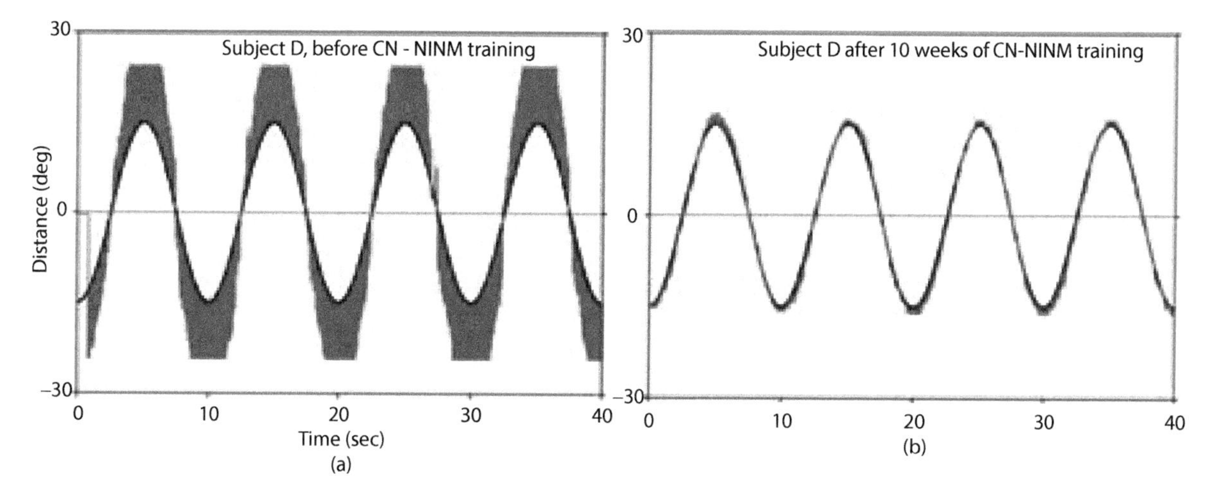

FIGURE 44.11 Eye movement before and after CN-NINM training. Smooth (sinusoidal) pursuit test, at target speed 0.1 Hz. Tested before and after CN-NINM. Subject D (stroke), 10 weeks apart, right eye presented (both eyes had similar results). Black line, trajectory of the visual target; red line, the right eye trajectory. Y-axis is target or eye horizontal distance (degree) from the center of display. Spatiotemporal difference between target and eye position was colored in red. Upper and lower cut on left picture, physical limits of recording area (edges of video display).

functional changes in physical, cognitive, and sensory domains in subjects with significant dysfunction from TBI.

Examples, presented in Section 44.8, demonstrate the efficiency of our neurorehabilitation approach when applied to very complex sensory-motor functional systems such as balance, gait, movement control, and static and dynamic eye movement control. Improvement of these functional systems demonstrates recovery of appropriate spatiotemporal activation patterns in affected neuromuscular systems, which requires optimization and exquisite synchronization of multiple movement control centers and systems to function correctly and automatically.

We believe that CN-NINM technology can help to reinforce existing methods of rehabilitation therapy and approach new areas in neurorehabilitation that many considered untreatable, especially in the domain of chronic TBI.

From a future perspective, evidence of extended brain activation and simultaneous improvement of multiple symptoms typical of TBI (as well as in multiple sclerosis, Parkinson disease, and stroke subjects; unpublished results), allow us to consider CN-NINM as a core technology that can be strengthened by integration with existing methods (such as neurofeedback technology) and potentially be a complementary component of new methods not yet developed.

Though our current aim in the field of neurorehabilitation is to demonstrate efficiency of CN-NINM technology in functional improvement of acquired brain injury symptoms, we believe that future work remains to uncover the underlying mechanisms for the remarkable phenomenon we have discovered.

REFERENCES

Abraham, W.C., and M.F. Bear. 1996. Metaplasticity: The plasticity of synaptic plasticity. *Trends Neurosci.* 19:126–130.

Bach-y-Rita, P. 2003a. Late postacute neurologic rehabilitation: Neuroscience, engineering, and clinical programs. *Arch Phys Med Rehabil.* 84:1100–1108.

Bach-y-Rita, P. 2003b. Theoretical basis for brain plasticity after a TBI. *Brain Inj.* 17:643–651.

Bach-y-Rita, P. 2001. Theoretical and practical considerations in the restoration of function after stroke. *Top Stroke Rehabil.* 8:1–15.

Bach-y-Rita, P. 1990. Brain plasticity as a basis for recovery of function in humans. *Neuropsychologia.* 28:547–554.

Bach-y-Rita, P. 1983. Tactile vision substitution: Past and future. *Int J Neurosci.* 19:29–36.

Bach-y-Rita, P. 1972. *Brain Mechanisms in Sensory Substitution.* Academic Press, New York.

Bach-y-Rita, P., C.C. Collins, F.A. Saunders, B. White, and L. Scadden. 1969a. Vision substitution by tactile image projection. *Nature.* 221:963–964.

Bach-y-Rita, P., C.C. Collins, B. White, F.A. Saunders, L. Scadden, and R. Blomberg. 1969b. A tactile vision substitution system. *Am J Optom Arch Am Acad Optom.* 46:109–111.

Bach-y-Rita, P., and K.A. Kaczmarek. 2002. Tongue placed tactile output device, from http://www.warf.org/technologies/medical-devices/adaptive-design/summary/tongue-placed-tactile-output-device-p98100us.cmsx. Wisconsin Alumni Research Foundation, Madison, WI.

Bach-y-Rita, P., K.A. Kaczmarek, M.E. Tyler, and J. Garcia-Lara. 1998. Form perception with a 49-point electrotactile stimulus array on the tongue: A technical note. *J Rehabil Res Dev.* 35:427–430.

Bach-y-Rita, P., and W.K.S. 2003. Sensory substitution and the human-machine interface. *Trends Cogn Sci.* 7:541–546.

Bach, Y.R.P. 2004. Is it possible to restore function with two percent surviving neural tissue? *J Integr Neurosci.* 3:3–6.

Bach, Y.R.P. 2005. Emerging concepts of brain function. *J Integr Neurosci.* 4:183–205.

Badke, M.B., J. Sherman, P. Boyne, S. Page, and K. Dunning. 2011. Tongue-based biofeedback for balance in stroke: Results of an 8-week pilot study. *Arch Phys Med Rehabil.* 92:1364–1370.

Baker, R.S., and A.D. Epstein. 1991. Ocular motor abnormalities from head trauma. *Surv Ophthalmol.* 35:245–267.

Barros, C.G., R.S. Bittar, and Y. Danilov. 2010. Effects of electrotactile vestibular substitution on rehabilitation of patients with bilateral vestibular loss. *Neurosci Lett.* 476:123–126.

Barroso-Chinea, P., and E. Bezard. 2010. Basal ganglia circuits underlying the pathophysiology of levodopa-induced dyskinesia. *Front Neuroanatomy.* 4.

Baugh, C.M., J.M. Stamm, D.O. Riley, B.E. Gavett, M.E. Shenton, A. Lin et al. 2012. Chronic traumatic encephalopathy: Neurodegeneration following repetitive concussive and subconcussive brain trauma. *Brain Imaging Behav.* 6:244–254.

Bear, M.F., and R.C. Malenka. 1994. Synaptic plasticity: LTP and LTD. *Curr Opin Neurobiol.* 4:389–399.

Bédard, M.-A. 2003. *Mental and Behavioral Dysfunction in Movement Disorders.* Humana Press, Totowa, NJ.

Bi, G.Q., V. Bolshakov, G. Bu, C.M. Cahill, Z.F. Chen, G.L. Collingridge et al. 2006. Recent advances in basic neurosciences and brain disease: From synapses to behavior. *Molecul Pain.* 2:38.

Binder, S., J.D. Corrigan, and J.A. Langlois. 2005. The public health approach to traumatic brain injury: An overview of CDC's research and programs. *J Head Trauma Rehabil.* 20:189–195.

Bittar, R.S., and G. Barros Cde. 2011. Vestibular rehabilitation with biofeedback in patients with central imbalance. *Braz J Otorhinolaryngol.* 77:356–361.

Bliss, T.V., and A.R. Gardner-Medwin. 1973. Long-lasting potentiation of synaptic transmission in the dentate area of the unanaestetized rabbit following stimulation of the perforant path. *J Physiol.* 232:357–374.

Bliss, T.V., and T. Lomo. 1973. Long-lasting potentiation of synaptic transmission in the dentate area of the anaesthetized rabbit following stimulation of the perforant path. *J Physiol.* 232:331–356.

Boringa, J., R. Lazeron, I. Reuling, H. Adèr, L. Pfennings, J. Lindeboom et al. 2001. The brief repeatable battery of neuropsychological tests: Normative values allow application in multiple sclerosis clinical practice. *Mult Scler.* 7:263–267.

Borisoff, J.F., S.L. Elliott, S. Hocaloski, and G.E. Birch. 2010. The development of a sensory substitution system for the sexual rehabilitation of men with chronic spinal cord injury. *J Sex Med.* 7:3647–3658.

Buchs, P.A., and D. Muller. 1996. Induction of long-term potentiation is associated with major ultrastructural changes of activated synapses. *Proc Natl Acad Sci U S A.* 93:8040–8045.

Calverley, R.K., and D.G. Jones. 1990. Contributions of dendritic spines and perforated synapses to synaptic plasticity. *Brain Res.* 15:215–249.

Chebat, D.R., C. Rainville, R. Kupers, and M. Ptito. 2007. Tactile-'visual' acuity of the tongue in early blind individuals. *Neuroreport.* 18:1901–1904.

Chebat, D.R., F.C. Schneider, R. Kupers, and M. Ptito. 2011. Navigation with a sensory substitution device in congenitally blind individuals. *Neuroreport.* 22:342–347.

Chen, Y.H., J.H. Kang, and H.C. Lin. 2011. Patients with traumatic brain injury: Population-based study suggests increased risk of stroke. *Stroke.* 42:2733–2739.

Cicerone, K.D., C. Dahlberg, K. Kalmar, D.M. Langenbahn, J.F. Malec, T.F. Bergquist et al. 2000. Evidence-based cognitive rehabilitation: Recommendations for clinical practice. *Arch Phys Med Rehabil.* 81:1596–1615.

Cifu, D.X., M.E. Huang, S.A. Kolakowsky-Hayner, and R.T. Seel. 1999. Age, outcome, and rehabilitation costs after paraplegia caused by traumatic injury of the thoracic spinal cord, conus medullaris, and cauda equina. *J Neurotrauma.* 16:805–815.

Collins, C.C., and J.M.J. Madey. 1974. Tactile sensory replacement. Proc San Diego Biomed Symp. 13:15–26.

Coronado, V.G., K.E. Thomas, R.W. Sattin, and R.L. Johnson. 2005. The CDC traumatic brain injury surveillance system: Characteristics of persons aged 65 years and older hospitalized with a TBI. *J Head Trauma Rehabil.* 20:215–228.

Coronado, V.G., L. Xu, S.V. Basavaraju, L.C. McGuire, M.M. Wald et al., Centers for Disease Control and Prevention. 2011. Surveillance for traumatic brain injury-related deaths—United States, 1997–2007. *MMWR.* 60:1–32.

Corso, P., E. Finkelstein, T. Miller, I. Fiebelkorn, and E. Zaloshnja. 2006. Incidence and lifetime costs of injuries in the United States. *Inj Prevent.* 12:212–218.

D'Angelo, E. 2011. Neural circuits of the cerebellum: Hypothesis for function. *J Integ Neurosci.* 10:317–352.

Danilov YP, T.M., Bach-Y-Rita P. 2004. Vestibular substitution for postural control. In R.J.-F. Lofaso F, Roby-Brami A, eds.. *Innovations Technologiques et Handicap: Actes des 17e[s] Entretiens de l'Institut Garches.* Vol. 17. Frison-Roche, Paris, France. pp. 216–224.

Danilov, Y.P., M.E. Tyler, K.L. Skinner, and P. Bach-y-Rita. 2006. Efficacy of electrotactile vestibular substitution in patients with bilateral vestibular and central balance loss. *Conf Proc IEEE Eng Med Biol Soc.* Suppl:6605–6609.

Danilov, Y.P., M.E. Tyler, K.L. Skinner, R.A. Hogle, and P. Bach-y-Rita. 2007. Efficacy of electrotactile vestibular substitution in patients with peripheral and central vestibular loss. *J Vestib Res.* 17:119–130.

DeKosky, S.T., M.D. Ikonomovic, and S. Gandy. 2010. Traumatic brain injury: Football, warfare, and long-term effects. *Minn Medic.* 93:46–47.

Doidge, N. 2007. *The Brain That Changes Itself: Stories of Personal Triumph from the Frontiers of Brain Science*. Viking, New York.

Dudek, S.M., and M.F. Bear. 1993. Bidirectional long-term modification of synaptic effectiveness in the adult and immature hippocampus. *J Neurosci*. 13:2910–2918.

Engert, F., and T. Bonhoeffer. 1999. Dendritic spine changes associated with hippocampal long-term synaptic plasticity. *Nature*. 399:66–70.

Gavett, B.E., R.A. Stern, and A.C. McKee. 2011. Chronic traumatic encephalopathy: A potential late effect of sport-related concussive and subconcussive head trauma. *Clin Sports Med*. 30:179–188, xi.

Geinisman, Y., R.W. Berry, J.F. Disterhoft, J.M. Power, and E.A. Van der Zee. 2001. Associative learning elicits the formation of multiple-synapse boutons. *J Neurosci*. 21:5568–5573.

Geinisman, Y., L. deToledo-Morrell, and F. Morrell. 1991. Induction of long-term potentiation is associated with an increase in the number of axospinous synapses with segmented postsynaptic densities. *Brain Res*. 566:77–88.

Geinisman, Y., L. deToledo-Morrell, F. Morrell, R.E. Heller, M. Rossi, and R.F. Parshall. 1993. Structural synaptic correlate of long-term potentiation: Formation of axospinous synapses with multiple, completely partitioned transmission zones. *Hippocampus*. 3:435–445.

Geinisman, Y., J.F. Disterhoft, H.J. Gundersen, M.D. McEchron, I.S. Persina, J.M. Poweret al. 2000. Remodeling of hippocampal synapses after hippocampus-dependent associative learning. *J Comp Neurol*. 417:49–59.

Gibson, R.H. 1968. Electrical stimulation of pain and touch. In D.R. Kenshalo, editor. *The Skin Senses*. Charles C Thomas, Springfield, IL. pp. 223–261.

Gisabella, B., V.Y. Bolshakov, and F.M. Benes. 2005. Regulation of synaptic plasticity in a schizophrenia model. *Proc Natl Acad Scie U S A*. 102:13301–13306.

Goldstein, L.E., A.M. Fisher, C.A. Tagge, X.L. Zhang, L. Velisek, J.A. Sullivan et al. 2012. Chronic traumatic encephalopathy in blast-exposed military veterans and a blast neurotrauma mouse model. *Sci Trans Med*. 4:134ra160.

Guerrero, J.L., D.J. Thurman, and J.E. Sniezek. 2000. Emergency department visits associated with traumatic brain injury: United States, 1995–1996. *Brain Inj*. 14:181–186.

Haase, S.J., and K.A. Kaczmarek. 2005. Electrotactile perception of scatterplots on the fingertips and abdomen. *Med Biol Eng Comput*. 43:283–289.

Heasman, P.A., and A.D. Beynon. 1986. Quantitative diameter analysis of lingual nerve axons in man. *J Dental Res*. 65:1016–1019.

Johansson, R.S., and A.B. Vallbo. 1979a. Detection of tactile stimuli. Thresholds of afferent units related to psychophysical thresholds in the human hand. *J Physiol*. 297:405–422.

Johansson, R.S., and A.B. Vallbo. 1979b. Tactile sensibility in the human hand: Relative and absolute densities of four types of mechanoreceptive units in glabrous skin. *J Physiol*. 286:283–300.

Jones, D.G., and R.K. Calverley. 1991. Frequency of occurrence of perforated synapses in developing rat neocortex. *Neurosci Lett*. 129:189–192.

Jones, D.G., W. Itarat, and R.K. Calverley. 1991. Perforated synapses and plasticity. A developmental overview. *Molec Neurobiol*. 5:217–228.

Kaczmarek, K., P. Bach-y-Rita, W.J. Tompkins, and J.G. Webster. 1985. A tactile vision-substitution system for the blind: Computer-controlled partial image sequencing. *IEEE Trans Biomed Eng*. 32:602–608.

Kaczmarek, K.A. 2011. The tongue display unit (TDU) for electrotactile spatiotemporal pattern presentation. *Scientia Iran*. 18:1476–1485.

Kaczmarek, K.A., and P. Bach-y-Rita. 1995. Tactile displays. In W. Barfield and T. Furness, eds. *Virtual Environments and Advanced Interface Design*. Oxford University Press, New York. pp. 349–414.

Kaczmarek, K.A., and S.J. Haase. 2003a. Pattern identification and perceived stimulus quality as a function of stimulation waveform on a fingertip-scanned electrotactile display. *IEEE Trans Neural Sys Rehab Eng*. 11:9–16.

Kaczmarek, K.A., and S.J. Haase. 2003b. Pattern identification as a function of stimulation current on a fingertip-scanned electrotactile display. *IEEE Trans Neural Sys Rehab Eng*. 11:269–275.

Kaczmarek, K.A., and M.E. Tyler. 2000. Effect of electrode geometry and intensity control method on comfort of electrotactile stimulation on the tongue. *Proc ASME Dyn Sys Contr Div*. Vol. DSC 69-2. ASME, Orlando, FL. pp. 1239–1243.

Kaczmarek, K.A., M.E. Tyler, and P. Bach-y-Rita. 1994. Electrotactile haptic display on the fingertips: Preliminary results. *Proc 16th Annu Int Conf IEEE Eng Med Biol Soc*.IEEE, Baltimore, MD. 940–941.

Kaczmarek, K.A., M.E. Tyler, A.J. Brisben, and K.O. Johnson. 2000. The afferent neural response to electrotactile stimuli: Preliminary results. *IEEE Trans Rehab Eng*. 8:268–270.

Kaczmarek, K.A., J.G. Webster, P. Bach-y-Rita, and W.J. Tompkins. 1991. Electrotactile and vibrotactile displays for sensory substitution systems. *IEEE Trans Biomed Eng*. 38:1–16.

Kaczmarek, K.A., J.G. Webster, and R.G. Radwin. 1992. Maximal dynamic range electrotactile stimulation waveforms. *IEEE Trans Biomed Eng*. 39:701–715.

Kirkwood, A., H.K. Lee, and M.F. Bear. 1995. Co-regulation of long-term potentiation and experience-dependent synaptic plasticity in visual cortex by age and experience. *Nature*. 375:328–331.

Kupers, R., and M. Ptito. 2013. Compensatory plasticity and cross-modal reorganization following early visual deprivation. *Neurosci Biobehav Rev*. 41:36–52.

Lambert, S., E. Sampaio, Y. Mauss, and C. Scheiber. 2004. Blindness and brain plasticity: Contribution of mental imagery? An fMRI study. *Brain Res*. 20:1–11.

Langlois Orman, J.A.K., J.F. Zaloshnja, and E. Miller, T. 2011. Epidemiology. In: J.M. Silver, T.W. McAllister, S.C. Yudofsk, eds. *Textbook of Traumatic Brain Injury*. 2nd ed. American Psychiatric Association, Washington, DC. pp. 3–22.

Larkman, A.U., and J.J. Jack. 1995. Synaptic plasticity: Hippocampal LTP. *Curr Opin Neurobiol*. 5:324–334.

Levenes, C., H. Daniel, and F. Crepel. 1998. Long-term depression of synaptic transmission in the cerebellum: Cellular and molecular mechanisms revisited. *Progr Neurobiol*. 55:79–91.

Li, Y., E.G. Meloni, W.A. Carlezon, Jr., M.R. Milad, R.K. Pitman, K. Nader et al. 2013. Learning and reconsolidation implicate different synaptic mechanisms. *Proc Natl Acad Sci U S A*. 110:4798–4803.

Machts, L. 1920. Device for converting light effects into effects perceptible by blind persons. Germany.

Maletic-Savatic, M., R. Malinow, and K. Svoboda. 1999. Rapid dendritic morphogenesis in CA1 hippocampal dendrites induced by synaptic activity. *Science*. 283:1923–1927.

Matteau, I., R. Kupers, E. Ricciardi, P. Pietrini, and M. Ptito. 2010. Beyond visual, aural and haptic movement perception: hMT+ is activated by electrotactile motion stimulation of the tongue in sighted and in congenitally blind individuals. *Brain Res Bull*. 82:264–270.

Max W, M.E., Rice DP. 1991. Head injuries: Costs and consequences. *J Head Trauma Rehabil*. 76–91.

McGregor, K., and B. Pentland. 1997. Head injury rehabilitation in the U.K.: An economic perspective. *Social Scie Med*. 45:295–303.

McKee, A.C., R.C. Cantu, C.J. Nowinski, E.T. Hedley-Whyte, B.E. Gavett, A.E. Budson et al. 2009. Chronic traumatic encephalopathy in athletes: Progressive tauopathy after repetitive head injury. *J Neuropathol Exp Neurol*. 68:709–735.

McKee, A.C., T.D. Stein, C.J. Nowinski, R.A. Stern, D.H. Daneshvar, V.E. Alvarez et al. 2013. The spectrum of disease in chronic traumatic encephalopathy. *Brain*. 136:43–64.

McKinlay, W.W., D.N. Brooks, M.R. Bond, D.P. Martinage, and M.M. Marshall. 1981. The short-term outcome of severe blunt head injury as reported by relatives of the injured persons. *J Neurol Neurosurg Psychiatry*. 44:527–533.

McMillan, T.M., G.M. Teasdale, C.J. Weir, and E. Stewart. 2011. Death after head injury: The 13 year outcome of a case control study. *J Neurol Neurosurg Psychiatry*. 82:931–935.

Morice, E., L.C. Andreae, S.F. Cooke, L. Vanes, E.M. Fisher, V.L. Tybulewicz et al. 2008. Preservation of long-term memory and synaptic plasticity despite short-term impairments in the Tc1 mouse model of Down syndrome. *Learn Memory*. 15:492–500.

Neuman, D.A. 2009. *Kinesiology of the Musculoskeletal System: Foundations for Rehabilitation*. Mosby/Elsevier, Philadelphia, PA.

Nudo, R.J., E.J. Plautz, and S.B. Frost. 2001. Role of adaptive plasticity in recovery of function after damage to motor cortex. *Muscle Nerve*. 24:1000–1019.

Omalu, B., J.L. Hammers, J. Bailes, R.L. Hamilton, M.I. Kamboh, G. Webster et al. 2011. Chronic traumatic encephalopathy in an Iraqi war veteran with posttraumatic stress disorder who committed suicide. *Neurosurg Focus*. 31:E3.

Pai, A.B., N.R. Jasper, and D.X. Cifu. 2012. Rehabilitation of injured U.S. servicemember with traumatic brain injury, stroke, spinal cord injury, and bilateral amputations: A case report. *J Rehabil Res Dev*. 49:1191–1196.

Perry, J. 1994. Gait analysis: Technology and the clinician. *J Rehabil Res Dev*. 31:vii.

Ponsford, J., M. Downing, J. Olver, M. Ponsford, R. Acher, M. Carty et al. 2013. Longitudinal follow-up of patients with traumatic brain injury: Outcome at 2, 5, and 10-years post-injury. *J Neurotrauma*. 31:64–77.

Portaccio, E., B. Goretti, V. Zipoli, A. Iudice, D.D. Pina, G.M. Malentacchi et al. 2010. Reliability, practice effects, and change indices for Rao's Brief Repeatable Battery. *Mult Scler*. 16:611–617.

Price, J., and W. Drevets. 2012. Neural circuits underlying the pathophysiology of mood disorders. *Trends Cogni Sci*. 16:61–71.

Ptito, M., S.M. Moesgaard, A. Gjedde, and R. Kupers. 2005. Cross-modal plasticity revealed by electrotactile stimulation of the tongue in the congenitally blind. *Brain*. 128:606–614.

Reilly, J.P. 1998. *Applied Bioelectricity*. Springer, New York.

Reis, J., H.M. Schambra, L.G. Cohen, E.R. Buch, B. Fritsch, E. Zarahn et al. 2009. Noninvasive cortical stimulation enhances motor skill acquisition over multiple days through an effect on consolidation. *Proc Natl Acad Sci U S A*. 106:1590–1595.

Rigg, J.L., and S.R. Mooney. 2011. Concussions and the military: Issues specific to service members. *PM & R*. 3:S380–386.

Robinson, B.S., J.L. Cook, C.M. Richburg, and S.E. Price. 2009. Use of an electrotactile vestibular substitution system to facilitate balance and gait of an individual with gentamicin-induced bilateral vestibular hypofunction and bilateral transtibial amputation. *J Neurol Phys Therapy*. 33:150–159.

Rohling, M.L., M.E. Faust, B. Beverly, and G. Demakis. 2009. Effectiveness of cognitive rehabilitation following acquired brain injury: A meta-analytic re-examination of Cicerone et al.'s (2000, 2005) systematic reviews. *Neuropsychology*. 23:20–39.

Rollman, G.B. 1973. Electrocutaneous stimulation. *Proc Conf Cutan Comm Sys*. Dev. F.A. Geldard, editor. Psychonomic Society. 38–51.

Rutland-Brown, W., J.A. Langlois, J.J. Bazarian, and D. Warden. 2008. Improving identification of traumatic brain injury after nonmilitary bomb blasts. *J Head Trauma Rehabil*. 23:84–91.

Rutland-Brown, W., J.A. Langlois, K.E. Thomas, and Y.L. Xi. 2006. Incidence of traumatic brain injury in the United States, 2003. *J Head Trauma Rehabil*. 21:544–548.

Sampaio, E., S. Maris, and P. Bach-y-Rita. 2001. Brain plasticity: 'visual' acuity of blind persons via the tongue. *Brain Res*. 908:204–207.

Schaning, M.A., and K.A. Kaczmarek. 2008. A high-voltage bipolar transconductance amplifier for electrotactile stimulation. *IEEE Trans Biomed. Eng*. 55:2433–2443.

Schneier, A.J., B.J. Shields, S.G. Hostetler, H. Xiang, and G.A. Smith. 2006. Incidence of pediatric traumatic brain injury and associated hospital resource utilization in the United States. *Pediatrics*. 118:483–492.

Segond, H., D. Weiss, M. Kawalec, and E. Sampaio. 2013. Perceiving space and optical cues via a visuo-tactile sensory substitution system: A methodological approach for training of blind subjects for navigation. *Perception*. 42:508–528.

Segond, H., D. Weiss, and E. Sampaio. 2005. Human spatial navigation via a visuo-tactile sensory substitution system. *Perception*. 34:1231–1249.

Selassie, A.W., E. Zaloshnja, J.A. Langlois, T. Miller, P. Jones, and C. Steiner. 2008. Incidence of long-term disability following traumatic brain injury hospitalization, United States, 2003. *J Head Trauma Rehabil*. 23:123–131.

Shin, R.M., E. Tsvetkov, and V.Y. Bolshakov. 2006. Spatiotemporal asymmetry of associative synaptic plasticity in fear conditioning pathways. *Neuron*. 52:883–896.

Shin, R.M., K. Tully, Y. Li, J.H. Cho, M. Higuchi, T. Suhara et al. 2010. Hierarchical order of coexisting pre- and postsynaptic forms of long-term potentiation at synapses in amygdala. *Proc Natl Acad Sci U S A*. 107:19073–19078.

Silver, J.M., R. Kramer, S. Greenwald, and M. Weissman. 2001. The association between head injuries and psychiatric disorders: Findings from the New Haven NIMH Epidemiologic Catchment Area Study. *Brain Inj*. 15:935–945.

Spitz, G., J.J. Maller, R. O'Sullivan, and J.L. Ponsford. 2013. White Matter integrity following traumatic brain injury: The association with severity of injury and cognitive functioning. *Brain Topogr*. 26:648–660.

Strauss, E.H. 2006. *A Compendium of Neuropsychological Tests: Administration, Norms, and Commentary*. Oxford University Press, Oxford, UK.

Szeto, A.Y., and F.A. Saunders. 1982. Electrocutaneous stimulation for sensory communication in rehabilitation engineering. *IEEE Trans Biomed Eng*. 29:300–308.

Szeto, A.Y.J., and R.R. Riso. 1990. Sensory feedback using electrical stimulation of the tactile sense. In R.V. Smith and J.H. Leslie Jr., eds. *Rehabilitation Engineering*. CRC Press, Boca Raton, FL. pp. 29–78.

Tate, D.F., M.E. Shenton, and E.D. Bigler. 2012. Introduction to the brain imaging and behavior special issue on neuroimaging findings in mild traumatic brain injury. *Brain Imaging Behav*. 6:103–107.

Taub, E., G. Uswatte, and T. Elbert. 2002. New treatments in neurorehabilitation founded on basic research. *Nat Rev*. 3:228–236.

Taylor, B.C., E.M. Hagel, K.F. Carlson, D.X. Cifu, A. Cutting, D.E. Bidelspach et al. 2012. Prevalence and costs of co-occurring traumatic brain injury with and without psychiatric disturbance and pain among Afghanistan and Iraq War Veteran V.A. users. *Med Care*. 50:342–346.

Thurman, D. 2001. The epidemiology and economics of head trauma. In L.P. Miller, R.L. Hayes, and J.K. Newcomb, eds. *Head Trauma: Basic, Preclinical, and Clinical Directions*. John Wiley & Sons, New York.

Thurman, D., and J. Guerrero. 1999. Trends in hospitalization associated with traumatic brain injury. *JAMA*. 282:954–957.

Thurman, D.J., C.M. Branche, and J.E. Sniezek. 1998. The epidemiology of sports-related traumatic brain injuries in the United States: Recent developments. *J Head Trauma Rehabil*. 13:1–8.

Toni, N., P.A. Buchs, I. Nikonenko, C.R. Bron, and D. Muller. 1999. LTP promotes formation of multiple spine synapses between a single axon terminal and a dendrite. *Nature*. 402:421–425.

Tschiriew, S., and A.D. Watteville. 1879. On the electrical excitability of the skin. *Brain*. 2:163–180.

Tully, K., and V.Y. Bolshakov. 2010. Emotional enhancement of memory: How norepinephrine enables synaptic plasticity. *Molec Brain*. 3:15.

Tully, K., Y. Li, and V.Y. Bolshakov. 2007a. Keeping in check painful synapses in central amygdala. *Neuron*. 56:757–759.

Tully, K., Y. Li, E. Tsvetkov, and V.Y. Bolshakov. 2007b. Norepinephrine enables the induction of associative long-term potentiation at thalamo-amygdala synapses. *Proc Natl Acad Sci U S A*. 104:14146–14150.

Tyler, M., Y. Danilov, and Y.R.P. Bach. 2003. Closing an open-loop control system: Vestibular substitution through the tongue. *J Integr Neurosci*. 2:159–164.

Tyler, M.E., J.G. Braun, and Y.P. Danilov. 2009. Spatial mapping of electrotactile sensation threshold and intensity range on the human tongue: Initial results. *Proc IEEE Eng Med Bio. Soc.*. 559–562.

Vallbo, A.B., and R.S. Johansson. 1984. Properties of cutaneous mechanoreceptors in the human hand related to touch sensation. *Human Neurobiol*. 3:3–14.

Vallbo, A.B., K.A. Olsson, K.G. Westberg, and F.J. Clark. 1984. Microstimulation of single tactile afferents from the human hand. Sensory attributes related to unit type and properties of receptive fields. *Brain*. 107(Pt 3):727–749.

Vuillerme, N., and R. Cuisinier. 2008. Head position-based electrotactile tongue biofeedback affects postural responses to Achilles tendon vibration in humans. *Exp Brain Res*. 186:503–508.

Vuillerme, N., N. Pinsault, O. Chenu, J. Demongeot, Y. Payan, and Y. Danilov. 2008. Sensory supplementation system based on electrotactile tongue biofeedback of head position for balance control. *Neurosci Lett*. 431:206–210.

Watson, T., S. Koutsikou, N. Cerminara, C. Flavell, J. Crook, B. Lumb et al. 2013. The olivo-cerebellar system and its relationship to survival circuits. *Front Neural Circuits*. 7:72.

Wehman, P., J. Kregel, L. Keyser-Marcus, P. Sherron-Targett, L. Campbell, M. West et al. 2003. Supported employment for persons with traumatic brain injury: A preliminary investigation of long-term follow-up costs and program efficiency. *Arch Phys Med Rehabil*. 84:192–196.

Wehman, P., J. Kregel, M. West, and D. Cifu. 1994. Return to work for patients with traumatic brain injury. Analysis of costs. *Am J Phys Med Rehabil*. 73:280–282.

Weiller, C., S.C. Ramsay, R.J. Wise, K.J. Friston, and R.S. Frackowiak. 1993. Individual patterns of functional reorganization in the human cerebral cortex after capsular infarction. *Ann Neurol*. 33:181–189.

Wilde, E.A., S.R. McCauley, J.V. Hunter, E.D. Bigler, Z. Chu, Z.J. Wang et al. 2008. Diffusion tensor imaging of acute mild traumatic brain injury in adolescents. *Neurology*. 70:948–955.

Wildenberg, J.C., M.E. Tyler, Y.P. Danilov, K.A. Kaczmarek, and M.E. Meyerand. 2010. Sustained cortical and subcortical neuromodulation induced by electrical tongue stimulation. *Brain Imaging Behav*. 4:199–211.

Wildenberg, J.C., M.E. Tyler, Y.P. Danilov, K.A. Kaczmarek, and M.E. Meyerand. 2011a. Electrical tongue stimulation normalizes activity within the motion-sensitive brain network in balance-impaired subjects as revealed by group independent component analysis. *Brain Connect*. 1:255–265.

Wildenberg, J.C., M.E. Tyler, Y.P. Danilov, K.A. Kaczmarek, and M.E. Meyerand. 2011b. High-resolution fMRI detects neuromodulation of individual brainstem nuclei by electrical tongue stimulation in balance-impaired individuals. *NeuroImage*. 56:2129–2137.

Wildenberg, J.C., M.E. Tyler, Y.P. Danilov, K.A. Kaczmarek, and M.E. Meyerand. 2013. Altered connectivity of the balance processing network after tongue stimulation in balance-impaired individuals. *Brain Connect*. 3:87–97.

Zaloshnja, E., T. Miller, J.A. Langlois, and A.W. Selassie. 2008. Prevalence of long-term disability from traumatic brain injury in the civilian population of the United States, 2005. *J Head Trauma Rehabil*. 23:394–400.

Zikopoulos, B., and H. Barbas. 2013. Altered neural connectivity in excitatory and inhibitory cortical circuits in autism. *Front Human Neurosci*. 7:609.

Zucker, R.S. 1999. Calcium- and activity-dependent synaptic plasticity. *Curr Opin Neurobiol*. 9:305–313.

Zucker, R.S., and W.G. Regehr. 2002. Short-term synaptic plasticity. *Annu Revi Physiol*. 64:355–405.

Section VII

Mild Brain Injury and Sport Concussion

45 Blast Injuries and Blast-Induced Neurotrauma

Overview of Pathophysiology and Experimental Knowledge Models and Findings

Ibolja Cernak

CONTENTS

45.1 INTRODUCTION

Explosions are physical phenomena that result in the sudden release of energy; they may be chemical, nuclear, or mechanical. This process results in a near-instantaneous pressure rise above atmospheric pressure. The positive pressure rise ("overpressure") compresses the surrounding medium (air or water) and results in the propagation of a blast wave, which extends outward from the explosion in a radial fashion. As the front or leading edge of the blast wave expands, the positive phase is followed by a decrease in pressure and the development of a negative wave ("underpressure") before subsequently returning to baseline. Figure 45.1 shows an idealized form of a shock wave (Friedländer wave) (Friedlander, 1955) generated by a spherical, uncased explosive in the air in free field conditions. The extent of damage from the blast wave mainly depends on five factors: (1) the peak of the initial positive-pressure wave (an overpressure of 690–1,724 kPa, for example, 100–250 psi, is considered potentially lethal) (Champion et al., 2009); (2) the duration of overpressure; (3) the medium of explosion; (4) the distance from the incident blast wave; and (5) the degree of focusing because of a confined area or walls. Intensity of an explosion pressure wave declines with the cubed root of the distance from the explosion. Thus, a person 3 m (10 ft) from an explosion experiences nine times more overpressure than a person 6 m (20 ft) away. Additionally, explosions near or within hard solid surfaces can be amplified two to nine times because of shock wave reflection (Rice and Heck, 2000). Indeed, it was observed that victims positioned between a blast and a building often suffer injuries two to three times the degree of the injury of a person in an open space. People exposed to explosion rarely experience the idealized pressure-wave form. Even in open-field conditions, the blast wave reflects from the ground, generating reflective waves that interact with the primary wave, thus changing its characteristics. In a closed environment (such as a building, an urban setting, or a vehicle), the blast wave interacts with the surrounding structures and creates multiple wave reflections, which, interacting with the primary wave and between each other, generate a complex wave (Ben-Dor et al., 2001; Mainiero and Sapko, 1996).

Blast injuries are characterized by interwoven mechanisms of systemic, local, and cerebral responses to blast exposure (Cernak, 2010). When a blast generated by explosion strikes

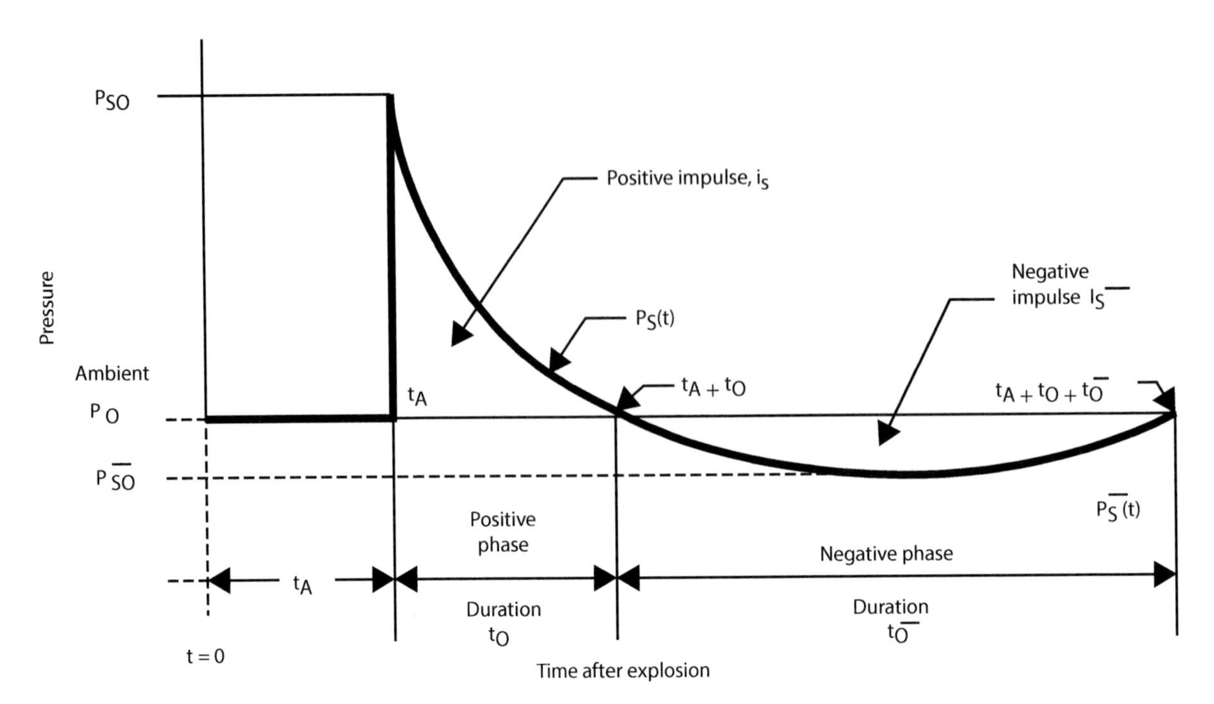

FIGURE 45.1 The Friedländer wave describing an ideal blast from a spherical source in an open environment. t_0 is the time at which the pressure began to rise above ambient pressure. Positive magnitude is the difference between peak pressure and ambient pressure. Positive duration is the time between t_0 and when the pressure goes below ambient pressure. Positive impulse is the integral of the pressure-time trace during the positive phase. Negative magnitude is the difference between ambient and peak negative pressure. (From Ngo, T. et al., *EJSE Special Issue: Loading on Structures*, 76–91, 2007.)

a living body, part of the shock wave is reflected and another fraction is absorbed becoming a tissue-transmitted shock wave. The transferred kinetic energy causes low-frequency stress waves that accelerate a medium from its resting state, leading to rapid physical movement, displacement, deformation, or rupture of the medium (Clemedson, 1956; Clemedson and Criborn, 1955). Thus, a militarily relevant blast injury model should be able to capture and measure these phenomena based on sufficient knowledge of shock wave physics, the characteristics of the injurious environment generated by an explosion, and the clinical manifestations and sequelae of the injuries. The purpose of this chapter is to outline the pathophysiology of blast-body/blast-brain interactions and to summarize the scientific evidence to date for the selection of appropriate experimental models for characterizing and understanding these interactions.

45.2 BLAST-BODY AND BLAST-BRAIN INTERACTIONS

Conceptually, explosive blast may have five distinct effects on the body (Figure 45.2): (1) primary blast effects causing injuries as sole consequences of the shock wave–body interaction; (2) secondary blast effects from the fragments of debris propelled by the explosion and connecting with the body, causing penetrating and/or blunt trauma; (3) tertiary blast effects from acceleration/deceleration of the body or part of the body (Richmond et al., 1961); (4) quaternary blast effects caused by the transient but intense heat of the explosion (flash burns) (Mellor, 1988); and (5) quinary blast effects caused by "post-detonation environmental contaminants," such as bacteria and radiation from dirty bombs, and tissue reactions to fuel and metal residues, among others (Kluger et al., 2007). Often, especially in the case of moderate-to-severe blast injuries, the multiple blast effects interact with the body simultaneously. In some literature sources, such an injurious environment and related injuries are referred to as "blast plus" scenarios (Moss et al., 2009).

When a shock wave generated by detonating a high-energy explosive strikes a living body, several physical events take place: a fraction of the shock wave is reflected, whereas another fraction of the shock wave energy is absorbed and propagates through the body as a tissue-transmitted shock wave (Clemedson and Criborn, 1955). Different organ and body structures differ in their reaction. Nevertheless, tissues typically respond (1) either on the impulse of the shock wave—this response is of longer duration—or (2) on the pressure variations of the shock wave, and this response is in a form of oscillations or pressure deflections of shorter duration (Clemedson and Pettersson, 1956). For example, basic experiments showed that tissues in the abdomen and costal interspaces react with typical impulse response, whereas the rib and the hind leg responded with a more or less pure maximum pressure type curve (Clemedson and Granstom, 1950; Clemedson et al., 1956, 1969).

The energy of the primary blast shock wave is either absorbed or transformed into the kinetic energy of a medium, which could be solid, liquid, gas, or plasma, when the interaction between them occurs (Clemedson and Jonsson, 1961). The transferred kinetic energy, then, moves and

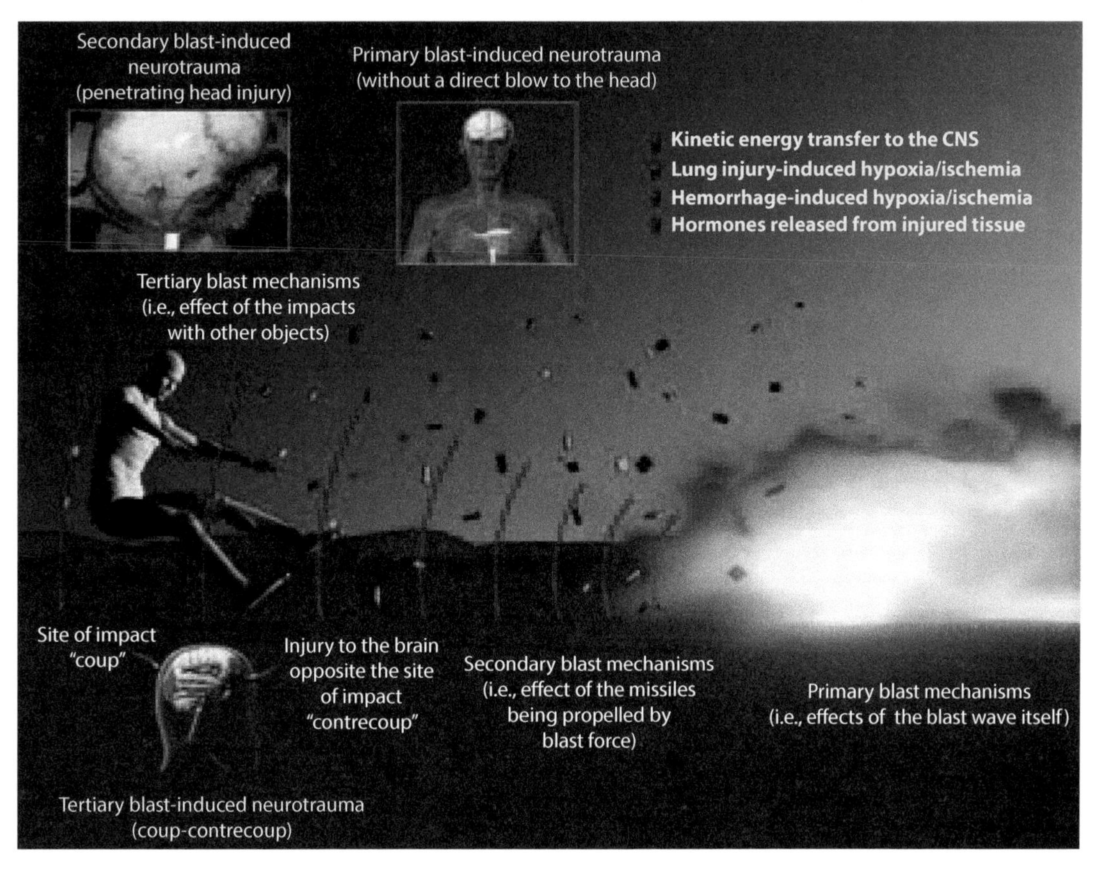

FIGURE 45.2 Complex injurious environment resulting from blast. Primary blast effects are caused by the blast wave itself (excludes penetrating and blunt force injury); secondary blast effects are caused by particles propelled by the blast (penetrating or blunt force injury); tertiary blast effects are caused by acceleration and deceleration of the body and its impact with other objects (penetrating or blunt force [including "coup-contrecoup"] injury). (From Cernak I. and L.J. Noble-Haeusslein. *J Cereb Blood Flow Metab.* 30:255–66. 2010. With permission.)

accelerates elements of the medium from their resting state with a speed that depends on the density of the medium; this leads to the medium's rapid physical movement, displacement, deformation, or rupture (Clemedson and Pettersson, 1956). As a result, the main physical mechanisms of the blast-body interaction and subsequent tissue damage include spalling, implosion, and inertia (Benzinger, 1950). Spallation occurs at the boundary between two media of different densities when a compression wave in the denser medium is reflected at the interface. Implosion happens in a liquid medium containing a dissolved gas. Because the shock wave penetrates such a medium, it compresses the gas bubbles, raising the pressure in the bubbles much higher than the initial shock pressure; after the pressure wave passes, the bubbles can rebound explosively and damage surrounding tissue. Inertial effects also occur at the interface of the different densities; the lighter object will be accelerated more than the heavier one, creating a large stress at the boundary (Sanborn et al., 2012)

Recent results suggest a frequency dependence of the primary blast effects. High-frequency (0.5–1.5 kHz) low-amplitude stress waves have been observed to target mostly organs that contain abrupt density changes from one medium to another (for example, the air–blood interface in the lungs or the blood–parenchyma interface in the brain). On the other hand, low-frequency (<0.5 kHz), high-amplitude shear waves show a tendency to disrupt a tissue by generating local motions that overcome natural tissue elasticity (for example, at the contact of gray and white brain matter) (Cooper et al., 1991; Gorbunov et al., 2004).

45.3 MODIFYING POTENTIAL OF SYSTEMIC CHANGES CAUSED BY BLAST

Because of the complexity of the injurious environment (i.e., multiple blast effects that may interact with the body in parallel), blast injuries involve interwoven mechanisms of systemic, local, and cerebral responses to blast exposure (Cernak et al., 1991, 1996b) (Figure 45.3). Even when the multiorgan responses are mild, systemic changes significantly extend the original organ damage and influence their severity and functional outcome. Air emboli, activation of the autonomous nervous system, vascular mechanisms, and systemic inflammation are among the most important deleterious systemic alterations that could modify the initial injuries due to blast.

45.3.1 Air Emboli

Air emboli develop as a consequence of the shock wave passing through the body and organs containing media of different densities and constituent states, that is, gas–air, fluid–blood, and solid–parenchyma (Clemedson and Hultman, 1954). Using

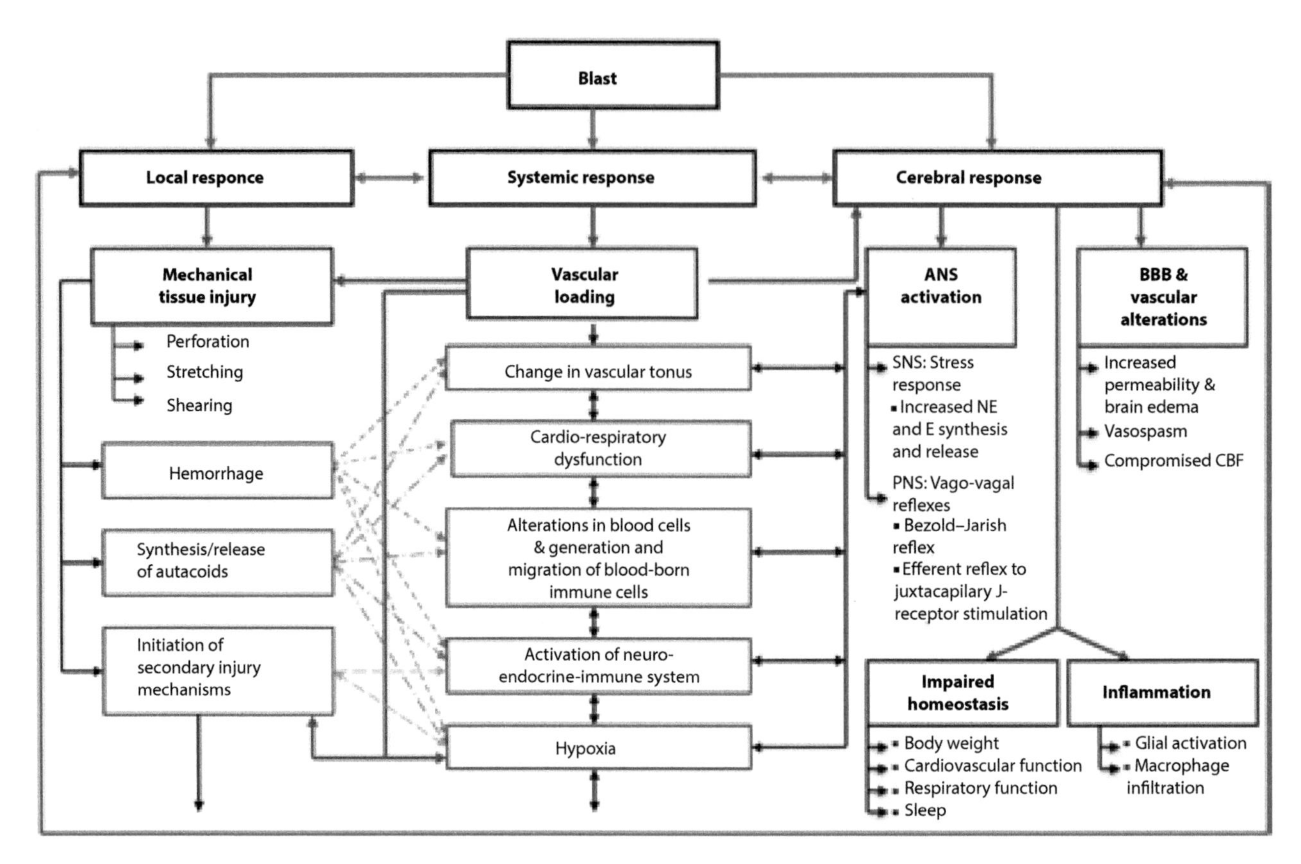

FIGURE 45.3 Simultaneous activation of systemic, local, and cerebral responses to blast exposure and interactive mechanisms causing or contributing to the pathobiology of BINT. ANS, autonomous nervous system; BBB, blood–brain barrier; CBF, cerebral blood flow; E, epinephrine; NE, norepinephrine; SNS, sympathetic nervous system; PNS, parasympathetic nervous system. (From Cernak, I. *Front Neurol.* 1:151, 2010.)

an ultrasonic Doppler blood-flow detector, Nevison, Mason, and colleagues have clearly shown air emboli passing through the carotid artery in dogs subjected to blast in a shock tube (Mason et al., 1971; Nevison et al., 1971). Interestingly, the dynamics of the air emboli release as recorded by the embolus detector showed a cyclic pattern, initially occurring over the first 10 seconds and again about 2 and 12 minutes after the blast. It is noteworthy that the air emboli release occurred parallel with a dramatic decrease in blood flow velocity and tissue convulsion likely from hypoxia/anoxia. Similar experimental findings have been described by others (Chu et al., 2005; Clemedson and Hultman, 1954; Kirkman and Watts, 2011) and supported by clinical studies (Freund et al., 1980; Tsokos et al., 2003a, 2003b). Indeed, a massive compressed-air embolism of the aorta and multiple air spaces in the interstitium, compressing the collecting tubules in the kidneys (Freund et al., 1980) and venous air embolism in the lungs (Tsokos et al., 2003a), have been reported in victims of severe blast injury. It is expected that the rate of the air emboli release is dependent on the intensity of blast, and the subsequent changes in blood flow and oxygenation level are also graded.

45.3.2 Activation of the Autonomous Nervous System

When the incident overpressure wave is transmitted through the body, it increases the pressure inside organs (Clemedson and Pettersson, 1956). The subsequent sudden hyperinflation of the lungs (Cernak et al., 1996b; Zuckerman, 1940) stimulates the juxtacapillary J receptors located in the alveolar interstitium and innervated by vagal fibers (Paintal, 1969). The resulting vagovagal reflex leads to apnea followed by tachypnea, bradycardia, and hypotension (i.e., symptoms frequently observed immediately after blast exposure). Moreover, hypoxia/ischemia due to damaged alveoli, air emboli, or triggered pulmonary vagal reflex can activate a cardiovascular decompressor Bezold–Jarish reflex, which markedly increases vagal (parasympathetic) efferent discharge to the heart (Zucker, 1986). This reflex slows the heart rate and dilates the peripheral blood vessels that precipitates a drop in blood pressure and could potentiate cerebral hypoxia (Cernak et al., 1996a, 1996b). Indeed, Axelsson and colleagues in their experimental study using pigs showed that the blast-induced short-lasting apnea correlated to diminished electrical activity of the brain (Axelsson et al., 2000). A substantial number of experimental studies further demonstrated the importance of vagally mediated cerebral effects of blast (Cernak et al., 1996b; Irwin et al., 1999; Ohnishi et al., 2001).

The explosive environment is a dramatic one, and may initiate endocrine mechanisms known as the classical "flight-or-fight" stress response (Selye, 1976). This is demonstrated in a recent experimental study (Tumer et al., 2013) showing increased expression of the catecholamine-biosynthesizing enzymes, tyrosine hydroxylase and dopamine hydroxylase, in the rat adrenal medulla, along with elevated plasma levels of norepinephrine at 6 hours after blast injury. Thus, accumulating experimental and clinical evidence suggests the following sequence of blast-induced alterations in autonomous nervous system activity: instantaneous triggering of parasympathetic reflexes followed by neuroendocrine changes from sympathetic nervous system activation.

45.3.3 Vascular Mechanisms

Bearing in mind the previously described mechanisms of blast effects, it becomes obvious that one of the most important media for the shock wave's energy transfer is blood. Veins contain approximately 70% of total blood volume. From this, the splanchnic system receives approximately 25% of cardiac output (translating into approximately 20% of total blood volume) compared with 18% in arteries and only 3% in terminal arteries and arterioles (Gelman, 2008). In general, veins are 30 times more compliant than the arteries, and splanchnic and cutaneous veins represent the most compliant venous vessels of all. Thus, these venous systems form the largest blood volume reservoirs in the human body. Figure 45.4 shows a simplified schematic representation of the consequences of blast-induced pressure changes and their extremely complex interactions, which form several interconnected loops. The shock wave's energy transferred to the body leads not only to a sudden increase in both abdominal and thoracic pressures (ThorP), but also causes increase in intramural central venous pressure (CVP). Hypoxia caused by alveolar damage and subsequently reduced surface area for gas exchange, impaired ventilation/perfusion caused by J-receptor activation, or decreased cardiac output from activation of Bezold–Jarish reflex, among others, increases pulmonary arterial resistance, which might also increase ThorP (Gelman, 2008). The elevated ThorP further amplifies the increase in central venous pressure.

Having a high density of α_1- and α_2-adrenergic receptors and hence high sensitivity to adrenergic stimulation, the venoconstriction and mobilization of blood volume mainly depend on splanchnic circulation (Pang, 2001; Rutlen et al., 1979). Thus, it is highly feasible that the initial sudden drop in systemic arterial pressure initiates a compensatory increase in sympathetic outflow through vagovagal reflexes with reduction in the inhibitory influences of the baroreceptors of the carotid sinus and aortic area on the vasomotor center. Consequently, the increased sympathetic stimuli induced by blast constrict venous smooth musculature and lead to mobilization of splanchnic blood toward the heart (Rutlen et al., 1979).

Cerebrovascular vasospasm has been found frequently in casualties with moderate or severe blast-induced traumatic brain injury (TBI), more often than in patients with TBI of other origins (i.e., impact, fall, or acceleration/deceleration) (Armonda et al., 2006; Ling et al., 2009). Vasospasm can develop early, often within 48 hours of injury, and can also manifest later, typically between 10–14 days postexposure. It is noteworthy that although cerebral vasospasm is usually prompted by subarachnoid hemorrhage, subarachnoid

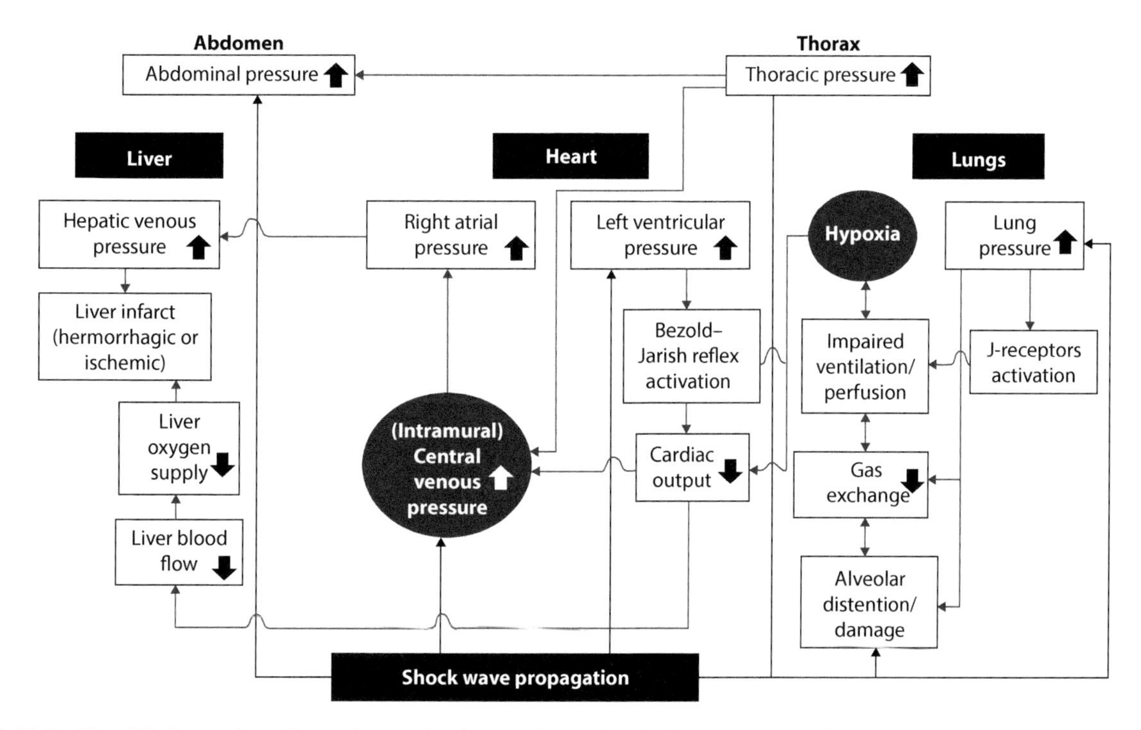

FIGURE 45.4 Simplified overview of vascular mechanisms activated by shock wave propagation through the body leading to alterations in functions of multiple organs and organ systems, which significantly influence the brain's response to blast. (Created by Ibolja Cernak for the Committee on Gulf War and Health: Long-Term Effects of Blast Exposures, Institute of Medicine, U.S. National Academies, Copyright 2014.)

hemorrhage is not required for vasospasm in blast-induced TBI (Magnuson et al., 2012). A recent experimental study using theoretical and in vitro models demonstrated that a single rapid mechanical insult is capable of inducing vascular hypercontractility and remodeling, indicative of vasospasm initiation. Furthermore, the results implied that the prolonged hypercontraction is linked to switching of the vascular smooth muscle phenotype in tissues exposed to simulated blast (Alford et al., 2011). These findings suggest a hypothetical scenario that when the shock wave passes through the vasculature, it interacts with cellular elements of vascular wall such as endothelium and vascular smooth muscle. As a consequence of this interaction, various mediators and modulators are released, which cause hypercontraction and subsequent genetic switch that potentiates vascular remodeling and cerebral vasospasm (Hald and Alford, 2013).

45.3.4 Systemic Inflammation

Blast exposure can activate multiple inflammatory mechanisms (Cernak, 2010). Tissue disruption stimulates synthesis and release of autacoids (i.e., biological factors acting like local hormones near the site of their synthesis with a brief duration of action). Indeed, increased concentrations of various prostaglandins, leukotrienes, and cytokines have been found in the blood of blast casualties (Cernak et al., 1999a, 1999b; Surbatovic et al., 2007). These autacoids directly affect a number of stages of cellular and humoral immunity, and also act as feedback modifiers connecting the early and late phases of the immune response (Melmon et al., 1981). Indeed, they can stimulate selected migration of cells to the injury site, and directly or indirectly modify the turnover of T- and B-lymphocytes, the production or release of lymphokines, and the activity of T helper or T-suppressor cells (Khan and Melmon, 1985; Melmon et al., 1981). It has been suggested that inflammatory cells of systemic origin that have been induced by shock wave propagation through the body significantly contribute to blast-induced inflammation in the brain and related neurodegeneration (Cernak, 2010), which was supported by additional experimental data (Valiyaveettil et al., 2013).

Blast exposures have been reported to cause significant alterations in neuroendocrine system involving multiple hypothalamopituitary end axes (Baxter et al., 2013; Cernak et al., 1999c; Wilkinson et al., 2012). The importance of the immunoneuroendocrine network in injury response and inflammation control is well established (Besedovsky and del Rey, 1996, 2002; Chrousos, 1995). Thus, it is highly likely that blast exposure, through multiple interwoven mechanisms, causes a massive stimulation of the central nervous system (CNS) with broad consequences on all aspects of vital functions.

45.4 REQUIREMENTS FOR BLAST-INDUCED INJURY MODELS

Regardless of the research questions to be addressed, clinically and militarily relevant blast injury models should satisfy all of the following four criteria: (1) the injurious component of the blast should be clearly identified and reproduced in controlled, reproducible, and quantifiable manner; (2) the inflicted injury should be reproducible, quantifiable, and mimic components of human blast injuries; (3) the injury outcome established, based on morphological, physiological, biochemical, and/or behavioral parameters, should be related to the chosen injurious component of the blast; and (4) the mechanical properties (intensity, complexity of blast signature, and/or its duration) of the injurious factor should predict the outcome severity (Cernak and Noble-Haeusslein, 2010). The mechanistic factors underlying blast injuries are extremely complex as compared with the injuries caused by an impact or acceleration/deceleration force. Hence, an appropriate and clinically relevant blast injury model should be based on sufficient knowledge of shock wave physics, the characteristics of the injurious environment generated by an explosion, and clinical manifestations of resultant injuries.

45.5 CHOICE OF MODELS

The purpose of experimental models of CNS damage such as TBI is to replicate certain pathological components or phases of clinical trauma in experimental animals, aiming to address pathology and/or treatment. The goal of research specifies the design and choice of the experimental model (Cernak, 2005; Risling and Davidsson, 2012). The extremely complex nature of blast injuries requires full understanding of blast physics, and a model reproducing multiple aspects of blast injuries should be defined with particular scientific fidelity to conditions observed in theater. Otherwise, a model will lack military and clinical relevance and the obtained results might be dangerously misleading. Because of the widely varying experimental conditions the currently existing models use, the results across studies are very difficult to compare or summarize (Panzer et al., 2012).

There are several decision-making steps in the process of choosing a model for blast research (Figure 45.5). Most importantly, the researcher should identify which of the blast effects should be reproduced. If the choice is primary blast, the experimenter should ensure the animals are restrained so that there will be no blast-induced acceleration of the body/head during the exposure. Namely, in a situation where the body/head is allowed to move, the injury mechanisms would involve both primary and tertiary blast effects; this would make the interpretation of the results difficult. Next, a decision should be made about the biological complexity of the research study. This factor will dictate the choice of the biological surrogate used to reproduce blast-induced pathologies seen in humans (e.g., cell culture, tissue, small or large experimental animals, nonhuman primates); positioning of the biological surrogates; means of generating a shock wave (open field, shock or blast tubes); and length of the experiment, among others. Thus, based on the research question and the scale of complexity, a choice is made between nonbiological models and biological models.

Nonbiological models such as in silico and surrogate physical models provide an experimental platform for analyzing

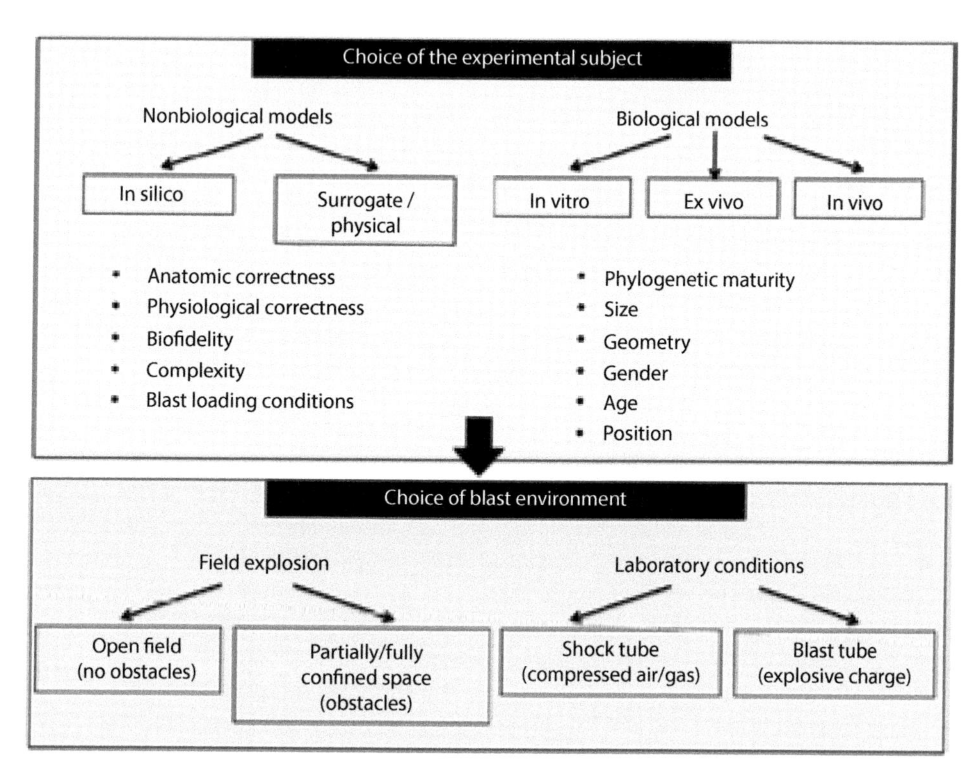

FIGURE 45.5 Factors influencing the choice of blast injury and BINT models. (Created by Ibolja Cernak for the Committee on Gulf War and Health: Long-Term Effects of Blast Exposures Institute of Medicine, U.S. National Academies, Copyright 2014.)

interactions between blast loading and different types of materials; the obtained information then is extrapolated to biological materials at different levels of scaling. The biofidelic computer (i.e., in silico) models provide spatially and temporally resolved descriptions of stress, strain, and acceleration that blast generates; as such, they are helpful tools in characterizing the physics of the blast-induced mechanical changes in the brain tissue (Chafi et al., 2010; Nyein et al., 2010; Zhu et al., 2010). The physical surrogate models use a human surrogate torso or head, which are made from synthetic materials (such as glass/epoxy or polyurethane) with biofidelic properties and incorporate multiple displacement and pressure sensors molded into the organs' material to record the biomechanical parameters such as linear/angular acceleration, velocity, displacement, force, torque, and pressure (Desmoulin and Dionne, 2009; Ganpule et al., 2012; Roberts et al., 2012).

The nonbiological models can be valuable in identifying blast-induced biomechanical alterations and suggest potential consequences in biological systems. Nevertheless, they are unable to give explanations for functional and physiological changes in a living system caused by blast exposure. Hence the need for biological (in vitro, ex vivo, and in vivo) models, which use biological systems of differing complexity. The in vitro models use cell cultures and can be helpful in characterizing the cell-response mechanisms to blast loading in a highly controlled experimental environment (Effgen et al., 2012; Panzer et al., 2012). The ex vivo models use an organ or a segment of a specific tissue such as brain or spinal cord, taken outside the organism into an artificial environment, which is more controlled than is possible with in vivo experiments. As for all blast injury models, applying operationally relevant loading histories is critical for the in vitro and ex vivo models. Namely, mechanisms of the energy transfer to the tissue and the resultant biological response can be reliably analyzed only when blast-loading conditions are realistic and would happen at the cellular or tissue level of an individual who had been exposed to and survived a militarily relevant blast environment (Effgen et al., 2012).

The success of a research study using biological models, especially at the whole-animal level, depends on rigorous selection of animal species used as experimental models. Rodents are the most frequently used experimental animals in trauma research. The relatively small size and low cost of rodents permit repetitive measurements of morphological, biochemical, cellular, and behavioral parameters that require relatively large numbers of animals; this, because of ethical, technical, and/or financial limitations, is less achievable in phylogenetically higher species (Cernak, 2005). Nevertheless, the anatomical and physiological differences between humans and rodents, especially in circulatory and nervous systems, limit the utilization of small experimental animals in blast-injury research.

Extensive studies conducted in Albuquerque, New Mexico, confirmed by German, Swedish, and British findings, demonstrated significant interspecies differences in blast tolerance among 15 mammals (Bowen et al., 1967; Richmond et al., 1967, 1968, 2005). The size-dependent differences in blast tolerance have been explained by reference to variation of interspecies lung densities and volumes. Namely, lung density in larger species, including man and monkey, cat, and dog, is only about half of the lung density of smaller species (e.g., rodents), whereas lung volumes normalized to body mass are three times bigger in large species than in smaller animals (White et al., 1965). Also, there are significant interspecies differences in body geometry that influence the blast-body blast-head interactions (Bass et al., 2012). The body position of the animal in the shock tube also has an important impact on blast injury severity. Animals facing the incoming shock wave front with their chest and abdomen (i.e., supine position) provide the best conditions for the shock wave's energy transfer; consequently, they have the highest mortality rate and most severe injuries (Cernak et al., 2011). Furthermore, in blast injury modeling, especially when acceleration is included as one of the mechanistic factors, scaling laws should be taken into careful consideration (Bass et al., 2008, 2012). Taking a blast-head scenario as an example, for a given blast, when calculating the net loading scales for cross-sectional area of the skull, even if other parameters would be identical, a specimen 20-fold the size would experience 20-fold less acceleration for the same blast. Nevertheless, other anatomical differences between the human and animal heads, such as bone volume fraction, trabecular separation and number, and connectivity density, among others (Holzer et al., 2012; Pietschmann et al., 2010), should also be considered. Evolutionary and developmental changes in the structure and arrangement of blood vessels (Vries, 1904) are also important factors that should be taken into account when choosing models for reproducing blast injury. For example, the internal carotid artery is the main blood supply both in humans/nonhuman primates and rodents (rats and mice). Nevertheless, although in lower vertebrates the internal carotid artery directs the blood to the brain parenchyma through the posterior branch without contribution from the basilar artery, in higher vertebrates, there are the two posterior branches that stem from a single and central branch turning into the branch of the basilar artery (Casals et al., 2011). These anatomical differences could significantly influence the shock wave propagation through the cerebrovasculature.

Previously, it has been shown that the level of phylogenetic maturity has a decisive role in brain's response to a high-pressure environment (Brauer et al., 1979), a factor that should be taken into account in planning blast-induced neurotrauma (BINT) experiments. Because basic molecular and gene injury-response mechanisms are conserved through evolution, phylogenetically lower species such as rodents can be used for studies studying blast-induced changes at cellular and subcellular levels. However, establishing the pathogenesis of impaired higher brain functions would require larger animals with a gyrencephalic brain.

45.5.1 Experimental Environment Generating Blast

Experimental studies on primary blast-induced biological responses are performed either in an open environment or laboratory conditions. The open field exposure studies use animals exposed to a blast wave that is generated by detonation of an explosive (Axelsson et al., 2000; Bauman et al., 2009; Lu et al., 2012; Richmond, 1991; Saljo and Hamberger, 2004; Savic et al., 1991). Although such an experimental setting is more comparable with in-theater conditions, the physical characteristics (such as homogeneity of the blast wave) are less controllable, so a broader range of biologic response should be expected.

Experiments performed in laboratory conditions use either shock tubes (which use compressed air or gas to generate a shock wave) or blast tubes (which use explosive charges) (Nishida, 2001; Robey, 2001). Both of these tubes focus the blast wave energy in a linear direction from the source to the subject, maximizing the amount of blast energy (Reneer et al., 2011) without the exponential decay of the shock wave's velocity and pressure seen in free-field explosions (Celander et al., 1955). The induction system routinely used in blast exposure models consists of a cylindrical metal tube divided by a plastic or metal diaphragm into two main, driver and driven, sections. The anesthetized animals are fixed individually in special holders designed to prevent any movement of their body in response to the blast. The high pressure in the driver section is generated by either explosive charge or compressed gas, which ruptures the diaphragm when reaching the material's tolerance to pressure. After the diaphragm ruptures, the resultant shock wave travels along the driven section with supersonic velocity, and interacts with the animal positioned inside the driven section. The duration of the overpressure can be varied by changing the length and/or diameter of the high-pressure chamber (Celander et al., 1955).

In the case of shock tubes, either compressed atmospheric air or another gas is used. The compressed air fails to expand as quickly as would an ideal gas when the membrane is ruptured and also generates a broad range of overpressure peaks. Use of a lighter gas, such as helium, improves the performance of shock tubes because of the increased speed of sound within a helium environment (Celander et al., 1955). Although shock/blast tubes are convenient means of generating shock waves, they lack the ability to generate the acoustic, thermal, optical, and electromagnetic components found in actual blast environments (Ling et al., 2009). Moreover, although the positive phase of free-field explosive blast can be reproduced by careful adjustment of the driver's length, driver's gas, and the specimen location, the negative phase and recompression shock are often artifacts of the rarefaction from the open end of the tube; thus, the simulated blast wave is incorrect when compared with the signature of the militarily relevant blast (Ritzel et al., 2012). Because of this, it is recommended that shock tubes be fitted with a reflection-eliminator at their end to eliminate the waves reverberating the length of the tube.

In the case of blast tubes, combustion of an explosive generates high pressures and volumes in the driver section without a diaphragm. Blast tubes use high-energy explosives placed within a heavy-walled, small-diameter driver section (often a gun barrel), expanding into the wider diameter driven (i.e., test) section (Ritzel et al., 2012). The cons for using a blast tube include (1) dispersion of the combustion products and residue in the test section; (2) generation of strong transverse waves either within the driver or in the wider test section, caused by the charge and charge and driver configuration; and (3) introduction of additional operating costs and more complex environmental control for safe handling, setting, and firing of the charges.

A wide range of blast overpressure sustained for various durations has been used in single-exposure experimental studies. In most studies, the animals are subjected to a shock or blast wave with a mean peak overpressure of 52–340 kPa (7.54–49.31 psi) on the nearest surface of an animal's body (Cernak et al., 2001b; Chavko et al., 2007; Clemedson et al., 1969; Saljo et al., 2000). Most experiments have used rodents (mice and rats) (Cernak et al., 2001a; Long et al., 2009), but some have subjected rabbits (Cernak et al., 1997), sheep (Savic et al., 1991), pigs (Bauman et al., 2009), or nonhuman primates to blast (Bogo et al., 1971; Damon et al., 1968; Lu et al., 2012; Richmond et al., 1967).

Special consideration should be given to positioning of the specimen in the shock/blast tubes, and its orientation in relation to the incident shock wave. There is an ongoing debate about whether the specimen should be positioned inside or outside shock/blast tubes. Namely, the biomechanical response of the animal significantly depends on the placement location in the tube (Sundaramurthy et al., 2012) as well as on the orientation of the specimen as compared with the propagating incident shock wave (Cernak et al., 2011; Varas et al., 2011). The majority of the currently existing literature supports the need of placing the specimen inside the shock/blast tube (Sundaramurthy et al., 2012; Varas et al., 2011). Experimental studies have shown that when the animal is positioned inside the shock tube, it is subjected to a load that is due to the pure blast wave comparable to the shock wave generated in free-field conditions (near-optimal, so-called Friedlander-type shock wave). When the animal was positioned at the exit, there was a sharp decay in pressure after the initial shock front, which was caused by the expansion wave from the exit of the shock tube eliminating the exponentially decaying blast wave (Sundaramurthy et al., 2012). This phenomenon led to significant decrease of the positive blast impulse (Figure 45.4) and conversion of most of the blast energy from supersonic blast wave to subsonic jet wind (Haselbacher et al., 2007), which has effects that are significantly different from those generated by a blast wave. Namely, because of the jet wind, the restrained animal experiences more severe compression of the head and neck and the thoracic cavity is exposed to higher pressure of longer positive-phase duration. The importance of animal's positioning in relation to the shock tube (inside versus outside) was further demonstrated by Svetlov and colleagues (Svetlov et al., 2010, 2012) who exposed the rats to blast loading by placing the rats 50 mm outside the shock tube. Their results suggested that the subsonic jet wind represented the bulk of the blast impulse, and the injuries were caused by the combination of blast wave and subsonic jet wind, as opposed to a pure blast wave injury. Experiments with surrogate physical models (dummy heads) placed at the exit of the shock tube supported the findings with animal models about the subsonic jet wind effects (Desmoulin and Dionne, 2009). Interestingly, a recent article argues that the environment inside a shock/blast tube induces artificially enhanced injuries because of reflected shock waves and rarefaction waves; thus, exposure of animals to shock waves in a shock/blast tube will cause severe, rare, and complex blast injuries, which are not comparable to injuries acquired in real-life, open-field blast conditions (Chen and Constantini, 2013).

Prone position with head and body oriented along the direction of shock wave propagation (perpendicular to the shock front) is the most commonly used orientation in current rodent model studies with shock tubes (Dal Cengio Leonardi et al., 2012; Saljo et al., 2010; Vandevord et al., 2012; Zhu et al., 2010). However, although this position is natural for quadrupeds, it does not reproduce the most frequent human scenario when the soldiers are in upright position and facing with torso toward the front of an incoming shock wave. It has been shown that both the pattern and severity of organ damage caused by blast depend on the orientation of the body toward the shock wave front (Cernak et al., 2011). The lung seems to be the most vulnerable organ to the effects of blast across injury severity and in both prone and supine body positions. Similarly, blast severity seemed to be positively correlated with lesion frequency and severity in both prone and supine positions (Koliatsos et al., 2011). As in the lung, supine position was associated with more severe findings in the heart such as dilation of ventricles and atria, right more than left. In contrast to lung injuries, prone disposition caused more severe liver pathology such as congestion, mottling, and white discoloration adjacent to apparently hemorrhagic sites, compared with supine positioning. In the prone, but not the supine position, there was some association between blast severity and liver infarct rate (Koliatsos et al., 2011). The prone position was also linked to more damage in kidneys and spleen. Recently, Ahlers and associates showed that low-intensity blast exposure produced an impairment of spatial memory that was specific to the orientation of the animal (Ahlers et al., 2012).

The choice of the animal holder is another important component in shock/blast tube experiments. Namely, if the animal is fixed on a solid platform, the waves reflecting from it will amplify the primary shock wave and increase the complexity and severity of blast injuries. Furthermore, a bulky animal holder when placed inside the tube could obstruct the central flow of the shock wave propagating along the driver and contribute to nonhomogeneous field conditions.

When choosing the system for generating blast, besides the physics, the physiological effects of the driver conditions also should be taken into account. Namely, when compressed helium was used to generate a shock wave, the oxygen content was reduced by approximately 75% in the driven section (Reneer et al., 2011). The oxygen content in the driven section was moderately reduced and the carbon monoxide content very high after oxyhydrogen-driven blasts. The carbon monoxide content was above the acceptable levels in the driver of an explosive-generated blast tube as a result of combustion (Reneer et al., 2011).

Taken together, when deciding which system to use to generate blast conditions, the blast and shock wave physics and the pros and cons of engineering solutions should be carefully weighed, especially when considering ultrasound or laser as a source for a shock wave (Takeuchi et al., 2013). The basic tenets of physics should always be remembered: that shock waves are single, mainly positive pressure pulses that are followed by comparatively small tensile wave components and their effects are mainly from forward-directed energy (i.e., in the direction of the shock wave propagation).

45.6 Blast-Induced Neurotrauma

Blast can interact with the brain by means of a (1) direct interaction with the head via direct passage of the blast wave through the skull (primary blast), causing acceleration and/or rotation of the head (tertiary blast), or through impacting particles accelerated by the energy released during explosion (secondary blast) (Goldstein et al., 2012; Mellor, 1988), and (2) kinetic energy transfer of the primary blast wave to organs and organ systems, including fluid medium (blood) in large blood vessels in the abdomen and chest, and the CNS (Cernak and Noble-Haeusslein, 2010; Cernak et al., 2001b). Namely, during the interaction with the body surface, the shock wave compresses the abdomen and chest and transfers its kinetic energy to the body's internal structures, including the blood as fluid medium. The resulting oscillating waves traverse the body at about the speed of sound in water and deliver the shock wave's energy to the brain. Clemedson, based on his extensive experimental work on shock wave propagation through the body (Clemedson and Jonsson, 1961), was among the first scientists to suggest the possibility of shock wave transmission to the CNS (Clemedson, 1956). These two potential avenues of interaction do not exclude each other (Cernak and Noble-Haeusslein, 2010). Most recent experimental data suggest both the importance of the blast's direct interaction with the head (Moss et al., 2009) and the role of shock wave–induced vascular load (Cernak, 2010) in the pathogenesis of BINT.

45.6.1 Animal Models of BINT

The majority of currently used experimental models of BINT use rodents exposed to a shock wave generated in laboratory conditions using a compressed gas–driven shock tube (Baalman et al., 2013; Cernak et al., 2011; Goldstein et al., 2012; Kamnaksh et al., 2011; Pun et al., 2011; Readnower et al., 2010; Reneer et al., 2011; Svetlov et al., 2012; Valiyaveettil et al., 2012a; Vandevord et al., 2012). Experiments with larger animals mainly involve pigs (Ahmed et al., 2012; Bauman et al., 2009) or nonhuman primates (Bogo et al., 1971; Lu et al., 2012).

Accumulating evidence suggests that primary blast causes significant behavioral impairments and cognitive deficits in multiple animal models (Bogo et al., 1971; Cernak et al., 2001a; Lu et al., 2012). These deficits show a dose-dependence from primary blast intensity as well as related to degenerative processes in the brain. The wide range of molecular changes starts early with metabolic impairments, including altered glucose metabolism shifting from aerobic toward anaerobic pathway measured as elevated lactate concentration and increased lactate/pyruvate ration (Cernak et al., 1996b), decline in energy reserve (Cernak et al., 1995, 1996b), and increased oxidative stress (Readnower et al., 2010) in parallel with ultrastructural changes in brainstem and hippocampus (Cernak et al., 2001b; Saljo et al., 2000), and activation of early immediate genes (Saljo et al., 2002). Later, the mechanisms include inflammation (Cernak et al., 2011; Kaur et al., 1997, 1995; Kwon et al., 2011; Readnower et al., 2010; Saljo et al., 2001), diffuse axonal injury (Garman et al., 2011; Risling et al., 2011), and apoptotic and nonapoptotic cascades leading to neurodegeneration (Svetlov et al., 2010; Vandevord et al., 2012; Wang et al., 2010). Emerging evidence suggests that certain brain structures might have a more pronounced sensitivity toward blast effects either from anatomical features and localization or functional properties of neuronal pathways and/or cells (Koliatsos et al., 2004; Valiyaveettil et al., 2012a, 2012b). Indeed, higher sensitivity of the cerebellum/brainstem, the corticospinal system, and the optic tract has been found (Koliatsos et al., 2011) based on the extent of multifocal axonal and neuronal cell degeneration. Additionally, based on regional specific alterations in the activity of acetylcholinesterase, the vulnerability of the frontal cortex and medulla has been observed in mice exposed to blast overpressure (Valiyaveettil et al., 2012a, 2012b). These changes showed a tendency toward chronicity.

The mechanisms involved in the pathobiology of BINT show some similarities with blunt TBI, although with earlier onset of brain edema and later onset of cerebral vasospasm (Agoston et al., 2009). Using a pig model of blast exposure, Ahmed and colleagues have shown that protein biomarker levels in cerebrospinal fluid can provide insight into the pathobiology of BINT (Ahmed et al., 2012). Their findings implicated neuronal and glial cell damage, compromised vascular permeability, and inflammation induced by blast. The early-phase biomarker included claudin-5, vascular endothelial growth factor, and von Willebrand factor, whereas neurofilament-heavy chain, neuron-specific enolase, vascular endothelial growth factor, and glial fibrillary acidic protein levels remained significantly elevated compared to baseline at 2 weeks postinjury.

Despite the progressively growing experimental data on BINT, it is very difficult to compare the results and consolidate the findings into one comprehensive pool of knowledge. Namely, because of the lack of well-defined criteria for reliable animal models of BINT, the current experimental models used in an attempt to study BINT vary widely, and include classical direct impact TBI models such as controlled cortical impact and fluid percussion injury models, air gun–type compressed air-delivered impact models, shock and blast tube models, and open-field explosion experiments, among others. Some experimental models expose only the head of the animal to an extremely focused overpressure field (Kuehn et al., 2011; Prima et al., 2013) causing brain damage comparable to those seen in impact TBIs rather than to structural alterations seen in soldiers exposed to blast and with diagnosed BINT. There are also models where the animals are exposed to whole body blast and their body protected aiming to prevent any blast-induced systemic effects. The utilization of different materials such as cardboard (Chavko et al., 2011) or Kevlar fabric without interceptive plate (Long et al., 2009; Reneer et al., 2011) further complicates the comparison of experimental findings.

Taken together, the lack of understanding of the physics of blast, unfocused rationale of experiments, and the broad variety of methods used to inflict head injury in the context of BINT research are significantly reducing the reliability of the published literature and slowing down the progress of this research field. Hence, there is a dire need for a well-coordinated, multidisciplinary research approach to clarify injury tolerance levels for animal models relevant for military experience and to define the injury mechanisms underlying acute and chronic consequences of blast exposure(s).

45.7 CONCLUSION

The problem of BINT and related long-term neurological deficits has been gradually increasing with the progress of military warfare and the pathological experience of returning veterans of Operation Enduring Freedom/Operation Iraqi Freedom. The long-term health problems manifesting in growing number of veterans have triggered intensified research efforts aiming to clarify the vital mechanisms underlying blast injuries and blast-induced brain damage. This is an extremely challenging task, which requires a unified front of physicists, military scientists, biomedical researchers, and clinicians applying out-of-the-box thinking and novel research approaches. Clear guidelines about experimental models that are acceptable for blast injury research should be established and strict adherence to those guidelines enforced. Without such consensus among blast researchers and close cooperation between the researchers and those with military operational experience, this research field will remain contradictory and misleading, and the soldiers with blast injuries and/or BINT left without improvement in their treatment.

REFERENCES

Agoston, D.V., A. Gyorgy, O. Eidelman, and H.B. Pollard. 2009. Proteomic biomarkers for blast neurotrauma: targeting cerebral edema, inflammation, and neuronal death cascades. *J Neurotrauma*. 26:901–11.

Ahlers, S.T., E. Vasserman-Stokes, M.C. Shaughness, A.A. Hall, D.A. Shear, M. Chavko et al. 2012. Assessment of the effects of acute and repeated exposure to blast overpressure in rodents: toward a greater understanding of blast and the potential ramifications for injury in humans exposed to blast. *Front Neurol*. 3:32.

Ahmed, F., A. Gyorgy, A. Kamnaksh, G. Ling, L. Tong, S. Parks et al. 2012. Time-dependent changes of protein biomarker levels in the cerebrospinal fluid after blast traumatic brain injury. *Electrophoresis*. 33:3705–11.

Alford, P.W., B.E. Dabiri, J.A. Goss, M.A. Hemphill, M.D. Brigham, and K.K. Parker. 2011. Blast-induced phenotypic switching in cerebral vasospasm. *Proc Natl Acad Sci U S A*. 108:12705–10.

Armonda, R.A., R.S. Bell, A.H. Vo, G. Ling, T.J. DeGraba, B. Crandall et al. 2006. Wartime traumatic cerebral vasospasm: recent review of combat casualties. *Neurosurgery*. 59:1215–25; discussion 1225.

Axelsson, H., H. Hjelmqvist, A. Medin, J.K. Persson, and A. Suneson. 2000. Physiological changes in pigs exposed to a blast wave from a detonating high-explosive charge. *Mil Med*. 165:119–26.

Baalman, K.L., R.J. Cotton, S.N. Rasband, and M.N. Rasband. 2013. Blast wave exposure impairs memory and decreases axon initial segment length. *J Neurotrauma*. 30:741–51.

Bass, C.R., M.B. Panzer, K.A. Rafaels, G. Wood, J. Shridharani, and B. Capehart. 2012. Brain injuries from blast. *Ann Biomed Eng*. 40:185–202.

Bass, C.R., K.A. Rafaels, and R.S. Salzar. 2008. Pulmonary injury risk assessment for short-duration blasts. *J Trauma*. 65:604–15.

Bauman, R.A., G. Ling, L. Tong, A. Januszkiewicz, D. Agoston, N. Delanerolle et al. 2009. An introductory characterization of a combat-casualty-care relevant swine model of closed head injury resulting from exposure to explosive blast. *J Neurotrauma*. 26:841–60.

Baxter, D., D.J. Sharp, C. Feeney, D. Papadopoulou, T.E. Ham, S. Jilka et al. 2013. Pituitary dysfunction after blast traumatic brain injury: the UK BIOSAP study. *Ann Neurol*. 74:527–36.

Ben-Dor, C., O. Igra, and T. Elperin. 2001. *Handbook of Shock Waves*. Academic Press, San Diego, CA.

Benzinger, T. 1950. Physiological effects of blast in air and water. In *German Aviation Medicine, World War II*. Vol. 2. Department of the Air Force, Washington, DC. pp. 1225–1229.

Besedovsky, H.O., and A. del Rey. 2002. Introduction: immune-neuroendocrine network. *Front Horm Res*. 29:1–14.

Besedovsky, H.O., and A. del Rey. 1996. Immune-neuro-endocrine interactions: facts and hypotheses. *Endocr Rev*. 17:64–102.

Bogo, V., R.A. Hutton, and A. Bruner. 1971. The effects of air-blast on discriminated avoidance behavior in rhesus monkeys. In *Technical Progress Report on Contract No. DA-49-146-XZ-372*. Vol. DASA 2659. Defense Nuclear Agency, Washington DC. pp. 1–32.

Bowen, I.G., E.R. Fletcher, D.R. Richmond, F.G. Hirsch, and C.S. White. 1967. Biophysical mechanisms and scaling procedures applicable in assessing responses of the thorax energized by air-blast overpressures or by non-penetrating missiles. Technical Progress Report DASA 1857. *Fission Prod Inhal Proj*. 1–46.

Brauer, R.W., W.M. Mansfield, Jr., R.W. Beaver, and H.W. Gillen. 1979. Stages in development of high-pressure neurological syndrome in the mouse. *J Appl Physiol Respir Environ Exerc Physiol*. 46:756–65.

Casals, J.B., N.C. Pieri, M.L. Feitosa, A.C. Ercolin, K.C. Roballo, R.S. Barreto et al. 2011. The use of animal models for stroke research: a review. *Comp Med*. 61:305–13.

Celander, H., C.J. Clemedson, U.A. Ericsson, and H.I. Hultman. 1955. A study on the relation between the duration of a shock wave and the severity of the blast injury produced by it. *Acta Physiol Scand*. 33:14–8.

Cernak, I. 2005. Animal models of head trauma. *NeuroRx*. 2:410–22.

Cernak, I. 2010. The importance of systemic response in the pathobiology of blast-induced neurotrauma. *Front Neurol*. 1(151):1–9.

Cernak, I., D. Ignjatovic, G. Andelic, and J. Savic. 1991. [Metabolic changes as part of the general response of the body to the effect of blast waves]. *Vojnosanit Pregl.* 48:515–22.

Cernak, I., Z. Malicevic, V. Prokic, G. Zunic, D. Djurdjevic, S. Ilic, and J. Savic. 1997. Indirect neurotrauma caused by pulmonary blast injury: Development and prognosis. *Int Rev. Armed Forces Med Serv.* 52:114–120.

Cernak, I., A.C. Merkle, V.E. Koliatsos, J.M. Bilik, Q.T. Luong, T.M. Mahota, L. Xu, N. Slack, D. Windle, and F.A. Ahmed. 2011. The pathobiology of blast injuries and blast-induced neurotrauma as identified using a new experimental model of injury in mice. *Neurobiol Dis.* 41:538–51.

Cernak, I., and L.J. Noble-Haeusslein. 2010. Traumatic brain injury: an overview of pathobiology with emphasis on military populations. *J Cereb Blood Flow Metab.* 30:255–66.

Cernak, I., P. Radosevic, Z. Malicevic, and J. Savic. 1995. Experimental magnesium depletion in adult rabbits caused by blast overpressure. *Magnes Res.* 8:249–59.

Cernak, I., J. Savic, D. Ignjatovic, and M. Jevtic. 1999a. Blast injury from explosive munitions. *J Trauma.* 47:96–103; discussion 103–4.

Cernak, I., J. Savic, Z. Malicevic, G. Zunic, D. Djurdjevic, and V. Prokic. 1996a. The pathogenesis of pulmonary blast injury: our point of view. *Chin J Traumatol.* 12:28–31.

Cernak, I., J. Savic, Z. Malicevic, G. Zunic, P. Radosevic, I. Ivanovic et al. 1996b. Involvement of the central nervous system in the general response to pulmonary blast injury. *J Trauma.* 40:S100–4.

Cernak, I., J. Savic, G. Zunic, N. Pejnovic, O. Jovanikic, and V. Stepic. 1999b. Recognizing, scoring, and predicting blast injuries. *World J Surg.* 23:44–53.

Cernak, I., V.J. Savic, A. Lazarov, M. Joksimovic, and S. Markovic. 1999c. Neuroendocrine responses following graded traumatic brain injury in male adults. *Brain Inj.* 13:1005–15.

Cernak, I., Z. Wang, J. Jiang, X. Bian, and J. Savic. 2001a. Cognitive deficits following blast injury-induced neurotrauma: possible involvement of nitric oxide. *Brain Inj.* 15:593–612.

Cernak, I., Z. Wang, J. Jiang, X. Bian, and J. Savic. 2001b. Ultrastructural and functional characteristics of blast injury-induced neurotrauma. *J Trauma.* 50:695–706.

Chafi, M.S., G. Karami, and M. Ziejewski. 2010. Biomechanical assessment of brain dynamic responses due to blast pressure waves. *Ann Biomed Eng.* 38:490–504.

Champion, H.R., J.B. Holcomb, and L.A. Young. 2009. Injuries from explosions: physics, biophysics, pathology, and required research focus. *J Trauma.* 66:1468–77; discussion 1477.

Chavko, M., W.A. Koller, W.K. Prusaczyk, and R.M. McCarron. 2007. Measurement of blast wave by a miniature fiber optic pressure transducer in the rat brain. *J Neurosci Methods.* 159:277–81.

Chavko, M., T. Watanabe, S. Adeeb, J. Lankasky, S.T. Ahlers, and R.M. McCarron. 2011. Relationship between orientation to a blast and pressure wave propagation inside the rat brain. *J Neurosci Methods.* 195:61–6.

Chen, Y., and S. Constantini. 2013. Caveats for using shock tube in blast-induced traumatic brain injury research. *Front Neurol.* 4:117.

Chrousos, G.P. 1995. The hypothalamic-pituitary-adrenal axis and immune-mediated inflammation. *N Engl J Med.* 332:1351–62.

Chu, S.J., T.Y. Lee, H.C. Yan, S.H. Lin, and M.H. Li. 2005. L-Arginine prevents air embolism-induced acute lung injury in rats. *Crit Care Med.* 33:2056–60.

Clemedson, C.J. 1956. Shock wave transmission to the central nervous system. *Acta Physiol Scand.* 37:204–14.

Clemedson, C.J., and C.O. Criborn. 1955. Mechanical response of different parts of a living body to a high explosive shock wave impact. *Am J Physiol.* 181:471–6.

Clemedson, C.J., L. Frankenberg, A. Jonsson, H. Pettersson, and A.B. Sundqvist. 1969. Dynamic response of thorax and abdomen of rabbits in partial and whole-body blast exposure. *Am J Physiol.* 216:615–20.

Clemedson, C.J., and S.A. Granstom. 1950. Studies on the genesis of "rib markings" in lung blast injury. *Acta Physiol Scand.* 21:131–44.

Clemedson, C.J., and H.I. Hultman. 1954. Air embolism and the cause of death in blast injury. *Mil Surg.* 114:424–37.

Clemedson, C.J., and A. Jonsson. 1961. Transmission of elastic disturbances caused by air shock waves in a living body. *J Appl Physiol.* 16:426–30.

Clemedson, C.J., A. Jonsson, and H. Pettersson. 1956. Propagation of an air-transmitted shock wave in muscular tissue. *Nature.* 177:380–1.

Clemedson, C.J., and H. Pettersson. 1956. Propagation of a high explosive air shock wave through different parts of an animal body. *Am J Physiol.* 184:119–26.

Cooper, G.J., D.J. Townend, S.R. Cater, and B.P. Pearce. 1991. The role of stress waves in thoracic visceral injury from blast loading: modification of stress transmission by foams and high-density materials. *J Biomech.* 24:273–85.

Dal Cengio Leonardi, A., N.J. Keane, C.A. Bir, A.G. Ryan, L. Xu, and P.J. Vandevord. 2012. Head orientation affects the intracranial pressure response resulting from shock wave loading in the rat. *J Biomech.* 45:2595–602.

Damon, E.G., C.S. Gaylord, J.T. Yelverton, D.R. Richmond, I.G. Bowen, R.K. Jones et al. 1968. Effects of ambient pressure on tolerance of mammals to air blast. *Aerosp Med.* 39:1039–47.

Desmoulin, G.T., and J.P. Dionne. 2009. Blast-induced neurotrauma: surrogate use, loading mechanisms, and cellular responses. *J Trauma.* 67:1113–22.

Effgen, G.B., C.D. Hue, E. Vogel, 3rd, M.B. Panzer, D.F. Meaney, C.R. Bass et al. 2012. A multiscale approach to blast neurotrauma modeling: part II: methodology for inducing blast injury to in vitro models. *Front Neurol.* 3:23.

Freund, U., J. Kopolovic, and A.L. Durst. 1980. Compressed air emboli of the aorta and renal artery in blast injury. *Injury.* 12:37–8.

Friedlander, F.G. 1955. *Propagation of a Pulse in an Inhomogeneous Medium.* Ulan Press.

Ganpule, S., A. Alai, E. Plougonven, and N. Chandra. 2012. Mechanics of blast loading on the head models in the study of traumatic brain injury using experimental and computational approaches. *Biomech Model Mechanobiol*

Garman, R.H., L.W. Jenkins, R.C. Switzer, 3rd, R.A. Bauman, L.C. Tong, P.V. Swauger et al. 2011. Blast exposure in rats with body shielding is characterized primarily by diffuse axonal injury. *J Neurotrauma.* 28:947–59.

Gelman, S. 2008. Venous function and central venous pressure: a physiologic story. *Anesthesiology.* 108:735–48.

Goldstein, L.E., A.M. Fisher, C.A. Tagge, X.L. Zhang, L. Velisek, J.A. Sullivan et al. 2012. Chronic traumatic encephalopathy in blast-exposed military veterans and a blast neurotrauma mouse model. *Sci Transl Med.* 4:134ra60.

Gorbunov, N.V., S.J. McFaul, S. Van Albert, C. Morrissette, G.M. Zaucha, and J. Nath. 2004. Assessment of inflammatory response and sequestration of blood iron transferrin complexes in a rat model of lung injury resulting from exposure to low-frequency shock waves. *Crit Care Med.* 32:1028–34.

Hald, E.S., and P.W. Alford. 2013. Smooth muscle phenotype switching in blast traumatic brain injury-induced cerebral vasospasm. *Transl Stroke Res.* 5:385–93.

Haselbacher, A., S. Balachandar, and S.W. Kieffer. 2007. Open-ended shock tube flows: Influence of pressure ration and diaphragm position. *AIAAA J.* 45:1917–1929.

Holzer, A., M.F. Pietschmann, C. Rosl, M. Hentschel, O. Betz, M. Matsuura et al. 2012. The interrelation of trabecular microstructural parameters of the greater tubercle measured for different species. *J Orthop Res.* 30:429–34.

Irwin, R.J., M.R. Lerner, J.F. Bealer, P.C. Mantor, D.J. Brackett, and D.W. Tuggle. 1999. Shock after blast wave injury is caused by a vagally mediated reflex. *J Trauma.* 47:105–10.

Kamnaksh, A., E. Kovesdi, S.K. Kwon, D. Wingo, F. Ahmed, N.E. Grunberg et al. 2011. Factors affecting blast traumatic brain injury. *J Neurotrauma.* 28:2145–53.

Kaur, C., J. Singh, M.K. Lim, B.L. Ng, and E.A. Ling. 1997. Macrophages/microglia as 'sensors' of injury in the pineal gland of rats following a non-penetrative blast. *Neurosci Res.* 27:317–22.

Kaur, C., J. Singh, M.K. Lim, B.L. Ng, E.P. Yap, and E.A. Ling. 1995. The response of neurons and microglia to blast injury in the rat brain. *Neuropathol Appl Neurobiol.* 21:369–77.

Khan, M.M., and K.L. Melmon. 1985. Are autacoids more than theoretic modulators of immunity? *Clin Immunol Rev.* 4:1–30.

Kirkman, E., and S. Watts. 2011. Characterization of the response to primary blast injury. *Philos Trans R Soc Lond B Biol Sci.* 366:286–90.

Kluger, Y., A. Nimrod, P. Biderman, A. Mayo, and P. Sorkin. 2007. The quinary pattern of blast injury. *Am J Disaster Med.* 2:21–5.

Koliatsos, V.E., I. Cernak, L. Xu, Y. Song, A. Savonenko, B.J. Crain et al. 2011. A mouse model of blast injury to brain: initial pathological, neuropathological, and behavioral characterization. *J Neuropathol Exp Neurol.* 70:399–416.

Koliatsos, V.E., T.M. Dawson, A. Kecojevic, Y. Zhou, Y.F. Wang, and K.X. Huang. 2004. Cortical interneurons become activated by deafferentation and instruct the apoptosis of pyramidal neurons. *Proc Natl Acad Sci U S A.* 101:14264–9. Epub 2004 Sep 20.

Kuehn, R., P.F. Simard, I. Driscoll, K. Keledjian, S. Ivanova, C. Tosun et al. 2011. Rodent model of direct cranial blast injury. *J Neurotrauma.* 28:2155–69.

Kwon, S.-K.C., E. Kovesdi, A.B. Gyorgy, D. Wingo, A. Kamnaksh, J. Walker, J.B. Long, and D.V. Agoston. 2011. Stress and traumatic brain injury; a behavioral, proteomics and histological study. *Front Neurol.* 2.

Ling, G., F. Bandak, R. Armonda, G. Grant, and J. Ecklund. 2009. Explosive blast neurotrauma. *J Neurotrauma.* 26:815–25.

Long, J.B., T.L. Bentley, K.A. Wessner, C. Cerone, S. Sweeney, and R.A. Bauman. 2009. Blast overpressure in rats: recreating a battlefield injury in the laboratory. *J Neurotrauma.* 26:827–40.

Lu, J., K.C. Ng, G. Ling, J. Wu, D.J. Poon, E.M. Kan et al. 2012. Effect of blast exposure on the brain structure and cognition in Macaca fascicularis. *J Neurotrauma.* 29:1434–54.

Magnuson, J., F. Leonessa, and G.S. Ling. 2012. Neuropathology of explosive blast traumatic brain injury. *Curr Neurol Neurosci Rep.* 12:570–9.

Mainiero, R., and M. Sapko. 1996. Blast and fire propagation in underground facilities, from http://www.dtic.mil/get-tr-doc/pdf?AD=ADA314855, Defense Nuclear Agency, Alexandria, VA.

Mason, W., T.G. Damon, A.R. Dickinson, and T.O. Nevison, Jr. 1971. Arterial gas emboli after blast injury. *Proc Soc Exp Biol Med.* 136:1253–5.

Mellor, S.G. 1988. The pathogenesis of blast injury and its management. *Br J Hosp Med.* 39:536–9.

Melmon, K.L., R.E. Rocklin, and R.P. Rosenkranz. 1981. Autacoids as modulators of the inflammatory and immune response. *Am J Med.* 71:100–6.

Moss, W.C., M.J. King, and E.G. Blackman. 2009. Skull flexure from blast waves: a mechanism for brain injury with implications for helmet design. *Phys Rev Lett.* 103:108702.

Nevison, T.O., W.V. Mason, and A.R. Dickinson. 1971. Measurement of blood velocity and detection of emboli by ultrasonic Doppler technique. In 11th Annual Symposium on Experimental Mechanics. University of New Mexico Press, Albuquerque, NM. Pp. 63–68.

Ngo, T., P. Mendis, A. Gupta, J. Ramsay. 2007. Blast loading and blast effects on structures—An overview. *EJSE Special Issue: Loading on Structures*, 76–91.

Nishida, M. 2001. Shock tubes and tunnels: facilities, instrumentation, and techniques. Shock tubes. In C. Ben-Dor, O. Igra, and T. Elperin, eds. *Handbook of Shock Waves.* Vol. 1. Academic Press, San Diego, CA. pp. 553–585.

Nyein, M.K., A.M. Jason, L. Yu, C.M. Pita, J.D. Joannopoulos, D.F. Moore et al. 2010. In silico investigation of intracranial blast mitigation with relevance to military traumatic brain injury. *Proc Natl Acad Sci U S A.* 107:20703–8.

Ohnishi, M., E. Kirkman, R.J. Guy, and P.E. Watkins. 2001. Reflex nature of the cardiorespiratory response to primary thoracic blast injury in the anaesthetised rat. *Exp Physiol.* 86:357–64.

Paintal, A.S. 1969. Mechanism of stimulation of type J pulmonary receptors. *J Physiol.* 203:511–32.

Pang, C.C. 2001. Autonomic control of the venous system in health and disease: effects of drugs. *Pharmacol Ther.* 90:179–230.

Panzer, M.B., K.A. Matthews, A.W. Yu, B. Morrison, 3rd, D.F. Meaney, and C.R. Bass. 2012. A multiscale approach to blast neurotrauma modeling: part I - development of novel test devices for in vivo and in vitro blast injury models. *Front Neurol.* 3:46.

Pietschmann, M.F., A. Holzer, C. Rosl, A. Scharpf, T. Niethammer, V. Jansson, and P.E. Muller. 2010. What humeri are suitable for comparative testing of suture anchors? An ultrastructural bone analysis and biomechanical study of ovine, bovine and human humeri and four different anchor types. *J Biomech.* 43:1125–30.

Prima, V., V.L. Serebruany, A. Svetlov, R.L. Hayes, and S.I. Svetlov. 2013. Impact of moderate blast exposures on thrombin biomarkers assessed by calibrated automated thrombography in rats. *J Neurotrauma.* 30:1881–7.

Pun, P.B., E.M. Kan, A. Salim, Z. Li, K.C. Ng, S.M. Moochhala et al. 2011. Low level primary blast injury in rodent brain. *Front Neurol.* 2:19.

Readnower, R.D., M. Chavko, S. Adeeb, M.D. Conroy, J.R. Pauly, R.M. McCarron et al. 2010. Increase in blood-brain barrier permeability, oxidative stress, and activated microglia in a rat model of blast-induced traumatic brain injury. *J Neurosci Res.* 88:3530–9.

Reneer, D.V., R.D. Hisel, J.M. Hoffman, R.J. Kryscio, B.T. Lusk, and J.W. Geddes. 2011. A multi-mode shock tube for investigation of blast-induced traumatic brain injury. *J Neurotrauma.* 28:95–104.

Rice, D., and J. Heck. 2000. Terrorist bombings: ballistics, patterns of blast injury and tactical emergency care. *Tactical Edge J.* Summer:53–55.

Richmond, D.R. 1991. Blast criteria for open spaces and enclosures. *Scand Audiol Suppl.* 34:49–76.

Richmond, D.R., I.G. Bowen, and C.S. White. 1961. Tertiary blast effects. Effects of impact on mice, rats, guinea pigs and rabbits. *Aerosp Med.* 32:789–805.

Richmond, D.R., E.G. Damon, I.G. Bowen, E.R. Fletcher, and C.S. White. 1967. Air-blast studies with eight species of mammals. Techn Progr Rep DASA 1854. *Fission Prod Inhal Proj*:1–44.

Richmond, D.R., E.G. Damon, E.R. Fletcher, I.G. Bowen, and C.S. White. 1968. The relationship between selected blast-wave parameters and the response of mammals exposed to air blast. *Ann N Y Acad Sci*. 152:103–21.

Risling, M., and J. Davidsson. 2012. Experimental animal models for studies on the mechanisms of blast-induced neurotrauma. *Front Neurol*. 3:30.

Risling, M., S. Plantman, M. Angeria, E. Rostami, B.M. Bellander, M. Kirkegaard et al. 2011. Mechanisms of blast induced brain injuries, experimental studies in rats. *Neuroimage*. 54 Suppl 1:S89–97.

Ritzel, D.V., S.A. Parks, J. Roseveare, G. Rude, and T.W. Sawyer. 2012. Experimental blast simulation for injury studies. RTO-MP-HFM-207, Halifax, Canada. 1–20.

Roberts, J.C., T.P. Harrigan, E.E. Ward, T.M. Taylor, M.S. Annett, and A.C. Merkle. 2012. Human head-neck computational model for assessing blast injury. *J Biomech*. 45:2899–906.

Robey, R. 2001. Shock tubes and tunnels: facilities, instrumentation, and techniques. Blast tubes. In C. Ben-Dor, O. Igra, and T. Elperin, eds. *Handbook of Shock Waves*. Vol. 1. Academic Press, San Diego, CA. pp. 623–650.

Rutlen, D.L., E.W. Supple, and W.J. Powell, Jr. 1979. The role of the liver in the adrenergic regulation of blood flow from the splanchnic to the central circulation. *Yale J Biol Med*. 52:99–106.

Saljo, A., F. Bao, K.G. Haglid, and H.A. Hansson. 2000. Blast exposure causes redistribution of phosphorylated neurofilament subunits in neurons of the adult rat brain. *J Neurotrauma*. 17:719–26.

Saljo, A., F. Bao, A. Hamberger, K.G. Haglid, and H.A. Hansson. 2001. Exposure to short-lasting impulse noise causes microglial and astroglial cell activation in the adult rat brain. *Pathophysiology*. 8:105–111.

Saljo, A., F. Bao, J. Shi, A. Hamberger, H.A. Hansson, and K.G. Haglid. 2002. Expression of c-Fos and c-Myc and deposition of beta-APP in neurons in the adult rat brain as a result of exposure to short-lasting impulse noise. *J Neurotrauma*. 19:379–85.

Saljo, A., H. Bolouri, M. Mayorga, B. Svensson, and A. Hamberger. 2010. Low-level blast raises intracranial pressure and impairs cognitive function in rats: prophylaxis with processed cereal feed. *J Neurotrauma*. 27:383–9.

Saljo, A., and A. Hamberger. 2004. Intracranial sound pressure levels during impulse noise exposure *In* 7th International Neurotrauma Symposium. Vol. CD E912. Medimond, Monduzzi Editore, Adelaide, Australia.

Sanborn, B., X. Nie, W. Chen, and T. Weerasooriya. 2012. Inertia effects on characterization of dynamic response of brain tissue. *J Biomech*. 45:434–9.

Savic, J., V. Tatic, D. Ignjatovic, V. Mrda, D. Erdeljan, I. Cernak et al. 1991. [Pathophysiologic reactions in sheep to blast waves from detonation of aerosol explosives]. *Vojnosanit Pregl*. 48:499–506.

Selye, H. 1976. Forty years of stress research: principal remaining problems and misconceptions. *Can Med Assoc J*. 115:53–6.

Sundaramurthy, A., A. Alai, S. Ganpule, A. Holmberg, E. Plougonven, and N. Chandra. 2012. Blast-induced biomechanical loading of the rat: an experimental and anatomically accurate computational blast injury model. *J Neurotrauma*. 29:2352–64.

Surbatovic, M., N. Filipovic, S. Radakovic, N. Stankovic, and Z. Slavkovic. 2007. Immune cytokine response in combat casualties: blast or explosive trauma with or without secondary sepsis. *Mil Med*. 172:190–5.

Svetlov, S.I., V. Prima, O. Glushakova, A. Svetlov, D. Kirk, H. Gutierrez et al. 2012. Neuro-glial and systemic mechanisms of pathological responses in rat models of primary blast overpressure compared to "composite" blast. *Front Neurol*. 3:15.

Svetlov, S.I., V. Prima, D.R. Kirk, H. Gutierrez, K.C. Curley, R.L. Hayes et al. 2010. Morphologic and biochemical characterization of brain injury in a model of controlled blast overpressure exposure. *J Trauma*. 69:795–804.

Takeuchi, S., H. Nawashiro, S. Sato, S. Kawauchi, K. Nagatani, H. Kobayashi et al. 2013. A better mild traumatic brain injury model in the rat. *Acta Neurochir Suppl*. 118:99–101.

Tsokos, M., F. Paulsen, S. Petri, B. Madea, K. Puschel, and E.E. Turk. 2003a. Histologic, immunohistochemical, and ultrastructural findings in human blast lung injury. *Am J Respir Crit Care Med*. 168:549–55.

Tsokos, M., E.E. Turk, B. Madea, E. Koops, F. Longauer, M. Szabo et al. 2003b. Pathologic features of suicidal deaths caused by explosives. *Am J Forensic Med Pathol*. 24:55–63.

Tumer, N., S. Svetlov, M. Whidden, N. Kirichenko, V. Prima, B. Erdos et al. 2013. Overpressure blast-wave induced brain injury elevates oxidative stress in the hypothalamus and catecholamine biosynthesis in the rat adrenal medulla. *Neurosci Lett*. 544:62–7.

Valiyaveettil, M., Y. Alamneh, Y. Wang, P. Arun, S. Oguntayo, Y. Wei et al. 2013. Contribution of systemic factors in the pathophysiology of repeated blast-induced neurotrauma. *Neurosci Lett*. 539:1–6.

Valiyaveettil, M., Y. Alamneh, S. Oguntayo, Y. Wei, Y. Wang, P. Arun et al. 2012a. Regional specific alterations in brain acetylcholinesterase activity after repeated blast exposures in mice. *Neurosci Lett*. 506:141–5.

Valiyaveettil, M., Y.A. Alamneh, S.A. Miller, R. Hammamieh, P. Arun, Y. Wang et al. 2012b. Modulation of cholinergic pathways and inflammatory mediators in blast-induced traumatic brain injury. *Chem Biol Interact*. 203:371–5.

Vandevord, P.J., R. Bolander, V.S. Sajja, K. Hay, and C.A. Bir. 2012. Mild neurotrauma indicates a range-specific pressure response to low level shock wave exposure. *Ann Biomed Eng*. 40:227–36.

Varas, J.M., M. Phillipens, S.R. Meijer, A. van den Berg, C., P.C. Sibma, van Bree, J.L.M., and D.V.W.M. de Vries. 2011. Physics of IED blast shock tube simulations for mTBI research. *Front Neurol*. 2:1–14.

Vries, H.D. 1904. The evidence of evolution. *Science*. 20:395–401.

Wang, Y., Z. Ye, X. Hu, J. Huang, and Z. Luo. 2010. Morphological changes of the neural cells after blast injury of spinal cord and neuroprotective effects of sodium beta-aescinate in rabbits. *Injury*. 41:707–16.

White, C.S., I.G. Bowen, and D.R. Richmond. 1965. Biological tolerance to air blast and related biomedical criteria. CEX-65.4. *CEX Rep Civ Eff Exerc*. 1–239.

Wilkinson, C.W., K.F. Pagulayan, E.C. Petrie, C.L. Mayer, E.A. Colasurdo, J.B. Shofer et al. 2012. High prevalence of chronic pituitary and target-organ hormone abnormalities after blast-related mild traumatic brain injury. *Front Neurol*. 3:11.

Zhu, F., H. Mao, A. Dal Cengio Leonardi, C. Wagner, C. Chou, X. Jin et al. 2010. Development of an FE model of the rat head subjected to air shock loading. *Stapp Car Crash J*. 54:211–25.

Zucker, I.H. 1986. Left ventricular receptors: physiological controllers or pathological curiosities? *Basic Res Cardiol*. 81:539–57.

Zuckerman, S. 1940. Experimental study of blast injuries to the lungs. *Lancet*. 236:219–24.

46 Animal Models for Concussion

Molecular and Cognitive Assessments—Relevance to Sport and Military Concussions

Hayde Bolouri and Henrik Zetterberg

CONTENTS

46.1 INTRODUCTION: BACKGROUND AND DEFINITIONS

Mild traumatic brain injury (mTBI) or concussion, the most common form of brain injury, results in a complex cascade of injurious and reparative events in the brain, and is not always as mild in nature as the mTBI term would imply. Over the last decades it has become clear that repeated mTBIs may give rise to chronic and sometimes progressive brain changes that may lead to a broad range of psychiatric and neurological symptoms. Presently, there is a convention to categorize TBI into three groups: mild, moderate, and severe, based on initial presentation. At the more severe end of the injury spectrum, the correlation between initial injury severity rating and various outcome measures is relatively robust. At the milder end of the spectrum, this correlation is less tight, and over the last 100 years this has generated confusion with regards to the typical presentation and outcome of milder injuries. For a successful translation of basic science knowledge to the clinic to occur, further techniques and models are needed that better reflect mTBI in humans. The purpose of this chapter is to overview the underlying evidence for the necessity of animal models for mTBI in sports and other high risk activities such as military service.

Traumatic brain injury (TBI) occurs when an external physical force impacts the head, either causing the brain to move within the intact skull or damaging the brain by fracturing the skull (McCrory et al., 2005). Various types and levels of impact damage the brain differently. TBI may acutely alter the state of consciousness and, with time, impair cognitive abilities, behavior, and/or physical function. Annually, around 1.7 million new cases of TBI are reported in the United States (www.cdc.gov/TraumaticBrainInjury, September 2013). Mild TBI constitutes most of these; an estimated 300,000 cases take place in the setting of sports and recreation, 95% being mTBI or concussion. In Europe, these cases have reached 60,000 deaths annually, and hospitalized TBI combined was estimated to be 235/100,000 inhabitants (Marklund and Hillered, 2011); the global magnitude of TBI is unknown, but available data suggest that the number of TBI victims globally is rising sharply (Corrigan et al., 2010; Maas et al., 2008; Tagliaferri et al., 2006).

TBI is not only a single pathophysiological phenomenon, but rather a complex disease process that gives rise to structural and functional damage from both primary and secondary injury mechanisms (Masel and DeWitt, 2010). The primary injury is the result of the immediate mechanical disruption of brain tissue that occurs at the time of exposure to the external force and includes contusion, damage to blood vessels (hemorrhage), and axonal shearing, in which the axons of neurons are stretched and wavering (Cernak, 2005; Gaetz, 2004). Secondary injury develops over minutes to months after the primary injury and is the result of cascades of metabolic, cellular, and molecular events that ultimately lead to brain cell death, tissue damage, and atrophy (Bramlett and Dietrich, 2007; Thompson et al., 2005).

Acute assessment of injury severity is critical for the diagnosis, management, and prognosis of TBI. Currently, in TBI clinical trials, the Glasgow Coma Scale is the primary means for initial patient classification, and the Glasgow Outcome Scale or its eight-point extended version remains a primary method for assessing outcomes (Lu et al., 2010; Maas and Lingsma, 2008). In contrast, much less is known about the pathophysiology of mTBI.

MTBI is characterized by a short deterioration of neural function that may or may not involve loss of consciousness (Kelly, 1997). It may result from neuropathological changes, but some believe that the acute clinical symptoms reflect a functional disorder rather than a structural damage. Generally, mTBI is associated with normal structural neuroimaging (Aubry et al., 2002; McCrory et al., 2005), but recent biomarker studies challenge the view that neurons and axons often stay intact during mTBI (Neselius et al., 2012, 2013; Zetterberg et al., 2006, 2009).

In the textbook *Traumatic Brain Injury* (Silver, 2010) the terms "concussion" and "mTBI" are used interchangeably. At the fourth concussion conference (McCrory et al., 2013), however, a distinction between concussion and mTBI was proposed, noting that mTBI refers to "different constructs and should not be used interchangeably" suggesting that the concussion may be followed by complete recovery, whereas mTBI may manifest persistent symptoms (McCrory et al., 2013).

Regardless of the different proposals around the terminology of concussion, the most important reason for developing animal models of TBI is to open new therapeutic windows. Developments and/or modifications of new and existing animal models of TBI provide opportunities for new therapeutic strategies and to cross the therapeutic gap between preclinical studies and patient care.

However, promising results from preclinical studies of potential TBI treatments have not been interpreted into successful outcomes in clinical trials. The pathophysiological heterogeneity observed in patients with TBI, the lack of adequate pharmacokinetic analyses to determine optimal doses of potential therapies, and the administration of compounds outside the therapeutic window may have led to the clinical trial failures (Schouten, 2007).

The pathophysiological heterogeneity observed in patients with TBI may arise from the location, nature, and severity of the primary injury and the effects of other factors and preexisting conditions, including but not restricted to age, health, sex, medication, alcohol and drug use, and genetics (Margulies and Hicks, 2009). Animal models of TBI are each designed to produce a relatively homogeneous type of injury, with age, sex, genetic background, and the injury parameters all well controlled. Thus, any one animal model may not be able to fully recapitulate all the aspects of secondary injury development that are observed in human TBI, and this may in part explain why drugs that showed promise in preclinical studies failed in clinical studies (Marklund et al., 2006). Unquestionably, however, animal models have been fundamental for studying the biomechanical, cellular, and molecular aspects of human TBI because of the limitation of clinical setting as well as for developing and characterizing novel therapeutic interventions. To achieve new therapeutic finding and based on the study design, appropriate animal models should be selected or modified.

46.2 SPORTS-RELATED CONCUSSION

Recently, there has been great interest in the long-term outcome of single and repetitive mTBI in contact sports, such as boxing, American football, rugby, and ice hockey. Although the symptoms of most athletes with mTBI resolve within days to weeks, some will experience long-term sequelae and a few will even develop chronic and progressive symptoms. Such symptoms may include changes in cognition (especially memory and executive functioning), mood (especially depression, apathy, and suicidality), personality and behavior (especially poor impulse control and behavioral disinhibition), and movement (including parkinsonism and signs of motor neuron disease) (Stern et al., 2011). These constellations of symptoms are often described as chronic traumatic encephalopathy (CTE) (Gavett et al., 2011; Jordan, 2000; Mendez, 1995), but this term is still debatable from both neuropsychiatric and neuropathological perspectives. Nonetheless, in 1928, a form of TBI-induced early dementia was first observed in professional boxers years after their careers had ended (Martland, 1928). Termed "dementia pugilistica" (DP) or "punch drunk syndrome," the prevalence of this neuropsychiatric manifestation has been estimated to about 20% in former pro boxers (Jordan et al., 1997; Roberts, 1969).

The brains of former boxers were also found to display hallmark pathologies of Alzheimer disease, neurofibrillary tangles composed of hyperphosphorylated tau and Aβ plaques (Roberts et al., 1990; Tokuda et al., 1991). Trauma-induced axonopathy provides a mechanism for the rapid production of Aβ after TBI (Johnson et al., 2010). Risk factors for CTE/DP are a long boxing career, many bouts, high sparring exposure, many knockouts, poor performance as a boxer, and being able to tolerate many blows without being knocked out, all more or less directly associated with accumulated exposure to repetitive brain trauma (Jordan, 2000). A positive APOE ε4 status, commonly associated with Alzheimer disease, may also be a genetic risk factor for CTE/DP in these individuals (Jordan et al., 1997). Recently, very similar tau and Aβ neuropathology as well as TDP-43 deposition have been found years after competition in other contact sports, such as American football (McKee et al., 2013).

Sport injuries are the most prevalent causes of repeated concussions and American football accounts for the highest proportion (Powell and Barber-Foss, 1999). In recent decades, a mounting body of new pathological findings has shown significant brain degeneration in professional athletes with history of repetitive concussions (Dietrich et al., 1994; Dixon et al., 1987; Graham et al., 2000; McIntosh et al., 1987; Pettus and Povlishock, 1996; Povlishock and Kontos, 1985; Schmidt and Grady, 1993; Wang et al., 1997).

More than 60% of the players in professional football (National Football League [NFL]) have clinical signs and symptoms of mTBI (Pellman et al., 2004, 2005). Headache is most common and found in 60% of NFL players (Sallis and Jones, 2000). Other symptoms are memory problems (40%), dizziness (40%), cognitive dysfunction (25%), tremor, irritability, anxiety, depression, tinnitus, other auditory symptoms, neck and back pain, anosmia, and visual problems (Collins et al., 1999; McCrea et al., 2002; Ponsford et al., 2008). Concussed football players are mostly free of symptoms after 1 week and return to the game, although it is not known whether this is sufficient time for recovery (Bleiberg et al., 2004; Collins et al., 1999; Macciocchi et al., 1996; Makdissi et al., 2001). Although concussion is classified as a milder form of severity, TBI may lead to lasting neurological effects that are not immediately evident (Mortimer et al., 1991). There is an increased risk of depression (Holsinger et al., 2002), Alzheimer disease, and other dementias (Plassman et al., 2000) many years after TBI.

One of the major problems being addressed in mTBI research is how to predict when a concussion or mTBI is so severe that there is risk of long-lasting or chronic brain changes. It appears as though repeated mTBIs are the most dangerous for this aspect, but how long after a concussion is the brain affected and how can we tell when it is safe to return to play or other activities where there is high risk of new brain impacts? To increase our understanding of mTBI, we need models to study the disease process. As will be clear in the following sections, there are well-established models for more severe forms of TBI. There are also very encouraging new models that may translate well to human mTBI processes.

46.3 ANIMAL MODELS OF TBI

In recent decades, several animal models have been developed and modified to mimic the clinical consequences of TBI in view of the heterogeneous nature of the clinical situation of such injury. Whereas early models of TBI addressed the biomechanical aspects of brain injury (Denny-Brown and Russell, 1941; Gurdjian et al., 1954), more recent models have been targeted at improving our understanding of the detrimental, complex molecular cascades that are initiated by head trauma.

However, not all available animal models of TBI do fulfill the requirement of high-impact velocity and rapid change in head velocity without injuring the skull bone as occurs in sport-related concussion. Simulating all aspects of TBI in a single animal model is impossible and for that reason, several TBI models are being used in animals of various ages and injury severity levels. Larger animals are closer in size and physiology to human; however, rodent models are the most common in TBI research for ethical reasons, their modest cost, and their small size, among other reasons (Finnie and Blumbergs, 2002). In addition to the heterogeneity of TBI, the difficulty in evaluating subtle cognitive and psychiatric impairments in small animal species is a major challenge in the preclinical evaluation of neuroprotective drug candidates. Clinically relevant experimental models require the following characteristics since the majority of repeated concussions are (1) closed and (2) of a mild level of severity, (3) with low mortality rates that (4) exhibit symptoms in the absence of gross neuropathology (Prins et al., 2010). Although the features of a useful preclinical TBI model are reproducibility, low costs, applicability to both rats and mice, and being technically easy to perform, perhaps most important is production of long-lasting behavioral deficits (Morales et al., 2005).

Although most TBI models show a substantial degree of diffuse injury, the animal models in the following are categorized into "head impact," "direct brain impacts," "fluid percussion," and "momentum-exchange model" (Table 46.1).

46.3.1 HEAD IMPACTS

Denny-Brown and Russel (1941), in an early experimental study, observed that the impact of the free-moving head better simulated cerebral concussion than when the head is fixed. Later on, Gurdjian et al. (1954, 1966) used a 1-kg piston in primate head impacts and caused concussion through coma and coup and contrecoup contusions. To produce moderate concussion, Goldman et al. (1991) used a pendulum drop. Special efforts were adopted to prevent skull fracture such as using a protective helmet. The concern for skull fractures increased with smaller animals because of proportionately higher head accelerations needed to produce brain injury (Ommaya et al., 1967). Marmarou et al. (1994) (Figure 46.1) established the drop-weight technique, one of the most widely used and modified techniques for different animal models for brain injury research. In this model, the skull is exposed to a free-falling, guided weight. The technique uses a 400–500 g weight dropped from heights of 1.0, 1.5, and 2.0 m (4.4, 5.4, and 6.3 m/s, respectively) on top of the head. In this model, injury severity can be altered by adjusting the mass of the weight and the height from which it falls. The head of the animal is supported on a thick layer of foam while the test causes an impact followed by the head being pushed into the foam until the impactor and head rebound. The 2-m impacts cause skull fracture, brain and cervical injuries, and death (Foda and Marmarou, 1994; Kallakuri et al., 2003; Marmarou et al., 1994).

DeFord et al. (2002) used the drop-weight impact on the mouse to produce mild brain injuries. The technique used a 12.5-mm diameter Teflon-coated guide tube, which was a modification of an existing procedure (Hall, 1985). The impact mass was dropped in the tube, and loaded into a 35-mm-diameter impactor placed on the vertex of the head. To load a lever-attached offset to the head, a pneumatic (Gutierrez et al., 2001) impactor and the drop-weight technique (Ellingson et al., 2005) were used to produce controlled head rotational acceleration in the rat. Tornheim et al.

TABLE 46.1
Existing Animal Models of TBI Corresponding to Each Category in Sections 46.3.1 through 46.3.4

Model	Type of Injury	Species	Comments	References
Head impact	Mainly focal	Rat, mouse	Device is easy to operate; biomechanics of injury mechanism are similar to those seen in human TBI. Not highly reproducible; high mortality rate.	Ellingson et al., 2005; DeFord et al., 2002; Foda and Marmarou, 1994; Gurdjian et al., 1954, 1966; Gutierrez et al., 2001; Marmarou et al., 1994; Nilsson et al., 1977; Tornheim et al. 1981
Direct brain impact	Mainly focal	Rat, mouse	Biomechanics of injury mechanism are similar to those seen in human TBI. Not highly reproducible; need for craniotomy with high mortality rate.	Dixon et al., 1991; Feeney et al., 1981; Gurdjian et al., 1954; Lewen et al., 1999; Lighthall, 1988; Meyer, 1956; Meyer and Denny-Brown, 1955; Viano and Lau, 1988
Fluid percussion	Mixed	Rat, rabbit, cat, dog, sheep, swine	Highly reproducible; allows fine-tuning of injury severity. Need for craniotomy with high mortality rate.	Dixon et al., 1987, 1988; Gurdjian, 1975; Gurdjian et al., 1968; Lindgren and Rinder, 1966, 1967; Lighthall et al., 1989, 1990; Sullivan et al., 1976
Momentum exchange	Mainly diffuse	Rat	Biomechanics of injury mechanism are similar to those seen in human concussion. Highly reproducible. No need for craniotomy with very low mortality rate.	Adelson et al., 1996; Hamberger et al., 2009; Lighthall, 1988; Nilsson et al., 1977; Viano et al., 2009

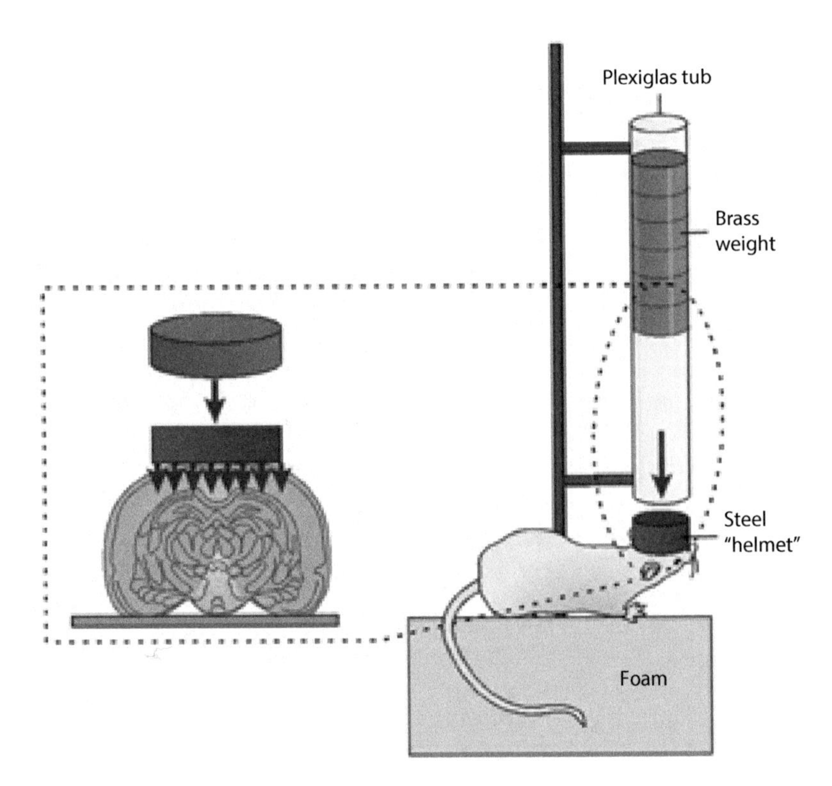

FIGURE 46.1 In Marmarou's weight-drop model, a metal disk is placed over the skull to prevent bone fracture.

(1981, 1983) used a captive bolt pistol to impact the head of anesthetized cats. Similar tests were conducted on sheep (Lewis et al., 1996) and other impact techniques were used (Anderson et al., 2003) to study diffuse axonal injury (DAI).

Another animal model of head impact was developed by Nilsson et al. (1977), involving the momentum-exchange impact with velocities of 6.0–11.0 m/s to the vertex of the supine rat. The tests involved a 600 g mass, but notable was the control of the severity by limiting the impact to 10-mm displacement after head contact while the animal was free to move in response to the head impact. Experiments by Nilsson et al. include a range of outcomes such as degrees of alertness, concussion, and death.

46.3.2 Direct Brain Impacts

In weight-drop models, another technique involves dropping a weight on the exposed dura (with or without a craniotomy) causing cortical compression (Figure 46.2) (Feeney et al., 1981; Gurdjian et al., 1954; Meyer, 1956; Meyer and Denny-Brown, 1955). The mechanics of the drop-weight procedure are based on a stationary mass over the dura and a softly supported head to limit the kinetic energy delivered to skull motion and brain deformation. A pneumatic impactor was used to load the ferret brain through a trephined opening in the skull to apply controlled loads on the brain (Lighthall, 1988). The technique showed repeatable moderate-to-severe injury in rats by controlling the speed of impact and depth of deformation (Dixon et al., 1991; Viano and Lau, 1988; Viano and Lovsund, 1999) and functional coma lasting 1.5 days was related to 8-m/s impacts with only a 2-mm brain deformation. A modification of the existing techniques was used (Dixon et al., 1999; Lighthall, 1988) (Figure 46.3) to study spatial memory changes after direct brain impact of rats by involving a 5-mm metal tip vertically driven and a stroke-constrained pneumatic cylinder (Lewen et al., 1999). The impact was at 4 m/s with 2.5-mm depth and approximately of 50-ms duration. Functional tests with morphological outcomes were studied from mild brain cortical contusion of rat brain (Lewen et al., 1999). The head was placed on a stereotaxic frame, and a craniotomy was made over the right parietal midline. Trauma was induced by a 21-g free falling mass dropped from 35 cm over a piston of 4.5 mm resting on the exposed dura. The impact was limited to 1.5-mm displacement of the cortex.

46.3.3 Fluid Percussion

In fluid percussion models, an insult is inflicted by a pendulum striking the piston of a reservoir of fluid to generate a fluid pressure pulse to intact dura through craniotomy, which is made either centrally around the midline between bregma and lambda (Sullivan et al., 1976) (Figure 46.4). However, the highest speed of impact used in lateral fluid percussion is lower than that required for concussion that occurs in contact sports, and the impact in football is mostly in the frontal or temporal region.

This model is one of the earliest and the most prominent head injury procedures, which Gurdjian (1975) attributed to Allen (1911), who simulated spinal cord crush from cervical-fracture dislocations. The technique has been modified over the years for transdural loading to study concussion, contusion, and more serious brain injuries (Denny-Brown, 1961; Denny-Brown and Russell, 1941; Gurdjian et al., 1968, Gurdjian et al., 1953; (Lindgren and Rinder, 1966, 1967; Walker, 1944).

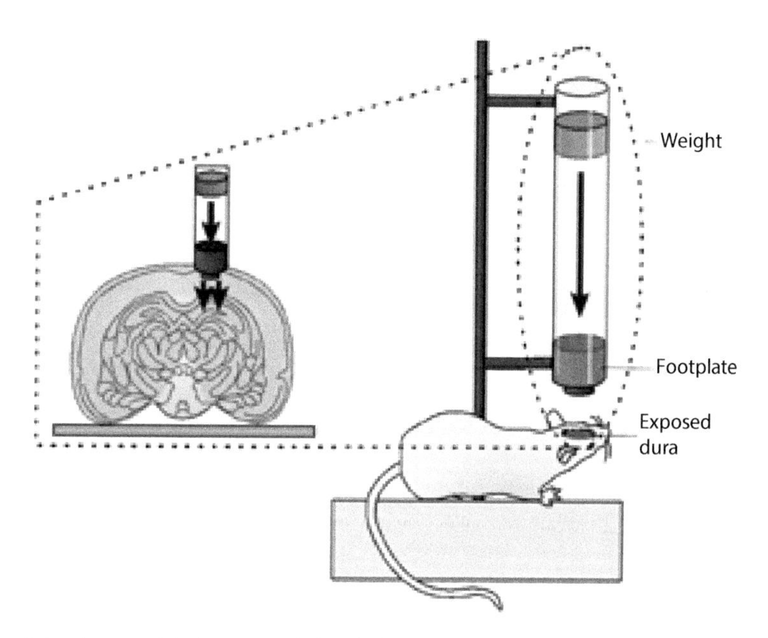

FIGURE 46.2 In Feeney's weight-drop model, the exposed dura is impacted by a known weight and distance.

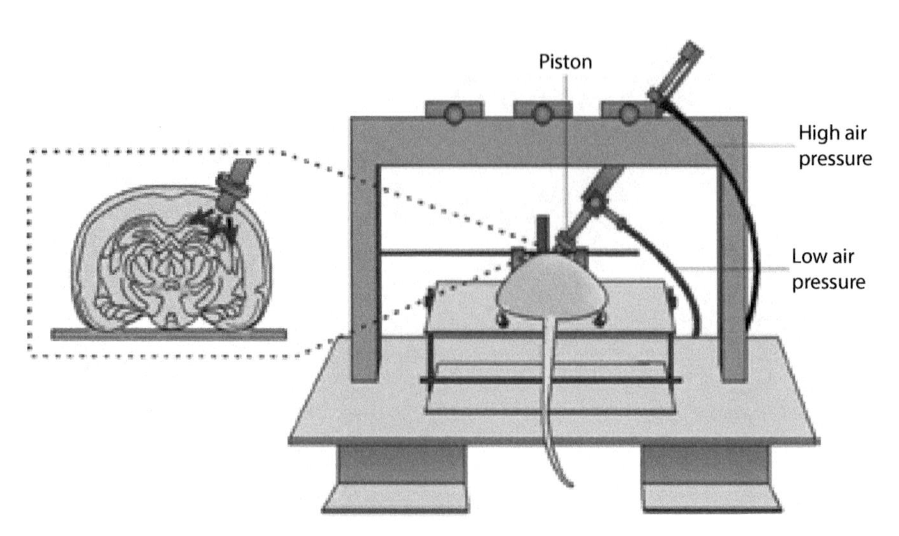

FIGURE 46.3 The controlled cortical impact injury model uses an air or electromagnetic driven piston to penetrate the brain through a trephined opening in the skull at a known distance and velocity.

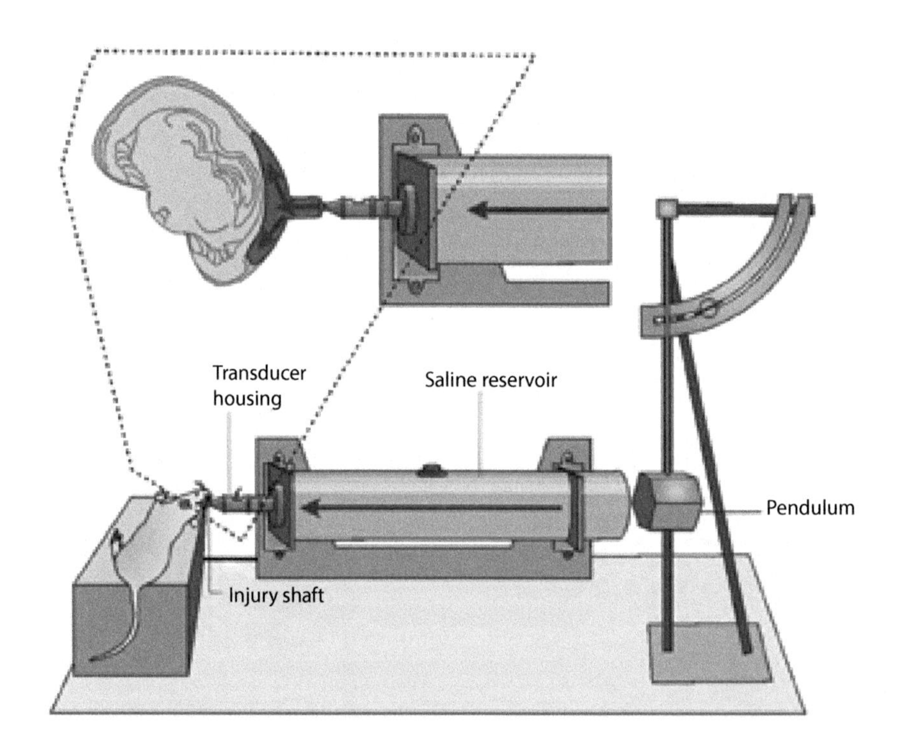

FIGURE 46.4 In fluid percussion models, an insult is inflicted by a rapid injection of fluid to generate a fluid pressure pulse into the intact epidural space.

Gurdjian refined the technique by developing an air-pulse impact procedure (Gurdjian, 1975), which controlled the amplitude and duration of pressure. The fluid percussion procedure provided data for the development of the Wayne State Tolerance Criterion (Gurdjian et al., 1966) by simulating brain injuries in the mid-range of head impact durations, greater than the short (2–4 ms) durations associated with skull fracture but lower than the longer (>10 ms) durations from the human volunteer exposures.

The fluid percussion procedure was further developed during the 1970–1980s (Dixon et al., 1987; Sullivan et al., 1976). However, it is invasive and involves direct brain loading by a volume of transdural fluid typically injected by a 450-g weight dropped from a height of 25–55 cm on the fluid volume. The injected fluid in the skull involves a loading duration of 5–20 ms at a velocity of 2.2–3.3 m/s. High-speed X-ray studies of the fluid infusion in rats (Dixon et al., 1988) and ferrets (Lighthall et al., 1989, 1990) showed complex movement of the fluid in the cranium around the epidural space and gross movement of the brain, which displaced the spinal cord down the cervical column. In more severe exposures, the displacement caused vascular injury of the cervical cord.

46.3.4 Momentum-Exchange Model

Techniques that involve air-driven pistons have the versatility to simulate concussion because they offer a wide range of impact velocity. The technique of Nilsson et al. (1977) combines an air-driven piston with the freely moving intact head of an animal, which is impacted on the top of the head. Other techniques involving air-driven pistons require surgical preparation (Adelson et al., 1996; Lighthall, 1988). In a later animal model for concussion, the technique involves a momentum-exchange impact to the side of the helmet-protected head of the rat. Level of injury is controlled from mild to severe (Hamberger et al., 2009; Viano et al., 2009) (Figure 46.5) by impacting the animal with a 50-g mass at 7.4–11.2 m/s velocity that mimics helmet impacts associated with concussion in NFL players (Pellman et al., 2003a, 2003b).

The animal's head, neck, and body are free to move in the impact. This model simulates the translational, high-speed head impact to a freely moving subject with techniques for scaling impact conditions in the game videos from man to animal (Gutierrez et al., 2001; Ommaya et al., 1967). The model does not require surgical preparation and does not cause musculoskeletal injuries. With this approach, impacts causing concussion are at higher speeds than with the weight-drop technique (Foda and Marmarou, 1994; Marmarou et al., 1994).

A clearer understanding of the similarities and differences between different techniques (Foda and Marmarou, 1994; Hamberger et al., 2009; Marmarou et al., 1994; Nilsson et al., 1977; Viano et al., 2009) is fundamental to making an informed decision on an animal model to study sport concussion, brain injuries, treatments, and long-term consequences of head impacts.

46.4 EFFECT OF REPEATED CONCUSSION/mTBI

Repeated concussion commonly occurs in contact sports (for example, in boxing, hockey, rugby, and American football), child abuse victims, and military personnel (Centers for Disease Control and Prevention, 1997; Weber, 2007). An increasing body of evidence indicates that repeated brain concussion can result in cumulative and long-term behavioral symptoms, neuropathological changes, and neurodegeneration (Daneshvar et al., 2011a, 2011b).

Several models have been developed to mimic the clinical consequences of repeated mTBI (Weber, 2007), including the weight-drop model (Creeley et al., 2004; DeFord et al., 2002), the mouse blast TBI model (Wang et al., 2011), the rat fluid percussion model (DeRoss et al., 2002), and the swine closed-head injury model (Raghupathi and Margulies, 2002).

Marmarou's impact acceleration model has been modified to allow repeated head impacts in lightly anesthetized mice (Kane et al., 2012). This method does not require scalp incision and protective skull helmets. Mice spontaneously recover the righting reflex without evidence of seizures and paralysis, and skull fractures and intracranial bleeding are rare. Minor deficits in motor coordination and locomotor hyperactivity recover over time. Mild astrocytic reactivity and increased phospho-tau levels occur without blood–brain barrier disruption, edema, and microglial activation. This new animal model is suitable for the screening of new therapies for mild concussive injuries.

A single mild fluid percussion injury induces short-term behavioral and neuropathological changes in the rat (Shultz et al., 2011), whereas repeated mild fluid percussion injury in rats causes cumulative long-term behavioral impairments, neuroinflammation, and cortical neuron loss (Shultz et al., 2012). Interestingly, subconcussive brain injury induces acute

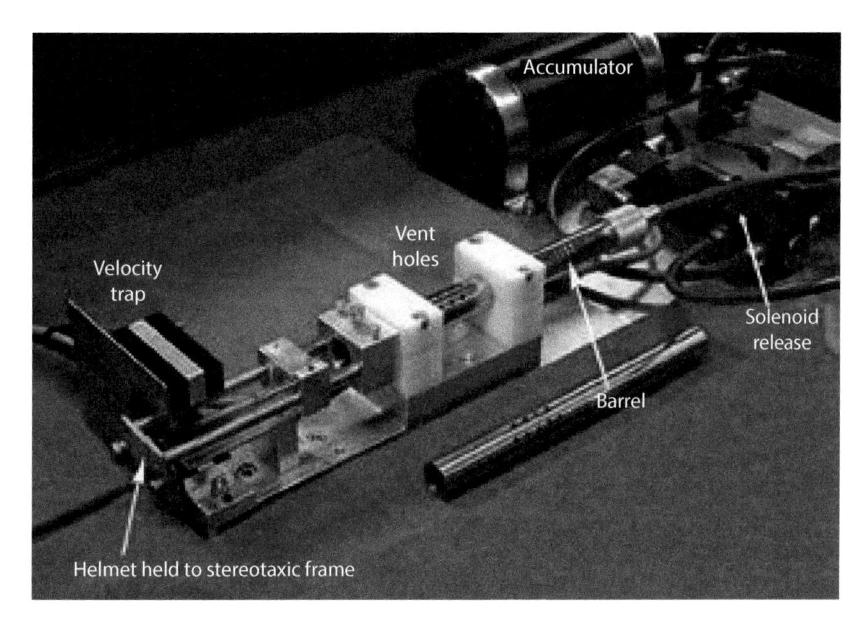

FIGURE 46.5 Photograph of the test setup including support for the ballistic impact barrel, solenoid release of accumulated pressure, and metal helmet.

neuroinflammation in the absence of behavioral impairments in rats after TBI (Shultz et al., 2012). In an immature large animal model of TBI in neonatal piglets, two head rotations after injury led to poorer outcomes, as assessed by neuropathology and neurobehavioral functional outcomes, than did a single rotation (Friess et al., 2009). White matter injury increased in the repeated rotation group compared with the single injury group. More importantly, an increase in injury severity and mortality was observed when the head rotations occurred 24 hours apart compared with 7 days apart. Worsening performance on the cognitive composite score was associated with increasing severity of white matter axonal injury.

Consequently, these findings of animal models suggest that repeated mTBI occurring within a short period might be devastating or fatal and are consistent with findings in human patients who have experienced repeated brain concussions. These models will provide further insights into sports- and combat-related repeated concussions and should help physicians make better decisions about allowing individuals with TBI to return to their normal lives and identify people who may be at enhanced risk for CTE.

In 2008, a center for the study of traumatic encephalopathy in collaboration between Boston University School of Medicine and Sports Legacy Institute was formed. They presented findings on the brain pathology of former football players, other athletes, and military personnel who wished to donate their brains upon death for scientists to examine the long-term effects of repetitive concussions. Examined brains exhibited pronounced CTE with axonal loss and sometimes extensive tau pathology (McKee et al., 2013). CTE is a progressive degenerative disease of the brain found in retired athletes with a history of repetitive concussions. It is clinically associated with the development of memory loss, confusion, impaired judgment, paranoid and aggressive behavior, depression, dementia, and parkinsonism (McKee et al., 2013). At the same time, the Center for the Study of Traumatic Encephalopathy Brain Bank at the Belford Veterans Administration hospital was established to analyze the brain and spinal cords after death of athletes, military veterans, and civilians who experienced repetitive concussions. Translational research at these and other centers will most likely shed more light on disease mechanisms in CTE, in particular how repeated mTBIs may initiate spreading of pathology over the brain and how this process can be halted.

46.4.1 Effects of Concussion on Brain Structure

Macroscopically, the brain is composed of a large number of regions, each with its specific function. Microscopically, the brain contains approximately 50 billion nerve cells or neurons, which are specialized to receive, conduct, and transmit impulses. Neurons have certain morphological features in common, which reflect that nervous tissue functions as a communication system (Heimer, 1995). The cell body generally comprises a minor portion of the neuron, whereas its axonal process often dominates the cell volume. There is a continuously ongoing transport of proteins and waste products into the axon (Droz and Leblond, 1962) from the cell body, which contains the normal setup of cell organelles. The intermediate filaments (neurofilaments), which are components of the cytoskeleton, contain different subunits with a molecular weight in the range of 61–115 kDa. The definition of DAI is microscopic evidence of damage to the axons, such as dissolution of microtubules, accumulation or modification of neurofilaments, and proteolysis of spectrin networks (Serbest et al., 2007). DAI was initially detected with silver impregnation or Nissl staining (Huh et al., 2002; Kallakuri et al., 2003; Marmarou et al., 1994) but is now often revealed with immunostaining of the neurofilaments (Inglese et al., 2005; Smith et al., 2003). Axonal swelling, retraction balls, microglial scars, long tract degeneration, and diffuse gliosis are also hallmarks of DAI (Adams et al., 1985; Simpson et al., 1985). Axonal swelling and retraction balls have been demonstrated with different antisera against either phosphorylated + unphosphorylated epitopes (Hoshino et al., 2003; Raghupathi et al., 2004; Serbest et al., 2007) or against the phosphorylated epitope of both heavy (NF-200) (Wang et al., 1994), medium (NF-180) (Ross et al., 1994), and light (NF-68) (Li et al., 1998) neurofilament proteins. The phosphorylation state of NF-200 is affected by TBI (Posmantur et al., 2000).

The brain also contains cells with a supportive function, the glial cells (i.e., astrocytes, oligodendrocytes, and microglial cells). The most abundant supporting cell is the astrocyte, which is physically supportive to neurons and performs essential tasks for their function. Its intermediate filaments have a major structural component, the glial fibrillary acidic protein.

Macroscopically, a TBI-induced brain injury may be seen as hemorrhages in the meninges and/or in the brain tissue. It may also involve destruction of brain tissue.

Evaluation of injury in various brain regions at the microscopical level is done with histological, immunohistochemical, autoradiographic, or similar techniques. Neuronal death may be visualized with various staining techniques (Gavrieli et al., 1992; Hall et al., 2008). TBI may involve injury to specific cell organelles, such as mitochondria (Lifshitz et al., 2004) and neurofilaments (Droz and Leblond, 1962). Neurodegeneration (Hall et al., 2008) and DAI, a frequent injury in TBI, have distinctive pathological features (Love et al. 2008; Gennarelli et al., 1982).

Astrocytes are activated in response to neuronal injury or death, inflammation, or the presence of plasma components in the brain tissue (Raivich et al., 1999; Stoll et al., 1998). Reactive gliosis is characterized by hyperplasia and hypertrophy of the astrocytes. Activated astrocytes are also characterized by an increase of the immunodetectable glial fibrillary acidic protein (Yu et al., 1993). Activated astrocytes migrate toward the injury and surround it with a scar, which can be detrimental for neurite growth (Davies et al., 1997; Fawcett and Asher, 1999; Reire, 1986).

46.4.2 Effects of Concussion on Brain Function

There are at least as many methods for the study of brain function as for structural studies. They test the function of specific sensory and/or motor systems, each having its site in the brain. The brain region that is essential for cognitive function is the hippocampal formation (Eichenbaum, 1997). Cognitive dysfunction and hippocampal damage (Bolouri et al., 2011, 2012; Smith and Mizumori, 2006) with loss of pyramidal neurons have been reported after TBI (Hamm et al., 1993; Goodman et al., 1994). One of the most frequently used tools in behavioral neuroscience is the Morris Water Maze (Morris, 1984), explicitly memory and learning in the laboratory rat. However, motor and other noncognitive impairments may bias the results. This is particularly critical in brain-injured animals, among which a decrease in spontaneous exploratory activity reflects increased stress/anxiety. In a recorded open field activity (Bolouri et al., 2011, 2012; Larsson et al., 2002), the rationale for using the widely used open field test in combination with MWM, detects and evaluates such behavior because rodents usually explore their environment and brain damage may disturb this behavior (Mikulecka et al., 2004).

46.4.3 Alteration in Intracranial Pressure

The time course and amplitude of the intracranial pressure (ICP) increase after TBI (Hamberger et al., 2009; Rooker et al., 2003; Vink et al., 2003). The ICP peaks 10 hours after impact, at which time the blood–brain barrier is maximally permeable (Hanstock et al., 1994). High ICP is also accompanied by decreases in cerebral blood flow and oxygen consumption, which in turn may be secondary to brain edema (Engelborghs et al., 1998; Fritz et al., 2005). However, at 48 hours after concussion, the ICP was close to control level (Bolouri et al., 2011, 2012).

46.5 CONCLUSION AND PERSPECTIVES

Brain injury is a considerable challenge to individuals and society. Concussion models are necessary to characterize the effects of the brain injuries. The knowledge in the field of neuronal and glial reactions has increased dramatically in recent decades because of studies of the different kinds of brain injuries in models, which has given new hope to the prospect of structural and functional repair. A long-term goal of research in this field is to develop a treatment customized for concussion, which has both preventive and therapeutic effects. We need to explore the underlying mechanisms and biomechanics of concussion extensively and should also make use of the possibility of studying human concussion in greater detail, using biomarkers for brain pathologies and pathological and physiological reactions to the head impact. The understanding of adaptation and recovery of injured neurons may improve the neurological outcome after concussion. Factors that influence the susceptibility of neurons after TBI should be identified, if possible, because prevention of the detrimental, complex molecular cascades in the early stage may protect from degeneration.

TBI leads frequently to cognitive dysfunction in experimental studies (Raymont et al., 2008). Athletes with a history of concussions are more likely to have future concussive injuries than those with no history. One of fifteen players in college football with a concussion is likely to have additional concussions in the same playing season (Guskiewicz et al., 2003; Slobounov et al., 2007). Athletes who have experienced three or more concussions are nine times more likely to suffer from headache when experiencing another concussion. This may be associated with slower recovery of neurological function (Guskiewicz et al., 2003). Although there is a high frequency of repeated concussions in a football season (Guskiewicz et al., 2003), cognitive function after repeated concussions has not often been studied in experimental work.

In recent decades, the clinical setting and sport medicine have been struggling with an issue of fundamental importance for athletes who have experienced concussion: when to return to play (Vagnozzi et al., 2007)? Key questions on when the brain has fully recovered from a concussive injury and what parameters are suitable for assessing such recovery need to be studied and fully understood (Vagnozzi et al., 2007). A continued translation of new experiences from the bench to the bedside and then back to the bench will eventually teach us a lot about the relevance of the animal models.

46.6 LIMITATIONS

The main limitation of animal models is that no animal model simulates perfectly the conditions of sports-related concussion. There are known anatomic differences and geometric orientation of the neuraxis in the brains, and the reliance on a single scale factor based on the radius of the brain can be questioned. Furthermore, the rat brain has no gyri and little white matter.

Rodent models are not able to mimic the clinical problems such as emotional and language difficulties or produce long-lasting periods with a decreased level of consciousness or coma, commonly observed in severely injured TBI patients (Marklund and Hillered, 2011).

Animals must be anesthetized at the time of impact to the head based on ethical requirements; however, anesthetic agents have almost invariably a neuroprotective potential (Head and Patel, 2007; Koerner and Brambrink, 2006; Zhang et al., 2006). The scaling of time between man and animal is not precise. Therefore, to design new animal models, the use of equal stress–equal velocity scaling has been important because this relies on a rich history in the literature.

REFERENCES

Adams, J. H., D. Doyle, et al. (1985). Microscopic diffuse axonal injury in cases of head injury. *Med Sci Law* 25(4):265–269.

Adelson, P. D., P. Robichaud, et al. (1996). A model of diffuse traumatic brain injury in the immature rat. *J Neurosurg* 85(5):877–884.

Allen, A. R. (1911). Surgery of experimental lesion on spinal cord equivalent to crush injury of fracture dislocation of spinal column. *JAMA* 57:878–880.

Anderson, R. W., C. J. Brown, et al. (2003). Impact mechanics and axonal injury in a sheep model. *J Neurotrauma* 20(10):961–974.

Aubry, M., R. Cantu, et al. (2002). Summary and agreement statement of the 1st international symposium on concussion in sport, Vienna 2001. *Clin J Sport Med* 12:6–11.

Bleiberg, J., A. N. Cernich, et al. (2004). Duration of cognitive impairment after sports concussion. *Neurosurgery* 54(5):1073–1078; discussion 1078–1080.

Bolouri, H., A. Saljo, et al. (2012). Animal model for sport-related concussion; ICP and cognitive function. *Acta Neurol Scand* 125(4):241–247.

Bramlett, H. M. and W. D. Dietrich (2007). Progressive damage after brain and spinal cord injury: Pathomechanisms and treatment strategies. *Prog Brain Res* 161:125–141.

Centers for Disease Control and Prevention (1997). Sports-related recurrent brain injuries—United States. Centers for Disease Control and Prevention. *Int J Trauma Nurs* 3(3):88–90.

Cernak, I. (2005). Animal models of head trauma. *NeuroRx* 2(3):410–422.

Collins, M. W., S. H. Grindel, et al. (1999). Relationship between concussion and neuropsychological performance in college football players. *JAMA* 282(10):964–970.

Corrigan, J. D., A. W. Selassie, et al. (2010). The epidemiology of traumatic brain injury. *J Head Trauma Rehabil* 25(2):72–80.

Creeley, C. E., D. F. Wozniak, et al. (2004). Multiple episodes of mild traumatic brain injury result in impaired cognitive performance in mice. *Acad Emerg Med* 11(8):809–819.

Daneshvar, D. H., C. J. Nowinski, et al. (2011a). The epidemiology of sport-related concussion. *Clin Sports Med* 30(1):1–17, vii.

Daneshvar, D. H., D. O. Riley, et al. (2011b). Long-term consequences: Effects on normal development profile after concussion. *Phys Med Rehabil Clin N Am* 22(4):683–700, ix.

Davies, S. J., M. T. Fitch, et al. (1997). Regeneration of adult axons in white matter tracts of the central nervous system. *Nature* 390(6661):680–683.

DeFord, S. M., M. S. Wilson, et al. (2002). Repeated mild brain injuries result in cognitive impairment in B6C3F1 mice. *J Neurotrauma* 19(4):427–438.

Denny-Brown, D. (1961). Brain trauma and concussion. *Arch Neurol* 5:1–3.

Denny-Brown, D. and W. R. Russell (1941). Experimental cerebral concussion. *J Physiol* 99(1):153.

DeRoss, A. L., J. E. Adams, et al. (2002). Multiple head injuries in rats: Effects on behavior. *J Trauma* 52(4):708–714.

Dietrich, W. D., O. Alonso, et al. (1994). Early microvascular and neuronal consequences of traumatic brain injury: A light and electron microscopic study in rats. *J Neurotrauma* 11(3):289–301.

Dixon, C. E., G. L. Clifton, et al. (1991). A controlled cortical impact model of traumatic brain injury in the rat. *J Neurosci Methods* 39(3):253–262.

Dixon, C. E., P. M. Kochanek, et al. (1999). One-year study of spatial memory performance, brain morphology, and cholinergic markers after moderate controlled cortical impact in rats. *J Neurotrauma* 16(2):109–122.

Dixon, C. E., J. W. Lighthall, et al. (1988). Physiologic, histopathologic, and cineradiographic characterization of a new fluid-percussion model of experimental brain injury in the rat.*J Neurotrauma* 5(2):91–104.

Dixon, C. E., B. G. Lyeth, et al. (1987). A fluid percussion model of experimental brain injury in the rat. *J Neurosurg* 67(1):110–119.

Droz, B. and C. P. Leblond (1962). Migration of proteins along the axons of the sciatic nerve. *Science* 137:1047–1048.

Eichenbaum, H. (1997). Declarative memory: Insights from cognitive neurobiology. *Annu Rev Psychol* 48:547–572.

Ellingson, B. M., R. J. Fijalkowski, et al. (2005). New mechanism for inducing closed head injury in the rat. *Biomed Sci Instrum* 41:86–91.

Engelborghs, K., J. Verlooy, et al. (1998). Temporal changes in intracranial pressure in a modified experimental model of closed head injury. *J Neurosurg* 89(5):796–806.

Fawcett, J. W. and R. A. Asher (1999). The glial scar and central nervous system repair. *Brain Res Bull* 49(6):377–391.

Feeney, D. M., M. G. Boyeson, et al. (1981). Responses to cortical injury: I. Methodology and local effects of contusions in the rat. *Brain Res* 211(1):67–77.

Finnie, J. W. and P. C. Blumbergs (2002). Traumatic brain injury. *Vet Pathol* 39(6):679–689.

Foda, M. A. and A. Marmarou (1994). A new model of diffuse brain injury in rats. Part II: Morphological characterization. *J Neurosurg* 80(2):301–313.

Friess, S. H., R. N. Ichord, et al. (2009). Repeated traumatic brain injury affects composite cognitive function in piglets. *J Neurotrauma* 26(7):1111–1121.

Fritz, H. G., B. Walter, et al. (2005). A pig model with secondary increase of intracranial pressure after severe traumatic brain injury and temporary blood loss. *J Neurotrauma* 22(7):807–821.

Gaetz, M. (2004). The neurophysiology of brain injury. *Clin Neurophysiol* 115(1):4–18.

Gavett, B. E., R. C. Cantu, et al. (2011). Clinical appraisal of chronic traumatic encephalopathy: Current perspectives and future directions. *Curr Opin Neurol* 24(6):525–531.

Gavrieli, Y., Y. Sherman, et al. (1992). Identification of programmed cell death in situ via specific labeling of nuclear DNA fragmentation. *J Cell Biol* 119(3):493–501.

Gennarelli, T. A., L. E. Thibault, et al. (1982). Diffuse axonal injury and traumatic coma in the primate. *Ann Neurol* 12(6):564–574.

Goldman, H., V. Hodgson, et al. (1991). Cerebrovascular changes in a rat model of moderate closed-head injury. *J Neurotrauma* 8(2):129–144.

Goodman, J. C., L. Cherian, et al. (1994). Lateral cortical impact injury in rats: Pathologic effects of varying cortical compression and impact velocity. *J Neurotrauma* 11(5):587–597.

Graham, D. I., R. Raghupathi, et al. (2000). Tissue tears in the white matter after lateral fluid percussion brain injury in the rat: Relevance to human brain injury. *Acta Neuropathol (Berl)* 99(2):117–124.

Gurdjian, E. S. (1975). *Impact Head Injury: Mechanistic, Clinical and Preventative Correlations*. Charles C Thomas, Springfield, IL.

Gurdjian, E. S. (1975). Re-evaluation of the biomechanics of blunt impact injury of the head. *Surg Gynecol Obstet* 140(6):845–850.

Gurdjian, E. S., V. R. Hodgson, et al. (1968). Significance of relative movements of scalp, skull, and intracranial contents during impact injury of the head. *J Neurosurg* 29(1):70–72.

Gurdjian, E. S., H. R. Lissner, et al. (1953). Quantitative determination of acceleration and intracranial pressure in experimental head injury; preliminary report. *Neurology* 3(6):417–423.

Gurdjian, E. S., H. R. Lissner, et al. (1954). Studies on experimental concussion: Relation of physiologic effect to time duration of intracranial pressure increase at impact. *Neurology* 4(9):674–681.

Gurdjian, E. S., V. L. Roberts, et al. (1966). Tolerance curves of acceleration and intracranial pressure and protective index in experimental head injury. *J Trauma* 6(5):600–604.

Guskiewicz, K. M., M. McCrea, et al. (2003). Cumulative effects associated with recurrent concussion in collegiate football players: The NCAA Concussion Study. *JAMA* 290(19):2549–2555.

Gutierrez, E., Y. Huang, et al. (2001). A new model for diffuse brain injury by rotational acceleration: I model, gross appearance, and astrocytosis. *J Neurotrauma* 18(3):247–257.

Hall, E. D. (1985). High-dose glucocorticoid treatment improves neurological recovery in head-injured mice. *J Neurosurg* 62(6):882–887.

Hall, E. D., Y. D. Bryant, et al. (2008). Evolution of post-traumatic neurodegeneration after controlled cortical impact traumatic brain injury in mice and rats as assessed by the de Olmos silver and fluorojade staining methods. *J Neurotrauma* 25(3):235–247.

Hamberger, A., D. C. Viano, et al. (2009). Concussion in professional football: Morphology of brain injuries in the NFL concussion model—Part 16. *Neurosurgery* 64(6):1174–1182.

Hamm, R. J., B. G. Lyeth, et al. (1993). Selective cognitive impairment following traumatic brain injury in rats. *Behav Brain Res* 59(1–2):169–173.

Hanstock, C. C., A. I. Faden, et al. (1994). Diffusion-weighted imaging differentiates ischemic tissue from traumatized tissue. *Stroke* 25(4):843–848.

Head, B. P. and P. Patel (2007). Anesthetics and brain protection. *Curr Opin Anaesthesiol* 20(5):395–399.

Heimer, L. (1995). *The Human Brain and Spinal Cord, Functional Neuroanatomy and Dissection Guide.* Springer-Verlag, New York.

Holsinger, T., D. C. Steffens, et al. (2002). Head injury in early adulthood and the lifetime risk of depression. *Arch Gen Psychiatry* 59(1):17–22.

Hoshino, S., S. Kobayashi, et al. (2003). Multiple immunostaining methods to detect traumatic axonal injury in the rat fluid-percussion brain injury model. *Neurol Med Chir (Tokyo)* 43(4):165–173; discussion 174.

Huh, J. W., H. L. Laurer, et al. (2002). Rapid loss and partial recovery of neurofilament immunostaining following focal brain injury in mice. *Exp Neurol* 175(1):198–208.

Inglese, M., S. Makani, et al. (2005). Diffuse axonal injury in mild traumatic brain injury: A diffusion tensor imaging study. *J Neurosurg* 103(2):298–303.

Johnson, V. E., W. Stewart, et al. (2010). Traumatic brain injury and amyloid-beta pathology: A link to Alzheimer's disease? *Nat Rev Neurosci* 11(5):361–370.

Jordan, B. D. (2000). Chronic traumatic brain injury associated with boxing. *Semin Neurol* 20(2):179–185.

Jordan, B. D., N. R. Relkin, et al. (1997). Apolipoprotein E epsilon4 associated with chronic traumatic brain injury in boxing. *JAMA* 278(2):136–140.

Kallakuri, S., J. M. Cavanaugh, et al. (2003). The effect of varying impact energy on diffuse axonal injury in the rat brain: A preliminary study. *Exp Brain Res* 148(4):419–424.

Kane, M. J., M. Angoa-Perez, et al. (2012). A mouse model of human repetitive mild traumatic brain injury. *J Neurosci Methods* 203(1):41–49.

Kelly, J. (1997). Sports-related recurrent brain injuries—United States. *MMWR* Report 46(10):224.

Koerner, I. P. and A. M. Brambrink (2006). Brain protection by anesthetic agents. *Curr Opin Anaesthesiol* 19(5):481–486.

Larsson, F., B. Winblad, et al. (2002). Psychological stress and environmental adaptation in enriched vs. impoverished housed rats. *Pharmacol Biochem Behav* 73(1):193–207.

Lewen, A., A. Fredriksson, et al. (1999). Behavioural and morphological outcome of mild cortical contusion trauma of the rat brain: Influence of NMDA-receptor blockade. *Acta Neurochir (Wien)* 141(2):193–202.

Lewis, S. B., J. W. Finnie, et al. (1996). A head impact model of early axonal injury in the sheep. *J Neurotrauma* 13(9):505–514.

Li, R., N. Fujitani, et al. (1998). Immunohistochemical indicators of early brain injury: An experimental study using the fluid-percussion model in cats. *Am J Forensic Med Pathol* 19(2):129–136.

Lifshitz, J., P. G. Sullivan, et al. (2004). Mitochondrial damage and dysfunction in traumatic brain injury. *Mitochondrion* 4(5–6):705–713.

Lighthall, J. W. (1988). Controlled cortical impact: A new experimental brain injury model. *J Neurotrauma* 5(1):1–15.

Lighthall, J. W., C. E. Dixon, et al. (1989). Experimental models of brain injury. *J Neurotrauma* 6(2):83–97.

Lighthall, J. W., H. G. Goshgarian, et al. (1990). Characterization of axonal injury produced by controlled cortical impact. *J Neurotrauma* 7(2):65–76.

Lindgren, S. and L. Rinder (1966). Experimental studies in head injury. II. Pressure propagation in "percussion concussion". *Biophysik* 3(2):174–180.

Lindgren, S. and L. Rinder (1967). Decompression in percussion concussion: Effects on "concussive response" in rabbits. J*Trauma* 7(4):493–499.

Love, S., D.N. Louis, and D.W Ellison (2008). *Greenfield's Neuropathology*. 8th edition. Taylor & Francis Group, Boca Raton, FL.

Lu, J., A. Marmarou, et al. (2010). A method for reducing misclassification in the extended Glasgow Outcome Score. *J Neurotrauma* 27(5):843–852.

Maas, A. I. and H. F. Lingsma (2008). New approaches to increase statistical power in TBI trials: Insights from the IMPACT study. *Acta Neurochir Suppl* 101:119–124.

Maas, A. I., N. Stocchetti, et al. (2008). Moderate and severe traumatic brain injury in adults. *Lancet Neurol* 7(8):728–741.

Macciocchi, S. N., J. T. Barth, et al. (1996). Neuropsychological functioning and recovery after mild head injury in collegiate athletes. N*eurosurgery* 39(3):510–514.

Makdissi, M., A. Collie, et al. (2001). Computerised cognitive assessment of concussed Australian Rules footballers. *Br J Sports Med* 35(5):354–360.

Margulies, S. and R. Hicks (2009). Combination therapies for traumatic brain injury: Prospective considerations. *J Neurotrauma* 26(6):925–939.

Marklund, N., A. Bakshi, et al. (2006). Evaluation of pharmacological treatment strategies in traumatic brain injury. *Curr Pharm Des* 12(13):1645–1680.

Marklund, N. and L. Hillered (2011). Animal modelling of traumatic brain injury in preclinical drug development: Where do we go from here? *Br J Pharmacol* 164(4):1207–1229.

Marmarou, A., M. A. Foda, et al. (1994). A new model of diffuse brain injury in rats. Part I: Pathophysiology and biomechanics. *J Neurosurg* 80(2):291–300.

Martland, H. (1928). Punch drunk. *JAMA* 91:1103–1107

Masel, B. E. and D. S. DeWitt (2010). Traumatic brain injury: A disease process, not an event. *J Neurotrauma* 27(8):1529–1540.

McCrea, M., J. P. Kelly, et al. (2002). Immediate neurocognitive effects of concussion. *Neurosurgery* 50(5):1032–1040; discussion 1040–1032.

McCrory, P., K. Johnston, et al. (2005). Summary and agreement statement of the 2nd International Conference on Concussion in Sport, Prague 2004. *Br J Sports Med* 39(4):196–204.

McCrory, P., W. Meeuwisse, et al. (2013). Consensus statement on Concussion in Sport - The 4th International Conference on Concussion in Sport held in Zurich, November 2012. *Phys Therapy Sport* 14(2):e1-e13.

McIntosh, T. K., L. Noble, et al. (1987). Traumatic brain injury in the rat: Characterization of a midline fluid-percussion model. *Cent Nerv Syst Trauma* 4(2):119–134.

McKee, A. C., T. D. Stein, et al. (2013). The spectrum of disease in chronic traumatic encephalopathy. *Brain* 136(Pt 1):43–64.

Mendez, M. F. (1995). The neuropsychiatric aspects of boxing. *Int J Psychiatry Med* 25(3):249–262.

Meyer, J. S. (1956). Studies of cerebral circulation in brain injury. III. Cerebral contusion, laceration and brain stem injury. *Electroencephalogr Clin Neurophysiol* 8(1):107–116.

Meyer, J. S. and D. Denny-Brown (1955). Studies of cerebral circulation in brain injury. II. Cerebral concussion. *Electroencephalogr Clin Neurophysiol* 7(4):529–544.

Mikulecka, A., H. Kubova, et al. (2004). Lamotrigine does not impair motor performance and spontaneous behavior in developing rats. *Epilepsy Behav* 5(4):464–471.

Morales, D. M., N. Marklund, et al. (2005). Experimental models of traumatic brain injury: Do we really need to build a better mousetrap? *Neuroscience* 136(4):971–989.

Morris, R. (1984). Developments of a water maze procedure for studying spatial learninng in the rat. *J Neurosci Methods* 11(1):47–60.

Mortimer, J. A., C. M. van Duijn, et al. (1991). Head trauma as a risk factor for Alzheimer's disease: A collaborative re-analysis of case-control studies. EURODEM Risk Factors Research Group. *Int J Epidemiol* 20 Suppl 2:S28–35.

Neselius, S., H. Brisby, et al. (2012). CSF-biomarkers in Olympic boxing: Diagnosis and effects of repetitive head trauma. *PLoS One* 7(4):e33606.

Neselius, S., H. Zetterberg, et al. (2013). Olympic boxing is associated with elevated levels of the neuronal protein tau in plasma. *Brain Inj* 27(4):425–433.

Nilsson, B., U. Ponten, et al. (1977). Experimental head injury in the rat. Part 1: Mechanics, pathophysiology, and morphology in an impact acceleration trauma model. *J Neurosurg* 47(2):241–251.

Ommaya, A. K., P. Yarnell, et al. (1967). Scaling of experimental data on cerebral concussion in sub-human primates to concussion threshold for man. 11th Stapp Car Crash Conference, SAE 670906. Society of Automotive Engineers, Warrendale, PA, 73–81.

Pellman, E. J., M. R. Lovell, et al. (2004). Concussion in professional football: Neuropsychological testing—Part 6. *Neurosurgery* 55(6):1290–1303; discussion 1303–1295.

Pellman, E. J., J. W. Powell, et al. (2004). Concussion in professional football: Epidemiological features of game injuries and review of the literature—Part 3. *Neurosurgery* 54(1):81–94; discussion 94–86.

Pellman, E. J., D. C. Viano, et al. (2005). Concussion in professional football: Players returning to the same game—Part 7. *Neurosurgery* 56(1):79–90; discussion 90–72.

Pellman, E. J., D. C. Viano, et al. (2003a). Concussion in professional football: Location and direction of helmet impacts-Part 2. *Neurosurgery* 53(6):1328–1340; discussion 1340–1321.

Pellman, E. J., D. C. Viano, et al. (2003b). Concussion in professional football: Reconstruction of game impacts and injuries. *Neurosurgery* 53(4):799–814.

Pettus, E. H. and J. T. Povlishock (1996). Characterization of a distinct set of intra-axonal ultrastructural changes associated with traumatically induced alteration in axolemmal permeability. *Brain Res* 722(1–2):1–11.

Plassman, B. L., R. J. Havlik, et al. (2000). Documented head injury in early adulthood and risk of Alzheimer's disease and other dementias. *Neurology* 55(8):1158–1166.

Ponsford, J., K. Draper, et al. (2008). Functional outcome 10 years after traumatic brain injury: Its relationship with demographic, injury severity, and cognitive and emotional status. *J Int Neuropsychol Soc* 14(2):233–242.

Posmantur, R. M., J. K. Newcomb, et al. (2000). Light and confocal microscopic studies of evolutionary changes in neurofilament proteins following cortical impact injury in the rat. *Exp Neurol* 161(1):15–26.

Povlishock, J. T. and H. A. Kontos (1985). Continuing axonal and vascular change following experimental brain trauma. *Cent Nerv Syst Trauma* 2(4):285–298.

Powell, J. W. and K. D. Barber-Foss (1999). Traumatic brain injury in high school athletes. *JAMA* 282(10):958–963.

Prins, M. L., A. Hales, et al. (2010). Repeat traumatic brain injury in the juvenile rat is associated with increased axonal injury and cognitive impairments. *Dev Neurosci* 32(5–6):510–518.

Raghupathi, R. and S. S. Margulies (2002). Traumatic axonal injury after closed head injury in the neonatal pig. *J Neurotrauma* 19(7):843–853.

Raghupathi, R., M. F. Mehr, et al. (2004). Traumatic axonal injury is exacerbated following repetitive closed head injury in the neonatal pig. *J Neurotrauma* 21(3):307–316.

Raivich, G., M. Bohatschek, et al. (1999). Neuroglial activation repertoire in the injured brain: Graded response, molecular mechanisms and cues to physiological function. *Brain Res Brain Res Rev* 30(1):77–105.

Raymont, V., A. Greathouse, et al. (2008). Demographic, structural and genetic predictors of late cognitive decline after penetrating head injury. *Brain* 131(Pt 2):543–558.

Reire, P. J. (1986). Gliosis following CNS injury: The anatomy of astrocytic scars and their influences on axonal elongation. Celluar Neurobiology: A series, Astrocytes, Cell Biology and Pathology of Astrocytes. Segay Fedoroff and Antonia Vernadokis. United States of America, Academic press, INC. Orlando, Florida 32887. 3:263–324.

Reire, P. L. (1986). Gliosis following CNS injury: The anatomy of astrocytic scars and their influences on axonal elongation. In: *Celluar Neurobiology: A Series, Astrocytes, Cell Biology and Pathology of Astrocytes*, S. Fedoroff and A. Vernadokis (eds.). Academic Press, Orlando, FL. pp. 263–324.

Roberts, A. J. (1969). *Brain Damage in Boxers*. London: Pitman Medical Scientific Publications.

Roberts, G. W., D. Allsop, et al. (1990). The occult aftermath of boxing. *J Neurol Neurosurg Psychiatry* 53(5):373–378.

Rooker, S., P. G. Jorens, et al. (2003). Continuous measurement of intracranial pressure in awake rats after experimental closed head injury. *J Neurosci Methods* 131(1–2):75–81.

Ross, D. T., D. F. Meaney, et al. (1994). Distribution of forebrain diffuse axonal injury following inertial closed head injury in miniature swine. *Exp Neurol*126(2):291–299.

Sallis, R. E. and K. Jones (2000). Prevalence of headaches in football players. *Med Sci Sports Exerc* 32(11):1820–1824.

Schmidt, R. H. and M. S. Grady (1993). Regional patterns of blood-brain barrier breakdown following central and lateral fluid percussion injury in rodents. *J Neurotrauma* 10(4):415–430.

Schouten, J. W. (2007). Neuroprotection in traumatic brain injury: A complex struggle against the biology of nature. *Curr Opin Crit Care* 13(2):134–142.

Serbest, G., M. F. Burkhardt, et al. (2007). Temporal profiles of cytoskeletal protein loss following traumatic axonal injury in mice. *Neurochem Res* 32(12):2006–2014.

Shultz, S. R., F. Bao, et al. (2012). Repeated mild lateral fluid percussion brain injury in the rat causes cumulative long-term behavioral impairments, neuroinflammation, and cortical loss in an animal model of repeated concussion. *J Neurotrauma* 29(2):281–294.

Shultz, S. R., D. F. MacFabe, et al. (2011). A single mild fluid percussion injury induces short-term behavioral and neuropathological changes in the Long-Evans rat: Support for an animal model of concussion. *Behav Brain Res* 224(2):326–335.

Shultz, S. R., D. F. MacFabe, et al. (2012). Sub-concussive brain injury in the Long-Evans rat induces acute neuroinflammation in the absence of behavioral impairments. *Behav Brain Res* 229(1):145–152.

Silver, J. M., T. W. M., Stuart C. Yudofsky (2010). *Textbook of Traumatic Brain Injury*. American Psychiatric Publishing, Arlington, VA.

Simpson, R. H., D. S. Berson, et al. (1985). The diagnosis of diffuse axonal injury in routine autopsy practice. *Forensic Sci Int* 27(4):229–235.

Slobounov, S., E. Slobounov, et al. (2007). Differential rate of recovery in athletes after first and second concussion episodes. *Neurosurgery* 61(2):338–344; discussion 344.

Smith, D. H., D. F. Meaney, et al. (2003). Diffuse axonal injury in head trauma. *J Head Trauma Rehabil* 18(4):307–316.

Smith, D. M. and S. J. Mizumori (2006). Learning-related development of context-specific neuronal responses to places and events: The hippocampal role in context processing. *J Neurosci* 26(12):3154–3163.

Stern, R. A., D. O. Riley, et al. (2011). Long-term consequences of repetitive brain trauma: Chronic traumatic encephalopathy. *PM & R* 3(10 Suppl 2):S460–467.

Stoll, G., S. Jander, et al. (1998). Inflammation and glial responses in ischemic brain lesions. *Prog Neurobiol* 56(2):149–171.

Sullivan, H. G., J. Martinez, et al. (1976). Fluid-percussion model of mechanical brain injury in the cat. *J Neurosurg* 45(5):521–534.

Tagliaferri, F., C. Compagnone, et al. (2006). A systematic review of brain injury epidemiology in Europe. *Acta Neurochir (Wien)* 148(3): 55–268; discussion 268.

Thompson, H. J., J. Lifshitz, et al. (2005). Lateral fluid percussion brain injury: A 15-year review and evaluation. *J Neurotrauma* 22(1):42–75.

Tokuda, T., S. Ikeda, et al. (1991). Re-examination of ex-boxers' brains using immunohistochemistry with antibodies to amyloid beta-protein and tau protein. *Acta Neuropathol* 82(4):280–285.

Tornheim, P. A., B. H. Liwnicz, et al. (1983). Acute responses to blunt head trauma. Experimental model and gross pathology. *J Neurosurg* 59(3):431–438.

Tornheim, P. A. and R. L. McLaurin (1981). Acute changes in regional brain water content following experimental closed head injury. *J Neurosurg* 55(3):407–413.

Vagnozzi, R., B. Tavazzi, et al. (2007). Temporal window of metabolic brain vulnerability to concussions: Mitochondrial-related impairment—Part I. *Neurosurgery* 61(2):379–388; discussion 388–379.

Viano, D. C., and P. Lovsund (1999). Biomechanics of brain and spinal cord injury: Analysis of neurophysiological experiments. *Crash Prev Inj Control* 1(1):35–43.

Viano, D. C., A. Hamberger, et al. (2009). Concussion in professional football: Animal model of brain injury—Part 15. *Neurosurgery* 64(6):1162–1173.

Viano, D. C. and I. V. Lau (1988). A viscous tolerance criterion for soft tissue injury assessment. *J Biomech* 21(5):387–399.

Vink, R., A. Young, et al. (2003). Neuropeptide release influences brain edema formation after diffuse traumatic brain injury. *Acta Neurochir Suppl* 86:257–260.

Walker, A. E., J. J. Kollross, and T. J. Case (1944). The physiological basis of concussion. *J Neurosurg* 1:103–116.

Wang, S., A. Hamberger, et al. (1994). Changes in neurofilament protein NF-L and NF-H immunoreactivity following kainic acid-induced seizures. *J Neurochem* 62(2):739–748.

Wang, Y., Y. Wei, et al. (2011). Tightly coupled repetitive blast-induced traumatic brain injury: Development and characterization in mice. *J Neurotrauma* 28(10):2171–2183.

Wang, Y. J., T. Shimura, et al. (1997). [A lateral fluid percussion model for the experimental severe brain injury and a morphological study in the rats]. *Nihon Ika Daigaku Zasshi* 64(2):172–175.

Weber, J. T. (2007). Experimental models of repetitive brain injuries. *Prog Brain Res* 161:253–261.

Yu, A. C., Y. L. Lee, et al. (1993). Astrogliosis in culture: I. The model and the effect of antisense oligonucleotides on glial fibrillary acidic protein synthesis. *J Neurosci Res* 34(3):295–303.

Zetterberg, H., M. A. Hietala, et al. (2006). Neurochemical aftermath of amateur boxing. *Arch Neurol* 63(9):1277–1280.

Zetterberg, H., F. Tanriverdi, et al. (2009). Sustained release of neuron-specific enolase to serum in amateur boxers. *Brain Inj* 23(9):723–726.

Zhang, Y. L., P. B. Zhang, et al. (2006). Effects of ketamine-midazolam anesthesia on the expression of NMDA and AMPA receptor subunit in the peri-infarction of rat brain. *Chin Med J (Engl)* 119(18):1555–1562.

47 The Problem of Neurodegeneration in Cumulative Sports Concussions

Emphasis on Neurofibrillary Tangle Formation

Vassilis E. Koliatsos and Leyan Xu

CONTENTS

47.1 INTRODUCTION

Traumatic brain injury (TBI) has been a common cause of morbidity and mortality throughout history, but has had interesting twists in the industrial era and in the world of aggressive sports or modern warfare. The American public is especially troubled by the problem of hundreds of thousands of injured veterans returning from the lengthy post-9/11 campaigns and is also fearful that repeat concussions experienced in contact sports played by millions of young athletes may result in progressive neurodegenerative disease and dementia. The problem of traumatic degeneration, termed chronic traumatic encephalopathy (CTE), has been long recognized in the world of boxing, but has recently come to the forefront because of several highly publicized deaths of popular NFL protagonists. Despite extensive publicity, the real risk of CTE among amateur and professional players has not been measured or adequately characterized and notions derived from autopsy studies, although useful for understanding mechanisms, cannot give an accurate picture of the range of outcomes after repeat concussions and are limited because of ascertainment bias. It is imperative that prospective studies are designed and deployed to the problem, but such approaches will take a long time to bear fruits in view of the long incubation time of key pathologies. Meanwhile, emphasis on molecular and cellular mechanisms of deposition of a key protein, the microtubule-associated protein tau, as well as the construction of relevant animal models, may help define cause-and-effect relationships, shed light into mechanisms, and generate ideas about biomarkers and therapeutic targets. Imports from extensive work in other degenerative tauopathies such as frontotemporal degeneration may greatly facilitate innovation in this area.

47.1.1 Clinico-Pathological Considerations

47.1.1.1 The Pathological Signature of CTE: Differentiation from Other Common Types of Traumatic Brain Injury

There are between 2 and 3 million new cases of TBI in the United States every year, most of them from motor vehicle accidents (MVA) and falls. The majority of cases of TBI are concussions that cause brief or no changes in mental status and a constellation of symptoms that usually improve over a few weeks or months. However, a substantial number of TBI events are associated with chronic disability because of permanent damage to the brain or, in some cases, progressive disease.

Classical forms of severe TBI include focal contusions often caused by falls (Courville, 1942) and diffuse axonal lesions (diffuse axonal injury, DAI) usually associated with dynamic stretching of axons with rotational acceleration as it occurs in motor vehicle crashes (Adams et al., 1982; Strich, 1961). The association of chronic progressive TBI with boxing (dementia pugilistica) has also been well recognized since the 1930s (Corsellis et al., 1973; Martland, 1928; Millspaugh, 1937). The U.S. military has had a major exposure to TBI risk in the post-9/11 deployments, although exact numbers are difficult to establish and figures vary among estimates (Congressional Research Service, http://www.fas.org/sgp/crs/natsec/RS22452.pdf; Defense and Veterans Brain Injury Center, http://www.dvbic.org/dod-worldwide-numbers-tbi). Much of this exposure relates to explosion (blast) from mines or improvised explosive devices in the Iraq and Afghanistan war theaters (Warden, 2006). A 100-year-old problem recognized since World War I (Mott, 1916), blast injury to brain has become the signature health problem of recent wars (Warden, 2006).

A recent trend that has attracted much attention is an increasing number of cases of progressive tauopathy in professional and amateur athletes with careers in collision sports, especially football. These cases have been linked to repeat concussions and are identical to cases of dementia pugilistica, with which they are classified under the rubric of CTEs (McKee et al., 2009; Omalu et al., 2005, 2006). The recent discovery of CTE lesions in younger subjects with histories of subconcussive TBI (Schwarz, 2010) suggests a broader risk in all scenarios where repeat mild TBI takes place, including the military. Repeat mild blast TBI is also very common in the Iraq and Afghanistan war theaters and the Defense and Veterans Brain Injury Center has recently developed return-to-duty guidelines to prevent additional TBI in soldiers with concussive histories. The public concern over repeat concussions is growing not only because of the risk among National Football League (NFL) professionals and active-duty soldiers or veterans, but also because of the exposure of millions of nonprofessional athletes playing contact and collision sports such as football, mixed martial arts, hockey, and rugby. It has been argued that high school or college football is even more aggressive than professional NFL play (Broglio et al., 2009).

Neuropathologies associated with various types of TBI present important differences, but also overlap with each other. For example, MVA- and fall-associated TBI present primarily as static encephalopathies due to diffuse axonal or focal parenchymal lesions, although there is some evidence of limited progression, especially with MVA-associated TBI (Buki and Povlishock, 2006). With contusions, more so than with DAI, there are some generally accepted correlations between neuropathology and neuropsychiatric morbidity, although patterns are far from linear (Lishman et al., 1968).

Axonal injury is a common denominator in many types of TBI including, besides DAI, focal contusions (Mac Donald et al., 2007), blast injuries, and the early stages of CTE (Dr. Juan Troncoso, personal communication) (Figure 47.1). Importantly, the prevalence of DAI is increasingly recognized across various types of TBI, including mild injuries (Mittl et al., 1994).

In contrast to focal contusions or DAI that are generally viewed as static encephalopathies and the generally benign course of single concussions, CTE is a progressive tauopathy that more closely resembles other neurodegenerative tauopathies such as frontotemporal degeneration (FTD) and Alzheimer disease (AD) or progressive supranuclear palsy (PSP) (McKee et al., 2009). This is perhaps the most characteristic signature of CTE and is the source of some confusion and controversy, but also creates great opportunities for research into mechanisms.

47.1.1.2 The Clinical Signature of CTE: The Predominance of Neuropsychiatric Symptoms

Although it is still too early to have a full picture of the clinical signature of CTE, existing reports from cases that have come to autopsy (McKee et al., 2009, 2012) and a few recent highly publicized cases have emphasized the early presence of psychiatric symptoms. Clinical presentation in CTE is an important matter because, in the absence of biological markers, it may help settle issues of diagnosis and prognosis and facilitate early interventions. Common symptoms attributed to CTE are mood changes (mainly depression), irritability with aggressive/violent outbursts, paranoia, poor insight/judgment, apathy, memory loss and other cognitive deficits, parkinsonism, dysarthria and other speech abnormalities, and gait ataxia. Remarkably, causes of death include suicide and the often-violent result of poor judgment, such as an accidental gunshot while cleaning a gun (J. Grimsley) or falling from the back of a moving truck on which the patient jumped in the course of chasing his fiancé (C. Henry).

The prevalence, in the young adult population, of psychiatric symptoms, especially those related to mood and even personality disorders, is relatively high. If we accept, based on existing clinical information, that a long period in the early stage of the CTE lifecycle is featured almost exclusively by mood and behavioral symptoms. A key question is whether the clinical presentation is merely the corresponding common idiopathic psychiatric illness or the relatively uncommon organic psychiatric illness that will prove, in the end, to be the harbinger of CTE. For example, if a young NFL player or a young/middle-aged NFL veteran presents with depression, how do we know that the problem is a sign of CTE versus common major depression that prevails in a community sample?

In general, clinical neuropsychiatric presentations on the left column of Table 47.1 are consistent with idiopathic psychiatric (i.e., mental) illness, whereas presentations and symptoms on the right column are more typical of neurological illness. Based on the high prevalence of non-manic impulsivity, aggressiveness, and cognitive deficits in several studied or publicized cases of CTE, it appears that many patients have "organic" features that are common

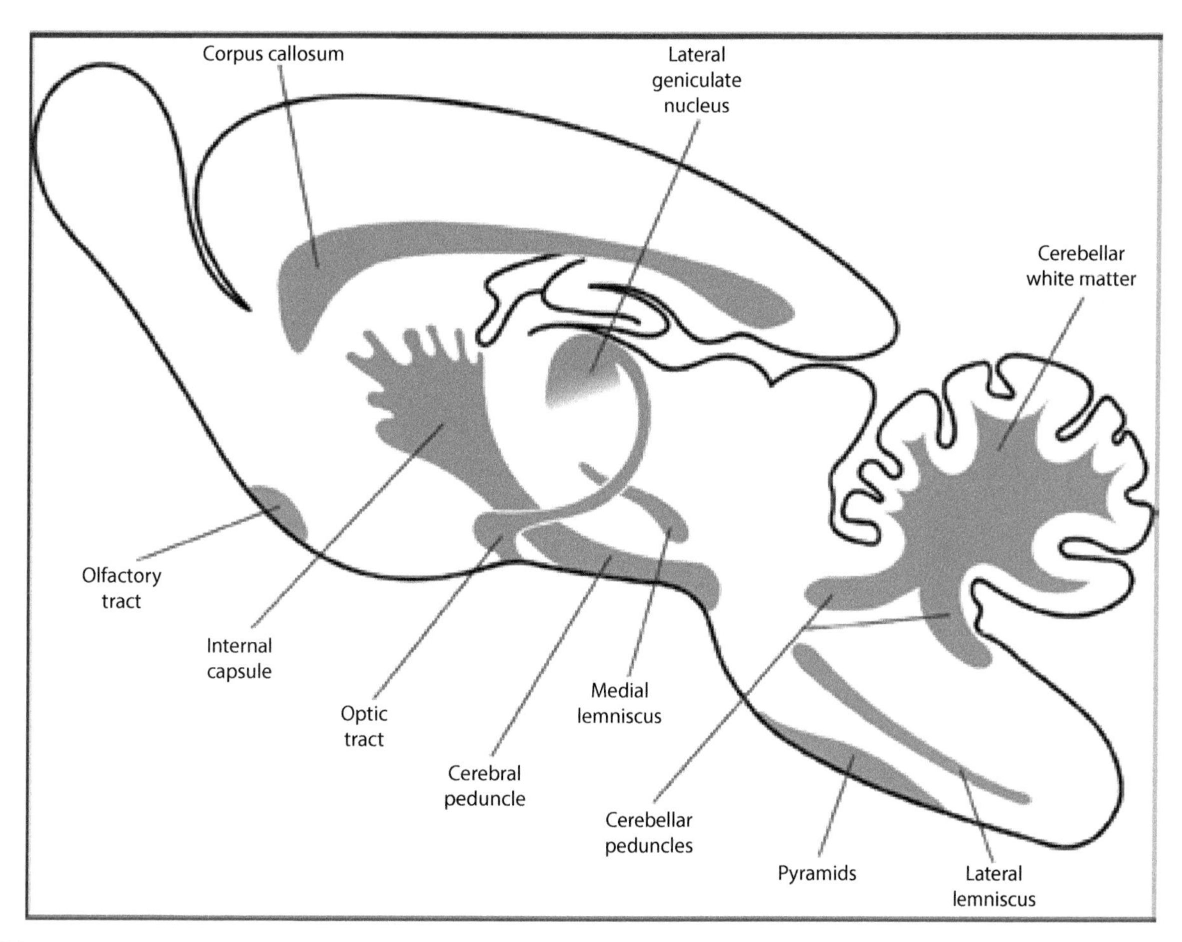

FIGURE 47.1 A sketch of fiber pathways affected in blast injury (gray) that recapitulate pathways at risk in many forms of DAI. Pathways include corticospinal, callosal, visual, somatosensory, and cerebellar circuits. (From Koliatsos et al., *J. Neuropathol. Exp. Neurol.* 70(5), 399–416, 2011. With permission.)

TABLE 47.1
Psychiatric Presentations Differentiating Idiopathic (Mental) from Organic (Neurological) Illness

Idiopathic Illness	Secondary (Organic) Illness
Depression	Apathy without depression
Short-lived mania	Pervasive impulsivity without mania
Psychotic episodes with organized delusions or bizarre hallucinations	Pervasive aggressiveness without delusional thinking
Relative paucity of cognitive deficits	Regression of social comportment without thought disorganization
Lack or paucity of neurological signs	Cognitive deficits
	Focal or diffuse neurological signs

with neuropsychiatric illnesses arising from frontal lobe lesions and neurodegenerative dementias (especially some variants of FTD).

47.1.1.3 CTE as a Distinct Illness: Some Challenges

One of the features that distinguish CTE from other forms of TBI is its very nature as a neurodegenerative disease. A challenging attribute in this regard is the delay between the occurrence of injury and the onset of symptoms. It has been noted that the less-than-decade incubation time of football-related CTE is shorter than that of dementia pugilistica, which is much as 15 years (McKee et al., 2009). The incubation time in the younger athletes may be a separate matter: for example, the brain of the 21-year-old University of Pennsylvania defensive lineman Owen Thomas showed some evidence of tau aggregation (Schwarz, 2010), although the diagnosis of CTE was far from certain in this young man. Whatever the scenario, there are several years intervening between the suspected cause and the degenerative outcome in all these cases.

The delay between concussive or even subconcussive hits and the appearance of symptoms raises classical cause-and-effect questions. Mood changes overlap with extremely common idiopathic psychiatric disorders such as major depression. Cognitive/personality symptoms overlap with other neurodegenerative diseases, especially presenile forms of FTD (Figure 47.2).

The neuropathological profile also raises issues of distinction. Although CTE does not resemble brain contusions or DAI, it overlaps extensively with other neurodegenerative tauopathies such as AD, FTD, and PSP. As we will see later, this overlap also creates enormous opportunities of importing possible mechanisms as well as models of disease from better-studied disorders. Meanwhile, the question that arises is whether trauma is both a necessary and sufficient condition for tauopathy or merely shortens the incubation of other presenile tauopathies whose course has been already initiated by genetic or other, nontrauma related, causes. In the same vein, does CTE also increase the risk for depression? The sketch in Figure 47.2 is meant to illustrate the issue of overlap, but arrows in the blue and purple sets depict the potential role of CTE in altering the prevalence and/or time course of more common disorders.

On a related topic, what is the role of genes? Is it possible that the same genes that cause familial FTD also play a causative role in CTE? And what is the role of severity of exposure measured by frequency of hits or hit (impact) nature and intensity? All these issues must be examined in light of the fact that cause-and-effect models for accumulating exposure to an offending agent or condition can be quite complex (Bandeen-Roche et al., 1999).

There are several well-established epidemiological methods to address cause-and-effect questions in medicine. From the weakest to the strongest level of evidence, these methods include case series, case-control studies, and cohort studies. All these methodologies have shortcomings and practical limitations when applied to the problem of CTE. Case series have well-known selection bias and they can best generate Oxford level-3 evidence. One such study has shown an earlier onset of AD in NFL players, although there was no association between recurrent concussions and incidence of AD (Guskiewicz et al., 2005). Another study from the University of Michigan (Institute for Social Research) based on a phone survey has shown a significant increase in memory complaints and dementia in NFL retirees, especially in the 30–49-year-old range (Weir et al., 2009). Case-control studies also have well-characterized confounding problems (i.e., problems of separation of the chooser from the choice) and cannot prove causation, although they have been traditionally associated with breakthroughs in medicine such as the first evidence that smoking may cause lung cancer. At best, case-control studies can generate level 2 evidence. Cohort (prospective) studies can generate level 1 evidence, but in the case of CTE, they would be hard-pressed to resolve between a relatively uncommon condition such as CTE and common chronic conditions such as idiopathic mood disorders (major depression) and, in the older age spectrum, AD and FTD, or other tauopathies. An earlier prospective NFL study drew severe criticism and was terminated; there are ongoing efforts to restart this work on new foundations.

47.1.2 Neurobiological Approaches to CTE

47.1.2.1 Lessons from Animal Models and Cellular/Molecular Neuropathology

In view of the several challenges of clinical-epidemiological studies, the question should turn to modeling. Can basic research from animal models and cellular and molecular biology provide insights and explanations that may help resolve issues critical to CTE, especially the problem of the link between trauma and degeneration? There are several steps in the sequence leading from trauma to cytoskeletal pathology to neuronal degeneration/cell death and here we will briefly examine each one of them in a logical order.

47.1.2.1.1 Can Injury Cause Cytoskeletal Abnormalities?

The neuronal cytoskeleton is the backbone that stabilizes and maintains the unusual geometry of neurons, including a complex dendritic tree and axons whose cytoplasm (axoplasm) often has volume exceeding multiple times that of the cell body. As shown by studies on the retrograde axonal transport of trophic peptides (Ehlers et al., 1995), cytoskeletal integrity is not only important for the structure and function of neurons

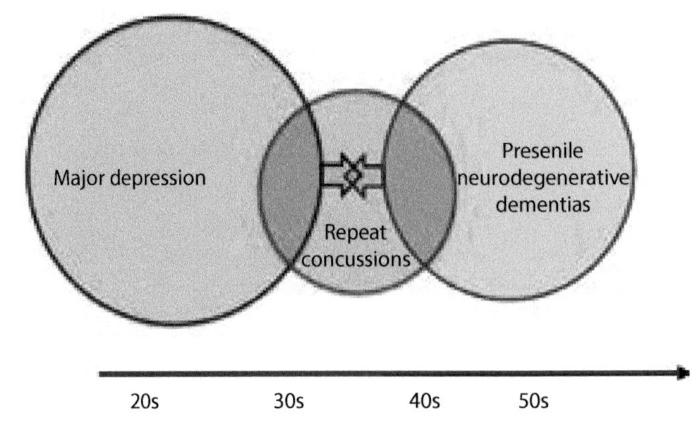

FIGURE 47.2 A sketch illustrating the problem of ascribing all or most mood or cognitive changes in patients with histories of multiple concussions to entities other than common mood disorders or neurodegenerative diseases. The problem is especially serious with major depression that is prevalent in young age and is very likely to coexist in individuals playing football and other contact sports.

as components of particular networks, but also for neuronal survival. In earlier studies that were first performed starting in the late 1980s, it became clear that various cytoskeletal constituents, especially neurofilaments, undergo alterations after axonal lesions, primarily in the form of aberrant or ectopic phosphorylation (Koliatsos et al., 1989). Tau abnormalities as a result of lesions are less well established, although some evidence for accumulation and phosphorylation exists with both repetitive mild injury (Kanayama et al., 1996) and a single lateral fluid percussion impact (Hoshino et al., 1998).

47.1.2.1.2 *Can Injury Cause Neuronal Cell Death?*

This association is very well established with proximal axonal lesions that cause retrograde degeneration of neurons in the peripheral nervous system, a type of degeneration that is reversible by trophic factors (Koliatsos and Price, 1996), but is also seen in the CNS with denervation lesions (Koliatsos et al., 2004). A correlation between injury and neuronal cell death is also established in focal models of traumatic brain injury, although it is less well shown in models of DAI.

47.1.2.1.3 *Is Tau Pathology Toxic to Neurons?*

There is some evidence that accumulation of hyperphosphorylated tau within neurons is associated with activation of caspase 3 (Gastard et al., 2003) and various enzymatic downstream events, including cleavage of alpha II spectrin (fodrin) to 120-kDa species (Koliatsos et al., 2006). Recent studies in Tg4510 mice that used cranial windows to image cortical neurons have indicated that the caspase-3 activation step comes first and tau cleavage and aggregation with tangle formation follows (de Calignon et al., 2010). The breakthrough finding here is that tangle formation is "off cycle" with respect to the apoptotic pathway and a "marker" or a consequence, rather than a direct cause, of neuronal degeneration.

47.1.2.1.4 *Can Repeat Mild Closed Injury Cause a Progressive Tauopathy with Brain Atrophy and Neuronal Cell Death?*

The answer is unknown. There have been very few attempts to address this issue in its entirety. In one study that used a model of repetitive closed-head injury on transgenic mice expressing the shortest human tau isoform (T44), no differences were found in severity of tauopathy and degeneration between noninjured and injured subjects (Yoshiyama et al., 2005). In more recent studies, there is evidence that repeat injury may accelerate some aspects of tauopathy in genetically predisposed subjects (Ojo et al., 2013), but the role of concussion in initiating tauopathy in genetically normal subjects remains elusive.

47.1.2.2 The Microtubule-Associated Protein Tau: Essential Biochemistry and Role in Neurodegenerative Disease

Tau is a microtubule-associated protein that occurs in six alternatively spliced isoforms differing by number of inserts at the N terminus (one or two) and the microtubule-binding domain (three or four) (Spillantini and Goedert, 1998). Normal tau is soluble and serves to stabilize microtubules (i.e., organelles critical for axonal transport). The physiological role of tau depends on a perpetual cycle between phosphorylated and nonphosphorylated states, a mechanism allowing for the smooth movement of cargo, comprising the motor proteins and associated vesicles, down the axon (Ballatore et al., 2007).

Perhaps as a result of its critical importance in axonal transport, tau pathology has been found in more than 20 neurodegenerative diseases including AD, FTD, parkinsonism-plus diseases such as PSP, and traumatic encephalopathies (Ludolph et al., 2009). In all of these disorders ("tauopathies"), tau protein displays misfolding, hyperphosphorylation, and aggregation, events leading to its critical transformation from the normal soluble state to a fibrillar insoluble state featured by beta conformation.

47.1.2.3 Molecular Mechanisms of Tauopathies: Tau as a Protein Prone to Aggregation

A number of studies in transgenic animals harboring mutations associated with FTD with parkinsonism mapping to chromosome 17 (P301L, V337M, others) and molecular studies summarized in the next section have indicated that tau is a protein with great proneness to aggregation (Wolfe, 2009). Aggregation may be the result of mutation, a particular type of enzymatic processing of tau, or both.

Transgenic tau associated with FTD-17 aggregates much more than normal tau. The microtubule-binding (repeat-domain) region aggregates faster than full-length tau and the pathogenic role of this region is highlighted by the fact that nearly all mutations associated with FTD-17 cluster in this region (Wolfe, 2009). The proneness of the repeat-domain tau to aggregation is increased substantially when the flanking regions are removed (see the following section).

47.1.2.4 Mechanisms of Tau Aggregation Are Being Increasingly Understood

As mentioned previously, repeat-domain tau fragments with FTD-17 mutations are especially predisposed to aggregation and beta structure. Cleavage of whole tau to repeat-domain fragments may be initiated by the action of a thrombin-like endogenous protease and then N- and C-terminal truncation of repeat-domain tau in an orderly fashion uncovered by Mandelkow and colleagues in seminal in vitro experiments (Khlistunova et al., 2006; Wang et al., 2007). These repeat-domain fragments nucleate full-length tau with a speed that is substantially higher in mutant than wild-type tau and this causes a slow aggregation process that is toxic to cells.

Aggregation is modulated by, but may not be dependent on, tau phosphorylation. Importantly, the transition from soluble to aggregated tau is a dynamic process that can be restored to normal with antiaggregation agents, for example N-phenylamine compounds and others (Khlistunova et al., 2006).

47.1.2.5 Aggregated Tau Enters Neurons and May Spread from One to Another in a Prion-Like Fashion

The work of Mandelkow and colleagues showing mechanisms of aggregation and the toxicity of aggregated tau has been nicely complemented by additional in vitro and in vivo work. Frost and Diamond have shown that tau aggregates are taken up from outside of cells and, acting as fibrillogenic nuclei, recruit normal tau onto their mass. Newly aggregated tau can be transferred from one cell to another (Frost and Diamond, 2010; Frost et al., 2009).

Clavaguera and colleagues have used transgenic mice that express FTD-P301S tau and develop fibrillar tau inclusions and neurodegeneration (2009). Mice expressing wild-type human tau have normal brains. These investigators were able to transmit tauopathy by injecting brain extracts from transgenic mice into the brains of wild-type mice. Importantly, they showed that pathology can spread from the injection site to other brain regions, presumably via synaptic contacts. Some of these findings have been confirmed using site-specific expression of tau transgenes (de Calignon et al., 2012; Liu et al., 2012)

47.1.3 Animal Modeling of CTE: Shedding Light Where Other Approaches Stumble

47.1.3.1 Animal Models of CTE Require a Central Hypothesis

Animal models can be extremely powerful tools to explore cause-and-effect relationships between traumatic insults and progressive neurodegeneration and, eventually, to serve as vehicles for therapeutic targeting and experimental therapeutics or for biomarker discovery and validation. To construct these models, we need a basic hypothesis or schema of disease cause and progression.

Based on various lines of clinical, epidemiological, and neurobiological data presented above we propose that CTE may arise from the effects of mild, yet repetitive head injury that occurs during contact sports, but that these outcomes are enabled or facilitated by a genetic predisposition of exposed individuals to form toxic tau aggregates within nerve cells. In other words, we propose a diathesis-stress approach in which diathesis (genetic risk) is configured as the genetic propensity of the subject's tau to form toxic aggregates in the brain and stress is defined as repeated, mild, noncontusive TBI. The genetic propensity to aggregation may be conferred by the expression of FTD with parkinsonism mapping to chromosome 17–linked mutations such as mutations P301L, P301S, V337M, the deletion mutation ΔK280, or the expression of particular tau isoforms (Caffrey and Wade-Martins, 2012; Wolfe, 2009). The stress contribution may be mediated by the increased likelihood of brain tau to be cleaved into multiple-repeat domain fragments under conditions of repeat concussive injury.

Repetitive concussive injury can cause axonal disruption/damage that is a very common phenomenon across various types of TBI (Figure 47.3). Such injury may allow the "leaking" of tau in the extracellular space (Figure 47.3a, step 1) and, in conjunction with microvascular damage or microhemorrhage or mere disruptions of the blood–brain barrier (see "leaky" capillary in sketch), may expose entire tau molecules to endogenous protease (thrombin) cleavage and truncation to repeat-domain, pro-aggregation tau (Figure 47.3a, step 2), as

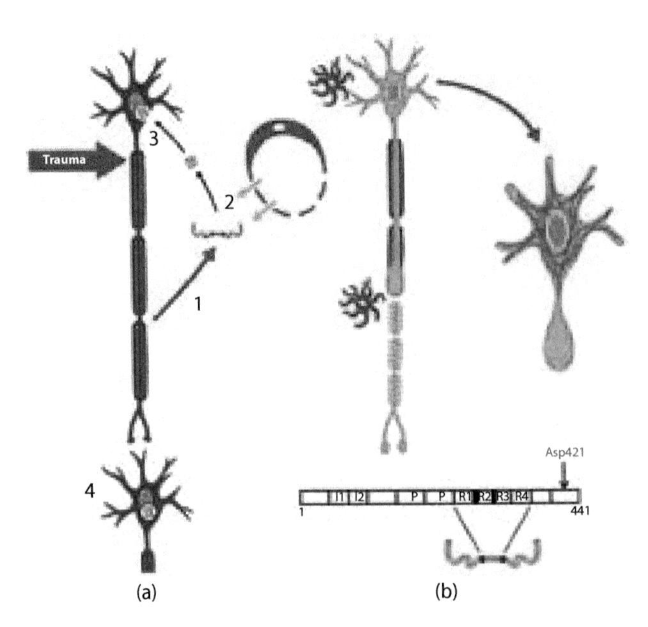

FIGURE 47.3 Two possible tau cascades initiated by repeat mild injury involve externalization of intact tau from its microtubular binding sites and digestion by endogenous proteases into multiple-repeat, pro-aggregation fragments (a). Axotomy followed by retrograde accumulation of tau (green) and subsequent phosphorylation-aggregation via fractalkine-IL1 signaling between axotomized neurons and activated microglia (b); caspase digestion at *Asp421* (bottom) may be common in both scenarios. For further explanation, refer to the main text.

described by Mandelkow and colleagues (Khlistunova et al., 2006; Wang et al., 2007). In this "vascular" model of tauopathy, tau mutations such as the ones associated with FTD can further facilitate aggregation. These mutations are the best characterized genetic causes of progressive tauopathy and appear to exert their pathogenic effect by cleaving tau to multiple-repeat fragments with high aggregation potential that self-assemble in fibrils and subsequently recruit and corrupt normal tau (Wang et al., 2007). Once tau aggregates form, they can be internalized from the extracellular to intracellular compartment (Figure 47.3a, step 3) or become propagated from one cell to another via prion-like mechanisms long after the injurious insults have ceased. Tau tangles may spread along synaptic circuits or via extrusion from dying neurons: as entire neural systems become involved, individuals develop symptoms (Figure 47.3a, step 4). The lengthy interval associated with the transition from aggregation to toxicity (Wang et al., 2007) or the propagation along circuits (Clavaguera et al., 2009) is consistent with the long incubation time required for clinically manifest CTE (McKee et al., 2009).

Another mechanism of tauopathy may be related to traumatic axonal injury (Figure 47.3b). Tau accumulates retrogradely in neuronal cell body. Neuronal injury activates adjacent microglia via fractalkine signaling and, in turn, activated microglia contributes to the phosphorylation of accumulated tau via IL1-p38 mitogen-activated protein K signaling (Bhaskar et al., 2010). Hyperphosphorylation increases the probability of fibrillization and, once this happens, steps 3 and 4 of Figure 47.3a are initiated. Asp 421 cleavage by caspases (Figure 47.3, bottom right) can contribute to either one of these two mechanisms.

In our laboratory, we have modified a well-characterized model of closed head injury (i.e. the impact acceleration model of Marmarou et al., 1994). Besides strictly replicating closed-head injury conditions, this model has been extensively characterized, during the past 15 years, to define mechanisms and treatments for TBI and can be modified to encompass a range of injury severity (Beaumont et al., 1999). Although the impact acceleration model can produce a graded, widespread injury involving neurons, astrocytes, axons, and the microvasculature, it does not cause focal damage regardless of injury severity. The standard impact acceleration model can be altered to generate repetitive mild injury and imitate the conditions of repeat concussion in contact sports.

47.2 CONCLUSIONS: CRITICAL ISSUES IN CTE AND THE ROLE OF EXPERIMENTAL NEUROBIOLOGY

As outlined in the beginning of the chapter, CTE may have neuropsychiatric symptoms and signs consistent with a frontal-limbic dementia with similarities to FTD, but also has a substantial early overlap with mood disorders. The neuropathological signature of CTE is that of a neurodegenerative tauopathy rather than other classical forms of focal or diffuse TBI (i.e., contusions and traumatic axonal injury). A key problem is that only a subpopulation of exposed individuals appears to develop CTE. The size of the exposed population is uncertain. The character and severity of injury is still being defined. The nature of genetic predisposition to the disorder is unclear. Epidemiological studies could easily stumble into problems of overlap with other diseases such as common forms of major depression and early-onset neurodegenerative disorders.

The CTE risk is a real threat to the community because of the popularity of contact and collision sports and is reminiscent of another era in medicine when scientists debated the nature and cause of transmission of contagious diseases (i.e., "miasma" versus some sort of microorganism). Robert Koch, one of the towering figures in modern medicine, while endorsing and actually doing clinical studies, also encouraged people to use experimental models in his famous third postulate: the suspected cause (microorganism) shall cause disease when introduced into a healthy organism (1884). A more precise clinical and neuropathological characterization can be certainly useful in clarifying the nature of CTE and well-designed cohort studies may sometimes determine the size of risk in the community. However, it is the animal models that will provide proof of principle that trauma, under certain circumstances, can cause degeneration and allow early insights into mechanisms. Such mechanistic insights will then facilitate the establishment of biomarkers and the formulation of appropriate treatments.

REFERENCES

Adams JH, Graham DI, Murray LS, Scott G (1982) Diffuse axonal injury due to nonmissile head injury in humans: An analysis of 45 cases. *Ann Neurol* 12:557–563.

Ballatore C, Lee VMY, Trojanowski JQ (2007) Tau-mediated neurodegeneration in Alzheimer's disease and related disorders. *Nature Reviews Neuroscience* 8:663–672.

Bandeen-Roche K, Hall CB, Stewart WF, Zeger SL (1999) Modelling disease progression in terms of exposure history. *Statistics in Medicine* 18:2899–2916.

Beaumont A, Marmarou A, Czigner A, Yamamoto M, Demetriadou K, Shirotani T et al. (1999) The impact-acceleration model of head injury: Injury severity predicts motor and cognitive performance after trauma. *Neurological Research* 21:742–754.

Bhaskar K, Konerth M, Kokiko-Cochran ON, Cardona A, Ransohoff RM, Lamb BT (2010) Regulation of tau pathology by the microglial fractalkine receptor. *Neuron* 68:19–31.

Broglio SP, Sosnoff JJ, Shin S, He XM, Alcaraz C, Zimmerman J (2009) Head impacts during high school football: A biomechanical assessment. *Journal of Athletic Training* 44:342–349.

Buki A, Povlishock JT (2006) All roads lead to disconnection? Traumatic axonal injury revisited. *Acta Neurochirurgica* 148:181–193.

Caffrey TM, Wade-Martins R (2012) The role of MAPT sequence variation in mechanisms of disease susceptibility. *Biochemical Society Transactions* 40:687–692.

Clavaguera F, Bolmont T, Crowther RA, Abramowski D, Frank S, Probst A et al. (2009) Transmission and spreading of tauopathy in transgenic mouse brain. *Nature Cell Biology* 11:909-U325.

Corsellis JAN, Bruton CJ, Freeman-Browne D (1973) The aftermath of boxing. *Psychological Medicine* 3:270–303.

Courville CB (1942) Coup-contrecoup mechanism of craniocerebral injuries — some observations. *Archives of Surgery* 45:19–43.

de Calignon A, Polydoro M, Suarez-Calvet M, William C, Adamowicz DH, Kopeikina KJet al. (2012) Propagation of tau pathology in a model of early Alzheimer's disease. *Neuron* 73:685–697.

de Calignon A, Fox LM, Pitstick R, Carlson GA, Bacskai BJ et al. (2010) Caspase activation precedes and leads to tangles. *Nature*464:1201–1204.

Ehlers MD, Kaplan DR, Price DL, Koliatsos VE (1995) NGF-stimulated retrograde transport of trkA in the mammalian nervous system. *The Journal of Cell Biology* 130:149–156.

Frost B, Diamond MI (2010) Prion-like mechanisms in neurodegenerative diseases. *Nature Reviews. Neuroscience* 11:155–159.

Frost B, Jacks RL, Diamond MI (2009) Propagation of tau misfolding from the outside to the inside of a cell. *The Journal of Cell Biology* 284:12845–12852.

Gastard MC, Troncoso JC, Koliatsos VE (2003) Caspase activation in the limbic cortex of subjects with early Alzheimer's disease. *Annals of Neurology* 54:393–398.

Guskiewicz KM, Marshall SW, Bailes J, McCrea M, Cantu RC, Randolph C, Jordan BD (2005) Association between recurrent concussion and late-life cognitive impairment in retired professional football players. *Neurosurgery* 57:719–724.

Hoshino S, Tamaoka A, Takahashi M, Kobayashi S, Furukawa T, Oaki Y et al. (1998) Emergence of immunoreactivities for phosphorylated tau and amyloid-beta protein in chronic stage of fluid percussion injury in rat brain. *Neuroreport* 9:1879–1883.

Kanayama G, Takeda M, Niigawa H, Ikura Y, Tamii H, Taniguchi N et al. (1996) The effects of repetitive mild brain injury on cytoskeletal protein and behavior. *Methods and Findings in Experimental and Clinical Pharmacology* 18:105–115.

Khlistunova I, Biernat J, Wang YP, Pickhardt M, von Bergen M, Gazova Z, Mandelkow E, Mandelkow M (2006) Inducible expression of tau repeat domain in cell models of tauopathy — Aggregation is toxic to cells but can be reversed by inhibitor drugs. *Journal of Biological Chemistry* 281:1205–1214.

Koliatsos VE, Applegate MD, Kitt CA, Walker LC, DeLong MR, Price DL (1989) Aberrant phosphorylation of neurofilaments accompanies transmitter-related changes in rat septal neurons following transection of the fimbria-fornix. *Brain Research* 482:205–218.

Koliatsos VE, Dawson TM, Kecojevic A, Zhou Y, Wang YF, Huang KX (2004) Cortical interneurons become activated by deafferentation and instruct the apoptosis of pyramidal neurons. *Proceedings of the National Academy of Sciences of the United States of America* 101:14264–14269.

Koliatsos VE, Kecojevic A, Troncoso JC, Gastard MC, Bennett DA, Schneider JA (2006) Early involvement of small inhibitory cortical interneurons in Alzheimer's disease. *Acta Neuropathologica* 112:147–162.

Koliatsos VE, Price DL (1996) Axotomy as an experimental model of neuronal injury and cell death. *Brain Pathology* 6:447–465.

Lishman WA, Eayrs JT, Chumbley CC (1968) Cortical and subcortical mechanisms in conditioned avoidance response to light in albino rat. *Physiology & Behavior* 3:739–744.

Liu L, Drouet V, Wu JW, Witter MP, Small SA, Clelland C, Duff K (2012) Trans-synaptic spread of tau pathology in vivo. *PLoS One* 7:e31302.

Ludolph AC, Kassubek J, Landwehrmeyer BG, Mandelkow E, Mandelkow EM, Burn DJ et al. (2009) Tauopathies with parkinsonism: Clinical spectrum, neuropathologic basis, biological markers, and treatment options. *European Journal of Neurology* 16:297–309.

Mac Donald CL, Dikranian K, Song SK, Bayly PV, Holtzman DM, Brody DL (2007) Detection of traumatic axonal injury with diffusion tensor imaging in a mouse model of traumatic brain injury. *Experimental Neurology* 205:116–131.

Marmarou A, Foda MAA, Vandenbrink W, Campbell J, Kita H, Demetriadou K (1994) A new model of diffuse brain injury in rats. 1. Pathophysiology and biomechanics. *Journal of Neurosurgery* 80:291–300.

Martland HS (1928) Punch drunk. *JAMA* 91:1103–1107.

McKee AC, Cantu RC, Nowinski CJ, Hedley-Whyte ET, Gavett BE, Budson AE et al. (2009) Chronic traumatic encephalopathy in athletes: Progressive tauopathy after repetitive head injury. *Journal of Neuropathology and Experimental Neurology* 68:709–735.

McKee AC, Stein TD, Nowinski CJ, Stern RA, Daneshvar DH, Alvarez VE et al (2012) The spectrum of disease in chronic traumatic encephalopathy. *Brain* 136:43–64.

Millspaugh (1937) Dementia pugilistica. *United States Naval Medical Bulletin* 35:297–303.

Mittl RL, Grossman RI, Hiehle JF, Hurst RW, Kauder DR, Gennarelli TA et al. (1994) Prevalence of MR evidence of diffuse axonal injury in patients with mild head injury and normal head CT findings. *AJNR. American Journal of Neuroradiology* 15:1583–1589.

Mott FW (1916) The effects of high explosives upon the central nervous system. *Lancet* 1:331–338.

Ojo JO, Mouzon B, Greenberg MB, Bachmeier C, Mullan M, Crawford F (2013) Repetitive mild traumatic brain injury augments tau pathology and glial activation in aged hTau mice. *Journal of Neuropathology and Experimental Neurology* 72:137–151.

Omalu BI, DeKosky ST, Hamilton RL, Minster RL, Kamboh MI, Shakir AM, Wecht CH (2006) Chronic traumatic encephalopathy in a national football league player: Part II. *Neurosurgery* 59:1086–1092.

Omalu BI, DeKosky ST, Minster RL, Kamboh MI, Hamilton RL, Wecht CH (2005) Chronic traumatic encephalopathy in a National Football League player. *Neurosurgery* 57:128–133.

Schwarz A. 2010. Suicide reveals signs of a disease seen in N.F.L., from http://www.nytimes.com/2010/09/14/sports/14football.html?pagewanted=all&_r=0. *The New York Times*

Spillantini MG, Goedert M (1998) Tau protein pathology in neurodegenerative diseases. *Trends in Neuroscience* 21:428–433.

Strich SJ (1961) Shearing of nerve fibres as a cause of brain damage due to head injury. A pathological study of twenty cases. *Lancet* 2:443–448.

Wang YP, Biernat J, Pickhardt M, Mandelkow E, Mandelkow EM (2007) Stepwise proteolysis liberates tau fragments that nucleate the Alzheimer-like aggregation of full-length tau in a neuronal cell model. *Proceedings of the National Academy of Sciences of the United States of America* 104:10252–10257.

Warden D (2006) Military TBI during the Iraq and Afghanistan wars. *Journal of Head Trauma Rehabilitation* 21:398–402.

Weir DR, Jackson JS, Sonnega A. 2009. Study of NFL retired players, from http://www.umich.edu/news/Releases/2009/Sep09/FinalReport.pdf. University of Michigan, Institute for Social Research.

Wolfe MS (2009) Tau mutations in neurodegenerative diseases. *Journal of Biological Chemistry* 284:6021–6025.

Yoshiyama Y, Uryu K, Higuchi M, Longhi L, Hoover R, Fujimoto S et al (2005) Enhanced neurofibrillary tangle formation, cerebral atrophy, and cognitive deficits induced by repetitive mild brain injury in a transgenic tauopathy mouse model. *Journal of Neurotrauma* 22:1134–1141.

Section VIII

Substance Abuse and Comorbid Conditions

48 Evidence for Beneficial and Adverse Effects of Alcohol in Animal Models and Clinical Studies of Traumatic Brain Injury

Anna N. Taylor and Richard L. Sutton

CONTENTS

48.1 INTRODUCTION

There is no doubt that acute and/or chronic alcohol consumption contributes to the incidence of traumatic brain injury (TBI). However, more than three decades of research on the impact of alcohol on the pathophysiological, neuroanatomical and functional outcomes of TBI have produced inconsistent results. In this chapter, the authors summarize the primary outcomes reported in the majority of animal studies and a great number of clinical studies that have been conducted in these areas. Studies reporting beneficial or adverse effects of acute alcohol treatment in animal models of experimental TBI are reviewed first, followed by those animal studies reporting on the effects of chronic or repeated alcohol exposures. Clinical studies on short-term outcomes such as morbidity, complications or mortality during acute care that have been associated with a positive admission blood alcohol content are then summarized, where the majority of studies have indicated beneficial or null effects of alcohol while fewer indicate injury exacerbation or worsened outcomes. Finally, clinical studies that have examined the impact of a history of preinjury alcohol abuse on TBI outcomes and those clinical studies that have focused more on functional outcomes are reviewed.

Alcohol intoxication increases the risk of occurrence of multiple types of injury, including TBI, and may affect the morbidity and mortality associated with head injury. Reviews of the literature have indicated that between 25% and 51% of TBI patients were intoxicated at the time of injury (Corrigan, 1995; Parry-Jones et al., 2006; Tien et al., 2006). In animal models of TBI, alcohol has been shown to produce both adverse and neuroprotective effects, whereas clinical studies assessing acute and the long-term neurological and behavioral impact of intoxication demonstrate no consistent effects (Asmaro et al., 2013; Chen et al., 2012; Opreanu et al., 2010). Such discrepancies may be due to the numerous dose-related effects of alcohol, the time of exposure in relation to the injury, and the nature of the injury and its outcome(s). The aim of this chapter is to systemically review animal and clinical studies on ethanol (EtOH) and TBI focusing on these elements to determine whether any specific pattern of interaction emerges.

48.2 ANIMAL STUDIES: ACUTE EtOH EXPOSURE

Neuroprotection in TBI with acute EtOH exposure has been observed in various studies. Kelly et al. administered EtOH (1, 2.5, or 3 g/kg, intraperitoneally [ip]) 40 minutes before lateral (parietal cortex) controlled cortical impact (CCI) injury in rats and found that preinjury EtOH was associated with acute postinjury mortality, the highest mortality rate occurring in the high-dose EtOH group (Kelly et al., 1997a). At 17 days postinjury, low- and moderate-dose groups had significantly less severe beam-walking impairment initially and a more rapid return to normal beam-walking ability, compared with the no- and high-dose EtOH groups. At 4 weeks postinjury, mean lesion volumes were significantly smaller in the low- and moderate-dose groups compared with the no- and high-dose groups.

In a subsequent study, Kelly et al. examined regional cerebral blood flow (CBF) and cerebral metabolic rates of glucose at various times up to 72 hours after CCI injury as a mechanism for the neuroprotective effect of EtOH (1 g/kg, ip) (Kelly et al., 2000). Their results indicated that low-dose EtOH is associated with attenuation of immediate postinjury hyperglycolysis and with more normal glucose metabolism in the injury penumbra over the ensuing 3 days. Simultaneously, the reduction in CBF typically seen within the contusion core and penumbra after CCI is less severe when EtOH is present. The net effect of these changes is a decreased degree of uncoupling between glucose metabolism and CBF that otherwise occurs in the absence of EtOH.

Janis et al. administered EtOH orally (blood alcohol content, BAC of 100 mg/dL) 2 hours before medial prefrontal cortex CCI. Commencing 7 days postinjury, the rats were tested for acquisition of spatial localization in a Morris Water Maze (MWM) task (Janis et al., 1998). Rats intoxicated at the time of injury performed better than the lesion/water controls, with EtOH reducing the severity of cognitive impairment caused by the CCI injury. Additionally, acute EtOH treatment reduced some of the histopathology that typically occurs after severe contusion of the medial frontal cortex but did not attenuate posttraumatic formation of cerebral edema.

Gottesfeld et al. studied the effects of low- (BAC of 100 mg/dL) and high-dose EtOH (BAC = 200 mg/dL) at 1 hour before lateral CCI injury on proinflammatory cytokine levels at 4 hours postinjury in hypothalamus, cortex, and hippocampus ipsilateral to the CCI (Gottesfeld et al., 2002). Their results showed that CCI injury elicits a marked increase in interleukin (IL)-1β and tumor necrosis factor (TNF)-α at these sites, as had previously been reported by Shohami et al. (1994). In contrast, EtOH pretreatment lowered the proinflammatory cytokine levels in a dose-dependent manner after TBI compared with the untreated injured rats. Given that posttraumatic cerebral and systemic inflammatory responses are increasingly recognized as playing an integral role in the progression of cerebral cellular damage after head injury, attenuation of the posttraumatic inflammatory cascade may diminish the severity and mortality of cerebral injury (Morganti-Kossmann et al., 2007; Schmidt et al., 2005).

Asa et al. examined effects of EtOH (2 g/kg, ip; peak BAC of 280 mg/dL) administered 1 hour after closed-head injury in rats (Asa et al., 2003). The EtOH treatment was found to not affect cerebral edema or neurological outcome 24 hours postinjury, although prostaglandin E2 release was reduced compared to saline-treated controls with TBI.

Tureci et al. designed a study to understand and modulate the pathophysiology of TBI (Tureci et al., 2004). Acute EtOH intoxication at moderate doses was performed 2 hours before TBI using an impact acceleration model in rats. At 2 days postinjury, there were slight increases in immunoreactivity for synaptophysin, a marker for injury severity, in the CA1 subfield of hippocampus when EtOH was administered before injury compared with moderate increases in the injury-only group. Histological ratings of neuronal damage in hippocampus were also decreased in the EtOH-treated group. The authors suggest that this neuroprotective effect of EtOH may be due to its inhibition of N-methyl-D-aspartate–mediated excitotoxicity. In a later study, this same group administered EtOH (1 or 2.5 mg/kg, ip) 40 minutes before severe TBI induced by impact acceleration injury (Is et al., 2005). Ratings of histological damage 24 hours postinjury suggested a dose-response neuroprotection in the hippocampus, with the 2.5 mg/kg dose of EtOH being most effective for reduction of red neuron formation and vacuolar degeneration.

Dash et al. studied the effects of combined caffeine (10 mg/kg, ip) and EtOH (0.65 g/kg, ip) given 15 minutes after a moderate lateral CCI (with bilateral skull openings) injury (Dash et al., 2004). This "caffeinol" treatment did not alter TBI-induced deficits in motor functions or cognitive functions assessed using trace fear conditioning or spatial memory in an MWM task, although MWM working memory was improved and injury volume was reduced by early caffeinol treatment. A higher dose of caffeinol (10 mg/kg caffeine with 1 g/kg EtOH, ip) or EtOH alone (1 g/kg, ip) given at 6 and 24 hours after TBI were found to have minimal effects on motor or cognitive functions and no effects on injury volume. Considering the previous results of Kelly et al. (1997a, 2000), protective effects of low-dose EtOH administered preinjury are apparently reduced or absent with delayed administration.

Goodman et al. recently reported that when mice were gavaged with either EtOH (5 g/kg, BAC of 250–300 mg/dL) or water 1 hour before TBI induced by weight-drop, the EtOH treatment reduced the local neuroinflammatory response to injury within 24 hours postinjury (Goodman et al., 2013). Preinjury EtOH treatment was associated with reduced levels of serum IL-6 at 3 hours and reduced levels of proinflammatory cytokines and chemokines (IL-6, KC, monocyte chemotactic protein (MCP)-1, and macrophage-inflammatory protein-1α) in ipsilateral cortex at 9 and/or 18 hours after TBI. This neuroprotective effect of acute EtOH exposure extends the findings of Gottesfeld et al. as presented previously (Gottesfeld et al., 2002).

Wang et al. investigated the role of EtOH on the well-known phenomenon that TBI causes brain edema by allowing excessive water passage through aquaporin (AQP) proteins (Wang et al., 2013). Rats received a closed-head TBI in which Maramarou's impact-acceleration method was used and were injected with 0.5 g/kg or 1.5 g/kg EtOH (ip) at 60 minutes post-TBI. EtOH administration at 1.5 g/kg post-TBI led to significantly lower levels of brain edema as measured by brain water content, associated with significantly reduced levels of AQP messenger RNA (mRNA) and protein expression as compared with the nontreated TBI group at 24 hours postinjury. Both doses of EtOH improved cognitive outcome assessed in a radial maze task and the higher dose of EtOH was also found to improve recovery of motor functions.

Kanbak et al. used the weight drop technique to induce lateral cortical contusion in adult male rats that received a single injection of EtOH (2.5 g/kg, ip) immediately after the injury to study apoptosis, one of the pathologic features of TBI, at 24 hours postinjury (Kanbak et al., 2013). They observed that this moderate EtOH dose caused inhibition of lysosomal cysteine protease release and nitric oxide production. Thus, the cytoprotective effect of moderate EtOH treatment in TBI might depend on the prevention of caspase-3–mediated neuronal apoptosis via decreased lysosomal release and low concentrations of nitric oxide. In contrast to the previous studies showing neuroprotective effects of acute EtOH exposure, various studies have reported disruptive or no effects.

Flamm et al. infused EtOH (5 mg/kg, iv) for 1 hour (BAC of 448 mg/dL) before impact of the exposed dura mater to induce cortical contusions of various severity in cats. At 3 and 24 hours after more severe injuries, they observed a considerable increase in the extent of blood–brain barrier (BBB) permeability to Evans blue-labeled albumin (EBA) and hemorrhagic necrosis with this high-dose EtOH treatment (Flamm et al., 1977). The authors suggested these outcomes may be related to EtOH alterations in membrane-bound enzymes, clotting mechanisms, and cell membranes through abnormal free radical reactions.

Persson and Rosengren have created small cerebral stab wounds and studied the extravasation of EBA in EtOH intoxicated rats given an ip injection (1.5 mL 20% w/v solution/100g BWt) at 1, 3, and 24 hours preinjury (Persson and Rosengren, 1977). EtOH-intoxicated rats showed a greater leakage of EBA as well as a wider area of leaking blood vessels around the stab wound at both time points, indicating increased BBB permeability from endothelial and neuronal cell injury.

Kim et al. induced midline fluid percussion injury in rats that received EtOH (4 g/kg, ip, BAC of 320 mg/dL) 10 minutes preinjury (Kim et al., 1989). They found no significant effect of this high-dose EtOH treatment on postinjury apnea or mortality, but grading of the extent of subarachnoid hemorrhage was significantly increased compared to their nontreated controls with TBI.

Zink's group in a porcine lateral fluid percussion (LFP) model of TBI administered 3.5 gm/kg EtOH intragastrically 1 hour before injury (Zink and Feustel, 1995; Zink et al., 1993). BAC levels were 136 mg/dL at the time of injury and 175 mg/dL at 2 hours after injury. Whereas EtOH exposure had no significant effect on pH, partial pressure of carbon dioxide, intracranial pressure, heart rate, brain temperature, or glucose levels at 10-minute and hourly intervals thereafter up to 3 hours postinjury, EtOH intoxication led to significantly higher blood lactate levels, longer postinjury apneas, and impairment of respiratory control (ventilation and hypercapnic response sensitivity) after TBI.

Yamakami et al. gavaged rats with low (1.5 g/kg) or high dose (3.0 g/kg) EtOH, or saline 2 hours before moderate or severe LFP injury (Yamakami et al., 1995). A significant and prolonged hypotension (up to 1 hour) was observed in animals pretreated with either low- or high-dose EtOH. Neither low-dose (BAC = 110 mg/dL) or high-dose (BAC of 340 mg/dL) EtOH had any effect on survival or neurological motor function up to 7 days after moderate injury. After severe brain injury, animals pretreated with high-dose EtOH showed a significantly increased mortality and markedly worsened neurological deficits at 24 hours postinjury. By 4 hours after severe TBI, free magnesium (Mg^{++}) and the cytosolic phosphorylation potential determined by phorphorus-31 magnetic resonance spectroscopy declined in both EtOH and control groups by approximately 60%. In contrast, brain intracellular pH in the high-dose EtOH-treated animals was consistently higher than in the control group after severe injury. It therefore appears that prior exposure to EtOH, particularly at high concentrations, may have detrimental effects on neurobehavioral function and survival in the acute period (up to 24 hours) after severe brain injury, and may be associated with posttraumatic cerebral alkalosis.

Shapira et al. exposed rats to EtOH (1.5 and 3.0 g/kg, ip) 2 hours before TBI induced by weight drop onto the left cranium (Shapira et al., 1997). Their data showed that in this closed-head injury model acute EtOH treatment increases mortality, neurological deficit, hemorrhagic necrosis volume, and brain edema, even at the low dose. Their mortality findings were similar to those of Franco et al., who studied mice given saline or EtOH (0.2 mL 50% EtOH, ip; BAC of 720 mg/dL) 30 minutes before closed-head injury and reported a significant increase in mortality of the EtOH-treated groups within the first week postinjury (Franco et al., 1988).

Zink's group in a subsequent series of studies added hemorrhagic-shock (HS) to their porcine LFP injury model described previously (Zink et al., 1998, 1999, 2006). They concluded that EtOH intoxication in the presence of HS potentiates the physiological and metabolic alterations that may contribute to secondary brain injury (Zink et al., 1998). To follow up on the observed increases in lactate concentrations in brain tissue and cerebral venous blood, Zink et al. gave a preinjury bolus of 2 g/kg EtOH (intravenously) over 30 minutes and an infusion of 0.4 g/kg EtOH 1 hour preinjury (BAC of 219 mg/dL) and found that these metabolic changes were associated with respiratory depression and reduced organ blood flow (Zink et al., 1999). More recently, Zink et al. extended these adverse effects of preinjury EtOH to the otherwise salutary outcomes of limited resuscitation in the LFP plus HS injury model (Zink et al., 2006).

Fabian and Proctor have also examined effects of EtOH in a porcine model of severe LFP injury combined with HS. EtOH administered by gavage (3.5 g/kg) 1 hour before injury raised the BAC to 200 mg/dL at the time of LFP and to 300–350 mg/dL at 30 minutes post-HS (Fabian and Proctor, 2002). Their data showed that acute EtOH intoxication did not alter mortality as long as cardiopulmonary support was provided, cerebral compliance and cerebral perfusion pressure were not aggravated, and CBF was adequate for cerebral metabolic demands. Although systemic and cerebral venous lactate levels were elevated, this likely reflected a decrease in lactate metabolism during EtOH intoxication.

Katada et al. studied the effects of administering 3 g/kg EtOH (ip) and 1 hour later subjecting rats to lateral brain contusion induced by weight-drop (Katada et al., 2009). The EtOH-pretreated group had a significantly decreased survival rate. Magnetic resonance imaging showed that EtOH pretreatment significantly augmented the volume of cytotoxic brain edema after contusion. In the EtOH-pretreated rat, the activities of NF-κB and AP-1 were reduced 6 hours after contusion and COX-2 mRNA expression was increased 24 hours after contusion. These findings suggest that EtOH augmented cerebral edema and mortality in rats with brain contusion, possibly through actions on cell survival pathways, COX-2 expression, and oxidative stress, because they were able to show reduced mortality and edema with antioxidant treatment at 3 hours postinjury. It should be noted, however, that at a lower dose of EtOH (1.5 g/kg postinjury), Wang et al. (2013) observed lower levels of brain edema as measured by brain water content, as described previously.

Katada et al., in the same preparation as in their 2009 study, described previously, found that acute EtOH administration (3 g/kg) increased expression of AQP4, a water channel, particularly in astrocyte end feet, at 24 hours after TBI (Katada et al., 2012). These findings suggest that EtOH induces upregulation of AQP4 leading to an augmentation of brain edema after TBI. The discrepancy in these results with those of Wang et al., as described previously, who observed less brain edema associated with significantly reduced levels of AQP mRNA and protein expression in their EtOH-TBI group (1.5 g/kg) as compared with the nontreated TBI controls, may well be an EtOH-dose related phenomenon (Wang et al., 2013).

Vink and Byard gave rats three equal-volume injections (ip) of 50% EtOH/water aliquots totaling 3 g/kg over 20 minutes (BAC of 300 mg/dL) before LFP injury (Vink and Byard, 2012). All animals were monitored by phorphorus-31 magnetic resonance spectroscopy before and for 3 hours after induction of TBI. TBI in untreated animals caused a gradual decline of free Mg^{++} over the subsequent 3 hours. EtOH not only caused preinjury depletion of Mg^{++} but also exacerbated the rate and degree of the decline in Mg^{++} levels after TBI. This combination could predispose the brain to a worse outcome, given the multiple adverse effects that Mg^{++} decline could have on a number of cellular processes. These include facilitating activity of N-methyl-D-aspartate channels, thus upregulating glutamate-induced excitotoxicity after trauma, although this effect may relate to EtOH dosage (Kelly et al., 1997a); adverse effects on all energy-producing and energy-consuming reactions given the essential role Mg^{++} plays in these processes; inhibiting cellular ion homeostasis, especially that of Na^+, K^+ and Ca^{++}, given that they all require the activity of Mg^{++} dependent ATPase, thus promoting posttraumatic edema formation which in itself has also been associated with repeated EtOH exposure (Collins et al., 1998); and adverse effects on membrane structure itself including a disruption of BBB permeability with consequent neuroinflammation.

Kong et al. gavaged rats with EtOH (wine, 56% vol, 15 mL/kg) 30 minutes before severe impact acceleration injury, with magnetic resonance imaging and diffusion tensor imaging (DTI) being performed 1, 3, 6, 12, and 24 hours after TBI, followed by histological analyses (Kong et al., 2013). Their data indicated that the relative proportion of cytotoxic and vasogenic (decrease or increase in the DTI-apparent diffusion coefficient values) edema in brainstem over time differed between the EtOH and trauma control group, whereas the EtOH treatment enhanced axonal injury based on lowering of brainstem DTI-FA values. Although not quantified, morphological images suggest increased cellular, vascular, and axonal damage with EtOH treatment, and intensity of AQP4 immunostaining was found to be lower 12 hours postinjury in the EtOH-TBI group compared with TBI only.

In summary, irrespective of differences in animal models of TBI, the bulk of the evidence indicates that acute exposure to low and moderate doses of EtOH (BAC <300 mg/dL) is neuroprotective on multiple outcomes if EtOH exposure occurred within 2 hours or less before TBI in rats and mice, but not in pigs. Neuroprotection was also obtained with these low doses when EtOH was administered immediately after the injury but not if delayed for more than 6 hours. Higher doses of EtOH (BAC >300 mg/dL) in rodent and feline models generally produced adverse effects on neurobehavioral and morphological outcomes of TBI.

48.3 ANIMAL STUDIES: CHRONIC EtOH EXPOSURE

Chronic EtOH exposure before and after TBI has been found to produce either beneficial or adverse effects, depending upon the outcomes being assessed. Taylor et al. reported that continuous consumption of EtOH before and after lateral CCI attenuated TBI-induced hyperthermia and deficits in spatial learning in the MWM (Taylor et al., 2002). Adult male rats were fed an EtOH-containing liquid diet (5% w/v, 35% EtOH-derived calories) for 2 weeks before CCI and for an additional week postinjury; controls were pair-fed the isocaloric liquid diet. BAC at the time of injury was approximately 150 mg/dL. Continuous biotelemetric recording of core body temperature revealed that CCI produced a significant febrile response that persisted for at least 6 days in the control group and in rats consuming EtOH only preinjury, but lasted for only 2 days in rats consuming EtOH both pre- and postinjury. The detrimental effects of TBI-induced hyperthermia on neurologic recovery in humans (Dietrich et al., 1996) and the improved neurological and behavioral recovery seen after application of mild to moderate hypothermia have been well documented (Bramlett et al., 1995; Clifton et al., 1991; Dixon et al., 1998; Marion et al., 1997). Our finding that the TBI-induced febrile response can be attenuated by EtOH consumption pre- and post-TBI suggests that such a regimen may be neuroprotective. In this study we also found that when tested at 3–4 weeks after CCI, rats consuming EtOH pre- and postinjury required significantly fewer trials to reach criterion in the MWM than rats consuming EtOH only preinjury.

Taylor et al. subsequently studied various neuroendocrine-immune effects of lateral CCI at 4–8 weeks postinjury and the development of EtOH dependence and withdrawal after feeding their EtOH-containing liquid diet for 2 weeks commencing at 9 weeks postinjury (Taylor et al., 2006). Compared with the sham-injured control diet-fed group, the TBI rats became hyperthermic during ethanol dependence and responded to EtOH withdrawal with a further increase in body temperature. Light- and dark-phase biotelemetric activity analysis indicated that TBI rats were significantly more active than the sham group, and that EtOH and withdrawal differentially affected their activity. These long-term post-TBI adverse reactions to EtOH may contribute to the high incidence of alcoholism in this population.

Baratz et al. examined the effects of binge EtOH drinking in mice subjected to mild TBI by closed-head weight-drop. Mice had access to 0%, 7.5%, 15%, or 30% EtOH solutions for 48 consecutive hours once a week for 4 weeks as the sole source of fluids (the remaining time they drank water) either before and/or post TBI (Baratz et al., 2010). The mild TBI mice exhibited lower memory ability in the Y-maze, higher anxiety in the elevated plus maze, and lower retention in a passive avoidance test than did sham animals. EtOH reversed these effects at all doses if consumed pre- and post-TBI, but not if consumed only post-TBI. The results suggest that EtOH drinking before trauma might have a protective effect on recovery from brain trauma, but not if consumed after the trauma.

Shapira et al., in contrast, found that chronic exposure (6% EtOH in drinking water) for 40 days before TBI with a weight-drop device had little effect on mortality, neurological deficit, hemorrhagic necrosis volume, and brain edema at 24 hours postinjury compared with the no EtOH or acute EtOH-treated groups (described previously) (Shapira et al., 1997). The lack of an apparent effect of chronic EtOH exposure on BBB integrity was subsequently confirmed by Arican et al. in a closed-head model of TBI in rats (Arican et al., 2006).

Biros et al. likewise did not find that EtOH intoxication provided neuroprotection in a repeated TBI model (Biros et al., 1999). They administered EtOH (3 g/kg) orogastrically every other day for 2 weeks and 2 h before LFP injury, followed by three episodes of LFP injury once every 4 days. The injured chronic EtOH exposed (CA) animals had a more rapid recovery of reflexes compared with the non-EtOH exposed (NA) animals with repeated injury. Injured NA rats eventually began to learn the MWM (days: 11–19 postinjury), whereas injured CA rats never learned the maze. Acute EtOH intoxication at the time of multiple episodes of minor head trauma did not provide neuroprotection for NA or CA rodents.

Prasad's group has conducted a series of studies in rats that received an EtOH-containing liquid diet for either 6 weeks (Prasad et al., 1997; Zhang et al., 1999) or 3 months (Masse et al., 2000) before LFP injury, or as a 4-week binge before injury (Prasad et al., 2001). Six weeks of EtOH administration did not alter TBI-induced lactate accumulation or high-energy phosphates (adenosine triphosphate + phosphocreatine) concentrations. Neither 6 weeks nor 3 months of EtOH exposure produced any significant alterations in behavioral and morphologic outcomes of the brain injury. However, 4 weeks of biweekly EtOH exposure (3.0 g/kg intragastrically, BAC of 100–150 mg/dL) before injury worsened some aspects of spatial learning ability (but not memory) in the MWM, but did not differentially affect the extent of ipsilateral cortical and hippocampal damage.

Lai et al. gave rats EtOH twice daily (4 g/kg for 2 weeks and 6 g/kg for another 2 weeks) before a minor strike on the occipital tuberosity with an iron pendulum. Compared with rats in the control group, rats in the chronic alcoholism group showed minor axonal degeneration, a significant decrease in the numerical density of synapses, and compensatory increase in postsynaptic density-95 expression in the medulla oblongata (Lai et al., 2013). Rats in the EtOH-TBI group showed high mortality (50%), inhibited respiration before death, severe axonal injury, and a decrease in postsynaptic density-95 expression in the medulla oblongata. The investigators conclude that chronic alcoholism induces significant synapse loss and axonal impairment in the medulla oblongata and renders the brain more susceptible to TBI. The combined effects of chronic alcoholism and TBI induce significant synapse and axon impairment and result in high mortality.

Teng and Molina studied the effects of a 15-hour intragastric infusion of EtOH (7 g/kg; BAC of 166 mg/dL) or isocaloric/isovolumic dextrose infusion, with a mild LFP injury induced in adult rats 30 minutes after discontinuation of the infusion (Teng and Molina, 2013). Mild TBI induced apnea and a delay in the acute righting reflex. EtOH intoxication at the time of injury increased delays in the righting reflex without altering apnea duration. Neurological and behavioral dysfunction was observed at 6 and 24 hours post-TBI, and this was not exacerbated by EtOH. TBI induced a transient upregulation of cortical IL-6 and MCP-1 mRNA expression at 6 hours, which was resolved at 24 hours. EtOH did not modulate the inflammatory response at 6 hours, but prevented resolution of inflammation (IL-1, IL-6, TNFα, and MCP-1 expression) at 24 hours post-TBI. Therefore, intoxication with EtOH at the time of mild TBI did not delay the recovery of neurological and neurobehavioral function but prevented the resolution of neuroinflammation post-TBI.

Although there is clear evidence that TBI initiates pro- and anti-inflammatory responses (e.g., production and release of IL-1β, TNFα, IL-6, IL-10) that evolve over hours to days and contribute to secondary damage to subcortical sites involved in thermoregulation and cognition, such as hypothalamus and hippocampus (Knoblach and Faden, 1998; McClain et al., 1987, 1991; Morganti-Kossmann et al., 2007; Perez-Polo et al., 2013; Shohami et al., 1994, 1999; Toulmond and Rothwell, 1995), the effects of EtOH on these factors in the models of experimental TBI have been somewhat inconsistent. Whereas the preceding manuscript by Teng and Molina indicates no EtOH effect at 6 hours and prevention of resolution of inflammation at 24 hours after mild LFP in rats (Teng and Molina, 2013), Gottesfeld et al. found EtOH lowered pro-inflammatory cytokines at 4 hours after moderate lateral CCI injury in rats, and Gottesfeld et al. report that EtOH reduced IL-6 and chemokines at 9 and/or 18 hours after moderately severe closed-head injury in mice (Goodman et al., 2013; Gottesfeld et al., 2002). Thus, EtOH effects on TBI-induced inflammation may be related to injury severity and not because of differing species. Time factors may also impact on these outcomes, and our unpublished data obtained in rats exposed to a 4-day EtOH binge immediately before moderately severe lateral CCI (once daily gavage with 5 g/kg EtOH and 2.5 g/kg at 1 hour before CCI) show that the preinflammatory cytokines, IL-1β, and TNFα, remain elevated in ipsilateral cortex and hypothalamus at 7 days after TBI with or without preinjury EtOH exposure.

In summary, chronic exposure to EtOH produced variable effects on outcomes of TBI, ranging from neuroprotective, to null, or to adverse effects. Also critical is the duration and timing of pre- and/or postinjury exposure as well as the level of intoxication.

48.4 HUMAN STUDIES: ACUTE EFFECTS OF EtOH EXPOSURE

In contrast with animal studies that have focused primarily on isolated TBI, human investigations on the effects of alcohol are most often conducted with patients suffering from polytrauma induced by multiple mechanisms of injury. Multiple studies have examined EtOH effects in trauma patients who did not necessarily have TBI, whereas other investigations have examined effects of EtOH exposure on outcomes of patients with clear evidence of head trauma or TBI. Both categories of studies are included in the following section, although more details and results are provided for the latter studies. It is well known that alcohol intoxication alters the level of consciousness, and an ongoing concern for clinicians is whether this confounds measures of the initial severity of TBI as assessed by the Glasgow Coma Score (GCS). The association between depressed GCS and a positive BAC in patients with moderate-to-severe TBI has proven to be variable (Alexander et al., 2004; Jagger et al., 1984; Nath et al., 1986; Pories et al., 1992; Sperry et al., 2006; Stuke et al., 2007; Talving et al., 2010). This may be in part due to the fact that BAC determinations are most often made at admission, with underestimations of the actual BAC at time of injury, and elapsed time from injury to admission and rates of EtOH metabolism are rarely examined. Nonetheless, clinical studies have increasingly incorporated other indices of overall injury severity or region-specific injury severity, such as those provided by the Abbreviated Injury Score (AIS; scored for head, chest, abdomen, and extremity) or the Injury Severity Score (ISS) in conjunction with the GCS, as noted within several of the studies that follow.

Research on the effects of alcohol in human trauma and TBI has most often been focused on whether or not a positive BAC at time of admission to the emergency room or trauma center impacts on mortality, morbidity, or complications during hospitalization or while in the intensive care unit (ICU). As reviewed next, most of these studies have reported beneficial or no effects, whereas fewer studies have reported negative consequences in patients with positive BAC at the time of injury or admission.

Several clinical studies have indicated that a positive BAC may exert beneficial effects. Ward et al. reported no difference in the overall ISS in multisystem trauma patients with or without detectable EtOH in blood, although mortality in patients with a positive BAC was significantly lower, particularly at high levels (≥300 mg/dL) (Ward et al., 1982). Blondell et al. found that BAC-positive patients with multiple mechanisms of traumatic injury were less likely to die and had shorter ICU length of stay than BAC-negative patients, even though they had comparable ISS and similar hospital length of stay (Blondell et al., 2002). Plurad et al. categorized victims of motor vehicle crashes into groups of no (BAC ≤ 5 mg/dL), low (BAC > 5 to < 80 mg/dL) and high (BAC ≥ 80 mg/dL) based on alcohol screening upon admission (Plurad et al., 2010). They found that BAC was not related to overall injury severity, hypotension, or ICU length of stay, but patients with high BAC had a higher incidence of severe head trauma (head AIS > 3) and sepsis. However, the high BAC group also had significantly better survival than did patients in the no BAC group. These findings are not incompatible with those of Tien et al. who studied patients with severe TBI from blunt head trauma and found lower mortality in patients with BAC < 230 mg/dL compared with those having no BAC, whereas BAC ≥ 230 mg/dL was found to increase in-hospital death compared with no BAC (Tien et al., 2006).

O'Phelan et al. studied patients with severe TBI who had a positive drug toxicology screen, and reported that both alcohol and methamphetamine were associated with decreased mortality (O'Phelan et al., 2008). Salim et al. also found that severe TBI patients testing positive for BAC had lower in-hospital mortality (Salim et al., 2009a). Their BAC-positive patients also required less ventilator time and had reduced ICU but not overall hospital length of stay, although more complications were found compared with BAC-negative patients. They also reported reduced in-hospital mortality in severely injured TBI patients who tested positive for BAC, although these patients had higher rates of sepsis compared with their BAC-negative counterparts (Salem et al., 2009b). Talving et al. reported that BAC (<0.08 mg/dL versus>0.08 mg/dL) in patients with isolated severe TBI was not associated with overall injury severity, head injury severity, major in-hospital complications, or with length of stay in the ICU or hospital (Talving et al., 2010). Moreover, these authors also report that in-hospital mortality was significantly lower in BAC-positive patients. Berry et al. studied patients with moderate to severe TBI (head AIS ≥3, all other AIS ≤3) and found that those testing positive for alcohol had significantly lower mortality compared with alcohol-negative patients (Berry et al., 2010). Their alcohol-positive patients were more severely injured (>ISS) and had a lower GCS, with a shorter hospital length of stay compared with alcohol-negative patients. Lustenberger et al. also studied patients with TBI (head AIS ≥ 3, all other AIS ≤ 3) and found that admission coagulopathy, in-hospital sepsis, and mortality were reduced in BAC-positive compared with BAC-negative patients (Lustenberger et al., 2011). The ICU and hospital length of stay were not affected by presence of alcohol in their study. Berry et al. stratified moderate to severe TBI patients (head AIS ≥3, all other AIS ≤3) by four levels of admission BAC (0, 0–100, 100–230, and ≥230 mg/dL), and found that mortality rates were decreased with increasing BAC (Berry et al., 2011). Although head AIS were equivalent between EtOH-positive and EtOH-negative groups, patients with positive BAC had reduced ISS and GCS indices of injury severity and their ICU and hospital lengths of stay were reduced compared with BAC-negative patients. Finally, Hadjibashi et al. studied TBI patients (head AIS ≥3, all other AIS ≤3) and reported that a positive BAC was associated with less severe injury (lower ISS) and lower GCS, but these patients had similar severity of head injury (head AIS) compared with those with no BAC detected (Hadjibashi et al., 2012). However, the pneumonia rate was significantly lower in TBI patients with positive BAC, with a significant linear trend for BAC levels (0, 0–100, 100–230 and ≥230 mg/dL).

Several published reports have also shown no apparent effect of acute alcohol intoxication on outcomes related to head injury in the setting of multisystem trauma or polytrauma, as discussed later. The study by the Ward group found that, although patients with a positive BAC had a higher incidence of head injuries, neurologic injury as a cause of death was not increased by alcohol (Ward et al., 1982). Several investigators have found no benefits or detrimental effects of acute alcohol intoxication on mortality or posttraumatic complications in multiple trauma patients (Huth et al., 1983; Jurkovich et al., 1993; Mann et al., 2011; Rootman et al., 2007; Scheyerer et al., 2013; Shih et al., 2003; Thal et al., 1985; Zeckey et al., 2011), although Shih et al. (2003) did find increased morbidity associated with alcohol intoxication. Ruff et al. found no effects of acute EtOH intoxication on in-hospital outcomes of patients with severe head injury (Ruff et al., 1990). Christensen et al. found no significant contributions of EtOH intoxication on admission GCS, pulmonary complications, or length of stay in head injury patients without significant intracranial bleeding (Christensen et al., 2001). Alexander et al. reported that a positive BAC (>100 mg/dL) correlated with reduced CBF in patients with severe TBI acutely after admission (Alexander et al., 2004), although no effects of serum EtOH levels on mortality were found. Shandro et al. found that admission BAC had no significant effect on in-hospital, 90 day, or 1 year post-TBI mortality (Shandro et al., 2009), although a trend towards lower mortality in patients with higher BAC (>230 mg/dL) was reported. Chen et al., in a large database study, matched alcohol-positive and alcohol-negative groups of TBI patients by sex, age, ethnicity, and research facility (Chen et al., 2012). These researchers found that a positive BAC at time of injury increased the risk for TBI across all ages (from 16 to 66 years). In their analyses of the relationship between BAC and mortality, which included covariates of ISS and the various causes and intents of TBI, Chen et al. (2012) found no protective effects of alcohol against TBI mortality for all categories of TBI as well as no protection in patients with isolated moderate to severe TBI.

In contrast to the preceding reports of null or beneficial effects, a few investigations have indicated that acute alcohol intoxication can exacerbate injury or worsen outcomes after multitrauma or head injury. Although Kraus et al. found that TBI severity and in-hospital mortality were inversely related to the admission BAC (Kraus et al., 1989), they provide evidence that selection bias related to differential rates of BAC testing underlie this effect. Rates of neurologic limitations at discharge for individuals with moderate or severe TBI with BAC ≥100 mg/dL were increased compared with patients with no detectable BAC, and a positive BAC resulted in longer durations of hospital stay. Luna et al. found increased incidence of severe head injury and mortality in intoxicated compared to nonintoxicated motorcyclists (Luna et al., 1984), although the overall ISS levels were similar in EtOH-positive and -negative groups. Pories et al. studied patients with polytrauma (motor vehicle crash, fall, sports injury) and found a positive BAC was associated with increased injury severity, mortality, and incidence of head injuries, with no alcohol effects found for admission GCS compared with GCS values 24 hours after admission (Pories et al., 1992). Gurney et al. reported increased likelihood for mechanical intubation and ICP monitoring and increased risk for respiratory distress and pneumonia in intoxicated (BAC ≥100 mg/dL) versus nonintoxicated (BAC <100 mg/dL) head-injured patients (Gurney et al., 1992). Cunningham et al. provided some evidence that alcohol potentiated the severity of TBI as judged from initial head computed tomography ratings (Cunningham et al., 2002), although their study was underpowered with small sample sizes. Hadjizacharia et al. studied a mixed population of trauma patients, matching patients with a positive BAC with alcohol-negative patients on demographics and injury severity indices (Hadjizacharia et al., 2011). Their alcohol-positive patients were found to have increased overall mortality as well as increased mortality within low, moderate, and high ISS stratifications, and a higher incidence of systolic blood pressure <90 or GCS ≤8 at admission. No group differences in ICU or hospital length of stay were found.

48.5 HUMAN STUDIES: FUNCTIONAL OUTCOMES WITH TBI AND EtOH EXPOSURE AND EFFECTS OF CHRONIC CONSUMPTION

The majority of the preceding clinical studies have examined relationships between a positive admission BAC and mortality, morbidity, or complications during acute care of TBI patients. Less frequently studied are the effects of chronic EtOH or the effects of EtOH on functional outcomes. Several studies have noted the substantial relationship between a positive admission BAC and a history of heavy EtOH use or abuse, and chronic preinjury EtOH use has been reported at 16% to ≥50% in TBI patients (Corrigan, 1995; Dikmen et al., 1995; Kreutzer et al., 1996; Parry-Jones et al., 2006; Sparadeo and Gill., 1989). Although adding potentially confounding factors such as alcohol tolerance and increased rate of EtOH metabolism to studies of acutely elevated BAC, a history of alcohol abuse with attendant damaging effects on organ systems including central nervous system functions can obviously complicate interpretations of TBI effects from any separate or additive effects of acute intoxication at time of injury. In this next section, we will review studies that have examined acute or chronic EtOH effects on acute or chronic functional status after TBI.

Studies of functional status post-TBI have also reported somewhat inconsistent findings for the impact of a positive admission BAC or history of EtOH use. Sparadeo and Gill reported lower global cognitive and neurobehavioral status at hospital discharge in persons with TBI who were intoxicated at admission compared to those not intoxicated (Sparadeo and Gill, 1989). Brooks et al., using relative's report of alcohol consumption at the time or injury or habitually, found that alcohol intoxication at the time of severe head injury was related to long-term (2–7 years postinjury) verbal learning and visual memory impairment, but habitual alcohol abuse conferred no additional disadvantage on cognitive outcome (Brooks et al., 1989). Ruff et al. reported that mortality in head injury patients with mass lesions and a history of chronic EtOH abuse was nearly doubled compared with similarly injured patients who did not drink excessively, and they had poorer global outcome at time of hospital discharge (Ruff et al., 1990). Chronic alcohol use (not admission BAC) was also found to be associated with increased pneumonia in trauma patients (Jurkovich et al., 1991). Kaplan and Corrigan reported that an increasing admission BAC level (0–0.099, ≥0.10 mg/dL) was associated with increased time to admission to rehabilitation after severe TBI, but neuropsychological test scores 2–3 months post-TBI did not vary as a function of admission BAC levels (Kaplan and Corrigan, 1992). Ronty et al. found that TBI patients with a history of alcohol abuse had increased intracranial hemorrhage volumes compared with nonalcoholic patients with similar TBI severity (Ronty et al., 1993). At 1 year after TBI, the alcohol-abusing individuals had increased cerebral atrophy, electroencephalogram impairments, and reduced rates of return to work. Dikmen et al. also reported that neuropsychological performance in TBI patients was related to the degree of preinjury EtOH use, with verbal intelligence and attention/visuomotor tracking impairments at 1 month and verbal intelligence impairments at 1 year in those individuals with more chronic pre-TBI EtOH use (Dikmen et al., 1995). However, they also found TBI severity was more strongly associated with the extent of neuropsychological impairment than was a history of alcohol use.

Kelly et al. studied severe TBI patients who had a positive BAC at the time of hospital admission during inpatient rehabilitation (mean 48.8 days postinjury) and found them to be worse on measures of intelligence, verbal and visual memory, attention and concentration, and delayed recall compared with nonintoxicated TBI patients (Kelly et al., 1997b). Head injury severity (GCS) was greater in the intoxicated TBI group, but not used as a covariate in the analyses of cognitive outcomes. Duration of posttraumatic amnesia was not different between their study groups. Importantly, this study also found that patients intoxicated with illicit drugs (primarily cocaine and opiates) at the time of TBI, with or without concurrent EtOH use, performed more poorly on the tests of intelligence, verbal memory, attention, and concentration compared with nonintoxicated persons with TBI. Cointoxication with EtOH and illicit drugs in trauma patients is also reported to increase in-hospital complications, particularly pneumonia (Rootman et al., 2007), and these studies (De Guise et al., 2009) indicate a need for increased screening and attention to the drug abuse and concomitant drug and EtOH intoxication on outcomes of TBI. Bombardier and Thurber (1998) tested effects of admission BAC on neuropsychological outcomes 2–3 months after mild to severe TBI, finding that higher BAC was predictive of poorer performance on orientation, concentration/mental speed, naming ability, verbal memory, and verbal abstract reasoning. The relationship between BAC and neuropsychological outcomes tended to diminish from 1 to 2 months post TBI. Tate et al. studied mild to severe TBI patients an average of 2.3 months postinjury (Tate et al., 1999). They reported that admission BAC (0, 0.01–0.099, 0.10–0.199, ≥0.20 mg/dL) was associated with the incidence of history of alcohol abuse. The admission BAC predicted poorer delayed recall of prose material, greater verbal memory deficits and poorer planning, organization, and visuospatial ability. Their results indicated that a history of alcohol abuse before TBI was not a significant predictor of cognitive performance in the post-acute stage of recovery. The previously cited study of Alexander et al. found no effects of admission BAC levels on functional outcome as measured using the Glasgow Outcome Scale at 3, 6, and 12 months after severe TBI (Alexander et al., 2004). Wilde et al. compared TBI patients who were BAC-positive or BAC-negative at the time of injury and found neuroimaging evidence (91–2,729 days after injury) for greater brain atrophy in patients with a positive BAC at the time of injury and/or a history of moderate to heavy alcohol use (Wilde et al., 2004). Although some trends toward worsened neuropsychological outcomes were apparent in patients with positive BAC or history of moderate to heavy EtOH use before TBI, cognitive outcome was not significantly decreased. Turner et al. also found that preinjury consumption of alcohol and admission BAC were not consistently related to any post-TBI cognitive outcomes (Turner et al., 2006).

Lange et al. have reported that preinjury alcohol abuse, compared with alcohol intoxication on the day of injury, had the greatest impact on short-term (within 7 days of injury) neuropsychological outcome after uncomplicated mild TBI (Lange et al., 2007). In a follow-up study this group (Lange et al., 2008) examined cognitive outcome within 10 days of mild to severe TBI in individuals with no history of preinjury heavy alcohol use or dependence who were either intoxicated (BAC ≥100 mg/dL) or had no positive BAC at the time of admission to the trauma center. Using matched samples of patients to control for demographic and injury-related variables, they found that sober patients with TBI performed worse on neuropsychological tests (immediate and delayed verbal memory, visual memory, executive functions) than did patients who were intoxicated at the time of injury. As the authors discuss, their findings of improved cognitive outcome in intoxicated TBI patients may be related to exclusion of patients with preinjury EtOH abuse. Vickery et al. studied the effects of trauma center admission BAC (0, 0.01–0.79, 0.80–1.59, ≥1.60 mg/dL) or EtOH use history (bingeing or drinking frequency of none, infrequent/light, moderate, or heavy) on functional status of mild to severe TBI patients at the time of admission to inpatient rehabilitation (Vickery et al., 2008). A higher BAC was associated with greater disability as assessed using the Disability Rating Scale (DRS) at admission, but neither BAC nor patterns of pretrauma drinking were predictive of Functional Independence Measure (FIM) cognitive or motor scores. A history of binge drinking was associated with better global outcome on the DRS. Weak, but significant, correlations between injury severity and BAC and patterns of EtOH use were found. Given that injury severity was more strongly associated with TBI functional status, these authors concluded that chronic alcohol use or use at the time of injury is only weakly associated with early post-TBI status. De Guise et al. also examined the relationship between preinjury alcohol or drug abuse and a positive BAC on hospital admission in mild to severe TBI patients (De Guise et al., 2009). In their study, preinjury EtOH abuse was shown to significantly increase the duration of posttraumatic amnesia and hospital length of stay, regardless of alcohol state at time of admission. A history of drug abuse was more predictive of longer length of stay than was preinjury alcohol. Assessments at the time of discharge from acute care indicated that a history of preinjury abuse (alcohol or drug) or intoxication at time of injury did not adversely affect physical domains of activities of daily living or on cognitive, language, memory, or emotional state and behavior. In fact, FIM total and cognitive scores of patients intoxicated at time of injury were higher than those of patients who were sober at the time of TBI.

Schutte and Hanks reported that an admission BAC ≥0.08 mg/dL at the time of sustaining a mild complicated to severe TBI did lower the GCS and that admission BAC (as well as age and GCS) did impact functional outcome as measured by the FIM total score at the time of entry to inpatient rehabilitation (Schutte and Hanks, 2010). However, admission BAC did not predict FIM scores at discharge from rehabilitation or at 1 year follow-up. These authors also found no relationship between BAC and cognitive outcome (processing speed, memory, or executive function) 1 year after TBI. O'Dell et al. have reported no ethnicity differences for frequency of a positive admission BAC in cases of severe TBI (GCS motor score ≤5), but admission BAC was higher in Hispanics compared with Anglo-Caucasians and African-Americans (O'Dell et al., 2012). The admission BAC did not predict 6-month global outcome (DRS) outcome after controlling for injury severity and demographic variables and chronic preinjury alcohol use. Ethnicity and preinjury alcohol use were found to interact, such that Anglo-Caucasians and African-Americans with a history of alcohol use had worse DRS outcomes, whereas Hispanics with chronic preinjury alcohol use had more favorable outcomes. Ponsford et al. reported that harmful or hazardous preinjury EtOH use was associated with poor memory performance and slower processing speed at 6–9 and 12–15 months after moderate to severe TBI (Ponsford et al., 2013). A history of alcohol use before TBI did not affect executive functions in these time frames. However, if individuals with TBI were consuming any alcohol in the month before testing, their executive functioning was poorer than those not consuming alcohol at 6–9 months post-TBI. This latter finding is compatible with a report that individuals with TBI who resumed alcohol use within 1 year of injury had impairments on executive tasks (Jorge et al., 2005).

In summary, as is the case for research on the impact of alcohol on mortality, morbidity, or complications in TBI, the preceding work examining functional outcomes in TBI patients with acute or chronic EtOH exposure has produced a wide range of results. These variable results may depend in part on the types of outcomes that were assessed, the time of the assessments relative to TBI onset, and possibly ethnicity. Clinical studies examining the impact of day of injury EtOH use on functional outcomes after TBI have provided evidence for null effects (Alexander et al., 2004; Kaplan and Corrigan, 1992; O'Dell et al., 2012; Schutte and Hanks, 2010; Turner et al., 2006; Vickery et al., 2008; Wilde et al., 2004), deleterious effects (Bombardier and Thurber, 1998; Brooks et al., 1989; Kaplan and Corrigan, 1992; Kelly et al., 1997a; Schutte and Hanks, 2010; Sparadeo and Gill, 1989; Tate et al., 1999; Vickery et al., 2008), and rarely some beneficial effects (De Guise et al., 2009; Lange et al., 2008). A history of heavy or abusive alcohol use before TBI has been reported to have no effect on functional outcomes (Tate et al., 1999; Turner et al., 2006; Vickery et al., 2008; Wilde et al., 2004), to confer no additional disadvantage for outcomes compared with having a positive admission BAC (Brooks et al., 1989), to be equally or more important as having a positive admission BAC for impairing outcomes (Lange et al., 2007; Wilde et al., 2004), to worsen outcomes (Dikmen et al., 1995; O'Dell et al., 2012; Ponsford et al., 2013; Ronty et al., 1993; Ruff et al., 1990), or to improve functional outcomes after TBI (De Guise et al., 2009; O'Dell et al., 2012; Vickery et al., 2008).

REFERENCES

Alexander, S., M.E. Kerr, H. Yonas, and D.W. Marion. 2004. The effects of admission alcohol level on cerebral blood flow and outcomes after severe traumatic brain injury. *J Neurotrauma*. 21:575–83.

Arican, N., M. Kaya, C. Yorulmaz, R. Kalayci, H. Ince, M. Kucuk et al. 2006. Effect of hypothermia on blood-brain barrier permeability following traumatic brain injury in chronically ethanol-treated rats. *Int J Neurosci*. 116:1249–61.

Asa, I., Y. Ivashkova, A.A. Artru, M. Lifshitz, V. Gavrilov, A.N. Azab et al. 2003. LF 16–0687 Ms, a new bradykinin B2 receptor antagonist, improves neurologic outcome but not brain tissue prostaglandin E2 release in a rat model of closed head trauma combined with ethanol intoxication. *J Trauma*. 54:881–7.

Asmaro, K., P. Fu, and Y. Ding. 2013. Neuroprotection & mechanism of ethanol in stroke and traumatic brain injury therapy: New prospects for an ancient drug. *Curr Drug Targets*. 14:74–80.

Baratz, R., V. Rubovitch, H. Frenk, and C.G. Pick. 2010. The influence of alcohol on behavioral recovery after mTBI in mice. *J Neurotrauma*. 27:555–63.

Berry, C., E.J. Ley, D.R. Margulies, J. Mirocha, M. Bukur, D. Malinoski et al. 2011. Correlating the blood alcohol concentration with outcome after traumatic brain injury: Too much is not a bad thing. *Am Surg*. 77:1416–9.

Berry, C., A. Salim, R. Alban, J. Mirocha, D.R. Margulies, and E.J. Ley. 2010. Serum ethanol levels in patients with moderate to severe traumatic brain injury influence outcomes: A surprising finding. *Am Surg*. 76:1067–70.

Biros, M.H., D. Kukielka, R.L. Sutton, G.L. Rockswold, and T.A. Bergman. 1999. The effects of acute and chronic alcohol ingestion on outcome following multiple episodes of mild traumatic brain injury in rats. *Acad Emerg Med*. 6:1088–97.

Blondell, R.D., S.W. Looney, C.L. Krieg, and D.A. Spain. 2002. A comparison of alcohol-positive and alcohol-negative trauma patients. *J Stud Alcohol*. 63:380–3.

Bombardier, C.H., and C.A. Thurber. 1998. Blood alcohol level and early cognitive status after traumatic brain injury. *Brain Inj*. 12:725–34.

Bramlett, H.M., E.J. Green, W.D. Dietrich, R. Busto, M.Y. Globus, and M.D. Ginsberg. 1995. Posttraumatic brain hypothermia provides protection from sensorimotor and cognitive behavioral deficits. *J Neurotrauma*. 12:289–98.

Brooks, N., C. Symington, A. Beattie, L. Campsie, J. Bryden, and W. McKinlay. 1989. Alcohol and other predictors of cognitive recovery after severe head injury. *Brain Inj*. 3:235–46.

Chen, C.M., H.Y. Yi, Y.H. Yoon, and C. Dong. 2012. Alcohol use at time of injury and survival following traumatic brain injury: Results from the National Trauma Data Bank. *J Stud Alcohol Drugs*. 73:531–41.

Christensen, M.A., S. Janson, and J.A. Seago. 2001. Alcohol, head injury, and pulmonary complications. *J Neurosci Nurs*. 33:184–9.

Clifton, G.L., J.Y. Jiang, B.G. Lyeth, L.W. Jenkins, R.J. Hamm, and R.L. Hayes. 1991. Marked protection by moderate hypothermia after experimental traumatic brain injury. *J Cereb Blood Flow Metab*. 11:114–21.

Collins, M.A., J.Y. Zou, and E.J. Neafsey. 1998. Brain damage due to episodic alcohol exposure in vivo and in vitro: Furosemide neuroprotection implicates edema-based mechanism. *FASEB J*. 12:221–30.

Corrigan, J.D. 1995. Substance abuse as a mediating factor in outcome from traumatic brain injury. *Arch Phys Med Rehabil*. 76:302–9.

Cunningham, R.M., R.F. Maio, E.M. Hill, and B.J. Zink. 2002. The effects of alcohol on head injury in the motor vehicle crash victim. *Alcohol Alcohol*. 37:236–40.

Dash, P.K., A.N. Moore, M.R. Moody, R. Treadwell, J.L. Felix, and G.L. Clifton. 2004. Post-trauma administration of caffeine plus ethanol reduces contusion volume and improves working memory in rats. *J Neurotrauma*. 21:1573–83.

De Guise, E., J. Leblanc, J. Dagher, J. Lamoureux, A.A. Jishi, M. Maleki et al. 2009. Early outcome in patients with traumatic brain injury, pre-injury alcohol abuse and intoxication at time of injury. *Brain Inj*. 23:853–65.

Dietrich, W.D., O. Alonso, M. Halley, and R. Busto. 1996. Delayed posttraumatic brain hyperthermia worsens outcome after fluid percussion brain injury: A light and electron microscopic study in rats. *Neurosurgery*. 38:533–41; discussion 541.

Dikmen, S., T. Donovan. T. Loberg, J. Machamer, and N. Tempkin. 1993. Alcohol use and its effects on neuropsychological outcome in head injury. *Neuropsychology* 7:296–305.

Dikmen, S.S., J.E. Machamer, D.M. Donovan, H.R. Winn, and N.R. Temkin. 1995. Alcohol use before and after traumatic head injury. *Ann Emerg Med*. 26:167–76.

Dixon, C.E., C.G. Markgraf, F. Angileri, B.R. Pike, B. Wolfson, J.K. Newcomb et al. 1998. Protective effects of moderate hypothermia on behavioral deficits but not necrotic cavitation following cortical impact injury in the rat. *J Neurotrauma*. 15:95–103.

Fabian, M.J., and K.G. Proctor. 2002. Hemodynamic actions of acute ethanol after resuscitation from traumatic brain injury. *J Trauma*. 53:864–75.

Flamm, E.S., H.B. Demopoulos, M.L. Seligman, J.J. Tomasula, V. De Crescito, and J. Ransohoff. 1977. Ethanol potentiation of central nervous system trauma. *J Neurosurg*. 46:328–35.

Franco, C.D., C.R. Spillert, K.R. Spillert, and E.J. Lazaro. 1988. Alcohol increases mortality in murine head injury. *J Natl Med Assoc*. 80:63–5.

Goodman, M.D., A.T. Makley, E.M. Campion, L.A. Friend, A.B. Lentsch, and T.A. Pritts. 2013. Preinjury alcohol exposure attenuates the neuroinflammatory response to traumatic brain injury. *J Surg Res*. 184:1053–8.

Gottesfeld, Z., A.N. Moore, and P.K. Dash. 2002. Acute ethanol intake attenuates inflammatory cytokines after brain injury in rats: A possible role for corticosterone. *J Neurotrauma*. 19:317–26.

Gurney, J.G., F.P. Rivara, B.A. Mueller, D.W. Newell, M.K. Copass, and G.J. Jurkovich. 1992. The effects of alcohol intoxication on the initial treatment and hospital course of patients with acute brain injury. *J Trauma*. 33:709–13.

Hadjibashi, A.A., C. Berry, E.J. Ley, M. Bukur, J. Mirocha, D. Stolpner, and A. Salim. 2012. Alcohol is associated with a lower pneumonia rate after traumatic brain injury. *J Surg Res*. 173:212–5.

Hadjizacharia, P., T. O'Keeffe, D.S. Plurad, D.J. Green, C.V. Brown, L.S. Chan et al. 2011. Alcohol exposure and outcomes in trauma patients. *Eur J Trauma Emerg Surg*. 37:169–175.

Huth, J.F., R.V. Maier, D.A. Simonowitz, and C.M. Herman. 1983. Effect of acute ethanolism on the hospital course and outcome of injured automobile drivers. *J Trauma*. 23:494–8.

Is, M., T. Tanriverdi, F. Akyuz, M.O. Ulu, N. Ustundag, F. Gezen et al. 2005. Yings and Yangs of acute ethanol intoxication in experimental traumatic brain injury. *Neurosurg Q*15:60–64.

Jagger, J., D. Fife, K. Vernberg, and J.A. Jane. 1984. Effect of alcohol intoxication on the diagnosis and apparent severity of brain injury. *Neurosurgery*. 15:303–6.

Janis, L.S., M.R. Hoane, D. Conde, Z. Fulop, and D.G. Stein. 1998. Acute ethanol administration reduces the cognitive deficits associated with traumatic brain injury in rats. *J Neurotrauma*. 15:105–15.

Jorge, R.E., S.E. Starkstein, S. Arndt, D. Moser, B. Crespo-Facorro, and R.G. Robinson. 2005. Alcohol misuse and mood disorders following traumatic brain injury. *Arch Gen Psychiatry*. 62:742–9.

Jurkovich, G.J., F.P. Rivara, J.G. Gurney, C. Fligner, R. Ries, B.A. Mueller et al. 1993. The effect of acute alcohol intoxication and chronic alcohol abuse on outcome from trauma. *JAMA*. 270:51–6.

Kanbak, G., K. Kartkaya, E. Ozcelik, A.B. Guvenal, S.C. Kabay, G. Arslan et al. 2013. The neuroprotective effect of acute moderate alcohol consumption on caspase-3 mediated neuroapoptosis in traumatic brain injury: The role of lysosomal cathepsin L and nitric oxide. *Gene*. 512:492–5.

Kaplan, C.P., and J.D. Corrigan. 1992. Effect of blood alcohol level on recovery from severe closed head injury. *Brain Inj*. 6:337–49.

Katada, R., Y. Nishitani, O. Honmou, K. Mizuo, S. Okazaki, K. Tateda et al. 2012. Expression of aquaporin-4 augments cytotoxic brain edema after traumatic brain injury during acute ethanol exposure. *Am J Pathol*. 180:17–23.

Katada, R., Y. Nishitani, O. Honmou, S. Okazaki, K. Houkin, and H. Matsumoto. 2009. Prior ethanol injection promotes brain edema after traumatic brain injury. *J Neurotrauma*. 26:2015–25.

Kelly, D.F., D.A. Kozlowski, E. Haddad, A. Echiverri, D.A. Hovda, and S.M. Lee. 2000. Ethanol reduces metabolic uncoupling following experimental head injury. *J Neurotrauma*. 17:261–72.

Kelly, D.F., S.M. Lee, P.A. Pinanong, and D.A. Hovda. 1997a. Paradoxical effects of acute ethanolism in experimental brain injury. *J Neurosurg*. 86:876–82.

Kelly, M.P., C.T. Johnson, N. Knoller, D.A. Drubach, and M.M. Winslow. 1997b. Substance abuse, traumatic brain injury and neuropsychological outcome. *Brain Inj*. 11:391–402.

Kim, H.J., J.E. Levasseur, J.L. Patterson, Jr., G.F. Jackson, G.E. Madge, J.T. Povlishock et al. 1989. Effect of indomethacin pretreatment on acute mortality in experimental brain injury. *J Neurosurg*. 71:565–72.

Knoblach, S.M., and A.I. Faden. 1998. Interleukin-10 improves outcome and alters proinflammatory cytokine expression after experimental traumatic brain injury. *Exp Neurol*. 153:143–51.

Kong, L., G. Lian, W. Zheng, H. Liu, H. Zhang, and R. Chen. 2013. Effect of alcohol on diffuse axonal injury in rat brainstem: Diffusion tensor imaging and aquaporin-4 expression study. *Biomed Res Int*. 2013:798261.

Kraus, J.F., H. Morgenstern, D. Fife, C. Conroy, and P. Nourjah. 1989. Blood alcohol tests, prevalence of involvement, and outcomes following brain injury. *Am J Public Health*. 79:294–9.

Kreutzer, J.S., J.H. Marvitz, R. Seel, and C.D. Serio. 1996. Validation of a neurobehavioral functioning inventory for adults with traumatic brain injury. *Arch. Phys. Med. Rehabil*. 77:116–24.

Lai, X.P., X.J. Yu, H. Qian, L. Wei, J.Y. Lv, and X.H. Xu. 2013. Chronic alcoholism-mediated impairment in the medulla oblongata: A mechanism of alcohol-related mortality in traumatic brain injury? *Cell Biochem Biophys*. 67:1049–57.

Lange, R.T., G.L. Iverson, and M.D. Franzen. 2007. Short-term neuropsychological outcome following uncomplicated mild TBI: Effects of day-of-injury intoxication and pre-injury alcohol abuse. *Neuropsychology*. 21:590–8.

Lange, R.T., G.L. Iverson, and M.D. Franzen. 2008. Effects of day-of-injury alcohol intoxication on neuropsychological outcome in the acute recovery period following traumatic brain injury. *Arch Clin Neuropsychol*. 23:809–22.

Luna, G.K., R.V. Maier, L. Sowder, M.K. Copass, and M.R. Oreskovich. 1984. The influence of ethanol intoxication on outcome of injured motorcyclists. *J Trauma*. 24:695–700.

Lustenberger, T., K. Inaba, G. Barmparas, P. Talving, D. Plurad, L. Lam et al. 2011. Ethanol intoxication is associated with a lower incidence of admission coagulopathy in severe traumatic brain injury patients. *J Neurotrauma*. 28:1699–706.

Mann, B., E. Desapriya, T. Fujiwara, and I. Pike. 2011. Is blood alcohol level a good predictor for injury severity outcomes in motor vehicle crash victims? *Emerg Med Int*. 2011:616323.

Marion, D.W., L.E. Penrod, S.F. Kelsey, W.D. Obrist, P.M. Kochanek, A.M. Palmer et al. 1997. Treatment of traumatic brain injury with moderate hypothermia. *N Engl J Med*. 336:540–6.

Masse, J., B. Billings, H.S. Dhillon, D. Mace, R. Hicks, S. Barron et al. 2000. Three months of chronic ethanol administration and the behavioral outcome of rats after lateral fluid percussion brain injury. *J Neurotrauma*. 17:421–30.

McClain, C., D. Cohen, R. Phillips, L. Ott, and B. Young. 1991. Increased plasma and ventricular fluid interleukin-6 levels in patients with head injury. *J Lab Clin Med*. 118:225–31.

McClain, C.J., D. Cohen, L. Ott, C.A. Dinarello, and B. Young. 1987. Ventricular fluid interleukin-1 activity in patients with head injury. *J Lab Clin Med*. 110:48–54.

Morganti-Kossmann, M.C., L. Satgunaseelan, N. Bye, and T. Kossmann. 2007. Modulation of immune response by head injury. *Injury*. 38:1392–400.

Nath, F.P., G. Beastal, and G.M. Teasdale. 1986. Alcohol and traumatic brain damage. *Injury*. 17:150–3.

O'Dell, K.M., H.J. Hannay, F.O. Biney, C.S. Robertson, and T.S. Tian. 2012. The effect of blood alcohol level and preinjury chronic alcohol use on outcome from severe traumatic brain injury in Hispanics, anglo-Caucasians, and African-Americans. *J Head Trauma Rehabil*. 27:361–9.

O'Phelan, K., D.L. McArthur, C.W. Chang, D. Green, and D.A. Hovda. 2008. The impact of substance abuse on mortality in patients with severe traumatic brain injury. *J Trauma*. 65:674–7.

Opreanu, R.C., D. Kuhn, and M.D. Basson. 2010. Influence of alcohol on mortality in traumatic brain injury. *J Am Coll Surg*. 210:997–1007.

Parry-Jones, B.L., F.L. Vaughan, and W. Miles Cox. 2006. Traumatic brain injury and substance misuse: A systematic review of prevalence and outcomes research (1994–2004). *Neuropsychol Rehabil*. 16:537–60.

Perez-Polo, J.R., H.C. Rea, K.M. Johnson, M.A. Parsley, G.C. Unabia, G. Xu et al. 2013. Inflammatory consequences in a rodent model of mild traumatic brain injury. *J Neurotrauma*. 30:727–40.

Persson, L. and L. Rosengren. 1977. Increased blood-brain barrier permeability around cerebral stab wounds, aggravated by acute ethanol intoxication. *Acta Neurol Scand*. 56:7–16.

Plurad, D., D. Demetriades, G. Gruzinski, C. Preston, L. Chan, D. Gaspard et al. 2010. Motor vehicle crashes: The association of alcohol consumption with the type and severity of injuries and outcomes. *J Emerg Med*. 38:12–7.

Ponsford, J., L. Tweedly, and J. Taffe. 2013. The relationship between alcohol and cognitive functioning following traumatic brain injury. *J Clin Exp Neuropsychol*. 35:103–12.

Pories, S.E., R.L. Gamelli, P. Vacek, G. Goodwin, T. Shinozaki, and F. Harris. 1992. Intoxication and injury. *J Trauma*. 32:60–4.

Prasad, R.M., I. Doubinskaia, D.K. Singh, G. Campbell, D. Mace, A. Fletcher et al. 2001. Effects of binge ethanol administration on the behavioral outcome of rats after lateral fluid percussion brain injury. *J Neurotrauma*. 18:1019–29.

Prasad, R.M., A. Laabich, H.S. Dhillon, L. Zhang, A. Maki, W.J. Clerici et al. 1997. Effects of six weeks of chronic ethanol administration on lactic acid accumulation and high energy phosphate levels after experimental brain injury in rats. *J Neurotrauma*. 14:919–30.

Ronty, H., A. Ahonen, U. Tolonen, J. Heikkila, and O. Niemela. 1993. Cerebral trauma and alcohol abuse. *Eur J Clin Invest*. 23:182–7.

Rootman, D.B., R. Mustard, V. Kalia, and N. Ahmed. 2007. Increased incidence of complications in trauma patients cointoxicated with alcohol and other drugs. *J Trauma*. 62:755–8.

Ruff, R.M., L.F. Marshall, M.R. Klauder, B.A. Blunt, I. Grant, M.A. Foulkes et al. 1990. Alcohol abuse and neurological outcome of the severely head injured patient. *J Head Trauma Rehab*. 521–31.

Salim, A., E.J. Ley, H.G. Cryer, D.R. Margulies, E. Ramicone, and A. Tillou. 2009a. Positive serum ethanol level and mortality in moderate to severe traumatic brain injury. *Arch Surg*. 144:865–71.

Salim, A., P. Teixeira, E.J. Ley, J. DuBose, K. Inaba, and D.R. Margulies. 2009b. Serum ethanol levels: Predictor of survival after severe traumatic brain injury. *J Trauma*. 67:697–703.

Scheyerer, M.J., J. Dutschler, A. Billeter, S.M. Zimmermann, K. Sprengel, C.M. Werner et al. 2013. Effect of elevated serum alcohol level on the outcome of severely injured patients. *Emerg Med J*. July 12.

Schmidt, O.I., C.E. Heyde, W. Ertel, and P.F. Stahel. 2005. Closed head injury—an inflammatory disease? *Brain Res Brain Res Rev*. 48:388–99.

Schutte, C., and R. Hanks. 2010. Impact of the presence of alcohol at the time of injury on acute and one-year cognitive and functional recovery after traumatic brain injury. *Int J Neurosci*. 120:551–6.

Shandro, J.R., F.P. Rivara, J. Wang, G.J. Jurkovich, A.B. Nathens, and E.J. MacKenzie. 2009. Alcohol and risk of mortality in patients with traumatic brain injury. *J Trauma*. 66:1584–90.

Shapira, Y., A.M. Lam, A. Paez, A.A. Artru, V. Laohaprasit, and T. Donato. 1997. The influence of acute and chronic alcohol treatment on brain edema, cerebral infarct volume and neurological outcome following experimental head trauma in rats. *J Neurosurg Anesthesiol*. 9:118–27.

Shih, H.C., S.C. Hu, C.C. Yang, T.J. Ko, J.K. Wu, and C.H. Lee. 2003. Alcohol intoxication increases morbidity in drivers involved in motor vehicle accidents. *Am J Emerg Med*. 21:91–4.

Shohami, E., I. Ginis, and J.M. Hallenbeck. 1999. Dual role of tumor necrosis factor alpha in brain injury. *Cytokine Growth Factor Rev*. 10:119–30.

Shohami, E., M. Novikov, R. Bass, A. Yamin, and R. Gallily. 1994. Closed head injury triggers early production of TNF alpha and IL-6 by brain tissue. *J Cereb Blood Flow Metab*. 14:615–9.

Sparadeo, F.R., and D. Gill. 1989. Effects of prior alcohol use on head injury recovery. *J. Head Trauma Rehabil*. 4:75–81.

Sperry, J.L., L.M. Gentilello, J.P. Minei, R.R. Diaz-Arrastia, R.S. Friese, and S. Shafi. 2006. Waiting for the patient to "sober up": Effect of alcohol intoxication on Glasgow Coma Scale score of brain injured patients. *J Trauma*. 61:1305–11.

Stuke, L., R. Diaz-Arrastia, L.M. Gentilello, and S. Shafi. 2007. Effect of alcohol on Glasgow Coma Scale in head-injured patients. *Ann Surg*. 245:651–5.

Talving, P., D. Plurad, G. Barmparas, J. Dubose, K. Inaba, L. Lam et al. 2010. Isolated severe traumatic brain injuries: Association of blood alcohol levels with the severity of injuries and outcomes. *J Trauma*. 68:357–62.

Tate, P.S., D.M. Freed, C.H. Bombardier, S.L. Harter, and S. Brinkman. 1999. Traumatic brain injury: Influence of blood alcohol level on post-acute cognitive function. *Brain Inj*. 13:767–84.

Taylor, A.N., S.U. Rahman, D.L. Tio, M.J. Sanders, J.K. Bando, A.H. Truong et al. 2006. Lasting neuroendocrine-immune effects of traumatic brain injury in rats. *J Neurotrauma*. 23:1802–13.

Taylor, A.N., H.E. Romeo, A.V. Beylin, D.L. Tio, S.U. Rahman, and D.A. Hovda. 2002. Alcohol consumption in traumatic brain injury: Attenuation of TBI-induced hyperthermia and neurocognitive deficits. *J Neurotrauma*. 19:1597–608.

Teng, S.X., and P.E. Molina. 2013. Acute alcohol intoxication prolongs neuroinflammation without exacerbating neurobehavioral dysfunction following mild traumatic brain injury. *J Neurotrauma*. 31:378–86.

Thal, E.R., R.O. Bost, and R.J. Anderson. 1985. Effects of alcohol and other drugs on traumatized patients. *Arch Surg*. 120:708–12.

Tien, H.C., L.N. Tremblay, S.B. Rizoli, J. Gelberg, T. Chughtai, P. Tikuisis et al. 2006. Association between alcohol and mortality in patients with severe traumatic head injury. *Arch Surg*. 141:1185–91; discussion 1192.

Toulmond, S., and N.J. Rothwell. 1995. Interleukin-1 receptor antagonist inhibits neuronal damage caused by fluid percussion injury in the rat. *Brain Res*. 671:261–6.

Tureci, E., R. Dashti, T. Tanriverdi, G.Z. Sanus, B. Oz, and M. Uzan. 2004. Acute ethanol intoxication in a model of traumatic brain injury: The protective role of moderate doses demonstrated by immunoreactivity of synaptophysin in hippocampal neurons. *Neurol Res*. 26:108–12.

Turner, A.P., R.M. Williams, J.D. Bowen, D.R. Kivlahan, and J.K. Haselkorn. 2006. Suicidal ideation in multiple sclerosis. *Arch Phys Med Rehabil*. 87:1073–8.

Vickery, C.D., M. Sherer, T.G. Nick, R. Nakase-Richardson, J.D. Corrigan, F. Hammond et al. 2008. Relationships among premorbid alcohol use, acute intoxication, and early functional status after traumatic brain injury. *Arch Phys Med Rehabil*. 89:48–55.

Vink, R., and R.W. Byard. 2012. Alcohol intoxication may exacerbate the effects of blunt cranial trauma through changes in brain free magnesium levels. *J Forensic Sci*. 57:979–82.

Wang, T., D.Y. Chou, J.Y. Ding, V. Fredrickson, C. Peng, S. Schafer et al. 2013. Reduction of brain edema and expression of aquaporins with acute ethanol treatment after traumatic brain injury. *J Neurosurg*. 118:390–6.

Ward, R.E., T.C. Flynn, P.W. Miller, and W.F. Blaisdell. 1982. Effects of ethanol ingestion on the severity and outcome of trauma. *Am J Surg*. 144:153–7.

Wilde, E.A., E.D. Bigler, P.V. Gandhi, C.M. Lowry, D.D. Blatter, J. Brooks et al. 2004. Alcohol abuse and traumatic brain injury: Quantitative magnetic resonance imaging and neuropsychological outcome. *J Neurotrauma*. 21:137–47.

Yamakami, I., R. Vink, A.I. Faden, T.A. Gennarelli, R. Lenkinski, and T.K. McIntosh. 1995. Effects of acute ethanol intoxication on experimental brain injury in the rat: Neurobehavioral and phosphorus-31 nuclear magnetic resonance spectroscopy studies. *J Neurosurg*. 82:813–21.

Zeckey, C., S. Dannecker, F. Hildebrand, P. Mommsen, R. Scherer, C. Probst et al. 2011. Alcohol and multiple trauma: Is there an influence on the outcome? *Alcohol*. 45:245–51.

Zhang, L., A. Maki, H.S. Dhillon, S. Barron, W.J. Clerici, R. Hicks et al. 1999. Effects of six weeks of chronic ethanol administration on the behavioral outcome of rats after lateral fluid percussion brain injury. *J Neurotrauma*. 16:243–54.

Zink, B.J., and P.J. Feustel. 1995. Effects of ethanol on respiratory function in traumatic brain injury. *J Neurosurg*. 82:822–8.

Zink, B.J., C.H. Schultz, X. Wang, M. Mertz, S.A. Stern, and A.L. Betz. 1999. Effects of ethanol on brain lactate in experimental traumatic brain injury with hemorrhagic shock. *Brain Res*. 837:1–7.

Zink, B.J., M.A. Sheinberg, X. Wang, M. Mertz, S.A. Stern, and A.L. Betz. 1998. Acute ethanol intoxication in a model of traumatic brain injury with hemorrhagic shock: Effects on early physiological response. *J Neurosurg*. 89:983–90.

Zink, B.J., S.A. Stern, B.D. McBeth, X. Wang, and M. Mertz. 2006. Effects of ethanol on limited resuscitation in a model of traumatic brain injury and hemorrhagic shock. *J Neurosurg*. 105:884–93.

Zink, B.J., R.F. Walsh, and P.J. Feustel. 1993. Effects of ethanol in traumatic brain injury. *J Neurotrauma*. 10:275–86.

49 Modeling Traumatic Brain Injury-Induced Alterations in Alcohol Use Behaviors

James P. Caruso, Jennifer L. Lowing, and Alana C. Conti

CONTENTS

49.1 INTRODUCTION

Traumatic brain injury (TBI) is a heterogeneous condition that often results in long-term consequences, such as the development of aberrant alcohol use patterns and progression to dependence. Despite this observation, limited data exists from use of preclinical TBI models to explore injury-induced alterations in alcohol-related behaviors, which has hampered progress in the field. Here we describe a subset of paradigms useful for evaluating features of alcohol use in rodents and their compatibility with existing models of TBI. The combination of TBI and alcohol-use models comes with considerations, but can be an effective means to further our understanding of TBI as a risk factor for alcohol use behaviors, identify underlying mechanisms of this clinical observation and develop therapies for use in the affected population.

49.1.1 Traumatic Brain Injury and Alcohol Misuse: The Clinical Perspective

Although injury symptoms may be subtle or even undetectable, as in the case of mild TBI, long-lasting consequences often develop in the brain-injured patient. The heterogeneity of TBI results in a wide spectrum of behavioral dysfunction. Numerous reports demonstrate that TBI results in cognitive impairments and increased risk of both affective disorder and Alzheimer disease (Hibbard et al., 1998; McDonald et al., 2002; Salmond and Sahakian, 2005; Rogers and Read, 2007; VanDenHeuvel et al., 2007). Additional reports indicate that TBI survivors often acquire behavioral disturbances, including development of substance and/or alcohol misuse behaviors (Adams et al., 2012a; Corrigan and Cole, 2008). However, data fully describing the propensity for TBI victims to develop substance use disorders (SUD), specifically alcohol use disorders, especially in the absence of additional risk factors for SUD at the time of injury, are limited.

Many studies have focused on a role for drug and alcohol use as a precipitating factor in causing TBI (motor vehicle accidents, falls, or violent acts) (Kolakowsky-Hayner et al., 1999; Taylor et al., 2010). Likewise, alcohol intoxication has been shown to impair recovery from TBI (Corrigan, 1995; Jorge et al., 2005) and increase the risk of SUD in psychologically vulnerable individuals after TBI (Horger et al., 1990). However, despite the finding that alcohol abuse/dependence is the second most common axis I disorder in persons with TBI, examination of alcohol use behaviors after TBI have received surprisingly little attention in the literature (Koponen et al., 2002) and have been largely focused on clinical studies. Limited data suggest increased rates of SUD in a cohort of 100 TBI patients (in the absence of axis I disorders before injury) compared with controls (Hibbard et al., 1998). Similar results were observed in self-reported severe TBI victims after controlling for alcohol use before injury (Silver et al., 2001). In a longitudinal study, McKinlay et al. identified significant increases in substance abuse after TBI (McKinlay et al., 2009). Similarly, alcohol misuse has been reported after TBI in civilians (Corrigan et al., 1995, 1998b; Kolakowsky-Hayner et al., 2002; Ponsford et al., 2007) and military populations (Adams et al., 2012a, 2012b; Carlson et al., 2010; Rona et al., 2012). Among the consequences of TBI and subsequent alcohol misuse are prominent health concerns with staggering associated psychosocial complications and financial costs. Studies have shown that excessive substance abuse leads to increased mortality post-TBI, poorer neuropsychological performances at 1 month and 1 year post-TBI, and greater likelihood of repeat injury (Corrigan, 1995). These consequences contribute to the estimated $9–10 billion cost of care resulting from 1.5–2 million new TBI cases each year (Graham and Cardon, 2008).

Because prior patterns of alcohol use can be risk factor for future TBI, it is important to consider preinjury alcohol use patterns. Generally, preinjury heavy alcohol use largely predicts postinjury heavy alcohol use (Bombardier et al., 2003; Horner et al., 2005). Although there are reports of reduced drinking after TBI, those who continue to drink are more often the moderate/heavy preinjury drinkers (Bombardier et al., 2002; Kolakowsky-Hayner et al., 2002). The initial reduction in drinking after TBI may be attributed to reduced mobility of injured patients or the inability to access alcohol while under medical supervision. Importantly, as many as 20% of light or nondrinkers preinjury report high-volume drinking after TBI, suggesting that TBI is a risk factor for alcohol use disorders in nondependent subjects (Corrigan et al., 1995).

Considering the history of pervasive alcohol abuse among military populations, the consequences of combat-related TBI prove especially foreboding. The ratio of wounded to killed service members has increased in recent conflicts such as Operation Iraqi Freedom (OIF) and Operation Enduring Freedom (OEF), because of improvements in body armor, war zone evacuation techniques, and combat-zone medical treatment (Adams et al., 2012a; IOM, 2008). Approximately 20%–22% of OIF and OEF veterans screen positive for TBI, which may exacerbate the effects of alcohol use disorders in these veterans (Adams et al., 2012a; Carlson et al., 2010; Terrio et al., 2009). This increase in the number of injured survivors, combined with reports that one-third of military personnel meet the criteria for hazardous drinking (Fiellin et al., 2000), highlight the vulnerability of the military population.

Among OIF/OEF personnel, those that have sustained TBI are significantly more likely to report frequent binge drinking and the odds of binge drinking increase with TBI severity (Adams et al., 2012a). Considering that 50%–60% of persons with TBI endure struggles with substance misuse (West, 2011), investigation of the relationship between TBI and alcohol use behaviors proves imperative.

Animal models can provide a means to further understand the neurobiology and molecular mechanisms associated with injury-induced alterations in alcohol sensitivity and use. It is most important, however, to combine models of TBI that are compatible with behavioral paradigms that assess alcohol use and sensitivity while maintaining high levels of validity. Only then can these combined models be further explored to identify the molecular underpinnings of alcohol behaviors after TBI. Clearly, no single rodent model can sufficiently represent the complexities associated with alcohol abuse disorders and dependence (Rhodes et al., 2005; Crabbe et al., 2011). However, several well-characterized paradigms address individual features of these clinical conditions, including alcohol preference, motivation to consume alcohol, voluntary alcohol consumption/binge patterns, and alcohol sensitivity. Here, some of the key rodent models of alcohol use and the facets of alcoholism that they represent will be introduced. This chapter will focus on those alcohol-related outcomes that have been examined in the brain-injured population.

49.2 ETHANOL BEHAVIOR MODELS

49.2.1 Binge Drinking

Early-life binge drinking increases the risk for developing alcohol dependence (Hingson et al., 2005, 2006; Miller et al., 2007) and may represent a behavioral transition to alcoholism. Binge drinking is defined by the National Institute on Alcohol Abuse and Alcoholism as a pattern of consumption that results in blood ethanol concentrations (BECs) that exceed 0.08% (80 mg/dL) (NIAAA, 2004). This is generally achieved by ingesting four or five drinks within a 2-hour period for adult women or men, respectively. It should follow then that models of binge drinking should incorporate voluntary oral administration of ethanol for intoxicating effects over a fixed duration that results in pharmacologically relevant BECs. This is most readily and recently modeled using the drinking-in-the-dark model (DID) (Rhodes et al., 2005). In this paradigm, mice (typically C57Bl/6) are provided 20% (v/v) ethanol for 2 hours, 3 hours after the start of the dark cycle for 3 days.

On day 4, the ethanol is available for 4 hours to complete the 4-day DID cycle. A distinct advantage of DID is that ethanol consumption is voluntary, as opposed to investigator-administered, although the mechanisms responsible for the propensity of C57Bl/6 mice to consume ethanol over other strains are less clear. The inherent appetitive behavior of mice during the dark cycle facilitates binge drinking of C57Bl/6 mice early in the dark cycle, overcoming the innate tendency for mice to avoid or limit alcohol consumption to levels that do not result in intoxication. No fluid restriction is needed to encourage ethanol consumption, as is required in some scheduled high-alcohol consumption paradigms. In addition, alcohol access is limited, as opposed to unlimited two-bottle choice access. Limited access allows for more precise regulation of consumption volume to determine whether alcohol intake reaches physiologically relevant levels. In the DID paradigm, mice readily consume ethanol in the early phase of the dark cycle, resulting in BECs that exceed 80 mg/dL, typically approaching 100 mg/dL. Of note, higher BECs will be achieved if no water is available during this ethanol exposure period, but animals will readily drink whether or not water is present concurrently with ethanol (Rhodes et al., 2007), suggesting animals are not motivated by thirst to consume ethanol. Consumption of high concentrations of ethanol (20%) is observed in this paradigm without training or sucrose fading and does not appear to be a result of dietary demand (Lyons et al., 2008), but does result in physical signs of intoxication (Rhodes et al., 2007). Since its initial characterization, modifications of the DID paradigm have been reported. For example, the number of days of ethanol exposure per cycle can be increased, which can increase rates of consumption, as can the number of DID cycles that can be presented (Linsenbardt et al., 2009, 2011; Sajja and Rahman, 2011). Likewise, the duration of ethanol access can be reduced from 2 hours to better adapt to the overall experimental design, as long as ethanol is presented during the active period of the dark cycle (3 hours after dark onset) (Sajja and Rahman, 2011). In sum, when used with ethanol-preferring mice given limited access at 3 hours into the dark cycle, the DID paradigm demonstrates high face validity as a model of human binge alcohol drinking. The straightforward experimental design also makes it amenable for testing pharmacological agents or therapies.

49.2.2 Chronic High-Level Consumption

Chronic high-level alcohol consumption that extends beyond binge drinking patterns is facilitated using chronic intermittent access to alcohol. Animals that have extended daily access to alcohol demonstrate a transient, but robust, increase in voluntary ethanol consumption upon reintroduction to alcohol after a period of deprivation (alcohol deprivation effect), although this enhancement in consumption is not consistently observed and may be nonsustained (Agabio et al., 2000; Sinclair and Li, 1989; Sinclair and Tiihonen, 1988). Some of these disadvantages have been overcome with multiple deprivation periods and simultaneous access to multiple concentrations of alcohol (Rodd et al., 2009; Rodd-Henricks et al., 2000; Spanagel and Holter, 1999, 2000).

A modification of this paradigm has recently regained popularity. Free access to alcohol, in the presence of water, is interspersed with alcohol abstinence periods as described; however, both access and abstinence periods are shorter (Simms et al., 2008; Wayner et al., 1972; Wise, 1973) relative to the described scheduled alcohol deprivation models (Sinclair and Li, 1989; Salimov and Salimova, 1993; Sinclair and Senter, 1968). For example, Simms and colleagues provided 24-hour free access to 20% ethanol in the presence of water for 3 days per week (with no more than 2 days between ethanol access), which increased consumption from 2 g/kg to nearly 6 g/kg after five to six drinking sessions in Long-Evans rats (Simms et al., 2008). Using this exposure paradigm, most subjects reached BECs exceeding 0.08 g/dL in the first 30 minutes of drinking (Carnicella et al., 2009; Simms et al., 2008). The strength of chronic intermittent access paradigms is the escalation of intake in the absence of training, sucrose fading, fluid restriction, or schedule-induced polydipsia. Similarly, chronic intermittent access has been coupled with oral alcohol self-administration using operant conditioning paradigms (Simms et al., 2010).

49.2.3 Alcohol Dependence, Withdrawal, and Consumption

There is a long-standing association between alcohol dependence and promotion of excessive drinking (Cappell, 1981; Grant, 1995). Recent studies have explored this relationship by establishing the ability of dependence to alter consumption behavior in rodent models (Becker, 2008, 2011). Dependence is most commonly induced by intermittently interspersing alcohol vapor exposure using inhalation chambers (16 hours/day for 4 days) with withdrawal periods. This paradigm has superior efficacy for establishing dependence as rodents typically will not orally consume sufficient amounts of alcohol to produce dependence. Consumption has subsequently been measured in dependent animals using unlimited- or limited-access schedules (Becker and Lopez, 2004; Dhaher et al., 2008; Finn et al., 2007; Lopez and Becker, 2005; Rimondini et al., 2002, 2003; Sommer et al., 2008). Similarly, dependent animals have been evaluated for alcohol self-administration (Chu et al., 2007; Funk and Koob, 2007; Gilpin et al., 2008, 2009; Lopez, 2008; O'Dell et al., 2004; Richardson et al., 2008; Roberts et al., 1996, 2000).

Repeated withdrawal experiences have been demonstrated to be critical to increased alcohol self-administration and are proportional to volume and duration of intake (Lopez and Becker, 2005). Although the induction of dependence using this model suffers from weak face validity (use of inhalation compared with oral administration), it is effective in establishing a history of chronic alcohol exposure that enhances alcohol consumption. This is an appealing feature when probing the effects of TBI on alcohol-dependent subjects compared with those that are alcohol-naive.

49.3 COMPATIBILITY OF TBI MODELS WITH ETHANOL PARADIGMS

49.3.1 Species Considerations

The discussed ethanol paradigms are largely applicable to rats and mice alike, making them amenable for use with common models of TBI. For example, although traditionally used in mice, the DID model has been modified for use in alcohol-preferring rats (a strain akin to C57Bl/6 mice that readily drink ethanol) by introducing ethanol for three 1-hour sessions, 2 hours apart, immediately upon dark cycle onset (Bell et al., 2006, 2011; McBride et al., 2010), termed the DID-multiple-schedule-access paradigm. Similarly, when using the DID model in mice, choice of mouse strain is paramount. C57Bl/6 mice exhibit high levels of ethanol drinking, whereas 129 mice drink considerably less ethanol (Belknap et al., 1993). This consideration is especially important when using genetically engineered mouse models that are commonly generated using embryonic stem cells from 129 substrains. This may require multiple generations of backcrossing to C57Bl/6 mice to achieve transgenic or knockout mice that are on an ethanol-preferring background strain. Conversely, chronic intermittent access, a model initially characterized in rats (Simms et al., 2008; Wise, 1973), has been adapted for use in mice using 15–20% ethanol with similar patterns of consumption escalation compared to continuous access (Hwa et al., 2011; Melendez, 2011).

49.3.2 TBI Outcomes

A factor to consider when using TBI models is the dependence of many of these alcohol consumption paradigms on intact motor performance. This prohibits the use of most severe TBI models and limits the extent to which moderate

TBI models can be used, at least during the acute post-TBI period when neuromotor performance is most affected. Alternatively, models of mild, repetitive, or blast TBI would be highly compatible with these ethanol paradigms and parallel clinical reports that document increased alcohol use after these types of brain injuries (Rona et al., 2012). Similarly, use of operant self-administration may be challenging with use of TBI models that result in significant cognitive impairment, demanding careful interpretation of results. In general, a variety of TBI models would be suitable for use with ethanol behavior paradigms, including fluid percussion injury, controlled cortical impact, blast neurotrauma, and weight-drop models. Of these, controlled cortical impact has been well-characterized for use in both mice and rats (Dixon et al., 1991; Smith et al., 1995).

49.4 CONCLUSION

Numerous experimental approaches have been developed to represent facets of alcohol consumption and patterns of alcohol use because the complexities associated with alcoholism cannot be fully represented in a single model. Although each approach is unique, they share the common goal of exploring the neurophysiology underlying alcohol use disorders. The application of these paradigms for use in TBI comes with considerations; however, they are largely amenable for combination with most TBI rodent models. Together, these approaches will further our understanding of TBI as a risk factor for alcohol use behaviors, the underlying mechanisms of this observation, and the development of therapies for use in the affected population.

REFERENCES

Adams, R. S., J. D. Corrigan, and M. J. Larson (2012a). Alcohol use after combat-acquired traumatic brain injury: What we know and don't know. *J Soc Work Pract Addict* 12(1):28–51.

Adams, R. S., M. J. Larson, J. D. Corrigan, C. M. Horgan, and T. V. Williams (2012b). Frequent binge drinking after combat-acquired traumatic brain injury among active duty military personnel with a past year combat deployment. *J Head Trauma Rehabil* 27(5):349–360.

Agabio, R., M. A. Carai, C. Lobina, M. Pani, R. Reali, G. Vacca et al. (2000). Development of short-lasting alcohol deprivation effect in sardinian alcohol-preferring rats. *Alcohol* 21(1):59–62.

Becker, H. C. (2008). Alcohol dependence, withdrawal and relapse. *Alcohol Res Health* 31:348–361.

Becker, H. C. and M. F. Lopez (2004). Increased ethanol drinking after repeated chronic ethanol exposure and withdrawal experience in C57BL/6 mice. *Alcohol Clin Exp Res* 28(12):1829–1838.

Becker, H. C., Lopez, M. F., and Doremus-Fitzwater, T. L. (2011). Effects of stress on alcohol drinking: A review of animal studies. *Psychopharmacology (Berl)* 218:131–156.

Belknap, J. K., J. C. Crabbe, and E. R. Young (1993). Voluntary consumption of ethanol in 15 inbred mouse strains. *Psychopharmacology (Berl)* 112(4):503–510.

Bell, R. L., Z. A. Rodd, L. Lumeng, J. M. Murphy, and W. J. McBride (2006). The alcohol-preferring P rat and animal models of excessive alcohol drinking. *Addict Biol* 11(3–4):270–288.

Bell, R. L., Z. A. Rodd, R. J. Smith, J. E. Toalston, K. M. Franklin, and W. J. McBride (2011). Modeling binge-like ethanol drinking by peri-adolescent and adult P rats. *Pharmacol Biochem Behav* 100(1):90–97.

Bombardier, C. H., C. T. Rimmele, and H. Zintel (2002). The magnitude and correlates of alcohol and drug use before traumatic brain injury. *Arch Phys Med Rehabil* 83(12):1765–1773.

Bombardier, C. H., N. R. Temkin, J. Machamer, and S. S. Dikmen (2003). The natural history of drinking and alcohol-related problems after traumatic brain injury. *Arch Phys Med Rehabil*84(2):185–191.

Cappell, H. and A. E. LeBlanc (1981). Tolerance and physical dependence: Do they play a role in alcohol and drug self-administration? In Israel, Y., Glaser, F.B., Kalant, H., Popham, R.E., Schmidt, W., and Smart, R.G., eds. *Research Advances in Alcohol and Drug Problems*. Plenum, New York. pp. 159–196.

Carlson, K. F., D. Nelson, R. J. Orazem, S. Nugent, D. Cifu, and N. A. Sayer (2010). Psychiatric diagnoses among Iraq and Afghanistan war veterans screened for deployment-related traumatic brain injury. *J Trauma Stress* 23(1):17–24.

Carnicella, S., R. Amamoto, and D. Ron (2009). Excessive alcohol consumption is blocked by glial cell line-derived neurotrophic factor. *Alcohol* 43(1):35–43.

Chu, K., G. F. Koob, M. Cole, E. P. Zorrilla, and A. J. Roberts (2007). Dependence-induced increases in ethanol self-administration in mice are blocked by the CRF1 receptor antagonist antalarmin and by CRF1 receptor knockout. *Pharmacol Biochem Behav* 86(4):813–821.

Corrigan, J. D. (1995). Substance abuse as a mediating factor in outcome from traumatic brain injury. *Arch Phys Med Rehabil*76(4):302–309.

Corrigan, J. D. and T. B. Cole (2008). Substance use disorders and clinical management of traumatic brain injury and posttraumatic stress disorder. *JAMA* 300(6):720–721.

Corrigan, J. D., G. L. Lamb-Hart, and E. Rust (1995). A programme of intervention for substance abuse following traumatic brain injury. *Brain Inj* 9(3):221–236.

Corrigan, J. D., K. Smith-Knapp, and C. V. Granger (1998). Outcomes in the first 5 years after traumatic brain injury. *Arch Phys Med Rehabil* 79(3):298–305.

Crabbe, J. C., R. A. Harris, and G. F. Koob (2011). Preclinical studies of alcohol binge drinking. *Ann N Y Acad Sci*1216:24–40.

Dhaher, R., D. Finn, C. Snelling, and R. Hitzemann (2008). Lesions of the extended amygdala in C57BL/6J mice do not block the intermittent ethanol vapor-induced increase in ethanol consumption. *Alcohol Clin Exp Res* 32(2):197–208.

Dixon, C. E., G. L. Clifton, J. W. Lighthall, A. A. Yaghmai, and R. L. Hayes (1991). A controlled cortical impact model of traumatic brain injury in the rat. *J Neurosci Methods* 39(3):253–262.

Fiellin, D. A., M. C. Reid, and P. G. O'Connor (2000). Screening for alcohol problems in primary care: A systematic review. *Arch Intern Med*160(13):1977–1989.

Finn, D. A., C. Snelling, A. M. Fretwell, M. A. Tanchuck, L. Underwood, M. Cole et al. (2007). Increased drinking during withdrawal from intermittent ethanol exposure is blocked by the CRF receptor antagonist D-Phe-CRF(12–41). *Alcohol Clin Exp Res* 31(6):939–949.

Funk, C. K. and G. F. Koob (2007). A CRF(2) agonist administered into the central nucleus of the amygdala decreases ethanol self-administration in ethanol-dependent rats. *Brain Res* 1155:172–178.

Gilpin, N. W., H. N. Richardson, L. Lumeng, and G. F. Koob (2008). Dependence-induced alcohol drinking by alcohol-preferring (P) rats and outbred Wistar rats. *Alcohol Clin Exp Res* 32(9):1688–1696.

Gilpin, N. W., A. D. Smith, M. Cole, F. Weiss, G. F. Koob, and H. N. Richardson (2009). Operant behavior and alcohol levels in blood and brain of alcohol-dependent rats. *Alcohol Clin Exp Res* 33(12):2113–2123.

Graham, D. P. and A. L. Cardon (2008). An update on substance use and treatment following traumatic brain injury. *Ann N Y Acad Sci*1141:148–162.

Grant, K. (1995). Animal models of the alcohol addiction process. In H. Kransler, ed. *The Pharmacology of Alcohol Abuse*. Springer, Berlin. pp. 185–230.

Hibbard, M. R., S. Uysal, K. Kepler, J. Bogdany, and J. Silver (1998). Axis I psychopathology in individuals with traumatic brain injury. *J Head Trauma Rehabil* 13(4):24–39.

Hingson, R., T. Heeren, M. Winter, and H. Wechsler (2005). Magnitude of alcohol-related mortality and morbidity among U.S. college students ages 18–24: Changes from 1998 to 2001. *Annu Rev Public Health* 26:259–279.

Hingson, R. W., T. Heeren, and M. R. Winter (2006). Age at drinking onset and alcohol dependence: Age at onset, duration, and severity. *Arch Pediatr Adolesc Med* 160(7):739–746.

Horger, B. A., K. Shelton, and S. Schenk (1990). Preexposure sensitizes rats to the rewarding effects of cocaine. *Pharmacol Biochem Behav* 37(4):707–711.

Horner, M. D., P. L. Ferguson, A. W. Selassie, L. A. Labbate, K. Kniele, and J. D. Corrigan (2005). Patterns of alcohol use 1 year after traumatic brain injury: A population-based, epidemiological study. *J Int Neuropsychol Soc* 11(3):322–330.

Hwa, L. S., A. Chu, S. A. Levinson, T. M. Kayyali, J. F. DeBold, and K. A. Miczek (2011). Persistent escalation of alcohol drinking in C57BL/6J mice with intermittent access to 20% ethanol. *Alcohol Clin Exp Res* 35(11):1938–1947.

IOM (2008). Institute of Medicine. Gulf War and Health: Vol. 7. Long-term consequences of traumatic brain injury. National Academies Press, Washington, DC.

Jorge, R. E., S. E. Starkstein, S. Arndt, D. Moser, B. Crespo-Facorro, and R. G. Robinson (2005). Alcohol misuse and mood disorders following traumatic brain injury. *Arch Gen Psychiatry* 62(7):742–749.

Kolakowsky-Hayner, S. A., E. V. Gourley, 3rd, J. S. Kreutzer, J. H. Marwitz, D. Cifu, and W. O. McKinley (1999). Pre-injury substance abuse among persons with brain injury and persons with spinal cord injury. *Brain Inj* 13(8):571–581.

Kolakowsky-Hayner, S. A., E. V. Gourley, 3rd, J. S. Kreutzer, J. H. Marwitz, M. A. Meade, and D. Cifu (2002). Post-injury substance abuse among persons with brain injury and persons with spinal cord injury. *Brain Inj* 16(7):583–592.

Koponen, S., T. Taiminen, R. Portin, L. Himanen, H. Isoniemi, H. Heinonen et al. (2002). Axis I and II psychiatric disorders after traumatic brain injury: A 30-year follow-up study. *Am J Psychiatry* 159(8):1315–1321.

Linsenbardt, D. N., E. M. Moore, K. D. Griffin, E. D. Gigante, and S. L. Boehm, 2nd (2011). Tolerance to ethanol's ataxic effects and alterations in ethanol-induced locomotion following repeated binge-like ethanol intake using the DID model. *Alcohol Clin Exp Res* 35(7):1246–1255.

Linsenbardt, D. N., E. M. Moore, C. D. Gross, K. J. Goldfarb, L. C. Blackman, and S. L. Boehm, 2nd (2009). Sensitivity and tolerance to the hypnotic and ataxic effects of ethanol in adolescent and adult C57BL/6J and DBA/2J mice. *Alcohol Clin Exp Res* 33(3):464–476.

Lopez, M. F., Anderson, R. I., and Becker, H. C. (2008). Repeated cycles of chronic intermittent ethanol exposure leads to the development of tolerance to aversive effects of ethanol in C57BL/6J mice. *Alcohol Clin Exp Res* 32:163A.

Lopez, M. F. and H. C. Becker (2005). Effect of pattern and number of chronic ethanol exposures on subsequent voluntary ethanol intake in C57BL/6J mice. *Psychopharmacology (Berl)* 181(4):688–696.

Lyons, A. M., E. G. Lowery, D. R. Sparta, and T. E. Thiele (2008). Effects of food availability and administration of orexigenic and anorectic agents on elevated ethanol drinking associated with drinking in the dark procedures. *Alcohol Clin Exp Res* 32(11):1962–1968.

McBride, W. J., M. W. Kimpel, J. A. Schultz, J. N. McClintick, H. J. Edenberg, and R. L. Bell (2010). Changes in gene expression in regions of the extended amygdala of alcohol-preferring rats after binge-like alcohol drinking. *Alcohol* 44(2):171–183.

McDonald, B. C., L. A. Flashman, and A. J. Saykin (2002). Executive dysfunction following traumatic brain injury: Neural substrates and treatment strategies. *NeuroRehabilitation* 17(4):333–344.

McKinlay, A., R. Grace, J. Horwood, D. Fergusson, and M. MacFarlane (2009). Adolescent psychiatric symptoms following preschool childhood mild traumatic brain injury: Evidence from a birth cohort. *J Head Trauma Rehabil* 24(3):221–227.

Melendez, R. I. (2011). Intermittent (every-other-day) drinking induces rapid escalation of ethanol intake and preference in adolescent and adult C57BL/6J mice. *Alcohol Clin Exp Res* 35(4):652–658.

Miller, J. W., T. S. Naimi, R. D. Brewer, and S. E. Jones (2007). Binge drinking and associated health risk behaviors among high school students. *Pediatrics* 119(1):76–85.

NIAAA (2004). National Institute on Alcohol Abuse and Alcoholism Council approves definition of binge drinking. NIAAA Newsletter. 3.

O'Dell, L. E., A. J. Roberts, R. T. Smith, and G. F. Koob (2004). Enhanced alcohol self-administration after intermittent versus continuous alcohol vapor exposure. *Alcohol Clin Exp Res* 28(11):1676–1682.

Ponsford, J., R. Whelan-Goodinson, and A. Bahar-Fuchs (2007). Alcohol and drug use following traumatic brain injury: A prospective study. *Brain Inj* 21(13–14):1385–1392.

Rhodes, J. S., K. Best, J. K. Belknap, D. A. Finn, and J. C. Crabbe (2005). Evaluation of a simple model of ethanol drinking to intoxication in C57BL/6J mice" *Physiol Behav* 84(1):53–63.

Rhodes, J. S., M. M. Ford, C. H. Yu, L. L. Brown, D. A. Finn, T. Garland, Jr et al. (2007). Mouse inbred strain differences in ethanol drinking to intoxication. *Genes Brain Behav* 6(1):1–18.

Richardson, H. N., S. Y. Lee, L. E. O'Dell, G. F. Koob, and C. L. Rivier (2008). Alcohol self-administration acutely stimulates the hypothalamic-pituitary-adrenal axis, but alcohol dependence leads to a dampened neuroendocrine state. *Eur J Neurosci* 28(8):1641–1653.

Rimondini, R., C. Arlinde, W. Sommer, and M. Heilig (2002). Long-lasting increase in voluntary ethanol consumption and transcriptional regulation in the rat brain after intermittent exposure to alcohol. *FASEB J* 16(1):27–35.

Rimondini, R., W. Sommer, and M. Heilig (2003). A temporal threshold for induction of persistent alcohol preference: Behavioral evidence in a rat model of intermittent intoxication. *J Stud Alcohol* 64(4):445–449.

Roberts, A. J., M. Cole, and G. F. Koob (1996). Intra-amygdala muscimol decreases operant ethanol self-administration in dependent rats. *Alcohol Clin Exp Res* 20(7):1289–1298.

Roberts, A. J., C. J. Heyser, M. Cole, P. Griffin, and G. F. Koob (2000). Excessive ethanol drinking following a history of dependence: Animal model of allostasis. *Neuropsychopharmacology* 22(6):581–594.

Rodd-Henricks, Z. A., D. L. McKinzie, J. M. Murphy, W. J. McBride, L. Lumeng, and T. K. Li (2000). The expression of an alcohol deprivation effect in the high-alcohol-drinking replicate rat lines is dependent on repeated deprivations. *Alcohol Clin Exp Res* 24(6):747–753.

Rodd, Z. A., R. L. Bell, K. A. Kuc, J. M. Murphy, L. Lumeng, and W. J. McBride (2009). Effects of concurrent access to multiple ethanol concentrations and repeated deprivations on alcohol intake of high-alcohol-drinking (HAD) rats. *Addict Biol* 14(2):152–164.

Rogers, J. M. and C. A. Read (2007). Psychiatric comorbidity following traumatic brain injury. *Brain Inj* 21(13–14):1321–1333.

Rona, R. J., M. Jones, N. T. Fear, L. Hull, D. Murphy, L. Machell et al. (2012). Mild traumatic brain injury in UK military personnel returning from Afghanistan and Iraq: Cohort and cross-sectional analyses. *J Head Trauma Rehabil* 27(1):33–44.

Sajja, R. K. and S. Rahman (2011). Lobeline and cytisine reduce voluntary ethanol drinking behavior in male C57BL/6J mice. *Prog Neuropsychopharmacol Biol Psychiatry* 35(1):257–264.

Salimov, R. M. and N. B. Salimova (1993). The alcohol-deprivation effect in hybrid mice. *Drug Alcohol Depend* 32(2):187–191.

Salmond, C. H. and B. J. Sahakian (2005). Cognitive outcome in traumatic brain injury survivors. *Curr Opin Crit Care* 11(2):111–116.

Silver, J. M., R. Kramer, S. Greenwald, and M. Weissman (2001). The association between head injuries and psychiatric disorders: Findings from the New Haven NIMH Epidemiologic Catchment Area Study. *Brain Inj* 15(11):935–945.

Simms, J. A., J. J. Bito-Onon, S. Chatterjee and S. E. Bartlett (2010). Long-Evans rats acquire operant self-administration of 20% ethanol without sucrose fading. *Neuropsychopharmacology* 35(7):1453–1463.

Simms, J. A., P. Steensland, B. Medina, K. E. Abernathy, L. J. Chandler, R. Wise et al. (2008). Intermittent access to 20% ethanol induces high ethanol consumption in Long-Evans and Wistar rats. *Alcohol Clin Exp Res* 32(10):1816–1823.

Sinclair, J. D. and T. K. Li (1989). Long and short alcohol deprivation: Effects on AA and P alcohol-preferring rats. *Alcohol* 6(6):505–509.

Sinclair, J. D. and R. J. Senter (1968). Development of an alcohol-deprivation effect in rats. *Q J Stud Alcohol* 29(4):863–867.

Sinclair, J. D. and K. Tiihonen (1988). Lack of alcohol-deprivation effect in AA rats. *Alcohol* 5(1):85–87.

Smith, D. H., H. D. Soares, J. S. Pierce, K. G. Perlman, K. E. Saatman, D. F. Meaney et al. (1995). A model of parasagittal controlled cortical impact in the mouse: Cognitive and histopathologic effects. *J Neurotrauma* 12(2):169–178.

Sommer, W. H., R. Rimondini, A. C. Hansson, P. A. Hipskind, D. R. Gehlert, C. S. Barr et al. (2008). Upregulation of voluntary alcohol intake, behavioral sensitivity to stress, and amygdala crhr1 expression following a history of dependence. *Biol Psychiatry* 63(2):139–145.

Spanagel, R. and S. M. Holter (1999). Long-term alcohol self-administration with repeated alcohol deprivation phases: An animal model of alcoholism? *Alcohol Alcohol*34(2):231–243.

Spanagel, R. and S. M. Holter (2000). Pharmacological validation of a new animal model of alcoholism. *J Neural Transm* 107(6):669–680.

Taylor, B., H. M. Irving, F. Kanteres, R. Room, G. Borges, C. Cherpitel et al. (2010). The more you drink, the harder you fall: A systematic review and meta-analysis of how acute alcohol consumption and injury or collision risk increase together. *Drug Alcohol Depend* 110(1–2):108–116.

Terrio, H., L. A. Brenner, B. J. Ivins, J. M. Cho, K. Helmick, K. Schwab et al. (2009). Traumatic brain injury screening: Preliminary findings in a US Army Brigade Combat Team. *J Head Trauma Rehabil* 24(1):14–23.

Van Den Heuvel, C., E. Thornton and R. Vink (2007). Traumatic brain injury and Alzheimer's disease: A review. *Prog Brain Res* 161:303–316.

Wayner, M. J., I. Greenberg, R. Tartaglione, D. Nolley, S. Fraley and A. Cott (1972). A new factor affecting the consumption of ethyl alcohol and other sapid fluids. *Physiol Behav* 8(2):345–362.

West, S. L. (2011). Substance use among persons with traumatic brain injury: A review. *NeuroRehabilitation* 29(1):1–8.

Wise, R. A. (1973). Voluntary ethanol intake in rats following exposure to ethanol on various schedules. *Psychopharmacologia* 29(3):203–210.

Index

C

D

E

N

O

P

Q

R

S

T

U

V

W

X

Y

Z